Special Features

P9-BXW-752

MOLECULAR BIOLOGY OF
THE CELL
THIRD EDITION

MOLECULAR BIOLOGY OF
THE CELL
THIRD EDITION

Bruce Alberts • Dennis Bray
Julian Lewis • Martin Raff • Keith Roberts
James D. Watson

GARLAND PUBLISHING
ALERE FLAMMAM
Taylor & Francis Group

GARLAND STAFF

Text Editor: Miranda Robertson
Managing Editor: Ruth Adams
Illustrator: Nigel Orme
Molecular Model Drawings: Kate Hesketh-Moore
Director of Electronic Publishing: John M-Roblin
Computer Specialist: Chuck Bartelt
Disk Preparation: Carol Winter
Copy Editor: Shirley M. Cobert
Production Editor: Douglas Goertzen
Production Coordinator: Perry Bessas
Indexer: Maija Hinkle

Bruce Alberts received his Ph.D. from Harvard University and is currently President of the National Academy of Sciences and Professor of Biochemistry and Biophysics at the University of California, San Francisco. *Dennis Bray* received his Ph.D. from the Massachusetts Institute of Technology and is currently a Medical Research Council Fellow in the Department of Zoology, University of Cambridge. *Julian Lewis* received his D.Phil. from the University of Oxford and is currently a Senior Scientist in the Imperial Cancer Research Fund Developmental Biology Unit, University of Oxford. *Martin Raff* received his M.D. from McGill University and is currently a Professor in the MRC Laboratory for Molecular Cell Biology and the Biology Department, University College, London. *Keith Roberts* received his Ph.D. from the University of Cambridge and is currently Head of the Department of Cell Biology, the John Innes Institute, Norwich. *James D. Watson* received his Ph.D. from Indiana University and is currently Director of the Cold Spring Harbor Laboratory. He is the author of *Molecular Biology of the Gene* and, with Francis Crick and Maurice Wilkins, won the Nobel Prize in Medicine and Physiology in 1962.

Library of Congress Cataloging-in-Publication Data
Molecular biology of the cell / Bruce Alberts . . . [et al.].—3rd ed.
     p. cm.
    Includes bibliographical references and index.
    ISBN 0-8153-1619-4 (hard cover).—ISBN 0-8153-1620-8 (pbk.)
    1. Cytology. 2. Molecular biology. I. Alberts, Bruce.
    [DNLM: 1. Cells. 2. Molecular Biology.   QH 581.2 M718 1994]
QH581.2.M64 1994
574.87—dc20
DNLM/DLC
for Library of Congress   93-45907
                              CIP

Published by Garland Publishing
A member of the Taylor & Francis Group
19 Union Square West, New York, NY 10003

Printed in the United States of America

15 14 13 12 10 9 8

Front cover: The photograph shows a rat nerve cell in culture. It is labeled (*yellow*) with a fluorescent antibody that stains its cell body and dendritic processes. Nerve terminals (*green*) from other neurons (not visible), which have made synapses on the cell, are labeled with a different antibody. (Courtesy of Olaf Mundigl and Pietro de Camilli.)

Dedication page: Gavin Borden, late president of Garland Publishing, weathered in during his mid-1980s climb near Mount McKinley with MBoC author Bruce Alberts and famous mountaineer guide Mugs Stump (1940–1992).

Back cover: The authors, in alphabetical order, crossing Abbey Road in London on their way to lunch. Much of this third edition was written in a house just around the corner. (Photograph by Richard Olivier.)

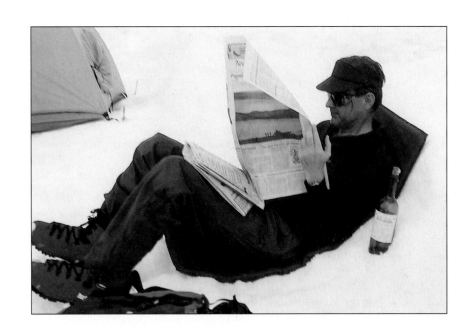

Gavin Borden
May 1, 1939–December 20, 1991

Preface to the Third Edition

Biological research is in an explosive phase, driven by a new awareness of the basic unity of all life forms. Today, the key to a problem about neurons, plant cells, or cancer may lie in research on yeasts, frogs, or flies. More than ever before, molecular genetics is showing us how to recognize and exploit such connections and is forcing us to reflect on the ancient origins of the components from which we are made. In surveying cell biology, one does not know whether to marvel more at the endless variety of living systems or at the fundamental similarities in the mechanisms by which all cells operate.

The challenge in writing a textbook is to pick out the concepts that are most important for our understanding. The problem in revising it is to see where these concepts have changed or new ones have emerged. This is a very different matter from summarizing the new facts: many new facts may change our view of cells very little, whereas a few can transform the whole picture. Recognizing that it is impossible to be comprehensive, we have struggled to ensure that our book should provide a survey of cell biology that students can readily read and digest. We have therefore devoted as much thought to the problem of what to leave out as to the problem of what to add.

In this third edition, large parts of the text have been radically rewritten to take account of recent advances. Two chapters—those on plants and on neurobiology—have been dismantled and the material from them integrated into other chapters, emphasizing that these topics are not mere side issues for specialists, but belong to the core of cell biology. Several of the central chapters of the book have been subdivided into more manageable parts, and a new chapter devoted to recombinant DNA technology has been added. We have also inserted a new chapter on the principles of protein function, showing how proteins operate as molecules with moving parts to form the dynamic gadgetry of the living cell—a fundamental topic that is only now gaining its rightful prominence, with the rapid growth in knowledge of protein structures. To bring protein molecules to life, we have selected an anthology for inclusion on a computer diskette together with the Kinemage program, which allows one to view the molecular structures three-dimensionally, rotating and manipulating them on the computer screen and watching as they undergo their conformational changes. The diskette is available from the publisher for a small charge. At the end of the book, we have added a glossary of essential technical terms. Tim Hunt and John Wilson have updated and enlarged their problems book to accompany the main text and help students appreciate the experiments and the reasoning on which our understanding of cells is based.

A special acknowledgment is due to four contributors (all from the University of California, San Francisco) who have had a major hand in writing specific chapters: Sandy Johnson (Chapters 8 and 9), Peter Walter (Chapters 12 and 13), Andrew Murray (Chapter 17), and Tim Mitchison (Chapters 16 and 18); also to Michael Klymkowsky (University of Colorado, Boulder) for his contribution to Chapter 16. We once again owe a debt also to the many experts who have generously given advice, comments, and corrections; they are acknowledged individually on pages xxxix–xli. As editor, Miranda Robertson, as always, had a crucial role in making the text clear, coherent, and accessible to students.

With this edition, we have ventured into full color. Nigel Orme converted drawings into printable color artwork with magical promptness and skill. Carol Winter again typed the entire book with painstaking care and prepared the final disks. John M-Roblin wove text and figures together into pages for printing and made the transition to computerized book production seem easy. Brian Rubin helped with references, Fran Dependahl helped with the typing, and Emily Preece fed us in style while we were writing. We give our thanks to all of them, and to many others, especially the staff of Garland Publishing, who contributed in ways too numerous to list. We are particularly indebted to Ruth Adams, who once again oversaw the production of the book with unfailing efficiency and good humor, and to Elizabeth Borden, now head of Garland Publishing, whose warm hospitality, good sense, and friendly encouragement have carried us through to the end of a daunting task for the third time. Thanks, above all, to our long-suffering families, students, and colleagues, who patiently put up with our absences and preoccupations.

Lastly, on a sad note, we record our debt to Gavin Borden, our publisher, who became our close friend and who died of cancer in December 1991. Without him, there would have been no book. We, and the book itself, owe more than we can say to his spirit of adventure, his generosity, his love of life, and his friendship. We dedicate this edition to his memory.

Preface to the Second Edition

More than 50 years ago, E.B. Wilson wrote that "the key to every biological problem must finally be sought in the cell." Yet, until very recently, cell biology was usually taught to biology majors as a specialized upper-level course based largely on electron microscopy. And in most medical schools, many cell-biological topics, such as the mechanisms of endocytosis, chemotaxis, cell movement, and cell adhesion, were hardly taught at all—being regarded as too cellular for biochemistry and too molecular for histology. With the recent dramatic advances in our understanding of cells, however, cell biology is beginning to take its rightful place at the center of biological and medical teaching. In an increasing number of universities, it is a required full-year course for all undergraduate biology and biochemistry majors and is becoming the organizing theme for much of the first-year medical curriculum. The first edition of *Molecular Biology the Cell* was written in anticipation of these much needed curriculum reforms and in the hope of catalyzing them. We will be gratified if the second edition helps to accelerate and spread these reforms.

In revising the book, we have found few cases where recent discoveries have proved old conclusions wrong. But in the six years since the first edition appeared, a fantastically rich harvest of new information about cells has uncovered new connections and in many places forced a radical change of emphasis. The present revision, therefore, goes deep: every chapter has undergone substantial changes, many have been almost entirely rewritten, and two new chapters—on the control of gene expression and on cancer—have been added.

Some commentators on the first edition, especially teachers, wanted more detailed discussion of the experimental evidence that supports the concepts discussed. We did not wish to disrupt the flow or to enlarge an already very large book; but we agree that it is crucial to give students a sense of how advances are made. To this end, John Wilson and Tim Hunt have written a Problems Book to accompany the central portion of the main text (Chapters 5–14). As explained in the Note to the Reader on page xxxv, each section of *The Problems Book* covers a section of *Molecular Biology of the Cell* and takes as its principal focus a set of experiments drawn from the relevant research literature. These are the basis for a series of questions—some easy and some hard—that are meant to involve the reader actively in the reasoning that underlies the process of discovery.

The second edition, like the first, has been a long time in the making. As before, each chapter has been passed back and forth between the author who wrote the first draft and the other authors for criticism and extensive revision, so that every part of the book represents a joint composition; Tim Hunt and John Wilson often helped in this process. In addition, outside experts were invited to make suggestions for revision and, in a few cases, to contribute material for the text, which the authors reworked to fit with the rest of the book. We are especially indebted to James Rothman (Princeton University) for his contribution to Chapter 8 and to Jeremy Hyams (University College London), Tim Mitchison (University of California, San Francisco), and Paul Nurse (University of Oxford) for their contributions to Chapter 13. All sections of the revised text were read by outside experts, whose comments and suggestions were invaluable; a list of acknowledgments is on page xxxiii.

Miranda Robertson again played a major part in the creation of a readable book, by insisting that every page be lucid and coherent and rewriting many pages that were not. We are also indebted to the staff at Garland Publishing, and in particular to Ruth Adams, Alison Walker, and Gavin Borden, for their kindness, good humour, efficiency, and unfailingly generous support during the four years that it has taken to prepare this edition. Special thanks go to Carol Winter, for her painstaking care in typing the entire book and preparing the disks for the printer. Finally, to our wives, families, colleagues, and students we again offer our gratitude and apologies for several years of impositions and neglect; without their help and tolerance this book could not have been written.

Preface to the First Edition

There is a paradox in the growth of scientific knowledge. As information accumulates in ever more intimidating quantities, disconnected facts and impenetrable mysteries give way to rational explanations, and simplicity emerges from chaos. Gradually the essential principles of a subject come into focus. This is true of cell biology today. New techniques of analysis at the molecular level are revealing an astonishing elegance and economy in the living cell and a gratifying unity in the principles by which cells function. This book is concerned with those principles. It is not an encyclopedia but a guide to understanding. Admittedly, there are still large areas of ignorance in cell biology and many facts that cannot yet be explained. But these unsolved problems provide much of the excitement, and we have tried to point them out in a way that will stimulate readers to join in the enterprise of discovery. Thus, rather than simply present disjointed facts in areas that are poorly understood, we have often ventured hypotheses for the reader to consider and, we hope, to criticize.

Molecular Biology of the Cell is chiefly concerned with eucaryotic cells, as opposed to bacteria, and its title reflects the prime importance of the insights that have come from the molecular approach. Part I and Part II of the book analyze cells from this perspective and cover the traditional material of cell biology courses. But molecular biology by itself is not enough. The eucaryotic cells that form multicellular animals and plants are social organisms to an extreme degree: they live by cooperation and specialization. To understand how they function, one must study the ways of cells in multicellular communities, as well as the internal workings of cells in isolation. These are two very different levels of investigation, but each depends on the other for focus and direction. We have therefore devoted Part III of the book to the behavior of cells in multicellular animals and plants. Thus developmental biology, histology, immunobiology, and neurobiology are discussed at much greater length than in other cell biology textbooks. While this material may be omitted from a basic cell biology course, serving as optional or supplementary reading, it represents an essential part of our knowledge about cells and should be especially useful to those who decide to continue with biological or medical studies. The broad coverage expresses our conviction that cell biology should be at the center of a modern biological education.

This book is principally for students taking a first course in cell biology, be they undergraduates, graduate students, or medical students. Although we assume that most readers have had at least an introductory biology course, we have attempted to write the book so that even a stranger to biology could follow it by starting at the beginning. On the other hand, we hope that it will also be useful to working scientists in search of a guide to help them pick their way through a vast field of knowledge. For this reason, we have provided a much more thorough list of references than the average undergraduate is likely to require, at the same time making an effort to select mainly those that should be available in most libraries.

This is a large book, and it has been a long time in gestation—three times longer than an elephant, five times longer than a whale. Many people have had a hand in it. Each chapter has been passed back and forth between the author who wrote the first draft and the other authors for criticism and revision, so that each chapter represents a joint composition. In addition, a small number of outside experts contributed written material, which the authors reworked to fit with the rest of the book, and all the chapters were read by experts, whose comments and corrections were invaluable. A full list of acknowledgments to these contributors and readers for their help with specific chapters is appended. Paul R. Burton (University of Kansas), Douglas Chandler (Arizona State University), Ursula Goodenough (Washington University), Robert E. Pollack (Columbia University), Robert E. Savage (Swarthmore College), and Charles F. Yocum (University of Michigan) read through all or some of the manuscript and made many helpful suggestions. The manuscript was also read by undergraduate students, who helped to identify passages that were obscure or difficult.

Most of the advice obtained from students and outside experts was collated and digested by Miranda Robertson. By insisting that every page be lucid and coherent, and by rewriting many of those that were not, she has played a major part in the creation of a textbook that undergraduates will read with ease. Lydia Malim drew many of the figures for Chapters 15 and 16, and a large number of scientists very generously provided us with photographs: their names are given in the figure credits. To our families, colleagues, and students we offer thanks for forbearance and apologies for several years of imposition and neglect. Finally, we owe a special debt of gratitude to our editors and publisher. Tony Adams played a large part in improving the clarity of the exposition, and Ruth Adams, with a degree of good-humored efficiency that put the authors to shame, organized the entire production of the book. Gavin Borden undertook to publish it, and his generosity and hospitality throughout have made the enterprise of writing a pleasure as well as an education for us.

Contents in Brief

Special Features

List of Topics

Introduction to the Cell

Part I

The Evolution of the Cell

Chapter 1

Small Molecules, Energy, and Biosynthesis

Chapter 2

Macromolecules: Structure, Shape, and Information

Chapter **3**

How Cells Are Studied

Chapter
4

Molecular Genetics

Part
II

Protein Function

Chapter
5

Basic Genetic Mechanisms

Chapter
6

Recombinant DNA Technology

Chapter
7

The Cell Nucleus

Chapter
8

Control of Gene Expression

Chapter
9

Internal Organization of the Cell

Part **III**

Membrane Structure

Chapter **10**

Membrane Transport of Small Molecules and the Ionic Basis of Membrane Excitability

Chapter
11

Intracellular Compartments and Protein Sorting

Vesicular Traffic in the Secretory and Endocytic Pathways

Energy Conversion: Mitochondria and Chloroplasts

Chapter **14**

Cell Signaling

The Cytoskeleton

Chapter
16

The Cell-Division Cycle

Chapter **17**

The Mechanics of Cell Division

Chapter
18

Cells in Their Social Context

Part
IV

Cell Junctions, Cell Adhesion, and the Extracellular Matrix

Chapter
19

Germ Cells and Fertilization

Chapter
20

Cellular Mechanisms of Development

Chapter **21**

Differentiated Cells and the Maintenance of Tissues

Chapter
22

The Immune System

Chapter **23**

Cancer

Chapter
24

Acknowledgments

In writing this book we have benefited greatly from the advice of many biologists and biochemists. In addition to those who advised us by telephone, we would like to thank the following for their written advice in preparing this edition, as well as those who helped in preparing the first and second editions. (Those who helped on this edition are listed first, and those who helped with the first and second editions follow.)

Chapter 1   Alan Grafen (University of Oxford), David Haig (Harvard University), Laurence Hurst (University of Cambridge), Jack Szostak (Massachusetts General Hospital, Boston).

Chapter 2   Carl Branden (European Synchrotron Facility, Grenoble, France), Greg Petsko (Brandeis University).

Chapter 3   David Agard (University of California, San Francisco), Greg Petsko (Brandeis University), Richard Wolfenden (University of North Carolina).

Chapter 4   Tim Mitchison (University of California, San Francisco).

Chapter 5   Henry Bourne (University of California, San Francisco), Steven Harrison (Harvard University), David Morgan (University of California, San Francisco), Greg Petsko (Brandeis University), Martin Rechsteiner (University of Utah, Salt Lake City), Howard Schachman (University of California, Berkeley), Mathias Sprinzl (University of Bayreuth), Alex Varshavsky (California Institute of Technology, Pasadena).

Chapter 6   Nancy Craig (Johns Hopkins University), Sandy Johnson (University of California, San Francisco), Kelly Komachi (University of California, San Francisco), Tomas Lindahl (ICRF Clare Hall Laboratories), Harry Noller (University of California, Santa Cruz), John Wilson (Baylor University), Rick Wood (ICRF Clare Hall Laboratories).

Chapter 7   Martha Arnaud (University of California, San Francisco), Mario Capecchi (University of Utah), Walter Gehring (Biozentrum, University of Basel), Tim Hunt (ICRF Clare Hall Laboratories), Barbara Meyer (University of California, Berkeley), Richard Myers (Stanford University), John Wilson (Baylor University).

Chapter 8   Sandy Johnson [major contribution] (University of California, San Francisco), Larry Gerace (Scripps Clinic, La Jolla, CA), Christine Guthrie (University of California, San Francisco), Martha Stark (University of California, San Francisco), Peter Walter (University of California, San Francisco), Keith Yamamoto (University of California, San Francisco).

Chapter 9   Sandy Johnson [major contribution] (University of California, San Francisco), Tanya Awabdy (University of California, San Francisco), Steve Burley (The Rockefeller University), Beverly Emerson (The Salk Institute), Frank Grosveld (Erasmus Universiteit, Rotterdam), Alan Hinnebusch (National Institutes of Health, Bethesda), Nancy Hollingsworth (University of California, San Francisco), Mike Levine (University of California, San Diego), Richard Losick (Harvard University), Stuart Orkin (Children's Hospital, Boston), Roy Parker (University of Arizona, Tucson), Alan Sachs (University of California, Berkeley), Madhu Wahi (University of California, San Francisco), Peter Walter (University of California, San Francisco), Harold Weintraub (Fred Hutchinson Cancer Research Center), Sandra Wolin (Yale University), Keith Yamamoto (University of California, San Francisco).

Chapter 10   Mark Bretscher (MRC Laboratory of Molecular Biology, Cambridge, UK), Richard Henderson (MRC Laboratory of Molecular Biology, Cambridge, UK), Kai Simons (EMBL, Heidelberg, Germany), Tim Springer (Center for Blood Research, Boston).

Chapter 11   Barbara Barres (Stanford University), Bertil Hille (University of Washington, Seattle), Ron Kaback (University of California, Los Angeles), Chuck Stevens (The Salk Institute, La Jolla, CA), Roger Thomas (University of Bristol, Bristol, UK).

Chapter 12   Peter Walter [major contribution] (University of California, San Francisco), Larry Gerace (Scripps Clinic, La Jolla, CA), Reid Gilmore (University of Massachusetts), Walter Neupert (University of Munchen, Germany), George Palade (University of California, San Diego), Gottfried Schatz (Biozentrum, University of Basel).

Chapter 13   Peter Walter [major contribution] (University of California, San Francisco), Ari Helenius (Yale University), Ira Mellman (Yale University), Keith Mostov (University of California, San Francisco), Jim Rothman (Memorial Sloan-Kettering Cancer Center), Randy Schekman (University of California, Berkeley).

Chapter 14   Martin Brand (University of Cambridge), Richard Henderson (MRC Laboratory of Molecular Biology, Cambridge, UK), William W. Parson (University of Washington, Seattle), Gottfried Schatz (University of Basel), Alison Smith (John Innes Institute, Norfolk, UK).

Chapter 15   Michael J. Berridge (University of Cambridge), Henry Bourne (University of California, San Francisco), Ernst Hafen (Universitat Zurich), Robert J. Lefkowitz (Duke University, Durham, NC), Mark E. Nelson (University of Illinois, Urbana), Melvin I. Simon (California Institute of Technology, Pasadena), Lewis T. Williams (University of California, San Francisco).

Chapter 16   Tim Mitchison [major contribution] (University of California, San Francisco), Mike Klymkowsky [substantial contribution] (University of Colorado, Boulder), Douglas Kellogg (University of California, San Francisco), Michelle Moritz (University of California, San Francisco), Jordan Raff (University of California, San Francisco), Murray Stewart (MRC Laboratory of Molecular Biology, Cambridge, UK).

Chapter 17 Andrew Murray [major contribution] (University of California, San Francisco), Gerard Evan (Imperial Cancer Research Fund, London), Tim Hunt (Imperial Cancer Research Fund, Clare Hall Laboratories), Paul Nurse (Imperial Cancer Research Fund, London).

Chapter 18 Tim Mitchison [major contribution] (University of California, San Francisco).

Chapter 19 David Birk (UMNDJ—Robert Wood Johnson Medical School), Robert Cohen (University of California, San Francisco), Benny Geiger (Weizmann Institute of Science, Rehovot, Israel), Dan Goodenough (Harvard University), Barry Gumbiner (Memorial Sloan-Kettering Cancer Center), Richard Hynes (Massachusetts Institute of Technology), Louis Reichardt (University of California, San Francisco), Erkki Ruoslahti (La Jolla Cancer Research Foundation), Robert Trelstad (UMDNJ—Robert Wood Johnson Medical School), Frank Walsh (Guy's Hospital, London), Peter Yurchenco (UMDNJ—Robert Wood Johnson Medical School).

Chapter 20 Nancy Kleckner (Harvard University), Daniel Szollosi (Institut National de la Recherche Agronomique, Jouy-en-Josas, France), Paul Wassarman (Roche Institute of Molecular Biology, Nutley, NJ), Keith Willison (Chester Beatty Laboratories, London).

Chapter 21 Michael Akam (University of Cambridge), Enrico Coen (John Innes Institute, Norwich, UK), David Ish-Horowicz (Imperial Cancer Research Fund, Oxford), Roger Keynes (University of Cambridge), Jonathan Slack (Imperial Cancer Research Fund, Oxford), Paul Sternberg (California Institute of Technology, Pasadena).

Chapter 22 John Harris (University of Otago, Dunedin, New Zealand).

Chapter 23 N.A. Mitchison (Deutsches Rheuma-Forschungszentrum Berlin), Klaus Rajewsky (Institut für Genetik der Universität, Cologne, Germany), Ronald Schwartz (National Institutes of Health, Bethesda), Alain Townsend (Institute of Molecular Medicine, John Radcliffe Hospital, Oxford, UK).

Chapter 24 Michael Bishop (University of California, San Francisco), Julian Downward (Imperial Cancer Research Fund, London), David Lane (University of Dundee), Andrew Murray (University of California, San Francisco), Bruce Ponder (University of Cambridge), Harold Varmus (National Institutes of Health).

First and second editions Fred Alt (Columbia University), Michael Ashburner (University of Cambridge), Jonathan Ashmore (University of Sussex, UK), Peter Baker (King's College London), Michael Banda (University of California, San Francisco), Michael Bennett (Albert Einstein College of Medicine), Darwin Berg (University of California, San Diego), Tim Bliss (National Institute for Medical Research, London), Hans Bode (University of California, Irvine), Piet Borst (Jan Swammerdam Institute, University of Amsterdam), Alan Boyde (University College London), Marianne Bronner-Fraser (University of California, Irvine), Robert Brooks (King's College London), Barry Brown (King's College London), Michael Brown (University of Oxford), Stephen Burden (Harvard Medical School), Steven Burden (Massachusetts Institute of Technology), Max Burger (University of Basel), John Cairns (Harvard School of Public Health), Zacheus Cande (University of California, Berkeley), Lewis Cantley (Harvard University), Charles Cantor (Columbia University), Roderick Capaldi (University of Oregon), Adelaide Carpenter (University of California, San Diego), Tom Cavalier-Smith (King's College London), Pierre Chambon (University of Strasbourg), Philip Cohen (University of Dundee, Scotland), Roger Cooke (University of California, San Francisco), James Crow (University of Wisconsin, Madison), Stuart Cull-Candy (University College London), Michael Dexter (Paterson Institute for Cancer Research, Manchester, UK), Russell Doolittle (University of California, San Diego), Graham Dunn (MRC Cell Biophysics Unit, London), Jim Dunwell (John Innes Institute, Norwich, UK), Sarah Elgin (Washington University, St. Louis), Ruth Ellman (Institute of Cancer Research, Sutton, UK), Charles Emerson (University of Virginia), David Epel (Stanford University), Ray Evert (University of Wisconsin, Madison), Gary Felsenfeld (National Institutes of Health, Bethesda), Gary Firestone (University of California, Berkeley), Gerald Fischbach (Washington University, St. Louis), Robert Fletterick (University of California, San Francisco), Judah Folkman (Harvard Medical School), Larry Fowke (University of Saskatchewan, Saskatoon), Daniel Friend (University of California, San Francisco), Joseph Gall (Yale University), Anthony Gardner-Medwin (University College London), Peter Garland (University of Dundee, Scotland), John Gerhart (University of California, Berkeley), Günther Gerisch (Max Planck Institute for Biochemistry, Martinsried), Bernie Gilula (Baylor University), Charles Gilvarg (Princeton University), Bastien Gomperts (University College Hospital Medical School, London), Peter Gould (Middlesex Hospital Medical School, London), Walter Gratzer (King's College London), Howard Green (Harvard University), Leslie Grivell (University of Amsterdam), Brian Gunning (Australian National University, Canberra), Jeffrey Hall (Brandeis University), John Hall (University of Southampton, UK), Zach Hall (University of California, San Francisco), David Hanke (University of Cambridge, UK), Graham Hardie (University of Dundee, Scotland), Leland Hartwell (University of Washington, Seattle), John Heath (University of Oxford), Glenn Herrick (University of Utah), Ira Herskowitz (University of California, San Francisco), Leroy Hood (California Institute of Technology), Robert Horvitz (Massachusetts Institute of Technology), David Housman (Massachusetts Institute of Technology), James Hudspeth (University of California, San Francisco), Jeremy Hyams (University College London), Philip Ingham (Imperial Cancer Research Fund, Oxford), Norman Iscove (Ontario Cancer Institute, Toronto), Tom Jessell (Columbia University), Andy Johnston (John Innes Institute, Norwich, UK), E.G. Jordan (Queen Elizabeth College, London), Ray Keller (University of California, Berkeley), Regis Kelly (University of California, San Francisco), John Kendrick-Jones (MRC Laboratory of Molecular Biology, Cambridge, UK), Cynthia Kenyon (University of California, San Francisco), Judith Kimble (University of Wisconsin, Madison), Marc Kirschner (University of California, San Francisco), Juan Korenbrot (University of California, San Francisco), Tom Kornberg (University of California, San Francisco), Stuart Kornfeld (Washington University, St. Louis), Daniel Koshland (University of California, Berkeley), Marilyn Kozak (University of Pittsburgh), Mark Krasnow (Stanford University), Peter Lachmann (MRC

Center, Cambridge, UK), Trevor Lamb (University of Cambridge), Hartmut Land (Imperial Cancer Research Fund, London), Jay Lash (University of Pennsylvania), Peter Lawrence (MRC Laboratory of Molecular Biology, Cambridge, UK), Alex Levitzki (Hebrew University), Vishu Lingappa (University of California, San Francisco), Clive Lloyd (John Innes Institute, Norwich, UK), Brian McCarthy (University of California, Irvine), Richard McCarty (Cornell University), Anne McLaren (University College London), James Maller (University of Colorado Medical School), Colin Manoil (Harvard Medical School), Mark Marsh (Institute of Cancer Research, London), Gail Martin (University of California, San Francisco), Freiderick Meins (Freiderich Miescher Institut, Basel), Robert Mishell (University of Birmingham, UK), Avrion Mitchison (University College London), J. Murdoch Mitchison (University of Edinburgh), Mark Mooseker (Yale University), Montrose Moses (Duke University), Anne Mudge (University College London), Hans Müller-Eberhard (Scripps Clinic and Research Institute), Alan Munro (University of Cambridge), David Nicholls (University of Dundee, Scotland), Duncan O'Dell (University College London), Patrick O'Farrell (University of California, San Francisco), John Owen (University of Birmingham, UK), Dale Oxender (University of Michigan), William Paul (National Institutes of Health, Bethesda), Robert Perry (Institute of Cancer Research, Philadelphia), David Phillips (Rockefeller University), Jeremy Pickett-Heaps (University of Colorado), Tom Pollard (Johns Hopkins University), Darwin Prockop (Rutgers Medical School), Dale Purves (Washington University, St. Louis), Efraim Racker (Cornell University), George Ratcliffe (University of Oxford), David Rees (National Institute for Medical Research, Mill Hill, London), Fred Richards (Yale University), Phillips Robbins (Massachusetts Institute of Technology), Elaine Robson (Univelsity of Reading, UK), Robert Roeder (Rockefeller University), Joel Rosenbaum (Yale University), Jesse Roth (National Institutes of Health, Bethesda), David Sabatini (Rockefeller University), Michael Schramm (Hebrew University), Robert Schreiber (Scripps Clinic and Research Institute), James Schwartz (Columbia University), John Scott (University of Manchester), John Sedat (University of California, San Francisco), Zvi Sellinger (Hebrew University), Philippe Sengel (University of Grenoble), Peter Shaw (John Innes Institute, Norwich, UK), Samuel Silverstein (Columbia University), John Maynard Smith (University of Sussex), Michael Solursh (University of Iowa), Scott Stachel (University of California, Berkeley), Andrew Staehelin (University of Colorado, Boulder), David Standring (University of California, San Francisco), Wilfred Stein (Hebrew University), Malcolm Steinberg (Princeton University), Monroe Strickberger (University of Missouri, St. Louis), Michael Stryker (University of California, San Francisco), Masatoshi Takeichi (Kyoto University), Vernon Thornton (King's College London), Cheryll Tickle (Middlesex Hospital Medical School, London), Jim Till (Ontario Cancer Institute, Toronto), Lewis Tilney (University of Pennsylvania), Anthony Trewavas (Edinburgh University), Victor Vacquier (University of California, San Diego), Tom Vanaman (University of Kentucky), Harry van der Westen (Wageningen, The Netherlands), Virginia Walbot (Stanford University), Trevor Wang (John Innes Institute, Norwich, UK), Anne Warner (University College London), Fiona Watt (Imperial Cancer Research Fund, London), John Watts (John Innes Institute, Norwich, UK), Klaus Weber (Max Planck Institute for Biophysical Chemistry, Göttingen), Martin Weigert (Institute of Cancer Research, Philadelphia), Norman Wessells (Stanford University), Judy White (University of California, San Francisco), William Wickner (University of California, Los Angeles), Michael Wilcox (MRC Laboratory of Molecular Biology, Cambridge, UK), Richard Wolfenden (University of North Carolina), Lewis Wolpert (Middlesex Hospital Medical School, London), Abraham Worcel (University of Rochester), John Wyke (Imperial Cancer Research Fund, London), Charles Yocum (University of Michigan), Rosalind Zalin (University College London), Patricia Zambryski (University of California, Berkeley).

Acknowledgments

A Note to the Reader

Although the chapters of this book can be read independently of one another, they are arranged in a logical sequence of four parts. The first three chapters of **Part I** cover elementary principles and basic biochemistry. They can serve either as an introduction for those who have not studied biochemistry or as a refresher course for those who have. Chapter 4, which concludes Part I, deals with the principles of the main experimental methods for investigating cells. It is not necessary to read this chapter in order to understand the later chapters, but a reader will find it a useful reference.

Parts II and III represent the central core of cell biology and are concerned mainly with those properties that are common to most eukaryotic cells. Part II deals with the expression and transmission of genetic information, while Part III discusses the internal organization of the cell.

Part IV follows the behavior of cells in multicellular organisms, starting with cell-cell junctions and extracellular matrix and ending with the disruption of multicellular organization that occurs in cancer.

Chapter 4 includes several tables giving the dates of crucial developments along with the names of the scientists involved. Elsewhere in the book the policy has been to avoid naming individual scientists. The authors of major discoveries, however, can often be identified by consulting the **list of references** at the end of each chapter. These references frequently include the original papers in which important discoveries were first reported. **Superscript numbers** that accompany the text headings refer to the numbered citations in the reference lists, providing a convenient means of following up specific topics.

Throughout the book, **boldface type** has been used to highlight key terms at the point in a chapter where the main discussion of them occurs. This may or may not coincide with the first appearance of the term in the text. *Italics* are used to set off important terms with a lesser degree of emphasis. We have also adopted the convention that the names of **genes** are set in italics (for example, *ras, Notch*), while the names of the corresponding **proteins** are set in roman type with the first letter capitalized (for example, Ras, Notch).

At the end of the book we have added a **glossary**, covering technical terms that are part of the common currency of cell biology; it is intended as a first resort for a reader who encounters an unfamiliar term used without explanation.

The Problems Book is designed as a companion volume that will help the reader appreciate the elegance, the ingenuity, and the surprises of research. It provides problems to accompany the central portion of this book (Chapters 6–19). Each chapter of problems is divided into sections that correspond to the sections of the main textbook, the principal focus of each section being a set of research-oriented problems derived from the scientific literature. Most of the research problems illustrate points in the main text. In addition, each section of *The Problems Book* begins with a set of short fill-in-the-blank and true-false questions intended to help the reader review the vocabulary and main concepts of the relevant topic. *The Problems Book* should be useful for homework assignments and as a basis for class discussion. It could even provide ideas for exam questions.

Introduction
to the Cell

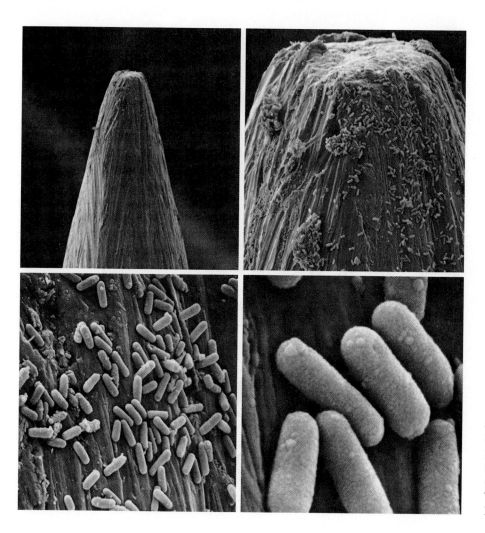

A sense of scale. These scanning electron micrographs, taken at progressively higher magnifications, show bacterial cells on the point of an ordinary domestic pin. (Courtesy of Tony Brain and the Science Photo Library.)

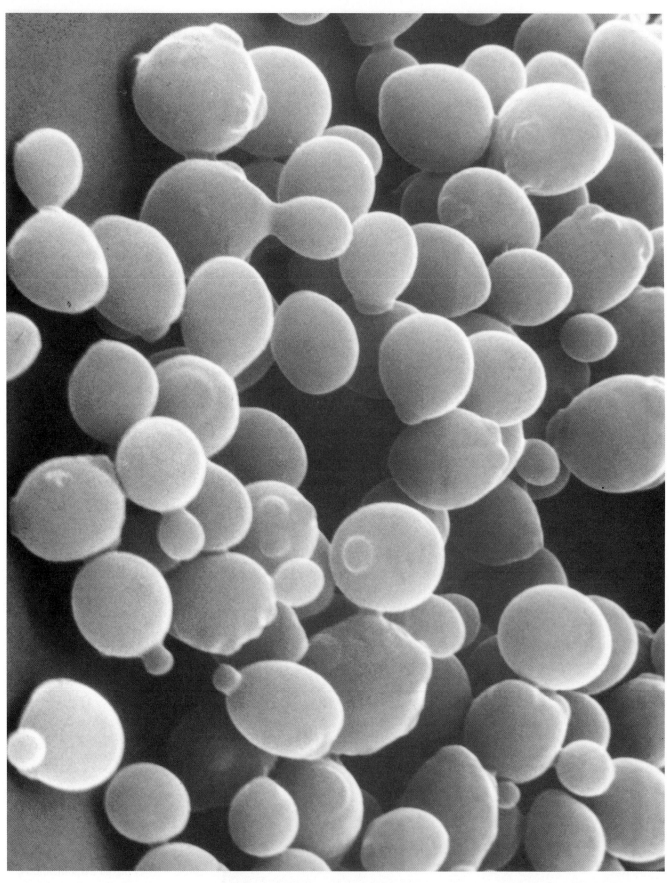

Scanning electron micrograph of growing yeast cells. These unicellular eucaryotes bud off small daughter cells as they multiply. (Courtesy of Ira Herskowitz and Eric Schabatach.)

The Evolution of the Cell

All living creatures are made of cells—small membrane-bounded compartments filled with a concentrated aqueous solution of chemicals. The simplest forms of life are solitary cells that propagate by dividing in two. Higher organisms, such as ourselves, are like cellular cities in which groups of cells perform specialized functions and are linked by intricate systems of communication. Cells occupy a halfway point in the scale of biological complexity. We study them to learn, on the one hand, how they are made from molecules and, on the other, how they cooperate to make an organism as complex as a human being.

All organisms, and all of the cells that constitute them, are believed to have descended from a common ancestor cell through *evolution by natural selection.* This involves two essential processes: (1) the occurrence of random *variation* in the genetic information passed from an individual to its descendants and (2) *selection* in favor of genetic information that helps its possessors to survive and propagate. Evolution is the central principle of biology, helping us to make sense of the bewildering variety in the living world.

This chapter, like the book as a whole, is concerned with the progression from molecules to multicellular organisms. It discusses the evolution of the cell, first as a living unit constructed from smaller parts and then as a building block for larger structures. Through evolution, we introduce the cell components and activities that are to be treated in detail, in broadly similar sequence, in the chapters that follow. Beginning with the origins of the first cell on earth, we consider how the properties of certain types of large molecules allow hereditary information to be transmitted and expressed and permit evolution to occur. Enclosed in a membrane, these molecules provide the essentials of a self-replicating cell. Following this, we describe the major transition that occurred in the course of evolution, from small bacteriumlike cells to much larger and more complex cells such as are found in present-day plants and animals. Lastly, we suggest ways in which single free-living cells might have given rise to large multicellular organisms, becoming specialized and cooperating in the formation of such intricate organs as the brain.

Clearly, there are dangers in introducing the cell through its evolution: the large gaps in our knowledge can be filled only by speculations that are liable to be wrong in many details. We cannot go back in time to witness the unique molecular events that took place billions of years ago. But those ancient events have left many traces for us to analyze. Ancestral plants, animals, and even bacteria are preserved as fossils. Even more important, every modern organism provides evidence of the character of living organisms in the past. Present-day biological molecules, in particular, are a rich source of information about the course of evolution, revealing fundamental similarities between the most disparate of living organisms and allowing us to map out the differences between them on an objective universal scale. These molecular similarities and differences present us with a problem like that which confronts the literary scholar who seeks to establish the original text of an ancient author by comparing a mass of variant manuscripts that have been corrupted through repeated copying and editing. The task is hard, and the evidence is incomplete, but it is possible at least to make intelligent guesses about the major stages in the evolution of living cells.

From Molecules to the First Cell [1]

Simple Biological Molecules Can Form Under Prebiotic Conditions [1, 2]

The conditions that existed on the earth in its first billion years are still a matter of dispute. Was the surface initially molten? Did the atmosphere contain ammonia, or methane? Everyone seems to agree, however, that the earth was a violent place with volcanic eruptions, lightning, and torrential rains. There was little if any free oxygen and no layer of ozone to absorb the ultraviolet radiation from the sun. The radiation, by its photochemical action, may have helped to keep the atmosphere rich in reactive molecules and far from chemical equilibrium.

Simple organic molecules (that is, molecules containing carbon) are likely to have been produced under such conditions. The best evidence for this comes from laboratory experiments. If mixtures of gases such as CO_2, CH_4, NH_3, and H_2 are heated with water and energized by electrical discharge or by ultraviolet radiation, they react to form small organic molecules—usually a rather small selection, each made in large amounts (Figure 1–1). Among these products are compounds, such as hydrogen cyanide (HCN) and formaldehyde (HCHO), that readily undergo further reactions in aqueous solution (Figure 1–2). Most important, representatives of most of the major classes of small organic molecules found in cells are generated, including *amino acids, sugars,* and the *purines* and *pyrimidines* required to make *nucleotides.*

Although such experiments cannot reproduce the early conditions on the earth exactly, they make it plain that the formation of organic molecules is surprisingly easy. And the developing earth had immense advantages over any human experimenter; it was very large and could produce a wide spectrum of conditions. But above all, it had much more time—tens to hundreds of millions of years. In such circumstances it seems very likely that, at some time and place, many of the simple organic molecules found in present-day cells accumulated in high concentrations.

Complex Chemical Systems Can Develop in an Environment That Is Far from Chemical Equilibrium

Simple organic molecules such as amino acids and nucleotides can associate to form *polymers.* One amino acid can join with another by forming a peptide bond, and two nucleotides can join together by a phosphodiester bond. The repetition of these reactions leads to linear polymers known as polypeptides and polynucleotides, respectively. In present-day living cells, large polypeptides—known as **proteins**—and polynucleotides—in the form of both **ribonucleic acids (RNA)** and **deoxyribonucleic acids (DNA)**—are commonly viewed as the most important constituents. A restricted set of 20 amino acids constitute the universal building blocks of the proteins, while RNA and DNA molecules are constructed from just

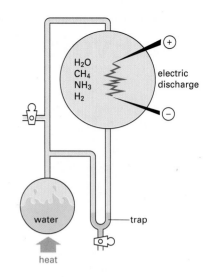

Figure 1–1 A typical experiment simulating conditions on the primitive earth. Water is heated in a closed apparatus containing CH_4, NH_3, and H_2, and an electric discharge is passed through the vaporized mixture. Organic compounds accumulate in the U-tube trap.

HCHO	formaldehyde
HCOOH	formic acid
HCN	hydrogen cyanide
CH₃COOH	acetic acid
NH₂CH₂COOH	glycine
CH₃CHCOOH OH	lactic acid
NH₂CHCOOH CH₃	alanine
NH—CH₂COOH CH₃	sarcosine
NH₂—C—NH₂ O	urea
NH₂CHCOOH CH₂ COOH	aspartic acid

Figure 1–2 A few of the compounds that might form in the experiment described in Figure 1–1. Compounds shown in color are important components of present-day living cells.

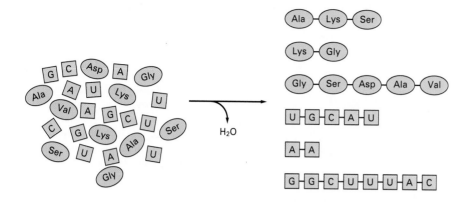

Figure 1–3 Formation of polynucleotides and polypeptides. Nucleotides of four kinds (here represented by the single letters A, U, G, and C) can undergo spontaneous polymerization with the loss of water. The product is a mixture of polynucleotides that are random in length and sequence. Similarly, amino acids of different types, symbolized here by three-letter abbreviated names, can polymerize with one another to form polypeptides. Present-day proteins are built from a standard set of 20 types of amino acids.

four types of nucleotides each. Although it is uncertain why these particular sets of monomers were selected for biosynthesis in preference to others that are chemically similar, we shall see that the chemical properties of the corresponding polymers suit them especially well for their specific roles in the cell.

The earliest polymers may have formed in any of several ways—for example, by the heating of dry organic compounds or by the catalytic activity of high concentrations of inorganic polyphosphates or other crude mineral catalysts. Under laboratory conditions the products of similar reactions are polymers of variable length and random sequence in which the particular amino acid or nucleotide added at any point depends mainly on chance (Figure 1–3). Once a polymer has formed, however, it can itself influence subsequent chemical reactions by acting as a catalyst.

The origin of life requires that in an assortment of such molecules there must have been some possessing, if only to a small extent, a crucial property: the ability to catalyze reactions that lead, directly or indirectly, to production of more molecules of the catalyst itself. Production of catalysts with this special self-promoting property would be favored, and the molecules most efficient in aiding their own production would divert raw materials from the production of other substances. In this way one can envisage the gradual development of an increasingly complex chemical system of organic monomers and polymers that function together to generate more molecules of the same types, fueled by a supply of simple raw materials in the environment. Such an **autocatalytic** system would have some of the properties we think of as characteristic of living matter: it would comprise a far from random selection of interacting molecules; it would tend to reproduce itself; it would compete with other systems dependent on the same feedstocks; and if deprived of its feedstocks or maintained at a wrong temperature that upsets the balance of reaction rates, it would decay toward chemical equilibrium and "die."

But what molecules could have had such autocatalytic properties? In present-day living cells the most versatile catalysts are polypeptides, composed of many different amino acids with chemically diverse side chains and, consequently, able to adopt diverse three-dimensional forms that bristle with reactive sites. But although polypeptides are versatile as catalysts, there is no known way in which one such molecule can reproduce itself by directly specifying the formation of another of precisely the same sequence.

Polynucleotides Are Capable of Directing Their Own Synthesis [3]

Polynucleotides have properties that contrast with those of polypeptides. They have more limited capabilities as catalysts, but they can directly guide the formation of exact copies of their own sequence. This capacity depends on **complementary pairing** of nucleotide subunits, which enables one polynucleotide to act

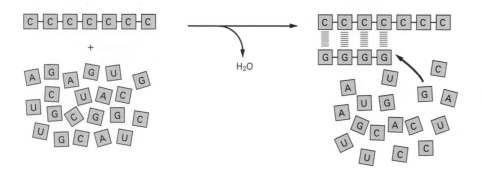

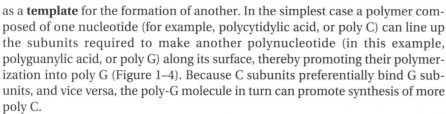

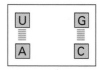

Figure 1–4 **Polynucleotides as templates.** Preferential binding occurs between pairs of nucleotides (G with C and U with A) by relatively weak chemical bonds (*above*). This pairing enables one polynucleotide to act as a template for the synthesis of another (*left*).

as a **template** for the formation of another. In the simplest case a polymer composed of one nucleotide (for example, polycytidylic acid, or poly C) can line up the subunits required to make another polynucleotide (in this example, polyguanylic acid, or poly G) along its surface, thereby promoting their polymerization into poly G (Figure 1–4). Because C subunits preferentially bind G subunits, and vice versa, the poly-G molecule in turn can promote synthesis of more poly C.

Consider now a polynucleotide with a more complex sequence of subunits—specifically, a molecule of RNA strung together from four types of nucleotides, containing the bases uracil (U), adenine (A), cytosine (C), and guanine (G), arranged in some particular sequence. Because of complementary pairing between the bases A and U and between the bases G and C, this molecule, when added to a mixture of activated nucleotides under suitable conditions, will line them up for polymerization in a sequence complementary to its own. The resulting new RNA molecule will be rather like a mold of the original, with each A in the original corresponding to a U in the copy and so on. The sequence of nucleotides in the original RNA strand contains information that is, in essence, preserved in the newly formed complementary strands: a second round of copying, with the complementary strand as a template, restores the original sequence (Figure 1–5).

Such **complementary templating** mechanisms are elegantly simple, and they lie at the heart of information transfer processes in biological systems. Genetic information contained in every cell is encoded in the sequences of nucleotides in its polynucleotide molecules, and this information is passed on (inherited) from generation to generation by means of complementary base-pairing interactions.

Templating mechanisms, however, require additional catalysts to promote polymerization; without these the process is slow and inefficient and other, competing reactions prevent the formation of accurate replicas. Today, the catalytic functions that polymerize nucleotides are provided by highly specialized catalytic proteins—that is, by *enzymes*. In the "prebiotic soup" primitive polypeptides might perhaps have provided some catalytic help. But molecules with the appropriate catalytic specificity would have remained rare unless the RNA itself were able somehow to reciprocate and favor *their* production. We shall come back to the reciprocal relationship between RNA synthesis and protein synthesis, which

Figure 1–5 **Replication of a polynucleotide sequence (here an RNA molecule).** In step 1 the original RNA molecule acts as a template to form an RNA molecule of complementary sequence. In step 2 this complementary RNA molecule itself acts as a template, forming RNA molecules of the original sequence. Since each templating molecule can produce many copies of the complementary strand, these reactions can result in the "multiplication" of the original sequence.

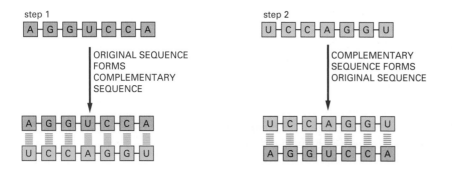

is crucially important in all living cells. But let us first consider what could be done with RNA itself, for RNA molecules can have a variety of catalytic properties, besides serving as templates for their own replication. In particular, an RNA molecule with an appropriate nucleotide sequence can act as catalyst for the accurate replication of another RNA molecule—the template—whose sequence can be arbitrary. The special versatility of RNA molecules is thought to have enabled them to play a central role in the origin of life.

Self-replicating Molecules Undergo Natural Selection [3, 4]

RNA molecules are not just strings of symbols that carry information in an abstract way. They also have chemical personalities that affect their behavior. In particular, the specific sequence of nucleotides governs how the molecule folds up in solution. Just as the nucleotides in a polynucleotide can pair with free complementary nucleotides in their environment to form a new polymer, so they can pair with complementary nucleotide residues within the polymer itself. A sequence GGGG in one part of a polynucleotide chain can form a relatively strong association with a CCCC sequence in another region of the same molecule. Such associations produce complex three-dimensional patterns of folding, and the molecule as a whole takes on a specific shape that depends entirely on the sequence of its nucleotides (Figure 1–6).

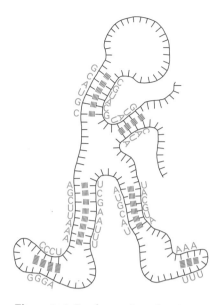

Figure 1–6 Conformation of an RNA molecule. Nucleotide pairing between different regions of the same polynucleotide (RNA) chain causes the molecule to adopt a distinctive shape.

The three-dimensional folded structure of a polynucleotide affects its stability, its actions on other molecules, and its ability to replicate, so that not all polynucleotide shapes will be equally successful in a replicating mixture. Moreover, errors inevitably occur in any copying process, and imperfect copies of the originals will be propagated. With repeated replication, therefore, new variant sequences of nucleotides will be continually generated. Thus, in laboratory studies, replicating systems of RNA molecules have been shown to undergo a form of natural selection in which different favorable sequences eventually predominate, depending on the exact conditions. Most important, RNA molecules can be selected for the ability to bind almost any other molecule specifically. This too has been shown, in experiments *in vitro* that begin with a preparation of short RNA molecules with random nucleotide sequences manufactured artificially. These are passed down a column packed with beads to which some chosen substance is bonded. RNA molecules that fail to bind to the chosen substance are washed through the column and discarded; those few that bind are retained and used as templates to direct production of multiple copies of their own sequences. This new RNA preparation, enriched in sequences that bind the chosen substance, is then used as the starting material for a repetition of the procedure. After several such cycles of selection and reproduction, the RNA is found to consist of multiple copies of a relatively small number of sequences, each of which binds the test substance quite specifically.

An RNA molecule therefore has two special characteristics: it carries information encoded in its nucleotide sequence that it can pass on by the process of replication, and it has a specific folded structure that enables it to interact selectively with other molecules and determines how it will respond to the ambient conditions. These two features—one informational, the other functional—are the two properties essential for evolution. The nucleotide sequence of an RNA molecule is analogous to the **genotype**—the hereditary information—of an organism. The folded three-dimensional structure is analogous to the **phenotype**—the expression of the genetic information on which natural selection operates.

Specialized RNA Molecules Can Catalyze Biochemical Reactions [5]

Natural selection depends on the environment, and for a replicating RNA molecule a critical component of the environment is the set of other RNA molecules in the mixture. Besides acting as templates for their own replication, these can catalyze the breakage and formation of covalent bonds between nucleotides. For example, some specialized RNA molecules can catalyze a change in other RNA molecules, cutting the nucleotide sequence at a particular point; and other types of RNA molecules spontaneously cut out a portion of their own nucleotide sequence and rejoin the cut ends (a process known as self-splicing). Each RNA-catalyzed reaction depends on a specific arrangement of atoms that forms on the surface of the catalytic RNA molecule (the *ribozyme*), causing particular chemical groups on one or more of its nucleotides to become highly reactive.

Certain catalytic activities would have had a cardinal importance in the primordial soup. Consider in particular an RNA molecule that helps to catalyze the process of templated polymerization, taking any given RNA molecule as template. (This ribozyme activity has been directly demonstrated *in vitro*, albeit in a rudimentary form.) Such a molecule, by acting on copies of itself, can replicate with heightened speed and efficiency (Figure 1–7A). At the same time, it can promote the replication of any other type of RNA molecules in its neighborhood (Figure 1–7B). Some of these may have catalytic actions that help or hinder the survival or replication of RNA in other ways. If beneficial effects are reciprocated, the different types of RNA molecules, specialized for different activities, may evolve into a cooperative system that replicates with unusually great efficiency.

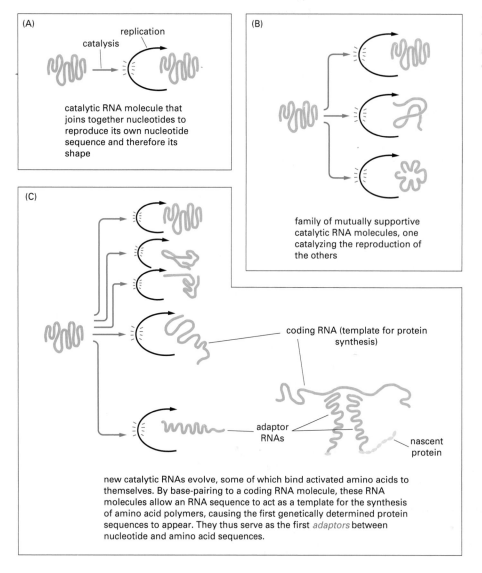

(A)
replication
catalysis

catalytic RNA molecule that joins together nucleotides to reproduce its own nucleotide sequence and therefore its shape

(B)

family of mutually supportive catalytic RNA molecules, one catalyzing the reproduction of the others

(C)

coding RNA (template for protein synthesis)

adaptor RNAs

nascent protein

new catalytic RNAs evolve, some of which bind activated amino acids to themselves. By base-pairing to a coding RNA molecule, these RNA molecules allow an RNA sequence to act as a template for the synthesis of amino acid polymers, causing the first genetically determined protein sequences to appear. They thus serve as the first *adaptors* between nucleotide and amino acid sequences.

Figure 1–7 Three successive steps in the evolution of a self-replicating system of RNA molecules capable of directing protein synthesis.

Information Flows from Polynucleotides to Polypeptides [6]

There are strong suggestions, therefore, that between 3.5 and 4 billion years ago, somewhere on earth, self-replicating systems of RNA molecules, mixed with other organic molecules including simple polypeptides, began the process of evolution. Systems with different sets of polymers competed for the available precursor materials to construct copies of themselves, just as organisms now compete; success depended on the accuracy and the speed with which the copies were made and on the stability of those copies.

However, as we emphasized earlier, while the structure of polynucleotides is well suited for information storage and replication, their catalytic abilities are limited by comparison with those of polypeptides, and efficient replication of polynucleotides in modern cells is absolutely dependent on proteins. At the origin of life any polynucleotide that helped guide the synthesis of a useful polypeptide in its environment would have had a great advantage in the evolutionary struggle for survival.

But how could the information encoded in a polynucleotide specify the sequence of a polymer of a different type? Clearly, the polynucleotides must act as catalysts to join selected amino acids together. In present-day organisms a collaborative system of RNA molecules plays a central part in directing the synthesis of polypeptides—that is, **protein synthesis**—but the process is aided by other proteins synthesized previously. The biochemical machinery for protein synthesis is remarkably elaborate. One RNA molecule carries the genetic information for a particular polypeptide in the form of a code, while other RNA molecules act as adaptors, each binding a specific amino acid. These two types of RNA molecules form complementary base pairs with one another to enable sequences of nucleotides in the coding RNA molecule to direct the incorporation of specific amino acids held on the adaptor RNAs into a growing polypeptide chain. Precursors to these two types of RNA molecules presumably directed the first protein synthesis without the aid of proteins (Figure 1–7C).

Today, these events in the assembly of new proteins take place on the surface of *ribosomes*—complex particles composed of several large RNA molecules of yet another class, together with more than 50 different types of protein. In Chapter 5 we shall see that the ribosomal RNA in these particles plays a central catalytic role in the process of protein synthesis and forms more than 60% of the ribosome's mass. At least in evolutionary terms, it appears to be the fundamental component of the ribosome.

It seems likely, then, that RNA guided the primordial synthesis of proteins, perhaps in a clumsy and primitive fashion. In this way RNA was able to create tools—in the form of proteins—for more efficient biosynthesis, and some of these could have been put to use in the replication of RNA and in the process of tool production itself.

The synthesis of specific proteins under the guidance of RNA required the evolution of a code by which the polynucleotide sequence specifies the amino acid sequence that makes up the protein. This code—the *genetic code*—is spelled out in a "dictionary" of three-letter words: different triplets of nucleotides encode specific amino acids. The code seems to have been selected arbitrarily (subject to some constraints, perhaps); yet it is virtually the same in all living organisms. This strongly suggests that all present-day cells have descended from a single line of primitive cells that evolved the mechanism of protein synthesis.

Membranes Defined the First Cell [7]

One of the crucial events leading to the formation of the first cell must have been the development of an outer membrane. For example, the proteins synthesized under the control of a certain species of RNA would not facilitate reproduction of that species of RNA unless they remained in the neighborhood of the RNA;

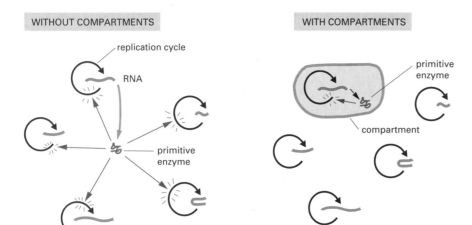

WITHOUT COMPARTMENTS

replication cycle

RNA

primitive enzyme

WITH COMPARTMENTS

primitive enzyme

compartment

Figure 1–8 Evolutionary significance of cell-like compartments. In a mixed population of self-replicating RNA molecules capable of influencing protein synthesis (as illustrated in Figure 1–7), any improved form of RNA that is able to promote formation of a more useful protein must share this protein with its neighboring competitors. However, if the RNA is enclosed within a compartment, such as a lipid membrane, then any protein the RNA causes to be made is retained for its own use; the RNA can therefore be selected on the basis of its guiding production of a better protein.

moreover, as long as these proteins were free to diffuse among the population of replicating RNA molecules, they could benefit equally any competing species of RNA that might be present. If a variant RNA arose that made a superior type of enzyme, the new enzyme could not contribute *selectively* to the survival of the variant RNA in its competition with its fellows. Selection of RNA molecules according to the quality of the proteins they generated could not occur efficiently until some form of compartment evolved to contain the proteins made by an RNA molecule and thereby make them available only to the RNA that had generated them (Figure 1–8).

The need for containment is easily fulfilled by another class of molecules that has the simple physicochemical property of being *amphipathic,* that is, consisting of one part that is hydrophobic (water insoluble) and another part that is hydrophilic (water soluble). When such molecules are placed in water, they aggregate, arranging their hydrophobic portions as much in contact with one another as possible and their hydrophilic portions in contact with the water. Amphipathic molecules of appropriate shape spontaneously aggregate to form *bilayers,* creating small closed vesicles whose aqueous contents are isolated from the external medium (Figure 1–9). The phenomenon can be demonstrated in a test tube by simply mixing phospholipids and water together: under appropriate conditions, small vesicles will form. All present-day cells are surrounded by a **plasma membrane** consisting of amphipathic molecules—mainly phospholipids—in this configuration; in cell membranes, the lipid bilayer also contains amphipathic proteins. In the electron microscope such membranes appear as sheets about 5 nm thick, with a distinctive three-layered appearance due to the tail-to-tail packing of the phospholipid molecules.

Presumably, the first membrane-bounded cells were formed by spontaneous assembly of phospholipid molecules from the prebiotic soup, enclosing a self-replicating mixture of RNA and other molecules. It is not clear at what point in the evolution of biological catalysts and protein synthesis this first occurred. In any case, once RNA molecules were sealed within a closed membrane, they could begin to evolve in earnest as carriers of genetic instructions: they could be selected not merely on the basis of their own structure, but also according to their effect on the other molecules in the same compartment. The nucleotide sequences of the RNA molecules could now be expressed in the character of a unitary living cell.

All Present-Day Cells Use DNA as Their Hereditary Material [3, 6, 8]

The picture we have presented is, of course, speculative: there are no fossil records that trace the origins of the first cell. Nevertheless, there is persuasive

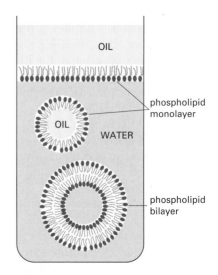

OIL

phospholipid monolayer

OIL

WATER

phospholipid bilayer

Figure 1–9 Formation of membranes by phospholipids. Because these molecules have hydrophilic heads and lipophilic tails, they will align themselves at an oil-water interface with their heads in the water and their tails in the oil. In water they will associate to form closed bilayer vesicles in which the lipophilic tails are in contact with one another and the hydrophilic heads are exposed to the water.

evidence from present-day organisms and from experiments that the broad features of this evolutionary story are correct. The prebiotic synthesis of small molecules, the self-replication of catalytic RNA molecules, the translation of RNA sequences into amino acid sequences, and the assembly of lipid molecules to form membrane-bounded compartments—all presumably occurred to generate primitive cells 3.5 to 4 billion years ago.

It is useful to compare these early cells with the simplest and smallest present-day cells, the **mycoplasmas.** Mycoplasmas are small bacteria of a degenerate type that normally lead a parasitic existence in close association with animal or plant cells (Figure 1–10). Some have a diameter of about 0.3 μm and contain only enough nucleic acid to direct the synthesis of about 400 different proteins. Some of these proteins are enzymes, some are structural; some lie in the cell's interior, others are embedded in its membrane. Together they synthesize essential small molecules that are not available in the environment, redistribute the energy needed to drive biosynthetic reactions, and maintain appropriate conditions inside the cell.

The first cells on the earth were presumably less sophisticated than a mycoplasma and less efficient in reproducing themselves. There was, however, a more fundamental difference between these primitive cells and a mycoplasma, or indeed any other present-day cell: the hereditary information in all cells alive today is stored in DNA rather than in the RNA that is thought to have stored the hereditary information during the earliest stages of evolution. Both types of polynucleotides are found in present-day cells, but they function in a collaborative manner, each having evolved to perform specialized tasks. Small chemical differences fit the two kinds of molecules for distinct functions. DNA acts as the permanent repository of genetic information, and, unlike RNA, it is found in cells principally in a double-stranded form, composed of a pair of complementary polynucleotide molecules. This double-stranded structure makes DNA in cells more robust and stable than RNA; it also makes DNA relatively easy to replicate (as will be explained in Chapter 3) and permits a repair mechanism to operate that uses the intact strand as a template for the correction or repair of the associated damaged strand. DNA guides the synthesis of specific RNA molecules, again by the principle of complementary base-pairing, though now this pairing is between slightly different types of nucleotides. The resulting single-stranded RNA molecules then perform two primeval functions: they direct protein synthesis both as coding RNA molecules (*messenger* RNAs) and as RNA catalysts (*ribosomal* and other nonmessenger RNAs).

The suggestion, in short, is that RNA preceded DNA in evolution, having both genetic and catalytic properties; eventually, DNA took over the primary genetic function and proteins became the major catalysts, while RNA remained primarily as the intermediary connecting the two (Figure 1–11). With the advent of DNA cells were enabled to become more complex, for they could then carry and transmit an amount of genetic information greater than that which could be stably maintained in RNA molecules.

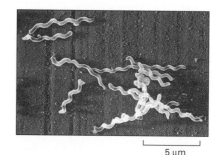

Figure 1–10 *Spiroplasma citrii*, a mycoplasma that grows in plant cells. (Courtesy of Jeremy Burgess.)

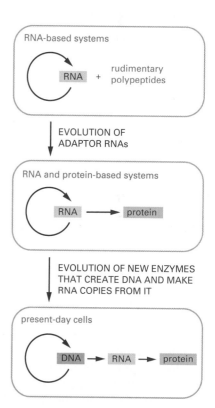

Figure 1–11 **Suggested stages of evolution from simple self-replicating systems of RNA molecules to present-day cells.** Today, DNA is the repository of genetic information and RNA acts largely as a go-between to direct protein synthesis.

Summary

Living cells probably arose on earth about 3.5 billion years ago by spontaneous reactions between molecules in an environment that was far from chemical equilibrium. From our knowledge of present-day organisms and the molecules they contain, it seems likely that the development of the directly autocatalytic mechanisms fundamental to living systems began with the evolution of families of RNA molecules that could catalyze their own replication. With time, one of these families of cooperating RNA catalysts developed the ability to direct synthesis of polypeptides. Finally, as the accumulation of additional protein catalysts allowed more efficient and complex cells to evolve, the DNA double helix replaced RNA as a more stable molecule for storing the increased amounts of genetic information required by such cells.

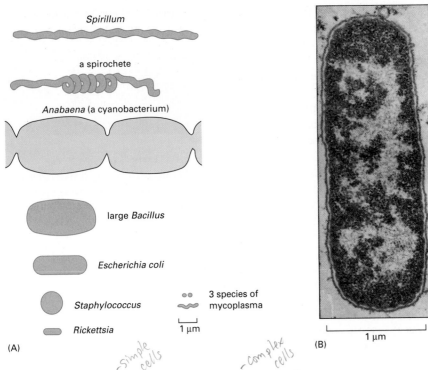

Spirillum

a spirochete

Anabaena (a cyanobacterium)

large *Bacillus*

Escherichia coli

Staphylococcus

3 species of mycoplasma

1 μm

Rickettsia

1 μm

(A)

(B)

1 μm

~Simple cells

~complex cells

From Procaryotes to Eucaryotes [9]

It is thought that all organisms living now on earth derive from a single primordial cell born more than 3 billion years ago. This cell, out-reproducing its competitors, took the lead in the process of cell division and evolution that eventually covered the earth with green, changed the composition of its atmosphere, and made it the home of intelligent life. The family resemblances among all organisms seem too strong to be explained in any other way. One important landmark along this evolutionary road occurred about 1.5 billion years ago, when there was a transition from small cells with relatively simple internal structures—the so-called **procaryotic** cells, which include the various types of bacteria—to a flourishing of larger and radically more complex *eucaryotic* cells such as are found in higher animals and plants.

Procaryotic Cells Are Structurally Simple but Biochemically Diverse [10]

Bacteria are the simplest organisms found in most natural environments. They are spherical or rod-shaped cells, commonly several micrometers in linear dimension (Figure 1–12). They often possess a tough protective coat, called a *cell wall*, beneath which a plasma membrane encloses a single cytoplasmic compartment containing DNA, RNA, proteins, and small molecules. In the electron microscope this cell interior appears as a matrix of varying texture without any obvious organized internal structure (see Figure 1–12B).

Bacteria are small and can replicate quickly, simply dividing in two by *binary fission.* When food is plentiful, "survival of the fittest" generally means survival of those that can divide the fastest. Under optimal conditions a single procaryotic cell can divide every 20 minutes and thereby give rise to 5 billion cells (approximately equal to the present human population on earth) in less than 11 hours. The ability to divide quickly enables populations of bacteria to adapt rapidly to changes in their environment. Under laboratory conditions, for example, a population of bacteria maintained in a large vat will evolve within a few weeks by spontaneous mutation and natural selection to utilize new types of sugar molecules as carbon sources.

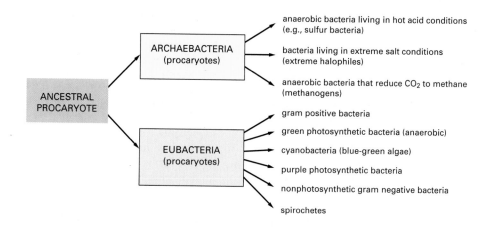

Figure 1–13 **Family relationships between present-day bacteria.** Arrows indicate probable paths of evolution. The origin of eucaryotic cells is discussed later in the text.

In nature bacteria live in an enormous variety of ecological niches, and they show a corresponding richness in their underlying biochemical composition. Two distantly related groups can be recognized: the *eubacteria,* which are the commonly encountered forms that inhabit soil, water, and larger living organisms; and the *archaebacteria,* which are found in such incommodious environments as bogs, ocean depths, salt brines, and hot acid springs (Figure 1–13).

There are species of bacteria that can utilize virtually any type of organic molecule as food, including sugars, amino acids, fats, hydrocarbons, polypeptides, and polysaccharides. Some are even able to obtain their carbon atoms from CO_2 and their nitrogen atoms from N_2. Despite their relative simplicity, bacteria have existed for longer than any other organisms and still are the most abundant type of cell on earth.

Metabolic Reactions Evolve [10, 11]

A bacterium growing in a salt solution containing a single type of carbon source, such as glucose, must carry out a large number of chemical reactions. Not only must it derive from the glucose the chemical energy needed for many vital processes, it must also use the carbon atoms of glucose to synthesize every type of organic molecule that the cell requires. These reactions are catalyzed by hundreds of enzymes working in reaction "chains" so that the product of one reaction is the substrate for the next; such enzymatic chains, called *metabolic pathways,* will be discussed in the following chapter.

Originally, when life began on earth, there was probably little need for such elaborate metabolic reactions. Cells with relatively simple chemistry could survive and grow on the molecules in their surroundings. But as evolution proceeded, competition for these limited natural resources would have become more intense. Organisms that had developed enzymes to manufacture useful organic molecules more efficiently and in new ways would have had a strong selective advantage. In this way the complement of enzymes possessed by cells is thought to have gradually increased, generating the metabolic pathways of present organisms. Two plausible ways in which a metabolic pathway could arise in evolution are illustrated in Figure 1–14.

If metabolic pathways evolved by the sequential addition of new enzymatic reactions to existing ones, the most ancient reactions should, like the oldest rings in a tree trunk, be closest to the center of the "metabolic tree," where the most fundamental of the basic molecular building blocks are synthesized. This position in metabolism is firmly occupied by the chemical processes that involve sugar phosphates, among which the most central of all is probably the sequence of reactions known as **glycolysis,** by which glucose can be degraded in the absence of oxygen (that is, *anaerobically*). The oldest metabolic pathways would have had to be anaerobic because there was no free oxygen in the atmosphere

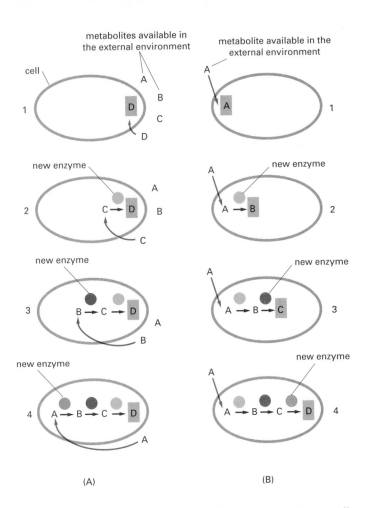

Figure 1–14 **Two possible ways in which metabolic pathways might have evolved.** (A) The cell on the left is provided with a supply of related substances (A, B, C, and D) produced by prebiotic synthesis. One of these, substance D, is metabolically useful. As the cell exhausts the available supply of D, a selective advantage is obtained by the evolution of a new enzyme that is able to produce D from the closely related substance C. Fundamentally important metabolic pathways may have evolved by a series of similar steps. (B) On the right, a metabolically useful compound A is available in abundance. An enzyme appears in the course of evolution that, by chance, has the ability to convert substance A to substance B. Other changes then occur within the cell that enable it to make use of the new substance. The appearance of further enzymes can build up a long chain of reactions.

of the primitive earth. Glycolysis occurs in virtually every living cell and drives the formation of the compound *adenosine triphosphate*, or *ATP*, which is used by all cells as a versatile source of chemical energy. Certain *thioester* compounds play a fundamental role in the energy-transfer reactions of glycolysis and in a host of other basic biochemical processes in which two organic molecules (a thiol and a carboxylic acid) are joined by a high-energy bond involving sulfur (Figure 1–15). It has been argued that this simple but powerful chemical device is a relic of prebiotic processes, reflecting the reactions that occurred in the sulfurous, volcanic environment of the early earth, before even RNA had begun to evolve.

Linked to the core reactions of glycolysis are hundreds of other chemical processes. Some of these are responsible for the synthesis of small molecules, many of which in turn are utilized in further reactions to make the large polymers specific to the organism. Other reactions are used to degrade complex molecules, taken in as food, into simpler chemical units. One of the most striking features of these metabolic reactions is that they take place similarly in all kinds of organisms, suggesting an extremely ancient origin.

Evolutionary Relationships Can Be Deduced by Comparing DNA Sequences [12]

The enzymes that catalyze the fundamental metabolic reactions, while continuing to serve the same essential functions, have undergone progressive modifica-

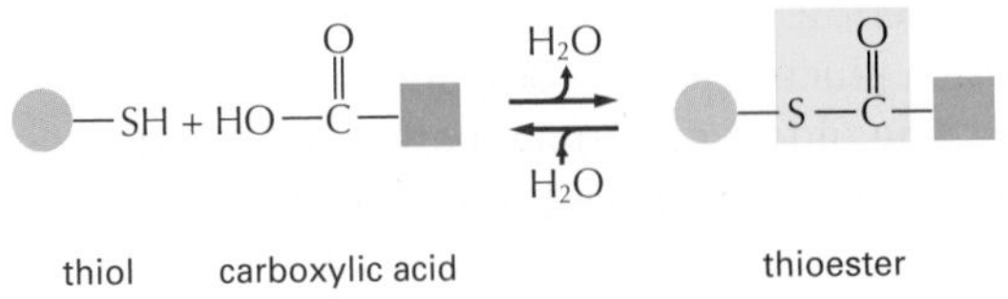

Figure 1–15 **The thioester bond.**

thiol   carboxylic acid   thioester

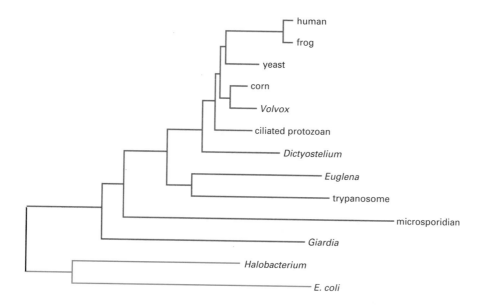

Figure 1–16 Evolutionary relationships of organisms deduced from the nucleotide sequences of their small-subunit ribosomal RNA genes. These genes contain highly conserved sequences, which change so slowly that they can be used to measure phylogenetic relationships spanning the entire range of living organisms. The data suggest that the plant, animal, and fungal lineages diverged from a common ancestor relatively late in the history of eucaryotic cells. *Halobacterium* and *E. coli* are procaryotes; the rest are eucaryotes. *Giardia*, microsporidians, trypanosomes, *Euglena*, and ciliated protozoans are protists (single-cell eucaryotes). (Adapted from M.L. Sogin, J.H. Gunderson, H.J. Elwood, R.A. Alonso, and D.A. Peattie, *Science* 243:75–77, 1989. © 1989 the AAAS.)

tions as organisms have evolved into divergent forms. For this reason the amino acid sequence of the same type of enzyme in different living species provides a valuable indication of the evolutionary relationship between these species. The evidence obtained closely parallels that from other sources, such as the fossil record. An even richer source of information is locked in the living cell in the sequences of nucleotides in DNA, and modern methods of analysis allow these *DNA sequences* to be determined in large numbers and compared between species. Comparisons of highly conserved sequences, which have a central function and therefore change only slowly during evolution, can reveal relationships between organisms that diverged long ago (Figure 1–16), while very rapidly evolving sequences can be used to determine how more closely related species evolved. It is expected that continued application of these methods will enable the course of evolution to be followed with unprecedented accuracy.

Cyanobacteria Can Fix CO_2 and N_2 [13]

As competition for the raw materials for organic syntheses intensified, a strong selective advantage would have been gained by any organisms able to utilize carbon and nitrogen atoms (in the form of CO_2 and N_2) directly from the atmosphere. But while they are abundantly available, CO_2 and N_2 are also very stable. It therefore requires a large amount of energy as well as a number of complicated chemical reactions to convert them to a usable form—that is, into organic molecules such as simple sugars.

In the case of CO_2 the major mechanism that evolved to achieve this transformation was **photosynthesis,** in which radiant energy captured from the sun drives the conversion of CO_2 into organic compounds. The interaction of sunlight with a pigment molecule, *chlorophyll,* excites an electron to a more highly energized state. As the electron drops back to a lower energy level, the energy it gives up drives chemical reactions that are facilitated and directed by protein molecules.

One of the first sunlight-driven reactions was probably the generation of "reducing power." The carbon and nitrogen atoms in atmospheric CO_2 and N_2 are in an oxidized and inert state. One way to make them more reactive, so that they participate in biosynthetic reactions, is to reduce them—that is, to give them a larger number of electrons. This is achieved in several steps. In the first step electrons are removed from a poor electron donor and transferred to a strong electron donor by chlorophyll in a reaction that is driven by sunlight. The strong electron donor is then used to reduce CO_2 or N_2. Comparison of the mechanisms of photosynthesis in various present-day bacteria suggests that one of the first

From Procaryotes to Eucaryotes

sources of electrons was H_2S, from which the primary waste product would have been elemental sulfur. Later the more difficult but ultimately more rewarding process of obtaining electrons from H_2O was accomplished, and O_2 was released in large amounts as a waste product.

Cyanobacteria (also known as blue-green algae) are today a major route by which both carbon and nitrogen are converted into organic molecules and thus enter the *biosphere*. They include the most self-sufficient organisms that now exist. Able to "fix" both CO_2 and N_2 into organic molecules, they are, to a first approximation, able to live on water, air, and sunlight alone; the mechanisms by which they do this have probably remained essentially constant for several billion years. Together with other bacteria that have some of these capabilities, they created the conditions in which more complex types of organisms could evolve: once one set of organisms had succeeded in synthesizing the whole gamut of organic cell components from inorganic raw materials, other organisms could subsist by feeding on the primary synthesizers and on their products.

Bacteria Can Carry Out the Aerobic Oxidation of Food Molecules [13]

Many people today are justly concerned about the environmental consequences of human activities. But in the past other organisms have caused revolutionary changes in the earth's environment (although very much more slowly). Nowhere is this more apparent than in the composition of the earth's atmosphere, which through oxygen-releasing photosynthesis was transformed from a mixture containing practically no molecular oxygen to one in which oxygen constitutes 21% of the total (Figure 1–17).

Since oxygen is an extremely reactive chemical that can interact with most cytoplasmic constituents, it must have been toxic to many early organisms, just as it is to many present-day anaerobic bacteria. However, this reactivity also provides a source of chemical energy, and, not surprisingly, this has been exploited by organisms during the course of evolution. By using oxygen, organisms are able to oxidize more completely the molecules they ingest. For example, in the absence of oxygen glucose can be broken down only to lactic acid or ethanol, the end products of anaerobic glycolysis. But in the presence of oxygen glucose can be completely degraded to CO_2 and H_2O. In this way much more energy can be derived from each gram of glucose. The energy released in **respiration**— the aerobic oxidation of food molecules—is used to drive the synthesis of ATP in much the same way that photosynthetic organisms produce ATP from the energy

Figure 1–17 Atmospheric oxygen and the course of evolution. The relationship between changes in atmospheric oxygen levels and some of the major stages that are believed to have occurred during the evolution of living organisms on earth. As indicated, geological evidence suggests that there was more than a billion-year delay between the rise of cyanobacteria (thought to be the first organisms to release oxygen) and the time that high oxygen levels began to accumulate in the atmosphere. This delay is thought to have been due largely to the rich supply of dissolved ferrous iron in the oceans, which reacted with the released oxygen to form enormous iron oxide deposits.

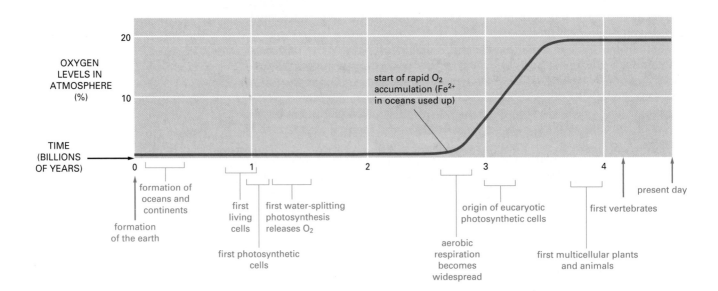

of sunlight. In both processes there is a series of electron-transfer reactions that generates an H⁺ gradient between the outside and inside of a membrane-bounded compartment; the H⁺ gradient then serves to drive the synthesis of the ATP. Today, respiration is used by the great majority of organisms, including most procaryotes.

Eucaryotic Cells Contain Several Distinctive Organelles [14]

As molecular oxygen accumulated in the atmosphere, what happened to the remaining anaerobic organisms with which life had begun? In a world that was rich in oxygen, which they could not use, they were at a severe disadvantage. Some, no doubt, became extinct. Others either developed a capacity for respiration or found niches in which oxygen was largely absent, where they could continue an anaerobic way of life. Others became predators or parasites on aerobic cells. And some, it seems, hit upon a strategy for survival more cunning and vastly richer in implications for the future: they are believed to have formed an intimate association with an aerobic type of cell, living with it in *symbiosis*. This is the most plausible explanation for the metabolic organization of present-day cells of the **eucaryotic** type (Panel 1–1, pp. 18–19) with which this book will be chiefly concerned.

Eucaryotic cells, by definition and in contrast to procaryotic cells, have a *nucleus* (*caryon* in Greek), which contains most of the cell's DNA, enclosed by a double layer of membrane (Figure 1–18). The DNA is thereby kept in a compartment separate from the rest of the contents of the cell, the **cytoplasm,** where most of the cell's metabolic reactions occur. In the cytoplasm, moreover, many distinctive *organelles* can be recognized. Prominent among these are two types of small bodies, the *chloroplasts* and *mitochondria* (Figures 1–19 and 1–20). Each of these is enclosed in its own double layer of membrane, which is chemically different from the membranes surrounding the nucleus. Mitochondria are an almost universal feature of eucaryotic cells, whereas chloroplasts are found only in those eucaryotic cells that are capable of photosynthesis—that is, in plants but not in animals or fungi. Both organelles almost certainly have a symbiotic origin.

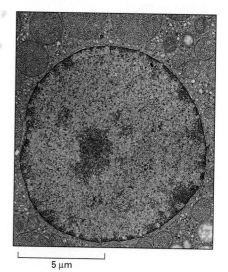

Figure 1–18 The cell nucleus. The nucleus contains most of the DNA of the eucaryotic cell. It is seen here in a thin section of a mammalian cell examined in the electron microscope. How and why the nucleus originated is uncertain; some speculations on its origin are presented in Figure 12–5. (Courtesy of Daniel S. Friend.)

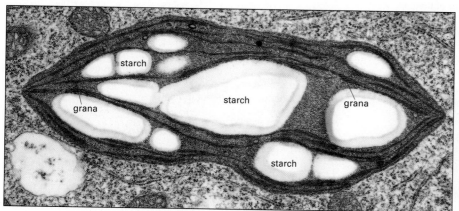

Figure 1–19 A chloroplast. The extensive system of internal membranes can be seen in this electron micrograph of a chloroplast in a moss cell. The flattened sacs of membrane contain chlorophyll and are arranged in stacks, or *grana*. This chloroplast also contains large accumulations of starch. (Courtesy of Jeremy Burgess.)

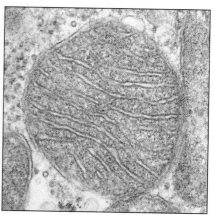

Figure 1–20 A mitochondrion. Mitochondria carry out the oxidative degradation of nutrient molecules in almost all eucaryotic cells. As seen in this electron micrograph, they possess a smooth outer membrane and a highly convoluted inner membrane. (Courtesy of Daniel S. Friend.)

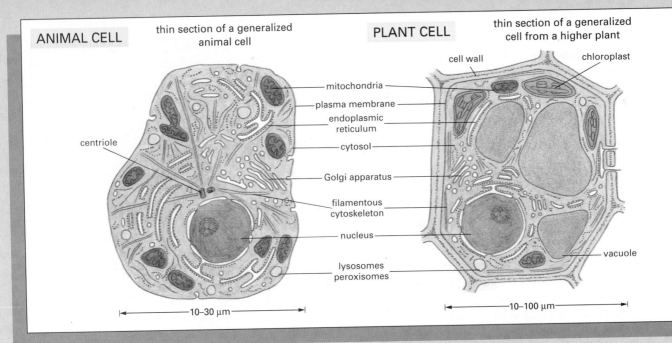

ANIMAL CELL — thin section of a generalized animal cell

PLANT CELL — thin section of a generalized cell from a higher plant

- cell wall
- chloroplast
- mitochondria
- plasma membrane
- endoplasmic reticulum
- centriole
- cytosol
- Golgi apparatus
- filamentous cytoskeleton
- nucleus
- lysosomes peroxisomes
- vacuole

10–30 μm

10–100 μm

THE MEMBRANE SYSTEM OF THE CELL

PLASMA MEMBRANE

The outer boundary of the cell is the plasma membrane, a continuous sheet of phospholipid molecules about 4–5 nm thick in which various proteins are embedded.

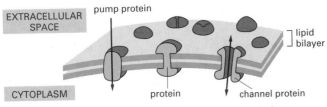

- EXTRACELLULAR SPACE
- pump protein
- lipid bilayer
- CYTOPLASM
- protein
- channel protein

Some of these proteins serve as pumps and channels for transporting specific molecules into and out of the cell.

ENDOPLASMIC RETICULUM

Flattened sheets, sacs, and tubes of membrane extend throughout the cytoplasm of eucaryotic cells, enclosing a large intracellular space. The ER membrane is in structural continuity with the outer membrane of the nuclear envelope, and it specializes in the synthesis and transport of lipids and membrane proteins.

The rough endoplasmic reticulum (rough ER) generally occurs as flattened sheets and is studded on its outer face with ribosomes engaged in protein synthesis.

- ribosomes
- nucleus
- lumen

The smooth endoplasmic reticulum (smooth ER) is generally more tubular and lacks attached ribosomes. It has a major function in lipid metabolism.

- lumen

GOLGI APPARATUS

A system of stacked, membrane-bounded, flattened sacs involved in modifying, sorting, and packaging macromolecules for secretion or for delivery to other organelles.

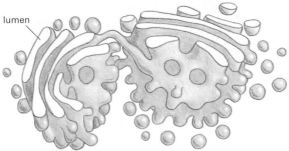

- lumen

Around the Golgi apparatus are numerous small membrane-bounded vesicles (50 nm and larger) that carry material between the Golgi apparatus and different compartments of the cell.

LYSOSOMES

0.2–0.5 μm

membrane-bounded vesicles that contain hydrolytic enzymes involved in intracellular digestions

PEROXISOMES

0.2–0.5 μm

membrane-bounded vesicles containing oxidative enzymes that generate and destroy hydrogen peroxide

NUCLEUS

The nucleus is the most conspicuous organelle in the cell. It is separated from the cytoplasm by an envelope consisting of two membranes. All of the chromosomal DNA is held in the nucleus, packed into chromatin fibers by its association with an equal mass of histone proteins. The nuclear contents communicate with the cytosol by means of openings in the nuclear envelope called nuclear pores.

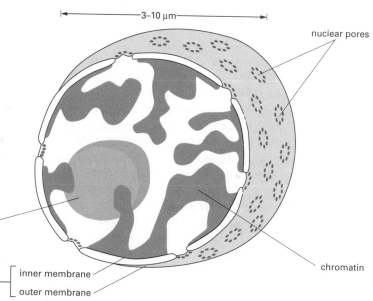

←——— 3–10 µm ———→

nuclear pores

nucleolus: a factory in the nucleus where the cell's ribosomes are assembled

chromatin

nuclear envelope — inner membrane
outer membrane

CYTOSKELETON

In the cytosol, arrays of protein filaments form networks that give the cell its shape and provide a basis for its movements. Three main kinds of cytoskeletal filaments are

1. microtubules

25-nm diameter

2. actin filaments

8-nm diameter

3. intermediate filaments

10-nm diameter

MITOCHONDRIA

About the size of bacteria, mitochondria are the power plants of all eucaryotic cells, harnessing energy obtained by combining oxygen with food molecules to make ATP.

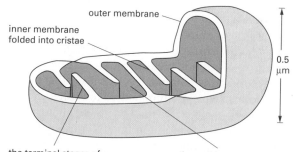

outer membrane

inner membrane folded into cristae

0.5 µm

the terminal stages of oxidation occur at the inner membrane

the matrix space contains a concentrated solution of many different enzymes

SPECIAL PLANT CELL ORGANELLES

Chloroplasts: These chlorophyll-containing plastids are double-membrane-bounded organelles found in all higher plants. An elaborate membrane system in the interior of the chloroplast contains the photosynthetic apparatus.

Vacuole: A very large single-membrane-bounded vesicle occupying up to 90% of the cell volume, the vacuole functions in space-filling and also intracellular digestion.

vacuole

plasma membrane

vacuole membrane (tonoplast)

Cell wall: Plant cells are surrounded by a rigid wall composed of tough fibrils of cellulose laid down in a matrix of other polysaccharides.

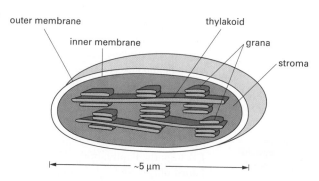

outer membrane

inner membrane

thylakoid

grana

stroma

←——— ~5 µm ———→

cell wall 0.1–10 µm

plasma membrane

Eucaryotic Cells Depend on Mitochondria for Their Oxidative Metabolism [15]

Mitochondria show many similarities to free-living procaryotic organisms: for example, they often resemble bacteria in size and shape, they contain DNA, they make protein, and they reproduce by dividing in two. By breaking up eucaryotic cells and separating their component parts, it is possible to show that mitochondria are responsible for respiration and that this process occurs nowhere else in the eucaryotic cell. Without mitochondria the cells of animals and fungi would be anaerobic organisms, depending on the relatively inefficient and antique process of glycolysis for their energy. Many present-day bacteria respire like mitochondria, and it seems probable that eucaryotic cells are descendants of primitive anaerobic organisms that survived, in a world that had become rich in oxygen, by engulfing aerobic bacteria—keeping them in symbiosis for the sake of their capacity to consume atmospheric oxygen and produce energy. Certain present-day microorganisms offer strong evidence of the feasibility of such an evolutionary sequence. There are several hundred species of single-celled eucaryotes that resemble the hypothetical ancestral eucaryote in that they live in oxygen-poor conditions (in the guts of animals, for example) and lack mitochondria altogether. Comparative nucleotide sequence analyses have revealed that at least two groups of these organisms, the *diplomonads* and the *microsporidia*, diverged very early from the line leading to other eucaryotic cells (Figure 1–21). There is another eucaryote, the amoeba *Pelomyxa palustris*, that, while lacking mitochondria, nevertheless carries out oxidative metabolism by harboring aerobic bacteria in its cytoplasm in a permanent symbiotic relationship. Diplomonads and microsporidia, on the one hand, and *Pelomyxa*, on the other, therefore resemble two proposed stages in the evolution of eucaryotes such as ourselves.

Acquisition of mitochondria must have had many repercussions. The plasma membrane, for example, is heavily committed to energy metabolism in procaryotic cells but not in eucaryotic cells, where this crucial function has been relegated to the mitochondria. It seems likely that the separation of functions left the eucaryotic plasma membrane free to evolve important new features. In particular, because eucaryotic cells need not maintain a large H^+ gradient across their plasma membrane, as required for ATP production in procaryotes, it became possible to use controlled changes in the ion permeability of the plasma membrane for cell-signaling purposes. Thus, a variety of ion channels appeared in the eucaryotic plasma membrane. Today, these channels mediate the elaborate electrical signaling processes in higher organisms—notably in the nervous system—

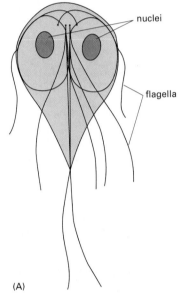

(A)

Figure 1–21 The diplomonad *Giardia.* (A) Drawing, as seen in the light microscope. (B) Electron micrograph of a cross-section through the broad, flattened body of the cell. *Giardia* is thought to be one of the most primitive types of eucaryotic cell. It is nucleated (in fact, it has, strangely, two identical nuclei), it possesses a cytoskeleton with actin and tubulin, and it moves by means of typical eucaryotic flagella containing microtubules; but it has no mitochondria or chloroplasts and no normal endoplasmic reticulum or Golgi apparatus. Nucleotide sequencing studies indicate that it is related almost as closely to bacteria as it is to other eucaryotes, from which it must have diverged very early in evolution. *Giardia* lives as a parasite in the gut and can cause disease in humans. (A, after G.D. Schmidt and L.S. Roberts, Foundations of Parasitology, 4th Ed. St Louis: Times Mirror/Mosby, 1989; B, courtesy of Dennis Feely.)

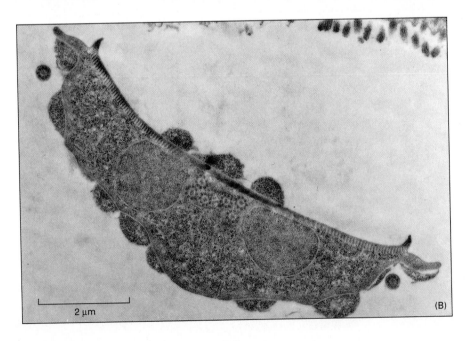

2 μm   (B)

and they control much of the behavior of single-celled free-living eucaryotes such as protozoa (see below).

Chloroplasts Are the Descendants of an Engulfed Procaryotic Cell [16]

Chloroplasts carry out photosynthesis in much the same way as procaryotic cyanobacteria, absorbing sunlight in the chlorophyll attached to their membranes. Some bear a close structural resemblance to the cyanobacteria, being similar in size and in the way that their chlorophyll-bearing membranes are stacked in layers (see Figure 1–20). Moreover, chloroplasts reproduce by dividing, and they contain DNA that is nearly indistinguishable in nucleotide sequence from portions of a bacterial chromosome. All this strongly suggests that chloroplasts share a common ancestry with cyanobacteria and evolved from procaryotes that made their home inside eucaryotic cells. These procaryotes performed photosynthesis for their hosts, who sheltered and nourished them. Symbiosis of photosynthetic cells with other cell types is, in fact, a common phenomenon, and some present-day eucaryotic cells contain authentic cyanobacteria (Figure 1–22).

Figure 1–23 shows the evolutionary origins of the eucaryotes according to the symbiotic theory. It must be stressed, however, that mitochondria and chloroplasts show important differences from, as well as similarities to, present-day aerobic bacteria and cyanobacteria. Their quantity of DNA is very small, for example, and most of the molecules from which they are constructed are synthesized elsewhere in the eucaryotic cell and imported into the organelle. Although there is good evidence that they originated as symbiotic bacteria, they have undergone large evolutionary changes and have become greatly dependent on—and subject to control by—their host cells.

The major existing eucaryotes have in common both mitochondria and a whole constellation of other features that distinguish them from procaryotes (Table 1–1). These function together to give eucaryotic cells a wealth of different capabilities, and it is debatable which of them evolved first. But the acquisition of mitochondria by an anaerobic eucaryotic cell must have been a crucial step in the success of the eucaryotes, providing them with the means to tap an abundant source of energy to drive all their complex activities.

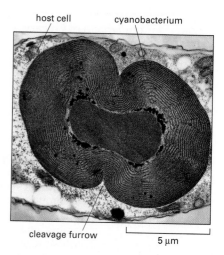

Figure 1–22 A close relative of present-day cyanobacteria that lives in a permanent symbiotic relationship inside another cell. The two organisms are known jointly as *Cyanophora paradoxa*. The "cyanobacterium" is in the process of dividing. (Courtesy of Jeremy D. Pickett-Heaps.)

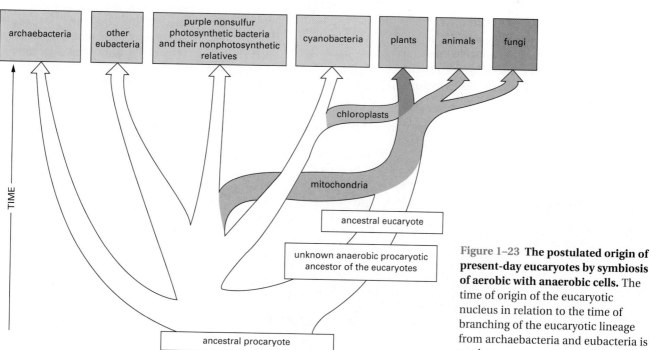

Figure 1–23 The postulated origin of present-day eucaryotes by symbiosis of aerobic with anaerobic cells. The time of origin of the eucaryotic nucleus in relation to the time of branching of the eucaryotic lineage from archaebacteria and eubacteria is not known.

Table 1–1 Comparison of Procaryotic and Eucaryotic Organisms

	Procaryotes	Eucaryotes
Organisms	bacteria and cyanobacteria	protists, fungi, plants, and animals
Cell size	generally 1 to 10 μm in linear dimension	generally 5 to 100 μm in linear dimension
Metabolism	anaerobic or aerobic	aerobic
Organelles	few or none	nucleus, mitochondria, chloroplasts, endoplasmic reticulum, etc.
DNA	circular DNA in cytoplasm	very long linear DNA molecules containing many noncoding regions; bounded by nuclear envelope
RNA and protein	RNA and protein synthesized in same compartment	RNA synthesized and processed in nucleus; proteins synthesized in cytoplasm
Cytoplasm	no cytoskeleton: cytoplasmic streaming, endocytosis, and exocytosis all absent	cytoskeleton composed of protein filaments; cytoplasmic streaming; endocytosis and exocytosis
Cell division	chromosomes pulled apart by attachments to plasma membrane	chromosomes pulled apart by cytoskeletal spindle apparatus
Cellular organization	mainly unicellular	mainly multicellular, with differentiation of many cell types

Eucaryotic Cells Contain a Rich Array of Internal Membranes

Eucaryotic cells are usually much larger in volume than procaryotic cells, commonly by a factor of 1000 or more, and they carry a proportionately larger quantity of most cellular materials; for example, a human cell contains about 1000 times as much DNA as a typical bacterium. This large size creates problems. Since all the raw materials for the biosynthetic reactions occurring in the interior of a cell must ultimately enter and leave by passing through the plasma membrane covering its surface, and since the membrane is also the site of many important reactions, an increase in cell volume requires an increase in cell surface. But it is a fact of geometry that simply scaling up a structure increases the volume as the cube of the linear dimension while the surface area increases only as the square. Therefore, if the large eucaryotic cell is to keep as high a ratio of surface to volume as the procaryotic cell, it must supplement its surface area by means of convolutions, infoldings, and other elaborations of its membrane.

This probably explains in part the complex profusion of **internal membranes** that is a basic feature of all eucaryotic cells. Membranes surround the nucleus, the mitochondria, and (in plant cells) the chloroplasts. They form a labyrinthine compartment called the **endoplasmic reticulum** (Figure 1–24), where lipids and proteins of cell membranes, as well as materials destined for export from the cell,

Figure 1–24 Endoplasmic reticulum. Electron micrograph of a thin section of a mammalian cell showing both smooth and rough regions of the endoplasmic reticulum (ER). The smooth regions are involved in lipid metabolism; the rough regions, studded with ribosomes, are sites of synthesis of proteins that are destined to leave the cytosol and enter certain other compartments of the cell. (Courtesy of George Palade.)

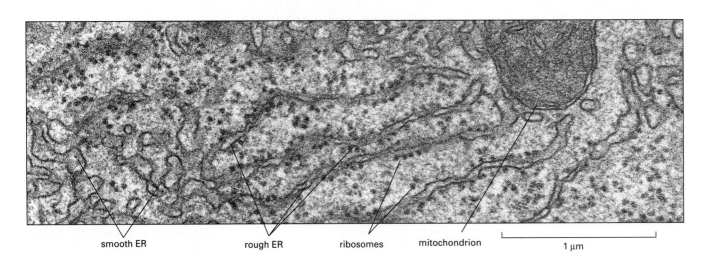

smooth ER   rough ER   ribosomes   mitochondrion   1 μm

are synthesized. They also form stacks of flattened sacs constituting the **Golgi apparatus** (Figure 1–25), which is involved in the modification and transport of the molecules made in the endoplasmic reticulum. Membranes surround **lysosomes,** which contain stores of enzymes required for intracellular digestion and so prevent them from attacking the proteins and nucleic acids elsewhere in the cell. In the same way membranes surround **peroxisomes,** where dangerously reactive hydrogen peroxide is generated and degraded during the oxidation of various molecules by O_2. Membranes also form small vesicles and, in plants, a large liquid-filled *vacuole.* All these membrane-bounded structures correspond to distinct internal compartments within the cytoplasm. In a typical animal cell these compartments (or organelles) occupy nearly half the total cell volume. The remaining compartment of the cytoplasm, which includes everything other than the membrane-bounded organelles, is usually referred to as the **cytosol.**

All of the aforementioned membranous structures lie in the interior of the cell. How, then, can they help to solve the problem we posed at the outset and provide the cell with a surface area that is adequate to its large volume? The answer is that there is a continual exchange between the internal membrane-bounded compartments and the outside of the cell, achieved by *endocytosis* and *exocytosis,* processes unique to eucaryotic cells. In endocytosis portions of the external surface membrane invaginate and pinch off to form membrane-bounded cytoplasmic vesicles that contain both substances present in the external medium and molecules previously adsorbed on the cell surface. Very large particles or even entire foreign cells can be taken up by *phagocytosis*—a special form of endocytosis. Exocytosis is the reverse process, whereby membrane-bounded vesicles inside the cell fuse with the plasma membrane and release their contents into the external medium. In this way membranes surrounding compartments deep inside the cell serve to increase the effective surface area of the cell for exchanges of matter with the external world.

As we shall see in later chapters, the various membranes and membrane-bounded compartments in eucaryotic cells have become highly specialized—some for secretion, some for absorption, some for specific biosynthetic processes, and so on.

Eucaryotic Cells Have a Cytoskeleton

The larger a cell is, and the more elaborate and specialized its internal structures, the greater is its need to keep these structures in their proper places and to control their movements. All eucaryotic cells have an internal skeleton, the **cytoskeleton,** that gives the cell its shape, its capacity to move, and its ability to arrange its organelles and transport them from one part of the cell to another. The cytoskeleton is composed of a network of protein filaments, two of the most important of which are *actin filaments* (Figure 1–26) and *microtubules.* These two must date from a very early epoch in evolution since they are found almost un-

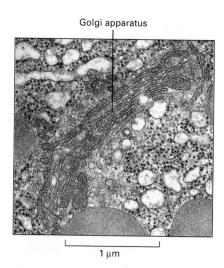

Golgi apparatus

1 μm

Figure 1–25 The Golgi apparatus. Electron micrograph of a thin section of a mammalian cell showing the Golgi apparatus, which is composed of flattened sacs of membrane arranged in multiple layers (see also Panel 1–1, pp. 18–19). The Golgi apparatus is involved in the synthesis and packaging of molecules destined to be secreted from the cell, as well as in the routing of newly synthesized proteins to the correct cellular compartments. (Courtesy of Daniel S. Friend.)

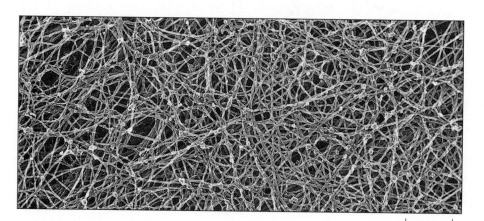

100 nm

Figure 1–26 Actin. A network of actin filaments underlying the plasma membrane of an animal cell is seen in this electron micrograph prepared by the deep-etch technique. (Courtesy of John Heuser.)

From Procaryotes to Eucaryotes

changed in all eucaryotes. Both are involved in the generation of cellular movements. Actin filaments enable individual eucaryotic cells to crawl about, for example, and they participate in the contraction of muscle in animals; microtubules are the main structural and force-generating elements in *cilia* and *flagella*—the long projections on some cell surfaces that beat like whips and serve as instruments of propulsion.

Actin filaments and microtubules are also essential for the internal movements that occur in the cytoplasm of all eucaryotic cells. Thus microtubules in the form of a *mitotic spindle* are a vital part of the usual machinery for partitioning DNA equally between the two daughter cells when a eucaryotic cell divides. Without microtubules, therefore, the eucaryotic cell could not reproduce. In this and other examples movement by free diffusion would be either too slow or too haphazard to be useful. In fact, most of the organelles in a eucaryotic cell appear to be attached, directly or indirectly, to the cytoskeleton and, when they move, to be propelled along cytoskeletal tracks.

Protozoa Include the Most Complex Cells Known [17]

The complexity that can be achieved by a single eucaryotic cell is nowhere better illustrated than in the free-living, single-celled eucaryotes known as *protists* (Figure 1–27). These are evolutionarily diverse (see Figure 1–16) and exhibit a bewildering variety of different forms and behaviors: they can be photosynthetic or carnivorous, motile or sedentary. Their anatomy is often complex and includes such structures as sensory bristles, photoreceptors, flagella, leglike appendages, mouth parts, stinging darts, and musclelike contractile bundles. Although they are single cells, protists, especially the larger and more active types known as **protozoa,** can be as intricate and versatile as many multicellular organisms. This is particularly well illustrated by the group known as **ciliates.**

Figure 1–27 An assortment of protists, illustrating some of the enormous variety to be found among this class of single-celled organisms. These drawings are done to different scales, but in each case the bar denotes 10 μm. The organisms in (A), (B), (E), (F), and (I) are ciliates; (C) is an euglenoid; (D) is an amoeba; (G) is a dinoflagellate; (H) is a heliozoan. (From M.A. Sleigh, The Biology of Protozoa. London: Edward Arnold, 1973.)

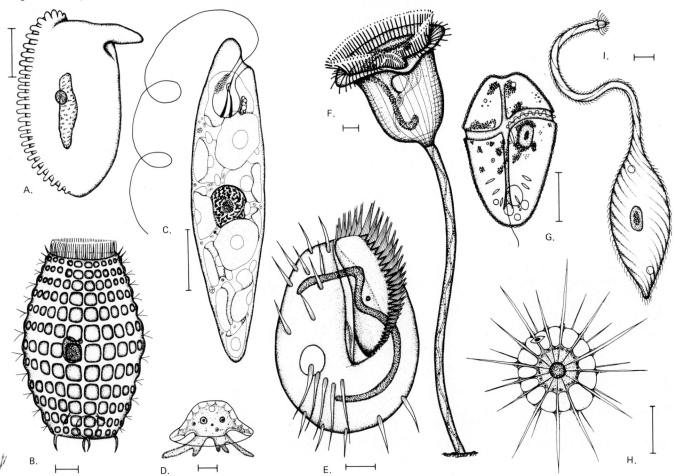

Didinium is a carnivorous ciliate. It has a globular body, about 150 μm in diameter, encircled by two fringes of cilia; its front end is flattened except for a single protrusion rather like a snout (Figure 1–28). *Didinium* swims around in the water at high speed by means of the synchronous beating of its cilia. When it encounters a suitable prey, usually another type of protozoan, such as a *Paramecium*, it releases numerous small paralyzing darts from its snout region. Then the *Didinium* attaches to and devours the *Paramecium*, inverting like a hollow ball to engulf the other cell, which is as large as itself. Most of this complex behavior—swimming, and paralyzing and capturing its prey—is generated by the cytoskeletal structures lying just beneath the plasma membrane. Included in this *cell cortex*, for example, are the parallel bundles of microtubules that form the core of each cilium and enable it to beat.

Predatory behavior of this sort and the set of features on which it depends—large size, the capacity for phagocytosis, and the ability to move in pursuit of prey—are peculiar to eucaryotes. Indeed, it is probable that these features came very early in eucaryotic evolution, making possible the subsequent capture of bacteria and their domestication as mitochondria and chloroplasts.

In Eucaryotic Cells the Genetic Material Is Packaged in Complex Ways

Eucaryotic cells contain a very large quantity of DNA. In human cells, for example, there is about 1000 times more DNA than in typical bacteria. The length of DNA in eucaryotic cells is so great that the risk of entanglement and breakage becomes severe. Probably for this reason, proteins unique to eucaryotes, the *histones*, have evolved to bind to the DNA and wrap it up into compact and manageable **chromosomes** (Figure 1–29). Tight packaging of the DNA in chromosomes is an essential part of the preparation for cell division in eucaryotes (Figure 1–30). All eucaryotes (with minor exceptions) have histones bound to their DNA, and the importance of these proteins is reflected in their remarkable conservation in evolution: several of the histones of a pea plant are almost exactly the same, amino acid for amino acid, as those of a cow.

The membranes enclosing the nucleus in eucaryotic cells further protect the structure of the DNA and its associated control machinery, sheltering them from entanglement with the moving cytoskeleton and from many of the chemical changes that take place in the cytoplasm. They also allow the segregation of two crucial steps in the expression of genetic information: (1) the copying of DNA sequences into RNA sequences (*DNA transcription*) and (2) the use of these RNA sequences, in turn, to direct the synthesis of specific proteins (*RNA translation*). In procaryotic cells there is no compartmentalization of these processes—the translation of RNA sequences into protein begins as soon as they are transcribed, even before their synthesis is completed. In eucaryotes, however (except in mitochondria and chloroplasts, which in this respect as in others are closer to bacteria), the two steps in the path from gene to protein are kept strictly separate: transcription occurs in the nucleus, translation in the cytoplasm. The RNA has to leave the nucleus before it can be used to guide protein synthesis. While in the nucleus it undergoes elaborate changes in which some parts of the RNA molecule are discarded and other parts are modified (*RNA processing*).

Because of these complexities, the genetic material of a eucaryotic cell offers many more opportunities for control than are present in bacteria.

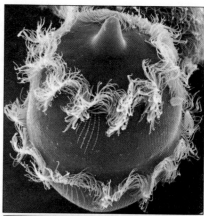
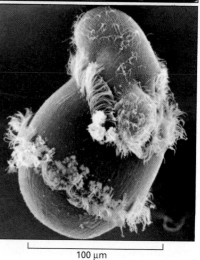

100 μm

Figure 1–28 One protozoan eating another. Ciliates are single-cell animals that show an amazing diversity of form and behavior. The top micrograph shows *Didinium*, a ciliated protozoan with two circumferential rings of motile cilia and a snoutlike protuberance at its leading end, with which it captures its prey. In the bottom micrograph *Didinium* is shown engulfing another protozoan, *Paramecium*. (Courtesy of D. Barlow.)

Summary

Present-day living cells are classified as procaryotic (bacteria and their close relatives) or eucaryotic. Although they have a relatively simple structure, procaryotic cells are biochemically versatile and diverse: for example, all of the major metabolic pathways can be found in bacteria, including the three principal energy-yielding processes of glycolysis, respiration, and photosynthesis. Eucaryotic cells are larger and more com-

plex than procaryotic cells and contain more DNA, together with components that allow this DNA to be handled in elaborate ways. The DNA of the eucaryotic cell is enclosed in a membrane-bounded nucleus, while the cytoplasm contains many other membrane-bounded organelles, including mitochondria, which carry out the oxidation of food molecules, and, in plant cells, chloroplasts, which carry out photosynthesis. Mitochondria and chloroplasts are almost certainly the descendants of earlier procaryotic cells that established themselves as internal symbionts of a larger anaerobic cell. Eucaryotic cells are also unique in containing a cytoskeleton of protein filaments that helps organize the cytoplasm and provides the machinery for movement.

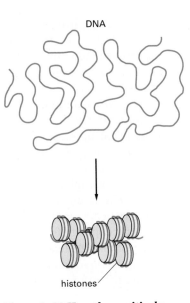

Figure 1–29 **How the positively charged proteins called histones mediate the folding of DNA in chromosomes.**

From Single Cells to Multicellular Organisms [18]

Single-cell organisms, such as bacteria and protozoa, have been so successful in adapting to a variety of different environments that they comprise more than half of the total biomass on earth. Unlike animals, many of these unicellular organisms can synthesize all of the substances they need from a few simple nutrients, and some of them divide more than once every hour. What, then, was the selective advantage that led to the evolution of **multicellular organisms**?

A short answer is that by collaboration and by division of labor it becomes possible to exploit resources that no single cell could utilize so well. This principle, applying at first to simple associations of cells, has been taken to an extreme in the multicellular organisms we see today. Multicellularity enables a plant, for example, to become physically large; to have roots in the ground, where one set of cells can take up water and nutrients; and to have leaves in the air, where another set of cells can efficiently capture radiant energy from the sun. Specialized cells in the stem of the plant form channels for transporting water and nutrients between the roots and the leaves. Yet another set of specialized cells forms a layer of epidermis to prevent water loss and to provide a protected internal environment (see Panel 1–2, pp. 28–29). The plant as a whole does not compete directly with unicellular organisms for its ecological niche; it has found a radically different way to survive and propagate.

As different animals and plants appeared, they changed the environment in which further evolution occurred. Survival in a jungle calls for different talents than survival in the open sea. Innovations in movement, sensory detection, communication, social organization—all enabled eucaryotic organisms to compete, propagate, and survive in ever more complex ways.

Single Cells Can Associate to Form Colonies

It seems likely that an early step in the evolution of multicellular organisms was the association of unicellular organisms to form colonies. The simplest way of achieving this is for daughter cells to remain together after each cell division. Even some procaryotic cells show such social behavior in a primitive form. Myxobacteria, for example, live in the soil and feed on insoluble organic molecules that they break down by secreting degradative enzymes. They stay together in loose colonies in which the digestive enzymes secreted by individual cells are pooled, thus increasing the efficiency of feeding (the "wolf-pack" effect). These cells indeed represent a peak of social sophistication among procaryotes, for when food supplies are exhausted, the cells aggregate tightly together and form a multicellular *fruiting body* (Figure 1–31), within which the bacteria differentiate into spores that can survive even in extremely hostile conditions. When conditions are more favorable, the spores in a fruiting body germinate to produce a new swarm of bacteria.

Green algae (not to be confused with the procaryotic "blue-green algae" or cyanobacteria) are eucaryotes that exist as unicellular, colonial, or multicellular

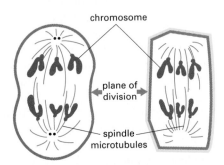

Figure 1–30 **Schematic drawing of eucaryotic cells in mitosis.** An animal cell is shown on the left and a plant cell on the right. The nuclear envelope has broken down, and the DNA, having replicated, has condensed into two complete sets of chromosomes. One set is distributed to each of the two newly forming cells by a mitotic spindle composed largely of microtubules.

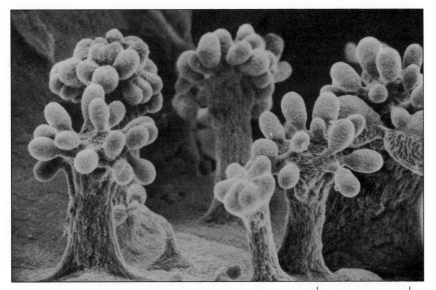

0.1 mm

Figure 1–31 **Fruiting bodies formed by a myxobacterium (*Chondromyces crocatus*), seen by scanning electron microscopy.** Each fruiting body, packed with spores, is created by the aggregation and differentiation of about a million myxobacteria. (From P.L. Grilione and J. Pangborn, *J. Bacteriol.* 124:1558–1565, 1975.)

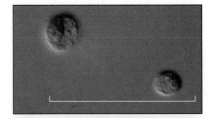

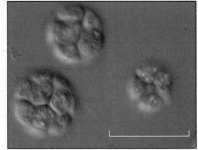

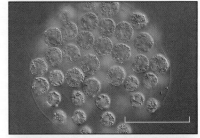

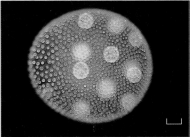

the scale bar shown represents 50 μm in each case

Figure 1–32 **Four closely related genera of green algae, showing a progression from unicellular to colonial and multicellular organization.** (Courtesy of David Kirk.)

forms (Figure 1–32). Different species of green algae can be arranged in order of complexity, illustrating the kind of progression that probably occurred in the evolution of higher plants and animals. Unicellular green algae, such as *Chlamydomonas,* resemble flagellated protozoa except that they possess chloroplasts, which enable them to carry out photosynthesis. In closely related genera, groups of flagellated cells live in colonies held together by a matrix of extracellular molecules secreted by the cells themselves. The simplest species (those of the genus *Gonium*) have the form of a concave disc made of 4, 8, 16, or 32 cells. Their flagella beat independently, but since they are all oriented in the same direction, they are able to propel the colony through the water. Each cell is equivalent to every other, and each can divide to give rise to an entirely new colony. Larger colonies are found in other genera, the most spectacular being *Volvox,* some of whose species have as many as 50,000 or more cells linked together to form a hollow sphere. In *Volvox* the individual cells forming a colony are connected by fine cytoplasmic bridges so that the beating of their flagella is coordinated to propel the entire colony along like a rolling ball (see Figure 1–32). Within the *Volvox* colony there is some division of labor among cells, with a small number of cells being specialized for reproduction and serving as precursors of new colonies. The other cells are so dependent on one another that they cannot live in isolation, and the organism dies if the colony is disrupted.

The Cells of a Higher Organism Become Specialized and Cooperate

In some ways *Volvox* is more like a multicellular organism than a simple colony. All of its flagella beat in synchrony as it spins through the water, and the colony is structurally and functionally polarized and can swim toward a distant source of light. The reproductive cells are usually confined to one end of the colony, where they divide to form new miniature colonies, which are initially sheltered inside the parent sphere. Thus, in a primitive way, *Volvox* displays the two essential features of all multicellular organisms: its cells become *specialized,* and they *cooperate.* By specialization and cooperation the cells combine to form a coordinated single organism with more capabilities than any of its component parts.

Organized patterns of cell differentiation occur even in some procaryotes. For example, many kinds of cyanobacteria remain together after cell division, forming filamentous chains that can be as much as a meter in length. At regular intervals

THE PLANT

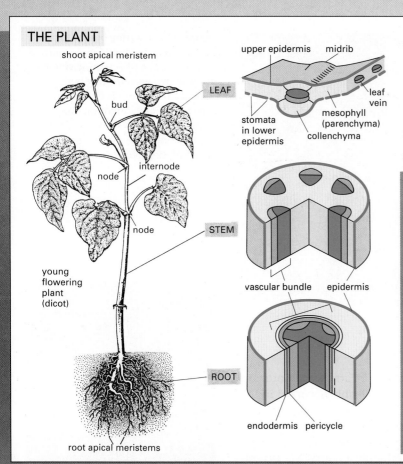

- shoot apical meristem
- bud
- LEAF
- internode
- node
- node
- node
- STEM
- young flowering plant (dicot)
- ROOT
- root apical meristems

Leaf labels: upper epidermis, midrib, leaf vein, mesophyll (parenchyma), collenchyma, stomata in lower epidermis

Stem labels: vascular bundle, epidermis

Root labels: endodermis, pericycle

The young flowering plant shown on the *left* is constructed from three main types of organs: leaves, stems, and roots. Each plant organ in turn is made from three tissue systems: ground, dermal, and vascular.

All three tissue systems derive ultimately from the cell proliferative activity of the shoot or root apical meristems, and each contains a relatively small number of specialized cell types. These three common tissue systems, and the cells that comprise them, are described in this panel.

THE THREE TISSUE SYSTEMS

Cell division, growth, and differentiation give rise to tissue systems with specialized functions.

DERMAL TISSUE (▬): This is the plant's protective outer covering in contact with the environment. It facilitates water and ion uptake in roots and regulates gas exchange in leaves and stems.

VASCULAR TISSUE: Together the phloem (▬) and the xylem (▬) form a continuous vascular system throughout the plant. This tissue conducts water and solutes between organs and also provides mechanical support.

GROUND TISSUE (▭): This packing and supportive tissue accounts for much of the bulk of the young plant. It also functions in food manufacture and storage.

GROUND TISSUE

The ground tissue system contains three main cell types called parenchyma, collenchyma, and sclerenchyma.

Parenchyma cells are found in all tissue systems. They are living cells, generally capable of further division, and have a thin primary cell wall. These cells have a variety of functions. The apical and lateral meristematic cells of shoots and roots provide the new cells required for growth. Food production and storage occur in the photosynthetic cells of the leaf and stem (called mesophyll cells); storage parenchyma cells form the bulk of most fruits and vegetables. Because of their proliferative capacity, parenchyma cells also serve as stem cells for wound healing and regeneration.

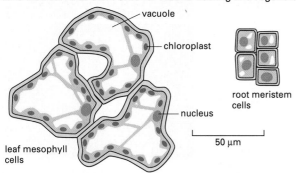

- vacuole
- chloroplast
- nucleus
- leaf mesophyll cells
- root meristem cells
- 50 μm

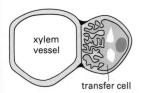

- xylem vessel
- transfer cell

A transfer cell, a specialized form of the parenchyma cell, is readily identified by elaborate ingrowths of the primary cell wall. The increase in the area of the plasma membrane beneath these walls facilitates the rapid transport of solutes to and from cells of the vascular system.

Collenchyma are living cells similar to parenchyma cells except that they have much thicker cell walls and are usually elongated and packed into long ropelike fibers. They are capable of stretching and provide mechanical support in the ground tissue system of the elongating regions of the plant. Collenchyma cells are especially common in subepidermal regions of stems.

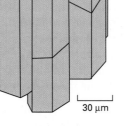

30 μm

- typical locations of supporting groups of cells in a stem
- sclerenchyma fibers
- vascular bundle
- collenchyma

Sclerenchyma, like collenchyma, have strengthening and supporting functions. However, they are usually dead cells with thick, lignified secondary cell walls that prevent them from stretching as the plant grows. Two common types are fibers, which often form long bundles, and sclereids, which are shorter branched cells found in seed coats and fruit.

fiber bundle

10 μm

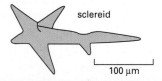

sclereid

100 μm

DERMAL TISSUE

The epidermis is the primary outer protective covering of the plant body. Cells of the epidermis are also modified to form stomata and hairs of various kinds.

Epidermis

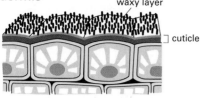

waxy layer

cuticle

The epidermis (usually one layer of cells deep) covers the entire stem, leaf, and root of the young plant. The cells are living, have thick primary cell walls, and are covered on their outer surface by a special cuticle with an outer waxy layer. The cells are tightly interlocked in different patterns.

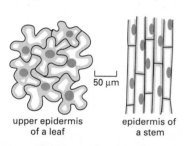

50 μm

upper epidermis of a leaf

epidermis of a stem

Stomata

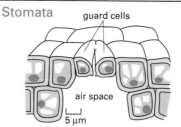

guard cells

air space

5 μm

Stomata are openings in the epidermis, mainly on the lower surface of the leaf, that regulate gas exchange in the plant. They are formed by two specialized epidermal cells called *guard cells,* which regulate the diameter of the pore. Stomata are distributed in a distinct species-specific pattern within each epidermis.

Vascular bundles

Roots usually have a single vascular bundle, but stems have several bundles. These are arranged with strict radial symmetry in dicots, but they are more irregularly dispersed in monocots.

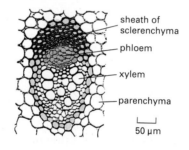

sheath of sclerenchyma

phloem

xylem

parenchyma

50 μm

a typical vascular bundle from the young stem of a buttercup

Hairs (or trichomes) are appendages derived from epidermal cells. They exist in a variety of forms and are commonly found in all plant parts. Hairs function in protection, absorption, and secretion; for example,

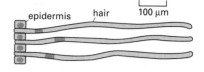

epidermis   hair   100 μm

young, single-celled hairs in the epidermis of the cotton seed. When these grow, the walls will be secondarily thickened with cellulose to form cotton fibers.

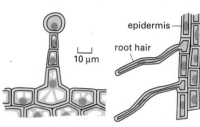

epidermis

root hair

10 μm

a multicellular secretory hair from a geranium leaf

Single-celled root hairs have an important function in water and ion uptake.

VASCULAR TISSUE

The phloem and the xylem together form a continuous vascular system throughout the plant. In young plants they are usually associated with a variety of other cell types in *vascular bundles*. Both phloem and xylem are complex tissues. Their conducting elements are associated with parenchyma cells that maintain and exchange materials with the elements. Also, groups of collenchyma and sclerenchyma cells provide mechanical support.

Phloem

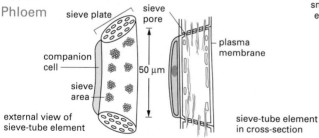

sieve plate

sieve pore

companion cell

50 μm

sieve area

plasma membrane

external view of sieve-tube element

sieve-tube element in cross-section

Phloem is involved in the transport of organic solutes in the plant. The main conducting cells (elements) are aligned to form tubes called *sieve tubes*. The sieve-tube elements at maturity are living cells, interconnected by perforations in their end walls formed from enlarged and modified plasmodesmata (sieve plates). These cells retain their plasma membrane, but they have lost their nuclei and much of their cytoplasm; they therefore rely on associated *companion cells* for their maintenance. These companion cells have the additional function of actively transporting soluble food molecules into and out of sieve-tube elements through porous sieve areas in the wall.

Xylem

Xylem carries water and dissolved ions in the plant. The main conducting cells are the vessel elements shown here, which are dead cells at maturity that lack a plasma membrane. The cell wall has been secondarily thickened and heavily lignified. As shown below, its end wall is largely removed, enabling very long, continuous tubes to be formed.

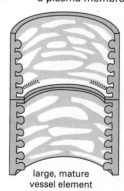

small vessel element in root tip

large, mature vessel element

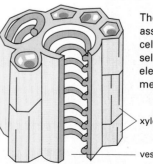

The vessel elements are closely associated with xylem parenchyma cells, which actively transport selected solutes into and out of the elements across their plasma membrane.

xylem parenchyma cells

vessel element

along the filament, individual cells take on a distinctive character and become able to incorporate atmospheric nitrogen into organic molecules. These few specialized cells perform nitrogen fixation for their neighbors and share the products with them. But eucaryotic cells appear to be very much better at this sort of organized division of labor; they, and not procaryotes, are the living units from which all the more complex multicellular organisms are constructed.

Multicellular Organization Depends on Cohesion Between Cells

To form a multicellular organism, the cells must be somehow bound together, and eucaryotes have evolved a number of different ways to satisfy this need. In *Volvox*, as noted above, the cells do not separate entirely at cell division but remain connected by cytoplasmic bridges. In higher plants the cells not only remain connected by cytoplasmic bridges (called *plasmodesmata*), they also are imprisoned in a rigid honeycomb of chambers walled with cellulose that the cells themselves have secreted (*cell walls*).

The cells of most animals do not have rigid walls, and cytoplasmic bridges are unusual. Instead, the cells are bound together by a relatively loose meshwork of large extracellular organic molecules (called the *extracellular matrix*) and by adhesions between their plasma membranes. Very often, side-to-side attachments between the cells hold them together to form a multicellular sheet, or **epithelium.**

Epithelial Sheets of Cells Enclose a Sheltered Internal Environment

Of all the ways in which animal cells are woven together into multicellular tissues, the epithelial arrangement is perhaps the most fundamentally important. The epithelial sheet has much the same significance for the evolution of complex multicellular organisms that the cell membrane has for the evolution of complex single cells.

The importance of epithelial sheets is well illustrated in the lowly group of animals known as *coelenterates*. The group includes sea anemones, jellyfish, and

Figure 1–33 The body plan of *Hydra.* (A) A living specimen of *Hydra vulgaris*, showing its transparent ectoderm and numerous sting cells, glittering on its tentacles. The projections budding from the side of the body are progeny that will eventually detach from their parent. (B) Diagram of the cellular architecture of the body of a typical *Hydra*. The outer layer of cells (ectoderm) has protective, predatory, and sensory functions, while cells of the inner layer (endoderm) function principally in digestion. Both epithelial sheets also have a contractile or muscular function, enabling the animal to move. The movements are coordinated by nerve cells that occupy a deep, protected position within each epithelium, forming an interconnected network. (A, courtesy of Timm Nüchter and Thomas Holstein.)

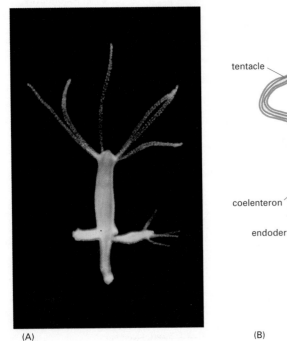

(A)

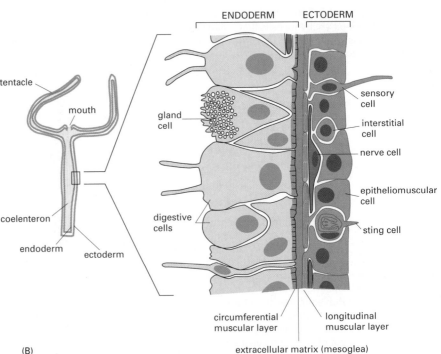

(B)

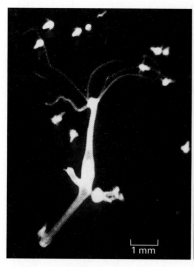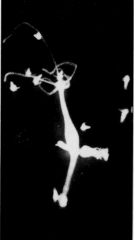

Figure 1–34 *Hydra* **feeding.** Feeding is one of a range of fairly complex activities this animal can perform. A single *Hydra* is photographed catching small water fleas in its tentacles; in the last panel it is stuffing these prey into its coelenteron for digestion. (Courtesy of Amata Hornbruch.)

corals, as well as the small freshwater organism *Hydra.* Coelenterates are constructed from two layers of epithelium, the outer layer being the *ectoderm,* the inner being the *endoderm.* The endodermal layer surrounds a cavity, the *coelenteron,* in which food is digested (Figure 1–33). Among the endodermal cells are some that secrete digestive enzymes into the coelenteron, while other cells absorb and further digest the nutrient molecules that these enzymes release. By forming a tightly coherent epithelial sheet that prevents all these molecules from being lost to the exterior, the endodermal cells create for themselves an environment in the coelenteron that is suited to their own digestive tasks. Meanwhile, the ectodermal cells, facing the exterior, remain specialized for encounters with the outside world. In the ectoderm, for example, are cells that contain a poison capsule with a coiled dart that can be unleashed to kill the small animals on which *Hydra* feeds. The majority of other ectodermal and endodermal cells have musclelike properties, enabling *Hydra* to move, as a predator must.

Within the double layer of ectoderm and endoderm is another compartment, separate both from the coelenteron and from the outside world. Here *nerve cells* lie, occupying narrow enclosed spaces between the epithelial cells, below the external surface where the specialized *cell junctions* between the epithelial cells form an impermeable barrier. The animal can change its shape and move by contractions of the musclelike cells in the epithelia, and it is the nerve cells that convey electrical signals to control and coordinate these contractions (Figures 1–33, 1–34, and 1–35). As we shall see later, the concentrations of simple inorganic ions in the medium surrounding a nerve cell are crucial for its function. Most nerve cells—our own included—are designed to operate when bathed in a solution with an ionic composition roughly similar to that of seawater. This may well reflect the conditions under which the first nerve cells evolved. Most coelenterates still live in the sea, but not all. *Hydra,* in particular, lives in fresh water. It has evidently been able to colonize this new habitat only because its nerve cells are contained in a space that is sealed and isolated from the exterior within sheets of epithelial cells that maintain the internal environment necessary for nerve cell function.

Cell-Cell Communication Controls the Spatial Pattern of Multicellular Organisms [19]

The cells of *Hydra* are not only bound together mechanically and connected by junctions that seal off the interior from the exterior environment, they also communicate with one another along the length of the body. If one end of a *Hydra* is cut off, the remaining cells react to the absence of the amputated part by adjusting their characters and rearranging themselves so as to regenerate a com-

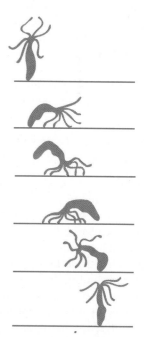

Figure 1–35 *Hydra* **traveling.** A *Hydra* can swim, glide on its base, or, as shown here, travel by somersaulting.

plete animal. Evidently, signals pass from one part of the organism to the other, governing the development of its body pattern—with tentacles and a mouth at one end and a foot at the other. Moreover, these signals are independent of the nervous system. If a developing *Hydra* is treated with a drug that prevents nerve cells from forming, the animal is unable to move about, catch prey, or feed itself. Its digestive system still functions normally, however, so that it can be kept alive by anyone with the patience to stuff its normal prey into its mouth. In such force-fed animals the body pattern is maintained, and lost parts are regenerated just as well as in an animal that has an intact nervous system.

The vastly more complex higher animals have evolved from simpler ancestors resembling coelenterates, and these higher animals owe their complexity to more sophisticated exploitation of the same basic principles of cell cooperation that underlie the construction of *Hydra*. Epithelial sheets of cells line all external and internal surfaces in the body, creating sheltered compartments and controlled internal environments in which specialized functions are performed by differentiated cells. Specialized cells interact and communicate with one another, setting up signals to govern the character of each cell according to its place in the structure as a whole. To show how it is possible to generate multicellular organisms of such size, precision, and complexity as a tree, a fly, or a mammal, however, it is necessary to consider more closely the sequence of events in **development.**

Cell Memory Permits the Development of Complex Patterns

The cells of almost every multicellular organism are generated by repeated division from a single precursor cell; they constitute a *clone.* As proliferation continues and the clone grows, some of the cells, as we have seen, become differentiated from others, adopting a different structure, a different chemistry, and a different function, usually in response to cues from their neighbors. It is remarkable that eucaryotic cells and their progeny will usually persist in their differently specialized states even after the influences that originally directed their differentiation have disappeared—in other words, these cells have a *memory.* Consequently, their final character is not determined simply by their final environment, but rather by the entire sequence of influences to which the cells have been exposed in the course of development. Thus as the body grows and matures, progressively finer details of the adult body pattern become specified, creating an organism of gradually increasing complexity whose ultimate form is the expression of a long developmental history.

Basic Developmental Programs Tend to Be Conserved in Evolution [20]

The final structure of an animal or plant reflects its evolutionary history, which, like development, presents a chronicle of progress from the simple to the complex. What then is the connection between the two perspectives, of evolution on the one hand and development on the other?

During evolution many of the developmental devices that evolved in the simplest multicellular organisms have been conserved as basic principles for the construction of their more complex descendants. We have already mentioned, for example, the organization of cells into epithelia. Some specialized cell types, such as nerve cells, are found throughout nearly the whole of the animal kingdom, from *Hydra* to humans. Molecular studies, to be discussed later in this book, reveal an astonishing number of developmental resemblances at a fundamental genetic level, even between species as remotely related as mammals and insects. In terms of anatomy, furthermore, early developmental stages of animals whose adult forms appear radically different are often surprisingly similar; it takes

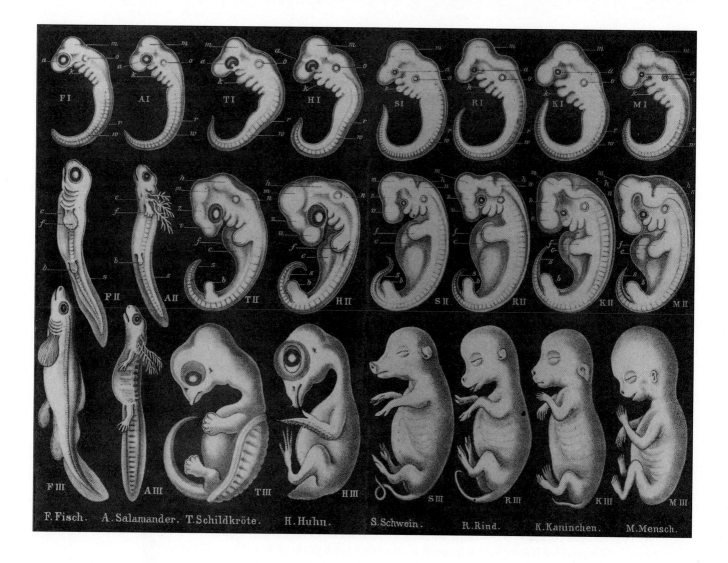

an expert eye to distinguish, for example, a young chick embryo from a young human embryo (Figure 1–36).

Such observations are not difficult to understand. Consider the process by which a new anatomical feature—say, an elongated beak—appears in the course of evolution. A random mutation occurs that changes the amino acid sequence of a protein or the timing of its synthesis and hence its biological activity. This alteration may, by chance, affect the cells responsible for the formation of the beak in such a way that they make one that is longer. But the mutation must also be compatible with the development of the rest of the organism; only then will it be propagated by natural selection. There would be little selective advantage in forming a longer beak if, in the process, the tongue was lost or the ears failed to develop. A catastrophe of this type is more likely if the mutation affects events occurring early in development than if it affects those near the end. The early cells of an embryo are like cards at the bottom of a house of cards—a great deal depends on them, and even small changes in their properties are likely to result in disaster. Fundamental steps appear to have been "frozen" into developmental processes, just as the genetic code or protein synthesis mechanisms have become frozen into the basic biochemical organization of the cell. In contrast, cells produced near the end of development (or produced early but forming accessory structures such as the placenta that are not incorporated in the adult body) have more freedom to change. It is presumably for this reason that the embryos of different species so often resemble each other in their early stages and, as they develop, seem sometimes to replay the steps of evolution.

Figure 1–36 Comparison of the embryonic development of a fish, an amphibian, a reptile, a bird, and a selection of mammals. The early stages (*above*) are very similar; the later stages (*below*) are more divergent. The earliest stages are drawn roughly to scale; the later stages are not. (From E. Haeckel, Anthropogenie, oder Entwickelungsgeschichte des Menschen. Leipzig: Engelmann, 1874. Courtesy of the Bodleian Library, Oxford.)

From Single Cells to Multicellular Organisms

The Cells of the Vertebrate Body Exhibit More Than 200 Different Modes of Specialization

The wealth of diverse specializations to be found among the cells of a higher animal is far greater than any procaryote can show. In a vertebrate more than 200 distinct **cell types** are plainly distinguishable, and many of these types of cells certainly include, under a single name, a large number of more subtly different varieties. Panel 1–3 (pp. 36–37) shows a small selection. In this profusion of specialized behaviors one can see displayed, in a single organism, the astonishing versatility of the eucaryotic cell. Much of our current knowledge of the general properties of eucaryotic cells has depended on the study of such specialized types of cells, because they demonstrate exceptionally well particular features on which all cells depend in some measure. Each feature and each organelle of the prototype that we have outlined in Panel 1–1 (pp. 18–19) is developed to an unusual degree or revealed with special clarity in one cell type or another. To take one arbitrary example, consider the *neuromuscular junction*, where just three types of cells are involved: a muscle cell, a nerve cell, and a Schwann cell. Each has a very different role (Figure 1–37):

1. The muscle cell has made contraction its specialty. Its cytoplasm is packed with organized arrays of protein filaments, including vast numbers of actin filaments. There are also many mitochondria interspersed among the protein filaments, supplying ATP as fuel for the contractile apparatus.

2. The nerve cell stimulates the muscle to contract, conveying an excitatory signal to the muscle from the brain or spinal cord. The nerve cell is therefore extraordinarily elongated: its main body, containing the nucleus, may lie a meter or more from the junction with the muscle. The cytoskeleton is consequently well developed so as to maintain the unusual shape of the cell and to transport materials efficiently from one end of the cell to the other. The most crucial specialization of the nerve cell, however, is its plasma membrane, which contains proteins that act as ion *pumps* and ion *channels*, causing a movement of ions that is equivalent to a flow of electricity. Whereas all cells contain such pumps and channels in their plasma membranes, the nerve cell has exploited them in such a way that a pulse of electricity can propagate in a fraction of a second from one end of the cell to the other, conveying a signal for action.

3. Lastly, Schwann cells are specialists in the mass production of plasma membrane, which they wrap around the elongated portion of the nerve cell, laying down layer upon layer of membrane like a roll of tape, to form a *myelin sheath* that serves as insulation.

Genes Can Be Switched On and Off

The various specialized cell types in a single higher plant or animal appear as different from one another as any cells could be. This seems paradoxical, since all of the cells in a multicellular organism are closely related, having recently descended from the same precursor cell—the fertilized egg. Common lineage implies similar genes; how then do the differences arise? In a few cases cell specialization involves the loss of genetic material. An extreme example is the mammalian red blood cell, which loses its entire nucleus in the course of differentiation. But the overwhelming majority of cells in most plant and animal species retain all of the genetic information contained in the fertilized egg. Specialization depends on changes in *gene expression*, not on the loss or acquisition of genes.

Even bacteria do not make all of their types of protein all of the time but are able to adjust the level of synthesis according to external conditions. Proteins required specifically for the metabolism of lactose, for example, are made by

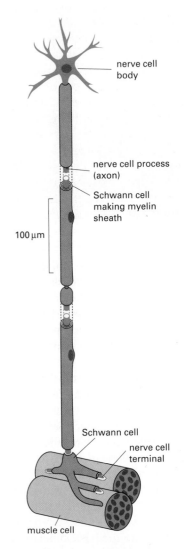

Figure 1–37 A nerve cell, with its associated Schwann cells, contacting a muscle cell at a neuromuscular junction. Schematic diagram.

many bacteria only when this sugar is available for use; and when conditions are unfavorable for cell proliferation, some bacteria arrest most of their normal metabolic processes and form *spores,* which have tough, impermeable outer walls and a cytoplasm of altered composition.

Eucaryotic cells have evolved far more sophisticated mechanisms for controlling gene expression, and these affect entire systems of interacting gene products. Groups of genes are activated or repressed in response to both external and internal signals. Membrane composition, cytoskeleton, secretory products, even metabolism—all these and other features must change in a coordinated manner when cells become differentiated. The radical differences of character between cell types reflect stable changes in gene expression. The controls that bring about these changes have evolved in eucaryotes to a degree unmatched in procaryotes, defining the complex rules of cell behavior that can generate an organized multicellular organism from a single egg.

Sequence Comparisons Reveal Hundreds of Families of Homologous Genes [12, 21]

To outward appearances, evolution has transformed the universe of living things to such a degree that they are no longer recognizable as relatives. A human being, a fly, a daisy, a yeast, a bacterium—they seem so different that it scarcely makes sense to compare them. Yet all are descendants of one ancestor, and as we probe their inner workings more and more deeply, we find more and more evidence of their common origins. We now know that the basic molecular machinery of life has been conserved to an extent that would surely have astonished the originators of the theory of evolution. As we have seen, all life forms have essentially the same chemistry, based on amino acids, sugars, fatty acids, and nucleotides; all synthesize these chemical constituents in an essentially similar way; all store their genetic information in DNA and express it through RNA and protein. But the degree of evolutionary conservatism becomes even more striking when we examine the detailed sequences of nucleotides in specific genes and of amino acids in specific proteins. The chances are that the bacterial enzyme catalyzing any particular common reaction, such as the splitting of a six-carbon sugar into two three-carbon sugars in glycolysis, will have an amino acid sequence (and a three-dimensional structure) unmistakably similar to the enzyme catalyzing the same reaction in human beings. The two enzymes—and, equivalently, the genes that specify them—not only have a similar function, but also almost certainly a common evolutionary origin. One can exploit these relationships to trace ancient evolutionary pathways; and by comparing gene sequences and recognizing homologies, one discovers hidden parallels and similarities between different organisms.

Family resemblances are also often found among genes coding for proteins that carry out related functions within a single organism. These genes are also evolutionarily related, and their existence reveals a basic strategy by which increasingly complex organisms have arisen: genes and portions of genes become duplicated, and the new copies then diverge from the old by mutation and recombination to serve new, additional purposes. In this way, starting from a relatively small set of genes in primitive cells, the more complex life forms have been able to evolve the more than 50,000 genes thought to be present in a higher animal or plant. From an understanding of one gene or protein, we consequently gain insight into a whole family of others homologous to it. Thus molecular biology both underscores the unity of the living world and gives us tools to discover the general mechanisms that underlie its endless variety of inventions.

In the next chapter we begin our account of these mechanisms with a discussion of the most basic components of the biological construction kit—the small molecules from which all larger components of living cells are made.

CELL TYPES

There are over 200 types of cells in the human body. These are assembled into a variety of types of tissue such as

epithelia

connective tissue

muscle

nervous tissue

Most tissues contain a mixture of cell types.

EPITHELIA

Epithelial cells form coherent cell sheets called epithelia, which line the inner and outer surfaces of the body. There are many specialized types of epithelia.

Absorptive cells have numerous hairlike microvilli projecting from their free surface to increase the area for absorption.

Ciliated cells have cilia on their free surface that beat in synchrony to move substances (such as mucus) over the epithelial sheet.

Secretory cells are found in most epithelial layers. These specialized cells secrete substances onto the surface of the cell sheet.

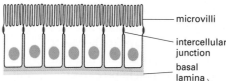

microvilli

intercellular junction

basal lamina

Adjacent epithelial cells are bound together by junctions that give the sheet mechanical strength and also make it impermeable to small molecules. The sheet rests on a basal lamina.

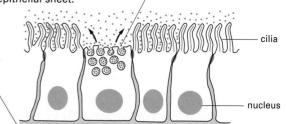

cilia

nucleus

CONNECTIVE TISSUE

The spaces between organs and tissues in the body are filled with connective tissue made principally of a network of tough protein fibers embedded in a polysaccharide gel. This extracellular matrix is secreted mainly by fibroblasts.

Two main types of extracellular protein fiber are collagen and elastin.

fibroblasts in loose connective tissue

Bone is made by cells called osteoblasts. These secrete an extracellular matrix in which crystals of calcium phosphate are later deposited.

Calcium salts are deposited in the extracellular matrix.

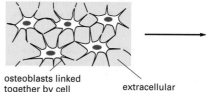

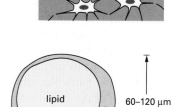

osteoblasts linked together by cell processes

extracellular matrix

Adipose cells, among the largest cells in the body, are responsible for the production and storage of fat. The nucleus and cytoplasm are squeezed by a large lipid droplet.

lipid

60–120 μm

NERVOUS TISSUE

dendrites

OUTPUT

axon

INPUTS

The axon conducts electrical signals away from the cell body. These signals are produced by a flux of ions across the nerve cell membrane.

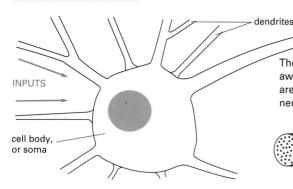

cell body, or soma

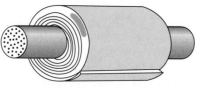

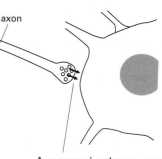

Nerve cells, or neurons, are specialized for communication. The brain and spinal cord, for example, are composed of a network of neurons among supporting glial cells.

Specialized cells, called Schwann cells, or oligodendrocytes, wrap around an axon to form a multilayered membrane sheath.

A synapse is where a neuron forms a specialized junction with another neuron (or with a muscle cell). At synapses, signals pass from one neuron to another (or from a neuron to a muscle cell).

Panel 1–3 Some of the different types of cells present in the vertebrate body.

Secretory epithelial cells are often collected together to form a gland that specializes in the secretion of a particular substance. As illustrated, exocrine glands secrete their products (such as tears, mucus, and gastric juices) into ducts. Endocrine glands secrete hormones into the blood.

secreted material

duct of gland

secretory cells of gland

BLOOD

Erythrocytes (red blood cells) are very small cells, usually with no nucleus or internal membranes, and are stuffed full of the oxygen-binding protein hemoglobin.

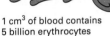

1 cm³ of blood contains 5 billion erythrocytes

their normal shape is a biconcave disc

Leucocytes (white blood cells) protect against infections. Blood contains about one leucocyte for every 100 red blood cells. Although leucocytes travel in the circulation, they can pass through the walls of blood vessels to do their work in the surrounding tissues. There are several different kinds, including

lymphocytes—responsible for immune responses such as the production of antibodies.
macrophages and neutrophils—move to sites of infection, where they ingest bacteria and debris.

wall of small blood vessel

bacterial infection in connective tissue

GERM CELLS

Both sperm and egg are *haploid,* that is, they carry only one set of chromosomes. A sperm from the male fuses with an egg from the female, which then forms a new diploid organism by successive cell divisions.

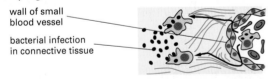

egg with sperm drawn to scale

sperm

MUSCLE

Muscle cells produce mechanical force by their contraction. In vertebrates there are three main types:

skeletal muscle—this moves joints by its strong and rapid contraction. Each muscle is a bundle of muscle fibers, each of which is an enormous multinucleated cell.

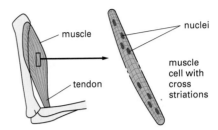

muscle

nuclei

muscle cell with cross striations

tendon

smooth muscle—present in digestive tract, bladder, arteries, and veins. It is composed of thin elongated cells (not striated), each of which has one nucleus.

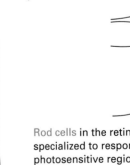

cardiac muscle—intermediate in character between skeletal and smooth muscle. It produces the heart beat. Adjacent cells are linked by electrically conducting junctions that cause the cells to contract in synchrony.

SENSORY CELLS

Among the most strikingly specialized cells in the vertebrate body are those that detect external stimuli. Hair cells of the inner ear are primary detectors of sound. Modified epithelial cells, they carry special microvilli (stereocilia) on their surface. The movement of these in response to sound vibrations causes an electrical signal to pass to the brain.

stereocilia are very rigid because they are packed with actin filaments

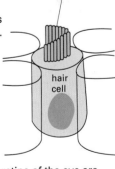

hair cell

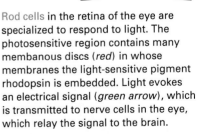

Rod cells in the retina of the eye are specialized to respond to light. The photosensitive region contains many membanous discs (*red*) in whose membranes the light-sensitive pigment rhodopsin is embedded. Light evokes an electrical signal (*green arrow*), which is transmitted to nerve cells in the eye, which relay the signal to the brain.

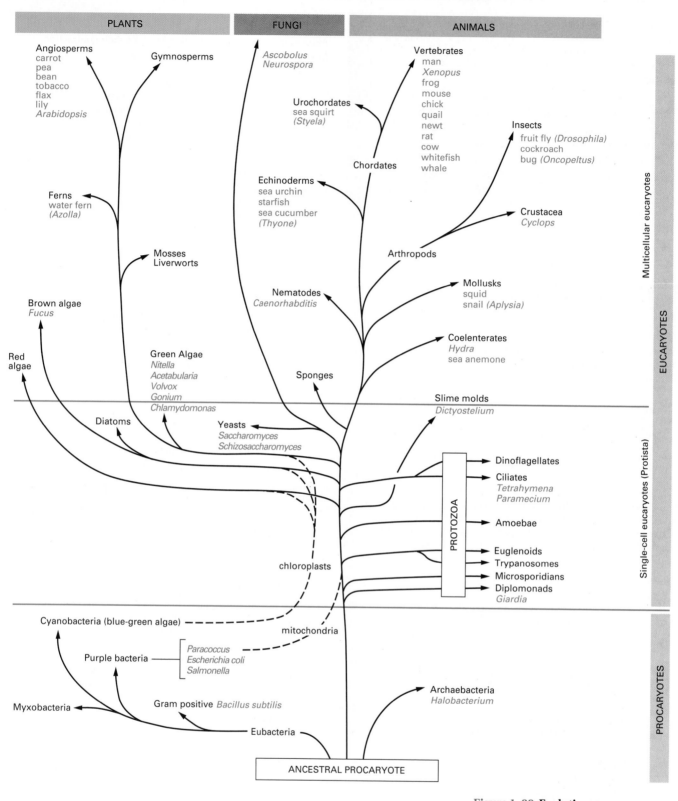

PLANTS | **FUNGI** | **ANIMALS**

Angiosperms
carrot
pea
bean
tobacco
flax
lily
Arabidopsis

Gymnosperms

Ascobolus
Neurospora

Vertebrates
man
Xenopus
frog
mouse
chick
quail
newt
rat
cow
whitefish
whale

Urochordates
sea squirt
(Styela)

Insects
fruit fly *(Drosophila)*
cockroach
bug *(Oncopeltus)*

Chordates

Ferns
water fern
(Azolla)

Echinoderms
sea urchin
starfish
sea cucumber
(Thyone)

Crustacea
Cyclops

Mosses
Liverworts

Arthropods

Brown algae
Fucus

Nematodes
Caenorhabditis

Mollusks
squid
snail *(Aplysia)*

Red
algae

Green Algae
Nitella
Acetabularia
Volvox
Gonium
Chlamydomonas

Sponges

Coelenterates
Hydra
sea anemone

Slime molds
Dictyostelium

Diatoms

Yeasts
Saccharomyces
Schizosaccharomyces

Dinoflagellates

Ciliates
Tetrahymena
Paramecium

PROTOZOA

Amoebae

chloroplasts

Euglenoids

Trypanosomes

Microsporidians

Diplomonads
Giardia

Cyanobacteria (blue-green algae)

mitochondria

Purple bacteria

Paracoccus
Escherichia coli
Salmonella

Archaebacteria
Halobacterium

Myxobacteria

Gram positive *Bacillus subtilis*

Eubacteria

ANCESTRAL PROCARYOTE

Multicellular eucaryotes

EUCARYOTES

Single-cell eucaryotes (Protista)

PROCARYOTES

**Figure 1–38 Evolutionary
relationships among some of the
organisms mentioned in this book.**
The branches of the evolutionary tree
show paths of descent but do not
indicate by their length the passage
of time. (Note, similarly, that the
vertical axis of the diagram shows
major categories of organisms and
not time.)

Summary

The evolution of large multicellular organisms depended on the ability of eucaryotic cells to express their hereditary information in many different ways and to function cooperatively in a single collective. In animals one of the earliest developments was probably the formation of epithelial cell sheets, which separate the internal space of the body from the exterior. In addition to epithelial cells, primitive differentiated cell types would have included nerve cells, muscle cells, and connective tissue cells, all of which can be found in very simple present-day animals. The evolution of higher animals and plants (Figure 1–38) depended on production of an increasing number of specialized cell types and more sophisticated methods of coordination among them, reflecting an increasingly elaborate system of controls over gene expression in the individual component cells.

References

General

Bendall, D.S., ed. Evolution from Molecules to Men. Cambridge, UK: Cambridge University Press, 1983.

Evolution of Catalytic Function. *Cold Spring Harbor Symp. Quant. Biol.* 52, 1987.

Curtis, H.; Barnes, N.S. Biology, 5th ed. New York: Worth, 1989.

Darnell, J.E.; Lodish, H.F.; Baltimore, D. Molecular Cell Biology, 2nd ed., Chapter 26. New York: Scientific American Books, 1990.

Darwin, C. On the Origin of Species. London: Murray, 1859. Reprinted, New York: Penguin, 1984.

Dawkins, R. The Blind Watchmaker. New York: Viking Penguin, 1988.

Margulis, L.M.; Schwartz, K.V. Five Kingdoms: An Illustrated Guide to the Phyla of Life on Earth, 2nd ed. New York: W.H. Freeman, 1988.

Watson, J.D.; Hopkins, N.H.; Roberts, J.W.; Steitz, J.A.; Weiner, A.M. Molecular Biology of the Gene, 4th ed., Chapter 28. Menlo Park, CA: Benjamin-Cummings, 1987.

Cited

1. Lazcano, A.; Fox, G.E.; Oró, J.F. Life before DNA: the origin and evolution of early archaean cells. In The Evolution of Metabolic Function (R.P. Mortlock, ed.), pp. 237–295. Boca Raton, FL: CRC Press, 1992.

   Schopf, J.W.; Hayes, J.M.; Walter, M.R. Evolution of earth's earliest ecosystems: recent progress and unsolved problems. In Earth's Earliest Biosphere: Its Origin and Evolution (J.W. Schopf, ed.), pp. 361–384. Princeton, NJ: Princeton University Press, 1983.

2. Miller, S.L. Which organic compounds could have occurred on the prebiotic earth? *Cold Spring Harbor Symp. Quant. Biol.* 52:17–27, 1987.

3. Orgel, L.E. Molecular replication. *Nature* 358:203–209, 1992.

4. Ellington, A.D.; Szostak, J.W. *In vitro* selection of RNA molecules that bind specific ligands. *Nature* 346:818–822, 1990.

   Joyce, G.F. Directed molecular evolution. *Sci. Am.* 267(6):90–97, 1992.

5. Bartel, D.P.; Szostak, J.W. Isolation of new ribozymes from a large pool of random sequences. *Science* 261:1411–1418, 1993.

   Cech, T.R. RNA as an enzyme. *Sci. Am.* 255(5):64–75, 1986.

6. Alberts, B.M. The function of the hereditary materials: biological catalyses reflect the cell's evolutionary history. *Am. Zool.* 26:781–796, 1986.

   Maizels, N.; Weiner, A.M. Peptide-specific ribosomes, genomic tags, and the origin of the genetic code. *Cold Spring Harbor Symp. Quant. Biol.* 52:743–749, 1987.

7. Cavalier-Smith, T. The origin of cells: a symbiosis between genes, catalysts, and membranes. *Cold Spring Harbor Symp. Quant. Biol.* 52:805–824, 1987.

8. Muto, A.; Andachi, Y.; Yuzawa, H.; Yamao, F.; Osawa, S. The organization and evolution of transfer RNA genes in *Mycoplasma capricolum. Nucl. Acid Res.* 18:5037–5043, 1990.

9. Sogin, M.L. Early evolution and the origin of eukaryotes. *Curr. Opin. Genet. Devel.* 1:457–463, 1991.

   Vidal, G. The oldest eukaryotic cells. *Sci. Am.* 250(2):48–57, 1984.

10. Woese, C.R. Bacterial evolution. *Microbiol. Rev.* 51:221–271, 1987.

    Zillig, W. Comparative biochemistry of *Archaea* and *Bacteria. Curr. Opin. Genet. Devel.* 1:544–551, 1991.

11. Clarke, P.H. Enzymes in bacterial populations. In Biochemical Evolution (H. Gutfreund, ed.), pp. 116–149. Cambridge, UK: Cambridge University Press, 1981.

    De Duve, C. Blueprint for a Cell: The Nature and Origin of Life. Burlington, NC: Neil Patterson Publishers, 1991.

12. Li, W.-H.; Graur, D. Fundamentals of Molecular Evolution. Sunderland, MA: Sinauer Associates, 1991.

    Sidow, A.; Bowman, B.H. Molecular phylogeny. *Curr. Opin. Genet. Devel.* 1:451–456, 1991.

13. Dickerson, R.E. Cytochrome c and the evolution of energy metabolism. *Sci. Am.* 242(3):136–153, 1980.

14. Cavalier-Smith, T. The origin of eukaryote and archaebacterial cells. *Ann. N.Y. Acad. Sci.* 503:17–54, 1987.

    Gray, M.W. The evolutionary origins of organelles. *Trends Genet.* 5:294–299, 1989.

    Margulis, L. Symbiosis in Cell Evolution. New York: W.H. Freeman, 1981.

15. Gray, M.W. Origin and evolution of mitochondrial DNA. *Annu. Rev. Cell Biol.* 5:25–50, 1989.

    Sogin, M.L.; Gunderson, J.H.; Elwood, H.J.; Alonso, R.A.; Peattie, D.A. Phylogenetic meaning of the kingdom concept: an unusual ribosomal RNA from *Giardia lamblia. Science* 243:75–77, 1989.

Vossbrinck, C.R.; Maddox, J.V.; Friedman, S.; Debrunner-Vossbrinck, B.A.; Woese, C.R. Ribosomal RNA sequence suggests microsporidia are extremely ancient eukaryotes. *Nature* 326:411–414, 1987.

16. Bryant, D.A. Puzzles of chloroplast ancestry. *Curr. Biol.* 2:240–242, 1992.

17. Sleigh, M.A. Protozoa and Other Protists. London: Edward Arnold, 1989.

18. Buchsbaum, R., et al. Animals Without Backbones, 3rd ed. Chicago: University of Chicago Press, 1987.

    Field, K.G., et al. Molecular phylogeny of the animal kingdom. *Science* 239:748–753, 1988.

    Knoll, A.H. The end of the proterozoic eon. *Sci. Am.* 265(4):64–73, 1991.

    Levinton, J.S. The big bang of animal evolution. *Sci. Am.* 267(5):84–91, 1992

    Shapiro, J.A. Bacteria as multicellular organisms. S*ci. Am.* 258(6):82–89, 1988.

    Valentine, J.W. The evolution of multicellular plants and animals. *Sci. Am.* 239(3):140–158, 1978.

19. Bode, P.M.; Bode, H.R. Patterning in Hydra. In Pattern Formation (G.M. Malacinski, S.V. Bryant, eds.), pp. 213–244. New York: Macmillan, 1984.

20. Raff, R.A.; Kaufman, T.C. Embryos, Genes, and Evolution. New York: Macmillan, 1983.

    Slack, J.M.W.; Holland, P.W.H.; Graham, C.F. The zootype and the phylotypic stage. *Nature* 361:490–492, 1993.

21. Chothia, C. One thousand families for the molecular biologist. *Nature* 357:543–544, 1992.

    Miklos, G.L.; Campbell, H.D. The evolution of protein domains and the organizational complexities of metazoans. *Curr. Opin. Genet. Devel.* 2:902–906, 1992.

Small Molecules, Energy, and Biosynthesis

"I must tell you that I can prepare urea without requiring a kidney or an animal, either man or dog." This sentence, written 165 years ago by the young German chemist Wöhler, signaled an end to the belief in a special *vital force* that exists in living organisms and gives rise to their distinctive properties and products. But what was a revelation in Wöhler's time is common knowledge today—living creatures are made of chemicals, obedient simply to the laws of chemistry and physics. This is not to say that no mysteries remain in biology: there are many areas of ignorance, as will become apparent in later chapters. But we should begin by emphasizing the enormous amount that is known.

We now have detailed information about the essential molecules of the cell—not just a small number of molecules, but thousands of them. In many cases we know their precise chemical structures and exactly how they are made and broken down. We know in general terms how chemical energy drives the biosynthetic reactions of the cell, how thermodynamic principles operate in cells to create molecular order, and how the myriad chemical changes occurring continuously within cells are controlled and coordinated.

In this and the next chapter we briefly survey the chemistry of the living cell. Here we deal with the processes involving small molecules: those mechanisms by which the cell synthesizes its fundamental chemical ingredients and by which it obtains its energy. Chapter 3 describes the giant molecules of the cell, which are polymers of a subset of the small molecules; these polymers are responsible both for the specificity of biological processes and for the transfer of biological information.

The Chemical Components of a Cell

Cell Chemistry Is Based on Carbon Compounds [1]

A living cell is composed of a restricted set of elements, four of which (C, H, N, and O) make up nearly 99% of its weight. This composition differs markedly from that of the earth's crust and is evidence of a distinctive type of chemistry (Figure 2–1). What is this special chemistry, and how did it evolve?

The most abundant substance of the living cell is water. It accounts for about 70% of a cell's weight, and most intracellular reactions occur in an aqueous environment. Life on this planet began in the ocean, and the conditions in that

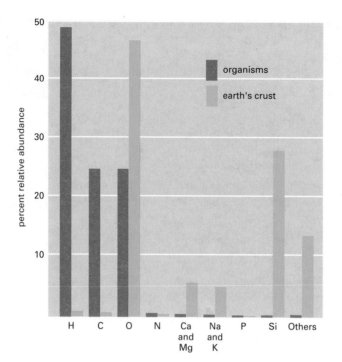

Figure 2–1 The relative abundance of chemical elements found in the earth's crust (the nonliving world) compared to that in the soft tissues of living organisms. The relative abundance is expressed as a percentage of the total *number* of atoms present. Thus, for example, nearly 50% of the atoms in living organisms are hydrogen atoms.

primeval environment put a permanent stamp on the chemistry of living things. All organisms have been designed around the special properties of water, such as its polar character, its ability to form hydrogen bonds, and its high surface tension. Water will completely surround polar molecules, for example, while tending to push nonpolar molecules together into larger assemblies. Some important properties of water are summarized in Panel 2–1 (pp. 48–49).

If we disregard water, nearly all of the molecules in a cell are carbon compounds, which are the subject matter of **organic chemistry.** Carbon is outstanding among all the elements on earth for its ability to form large molecules; silicon is a poor second. The carbon atom, because of its small size and four outer-shell electrons, can form four strong covalent bonds with other atoms. Most important, it can join to other carbon atoms to form chains and rings and thereby generate large and complex molecules with no obvious upper limit to their size. The other abundant atoms in the cell (H, N, and O) are also small and able to make very strong covalent bonds (Panel 2–2, pp. 50–51).

A typical covalent bond in a biological molecule has an energy of 15 to 170 Kcal/mole, depending on the atoms involved. Since the average thermal energy at body temperature is only 0.6 kcal/mole, even an unusually energetic collision with another molecule will leave a covalent bond intact. Specific catalysts, however, can rapidly break or rearrange covalent bonds. Biology is made possible by the combination of the stability of covalent bonds under physiological conditions and the ability of biological catalysts (called *enzymes*) to break and rearrange these bonds in a controlled way in selected molecules.

In principle, the simple rules of covalent bonding between carbon and other elements permit an infinitely large number of compounds. Although the number of different carbon compounds in a cell is very large, it is only a tiny subset of what is theoretically possible. In some cases we can point to good reasons why this compound or that performs a given biological function; more often it seems that the actual "choice" was one among many reasonable alternatives and therefore something of an accident (Figure 2–2). Once established in an ancient cell, certain chemical themes and patterns of reaction were preserved, with variations, during billions of years of cellular evolution. Apparently, the development of new classes of compounds was only rarely necessary or useful.

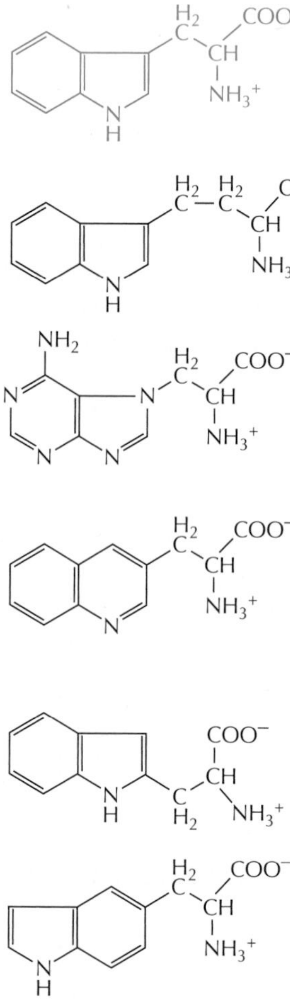

Figure 2–2 Living organisms synthesize only a small number of the organic molecules that they in principle could make. Of the six amino acids shown, only the top one (tryptophan) is made by cells.

Cells Use Four Basic Types of Small Molecules [2]

Certain simple combinations of atoms—such as the methyl ($-CH_3$), hydroxyl ($-OH$), carboxyl ($-COOH$), and amino ($-NH_2$) groups—recur repeatedly in biological molecules. Each such group has distinct chemical and physical properties that influence the behavior of whatever molecule the group occurs in. The main types of chemical groups and some of their salient properties are summarized in Panel 2–2 (pp. 50–51).

The atomic weights of H, C, N, and O are 1, 12, 14, and 16, respectively. The **small organic molecules** of the cell have molecular weights in the range 100 to 1000 and contain up to 30 or so carbon atoms. They are usually found free in solution, where some of them form a pool of intermediates from which large polymers, called **macromolecules,** are made. They are also essential intermediates in the chemical reactions that transform energy derived from food into usable forms (discussed below).

The small molecules amount to about one-tenth of the total organic matter in a cell, and (at a rough estimate) only on the order of a thousand different kinds are present (Table 2–1). All biological molecules are synthesized from and broken down to the same simple compounds. Both synthesis and breakdown occur through sequences of chemical changes that are limited in scope and follow definite rules. As a consequence, the compounds in a cell are chemically related and can be classified into a small number of distinct families. Since the macromolecules in a cell, which form the subject of Chapter 3, are assembled from these small molecules, they belong to corresponding families.

Broadly speaking, cells contain just four major families of small organic molecules: the simple **sugars,** the **fatty acids,** the **amino acids,** and the **nucleotides.** Each of these families contains many different members with common chemical features. Although some cellular compounds do not fit into these categories, the four families, and especially the macromolecules made from them, account for a surprisingly large fraction of the mass of every cell (Table 2–1).

Sugars Are Food Molecules of the Cell [3]

The simplest sugars—the **monosaccharides**—are compounds with the general formula $(CH_2O)_n$, where n is an integer from 3 through 7. *Glucose,* for example, has the formula $C_6H_{12}O_6$ (Figure 2–3). As shown in Figure 2–3, sugars can exist in either a ring or an open-chain form. In their open-chain form sugars contain a number of hydroxyl groups and either one aldehyde ($_H{>}C{=}O$) or one ketone (${>}C{=}O$) group. The aldehyde or ketone group plays a special role. First, it can react with a hydroxyl group in the same molecule to convert the molecule into

---

Table 2–1 The Approximate Chemical Composition of a Bacterial Cell

	Percent of Total Cell Weight	Number of Types of Each Molecule
Water	70	1
Inorganic ions	1	20
Sugars and precursors	1	250
Amino acids and precursors	0.4	100
Nucleotides and precursors	0.4	100
Fatty acids and precursors	1	50
Other small molecules	0.2	~300
Macromolecules (proteins, nucleic acids, and polysaccharides)	26	~3000

(handwritten annotation: "organic small molecules" bracketing Sugars through Fatty acids; "organic" next to Macromolecules)

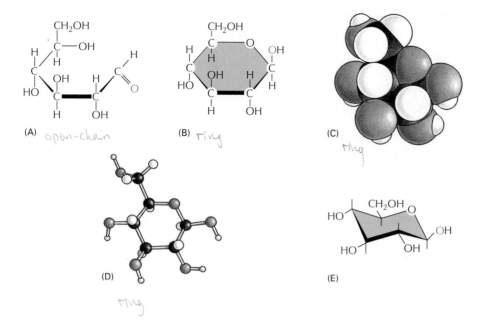

(A) open-chain   (B) ring   (C) ring

(D) ring   (E)

Figure 2–3 The structure of the monosaccharide glucose, a common hexose sugar. (A) is the open-chain form of this sugar, which is in equilibrium with the more stable cyclic or ring form in (B). (C) and (D) are space-filling and ball-and-stick models, respectively, of this cyclic form (β-D-glucose). The chair form (E) is an alternative representation of the cyclic form that is frequently used because it more accurately reflects the structure. In (A), (B), and (E) the *red* O denotes the oxygen atom of the aldehyde group. For an outline of sugar structures and chemistry, see Panel 2–3 (pp. 52–53).

a ring; in the ring form the carbon of the original aldehyde or ketone group can be recognized as the only one that is bonded to two oxygens. Second, once the ring is formed, this carbon can become further linked to one of the carbons bearing a hydroxyl group on another sugar molecule, creating a *disaccharide* (Panel 2–3, pp. 52–53). The addition of more monosaccharides in the same way results in **oligosaccharides** of increasing length (trisaccharides, tetrasaccharides, and so on) up to very large **polysaccharide** molecules with thousands of monosaccharide units. Because each monosaccharide has several free hydroxyl groups that can form a link to another monosaccharide (or to some other compound), the number of possible polysaccharide structures is extremely large. Even a simple disaccharide consisting of two glucose residues can exist in eleven different varieties (Figure 2–4), while three different hexoses ($C_6H_{12}O_6$) can join together to

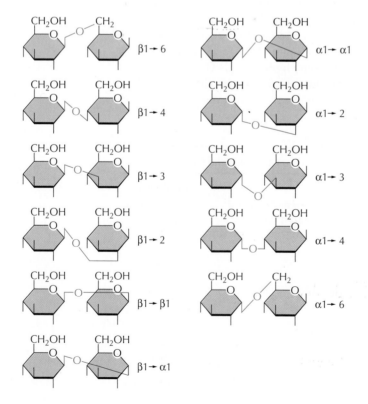

Figure 2–4 Eleven disaccharides consisting of two D-glucose units. Although these differ only in the type of linkage between the two glucose units, they are chemically distinct. Since the oligosaccharides associated with proteins and lipids may have six or more different kinds of sugar joined in both linear and branched arrangements through linkages such as those illustrated here, the number of distinct types of oligosaccharides that can be used in cells is extremely large.

make several thousand different trisaccharides. It is very difficult to determine the structure of any particular polysaccharide because one needs to determine the sites of linkage between each sugar unit and its neighbors. With present methods, for instance, it takes longer to determine the arrangement of half a dozen linked sugars (those in a glycoprotein, for example) than to determine the nucleotide sequence of a DNA molecule containing many thousands of nucleotides (where each unit is joined to the next in exactly the same way).

Glucose is the principal food compound of many cells. A series of oxidative reactions (see p. 62) leads from this hexose to various smaller sugar derivatives and eventually to CO_2 and H_2O. The net result can be written

$$C_6H_{12}O_6 + 6O_2 \rightarrow 6CO_2 + 6H_2O + energy$$

In the course of glucose breakdown, energy and "reducing power," both of which are essential in biosynthetic reactions, are salvaged and stored, mainly in the form of ATP in the case of energy and NADH for reducing power. We discuss the structures and functions of these two crucial molecules later in the chapter.

Simple polysaccharides composed only of glucose residues—principally _glycogen_ in animal cells and _starch_ in plant cells—are used to store energy for future use. But sugars have functions in addition to the production and storage of energy. Important extracellular structural materials (such as cellulose) are composed of simple polysaccharides, and smaller but more complex chains of sugar molecules are often covalently linked to proteins in _glycoproteins_ and to lipids in _glycolipids_.

Fatty Acids Are Components of Cell Membranes [4]

A fatty acid molecule, such as _palmitic acid_ (Figure 2–5), has two distinct regions: a long hydrocarbon chain, which is hydrophobic (water insoluble) and not very reactive chemically, and a carboxylic acid group, which is ionized in solution (COO^-), extremely hydrophilic (water soluble), and readily reacts with a hydroxyl or an amino group on a second molecule to form esters and amides. In fact, almost all of the fatty acid molecules in a cell are covalently linked to other molecules by their carboxylic acid group. The many different fatty acids found in cells differ in the length of their hydrocarbon chains and the number and position of the carbon-carbon double bonds they contain (Panel 2–4, pp. 54–55).

Fatty acids are a valuable source of food since they can be broken down to produce more than twice as much usable energy, weight for weight, as glucose. They are stored in the cytoplasm of many cells in the form of droplets of _triglyceride_ molecules, which consist of three fatty acid chains, each joined to a glycerol molecule (Panel 2–4, pp. 54–55); these molecules are the animal fats familiar from everyday experience. When required to provide energy, the fatty acid chains can be released from triglycerides and broken down into two-carbon units. These two-carbon units, present as the acetyl group in a water-soluble molecule called _acetyl CoA_, are then further degraded in various energy-yielding reactions, which we describe below.

But the most important function of fatty acids is in the construction of cell membranes. These thin, impermeable sheets that enclose all cells and surround their internal organelles are composed largely of **phospholipids,** which are small molecules that resemble triglycerides in that they are constructed mostly from fatty acids and glycerol. In phospholipids, however, the glycerol is joined to two rather than three fatty acid chains. The remaining site on the glycerol is coupled to a negatively charged phosphate group, which is in turn attached to another small hydrophilic compound, such as _ethanolamine, choline,_ or _serine._

Each phospholipid molecule, therefore, has a hydrophobic tail—composed of the two fatty acid chains—and a hydrophilic polar head group, where the phosphate is located. A small amount of phospholipid will spread over the surface of water to form a _monolayer_ of phospholipid molecules; in this thin film,

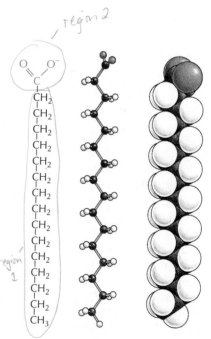

Figure 2–5 **Palmitic acid.** The carboxylic acid group (_red_) is shown in its ionized form. A ball-and-stick model (_center_) and a space-filling model (_right_) are also shown.

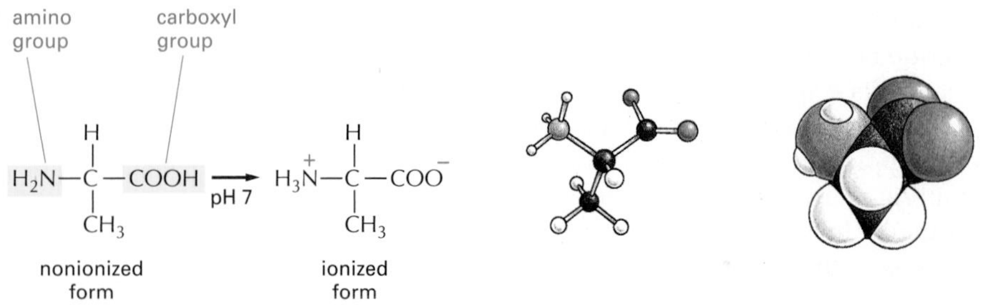

Figure 2–6 **The amino acid alanine.** In the cell, where the pH is close to 7, the free amino acid exists in its ionized form; but when it is incorporated into a polypeptide chain, the charges on the amino and carboxyl groups disappear. A ball-and-stick model and a space-filling model are shown to the right of the structural formulas. For alanine, the side chain is a $-CH_3$ group.

the hydrophobic tail regions pack together very closely facing the air and the hydrophilic head groups are in contact with the water (Panel 2–4, pp. 54–55). Two such films can combine tail to tail in water to make a phospholipid sandwich, or **lipid bilayer,** an extremely important assembly that is the structural basis of all cell membranes (discussed in Chapter 10).

Amino Acids Are the Subunits of Proteins [5]

The common amino acids are chemically varied, but they all contain a carboxylic acid group and an amino group, both linked to a single carbon atom (called the α-carbon; Figure 2–6). They serve as subunits in the synthesis of **proteins,** which are long linear polymers of amino acids joined head to tail by a *peptide bond* between the carboxylic acid group of one amino acid and the amino group of the next (Figure 2–7). Although there are many different possible amino acids, only 20 are common in proteins, each with a different *side chain* attached to the α-carbon atom (Panel 2–5, pp. 56–57). The same 20 amino acids occur over and over again in all proteins, including those made by bacteria, plants, and animals. Although the choice of these particular 20 amino acids probably occurred by chance in the course of evolution, the chemical versatility they provide is vitally important. For example, 5 of the 20 amino acids have side chains that can carry a charge (Figure 2–8), whereas the others are uncharged but reactive in specific ways (Panel 2–5, pp. 56–57). As we shall see, the properties of the amino acid side chains, in aggregate, determine the properties of the proteins they constitute and underlie all of the diverse and sophisticated functions of proteins.

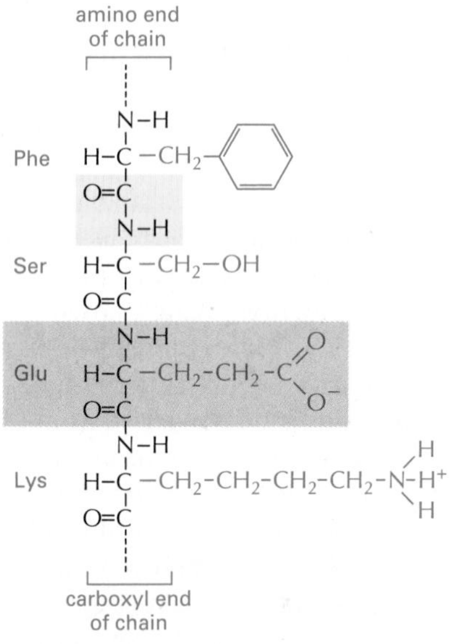

Nucleotides Are the Subunits of DNA and RNA [6]

In nucleotides one of several different nitrogen-containing ring compounds (often referred to as *bases* because they can combine with H^+ in acidic solutions) is linked to a five-carbon sugar (either *ribose* or *deoxyribose*) that carries a phosphate group. There is a strong family resemblance between the different nitrogen-containing rings found in nucleotides. *Cytosine* (C), *thymine* (T), and *uracil* (U) are called *pyrimidine* compounds because they are all simple derivatives of a six-membered pyrimidine ring; *guanine* (G) and *adenine* (A) are *purine* compounds, with a second five-membered ring fused to the six-membered ring. Each nucleotide is named by reference to the unique base that it contains (Panel 2–6, pp.58–59).

Nucleotides can act as carriers of chemical energy. The triphosphate ester of adenine, **ATP** (Figure 2–9), above all others, participates in the transfer of energy in hundreds of individual cellular reactions. Its terminal phosphate is added using energy from the oxidation of foodstuffs, and this phosphate can be split off readily by hydrolysis to release energy that drives energetically unfavorable biosynthetic reactions elsewhere in the cell. As we discuss later, other nucleotide derivatives serve as carriers for the transfer of particular chemical groups, such as hydrogen atoms or sugar residues, from one molecule to another. And a cyclic phosphate-

Figure 2–7 **A small part of a protein molecule, showing four amino acids.** Each amino acid is linked to the next by a covalent *peptide bond*, one of which is shaded *yellow*. A protein is therefore also sometimes referred to as a *polypeptide*. The amino acid *side chains* are shown in *red*, and the atoms of one amino acid (glutamic acid) are outlined by the *gray box*.

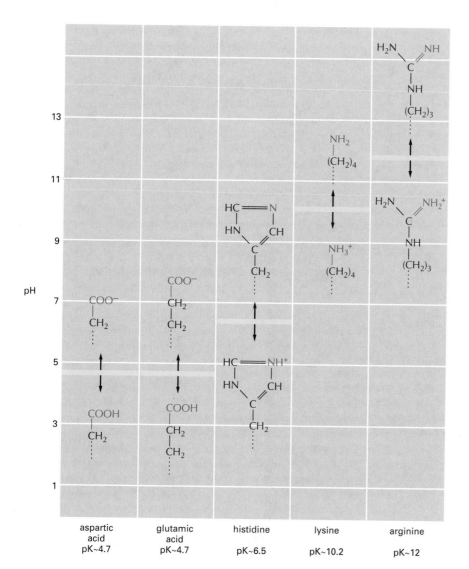

Figure 2–8 The charge on amino acid side chains depends on the pH. Carboxylic acids readily lose H⁺ in aqueous solution to form a negatively charged ion, which is denoted by the suffix "-ate," as in aspart*ate* or glutam*ate*. A comparable situation exists for amines, which in aqueous solution take up H⁺ to form a positively charged ion (which does not have a special name). These reactions are rapidly reversible, and the amounts of the two forms, charged and uncharged, depend on the pH of the solution. At a high pH, carboxylic acids tend to be charged and amines uncharged. At a low pH, the opposite is true—the carboxylic acids are uncharged and amines are charged. The pH at which exactly *half* of the carboxylic acid or amine residues are charged is known as the pK of that amino acid side chain.

In the cell the pH is close to 7, and almost all carboxylic acids and amines are in their fully charged form.

containing adenine derivative, *cyclic AMP*, serves as a universal signaling molecule within cells.

The special significance of nucleotides is in the storage of biological information. Nucleotides serve as building blocks for the construction of **nucleic acids,** long polymers in which nucleotide subunits are covalently linked by the formation of a phosphate ester between the 3′-hydroxyl group on the sugar residue of one nucleotide and the 5′-phosphate group on the next nucleotide (Fig-

Figure 2–9 Chemical structure of adenosine triphosphate (ATP). A space-filling model (A), a ball-and-stick model (B), and the structural formula (C) are shown. Note the negative charges on each of the three phosphates.

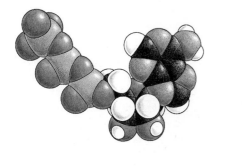

(A)

(B)

(C)

The Chemical Components of a Cell

HYDROGEN BONDS

Because they are polarized, two adjacent H_2O molecules can form a linkage known as a hydrogen bond. Hydrogen bonds have only about 1/20 the strength of a covalent bond.

Hydrogen bonds are strongest when the three atoms lie in a straight line.

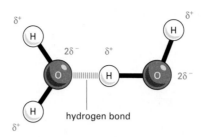

bond lengths

hydrogen bond
0.28 nm

0.104 nm
covalent bond

WATER

Two atoms, connected by a covalent bond, may exert different attractions for the electrons of the bond. In such cases the bond is dipolar, with one end slightly negatively charged (δ^-) and the other slightly positively charged (δ^+). A bond in which both atoms are the same, or in which they attract electrons equally, is called nonpolar.

electropositive
region

electronegative
region

Although a water molecule has an overall neutral charge (having the same number of electrons and protons), the electrons are asymmetrically distributed, which makes the molecule polar. The oxygen nucleus draws electrons away from the hydrogen nuclei, leaving these nuclei with a small net positive charge. The excess of electron density on the oxygen atom creates weakly negative regions at the other two corners of an imaginary tetrahedron.

WATER STRUCTURE

Molecules of water join together transiently in a hydrogen-bonded lattice. Even at 37°C, 15% of the water molecules are joined to four others in a short-lived assembly known as a "flickering cluster."

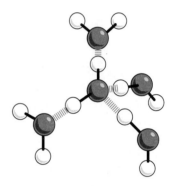

The cohesive nature of water is responsible for many of its unusual properties, such as high surface tension, specific heat, and heat of vaporization.

HYDROPHILIC AND HYDROPHOBIC MOLECULES

Because of the polar nature of water molecules, they will cluster around ions and other polar molecules.

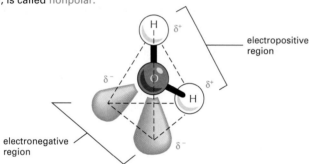

Molecules that can thereby be accommodated in water's hydrogen-bonded structures are hydrophilic and relatively water-soluble.

Nonpolar molecules interrupt the H-bonded structure of water without forming favorable interactions with water molecules. They are therefore hydrophobic and quite insoluble in water.

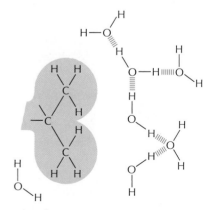

HYDROPHOBIC MOLECULES AND CLATHRATE WATER STRUCTURES

Molecules that are nonpolar and cannot form hydrogen bonds—such as hydrocarbons—have only limited solubility in water and are called hydrophobic. In water, ordered cages of water molecules are formed around hydrocarbons. These icelike cages, called "clathrate structures," are relatively more ordered than water and cause an entropy decrease of the mixture. Part of a clathrate cage (*red*) surrounding a hydrocarbon (*black*) is shown. In the intact cage, each oxygen atom (*red circles*) would be tetrahedrally coordinated to four others.

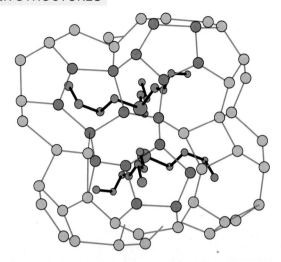

ACIDS AND BASES

An acid is a molecule that releases an H^+ ion (proton) in solution. For example,

$$CH_3-C\underset{OH}{\overset{O}{\big|}} \rightleftharpoons CH_3-C\underset{O^-}{\overset{O}{\big|}} + H^+$$

acid   base   proton

A base is a molecule that accepts an H^+ ion (proton) in solution. For example,

$$CH_3-NH_2 + H^+ \rightleftharpoons CH_3-NH_3^+$$

base   proton   acid

Water itself has a slight tendency to ionize and can act both as a weak acid and as a weak base. When it acts as an acid, it releases a proton to form a hydroxyl ion. When it acts as a base, it accepts a proton to form a hydronium ion. Most protons in aqueous solutions exist as hydronium ions.

hydroxyl ion   hydronium ion

pH

The acidity of a solution is defined by the concentration of H^+ ions it possesses. For convenience we use the pH scale where

$$pH = -\log_{10}[H^+]$$

For pure water

$$[H^+] = 10^{-7} \text{ moles/liter}$$

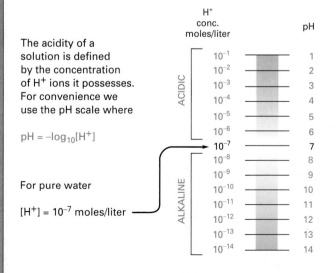

	H⁺ conc. moles/liter	pH
ACIDIC	10^{-1}	1
	10^{-2}	2
	10^{-3}	3
	10^{-4}	4
	10^{-5}	5
	10^{-6}	6
	10^{-7}	7
ALKALINE	10^{-8}	8
	10^{-9}	9
	10^{-10}	10
	10^{-11}	11
	10^{-12}	12
	10^{-13}	13
	10^{-14}	14

OSMOSIS

If two aqueous solutions are separated by a membrane that allows only water molecules to pass, water will move into the solution containing the greatest concentration of solute molecules by a process known as osmosis.

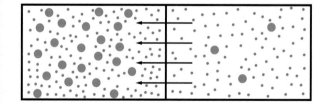

This movement of water from a hypotonic to a hypertonic solution can cause an increase in hydrostatic pressure in the hypertonic compartment. Two solutions that have identical solute concentrations and are therefore osmotically balanced are said to be isotonic.

CARBON SKELETONS

The unique role of carbon in the cell comes from its ability to form strong covalent bonds with other carbon atoms. Thus carbon atoms can join to form chains.

or branched trees

or rings

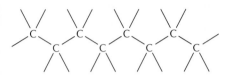

also written as

also written as

also written as

COVALENT BONDS

A covalent bond forms when two atoms come very close together and share one or more of their electrons. In a single bond one electron from each of the two atoms is shared; in a double bond a total of four electrons are shared.

Each atom forms a fixed number of covalent bonds in a defined spatial arrangement. For example, carbon forms four single bonds arranged tetrahedrally, whereas nitrogen forms three single bonds and oxygen forms two single bonds arranged as shown below.

Double bonds exist and have a different spatial arrangement.

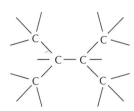

Atoms joined by two or more covalent bonds cannot rotate freely about the bond axis. This restriction is a major influence on the three-dimensional shape of many macromolecules.

HYDROCARBONS

Carbon and hydrogen together make stable compounds called hydrocarbons. These are nonpolar, do not form hydrogen bonds, and are generally insoluble in water.

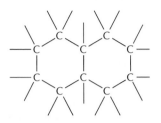

methane   methyl group

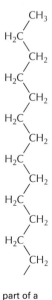

part of a
fatty acid chain

RESONANCE AND AROMATICITY

The carbon chain can include double bonds. If these are on alternate carbon atoms, the bonding electrons move within the molecule, stabilizing the structure by a phenomenon called resonance.

the truth is somewhere between these two structures

When resonance occurs throughout a ring compound, an aromatic ring is generated.

often written as

Panel 2–2 Chemical bonds and groups commonly encountered in biological molecules.

C—O COMPOUNDS

Many biological compounds contain a carbon bonded to an oxygen. For example,

alcohol

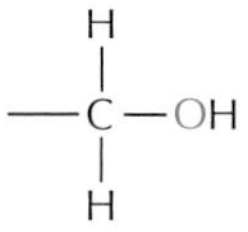

The —OH is called a hydroxyl group.

aldehyde

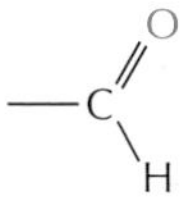

ketone

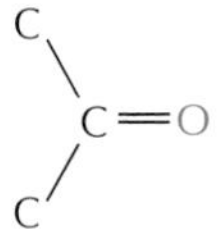

The C=O is called a carbonyl group.

carboxylic acid

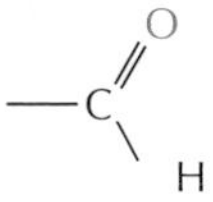

The —COOH is called a carboxyl group. In water this loses an H^+ ion to become —COO⁻.

esters Esters are formed by combining an acid and an alcohol.

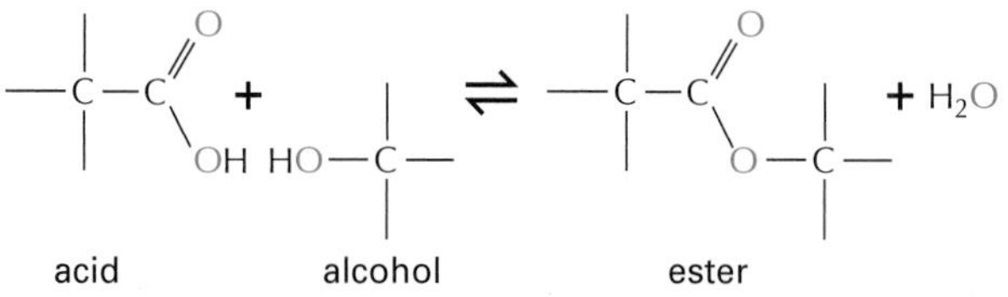

acid alcohol ester

C—N COMPOUNDS

Amines and amides are two important examples of compounds containing a carbon linked to a nitrogen.

Amines in water combine with an H^+ ion to become positively charged.

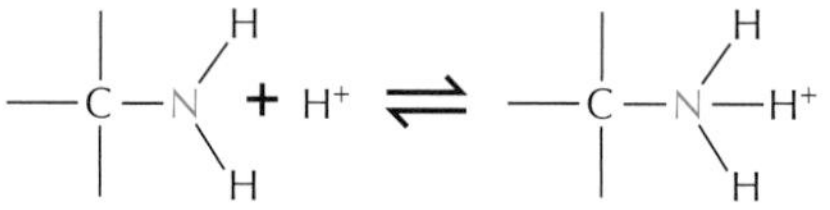

They are therefore basic.

Amides are formed by combining an acid and an amine. They are more stable than esters. Unlike amines, they are uncharged in water. An example is the peptide bond.

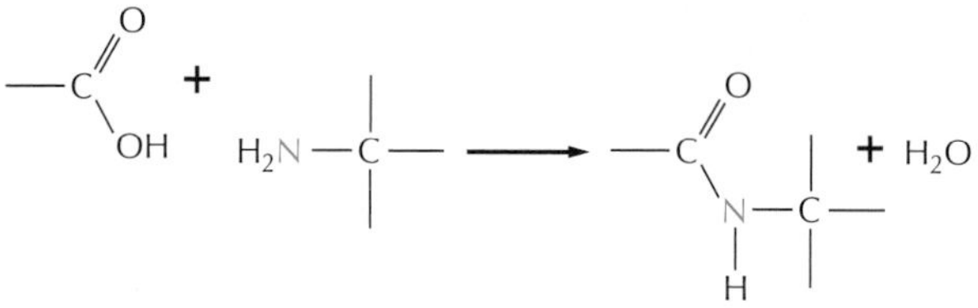

Nitrogen also occurs in several ring compounds, including important constituents of nucleic acids: purines and pyrimidines.

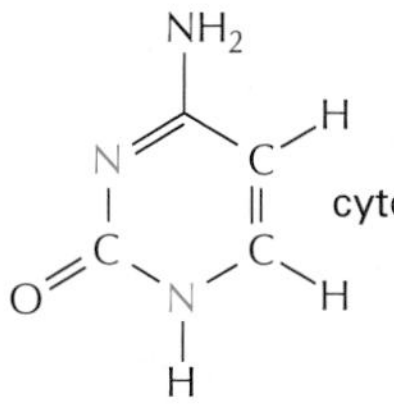

cytosine (a pyrimidine)

PHOSPHATES

Inorganic phosphate is a stable ion formed from phosphoric acid, H_3PO_4. It is often written as P_i.

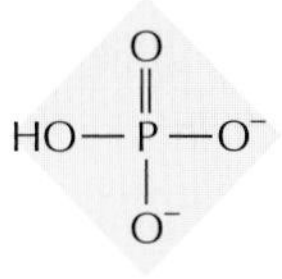

Phosphate esters can form between a phosphate and a free hydroxyl group.

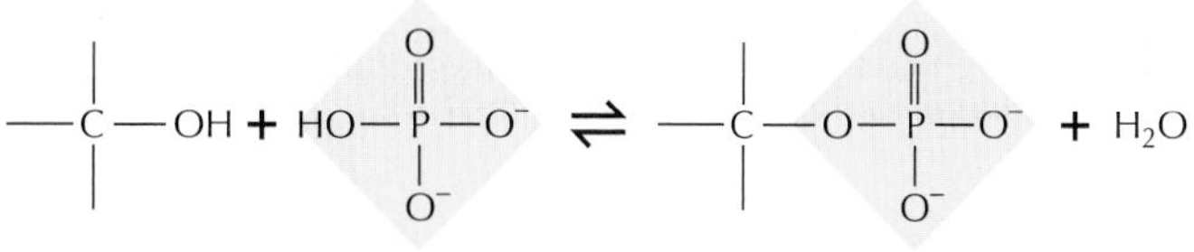

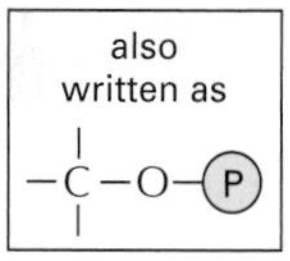

also written as

The combination of a phosphate and a carboxyl group, or two or more phosphate groups, gives an acid anhydride.

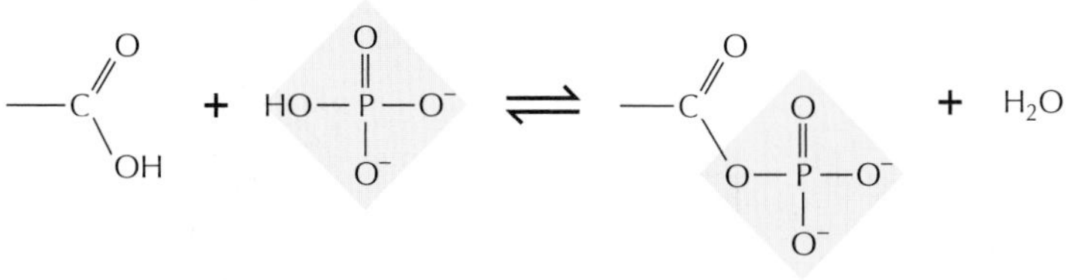

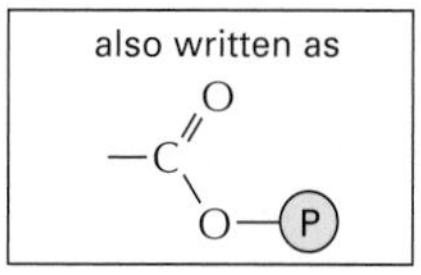

also written as

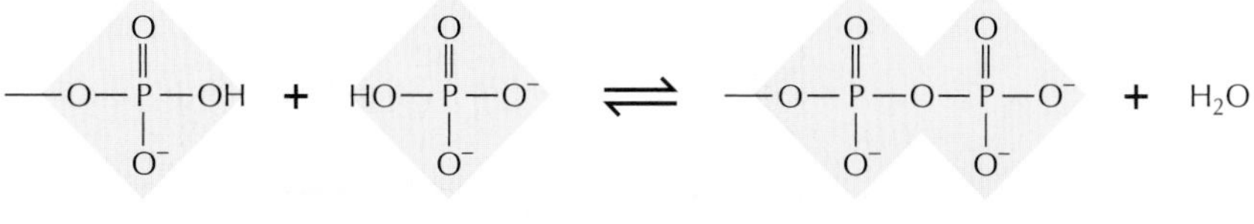

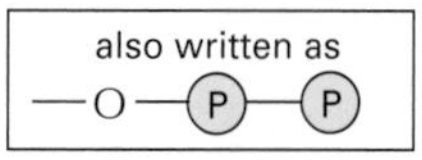

also written as

HEXOSES $n = 6$

Two common hexoses are

glucose   fructose

MONOSACCHARIDES

Monosaccharides are
aldehydes or ketones

that also have two or more hydroxyl groups.
Their general formula is $(CH_2O)n$. The
simplest are trioses ($n = 3$) such as

glyceraldehyde (an aldose)

dihydroxyacetone (a ketose)

PENTOSES $n = 5$

A common pentose is

ribose

D-glucose (open-chain form)

D-ribose (open-chain form)

RING FORMATION

The aldehyde or ketone group of a
sugar can react with a hydroxyl group

For the larger sugars ($n > 4$) this
happens within the same molecule
to form a 5- or 6-membered ring.

NUMBERING

The carbon atoms of a sugar are
numbered from the end closest
to the aldehyde or ketone.

β-D-glucose   α-D-glucose

STEREOISOMERS

β-D-ribose   α-D-ribose

STEREOISOMERS

ISOMERS

Monosaccharides have many isomers that differ only in the
orientation of their hydroxyl groups—e.g., glucose, mannose,
and galactose are isomers of each other.

glucose   mannose

galactose

D AND L FORMS

Two isomers that are mirror images of each other have the
same chemistry and therefore are given the same name and
distinguished by the prefix D or L.

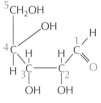
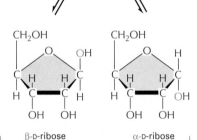

D-glucose

mirror plane

L-glucose

    Panel 2–3 An outline of some of the types of sugars commonly found in cells.

α- AND β-LINKS

The hydroxyl group on the carbon that carries the aldehyde or ketone can rapidly change from one position to another. These two positions are called α- and β-.

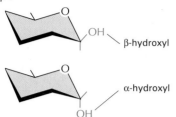

β-hydroxyl

α-hydroxyl

As soon as one sugar is linked to another, the α- or β-form is frozen.

SUGAR DERIVATIVES

The hydroxyl groups of a simple monosaccharide can be replaced by other groups. For example,

D-glucuronic acid

D-glucosamine

N-acetyl-D-glucosamine

DISACCHARIDES

The carbon that carries the aldehyde or the ketone can react with any hydroxyl group on a second sugar molecule to form a glycosidic bond. Three common disaccharides are maltose (glucose α1,4 glucose), lactose (galactose β1,4 glucose), and sucrose (glucose α1,2 fructose). Sucrose is shown here.

α-D-glucose

β-D-fructose

$+$

H_2O

sucrose (glucose α1,2 fructose)

OLIGOSACCHARIDES AND POLYSACCHARIDES

Large linear and branched molecules can be made from simple repeating units. Short chains are called oligosaccharides, while long chains are called polysaccharides. Glycogen, for example, is a polysaccharide made entirely of glucose units joined together.

α1,6 links occur at branch points

glycogen

all other links are α1,4

COMPLEX OLIGOSACCHARIDES

In many cases a sugar sequence is nonrepetitive. Many different molecules are possible. Such complex oligosaccharides are usually linked to proteins or to lipids.

a blood group oligosaccharide

COMMON FATTY ACIDS

These are carboxylic acids with long hydrocarbon tails.

Hundreds of different kinds of fatty acids exist. Some have one or more double bonds and are said to be unsaturated.

```
COOH    COOH    COOH
 |       |       |
CH2     CH2     CH2
 |       |       |
CH2     CH2     CH2
 |       |       |
CH2     CH2     CH2
 |       |       |
CH2     CH2     CH2
 |       |       |
CH2     CH2     CH2
 |       |       |
CH2     CH2     CH2
 |       |       |
CH2     CH2     CH2
 |       |       |
CH2     CH2     CH
 |       |       ||
CH2     CH2     CH
 |       |       |
CH2     CH2     CH2
 |       |       |
CH2     CH2     CH2
 |       |       |
CH2     CH2     CH2
 |       |       |
CH2     CH2     CH2
 |       |       |
CH2     CH3     CH2
 |     palmitic  |
CH2     acid    CH2
 |      (C16)    |
CH3             CH3
stearic        oleic
acid           acid
(C18)          (C18)
```

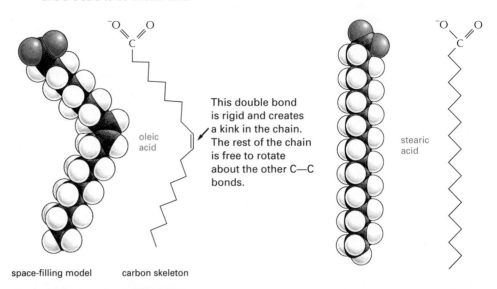

This double bond is rigid and creates a kink in the chain. The rest of the chain is free to rotate about the other C—C bonds.

oleic acid

stearic acid

space-filling model   carbon skeleton

TRIGLYCERIDES

Fatty acids are stored as an energy reserve (fat) through an ester linkage to glycerol to form triglycerides.

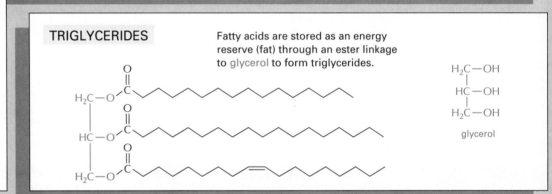

```
H2C—OH
 |
HC—OH
 |
H2C—OH
```
glycerol

CARBOXYL GROUP

If free, the carboxyl group of a fatty acid will be ionized.

But more usually it is linked to other groups to form either esters

or amides.

PHOSPHOLIPIDS

Phospholipids are the major constituents of cell membranes.

a phospholipid

polar head group

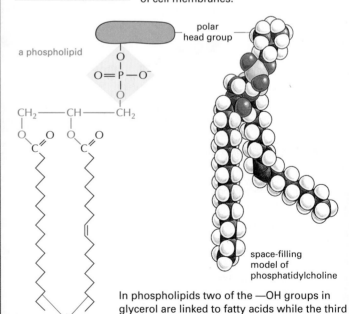

space-filling model of phosphatidylcholine

hydrophobic fatty acid "tails"

In phospholipids two of the —OH groups in glycerol are linked to fatty acids while the third —OH group is linked to phosphoric acid. The phosphate is further linked to one of a variety of small polar head groups (alcohols).

Panel 2–4 An outline of some of the types of fatty acids commonly encountered in cells and the structures that they form.

LIPID AGGREGATES

Fatty acids have a hydrophilic head and a hydrophobic tail.

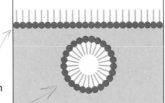

In water they can form a surface film or form small micelles.

Their derivatives can form larger aggregates held together by hydrophobic forces:

Triglycerides form large spherical fat droplets in the cell cytoplasm.

Phospholipids and glycolipids form self-sealing lipid bilayers that are the basis for all cellular membranes.

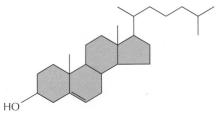

200 nm or more

cell membrane

4 nm

OTHER LIPIDS

Lipids are defined as the water-insoluble molecules in cells that are soluble in organic solvents. Two other common types of lipids are steroids and polyisoprenoids. Both are made from isoprene units.

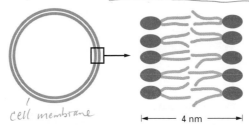

CH_3
$C-CH=CH_2$
CH_2 isoprene

STEROIDS

Steroids have a common multiple-ring structure.

HO

cholesterol—found in many membranes

OH

O

testosterone—male steroid hormone

GLYCOLIPIDS

Like phospholipids, these compounds are composed of a hydrophobic region, containing two long hydrocarbon tails, and a polar region, which now contains one or more sugar residues and no phosphate.

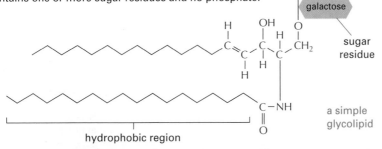

galactose

sugar residue

a simple glycolipid

hydrophobic region

POLYISOPRENOIDS

long chain polymers of isoprene

dolichol phosphate—used to carry activated sugars in the membrane-associated synthesis of glycoproteins and some polysaccharides

THE AMINO ACID

The general formula of an amino acid is

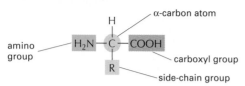

amino group

α-carbon atom

carboxyl group

side-chain group

R is commonly one of 20 different side chains. At pH 7 both the amino and carboxyl groups are ionized.

$$H_3N^{\oplus}-\overset{\overset{\displaystyle H}{|}}{\underset{\underset{\displaystyle R}{|}}{C}}-COO^{\ominus}$$

OPTICAL ISOMERS

The α-carbon atom is asymmetric, which allows for two mirror image (or stereo-) isomers, D and L.

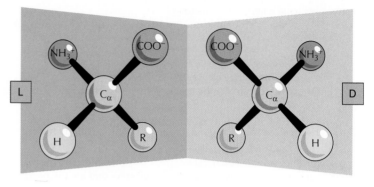

L   D

Proteins consist exclusively of L-amino acids.

PEPTIDE BONDS

Amino acids are commonly joined together by an amide linkage, called a peptide bond.

peptide bond: The four atoms in each *gray box* form a rigid planar unit. There is no freedom of rotation about the C—N bond.

Proteins are long polymers of amino acids linked by peptide bonds, and they are always written with the *N*-terminus toward the left. The sequence of this tripeptide is His Cys Val.

amino or *N*-terminus

carboxyl or *C*-terminus

These two single bonds, on either side of the rigid peptide unit, exhibit a high degree of rotational freedom.

FAMILIES OF AMINO ACIDS

The common amino acids are grouped according to whether their side chains are

acidic
basic
uncharged polar
nonpolar

These 20 amino acids are given both three-letter and one-letter abbreviations.

Thus: alanine = Ala = A

BASIC SIDE CHAINS

lysine
(Lys, or K)

This group is very basic because its positive charge is stabilized by resonance.

arginine
(Arg, or R)

histidine
(His, or H)

These nitrogens have a relatively weak affinity for an H⁺ and are only partly positive at neutral pH.

ACIDIC SIDE CHAINS

aspartic acid
(Asp, or D)

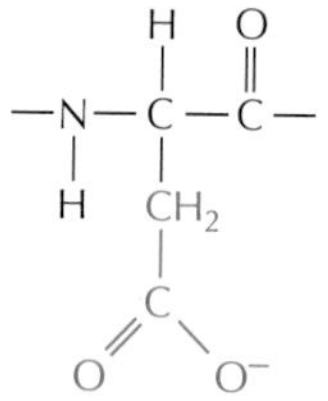

glutamic acid
(Glu, or E)

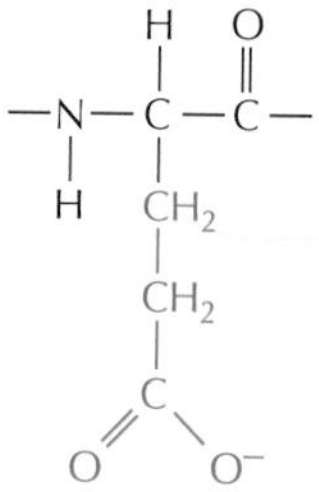

Amino acids with uncharged polar side chains are relatively hydrophilic and are usually on the outside of proteins, while the side chains on nonpolar amino acids tend to cluster together on the inside. Amino acids with basic or acidic side chains are very polar, and they are nearly always found on the outside of protein molecules.

The one letter code in alphabetical order:

A = Ala	G = Gly	M = Met	S = Ser
C = Cys	H = His	N = Asn	T = Thr
D = Asp	I = Ile	P = Pro	V = Val
E = Glu	K = Lys	Q = Gln	W = Trp
F = Phe	L = Leu	R = Arg	Y = Tyr

UNCHARGED POLAR SIDE CHAINS

asparagine
(Asn, or N)

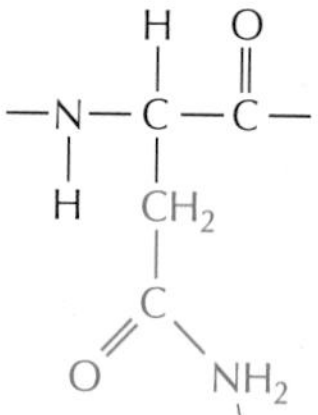

glutamine
(Gln, or Q)

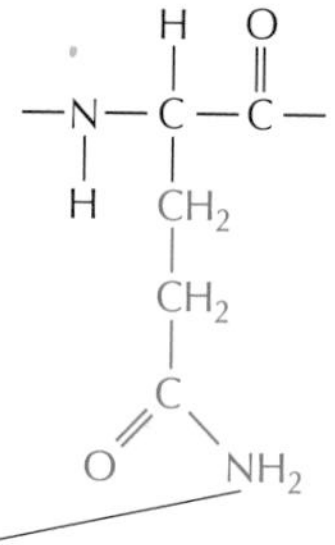

Although the amide N is not charged at neutral pH, it is polar.

serine
(Ser, or S)

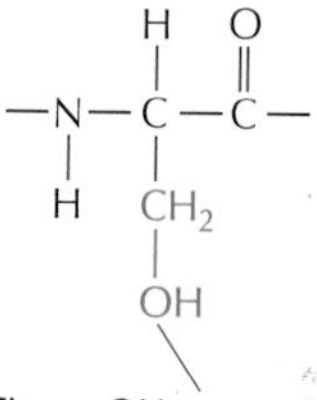

threonine
(Thr, or T)

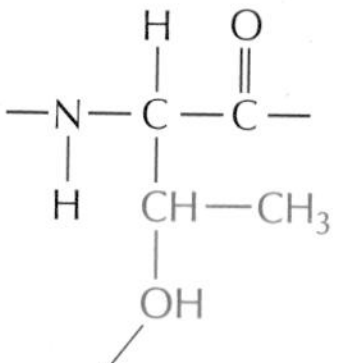

tyrosine
(Tyr, or Y)

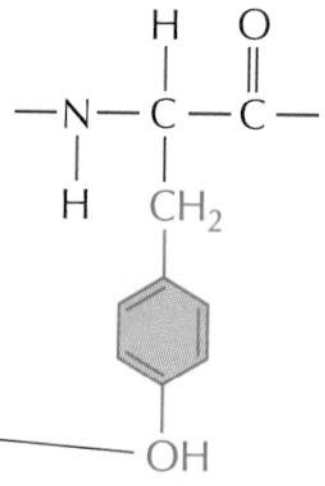

The —OH group is polar.

NONPOLAR SIDE CHAINS

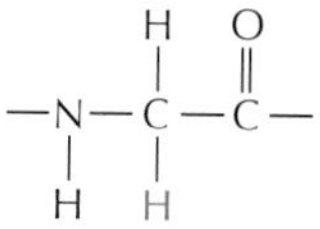

glycine
(Gly, or G)

alanine
(Ala, or A)

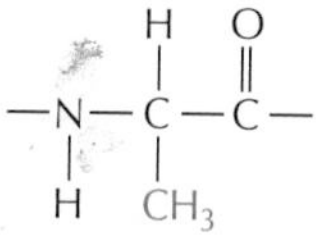

valine
(Val, or V)

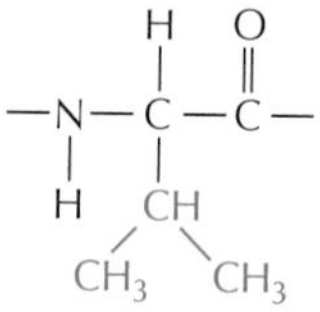

leucine
(Leu, or L)

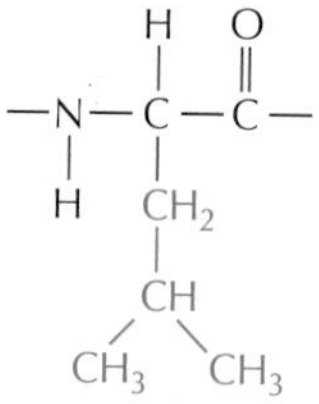

isoleucine
(Ile, or I)

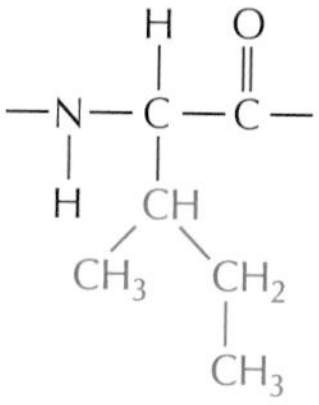

proline
(Pro, or P)

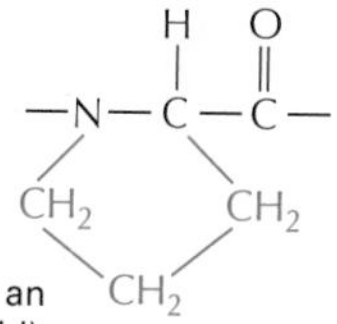

(actually an imino acid)

phenylalanine
(Phe, or F)

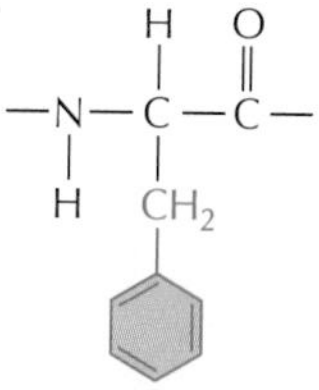

methionine
(Met, or M)

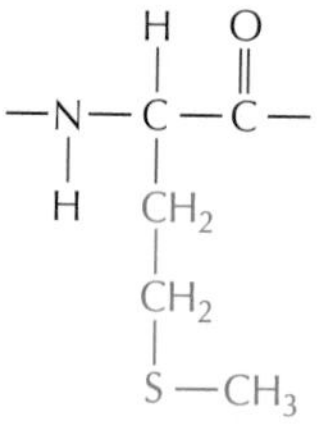

tryptophan
(Trp, or W)

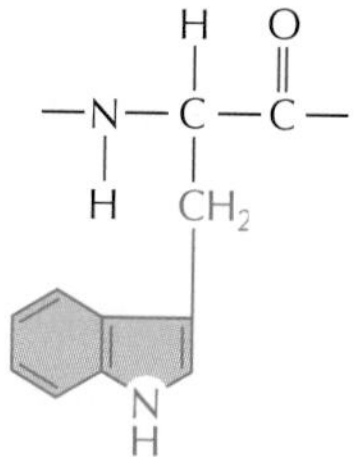

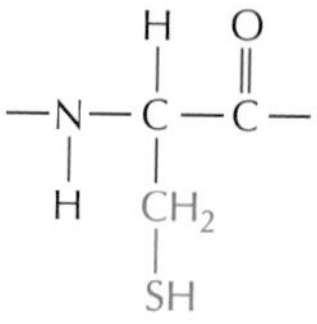

cysteine
(Cys, or C)

Paired cysteines allow disulfide bonds to form in proteins.

$$--CH_2-S-S-CH_2--$$

BASES

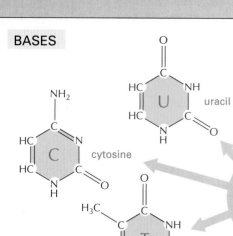

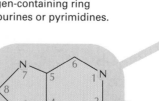

The bases are nitrogen-containing ring compounds, either purines or pyrimidines.

PYRIMIDINE

PURINE

adenine

A

guanine

G

cytosine

uracil

thymine

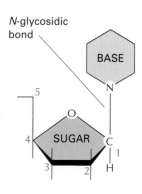

PHOSPHATES

The phosphates are normally joined to the C5 hydroxyl of the ribose or deoxyribose sugar. Mono-, di-, and triphosphates are common.

as in AMP

as in ADP

as in ATP

The phosphate makes a nucleotide negatively charged.

NUCLEOTIDES

A nucleotide consists of a nitrogen-containing base, a 5-carbon sugar, and one or more phosphate groups.

BASE

PHOSPHATE

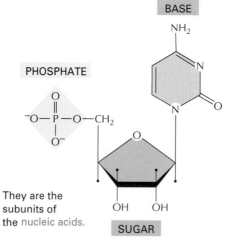

They are the subunits of the nucleic acids.

SUGAR

BASIC SUGAR LINKAGE

N-glycosidic bond

BASE

SUGAR

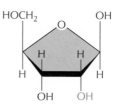

The base is linked to the same carbon (C1) used in sugar-sugar bonds.

SUGARS

PENTOSE

a 5-carbon sugar

two kinds are used

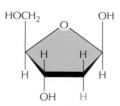

β-D-RIBOSE
used in ribonucleic acid

β-D-2-DEOXYRIBOSE
used in deoxyribonucleic acid

Each numbered carbon on the sugar of a nucleotide is followed by a prime mark; therefore, one speaks of the "5-prime carbon," etc.

The names can be confusing, but the abbreviations are clear.

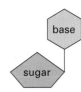

BASE	NUCLEOSIDE	ABBR.
adenine	adenosine	A
guanine	guanosine	G
cytosine	cytidine	C
uracil	uridine	U
thymine	thymidine	T

Nucleotides are abbreviated by three capital letters. Some examples follow:

AMP = adenosine monophosphate
dAMP = deoxyadenosine monophosphate
UDP = uridine diphosphate
ATP = adenosine triphosphate

BASE + SUGAR = NUCLEOSIDE

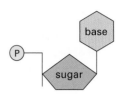

BASE + SUGAR + PHOSPHATE = NUCLEOTIDE

NUCLEIC ACIDS

Nucleotides are joined together by a phosphodiester linkage between 5' and 3' carbon atoms to form nucleic acids. The linear sequence of nucleotides in a nucleic acid chain is commonly abbreviated by a one-letter code, A—G—C—T—T—A—C—A, with the 5' end of the chain at the left.

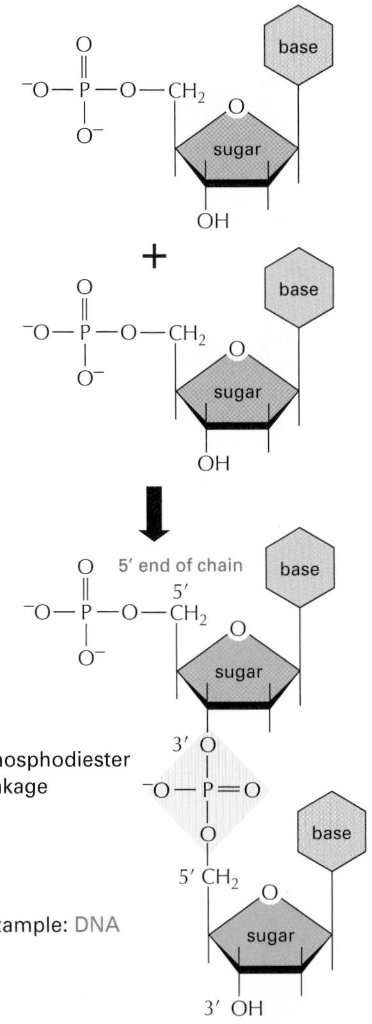

+

phosphodiester linkage

5' end of chain

3' end of chain

example: DNA

NUCLEOTIDES HAVE MANY OTHER FUNCTIONS

1 They carry chemical energy in their easily hydrolyzed acid-anhydride bonds.

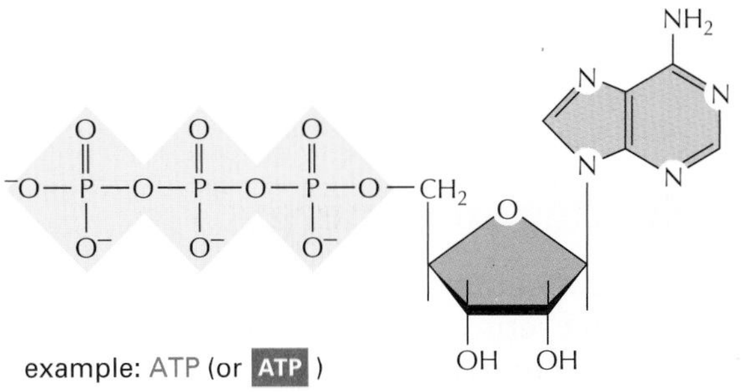

example: ATP (or ATP)

2 They combine with other groups to form enzymes.

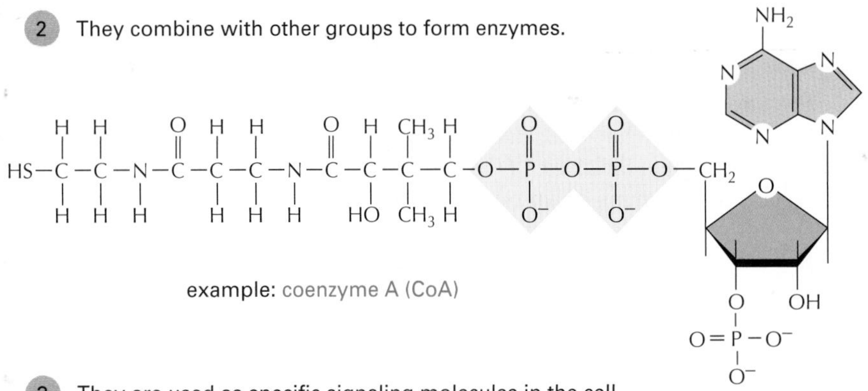

example: coenzyme A (CoA)

3 They are used as specific signaling molecules in the cell.

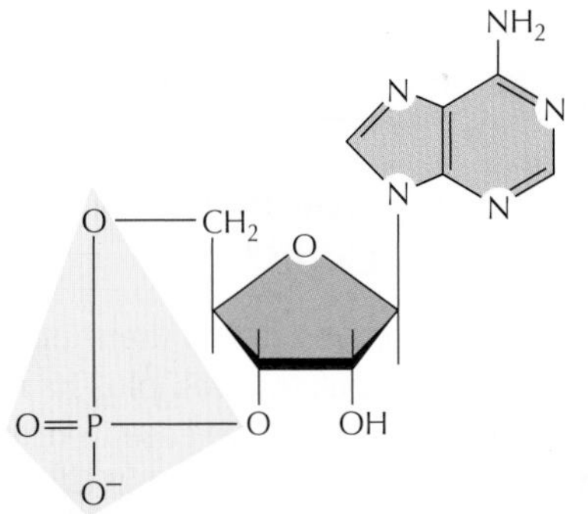

example:
cyclic AMP (cAMP)

ure 2–10). There are two main types of nucleic acids, differing in the type of sugar that forms their polymeric backbone. Those based on the sugar *ribose* are known as **ribonucleic acids,** or **RNA,** and contain the four bases A, U, G, and C. Those based on *deoxyribose* (in which the hydroxyl at the 2′ position of ribose is replaced by a hydrogen) are known as **deoxyribonucleic acids,** or **DNA,** and contain the four bases A, T, G, and C. The sequence of bases in a DNA or RNA polymer represents the genetic information of the living cell. The ability of the bases from different nucleic acid molecules to recognize each other by noncovalent interactions (called **base-pairing**)—G with C, and A with either T (in DNA) or U (in RNA)—underlies all of heredity and evolution, as explained in Chapter 3.

Summary

Living organisms are autonomous, self-propagating chemical systems. They are made from a distinctive and restricted set of small carbon-based molecules that are essentially the same for every living species. The main categories are sugars, fatty acids, amino acids, and nucleotides. Sugars are a primary source of chemical energy for cells and can be incorporated into polysaccharides for energy storage. Fatty acids are also important for energy storage, but their most significant function is in the formation of cell membranes. Polymers consisting of amino acids constitute the remarkably diverse and versatile macromolecules known as proteins. Nucleotides play a central part in energy transfer and also are the subunits from which the informational macromolecules, RNA and DNA, are made.

Biological Order and Energy [7]

Cells must obey the laws of physics and chemistry. The rules of mechanics and of the conversion of one form of energy to another apply just as much to a cell as to a steam engine. There are, however, puzzling features of a cell that, at first sight, seem to place it in a special category. It is common experience that things left to themselves eventually become disordered: buildings crumble, dead organisms become oxidized, and so on. This general tendency is expressed in the *second law of thermodynamics,* which states that the degree of disorder in the universe (or in any isolated system in the universe) can only increase.

The puzzle is that living organisms maintain, at every level, a very high degree of order; and as they feed, develop, and grow, they appear to create this order out of raw materials that lack it. Order is strikingly apparent in large structures such as a butterfly wing or an octopus eye, in subcellular structures such as a mitochondrion or a cilium, and in the shape and arrangement of molecules from which these structures are built. The constituent atoms have been captured, ultimately, from a relatively disorganized state in the environment and locked together into a precise structure. Even a nongrowing cell requires constant ordering processes for survival since all of its organized structures are subject to spontaneous accidents and must be repaired continually. How is this possible thermodynamically? The answer is that the cell draws in fuel from its environment and releases heat as a waste product. The cell is therefore not an isolated system in the thermodynamic sense.

Biological Order Is Made Possible by the Release of Heat Energy from Cells [8]

As already mentioned, the second law of thermodynamics states that the amount of order in the universe (that is, in a cell plus its environment) must always decrease. Therefore, the continuous increase in order inside a living cell must be accompanied by an even greater increase in *dis*order in the cell's environment. Heat is energy in its most disordered form—the random commotion of mol-

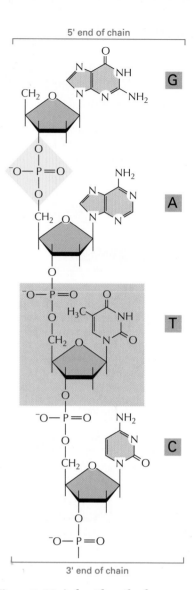

Figure 2–10 A short length of deoxyribonucleic acid (DNA), showing four nucleotides. One of the phosphodiester bonds that link adjacent nucleotides is shaded *yellow,* and one of the nucleotides is enclosed in a *gray box.* DNA and its close relative RNA are the nucleic acids of the cell.

ecules—and heat is released from the cell by the reactions that order the molecules it contains. The increase in random motion, including bond distortions, of the molecules in the rest of the universe creates a disorder that more than compensates for the increased order in the cell, as required by the laws of thermodynamics for spontaneous processes. In this way the release of heat by a cell to its surroundings allows it to become more highly ordered internally at the same time that the universe as a whole becomes more disordered (Figure 2–11).

It is important to note that the cell will achieve nothing by producing heat unless the heat-generating reactions are directly linked with the processes that generate molecular order in the cell. Such linked reactions are said to be *coupled*, as we explain later. It is the tight coupling of heat production to an increase in order that distinguishes the metabolism of the cell from the wasteful burning of fuel in a fire.

The creation of order inside the cell is at the expense of the degradation of fuel energy. For plants this fuel energy is initially derived from the electromagnetic radiation of the sun; for animals it is derived from the energy stored in the covalent bonds of the organic molecules that animals eat. Since these organic nutrients are themselves produced by photosynthetic organisms such as green plants, however, the sun is in fact the ultimate energy source for animals also.

Photosynthetic Organisms Use Sunlight to Synthesize Organic Compounds [9]

Solar energy enters the living world (the *biosphere*) by means of the **photosynthesis** carried out by photosynthetic organisms—either plants or bacteria. In photosynthesis electromagnetic energy is converted into chemical bond energy. At the same time, however, part of the energy of sunlight is converted into heat energy, and the release of this heat to the environment increases the disorder of the universe and thereby drives the photosynthetic process.

The reactions of photosynthesis are described in detail in Chapter 14. In broad terms, they occur in two distinct stages. In the first (the *light-activated reactions*) the visible radiation impinging on a pigment molecule drives the transfer of electrons from water to NADPH and at the same time provides the energy needed for the synthesis of ATP. In the second (the *dark reactions*) the ATP and NADPH are used to drive a series of "carbon-fixation" reactions in which CO_2 from the air is used to form sugar molecules (Figure 2–12).

The *net* result of photosynthesis, so far as the green plant is concerned, can be summarized by the equation

organic compound

$$\text{energy} + CO_2 + H_2O \rightarrow \text{sugar} + O_2$$

This simple equation hides the complex nature of the reactions, which involve many linked reaction steps. Furthermore, although the initial fixation of CO_2 results in sugars, subsequent metabolic reactions soon convert these into the many other small and large molecules essential to the plant cell.

Chemical Energy Passes from Plants to Animals

Animals and other nonphotosynthetic organisms cannot capture energy from sunlight directly and so have to survive on "secondhand" energy obtained by eating plants or on "thirdhand" energy obtained by eating other animals. The organic molecules made by plant cells provide both building blocks and fuel to the organisms that feed on them. All types of plant molecules can serve this purpose—sugars, proteins, polysaccharides, lipids, and many others.

The transactions between plants and animals are not all one-way. Plants, animals, and microorganisms have existed together on this planet for so long that many of them have become an essential part of the others' environment. The oxygen released by photosynthesis, for example, is consumed in the combustion

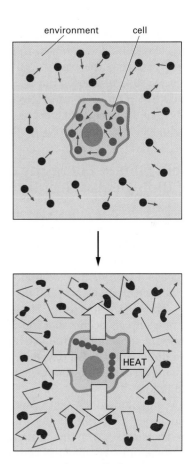

environment   cell

Figure 2–11 A simple thermodynamic analysis of a living cell. In the upper diagram, the molecules of both the cell and the rest of the universe (its environment) are depicted in a relatively disordered state. In the lower diagram, heat has been released from the cell by a reaction that orders the molecules that the cell contains (*green*). Because the heat increases the disorder in the environment around the cell, the second law of thermodynamics is satisfied as the cell grows and divides.

of organic molecules by nearly all organisms, and some of the CO_2 molecules that are "fixed" today into larger organic molecules by photosynthesis in a green leaf were yesterday released into the atmosphere by the respiration of an animal. Thus, carbon utilization is a cyclic process that involves the biosphere as a whole and crosses boundaries between individual organisms (Figure 2–13). Similarly, atoms of nitrogen, phosphorus, and sulfur can, in principle, be traced from one biological molecule to another in a series of similar cycles.

Cells Obtain Energy by the Oxidation of Biological Molecules [10]

The carbon and hydrogen atoms in the molecules taken up as food materials by a cell can serve as fuel because they are not in their most stable form. The earth's atmosphere contains a great deal of oxygen, and in the presence of oxygen the most energetically stable form of carbon is as CO_2 and that of hydrogen is as H_2O. A cell is therefore able to obtain energy from sugars or other organic molecules by allowing their carbon and hydrogen atoms to combine with oxygen to produce CO_2 and H_2O, respectively. The cell, however, does not oxidize organic molecules in one step, as occurs in a fire. Through the use of specific enzyme catalysts, it takes the molecules through a large number of reactions that only rarely involve the direct addition of oxygen. Before we can consider these reactions and the driving force behind them, we need to discuss what is meant by the process of oxidation.

Oxidation, in the sense used above, does not mean only the addition of oxygen atoms; rather, it applies more generally to any reaction in which electrons are transferred from one atom to another. **Oxidation** in this sense refers to the removal of electrons, and **reduction**—the converse of oxidation—means the addition of electrons. Thus, Fe^{2+} is oxidized if it loses an electron to become Fe^{3+}, and a chlorine atom is reduced if it gains an electron to become Cl^-. The same terms are used when there is only a partial shift of electrons between atoms linked by a covalent bond (Figure 2–14A). When a carbon atom becomes covalently bonded to an atom with a strong affinity for electrons, such as oxygen, chlorine, or sulfur, for example, it gives up more than its equal share of electrons; it therefore acquires a partial positive charge and is said to be oxidized. Conversely, a carbon atom in a C–H linkage has more than its share of electrons, and so it is said to be reduced (Figure 2–14B).

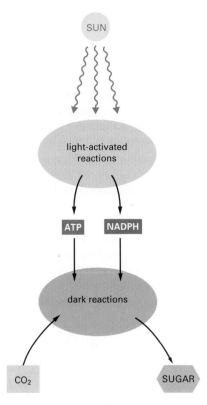

Figure 2–12 Photosynthesis. The two stages of photosynthesis in a green plant.

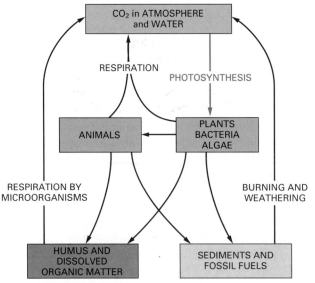

Figure 2–13 The carbon cycle. Individual carbon atoms are incorporated into organic molecules of the living world by the photosynthetic activity of plants, bacteria, and marine algae. They pass to animals, microorganisms, and organic material in soil and oceans in cyclic paths. CO_2 is restored to the atmosphere when organic molecules are oxidized by cells or burned by humans as fossil fuels.

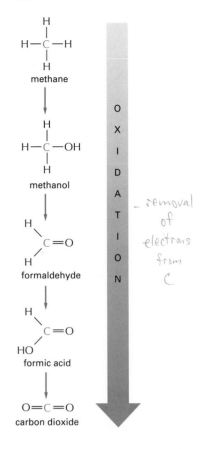

(A)

FORMATION OF COVALENT BOND

ATOM 1 + ATOM 2 → MOLECULE

this end has partial positive charge *(oxidized)*

this end has partial negative charge due to excess of electrons *(reduced)*

(B)

H—C—H (H above and below) methane

↓

H—C—OH (H above and below) methanol

↓

C=O (H above, H below) formaldehyde

↓

C=O (H above, HO below) formic acid

↓

O=C=O carbon dioxide

O X I D A T I O N

— removal of electrons from C

Figure 2–14 Oxidation and reduction. (A) When two atoms form a covalent bond, an atom ending up with a greater share of electrons acquires a partial negative charge and is said to be *reduced*, while the other atom acquires a partial positive charge and is said to be *oxidized*. (B) The carbon atom of methane can be converted to that of carbon dioxide by the successive removal of its hydrogen atoms. With each step, electrons are shifted away from the carbon, and the carbon atom becomes progressively more oxidized. Each of these steps is energetically favorable inside a cell.

Often, when a molecule picks up an electron (e^-), it picks up a proton (H^+) at the same time (protons being freely available in an aqueous solution). The net effect in this case is to add a hydrogen atom to the molecule

$$A + e^- + H^+ \rightarrow AH$$

Even though a proton plus an electron is involved (instead of just an electron), such *hydrogenation* reactions are reductions, and the reverse, *dehydrogenation* reactions, are oxidations.

The combustion of food materials in a cell converts the C and H atoms in organic molecules (where they are both in a relatively electron-rich, or reduced, state) to CO_2 and H_2O, where they have given up electrons and are therefore highly oxidized. The shift of electrons from carbon and hydrogen to oxygen allows these atoms to achieve a more stable state and hence is energetically favorable.

The Breakdown of an Organic Molecule Takes Place in a Sequence of Enzyme-catalyzed Reactions [11]

Although the most energetically favorable form of carbon is as CO_2 and that of hydrogen is as H_2O, a living organism does not disappear in a puff of smoke for the same reason that the book in your hands does not burst into flame: the molecules of both exist in metastable energy troughs and require **activation energy** (Figure 2–15) before they can pass to more stable configurations. In the case of the book, the activation energy can be provided by a lighted match. For a living cell the combustion is achieved molecule by molecule in a much more controlled way. The place of the match is taken by an unusually energetic collision of one molecule with another. Moreover, the only molecules that react are those that are bound to the surface of *enzymes*.

As explained in Chapter 3, **enzymes** are highly specific protein *catalysts*. Like all other types of catalysts, they speed up reactions by reducing the activation energy for a particular chemical change. Enzymes bind tightly to their substrate molecules and hold them in a way that greatly reduces the activation energy of one particular reaction that rearranges covalent bonds. By selectively lowering the activation energy of only one reaction path for the bound molecule, enzymes determine which of several alternative bond-breaking and bond-forming reactions occurs (Figure 2–16). After the product of one enzyme is released, it can

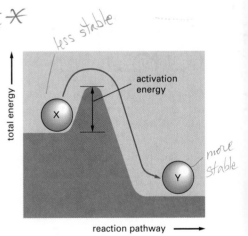

less stable

activation energy

X

Y

more stable

total energy

reaction pathway

Figure 2–15 The principle of activation energy. Compound X is in a metastable state because energy is released when it is converted to compound Y. This transition will not take place, however, unless X can acquire enough *activation energy* from its surroundings (by means of an unusually energetic collision with other molecules) to undergo the reaction.

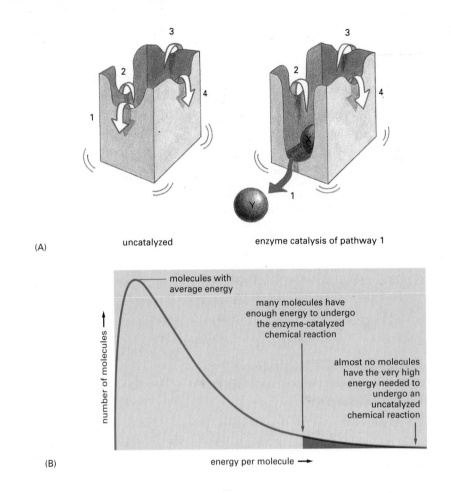

(A)   uncatalyzed   enzyme catalysis of pathway 1

(B)

molecules with average energy

many molecules have enough energy to undergo the enzyme-catalyzed chemical reaction

almost no molecules have the very high energy needed to undergo an uncatalyzed chemical reaction

number of molecules →

energy per molecule →

Figure 2–16 Enzyme catalysis. (A) A "jiggling-box" model illustrates how enzymes direct molecules along desired reaction pathways. In this model the *green ball* represents a potential enzyme substrate (compound X) that is bouncing up and down in energy level due to the constant bombardment of colliding water molecules. The four walls of the box represent the activation energy barriers for four different chemical reactions that are energetically favorable. In the left-hand box none of these reactions occurs because the energy available from even the most energetic collisions is insufficient to surmount any of the energy barriers. In the right-hand box enzyme catalysis lowers the activation energy for reaction number 1 only. It thereby allows this reaction to proceed with available energies, causing compound Y to form from compound X. Compound Y may then bind to a different enzyme that converts it to compound Z, and so on. (B) The distribution of energy in a population of identical molecules. In order to undergo a chemical reaction, the energy of the molecule (as translational, vibrational, and rotational motions) must exceed the activation energy; for most biological reactions, this never happens without enzyme catalysis.

bind to a second enzyme that catalyzes an additional change. In this way each of the many different molecules in a cell moves from enzyme to enzyme along a *specific* reaction pathway, and it is the sum of all of these pathways that determines the cell's chemistry. We discuss a few central pathways of energy metabolism later.

The success of living organisms is attributable to their cells' ability to make enzymes of many different types, each with precisely specified properties. Each enzyme has a unique shape and binds a particular set of other molecules (called *substrates*) in such a way as to speed up a particular chemical reaction enormously, often by a factor of as much as 10^{14}. Like all other catalysts, enzyme molecules themselves are not changed after participating in a reaction and therefore can function over and over again.

Part of the Energy Released in Oxidation Reactions Is Coupled to the Formation of ATP [12]

Cells derive useful energy from the "burning" of glucose only because they burn it in a very complex and controlled way. By means of enzyme-directed reaction paths, the synthetic, or *anabolic*, chemical reactions that create biological order are closely coupled to the degradative, or *catabolic*, reactions that provide the energy. The crucial difference between a **coupled reaction** and an uncoupled reaction is illustrated by the mechanical analogy shown in Figure 2–17, where an energetically favorable chemical reaction is represented by rocks falling from a cliff. The kinetic energy of falling rocks would normally be entirely wasted in the form of heat generated when they hit the ground (section A). But, by careful design, part of the kinetic energy could be used to drive a paddle wheel that lifts a bucket of water (section B). Because the rocks can reach the ground only by

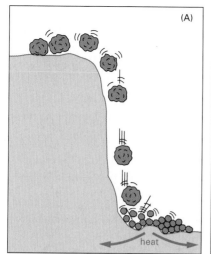

(A)

kinetic energy transformed into heat energy only

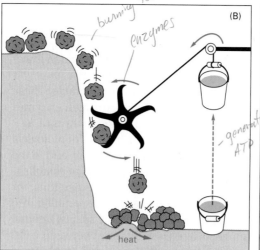

(B)

burning foodstuff

enzymes

generate ATP

part of the kinetic energy is used to lift a bucket of water, and a correspondingly smaller amount is transformed into heat

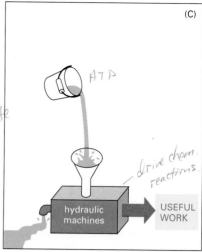

(C)

ATP

drive chem. reactions.

hydraulic machines

USEFUL WORK

the potential kinetic energy stored in the elevated bucket of water can be used to drive a wide variety of different hydraulic machines

moving the paddle wheel, we say that the spontaneous reaction of rock falling has been directly coupled to the nonspontaneous reaction of lifting the bucket of water. Note that because part of the energy is now used to do work in section B, the rocks hit the ground with less velocity than in section A, and therefore correspondingly less energy is wasted as heat.

In cells enzymes play the role of paddle wheels in our analogy and couple the spontaneous burning of foodstuffs to reactions that generate ATP. Just as the energy stored in the elevated bucket of water in Figure 2–17 can be dispensed in small doses to drive a wide variety of hydraulic machines (section C), ATP serves as a convenient and versatile store, or currency, of energy to drive many different chemical reactions that the cell needs (Figure 2–18).

The Hydrolysis of ATP Generates Order in Cells [13]

How does ATP act as a carrier of chemical energy? Under the conditions existing in the cytoplasm, the breakdown of ATP by hydrolysis to release inorganic phosphate (P_i) requires catalysis by an enzyme, but whenever it occurs, it releases a great deal of usable energy. A chemical group that is linked by such a reactive bond is readily transferred to another molecule; for this reason the terminal

Figure 2–17 A mechanical model illustrating the principle of coupled chemical reactions. The spontaneous reaction shown in (A) might serve as an analogy for the direct oxidation of glucose to CO_2 and H_2O, which produces heat only. In (B) the same reaction is coupled to a second reaction; the second reaction might serve as an analogy for the synthesis of ATP. The more versatile form of energy produced in (B) can be used to drive other cellular processes, as in (C). ATP is the most versatile form of energy in cells.

energy from catabolism or photosynthesis

oxidation (rocks falling)

ADENINE

RIBOSE

ATP

hydrolysis

energy available for cellular work and for chemical synthesis

ADENINE

RIBOSE

ADP

Figure 2–18 The ATP molecule serves as a convenient energy store in cells. As indicated, the energetically unfavorable formation of ATP from ADP and inorganic phosphate is coupled to the energetically favorable oxidation of foodstuffs (see Figure 2–17B). The hydrolysis of this ATP back to ADP and inorganic phosphate in turn provides the energy needed to drive many important cellular reactions.

Biological Order and Energy

phosphate in ATP can be considered to exist in an _activated_ state. The bond broken in this hydrolysis reaction is sometimes described as a _high-energy bond_. There is nothing special about the covalent bond itself, however; it is simply that in aqueous solution the hydrolysis of ATP creates two molecules of much lower energy (ADP and P_i).

Many of the chemical reactions in cells are energetically unfavorable. These reactions are driven by the energy released by ATP hydrolysis through enzymes that directly couple the unfavorable reaction to the favorable reaction of ATP hydrolysis. Among these reactions are those involved in the synthesis of biological molecules, in the active transport of molecules across cell membranes, and in the generation of force and movement. These processes play a vital part in establishing biological order. The macromolecules formed in biosynthetic reactions, for example, carry information, catalyze specific reactions, and are assembled into highly ordered structures. Membrane-bound pumps maintain the special internal composition of cells and permit many signals to pass within and between cells. And the production of force and movement enables the cytoplasmic contents of cells to become organized and the cells themselves to move about and assemble into tissues.

Summary

Living cells are highly ordered and must create order within themselves in order to survive and grow. This is thermodynamically possible only because of a continual input of energy, part of which is released from the cells to their environment as heat. The energy comes ultimately from the electromagnetic radiation of the sun, which drives the formation of organic molecules in photosynthetic organisms such as green plants. Animals obtain their energy by eating these organic molecules and oxidizing them in a series of enzyme-catalyzed reactions that are coupled to the formation of ATP. ATP is a common currency of energy in all cells, and its energetically favorable hydrolysis is coupled to other reactions to drive a variety of energetically unfavorable processes that create the high degree of order essential for life.

Food and the Derivation of Cellular Energy [14]

Food Molecules Are Broken Down in Three Stages to Give ATP

The proteins, lipids, and polysaccharides that make up the major part of the food we eat must be broken down into smaller molecules before our cells can use them. The enzymatic breakdown, or catabolism, of these molecules may be regarded as proceeding in three stages (Figure 2–19). We shall give a short outline of these stages before discussing the last two of them in more detail.

Stage 1, called **digestion,** occurs mainly in our intestine. Here, large polymeric molecules are broken down into their monomeric subunits—proteins into amino acids, polysaccharides into sugars, and fats into fatty acids and glycerol—through the action of secreted enzymes. Stage 2 occurs in the cytoplasm after the small molecules generated in stage 1 enter cells, where they are further degraded. Most of the carbon and hydrogen atoms of sugars are converted into _pyruvate_, which then enters mitochondria, where it is converted to the acetyl groups of the chemically reactive compound _acetyl coenzyme A (acetyl CoA)_ (Figure 2–20). Major amounts of acetyl CoA are also produced by the oxidation of fatty acids. In stage 3 the acetyl group of acetyl CoA is completely degraded to CO_2 and H_2O in the mitochondrion. It is in this final stage that most of the ATP is generated. Through a series of coupled chemical reactions, about half of the energy theo-

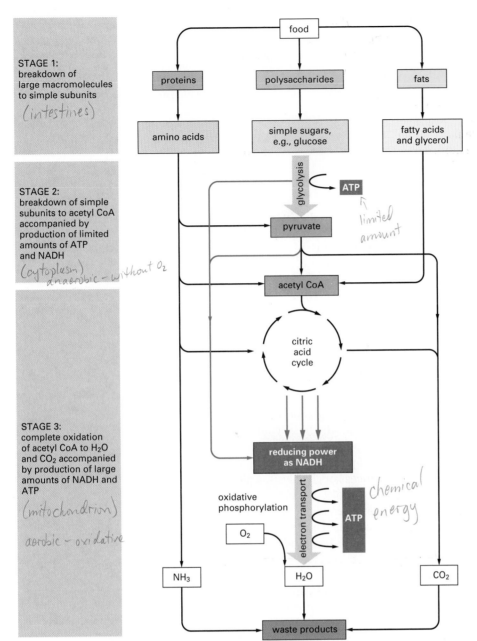

STAGE 1:
breakdown of
large macromolecules
to simple subunits

(intestines)

STAGE 2:
breakdown of simple
subunits to acetyl CoA
accompanied by
production of limited
amounts of ATP
and NADH

(cytoplasm)
anaerobic - without O_2

STAGE 3:
complete oxidation
of acetyl CoA to H_2O
and CO_2 accompanied
by production of large
amounts of NADH and
ATP

(mitochondrion)

aerobic - oxidative

food

proteins   polysaccharides   fats

amino acids   simple sugars,   fatty acids
                  e.g., glucose   and glycerol

glycolysis   ATP

limited
amount

pyruvate

acetyl CoA

citric
acid
cycle

reducing power
as NADH

oxidative
phosphorylation

electron transport

chemical
energy

ATP

O_2

NH_3 H_2O CO_2

waste products

Figure 2–19 **Simplified diagram of
the three stages of catabolism that
lead from food to waste products.**
This series of reactions produces ATP,
which is then used to drive
biosynthetic reactions and other
energy-requiring processes in the cell.

retically derivable from the combustion of carbohydrates and fats to H_2O and CO_2
is channeled into driving the energetically unfavorable reaction P_i + ADP → ATP.
Because the rest of the combustion energy is released by the cell as heat, this
generation of ATP creates net disorder in the universe, in conformity with the
second law of thermodynamics.

Through the production of ATP, the energy originally derived from the combustion of carbohydrates and fats is redistributed as a conveniently packaged
form of chemical energy. Roughly 10^9 molecules of ATP are in solution throughout the intracellular space in a typical cell, where their energetically favorable
hydrolysis back to ADP and phosphate in coupled reactions provides the driving energy for a very large number of different coupled reactions that would
otherwise not occur.

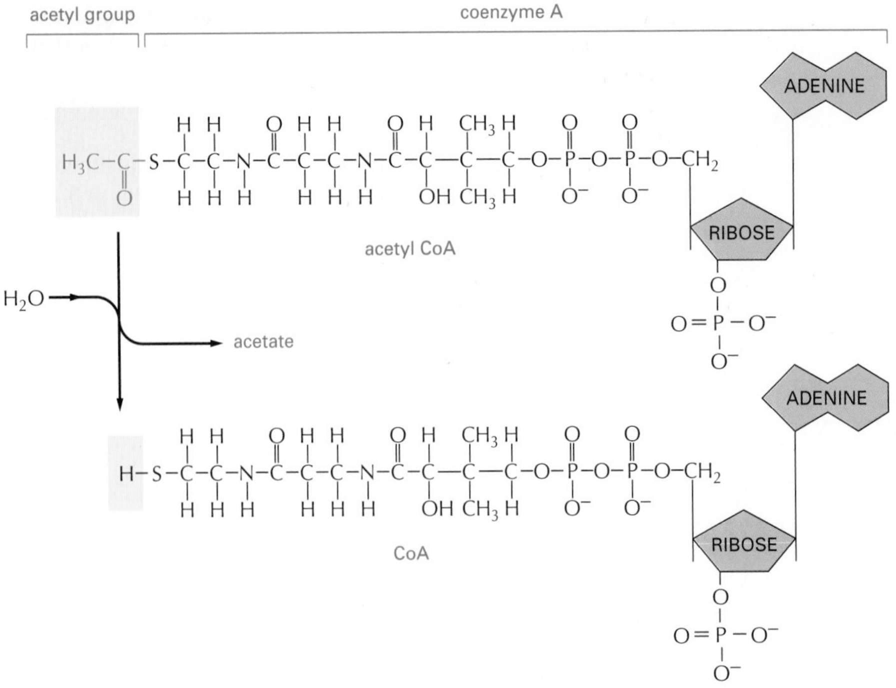

acetyl group   coenzyme A

acetyl CoA

CoA

Figure 2–20 **Acetyl coenzyme A (acetyl CoA).** This crucial metabolic intermediate is generated when acetyl groups, produced in stage 2 of catabolism, are covalently linked to coenzyme A (CoA).

Glycolysis Can Produce ATP Even in the Absence of Oxygen

The most important process in stage 2 of catabolism is the degradation of carbohydrates in a sequence of reactions known as **glycolysis**—the lysis (splitting) of glucose. Glycolysis can produce ATP in the absence of oxygen, and it probably evolved early in the history of life, before the activities of photosynthetic organisms introduced oxygen into the atmosphere. In the process of glycolysis, a glucose molecule with six carbon atoms is converted into two molecules of pyruvate, each with three carbon atoms. This conversion involves a sequence of nine enzymatic steps that create phosphate-containing intermediates (Figure 2–21). The cell hydrolyzes two molecules of ATP to drive the early steps but produces four molecules of ATP in the later steps, so that there is a net gain of ATP by the end of glycolysis.

Logically, the sequence of reactions that constitute glycolysis can be divided into three parts: (1) in steps 1 to 4, glucose is converted to two molecules of the three-carbon aldehyde *glyceraldehyde 3-phosphate*—a conversion that requires an investment of energy in the form of ATP hydrolysis to provide the two phosphates; (2) in steps 5 and 6, the aldehyde group of each glyceraldehyde 3-phosphate molecule is oxidized to a carboxylic acid, and the energy from this reaction is coupled to the synthesis of ATP from ADP and inorganic phosphate; and (3) in steps 7, 8, and 9, the same two phosphate molecules that were added to sugars in the first reaction sequence are transferred back to ADP to form ATP, thereby repaying the original investment of two ATP molecules hydrolyzed in the first reaction sequence (see Figure 2–21).

At the end of glycolysis, therefore, the ATP balance sheet shows a net profit of the two molecules of ATP (per glucose molecule) that were produced in steps 5 and 6. As the only reactions in the sequence in which a high-energy phosphate linkage is created from inorganic phosphate, these two steps lie at the heart of glycolysis. They also provide an excellent illustration of the way in which reactions in the cell can be coupled together by enzymes to harvest the energy released by oxidations (Figure 2–22). The overall result is that an aldehyde group on a sugar is oxidized to a carboxylic acid and an inorganic phosphate group is transferred to a high-energy linkage on ATP; in addition, a molecule of NAD$^+$ is

9 step process

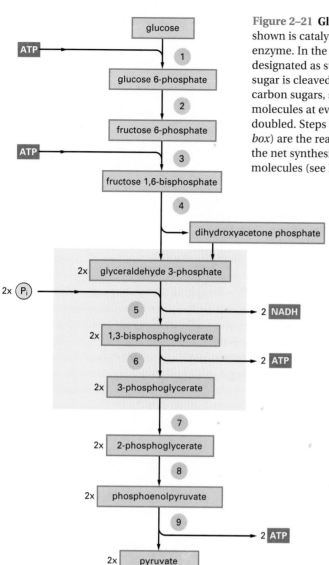

Figure 2–21 Glycolysis. Each reaction shown is catalyzed by a different enzyme. In the series of reactions designated as step 4, a six-carbon sugar is cleaved to give two three-carbon sugars, so that the number of molecules at every step after this is doubled. Steps 5 and 6 (in the *yellow box*) are the reactions responsible for the net synthesis of ATP and NADH molecules (see Figure 2–22).

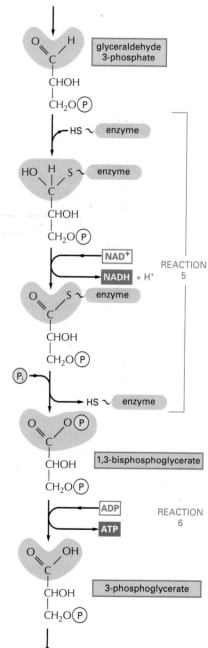

Figure 2–22 Steps 5 and 6 of glycolysis. In these steps the oxidation of an aldehyde to a carboxylic acid is coupled to the formation of ATP and NADH (see also Figure 2–21). Step 5, shown here as a series of three steps, begins when the enzyme *glyceraldehyde 3-phosphate dehydrogenase* (see Figure 3–42) forms a covalent bond to the carbon carrying the aldehyde group on glyceraldehyde 3-phosphate. Next, hydrogen (as a hydride ion—a proton plus two electrons) is removed from the enzyme-linked aldehyde group in glyceraldehyde 3-phosphate and transferred to the important hydrogen carrier NAD+ (see Figure 2–24). This oxidation step creates a sugar carbonyl group attached to the enzyme in a high-energy linkage (shown as a *red bond*). This linkage is then broken by a phosphate ion (Pi) from solution, creating a high-energy sugar-phosphate bond instead (*red bond*). In these last two reactions, the enzyme has *coupled* the energetically favorable process of oxidizing an aldehyde to the energetically unfavorable formation of a high-energy phosphate bond, allowing the second step to be driven by the first. Finally, in step 6 of glycolysis, the newly created reactive phosphate group is transferred to ADP to form ATP, leaving a free carboxylic acid group on the oxidized sugar.

reduced to NADH, a molecule with a central role in energy metabolism, as we discuss next. This elegant set of coupled reactions was probably among the earliest metabolic steps to appear in the evolving cell.

For most animal cells glycolysis is only a prelude to stage 3 of catabolism, since the pyruvic acid that is formed at the last step quickly enters the mitochondria to be completely oxidized to CO_2 and H_2O. In the case of anaerobic organisms (those that do not utilize molecular oxygen), however, and for tissues, such as skeletal muscle, that can function under anaerobic conditions, glycolysis can become a major source of the cell's ATP. Anaerobic energy-yielding reactions of this type are called *fermentations*. Here, instead of being degraded in mitochondria, the pyruvate molecules stay in the cytosol and, depending on the organism, can be converted into ethanol plus CO_2 (as in yeast) or into lactate (as in muscle), which is then excreted from the cell. These further reactions of pyruvate use up the reducing power produced in reaction 5 of glycolysis, thereby regenerating the NAD^+ required for glycolysis to continue, as we discuss in Chapter 14.

NADH Is a Central Intermediate in Oxidative Catabolism

The anaerobic generation of ATP from glucose through the reactions of glycolysis is relatively inefficient. The end products of anaerobic glycolysis still contain a great deal of chemical energy that can be released by further oxidation. The evolution of *oxidative catabolism* (cellular respiration) became possible only after molecular oxygen had accumulated in the earth's atmosphere as a result of photosynthesis by the cyanobacteria. Earlier, anaerobic processes had dominated life on earth. The addition of an oxygen-requiring stage to the catabolic process (stage 3 in Figure 2–19) provided cells with a much more powerful and efficient method for extracting energy from food molecules. This third stage begins with a series of reactions called the *citric acid cycle* (also called the tricarboxylic acid cycle, or the Krebs cycle) and ends with *oxidative phosphorylation*, both of which occur in aerobic bacteria and the mitochondria of eucaryotic cells.

A simplified version of the two central processes of oxidative catabolism is given in Figure 2–23. First, in the citric acid cycle, the acetyl groups from acetyl CoA are oxidized to produce CO_2 and NADH. Next, in the process of oxidative phosphorylation, the NADH generated reacts with molecular oxygen (O_2) to produce ATP and H_2O in a complicated series of steps that relies on electron transport in a membrane.

NADH, which serves as a central intermediate in the above process, was previously encountered as a product of glycolysis (see Figure 2–22). It is an important carrier of reducing power in cells. As illustrated in Figure 2–24, it is formed by the addition of a hydrogen nucleus and two electrons (a hydride ion, H^-) to nicotinamide adenine dinucleotide (NAD). Because this addition occurs in a way that leaves the hydride ion held in a high-energy linkage, NADH acts as a convenient source of readily transferable electrons in cells, in much the same way that ATP acts as a convenient source of readily transferable phosphate groups.

Figure 2–23 **A simplified outline of stage 3 of catabolism.** The process is primarily designed to produce large amounts of ATP from ADP and inorganic phosphate (P_i). NADH is produced by the citric acid cycle and is then used to drive the production of ATP during oxidative phosphorylation. NADH thus serves as a central intermediate in the oxidation of acetyl groups to CO_2 and H_2O (lesser amounts of $FADH_2$ play a similar part).

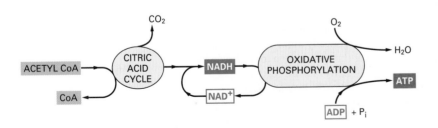

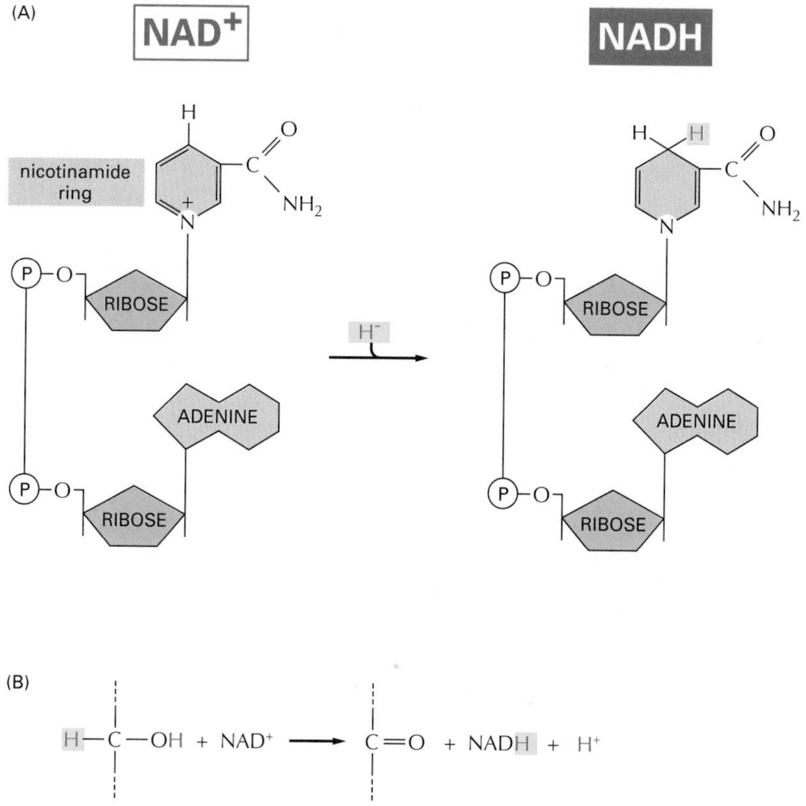

(A) NAD⁺ → NADH

(B)

$$H—C—OH + NAD^+ \longrightarrow C{=}O + NADH + H^+$$

Figure 2–24 NAD⁺ and NADH. These two molecules are the most important carriers of readily transferable electrons in catabolic reactions. Their structures are shown in (A). NAD is an abbreviation for *nicotinamide adenine dinucleotide*, reflecting the fact that the bottom half of the molecule, as drawn, is adenosine monophosphate (AMP). The part of the NAD⁺ molecule known as the nicotinamide ring (labeled in *gray box*) is able to accept two electrons together with a proton (in sum, a hydride ion, H⁻), forming NADH. In this reduced form, the nicotinamide ring has a reduced stability because it is no longer stabilized by resonance. As a result, the added hydride ion is *activated* in the sense that it can be easily transferred to other molecules.

(B) An example of a reaction involving NAD⁺ and NADH. In the biological oxidation of a substrate molecule such as an alcohol, two hydrogen atoms are lost from the substrate. One of these is added as a hydride ion to NAD⁺, producing NADH, while the other is released into solution as a proton (H⁺).

Metabolism Is Dominated by the Citric Acid Cycle [15]

The primary function of the citric acid cycle is to oxidize acetyl groups that enter the cycle in the form of acetyl CoA molecules. The reactions form a cycle because the acetyl group is not oxidized directly, but only after it has been covalently added to a larger molecule, *oxaloacetate*, which is regenerated at the end of one turn of the cycle. As illustrated in Figure 2–25, the cycle begins with the reaction between acetyl CoA and oxaloacetate to form the tricarboxylic acid molecule called *citric acid* (or *citrate*). A series of enzymatically catalyzed reactions then occurs in which two of the six carbons of citrate are oxidized to CO_2, forming another molecule of oxaloacetate to repeat the cycle. (Because the two carbons that are newly added in each cycle enter a different part of the citrate molecule from the part oxidized to CO_2, it is only after several cycles that their turn comes to be oxidized.) The CO_2 produced in these reactions then diffuses from the mitochondrion (or from the bacterium) and leaves the cell.

The energy made available when the C—H and C—C bonds in citrate are oxidized is captured in several ways in the course of the citric acid cycle. At one step in the cycle (succinyl CoA to succinate), a high-energy phosphate linkage is created by a mechanism resembling that described for glycolysis above. All of the remaining energy of oxidation that is captured is channeled into the conversion of hydrogen—or hydride ion—carrier molecules to their reduced forms; for each turn of the cycle, three molecules of NAD⁺ are converted to NADH and one *flavin adenine nucleotide* (FAD) is converted to $FADH_2$. The energy that is stored in the readily transferred electrons on these carrier molecules will subsequently be harnessed through the reactions of *oxidative phosphorylation* (considered in more detail below), which are the only reactions described here that require molecular oxygen from the atmosphere.

The additional oxygen atoms required to make CO_2 from the acetyl groups entering the citric acid cycle are supplied not by molecular oxygen but by water. Three molecules of water are split in each cycle, and their oxygen atoms are used

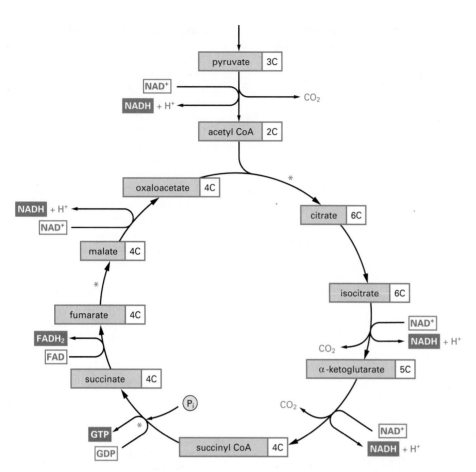

Figure 2–25 The citric acid cycle. In mitochondria and in aerobic bacteria the acetyl groups produced from pyruvate are further oxidized. The carbon atoms of the acetyl groups are converted to CO_2, while the hydrogen atoms are transferred to the carrier molecules NAD^+ and FAD. Additional oxygen and hydrogen atoms enter the cycle in the form of water at the steps marked with an asterisk (*). The number of carbon atoms in each molecule is indicated by a *white box*. For details, see Figure 14–14.

to make CO_2. Some of their hydrogen atoms enter substrate molecules and, like the hydrogen atoms of the acetyl groups, are ultimately removed to carrier molecules such as NADH.

In the eucaryotic cell the mitochondrion is the center toward which all catabolic processes lead, whether they begin with sugars, fats, or proteins. For, in addition to pyruvate, fatty acids and some amino acids also pass from the cytosol into mitochondria, where they are converted into acetyl CoA or one of the other intermediates of the citric acid cycle. The mitochondrion also functions as the starting point for some biosynthetic reactions by producing vital carbon-containing intermediates, such as *oxaloacetate* and *α-ketoglutarate*. These substances are transferred back from the mitochondrion to the cytosol, where they serve as precursors for the synthesis of essential molecules, such as amino acids.

In Oxidative Phosphorylation the Transfer of Electrons to Oxygen Drives ATP Formation [10, 16]

Oxidative phosphorylation is the last step in catabolism and the point at which the major portion of metabolic energy is released. In this process molecules of NADH and $FADH_2$ transfer the electrons that they have gained from the oxidation of food molecules to molecular oxygen, O_2, forming H_2O. The reaction, which is formally equivalent to the burning of hydrogen in air to form water, releases a great deal of chemical energy. Part of this energy is used to make the major portion of the cell's ATP; the rest is liberated as heat.

Although the overall chemistry of NADH and $FADH_2$ oxidation involves a transfer of hydrogen to oxygen, each hydrogen atom is transferred as an *electron* plus a proton (the hydrogen nucleus, H^+). This is possible because a hydrogen atom can be readily dissociated into its constituent electron and proton (H^+). The electron can then be transferred separately to a molecule that accepts only electrons, while the proton remains in aqueous solution. Conversely, if an electron alone is donated to a molecule with a strong affinity for hydrogen, then a hydrogen atom will be reconstituted automatically by the capture of a proton from solution. In the course of oxidative phosphorylation, electrons from NADH and $FADH_2$ pass down a long chain of carrier molecules that are known as the **electron-transport chain**. The presence or absence of intact hydrogen atoms at each step of the electron-transfer process depends on the nature of the carrier.

In a eucaryotic cell this series of electron transfers along the electron-transport chain takes place on the inner membrane of the mitochondrion, in which all of the electron carrier molecules are embedded. At each step of the transfer, the electrons fall to a lower energy state, until at the end they are transferred to oxygen molecules. Each oxygen molecule (O_2) picks up four electrons from the electron-transport chain plus four protons from aqueous solution to form two molecules of water. Oxygen molecules have a high affinity for electrons, and electrons bound to oxygen are thus in a low energy state.

The electron-transport chain is important for the cell because the energy released as the electrons fall to lower energy states is harnessed in a remarkable way. As described in Chapter 14, particular electron transfers cause protons to be pumped across the membrane from the inner mitochondrial compartment to the outside (Figure 2–26). An *electrochemical proton gradient* is thereby generated across the inner mitochondrial membrane. This gradient in turn drives a flux of protons back through a special enzyme complex in the same membrane, causing the enzyme (*ATP synthase*) to add a phosphate group to ADP and thereby generating ATP inside the mitochondrion. The newly made ATP is then transferred from the mitochondrion to the rest of the cell.

Amino Acids and Nucleotides Are Part of the Nitrogen Cycle

In our discussion so far we have concentrated mainly on carbohydrate metabolism. We have not yet considered the metabolism of nitrogen or sulfur. These two elements are constituents of proteins and nucleic acids, which are the two most important classes of macromolecules in the cell and make up approximately two-thirds of its dry weight. Atoms of nitrogen and sulfur pass from compound to compound and between organisms and their environment in a series of reversible cycles.

Although molecular nitrogen is abundant in the earth's atmosphere, nitrogen is chemically unreactive as a gas. Only a few living species are able to incorporate it into organic molecules, a process called **nitrogen fixation.** Nitrogen fixation occurs in certain microorganisms and by some geophysical processes, such as lightning discharge. It is essential to the biosphere as a whole, for without it life would not exist on this planet. Only a small fraction of the nitrogenous compounds in today's organisms, however, represents fresh products of nitrogen fixation from the atmosphere. Most organic nitrogen has been in circulation for some time, passing from one living organism to another. Thus present-day nitrogen-fixing reactions can be said to perform a "topping-up" function for the total nitrogen supply.

Vertebrates receive virtually all of their nitrogen in their dietary intake of proteins and nucleic acids. In the body these macromolecules are broken down to component amino acids and nucleotides, which are then repolymerized into new proteins and nucleic acids or utilized to make other molecules. About half of the 20 amino acids found in proteins are *essential amino acids* (Figure 2–27) for vertebrates, which means that they cannot be synthesized from other ingredients of the diet. The others can be so synthesized, using a variety of raw materials, including intermediates of the citric acid cycle. The essential amino acids are made by nonvertebrate organisms, usually by long and energetically expensive pathways that have been lost in the course of vertebrate evolution.

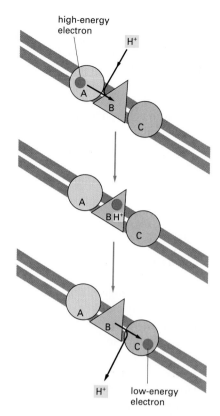

high-energy electron

H^+

A

B

C

A

B H^+

C

A

B

C

H^+

low-energy electron

Figure 2–26 The generation of an H^+ gradient across a membrane by electron-transport reactions. A high-energy electron (derived, for example, from the oxidation of a metabolite) is passed sequentially by carriers A, B, and C to a lower energy state. In this diagram carrier B is arranged in the membrane in such a way that it takes up H^+ from one side and releases it to the other as the electron passes. The resulting H^+ gradient represents a form of stored energy that is harnessed by other membrane proteins in the mitochondrion to drive the formation of ATP, as discussed in Chapter 14.

The nucleotides needed to make RNA and DNA can be synthesized using specialized biosynthetic pathways: there are no "essential nucleotides" that must be provided in the diet. All of the nitrogens in the purine and pyrimidine bases (as well as some of the carbons) are derived from the plentiful amino acids glutamine, aspartic acid, and glycine, whereas the ribose and deoxyribose sugars are derived from glucose.

Amino acids that are not utilized in biosynthesis can be oxidized to generate metabolic energy. Most of their carbon and hydrogen atoms eventually form CO_2 or H_2O, whereas their nitrogen atoms are shuttled through various forms and eventually appear as urea, which is excreted. Each amino acid is processed differently, and a whole constellation of enzymatic reactions exists for their catabolism.

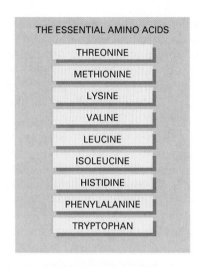

Figure 2–27 **The nine essential amino acids.** These cannot be synthesized by human cells and so must be supplied in the diet.

Summary

Animal cells derive energy from food in three stages. In stage 1, called digestion, proteins, polysaccharides, and fats are broken down by extracellular reactions to small molecules. In stage 2, these small molecules are degraded within cells to produce acetyl CoA and a limited amount of ATP and NADH. These are the only reactions that can yield energy in the absence of oxygen. In stage 3, the acetyl CoA molecules are degraded in mitochondria to give CO_2 and hydrogen atoms that are linked to carrier molecules such as NADH. Electrons from the hydrogen atoms are passed through a complex chain of membrane-bound carriers, finally being passed to molecular oxygen to form water. Driven by the energy released in these electron-transfer steps, protons (H^+) are transported out of the mitochondria. The resulting electrochemical proton gradient across the inner mitochondrial membrane is harnessed to drive the synthesis of most of the cell's ATP.

Biosynthesis and the Creation of Order [17]

Thousands of different chemical reactions are occurring in a cell at any instant of time. The reactions are all linked together in chains and networks in which the product of one reaction becomes the substrate of the next. Most of the chemical reactions in cells can be roughly classified as being concerned either with catabolism or with biosynthesis (anabolism). Having discussed the catabolic reactions, we now turn to the reactions of biosynthesis. These begin with the intermediate products of glycolysis and the citric acid cycle (and closely related compounds) and generate the larger and more complex molecules of the cell.

The Free-Energy Change for a Reaction Determines Whether It Can Occur [18]

Although enzymes speed up energetically favorable reactions, they cannot force energetically unfavorable reactions to occur. In terms of a water analogy, enzymes by themselves cannot make water run uphill. Cells, however, must do just that in order to grow and divide; they must build highly ordered and energy-rich molecules from small and simple ones. We have seen that, in a general way, this is done through enzymes that directly couple energetically favorable reactions, which consume energy (derived ultimately from the sun) and produce heat, to energetically unfavorable reactions, which produce biological order. Let us examine in greater detail how such coupling is achieved.

First, we must consider more carefully the term "energetically favorable," which we have so far used loosely without giving it a definition. As explained earlier, a chemical reaction can proceed spontaneously only if it results in a net increase in the disorder of the universe. Disorder increases when useful energy

(energy that could be harnessed to do work) is dissipated as heat; and the criterion for an increase of disorder can be expressed conveniently in terms of a quantity called the **free energy, G.** This is defined in such a way that changes in its value, denoted by ΔG, measure the amount of disorder created in the universe when a reaction takes place. *Energetically favorable reactions*, by definition, are those that release a large quantity of free energy, or, in other words, have a large *negative* ΔG and create much disorder. A familiar example on a macroscopic scale would be the "reaction" by which a compressed spring relaxes to an expanded state, releasing its stored elastic energy as heat to its surroundings. Energetically favorable reactions, with $\Delta G <$ O, have a strong tendency to occur spontaneously, although their rate will depend on other factors, such as the availability of specific enzymes (discussed below). Conversely, *energetically unfavorable reactions*, with a *positive* ΔG, such as those in which two amino acids are joined together to form a peptide bond, by themselves create order in the universe and therefore do not occur spontaneously. Reactions of this kind can take place only if they are coupled to a second reaction with a negative ΔG so large that the ΔG of the entire process is negative.

The course of most reactions can be predicted quantitatively. A large body of thermodynamic data has been collected that makes it possible to calculate the change in free energy for most of the important metabolic reactions of the cell. The overall free-energy change for a pathway is then simply the sum of the free-energy changes in each of its component steps. Consider, for example, two reactions

$$X \rightarrow Y \text{ and } C \rightarrow D$$

where the ΔG values are +1 and –13 kcal/mole, respectively. (Recall that a mole is 6×10^{23} molecules of a substance.) If these two reactions can be coupled together, the ΔG for the coupled reaction will be –12 kcal/mole. Thus, the unfavorable reaction $X \rightarrow Y$, which will not occur spontaneously, can be driven by the favorable reaction $C \rightarrow D$, provided that a mechanism exists by which the two reactions can be coupled together.

Biosynthetic Reactions Are Often Directly Coupled to ATP Hydrolysis

Consider a typical biosynthetic reaction in which two monomers, A and B, are to be joined in a **dehydration** (also called *condensation*) reaction, in which water is released:

$$A\text{-}H + B\text{-}OH \rightarrow A\text{-}B + H_2O$$

Almost invariably the reverse reaction (called **hydrolysis**), in which water breaks the covalently linked compound A-B, will be the energetically favorable one. This is the case, for example, in the hydrolysis of proteins, nucleic acids, and polysaccharides into their subunits.

The general strategy that allows the cell to make A-B from A-H and B-OH involves a sequence of steps through which the energetically unfavorable synthesis of the desired compound is coupled to an even more energetically favorable reaction (see Figure 2–17). ATP hydrolysis (Figure 2–28) has a large negative ΔG, and it is the usual source of the free energy used to drive the biosynthetic reactions in a cell. In the coupled pathway from A-H and B-OH to A-B, energy from ATP hydrolysis is first used to convert B-OH to a higher-energy intermediate compound, which then reacts directly with A-H to give A-B. The simplest mechanism involves the transfer of a phosphate from ATP to B-OH to make $B\text{-}OPO_3$ (or B-O-Ⓟ), in which case the reaction pathway contains only two steps:

1. $B\text{-}OH + ATP \rightarrow B\text{-}O\text{-}Ⓟ + ADP$
2. $A\text{-}H + B\text{-}O\text{-}Ⓟ \rightarrow A\text{-}B + P_i$

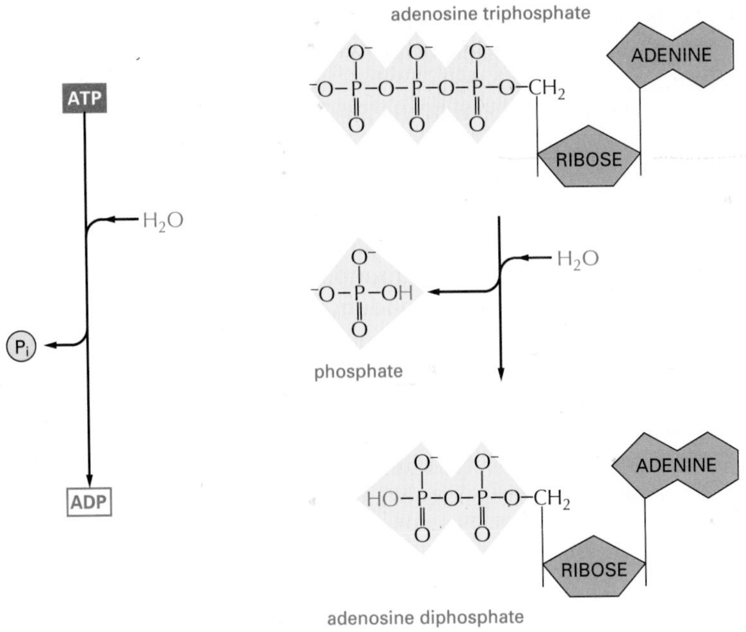

adenosine triphosphate

phosphate

adenosine diphosphate

Figure 2–28 **ATP hydrolysis.** The hydrolysis of the terminal phosphate of ATP yields between 11 and 13 kcal/mole of usable energy, depending on the intracellular conditions. The large negative ΔG of this reaction arises from a number of factors. Release of the terminal phosphate group removes an unfavorable repulsion between adjacent negative charges. In addition, the inorganic phosphate ion (P_i) released is stabilized by resonance and by favorable hydrogen bond formation with water.

Since the intermediate B-O-(P) is formed only transiently, the overall reactions that occur are

$$A\text{-}H + B\text{-}OH \rightarrow A\text{-}B \text{ and } ATP \rightarrow ADP + P_i$$

The first reaction, which by itself is energetically unfavorable, is forced to occur by being directly coupled to the second energetically favorable reaction (ATP hydrolysis). An example of a coupled biosynthetic reaction of this kind, the synthesis of the amino acid glutamine, is shown in Figure 2–29.

The ΔG for the hydrolysis of ATP to ADP and inorganic phosphate (P_i) depends on the concentrations of all of the reactants, and under the usual conditions in a cell it is between -11 and -13 kcal/mole. In principle, this hydrolysis reaction can be used to drive an unfavorable reaction with a ΔG of, perhaps, $+10$ kcal/mole, provided that a suitable reaction path is available. For some biosynthetic reactions, however, even -13 kcal/mole may not be enough. In these cases the path of ATP hydrolysis can be altered so that it initially produces AMP and pyrophosphate (PP_i), which is itself then hydrolyzed in a subsequent step (Figure 2–30). The whole process makes available a total free-energy change of about -26 kcal/mole.

How is the energy of pyrophosphate hydrolysis coupled to a biosynthetic reaction? One way can be illustrated by considering again the synthesis of compound A-B from A-H and B-OH. By an appropriate enzyme, B-OH can be converted to the higher-energy intermediate B-O-(P)-(P) by its reaction with ATP. The complete reaction now contains three steps:

1. $B\text{-}OH + ATP \rightarrow B\text{-}O\text{-}\text{(P)}\text{-}\text{(P)} + AMP$
2. $A\text{-}H + B\text{-}O\text{-}\text{(P)}\text{-}\text{(P)} \rightarrow A\text{-}B + PP_i$
3. $PP_i + H_2O \rightarrow 2P_i$

And the overall reactions are

$$A\text{-}H + B\text{-}OH \rightarrow A\text{-}B \text{ and } ATP + H_2O \rightarrow AMP + 2P_i$$

Since an enzyme always facilitates equally the forward and backward directions of the reaction it catalyzes, the compound A-B can be destroyed by recombining it with pyrophosphate (a reversal of step 2). But the energetically favorable reaction of pyrophosphate hydrolysis (step 3) greatly stabilizes compound A-B

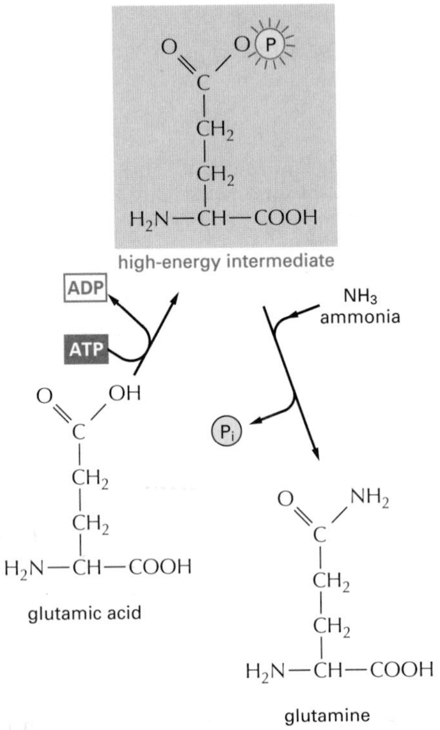

high-energy intermediate

glutamic acid

glutamine

Figure 2–29 **An example of a biosynthetic reaction of the dehydration type driven by ATP hydrolysis.** Glutamic acid is first converted to a high-energy phosphorylated intermediate (corresponding to the compound B-O-(P) described in the text), which then reacts with ammonia to form glutamine. In this example both steps occur on the surface of the same enzyme, *glutamine synthase*. Note that, for clarity, these molecules are shown in their uncharged forms.

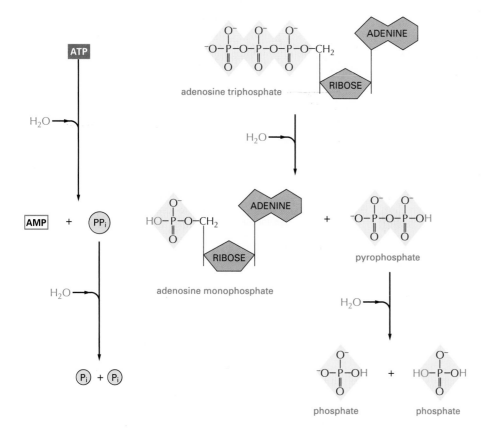

Figure 2–30 **An alternative route for the hydrolysis of ATP, in which pyrophosphate is first formed and then hydrolyzed.** This route releases about twice as much free energy as the reaction shown in Figure 2–28. The hydrogen atoms derived from water are shown attached to the phosphate groups following hydrolysis. At the pH of the cytoplasm, however, most of these dissociate to form free hydrogen ions, H+.

by keeping the concentration of pyrophosphate very low, essentially preventing the reversal of step 2. In this way the energy of pyrophosphate hydrolysis is used to drive this reaction in the forward direction. An example of an important biosynthetic reaction of this kind, polynucleotide synthesis, is illustrated in Figure 2–31.

Coenzymes Are Involved in the Transfer of Specific Chemical Groups

Because the terminal phosphate linkage in ATP is easily cleaved, with release of free energy, ATP acts as an efficient donor of a phosphate group in a large number of phosphorylation reactions. A wide variety of other chemically labile linkages also function in this way, and molecules bearing them often bind tightly to the surface of enzymes so that they can be used efficiently as donors of their reactive group in enzymatically catalyzed reactions. Such molecules are called **coenzymes** because they are essential for the activity of the enzyme; the same coenzyme can participate in many different biosynthetic reactions in which its group is needed.

Some examples of coenzymes are listed in Table 2–2. Among them is acetyl coenzyme A (acetyl CoA), which we encountered earlier. It carries an acetyl group linked to CoA through a reactive thioester bond (see Figure 2–20). This acetyl group is readily transferred to another molecule, such as a growing fatty acid. Other important coenzymes are NADH, which carries a hydride ion (see Figure 2–24), and *biotin*, which transfers a carboxyl group in many biosynthetic reactions (Figure 2–32).

Many coenzymes cannot be synthesized by animals and must be obtained from plants or microorganisms in the diet. *Vitamins*—essential nutritional factors that animals need in trace amounts—are often the precursors of required coenzymes.

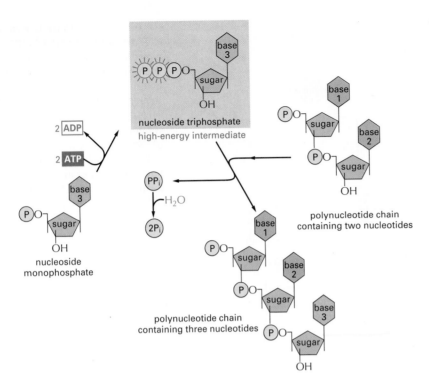

Figure 2–31 **Synthesis of a polynucleotide, RNA or DNA, is a multistep process driven by ATP hydrolysis.** In the first step a nucleoside monophosphate is activated by the sequential transfer of the terminal phosphate groups from two ATP molecules. The high-energy intermediate formed—a nucleoside triphosphate—exists free in solution until it reacts with the growing end of an RNA or DNA chain with release of pyrophosphate. Hydrolysis of the latter to inorganic phosphate is highly favorable and helps drive the overall reaction in the direction of polynucleotide synthesis.

The Structure of Coenzymes Suggests That They May Have Originated in an RNA World

As discussed in Chapter 1, the recent discovery that RNA molecules can fold up to form highly specific catalytic surfaces has led to the view that RNAs (or their close relatives) were probably the main catalysts for early life forms and that proteins were a later evolutionary addition. To allow efficient catalysis of the many types of reactions needed in cells, it seems likely that these RNAs would have required a large variety of coenzymes to compensate for their own chemical monotony (being made from only four different nucleotide subunits). Some present-day RNAs contain highly specific binding sites for nucleotides, which utilize non-Watson-Crick, hydrogen-bonded base-base contacts (see Figure 3–21). It is conceivable that coenzymes bound to early RNAs by means of the same type of nucleotide "handle" that is present on the back side of many present-day coenzymes, such as ATP, acetyl CoA, and NADH. According to this view, these molecules evolved with a covalently attached nucleotide because of the nucleotide's earlier usefulness for coenzyme binding in an RNA world.

Table 2–2 **Some Coenzymes Involved in Group-Transfer Reactions**

Coenzyme*	Group Transferred
ATP	phosphate
NADH, NADPH	hydrogen and electron (hydride ion)
Coenzyme A	acetyl
Biotin	carboxyl
S-Adenosylmethionine	methyl

*Coenzymes are small molecules that are associated with some enzymes and are essential for their activity. Each one listed is a carrier molecule for a small chemical group, and it participates in various reactions in which that group is transferred to another molecule. Some coenzymes are covalently linked to their enzyme; others are less tightly bound.

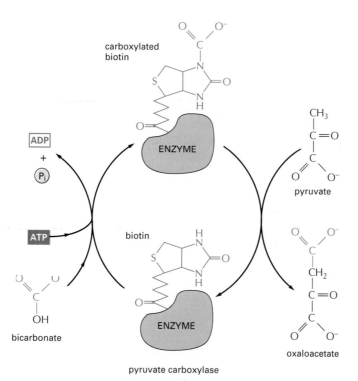

carboxylated
biotin

ADP
+
P_i

ATP

biotin

bicarbonate

ENZYME

ENZYME

pyruvate carboxylase

CH_3

pyruvate

oxaloacetate

Figure 2–32 **Transfer of a carboxyl group by the coenzyme biotin.** Biotin (shown in *green*) acts as a carrier molecule for the carboxyl group (shown in *red*). In the sequence of reactions shown, biotin is covalently bound to the enzyme pyruvate carboxylase. An activated carboxyl group derived from a bicarbonate ion (HCO_3^-) is coupled to biotin in a reaction that requires an input of energy from the hydrolysis of an ATP molecule. Subsequently, this carboxyl group is transferred to the methyl group of pyruvate to form oxaloacetate.

Biosynthesis Requires Reducing Power

We have seen that oxidation and reduction reactions occur continuously in cells. The chemical energy in food molecules is released by catabolic oxidation reactions, while, in order to make biological molecules, the cell needs (among other things) to carry out a series of reduction reactions that require an input of chemical energy. By using the principle of coupled reactions described previously, cells directly channel chemical energy derived from catabolism into the synthesis of NADH (see Figure 2–22, for example). The high-energy bond between hydrogen and the nicotinamide ring in NADH then provides the energy for otherwise unfavorable enzyme reactions that transfer two electrons plus a proton (as a hydride ion) to another molecule. NADH, and the closely related NADPH to which it can be readily converted, are therefore said to carry "reducing power"; both are used as coenzymes in many types of reduction reactions.

To see how this hydrogen transfer works in practice, consider just one biosynthetic step: the last reaction in a pathway for the synthesis of the lipid molecule *cholesterol*. In this reaction two hydrogen atoms are added to the polycyclic steroid ring in order to reduce a carbon-carbon double bond. As in most biosynthetic reactions, the constituents of the two hydrogen atoms required in this reaction are supplied as a hydride ion from NADPH and a proton (H^+) from the solution ($H^- + H^+ = 2H$) (Figure 2–33). As in NADH, the hydride ion to be transferred from NADPH is part of a nicotinamide ring and is easily lost because the ring can achieve a more stable aromatic state without it (see Figure 2–24). Therefore, NADH and NADPH both hold a hydride ion in a high-energy linkage from which it can be transferred to another molecule when a suitable enzyme is available to catalyze the transfer.

The difference between NADH and NADPH is trivial in chemical terms, but it is crucial for their distinctive functions. The extra phosphate group on NADPH is far from the active region (Figure 2–34) and is of no importance to the hydride ion transfer reaction; but it determines the enzymes to which NADPH can bind as a coenzyme. As a general rule, NADH operates with enzymes catalyzing catabolic reactions, whereas NADPH operates with enzymes that catalyze biosyn-

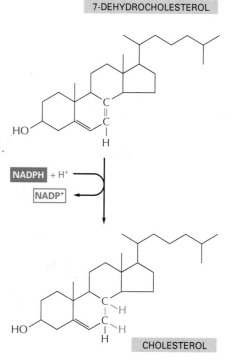

7-DEHYDROCHOLESTEROL

HO

NADPH + H^+

NADP$^+$

HO

CHOLESTEROL

Figure 2–33 **The final stage in one of the biosynthetic routes leading to cholesterol.** The reduction of the C=C bond is achieved by the transfer of a hydride ion from the carrier molecule NADPH plus a proton (H^+) from the solution.

Figure 2–34 The structure of NADPH. It differs from NADH (see Figure 2–24) only in the presence of an extra phosphate group that allows it to be recognized selectively by the enzymes involved in biosynthesis.

thetic reactions. By having the two coenzymes act in different pathways, the cell can keep NADPH:NADP$^+$ ratios high to provide the reducing power necessary for biosynthetic pathways, while at the same time it can keep NADH:NAD$^+$ ratios low to provide the NAD$^+$ required to accept electrons during catabolism.

Biological Polymers Are Synthesized by Repetition of Elementary Dehydration Reactions

The principal macromolecules synthesized by cells are polynucleotides (DNA and RNA), polysaccharides, and proteins. They are enormously diverse in structure and include the most complex molecules known. Despite this, they are synthe-

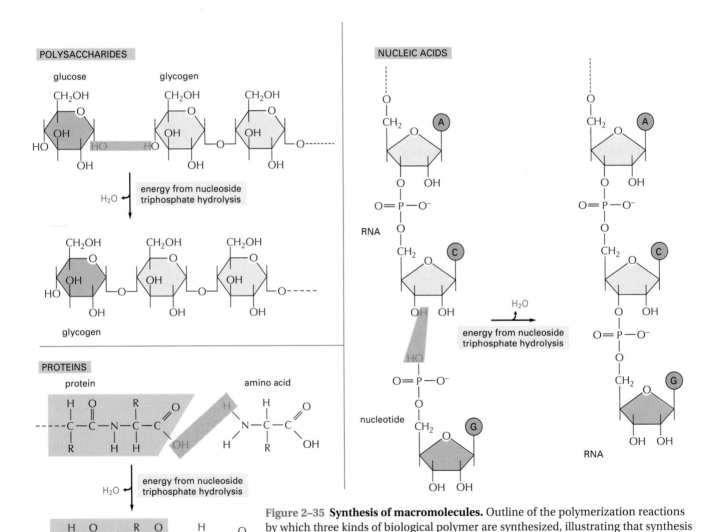

Figure 2–35 Synthesis of macromolecules. Outline of the polymerization reactions by which three kinds of biological polymer are synthesized, illustrating that synthesis in every case involves the loss of water (dehydration). Not shown is the consumption of high-energy nucleoside triphosphates that is required to activate each monomer prior to its addition. In contrast, the reverse reaction—the breakdown of all three types of polymer—occurs by the simple addition of water (hydrolysis).

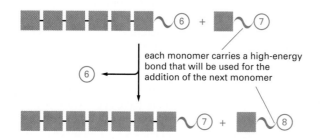

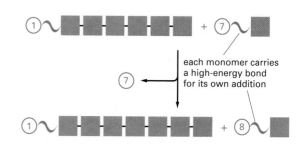

sized from relatively few kinds of small molecules (referred to as either *monomers* or *subunits*) by a restricted repertoire of chemical reactions.

An oversimplified outline of the mechanism of addition of monomers to proteins, polynucleotides, and polysaccharides is shown in Figure 2–35. Although the synthetic reactions for each polymer involve a different kind of covalent bond and different enzymes and cofactors, there are strong underlying similarities. As indicated by the red shading, the addition of subunits in each case occurs by a dehydration reaction, involving the removal of a molecule of water from the two reactants.

As in the general case discussed on page 76, the formation of these polymers requires the input of chemical energy, which is achieved by the standard strategy of coupling the biosynthetic reaction to the energetically favorable hydrolysis of a nucleoside triphosphate. For all three types of macromolecules, at least one nucleoside triphosphate is cleaved to produce pyrophosphate, which is subsequently hydrolyzed so that the driving force for the reaction is large. The mechanism used for polynucleotide synthesis was illustrated earlier in Figure 2–31.

The activated intermediates in the polymerization reactions can be oriented in one of two ways, giving rise to either the head polymerization or the tail polymerization of monomers. In *head polymerization* the reactive bond is carried on the end of the growing polymer and must therefore be regenerated each time a monomer is added. In this case each monomer brings with it the reactive bond that will be used to react with the *next* monomer in the series (Figure 2–36). In *tail polymerization* the reactive bond carried by each monomer is used instead for its own addition. Both types of polymerization are used for the synthesis of biological macromolecules. The synthesis of polynucleotides and some simple polysaccharides occurs by tail polymerization, for example, whereas the synthesis of proteins occurs by head polymerization.

Figure 2–36 Activated intermediates in polymerization reactions. Head growth of polymers is compared to tail growth.

Summary

The hydrolysis of ATP is commonly coupled to energetically unfavorable reactions, such as the biosynthesis of macromolecules, by the transfer of phosphate groups to form reactive phosphorylated intermediates. Because the energetically unfavorable reaction now becomes energetically favorable, ATP hydrolysis is said to drive the reaction. Polymeric molecules such as proteins, nucleic acids, and polysaccharides are assembled from small activated precursor molecules by repetitive dehydration reactions that are driven in this way. Other reactive molecules, called coenzymes, transfer other chemical groups in the course of biosynthesis: NADPH transfers hydrogen as a proton plus two electrons (a hydride ion), for example, whereas acetyl CoA transfers acetyl groups.

The Coordination of Catabolism and Biosynthesis [19]

Metabolism Is Organized and Regulated

Some idea of how intricate a cell is when viewed as a chemical machine can be obtained from Figure 2–37, which is a chart showing only some of the enzymatic pathways in a cell. All of these reactions occur in a cell that is less than 0.1 mm in diameter, and each requires an enzyme that is itself the product of a whole series of information-transfer and protein-synthesis reactions. For a typical small molecule—the amino acid *serine*, for example—there are half a dozen or more enzymes that can modify it chemically in different ways: it can be linked to AMP (adenylated) in preparation for protein synthesis, or degraded to glycine, or converted to pyruvate in preparation for oxidation; it can be acetylated by acetyl CoA or transferred to a fatty acid to make phosphatidyl serine. All of these different pathways compete for the same serine molecule, and similar competitions for thousands of other small molecules go on at the same time. One might think that the whole system would need to be so finely balanced that any minor upset, such as a temporary change in dietary intake, would be disastrous.

In fact, the cell is amazingly stable. Whenever it is perturbed, the cell reacts so as to restore its initial state. It can adapt and continue to function during starvation or disease. Mutations of many kinds can eliminate particular reaction pathways, and yet—provided that certain minimum requirements are met—the cell survives. It does so because an elaborate network of control mechanisms regulates and coordinates the rates of its reactions. Some of the higher levels of control will be considered in later chapters. Here we are concerned only with the simplest mechanisms that regulate the flow of small molecules through the various metabolic pathways.

Metabolic Pathways Are Regulated by Changes in Enzyme Activity [20]

The concentrations of the various small molecules in a cell are buffered against major changes by a process known as **feedback regulation,** which fine-tunes the flux of metabolites through a particular pathway by temporarily increasing or decreasing the activity of crucial enzymes. The first enzyme of a series of reactions, for example, is usually inhibited by a *negative feedback* effect of the final product of that pathway: if large quantities of the final product accumulate, further entry of precursors into the reaction pathway is automatically inhibited (Figure 2–38). Where pathways branch or intersect, as they often do, there are usually multiple points of control by different final products. The complexity of such feedback control processes is illustrated in Figure 2–39, which shows the pattern of enzyme regulation observed in a set of related amino acid pathways.

Feedback regulation can work almost instantaneously and is reversible; in addition, a given end product may activate enzymes leading along other pathways, as well as inhibit enzymes that cause its own synthesis. The molecular basis for this type of control in cells is well understood, but since an explanation requires some knowledge of protein structure, it will be deferred until Chapter 5.

Catabolic Reactions Can Be Reversed by an Input of Energy [21]

By regulating a few enzymes at key points in a metabolic network, a cell can effect large-scale changes in its general metabolism. A special pattern of feedback regulation enables a cell to switch, for example, from glucose degradation to glucose biosynthesis (denoted *gluconeogenesis*). The need for gluconeogenesis is especially acute in periods of violent exercise, when the glucose needed for muscle

Figure 2–37 (*opposite page*) **Some of the chemical reactions occurring in a cell.** (A) About 500 common metabolic reactions are shown diagrammatically, with each chemical species represented by a filled circle. The centrally placed reactions of the glycolytic pathway and the citric acid cycle are shown in *red*. A typical mammalian cell synthesizes more than 10,000 different proteins, a major proportion of which are enzymes. In the arbitrarily selected segment of this metabolic maze (shaded *yellow*), cholesterol is synthesized from acetyl CoA. To the right and below the maze, this segment is shown in detail in an enlargement (B).

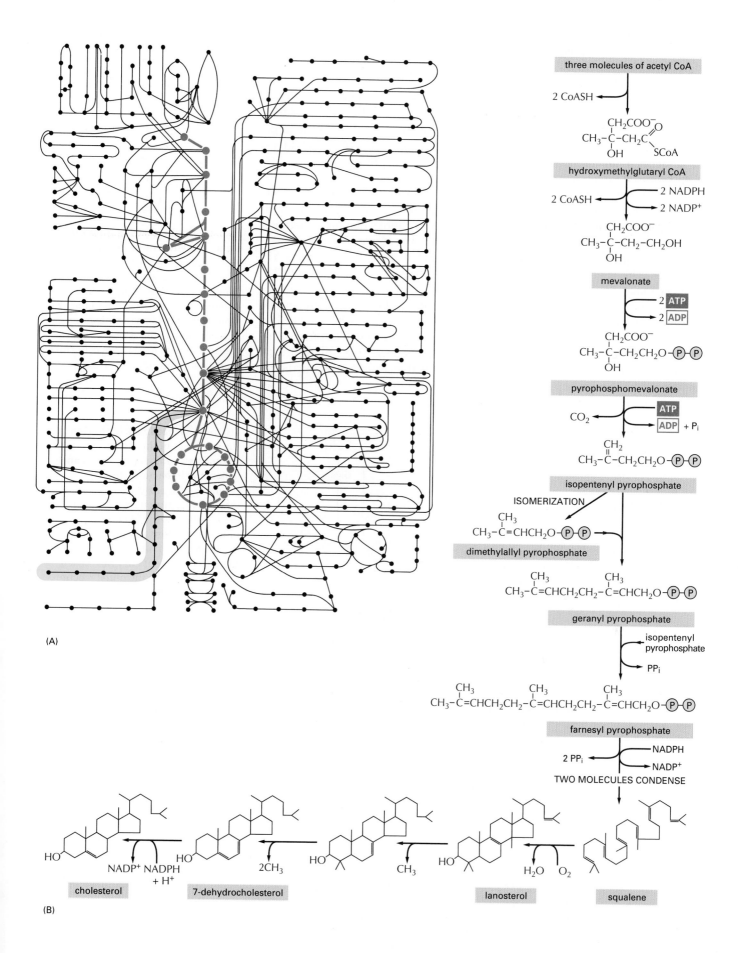

three molecules of acetyl CoA

2 CoASH

$$CH_3-\underset{\underset{OH}{|}}{C}-CH_2\underset{\underset{SCoA}{|}}{C}\overset{\overset{CH_2COO^-}{|}}{}=O$$

hydroxymethylglutaryl CoA

2 CoASH   2 NADPH

2 NADP$^+$

$$CH_3-\underset{\underset{OH}{|}}{C}-CH_2-CH_2OH \quad\overset{CH_2COO^-}{|}$$

mevalonate

2 ATP

2 ADP

$$CH_3-\underset{\underset{OH}{|}}{C}-CH_2CH_2O-\text{P}-\text{P} \quad\overset{CH_2COO^-}{|}$$

pyrophosphomevalonate

CO_2 ATP

ADP + P$_i$

$$CH_3-\underset{}{C}-CH_2CH_2O-\text{P}-\text{P} \quad\overset{CH_2}{\|}$$

isopentenyl pyrophosphate

ISOMERIZATION

$$CH_3-\underset{\underset{CH_3}{|}}{C}=CHCH_2O-\text{P}-\text{P}\longrightarrow$$

dimethylallyl pyrophosphate

$$CH_3-\underset{\underset{}{\|}}{\overset{CH_3}{C}}=CHCH_2CH_2-\overset{CH_3}{\underset{}{C}}=CHCH_2O-\text{P}-\text{P}$$

geranyl pyrophosphate

isopentenyl
pyrophosphate

PP$_i$

$$CH_3-\overset{CH_3}{C}=CHCH_2CH_2-\overset{CH_3}{C}=CHCH_2CH_2-\overset{CH_3}{C}=CHCH_2O-\text{P}-\text{P}$$

farnesyl pyrophosphate

2 PP$_i$ NADPH

NADP$^+$

TWO MOLECULES CONDENSE

cholesterol   NADP$^+$ NADPH + H$^+$ 7-dehydrocholesterol   2CH$_3$ CH$_3$ lanosterol   H$_2$O   O$_2$ squalene

(A)

(B)

83

contraction is generated from lactic acid by liver cells, and also in periods of starvation, when glucose must be formed from the glycerol portion of fats and from amino acids for survival.

The normal breakdown of glucose to pyruvate during glycolysis is catalyzed by a number of enzymes acting in series. The reactions catalyzed by most of these enzymes are readily reversible, but three reaction steps (numbers 1, 3, and 9 in the sequence of Figure 2–21) are effectively irreversible. In fact, it is the large negative free-energy change that occurs in these reactions that normally drives the sequence in the direction of glucose breakdown. For the reactions to proceed in the opposite direction and make glucose from pyruvate, each of these three reactions must be bypassed. This is achieved by substituting three enzyme-catalyzed bypass reactions that are driven in the uphill direction by an input of chemical energy (Figure 2–40). Thus, whereas two ATP molecules are generated as each molecule of glucose is degraded to two molecules of pyruvate, the reverse reaction during gluconeogenesis requires the hydrolysis of four ATP and two GTP molecules. This is equivalent, in total, to the hydrolysis of six molecules of ATP for every molecule of glucose synthesized.

The bypass reactions in Figure 2–40 must be controlled so that glucose is broken down rapidly when energy is needed but synthesized when the cell is nutritionally replete. If both forward and reverse reactions were allowed to proceed at the same time without restraint, they would shuttle large quantities of metabolites backward and forward in futile cycles that would consume large amounts of ATP and generate heat for no purpose.

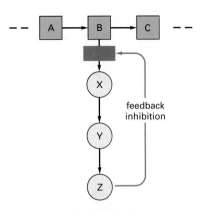

Figure 2–38 Feedback inhibition of a single biosynthetic pathway. Each letter represents a different small molecule, and each *black arrow* denotes a reaction catalyzed by a different enzyme. The end product Z inhibits the first enzyme that is unique to its synthesis and thereby controls its own level in the cell. This is an example of *negative feedback.*

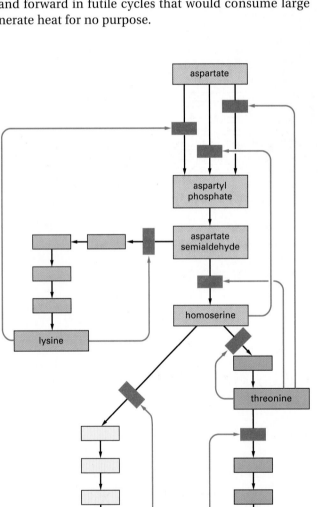

Figure 2–39 Feedback inhibition in the synthesis of the amino acids lysine, methionine, threonine, and isoleucine in bacteria. In this diagram, each enzyme-catalyzed reaction is represented by a *black arrow,* whereas the *red arrows* indicate positions at which products "feed back" to inhibit enzymes. Note that three different enzymes (called *isozymes*) catalyze the initial reaction, each inhibited by a different product.

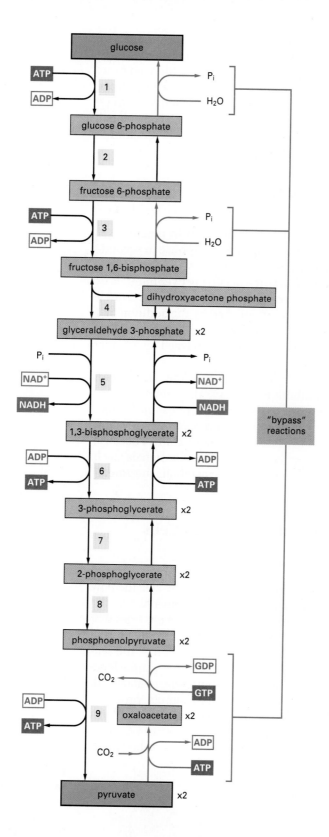

Figure 2–40 **Comparison of the reactions that produce glucose during gluconeogenesis with those that degrade glucose during glycolysis.** The degradative (glycolytic) reactions are energetically favorable (the free-energy change is less than zero), whereas the synthetic reactions require an input of energy. To synthesize glucose, different "bypass enzymes" are needed that bypass reactions 1, 3, and 9 of glycolysis. The overall flux of reactants between glucose and pyruvate is determined by feedback control mechanisms that regulate the enzymes that participate in these three steps.

The elegance of the control mechanisms involved can be illustrated by a single example. Step 3 of glycolysis is one of the reactions that must be bypassed during glucose formation (see Figure 2–40). Normally, this step involves the addition of a second phosphate group to fructose 6-phosphate from ATP and is catalyzed by the enzyme *phosphofructokinase.* This enzyme is activated by AMP, ADP, and inorganic phosphate, whereas it is inhibited by ATP, citrate, and fatty acids. Therefore, the enzyme is activated by the accumulation of the products of ATP hydrolysis when energy supplies are low, and it is inactivated when energy (in the form of ATP) or food supplies such as fatty acids or citrate (derived from amino acids) are abundant. *Fructose bisphosphatase* is the enzyme that catalyzes the reverse bypass reaction (the hydrolysis of fructose 1,6-bisphosphate to fructose 6-phosphate, leading to the formation of glucose); this enzyme is regulated in the opposite way by the same feedback control molecules so that it is stimulated when the phosphofructokinase is inhibited.

Enzymes Can Be Switched On and Off by Covalent Modification [22]

The types of feedback control just described permit the rates of reaction sequences to be regulated continuously and automatically in response to second-by-second fluctuations in metabolism. Cells have different devices for regulating enzymes when longer-lasting changes in activity, occurring over minutes or hours, are required. These involve reversible covalent modification of enzymes. This modification is usually, but not always, accomplished by the addition of a phosphate group to a specific serine, threonine, or tyrosine residue in the enzyme. The phosphate comes from ATP, and its transfer is catalyzed by a family of enzymes known as *protein kinases.*

In Chapter 5 we describe how phosphorylation can alter the shape of an enzyme in such a way as to increase or inhibit its activity. The subsequent removal of the phosphate group, which reverses the effect of the phosphorylation, is achieved by a second type of enzyme, called a *protein phosphatase.* Covalent modification of enzymes adds another dimension to metabolic control because it allows specific reaction pathways to be regulated by extracellular signals (such as hormones and growth factors) that are unrelated to the metabolic intermediates themselves.

Reactions Are Compartmentalized Both Within Cells and Within Organisms [23]

Not all of a cell's metabolic reactions occur within the same subcellular compartment. Because different enzymes are found in different parts of the cell, the flow of chemical components is channeled physically as well as chemically.

The simplest form of such spatial segregation occurs when two enzymes that catalyze sequential reactions form an enzyme complex and the product of the first enzyme does not have to diffuse through the cytosol to encounter the second enzyme. The second reaction begins as soon as the first is over. Some large enzyme aggregates carry out whole series of reactions without losing contact with the substrate. The conversion of pyruvate to acetyl CoA, for example, proceeds in three chemical steps, all of which take place on the same large enzyme complex (Figure 2–41). In fatty acid synthesis an even longer sequence of reactions is catalyzed by a single enzyme assembly. Not surprisingly, some of the largest

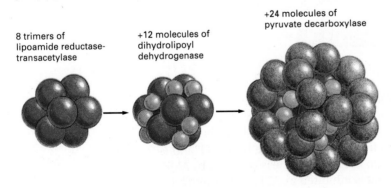

8 trimers of lipoamide reductase-transacetylase

+12 molecules of dihydrolipoyl dehydrogenase

+24 molecules of pyruvate decarboxylase

Figure 2–41 The structure of pyruvate dehydrogenase. This enzyme complex catalyzes the conversion of pyruvate to acetyl CoA. It is an example of a large multienzyme complex in which reaction intermediates are passed directly from one enzyme to another.

enzyme complexes are concerned with the synthesis of macromolecules such as proteins and DNA.

The next level of spatial segregation in cells involves the confinement of functionally related enzymes within the same membrane or within the aqueous compartment of an organelle that is bounded by a membrane. The oxidative metabolism of glucose is a good example. After glycolysis, pyruvate is actively taken up from the cytosol into the inner compartment of the mitochondrion, which contains all of the enzymes and metabolites involved in the citric acid cycle (Figure 2–42). Moreover, the inner mitochondrial membrane itself contains all of the enzymes that catalyze the subsequent reactions of oxidative phosphorylation, including those involved in the transfer of electrons from NADH to O_2 and in the synthesis of ATP. The entire mitochondrion can therefore be regarded as a small ATP-producing factory. In the same way other cellular organelles, such as the nucleus, the Golgi apparatus, and the lysosomes, can be viewed as specialized compartments where functionally related enzymes are confined to perform a specific task. In a sense, the living cell is like a city, with many specialized services concentrated in different areas that are extensively interconnected by various paths of communication.

Spatial organization in a multicellular organism extends beyond the individual cell. The different tissues of the body have different sets of enzymes and make distinct contributions to the chemistry of the organism as a whole. In addition to differences in specialized products such as hormones or antibodies, there are significant differences in the "common" metabolic pathways among various types of cells in the same organism. Although virtually all cells contain the enzymes of glycolysis, the citric acid cycle, lipid synthesis and breakdown, and amino acid metabolism, the levels of these processes in different tissues are differently regulated. Nerve cells, which are probably the most fastidious cells in the body, maintain almost no reserves of glycogen or fatty acids and rely almost entirely on a supply of glucose from the bloodstream. Liver cells supply glucose to actively contracting muscle cells and recycle the lactic acid produced by muscle cells back into glucose (Figure 2–43). All types of cells have their distinctive metabolic traits and cooperate extensively in the normal state as well as in response to stress and starvation.

Summary

The many thousands of different chemical reactions carried out simultaneously by a cell are closely coordinated. A variety of control mechanisms regulate the activities of key enzymes in response to the changing conditions in the cell. One very common form of regulation is a rapidly reversible feedback inhibition exerted on the first enzyme of a pathway by the final product of that pathway. A longer-lasting form of regulation involves the chemical modification of one enzyme by another, usually by phosphorylation. Combinations of regulatory mechanisms can produce major and long-lasting changes in the metabolism of the cell. Not all cellular reactions occur within the same intracellular compartment, and spatial segregation by internal membranes permits organelles to specialize in their biochemical tasks.

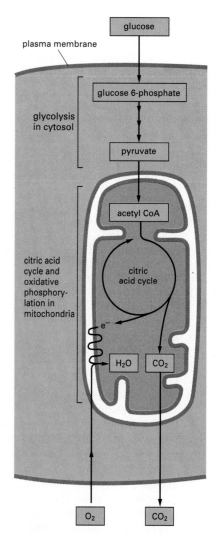

Figure 2–42 Segregation of the various steps in the breakdown of glucose in the eucaryotic cell. Glycolysis occurs in the cytosol, whereas the reactions of the citric acid cycle and oxidative phosphorylation take place only in mitochondria.

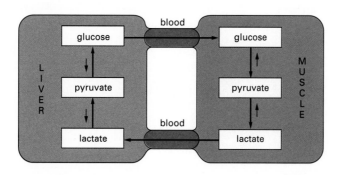

Figure 2–43 Schematic view of the metabolic cooperation between liver and muscle cells. The principal fuel of actively contracting muscle cells is glucose, much of which is supplied by liver cells. Lactic acid, the end product of anaerobic glucose breakdown by glycolysis in muscle, is converted back to glucose in the liver by the process of gluconeogenesis.

References

General

Becker, W.M.; Deamer, D.W. The World of the Cell, 2nd ed. Redwood City, CA: Benjamin-Cummings, 1991.

Herriott, J.; Jacobson, G.; Marmur, J.; Parsom, W. Papers in Biochemistry. Reading, MA: Addison-Wesley, 1984.

Lehninger, A.L.; Nelson, D.L.; Cox, M.M. Principles of Biochemistry, 2nd ed. New York: Worth, 1993.

Mathews, C.K.; van Holde, K.E. Biochemistry. Menlo Park, CA: Benjamin-Cummings, 1990.

Stryer, L. Biochemistry, 3rd ed. New York: W.H. Freeman, 1988.

Cited

1. Atkins, P.W. Molecules. New York: Scientific American Library, 1987.

   Henderson, L.J. The Fitness of the Environment. Boston: Beacon, 1927; reprinted 1958. (A classical and readable analysis.)

2. Ingraham, J.L.; Maaløe, O.; Neidhardt, F.C. Growth of the Bacterial Cell. Sunderland, MA: Sinauer, 1983.

3. Sharon, N. Carbohydrates. *Sci. Am.* 243(5):90–116, 1980.

4. Robertson, R.N. The Lively Membranes. Cambridge, UK: Cambridge University Press, 1983.

5. Branden, C.; Tooze, J. Introduction to Protein Structure. New York: Garland, 1991.

6. Saenger, W. Principles of Nucleic Acid Structure. New York: Springer, 1984.

7. Hess, B.; Markus, M. Order and chaos in biochemistry. *Trends Biochem. Sci.* 12:45–48, 1987.

   Schrödinger, E. What Is Life? Mind and Matter. Cambridge, UK: Cambridge University Press, 1969.

8. Atkins, P.W. The Second Law. New York: Scientific American Books, 1984.

   Dickerson, R.E. Molecular Thermodynamics. Menlo Park, CA: Benjamin, 1969.

   Klotz, I.M. Energy Changes in Biochemical Reactions. New York: Academic Press, 1967.

9. Youvan, D.C.; Marrs, B.L. Molecular mechanisms of photosynthesis. *Sci. Am.* 256(6):42–49, 1987.

10. Lehninger, A.L. Bioenergetics: The Molecular Basis of Biological Energy Transformations, 2nd ed. Menlo Park, CA: Benjamin-Cummings, 1971.

    Racker, E. From Pasteur to Mitchell: a hundred years of bioenergetics. *Fed. Proc.* 39:210–215, 1980.

11. Fothergill-Gilmore, L.A. The evolution of the glycolytic pathway. *Trends Biochem. Sci.* 11:47–51, 1986.

12. Lipmann, F. Wanderings of a Biochemist. New York: Wiley, 1971.

13. Schlenk, F. The ancestry, birth and adolescence of ATP. *Trends Biochem. Sci.* 12:367–368, 1987.

14. Abeles, R.H.; Frey, P.A.; Jencks, W.P. Biochemistry. Boston: Jones and Bartlett, 1992.

    McGilvery, R.W. Biochemistry: A Functional Approach, 3rd ed. Philadelphia: Saunders, 1983.

15. Kornberg, H.L. Tricarboxylic acid cycles. *Bioessays* 7:236–238, 1987.

    Krebs, H.A. The history of the tricarboxylic acid cycle. *Perspect. Biol. Med.* 14:154–170, 1970.

    Krebs, H.A.; Martin, A. Reminiscences and Reflections. Oxford, UK: Clarendon Press; New York: Oxford University Press, 1981.

16. Pullmann, M.E., et al. Partial resolution of the enzymes catalyzing oxidative phosphorylation. *J. Biol. Chem.* 235:3322–3329, 1960.

17. Prigogine, I.; Stengers, I. Order out of Chaos: Man's New Dialogue with Nature. New York: Bantam Books, 1984.

18. Eisenberg, D.; Crothers, D. Physical Chemistry with Applications to the Life Sciences. Menlo Park, CA: Benjamin-Cummings, 1979.

    Kauzmann, W. Thermodynamics and Statistics: With Applications to Gases. New York: Benjamin, 1967.

19. Martin, B.R. Metabolic Regulation: A Molecular Approach. Oxford, UK; Blackwell Scientific, 1987.

    Newsholme, E.A.; Start, C. Regulation in Metabolism. New York: Wiley, 1973.

20. Pardee, A.B. Molecular basis of biological regulation: origins of feedback inhibition and allostery. *Bioessays* 2:37–40, 1985.

21. Hess, B. Oscillating reactions. *Trends Biochem. Sci.* 2:193–195, 1977.

22. Cohen, P. Control of Enzyme Activity, 2nd ed. London: Chapman and Hall, 1983.

    Koshland, D.E., Jr. Switches, thresholds and ultra-sensitivity. *Trends Biochem. Sci.* 12:225–229, 1987.

23. Banks, P.; Bartley, W.; Birt, M. The Biochemistry of the Tissues, 2nd ed. New York: Wiley, 1976.

    Sies, H. Metabolic Compartmentation. New York: Academic Press, 1982.

Macromolecules: Structure, Shape, and Information

In moving from the small molecules of the cell to the giant macromolecules, we encounter a transition of more than size alone. Even though proteins, nucleic acids, and polysaccharides are made from a limited repertoire of amino acids, nucleotides, and sugars, respectively, they can have unique and truly astounding properties that bear little resemblance to those of their simple chemical precursors. Biological macromolecules are composed of many thousands—sometimes millions—of atoms linked together in precisely defined spatial arrangements. Each of these macromolecules carries specific information. Incorporated in its structure is a series of biological messages that can be "read" in its interactions with other molecules, enabling it to perform a precise function.

In this chapter we examine the structures of macromolecules, emphasizing proteins and nucleic acids, and explain how they have adapted in the course of evolution to perform specific functions. We consider the principles by which these molecules catalyze chemical transformations, build complex multimolecular structures, generate movement, and—most fundamental of all—store and transmit hereditary information.

Molecular Recognition Processes [1]

Macromolecules typically have molecular weights between about 10,000 and 1 million and are intermediate in size between the organic molecules of the cell discussed in Chapter 2 and the large macromolecular assemblies and organelles that will be discussed in subsequent chapters (Figure 3–1). One small molecule, water, constitutes 70% of the total mass of a cell; nearly all of the remaining cell mass is due to macromolecules (Table 3–1).

As described in Chapter 2, a macromolecule is assembled from low-molecular-weight subunits that are repeatedly added to one end to form a long, chainlike polymer. Usually only one family of subunits is used to construct each chain: amino acids are linked to other amino acids to form proteins, nucleotides are linked to other nucleotides to form nucleic acids, and sugars are linked to other sugars to form polysaccharides. Because the precise sequence of subunits is crucial to the function of a macromolecule, its biosynthesis requires mechanisms to ensure that the correct subunit goes into the polymer at each position in the chain.

sugars, amino acids, and nucleotides ~0.5–1 nm

globular proteins ~2–10 nm

ribosome ~30 nm

Figure 3–1 The size of protein molecules compared to some other cell components. The ribosome is an important macromolecular assembly composed of about 60 protein and RNA molecules.

Table 3–1 Approximate Chemical Compositions of a Typical Bacterium and a Typical Mammalian Cell

Component	Percent of Total Cell Weight	
	E. Coli *Bacterium*	*Mammalian Cell*
H_2O	70	70
Inorganic ions (Na^+, K^+, Mg^{2+}, Ca^{2+}, Cl^-, etc.)	1	1
Miscellaneous small metabolites	3	3
Proteins	15	18
RNA	6	1.1
DNA	1	0.25
Phospholipids	2	3
Other lipids	—	2
Polysaccharides	2	2
Total cell volume:	2×10^{-12} cm^3	4×10^{-9} cm^3
Relative cell volume:	1	2000

Proteins, polysaccharides, DNA, and RNA are macromolecules. Lipids are not generally classed as macromolecules even though they share some of their features; for example, most are synthesized as linear polymers of a smaller molecule (the acetyl group on acetyl CoA), and they self-assemble into larger structures (membranes). Note that water and protein comprise most of the mass of both mammalian and bacterial cells.

The Specific Interactions of a Macromolecule Depend on Weak, Noncovalent Bonds [2]

A macromolecular chain is held together by *covalent* bonds, which are strong enough to preserve the sequence of subunits for long periods of time. Although the sequence of subunits determines the information content of a macromolecule, utilizing that information depends largely on much weaker, *noncovalent* bonds. These weak bonds form between different parts of the same macromolecule and between different macromolecules. They therefore play a major part in determining both the three-dimensional structure of macromolecular chains and how these structures interact with one another.

The noncovalent bonds encountered in biological molecules are usually classified into three types: **ionic bonds, hydrogen bonds,** and **van der Waals attractions.** Another important weak force is created by the three-dimensional structure of water, which forces exposed hydrophobic groups together in order to minimize their disruptive effect on the hydrogen-bonded network of water molecules (see Panel 2–1, pp. 48–49). This expulsion from the aqueous solution generates what is sometimes thought of as a fourth kind of weak, noncovalent bond. These four types of weak attractive forces are the subject of Panel 3–1, pages 92–93.

In an aqueous environment each noncovalent bond is 30 to 300 times weaker than the typical covalent bonds that hold biological molecules together (Table 3–2) and only slightly stronger than the average energy of thermal collisions at

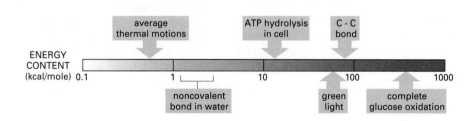

Figure 3–2 Comparative energies of some important molecular events in cells. Note that energy is displayed on a logarithmic scale.

Table 3–2 Covalent and Noncovalent Chemical Bonds

Bond Type	Length (nm)	Strength (kcal/mole)* In Vacuum	Strength (kcal/mole)* In Water
Covalent	0.15	90	90
Ionic	0.25	80	3
Hydrogen	0.30	4	1
van der Waals attraction (per atom)	0.35	0.1	0.1

*The strength of a bond can be measured by the energy required to break it, here given in kilocalories per mole (kcal/mole). (*One kilocalorie* is the quantity of energy needed to raise the temperature of 1000 g of water by 1°C. An alternative unit in wide use is the kilojoule, kJ, equal to 0.24 kcal.) Individual bonds vary a great deal in strength, depending on the atoms involved and their precise environment, so that the above values are only a rough guide. Note that the aqueous environment in a cell will greatly weaken both the ionic and the hydrogen bonds between nonwater molecules (Panel 3–1, pp. 92–93). The bond length is the center-to-center distance between the two interacting atoms; the length given here for a hydrogen bond is that between its two nonhydrogen atoms.

37°C (Figure 3–2). A single noncovalent bond—unlike a single covalent bond—is therefore too weak to withstand the thermal motions that tend to pull molecules apart. Large numbers of noncovalent bonds are needed to hold two molecular surfaces together, and these can form between two surfaces only when large numbers of atoms on the surfaces are precisely matched to each other (Figure 3–3). The exacting requirements for matching account for the specificity of biological recognition, such as occurs between an enzyme and its substrate.

As explained at the top of Panel 3–1, atoms behave almost as if they were hard spheres with a definite radius (their van der Waals radius). The requirement that no two atoms overlap limits the possible bond angles in a polypeptide chain (Figure 3–4). These and other steric interactions severely constrain the number of three-dimensional arrangements of atoms (or **conformations**) that are possible. Nevertheless, a long flexible chain such as a protein can still fold in an enormous number of ways. Each conformation will have a different set of weak intrachain interactions, and it is the total strength of these interactions that determines which conformations will form.

Most proteins in a cell fold stably in only one way: during the course of evolution the sequence of amino acid subunits in each protein has been selected so that one conformation is able to form many more favorable intrachain interactions than any other.

Figure 3–3 Noncovalent bonds. How weak bonds mediate recognition between macromolecules.

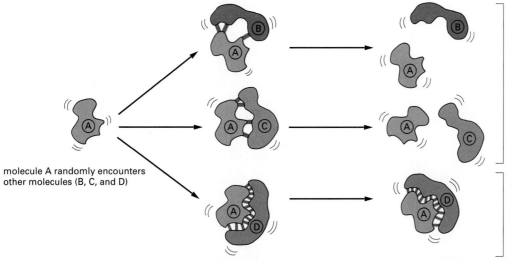

molecule A randomly encounters other molecules (B, C, and D)

the surfaces of molecules A and B, and A and C, are a poor match and are capable of forming only a few weak bonds; thermal motion rapidly breaks them apart

the surfaces of molecules A and D match well and therefore can form enough weak bonds to withstand thermal motion; they therefore stay bound to each other

VAN DER WAALS FORCES

At very short distances any two atoms show a weak bonding interaction due to their fluctuating electrical charges. This force is known as van der Waals attraction. However, two atoms will very strongly repel each other if they are brought too close together. This van der Waals repulsion plays a major part in limiting the possible conformations of a molecule.

Each type of atom has a radius, known as its van der Waals radius at which van der Waals forces are in equilibrium.

H	C	N	O
1.2 Å (0.12 nm)	2.0 Å (0.2 nm)	1.5 Å (0.15 nm)	1.4 Å (0.14 nm)

Two atoms will be attracted to each other by van der Waals forces until the distance between them equals the sum of their van der Waals radii. Although they are individually very weak, these van der Waals attractions can become important when two macromolecular surfaces fit very close together.

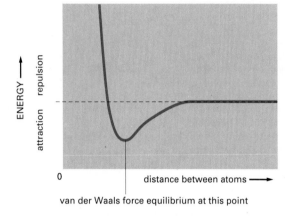

van der Waals force equilibrium at this point

WEAK CHEMICAL BONDS

Organic molecules can interact with other molecules through short-range noncovalent forces.

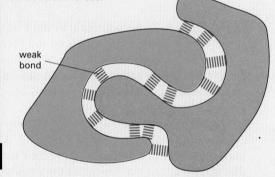

weak bond

Weak chemical bonds have less than 1/20 the strength of a strong covalent bond. They are strong enough to provide tight binding only when many of them are formed simultaneously.

HYDROGEN BONDS

A hydrogen atom is shared between two other atoms (both electronegative, such as O and N) to give a hydrogen bond.

hydrogen bond, ~ 0.3 nm long

Hydrogen bonds are strongest when the three atoms are in a straight line:

O—H |||||||||O   N—H |||||||||O

Examples in macromolecules:

Amino acids in polypeptide chains hydrogen-bonded together.

Two bases, G and C, hydrogen-bonded in DNA or RNA.

HYDROGEN BONDS IN WATER

Any molecules that can form hydrogen bonds to each other can alternatively form hydrogen bonds to water molecules. Because of this competition with water molecules, the hydrogen bonds formed between two molecules dissolved in water are relatively weak.

peptide bond

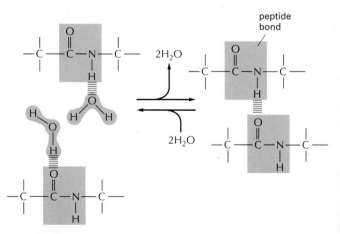

$2H_2O$

HYDROPHOBIC FORCES

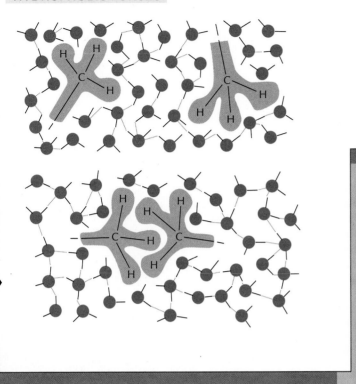

Water forces hydrophobic groups together in order to minimize their disruptive effects on the hydrogen-bonded water network. Hydrophobic groups held together in this way are sometimes said to be held together by "hydrophobic bonds," even though the attraction is actually caused by a repulsion from the water.

IONIC BONDS IN AQUEOUS SOLUTIONS

Charged groups are shielded by their interactions with water molecules. Ionic bonds are therefore quite weak in aqueous solution.

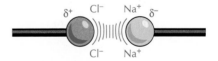

Ionic bonds are further weakened by the presence of salts, whose atoms form the counterions that cluster around ions of opposite charge.

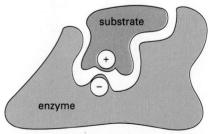

Measurement of the extent of destabilization of an interaction by salt provides a quantitative estimate of the total number of ionic bonds involved.

Despite being weakened by water and salt, ionic bonds are very important in biological systems; an enzyme that binds a positively charged substrate will often have a negatively charged amino acid side chain at the appropriate place.

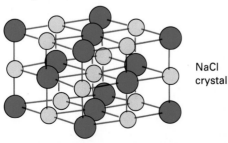

IONIC BONDS

Ionic interactions occur either between fully charged groups (ionic bond) or between partially charged groups.

The force of attraction between the two charges δ^+ and δ^- is

$$\text{force} = \frac{\delta^+ \delta^-}{r^2 D} \quad \text{(Coulomb's law)}$$

where D = dielectric constant
(1 for vacuum; 80 for water)

r = distance of separation

In the absence of water, ionic forces are very strong. They are responsible for the strength of such minerals as marble and agate.

NaCl crystal

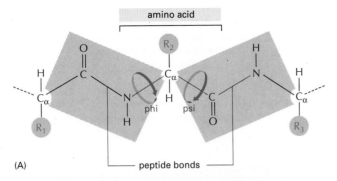

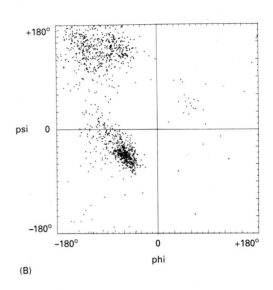

(A)

(B)

Figure 3–4 Steric limitations on the bond angles in a polypeptide chain. (A) Each amino acid contributes three bonds (colored *red*) to its polypeptide chain. The peptide bond is planar (*gray* shading) and does not permit rotation. By contrast, rotation can occur about the C$_\alpha$—C bond, whose angle of rotation is called psi (ψ), and about the N—C$_\alpha$ bond, whose angle of rotation is called phi (φ). The R group denotes an amino acid side chain. (B) The conformation of the main-chain atoms in a protein is determined by one pair of phi and psi angles for each amino acid; because of steric collisions within each amino acid, most pairs of phi and psi angles do not occur. In this so-called Ramachandran plot, each dot represents an observed pair of angles in a protein. (B, from J. Richardson, *Adv. Prot. Chem.* 34:174–175, 1981.)

A Helix Is a Common Structural Motif in Biological Structures Made from Repeated Subunits [3]

Biological structures are often formed by linking subunits that are very similar to each other—such as amino acids or nucleotides—into a long, repetitive chain. If all the subunits are identical, neighboring subunits in the chain will often fit together in only one way, adjusting their relative positions so as to minimize the

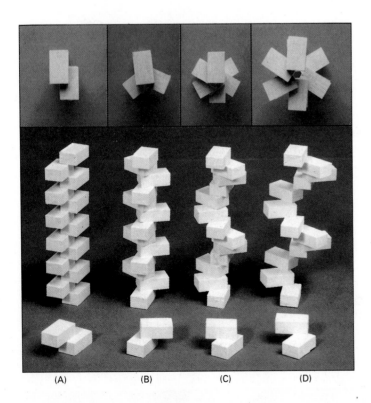

(A)   (B)   (C)   (D)

Figure 3–5 A helix will form when a series of subunits bind to each other in a regular way. In the foreground the interaction between two subunits is shown; behind it are the helices that result. These helices have two (A), three (B), and six (C and D) subunits per turn. At the top, the arrangement of subunits has been photographed from directly above the helix. Note that the helix in (D) has a wider path than that in (C).

free energy of the contact between them. In this case, each subunit will be positioned in exactly the same way in relation to its neighboring subunits, so that subunit 3 will fit onto subunit 2 in the same way that subunit 2 fits onto subunit 1, and so on. Because it is very rare for subunits to join up in a straight line, this arrangement will generally result in a **helix**—a regular structure that resembles a spiral staircase, as illustrated in Figure 3–5. Depending on the twist of the staircase, a helix is said to be either right-handed or left-handed (Figure 3–6). Handedness is not affected by turning the helix upside down, but it is reversed if the helix is reflected in a mirror.

Helices occur commonly in biological structures, whether the subunits are small molecules that are covalently linked together (as in DNA) or large protein molecules that are linked by noncovalent forces (as in actin filaments). This is not surprising. A helix is an unexceptional structure, generated simply by placing many similar subunits next to each other, each in the same strictly repeated relationship to the one before.

Diffusion Is the First Step to Molecular Recognition [4]

Before two molecules can bind to each other, they must come into close contact. This is achieved by the thermal motions that cause molecules to wander, or *diffuse*, from their starting positions. As the molecules in a liquid rapidly collide and bounce off one another, an individual molecule moves first one way and then another, its path constituting a "random walk" (Figure 3–7). The average distance that each type of molecule travels from its starting point is proportional to the square root of the time involved: that is, if it takes a particular molecule 1 second on average to go 1 μm, it will go 2 μm in 4 seconds, 10 μm in 100 seconds, and so on. Diffusion is therefore an efficient way for molecules to move limited distances but an inefficient way for molecules to move long distances.

Experiments performed by injecting fluorescent dyes and other labeled molecules into cells show that the diffusion of small molecules through the cytoplasm is nearly as rapid as it is in water. A molecule the size of ATP, for example, requires only about 0.2 second to diffuse an average distance of 10 μm—the diameter of a small animal cell. Large macromolecules, however, move much more slowly. Not only is their diffusion rate intrinsically slower, but their movement is retarded by frequent collisions with many other macromolecules that are held in place by molecular associations in the cytoplasm (Figure 3–8).

Thermal Motions Bring Molecules Together and Then Pull Them Apart [4]

Encounters between two macromolecules or between a macromolecule and a small molecule occur randomly through simple diffusion. An encounter may lead immediately to the formation of a complex between the two molecules, in which case the rate of complex formation is said to be *diffusion-limited*. Alternatively, the rate of complex formation may be slower, requiring some adjustment of the structure of one or both molecules before the interacting surfaces can fit together, so that most often the two colliding molecules will bounce off each other without sticking. In either case once the two interacting surfaces have come sufficiently close together, they will form multiple weak bonds with each other that persist until random thermal motions cause the molecules to dissociate again (see Figure 3–3).

In general, the stronger the binding of the molecules in the complex, the slower their rate of dissociation. At one extreme the total energy of the bonds formed is negligible compared with that of thermal motion, and the two molecules dissociate as rapidly as they came together. At the other extreme the total bond energy is so high that dissociation rarely occurs. Strong interactions occur in cells whenever a biological function requires that two macromolecules remain tightly associated for a long time—for example, when a gene regulatory

left-handed   right-handed
helix   helix

Figure 3–6 Comparison of a left-handed and a right-handed helix. As a reference, it is useful to remember that standard screws, which insert when turned clockwise, are right-handed. Note that a helix preserves the same handedness when it is turned upside down.

Figure 3–7 A random walk. Molecules in solution move in a random fashion due to the continual buffeting they receive in collisions with other molecules. This movement allows small molecules to diffuse from one part of the cell to another in a surprisingly short time: such molecules will generally diffuse across a typical animal cell in less than a second.

protein binds to DNA to turn off a gene. Weaker interactions occur when the function demands a rapid change in the structure of a complex—for example, when two interacting proteins change partners during the movements of a protein machine.

The Equilibrium Constant Is a Measure of the Strength of an Interaction Between Two Molecules [5]

The precise strength of the bonding between two molecules is a useful index of the specificity of their interaction. To illustrate how the binding strength is measured, let us consider a reaction in which molecule A binds to molecule B. The reaction will proceed until it reaches an *equilibrium point*, at which the rates of formation and dissociation are equal (Figure 3–9). The concentrations of A, B, and the complex AB at this point can be used to determine an **equilibrium constant** (**K**) for the reaction, as explained in Figure 3–9. This constant is sometimes termed the **affinity constant** and is commonly employed as a measure of the strength of binding between two molecules: the *stronger* the binding, the *larger* is the value of the affinity constant.

The equilibrium constant of a reaction in which two molecules bind to each other is related directly to the standard free-energy change for the binding ($\Delta G°$) by the equation described in Table 3–3. The table also lists the $\Delta G°$ values corresponding to a range of K values. Affinity constants for simple binding interactions in biological systems often range between 10^3 and 10^{12} liters/mole; this corresponds to binding energies in the range 4–17 kcal/mole, which could arise from 4 to 17 average hydrogen bonds.

100 nm

Figure 3–8 **Macromolecules in the cell cytoplasm.** The drawing is approximately to scale and emphasizes the crowding in the cytoplasm. Only the macromolecules are shown: RNAs are shown in *blue*, ribosomes in *green*, and proteins in *red*. Macromolecules diffuse relatively slowly in the cytoplasm because they interact with many other macromolecules; small molecules, by contrast, diffuse nearly as rapidly as they do in water. (Adapted from D.S. Goodsell, *Trends in Biochem. Sci.* 16:203–206, 1991.)

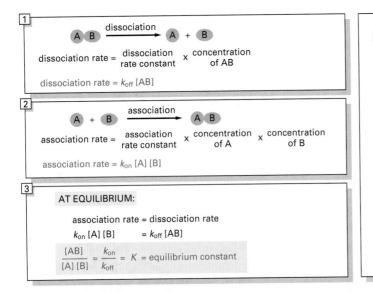

Figure 3–9 **The principle of equilibrium.** The equilibrium between molecules A and B and the complex AB is maintained by a balance between the two opposing reactions shown in (1) and (2). As shown in (3), the ratio of the rate constants for the association and the dissociation reactions is equal to the *equilibrium constant (K)* for the reaction. Molecules A and B must collide in order to react, and the rate in reaction (2) is therefore proportional to the product of their individual concentrations. As a result, the product [A] × [B] appears in the final expression for K, where [ ] indicates concentration.

As traditionally defined, the concentrations of products appear in the numerator and the concentrations of reactants appear in the denominator of the equation for an equilibrium constant. Thus the equilibrium constant in (3) is that for the *association* reaction A + B → AB. For simple binding interactions this constant is called the *affinity constant* or *association constant* (in units of liters per mole); the larger the value of the association constant (K_a), the stronger is the binding between A and B. The reciprocal of K_a is the *dissociation constant* (in units of moles per liter); the smaller the value of the dissociation constant (K_d), the stronger is the binding between A and B.

Atoms and Molecules Move Very Rapidly [6]

The chemical reactions in a cell occur at amazingly fast rates. A typical enzyme molecule, for example, will catalyze on the order of 1000 reactions per second, and rates of more than 10^6 reactions per second are achieved by some enzymes. Since each reaction requires a separate encounter between an enzyme and a substrate molecule, such rates are possible only because the molecules are moving so rapidly. Molecular motions can be classified broadly into three kinds: (1) the movement of a molecule from one place to another (translational motion), (2) the rapid back-and-forth movement of covalently linked atoms with respect to one another (vibrations), and (3) rotations. All of these motions are important in bringing the surfaces of interacting molecules together.

The rates of molecular motions can be measured by a variety of spectroscopic techniques. These indicate that a large globular protein is constantly tumbling, rotating about its axis about a million times per second. The rates of diffusional encounters due to translational movements are proportional to the concentration of the diffusing molecule. If ATP is present at its typical intracellular concentration of about 1 mM, for instance, each site on a protein molecule will be bombarded by about 10^6 random collisions with ATP molecules per second; for an ATP concentration tenfold lower, the number of collisions would drop to 10^5 per second and so on.

Once two molecules have collided and are in the correct relative orientation, a chemical reaction can occur between them extremely rapidly. When one appreciates how quickly molecules move and react, the observed rates of enzymatic catalysis do not seem so amazing.

Molecular Recognition Processes Can Never Be Perfect [7]

All molecules possess energy—the kinetic energy of their translational movements, vibrations, and rotations and the potential energy stored in their electron distributions. Through molecular collisions this energy is randomly distributed to all of the atoms present, so that most atoms will have energy levels close to the average, with only a small proportion possessing very high energy. Although the favored conformations or states for a molecule will be those of lowest free energy (see p. 75), states of higher energy occur through unusually violent collisions. Given the temperature, it is possible to calculate the probability that an atom or a molecule will be in a particular energy state (see Table 3–3). The probability of a high-energy state becomes smaller relative to a low-energy state as the difference in free energy between the two increases. It reaches zero, however, only when this energy difference becomes infinite.

Because of the random factor in molecular interactions, minor "side reactions" are bound to occur occasionally. As a consequence, a cell continually makes errors. Even reactions that are very energetically unfavorable will take place occasionally. Two atoms joined to each other by a covalent bond, for example, will eventually be subjected to an especially energetic collision and fall apart. Similarly, the specificity of an enzyme for its substrate cannot be absolute because the recognition of one molecule as distinct from another can never be perfect. Mistakes could be avoided completely only if the cell could evolve mechanisms with infinite energy differences between alternatives. Since this is not possible, cells are forced to tolerate a certain level of failure and have instead evolved a variety of repair reactions to correct those errors that are the most damaging.

On the other hand, errors are essential to life as we know it. If it were not for occasional mistakes in the maintenance of DNA sequences, evolution could not occur.

Table 3–3 The Relationship Between Free-Energy Differences and Equilibrium Constants

Equilibrium Constant $\dfrac{[AB]}{[A][B]} = K$ (liters/mole)	Free Energy of AB minus Free Energy of A + B (kcal/mole)
10^5	−7.1
10^4	−5.7
10^3	−4.3
10^2	−2.8
10	−1.4
1	0
10^{-1}	1.4
10^{-2}	2.8
10^{-3}	4.3
10^{-4}	5.7
10^{-5}	7.1

If the reaction $A + B \rightleftharpoons AB$ is allowed to come to equilibrium, the relative amounts of A, B, and AB will depend on the free-energy difference, $\Delta G°$, between them. The above values are given for 37°C (310°K) and are calculated from the equation

$$\Delta G° = -RT \ln \frac{[AB]}{[A][B]}$$

or

$$\frac{[AB]}{[A][B]} = e^{-\Delta G°/RT} = e^{-\Delta G°/0.6138}$$

Here $\Delta G°$ is in kilocalories per mole and represents the free-energy difference under standard conditions (where all components are present at a concentration of 1.0 mole/liter); T is the temperature in kelvins (°K).

Similar principles apply to the even simpler case of a reaction $A \rightleftharpoons A^*$, where a molecule is interconvertible between two states A and A^* differing in free energy by an amount $\Delta G°$. The quantities of molecules in the two states at equilibrium will be in the ratio

$$\frac{[A^*]}{[A]} = e^{-\Delta G°/RT}$$

Thus, we see from this table that, if there is a favorable free-energy change of 4.3 kcal/mole for the transition $A \rightarrow A^*$, there will be 1000 times more molecules in state A^* than state A.

Summary

The sequence of subunits in a macromolecule contains information that determines the three-dimensional contours of its surface. These contours in turn govern the recognition between one molecule and another, or between different parts of the same molecule, by means of weak, noncovalent bonds. The attractive forces are of four types: ionic bonds, van der Waals attractions, hydrogen bonds, and an interaction between nonpolar groups caused by their hydrophobic expulsion from water. Two molecules will recognize each other by a process in which they meet by random diffusion, stick together for a while, and then dissociate. The strength of this interaction is generally expressed in terms of an equilibrium constant. Since the only way to make recognition infallible is to make the energy of binding infinitely large, living cells constantly make errors; those that are intolerable are corrected by specific repair processes.

Nucleic Acids [8]

Genes Are Made of DNA [9]

It has been obvious for as long as humans have sown crops or raised animals that each seed or fertilized egg must contain a hidden plan, or design, for the development of the organism. In modern times the science of genetics grew up around the premise of invisible information-containing elements, called **genes,** that are distributed to each daughter cell when a cell divides. Therefore, before dividing, a cell has to make a copy of its genes in order to give a complete set to each daughter cell. The genes in the sperm and egg cells carry the hereditary information from one generation to the next.

The inheritance of biological characteristics must involve patterns of atoms that follow the laws of physics and chemistry: in other words, genes must be formed from molecules. At first the nature of these molecules was hard to imagine. What kind of molecule could be stored in a cell and direct the activities of a developing organism and also be capable of accurate and almost unlimited replication?

By the end of the nineteenth century biologists had recognized that the carriers of inherited information were the *chromosomes* that become visible in the nucleus as a cell begins to divide. But the evidence that the deoxyribonucleic acid (DNA) in these chromosomes is the substance of which genes are made came only much later, from studies on bacteria. In 1944 it was shown that adding purified DNA from one strain of bacteria to a second, slightly different bacterial strain conferred heritable properties characteristic of the first strain upon the second. Because it had been commonly believed that only proteins have enough conformational complexity to carry genetic information, this discovery came as a surprise, and it was not generally accepted until the early 1950s. Today the idea that DNA carries genetic information in its long chain of nucleotides is so fundamental to biological thought that it is sometimes difficult to realize the enormous intellectual gap that it filled.

DNA Molecules Consist of Two Long Chains Held Together by Complementary Base Pairs [10]

The difficulty that geneticists had in accepting DNA as the substance of genes is understandable, considering the simplicity of its chemistry. A DNA chain is a long, unbranched polymer composed of only four types of subunits. These are the deoxyribonucleotides containing the bases adenine (A), cytosine (C), guanine (G), and thymine (T). The nucleotides are linked together by covalent phosphodiester bonds that join the 5′ carbon of one deoxyribose group to the 3′ carbon of the next (see Panel 2–6, pp. 58–59). The four kinds of bases are attached to this

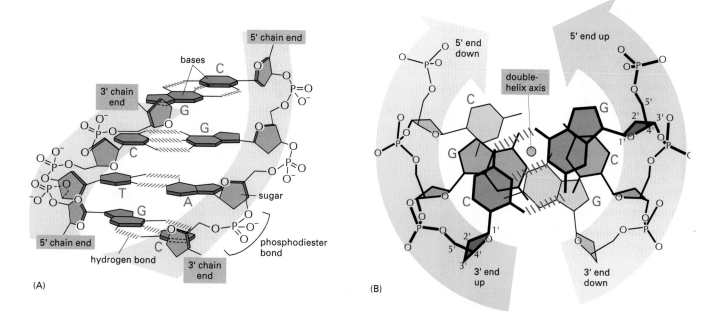

(A)

(B)

repetitive sugar-phosphate chain almost like four kinds of beads strung on a necklace.

How can a long chain of nucleotides encode the instructions for an organism or even a cell? And how can these messages be copied from one generation of cells to the next? The answers lie in the structure of the DNA molecule.

Early in the 1950s x-ray diffraction analyses of specimens of DNA pulled into fibers suggested that the DNA molecule is a helical polymer composed of two strands. The helical structure of DNA was not surprising since, as we have seen, a helix will often form if each of the neighboring subunits in a polymer is regularly oriented. But the finding that DNA is two-stranded was of crucial significance. It provided the clue that led, in 1953, to the construction of a model that fitted the observed x-ray diffraction pattern and thereby solved the puzzle of DNA structure and function.

An essential feature of the model was that all of the bases of the DNA molecule are on the *inside* of the double helix, with the sugar phosphates on the outside. This demands that the bases on one strand be extremely close to those on the other, and the fit proposed required specific *base-pairing* between a large purine base (A or G, each of which has a double ring) on one chain and a smaller pyrimidine base (T or C, each of which has a single ring) on the other chain (Figure 3–10).

Both evidence from earlier biochemical experiments and conclusions derived from model building suggested that **complementary base pairs** (also called *Watson-Crick base pairs*) form between A and T and between G and C. Biochemical analyses of DNA preparations from different species had shown that, although the nucleotide composition of DNA varies a great deal (for example, from 13% A residues to 36% A residues in the DNA of different types of bacteria), there is a general rule that quantitatively [G] = [C] and [A] = [T]. Model building revealed that the numbers of effective hydrogen bonds that could be formed between G and C or between A and T were greater than for any other combinations (see Panel 3–2, pp. 100–101). The double-helical model for DNA thus neatly explained the quantitative biochemistry.

The Structure of DNA Provides an Explanation for Heredity [11]

A gene carries biological information in a form that must be precisely copied and transmitted from each cell to all of its progeny. The implications of the discov-

Figure 3–10 The DNA double helix. (A) A short section of the helix viewed from its side. Four complementary base pairs are shown. The bases are shown in *green*, while the deoxyribose sugars are *blue*. (B) The helix viewed from an end. Note that the two DNA strands run in opposite directions and that each base pair is held together by either two or three hydrogen bonds (see also Panel 3–2, pp. 100–101).

DNA AND RNA

The structure of RNA is shown in this half of the panel, while the structure of DNA is shown in the other half. Both DNA and RNA are linear polymers of nucleotides (see Panel 2–6, pp. 58–59). RNA differs from DNA in three ways:

1. the sugar phosphate backbone contains ribose rather than deoxyribose

2. it contains the base uracil (U) instead of thymine (T)

3. it exists as a single strand rather than a double-stranded helix

SUGAR-PHOSPHATE BACKBONE OF RNA

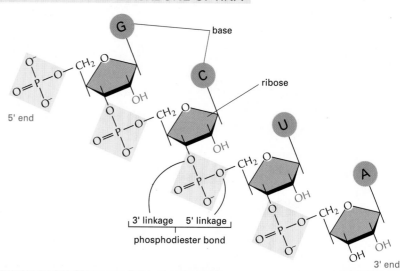

FOUR BASES OF RNA

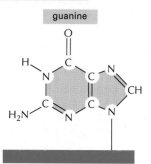 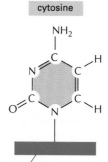 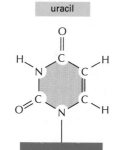 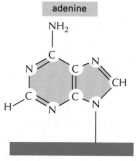

guanine   cytosine   uracil   adenine

sugar-phosphate backbone

RNA SINGLE STRAND

RNA is single-stranded, but it contains local regions of short complementary base-pairing that can form from a random matching process. Regions of base-pairing can be seen in the electron micrograph as branches off the stretched-out chain.

hydrogen bond

bases

sugar-phosphate backbone

ELECTRON MICROGRAPH OF RNA

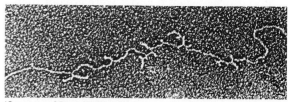

(Courtesy of Peter Wellauer.)

SUGAR-PHOSPHATE BACKBONE OF DNA

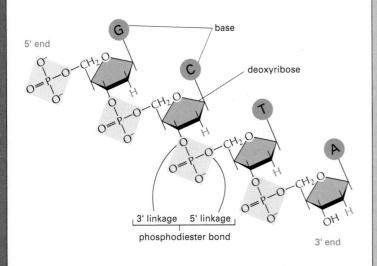

5' end

base

deoxyribose

3' linkage 5' linkage

phosphodiester bond

3' end

FOUR BASES AS BASE PAIRS OF DNA

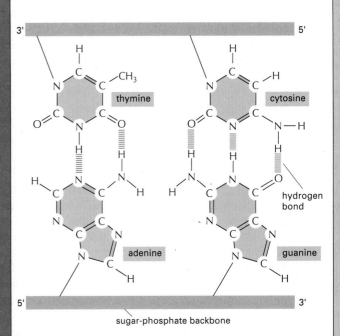

thymine

cytosine

hydrogen bond

adenine

guanine

sugar-phosphate backbone

Specific hydrogen bonding between G and C and between A and T (A and U in RNA) generates complementary base-pairing.

ELECTRON MICROGRAPH OF DNA

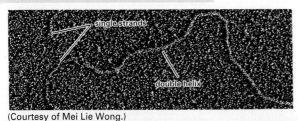

single strands

double helix

(Courtesy of Mei Lie Wong.)

DNA DOUBLE HELIX

In a DNA molecule two antiparallel strands that are complementary in their nucleotide sequence are paired in a right-handed double helix with about 10 nucleotide pairs per helical turn. A schematic representation (*top*) and a space-filling model (*bottom*) are illustrated here.

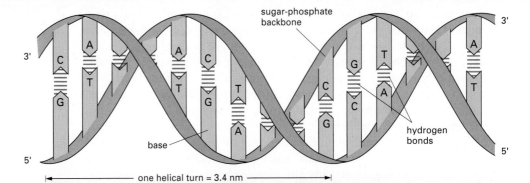

sugar-phosphate backbone

base

hydrogen bonds

one helical turn = 3.4 nm

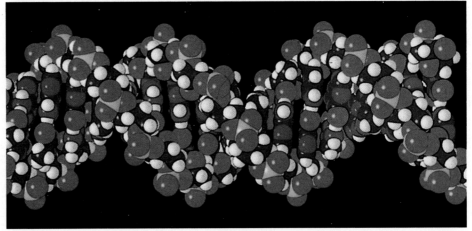

major groove   minor groove

101

ery of the DNA double helix were profound because the structure immediately suggested how information transfer could be accomplished. Since each strand contains a nucleotide sequence that is exactly complementary to the nucleotide sequence of its partner strand, both strands actually carry the same genetic information. If we designate the two strands A and A', strand A can serve as a mold or *template* for making a new strand A', while strand A' can serve in the same way to make a new strand A. Thus genetic information can be copied by a process in which strand A separates from strand A' and each separated strand then serves as a template for the production of a new complementary partner strand.

As a direct consequence of the base-pairing mechanism, it becomes evident that DNA carries information by means of the linear sequence of its nucleotides. Each nucleotide—A, C, T, or G—can be considered a letter in a four-letter alphabet that is used to write out biological messages in a linear "ticker-tape" form. Organisms differ because their respective DNA molecules carry different nucleotide sequences and therefore different biological messages.

Since the number of possible sequences in a DNA chain n nucleotides long is 4^n, the biological variety that could in principle be generated using even a modest length of DNA is enormous. A typical animal cell contains a meter of DNA (3×10^9 nucleotides). Written in a linear alphabet of four letters, an unusually small human gene would occupy a quarter of a page of text (Figure 3–11), while the genetic information carried in a human cell would fill a book of more than 500,000 pages.

Although the principle underlying gene replication is both elegant and simple, the actual machinery by which this copying is carried out in the cell is complicated and involves a complex of proteins that form a "replication machine." The fundamental reaction is that shown in Figure 3–12, in which the enzyme *DNA polymerase* catalyzes the addition of a deoxyribonucleotide to the 3' end of a DNA chain. Each nucleotide added to the chain is a *deoxyribonucleoside triphosphate;* the release of pyrophosphate from this activated nucleotide and its subsequent hydrolysis provide the energy for the **DNA replication** reaction and make it effectively irreversible.

Replication of the DNA helix begins with the local separation of its two complementary DNA strands. Each strand then acts as a template for the formation of a new DNA molecule by the sequential addition of deoxyribonucleoside triphosphates. The nucleotide to be added at each step is selected by a process that requires it to form a complementary base pair with the next nucleotide in the parental template strand, thereby generating a new DNA strand that is complementary in sequence to the template strand (see Figure 3–12). Eventually, the genetic information is duplicated in its entirety, so that two complete DNA double helices are formed, each identical in nucleotide sequence to the parental DNA helix that served as the template. Since each daughter DNA molecule ends up with one of the original strands plus one newly synthesized strand, the mechanism of DNA replication is said to be *semiconservative* (Figure 3–13).

Errors in DNA Replication Cause Mutations [12]

One of the most impressive features of DNA replication is its accuracy. Several proofreading mechanisms are used to eliminate incorrectly positioned nucleotides; as a result, the sequence of nucleotides in a DNA molecule is copied with

Figure 3–11 **The DNA sequence of the human β-globin gene.** The gene encodes one of the two subunits of the hemoglobin molecule, which carries oxygen in the blood. Only one of the two DNA strands is shown (the "coding strand"), since the other strand has a precisely complementary sequence. The sequence should be read from left to right in successive lines down the page, as if it were normal English text.

```
CCCTGTGGAGCCACACCCTAGGGTTGGCCA
ATCTACTCCCAGGAGCAGGGAGGGCAGGAG
CCAGGGCTGGGCATAAAAGTCAGGGCAGAG
CCATCTATTGCTTACATTTGCTTCTGACAC
AACTGTGTTCACTAGCAACTCAAACAGACA
CCATGGTGCACCTGACTCCTGAGGAGAAGT
CTGCCGTTACTGCCCTGTGGGGCAAGGTGA
ACGTGGATGAAGTTGGTGGTGAGGCCCTGG
GCAGGTTGGTATCAAGGTTACAAGACAGGT
TTAAGGAGACCAATAGAAACTGGGCATGTG
GAGACAGAGAAGACTCTTGGGTTTCTGATA
GGCACTGACTCTCTCTGCCTATTGGTCTAT
TTTCCCACCCTTAGGCTGCTGGTGGTCTAC
CCTTGGACCCAGAGGTTCTTTGAGTCCTTT
GGGGATCTGTCCACTCCTGATGCTGTTATG
GGCAACCCTAAGGTGAAGGCTCATGGCAAG
AAAGTGCTCGGTGCCTTTAGTGATGGCCTG
GCTCACCTGGACAACCTCAAGGGCACCTTT
GCCACACTGAGTGAGCTGCACTGTGACAAG
CTGCACGTGGATCCTGAGAACTTCAGGGTG
AGTCTATGGGACCCTTGATGTTTTCTTTCC
CCTTCTTTTCTATGGTTAAGTTCATGTCAT
AGGAAGGGGAGAAGTAACAGGGTACAGTTT
AGAATGGGAAACAGACGAATGATTGCATCA
GTGTGGAAGTCTCAGGATCGTTTTAGTTTC
TTTTATTTGCTGTTCATAACAATTGTTTTC
TTTTGTTTAATTCTTGCTTTCTTTTTTTTT
CTTCTCCGCAATTTTTACTATTATACTTAA
TGCCTTAACATTGTGTATAACAAAAGGAAA
TATCTCTGAGATACATTAAGTAACTTAAAA
AAAAACTTTACACAGTCTGCCTAGTACATT
ACTATTTGGAATATATGTGTGCTTATTTGC
ATATTCATAATCTCCCTACTTTATTTTCTT
TTATTTTTAATTGATACATAATCATTATAC
ATATTTATGGGTTAAAGTGTAATGTTTTAA
TATGTGTACACATATTGACCAAATCAGGGT
AATTTTGCATTTGTAATTTTAAAAAATGCT
TTCTTCTTTTAATATACTTTTTTGTTTATC
TTATTTCTAATACTTTCCCTAATCTCTTTC
TTTCAGGGCAATAATGATACAATGTATCAT
GCCTCTTTGCACCATTCTAAAGAATAACAG
TGATAATTTCTGGGTTAAGGCAATAGCAAT
ATTTCTGCATATAAATATTTCTGCATATAA
ATTGTAACTGATGTAAGAGGTTTCATATTG
CTAATAGCAGCTACAATCCAGCTACCATTC
TGCTTTTATTTTATGGTTGGGATAAGGCTG
GATTATTCTGAGTCCAAGCTAGGCCCTTTT
GCTAATCATGTTCATACCTCTTATCTTCCT
CCCACAGCTCCTGGGCAACGTGCTGGTCTG
TGTGCTGGCCCATCACTTTGGCAAAGAATT
CACCCCACCAGTGCAGGCTGCCTATCAGAA
AGTGGTGGCTGGTGTGGCTAATGCCCTGGC
CCACAAGTATCACTAAGCTCGCTTTCTTGC
TGTCCAATTTCTATTAAAGGTTCCTTTGTT
CCCTAAGTCCAACTACTAAACTGGGGGATA
TTATGAAGGGCCTTGAGCATCTGGATTCTG
CCTAATAAAAAACATTTATTTTCATTGCAA
TGATGTATTTAAATTATTTCTGAATATTTT
ACTAAAAAGGGAATGTGGGAGGTCAGTGCA
TTTAAAACATAAAGAAATGATGAGCTGTTC
AAACCTTGGGAAAATACACTATATCTTAAA
CTCCATGAAAGAAGGTGAGGCTGCAACCAG
CTAATGCACATTGGCAACAGCCCCTGATGC
CTATGCCTTATTCATCCCTCAGAAAAGCAT
TCTTGTAGAGGCTTGATTTGCAGGTTAAAG
TTTTGCTATGCTGTATTTTACATTACTTAT
TGTTTTAGCTGTCCTCATGAATGTCTTTTC
```

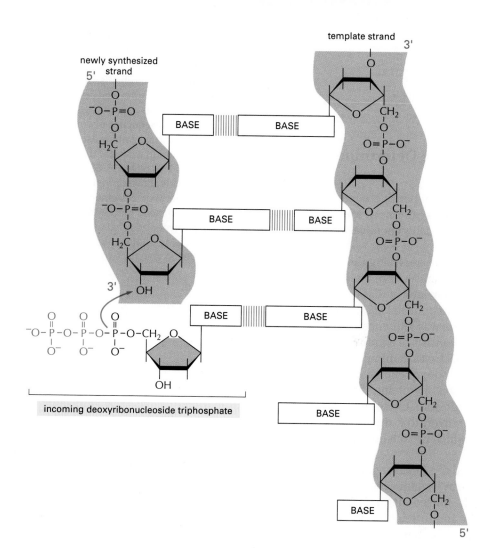

Figure 3–12 **DNA synthesis.** The addition of a deoxyribonucleotide to the 3′ end of a polynucleotide chain is the fundamental reaction by which DNA is synthesized. As shown, base-pairing between this incoming deoxyribonucleotide and an existing strand of DNA (the *template* strand) guides the formation of a new strand of DNA with a complementary nucleotide sequence.

fewer than one mistake in 10^9 nucleotides added. Very rarely, however, the replication machinery skips or adds a few nucleotides, or puts a T where it should have put a C, or an A instead of a G. Any change of this kind in the DNA sequence constitutes a genetic mistake, called a **mutation,** which will be copied in all future cell generations since "wrong" DNA sequences are copied as faithfully as "correct" ones. The consequence of such an error can be great, for even a single nucleotide change can have important effects on the cell, depending on where the mutation has occurred.

Geneticists demonstrated conclusively in the early 1940s that genes specify the structure of individual proteins. Thus a mutation in a gene, caused by an alteration in its DNA sequence, may lead to the inactivation of a crucial protein and result in cell death, in which case the mutation will be lost. On the other hand, a mutation may be *silent* and not affect the function of the protein. Very rarely, a mutation will create a gene with an improved or novel useful function. In this case organisms carrying the mutation will have an advantage, and the mutated gene may eventually replace the original gene in the population through natural selection.

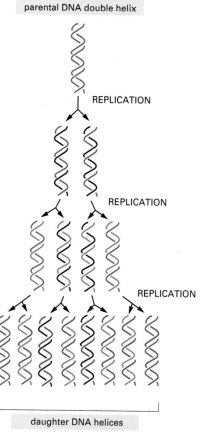

Figure 3–13 **The semiconservative replication of DNA.** In each round of replication each of the two strands of DNA is used as a template for the formation of a complementary DNA strand. The original strands therefore remain intact through many cell generations.

Nucleic Acids

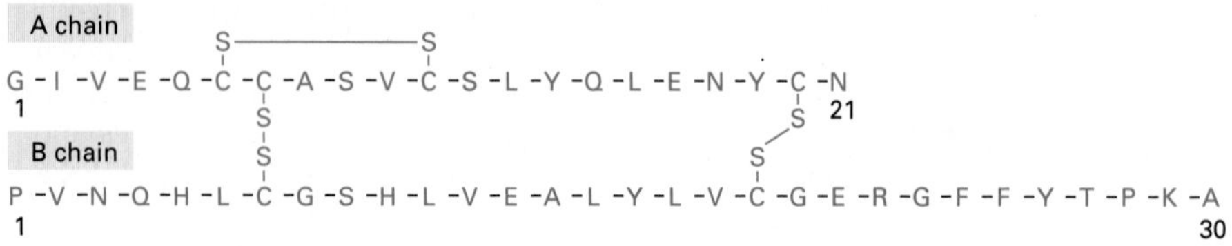

A chain

```
S ——————— S
|         |
G - I - V - E - Q - C - C - A - S - V - C - S - L - Y - Q - L - E - N - Y - C - N
1                 |                   |                               |    21
                  S                   S                               S
                  |                   |
```

B chain

```
                  S                   S
                  |                   |
P - V - N - Q - H - L - C - G - S - H - L - V - E - A - L - Y - L - V - C - G - E - R - G - F - F - Y - T - P - K - A
1                                                                                                              30
```

Figure 3–14 The amino acid sequence of bovine insulin. Insulin is a very small protein that consists of two polypeptide chains, one 21 and the other 30 amino acid residues long. Each chain has a unique, genetically determined sequence of amino acids. The one-letter symbols used to specify amino acids are those listed in Panel 2–5, pages 56–57; the S—S bonds shown in *red* are disulfide bonds between cysteine residues. The protein is made initially as a single long polypeptide chain (encoded by a single gene) that is subsequently cleaved to give the two chains.

The Nucleotide Sequence of a Gene Determines the Amino Acid Sequence of a Protein [13]

DNA is relatively inert chemically. The information it contains is expressed indirectly via other molecules: DNA directs the synthesis of specific RNA and protein molecules, which in turn determine the cell's chemical and physical properties.

At about the time that biophysicists were analyzing the three-dimensional structure of DNA by x-ray diffraction, biochemists were intensively studying the chemical structure of proteins. It was already known that proteins are chains of amino acids joined together by sequential peptide linkages; but it was only in the early 1950s, when the small protein *insulin* was sequenced (Figure 3–14), that it was discovered that each type of protein consists of a unique sequence of amino acids. Just as solving the structure of DNA was seminal in understanding the molecular basis of genetics and heredity, so sequencing insulin provided a key to understanding the structure and function of proteins. If insulin had a definite, genetically determined sequence, then presumably so did every other protein. It seemed reasonable to suppose, moreover, that the properties of a protein would depend on the precise order in which its constituent amino acids are arranged.

Both DNA and protein are composed of a linear sequence of subunits; eventually, the analysis of the proteins made by mutant genes demonstrated that the two sequences are *co-linear*—that is, the nucleotides in DNA are arranged in an order corresponding to the order of the amino acids in the protein they specify. It became evident that the DNA sequence contains a coded specification of the protein sequence. The central question in molecular biology then became how a cell translates a nucleotide sequence in DNA into an amino acid sequence in a protein.

Portions of DNA Sequence Are Copied into RNA Molecules That Guide Protein Synthesis [14]

The synthesis of proteins involves copying specific regions of DNA (the *genes*) into polynucleotides of a chemically and functionally different type known as **ribonucleic acid**, or **RNA**. RNA, like DNA, is composed of a linear sequence of nucleotides, but it has two small chemical differences: (1) the sugar-phosphate backbone of RNA contains ribose instead of a deoxyribose sugar and (2) the base thymine (T) is replaced by uracil (U), a very closely related base that likewise pairs with A (see Panel 3–2, pp. 100–101).

RNA retains all of the information of the DNA sequence from which it was copied, as well as the base-pairing properties of DNA. Molecules of RNA are synthesized by a process known as **DNA transcription,** which is similar to DNA replication in that one of the two strands of DNA acts as a template on which the base-pairing abilities of incoming nucleotides are tested. When a good match is achieved with the DNA template, a ribonucleotide is incorporated as a covalently bonded unit. In this way the growing RNA chain is elongated one nucleotide at a time.

DNA transcription differs from DNA replication in a number of ways. The RNA product, for example, does not remain as a strand annealed to DNA. Just behind the region where the ribonucleotides are being added, the original DNA helix re-forms and releases the RNA chain. Thus RNA molecules are single-

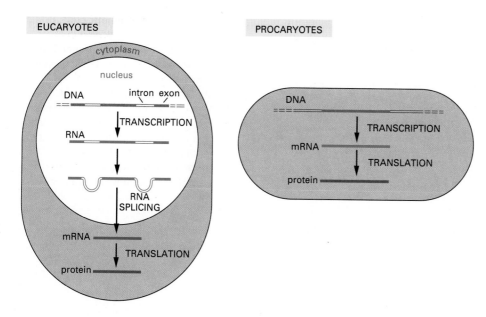

Figure 3–15 The transfer of information from DNA to protein. The transfer proceeds by means of an RNA intermediate called messenger RNA (mRNA). In procaryotic cells the process is simpler than in eucaryotic cells. In eucaryotes the coding regions of the DNA (in the *exons,* shown in color) are separated by noncoding regions (the *introns*). As indicated, these introns must be removed by an enzymatically catalyzed RNA-splicing reaction to form the mRNA.

stranded. Moreover, RNA molecules are relatively short compared to DNA molecules since they are copied from a limited region of the DNA—enough to make one or a few proteins (Figure 3–15). RNA transcripts that direct the synthesis of protein molecules are called **messenger RNA (mRNA)** molecules, while other RNA transcripts serve as *transfer RNAs (tRNAs)* or form the RNA components of ribosomes (rRNA) or smaller ribonucleoprotein particles.

The amount of RNA made from a particular region of DNA is controlled by *gene regulatory proteins* that bind to specific sites on DNA close to the coding sequences of a gene. In any cell at any given time, some genes are used to make RNA in very large quantities while other genes are not transcribed at all. For an active gene thousands of RNA transcripts can be made from the same DNA segment in each cell generation. Because each mRNA molecule can be translated into many thousands of copies of a polypeptide chain, the information contained in a small region of DNA can direct the synthesis of millions of copies of a specific protein. The protein *fibroin*, for example, is the major component of silk. In each silk gland cell a single fibroin gene makes 10^4 copies of mRNA, each of which directs the synthesis of 10^5 molecules of fibroin—producing a total of 10^9 molecules of fibroin in just 4 days.

Eucaryotic RNA Molecules Are Spliced to Remove Intron Sequences [15]

In bacterial cells most proteins are encoded by a single uninterrupted stretch of DNA sequence that is copied without alteration to produce an mRNA molecule. In 1977 molecular biologists were astonished by the discovery that most eucaryotic genes have their coding sequences (called *exons*) interrupted by noncoding sequences (called *introns*). To produce a protein, the entire length of the gene, including both its introns and its exons, is first transcribed into a very large RNA molecule—the *primary transcript.* Before this RNA molecule leaves the nucleus, a complex of RNA-processing enzymes removes all of the intron sequences, thereby producing a much shorter RNA molecule. After this RNA-processing step, called **RNA splicing,** has been completed, the RNA molecule moves to the cytoplasm as an mRNA molecule that directs the synthesis of a particular protein (see Figure 3–15).

This seemingly wasteful mode of information transfer in eucaryotes is presumed to have evolved because it makes protein synthesis much more versatile. The primary RNA transcripts of some genes, for example, can be spliced in various ways to produce different mRNAs, depending on the cell type or stage of

Nucleic Acids

1st position (5' end) ↓	2nd position				3rd position (3' end) ↓
	U	C	A	G	
U	Phe Phe Leu Leu	Ser Ser Ser Ser	Tyr Tyr STOP STOP	Cys Cys STOP Trp	U C A G
C	Leu Leu Leu Leu	Pro Pro Pro Pro	His His Gln Gln	Arg Arg Arg Arg	U C A G
A	Ile Ile Ile Met	Thr Thr Thr Thr	Asn Asn Lys Lys	Ser Ser Arg Arg	U C A G
G	Val Val Val Val	Ala Ala Ala Ala	Asp Asp Glu Glu	Gly Gly Gly Gly	U C A G

Figure 3–16 The genetic code. Sets of three nucleotides (*codons*) in an mRNA molecule are translated into amino acids in the course of protein synthesis according to the rules shown. The codons GUG and GAG, for example, are translated into valine and glutamic acid, respectively. Note that those codons with U or C as the second nucleotide tend to specify the more hydrophobic amino acids (compare with Panel 2–5, pp. 56–57).

development. This allows different proteins to be produced from the same gene. Moreover, because the presence of numerous introns facilitates genetic recombination events between exons, this type of gene arrangement is likely to have been profoundly important in the early evolutionary history of genes, speeding up the process whereby organisms evolve new proteins from parts of preexisting ones instead of evolving totally new amino acid sequences.

Sequences of Nucleotides in mRNA Are "Read" in Sets of Three and Translated into Amino Acids [16]

The rules by which the nucleotide sequence of a gene is translated into the amino acid sequence of a protein, the so-called **genetic code,** were deciphered in the early 1960s. The sequence of nucleotides in the mRNA molecule that acts as an intermediate was found to be read in serial order in groups of three. Each triplet of nucleotides, called a **codon,** specifies one amino acid. Since RNA is a linear polymer of four different nucleotides, there are $4^3 = 64$ possible codon triplets (remember that it is the *sequence* of nucleotides in the triplet that is important). However, only 20 different amino acids are commonly found in proteins, so that most amino acids are specified by several codons; that is, the genetic code is *degenerate*. The code (shown in Figure 3–16) has been highly conserved during evolution: with a few minor exceptions, it is the same in organisms as diverse as bacteria, plants, and humans.

In principle, each RNA sequence can be translated in any one of three different *reading frames* depending on where the decoding process begins (Figure 3–17). In almost every case only one of these reading frames will produce a functional protein. Since there are no punctuation signals except at the beginning and end of the RNA message, the reading frame is set at the initiation of the translation process and is maintained thereafter.

tRNA Molecules Match Amino Acids to Groups of Nucleotides [17]

The codons in an mRNA molecule do not directly recognize the amino acids they specify in the way that an enzyme recognizes a substrate. The **translation** of mRNA into protein depends on "adaptor" molecules that recognize both an

Figure 3–17 The three possible reading frames in protein synthesis. In the process of translating a nucleotide sequence (*blue*) into an amino acid sequence (*green*), the sequence of nucleotides in an mRNA molecule is read from the 5′ to the 3′ end in sequential sets of three nucleotides. In principle, therefore, the same RNA sequence can specify three completely different amino acid sequences, depending on the "reading frame."

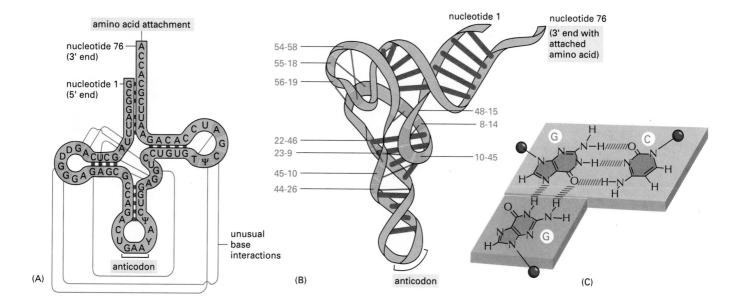

amino acid and a group of three nucleotides. These adaptors consist of a set of small RNA molecules known as **transfer RNAs (tRNAs)**, each about 80 nucleotides in length.

A tRNA molecule has a folded three-dimensional conformation that is held together in part by noncovalent base-pairing interactions like those that hold together the two strands of the DNA helix. In the single-stranded tRNA molecule, however, the complementary base pairs form between nucleotide residues in the *same* chain, which causes the tRNA molecule to fold up in a unique way that is important for its function as an adaptor. Four short segments of the molecule contain a double-helical structure, producing a molecule that looks like a "cloverleaf" in two dimensions. This cloverleaf is in turn further compacted into a highly folded, L-shaped conformation that is held together by more complex hydrogen-bonding interactions (Figure 3–18). Two sets of unpaired nucleotide residues at either end of the "L" are especially important for the function of the tRNA molecule in protein synthesis: one forms the *anticodon* that base-pairs to a complementary triplet in an mRNA molecule (the codon), while the *CCA sequence* at the 3′ end of the molecule is attached covalently to a specific amino acid (see Figure 3–18A).

The RNA Message Is Read from One End to the Other by a Ribosome [18]

The codon recognition process by which genetic information is transferred from mRNA via tRNA to protein depends on the same type of base-pair interactions that mediate the transfer of genetic information from DNA to DNA and from DNA to RNA (Figure 3–19). But the mechanics of ordering the tRNA molecules on the mRNA are complicated and require a **ribosome,** a complex of more than 50 different proteins associated with several structural RNA molecules (rRNAs). Each ribosome is a large protein-synthesizing machine on which tRNA molecules position themselves so as to read the genetic message encoded in an mRNA molecule. The ribosome first finds a specific start site on the mRNA that sets the reading frame and determines the amino-terminal end of the protein. Then, as the ribosome moves along the mRNA molecule, it translates the nucleotide sequence into an amino acid sequence one codon at a time, using tRNA molecules to add amino acids to the growing end of the polypeptide chain (Figure 3–20). When a ribosome reaches the end of the message, both it and the freshly made carboxyl end of the protein are released from the 3′ end of the mRNA molecule into the cytoplasm.

Figure 3–18 Phenylalanine tRNA of yeast. (A) The molecule is drawn with a cloverleaf shape to show the complementary base-pairing (*short gray bars*) that occurs in the helical regions of the molecule. (B) The actual shape of the molecule, based on x-ray diffraction analysis, is shown schematically. Complementary base pairs are indicated as *long gray bars*. In addition, the nucleotides involved in unusual base-pair interactions that hold different parts of the molecule together are colored *red* and are connected by a *red line* in both (A) and (B). The pairs are numbered in (B). (C) One of the unusual base-pair interactions. Here one base forms hydrogen-bond interactions with two others; several such "base triples" help fold up this tRNA molecule.

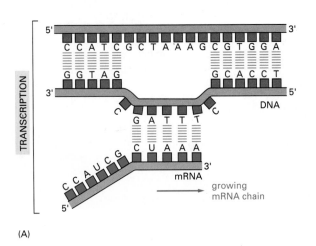

(A)

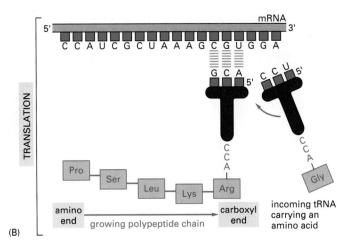

(B)

Ribosomes operate with remarkable efficiency: in one second a single bacterial ribosome adds about 20 amino acids to a growing polypeptide chain. Ribosome structure and the mechanism of protein synthesis are discussed in Chapter 6.

Some RNA Molecules Function as Catalysts [19]

RNA molecules have commonly been viewed as strings of nucleotides with a relatively uninteresting chemistry. In 1981 this view was shattered by the discovery of a catalytic RNA molecule with the type of sophisticated chemical reactivity that biochemists had previously associated only with proteins. The ribosomal RNA molecules of the ciliated protozoan *Tetrahymena* are initially synthesized as a large precursor from which one of the rRNAs is produced by an RNA-splicing reaction. The surprise came with the discovery that this splicing can occur *in vitro* in the absence of protein. It was subsequently shown that the intron sequence itself has an enzymelike catalytic activity that carries out the two-step reaction illustrated in Figure 3–21. The 400-nucleotide-long intron sequence was then synthesized in a test tube and shown to fold up to form a complex surface that can function like an enzyme in reactions with other RNA molecules. For example, it can bind two specific substrates tightly—a guanine nucleotide and an RNA chain—and catalyze their covalent attachment so as to sever the RNA chain at a specific site (Figure 3–22).

In this model reaction, which mimics the first step in Figure 3–21, the same intron sequence acts repeatedly to cut many RNA chains. Although RNA splicing is most commonly achieved by means that are not autocatalytic (discussed

Figure 3–19 Information flow in protein synthesis. (A) The nucleotides in an mRNA molecule are joined together to form a complementary copy of a segment of one strand of DNA. (B) They are then matched three at a time to complementary sets of three nucleotides in the anticodon regions of tRNA molecules. At the other end of each type of tRNA molecule, a specific amino acid is held in a high-energy linkage, and when matching occurs, this amino acid is added to the end of the growing polypeptide chain. Thus translation of the mRNA nucleotide sequence into an amino acid sequence depends on complementary base-pairing between codons in the mRNA and corresponding tRNA anticodons. The molecular basis of information transfer in translation is therefore very similar to that in DNA replication and transcription. Note that the mRNA is both synthesized and translated starting from its 5′ end.

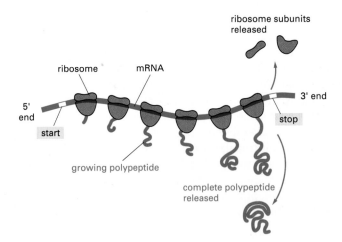

Figure 3–20 Synthesis of a protein by ribosomes attached to an mRNA molecule. Ribosomes become attached to a start signal near the 5′ end of the mRNA molecule and then move toward the 3′ end, synthesizing protein as they go. A single mRNA will usually have a number of ribosomes traveling along it at the same time, each making a separate but identical polypeptide chain; the entire structure is known as a *polyribosome*.

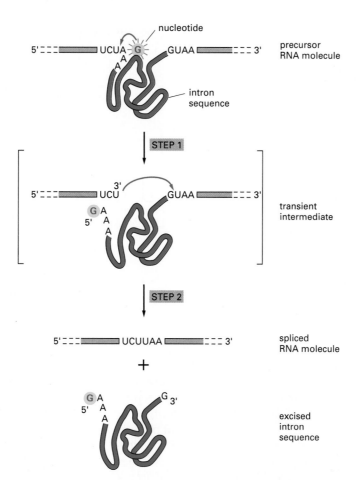

Figure 3–21 **A self-splicing RNA molecule.** The diagram shows the self-splicing reaction in which an intron sequence catalyzes its own excision from a *Tetrahymena* ribosomal RNA molecule. As shown, the reaction is initiated when a G nucleotide is added to the intron sequence, cleaving the RNA chain in the process; the newly created 3′ end of the RNA chain then attacks the other side of the intron to complete the reaction.

in Chapter 8), self-splicing RNAs with intron sequences related to that in *Tetrahymena* have been discovered in other types of cells, including fungi and bacteria. This suggests that these RNA sequences may have arisen before the eucaryotic and procaryotic lineages diverged about 1.5 billion years ago.

Several other families of catalytic RNAs have recently been discovered. Most tRNAs, for example, are initially synthesized as a larger precursor RNA, and an RNA molecule has been shown to play the major catalytic role in an RNA-protein complex that recognizes these precursors and cleaves them at specific sites. A catalytic RNA sequence also plays an important part in the life cycle of many plant viroids. Most remarkably, ribosomes are now suspected to function largely by RNA-based catalysis, with the ribosomal proteins playing a supporting role to the ribosomal RNAs (rRNAs), which make up more than half the mass of the

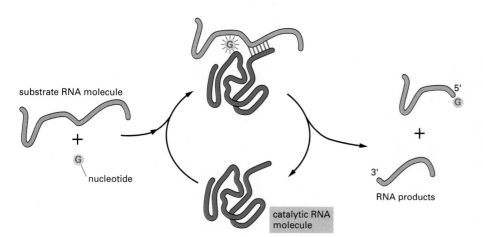

Figure 3–22 **An enzymelike reaction catalyzed by the purified *Tetrahymena* intron sequence.** In this reaction, which corresponds to the first step in Figure 3–21, both a specific substrate RNA molecule and a G nucleotide become tightly bound to the surface of the catalytic RNA molecule. The nucleotide is then covalently attached to the substrate RNA molecule, cleaving it at a specific site. The release of the resulting two RNA chains frees the intron sequence for further cycles of reaction.

Nucleic Acids

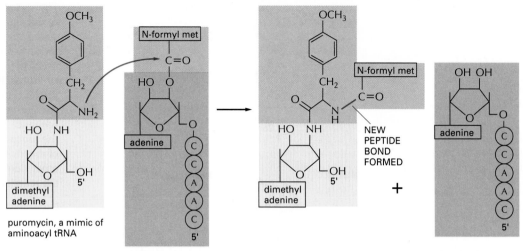

puromycin, a mimic of
aminoacyl tRNA

mimic of tRNA (*red*)
linked to C-terminus
of growing polypeptide
(*blue*)

NEW
PEPTIDE
BOND
FORMED

Figure 3–23 A peptidyl transferase reaction catalyzed by a deproteinized ribosomal RNA molecule. The puromycin molecule mimics a tRNA charged with the amino acid tyrosine, and it acts as a powerful inhibitor of protein synthesis in cells by adding to the growing end of a polypeptide chain on a ribosome. In this model reaction the growing polypeptide chain end is mimicked by a hexanucleotide (*red*, representing a tRNA) that is covalently linked to N-formyl methionine (representing the polypeptide). A highly purified large rRNA molecule catalyzes the addition of the puromycin to the N-formyl methionine, forming a new peptide bond and releasing the hexanucleotide.

ribosome. The large rRNA by itself, for example, has peptidyl transferase activity and will catalyze the formation of new peptide bonds (Figure 3–23).

How is it possible for an RNA molecule to act like an enzyme? The example of tRNA indicates that RNA molecules can fold up in highly specific ways. A proposed three-dimensional structure for the core of the self-splicing *Tetrahymena* intron sequence is shown in Figure 3–24. Interactions between different parts of this RNA molecule (analogous to the unusual hydrogen bonds in tRNA molecules—see Figure 3–18) are responsible for folding it to create a complex three-dimensional surface with catalytic activity. An unusual juxtaposition of atoms presumably strains covalent bonds and thereby makes selected atoms in the folded RNA chain unusually reactive.

As explained in Chapter 1, the discovery of catalytic RNA molecules has profoundly changed our views of how the first living cells arose.

Summary

Genetic information is carried in the linear sequence of nucleotides in DNA. Each molecule of DNA is a double helix formed from two complementary strands of nucleotides held together by hydrogen bonds between G-C and A-T base pairs. Duplication of the genetic information occurs by the polymerization of a new complementary strand onto each of the old strands of the double helix during DNA replication.

The expression of the genetic information stored in DNA involves the translation of a linear sequence of nucleotides into a co-linear sequence of amino acids in proteins. A limited segment of DNA is first copied into a complementary strand of RNA.

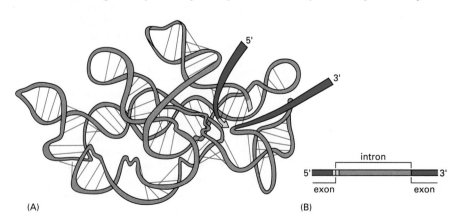

(A) (B)

Figure 3–24 A three-dimensional view of the catalytic core of the type of intron RNA sequence illustrated in Figures 3–21 and 3–22. (A) The folded molecule, with hydrogen-bond interactions shown in *red*. This molecule, which is about 240 nucleotides long, is shown immediately after the initial cut at the 5′ side of the intron (*yellow*). (B) Schematic of the molecule in (A) in its unfolded form. (Adapted from L. Jaeger, E. Westhof, and F. Michel, *J. Mol. Biol.* 221:1153–1164, 1991.)

This primary RNA transcript is spliced to remove intron sequences, producing an mRNA molecule. Finally, the mRNA is translated into protein in a complex set of reactions that occur on a ribosome. The amino acids used for protein synthesis are first attached to a family of tRNA molecules, each of which recognizes, by complementary base-pairing interactions, particular sets of three nucleotides in the mRNA (codons). The sequence of nucleotides in the mRNA is then read from one end to the other in sets of three, according to a universal genetic code.

Other RNA molecules in cells function as enzymelike catalysts. These RNA molecules fold up to create a surface containing nucleotides that have become unusually reactive. One of these catalysts is the large rRNA of the ribosome, which catalyzes the formation of peptide bonds during protein synthesis.

Protein Structure [20]

To a large extent, cells are made of protein, which constitutes more than half of their dry weight (see Table 3–1). Proteins determine the shape and structure of the cell and also serve as the main instruments of molecular recognition and catalysis. Although DNA stores the information required to make a cell, it has little direct influence on cellular processes. The gene for hemoglobin, for example, cannot carry oxygen; that is a property of the protein specified by the gene.

DNA and RNA are chains of nucleotides that are chemically very similar to one another. In contrast, proteins are made from an assortment of 20 very different amino acids, each with a distinct chemical personality (see Panel 2–5, pp. 56–57). This variety allows for enormous versatility in the chemical properties of different proteins, and it presumably explains why evolution eventually selected proteins rather than RNA molecules to catalyze most cellular reactions.

The Shape of a Protein Molecule Is Determined by Its Amino Acid Sequence [21]

Many of the bonds in a long polypeptide chain allow free rotation of the atoms they join, giving the protein backbone great flexibility. In principle, then, any protein molecule could adopt an almost unlimited number of shapes (*conformations*). Most polypeptide chains, however, fold into only one particular conformation determined by their amino acid sequence. This is because the backbones and side chains of the amino acids associate with one another and with water to form various weak noncovalent bonds (see Panel 3–1, pp. 92–93). Provided that the appropriate side chains are present at crucial positions in the chain, large forces are developed that make one particular conformation especially stable.

Most proteins can fold spontaneously into their correct shape. By treatment with certain solvents, a protein can be unfolded, or *denatured,* to give a flexible polypeptide chain that has lost its native conformation. When the denaturing solvent is removed, the protein will usually refold spontaneously into its original conformation, indicating that all the information necessary to specify the shape of a protein is contained in the amino acid sequence itself.

One of the most important factors governing the folding of a protein is the distribution of its polar and nonpolar side chains. The many hydrophobic side chains in a protein tend to be pushed together in the interior of the molecule, which enables them to avoid contact with the aqueous environment (just as oil droplets coalesce after being mechanically dispersed in water). By contrast, the polar side chains tend to arrange themselves near the outside of the protein molecule, where they can interact with water and with other polar molecules (Figure 3–25). Since the peptide bonds are themselves polar, they tend to interact both with one another and with polar side chains to form hydrogen bonds (Figure 3–26); nearly all polar residues buried within the protein are paired in this way (Figure 3–27). Hydrogen bonds thus play a major part in holding together differ-

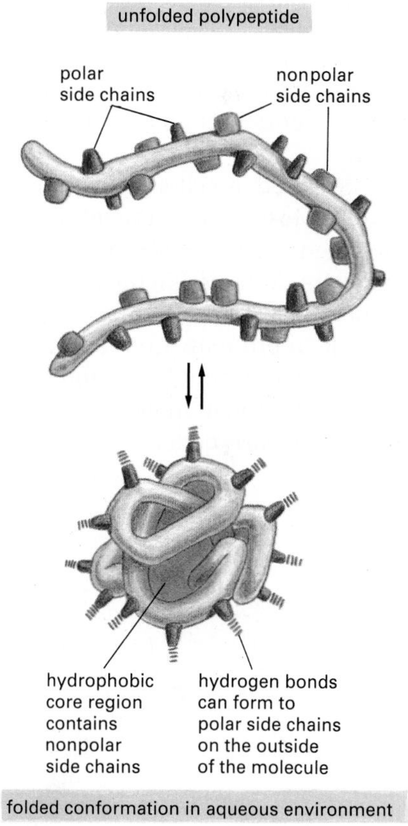

unfolded polypeptide

polar side chains

nonpolar side chains

hydrophobic core region contains nonpolar side chains

hydrogen bonds can form to polar side chains on the outside of the molecule

folded conformation in aqueous environment

Figure 3–25 How a protein folds into a globular conformation. The polar amino acid side chains tend to gather on the outside of the protein, where they can interact with water. The nonpolar amino acid side chains are buried on the inside to form a hydrophobic core that is "hidden" from water.

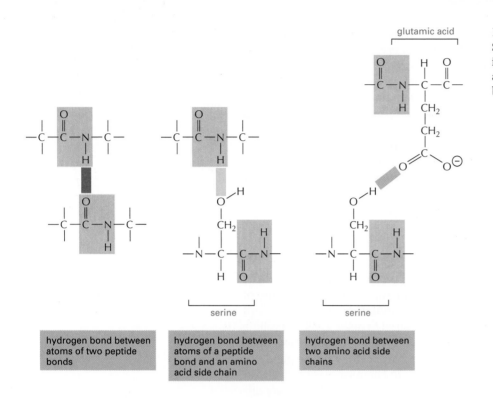

glutamic acid

serine   serine

hydrogen bond between atoms of two peptide bonds

hydrogen bond between atoms of a peptide bond and an amino acid side chain

hydrogen bond between two amino acid side chains

Figure 3–26 Hydrogen bonding. Some of the hydrogen bonds (shown in color) that can form between the amino acids in a protein. The peptide bonds are shaded in *gray*.

ent regions of polypeptide chain in a folded protein molecule. They are also crucially important for many of the binding interactions that occur on protein surfaces.

Secreted or cell-surface proteins often form additional *covalent* intrachain bonds. Most notably, the formation of **disulfide bonds** (also called S—S bonds) between the two —SH groups of neighboring cysteine residues in a folded polypeptide chain (Figure 3–28) often serves to stabilize the three-dimensional structure of extracellular proteins. These bonds are not required for the specific folding of proteins, since folding occurs normally in the presence of reducing agents that prevent S—S bond formation. In fact, S—S bonds are rarely, if ever, formed in protein molecules in the cytosol because the high cytosolic concentration of —SH reducing agents breaks such bonds.

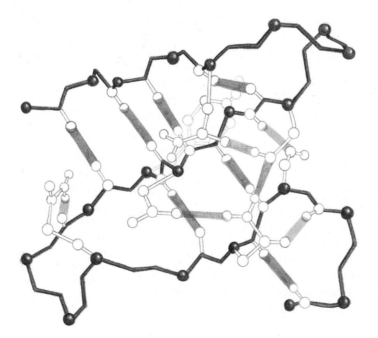

Figure 3–27 Details of intramolecular hydrogen bonds in a protein. In this region of the enzyme lysozyme, hydrogen bonds form between two side chains (*blue*), between a side chain and an atom in a peptide bond (*yellow*), or between atoms in two peptide bonds (*red*). For reference, see Figure 3–26. (After C.K. Mathews and K.E. van Holde, Biochemistry. Redwood City, CA: Benjamin/Cummings, 1990.)

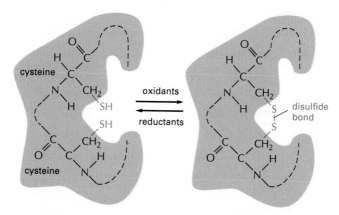

Figure 3–28 Disulfide-bond formation. The drawing illustrates the formation of a covalent disulfide bond between the side chains of neighboring cysteine residues in a protein.

The net result of all the individual amino acid interactions is that most protein molecules fold up spontaneously into precisely defined conformations. Those that are compact and globular have an inner core composed of clustered hydrophobic side chains—packed into a tight, nearly crystalline arrangement—while a very complex and irregular exterior surface is formed by the more polar side chains. The positioning and chemistry of the different atoms on this intricate surface make each protein unique and enable it to bind specifically to other macromolecular surfaces and to certain small molecules (discussed below). From both a chemical and a structural standpoint, proteins are the most sophisticated molecules known.

Common Folding Patterns Recur in Different Protein Chains [22]

Although all the information required for the folding of a protein chain is contained in its amino acid sequence, we have not yet learned how to "read" this information so as to predict the detailed three-dimensional structure of a protein whose sequence is known. Consequently, the folded conformation can be determined only by an elaborate *x-ray diffraction analysis* performed on crystals of the protein or, if the protein is very small, by nuclear magnetic resonance techniques (see Chapter 4). So far, more than 100 types of protein folds have been discovered by this technique. Each protein has a specific conformation so intricate and irregular that it would require a chapter to describe it in full three-dimensional detail.

When the three-dimensional structures of different protein molecules are compared, it becomes clear that, although the overall conformation of each protein is unique, several structural patterns recur repeatedly in parts of these macromolecules. Two patterns are particularly common because they result from regular hydrogen-bonding interactions between the peptide bonds themselves rather than between the side chains of particular amino acids. Both patterns were correctly predicted in 1951 from model-building studies based on the different x-ray diffraction patterns of silk and hair. The two regular patterns discovered are now known as the *β sheet*, which occurs in the protein fibroin, found in silk, and the *α helix*, which occurs in the protein α-keratin, found in skin and its appendages, such as hair, nails, and feathers.

The core of most (but not all) globular proteins contains extensive regions of **β sheet**. In the example illustrated in Figure 3–29, which shows part of an antibody molecule, an *antiparallel β sheet* is formed when an extended polypeptide chain folds back and forth upon itself, with each section of the chain running in the direction opposite to that of its immediate neighbors. This gives a very rigid structure held together by hydrogen bonds that connect the peptide bonds in neighboring chains. The antiparallel β sheet and the closely related *parallel β sheet* (which is formed by regions of polypeptide chain that run in the same direction) frequently serve as the framework around which globular proteins are constructed.

Protein Structure

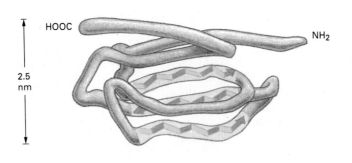

2.5 nm

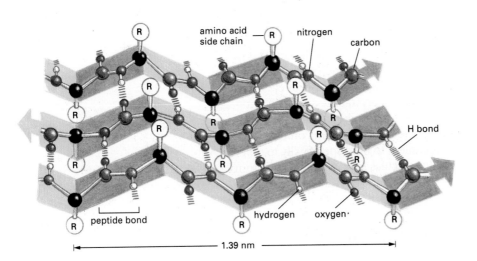

amino acid side chain — R   nitrogen
carbon
R
H bond
peptide bond
hydrogen   oxygen·
1.39 nm

Figure 3–29 A β sheet is a common structure formed by parts of the polypeptide chain in globular proteins. At the top, a domain of 115 amino acids from an immunoglobulin molecule is shown; it consists of a sandwichlike structure of two β sheets, one of which is drawn in color. At the bottom, a perfect antiparallel β sheet is shown in detail, with the amino acid side chains denoted R. Note that every peptide bond is hydrogen-bonded to a neighboring peptide bond. The actual sheet structures in globular proteins are usually less regular than the β sheet shown here, and most sheets are slightly twisted (see Figure 3–31).

Figure 3–30 An α helix is another common structure formed by parts of the polypeptide chain in proteins. (A) The oxygen-carrying molecule myoglobin (153 amino acids long) is shown, with one region of α helix outlined in color. (B) A perfect α helix is shown in outline. (C) As in the β sheet, every peptide bond in an α helix is hydrogen-bonded to a neighboring peptide bond. Note that for clarity in (B) both the side chains [which protrude radially along the outside of the helix and are denoted by R in (C)] and the hydrogen atom are omitted on the α-carbon atom of each amino acid (see also Figure 3–31).

An **α helix** is generated when a single polypeptide chain turns regularly about itself to make a rigid cylinder in which each peptide bond is regularly hydrogen-bonded to other peptide bonds nearby in the chain. Many globular proteins contain short regions of such α helices (Figure 3–30), and those portions of a transmembrane protein that cross the lipid bilayer are usually α helices because of the constraints imposed by the hydrophobic lipid environment (discussed in Chapter 10).

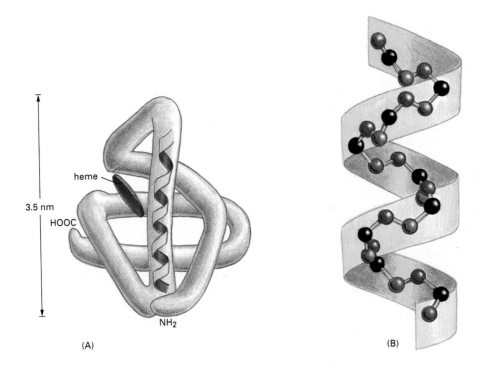

heme

3.5 nm

HOOC

NH₂

(A)

(B)

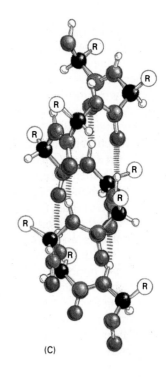

(C)

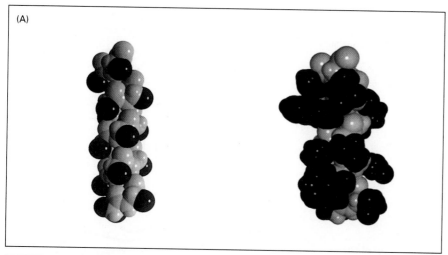

(A)

(B)

Figure 3–31 **Space-filling models of an α helix and a β sheet with (*right*) and without (*left*) their amino acid side chains.** (A) An α helix (part of the structure of myoglobin). (B) A region of β sheet (part of the structure of an immunoglobulin domain). In the photographs on the left, each side chain is represented by a single darkly shaded atom (the R groups in Figures 3–29 and 3–30), while the entire side chain is shown on the right. (Courtesy of Richard J. Feldmann.)

In aqueous environments an isolated α helix is usually not stable on its own. Two identical α helices that have a repeating arrangement of nonpolar side chains, however, will twist around each other gradually to form a particularly stable structure known as a *coiled-coil* (see p. 125). Long rodlike coiled-coils are found in many fibrous proteins, such as the intracellular α-keratin fibers that reinforce skin and its appendages.

Space-filling representations of an α helix and a β sheet from actual proteins are shown with and without their side chains in Figure 3–31.

Proteins Are Amazingly Versatile Molecules [23]

Because of the variety of their amino acid side chains, proteins are remarkably versatile with respect to the types of structures they can form. Contrast, for example, two abundant proteins secreted by cells in connective tissue—collagen and elastin—both present in the extracellular matrix. In **collagen** molecules three separate polypeptide chains, each rich in the amino acid proline and containing the amino acid glycine at every third residue, are wound around one another to generate a regular triple helix. These collagen molecules are packed together into fibrils in which adjacent molecules are tied together by covalent cross-links between neighboring lysine residues, giving the fibril enormous tensile strength (Figure 3–32).

Elastin is at the opposite extreme. Its relatively loose and unstructured polypeptide chains are cross-linked covalently to generate a rubberlike elastic meshwork that enables tissues such as arteries and lungs to deform and stretch without damage. As illustrated in Figure 3–32, the elasticity is due to the ability of individual protein molecules to uncoil reversibly whenever a stretching force is applied.

It is remarkable that the same basic chemical structure—a chain of amino acids—can form so many different structures: a rubberlike elastic meshwork

Protein Structure

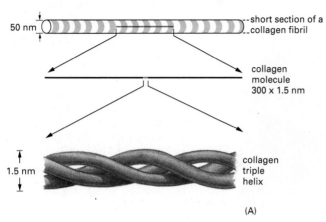

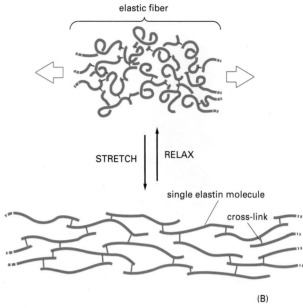

Figure 3–32 **Contrast between collagen and elastin.**
(A) *Collagen* is a triple helix formed by three extended protein chains that wrap around each other. Many rodlike collagen molecules are cross-linked together in the extracellular space to form inextensible collagen fibrils (*top*) that have the tensile strength of steel. (B) *Elastin* polypeptide chains are cross-linked together to form elastic fibers. Each elastin molecule uncoils into a more extended conformation when the fiber is stretched. The striking contrast between the physical properties of elastin and collagen is due entirely to their very different amino acid sequences.

(elastin), an inextensible cable with the tensile strength of steel (collagen), or any of the wide variety of catalytic surfaces on the globular proteins that function as enzymes. Figure 3–33 illustrates and compares the range of shapes that could, in theory, be adopted by a polypeptide chain 300 amino acids long. As we have already emphasized, the conformation actually adopted depends on the amino acid sequence.

Proteins Have Different Levels of Structural Organization [24]

In describing the structure of a protein, it is helpful to distinguish various levels of organization. The amino acid sequence is called the **primary structure** of the protein. Regular hydrogen-bond interactions within contiguous stretches of polypeptide chain give rise to α helices and β sheets, which constitute the protein's **secondary structure.** Certain combinations of α helices and β sheets pack together to form compactly folded globular units, each of which is called a protein **domain.** Domains are usually constructed from a section of polypeptide chain that contains between 50 and 350 amino acids, and they seem to be the modular units from which proteins are constructed (see below). While small proteins may contain only a single domain, larger proteins contain a number of domains, which are often connected by relatively open lengths of polypeptide chain. Finally, individual polypeptides often serve as subunits for the formation of larger molecules, sometimes called *protein assemblies* or *protein complexes,* in which the subunits are bound to one another by a large number of weak, noncovalent interactions; in extracellular proteins these interactions are often stabilized by disulfide bonds.

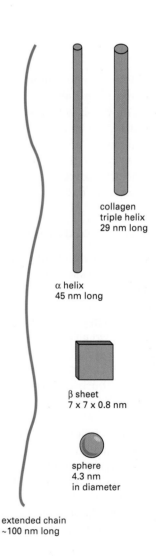

collagen triple helix 29 nm long

α helix 45 nm long

β sheet 7 x 7 x 0.8 nm

sphere 4.3 nm in diameter

extended chain ~100 nm long

Figure 3–33 **Some possible sizes and shapes of a protein molecule 300 amino acid residues long.** The structure formed is determined by the amino acid sequence. (Adapted from D.E. Metzler, Biochemistry. New York: Academic Press, 1977.)

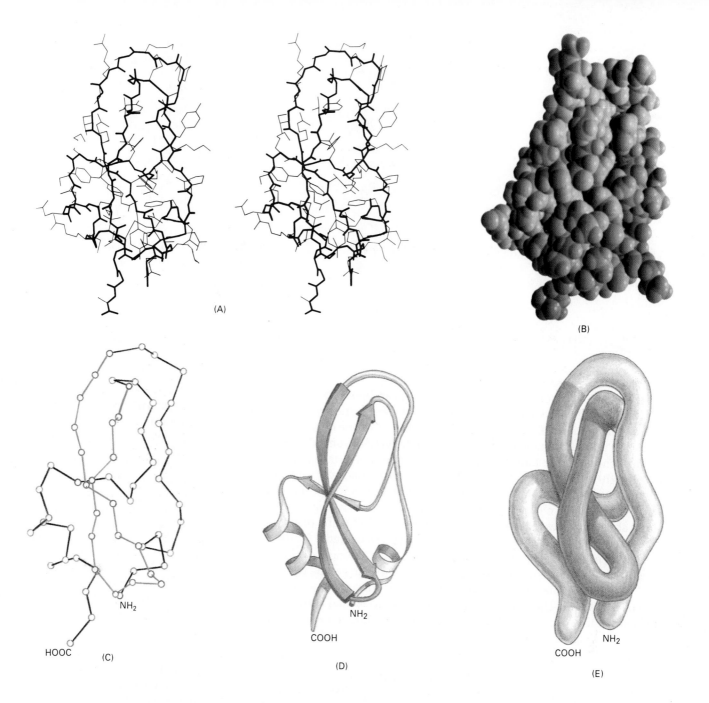

Figure 3–34 Basic pancreatic trypsin inhibitor (BPTI). The three-dimensional conformation of this small protein is shown in five commonly used representations. (A) A stereo pair illustrating the positions of all nonhydrogen atoms. The main chain is shown with heavy lines and the side chains with thin lines. (B) Space-filling model showing the van der Waals radii of all atoms (see Panel 3–1, pp. 92–93). (C) Backbone wire model composed of lines that connect each α carbon along the polypeptide backbone. (D) "Ribbon model," which represents all regions of regular hydrogen-bonded interactions as either helices (α helices) or sets of arrows (β sheets) pointing toward the carboxyl-terminal end of the chain. (E) "Sausage model," which shows the course of the polypeptide chain but omits all detail. In the bottom three panels the *hairpin beta motif* is colored *green*; this motif is also found in many other proteins (see text). Note that the core of all globular proteins is densely packed with atoms. Thus the impression of an open structure produced by models (C), (D), and (E) is misleading. (B and C, courtesy of Richard J. Feldmann; A and D, courtesy of Jane Richardson.)

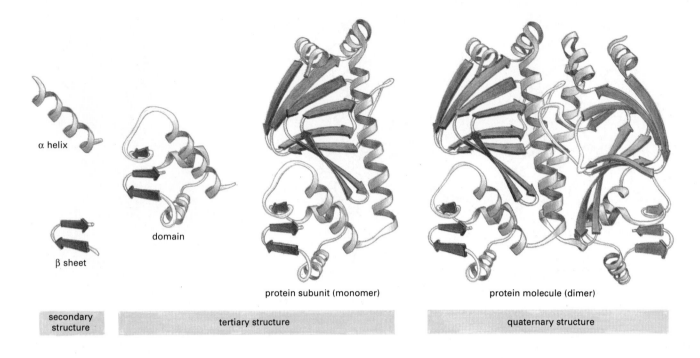

α helix

domain

β sheet

protein subunit (monomer)

protein molecule (dimer)

secondary structure

tertiary structure

quaternary structure

The three-dimensional structure of a protein can be illustrated in various ways. Consider the unusually small protein basic pancreatic trypsin inhibitor (BPTI), which contains 58 amino acid residues folded into one domain. BPTI can be shown as a stereo pair displaying all of its nonhydrogen atoms (Figure 3–34A) or as an accurate space-filling model, where most of the details are obscured (Figure 3–34B). Alternatively, it can be shown more schematically, with all of the side chains and actual atoms omitted so that it is easier to follow the course of the main polypeptide chain (Figures 3–34C, D, and E). An average-size protein contains about six times more amino acid residues than BPTI, and many proteins are more than 20 times its size. Schematic drawings are essential for displaying the structure of these larger proteins, and we use them throughout this text.

Figure 3–35 shows how the structure of a large protein can be resolved into several levels of organization, each level constructed from the one below it in a hierarchical fashion. These levels of increased organizational complexity may correspond to the steps by which a newly synthesized protein folds into its final native structure inside the cell.

Domains Are Formed from a Polypeptide Chain That Winds Back and Forth, Making Sharp Turns at the Protein Surface [24]

A protein domain can be viewed as the basic structural unit of a protein structure. The core of each domain is largely composed of a set of interconnected β sheets or α helices or both. These regular secondary structures are favored because they permit an extensive hydrogen bonding between the backbone atoms, which is essential for stabilizing the interior of the domain, where water is not available to form hydrogen bonds with the polar carbonyl oxygen or amide hydrogen of the peptide bond.

Because there are only a limited number of ways of combining α helices and β sheets to make a globular structure, certain combinations of these elements, called **motifs,** occur repeatedly in the core of many unrelated proteins. One example is the *hairpin beta motif* found in BPTI (colored *green* in Figure 3–34D), which consists of two antiparallel β strands joined by a sharp turn formed by a loop of polypeptide chain. Another example is the *beta-alpha-beta motif,* in

Figure 3–35 **Three levels of organization of a protein.** The three-dimensional structure of a protein can be described in terms of different levels of folding, each of which is constructed from the preceding one in hierarchical fashion. These levels are illustrated here using the catabolite activator protein (CAP), a bacterial gene regulatory protein with two domains. When the large domain binds cyclic AMP, it causes a conformational change in the protein that enables the small domain to bind to a specific DNA sequence. The amino acid sequence is termed the *primary structure* and the first folding level the *secondary structure.* As indicated under the brackets at the bottom of this figure, the combination of the second and third folding levels shown here is commonly termed the *tertiary structure,* and the fourth level (the assembly of subunits) the *quaternary structure* of a protein. (Modified from a drawing by Jane Richardson.)

which two adjacent parallel β strands are connected by a length of α helix (Figure 3–36). Several other common motifs are discussed in Chapter 9, where we consider the various DNA-binding motifs found in several families of gene regulatory proteins.

Various combinations of motifs form the protein domain itself, in which the polypeptide chain tends to wind its way back and forth across the entire structure, either as a β sheet or an α helix, reversing direction suddenly by making a tight turn when it reaches the surface of the domain. As a result, a typical domain is a compact structure whose surface is covered by protruding loops of polypeptide chain (Figure 3–37). The *loop regions*, which vary in length and have an irregular shape, often form the binding sites for other molecules. Because the loop regions are exposed to water, they are rich in hydrophilic amino acids, and on this basis their positions can frequently be predicted from a careful examination of the amino acid sequence of a protein.

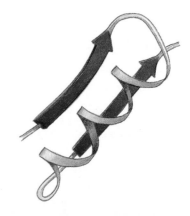

Figure 3–36 Example of a common protein motif. In the beta-alpha-beta motif two adjacent parallel strands that form a β sheet structure are connected by an α helix. Like the hairpin beta motif highlighted in Figure 3–34, this motif is found in many different proteins.

Relatively Few of the Many Possible Polypeptide Chains Would Be Useful

Since each of the 20 amino acids is chemically distinct and each can, in principle, occur at any position in a protein chain, there are $20 \times 20 \times 20 \times 20 = 160,000$ different possible polypeptide chains 4 amino acids long, or 20^n different possible polypeptide chains n amino acids long. For a typical protein length of about 300 amino acids, more than 10^{390} different proteins can be made.

We know, however, that only a very small fraction of these possible proteins would adopt a stable three-dimensional conformation. The vast majority would have many different conformations of roughly equal energy, each with different chemical properties. Proteins with such variable properties would not be useful and would therefore be eliminated by natural selection in the course of evolution. Present-day proteins have an amazingly sophisticated structure and chemistry because of their unique folding properties. Not only is the amino acid sequence such that a single conformation is extremely stable, but this conformation has the precise chemical properties that enable the protein to perform a specific catalytic or structural function in the cell. Proteins are so precisely built that the change of even a few atoms in one amino acid can sometimes disrupt the structure and cause a catastrophic change in function.

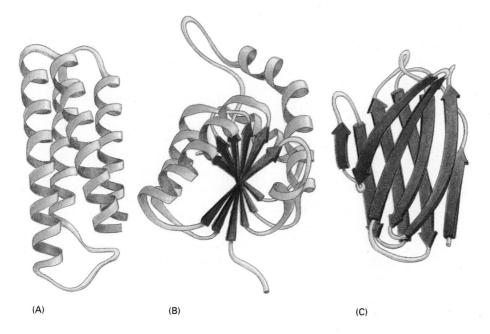

(A)   (B)   (C)

Figure 3–37 Ribbon models of the three-dimensional structure of several differently organized protein domains. (A) Cytochrome b$_{562}$, a single-domain protein composed almost entirely of α helices. (B) The NAD-binding domain of lactic dehydrogenase, composed of a mixture of α helices and β sheets. (C) The variable domain of an immunoglobin light chain, composed of a sandwich of two β sheets. In these examples the α helices are shown in *green*, while strands organized as β sheets are denoted by *red arrows*. Note that the polypeptide chain generally traverses back and forth across the entire domain, making sharp turns only at the protein surface. The protruding *loop regions* (*yellow*) often form the binding sites for other molecules. (Drawings courtesy of Jane Richardson.)

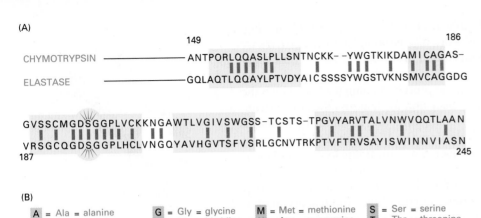

(A)

149   186
CHYMOTRYPSIN ——————— ANTPORLQQASLPLLSNTNCKK- -YWGTKIKDAMICAGAS-
                                   |||| || | ||| | | |||
ELASTASE ——————— GQLAQTLQQAYLPTVDYAICSSSSYWGSTVKNSMVCAGGDG

GVSSCMGDSGGPLVCKKNGAWTLVGIVSWGSS-TCSTS-TPGVYARVTALVNWVQQTLAAN
| | |||||| || || | | | | | || | | | |
VRSGCQGDSGGPLHCLVNGQYAVHGVTSFVSRLGCNVTRKPTVFTRVSAYISWINNVIASN
187   245

(B)

A = Ala = alanine   G = Gly = glycine   M = Met = methionine   S = Ser = serine
C = Cys = cysteine   H = His = histidine   N = Asn = asparagine   T = Thr = threonine
D = Asp = aspartic acid I = Ile = isoleucine   P = Pro = proline   V = Val = valine
E = Glu = glutamic acid K = Lys = lysine   Q = Gln = glutamine   W = Trp = tryptophan
F = Phe = phenylalanine L = Leu = leucine   R = Arg = arginine   Y = Tyr = tyrosine

Figure 3–38 (A) Comparison of the amino acid sequences of two members of the serine protease family of enzymes. The carboxyl-terminal portions of the two proteins are shown (amino acids 149 to 245). Identical amino acids are connected by colored bars, and the serine residue in the active site at position 195 is highlighted. In the *yellow* boxed sections of the polypeptide chains, each amino acid occupies a closely equivalent position in the three-dimensional structures of the two enzymes (see Figure 3–39). (B) The standard one-letter and three-letter codes for amino acids. (Modified from J. Greer, *Proc. Natl. Acad. Sci. USA* 77:3393–3397, 1980.)

New Proteins Usually Evolve by Alterations of Old Ones [25]

Cells have genetic mechanisms that allow genes to be duplicated, modified, and recombined in the course of evolution. Consequently, once a protein with useful surface properties has evolved, its basic structure can be incorporated in many other proteins. Proteins of different but related function in present-day organisms often have similar amino acid sequences. Such families of proteins are believed to have evolved from a single ancestral gene that duplicated in the course of evolution to give rise to other genes in which mutations gradually accumulated to produce related proteins with new functions.

Consider the **serine proteases,** a family of protein-cleaving (proteolytic) enzymes that includes the digestive enzymes chymotrypsin, trypsin, and elastase and some of the proteases in the blood-clotting and complement enzymatic cascades. When two of these enzymes are compared, about 40% of the positions in their amino acid sequences are found to be occupied by the same amino acid (Figure 3–38). The similarity of their three-dimensional conformations as determined by x-ray crystallography is even more striking: most of the detailed twists and turns in their polypeptide chains, which are several hundred amino acids long, are identical (Figure 3–39).

The story that we have told for the serine proteases could be repeated for hundreds of other protein families. In many cases the amino acid sequences have diverged much further than for the serine proteases, so that one cannot be sure of a family relationship between two proteins without determining their three-

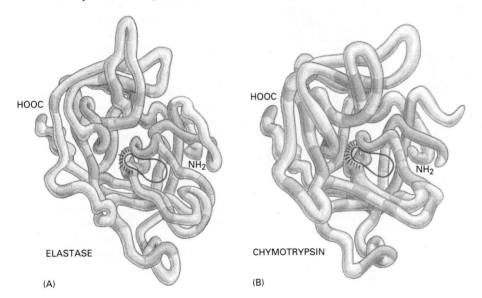

Figure 3–39 Comparison of the conformations of the two serine proteases shown in Figure 3–38. Elastase is shown in (A) and chymotrypsin in (B). Although only those amino acid residues in the polypeptide chain shaded in *green* are the same in the two proteins, their conformations are very similar everywhere. The active site, which is circled in *red,* contains an activated serine residue (see Figure 3–57). Chymotrypsin contains more than two chain termini because it is formed by the proteolytic cleavage of chymotrypsinogen, an inactive precursor.

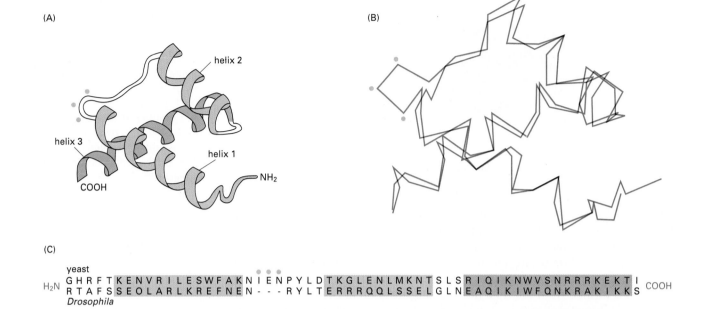

(A)

helix 2

helix 3

helix 1

COOH

NH₂

(B)

(C)

yeast

H₂N G H R F T **K E N V R I L E S W F A K N** I E N P Y L **D T K G L E N L M K N T** S L S **R I Q I K N W V S N R R R K E K T** I COOH
      R T A F S **S E O L A R L K R E F N E N** - - - R Y L **T E R R R Q Q L S S E L** G L N **E A Q I K I W F Q N K R A K I K K** S

Drosophila

dimensional structures. The yeast α2 protein and the *Drosophila* engrailed protein, for example, are both gene regulatory proteins in the homeodomain family. Because they are identical in only 17 of their 60 amino acid residues, their relationship became certain only when their three-dimensional structures were compared (Figure 3–40).

The various members of a large protein family will often have distinct functions. Some of the amino acid changes that make these proteins different were no doubt selected in the course of evolution because they resulted in changes in biological activity, giving the individual family members the different functional properties that they have today. Other amino acid changes are likely to be "neutral," having neither a beneficial nor a damaging effect on the basic structure and function of the protein. Since mutation is a random process, there must also have been many deleterious changes that altered the three-dimensional structure of these proteins sufficiently to inactivate them. Such inactive proteins would have been lost whenever the individual organisms making them were at enough of a disadvantage to be eliminated by natural selection. It is not surprising, then, that cells contain whole sets of structurally related polypeptide chains that have a common ancestry but different functions.

New Proteins Can Evolve by Recombining Preexisting Polypeptide Domains [26]

Once a number of stable protein surfaces have been made in a cell, new surfaces with different binding properties can be generated by joining two or more proteins together by noncovalent interactions between them, producing a *protein complex*. This combining of proteins to make larger, functional protein assemblies is common. Many protein complexes have molecular weights of a million or more, even though an average polypeptide chain has a molecular weight of 40,000 (about 300 to 400 amino acids), and relatively few polypeptide chains are more than three times this size.

An alternative way of making a new protein from existing chains is to join the corresponding DNA sequences to make a gene that encodes a single large polypeptide chain. Proteins in which different parts of the polypeptide chain fold independently into separate globular domains are believed to have evolved in this way, perhaps after existing for a prolonged period as a protein complex formed from separate polypeptides. Many proteins have such "multidomain" structures, and, as might be expected from the evolutionary considerations discussed above,

Figure 3–40 Comparison of DNA-binding homeodomains from two organisms separated by more than a billion years of evolution.
(A) Schematic of structure. (B) Trace of the α-carbon positions. The three-dimensional structures shown were determined by x-ray crystallography for the yeast α2 protein (*green*) and the *Drosophila* engrailed protein (*red*). (C) Comparison of amino acid sequences for the region of the proteins shown in (A) and (B). *Orange dots* demark the position of a three amino acid insert in the α2 protein. (Adapted from C. Wolberger, et al., *Cell* 67:517–528, 1991.)

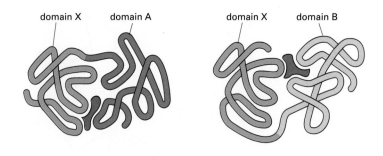

domain X   domain A   domain X   domain B

the binding sites for substrate molecules frequently lie where the separate domains are juxtaposed (Figure 3–41). Thus, for the multidomain protein whose three-dimensional structure is shown in Figure 3–42, a protein surface on one domain that binds NAD$^+$ was apparently combined with a surface on a second domain that binds a sugar, as part of the process of evolving an active site that uses the NAD$^+$ to catalyze sugar oxidation.

Another way of reutilizing an amino acid sequence is especially widespread among long fibrous proteins such as collagen (see Figure 3–32). In these cases a structure is formed from multiple internal repeats of an ancestral amino acid sequence. Putting together amino acid sequences by joining preexisting coding DNA sequences is clearly a much more efficient strategy for a cell than the alternative of deriving new protein sequences from scratch by random DNA mutation.

Structural Homologies Can Help Assign Functions to Newly Discovered Proteins [27]

The development of techniques for rapidly sequencing DNA molecules has made it possible to determine the amino acid sequences of many thousands of proteins from the nucleotide sequences of their genes. A rapidly enlarging *protein data base* is therefore available that biologists routinely scan by computer to search for possible sequence homologies between a newly sequenced protein and previously studied ones. Although sequences have so far been determined for only

Figure 3–41 The evolution of new ligand-binding sites. The general principle by which the juxtaposition of separate protein surfaces in the course of evolution has given rise to proteins that contain new binding sites for other molecules (*ligands*—see p. 129). As indicated here, the ligand-binding sites often lie at the interface between two protein domains and are formed from loop regions on the protein surface (see also Figure 3–42).

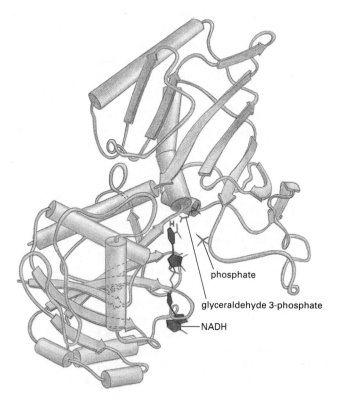

phosphate

glyceraldehyde 3-phosphate

NADH

Figure 3–42 The structure of the glycolytic enzyme glyceraldehyde 3-phosphate dehydrogenase. The protein is composed of two domains, each shown in a different color, with regions of α helix represented by cylinders and regions of β sheet represented by arrows. The details of the reaction catalyzed by the enzyme are shown in Figure 2–22. Note that the three bound substrates lie at an interface between the two domains. (Courtesy of Alan J. Wonacott.)

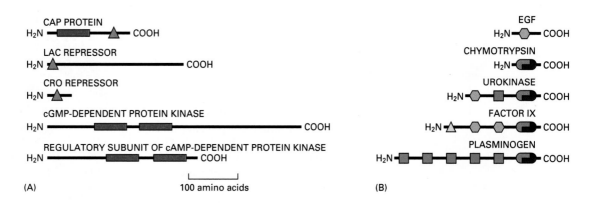

(A)

100 amino acids

(B)

a few percent of the proteins in eucaryotic organisms, it is common to find that a newly sequenced protein is homologous to some other, known protein over part of its length, indicating that most proteins may have descended from relatively few ancestral types. As expected, the sequences of many large proteins often show signs of having evolved by the joining of preexisting domains in new combinations—a process called *domain shuffling* (Figure 3–43).

These protein comparisons are important because related structures often imply related functions. Many years of experimentation can be saved by discovering an amino acid sequence homology with a protein of known function. Such sequence homologies, for example, first indicated that certain cell-cycle regulatory genes in yeast cells and certain genes that cause mammalian cells to become cancerous are protein kinases. In the same way many of the proteins that control pattern formation in the fruit fly *Drosophila* were recognized to be gene regulatory proteins, while another protein involved in pattern formation was identified as a serine protease.

The discovery of domain homologies can also be useful in another way. It is much more difficult to determine the three-dimensional structure of a protein than to determine its amino acid sequence. But the conformation of a newly sequenced protein domain can be guessed if it is homologous to a domain of a protein whose conformation has already been determined by x-ray diffraction analysis. By assuming that the twists and turns of the polypeptide chain will be conserved in the two proteins despite the presence of discrepancies in amino acid sequence, one can often sketch the structure of the new protein with reasonable accuracy (see Figure 3–40).

Many new protein sequences are being added to the data base each year, each one increasing the chance of finding useful homologies. Protein-sequence comparisons have therefore become a very important tool in cell biology.

Protein Subunits Can Assemble into Large Structures [28]

The same principles that enable several protein domains to associate to form binding sites for small molecules operate to generate much larger structures in the cell. Supramolecular structures such as enzyme complexes, ribosomes, protein filaments, viruses, and membranes are not made as single, giant, covalently linked molecules; instead they are formed by the noncovalent assembly of many preformed molecules, which are called *subunits* of the final structure.

There are several advantages to the use of smaller subunits to build larger structures: (1) building a large structure from one or a few repeating smaller subunits reduces the amount of genetic information required; (2) both assembly and disassembly can be readily controlled, since the subunits associate through multiple bonds of relatively low energy; and (3) errors in the synthesis of the structure can be more easily avoided, since correction mechanisms can operate during the course of assembly to exclude malformed subunits.

Figure 3–43 Domain shuffling. An extensive shuffling of blocks of protein sequence (protein *modules*) has occurred during the evolution of proteins. Those portions of a protein denoted by the same shape and color are evolutionarily related but not identical. (A) The bacterial catabolite gene activator protein (CAP) contains one domain (*blue triangle*) that binds a specific DNA sequence and a second domain (*red rectangle*) that binds cyclic AMP (see Figure 3–35). The DNA-binding domain here is related to the DNA-binding domains of many other gene regulatory proteins, including the lac repressor and cro repressor proteins. In addition, two copies of the cyclic-AMP-binding domain are found in eucaryotic protein kinases regulated by the binding of cyclic nucleotides. (B) Serine proteases like chymotrypsin are formed from two domains (*brown*). In some related proteases that are highly regulated and more specialized, the two protease domains are connected to one or more domains homologous to domains found in epidermal growth factor (*green hexagon*), to a calcium-binding protein (*yellow triangle*), or to a "kringle" domain (*blue square*) that contains three internal disulfide bridges.

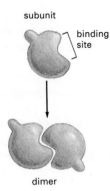

subunit

binding site

dimer

Figure 3–44 The formation of a dimer from a single type of protein subunit. A protein with a binding site that recognizes itself will often form symmetrical dimers. These may then pair with other subunits to form tetramers and larger assemblies (not shown).

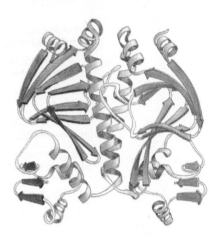

Figure 3–45 Ribbon model of a dimer formed from two identical protein subunits (monomers). The protein shown is the bacterial catabolite gene activator protein (CAP) illustrated previously in Figure 3–35. (Courtesy of Jane Richardson.)

A Single Type of Protein Subunit Can Interact with Itself to Form Geometrically Regular Assemblies [29]

If a protein has a binding site that is complementary to a region of its own surface, it will assemble spontaneously to form a larger structure. In the simplest case, a binding site recognizes itself and forms a symmetrical *dimer*. Many enzymes and other proteins form dimers of this kind, which frequently act as subunits in the formation of larger assemblies (Figures 3–44 and 3–45).

If the binding site of a protein is complementary to a region of its surface that does not include the binding site itself, a chain of subunits will be formed. For certain special orientations of the two binding sites, the chain will soon run into itself and terminate, forming a closed ring of subunits (Figure 3–46). More commonly, an extended polymer of subunits will result, and provided that each subunit is bound to its neighbor in an identical way, the subunits in the polymer will be arranged in a helix that can be extended indefinitely (see Figure 3–5). An *actin filament*, for example, is a helical structure formed from a single globular protein subunit called *actin*; actin filaments are major components in the cytosol of most eucaryotic cells (Figure 3–47). As we discuss below, globular proteins may also associate with like neighbors to form extended sheets or tubes (see Figure 3–49).

Coiled-Coil Proteins Help Build Many Elongated Structures in Cells [30]

Where mechanical strength is of major importance, supramolecular assemblies are usually made from fibrous rather than globular subunits. Such assemblies can be stabilized by extensive regions of protein-protein contact when the subunits

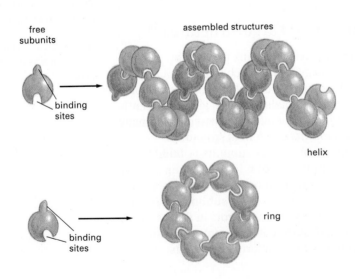

free subunits

assembled structures

binding sites

helix

binding sites

ring

Figure 3–46 Rings or helices can form if a single type of protein subunit interacts with itself repeatedly. The formation of a helix was illustrated in Figure 3–5; a ring forms instead of a helix if the subunits run into one another, stopping further growth of the chain.

actin helix

are wound around one another as a multistranded helix. A particularly stable structural unit that is used repeatedly for this purpose is known as the **coiled-coil.** It forms by the pairing of two α-helical subunits that have a repeating arrangement of nonpolar side chains. The two α-helical subunits are usually identical and run in parallel (that is, in the same direction from amino to carboxyl terminal). They coil gradually around each other to produce a stiff filament with a diameter of about 2 nm (Figure 3–48). Whereas short coiled-coils serve as dimerization domains in several families of gene regulatory proteins, more commonly a coiled-coil will extend for more than 100 nm and serve as a building block for a large fibrous structure, such as the thick filaments in a muscle cell.

Proteins Can Assemble into Sheets, Tubes, or Spheres [31]

Some protein subunits assemble into flat sheets in which the subunits are arranged in hexagonal arrays. Specialized membrane proteins are sometimes arranged in this way in lipid bilayers. With a slight change in the geometry of the individual subunits, a hexagonal sheet can be converted into a tube (Figure 3–49) or, with more changes, into a hollow sphere. Protein tubes and spheres that bind specific RNA and DNA molecules form the coats of viruses.

The formation of closed structures, such as rings, tubes, or spheres, provides additional stability because it increases the number of bonds that can form between the protein subunits. Moreover, because such a structure is formed by mutually dependent, cooperative interactions between subunits, it can be driven

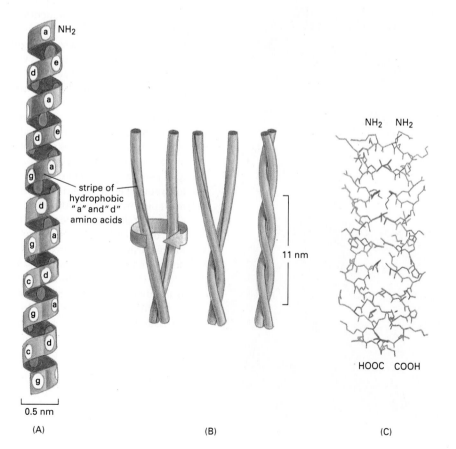

(A)

(B)

(C)

Figure 3–48 **The structure of a coiled-coil.** In (A) a single α helix is shown, with successive amino acid side chains labeled in a sevenfold sequence "abcdefg" (from bottom to top). Amino acids "a" and "d" in such a sequence lie close together on the cylinder surface, forming a "stripe" (shaded in *red*) that winds slowly around the α helix. Proteins that form coiled-coils typically have hydrophobic amino acids at positions "a" and "d." Consequently, as shown in (B), the two α helices can wrap around each other with the hydrophobic side chains of one α helix interacting with the hydrophobic side chains of the other, while the more hydrophilic amino acid side chains are left exposed to the aqueous environment. (C) The atomic structure of a coiled-coil determined by x-ray crystallography. The *red* side chains are hydrophobic. (C, from T. Alber, *Curr. Opin. Genet. Devel.* 2:205–210, 1992. © Current Science.)

Protein Structure

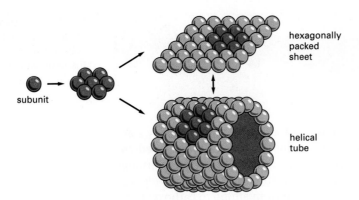

Figure 3–49 **Hexagonally packed globular protein subunits can form either a flat sheet or a tube.**

hexagonally packed sheet

helical tube

subunit

to assemble or disassemble by a relatively small change that affects the subunits individually. These principles are dramatically illustrated in the protein *capsid* of many simple viruses, which takes the form of a hollow sphere. These coats are often made of hundreds of identical protein subunits that enclose and protect the viral nucleic acid (Figure 3–50). The protein in such a capsid must have a particularly adaptable structure, since it must make several different kinds of contacts and also change its arrangement to let the nucleic acid out to initiate viral replication once the virus has entered a cell.

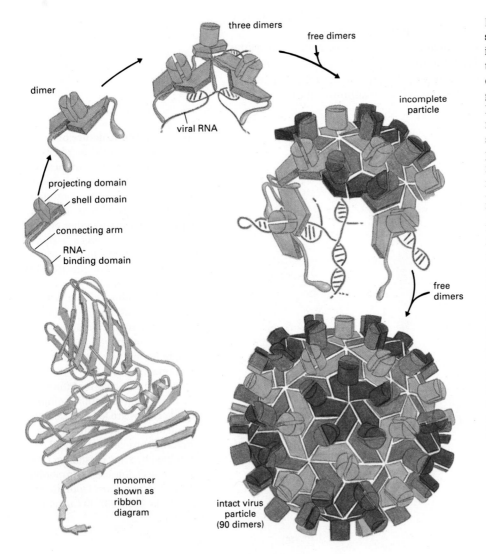

three dimers

free dimers

dimer

viral RNA

incomplete particle

projecting domain

shell domain

connecting arm

RNA-binding domain

free dimers

monomer shown as ribbon diagram

intact virus particle (90 dimers)

Figure 3–50 **The structure of a spherical virus.** In many viruses, identical protein subunits pack together to create a spherical shell (a capsid) that encloses the viral genome, composed of either RNA or DNA (see Figure 6–72). For geometric reasons, no more than 60 identical subunits can pack together in a precisely symmetrical way. If slight irregularities are allowed, however, more subunits can be used to produce a larger capsid. The tomato bushy stunt virus (TBSV) shown here, for example, is a spherical virus about 33 nm in diameter that is formed from 180 identical copies of a 386 amino acid capsid protein plus an RNA genome of 4500 nucleotides. To form such a large capsid, the protein must be able to fit into three somewhat different environments, each of which is differently colored in the particle shown here. The postulated pathway of assembly is shown; the precise three-dimensional structure has been determined by x-ray diffraction. (Courtesy of Steve Harrison.)

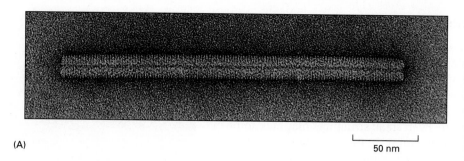

(A)

50 nm

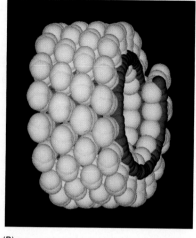

(B)

Many Structures in Cells Are Capable of Self-assembly [32]

The information for forming many of the complex assemblies of macromolecules in cells must be contained in the subunits themselves, since under appropriate conditions the isolated subunits can spontaneously assemble in a test tube into the final structure. The first large macromolecular aggregate shown to be capable of self-assembly from its component parts was *tobacco mosaic virus (TMV)*. This virus is a long rod in which a cylinder of protein is arranged around a helical RNA core (Figure 3–51). If the dissociated RNA and protein subunits are mixed together in solution, they recombine to form fully active virus particles. The assembly process is unexpectedly complex and includes the formation of double rings of protein, which serve as intermediates that add to the growing virus coat.

Another complex macromolecular aggregate that can reassemble from its component parts is the bacterial ribosome. These ribosomes are composed of about 55 different protein molecules and 3 different rRNA molecules. If the individual components are incubated under appropriate conditions in a test tube, they spontaneously re-form the original structure. Most important, such reconstituted ribosomes are able to carry out protein synthesis. As might be expected, the reassembly of ribosomes follows a specific pathway: certain proteins first bind to the RNA, and this complex is then recognized by other proteins, and so on until the structure is complete.

It is still not clear how some of the more elaborate self-assembly processes are regulated. Many structures in the cell, for example, appear to have a precisely defined length that is many times greater than that of their component macromolecules. How such length determination is achieved is in most cases a mystery. Three possible mechanisms are illustrated in Figure 3–52. In the simplest case a long core protein or other macromolecule provides a scaffold that determines the extent of the final assembly. This is the mechanism that determines the length of the TMV particle, where the RNA chain provides the core. Similarly, a core protein is thought to determine the length of the thin filaments in muscle, as well as the long tails of some bacterial viruses (Figure 3–53).

Figure 3–51 The structure of tobacco mosaic virus (TMV). (A) Electron micrograph of a tobacco mosaic virus (TMV), which consists of a single long RNA molecule enclosed in a cylindrical protein coat composed of a tight helical array of identical protein subunits. (B) A model showing part of the structure of TMV. A single-stranded RNA molecule of 6000 nucleotides is packaged in a helical coat constructed from 2130 copies of a coat protein 158 amino acids long. Fully infective virus particles can self-assemble in a test tube from purified RNA and protein molecules. (A, courtesy of Robley Williams; B, courtesy of Richard J. Feldmann.)

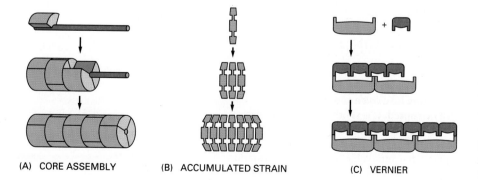

(A) CORE ASSEMBLY   (B) ACCUMULATED STRAIN   (C) VERNIER

Figure 3–52 Three ways in which a large protein assembly can be made to a fixed length. (A) Coassembly along an elongated core protein or other macromolecule that acts as a measuring device. (B) Termination of assembly because of strain that accumulates in the polymeric structure as additional subunits are added, so that beyond a certain length the energy required to fit another subunit onto the chain becomes excessively large. (C) A vernier type of assembly, in which two sets of rodlike molecules differing in length form a staggered complex that grows until their ends exactly match.

Not All Biological Structures Form by Self-assembly [33]

Some cellular structures held together by noncovalent bonds are not capable of self-assembly. A mitochondrion, a cilium, or a myofibril, for example, cannot form spontaneously from a solution of their component macromolecules because part of the information for their assembly is provided by special enzymes and other cellular proteins that perform the function of jigs or templates but do not appear in the final assembled structure. Even small structures may lack some of the ingredients necessary for their own assembly. In the formation of some bacterial viruses, for example, the head structure, which is composed of a single protein subunit, is assembled on a temporary scaffold composed of a second protein. The second protein is absent from the final virus particle, and so the head structure cannot spontaneously reassemble once it is taken apart. Other examples are known in which proteolytic cleavage is an essential and irreversible step in the assembly process. This is the case for the coats of some bacterial viruses and even for some simple protein assemblies, including the structural protein collagen and the hormone insulin (Figure 3–54). From these relatively simple examples, it seems very likely that the assembly of a structure as complex as a mitochondrion or a cilium will involve both temporal and spatial ordering imparted by other cellular components, as well as irreversible processing steps catalyzed by degradative enzymes.

Summary

The three-dimensional conformation of a protein molecule is determined by its amino acid sequence. The folded structure is stabilized by noncovalent interactions between different parts of the polypeptide chain. The amino acids with hydrophobic side chains tend to cluster in the interior of the molecule, and local hydrogen-bond interactions between neighboring peptide bonds give rise to α helices and β sheets. Globular regions known as domains are the modular units from which many proteins are constructed; small proteins typically contain only a single domain, while large proteins contain several domains linked together by short lengths of polypeptide chain. As proteins evolved, domains were modified and combined with other domains to construct new proteins.

Proteins are brought together into larger structures by the same noncovalent forces that determine protein folding. Proteins with binding sites for their own surface can assemble into dimers, closed rings, spherical shells, or helical polymers. Although mixtures of proteins and nucleic acids can assemble spontaneously into complex structures in the test tube, many assembly processes involve irreversible steps. Consequently, not all structures in the cell are capable of spontaneous reassembly after they are dissociated into their component parts.

Proteins as Catalysts [34]

The chemical properties of a protein molecule depend almost entirely on its exposed surface residues, which are able to form weak, noncovalent bonds with other molecules. When a protein molecule binds to another molecule, the second molecule is commonly referred to as a **ligand.** Because an effective interaction between a protein molecule and a ligand requires that many weak bonds be

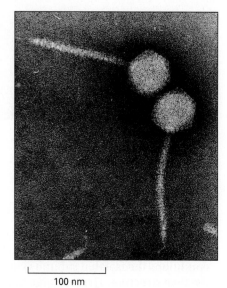

Figure 3–53 Electron micrograph of bacteriophage lambda. The tip of the virus tail attaches to a specific protein on the surface of a bacterial cell, following which the tightly packaged DNA in the head is injected through the tail into the cell. The tail has a precise length, which is determined by the mechanism shown in Figure 3–52A.

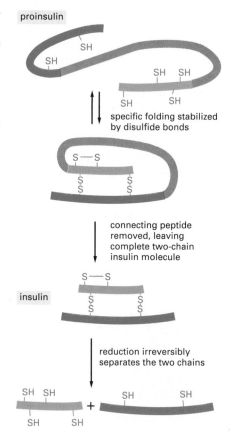

Figure 3–54 **The polypeptide hormone insulin cannot spontaneously re-form if its disulfide bonds are disrupted.** It is synthesized as a larger protein (*proinsulin*) that is cleaved by a proteolytic enzyme after the protein chain has folded into a specific shape. Excision of part of the proinsulin polypeptide chain causes an irretrievable loss of the information needed for the protein to fold spontaneously into its normal conformation.

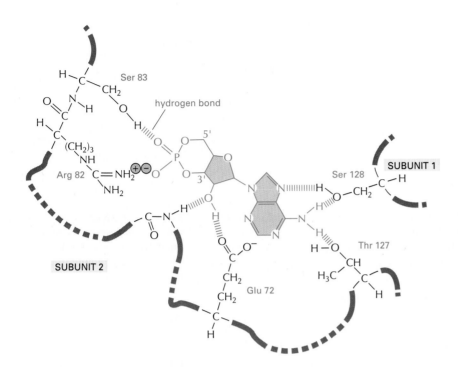

Figure 3–55 The ligand-binding site of the catabolite gene activator protein (CAP). Hydrogen bonding between CAP and its ligand, cyclic AMP (*green*), was determined by x-ray crystallographic analysis of the complex. As indicated, the two identical subunits of the dimer cooperate to form this binding site (see also Figure 3–45). (Courtesy of Tom Steitz.)

formed simultaneously between them, the only ligands that can bind tightly to a protein are those that fit precisely onto its surface.

The region of a protein that associates with a ligand, known as its **binding site,** usually consists of a cavity formed by a specific arrangement of amino acids on the protein surface. These amino acids often belong to widely separated regions of the polypeptide chain (Figure 3–55), and they represent only a minor fraction of the total amino acids present. The rest of the protein molecule is presumably necessary to maintain the polypeptide chain in the correct position and to provide additional binding sites for regulatory purposes; the interior of the protein is often important only insofar as it gives the surface of the molecule the appropriate shape and rigidity.

A Protein's Conformation Determines Its Chemistry [20]

Neighboring surface residues on a protein often interact in a way that alters the chemical reactivity of selected amino acid side chains. These interactions are of several types.

First, neighboring parts of the polypeptide chain may interact in a way that restricts the access of water molecules to other parts of the protein surface. Because water molecules tend to form hydrogen bonds, they compete with ligands for selected side chains on the protein surface (Figure 3–56). The tightness of hydrogen bonds (and ionic interactions) between proteins and their ligands is therefore greatly increased if water molecules are excluded. At first sight it is hard to imagine a mechanism that would exclude a molecule as small as water from a protein surface without affecting the access of the ligand itself. Because of their strong tendency for hydrogen bonding, however, water molecules exist in a large hydrogen-bonded network (see Panel 2–1, pp. 48–49), and it is often energetically unfavorable for individual molecules to break away from this network to reach into a crevice on the protein surface.

Second, the clustering of neighboring polar amino acid side chains can alter their reactivity. If a number of negatively charged side chains are forced together against their mutual repulsion by the way the protein folds, for example, the affinity of the site for a positively charged ion is greatly increased. Selected amino acid side chains can also interact with one another through hydrogen bonds, which can activate normally unreactive side groups (such as the $-CH_2OH$

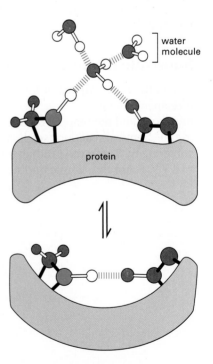

Figure 3–56 Competition for hydrogen bonding. The ability of water molecules to make favorable hydrogen bonds with groups on the protein surface greatly reduces the tendency of these groups to pair with each other.

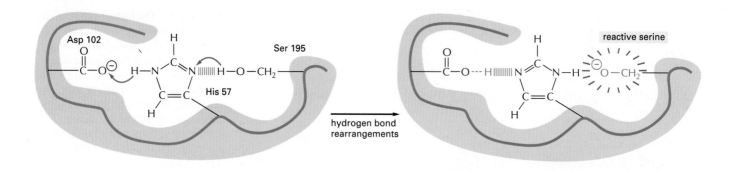

Figure 3–57 **An unusually reactive amino acid at the active site of an enzyme.** The example shown is the "catalytic triad" found in chymotrypsin, elastase, and other serine proteases (see Figure 3–39). The aspartic acid side chain induces the histidine to remove the proton from serine 195; this activates the serine to form a covalent bond with the enzyme substrate, hydrolyzing a peptide bond as illustrated later in Figure 3–64.

on the serine shown in Figure 3–57) so that they are able to enter into reactions that make or break selected covalent bonds.

The surface of each protein molecule therefore has a unique chemical reactivity that depends not only on which amino acid side chains are exposed, but also on their exact orientation relative to one another. For this reason even two slightly different conformations of the same protein molecule may differ greatly in their chemistry.

Where side-chain reactivities are insufficient, proteins often enlist the help of selected nonpolypeptide molecules that the proteins bind to their surface. These ligands serve as **coenzymes** in enzyme-catalyzed reactions, and they may be so tightly bound to the protein that they are effectively part of the protein itself. Examples are the iron-containing *hemes* in hemoglobin and cytochromes, *thiamine pyrophosphate* in enzymes involved in aldehyde-group transfers, and *biotin* in enzymes involved in carboxyl-group transfers. Most coenzymes are very complex organic molecules that have been selected for the unique chemical reactivity they acquire when bound to a protein surface. Besides its reactive center such a coenzyme has other residues designed to bind it to its host protein (Figure 3–58). A space-filling model of an enzyme bound to a coenzyme is shown in Figure 3–59A.

Substrate Binding Is the First Step in Enzyme Catalysis [35]

One of the most important functions of proteins is to act as enzymes that catalyze specific chemical reactions. The ligand in this case is called a **substrate** molecule, and the binding of the substrate to the enzyme is an essential prelude to the chemical reaction (see Figure 3–59B). If we denote the enzyme by E, the substrate by S, and the product by P, the basic reaction path is $E + S \rightleftharpoons ES \rightleftharpoons EP \rightleftharpoons E + P$. From this simple outline of an enzyme-catalyzed reaction, we see that

Figure 3–58 **Coenzymes.** Coenzymes, such as thiamine pyrophosphate (TPP), shown here in *gray,* are small molecules that bind to an enzyme's surface and enable it to catalyze specific reactions. The reactivity of TPP depends on its "acidic" carbon atom, which readily exchanges its hydrogen atom for a carbon atom of a substrate molecule. Other regions of the TPP molecule act as "handles" by which the enzyme holds the coenzyme in the correct position. Coenzymes presumably evolved first in an "RNA world," where they were bound to RNA molecules to help with catalysis (discussed in Chapter 1).

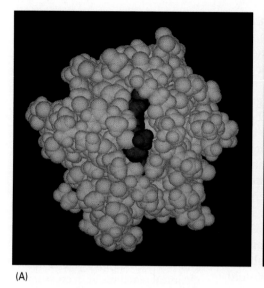
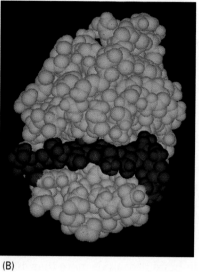

(A)   (B)

Figure 3–59 Computer-generated space-filling models of two enzymes. In (A) cytochrome c is shown with its bound heme coenzyme. In (B) egg-white lysozyme is shown with a bound oligosaccharide substrate. In both cases the bound ligand is *red*. (Courtesy of Richard J. Feldmann.)

there is a limit to the amount of substrate that a single enzyme molecule can process in a given time. If the concentration of substrate is increased, the rate at which product is formed also increases, up to a maximum value (Figure 3–60). At that point the enzyme molecule is saturated with substrate and the rate of reaction (denoted V_{max}) depends only on how rapidly the substrate molecule can be processed. This rate divided by the enzyme concentration is called the **turnover number.** The turnover number is often about 1000 substrate molecules processed per second per enzyme molecule, but it can be much greater in extreme cases.

The other kinetic parameter frequently used to characterize an enzyme is its K_M, which is the substrate concentration that allows the reaction to proceed at one-half its maximum rate (see Figure 3–60). A *low* K_M value means that the enzyme reaches its maximum catalytic rate at a *low concentration* of substrate and generally indicates that the enzyme binds its substrate very tightly.

Enzymes Speed Reactions by Selectively Stabilizing Transition States [36]

Extremely high rates of chemical reaction are achieved by enzymes—far higher than for any synthetic catalysts. This efficiency is attributable to several factors. The enzyme serves, first, to increase the local concentration of substrate molecules at the catalytic site and to hold all of the appropriate atoms in the correct orientation for the reaction that is to follow. More important, however, some of the binding energy contributes directly to the catalysis. Substrate molecules pass through a series of intermediate forms of altered geometry and electron distribution before they form the ultimate products of the reaction, and the free energies of these intermediate forms—especially of those in the most unstable

Figure 3–60 Enzyme kinetics. The rate of an enzyme reaction (V) increases as the substrate concentration increases until a maximum value (V_{max}) is reached. At this point all substrate-binding sites on the enzyme molecules are fully occupied, and the rate of reaction is limited by the rate of the catalytic process on the enzyme surface. For most enzymes the concentration of substrate at which the reaction rate is half-maximal (K_M) is a measure of how tightly the substrate is bound, with a large value of K_M corresponding to weak binding.

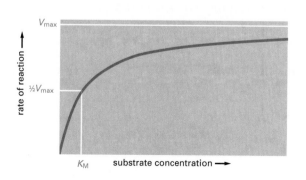

transition states—are the major determinants of the rate of reaction. Enzymes have a much greater affinity for these transition states of the substrate than they have for the stable forms. Because this binding interaction lowers the energies of crucial transition states, the enzyme greatly accelerates one particular reaction (Figure 3–61).

A dramatic demonstration of how stabilizing a transition state can greatly increase reaction rates is provided by the intentional production of antibodies that act like enzymes. Consider, for example, the hydrolysis of an amide bond, which is similar to the peptide bond that joins adjacent amino acids in a protein. In an aqueous solution an amide bond hydrolyzes very slowly by the mechanism illustrated in Figure 3–62A. In the central intermediate, or transition state, the carbonyl carbon is bonded to four atoms that are arranged at the corners of a tetrahedron. By generating monoclonal antibodies that bind tightly to a stable analogue of this very unstable *tetrahedral intermediate*, as illustrated in Figure 3–62B, an antibody that functions like an enzyme can be obtained. This *catalytic antibody* binds to and stabilizes the tetrahedral intermediate and thereby increases the spontaneous rate of amide-bond hydrolysis more than 10,000-fold.

Enzymes Can Promote the Making and Breaking of Covalent Bonds Through Simultaneous Acid and Base Catalysis [37]

Enzymes are better catalysts than catalytic antibodies. In addition to binding tightly to the transition state, the active site of an enzyme contains precisely positioned atoms that speed up the reaction by altering the distribution of electrons in those atoms involved in the making and breaking of covalent bonds. Peptide bonds, for example, can be hydrolyzed in the absence of an enzyme by exposing a polypeptide to either a strong acid or a strong base, as explained in Figure 3–63B and C. Enzymes are unique, however, in being able to use acid and base catalysis simultaneously, since the acidic and basic residues required are prevented from combining with each other (as they would do in solution) by being tied to the rigid framework of the protein itself (Figure 3–63D).

The fit between an enzyme and its substrate needs to be precise. A small change introduced by genetic engineering in the active site of an enzyme can have a profound effect. Replacing a glutamic acid with an aspartic acid in one enzyme, for example, shifts the position of the catalytic carboxylate ion by only 1 Å (about the radius of a hydrogen atom), and yet this is enough to reduce the activity of the enzyme a thousandfold.

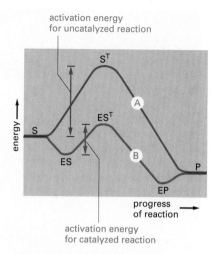

Figure 3–61 Enzymes accelerate chemical reactions by decreasing the activation energy. Often both the uncatalyzed reaction (A) and the enzyme-catalyzed reaction (B) go through several transition states. It is the transition state with the highest energy (S^T and ES^T) that determines the activation energy and limits the rate of the reaction. (S = substrate; P = product of the reaction.)

(A) HYDROLYSIS OF AN AMIDE BOND

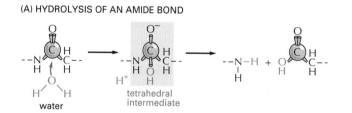

(B) TRANSITION-STATE ANALOGUE FOR AMIDE HYDROLYSIS

analogue   amide

Figure 3–62 Catalytic antibodies. The stabilization of a transition state by an antibody creates an enzyme. (A) The reaction path for hydrolysis of an amide bond goes through a *tetrahedral intermediate*, which is the high-energy transition state for the reaction. (B) The molecule shown on the left was covalently linked to a protein and used as an antigen to generate an antibody that binds tightly to the region of the molecule shown in *yellow*. Because this antibody also bound tightly to the transition state in (A), it was found to function as an enzyme that efficiently catalyzed the hydrolysis of the amide bond in the molecule shown on the right.

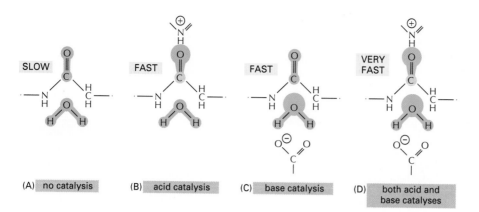

(A) no catalysis (B) acid catalysis (C) base catalysis (D) both acid and base catalyses

Figure 3–63 Acid catalysis and base catalysis. (A) The start of the uncatalyzed reaction shown in Figure 3–62A is diagrammed, with *blue* shading as a schematic indicator of electron distribution in the water and carbonyl bonds. (B) An acid likes to donate a proton (H^+) to other atoms. By pairing with the carbonyl oxygen, an acid causes electrons to move away from the carbonyl carbon, making this atom much more attractive to the electronegative oxygen of an attacking water molecule. (C) A base likes to take up H^+; by pairing with a hydrogen of the attacking water molecule, a base causes electrons to move toward the water oxygen, making it a better attacking group for the carbonyl carbon. (D) By having appropriately positioned atoms on its surface, an enzyme can carry out both acid catalysis and base catalysis at the same time.

Enzymes Can Further Increase Reaction Rates by Forming Covalent Intermediates with Their Substrates [38]

In addition to the above roles, many enzymes further speed the reaction they catalyze by interacting covalently with one of their substrates, thereby temporarily attaching the substrate to an amino acid or to a coenzyme molecule. Generally, one substrate enters the binding site, becomes covalently bound, and then reacts with a second molecule on the enzyme surface that breaks the covalent attachment just made. At the end of each reaction cycle, the free enzyme is regenerated.

Consider, for example, the mechanism of action of the serine proteases. The reaction they catalyze, the hydrolysis of a peptide bond, is greatly accelerated by the enzymes' affinity for the tetrahedral intermediate of the reaction. But a serine protease does more than a typical catalytic antibody: instead of waiting for an oxygen from a water molecule to attack the carbonyl carbon, it makes the reaction go much more quickly by first using a precisely positioned amino acid side chain for this purpose (the activated serine in Figure 3–57). This step breaks the peptide bond, but it leaves the enzyme covalently linked to the carboxyl group. Then, in a rapid second step, this covalent intermediate is destroyed by the enzyme-catalyzed addition of water, completing the reaction and regenerating the free enzyme (Figure 3–64). Even though this two-step reaction is less direct than a one-step reaction (in which water is added to the peptide bond), it is faster because each step has a relatively low activation energy.

Enzymes Accelerate Chemical Reactions but Cannot Make Them Energetically More Favorable

No matter how sophisticated an enzyme is, it cannot make the chemical reaction it catalyzes either more or less energetically favorable. It cannot alter the free-energy difference between the initial substrates and the final products of the reaction. Like the simple binding interactions already discussed, any given chemical reaction has an *equilibrium point*, at which the backward and forward reaction fluxes are equal, so that no net change occurs (see Figure 3–9). If an enzyme speeds up the rate of the forward reaction, $A + B \rightarrow AB$, by a factor of 10^8, it must speed up the rate of the backward reaction, $AB \rightarrow A + B$, by a factor of 10^8 as well. The *ratio* of the forward to the backward rates of reaction depends only on the concentrations of A, B, and AB. The equilibrium point remains precisely the same whether or not the reaction is catalyzed by an enzyme.

Enzymes Determine Reaction Paths by Coupling Selected Reactions to ATP Hydrolysis [39]

The living cell is a chemical system that is far from equilibrium. The product of each enzyme usually serves as a substrate for another enzyme in the metabolic

Proteins as Catalysts

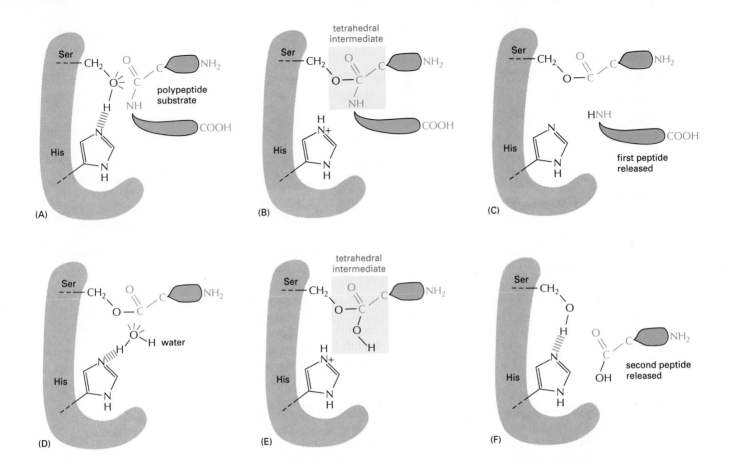

pathway and is rapidly consumed. More important, by means of a reaction pathway that is determined by enzymes, many reactions are driven in one direction by being *coupled* to the energetically favorable hydrolysis of ATP to ADP and inorganic phosphate, as previously described in Chapter 2. To make this strategy effective, the ATP pool is itself maintained at a level far from its equilibrium point, with a high ratio of ATP to its hydrolysis products (discussed in Chapter 14). This ATP pool thereby serves as a "storage battery" that keeps energy and atoms continually passing through the cell directed along pathways determined by the enzymes present. For a living system, approaching chemical equilibrium means decay and death.

Multienzyme Complexes Help to Increase the Rate of Cell Metabolism [40]

The efficiency of enzymes in accelerating chemical reactions is crucial to the maintenance of life. Cells, in effect, must race against the unavoidable processes of decay, which run downhill toward chemical equilibrium. If the rates of desirable reactions were not greater than the rates of competing side reactions, a cell would soon die. Some idea of the rate at which cellular metabolism proceeds can be obtained by measuring the rate of ATP utilization. A typical mammalian cell turns over (that is, completely degrades and replaces) its entire ATP pool once every 1 or 2 minutes. For each cell this turnover represents the utilization of roughly 10^7 molecules of ATP per second (or, for the human body, about a gram of ATP every minute).

The rates of cellular reactions are rapid because of the effectiveness of enzyme catalysis. Many important enzymes have become so efficient that there is no possibility of further useful improvement: the factor limiting the reaction rate is no longer the intrinsic speed of action of the enzyme, rather it is the frequency

Figure 3–64 Some enzymes form covalent bonds with their substrates. In the example shown here, a carbonyl group in a polypeptide chain (shown in *green*) forms a covalent bond with a specially activated serine residue (see Figure 3–57) of a serine protease (shown in *gray*), which cleaves the polypeptide chain. When the unbound portion of the polypeptide chain has diffused away, a second step occurs in which a water molecule hydrolyzes the newly formed covalent bond, thereby releasing the portion of the polypeptide bound to the enzyme surface and freeing the serine for another cycle of reaction. Note that two unstable tetrahedral intermediates (shaded in *yellow*) serve as transition states in this reaction and both are stabilized by the enzyme.

with which the enzyme collides with its substrate. Such a reaction is said to be *diffusion-limited.*

If a reaction is diffusion-limited, its rate will depend on the concentration of both the enzyme and its substrate. For a sequence of reactions to occur very rapidly, each metabolic intermediate and enzyme involved must therefore be present in high concentration. Given the enormous number of different reactions carried out by a cell, there are limits to the concentrations of substrates that can be achieved. In fact, most metabolites are present in micromolar (10^{-6} M) concentrations, and most enzyme concentrations are much lower. How is it possible, therefore, to maintain very fast metabolic rates?

The answer lies in the spatial organization of cell components. Reaction rates can be increased without raising substrate concentrations by bringing the various enzymes involved in a reaction sequence together to form a large protein assembly known as a **multienzyme complex.** In this way the product of enzyme A is passed directly to enzyme B and so on to the final product, and diffusion rates need not be limiting even when the concentration of substrate in the cell as a whole is very low. Such enzyme complexes are very common (the structure of one, pyruvate dehydrogenase, was shown in Figure 2–41), and they are involved in nearly all aspects of metabolism, including the central genetic processes of DNA, RNA, and protein synthesis. In fact, it may be that few enzymes in eucaryotic cells diffuse freely in solution; instead, most may have evolved binding sites that concentrate them with other proteins of related function in particular regions of the cell, thereby increasing the rate and efficiency of the reactions that they catalyze.

Cells have another way of increasing the rate of metabolic reactions. It depends on the extensive intracellular membrane systems of eucaryotic cells. These membranes can segregate certain substrates and the enzymes that act on them into the same membrane-bounded compartment, such as the endoplasmic reticulum or the cell nucleus. If, for example, the compartment occupies a total of 10% of the volume of the cell, the concentration of reactants in the compartment can be 10 times greater than in a similar cell with no compartmentalization (Figure 3–65). Reactions that would otherwise be limited by the speed of diffusion can thereby be speeded up by the same factor.

Further details of protein structure and function will be presented in Chapter 5, where we discuss how cells construct tiny machines out of proteins.

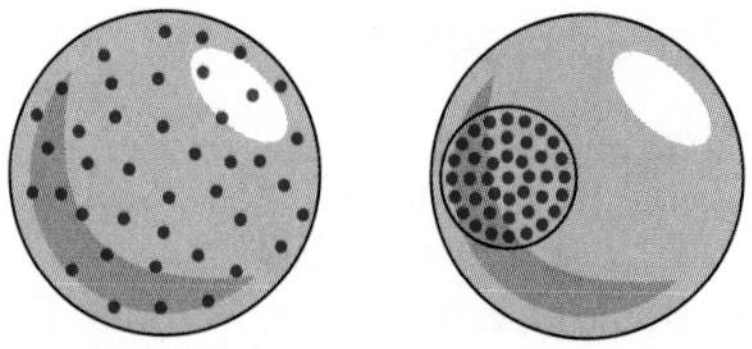

Figure 3–65 Compartmentalization. A large increase in the concentration of interacting molecules can be achieved by confining them to the same membrane-bounded compartment in a eucaryotic cell.

Summary

The biological function of a protein depends on the detailed chemical properties of its surface. Binding sites for ligands are formed as surface cavities in which precisely positioned amino acid side chains are brought together by protein folding. In this way, normally unreactive amino acid side chains can be activated. Enzymes greatly speed up reaction rates by binding the high-energy transition states in a reaction especially tightly; they also carry out acid catalysis and base catalysis simultaneously. The rates of enzyme reactions are often so fast that they are limited only by diffusion; rates can be further increased if enzymes that act sequentially on a substrate are joined into a single multienzyme complex or if the enzymes and their substrates are confined to the same compartment of the cell.

References

General

Branden, C.; Tooze, J. Introduction to Protein Structure. New York: Garland, 1991.

Gesteland, R.F.; Atkins, J.F., eds. The RNA World. Cold Spring Harbor, NY: Cold Spring Harbor Laboratory Press, 1993.

Judson, H.F. The Eighth Day of Creation: Makers of the Revolution in Biology. New York: Simon & Schuster, 1979.

Lehninger, A.L.; Nelson, D.L.; Cox, M.M. Principles of Biochemistry, 2nd ed. New York: Worth, 1993.

Mathews, C.K.; van Holde, K.E. Biochemistry. Menlo Park, CA: Benjamin/Cummings, 1990.

Schulz, G.E.; Schirmer, R.H. Principles of Protein Structure. New York: Springer, 1979.

Stryer, L. Biochemistry, 3rd ed. New York: W.H. Freeman, 1988.

Voet, D.; Voet, J.G. Biochemistry. New York: Wiley, 1990.

Watson, J.D., et al. Molecular Biology of the Gene, 3rd ed. Menlo Park, CA: Benjamin-Cummings, 1987.

Cited

1. Cantor, C.R.; Schimmel, P.R. Biophysical Chemistry, Part I and Part III. New York: W.H. Freeman, 1980.

   Eisenberg, D.; Crothers, D. Physical Chemistry with Applications to the Life Sciences. Menlo Park, CA: Benjamin-Cummings, 1979.

   Pauling, L. The Nature of the Chemical Bond, 3rd ed. Ithaca, NY: Cornell University Press, 1960.

   Whitesides, G.M.; Mathias, J.P.; Seto, C.T. Molecular self-assembly and nanochemistry: a chemical strategy for the synthesis of nanostructures. *Science* 254:1312–1319, 1991.

2. Abeles, R.H.; Frey, P.A.; Jencks, W.P. Biochemistry. Boston: Jones and Bartlett, 1992.

   Burley, S.K.; Petsko, G.A. Weakly polar interactions in proteins. *Adv. Prot. Chem.* 39:125–189, 1988.

   Fersht, A.R. The hydrogen bond in molecular recognition. *Trends Biochem. Sci.* 12:301–304, 1987.

3. Cohen, C.; Parry, D.A.D. α-helical coiled coils—a widespread motif in proteins. *Trends Biochem. Sci.* 11:245–248, 1986.

   Dickerson, R.E. The DNA helix and how it is read. *Sci. Am.* 249(6):94–111, 1983.

4. Berg, H.C. Random Walks in Biology. Princeton, NJ: Princeton University Press, 1983.

   Einstein, A. Investigations on the Theory of Brownian Movement. New York: Dover, 1956.

   Lavenda, B.H. Brownian motion. *Sci. Am.* 252(2):70–85, 1985.

5. Lehninger, A.L. Bioenergetics: The Molecular Basis of Biological Energy Transformations, 2nd ed. Menlo Park, CA: Benjamin-Cummings, 1971.

6. Karplus, M.; McCammon, J.A. The dynamics of proteins. *Sci. Am.* 254(4):42–51, 1986.

   Karplus, M.; Petsko, G.A. Molecular dynamics simulations in biology. *Nature* 347:631–639, 1990.

   McCammon, J.A.; Harvey, S.C. Dynamics of Proteins and Nucleic Acids. Cambridge, UK: Cambridge University Press, 1987.

7. Kirkwood, T.B.; Rosenberger, R.F.; Galas, D.J., eds. Accuracy in Molecular Processes: Its Control and Relevance to Living Systems. London: Chapman and Hall, 1986.

8. Berg, P.; Singer, M. Dealing with Genes. The Language of Heredity. Mill Valley, CA: University Science Books, 1992.

   Rosenfield, I.; Ziff, E.; Van Loon, B. DNA for Beginners. London: Writers and Readers Publishing Cooperative. New York: Distributed in the USA by Norton, 1983.

   Saenger, W. Principles of Nucleic Acid Structure. Berlin: Springer, 1984.

9. Moore, J. Heredity and Development, 2nd ed. New York: Oxford University Press, 1992.

   Olby, R. The Path to the Double Helix. Seattle: University of Washington Press, 1974.

   Stent, G.S.; Calendar, A.Z. Molecular Genetics: An Introductory Narrative, 2nd ed. San Francisco: W.H. Freeman, 1978.

10. Watson, J.D.; Crick, F.H.C. Molecular structure of nucleic acids. A structure for deoxyribose nucleic acid. *Nature* 171:737–738, 1953.

11. Felsenfeld, G. DNA. *Sci. Am.* 253(4):58–66, 1985.

    Meselson, M.; Stahl, F.W. The replication of DNA in *E. coli. Proc. Natl. Acad. Sci. USA* 44:671–682, 1958.

    Watson, J.D.; Crick, F.H.C. Genetic implications of the structure of deoxyribonucleic acid. *Nature* 171:964–967, 1953.

12. Drake, J.W. Spontaneous mutation. *Annu. Rev. Genet.* 25:125–146, 1991.

    Lindahl, T. Instability and decay of the primary structure of DNA. *Nature* 362:709–715, 1993.

    Wilson, A.C. Molecular basis of evolution. *Sci. Am.* 253(4):164–173, 1985.

13. Sanger, F. Sequences, sequences, and sequences. *Annu. Rev. Biochem.* 57:1–28, 1988.

    Thompson, E.O.P. The insulin molecule. *Sci. Am.* 192(5):36–41, 1955.

    Yanofsky, C. Gene structure and protein structure. *Sci. Am.* 216(5):80–94, 1967.

14. Brenner, S.; Jacob, F.; Meselson, M. An unstable intermediate carrying information from genes to ribosomes for protein synthesis. *Nature* 190:576–581, 1961.

    Darnell, J.E., Jr. RNA. *Sci. Am.* 253(4):68–78, 1985.

15. Chambon, P. Split genes. *Sci. Am.* 244(5):60–71, 1981.

    Steitz, J.A. Snurps. *Sci. Am.* 258(6):58–63, 1988.

    Witkowski, J.A. The discovery of "split" genes: a scientific revolution. *Trends Biochem. Sci.* 13:110–113, 1988.

16. Crick, F.H.C. The genetic code: III. *Sci. Am.* 215(4):55–62, 1966.

    The Genetic Code. *Cold Spring Harbor Symp. Quant. Biol.* 31, 1965.

17. Rich, A.; Kim, S.H. The three-dimensional structure of transfer RNA. *Sci. Am.* 238(1):52–62, 1978.

18. Lake, J.A. The ribosome. *Sci. Am.* 245(2):84–97, 1981.

    Watson, J.D. Involvement of RNA in the synthesis of proteins. *Science* 140:17–26, 1963.

    Zamecnik, P. The machinery of protein synthesis. *Trends Biochem. Sci.* 9:464–466, 1984.

19. Altman, S.; Baer, M.; Guerrier-Takada, C.; Viogue, A. Enzymatic cleavage of RNA by RNA. *Trends Biochem. Sci.* 11:515–518, 1986.

    Cech, T. RNA as an enzyme. *Sci. Am.* 255(5):64–75, 1986.

Cech, T. Fishing for fresh catalysts. *Nature* 365:204–205, 1993.

Michel, F.; Westhof, E. Modelling of the three-dimensional architecture of group I catalytic introns based on comparative sequence analysis. *J. Mol. Biol.* 216:585–610, 1990.

Noller, H.F. Ribosomal RNA and translation. *Annu. Rev. Biochem.* 60:191–227, 1991.

Noller, H.F.; Hoffarth, V.; Zimniak, L. Unusual resistance of peptidyl transferase to protein extraction procedures. *Science* 256:1416–1419, 1992.

20. Branden, C.; Tooze, J. Introduction to Protein Structure. New York: Garland, 1991.

Creighton, T.E. Proteins: Structure and Molecular Properties, 2nd ed. New York: W.H. Freeman, 1993.

Dickerson, R.E.; Geis, I. The Structure and Action of Proteins. New York: Harper & Row, 1969.

Schulz, G.E.; Schirmer, R.H. Principles of Protein Structure. New York: Springer, 1990.

21. Anfinsen, C.B. Principles that govern the folding of protein chains. *Science* 181:223–230, 1973.

Baldwin, R.L. Seeding protein folding. *Trends Biochem. Sci.* 11:6–9, 1986.

Creighton, T.E. Disulphide bonds and protein stability. *Bioessays* 8:57–63, 1988.

Richards, F.M. The protein folding problem. *Sci. Amer.* 264(1):54–63, 1991.

Rupley, J.A.; Gratton, E.; Careri, G. Water and globular proteins. *Trends Biochem. Sci.* 8:18–22, 1983.

22. Doolittle, R.F. Proteins. *Sci. Am.* 253(4):88–99, 1985.

Milner-White, E.J.; Poet, R. Loops, bulges, turns and hairpins in proteins. *Trends Biochem Sci.* 12:189–192, 1987.

Pauling, L.; Corey, R.B. Configurations of polypeptide chains with favored orientations around single bonds: two new pleated sheets. *Proc. Natl. Acad. Sci. USA* 37:729–740, 1951.

Pauling, L.; Corey, R.B.; Branson, H.R. The structure of proteins: two hydrogen-bonded helical configurations of the polypeptide chain. *Proc. Natl. Acad. Sci. USA* 27:205–211, 1951.

Richardson, J.S. The anatomy and taxonomy of protein structure. *Adv. Protein Chem.* 34:167–339, 1981.

23. Scott, J.E. Molecules for strength and shape. *Trends Biochem. Sci.* 12:318–321, 1987.

24. Branden, C.; Tooze, J. Introduction to Protein Structure. New York: Garland, 1991.

Hardie, D.G.; Coggins, J.R. Multidomain Proteins—Structure and Evolution. Amsterdam: Elsevier, 1986.

25. Doolittle, R.F. The genealogy of some recently evolved vertebrate proteins. *Trends Biochem. Sci.* 10:233–237, 1985.

Neurath, H. Evolution of proteolytic enzymes. *Science* 224:350–357, 1984.

26. Blake, C. Exons and the evolution of proteins. *Trends Biochem. Sci.* 8:11–13, 1983.

Gilbert, W. Genes-in-pieces revisited. *Science* 228:823–824, 1985.

McCarthy, A.D.; Hardie, D.G. Fatty acid synthase: an example of protein evolution by gene fusion. *Trends Biochem. Sci.* 9:60–63, 1984.

Rossmann, M.G.; Argos, P. Protein folding. *Annu. Rev. Biochem.* 50:497–532, 1981.

27. Gehring, W.J. On the homeobox and its significance. *Bioessays* 5:3–4, 1986.

Hanks, S.K.; Quinn, A.M.; Hunter, T. The protein kinase family: conserved features and deduced phylogeny of the catalytic domains. *Science* 241:42–52, 1988.

Shabb, J.B.; Corbin, J.D. Cyclic nucleotide-binding domains in proteins having diverse functions. *J. Biol. Chem.* 267:5723–5726, 1992.

Thornton, J.M.; Gardner, S.P. Protein motifs and database searching. *Trends Biochem. Sci.* 14:300–304, 1989.

28. Bajaj, M.; Blundell, T. Evolution and the tertiary structure of proteins. *Annu. Rev. Biophys. Bioeng.* 13:453–492, 1984.

Klug, A. From macromolecules to biological assemblies. *Biosci. Rep.* 3:395–430, 1983.

Metzler, D.E. Biochemistry. New York: Academic Press, 1977. (Chapter 4 describes how macromolecules pack together into large assemblies.)

29. Caspar, D.L.D.; Klug, A. Physical principles in the construction of regular viruses. *Cold Spring Harbor Symp. Quant. Biol.* 27:1–24, 1962.

Goodsell, D.S. Inside a living cell. *Trends Biochem. Sci.* 16:203–206, 1991.

30. Cohen, C.; Parry, D.A.D. α-helical coiled coils—a widespread motif in proteins. *Trends Biochem. Sci.* 11:245–248, 1986.

31. Harrison, S.C. Multiple modes of subunit association in the structures of simple spherical viruses. *Trends Biochem. Sci.* 9:345–351, 1984.

Harrison, S.C. Viruses. *Curr. Opin. Struc. Biol.* 2:293–299, 1992.

Hogle, J.M.; Chow, M.; Filman, D.J. The structure of polio virus. *Sci. Am.* 256(3):42–49, 1987.

Rossmann, M.G.; Johnson, J.E. Icosahedral RNA virus structure. *Annu. Rev. Biochem.* 58:533–573, 1989.

32. Fraenkel-Conrat, H.; Williams, R.C. Reconstitution of active tobacco mosaic virus from its inactive protein and nucleic acid components. *Proc. Natl. Acad. Sci. USA* 41:690–698, 1955.

Hendrix, R.W. Tail length determination in double-stranded DNA bacteriophages. *Curr. Top. Microbiol. Immunol.* 136:21–29, 1988.

Namba, K.; Caspar, D.L.D.; Stubbs, G.J. Computer graphics representation of levels of organization in tobacco mosaic virus structure. *Science* 227:773–776, 1985.

Nomura, M. Assembly of bacterial ribosomes. *Science* 179:864–873, 1973.

Trinick, J. Understanding the functions of titin and nebulin. *FEBS Lett.* 307:44–48, 1992.

33. Mathews, C.K., Kutter, E.M.; Mosig, G.; Berget, P.B. Bacteriophage T4, Chaps. 1 and 4. Washington, DC: American Society of Microbiologists, 1983.

Steiner, D.F.; Kemmler, W.; Tager, H.S.; Peterson, J.D. Proteolytic processing in the biosynthesis of insulin and other proteins. *Fed. Proc.* 33:2105–2115, 1974.

34. Dressler, D.; Potter, H. Discovering Enzymes. New York: Scientific American Library, 1991.

Fersht, A. Enzyme Structure and Mechanism, 2nd ed. New York: W.H. Freeman, 1985.

35. Fersht, A.R.; Leatherbarrow, R.J.; Wells, T.N.C. Binding energy and catalysis. *Trends Biochem. Sci.* 11:321–325, 1986.

Hansen, D.E.; Raines, R.T. Binding energy and enzymatic catalysis. *J. Chem. Educ.* 67:483–489, 1990.

36. Lerner, R.A.; Benkovic, S.J.; Schultz, P.G. At the crossroads of chemisry and immunology: catalytic antibodies. *Science* 252:659–667, 1991.

Lerner, R.A.; Tramontano, A. Catalytic antibodies. *Sci. Am.* 258(3):58–70, 1988.

Wolfenden, R. Analog approaches to the structure of the transition state in enzyme reactions. *Accounts Chem. Res.* 5:10–18, 1972.

37. Knowles, J.R. Tinkering with enzymes: what are we learning? *Science* 236:1252–1258, 1987.

Knowles, J.R. Enzyme catalysis: not different, just better. *Nature* 350:121–124, 1991.

38. Stroud, R.M. A family of protein-cutting proteins. *Sci. Am.* 231(1):74–88, 1974.

39. Wood, W.B.; Wilson, J.H.; Benbow, R.M.; Hood, L.E. Biochemistry, A Problems Approach, 2nd ed. Menlo Park, CA: Benjamin-Cummings, 1981. (Chapters 9 and 15 and associated problems.)

40. Barnes, S.J.; Weitzman, P.D.J. Organization of citric acid cycle enzymes into a multienzyme cluster. *FEBS Lett.* 201:267–270, 1986.

Berg, O.G.; von Hippel, P.H. Diffusion-controlled macromolecular interactions. *Annu. Rev. Biophys. Biophys. Biochem.* 14:131–160, 1985.

Reed, L.J.; Cox, D.J. Multienzyme complexes. In The Enzymes, 3rd ed. (P.D. Boyer, ed.), Vol. 1, pp. 213–240. New York: Academic Press, 1970.

How Cells Are Studied

4

- Looking at the Structure of Cells in the Microscope
- Isolating Cells and Growing Them in Culture
- Fractionation of Cells and Analysis of Their Molecules
- Tracing and Assaying Molecules Inside Cells

Cells are small and complex. It is hard to see their structure, hard to discover their molecular composition, and harder still to find out how their various components function. What we can learn about cells depends on the tools at our disposal, and major advances in cell biology have frequently sprung from the introduction of new techniques. To understand contemporary cell biology, therefore, it is necessary to know something of its methods.

In this chapter we briefly review some of the principal methods used to study cells. We start with techniques for examining the cell as a whole and then proceed to techniques for analyzing its constituent macromolecules. Microscopy will be our starting point, for cell biology began with the light microscope, and this is still an essential tool in the field, along with more recent imaging devices based on beams of electrons and other forms of radiation. From passive observation we move to active intervention: we consider how cells of different types can be separated from tissues and grown outside the body and how cells can be disrupted and their organelles and constituent macromolecules isolated in pure form. Finally, we describe how we can detect, follow, and quantify individual types of molecules and ions within the cell. A revolution in our understanding of cellular function has come from recombinant DNA technology, but because it is a complex subject in itself and depends on an understanding of basic genetic mechanisms, this powerful array of methods will be considered in detail in Chapter 7.

Although methods are of basic importance, it is what we discover with them that makes them interesting. The present chapter, therefore, is meant to be used for reference and to be read in conjunction with the later chapters of the book, rather than as an introduction to them.

Looking at the Structure of Cells in the Microscope [1]

A typical animal cell is 10 to 20 µm in diameter, which is about five times smaller than the smallest particle visible to the naked eye. It was not until good light microscopes became available in the early part of the nineteenth century that all plant and animal tissues were discovered to be aggregates of individual cells. This discovery, proposed as the **cell doctrine** by Schleiden and Schwann in 1838, marks the formal birth of cell biology.

139

Animal cells are not only tiny, they are also colorless and translucent. Consequently, the discovery of their main internal features depended on the development, in the latter part of the nineteenth century, of a variety of stains that provided sufficient contrast to make those features visible. Likewise, introduction of the far more powerful electron microscope in the early 1940s required the development of new techniques for preserving and staining cells before the full complexities of their internal fine structure could begin to emerge. To this day, microscopy depends as much on techniques for preparing the specimen as on the performance of the microscope itself. In the discussions that follow, we therefore consider both instruments and specimen preparation, beginning with the light microscope.

Figure 4–1 shows the fineness of detail that can be resolved with modern light microscopes, in comparison with electron microscopes. Some of the landmarks in the development of light microscopy are outlined in Table 4–1.

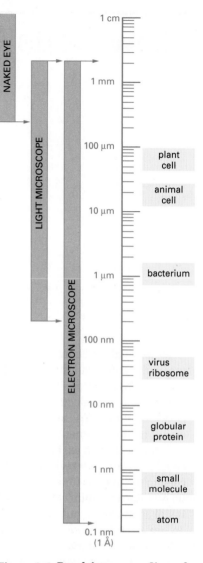

Figure 4–1 Resolving power. Sizes of cells and their components drawn on a logarithmic scale, indicating the range of objects that can be readily resolved by the naked eye and in the light and electron microscopes. The following units of length are commonly employed in microscopy:
μm (micrometer) $= 10^{-6}$ m
nm (nanometer) $= 10^{-9}$ m
Å (Ångström unit) $= 10^{-10}$ m

Table 4–1 Some Important Discoveries in the History of Light Microscopy

1611 **Kepler** suggested a way of making a compound microscope.

1655 **Hooke** used a compound microscope to describe small pores in sections of cork that he called "cells."

1674 **Leeuwenhoek** reported his discovery of protozoa. He saw bacteria for the first time nine years later.

1833 **Brown** published his microscopic observations of orchids, clearly describing the cell nucleus.

1838 **Schleiden** and **Schwann** proposed the cell theory, stating that the nucleated cell is the unit of structure and function in plants and animals.

1857 **Kolliker** described mitochondria in muscle cells.

1876 **Abbé** analyzed the effects of diffraction on image formation in the microscope and showed how to optimize microscope design.

1879 **Flemming** described with great clarity chromosome behavior during mitosis in animal cells.

1881 **Retzius** described many animal tissues with a detail that has not been surpassed by any other light microscopist. In the next two decades he, **Cajal**, and other histologists developed staining methods and laid the foundations of microscopic anatomy.

1882 **Koch** used aniline dyes to stain microorganisms and identified the bacteria that cause tuberculosis and cholera. In the following two decades other bacteriologists, such as **Klebs** and **Pasteur**, identified the causative agents of many other diseases by examining stained preparations under the microscope.

1886 **Zeiss** made a series of lenses, to the design of **Abbé**, that enabled microscopists to resolve structures at the theoretical limits of visible light.

1898 **Golgi** first saw and described the Golgi apparatus by staining cells with silver nitrate.

1924 **Lacassagne** and collaborators developed the first autoradiographic method to localize radioactive polonium in biological specimens.

1930 **Lebedeff** designed and built the first interference microscope. In 1932 **Zernicke** invented the phase-contrast microscope. These two developments allowed unstained living cells to be seen in detail for the first time.

1941 **Coons** used antibodies coupled to fluorescent dyes to detect cellular antigens.

1952 **Nomarski** devised and patented the system of differential interference contrast for the light microscope that still bears his name.

1981 **Allen** and **Inoué** perfected video-enhanced-contrast light microscopy.

1988 Commercial confocal scanning microscopes came into widespread use.

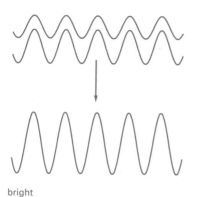

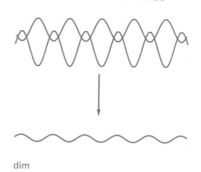

bright

dim

Figure 4–2 Interference between light waves. When two light waves combine *in phase*, the amplitude of the resultant wave is larger and the brightness is increased. Two light waves that are *out of phase* partially cancel each other and produce a wave whose amplitude, and therefore brightness, is decreased.

The Light Microscope Can Resolve Details 0.2 μm Apart [2]

In general, a beam of a given type of radiation cannot be used to probe structural details much smaller than its own wavelength. This is a fundamental limitation of microscopes. The limit to the resolution of a light microscope, therefore, is set by the wavelength of visible light, which ranges from about 0.4 μm (for violet) to 0.7 μm (for deep red). In practical terms, bacteria and mitochondria, which are about 500 nm (0.5 μm) wide, are generally the smallest objects whose shape can be clearly discerned in the light microscope; details smaller than this are obscured by effects resulting from the wave nature of light. To understand why this occurs, we must follow what happens to a beam of light waves as it passes through the lenses of a microscope.

Because of its wave nature, light does not follow exactly the idealized straight ray paths predicted by geometrical optics. Instead, light waves travel through an optical system by a variety of slightly different routes, so that they interfere with one another and cause *optical diffraction* effects. If two trains of waves reaching the same point by different paths are precisely in phase, with crest matching crest and trough matching trough, they will reinforce each other so as to increase brightness. On the other hand, if the trains of waves are out of phase, they will interfere with each other in such a way as to cancel each other partially or entirely (Figure 4–2). The interaction of light with an object will change the phase relationships of the light waves in a way that produces complex interference effects. At high magnification, for example, the shadow of a straight edge that is evenly illuminated with light of uniform wavelength appears as a set of parallel lines, whereas that of a circular spot appears as a set of concentric rings (Figure 4–3). For the same reason, a single point seen through a microscope appears as a blurred disc, and two point objects close together give overlapping images and may merge into one. No amount of refinement of the lenses can overcome this limitation imposed by the wavelike nature of light.

The limiting separation at which two objects can still be seen as distinct—the so-called **limit of resolution**—depends on both the wavelength of the light and the numerical aperture of the lens system used (Figure 4–4). Under the best conditions, with violet light (wavelength, $\lambda = 0.4$ μm) and a numerical aperture of 1.4, a limit of resolution of just under 0.2 μm can theoretically be obtained in the light microscope. This resolution was achieved by microscope makers at the end of the nineteenth century and is only rarely matched in contemporary, factory-produced microscopes. Although it is possible to *enlarge* an image as much as one wants—for example, by projecting it onto a screen—it is never possible to resolve two objects in the light microscope that are separated by less than about 0.2 μm: such objects will appear as one.

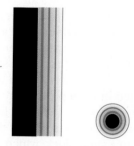

Figure 4–3 Edge effects. The interference effects observed at high magnification when light passes the edges of a solid object placed between the light source and the observer.

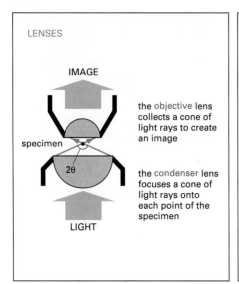

LENSES

IMAGE

the objective lens collects a cone of light rays to create an image

specimen

2θ

the condenser lens focuses a cone of light rays onto each point of the specimen

LIGHT

RESOLUTION: the resolving power of the microscope depends on the width of the cone of illumination and therefore on both the condenser and the objective lens. It is calculated using the formula

$$resolution = \frac{0.61\,\lambda}{n\,\sin\theta}$$

where:

θ = half the angular width of the cone of rays collected by the objective lens from a typical point in the specimen (since the maximum width is 180°, sin θ has a maximum value of 1)

n = the refractive index of the medium (usually air or oil) separating the specimen from the objective and condenser lenses

λ = the wavelength of light used (for white light, a figure of 0.53 μm is commonly assumed)

NUMERICAL APERTURE: $n \sin\theta$ in the equation above is called the numerical aperture of the lens (NA) and is a function of its light-collecting ability. For dry lenses this cannot be more than 1, but for oil-immersion lenses it can be as high as 1.4. The higher the numerical aperture, the greater the resolution and the brighter the image (brightness is important in fluorescence microscopy). However, this advantage is obtained at the expense of very short working distances and a very small depth of field.

Figure 4–4 **Numerical aperture.** The path of light rays passing through a transparent specimen in a microscope, illustrating the concept of numerical aperture and its relation to the limit of resolution.

We shall see later how interference and diffraction can be exploited to study unstained cells in the living state. First we discuss how permanent preparations of cells are made for viewing in the light microscope and how chemical stains are used to enhance the visibility of the cell structures in such preparations.

Tissues Are Usually Fixed and Sectioned for Microscopy

To make a permanent preparation that can be stained and viewed at leisure in the microscope, one first must treat cells with a fixative so as to immobilize, kill, and preserve them. In chemical terms, **fixation** makes cells permeable to staining reagents and cross-links their macromolecules so that they are stabilized and locked in position. Some of the earliest fixation procedures involved immersion in acids or in organic solvents, such as alcohol. Current procedures usually include treatment with reactive aldehydes, particularly formaldehyde and glutaraldehyde, which form covalent bonds with the free amino groups of proteins and thereby cross-link adjacent proteins.

Most tissue samples are too thick for their individual cells to be examined directly at high resolution. After fixation, therefore, the tissues are usually cut into very thin slices (**sections**) with a *microtome*, a machine with a sharp metal blade that operates like a meat slicer (Figure 4–5). The sections (typically 1 to 10 μm thick) are then laid flat on the surface of a glass microscope slide.

Tissues are generally soft and fragile, even after fixation, and need to be **embedded** in a supporting medium before sectioning. The usual embedding media are waxes or resins. In liquid form these media will both permeate and surround the fixed tissue; they then can be hardened (by cooling or by polymerization) to a solid block, which is readily sectioned by the microtome.

There is a serious danger that any treatment used for fixation and embedding may alter the structure of the cell or its constituent molecules in undesirable ways. Rapid freezing provides an alternative method of preparation that to some extent avoids this problem by eliminating the need for fixation and embedding. The frozen tissue can be cut directly with a cryostat—a special microtome that is maintained in a cold chamber. Although **frozen sections** produced in this

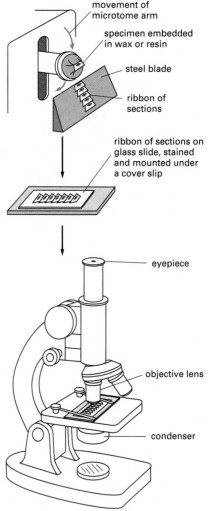

movement of microtome arm

specimen embedded in wax or resin

steel blade

ribbon of sections

ribbon of sections on glass slide, stained and mounted under a cover slip

eyepiece

objective lens

condenser

EXAMINATION WITH LIGHT MICROSCOPE

Figure 4–5 **Making tissue sections.** How an embedded tissue is sectioned with a microtome in preparation for examination in the light microscope.

way avoid some artifacts, they suffer from others: the native structures of individual molecules such as proteins are well preserved, but the fine structure of the cell is often disrupted by ice crystals.

Once sections have been cut, by whatever method, the next step is usually to stain them.

Different Components of the Cell Can Be Selectively Stained [3]

There is little in the contents of most cells (which are 70% water by weight) to impede the passage of light rays. Thus most cells in their natural state, even if fixed and sectioned, are almost invisible in an ordinary light microscope. One way to make them visible is to stain them with dyes.

In the early nineteenth century the demand for dyes to stain textiles led to a fertile period for organic chemistry. Some of the dyes were found to stain biological tissues and, unexpectedly, often showed a preference for particular parts of the cell—the nucleus or mitochondria, for example—making these internal structures clearly visible. Today a rich variety of organic dyes is available, with such colorful names as *Malachite green, Sudan black,* and *Coomassie blue,* each of which has some specific affinity for particular subcellular components. The dye *hematoxylin,* for example, has an affinity for negatively charged molecules and therefore reveals the distribution of DNA, RNA, and acidic proteins in a cell (Figure 4–6). The chemical basis for the specificity of many dyes, however, is not known.

The relative lack of specificity of these dyes at the molecular level has stimulated the design of more rational and selective staining procedures and, in particular, of methods that reveal specific proteins or other macromolecules in cells. It is a problem, however, to achieve adequate sensitivity for this purpose. Since relatively few copies of most macromolecules are present in any given cell, one or two molecules of stain bound to each macromolecule will often be invisible. One way to solve this problem is to increase the number of stain molecules associated with a single macromolecule. Thus some enzymes can be located in cells through their catalytic activity: when supplied with appropriate substrate molecules, each enzyme molecule generates many molecules of a localized, visible reaction product. An alternative and much more generally applicable approach to the problem of sensitivity depends on using dyes that are fluorescent, as we explain next.

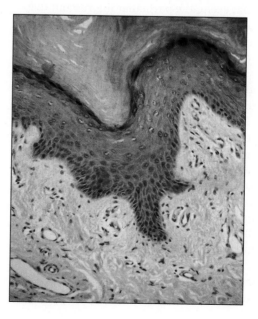

Figure 4–6 **A stained tissue section.** A section of thick human skin, stained with a combination of dyes, hematoxylin and eosin, that is commonly used in histology.

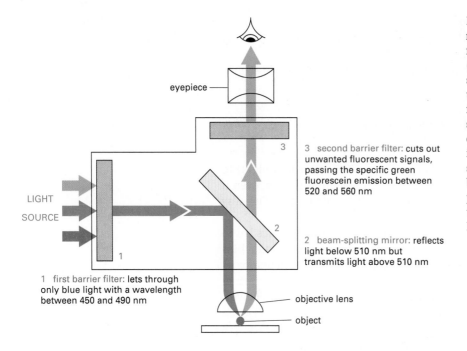

eyepiece

3 second barrier filter: cuts out unwanted fluorescent signals, passing the specific green fluorescein emission between 520 and 560 nm

LIGHT SOURCE

2 beam-splitting mirror: reflects light below 510 nm but transmits light above 510 nm

1 first barrier filter: lets through only blue light with a wavelength between 450 and 490 nm

objective lens

object

Figure 4–7 The optical system of a modern fluorescence microscope. A filter set consists of two barrier filters (1 and 3) and a dichroic (beam-splitting) mirror (2). In this example the filter set for detection of the fluorescent molecule fluorescein is shown. High-numerical-aperture objective lenses are especially important in this type of microscopy since, for a given magnification, the brightness of the fluorescent image is proportional to the fourth power of the numerical aperture (see also Figure 4–4).

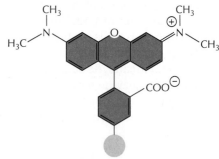

fluorescein (green)

tetramethylrhodamine (red)

Figure 4–8 Fluorescent dyes. The structures of fluorescein and tetramethylrhodamine, two dyes that are commonly used for fluorescence microscopy. Fluorescein emits green light when activated by light of the appropriate wavelength, whereas the rhodamine dye emits red light. The portion of each molecule shown in *orange* denotes the position of a chemically reactive group; at this position a covalent bond is commonly formed between the dye and a protein (or other molecule). Commercially available versions of these dyes with different types of reactive groups allow the dye to be coupled either to an —SH group or to an —NH$_2$ group on a protein.

Specific Molecules Can Be Located in Cells by Fluorescence Microscopy [4]

Fluorescent molecules absorb light at one wavelength and emit it at another, longer wavelength. If such a compound is illuminated at its absorbing wavelength and then viewed through a filter that allows only light of the emitted wavelength to pass, it is seen to glow against a dark background. Because the background is dark, even a minute amount of the glowing fluorescent dye can be detected. The same number of molecules of an ordinary stain viewed conventionally would be practically invisible because they would give only the faintest tinge of color to the light transmitted through this stained part of the specimen.

The fluorescent dyes used for staining cells are detected with the help of a **fluorescence microscope.** This microscope is similar to an ordinary light microscope except that the illuminating light, from a very powerful source, is passed through two sets of filters—one to filter the light before it reaches the specimen and one to filter the light obtained from the specimen. The first filter is selected so that it passes only the wavelengths that excite the particular fluorescent dye, while the second filter blocks out this light and passes only those wavelengths emitted when the dye fluoresces (Figure 4–7).

Fluorescence microscopy is most often used to detect specific proteins or other molecules in cells and tissues. A very powerful and widely used technique is to couple fluorescent dyes to antibody molecules, which then serve as highly specific and versatile staining reagents that bind selectively to the particular macromolecules that they recognize in cells or in the extracellular matrix. Two fluorescent dyes that are commonly used for this purpose are *fluorescein,* which emits an intense green fluorescence when excited with blue light, and *rhodamine,* which emits a deep red fluorescence when excited with green-yellow light (Figure 4–8). By coupling one antibody to fluorescein and another to rhodamine, the distributions of different molecules can be compared in the same cell; the two molecules are visualized separately in the microscope by switching back and forth between two sets of filters, each specific for one dye. As shown in Figure 4–9, three fluorescent dyes can be used in the same way to distinguish three types of molecules in the same cell.

Important new methods, to be discussed later, enable fluorescence microscopy to be used to monitor changes in the concentration and location of specific molecules inside *living* cells (see p. 183).

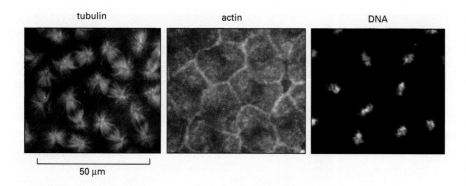

tubulin   actin   DNA

50 μm

Figure 4–9 Fluorescence microscopy. Micrographs of a portion of the surface of an early *Drosophila* embryo in which the microtubules have been labeled with an antibody coupled to fluorescein (*left panel*) and the actin filaments have been labeled with an antibody coupled to rhodamine (*middle panel*). In addition, the chromosomes have been labeled with a third dye that fluoresces only when it binds to DNA (*right panel*). At this stage, all the nuclei of the embryo share a common cytoplasm, and they are in the metaphase stage of mitosis. The three micrographs were taken of the same region of a fixed embryo using three different filter sets in the fluorescence microscope (see also Figure 4–7). (Courtesy of Tim Karr.)

Living Cells Are Seen Clearly in a Phase-Contrast or a Differential-Interference-Contrast Microscope [2, 5]

The possibility that some components of the cell may be lost or distorted during specimen preparation has always worried microscopists. The only certain way to avoid the problem is to examine cells while they are alive, without fixing or freezing. For this purpose light microscopes with special optical systems are especially useful.

When light passes through a living cell, the phase of the light wave is changed according to the cell's refractive index: light passing through a relatively thick or dense part of the cell, such as the nucleus, is retarded; its phase, consequently, is shifted relative to light that has passed through an adjacent thinner region of the cytoplasm. Both the **phase-contrast microscope** and the **differential-interference-contrast microscope** exploit the interference effects produced when these two sets of waves recombine, thereby creating an image of the cell's structure (Figure 4–10). Both types of light microscopy are widely used to visualize living cells.

A simpler way to see some of the features of a living cell is to observe the light that is scattered by its various components. In the **dark-field microscope** the illuminating rays of light are directed from the side so that only scattered light enters the microscope lenses. Consequently, the cell appears as an illuminated object against a black background. Images of the same cell obtained by four kinds of light microscopy are shown in Figure 4–11.

One of the great advantages of phase-contrast, differential-interference-contrast, and dark-field microscopy is that each makes it possible to watch the movements involved in such processes as mitosis and cell migration. Since many cel-

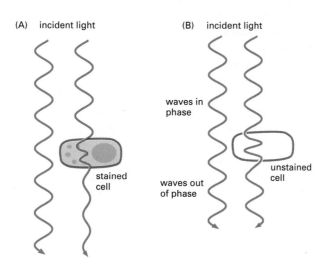

(A) incident light   (B) incident light

waves in phase

waves out of phase

stained cell

unstained cell

Figure 4–10 Two ways to obtain contrast in light microscopy. The stained portions of the cell in (A) reduce the amplitude of light waves of particular wavelengths passing through them. A colored image of the cell is thereby obtained that is visible in the ordinary way. Light passing through the unstained, living cell (B) undergoes very little change in amplitude, and the structural details cannot be seen even if the image is highly magnified. The *phase* of the light, however, is altered by its passage through the cell, and small phase differences can be made visible by exploiting interference effects using a phase-contrast or a differential-interference-contrast microscope.

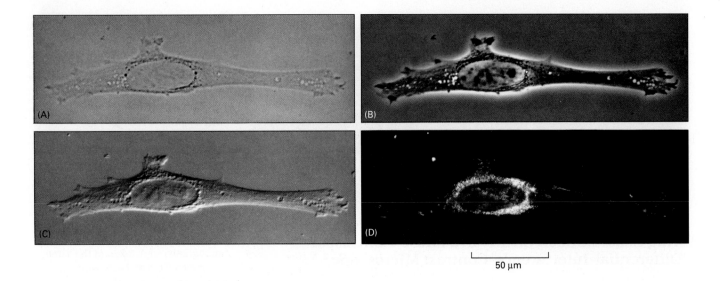

50 μm

lular motions are too slow to be seen in real time, it is often helpful to take time-lapse motion pictures (*microcinematography*) or video recordings. Here, successive frames separated by a short time delay are recorded, so that when the resulting film or videotape is projected or played at normal speed, events appear greatly speeded up.

Images Can Be Enhanced and Analyzed by Electronic Techniques [6]

In recent years **electronic imaging systems** and the associated technology of **image processing** have had a major impact on light microscopy. They have enabled certain practical limitations of microscopes (due to imperfections in the optical system) to be largely overcome. They have also circumvented two fundamental limitations of the human eye: the eye cannot see well in extremely dim light, and it cannot perceive small differences in light intensity against a bright background. The first limitation can be overcome by attaching highly light-sensitive video cameras (of the kind used in night surveillance) to a microscope. It is then possible to observe cells for long periods at very low light levels, thereby avoiding the damaging effects of prolonged bright light (and heat). Such *image-intensification systems* are especially important for viewing fluorescent molecules in living cells.

Because images produced by video cameras are in electronic form, they can be readily digitized, fed to a computer, and processed in various ways to extract latent information. Such image processing makes it possible to compensate for various optical faults in microscopes in order to attain the theoretical limit of resolution. Moreover, by using video systems linked to image processors, contrast can be greatly enhanced so that the eye's limitations in detecting small differences in light intensity are overcome. Although this processing also enhances the effects of random background irregularities in the optical system, this "noise" can be removed by electronically subtracting an image of a blank area of the field. Small transparent objects then become visible that were previously impossible to distinguish from the background.

The high contrast attainable by computer-assisted, differential-interference-contrast microscopy makes it possible to see even very small objects such as single microtubules (Figure 4–12), which have a diameter of 0.025 μm, less than one-tenth the wavelength of light. Individual microtubules can also be seen in a fluorescence microscope if they are fluorescently labeled (see Figure 4–62). In both cases, however, the unavoidable diffraction effects badly blur the image so that the microtubules appear at least 0.2 μm wide, making it impossible to distinguish a single microtubule from a bundle of several microtubules.

Figure 4–11 Four types of light microscopy. (A) The image of a fibroblast in culture obtained by the simple transmission of light through the cell, a technique known as *bright-field microscopy*. The other images were obtained by techniques discussed in the text: (B) phase-contrast microscopy, (C) Nomarski differential-interference-contrast microscopy, and (D) dark-field microscopy. All four types of image can be obtained with most modern microscopes simply by interchanging optical components.

Imaging of Complex Three-dimensional Objects Is Possible with the Confocal Scanning Microscope [7]

For ordinary light microscopy, as we have seen, a tissue has to be sliced into thin sections in order to be examined; the thinner the section, the crisper the image. In the process of sectioning, information about the third dimension is lost. How then can one get a picture of the three-dimensional architecture of a cell or tissue, and how can one view the microscopic structure of a specimen that, for one reason or another, cannot first be sliced into sections? If a thick specimen is viewed with a conventional light microscope, the image obtained by focusing at any one level is degraded by blurred, out-of-focus information from the parts of the specimen that lie above and below the plane of focus. Although this problem can be overcome by complex computer-based image processing applied to a series of images in different focal planes, the method is slow and costly in computing power. The **confocal scanning microscope** provides another, more direct way of achieving the same end result: electronic-imaging methods make it possible to focus on a chosen plane in a thick specimen while rejecting the light that comes from out-of-focus regions above and below that plane. Thus one sees a crisp, thin *optical section*. From a series of such optical sections taken at different depths and stored in a computer, it is easy to reconstruct a three-dimensional image. The confocal scanning microscope does for the microscopist what the CAT scanner does (by different means) for the radiologist investigating a human body: both machines give detailed sectional views of the interior of an intact structure.

The optical details of the confocal scanning microscope are complex, but the basic idea is simple, as illustrated in Figure 4–13. The microscope is generally used with fluorescence optics (see Figure 4–7), but instead of illuminating the whole specimen at once, in the usual way, the optical system at any instant focuses a spotlight onto a single point at a specific depth in the specimen. A very bright source of pinpoint illumination is required; this is usually supplied by a laser whose light has been passed through a pinhole. The fluorescence emitted from the illuminated material is collected and brought to an image at the entry port of a suitable light detector. A pinhole aperture is placed at the detector, at the site that is *confocal* with the illuminating pinhole—that is, precisely where the rays emitted from the illuminated point in the specimen come to a focus. Thus the light from this point in the specimen converges on this aperture and enters the detector. By contrast, the light from regions out of the plane of focus of the

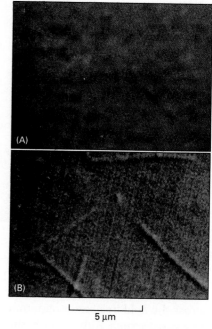

Figure 4–12 Extending the limits of detection. Light-microscope images of unstained microtubules that have been visualized by differential-interference-contrast microscopy followed by electronic image processing. (A) The original unprocessed image. (B) The final result of an electronic process that greatly enhances contrast and reduces "noise." Microtubules are only 0.025 μm in diameter and therefore in this image should appear only 0.1 mm wide. Instead, they appear much wider because of diffraction effects. (Courtesy of Bruce Schnapp.)

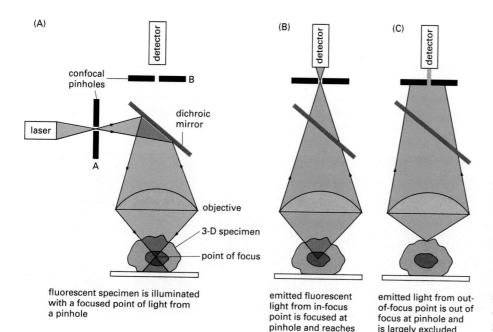

fluorescent specimen is illuminated with a focused point of light from a pinhole

emitted fluorescent light from in-focus point is focused at pinhole and reaches detector

emitted light from out-of-focus point is out of focus at pinhole and is largely excluded from detector

Figure 4–13 The confocal scanning fluorescence microscope. This simplified diagram shows that the basic arrangement of optical components is similar to that of the standard fluorescence microscope shown in Figure 4–7 except that a laser is used to illuminate a small pinhole whose image is focused at a single point in the specimen (A). Emitted fluorescence from this focal point in the specimen is focused at a second (confocal) pinhole (B). Emitted light from elsewhere in the specimen is not focused here and therefore does not contribute to the final image (C). By scanning the beam of light across the specimen, a very sharp two-dimensional image of the exact plane of focus is built up that is not significantly degraded by light from other regions of the specimen.

Looking at the Structure of Cells in the Microscope

147

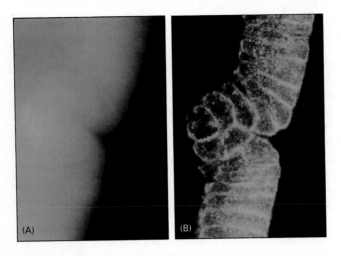

Figure 4–14 **Comparison of conventional and confocal fluorescence microscopy.** These two micrographs are of the same intact gastrula-stage *Drosophila* embryo that has been stained with a fluorescent probe for actin filaments. The conventional, unprocessed image (A) is blurred by the presence of fluorescent structures above and below the plane of focus. In the confocal image (B), this out-of-focus information is removed, which results in a crisp optical section of the cell in the embryo. (Courtesy of Richard Warn and Peter Shaw.)

spotlight is also out of focus at the pinhole aperture and is therefore largely excluded from the detector (Figure 4–14). To build up a two-dimensional image, data from each point in the plane of focus are collected sequentially by scanning across the field in a raster pattern (as on a television screen) and are displayed on a video screen. Although not shown in Figure 4–13, the scanning is done by deflecting the beam with an oscillating mirror placed between the dichroic mirror and the objective lens in such a way that the illuminating spotlight and the confocal pinhole at the detector remain strictly in register.

The confocal scanning microscope has been used to resolve the structure of numerous complex three-dimensional objects (Figure 4–15), including the networks of cytoskeletal fibers in the cytoplasm and the arrangements of chromosomes and genes in the nucleus.

The Electron Microscope Resolves the Fine Structure of the Cell [8]

The relationship between the limit of resolution and the wavelength of the illuminating radiation (see Figure 4–4) holds true for any form of radiation, whether it is a beam of light or a beam of electrons. With electrons, however, the limit of resolution can be made very small. The wavelength of an electron decreases as its velocity increases. In an electron microscope with an accelerating voltage of 100,000 V, the wavelength of an electron is 0.004 nm. In theory the resolution of such a microscope should be about 0.002 nm, which is 10,000 times greater than that of the light microscope. Because the aberrations of an electron lens are considerably harder to correct than those of a glass lens, however, the practical resolving power of most modern electron microscopes is, at best, 0.1 nm (1 Å) (Figure 4–16). Furthermore, problems of specimen preparation, contrast, and radiation damage effectively limit the normal resolution for biological objects to 2 nm (20 Å). This is nonetheless about 100 times better than the resolution of the

Figure 4–15 **Three-dimensional reconstruction from confocal scanning microscope images.** Pollen grains, in this case from a passion flower, have a complex sculptured cell wall that contains fluorescent compounds. Images obtained at different depths through the grain, using a confocal scanning microscope, can be recombined to give a three-dimensional view of the whole grain, shown on the right. Three selected individual optical sections from the full set of 30, each of which shows little contribution from its neighbors, are shown on the left. (Courtesy of John White.)

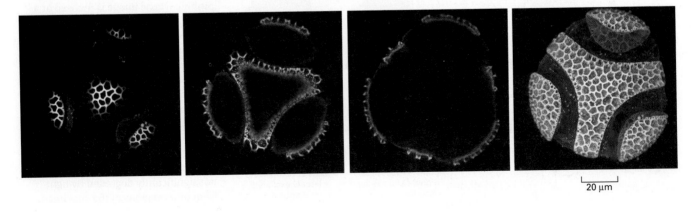

20 μm

Table 4–2 Major Events in the Development of the Electron Microscope and Its Applications to Cell Biology

1897 **J.J. Thomson** announced the existence of negatively charged particles, later termed *electrons*.

1924 **de Broglie** proposed that a moving electron has wavelike properties.

1926 **Busch** proved that it was possible to focus a beam of electrons with a cylindrical magnetic lens, laying the foundations of electron optics.

1931 **Ruska** and colleagues built the first transmission electron microscope.

1935 **Knoll** demonstrated the feasibility of the scanning electron microscope; three years later a prototype instrument was built by **Von Ardenne**.

1939 **Siemens** produced the first commercial transmission electron microscope.

1944 **Williams** and **Wyckoff** introduced the metal shadowing technique.

1945 **Porter, Claude**, and **Fullam** used the electron microscope to examine cells in tissue culture after fixing and staining them with OsO_4.

1948 **Pease** and **Baker** reliably prepared thin sections (0.1 to 0.2 µm thick) of biological material.

1952 **Palade, Porter**, and **Sjöstrand** developed methods of fixation and thin sectioning that enabled many intracellular structures to be seen for the first time. In one of the first applications of these techniques, **H.E. Huxley** showed that skeletal muscle contains overlapping arrays or protein filaments, supporting the "sliding filament" hypothesis of muscle contraction.

1953 **Porter and Blum** developed the first widely accepted ultramicrotome, incorporating many features introduced by **Claude** and **Sjöstrand** previously.

1956 **Glauert** and associates showed that the epoxy resin Araldite was a highly effective embedding agent for electron microscopy. **Luft** introduced another embedding resin, Epon, five years later.

1957 **Robertson** described the trilaminar structure of the cell membrane, seen for the first time in the electron microscope.

1957 Freeze-fracture techniques, initially developed by **Steere**, were perfected by **Moor and Mühlethaler**. Later (1966), **Branton** demonstrated that freeze-fracture allows the interior of the membrane to be visualized.

1959 **Singer** used antibodies coupled to ferritin to detect cellular molecules in the electron microscope.

1959 **Brenner and Horne** developed the negative staining technique, invented four years previously by **Hall**, into a generally useful technique for visualizing viruses, bacteria, and protein filaments.

1963 **Sabatini, Bensch**, and **Barrnett** introduced glutaraldehyde (usually followed by OsO_4) as a fixative for electron microscopy.

1965 **Cambridge Instruments** produced the first commercial scanning electron microscope.

1968 **de Rosier and Klug** described techniques for the reconstruction of three-dimensional structures from electron micrographs.

1975 **Henderson and Unwin** determined the first structure of a membrane protein by computer-based reconstruction from electron micrographs of unstained samples.

1979 **Heuser, Reese**, and colleagues developed a high-resolution, deep-etching technique using very rapidly frozen specimens.

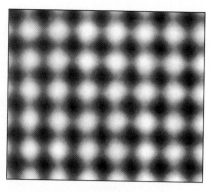

Figure 4–16 Limit of resolution of the electron microscope. Electron micrograph of a thin layer of gold showing the individual files of atoms in the crystal as bright spots. The distance between adjacent files of gold atoms is about 0.2 nm (2 Å). (Courtesy of Graham Hills.)

light microscope. Some of the landmarks in the development of electron microscopy are outlined in Table 4–2.

In overall design the **transmission electron microscope (TEM)** is similar to a light microscope, although it is much larger and upside down (Figure 4–17). The source of illumination is a filament or cathode that emits electrons at the top of a cylindrical column about 2 meters high. Since electrons are scattered by collisions with air molecules, air must first be pumped out of the column to create a vacuum. The electrons are then accelerated from the filament by a nearby

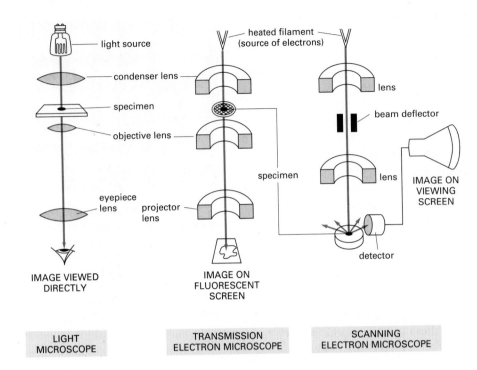

light source — light source
condenser lens — condenser lens
specimen — specimen
objective lens — objective lens
eyepiece lens — eyepiece lens
projector lens — projector lens
heated filament (source of electrons)
lens
beam deflector
lens
IMAGE ON VIEWING SCREEN
specimen
detector
IMAGE VIEWED DIRECTLY
IMAGE ON FLUORESCENT SCREEN

LIGHT MICROSCOPE

TRANSMISSION ELECTRON MICROSCOPE

SCANNING ELECTRON MICROSCOPE

Figure 4–17 Principal features of a light microscope, a transmission electron microscope, and a scanning electron microscope. These drawings emphasize the similarities of overall design. Whereas the lenses in the light microscope are made of glass, those in the electron microscope are magnetic coils. The two types of electron microscopes require that the specimen be placed in a vacuum.

anode and allowed to pass through a tiny hole to form an electron beam that travels down the column. Magnetic coils placed at intervals along the column focus the electron beam, just as glass lenses focus the light in a light microscope. The specimen is put into the vacuum, through an airlock, into the path of the electron beam. As in the case of light microscopy, the specimen is usually stained, in this case with *electron-dense* material, as we see in the next section. Some of the electrons passing through the specimen are scattered by structures stained with the electron-dense material; the remainder are focused to form an image—in a manner analogous to the way an image is formed in a light microscope—either on a photographic plate or on a phosphorescent screen. Because the scattered electrons are lost from the beam, the dense regions of the specimen show up in the image as areas of reduced electron flux, which look dark.

Biological Specimens Require Special Preparation for the Electron Microscope [9]

In the early days of its application to biological materials, the electron microscope revealed many previously unimagined structures in cells. But before these discoveries could be made, electron microscopists had to develop new procedures for embedding, cutting, and staining tissues.

Since the specimen is exposed to a very high vacuum in the electron microscope, there is no possibility of viewing it in the living, wet state. Tissues are usually preserved by fixation—first with *glutaraldehyde*, which covalently cross-links protein molecules to their neighbors, and then with *osmium tetroxide*, which binds to and stabilizes lipid bilayers as well as proteins (Figure 4–18). Since electrons have very limited penetrating power, the fixed tissues normally have to be cut into extremely thin sections (50 to 100 nm thick—about 1/200 of the thickness of a single cell) before they are viewed. This is achieved by dehydrating the specimen and permeating it with a monomeric resin that polymerizes to form a solid block of plastic; the block is then cut with a fine glass or diamond knife on a special microtome. These *thin sections*, free of water and other volatile solvents, are placed on a small circular metal grid for viewing in the microscope (Figure 4–19).

Contrast in the electron microscope depends on the atomic number of the atoms in the specimen: the higher the atomic number, the more electrons are

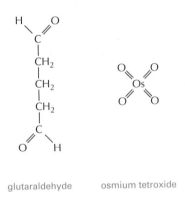

glutaraldehyde   osmium tetroxide

Figure 4–18 Two common chemical fixatives used for electron microscopy. The two reactive aldehyde groups of glutaraldehyde enable it to cross-link various types of molecules, forming covalent bonds between them. Osmium tetroxide is reduced by many organic compounds with which it forms cross-linked complexes. It is especially useful for fixing cell membranes, since it reacts with the C=C double bonds present in many fatty acids.

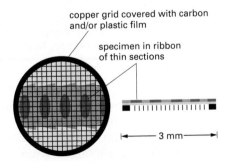

copper grid covered with carbon and/or plastic film

specimen in ribbon of thin sections

3 mm

Figure 4–19 Diagram of the copper grid used to support the thin sections of a specimen in the transmission electron microscope.

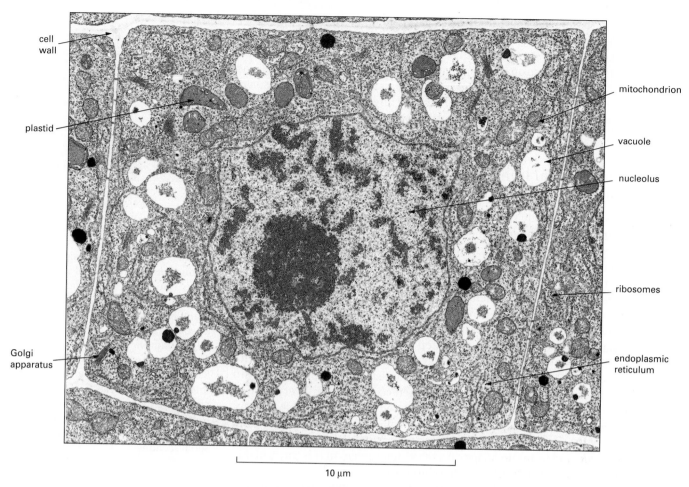

cell wall

plastid

Golgi apparatus

mitochondrion

vacuole

nucleolus

ribosomes

endoplasmic reticulum

10 μm

scattered and the greater is the contrast. Biological molecules are composed of atoms of very low atomic number (mainly carbon, oxygen, nitrogen, and hydrogen). To make them visible, they are usually impregnated (before or after sectioning) with the salts of heavy metals such as uranium and lead. Different cellular constituents are revealed with various degrees of contrast according to their degree of impregnation, or "staining," with these salts. Lipids, for example, tend to stain darkly following osmium fixation, revealing the location of cell membranes (Figure 4–20).

Figure 4–20 Electron micrograph of a root-tip cell stained with osmium and other heavy metal ions. The cell wall, nucleus, vacuoles, mitochondria, endoplasmic reticulum, Golgi apparatus, and ribosomes are easily seen. (Courtesy of Brian Gunning.)

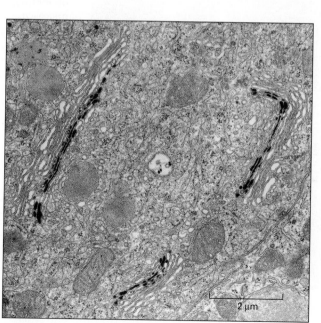

2 μm

Figure 4–21 Electron micrograph of a cell showing the location of a particular enzyme (nucleotide diphosphatase) in the Golgi apparatus. A thin section of the cell was incubated with a substrate that formed an electron-dense precipitate upon reaction with the enzyme. (Courtesy of Daniel S. Friend.)

Looking at the Structure of Cells in the Microscope

In some cases specific macromolecules can be located in thin sections by techniques adapted from light microscopy. Certain enzymes in cells can be detected by incubating the specimen with a substrate whose reaction leads to the local deposition of an electron-dense precipitate (Figure 4–21). Alternatively, as discussed on page 186, antibodies can be coupled to an indicator enzyme (usually peroxidase) or to an electron-dense marker (usually tiny spheres of metallic gold, which are referred to as *colloidal gold* particles) and then used to locate the macromolecules that the antibodies recognize (see Figure 4–63).

Three-dimensional Images of Surfaces Can Be Obtained by Scanning Electron Microscopy [10]

Thin sections are effectively two-dimensional slices of tissue and fail to convey the three-dimensional arrangement of cellular components. Although the third dimension can be reconstructed from serial sections (Figure 4–22), this is a lengthy and tedious process.

Fortunately, there are more direct means to obtain a three-dimensional image. One is to examine a specimen in a **scanning electron microscope (SEM),** which is usually a smaller, simpler, and cheaper device than a transmission electron microscope. Whereas the transmission electron microscope uses the electrons that have passed through the specimen to form an image, the scanning electron microscope uses electrons that are scattered or emitted from the specimen's surface. The specimen to be examined is fixed, dried, and coated with a thin layer of heavy metal. The specimen is then scanned with a very narrow beam of electrons. The quantity of electrons scattered or emitted as this primary beam bombards each successive point of the metallic surface is measured and used to control the intensity of a second beam, which moves in synchrony with the primary beam and forms an image on a television screen. In this way a highly enlarged image of the surface as a whole is built up.

The SEM technique provides great depth of focus; moreover, since the amount of electron scattering depends on the angle of the surface relative to the

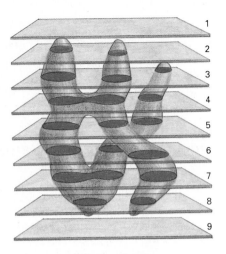

Figure 4–22 Three-dimensional reconstruction from serial sections. Single thin sections sometimes give misleading impressions. In this example most sections through a cell containing a branched mitochondrion will appear to contain two or three separate mitochondria. Sections 4 and 7, moreover, might be interpreted as showing a mitochondrion in the process of dividing. The true three-dimensional shape, however, can be reconstructed from serial sections.

Figure 4–23 Scanning electron microscopy. Scanning electron micrograph of the stereocilia projecting from a hair cell in the inner ear of a bullfrog (A). For comparison, the same structure is shown by differential-interference-contrast light microscopy (B) and by thin-section electron microscopy (C). (Courtesy of Richard Jacobs and James Hudspeth.)

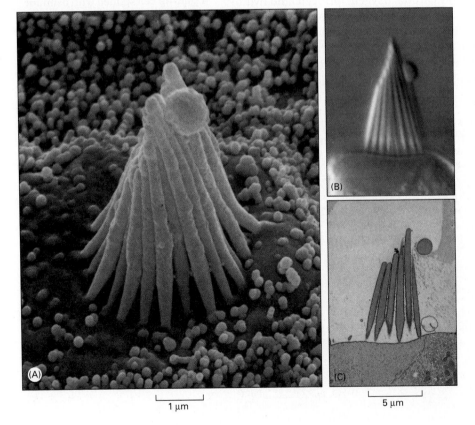

beam, the image has highlights and shadows that give it a three-dimensional appearance (Figure 4–23). Only surface features can be examined, however, and in most forms of SEM the resolution attainable is not very high (about 10 nm, with an effective magnification of up to 20,000 times). As a result, the technique is usually used to study whole cells and tissues rather than subcellular organelles (see also Figure 4–32).

Metal Shadowing Allows Surface Features to Be Examined at High Resolution by Transmission Electron Microscopy [11]

The transmission electron microscope can also be used to study the surface of a specimen—and at a generally higher resolution than in a scanning electron microscope—such that individual macromolecules can be seen. As for scanning electron microscopy, a thin film of a heavy metal such as platinum is evaporated onto the dried specimen. The metal is sprayed from an oblique angle in order to deposit a coating that is thicker in some places than others—a process known as **shadowing** because a shadow effect is created that gives the image a three-dimensional appearance.

Some specimens coated in this way are thin enough or small enough for the electron beam to penetrate them directly; this is the case for individual molecules, viruses, and cell walls (Figure 4–24). For thicker specimens the organic material of the cell must be dissolved away after shadowing so that only the thin metal *replica* of the surface of the specimen is left. The replica is reinforced with a film of carbon so that it can be placed on a grid and examined in the transmission electron microscope in the ordinary way (Figure 4–25).

Freeze-Fracture and Freeze-Etch Electron Microscopy Provide Unique Views of the Cell Interior [12]

Two methods that use metal replicas have been particularly useful in cell biology. One of these, **freeze-fracture** electron microscopy, provides a way of visualizing the interior of cell membranes. Cells are frozen at the temperature of liquid nitrogen (–196°C) in the presence of a *cryoprotectant* (antifreeze) to prevent distortion from ice crystal formation, and then the frozen block is cracked with a knife blade. The fracture plane often passes through the hydrophobic middle of lipid bilayers, thereby exposing the interior of cell membranes. The resulting fracture faces are shadowed with platinum, the organic material is dissolved away, and the replicas are floated off and viewed in the electron microscope (as in Figure 4–25). Such replicas are studded with small bumps, called *intramembrane particles,* which represent large transmembrane proteins. The technique provides a convenient and dramatic way to visualize the distribution of such proteins in the plane of a membrane (Figure 4–26).

Another important and related replica method is **freeze-etch** electron microscopy, which can be used to examine either the exterior or interior of cells. In this technique the cells are frozen extremely rapidly—using a special device to slam the sample against a copper block cooled with liquid helium, for example—and the frozen block is cracked with a knife blade as just described. But now the ice level is lowered around the cells (and to a lesser extent within the cells) by the sublimation of ice in a vacuum as the temperature is raised (a process called *freeze-drying*) (Figure 4–27). The parts of the cell exposed by this *etching* process are then shadowed as before to make a platinum replica. This technique exposes structures in the interior of the cell and can reveal their three-dimensional organization with exceptional clarity (Figure 4–28).

Because a metal-shadowed replica rather than the sample itself is viewed under vacuum in the microscope, both freeze-fracture and freeze-etch microscopy can be used to study frozen unfixed cells, thereby avoiding the risk of artifacts caused by fixation.

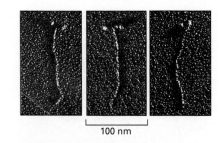

100 nm

Figure 4–24 Electron micrographs of individual myosin protein molecules that have been shadowed with platinum. Myosin is a major component of the contractile apparatus of muscle. As shown here, it is composed of two globular head regions linked to a common rodlike tail. (Courtesy of Arthur Elliot.)

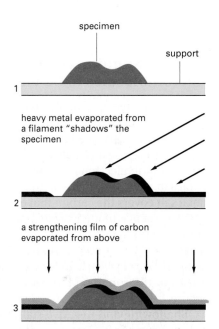

Figure 4–25 Preparation of a metal-shadowed replica of the surface of a specimen. Note that the thickness of the metal reflects the surface contours of the original specimen.

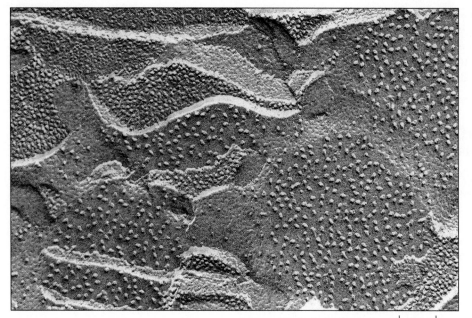

Figure 4–26 **Freeze-fracture electron micrograph of the thylakoid membranes from the chloroplast of a plant cell.** These membranes, which carry out photosynthesis, are stacked up in multiple layers (see Figure 14–39). The plane of the fracture has moved from layer to layer, passing through the middle of each lipid bilayer and exposing transmembrane proteins that have sufficient bulk in the interior of the bilayer to cast a shadow and show up as intramembrane particles in this platinum replica. The largest particles seen in the membrane are the complete photosystem II—a complex of multiple proteins. (Courtesy of L.A. Staehelin.)

0.1 μm

Negative Staining and Cryoelectron Microscopy Allow Macromolecules to Be Viewed at High Resolution [13]

Although isolated macromolecules, such as DNA or large proteins, can be visualized readily in the electron microscope if they are shadowed with a heavy metal to provide contrast (see Figure 4–24), finer detail can be seen by using **negative staining.** Here, the molecules, supported on a thin film of carbon (which is nearly transparent to electrons), are washed with a concentrated solution of a heavy-metal salt such as uranyl acetate. After the sample has dried, a very thin film of metal salt covers the carbon film everywhere except where it has been excluded by the presence of an adsorbed macromolecule. Because the macromolecule allows electrons to pass much more readily than does the surrounding heavy-metal stain, a reversed or negative image of the molecule is created. Negative staining is especially useful for viewing large macromolecular aggregates such as viruses or ribosomes and for seeing the subunit structure of protein filaments (Figure 4–29).

Shadowing and negative staining are capable of providing high-contrast surface views of small macromolecular assemblies, but both are limited in reso-

Figure 4–27 **Freeze-etch electron microscopy.** The specimen is rapidly frozen, and the block of ice is fractured with a knife (A). The ice level is then lowered by sublimation in a vacuum, exposing structures in the cell that were near the fracture plane (B). Following these steps, a replica of the still frozen surface is prepared (as described in Figure 4–25), and this is examined in a transmission electron microscope.

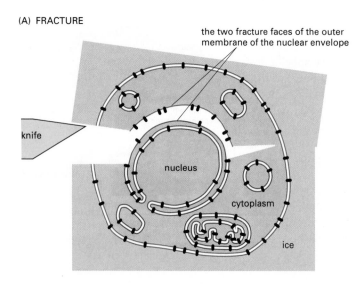

(A) FRACTURE

the two fracture faces of the outer membrane of the nuclear envelope

knife

nucleus

cytoplasm

ice

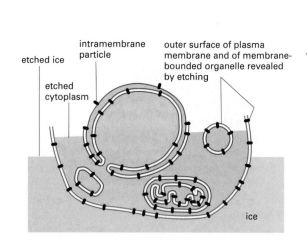

(B) ETCH

intramembrane particle

etched ice

etched cytoplasm

outer surface of plasma membrane and of membrane-bounded organelle revealed by etching

ice

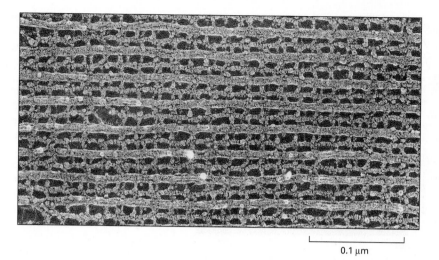

Figure 4–28 **Regular array of protein filaments in an insect muscle.** To obtain this image, the muscle cells were rapidly frozen to liquid helium temperature, fractured through the cytoplasm, and subjected to deep etching. A metal-shadowed replica was then prepared and examined at high magnification. (Courtesy of Roger Cooke and John Heuser.)

0.1 µm

lution by the size of the smallest metal particles in the shadow or stain employed. Recent methods provide an alternative that has allowed even the interior features of three-dimensional structures such as viruses to be visualized directly at high resolution. In this technique, called **cryoelectron microscopy,** a very thin (~100 nm) layer of rapidly frozen hydrated sample is prepared on a microscope grid. A special sample holder is required to keep this hydrated specimen at –160°C in the vacuum of the microscope, where it can be viewed directly without fixation, staining, or drying. Surprisingly for unstained material, these specimens can be imaged with a considerable degree of contrast (Figure 4–30).

Regardless of the method used, a single protein molecule gives only a weak and ill-defined image in the electron microscope. Efforts to get better information by prolonging the time of inspection or by increasing the intensity of the illuminating beam are self-defeating because they damage and disrupt the object under examination. Therefore, to discover the details of molecular structure, it is necessary to combine the information obtained from many molecules in such a way as to average out the random errors in the individual images. This is possible for viruses or protein filaments, in which the individual subunits are present in regular repeating arrays; it is also possible for any substance that can be made to form a crystalline array in two dimensions in which large numbers of molecules are held in identical orientation and in regularly spaced positions. Given an electron micrograph of either type of array, one can use image-processing techniques to compute the average image of an individual molecule, revealing details obscured by the random "noise" in the original picture.

Image reconstructions of this type have allowed the interior structure of an enveloped virus to be obtained to a resolution of 3.5 nm and have revealed the

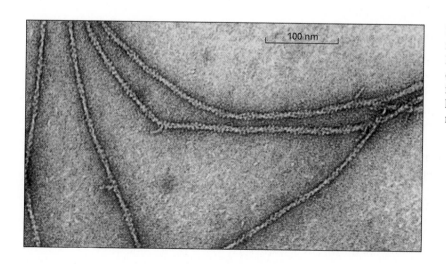

100 nm

Figure 4–29 **Electron micrograph of negatively stained actin filaments.** Each filament is about 8 nm in diameter and is seen, on close inspection, to be composed of a helical chain of globular actin molecules. (Courtesy of Roger Craig.)

Looking at the Structure of Cells in the Microscope

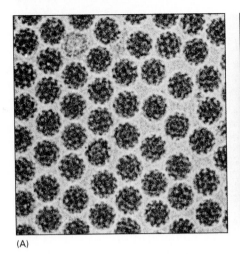

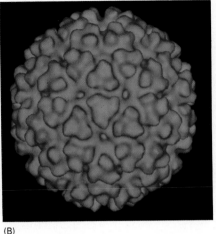

(A)

(B)

Figure 4–30 **Electron microscopy of a virus.** (A) Unstained Semliki forest virus in a thin layer of vitrified water viewed by cryoelectron microscopy at −160°C. As in light microscopy, phase contrast can be used to get an image of the unstained specimen. A large number of these images can then be combined by image-processing methods to produce a three-dimensional image of the virus at high resolution (B). (Courtesy of Stephen Fuller.)

shape of an individual protein molecule to the remarkable resolution of 0.35 nm (see Figure 10–31). But even in its most sophisticated forms, electron microscopy falls short of providing a full description of molecular structure because the atoms in a molecule are separated by distances of only 0.1 or 0.2 nm. Resolving molecular structure in atomic detail takes us beyond microscopy to techniques such as x-ray diffraction, which are described in a later section.

Summary

Many light-microscope techniques are available for observing cells. Cells that have been fixed and stained can be studied in a conventional light microscope, while antibodies coupled to fluorescent dyes can be used to locate specific molecules in cells in a fluorescence microscope. The confocal scanning microscope provides thin optical sections and can be used to reconstruct a three-dimensional image. Living cells can be seen with phase-contrast, differential-interference-contrast, or dark-field optics. All forms of light microscopy are facilitated by electronic image-processing techniques, which enhance sensitivity and refine the image.

Determining the detailed structure of the membranes and organelles in cells requires the higher resolution attainable in a conventional transmission electron microscope. Three-dimensional views of the surfaces of cells and tissues can be obtained by scanning electron microscopy, while the interior of membranes and cells can be visualized by freeze-fracture and freeze-etch electron microscopy, respectively. The shapes of isolated macromolecules that have been shadowed with a heavy metal or outlined by negative staining can also be readily visualized by electron microscopy.

Isolating Cells and Growing Them in Culture [14]

Although the structure of organelles and large molecules in a cell can be seen with microscopes, a molecular understanding of a cell requires detailed biochemical analysis. Unfortunately, most biochemical procedures require large numbers of cells and begin by disrupting them. If the sample is a piece of tissue, fragments of all of its cells will be mixed together, creating confusion if the cells are of several types, which is almost always the case. In order to preserve as much information as possible about each individual type of cell, cell biologists have developed ways of dissociating cells from tissues and separating the various types. The resulting, relatively homogenous population of cells then can be analyzed—either directly or after their number has been greatly increased by allowing them to proliferate in culture.

Cells Can Be Isolated from a Tissue and Separated into Different Types [15]

The first step in isolating cells of a uniform type from a tissue that contains a mixture of cell types is to disrupt the extracellular matrix and intercellular junctions that hold the cells together. The best yields of viable dissociated cells are usually obtained from fetal or neonatal tissues, typically by treating them with proteolytic enzymes (such as trypsin and collagenase) and with agents (such as ethylenediaminetetraacetic acid, or EDTA) that bind, or *chelate*, the Ca^{2+} on which cell-cell adhesion depends. The tissue can then be dissociated into single viable cells by gentle agitation.

Several approaches are used to separate the different cell types from a mixed cell suspension. One involves exploiting differences in physical properties. Large cells can be separated from small cells and dense cells from light cells by centrifugation, for example. These techniques will be described in connection with the separation of organelles and macromolecules, for which they were originally developed. Another approach is based on the tendency of some cell types to adhere strongly to glass or plastic, which allows them to be separated from cells that adhere less strongly.

An important refinement of this last technique depends on the specific binding properties of antibodies. Antibodies that bind specifically to the surface of only one cell type in a tissue can be coupled to various matrices—such as collagen, polysaccharide beads, or plastic—to form an *affinity surface* to which only cells recognized by the antibodies will adhere. The bound cells are then recovered by gentle shaking, by treatment with trypsin to digest the proteins that mediate the adhesion, or, in the case of a digestible matrix (such as collagen), by degrading the matrix itself with enzymes (such as collagenase).

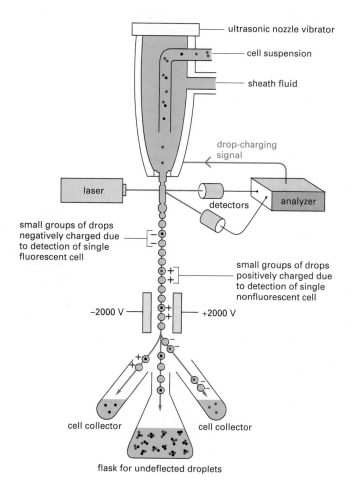

Figure 4–31 **A fluorescence-activated cell sorter.** When a cell passes through the laser beam, it is monitored for fluorescence. Droplets containing single cells are given a negative or positive charge, depending on whether the cell is fluorescent or not. The droplets are then deflected by an electric field into collection tubes according to their charge. Note that the cell concentration must be adjusted so that most droplets contain no cells and flow to a waste container together with any cell clumps. The same apparatus can also be used to separate fluorescently labeled chromosomes from one another, providing valuable starting material for the isolation and mapping of genes.

Figure labels:
- ultrasonic nozzle vibrator
- cell suspension
- sheath fluid
- drop-charging signal
- laser
- detectors
- analyzer
- small groups of drops negatively charged due to detection of single fluorescent cell
- small groups of drops positively charged due to detection of single nonfluorescent cell
- −2000 V
- +2000 V
- cell collector
- cell collector
- flask for undeflected droplets

The most sophisticated cell-separation technique involves labeling specific cells with antibodies coupled to a fluorescent dye and then separating the labeled cells from the unlabeled ones in an electronic **fluorescence-activated cell sorter.** Here, individual cells traveling in single file in a fine stream are passed through a laser beam and the fluorescence of each cell is measured. Slightly farther downstream, tiny droplets, most containing either one cell or no cells, are formed by a vibrating nozzle. The droplets containing a single cell are automatically given a positive or a negative charge at the moment of formation, depending on whether the cell they contain is fluorescent; they are then deflected by a strong electric field into an appropriate container. Occasional clumps of cells, detected by their increased light scattering, are left uncharged and are discarded into a waste container. Such machines can select 1 cell in 1000 and sort about 5000 cells each second (Figure 4–31).

When a uniform population of cells has been obtained by any of these methods, it can be used directly for biochemical analysis. Alternatively, it provides a suitable starting material for cell culture, allowing the complex behavior of cells to be studied under the strictly defined conditions of a culture dish.

Cells Can Be Grown in a Culture Dish [16]

Given appropriate conditions, most kinds of plant and animal cells will live, multiply, and even express differentiated properties in a tissue-culture dish. The cells can be watched under the microscope or analyzed biochemically, and the effects of adding or removing specific molecules, such as hormones or growth factors, can be explored. In addition, in a mixed culture the interactions between one cell type and another can be studied. Experiments on cultured cells are sometimes said to be carried out *in vitro* (literally, "in glass") to contrast them with experiments on intact organisms, which are said to be carried out *in vivo* (literally, "in the living organism"). The terms can be confusing because they are often used in a different sense by biochemists, for whom *in vitro* refers to biochemical reactions occurring outside living cells, while *in vivo* refers to any reaction taking place inside a living cell.

Tissue culture began in 1907 with an experiment designed to settle a controversy in neurobiology. The hypothesis under examination was known as the *neuronal doctrine*, which states that each nerve fiber is the outgrowth of a single nerve cell and not the product of the fusion of many cells. To test this contention, small pieces of spinal cord were placed on clotted tissue fluid in a warm, moist chamber and observed at regular intervals under the microscope. After a day or so, individual nerve cells could be seen extending long, thin processes into the clot. Thus the neuronal doctrine was validated, and the foundations for the cell-culture revolution were laid.

The original experiments in 1907 involved the culture of small tissue fragments, or **explants.** Today, cultures are more commonly made from suspensions of cells dissociated from tissues as described above. Unlike bacteria, most tissue cells are not adapted to living in suspension and require a solid surface on which to grow and divide, which is now usually the surface of a plastic tissue-culture dish (Figure 4–32). Cells vary in their requirements, however, and some will not grow or differentiate unless the culture dish is coated with specific extracellular matrix components, such as collagen or laminin.

Cultures prepared directly from the tissues of an organism, either with or without an initial cell-fractionation step, are called **primary cultures.** In most cases cells in primary cultures can be removed from the culture dish and used to form a large number of **secondary cultures;** they may be repeatedly subcultured in this way for weeks or months. Such cells often display many of the differentiated properties appropriate to their origin: fibroblasts continue to secrete

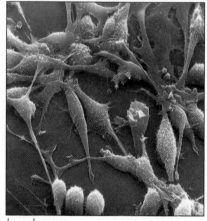

⊢——⊣
10 μm

Figure 4–32 Cells in culture. Scanning electron micrograph of rat fibroblasts growing on the plastic surface of a tissue-culture dish. (Courtesy of Guenter Albrecht-Buehler.)

collagen; cells derived from embryonic skeletal muscle fuse to form giant muscle fibers that spontaneously contract in the culture dish; nerve cells extend axons that are electrically excitable and make synapses with other nerve cells; and epithelial cells form extensive sheets with many of the properties of an intact epithelium. Since these phenomena occur in culture, they are accessible to study in ways that are not possible in intact tissues.

Serum-free, Chemically Defined Media Permit Identification of Specific Growth Factors [17]

Until the early 1970s tissue culture was something of a blend of science and witchcraft. Although tissue fluid clots were replaced by dishes of liquid media containing specified quantities of small molecules such as salts, glucose, amino acids, and vitamins, most media also included a poorly defined mixture of macromolecules in the form of horse serum or fetal calf serum or a crude extract made from chick embryos. Such media are still used today for most routine cell culture (Table 4–3), but they make it difficult for the investigator to know which specific macromolecules a particular type of cell needs to have in the medium in order to thrive and function normally.

This difficulty led to the development of various *serum-free, chemically defined media.* In addition to the usual small molecules, such defined media contain one or more specific proteins that most cells require in order to survive and proliferate in culture. These include **growth factors,** which stimulate cell proliferation, and *transferrin,* which carries iron into cells. Many of the extracellular protein signaling molecules essential for the survival, development, and proliferation of specific cell types have been discovered by studies in cell culture, and the search for new ones has been made very much easier by the availability of serum-free, chemically defined media.

Table 4–3 **Composition of a Typical Medium Suitable for the Cultivation of Mammalian Cells**

Amino Acids	Vitamins	Salts	Miscellaneous	Proteins (required in serum-free, chemically defined media)
Arginine	Biotin	NaCl	Glucose	Insulin
Cystine	Choline	KCl	Penicillin	Transferrin
Glutamine	Folate	NaH_2PO_4	Streptomycin	Specific growth factors
Histidine	Nicotinamide	$NaHCO_3$	Phenol red	
Isoleucine	Pantothenate	$CaCl_2$	Whole serum	
Leucine	Pyridoxal	$MgCl_2$		
Lysine	Thiamine			
Methionine	Riboflavin			
Phenylalanine				
Threonine				
Tryptophan				
Tyrosine				
Valine				

Glucose is used at a concentration of 5 to 10 mM. The amino acids are all in the L form and, with one or two exceptions, are used at concentrations of 0.1 or 0.2 mM; vitamins are used at a 100-fold lower concentration, that is, about 1 μM. Serum, which is usually from horse or calf, is added to make up 10% of the total volume. Penicillin and streptomycin are antibiotics added to suppress the growth of bacteria. Phenol red is a pH indicator dye whose color is monitored to assure a pH of about 7.4.

Cultures are usually grown in a plastic or glass container with a suitably prepared surface that allows the attachment of cells. The containers are kept in an incubator at 37°C in an atmosphere of 5% CO_2, 95% air.

Eucaryotic Cell Lines Are a Widely Used Source of Homogeneous Cells [18]

Most vertebrate cells die after a finite number of divisions in culture. Human skin cells, for example, typically last for several months in culture, dividing only 50 to 100 times before they die out. It has been suggested that this limited life-span is related to the limited life-span of the animal from which the cells are derived. Occasionally, however, some cells in a culture will undergo a genetic change that makes them effectively immortal. Such cells will proliferate indefinitely and can be propagated as a **cell line** (Table 4–4).

Cells lines can also be prepared from cancer cells, but they differ from those prepared from normal cells in several ways. Cancer cell lines often grow without attaching to a surface, for example, and they proliferate to a very much higher density in a culture dish. Similar properties can be experimentally induced in normal cells by *transforming* them with a tumor-inducing virus or chemical. The resulting **transformed cell lines,** in reciprocal fashion, can often cause tumors if injected into a susceptible animal. Both transformed and untransformed cell lines are extremely useful in cell research as sources of very large numbers of cells of a uniform type, especially since they can be stored in liquid nitrogen at −196°C for an indefinite period and are still viable when thawed. It is important to recognize, however, that the cells in both types of cell lines nearly always differ in important ways from their normal progenitors in the tissues from which they were derived.

Although all the cells in a cell line are very similar, they are often not identical. The genetic uniformity of a cell line can be improved by **cell cloning,** in which a single cell is isolated and allowed to proliferate to form a large colony. A *clone* is any such collection of cells that are all descendants of a single ancestor cell. One of the most important uses of cell cloning has been the isolation of mutant cell lines with defects in specific genes. Studying cells that are defective in a specific protein often reveals a good deal about the function of that protein in normal cells.

Cells Can Be Fused Together to Form Hybrid Cells [19]

It is possible to fuse one cell with another to form a combined cell with two separate nuclei, called a **heterocaryon.** Typically, a suspension of cells is treated with certain inactivated viruses or with polyethylene glycol, either of which alters the plasma membranes of cells in a way that induces them to fuse with each other. Heterocaryons provide a way of mixing the components of two separate cells in order to study their interactions. The inert nucleus of a chicken red blood cell,

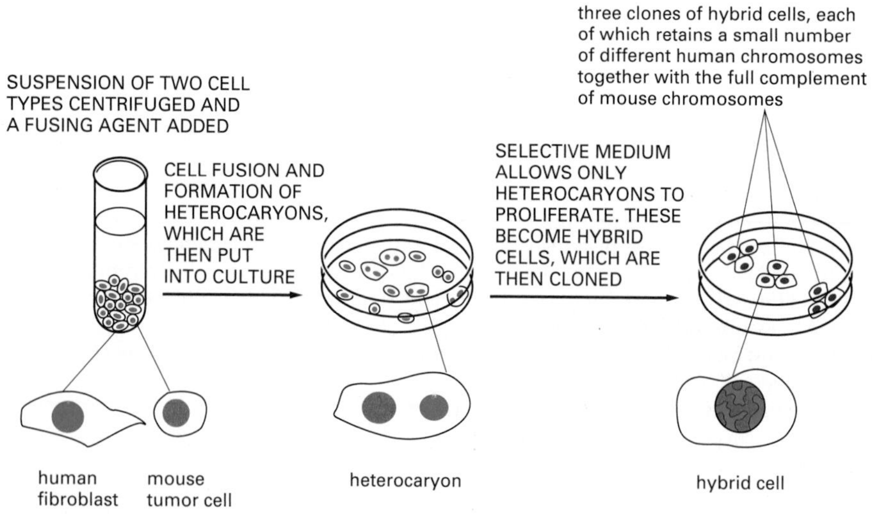

Table 4–4 Some Commonly Used Cell Lines

Cell Line*	Cell Type and Origin
3T 3	fibroblast (mouse)
BHK 21	fibroblast (Syrian hamster)
MDCK	epithelial cell (dog)
HeLa	epithelial cell (human)
PtK 1	epithelial cell (rat kangaroo)
L 6	myoblast (rat)
PC 12	chromaffin cell (rat)
SP 2	plasma cell (mouse)

*Many of these cell lines were derived from tumors. All of them are capable of indefinite replication in culture and express at least some of the differentiated properties of their cell of origin. BHK 21 cells, HeLa cells, and SP 2 cells are capable of growth in suspension; the other cell lines require a solid culture substratum in order to multiply.

three clones of hybrid cells, each of which retains a small number of different human chromosomes together with the full complement of mouse chromosomes

SUSPENSION OF TWO CELL TYPES CENTRIFUGED AND A FUSING AGENT ADDED

CELL FUSION AND FORMATION OF HETEROCARYONS, WHICH ARE THEN PUT INTO CULTURE

SELECTIVE MEDIUM ALLOWS ONLY HETEROCARYONS TO PROLIFERATE. THESE BECOME HYBRID CELLS, WHICH ARE THEN CLONED

human fibroblast   mouse tumor cell   heterocaryon   hybrid cell

Figure 4–33 The production of hybrid cells. Human cells and mouse cells are fused to produce heterocaryons (each with two or more nuclei), which eventually form hybrid cells (each with one fused nucleus). These particular hybrid cells are useful for mapping human genes on specific human chromosomes because most of the human chromosomes are quickly lost in a random manner, leaving clones that retain only one or a few. The hybrid cells produced by fusing other types of cells often retain most of their chromosomes.

Table 4–5 Some Landmarks in the Development of Tissue and Cell Culture

1885 **Roux** showed that embryonic chick cells could be maintained alive in a saline solution outside the animal body.

1907 **Harrison** cultivated amphibian spinal cord in a lymph clot, thereby demonstrating that axons are produced as extensions of single nerve cells.

1910 **Rous** induced a tumor by using a filtered extract of chicken tumor cells, later shown to contain an RNA virus (Rous sarcoma virus).

1913 **Carrel** showed that cells could grow for long periods in culture provided they were fed regularly under aseptic conditions.

1948 **Earle** and colleagues isolated single cells of the L cell line and showed that they formed clones of cells in tissue culture.

1952 **Gey** and colleagues established a continuous line of cells derived from human cervical carcinoma, which later became the well-known HeLa cell line.

1954 **Levi-Montalcini** and associates showed that nerve growth factor (NGF) stimulated the growth of axons in tissue culture.

1955 **Eagle** made the first systematic investigation of the essential nutritional requirements of cells in culture and found that animal cells could propagate in a defined mixture of small molecules supplemented with a small proportion of serum proteins.

1956 **Puck** and associates selected mutants with altered growth requirements from cultures of HeLa cells.

1958 **Temin** and **Rubin** developed a quantitative assay for the infection of chick cells in culture by purified Rous sarcoma virus. In the following decade the characteristics of this and other types of viral transformation were established by **Stoker**, **Dulbecco**, **Green**, and other virologists.

1961 **Hayflick and Moorhead** showed that human fibroblasts die after a finite number of divisions in culture.

1964 **Littlefield** introduced HAT medium for the selective growth of somatic cell hybrids. Together with the technique of cell fusion, this made somatic-cell genetics accessible.

     Kato and Takeuchi obtained a complete carrot plant from a single carrot root cell in tissue culture.

1965 **Ham** introduced a defined, serum-free medium able to support the clonal growth of certain mammalian cells.

     Harris and Watkins produced the first heterocaryons of mammalian cells by the virus-induced fusion of human and mouse cells.

1968 **Augusti-Tocco and Sato** adapted a mouse nerve cell tumor (neuroblastoma) to tissue culture and isolated clones that were electrically excitable and that extended nerve processes. A number of other differentiated cell lines were isolated at about this time, including skeletal muscle and liver cell lines.

1975 **Köhler and Milstein** produced the first monoclonal antibody-secreting hybridoma cell lines.

1976 **Sato** and associates published the first of a series of papers showing that different cell lines require different mixtures of hormones and growth factors to grow in serum-free medium.

1977 **Wigler and Axel** and their associates developed an efficient method for introducing single-copy mammalian genes into cultured cells, adapting an earlier method developed by **Graham and van der Eb**.

for example, is reactivated to make RNA and eventually to replicate its DNA when it is exposed to the cytoplasm of a growing tissue-culture cell by fusion. The first direct evidence that membrane proteins are able to move in the plane of the plasma membrane came from an experiment in which mouse cells and human cells were fused: although the mouse and human cell-surface proteins were initially confined to their own halves of the heterocaryon plasma membrane, they quickly diffused and mixed over the entire surface of the cell.

Isolating Cells and Growing Them in Culture

Eventually, a heterocaryon will proceed to mitosis and produce a **hybrid cell** in which the two separate nuclear envelopes have been disassembled, allowing all the chromosomes to be brought together in a single large nucleus (Figure 4–33). Although such hybrid cells can be cloned to produce hybrid cell lines, the cells tend to be unstable and lose chromosomes. For unknown reasons, mouse-human hybrid cells predominantly lose human chromosomes. These chromosomes are lost at random, giving rise to a variety of mouse-human hybrid cell lines, each of which contains only one or a few human chromosomes. This phenomenon has been put to good use in mapping the locations of genes in the human genome: only hybrid cells containing human chromosome 11, for example, synthesize human insulin, indicating that the gene encoding insulin is located on chromosome 11. The same hybrid cells are also used as a source of human DNA for preparing chromosome-specific human DNA libraries. Later in this chapter we shall learn how hybrid cells have been useful in the production of monoclonal antibodies.

Some important steps in the development of cell culture are outlined in Table 4–5.

Summary

Tissues, usually from very young organisms, can be dissociated into their component cells, from which individual cell types can be purified and used for biochemical analysis or for the establishment of cell cultures. Many animal and plant cells survive and proliferate in a culture dish if they are provided with a suitable medium containing nutrients and specific protein growth factors. Although most animal cells die after a finite number of divisions, rare immortal variant cells arise spontaneously in culture and can be maintained indefinitely as cell lines. Clones derived from a single ancestor cell make it possible to isolate uniform populations of mutant cells with defects in a single protein. Two types of cell can be fused to produce heterocaryons with two nuclei, which can be used to study interactions between the components of the original two cells. Heterocaryons eventually form hybrid cells with one fused nucleus; such cells provide a convenient method for assigning genes to specific chromosomes.

Fractionation of Cells and Analysis of Their Molecules [20]

Although biochemical analysis requires disruption of the anatomy of the cell, gentle separation techniques have been devised that preserve the functions of the various cell components. Just as a tissue can be separated into its living constituent cell types, so the cell can be separated into its functioning organelles and macromolecules. In this section we consider the methods that allow organelles and proteins to be purified and analyzed biochemically. The powerful recombinant DNA techniques used to analyze genes and proteins are discussed in Chapter 7.

Organelles and Macromolecules Can Be Separated by Ultracentrifugation [21]

Cells can be disrupted in various ways: they can be subjected to osmotic shock or ultrasonic vibration, or forced through a small orifice, or ground up. These procedures break many of the membranes of the cell (including the plasma membrane and membranes of the endoplasmic reticulum) into fragments that immediately reseal to form small closed vesicles. If carefully applied, however, the disruption procedures leave organelles such as nuclei, mitochondria, the

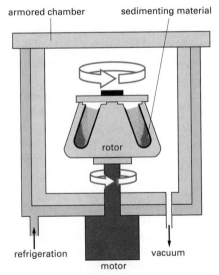

Figure 4–34 The preparative ultracentrifuge. The sample is contained in tubes that are inserted into a ring of cylindrical holes in a metal *rotor*. Rapid rotation of the rotor generates enormous centrifugal forces, which cause particles in the sample to sediment. The vacuum reduces friction, preventing heating of the rotor and allowing the refrigeration system to maintain the sample at 4°C.

Golgi apparatus, lysosomes, and peroxisomes largely intact. The suspension of cells is thereby reduced to a thick soup (called a *homogenate* or *extract*) containing a variety of membrane-bounded particles, each with a distinctive size, charge, and density. Provided that the homogenization medium has been carefully chosen (by trial and error for each organelle), the various particles—including the vesicles derived from the endoplasmic reticulum, called *microsomes*—retain most of their original biochemical properties.

The various components of the homogenate must then be separated. This became possible only after the commercial development in the early 1940s of an instrument known as the **preparative ultracentrifuge,** in which preparations of broken cells are rotated at high speeds (Figure 4–34). This treatment separates cell components on the basis of size and density: in general, the largest units experience the largest centrifugal force and move the most rapidly. At relatively low speed, large components such as nuclei and unbroken cells sediment to form a pellet at the bottom of the centrifuge tube; at slightly higher speed, a pellet of mitochondria is deposited; and at even higher speeds and with longer periods of centrifugation, first the small closed vesicles and then the ribosomes can be collected (Figure 4–35). All of these fractions are impure, but many of the contaminants can be removed by resuspending the pellet and repeating the centrifugation procedure several times.

Centrifugation is the first step in most fractionations, but it separates only components that differ greatly in size. A finer degree of separation can be achieved by layering the homogenate as a narrow band on top of a dilute salt solution that fills a centrifuge tube. When centrifuged, the various components in the mixture move as a series of distinct bands through the salt solution, each at a different rate, in a process called **velocity sedimentation** (Figure 4–36). For the procedure to work effectively, the bands must be protected from convective mixing, which would normally occur whenever a denser solution (for example, one containing organelles) finds itself on top of a lighter one (the salt solution). This is achieved by filling the centrifuge tube with a shallow gradient of sucrose prepared by a special mixing device; the resulting *density gradient*, with the dense end at the bottom of the tube, keeps each region of the salt solution denser than any solution above it and thereby prevents convective mixing from distorting the separation.

When sedimented through such dilute sucrose gradients, different cell components separate into distinct bands that can be collected individually. The rate at which each component sediments depends primarily on its size and shape and is normally described in terms of its *sedimentation coefficient*, or *s value*. Present-day ultracentrifuges rotate at speeds of up to 80,000 rpm and produce forces as high as 500,000 times gravity. With these enormous forces, even small macromolecules, such as tRNA molecules and simple enzymes, can be driven to sediment at an appreciable rate and so can be separated from one another on the basis of their size. Measurements of sedimentation coefficients are routinely used to help determine the size and subunit composition of the organized assemblies of macromolecules found in cells.

The ultracentrifuge is also used to separate cellular components on the basis of their *buoyant density*, independently of their size and shape. In this case the sample is usually sedimented through a steep density gradient that contains a very high concentration of sucrose or cesium chloride. Each cellular component begins to move down the gradient as in Figure 4–36, but it eventually reaches a position where the density of the solution is equal to its own density. At this point the component floats and can move no farther. A series of distinct bands is thereby produced in the centrifuge tube, with the bands closest to the bottom of the tube containing the components of highest buoyant density. This method, called **equilibrium sedimentation,** is so sensitive that it is capable of separating macromolecules that have incorporated heavy isotopes, such as ^{13}C or ^{15}N, from the same macromolecules that have not. In fact, the cesium-chloride method was

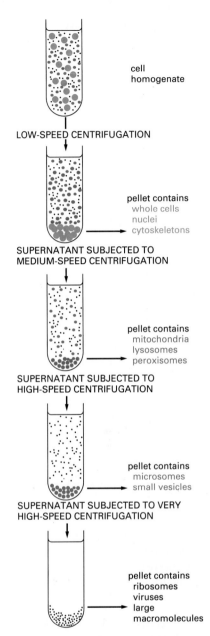

cell homogenate

LOW-SPEED CENTRIFUGATION

pellet contains
whole cells
nuclei
cytoskeletons

SUPERNATANT SUBJECTED TO
MEDIUM-SPEED CENTRIFUGATION

pellet contains
mitochondria
lysosomes
peroxisomes

SUPERNATANT SUBJECTED TO
HIGH-SPEED CENTRIFUGATION

pellet contains
microsomes
small vesicles

SUPERNATANT SUBJECTED TO VERY
HIGH-SPEED CENTRIFUGATION

pellet contains
ribosomes
viruses
large
macromolecules

Figure 4–35 Cell fractionation by centrifugation. Repeated centrifugation at progressively higher speeds will fractionate homogenates of cells into their components. In general, the smaller the subcellular component, the greater is the centrifugal force required to sediment it. Typical values for the various centrifugation steps referred to in the figure are

low speed: 1,000 times gravity
for 10 minutes
medium speed: 20,000 times gravity
for 20 minutes
high speed: 80,000 times gravity
for 1 hour
very high speed: 150,000 times gravity
for 3 hours

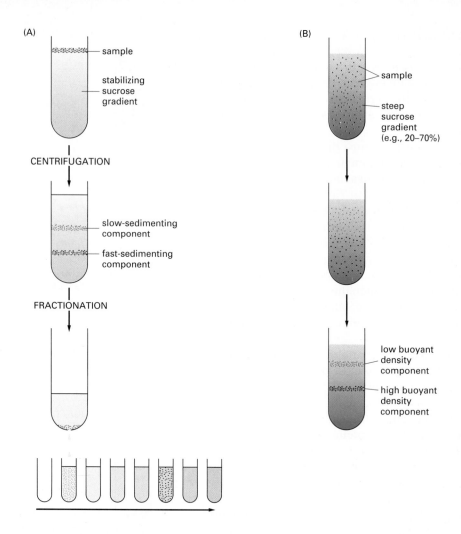

(A)

sample

stabilizing sucrose gradient

CENTRIFUGATION

slow-sedimenting component

fast-sedimenting component

FRACTIONATION

(B)

sample

steep sucrose gradient (e.g., 20–70%)

low buoyant density component

high buoyant density component

Figure 4–36 Comparison of methods of velocity sedimentation and equilibrium sedimentation. In velocity sedimentation (A) subcellular components sediment at different speeds according to their size when layered over a dilute sucrose-containing solution. In order to stabilize the sedimenting bands against convective mixing caused by small differences in temperature or solute concentration, the tube contains a continuous shallow gradient of sucrose that increases in concentration toward the bottom of the tube (typically from 5% to 20% sucrose). Following centrifugation, the different components can be collected individually, most simply by puncturing the plastic centrifuge tube and collecting drops from the bottom, as illustrated here. In equilibrium sedimentation (B) subcellular components may move up or down when centrifuged in a gradient until they reach a position where their density matches their surroundings. Although a sucrose gradient is shown here, denser gradients, which are especially useful for protein and nucleic acid separation, can be formed from cesium chloride. The final bands, at equilibrium, can be collected as in (A).

developed in 1957 to separate the labeled from the unlabeled DNA produced after exposure of a growing population of bacteria to nucleotide precursors containing ^{15}N; this classic experiment provided direct evidence for the semiconservative replication of DNA.

The Molecular Details of Complex Cellular Processes Can Be Deciphered in Cell-free Systems [22]

Studies of organelles and other large subcellular components isolated in the ultracentrifuge have played an important part in determining the *functions* of different components of the cell. Experiments on mitochondria and chloroplasts purified by centrifugation, for example, demonstrated the central function of these organelles in energy interconversions. Similarly, resealed vesicles formed from fragments of rough and smooth endoplasmic reticulum (microsomes) have been separated from each other and analyzed as functional models of these compartments of the intact cell. An extension of this approach makes it possible to study other biological processes free from all of the complex side reactions that occur in a cell and without being constrained by the need to keep the cell as a whole alive. Fractionated cell homogenates that maintain a biological function (called **cell-free systems**) are widely used in this way.

An early triumph was the elucidation of the mechanisms of protein synthesis. The starting point was a crude cell homogenate that could translate RNA molecules into protein. Fractionation of this homogenate, step by step, produced in turn the ribosomes, tRNA, and various enzymes that together constitute the protein-synthetic machinery. Once individual pure components were available, they could be added separately or withheld and their exact role in the process defined. The same *in vitro translation system* was later used to decipher the genetic code by using synthetic polyribonucleotides of known sequence as the "messenger RNA" (mRNA) to be translated. Today, *in vitro* translation systems are used to determine how newly made proteins are sorted into various intracellular compartments, as well as to identify the proteins encoded by purified preparations of mRNA—an important step in gene-cloning procedures. Some landmarks in the development of methods for the preparation of fractionated cell homogenates are outlined in Table 4–6.

Much of what we know about the molecular biology of the cell has been discovered by studying cell-free systems. As a few of many examples, they have been used to analyze the molecular details of DNA replication and transcription, RNA splicing, muscle contraction, and particle transport along microtubules. Cell-free systems have even been used to study such complex and highly organized processes as the cell-division cycle, the separation of chromosomes on the mitotic spindle, and the vesicular-transport steps involved in the movement of proteins from the endoplasmic reticulum through the Golgi apparatus to the plasma membrane. Cell-free extracts of this kind then provide the starting material for the complete separation of all of the individual macromolecular components of the extract, especially all of its proteins. We now consider how this is achieved.

Table 4–6 Some Major Events in the Development of the Ultracentrifuge and the Preparation of Cell-free Extracts

1897 **Buchner** showed that cell-free extracts of yeast can ferment sugars to form carbon dioxide and ethanol, laying the foundations of enzymology.

1926 **Svedberg** developed the first analytical ultracentrifuge and used it to estimate the molecular weight of hemoglobin as 68,000.

1935 **Pickels and Beams** introduced several new features of centrifuge design that led to its use as a preparative instrument.

1938 **Behrens** employed differential centrifugation to separate nuclei and cytoplasm from liver cells, a technique further developed for the fractionation of cell organelles by **Claude, Brachet, Hogeboom,** and others in the 1940s and early 1950s.

1939 **Hill** showed that isolated chloroplasts, when illuminated, could perform the reactions of photosynthesis.

1949 **Szent-Györgyi** showed that isolated myofibrils from skeletal muscle cells contract upon the addition of ATP. In 1955 a similar cell-free system was developed for ciliary beating by **Hofmann-Berling.**

1951 **Brakke** used density-gradient centrifugation in sucrose solutions to purify a plant virus.

1954 **de Duve** isolated lysosomes and, later, peroxisomes by centrifugation.

1954 **Zamecnik** and colleagues developed the first cell-free system to carry out protein synthesis. This was followed by a decade of intense research activity, during which the genetic code was elucidated.

1957 **Meselson, Stahl, and Vinograd** developed equilibrium density-gradient centrifugation in cesium chloride solutions for separating nucleic acids.

1975 **Dobberstein and Blobel** demonstrated protein translocation across membranes in a cell-free system.

1976 **Neher and Sakmann** developed patch-clamp recording to measure the activity of single ion channels.

1983 **Lohka and Masui** made concentrated extracts from frog eggs that perform the entire cell cycle *in vitro.*

1984 **Rothman** and colleagues reconstituted Golgi vesicle trafficking *in vitro* using a cell-free system.

Fractionation of Cells and Analysis of Their Molecules

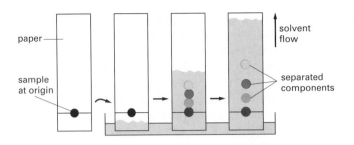

Figure 4–37 The separation of small molecules by paper chromatography. After the sample has been applied to one end of the paper (the "origin") and dried, a solution containing a mixture of two or more solvents is allowed to flow slowly through the paper by capillary action. Different components in the sample move at different rates in the paper according to their relative solubility in the solvent that is preferentially adsorbed onto the fibers of the paper. The development of this technique revolutionized biochemical analyses in the 1940s.

Proteins Can Be Separated by Chromatography [23]

One of the most generally useful methods for protein fractionation is **chromatography,** a technique originally developed to separate small molecules such as sugars and amino acids. A common type of chromatography, still widely used to separate small molecules, is *partition chromatography*. Typically, a drop of the sample is applied as a spot to a sheet of absorbent material, which may be paper (*paper chromatography*) or a sheet of plastic or glass covered with a thin layer of inert absorbent material, such as cellulose or silica gel (*thin-layer chromatography*). A mixture of solvents, such as water and an alcohol, is allowed to permeate the sheet from one edge; as the liquid moves across the sheet, the molecules in the sample become separated according to their relative solubilities in the two solvents. To achieve this, the solvents are selected so that one of them is held more strongly than the other by the absorbent material and forms a stationary solvent layer adsorbed on its surface. In each region of the sheet molecules equilibrate between the stationary and moving solvents: those that are most soluble in the strongly adsorbed solvent are relatively retarded because they spend more time in the stationary layer, while those that are most soluble in the other solvent move more quickly. After a number of hours the sheet is dried and stained to determine the location of the various molecules (Figure 4–37).

Proteins are most often fractionated by *column chromatography*, in which a mixture of proteins in solution is passed through a column containing a porous solid matrix. The different proteins are retarded to different extents by their interaction with the matrix, and they can be collected separately as they flow out

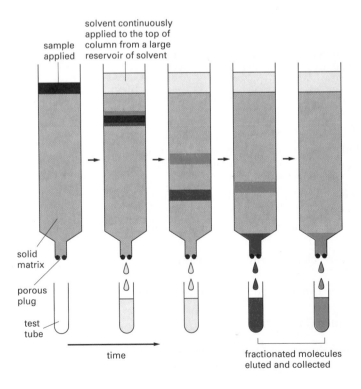

Figure 4–38 The separation of molecules by column chromatography. The sample is applied to the top of a cylindrical glass or plastic column filled with a permeable solid matrix, such as cellulose, immersed in solvent. Then a large amount of solvent is pumped slowly through the column and is collected in separate tubes as it emerges from the bottom. Various components of the sample travel at different rates through the column and are thereby fractionated into different tubes.

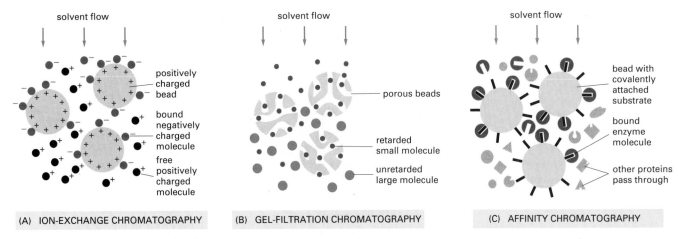

solvent flow

positively charged bead

bound negatively charged molecule

free positively charged molecule

(A) ION-EXCHANGE CHROMATOGRAPHY

solvent flow

porous beads

retarded small molecule

unretarded large molecule

(B) GEL-FILTRATION CHROMATOGRAPHY

solvent flow

bead with covalently attached substrate

bound enzyme molecule

other proteins pass through

(C) AFFINITY CHROMATOGRAPHY

Figure 4–39 Three kinds of matrix used for chromatography. In ion-exchange chromatography (A) the insoluble matrix carries ionic charges that retard molecules of opposite charge. Matrices commonly used for separating proteins are diethylaminoethylcellulose (DEAE-cellulose), which is positively charged, and carboxymethylcellulose (CM-cellulose) and phosphocellulose, which are negatively charged. The strength of the association between the dissolved molecules and the ion-exchange matrix depends on both the ionic strength and the pH of the solution that is passing down the column, which may therefore be varied in a systematic fashion (as in Figure 4–40) to achieve an effective separation. In gel-filtration chromatography (B) the matrix is inert but porous. Molecules that are small enough to penetrate into the matrix are thereby delayed and travel more slowly through the column. Beads of cross-linked polysaccharide (dextran or agarose) are available commercially in a wide range of pore sizes, making them suitable for the fractionation of molecules of various molecular weights, from less than 500 to more than 5×10^6. Affinity chromatography (C) utilizes an insoluble matrix that is covalently linked to a specific ligand, such as an antibody molecule or an enzyme substrate, that will bind a specific protein. Enzyme molecules that bind to immobilized substrates on such columns can be eluted with a concentrated solution of the free form of the substrate molecule, while molecules that bind to immobilized antibodies can be eluted by dissociating the antibody-antigen complex with concentrated salt solutions or solutions of high or low pH. High degrees of purification are often achieved in a single pass through an affinity column.

of the bottom of the column (Figure 4–38). According to the choice of matrix, proteins can be separated according to their charge (ion-exchange chromatography), their hydrophobicity (hydrophobic chromatography), their size (gel-filtration chromatography), or their ability to bind to particular chemical groups (affinity chromatography).

Many types of matrices are commercially available for these purposes (Figure 4–39). *Ion-exchange columns* are packed with small beads that carry either a positive or negative charge, so that proteins are fractionated according to the arrangement of charges on their surface. *Hydrophobic columns* are packed with beads from which hydrophobic side chains protrude, so that proteins with exposed hydrophobic regions are retarded. *Gel-filtration columns*, which separate proteins according to their size, are packed with tiny porous beads: molecules that are small enough to enter the pores linger inside successive beads as they pass down the column, while larger molecules remain in the solution flowing between the beads and therefore move more rapidly, emerging from the column first. Besides providing a means of separating molecules, gel-filtration chromatography is a convenient way to determine their size.

These types of column chromatography do not produce very highly purified fractions if one starts with a complex mixture of proteins: a single passage through the column generally increases the proportion of a given protein in the mixture by no more than twentyfold. Since most individual proteins represent less than 1/1000 of the total cellular protein, it is usually necessary to use several different types of column in succession to attain sufficient purity (Figure 4–40). A more efficient procedure, known as **affinity chromatography,** takes advantage of the biologically important binding interactions that occur on protein surfaces. If an enzyme substrate is covalently coupled to an inert matrix such as a polysaccharide bead, for example, the enzyme that operates on that substrate will often

Fractionation of Cells and Analysis of Their Molecules

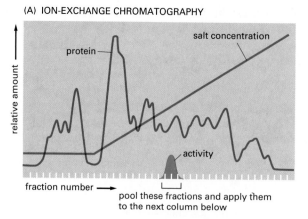

(A) ION-EXCHANGE CHROMATOGRAPHY

salt concentration

protein

relative amount

activity

fraction number →

pool these fractions and apply them
to the next column below

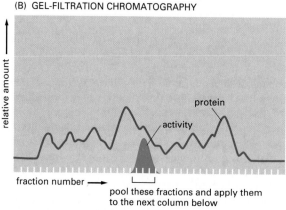

(B) GEL-FILTRATION CHROMATOGRAPHY

relative amount

protein

activity

fraction number →

pool these fractions and apply them
to the next column below

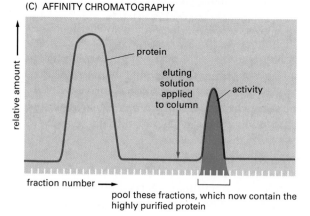

(C) AFFINITY CHROMATOGRAPHY

relative amount

protein

eluting
solution
applied
to column

activity

fraction number →

pool these fractions, which now contain the
highly purified protein

Figure 4–40 Protein purification by chromatography. Typical results obtained when three different chromatographic steps are used in succession to purify a protein. In this example a homogenate of cells was first fractionated by allowing it to percolate through an ion-exchange resin packed into a column (A). The column was washed, and the bound proteins were then eluted by passing a solution containing a gradually increasing concentration of salt onto the top of the column. Proteins with the lowest affinity for the ion-exchange resin passed directly through the column and were collected in the earliest fractions eluted from the bottom of the column. The remaining proteins were eluted in sequence according to their affinity for the resin—those proteins binding most tightly to the resin requiring the highest concentration of salt to remove them. The protein of interest eluted in a narrow peak and was detected by its enzymatic activity. The fractions with activity were pooled and then applied to a second, gel-filtration column (B). The elution position of the still-impure protein was again determined by its enzymatic activity and the active fractions pooled and purified to homogeneity on an affinity column (C) that contained an immobilized substrate of the enzyme.

be specifically retained by the matrix and can then be eluted (washed out) in nearly pure form. Likewise, short DNA oligonucleotides of a specifically designed sequence can be immobilized in this way and used to purify DNA-binding proteins that normally recognize this sequence of nucleotides in chromosomes (see Figure 9–23). Alternatively, specific antibodies can be coupled to a matrix in order to purify protein molecules recognized by the antibodies. Because of the great specificity of all such *affinity columns,* 1000- to 10,000-fold purifications can sometimes be achieved in a single pass.

The resolution of conventional column chromatography is limited by inhomogeneities in the matrices (such as cellulose), which cause an uneven flow of solvent through the column. Newer chromatography resins (usually silica-based) have been developed in the form of tiny spheres (3 to 10 μm in diameter) that can be packed with a special apparatus to form a uniform column bed. A high degree of resolution is attainable on such **high-performance liquid chromatography (HPLC)** columns. Because they contain such tightly packed particles, HPLC columns have negligible flow rates unless high pressures are applied.

For this reason these columns are typically packed in steel cylinders and require an elaborate system of pumps and valves to force the solvent through them at sufficient pressure to produce the desired rapid flow rates of about one column volume per minute. In conventional column chromatography, flow rates must be kept slow (often about one column volume per hour) to give the solutes being fractionated time to equilibrate with the interior of the large matrix particles. In HPLC the solutes equilibrate very rapidly with the interior of the tiny spheres, so solutes with different affinities for the matrix are efficiently separated from one another even at fast flow rates. This allows most fractionations to be carried out in minutes, whereas hours are required to obtain a poorer separation by conventional chromatography. HPLC has therefore become the method of choice for separating many proteins and small molecules.

The Size and Subunit Composition of a Protein Can Be Determined by SDS Polyacrylamide-Gel Electrophoresis [24]

Proteins usually have a net positive or negative charge that reflects the mixture of charged amino acids they contain. If an electric field is applied to a solution containing a protein molecule, the protein will migrate at a rate that depends on its net charge and on its size and shape. This technique, known as **electrophoresis,** was originally used to separate mixtures of proteins either in free aqueous solution or in solutions held in a solid porous matrix such as starch.

In the mid-1960s a modified version of this method—which is known as **SDS polyacrylamide-gel electrophoresis (SDS-PAGE)**—was developed that has revolutionized the way proteins are routinely analyzed. It uses a highly cross-linked gel of polyacrylamide as the inert matrix through which the proteins migrate. The gel is usually prepared immediately before use by polymerization from monomers; the pore size of the gel can be adjusted so that it is small enough to retard the migration of the protein molecules of interest. The proteins themselves are not in a simple aqueous solution but in one that includes a powerful negatively charged detergent, **sodium dodecyl sulfate,** or **SDS** (Figure 4–41). Because this detergent binds to hydrophobic regions of the protein molecules, causing them to unfold into extended polypeptide chains, the individual protein molecules are released from their associations with other proteins or lipid molecules and rendered freely soluble in the detergent solution. In addition, a reducing agent such as *mercaptoethanol* (Figure 4–41) is usually added to break any S—S linkages in the proteins so that all of the constituent polypeptides in multisubunit molecules can be analyzed separately.

What happens when a mixture of SDS-solubilized proteins is electrophoresed through a slab of polyacrylamide gel? Each protein molecule binds large numbers of the negatively charged detergent molecules, which overwhelm the protein's intrinsic charge and cause it to migrate toward the positive electrode when a voltage is applied. Proteins of the same size tend to behave identically because (1) their native structure is completely unfolded by the SDS, so that their shapes are the same, and (2) they bind the same amount of SDS and therefore have the same amount of negative charge. Larger proteins, with more charge, will be subjected to larger electrical forces and also to a larger drag. In free solution the two effects would cancel out, but in the meshes of the polyacrylamide gel, which acts as a molecular sieve, large proteins are retarded much more severely than small ones. As a result, a complex mixture of proteins is fractionated into a series of discrete protein bands arranged in order of molecular weight (Figure 4–42). The major proteins are readily detected by staining the gel with a dye such as Coomassie blue, and even minor proteins are seen in gels treated with a silver stain (where as little as 10 ng of protein can be detected in a band).

SDS polyacrylamide-gel electrophoresis is a more powerful procedure than any previous method of protein analysis principally because it can be used to separate all types of proteins, including those that are insoluble in water. Membrane proteins, protein components of the cytoskeleton, and proteins that are

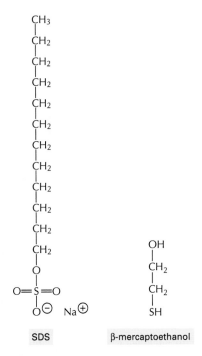

Figure 4–41 The detergent sodium dodecyl sulfate (SDS) and the reducing agent β-mercaptoethanol. These two chemicals are used to solubilize proteins for SDS polyacrylamide-gel electrophoresis. The SDS is shown here in its ionized form.

(A)

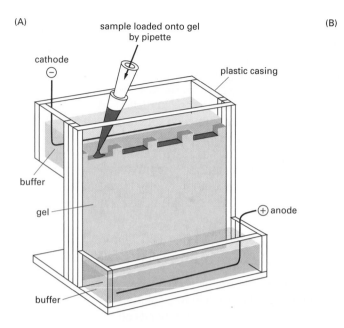

(B)

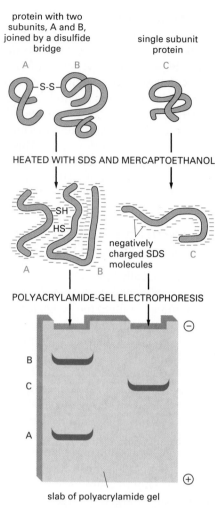

Figure 4–42 SDS polyacrylamide-gel electrophoresis (SDS-PAGE). (A) Apparatus. (B) Individual polypeptide chains form a complex with negatively charged molecules of sodium dodecyl sulfate (SDS) and therefore migrate as a negatively charged SDS-protein complex through a porous gel of polyacrylamide. Since the speed of migration under these conditions is greater the smaller the polypeptide, this technique can be used to determine the approximate molecular weight of a polypeptide chain as well as the subunit composition of a complex protein. If the protein contains a large amount of carbohydrate, however, it will move anomalously on the gel and its apparent molecular weight estimated by SDS-PAGE will be misleading.

part of large macromolecular aggregates can all be resolved. Since the method separates polypeptides according to size, it also provides information about the molecular weight and the subunit composition of any protein complex. A photograph of a gel that has been used to analyze each of the successive stages in the purification of a protein is shown in Figure 4–43.

More Than 1000 Proteins Can Be Resolved on a Single Gel by Two-dimensional Polyacrylamide-Gel Electrophoresis [25]

Since closely spaced protein bands or peaks tend to overlap, one-dimensional separation methods, such as SDS polyacrylamide-gel electrophoresis or chromatography, can resolve only a relatively small number of proteins (generally fewer than 50). In contrast, **two-dimensional gel electrophoresis,** which combines two different separation procedures, can be used to resolve more than 1000 proteins in the form of a two-dimensional protein map.

Figure 4–43 Analysis of protein samples by SDS polyacrylamide-gel electrophoresis. The photograph shows a gel that has been used to detect the proteins present at successive stages in the purification of an enzyme. The leftmost lane (lane 1) contains the complex mixture of proteins in the starting cell extract, and each succeeding lane analyzes the proteins obtained after a chromatographic fractionation of the protein sample analyzed in the previous lane (see Figure 4–40). The same total amount of protein (10 μg) was loaded onto the gel at the top of each lane. Individual proteins normally appear as sharp, dye-stained bands; a band broadens, however, when it contains too much protein. (Courtesy of Tim Formosa.)

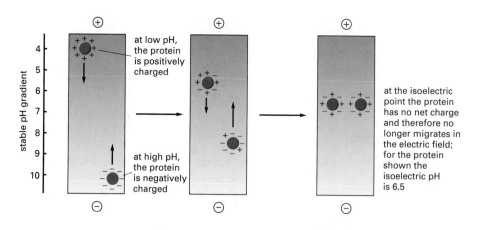

In the first step the proteins are separated on the basis of their intrinsic charge. The sample is dissolved in a small volume of a solution containing a nonionic (uncharged) detergent, together with mercaptoethanol and the denaturing reagent urea. This solution solubilizes, denatures, and dissociates all the polypeptide chains but leaves their intrinsic charge unchanged. The polypeptide chains are then separated by a procedure called **isoelectric focusing,** which depends on the fact that the net charge on a protein molecule varies with the pH of the surrounding solution. For any protein there is a characteristic pH, called its *isoelectric point,* at which the protein has no net charge and therefore will not migrate in an electric field. In isoelectric focusing, proteins are electrophoresed in a narrow tube of polyacrylamide gel in which a gradient of pH is established by a mixture of special buffers. Each protein moves to a position in the gradient that corresponds to its isoelectric point and stays there (Figure 4–44). This is the first dimension of two-dimensional gel electrophoresis.

In the second step the narrow gel containing the separated proteins is again subjected to electrophoresis but in a direction at right angles to that used in the first step. This time SDS is added, and the proteins are separated according to their size, as in one-dimensional SDS-PAGE: the original narrow gel is soaked in SDS and then placed on one edge of an SDS polyacrylamide-gel slab, through which each polypeptide chain migrates to form a discrete spot. This is the second dimension of two-dimensional polyacrylamide-gel electrophoresis. The only proteins left unresolved will be those that have both an identical size and an identical isoelectric point, a relatively rare situation. Even trace amounts of each

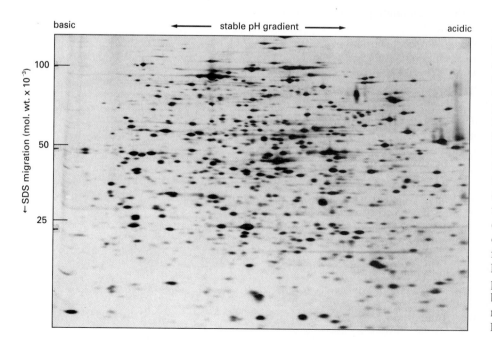

Figure 4–45 **Two-dimensional polyacrylamide-gel electrophoresis.** All the proteins in an *E. coli* bacterial cell are separated in this gel, in which each spot corresponds to a different polypeptide chain. The proteins were first separated according to their isoelectric points by isoelectric focusing from left to right. They were then further fractionated according to their molecular weights by electrophoresis from top to bottom in the presence of SDS. Note that different proteins are present in very different amounts. The bacteria were fed with a mixture of radioisotope-labeled amino acids so that all of their proteins were radioactive and could be readily detected by autoradiography (see p. 180). (Courtesy of Patrick O'Farrell.)

Fractionation of Cells and Analysis of Their Molecules

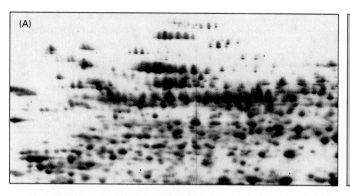

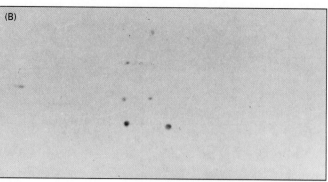

polypeptide chain can be detected on the gel by various staining procedures—or by autoradiography if the protein sample was initially labeled with a radioisotope. Up to 2000 individual polypeptide chains—almost the number of different proteins in a bacterium—have been resolved on a single two-dimensional gel (Figure 4–45). The resolving power is so great that two proteins that differ in only a single charged amino acid can be readily distinguished.

A specific protein can be identified after its fractionation on either one-dimensional or two-dimensional gels by exposing all the proteins present to a specific antibody that has been coupled to a radioactive isotope, to an easily detectable enzyme, or to a fluorescent dye. For convenience, this is normally done after all the separated proteins present in the gel have been transferred (by "blotting") onto a sheet of nitrocellulose paper, as will be described later for nucleic acids (see Figure 7–13). This protein-detection method is called *Western blotting* (Figure 4–46).

Figure 4–46 Western blotting or immunoblotting. The total proteins from dividing tobacco cells in culture are first separated by two-dimensional polyacrylamide-gel electrophoresis as shown in Figure 4–45 and their positions revealed by a sensitive protein stain (A). The separated proteins on an identical gel were then transferred to a sheet of nitrocellulose and incubated with an antibody that recognizes those proteins that, during mitosis, are phosphorylated on threonine residues. The positions of the dozen or so proteins that are recognized by this antibody are revealed by an enzyme-linked second antibody (B). (From J.A. Traas, A.F. Bevan, J.H. Doonan, J. Cordewener, and P.J. Shaw. *Plant Journal* 2:723–732, 1992, by permission of Blackwell Scientific Publications.)

Table 4–7 Landmarks in the Development of Chromatography and Electrophoresis and Their Application to Protein Molecules

1833 **Faraday** described the fundamental laws concerning the passage of electricity through ionic solutions.

1850 **Runge** separated inorganic chemicals by their differential adsorption to paper, a forerunner of later chromatographic separations.

1906 **Tswett** invented column chromatography, passing petroleum extracts of plant leaves through columns of powdered chalk.

1933 **Tiselius** introduced electrophoresis for separating proteins in solution.

1942 **Martin and Synge** developed partition chromatography, leading to paper chromatography two years later.

1946 **Stein and Moore** determined for the first time the amino acid composition of a protein, initially using column chromatography on starch and later developing chromatography on ion-exchange resins.

1955 **Smithies** used gels made of starch to separate serum proteins by electrophoresis.

Sanger completed the analysis of the amino acid sequence of bovine insulin, the first protein to be sequenced.

1956 **Ingram** produced the first protein fingerprints, showing that the difference between sickle-cell hemoglobin and normal hemoglobin is due to a change in a single amino acid.

1959 **Raymond** introduced polyacrylamide gels, which are superior to starch gels for separating proteins by electrophoresis; improved buffer systems allowing high-resolution separations were developed in the next few years by **Ornstein and Davis.**

1966 **Maizel** introduced the use of sodium dodecyl sulfate (SDS) for improving polyacrylamide-gel electrophoresis of proteins.

1975 **O'Farrell** devised a two-dimensional gel system for analyzing protein mixtures in which SDS polyacrylamide-gel electrophoresis is combined with separation according to isoelectric point.

Table 4–8 Some Reagents Commonly Used to Cleave Peptide Bonds in Proteins

	Amino Acid 1	Amino Acid 2
Enzyme		
Trypsin	Lys or Arg	any
Chymotrypsin	Phe, Trp, or Tyr	any
V8 protease	Glu	any
Chemical		
Cyanogen bromide	Met	any
2-Nitro-5-thiocyanobenzoate	any	Cys

The specificity for the amino acids on either side of the cleaved bond is indicated. The carboxyl group of amino acid 1 is released by the cleavage; this amino acid is to the left of the peptide bond as normally written (see Panel 2–5, pp. 56–57).

Some landmarks in the development of chromatography and electrophoresis are outlined in Table 4–7.

Selective Cleavage of a Protein Generates a Distinctive Set of Peptide Fragments [26]

Although the molecular weight and isoelectric point are distinctive features of a protein, unambiguous identification ultimately depends on determining the amino acid sequence. The first stage of this process, which involves cleaving the protein into smaller fragments, can itself provide information that helps to characterize the molecule. Proteolytic enzymes and chemical reagents are available that will cleave proteins between specific amino acid residues (Table 4–8). The enzyme *trypsin*, for instance, cuts on the carboxyl side of lysine or arginine residues, whereas the chemical *cyanogen bromide* cuts peptide bonds next to methionine residues. Since these enzymes and chemicals cleave at relatively few sites in a protein, they tend to produce relatively large and relatively few peptides. If such a mixture of peptides is separated by chromatographic or electrophoretic procedures, the resulting pattern, or **peptide map,** is diagnostic of the protein from which the peptides were generated and is sometimes referred to as the protein's "fingerprint" (Figure 4–47).

Protein fingerprinting was developed in 1956 as a way of comparing normal hemoglobin with the mutant form of the protein found in patients suffering from *sickle-cell anemia.* A single peptide difference was found and eventually traced to a single amino acid change, providing the first demonstration that a mutation can change a single amino acid in a protein.

Short Amino Acid Sequences Can Be Analyzed by Automated Machines [27]

Once a protein has been cleaved into smaller peptides, the next logical step in the analysis is to determine the amino acid sequence of each isolated peptide fragment. This is accomplished by a repeated series of chemical reactions originally devised in 1967. First the peptide is exposed to a chemical that forms a covalent bond only with the free amino group at the amino terminus of the peptide. This chemical is then further activated by exposure to a weak acid so that it specifically cleaves the peptide bond that attaches the amino-terminal amino acid to the peptide chain; the released amino acid is then identified by chromatographic methods. The remaining peptide, which is shorter by one amino acid, is then submitted to the same sequence of reactions, and so on, until every amino acid in the peptide has been determined.

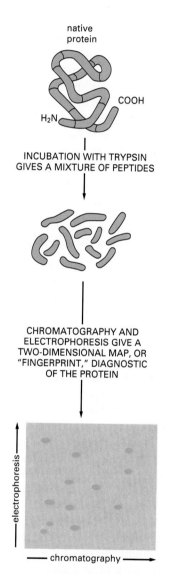

native protein

H₂N COOH

INCUBATION WITH TRYPSIN GIVES A MIXTURE OF PEPTIDES

CHROMATOGRAPHY AND ELECTROPHORESIS GIVE A TWO-DIMENSIONAL MAP, OR "FINGERPRINT," DIAGNOSTIC OF THE PROTEIN

← electrophoresis →

← chromatography →

Figure 4–47 Production of a peptide map, or fingerprint, of a protein. Here, the protein was digested with trypsin to generate a mixture of polypeptide fragments, which was then fractionated in two dimensions by electrophoresis and partition chromatography. The pattern of spots obtained is diagnostic of the protein analyzed.

Fractionation of Cells and Analysis of Their Molecules

The reiterative nature of these reactions lends itself to automation, and machines called **amino acid sequenators** are commercially available for automatic determination of the amino acid sequence of peptide fragments. The final step is to arrange the sequences of the various peptide fragments in the order in which they occur in the intact polypeptide chain. This was traditionally achieved by comparing the sequences of different sets of overlapping peptide fragments obtained by cleaving the same protein with different proteolytic enzymes.

Improvements in protein-sequencing technology have greatly increased its speed and sensitivity, allowing analysis of minute samples; the sequence of several dozen amino acids at the amino-terminal end of a peptide can be obtained overnight from a few micrograms of protein—the amount available from a single band on an SDS polyacrylamide gel. This has been important for characterizing many minor cell proteins, such as the receptors for steroid and polypeptide hormones. Knowing the sequence of as few as 20 amino acids of a protein is frequently enough to allow a DNA probe to be designed so that the gene encoding the protein can be cloned (Figure 7–27). Once the gene has been isolated, the rest of the protein's amino acid sequence can be deduced from the DNA sequence by reference to the genetic code. This is a major advantage because, even with automation, the direct determination of the entire amino acid sequence of a protein is a major undertaking. A protein of 100 residues can often be sequenced in a month of hard work. But the difficulty increases steeply with the length of the polypeptide chain, and the chemical peculiarities of individual peptide fragments prevent the process from being routine. Since DNA sequencing can be done so quickly and simply (see Chapter 7), the sequences of most proteins are now determined largely from the nucleotide sequences of their genes.

The Diffraction of X-rays by Protein Crystals Can Reveal a Protein's Exact Structure [28]

From the amino acid sequence of a protein, one can often predict secondary structural elements in the protein, such as membrane-spanning α helices, as well as the protein's resemblance to other known proteins. It is presently not possible, however, to deduce reliably the three-dimensional folded structure of a protein from its amino acid sequence, and without knowing its detailed folded structure, it is not possible to understand the molecular basis of a protein's function. The main technique that has been used to discover the three-dimensional structure of molecules, including proteins, at atomic resolution is **x-ray crystallography.**

X-rays, like light, are a form of electromagnetic radiation, but they have a much smaller wavelength, typically around 0.1 nm (the diameter of a hydrogen atom). If a narrow parallel beam of x-rays is directed at a sample of a pure protein, most of the x-rays will pass straight through it. A small fraction, however, will be scattered by the atoms in the sample. If the sample is a well-ordered crystal, the scattered waves will reinforce one another at certain points (see Figure 4–2) and will appear as diffraction spots when the x-rays are recorded by a suitable detector (Figure 4–48).

The position and intensity of each spot in the **x-ray diffraction pattern** contain information about the positions of the atoms in the crystal that gave rise to it. Deducing the three-dimensional structure of a large molecule from the diffraction pattern of its crystal is a complex task and was not achieved for a protein molecule until 1960. In recent years x-ray diffraction analysis has become increasingly automated, and now the slowest step is likely to be the production of suitable protein crystals. This requires large amounts of very pure protein and often involves years of trial-and-error searching for the proper crystallization conditions. There are still many proteins, especially membrane proteins, that have so far resisted all attempts to crystallize them.

The immediate product of a diffraction-pattern analysis is a complex three-dimensional electron-density map. It is then a complicated matter to interpret

(A)

Figure 4–48 X-ray crystallography. (A) Protein crystal of ribulose bisphosphate carboxylase, an enzyme that plays a central role in CO_2 fixation during photosynthesis. (B) X-ray diffraction pattern obtained from the crystal. (C) Simplified model of the protein structure derived from the x-ray diffraction data. The complete atomic model is hard to interpret, but this version shows its structural features clearly (α helices, *green*; β strands, *red*). (A, courtesy of C. Branden; B, courtesy of J. Hajdu and I. Andersson; C, adapted from original provided by B. Furugren.)

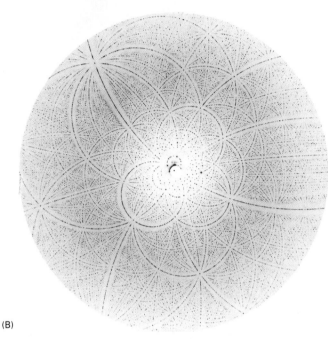

(B)

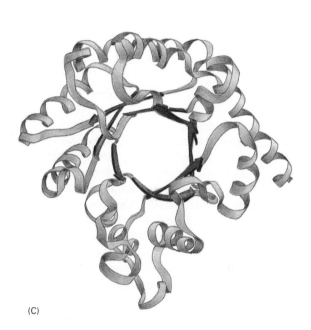

(C)

this map, and to do so requires the amino acid sequence of the protein. Largely by trial and error, the sequence and the electron-density map are correlated by computer to give the best possible fit. The reliability of the final atomic model will depend on the resolution of the original crystallographic data: 0.5 nm resolution might produce a low-resolution map of the polypeptide backbone, whereas a resolution of 0.15 nm allows all of the non-hydrogen atoms in the molecule to be reliably positioned. A complete atomic model is often too complex to appreciate directly, but simplified versions that show the essential structural features of the structure can be readily derived from it (see Figure 3–34). The structures of several hundred proteins have been determined by x-ray crystallography— enough to begin to see families of common structures emerging. These structures have often been more conserved in evolution than the amino acid sequences that form them.

Molecular Structure Can Also Be Determined Using Nuclear Magnetic Resonance (NMR) Spectroscopy [29]

Nuclear magnetic resonance (NMR) spectroscopy has been used in the past to analyze the structure of small molecules and now is increasingly used to study the structure of small proteins or protein domains. Unlike x-ray crystallography, NMR does not depend on having a crystalline sample; it simply requires a small volume of concentrated protein solution that is placed in a strong magnetic field.

Certain atomic nuclei, and in particular those of hydrogen, have a magnetic moment or *spin:* that is, they have an intrinsic magnetization, like a bar magnet.

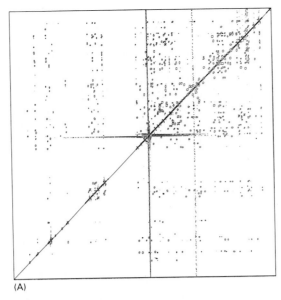

(A)

(B)

The spin aligns along the strong magnetic field, but it can be changed to a mis-aligned excited state in response to applied radiofrequency (RF) pulses of electromagnetic radiation. When the excited hydrogen nuclei relax to their aligned state, they emit RF radiation, which can be measured and displayed as a spectrum. The nature of the emitted radiation depends on the environment of each hydrogen nucleus, and if one nucleus is excited, it will influence the absorption and emission of radiation by other nuclei that lie close to it. It is consequently possible, by an ingenious elaboration of the basic NMR technique known as *two-dimensional NMR*, to distinguish the signals from hydrogen nuclei in different amino acid residues and to identify and measure the small shifts in these signals that occur when these hydrogen nuclei lie close enough together to interact: the size of such a shift reveals the distance between the interacting pair of hydrogen atoms. In this way NMR can give information about the distances between the parts of the protein molecule. By combining this information with a knowledge of the amino acid sequence, it is possible in principle to compute the three-dimensional structure of the protein (Figure 4–49).

For technical reasons only the structure of small proteins of about 15,000–20,000 daltons or less can currently be determined by NMR spectroscopy, and it is very unlikely that the method will ever be able to tackle molecules larger than about 30,000–40,000 daltons. Many functional domains of proteins are much smaller than this, however, and can often be obtained on their own as stable structures amenable to analysis by NMR. This is especially useful when the protein has resisted attempts at crystallization. The structures of the DNA-binding domains of various gene regulatory proteins, for example, were first determined in this way. NMR is also used widely to investigate molecules other than proteins and is valuable, for example, as a method to discover the structures of the complex carbohydrate side chains of glycoproteins.

Some landmarks in the development of x-ray crystallography and NMR are outlined in Table 4–9.

Figure 4–49 **NMR spectroscopy.** (A) An example of the data from an NMR machine. This is a two-dimensional NMR spectrum derived from the carboxyl-terminal domain of the enzyme cellulase. The spots represent interactions between hydrogen atoms that are near neighbors in the protein and hence their distance apart. Complex computing methods, in conjunction with the known amino acid sequence, enable possible compatible structures to be derived. In (B) 10 structures, which all satisfy the distance constraints equally well, are shown superimposed on one another, giving a good indication of the probable three-dimensional structure. (Courtesy of P. Kraulis.)

Summary

Populations of cells can be analyzed biochemically by disrupting them and fractionating their contents by ultracentrifugation. Further fractionations allow functional cell-free systems to be developed; such systems are required to determine the molecular details of complex cellular processes. Protein synthesis, DNA replication, RNA splicing, the cell cycle, mitosis, and various types of intracellular transport are all currently being studied in this way. The molecular weight and subunit composition of even very small amounts of a protein can be determined by SDS polyacrylamide-gel electrophoresis. In two-dimensional gel electrophoresis proteins are resolved as

Table 4–9 Landmarks in the Development of X-ray Crystallography and NMR and Their Application to Biological Molecules

1864 **Hoppe-Seyler** crystallized, and named, the protein hemoglobin.

1895 **Röntgen** observed that a new form of penetrating radiation, which he named x-rays, was produced when cathode rays (electrons) hit a metal target.

1912 **Von Laue** obtained the first x-ray diffraction patterns by passing x-rays through a crystal of zinc sulfide.

  W.L. Bragg proposed a simple relationship between an x-ray diffraction pattern and the arrangement of atoms in a crystal that produced the pattern.

1926 **Summer** obtained crystals of the enzyme urease from extracts of jack beans and demonstrated that proteins possess catalytic activity.

1931 **Pauling** published his first essays on "The Nature of the Chemical Bond," detailing the rules of covalent bonding.

1934 **Bernal and Crowfoot** presented the first detailed x-ray diffraction patterns of a protein obtained from crystals of the enzyme pepsin.

1935 **Patterson** developed an analytical method for determining interatomic spacings from x-ray data.

1941 **Astbury** obtained the first x-ray diffraction pattern of DNA.

1951 **Pauling and Corey** proposed the structure of a helical conformation of a chain of L-amino acids—the α helix—and the structure of the β sheet, both of which were later found in many proteins.

1953 **Watson and Crick** proposed the double-helix model of DNA, based on x-ray diffraction patterns obtained by **Franklin and Wilkins.**

1954 **Perutz** and colleagues developed heavy-atom methods to solve the phase problem in protein crystallography.

1960 **Kendrew** described the first detailed structure of a protein (sperm whale myoglobin) to a resolution of 0.2 nm, and **Perutz** proposed a lower-resolution structure of the larger protein hemoglobin.

1966 **Phillips** described the structure of lysozyme, the first enzyme to be analyzed in detail.

1971 **Jeener** proposed the use of two-dimensional NMR, and **Wuthrich** and colleagues first used the method to solve a protein structure in the early 1980s.

1976 **Kim and Rich** and **Klug** and colleagues described the detailed three-dimensional structure of tRNA determined by x-ray diffraction.

1977–1978 **Holmes and Klug** determined the structure of tobacco mosaic virus (TMV), and **Harrison** and **Rossman** determined the structure of two small spherical viruses.

1985 **Michel** and colleagues determined the first structure of a transmembrane protein (a bacterial photosynthetic reaction center) by x-ray crystallography. **Henderson** and colleagues obtained the structure of bacteriorhodopsin, a transmembrane protein, by electron-microscopy methods between 1975 and 1990.

---

separate spots by isoelectric focusing in one dimension, followed by SDS polyacrylamide-gel electrophoresis in a second dimension. These electrophoretic separations can be applied even to proteins that are normally insoluble in water.

The major proteins in soluble cell extracts can be purified by column chromatography; depending on the type of column matrix, biologically active proteins can be separated according to their molecular weight, hydrophobicity, charge characteristics, or affinity for other molecules. In a typical purification the sample is passed through several different columns in turn—the enriched fractions obtained from one column being applied to the next. Once a protein has been purified to homogeneity, its biological activities can be examined in detail. In addition, a small part of the protein's amino acid sequence can be determined, which then allows the DNA sequence that encodes the entire protein to be cloned; the remaining amino acid sequence is then deduced from the nucleotide sequence of the cloned DNA. While the

Fractionation of Cells and Analysis of Their Molecules

amino acid sequence is required to determine the three-dimensional structure of a protein, it is not sufficient on its own. To determine this structure at atomic resolution, large proteins have to be crystallized and studied by x-ray diffraction. The structure of small proteins in solution can be determined using a nuclear magnetic resonance analysis.

Tracing and Assaying Molecules Inside Cells

The classical methods of microscopy give good views of cell architecture, but they provide little information about cell chemistry. In cell biology it is often important to determine the quantities of specific molecules and to know where they are in the cell and how their level or location changes in response to extracellular signals. The molecules of interest range from small inorganic ions, such as Ca^{2+} or H^+, to large macromolecules, such as specific proteins, RNAs, or DNA sequences.

Sensitive methods have been developed for assaying each of these types of molecules, as well as for following the dynamic behavior of many of them in living cells. In this section we describe how specific probes can be introduced into living cells in order to monitor the chemical conditions in the cytosol. In addition, two other detection methods that are widely used in cell biology are discussed: those involving *radioisotopes* and those utilizing *antibodies*. Both of these methods are capable of detecting a specific molecule in a complex mixture with great sensitivity: under optimal conditions, they can detect fewer than 1000 copies of a molecule in a sample.

Radioactive Atoms Can Be Detected with Great Sensitivity [30]

Most naturally occurring elements are a mixture of slightly different *isotopes.* These differ from one another in the mass of their atomic nuclei, but because they have the same number of electrons, they have the same chemical properties. In radioactive isotopes, or radioisotopes, the nucleus is unstable and undergoes random disintegration to produce a different atom. In the course of these disintegrations, either energetic subatomic particles, such as electrons, or radiations, such as γ-rays, are given off. By using chemical synthesis to incorporate one or more radioactive atoms into a small molecule of interest, such as a sugar or an amino acid, the fate of that molecule can be traced during any biological reaction.

Although naturally occurring radioisotopes are rare (because of their instability), they can be produced in large amounts in nuclear reactors, where stable atoms are bombarded with high-energy particles. As a result, radioisotopes of many biologically important elements are readily available (Table 4–10). The radiation they emit is detected in various ways. Electrons (β particles) can be detected in a *Geiger counter* by the ionization they produce in a gas, or they can be measured in a *scintillation counter* by the small flashes of light they induce in a scintillation fluid. These methods make it possible to measure accurately the quantity of a particular radioisotope present in a biological specimen. It is also possible to determine the location of a radioisotope in a specimen by *autoradiography,* as we describe below. All of these methods of detection are extremely sensitive: in favorable circumstances, nearly every disintegration—and therefore every radioactive atom that decays—can be detected.

Radioisotopes Are Used to Trace Molecules in Cells and Organisms [31]

One of the earliest uses of radioactivity in biology was to trace the chemical pathway of carbon during photosynthesis. Unicellular green algae were maintained

Table 4–10 Some Radioisotopes in Common Use in Biological Research

Isotope	Half-life
^{32}P	14 days
^{131}I	8.1 days
^{35}S	87 days
^{14}C	5570 years
^{45}Ca	164 days
^3H	12.3 years

The isotopes are arranged in decreasing order of the energy of the β radiation (electrons) they emit. ^{131}I also emits γ radiation. The *half-life* is the time required for 50% of the atoms of an isotope to disintegrate.

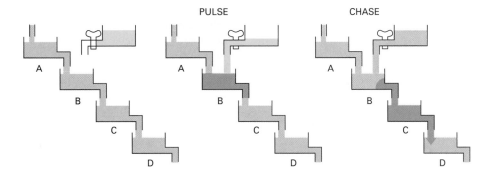

PULSE   CHASE

Figure 4–50 The logic of a typical pulse-chase experiment using radioisotopes. The chambers labeled A, B, C, and D represent either different compartments in the cell (detected by autoradiography or by cell-fractionation experiments) or different chemical compounds (detected by chromatography or other chemical methods). The results of a real pulse-chase experiment can be seen in Figure 4–51.

Figure 4–51 Electron-microscopic autoradiography. The results of a pulse-chase experiment in which pancreatic B cells were fed ^{3}H-leucine for 5 minutes (the pulse) followed by excess unlabeled leucine (the chase). The amino acid is largely incorporated into insulin, which is destined for secretion. After a 10-minute chase the labeled protein has moved from the rough ER to the Golgi stacks (A), where its position is revealed by the black silver grains in the photographic emulsion. After a further 45-minute chase the labeled protein is found in electron-dense secretory granules (B). The small round silver grains seen here are produced by using a special photographic developer and should not be confused with the similar-looking black dots seen with immunogold labeling methods (e.g., Figure 4–63). Experiments similar to this one were important in establishing the intracellular pathway taken by newly synthesized secretory proteins. (Courtesy of L. Orci, from *Diabetes* 31:538–565, 1982. © 1982 American Diabetes Association, Inc.)

in an atmosphere containing radioactively labeled CO_2 ($^{14}CO_2$), and at various times after they had been exposed to sunlight, their soluble contents were separated by paper chromatography. Small molecules containing ^{14}C atoms derived from CO_2 were detected by a sheet of photographic film placed over the dried paper chromatogram. In this way most of the principal components in the photosynthetic pathway from CO_2 to sugar were identified.

Radioactive molecules can be used to follow the course of almost any process in cells. In a typical experiment the cells are supplied with a precursor molecule in radioactive form. The radioactive molecules mix with the preexisting unlabeled ones; both are treated identically by the cell as they differ only in the weight of their atomic nuclei. Changes in the location or chemical form of the radioactive molecules can be followed as a function of time. The resolution of such experiments is often sharpened by using a **pulse-chase** labeling protocol, in which the radioactive material (the *pulse*) is added for only a very brief period and then washed away and replaced by nonradioactive molecules (the *chase*). Samples are taken at regular intervals, and the chemical form or location of the radioactivity is identified for each sample (Figure 4–50). Pulse-chase experiments, combined with autoradiography (see below), have been important, for example, in elucidating the pathway taken by secreted proteins from the ER to the cell exterior (Figure 4–51).

Radioisotopic labeling is a uniquely valuable way of distinguishing between molecules that are chemically identical but have different histories—for example, those that differ in their time of synthesis. In this way, for example, it was shown that almost all of the molecules in a living cell are continually being degraded and replaced, even when the cell is not growing and is apparently in a steady state. This "turnover," which sometimes takes place very slowly, would be almost impossible to detect without radioisotopes.

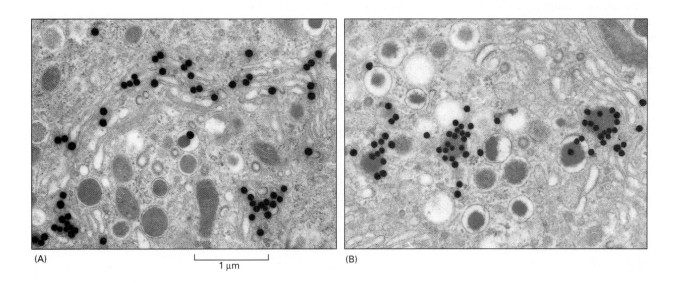

(A)   1 μm   (B)

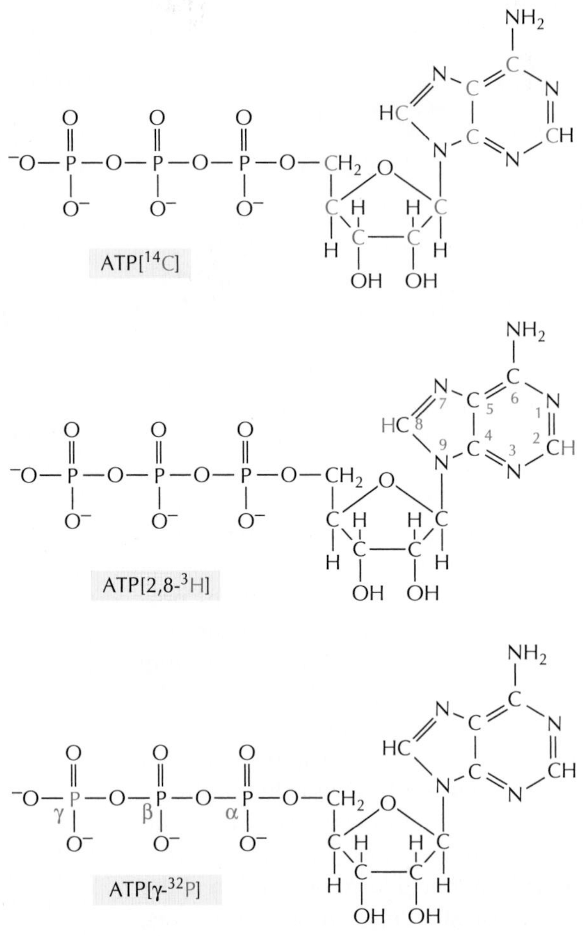

Figure 4–52 Radioisotopically labeled molecules. Three commercially available radioactive forms of ATP, with the radioactive atoms shown in *red*. The nomenclature used to identify the position and type of the radioactive atoms is also shown.

Today, nearly all common small molecules are available in radioactive form from commercial sources, and virtually any biological molecule, no matter how complicated, can be radioactively labeled. Compounds can be made with radioactive atoms incorporated at particular positions in their structure, enabling the separate fates of different parts of the same molecule to be followed during biological reactions (Figure 4–52).

One of the important uses of radioactivity in cell biology is to localize a radioactive compound in sections of whole cells or tissues by **autoradiography**. In this procedure living cells are briefly exposed to a pulse of a specific radioactive compound and then incubated for a variable period—to allow them time to incorporate the compound—before being fixed and processed for light or electron microscopy. Each preparation is then overlaid with a thin film of photographic emulsion and left in the dark for a number of days, during which the radioisotope decays. The emulsion is then developed, and the position of the radioactivity in each cell is indicated by the position of the developed silver grains (see Figure 4–51). If cells are exposed to ^3H-*thymidine*, a radioactive precursor of DNA, for example, it can be shown that DNA is made in the nucleus and remains there. By contrast, if cells are exposed to ^3H-*uridine*, a radioactive precursor of RNA, it is found that RNA is initially made in the nucleus and then moves rapidly into the cytoplasm. Radiolabeled molecules can also be detected by autoradiography after they are separated from other molecules by gel electrophoresis: the positions of both proteins (see Figure 4–45) and nucleic acids (see Figure 7–5) are commonly detected on gels in this way. In addition, when radioactive nucleic acid molecules are used as probes to find complementary nucleic acid molecules, autoradiography serves to detect both the positions and amounts of those molecules hybridizing to the probe, either on a gel or in a tissue section (see Figure 7–20).

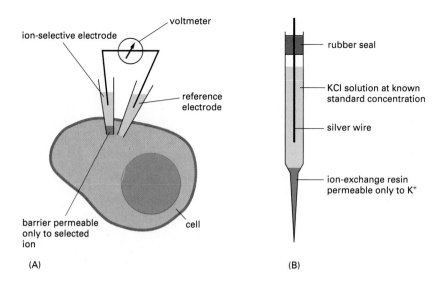

Figure 4–53 An ion-selective microelectrode used to measure the intracellular concentration of a specific ion. The experimental arrangement is shown in (A). The construction of a microelectrode selectively permeable to K^+ is shown in (B). In general, the tip of the ion-sensitive intracellular microelectrode is either constructed from special glass or filled with a special organic compound to make it permeable only to a chosen ion. The rest of the tube is filled with an aqueous solution of the ion at a known concentration, with a metal conductor connected to one terminal of a voltmeter dipping into it. The other terminal of the voltmeter is connected in a similar way to a glass reference microelectrode that has an open tip and contains a simple conducting solution. The two microelectrodes are both poked through the plasma membrane of the cell to be studied. The voltage measured by the voltmeter, which equals the potential difference across the selectively permeable barrier, reveals the concentration of the ion in the cell.

Ion Concentrations Can Be Measured with Intracellular Electrodes [32]

One way to study the chemistry of a single living cell is to insert the tip of a fine glass pipette directly into the cell interior, a technique developed by electrophysiologists to study voltages and current flows across the plasma membrane. For this purpose intracellular *microelectrodes* are made from pieces of fine glass tubing that are pulled to a tip diameter of a fraction of a micrometer and filled with a conducting solution (usually a simple salt, such as KCl, in water). The tip of the microelectrode can be poked into the cytoplasm through the plasma membrane, which seals around the shaft, adhering tightly to the glass so that the cell is left relatively undisturbed.

Microelectrodes are useful for studying the cell interior in two ways: they can be converted into probes for measuring intracellular concentrations of common inorganic ions, such as H^+, Na^+, K^+, Cl^-, Ca^{2+}, and Mg^{2+}, and they can be used as micropipettes to inject molecules into cells. The principle that allows the measurement of ion concentrations with a microelectrode is the same as that of the familiar pH meter. The tendency of ions to diffuse down their concentration gradient can be balanced by an opposing electric field: the greater the concentration gradient, the greater is the electric field required. The magnitude of the electric field required to maintain the concentration gradient at a stable equilibrium therefore gives a measure of the ion concentration gradient. To find the concentration of a specific ion, it is necessary to use a material that is permeable only to that ion, formed into a sheet or barrier, which is placed between the test solution and a solution that contains a known concentration of the ion. The electrical potential difference across the selectively permeable barrier when no current flows will then be a direct measure of the ratio of concentrations of the specific ion on the two sides. (The theory is discussed in Panel 11–2, in connection with ion transport across a cell membrane.) In practice, the tip of a microelectrode is filled with an appropriate organic compound to make a barrier that is selectively permeable to the chosen ion. The microelectrode is then inserted, together with a reference microelectrode, into the interior of a cell, as shown in Figure 4–53.

More recently, the microelectrode technique has been adapted to study the movement of ions through specialized channel proteins (called ion channels) contained in a small patch of plasma membrane. Here, a glass microelectrode with a somewhat larger tip is pressed gently against the plasma membrane instead of being poked through it (Figure 4–54). It is then possible to study the electrical behavior of the small patch of membrane covering the tip of the electrode;

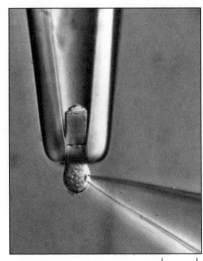

Figure 4–54 **Micropipettes used for patch-clamp recording.** A rod cell from the eye of a salamander is shown held by a suction pipette while a fine-tipped glass pipette, pressed against the cell so that the glass is sealed tightly to the plasma membrane, serves as a microelectrode. The term "clamp" is used because an electronic device is generally utilized to "clamp" the voltage across the patch so that the voltage is maintained at a fixed value. (From T.D. Lamb, H.R. Matthews, and V. Torre, *J. Physiol.* 37:315–349, 1986.)

20 μm

the patch either can be left attached to the cell or can be pulled free of it (Figure 4–55). This technique, known as patch-clamp recording, has revolutionized the study of ion channels. It is one of the few techniques in cell biology that allows one to study the function of a single protein molecule in real time, as we discuss in Chapter 11.

Rapidly Changing Intracellular Ion Concentrations Can Be Measured with Light-emitting Indicators [33]

Ion-sensitive electrodes reveal the ion concentration only at one point in a cell, and for an ion, such as Ca^{2+}, that is present at very low concentration, their responses are slow and somewhat erratic. Thus these electrodes are not ideally suited to record the rapid and transient changes in the concentration of cytosolic Ca^{2+} that play an important part in allowing cells to respond to extracellular signals. Such changes can be analyzed with the use of ion-sensitive indicators, whose light emission reflects the local concentration of the ion. Some of these indicators are luminescent (emitting light spontaneously), while others are fluorescent (emitting light on exposure to light). *Aequorin* is a luminescent protein isolated from certain marine jellyfish; it emits light in the presence of Ca^{2+} and responds to changes in Ca^{2+} concentration in the range 0.5 to 10 mM. If microinjected into an egg, for example, aequorin emits a flash of light in response to

Figure 4–55 **The four standard configurations used for patch-clamp recording.** The mouth of the glass recording pipette is first pressed against the cell membrane so that a tight seal forms (*top*). Recordings of the current entering the pipette through the patch of membrane can then be made with the patch still attached to the cell (A) or pulled free from it, exposing the cytoplasmic surface of the plasma membrane (B). Alternatively, the patch can be ruptured by gentle suction so that the interior of the electrode communicates directly with the interior of the cell (C); in this latter, "whole-cell" configuration, one can record the electrical behavior of the cell in the same way as with an intracellular electrode, with the added option that the internal chemistry of the cell can be altered by allowing substances to diffuse out of the relatively wide recording pipette into the cytoplasm. Configuration (D) is reached via configuration (C) by pulling the pipette away from the cell, thereby causing a fragment of the adjacent plasma membrane to fold back over the tip to form a seal. In (D) the exterior surface rather than the cytoplasmic surface of the membrane is exposed [compare with (B)].

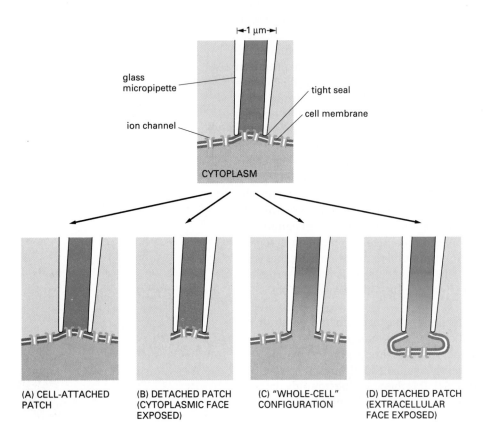

(A) CELL-ATTACHED PATCH

(B) DETACHED PATCH (CYTOPLASMIC FACE EXPOSED)

(C) "WHOLE-CELL" CONFIGURATION

(D) DETACHED PATCH (EXTRACELLULAR FACE EXPOSED)

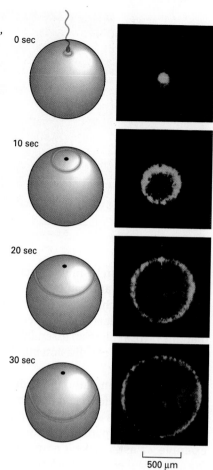

Figure 4–56 **The luminescent protein aequorin emits light in the presence of free Ca²⁺.** Here an egg of the medaka fish has been injected with aequorin, which has diffused throughout the cytosol, and the egg has then been fertilized with a sperm and examined with the help of an image intensifier. The four photographs shown were taken looking down on the site of sperm entry at intervals of 10 seconds and reveal a wave of release of free Ca²⁺ into the cytosol from internal stores just beneath the plasma membrane. This wave sweeps across the egg starting from the site of sperm entry, as indicated in the diagrams on the left. (Photographs reproduced from J.C. Gilkey, L.F. Jaffe, E.B. Ridgway, and G.T. Reynolds, *J. Cell Biol.* 76:448–466, 1978, by copyright permission of the Rockefeller University Press.)

0 sec

10 sec

20 sec

30 sec

500 μm

the sudden localized release of free Ca²⁺ into the cytoplasm that occurs when the egg is fertilized (Figure 4–56).

Fluorescent Ca²⁺ indicators have recently been synthesized that bind Ca²⁺ tightly and are excited at slightly longer wavelengths when they are free of Ca²⁺ than when in their Ca²⁺-bound form. By measuring the ratio of fluorescence intensity at two excitation wavelengths, the concentration ratio of the Ca²⁺-bound indicator to the Ca²⁺-free indicator can be determined; this provides an accurate measurement of the free Ca²⁺ concentration. Two indicators of this type, called *quin-2* and *fura-2*, are widely used for second-by-second monitoring of changes in intracellular Ca²⁺ concentrations in the different parts of a cell viewed in a fluorescence microscope (Figure 4–57). Similar fluorescent indicators are available for measuring other ions; some are used for measuring H⁺, for example, and hence intracellular pH. Some of these indicators can enter cells by diffusion and so need not be microinjected; this makes it possible to monitor large numbers of individual cells simultaneously in a fluorescence microscope. By constructing new types of indicators and using them in conjunction with modern image-processing methods, it should be possible to develop similarly rapid and precise methods for analyzing changes in the concentrations of many types of small molecules in cells.

There Are Several Ways of Introducing Membrane-impermeant Molecules into Cells [34]

It is often useful to be able to introduce membrane-impermeant molecules into a living cell, whether they are antibodies that recognize intracellular proteins, normal cell proteins tagged with a fluorescent label, or molecules that influence cell behavior. One approach is to microinject the molecules into the cell through a glass micropipette. An especially useful technique is called **fluorescent analogue cytochemistry**, in which a purified protein is coupled to a fluorescent dye and microinjected into a cell; in this way the fate of the injected protein can be followed in a fluorescence microscope as the cell grows and divides. If tubulin (the subunit of microtubules) is labeled with a dye that fluoresces red, for example, microtubule dynamics can be followed second by second in a living cell (Figure 4–58).

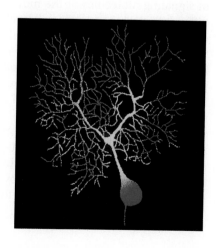

Figure 4–57 **Visualizing intracellular Ca²⁺ concentrations using a fluorescent indicator.** The branching tree of dendrites of the Purkinje cell in the cerebellum receives more than 100,000 synapses from other neurons. The output from the cell is conveyed along the single axon seen leaving the cell body at the bottom of the picture. This image of the intracellular calcium concentration in a single Purkinje cell (from the brain of a guinea pig) was taken using a low-light camera and the Ca²⁺-sensitive fluorescent indictor fura-2. The concentration of free Ca²⁺ is represented by different colors, *red* being the highest and *blue* the lowest. The highest Ca²⁺ levels are present in the thousands of dendritic branches. (Courtesy of D.W. Tank, J.A. Connor, M. Sugimori, and R.R. Llinas.)

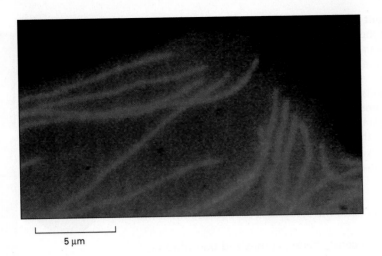

L————5 µm————I

Figure 4–58 **Fluorescent analogue cytochemistry.** Fluorescence micrograph of the leading edge of a living fibroblast that has been injected with rhodamine-labeled tubulin. The microtubules throughout the cell have incorporated the labeled tubulin molecules. Thus individual microtubules can be detected and their dynamic behavior followed using computer-enhanced imaging, as shown here. Although the microtubules appear to be about 0.25 µm thick, this is an optical effect; they are, in reality, only one-tenth this diameter. (Courtesy of P. Sammeh and G. Borisy.)

Antibodies can be microinjected into a cell in order to block the function of the molecule the antibodies recognize. Anti-myosin-II antibodies injected into a fertilized sea urchin egg, for example, prevent the egg cell from dividing in two, even though nuclear division occurs normally. This observation demonstrates that this myosin plays an essential part in the contractile process that divides the cytoplasm during cell division but that it is not required for nuclear division (see Figure 18–34).

Microinjection, although widely used, demands that each cell be injected individually; therefore it is possible to study at most a few hundred cells at a time. Other approaches allow large populations of cells to be permeabilized simultaneously. One can partially disrupt the structure of the cell plasma membrane, for example, so as to make it more permeable; this is usually accomplished by using a powerful electric shock or a chemical such as a low concentration of detergent. The electrical technique has the advantage of creating large pores in the plasma membrane without damaging intracellular membranes. The pores remain open for minutes or hours, depending on the cell type and size of the electric shock, and allow even macromolecules to enter (and leave) the cytosol rapidly. With a limited treatment a large fraction of the cells repair their plasma membrane and survive. A third method for introducing large molecules into cells is to cause membranous vesicles that contain these molecules to fuse with the cell's plasma membrane. These three methods are used widely in cell biology and are illustrated in Figure 4–59.

The Light-induced Activation of "Caged" Precursor Molecules Facilitates Studies of Intracellular Dynamics [35]

The complexity and rapidity of many intracellular processes, such as the action of signaling molecules or the movements of cytoskeletal proteins, make them difficult to study at a single-cell level. Ideally, one would like to be able to introduce any molecule of interest into a living cell at a precise time and location and follow its subsequent behavior, as well as the response of the cell. Microinjection is limited by the difficulty of controlling the place and time of delivery. A more powerful approach involves synthesizing an inactive form of the molecule of interest, introducing it into the cell and then activating it suddenly at a chosen site in the cell by focusing a spot of light on it. Inactive photosensitive precursors of this type, called **caged molecules**, have been made for a variety of small molecules, including Ca^{2+}, cyclic AMP, GTP, and inositol trisphosphate. The caged molecules can be introduced into living cells by any of the methods described in Figure 4–59 and then activated by a strong pulse of light from a laser (Figure 4–60). A microscope can be used to focus the light pulse on any tiny region of the

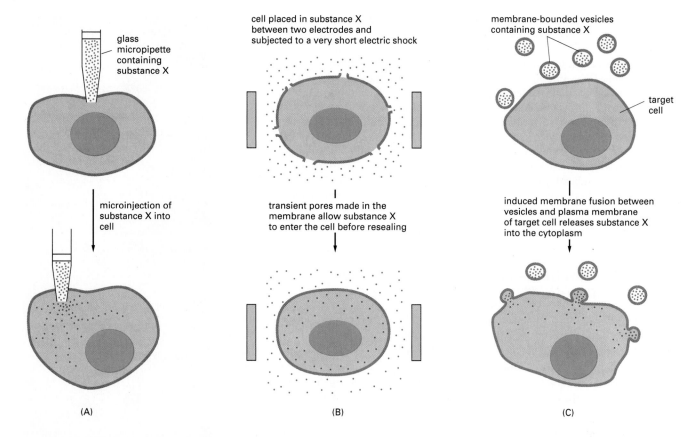

cell, so that the experimenter can control exactly where and when a molecule is delivered. In this way, for example, one can study the instantaneous effects of releasing an intracellular signaling molecule into the cytosol.

Caged fluorescent molecules are also tools with great promise. They are made by attaching a photoactivatable fluorescent dye to a purified protein. It is important that the modified protein remain biologically active: unlike labeling with radioisotopes (which changes only the number of neutrons in the nuclei of the labeled atoms), labeling with a caged fluorescent dye adds a large bulky group to the surface of a protein, which can easily change the protein's properties. A satisfactory labeling protocol is usually found by trial and error. Once a biologically active labeled protein has been produced, its behavior can be followed inside living cells. Tubulin labeled with caged fluorescein, for example, has been incorporated into microtubules of the mitotic spindle: when a small region of the spindle was illuminated with a laser, the labeled tubulin became fluorescent, so that its movement along the spindle microtubules could be readily followed (Figure 4–61). In principle, the same technique can be applied to any protein.

Figure 4–59 Methods to introduce a membrane-impermeant substance into a cell. In (A) the substance is injected through a micropipette, either by applying pressure or, if the substance is electrically charged, by applying a voltage that drives the substance into the cell as an ionic current (a technique called *iontophoresis*). In (B) the cell membrane is made transiently permeable to the substance by disrupting the membrane structure with a brief but intense electric shock (2000 volts per centimeter for 200 microseconds, for example). In (C) membrane-bounded vesicles are loaded with the desired substance and then induced to fuse with the target cells.

Figure 4–60 Caged molecules. Generalized scheme to show how a light-sensitive caged derivative of a molecule (here designated as X) can be converted by a flash of UV light to its free, active form. Small molecules such as ATP can be caged in this way. Even ions like Ca^{2+} can be indirectly caged; in this case a Ca^{2+}-binding chelator is used, which is inactivated by photolysis, thus releasing its Ca^{2+}.

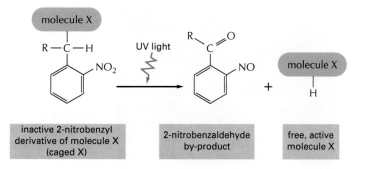

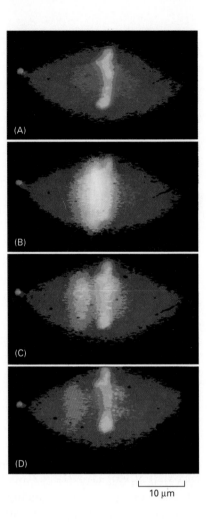

Figure 4–61 **Determining microtubule flux in the mitotic spindle using caged fluorescein linked to tubulin.** (A) A metaphase spindle formed *in vitro* from an extract of *Xenopus* eggs has incorporated three fluorescent markers: rhodamine-labeled tubulin (*red*) to mark all of the microtubules, a *blue* DNA-binding dye that labels the chromosomes, and caged-fluorescein-labeled tubulin, which is also incorporated into all of the microtubules but is invisible because it is nonfluorescent until activated by ultraviolet light. In (B) a beam of ultraviolet light is used to uncage the caged-fluorescein-labeled tubulin locally, mainly just to the left side of the metaphase plate. Over the next few minutes (after 1½ minutes in C, after 2½ minutes in D) the uncaged fluorescein-tubulin signal is seen to move toward the left spindle pole, indicating that tubulin is continuously moving poleward even though the spindle (visualized by the *red* rhodamine-labeled tubulin fluorescence) remains largely unchanged. (From K.E. Sawin and T.J. Mitchison, *J. Cell Biol.* 112:941–954, 1991, by copyright permission of the Rockefeller University Press.)

Antibodies Can Be Used to Detect and Isolate Specific Molecules [36]

Antibodies are proteins produced by the vertebrate immune system as a defense against infection (see Chapter 23). They are unique among proteins because they are made in billions of different forms, each with a different binding site that recognizes a specific target molecule (or *antigen*). The precise antigen specificity of antibodies makes them powerful tools for the cell biologist. Labeled with fluorescent dyes, they are invaluable for locating specific molecules in cells by fluorescence microscopy (Figure 4–62); labeled with electron-dense particles such as colloidal gold spheres, they are used for similar purposes in the electron microscope (Figure 4–63). As biochemical tools, they are used to detect and quantify molecules in cell extracts and to identify specific proteins after they have been fractionated by electrophoresis in polyacrylamide gels (see Figure 4–46). When coupled to an inert matrix to produce an affinity column, antibodies can be used either to purify a specific molecule from a crude cell extract or, if the molecule is on the cell surface, to pick out specific types of living cells from a heterogeneous population.

The sensitivity of antibodies as probes for detecting and assaying specific molecules in cells and tissues is frequently enhanced by a signal-amplification method. For example, although a marker molecule such as a fluorescent dye can be linked directly to an antibody used for specific recognition (the *primary an-*

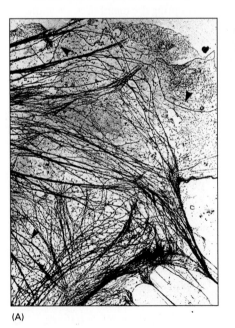
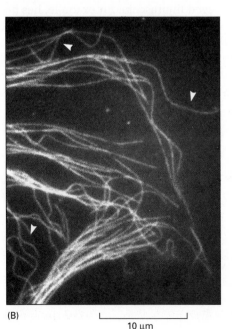

Figure 4–62 **Immunofluorescence.** (A) An electron micrograph of the periphery of a cultured epithelial cell showing the distribution of microtubules and other filaments. (B) The same area stained with fluorescent antibodies to tubulin, the protein subunit of microtubules, using the technique of indirect immunocytochemistry (see Figure 4–64). Arrows indicate individual microtubules that are readily recognizable in the two figures. (From M. Osborn, R. Webster, and K. Weber, *J. Cell Biol.* 77:R27–R34, 1978, by copyright permission of the Rockefeller University Press.)

tibody), a stronger signal is achieved by using an unlabeled primary antibody and then detecting it with a group of labeled *secondary antibodies* that bind to it (Figure 4–64).

The most sensitive and versatile amplification methods use an enzyme as a marker molecule attached to the secondary antibody. The enzyme alkaline phosphatase, for example, in the presence of appropriate chemicals, produces inorganic phosphate and leads to the local formation of a colored precipitate. This reveals the location of the secondary antibody that is coupled to the enzyme and hence the location of the antibody-antigen complex to which the secondary antibody is bound. Since each enzyme molecule acts catalytically to generate many thousands of molecules of product, even tiny amounts of antigen can be detected. Enzyme-linked immunosorbent assays (ELISA) based on this principle are frequently used in medicine as a sensitive test—for pregnancy or for various types of infections, for example.

Antibodies are made most simply by injecting a sample of the antigen several times into an animal such as a rabbit or a goat and then collecting the antibody-rich serum. This *antiserum* contains a heterogeneous mixture of antibodies, each produced by a different antibody-secreting cell (a B lymphocyte). The different antibodies recognize various parts of the antigen molecule as well as impurities in the antigen preparation. The specificity of an antiserum for a particular antigen sometimes can be sharpened by removing the unwanted antibody molecules that bind to other molecules; an antiserum produced against protein X, for example, can be passed through an affinity column of antigens Y and Z to remove any contaminating anti-Y and anti-Z antibodies. Even so, the heterogeneity of such antisera sometimes limits their usefulness.

Hybridoma Cell Lines Provide a Permanent Source of Monoclonal Antibodies [37]

In 1976 the problem of antiserum heterogeneity was overcome by the development of a technique that revolutionized the use of antibodies as tools in cell biology. The principle is to propagate a clone of cells from a single antibody-secreting B lymphocyte so that a homogeneous preparation of antibodies can be obtained in large quantities. The practical problem, however, is that B lymphocytes normally have a limited life-span in culture. To overcome this limitation, individual antibody-producing B lymphocytes from an immunized mouse or rat are fused with cells derived from an "immortal" B lymphocyte tumor. From the resulting heterogeneous mixture of hybrid cells, those hybrids that have both the ability to make a particular antibody and the ability to multiply indefinitely in culture are selected. These **hybridomas** are propagated as individual clones, each of which provides a permanent and stable source of a single type of **monoclonal antibody** (Figure 4–65). This antibody will recognize a single type of antigenic site—for example, a particular cluster of five or six amino acid side chains on the surface of a protein. Their uniform specificity makes monoclonal antibodies much more useful for most purposes than conventional antisera, which usually contain a mixture of antibodies that recognize a variety of different antigenic sites on even a small macromolecule.

But the most important advantage of the hybridoma technique is that monoclonal antibodies can be made against molecules that constitute only a minor

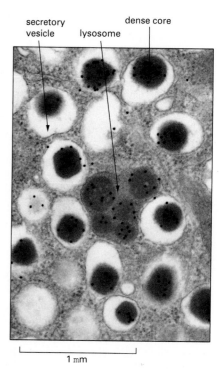

secretory vesicle   lysosome   dense core

Figure 4–63 Immunogold electron microscopy. Electron micrograph of an insulin-secreting cell in which the insulin molecules have been labeled with anti-insulin antibodies bound to tiny colloidal gold spheres (each seen as a black dot). Most of the insulin is stored in the dense cores of secretory vesicles; in addition, some cores are being degraded in lysosomes. (From L. Orci, *Diabetologia* 28:528–546, 1985.)

Figure 4–64 Indirect immunocytochemistry. The method is very sensitive because the primary antibody is itself recognized by many molecules of the secondary antibody. The secondary antibody is covalently coupled to a marker molecule that makes it readily detectable. Commonly used marker molecules include fluorescein or rhodamine dyes (for fluorescence microscopy), the enzyme horseradish peroxidase (for either conventional light microscopy or electron microscopy), the iron-containing protein ferritin or colloidal gold spheres (for electron microscopy), and the enzymes alkaline phosphatase or peroxidase (for biochemical detection).

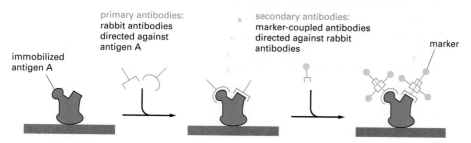

immobilized antigen A

primary antibodies: rabbit antibodies directed against antigen A

secondary antibodies: marker-coupled antibodies directed against rabbit antibodies

marker

Tracing and Assaying Molecules Inside Cells

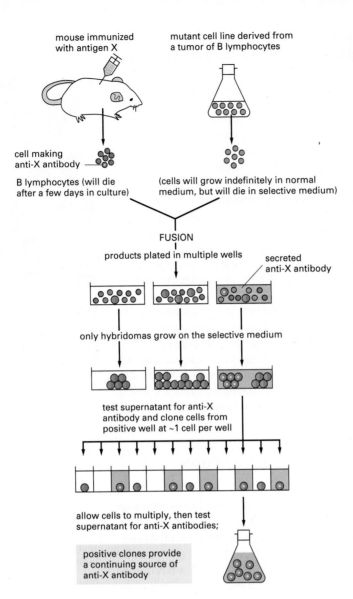

mouse immunized
with antigen X

mutant cell line derived from
a tumor of B lymphocytes

cell making
anti-X antibody

B lymphocytes (will die
after a few days in culture)

(cells will grow indefinitely in normal
medium, but will die in selective medium)

FUSION

products plated in multiple wells

secreted
anti-X antibody

only hybridomas grow on the selective medium

test supernatant for anti-X
antibody and clone cells from
positive well at ~1 cell per well

allow cells to multiply, then test
supernatant for anti-X antibodies;

positive clones provide
a continuing source of
anti-X antibody

Figure 4–65 Preparation of hybridomas that secrete monoclonal antibodies against a particular antigen (X). The selective growth medium used contains an inhibitor (aminopterin) that blocks the normal biosynthetic pathways by which nucleotides are made. The cells must therefore use a bypass pathway to synthesize their nucleic acids, and this pathway is defective in the mutant cell line to which the normal B lymphocytes are fused. Because neither cell type used for the initial fusion can grow on its own, only the hybrid cells survive.

component of a complex mixture. In an ordinary antiserum made against such a mixture, the proportion of antibody molecules that recognize the minor component would be too small to be useful. But if the B lymphocytes that produce the various components of this antiserum are made into hybridomas, it becomes possible to screen individual hybridoma clones from the large mixture to select one that produces the desired type of monoclonal antibody and to propagate the selected hybridoma indefinitely so as to produce that antibody in unlimited quantities. In principle, therefore, a monoclonal antibody can be made against any protein in a biological sample. Once the antibody is made, it can be used as a specific probe—both to track down and localize the protein that induced its formation and to purify the protein in order to study its structure and function. Since fewer than 5% of the estimated 10,000 proteins in a typical mammalian cell have thus far been isolated, many monoclonal antibodies made against impure protein mixtures in fractionated cell extracts identify new proteins. Using monoclonal antibodies and gene-cloning technology, it is no longer difficult to identify and characterize novel proteins and genes. The problem is to determine their function, and the most powerful way of doing this is often by the use of recombinant DNA technology, as we discuss in Chapter 7.

Summary

A large number of techniques are now available for detecting, measuring, and following almost any chosen molecule in a cell. Any type of molecule can, for example, be "labeled" by the incorporation of one or more radioactive atoms. The unstable nuclei of these atoms disintegrate, emitting radiation that allows the molecule to be detected and its movements and metabolism to be traced in the cell. Applications of radioisotopes in cell biology include the analysis of metabolic pathways by pulse-chase methods and the determination of the location of individual molecules in a cell or on a gel by autoradiography.

Fluorescent indicator dyes can be used to measure the concentration of specific ions in individual cells and even in different parts of a cell. Glass microelectrodes, besides being indispensable for studying the electrical potentials and ionic currents across the plasma membrane, provide an alternative way of measuring the concentrations of specific intracellular ions. They can also be used as micropipettes to inject membrane-impermeant molecules, such as fluorescently labeled proteins and antibodies, into living cells. The dynamic behavior and movements of many types of molecules can be followed in a living cell by constructing an inactive "caged" precursor, which can be introduced into a cell and then activated in a selected region of the cell by a light-stimulated reaction.

Antibodies are also versatile and sensitive tools for detecting and localizing specific biological molecules. Vertebrates make billions of different antibody molecules, each with a binding site that recognizes a specific region of a macromolecule. The hybridoma technique allows monoclonal antibodies of a single specificity to be obtained in virtually unlimited amounts. In principle, monoclonal antibodies can be made against any cell macromolecule and so can be used to locate and purify the molecule and, in some cases, to analyze its function.

References

General

Cantor, C.R.; Schimmel, P.R. Biophysical Chemistry, 3 vols. New York: W.H. Freeman, 1980. (A comprehensive account of the physical principles underlying many biochemical and biophysical techniques.)

Freifelder, D.M. Physical Biochemistry, 2nd ed. New York: W.H. Freeman, 1982.

Van Holde, K.E. Physical Biochemistry. Englewood Cliffs, NJ: Prentice-Hall, 1985.

Cited

1. Bradbury, S. An Introduction to the Optical Microscope, 2nd ed. Oxford, UK: Oxford University Press, 1989.

   Fawcett, D.W. A Textbook of Histology, 11th ed. Philadelphia: Saunders, 1986.

   Lacey, A.J., ed. Light Microscopy in Biology. A Practical Approach. Oxford, UK: IRL Press, 1989.

2. Hecht, E. Optics, 2nd ed. Reading, MA: Addison-Wesley, 1987.

   Slayter, E.M.; Slayter, H.S. Light and Electron Microscopy. Cambridge, UK: Cambridge University Press, 1992.

   Spencer, M. Fundamentals of Light Microscopy. Cambridge, UK: Cambridge University Press, 1982.

3. Boon, M.E.; Drijver, J.S. Routine Cytological Staining Methods. London: Macmillan, 1986.

   Hayat, M.A. Cytochemical Methods. New York: Wiley-Liss, 1989.

4. Haugland, R.P. Handbook of Fluorescent Probes and Research Chemicals, 5th ed. Eugene, OR: Molecular Probes, Inc., 1992.

Ploem, J.S.; Tanke, H.J. Introduction to Fluorescence Microscopy. Royal Microscopical Society Handbook No. 10. Oxford, UK: Oxford Scientific Publications, 1987.

Wang, Y.-L.; Taylor, D.L., eds. Fluorescence Microscopy of Living Cells in Culture, Parts A & B (Methods in Cell Biology, Vols. 29 and 30). San Diego, CA: Academic Press, 1989. (A comprehensive guide that extends far beyond basic fluorescence microscopy.)

Willingham, M.C.; Pastan, I.H. An Atlas of Immunofluorescence in Cultured Cells. Orlando, FL: Academic Press, 1985.

5. Zernike, F. How I discovered phase contrast. *Science* 121:345–349, 1955.

6. Allen, R.D. New observations on cell architecture and dynamics by video-enhanced contrast optical microscopy. *Annu. Rev. Biophys. Biophys. Chem.* 14:265–290, 1985.

   Herman, B.; Jacobson, K., eds. Optical Microscopy for Biology. New York: Wiley-Liss, 1990.

   Inoue, S. Video Microscopy. New York: Plenum Press, 1986.

   Shotton, D., ed. Electronic Light Microscopy: Techniques in Modern Biomedical Microscopy. New York: Wiley-Liss, 1993.

7. Minsky, M. Memoir on inventing the confocal scanning microscope. *Scanning* 10:128–138, 1988.

   Pawley, J.B., ed. Handbook of Confocal Microscopy, 2nd ed. New York: Plenum Press, 1990.

   White, J.G.; Amos, W.B.; Fordham, M. An evaluation of confocal versus conventional imaging of biological structures by fluorescence light microscopy. *J. Cell Biol.* 105:41–48, 1987.

8. Pease, D.C.; Porter, K.R. Electron microscopy and ultramicrotomy. *J. Cell Biol.* 91:287s–292s, 1981.

9. Hayat, M.A. Principles and Techniques of Electron Microscopy, 3rd ed. Boca Raton, FL: CRC Press, 1989.

   Wischnitzer, S. Introduction to Electron Microscopy. Oxford: Pergamon, 1981.

10. Everhart, T.E.; Hayes, T.L. The scanning electron microscope. *Sci. Am.* 226(1):54–69, 1972.

    Hayat, M.A. Introduction to Biological Scanning Electron Microscopy. Baltimore: University Park Press, 1978.

    Kessel, R.G.; Kardon, R.H. Tissues and Organs. New York: W.H. Freeman, 1979. (An atlas of vertebrate tissues seen by scanning electron microscopy.)

    Pawley, J.B.; Walther, P.; Shih, S.-J.; Malecki, M. Early results using high-resolution, low-voltage, low-temperature SEM. *J. Microsc.* 161:327–335, 1991.

11. Sommerville, J.; Scheer, U., eds. Electron Microscopy in Molecular Biology: A Practical Approach. Oxford: IRL Press, 1987.

12. Heuser, J. Quick-freeze, deep-etch preparation of samples for 3-D electron microscopy. *Trends Biochem. Sci.* 6:64–68, 1981.

    Pinto da Silva, P.; Branton, D. Membrane splitting in freeze-etching. *J. Cell Biol.* 45:598–605, 1970.

13. Chiu, W. Electron microscopy of frozen, hydrated biological specimens. *Annu. Rev. Biophys. Biophys. Chem.* 15:237–257, 1986.

    Darst, S.A.; Edwards, A.M.; Kubalek, E.W.; Kornberg, R.D. Three-dimensional structure of yeast RNA polymerase II at 16 A resolution. *Cell* 66:121–128, 1991.

    Unwin, N. Nicotinic acetylcholine receptor at 9 A resolution. *J. Mol. Biol.* 229:1101–1124, 1993.

    Unwin, P.N.T.; Henderson, R. Molecular structure determination by electron microscopy of unstained crystal specimens. *J. Mol. Biol.* 94:425–440, 1975.

14. Freshney, R.I. Culture of Animal Cells: A Manual of Basic Techniques. New York: Liss, 1987.

15. Herzenberg, L.A.; Sweet, R.G.; Herzenberg, L.A. Fluorescence-activated cell sorting. *Sci. Am.* 234(3):108–116, 1976.

    Kamarck, M.E. Fluorescence-activated cell sorting of hybrid and transfected cells. *Methods Enzymol.* 151:150–165, 1987.

    Nolan, G.P.; Fiering, S.; Nicolas, J.-F.; Herzenberg, L.A. Fluorescence-activated cell analysis and sorting of viable mammalian cells based on beta-D-galactosidase activity after transduction of *E. coli lac Z. Proc. Natl. Acad. Sci. USA* 85:2603–2607, 1988.

16. Harrison, R.G. The outgrowth of the nerve fiber as a mode of protoplasmic movement. *J. Exp. Zool.* 9:787–848, 1910.

17. Ham, R.G. Clonal growth of mammalian cells in a chemically defined, synthetic medium. *Proc. Natl. Acad. Sci. USA* 53:288–293, 1965.

    Levi-Montalcini, R. The nerve growth factor thirty-five years later. *Science* 237:1154–1162, 1987.

    Loo, D.T.; Fuquay, J.I.; Rawson, C.L.; Barnes, D.W. Extended culture of mouse embryo cells without senescence: inhibition by serum. *Science* 236:200–202, 1987.

    Sato, G.H.; Pardee, A.B.; Sirbasku, D.A., eds. Growth of Cells in Hormonally Defined Media. Cold Spring Harbor, NY: Cold Spring Harbor Laboratory, 1982.

18. Hay, R., et al., eds. American Type Culture Collection Catalogue of Cell Lines and Hybridomas, 6th ed. Rockville, MD: American Type Culture Collection, 1988.

19. Ruddle, F.H.; Creagan, R.P. Parasexual approaches to the genetics of man. *Annu. Rev. Genet.* 9:407–486, 1975.

20. Colowick, S.P.; Kaplan, N.O., eds. Methods in Enzymology, Vols. 1–.... San Diego, CA: Academic Press, 1955–1994. (A multivolume series containing general and specific articles on many procedures.)

    Cooper, T.G. The Tools of Biochemistry. New York: Wiley, 1977.

    Deutscher, M.P., ed. Guide to Protein Purification (Methods in Enzymology, Vol. 182). San Diego, CA: Academic Press, 1990.

    Scopes, R.K. Protein Purification, Principles and Practice, 2nd ed. New York: Springer-Verlag, 1987.

21. Claude, A. The coming of age of the cell. *Science* 189:433–435, 1975.

    de Duve, C.; Beaufay, H. A short history of tissue fractionation. *J. Cell Biol.* 91:293s–299s, 1981.

    Meselson, M.; Stahl, F.W. The replication of DNA in *Escherichia coli. Proc. Natl. Acad. Sci. USA* 47:671–682, 1958. (Density gradient centrifugation was used to show the semiconservative replication of DNA.)

    Palade, G. Intracellular aspects of the process of protein synthesis. *Science* 189:347–358, 1975.

    Sheeler, P. Centrifugation in Biology and Medical Science. New York: Wiley, 1981.

22. Morre, D.J.; Howell, K.E.; Cook, G.M.W.; Evans, W.H., eds. Cell Free Analysis of Membrane Traffic. New York: Liss, 1986.

    Nirenberg, M.W.; Matthaei, J.H. The dependence of cell-free protein synthesis in *E. coli* on naturally occurring or synthetic polyribonucelotides. *Proc. Natl. Acad. Sci. USA* 47:1588–1602, 1961.

    Racker, E. A New Look at Mechanisms in Bioenergetics. New York: Academic Press, 1976. (Cell-free systems in the working out of energy metabolism.)

    Zamecnik, P.C. An historical account of protein synthesis, with current overtones—a personalized view. *Cold Spring Harbor Symp. Quant. Biol.* 34:1–16, 1969.

23. Dean, P.D.G.; Johnson, W.S.; Middle, F.A. Affinity Chromatography: A Practical Approach. Arlington, VA: IRL Press, 1985.

    Gilbert, M.T. High Performance Liquid Chromatography. Littleton, MA: John Wright-PSG, 1987.

24. Andrews, A.T. Electrophoresis, 2nd ed. Oxford, UK: Clarendon Press, 1986.

    Laemmli, U.K. Cleavage of structural proteins during the assembly of the head of bacteriophage T4. *Nature* 227:680–685, 1970.

    Hames, B.D.; Rickwood, D., eds. Gel Electrophoresis of Proteins: A Practical Approach, 2nd ed. New York: Oxford University Press, 1990.

25. Celis, J.E.; Bravo, R., eds. Two-Dimensional Gel Electrophoresis of Proteins. New York: Academic Press, 1984.

    O'Farrell, P.H. High-resolution two-dimensional electrophoresis of proteins. *J. Biol. Chem.* 250:4007–4021, 1975.

26. Cleveland, D.W.; Fischer, S.G.; Kirschner, M.W.; Laemmli, U.K. Peptide mapping by limited proteoly-

sis in sodium dodecyl sulfate and analysis by gel electrophoresis. *J. Biol. Chem.* 252:1102–1106, 1977.

Ingram, V.M. A specific chemical difference between the globins of normal human and sickle-cell anemia hemoglobin. *Nature* 178:792–794, 1956.

27. Edman, P.; Begg, G. A protein sequenator. *Eur. J. Biochem.* 1:80–91, 1967.

Hewick, R.M.; Hunkapiller, M.W.; Hood, L.E.; Dreyer, W.J. A gas-liquid solid phase peptide and protein sequenator. *J. Biol. Chem.* 256:7990–7997, 1981.

McCloskey, J.A., ed. Mass Spectrometry (Methods in Enzymology, Vol. 193). San Diego, CA: Academic Press, 1990.

Sanger, F. The arrangement of amino acids in proteins. *Adv. Protein Chem.* 7:1–67, 1952.

Walsh, K.A.; Ericsson, L.H.; Parmelee, D.C.; Titani, K. Advances in protein sequencing. *Annu. Rev. Biochem.* 50:261–284, 1981.

28. Branden, C.; Tooze, J. Introduction to Protein Structure. New York: Garland, 1991.

Glusker, J.P.; Trueblood, K.N. Crystal Stucture Analysis: A Primer. Oxford, UK: Oxford University Press, 1985.

Kendrew, J.C. The three-dimensional structure of a protein molecule. *Sci. Am.* 205(6):96–111, 1961.

Perutz, M.F. The hemoglobin molecule. *Sci. Am.* 211(5):64–76, 1964.

29. Cooke, R.M.; Cambell, I.D. Protein structure determination by nuclear magnetic resonance. *Bioessays* 8:52–56, 1988.

Shulman, R.G. NMR spectroscopy of living cells. *Sci. Am.* 248(1):86–93, 1983.

Wuthrich, K. Protein structure determination in solution by nuclear magnetic resonance spectroscopy. *Science* 243:45–50, 1989.

30. Chase, G.D.; Rabinowitz, J.L. Principles of Radioisotope Methodology, 2nd ed. Minneapolis: Burgess, 1962.

Dyson, N.A. An Intoduction to Nuclear Physics with Applications in Medicine and Biology. Chichester, UK: Horwood, 1981.

Yalow, R.S. Radioimmunoassay: a probe for the fine structure of biologic systems. *Science* 200:1236–1245, 1978.

31. Calvin, M. The path of carbon in photosynthesis. *Science* 135:879–889, 1962. (Description of one of the earliest uses of radioisotopes in biology.)

Hershey, A.D.; Chase, M. Independent functions of viral protein and nucleic acid in growth of bacteriophage. *J. Gen. Physiol.* 36:39–56, 1952.

Rogers, A.W. Techniques of Autoradiography, 3rd ed. New York: Elsevier/North Holland, 1979.

Slater, R.J., ed. Radioisotopes in Biology: A Practical Approach. New York: Oxford University Press, 1990.

32. Ammann, D. Ion-Selective Microelectrodes: Principles, Design and Application. Berlin: Springer-Verlag, 1986.

Auerbach, A.; Sachs, F. Patch clamp studies of single ionic channels. *Annu. Rev. Biophys. Bioeng.* 13:269–302, 1984.

Neher, E. Ion channels for communication between and within cells. *Science* 256:498–502, 1992.

Sakmann, B. Elementary steps in synaptic transmission revealed by currents through single ion channels. *Science* 256:503–512, 1992.

33. Grynkiewicz, G.; Poenie, M.; Tsien, R.Y. A new generation of Ca^{2+} indicators with greatly improved fluorescence properties. *J. Biol. Chem.* 260:3440–3450, 1985.

Tsien, R.Y. Fluorescent probes of cell signaling. *Annu. Rev. Neurosci.* 12:227–253, 1989.

34. Celis, J.E.; Graessman, A.; Loyter, A., eds. Microinjection and Organelle Transplantation Techniques. London: Academic Press, 1986.

Chang, D.C.; Chassy, B.M.; Saunders, J.A.; Sowers, A.E. Guide to Electroporation and Electrofusion. New York: Academic Press, 1991.

Gomperts, B.D.; Fernandez, J.M. Techniques for membrane permeabilization. *Trends Biochem. Sci.* 10:414–417, 1985.

Ostro, M.J. Liposomes. *Sci. Am.* 256(1):102–111, 1987.

Ureta, T.; Radojkovic, J. Microinjected frog oocytes: a first-rate test tube for studies on metabolism and its control. *Bioessays* 2:221–226, 1985.

35. Adams, S.R.; Tsien, R.Y. Controlling cell chemistry with caged compounds. *Annu. Rev. Physiol.* 55:755–784, 1993.

Gurney, A.M. Photolabile caged compounds. In Fluorescent and Luminescent Probes for Biological Activity (W.T. Mason, ed.). London: Academic Press, 1993.

Sawin, K.E.; Theriot, J.A.; Mitchison, T.J. Photoactivation of fluorescence as a probe for cytoskeletal dynamics in mitosis and cell motility. In Fluorescent and Luminescent Probes for Biological Activity (W.T. Mason, ed.). London: Academic Press, 1993.

Walker, J.W.; Somlyo, A.V.; Goldman, Y.E.; Somlyo, A.P.; Trentham, D.R. Kinetics of smooth and skeletal muscle activation by laser pulse photolysis of caged inositol 1,4,5–trisphosphate. *Nature* 327:249–252, 1987.

36. Anderton, B.H.; Thorpe, R.C. New methods of analysing for antigens and glycoproteins in complex mixtures. *Immunol. Today.* 1:122–127, 1980.

Coons, A.H. Histochemistry with labeled antibody. *Int. Rev. Cytol.* 5:1–23, 1956.

Harlow, E.; Lane, D. Antibodies. A Laboratory Manual. Cold Spring Harbor, NY: Cold Spring Harbor Laboratory, 1988.

Wilson, L.; Matsudaira, P., eds. Antibodies in Cell Biology (Methods in Cell Biology, Vol. 37). New York: Academic Press, 1993.

37. Clackson, T.; Hoogenboom, H.R.; Griffiths, A.D.; Winter, G. Making antibody fragments using phage display libraries. *Nature* 352:624–628, 1991.

Kang, A.S.; Barbas, C.F.; Janda, K.D.; Benkovic, S.J.; Lerner, R.A. Linkage of recognition and replication functions by assembling combinatorial antibody Fab libraries along phage surfaces. *Proc. Natl. Acad. Sci. USA* 88:4363–4366, 1991.

Milstein, C. Monoclonal antibodies. *Sci. Am.* 243(4):66–74, 1980.

Yelton, D.E.; Scharff, M.D. Monoclonal antibodies: a powerful new tool in biology and medicine. *Annu. Rev. Biochem.* 50:657–680, 1981.

Molecular Genetics

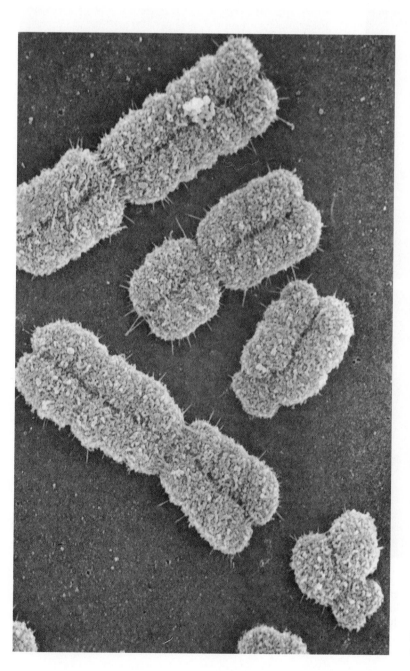

Scanning electron micrograph of human metaphase chromosomes. (Courtesy of Terry D. Allen.)

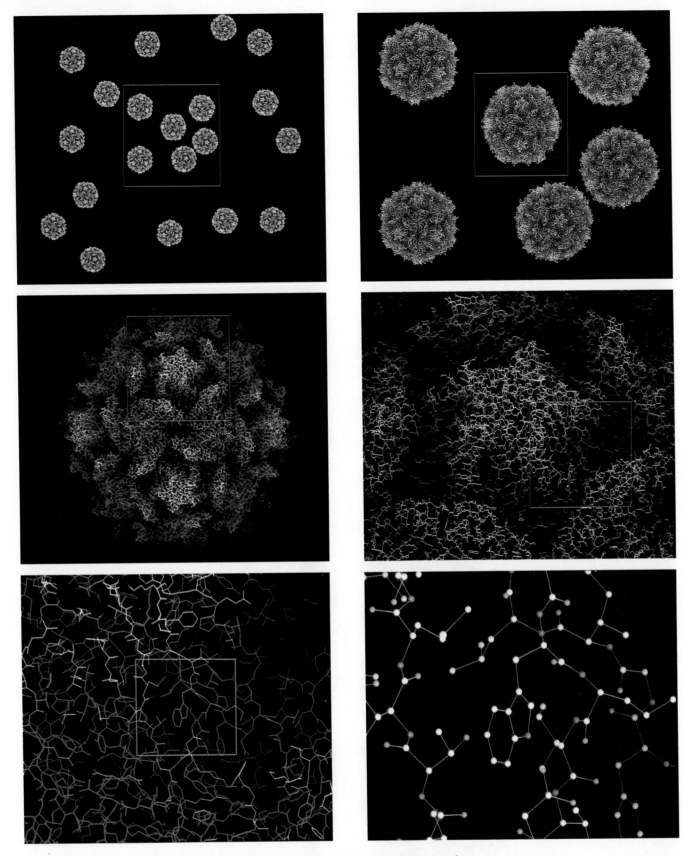

A series of six views of an atomic model of a poliovirus, whose exact three-dimensional structure is known from x-ray crystallography. Each view is magnified three times over the one before until, in the final view, we can see a region of the virus at atomic resolution. In the center of the final view you can clearly see the amino acid side chain of tryptophan. (Courtesy of James Hogle.)

Protein Function

Proteins make up most of the dry mass of a cell, and they play the predominant part in most biological processes. One must understand proteins, therefore, before one can hope to understand the cell. An elementary introduction to the structure of proteins was provided in Chapter 3, where we presented a general overview of biological macromolecules and discussed their shapes and chemistry. But proteins are not just rigid lumps of material with chemically reactive surfaces. They have precisely engineered moving parts whose mechanical actions are coupled to chemical events. It is this coupling of chemistry and movement that gives proteins the extraordinary capabilities that underlie all the dynamic processes in living cells. Without a grasp of how proteins operate as molecules with moving parts, it is hard to appreciate the rest of cell biology.

In this chapter, which begins the more advanced sections of the book, we use selected examples to show how proteins function not only as catalysts but also as sophisticated transducers of motion, signal integrators, and components of multisubunit protein machines. The discussion relies on advances that have revealed the detailed three-dimensional structures of many proteins; it will emphasize general principles and is intended to set the stage for the descriptions of specific cell structures and processes in subsequent chapters.

In the last part of the chapter we describe the life and death of proteins—from their folding, guided by molecular chaperones, to their destruction by targeted proteolysis—emphasizing the modular construction of most proteins and protein complexes.

Making Machines Out of Proteins

We begin by considering how the shape of a protein can be altered by the binding of another molecule, called a *ligand*. We then demonstrate the profound implications of this apparently simple phenomenon by describing a few of the many ways in which ligand-driven alterations in protein shape are exploited by cells.

The Binding of a Ligand Can Change the Shape of a Protein [1]

The first example involves the enzyme *hexokinase*, which is present in nearly all cells. This enzyme catalyzes an early step in sugar metabolism—the transfer of the terminal phosphate of an ATP molecule to glucose, forming glucose 6-phos-

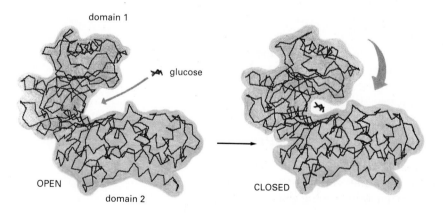

Figure 5–1 **The reaction catalyzed by hexokinase.** As the first step in the breakdown of glucose, a phosphate group is transferred from ATP to glucose to form glucose 6-phosphate. The glucose 6-phosphate is then processed by a series of other enzymes, which catalyze the chain of reactions known as glycolysis. Glycolysis converts glucose to pyruvate and produces a net gain of ATP molecules for the cell (see Figure 2–21).

phate (discussed in Chapter 2). Hexokinase binds glucose tightly, and this greatly increases the affinity of the enzyme for ATP, which binds to a neighboring site on the protein. Specific amino acid side chains on the protein then catalyze the phosphate transfer, and the two products—glucose 6-phosphate and ADP—are released to finish the reaction cycle (Figure 5–1).

The hexokinase from yeast is composed of two domains. The binding sites for glucose and ATP lie in a cleft between these domains, and the domains move toward each other to narrow the cleft when glucose binds (Figure 5–2).

This type of domain movement in response to ligand binding is common and is easily explained. In the case of hexokinase there are binding sites for different parts of the glucose molecule on the inside face of each domain. The unfavorable change in the free energy of the protein that occurs when the domains move relative to each other to close the cleft is more than compensated for by the free energy released when the cleft clamps down on the glucose; in other words, the noncovalent bonds that glucose forms with the protein serve to "glue" the two domains together, causing the protein to shift from an *open* to a *closed* conformation.

Two Ligands That Bind to the Same Protein Often Affect Each Other's Binding [2]

The binding of glucose to hexokinase causes a fiftyfold increase in the affinity of the enzyme for ATP. The reason is easy to see. Like glucose, ATP can form noncovalent bonds with amino acids on the inside faces of the two domains if the cleft closes. When ATP alone binds to hexokinase, some of the binding energy must be used to close up the cleft; this energy is not required, however, if glucose binding has already induced this shape change (Figure 5–3). By the same

Figure 5–2 **The conformational change in hexokinase caused by glucose binding.** The lines trace the course of the polypeptide backbone of hexokinase. These structures were determined by x-ray diffraction analysis of crystals of the protein with and without glucose bound. Glucose binding shifts the protein from an *open* to a *closed* conformation.

reasoning, one would predict that glucose would bind more tightly to hexokinase when ATP is present than when it is absent, and this is what one observes (Figure 5–4).

ATP and glucose bind to neighboring sites in hexokinase. But the binding of one ligand to a protein's surface can sometimes affect the binding of a second ligand even if the two binding sites are far apart. Suppose, for example, that a protein that binds glucose in the same way as hexokinase also binds another molecule, X, at a distant site on the protein's surface. If the binding site for X changes shape as part of the large conformational change induced by glucose binding, one would say that the binding sites for X and for glucose are *coupled*. If the shift to the closed conformation, for example, causes the binding site for X to fit X better, then glucose binding will increase the affinity of the protein for X, just as glucose binding increases the affinity of hexokinase for ATP (Figure 5–5).

As we discuss next, proteins in which conformational changes couple two widely separated binding sites have been selected in evolution because they enable a cell to link the fate of one molecule to the presence or absence of any other. This type of conformational coupling is known as *allostery*. A protein whose activity is regulated in this way is said to undergo an **allosteric transition,** and the protein is called an *allosteric protein*.

Two Ligands Whose Binding Sites Are Coupled Must Reciprocally Affect Each Other's Binding [2]

Whenever two ligands prefer to bind to the *same* conformation of an allosteric protein, it follows from basic thermodynamic considerations that each ligand

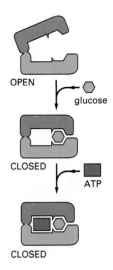

Figure 5–3 Glucose helps ATP bind to hexokinase. Like glucose, ATP binds best to the *closed* conformation of the enzyme and therefore binds best if glucose has already bound. For simplicity, the actual structure of the protein shown in Figure 5–2 has been replaced (both here and in Figure 5–4) by a schematic diagram.

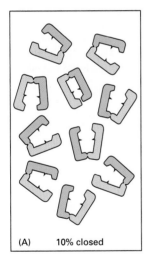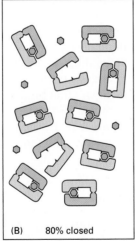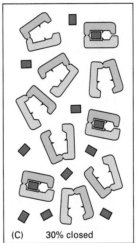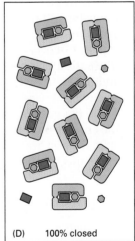

(A) 10% closed (B) 80% closed (C) 30% closed (D) 100% closed

Figure 5–4 The conformational equilibrium in hexokinase. Because ATP and glucose both individually drive hexokinase toward its *closed* conformation, each ligand helps the other to bind. To help make this clear, each panel has been drawn to represent a test tube containing 10 molecules of hexokinase in an aqueous solution. Panel A shows how the protein behaves with no ligand present; although a small fraction of the molecules spontaneously adopt the closed form, most are in the open configuration. The other panels show how the 10 molecules of protein behave with 12 molecules of glucose (panel B), with 12 molecules of ATP (panel C), and with 12 molecules of glucose and 12 molecules of ATP (panel D). The symbols for glucose and ATP are the same as in Figure 5–3. A comparison of the amount of free (unbound) glucose in panels B and D shows that the addition of ATP helps glucose to bind, whereas a comparison of the amount of free ATP in panels C and D shows that the addition of glucose helps ATP to bind.

Making Machines Out of Proteins

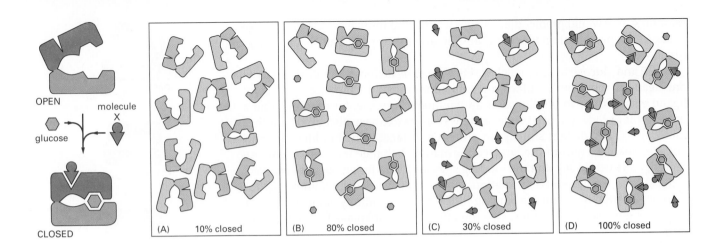

OPEN

molecule X

glucose

CLOSED

(A) 10% closed   (B) 80% closed   (C) 30% closed   (D) 100% closed

must increase the affinity of the protein for the other. This concept is called **linkage**. It is well illustrated by the example already considered in Figure 5–5, where the binding of glucose to hexokinase increases the enzyme's affinity for molecule X and vice versa. The linkage relationship is quantitatively reciprocal, so that, for example, if glucose has a very large effect on the binding of X, X will have a very large effect on the binding of glucose.

Linkage will operate in a negative way if two ligands bind to *different* conformations of an allosteric protein. As a general rule, a ligand will act to stabilize the particular conformation of the protein to which it binds; if this is different from the conformation favored by a second ligand, the binding of the first will discourage the binding of the second. Thus, if a shape change caused by glucose binding reduces the affinity of a protein for molecule X, the binding of X must decrease the protein's affinity for glucose (Figure 5–6).

The relationships shown schematically in Figures 5–5 and 5–6 underlie all of cell biology. They seem so obvious in retrospect that we now take them for granted. But their discovery in the 1950s, followed by a general description of allostery in the early 1960s, was revolutionary at the time. Since the X in these examples binds at a site that is distinct from the site where catalysis occurs, it need have no chemical relationship to glucose or to any other ligand that binds at the active site. For enzymes that are regulated in this way, molecule X could either turn the enzyme on (see Figure 5–5) or turn it off (see Figure 5–6). By such a mechanism, allosteric proteins serve as general switches that allow one molecule in a cell to affect the fate of another.

Allosteric Transitions Help Regulate Metabolism [3]

As described in Chapter 2, the end product of a metabolic pathway often inhibits the enzyme that starts the pathway. Because of this *negative feedback* on the flux through a pathway, the intracellular concentration of the end product is kept approximately constant, despite large changes in the chemical conditions in the cell. Allosteric transitions are essential to this type of *feedback regulation*. Enzymes that act early in a pathway, for example, generally exist in two conformations. One is an active conformation that binds substrate at its **active site** and catalyzes its conversion to the next substance in the pathway. The other is an inactive conformation that binds the final product of the pathway at a different, **regulatory site.** As the final product accumulates, it binds to the enzyme and converts it to its inactive conformation (see Figure 2–38).

An enzyme involved in a metabolic pathway can also be *activated* by an allosteric transition induced by ligand binding. In this case the ligand is a molecule that accumulates when the cell is deficient in a product of the pathway; because the ligand binds preferentially to the active form of the protein, it drives the enzyme from an inactive to an active conformation. Examples of this type of *positive*

Figure 5–5 Cooperative binding caused by conformational coupling between two distant binding sites. In this example both glucose and molecule X bind best to the *closed* conformation of a protein with two domains. Because both glucose and molecule X drive the protein toward its closed conformation, each ligand helps the other to bind. Glucose and molecule X are therefore said to bind cooperatively to the protein.

This figure is very similar to Figures 5–3 and 5–4; the only difference is that whereas the binding site for ATP lies in the cleft of hexokinase, the binding site for molecule X lies outside the cleft.

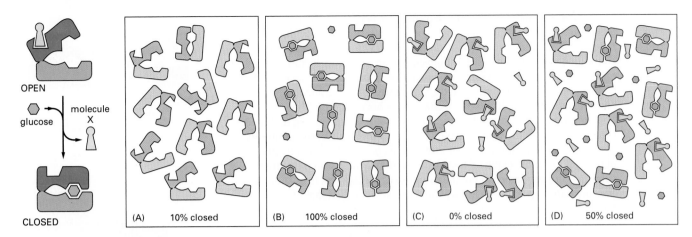

OPEN

glucose ⬡ → molecule X

CLOSED

(A) 10% closed   (B) 100% closed   (C) 0% closed   (D) 50% closed

Figure 5–6 **Competitive binding caused by conformational coupling between two distant binding sites.** The design of this figure is the same as that described previously for Figure 5–5, but here molecule X prefers the *open* conformation, while glucose prefers the *closed* conformation. Because glucose and molecule X drive the protein toward opposite conformations (*closed* and *open*, respectively), the presence of either ligand interferes with the binding of the other.

feedback are provided by many of the enzymes involved in the catabolic pathways that produce ATP: they are stimulated by the rise in ADP concentration that occurs when ATP levels drop. For these enzymes the ADP has a purely regulatory role, in contrast to the substrate role played by ATP in the function of hexokinase.

Proteins Often Form Symmetrical Assemblies That Undergo Cooperative Allosteric Transitions [4]

An enzyme that is regulated by negative feedback and that consists of only one subunit with one regulatory site can at most decrease from 90% to about 10% activity in response to a 100-fold increase in the concentration of the inhibitory ligand (Figure 5–7, *red line*). Responses of this type are apparently not sharp enough for optimal cell regulation, and most enzymes that are turned on or off by ligand binding consist of symmetrical assemblies of identical subunits. With this arrangement the binding of a molecule of ligand to a single site on one subunit can trigger an allosteric change in the subunit that can be transmitted to the neighboring subunits, helping them to bind the same ligand. As a result of this *cooperative* allosteric transition, a relatively small change in ligand concentration in the cell can switch the whole assembly from an almost fully active to an almost fully inactive conformation or vice versa (Figure 5–7, *blue line*).

Figure 5–7 **A plot of enzyme activity versus the concentration of inhibitory ligand for monomeric and multisubunit allosteric enzymes.** For an enzyme with a single subunit (*red line*) a drop from 90% enzyme activity to 10% activity (indicated by *dots* on the curve) requires a 100-fold increase in the concentration of inhibitor. The enzyme activity is calculated from the simple equilibrium relationship K = [I][P]/[IP], where P is active protein, I is inhibitor, and IP is the inactive protein bound to inhibitor. An identical curve applies to any simple binding interaction between two molecules, A and B (see Figure 3–9). In contrast, a multisubunit allosteric enzyme can respond in a switchlike manner to a change in ligand concentration: the steep response is caused by a cooperative binding of the ligand molecules, as explained in Figure 5–8. The *green line* represents the idealized result expected for the cooperative binding of 2 inhibitory ligand molecules to an allosteric enzyme with 2 subunits, and the *blue line* shows the idealized response of an enzyme with 4 subunits. As indicated by the dots on the curves, the more complex enzymes drop from 90% to 10% activity over a much narrower range of inhibitor concentration than does the enzyme composed of a single subunit.

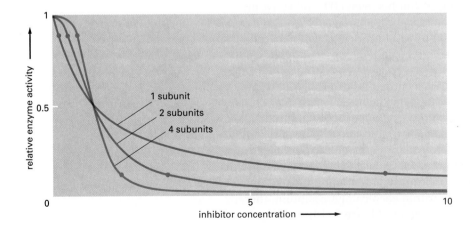

1 subunit
2 subunits
4 subunits

relative enzyme activity

inhibitor concentration

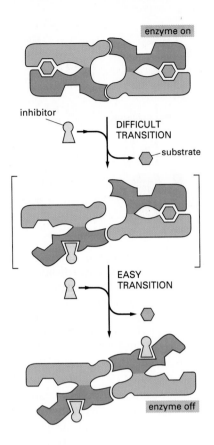

enzyme on

inhibitor

DIFFICULT
TRANSITION

substrate

EASY
TRANSITION

enzyme off

Figure 5–8 **A cooperative allosteric transition.** Schematic diagram illustrating how the conformation of one subunit can influence that of its neighbor in a symmetrical protein composed of two identical allosteric subunits. The binding of a single molecule of an inhibitory ligand (*yellow*) to one subunit of the enzyme occurs with difficulty because it changes the conformation of this subunit and thereby destroys the symmetry of the enzyme; once this conformational change has been accomplished, however, the energy gained by restoring the symmetrical pairing makes it especially easy for the second subunit to bind a molecule of the inhibitory ligand and undergo the same conformational change. Because the binding of the first molecule of ligand increases the affinity with which the other subunit binds the same ligand, the response of the enzyme to changes in the concentration of the ligand will be much steeper than that of a monomeric enzyme (see Figure 5–7).

The principles involved in a cooperative "all-or-none" transition are easiest to visualize for an enzyme that forms a symmetrical dimer. In the example shown in Figure 5–8, the first molecule of an inhibitory ligand binds with great difficulty since its binding destroys an energetically favorable interaction between the two identical monomers in the dimer. A second ligand molecule now binds more easily, however, because its binding restores the monomer-monomer contacts of a symmetrical dimer (and also completely inactivates the enzyme). An even sharper response to a ligand can be obtained with larger assemblies, such as the enzyme formed from 12 polypeptide chains discussed next.

The Allosteric Transition in Aspartate Transcarbamoylase Is Understood in Atomic Detail [5]

One enzyme used in the early studies of negative feedback, allosteric regulation was aspartate transcarbamoylase from *E. coli*. It catalyzes the important reaction carbamoylphosphate + aspartate → N-carbamoylaspartate, which begins the synthesis of the pyrimidine ring of C, U, and T nucleotides. One of the final products of this pathway, cytosine triphosphate (CTP), binds to the enzyme to turn it off whenever CTP is plentiful.

Aspartate transcarbamoylase is a large complex of six regulatory and six catalytic subunits. The catalytic subunits are present as two trimers, each arranged like an equilateral triangle; the two trimers face each other and are held together by three regulatory dimers that form a bridge between them. The entire molecule is poised to undergo a concerted, all-or-none allosteric transition between two conformations, designated T ("tense") and R ("relaxed") states (Figure 5–9).

The binding of substrates (carbamoylphosphate and aspartate) to the catalytic trimers drives aspartate transcarbamoylase into its catalytically active R state, from which the regulatory CTP molecules dissociate. By contrast, the binding of CTP to the regulatory dimers converts the enzyme to the inactive T state, from which the substrates dissociate. This tug-of-war between CTP and substrates is identical in principle to that described previously in Figure 5–6 for a simpler allosteric protein. But because here the tug-of-war occurs in a symmetrical molecule with multiple binding sites, the effect is a cooperative allosteric transition that can either turn the enzyme on suddenly as substrates accumulate (forming the R state) or shut it off rapidly when CTP accumulates (forming the T state).

A combination of biochemistry and x-ray crystallography has revealed many fascinating details of this allosteric transition. Each regulatory subunit has two domains, and the binding of CTP causes the two domains to move relative to each other, so that they function like a lever that rotates the two catalytic trimers and pulls them closer together into the T state (see Figure 5–9). When this occurs, hydrogen bonds form between opposing catalytic subunits that help to widen the cleft that forms the active site within each catalytic subunit, thereby destroying the binding sites for the substrates (Figure 5–10). Adding large

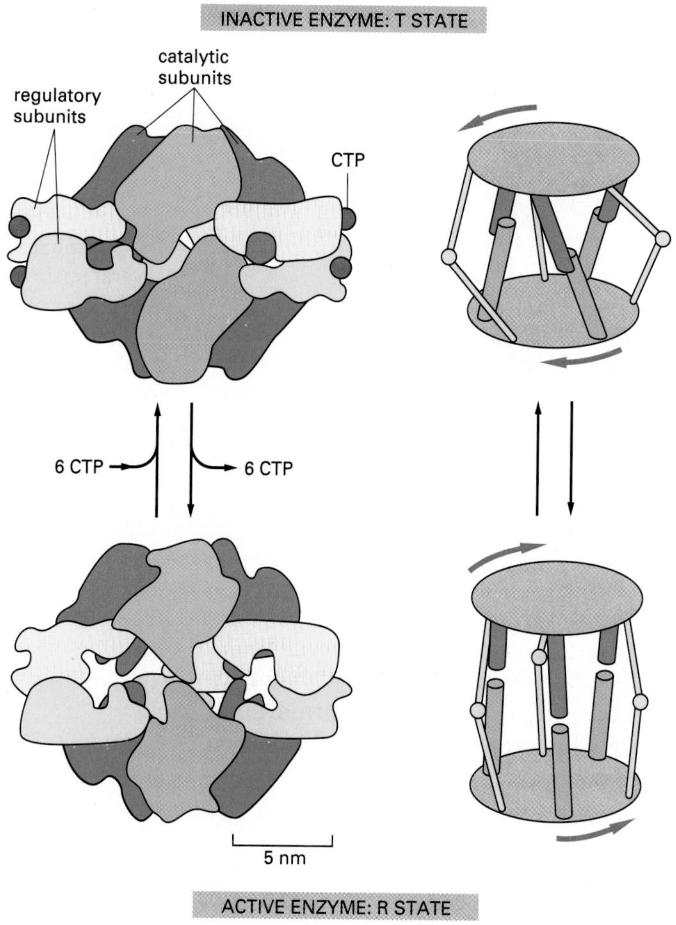

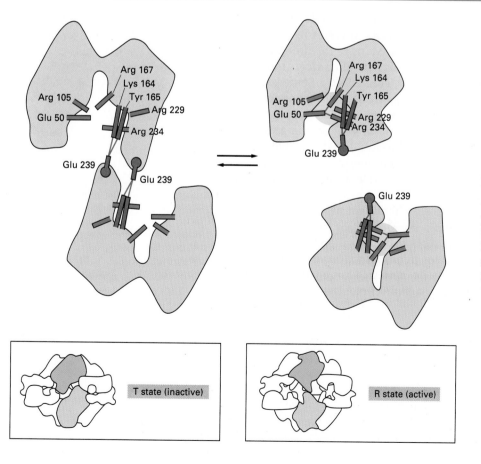

INACTIVE ENZYME: T STATE

regulatory subunits

catalytic subunits

CTP

6 CTP ⟶   ⟵ 6 CTP

5 nm

ACTIVE ENZYME: R STATE

Figure 5–9 The transition between the R and T states in the enzyme aspartate transcarbamoylase. The enzyme consists of a complex of six catalytic subunits and six regulatory subunits, and the structures of its inactive (*T state*) and active (*R state*) forms have been determined by x-ray crystallography. The enzyme is turned off when CTP concentrations rise. Each of the regulatory subunits can bind one molecule of CTP, which is one of the final products in the pathway. By means of this negative feedback regulation, the pathway is prevented from producing more CTP than the cell needs. (Based on K.L. Krause, K.W. Volz, and W.N. Lipscomb, *Proc. Natl. Acad. Sci. USA* 82:1643–1647, 1985.)

Arg 167
Lys 164
Tyr 165
Arg 105
Glu 50
Arg 229
Arg 234
Glu 239
Glu 239

Arg 167
Lys 164
Tyr 165
Arg 105
Glu 50
Arg 229
Arg 234
Glu 239

Glu 239

T state (inactive)

R state (active)

Figure 5–10 Part of the on-off switch in the catalytic subunits of aspartate transcarbamoylase. Changes in the indicated hydrogen-bonding interactions are partly responsible for switching this enzyme's active site between active (*yellow*) and inactive conformations. Hydrogen bonds are indicated by *thin red lines*. The amino acids involved in the subunit-subunit interaction are shown in *red*, while those that form the active site of the enzyme are shown in *blue*. The upper pair of pictures show the catalytic site in the interior of the enzyme; the lower pictures show the external surface of the enzyme. (Adapted from E.R. Kantrowitz and W.N. Lipscomb, *Trends Biochem. Sci.* 15:53–59, 1990.)

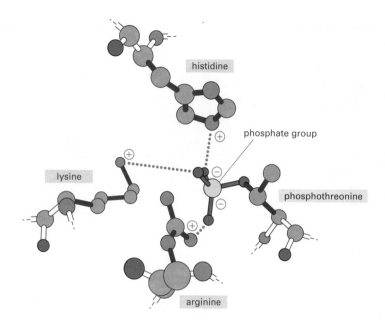

Figure 5–11 **The influence of a phosphate group on a protein.** The negatively charged phosphate group shown here is covalently attached to a threonine side chain of the protein cyclic AMP-dependent protein kinase, which is discussed in Chapter 15. As determined by x-ray crystallography, the phosphate is surrounded by several positively charged amino acid side chains of the same protein. (Adapted from S.S. Taylor et al., *Annu. Rev. Cell Biol.* 8:429–462, 1992. ©1992 Annual Reviews Inc.)

amounts of substrate has the opposite effect, favoring the R state by binding in the cleft of each catalytic subunit and opposing the above conformational change. Conformations that are intermediate between R and T are unstable, so that the enzyme mostly clicks back and forth between its R and T forms, producing a mixture of these two species, whose composition varies depending on the relative concentrations of CTP and substrates.

Protein Phosphorylation Is a Common Way of Driving Allosteric Transitions in Eucaryotic Cells [6]

The activity of proteins in a bacterium such as *E. coli* is regulated mainly by the myriad small molecules in the cell that bind to specific proteins to cause allosteric transitions that control the protein's activity. Many of the proteins regulated in this way are enzymes that catalyze metabolic reactions; others transduce signals or turn genes on and off (see, for example, Figure 9–27). Some bacterial proteins are controlled in a different way, however—by the covalent attachment of a phosphate group to an amino acid side chain. Because each phosphate group carries two negative charges, its addition to a protein can cause a structural change, for example, by attracting a cluster of positively charged side chains (Figure 5–11). Such a change occurring at one site in a protein can in turn alter the protein's conformation elsewhere—to control allosterically the activity of a distant ligand-binding site, for instance.

Reversible protein phosphorylation is the predominant strategy used to control the activity of proteins in eucaryotic cells. More than 10% of the 10,000 proteins in a typical mammalian cell are thought to be phosphorylated. The phosphates are transferred from ATP molecules by *protein kinases* and are taken off by *protein phosphatases*. Eucaryotic cells contain a large variety of these enzymes, many of which play a central role in intracellular signaling (discussed in Chapter 15).

A Eucaryotic Cell Contains Many Protein Kinases and Phosphatases [7]

The **protein kinases** that phosphorylate proteins in eucaryotic cells belong to a large family of enzymes, which contain a similar 250 amino acid catalytic (kinase) domain (Figure 5–12). The various family members contain different amino acid sequences on either side of the kinase domain, and often have short amino acid

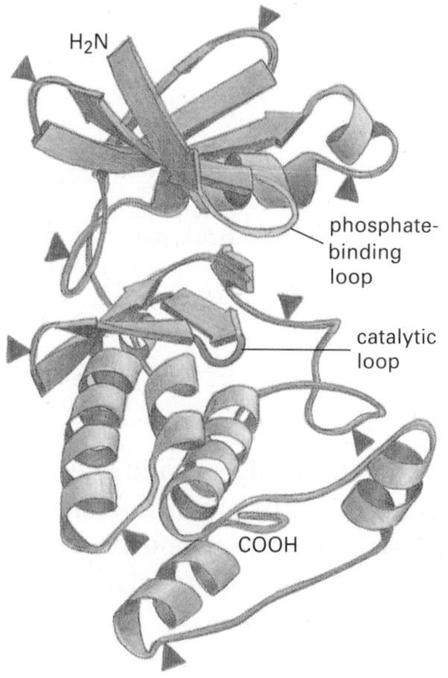

Figure 5–12 The three-dimensional structure of a protein kinase domain.
Superimposed on this structure of the kinase domain of cyclic AMP-dependent kinase are *red arrowheads* to indicate sites where insertions of 5 to 100 amino acids are found in some other members of the protein kinase family. These insertions are located in loops on the surface of the enzyme where other ligands interact with the protein. Thus they distinguish different kinases and confer on them distinctive interactions with other proteins. The ATP (which will donate a phosphate group) and the peptide to be phosphorylated are held in the active site, which extends between the phosphate-binding loop (*yellow*) and the catalytic loop (*orange*). (Adapted from D.R. Knighton et al., *Science* 253:407–414, 1991. © 1991 the AAAS.)

sequences inserted into loops within it (see *red arrowheads* in Figure 5–12). Some of these additional amino acid sequences enable each kinase to recognize the specific set of proteins that it phosphorylates. Other unique sequences allow the activity of each enzyme to be tightly regulated, so that it can be turned on and off in response to different specific signals, as described below.

By comparing the numbers of amino acid sequence differences between the members of a protein family, one can construct an "evolutionary tree" that is thought to reflect the pattern of gene duplication and divergence that gave rise to the family (see Figure 8–76). An evolutionary tree of protein kinases is shown in Figure 5–13. Not surprisingly, kinases with related functions are often located on nearby branches of the tree: the protein kinases involved in cell signaling that phosphorylate tyrosine side chains, for example, are all clustered at the upper left corner of the tree. The other kinases shown phosphorylate either a serine or a threonine side chain, and many are organized into clusters that seem to reflect their function—in transmembrane signaling, intracellular amplification of signals, cell-cycle control, and so on.

The basic reaction catalyzed by a protein kinase is illustrated in Figure 5–14. A phosphate group is transferred from an ATP molecule to a hydroxyl group on a serine, threonine, or tyrosine side chain of a protein. This reaction is essentially unidirectional because of the large amount of free energy released when the phosphate-phosphate bond in ATP is broken to produce ADP (see Figure 2–28). The phosphorylations catalyzed by protein kinases can nevertheless be reversed by a second group of enzymes, called **protein phosphatases,** which remove the phosphate (see Figure 5–14). There are several families of protein phosphatases:

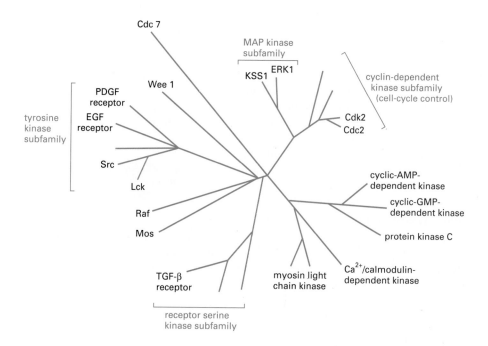

Figure 5–13 An evolutionary tree of selected protein kinases. Although a higher eucaryotic cell contains hundreds of such enzymes, only some of those discussed in this book are shown.

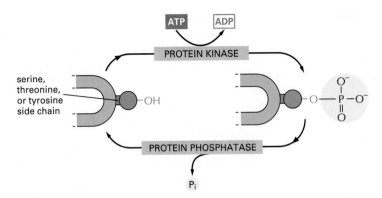

Figure 5–14 **The enzymes that control the phosphorylation of proteins in cells.** The reaction catalyzed by a protein kinase puts a phosphate onto an amino acid side chain, whereas the reaction catalyzed by a protein phosphatase removes this phosphate.

some are highly specific and remove phosphate groups from only one or a few proteins, while others are relatively nonspecific and act on a broad range of proteins. The extent of phosphorylation of a particular protein in a cell at a particular time depends on the relative activities of the protein kinases and phosphatases that act on it.

The Structure of Cdk Protein Kinase Shows How a Protein Can Function as a Microchip [8]

The hundreds of different protein kinases in a eucaryotic cell are organized into complex networks of signaling pathways that help coordinate the cell's activities, drive the cell cycle, and relay signals into the cell from the cell's environment. Many of the signals involved need to be both integrated and amplified. Individual protein kinases (and other signaling proteins) serve as processing devices, or "microchips," in the integration process. An important part of the input to these proteins comes from the control that is exerted by phosphates added to them by other protein kinases in the network: specific sets of phosphate groups serve to activate the protein, while other sets inactivate it.

A **cyclin-dependent protein kinase (Cdk)** represents a good example of such a processing device. Kinases in this class are central components of the cell-division-cycle control system in eucaryotic cells (discussed in Chapter 17). In a vertebrate cell, individual Cdk enzymes turn on and off in succession as a cell proceeds through the different phases of its division cycle, and when they are on, they influence various aspects of cell behavior through their effects on the proteins they phosphorylate. The three-dimensional structure of this important class of protein kinases is now known, and we shall use it to demonstrate how a protein can function as a microchip.

A Cdk protein is active as a protein kinase only when it is bound to a second protein called a *cyclin*. But, as illustrated in Figure 5–15, the binding of cyclin is only one of three distinct "inputs" required to activate the Cdk: in addition, a phosphate must be added to a specific threonine side chain and a phosphate elsewhere in the protein (covalently bound to a specific tyrosine side chain) must be removed. Cdk thus monitors a specific set of cell components—a cyclin, a protein kinase, and a protein phosphatase—and turns on if, and only if, each of these components has attained its appropriate activity state. Some cyclins, for example, rise and fall in concentration in step with the cell cycle, increasing gradually in amount until they are suddenly destroyed at a particular point in the cycle. The sudden destruction of a cyclin (by targeted proteolysis) will immediately shut off its partner Cdk enzyme, and this is an important way of controlling intracellular events such as mitosis.

The three-dimensional structure of Cdk (Figure 5–16A) suggests a likely molecular explanation for the regulation of this enzyme. The Cdk protein on its own is inactive for two reasons: its ATP-binding site is distorted, and a flexible

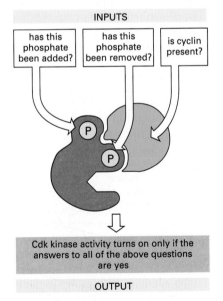

INPUTS

| has this phosphate been added? | has this phosphate been removed? | is cyclin present? |

Cdk kinase activity turns on only if the answers to all of the above questions are yes

OUTPUT

Figure 5–15 **How a Cdk acts as an integrating device.** The function of these central regulators of the cell cycle is discussed in Chapter 17.

204   Chapter 5 : Protein Function

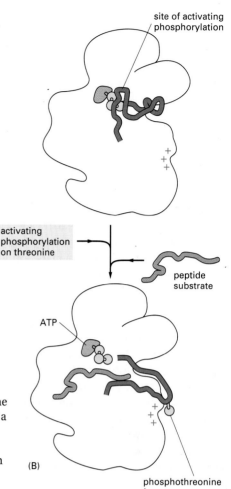

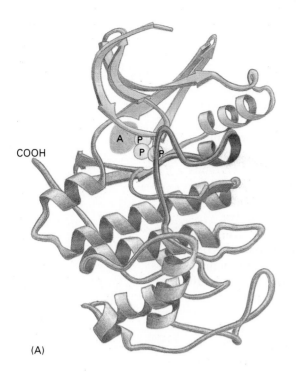

(A)

Figure 5–16 The three-dimensional structure of a Cdk. (A) A diagram of the detailed structure, as determined by x-ray diffraction analysis. Bound ATP is shown in *light red,* with its three phosphate groups in *yellow.* (B) The suggested pathway for enzyme activation includes the phosphorylation of a specific threonine located at the tip of a flexible loop (*red*) that otherwise blocks access of the protein substrate to the active site in the kinase domain. This activation also requires the binding of cyclin, as illustrated in Figure 5–17. (A, adapted from H.L. DeBondt et al., *Nature* 363:595–602, 1993. © 1993 Macmillan Magazines Ltd.)

loop of about 20 amino acids blocks access of the protein substrate to the active site. Cyclin binding both removes the distortion and permits the addition of the activating phosphate group to the tip of the flexible loop; this phosphate is then thought to be attracted to a pocket formed by positively charged amino acids, pulling down the loop so as to permit access to the active site (Figure 5–16B). Cyclin binding also allows the rapid addition of the inhibitory phosphate, however, which interferes with the ATP site, and this keeps the Cdk protein in an inactive state. The kinase is finally activated when a specific phosphatase removes the inhibiting phosphate (Figure 5–17).

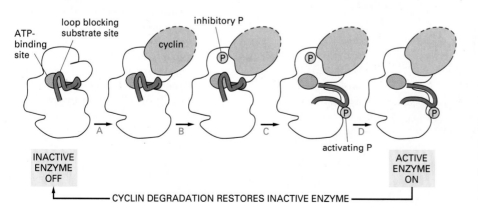

Figure 5–17 A detailed model for Cdk activation. This model, based on the three-dimensional structure of Cdk, explains why Cdk is turned on only if the three separate conditions specified in Figure 5–15 are satisfied. In step A cyclin binds, leading to the addition of the inhibitory phosphate in step B. The activating phosphorylation occurs in step C, but the enzyme turns on only after the inhibitory phosphate is removed in step D. The sudden degradation of cyclin after step D causes enzyme inactivation, including the loss of the activating phosphate, which resets the system to its initial inactive state.

Making Machines Out of Proteins

Proteins That Bind and Hydrolyze GTP Are Ubiquitous Cellular Regulators [9]

We have described how the addition or removal of phosphate groups on a protein can be used by a cell to control the protein's activity. In the examples discussed so far, the phosphate is transferred from an ATP molecule to an amino acid side chain of the protein in a reaction that is catalyzed by a specific protein kinase. Eucaryotic cells also use another way to control protein activity by phosphate addition and removal. In this case the phosphate is not attached directly to the protein; instead, it is a part of the guanine nucleotide GTP, which binds tightly to the protein. With GTP bound the protein is active. The loss of a phosphate group occurs when the bound GTP is hydrolyzed to GDP in a reaction that is catalyzed by the protein itself; with GDP bound the protein is inactive.

GTP-binding proteins (also called *GTPases* because of the GTP hydrolysis that they catalyze) constitute a large family of proteins that all have a similar GTP-binding globular domain. When its bound GTP is hydrolyzed to GDP, this domain undergoes a conformational change that inactivates the protein. The three-dimensional structure of a small GTP-binding protein called *Ras* is illustrated in Figure 5–18.

The **Ras protein** plays a crucial role in cell signaling (as discussed in Chapter 15). In its GTP-bound form it is active and stimulates a cascade of protein phosphorylations in the cell. Most of the time, however, the protein is in its inactive, GDP-bound form. It is activated when it exchanges its GDP for a GTP molecule in response to extracellular signals, such as growth factors, that bind to receptors in the plasma membrane (see Figure 15–53). Thus the Ras protein acts as an on-off switch whose activity is determined by the presence or absence of an additional phosphate on a bound GDP molecule, just as the activity of a Cdk protein is controlled by the presence of one or more phosphate groups on amino acid side chains (see Figure 5–17).

Other Proteins Control the Activity of GTP-binding Proteins by Determining Whether GTP or GDP Is Bound [10]

The activity of Ras and other GTP-binding proteins is controlled by regulatory proteins that determine whether GTP or GDP is bound, just as the activity of a Cdk protein is controlled by cyclins, protein kinases, and protein phosphatases. Ras is inactivated by a **GTPase-activating protein** (or **GAP**), which binds to the Ras protein and induces it to hydrolyze its bound GTP molecule to GDP—which

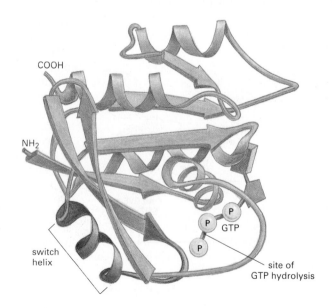

Figure 5–18 The structure of the Ras protein in its GTP-bound form. This relatively small protein illustrates the structure of a GTP-binding domain, which is present in other GTP-binding proteins (see Figure 5–20, for example). The regions shown in *red* change their conformation when the GTP molecule is hydrolyzed to GDP and inorganic phosphate by the protein; the GDP remains bound to the protein, while the inorganic phosphate is released. The special role of the "switch helix" in proteins related to Ras is explained below (see Figure 5–20).

COOH

NH₂

switch helix

P
P
GTP
P

site of GTP hydrolysis

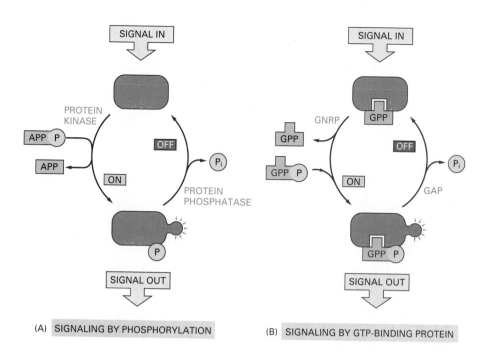

Figure 5–19 **A comparison of the two major intracellular signaling mechanisms in eucaryotic cells.** In both cases a signaling protein is activated by the addition of a phosphate group and inactivated by removal of this phosphate. To emphasize the similarities in the two pathways, ATP and GTP are drawn as APPP and GPPP, and ADP and GDP are drawn as APP and GPP, respectively. As shown in Figure 5–17, addition of a phosphate to a protein can also be inhibitory.

remains tightly bound—and inorganic phosphate (P_i), which is rapidly released. The Ras protein will stay in its inactive, GDP-bound conformation until it encounters a **guanine nucleotide releasing protein (GNRP),** which binds to GDP-Ras and causes it to release its GDP. Because the empty nucleotide-binding site is immediately filled by a GTP molecule (GTP is present in large excess over GDP in cells), the GNRP activates Ras by *indirectly* adding back the phosphate removed by GTP hydrolysis. Thus, in a sense, the roles of GAP and GNRP are analogous to those of a protein phosphatase and a protein kinase, respectively (Figure 5–19).

The Allosteric Transition in EF-Tu Protein Shows How Large Movements Can Be Generated from Small Ones [11]

The Ras protein is a member of a family of *monomeric regulatory GTPases*, each of which consists of a single GTP-binding domain of about 200 amino acids. During the course of evolution this domain has also become joined to other protein domains to create a large family of GTP-binding proteins, whose members include the receptor-associated trimeric G proteins (discussed in Chapter 15), proteins regulating the traffic of vesicles between intracellular compartments (discussed in Chapter 13), and proteins that bind to transfer RNA and are required for protein synthesis on the ribosome (discussed in Chapter 6). In each case, an important biological activity is controlled by a change in the protein's conformation caused by GTP hydrolysis in a Ras-like domain.

The *EF-Tu protein* provides a good example of how this family of proteins works. EF-Tu is an abundant molecule in bacterial cells, where it serves as an elongation factor in protein synthesis, loading each amino-acyl tRNA molecule onto the ribosome. The tRNA molecule forms a tight complex with the GTP-bound form of EF-Tu. In this complex, the amino acid attached to the tRNA is masked; its unmasking, which is required for protein synthesis, occurs on the ribosome when the tRNA is released following hydrolysis of the GTP bound to EF-Tu (see Figure 6–31 for an illustration of the clock-like function of EF-Tu).

The three-dimensional structure of EF-Tu, in both its GTP- and GDP-bound forms, has been determined by x-ray crystallography. These studies reveal how the unmasking of the tRNA occurs. The dissociation of the inorganic phosphate group (P_i), which follows the reaction GTP → GDP + P_i, causes a shift of a few tenths of a nanometer at the GTP-binding site, just as it does in the Ras protein.

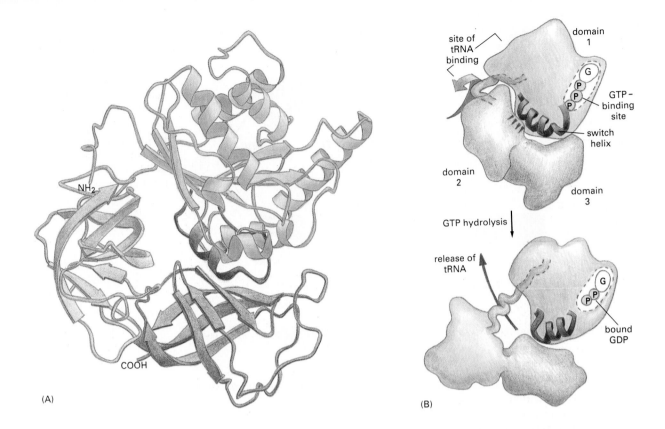

(A)

(B)

This tiny movement, equivalent to a few times the diameter of a hydrogen atom, causes a conformational change to propagate along a crucial piece of α helix, called the *switch helix*, in the Ras-like domain of the protein. The switch helix seems to serve as a latch that adheres to a specific site in another domain of the molecule, holding the protein in a "shut" conformation. The conformational change triggered by GTP hydrolysis causes the switch helix to detach, allowing separate domains of the protein to swing apart, through a distance of about 4 nanometers, thereby releasing the bound tRNA (Figure 5–20).

One can see from this example how cells can exploit simple chemical changes that occur on the surface of a small protein domain to evolve larger proteins with sophisticated functions. In the transition from Ras to EF-Tu we have entered a world that begins to feel like biology.

Proteins That Hydrolyze ATP Do Mechanical Work in Cells [12]

Allosteric shape changes can be used to generate orderly movements in cells as well as to regulate chemical reactions. Suppose, for example, that a protein is required that can "walk" along a narrow thread, such as a DNA molecule. Figure 5–21 shows schematically how an allosteric protein might do this by undergoing a series of conformational changes. With nothing to drive these changes in an orderly sequence, however, they will be perfectly reversible, and the protein will wander randomly back and forth along the thread.

We can look at this situation another way. Since the directional movement of a protein does work, the laws of thermodynamics demand that such movement depletes free energy from some other source (otherwise the protein could be used to make a perpetual motion machine). Therefore, no matter what modifications we make to the model shown in Figure 5–21, such as adding ligands that favor particular conformations, without an input of energy the protein molecule shown could only wander aimlessly.

How can one make the series of conformational changes unidirectional? To make the entire cycle proceed in one direction, it is enough to make any one of the steps irreversible. One way to do this is to use the mechanism just dis-

Figure 5–20 **The large conformational change in EF-Tu caused by GTP hydrolysis.** (A) The three-dimensional structure of EF-Tu with GTP bound. The domain at the top is homologous to the Ras protein, and its *red* α helix is the "switch helix," which moves after GTP hydrolysis, as shown in Figure 5–18. (B) The change in the conformation of the switch helix in domain 1 causes domains 2 and 3 to rotate as a single unit by about 90° toward the viewer, which releases the tRNA. (A, adapted from Berchtold et al., *Nature* 365:126–132, 1993. © 1993 Macmillan Magazines, Ltd.; B, courtesy of Mathias Sprinzl and Rolf Hilgenfeld.)

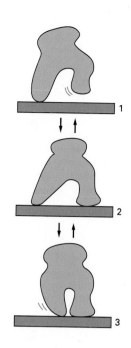

Figure 5–21 **An allosteric "walking" protein.** Although its three different conformations allow it to wander randomly back and forth while bound to the thread, the protein cannot move uniformly in a single direction.

cussed for driving allosteric changes in a protein molecule by GTP hydrolysis. For example, because a great deal of free energy is released when GTP is hydrolyzed, it is very unlikely that the EF-Tu protein will directly add a phosphate molecule to GDP to reverse the hydrolysis of its GTP. Precisely the same principle applies to ATP hydrolysis, and most proteins that are able to walk in one direction for long distances (the so-called *motor proteins*) do so by hydrolyzing ATP.

In the highly schematic model shown in Figure 5–22, ATP binding shifts a motor protein from conformation 1 to conformation 2. The bound ATP is then hydrolyzed to produce ADP and inorganic phosphate (P_i), causing a change from conformation 2 to conformation 3. Finally, the release of the bound ADP and P_i drives the protein back to conformation 1. Because the transitions $1 \rightarrow 2 \rightarrow 3 \rightarrow 1$ are driven by the energy provided by ATP hydrolysis, this series of conformational changes will be effectively irreversible under physiological conditions (that is, the probability that ADP will recombine with P_i to form ATP by the route $1 \rightarrow 3 \rightarrow 2 \rightarrow 1$ is extremely low). Thus the entire cycle will go in only one direction, causing the protein molecule to move continuously to the right in this example. Many proteins generate directional movement in this way, including *DNA helicase* enzymes that propel themselves along DNA at rates as high as 1000 nucleotides per second.

The Structure of Myosin Reveals How Muscles Exert Force [13]

In Chapter 16 we discuss how various cell movements are produced by motor proteins that move rapidly along protein filaments, driven by energy derived from repeated cycles of ATP hydrolysis (see Figure 5–22). The best understood of these motor proteins is **myosin,** whose directed movement along actin filaments causes both intracellular movements and muscle contraction. The three-dimensional structures of myosin (and actin) have been determined by x-ray diffraction analyses, providing a glimpse of the inner workings of a biological motor. The structure of the myosin head domain (Figure 5–23) suggests how ATP hydrolysis may be coupled to force generation. ATP binding and hydrolysis are thought to cause an ordered series of conformational changes that move the tip of the head by about 5 nanometers, as illustrated schematically in Figure 5–24. This movement, coupled to the making and breaking of interactions with actin and repeated with each round of ATP hydrolysis, propels the myosin molecule unidirectionally along an actin filament (see Figure 16–91). Thus in myosin, as in the EF-Tu protein discussed earlier, a small perturbation in the nucleotide-binding site is translated, via allosteric transitions that magnify the effect, to create the much more extensive, orderly protein motions that underlie much of cell biology.

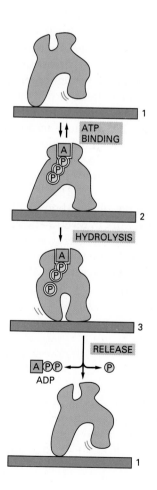

ATP-driven Membrane-bound Allosteric Proteins Can Either Act as Ion Pumps or Work in Reverse to Synthesize ATP [14]

Besides generating mechanical force, allosteric proteins can use the energy of ATP hydrolysis to do other forms of work, such as pumping specific ions into or out of the cell. An important example is the **Na$^+$-K$^+$ ATPase** found in the plasma

Figure 5–22 **An allosteric motor protein.** An orderly transition among three conformations is driven by the hydrolysis of a bound ATP molecule. Because one of these transitions is coupled to the hydrolysis of ATP, the cycle is essentially irreversible. By repeated cycles the protein moves continuously to the right along the thread.

Making Machines Out of Proteins

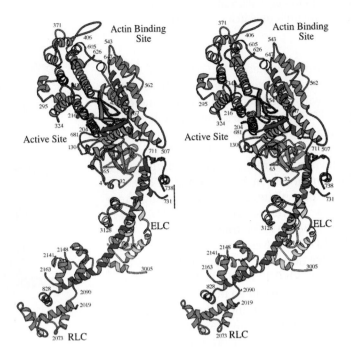

Figure 5–23 **The structure of the myosin head.** In this stereo diagram of the myosin head domain, ATP hydrolysis occurs at the *active site*. ELC denotes the essential light chain and the RLC the regulatory light chain, both of which contribute, along with the myosin heavy chain, to the head domain. (From I. Rayment et al., *Science* 261:50–58, 1993. © 1993 the AAAS.)

membrane of all animal cells, which pumps 3 Na⁺ out of the cell and 2 K⁺ in during each cycle of conformational changes driven by ATP hydrolysis (see Figure 11–11). This ATP-driven pump consumes more than 30% of the total energy requirement of most cells. By continuously pumping Na⁺ out and K⁺ in, it keeps the Na⁺ concentration much lower inside the cell than outside and the K⁺ concentration much higher inside than outside, thereby generating two ion gradients (in opposite directions) across the plasma membrane. These and other ion gradients across various cell membranes can store energy, just as the differences of water pressure on either side of a dam can. The energy is used to drive conformational changes in a variety of membrane-bound allosteric proteins that do useful work. The large Na⁺ gradient across the plasma membrane, for example, drives many other plasma-membrane-bound protein pumps that transport glucose or specific amino acids into the cell; the glucose and amino acids are dragged in by the simultaneous influx of Na⁺ that occurs as Na⁺ moves down its concentration gradient.

The membrane-bound allosteric pumps that are driven by ATP hydrolysis can also work in reverse and employ the energy in the ion gradient to synthesize ATP. In fact, the energy available in the H⁺ gradient across the inner mitochondrial membrane is used in this way by the membrane-bound allosteric protein complex, **ATP synthase,** which synthesizes most of the ATP required by animal cells, as we discuss in Chapter 14.

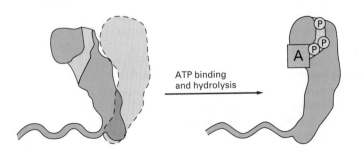

ATP binding and hydrolysis

Figure 5–24 **A conceptual view of a major conformational change in myosin that is postulated to be caused by ATP binding and hydrolysis.** This model is based on the structure shown in Figure 5–23. At the next step in the hydrolysis process, the inorganic phosphate molecule produced (*top*) will be released into solution. (After I. Rayment et al., *Science* 261:58–65, 1993. © 1993 the AAAS.)

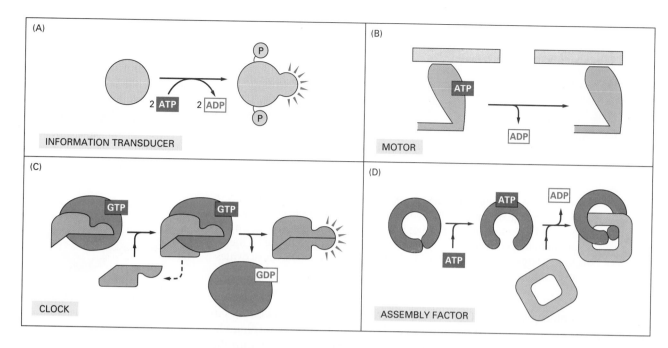

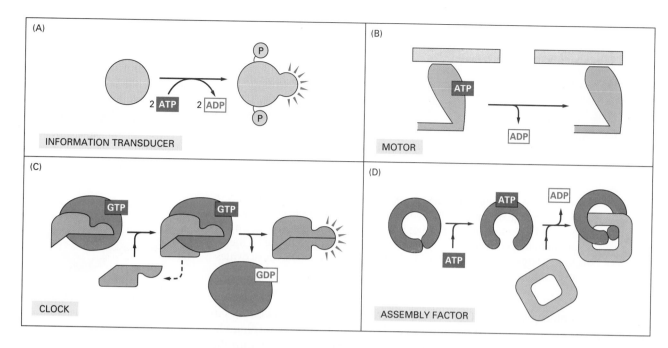

(A) INFORMATION TRANSDUCER

(B) MOTOR

(C) CLOCK

(D) ASSEMBLY FACTOR

Energy-coupled Allosteric Transitions in Proteins Allow the Proteins to Function as Motors, Clocks, Assembly Factors, or Transducers of Information [15]

Many proteins undergo ordered conformational changes that are coupled to the energy released when a nucleoside triphosphate (either ATP or GTP) is hydrolyzed to a nucleoside diphosphate (ADP or GDP, respectively). Some of these changes involve the covalent attachment of a phosphate group to the protein (protein phosphorylation), but many others, as for myosin, do not. Each change is generally triggered by a specific event (the binding of myosin to an actin filament, for example, triggers ATP hydrolysis by myosin), imparting directionality and order to the interactions of macromolecules in the cell.

The ability to harness the energy in nucleoside triphosphates to drive allosteric changes in proteins has been crucial for the evolution of cells in much the same way that the ability to harness electrical energy has been crucial for the development of modern technology. In both cases rich opportunities have opened up for the development of useful devices. Proteins like Cdk, for example, act as sophisticated integrating switches (see Figure 5–15), receiving information about a cell's environment and the stage of the cell cycle and using it to coordinate the behavior of the cell. Motor proteins like myosin move unidirectionally along filaments to generate various movements and create order inside the cell. Proteins such as EF-Tu serve as timing devices that improve the fidelity of important biological reactions (see Figure 6–31). Other proteins use the energy released by nucleoside triphosphate hydrolysis to catalyze the assembly of specific protein complexes. A summary is presented in Figure 5–25.

Proteins Often Form Large Complexes That Function as Protein Machines [16]

As one progresses from small proteins to large proteins formed from many domains, the functions that a protein can perform become more elaborate. The most impressive tasks, however, are carried out by large protein assemblies formed from multiple individual subunits. Now that it is possible to reconstruct most biological processes in cell-free systems in a test tube, one can see that each central process in a cell—such as DNA replication, RNA or protein synthesis, vesicle budding, or transmembrane signaling—is catalyzed by a complex of 10 or more proteins. In such **protein machines** the hydrolysis of bound nucleoside

Figure 5–25 **Some devices made from proteins.** In these examples the energy of nucleoside triphosphate hydrolysis is used to drive conformational changes in allosteric proteins. (A) A transducer of information, such as a protein kinase. (B) A motor, such as myosin. (C) A clock, such as EF-Tu, that delays assembly of an active complex to insure that incorrect complexes dissociate (*dotted line*). (D) An assembly factor that builds larger structures.

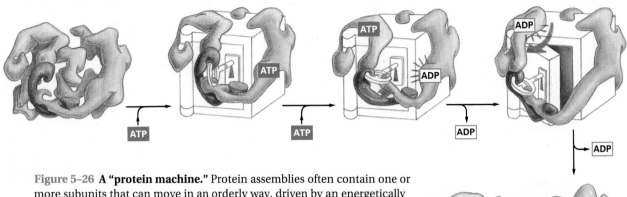

Figure 5–26 **A "protein machine."** Protein assemblies often contain one or more subunits that can move in an orderly way, driven by an energetically favorable change that occurs in a bound substrate molecule (see Figure 5–22). Protein movements of this type are especially useful to the cell if they occur in a large protein assembly in which, as illustrated here, the activities of several subunits can be coordinated.

triphosphate molecules (ATP or GTP) drives ordered conformational changes in the individual proteins, enabling the ensemble of proteins to move coordinately. In this way, for example, the appropriate enzymes are moved directly into the positions where they are needed to carry out each reaction in a series instead of waiting for the random collision of each separate component that would otherwise be required. A simple mechanical analogy is illustrated in Figure 5–26.

Cells have evolved protein machines for the same reason that humans have invented mechanical and electronic machines: manipulations that are spatially and temporally coordinated through linked processes are much more efficient for accomplishing almost any task than is the sequential use of individual tools.

Summary

Allosteric proteins reversibly change their shape when ligands bind to their surface. The changes produced by one ligand often affect the binding of a second ligand, and this type of linkage between two ligand-binding sites provides a crucial mechanism for regulating cell processes. Metabolic pathways, for example, are controlled by feedback regulation: some small molecules will inhibit and other small molecules activate enzymes early in a pathway. Enzymes regulated in this way generally form symmetrical assemblies, allowing cooperative conformational changes to create a steep response to ligands.

Changes in protein shape can be driven in a unidirectional manner by the expenditure of chemical energy. By coupling allosteric shape changes to ATP hydrolysis, for example, proteins can do useful work, such as generating a mechanical force or pumping ions across a membrane. The three-dimensional structures of several proteins, determined by x-ray crystallography, have revealed how a small local change caused by nucleoside triphosphate hydrolysis is amplified to create major changes elsewhere in the protein; by such means these proteins are able to serve as transducers of information, motors, clocks, or assembly factors. Highly efficient "protein machines" are formed by incorporating many different protein subunits into larger assemblies in which allosteric movements of the individual components are coordinated to carry out many, if not most, biological reactions.

The Birth, Assembly, and Death of Proteins

Having described some of the remarkable devices that cells make out of proteins, we now consider how these devices are produced and how they are destroyed. The mechanism of protein synthesis is discussed elsewhere. We begin by con-

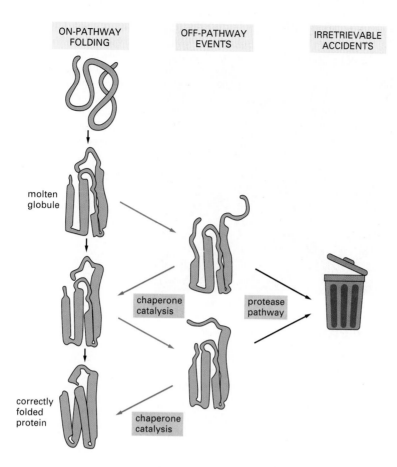

ON-PATHWAY FOLDING   OFF-PATHWAY EVENTS   IRRETRIEVABLE ACCIDENTS

molten globule

chaperone catalysis

protease pathway

correctly folded protein

chaperone catalysis

Figure 5–27 **A current view of protein folding.** A newly synthesized protein rapidly attains a "molten globule" state (see Figure 5–28). Subsequent folding occurs more slowly and by multiple pathways, some of which reach dead ends without the help of a molecular chaperone. Some molecules may still fail to fold correctly; these are recognized and degraded by proteolytic enzymes (see Figure 5–39).

sidering how a protein folds and assembles once it leaves the ribosome as a finished polypeptide chain.

Proteins Are Thought to Fold Through a Molten Globule Intermediate [17]

Because many purified proteins will refold properly on their own after being unfolded *in vitro,* for many years it was thought that a protein will try out every conceivable conformation as it folds until it attains the one conformation with the lowest free energy, which was assumed to be its correctly folded state. We now know that this view is incorrect: despite the high speed of molecular motions in a protein (see p. 97), there are vastly more possible conformations for any large protein than can be explored in the few seconds that are typically required for folding. Moreover, the existence of mutant proteins that have specific defects in folding indicates that a protein's amino acid sequence has been selected during evolution, not only for the properties of its final structure, but also for the ability to fold rapidly into its native conformation.

The ability of pure, denatured proteins to reform their native structures on their own has made it possible to dissect the process of protein folding experimentally. Proteins appear to fold rapidly into a structure in which most (but not all) of the final secondary structure (α helices and β sheets) has formed and in which these elements of structure are aligned in roughly the right way (Figure 5–27). This unusually open and flexible conformation, which is called a *molten globule* (Figure 5–28), is the starting point for a relatively slow process in which many side-chain adjustments occur in order to form the correct tertiary structure. In the latter process a variety of pathways can be taken toward the final conformation. Some of these may be nonproductive dead ends without the help of a *molecular chaperone,* special proteins in cells whose function is to help other proteins fold and assemble into stable, active structures (see Figure 5–27).

The Birth, Assembly, and Death of Proteins

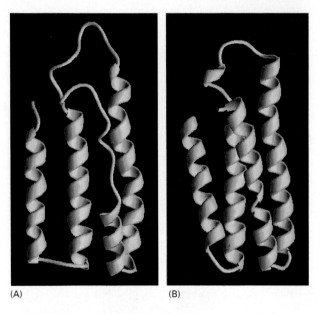

(A)   (B)

Figure 5–28 The structure of a molten globule. (A) A molten globule form of cytochrome b562 is more open and less highly ordered than the native protein, shown in (B). Note that the molten globule contains most of the secondary structure of the native form, although the ends of the α helices are frayed and one of these helices is only partly formed. (Courtesy of Joshua Wand.)

Molecular Chaperones Facilitate Protein Folding [18]

Molecular chaperones were first identified in bacteria when *E. coli* mutants that failed to allow bacteriophage lambda to replicate in them were studied. These mutants produce slightly altered versions of two components of the chaperone machinery, related to heat-shock proteins 60 and 70 (hsp60 and hsp70), and as a result are defective in specific steps in the assembly of the viral proteins.

Eucaryotic cells have families of hsp60 and hsp70 proteins, and different family members function in different organelles. Thus, as discussed in Chapter 12, mitochondria contain their own hsp60 and hsp70 molecules that are distinct from those that function in the cytosol, and a special hsp70 (called *BIP*) helps to fold proteins in the endoplasmic reticulum.

Both hsp60-like and hsp70 proteins work with a small set of associated proteins when they help other proteins to fold. They share an affinity for the exposed hydrophobic patches on incompletely folded proteins, and they hydrolyze ATP, possibly binding and releasing their protein with each cycle of ATP hydrolysis. Originally, molecular chaperones were thought to act only by preventing the promiscuous aggregation of still unfolded proteins (hence their name). It is now thought, however, that they also interact more intimately with their clients, producing effects that can be likened to a "protein massage." By binding to exposed hydrophobic regions, the chaperone massages those regions of a protein that are likely to have misfolded from the molten globule state, changing their structure in a way that gives the protein another chance to fold (see Figure 5–27).

In some other respects the two types of hsp proteins function differently. The hsp70 machinery is thought to act early in the life of a protein, binding to a string of about seven hydrophobic amino acids before the protein leaves the ribosome (Figure 5–29). In contrast, hsp60-like proteins form a large barrel-shaped structure (Figure 5–30) that acts later in a protein's life; this chaperone is thought to form an "isolation chamber" into which misfolded proteins are fed, providing them with a favorable environment in which to attempt to refold (see Figure 5–29).

These molecular chaperones are called *heat-shock proteins* because they are synthesized in dramatically increased amounts following a brief exposure of cells to an elevated temperature (for example, 42°C). This seems to reflect the operation of a feedback system that responds to any increase in misfolded proteins (such as those produced by elevated temperatures) by boosting the synthesis of the chaperones that help the protein refold.

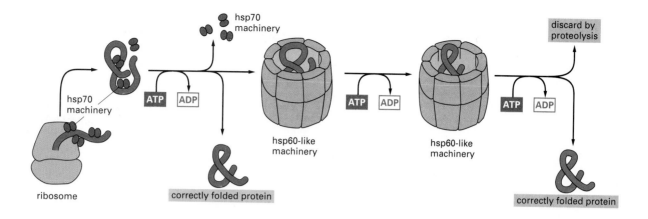

ribosome

correctly folded protein

hsp60-like machinery

hsp60-like machinery

correctly folded protein

discard by proteolysis

Many Proteins Contain a Series of Independently Folded Modules [19]

The folding of a newly synthesized protein often begins with the formation of a number of distinct structurally stable domains that correspond to functional units, which seem to have ancient evolutionary origins. Elsewhere we discuss the pathways by which proteins are thought to have evolved, emphasizing how new proteins have been created by the shuffling of exons that code for conserved domains with useful properties (see pp. 386–394). Evolution has preserved some of these domains as folding units that retain their structure even when cut out of the protein—either by selected proteolysis or, more efficiently, by genetic engineering techniques. Protein domains of this type that are very frequently involved in evolutionary exon shuffling are called **modules;** their importance has become clear now that DNA sequences are available for thousands of genes.

Protein modules are typically 40 to 100 amino acids in length. Their small size and ability to fold independently has made it possible to determine many of their three-dimensional structures in solution by high-resolution NMR techniques, which is a convenient alternative to x-ray crystallography. Some typical modules are illustrated in Figure 5–31. Each of these modules has a stable core structure formed from strands of β sheet, from which less-ordered loops of polypeptide chain protrude (shown in *green*). The loops are ideally situated to form binding sites for other molecules, as well demonstrated for the immunoglobulin fold, which was first recognized in antibody molecules (see Figure 23–35). The evolutionary success of β-sheet-based modules is likely to have been due to their forming a convenient framework for the generation of new binding sites for ligands through changes to these protruding loops.

Figure 5–29 Two families of molecular chaperones. The hsp70 proteins act early, recognizing small patches on a protein's surface. The hsp60-like proteins appear to act later and form a container into which proteins that have still failed to fold are transferred. In both cases repeated cycles of ATP hydrolysis by the hsp proteins contribute to a cycle of binding and release of the client protein that helps this protein to fold.

Figure 5–30 The structure of an hsp60-like chaperone, as determined by electron microscopy. A large number of negatively stained particles is shown in (A) and a 3-D model of a single particle, derived by computer-based image processing methods, is shown in (B). A similar large barrel-shaped structure is found in both eucaryotes and procaryotes. This type of protein is called hsp60 in mitochondria, groEL in bacteria, and TCP-1 in the cytosol of vertebrate cells. (A, from B.M. Phipps et al., *EMBO J.* 10:1711–1722, 1991; B, from B.M. Phipps et al., *Nature* 361:475–477, 1993. © 1993 Macmillan Magazines Ltd.)

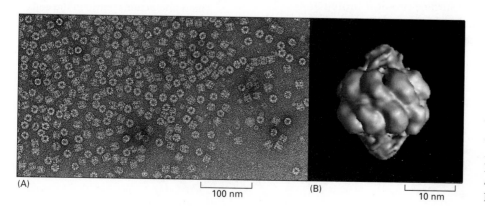

(A) 100 nm

(B) 10 nm

The Birth, Assembly, and Death of Proteins

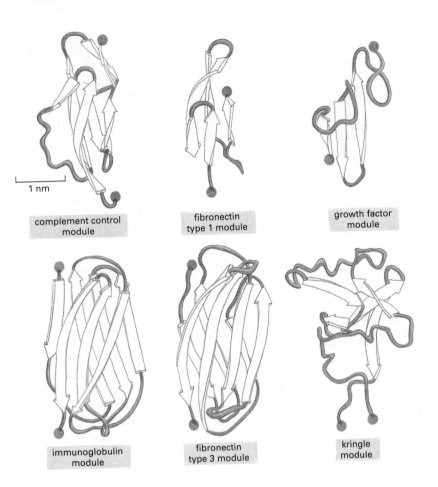

complement control
module

fibronectin
type 1 module

growth factor
module

immunoglobulin
module

fibronectin
type 3 module

kringle
module

1 nm

Figure 5–31 The three-dimensional structures of some protein modules. In these ribbon diagrams, β-sheet strands are shown as *arrows*, and the N- and C-termini are marked with *red balls*. (Adapted from M. Baron, D.G. Norman, and I.D. Campbell, *Trends Biochem. Sci.* 16:13–17, 1991, and D.J. Leahy et al., *Science* 258:987–991, 1992. © by AAAS.)

Modules Confer Versatility and Often Mediate Protein-Protein Interactions [19, 20]

A second feature of protein modules that explains their utility is the ease with which they can be integrated into other proteins. Five of the six modules illustrated in Figure 5–31 have their N- and C-terminal ends (marked with *red balls*) at opposite ends of the module. This "in-line" arrangement means that when the DNA encoding such a module undergoes tandem duplication, which is not unusual in the evolution of genomes (discussed in Chapter 8), the duplicated modules can be readily accommodated in the protein. In this way such modules can become linked in series to form extended structures, either with themselves (Figure 5–32) or with other in-line modules. Stiff extended structures composed of a series of modules are commonly found both in extracellular matrix molecules and in the extracellular portions of cell surface receptor proteins.

Other modules, like the kringle module in Figure 5–31, are of a "plug-in" type. After genomic rearrangements, they can be easily accommodated as an insertion

Figure 5–32 An extended structure formed from a series of in-line protein modules. Here, five fibronectin type 3 modules are shown forming a repeating array. Similar structures are found in several extracellular matrix molecules. Side-chain interactions between the ends of modules are thought to impart rigidity to such structures.

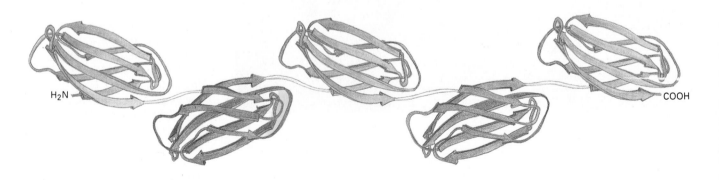

H₂N

COOH

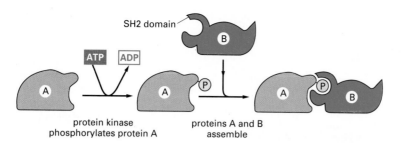

Figure 5–33 **SH2 domains mediate protein assembly reactions that depend on protein phosphorylations.** The structure of an SH2 domain, which has the form of a plug-in module, is illustrated in Figure 15–49.

into a loop region of a second protein. Some of these modules act as specific binding sites for other proteins or structures in the cell. An important example is the SH2 domain, which can bind tightly to a region of polypeptide chain that contains a phosphorylated tyrosine side chain. Because each SH2 domain also recognizes other features of the polypeptide, it binds only to a subset of proteins that contains phosphorylated tyrosines. The presence of an SH2 domain in a protein allows it to form complexes with proteins that become phosphorylated on tyrosines in response to cell-signaling events (Figure 5–33). Such protein complexes that form and break up as a result of changes in protein phosphorylation play a central part in transducing extracellular signals into intracellular ones, as described in Chapter 15.

Proteins Can Bind to Each Other Through Several Types of Interfaces

Proteins can bind to other proteins in at least three ways. In many cases a portion of the surface of one protein contacts an extended loop of polypeptide chain (a "string") on a second protein (Figure 5–34A). Such a surface-string interaction, for example, allows the SH2 domain to recognize a phosphorylated loop of another protein, and it also enables a protein kinase to recognize the proteins that it will phosphorylate (see Figure 5–16B).

A second type of protein-protein interface is formed when two α helices, one from each protein, pair together to form a coiled-coil (Figure 5–34B). This type of protein interface is found in several families of gene regulatory proteins, as discussed in Chapter 9.

The most common way for proteins to interact, however, is by the precise matching of one rigid surface with that of another (Figure 5–34C). Such interactions can be very tight, since a large number of weak bonds can form between two surfaces that match well. For the same reason such surface-surface interactions can be extremely specific, allowing one protein to select a specific partner from the many thousands of different proteins found in a higher eucaryotic cell.

Linkage and Selective Proteolysis Ensure All-or-None Assembly

Many proteins are present in large complexes with other proteins. This requires that the protein bind to several other proteins at the same time. It is crucial for

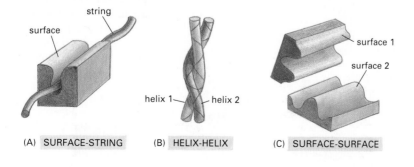

(A) SURFACE-STRING   (B) HELIX-HELIX   (C) SURFACE-SURFACE

Figure 5–34 **Three ways that two proteins can bind to each other.** Only the interacting parts of the two proteins are shown. (A) A rigid surface on one protein can bind to an extended loop of polypeptide chain (a "string") on a second protein. (B) Two α helices can bind together to form a coiled-coil. (C) Two complementary rigid surfaces often link two proteins together.

The Birth, Assembly, and Death of Proteins

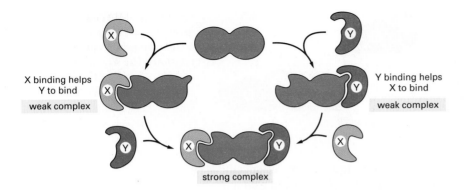

Figure 5–35 Linkage facilitates an efficient all-or-none assembly of protein complexes. As indicated, proteins X and Y each induce an allosteric shape change in a third protein (shown in *blue*) that helps the other protein to bind. As a result, the complex of all three proteins may be the only one that is strong enough to exist in the cell, resulting effectively in all-or-none assembly.

the cell that each protein complex form efficiently and that the formation of partial complexes, which can interfere with the function of complete complexes, be kept to a minimum. There must be mechanisms, therefore, for ensuring that assembly is an all-or-none process.

One important mechanism relies on the phenomenon of *linkage*, which we described earlier. Because of linkage, if a ligand changes the shape of an allosteric protein so that the protein binds a second ligand more tightly, the second ligand must similarly increase the affinity of the protein for the first ligand (see Figure 5–5). The same principle applies to protein-protein interactions. When two proteins bind to each other, they often increase the affinity of one of the partners for a third protein. Because of linkage, the complex of all three proteins will be much more stable than a complex containing only two. A mechanism of this type can produce all-or-none assembly (Figure 5–35).

Even if an all-or-none assembly mechanism drives the formation of complete protein complexes, unless the cell contains exactly the right proportions of each protein in the complex, unassembled proteins will be left over. In fact, cells do not always produce their components in precise amounts and are instead able to degrade selectively any protein component that is left unassembled (Figure 5–36). Cells therefore require a sophisticated system to identify abnormally assembled proteins and destroy them. Indeed, the eucaryotic cell contains an elaborate set of proteins that enables such incomplete assemblies to be selectively directed to its protein-degradation machinery, as we now discuss.

Ubiquitin-dependent Proteolytic Pathways Are Largely Responsible for Selective Protein Turnover in Eucaryotes [21]

One function of intracellular proteolytic mechanisms is to recognize and eliminate unassembled proteins, as just described. Another is to dispose of damaged or misfolded proteins (see Figure 5–27). Yet another is to confer short half-lives on certain normal proteins whose concentrations must change promptly with

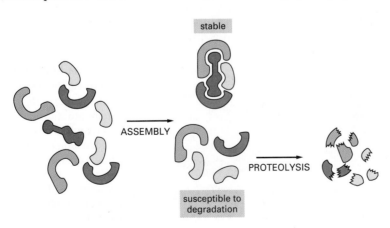

Figure 5–36 Proteolysis of the extra components of a protein complex prevents them from accumulating in a cell. The degradation shown here requires that an unassembled protein be recognized by enzymes that covalently add ubiquitin to it, as discussed in the text.

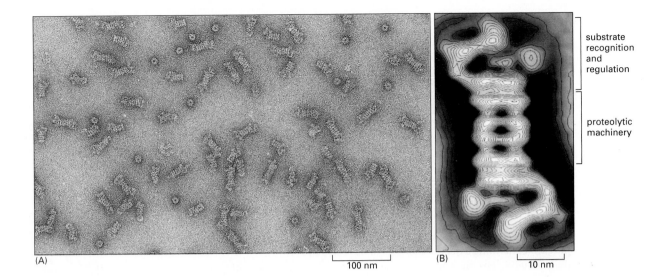

alterations in the state of a cell; many of these short-lived proteins are degraded rapidly at all times, while others, most notably the cyclins, are stable until they are suddenly degraded at one particular point in the cell cycle. Although here we mainly discuss how proteins are degraded in the cytosol, important degradation pathways also operate in the endoplasmic reticulum (ER) and, as discussed in Chapter 13, in lysosomes.

Most of the proteins that are degraded in the cytosol are delivered to large protein complexes called **proteasomes,** which are present in many copies and are dispersed throughout the cell. Each proteasome consists of a central cylinder formed from multiple distinct proteases, whose active sites are thought to face an inner chamber. Each end of the cylinder is "stoppered" by a large protein complex formed from at least 10 types of polypeptides, some of which hydrolyze ATP (Figure 5–37). These protein stoppers are thought to select the proteins for destruction by binding to them and feeding them into the inner chamber of the cylinder, where multiple proteases degrade the proteins to short peptides that are then released.

Proteasomes act on proteins that have been specifically marked for destruction by the covalent attachment of a small protein called **ubiquitin** (Figure 5–38). Ubiquitin exists in cells either free or covalently linked to proteins. Most ubiquinated proteins have been tagged for degradation. (Some long-lived proteins such as histones are also ubiquinated, but in these cases the function of ubiquitin is not understood.) Different *ubiquitin-dependent* proteolytic pathways employ structurally similar but distinct ubiquitin-conjugating enzymes that are associated with recognition subunits that direct them to proteins carrying a particular degradation signal. The conjugating enzyme adds ubiquitin to a lysine residue of a target protein and thereafter adds a series of additional ubiquitin moieties, forming a multiubiquitin chain (Figure 5–39) that is thought to be recognized by a specific receptor protein in the proteasome.

Denatured or misfolded proteins, as well as proteins containing oxidized or otherwise abnormal amino acids, are recognized and degraded by ubiquitin-dependent proteolytic systems. The ubiquitin-conjugating enzymes presumably recognize signals that are exposed on these proteins as a result of their misfolding or chemical damage; such signals are likely to include amino acid sequences or

Figure 5–37 **A proteasome.** A large number of negatively stained particles is shown in (A). A 3-D model of a single complete proteasome complex, derived by computer-based image processing of such images, is shown in (B). Many copies of this structure are present throughout the cell, where they serve as trash cans for the cell's unwanted proteins. (Electron micrographs courtesy of Wolfgang Baumeister, from J.M. Peters et al. *J. Mol Biol.* 234: 932–937, 1993.)

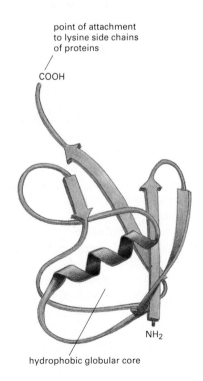

point of attachment to lysine side chains of proteins

COOH

NH₂

hydrophobic globular core

Figure 5–38 **The three-dimensional structure of ubiquitin.** This protein contains 76 amino acid residues. The addition of a chain of ubiquitin molecules to a protein results in the degradation of this protein by the proteasome (see Figure 5–39). (Based on S. Vijay-Kumar, C.E. Bugg, K.D. Wilkinson, and W.J. Cook, *Proc. Natl. Acad. Sci. USA* 82:3582–3585, 1985.)

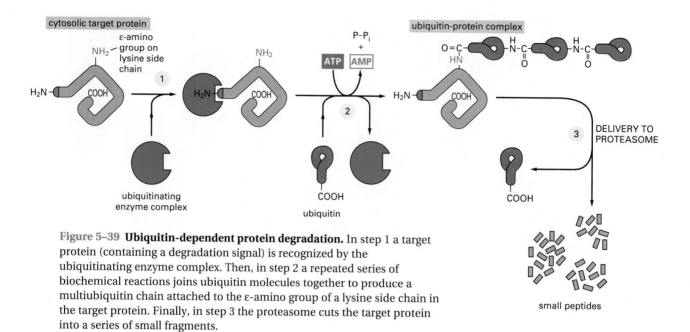

Figure 5–39 Ubiquitin-dependent protein degradation. In step 1 a target protein (containing a degradation signal) is recognized by the ubiquitinating enzyme complex. Then, in step 2 a repeated series of biochemical reactions joins ubiquitin molecules together to produce a multiubiquitin chain attached to the ε-amino group of a lysine side chain in the target protein. Finally, in step 3 the proteasome cuts the target protein into a series of small fragments.

conformational motifs that are buried and therefore inaccessible in the normal counterparts of these proteins.

A proteolytic pathway that recognizes and destroys abnormal proteins must be able to distinguish between *completed* proteins that have "wrong" conformations and the many growing polypeptides on ribosomes (as well as polypeptides just released from ribosomes) that have not yet achieved their normal folded conformation. That this is not a trivial problem can be demonstrated experimentally: if puromycin—an inhibitor of protein synthesis—is added to cells, the prematurely terminated proteins that are formed are rapidly degraded by a ubiquitin-dependent pathway. One possibility is that the normally forming proteins are temporarily protected by the translation machinery or by chaperone molecules. Another is that nascent and newly completed proteins are actually vulnerable to proteolysis but manage to fold up into their native conformations fast enough to escape being targeted for destruction by proteolysis.

The Lifetime of a Protein Can Be Determined by Enzymes That Alter Its N-Terminus [22]

One feature that has an important influence on the stability of a protein is the nature of the first (N-terminal) amino acid in the polypeptide chain. There is a strong relation, called the *N-end rule*, between the *in vivo* half-life of a protein and the identity of its N-terminal amino acid. Distinct versions of the N-end rule operate in all organisms examined, from bacteria to mammals. The amino acids Met, Ser, Thr, Ala, Val, Cys, Gly, or Pro, for example, protect proteins in the yeast *S. cerevisiae* when present at the N-terminus; these amino acids are not recognized by targeting components of the N-end rule pathway, while the remaining 12 amino acids attract a proteolytic attack. Most of the proteins that are rapidly degraded by the N-end rule pathway (which operates in both the cytosol and the nucleus) remain to be identified. Since destabilizing amino acids, however, are rare at the N-termini of cytosolic proteins but are frequently present at the N-terminus of proteins that have been transported to other compartments, one hypothetical function of the N-end rule pathway is to degrade proteins that normally function in the ER, the Golgi apparatus, or another membrane-bounded compartment but for some reason have leaked back into the cytosol.

It is not known how destabilizing amino acids become exposed at the N-terminus of a newly formed protein. As discussed in Chapter 6, all proteins are initially synthesized with methionine (or formyl-methionine in bacteria) as their N-terminal amino acid. This methionine, which is a stabilizing amino acid in the N-end rule, is often removed by a specific aminopeptidase. The presently known methionine aminopeptidases, however, will remove the N-terminal methionine if and only if the second amino acid is *also* stabilizing in the N-end rule. The proteases that produce physiological substrates of the N-end rule pathway, and the sequences they recognize as signals for cleavage, remain to be discovered.

Certain destabilizing N-terminal amino acids, such as aspartate and glutamate, are not recognized directly by the targeting component of the N-end rule pathway. Instead, they are modified by the enzyme arginyl-tRNA-protein transferase, which links arginine, one of the directly recognized destabilizing amino acids, to the N-terminus of proteins bearing N-terminal aspartate or glutamate. Arginine is thus one of the *primary* destabilizing amino acids in the N-end rule, while aspartate and glutamate are *secondary* destabilizing amino acids. In eucaryotes there are also *tertiary* destabilizing N-terminal amino acids—asparagine and glutamine—which are destabilizing through their conversion, by a specific amidase, into the secondary destabilizing amino acids aspartate and glutamate.

The N-terminal amino acid of a protein is often found to be resistant to hydrolysis by the reagents used in protein sequenators. Such proteins have a chemically modified ("blocked") N-terminus, the most frequent modification being acetylation. This modification was believed to play a role in protecting long-lived proteins from degradation. However, recent experiments with yeast mutants that lack the major species of N-terminal acetylase, so that the bulk of the normally acetylated proteins are unacetylated, show that most of these unacetylated proteins remain long-lived. The function of N-terminal acetylation in these proteins remains to be deciphered.

Summary

From the moment of its birth on a ribosome to its death by targeted proteolysis, a protein is accompanied by molecular chaperones and other surveying devices whose purpose is to massage it into shape, repair it, or eliminate it. Misfolded proteins are first induced to refold correctly by hsp70 or hsp60 chaperone molecules; if this fails, they are coupled to ubiquitin and thereby targeted for digestion in proteasomes.

Proteins are often composed of discrete modular domains that have been juxtaposed during evolution by duplication and shuffling of the DNA sequences that encode the modules. The modules often contain specific binding sites for other molecules, including other proteins, and they often enable proteins to assemble into large complexes. The principle of linkage explains how cells manage to use allosteric transitions to assemble such protein complexes in an all-or-none fashion.

References

General

Branden, C.; Tooze, J. Introduction to Protein Structure. New York: Garland, 1991.

Creighton, T.E. Proteins: Structures and Molecular Properties, 2nd ed. New York: W.H. Freeman, 1993.

Cited

1. Steitz, T.A.; Shoham, M.; Bennett, W.S., Jr. Structural dynamics of yeast hexokinase during catalysis. *Philos. Trans. R. Soc. Lond. (Biol.)* 293:43–52, 1981.

2. Monod, J.; Changeux, J.-P.; Jacob, F. Allosteric proteins and cellular control systems. *J. Mol. Biol.* 6:306–329, 1963.

Perutz, M. Cooperativity and Allosteric Regulation in Proteins. Cambridge, UK: Cambridge University Press, 1990.

3. Koshland, D.E., Jr. Control of enzyme activity and metabolic pathways. *Trends Biochem. Sci.* 9:155–159, 1984.

Umbarger, H.E. The origin of a useful concept—feedback inhibition. *Protein Sci.* 1:1392–1395, 1992.

4. Cantor, C.R.; Schimmel, P.R. Biophysical Chemistry. Part III: The Behavior of Biological Macromolecules, Chaps. 15 and 17. New York: W.H. Freeman, 1980.

Dickerson, R.E.; Geis, I. Hemoglobin: Structure, Function, Evolution and Pathology. Menlo Park, CA: Benjamin-Cummings, 1983.

Monod, J.; Changeux, J.-P.; Jacob, F. Allosteric proteins and cellular control systems. *J. Mol. Biol.* 6:306–329, 1963.

5. Kantrowitz, E.R.; Lipscomb, W.N. *Escherichia coli* aspartate transcarbamylase: the relation between structure and function. *Science* 241:669–674, 1988.

Kantrowitz, E.R.; Lipscomb, W.N. *Escherichia coli* aspartate transcarbamoylase: the molecular basis for a concerted allosteric transition. *Trends. Biochem. Sci.* 15:53–59, 1990.

Schachman, H.K. Can a simple model account for the allosteric transition of aspartate transcarbamoylase? *J. Biol. Chem.* 263:18583–18586, 1988.

6. Fischer, E.H.; Krebs, E.G. Conversion of phosphorylase b to phosphorylase a in muscle extracts. *J. Biol. Chem.* 216:121–132, 1955.

Johnson, L.N.; Barford, D. The effects of phosphorylation on the structure and function of proteins. *Annu. Rev. Biophys. Biomol. Struct.* 22:199–232, 1993.

7. Cohen, P. The structure and regulation of protein phosphatases. *Annu. Rev. Biochem.* 58:453–508, 1989.

Hanks, S.K.; Quinn, A.M.; Hunter, T. The protein kinase family: conserved features and deduced phylogeny of the catalytic domains. *Science* 241:42–52, 1988.

Taylor, S.S.; Knighton, D.R.; Zheng, J.; Ten Eyck, L.F.; Sowadski, J.M. Structural framework for the protein kinase family. *Annu. Rev. Cell Biol.* 8:429–462, 1992.

8. DeBondt, H.L.; Rosenblatt, J.; Jancarik, J.; et al. Crystal structure of cyclin-dependent kinase 2. *Nature* 363:595–602, 1993.

9. Bourne, H.R.; Sanders, D.A.; McCormick, F. The GTPase superfamily: conserved structure and molecular mechanism. *Nature* 349:117–127, 1991.

Wittinghofer, A.; Pai, E.F. The structure of ras protein: a model for a universal molecular switch. *Trends Biochem. Sci.* 16:382–387, 1991.

10. McCormick, F. *ras* GTPase activating protein: signal transmitter and signal terminator. *Cell* 56:5–8, 1989.

11. Berchtold, H.; Reshetnikova, L.; Reiser, C.O.A.; et al. Crystal structure of active elongation factor Tu reveals major domain rearrangements. *Nature* 365:126–132, 1993.

Kjeldgaard, M.; Nissen, P.; Thirup, S.; Nyborg, J. The crystal structure of elongation factor EF-Tu from *Thermus aquaticus* in the GTP conformation. *Structure* 1:35–50, 1993.

12. Hill, T.L. Biochemical cycles and free energy transduction. *Trends Biochem. Sci.* 2:204–207, 1977.

13. Rayment, I.; Holden, H.M.; Whittaker, M.; et al. Structure of the actin-myosin complex and its implications for muscle contraction. *Science* 261:58–65, 1993.

14. Hokin, L.E. The molecular machine for driving the coupled transports of Na$^+$ and K$^+$ is an (Na$^+$ + K$^+$)-activated ATPase. *Trends Biochem. Sci.* 1:233–237, 1976.

Jencks, W.P. How does a calcium pump pump calcium? *J. Biol. Chem.* 264:18855–18858, 1989.

15. Alberts, B.; Miake-Lye, R. Unscrambling the puzzle of biological machines: the importance of the details. *Cell* 68:415–420, 1992.

16. Alberts, B.M. Protein machines mediate the basic genetic processes. *Trends Genet.* 1:26–30, 1985.

17. Udgaonkar, J.B.; Baldwin, R.L. NMR evidence for an early framework intermediate on the folding pathway of ribonuclease A. *Nature* 335:694–699, 1988.

Kuwajima, K. The molten globular state as a clue for understanding the folding and cooperativity of globular-protein structure. *Proteins* 6:87–103, 1989.

18. Agard, D.A. To fold or not to fold. *Science* 260:1903–1904, 1993.

Georgopoulos, C.; Welch, W.J. Role of the major heat shock proteins as molecular chaperones. *Annu. Rev. Cell Biol.* 9:601–634, 1993.

Martin, J.; Langer, T.; Boteva, R.; et al. Chaperonin-mediated protein folding at the surface of groEL through a "molten globule"-like intermediate. *Nature* 352:36–42, 1991.

19. Baron, M.; Norman, D.G.; Campbell, I.D. Protein modules. *Trends Biochem. Sci.* 16:13–17, 1991.

20. Koch, C.A.; Anderson, D.; Moran, M.F.; Ellis, C.; Pawson, T. SH2 and SH3 domains: elements that control interactions of cytoplasmic signaling proteins. *Science* 252:668–674, 1991.

21. Hershko, A.; Ciechanover, A. The ubiquitin system for protein degradation. *Annu. Rev. Biochem.* 61:761–807, 1992.

Rechsteiner, M.; Hoffman, L.; Dubiel, W. The multicatalytic and 26S proteases. *J. Biol. Chem.* 268:6065–6068, 1993.

22. Varshavsky, A. The N-end rule. *Cell* 69:725–735, 1992.

Basic Genetic Mechanisms

The ability of cells to maintain a high degree of order in a chaotic universe depends on the genetic information that is expressed, maintained, replicated, and occasionally improved by the basic genetic processes—*RNA and protein synthesis, DNA repair, DNA replication,* and *genetic recombination.* In these processes, which produce and maintain the proteins and nucleic acids of a cell (Figure 6–1), the information in a linear sequence of nucleotides is used to specify either another linear chain of nucleotides (a DNA or an RNA molecule) or a linear chain of amino acids (a protein molecule). The framework underlying genetic events is therefore one-dimensional and conceptually simple. In contrast, most other processes in cells result solely from information expressed in the complex three-dimensional surfaces of protein molecules. Perhaps that is why we understand more about genetic mechanisms than about most other biological processes.

In this chapter we examine the molecular machinery that repairs, replicates, and alters on occasion the DNA of the cell. We shall see that the machinery depends on enzymes that cut, copy, and recombine nucleotide sequences. We shall also see that these and other enzymes can be parasitized by viruses, plasmids, and transposable genetic elements, which not only direct their own replication, but also can alter the cell genome by genetic recombination events.

First, however, we reconsider a central topic mentioned briefly in Chapter 3—the mechanisms of RNA and protein synthesis.

RNA and Protein Synthesis

Proteins constitute more than half the total dry mass of a cell, and their synthesis is central to cell maintenance, growth, and development. Protein synthesis occurs on *ribosomes*. It depends on the collaboration of several classes of RNA molecules and begins with a series of preparatory steps. First, a molecule of *messenger RNA (mRNA)* must be copied from the DNA that encodes the protein. Meanwhile, in the cytoplasm, each of the 20 amino acids from which the protein is to be built must be attached to its specific *transfer RNA (tRNA)* molecule, and the subunits of the ribosome on which the new protein is to be made must be preloaded with auxiliary protein factors. Protein synthesis begins when all of these components come together in the cytoplasm to form a functioning ribosome. As a single molecule of mRNA moves stepwise through a ribosome, the sequence of nucleotides in the mRNA molecule is translated into a corresponding sequence of amino acids to produce a distinctive protein chain, as specified by the DNA sequence of

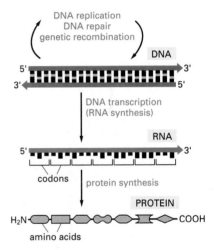

Figure 6–1 The basic genetic processes. The processes shown here are thought to occur in all present-day cells. Very early in the evolution of life, however, much simpler cells probably existed that lacked both DNA and proteins (see Figure 1–11). Note that a sequence of three nucleotides (a codon) in an RNA molecule codes for a specific amino acid in a protein.

its gene. We begin by considering how the many different RNA molecules in a cell are made.

RNA Polymerase Copies DNA into RNA: The Process of DNA Transcription [1]

RNA is synthesized on a DNA template by a process known as **DNA transcription.** Transcription generates the mRNAs that carry the information for protein synthesis, as well as the transfer, ribosomal, and other RNA molecules that have structural or catalytic functions. All of these RNA molecules are synthesized by **RNA polymerase** enzymes, which make an RNA copy of a DNA sequence. In eucaryotes three kinds of RNA polymerase molecules synthesize different types of RNA, as described in Chapter 8. These RNA polymerases are thought to have derived during evolution from the single enzyme present in bacteria that mediates all bacterial RNA synthesis.

The bacterial RNA polymerase is a large multisubunit enzyme associated with several additional protein subunits that enter and leave the polymerase-DNA complex at different stages of transcription. Free RNA polymerase molecules collide randomly with the bacterial chromosome, sliding along it but sticking only weakly to most DNA. The polymerase binds very tightly, however, when it con-

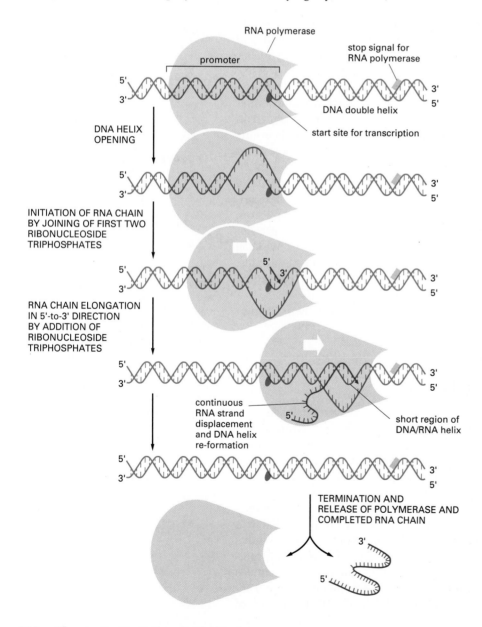

Figure 6–2 The synthesis of an RNA molecule by RNA polymerase. The enzyme binds to the promoter sequence on the DNA and begins its synthesis at a start site within the promoter. It completes its synthesis at a stop (termination) signal, whereupon both the polymerase and its completed RNA chain are released. During RNA chain elongation, polymerization rates average about 30 nucleotides per second at 37°C. Therefore, an RNA chain of 5000 nucleotides takes about 3 minutes to complete.

RNA polymerase

promoter

stop signal for RNA polymerase

5'
3'

3'
5'

DNA double helix

start site for transcription

DNA HELIX OPENING

INITIATION OF RNA CHAIN BY JOINING OF FIRST TWO RIBONUCLEOSIDE TRIPHOSPHATES

RNA CHAIN ELONGATION IN 5'-to-3' DIRECTION BY ADDITION OF RIBONUCLEOSIDE TRIPHOSPHATES

continuous RNA strand displacement and DNA helix re-formation

short region of DNA/RNA helix

TERMINATION AND RELEASE OF POLYMERASE AND COMPLETED RNA CHAIN

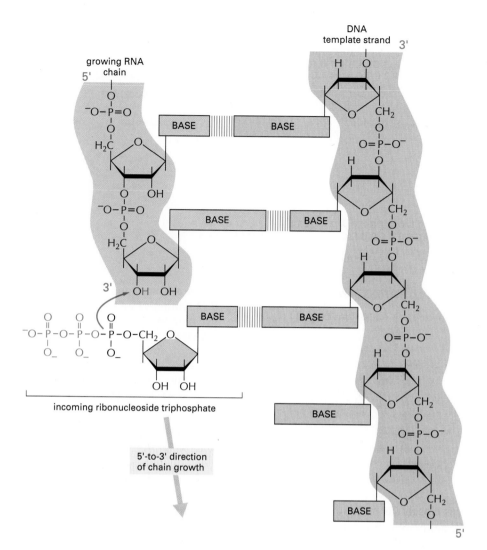

Figure 6–3 The chain elongation reaction catalyzed by an RNA polymerase enzyme. In each step an incoming ribonucleoside triphosphate is selected for its ability to base-pair with the exposed DNA template strand; a ribonucleoside monophosphate is then added to the growing, 3′-OH end of the RNA chain (*red arrow*), and pyrophosphate is released (*red atoms*). The new RNA chain therefore grows by one nucleotide at a time in the 5′-to-3′ direction, and it is complementary in sequence to the DNA template strand. The reaction is driven both by the favorable free-energy change that accompanies the release of pyrophosphate and by the subsequent hydrolysis of the pyrophosphate to inorganic phosphate (see Figure 2–30).

tacts a specific DNA sequence, called the **promoter,** that contains the *start site* for RNA synthesis and signals where RNA synthesis should begin. The reactions that ensue are outlined in Figure 6–2. After binding to the promoter, the RNA polymerase opens up a local region of the double helix to expose the nucleotides on a short stretch of DNA on each strand. One of the two exposed DNA strands acts as a template for complementary base-pairing with incoming ribonucleoside triphosphate monomers, two of which are joined together by the polymerase to begin an RNA chain. The RNA polymerase molecule then moves stepwise along the DNA, unwinding the DNA helix just ahead to expose a new region of the template strand for complementary base-pairing. In this way the growing RNA chain is extended by one nucleotide at a time in the 5′-to-3′ direction (Figure 6–3). The chain elongation process continues until the enzyme encounters a second special sequence in the DNA, the **stop (termination) signal,** where the polymerase halts and releases both the DNA template and the newly made RNA chain.

By convention, when a DNA sequence associated with a gene is specified, it is the sequence of the nontemplate strand that is given, and it is written in the 5′-to-3′ direction. This convention is adopted because the sequence of the nontemplate strand corresponds to the sequence of the RNA that is made.

Nucleotide sequences that act as start sites and stop signals for the bacterial RNA polymerase are illustrated in Figure 6–4. Nucleotide sequences that are found in many examples of a particular type of region in DNA (such as a promoter) are called **consensus sequences.** In bacteria strong promoters (those associated with genes that produce large amounts of mRNA) have sequences that

RNA and Protein Synthesis

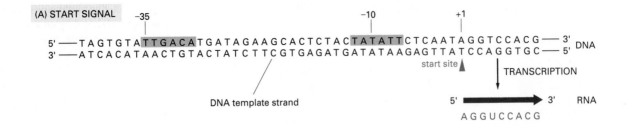

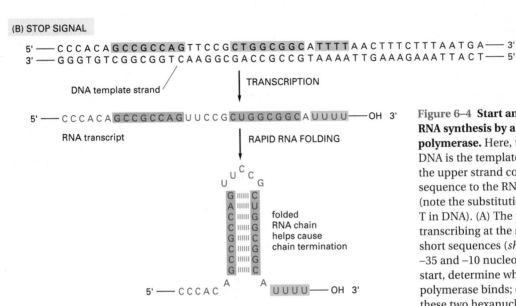

Figure 6–4 Start and stop signals for RNA synthesis by a bacterial RNA polymerase. Here, the lower strand of DNA is the template strand, whereas the upper strand corresponds in sequence to the RNA that is made (note the substitution of U in RNA for T in DNA). (A) The polymerase begins transcribing at the start site. Two short sequences (*shaded red*), about −35 and −10 nucleotides from the start, determine where the polymerase binds; close relatives of these two hexanucleotide sequences, properly spaced from each other, specify the promoter for most *E. coli* genes. (B) A stop (termination) signal. The *E. coli* RNA polymerase stops when it synthesizes a run of U residues (*shaded blue*) from a complementary run of A residues on the template strand, provided that it has just synthesized a self-complementary RNA nucleotide sequence (*shaded green*), which rapidly forms a hairpin helix that is crucial for stopping transcription. The sequence of nucleotides in the self-complementary region can vary widely.

match the promoter consensus sequences closely (as in Figure 6–4A), whereas weak promoters (those associated with genes that produce relatively small amounts of mRNA) match these sequences less well.

Only Selected Portions of a Chromosome Are Used to Produce RNA Molecules [2]

As an RNA polymerase molecule moves along the DNA, an RNA/DNA double helix is formed at the enzyme's active site. This helix is very short because the RNA just made is displaced, allowing the DNA/DNA helix immediately at the rear of the polymerase to rewind (Figure 6–5). As a result, each completed RNA chain is released from the DNA template as a free, single-stranded RNA molecule, typically between 70 and 10,000 nucleotides long.

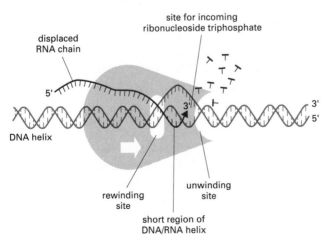

Figure 6–5 DNA unwinding and rewinding by RNA polymerase. A moving RNA polymerase molecule is continuously unwinding the DNA helix ahead of the polymerization site while rewinding the two DNA strands behind this site to displace the newly formed RNA chain. A short region of DNA/RNA helix is therefore formed only transiently, and the final RNA product is released as a single-stranded copy of one of the two DNA strands.

In principle, any region of the DNA double helix could be copied into two different RNA molecules—one from each of the two DNA strands. In reality, only one DNA strand is used as a template in each region. The RNA made is equivalent in nucleotide sequence to the opposite, nontemplate DNA strand. Which of the two strands is copied varies along the length of a single DNA molecule and is determined by the promoter of each gene. As illustrated in Figure 6–4, a promoter is an oriented DNA sequence that points the RNA polymerase in one direction or the other, and this orientation determines which DNA strand is copied (Figure 6–6). The DNA strand that is copied into RNA can be either different or the same for neighboring genes (Figure 6–7).

Both bacterial and eucaryotic RNA polymerases are large, complicated molecules, with multiple subunits and a total mass of more than 500,000 daltons. Some bacterial viruses, in contrast, encode single-chain RNA polymerases of one-fifth this mass that catalyze RNA synthesis at least as well as the host-cell enzyme. Presumably, the multiple subunit composition of the cellular RNA polymerases is important for various regulatory aspects of cellular RNA synthesis that have not yet been well defined.

This brief outline of DNA transcription omits many details. Other complex steps usually must occur before an mRNA molecule is produced. *Gene regulatory proteins*, for example, help to determine which regions of DNA are transcribed by the RNA polymerase and thereby play a major part in determining which proteins are made by a cell. Moreover, although mRNA molecules are produced directly by DNA transcription in procaryotes, in higher eucaryotic cells most RNA transcripts are altered extensively—by a process called *RNA splicing*—before they leave the cell nucleus and enter the cytoplasm as mRNA molecules. All of these aspects of mRNA production are discussed in Chapters 8 and 9, where we consider the cell nucleus and the control of gene expression, respectively. For now, let us assume that functional mRNA molecules have been produced and proceed to examine how they direct protein synthesis.

Transfer RNA Molecules Act as Adaptors That Translate Nucleotide Sequences into Protein Sequences [3]

All cells contain a set of **transfer RNAs (tRNAs),** each of which is a small RNA molecule (most have a length between 70 and 90 nucleotides). The tRNAs, by binding at one end to a specific codon in the mRNA and at their other end to the amino acid specified by that codon, enable amino acids to line up according to the sequence of nucleotides in the mRNA. Each tRNA is designed to carry only one of the 20 amino acids used for protein synthesis: a tRNA that carries glycine is designated tRNA^{Gly} and so on. Each of the 20 amino acids has at least one type of tRNA assigned to it, and most have several tRNAs. Before an amino acid is incorporated into a protein chain, it is attached by its carboxyl end to the 3′ end of an appropriate tRNA molecule. This attachment serves two purposes. First, and most important, it covalently links the amino acid to a tRNA containing the correct **anticodon**—the sequence of three nucleotides that is complementary to the three-nucleotide *codon* that specifies that amino acid on an mRNA molecule. Codon-anticodon pairings enable each amino acid to be inserted into a growing protein chain according to the dictates of the sequence of nucleotides in the mRNA, thereby allowing the genetic code to be used to translate nucleotide se-

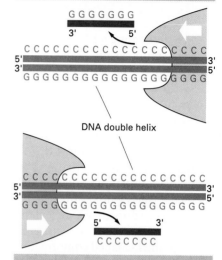

an RNA polymerase that moves from right to left makes RNA by using the top strand as a template

DNA double helix

an RNA polymerase that moves from left to right makes RNA by using the bottom strand as a template

Figure 6–6 RNA polymerase orientation determines which DNA strand serves as template. The DNA strand serving as template must be traversed from its 3′ end to its 5′ end, as illustrated in Figure 6–3. Thus the direction of RNA polymerase movement determines which of the two DNA strands will serve as a template for the synthesis of RNA, as shown here. Polymerase direction is, in turn, determined by the orientation of the promoter sequence, where the RNA polymerase initially binds.

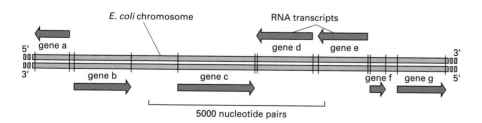

Figure 6–7 Directions of transcription along a short portion of a bacterial chromosome. Note that some genes are transcribed from one DNA strand, while others are transcribed from the other DNA strand. Approximately 0.2% of the *E. coli* chromosome is depicted here. (Adapted from D.L. Daniels et al., *Science* 257:771–777, 1992.)

RNA and Protein Synthesis

quences into protein sequences. This is the essential "adaptor" function of the tRNA molecule: with one end attached to an amino acid and the other paired to a codon, the tRNA converts sequences of nucleotides into sequences of amino acids.

The second function of the amino acid attachment is to activate the amino acid by generating a high-energy linkage at its carboxyl end so that it can react with the amino group of the next amino acid in the protein sequence to form a *peptide bond*. The activation process is necessary for protein synthesis because nonactivated amino acids cannot be added directly to a growing polypeptide chain. (In contrast, the reverse process, in which a peptide bond is hydrolyzed by the addition of water, is energetically favorable and can occur spontaneously.)

The function of a tRNA molecule depends on its precisely folded three-dimensional structure. A few tRNAs have been crystallized and their complete structures determined by x-ray diffraction analyses. Both intramolecular complementary base-pairings and unusual base interactions are required to fold a tRNA molecule (see Figure 3–18). The nucleotide sequences of tRNA molecules from many types of organisms reveal that tRNAs can form the loops and base-paired stems of a "cloverleaf" structure (Figure 6–8), and all are thought to fold further to adopt the L-shaped conformation detected in crystallographic analyses. In the native structure the amino acid is attached to one end of the "L," while the anticodon is located at the other (Figure 6–9).

The nucleotides in a completed nucleic acid chain (like the amino acids in proteins) can be covalently modified to modulate the biological activity of the nucleic acid molecule. Such posttranscriptional modifications are especially common in tRNA molecules, which contain a variety of modified nucleotides (Figure 6–10). Some of the modified nucleotides affect the conformation and base-pairing of the anticodon and thereby facilitate the recognition of the appropriate mRNA codon by the tRNA molecule.

Specific Enzymes Couple Each Amino Acid to Its Appropriate tRNA Molecule [4]

Only the tRNA molecule, and not its attached amino acid, determines where the amino acid is added during protein synthesis. This was established by an ingenious experiment in which an amino acid (cysteine) was chemically converted into a different amino acid (alanine) after it was already attached to its specific tRNA. When such "hybrid" tRNA molecules were used for protein synthesis in a cell-free system, the wrong amino acid was inserted at every point in the protein chain where that tRNA was used. Thus the accuracy of protein synthesis is crucially dependent on the accuracy of the mechanism that normally links each activated amino acid specifically to its corresponding tRNA molecules.

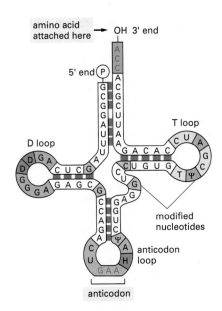

Figure 6–8 The "cloverleaf" structure of tRNA. This is a view of the molecule shown in Figure 6–9 after it has been partially unfolded. There are many different tRNA molecules, including at least one for each kind of amino acid. Although they differ in nucleotide sequence, they all have the three stem loops shown plus an amino acid-accepting arm. The particular tRNA molecule shown binds phenylalanine and is therefore denoted tRNA[Phe]. In all tRNA molecules the amino acid is attached to the A residue of a CCA sequence at the 3′ end of the molecule. Complementary base-pairings are shown by *red bars*.

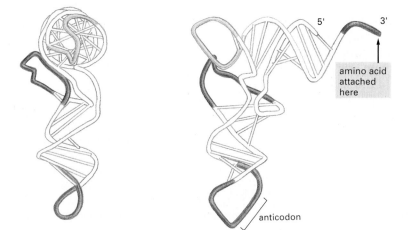

Figure 6–9 The folded structure of a typical tRNA molecule. Two views of the three-dimensional conformation determined by x-ray diffraction are shown. Note that the molecule is L-shaped; one end is designed to accept the amino acid, while the other end contains the three nucleotides of the anticodon. Each loop is colored to match Figure 6–8.

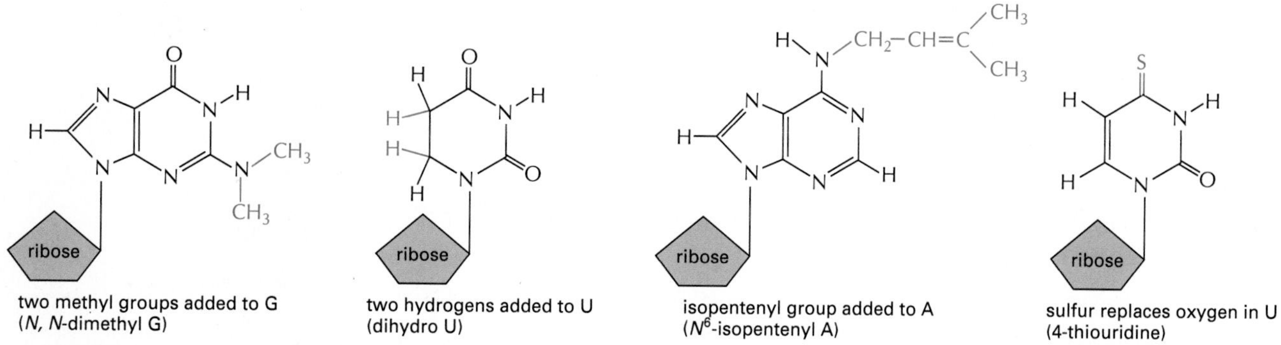

The top of the page shows four chemical structures with labels:

- two methyl groups added to G (*N, N*-dimethyl G)
- two hydrogens added to U (dihydro U)
- isopentenyl group added to A (*N*6-isopentenyl A)
- sulfur replaces oxygen in U (4-thiouridine)

How does a tRNA molecule become covalently linked to the one amino acid in 20 that is its appropriate partner? The mechanism depends on enzymes called **aminoacyl-tRNA synthetases,** which couple each amino acid to its appropriate set of tRNA molecules. There is a different synthetase enzyme for every amino acid (20 synthetases in all): one attaches glycine to all tRNA^Gly molecules, another attaches alanine to all tRNA^Ala molecules, and so on. The coupling reaction that creates an **aminoacyl-tRNA** molecule is catalyzed in two steps, as illustrated in Figure 6–11. The structure of the amino acid-RNA linkage is shown in Figure 6–12.

Although the tRNA molecules serve as the final adaptors in converting nucleotide sequences into amino acid sequences, the aminoacyl-tRNA synthetase

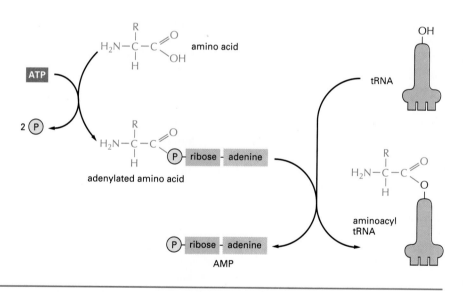

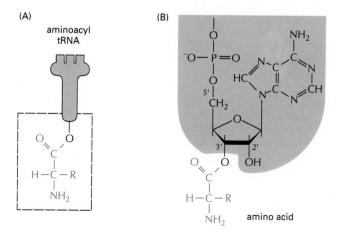

RNA and Protein Synthesis

enzymes are adaptors of equal importance to the decoding process (Figure 6–13). Thus the genetic code is translated by two sets of adaptors that act sequentially, each matching one molecular surface to another with great specificity; it is their combined action that associates each sequence of three nucleotides in the mRNA molecule—that is, each codon—with its particular amino acid (Figure 6–14).

Amino Acids Are Added to the Carboxyl-Terminal End of a Growing Polypeptide Chain

The fundamental reaction of protein synthesis is the formation of a peptide bond between the carboxyl group at the end of a growing polypeptide chain and a free amino group on an amino acid. Consequently, a protein is synthesized stepwise from its amino-terminal end to its carboxyl-terminal end. Throughout the entire process the growing carboxyl end of the polypeptide chain remains activated by its covalent attachment to a tRNA molecule (a *peptidyl-tRNA* molecule). This high-energy covalent linkage is disrupted in each cycle but is immediately replaced by the identical linkage on the most recently added amino acid (Figure 6–15). In this way each amino acid added carries with it the activation energy for the addition of the *next* amino acid rather than the energy for its own addition—an example of the "head growth" type of polymerization described in Chapter 2 (see Figure 2–36).

The Genetic Code Is Degenerate [5]

In the course of protein synthesis, the translation machinery moves in the 5′-to-3′ direction along an mRNA molecule and the mRNA sequence is read three nucleotides at a time. As we have seen, each amino acid is specified by the triplet of nucleotides (**codon**) in the mRNA molecule that pairs with a sequence of three complementary nucleotides at the anticodon tip of a particular tRNA. Because only one of the many types of tRNA molecules in a cell can base-pair with each codon, the codon determines the specific amino acid residue to be added to the growing polypeptide chain end (Figure 6–16).

Since RNA is constructed from four types of nucleotides, there are 64 possible sequences composed of three nucleotides ($4 \times 4 \times 4$). Three of these 64 sequences do not code for amino acids but instead specify the termination of a polypeptide chain; they are known as *stop codons*. That leaves 61 codons to specify only 20 different amino acids. For this reason, most of the amino acids are represented by more than one codon (Figure 6–17) and the genetic code is

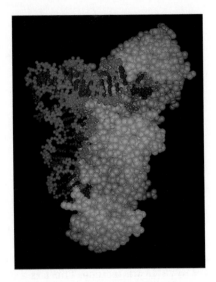

Figure 6–13 The recognition of a tRNA molecule by its aminoacyl-tRNA synthetase. For this tRNA (tRNA^{Gln}), specific nucleotides in both the anticodon (*bottom*) and the amino acid-accepting arm allow the correct tRNA to be recognized by the synthetase enzyme (*blue*). (Courtesy of Tom Steitz.)

Figure 6–14 The genetic code is translated by means of two sequential "adaptors." The first adaptor is the aminoacyl-tRNA synthetase enzyme, which couples a particular amino acid to its corresponding tRNA; the second adaptor is the tRNA molecule, whose *anticodon* forms base pairs with the appropriate nucleotide sequence (*codon*) on the mRNA. An error in either step will cause the wrong amino acid to be incorporated into a protein chain.

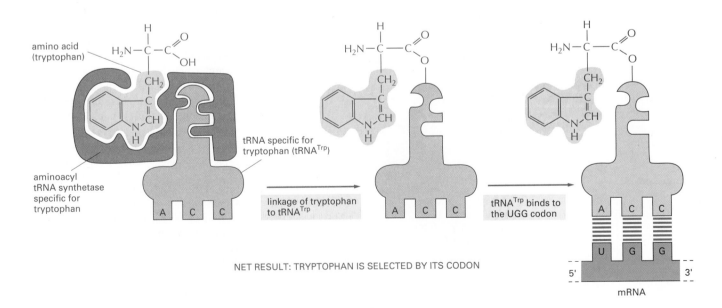

NET RESULT: TRYPTOPHAN IS SELECTED BY ITS CODON

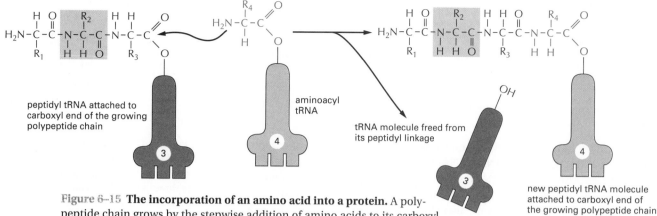

Figure 6–15 **The incorporation of an amino acid into a protein.** A polypeptide chain grows by the stepwise addition of amino acids to its carboxyl-terminal end. The formation of each peptide bond is energetically favorable because the growing carboxyl terminus has been activated by the covalent attachment of a tRNA molecule. The peptidyl-tRNA linkage that activates the growing end is regenerated in each cycle. The amino acid side chains have been abbreviated as R_1, R_2, R_3, and R_4; as a reference point, all of the atoms in the second amino acid in the polypeptide chain are shaded *gray*.

said to be *degenerate.* Two amino acids, methionine and tryptophan, have only one codon each, and they are the least abundant amino acids in proteins.

The degeneracy of the genetic code implies either that there is more than one tRNA for each amino acid or that a single tRNA molecule can base-pair with more than one codon. In fact, both situations occur. For some amino acids there is more than one tRNA molecule, and some tRNA molecules are constructed so that they require accurate base-pairing only at the first two positions of the codon and can tolerate a mismatch (or wobble) at the third. This *wobble base-pairing* explains why so many of the alternative codons for an amino acid differ only in their third nucleotide (see Figure 6–17). The standard wobble pairings make it possible to fit the 20 amino acids to 61 codons with as few as 31 kinds of tRNA molecules; in animal mitochondria a more extreme wobble allows protein synthesis with only 22 tRNAs (discussed in Chapter 14).

The Events in Protein Synthesis Are Catalyzed on the Ribosome [6]

The protein synthesis reactions just described require a complex catalytic machinery to guide them. The growing end of the polypeptide chain, for example, must be kept in register with the mRNA molecule to ensure that each successive codon in the mRNA engages precisely with the anticodon of a tRNA molecule and does not slip by one nucleotide, thereby changing the reading frame (see Figure 3–17). This precise movement and the other events in protein synthesis are catalyzed by **ribosomes,** which are large complexes of RNA and protein molecules. Eucaryotic and procaryotic ribosomes are very similar in design and function. Both are composed of one large and one small subunit that fit together to form a complex with a mass of several million daltons (Figure 6–18). The small subunit binds the mRNA and tRNAs, while the large subunit catalyzes peptide bond formation.

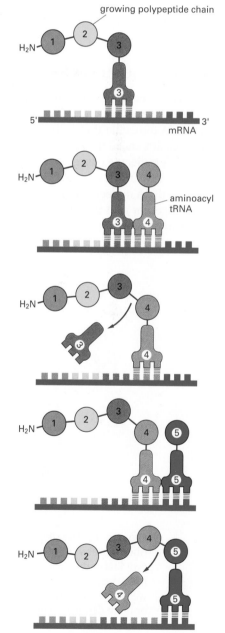

Figure 6–16 **Decoding an mRNA molecule.** Each amino acid added to the growing end of a polypeptide chain is selected by complementary base-pairing between the anticodon on its attached tRNA molecule and the next codon on the mRNA chain.

	AGA						GGA			UUA					AGC					
	AGG						GGC			UUG					AGU					
GCA	CGA				GGA		GGG			CUA		CCA	UCA	ACA		GUA	UAA
GCA	CGA					AUA	CUC		CCA	UCA	ACA		GUA	UAA
GCC	CGC								AUC	CUG		CCC	UCC	ACC		GUC	UAG
GCG	CGG	GAC	AAC	UGC	GAA	CAA	GGG	CAC	AUC	CUG	AAA		UUC	CCG	UCG	ACG		UAC	GUG	UGA
GCU	CGU	GAU	AAU	UGU	GAG	CAG	GGU	CAU	AUU	CUU	AAG	AUG	UUU	CCU	UCU	ACU	UGG	UAU	GUU	
Ala	Arg	Asp	Asn	Cys	Glu	Gln	Gly	His	Ile	Leu	Lys	Met	Phe	Pro	Ser	Thr	Trp	Tyr	Val	stop
A	R	D	N	C	E	Q	G	H	I	L	K	M	F	P	S	T	W	Y	V	

More than half of the weight of a ribosome is RNA, and there is increasing evidence that the **ribosomal RNA (rRNA)** molecules play a central part in its catalytic activities. Although the rRNA molecule in the small ribosomal subunit varies in size depending on the organism, its complicated folded structure is highly conserved (Figure 6–19); there are also close homologies between the rRNAs of the large ribosomal subunits in different organisms. Ribosomes contain a large number of proteins (Figure 6–20), but many of these have been relatively poorly conserved in sequence during evolution, and a surprising number seem not to be essential for ribosome function. Therefore, it has been suggested that the ribosomal proteins mainly enhance the function of the rRNAs and that the RNA molecules rather than the protein molecules catalyze many of the reactions on the ribosome.

A Ribosome Moves Stepwise Along the mRNA Chain [7]

A ribosome contains three binding sites for RNA molecules: one for mRNA and two for tRNAs. One site, called the **peptidyl-tRNA-binding site,** or **P-site,** holds the tRNA molecule that is linked to the growing end of the polypeptide chain. Another site, called the **aminoacyl-tRNA-binding site,** or **A-site,** holds the incoming tRNA molecule charged with an amino acid. A tRNA molecule is held tightly at either site only if its anticodon forms base pairs with a complementary codon on the mRNA molecule that is bound to the ribosome. The A- and P-sites are so close together that the two tRNA molecules are forced to form base pairs with adjacent codons in the mRNA molecule (Figure 6–21).

Figure 6–17 The genetic code. The standard one-letter abbreviation for each amino acid is presented below its three-letter abbreviation. Codons are written with the 5'-terminal nucleotide on the left. Note that most amino acids are represented by more than one codon and that variation is common at the third nucleotide (see also Figure 3–16).

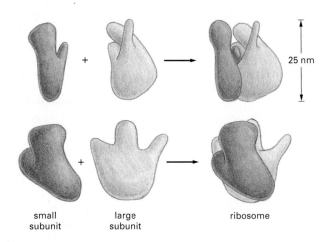

25 nm

small subunit   large subunit   ribosome

Figure 6–18 The ribosome. A three-dimensional model of the bacterial ribosome as viewed from two angles. The positions of many ribosomal proteins in this structure have been determined by using an electron microscope to visualize the positions where specific antibodies bind, as well as by measuring the neutron scattering from ribosomes containing one or more deuterated proteins. (After J.A. Lake, *Annu. Rev. Biochem.* 54:507–530, 1985. © 1985 by Annual Reviews Inc.)

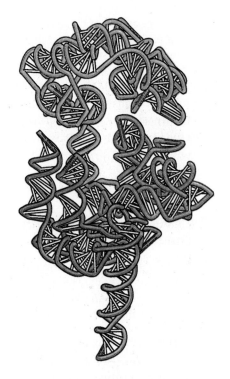

Figure 6–19 The structure of the rRNA in the small subunit. This model of *E. coli* 16S rRNA is indicative of the complex folding that underlies the catalytic activities of the RNAs in the ribosome. The 16S rRNA molecule contains 1540 nucleotides, and it is folded into three domains: 5' (*blue*), central (*red*), and 3' (*green*). (Adapted from S. Stern, B. Weiser, and H.F. Noller, *J. Mol. Biol.* 204:447–481, 1988.)

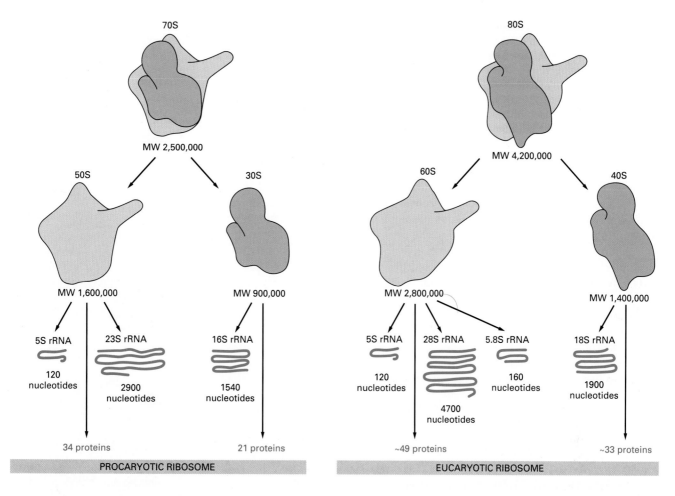

Figure 6–20 **A comparison of the structures of procaryotic and eucaryotic ribosomes.** Ribosomal components are commonly designated by their "S values," which indicate their rate of sedimentation in an ultracentrifuge. Despite the differences in the number and size of their rRNA and protein components, both types of ribosomes have nearly the same structure and they function in very similar ways. Although the 18S and 28S rRNAs of the eucaryotic ribosome contain many extra nucleotides not present in their bacterial counterparts, these nucleotides are present as multiple insertions that are thought to protrude as loops and leave the basic structure of each rRNA largely unchanged.

The process of polypeptide chain elongation on a ribosome can be thought of as a cycle with three discrete steps (Figure 6–22):

1. In step 1, an aminoacyl-tRNA molecule becomes bound to a vacant ribosomal A-site (adjacent to an occupied P-site) by forming base pairs with the three mRNA nucleotides (*codon*) exposed at the A-site.

2. In step 2, the carboxyl end of the polypeptide chain is uncoupled from the tRNA molecule in the P-site and joined by a peptide bond to the amino acid linked to the tRNA molecule in the A-site. This central reaction of protein synthesis (see Figure 6–15) is catalyzed by a **peptidyl transferase** enzyme. Recent experiments with ribosomes that have been experimentally stripped of proteins show that this catalysis is mediated not by a protein but by a specific region of the major rRNA molecule in the large subunit (see Figure 3–23).

3. In step 3, the new peptidyl-tRNA in the A-site is translocated to the P-site as the ribosome moves exactly three nucleotides along the mRNA molecule. This step requires energy and is driven by a series of conformational changes induced in one of the ribosomal components by the hydrolysis of a GTP molecule.

Figure 6–21 **The three major RNA-binding sites on a ribosome.** An empty ribosome is shown on the left and a loaded ribosome on the right. The representation of a ribosome used here and in the next three figures is highly schematic; for a more accurate view, see Figures 6–18 and 6–25.

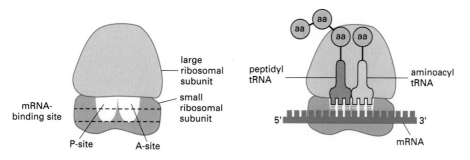

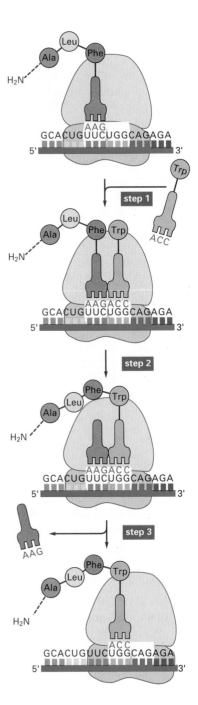

Figure 6–22 The elongation phase of protein synthesis on a ribosome.
The three-step cycle shown is repeated over and over during the synthesis of a protein chain. An aminoacyl-tRNA molecule binds to the A-site on the ribosome in step 1, a new peptide bond is formed in step 2, and the ribosome moves a distance of three nucleotides along the mRNA chain in step 3, ejecting an old tRNA molecule and "resetting" the ribosome so that the next aminoacyl-tRNA molecule can bind. As indicated in Figure 6–21, the P-site is drawn on the left side of the ribosome, with the A-site on the right.

As part of the translocation process of step 3, the free tRNA molecule that was generated in the P-site during step 2 is released from the ribosome to reenter the cytoplasmic tRNA pool. Upon completion of step 3, the unoccupied A-site is free to accept a new tRNA molecule linked to the next amino acid, which starts the cycle again. In a bacterium each cycle requires about 1/20th of a second under optimal conditions, so that the complete synthesis of an average-sized protein of 400 amino acids is accomplished in about 20 seconds. Ribosomes move along an mRNA molecule in the 5′-to-3′ direction, which is also the direction of RNA synthesis (see Figure 6–3).

In most cells protein synthesis consumes more energy than any other biosynthetic process. At least four high-energy phosphate bonds are split to make each new peptide bond: two of these are required to charge each tRNA molecule with an amino acid (see Figure 6–11), and two more drive steps in the cycle of reactions occurring on the ribosome during synthesis itself—one for the aminoacyl-tRNA binding in step 1 (see Figure 6–31) and one for the ribosome translocation in step 3.

A Protein Chain Is Released from the Ribosome When Any One of Three Stop Codons Is Reached [8]

Of the 64 possible codons in an mRNA molecule, 3 (UAA, UAG, and UGA) are **stop codons,** which terminate the translation process. Cytoplasmic proteins called *release factors* bind directly to any stop codon that reaches the A-site on the ribosome. This binding alters the activity of the peptidyl transferase, causing it to catalyze the addition of a water molecule instead of an amino acid to the peptidyl-tRNA. This reaction frees the carboxyl end of the growing polypeptide chain from its attachment to a tRNA molecule, and since only this attachment normally holds the growing polypeptide to the ribosome, the completed protein chain is immediately released into the cytoplasm. The ribosome releases the mRNA and dissociates into its two separate subunits (Figure 6–23), which can assemble on another mRNA molecule to begin a new round of protein synthesis by the process to be described next.

The Initiation Process Sets the Reading Frame for Protein Synthesis [9]

In principle, an RNA sequence can be translated in any one of three **reading frames,** each of which will specify a completely different polypeptide chain (see Figure 3–17). Which of the three frames is actually read is determined by the RNA sequence, which determines how the ribosome assembles. During the *initiation* phase of protein synthesis, the two subunits of the ribosome are brought together at the exact spot on the mRNA where the polypeptide chain is to begin.

The initiation process is complicated, involving a number of steps catalyzed by proteins called **initiation factors (IFs),** many of which are themselves composed of several polypeptide chains. Because the process is so complex, many of the details of initiation are still uncertain. It is clear, however, that each ribosome is assembled onto an mRNA chain in two steps: only after the small ribosomal

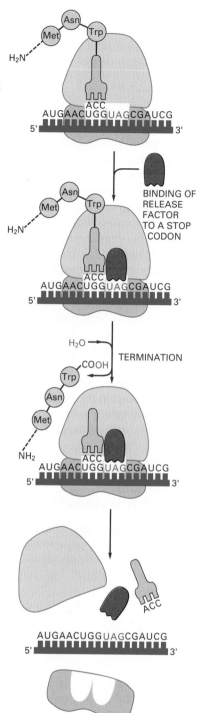

Figure 6–23 The final phase of protein synthesis. The binding of release factor to a stop codon terminates translation. The completed polypeptide is released, and the ribosome dissociates into its two separate subunits.

BINDING OF RELEASE FACTOR TO A STOP CODON

TERMINATION

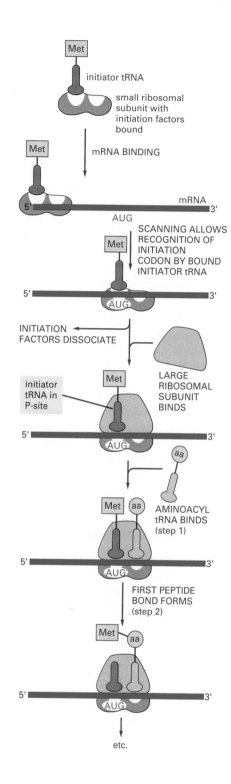

Figure 6–24 The initiation phase of protein synthesis in eucaryotes. Step 1 and step 2 refer to steps in the elongation reaction shown in Figure 6–22.

subunit loaded with initiation factors finds the start codon (AUG, see below) does the large subunit bind.

Before a ribosome can begin a new protein chain, it must bind an aminoacyl tRNA molecule in its P-site, where normally only peptidyl tRNA molecules are bound. (As explained previously, the peptidyl tRNA is translocated to the P-site during step 3 of the elongation reaction.) A special tRNA molecule is required for this purpose. This **initiator tRNA** provides the amino acid that starts a protein chain, and it always carries methionine (aminoformyl methionine in bacteria). In eucaryotes the initiator tRNA molecule must be loaded onto the small ribosomal subunit before this subunit can bind to an mRNA molecule. An initiation factor called *eucaryotic initiation factor 2 (eIF-2)* is required to position the initiator tRNA on the small subunit. One molecule of eIF-2 becomes tightly bound to each initiator tRNA molecule as soon as this tRNA acquires its methionine, and in some cells the overall rate of protein synthesis is controlled by this factor (see Figure 9–82).

As described in more detail in the next section, the small ribosomal subunit helps its bound initiator tRNA molecule find a special AUG codon (the *start codon*) on an mRNA molecule. Once this has occurred, the several initiation factors that were previously associated with the small ribosomal subunit are discharged to make way for the binding of a large ribosomal subunit to the small one. Because the initiator tRNA molecule is bound to the P-site of the ribosome, the synthesis of a protein chain can begin directly with the binding of a second aminoacyl-tRNA molecule to the A-site of the ribosome (Figure 6–24). Thus a complete functional ribosome is assembled, with the mRNA molecule threaded through it (Figure 6–25). Further steps in the elongation phase of protein synthesis then proceed as described previously (see Figure 6–22). Because an initiator tRNA molecule has begun each polypeptide chain, all newly made proteins have a methionine (or the aminoformyl derivative of methionine in bacteria) as their amino-terminal residue. The methionine is often removed shortly after its incorporation by a specific aminopeptidase; this trimming process is important because the amino acid left at the amino terminus can determine the protein's lifetime in the cell by its effects on a ubiquitin-dependent protein-degradation pathway (see Figure 5–39).

Evidently the correct initiation site on the mRNA molecule must be selected by the small subunit acting in concert with initiation factors but in the absence of the large subunit. This requirement helps to explain why all ribosomes are formed from two separate subunits. We shall now consider how the correct start codon is selected.

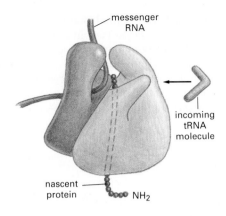

Figure 6–25 A three-dimensional model of a functioning bacterial ribosome. The small (*dark green*) subunit and the large (*light green*) subunit form a complex through which the messenger RNA is threaded. Although the exact paths of the mRNA and the nascent polypeptide chain are unknown, the addition of amino acids occurs in the general region shown, with the tRNAs held in the pocket formed between the large and small subunit. (Modified from J.A. Lake, *Annu. Rev. Biochem.* 54:507–530, 1985. © 1985 by Annual Reviews Inc.)

Only One Species of Polypeptide Chain Is Usually Synthesized from Each mRNA Molecule in Eucaryotes [10]

A messenger RNA molecule will typically contain many AUG sequences, each of which can code for methionine. In eucaryotes, however, only one of these AUG sequences will normally be recognized by the initiator tRNA and thereby serve as a **start codon.** How does the ribosome distinguish this start codon?

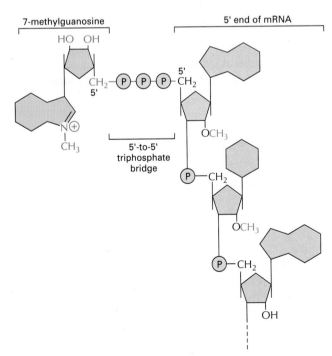

Figure 6–26 The structure of the cap at the 5′ end of eucaryotic mRNA molecules. Note the unusual 5′-to-5′ linkage to the positively charged 7-methylguanosine and the methylation of the 2′ hydroxyl group on the first ribose sugar in the RNA. (The second sugar may or may not be methylated.)

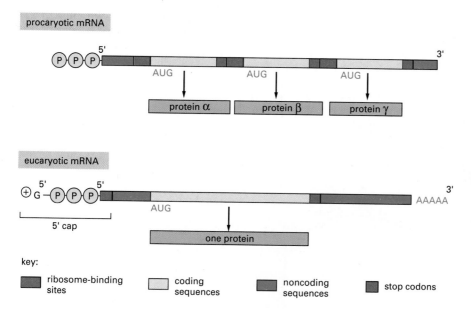

procaryotic mRNA

AUG → protein α

AUG → protein β

AUG → protein γ

eucaryotic mRNA

5' cap

AUG → one protein

key:
- ribosome-binding sites
- coding sequences
- noncoding sequences
- stop codons

Figure 6–27 A comparison of the structures of procaryotic and eucaryotic messenger RNA molecules. Although both mRNAs are synthesized with a triphosphate group at the 5' end, the eucaryotic RNA molecule immediately acquires a 5' cap, which is part of the structure recognized by the small ribosomal subunit. Protein synthesis therefore begins at a start codon near the 5' end of the mRNA (see Figure 6–24). In procaryotes, by contrast, the 5' end has no special significance, and there can be multiple ribosome-binding sites (called *Shine-Dalgarno sequences*) in the interior of an mRNA chain, each resulting in the synthesis of a different protein.

Eucaryotic RNAs (except those that are synthesized in mitochondria and chloroplasts) are extensively modified in the nucleus immediately after their transcription (discussed in Chapter 8). Two general modifications are the addition of a unique "cap" structure, composed of a 7-methylguanosine residue linked to a triphosphate at the 5' end (Figure 6–26) and the addition of a run of about 200 adenylic residues ("poly A") at the 3' end. What part the poly A plays in the translation process is uncertain (see Figure 9–87), but the 5' cap structure is essential for efficient protein synthesis. Experiments carried out with extracts of eucaryotic cells have shown that the small ribosomal subunit first binds at the 5' end of an mRNA chain, aided by recognition of the 5' cap (see Figure 6–24). This subunit then propels itself along the mRNA chain in a scanning mode, carrying its bound initiator tRNA in a search of an AUG start codon. The requirements for a start codon apparently are not very stringent, since the small subunit usually selects the first AUG it encounters; however, a few nucleotides in addition to the AUG are also important for the selection process. For most eucaryotic RNAs, once a start codon near the 5' end has been selected, none of the many other AUG codons farther down the chain will serve as initiation sites. As a result, only a single species of polypeptide chain is usually synthesized from an mRNA molecule (for exceptions see p. 467).

The mechanism for selecting a start codon in bacteria is different. Bacterial mRNAs have no 5' cap structure. Instead, they contain a specific ribosome-binding site sequence, up to six nucleotides long, which can occur at several places in the same mRNA molecule. These sequences are located four to seven nucleotides upstream from an AUG, and they form base pairs with a specific region of the rRNA in a ribosome to signal the initiation of protein synthesis at this nearby start codon. Bacterial ribosomes, unlike eucaryotic ribosomes, bind directly to start codons in the interior of an mRNA molecule to initiate protein synthesis. As a result, bacterial messenger RNAs are commonly *polycistronic*—that is, they encode multiple proteins that are separately translated from the same mRNA molecule. Eucaryotic mRNAs, in contrast, are typically *monocistronic,* with only one species of polypeptide chain being translated per messenger molecule (Figure 6–27).

The Binding of Many Ribosomes to an Individual mRNA Molecule Generates Polyribosomes [11]

The complete synthesis of a protein takes 20 to 60 seconds on average. But even during this very short period, multiple initiations usually take place on each

RNA and Protein Synthesis

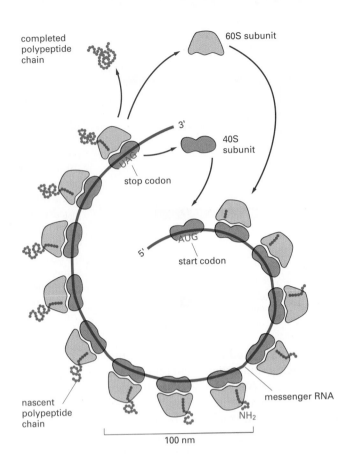

completed polypeptide chain

60S subunit

3'

40S subunit

stop codon

UAG

start codon

AUG

5'

nascent polypeptide chain

NH₂

messenger RNA

100 nm

Figure 6–28 A polyribosome.
Schematic drawing showing how a series of ribosomes can simultaneously translate the same mRNA molecule.

mRNA molecule being translated. A new ribosome hops onto the 5' end of the mRNA molecule almost as soon as the preceding ribosome has translated enough of the amino acid sequence to be out of the way. Such mRNA molecules are thus found in the cell as **polyribosomes,** or **polysomes,** formed by several ribosomes spaced as close as 80 nucleotides apart along a single messenger molecule (Figures 6–28 and 6–29). Polyribosomes are a common feature of cells. They can be isolated and separated from single ribosomes in the cytosol by ultracentrifugation after cell lysis (Figure 6–30). The mRNA purified from these polyribosomes can be used to determine if the protein encoded by a particular DNA sequence is being actively synthesized in the cells used to prepare the polyribosomes. These mRNA molecules can also serve as the starting material for the preparation of specialized cDNA libraries (discussed in Chapter 7).

In eucaryotes the nuclear envelope keeps transcription and protein synthesis separate. But in procaryotes, RNA is accessible to ribosomes as soon as it is made. Thus, ribosomes will begin synthesizing a polypeptide chain at the 5' end of a nascent mRNA molecule and then follow behind the RNA polymerase as it completes an mRNA chain.

The Overall Rate of Protein Synthesis in Eucaryotes Is Controlled by Initiation Factors [12]

As we discuss in Chapter 17, the cells in a multicellular organism proliferate only when they are stimulated to do so by specific growth factors. Although the

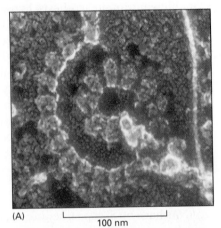

(A)

100 nm

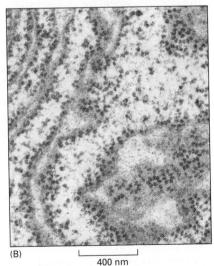

(B)

400 nm

Figure 6–29 Freeze-etch (A) and transmission (B) electron micrographs of typical polyribosomes in a eucaryotic cell. The cell cytoplasm is generally crowded with such polyribosomes, some free in the cytosol and some membrane-bound. (A, courtesy of John Heuser; B, courtesy of George Palade.)

mechanisms by which growth factors act are incompletely understood, one of their major effects must be to increase the overall rate of protein synthesis, for cells must double their contents before they divide. What determines the rate of protein synthesis? When eucaryotic cells in culture are starved of nutrients, there is a marked reduction in the rate of polypeptide chain *initiation*. This is the result of inactivation of the protein synthesis initiation factor eIF-2 (see Figure 9–82). The initiation factors required for protein synthesis are much more numerous and complex in eucaryotes than in procaryotes, even though they perform the same basic functions. Many of the extra components may be regulatory proteins that respond to growth factors and help coordinate cell growth and proliferation in multicellular organisms. Less complex controls are needed in bacteria, which generally grow as fast as the nutrients in their environment allow.

The Fidelity of Protein Synthesis Is Improved by Two Proofreading Mechanisms [13]

The error rate in protein synthesis can be estimated by monitoring the frequency of incorporation of an amino acid into a protein that normally lacks that amino acid. Error rates of about 1 amino acid misincorporated for every 10^4 amino acids polymerized are observed, which means that only about 1 in every 25 protein molecules of average size (400 amino acids) should contain an error. The fidelity of protein synthesis depends on the accuracy of the two adaptor mechanisms previously discussed: the linking of each amino acid to its corresponding tRNA molecule and the base-pairing of the codons in mRNA to the anticodons in tRNA (see Figure 6–14). Not surprisingly, cells have evolved "proofreading" mechanisms to reduce the number of errors in both these crucial steps of protein synthesis.

Two fundamentally different proofreading mechanisms are used, each representative of strategies used in other processes in the cell. Both involve expenditure of free energy, since, as discussed in Chapter 2, a price must be paid for any increase in order in the cell. A relatively simple mechanism is used to improve the accuracy of amino acid attachment to tRNA. Many aminoacyl tRNA synthetases have two active sites, one that carries out the loading reaction shown earlier (Figure 6–11) and one that recognizes an incorrect amino acid attached to its tRNA molecule and removes it by hydrolysis. The correction process is energetically costly because to be effective it must remove an appreciable fraction of correctly attached amino acids as well. The same type of costly two-step proofreading process is used in DNA replication (see Figure 6–42).

A more subtle "kinetic proofreading" mechanism is used to improve the fidelity of codon-anticodon pairing. Thus far we have given a simplified account of this pairing. In fact, once tRNA molecules have acquired an amino acid, they form a complex with an abundant protein called an *elongation factor (EF)*, which binds tightly to both the amino acid end of a tRNA and to a molecule of GTP. It is this complex, and not free tRNA, that pairs with the appropriate codon in an mRNA molecule. The bound elongation factor allows correct codon-anticodon pairing to occur but prevents the amino acid from being incorporated into the growing polypeptide chain. The initial codon recognition, however, triggers the elongation factor to hydrolyze its bound GTP (to GDP and inorganic phosphate), whereupon the factor dissociates from the ribosome without its tRNA, allowing protein synthesis to proceed. The elongation factor thereby introduces a short delay between codon-anticodon base-pairing and polypeptide chain elongation, which provides an opportunity for the bound tRNA molecule to exit from the ribosome. An incorrect tRNA molecule forms a smaller number of codon-anticodon hydrogen bonds than a correct one; it therefore binds more weakly to the ribosome and is more likely to dissociate during this period. Because the delay introduced by the elongation factor causes most incorrectly bound tRNA molecules to leave the ribosome without being used for protein synthesis, this factor increases the ratio of correct to incorrect amino acids incorporated into protein (Figure 6–31).

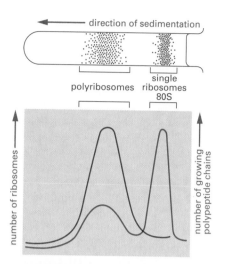

Figure 6–30 The isolation of polyribosomes. Polyribosomes are separated from single ribosomes (and their subunits) by sedimentation in a centrifuge. This method is based on the fact that large molecular aggregates move faster than small ones in a strong gravitational field. Generally, the sedimentation is done through a gradient of sucrose to stabilize the solution against convective mixing. Note that most of the growing polypeptide chains (*red line*) are associated with the polyribosomes.

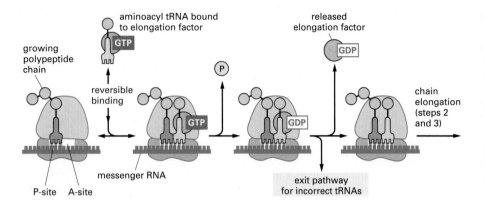

growing polypeptide chain

aminoacyl tRNA bound to elongation factor

GTP

reversible binding

released elongation factor

GDP

P

GTP

GDP

chain elongation (steps 2 and 3)

messenger RNA

P-site A-site

exit pathway for incorrect tRNAs

Figure 6–31 Kinetic proofreading selects for the correct tRNA molecule on the ribosome. This more detailed view of step 1 of the elongation phase of protein synthesis shows how, in the initial binding event, an aminoacyl-tRNA molecule that is tightly bound to an elongation factor pairs transiently with the codon at the A-site. This pairing triggers GTP hydrolysis by the elongation factor, enabling the factor to dissociate from the aminoacyl-tRNA molecule, which can now participate in chain elongation (see Figure 6–22). A delay between aminoacyl tRNA binding and its availability for protein synthesis is thereby inserted into the protein synthesis mechanism. As a result, only those tRNAs with the correct anticodon are likely to remain paired to the mRNA long enough to be added to the growing polypeptide chain.

The elongation factor, which is an abundant protein, is called EF-Tu in procaryotes and EF-1 in eucaryotes. The dramatic change in the three-dimensional structure of EF-Tu that is caused by GTP hydrolysis is illustrated in Figure 5–20.

Many Inhibitors of Procaryotic Protein Synthesis Are Useful as Antibiotics [14]

Many of the most effective antibiotics used in modern medicine are compounds made by fungi that act by inhibiting bacterial protein synthesis. A number of these drugs exploit the structural and functional differences between procaryotic and eucaryotic ribosomes so as to interfere with the function of procaryotic ribosomes preferentially. Thus some of these compounds can be taken in high doses without undue toxicity to humans. Because different antibiotics bind to different regions of bacterial ribosomes, they often inhibit different steps in the synthetic process. Some of the more common antibiotics of this kind are listed in Table 6–1 along with several other commonly used inhibitors of protein synthesis, some of which act on eucaryotic cells and therefore cannot be used as antibiotics.

Table 6–1 Inhibitors of Protein or RNA Synthesis

Inhibitor	Specific Effect
*Acting Only on Procaryotes**	
Tetracycline	blocks binding of aminoacyl-tRNA to A-site of ribosome
Streptomycin	prevents the transition from initiation complex to chain-elongating ribosome and also causes miscoding
Chloramphenicol	blocks the peptidyl transferase reaction on ribosomes (step 2 in Figure 6–22)
Erythromycin	blocks the translocation reaction on ribosomes (step 3 in Figure 6–22)
Rifamycin	blocks initiation of RNA chains by binding to RNA polymerase (prevents RNA synthesis)
Acting on Procaryotes and Eucaryotes	
Puromycin	causes the premature release of nascent polypeptide chains by its addition to growing chain end
Actinomycin D	binds to DNA and blocks the movement of RNA polymerase (prevents RNA synthesis)
Acting Only on Eucaryotes	
Cycloheximide	blocks the translocation reaction on ribosomes (step 3 in Figure 6–22)
Anisomycin	blocks the peptidyl transferase reaction on ribosomes (step 2 in Figure 6–22)
α-Amanitin	blocks mRNA synthesis by binding preferentially to RNA polymerase II

*The ribosomes of eucaryotic mitochondria (and chloroplasts) often resemble those of procaryotes in their sensitivity to inhibitors. Therefore, some of these antibiotics can have a deleterious effect on human mitochondria.

Because they block specific steps in the processes that lead from DNA to protein, many of the compounds listed in Table 6–1 are useful for cell biological studies. Among the most commonly used drugs in such experimental studies are *chloramphenicol, cycloheximide,* and *puromycin,* all of which specifically inhibit protein synthesis. In a eucaryotic cell, for example, chloramphenicol inhibits protein synthesis on ribosomes only in mitochondria (and in chloroplasts in plants), presumably reflecting the procaryotic origins of these organelles (discussed in Chapter 14). Cycloheximide, on the other hand, affects only ribosomes in the cytosol. The difference in the sensitivity of protein synthesis to these two drugs provides a powerful way to determine in which cell compartment a particular protein is translated. Puromycin is especially interesting because it is a structural analogue of a tRNA molecule linked to an amino acid; the ribosome mistakes it for an authentic amino acid and covalently incorporates it at the carboxyl terminus of the growing polypeptide chain, thereby causing the premature termination and release of the polypeptide (see Figure 3–23). As might be expected, puromycin inhibits protein synthesis in both procaryotes and eucaryotes.

How Did Protein Synthesis Evolve? [15]

The molecular processes underlying protein synthesis seem inexplicably complex. Although we can describe many of them, they do not make conceptual sense in the way that DNA transcription, DNA repair, and DNA replication do. As we have seen, protein synthesis in present-day organisms centers on the ribosome, which consists of proteins arranged around a core of rRNA molecules. Why should rRNA molecules exist at all, and how did they come to play such a dominant part in the structure and function of the ribosome?

Before the discovery of mRNA in the early 1960s, it was suspected that the large amount of RNA in ribosomes served a "messenger" function, carrying genetic information from DNA to proteins. Now we know, however, that all of the ribosomes in a cell contain an identical set of rRNA molecules that have no such informational role. In bacterial ribosomes, rRNA molecules have been shown to have catalytic functions in protein synthesis. As mentioned earlier, the major rRNA of the large ribosomal subunit appears to be the peptidyl transferase; in addition, the rRNA of the small ribosomal subunit forms a short base-paired helix with the initiation site sequence on bacterial mRNA molecules, positioning the neighboring AUG start codon at the P-site. A variety of specific base-pair interactions likewise form between tRNA molecules and bacterial rRNAs, although these interactions involve individual bases on the rRNA that are far apart in the nucleotide sequence, suggesting complex sets of interactions that depend on the tertiary structure of the rRNA.

Protein synthesis also relies heavily on a large number of proteins that are bound to the rRNAs in a ribosome (see Figure 6–20). The complexity of a process with so many interacting components has made many biologists despair of ever understanding the pathway by which protein synthesis evolved. The discovery that RNA molecules can act as enzymes, however, has provided a new way of viewing the pathway. As discussed in Chapter 1, early biological reactions probably used RNA molecules rather than protein molecules as catalysts. In the earliest cells tRNA molecules on their own may have formed catalytic surfaces that allowed them to bind and activate specific amino acids without requiring aminoacyl-tRNA synthetase enzymes. Likewise, rRNA molecules may have served by themselves as the entire "ribosome," folding up in complex ways to generate an intricate set of surfaces that both guided tRNA pairings with mRNA codons and catalyzed the polymerization of the tRNA-linked amino acids (see Figure 1–7). Over the course of evolution individual proteins have been added to this machinery, each one making the process a little more accurate and efficient, or adding regulatory controls. In this view the large amount of RNA in present-day ribosomes is a remnant of a very early stage in evolution, before proteins dominated biological catalysis.

RNA and Protein Synthesis

Summary

Before the synthesis of a particular protein can begin, the corresponding mRNA molecule must be produced by DNA transcription. Then a small ribosomal subunit binds to the mRNA molecule at a start codon (AUG) that is recognized by a unique initiator tRNA molecule. A large ribosomal subunit binds to complete the ribosome and initiate the elongation phase of protein synthesis. During this phase aminoacyl tRNAs, each bearing a specific amino acid, sequentially bind to the appropriate codon in mRNA by forming complementary base pairs with the tRNA anticodon. Each amino acid is added to the carboxyl-terminal end of the growing polypeptide by means of a cycle of three sequential steps: aminoacyl-tRNA binding, followed by peptide bond formation, followed by ribosome translocation. The ribosome progresses from codon to codon in the 5′-to-3′ direction along the mRNA molecule until one of three stop codons is reached. A release factor then binds to the stop codon, terminating translation and releasing the completed polypeptide from the ribosome.

Eucaryotic and procaryotic ribosomes are highly homologous, despite substantial differences in the number and size of their rRNA and protein components. The predominant role of rRNA in ribosome structure and function is likely to reflect the ancient origin of protein synthesis, which is thought to have evolved in an environment dominated by RNA-mediated catalysis.

DNA Repair [16]

The long-term survival of a species may be enhanced by genetic changes, but the survival of the individual demands genetic stability. Maintaining genetic stability requires not only an extremely accurate mechanism for replicating the DNA before a cell divides, but also mechanisms for repairing the many accidental lesions that occur continually in DNA. Most such spontaneous changes in DNA are temporary because they are immediately corrected by processes collectively called *DNA repair*. Only rarely do the cell's DNA maintenance processes fail and allow a permanent change in the DNA. Such a change is called a **mutation,** and it can destroy an organism if the change occurs in a vital position in the DNA sequence.

Before examining the mechanisms of DNA repair, we briefly discuss the maintenance of DNA sequences from one generation to the next.

DNA Sequences Are Maintained with Very High Fidelity [17]

The rate at which stable changes occur in DNA sequences (the *mutation rate*) can be estimated only indirectly. One way is to compare the amino acid sequence of the same protein in several species. The fraction of the amino acids that are different can then be compared with the estimated number of years since each pair of species diverged from a common ancestor, as determined from the fossil record. In this way one can calculate the number of years that elapse, on average, before an inherited change in the amino acid sequence of a protein becomes fixed in the species. Because each such change will commonly reflect the alteration of a single nucleotide in the DNA sequence of the gene encoding that protein, this value can be used to estimate the average number of years required to produce a single, stable mutation in the gene.

Such calculations always will substantially underestimate the actual mutation rate because most mutations will spoil the function of the protein and vanish from the population through natural selection. But there is one family of proteins whose sequence does not seem to matter, and so the genes that encode them can accumulate mutations without being selected against. These proteins are the **fibrinopeptides**—20-residue-long fragments that are discarded from the protein *fibrinogen* when it is activated to form *fibrin* during blood clotting. Since the function of fibrinopeptides apparently does not depend on their amino acid sequence, they can tolerate almost any amino acid change. Sequence analysis of the fibrinopeptides indicates that an average-sized protein 400 amino acids long would be randomly altered by an amino acid change roughly once every 200,000 years. More recently, DNA sequencing technology has made it possible to compare corresponding nucleotide sequences in regions of the genome that do not code for protein. Comparisons of such sequences in several mammalian species produce estimates of the mutation rate during evolution that are in excellent agreement with those obtained from the fibrinopeptide studies.

The Observed Mutation Rates in Proliferating Cells Are Consistent with Evolutionary Estimates [18]

The mutation rate can be estimated more directly by observing the rate at which spontaneous genetic changes arise in a large population of cells followed over a relatively short period of time. This can be done either by estimating the frequency with which new mutants arise in very large animal populations (in a colony of fruit flies or mice, for example) or by screening for changes in specific proteins in cells growing in culture. Although they are only approximate, the numbers obtained in both cases are consistent with an error frequency of 1 base-pair change in roughly 10^9 base pairs for each cell generation. Consequently, a single gene that encodes an average-sized protein (containing about 10^3 coding base pairs) would suffer a mutation once in about 10^6 cell generations. This number is at least roughly consistent with the evolutionary estimate described above, in which one mutation appears in an average gene in the germ line every 200,000 years.

Most Mutations in Proteins Are Deleterious and Are Eliminated by Natural Selection [19]

When the number of amino acid differences in a particular protein is plotted for several pairs of species against the time since the species diverged, the result is a reasonably straight line. That is, the longer the period since divergence, the larger the number of differences. For convenience, the slope of this line can be expressed in terms of the "unit evolutionary time" for that protein, which is the average time required for 1 amino acid change to appear in a sequence of 100 amino acid residues. When various proteins are compared, each shows a different but characteristic rate of evolution (Figure 6–32). Since all DNA base pairs are thought to be subject to roughly the same rate of random mutation, these different rates must reflect differences in the probability that an organism with a random mutation over the given protein will survive and propagate. Changes in amino acid sequence are evidently much more harmful for some proteins than for others. From Table 6–2 we can estimate that about 6 of every 7 random amino acid changes are harmful over the long term in hemoglobin, about 29 of every 30 amino acid changes are harmful in cytochrome c, and virtually all amino acid changes are harmful in histone H4. We assume that individuals who carried such harmful mutations have been eliminated from the population by natural selection.

Low Mutation Rates Are Necessary for Life as We Know It [19]

Since most mutations are deleterious, no species can afford to allow them to accumulate at a high rate in its germ cells. We discuss later why the observed mutation frequency, low though it is, nevertheless, is thought to limit the number of essential proteins that any organism can encode in its germ line to about 60,000. By an extension of the same arguments, a mutation frequency tenfold higher would limit an organism to about 6000 essential proteins. In this case evolution would probably have stopped at an organism no more complex than a fruit fly.

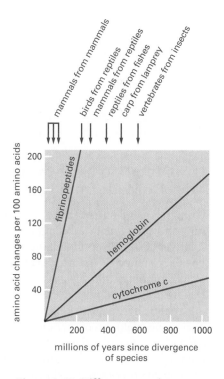

Figure 6–32 **Different proteins evolve at very different rates.** A comparison of the rates of amino acid change found in hemoglobin, cytochrome c, and the fibrinopeptides. Hemoglobin and cytochrome c have changed much more slowly during evolution than the fibrinopeptides. In determining rates of change per year (as in Table 6–2), it is important to realize that two species that diverged from a common ancestor 100 million years ago are separated by 200 million years of evolutionary time.

Table 6–2 Observed Rates of Change of the Amino Acid Sequences in Four Proteins over Evolutionary Time

Protein	Unit Evolutionary Time* (in millions of years)
Fibrinopeptide	0.7
Hemoglobin	5
Cytochrome c	21
Histone H4	500

*The "unit evolutionary time" is defined as the average time required for one *acceptable* amino acid change to appear in the indicated protein for every 100 amino acids that it contains.

While *germ* cells must be protected against high rates of mutation in order to maintain the species, the other cells of a multicellular organism (its *somatic* cells) must be protected from genetic change to safeguard the individual. Nucleotide changes in somatic cells can give rise to variant cells, some of which, through a process of natural selection, grow rapidly at the expense of the rest of the organism. In the extreme case the uncontrolled cell proliferation known as cancer results, which is responsible for about 30% of the deaths that occur in Europe and North America. These deaths are due largely to the accumulation of changes in the DNA sequences of somatic cells (discussed in Chapter 24). A tenfold increase in the mutation frequency would presumably cause a disastrous increase in the incidence of cancer by accelerating the rate at which somatic cell variants arise. Thus, both for the perpetuation of a species with 60,000 proteins (germ cell stability) and for the prevention of cancer resulting from mutations in somatic cells (somatic cell stability), eucaryotes depend on the remarkably high fidelity with which DNA sequences are maintained.

Low Mutation Rates Mean That Related Organisms Must Be Made from Essentially the Same Proteins [20]

Humans, as a genus distinct from the great apes, have existed for only a few million years. Each human gene has therefore had the chance to accumulate relatively few nucleotide changes since our inception, and most of these have been eliminated by natural selection. A comparison of humans and monkeys, for example, shows that their cytochrome c molecules differ in about 1% and their hemoglobins in about 4% of their amino acid positions. Clearly, a great deal of our genetic heritage must have been formed long before *Homo sapiens* appeared, during the evolution of mammals (which started about 300 million years ago) and even earlier. Because the proteins of mammals as different as whales and humans are very similar, the evolutionary changes that have produced such striking morphological differences must involve relatively few changes in the molecules from which we are made. Instead, it is thought that the morphological differences arise from differences in the temporal and spatial pattern of gene expression during embryonic development, which then determine the size, shape, and other characteristics of the adult. At the end of Chapter 8 we discuss the mechanisms that are thought to underlie such evolutionary changes in gene expression.

If Left Uncorrected, Spontaneous DNA Damage Would Rapidly Change DNA Sequences [21]

The physicist Erwin Schroedinger pointed out in 1945 that, whatever its chemical nature (at that time unknown), a gene must be extremely small and composed

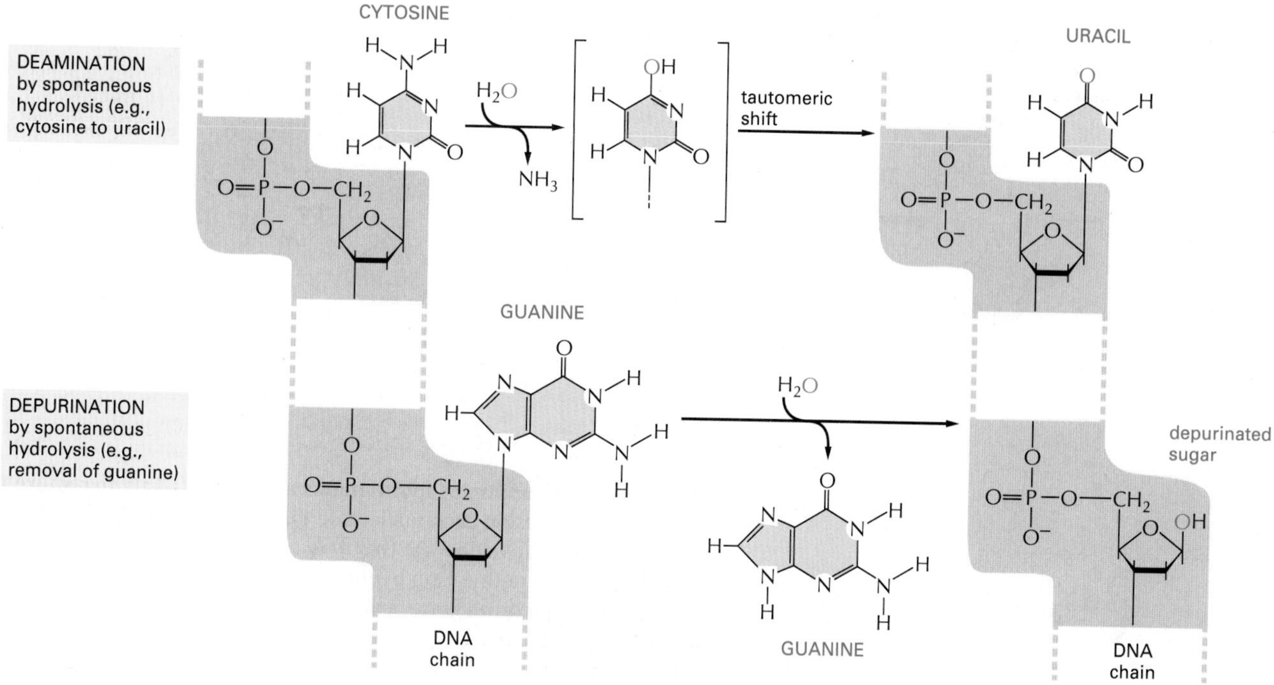

Figure 6–33 Deamination and depurination. These hydrolytic reactions are the two most frequent spontaneous chemical reactions known to create serious DNA damage in cells. Only a single example is shown for each type of reaction. (See also Figure 6–39.)

of few atoms. Otherwise the very large number of genes thought to be necessary to generate an organism would not fit in the cell nucleus. On the other hand, because it was so small, a gene would be expected to undergo significant changes as a result of spontaneous reactions induced by random thermal collisions with solvent molecules. This poses a serious dilemma, since genetic data imply that genes are composed of a remarkably stable substance in which spontaneous changes (mutations) occur rarely.

This dilemma is real. DNA does undergo major changes as a result of thermal fluctuations. We now know, for example, that about 5000 purine bases (adenine and guanine) are lost per day from the DNA of each human cell because of the thermal disruption of their *N*-glycosyl linkages to deoxyribose (*depurination*). Similarly, spontaneous *deamination* of cytosine to uracil in DNA is estimated to occur at a rate of 100 bases per genome per day (Figure 6–33). DNA bases are also subject to change by reactive metabolites (including reactive forms of oxygen) that can alter their base-pairing abilities and by ultraviolet light from the sun, which promotes a covalent linkage of two adjacent pyrimidine bases in DNA (forming, for example, the *thymine dimers* shown in Figure 6–34B). These are only a few of many changes that can occur in our DNA (Figure 6–34A). Most of them would be expected to lead either to deletion of one or more base pairs in the daughter DNA chain after DNA replication or to a base-pair substitution (each C → U deamination, for example, would eventually change a C-G base pair to a T-A base pair, since U closely resembles T and forms a complementary base pair with A). As we have seen, a high rate of such random changes would have disastrous consequences for an organism.

The Stability of Genes Depends on DNA Repair [22]

Despite the thousands of random changes created every day in the DNA of a human cell by heat energy and metabolic accidents, only a few stable changes (mutations) accumulate in the DNA sequence of an average cell in a year. We now know that fewer than one in a thousand accidental base changes in DNA causes a mutation; the rest are eliminated with remarkable efficiency by **DNA repair.**

(A)

(B)

Figure 6–34 A summary of spontaneous alterations likely to require DNA repair. (A) The sites on each nucleotide that are known to be modified by spontaneous oxidative damage (*red arrows*), hydrolytic attack (*blue arrows*), and uncontrolled methylation by the methyl group donor S-adenosyl-methionine (*green arrows*) are indicated, with the size of each arrow indicating the relative frequency of each event. The two most frequent types of hydrolytic events are illustrated in more detail in Figure 6–33. (B) The thymine dimer, a type of damage introduced into DNA in cells that are exposed to ultraviolet irradiation (as in sunlight). A similar dimer will form between any two neighboring pyrimidine bases (C or T residues) in DNA. (A, after T. Lindahl, *Nature* 362:709–715, 1993. © 1993 Macmillan Magazines Ltd.)

There are a variety of repair mechanisms, each catalyzed by a different set of enzymes. Nearly all of these mechanisms depend on the existence of two copies of the genetic information, one in each strand of the DNA double helix: if the sequence in one strand is accidentally changed, information is not lost irretrievably because a complementary copy of the altered strand remains in the sequence of nucleotides in the other strand. The basic pathway for DNA repair is illustrated schematically in Figure 6–35. As indicated, it involves three steps:

1. The altered portion of a damaged DNA strand is recognized and removed by enzymes called *DNA repair nucleases*, which hydrolyze the phosphodiester bonds that join the damaged nucleotides to the rest of the DNA molecule, leaving a small gap in the DNA helix in this region.

Figure 6–35 DNA repair. The three steps common to most types of repair are excision (step 1), resynthesis (step 2), and ligation (step 3). In step 1 the damage is excised; in steps 2 and 3 the original DNA sequence is restored. *DNA polymerase* fills in the gap created by the excision events, and *DNA ligase* seals the nick left in the repaired strand. Nick sealing consists of the re-formation of a broken phosphodiester bond (see Figure 6–37).

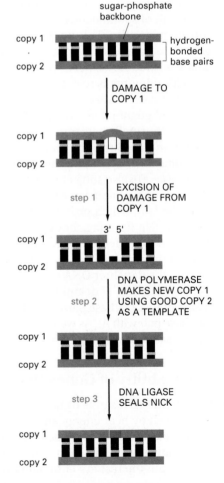

sugar-phosphate backbone

copy 1

copy 2

hydrogen-bonded base pairs

DAMAGE TO COPY 1

copy 1

copy 2

step 1

EXCISION OF DAMAGE FROM COPY 1

3' 5'

copy 1

copy 2

step 2

DNA POLYMERASE MAKES NEW COPY 1 USING GOOD COPY 2 AS A TEMPLATE

copy 1

copy 2

step 3

DNA LIGASE SEALS NICK

copy 1

copy 2

NET RESULT: RESTORATION OF TWO GOOD COPIES

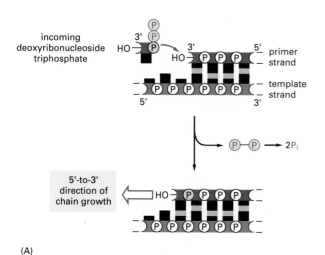

(A)

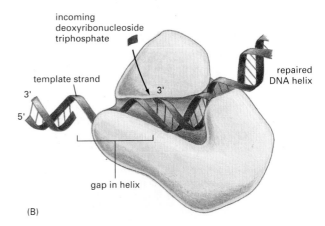

(B)

2. Another enzyme, **DNA polymerase,** binds to the 3′-OH end of the cut DNA strand and fills in the gap by making a complementary copy of the information stored in the "good" (*template*) strand.

3. The break or "nick" in the damaged strand left when the DNA polymerase has filled in the gap is sealed by a third type of enzyme, **DNA ligase,** which completes the restoration process.

Both DNA polymerase and DNA ligase have important general roles in DNA metabolism; both function in DNA replication as well as in DNA repair, for example. The reactions that these two enzymes catalyze are illustrated in Figures 6–36 and 6–37, respectively.

DNA Damage Can Be Removed by More Than One Pathway [23]

The details of the excision step in DNA repair depend on the type of damage. Depurination, for example, which is by far the most frequent lesion that occurs in DNA, leaves a deoxyribose sugar with a missing base (see Figure 6–33). This exposed sugar is rapidly recognized by the enzyme *AP endonuclease,* which cuts the DNA phosphodiester backbone at the 5′ side of the altered site. After excision of the sugar phosphate residue by a phosphodiesterase enzyme, an undamaged DNA sequence is restored by DNA polymerase and DNA ligase (see Figure 6–35).

Figure 6–36 The DNA polymerase enzyme. (A) The reaction catalyzed by DNA polymerase. This enzyme catalyzes the stepwise addition of a deoxyribonucleotide to the 3′-OH end of a polynucleotide chain (the *primer strand*) that is paired to a second, *template strand.* The new DNA strand therefore grows in the 5′-to-3′ direction. Because each incoming deoxyribonucleoside triphosphate must pair with the template strand in order to be recognized by the polymerase, this strand determines which of the four possible deoxyribonucleotides (A, C, G, or T) will be added. As in the case of RNA polymerase, the reaction is driven by a large favorable free-energy change (see Figure 6–3). (B) The structure of an *E. coli* DNA polymerase molecule has been determined by x-ray crystallography. This drawing illustrates how the polymerase is thought to function during the DNA synthesis involved in DNA repair. (B, adapted from L.S. Beese, V. Derbyshire, and T.A. Steitz, *Science* 260:352–355, 1993. © 1993 the AAAS.)

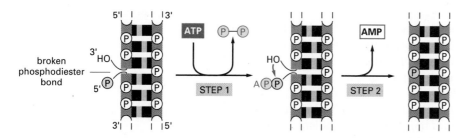

Figure 6–37 The reaction catalyzed by DNA ligase. This enzyme seals a broken phosphodiester bond. As shown, DNA ligase uses a molecule of ATP to activate the 5′ end at the nick (step 1) before forming the new bond (step 2). In this way the energetically unfavorable nick-sealing reaction is driven by being coupled to the energetically favorable process of ATP hydrolysis. In *Bloom's syndrome,* an inherited human disease, individuals are partially defective in DNA ligation and consequently are deficient in DNA repair; as a consequence, they have a dramatically increased incidence of cancer.

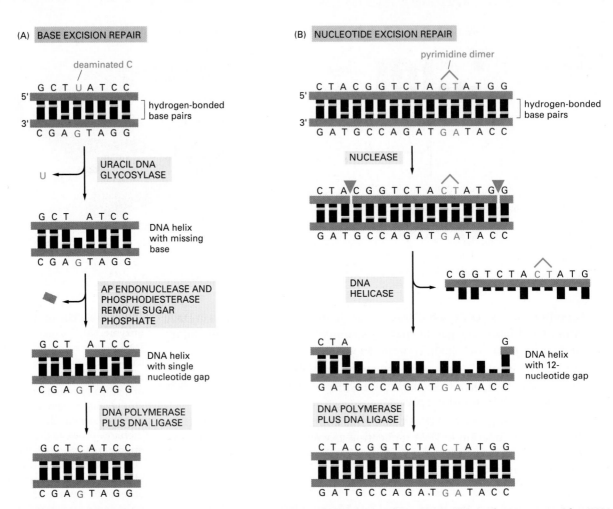

Figure 6–38 Comparison of two major DNA repair pathways. (A) *Base excision repair.* This pathway starts with a DNA glycosylase. Here the enzyme uracil DNA glycosylase removes an accidentally deaminated cytosine in DNA. After the action of this glycosylase (or another DNA glycosylase that recognizes a different kind of damage) the sugar phosphate with the missing base is cut out by the sequential action of AP endonuclease and a phosphodiesterase, the same enzymes that initiate the repair of depurinated sites. The gap of a single nucleotide is then filled by DNA polymerase and DNA ligase. The net result is that the U that was created by accidental deamination is restored to a C. The AP endonuclease derives its name from the fact that it recognizes any site in the DNA helix that contains a deoxyribose sugar with a missing base; such sites can arise either by the loss of a purine (*ap*urinic sites) or by the loss of a pyrimidine (*ap*yriminic sites). (B) *Nucleotide excision repair.* After a multienzyme complex recognizes a bulky lesion such as a pyrimidine dimer (see Figure 6–34B), one cut is made on each side of the lesion, and an associated DNA helicase then removes the entire portion of the damaged strand. The multienzyme complex in bacteria leaves the gap of 12 nucleotides shown; the gap produced in human DNA is more than twice this size.

A related repair pathway, called **base excision repair,** involves a battery of enzymes called *DNA glycosylases.* Each DNA glycosylase recognizes an altered base in DNA and catalyzes its hydrolytic removal. There are at least six types of these enzymes, including those that remove deaminated Cs, deaminated As, different types of alkylated or oxidized bases, bases with opened rings, and bases in which a carbon-carbon double bond has been accidentally converted to a carbon-carbon single bond. As an example of the general mechanism that operates in all cases, the removal of a deaminated C by uracil DNA glycosylase is shown in Figure 6–38A. The DNA glycosylase reaction produces a deoxyribose sugar with a missing base. Because this sugar phosphate is the same substrate recognized by the AP endonuclease, the subsequent steps in the repair process proceed in the same way as for depurinated sites. The importance of removing accidentally deaminated DNA bases has been directly demonstrated. In mutant bacteria that lack the enzyme uracil DNA glycosylase, the normally low spontaneous rate of change of a C-G to a T-A base pair is increased about twentyfold.

Cells have a separate **nucleotide excision repair** pathway capable of removing almost any type of DNA damage that creates a large change in the DNA double helix. Such "bulky lesions" include those created by the covalent reaction of DNA bases with large hydrocarbons (such as the carcinogen benzopyrene), as well as the various pyrimidine dimers (T-T, T-C, and C-C) caused by sunlight. In these cases a large multienzyme complex scans the DNA for a distortion in the double helix rather than for a specific base change. Once a bulky lesion is found, the phosphodiester backbone of the abnormal strand is cleaved on both sides of the distortion, and the portion of the strand containing the lesion (an oligonucleotide) is peeled away from the DNA double helix by a *DNA helicase* enzyme (discussed later). The gap produced in the DNA helix is then repaired in the usual manner by DNA polymerase and DNA ligase (Figure 6–38B).

The importance of these repair processes is indicated by the large investment that cells make in DNA repair enzymes. A comprehensive genetic analysis of a yeast suggests that these cells contain more than 50 different genes that code for DNA repair functions. DNA repair pathways are likely to be at least as complex in humans. Individuals with the genetic disease *xeroderma pigmentosum*, for example, are defective in a nucleotide excision repair process that can be shown by genetic analysis to require at least seven different gene products. Such individuals develop severe skin lesions, including skin cancer, because of the accumulation of pyrimidine dimers in cells that are exposed to sunlight.

Cells Can Produce DNA Repair Enzymes in Response to DNA Damage [24]

Cells have evolved a number of mechanisms to help them survive in a hazardous world. Often an extreme environmental insult activates a battery of genes whose products protect the cell from its effects. One such mechanism shared by all cells is the *heat-shock response,* which is evoked by the exposure of cells to unusually high temperatures. The induced "heat-shock proteins" include some that are thought to help stabilize and repair partially denatured cell proteins (see Figure 5–29).

Many cells also have mechanisms that enable them to synthesize DNA repair enzymes as an emergency response to severe DNA damage. The best-studied example is the **SOS response** in *E. coli.* In this bacterium any block to DNA replication caused by DNA damage produces a signal (thought to be an excess of single-stranded DNA) that induces an increase in the transcription of more than 15 genes, many of which code for proteins that function in DNA repair. The signal first activates the *E. coli* RecA protein (discussed later), which then destroys a negatively acting gene regulatory protein (a *repressor*) that normally suppresses the transcription of the entire set of SOS response genes. Studies of mutant bacteria deficient in different parts of the response indicate that the newly synthesized proteins have two effects. First, as would be expected, the induction of new DNA repair enzymes increases cell survival. When the mutants deficient in this part of the SOS response are treated with a DNA-damaging agent such as ultraviolet radiation, an unusually high proportion of them die. Second, several of the induced proteins transiently increase the mutation rate by greatly increasing the number of errors made in copying DNA sequences. While this has little effect on short-term survival, it is presumably advantageous in the long term because it produces a burst of genetic variability in the bacterial population and hence increases the chance that a mutant cell with increased fitness will arise.

The DNA repair system activated in the SOS response is not the only inducible DNA repair system known. Bacteria have another system that is activated specifically by the presence of methylated nucleotides in DNA, and there is at least one inducible DNA repair system in yeast cells. Some higher eucaryotic cells have been reported to adapt to DNA damage in similar ways.

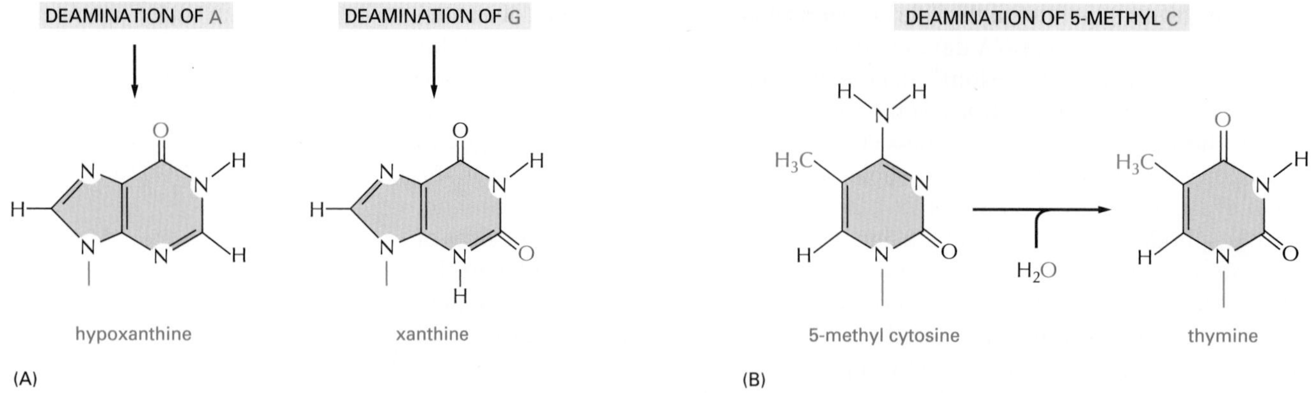

DEAMINATION OF A   DEAMINATION OF G   DEAMINATION OF 5-METHYL C

hypoxanthine   xanthine   5-methyl cytosine   thymine

(A)   (B)

The Structure and Chemistry of the DNA Double Helix Make It Easy to Repair [25]

The DNA double helix seems to be optimally constructed for repair. As discussed in Chapter 1, RNA is thought to have evolved before DNA, and it seems likely that the genetic code was initially carried in the four nucleotides A, C, G, and U. This raises the question of why the U in RNA has been replaced in DNA by T (which is 5-methyl U). We have seen that spontaneous C deamination converts C to U but that this event is rendered harmless by uracil DNA glycosylase (see Figure 6–38A). One can imagine how any repair enzyme designed to recognize and excise such accidents would be confused by the normal U nucleotides in a U-containing DNA molecule. Thus it is not surprising that U is not used in DNA.

This line of argument is strengthened by the observation that every possible deamination event in DNA yields an unnatural base, which can therefore be directly recognized and removed by a specific DNA glycosylase. Hypoxanthine, for example, is the simplest purine base capable of pairing specifically with C, but hypoxanthine is the direct deamination product of A. The addition of a second amino group to hypoxanthine produces G, which cannot be formed from A by spontaneous deamination and whose deamination product is likewise unique (Figure 6–39A).

A special situation occurs in vertebrate DNA, where selected C nucleotides are methylated at specific CG sequences associated with inactive genes (discussed in Chapter 9). As illustrated in Figure 6–39B, the accidental deamination of these methylated C nucleotides produces the natural nucleotide T, which forms a mismatched base pair with a G on the opposite DNA strand. To help protect methylated C nucleotides against such mutations, a special DNA glycosylase recognizes a mismatched base pair involving T in the sequence TG and removes the T. This DNA repair mechanism must be relatively ineffective, however, as methylated C nucleotides are common sites for mutations in vertebrate DNA. Even though only about 3% of the C nucleotides in human DNA are methylated, mutations in these methylated nucleotides account for about one-third of the single-base mutations that have been observed in inherited human diseases (see also Figure 9–71).

Whereas the chemistry of the bases ensures that deamination will be detected, accurate repair—and the fundamental answer to Schroedinger's dilemma—depends on the existence of separate copies of the genetic information in the two strands of the double helix. Only in the very unlikely event that both strands are damaged simultaneously at the same base pair is the cell left without one good copy to serve as a template for DNA repair. Even in this case mechanisms have evolved that are sometimes able to repair the damage. These repair mechanisms require that a second DNA helix of the same sequence be present in the cell, and they use genetic recombination mechanisms to transfer the missing information from one DNA helix to another—a process called *gene conversion*, which we discuss later.

Figure 6–39 **The deamination of DNA nucleotides.** In each case the oxygen atom added from the reaction with water is colored *red*. (A) The spontaneous deamination products of A and G are recognizable as unnatural when they occur in DNA and thus are readily recognized and repaired. The deamination of C to U was illustrated in Figure 6–33, and T has no amino group to deaminate. (B) A few percent of the C nucleotides in vertebrate DNAs are methylated to help control gene expression. When these 5-methyl C nucleotides are accidentally deaminated, they form T. This T will be paired with a G on the opposite strand, forming a mismatched base pair.

Genetic information is stored in single-stranded DNA or RNA molecules only in some very small viruses with genomes of a few thousand nucleotides. The types of repair processes that we have described cannot operate on such nucleic acids, and the chance of a nucleotide change occurring in these viruses is very high. It seems that only organisms with tiny genomes can afford to encode their genetic information in a structure other than a DNA double helix.

Summary

The fidelity with which DNA sequences are maintained in higher eucaryotes can be estimated from the rates at which changes have occurred in nonessential protein and DNA sequences over evolutionary time. This fidelity is so high that a mammalian germ-line cell with a genome of 3×10^9 base pairs is subjected on average to only about 10 to 20 base-pair changes per year. But unavoidable chemical processes damage thousands of DNA nucleotides in a typical mammalian cell every day. Genetic information can be stored stably in DNA sequences only because a large variety of DNA repair enzymes continuously scan the DNA and replace the damaged nucleotides.

The process of DNA repair depends on the presence of a separate copy of the genetic information in each strand of the DNA double helix. An accidental lesion on one strand can therefore be cut out by a repair enzyme and a good strand resynthesized from the information in the undamaged strand. Most of the damage to DNA bases is excised by one of two major pathways. In base excision repair an altered base is removed by a DNA glycosylase enzyme, followed by excision of the resulting sugar phosphate. In nucleotide excision repair a small region of the strand surrounding the damage is removed from the DNA helix as an oligonucleotide. In both cases the small gap left in the DNA helix is filled in by the sequential action of DNA polymerase and DNA ligase.

DNA Replication [26]

Besides maintaining the integrity of DNA sequences by DNA repair, all organisms must duplicate their DNA accurately before every cell division. *DNA replication* occurs at polymerization rates of about 500 nucleotides per second in bacteria and about 50 nucleotides per second in mammals. Clearly, the proteins that catalyze this process must be both accurate and fast. Speed and accuracy are achieved by means of a multienzyme complex that guides the process and constitutes an elaborate "replication machine."

Base-pairing Underlies DNA Replication as well as DNA Repair [27]

DNA templating is the process in which the nucleotide sequence of a DNA strand (or selected portions of a DNA strand) is copied by complementary base-pairing (A with T or U, and G with C) into a complementary nucleic acid sequence (either DNA or RNA). The process entails the recognition of each nucleotide in the DNA strand by an unpolymerized complementary nucleotide and requires that the two strands of the DNA helix be separated, at least transiently, so that the hydrogen bond donor and acceptor groups on each base become exposed for base-pairing. The appropriate incoming single nucleotides are thereby aligned for their enzyme-catalyzed polymerization into a new nucleic acid chain. In 1957 the first such nucleotide polymerizing enzyme, *DNA polymerase,* was discovered. The substrates for this enzyme were found to be deoxyribonucleoside triphosphates, which are polymerized on a single-stranded DNA template. The stepwise mechanism of this reaction is the one previously illustrated in Figure 6–36 in connection with DNA repair. The discovery of DNA polymerase led to the isolation of *RNA polymerase*, which was correctly inferred to use ribonucleoside tri-phosphates as its substrates.

During DNA replication each of the two old DNA strands serves as a template for the formation of an entire new strand. Because each of the two daughters of a dividing cell inherits a new DNA double helix containing one old and one new strand (see Figure 3–13), DNA is said to be replicated "semiconservatively" by DNA polymerase.

The DNA Replication Fork Is Asymmetrical [28]

Autoradiographic analyses carried out in the early 1960s on whole replicating chromosomes labeled with a short pulse of the radioactive DNA precursor ^{3}H-thymidine revealed a localized region of replication that moves along the parental DNA double helix. Because of its Y-shaped structure, this active region is called a **DNA replication fork.** At a replication fork the DNA of both new daughter strands is synthesized by a multienzyme complex that contains the DNA polymerase.

Initially, the simplest mechanism of DNA replication appeared to be continuous growth of both new strands, nucleotide by nucleotide, at the replication fork as it moves from one end of a DNA molecule to the other. But because of the antiparallel orientation of the two DNA strands in the DNA double helix (see Figure 3–10 and Panel 3–2, pp. 100–101), this mechanism would require one daughter strand to grow in the 5′-to-3′ direction and the other in the 3′-to-5′ direction. Such a replication fork would require two different DNA polymerase enzymes. One would polymerize in the 5′-to-3′ direction (see Figure 6–36), where each incoming deoxyribonucleoside triphosphate carries the triphosphate activation needed for its own addition. The other would move in the 3′-to-5′ direction and work by so-called "head growth," in which the end of the growing DNA chain carries the triphosphate activation required for the addition of each subsequent nucleotide. Although head-growth polymerization occurs elsewhere in biochemistry (see Figure 2–36), it does not occur in DNA synthesis; no 3′-to-5′ DNA polymerase has ever been found (Figure 6–40).

How, then, is 3′-to-5′ DNA synthesis achieved? The answer was first suggested in the late 1960s by experiments in which highly radioactive ^{3}H-thymidine was added to dividing bacteria for a few seconds so that only the most recently replicated DNA, just behind the replication fork, became radiolabeled. This selective labeling method revealed the transient existence of pieces of DNA that were 1000 to 2000 nucleotides long, now commonly known as *Okazaki fragments,* at the bacterial growing fork. (Such replication intermediates were later found in eucaryotes, where they are only 100 to 200 nucleotides long.) The Okazaki fragments were shown to be synthesized only in the 5′-to-3′ chain direction and to

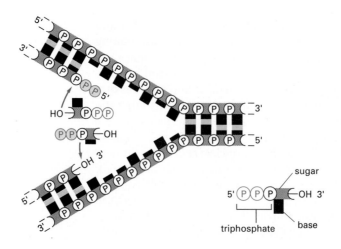

Figure 6–40 **An incorrect model for DNA replication.** Although it might appear to be the simplest mechanism for DNA replication, the mechanism illustrated here is not the one that cells use. Note that in this scheme both daughter DNA strands would grow continuously, using the energy of hydrolysis of the *yellow* phosphates to add the next nucleotide on each strand. This would require chain growth in both the 5′-to-3′ direction (*bottom*) and the 3′-to-5′ direction (*top*). No enzyme that catalyzes 3′-to-5′ nucleotide polymerization has ever been found.

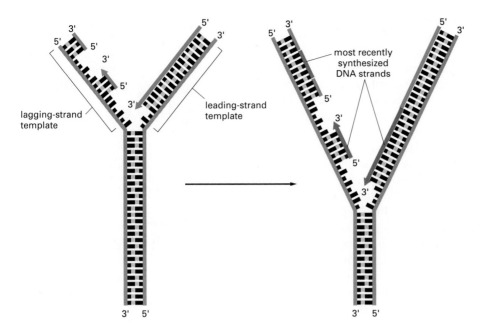

be joined together after their synthesis to create long DNA chains by the same *DNA ligase* enzyme that seals nicks during DNA repair (see Figure 6–37).

A replication fork has an asymmetric structure. The DNA daughter strand that is synthesized continuously is known as the **leading strand,** and its synthesis slightly precedes the synthesis of the daughter strand that is synthesized discontinuously, which is known as the **lagging strand.** The synthesis of the lagging strand is delayed because it must wait for the leading strand to expose the template strand on which each Okazaki fragment is synthesized (Figure 6–41). The synthesis of the lagging strand by a discontinuous, "backstitching" mechanism means that only the 5′-to-3′ type of DNA polymerase is needed for DNA replication.

The High Fidelity of DNA Replication Requires a Proofreading Mechanism [29]

The fidelity of copying that is observed after DNA replication has occurred is such that only about 1 error is made in every 10^9 base-pair replications, as required to maintain the mammalian genome of 3×10^9 DNA base pairs. This fidelity is much higher than expected, given that the standard complementary base pairs are not the only ones possible. With small changes in helix geometry, for example, two hydrogen bonds will form between G and T in DNA. In addition, rare tautomeric forms of the four DNA bases occur transiently in ratios of 1 part to 10^4 or 10^5. These forms will mispair without a change in helix geometry: the rare tautomeric form of C pairs with A instead of G, for example. If the DNA polymerase accepts a mispairing that occurs between an incoming deoxyribonucleoside triphosphate and the DNA template, the wrong nucleotide can be incorporated into the new DNA chain, producing a mutation. The high fidelity of DNA replication depends on several "proofreading" mechanisms that act sequentially to remove errors brought about in these ways.

One important proofreading process depends on special properties of the DNA polymerase enzyme. Unlike RNA polymerases, DNA polymerases do not begin a new polynucleotide chain by linking two nucleoside triphosphates together. They absolutely require the 3′-OH end of a base-paired *primer strand* on which to add further nucleotides (see Figure 6–36). Moreover, DNA molecules with a mismatched (not base-paired) nucleotide at the 3′-OH end of the primer strand are not effective as templates. DNA polymerase molecules are able to deal with such mismatched DNAs by means of either a separate catalytic subunit or a covalently linked, separate catalytic site that clips off any unpaired residues at the primer terminus. Clipping by this *3'-to-5' proofreading exonuclease* activity continues until enough nucleotides have been removed from the 3' end to regenerate a base-paired terminus that can prime DNA synthesis. In this way DNA polymerase functions as a "self-correcting" enzyme that removes its own polymerization errors as it moves along the DNA. Figure 6–42 illustrates how this proofreading process can correct a base-pairing error.

The requirement for a perfectly base-paired terminus is essential to the self-correcting properties of the DNA polymerase. For such an enzyme to start synthesis in the complete absence of a primer without losing any of its discrimination between base-paired and unpaired growing 3'-OH termini is apparently not possible. By contrast, the RNA polymerase enzymes involved in gene transcription need not be self-correcting: errors in making RNA are not passed on to the next generation, and an occasional defective molecule has no significance. RNA polymerases are able to start new polynucleotide chains without a primer, and an error frequency of about 1 in 10^4 is found both in RNA synthesis and in the separate process of translating mRNA sequences into protein sequences.

Only DNA Replication in the 5'-to-3' Direction Allows Efficient Error Correction

The need for accuracy probably explains why DNA replication occurs only in the 5'-to-3' direction. If there were a DNA polymerase that added deoxyribonucleoside triphosphates in such a way as to cause chains to grow in the 3'-to-5' chain direction, the growing 5'-chain end rather than the incoming mononucleotide would carry the activating triphosphate. In this case the mistakes in polymerization could not be simply hydrolyzed away, since the bare 5'-chain end thus created would immediately terminate DNA synthesis. It is much easier, therefore, to correct a mismatched base that has just been added to the 3' end than one that has just been added to the 5' end of a DNA chain. Although the type of mechanism for DNA replication shown in Figure 6–41 seems at first sight much more complex than the incorrect mechanism depicted in Figure 6–40, it is much more accurate because it involves DNA synthesis only in the 5'-to-3' direction.

A Special Nucleotide Polymerizing Enzyme Synthesizes Short RNA Primer Molecules on the Lagging Strand [30]

For the leading strand a special primer is needed only at the start of replication; once a replication fork is established, the DNA polymerase is continuously presented with a base-paired chain end on which to add new nucleotides. But the DNA polymerase on the lagging side of the fork requires only about 4 seconds to complete each short DNA fragment, after which it must start synthesizing a completely new fragment at a site farther along the template strand (see Figure 6–41). A special mechanism is needed to produce the base-paired primer strand required by this DNA polymerase molecule. The mechanism involves an enzyme called *DNA primase*, which uses ribonucleoside triphosphates to synthesize short **RNA primers** (Figure 6–43). These primers are about 10 nucleotides long in eucaryotes, and they are made at intervals on the lagging strand, where they are elongated by the DNA polymerase to begin each Okazaki fragment. The synthesis of each Okazaki fragment ends when this DNA polymerase runs into the RNA primer attached to the 5' end of the previous fragment. To produce a continuous DNA chain from the many DNA fragments made on the lagging strand, a special DNA repair system acts quickly to erase the old RNA primer and replace it with DNA. DNA ligase then joins the 3' end of the new DNA fragment to the 5' end of the previous one to complete the process (Figure 6–44).

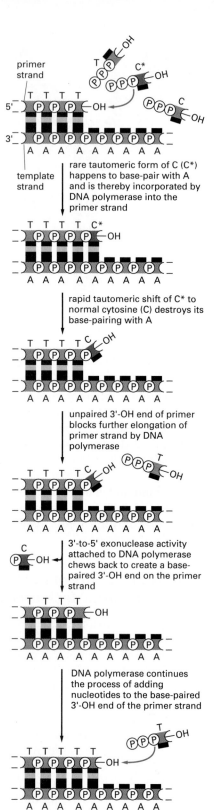

rare tautomeric form of C (C*) happens to base-pair with A and is thereby incorporated by DNA polymerase into the primer strand

rapid tautomeric shift of C* to normal cytosine (C) destroys its base-pairing with A

unpaired 3'-OH end of primer blocks further elongation of primer strand by DNA polymerase

3'-to-5' exonuclease activity attached to DNA polymerase chews back to create a base-paired 3'-OH end on the primer strand

DNA polymerase continues the process of adding nucleotides to the base-paired 3'-OH end of the primer strand

Figure 6–42 Proofreading during DNA replication.

Why might an erasable RNA primer be preferred to a DNA primer that need not be erased? The argument that a self-correcting polymerase cannot start chains *de novo* also implies its converse: an enzyme that starts chains *de novo* cannot be efficient at self-correction. Thus any enzyme that primes the synthesis of Okazaki fragments will of necessity make a relatively inaccurate copy (at least 1 error in 10^5). Even if the copies retained in the final product constituted as little as 5% of the total genome (for example, 10 nucleotides per 200-nucleotide DNA fragment), the resulting increase in overall mutation rate would be enormous. It therefore seems likely that the evolution of RNA rather than DNA for priming entailed a powerful advantage, since the ribonucleotides in the primer automatically mark these sequences as "bad copy" to be removed.

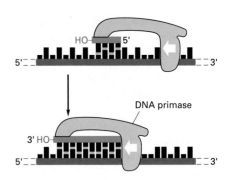

Figure 6–43 RNA primer synthesis. A schematic view of the reaction catalyzed by DNA primase, the enzyme that synthesizes the short RNA primers made on the lagging strand. Unlike DNA polymerase, this enzyme can start a new polynucleotide chain by joining two nucleoside triphosphates together. The primase stops after a short polynucleotide has been synthesized and makes the 3' end of this primer available for the DNA polymerase.

Special Proteins Help Open Up the DNA Double Helix in Front of the Replication Fork [31]

The DNA double helix must be opened up ahead of the replication fork so that the incoming deoxyribonucleoside triphosphates can form base pairs with the template strand. The DNA double helix is very stable under normal conditions: the base pairs are locked in place so strongly that temperatures approaching that of boiling water are required to separate the two strands in a test tube. For this reason most DNA polymerases can copy DNA only when the template strand has already been separated from its complementary strand. Additional proteins are needed to help open the double helix and thus provide the appropriate exposed DNA template for the DNA polymerase to copy. Two types of replication proteins contribute to this process—DNA helicases and single-strand DNA-binding proteins.

DNA helicases were first isolated as proteins that hydrolyze ATP when they are bound to single strands of DNA. As described in Chapter 5, the hydrolysis of ATP can change the shape of a protein molecule in a cyclical manner that allows the protein to perform mechanical work. DNA helicases utilize this principle to move rapidly along a DNA single strand; where they encounter a region of double helix, they continue to move along their strand, thereby prying apart the helix (Figure 6–45). We have previously described how a special DNA repair helicase functions in nucleotide excision repair (see Figure 6–38B).

The unwinding of the template DNA helix at a replication fork could in principle be catalyzed by two DNA helicases acting in concert—one running along the leading strand and one along the lagging strand. These two helicases would need to move in opposite directions along a DNA single strand and therefore would have to be different enzymes. Both types of DNA helicase, in fact, do exist, although in bacteria the DNA helicase on the lagging strand plays the predominant role, for reasons that will become clear shortly.

Single-strand DNA-binding (SSB) proteins—also called *helix-destabilizing proteins*—bind to exposed DNA strands without covering the bases, which therefore remain available for templating. These proteins are unable to open a long DNA helix directly, but they aid helicases by stabilizing the unwound, single-stranded conformation. In addition, their cooperative binding completely coats the regions of single-stranded DNA on the lagging strand, thereby preventing for-

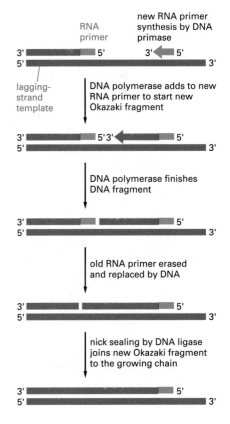

Figure 6–44 The synthesis of one of the many DNA fragments on the lagging strand. In eucaryotes the RNA primers are made at intervals spaced by about 200 nucleotides on the lagging strand, and each RNA primer is 10 nucleotides long. This primer is erased by a special DNA repair enzyme that recognizes an RNA strand in an RNA/DNA helix and excises it; this leaves a gap that is filled in by DNA polymerase and DNA ligase, as we saw for the DNA repair process (see Figure 6–35).

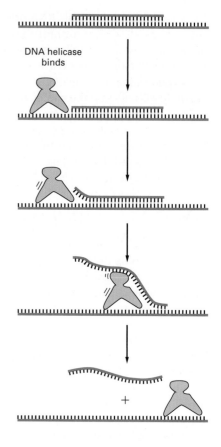

Figure 6–45 **The assay used to test for DNA helicase enzymes.** A short DNA fragment is annealed to a long DNA single strand to form a region of DNA double helix. The double helix is melted as the helicase runs along the DNA single strand, releasing the short DNA fragment in a reaction that requires the presence of both the helicase protein and ATP. The movement of the helicase is powered by its ATP hydrolysis (see Figure 5–22).

mation of the short hairpin helices that would otherwise impede synthesis by the DNA polymerase (Figure 6–46).

A Moving DNA Polymerase Molecule Is Kept Tethered to the DNA by a Sliding Ring [32]

On their own, most DNA polymerase molecules will synthesize only a short string of nucleotides before falling off a DNA template. This tendency to leave a DNA molecule quickly allows the DNA polymerase molecule that has just finished synthesizing one Okazaki fragment on the lagging strand to be recycled quickly to begin the synthesis of the next Okazaki fragment on the same strand. This rapid dissociation, however, would make it difficult for the polymerase to synthesize long DNA strands at a replication fork were it not for an accessory protein that functions as a regulated clamp. This clamp keeps the polymerase firmly on the DNA when it is moving, but releases it as soon as the polymerase stops.

How can a clamp prevent the polymerase from dissociating without at the same time impeding the polymerase's rapid movement along the DNA molecule? The three-dimensional structure of a clamp protein, determined by x-ray diffraction, indicates that it forms a large ring around the DNA helix. One side of the ring binds to the back of the DNA polymerase, and the whole ring slides freely as the polymerase moves along a DNA strand (Figure 6–47). The assembly of the clamp around DNA requires ATP hydrolysis by special accessory proteins that bind both to the clamp protein and to DNA; it is not known how the clamp is disassembled to remove it from the DNA.

The Proteins at a Replication Fork Cooperate to Form a Replication Machine [33]

Although we have discussed DNA replication as though it were carried out by a mixture of replication proteins that act independently, in reality most of the proteins are held together in a large multienzyme complex that moves rapidly along

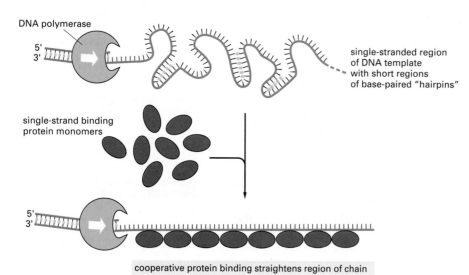

cooperative protein binding straightens region of chain

Figure 6–46 **The effect of single-strand binding proteins on the structure of single-stranded DNA.** Because each protein molecule prefers to bind next to a previously bound molecule (*cooperative binding*) long rows of this protein will form on a DNA single strand. This cooperative binding straightens out the DNA template and facilitates the DNA polymerization process. The "hairpin helices" shown in the bare single-stranded DNA result from a chance matching of short regions of complementary nucleotide sequence; they are similar to the short helices that typically form in RNA molecules.

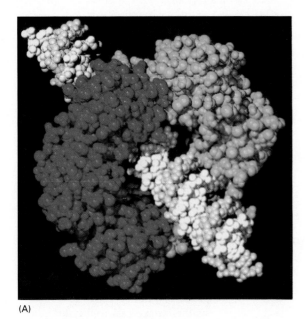

(A)

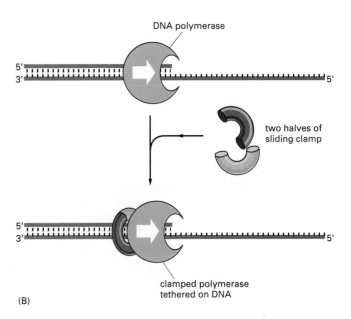

DNA polymerase

5'
3'

two halves of
sliding clamp

5'
3'

5'

5'

clamped polymerase
tethered on DNA

(B)

the DNA. This complex can be likened to a tiny sewing machine composed of protein parts and powered by nucleoside triphosphate hydrolyses. Although the replication complex has been best characterized in *E. coli* and several of its viruses, a very similar complex operates in eucaryotes (see p. 358).

The functions of the subunits of the replication machine are summarized in the two-dimensional diagram of the complete replication fork shown in Figure 6–48. Two identical DNA polymerase molecules work at the fork, one on the leading strand and one on the lagging strand. The DNA helix is opened by a DNA polymerase molecule clamped on the leading strand, acting in concert with a DNA helicase molecule running along the lagging strand; helix opening is aided by cooperatively bound molecules of single-strand DNA-binding protein. While the DNA polymerase molecule on the leading strand can operate in a continuous fashion, the DNA polymerase molecule on the lagging strand must restart at short intervals, using a short RNA primer made by a DNA primase molecule.

The efficiency of replication is greatly increased by the close association of all these protein components. The primase molecule is linked directly to the DNA

Figure 6–47 The regulated sliding clamp that holds DNA polymerase on the DNA. (A) The structure of the sliding clamp from *E. coli*, with a DNA helix added to indicate how the protein fits around DNA. A similar protein is present in eucaryotic cells. (B) Schematic illustration of how the clamp is thought to hold a moving DNA polymerase molecule on the DNA. (A, from X.-P. Kong et al., *Cell* 69:425–437, 1992. © Cell Press.)

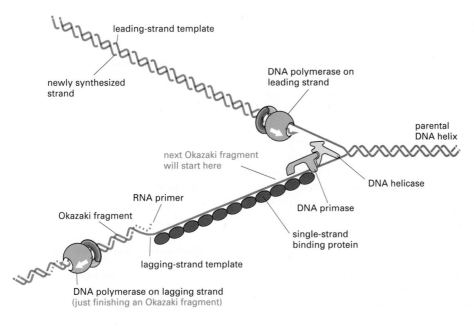

leading-strand template

newly synthesized strand

DNA polymerase on leading strand

parental DNA helix

next Okazaki fragment will start here

DNA helicase

RNA primer

DNA primase

Okazaki fragment

single-strand binding protein

lagging-strand template

DNA polymerase on lagging strand
(just finishing an Okazaki fragment)

Figure 6–48 The proteins at a DNA replication fork. The major types of proteins that act at a DNA replication fork are illustrated, showing their positions on the DNA.

DNA Replication

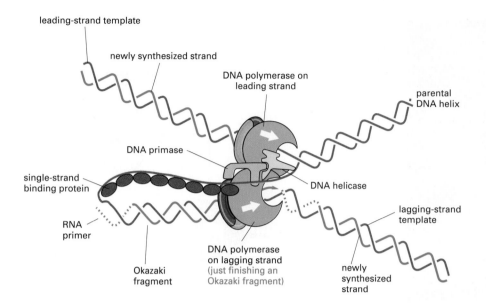

leading-strand template

newly synthesized strand

DNA polymerase on leading strand

parental DNA helix

DNA primase

single-strand binding protein

DNA helicase

RNA primer

lagging-strand template

Okazaki fragment

DNA polymerase on lagging strand (just finishing an Okazaki fragment)

newly synthesized strand

Figure 6–49 A replication fork in three dimensions. This diagram shows a current view of how the replication proteins are arranged at a replication fork when the fork is moving. The two-dimensional structure of Figure 6–48 has been altered by folding the DNA on the lagging strand to bring the lagging-strand DNA polymerase molecule into a complex with the leading-strand DNA polymerase molecule. This folding process also brings the 3′ end of each completed Okazaki fragment close to the start site for the next Okazaki fragment (compare with Figure 6–48). Because the lagging-strand DNA polymerase molecule is held to the rest of the replication proteins, it can be reused to synthesize successive Okazaki fragments; thus it is about to let go of its completed DNA fragment and move to the RNA primer that will be synthesized nearby, as required to start the next DNA fragment. Note that one daughter DNA helix extends toward the bottom right and the other toward the top left in this diagram.

helicase to form a unit on the lagging strand called a **primosome**. Powered by the DNA helicase, the primosome moves with the fork, synthesizing RNA primers as it goes. Similarly, the DNA polymerase molecule that synthesizes DNA on the lagging strand moves in concert with the rest of the proteins, synthesizing a succession of new Okazaki fragments. To accommodate this arrangement, its DNA template strand is thought to be folded back in the manner shown in Figure 6–49. The replication proteins are thus linked together into a single large unit (total mass > 10^6 daltons) that moves rapidly along the DNA, enabling DNA to be synthesized on both sides of the fork in a coordinated and efficient manner.

This DNA replication machine leaves behind on the lagging strand a series of unsealed Okazaki fragments, which still contain the RNA that primed their synthesis at their 5′ ends. This RNA must be removed and the fragments joined up by DNA repair enzymes that operate behind the replication fork (see Figure 6–44).

A Mismatch Proofreading System Removes Replication Errors That Escape from the Replication Machine [34]

Bacteria such as *E. coli* are capable of dividing once every 30 minutes, so it is relatively easy to screen large populations to find rare mutants that are altered in a specific process. One interesting class of mutants contains alterations in so-called *mutator genes,* which greatly increase the rate of spontaneous mutation. Not surprisingly, one such mutant encodes a defective form of the 3′-to-5′ proofreading exonuclease (discussed earlier) that is a subunit of the DNA polymerase enzyme (see Figure 6–42). When this protein is defective, the DNA poly-

Figure 6–50 A model for mismatch proofreading in eucaryotes. The two proteins shown are present in both bacteria and eucaryotic cells: MutS binds specifically to a mismatched base pair, while MutL scans the nearby DNA for a nick. Once a nick is found, MutL triggers the degradation of the nicked strand all the way back through the mismatch. Because nicks are largely confined to newly replicated strands in eucaryotes, replication errors are selectively removed. In bacteria the mechanism is the same except that an additional protein in the complex (MutH) nicks unmethylated (and therefore newly replicated) GATC sequences and thereby begins the process that is illustrated here. We know the mechanism because these reactions have been reconstituted in a cell-free system containing purified bacterial proteins and DNA.

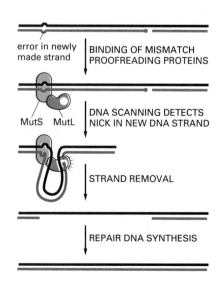

error in newly made strand

BINDING OF MISMATCH PROOFREADING PROTEINS

MutS   MutL

DNA SCANNING DETECTS NICK IN NEW DNA STRAND

STRAND REMOVAL

REPAIR DNA SYNTHESIS

merase no longer proofreads effectively, and many replication errors that would otherwise have been removed accumulate in the DNA.

The study of other *E. coli* mutants that exhibit abnormally high mutation rates has uncovered another proofreading system that removes replication errors missed by the proofreading exonuclease. This **mismatch proofreading** system (also called a *mismatch repair* system) differs from most DNA repair systems in that it does not depend on the presence in the DNA of abnormal nucleotides that can be recognized and excised. Instead, it detects the distortion on the outside of the helix that results from the misfit between noncomplementary base pairs. But if the proofreading system simply recognized a mismatch in newly replicated DNA and randomly excised one of the two mismatched nucleotides, it would make the mistake of "correcting" the original template strand to match the error exactly half the time and would not therefore lower the overall error rate. To be effective, the proofreading system must be able to distinguish and remove the mismatched nucleotide only on the new strand, where the replication error occurred.

The recognition system used by the mismatch proofreading system in *E. coli* depends on the methylation of selected A residues in the DNA. Methyl groups are added to all A residues in the sequence GATC, but not until some time after the A has been incorporated into a newly synthesized DNA chain. Because only the new strands just behind a replication fork will contain GATC sequences that have not yet been methylated, these new DNA strands can be distinguished from old ones.

More recently, eucaryotic proteins have been discovered that are homologous in their amino acid sequence to several of the bacterial proteins that catalyze mismatch proofreading. As expected, when the genes that encode these proteins are deleted in a yeast cell, mutation rates can increase by 100-fold or more. There must, however, be some important differences between the bacterial and eucaryotic proofreading mechanisms, as the mechanism for distinguishing the newly synthesized strand from the parental template strand at the site of a mismatch cannot depend on DNA methylation as in bacteria, since some eucaryotes, such as yeasts and *Drosophila*, do not methylate any of their DNA. Newly synthesized DNA strands are known to be preferentially *nicked*, and it has been suggested that such nicks (single-strand breaks) provide the signal that directs mismatch proofreading to the appropriate strand in a eucaryotic cell (Figure 6–50).

Replication Forks Initiate at Replication Origins [35]

In both bacteria and mammals replication forks originate at a structure called a *replication bubble*, a local region where the two strands of the parental DNA helix have been separated from each other to serve as templates for DNA synthesis (Figure 6–51). For bacteria, yeasts, and several viruses that grow in mammalian cells, replication bubbles have been shown to form at special DNA sequences called **replication origins**, which can be as long as 300 nucleotides. For reasons that are not clear, the replication origins in mammalian chromosomes have thus far been very difficult to characterize at the molecular level.

For several well-defined replication origins, it has been possible to reproduce the fork initiation reaction *in vitro*. The *in vitro* studies reveal that fork initiation in bacteria and bacterial viruses starts in the manner indicated in Figure 6–52. *Initiator proteins* bind in multiple copies to specific sites at the replication origin, wrapping the DNA around them to form a large protein-DNA complex. This complex then binds the DNA helicase and loads it onto an exposed DNA single strand in an adjacent region of helix. The DNA primase also binds, forming the primosome, which moves away from the origin and makes an RNA primer that starts the first DNA chain. This quickly leads to assembly of the remaining proteins to create two replication protein complexes moving away from the origin in opposite directions (see Figure 6–51); these continue to synthesize DNA until all of the DNA template downstream of each fork has been replicated.

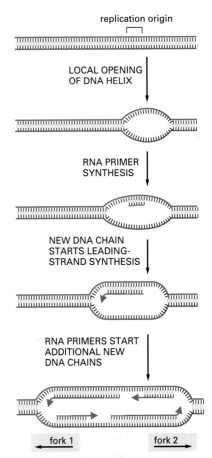

replication origin

LOCAL OPENING OF DNA HELIX

RNA PRIMER SYNTHESIS

NEW DNA CHAIN STARTS LEADING-STRAND SYNTHESIS

RNA PRIMERS START ADDITIONAL NEW DNA CHAINS

fork 1   fork 2

two complete replication forks are formed, one moving leftward with leading (*top*) and lagging (*bottom*) strands, and one moving rightward with leading (*bottom*) and lagging (*top*) strands

Figure 6–51 Replication fork initiation. The figure outlines the processes involved in the initiation of replication forks at replication origins. (See also Figure 6–52.)

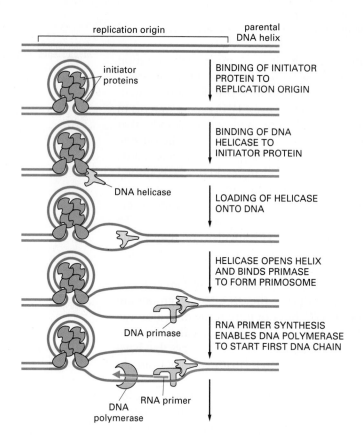

initiator proteins

BINDING OF INITIATOR PROTEIN TO REPLICATION ORIGIN

BINDING OF DNA HELICASE TO INITIATOR PROTEIN

DNA helicase

LOADING OF HELICASE ONTO DNA

HELICASE OPENS HELIX AND BINDS PRIMASE TO FORM PRIMOSOME

DNA primase

RNA PRIMER SYNTHESIS ENABLES DNA POLYMERASE TO START FIRST DNA CHAIN

DNA polymerase

RNA primer

replication origin

parental DNA helix

Figure 6–52 The proteins that initiate DNA replication. The major types of proteins involved in the formation of replication forks at the *E. coli* and bacteriophage lambda replication origins are indicated. The mechanism shown was established by *in vitro* studies utilizing a mixture of highly purified proteins. Subsequent steps result in the initiation of three more DNA chains (see Figure 6–51) by a pathway that is not yet clear. For *E. coli* DNA replication, the major initiator protein is the dnaA protein; for both lambda and *E. coli*, the primosome is composed of the dnaB (DNA helicase) and dnaG (DNA primase) proteins.

Replication fork initiation in eucaryotic chromosomes is discussed in detail in Chapter 8.

DNA Topoisomerases Prevent DNA Tangling During Replication [36]

When we draw the DNA helix (incorrectly) as a flat, ladderlike structure, we are ignoring the "winding problem" that arises during DNA replication. Every 10 base pairs replicated at the fork correspond to one complete turn about the axis of the parental double helix. Therefore, for a replication fork to move, the entire chromosome ahead of the fork would normally have to rotate rapidly (Figure 6–53), which would require large amounts of energy for long chromosomes. An alternative strategy is used during DNA replication: a swivel is formed in the DNA helix by proteins known as **DNA topoisomerases.**

A DNA topoisomerase can be viewed as a reversible nuclease that adds itself covalently to a DNA phosphate, thereby breaking a phosphodiester bond in a DNA strand. Because the covalent linkage that joins a topoisomerase to a DNA phosphate retains the energy of the cleaved phosphodiester bond, the cleavage reaction is reversible; resealing is rapid and does not require additional energy input. The rejoining mechanism is different in this respect from that of the enzyme DNA ligase, discussed previously (see Figure 6–37).

One type of topoisomerase (*topoisomerase I*) causes a *single-strand break* (or *nick*), which can allow the two sections of DNA helix on either side of the nick to rotate freely relative to each other, using the phosphodiester bond in the strand opposite the nick as a swivel point (Figure 6–54). Any tension in the DNA helix will drive this rotation in the direction that relieves the tension. As a result, DNA replication can occur with the rotation of only a short length of helix—the part just ahead of the fork. The analogous problem that arises during DNA transcription is solved in a similar way.

A second type of DNA topoisomerase (*topoisomerase II*) forms a covalent linkage to both strands of the DNA helix at the same time, making a transient

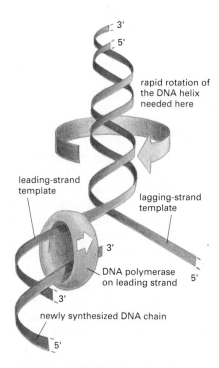

rapid rotation of the DNA helix needed here

leading-strand template

lagging-strand template

DNA polymerase on leading strand

newly synthesized DNA chain

Figure 6–53 The "winding problem" that arises during DNA replication. For a bacterial replication fork moving at 500 nucleotides per second, the parental DNA helix ahead of the fork must rotate at 50 revolutions per second.

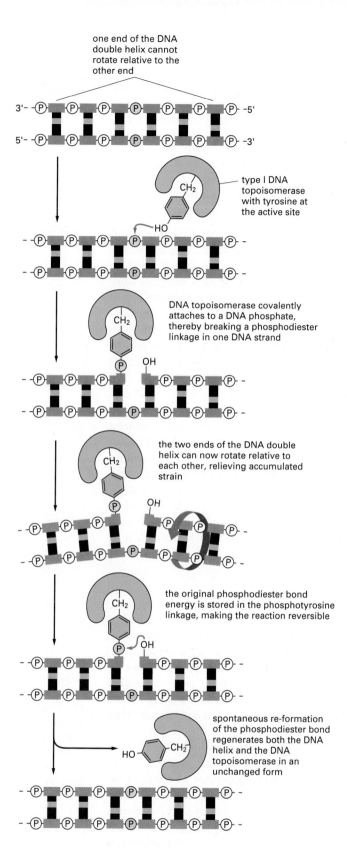

one end of the DNA double helix cannot rotate relative to the other end

3'- (P) (P) (P) (P) (P) (P) (P) -5'

5'- (P) (P) (P) (P) (P) (P) (P) -3'

type I DNA topoisomerase with tyrosine at the active site

CH₂

HO

DNA topoisomerase covalently attaches to a DNA phosphate, thereby breaking a phosphodiester linkage in one DNA strand

CH₂

(P) OH

the two ends of the DNA double helix can now rotate relative to each other, relieving accumulated strain

CH₂

(P) OH

the original phosphodiester bond energy is stored in the phosphotyrosine linkage, making the reaction reversible

CH₂

(P) OH

spontaneous re-formation of the phosphodiester bond regenerates both the DNA helix and the DNA topoisomerase in an unchanged form

HO CH₂

Figure 6–54 The reversible nicking reaction catalyzed by a eucaryotic DNA topoisomerase I enzyme. As indicated, these enzymes form a transient covalent bond with DNA so as to allow free rotation about the covalent bonds linked to the *blue* phosphate.

double-strand break in the helix. These enzymes are activated by sites on chromosomes where two double helices cross over each other. When the topoisomerase binds to such a crossing site, it (1) breaks one double helix reversibly to create a DNA "gate," (2) causes the second, nearby double helix to pass through this break, and (3) reseals the break and dissociates from the DNA. In this way type II DNA topoisomerases can efficiently separate two interlocked DNA circles

DNA Replication

(Figure 6–55). The same reaction prevents the severe DNA tangling problems that would otherwise arise during DNA replication. For example, mutant yeast cells have been isolated that produce, in place of the normal topoisomerase II, a version that is inactive at 37°C. When the mutant yeast cells are warmed to this temperature, their chromosomes remain intertwined at mitosis and are unable to separate. The usefulness of topoisomerase II for untangling chromosomes can readily be appreciated by anyone who has struggled to remove a tangle from a fishing line without the aid of scissors.

DNA Replication Is Basically Similar in Eucaryotes and Procaryotes [37]

Much of what we know about DNA replication comes from studies of purified bacterial and bacteriophage multienzyme systems capable of DNA replication *in vitro*. The development of these systems in the 1970s was greatly facilitated by the prior isolation of mutants in a variety of replication genes; these mutants were exploited to identify and purify the corresponding replication proteins.

Less is known about the detailed enzymology of DNA replication in eucaryotes, largely because it is difficult to obtain replication-deficient mutants. Nevertheless, the basic mechanisms of DNA replication, including both the geometry of the replication fork and the protein components of the multiprotein replication machine, are similar for procaryotes and eucaryotes (see Figure 8–35). The major difference is that eucaryotic DNA is replicated not as bare DNA but as *chromatin*, in which the DNA is complexed with tightly bound proteins called *histones*. As described in Chapter 8, histones form disclike structures around which the eucaryotic DNA is wound, creating a repeating structural unit called a *nucleosome*. Nucleosomes are spaced at intervals of about 200 base pairs along the DNA, which may be why new Okazaki fragments are synthesized on the lagging strand at intervals of 100 to 200 nucleotides in eucaryotes instead of at intervals of 1000 to 2000 nucleotides as in bacteria. Nucleosomes may also act as barriers that slow down the movement of DNA polymerase molecules, which could explain why eucaryotic replication forks move only one-tenth as fast as bacterial replication forks.

Summary

A self-correcting DNA polymerase catalyzes nucleotide polymerization in a 5′-to-3′ direction, copying a DNA template with remarkable fidelity. Since the two strands of a DNA double helix are antiparallel, this 5′-to-3′ DNA synthesis can take place continuously on only one of the strands at a replication fork (the leading strand). On the lagging strand short DNA fragments are made by a "backstitching" process. Because the self-correcting DNA polymerase cannot start a new chain, these lagging-strand DNA fragments are primed by short RNA primer molecules that are subsequently erased and replaced with DNA.

DNA replication requires the cooperation of many proteins, including (1) DNA polymerase and DNA primase to catalyze nucleoside triphosphate polymerization, (2) DNA helicases and single-strand binding proteins to help open up the DNA helix so that it can be copied, (3) DNA ligase and an enzyme that degrades RNA primers to seal together the discontinuously synthesized lagging-strand DNA fragments, (4) DNA topoisomerases to help relieve helical winding and tangling problems, and (5) initiator proteins that bind to specific DNA sequences at a replication origin and catalyze the formation of a replication fork at that site. At a replication origin a specialized protein-DNA structure is formed that subsequently loads a DNA helicase onto the DNA template; other proteins are then added to form the multienzyme "replication machine" that catalyzes DNA synthesis.

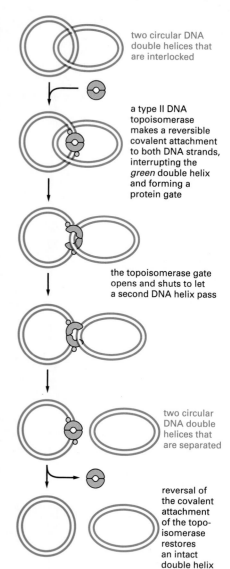

two circular DNA double helices that are interlocked

a type II DNA topoisomerase makes a reversible covalent attachment to both DNA strands, interrupting the *green* double helix and forming a protein gate

the topoisomerase gate opens and shuts to let a second DNA helix pass

two circular DNA double helices that are separated

reversal of the covalent attachment of the topoisomerase restores an intact double helix

Figure 6–55 DNA topoisomerase II. An example of a DNA-helix-passing reaction catalyzed by a type II DNA topoisomerase. Unlike type I topoisomerases, these enzymes require ATP hydrolysis for their function, and some of the bacterial versions can introduce superhelical tension into DNA (see p. 438). Type II topoisomerases are largely confined to proliferating cells in eucaryotes; partly for that reason, they have been popular targets for anticancer drugs.

Genetic Recombination [38]

In the two preceding sections we discussed the mechanisms by which DNA sequences in cells are maintained from generation to generation with very little change. Although such genetic stability is crucial for the survival of individuals, in the longer term the survival of organisms may depend on genetic variation, through which they can adapt to a changing environment. Thus an important property of the DNA in cells is its ability to undergo rearrangements that can vary the particular combination of genes present in any individual genome, as well as the timing and the level of expression of these genes. These DNA rearrangements are caused by **genetic recombination.** Two broad classes of genetic recombination are commonly recognized—general recombination and site-specific recombination.

In *general recombination,* genetic exchange takes place between any pair of homologous DNA sequences, usually located on two copies of the same chromosome. One of the most important examples is the exchange of sections of homologous chromosomes (homologues) in the course of *meiosis.* This "crossing-over" occurs between tightly apposed chromosomes early in the development of eggs and sperm (discussed in Chapter 20), and it allows different versions (*alleles*) of the same gene to be tested in new combinations with other genes, increasing the chance that at least some members of a mating population will survive in a changing environment. Although meiosis occurs only in eucaryotes, the advantage of this type of gene mixing is so great that mating and the reassortment of genes by general recombination is also widespread in bacteria.

DNA homology is not required in *site-specific recombination.* Instead, exchange occurs at short, specific nucleotide sequences (on either one or both of the two participating DNA molecules) that are recognized by a variety of site-specific recombination enzymes. Site-specific recombination therefore alters the relative positions of nucleotide sequences in genomes. In some cases these changes are scheduled and organized, as when an integrated bacterial virus is induced to leave a chromosome of a bacterium under stress (see Figure 6–80); in others they are haphazard, as when the DNA sequence of a transposable element is inserted at a randomly selected site in a chromosome.

As for DNA replication, most of what we know about the biochemistry of genetic recombination has come from studies of procaryotic organisms, especially of *E. coli* and its viruses.

General Recombination Is Guided by Base-pairing Interactions Between Complementary Strands of Two Homologous DNA Molecules [39]

General recombination involves DNA strand-exchange intermediates that require some effort to understand. Although the exact pathway followed is likely to be different in different organisms, detailed genetic analyses of viruses, bacteria, and fungi suggest that the major outcome of general recombination is always the same. (1) Two homologous DNA molecules "cross over"; that is, their double helices break and the two broken ends join to their opposite partners to re-form two intact double helices, each composed of parts of the two initial DNA molecules (Figure 6–56). (2) The site of exchange (that is, where a *red* double helix is joined to a *green* double helix in Figure 6–56) can occur anywhere in the homologous nucleotide sequences of the two participating DNA molecules. (3) At the site of exchange, a strand of one DNA molecule becomes base-paired to a strand of the second DNA molecule to create a *staggered joint* (usually called a *heteroduplex joint*) between the two double helices (Figure 6–57). The heteroduplex region can be thousands of base pairs long; we shall explain later how it

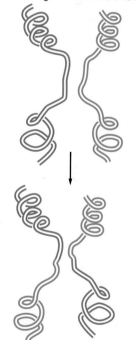

two homologous DNA double helices

DNA molecules that have crossed over

Figure 6–56 General recombination. The breaking and rejoining of two homologous DNA double helices creates two DNA molecules that have "crossed over."

forms. (4) No nucleotide sequences are altered at the site of exchange; the cleavage and rejoining events occur so precisely that not a single nucleotide is lost or gained. Despite this precision, general recombination creates DNA molecules of novel sequence: the heteroduplex joint can contain a small number of mismatched base pairs, and, more important, the two DNAs that cross over are usually not exactly the same on either side of the joint.

The mechanism of general recombination ensures that two regions of DNA double helix undergo an exchange reaction only if they have extensive sequence homology. The formation of a heteroduplex joint requires that such homology be present because it involves a long region of complementary base-pairing between a strand from one of the two original double helices and a complementary strand from the other. But how does this heteroduplex joint arise, and how do the two homologous regions of DNA at the site of crossing-over recognize each other? As we shall see, recognition takes place by means of a direct base-pairing interaction. The formation of base pairs between complementary strands from the two DNA molecules then guides the general recombination process, allowing it to occur only between long regions of matching DNA sequence.

General Recombination Can Be Initiated at a Nick in One Strand of a DNA Double Helix [40]

Each of the two strands in a DNA molecule is helically wound around the other. As a result, extensive base-pair interactions can occur between two homologous DNA double helices only if a nick is first made in a strand of one of them, freeing that strand for the unwinding and rewinding events required to form a heteroduplex with another DNA molecule. For the same reason, any *exchange* of strands between two DNA double helices requires at least two nicks, one in a

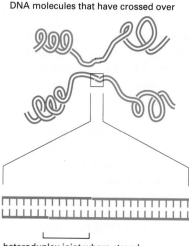

Figure 6–57 **A heteroduplex joint.** This structure unites two DNA molecules where they have crossed over. Such a joint is often thousands of nucleotides long.

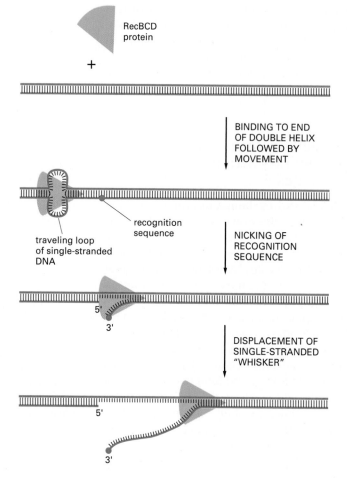

Figure 6–58 **One way to start a recombination event.** The RecBCD protein is an enzyme required for general genetic recombination in *E. coli*. The protein enters the DNA from one end of the double helix and then uses energy derived from the hydrolysis of bound ATP molecules to propel itself in one direction along the DNA at a rate of about 300 nucleotides per second. A special recognition site (a DNA sequence of eight nucleotides scattered throughout the *E. coli* chromosome) is cut in the traveling loop of DNA created by the RecBCD protein, and thereafter a single-stranded whisker is displaced from the helix, as shown. This whisker is thought to initiate genetic recombination by pairing with a homologous helix, as in Figure 6–59.

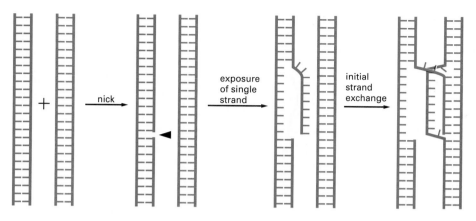

Figure 6–59 The initial strand exchange in general recombination. A nick in a single DNA strand frees the strand, which then invades a homologous DNA double helix to form a short pairing region with one of the strands in the second helix. Only two DNA molecules that are complementary in nucleotide sequence can base-pair in this way and thereby initiate a general recombination event. All of the steps shown here can be catalyzed by known enzymes (see Figures 6–58 and 6–62).

strand of each interacting double helix. Finally, to produce the heteroduplex joint illustrated in Figure 6–57, each of the four strands present must be cut to allow each to be joined to a different partner. In general recombination, these nicking and resealing events are coordinated so that they occur only when two DNA helices share an extensive region of matching DNA sequence.

There is evidence from a number of sources that a single nick in only one strand of a DNA molecule is sufficient to initiate general recombination. Chemical agents or types of irradiation that introduce single strand nicks, for example, will trigger a genetic recombination event. Moreover, one of the special proteins required for general recombination in *E. coli*—the *RecBCD protein*—has been shown to make single strand nicks in DNA molecules. The RecBCD protein is also a DNA helicase, hydrolyzing ATP and traveling along a DNA helix transiently exposing its strands. By combining its nuclease and helicase activities, the RecBCD protein will create a single-stranded "whisker" on the DNA double helix (Figure 6–58). Figure 6–59 shows how such a whisker could initiate a base-pairing interaction between two complementary stretches of DNA double helix.

DNA Hybridization Reactions Provide a Simple Model for the Base-pairing Step in General Recombination [41]

In its simplest form, the type of base-pairing interaction central to general recombination can be mimicked in a test tube by allowing a DNA double helix to re-form from its separated single strands. This process, called **DNA renaturation** or **hybridization,** occurs when a rare random collision juxtaposes complementary nucleotide sequences on two matching DNA single strands, allowing the formation of a short stretch of double helix between them. This relatively slow *helix nucleation* step is followed by a very rapid "zippering" step as the region of double helix is extended to maximize the number of base-pairing interactions (Figure 6–60).

Formation of a new double helix in this way requires that the annealing strands be in an open, unfolded conformation. For this reason *in vitro* hybridization reactions are carried out at high temperature or in the presence of an organic solvent such as formamide; these conditions "melt out" the short hairpin helices formed where base-pairing interactions occur within a single strand that folds back on itself. Bacterial cells could not survive such harsh conditions and instead use a single-strand binding protein, the *SSB protein*, to open their helices. This protein is essential for DNA replication as well as for general recombination in *E. coli*; it binds tightly and cooperatively to the sugar-phosphate backbone of all single-stranded regions of DNA, holding them in an extended conformation with their bases exposed (see Figure 6–46). In this extended conformation a DNA single strand can base-pair efficiently with either a nucleoside triphosphate molecule (in DNA replication) or a complementary section of another DNA

Genetic Recombination

nonpairing interactions

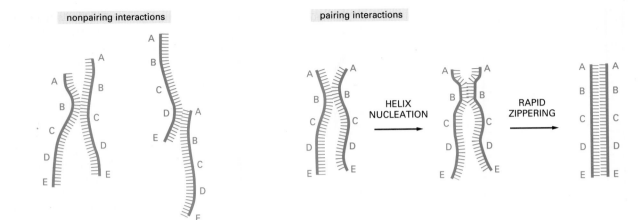

pairing interactions

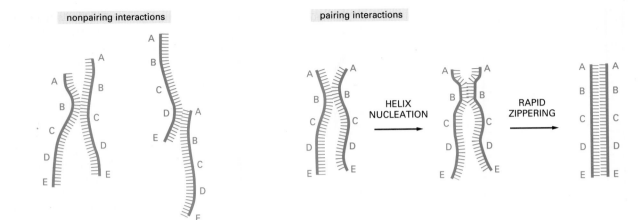

single strand (in genetic recombination). When hybridization reactions are carried out *in vitro* under conditions that mimic those inside a cell, the SSB protein speeds up the rate of DNA helix nucleation and thereby the overall rate of strand annealing by a factor of more than 1000.

The RecA Protein Enables a DNA Single Strand to Pair with a Homologous Region of DNA Double Helix in *E. coli* [42]

General recombination is more complex than the simple hybridization reactions just described. In the course of general recombination, a single DNA strand from one DNA double helix must invade another double helix (see Figure 6–59). In *E. coli* this requires the **RecA protein,** produced by the *recA* gene, which was identified in 1965 as being essential for recombination between chromosomes. Long sought by biochemists, this elusive gene product was finally purified to homogeneity in 1976, a feat that allowed its detailed characterization (Figure 6–61). Like a single-strand binding (SSB) protein, the RecA protein binds tightly and in large cooperative clusters to single-stranded DNA to form a nucleoprotein filament. This filament has several distinctive properties. The RecA protein has more than one DNA-binding site, for example, and it can therefore hold a single strand and a double helix together. These sites allow the RecA protein to catalyze a multistep reaction (called **synapsis**) between a DNA double helix and a homologous region of single-stranded DNA. The crucial step in synapsis occurs when a region of homology is identified by an initial base-pairing between complementary

Figure 6–60 DNA hybridization. DNA double helices re-form from their separated strands in a reaction that depends on the random collision of two complementary strands (see p. 300). Most such collisions are not productive, as shown at the left, but a few result in a short region where complementary base pairs have formed (helix nucleation). A rapid zippering then leads to the formation of a complete double helix. A DNA strand can use this trial-and-error process to find its complementary partner in the midst of millions of nonmatching DNA strands. Trial-and-error recognition of a complementary partner DNA sequence appears to initiate all general recombination events.

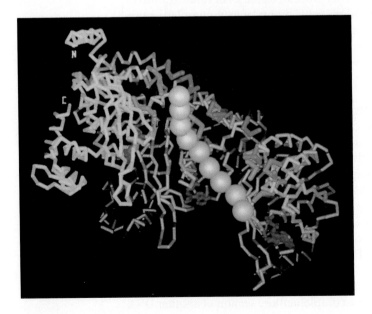

Figure 6–61 The structure of the RecA protein. A string of three RecA monomers is shown, with the position of each ATP in *red*. The *white spheres* show the putative position of the single-strand DNA in the filament, with three nucleotides (each shown as a sphere) bound per monomer. (From R.M. Story, I.T. Weber, and T.A. Steitz, *Nature* 256:318–325, 1992. © 1992 Macmillan Magazines Ltd.)

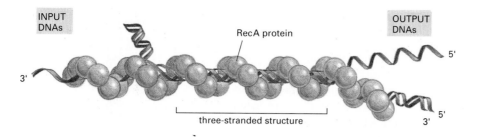

INPUT DNAs

RecA protein

OUTPUT DNAs

5'

3'

5'

three-stranded structure

3' 5'

Figure 6–62 DNA synapsis catalyzed by the RecA protein. *In vitro* experiments show that several types of complexes are formed between a DNA single strand covered with RecA protein (*red*) and a DNA double helix (*green*). First a non-base-paired complex is formed, which is converted to a three-stranded structure as soon as a region of homologous sequence is found. This complex is presumably unstable because it involves an unusual form of DNA, and it spins out a DNA heteroduplex (one strand *green* and the other strand *red*) plus a displaced single strand from the original helix (*green*); thus the structure shown in this diagram migrates to the left, reeling in the "input DNAs" while producing the "output DNAs." The net result is a DNA strand exchange identical to that diagrammed earlier in Figure 6–59. (Adapted from S.C. West, *Annu. Rev. Biochem.* 61:603–640, 1992. © Annual Reviews Inc.)

nucleotide sequences. The nucleation step in this case appears to involve a three-stranded structure, in which the DNA single strand forms nonconventional base pairs in the major groove of the DNA double helix (Figure 6–62). This begins the pairing shown previously in Figure 6–59 and so initiates the exchange of strands between two recombining DNA double helices. Studies *in vitro* suggest that the *E. coli* SSB protein cooperates with the RecA protein to facilitate these reactions.

Once synapsis has occurred, a short heteroduplex region where the strands from two different DNA molecules have begun to pair is enlarged through *protein-directed branch migration*, which can also be catalyzed by the RecA protein. **Branch migration** can take place at any point where two single DNA strands with the same sequence are attempting to pair with the same complementary strand; an unpaired region of one of the single strands will displace a paired region of the other single strand, moving the branch point without changing the total number of DNA base pairs. Spontaneous branch migration proceeds equally in both directions, and so it makes little progress and is unlikely to complete recombination efficiently (Figure 6–63A). Because the RecA protein catalyzes unidirectional branch migration, it readily produces a region of heteroduplex that is thousands of base pairs long (Figure 6–63B).

The catalysis of branch migration depends on a further property of the RecA protein. In addition to having two DNA-binding sites, the RecA protein is a DNA-dependent ATPase, with an additional site for binding and hydrolyzing ATP. The protein associates much more tightly with DNA when it has ATP bound than when it has ADP bound. Moreover, new RecA molecules with ATP bound are preferentially added at one end of the RecA protein filament, and the ATP is then hydrolyzed to ADP. The RecA protein filaments that form on DNA may therefore share many of the dynamic assembly properties displayed by the cytoskeletal filaments formed from actin or tubulin (discussed in Chapter 16); an ability of the protein to "treadmill" unidirectionally along a DNA strand, for example, could drive the branch migration reaction shown in Figure 6–63B.

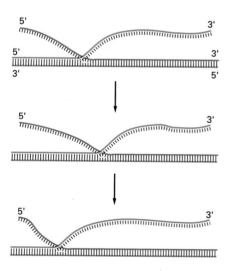

(A) SPONTANEOUS BRANCH MIGRATION

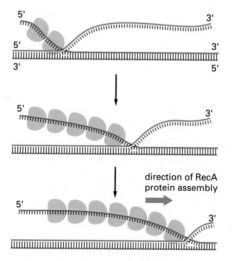

(B) PROTEIN-DIRECTED BRANCH MIGRATION

direction of RecA protein assembly

Figure 6–63 Two types of DNA branch migration observed in experiments *in vitro*. (A) Spontaneous branch migration is a back-and-forth, random-walk type of process, and it therefore makes little progress over long distances. (B) RecA-protein-directed branch migration proceeds at a uniform rate in one direction, and it may be driven by the polarized assembly of the RecA protein filament on a DNA single strand, which occurs in the direction indicated. In addition, special DNA helicases that catalyze protein-directed branch migration even more efficiently are involved in recombination.

Genetic Recombination

General Genetic Recombination Usually Involves a Cross-Strand Exchange [43]

Exchanging a single strand between two double helices is presumed to be the slow and difficult step in a general recombination event (see Figure 6–59). After this initial exchange, extending the region of pairing and establishing further strand exchanges between the two closely apposed helices is thought to proceed rapidly. During these events a limited amount of nucleotide excision and local DNA resynthesis often occurs, resembling some of the events in DNA repair. Because of the large number of possibilities, different organisms are likely to follow different pathways at this stage. In most cases, however, an important intermediate structure, the **cross-strand exchange,** will be formed by the two participating DNA helices. One of the simplest ways in which this structure can form is shown in Figure 6–64.

In the cross-strand exchange (also called a *Holliday junction*) the two homologous DNA helices that initially paired are held together by mutual exchange of two of the four strands present, one originating from each of the helices. No disruption of base-pairing is necessary to maintain this structure, which has two important properties—(1) the point of exchange between the two homologous DNA double helices (where the two strands cross in Figure 6–64) can migrate rapidly back and forth along the helices by a double branch migration; (2) the cross-strand exchange contains two pairs of strands: one pair of crossing strands and one pair of noncrossing strands. The structure can *isomerize,* however, by undergoing a series of rotational movements, so that the two original noncrossing strands become crossing strands and vice versa (Figure 6–65).

In order to regenerate two separate DNA helices and thus terminate the pairing process, the two crossing strands must be cut. If the crossing strands are cut *before* isomerization, the two original DNA helices separate from each other nearly unaltered, with only a very short piece of single-stranded DNA exchanged. If the crossing strands are cut *after* isomerization, however, one section of each original DNA helix is joined to a section of the other DNA helix; in other words, the two DNA helices have crossed over (see Figure 6–65).

The isomerization of the cross-strand exchange should occur spontaneously at some rate, but it may also be enzymatically driven or otherwise regulated by cells. Some kind of control probably operates during meiosis, when the two DNA double helices that pair are constrained in an elaborate structure called the *synaptonemal complex* (discussed in Chapter 20).

Gene Conversion Results from Combining General Recombination and Limited DNA Synthesis [44]

It is a fundamental law of genetics that each parent makes an equal genetic contribution to the offspring, one complete set of genes being inherited from the father and one from the mother. Thus, when a diploid cell undergoes meiosis to produce four haploid cells (discussed in Chapter 20), exactly half of the genes in these cells should be maternal (genes that the diploid cell inherited from its mother) and the other half paternal (genes that the diploid cell inherited from its father). In a complex animal, such as a human, it is not possible to check this prediction directly. But in other organisms, such as fungi, where it is possible to recover and analyze all four of the daughter cells produced from a single cell by meiosis, one finds cases in which the standard genetic rules have apparently been violated. Occasionally, for example, meiosis yields three copies of the maternal version of a gene (allele) and only one copy of the paternal allele, indicating that one of the two copies of the paternal allele has been changed to a copy of the maternal allele. This phenomenon is known as **gene conversion.** It often occurs in association with general genetic recombination events, and it is thought to be important in the evolution of certain genes (see Figure 8–74). Gene conversion is believed to be a straightforward consequence of the mechanisms of general recombination and DNA repair.

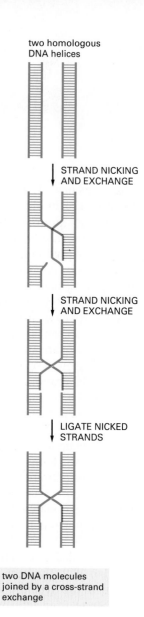

two homologous DNA helices

STRAND NICKING AND EXCHANGE

STRAND NICKING AND EXCHANGE

LIGATE NICKED STRANDS

two DNA molecules joined by a cross-strand exchange

Figure 6–64 The formation of a cross-strand exchange. There are many possible pathways that can lead from a single-strand exchange (see Figure 6–59) to a cross-strand exchange, but only one is shown.

Figure 6–65 The isomerization of a cross-strand exchange. Without isomerization, cutting the two crossing strands would terminate the exchange and crossing over would not occur. With isomerization (steps B and C), cutting the two crossing strands creates two DNA molecules that have crossed over (*bottom*). Isomerization is therefore thought to be required for the breaking and rejoining of two homologous DNA double helices that result from general genetic recombination. Step A was illustrated previously (see Figure 6–64).

During meiosis heteroduplex joints are formed at the sites of crossing-over between homologous maternal and paternal chromosomes. If the maternal and paternal DNA sequences are slightly different, the heteroduplex joint may include some mismatched base pairs. The resulting mismatch in the double helix may then be corrected by the DNA repair machinery, which either can erase nucleotides on the paternal strand and replace them with nucleotides that match the maternal strand or vice versa. The consequence of this mismatch repair will be a gene conversion. Gene conversion can also take place by a number of other mechanisms, but they all require some type of general recombination event that

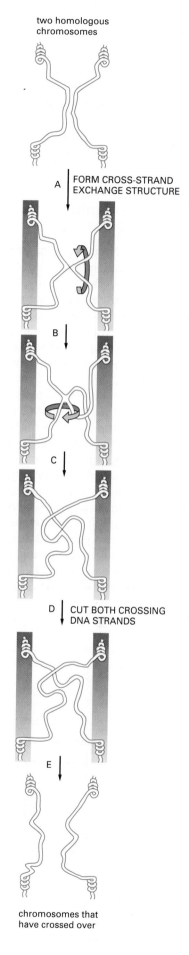

two homologous chromosomes

A | FORM CROSS-STRAND EXCHANGE STRUCTURE

B |

C |

D | CUT BOTH CROSSING DNA STRANDS

E |

chromosomes that have crossed over

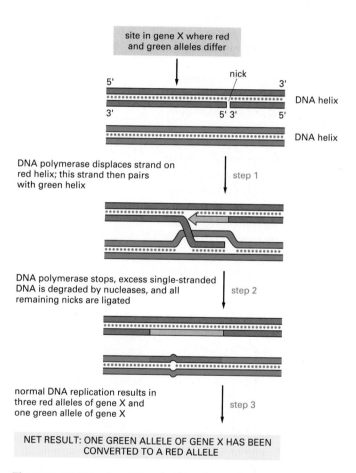

site in gene X where red and green alleles differ

nick

5'   3'
                                    DNA helix
3'   5' 3'   5'

                                    DNA helix

DNA polymerase displaces strand on red helix; this strand then pairs with green helix | step 1

DNA polymerase stops, excess single-stranded DNA is degraded by nucleases, and all remaining nicks are ligated | step 2

normal DNA replication results in three red alleles of gene X and one green allele of gene X | step 3

NET RESULT: ONE GREEN ALLELE OF GENE X HAS BEEN CONVERTED TO A RED ALLELE

Figure 6–66 One general recombination pathway that can cause gene conversion. The process begins when a nick is formed in one of the strands in the *red* DNA helix. In step 1 DNA polymerase begins the synthesis of an extra copy of a strand in the *red* helix, displacing the original copy as a single strand. This single strand then pairs with the homologous region of the *green* helix in the manner shown in Figure 6–59. In step 2 the short region of unpaired *green* strand produced in step 1 is degraded, completing the transfer of nucleotide sequences. The result is normally seen in the next cell cycle, after DNA replication has separated the two nonmatching strands (step 3). As described in the text, the repair of mismatched base pairs in a heteroduplex joint also causes gene conversion.

Genetic Recombination

brings two copies of a closely related DNA sequence together. Because an extra copy of one of the two DNA sequences is generated, a limited amount of DNA synthesis must also be involved. Genetic studies show that usually only small sections of DNA undergo gene conversion, and in many cases only part of a gene is changed.

Gene conversion can also occur in mitotic cells, but it does so more rarely. As in meiotic cells, some gene conversions in mitotic cells probably result from a mismatch repair process operating on heteroduplex DNA. Another likely mechanism in both meiotic and mitotic cells is illustrated in Figure 6–66.

Mismatch Proofreading Can Prevent Promiscuous Genetic Recombination Between Two Poorly Matched DNA Sequences [45]

As previously discussed, general recombination is triggered whenever two DNA strands of complementary sequence pair to form a heteroduplex joint between two double helices (see Figure 6–64). Experiments carried out *in vitro* with purified RecA protein show that pairing can occur efficiently even when the sequences of the two DNA strands do not match well—when, for example, only four out of every five nucleotides on average can form base pairs. How, then, do vertebrate cells avoid promiscuous general recombination between the many thousands of copies of closely related DNA sequences that are repeated in their genomes (see p. 395)?

Although the answer is not known, studies with bacteria and yeasts demonstrate that the same mismatch proofreading system that removes replication errors (see Figure 6–50) has the additional role of interrupting genetic recombination events between imperfectly matched DNA sequences. It has long been known, for example, that homologous genes in two closely related bacteria, *Escherichia coli* and *Salmonella typhimurium*, generally will not recombine, even though their nucleotide sequences are 80% identical; when the mismatch proofreading system is inactivated by mutation, however, there is a 1000-fold increase in the frequency of such interspecies recombination events. It is thought, then, that the mismatch proofreading system normally recognizes the mispaired bases in an initial strand exchange and prevents the subsequent steps required to break and rejoin the two paired DNA helices. This mechanism protects the bacterial genome from the sequence changes that would otherwise be caused by recom-

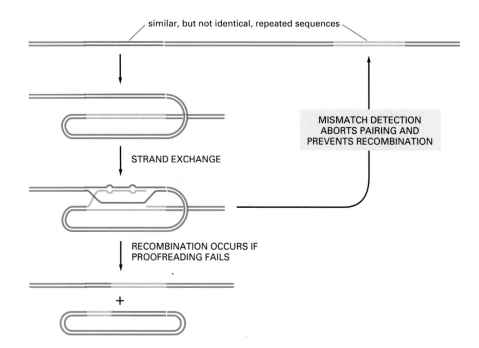

MISMATCH DETECTION ABORTS PAIRING AND PREVENTS RECOMBINATION

Figure 6–67 **Proofreading prevents general recombination from destabilizing genomes that contain repeated sequences.** Studies with bacterial and yeast cells suggest that the mismatch proofreading system diagrammed previously in Figure 6–50 has the additional function shown here.

bination with foreign DNA molecules that occasionally enter the cell. In vertebrate cells, which contain many closely related DNA sequences, the same type of proofreading is thought to help prevent promiscuous recombination events that would otherwise scramble the genome (Figure 6–67).

Site-specific Recombination Enzymes Move Special DNA Sequences into and out of Genomes [46]

Site-specific genetic recombination, unlike general recombination, is guided by a recombination enzyme that recognizes specific nucleotide sequences present on one or both of the recombining DNA molecules. Base-pairing between the recombining DNA molecules need not be involved, and even when it is, the heteroduplex joint that is formed is only a few base pairs long. By separating and joining double-stranded DNA molecules at specific sites, this type of recombination enables various types of mobile DNA sequences to move about within and between chromosomes.

Site-specific recombination was first discovered as the means by which a bacterial virus, bacteriophage *lambda,* moves its genome into and out of the *E. coli* chromosome. In its integrated state the virus is hidden in the bacterial chromosome and replicated as part of the host's DNA. When the virus enters a cell, a virus-encoded enzyme called *lambda integrase* is synthesized. This enzyme catalyzes a recombination process that begins when several molecules of the integrase protein bind tightly to a specific DNA sequence on the circular bacteriophage chromosome. The resulting DNA-protein complex can now bind to a related but different specific DNA sequence on the bacterial chromosome, bringing the bacterial and bacteriophage chromosomes close together. The integrase then catalyzes the required DNA cutting and resealing reactions, using a short region of sequence homology to form a tiny heteroduplex joint at the point of union (Figure 6–68). The integrase resembles a DNA topoisomerase in that it forms a reversible covalent linkage to DNA wherever it breaks a DNA chain.

The same type of site-specific recombination mechanism can also be carried out in *reverse* by the lambda bacteriophage, enabling it to exit from its integration site in the *E. coli* chromosome in order to multiply rapidly within the bac-

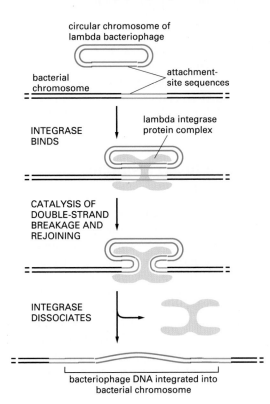

Figure 6–68 The insertion of bacteriophage lambda DNA into the bacterial chromosome. In this example of site-specific recombination, the lambda integrase enzyme binds to a specific "attachment site" DNA sequence on each chromosome, where it makes cuts that bracket a short homologous DNA sequence; the integrase thereby switches the partner strands and reseals them so as to form a heteroduplex joint 7 base pairs long. Each of the four strand-breaking and strand-joining reactions required resembles that made by a DNA topoisomerase, inasmuch as the energy of a cleaved phosphodiester bond is stored in a transient covalent linkage between the DNA and the enzyme (see Figure 6–64).

circular chromosome of lambda bacteriophage

bacterial chromosome

attachment-site sequences

INTEGRASE BINDS

lambda integrase protein complex

CATALYSIS OF DOUBLE-STRAND BREAKAGE AND REJOINING

INTEGRASE DISSOCIATES

bacteriophage DNA integrated into bacterial chromosome

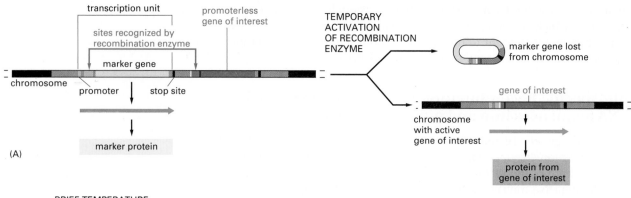

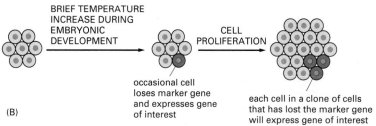

(A)

(B)

Figure 6–69 **Using a site-specific recombination enzyme to turn on a gene in a group of cells in a transgenic animal.** (A) The DNA molecule shown has been engineered so that the gene of interest is transcribed only when a site-specific recombination enzyme is activated, which both removes the marker gene and brings the promoter next to the gene of interest. The recombination enzyme is encoded by a second DNA molecule (not shown) that is engineered so that the enzyme is made only when the temperature is increased. Both DNA molecules are introduced into the chromosomes of the same transgenic animal. When the temperature of this animal is transiently increased, there is a brief burst of synthesis of the recombination enzyme, which causes a DNA rearrangement in an occasional cell such that the marker gene is removed and the gene of interest is simultaneously activated. (B) The strategy can be used to turn on a gene of interest permanently in small clones of cells in a developing animal. The clones can be identified by their loss of the marker gene product, which, for example, could cause a change in the pigmentation of the cells. This technique therefore allows one to study the effect of expressing any gene of interest in a group of cells in an intact animal.

terial cell. This excision reaction is catalyzed by a complex of the integrase enzyme with a second bacteriophage protein, which is produced by the virus only when its host cell is stressed. If the sites recognized by such a recombination enzyme are flipped, the DNA between them will be inverted rather than excised (see Figure 9–57).

Many other enzymes that catalyze site-specific recombination resemble lambda integrase in requiring a short region of identical DNA sequence on the two regions of DNA helix to be joined. Because of this requirement, each enzyme in this class is fastidious with respect to the DNA sequences that it recombines, and it can be expected to catalyze one particular DNA joining event that is useful to the virus, plasmid, transposable element, or cell that contains it. These enzymes can be exploited as tools in transgenic animals to study the influence of specific genes on cell behavior, as illustrated in Figure 6–69.

Site-specific recombination enzymes that break and rejoin two DNA double helices at specific sequences on each DNA molecule often do so in a reversible way: as for lambda bacteriophage, the same enzyme system that joins two DNA molecules can take them apart again, precisely restoring the sequences of the two original DNA molecules. This type of recombination is therefore called *conservative site-specific recombination* to distinguish it from the mechanistically distinct *transpositional site-specific recombination* that we discuss next.

Transpositional Recombination Can Insert a Mobile Genetic Element into Any DNA Sequence [47]

Many mobile DNA sequences, including many viruses and transposable elements, encode integrases that insert their DNA into a chromosome by a mechanism that is different from that used by bacteriophage lambda. Like the lambda integrase, each of these enzymes recognizes a specific DNA sequence in the particular mobile genetic element whose recombination it catalyzes. Unlike the lambda enzyme, however, these integrases do not require a specific DNA sequence in the "target" chromosome and they do not form a heteroduplex joint. Instead, they introduce cuts into both ends of the linear DNA sequence of the mobile genetic element and then catalyze a direct attack by these DNA ends on the target DNA molecule, breaking two closely spaced phosphodiester bonds in the target molecule. Because of the way that these breaks are made, two short single-stranded gaps are left in the recombinant DNA molecule, one at each end

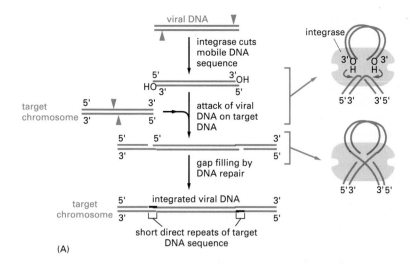

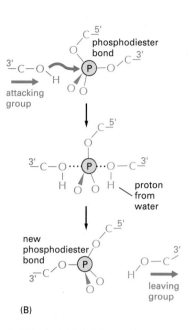

(A)

(B)

of the mobile element; these are filled in by DNA polymerase to complete the recombination process. As illustrated in Figure 6–70, this mechanism creates a short duplication of the adjacent target DNA sequence; such flanking duplications are the hallmark of a transpositional site-specific recombination event.

An integrase enzyme of this type was first purified in active form from bacteriophage Mu. Like the bacteriophage lambda integrase, it carries out all of its cutting and rejoining reactions without requiring an energy source (such as ATP). Very similar enzymes are present in organisms as diverse as bacteria, fruit flies, and humans—all of which contain mobile genetic elements, as we discuss next.

Summary

Genetic recombination mechanisms allow large sections of DNA double helix to move from one chromosome to another. There are two broad classes of recombination events. In general recombination the initial reactions rely on extensive base-pairing interactions between strands of the two DNA double helices that will recombine. As a result, general recombination occurs only between two homologous DNA molecules, and although it moves sections of DNA back and forth between chromosomes, it does not normally change the arrangement of the genes in a chromosome. Site-specific recombination, on the other hand, alters the relative positions of nucleotide sequences in chromosomes because the pairing reactions depend on a protein-mediated recognition of the two DNA sequences that will recombine, and extensive sequence homology is not required. Two site-specific recombination mechanisms are common: (1) conservative site-specific recombination, which produces a very short heteroduplex and therefore requires some DNA sequence that is the same on the two DNA molecules, and (2) transpositional site-specific recombination, which produces no heteroduplex and usually does not require a specific sequence on the target DNA.

Viruses, Plasmids, and Transposable Genetic Elements [48]

In our description of the basic genetic mechanisms, we have so far focused on their selective advantage for the cell. We saw that the short-term survival of the cell depends on the maintenance of genetic information by DNA repair, while the multiplication of the cell requires rapid and accurate DNA replication. On a longer time scale the appearance of genetic variants, on which evolution of the species depends, is greatly facilitated by the reassortment of genes and the occasional rearrangement of DNA sequences caused by genetic recombination. We shall now examine a group of genetic elements that seem to act as parasites,

Figure 6–70 Transpositional site-specific recombination. (A) Outline of the strand-breaking and -rejoining events that lead to integration of the linear double-stranded DNA of a retrovirus (*red*) into an animal cell chromosome (*blue*). In an initial endonuclease step the integrase enzyme makes a cut in one strand at each end of the viral DNA sequence, exposing a protruding 3′-OH group. Each of these 3′-OH ends then directly attacks a phosphodiester bond on opposite strands of a randomly selected site on a target chromosome. This inserts the viral DNA sequence into the target chromosome, leaving short gaps on each side that are filled in by DNA repair processes. Because of the gap filling, this type of mechanism leaves short repeats of target DNA sequence [3 to 12 nucleotides in length (*black*), depending on the integrase enzyme] on either side of the integrated DNA segment. (B) An atomic-level view of the attack by one DNA chain end in (A) on a phosphodiester bond of the target DNA (*blue*). This mechanism resembles that used in RNA splicing, and is distinctly different from the topoisomerase-like activity of lambda integrase. (Adapted from K. Mizuuchi, *J. Biol. Chem.* 267:21273–21276, 1992.)

subverting the genetic mechanisms of the cell for their own benefit. These genetic elements are interesting in their own right. In addition, because they must heavily exploit the metabolism of the host cell in order to multiply, they serve as powerful tools for investigating the normal cell machinery.

Many DNA sequences can replicate independently of the rest of the genome. Such sequences have widely different degrees of independence from their host cells. Of these, virus chromosomes are the most independent because they have a protein coat that allows them to move freely from cell to cell. To varying degrees, the *viruses* are closely related to *plasmids* and *transposable elements*, which are DNA sequences that lack a coat and are therefore more host-cell-dependent and confined to replicate within a single cell and its progeny. More primitive still are some DNA sequences that are suspected of being mobile because they are repeated many times in a cell's chromosome. They move or multiply so rarely, however, that it is not clear if they should be considered as separate genetic elements at all.

We begin our discussion with viruses, which are the best understood of the mobile genetic elements. Then we describe the properties of plasmids and transposable elements, some of which bear a remarkable resemblance to viruses and may in fact have been their ancestors. The many repetitive DNA sequences in vertebrate chromosomes are discussed in Chapter 8.

Viruses Are Mobile Genetic Elements [49]

Viruses were first described as disease-causing agents that can multiply only in cells and that by virtue of their tiny size pass through ultrafine filters that hold back even the smallest bacteria. Before the advent of the electron microscope, their nature was obscure, although it was suspected that they might be naked genes that had somehow acquired the ability to move from one cell to another. The use of ultracentrifuges in the 1930s made it possible to separate viruses from host cell components, and by the early 1940s the generalization emerged that all viruses contain nucleic acids. The idea that viruses and genes carry out similar functions was confirmed by studies on **bacteriophages,** which are bacterial viruses. In 1952 it was shown for the bacteriophage T4 that only the phage DNA, and not the phage protein, enters the bacterial host cell and initiates the replication events that lead to the production of several hundred progeny viruses in every infected cell.

These observations led to the notion of viruses as genetic elements enclosed by a protective coat that enables them to move from one cell to another. Virus multiplication per se is often lethal to the cells in which it occurs; in many cases the infected cell breaks open (*lyses*) and thereby allows the progeny viruses access to nearby cells. Many of the clinical manifestations of viral infection reflect this cytolytic effect of the virus. Both the cold sores formed by herpes simplex virus and the lesions caused by smallpox, for example, reflect the killing of the epithelial cells in a local area of the skin.

As we shall see, the type of nucleic acid in a virus, the structure of its coat, its mode of entry into the host cell, and its mechanism of replication once inside all vary from one type of virus to another.

The Outer Coat of a Virus May Be a Protein Capsid or a Membrane Envelope [50]

Initially, it was thought that the outer coat of a virus might be constructed from a single type of protein molecule. Viral infections were believed to start with the dissociation of the viral chromosome (its nucleic acid) from its protein coat, followed by replication of the chromosome inside the host cell, to form many identical copies. After the synthesis of new copies of the virus-specific coat protein from virally encoded messenger RNA molecules, formation of the progeny virus

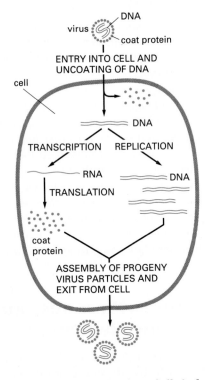

Figure 6–71 The simplest of all viral life cycles. The hypothetical virus shown consists of a small double-stranded DNA molecule that codes for only a single viral capsid protein. No known virus is this simple.

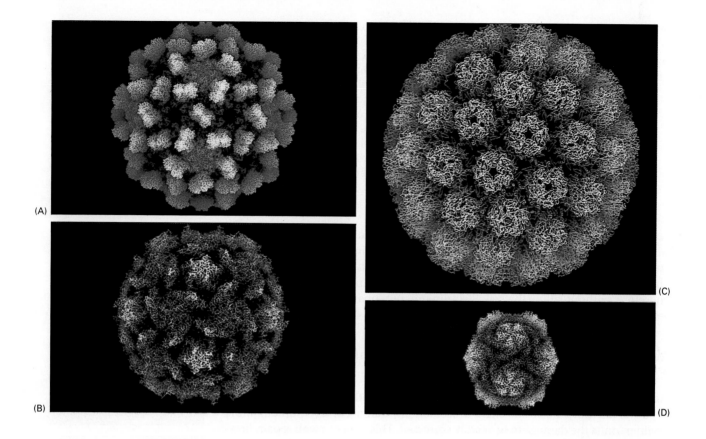

particles would occur by the spontaneous assembly of these coat protein molecules around the progeny viral chromosomes (Figure 6–71).

It is now known that these ideas vastly oversimplify the diversity of virus life cycles. The protein shell that surrounds the nucleic acid of most viruses (the **capsid**), for example, contains more than one type of polypeptide chain, often arranged in several layers (Figure 6–72). In many viruses, moreover, the protein

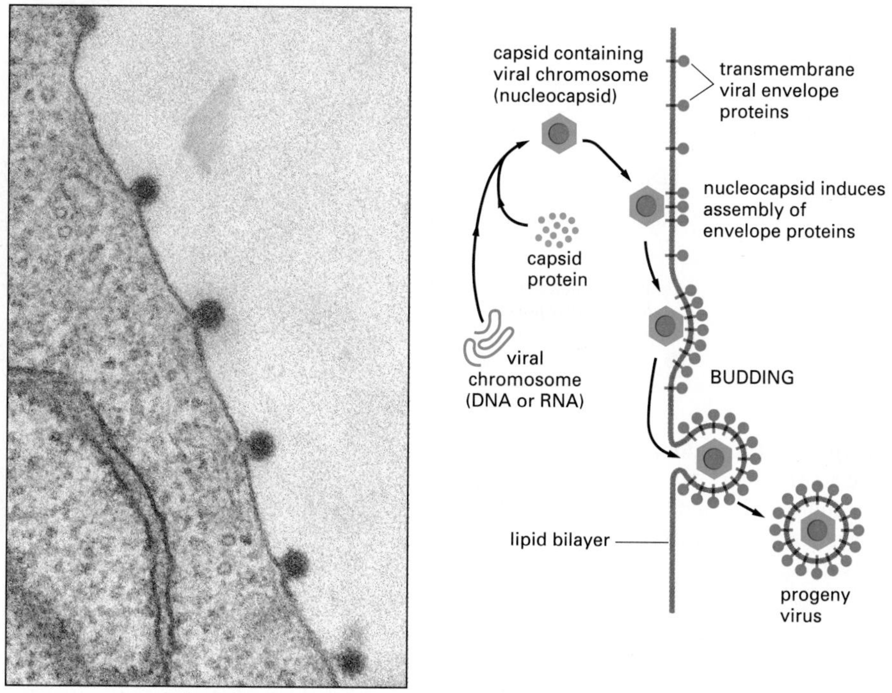

Figure 6–72 The capsids of some viruses, all shown at the same scale. (A) Tomato bushy stunt virus; (B) poliovirus; (C) simian virus 40 (SV40); (D) satellite tobacco necrosis virus. The structures of all of these capsids have been determined by x-ray crystallography and are known in atomic detail. (Courtesy of Robert Grant, Stephan Crainic, and James M. Hogle.)

capsid containing
viral chromosome
(nucleocapsid)

transmembrane
viral envelope
proteins

nucleocapsid induces
assembly of
envelope proteins

capsid
protein

viral
chromosome
(DNA or RNA)

BUDDING

lipid bilayer

progeny
virus

(A)

100 nm

(B)

Figure 6–73 Acquisition of a viral envelope. (A) Electron micrograph of a thin section of an animal cell from which several copies of an enveloped virus (Semliki forest virus) are budding. (B) Schematic view of the envelope assembly and budding process. Whereas the lipid bilayer that surrounds the capsid is parasitized directly from the plasma membrane of the host cell, the only proteins in this lipid bilayer are those encoded by the viral genome. (A, courtesy of M. Olsen and G. Griffiths.)

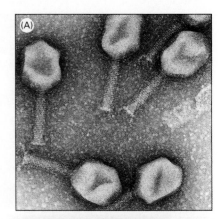

Figure 6–74 **The coats of viruses.** These electron micrographs of negatively stained virus particles are all at the same scale. (A) *Bacteriophage T4*, a large DNA-containing virus that infects *E. coli*. The DNA is stored in the bacteriophage head and injected into the bacterium through the cylindrical tail. (B) *Potato virus X*, a filamentous plant virus that contains an RNA genome. (C) *Adenovirus*, a DNA-containing virus that can infect human cells. The protein capsid forms the outer surface of this virus. (D) *Influenza virus*, a large RNA-containing animal virus whose protein capsid is further enclosed in a lipid-bilayer-based envelope containing protruding spikes of viral glycoprotein. (A, courtesy of James Paulson; B, courtesy of Graham Hills; C, courtesy of Mei Lie Wong; D, courtesy of R.C. Williams and H.W. Fisher.)

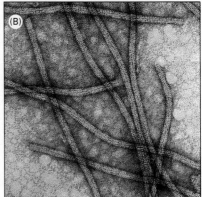

capsid is further enclosed by a lipid bilayer membrane that contains proteins. Many of these *enveloped viruses* acquire their envelope in the process of budding from the plasma membrane (Figure 6–73). This budding process allows the virus particles to leave the cell without disrupting the plasma membrane and, therefore, without killing the cell. Electron micrographs that emphasize the differences among viral coats are presented in Figure 6–74.

Viral Genomes Come in a Variety of Forms and Can Be Either RNA or DNA [51]

As discussed earlier, the DNA double helix has the advantages of stability and easy repair. If one polynucleotide chain is accidentally damaged, its complementary chain permits the damage to be readily corrected. This concern with repair, however, need not bother small viral chromosomes that contain only several thousand nucleotides. The chance of accidental damage is very small compared with the risk to a cell genome containing millions of nucleotides.

The genetic information of a virus can, therefore, be carried in a variety of unusual forms, including RNA instead of DNA. A viral chromosome may be a single-stranded RNA chain, a double-stranded RNA helix, a circular single-

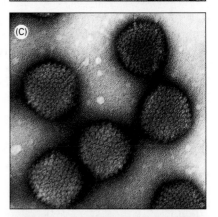

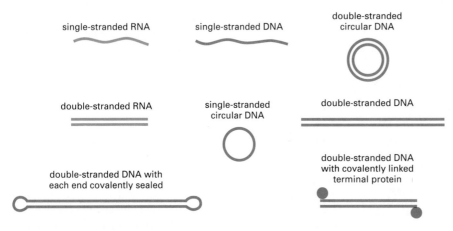

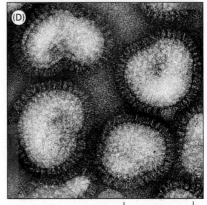

Figure 6–75 **Schematic drawings of several types of viral genomes.** The smallest viruses contain only a few genes and can have an RNA or a DNA genome; the largest viruses contain hundreds of genes and have a double-stranded DNA genome. Some examples of these types of viruses are as follows: *single-stranded RNA*—tobacco mosaic virus, bacteriophage R17, poliovirus; *double-stranded RNA*—reovirus; *single-stranded DNA*—parvovirus; *single-stranded circular DNA*—M13 and φX174 bacteriophages; *double-stranded circular DNA*—SV40 and polyomaviruses; *double-stranded DNA*—T4 bacteriophage, herpes virus; *double-stranded DNA with covalently linked terminal protein*—adenovirus; *double-stranded DNA with covalently sealed ends*—poxvirus. The peculiar ends (as well as the circular forms) overcome the difficulty of replicating the last few nucleotides at the end of a DNA chain (see pp. 338 and 364).

stranded DNA chain, or a linear single-stranded DNA chain. Moreover, although some viral chromosomes are simple linear DNA double helices, circular DNA double helices and more complex linear DNA double helices are also common. Several viruses have protein molecules covalently attached to the 5′ ends of their DNA strands, for example, and the DNA double helices from the very large poxviruses have their opposite strands at each end covalently joined through phosphodiester linkages (Figure 6–75).

A Viral Chromosome Codes for Enzymes Involved in the Replication of Its Nucleic Acid [52]

Each type of viral genome requires unique enzymatic tricks for its replication and thus must encode not only the viral coat protein but also one or more of the enzymes needed to replicate the viral nucleic acid. The amount of information that a virus brings into a cell to ensure its own selective replication varies greatly. The DNA of the relatively large bacteriophage T4, for example, contains about 300 genes, including at least 30 genes that ensure the rapid replication of the T4 chromosome in its *E. coli* host cell (Figure 6–76). T4 DNA replication has the unusual feature that 5-hydroxymethyl-C is incorporated in place of C in its DNA. The unusual base composition of the T4 DNA makes it readily distinguishable from host DNA and selectively protects it from nucleases encoded in the T4 genome that thus degrade only the *E. coli* chromosome. Still other proteins alter host cell RNA polymerase molecules so that they are unable to transcribe *E. coli* DNA and instead transcribe different sets of bacteriophage genes at different stages of infection, according to the needs of the phage.

Smaller DNA viruses, such as the monkey virus SV40 and the tiny bacteriophage M13, carry much less genetic information. They rely heavily on host-cell enzymes to carry out their DNA synthesis, parasitizing most of the host-cell DNA replication proteins. Most DNA viruses, however, code for proteins that selectively initiate the synthesis of their own DNA, recognizing a particular nucleotide sequence in the virus that serves as a *replication origin*. This is important because a virus must override the cellular control signals that would otherwise cause the viral DNA to replicate in pace with the host cell DNA, doubling only once in each cell cycle. We do not yet understand very much about how eucaryotic cells regulate their own DNA synthesis, and the mechanisms used by viruses to escape from this regulation—which are much more accessible to study—provide insights into the host mechanisms.

RNA viruses have particularly specialized requirements for replication, since to reproduce their genomes they must copy RNA molecules, which means polymerizing nucleoside triphosphates on an RNA template. Cells normally do not have enzymes to carry out this reaction, so even the smallest RNA viruses must encode their own RNA-dependent polymerase enzymes in order to replicate. We now look in more detail at the replication mechanisms of the various types of viruses.

Both RNA Viruses and DNA Viruses Replicate Through the Formation of Complementary Strands [53]

Like DNA replication, the replication of the genomes of RNA viruses occurs through the formation of complementary strands. For most RNA viruses this process is catalyzed by specific RNA-dependent RNA polymerase enzymes (*replicases*). These enzymes are encoded by the viral RNA chromosome and are often incorporated into the progeny virus particles, so that upon entry of the virus into a cell, they can immediately begin replicating the viral RNA. Replicases are always packaged into the capsid of the so-called *negative-strand RNA viruses*, such as influenza or vesicular stomatitis virus. Negative-strand viruses are so called because the infecting single strand does not code for protein; instead its complementary strand carries the coding sequences. Thus the infecting strand

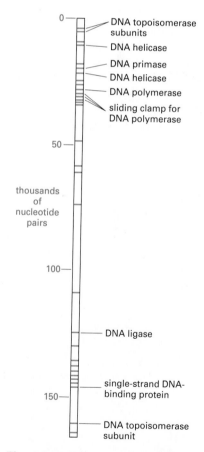

Figure 6–76 **The T4 bacteriophage chromosome, showing the positions of the more than 30 genes involved in T4 DNA replication.** The genome of bacteriophage T4 consists of 169,000 nucleotide pairs and encodes about 300 different proteins.

remains impotent without a preformed replicase. In contrast, the viral RNA of *positive-strand RNA viruses*, such as poliovirus, can serve as mRNA and produce a replicase once it enters the cell; therefore the naked genome itself is infectious.

The synthesis of viral RNA always begins at the 3′ end of the RNA template, starting with the synthesis of the 5′ end of the new viral RNA molecule and progressing in the 5′-to-3′ direction until the 5′ end of the template is reached. There are no error-correcting mechanisms for viral RNA synthesis, and error rates are similar to those in DNA transcription (about 1 error in 10^4 nucleotides synthesized). This is not a serious deficiency as long as the RNA chromosome is relatively short; for this reason the genomes of all RNA viruses are small relative to those of the large DNA viruses.

All DNA viruses begin their replication at a replication origin, where special initiator proteins bind and then attract the replication enzymes of the host cell (see Figure 8–34). There are many different replication pathways, however. The complexity of these diverse replication schemes reflects, in part, the problem of replicating the ends of a simple linear DNA molecule, given a DNA polymerase enzyme that cannot begin synthesis without a primer (see pp. 253–254). DNA viruses have solved this problem in a variety of ways: some have circular DNA genomes and thus no ends; others have linear DNA genomes that repeat their terminal sequences or end in loops; while still others have special terminal proteins that serve to prime the DNA polymerase directly (see Figure 6–75).

Viruses Exploit the Intracellular Traffic Machinery of their Host Cells [54]

All viruses have only a limited amount of nucleic acid in their genome, and so they must parasitize host-cell pathways for most of the steps in their reproduction. In fact, because viral products are usually synthesized in large amounts during infection, and because during its life cycle the virus follows a sequential route through the compartments of the host cell, virus-infected cells have served as important models for tracing the pathways of intracellular transport and for studying how essential biosynthetic reactions are compartmentalized in eucaryotic cells.

Enveloped animal viruses, in which the genome is enclosed in a lipid-bilayer membrane, have exploited the compartmentalization of the cell to an especially fine degree. To follow the life cycle of an enveloped virus is to take a tour through the cell. A well-studied example is *Semliki forest virus,* which consists of a single-stranded RNA genome surrounded by a *capsid* formed by a regularly arranged icosahedral (20-faced) shell composed of many copies of one protein (called C protein). The *nucleocapsid* (genome + capsid) is surrounded by a closely apposed lipid bilayer that contains only three types of polypeptide chains, each encoded

Figure 6–77 The structure of Semliki forest virus. Schematic drawings of a cross-section (A) and an exploded three-dimensional view (B) of the virus. (C) A three-dimensional reconstruction of the surface of the virus derived from cryoelectron micrographs of unstained specimens. The virus has a total mass of 46 million daltons. (B, adapted from S.C. Harrison, *Curr. Opin. Struct. Biol.* 2:293–299, 1992. Current Science; C, courtesy of Stephen Fuller.)

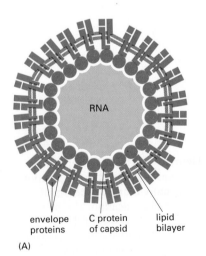

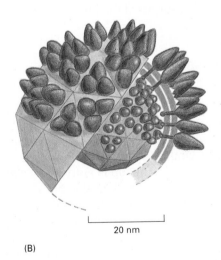

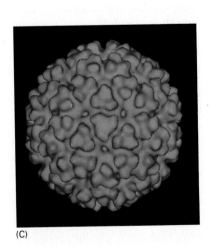

(A) envelope proteins | C protein of capsid | lipid bilayer

(B) 20 nm

(C)

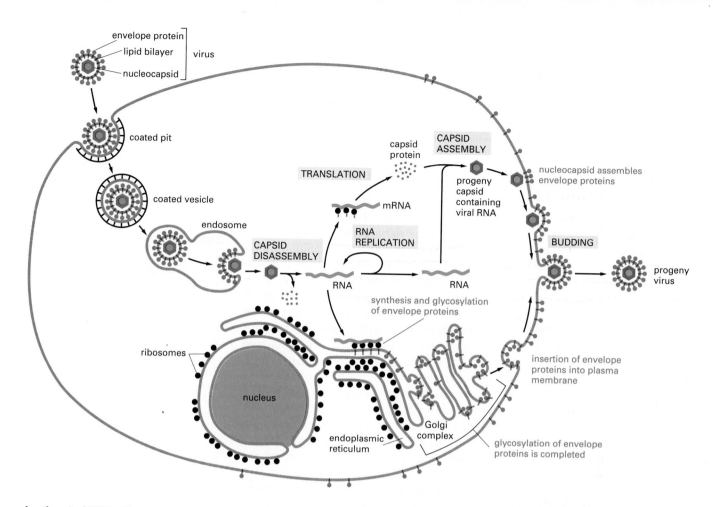

by the viral RNA. These **envelope proteins** form heterotrimers that span the lipid bilayer and interact with the C protein of the nucleocapsid, linking the membrane and nucleocapsid together (Figure 6–77). The glycosylated portions of the envelope proteins are always on the outside of the lipid bilayer, and each trimer forms a "spike" that can be seen in electron micrographs projecting outward from the surface of the virus (Figure 6–77C).

Infection is initiated when an envelope protein on the virus binds to a normal cell protein that serves as its receptor on the host-cell plasma membrane. The virus then uses the cell's normal endocytic pathway to enter the cell by receptor-mediated endocytosis and is delivered to early endosomes (discussed in Chapter 13). But instead of being transferred from endosomes to lysosomes, the virus escapes from the endosome by virtue of the special properties of one of its envelope proteins. At the acidic pH of the endosome, this protein causes the viral envelope to fuse with the endosome membrane, releasing the bare nucleocapsid into the cytosol. The nucleocapsid is "uncoated" in the cytosol, releasing the viral RNA, which is then translated by host-cell ribosomes to produce a virus-encoded RNA polymerase. This in turn makes many copies of viral RNA, some of which serve as mRNA molecules to direct the synthesis of the structural proteins of the virus—the capsid C protein and the three envelope proteins.

The newly synthesized capsid and envelope proteins follow separate pathways through the cytoplasm. The envelope proteins, like the plasma membrane proteins of the host cell, are synthesized by ribosomes that are bound to the rough ER; in contrast, the capsid protein, like the cytosolic proteins of the cell, is synthesized by ribosomes that are not membrane bound. The newly synthesized capsid protein binds to the recently replicated viral RNA to form new nucleocapsids. The envelope proteins, in contrast, are inserted into the membrane of the ER, where they are glycosylated, transported to the Golgi apparatus, and then delivered to the plasma membrane (Figure 6–78).

Figure 6–78 The life cycle of the Semliki forest virus. The virus parasitizes the host cell for most of its biosyntheses.

Viruses, Plasmids, and Transposable Genetic Elements

The viral nucleocapsids and envelope proteins finally meet at the plasma membrane. As a result of a specific interaction with a cluster of envelope proteins, the nucleocapsid forms a bud whose envelope contains the envelope proteins embedded in host-cell lipids. Finally, the bud pinches off and a free virus is released on the outside of the cell. The clustering of envelope proteins as they assemble around the nucleocapsid during viral budding excludes the host plasma membrane proteins from the final virus particle.

Different Enveloped Viruses Bud from Different Cellular Membranes [55]

Viral envelope proteins are all transmembrane proteins that are synthesized in the ER. Like other ER proteins, they carry sorting signals that direct them to a particular cell membrane (discussed in Chapter 13). Their final location determines the site of viral budding. Epithelial cell lines, for example, can form polarized cell sheets when they are cultured on an appropriate surface, such as a collagen-coated porous filter. When viruses infect such polarized cells, which maintain distinct domains of apical and basolateral plasma membrane, some of them (such as influenza virus) bud exclusively from the apical plasma membrane, whereas others (such as Semliki forest virus and vesicular stomatitis virus) bud only from the basolateral plasma membrane (Figure 6–79). This polarity of budding reflects the presence on the envelope proteins of distinct apical or basolateral sorting signals, which direct the proteins to only one cell-surface domain; the proteins in turn cause the virus to assemble in that domain.

Other viruses have envelope proteins with different kinds of sorting signals. *Herpes virus*, for example, is a DNA virus that replicates in the nucleus, where its nucleocapsid assembles, and then acquires an envelope by budding through the inner nuclear membrane into the ER lumen; the envelope proteins therefore must be specifically transported from the ER membrane to the inner nuclear membrane, probably via the lipid bilayer that surrounds the nuclear pores. *Flavivirus*, in contrast, buds directly into the ER lumen, and *bunyavirus* buds into the Golgi apparatus, indicating that their envelope proteins carry signals for retention in the ER and Golgi membranes, respectively. After budding, the enveloped herpes virus, flavivirus, and bunyavirus particles become soluble in the ER and Golgi lumen, and they move outward toward the cell surface exactly as if they were

Figure 6–79 Two enveloped viruses that bud from different domains of the plasma membrane. Electron micrographs showing that one type of enveloped virus buds from the apical plasma membrane while another type buds from the basolateral plasma membrane of the same epithelial cell line grown in culture. These cells grow with their basal surface attached to the culture dish. The boxed area in each schematic drawing corresponds to the indicated electron micrograph. (Micrographs courtesy of E. Rodriguez-Boulan and D.D. Sabatini.)

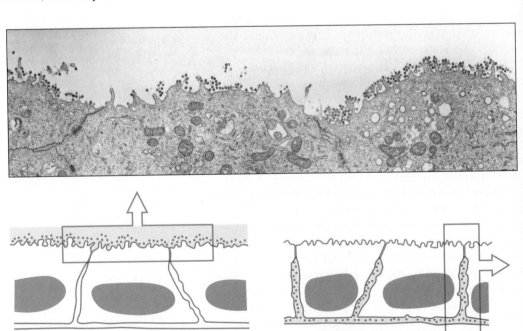

influenza virus buds only from the apical plasma membrane

vesicular stomatitis virus buds only from the basolateral plasma membrane secreted proteins; in the *trans* Golgi network they are incorporated into transport vesicles and secreted from the cell by the constitutive secretory pathway (discussed in Chapter 13).

Viral Chromosomes Can Integrate into Host Chromosomes [56]

The end result of the entry of a viral chromosome into a cell is not always its immediate multiplication to produce large numbers of progeny. Many viruses enter a *latent* state, in which their genomes are present but inactive in the cell and no progeny are produced. Viral latency was discovered when it was found that exposure to ultraviolet light induced many apparently uninfected bacteria to produce progeny bacteriophages. Subsequent experiments showed that these *lysogenic bacteria* carry in their chromosomes a dormant but complete viral chromosome. Such integrated viral chromosomes are called **proviruses.**

Bacteriophages that can integrate their DNA into bacterial chromosomes are known as *temperate bacteriophages.* The prototypic example is the bacteriophage lambda, discussed earlier. When lambda infects a suitable *E. coli* host cell, it normally multiplies to produce several hundred progeny particles, which are released when the bacterial cell lyses; this is called a *lytic infection.* More rarely, the free ends of the linear infecting DNA molecules join to form a DNA circle that becomes integrated into the circular host *E. coli* chromosome by a site-specific

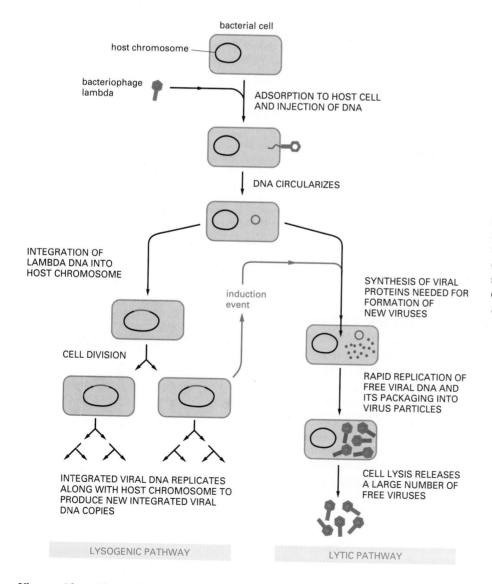

Figure 6–80 **The life cycle of bacteriophage lambda.** The lambda genome contains about 50,000 nucleotide pairs and encodes about 50 proteins. Its double-stranded DNA can exist in either linear or circular forms. As shown, the bacteriophage can multiply by either a lytic or a lysogenic pathway in the *E. coli* bacterium. When the bacteriophage is growing in the lysogenic state, damage to the cell causes the integrated viral DNA (provirus) to exit from the host chromosome and shift to lytic growth. The entrance and exit of the DNA from the chromosome are site-specific genetic recombination events catalyzed by the lambda *integrase* protein (see Figure 6–68).

recombination event. The resulting lysogenic bacterium, carrying the proviral lambda chromosome, multiples normally until it is subjected to an environmental insult, such as exposure to ultraviolet light or ionizing radiation. The resulting cell debilitation induces the integrated provirus to leave the host chromosome and begin a normal cycle of viral replication. In this way the integrated provirus need not perish with its damaged host cell but has a chance to escape to other *E. coli* cells (Figure 6–80).

The Continuous Synthesis of Some Viral Proteins Can Make Cells Cancerous [57]

Animal cells, like bacteria, can offer viruses an alternative to lytic growth. *Permissive* cells permit DNA viruses to multiply lytically and kill the cell. *Nonpermissive* cells may allow the DNA virus to enter but not to replicate lytically; in a small percentage of such cells the viral chromosome either becomes integrated into the host cell genome, where it is replicated along with the host chromosomes, or forms a plasmid—a circular DNA molecule—that replicates in a controlled fashion without killing the cell. Such nonpermissive infections sometimes result in a genetic change in the host cell, causing it to proliferate in an ill-controlled way and thus transforming it into its cancerous equivalent. In this case the DNA virus is called a *DNA tumor virus* and the process is called virus-mediated *neoplastic transformation*. The most extensively studied DNA tumor viruses are two papovaviruses, SV40 and polyoma. Their transforming ability has been traced to several viral proteins that cooperate to stimulate quiescent cells to proliferate—that is, they drive the cells from G_0 into S phase. In permissive cells the shift to S phase (the phase of the cell cycle where DNA is synthesized) provides the virus with all of the host-cell replication enzymes required for viral DNA synthesis. When a provirus happens to make these viral proteins in a nonpermissive cell, they can override some of the normal growth control mechanisms in the cell and its progeny. By this means some DNA tumor viruses that infect humans are known to contribute to the development of some types of human cancers (although the great majority of human cancers are thought not to involve tumor viruses).

RNA Tumor Viruses Are Retroviruses [58]

For one group of RNA viruses, the so-called *RNA tumor viruses,* the infection of a permissive cell often leads simultaneously to a nonlethal release of progeny virus from the cell surface by budding and a permanent genetic change in the infected cell that makes it cancerous. How RNA virus infection could lead to a permanent genetic alteration was unclear until the discovery of the enzyme *reverse transcriptase,* which transcribes the infecting RNA chains of these viruses into complementary DNA molecules that integrate into the host cell genome. RNA tumor viruses—which include the first well-known tumor virus, the Rous sarcoma virus—are members of a large class of viruses known as **retroviruses.** These viruses are so named because as part of their normal life cycle they reverse the normal process in which DNA is transcribed into RNA.

The enzyme **reverse transcriptase** is an unusual DNA polymerase that uses either RNA or DNA as a template (Figure 6–81); it is encoded by the retrovirus RNA and is packaged inside each viral capsid during the production of new virus particles. When the single-stranded RNA of the retrovirus enters a cell, the reverse transcriptase brought in with the capsid first makes a DNA copy of the RNA strand to form a DNA-RNA hybrid helix, which is then used by the same enzyme to make a double helix with two DNA strands. The two ends of the linear viral DNA molecule are recognized by a virus-encoded integrase that catalyzes the insertion of the viral DNA into virtually any site on a host-cell chromosome (see Figure 6–70). The next step in the infectious process is transcription of the integrated viral DNA by host-cell RNA polymerase, producing large num-

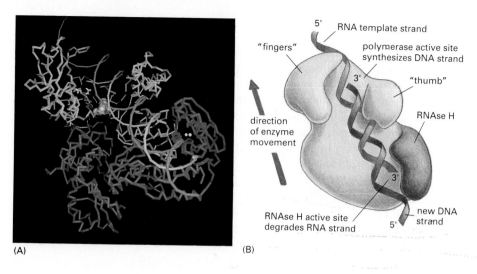

Figure 6–81 Reverse transcriptase. (A) The three-dimensional structure of the enzyme from HIV-1 (the AIDS virus), determined by x-ray crystallography; (B) a schematic view of a model for its activity on an RNA template. Note that the polymerase domain (*yellow*) has a covalently attached RNAse domain (*red*) that degrades an RNA strand in an RNA/DNA helix. This activity helps the polymerase convert the initial hybrid helix into a DNA double helix. (A, courtesy of Tom Steitz; B, adapted from L.A. Kohlstaedt et al., *Science* 256:1783–1790, 1992. © 1992 the AAAS.)

Figure labels: 5′, RNA template strand, "fingers", polymerase active site synthesizes DNA strand, 3′, "thumb", RNAse H, direction of enzyme movement, RNAse H active site degrades RNA strand, 3′, new DNA strand, 5′
(A)   (B)

bers of viral RNA molecules identical to the original infecting genome. Finally, these RNA molecules are translated to produce the capsid, envelope, and reverse transcriptase proteins that are assembled with the RNA into new enveloped virus particles, which bud from the plasma membrane (Figure 6–82).

Both RNA and DNA tumor viruses transform cells because the permanent presence of the viral DNA in the cell causes the synthesis of new proteins that alter the control of host-cell proliferation. The genes that code for such proteins are called *oncogenes*. Unlike DNA tumor viruses, whose oncogenes typically encode normal viral proteins essential for viral multiplication, the oncogenes carried by RNA tumor viruses are modified versions of normal host-cell genes that are not required for viral replication. Since only a limited amount of RNA can be packed into the capsid of a retrovirus, the acquired oncogene sequences often replace an essential part of the retroviral genome. In Chapters 15 and 24 we discuss how viral oncogenes have provided important clues to the causes and nature of cancer, as well as to the normal mechanisms that control cell growth and division in multicellular animals. We also discuss how the random integration of viral DNA into genomes can alter normal genes and thereby affect cell behavior (see Figure 24–24).

Figure 6–82 The life cycle of a retrovirus. The retrovirus genome consists of an RNA molecule of about 8500 nucleotides; two such molecules are packaged into each viral particle. The enzyme *reverse transcriptase* first makes a DNA copy of the viral RNA molecule and then a second DNA strand, generating a double-stranded DNA copy of the RNA genome. The integration of this DNA double helix into the host chromosome, catalyzed by the viral integrase, is required for the synthesis of new viral RNA molecules by the host-cell RNA polymerase.

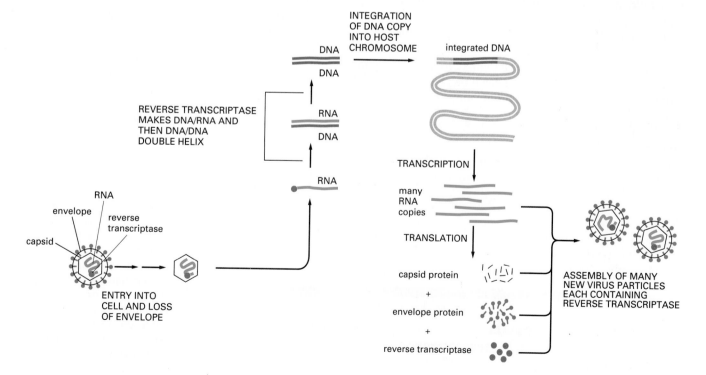

The Virus That Causes AIDS Is a Retrovirus [59]

In 1982 physicians first became aware of a new sexually transmitted disease that was associated with an unusual form of cancer (Kaposi's sarcoma) and a variety of unusual infections. Because both of these problems reflect a severe deficiency in the immune system—specifically in helper T lymphocytes—the disease was named *acquired immune deficiency syndrome (AIDS)*. By culturing lymphocytes from patients with an early stage of the disease, a retrovirus was isolated that is now known to be the causative agent of AIDS, which has become a rapidly spreading epidemic that threatens to kill millions of people worldwide.

The retrovirus, called *human immunodeficiency virus (HIV)*, enters helper T lymphocytes by first binding to a functionally important plasma membrane protein called CD4 (discussed in Chapter 23). There are two features of HIV that make it especially deadly. First, it eventually kills the helper T cells that it infects rather than living in symbiosis with them, as do most other retroviruses, and helper T cells are vitally important in defending us against infection. Second, the provirus tends to persist in a latent state in the chromosomes of an infected cell without producing virus until it is activated by an unknown rare event; this ability to hide greatly complicates any attempt to treat the infection with antiviral drugs.

Much current research on AIDS is aimed at understanding the life cycle of HIV. The complete nucleotide sequence of the viral RNA has been determined. This has made it possible to identify and study each of the proteins that it encodes. The three-dimensional structure of its reverse transcriptase (see Figure 6–81) is being used to help design new drugs that inhibit the enzyme. The nine genes of this retrovirus are displayed on the HIV genetic map in Figure 6–83.

Some Transposable Elements Are Close Relatives of Retroviruses [60]

Because many viruses can move into and out of their host chromosomes, any large genome is likely to contain a number of different proviruses. Most genomes are also likely to house a variety of mobile DNA sequences that do not form viral particles and cannot leave the cell. Such **transposable elements** range in length from a few hundred to tens of thousands of base pairs, and they are usually present in multiple copies per cell. One can consider these elements as tiny parasites hidden in chromosomes. Each transposable element is occasionally activated to move to another DNA site in the same cell by a process called *transposition*, catalyzed by its own site-specific recombination enzyme. These integrases, also referred to as *transposases*, are often encoded in the DNA of the element itself. Since most transposable elements move only very rarely (once in 10^5 cell generations for many elements in bacteria), it is often difficult to distinguish them from nonmobile parts of the chromosome. It is not known what suddenly triggers their movement.

Transposition can occur by a variety of mechanisms. One large family of transposable elements uses a mechanism that is indistinguishable from part of a retrovirus life cycle. These elements, called *retrotransposons,* are present in organisms as diverse as yeasts, flies, and mammals. One of the best-understood

Figure 6–83 A map of the HIV genome. The genome consists of about 9000 nucleotides and contains nine genes, whose locations are shown in *green* and *red*. Three of the genes (*green*) are common to all retroviruses: *gag* encodes capsid proteins, *env* encodes envelope proteins, and *pol* encodes both the reverse transcriptase (see Figure 6–81) and the integrase (see Figure 6–70) proteins. The HIV genome is unusually complex, since it contains six small genes (in *red*) in addition to the three (in *green*) that are normally required for the retrovirus life cycle. At least some of these small genes encode proteins that regulate viral gene expression, and it is tempting to speculate that it is this extra complexity that makes HIV so deadly. As indicated by the *red lines*, RNA splicing (see Figure 8–7) is required to produce the Rev and Tat proteins.

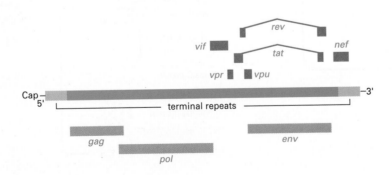

retrotransposons is the so-called Ty1 element of yeasts. The first step in its transposition is the transcription of the entire transposable element, producing an RNA copy of the element that is more than 5000 nucleotides long. This transcript encodes a reverse transcriptase enzyme that makes a double-stranded DNA copy of the RNA molecule via a RNA/DNA hybrid intermediate, precisely mimicking the early stages of infection by a retrovirus (see Figure 6–82). The analogy continues as the linear DNA molecule uses an integrase to integrate into a randomly selected site on the chromosome. Although the resemblance to a retrovirus is striking, unlike a retrovirus, the Ty1 element does not have a functional protein coat and therefore can only move within a single cell and its progeny.

Other Transposable Elements Transfer Themselves Directly from One Site in the Genome to Another [61]

Unlike retrotransposons, many transposable elements rarely exist free of the host chromosome; the transposases that catalyze their movement can act on the DNA of the element while it is still integrated in the host genome. The transposase binds to a short sequence that is repeated in reverse orientation at each end of the element, thereby holding these two ends close together while catalyzing the subsequent recombination event. The mechanism is closely related to that used by the retrovirus integrase (see Figure 6–70). For some transposable elements the transposition mechanism differs only in that the linear DNA molecule to be integrated must be cut out of a much longer DNA molecule, leaving a break in the vacated chromosome (Figure 6–84). This break is subsequently resealed, but in the process the DNA sequence is often altered, resulting in a mutation at the old chromosomal site.

Other transposable elements replicate when they move. In the best-studied example, a covalent connection is first made between the transposable element and a randomly selected target site; this connection then triggers a localized synthesis of DNA that results in one copy of the replicated transposable element being inserted at a new chromosomal site, while the other copy remains at the old one (Figure 6–85). The mechanism is closely related to the nonreplicative mechanism just described, and it starts in nearly the same way; indeed, some transposable elements can move by either pathway.

In addition to moving themselves, all types of transposable elements occasionally move or rearrange neighboring DNA sequences of the host genome. They frequently cause deletions of adjacent nucleotide sequences, for example, or carry them to another site. The presence of transposable elements makes the arrangement of the DNA sequences in chromosomes much less stable than previously thought, and it is likely that they have been responsible for many important evolutionary changes in genomes (discussed in Chapter 8).

Are the transposable elements also of evolutionary importance as the most ancient ancestors of viruses? Although the precursors of retroviruses were almost certainly retrotransposons, all present-day transposable elements rely heavily on DNA-based reaction mechanisms. But very early cells are thought to have had

Figure 6–84 The direct movement of a transposable element from one chromosomal site to another. Transposable elements of this type can be recognized by the "inverted repeat DNA sequences" (*orange*) at their ends. Experiments show that these sequences, which can be as short as 20 nucleotides, are all that is necessary for the DNA between them to be transposed by the particular transposase enzyme associated with the element. The mechanism shown here is closely related to that used by a retrovirus to integrate its double-stranded DNA into a chromosome (compare with Figure 6–70). Although the gap left in donor chromosome is resealed, the process often alters the DNA sequence, causing a mutation at the donor site (not shown).

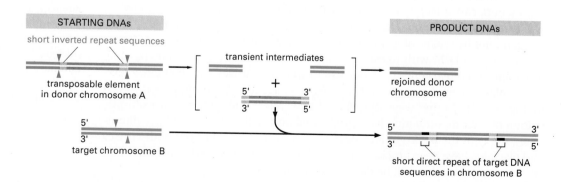

RNA rather than DNA genomes, so we must look to RNA-based mechanisms for the ultimate origin of viruses.

Most Viruses Probably Evolved from Plasmids [62]

Even the largest viruses depend heavily on their host cells for biosynthesis; no known virus makes its own ribosomes or generates the ATP it requires, for example. Clearly, therefore, cells must have evolved before viruses. The precursors of the first viruses were probably small nucleic acid fragments that developed the ability to multiply independently of the chromosomes of their host cells. Such independently replicating elements, called **plasmids,** can replicate indefinitely outside the host chromosome. Plasmids occur in both DNA and RNA forms, and, like viruses, they contain a special nucleotide sequence that serves as an origin of replication. Unlike viruses, however, they cannot make a protein coat and therefore cannot move from cell to cell in this way.

The first RNA plasmids may have resembled the *viroids* found in some plant cells. These small RNA circles, only 300 to 400 nucleotides long, are replicated despite the fact that they do not code for any protein. Having no protein coat, viroids exist as naked RNA molecules and pass from plant to plant only when the surfaces of both donor and recipient cells are damaged so that there is no membrane barrier for the viroid to pass. Under the pressure of natural selection, such independently replicating elements could be expected to acquire nucleotide sequences from the host cell that would facilitate their own multiplication, including sequences that code for proteins. Some present-day plasmids are indeed quite complex, encoding proteins and RNA molecules that regulate their replication, as well as proteins that control their partitioning into daughter cells. The largest known plasmids are double-stranded DNA circles more than 100,000 base pairs long.

The first virus probably appeared when an RNA plasmid acquired a gene coding for a capsid protein. But a capsid can enclose only a limited amount of nucleic acid; therefore a virus is limited in the number of genes it can contain. Forced to make optimal use of their limited genomes, some small viruses evolved *overlapping genes,* in which part of the nucleotide sequence encoding one protein is used (in the same or a different reading frame) to encode a second protein. Other viruses evolved larger capsids and consequently could accommodate more genes.

With their unique ability to transfer nucleic acid sequences across species barriers, viruses have almost certainly played an important part in the evolution of the organisms they infect. Many recombine frequently with their host-cell genome and with one another. In this way they can pick up small pieces of host chromosome at random and carry them to different cells or organisms. Moreover, integrated copies of viral DNA (proviruses) have become a normal part of the genome of most organisms. Examples of such proviruses include the lambda family of bacteriophages and the so-called endogenous retroviruses found in numerous copies in vertebrate genomes. The integrated viral DNA can become altered so that it cannot produce a complete virus but can still encode proteins, some of which may be useful to the host cell. Therefore, viruses, like sexual reproduction, can speed up evolution by promoting the mixing of gene pools.

The process in which DNA sequences are transferred between different host-cell genomes by means of a virus is called *DNA transduction,* and several viruses that transduce DNA with particularly high frequencies are commonly used by researchers to move genes from one cell to another. Viruses and their close relatives—plasmids and transposable elements—have also been important to cell biology in many other ways. Because of their relative simplicity, for example, studies of their reproduction have progressed unusually rapidly and have illuminated many of the basic genetic mechanisms in cells. In addition, both viruses and plasmids have been crucial elements in the development of the recombinant DNA technologies that will be described in Chapter 7.

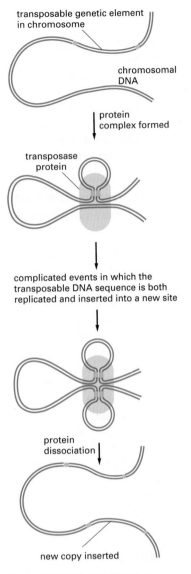

Figure 6–85 The replicative movement of a transposable element within a chromosome. The element shown replicates during transposition, its movement occurring without it being excised from its original site. The two inverted repeat DNA sequences that commonly flank the two ends of transposable elements are shown in *orange.* At the start of transposition the transposase cuts one of the two DNA strands at each end of the element, and the element then serves as a template for DNA synthesis, which begins by the addition of nucleotides to the 3′ ends of chromosomal DNA sequences. Many details are known, but the process is too complex to be illustrated here.

Summary

Viruses are infectious particles that consist of a DNA or an RNA molecule (the viral genome) packaged in a protein capsid, which in the enveloped viruses is surrounded by a lipid-bilayer-based membrane. Both the structure of the viral genome and its mode of replication vary widely among viruses. A virus can multiply only inside a host cell, whose genetic mechanisms it subverts for its own reproduction. A common outcome of a viral infection is the lysis of the infected cell and release of infectious viral particles. In some cases, however, the viral chromosome instead integrates into a host-cell chromosome, where it is replicated as a provirus along with the host genome. Many viruses are thought to have evolved from plasmids, which are self-replicating DNA or RNA molecules that lack the ability to wrap themselves in a protein coat.

Transposable elements are DNA sequences that differ from viruses in being able to multiply only in their host cell and its progeny; like plasmids, they cannot exist stably outside of cells. Unlike plasmids, they normally replicate only as an integral part of a chromosome. Some transposable elements, however, are closely related to retroviruses and can move from place to place in the genome by the reverse transcription of an RNA intermediate. Although both viruses and transposable elements can be viewed as parasites, many of the DNA sequence rearrangements they cause are important for the evolution of cells and organisms.

References

General

Judson, H.F. The Eighth Day of Creation: Makers of the Revolution in Biology. New York: Simon & Schuster, 1979.

Lewin, B. Genes IV. Oxford, NY: Oxford University Press, 1990.

Stent, G.S.; Calendar, R. Genetics: An Introductory Narrative, 2nd ed. San Francisco: W.H. Freeman, 1978.

Watson, J.D.; Hopkins, N.H.; Roberts, J.W.; Steitz, J.A.; Weiner, A.M. Molecular Biology of the Gene, 4th ed. Menlo Park, CA: Benjamin-Cummings, 1987.

Cited

1. Chamberlin, M. Bacterial DNA-dependent RNA polymerases. In The Enzymes, 3rd ed., Vol. 15B (P. Boyer, ed.), pp. 61–108. New York: Academic Press, 1982.

   Gralla, J.D. Promoter recognition and mRNA initiation by *Escherichia coli* Eσ⁷⁰. *Methods Enzymol.* 185:37–54, 1990.

   Kerppola, T.K.; Kane, C.M. RNA polymerase: regulation of transcript elongation and termination. *FASEB J.* 5(13):2833–2842, 1991.

   Sentenac, A. Eukaryotic RNA polymerases. *Crc Crit. Rev. Biochem.* 18(1):31–90, 1985.

2. Collado-Vides, J.; Magasanik, B.; Gralla, J.D. Control site location and transcriptional regulation in *Escherichia coli. Microbiol. Rev.* 55(3):371–394, 1991.

   Murphy, S.; Moorefield, B.; Pieler, T. Common mechanisms of promoter recognition by RNA polymerases II and III. *Trends Genet.* 5(4):122–126, 1989.

   Travers, A.A. Structure and function of E. coli promoter DNA. *Crc Crit. Rev. Biochem.* 22(3):181–219, 1987.

3. McClain, W.H. Transfer RNA identity. *FASEB J.* 7(1):72–78, 1993.

   Rich, A.; Kim, S.H. The three-dimensional structure of transfer RNA. *Sci. Am.* 238(1):52–62, 1978.

   Sprinzl, M.; Hartmann, T.; Weber, J.; Blank, J.; Zeidler, R. Compilation of tRNA sequences and sequences of tRNA genes. *Nucleic Acids Res.* 17 Suppl.:1–172, 1989.

4. Cavarelli, J.; Moras, D. Recognition of tRNAs by aminoacyl-tRNA synthetases. *FASEB J.* 7(1):79–86, 1993.

   Freist, W. Mechanisms of aminoacyl-tRNA synthetases: a critical consideration of recent results. *Biochemistry* 28(17):6787–6795, 1989.

   Schimmel, P. Aminoacyl tRNA synthetases: general scheme of structure-function relationships in the polypeptides and recognition of transfer RNAs. *Annu. Rev. Biochem.* 56:125–158, 1987.

5. Crick, F.H.C. The genetic code. III. *Sci. Am.* 215(4):55–62, 1966.

   The Genetic Code. *Cold Spring Harbor Symp. Quant. Biol.*, Vol. 31, 1966.

   Wong, J.T. Evolution of the genetic code. *Microbiol. Sci.* 5(6):174–181, 1988.

6. Brimacombe, R. RNA-protein interactions in the *Escherichia coli* ribosome. *Biochimie* 73(7–8):927–936, 1991.

   Stern, S.; Powers, T.; Changchien, L.-M.; Noller, H.F. RNA-protein interactions in 30S ribosomal subunits: folding and function of 16S rRNA. *Science* 244(4906):783–790, 1989.

   Yonath, A.; Wittman, H.G. Challenging the three-dimensional structure of ribosomes. *Trends Biochem. Sci.* 14(8):329–335, 1989.

7. Nierhaus, K.H. The allosteric three-site model for the ribosomal elongation cycle: features and future. *Biochemistry* 29(21):4997–5008, 1990.

   Noller, H.F. Peptidyl transferase: protein, ribonucleoprotein, or RNA? *J. Bacteriol.* 175(17):5297–5300, 1993.

   Watson, J.D. The involvement of RNA in the synthesis of proteins. *Science* 140:17–26, 1963.

8. Craigen, W.J.; Lee, C.C.; Caskey, C.T. Recent advances in peptide chain termination. *Mol. Microbiol.* 4(6):861–865, 1990.

   Tate, W.P.; Brown, C.M. Translational termination: "stop" for protein synthesis or "pause" for regulation of gene expression. *Biochemistry* 31(9):2443–2450, 1992.

9. Gualerzi, C.O.; Pon, C.L. Initiation of mRNA translation in prokaryotes. *Biochemistry* 29(25):5881–5889, 1990.

Hunt, T. The initiation of protein synthesis. *Trends Biochem. Sci.* 5:178–181, 1980.

Merrick, W.C. Overview: mechanism of translation initiation in eukaryotes. *Enzyme* 44(1–4):7–16, 1990.

10. Jacques, N.; Dreyfus, M. Translation initiation in *Escherichia coli*: old and new questions. *Mol. Microbiol.* 4(7):1063–1067, 1990.

Kozak, M. Comparison of initiation of protein synthesis in procaryotes, eucaryotes, and organelles. *Microbiol. Rev.* 47:1–45, 1983.

Kozak, M. Structural features in eukaryotic mRNAs that modulate the initiation of translation. *J. Biol. Chem.* 266(30):19867–19870, 1991.

Yoon, H.; Donohue, T.F. Control of translation initiation in *Saccharomyces cerevisiae. Mol. Microbiol.* 6(11):1413–1419, 1992.

11. Hesketh, J.E.; Pryme, I.F. Interaction between mRNA, ribosomes and the cytoskeleton. *Biochem. J.* 277(Pt 1):1–10, 1991.

Rich, A. Polyribosomes. *Sci. Am.* 209(6):44–53, 1963.

Ryazanov, A.G.; Ovchinnikov, L.P.; Spirin, A.S. Development of structural organization of protein-synthesizing machinery from prokaryotes to eukaryotes. *Biosystems* 20(3):275–288, 1987.

12. Hershey, J.W.B. Overview: phosphorylation and translation control. *Enzyme* 44(1–4):17–27, 1990.

Lindahl, L.; Hinnebusch, A. Diversity of mechanisms in the regulation of translation in prokaryotes and lower eukaryotes. *Curr. Opin. Genet. Dev.* 2(5):720–726, 1992.

Merrick, W.C. Mechanism and regulation of eukaryotic protein synthesis. *Microbiol. Rev.* 56(2):291–315, 1992.

13. Riis, B.; Rattan, S.I.; Clark, B.F.; Merrick, W.C. Eukaryotic protein elongation factors. *Trends Biochem. Sci.* 15(11):420–424, 1990.

Soll, D. The accuracy of aminoacylation—ensuring the fidelity of the genetic code. *Experientia* 46(11–12):1089–1096, 1990.

Thompson, R.C. EFTu provides an internal kinetic standard for translational accuracy. *Trends Biochem. Sci.* 13:91–93, 1988.

14. Jiménez, A. Inhibitors of translation. *Trends Biochem. Sci.* 1:28–30, 1988.

Lord, J.M.; Hartley, M.R.; Roberts, L.M. Ribosome inactivating proteins of plants. *Semin. Cell Biol.* 2(1):15–22, 1991.

Perentesis, J.P.; Miller, S.P.; Bodley, J.W. Protein toxin inhibitors of protein synthesis. *Biofactors* 3(3):173–184, 1992.

15. Campbell, J.H. An RNA replisome as the ancestor of the ribosome. *J. Mol. Evol.* 32(1):3–5, 1991.

Lake, J.A. Evolving ribosome structure: domains in archaebacteria, eubacteria, eocytes and eukaryotes. *Annu. Rev. Biochem.* 54:507–530, 1985.

Orgel, L.E. RNA catalysis and the origin of life. *J. Theor. Biol.* 123:127–149, 1983.

16. Barnes, D.E.; Lindahl, T.; Sedgwick, B. DNA repair. *Curr. Opin. Cell Biol.* 5(3):424–433, 1993.

Friedberg, E.C. DNA Repair. San Francisco: W.H. Freeman, 1985.

Wevrick, R.; Buchwald, M. Mammalian DNA-repair genes. *Curr. Opin. Genet. Dev.* 3(3):470–474, 1993.

17. Behe, M.J. Histone deletion mutants challenge the molecular clock hypothesis. *Trends Biochem. Sci.* 15(10):374–376, 1990.

Wilson, A.C.; Carlson, S.S.; White, T.J. Biochemical Evolution. *Annu. Rev. Biochem.* 46:573–639, 1977.

Wilson, A.C.; Ochman, H.; Prager, E.M. Molecular time scale for evolution. *Trends Genet.* 3:241–247, 1987.

18. Drake, J.W. Comparative rates of spontaneous mutation. *Nature* 221:1132, 1969.

Drake, J.W. Spontaneous mutation. *Annu. Rev. Genet.* 25:125–146, 1991.

19. Ohta, T.; Kimura, M. Functional organization of genetic material as a product of molecular evolution. *Nature* 233:118–119, 1971. (The mutation rate limits the maximum number of genes in an organism.)

Smith, K.C. Spontaneous mutagenesis: experimental, genetic and other factors. *Mutat. Res.* 277(2):139–162, 1992.

Vulliamy, T.; Mason, P.; Luzzatto, L. The molecular basis of glucose-6-phosphate dehydrogenase deficiency. *Trends Genet.* 8(4):138–143, 1992.

20. Loomis, W.F. Similarities in eukaryotic genomes. *Comp. Biochem. Physiol.* 95B(1):21–27, 1990.

21. Lindahl, T. Instability and decay of the primary structure of DNA. *Nature* 362(6422):709–715, 1993.

Schrodinger, E. What Is Life? Cambridge, UK: Cambridge University Press, 1945.

22. Kornberg, A.; Baker, T.A. DNA Replication, 2nd ed. New York: W.H. Freeman, 1992. (Chapters 4 and 6 describe DNA polymerases; Chapter 9 describes DNA ligase; Chapter 21 covers the enzymology of DNA repair.)

23. Hoeijmakers, J.H. Nucleotide excision repair I: from *E. coli* to yeast. *Trends Genet.* 9(5):173–177, 1993.

Hoeijmakers, J.H. Nucleotide excision repair II: from yeast to mammals. *Trends Genet.* 9(6):211–217, 1993.

Sancar, A.Z.; Sancar, G.B. DNA repair enzymes. *Annu. Rev. Biochem.* 57:29–67, 1988.

24. Elledge, S.J.; Zhou, Z.; Allen, J.B. Ribonucleotide reductase: regulation, regulation, regulation. *Trends Biochem. Sci.* 17(3):119–123, 1992.

Walker, G.C. Inducible DNA repair systems. *Annu. Rev. Biochem.* 54:425–457, 1985.

Witkin, E.M. RecA protein in the SOS response: milestones and mysteries. *Biochimie* 73(2–3):133–141, 1991.

25. Perutz, M.F. Frequency of abnormal human haemoglobins caused by C→T transitions in CpG dinucleotides. *J. Mol. Biol.* 213(2):203–206, 1990.

26. Baker, T.A.; Wickner, S.H. Genetics and enzymology of DNA replication in *Escherichia coli. Annu. Rev. Genet.* 26:447–477, 1992.

Kornberg, A.; Baker, T.A. DNA Replication, 2nd ed. New York: W.H. Freeman, 1992.

So, A.G.; Downey, K.M. Eukaryotic DNA replication. *Crit. Rev. Biochem. Mol. Biol.* 27(1–2):129–155, 1992.

27. Linn, S. How many pols does it take to replicate nuclear DNA? *Cell* 66(2):185–187, 1991.

Meselsohn, M.; Stahl, F.W. The replication of DNA in *E. coli. Proc. Natl. Acad. Sci. USA* 44:671–682, 1958.

Young, M.C.; Reddy, M.K.; von Hippel, P.H. Structure and function of the bacteriophage T4 DNA polymerase holoenzyme. *Biochemistry* 31(37):8675–8690, 1992.

28. Inman, R.B.; Schnos, M. Structure of branch points in replicating DNA: presence of single-stranded connections in lambda DNA branch points. *J. Mol. Biol.* 56:319–325, 1971.

Ogawa, T.; Okazaki, T. Discontinuous DNA replication. *Annu. Rev. Biochem.* 49:421–457, 1980.

Thommes, P.; Hubscher, U. Eukaryotic DNA replication. Enzymes and proteins acting at the fork. *Eur. J. Biochem.* 194(3):699–712, 1990.

29. Echols, H.; Goodman, M.F. Fidelity mechanisms in DNA replication. *Annu. Rev. Biochem.* 60:477–511, 1991.

Fersht, A.R. Enzymatic editing mechanisms in protein synthesis and DNA replication. *Trends Biochem. Sci.* 5:262–265, 1980.

Goodman, M.F.; Creighton, S.; Bloom, L.B.; Petruska, J. Biochemical basis of DNA replication fidelity. *Crit. Rev. Biochem. Mol. Biol.* 28(2):83–126, 1993.

30. Crouch, R.J. Ribonuclease H: from discovery to 3D structure. *New Biol.* 2(9):771–777, 1990.

Kaguni, L.S; Lehman, I.R. Eukaryotic DNA polymerase-primase: structure, mechanism and function. *Biochim. Biophys. Acta* 950(2):87–101, 1988.

Rowen, L.; Kornberg, A. Primase, the dnaG protein of *Escherichia coli*: an enzyme which starts DNA chains. *J. Biol. Chem.* 253:758–764, 1978.

31. Lohman, T.M. Helicase-catalyzed DNA unwinding. *J. Biol. Chem.* 268(4):2269–2272, 1993.

Meyer, R.R.; Laine, P.S. The single-stranded DNA-binding protein of *Escherichia coli*. *Microbiol. Rev.* 54(4):342–380, 1990.

Thommes, P.; Hubscher, U. Eukaryotic DNA helicases: essential enzymes for DNA transactions. *Chromosoma* 101(8):467–473, 1992.

32. Kong, X.-P.; Onrust, R.; O'Donnell, M.; Kuriyan, J. Three-dimensional structure of the beta subunit of *E. coli* DNA polymerase III holoenzyme: a sliding DNA clamp. *Cell* 69(3):425–437, 1992.

33. Alberts, B.M. Protein machines mediate the basic genetic processes. *Trends Genet.* 1:26–30, 1985.

Nossal, N.G. Protein-protein interactions at a DNA replication fork: bacteriophage T4 as a model. *FASEB J.* 6(3):871–878, 1992.

Stukenberg, P.T; Studwell-Vaughan, P.S.; O'Donnell, M. Mechanism of the sliding beta-clamp of DNA polymerase III holoenzyme. *J. Biol. Chem.* 266(17):11328–11334, 1991.

34. Barras, F.; Marinus, M.G. The great GATC: DNA methylation in *E. coli*. *Trends Genet.* 5(5):139–143, 1989.

Heywood, L.A.; Burke, J.F. Mismatch repair in mammalian cells. *Bioessays* 12(10):473–477, 1990.

Modrich, P. Methyl-directed DNA mismatch correction. *J. Biol. Chem.* 264(12):6597–6600, 1989.

35. Echols, H. Nucleoprotein structures initiating DNA replication, transcription, and site-specific recombination. *J. Biol. Chem.* 265(25):14697–14700, 1990.

Georgopoulos, C. The *E. coli* dnaA initiation protein, a protein for all seasons. *Trends Genet.* 5(10):319–321, 1989.

Salas, M. Protein-priming of DNA replication. *Annu. Rev. Biochem.* 60:39–71, 1991.

36. Drlica, K. Bacterial topoisomerases and the control of DNA supercoiling. *Trends Genet.* 6(12):433–437, 1990.

Sternglanz, R. DNA topoisomerases. *Curr. Opin. Cell Biol.* 1(3):533–535, 1989.

Wang, J.C. DNA topoisomerases: why so many? *J. Biol. Chem.* 266(11):6659–6662, 1991.

37. Gruss, C.; Sogo, J.M. Chromatin replication. *Bioessays* 14(1):1–8, 1992.

Wang, T.S.-F. Eukaryotic DNA polymerases. *Annu. Rev. Biochem.* 60:513–552, 1991.

38. Roeder, G.S. Chromosome synapsis and genetic recombination: their roles in meiotic chromosome segregation. *Trends Genet.* 6(12):385–389, 1990.

Sadowski, P.D. Site-specific genetic recombination: hops, flips and flops. *FASEB J.* 7(9):760–767, 1993.

Whitehouse, H.L.K. Genetic Recombination: Understanding the Mechanisms. New York: Wiley, 1982.

39. Camerini-Otero, R.D.; Hsieh, P. Parallel DNA triplexes, homologous recombination, and other homology-dependent DNA interactions. *Cell* 73(2):217–223, 1993.

Smith, G.R. Homologous recombination in *E. coli*: multiple pathways for multiple reasons. *Cell* 58(5):807–809, 1989.

West, S.C. Enzymes and molecular mechanisms of genetic recombination. *Annu. Rev. Biochem.* 61:603–640, 1992.

40. Lloyd, R.G.; Sharples, G.J. Genetic analysis of recombination in prokaryotes. *Curr. Opin. Genet. Dev.* 2(5):683–690, 1992.

Lohman, T.M. *Escherichia coli* DNA helicases: mechanisms of DNA unwinding. *Mol. Microbiol.* 6(1):5–14, 1992.

Weinstock, G.M. General recombination in *Escherichia coli*. In *Escherichia coli* and *Salmonella Typhimurium*: Cellular and Molecular Biology (F.C. Neidhardt, ed.), pp. 1034–1043. Washington, DC: American Society for Microbiology, 1987.

41. Bradley, S.G. DNA reassociation and base composition. *Society for Applied Bacteriology Symposium Series* 8:11–26, 1980.

Gotoh, O. Prediction of melting profiles and local helix stability for sequenced DNA. *Adv. Biophys.* 16:1–52, 1983.

Wetmur, J.G.; Davidson, N. Kinetics of renaturation of DNA. *J. Mol. Biol.* 31:349–370, 1968.

42. Eggleston, A.K.; Kowalczykowski, S.C. An overview of homologous pairing and DNA strand exchange proteins. *Biochimie* 73(2–3):163–176, 1991.

Kowalczykowski, S.C. Biochemical and biological function of *Escherichia coli* RecA protein: behavior of mutant RecA proteins. *Biochimie* 73(2–3):289–304, 1991.

Roca, A.I., Cox, M.M. The RecA protein: structure and function. *Crit. Rev. Biochem. Mol. Biol.* 25(6):415–456, 1990.

43. Holliday, R. The history of the DNA heteroduplex. *Bioessays* 12(3):133–142, 1990.

Kowalczykowski, S.C. Biochemistry of genetic recombination: energetics and mechanism of DNA strand exchange. *Annu. Rev. Biophys. Biophys. Chem.* 20:539–575, 1991.

Messelsohn, M.S.; Radding, C.M. A general model for genetic recombination. *Proc. Natl. Acad. Sci. USA* 72:358–361, 1975.

44. Fogel, S.; Mortimer, R.K.; Lusnak, K. Mechanisms of meiotic gene conversion or "wanderings on a foreign strand." In The Molecular Biology of the Yeast *Saccharomyces Cerevisiae*: Life Cycle and Inheritance (J.N. Strathern, ed.), pp. 289–340. Cold Spring Harbor, NY: Cold Spring Harbor Laboratory Press, 1981.

Kobayashi, I. Mechanisms for gene conversion and homologous recombination: the double-strand break repair model and the successive half crossing-over model. *Adv. Biophys.* 28:81–133, 1992.

Kourilsky, P. Molecular mechanisms of gene conversion in higher cells. *Trends Genet.* 2:60–62, 1986.

45. Radman, M. Mismatch repair and the fidelity of genetic recombination. *Genome* 31(1):68–73, 1989.

    Rayssiguier, C.; Thaler, D.S.; Radman, M. The barrier to recombination between *Escherichia coli* and *Salmonella typhimurium* is disrupted in mismatch repair mutants. *Nature* 342(6248):396–401, 1989.

46. Landy, A. Dynamic, structural, and regulatory aspects of lambda site-specific recombination. *Annu. Rev. Biochem.* 58:913–949, 1989.

    O'Gorman, S.; Fox, D.T.; Wahl, G.M. Recombinase-mediated gene activation and site-specific integration in mammalian cells. *Science* 251(4999):1351–1355, 1991.

    Stark, W.M.; Boocock, M.R.; Sherratt, D.J. Catalysis by site-specific recombinases. *Trends Genet.* 8(12):432–439, 1992.

47. Mizuuchi, K. Transpositional recombination: mechanistic insights from studies of mu and other elements. *Annu. Rev. Biochem.* 61:1011–1051, 1992.

48. Borg, D.E.; Howe, M.M., ed. Mobile DNA. Washington, DC: American Society for Microbiology, 1989.

    Joklik, W.K., ed. Virology, 2nd ed. Norwalk, CT: Appleton & Lange, 1985.

    Levine, A. Viruses. New York: Scientific American Library, 1992.

49. Hershey, A.D.; Chase, M. Independent functions of viral protein and nucleic acid in growth of bacteriophage. *J. Gen. Physiol.* 36:39–56, 1952.

    Brock, T.D. The Emergence of Bacterial Genetics. Cold Spring Harbor, NY: Cold Spring Harbor Laboratory Press, 1990. (Chapter 6 provides a historical overview of bacteriophage research.)

50. Fields, B.N., ed. Virology. New York: Raven Press, 1985. (Chapters 3 and 4 discuss virus capsid structure and viral membranes, respectively.)

    Simons, K.; Garoff, H.; Helenius, A. How an animal virus gets into and out its host cell. *Sci. Am.* 246(2):58–66, 1982.

51. Fields, B.N., ed. Virology. New York: Raven Press, 1985. (Chapter 5 summarizes types of viral genomes.)

    Gierer, A.; Schramm, G. Infectivity of ribonucleic acid from tobacco mosaic virus. *Nature* 177:702–703, 1956.

    Strauss, E.G.; Strauss, J.H. RNA viruses: genome structure and evolution. *Curr. Opin. Genet. Dev.* 1(4):485–493, 1991.

52. Borowiec, J.A.; Dean, F.B.; Bullock, P.A.; Hurwitz, J. Binding and unwinding—how T antigen engages the SV40 origin of DNA replication. *Cell* 60(2):181–184, 1990.

    Carlson, K.; Overvatn, A. Bacteriophage T4 endonucleases II and IV, oppositely affected by dCMP hydroxymethylase activity, have different roles in the degradation and in the RNA polymerase-dependent replication of T4 cytosine containing DNA. *Genetics* 114(3):669–685, 1986.

    Cohen, S.S. Virus-induced Enzymes. New York: Columbia University Press, 1968.

53. David, C.; Gargouri-Bouzid, R.; Haenni, A.-L. RNA replication of plant viruses containing an RNA genome. *Prog. Nucleic Acid Res. Mol. Biol.* 42:157–227, 1992.

Kornberg, A.; Baker, T.A. DNA Replication, 2nd ed. New York: W.H. Freeman, 1992. (Chapters 17 and 19 describe the replication of DNA phages and animal viruses.)

54. Kielian, M.; Jungerwirth, S. Mechanisms of enveloped virus entry into cells. *Mol. Biol. Med.* 7(1):17–31, 1990.

    White, J.M. Membrane fusion. *Science* 258(5084):917–924, 1992.

    Zhao, H.; Garoff, H. Role of cell surface spikes in alphavirus budding. *J. Virol.* 66(12):7089–7095, 1992.

55. Griffiths, G.; Rottier, P. Cell biology of viruses that assemble along the biosynthetic pathway. *Semin. Cell Biol.* 3(5):367–381, 1992.

56. Campbell, A.M. Thirty years ago in genetics: prophage insertion into bacterial chromosomes. *Genetics* 133(3):433–438, 1993.

    Friedman, D.I. Interaction between bacteriophage lambda and its *Escherichia coli* host. *Curr. Opin. Genet. Dev.* 2(5):727–738, 1992.

57. Tooze, J. DNA Tumor Viruses. Molecular Biology of Tumor Viruses, 2nd ed., Part 2. Cold Spring Harbor, NY: Cold Spring Harbor Laboratory Press, 1980. (Chapter 4 describes transformation by SV40 and polyoma viruses.)

    zur Hausen, H. Viruses in human cancers. *Science* 254(5035):1167–1173, 1991.

58. Baltimore, D. Viral RNA-dependent DNA polymerase. *Nature* 226:1209–1211, 1970.

    Varmus, H. Retroviruses. *Science* 240:1427–1435, 1988.

    Whitcomb, J.M.; Hughes, S.H. Retroviral reverse transcription and integration: progress and problems. *Annu. Rev. Cell Biol.* 8:275–306, 1992.

59. Gallo, R.C.; Montagnier, L. The chronology of AIDS research. *Nature* 326(6112):435–436, 1987.

60. Boeke, J.D.; Chapman, K.B. Retrotransposition mechanisms. *Curr. Opin. Cell Biol.* 3(3):502–507, 1991.

    Corces, V.G.; Geyer, P.K. Interactions of retrotransposons with the host genome: the case of the gypsy element of *Drosophila*. *Trends Genet.* 7(3):86–90, 1991.

    Sandmeyer, S.B. Yeast retrotransposons. *Curr. Opin. Genet. Dev.* 2(5):705–711, 1992.

61. Gierl, A.; Frey, M. Eukaryotic transposable elements with short terminal inverted repeats. *Curr. Opin. Genet. Dev.* 1(4):494–497, 1991.

    Haniford, D.B.; Chaconas, G. Mechanistic aspects of DNA transposition. *Curr. Opin. Genet. Dev.* 2(5):698–704, 1992.

    Mizuuchi, K. Mechanism of transposition of bacteriophage mu: polarity of the strand transfer reaction at the initiation of transposition. *Cell* 39:395–404, 1984.

62. Levine, A. Viruses. New York: Scientific American Library, 1992. (Chapter 10 discusses the evolution of viruses.)

    Novick, R.P. Plasmids. *Sci. Am.* 243(6):102–127, 1980.

    Symons, R.H. The intriguing viroids and virusoids: what is their information content and how did they evolve? *Mol. Plant-Microbe Interact.* 4(2):111–121, 1991.

Recombinant DNA Technology

Until the early 1970s DNA was the most difficult cellular molecule for the biochemist to analyze. Enormously long and chemically monotonous, the nucleotide sequence of DNA could be approached only by indirect means, such as by protein or RNA sequencing or by genetic analysis. Today the situation has changed entirely. From being the most difficult macromolecule of the cell to analyze, DNA has become the easiest. It is now possible to excise a specific region of DNA, to produce a virtually unlimited number of copies of it, and to determine the sequence of its nucleotides at a rate of hundreds of nucleotides a day. By variations of the same techniques, an isolated gene can be altered (engineered) at will and transferred back into cells in culture. With more difficulty, the redesigned gene can be inserted into the germ line of an animal or plant, so as to become a functional and heritable part of the organism's genome.

These technical breakthroughs have had a dramatic impact on all aspects of cell biology by allowing the study of cells and their macromolecules in previously unimagined ways. They have led to the discovery of whole new classes of genes and proteins and have revealed that many proteins have been much more highly conserved in evolution than had been suspected. They have provided new means to determine the functions of proteins and of individual domains within proteins, revealing a host of unexpected relationships between them. By making available large amounts of any protein, they have shown the way to efficient mass production of protein hormones and vaccines. Finally, by allowing the regulatory regions of genes to be dissected, they have provided biologists with an important tool for unraveling the complex mechanisms by which eucaryotic gene expression is regulated.

Recombinant DNA technology comprises a mixture of techniques, some new and some borrowed from other fields such as microbial genetics (Table 7–1). The most important of these techniques are the following:

1. Cleavage of DNA at specific sites by *restriction nucleases,* which greatly facilitates the isolation and manipulation of individual genes.

2. Rapid *sequencing* of all the nucleotides in a purified DNA fragment, which makes it possible to determine the boundaries of a gene and the amino acid sequence it encodes.

3. *Nucleic acid hybridization,* which makes it possible to find a specific sequence of DNA or RNA with great accuracy and sensitivity on the basis of its ability to bind a complementary nucleic acid sequence.

4. *DNA cloning*, whereby a single DNA molecule can be copied to generate many billions of identical molecules.

5. *DNA engineering*, by which DNA sequences are altered to make modified versions of genes, which are reinserted back into cells or organisms.

In this chapter we explain how recombinant DNA technology has generated the new experimental approaches that have revolutionized cell biology.

The Fragmentation, Separation, and Sequencing of DNA Molecules [1]

Before the 1970s the goal of isolating a single gene from a large chromosome seemed unattainable. Unlike a protein, a gene does not exist as a discrete entity in cells, but rather as a small region of a much larger DNA molecule. Although the DNA molecules in a cell can be randomly broken into small pieces by mechanical force, a fragment containing a single gene in a mammalian genome would still be only one among a hundred thousand or more DNA fragments, indistinguishable in their average size. How could such a gene be purified? Since all DNA molecules consist of an approximately equal mixture of the same four nucleotides, they cannot be readily separated, as proteins can, on the basis of their different charges and binding properties. Moreover, even if a purification scheme could be devised, vast amounts of DNA would be needed to yield enough of any particular gene to be useful for further experiments.

The solution to all of these problems began to emerge with the discovery of *restriction nucleases*. These enzymes, which can be purified from bacteria, cut the DNA double helix at specific sites defined by the local nucleotide sequence, producing double-stranded DNA fragments of strictly defined sizes. Different species of bacteria make restriction nucleases with different sequence specificities, and it is relatively simple to find a restriction nuclease that will create a DNA fragment that includes a particular gene. The size of the DNA fragment can then

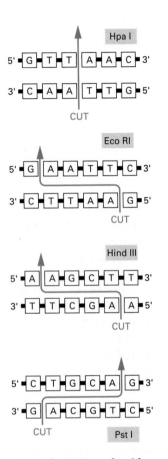

Figure 7–1 The DNA nucleotide sequences recognized by four widely used restriction nucleases. As in the examples shown, such sequences are often six base pairs long and "palindromic" (that is, the nucleotide sequence is the same if the helix is turned by 180 degrees around the center of the short region of helix that is recognized). The enzymes cut the two strands of DNA at or near the recognition sequence. For some enzymes, such as Hpa I, the cleavage leaves blunt ends; for others, such as Eco RI, Hind III, and Pst I, the cleavage is staggered and creates cohesive ends. Restriction nucleases are obtained from various species of bacteria: Hpa I is from *Hemophilus parainfluenzae*, Eco RI is from *Escherichia coli*, Hind III is from *Hemophilus influenzae*, and Pst I is from *Providencia stuartii*.

Table 7–1 **Some Major Steps in the Development of Recombinant DNA Technology**

1869	**Miescher** isolated DNA for the first time.
1944	**Avery** provided evidence that DNA, rather than protein, carries the genetic information during bacterial transformation.
1953	**Watson and Crick** proposed the double-helix model for DNA structure based on x-ray results of **Franklin and Wilkins.**
1957	**Kornberg** discovered DNA polymerase, the enzyme now used to produce labeled DNA probes.
1961	**Marmur and Doty** discovered DNA renaturation, establishing the specificity and feasibility of nucleic acid hybridization reactions.
1962	**Arber** provided the first evidence for the existence of DNA restriction nucleases, leading to their later purification and use in DNA sequence characterization by **Nathans and H. Smith.**
1966	**Nirenberg, Ochoa,** and **Khorana** elucidated the genetic code.
1967	**Gellert** discovered DNA ligase, the enzyme used to join DNA fragments together.
1972–1973	DNA cloning techniques were developed by the laboratories of **Boyer, Cohen, Berg,** and their colleagues at Stanford University and the University of California at San Francisco.
1975	**Southern** developed gel-transfer hybridization for the detection of specific DNA sequences.
1975–1977	**Sanger and Barrell** and **Maxam and Gilbert** developed rapid DNA-sequencing methods.
1981–1982	**Palmiter and Brinster** produced transgenic mice; **Spradling and Rubin** produced transgenic fruit flies.
1985	**Mullis** and co-workers invented the polymerase chain reaction (PCR).

be used as a basis for partially purifying the gene from a mixture. Most important, the DNA fragment usually serves as the starting material for the production of the highly purified gene in unlimited amounts by *DNA cloning.*

We begin this section by discussing how restriction nucleases are used to produce specific DNA fragments and how these (and other) DNA molecules are separated according to their size. We then explain how, after the purification and amplification of a DNA fragment by DNA cloning, this DNA can be *sequenced* to determine the order of its nucleotides.

Restriction Nucleases Hydrolyze DNA Molecules at Specific Nucleotide Sequences [2]

Many bacteria make **restriction nucleases,** which protect the bacterial cell from viruses by degrading the viral DNA. Each such nuclease recognizes a specific sequence of four to eight nucleotides in DNA. These sequences, where they occur in the genome of the bacterium itself, are protected from cleavage by methylation at an A or a C residue; where the sequences occur in foreign DNA, they are generally not methylated and so are cleaved by the restriction nucleases. Large numbers of restriction nucleases have been purified from various species of bacteria; more than 100, most of which recognize different nucleotide sequences, are now available commercially.

Some restriction nucleases produce staggered cuts, which leave short single-stranded tails at the two ends of each fragment (Figure 7–1). Ends of this type are known as *cohesive ends,* as each tail can form complementary base pairs with the tail at any other end produced by the same enzyme (Figure 7–2). The cohesive ends generated by restriction enzymes allow any two DNA fragments to be easily joined together, as long as the fragments were generated with the same restriction nuclease (or with another nuclease that produces the same cohesive ends). DNA molecules produced by splicing together two or more DNA fragments in this way are called *recombinant DNA molecules;* they have made possible many new types of cell biological studies.

Restriction Maps Show the Distribution of Short Marker Nucleotide Sequences Along a Chromosome [3]

A particular restriction nuclease will cut any double-helical DNA molecule extracted from a cell into a series of specific DNA fragments (known as *restriction fragments*). By comparing the sizes of the DNA fragments produced from a particular genetic region after treatment with a combination of different restriction nucleases, a **restriction map** of that region can be constructed showing the location of each cutting (restriction) site in relation to neighboring restriction sites (Figure 7–3). The different short DNA sequences recognized by different restriction nucleases serve as convenient markers, and the restriction map reflects their arrangement in the region. This allows one to compare the same region of DNA in different individuals (by comparing their restriction maps) without having to determine the nucleotide sequences in detail. By comparing the restriction maps illustrated in Figure 7–4, for example, we know that the chromosomal regions that code for hemoglobin chains in humans and various other primates have remained largely unchanged during the 5 to 10 million years since these species first diverged. Restriction maps are also used in DNA cloning and DNA engineering, where they make it possible to locate a gene of interest on a particular restriction fragment and thus facilitate its isolation.

Gel Electrophoresis Separates DNA Molecules of Different Sizes [4]

In the early 1970s it was found that the length and purity of DNA molecules could be accurately determined by the same types of gel electrophoresis methods that

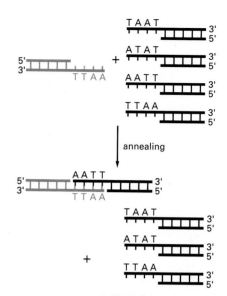

Figure 7–2 Many kinds of restriction nucleases produce DNA fragments with cohesive ends. DNA fragments with the same cohesive ends can readily join by complementary base-pairing between their cohesive ends as illustrated. The two DNA fragments that join in this example were both produced by the Eco RI restriction nuclease, whereas the three other fragments were produced by different restriction nucleases that generate different cohesive ends (see Figure 7–1).

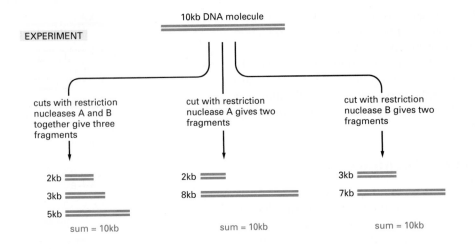

EXPERIMENT

10kb DNA molecule

cuts with restriction nucleases A and B together give three fragments

2kb
3kb
5kb

sum = 10kb

cut with restriction nuclease A gives two fragments

2kb
8kb

sum = 10kb

cut with restriction nuclease B gives two fragments

3kb
7kb

sum = 10kb

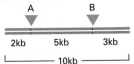

CONCLUSION: enzyme A cuts near one end of the molecule. Enzyme B must cut either near the same end or near the other end. The size of the fragments produced by both enzymes acting together rules out the first alternative and leads to the unambiguous order of restriction nuclease cutting sites shown below

A   B

2kb   5kb   3kb

10kb

had proved so useful in the analysis of proteins. The procedure is actually simpler than for proteins: because each nucleotide in a nucleic acid molecule already carries a single negative charge, there is no need to add the negatively charged detergent SDS that is required to make protein molecules move uniformly toward the positive electrode. For DNA fragments less than 500 nucleotides long, specially designed polyacrylamide gels allow molecules that differ in length by as little as a single nucleotide to be separated from each other (Figure 7–5A). The pores in polyacrylamide gels, however, are too small to permit very large DNA molecules to pass; to separate these by size, the much more porous gels formed by dilute solutions of agarose (a polysaccharide isolated from seaweed) are used (Figure 7–5B). These DNA separation methods are widely used for both analytical and preparative purposes.

Figure 7–3 Restriction mapping. A simple example illustrating how the positions of cutting sites for different restriction nucleases (known as *restriction sites*) are determined relative to one another on double-helical DNA molecules to create a *restriction map.* (kb = kilobases, an abbreviation designating either 1000 nucleotides or 1000 nucleotide pairs.)

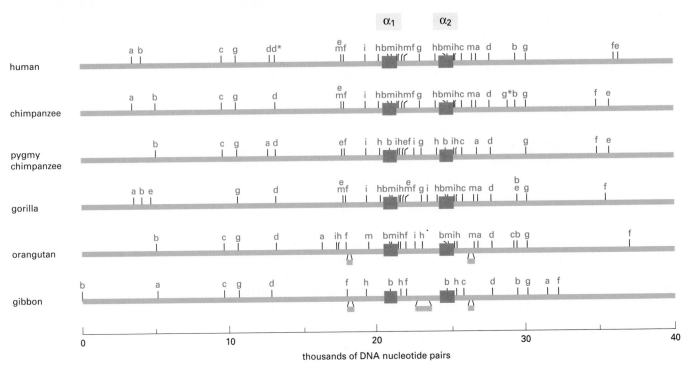

Figure 7–4 Restriction maps of human and various primate DNAs in a cluster of genes coding for hemoglobin. The two *red squares* in each map indicate the positions of the DNA corresponding to the two α-globin genes. Each letter stands for a site cut by a different restriction nuclease. As in Figure 7–3, the location of each cut was determined by comparing the sizes of the DNA fragments generated by treating the DNAs with the various restriction nucleases, individually and in combinations. Note that the chimpanzee, which is most closely related to humans, has the most similar restriction map, whereas the gibbon is more distantly related and has the most diverged map—including three DNA insertions. (Courtesy of Elizabeth Zimmer and Allan Wilson.)

Figure 7–5 Gel electrophoresis techniques for separating DNA molecules by size. In the three examples shown, electrophoresis is from top to bottom, so that the largest—and so slowest-moving—DNA molecules are near the top of the gel. In (A) a polyacrylamide gel with small pores is used to fractionate single-stranded DNA. In the size range 10 to 500 nucleotides, DNA molecules that differ in size by only a single nucleotide can be separated from each other. In the example the four lanes represent sets of DNA molecules synthesized in the course of a DNA-sequencing procedure. The DNA to be sequenced has been artificially replicated from a fixed start site up to a variable stopping point, producing a set of partial replicas of differing lengths. (Figure 7–8 explains how such sets of partial replicas are synthesized.) Lane 1 shows all the partial replicas that terminate in a G; lane 2, all those that terminate in an A; lane 3, all those that terminate in a T; and lane 4, all those that terminate in a C. Since the DNA molecules used in these reactions are radiolabeled, their positions can be determined by autoradiography, as shown. In (B) an agarose gel with medium-sized pores is used to separate double-stranded DNA molecules. This method is most useful in the size range 300 to 10,000 nucleotide pairs. These DNA molecules are restriction fragments produced from the genome of a bacterial virus, and they have been detected by their fluorescence when stained with the dye ethidium bromide. In (C) the technique of pulsed-field agarose gel electrophoresis has been used to separate 16 different yeast (*Saccharomyces cerevisiae*) chromosomes that range in size from 220,000 to 2.5 million nucleotide pairs. The DNA was stained as in (B). DNA molecules as large as 10^7 nucleotide pairs can be separated in this way. (A, courtesy of Leander Lauffer and Peter Walter; B, courtesy of Ken Kreuzer; C, from D. Vollrath and R.W. Davis, *Nucleic Acids Res.* 15:7876, 1987, by permission of Oxford University Press.)

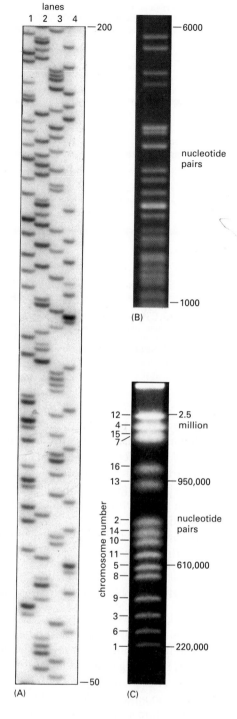

A variation of agarose gel electrophoresis, called *pulsed-field gel electrophoresis,* makes it possible to separate even extremely long DNA molecules. Ordinary gel electrophoresis fails to separate such molecules because the steady electric field stretches them out so that they travel end-first through the gel in snakelike configurations at a rate that is independent of their length. In pulsed-field gel electrophoresis, by contrast, the direction of the electric field is changed periodically, which forces the molecules to reorient before continuing to move snake-like through the gel. This reorientation takes much more time for larger molecules, so that progressively longer molecules move more and more slowly. As a consequence, even entire bacterial or yeast chromosomes separate into discrete bands in pulsed-field gels and so can be sorted and identified on the basis of their size (Figure 7–5C). Although a typical mammalian chromosome of 10^8 base pairs is too large to be sorted even in this way, large subregions of these chromosomes are readily separated and identified if the chromosomal DNA is first cut with a restriction nuclease selected to recognize sequences that occur only extremely rarely (once every 10^6 to 10^7 nucleotide pairs, for example).

The DNA bands on agarose or polyacrylamide gels are invisible unless the DNA is labeled or stained in some way. One sensitive method of staining DNA is to expose it to the dye *ethidium bromide,* which fluoresces under ultraviolet light when it is bound to DNA (see Figure 7–5B and C). An even more sensitive detection method involves incorporating a radioisotope into the DNA molecules before electrophoresis; ^{32}P is often used as it can be incorporated into DNA phosphates and emits an energetic β particle that is easily detected by autoradiography (Figure 7–5A).

Purified DNA Molecules Can Be Specifically Labeled with Radioisotopes or Chemical Markers *in Vitro* [5]

Two procedures are widely used to add distinct labels to isolated DNA molecules. In the first the DNA is copied by an *E. coli* enzyme, *DNA polymerase I,* in the presence of nucleotides that are either radioactive (usually labeled with ^{32}P) or

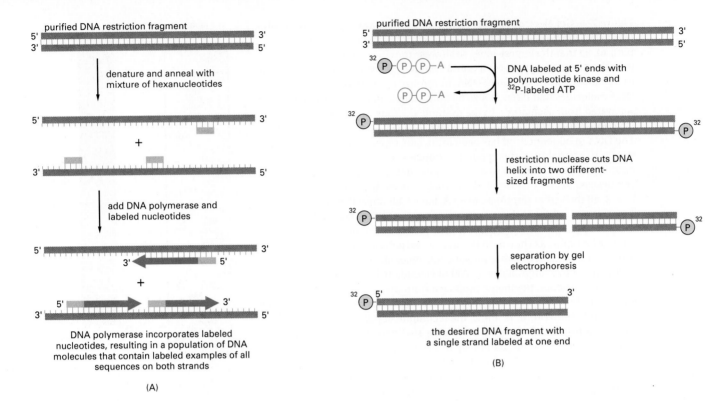

purified DNA restriction fragment

denature and anneal with
mixture of hexanucleotides

+

add DNA polymerase and
labeled nucleotides

+

DNA polymerase incorporates labeled
nucleotides, resulting in a population of DNA
molecules that contain labeled examples of all
sequences on both strands

(A)

purified DNA restriction fragment

DNA labeled at 5' ends with
polynucleotide kinase and
^{32}P-labeled ATP

restriction nuclease cuts DNA
helix into two different-
sized fragments

separation by gel
electrophoresis

the desired DNA fragment with
a single strand labeled at one end

(B)

chemically tagged (Figure 7–6A). In this way "DNA probes" containing many labeled nucleotides can be produced for nucleic acid hybridization reactions (see below). The second procedure uses the bacteriophage enzyme *polynucleotide kinase* to transfer a single ^{32}P-labeled phosphate from ATP to the 5' end of each DNA chain (Figure 7–6B). Because only one ^{32}P atom is incorporated by the kinase into each DNA strand, the DNA molecules labeled in this way are often not radioactive enough to be used as DNA probes; because they are labeled at only one end, however, they are invaluable for DNA sequencing and DNA footprinting, as we see next.

Isolated DNA Fragments Can Be Rapidly Sequenced [6]

In the late 1970s methods were developed that allow the nucleotide sequence of any purified DNA fragment to be determined simply and quickly. They have made it possible to determine the complete DNA sequences of thousands of genes, including those coding for such well-known proteins as insulin, hemoglobin, interferon, and cytochrome c. The volume of DNA sequence information is already so large (many tens of millions of nucleotides) that computers must be used to store and analyze it. Several continuous stretches of DNA sequence have been determined that each contain more than 10^5 nucleotide pairs; these include the entire genomes of the Epstein-Barr virus (which infects humans and causes infectious mononucleosis) and of a plant chloroplast, as well as an entire chromosome of yeast. Two powerful DNA sequencing methods were originally developed: the principle underlying the *chemical method* is illustrated in Figure 7–7, and the *enzymatic method* is explained in Figure 7–8. The latter method, which is based on *in vitro* DNA synthesis carried out in the presence of chain-terminating nucleoside triphosphates, has now become the standard procedure for sequencing DNA.

DNA sequencing methods are so rapid and reliable that the easiest and most accurate way to determine the amino acid sequence of a protein is to determine the nucleotide sequence of its gene: a cDNA clone is made from the appropriate mRNA (see below), its nucleotide sequence is determined, and the genetic

Figure 7–6 Two procedures are used routinely for making DNA molecules radioactive. (A) A purified DNA polymerase enzyme labels all the nucleotides in a DNA molecule and can thereby produce highly radioactive DNA probes. (B) Polynucleotide kinase labels only the 5' ends of DNA strands; therefore, when labeling is followed by restriction nuclease cleavage, as shown, DNA molecules containing a single 5'-end-labeled strand can be readily obtained. The method in (A) is also used to produce nonradioactive DNA molecules that carry a specific chemical marker that can be detected with an appropriate antibody (see Figure 7–18).

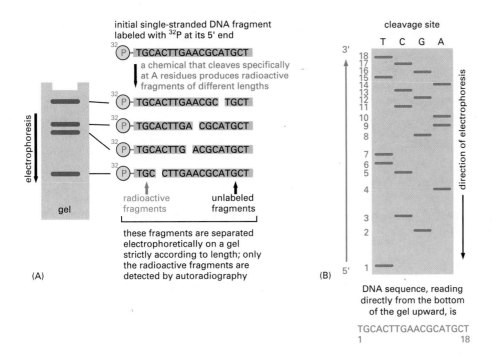

Figure 7–7 The chemical method for sequencing DNA. (A) The procedure starts with a set of identical end-labeled double-stranded DNA molecules produced by the method outlined in Figure 7–6B. In the first step the strands of the double helix are dissociated and exposed to mild treatment with a chemical that destroys one of the four bases (in this case, A residues) in the DNA. Because the treatment is mild, usually only one of the A residues in each molecule is destroyed at random. This generates a family of DNA fragments of different lengths, reflecting the different sites at which A residues occur in the original DNA. These fragments are separated on a gel and detected by autoradiography: only the fragments possessing a 5′-terminal 32P-phosphate group show up on the gel, and their sizes reveal the distances from the labeled end at which A residues occur. (B) To determine the full sequence, similar procedures are carried out simultaneously on four separate samples of the same 5′-end-labeled DNA molecule using chemicals that cleave DNA preferentially at T for the first sample, C for the second, G for the third, and A for the fourth. The resulting fragments are separated in parallel lanes of a gel like that shown in Figure 7–5A, giving a pattern of radioactive DNA bands from which the DNA sequence is read. The nucleotide closest to the 5′ end of the sequence is determined by looking across the gel at level 1 (at the bottom of the gel) and seeing in which lane a band appears (T). The same procedure is repeated for level 2, then level 3, and so on, to obtain the sequence. The method has been idealized here; the actual chemical treatments are less specific than shown.

Labels within figure:

initial single-stranded DNA fragment labeled with 32P at its 5′ end

a chemical that cleaves specifically at A residues produces radioactive fragments of different lengths

electrophoresis

gel

radioactive fragments

unlabeled fragments

these fragments are separated electrophoretically on a gel strictly according to length; only the radioactive fragments are detected by autoradiography

(A)

cleavage site

T C G A

direction of electrophoresis

(B)

DNA sequence, reading directly from the bottom of the gel upward, is

TGCACTTGAACGCATGCT
1   18

---

code is then used as a dictionary to convert this sequence back to an amino acid sequence. Although in principle there are six different reading frames in which a DNA sequence can be translated into protein (three on each strand), the correct one is generally recognizable as the only one lacking frequent *stop codons* (Figure 7–9). A limited amount of amino acid sequence can be determined from the purified protein to confirm the sequence determined from the DNA.

As techniques for DNA sequencing have improved, scientists have begun to envisage the possibility of determining the entire DNA sequence of a human genome of 3×10^9 nucleotides. Because this sequence specifies all of the possible RNA and protein molecules that are used to construct the body, knowledge of it will provide us with a "dictionary of the human being," which will greatly expedite future studies of human cells and tissues. DNA sequencing on such a scale will require the development of highly automated methods that will lower the current cost of DNA sequencing by at least fivefold.

DNA Footprinting Reveals the Sites Where Proteins Bind on a DNA Molecule [7]

A modification of the chemical method for DNA sequencing can be used to determine the nucleotide sequences recognized by DNA-binding proteins. Some of these proteins play a central part in determining which genes are active in a particular cell by binding to regulatory DNA sequences, which are usually located outside the coding regions of a gene. In analyzing how such a protein functions, it is important to identify the specific sequences to which it binds. A method used for this purpose is called **DNA footprinting.** First, a pure DNA fragment that is labeled at one end with 32P is isolated (see Figure 7–6B); this molecule is then cleaved with a nuclease or a chemical that makes random single-stranded cuts in the DNA. After the DNA molecule is denatured to separate its two strands, the resultant subfragments from the labeled strand are separated on a gel and detected by autoradiography. The pattern of bands from DNA cut in the presence of a DNA-binding protein is compared with that from DNA cut in its absence. When the protein is present, it covers the nucleotides at its binding site and protects their phosphodiester bonds from cleavage. As a result, the labeled fragments that terminate in the binding site will be missing, leaving a gap in the gel pattern called a "footprint" (Figure 7–10A). The footprint of a DNA-binding protein that activates the transcription of a eucaryotic gene is shown in Figure 7–10B.

(A)

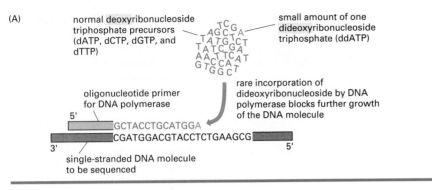

normal deoxyribonucleoside triphosphate precursors (dATP, dCTP, dGTP, and dTTP)

TCG
AGCGA
TATGCT
TATCGA
AACTCAT
GCCCAT
GTGGCT

small amount of one dideoxyribonucleoside triphosphate (ddATP)

rare incorporation of dideoxyribonucleoside by DNA polymerase blocks further growth of the DNA molecule

oligonucleotide primer for DNA polymerase

5'
GCTACCTGCATGGA
CGATGGACGTACCTCTGAAGCG

3'   5'

single-stranded DNA molecule to be sequenced

(B)

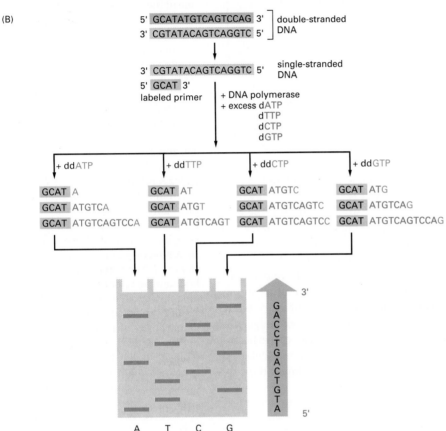

5' GCATATGTCAGTCCAG 3' } double-stranded
3' CGTATACAGTCAGGTC 5'   DNA

3' CGTATACAGTCAGGTC 5'   single-stranded DNA

5' GCAT 3'
labeled primer

+ DNA polymerase
+ excess dATP
         dTTP
         dCTP
         dGTP

+ ddATP   + ddTTP   + ddCTP   + ddGTP

GCAT A   GCAT AT   GCAT ATGTC   GCAT ATG
GCAT ATGTCA   GCAT ATGT   GCAT ATGTCAGTC   GCAT ATGTCAG
GCAT ATGTCAGTCCA GCAT ATGTCAGT   GCAT ATGTCAGTCC   GCAT ATGTCAGTCCAG

3'
G
A
C
C
T
G
A
C
T
G
T
A
5'

A   T   C   G

Summary

Recombinant DNA technology has revolutionized the study of the cell. The development of this technology was neither planned nor anticipated. Instead, steady advances in the ability of researchers to manipulate DNA molecules were made on many different fronts until the combination of techniques became powerful enough to allow researchers to pick out any gene at will and, after an amplification step, to determine the exact molecular structure of both the gene and its products. Crucial elements in this technology are the ability to fragment chromosomes into DNA molecules with specific ends and the means to separate and sequence the resulting DNA fragments. The fragmentation step is carried out by proteins that are normally produced by bacteria in order to destroy invading DNA molecules. These proteins, called restriction nucleases, are purified and used to cut the DNA double helix at a specific short nucleotide sequence. The resulting DNA fragments can be separated from one another according to their size by using electrophoresis to move the mixture of fragments through the pores of a gel. Under certain conditions, even DNA molecules that differ in length by only a single nucleotide can be detected as separate bands. DNA sequencing takes advantage of this powerful separation method: after electrophoresis,

Figure 7–8 The enzymatic method for sequencing DNA. (A) The DNA to be sequenced is used as template for the *in vitro* synthesis, by DNA polymerase, of a set of partial replicas, all beginning at the same place, but terminating at different points along the DNA chain. The key to this method is the use of dideoxyribonucleoside triphosphates in which the deoxyribose 3'-OH group present in normal nucleotides is missing; when such a modified nucleotide is incorporated into a DNA chain, it blocks the addition of the next nucleotide. In the example illustrated, dideoxy ATP (ddATP) competes with an excess of deoxy ATP (dATP), so that each newly synthesized DNA strand made in a test tube by DNA polymerase will stop at a randomly selected A in the sequence. This reaction therefore generates a ladder of DNA fragments similar to that shown previously for the chemical method (Figure 7–7); these fragments are detected by a label (chemical or radioactive) that is either incorporated into the oligonucleotide primer (*orange*), or into one of the deoxyribonucleoside triphosphates (*green*) used to extend the DNA chain. (B) To determine the full sequence, four different chain-terminating nucleoside triphosphates (*red*) are used in separate DNA synthesis reactions on the same primed single-stranded DNA template. When the products of these four reactions are analyzed by electrophoresis in four parallel lanes of a polyacrylamide gel, the DNA sequence can be derived in the manner illustrated for the chemical method in Figure 7–7B.

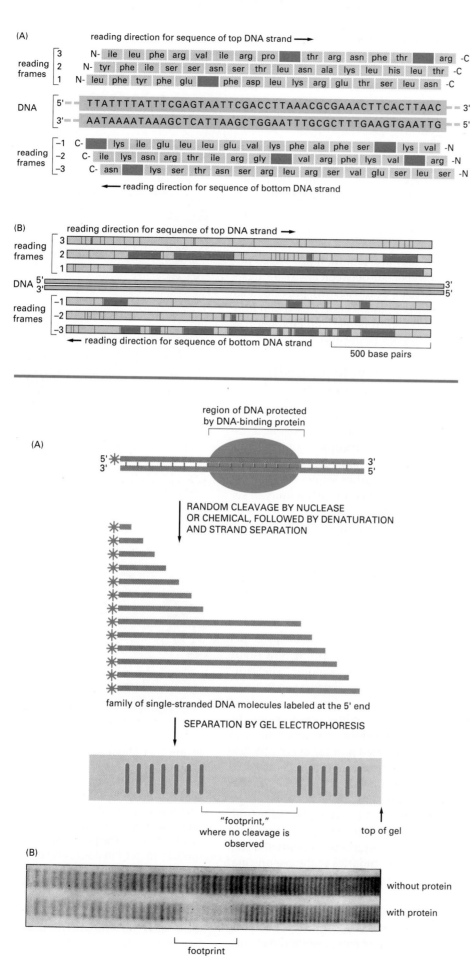

(A)

reading direction for sequence of top DNA strand ⟶

reading frames

3 | N- ile leu phe arg val ile arg pro ▨ thr arg asn phe thr ▨ arg -C
2 | N- tyr phe ile ser ser asn ser thr leu asn ala lys leu his leu thr -C
1 | N- leu phe tyr phe glu ▨ phe asp leu lys arg glu thr ser leu asn -C

DNA
5'—TTATTTTATTTCGAGTAATTCGACCTTAAACGCGAAACTTCACTTAAC—3'
3'—AATAAAATAAAGCTCATTAAGCTGGAATTTGCGCTTTGAAGTGAATTG—5'

reading frames

−1 | C- ▨ lys ile glu leu leu glu val lys phe ala phe ser ▨ lys val -N
−2 | C- ile lys asn arg thr ile arg gly ▨ val arg phe lys val ▨ arg -N
−3 | C- asn ▨ lys ser thr asn ser arg leu arg ser val glu ser leu ser -N

⟵ reading direction for sequence of bottom DNA strand

(B)

reading direction for sequence of top DNA strand ⟶

reading frames
3
2
1

DNA 5'
3'

reading frames
−1
−2
−3

⟵ reading direction for sequence of bottom DNA strand

|⸺ 500 base pairs ⸺|

(A)

region of DNA protected
by DNA-binding protein

5'✷
3'

RANDOM CLEAVAGE BY NUCLEASE
OR CHEMICAL, FOLLOWED BY DENATURATION
AND STRAND SEPARATION

family of single-stranded DNA molecules labeled at the 5' end

SEPARATION BY GEL ELECTROPHORESIS

"footprint,"
where no cleavage is
observed

top of gel

(B)

without protein

with protein

footprint

Figure 7–9 Finding the regions in a DNA sequence that encode a protein. (A) Any region of the DNA sequence can, in principle, code for six different amino acid sequences, because any one of three different reading frames can be used to interpret the nucleotide sequence on each strand. Note that a nucleotide sequence is always read in the 5'-to-3' chain direction and encodes a polypeptide from the amino (N) to the carboxyl (C) terminus. For a random nucleotide sequence, a stop signal for protein synthesis will be encountered, on average, about once every 21 amino acids (once every 63 nucleotides). In this sample sequence of 48 base pairs, each such signal (*stop codon*) is colored *green*, and only reading frame 2 lacks a stop signal. (B) Search of a 1700 base-pair DNA sequence for a possible protein-encoding sequence. The information is displayed as in (A), with each stop signal for protein synthesis denoted by a *green line*. In addition, all of the regions between possible start and stop signals for protein synthesis (see p. 234) are displayed as *red bars*. Only reading frame 1 actually encodes a protein, which is 475 amino acid residues long.

Figure 7–10 The DNA footprinting technique. (A) This technique requires a DNA molecule that has been radioactively labeled at one end (see Figure 7–6B). The protein shown binds tightly to a specific DNA sequence that is seven nucleotides long, thereby protecting these seven nucleotides from the cleaving agent. If the same reaction were carried out without the DNA-binding protein, a complete ladder of bands would be seen on the gel (not shown). (B) An actual footprint used to determine the binding site for a human protein that stimulates the transcription of specific eucaryotic genes. These results locate the binding site about 60 nucleotides upstream from the start site for RNA synthesis. The cleaving agent was a small, iron-containing organic molecule that normally cuts at every phosphodiester bond with nearly equal frequency. (B, courtesy of Michele Sawadogo and Robert Roeder.)

the sequence of nucleotides in a DNA molecule is determined from the pattern of labeled DNA bands observed when a fixed end of one of its DNA strands has been labeled and the other end terminates at a randomly selected nucleotide of a single type (A, G, C, or T).

Nucleic Acid Hybridization [8]

When an aqueous solution of DNA is heated at 100°C or exposed to a very high pH (pH ≥ 13), the complementary base pairs that normally hold the two strands of the double helix together are disrupted and the double helix rapidly dissociates into two single strands. This process, called *DNA denaturation*, was for many years thought to be irreversible. In 1961, however, it was discovered that complementary single strands of DNA will readily re-form double helices (a process called **DNA renaturation** or **hybridization**) if they are kept for a prolonged period at 65°C. Similar hybridization reactions will occur between any two single-stranded nucleic acid chains (DNA/DNA, RNA/RNA, or RNA/DNA), provided that they have a complementary nucleotide sequence. In this section we explain how these specific hybridization reactions can be used to detect and characterize specific nucleotide sequences in both RNA and DNA molecules.

Nucleic Acid Hybridization Reactions Provide a Sensitive Way of Detecting Specific Nucleotide Sequences [9]

The rate of double-helix formation during hybridization reactions is limited by the rate at which two complementary nucleic acid chains happen to collide. Since this collision rate is proportional to the concentration of complementary chains in the solution, hybridization rates can be used to determine the concentration of any desired DNA or RNA sequence in a mixture of other sequences. The assay requires a pure single-stranded DNA fragment that is complementary in sequence to the desired nucleic acid (DNA or RNA). This DNA fragment can be obtained by cloning, or, if the sequence is short, it can be synthesized by chemical means. In either case the DNA fragment must carry a unique marker (either a radioisotope or a chemical label), so that its incorporation into double-stranded molecules can be followed during the course of a hybridization reaction. A single-stranded DNA molecule used as an indicator in this way is known as a *DNA probe;* it can be anywhere from fifteen to thousands of nucleotides long.

Hybridization reactions using DNA probes are so sensitive and selective that complementary sequences present at a concentration as low as one molecule per cell can be detected (Figure 7–11). It is thus possible to determine how many copies of a particular DNA sequence are present in a cell's genome. The same technique can be used to search for related but nonidentical genes; once an interesting gene has been cloned from a mouse or a chicken, for example, part of its sequence can be used as a probe to find the corresponding gene in a human.

Alternatively, DNA probes can be used in hybridization reactions with RNA rather than DNA to find out whether a cell is *expressing* a given gene. In this case a DNA probe that contains part of the gene's sequence is hybridized with RNA purified from the cell in question to see whether the RNA includes molecules matching the probe DNA and, if so, in what quantities. In somewhat more elaborate procedures the DNA probe is treated with specific nucleases after the hybridization is complete to determine the exact regions of the DNA probe that have paired with cellular RNA molecules. One can thereby determine the start and stop sites for RNA transcription, as well as the precise boundaries of the regions that are cut out of the RNA transcripts by *RNA splicing* (the intron sequences) (Figure 7–12).

Large numbers of genes are switched on and off in elaborate patterns as an embryo develops. The hybridization of DNA probes to cellular RNAs allows one to determine whether a particular gene is off or on; moreover, when the expres-

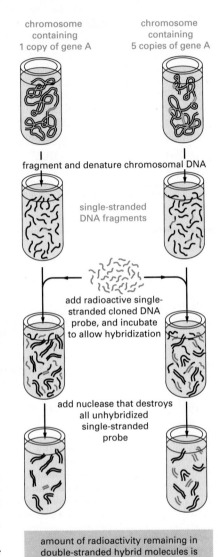

Figure 7–11 Measurement of the number of copies of a specific gene in a sample of DNA by means of DNA hybridization. The single-stranded DNA fragment used in such experiments is commonly referred to as a *DNA probe*. The probe must be specially marked to distinguish it from the vast excess of the chromosomal DNA. In this example the label is a radioactive isotope; alternatively, chemically labeled nucleotides can be used as the label (see Figure 7–18).

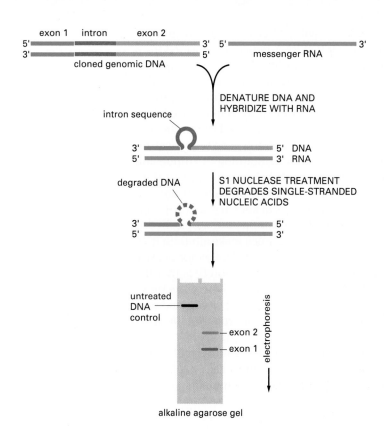

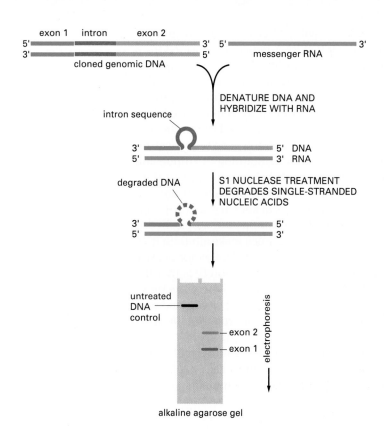

Figure 7–12 **The use of nucleic acid hybridization to determine the region of a cloned DNA fragment that is present in an mRNA molecule.** The method shown requires a nuclease that cuts the DNA chain only where it is not base-paired to a complementary RNA chain. The positions of the introns (intervening sequences) in eucaryotic genes are mapped by the method shown; the beginning and the end of an RNA molecule can be determined in the same way.

sion of a gene changes, one can determine whether the change is due to controls that act on the transcription of DNA, the splicing of the gene's RNA, or the translation of its mature mRNA molecules into protein. Hybridization methods are in such wide use in cell biology today that it is difficult to imagine what it would be like to study gene structure and expression without them.

Northern and Southern Blotting Facilitate Hybridization with Electrophoretically Separated Nucleic Acid Molecules [10]

DNA probes are often used following gel electrophoresis to detect the nucleic acid molecules with sequences that are complementary to all or part of the probe. The electrophoresis fractionates the many different RNA or DNA molecules in a crude mixture according to their size before the hybridization reaction is carried out; if molecules of only one or a few sizes become labeled with the probe, one can be certain that the hybridization was indeed specific. Moreover, the size information obtained can be invaluable in itself. An example will illustrate this point.

Suppose that one wishes to determine the nature of the defect in a mutant mouse that produces abnormally low amounts of *albumin,* a protein that liver cells normally secrete into the blood in large amounts. First, one collects identical samples of liver tissue from mutant and normal mice (the latter serving as controls) and disrupts the cells in a strong detergent to inactivate cellular nucleases that might otherwise degrade the nucleic acids. Next, one separates the RNA and DNA from all of the other cell components: the proteins present are completely denatured and removed by repeated extractions with phenol—a potent organic solvent that is partly miscible with water; the nucleic acids, which remain in the aqueous phase, are then precipitated with alcohol to separate them from the small molecules of the cell. Then one separates the DNA from the RNA by their different solubilities in alcohols and degrades any contaminating nucleic acid of the unwanted type by treatment with a highly specific enzyme—either an RNase or a DNase.

To analyze the albumin-encoding RNAs with a DNA probe, a technique called **Northern blotting** is used. First, the intact RNA molecules purified from mutant and control liver cells are fractionated according to their size into a series of bands by gel electrophoresis. Then, to make the RNA molecules accessible to DNA probes, a replica of the pattern of RNA bands on the gel is made by transferring ("blotting") the fractionated RNA molecules onto a sheet of nitrocellulose paper (nylon). The RNA molecules that hybridize to the labeled DNA probe (because they are complementary to part of the normal albumin gene sequence) are then located by incubating the paper with a solution containing the probe and detecting the hybridized probe by autoradiography or by chemical means (Figure 7–13). The size of the RNA molecules in each band that binds the probe can be determined by reference to bands of RNA molecules of known size (*RNA standards*) that are electrophoresed side by side with the experimental sample. In this way one might discover that liver cells from the mutant mice make albumin RNA in normal amounts and of normal size; alternatively, albumin RNA of normal size might be detected in greatly reduced amounts. Another possibility is that the mutant albumin RNA molecules might be abnormally short and therefore move unusually quickly through the gel; in this case the gel blot could be retested with a series of shorter DNA probes, each corresponding to small portions of the gene, to reveal which part of the normal RNA is missing.

An analogous gel-transfer hybridization method, called **Southern blotting,** analyzes DNA rather than RNA. Isolated DNA is first cut into readily separable fragments with restriction nucleases. The double-stranded fragments are then

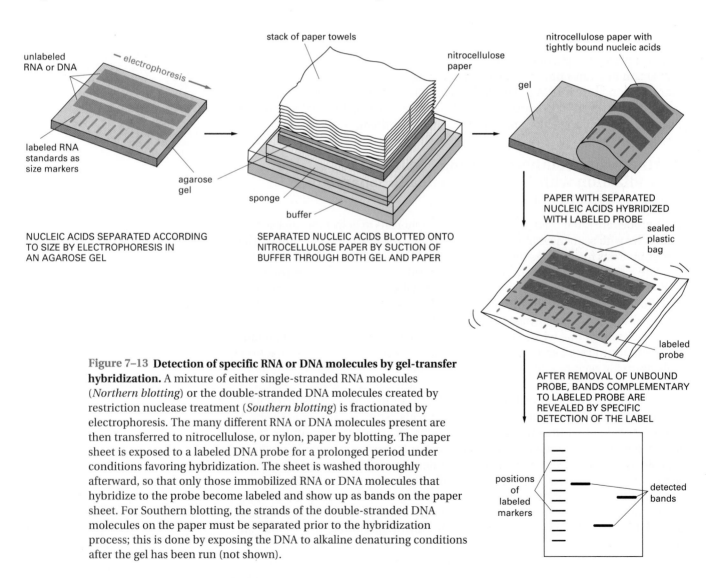

Figure 7–13 Detection of specific RNA or DNA molecules by gel-transfer hybridization. A mixture of either single-stranded RNA molecules (*Northern blotting*) or the double-stranded DNA molecules created by restriction nuclease treatment (*Southern blotting*) is fractionated by electrophoresis. The many different RNA or DNA molecules present are then transferred to nitrocellulose, or nylon, paper by blotting. The paper sheet is exposed to a labeled DNA probe for a prolonged period under conditions favoring hybridization. The sheet is washed thoroughly afterward, so that only those immobilized RNA or DNA molecules that hybridize to the probe become labeled and show up as bands on the paper sheet. For Southern blotting, the strands of the double-stranded DNA molecules on the paper must be separated prior to the hybridization process; this is done by exposing the DNA to alkaline denaturing conditions after the gel has been run (not shown).

separated according to their size by gel electrophoresis, and those complementary to a DNA probe are identified by blotting and hybridization, as just described for RNA (see Figure 7–13). To characterize the structure of the albumin gene in the mutant mice, an albumin-specific DNA probe would be used (after cutting the DNA with different restriction nucleases) to construct a detailed *restriction map* of the genome in the region of the albumin gene. From this map one could determine if the albumin gene has been rearranged in the defective animals—for example, by the deletion or the insertion of a short DNA sequence; most single base changes, however, could not be detected in this way.

RFLP Markers Greatly Facilitate Genetic Approaches to the Mapping and Analysis of Large Genomes [11]

Very large genomes can be mapped either physically or genetically. **Physical maps** are based on direct analysis of the DNA molecules that constitute each chromosome. They include restriction maps and ordered libraries of genomic DNA clones (see Figure 7–29), both of which can be viewed as incomplete representations of the ultimate physical map, which is the complete nucleotide sequence of the genome. **Genetic linkage maps,** by contrast, are based on the frequency of coinheritance of two or more features of the organism that serve as *genetic markers* when individuals interbreed. A genetic marker can be any site in the genome where there is variation in the DNA sequence that is detectable as differences between individuals in a population. If the difference is a rare one, it is called a *mutation;* if it is a common one, it is called a *polymorphism.* Genetic linkage maps are constructed by following the pattern of inheritance of such genetic variants.

Traditionally, linkage mapping has depended on the identification of mutations by their effects on characteristics such as eye color or blood groups. More recently, recombinant DNA methods have made it possible to use as genetic markers short DNA sequences that differ between normal individuals without having a detectable effect on the organism. A widely used genetic marker of this type depends on the way small differences in DNA sequence can alter restriction-enzyme cutting patterns. A single base-pair difference in a particular chromosomal position, for example, may eliminate a restriction-enzyme cutting site, giving rise to a large difference in the lengths of certain restriction fragments produced from the DNA at that position. Similarly, short deletions or insertions will change the size of any DNA restriction fragment in which they occur. Such small differences between individuals are known as **restriction fragment length polymorphisms (RFLPs).** An RFLP is most useful when it is very common in the population, so that there is a high probability that the parents of an individual will carry distinguishable markers. For this reason, short, tandemly repeated sequences whose exact length is highly variable in the population are the basis for the most useful RFLP markers (Figure 7–14).

Figure 7–14 Some DNA sequence changes that produce a restriction fragment length polymorphism (RFLP). (A) In many individuals in the population, a restriction nuclease cuts the chromosomal DNA shown at three sites, producing two DNA fragments; one of these DNA fragments is detected with a labeled DNA probe after gel electrophoresis and Southern blotting. (B) In other examples of the same chromosome, a single base-pair change in the short sequence recognized by the restriction nuclease prevents cutting at the central restriction site. The labeled probe now detects a much longer DNA fragment as an RFLP. (C) In yet other examples of the same chromosome, a sequence duplication increases the length of the DNA fragment that is detected by the labeled probe. This type of RFLP is commonly found in regions containing runs of short repeated sequences, such as GTGTGT . . . ; the number of such tandem repeats is found to be highly variable in the population, with each individual typically containing a different number of copies (generally 4 to 40) of the repeat at each particular locus. A repeat of this type is commonly referred to as either a *microsatellite* or a *VNTR (variable number of tandem repeat)* sequence.

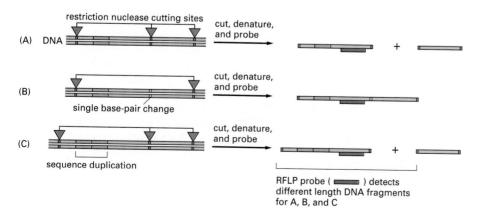

Nucleic Acid Hybridization

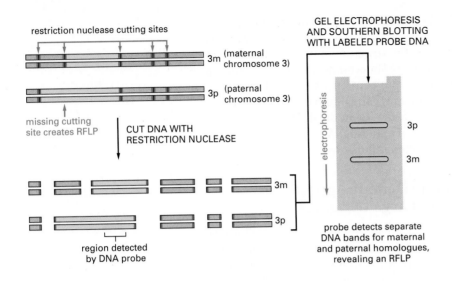

restriction nuclease cutting sites

3m (maternal chromosome 3)

3p (paternal chromosome 3)

missing cutting site creates RFLP

CUT DNA WITH RESTRICTION NUCLEASE

3m

3p

region detected by DNA probe

GEL ELECTROPHORESIS AND SOUTHERN BLOTTING WITH LABELED PROBE DNA

electrophoresis

3p

3m

probe detects separate DNA bands for maternal and paternal homologues, revealing an RFLP

Figure 7–15 Detection of an RFLP by Southern blotting. In this example, the technique is being used to distinguish the maternally and paternally inherited copies of a region of chromosome 3 that are present in a single individual. A variant of this method uses the PCR technique (see Figure 7–32) to amplify the appropriate DNA region selectively before the restriction nuclease treatment. The relevant restriction fragments are then vastly more abundant than other DNA fragments in the preparation and can be seen simply by staining the gel, avoiding the need for blotting and labeled probes.

RFLPs can be used for genetic mapping because the fragment size differences that an individual inherits are readily detected by Southern blotting with a DNA probe complementary to a specific DNA sequence in the region (Figure 7–15). RFLPs provide an immediate way to relate a genetic linkage map to a physical map: on the one hand, they serve as heritable genetic markers, like eye color; on the other hand, the DNA probes that detect them represent DNA sequences whose positions on the physical map are easily found by DNA hybridization.

If two genetic markers are on different chromosomes, the inheritance of a variation in either of them will be *unlinked*—that is, they will have only a 50-50 chance of being inherited together. The same is true for the variants of markers at opposite ends of a single chromosome because of the high probability that they will be separated during the frequent crossing-over that occurs during meiosis in the development of eggs and sperm. The closer together two markers are on the same chromosome, the greater the chance that the form of each marker in the parent will not be separated by crossover events and will therefore be

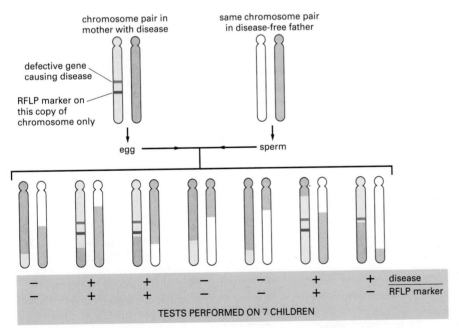

chromosome pair in mother with disease

same chromosome pair in disease-free father

defective gene causing disease

RFLP marker on this copy of chromosome only

egg   sperm

							disease
−	+	+	−	−	+	+	
−	+	+	−	−	+	−	RFLP marker

TESTS PERFORMED ON 7 CHILDREN

CONCLUSION: gene causing disease is coinherited with RFLP marker from diseased mother in 75% of the diseased progeny. If this same correlation is observed in other families that have been examined, the gene causing disease is mapped to this chromosome close to the RFLP marker

Figure 7–16 Genetic linkage analysis using an RFLP marker. In this procedure one studies the coinheritance of a specific human phenotype (here a genetic disease) with an RFLP marker. If individuals who inherit the disease nearly always inherit the RFLP marker, then the gene causing the disease and the RFLP marker are likely to be close together on the chromosome, as shown here. To prove that an observed linkage is statistically significant, hundreds of individuals may need to be examined. Note that the linkage will not be absolute unless the RFLP marker is located in the gene itself. Thus, occasionally the RFLP marker will be separated from the disease gene by meiotic crossing-over during the formation of the egg or sperm: this has happened in the case of the chromosome pair on the far right.

coinherited by each progeny (*linked*). By screening large family groups for the coinheritance of a gene of interest (such as one associated with a disease) and a large number of individual RFLPs, a few RFLP markers can be identified that are unambiguously coinherited with the gene (this requires the analysis of many individuals, as explained in Figure 7–16). DNA sequences that surround the gene can thereby be located. Eventually the DNA corresponding to the gene itself can be found by techniques to be described later (see Figure 7–30). Many genes that cause human diseases are being isolated in this way, allowing the proteins they encode to be analyzed in detail.

It is clear from Figure 7–16 that the closer an RFLP marker is to the mutant gene of interest, the easier it is to locate the gene unambiguously. To facilitate these and other studies, an intensive effort is underway to prepare a *high-resolution RFLP map* of the human genome, with thousands of RFLP markers spaced an average of 10^6 nucleotide pairs apart. On average, two markers separated by this distance will be coinherited by 99 of every 100 progeny. Such a map will make it relatively easy to use genetic linkage studies to locate a gene that has been identified only by the effect of a mutation in humans, allowing it to be mapped to one or a few large DNA clones in an ordered genomic DNA library. Isolation of the gene might then be accomplished relatively quickly.

Synthetic DNA Molecules Facilitate the Prenatal Diagnosis of Genetic Diseases [12]

At the same time that microbiologists were developing DNA cloning techniques, organic chemists were improving the methods for synthesizing short DNA chains. Today, such synthetic *DNA oligonucleotides* are routinely produced by machines that can automatically synthesize any DNA sequence up to 120 nucleotides long overnight. This ability to produce DNA molecules of a desired sequence makes it possible to redesign genes at will, an important aspect of genetic engineering, as explained later. Such synthetic oligonucleotides can also be used as labeled probes to detect corresponding genomic sequences by DNA hybridization. By varying the temperature at which the hybridization reaction is run, it is possible to vary the *stringency* of the hybridization: above a certain temperature, only perfectly matched sequences will hybridize, and this can make it possible, for example, to detect a mutant gene in the prenatal diagnosis of genetic disease.

More than 3000 human genetic diseases are attributable to single-gene defects. In most of these the mutation is recessive: that is, it shows its effect only when an individual inherits two defective copies of the gene, one from each parent. One goal of modern medicine is to identify those fetuses that carry two copies of the defective gene long before birth so that the mother, if she wishes, can have the pregnancy terminated. In sickle-cell anemia, for example, the exact nucleotide change in the mutant gene is known (the sequence GAG is changed to GTG at a specific point in the DNA strand that codes for the β chain of hemoglobin). For prenatal diagnosis, two DNA oligonucleotides are synthesized—one corresponding to the normal gene sequence in the region of the mutation and the other corresponding to the mutant sequence. If the oligonucleotides are kept short (about 20 nucleotides), they can be hybridized with DNA at a temperature selected so that only the perfectly matched helix will be stable. Such oligonucleotides can thus be used as labeled probes to distinguish between the two forms of the gene by Southern blotting on DNA isolated from fetal cells collected by amniocentesis. A fetus carrying two copies of the mutant β-chain gene can be readily recognized because its DNA will hybridize *only* with the oligonucleotide that is complementary to the mutant DNA sequence.

For many genetic abnormalities the exact nucleotide sequence change is not known. For an increasing number of these, prenatal diagnosis is still possible by using Southern blotting to assay for specific variations in the human genome (the *RFLPs* in Figure 7–16) that are known to be closely linked to the defective gene.

Nucleic Acid Hybridization

The same techniques can also be used to detect an individual's susceptibility to future disease. Individuals who have inherited abnormal copies of certain genes, for example, have a greatly increased risk of cancer. They need to take protective measures to improve their prospects for a healthy life.

Hybridization at Reduced Stringency Allows Even Distantly Related Genes to Be Identified [13]

New genes arise during evolution by the duplication and divergence of old genes and by the reutilization of portions of old genes in new combinations. For this reason, most genes have a family of close relatives elsewhere in the genome, many of which are likely to have a related function. Laborious methods are usually required to isolate a DNA clone corresponding to the first member of such a *gene family*. Other members of the family, however, can then often be isolated relatively easily by using sequences from the first gene as DNA probes. Because the new genes are unlikely to have *identical* sequences, hybridizations with the DNA probe are usually carried out under conditions of "reduced stringency"—that is, conditions that allow even an imperfect match with the probe sequence to form a stable double helix (Figure 7–17). Although using reduced stringency for hybridization carries the risk of obtaining a false signal from a chance region of short sequence homology in an unrelated DNA sequence, such false positives can be eliminated by further investigation.

This technique is one of the most powerful uses of recombinant DNA technology. It has led, for example, to the isolation of a whole family of DNA-binding proteins that function as master regulators of gene expression during embryonic development in *Drosophila* and also made it possible to isolate members of this same gene family from a variety of other organisms, including mice and humans.

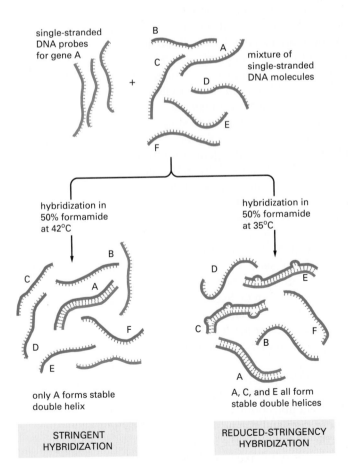

single-stranded DNA probes for gene A

mixture of single-stranded DNA molecules

hybridization in 50% formamide at 42°C

hybridization in 50% formamide at 35°C

only A forms stable double helix

A, C, and E all form stable double helices

STRINGENT HYBRIDIZATION

REDUCED-STRINGENCY HYBRIDIZATION

Figure 7–17 Comparison of stringent and reduced-stringency hybridization conditions. In the reaction on the left (stringent conditions), the solution is kept only a few degrees below the temperature at which a perfect DNA helix denatures (its *melting temperature*), so that the imperfect helices that can form under the conditions of reduced stringency are unstable. Only the hybridization conditions on the right can be used to find genes that are nonidentical but related to gene A.

Figure 7–18 **A chemical label for DNA probes.** The modified nucleotide shown can be incorporated into DNA by DNA polymerase so as to allow the DNA molecule to serve as a probe that can be readily detected. The base on the nucleoside triphosphate shown is an analogue of thymine, in which the methyl group on T has been replaced by a spacer arm linked to the plant steroid digoxygenin. To visualize the probe, the digoxygenin is detected by a specific antibody coupled to a visible marker such as a fluorescent dye. Other chemical labels such as biotin can be attached to nucleotides and used in essentially the same way.

digoxygenin

spacer

this region still available for base-pairing with A

modified nucleoside triphosphate

In Situ Hybridization Techniques Locate Specific Nucleic Acid Sequences in Cells or on Chromosomes [14]

Nucleic acids, no less than other macromolecules, occupy precise positions in cells and tissues, and a great deal of potential information is lost when these molecules are extracted by homogenization. For this reason, techniques have been developed in which nucleic acid probes are used in much the same way as labeled antibodies to locate specific nucleic acid sequences *in situ*, a procedure called *in situ* **hybridization.** This can now be done both for DNA in chromosomes and for RNA in cells. Labeled nucleic acid probes can be hybridized to chromosomes that have been exposed briefly to a very high pH to disrupt their DNA base pairs. The chromosomal regions that bind the probe during the hybridization step are then visualized. Originally, this technique was developed using highly radioactive DNA probes, which were detected by autoradiography. The spatial resolution of the technique, however, can be greatly improved by labeling the DNA probes chemically instead of radioactively. For this purpose the probes are synthesized with special nucleotides that contain a modified side chain (Figure 7–18), and the hybridized probes are detected with an antibody (or other ligand) that specifically recognizes this side chain (Figure 7–19).

In situ hybridization methods have also been developed that reveal the distribution of specific RNA molecules in cells in tissues. In this case the tissues are not exposed to a high pH, so the chromosomal DNA remains double-stranded and cannot bind the probe. Instead the tissue is gently fixed so that its RNA is retained in an exposed form that will hybridize when the tissue is incubated with a complementary DNA or RNA probe. In this way the patterns of differential gene expression can be observed in tissues. In the *Drosophila* embryo, for example, such patterns have provided new insights into the mechanisms that create distinctions between cells in different positions during development (Figure 7–20).

Summary

In nucleic acid hybridization reactions, single strands of DNA or RNA randomly collide with one another, rapidly testing out billions of possible alignments. Under stringent hybridization conditions (a combination of solvent and temperature where a perfect double helix is barely stable), two strands will pair to form a "hybrid" helix only if their nucleotide sequences are almost perfectly complementary. The enormous specificity of this hybridization reaction allows any single-stranded sequence of nucleotides to be labeled with a radioisotope or chemical and used as a probe to find

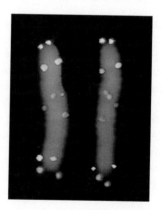

Figure 7–19 *In situ* **hybridization to locate specific genes on chromosomes.** Here, six different DNA probes have been used to mark the location of their respective nucleotide sequences on human chromosome 5 at metaphase. The probes have been chemically labeled and detected with fluorescent antibodies. Both copies of chromosome 5 are shown, aligned side by side. Each probe produces two dots on each chromosome, since a metaphase chromosome has replicated its DNA and therefore contains two identical DNA helices (see Figure 8–4). (Courtesy of David C. Ward.)

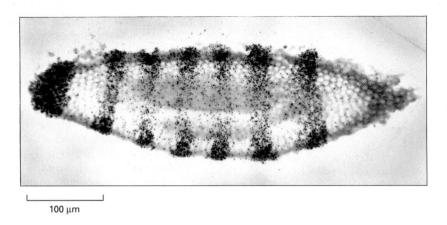

Figure 7–20 *In situ* hybridization for RNA localization in tissues.
Autoradiograph of a section of a very young *Drosophila* embryo that has been subjected to *in situ* hybridization using a radioactive DNA probe complementary to a gene involved in segment development. The probe has hybridized to RNA in the embryo, and the pattern of autoradiographic silver grains reveals that the RNA made by the gene (called *ftz*) is localized in alternating stripes across the embryo that are three or four cells wide. At this stage of development (cellular blastoderm), the embryo contains about 6000 cells. (From E. Hafen, A. Kuriowa, and W.J. Gehring, *Cell* 37:833–841, 1984. © Cell Press.)

100 µm

a complementary partner strand, even in a cell or cell extract that contains millions of different DNA and RNA sequences. Probes of this type are widely used to detect the nucleic acids corresponding to specific genes, both to facilitate the purification and characterization of the genes after cell lysis and to localize them in cells, tissues, and organisms. Moreover, by carrying out hybridization reactions under conditions of "reduced stringency," a probe prepared from one gene can be used to find its evolutionary relatives—both in the same organism, where the relatives form part of a gene family, and in other organisms, where the evolutionary history of the nucleotide sequence can be traced.

DNA Cloning [15]

In **DNA cloning,** a DNA fragment that contains a gene of interest is inserted into the purified DNA genome of a self-replicating genetic element—generally a virus or a plasmid. A DNA fragment containing a human gene, for example, can be joined in a test tube to the chromosome of a bacterial virus, and the new *recombinant DNA molecule* can then be introduced into a bacterial cell. Starting with only one such recombinant DNA molecule that infects a single cell, the normal replication mechanisms of the virus can produce more than 10^{12} identical virus DNA molecules in less than a day, thereby amplifying the amount of the inserted human DNA fragment by the same factor. A virus or plasmid used in this way is known as a *cloning vector,* and the DNA propagated by insertion into it is said to have been *cloned.*

A DNA Library Can Be Made Using Either Viral or Plasmid Vectors [16]

In order to clone a specific gene, one begins by constructing a *DNA library*—a comprehensive collection of cloned DNA fragments, including (one hopes) at least one fragment that contains the gene of interest. The library can be constructed using either a virus or a plasmid vector and is generally housed in a population of bacterial cells. The principles underlying the methods used for cloning genes are the same for either type of cloning vector, although the details may be different. For simplicity, in this chapter we ignore these differences and illustrate the methods with reference to plasmid vectors.

The **plasmid vectors** used for gene cloning are small circular molecules of double-stranded DNA derived from larger plasmids that occur naturally in bacterial cells. They generally account for only a minor fraction of the total host bacterial cell DNA, but they can easily be separated on the basis of their small size from chromosomal DNA molecules, which are large and precipitate as a pellet upon centrifugation. For use as cloning vectors, the purified plasmid DNA circles are first cut with a restriction nuclease to create linear DNA molecules. The cel-

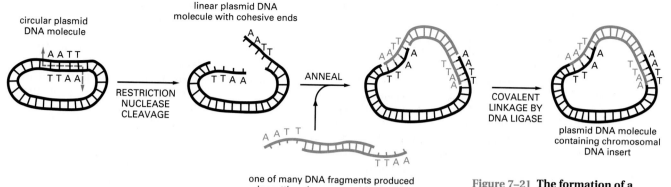

circular plasmid
DNA molecule

linear plasmid DNA
molecule with cohesive ends

RESTRICTION
NUCLEASE
CLEAVAGE

ANNEAL

COVALENT
LINKAGE BY
DNA LIGASE

plasmid DNA molecule
containing chromosomal
DNA insert

one of many DNA fragments produced
by cutting chromosomal DNA with
the same restriction nuclease

Figure 7–21 The formation of a recombinant DNA molecule. The cohesive ends produced by many kinds of restriction nucleases allow two DNA fragments to join by complementary base-pairing (see Figure 7–2). DNA fragments joined in this way can be covalently linked in a highly efficient reaction catalyzed by the enzyme DNA ligase. In this example a recombinant plasmid DNA molecule containing a chromosomal DNA insert is formed.

lular DNA to be used in constructing the library is cut with the same restriction nuclease, and the resulting restriction fragments (including those containing the gene to be cloned) are then added to the cut plasmids and annealed via their cohesive ends to form recombinant DNA circles. These recombinant molecules containing foreign DNA inserts are then covalently sealed with the enzyme DNA ligase (Figure 7–21).

In the next step in preparing the library, the recombinant DNA circles are introduced into bacterial cells that have been made transiently permeable to DNA; such cells are said to be *transfected* with the plasmids. As these cells grow and divide, doubling in number every 30 minutes, the recombinant plasmids also replicate to produce an enormous number of copies of DNA circles containing the foreign DNA (Figure 7–22). Many bacterial plasmids carry genes for antibiotic resistance, a property that can be exploited to select those cells that have been successfully transfected; if the bacteria are grown in the presence of the antibiotic, only cells containing plasmids will survive. Each original bacterial cell that was initially transfected will, in general, contain a different foreign DNA insert; this insert will be inherited by all of the progeny cells of that bacterium, which together form a small colony in a culture dish.

The mixture of many different surviving bacteria contains the DNA library, composed of a large number of different DNA inserts. The problem is that only a few of the bacteria will harbor the particular recombinant plasmids that contain the desired gene. One needs to be able to identify these rare cells in order to recover the DNA of interest in pure form and in useful quantities. Before discussing how this is achieved, we need to describe a second strategy for generating a DNA library that is commonly used in gene cloning.

Two Types of DNA Libraries Serve Different Purposes [17]

Cleaving the entire genome of a cell with a specific restriction nuclease as just described is sometimes called the "shotgun" approach to gene cloning. It produces a very large number of DNA fragments—on the order of a million for a mammalian genome—which will generate millions of different colonies of transfected bacterial cells. Each of these colonies will be composed of a *clone* derived from a single ancestor cell and therefore will harbor a recombinant plasmid with the same inserted genomic DNA sequence. Such a plasmid is said to contain a **genomic DNA clone,** and the entire collection of plasmids is said to constitute

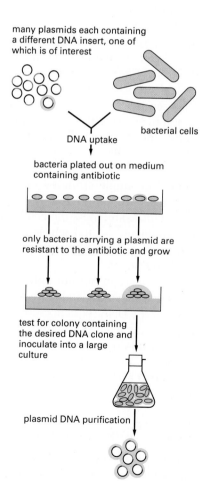

many plasmids each containing
a different DNA insert, one of
which is of interest

bacterial cells

DNA uptake

bacteria plated out on medium
containing antibiotic

only bacteria carrying a plasmid are
resistant to the antibiotic and grow

test for colony containing
the desired DNA clone and
inoculate into a large
culture

plasmid DNA purification

Figure 7–22 Purification and amplification of a specific DNA sequence by DNA cloning in a bacterium. Each bacterial cell carrying a recombinant plasmid develops into a colony of identical cells, visible as a spot on the nutrient agar. By inoculating a single colony of interest into a liquid culture, one can obtain a large number of identical plasmid DNA molecules, each containing the same DNA insert.

DNA Cloning

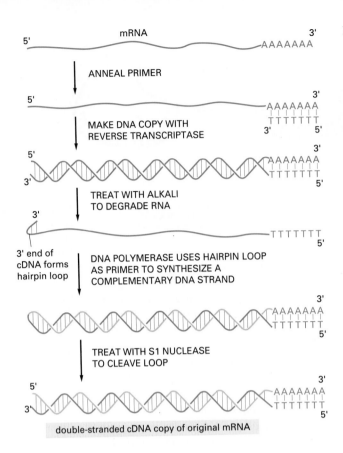

Figure 7–23 The synthesis of cDNA. A DNA copy (cDNA) of an mRNA molecule is produced by the enzyme reverse transcriptase (see p. 282), thereby forming a DNA/RNA hybrid helix. Treating the DNA/RNA hybrid with alkali selectively degrades the RNA strand into nucleotides. The remaining single-stranded cDNA is then copied into double-stranded cDNA by the enzyme DNA polymerase. As indicated, both reverse transcriptase and DNA polymerase require a primer to begin their synthesis. For reverse transcriptase a small oligonucleotide is used; in this example oligo(dT) has been annealed with the long poly-A tract at the 3′ end of most mRNAs. Note that the double-stranded cDNA molecule produced here lacks cohesive ends; such blunt-ended DNA molecules can be cloned by one of several procedures that are analogous to (but less efficient than) that shown in Figure 7–21.

Labels in figure:
- mRNA, 5′ ... 3′ AAAAAAA
- ANNEAL PRIMER
- MAKE DNA COPY WITH REVERSE TRANSCRIPTASE
- TREAT WITH ALKALI TO DEGRADE RNA
- 3′ end of cDNA forms hairpin loop
- DNA POLYMERASE USES HAIRPIN LOOP AS PRIMER TO SYNTHESIZE A COMPLEMENTARY DNA STRAND
- TREAT WITH S1 NUCLEASE TO CLEAVE LOOP
- double-stranded cDNA copy of original mRNA

a **genomic DNA library.** But because the genomic DNA is cut into fragments at random, only some fragments will contain genes; many will contain only a portion of a gene, while most of the genomic DNA clones obtained from the DNA of a higher eucaryotic cell will contain only noncoding DNA, which, as we shall discuss in Chapter 8, makes up most of the DNA in such genomes.

An alternative strategy is to begin the cloning process by selecting only those DNA sequences that are transcribed into RNA and thus are presumed to correspond to genes. This is done by extracting the mRNA (or a purified subfraction of the mRNA) from cells and then making a **complementary DNA (cDNA)** copy of each mRNA molecule present; this reaction is catalyzed by the *reverse transcriptase* enzyme of retroviruses, which synthesizes a DNA chain on an RNA template. The single-stranded DNA molecules synthesized by the reverse transcriptase are converted into double-stranded DNA molecules by DNA polymerase, and these molecules are inserted into a plasmid or virus vector and cloned (Figure 7–23). Each clone obtained in this way is called a **cDNA clone,** and the entire collection of clones derived from one mRNA preparation constitutes a **cDNA library.**

There are important differences between genomic DNA clones and cDNA clones, as illustrated in Figure 7–24. Genomic clones represent a random sample of all of the DNA sequences in an organism and, with very rare exceptions, will be the same regardless of the cell type used to prepare them. By contrast, cDNA clones contain only those regions of the genome that have been transcribed into mRNA; as the cells of different tissues produce distinct sets of mRNA molecules, a different cDNA library will be obtained for each type of cell used to prepare the library.

cDNA Clones Contain Uninterrupted Coding Sequences [18]

The use of a cDNA library for gene cloning has several advantages. First, some proteins are produced in very large quantities by specialized cells. In this case,

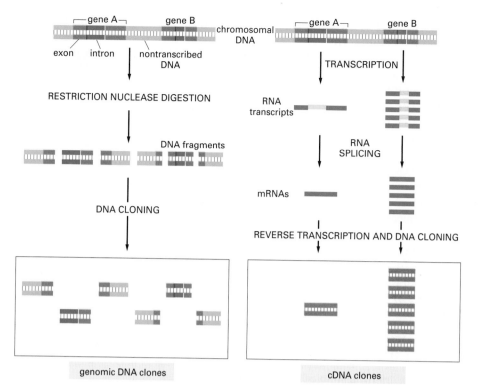

Figure 7–24 The differences between cDNA clones and genomic DNA clones. In this example gene A is infrequently transcribed while gene B is frequently transcribed, and both genes contain introns (*green*). In the genomic DNA clones both the introns and the nontranscribed DNA are included, and most clones will contain only part of the coding sequence of a gene. In the cDNA clones the intron sequences have been removed by RNA splicing during the formation of the mRNA, and a continuous coding sequence is therefore present.

the mRNA encoding the protein is likely to be produced in such large quantities that a cDNA library prepared from the cells will be highly enriched for the cDNA molecules encoding the protein, greatly reducing the problem of identifying the desired clone in the library (see Figure 7–24). Hemoglobin, for example, is made in large amounts by developing erythrocytes (red blood cells); for this reason the globin genes were among the first to be cloned.

By far the most important advantage of cDNA clones is that they contain the uninterrupted coding sequence of a gene. Eucaryotic genes usually consist of short coding sequences of DNA (exons) separated by longer noncoding sequences (introns); the production of mRNA entails the removal of the noncoding sequences from the initial RNA transcript and the splicing together of the coding sequences. Neither bacterial nor yeast cells will make these modifications to the RNA produced from a gene of a higher eucaryotic cell. Thus, if the aim of the cloning is either to deduce the amino acid sequence of the protein from the DNA or to produce the protein in bulk by expressing the cloned gene in a bacterial or yeast cell, it is much preferable to start with cDNA.

Genomic and cDNA libraries are inexhaustible resources that are widely shared among investigators. Today, many such libraries are also available from commercial sources.

cDNA Libraries Can Be Prepared from Selected Populations of mRNA Molecules [19]

When cDNAs are prepared from cells that express the gene of interest at extremely high levels, the majority of cDNA clones may contain the gene sequence, which can therefore be selected with minimal effort. For less abundantly transcribed genes, various methods can be used to enrich for particular mRNAs before making the cDNA library. If an antibody against the protein is available, for example, it can be used to precipitate selectively those polyribosomes (see pp. 237–238) that have the appropriate growing polypeptide chains attached to them. Since these polyribosomes will also have attached to them the mRNA coding for the protein, the precipitate may be enriched in the desired mRNA by as much as 1000-fold.

DNA Cloning

Subtractive hybridization provides a powerful alternative way of enriching for particular nucleotide sequences prior to cDNA cloning. This selection procedure can be used, for example, if two closely related cell types are available from the same organism, only one of which produces the protein or proteins of interest. It was first used to identify cell-surface receptor proteins present on T lymphocytes but not on B lymphocytes. It can also be used wherever a cell that expresses the protein has a mutant counterpart that does not. The first step is to synthesize cDNA molecules using the mRNA from the cell type that makes the protein of interest. These cDNAs are then hybridized with a large excess of mRNA molecules from the second cell type. Those rare cDNA sequences that fail to find a complementary mRNA partner are likely to represent mRNA sequences present only in the first cell type. Because these cDNAs remain unpaired after the hybridization, they can be purified by a simple biochemical procedure (a hydroxyapatite column) that separates single-stranded from double-stranded nucleic acids (Figure 7–25). Besides providing a powerful way to clone genes whose products are known to be restricted to a specific differentiated cell type, cDNA libraries prepared after subtractive hybridization are useful for defining the differences in gene expression between any two related types of cells.

Either a DNA Probe or a Test for Expressed Protein Can Be Used to Identify the Clones of Interest in a DNA Library [20]

The most difficult part of gene cloning is often the identification of the rare colonies in the library that contain the DNA fragment of interest. This is especially true in the case of a genomic library, where one has to identify one bacterial cell in a million to select a specific mammalian gene. The technique most frequently used is a form of *in situ* hybridization that takes advantage of the exquisite specificity of the base-pairing interactions between two complementary nucleic acid molecules. Culture dishes containing the growing bacterial colonies are blotted with a piece of filter paper, to which some members of each bacterial colony adhere. The adhering colonies, known as *replicas,* are treated with alkali to disrupt the cells and to separate the strands of their DNA molecules; the paper is then incubated with either a radioactive or a chemically labeled DNA probe containing part of the sequence of the gene being sought (Figure 7–26). If necessary, millions of bacterial clones can be screened in this way to find the one clone that hybridizes with the probe.

In order to find the clone of interest, a specific probe must be made. How this is done will depend on the information that is available about the gene to be cloned. In many cases the protein of interest has been identified by biochemical studies and purified in small amounts. Only a few micrograms of pure protein are often enough to determine the sequence of 30 or so amino acid residues. From this amino acid sequence the corresponding nucleotide sequence can be deduced using the genetic code (with some ambiguities corresponding to amino acids that can be represented by several alternative codons). Two sets of DNA oligonucleotides, chosen to match different parts of the predicted nucleotide sequence of the gene, are then synthesized by chemical methods (Figure 7–27). Colonies of cells that hybridize with both sets of DNA probes are strong candidates for containing the desired gene and are saved for further characterization (see below).

Probes can also be obtained in other ways. If an antibody is available that recognizes the protein produced by the gene, it can be labeled and used as a probe to find a clone that is producing the protein, which therefore contains the desired gene. Any other ligand that is known to bind to the protein encoded by the gene can also be used as a probe: if the gene encodes a receptor protein, for example, the ligand that normally binds to the receptor can, in principle, be used as a probe.

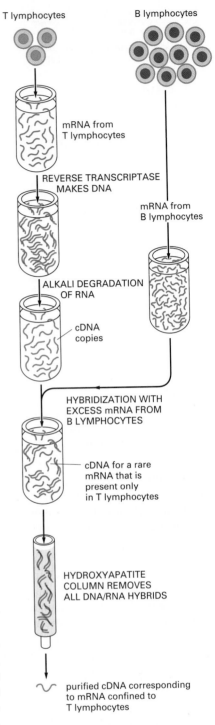

T lymphocytes

B lymphocytes

mRNA from
T lymphocytes

REVERSE TRANSCRIPTASE
MAKES DNA

mRNA from
B lymphocytes

ALKALI DEGRADATION
OF RNA

cDNA
copies

HYBRIDIZATION WITH
EXCESS mRNA FROM
B LYMPHOCYTES

cDNA for a rare
mRNA that is
present only
in T lymphocytes

HYDROXYAPATITE
COLUMN REMOVES
ALL DNA/RNA HYBRIDS

purified cDNA corresponding
to mRNA confined to
T lymphocytes

Figure 7–25 Subtractive hybridization. In this example the technique is used to purify rare cDNA clones corresponding to mRNA molecules present in T lymphocytes but not in B lymphocytes. Because the two cell types are very closely related, most of the mRNAs will be common to both cell types; subtractive hybridization is thus a powerful way to enrich for those specialized molecules that distinguish the two cells.

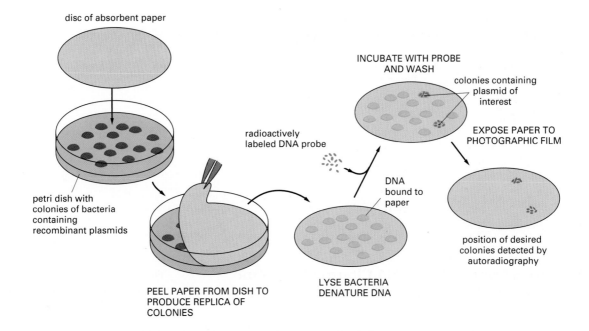

disc of absorbent paper

petri dish with
colonies of bacteria
containing
recombinant plasmids

PEEL PAPER FROM DISH TO
PRODUCE REPLICA OF
COLONIES

radioactively
labeled DNA probe

INCUBATE WITH PROBE
AND WASH

colonies containing
plasmid of
interest

LYSE BACTERIA
DENATURE DNA

DNA
bound to
paper

EXPOSE PAPER TO
PHOTOGRAPHIC FILM

position of desired
colonies detected by
autoradiography

Figure 7–26 An efficient technique commonly used to detect a bacterial colony carrying a particular DNA clone. A replica of the culture is made by pressing a piece of absorbent paper against the surface. This replica is treated with alkali (to disrupt the cells and denature the plasmid DNA) and then hybridized to a highly radioactive DNA probe. Those bacterial colonies that have bound the probe are identified by autoradiography. (See also Figure 7–22.)

Whenever the protein product of a gene is to be detected rather than the gene itself, a special type of cDNA library is required. It is prepared in a special plasmid or virus called an *expression vector*, which directs the transfected bacterium to synthesize large amounts of the protein encoded by the foreign DNA insert contained within the vector's DNA, as we shall discuss later.

In Vitro Translation Facilitates Identification of the Correct DNA Clone [21]

Any method that is used to find a specific clone from a cDNA or genomic DNA library will usually pick out many false positive clones. Further ingenuity is required to discriminate between these and the authentic clones desired. The task is easiest when the desired clone encodes a protein that has already been characterized by other means. In this case each candidate DNA can be tested by one of several methods for its ability to encode the appropriate protein. The cloned DNA can be inserted into an expression vector, for example, so that the protein that it encodes is produced in large amounts in a bacterium. Alternatively, the cloned DNA can be used to obtain a corresponding RNA molecule, either through *in vitro* synthesis with a purified RNA polymerase (see Figure 7–36) or by a technique called *hybrid selection*. In the latter method a mixture of cellular RNAs is added to an excess of single strands of the candidate DNA, and DNA/RNA hybridization is used to purify complementary mRNA molecules from the mixture. In either case the mRNA obtained is allowed to direct protein synthesis in a cell-free system using radioactive amino acids, and the radioactive protein produced is then characterized and compared with the expected protein product of the desired clone. A match in any of these tests allows one to conclude that a cloned DNA fragment encodes the correct protein.

DNA Cloning

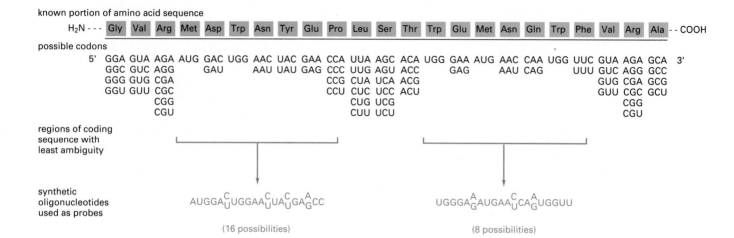

known portion of amino acid sequence

possible codons

regions of coding sequence with least ambiguity

synthetic oligonucleotides used as probes

(16 possibilities)　　　　　(8 possibilities)

The Selection of Overlapping DNA Clones Allows One to "Walk" Along the Chromosome to a Nearby Gene of Interest [22]

Many of the most interesting genes—for example, those that control development—are known only from genetic analysis of mutants in such organisms as the fruit fly *Drosophila* and the nematode *Caenorhabditis elegans*. The protein products of these genes are unknown and may be present in very small quantities in a few cells or produced only at one stage of development. A study of the genetic linkage between different mutations, however, can be used to generate *chromosome maps,* which give the relative locations of the genes (see Figure 7–16). Once one mapped gene has been cloned, the clones in a genomic DNA library that correspond to neighboring genes can be identified using a technique called **chromosome walking.** The methods described in this chapter can then be used to deduce the exact structure and function of the gene of interest and the protein that it encodes.

In chromosome walking one starts with a DNA clone corresponding to a gene or an RFLP marker that is known to be as close as possible to the gene of interest. One end of this clone is used to prepare a DNA probe, which is then used in DNA hybridization experiments to find an overlapping clone in a genomic DNA clone library. The DNA from this second DNA clone is purified, and its far end is used to prepare a second DNA probe, which is used to find a clone that is overlapping, and so on. In this way one can walk along a chromosome one clone at a time, in steps of 30,000 base pairs or more in either direction (Figure 7–28).

How does one know when the gene of interest (identified originally by a deleterious mutation) has been reached, given that the walk is generally too long for complete DNA sequencing to be practicable? For experimental organisms such as fruit flies, nematodes, *Arabidopsis*, yeast, and mice, the ultimate proof of the correct gene is to transfer the normal form of the gene (as a cloned DNA molecule) into a chromosome of the mutant organism, producing a transgenic organism (see Figures 7–45 and 7–49). If the original mutation was a recessive one, the correct DNA should reverse the original mutant phenotype. Other, less stringent criteria, however, are often used and are necessary in the case of human genes, as described later (see Figure 7–30).

Figure 7–27 Selecting regions of a known amino acid sequence to make synthetic oligonucleotide probes. Although only one nucleotide sequence will actually code for the protein, the degeneracy of the genetic code means that several different nucleotide sequences will give the same amino acid sequence, and it is impossible to tell in advance which is the correct one. Because it is desirable to have as large a fraction of the correct nucleotide sequence as possible in the mixture of oligonucleotides to be used as a probe, those regions with the fewest possibilities are chosen, as illustrated. In this example the mixture of 8 closely related oligonucleotides shown might be synthesized and used to probe a clone library, and the indicated mixture of 16 oligonucleotides would be used to reprobe all positive clones to find those that actually code for the desired protein. After the oligonucleotide mixture is synthesized by chemical means, the 5′ end of each oligonucleotide is radioactively labeled (see Figure 7–6B); alternatively, the probe can be marked with a chemical label by incorporating a modified nucleotide during its synthesis (see Figure 7–18).

Ordered Genomic Clone Libraries Are Being Produced for Selected Organisms [23]

The whole task of identifying mutant genes should become vastly easier as knowledge of the sequence of the normal genome becomes more complete and systematic. By using methods related to those described for chromosome walking, it has been possible to order (map) a complete, or nearly complete, set of large

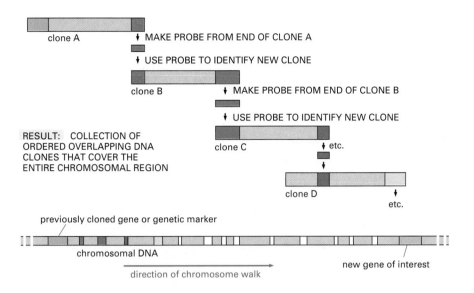

Figure 7–28 The use of overlapping DNA clones to find a new gene by "chromosome walking." To speed up the walk, genomic libraries containing very large cloned DNA molecules are optimal. To probe for the next clone in the walk by DNA hybridization, a short DNA fragment (labeled with a chemical or a radioisotope) from one end of the previously identified clone is purified: If a "right-handed" end is used, for example, the walk will go in the "rightward" direction, as shown in this example. Use of a small end fragment as a probe also reduces the probability that the probe will contain a repeated DNA sequence that would hybridize with many clones from different parts of the genome and thereby interrupt the walk.

In the figure:
clone A
↓ MAKE PROBE FROM END OF CLONE A
↓ USE PROBE TO IDENTIFY NEW CLONE
clone B
↓ MAKE PROBE FROM END OF CLONE B
↓ USE PROBE TO IDENTIFY NEW CLONE
RESULT: COLLECTION OF ORDERED OVERLAPPING DNA CLONES THAT COVER THE ENTIRE CHROMOSOMAL REGION
clone C
↓ etc.
↓
clone D
↓
etc.
previously cloned gene or genetic marker
chromosomal DNA
new gene of interest
direction of chromosome walk

genomic clones along the chromosomes of the *E. coli* bacterium, the yeast *Saccharomyces cerevisiae*, the fruit fly *Drosophila*, the plant *Arabidopsis*, and the nematode *C. elegans*. Such large clones, each about 30,000 base pairs in length, are usually prepared in bacteriophage lambda vectors called *cosmids,* which are specially designed to accept only large DNA inserts. It takes a few thousand such clones to cover the entire genome of an organism such as *C. elegans* or *Drosophila.* To map the entire human genome in this way would require ordering more than 100,000 clones in cosmids, which is very time consuming but technically feasible. DNA fragments that are more than 10 times larger than these clones (300,000 to 1.5 million base pairs) can be cloned in yeast cells as *YACs (yeast artificial chromosomes)* (Figure 7–29); in principle, the human genome could be represented as about 10,000 clones of this type (see Figure 8–5).

In the near future, ordered sets of genomic clones will no doubt be available from centralized DNA libraries for use by all research workers. Eventually, a complete library will be available for each commonly studied organism, with each DNA fragment catalogued according to its chromosome of origin and numbered sequentially with respect to the positions of all other DNA fragments derived from the same chromosome. One will then begin a "chromosome walk" simply by obtaining from the library all the clones covering the region of the genome that contains the mutant gene of interest.

Positional DNA Cloning Reveals Human Genes with Unanticipated Functions [24]

Thousands of human diseases are caused by alterations in single genes. Our understanding of these genetic diseases is being revolutionized by recombinant

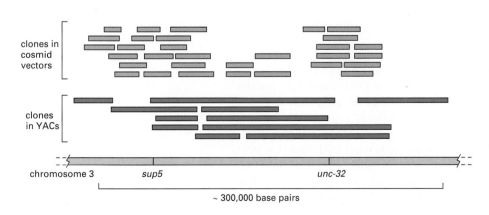

clones in cosmid vectors
clones in YACs
chromosome 3
sup5
unc-32
~ 300,000 base pairs

Figure 7–29 Overlapping genomic DNA clones. The collection of clones shown covers a small region of a chromosome of the nematode worm *Caenorhabditis elegans* and represents 0.3% of the total genome. (Adapted from J. Sulston et al., *Nature* 356:37–41, 1992. © 1992 Macmillan Magazines Ltd.)

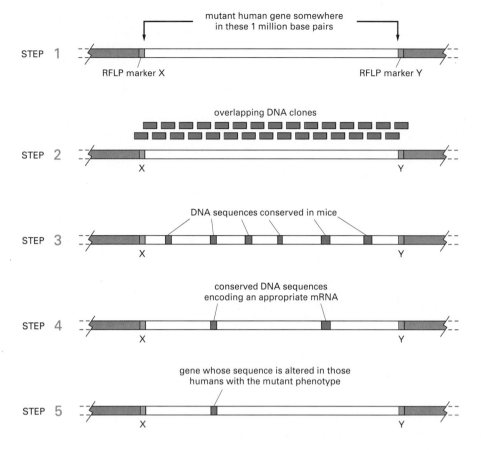

STEP 1

mutant human gene somewhere
in these 1 million base pairs

RFLP marker X   RFLP marker Y

overlapping DNA clones

STEP 2

X   Y

DNA sequences conserved in mice

STEP 3

X   Y

conserved DNA sequences
encoding an appropriate mRNA

STEP 4

X   Y

gene whose sequence is altered in those
humans with the mutant phenotype

STEP 5

X   Y

Figure 7–30 Positional cloning. The procedure requires a mutant human gene whose inheritance can be traced in many family groups by virtue of the phenotype that the mutation causes. Step 1, *genetic mapping:* RFLP markers that are coinherited with the phenotype are identified and used to position the gene within about 10^6 base pairs (one megabase, or about 1% the length of a typical human chromosome). Step 2, *assembly of an ordered clone library:* genomic DNA clones are obtained that cover the entire region between two RFLP markers that bracket the gene. Step 3, *search for conserved DNA sequences:* the portions of each DNA clone that hybridize with mouse DNA are identified; only those regions of the human chromosome whose nucleotide sequence is important will have been sufficiently conserved during evolution to form such a hybrid DNA helix (one strand human and the other mouse). Step 4, *search for appropriate mRNAs:* the subset of conserved DNA sequences that encode an mRNA in tissues where the mutant phenotype is expressed are the most likely to represent the mutant gene. Step 5, *finding a difference in the DNA sequence of mutated genes.* When the deleterious mutations in a typical human gene are analyzed, about 1 in 10 turns out to be a deletion that is easily detected as a change in the size of a restriction fragment detected by Southern blotting. For this reason one generally begins by screening the DNA of many human patients with the same disease using probes identified in step 4, looking for such a change. If the mutation is not detectable as a deletion, other, more laborious methods that are capable of detecting single base changes must be used to identify the gene of interest.

DNA methods, which allow the altered DNA to be cloned and sequenced, revealing the precise defect in each patient. In this way, for example, Duchenne's muscular dystrophy was shown to be due to an abnormal cytoskeletal protein in muscle cells, and cystic fibrosis to be an abnormal chloride channel in epithelial cells. This knowledge not only improves the accuracy of diagnosis but also makes it possible, in principle at least, to design treatments.

Although the techniques used to find human disease genes have differed depending on the disease, a standard approach has recently been developed that makes it possible to isolate any human gene that is responsible by itself for a specific trait or disease. It is called **positional cloning** because it starts with genetic linkage mapping to locate the gene in the genome (Figure 7–30). While the approach is straightforward, it presently requires 10 to 100 person-years to isolate a gene in this way. It will become much easier once DNA sequencing is highly automated and the full DNA sequence of the human genome is known: genetic linkage mapping will reveal immediately which genes are prime suspects, and the sequences of these genes can then be analyzed directly in individual patients. The functions of thousands of human genes are likely to be identified in this way.

Selected DNA Segments Can Be Cloned in a Test Tube by a Polymerase Chain Reaction [25]

The availability of purified DNA polymerases and chemically synthesized DNA oligonucleotides has made it possible to clone specific DNA sequences rapidly without the need for a living cell. The technique, called the **polymerase chain reaction (PCR),** allows the DNA from a selected region of a genome to be amplified a billionfold, provided that at least part of its nucleotide sequence is already known. First, the known part of the sequence is used to design two synthetic DNA oligonucleotides, one complementary to each strand of the DNA double helix and lying on opposite sides of the region to be amplified. These oligonucleotides serve as primers for *in vitro* DNA synthesis, which is catalyzed by a DNA polymerase, and they determine the ends of the final DNA fragment that is obtained (Figure 7–31).

The principle of the PCR technique is illustrated in Figure 7–32. Each cycle of the reaction requires a brief heat treatment to separate the two strands of the genomic DNA double helix (step 1). The success of the technique depends on the use of a special DNA polymerase isolated from a thermophilic bacterium that is stable at much higher temperatures than normal, so that it is not denatured by the repeated heat treatments. A subsequent cooling of the DNA in the presence of a large excess of the two primer DNA oligonucleotides allows these oligonucleotides to hybridize to complementary sequences in the genomic DNA (step 2). The annealed mixture is then incubated with DNA polymerase and the four deoxyribonucleoside triphosphates so that the regions of DNA downstream from each of the two primers are selectively synthesized (step 3). When the procedure is repeated, the newly synthesized fragments serve as templates in their turn, and within a few cycles the predominant product is a single species of DNA fragment whose length corresponds to the distance between the two original primers. In practice, 20 to 30 cycles of reaction are required for effective DNA amplification. Each cycle doubles the amount of DNA synthesized in the previous cycle. A single cycle requires only about 5 minutes, and an automated procedure permits "cell-free molecular cloning" of a DNA fragment in a few hours, compared with the several days required for standard cloning procedures.

The PCR method is extremely sensitive; it can detect a single DNA molecule in a sample. Trace amounts of RNA can be analyzed in the same way by first transcribing them into DNA with reverse transcriptase. The PCR cloning technique is rapidly replacing Southern blotting for the diagnosis of genetic diseases and for the detection of low levels of viral infection. It also has great promise in forensic medicine as a means of analyzing minute traces of blood or other tissues—even as little as a single cell—and identifying the person from whom they came by his or her genetic "fingerprint" (Figure 7–33).

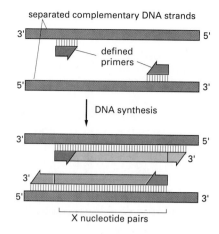

Figure 7–31 The start of the polymerase chain reaction (PCR) for amplifying specific nucleotide sequences *in vitro*. DNA isolated from cells is heated to separate its complementary strands. These strands are then annealed with an excess of two DNA oligonucleotides (each 15 to 20 nucleotides long) that have been chemically synthesized to match sequences separated by X nucleotides (where X is generally between 50 and 2000). The two oligonucleotides serve as specific primers for *in vitro* DNA synthesis catalyzed by DNA polymerase, which copies the DNA between the sequences corresponding to the two oligonucleotides.

Figure 7–32 PCR amplification. PCR produces an amount of DNA that doubles in each cycle of DNA synthesis and includes a uniquely sized DNA species. Three steps constitute each cycle, as described in the text. After many cycles of reaction, the population of DNA molecules becomes dominated by a single DNA fragment, X nucleotides long, provided that the original DNA sample contains the DNA sequence that was anticipated when the two oligonucleotides were designed. In the example illustrated, three cycles of reaction produce 16 DNA chains, 8 of which have this unique length (*yellow*); but after three more cycles, 240 of the 256 DNA chains would be X nucleotides long.

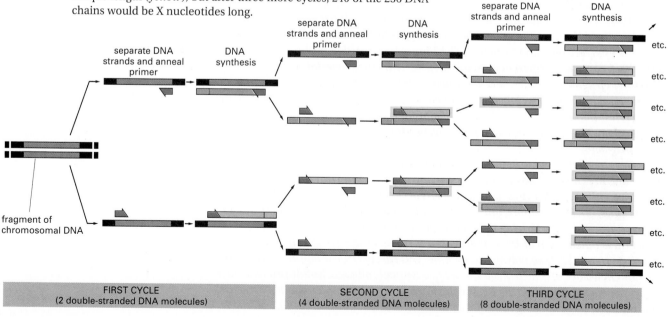

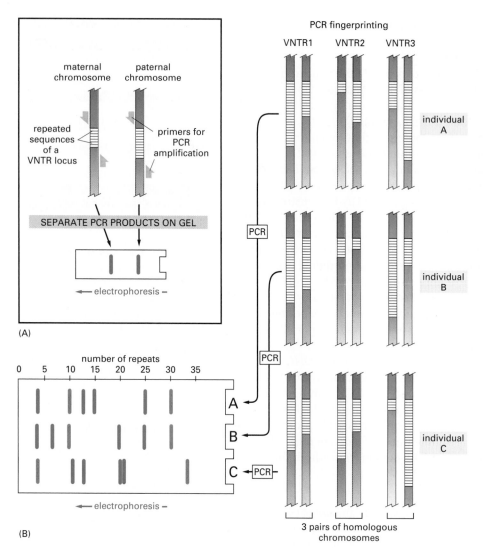

(A)

(B)

Figure 7–33 The use of PCR in forensic science. (A) A PCR reaction using two primers that bracket a particular microsatellite, or VNTR, sequence (see Figure 7–14C) produces a different pair of DNA bands from each individual. One of these bands contains the repeated VNTR sequence that was inherited from the individual's mother and the other contains the repeated VNTR sequence that was inherited from the individual's father. (B) The large set of DNA bands obtained from a set of different PCR reactions, each of which amplifies the DNA from a different VNTR sequence, can serve as a "fingerprint" to identify each individual nearly uniquely. The starting material for the PCR reaction can be a single hair that was left at the scene of a crime.

Summary

DNA cloning allows a copy of any specific part of a DNA or RNA sequence to be selected from the millions of other sequences in a cell and produced in unlimited amounts in pure form. DNA sequences are amplified after cutting chromosomal DNA with a restriction nuclease and inserting the resulting DNA fragments into the chromosome of a self-replicating genetic element (a plasmid or a virus). When a plasmid vector is used, the resulting "genomic DNA library" is housed in millions of bacterial cells, each carrying a different cloned DNA fragment. The bacterial colony containing a DNA fragment of interest is identified by hybridization using a DNA probe or, following expression of a cloned gene or gene fragment in the bacterial host cell, by using a test that detects the desired protein product. The cells in the identified bacterial colony are then allowed to proliferate, producing large amounts of the desired DNA fragment.

The procedure used to obtain DNA clones that correspond in sequence to mRNA molecules are the same except that the starting material is a DNA copy of the mRNA sequence, called cDNA, rather than fragments of chromosomal DNA. Unlike genomic DNA clones, cDNA clones lack intron sequences, making them the clones of choice for expressing and characterizing the protein product of a gene.

PCR is a new form of DNA cloning that is carried out outside cells using a purified, thermostable DNA polymerase enzyme. This type of DNA amplification requires a prior knowledge of gene sequence, since two synthetic oligonucleotide primers must be synthesized that bracket the DNA sequence to be amplified. PCR cloning, however, has the advantage of being much faster and easier than standard cloning methods.

DNA Engineering [26]

The methods described thus far in this chapter make it possible, in principle, to determine the exact organization and nucleotide sequence of any chromosome, including all of the chromosomes that constitute the human genome. In this section we describe how extensions of these methods have revolutionized all other aspects of cell biology by providing new ways to study the functions of genes, RNA molecules, and proteins.

New DNA Molecules of Any Sequence Can Be Formed by Joining Together DNA Fragments [27]

As we have seen, recombinant DNA molecules are generally constructed by using the enzyme DNA ligase to join together two DNA molecules with matching cohesive ends. For production of a DNA library, one of the two DNA molecules to be joined is the *vector*, derived from a bacterial plasmid or virus, while the other is either a fragment of a chromosome or a cDNA molecule (see Figure 7–21). A recombinant DNA molecule can in turn serve as a vector for cloning additional DNA molecules, and by the stepwise repetition of the cloning procedure, a series of DNA fragments can be joined together end to end. In this way new DNA molecules can be generated that are different from any molecule that occurs naturally (Figure 7–34).

Automated oligonucleotide synthesizers rapidly produce any DNA molecule containing up to about 100 nucleotides. The sequence of such synthetic DNA molecules is entirely determined by the experimenter, and they can be joined together by repeated DNA cloning steps in various combinations to produce long custom-designed DNAs of any sequence. But the DNA sequences of interest to cell biologists are mostly those that code for protein domains that already exist in nature. The polymerase chain reaction (PCR) can be used both to amplify any natural nucleotide sequence and to redesign its two ends, so that any two natu-

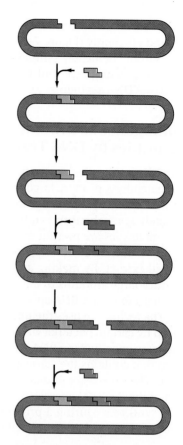

Figure 7–34 **Serial DNA cloning can be used to splice together a set of DNA fragments derived from different genes.** All of the DNA molecules shown here are double-stranded. After each DNA insertion step, the plasmid is cloned to purify and amplify the new recombinant DNA molecule. The recombinant molecule is then cut once with a restriction nuclease, as indicated, and used as a cloning vector for the next DNA fragment.

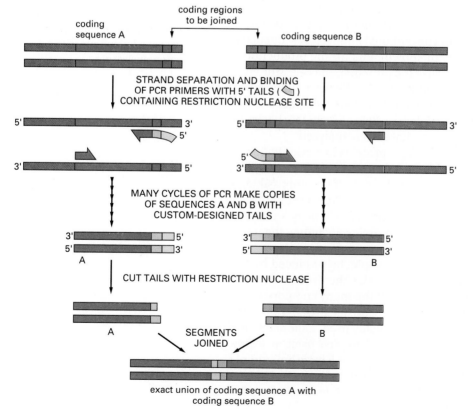

Figure 7–35 **Tailoring the ends of DNA fragments facilitates their precise joining.** The PCR reaction allows any DNA segment to be amplified with modified ends in preparation for cloning. The synthetic oligonucleotide primers used here contain 5′ tails whose sequence has been chosen to create a particular restriction nuclease cutting site.

rally occurring DNA sequences can be amplified and spliced together very rapidly and efficiently (Figure 7–35). Thus, rather than generating novel DNA molecules from chemically synthesized fragments, it is almost always easier to produce them using PCR and repeated cloning steps to link selected naturally occurring DNA segments.

Homogeneous RNA Molecules Can Be Produced in Large Quantities by DNA Transcription *in Vitro* [28]

Pure RNA molecules are useful for many types of cell biological studies. If the RNA encodes a protein, large amounts of the pure species will facilitate *in vitro* studies of RNA splicing or protein synthesis; if the RNA has a catalytic function, its catalytic mechanism can be tested in cell-free systems. But most species of RNA molecules are present in only tiny quantities in cells, and they are very difficult to purify away from the many thousands of other species present in a typical cell extract. By making available pure DNA templates that can be used to produce large amounts of any RNA molecule, genetic engineering provides the best path to a pure RNA species.

The technique uses the unusually efficient RNA polymerase molecules produced by certain bacterial viruses. To prepare a suitable DNA template, the DNA sequence encoding the RNA is cloned in a way that joins it to a second DNA segment that contains the start signal (promoter) for the viral RNA polymerase enzyme. The pure recombinant DNA molecule is then mixed with a pure preparation of the RNA polymerase plus the four ribonucleoside triphosphates used in RNA synthesis. During a prolonged incubation at 37°C, large amounts of the desired RNA are then generated by *in vitro* transcription (Figure 7–36).

Rare Cellular Proteins Can Be Made in Large Amounts Using Expression Vectors [29]

Until recently, the only proteins in a cell that could be studied easily were the relatively abundant ones. Starting with several hundred grams of cells, a major protein—one that constitutes 1% or more of the total cellular protein—can be purified by sequential chromatography steps to yield perhaps 0.1 g (100 mg) of pure protein. This amount is sufficient for conventional amino acid sequencing, for detailed analysis of biological activity (if known), and for the production of antibodies, which can then be used to localize the protein in the cell. Moreover, if suitable crystals can be grown (usually a difficult task), the three-dimensional structure of the protein can be determined by x-ray diffraction techniques. In this way the structure and function of many abundant proteins—including hemoglobin, trypsin, immunoglobulin, and lysozyme—have been analyzed.

The vast majority of the thousands of different proteins in a eucaryotic cell, however, including many with crucially important functions, are present in very small amounts. For most of them it is extremely difficult, if not impossible, to obtain more than a few micrograms of pure material. One of the most important contributions of DNA cloning and genetic engineering to cell biology is that they have made it possible to produce any of the cell's proteins, including the minor ones, in large amounts.

Large amounts of a pure protein could, in principle, be produced by first synthesizing a large amount of the mRNA that encodes the protein and then translating it in a cell-free system that contains all of the many components required for protein synthesis, including ribosomes, transfer RNAs, translation enzymes, and so on. While this approach is occasionally used, a more efficient way is to produce both the mRNA and the protein in a living cell, using an *expression vector* (Figure 7–37). The vector is designed to produce a large amount of a stable mRNA that will be efficiently translated into protein in the transfected bacterial, yeast, insect, or mammalian cell. To prevent the high level of the for-

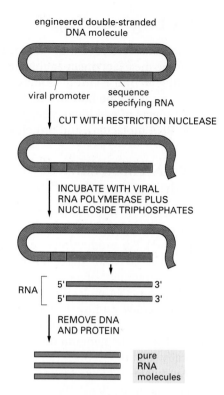

Figure 7–36 Large amounts of any RNA molecule can be prepared *in vitro* using a viral RNA polymerase. The use of this unusually efficient enzyme requires that the DNA template be engineered so that the short DNA sequence that specifies the viral promoter is adjacent to the DNA sequence that encodes the RNA molecule to be synthesized. As indicated, the 3′ end of the RNA is determined by cutting the DNA template at a unique site with a restriction nuclease.

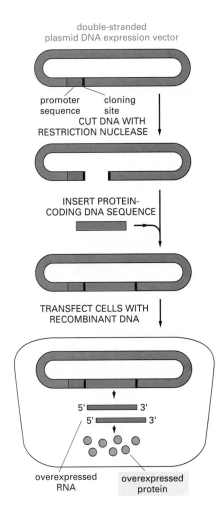

promoter   cloning
sequence   site
CUT DNA WITH
RESTRICTION NUCLEASE

INSERT PROTEIN-
CODING DNA SEQUENCE

TRANSFECT CELLS WITH
RECOMBINANT DNA

5'   3'
5'   3'

overexpressed   overexpressed
RNA   protein

Figure 7–37 Production of large amounts of a protein by cloning the protein-coding DNA sequence in a plasmid expression vector. The plasmid has been engineered to contain a highly active promoter, which causes unusually large amounts of mRNA encoding the protein to be synthesized in the transfected cells.

eign protein from interfering with the transfected cell's growth, the expression vector is often designed so that the synthesis of the foreign mRNA and protein can be delayed until shortly before the cells are harvested (Figure 7–38).

Because the desired protein made from an expression vector is produced inside a cell, it must be purified away from the host cell proteins by chromatography following cell lysis; but because it is such a plentiful species in the cell lysate (often 1% to 10% of the total cell protein), the purification is usually easy to accomplish in only a few steps. A variety of expression vectors are available, each engineered to function in the type of cell in which the protein is to be made. In this way cells can be induced to make vast quantities of medically useful proteins—such as human insulin and growth hormone, interferon, and viral antigens for vaccines. More generally, these methods make it possible to produce every protein in large enough amounts for the kinds of detailed structural and functional studies that were previously possible for only a rare few.

Reporter Genes Enable Regulatory DNA Sequences to Be Dissected [30]

The transcription of a gene is controlled by regulatory DNA sequences that are not transcribed. These sequences determine which cells will express the gene and under what conditions, as explained in Chapter 9. In higher eucaryotes these regulatory sequences can be located many thousands of base pairs upstream or downstream from the transcribed sequences, and they act in combinations to control gene transcription. The regulatory DNA sequences can be identified for any particular gene by a manipulation that involves replacing part of the gene's coding sequence with a different coding sequence (a so-called *reporter sequence*), selected because the protein it encodes is easily detected by either a simple cytological stain or an enzyme assay. The regulatory sequences for the gene of interest can then be identified by attaching the reporter sequence to various fragments of DNA sequence taken from upstream or downstream of the gene. When such recombinant DNA molecules are used to transfect cells, the level, timing, and cell specificity of reporter protein production will reflect the action of the regulatory sequences that each DNA construct contains (Figure 7–39).

Mutant Organisms Best Reveal the Function of a Gene [31]

Suppose that one has cloned a gene that codes for a protein whose function is unknown. How can one discover what the protein does in the cell? This has become a common problem in cell biology and one that is surprisingly difficult to solve, since neither the three-dimensional structure of the protein nor the com-

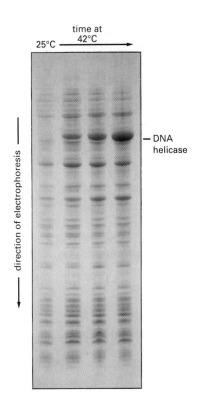

time at
42°C
25°C

— DNA
helicase

direction of electrophoresis

Figure 7–38 Production of large amounts of a protein using a plasmid expression vector. In this example bacterial cells have been transfected with the coding sequence for an enzyme, DNA helicase (*arrow*); transcription from this coding sequence is under the control of a viral promoter that becomes active only at temperatures of 37°C or higher. The total cell protein has been analyzed by SDS-polyacrylamide gel electrophoresis, either from bacteria grown at 25°C (no helicase protein made), or after a shift of the same bacteria to 42°C for up to two hours. (Photograph courtesy of Jack Barry.)

DNA Engineering

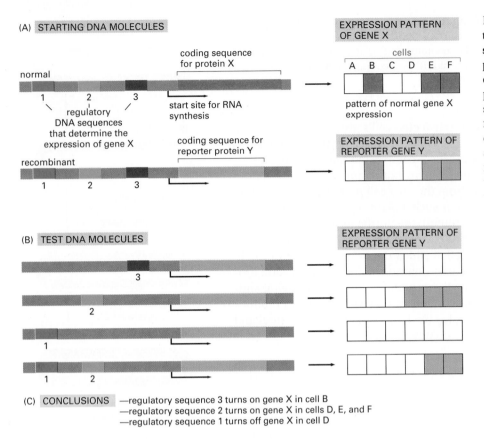

(A) STARTING DNA MOLECULES

coding sequence
for protein X

normal

1   2   3
regulatory
DNA sequences
that determine the
expression of gene X

start site for RNA
synthesis

coding sequence for
reporter protein Y

recombinant

1   2   3

EXPRESSION PATTERN
OF GENE X

cells
A B C D E F

pattern of normal gene X
expression

EXPRESSION PATTERN OF
REPORTER GENE Y

(B) TEST DNA MOLECULES

3

2

1

1   2

EXPRESSION PATTERN OF
REPORTER GENE Y

(C) CONCLUSIONS —regulatory sequence 3 turns on gene X in cell B
—regulatory sequence 2 turns on gene X in cells D, E, and F
—regulatory sequence 1 turns off gene X in cell D

Figure 7–39 Using a reporter protein to locate the regulatory DNA sequences that determine the pattern of a gene's expression. In this example the coding sequence for protein X is replaced by the coding sequence for protein Y, and various fragments of DNA containing candidate regulatory sequences are added in combinations. The recombinant DNA molecules are then tested for expression after their transfection into a variety of different types of mammalian cells. For experiments in eucaryotic cells, two commonly used reporters are the enzymes β-galactosidase (*β-gal*) and chloramphenicol acetyltransferase (*CAT*). Because these are bacterial enzymes, their presence can be monitored by simple and sensitive assays of enzyme activity, without any interference from host cell enzymes.

plete nucleotide sequence of its gene is usually sufficient to deduce the protein's function. Moreover, many proteins—such as those that have a structural role in the cell or normally form part of a large multienzyme complex—will have no obvious activity by themselves.

One approach, discussed in Chapter 4, is to inactivate the protein by injecting into a cell a specific antibody that recognizes the protein and then observing how the cell is affected. Although this provides a useful way to test protein function, the effect is transitory because the injected antibody is eventually diluted out during cell proliferation or destroyed by intracellular proteolytic enzymes. Moreover, many antibodies—even those that bind tightly to some part of the given protein—fail to block the protein's function.

Genetics provides a much more powerful solution to this problem. Mutants that lack a particular protein may quickly reveal the function of the normal molecule. Even more useful are mutants in which the abnormal protein is temperature sensitive, so that it is inactivated by a small increase or decrease in temperature, since in these mutants the abnormality can be switched on and off simply by changing the temperature. Before the advent of gene cloning technology, most genes were identified in this way, according to the processes disrupted when mutations occur. The genetic approach is most easily applicable to organisms that reproduce rapidly—such as bacteria, yeasts, nematode worms, and fruit flies. By treating these organisms with agents that alter their DNA (*mutagens*), very large numbers of mutants can be created quickly and then screened for a particular defect of interest. By screening populations of mutagen-treated bacteria for cells that stop making DNA when they are shifted from 30°C to 42°C, for example, many temperature-sensitive mutants were isolated in the genes that encode the bacterial proteins required for DNA replication. These mutants were later used to identify and characterize the corresponding DNA replication proteins. Similarly, the genetic approach has been used to demonstrate the function of enzymes involved in the principal metabolic pathways of bacteria, as well as to discover many of the gene products responsible for the orderly development of the *Drosophila* embryo.

Humans do not reproduce rapidly, and they are not, as a rule, intentionally treated with mutagens. Moreover, any human with a serious defect in an essential process, such as DNA replication, would die long before birth. Many mutations that are compatible with life, however—for example, tissue-specific defects in lysosomes or in cell-surface receptors—have arisen spontaneously in the human population. Analyses of the phenotypes of the affected individuals, together with studies of their cultured cells, have provided many unique insights into important cell functions. Although such mutants are rare, they are very efficiently discovered because of a unique human property: the mutant individuals call attention to themselves by seeking special medical care.

Cells Containing Mutated Genes Can Be Made to Order [32]

Although in rapidly reproducing organisms it is often not difficult to obtain mutants that are deficient in a particular process, such as DNA replication or eye development, it can take a long time to trace the defect to a particular altered protein. Recently, recombinant DNA technology has made possible a different type of genetic approach. Instead of starting with a randomly generated mutant and using it to identify a gene and its protein, one can start with a particular gene and proceed to make mutations in it, creating mutant cells or organisms so as to analyze the gene's function. Because the new approach reverses the traditional direction of genetic discovery—proceeding from genes and proteins to mutants, rather than vice versa—it is commonly referred to as *reverse genetics.*

Reverse genetics begins with a cloned gene or a protein with interesting features that has been isolated from a cell. If the starting point is a protein, the gene encoding it is first cloned and its nucleotide sequence is determined. The gene sequence is then altered by biochemical means to create a mutant gene that codes for an altered version of the protein. The mutant gene is transferred into a cell, where it can integrate into a chromosome by genetic recombination to become a permanent part of the cell's genome. If the gene is expressed, the cell and all of its descendants will now synthesize an altered protein.

If the original cell used for the gene transfer is a fertilized egg, whole multicellular organisms can be obtained that contain the mutant gene, and some of these *transgenic organisms* will pass the gene on to their progeny as a permanent part of their germ line. Such *genetic transformations* are now routinely performed with organisms as complex as fruit flies and mammals (see Figure 7–45). Technically, even humans could now be transformed in this way, although such procedures are not undertaken, even for therapeutic purposes, for fear of the unpredictable aberrations that might occur in such individuals.

Genes Can Be Redesigned to Produce Proteins of Any Desired Sequence [33]

To facilitate reverse genetic studies of protein function, both the coding sequence of a gene and its regulatory regions can be altered to change the functional properties of the protein product, the amount of protein made, or the particular cell type in which the protein is produced.

Special techniques are required to alter a gene in such subtle ways: it is often desirable, for example, to change the protein the gene encodes by one or a few amino acids. The first step in redesigning a gene in this way is the chemical synthesis of a short DNA molecule containing the altered portion of the gene's nucleotide sequence. This synthetic DNA oligonucleotide is hybridized with single-stranded plasmid DNA that contains the DNA sequence to be altered, using conditions that allow imperfectly matched DNA strands to pair (Figure 7–40). The synthetic oligonucleotide will now serve as a primer for DNA synthesis

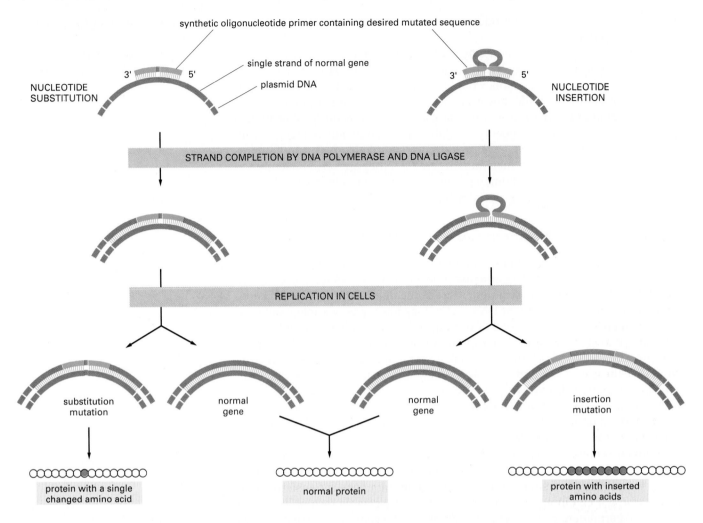

NUCLEOTIDE
SUBSTITUTION

synthetic oligonucleotide primer containing desired mutated sequence

single strand of normal gene

plasmid DNA

3' 5'

3' 5'

NUCLEOTIDE
INSERTION

STRAND COMPLETION BY DNA POLYMERASE AND DNA LIGASE

REPLICATION IN CELLS

substitution
mutation

normal
gene

normal
gene

insertion
mutation

protein with a single
changed amino acid

normal protein

protein with inserted
amino acids

by DNA polymerase, thereby generating a DNA double helix that incorporates the altered sequence into one of its two strands. After transfection, plasmids that carry the fully modified gene sequence are obtained, and the appropriate DNA is inserted into an expression vector so that the redesigned protein can be produced in the appropriate type of cells for detailed studies of its function. By changing selected amino acids in a protein in this way, one can analyze which parts of the polypeptide chain are important in such fundamental processes as protein folding, protein-ligand interactions, and enzymatic catalysis.

Fusion Proteins Are Often Useful for Analyzing Protein Function [34]

A protein will often contain short amino acid sequences that determine its location in a cell, or its stability after the protein has been synthesized. These special regions of the protein can be identified by fusing them to an easily detected reporter protein that lacks such regions and then following the behavior of the reporter protein in a cell (Figure 7–41). Such **fusion proteins** are produced by the recombinant DNA techniques discussed previously. Most nuclear proteins, for example, contain one or more specific short sequences of amino acids that serve as signals for their import into the nucleus after their synthesis in the cytosol. By artificially attaching different segments of such a nuclear protein to a cytoplasmic protein using gene-fusion techniques, the "signal peptides" responsible for nuclear import have been identified.

Not all of the important signals carried by proteins, however, can be transferred to a large reporter protein in the form of a short amino acid sequence. The

Figure 7–40 The use of synthetic oligonucleotides to modify the protein-coding regions of genes. Only two of the many types of changes that can be engineered in this way are shown. With an appropriate oligonucleotide, for example, more than one amino acid substitution can be made at a time, or one or more amino acids can be deleted. As indicated, because only one of the two DNA strands in the original recombinant plasmid is altered by this procedure, only half of the transfected cells will end up with a plasmid that contains the desired mutant gene. Note that most of the plasmid DNA sequence is not illustrated here.

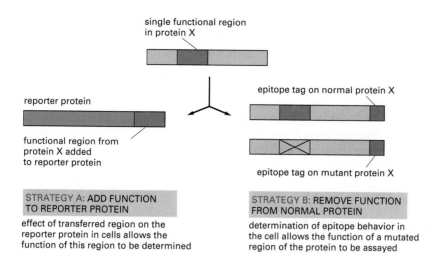

STRATEGY A: ADD FUNCTION TO REPORTER PROTEIN

effect of transferred region on the reporter protein in cells allows the function of this region to be determined

STRATEGY B: REMOVE FUNCTION FROM NORMAL PROTEIN

determination of epitope behavior in the cell allows the function of a mutated region of the protein to be assayed

Figure 7–41 Two strategies for the analysis of protein function that exploit genetic engineering. In strategy A, a new function is imparted to a reporter protein by the addition of a small segment of normal protein X. In strategy B, minor alterations are made in the normal protein X and its behavior is followed by detection of a short epitope tag and compared to that of the normal protein. Both strategies have been used to analyze the structure and function of the nuclear localization signals found on nuclear proteins (see Chapter 12).

information that moves newly synthesized lysosomal proteins to lysosomes from the Golgi apparatus, for example, resides in a "signal patch," whose formation requires that the lysosomal protein fold correctly so as to bring distant portions of the polypeptide chain into close proximity. In these cases a different strategy, called **epitope tagging,** can often be used. Again, a fusion protein is produced, but it contains the entire protein being analyzed plus a short peptide of 8 to 12 amino acids (the "epitope") that can be recognized by an antibody—hence its name. The fusion protein can therefore be specifically followed in cells with the anti-epitope antibody, even in the presence of a large excess of the normal protein, so that one can determine the effect of altering any of the amino acids in the fusion protein on its intracellular location (strategy B in Figure 7–41).

Normal Genes Can Be Easily Replaced by Mutant Ones in Bacteria and Some Lower Eucaryotes [35]

Unlike higher eucaryotes (which are multicellular and diploid), bacteria, yeasts, and the cellular slime mold *Dictyostelium* generally exist as haploid single cells. In these organisms an artificially introduced DNA molecule carrying a mutant gene will, with a relatively high frequency, replace the single copy of the normal gene by *homologous recombination* (see p. 263), so that it is easy to produce cells in which the mutant gene has replaced the normal gene (Figure 7–42A). In this way cells can be made to order that produce an altered form of any specific protein or RNA molecule instead of the normal form of the molecule. If the mutant gene is completely inactive and the gene product normally performs an essential function, the cell will die; but in this case a less severely mutated version of the gene can be used to replace the normal gene, so that the mutant cell survives

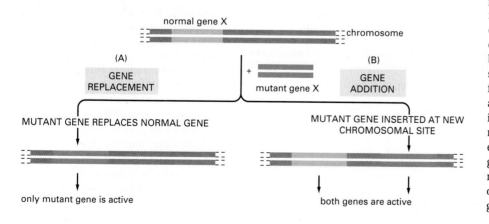

Figure 7–42 Gene replacement and gene addition. A gene whose nucleotide sequence has been altered can be inserted back into the chromosomes of an organism. In bacteria and some haploid eucaryotes such as yeast, the altered gene frequently replaces the normal gene, a process called *gene replacement* (A); in these cases, only the mutant gene remains in the cell. In higher eucaryotes, *gene addition* (B) generally occurs instead of gene replacement; the transformed cell or organism now contains the mutated gene in addition to the normal gene.

but is abnormal in the process for which the gene is required. Often the mutant of choice is one that produces a temperature-sensitive gene product, which functions normally at one temperature but is inactivated when cells are shifted to a higher or lower temperature.

The ability to perform direct *gene replacements* in lower eucaryotes, combined with the power of standard genetic analyses in these haploid organisms, in large part explains why studies in these types of cells have been so important for working out the details of those processes that are shared by all eucaryotes. Gene replacements occur more rarely in higher eucaryotes, for reasons that are not known.

Engineered Genes Can Be Used to Create Specific Dominant Mutations in Diploid Organisms [36]

Higher eucaryotes, such as mammals or fruit flies, are diploid and therefore have two copies of each chromosome. Moreover, transfection with an altered gene generally leads to *gene addition* rather than gene replacement: the altered gene inserts at a random location in the genome, so that the cell (or the organism) ends up with the mutated gene in addition to its normal gene copies (Figure 7–42B).

Because gene addition is much more easily accomplished than gene replacement in higher eucaryotic cells, it would be enormously useful to be able to create specific **dominant negative mutations** in which a mutant gene eliminates the activity of its normal counterparts in the cell. One ingenious and promising approach exploits the specificity of hybridization reactions between two complementary nucleic acid chains. Normally, only one of the two DNA strands in a given portion of double helix is transcribed into RNA, and it is always the same strand for a given gene. If a cloned gene is engineered so that the opposite DNA strand is transcribed instead, it will produce **antisense RNA molecules** that have a sequence complementary to the normal RNA transcripts. Antisense RNA, when synthesized in large enough amounts, will often hybridize with the "sense" RNA made by the normal genes and thereby inhibit the synthesis of the corresponding protein (Figure 7–43). A related method is to synthesize short antisense nucleic acid molecules by chemical or enzymatic means and then inject (or otherwise deliver) them into cells, again blocking (though only temporarily) production of the corresponding protein.

For unknown reasons the antisense RNA approach frequently fails to inactivate the desired gene. An alternative way of producing a dominant negative mutation takes advantage of the fact that most proteins function as part of a

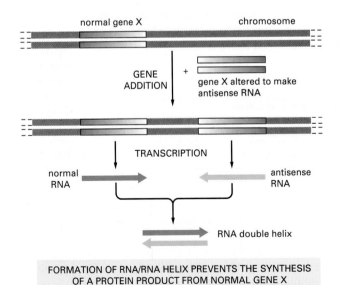

Figure 7–43 The antisense RNA strategy for generating dominant negative mutations. Mutant genes that have been engineered to produce antisense RNA, which is complementary in sequence to the RNA made by the normal gene X, can cause double-stranded RNA to form inside cells. If a large excess of the antisense RNA is produced, it can hybridize with—and thereby inactivate—most of the normal RNA produced by gene X. Although in the future it may become possible to inactivate any gene in this way, at present the technique seems to work for some genes but not others.

normal gene X   chromosome

GENE ADDITION   +   gene X altered to make antisense RNA

TRANSCRIPTION

normal RNA   antisense RNA

RNA double helix

FORMATION OF RNA/RNA HELIX PREVENTS THE SYNTHESIS OF A PROTEIN PRODUCT FROM NORMAL GENE X

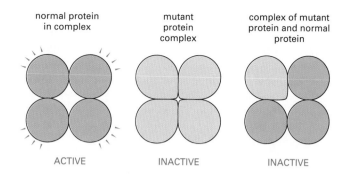

normal protein in complex   mutant protein complex   complex of mutant protein and normal protein

ACTIVE   INACTIVE   INACTIVE

Figure 7–44 A dominant negative effect of a protein. Here a gene is engineered to produce a mutant protein that prevents the normal copies of the same protein from performing their function. In this simple example the normal protein must form a multi-subunit complex to be active, and the mutant protein blocks function by forming a mixed complex that is inactive. In this way a single copy of a mutant gene located anywhere in the genome can inactivate the normal products produced by other gene copies.

larger protein complex. Such complexes can be inactivated by the inclusion of just one nonfunctional component. Therefore, by designing a gene that produces large quantities of a mutant protein that is inactive but still able to assemble into the complex, it is often possible to produce a cell in which all the complexes are inactivated despite the presence of both normal and mutant copies of the protein (Figure 7–44).

If a protein is required for the survival of the cell (or the organism), a dominant negative mutant will die, making it impossible to test the function of the protein. To avoid this problem, one can couple the mutant gene to control sequences that have been engineered to produce the gene product only on command—for example, in response to an increase in temperature or to the presence of a specific signaling molecule. Cells or organisms containing such an *inducible* dominant mutant gene can be deprived of a specific protein at a particular time, and the effect can then be followed. In the future, techniques for producing dominant negative mutations to inactivate specific genes are likely to be widely used to determine the functions of proteins in higher organisms.

Engineered Genes Can Be Permanently Inserted into the Germ Line of Mice or Fruit Flies to Produce Transgenic Animals [37]

The ultimate test of the function of an altered gene is to reinsert it into an organism and see what effect it has. Ideally one would like to be able to replace the normal gene with the altered one so that the function of the mutant protein can be analyzed in the absence of the normal protein. As discussed above, this can be readily accomplished in some haploid organisms, but in higher eucaryotic cells an integrative event leading to a gene replacement occurs only very rarely. Foreign DNA can, however, rather easily be randomly integrated into the genome. In mammals, for example, linear DNA fragments introduced into cells are rapidly ligated end to end by intracellular enzymes to form long tandem arrays, which usually become integrated into a chromosome at an apparently random site. Fertilized mammalian eggs behave like other mammalian cells in this respect. A mouse egg injected with 200 copies of a linear DNA molecule will often develop into a mouse containing, in many of its cells, a tandem array of copies of the injected gene integrated at a single random site in one of its chromosomes (Figure 7–45). If the modified chromosome is present in the germ line cells (eggs or sperm), the mouse will pass these foreign genes on to its progeny. Animals that have been permanently altered in this way are called **transgenic organisms,** and the foreign genes are called **transgenes.** Because the normal gene generally remains present, only dominant effects of the alteration will show up. Nevertheless, such transgenic animals have already provided important insights into how mammalian genes are regulated and how certain altered genes (called oncogenes) cause cancer.

It is also possible to produce transgenic fruit flies, in which single copies of a gene are inserted at random into the *Drosophila* genome. The trick in this case is first to insert the DNA fragment between the two terminal sequences of a par-

DNA Engineering

327

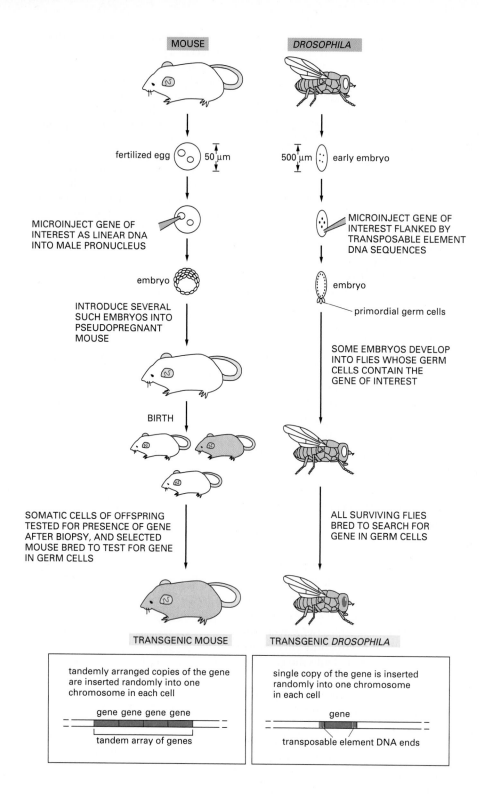

MOUSE | DROSOPHILA

fertilized egg — 50 µm

500 µm — early embryo

MICROINJECT GENE OF INTEREST AS LINEAR DNA INTO MALE PRONUCLEUS

MICROINJECT GENE OF INTEREST FLANKED BY TRANSPOSABLE ELEMENT DNA SEQUENCES

embryo

embryo

primordial germ cells

INTRODUCE SEVERAL SUCH EMBRYOS INTO PSEUDOPREGNANT MOUSE

SOME EMBRYOS DEVELOP INTO FLIES WHOSE GERM CELLS CONTAIN THE GENE OF INTEREST

BIRTH

SOMATIC CELLS OF OFFSPRING TESTED FOR PRESENCE OF GENE AFTER BIOPSY, AND SELECTED MOUSE BRED TO TEST FOR GENE IN GERM CELLS

ALL SURVIVING FLIES BRED TO SEARCH FOR GENE IN GERM CELLS

TRANSGENIC MOUSE | TRANSGENIC *DROSOPHILA*

tandemly arranged copies of the gene are inserted randomly into one chromosome in each cell

gene gene gene gene

tandem array of genes

single copy of the gene is inserted randomly into one chromosome in each cell

gene

transposable element DNA ends

Figure 7–45 Comparison of the standard procedures used to make transgenic mice and transgenic *Drosophila*. In these examples the gene injected into the mouse egg causes a change in coat color, whereas the gene injected into the fly embryo causes a change in eye color. In both organisms some of the transgenic animals are found to have DNA insertions at more than one chromosomal site.

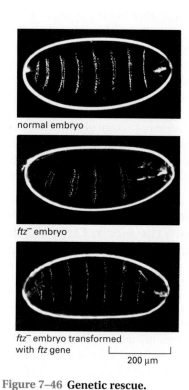

normal embryo

ftz⁻ embryo

ftz⁻ embryo transformed with *ftz* gene

200 µm

Figure 7–46 Genetic rescue. Comparison of a normal *Drosophila* larva and two mutant larvae that contain defective *ftz* genes. One of the defective (*ftz⁻*) larvae has been rescued by the injection of a DNA clone containing the normal *ftz* gene sequence into the egg of one of its ancestors. This added DNA sequence has become permanently integrated into one of the fly's chromosomes and is therefore faithfully inherited and expressed. The *ftz* gene is required for normal development, and the addition of this gene to the mutant genome restores the larval segments that are missing in the *ftz⁻* organism. For the technique used to make transgenic animals, see Figure 7–45. (Courtesy of Walter Gehring.)

ticular *Drosophila* transposable element, called the *P element*. The terminal sequences enable the P element to integrate into *Drosophila* chromosomes if the P element transposase enzyme is also present (see p. 285). To make transgenic fruit flies, therefore, the appropriately modified DNA fragment is injected into a very young fruit fly embryo along with a separate plasmid containing the gene encoding the transposase. When this is done, the injected gene often enters the germ line in a single copy as the result of a transposition event (Figures 7–45 and 7–46).

Gene Targeting Makes It Possible to Produce Transgenic Mice That Are Missing Specific Genes [38]

If a DNA molecule carrying a mutated mouse gene is transferred into a mouse cell, it usually inserts into the chromosomes at random, but about once in a thousand times, it will replace one of the two copies of the normal gene by homologous recombination. By exploiting these rare "gene targeting" events, any specific gene can be inactivated in a mouse cell by a direct gene replacement. This technique can be extended to produce a mouse with a missing gene by means of the following two-step pathway.

In the first step, a DNA fragment containing a desired mutant gene (or part of a gene) is inserted into a vector and then introduced into a special line of embryo-derived mouse stem cells (called *embryonic stem cells,* or *ES cells*) that grow in cell culture and are capable of producing cells of any tissue. After a period of cell proliferation, the rare colonies of cells in which a homologous recombination event is likely to have caused a gene replacement to occur are isolated by a double drug selection (Figure 7–47). The correct colonies among these are identified by PCR or by Southern blotting: they will contain recombinant DNA sequences in which the inserted fragment has replaced all or part of one copy of the normal gene. In the second step, individual cells from the identified colony are taken up into a fine micropipette and injected into an early mouse embryo. The transfected embryo-derived stem cells collaborate with the cells of the host embryo to produce a normal-looking mouse (see Figure 21–32); large parts of this chimeric animal, including in favorable cases cells of the germ line, will often derive from the artificially altered stem cells. Such mice are bred to produce both a male and a female animal, each heterozygous for the gene replacement (that is, they have one normal and one mutant copy of the gene). When these two mice are mated, one-fourth of their progeny will be homozygous for the altered gene. Studies of these homozygotes allow the function of the altered gene to be examined in the absence of the corresponding normal gene.

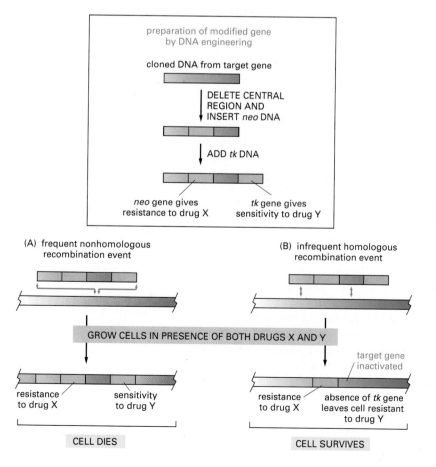

Figure 7–47 Selective gene knockout by homologous recombination in the mouse. Cloned DNA from the mouse gene to be mutated is modified by genetic engineering so that it contains a bacterial gene within it, called *neo*, whose integration into a mouse chromosome makes the mouse cells *resistant* to a drug that otherwise kills them (drug X). A viral gene, called *tk*, is also added, attached to one end of the mouse DNA; the integration of *tk* into a mouse chromosome makes the cells *sensitive* to a different drug (drug Y). Most insertions occur into random sites in the mouse chromosome, and these nearly always include both ends of the engineered DNA fragment, as shown (A). By selecting for those rare mouse cells that grow in the presence of *both* drugs, colonies of cells are obtained in which homologous recombination has incorporated the center of the engineered DNA fragment *without* the ends; most of these cells will turn out to carry the targeted gene replacement shown in (B).

The ability to prepare transgenic mice lacking a known normal gene has been a major advance, and the technique is now being widely used to dissect the functions of specific mammalian genes (Figure 7–48).

Transgenic Plants Are Important for Both Cell Biology and Agriculture [39]

When a plant is damaged, it can often repair itself by a process in which mature differentiated cells "dedifferentiate," proliferate, and then redifferentiate into other cell types. In some circumstances the dedifferentiated cells can even form an apical meristem, which can then give rise to an entire new plant, including gametes. This remarkable plasticity of plant cells can be exploited to generate transgenic plants from cells growing in culture.

When a piece of plant tissue is cultured in a sterile medium containing nutrients and appropriate growth regulators, many of the cells are stimulated to proliferate indefinitely in a disorganized manner, producing a mass of relatively undifferentiated cells called a *callus*. If the nutrients and growth regulators are carefully manipulated, one can induce the formation of a shoot and then root apical meristems within the callus, and, in many species, a whole new plant can be regenerated.

Callus cultures can also be mechanically dissociated into single cells, which will grow and divide as a suspension culture. In a number of plants—including tobacco, petunia, carrot, potato, and *Arabidopsis*—a single cell from such a suspension culture can be grown into a small clump (a clone) from which a whole plant can be regenerated. Just as mutant mice can be derived by genetic manipulation of embryonic stem cells in culture, so transgenic plants can be created from single plant cells transfected with DNA in culture (Figure 7–49).

The ability to produce transgenic plants has greatly accelerated progress in many areas of plant cell biology. It has played an important part, for example, in isolating receptors for growth regulators and in analyzing the mechanisms of morphogenesis and of gene expression in plants. It has also opened up many new possibilities in agriculture that could benefit both the farmer and the consumer. It has made it possible, for example, to modify the lipid, starch, and protein storage reserved in seeds, to impart pest and virus resistance to plants, and to create modified plants that tolerate extreme habitats such as salt marshes or waterlogged soil.

Many of the major advances in understanding animal development have come from studies on the fruit fly *Drosophila* and the nematode worm *Caenorhabditis elegans,* which are amenable to extensive genetic analysis as well as to experimental manipulation. Progress in plant developmental biology has been relatively slow by comparison. Many of the organisms that have proved most amenable to genetic analysis, such as maize and tomato, have long life cycles and very large genomes, which have made both classical and molecular genetic analysis time-consuming. Increasing attention is consequently being paid to a fast-growing small weed, the common wall cress (*Arabidopsis thaliana*), which has several major advantages as a "model plant" (see Figure 21–93).

Summary

Genetic engineering has revolutionized the study of both cells and organisms. DNA engineering techniques can be used to create any mutant gene and insert it into a cell's chromosomes so that it becomes a permanent part of the genome. If the cell used for this gene transfer is a fertilized egg (for an animal) or a totipotent plant cell in culture, transgenic organisms can be produced that express the mutant gene and will

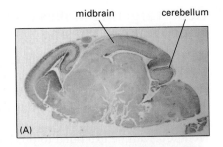

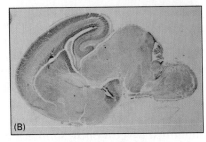

midbrain   cerebellum

(A)

(B)

Figure 7–48 A transgenic mouse in which both copies of the *Wnt-1* growth factor gene have been eliminated by homologous recombination. (A) Section of the brain from a normal embryo. (B) Section of the brain from a mutant embryo, which lacks a cerebellum and most of the midbrain, and dies *in utero*. (Photographs courtesy of Mario Capecchi.)

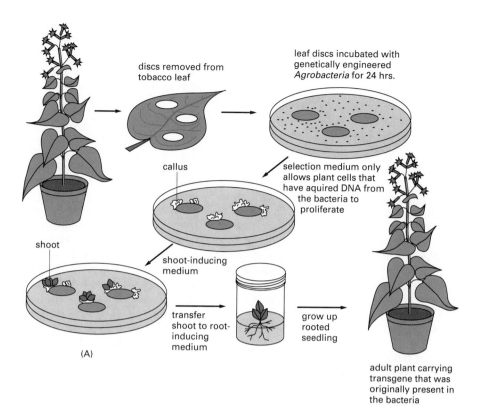

discs removed from
tobacco leaf

leaf discs incubated with
genetically engineered
Agrobacteria for 24 hrs.

callus

selection medium only
allows plant cells that
have aquired DNA from
the bacteria to
proliferate

shoot

shoot-inducing
medium

transfer
shoot to root-
inducing
medium

grow up
rooted
seedling

adult plant carrying
transgene that was
originally present in
the bacteria

(A)

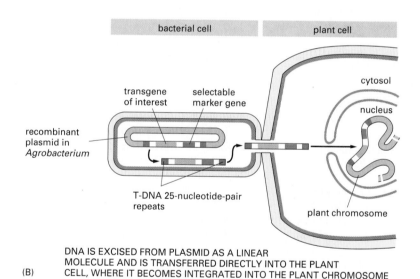

bacterial cell

plant cell

transgene
of interest

selectable
marker gene

recombinant
plasmid in
Agrobacterium

cytosol

nucleus

T-DNA 25-nucleotide-pair
repeats

plant chromosome

plant chromosome

DNA IS EXCISED FROM PLASMID AS A LINEAR
MOLECULE AND IS TRANSFERRED DIRECTLY INTO THE PLANT
CELL, WHERE IT BECOMES INTEGRATED INTO THE PLANT CHROMOSOME

(B)

Figure 7–49 A procedure used to make a transgenic plant. (A) Outline of the process. A disc is cut out of a leaf and incubated in culture with *Agrobacteria* that carry a recombinant plasmid with both a selectable marker and a desired transgene. The wounded cells at the edge of the disc release substances that attract the *Agrobacteria* and cause them to inject DNA into these cells. Only those plant cells that take up the appropriate DNA and express the selectable marker gene survive to proliferate and form a callus. The manipulation of growth factors supplied to the callus induces it to form shoots that subsequently root and grow into adult plants carrying the transgene. (B) The preparation of the recombinant plasmid and its transfer to plant cells. An *Agrobacterium* plasmid that normally carries the T-DNA sequence is modified by substituting a selectable marker (such as the kanamycin-resistance gene) and a desired transgene between the 25-nucleotide-pair T-DNA repeats. When the *Agrobacterium* recognizes a plant cell, it efficiently passes a DNA strand that carries these sequences into the plant cell, using the special machinery that normally transfers the plasmid's T-DNA sequence.

pass it on to their progeny. *Especially important for cell biology is the power that such technology gives the experimenter to alter cells and organisms in highly specific ways—allowing one to discern the effect on the cell or the organism of a designed change in a single protein that has been intentionally mutated by "reverse genetic" techniques.*

The consequences of genetic engineering are far-reaching in other ways as well. Plants can be altered in ways that increase their value as food crops. Bacteria, yeasts, or mammalian cells can be engineered to synthesize any desired protein in large quantities, making it possible to analyze the structure and function of the protein in detail or to use the protein as a drug or a vaccine for medical purposes.

References

Cited

1. Sambrook, J.; Fritsch, E.F.; Maniatis, T. Molecular Cloning: A Laboratory Manual, 2nd ed. Cold Spring Harbor, NY: Cold Spring Harbor Laboratory Press, 1989.

   Watson, J.D.; Tooze, J. The DNA Story: A Documentary History of Gene Cloning. New York: W.H. Freeman, 1981.

   Watson, J.D.; Gilman, M.; Witkowski, J.; Zoller, M. Recombinant DNA, 2nd ed. New York: W.H. Freeman, 1992.

2. Kessler, C.; Manta, V. Specificity of restriction endonucleases and DNA modification methyltransferases—a review (Edition 3). *Gene* 92:1–248, 1990.

3. Danna, K.J. Determination of fragment order through partial digests and multiple enzyme digests. *Methods Enzymol.* 65:449–467, 1980.

   Nathans, D.; Smith, H.O. Restriction endonucleases in the analysis and restructuring of DNA molecules. *Annu. Rev. Biochem.* 44:273–293, 1975.

4. Andrews, A.T. Electrophoresis, 2nd ed. Oxford, UK: Clarendon Press, 1986.

   Evans, G.A. Physical mapping of the human genome by pulsed field gel analysis. *Curr. Opin. Genet. Dev.* 1:75–81, 1991.

   Southern, E.M. Gel electrophoresis of restriction fragments. *Methods Enzymol.* 68:152–176, 1979.

5. Lehman, I.R. DNA polymerase I of *Escherichia coli.* In The Enzymes, Vol. 14A (P.D. Boyer, ed.), pp. 16–38. New York: Academic Press, 1981.

   Richardson, C.C. Polynucleotide kinase from *Escherichia coli* infected with bacteriophage T4. *Nucleic Acids Res.* 2:815, 1971.

   Rigby, P.W.; Dieckmann, M.; Rhodes, C.; Berg, P. Labeling deoxyribonucleic acid to high specific activity *in vitro* by nick translation with DNA polymerase I. *J. Mol. Biol.* 113:237–251, 1977.

6. Griffin, H.G.; Griffin, A.M. DNA sequencing. Recent innovations and future trends. *Appl. Biochem. Biotech.* 38:147–159, 1993.

   Maxam, A.M.; Gilbert, W. A new method for sequencing DNA. *Proc. Natl. Acad. Sci. USA* 74:560–564, 1977.

   Maxam, A.M.; Gilbert, W. Sequencing end-labeled DNA with base-specific chemical cleavages. *Methods Enzymol.* 65:499–560, 1980.

   Prober, J.M.; Trainor, G.; Dam, R.; et al. A system for rapid DNA sequencing with fluorescent chain terminating dideoxynucleotides. *Science* 238:336–341, 1987.

   Sanger, F.; Nicklen, S.; Coulson, A.R. DNA sequencing with chain-terminating inhibitors. *Proc. Natl. Acad. Sci. USA* 74:5463–5467, 1977.

7. Cartwright I.L.; Kelly, S.E. Probing the nature of chromosomal DNA-protein contacts by *in vivo* footprinting. *Biotechniques* 11:188–203, 1991.

   Galas, D.J.; Schmitz, A. DNAse footprinting: a simple method for the detection of protein-DNA binding specificity. *Nucleic Acids Res.* 5:3157–3170, 1978.

   Tullius, T.D. Physical studies of protein-DNA complexes by footprinting. *Annu. Rev. Biophys. Biophys. Chem.* 18:213–237, 1989.

8. Schildkraut, C.L.; Marmur, J.; Doty, P. The formation of hybrid DNA molecules and their use in studies of DNA homologies. *J. Mol. Biol.* 3:595–617, 1961.

   Wetmur, J.G. Hybridization and renaturation kinetics of nucleic acids. *Annu. Rev. Biophys. Bioeng.* 5:337–361, 1976.

   Wetmur, J.G. DNA probes: applications of the principles of nucleic acid hybridization. *Critical Reviews in Biochem. Mol. Biol.* 26:227–259, 1991.

9. Berk, A.J.; Sharp, P.A. Sizing and mapping of early adenovirus mRNAs by gel electrophoresis of S1 endonuclease digested hybrids. *Cell* 12:721–732, 1977.

   Gerhard, D.S.; Kawasaki, E.S.; Bancroft, F.C.; Szabo, P. Localization of a unique gene by direct hybridization. *Proc. Natl. Acad. Sci. USA* 78:3755–3759, 1981.

   Pardue, M.L.; Gall, J.G. Molecular hybridization of radioactive DNA to the DNA of cytological preparations. *Proc. Natl. Acad. Sci. USA* 64:600–604, 1969.

   Wallace, R.B.; Shaffer, J.; Murphy, R.F.; et al. Hybridization of synthetic oligodeoxyribonucleotides to ϕX174 DNA: the effect of a single base pair mismatch. *Nucleic Acids Res.* 6:3543–3557, 1979.

10. Alwine J.C.; Kemp, D.J.; Stark, G.R. Method for detection of specific RNAs in agarose gels by transfer to diabenzyloxymethyl-paper and hybridization with DNA probes. *Proc. Natl. Acad. Sci. USA* 74:5350–5354, 1977.

    Southern, E.M. Detection of specific sequences among DNA fragments separated by gel electrophoresis. *J. Mol. Biol.* 98:503–517, 1975.

    Thomas, P.S. Hybridization of denatured RNA and small DNA fragments transferred to nitrocellulose. *Proc. Natl. Acad. Sci. USA* 77:5201–5205, 1980.

11. Chang, C; Meyerowitz, E.M. Plant genome studies: restriction fragment length polymorphism and chromosome mapping information. *Curr. Opin. Genet. Dev.* 1:112–118, 1991.

    Kidd, K.K. Progress towards completing the human linkage map. *Curr. Opin. Genet. Dev.* 1:99–104, 1991.

    Pourzand, C.; Cerutti, P. Genotypic mutation analysis by RFLP/PCR. *Mutat. Res.* 288:113–121, 1993.

12. Gilbert, F.; Marinduque, B. DNA prenatal diagnosis. *Curr. Opin. Obstet. Gynecol.* 2:226–235, 1990.

    Lathe, R. Synthetic oligonucleotide probes deduced from amino acid sequence data. Theoretical and practical considerations. *J. Mol. Biol.* 183:1–12, 1985.

13. McGinnis, W.; Garber, R.L.; Wirz, J.; et al. A homologous protein-coding sequence in *Drosophila* homeotic genes and its conservation in other metazoans. *Cell* 37:403–408, 1984.

14. Stephenson, E.C.; Pokrywka, N.J. Localization of bicoid message during *Drosophila* oogenesis. *Curr. Top. Dev. Biol.* 26:23–34, 1992.

    Trask, B.J. Gene mapping by *in situ* hybridization. *Curr. Opin. Genet. Dev.* 1:82–87, 1991.

15. Ausubel, F.M.; Brent, R.; Kingston, R.E.; et al., eds. Current Protocols in Molecular Biology. New York: Wiley, 1993.

    Drlica, K. Understanding DNA and Gene Cloning. New York: Wiley, 1984.

16. Cohen, S.N. The manipulation of genes. *Sci. Am.* 233(1):24–33, 1975.

    Foster, T.J. Plasmid-determined resistance to antimicrobial drugs and toxic metal ions in bacteria. *Microbiol. Rev.* 47:361–409, 1983.

    Hanahan, D. Studies on transformation of *Escherichia coli* with plasmids. *J. Mol. Biol.* 166:557–580, 1983.

    Novick, R.P. Plasmids. *Sci. Am.* 243(6):102–127, 1980.

17. Lennon, G.G.; Lehrach, H. Hybridization analyses of arrayed cDNA libraries. *Trends Genet.* 7:314–317, 1991.

Maniatis, T., et al. The isolation of structural genes from libraries of eucaryotic DNA. *Cell* 15:687–701, 1978.

18. Okayama, H.; Berg, P. High-efficiency cloning of full-length cDNA. *Mol. Cell Biol.* 2:161–170, 1982.

Southern, E.M. Genome mapping: cDNA approaches. *Curr. Opin. Genet. Dev.* 2:412–416, 1992.

19. Calvet, J.P. Molecular approaches for analyzing differential gene expression: differential cDNA library construction and screening. *Pediat. Nephrol.* 5:751–757, 1991.

Hedrick, S.M.; Cohen, D.I.; Nielsen, E.A.; Davis, M.M. Isolation of cDNA clones encoding T cell-specific membrane-associated proteins. *Nature* 308:149–153, 1984.

20. Yang, J.; Ye, J.; Wallace, D.C. Computer selection of oligonucleotide probes from amino acid sequences for use in gene library screening. *Nucleic Acids Res.* 12:837–843, 1984.

Young, R.A.; Davis, R.W. Efficient isolation of genes using antibody probes. *Proc. Natl. Acad. Sci. USA* 80:1194–1198, 1983.

21. Hope, I.A.; Struhl, K. GCN4 protein, synthesized *in vitro*, binds HIS3 regulatory sequences: implications for general control of amino acid biosynthetic genes in yeast. *Cell* 43:177–188, 1985.

Ricciardi, R.P.; Miller, J.S.; Roberts, B.E. Purification and mapping of specific mRNAs by hybridization-selection and cell-free translation. *Proc. Natl. Acad. Sci. USA* 76:4927–4931, 1979.

22. Stubbs, L. Long-range walking techniques in positional cloning strategies. *Mammalian Genome* 3:127–142, 1992.

23. Burke, D.T. The role of yeast artificial chromosomes in generating genome maps. *Curr. Opin. Genet. Dev.* 1:69–74, 1991.

Carrano, A.V.; de Jong, P.J.; Branscomb, E.; et al. Constructing chromosome- and region-specific cosmid maps of the human genome. *Genome* 31:1059–1065, 1989.

Coulson, A.; Kozono, Y.; Lutterbach, B.; et al. YACs and the *C. elegans* genome. *Bioessays* 13:413–417, 1991.

Hauge, B.M.; Hanley, S.; Giraudat, J.; et al. Mapping the *Arabidopsis* genome. *Symp. Soc. Exper. Biol.* 45:45–56, 1991.

24. Harris, A.; Argent, B.E. The cystic fibrosis gene and its product CFTR. *Semin. Cell Biol.* 4:37–44, 1993.

Hyser, C.L. Unraveling the mysteries of Duchenne and Becker muscular dystrophy. *Mol. Chem. Neuropathol.* 10:15–20, 1989.

Wicking, C.; Williamson, B. From linked marker to gene. *Trends Genet.* 7:288–293, 1991.

25. Allen, R.W.; Wallhermfechtel, M.; Miller, W.V. The application of restriction fragment length polymorphism mapping to parentage testing. *Transfusion* 30:552–564, 1990.

Bottema, C.D.; Sommer, S.S. PCR amplification of specific alleles: rapid detection of known mutations and polymorphisms. *Mutat. Res.* 288:93–102, 1993.

Nelson, D.L. Applications of polymerase chain reaction methods in genome mapping. *Curr. Opin. Genet. Dev.* 1:62–68, 1991.

Pourzand, C.; Cerutti, P. Genotypic mutation analysis by RFLP/PCR. *Mutat. Res.* 288:113–121, 1993.

Saiki, R.K.; Gelfand, D.H.; Stoffel, S.; et al. Primer-directed enzymatic amplification of DNA with a thermostable DNA polymerase. *Science* 239:487–491, 1988.

26. Olson, M.V. The human genome project. *Proc. Natl. Acad. Sci. USA* 90:4338–4344, 1993.

Szostak, J.W. *In vitro* genetics. *Trends Biochem. Sci.* 17:89–93, 1992.

27. Itakura, K.; Rossi, J.J.; Wallace, R.B. Synthesis and use of synthetic oligonucleotides. *Annu. Rev. Biochem.* 53:323–356, 1984.

28. Melton, D.A.; Krieg, P.A.; Rebagliati, M.R.; et al. Efficient *in vitro* synthesis of biologically active RNA and RNA hybridization probes from plasmids containing a bacteriophage SP6 promoter. *Nucleic Acids Res.* 12:7035–7056, 1984.

29. Abelson, J.; Butz, E., eds. Recombinant DNA. *Science* 209:1317–1438, 1980.

Balbas, P.; Bolivar, F. Design and construction of expression plasmid vectors in *E. coli. Methods Enzymol.* 185:14–37, 1990.

Gilbert, W.; Villa-Komaroff, L. Useful proteins from recombinant bacteria. *Sci. Am.* 242(4):74–94, 1980.

Miller, L.K. Baculoviruses: high-level expression in insect cells. *Curr. Opin. Genet. Dev.* 3:97–101, 1993.

Rose, A.B.; Broach, J.R. Propagation and expression of cloned genes in yeast: 2-micron circle-based vectors. *Methods Enzymol.* 185:234–279, 1990.

30. Alam, J.; Cook, J.L. Reporter genes: application to the study of mammalian gene transcription. *Anal. Biochem.* 188:245–254, 1990.

Bellen, H.J.; Wilson, C.; Gehring, W.J. Dissecting the complexity of the nervous system by enhancer detection. *Bioessays* 12:199–204, 1990.

31. Jockusch, B.M.; Zurek, B.; Zahn, R.; Westmeyer, A.; Fuchtbauer, A. Antibodies against vertebrate microfilament proteins in the analysis of cellular motility and adhesion. *J. Cell Sci.* S14:41–47, 1991.

Lederberg, J.; Lederberg, E.M. Replica plating and indirect selection of bacterial mutants. *J. Bacteriol.* 63:399–406, 1952.

Nusslein-Volhard, C.; Weischaus, E. Mutations affecting segment number and polarity in *Drosophila. Nature* 287:795–801, 1980.

32. Kaiser, K. From gene to phenotype in *Drosophila* and other organisms. *Bioessays* 12:297–301, 1990.

Landel, C.P.; Chen, S.; Evans, G.A. Reverse genetics using transgenic mice. *Annu. Rev. Physiol.* 52:841–851, 1990.

Provost, G.S.; Kretz, P.L.; Hamner, R.T.; et al. Transgenic systems for *in vivo* mutation analysis. *Mutat. Res.* 288:133–149, 1993.

33. Baase, W.A.; Eriksson, A.E.; Zhang, X.J.; et al. Dissection of protein structure and folding by directed mutagenesis. *Faraday Discuss.* 93:173–181, 1992.

Sharon, J.; Kao C.Y.; Sompuram, S.R. Oligonucleotide-directed mutagenesis of antibody combining sites. *Int. Rev. Immunol.* 10:113–127, 1993.

34. Garoff, H. Using recombinant DNA techniques to study protein targeting in the eucaryotic cell. *Annu. Rev. Cell Biol.* 1:403–445, 1985.

Kelly, J.H.; Darlington, G.J. Hybrid genes: molecular approaches to tissue-specific gene regulation. *Annu. Rev. Genet.* 19:273–296, 1985.

35. Leclerc, D.; Brakier-Gingras, L. Study of the function of *Escherichia coli* ribosomal RNA through site-directed mutagenesis. *Biochem. Cell Biol.* 68:169–179, 1990.

Scherer, S.; Davis, R.W. Replacement of chromosome segments with altered DNA segments constructed *in vitro. Proc. Natl. Acad. Sci. USA* 76:4951–4955, 1979.

Struhl, K. The new yeast genetics. *Nature* 305:391–397, 1983.

36. Herskowitz, I. Functional inactivation of genes by dominant negative mutations. *Nature* 329:219–222, 1987.

Weintraub, H.; Izant, J.G.; Harland, R.M. Anti-sense RNA as a molecular tool for genetic analysis. *Trends Genet.* 1:22–25, 1985.

37. Boyd, A.L.; Samid, D. Review: molecular biology of transgenic animals. *J. Anim. Sci.* 71(S3):1–9, 1993.

Palmiter, R.D.; Brinster, R.L. Germ line transformation of mice. *Annu. Rev. Genet.* 20:465–499, 1986.

Rubin, G.M.; Sprading, A.C. Genetic transformation of *Drosophila* wth transposable element vectors. *Science* 218:348–353, 1982.

Smith, P.A.; Corces, V.G. *Drosophila* transposable elements: mechanisms of mutagenesis and interactions with the host genome. *Adv. Genet.* 29:229–300, 1991.

38. Babinet, C.; Morello, D.; Renard, J.P. Transgenic mice. *Genome* 31:938–949, 1989.

Gossen, J.; Vijg, J. Transgenic mice as model systems for studying gene mutations *in vivo. Trends Genet.* 9:27–31, 1993.

Gridley, T. Insertional versus targeted mutagenesis in mice. *New Biol.* 3:1025–1034, 1991.

39. Davey, M.R.; Rech, E.L.; Mulligan, B.J. Direct DNA transfer to plant cells. *Plant Mol. Biol.* 133:273–285, 1989.

Walden, R.; Schell, J. Techniques in plant molecular biology—progress and problems. *Eur. J. Biochem.* 192:563–576, 1990.

The Cell Nucleus

The DNA in a eucaryotic cell is sequestered in the nucleus, which occupies about 10% of the total cell volume. The nucleus is delimited by a *nuclear envelope* formed by two concentric membranes. These membranes are punctured at intervals by nuclear pores, which actively transport selected molecules to and from the cytosol. The envelope is directly connected to the extensive membranes of the endoplasmic reticulum, and it is supported by two networks of intermediate filaments: one called the *nuclear lamina* forms a thin shell just inside the nucleus underlying the inner nuclear membrane, while the other, less regularly organized, surrounds the outer nuclear membrane (Figure 8–1).

Like modern procaryotes, the ancestors of eucaryotic cells almost certainly lacked a nucleus, and one can only speculate on why a separate nuclear compartment evolved, segregating the DNA from the activities in the cytoplasm. Two special features of eucaryotic cells suggest possible reasons. One is the eucaryotic cytoskeleton, which is composed mainly of microtubules and actin filaments and mediates cell movements. Bacteria, whose DNA is in direct contact with the cytoplasm, lack a cytoskeleton and move by means of external structures such as flagella. One function of the nuclear envelope may therefore be to protect the long, fragile DNA molecules from the mechanical forces generated by the cytoplasmic filaments in eucaryotes.

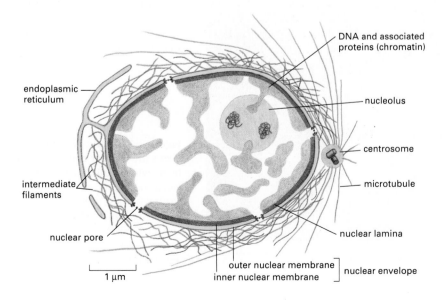

Figure 8–1 Cross-section of a typical cell nucleus. The nuclear envelope consists of two membranes, the outer one being continuous with the endoplasmic reticulum membrane (see also Figure 12–9). The lipid bilayers of the inner and outer nuclear membranes are connected at each nuclear pore. Two networks of intermediate filaments (*green*) provide mechanical support for the nuclear envelope; the intermediate filaments inside the nucleus form a sheetlike *nuclear lamina*. The space inside the endoplasmic reticulum (the ER lumen) is colored *yellow*.

335

A second special feature of eucaryotic cells is the extensive processing that RNA molecules undergo before they are translated into protein. In procaryotic cells RNA synthesis (*transcription*) and protein synthesis (*translation*) occur concurrently: ribosomes translate the 5′ end of an RNA molecule into protein while the 3′ end of the RNA molecule is still being synthesized. Consequently, there is relatively little opportunity to alter the RNA transcripts before they are translated into protein. In eucaryotes, by contrast, transcription (in the nucleus) is separated both temporally and spatially from translation (in the cytoplasm). The RNA transcripts in the nucleus are immediately packaged into ribonucleo-protein complexes and subjected to *RNA splicing*, in which certain portions of the nucleotide sequence are removed. Only when splicing is complete are the packaging proteins removed and the RNA molecules transported out of the nucleus to the cytosol, where ribosomes begin translating the RNA into protein (Figure 8–2). As we discuss later, RNA splicing is an important intermediate step in the transfer of genetic information in eucaryotes. It provides a number of advantages for the cell, including the potential for a single gene to make several different proteins. This may help explain why eucaryotic cells have a nucleus, where splicing can occur without interference from ribosomes (Figure 8–3).

In this chapter we describe how proteins package eucaryotic DNA into chromosomes, how chromosomes are folded and organized in the nucleus, and how they are replicated during each cell-division cycle. We then discuss RNA synthesis and RNA splicing, and, finally, we describe how genetic information is organized in the eucaryotic genome and how this organization may have arisen during evolution. The nuclear envelope is discussed in detail in Chapter 12 in connection with the selective transport of macromolecules into and out of the nucleus, and a hypothetical scheme of how the nuclear compartment might have evolved is presented there (see Figure 12–5).

Chromosomal DNA and Its Packaging [1]

For the first 40 years of this century, biologists tended to dismiss the possibility that DNA could carry the genetic information in chromosomes, partly because

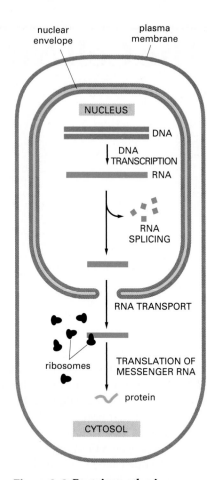

Figure 8–2 Protein synthesis (DNA → RNA → protein) in eucaryotes. Eucaryotic cells have evolved numerous membrane-bounded compartments that segregate their various chemical reactions so as to make them more efficient, and the nucleus is one such compartment. The nuclear envelope keeps functional ribosomes out of the nucleus, preventing RNA transcripts from being translated into protein until they have been extensively processed (spliced) and transported out of the nucleus into the cytosol. Thus RNA splicing and transport steps are interposed between DNA transcription and RNA translation.

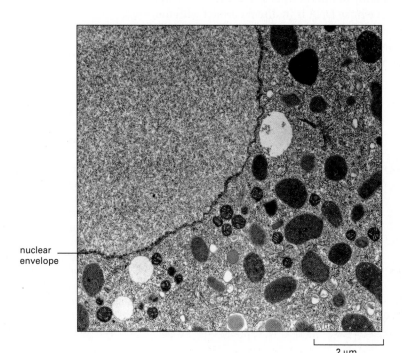

2 μm

Figure 8–3 The nuclear envelope keeps the nuclear compartment free from cytoplasmic organelles. This electron micrograph shows a thin section of a sea urchin egg, which has a nucleus that stains unusually evenly and a cytoplasm densely packed with organelles. (Courtesy of David Begg and Tim Hunt.)

nucleic acids were erroneously believed to contain only a simple repeating tetranucleotide sequence (such as AGCTAGCTAGCT...). We now know, however, that a DNA molecule is an enormously long, unbranched, linear polymer that can contain many millions of nucleotides arranged in an irregular but nonrandom sequence and that the genetic information of a cell is contained in the linear order of the nucleotides in its DNA. The genetic code, written in "words" of three nucleotides (*codons* that each specify an amino acid, discussed in Chapter 6), neatly solves the problem of storing a large amount of genetic information in a small amount of space: every million "letters" (nucleotides) take up a linear distance of only 3.4×10^5 nm (0.034 cm) and occupy a total volume of about 10^6 nm^3 (10^{-15} cm^3).

Each DNA molecule is packaged in a separate **chromosome,** and the total genetic information stored in the chromosomes of an organism is said to constitute its **genome**. The genome of the *E. coli* bacterium contains 4.7×10^6 nucleotide pairs of DNA, present in a single double-helical DNA molecule (one chromosome). The human genome, in contrast, contains about 3×10^9 nucleotide pairs, organized as 24 chromosomes (22 different autosomes and 2 different sex chromosomes), and thus consists of 24 different DNA molecules—each containing from 50×10^6 to 250×10^6 nucleotide pairs of DNA. DNA molecules of this size are 1.7 to 8.5 cm long when uncoiled, and even the slightest mechanical force will break them once the chromosomal proteins have been removed.

In diploid organisms such as ourselves, there are two copies of each different chromosome, one inherited from the mother and one from the father (except for the sex chromosomes in males, where a Y chromosome is inherited from the father and an X from the mother). A typical human cell thus contains a total of 46 chromosomes and about 6×10^9 nucleotide pairs of DNA. Other mammals have genomes of similar size. This amount of DNA could in theory be packed into a cube 1.9 μm on each side. By comparison, 6×10^9 letters in this book would occupy more than a million pages, thus requiring more than 10^{17} times as much space.

In this section we consider the relationship between DNA molecules, genes, and chromosomes, and we discuss how the DNA is folded into a compact and orderly structure—the chromosome—while still allowing access to its genetic information. Throughout the discussion, it is important to bear in mind that the chromosomes in a cell change their structure and activities according to the stage of the cell-division cycle: in mitosis, or *M phase*, they are very highly condensed and transcriptionally inactive; in the other, much longer part of the division cycle, called *interphase*, they are less condensed and are continuously active in directing RNA synthesis.

Each DNA Molecule That Forms a Linear Chromosome Must Contain a Centromere, Two Telomeres, and Replication Origins [2]

To form a functional chromosome, a DNA molecule must be able to do more than direct the synthesis of RNA: it must be able to propagate itself reliably from one cell generation to the next. This requires three types of specialized nucleotide sequences in the DNA, each of which serves to attach specific proteins that guide the machinery that replicates and segregates chromosomes. Experiments in yeasts, whose chromosomes are relatively small and easy to manipulate by recombinant DNA methods, have identified the minimal DNA sequence elements responsible for each of these functions. Two of the three elements were identified by studying small circular DNA molecules that can be propagated as *plasmids* in cells of the yeast *Saccharomyces cerevisiae*. In order to replicate, such a DNA molecule requires a specific nucleotide sequence to act as a **DNA replication origin;** as we discuss below, one can identify the many origins in each yeast chromosome by their ability to allow a test DNA molecule that contains one of them to replicate when free of the host chromosome. A second sequence element, called a **centromere,** attaches any DNA molecule that contains it to the mitotic spindle during cell division. Each yeast chromosome contains a single centromere; when this sequence is inserted into a plasmid, it guarantees that each daughter cell will receive one of the two copies of the newly replicated plasmid DNA molecule when the yeast cell divides.

The third required sequence element is a **telomere,** which is needed at each end of a linear chromosome. If a circular plasmid that contains a replication origin and a centromere is broken at a single site to create two free ends in the double helix, it will still replicate and attach to the mitotic spindle, but it will eventually be lost from the progeny cells. This is because replication on the lagging strand of a replication fork requires the presence of some DNA ahead of the sequence to be copied to serve as the template for an RNA primer (see Figure 6–44). Since there can never be such a template for the last few nucleotides of a linear DNA molecule, special mechanisms are required to prevent each such DNA strand from becoming shorter with each replication cycle. Bacteria and many viruses solve this "end-replication problem" by having a circular DNA molecule as their chromosome. Eucaryotic cells have instead evolved a specialized telomeric DNA sequence at each chromosome end. This simple repeating sequence is periodically extended by an enzyme, *telomerase*, thus compensating for the loss of a few nucleotides of telomeric DNA in each cycle and permitting a linear chromosome to be completely replicated.

Figure 8–4 summarizes the functions of the three DNA sequence elements that are required for a linear chromosome to propagate itself from generation to generation in a yeast cell. These sequence elements are relatively short (typically less than 1000 base pairs each) and therefore utilize only a tiny fraction of the information-carrying capacity of a chromosome. The same three types of sequence elements are thought to operate in human chromosomes, but to date only the human telomere sequences have been well defined. Although the yeast versions of these sequences do not function in higher eucaryotic cells, recombinant DNA methods allow the yeast sequence elements to be added to human DNA molecules, which can then replicate in yeast cells as **artificial chromosomes.** In this way yeast cells can be used to prepare human genomic DNA libraries (see p. 315) in which each DNA clone (propagated as an artificial chromosome) contains as many as a million nucleotide pairs of human DNA sequence (Figure 8–5).

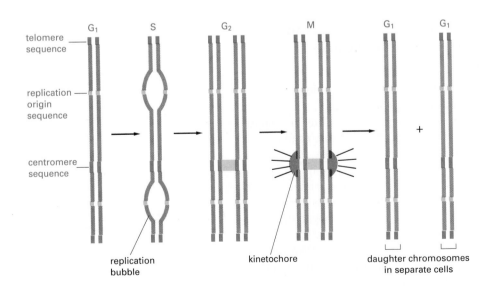

Figure 8–4 **The functions of the three DNA sequence elements needed to produce a stable linear eucaryotic chromosome.** Each chromosome has many origins of replication, one centromere, and two telomeres. The centromere serves to hold the two copies of the duplicated chromosome together and to attach them, via a protein complex called a kinetochore, to the mitotic spindle in such a way that one copy is distributed to each daughter cell at mitosis. The phases of the cell-division cycle corresponding to the events of chromosome replication and segregation are shown above the diagrams.

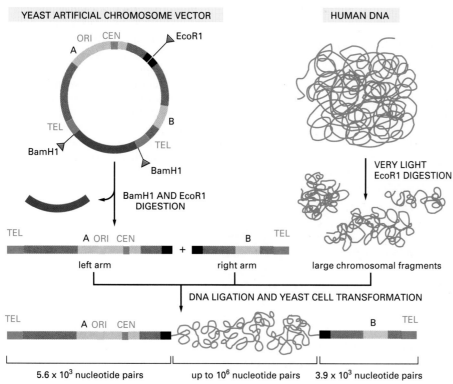

YEAST ARTIFICIAL CHROMOSOME VECTOR

HUMAN DNA

BamH1 AND EcoR1 DIGESTION

VERY LIGHT EcoR1 DIGESTION

TEL | A ORI CEN | B | TEL

left arm + right arm | large chromosomal fragments

DNA LIGATION AND YEAST CELL TRANSFORMATION

TEL A ORI CEN | B TEL

5.6 x 10³ nucleotide pairs | up to 10⁶ nucleotide pairs | 3.9 x 10³ nucleotide pairs

ARTIFICIAL YEAST CHROMOSOME WITH INSERTED HUMAN DNA

Figure 8–5 The making of a yeast artificial chromosome (YAC). A YAC vector allows the cloning of very large DNA molecules. TEL, CEN, and ORI are the telomere, centromere, and replication origin sequence elements, respectively, for the yeast *Saccharomyces cerevisiae*. BamH1 and EcoR1 are sites where the corresponding restriction nucleases cut the DNA double helix. The sequences denoted as A and B encode enzymes that serve as selectable markers to allow the easy isolation of yeast cells that have taken up the artificial chromosome. (Adapted from D.T. Burke, G.F. Carle, and M.V. Olson, *Science* 236:806–812, 1987. © 1987 AAAS.)

Most Chromosomal DNA Does Not Code for Proteins or RNAs [3]

The genomes of higher organisms seem to contain a large excess of DNA. Long before it was possible to examine the nucleotide sequences of chromosomal DNA directly, it was evident that the amount of DNA in the haploid genome of an organism has no systematic relationship to the complexity of the organism. Human cells, for example, contain about 700 times more DNA than the bacterium *E. coli*, but some amphibian and plant cells contain 30 times more DNA than human cells (Figure 8–6). Moreover, the genomes of different species of amphibians can vary 100-fold in their DNA content.

Population geneticists have tried to estimate how much of the DNA in higher organisms codes for essential proteins or RNA molecules on the basis of the following indirect argument. Each gene is inevitably subject to a small risk of accidental mutation, in which nucleotides in the DNA are altered at random. The greater the number of genes, the greater the probability that a mutation will occur in at least one of them. Since most mutations will impair the function of the gene in which they occur, the mutation rate sets an upper limit to the number of essential genes that an organism can depend on for its survival: if there are too many, disaster becomes almost a certainty, as with a complex machine dependent on too many components that are liable to fail. Using this argument and the observed mutation rate, it has been estimated that no more than a small percentage of the mammalian genome can be involved in regulating or encoding essential proteins or RNA molecules. We shall see later that other evidence supports this conclusion.

The most important implication of this estimate is that although the mammalian genome contains enough DNA, in principle, to code for nearly 3 million average-sized proteins (3×10^9 nucleotides), the limited fidelity with which DNA sequences can be maintained means that no mammal (or any other organism) is likely to be constructed from more than perhaps 60,000 essential proteins.

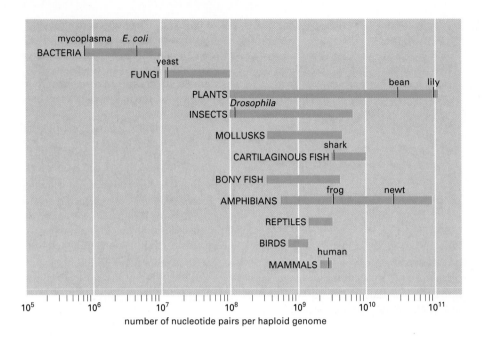

Figure 8–6 Lack of relationship between amount of DNA and organism complexity. The amount of DNA in a haploid genome varies over a 100,000-fold range from the smallest procaryotic cell—the mycoplasma—to the large cells of some plants and amphibia. Note that the genome size of humans (3×10^9 nucleotide pairs) is much smaller than that of many organisms that appear to be simpler.

Thus, from a genetic point of view, humans are unlikely to be more than about 10 times more complex than the fruit fly *Drosophila*, which is estimated to have about 5000 essential genes.

Whatever the remaining nonessential DNA in higher eucaryotic chromosomes may do (we discuss this later), the data shown in Figure 8–6 make it clear that it is not a great handicap for a higher eucaryotic cell to carry a large amount of extra DNA. Indeed, even the essential coding regions are often interrupted by long stretches of noncoding DNA.

Each Gene Produces an RNA Molecule [4]

The primary function of the genome is to specify RNA molecules. Selected portions of the DNA nucleotide sequence are copied into a corresponding RNA nucleotide sequence, which either encodes a protein (if it is an mRNA) or forms a "structural" RNA, such as a transfer RNA (tRNA) or ribosomal RNA (rRNA) molecule. Each region of the DNA helix that produces a functional RNA molecule constitutes a **gene.**

Table 8–1 The Size of Some Human Genes in Thousands of Nucleotides

	Gene Size	mRNA Size	Number of Introns
β-Globin	1.5	0.6	2
Insulin	1.7	0.4	2
Protein kinase C	11	1.4	7
Albumin	25	2.1	14
Catalase	34	1.6	12
LDL receptor	45	5.5	17
Factor VIII	186	9	25
Thyroglobulin	300	8.7	36
Dystrophin*	more than 2000	17	more than 50

*An altered form of this gene causes Duchenne muscular dystrophy.
The size specified here for a gene includes both its transcribed portion and nearby regulatory DNA sequences. (Compiled from data supplied by Victor McKusick.)

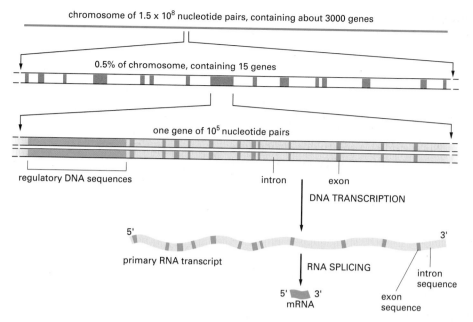

chromosome of 1.5 x 10^8 nucleotide pairs, containing about 3000 genes

0.5% of chromosome, containing 15 genes

one gene of 10^5 nucleotide pairs

regulatory DNA sequences

intron   exon

DNA TRANSCRIPTION

5'   3'
primary RNA transcript

RNA SPLICING

5' ▬ 3'
mRNA

intron
sequence

exon
sequence

Figure 8–7 The organization of genes on a typical vertebrate chromosome. Proteins that bind to the DNA in regulatory regions determine whether a gene is transcribed; although often located on the 5' side of a gene, as shown here, regulatory regions can also be located in introns, in exons, or on the 3' side of a gene. Intron sequences are removed from primary RNA transcripts to produce messenger RNA (mRNA) molecules. The figure given here for the number of genes per chromosome is a minimal estimate.

In higher eucaryotes genes that are more than 100,000 nucleotide pairs in length are common, and some contain more than 2 million nucleotide pairs (Table 8–1); yet only about 1000 nucleotide pairs are required to encode a protein of average size (one containing 300 to 400 amino acids). Most of the extra length consists of long stretches of noncoding DNA that interrupt the relatively short segments of coding DNA. The coding sequences are called **exons**; the intervening (noncoding) sequences are called **introns.** The RNA molecule (called a *primary RNA transcript*) synthesized from such a gene is altered to remove the intron sequences during its conversion to an mRNA molecule (see Figure 8–2) in the process of *RNA splicing*, as we discuss later.

Large genes consist of a long string of alternating exons and introns, with most of the gene consisting of introns. In addition, each gene is associated with *regulatory DNA sequences,* which are responsible for ensuring that the gene is transcribed at the proper time and in the appropriate cell type. We discuss in Chapter 9 how these regulatory sequences work. Many of them are located "upstream" (on the 5' side) of the site where RNA transcription begins, but they can also be located "downstream" (on the 3' side) of the site where RNA transcription ends, or even in introns or exons. A typical vertebrate chromosome is illustrated schematically in Figure 8–7, along with one of its many genes.

Comparisons Between the DNAs of Related Organisms Distinguish Conserved and Nonconserved Regions of DNA Sequence [5]

Technical improvements in DNA sequencing are expected to allow the routine sequencing of stretches of chromosomal DNA that are millions of nucleotide pairs long, so that in the foreseeable future the sequence of all 3×10^9 nucleotides of the human genome will be determined. As more than 90% of this sequence is probably unimportant, it will be crucial to have some way of identifying the small proportion of sequence that is important. One approach to this problem is based on the observation that important sequences are conserved during evolution, while unimportant ones are free to mutate randomly. The strategy, therefore, is to compare the human sequence with that of the corresponding regions of a related genome, such as that of the mouse. Humans and mice are thought to have diverged from a common mammalian ancestor about 80×10^6 years ago, which is long enough for roughly two out of every three nucleotides to have been changed by random mutational events. Consequently, the only regions that will

Chromosomal DNA and Its Packaging

have remained closely similar in the two genomes are those where mutations would impair function and put animals carrying them at a disadvantage, resulting in their elimination from the population by natural selection. Such closely similar regions are known as *conserved regions*. In general, *nonconserved* regions represent noncoding DNA—both between genes and in introns—whose sequence is not critical for function, whereas conserved regions represent functionally important exons and regulatory sequences. By revealing in this way the results of a very long natural "experiment," comparative DNA sequencing studies highlight the most interesting regions in genomes. Such studies also provide strong support for the conclusion that only about 10% of the vertebrate genome sequence is vitally important to the organism.

Histones Are the Principal Structural Proteins of Eucaryotic Chromosomes [6]

If chromosomes were composed simply of extended DNA, it is difficult to imagine how they could be replicated and segregated to daughter cells without becoming severely tangled or broken. In fact, the DNA of all chromosomes is packaged into a compact structure with the aid of specialized proteins. It is traditional to divide the DNA-binding proteins in eucaryotes into two general classes: the **histones** and the *nonhistone chromosomal proteins*. The complex of both classes of proteins with the nuclear DNA of eucaryotic cells is known as **chromatin.** Histones are unique to eucaryotes. They are present in such enormous quantities (about 60 million molecules of each type per cell, compared to 10,000 molecules per cell for a typical sequence-specific DNA-binding protein) that their total mass in chromatin is about equal to that of the DNA.

Histones are relatively small proteins with a very high proportion of positively charged amino acids (lysine and arginine); the positive charge helps the histones bind tightly to DNA (which is highly negatively charged), regardless of its nucleotide sequence. Histones probably only rarely dissociate from the DNA, and so they are likely to have an influence on any reaction that occurs on chromosomes.

The five types of histones fall into two main groups—the *nucleosomal histones* and the *H1 histones*. The **nucleosomal histones** are small proteins (102–135 amino acids) responsible for coiling the DNA into *nucleosomes*, as discussed later. These four histones are designated **H2A, H2B, H3,** and **H4.** H3 and H4 are among the most highly conserved of all known proteins (Figure 8–8). This evolutionary conservation suggests that their functions involve nearly all of their amino acids, so that a change in any position is deleterious to the cell. This suggestion has been tested in yeasts, where it is possible to mutate a given histone gene *in vitro* and introduce it into the yeast genome in place of the normal gene. As predicted, many mutations are found to be lethal; some that are not lethal cause changes in the normal pattern of gene expression (discussed in Chapter 9). The H1 histones are larger (containing about 220 amino acids) and have been less conserved during evolution than the nucleosomal histones.

Histones Associate with DNA to Form Nucleosomes, the Unit Particles of Chromatin [7]

If it were stretched out, the DNA double helix in each human chromosome would span the cell nucleus thousands of times. Histones play a crucial part in packing this very long DNA molecule in an orderly way into a nucleus only a few micrometers in diameter. Their role in DNA folding is also important for a second reason. As we shall see, not all the DNA is folded in exactly the same way, and the manner in which a region of the genome is packaged into chromatin in a particular cell seems to influence the activity of the genes the region contains.

A major advance in our understanding of chromatin structure came in 1974 with the discovery of the fundamental packing unit known as the **nucleosome,**

Figure 8–8 The amino acid sequence of histone H4. The amino acids are designated by their single-letter abbreviations, with the positively charged amino acids colored for emphasis. As in the three other nucleosomal histones, an elongated amino-terminal "tail" is reversibly modified in the cell by the acetylation of selected lysines, which removes the lysine's positive charge. The sequence from a cow is shown; the sequence is the same in peas except that one valine is changed to an isoleucine and one lysine is changed to an arginine.

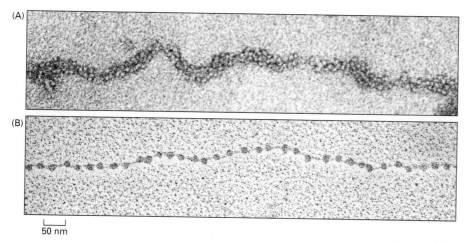

Figure 8–9 Nucleosomes as seen in the electron microscope. These electron micrographs show chromatin strands before and after treatments that unpack, or "decondense," the native structure to produce the "beads-on-a-string" form. The native structure, known as the 30-nm fiber (discussed later), is shown in (A). The decondensed, "beads-on-a-string" form of chromatin is shown at the same magnification in (B). For a schematic drawing of the relation between these two chromatin forms, see Figure 8–30. These electron micrographs were taken by modifications of the procedure outlined in Figure 8–45. (A, courtesy of Barbara Hamkalo; B, courtesy of Victoria Foe.)

50 nm

which gives chromatin a "beads-on-a-string" appearance in electron micrographs taken after treatments that unfold higher-order packing (Figure 8–9). The long DNA "string" can be broken into nucleosome "beads" by digestion with enzymes that degrade DNA, such as the bacterial enzyme micrococcal nuclease. (Enzymes that degrade both DNA and RNA are called *nucleases;* enzymes that degrade only DNA are *deoxyribonucleases,* or *DNases.*) After digestion for a short period with micrococcal nuclease, only the DNA between the nucleosome beads is degraded. The rest is protected from digestion and remains as double-helical DNA fragments 146 nucleotide pairs long bound to a specific complex of eight nucleosomal histones (the *histone octamer*). The nucleosome beads obtained in this way have been crystallized and analyzed by x-ray diffraction. Each is a disc-shaped

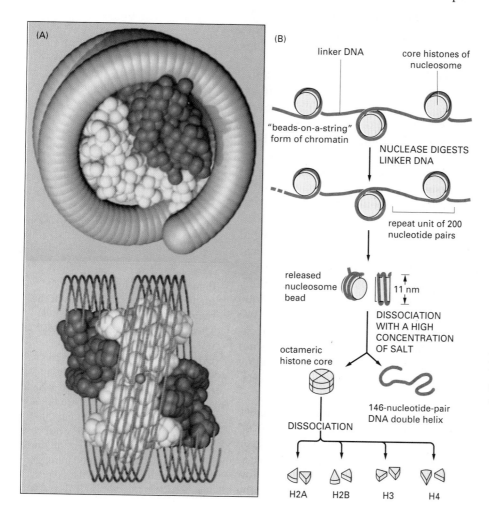

Figure 8–10 The nature of the nucleosome. (A) depicts two views of the three-dimensional structure of the histone octamer; the general path of the DNA wrapped around it is indicated by a coiled tube (*top*) and a series of parallel lines (*bottom*). Two H2A-H2B dimers (*blue*) flank an H3-H4 tetramer. The histone octamer is thus composed of two each of histones H2A, H2B, H3, and H4, with a total mass of about 100,000 daltons. (B) The nucleosome consists of two full turns of DNA (83 nucleotide pairs per turn) wound around an octameric histone core, plus the adjacent "linker DNA." The part of the nucleosome referred to here as the "nucleosome bead" is released from chromatin by digestion of the DNA with micrococcal nuclease. In each nucleosome bead 146 nucleotide pairs of DNA double helix (about 1.8 turns) remain wound around the octameric histone core. (A, courtesy of Evangelos Moudrianakis.)

Chromosomal DNA and Its Packaging

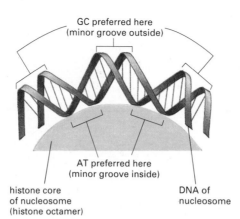

GC preferred here
(minor groove outside)

AT preferred here
(minor groove inside)

histone core
of nucleosome
(histone octamer)

DNA of
nucleosome

Figure 8–11 The bending of DNA in a nucleosome. The DNA helix makes two tight turns around the histone octamer. This diagram is drawn approximately to scale to illustrate how the minor groove is compressed on the inside of the turn. Due to certain structural features of the DNA molecule, A-T base pairs are preferentially accommodated in a narrow minor groove.

particle with a diameter of about 11 nm containing two copies of each of the four nucleosomal histones—H2A, H2B, H3, and H4. This **histone octamer** forms a protein core around which the double-stranded DNA helix is wound twice (Figure 8–10).

In undigested chromatin the DNA extends as a continuous double-helical thread from nucleosome to nucleosome. Each nucleosome is separated from the next by a region of *linker DNA*, which can vary in length from 0 to 80 nucleotide pairs. On average, nucleosomes repeat at intervals of about 200 nucleotide pairs (see Figure 8–10). Thus, a typical eucaryotic gene of 10,000 nucleotide pairs will be associated with 50 nucleosomes, and each human cell with 6×10^9 DNA nucleotide pairs contains 3×10^7 nucleosomes.

The Positioning of Nucleosomes on DNA Is Determined by the Propensity of the DNA to Form Tight Loops and by the Presence of Other DNA-bound Proteins [8]

Experiments performed *in vitro* with isolated chromatin suggest that the histone octamers generally remain fixed in one position under physiological conditions, inasmuch as their tight binding to DNA prevents them from sliding back and forth along the helix. There are two main influences that determine where nucleosomes form in the DNA. One is the difficulty of bending the DNA double helix into two tight turns around the outside of the histone octamer, a process that requires substantial compression of the minor groove of the helix. Because A-T-rich sequences in the minor groove are easier to compress than G-C-rich sequences, each histone octamer tends to position itself on the DNA so as to maximize A-T-rich minor grooves on the inside of the DNA coil (Figure 8–11). Thus a segment of DNA that contains short A-T-rich sequences spaced by integral numbers of DNA turns will be much easier to bend around the nucleosome than a segment of DNA lacking this feature. This probably explains some striking cases of very precise positioning of nucleosomes, such as those that bind to the tiny 5S rRNA genes, each of which has a single nucleosome bound to it in a unique location. If the DNA containing the 5S rRNA genes is added *in vitro* to a mixture of the four purified nucleosomal histones, nucleosomes will re-form at the exact position where they are located *in vivo*. For most of the DNA sequences found in chromosomes, however, there is no strongly preferred nucleosome binding site; instead, a nucleosome can occupy any one of a number of positions relative to the DNA sequence.

The second important influence on nucleosome positioning is the presence of other tightly bound proteins on the DNA that prevent nucleosomes from forming. For this reason some regions of DNA appear to lack a nucleosome even though they are hundreds of nucleotide pairs long. They can be detected by treating cell nuclei with trace amounts of a deoxyribonuclease (DNase I) that at low concentrations will digest long stretches of nucleosome-free DNA but not the

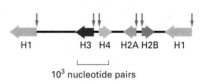

H1   H3 H4   H2A H2B   H1

10^3 nucleotide pairs

Figure 8–12 The location of nuclease-hypersensitive sites in the regulatory regions of active genes. The genes shown encode histones (H1, H2A, H2B, H3, and H4) in *Drosophila*. The *horizontal arrows* denoting each gene point in the direction of DNA transcription, which always proceeds from the 5′ to the 3′ end of the transcript. Although the nuclease-hypersensitive sites (*vertical red arrows*) are often present on the 5′ side of a gene, as illustrated here, they can also be located elsewhere (see Figure 9–34).

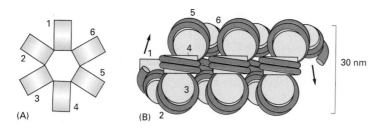

Figure 8–13 The 30-nm chromatin fiber. A model to explain how the "beads-on-a-string" form of nucleosomes is packed to form the 30-nm fiber seen in electron micrographs (see Figure 8–9A) in top (A) and side view (B). This type of packing requires one molecule of histone H1 per nucleosome (not shown). Although the position where the H1 attaches to the nucleosome has been defined (see Figure 8–15), the location of the H1 molecules in this fiber is unknown. (See also Figure 8–14.)

short stretches of linker DNA between nucleosomes. Such **nuclease-hypersensitive sites** often lie in the regulatory regions of genes (Figure 8–12). The first evidence for this idea came from studies of the monkey virus SV40, whose circular DNA chromosome binds to histones produced by its host cells. The SV40 chromosome often contains a single nucleosome-free region about 300 nucleotide pairs long very near the sequences at which viral DNA synthesis and RNA synthesis begin. Although several sequence-specific DNA-binding proteins are bound to this region, they do not protect long stretches of DNA against nuclease attack as do the nucleosomes, which is why the site is DNase-I sensitive.

The default state of the DNA in eucaryotic cells is to be fully covered with nucleosomes, and most nucleosome-free regions are specifically created by gene regulatory proteins as part of the process of activating DNA transcription (discussed in Chapter 9). Wherever nucleosomes are specifically positioned by the DNA sequence itself, there may have been evolutionary pressure to keep the adjacent linker DNA free of a nucleosome so as to facilitate its recognition by sequence-specific DNA-binding proteins.

Nucleosomes Are Usually Packed Together by Histone H1 to Form Regular Higher-Order Structures [9]

The linker DNA that connects adjacent nucleosomes can vary in length since nucleosomes position themselves according to the local flexibility of the DNA helix and the distribution of other proteins bound to specific DNA sequences. Although long strings of nucleosomes form on most chromosomal DNA, in the living cell chromatin probably rarely adopts the extended "beads-on-a-string" form. Instead, the nucleosomes are packed upon one another to generate regular arrays in which the DNA is even more highly condensed. Thus, when nuclei are very gently lysed onto an electron microscope grid, most of the chromatin is seen to be in the form of a fiber with a diameter of about 30 nm, which is considerably wider than chromatin in the "beads-on-a-string" form. One of several models proposed to explain how nucleosomes are packed in the 30-nm chromatin fiber is illustrated in Figure 8–13. Such models represent an idealized structure, since both the range of linker lengths that result from preferred nucleosome positioning and the presence of occasional nucleosome-free sequences will punctuate the 30-nm fiber with irregular features (Figure 8–14).

The histone H1 molecules, of which there are about six closely related subtypes in a mammalian cell, are thought to be responsible for pulling nucleosomes together to form the 30-nm fiber. The H1 molecule has an evolutionarily con-

Figure 8–14 Nucleosome-free regions in 30-nm fibers. A schematic section of chromatin illustrating the interruption of its regular nucleosomal structure by short regions where the chromosomal DNA is unusually vulnerable to digestion by DNase I. At each of these nuclease-hypersensitive sites, a nucleosome appears to have been excluded from the DNA by one or more sequence-specific DNA-binding proteins. How these proteins bind DNA tightly is discussed in Chapter 9.

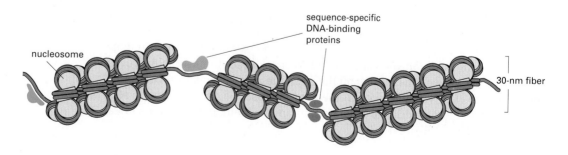

served globular central region linked to less conserved extended amino-terminal and carboxyl-terminal "arms." Each H1 molecule binds through its globular portion to a unique site on a nucleosome, and its arms extend to contact other sites on the histone cores of adjacent nucleosomes, so that the nucleosomes are pulled together into a regular repeating array (Figure 8–15).

Summary

A gene is defined as a nucleotide sequence in a DNA molecule that acts as a functional unit for the production of an RNA molecule. A chromosome is formed from a single, enormously long DNA molecule that contains a series of many genes. A chromosomal DNA molecule also contains three other types of functionally important nucleotide sequences: replication origins and telomeres allow the DNA molecule to be replicated, while a centromere is needed to attach the DNA molecule to the mitotic spindle, ensuring its accurate segregation to daughter cells. The human haploid genome contains 3×10^9 DNA nucleotide pairs, divided among 22 different autosomes and 2 sex chromosomes. Only a small percentage of this DNA is thought to code for proteins.

The DNA in eucaryotes is tightly bound to an equal mass of histones, which form a repeating array of DNA-protein particles called nucleosomes. The nucleosome is made up of an octameric core of histone proteins around which the DNA is wrapped twice. The ease with which a segment of DNA can undergo the severe bending required for this wrapping varies with its nucleotide sequence. Despite irregularities such as this, nucleosomes are usually packed together, with the aid of histone H1 molecules, into regular arrays to form a 30-nm fiber.

The Global Structure of Chromosomes

Having discussed the DNA and protein molecules from which chromosomes are made, we now turn to the organization of the chromosome on a more global scale. As a 30-nm fiber, the typical human chromosome would be 0.1 cm in length and could span the nucleus more than 100 times. Clearly, there must be a still higher level of folding. The DNA is not only packaged with histones into regularly repeating nucleosomes that are packed into 30-nm fibers; it is also elaborately folded and organized by other proteins into a series of subdomains of distinct character. This higher-order packaging is one of the most fascinating but also most poorly understood aspects of chromatin. Although its molecular basis is still a mystery, this packaging almost certainly plays a crucial part in the regulation of gene transcription. In this section we discuss what is known about higher-order chromatin structure and examine evidence that shows it is functionally important.

Lampbrush Chromosomes Contain Loops of Decondensed Chromatin [10]

Although packaged into chromatin, most chromosomes in interphase cells (cells not in mitosis) are too fine and too tangled to be visualized clearly. In a few exceptional cases, however, interphase chromosomes can be seen to have a precisely defined higher-order structure. The meiotically paired chromosomes in growing amphibian oocytes (immature eggs), for example, are highly active in RNA synthesis, and they form unusually stiff and extended chromatin loops that are covered with newly transcribed RNA packed into dense RNA-protein complexes. Because of this coating on the DNA, these so-called **lampbrush chromosomes** are clearly visible even in the light microscope, where they are seen to be organized into a series of large chromatin loops emanating from a linear chromosomal axis (Figure 8–16).

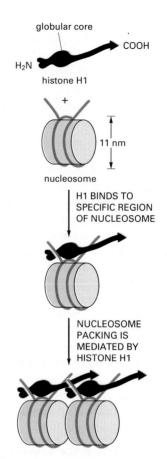

Figure 8–15 **The way histone H1 is thought to help pack adjacent nucleosomes together.** The globular core of H1 binds to each nucleosome near the site where the DNA helix enters and leaves the histone octamer. When H1 is present on the nucleosomes, 166 nucleotide pairs of DNA are protected from micrococcal nuclease digestion, compared with 146 nucleotide pairs for nucleosomes lacking H1 (see Figure 8–10).

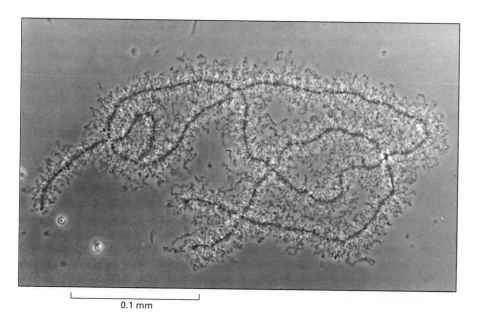

Figure 8–16 Light micrograph of lampbrush chromosomes in an amphibian oocyte. Early in oocyte differentiation each chromosome replicates to begin meiosis, and the homologous replicated chromosomes pair to form this highly extended structure containing a total of four replicated DNA molecules, or chromatids. The lampbrush chromosome stage persists for months or years as the oocyte builds up a supply of mRNA and other materials required for its ultimate development into a new individual. (Courtesy of Joseph G. Gall.)

0.1 mm

The organization of the lampbrush chromosome is illustrated schematically in Figure 8–17. Nucleic acid hybridization experiments show that a given loop always contains the same DNA sequence and that it remains extended in the same manner as the oocyte grows. Other experiments demonstrate that most of the looped DNA is being actively transcribed into RNA. The majority of the chro-

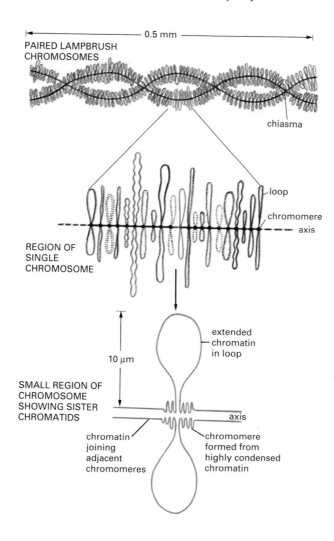

— 0.5 mm —

PAIRED LAMPBRUSH CHROMOSOMES

chiasma

loop

chromomere

axis

REGION OF SINGLE CHROMOSOME

10 μm

extended chromatin in loop

SMALL REGION OF CHROMOSOME SHOWING SISTER CHROMATIDS

axis

chromatin joining adjacent chromomeres

chromomere formed from highly condensed chromatin

Figure 8–17 Lampbrush chromosome structure. The set of lampbrush chromosomes in many amphibians contains a total of about 10,000 chromatin loops, although most of the DNA in each chromosome remains highly condensed in the *chromomeres*. Each loop corresponds to a particular DNA sequence. Four copies of each loop are present in each cell, since each of the two chromosomes shown at the top consists of two closely apposed sister chromatids. This four-stranded structure is characteristic of this stage of development of the oocyte (the diplotene stage of meiosis, see Figure 20–9).

The Global Structure of Chromosomes

matin, however, is not in loops but remains highly condensed in the *chromomeres*, which are generally not transcribed. A general chromosome model based on these studies has been proposed in which loops of 30-nm fibers extend at an angle from the main axis of the chromosome (Figure 8–18).

Lampbrush chromosomes illustrate the recurrent themes of this section—when chromatin is actively transcribed, it has an extended structure; when the chromatin is condensed, it is inactive; and the structural units of regulation are large, precisely defined domains.

Orderly Domains of Interphase Chromatin Also Can Be Seen in Insect Polytene Chromosomes [11]

Because of a specialization (which is different from that of the lampbrush chromosome), chromatin structure is also unusually visible in certain insect cells.

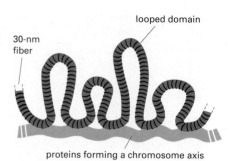

Figure 8–18 A model of chromosome structure. A section of a chromosome is shown folded into a series of looped domains, each containing perhaps 20,000 to 100,000 nucleotide pairs of double-helical DNA condensed in a 30-nm chromatin fiber.

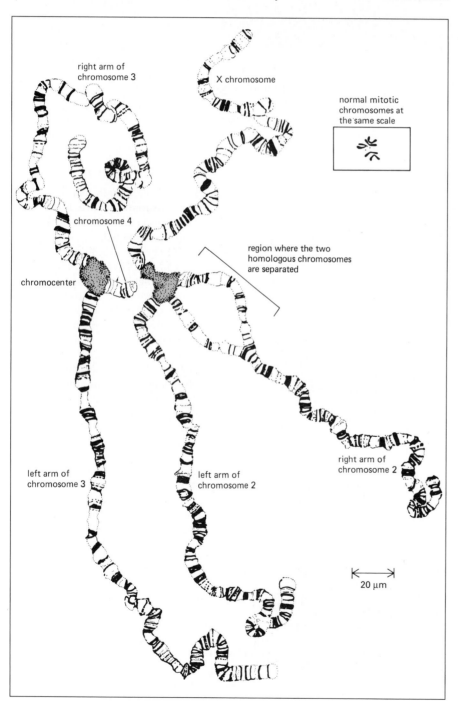

Figure 8–19 A detailed sketch of the entire set of polytene chromosomes in one *Drosophila* salivary cell. These chromosomes have been spread out for viewing by squashing them against a microscope slide. *Drosophila* has four chromosomes, and there are four different chromosome pairs present. But each chromosome is tightly paired with its homologue (so that each pair appears as a single structure), which is not the case in most nuclei (except in meiosis). The four polytene chromosomes are normally linked together by regions near their centromeres that aggregate to create a single large "chromocenter" (*colored region*); in this preparation, however, the chromocenter has been split into two halves by the squashing procedure used. (Modified from T.S. Painter, *J. Hered.* 25:465–476, 1934.)

Many of the cells of the larvae of flies grow to an enormous size through multiple cycles of DNA synthesis without cell division. The resulting giant cells contain as much as several thousand times the normal DNA complement. Cells with more than the normal DNA complement are said to be *polyploid* when, as is usually the case, they contain increased numbers of standard chromosomes. In several types of secretory cells of fly larvae, however, all the homologous chromosome copies remain side by side, creating a single giant *polytene chromosome*. The fact that some large insect cells can undergo a direct polytene-to-polyploid conversion demonstrates that these two chromosomal states are closely related and that the basic structure of a polytene chromosome must be similar to that of a normal chromosome.

Polytene chromosomes are easy to see in the light microscope because they are so large and because the precisely aligned side-to-side adherence of individual chromatin strands greatly elongates the chromosome axis and prevents tangling. Like lampbrush chromosomes, these chromosomes are active in RNA synthesis. Polyteny has been most studied in the salivary gland cells of *Drosophila* larvae, in which the DNA in each of the four *Drosophila* chromosomes has been replicated through 10 cycles without separation of the daughter chromosomes, so that 1024 (= 2^{10}) identical strands of chromatin are lined up side by side (Figure 8–19).

When polytene chromosomes are viewed in the light microscope, distinct alternating dark *bands* and light *interbands* are visible (Figure 8–20). Each band and interband represents a set of 1024 identical DNA sequences arranged in register. About 85% of the DNA in polytene chromosomes is in bands, and 15% is in interbands. The chromatin in each band stains darkly because it is much more condensed than the chromatin in the interbands (Figure 8–21). Depending on their size, individual bands are estimated to contain 3000 to 300,000 nucleotide pairs per chromatin strand. Since the bands can be recognized by their different thicknesses and spacings, each one has been given a number to generate a polytene chromosome "map." There are approximately 5000 bands and 5000 interbands in the total *Drosophila* genome.

Individual Chromatin Domains Can Unfold and Refold as a Unit [12]

Long before anything was known at the molecular level about chromatin structure, studies of polytene chromosomes suggested that a major change in DNA packing accompanies gene transcription, since individual chromosome bands often expand when the genes they contain become active and recondense when these genes become quiescent.

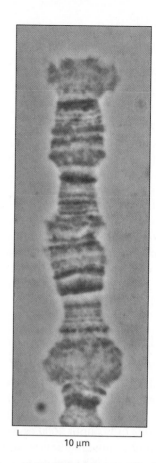

10 µm

Figure 8–20 Light micrograph of a portion of a polytene chromosome from *Drosophila* salivary glands. The distinct patterns recognizable in different chromosome bands are readily seen. The bands are regions of increased chromatin concentration. They occur in interphase chromosomes and are a special property of the giant polytene chromosomes. (Courtesy of Joseph G. Gall.)

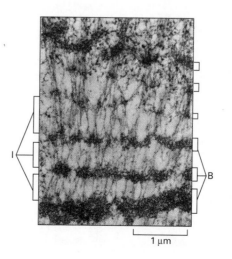

1 µm

Figure 8–21 Electron micrograph of a small section of a *Drosophila* polytene chromosome seen in thin section. Bands (B) of very different thickness can be readily distinguished, separated by interbands (I), which contain less condensed chromatin. (Courtesy of Viekko Sorsa.)

The Global Structure of Chromosomes

The regions on a polytene chromosome being transcribed at any instant can be identified by labeling the cells briefly with the radioactive RNA precursor ³H-uridine and locating the growing RNA transcripts by autoradiography (Figure 8–22). This analysis (see below) reveals that the most active chromosomal regions are decondensed, forming distinctive **chromosome puffs.**

One of the main factors controlling the activity of genes in polytene chromosomes of *Drosophila* is the insect steroid hormone *ecdysone,* the levels of which rise and fall periodically during larval development, inducing the transcription of various genes coding for proteins that the larva requires for each molt and for pupation. As the organism progresses from one developmental stage to another, new puffs arise and old puffs recede, as transcription units are activated and deactivated and different mRNAs and proteins are made (Figure 8–23). From inspection of each puff when it is relatively small and the banding pattern of the chromosome is still discernible, it seems that most puffs arise from the uncoiling of a single chromosome band (Figure 8–24). Electron microscopy of thin sections of such puffs shows that the DNA in the chromatin is much less condensed than it would be in the 30-nm chromatin fiber. These observations suggest that the chromatin in a band can decondense as a unit during transcription.

Both Bands and Interbands in Polytene Chromosomes Are Likely to Contain Genes [13]

The fixed pattern of bands and interbands in a *Drosophila* polytene chromosome suggested to early cytogeneticists that each band might correspond to a single gene. Mutational analyses that allowed geneticists to estimate that *Drosophila* contains only about 5000 essential genes, a number roughly equal to the number of chromosome bands, supported this view. Moreover, when an intensive effort was made to isolate as many mutants as possible in a small chromosomal region, about 50 essential genes were genetically identified in a region that contains about 50 visible bands. Although it is not possible with these techniques to determine whether a particular gene lies in a band or an interband region, these observations suggested that an average band might contain the DNA coding sequences for only one essential protein.

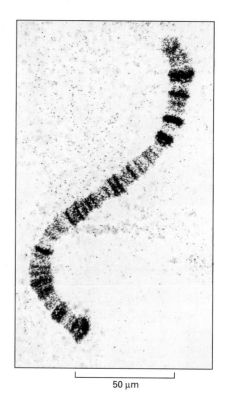

50 µm

Figure 8–22 **Synthesis of RNA along a giant polytene chromosome.** In this autoradiograph of a chromosome (from the salivary gland of the insect *Chironomus tentans*) labeled with ³H-uridine, sites of RNA synthesis are covered with dark silver grains in proportion to their activity. (From C. Pelling, *Chromosoma* 15:71–122, 1964.)

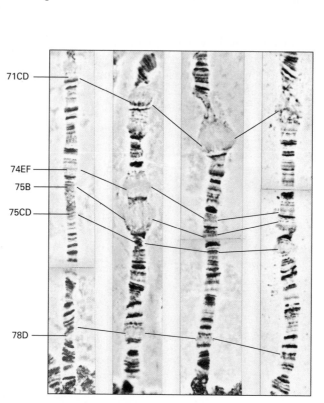

71CD
74EF
75B
75CD
78D

Figure 8–23 **Chromosome puffs.** A temporal series of photographs illustrating how puffs arise and recede in the polytene chromosomes of *Drosophila melanogaster* during larval development. A region of the left arm of chromosome 3 is shown. It exhibits five very large puffs in salivary gland cells, each active for only a short developmental period. The series of changes shown occur over a period of 22 hours, appearing in a reproducible pattern as the organism develops. (Courtesy of Michael Ashburner.)

Genes that are not essential for survival in the laboratory (where predators are absent and both food and mates are provided), however, would have been missed in these genetic analyses, and more recent data have made the simple "one-band, one-gene" hypothesis seem unlikely. A continuous 315,000-nucleotide-pair region of the *Drosophila* genome has been cloned, for example, and fragments have been used as DNA probes to catalogue the mRNAs produced from the region: three times as many separate mRNAs as bands were identified, indicating that each band probably contains several genes. In addition, it has now been shown directly that mRNAs are produced from both interbands and bands in polytene chromosomes.

Although still controversial, it has been proposed that the DNA in polytene chromosome bands is arranged in loops in a manner analogous to that of the lampbrush chromosomes. According to this model, puff formation would correspond to the decondensation of one or more looped domains. Whether or not this model is correct, the above findings show that at least some interphase chromosomes are arranged in a complex, precisely ordered sequence of structural domains, each typically containing a small number of genes whose transcription is regulated in a coordinated way. It may be that all interphase chromosomes are packaged into ordered structures according to these same principles.

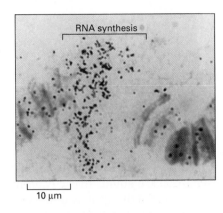

Figure 8–24 Autoradiogram of a single puff in a polytene chromosome. The portion of the chromosome indicated is undergoing RNA synthesis and has therefore become labeled with ³H-uridine. (Courtesy of Jose Bonner.)

Transcriptionally Active Chromatin Is Less Condensed [14]

The studies of puffs on insect polytene chromosomes suggest that the chromatin structure of transcribed genes is selectively decondensed. Experiments of a completely different type support this suggestion. When nuclei isolated from vertebrate cells are exposed to the enzyme DNase I, about 10% of the genome is preferentially degraded. Although a small subset of these degraded sequences corresponds to the hypersensitive sites described earlier, most of them represent genes that are being transcribed: different DNA sequences, for example, are digested by the nuclease in different cells of the same organism, depending on the genes that are being transcribed in the different cell types. Remarkably, even genes that are transcribed only a few times in every cell generation are sensitive

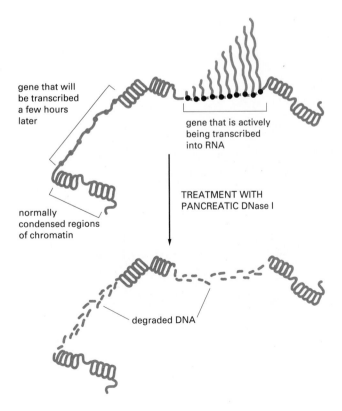

gene that will be transcribed a few hours later

normally condensed regions of chromatin

gene that is actively being transcribed into RNA

TREATMENT WITH PANCREATIC DNase I

degraded DNA

Figure 8–25 Digestion of chromatin with DNase I. The enzyme first cuts at nuclease-hypersensitive sites (not shown, but see Figure 8–12) and then selectively degrades the entire DNA sequence both of genes that are actively being transcribed and of genes that have been activated but are not yet being transcribed.

to DNase I, indicating that a special state of the chromatin, rather than the process of DNA transcription itself, makes these regions unusually accessible to digestion (Figure 8–25). Chromatin in this accessible state is often termed **active chromatin,** and its structure appears to be altered in a way that makes the packing of the nucleosomes less condensed.

Active Chromatin Is Biochemically Distinct [15]

What distinguishes active from inactive chromatin? The answer to this question will probably require purification and characterization of the chromosomal proteins that are unique to each chromatin conformation. Some progress has been made toward this goal by the development of biochemical methods designed to isolate active chromatin by taking advantage of its relatively decondensed state. The analysis of the chromosomal proteins in the active chromatin fraction has suggested the following: (1) Histone H1 seems to be less tightly bound to at least some active chromatin, and particular subtypes of this histone may be specific for active chromatin. (2) Although the four nucleosomal histones are present in normal amounts in active chromatin, they appear to be unusually highly acetylated when compared with the same histones in inactive chromatin. (3) The nucleosomal histone H2B in active chromatin appears to be less phosphorylated than it is in inactive chromatin. (4) Active chromatin is highly enriched in a minor variant form of histone H2A that is found in many species, including *Drosophila* and humans. (5) The nucleosomes in active chromatin selectively bind two closely related small chromosomal proteins, called HMG 14 and HMG 17. Consistent with their presence only in active chromatin, these "high-mobility-group" (HMG) proteins occur in roughly the quantities required to bind to 1 in every 10 nucleosomes in total chromatin; moreover, their amino acid sequences have been highly conserved during evolution, implying that they have important functions.

Any or all of these changes might play an important part in uncoiling the chromatin of active genes, helping to make the DNA available as a template for RNA synthesis, but more direct experiments are needed to test this idea. In Chapter 9 we discuss mechanisms by which the transition between inactive and active chromatin may be controlled.

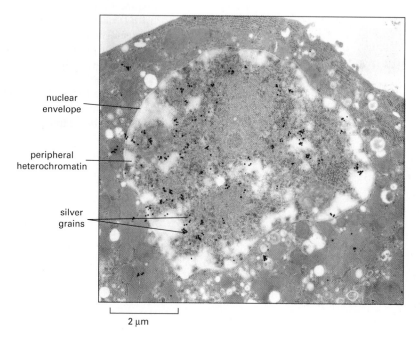

nuclear envelope

peripheral heterochromatin

silver grains

2 µm

Figure 8–26 Heterochromatin is transcriptionally inactive. An autoradiograph is shown of a thin section of a nucleus from a cell that has been pulse-labeled with ^{3}H-uridine in order to label sites of RNA synthesis (seen as black silver grains). The *white areas* are regions of heterochromatin, which tend to pack along the inside of the nuclear envelope. Normally, heterochromatin stains unusually darkly in electron micrographs (for example, see Figure 8–71), but because of the particular method used for sample preparation here, areas containing heterochromatin have been bleached. As can be seen, most of the RNA synthesis occurs in the euchromatin that borders these heterochromatic areas. (Courtesy of Stan Fakan.)

Heterochromatin Is Highly Condensed and Transcriptionally Inactive [16]

Light microscopic studies in the 1930s distinguished two types of chromatin in interphase nuclei of higher eucaryotic cells: a highly condensed form called **heterochromatin** and all the rest, which is less condensed, called **euchromatin.** Heterochromatin was originally identified because it remains unusually compact during interphase; it was later found to be transcriptionally inactive (Figure 8–26). In a typical cell in interphase, approximately 10% of the genome is packed into heterochromatin. It is now clear, as described above, that the euchromatin exists in at least two forms: about 10% is in the form of active chromatin, which is the least condensed, while the rest is inactive euchromatin, which is more condensed than active chromatin but less condensed than heterochromatin. Heterochromatin is therefore thought to be a special variety of transcriptionally inactive chromatin with distinctive functions. In mammals, for example, and in many other higher eucaryotes, the DNA surrounding each centromere is composed of relatively simple repeating nucleotide sequences. These so-called *satellite DNAs* constitute a major portion of the heterochromatin in these organisms.

Mitotic Chromosomes Are Formed from Chromatin in Its Most Condensed State [17]

With the exception of a few specialized cases, such as the lampbrush and polytene chromosomes discussed above, most *interphase* chromosomes are too extended, thin, and entangled to be readily detectable. In contrast, chromosomes from nearly all cells are visible during mitosis, when they coil up to form much more condensed structures. This coiling, which reduces a 5-cm length of DNA to about 5 μm, makes it possible for the mitotic spindle to segregate the chromosomes to separate daughter cells without breaking them. The condensation is accompanied by the phosphorylation of all of the histone H1 in the cell at five of its serine residues. Because histone H1 helps to pack nucleosomes together (see Figure 8–15), its phosphorylation during mitosis is suspected to play a role in chromosome condensation.

Figure 8–27 depicts a typical **mitotic chromosome** at the metaphase stage of mitosis. The two daughter DNA molecules produced by DNA replication during the S phase of the cell-division cycle are separately folded to produce two sister chromosomes (called *sister chromatids*) held together at their *centromeres*. These chromosomes are normally covered with a variety of molecules, including large amounts of ribonucleoproteins. Once this covering is stripped away, each chromatid can be seen in electron micrographs to be organized into loops of chromatin emanating from a central axis (Figures 8–28 and 8–29). Several types of experiments demonstrate that the order of visible features along a mitotic chromosome at least roughly reflects the order of the genes along the DNA molecule. To illustrate how such an organized folding of the long DNA helix might be achieved, we have drawn each chromatid as a closely packed series of looped domains wound in a tight helix in Figure 8–30, which also presents a schematic view of all the different folding processes that would contribute to this structure.

The condensed form of chromatin found in mitotic chromosomes is thought to resemble heterochromatin in its high degree of packing. Not surprisingly, mitotic chromosomes are found to be transcriptionally inactive: all RNA synthesis ceases as the chromosomes condense. Presumably, the condensation prevents RNA polymerase from gaining access to the DNA, although other control factors may also be involved.

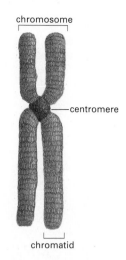

Figure 8–27 Drawing of a typical metaphase chromosome. Each sister chromatid contains one of two identical daughter DNA molecules generated earlier in the cell cycle by DNA replication.

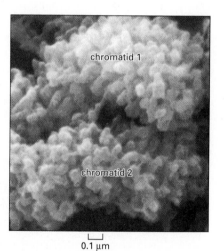

Figure 8–28 Scanning electron micrograph of a region near one end of a typical mitotic chromosome. Each knoblike projection is believed to represent the tip of a separate looped domain. Note that the two identical paired chromatids diagrammed in Figure 8–27 can be clearly distinguished. (From M.P. Marsden and U.K. Laemmli, *Cell* 17:849–858, 1979. © Cell Press.)

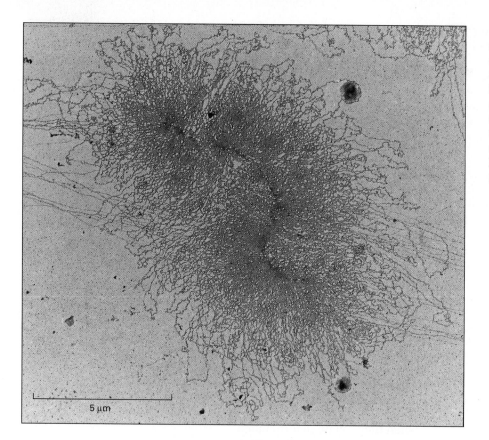

Figure 8–29 Electron micrograph of a single chromatid of a mitotic chromosome. The chromosome (from an insect) was treated to reveal loops of chromatin fibers that emanate from a central axis of the chromatid. Such micrographs support the idea that the chromatin in all chromosomes is folded into a series of looped domains (see Figure 8–18). (Courtesy of Victoria Foe.)

5 µm

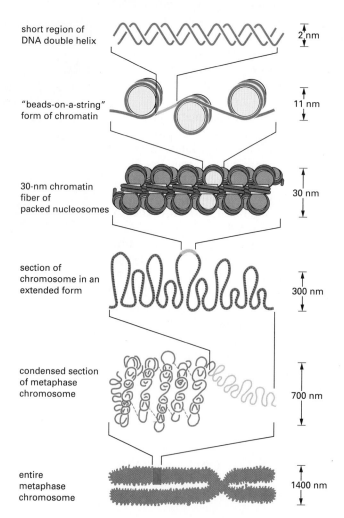

short region of DNA double helix

2 nm

"beads-on-a-string" form of chromatin

11 nm

30-nm chromatin fiber of packed nucleosomes

30 nm

section of chromosome in an extended form

300 nm

condensed section of metaphase chromosome

700 nm

entire metaphase chromosome

1400 nm

Figure 8–30 Model of chromatin packing. This schematic drawing shows some of the many orders of chromatin packing postulated to give rise to the highly condensed mitotic chromosome.

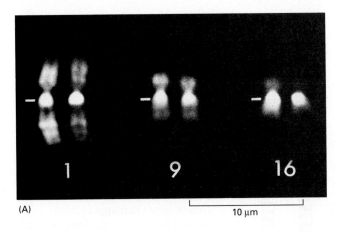

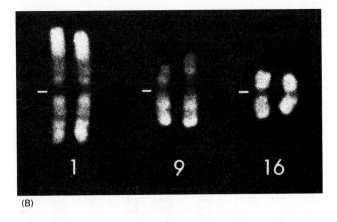

(A)

10 μm

(B)

Each Mitotic Chromosome Contains a Characteristic Pattern of Very Large Domains [18]

The display of the 46 human chromosomes at mitosis is called the human **karyotype.** Staining methods developed over the last 25 years permit unambiguous identification of each individual chromosome. Some of these methods involve staining mitotic chromosomes with dyes that fluoresce only when they bind to certain types of DNA sequences. Although these dyes have very low specificity and appear mainly to distinguish DNA that is rich in A-T nucleotide pairs (**G bands**) from DNA that is rich in G-C nucleotide pairs (**R bands**), they produce a striking and reproducible banding pattern along each mitotic chromosome (Figure 8–31). In this way each chromosome can be identified and numbered, as illustrated in Figure 8–32. Both the G bands and the R bands are known to contain genes. These bands are unrelated to those described earlier for the insect polytene chromosomes, which are thought to correspond to regions of condensed chromatin; in the human mitotic chromosomes, all of the chromatin is condensed and the bands are formed by the selective binding of specific dyes. Whereas a band in a polytene chromosome contains only a few genes, a typical band in a human karyotype contains several hundred.

By examining human chromosomes very early in mitosis, when they are less condensed than at metaphase, it has been possible to estimate that the total haploid genome contains about 2000 distinguishable A-T-rich bands. These progressively coalesce as condensation proceeds during mitosis to produce fewer and thicker bands.

Chromosome bands are detected in mitotic chromosomes from species as diverse as humans and flies. Moreover, the exact pattern of bands in a chromosome has remained unchanged over long periods of evolutionary time. Every human chromosome, for example, has a clearly recognizable counterpart with a nearly identical banding pattern in the chromosomes of the chimpanzee, gorilla, and orangutan (although there has been a single chromosome fusion that gives humans 46 chromosomes instead of the apes' 48). This conservation suggests that the long-range spatial organization of chromosomes may be important for chromosomal function. But why such bands exist at all is a mystery. Even the thinnest of the bands diagrammed in Figure 8–32 probably contain more than a million nucleotide pairs, which is nearly the size of a bacterial genome, and the average nucleotide-pair composition would be expected to be random over such long stretches of DNA sequence.

Figure 8–31 Fluorescence micrographs showing the banding patterns in three pairs of human mitotic chromosomes. In (A) the chromosomes were stained with the A-T-base-pair-specific dye Hoechst 33258, which stains the G bands, and in (B) with the G-C-base-pair-specific dye olivomycin, which stains the R bands. The bars indicate the position of the centromere. Note that these banding patterns are the reverse of each other: the bands that are bright in (A) are dark in (B), and vice versa. The G bands also stain with Giemsa stain—hence their name—while the R bands were so named because they formed the "reverse" of the G-band pattern. (From K.F. Jorgenson, J.H. van de Sande, and C.C. Lin, *Chromosoma* 68:287–302, 1978.)

Summary

Chromosomes are generally decondensed during interphase, so that their structure is difficult to discern. Notable exceptions are the specialized lampbrush chromosomes of vertebrate oocytes and the polytene chromosomes of insect giant secretory cells. Studies of these two types of interphase chromosomes suggest that each long DNA

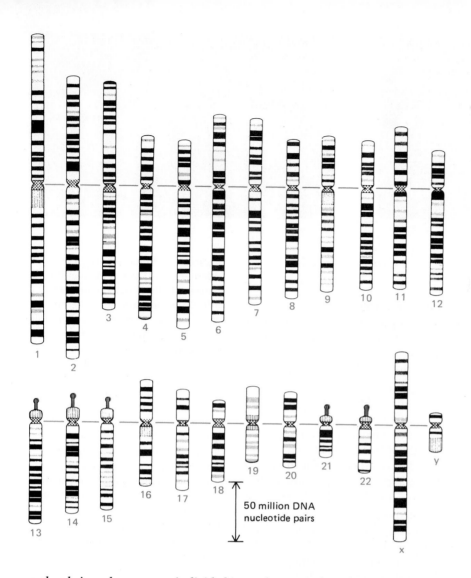

Figure 8–32 A standard map of the banding pattern of each chromosome in the human karyotype. This map was determined at the prometaphase stage of mitosis. Chromosomes 1 through 22 are labeled in the approximate order of their size; a diploid cell contains two of each of these autosomes plus two sex chromosomes—two X chromosomes (female) or an X and a Y chromosome (male). The 850 bands shown here are G bands, which stain with reagents that appear to be specific for A-T-rich DNA sequences. The *green knobs* on chromosomes 13, 14, 15, 21, and 22 indicate the positions of the genes that encode the large ribosomal RNAs; the *green lines* mark the centromere on each chromosome. (Adapted from U. Franke, *Cytogenet. Cell Genet.* 31:24–32, 1981.)

molecule in a chromosome is divided into a large number of discrete domains that are folded differently. In both lampbrush and polytene chromosomes the regions that are actively synthesizing RNA are least condensed. Likewise, as judged by nuclease sensitivity, about 10% of the DNA in interphase vertebrate cells is in a relatively uncondensed conformation that correlates with DNA transcription in these regions. Such "active chromatin" is biochemically distinct from the more condensed inactive regions of chromatin.

All chromosomes adopt a highly condensed conformation during mitosis. When they are specially stained, these mitotic chromosomes have a banded structure that allows each individual chromosome to be recognized unambiguously; these bands contain millions of DNA nucleotide pairs, and they reflect a coarse heterogeneity of chromosome structure that is not understood.

Chromosome Replication

Before a cell can divide, it must produce a new copy of each of its chromosomes, and it does this during a specific part of interphase called the DNA-synthesis phase, or **S phase,** of the cell-division cycle; the part of interphase preceding S phase is called Gap 1, or G_1, and the part following S phase is called Gap 2, or G_2 (see Figure 17–3). In a typical higher eucaryotic cell the S phase lasts for about 8 hours. By its end each chromosome has been replicated to produce two complete copies, which remain joined together at their centromeres until the M phase that soon follows (see Figure 8–27). Chromosome duplication requires both the replication of the long DNA molecule in each chromosome and the assembly of a new set of chromosomal proteins onto the DNA to form chromatin. In Chapter 6 we discussed the enzymology of DNA replication and described the structure of the *DNA replication fork*, where DNA synthesis occurs by a semi-conservative process (see Figures 6–48 and 6–49). In Chapter 17 we consider how entry into the S phase is regulated as part of a more general discussion of how the cell-division cycle is controlled. In this section we describe the timing and pattern of eucaryotic chromosome replication and its relation to chromosome structure.

Specific DNA Sequences Serve as Replication Origins [19]

We saw in Chapter 6 that replication origins have been precisely defined in bacteria as specific DNA sequences that allow the DNA replication machinery to assemble on the DNA double helix and move in opposite directions to produce *replication forks*. By analogy, one would expect the replication origins in eucaryotic chromosomes to be specific DNA sequences too. As mentioned earlier, the search for replication origins in the chromosomes of eucaryotic cells has been most successful in the yeast *Saccharomyces cerevisiae*. The powerful selection methods that have been devised to find them make use of mutant yeast cells that are defective for an essential gene: they will survive in a selective medium only if they are provided with a plasmid carrying a functional copy of the missing gene. If a bacterial plasmid that carries the essential yeast gene is introduced into the mutant yeast cells, it will not be able to replicate free of the host chromosome because the bacterial replication origin in the plasmid DNA cannot function in a yeast cell. If random pieces of yeast DNA are inserted into the plasmid, however, a small proportion of the plasmid DNA molecules will contain a yeast replication origin and therefore will be able to replicate in the yeast cells. The yeast cells that carry such plasmids will be able to proliferate since they will have been provided with the essential gene (Figure 8–33). A DNA sequence identified by its presence in a plasmid isolated from these cells is called an *autonomously replicating sequence (ARS)*. A number of ARSs recently have been shown to be authentic chromosomal origins of replication, thereby justifying the strategy used to obtain them.

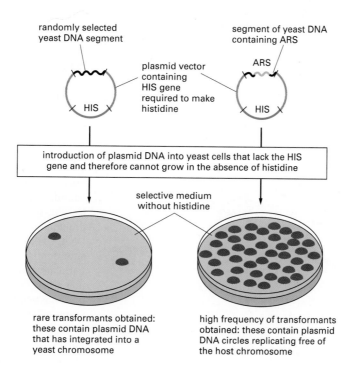

randomly selected yeast DNA segment

segment of yeast DNA containing ARS

ARS

plasmid vector containing HIS gene required to make histidine

HIS

HIS

introduction of plasmid DNA into yeast cells that lack the HIS gene and therefore cannot grow in the absence of histidine

selective medium without histidine

rare transformants obtained: these contain plasmid DNA that has integrated into a yeast chromosome

high frequency of transformants obtained: these contain plasmid DNA circles replicating free of the host chromosome

Figure 8–33 The strategy used to identify replication origins in yeast cells. A DNA sequence identified in this way was initially called an autonomously replicating sequence (ARS) since it enables a plasmid that contains it to replicate freely in the host cell without having to be incorporated into its chromosomes.

An 11-nucleotide "core consensus sequence" has been shown to be essential for yeast replication-origin function, and all known origins contain multiple near matches of this sequence, spaced throughout a region of about 100 nucleotides. These DNA sequences are recognized by a large protein complex that is thought to function in the initiation process. Although some replication-origin sequences in higher eucaryotic cells have been tentatively identified, the proteins that bind to them are not known.

A Mammalian Cell-free System Replicates the Chromosome of a Monkey Virus [20]

In bacteria a multienzyme complex that includes DNA polymerase, DNA primase, and DNA helicase moves the replication fork (see Figures 6–48 and 6–49); this complex assembles at a replication origin in a reaction involving an initiator protein that binds specifically to the origin DNA (see Figure 6–52).

Many attempts have been made to reconstitute mammalian DNA replication in a test tube. To do so would make it possible to identify all the enzymes required and to define their actions at a replication fork. The most successful *in vitro* system developed thus far replicates the small circular chromosome of the monkey virus SV40. This virus parasitizes its mammalian host cell for all but one of the proteins that the virus needs to replicate; the one exception is the *SV40 T-antigen*, a large multifunctional protein that allows the virus to bypass normal controls and thereby replicate faster than the host cell. Two hexamers of the T-antigen bind specifically to the SV40 replication origin, recognizing the outside of the DNA double helix. Then, in a process that is not well understood, the bound T-antigen undergoes a conformational change that pulls apart the DNA strands of the origin. This remarkable protein has a third function: utilizing the energy of ATP hydrolysis, it acts as a DNA helicase, moving along the DNA and opening the DNA double helix as it moves, forming a *replication bubble* (Figure 8–34). The cellular components of the replication machinery can now assemble in the bubble to form two **replication forks**, which then move away from the origin in opposite directions.

Although the replication forks are similar to those of procaryotes, the SV40 experiments have revealed that at least two distinct types of DNA polymerase are needed in eucaryotes: *DNA polymerase α (alpha)* on the lagging strand and *DNA polymerase δ (delta)* on the leading strand (Figure 8–35). In procaryotes, by contrast, a single type of DNA polymerase functions on both strands of the replication fork (see Figure 6–48).

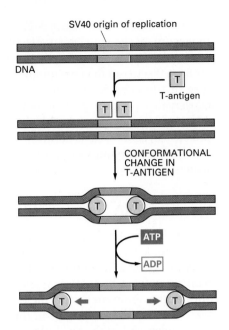

Figure 8–34 Initiation of DNA replication on the SV40 viral genome. The initiation protein (T-antigen) is encoded by the virus, and it specifically recognizes the viral origin of replication. After opening the DNA double helix at the origin, the T-antigen acts as a DNA helicase, traveling away from the origin and opening the double helix as it moves. Subsequent to the steps depicted, host cell components of the replication machinery assemble in the "replication bubble" to form two replication forks (see Figure 8–35).

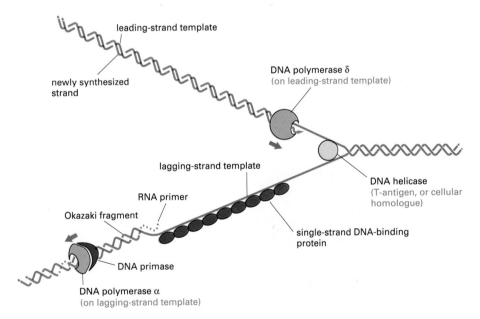

Figure 8–35 A mammalian replication fork. The fork is drawn to emphasize its similarity to the bacterial replication fork described in Chapter 6. Although both forks utilize the same basic components, the mammalian fork differs in two important respects. First, it makes use of two DNA polymerases, one for the leading strand and one for the lagging strand. It seems likely that the leading-strand polymerase is designed to keep a tight hold on the DNA, whereas that on the lagging strand must be able to release the template and then rebind each time that a new Okazaki fragment (see p. 256) is synthesized. Second, the mammalian DNA primase is a subunit of the lagging-strand DNA polymerase, while that of bacteria is associated with the DNA helicase.

The protein that normally initiates DNA replication of mammalian chromosomes has not yet been discovered. It is therefore not known whether it resembles the T-antigen in functioning as a DNA helicase at the fork or whether, as in procaryotes, a separate initiator protein and DNA helicase are involved.

Replication Origins Are Activated in Clusters on Higher Eucaryotic Chromosomes [21]

In Chapter 6 we describe how in bacteria two replication forks begin at each origin and proceed in opposite directions, moving away from the origin at a rate of about 500 nucleotides per second until all of the DNA in the single circular bacterial chromosome is replicated. The bacterial genome is so small that these two replication forks can duplicate it in less than 40 minutes. Because of the much larger size of most eucaryotic chromosomes, it seemed unlikely that their replication would begin from a single origin.

A method for determining the general pattern of eucaryotic chromosome replication was developed in the early 1960s. Human cells growing in culture are labeled for a short time with ^{3}H-thymidine so that the DNA synthesized during this period becomes highly radioactive. The cells are then gently lysed, and the DNA is streaked onto the surface of a glass slide, which is coated with a photographic emulsion, so that the pattern of labeled DNA can be determined by autoradiography. Since the time allotted for radioactive labeling is chosen to allow each replication fork to move several micrometers along the DNA, the replicated DNA can be detected in the light microscope as lines of silver grains, even though the DNA molecule itself is too thin to be visible. In this way both the rate and the direction of replication-fork movement can be determined (Figure 8–36). From the rate at which tracks of replicated DNA increase in length with increasing labeling time, the replication forks are estimated to travel at a speed of about 50 nucleotides per second. This is one-tenth the rate at which bacterial replication forks move, possibly reflecting the increased difficulty of replicating DNA that is packaged tightly in chromatin.

As discussed previously, an average human chromosome is composed of a single DNA molecule containing about 150 million nucleotide pairs. To replicate such a DNA molecule from end to end with a single replication fork moving at a rate of 50 nucleotides per second would require $0.02 \times 150 \times 10^6 = 3.0 \times 10^6$ seconds (about 800 hours). As expected, therefore, the autoradiographic experiments just described reveal that many forks are moving simultaneously on each eucaryotic chromosome. Moreover, many forks are often found close together in the same DNA region, while other regions of the same chromosome have none. Further experiments of this type have shown the following: (1) Replication origins tend to be activated in clusters (called *replication units*) of perhaps 20 to 80

Figure 8–36 **The experiments that demonstrated the pattern in which replication forks move during the S phase.** The new DNA made in human cells in culture was briefly labeled with a pulse of highly radioactive thymidine (^{3}H-thymidine). In the experiment illustrated in (A), the cells were lysed and the DNA was stretched out on a glass slide that was subsequently covered with a photographic emulsion. After several months the emulsion was developed, revealing a line of silver grains over the radioactive DNA. The experiment in (B) was the same except that a further incubation in unlabeled medium allowed additional DNA, with a lower level of radioactivity, to be replicated. The pairs of dark tracks in (B) were found to have silver grains tapering off in opposite directions, demonstrating bidirectional fork movement from a central replication origin (see Figure 6–51). The *red* DNA in this figure is shown only to help with the interpretation of the autoradiograph; the unlabeled DNA is invisible in such experiments. A replication fork is thought to stop only when it encounters a replication fork moving in the opposite direction or when it reaches the end of the chromosome; in this way all the DNA is eventually replicated.

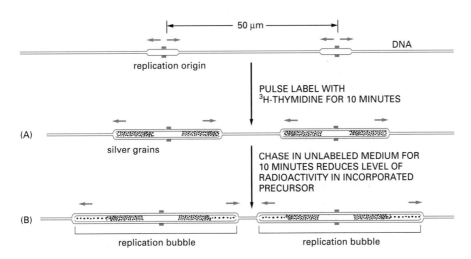

origins. (2) New replication units seem to be activated throughout the S phase until all of the DNA is replicated. (3) Within a replication unit, individual origins are spaced at intervals of 30,000 to 300,000 nucleotide pairs from one another. (4) As in bacteria, replication forks are formed in pairs and create a replication bubble as they move in opposite directions away from a common point of origin, stopping only when they collide head-on with a replication fork moving in the opposite direction (or reach a chromosome end). In this way many replication forks can operate independently on each chromosome and yet form two complete daughter DNA helices (see Figure 8–36).

Different Regions on the Same Chromosome Replicate at Distinct Times [22]

The replication of DNA in the region between one replication origin and the next should normally require only about an hour to complete, given the rate at which a replication fork moves and the largest distances measured between the replication origins in a replication unit. Yet the S phase usually lasts for about eight hours in a mammalian cell. This implies that the replication origins are not all activated simultaneously and that the DNA in each replication unit (which contains a cluster of perhaps 20 to 80 replication origins) is replicating for only a small part of the total S-phase interval.

Are different replication units activated at random, or is there a specified order in which different regions of the genome are replicated? One way to answer this question is to use the thymidine analogue bromodeoxyuridine (BrdU) to label the newly synthesized DNA in synchronized cell populations for different short periods throughout the S phase. Later, in M phase, those regions of the mitotic chromosomes that have incorporated BrdU into their DNA can be recognized by their altered staining properties or by means of anti-BrdU antibodies. The results show that different regions of each chromosome are replicated in a reproducible order during S phase (Figure 8–37). Moreover, as one would expect from the clusters of replication forks seen in DNA autoradiographs, the timing of replication is coordinated over large regions of the chromosome.

Highly Condensed Chromatin Replicates Late, While Genes in Active Chromatin Replicate Early [23]

It seems that the order in which replication origins are activated, at least in part, depends on the chromatin structure in which the origins reside. We have seen, for example, that heterochromatin remains in a highly condensed conformation (similar to that at mitosis) during interphase, while active chromatin has an especially decondensed conformation, which is apparently required to allow RNA synthesis. Heterochromatin is replicated very late in the S phase, suggesting that the timing of replication is related to the packing of the DNA in chromatin. This suggestion is supported by the timing of replication of the two X chromosomes in a female mammalian cell. While these two chromosomes contain essentially the same DNA sequences, one is active for DNA transcription and the other is not (discussed in Chapter 9). Nearly all of the inactive X chromosome is condensed into heterochromatin and its DNA replicates late in the S phase, whereas its active homologue is less condensed and replicates throughout the S phase. These findings support the hypothesis that those regions of the genome whose chromatin is least condensed during interphase, and therefore most accessible to the replication machinery, are replicated first. Autoradiography shows that replication forks move at comparable rates throughout the S phase, so that the extent of chromosome condensation appears to influence the initiation of replication forks but not their speed once formed.

The suggested relationship between chromatin structure and the time of DNA replication is also supported by studies in which the replication times of

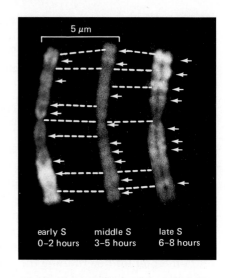

5 µm

early S   middle S   late S
0–2 hours 3–5 hours   6–8 hours

Figure 8–37 Different regions of a chromosome are replicated at different times in S phase. These light micrographs show stained mitotic chromosomes in which the replicating DNA has been differentially labeled during different defined intervals of the preceding S phase. In these experiments cells were first grown in the presence of BrdU (a thymidine analogue) and in the absence of thymidine to label the DNA uniformly. The cells were then briefly pulsed with thymidine in the absence of BrdU during early, middle, or late S phase. Because the DNA made during the thymidine pulse is a double helix with thymidine on one strand and BrdU on the other, it stains more darkly than the remaining DNA (which has BrdU on both strands) and shows up as a bright band (*arrows*) on these negatives. Dashed lines connect corresponding positions on the three copies of the chromosome shown. (Courtesy of Elton Stubblefield.)

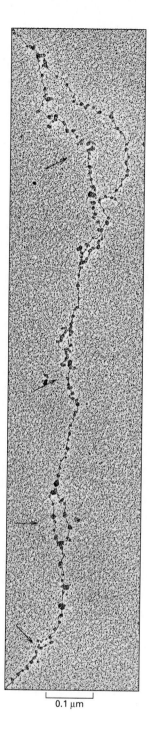

Figure 8–38 Large numbers of replication origins operate in rapidly dividing embryonic nuclei. This electron micrograph of spread chromatin from an early *Drosophila* embryo shows that replication bubbles (*arrows*) are extremely closely spaced. Only about 10 minutes elapse between some of the successive nuclear divisions in this embryo. Since the replication origins used here are so closely spaced (separated by a few thousand nucleotide pairs), it should require only about a minute to replicate all the DNA between them. (Courtesy of Victoria Foe.)

0.1 μm

specific genes are measured. The results show that so-called "housekeeping" genes, which are those active in all cells, replicate very early in S phase in all cells tested. Genes that are active in only a few cell types, in contrast, generally replicate early in the cells in which the genes are active and later in other types of cells.

The Late-replicating Replication Units Coincide with the A-T-rich Bands on Metaphase Chromosomes [24]

Many of the replication units seem to correspond to distinct chromosome bands made visible by the various fixation and staining procedures used for karyotyping. As discussed previously, as many as 2000 dark-staining *A-T-rich bands (G bands)* can be detected early in mitosis in the haploid set of mammalian chromosomes, and these are separated by an equal number of light-staining *G-C-rich bands (R bands)*. It is intriguing that the A-T-rich DNA and the G-C-rich DNA differ in the time of their replication during S phase. Experiments like the one shown in Figure 8–37 suggest that most G-C-rich bands replicate during the first half of S phase, while most A-T-rich bands replicate during the second half of S phase. It has therefore been suggested that housekeeping genes are located mostly in G-C-rich bands, while many cell-type-specific genes—the vast majority of which will be inactive in most cells—are located in A-T-rich bands. As stated earlier, it is a complete mystery why the mammalian genome should be segregated into such large alternating blocks of chromatin—many nearly equal in size to an entire bacterial genome. It is also not known how the many replication origins present in each replication unit are activated all at once. Perhaps the chromatin in a late-replicating unit remains condensed even after the end of M phase and decondenses only in mid S phase, making all the replication origins in the unit simultaneously accessible. If this is the case, the all-or-none coordinated replication of the DNA in a single replication unit could reflect the stepwise decondensation of large chromatin domains.

The Controlled Timing of DNA Replication May Contribute to Cell Memory [25]

The S phase is completed extremely rapidly in the cleaving eggs of many species, where large stores of chromatin components (such as histones) are present, as required for the rapid manufacture of new nuclei. As illustrated in Figure 8–38, a short S phase also requires the use of an exceptionally large number of replication origins spaced at intervals of only a few thousand nucleotide pairs (rather than the tens or hundreds of thousands of nucleotide pairs found between the replication origins later in development). Since any foreign DNA injected into a fertilized frog egg is replicated, a specific DNA sequence may not be required to form a replication origin in this cell.

Thus, DNA replication can be viewed as a potentially very rapid process that in most cells is subject to a complex system of regulation that restrains the initiation of replication forks in a way that causes different portions of the genome to replicate at very different times. It has been suggested, for example, that the chromatin that replicates early is assembled in part from a special store of chromosomal proteins produced during the G_1 phase and that once a chromosomal region becomes decondensed into active chromatin, its early replication causes the region to pick up proteins that help it to maintain a decondensed state from one cell generation to the next. In this view, the timing of DNA replication contributes to a cellular memory process that facilitates the continued transcription of expressed genes.

Chromatin-bound Factors Ensure That Each Region of the DNA Is Replicated Only Once [26]

In a normal S phase the whole genome must be replicated exactly once and no more. As we have just seen, DNA replication in most eucaryotic cells is an asynchronous process that takes a relatively long time to complete. Because the replication origins are used at different times in different chromosomal regions, in the middle of S phase some parts of a chromosome will not yet have begun replication, while other parts will have replicated completely. An enormous "bookkeeping" problem consequently arises during the middle and late stages of the S phase. Those replication origins already used have been duplicated and, at least with respect to their DNA sequences, are presumably identical to other replication origins not yet used. But each replication origin must be used only once in each S phase. How is this accomplished?

Cell-fusion experiments have provided an important clue. When an S-phase cell is fused with a G_1-phase cell (which has not yet begun S phase), DNA synthesis is induced in the G_1-phase nucleus, suggesting that the transition from G_1 to S phase is mediated by a diffusible activator of DNA synthesis. In contrast, when the S-phase cell is fused with a G_2-phase cell (that is, a cell that has just completed S phase), the G_2 nucleus is not stimulated to synthesize DNA. Since DNA synthesis continues undisturbed in the S-phase nucleus, this implies that the G_2 nucleus is prevented from entering further rounds of replication because every part of its DNA is somehow blocked. Some nondiffusible inhibitor of replication, for example, may have become tightly bound to its DNA. Such an inhibitor, if applied locally during S phase in the wake of each replication fork, would neatly solve the bookkeeping problem: by modifying the chromatin of freshly replicated DNA, it would ensure that once replicated DNA is not replicated again in the same S period (Figure 8–39A). An equally plausible alternative is that tightly bound initiator proteins, or "licensing factors," required for replication are de-

Figure 8–39 Two possible mechanisms to explain the "re-replication block." This block, which protects replicated DNA from further replication in the same cell cycle, is crucial to replication bookkeeping, but its molecular nature is not known. Normally, the block is removed at mitosis, but in a few specialized types of cells (the salivary gland cells of *Drosophila* larvae, for example), it is removed without mitosis, leading to the formation of giant *polytene* chromosomes. (A) A model based on the addition of an inhibitor to all newly replicated chromatin; (B) a model based on tightly bound initiator proteins, or "licensing factors," that act only once and that can be added to the DNA only during mitosis.

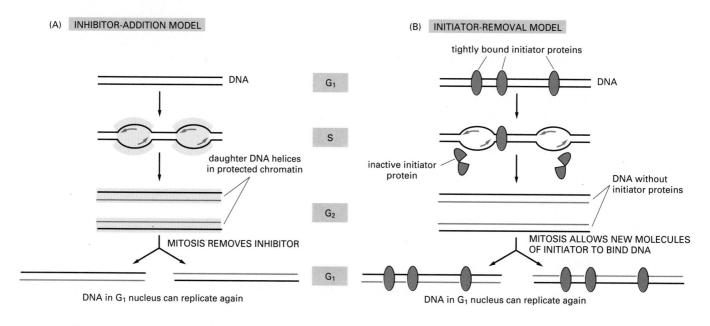

stroyed or inactivated by the passage of a replication fork (Figure 8–39B). Whatever its nature, the DNA re-replication block must be removed at or near the time of mitosis, since after cell division the DNA in the G$_1$ nuclei that emerges in the daughter cells is no longer protected.

The fragments of bacterial DNA that replicate when injected into a fertilized frog egg can be shown to be affected by the re-replication block. Therefore, the mechanism responsible for the block cannot require a highly specific replication origin. The block does not affect the SV40 virus, presumably because the T-antigen made by the virus supplies both initiator and DNA helicase functions for replication, substituting for analogous host cell components and thereby evading the controls to which they are subject.

New Histones Are Assembled into Chromatin as DNA Replicates [27]

A large amount of new histone, approximately equal in mass to the newly synthesized DNA, is required to make new chromatin in each cell cycle. For this reason most organisms possess multiple copies of the gene for each histone. Vertebrate cells, for example, have about 20 repeated sets, each set containing all five histone genes.

Unlike most proteins, which are made continuously throughout interphase, the histones are synthesized mainly in the S phase, when the level of histone mRNA increases about 50-fold as a result of both increased transcription and decreased mRNA degradation. By a mechanism that depends on special properties of their 3′ ends (discussed in Chapter 9), the major histone mRNAs become highly unstable and are degraded within minutes when DNA synthesis stops at the end of S phase (or when inhibitors are added to stop DNA synthesis prematurely). In contrast, the histone proteins themselves are remarkably stable and may survive for the entire life of a cell. The tight linkage between DNA synthesis and histone synthesis may be due, at least in part, to a feedback mechanism that monitors the level of free histone to ensure that the amount of histone made is appropriate for the amount of new DNA synthesized.

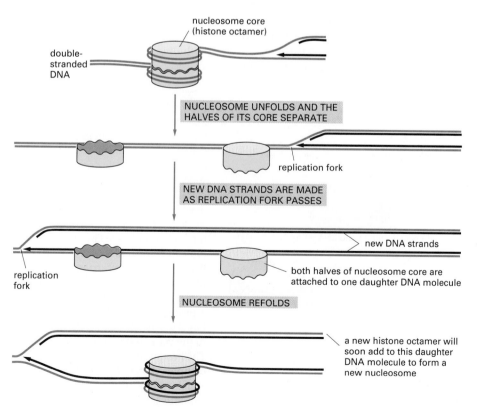

Figure 8–40 Speculative model showing how a nucleosome might open up to permit DNA replication. After the replication fork passes, the nucleosome reassembles. In this way the histones of the nucleosome core remain permanently bound to the DNA. Although in this diagram the old nucleosome has been inherited intact by the DNA helix made on the leading strand, there is evidence that an intact nucleosome can be inherited by either daughter DNA molecule. Moreover, this is only one of many possible models.

Chromosome Replication

Once they are assembled in nucleosomes, histone molecules rarely, if ever, leave the DNA to which they are bound. As a replication fork advances, therefore, it must somehow pass through the parental nucleosomes, which seem to have been designed to allow both transcription and replication to proceed past them. According to one hypothesis, each nucleosome transiently unfolds into two half-nucleosomes during DNA replication, thereby allowing the DNA polymerase to copy the uncoiled nucleosomal DNA (Figure 8–40).

The newly synthesized DNA behind a replication fork inherits some old histones, but since the amount of DNA has doubled, it also needs to bind an equal amount of new histones to complete its packaging into chromatin (see Figure 8–40). There is some evidence to suggest that a special nucleosome assembler may travel with the replication fork, packaging the newly synthesized DNA as soon as it emerges from the replication machinery. This newly formed chromatin, however, requires as much as an hour to become fully mature; until then, for example, the new histones are more susceptible than usual to histone-modifying enzymes. The chemical differences between mature and immature chromatin are unknown, although it is thought that maturation involves both covalent modifications of histones and changes in the binding of other proteins to the chromatin.

Telomeres Consist of Short G-rich Repeats That Are Added to Chromosome Ends by Telomerase [28]

As discussed earlier, it is not possible for DNA polymerase to replicate the end of a linear DNA molecule completely in the ordinary way, and this has led to the evolution of special DNA sequences, called **telomeres,** at the ends of eucaryotic chromosomes. These sequences, which are similar in organisms as diverse as protozoa, fungi, plants, and mammals, consist of many tandem repeats of a short sequence that contains a block of neighboring G nucleotides. In humans this sequence is GGGTTA.

The problem of replicating the ends of chromosomes is solved in an ingenious way by an enzyme called **telomerase.** This enzyme recognizes the G-rich strand of an existing telomere repeat sequence and elongates it in the 5′-to-3′ direction. In the absence of a complementary DNA strand, the telomerase synthesizes a new copy of the repeat using an RNA template that is a component of the enzyme itself. The enzyme thus contains the information used to maintain the characteristic telomere sequences. After several rounds of extension by telomerase, replication of the chromosome end can be completed using these extensions as a template for synthesis of the complementary strand by DNA

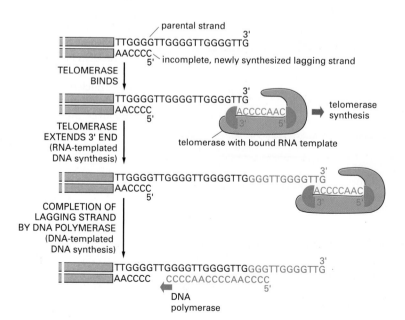

Figure 8–41 Telomere replication. The figure outlines the reactions involved in the formation of the repeating G-rich sequences that form the ends of chromosomes (telomeres) of diverse eucaryotic organisms, as suggested by experiments in the ciliate *Tetrahymena*. The incomplete, newly synthesized strand is the strand made on the lagging side of a replication fork (see Figure 6–41). As indicated, the telomerase is a protein-RNA complex that carries an RNA template for synthesizing a repeating, G-rich telomere DNA sequence. These repeats are GGGTTG in *Tetrahymena* but are GGGTTA in humans and $G_{1-3}A$ in the yeast *Saccharomyces cerevisiae*. The lagging strand is presumed to be completed by DNA polymerase α, which carries primase as one of its subunits (see Figure 8–35).

polymerase (Figure 8–41). Because the processes that shorten and restore the telomere sequence are only approximately balanced, each chromosome end contains a variable number of the tandem repeats, which generally extend for hundreds of nucleotide pairs.

Summary

In vitro studies using the monkey virus SV40 as a model system suggest that, in eucaryotes as in procaryotes, DNA replication begins with the loading of a DNA helicase onto the DNA by an initiator protein bound to a replication origin. A replication bubble forms at such an origin as two replication forks move away from each other. During S phase in higher eucaryotes, neighboring replication origins appear to be activated in clusters known as replication units, with the origins spaced an average of about 100,000 nucleotide pairs apart. Since the replication fork moves at about 50 nucleotides per second, only about an hour should be required to complete the DNA synthesis in a replication unit. Throughout a typical 8-hour S phase different replication units are activated in a sequence determined in part by the structure of the chromatin, the most condensed regions of chromatin being replicated last. The correspondence between replication units and the bands containing millions of nucleotide pairs seen on mitotic eucaryotic chromosomes suggests that replication units may correspond to structurally distinct domains in interphase chromatin.

After the replication fork passes, chromatin structure is re-formed by the addition of new histones and other chromosomal proteins to the old histones inherited on the daughter DNA molecules. A DNA re-replication block of unknown nature acts locally to prevent a second round of replication from occurring until a chromosome has passed through mitosis; this block is needed to ensure that each region of the DNA is replicated only once in each S phase.

The problem of replicating the ends of chromosomes is solved by a specialized end structure (the telomere) and an enzyme (telomerase) that extends this structure using an RNA template that is part of the telomerase.

RNA Synthesis and RNA Processing

We have thus far considered how chromosomes are organized as very large DNA-protein complexes and how they are duplicated before a cell divides. But the main function of a chromosome is to act as a template for the synthesis of RNA molecules, since only in this way does the genetic information stored in chromosomes become directly useful to the cell. There is a great deal of RNA synthesis in a cell: the total rate at which nucleotides are incorporated into RNA during interphase is about 20 times the rate at which nucleotides are incorporated into DNA during S phase.

RNA synthesis, which is also called **DNA transcription,** is a highly selective process. In most mammalian cells, for example, only about 1% of the DNA nucleotide sequence is copied into functional RNA sequences (mature messenger RNA or structural RNA). The selectivity occurs at two levels, which we discuss in turn in this section: (1) only part of the DNA sequence is transcribed to produce nuclear RNAs, and (2) only a minor proportion of the nucleotide sequences in nuclear RNAs survives the RNA processing steps that precede the export of RNA molecules to the cytoplasm. We begin by describing *RNA polymerases*, the enzymes that catalyze all DNA transcription.

RNA Polymerase Exchanges Subunits as It Begins Each RNA Chain [29]

As described in outline in Chapter 6, transcription begins when an RNA polymerase molecule binds to a **promoter** sequence on the DNA double helix. Next,

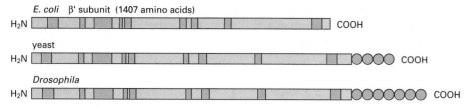

E. coli β' subunit (1407 amino acids)

H₂N [COOH

yeast

H₂N [COOH

Drosophila

H₂N [COOH

in a step that is not well understood, the two strands of the DNA are separated locally to form an *open complex*. At this stage the template strand is exposed, and synthesis of the complementary RNA chain can begin. The polymerase then moves along the template strand, extending its growing RNA chain in the 5'-to-3' direction by the stepwise addition of ribonucleoside triphosphates until it reaches a *stop (termination) signal*, at which point the newly synthesized RNA chain and the polymerase are released from the DNA. Each RNA molecule thus represents a single-strand copy of the nucleotide sequence of one DNA strand in a relatively short region of the genome (see Figure 6–2). This transcribed segment of DNA is called a *transcription unit*.

RNA polymerases are generally formed from multiple polypeptide chains and have masses of 500,000 daltons or more. The enzymes in bacteria and eucaryotes are evolutionarily related (Figure 8–42). Since the bacterial enzyme has been far easier to study, its properties provide a basis for understanding its eucaryotic relatives. The *E. coli* enzyme contains four different subunits, α, β, β', and σ, there being two copies of α and one each of the others. The complete amino acid sequence of each subunit has been determined from the nucleotide sequence of its gene.

The sigma (σ) subunit of the *E. coli* polymerase has a specific role in the initiation of transcription: it enables the enzyme to find promoter sequences to which it binds. After about eight nucleotides of an RNA molecule have been synthesized (step 4 in Figure 8–43), the σ subunit dissociates and a number of *elongation factors*—important for chain elongation and termination—become associated with the enzyme instead. The elongation factors include several proteins that function in ways that are incompletely understood. The initiation of transcription is an important control point where the cell can regulate the expression of a gene; for this reason it is discussed in more detail in Chapter 9.

Three Kinds of RNA Polymerase Make RNA in Eucaryotes [30]

Although the mechanism of DNA transcription is similar in eucaryotes and procaryotes such as *E. coli*, the machinery is considerably more complex in eu-

Figure 8–43 Schematic diagram of the steps in the initiation of RNA synthesis (DNA transcription) catalyzed by RNA polymerase. The steps indicated have been revealed by studies of the *E. coli* enzyme. A DNA molecule containing a promoter sequence for the *E. coli* polymerase is shown (see Figure 6–2). The enzyme first forms a *closed complex* in which the two DNA strands remain fully base-paired. In the next step the enzyme catalyzes the opening of a little more than one turn of the DNA helix to form an *open complex*, in which the template DNA strand is exposed for the initiation of an RNA chain. The polymerase containing the bound σ subunit, however, behaves as though it is tethered to the promoter site: it seems unable to proceed with the elongation of the RNA chain and on its own frequently synthesizes and releases short RNA chains. As indicated, the conversion to an actively elongating polymerase requires the release of initiation factors (the σ subunit in the case of the *E. coli* enzyme) and generally involves the binding of other proteins that serve as elongation factors.

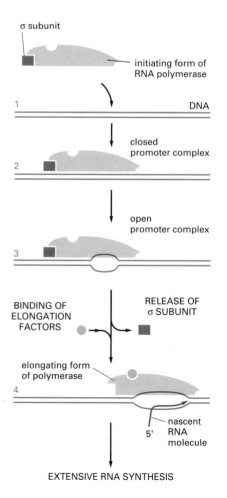

σ subunit

initiating form of RNA polymerase

1   DNA

closed promoter complex

2

open promoter complex

3

BINDING OF ELONGATION FACTORS   RELEASE OF σ SUBUNIT

elongating form of polymerase

4

nascent RNA molecule

5'

EXTENSIVE RNA SYNTHESIS

caryotes. In eucaryotes as diverse as yeasts and humans, for example, there are three types of RNA polymerases, each responsible for transcribing different sets of genes. These enzymes—denoted as *RNA polymerases I, II, and III*—are structurally similar to one another and have some common subunits, although other subunits are unique. Each is more complex than *E. coli* RNA polymerase and is thought to contain 10 or more polypeptide chains. Another important distinction between the bacterial and eucaryotic enzymes is that, whereas the purified bacterial enzyme can bind to promoters and initiate transcription on its own, the eucaryotic enzymes require the presence of additional initiation proteins that must bind to the promoter before the enzyme can bind. For this reason it was not until 1979 that systems with all the needed components became available so that eucaryotic initiation mechanisms could be analyzed *in vitro*. Because these initiation proteins and their interactions with the polymerases are intimately involved with the control of transcription initiation, we shall defer discussion of them until Chapter 9.

The three eucaryotic RNA polymerases were initially distinguished by their chemical differences during purification and by their sensitivity to α-*amanitin*, a poison isolated from the deadly toadstool *Amanita phalloides*. RNA polymerase I is unaffected by α-amanitin; RNA polymerase II is very sensitive to this poison; and RNA polymerase III is moderately sensitive to it. The sensitivity of RNA synthesis to α-amanitin is still used to determine which polymerase transcribes a gene. Such studies indicate that **RNA polymerase II** transcribes the genes whose RNAs will be translated into proteins. The other two polymerases synthesize RNAs that have structural or catalytic roles, chiefly as part of the protein synthetic machinery: **polymerase I** makes the large ribosomal RNAs, and **polymerase III** makes a variety of very small, stable RNAs—including the small 5S ribosomal RNA and the transfer RNAs. However, most of the small RNAs that form snRNPs, which we discuss later when we consider RNA processing, are made by polymerase II.

Mammalian cells typically contain 20,000 to 40,000 molecules of each of the RNA polymerases, and studies with cultured cells indicate that the concentrations of these enzymes are regulated individually according to the rate of cell growth.

RNA Polymerase II Transcribes Some DNA Sequences Much More Often Than Others [31]

Because RNA polymerase II makes all of the mRNA precursors and thus determines which proteins a cell will make, we shall focus most of our discussion on its activities and on the fate of its products. Although experiments with purified polymerases *in vitro* are essential for establishing the mechanism of transcription, much can also be learned about how the process occurs in a cell by using the electron microscope to examine genes in action, with their bound RNA polymerases caught in the act of transcription.

Ordinary thin-section electron micrographs of interphase nuclei show granular clumps of chromatin (see Figure 8–71) but reveal very little about how genes are transcribed. A much more detailed picture emerges if the nucleus is ruptured and its contents spilled out onto an electron microscope grid (Figures 8–44 and 8–45). At the farthest point from the center of the lysed nucleus, the chromatin is diluted sufficiently to make individual chromatin strands visible in the expanded, beads-on-a-string form shown previously in Figure 8–9B.

RNA polymerase molecules actively engaged in transcription appear as globular particles with a single RNA molecule trailing behind. Particles representing active RNA polymerase II molecules are usually seen as single units, without nearby neighbors. This observation indicates that most genes are transcribed into mRNA precursors only infrequently, so that one polymerase finishes transcription before another one begins. Occasionally, however, many polymerase molecules (and their associated RNA transcripts) are seen clustered together. These clusters occur on the relatively few genes that are transcribed at high frequency

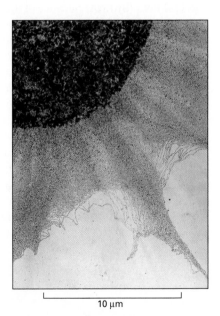

10 μm

Figure 8–44 A typical cell nucleus visualized by electron microscopy using the procedure shown in Figure 8–45. An enormous tangle of chromatin can be seen spilling out of the lysed nucleus; only the chromatin at the outermost edge of this tangle will be sufficiently dilute for meaningful examination at higher power. (Courtesy of Victoria Foe.)

(Figure 8–46). The length of the attached RNA molecules in such a cluster increases in the direction of transcription, producing a characteristic pattern. This pattern defines the RNA polymerase II start site and direction of transcription for a specific transcription unit (Figure 8–47).

Biochemical studies have confirmed and extended the results obtained by electron microscopy, leading to three major conclusions:

1. Eucaryotic RNA polymerase molecules, like those in procaryotes, begin at specific sites on the chromosome.

2. The average length of the complete RNA molecule produced by RNA polymerase II from a single transcription unit is about 7000 nucleotides, and RNA molecules 10,000 to 20,000 nucleotides long are common. These lengths, which are much longer than the 1200 nucleotides of RNA needed to code for an average protein of 400 amino acid residues, reflect the complex structure of eucaryotic genes and, in particular, the presence of long intron sequences, which, as we discuss later, are later removed from the RNA.

3. Although chain *elongation rates* of about 30 nucleotides per second are observed for all RNAs, different RNA polymerase II start sites have different *initiation frequencies*, so that some genes are transcribed at much higher rates than others. As indicated in Table 8–2, the majority of the genes that are transcribed give rise to very few mRNA molecules.

The Precursors of Messenger RNA Are Covalently Modified at Both Ends [32]

In eucaryotes mature mRNA is produced in several steps. The RNA molecules freshly synthesized by RNA polymerase II in the nucleus are known as *primary transcripts;* the collection of such transcripts was originally called **heterogeneous nuclear RNA (hnRNA)** because of the large variation in RNA size, contrasting with the more uniform and smaller size of the RNA sequences actually needed to encode proteins. We shall see shortly that much of this variation is due to the presence of long intron sequences in the primary transcripts. As they are being synthesized, these transcripts are covalently modified at both their 5′ end and their 3′ end in ways that clearly distinguish them from transcripts made by other RNA polymerases. These modifications will be used later in the cytoplasm as signals that these transcripts are to be translated into protein.

The 5′ end of the RNA molecule (which is the end synthesized first during transcription) is first *capped* by the addition of a methylated G nucleotide. Capping occurs almost immediately, after about 30 nucleotides of RNA have been synthesized, and it involves condensation of the triphosphate group of a molecule of GTP with a diphosphate left at the 5′ end of the initial transcript (Figure 8–48). This **5′ cap** will later play an important part in the initiation of protein synthesis; it also seems to protect the growing RNA transcript from degradation.

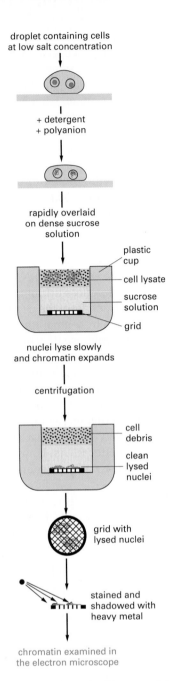

droplet containing cells
at low salt concentration

+ detergent
+ polyanion

rapidly overlaid
on dense sucrose
solution

plastic cup
cell lysate
sucrose solution
grid

nuclei lyse slowly
and chromatin expands

centrifugation

cell debris
clean lysed nuclei

grid with
lysed nuclei

stained and
shadowed with
heavy metal

chromatin examined in
the electron microscope

Figure 8–45 A method for examining chromatin in the electron microscope. The nuclei are first lysed, and then the chromatin is freed from cellular debris and spread out on a grid.

Figure 8–46 Electron micrograph of a region of chromatin containing a gene being transcribed at unusually high frequency. Many RNA polymerase II molecules with their growing RNA transcripts are visible. The direction of transcription is from left to right (see Figure 8–47). (From V.E. Foe, L.E. Wilkinson, and C.D. Laird, *Cell* 9:131–146, 1976. © 1976 Cell Press.)

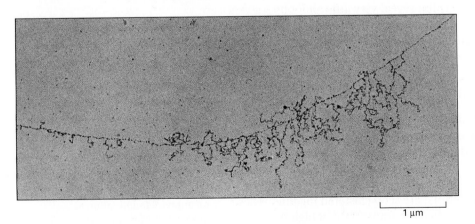

1 μm

Table 8–2 The Population of mRNA Molecules in a Typical Mammalian Cell

	Copies per Cell of Each mRNA Sequence		Number of Different mRNA Sequences in Each Class		Total Number of mRNA Molecules in Each Class
Abundant class	12,000	×	4	=	48,000
Intermediate class	300	×	500	=	150,000
Scarce class	15	×	11,000	=	165,000

This division of mRNAs into just three discrete classes is somewhat arbitrary, and in many cells a more continuous spread in abundances is seen. However, a total of 10,000 to 20,000 different mRNA species is normally observed in each cell, most species being present at a low level (5 to 15 molecules per cell). Most of the total cytoplasmic RNA is rRNA, and only 3% to 5% is mRNA, a ratio consistent with the presence of about 10 ribosomes per mRNA molecule. This particular cell type contains a total of about 360,000 mRNA molecules in its cytoplasm.

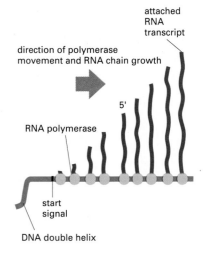

Figure 8–47 An idealized transcription unit. The drawing illustrates how the electron microscope appearance (see Figure 8–46) demonstrates the direction of transcription, as well as the start site of the unit.

The 3′ end of most polymerase II transcripts is defined not by the termination of transcription but by a second modification in which the growing transcript is cleaved at a specific site and a **poly-A tail** is added by a separate polymerase to the cut 3′ end. The signal for the cleavage is the appearance in the RNA chain of the sequence AAUAAA located 10 to 30 nucleotides upstream from the site of cleavage, plus a less well-defined downstream sequence. Immediately after cleavage, a *poly-A polymerase* enzyme adds 100 to 200 residues of adenylic acid (as *poly A*) to the 3′ end of the RNA chain to complete the **primary RNA transcript.** Meanwhile, the polymerase fruitlessly continues transcribing for hundreds or thousands of nucleotides until termination occurs at one of several later sites; the extra piece of functionless RNA transcript thus generated presumably lacks a 5′ cap and is rapidly degraded (Figure 8–49).

The poly-A tail appears to have several functions: (1) as described later, it aids in the export of mature mRNA from the nucleus; (2) it is thought to affect the stability of at least some mRNAs in the cytoplasm; and (3) it seems to serve as a recognition signal for the ribosome that is required for efficient translation of mRNA. The latter feature—in combination with the 5′ cap—would enable a ribosome to determine whether the mRNA was intact before expending energy and precursors to begin its translation.

Even though polymerase II transcripts comprise more than half of the RNA being synthesized by DNA transcription, we shall see below that most of the RNA in these transcripts is unstable and therefore short-lived. Consequently, the hnRNA in the cell nucleus and the cytoplasmic mRNA derived from it constitute only a minor fraction of the total RNA in a cell (Table 8–3). Despite their relative scarcity, these RNA molecules can be readily purified because of their poly-A tails. When the total cellular RNA is passed through a column containing poly dT linked to a solid support, the complementary base-pairing between T and A residues selectively binds the molecules with poly-A tails to the column; the bound molecules can then be released for further analysis. This procedure is widely used to separate the hnRNA and mRNA molecules from the ribosomal and transfer RNA molecules that predominate in cells.

Figure 8–48 The reactions that cap the 5′ end of each RNA molecule synthesized by RNA polymerase II. The final cap contains a novel 5′-to-5′ linkage between the positively charged 7-methyl G residue and the 5′ end of the RNA transcript (see Figure 6–26). At least some of the enzymes required for this process are thought to be bound to polymerase II, since polymerase I and III transcripts are not capped and the indicated reaction occurs almost immediately following initiation of each RNA chain. The letter N is used here to represent any one of the four ribonucleotides, although the nucleotide that starts an RNA chain is usually a purine (an A or a G). (After A.J. Shatkin, *Bioessays* 7:275–277, 1987. © ICSU Press.)

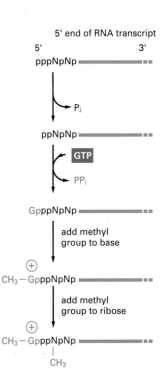

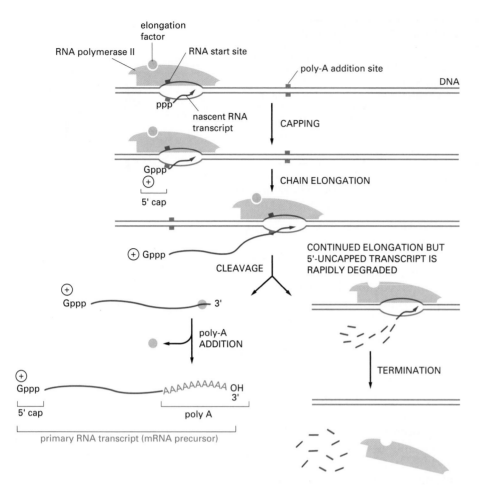

Figure 8–49 Synthesis of a primary RNA transcript (an mRNA precursor) by RNA polymerase II. This diagram starts with a polymerase that has just begun synthesizing an RNA chain (step 4 of Figure 8–43). Recognition of a poly-A addition signal in the growing RNA transcript causes the chain to be cleaved and then polyadenylated as shown. In yeasts the polymerase terminates its RNA synthesis almost immediately thereafter, but in higher eucaryotes it often continues transcription for thousands of nucleotides. It seems likely that the polymerase changes its properties once one RNA chain cleavage has occurred. Thus it cannot trigger poly-A addition to its downstream RNA transcript, and it seems to have a greater probability of responding to the sequences that cause chain termination and polymerase release. The simplest hypothesis, represented here, is that an elongation factor (or factors) is released from the polymerase after cleavage of the transcript.

Table 8–3 Selected Data on Amounts of RNA in a Typical Mammalian Cell

	Steady-State Amount (percent of total cell RNA)	Percent of Total RNA Synthesis
Nuclear rRNA precursors	4	39
↓		
Cytoplasmic rRNA	71	—
Nuclear hnRNA	7	58
↓		
Cytoplasmic mRNA	3	—
Small stable RNAs (mostly tRNAs)	15	3

The figures shown here were derived from the analysis of a mouse fibroblast cell line (L cells) in culture. Each cell contained 26 pg of RNA (5×10^{10} nucleotides of RNA), of which about 14% was located in the cell nucleus. (The cell nucleus thus contains about twice as much DNA as RNA.) An average of about 200×10^6 nucleotides is polymerized into RNA every minute during interphase. This is about 20 times the average rate at which DNA is synthesized during S phase. Note that although most of the RNA synthesized is hnRNA, most of this RNA is rapidly degraded in the nucleus. As a result, the mRNA produced from the hnRNA is only a minor fraction of the total RNA in the cell. (Modified from B.P. Brandhorst and E.H. McConkey, *J. Mol. Biol.* 85:451–563, 1974.)

Only RNA polymerase II transcripts have 5′ caps and 3′ poly-A tails. This seems to be because the capping and cleavage plus poly-A addition reactions are mediated by enzymes that bind selectively to polymerase II. Thus, if a gene that is normally transcribed by polymerase II is separated from its promoter by recombinant DNA methods and fused to a promoter recognized by polymerase I or by polymerase III, the RNA transcripts produced from it by these polymerases are neither capped nor polyadenylated. The requirement for a specific capping and polyadenylation of mRNA precursors may explain why these RNAs are synthesized by a separate type of RNA polymerase molecule in eucaryotes.

RNA Processing Removes Long Nucleotide Sequences from the Middle of RNA Molecules [33]

The discovery of interrupted genes in 1977 was entirely unexpected. Previous studies in bacteria had shown that their genes are composed of a continuous string of the nucleotides needed to encode the amino acids of a protein, and there seemed to be no obvious reason why a gene should be organized in any other way. The first indication that eucaryotic genes are not continuous like bacterial genes came when new methods allowing an accurate comparison of mRNA and DNA sequences were applied to mRNAs produced by a human *adenovirus* (a large DNA virus). The region of the viral DNA producing these RNAs turned out to contain sequences that are not present in the mature RNAs. The possibility that this situation was unique to viruses was quickly eliminated by the finding of similar interruptions in the ovalbumin and β-globin genes of vertebrates. As discussed earlier, the sequences present in the DNA but omitted from the mRNA are called *intron* sequences, while those present in the mRNA are called *exon* sequences (Figures 8–50 and 8–51).

Before the discovery of introns, the significance of hnRNA and its relationship to mRNA had seemed very mysterious. It had long been known that most of the RNA synthesized by RNA polymerase II is rapidly degraded in the nucleus. The hnRNA molecules of cultured cells can be radiolabeled by brief exposure to ³H-uridine and followed over a long period. This sort of experiment showed that the average length of the hnRNA molecules in the labeled population decreases rapidly, starting from about 7000 nucleotides, to reach the size of cytoplasmic mRNA molecules (an average of about 1500 nucleotides) after only about 30 minutes; at about the same time, radioactively labeled RNA molecules begin to leave the nucleus as mRNA molecules. Only about 5% of the mass of the labeled hnRNA ever reaches the cytoplasm, however; the remainder is degraded into small fragments in the nucleus over a period of about an hour. This seemed strangely wasteful. And the puzzle was deepened by the finding that even though the hnRNA molecules became progressively shorter, they retained their 5′ caps and their 3′ poly-A tails.

With the discovery of introns, the explanation became clear: the primary RNA transcript is a faithful copy of the gene, containing both exon and intron sequences, and the latter sequences are cut out of the middle of the RNA transcript to produce an mRNA molecule that codes directly for a protein (see Figure 3–15). Because the coding RNA sequences on either side of an intron sequence are joined to each other after the intron sequence has been cut out, this reaction is known as **RNA splicing.** RNA splicing occurs in the cell nucleus, out of reach of the ribosomes, and RNA is exported to the cytoplasm only when processing is complete.

Because most mammalian genes contain much more intron than exon sequence (see Table 8–1, p. 340), RNA splicing can account for the conversion of the very long nuclear hnRNA molecules to the much shorter cytoplasmic mRNA molecules.

Before discussing the distribution of introns in eucaryotic genes and some of their consequences for cell function, it is necessary to explain how intron sequences are recognized and removed by the splicing machinery.

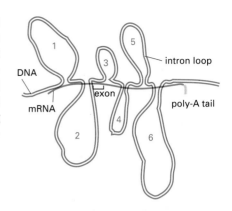

Figure 8–50 Early evidence for the existence of introns in eucaryotic genes. The evidence was provided by the "R-loop technique," in which a base-paired complex between mRNA and DNA molecules is visualized in the electron microscope. An unusually abundant mRNA molecule, such as β-globin mRNA or ovalbumin mRNA, is readily purified from the specialized cells that produce it. When this single-stranded mRNA preparation is annealed in a suitable solvent to a cloned double-stranded DNA molecule containing the gene that encodes the mRNA, the RNA can displace a DNA strand wherever the two sequences match and form regions of RNA-DNA helix. Regions of DNA where no match to the mRNA sequence is possible are clearly visible as large loops of double-stranded DNA. Each of these loops (numbered 1 to 6) represents an intron in the gene sequence.

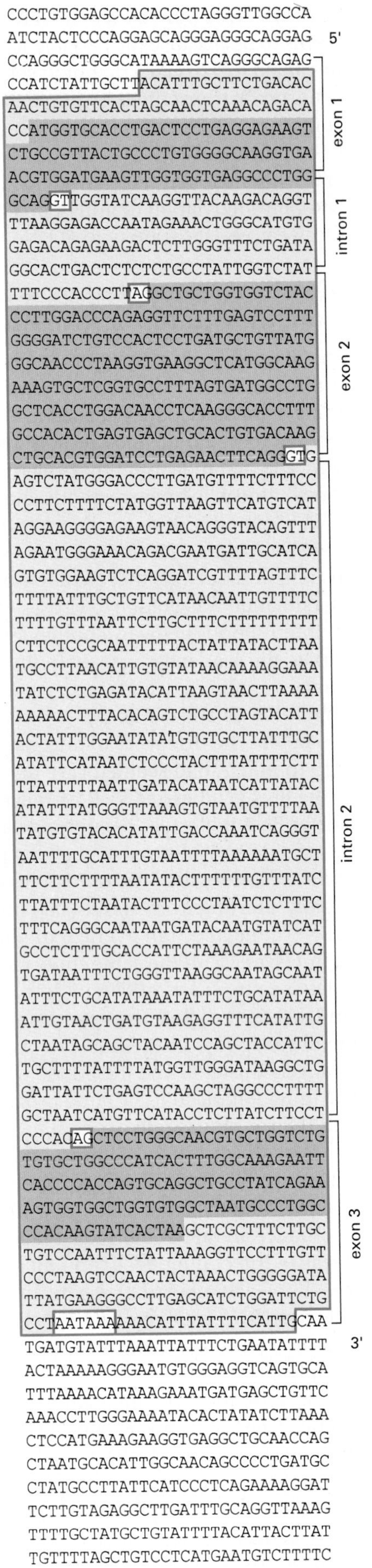

Figure 8–51 The transcribed portion of the human β-globin gene. The sequence of the DNA strand corresponding to the mRNA sequence is given, with the primary RNA transcript surrounded by a *green line* and the nucleotides in the three exons *shaded red*. Note that exon 1 includes a *5'-leader sequence* and that exon 3 includes a *3'-untranslated sequence*; although these sequences are included in the mRNA, they do not code for amino acids. The highly conserved GT and AG nucleotides at the ends of each intron are boxed (see Figure 8–53), along with the cleavage and polyadenylation signal near the 3' end of the gene (AATAAA, see Figure 8–49).

```
CCCTGTGGAGCCACACCCTAGGGTTGGCCA
ATCTACTCCCAGGAGCAGGGAGGGCAGGAG   5'
CCAGGGCTGGGCATAAAAGTCAGGGCAGAG
CCATCTATTGCTTACATTTGCTTCTGACAC
AACTGTGTTCACTAGCAACTCAAACAGACA   exon 1
CCATGGTGCACCTGACTCCTGAGGAGAAGT
CTGCCGTTACTGCCCTGTGGGGCAAGGTGA
ACGTGGATGAAGTTGGTGGTGAGGCCCTGG
GCAGGTTGGTATCAAGGTTACAAGACAGGT
TTAAGGAGACCAATAGAAACTGGGCATGTG
GAGACAGAGAAGACTCTTGGGTTTCTGATA   intron 1
GGCACTGACTCTCTCTGCCTATTGGTCTAT
TTTCCCACCCTTAGGCTGCTGGTGGTCTAC
CCTTGGACCCAGAGGTTCTTTGAGTCCTTT
GGGGATCTGTCCACTCCTGATGCTGTTATG
GGCAACCCTAAGGTGAAGGCTCATGGCAAG
AAAGTGCTCGGTGCCTTTAGTGATGGCCTG
GCTCACCTGGACAACCTCAAGGGCACCTTT   exon 2
GCCACACTGAGTGAGCTGCACTGTGACAAG
CTGCACGTGGATCCTGAGAACTTCAGGGTG
AGTCTATGGGACCCTTGATGTTTTCTTTCC
CCTTCTTTTCTATGGTTAAGTTCATGTCAT
AGGAAGGGGAGAAGTAACAGGGTACAGTTT
AGAATGGGAAACAGACGAATGATTGCATCA
GTGTGGAAGTCTCAGGATCGTTTTAGTTTC
TTTTATTTGCTGTTCATAACAATTGTTTTC
TTTTGTTTAATTCTTGCTTTCTTTTTTTTT
CTTCTCCGCAATTTTTACTATTATACTTAA
TGCCTTAACATTGTGTATAACAAAAGGAAA
TATCTCTGAGATACATTAAGTAACTTAAAA
AAAAACTTTACACAGTCTGCCTAGTACATT
ACTATTTGGAATATATGTGTGCTTATTTGC
ATATTCATAATCTCCCTACTTTATTTTCTT
TTATTTTTAATTGATACATAATCATTATAC
ATATTTATGGGTTAAAGTGTAATGTTTTAA
TATGTGTACACATATTGACCAAATCAGGGT
AATTTTGCATTTGTAATTTTAAAAAATGCT   intron 2
TTCTTCTTTTAATATACTTTTTTGTTTATC
TTATTTCTAATACTTTCCCTAATCTCTTTC
TTTCAGGGCAATAATGATACAATGTATCAT
GCCTCTTTGCACCATTCTAAAGAATAACAG
TGATAATTTCTGGGTTAAGGCAATAGCAAT
ATTTCTGCATATAAATATTTCTGCATATAA
ATTGTAACTGATGTAAGAGGTTTCATATTG
CTAATAGCAGCTACAATCCAGCTACCATTC
TGCTTTTATTTTATGGTTGGGATAAGGCTG
GATTATTCTGAGTCCAAGCTAGGCCCTTTT
GCTAATCATGTTCATACCTCTTATCTTCCT
CCCACAGCTCCTGGGCAACGTGCTGGTCTG
TGTGCTGGCCCATCACTTTGGCAAAGAATT
CACCCCACCAGTGCAGGCTGCCTATCAGAA
AGTGGTGGCTGGTGTGGCTAATGCCCTGGC   exon 3
CCACAAGTATCACTAAGCTCGCTTTCTTGC
TGTCCAATTTCTATTAAAGGTTCCTTTGTT
CCCTAAGTCCAACTACTAAACTGGGGGATA
TTATGAAGGGCCTTGAGCATCTGGATTCTG
CCTAATAAAAAACATTTATTTTCATTGCAA
TGATGTATTTAAATTATTTCTGAATATTTT   3'
ACTAAAAAGGGAATGTGGGAGGTCAGTGCA
TTTAAAACATAAAGAAATGATGAGCTGTTC
AAACCTTGGGAAAATACACTATATCTTAAA
CTCCATGAAAGAAGGTGAGGCTGCAACCAG
CTAATGCACATTGGCAACAGCCCCTGATGC
CTATGCCTTATTCATCCCTCAGAAAAGGAT
TCTTGTAGAGGCTTGATTTGCAGGTTAAAG
TTTTGCTATGCTGTATTTTACATTACTTAT
TGTTTTAGCTGTCCTCATGAATGTCTTTTC
```

hnRNA Transcripts Are Immediately Coated with Proteins and snRNPs [34]

Newly made RNA in eucaryotic cells, unlike that in bacteria, appears to become immediately condensed into a string of closely spaced protein-containing particles. Each particle consists of about 500 nucleotides of RNA wrapped around a protein complex that serves to condense and package each growing RNA transcript in a manner reminiscent of the DNA-protein complexes of nucleosomes. The resulting **hnRNP particles** (*heterogeneous nuclear ribonucleoprotein particles*) can be purified after nuclei have been treated with ribonucleases at levels just sufficient to destroy the linker RNA between them. Each particle has a diameter of about 20 nm, which is twice that of a nucleosome, and the protein core is more complex and less well characterized, being composed of a set of at least eight different proteins. Except for histones, the proteins in this core are the most abundant proteins in the cell nucleus. Several of them contain a conserved domain of about 80 amino acids, which is often repeated and shared by many other RNA-binding proteins.

The hnRNP particles are generally distorted by the standard spreading techniques used to view gene transcription in the electron microscope (see Figure 8–45). These micrographs, however, reveal especially stable particles of a less common type, whose position on the RNA strongly implicates them in RNA splicing. These particles form very quickly at specific RNA sequences—at or near the junctions between intron and exon sequences—and, as the RNA transcript elongates, they coalesce in pairs to form a larger assembly that is thought to be the *spliceosome* that catalyzes RNA splicing (Figure 8–52).

Biochemical analysis has revealed that the cell nucleus contains many complexes of proteins with small RNAs (generally RNAs of 250 nucleotides or less), which have arbitrarily been designated U1, U2, . . . , U12 RNAs. These complexes, called **small nuclear ribonucleoproteins** (snRNPs—pronounced "snurps"), resemble ribosomes in that each contains a set of proteins complexed to a stable RNA molecule. They are much smaller than ribosomes, however—only 250,000 daltons compared with 4.5 million daltons for a ribosome. Some proteins are present in several types of snRNPs, whereas others are unique to one type. This was first demonstrated using serum from patients with the autoimmune disease *systemic lupus erythematosus*, who make antibodies directed against one or more of their own snRNP proteins: a single antibody was found that binds the U1, U2, U5, and U4/U6 snRNPs, for example, and we now know that they all contain common proteins.

Individual snRNPs are believed to recognize specific nucleic acid sequences through RNA-RNA base-pairing. Some mediate RNA splicing, one is known to be involved in the cleavage reaction that generates the 3' ends of newly formed histone RNAs, while the function of others is unknown. The evidence for the role of snRNPs in RNA splicing comes from experiments on RNA splicing *in vitro*, as well as from analyses of yeast cells that are mutant in one of the snRNP components.

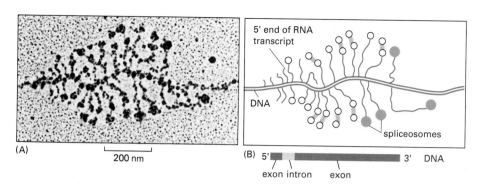

(A)

200 nm

(B)

5' end of RNA
transcript

DNA

spliceosomes

5' exon intron exon 3' DNA

Figure 8–52 Spliceosomes. (A) Electron micrograph of a chromatin spread showing large ribonucleoprotein particles assembling at the 5' and 3' splice site regions to form a *spliceosome*. The RNA transcripts are being produced from a gene encoding a *Drosophila* chorion protein, and the positions of the splice sites on the primary RNA transcript are known. As indicated in the drawing (B), most of the RNA transcripts have either one or two large RNP particles near their 5' ends; a schematic representation of the gene is shown below the drawing. When there are two particles on a transcript [*open circles* in (B)], they average 25 nm in diameter and occur at or very near the positions of the 5' and 3' splice sites for the single small intron sequence (228 nucleotides long) near the 5' end of the transcripts. The more mature, longer transcripts frequently display a single larger particle [*green circles* in (B)] in the region of the intron, which probably results from the stable association of the two smaller particles and represents the assembled spliceosome. Since in some cases (including the example shown in this figure) splicing occurs while the 3' end of the RNA chain is still being transcribed, the poly A at the 3' end of hnRNA molecules cannot be required for splicing. The hnRNP proteins have been removed from these transcripts by the spreading conditions used. (Adapted from Y.N. Osheim, O.L. Miller, and A.L. Beyer, *Cell* 43:143–151, 1985. © Cell Press.)

Intron Sequences Are Removed as Lariat-shaped RNA Molecules [35]

Introns range in size from about 80 nucleotides to 10,000 nucleotides or more. Unlike the sequence of an exon, the exact nucleotide sequence of an intron seems to be unimportant. Thus introns have accumulated mutations rapidly during evolution, and it is often possible to alter most of an intron's nucleotide sequence without greatly affecting gene function. This has led to the suggestion that intron sequences have no function at all and are largely genetic "junk," a proposition we shall examine at the end of the chapter. The only highly conserved sequences in introns are those required for intron removal, which are found at or near the ends of an intron and are very similar in all known intron sequences; they generally cannot be altered without affecting the splicing process that normally removes the intron sequence from the primary RNA transcript. These conserved boundary sequences at the **5' splice site (donor site)** and the **3' splice site (acceptor site)** of introns from higher eucaryotes are shown in Figure 8–53. The RNA breaking and rejoining reactions must be carried out precisely because an error of even one nucleotide would shift the reading frame in the resulting mRNA molecule and make nonsense of its message.

The pathway by which the intron sequences are removed from primary RNA transcripts has been elucidated by *in vitro* studies in which a pure RNA species containing a single intron is prepared by incubating an appropriately designed DNA fragment with an RNA polymerase (see Figure 7–36). When these RNA molecules are added to a cell extract, they become spliced in a two-step enzymatic reaction that requires prolonged incubation with ATP, the U1, U2, U5, and U4/U6 snRNPs, and a number of additional proteins; these components assemble into a large multicomponent ribonucleoprotein complex, or **spliceosome.** Characterization of the RNA species that appear as intermediates during the reaction, as well as the snRNPs required to produce them, led to the discovery that the intron is excised in the form of a *lariat*, according to the splicing pathway shown in Figures 8–54 and 8–55.

Individual roles have been defined for several of the snRNPs. The U1 snRNP, for example, binds to the 5' splice site, guided by a nucleotide sequence in the U1 RNA that forms base pairs complementary to the nine-nucleotide splice site consensus sequence (see Figure 8–53). Since RNA is capable of acting like an enzyme, either the RNA or the protein components of the spliceosome could be responsible for catalyzing the breakage and formation of covalent bonds required for RNA splicing.

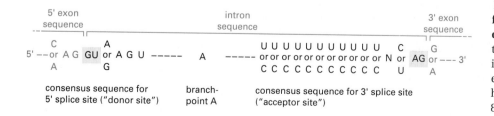

5' exon sequence

intron sequence

3' exon sequence

consensus sequence for 5' splice site ("donor site")

branch-point A

consensus sequence for 3' splice site ("acceptor site")

Figure 8–53 Consensus sequences for RNA splicing in higher eucaryotes. The sequence given is that for the RNA chain; the nearly invariant GU and AG dinucleotides at either end of the intron sequence are highlighted in *yellow* (see also Figure 8–51).

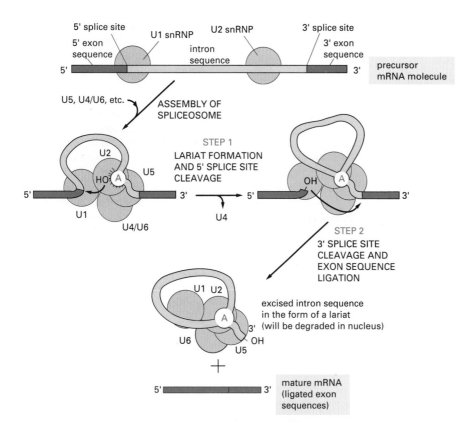

Figure 8–54 The RNA splicing mechanism. RNA splicing is catalyzed by a spliceosome formed from the assembly of U1, U2, U5, and U4/U6 snRNPs (shown as *green circles*) plus other components (not shown). After assembly of the spliceosome, the reaction occurs in two steps: in step 1 the branch-point A nucleotide in the intron sequence, which is located close to the 3′ splice site, attacks the 5′ splice site and cleaves it; the cut 5′ end of the intron sequence thereby becomes covalently linked to this A nucleotide, forming the branched nucleotide shown in Figure 8–55. In step 2 the 3′-OH end of the first exon sequence, which was created in the first step, adds to the beginning of the second exon sequence, cleaving the RNA molecule at the 3′ splice site; the two exon sequences are thereby joined to each other and the intron sequence is released as a lariat. The spliceosome complex sediments at 60S, indicating that it is nearly as large as a ribosome. These splicing reactions occur in the nucleus and generate mRNA molecules from primary RNA transcripts (mRNA precursor molecules).

Multiple Intron Sequences Are Usually Removed from Each RNA Transcript [36]

Because the spliceosome seems mainly to work by recognizing the consensus sequences that mark the two boundaries of an intron sequence (and for all intron sequences these consensus sequences are alike), the 5′ splice site (donor site) at the end of any one intron sequence can in principle be spliced to the 3′ splice site (acceptor site) of any other intron sequence. Indeed, when an RNA molecule is created artificially, with donor and acceptor splice sites from different intron sequences inserted into it, the intervening RNA is often recognized by the spliceosome and removed.

In view of this result, it is surprising that vertebrate genes can contain as many as 50 introns (see Table 8–1, p. 340). If *any* two 5′ and 3′ splice sites were mispaired for splicing, some functional mRNA sequences would be lost, with disastrous consequences. Somehow such mistakes are avoided: the RNA processing machinery normally guarantees that each 5′ splice site pairs only with the 3′ splice site that is closest to it in the downstream (5′-to-3′) direction of the linear RNA sequence (Figure 8–56). How this sequential pairing of splice sites is accomplished is not known, although the assembly of the spliceosome while the RNA transcript is still growing (see Figure 8–52) is presumed to play a major part in ensuring an orderly pairing of the appropriate splice sites. There is also evidence

Figure 8–55 Structure of the branched RNA chain that forms during nuclear RNA splicing. The nucleotide shown in *yellow* is the A nucleotide highlighted in Figure 8–54. The branch is formed in step 1 of the splicing reaction illustrated there, when the 5′ end of the intron sequence couples covalently to the 2′-OH ribose group of the A nucleotide, which is located about 30 nucleotides from the 3′ end of the intron sequence. The branched chain remains in the final excised intron sequence and is responsible for its lariat form (see Figure 8–54).

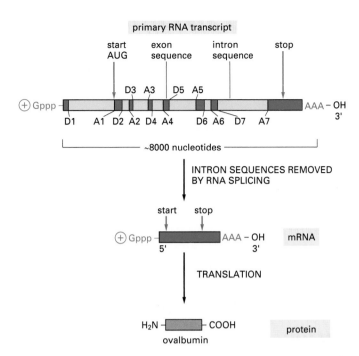

Figure 8–56 **Splicing the primary RNA transcript from the chicken ovalbumin gene.** The drawing shows the organized removal of seven introns required to obtain a functional mRNA molecule. The 5′ splice sites (donor sites) are denoted by D, and 3′ splice sites (acceptor sites) are denoted by A.

that the exact three-dimensional conformations adopted by the intron and exon sequences in the RNA transcript are important. We shall see in Chapter 9, however, that this simple 5′-to-3′ splicing can be altered by specialized control mechanisms that allow a single gene to produce several different mRNAs and hence several different proteins.

Studies of Thalassemia Reveal How RNA Splicing Can Allow New Proteins to Evolve [37]

Recombinant DNA techniques have made humans with inherited diseases an increasingly important source of material for genetic studies of cellular mechanisms. In a group of human genetic diseases called the **thalassemia syndromes,** for example, patients have an abnormally low level of hemoglobin—the oxygen-carrying protein in red blood cells. The change in the DNA sequence has been determined for more than 50 such mutants, and a large proportion of them have been found to have alterations in the pattern of splicing of globin RNA transcripts. Thus single nucleotide changes have been detected that inactivate a splice site. Surprisingly, analysis of the mRNAs produced in these mutant individuals reveals that the loss of a splice site does not prevent splicing but instead causes its normal partner site to seek out and become joined to a new "cryptic" site nearby; often a number of alternative splices are made in these mutants, causing the mutant gene to produce a set of altered proteins rather than just one (Figure 8–57). Other single nucleotide changes create new splice sites by changing a sequence in an intron or an exon into a consensus splice site. These results demonstrate that RNA splicing is a flexible process in higher eucaryotic cells, and they suggest that changes in the splicing pattern caused by random mutations could be an important pathway in the evolution of genes and organisms.

Spliceosome-catalyzed RNA Splicing Probably Evolved from Self-splicing Mechanisms [38]

When the lariat intermediate in nuclear RNA splicing was first discovered, it puzzled molecular biologists. Why was this bizarre pathway used rather than the apparently simpler alternative of bringing the 5′ and 3′ splice sites together in an initial step, followed by their direct cleavage and rejoining? The answer seems to lie in the way the spliceosome evolved.

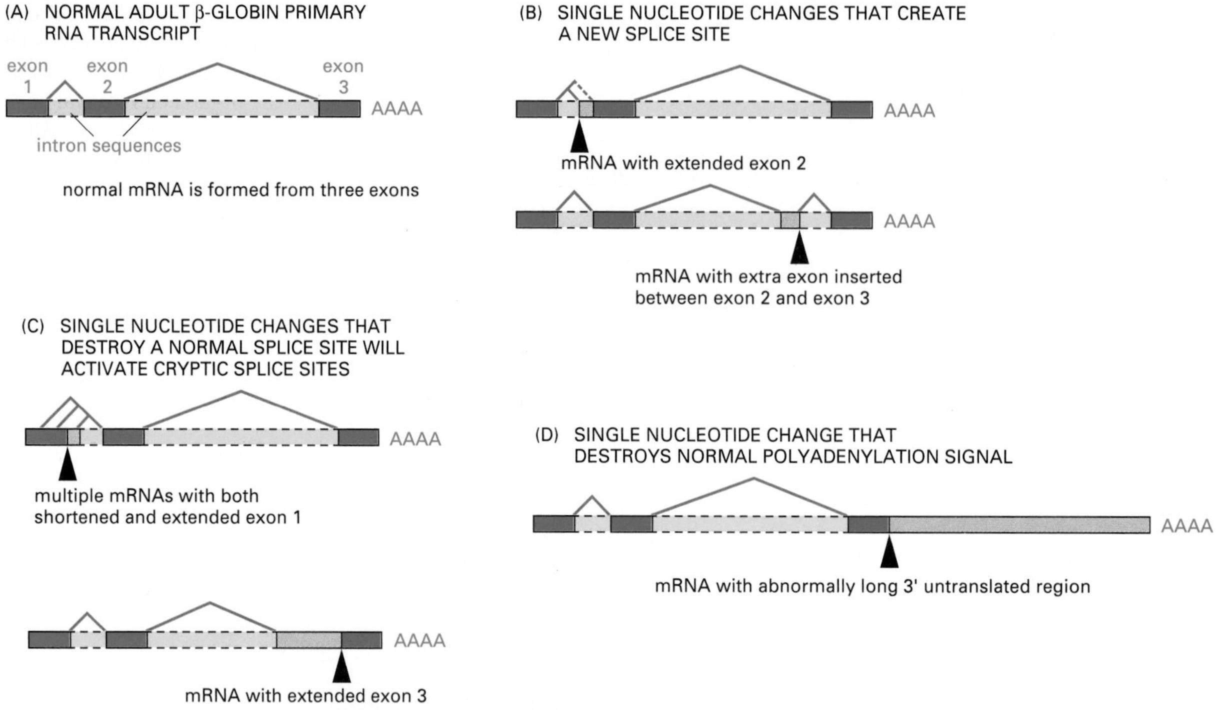

(A) NORMAL ADULT β-GLOBIN PRIMARY RNA TRANSCRIPT

exon 1   exon 2   exon 3

intron sequences

normal mRNA is formed from three exons

(B) SINGLE NUCLEOTIDE CHANGES THAT CREATE A NEW SPLICE SITE

mRNA with extended exon 2

mRNA with extra exon inserted between exon 2 and exon 3

(C) SINGLE NUCLEOTIDE CHANGES THAT DESTROY A NORMAL SPLICE SITE WILL ACTIVATE CRYPTIC SPLICE SITES

multiple mRNAs with both shortened and extended exon 1

mRNA with extended exon 3

(D) SINGLE NUCLEOTIDE CHANGE THAT DESTROYS NORMAL POLYADENYLATION SIGNAL

mRNA with abnormally long 3' untranslated region

Figure 8–57 **Abnormal processing of the β-globin primary RNA transcript in humans with β thalassemia.** The site of each mutation is denoted by a *black arrowhead*. The *dark blue boxes* represent the three normal exon sequences illustrated previously in Figure 8–51, and the *red lines* join the 5' and 3' splice sites utilized in splicing the primary RNA transcript produced by the gene. The *light blue boxes* depict new nucleotide sequences included in the final mRNA molecule as a result of a mutation. Note that when a mutation leaves a normal splice site without a partner, one or more abnormal "cryptic" splice sites nearby are used as the partner site, as in (C). (After S.H. Orkin, in The Molecular Basis of Blood Diseases [G. Stamatoyanno-poulos et al., eds.], pp. 106–126. Philadelphia: Saunders, 1987.)

As explained in Chapter 1, it is thought that early cells may have used RNA molecules rather than proteins as their major catalysts and stored their genetic information in RNA rather than DNA sequences. RNA-catalyzed splicing reactions presumably played important roles in these early cells, and some self-splicing RNA introns remain today—for example, in the nuclear rRNA genes of the ciliate *Tetrahymena*, in bacteriophage T4, and in some mitochondrial and chloroplast genes. A self-splicing intron sequence can be identified in a test tube by incubating a pure RNA molecule that contains the intron sequence and observing the splicing reaction; it can also be identified from the RNA sequence, inasmuch as large parts of the intron sequence need to be conserved in order to fold to create a catalytic surface in the RNA molecule. Two major classes of self-splicing intron sequences can be readily distinguished in this way. *Group I intron sequences* begin the splicing reaction by binding a G nucleotide to the intron sequence; the G is thereby activated to form the attacking group that will break the first of the phosphodiester bonds cleaved during splicing (the bond at the 5' splice site). In *group II intron sequences* a specially reactive A residue in the intron sequence is the attacking group, and a lariat intermediate is generated. Otherwise the reaction pathways for the two types of sequences are the same. Both are presumed to represent vestiges of very ancient mechanisms (Figure 8–58).

In the evolution of nuclear RNA splicing, the reaction pathway used by the group II self-splicing intron sequences seems to have been retained, with the catalytic role of the intron sequences being replaced by separate spliceosome components. Thus the small RNAs U1 and U2, for example, may well be remnants of catalytic RNA sequences that were originally present in intron sequences. Shifting the catalysis from intron sequence to spliceosome presumably lifted most of the constraints on the evolution of introns, allowing many new intron sequences to evolve.

The Transport of mRNAs to the Cytoplasm Is Delayed Until Splicing Is Complete [39]

Finished mRNA molecules are thought to be recognized by receptor proteins in the nuclear pore complex and to be transported actively to the cytoplasm (dis-

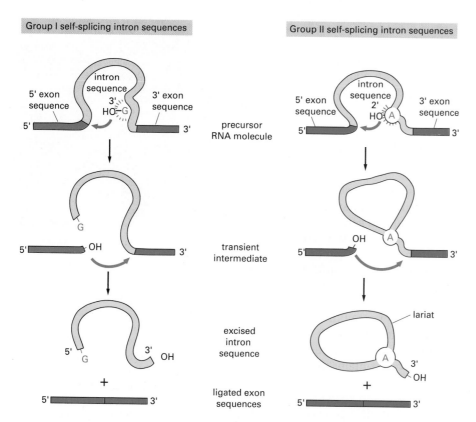

Group I self-splicing intron sequences	Group II self-splicing intron sequences	
		precursor RNA molecule
		transient intermediate
		excised intron sequence
		ligated exon sequences

Figure 8–58 The two known classes of self-splicing intron sequences. The group I intron sequences bind a free G nucleotide to a specific site to initiate splicing (see Figure 3–21), while the group II intron sequences use a specially reactive A nucleotide in the intron sequence itself for the same purpose. The two mechanisms have been drawn in a way that emphasizes their similarities. Both are normally aided by proteins that speed up the reaction, but the catalysis is nevertheless mediated by the RNA in the intron sequence. The mechanism used by group II intron sequences forms a lariat and resembles the pathway catalyzed by the spliceosome (compare to Figure 8–54). (After T.R. Cech, *Cell* 44:207–210, 1986. © Cell Press.)

cussed in Chapter 12). The major proteins of the hnRNP particles and various processing molecules bound to the RNA, by contrast, are largely confined to the nucleus, although some of them pass into the cytoplasm with the transported mRNA before being rapidly stripped from the RNA and returned to the nucleus (Figure 8–59).

Studies of mutant yeasts suggest that for RNAs that have splice sites transport out of the nucleus can occur only after the splicing reaction has been completed. When a mutation creates a defect in the splicing machinery so that splicing cannot occur, unspliced mRNA precursors remain in the nucleus, while those mRNAs that do not require splicing (which includes most of the mRNAs in this single-cell eucaryote) are transported normally to the cytosol. This observation suggests that RNAs may be retained in the nucleus by their bound spliceosome components, which seem to form numerous large aggregates throughout the nucleus of higher eucaryotes. These aggregates could serve as "splicing islands" (Figure 8–60); although it is not known how they form or function, they may be

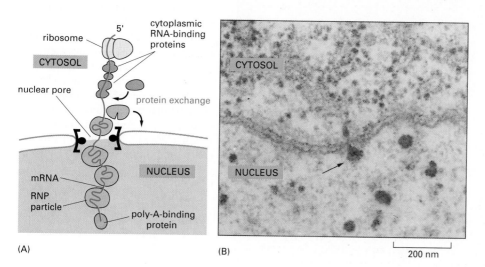

Figure 8–59 The transport of mRNA molecules through nuclear pores. (A) Schematic illustration of the change in the proteins bound to the RNA molecule as it moves out of the nucleus. (B) Electron micrograph of a large mRNA molecule produced in an insect salivary gland cell; this molecule (*arrow*) has apparently been caught in the process of moving to the cytosol. (B, from B.J. Stevens and H. Swift, *J. Cell Biol.* 31:55–77, 1966, by copyright permission of the Rockefeller University Press.)

RNA Synthesis and RNA Processing

analogous to the *nucleolus*, a much larger and more prominent structure in the nucleus, whose organization and function are better understood.

The nucleolus is the site where ribosomal RNA (rRNA) molecules are processed from a larger precursor RNA and assembled into ribosomes by the binding of ribosomal proteins. Before discussing nucleolar structure, however, we need to consider how the precursor rRNA molecules are synthesized from rRNA genes.

Ribosomal RNAs (rRNAs) Are Transcribed from Tandemly Arranged Sets of Identical Genes [40]

Many of the most abundant proteins of a differentiated cell, such as hemoglobin in the red blood cell and myoglobin in a muscle cell, are synthesized from genes that are present in only a single copy per haploid genome. These proteins are abundant because each of the many mRNA molecules transcribed from the gene can be translated into as many as 10 protein molecules per minute. This will normally produce more than 10,000 protein molecules per mRNA molecule in each cell generation. Such an amplification step is not available for the synthesis of the intrinsic RNA components of the ribosome, however, since they are the final gene products. Yet a growing higher eucaryotic cell must synthesize 10 million copies of each type of ribosomal RNA molecule in each cell generation in order to construct its 10 million ribosomes. Adequate quantities of ribosomal RNAs, in fact, can be produced only because the cell contains multiple copies of the **rRNA genes** that code for ribosomal RNAs.

Even *E. coli* needs seven copies of its rRNA genes to keep up with the cell's need for ribosomes. Human cells contain about 200 rRNA gene copies per haploid genome, spread out in small clusters on five different chromosomes, while cells of the frog *Xenopus* contain about 600 rRNA gene copies per haploid genome in a single cluster on one chromosome. In eucaryotes the multiple copies of the highly conserved rRNA genes on a given chromosome are located in a tandemly arranged series in which each gene (8000 to 13,000 nucleotide pairs long, depending on the organism) is separated from the next by a nontranscribed region known as *spacer DNA*, which can vary greatly in length and sequence. We shall see later that such multiple copies of tandemly arranged genes tend to co-evolve.

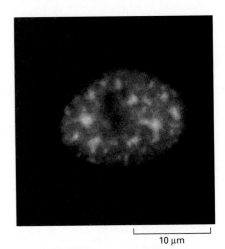

10 μm

Figure 8–60 Possible splicing islands. This immunofluorescence micrograph shows the staining of a human fibroblast nucleus with a monoclonal antibody that detects the snRNP particles involved in nuclear splicing of mRNA precursor molecules. The snRNP particles are present in large aggregates, which could function as "splicing islands." The antibody detects specific proteins that are present in several of the snRNPs that function in the spliceosome. (Courtesy of N. Ringertz.)

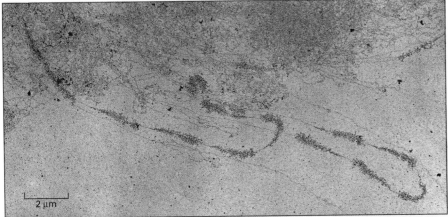

2 μm

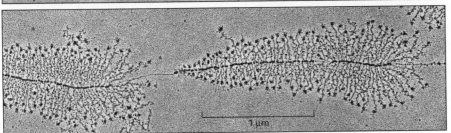

1 μm

Figure 8–61 Transcription from tandemly arranged rRNA genes, as visualized in the electron microscope. The pattern of alternating transcribed gene and nontranscribed spacer is readily seen in the lower-magnification view in the *upper panel*. The large particles at the 5′ end of each rRNA transcript (*lower panel*) are believed to reflect the beginning of ribosome assembly; RNA polymerase molecules are also clearly visible as a series of dots along the DNA. (Upper panel, from V.E. Foe, *Cold Spring Harbor Symp. Quant. Biol.* 42:723–740, 1978; lower panel, courtesy of Ulrich Scheer.)

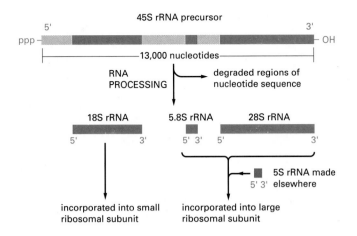

Because of their repeating arrangement, and because they are transcribed at a very high rate, the tandem arrays of rRNA genes can easily be seen in spread chromatin preparations. The RNA polymerase molecules and their associated transcripts are so densely packed (typically about 100 per gene) that the transcripts stick out perpendicularly from the DNA to give each transcription unit a "Christmas tree" appearance (Figure 8–61). As noted earlier (see Figure 8–47), the tip of each of these "trees" represents the point on the DNA at which transcription begins and where the transcripts are thus shortest, while the other end of the rRNA transcription unit is sharply demarcated by the sudden disappearance of RNA polymerase molecules and their transcripts.

The rRNA genes are transcribed by *RNA polymerase I*, and each gene produces the same primary RNA transcript. In humans this RNA transcript, known as *45S rRNA*, is about 13,000 nucleotides long. Before it leaves the nucleus in assembled ribosomal particles, the 45S rRNA is cleaved to give one copy each of the 28S rRNA (about 5000 nucleotides), the 18S rRNA (about 2000 nucleotides), and the 5.8S rRNA (about 160 nucleotides) of the final ribosome. The derivation of these three rRNAs from the same primary transcript ensures that they will be made in equal quantities. The remaining part of each primary transcript (about 6000 nucleotides) is degraded in the nucleus (Figure 8–62). Some of these extra RNA sequences are thought to play a transient part in ribosome assembly, which begins immediately as specific proteins bind to the growing 45S rRNA transcripts in the nucleus.

Another set of tandemly arranged genes with nontranscribed spacers codes for the 5S rRNA of the large ribosomal subunit (the only rRNA that is transcribed separately). The 5S rRNA genes are only about 120 nucleotide pairs in length, and like a number of other genes encoding small stable RNAs (most notably the transfer RNA [tRNA] genes), they are transcribed by *RNA polymerase III*. Humans have about 2000 5S rRNA genes tandemly arranged in a single cluster far from all the other rRNA genes. It is not known why this one type of rRNA is transcribed separately.

The Nucleolus Is a Ribosome-producing Machine [40]

The continuous transcription of multiple gene copies ensures an adequate supply of the rRNAs, which are immediately packaged with ribosomal proteins to form ribosomes. The packaging occurs in the nucleus, in a large, distinct structure called the **nucleolus.** The nucleolus contains large loops of DNA emanating from several chromosomes, each of which contains a cluster of rRNA genes. Each such gene cluster is known as a **nucleolar organizer** region. Here the rRNA genes are transcribed at a rapid rate by RNA polymerase I. The beginning of the rRNA packaging process can be seen in electron micrographs of these genes: the 5′ tail of each transcript is encased by a protein-rich granule (see Figure 8–61). These gran-

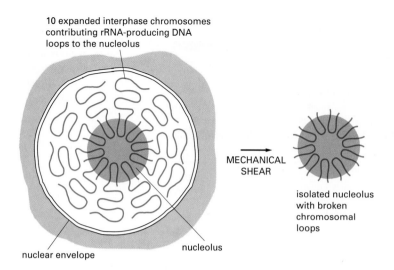

Figure 8–63 The nucleolus. This highly schematic view of a nucleolus in a human cell shows the contributions of loops of chromatin containing rRNA genes from 10 separate chromosomes. Purified nucleoli are very useful for biochemical studies of nucleolar function; to obtain such nucleoli, the loops of chromatin are mechanically sheared from their chromosomes, as shown.

10 expanded interphase chromosomes contributing rRNA-producing DNA loops to the nucleolus

MECHANICAL SHEAR

isolated nucleolus with broken chromosomal loops

nuclear envelope

nucleolus

ules, which do not appear on other types of RNA transcripts, presumably reflect the first of the protein-RNA interactions that take place in the nucleolus.

The biosynthetic functions of the nucleolus can be traced by briefly labeling newly made RNA with ^{3}H-uridine. After varying intervals of further incubation without ^{3}H-uridine, a cell fractionation procedure can be used to break the rRNA genes free of their chromosomes, thereby allowing the radioactive nucleoli to be isolated in relatively pure form (Figure 8–63). Such experiments show that the

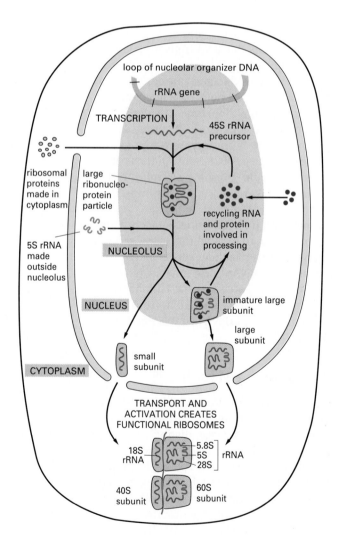

Figure 8–64 The function of the nucleolus in ribosome synthesis. The 45S rRNA transcript is packaged in a large ribonucleoprotein particle containing many ribosomal proteins imported from the cytoplasm. While this particle remains in the nucleolus, selected pieces are discarded as it is processed into immature large and small ribosomal subunits. These two subunits are thought to attain their final functional form only as each is individually transported through the nuclear pores into the cytoplasm.

loop of nucleolar organizer DNA

rRNA gene

TRANSCRIPTION

45S rRNA precursor

ribosomal proteins made in cytoplasm

large ribonucleo-protein particle

recycling RNA and protein involved in processing

5S rRNA made outside nucleolus

NUCLEOLUS

NUCLEUS

immature large subunit

large subunit

CYTOPLASM

small subunit

TRANSPORT AND ACTIVATION CREATES FUNCTIONAL RIBOSOMES

18S rRNA

5.8S
5S
28S

rRNA

40S subunit

60S subunit

intact 45S transcript is first packaged into a large complex containing many different proteins imported from the cytoplasm, where all proteins are synthesized. Most of the 80 different polypeptide chains that will make up the ribosome, as well as the 5S rRNAs, are incorporated at this stage. Other molecules are needed to process the 45S rRNA and to guide the assembly process. Thus the nucleolus also contains other RNA-binding proteins and certain small ribonucleoprotein particles (including U3 snRNP) that are believed to help catalyze the construction of ribosomes. These components remain in the nucleolus when the ribosomal subunits are exported to the cytoplasm in finished form. An especially notable component is *nucleolin*, an abundant, well-characterized RNA-binding protein that seems to coat only ribosomal transcripts; this protein stains with silver in the characteristic manner of the nucleolus itself.

As the 45S rRNA molecule is processed, it gradually loses some of its RNA and protein and then splits to form separate precursors of the large and small ribosomal subunits (Figure 8–64). Within 30 minutes of radioactive pulse labeling, the first mature small ribosomal subunits, containing their 18S rRNA, emerge from the nucleolus and appear in the cytoplasm. Assembly of the mature large ribosomal subunit, with its 28S, 5.8S, and 5S rRNAs, takes about an hour to complete. The nucleolus therefore contains many more incomplete large ribosomal subunits than small ones.

The last steps in ribosome maturation occur only as these subunits are transferred to the cytoplasm. This delay prevents functional ribosomes from gaining access to the incompletely processed hnRNA molecules in the nucleus.

The Nucleolus Is a Highly Organized Subcompartment of the Nucleus [41]

As seen in the light microscope, the large spheroidal nucleolus is the most obvious structure in the nucleus of a nonmitotic cell. Consequently, it was so closely scrutinized by early cytologists that an 1898 review could list some 700 references. By the 1940s cytologists had demonstrated that the nucleolus contains high concentrations of RNA and proteins, but its major function in ribosomal RNA synthesis and ribosome assembly was not discovered until the 1960s.

Figure 8–65 Electron micrograph of a thin section of a nucleolus in a human fibroblast, showing its three distinct zones. (A) View of entire nucleus. (B) High-power view of the nucleolus. (Courtesy of E.G. Jordan and J. McGovern.)

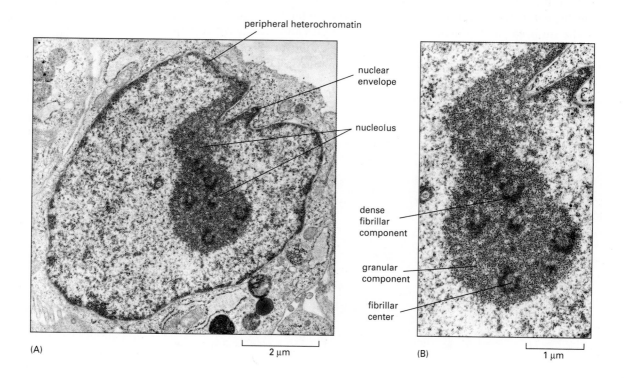

peripheral heterochromatin

nuclear envelope

nucleolus

dense fibrillar component

granular component

fibrillar center

(A)

2 μm

(B)

1 μm

Some of the details of nucleolar organization can be seen in the electron microscope. Unlike the cytoplasmic organelles, the nucleolus is not bounded by a membrane; instead, it seems to be constructed by the specific binding of unfinished ribosome precursors to one another to form a large network. In a typical electron micrograph three partially segregated regions can be distinguished (Figure 8–65): (1) a pale-staining *fibrillar center*, which contains DNA that is not being actively transcribed; (2) a *dense fibrillar component*, which contains RNA molecules in the process of being synthesized; and (3) *a granular component*, which contains maturing ribosomal precursor particles.

The size of the nucleolus reflects its activity. Its size therefore varies greatly in different cells and can change in a single cell. It is very small in some dormant plant cells, for example, but can occupy up to 25% of the total nuclear volume in cells that are making unusually large amounts of protein. The differences in size are due largely to differences in the amount of the granular component, which is probably controlled at the level of ribosomal gene transcription: electron microscopy of spread chromatin shows that both the fraction of activated ribosomal genes and the rate at which each gene is transcribed can vary according to circumstances.

The Nucleolus Is Reassembled on Specific Chromosomes After Each Mitosis [42]

The appearance of the nucleolus changes dramatically during the cell-division cycle. As the cell approaches mitosis, the nucleolus first decreases in size and then disappears as the chromosomes condense and all RNA synthesis stops, so that generally there is no nucleolus in a metaphase cell. When ribosomal RNA synthesis restarts at the end of mitosis (in telophase), tiny nucleoli reappear at the chromosomal locations of the ribosomal RNA genes (Figure 8–66).

In humans the ribosomal RNA genes are located near the tips of each of 5 different chromosomes, as shown previously in Figure 8–32 (that is, on 10 of the 46 chromosomes in a diploid cell). Correspondingly, 10 small nucleoli form after mitosis in a human cell, although they are rarely seen as separate entities because they quickly grow and fuse to form the single large nucleolus typical of many interphase cells (Figure 8–67).

What happens to the RNA and protein components of the disassembled nucleolus during mitosis? It seems that at least some of them become distributed over the surface of all of the metaphase chromosomes and are carried as cargo to each of the two daughter cell nuclei. As the chromosomes decondense at telophase, these "old" nucleolar components help reestablish the newly emerging nucleoli.

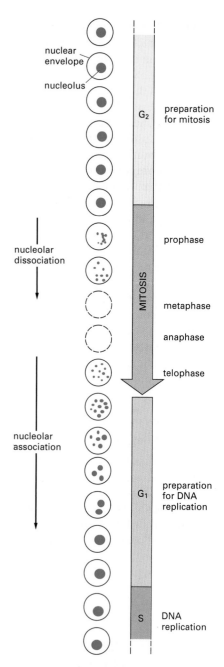

Figure 8–66 Changes in the appearance of the nucleolus in a human cell during the cell cycle. Only the cell nucleus is represented in this diagram. In most eucaryotic cells the nuclear membrane breaks down during mitosis, as indicated by the dashed circles.

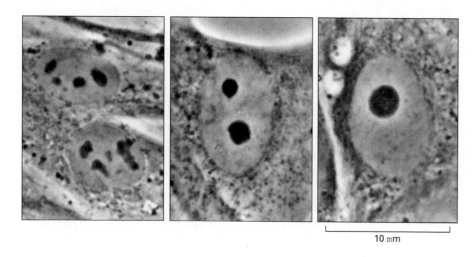

10 mm

Figure 8–67 Nucleolar fusion. These light micrographs of human fibroblasts grown in culture show various stages of nucleolar fusion. (Courtesy of E.G. Jordan and J. McGovern.)

Figure 8–68 **The polarized orientation of chromosomes in interphase cells of the early *Drosophila* embryo.** (A) Diagrams of the Rabl orientation, with all centromeres facing one nuclear pole and all telomeres pointing toward the opposite pole. In the embryo each nucleus is elongated as shown. (B) Low-magnification light micrograph of a *Drosophila* embryo at the cellular blastoderm stage. The chromosomes in each interphase nucleus have been stained with a fluorescent dye. Note that the most brightly staining region (the chromocenter), which is known to contain the centromeric regions of each of the four chromosomes (see Figure 8–19), is oriented toward the outer surface of the embryo and thus faces the apical plasma membrane of every cell. (Courtesy of John Sedat.)

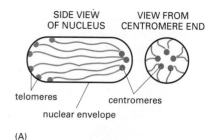

(A)

Individual Chromosomes Occupy Discrete Territories in the Nucleus During Interphase [43]

As we have just seen, specific genes from separate interphase chromosomes are brought together at a single site in the nucleus when the nucleolus forms. Are other parts of chromosomes also nonrandomly ordered in the nucleus? First raised by biologists in the late nineteenth century, this fundamental question still has not been answered satisfactorily.

A certain degree of chromosomal order results from the configuration that the chromosomes always have at the end of mitosis. Just before a cell divides, the condensed chromosomes are pulled to each spindle pole by microtubules attached to the centromeres; thus, as the chromosomes move, the centromeres lead the way and the distal arms (terminating in telomeres) lag behind. The chromosomes in many nuclei tend to retain this so-called *Rabl orientation* throughout interphase, with their centromeres facing one pole of the nucleus and their telomeres pointing toward the opposite pole (Figure 8–68A). In some cases the nucleus is specifically oriented in the cell: in the early *Drosophila* embryo, for example, all the centromeres face apically (Figure 8–68B). Such fixed nuclear orientations might have important effects on cell polarity, but it is difficult to design experiments to test this possibility.

In most cells the various chromosomes are indistinguishable from one another during interphase. Consequently, it is difficult to assess their arrangement in more detail than just described. The giant interphase chromosomes of the polytene cells of *Drosophila* larvae, however, are an exception. Here the individual chromosome bands can be resolved clearly enough to determine the precise positions of specific genes in intact nuclei by microscopic optical-sectioning and reconstruction techniques. The results of such analyses suggest that the interphase chromosome set is not highly ordered: although the Rabl orientation tends to be maintained, two apparently identical cells often have different chromosomes as nearest neighbors.

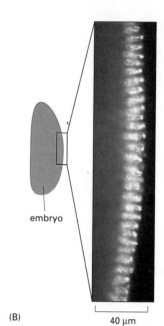

(B)

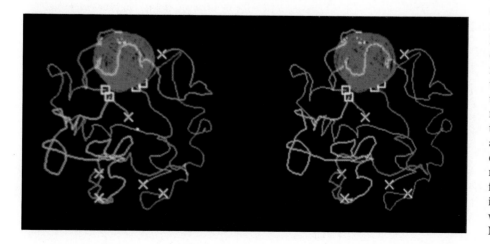

Figure 8–69 **A stereo pair that displays the three-dimensional arrangement of the polytene chromosomes in a single nucleus of a *Drosophila* larval gland cell.** The large ball is the nucleolus, and the course of each chromosome arm is represented by a line running along the chromosome axis. The telomeres tend to be on the surface of the nuclear envelope opposite the surface that is nearest the nucleolus, where all the centromeres are located. The chromosomes in such nuclei are never entangled, but their detailed foldings and neighbors are different in otherwise identical nuclei. (For viewing with crossed eyes; courtesy of Mark Hochstrasser and John Sedat.)

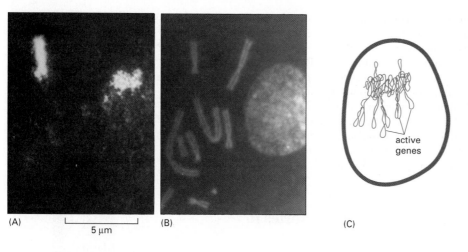

(A)   5 μm   (B)   (C)

active genes

Figure 8–70 **Selective labeling of a single chromosome in a cultured mammalian cell nucleus during interphase.** In (A) and (B), an interphase nucleus freed from its cytoplasm is shown at the right, with scattered mitotic chromosomes released from a second cell on the left. (A) The results of *in situ* hybridization (using a fluorescent probe) to outline the single human chromosome in a human-hamster hybrid cell line. The same preparation is shown with all of the DNA labeled with a second fluorescent dye in (B). (C) Schematic drawing of the human chromosome detected in the interphase nucleus in (A), shown at a somewhat larger scale. (A and B, courtesy of Joyce A. Kobori and David R. Cox.)

These analyses of polytene chromosomes have also indicated that each chromosome occupies its own territory in the interphase nucleus—that is, the individual chromosomes are not extensively intertwined (Figure 8–69). Other experiments have shown that nonpolytene chromosomes also tend to occupy discrete domains in interphase nuclei. *In situ* hybridization experiments with an appropriate DNA probe, for example, can outline a single chromosome in hybrid mammalian cells grown in culture (Figure 8–70). Most of the DNA of such a chromosome is seen to occupy only a small portion of the interphase nucleus, suggesting that each individual chromosome remains compact and organized while allowing selected portions of its DNA to be active in RNA synthesis.

How Well Ordered Is the Nucleus? [44]

The interior of the nucleus is not a random jumble of its many RNA, DNA, and protein components. We have seen that the nucleolus is organized as an efficient ribosome-construction machine, and clusters of spliceosome components are organized, possibly as discrete RNA-splicing islands (see Figure 8–60). Order is also seen in the electron microscope when one focuses on the regions around nuclear pores: the chromatin that lines the inner nuclear membrane (which is unusually condensed chromatin and therefore clearly visible in electron micrographs) is excluded from a considerable region beneath and around each nuclear pore, clearing a path between the cytoplasm and the nucleoplasm (Figure 8–71). In some special cases, moreover, the nuclear pores are found to be highly organized in the nuclear envelope (Figure 8–72), presumably reflecting a corresponding organization of the nuclear lamina to which the pores are attached.

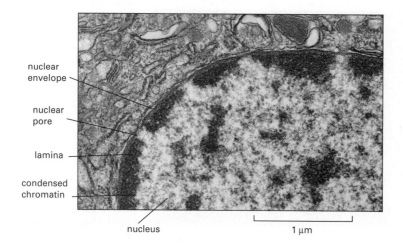

nuclear envelope

nuclear pore

lamina

condensed chromatin

nucleus   1 μm

Figure 8–71 **Electron micrograph of a mammalian cell nucleus.** Note that the condensed chromatin underlying the nuclear envelope is excluded from regions around the nuclear pores. (Courtesy of Larry Gerace.)

Is there an intranuclear framework, analogous to the cytoskeleton, on which nuclear components are organized? Many cell biologists believe there is. The *nuclear matrix*, or *scaffold*, has been defined as the insoluble material left in the nucleus after a series of biochemical extraction steps. Some of the proteins that constitute it can be shown to bind specific DNA sequences called *SARs* or *MARs* (for scaffold- or matrix-associated regions). Such DNA sequences have been postulated to form the base of chromosomal loops (see Figure 8–18). By means of such chromosomal attachment sites, the matrix might help organize chromosomes, localize genes, and regulate DNA transcription and replication within the nucleus. Because the structural components of the matrix have not yet been identified, however, it remains uncertain whether the matrix isolated by cell biologists represents a structure that is present in intact cells.

1 µm

Figure 8–72 Freeze-fracture electron micrograph of the elongated nuclear envelope of a fern spore. Note the ordered arrangement of the nuclear pore complexes in parallel rows. In other cells either concentrated clusters of nuclear pores or unusual areas free of nuclear pores have been detected in the nuclear envelope, and these are specifically oriented with respect to other structures in the cell. (Courtesy of Don H. Northcote; from K. Roberts and D.H. Northcote, *Microsc. Acta* 71:102–120, 1971.)

Summary

RNA polymerase, the enzyme that catalyzes DNA transcription, is a complex molecule containing many polypeptide chains. In eucaryotic cells there are three RNA polymerases, designated polymerases I, II, and III; they are evolutionarily related to one another and to bacterial RNA polymerase, and they have some subunits in common. After initiating transcription, each enzyme is thought to release one or more subunits and to bind other subunits that are required for RNA chain elongation and termination.

Most of the cell's mRNA is produced by a complex process beginning with the synthesis of heterogeneous nuclear RNA (hnRNA). The primary hnRNA transcript is made by RNA polymerase II. It is then capped by the addition of a special nucleotide to its 5′ end and is cleaved and then polyadenylated at its 3′ end. The modified RNA molecules are usually then subjected to one or more RNA splicing events, in which intron sequences are removed from the middle of the RNA molecule by a reaction catalyzed by a large ribonucleoprotein complex known as a spliceosome. In this process most of the mass of the primary RNA transcript is removed and degraded in the nucleus. As a result, although the rate of production of hnRNA typically accounts for about half of a cell's RNA synthesis, the mRNA produced represents only about 3% of the steady-state quantity of RNA in a cell.

Unlike genes that code for proteins, which are transcribed by polymerase II, the genes that code for most structural RNAs are transcribed by polymerase I and III. These genes are usually repeated many times in the genome and are often clustered in tandem arrays. RNA polymerase III makes a variety of small stable RNAs, including the tRNAs and the small 5S rRNA of the ribosome. RNA polymerase I makes the large rRNA precursor molecule (45S rRNA) containing the major rRNAs. Except for the ribosomes in mitochondria and chloroplasts, all the cell's ribosomes are assembled in the nucleolus—a distinct intranuclear organelle that is formed around the tandemly arranged rRNA genes, which are brought together from several chromosomes.

The Organization and Evolution of the Nuclear Genome

Much of evolutionary history is recorded in the genomes of present-day organisms and can be deciphered from a careful analysis of their DNA sequences. Tens of millions of DNA nucleotides have been sequenced thus far, and we can now see in outline how the genes coding for certain proteins have evolved over hundreds of millions of years. Studies of the occasional changes that occur in present-day chromosomes provide additional clues to the mechanisms that have brought about evolutionary change in the past. In this section we consider some of the general principles that have emerged from such molecular genetic studies, with emphasis on the organization and evolution of the nuclear genome in higher eucaryotes.

Genomes Are Fine-tuned by Point Mutation and Radically Remodeled or Enlarged by Genetic Recombination [45]

DNA nucleotide sequences must be accurately replicated and conserved. In Chapter 6 we discussed the elaborate DNA-replication and DNA-repair mechanisms that enable DNA sequences to be inherited with extraordinary fidelity: only about one nucleotide pair in a thousand is randomly changed every 200,000 years. Even so, in a population of 10,000 individuals, every possible nucleotide substitution will have been "tried out" on about 50 occasions in the course of a million years, which is a short span of time in relation to the evolution of species. Much of the variation created in this way will be disadvantageous to the organism and will be selected against in the population. When a rare variant sequence is advantageous, however, it will be rapidly propagated by natural selection. Consequently, it can be expected that in any given species the functions of most genes will have been optimized by random point mutation and selection.

While point mutation is an efficient mechanism for fine-tuning the genome, evolutionary progress in the long term must depend on more radical types of genetic change. Genetic recombination causes major rearrangements of the genome with surprising frequency: the genome can expand or contract by duplication or deletion, and its parts can be transposed from one region to another to create new combinations. Component parts of genes—their individual exons and regulatory elements—can be shuffled as separate modules to create proteins that have entirely new roles. In addition, duplicated copies of genes tend to diverge by further mutation and become specialized and individually optimized for subtly different functions. By these means the genome as a whole can evolve to become increasingly complex and sophisticated. In a mammal, for example, multiple variant forms of almost every gene exist—different actin genes for the different types of contractile cells, different opsin genes for the perception of lights of different colors, different collagen genes for the different types of connective tissues, and so on. The expression of each gene is regulated according to its own precise and specific rules. Moreover, DNA sequencing studies reveal that many genes share related modular segments but are otherwise very different: common sequence motifs are frequently found in otherwise unrelated proteins.

Genetic recombination, whereby one chromosome exchanges genetic material with another, is fundamental to the creation of such families of genes and gene segments. In Chapter 6 we discussed the molecular mechanisms of both general recombination and site-specific recombination. Here we consider some of their effects on the genome.

Tandemly Repeated DNA Sequences Tend to Remain the Same [46]

Gene duplications are usually attributed to rare accidents catalyzed by some of the enzymes that mediate normal recombination processes. Higher eucaryotes, however, contain an efficient enzymatic system that joins the two ends of a broken DNA molecule together, so that duplications (as well as inversions, deletions, and translocations of DNA segments) can also arise as a consequence of the erratic rejoining of fragments of chromosomes that have somehow become broken in more than one place. When duplicated DNA sequences are joined head to tail, they are said to be *tandemly repeated*. Once a single tandem repeat appears, it can be extended readily into a long series of tandem repeats by unequal crossover events between two sister chromosomes, inasmuch as the large amount of matching sequence provides an ideal substrate for general recombination (Figure 8–73). DNA duplication followed by sequential unequal crossing-over underlies *DNA amplification*, a process that often contributes to the formation of cancer by increasing the number of copies of genes (proto-oncogenes) that promote cancer (see Figure 24–27).

Tandemly repeated genes both increase and decrease in number due to unequal crossing-over (see Figure 8–73). They therefore would be expected to be maintained by natural selection in large numbers only if the extra copies were beneficial to the organism. We have already discussed the hundreds of tandemly repeated genes that code for the vertebrate large ribosomal RNA precursor; these are needed to keep up with a growing cell's demand for new ribosomes. Similarly, vertebrates have clusters of tandemly repeated genes that encode other structural RNAs, including 5S rRNA and the U1 and U2 snRNAs, as well as clusters of repeated histone genes, which produce the large amounts of histones required during each S phase.

One might expect that in the course of evolution the sequences of the genes in a tandem array—and of the nontranscribed *spacer* DNA between them—would drift apart. With many copies of the same gene there should be little selection against random mutations that alter one or a few of the copies, and most nucleotide changes in the long nontranscribed spacer regions would have no functional consequence. In fact, however, the sequences of the tandemly repeated genes and their spacer DNAs are generally almost identical. Two mechanisms are thought to account for this. First, recurring unequal crossing-over events will cause the continued expansion and contraction of tandem arrays, and computer simulations show that this will tend to keep the sequences the same (Figure 8–74A). Second, related DNA sequences can become homogenized through *gene conversion*—the process whereby a portion of the DNA sequence is changed by copying a closely similar sequence present at a different site in the genome, as described in Chapter 6 (Figure 8–74B). Although gene conversion does not require that the genes be tandemly repeated, in higher eucaryotes it seems to occur mainly between genes that are close to each other.

The movement of one gene copy in a tandem array to a new chromosomal location will protect it from both of the above homogenizing influences. Thus, in higher eucaryotes accidental gene translocation is an important step in the evolution of new genes: it allows the translocated DNA sequence to begin to evolve independently, so that it can acquire new functions that might benefit the organism.

The Evolution of the Globin Gene Family Shows How Random DNA Duplications Contribute to the Evolution of Organisms [47]

In addition to generating a number of sets of tandemly repeated genes, DNA duplications have played a more important general role in the evolution of new

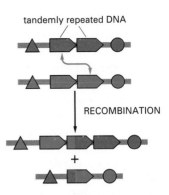

Figure 8–73 A family of tandemly repeated genes frequently loses and gains gene copies due to unequal crossing-over between sister chromosomes containing the genes. This type of event is frequent because the long regions of homologous DNA sequence are good substrates for the general genetic recombination process (discussed in Chapter 6).

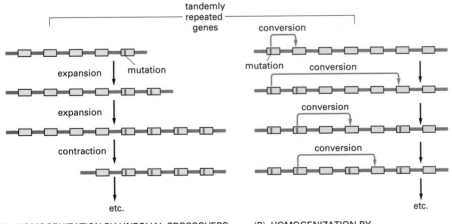

(A) HOMOGENIZATION BY UNEQUAL CROSSOVERS BETWEEN SISTER CHROMOSOMES

(B) HOMOGENIZATION BY GENE CONVERSION

Figure 8–74 Two types of events that help to keep all DNA sequences in a tandem array very similar to one another. (A) The continual expansion and contraction of the number of gene copies in a tandem array caused by unequal crossing-over (see Figure 8–73) tends to homogenize all of the gene sequences in the array. (B) In gene conversion one gene copy acts as a template that passes all or part of its DNA sequence to another gene copy. In higher eucaryotes this process seems to be confined largely to genes that are next to each other on the chromosome; in lower eucaryotes such as fungi, where gene conversion is most readily studied, it is not confined to neighboring genes.

proteins. The globin gene family provides a good example because its evolutionary history has been worked out particularly well. The unmistakable homologies in amino acid sequence and structure among the present-day globin genes indicate that they all must derive from a common ancestral gene, even though some now occupy widely separated locations in the mammalian genome.

We can reconstruct some of the past events that produced the various types of oxygen-carrying hemoglobin molecules by considering the different forms of the protein in organisms at different levels on the phylogenetic scale. A molecule like hemoglobin was necessary to allow multicellular animals to grow to a large size, since large animals could no longer rely on the simple diffusion of oxygen through the body surface to oxygenate their tissues adequately. Consequently, hemoglobinlike molecules are found in all vertebrates and in many invertebrates. The most primitive oxygen-carrying molecule in animals is a globin polypeptide chain of about 150 amino acids, which is found in many marine worms, insects, and primitive fish. The hemoglobin molecule in higher vertebrates, however, is composed of two kinds of globin chains. It appears that about 500 million years ago, during the evolution of higher fish, a series of gene mutations and duplications occurred. These events established two slightly different globin genes, coding for the α- and β-globin chains in the genome of each individual. In modern higher vertebrates each hemoglobin molecule is a complex of two α chains and two β chains (Figure 8–75). The four oxygen binding sites in the $\alpha_2\beta_2$ molecule interact, allowing a cooperative allosteric change in the molecule as it binds and releases oxygen, which enables hemoglobin to take up and to release oxygen more efficiently than the single-chain version.

Still later, during the evolution of mammals, the β-chain gene apparently underwent mutation and duplication to give rise to a second β-like chain that is synthesized specifically in the fetus. The resulting hemoglobin molecule has a higher affinity for oxygen than adult hemoglobin and thus helps in the transfer of oxygen from the mother to the fetus. The gene for the new β-like chain subsequently mutated and duplicated again to produce two new genes, ϵ and γ, the ϵ chain being produced earlier in development (to form $\alpha_2\epsilon_2$) than the fetal γ chain, which forms $\alpha_2\gamma_2$ (see Figure 9–52). A duplication of the adult β-chain gene occurred still later, during primate evolution, to give rise to a δ-globin gene and thus to a minor form of hemoglobin ($\alpha_2\delta_2$) found only in adult primates (Figure 8–76). Each of these duplicated genes has been modified by point mutations that affect the properties of the final hemoglobin molecule, as well as by changes in regulatory regions that determine the timing and level of expression of the gene.

The end result of the gene duplication processes that have given rise to the diversity of globin chains is seen clearly in the genes that arose from the original β gene, which are arranged as a series of homologous DNA sequences located within 50,000 nucleotide pairs of one another. A similar cluster of α-globin genes is located on a separate human chromosome. Because the α- and β-globin gene clusters are on separate chromosomes in birds and mammals but are together in the frog *Xenopus*, it is believed that a translocation event separated the two genes about 300 million years ago (see Figure 8–76). As previously discussed, such translocations probably help stabilize duplicated genes with distinct functions by protecting them from the homogenizing processes that act on closely linked genes of similar DNA sequence (see Figure 8–74).

There are several duplicated globin DNA sequences in the α- and β-globin gene clusters that are not functional genes. They are examples of *pseudogenes*. These have a close homology to the functional genes but have been disabled by mutations that prevent their expression. The existence of such pseudogenes should not be surprising since not every DNA duplication would be expected to lead to a new functional gene. Moreover, nonfunctional DNA sequences are not rapidly discarded, as indicated by the large excess of noncoding DNA in mammalian genomes, discussed previously.

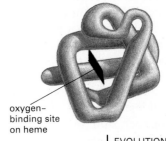

single-chain globin binds
one oxygen molecule

oxygen-
binding site
on heme

EVOLUTION OF A
SECOND GLOBIN
CHAIN BY
GENE DUPLICATION
FOLLOWED BY
MUTATION

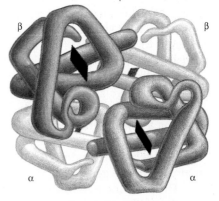

four-chain globin binds four
oxygen molecules in a
cooperative way

Figure 8–75 **A comparison of the structure of one-chain and four-chain globins.** The four-chain globin shown is hemoglobin, which is a complex of two α- and two β-globin chains. The one-chain globin in some primitive vertebrates forms a dimer that dissociates when it binds oxygen, representing an intermediate in the evolution of the four-chain globin.

A great deal of our evolutionary history will be discernible in our chromosomes once the DNA sequences of many gene families have been compared in different animals (see also Figure 7–4).

Genes Encoding New Proteins Can Be Created by the Recombination of Exons [45]

The role of DNA duplication in evolution is not confined to the generation of gene families. It can also be important in generating new single genes. The proteins encoded by genes generated in this way can be recognized by the presence of repeating, similar protein domains, which are covalently linked to one another in series. The immunoglobulins (Figure 8–77) and albumins, for example, as well as most fibrous proteins (such as the collagens) are encoded by genes that have evolved by repeated duplications of a primordial DNA sequence.

In genes that have evolved in this way, as well as in many other genes, each separate exon often encodes an individual protein folding unit, or domain. It is believed that the organization of DNA coding sequences as a series of such exons separated by long introns has greatly facilitated the evolution of new proteins. The duplications necessary to form a single gene coding for a protein with repeating domains, for example, can occur by breaking and rejoining the DNA anywhere in the long introns on either side of an exon encoding a useful protein domain; without introns there would be only a few sites in the original gene at which a recombinational exchange between sister DNA molecules could duplicate the domain. By enabling the duplication to occur at many potential recombination sites rather than at just a few, introns increase the probability of a favorable duplication event.

For the same reason, the presence of introns greatly increases the probability that a chance recombination event will generate a functional hybrid gene by joining two initially separated DNA sequences that code for different protein domains in such a way that both domains are preserved in the new protein that the hybrid gene encodes (see Figure 8–81, for example). The presumed results of such recombinations are seen in many present-day proteins (see Figure 3–43). Thus the large separation between the exons encoding individual domains in higher eucaryotes is thought to accelerate the process by which random genetic-recombination events generate useful new proteins. This could help to explain the successful evolution of these very complex organisms.

Most Proteins Probably Originated from Highly Split Genes [48]

The discovery in 1977 of genes split up by introns was unexpected. Previously all genes analyzed in detail were bacterial genes, which lack introns. Bacteria also lack nuclei and internal membranes and have smaller genomes than eucaryotic cells, and traditionally they were considered to resemble the simpler cells from which eucaryotic cells must have been derived. Not surprisingly, most biologists initially assumed that introns were a bizarre and late evolutionary addition to the eucaryotic line. It now seems likely, however, that split genes are the ancient condition and that bacteria lost their introns only after most of their proteins had evolved.

The idea that introns are very old is consistent with current concepts of protein evolution by the trial-and-error recombination of separate exons that encode distinct protein domains. Moreover, evidence for the ancient origin of introns has been obtained by examination of the gene that encodes the ubiquitous enzyme *triosephosphate isomerase*. Triosephosphate isomerase has an essential role in the metabolism of all cells, catalyzing the interconversion of glyceraldehyde-3 phosphate and dihydroxyacetone phosphate—a central step in glycolysis and gluconeogenesis (see Figure 2–21). By comparing the amino acid sequence of this en-

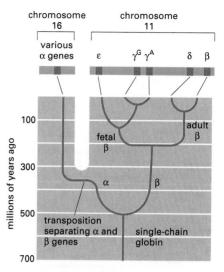

Figure 8–76 An evolutionary scheme for the globin chains that carry oxygen in the blood of animals. The scheme emphasizes the β-like globin gene family. A relatively recent gene duplication of the γ-chain gene produced γ^G and γ^A, which are fetal β-like chains of identical function. The location of the globin genes in the human genome is shown at the top of the figure.

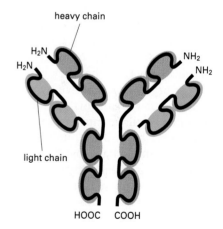

Figure 8–77 Schematic view of an antibody (immunoglobulin) molecule. This molecule is a complex of two identical heavy chains and two identical light chains. Each heavy chain contains four similar, covalently linked domains, while each light chain contains two such domains. Each domain is encoded by a separate exon, and all of the exons are thought to have evolved by the serial duplication of a single ancestral exon.

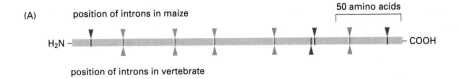

(A)

position of introns in maize

50 amino acids

H₂N — ▼ ▼ ▼ ▼ ▼ ▼ ▼ — COOH

position of introns in vertebrate

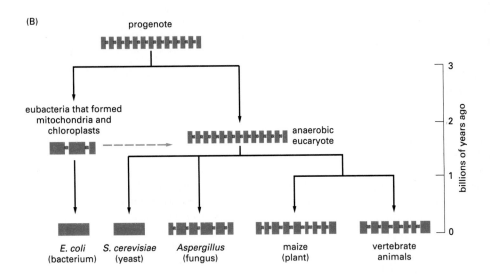

Figure 8–78 The ancient origin of split genes. (A) A comparison of the exon structure of the *triosephosphate isomerase* gene in plants and animals. The intron positions that are identical in maize (corn) and vertebrates are marked with *green arrowheads,* while the intron positions that differ are marked with *blue arrowheads.* Since plants and animals are thought to have diverged from a common ancestor about a billion years ago, the introns that they share must be of very ancient origin. (B) An outline of how a particular gene may have evolved. The exon sequences are shown in *red* and the intron sequences in *gray.* The gene illustrated here codes for a hypothetical protein that is required in all cells. Like *triosephosphate isomerase,* this protein is assumed to have evolved to its final three-dimensional structure before the eubacterial, archaebacterial, and eucaryotic lineages split off from a common ancestor cell—designated here as a "progenote." The dotted *green* line marks the approximate time of the endosymbiotic events that gave rise to mitochondria and chloroplasts (see pp. 20–21). (A, after W. Gilbert, M. Marchionni, and G. McKnight, *Cell* 46:151–154, 1987. © Cell Press.)

zyme in various organisms, it is possible to deduce that the enzyme evolved before the divergence of procaryotes and eucaryotes from a common ancestor; the human and bacterial amino acid sequences are 46% identical. The gene encoding the enzyme contains six introns in vertebrates (chickens and humans), and five of these are in precisely the same positions in maize. This implies that these five introns were present in the gene before plants and animals diverged in the eucaryotic lineage, an estimated 10^9 years ago (Figure 8–78).

In general, small unicellular organisms are under a strong selection pressure to reproduce by cell division at the maximum rate permitted by the levels of nutrients in the environment. To do this, they must minimize the amount of unnecessary DNA that they have to synthesize in each cell-division cycle. For larger organisms that live by predation, where size is an advantage, and for multicellular organisms in general, where rates of cell division are constrained by other requirements, there will not be such strong selection pressure to eliminate superfluous DNA from the genome. This argument may help to explain why bacteria should have lost their introns while eucaryotes have retained them. It also helps to explain why the multicellular fungus *Aspergillus* has five introns in its triosephosphate isomerase gene, whereas its unicellular relative, the yeast *Saccharomyces,* has none.

What is the mechanism by which introns are lost? Precise loss of introns would occur only rarely by piecemeal random deletions of short segments of DNA, yet precise and selective loss of introns seems not uncommon in eucaryotic cells (and perhaps was also frequent in the ancestors of bacteria). Whereas most vertebrates contain only a single insulin gene with two intron sequences, rats, for example, contain a second, neighboring insulin gene with only one intron. The second gene apparently arose by gene duplication relatively recently and subsequently lost one of its introns. Because intron loss requires the exact rejoining of DNA coding sequences, the most likely source of the information needed for such an event is an mRNA transcript of the original gene, from which the intron sequences will have been precisely removed. We know that messenger RNAs may be copied back into DNA through the activity of *reverse transcriptases* (see p. 282), and it is thought that recombination enzymes on occasion allow these DNA copies to become paired with the original sequence, which is then "corrected" to an intronless form by a gene-conversion type of event. This pathway of intron loss has been demonstrated in the laboratory using the powerful genetic tools available in the yeast *S. cerevisiae*.

Reverse transcriptases are not needed for the central genetic pathways, but they are produced in cells by specific transposable elements (see Table 8–4) as well as by all retroviruses. The generation of DNA copies of segments of the genome by reverse transcription has contributed in several ways to the evolution of the genomes of higher organisms, as we discuss later.

A Major Fraction of the DNA of Higher Eucaryotes Consists of Repeated, Noncoding Nucleotide Sequences [49]

Eucaryotic genomes contain not only introns but also large numbers of copies of other seemingly nonessential DNA sequences that do not code for protein. The presence of such repeated DNA sequences in higher eucaryotes was first revealed by a hybridization technique that measures the number of gene copies. In this procedure the genome is broken mechanically into short fragments of DNA double helix about 1000 nucleotide pairs long, and the fragments are then denatured to produce DNA single strands. The speed with which the single-stranded fragments in the mixture reanneal under conditions in which the double-helical conformation is stable depends on how long it takes each fragment to find a complementary fragment to pair with, which in turn depends on the concentration of suitable fragments in the mixture. For the most part, the reaction is very slow. The haploid genome of a mammalian cell, for example, is represented by about 6 million different 1000-nucleotide-long DNA fragments, and any fragment whose sequence is present in only one copy must randomly collide with 6 million noncomplementary strands for every complementary partner strand that it happens to find.

When the DNA from a human cell is analyzed in this way under conditions that require near perfect matching (high stringency conditions, see Figure 7–17), about 70% of the DNA strands reanneal as slowly as one would expect for a large collection of unique (nonrepeated) DNA sequences, requiring days for complete annealing. But most of the remaining 30% of the DNA strands anneal much more quickly. These strands contain sequences that are repeated many times in the genome, and they thus collide with a complementary partner relatively rapidly. Most of these highly repeated DNA sequences do not encode proteins, and they are of two types: about one-third are the tandemly repeated *satellite DNAs*, to be discussed next; the rest are *interspersed repeated DNAs*. As we shall see, most of the latter DNAs derive from a few transposable DNA sequences that have multiplied to especially high copy numbers in the human genome.

Satellite DNA Sequences Have No Known Function [50]

The most rapidly annealing DNA strands in an experiment of the type just described usually consist of very long tandem series of repetitions of a short nucleotide sequence (Figure 8–79). The repeat unit in a sequence of this type may be composed of only one or two nucleotides, but most repeats are longer, and in mammals they are typically composed of variants of a short sequence organized into a repeat of a few hundred nucleotides. These tandem repeats of a simple sequence are called **satellite DNAs** because the first DNAs of this type to be discovered had an unusual ratio of nucleotides that made it possible to separate them by density-gradient centrifugation from the bulk of the cell's DNA as a minor component (or "satellite"). Satellite DNA sequences generally are not transcribed and are located most often in the heterochromatin associated with the centromeric regions of chromosomes. In some mammals a single type of satellite DNA sequence constitutes 10% or more of the DNA and may even occupy a whole chromosome arm, so that the cell contains millions of copies of the basic repeated sequence.

The Organization and Evolution of the Nuclear Genome

Figure 8–79 Satellite DNA. A simple satellite DNA sequence from *Drosophila* is shown. It consists of many serially arranged repetitions of a sequence seven nucleotide pairs long, and it occurs millions of times in the *Drosophila* haploid genome.

Satellite DNA sequences seem to have changed unusually rapidly and even to have shifted their positions on chromosomes in the course of evolution. When two homologous mitotic chromosomes of any human are compared, for example, some of the satellite DNA sequences usually are found arranged in a strikingly different manner on the two chromosomes. Moreover, in contrast to the high degree of conservation of DNA sequences elsewhere in the genome, generally there are marked differences in the satellite DNA sequences of two closely related species. No function has yet been found for satellite DNA sequences: tests designed to demonstrate a role in chromosome pairing or nuclear organization have failed thus far to reveal any evidence for such a role. It has therefore been suggested that they are an extreme form of "selfish DNA" sequences, whose properties ensure their own retention in the genome but which do nothing to help the survival of the cells containing them. Other sequences that are commonly viewed as selfish are the *transposable elements*, which we discuss next.

The Evolution of Genomes Has Been Accelerated by Transposable Elements [51]

Genomes generally contain many varieties of **transposable elements.** These segments of DNA were first discovered in maize, where several have been sequenced and characterized. Eucaryotic transposable elements have been studied most extensively in *Drosophila*, where more than 30 varieties are known, varying in length between 2000 and 10,000 nucleotide pairs; most are present in 5 to 10 copies per diploid cell.

At least three broad classes of transposable elements can be distinguished by the peculiarities of their sequence organization (Table 8–4). Some elements move from place to place within chromosomes directly as DNA, while many others move via an RNA intermediate, as described in Chapter 6. In either case they

Table 8–4 Three Major Families of Transposable Elements

Structure	Genes in Complete Element	Mode of Movement	Examples
short inverted repeats at each end	encodes transposase	moves as DNA, either excising or following a replicative pathway	P element (*Drosophila*) Ac-Ds (maize) Tn3 and IS1 (*E. coli*) Tam3 (snapdragon)
directly repeated long terminal repeats (LTRs) at ends	encodes reverse transcriptase and resembles retrovirus	moves via an RNA intermediate produced by promoter in LTR	Copia (*Drosophila*) Ty (yeast) THE-1 (human) Bs1 (maize)
Poly A at 3′ end of RNA transcript; 5′ end is often truncated	encodes reverse transcriptase	moves via an RNA intermediate that is presumably produced from a neighboring promotor	F element (*Drosophila*) L1 (human) Cin4 (maize)

These elements range in length from 2000 to about 12,000 nucleotide pairs; each family contains many members, only a few of which are listed here.

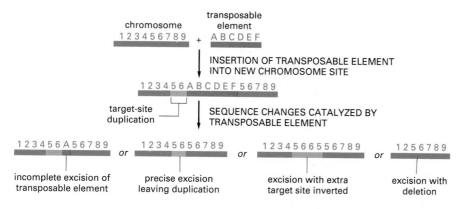

Figure 8–80 **Some changes in chromosomal DNA sequences caused by transposable elements.** The insertion of a transposable element always produces a short target-site duplication of the chromosomal sequence, which is generally 3 to 12 nucleotide pairs in length depending on the identity of the element. The site-specific recombination enzymes associated with the element can also cause its subsequent excision, which often fails to restore the original chromosomal DNA sequence, as in the four examples shown.

can multiply and spread from one site in a genome to a multitude of other sites, sometimes behaving as disruptive parasites.

Transposable elements seem to make up at least 10% of higher eucaryotic genomes. Although most of these elements move only very rarely, there are so many elements that their movement has a major effect on the variability of a species. More than half of the spontaneous mutations examined in *Drosophila*, for example, are due to the insertion of a transposable element in or near the mutant gene.

Mutations can occur either when an element inserts into a gene or when it exits to move elsewhere. All known transposable elements cause a short "target-site duplication" because of their mechanism of insertion (see Figure 6–70); when they exit, they generally leave behind part of this duplication—often with other local sequence changes as well (Figure 8–80). Thus, as transposable elements move in and out of chromosomes, they cause a variety of short additions and deletions of nucleotide sequences.

Transposable elements have also contributed to genome diversity in another way. When two transposable elements that are recognized by the same site-specific recombination enzyme (*transposase*) integrate into neighboring chromosomal sites, the DNA between them can become subject to transposition by the transposase. Because this provides a particularly effective pathway for the duplication and movement of exons (*exon shuffling*), these elements can help to create new genes (Figure 8–81).

Transposable Elements Often Affect Gene Regulation [52]

A DNA sequence rearrangement caused by a transposable element is often observed to alter the timing, level, or spatial pattern of expression of a nearby gene

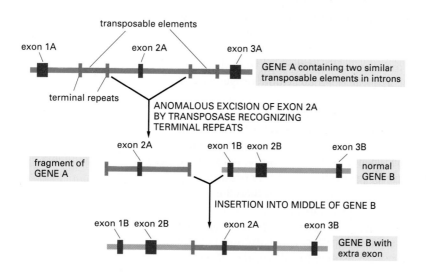

Figure 8–81 **An example of the exon shuffling that can be caused by transposable elements.** When two elements of the same type (*red* DNA) happen to insert near each other in a chromosome, the transposition mechanism may occasionally use the ends of two different elements (instead of the two ends of the same element) and thereby move the chromosomal DNA between them to a new chromosomal site. Since introns are very large relative to exons, the illustrated insertion of a new exon into a preexisting intron is a frequent outcome.

The Organization and Evolution of the Nuclear Genome

without affecting the sequence of the protein or RNA molecule that the gene encodes. This can change a subtle aspect of animal or plant development, such as the shape of an eye or a flower. While most of these changes in gene regulation would be expected to be detrimental to an organism, some of them will bring benefits and therefore tend to spread through the population by natural selection.

Effects on gene regulation are common, partly because the movement of a transposable element will generally bring with it new sequences that act as binding sites for sequence-specific DNA-binding proteins, including a transposase and the proteins that regulate the transcription of the transposable element DNA. These sequences can thereby act as regulatory sequences called **enhancers** (see p. 422) to affect the transcription of nearby genes. Similar effects commonly contribute to the evolution of cancer cells, where oncogenes can be created by the transposition of such regulatory sequences into the neighborhood of a proto-oncogene, as we discuss in Chapter 24.

The organization of higher eucaryotic genomes, with long noncoding DNA sequences interspersed with comparatively short coding sequences, provides an accommodating "playground" for the integration and excision of mobile DNA sequences. Because gene transcription can be regulated from distances that are tens of thousands of nucleotide pairs away from a promoter, many of the resulting changes in the genome would be expected to affect gene expression; by contrast, relatively few would be expected to disrupt the short exons that contain the coding sequences.

Might the vast excess of noncoding DNA in higher eucaryotes have been favored by selection during evolution because of the regulatory flexibility that it has provided to organisms with a large variety of transposable elements? What is known about the regulatory systems that control higher eucaryotic genes is consistent with this possibility. Enhancers, like exons, seem to function as separate modules, and the activity of a gene depends on a summation of the influences received at its promoter from a set of enhancers (see Figure 9–44). Transposable elements, by moving such enhancer modules around in a genome, may allow gene regulation to be optimized for the long-term survival of the organism.

Transposition Bursts Cause Cataclysmic Changes in Genomes and Increase Biological Diversity [53]

Another unique feature that distinguishes transposable elements as mutagens is their tendency to undergo long quiescent periods, during which they remain fixed in their chromosomal positions, followed by a period of intense movement. Their transposition, and therefore their mutagenic action, is activated from time to time in a few individuals in a population of organisms. Such cataclysmic changes in genomes, called *transposition bursts*, can involve near simultaneous transpositions of several types of transposable elements. Transposition bursts were first observed in developing maize plants that were subjected to repeated chromosome breakage. They also are observed in crosses between certain strains of flies—a phenomenon known as *hybrid dysgenesis*. When they occur in the germ line, they induce multiple changes in the genome of an individual progeny fly or plant.

By simultaneously changing several properties of an organism, transposition bursts increase the probability that two new traits that are useful together but of no selective value by themselves will appear in a single individual in a population. In several types of plants there is evidence that transposition bursts can be activated by a severe environmental stress, generating a variety of randomly modified progeny organisms, some of which may be better suited than the parent to survive in the new conditions. It seems that, at least in these plants, a mechanism has evolved to activate transposable elements to serve as mutagens that produce an enhanced range of variant organisms when this variation is most needed. Thus transposable elements are not necessarily just disruptive parasites;

rather, they may on occasion act as useful symbionts that aid the long-term survival of the species whose genomes they inhabit.

About 10% of the Human Genome Consists of Two Families of Transposable Elements [54]

Primate DNA is unusual in at least one respect: it contains a remarkably large number of copies of two transposable DNA sequences that seem to have overrun our chromosomes. Both of these sequences move by an RNA-mediated process that requires a reverse transcriptase. One is the **L1 transposable element,** which resembles the F element in *Drosophila* and the Cin4 element in maize and encodes a reverse transcriptase (see Table 8–4, p. 392). Transposable elements have generally evolved with feedback control systems that severely limit their numbers in each cell (thereby saving the cell from potential disaster); the L1 element in humans, however, constitutes about 4% of the mass of the genome.

Even more abundant is the ***Alu* sequence,** which is very short (about 300 nucleotide pairs) and moves like a transposable element, creating target-site duplications when it inserts. It was derived, however, from an internally deleted host-cell RNA gene (7SL), which encodes the RNA component of the signal-recognition particle (SRP) that functions in protein synthesis (see Figure 12–39); it is therefore not clear whether the *Alu* sequence should be considered a transposable element or an unusually mobile pseudogene. It is present in about 500,000 copies in the haploid genome and constitutes about 5% of human DNA; thus it is present on average about once every 5000 nucleotide pairs. The *Alu* DNA is transcribed from the 7SL RNA promoter, a polymerase-III promoter that is internal to the transcript, so that it carries the information necessary for its own transcription wherever it moves. It needs a reverse transcriptase encoded elsewhere, however, to transpose.

Comparisons of the sequence and locations of the L1- and *Alu*-like sequences in different mammals suggest that these sequences have multiplied to high copy numbers relatively recently (Figure 8–82). It is hard to imagine that these highly abundant sequences scattered throughout our genome have not had major effects on the expression of many nearby genes. How many of our uniquely human qualities, for example, do we owe to these parasitic elements?

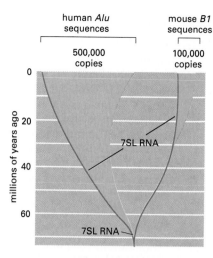

Figure 8–82 The proposed pattern of evolution of the abundant *Alu* sequence found in the human genome. A related transposable element, *B1*, is found in the mouse genome. Both of these transposable DNA sequences are thought to have evolved from the essential 7SL RNA gene. Based on the species distribution and sequence homology of these highly repeated elements, however, the major expansion in copy numbers seems to have occurred independently in mice and humans. (Adapted from P.L. Deininger and G.R. Daniels, *Trends Gen.* 2:76–80, 1986.)

Summary

The functional DNA sequences in the genomes of higher eucaryotes appear to be constructed from small genetic modules of at least two kinds. Modules of coding sequence are combined in many ways to produce proteins, whereas modules of regulatory sequences are scattered throughout long stretches of noncoding sequences and regulate the expression of genes. Both the coding sequences and the regulatory sequences are typically present in modules that are less than a few hundred nucleotide pairs long, which together account for only a small proportion of the total DNA.

A variety of genetic-recombination processes occur in genomes, causing the random duplication and translocation of DNA sequences. Some of these changes create duplicates of entire genes, which can then evolve new functions. Others produce new proteins by shuffling exons or alter the expression of old genes by exposing them to new regulatory sequences. This type of DNA sequence shuffling, which is of great importance for the evolution of organisms, is greatly facilitated by the split structure of higher eucaryotic genes and by the fact that these genes are often controlled by distant regulatory sequences.

Many types of transposable elements are present in genomes. Collectively, they constitute more than 10% of the mass of both Drosophila *and vertebrate genomes. Occasionally, transposition bursts occur in germ cells and cause many heritable changes in gene expression in the same individual. Transposable elements are thought to have had a special evolutionary role in the generation of organismal diversity.*

References

Cited

1. Adolph, K.W., ed. Chromosomes and Chromatin, Vols. 1–3. Boca Raton, FL: CRC Press, 1988.

   Felsenfeld, G. DNA. *Sci. Am.* 253(4):58–67, 1985.

   Hsu, T.C. Human and Mammalian Cytogenetics: A Historical Perspective. New York: Springer-Verlag, 1979.

   Stewart, A. The functional organization of chromosomes and the nucleus—a special issue. *Trends Genet.* 6:377–379, 1990.

2. Burke, D.T.; Carle, G.F.; Olson, M.V. Cloning of large segments of exogenous DNA into yeast by means of artificial chromosome vectors. *Science* 236:806–812, 1987.

   DePamphilis, M.L. Origins of DNA replication that function in eucaryotic cells. *Curr. Opin. Cell Biol.* 5:434–441, 1993.

   Murray, A.W. Chromosome structure and behavior. *Trends Biochem. Sci.* 10:112–115, 1985.

   Price, C.M. Centromeres and telomeres. *Curr. Opin. Cell Biol.* 4:379–384, 1992.

3. Gall, J.G. Chromosome structure and the C-value paradox. *J. Cell Biol.* 91:3s–14s, 1981.

   Ohta, T.; Kimura, M. Functional organization of genetic material as a product of molecular evolution. *Nature* 233:118–119, 1971.

4. Ruby, S.W.; Abelson, J. Pre-mRNA splicing in yeast. *Trends Genet.* 7:79–85, 1991.

5. Wilson, A.C.; Ochman, H.; Prager, E.M. Molecular time scale for evolution. *Trends Genet.* 3:241–247, 1987.

6. Grunstein, M. Histones as regulators of genes. *Sci. Am.* 267(4):68–74B, 1992.

   Isenberg, I. Histones. *Annu. Rev. Biochem.* 48:159–191, 1979.

   Smith, M.M. Histone structure and function. *Curr. Opin. Cell Biol.* 3:429–437, 1991.

   Wells, D.E. Compilation analysis of histones and histone genes. *Nucleic Acids Res.* 14:r119–r149, 1986.

7. Chromatin. *Cold Spring Harbor Symp. Quant. Biol.*, Vol. 42, 1978.

   Kornberg, R.D.; Klug, A. The nucleosome. *Sci. Am.* 244(2):52–64, 1981.

   McGhee, J.D.; Felsenfeld, G. Nucleosome structure. *Annu. Rev. Biochem.* 49:1115–1156, 1980.

   Richmond, T.J.; Finch, J.T.; Rushton, B.; Rhodes, D.; Klug, A. Structure of the nucleosome core particle at 7Å resolution. *Nature* 311:532–537, 1984.

8. Gross, D.S.; Garrard, W.T. Nuclease hypersensitive sites in chromatin. *Annu. Rev. Biochem.* 57:159–198, 1988.

   Simpson, R.T. Nucleosome position *in vivo* and *in vitro*. *Bioessays* 4:172–176, 1986.

   Svaren, J.; Chalkley, R. The structure and assembly of active chromatin. *Trends Genet.* 6:52–56, 1990.

   Travers, A.A. DNA bending and nucleosome positioning. *Trends Biochem. Sci.* 12:108–112, 1987.

   Zhang, L.; Gralla. J.D. *In situ* nucleoprotein structure involving origin-proximal SV40 DNA control elements. *Nucleic Acids Res.* 18:1797–1803, 1990.

9. Hansen, J.C.; Ausio, J. Chromatin dynamics and the modulation of genetic activity. *Trends Biochem. Sci.* 17:187–191, 1992.

   Pederson, D.S.; Thoma, F.; Simpson, R. Core particle, fiber, and transcriptionally active chromatin structure. *Annu. Rev. Cell Biol.* 2:117–147, 1986.

   Swedlow, J.R.; Agard, D.A.; Sedat, J.W. Chromosome structure inside the nucleus. *Curr. Opin. Cell Biol.* 5:412–416, 1993.

   Zlatanova, J.; Yaneva, J. Histone H1–DNA interactions and their relation to chromatin structure and function. *DNA Cell Biol.* 10:239–248, 1991.

10. Bellini, M.; Lacroix, J.-C.; Gall, J.G. A putative zinc-binding protein on lampbrush chromosome loops. *EMBO J.* 12:107–114, 1993.

    Bostock, C.J.; Sumner, A.T. The Eucaryotic Chromosome, pp. 347–374. Amsterdam: North-Holland, 1978.

    Callan, H.G. Lampbrush chromosomes. *Proc. R. Soc. Lond. (Biol.)* 214:417–448, 1982.

    Reznik, N.A.; Yampol, G.P.; Kiseleva, E.V.; Khristolyubova, N.B.; Gruzdev, A.D. Functional and structural units in the chromomere. *Genetica* 83:293–299, 1991.

11. Agard, D.A.; Sedat, J.W. Three-dimensional architecture of a polytene nucleus. *Nature* 302:676–681, 1983.

    Beermann, W. Chromosomes and genes. In Developmental Studies on Giant Chromosomes (W. Beermann, ed.), pp. 1–33. New York: Springer-Verlag, 1972.

    Zhimulev, I.F.; Belyaeva, E.S. Chromomeric organization of polytene chromosomes. *Genetica* 85:65–72, 1991.

12. Ashburner, M.; Chihara, C.; Meltzer, P.; Richards, G. Temporal control of puffing activity in polytene chromosomes. *Cold Spring Harbor Symp. Quant. Biol.* 38:655–662, 1974.

    Lamb, M.M.; Daneholt, B. Characterization of active transcription units in Balbiani rings of *Chironomous tentans*. *Cell* 17:835–848, 1979.

    Thummel, C.S. Puffs and gene regulation—molecular insights into the *Drosophila* ecdysone regulatory hierarchy. *Bioessays* 12:561–568, 1990.

13. Bossy, B.; Hall, L.M.; Spierer, P. Genetic activity along 315 kb of the *Drosophila* chromosome. *EMBO J.* 3:2537–2541, 1984.

    Friedman, T.B.; Owens, K.N.; Burnett, J.B.; Saura, A.O.; Wallrath, L.L. The faint band/interband region 28C2 to 28C4–5(−) of the *Drosophila melanogaster* salivary gland polytene chromosomes is rich in transcripts. *Mol. Gen. Genet.* 226:81–87, 1991.

    Judd, B.H.; Young, M.W. An examination of the one cistron: one chromomere concept. *Cold Spring Harbor Symp. Quant. Biol.* 38:573–579, 1974.

14. Croston, G.E.; Kadonaga, J.T. Role of chromatin structure in the regulation of transcription by RNA polymerase II. *Curr. Opin. Cell Biol.* 5:417–423, 1993.

    Nacheva, G.A.; Guschin, D.Y.; Preobrazhenskaya, O.V.; et al. Change in the pattern of histone binding to DNA upon transcriptional activation. *Cell* 58:27–36, 1989.

    Svaren, J.; Horz, W. Histones, nucleosomes and transcription. *Curr. Opin. Genet. Dev.* 3:219–225, 1993.

15. Allis, C.D., et al. hv1 is an evolutionarily conserved H2A variant that is preferentially associated with active genes. *J. Biol. Chem.* 261:1941–1948, 1986.

    Bradbury, E.M. Reversible histone modifications and the chromosome cell cycle. *Bioessays* 14:9–16, 1992.

Tremethick, D.J.; Drew, H.R. High mobility group proteins 14 and 17 can space nucleosomes *in vitro. J. Biol. Chem.* 268:11389–11393, 1993.

Turner, B.M. Histone acetylation and control of gene expression. *J. Cell Sci.* 99:13–20, 1991.

16. Brown, S.W. Heterochromatin. *Science* 151:417–425, 1966.

Pimpinelli, S.; Bonaccorsi, S.; Gatti, M.; Sandler, L. The peculiar genetic organization of *Drosophila* heterochromatin. *Trends Genet.* 2:17–20, 1986.

17. Georgiev, G.P.; Nedospasov, S.A.; Bakayev, V.V. Supranucleosomal levels of chromatin organization. In The Cell Nucleus (H. Busch, ed.), Vol. 6, pp. 3–34. New York: Academic Press, 1978.

Kitsberg, D.; Selig, S.; Cedar, H. Chromosome structure and eukaryotic gene organization. *Curr. Opin. Genet. Dev.* 1:534–537, 1991.

Marsden, M.; Laemmli, U.K. Metaphase chromosome structure: evidence for a radial loop model. *Cell* 17:849–858, 1979.

18. Bickmore, W.A.; Sumner, A.T. Mammalian chromosome banding—an expression of genome organization. *Trends Genet.* 5:144–148, 1989.

Holmquist, G. DNA sequences in G-bands and R-bands. In Chromosomes and Chromatin Structure (K.W. Adolph, ed.), Vol. 2, pp. 75–122. Boca Raton, FL: CRC Press, 1988.

Lewin, B. Gene Expression, Vol. 2: Eucaryotic Chromosomes, 2nd ed., pp. 428–440. New York: Wiley, 1980.

19. DePamphilis, M.L. Origins of DNA replication that function in eukaryotic cells. *Curr. Opin. Cell Biol.* 5:434–441, 1993.

Marahrens, Y.; Stillman, B. A yeast chromosomal origin of DNA replication defined by multiple functional elements. *Science* 255:817–823, 1992.

Struhl, K.; Stinchcomb, D.T.; Sherer, S.; Davis. R.W. High-frequency transformation of yeast: autonomous replication of hybrid DNA molecules. *Proc. Natl. Acad. Sci. USA* 76:1035–1039, 1979.

20. Dodson, M.; Dean, F.B.; Bullock, P.; Echols, H.; Hurwitz, J. Unwinding of duplex DNA from the SV40 origin of replication by T antigen. *Science* 238:964–967, 1987.

Stillman, B.; Bell, S.P.; Dutta, A.; Marahrens, Y. DNA replication and the cell cycle. *Ciba Found. Symp.* 170:147–156; discussion 156–160, 1992.

Tsurimoto, T.; Stillman, B. Replication factors required for SV40 DNA replication *in vitro.* I. DNA structure-specific recognition of a primer-template junction by eukaryotic DNA polymerases and their accessory proteins. *J. Biol. Chem.* 266:1950–1960, 1991.

21. Bonifer, C.; Hecht, A.; Saueressig, H.; Winter, D.M.; Sippel, A.E. Dynamic chromatin: the regulatory domain organization of eukaryotic gene loci. *J. Cell. Biochem.* 47:99–108, 1991.

Huberman, J.A.; Riggs, A.D. On the mechanism of DNA replication in mammalian chromosomes. *J. Mol. Biol.* 32:327–341, 1968.

22. Craig, J.M.; Bickmore, W.A. Chromosome bands—flavours to savour. *Bioessays* 15:349–354, 1993.

Stubblefield, E. Analysis of the replication pattern of Chinese hamster chromosomes using 5-bromodeoxyuridine suppression of 33258 Hoechst fluorescence. *Chromosoma* 53:209–221, 1975.

23. Lima-de-Faria, A.; Jaworska, H. Late DNA synthesis in heterochromatin. *Nature* 217:138–142, 1968.

24. Bickmore, W.A.; Sumner, A.T. Mammalian chromosome banding—an expression of genome organization. *Trends Genet.* 5:144–148, 1989.

25. Hofmann, A.; Montag, M.; Steinbeisser, H.; Trendelenburg, M.F. Plasmid and bacteriophage lambda-DNA show differential replication characteristics following injection into fertilized eggs of *Xenopus laevis*: dependence on period and site of injection. *Cell Differ. Dev.* 30:77–85, 1990.

Mechali, M.; Kearsey, S. Lack of specific sequence requirement for DNA replication in *Xenopus* eggs compared with high sequence specificity in yeast. *Cell* 38:55–64, 1984.

Riggs, A.D. DNA methylation and late replication probably aid cell memory, and type I DNA reeling could aid chromosome folding and enhancer function. *Phil. Trans. R. Soc. Lond. (Biol.)* 326:285–297, 1990.

26. Blow, J.J. Preventing re-replication of DNA in a single cell cycle: evidence for a replication licensing factor. *J. Cell Biol.* 122:993–1002, 1993.

Coverley, D.; Downes, C.S.; Romanowski, P.; Laskey, R.A. Reversible effects of nuclear membrane permeabilization on DNA replication: evidence for a positive licensing factor. *J. Cell Biol.* 122:985–992, 1993.

Rao, P.N., Johnson, R.T. Mammalian cell fusion: studies on the regulation of DNA synthesis and mitosis. *Nature* 225:159–164, 1970.

Wanka, F. Control of eukaryotic DNA replication at the chromosomal level. *Bioessays* 13:613–618, 1991.

27. Almouzni, G.; Clark, D.J.; Mechali, M.; Wolffe, A.P. Chromatin assembly on replicating DNA *in vitro*. *Nucleic Acids Res.* 18:5767–5774, 1990.

Randall, S.K.; Kelly, T.J. The fate of parental nucleosomes during SV40 DNA replication. *J. Biol. Chem.* 267:14259–14265, 1992.

Russev, G.; Hancock, R. Assembly of new histones into nucleosomes and their distribution in replicating chromatin. *Proc. Natl. Acad. Sci. USA* 79:3143–3147, 1982.

28. Blackburn, E.H. Telomeres. *Trends Biochem. Sci.* 16:378–381, 1991.

Blackburn, E.H.; Szostak, J.W. The molecular structure of centromeres and telomeres. *Annu. Rev. Biochem.* 53:163–194, 1984.

Vogt, P. Potential genetic functions of tandem repeated DNA sequence blocks in the human genome are based on a highly conserved "chromatin folding code." *Hum. Genet.* 84:301–336, 1990.

29. Chamberlin, M. Bacterial DNA-dependent RNA polymerases. In The Enzymes, 3rd ed. (P. Boyer, ed.), Vol. 15B, pp. 61–108. New York: Academic Press, 1982.

Gill, S.C.; Yager, T.D.; von Hippel, P.H. Thermodynamic analysis of the transcription cycle in *E. coli. Biophys. Chem.* 37:239–250, 1990.

Kerppola, T.K.; Kane, C.M. RNA polymerase: regulation of transcript elongation and termination. *FASEB J.* 5:2833–2842, 1991.

30. Chambon, P. Eucaryotic nuclear RNA polymerases. *Annu. Rev. Biochem.* 44:613–638, 1975.

Geiduschek, E.P.; Tocchini-Valentini, G.P. Transcription by RNA polymerase III. *Annu. Rev. Biochem.* 57:873–914, 1988.

References

Sentenac, A. Eucaryotic RNA polymerases. *CRC Crit. Rev. Biochem.* 18:31–91, 1985.

Young, R.A. RNA polymerase II. *Annu. Rev. Biochem.* 60:689–715, 1991.

31. Foe, V.E.; Wilkinson, L.E.; Laird, C.D. Comparative organization of active transcription units in *Oncopeltus fasciatus. Cell* 9:131–146, 1976.

Lewin, B. Gene Expression, Vol. 2: Eucaryotic Chromosomes, 2nd ed., pp. 708–719. New York: Wiley, 1980.

Miller, O.L. The nucleolus, chromosomes, and visualization of genetic activity. *J. Cell Biol.* 91:15s–27s, 1981.

32. Nevins, J.R. The pathway of eukaryotic mRNA formation. *Annu. Rev. Biochem.* 52:441–466, 1983.

Proudfoot, N.J. How RNA polymerase II terminates transcription in higher eukaryotes. *Trends Biochem. Sci.* 14:105–110, 1989.

Takgaki, V.; Ryner, L.C.; Manley, J.L. Separation and characterization of a poly(A) polymerase and a cleavage/specificity factor required for pre-mRNA polyadenylation. *Cell* 52:731–742, 1988.

Wickens, M. How the messenger got its tail: addition of poly(A) in the nucleus. *Trends Biochem. Sci.* 15:277–281, 1990.

33. Crick, F. Split genes and RNA splicing. *Science* 204:264–271, 1979.

Dreyfuss, G.; Matunis, M.J.; Piñol-Roma, S.; Burd, C.G. hnRNP proteins and the biogenesis of mRNA. *Annu. Rev. Biochem.* 62:289–321, 1993.

Hickey, D.A.; Benkel, B.F.; Abukashawa, S.M. A general model for the evolution of nuclear pre-mRNA introns. *J. Theor. Biol.* 137:41–53, 1989.

34. Dreyfuss, G.; Swanson, M.S.; Piñol-Roma, S. Heterogeneous nuclear ribonucleoprotein particles and the pathway of mRNA formation. *Trends Biochem. Sci.* 13:86–91, 1988.

Guthrie, C. Messenger RNA splicing in yeast: clues to why the spliceosome is a ribonucleoprotein. *Science* 253:157–163, 1991.

Guthrie, C.; Patterson, B. Spliceosomal snRNAs. *Annu. Rev. Genet.* 22:387–419, 1988.

Samarina, O.P.; Krichevskaya, A.A.; Georgiev, G.P. Nuclear ribonucleoprotein particles containing messenger ribonucleic acid. *Nature* 210:1319–1322, 1966.

Steitz, J.A. "Snurps." *Sci. Am.* 258(6):56–63, 1988.

35. Balvay, L.; Libri, D.; Fiszman, M.Y. Pre-mRNA secondary structure and the regulation of splicing. *Bioessays* 15:165–169, 1993.

Padgett, R.A.; Grabowski, P.J.; Konarska, M.M.; Seiler, S.; Sharp, P.A. Splicing of messenger RNA precursors. *Annu. Rev. Biochem.* 55:1119–1150, 1986.

36. Aebi, M.; Weissman, C. Precision and orderliness in splicing. *Trends Genet.* 3:102–107, 1987.

Goguel, V.; Liao, X.; Rymond, B.C.; Rosbash, M. U1 snRNP can influence 3′-splice site selection as well as 5′-splice site selection. *Genes Dev.* 5:1430–1438, 1991.

37. Orkin, S.H.; Kazazian, H.H. The mutation and polymorphism of the human β-globin gene and its surrounding DNA. *Annu. Rev. Genet.* 18:131–171, 1984.

Vidaud, M.; Gattoni, R.; Stevenin, J.; et al. A 5′ splice-region G→C mutation in exon 1 of the human beta-globin gene inhibits pre-mRNA splicing: a mecha-

nism for β$^+$-thalassemia. *Proc. Natl. Acad. Sci. USA* 86:1041–1045, 1989.

38. Cech, T.R. The generality of self-splicing RNA: relationship to nuclear mRNA splicing. *Cell* 44:207–210, 1986.

Saldanha, R.; Mohr, G.; Belfort, M.; Lambowitz, A.M. Group I and group II introns. *FASEB J.* 7:15–24, 1993.

39. Brown, J.D.; Plumpton, M.; Beggs, J.D. The genetics of nuclear pre-mRNA splicing: a complex story. *Antonie Van Leeuwenhoek* 62:35–46, 1992.

Green, M.R. Pre-mRNA processing and mRNA nuclear export. *Curr. Opin. Cell Biol.* 1:519–525, 1989.

Jimenez-Garcia, L.F.; Spector. D.L. *In vivo* evidence that transcription and splicing are coordinated by a recruiting mechanism. *Cell* 73:47–59, 1993.

Spector, D.L. Nuclear organization of pre-mRNA processing. *Curr. Opin. Cell Biol.* 5:442–447, 1993.

40. Long, E.O.; Dawid, I.B. Repeated genes in eucaryotes. *Annu. Rev. Biochem.* 49:727–764, 1980.

Miller, O.L. The nucleolus, chromosomes, and visualization of genetic activity. *J. Cell Biol.* 91:15s–27s, 1981.

Williams, S.M.; Robbins, L.G. Molecular genetic analysis of *Drosophila* rDNA arrays. *Trends Genet.* 8:335–340, 1992.

41. Fischer, D.; Weisenberger, D.; Scheer, U. Assigning functions to nucleolar structures. *Chromosoma* 101:133–140, 1991.

Hadjiolov, A.A. The Nucleolus and Ribosome Biogenesis. New York: Springer-Verlag, 1985.

Jordan, E.G.; Cullis, C.A., eds. The Nucleolus. Cambridge, UK: Cambridge University Press, 1982.

Larson, D.E.; Zahradka, P.; Sells, B.H. Control points in eucaryotic ribosome biogenesis. *Biochem. Cell Biol.* 69:5–22, 1991.

42. Anastassova-Kristeva, M. The nucleolar cycle in man. *J. Cell Sci.* 25:103–110, 1977.

McClintock, B. The relation of a particular chromosomal element to the development of the nucleoli in Zea Mays. *Z. Zellforsch. Mikrosk. Anat.* 21:294–323, 1934.

43. Cremer, T., et al. Rabl's model of the interphase chromosome arrangement tested in Chinese hamster cells by premature chromosome condensation and laser-UV-microbeam experiments. *Hum. Genet.* 60:46–56, 1980.

Haaf, T.; Schmid, M. Chromosome topology in mammalian interphase nuclei. *Exp. Cell Res.* 192:325–332, 1991.

Hochstrasser, M.; Sedat, J.W. Three-dimensional organization of *Drosophila melanogaster* interphase nuclei. II. Chromosome spatial organization and gene regulation. *J. Cell Biol.* 104:1471–1483, 1987.

Manuelidis, L. A view of interphase chromosomes. *Science* 250:1533–1540, 1990.

44. Dessev, G.N. Nuclear envelope structure. *Curr. Opin. Cell Biol.* 4:430–435, 1992.

Gasser, S.M.; Laemmli, U.K. A glimpse at chromosome order. *Trends Genet.* 3:16–22, 1987.

Gerace, L.; Burke, B. Functional organization of the nuclear envelope. *Annu. Rev. Cell Biol.* 4:335–374, 1988.

Newport, J.W.; Forbes, D.J. The nucleus: structure, function, and dynamics. *Annu. Rev. Biochem.* 56:535–566, 1987.

45. Doolittle, R.F. Proteins. *Sci. Am.* 253(4):88–99, 1985.

Holland, S.K.; Blake, C.C. Proteins, exons, and molecular evolution. *Biosystems* 20:181–206, 1987.

Maeda, N.; Smithies, O. The evolution of multigene families: human haptoglobin genes. *Annu. Rev. Genet.* 20:81–108, 1986.

46. Kourilsky, P. Molecular mechanisms for gene conversion in higher cells. *Trends Genet.* 2:60–63, 1986.

Roth, D.B.; Porter, T.N.; Wilson, J.H. Mechanisms of nonhomologous recombination in mammalian cells. *Mol. Cell. Biol.* 5:2599–2607, 1985.

Smith, G.P. Evolution of repeated DNA sequences by unequal crossover. *Science* 191:528–535, 1976.

Stark, G.R.; Wahl, G.M. Gene amplification. *Annu. Rev. Biochem.* 53:447–491, 1984.

47. Dickerson, R.E.; Geis, I. Hemoglobin: Structure, Function, Evolution, and Pathology. Menlo Park, CA: Benjamin-Cummings, 1983.

Efstratiadis, A., et al. The structure and evolution of the human β-globin gene family. *Cell* 21:653–668, 1980.

48. Anderson, C.L.; Carew, E.A.; Powell, J.R. Evolution of the Adh locus in the *Drosophila willistoni* group: the loss of an intron, and shift in codon usage. *Mol. Biol. Evol.* 10:605–618, 1993.

Doolittle, W.F. RNA-mediated gene conversion? *Trends Genet.* 1:64–65, 1985.

Nyberg, A.M.; Cronhjort, M.B. Intron evolution: a statistical comparison of two models. *J. Theor. Biol.* 157:175–190, 1992.

49. Britten, R.J.; Kohne, D.E. Repeated sequences in DNA. *Science* 161:529–540, 1968.

Jelinek, W.R.; Schmid, C.W. Repetitive sequences in eukaryotic DNA and their expression. *Annu. Rev. Biochem.* 51:813–844, 1982.

50. Craig-Holmes, A.P.; Shaw, M.W. Polymorphism of human constitutive heterochromatin. *Science* 174:702–704, 1971.

Hsu, T.C. Human and Mammalian Cytogenetics: A Historical Perspective. New York: Springer-Verlag, 1979.

John, B.; Miklos, G.L.G. Functional aspects of satellite DNA and heterochromatin. *Int. Rev. Cytol.* 58:1–114, 1979.

Orgel, L.E.; Crick, F.H.C. Selfish DNA: the ultimate parasite. *Nature* 284:604–607, 1980.

51. Berg, D.E.; Howe, M.M., eds. Mobile DNA. Washington, DC: American Society for Microbiology, 1989.

Döring, H.-P.; Starlinger, P. Molecular genetics of transposable elements in plants. *Annu. Rev. Genet.* 20:175–200, 1986.

Finnegan, D.J. Eukaryotic transposable elements and genome evolution. *Trends Genet.* 5:103–107, 1989.

McClintock, B. Controlling elements and the gene. *Cold Spring Harbor Symp. Quant. Biol.* 21:197–216, 1956.

52. Coen, E.S.; Carpenter, R. Transposable elements in *Antirrhinum majus*: generators of genetic diversity. *Trends Genet.* 2:292–296, 1986.

Georgiev, G.P. Mobile genetic elements in animal cells and their biological significance. *Eur. J. Biochem.* 145:203–220, 1984.

O'Kane, C.J.; Gehring, W.J. Detection *in situ* of genomic regulatory elements in *Drosophila. Proc. Natl. Acad. Sci. USA* 84:9123–9127, 1987.

53. Gerasimova, T.I.; Mizrokhi, L.J.; Georgiev, G.P. Transposition bursts in genetically unstable *Drosophila melanogaster. Nature* 309:714–716, 1984.

McClintock, B. The significance of responses of the genome to challenge. *Science* 226:792–801, 1984.

Walbot, V.; Cullis, C.A. Rapid genomic change in higher plants. *Annu. Rev. Plant Physiol.* 36:367–396, 1985.

54. Deininger, P.L.; Daniels, G.R. The recent evolution of mammalian repetitive DNA elements. *Trends Genet.* 2:76–80, 1986.

Ruffner, D.E.; Sprung, C.N.; Minghetti, P.P.; Gibbs, P.E.; Dugaiczyk, A. Invasion of the human albumin-α-fetoprotein gene family by Alu, Kpn, and two novel repetitive DNA elements. *Mol. Biol. Evol.* 4:1–9, 1987.

Weiner, A.M.; Deininger, P.L.; Estratiadis, A. Nonviral retroposons: genes, pseudogenes, and transposable elements generated by the reverse flow of genetic information. *Annu. Rev. Biochem.* 55:631–661, 1986.

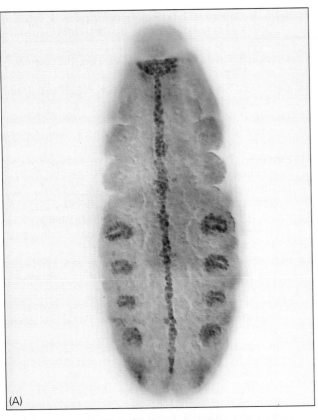

(A)

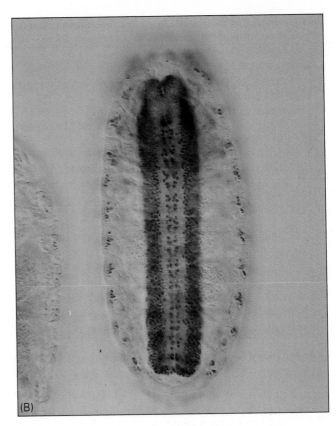

(B)

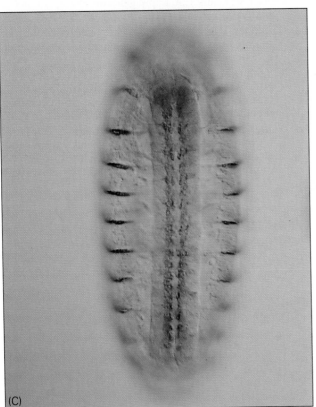

(C)

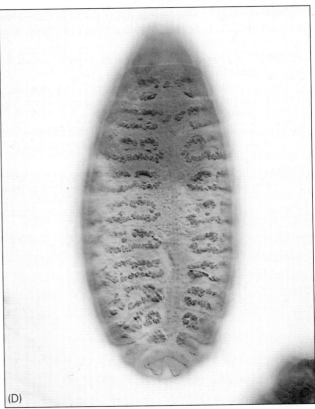

(D)

Position-dependent patterns of gene expression in four different transgenic *Drosophila* embryos. Each embryo has a single copy of a bacterial β-galactosidase gene inserted into its genome. Depending on the chromosomal insertion site, nearby *Drosophila* enhancers cause the gene to be expressed in a pattern that corresponds either to (A) trachea and mesectoderm, (B) nervous system, (C) glial cells of the central nervous system plus a segmentally repeating pattern, or (D) muscle. (Courtesy of Yuh-Nung Jan.)

Control of Gene Expression

An organism's DNA encodes all of the RNA and protein molecules required to construct its cells. Yet a complete description of the DNA sequence of an organism—be it the few million nucleotides of a bacterium or the few billion nucleotides of a human—would no more enable us to reconstruct the organism than a list of English words would enable us to reconstruct a play by Shakespeare. In both cases the problem is to know how the elements in the DNA sequence or the words on the list are used. Under what conditions is each gene product made, and, once made, what does it do?

In this chapter we discuss the first half of this problem—rules by which a subset of the genes are selectively expressed in each cell. The mechanisms that control the expression of genes operate at a variety of levels, and we discuss the different levels in turn. We begin, however, with an overview of some basic principles of gene control in multicellular organisms.

An Overview of Gene Control

The different cell types in a multicellular organism differ dramatically in both structure and function. If we compare a mammalian neuron with a lymphocyte, for example, the differences are so extreme that it is difficult to imagine that the two cells contain the same genome. For this reason, and because cell differentiation is usually irreversible, biologists originally suspected that genes might be selectively lost when a cell differentiates. We now know, however, that cell differentiation generally depends on changes in gene expression rather than on gene loss.

The Different Cell Types of a Multicellular Organism Contain the Same DNA [1]

The cell types in a multicellular organism become different from one another because they synthesize and accumulate different sets of RNA and protein molecules. They generally do this without altering the sequence of their DNA. The best evidence for the preservation of the genome during cell differentiation comes from a classic set of experiments in frogs. When the nucleus of a fully differentiated frog cell is injected into a frog egg whose nucleus has been removed, the injected donor nucleus is capable of programming the recipient egg to produce

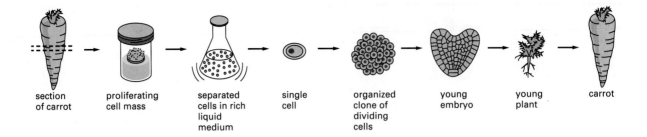

section of carrot → proliferating cell mass → separated cells in rich liquid medium → single cell → organized clone of dividing cells → young embryo → young plant → carrot

a normal tadpole. Because the tadpole contains a full range of differentiated cells that derived their DNA sequences from the nucleus of the original donor cell, it follows that the differentiated donor cell cannot have lost any important DNA sequences. A similar conclusion has been reached in experiments done with various plants. Here differentiated pieces of plant tissue are placed in culture and then dissociated into single cells. Often, one of these individual cells can regenerate an entire adult plant (Figure 9–1).

Further evidence that large blocks of DNA are not lost or rearranged during vertebrate development comes from comparing the detailed banding patterns detectable in condensed chromosomes at mitosis (see Figure 8–32). By this criterion the chromosome sets of all differentiated cells in the human body appear to be identical. Moreover, comparisons of the genomes of different cells based on recombinant DNA technology have shown, as a general rule, that the changes in gene expression that underlie the development of multicellular organisms are not accompanied by changes in the DNA sequences of the corresponding genes (for an important exception, however, see Figure 23–27).

Different Cell Types Synthesize Different Sets of Proteins [2]

As a first step in trying to understand cell differentiation, one would like to know how many differences there are between any one cell type and another. Although we still do not know the answer to this fundamental question, certain general statements can be made.

1. Many processes are common to all cells, and any two cells in a single organism therefore have many proteins in common. These include some abundant proteins that are easy to analyze, such as the major structural proteins of the cytoskeleton and of chromosomes, some of the proteins essential to the endoplasmic reticulum and Golgi membranes, ribosomal proteins, and so on. Many nonabundant proteins, such as various enzymes involved in the central reactions of metabolism, are also the same in all cell types.

2. Some proteins are abundant in the specialized cells in which they function and cannot be detected elsewhere, even by sensitive tests. Hemoglobin, for example, can be detected only in red blood cells.

3. If the 2000 or so most abundant proteins (those present in quantities of 50,000 or more copies per cell) are compared among different cell types of the same organism using two-dimensional polyacrylamide-gel electrophoresis, remarkably few differences are found. Whether the comparison is between two cell lines grown in culture (such as muscle and nerve cells lines) or between cells of two young rodent tissues (such as liver and lung), the great majority of the proteins detected are synthesized in both cell types and at rates that differ by less than a factor of five; only a few percent of the proteins are present in very different amounts in the two cell types.

Studies of the number of different mRNA sequences in a cell suggest that a typical higher eucaryotic cell synthesizes 10,000 to 20,000 different proteins. Most of these are too rare to be detected by two-dimensional gel electrophoresis of cell

Figure 9–1 **Regeneration of a whole plant from a single differentiated cell.** In many types of plants, differentiated cells retain the ability to "dedifferentiate" so that a single cell can form a clone of progeny cells that can later give rise to an entire plant.

extracts. If these minor cell proteins differ among cells to the same extent as the more abundant proteins, as is commonly assumed, only a small number of protein differences (perhaps several hundred) suffice to create very large differences in cell morphology and behavior.

A Cell Can Change the Expression of Its Genes in Response to External Signals [3]

Most of the specialized cells in a multicellular organism are capable of altering their patterns of gene expression in response to extracellular cues. If a liver cell is exposed to a glucocorticoid hormone, for example, the production of several specific proteins is dramatically increased. Glucocorticoids are released during periods of starvation or intense exercise and signal the liver to increase the production of glucose from amino acids and other small molecules; the set of proteins whose production is induced includes enzymes such as tyrosine aminotransferase, which helps to convert tyrosine to glucose. When the hormone is no longer present, the production of these proteins drops to its normal level.

Other cell types respond to glucocorticoids in different ways. In fat cells, for example, the production of tyrosine aminotransferase is reduced, while some other cell types do not respond to glucocorticoids at all. These examples illustrate a general feature of cell specialization—different cell types often respond in different ways to the same extracellular signal. Underlying this specialization are features that do not change, which give each cell type its permanently distinctive character. These features reflect the persistent expression of different sets of genes.

Gene Expression Can Be Regulated at Many of the Steps in the Pathway from DNA to RNA to Protein [4]

If differences between the various cell types of an organism depend on the particular genes that the cells express, at what level is the control of gene expression exercised? There are many steps in the pathway leading from DNA to protein, and all of them can in principle be regulated. Thus a cell can control the proteins it makes by (1) controlling when and how often a given gene is transcribed (**transcriptional control**), (2) controlling how the primary RNA transcript is spliced or otherwise processed (**RNA processing control**), (3) selecting which completed mRNAs in the cell nucleus are exported to the cytoplasm (**RNA transport control**), (4) selecting which mRNAs in the cytoplasm are translated by ribosomes (**translational control**), (5) selectively destabilizing certain mRNA molecules in the cytoplasm (**mRNA degradation control**), or (6) selectively activating, inactivating, or compartmentalizing specific protein molecules after they have been made (**protein activity control**) (Figure 9–2).

For most genes transcriptional controls are paramount. This makes sense because, of all the possible control points illustrated in Figure 9–2, only transcriptional control ensures that no superfluous intermediates are synthesized. In the

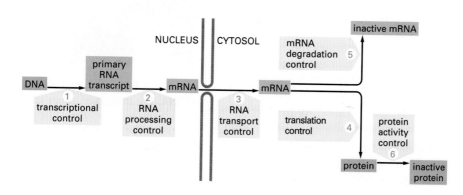

Figure 9–2 **Six steps at which eucaryote gene expression can be controlled.** Only controls that operate at steps 1 through 5 are discussed in this chapter. The regulation of protein activity (step 6) is discussed in Chapter 5; this includes reversible activation or inactivation by protein phosphorylation as well as irreversible inactivation by proteolytic degradation.

An Overview of Gene Control

following sections we discuss the DNA and protein components that regulate the initiation of gene transcription. We return at the end of the chapter to the other ways of regulating gene expression.

Summary

The genome of a cell contains in its DNA sequence the information to make many thousands of different protein and RNA molecules. A cell typically expresses only a fraction of its genes, and the different types of cells in multicellular organisms arise because different sets of genes are expressed. Moreover, cells can change the pattern of genes they express in response to changes in their environment, such as signals from other cells. Although all of the steps involved in expressing a gene can in principle be regulated, for most genes the initiation of RNA transcription is the most important point of control.

DNA-binding Motifs in Gene Regulatory Proteins [5]

How does a cell determine which of its thousands of genes to transcribe? As discussed in Chapter 8, the transcription of each gene is controlled by a regulatory region of DNA near the site where transcription begins. Some regulatory regions are simple and act as switches that are thrown by a single signal. Other regulatory regions are complex and act as tiny microprocessors, responding to a variety of signals that they interpret and integrate to switch the neighboring gene on or off. Whether complex or simple, these switching devices consist of two fundamental types of components: (1) short stretches of DNA of defined sequence and (2) *gene regulatory proteins* that recognize and bind to them.

We begin our discussion of gene regulatory proteins by describing how these proteins were discovered.

Gene Regulatory Proteins Were Discovered Using Bacterial Genetics [6]

Genetic analyses in bacteria carried out in the 1950s provided the first evidence of the existence of **gene regulatory proteins** that turn specific sets of genes on or off. One of these regulators, the *lambda repressor,* is encoded by a bacterial virus, *bacteriophage lambda.* The repressor shuts off the viral genes that code for the protein components of new virus particles and thereby enables the viral genome to remain a silent passenger in the bacterial chromosome, multiplying with the bacterium when conditions are favorable for bacterial growth (see Figure 6–80). The lambda repressor was among the first gene regulatory proteins to be characterized, and it remains one of the best understood, as we discuss later. Other bacterial regulators respond to nutritional conditions by shutting off genes encoding specific sets of metabolic enzymes when they are not needed. The *lac repressor,* for example, the first of these bacterial proteins to be recognized, turns off the production of the proteins responsible for lactose metabolism when this sugar is absent from the medium.

The first step toward understanding gene regulation was the isolation of mutant strains of bacteria and bacteriophage lambda that were unable to shut off specific sets of genes. It was proposed at the time, and later proved, that most of these mutants were deficient in proteins acting as specific repressors for these sets of genes. Because these proteins, like most gene regulatory proteins, are present in small quantities, it was difficult and time-consuming to isolate them. They were eventually purified by fractionating cell extracts on a series of standard chromatography columns (see pp. 166–169). Once isolated, the proteins were shown to bind to specific DNA sequences close to the genes that they

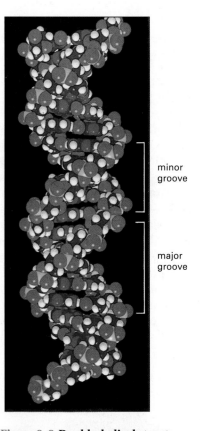

Figure 9–3 **Double-helical structure of DNA.** The major and minor grooves on the outside of the double helix are indicated. The atoms are colored as follows: carbon, *dark blue;* nitrogen, *light blue;* hydrogen, *white;* oxygen, *red;* phosphorus, *yellow.*

minor groove

major groove

regulate. The precise DNA sequences that they recognized were then determined by a combination of classical genetics, DNA sequencing, and DNA-footprinting experiments (discussed in Chapter 7).

The Outside of the DNA Helix Can Be Read by Proteins [7]

As discussed in Chapter 3, the DNA in a chromosome consists of a very long double helix (Figure 9–3). Gene regulatory proteins must recognize specific nucleotide sequences embedded within this structure. It was originally thought that these proteins might require direct access to the hydrogen bonds between base pairs in the interior of the double helix to distinguish between one DNA sequence and another. It is now clear, however, that the outside of the double helix is studded with DNA sequence information that gene regulatory proteins can recognize without having to open the double helix. The edge of each base pair is exposed at the surface of the double helix, presenting a distinctive pattern of hydrogen bond donors, hydrogen bond acceptors, and hydrophobic patches for proteins to recognize in both the major and minor groove (Figure 9–4). But only in the major groove are the patterns unique for each of the four base-pair arrangements (Figure 9–5). For this reason gene regulatory proteins generally bind to the major groove, as we shall see.

Although the patterns of hydrogen bond donor and acceptor groups are the most important features recognized by gene regulatory proteins, they are not the only ones: the nucleotide sequence also determines the overall geometry of the double helix.

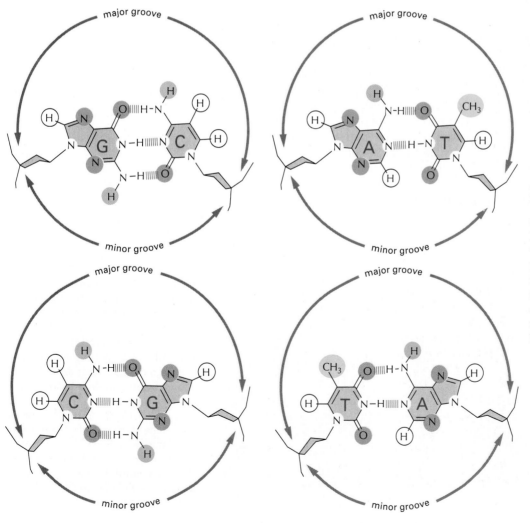

Figure 9–4 How the different base pairs in DNA can be recognized from their edges without the need to open the double helix. The four possible configurations of base pairs are shown, with potential hydrogen bond donors indicated in *blue*, potential hydrogen bond acceptors in *red*, and hydrogen bonds of the base pairs themselves as a series of short parallel *red* lines. Methyl groups, which form hydrophobic protuberances, are shown in *yellow*, and hydrogen atoms that are attached to carbons, and are therefore unavailable for hydrogen bonding, are *white*.

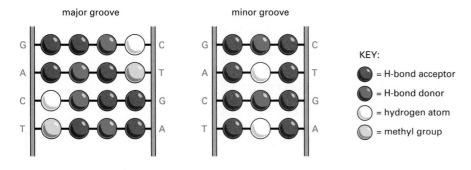

major groove   minor groove

KEY:

- = H-bond acceptor
- = H-bond donor
- = hydrogen atom
- = methyl group

Figure 9–5 A DNA recognition code. The edge of each base pair, seen here looking directly at the major or minor groove, contains a distinctive pattern of hydrogen bond donors, hydrogen bond acceptors, and methyl groups. From the major groove, each of the four base-pair configurations projects a unique pattern of features. From the minor groove, however, the patterns are similar for G-C and C-G as well as for A-T and T-A. The color code is the same as that in Figure 9–4.

The Geometry of the DNA Double Helix Depends on the Nucleotide Sequence [8]

For 20 years after the discovery of the DNA double helix in 1953, DNA was thought to have the same monotonous structure, with exactly 36° of helical twist between its adjacent nucleotide pairs (10 nucleotide pairs per helical turn) and a uniform helix geometry. This view was based on structural studies of heterogeneous mixtures of DNA molecules, however, and it changed once the three-dimensional structures of short DNA molecules of defined nucleotide sequence were solved by x-ray crystallography and NMR spectroscopy. Whereas the earlier studies provided a picture of an average, idealized DNA molecule, the later studies showed that any given nucleotide sequence had local irregularities, such as tilted nucleotide pairs or a helical twist angle larger or smaller than 36°. These unique features can be recognized by specific DNA-binding proteins.

An especially striking departure from the average structure is seen in the case of nucleotide sequences that cause the DNA double helix to bend. Some sequences (for example, AAAANNN, where N can be any base except A) form a double helix with a pronounced irregularity that causes a slight bend; if this sequence is repeated at 10-nucleotide-pair intervals in a long DNA molecule, the small bends add together so that the DNA molecule appears unusually curved when viewed in the electron microscope (Figure 9–6).

A related and equally important variable feature of DNA structure is the extent to which the double helix is deformable. For a protein to recognize and bind to a specific DNA sequence, there must be a tight fit between the DNA and the protein, and often the normal DNA conformation must be distorted to maximize this fit (Figure 9–7). The energetic cost of such distortion depends on the local nucleotide sequence. We encountered an example of this in the discussion of nucleosome assembly in Chapter 8: some DNA sequences can accommodate the

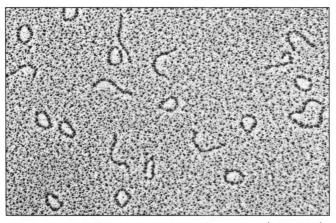

100 nm

Figure 9–6 Electron micrograph of fragments of a highly bent segment of DNA double helix. The DNA fragments are derived from the small, circular mitochondrial DNA molecules of a trypanosome. Although the fragments are only about 200 nucleotide pairs long, many of them have bent to form a complete circle. On average, a normal DNA helix of this length would bend only enough to produce one-fourth of a circle (one smooth right-angle turn). (From J. Griffith, M. Bleyman, C.A. Raugh, P.A. Kitchin, and P.T. Englund, *Cell* 46:717–724, 1986. © Cell Press.)

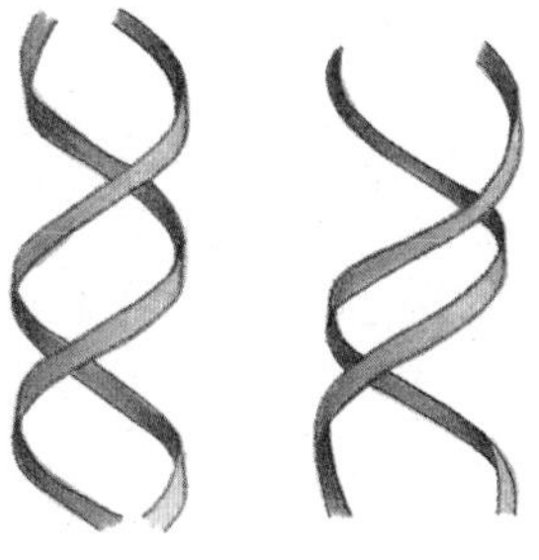

Figure 9–7 **DNA deformation induced by protein binding.** The figure shows the changes of DNA structure, from regular B-DNA (A) to a distorted version of B-DNA (B), that are observed when a well-studied gene regulatory protein (the repressor from bacteriophage 434) binds to specific sequences of DNA. The ease with which a DNA sequence can be deformed often affects the affinity with which a protein binds to it.

(A)   (B)

tight DNA wrapping required for nucleosome formation better than others. Similarly, a few gene regulatory proteins induce a striking bend in the DNA when they bind to it (Figure 9–8). In general, these proteins recognize DNA sequences that are easily bent.

Short DNA Sequences Are Fundamental Components of Genetic Switches [9]

We have seen how a specific nucleotide sequence can be detected as a pattern of structural features on the surface of the DNA double helix. Particular nucleotide sequences, each typically less than 20 nucleotide pairs in length, function as fundamental components of genetic switches by serving as recognition sites for the binding of specific gene regulatory proteins. Hundreds of such DNA sequences have been identified, each recognized by a different gene regulatory protein or by a set of related gene regulatory proteins. Some examples of such proteins are listed in Table 9–1 along with the DNA sequences that they recognize.

Table 9–1 Some Gene Regulatory Proteins and the DNA Sequences That They Recognize

	Name	DNA Sequence Recognized*
Bacteria	lac repressor	AATTGTGAGCGGATAACAATT TTAACACTCGCCTATTGTTAA
	CAP	TGTGAGTTAGCTCACT ACACTCAATCGAGTGA
	lambda repressor	TATCACCGCCAGAGGTA ATAGTGGCGGTCTCCAT
Yeast	GAL4	CGGAGGACTGTCCTCCG GCCTCCTGACAGGAGGC
	MAT α2	CATGTAATT GTACATTAA
	GCN4	ATGACTCAT TACTGAGTA
Drosophila	Krüppel	AACGGGTTAA TTGCCCAATT
	bicoid	GGGATTAGA CCCTAATCT
Mammals	Sp1	GGGCGG CCCGCC
	Oct-1	ATGCAAAT TACGTTTA
	GATA-1	TGATAG ACTATC

Figure 9–8 **The bending of DNA induced by the binding of the catabolite activator protein (CAP).** CAP is a gene regulatory protein from *E. coli*. In the absence of the bound protein, this DNA helix is straight.

*Each protein in this table can recognize a set of closely related DNA sequences; for convenience, only one recognition sequence is given for each protein.

DNA-binding Motifs in Gene Regulatory Proteins

We now turn to the gene regulatory proteins—the second fundamental component of genetic switches—that recognize short, specific DNA sequences contained in a much longer double helix.

Gene Regulatory Proteins Contain Structural Motifs That Can Read DNA Sequences [10]

Molecular recognition in biology generally relies on an exact fit between the surfaces of two molecules, and the study of gene regulatory proteins has provided some of the clearest examples of this principle. A gene regulatory protein recognizes a specific DNA sequence because the surface of the protein is extensively complementary to the special surface features of the double helix in that region. In most cases the protein makes a large number of contacts with the DNA, involving hydrogen bonds, ionic bonds, and hydrophobic interactions. Although each individual contact is weak, the 20 or so contacts that are typically formed at the protein-DNA interface add together to ensure that the interaction is both highly specific and very strong (Figure 9–9). In fact, DNA-protein interactions are among the tightest and most specific molecular interactions known in biology.

Although each example of protein-DNA recognition is unique in detail, x-ray crystallographic and NMR spectroscopic studies of about 30 gene regulatory proteins complexed with their specific DNA sequence have revealed that many of those proteins contain one of a small set of DNA-binding structural motifs. Each of these motifs uses either α helices or β sheets to bind to the major groove of DNA; this groove, as we have seen, contains sufficient information to distinguish one DNA sequence from any other. The fit is so good that it is tempting to speculate that the dimensions of the basic structural units of nucleic acids and proteins evolved together to permit these molecules to interlock.

The Helix-Turn-Helix Motif Is One of the Simplest and Most Common DNA-binding Motifs [11]

The first DNA-binding protein motif to be recognized was the **helix-turn-helix.** Originally identified in bacterial proteins, this motif has since been found in hundreds of DNA-binding proteins from both eucaryotes and procaryotes. It is

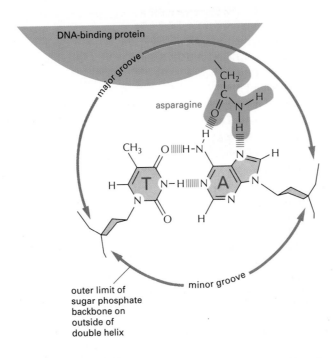

Figure 9–9 **The binding of a gene regulatory protein to the major groove of DNA.** Only a single type of contact is shown. Typically, the protein-DNA interface would consist of 10 to 20 such contacts, involving different amino acids, each contributing to the binding energy of the protein-DNA interaction.

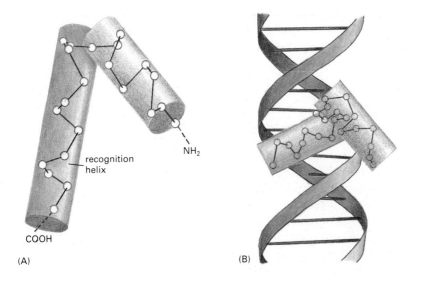

(A)

recognition helix

NH₂

COOH

(B)

Figure 9–10 **The DNA-binding helix-turn-helix motif.** The motif is shown in (A), where each *white* circle denotes the central carbon of an amino acid. The carboxyl-terminal α helix (*red*) is called the recognition helix because it participates in sequence-specific recognition of DNA. As shown in (B), this helix fits into the major groove of DNA, where it contacts the edges of the base pairs (see also Figure 9–4).

constructed from two α helices connected by a short extended chain of amino acids, which constitutes the "turn." The two helices are held at a fixed angle, primarily through interactions between the two helices. The more carboxyl-terminal helix is called the *recognition helix* because it fits into the major groove of DNA; its amino acid side chains, which differ from protein to protein, play an important part in recognizing the specific DNA sequence to which the protein binds (Figure 9–10).

Outside the helix-turn-helix region the structure of the various proteins that contain this motif can vary enormously (Figure 9–11). Thus each protein "presents" its helix-turn-helix motif to the DNA in a unique way, a feature thought to enhance the versatility of the helix-turn-helix motif by increasing the number of DNA sequences that the motif can be used to recognize. Moreover, in most of these proteins, parts of the polypeptide chain outside the helix-turn-helix domain also make important contacts with the DNA, helping to fine-tune the interaction.

The group of helix-turn-helix proteins shown in Figure 9–11 demonstrates a feature that is common to many sequence-specific DNA-binding proteins. They bind as symmetric dimers to DNA sequences that are composed of two very similar "half-sites," which are also arranged symmetrically (Figure 9–12). This arrangement allows each protein monomer to make a nearly identical set of contacts and enormously increases the binding affinity: as a first approximation, doubling the number of contacts doubles the free energy of the interaction but *squares* the affinity constant.

Figure 9–11 **Some helix-turn-helix DNA-binding proteins.** All of the proteins bind DNA as dimers in which the two copies of the recognition helix (*red cylinder*) are separated by exactly one turn of the DNA helix (3.4 nm). The second helix of the helix-turn-helix motif is colored *blue*, as in Figure 9–10. The lambda repressor and cro proteins control bacteriophage lambda gene expression, and the tryptophan repressor and the catabolite activator protein (CAP) control the expression of sets of *E. coli* genes.

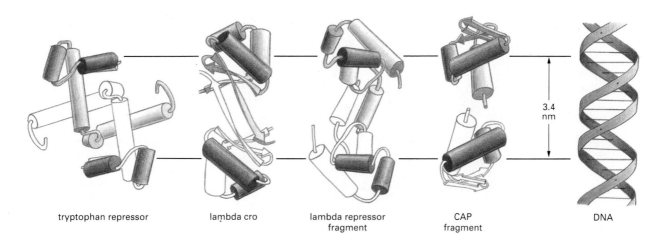

tryptophan repressor | lambda cro | lambda repressor fragment | CAP fragment | DNA

3.4 nm

DNA-binding Motifs in Gene Regulatory Proteins

```
5' T A A C A C C G T G C G T G T T G 3'
   | | | | | | | | | | | | | | | | | |
3' A T T G T G G C A C G C A C A A C 5'
```

Homeodomain Proteins Are a Special Class of Helix-Turn-Helix Proteins [12]

Not long after the first gene regulatory proteins were discovered in bacteria, genetic analyses in the fruit fly *Drosophila* led to the characterization of an important class of genes, the *homeotic selector genes,* that play a critical part in orchestrating fly development. As discussed in Chapter 21, they have since proved to have a fundamental role in the development of higher animals as well. Mutations in these genes cause one body part in the fly to be converted into another, showing that the proteins they encode control developmental switches.

When the nucleotide sequences of several homeotic selector genes were determined in the early 1980s, each proved to contain an almost identical stretch of 60 amino acids that defines this class of proteins and is termed the **homeodomain.** When the three-dimensional structure of the homeodomain was solved, it was seen to contain a helix-turn-helix motif related to that of the bacterial gene regulatory proteins, providing one of the first indications that the principles of gene regulation established in bacteria are relevant to higher organisms as well. More than 60 homeodomain proteins have now been discovered in *Drosophila* alone, and homeodomain proteins have been identified in virtually all eucaryotic organisms that have been studied, from yeasts to man.

The structure of a homeodomain bound to its specific DNA sequence is shown in Figure 9–13. Whereas the helix-turn-helix motif of bacterial gene regulatory proteins is often embedded in different structural contexts, the helix-turn-helix motif of homeodomains is always surrounded by the same structure (which forms the rest of the homeodomain), suggesting that the motif is always presented to DNA in the same basic manner. Indeed, structural studies show that a yeast homeodomain protein and a *Drosophila* homeodomain protein recognize DNA in almost exactly the same way, although they are identical at only 17 of 60 amino acid positions.

There Are Several Types of DNA-binding Zinc Finger Motifs [13]

The helix-turn-helix motif is composed solely of amino acids. A second important group of DNA-binding motifs, by contrast, utilizes one or more molecules of zinc as a structural component. Although all such zinc-coordinated DNA-bind-

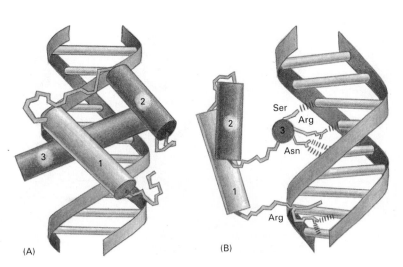

(A)   (B)

Figure 9–13 **A homeodomain bound to its specific DNA sequence.** The homeodomain is folded into three α helices, which are packed tightly together by hydrophobic interactions (A). The part containing helix 2 and 3 closely resembles the helix-turn-helix motif, with the recognition helix (*red*) making important contacts with the major groove (B). The Asn of helix 3, for example, contacts an adenine, as shown in Figure 9–9. Nucleotide pairs are also contacted in the minor groove by a flexible arm attached to helix 1. The homeodomain shown here is from a yeast gene regulatory protein, but it is nearly identical to two homeodomains from *Drosophila*, which interact with DNA in a similar fashion. (Adapted from C. Wolberger et al., *Cell* 67:517–528, 1991. © Cell Press.)

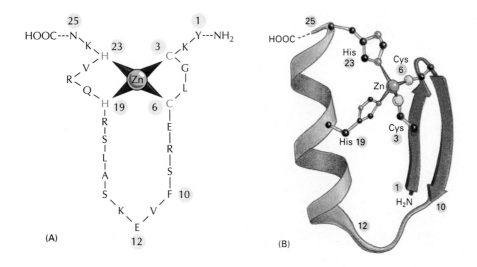

(A)

(B)

Figure 9–14 One type of zinc finger protein. This protein belongs to the Cys-Cys-His-His family of zinc finger proteins, named after the amino acids that grasp the zinc. This zinc finger is from a frog protein of unknown function. (A) Schematic drawing of the amino acid sequence of the zinc finger. (B) The three-dimensional structure of the zinc finger is constructed from an antiparallel β sheet (amino acids 1 to 10) followed by an α helix (amino acids 12 to 24). The four amino acids that bind the zinc (Cys 3, Cys 6, His 19, and His 23) hold one end of the α helix firmly to one end of the β sheet. (Adapted from M.S. Lee et al., *Science* 245:635–637, 1989. © 1989 the AAAS.)

ing motifs are called **zinc fingers,** this description refers only to their appearance in schematic drawings dating from their initial discovery (Figure 9–14A). Subsequent structural studies have shown that, in fact, they fall into several groups, only two of which are considered here. The first type was initially discovered in the protein that activates the transcription of a eucaryotic ribosomal RNA. It is a simple structure, consisting of an α helix and a β sheet held together by the zinc (Figure 9–14B). This type of zinc finger is often found in a cluster with additional zinc fingers, arranged one after the other so that the α helix of each can contact the major groove of the DNA, forming a nearly continuous stretch of α helix along the groove. In this way a strong and specific DNA-protein interaction is built up through a repeating basic structural unit (Figure 9–15). A particular advantage of this motif is that the strength and specificity of the DNA-protein interaction could be adjusted during evolution by changes in the number of zinc finger repeats. By contrast, it is difficult to imagine how any of the other motifs discussed in this section could be formed into repeating chains.

The other type of zinc finger is found in the large family of intracellular receptor proteins (discussed in Chapter 15). It forms a different structure (similar to the procaryotic helix-turn-helix motif) in which two α helices are packed together with two zinc atoms. Like the helix-turn-helix proteins, these proteins form dimers that allow one of the two α helices of each subunit to interact with the major groove of the DNA (Figure 9–16). Although the two types of zinc finger structures are distinct, they share two important features: both use zinc as a structural element, and both use an α helix to recognize the major groove of the DNA.

Figure 9–15 DNA binding by a zinc finger protein. (A) The structure of a fragment of a mouse gene regulatory protein bound to a specific DNA site. This protein recognizes DNA using three zinc fingers of the Cys-Cys-His-His type (see Figure 9–14) arranged as direct repeats. (B) The three fingers have similar amino acid sequences and contact the DNA in similar ways. In both (A) and (B) the zinc atom in each finger is represented by a small sphere. (Adapted from N. Pavletich and C. Pabo, *Science* 252:810–817, 1991. © 1991 the AAAS.)

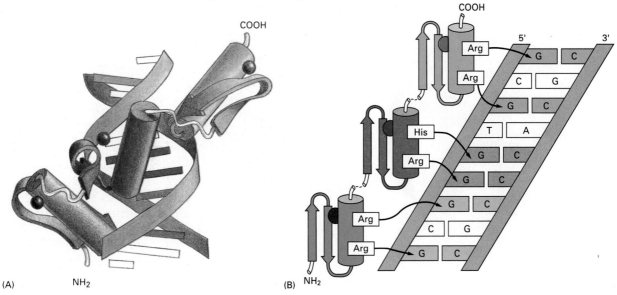

(A) NH₂

(B) NH₂

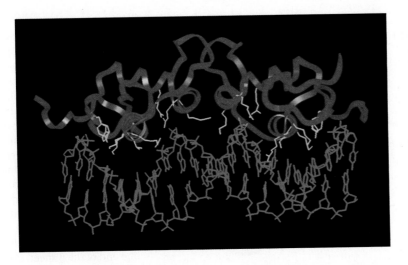

Figure 9–16 **A zinc finger protein of the intracellular receptor family bound to its specific DNA sequence.** This is an example of a Cys-Cys-Cys-Cys zinc finger protein, named after the amino acids that bind the zinc. A dimer of the DNA-binding domain (*red*) is shown, with all side chains omitted except those shown in *yellow*, which form contacts with the DNA (*green*). (Adapted from B.F. Luisi et al., *Nature* 352:497–505, 1991. © 1991 Macmillan Magazines Ltd.; photograph courtesy of Jay Thomas.)

β Sheets Can Also Recognize DNA [14]

In the DNA-binding motifs discussed so far, α helices are used as a primary mechanism to recognize specific DNA sequences. One group of gene regulatory proteins, however, has evolved an entirely different and no less ingenious recognition strategy. In this case the information on the surface of the major groove is read by a two-stranded β sheet, with side chains of the amino acids extending from the sheet toward the DNA as shown in Figure 9–17. As in the case of a recognition α helix, this β-sheet motif can be used to recognize many different DNA sequences; the exact DNA sequence recognized depends on the sequence of amino acids that make up the β sheet.

The Leucine Zipper Motif Mediates Both DNA Binding and Protein Dimerization [15]

Most gene regulatory proteins recognize DNA as dimers, probably because, as we have seen, this is a simple way of achieving strong specific binding (see Figure 9–12). Usually, the portion of the protein responsible for dimerization is distinct from the portion that is responsible for DNA binding (see Figure 9–11). One motif, however, combines these two functions in an elegant and economical way. It is called **the leucine zipper motif**, so named because of the way the two α helices, one from each monomer, are joined together to form a short coiled-coil (discussed in Chapter 3). The helices are held together by interactions between

Figure 9–17 **The bacterial *met* repressor protein.** The bacterial *met* repressor regulates the genes encoding the enzymes that catalyze methionine synthesis. When this amino acid is abundant, it binds to the repressor, causing a change in the structure of the protein that enables it to bind to DNA tightly, shutting off the synthesis of the enzymes. (A) In order to bind to DNA tightly, the *met* repressor must be complexed with S-adenosyl methionine, shown in *red*. One subunit of the dimeric protein is shown in *green*, while the other is shown in *blue*. The two-stranded β sheet that binds to DNA is formed by one strand from each subunit and is shown in *dark green* and *dark blue*. (B) Simplified diagram of the *met* repressor bound to DNA, to show how the two-stranded β sheet of the repressor binds to the major groove of DNA. For clarity, the other regions of the repressor are omitted. (A, adapted from S. Phillips, *Curr. Opin. Struct. Biol.* 1:89–98, 1991. © Current Sciences; B, adapted from W. Somers and S. Phillips, *Nature* 359:387–393, 1992. © 1992 Macmillan Magazines Ltd.)

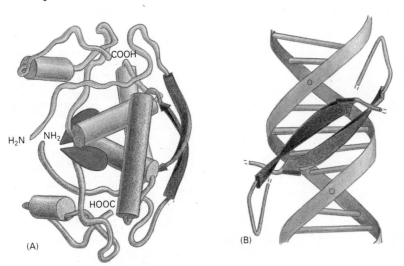

(A)   (B)

hydrophobic amino acid side chains (often on leucines) that extend from one side of each helix. Just beyond the dimerization interface the two α helices separate from each other to form a Y-shaped structure, which allows their side chains to contact the major groove of DNA. The dimer thus grips the double helix like a clothespin on a clothesline (Figure 9–18).

Gene regulatory proteins that contain a leucine zipper motif can form either homodimers, in which the two monomers are identical, or heterodimers, in which the monomers are different. Because heterodimers typically form from two proteins with distinct DNA-binding specificities, the ability of the leucine zipper proteins to form heterodimers greatly expands the repertoire of DNA-binding specificities that these proteins can display. As illustrated in Figure 9–19, for example, three distinct DNA-binding specificities could, in principle, be generated from two types of monomer, while six could be created from three types of monomer, and so on. This is an example of **combinatorial control,** in which combinations of proteins, rather than individual proteins, control a cellular process. It is one of the most important mechanisms used by eucaryotic cells to control gene expression. As we discuss later, the formation of heterodimeric gene regulatory complexes is only one of several combinatorial mechanisms for controlling gene expression.

There are many types of leucine zipper proteins, and they cannot all form heterodimers with one another. Otherwise the amount of cross-talk between the gene regulatory circuits of a cell would be so great as to cause chaos. Whether or not a particular heterodimer can form depends on how well the hydrophobic surfaces of the two leucine zipper α helices mesh with each other, which, in turn, depends on the exact amino acid sequences of the two zipper regions. Thus each leucine zipper protein in the cell can form dimers with only a small set of other leucine zipper proteins.

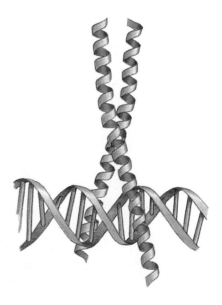

Figure 9–18 A leucine zipper dimer bound to DNA. Two α-helical DNA-binding domains (*bottom*) dimerize through their α-helical leucine zipper region (*top*) to form an inverted Y-shaped structure. Each arm of the Y is formed by a single α helix, one from each monomer, that mediates binding to a specific DNA sequence in the major groove of DNA. Each α helix binds to one-half of a symmetric DNA structure. (Adapted from T.E. Ellenberger et al., *Cell* 71:1223–1237, 1992. © Cell Press.)

The Helix-Loop-Helix Motif Also Mediates Dimerization and DNA Binding [16]

Another important DNA-binding motif, related to the leucine zipper, is the **helix-loop-helix (HLH) motif,** which should not be confused with the *helix-turn-helix* motif discussed earlier. An HLH motif consists of a short α helix connected by a loop to a second, longer α helix. The flexibility of the loop allows one helix to fold back and pack against the other. As shown in Figure 9–20, this two-helix structure binds both to DNA and to the HLH motif of a second HLH protein. As with leucine zipper proteins, the second HLH protein can be the same (resulting in a homodimer) or different (resulting in a heterodimer), and α helices extending from the dimerization interface make specific contacts with DNA.

Several HLH proteins lack the α-helical extension responsible for binding to DNA. These truncated proteins can form heterodimers with full-length HLH proteins, but the heterodimers are unable to bind DNA tightly because they can form only half of the necessary contacts. Thus, in addition to creating active dimers of hybrid DNA specificity, heterodimerization provides a useful control mechanism, enabling a cell to inactivate specific gene regulatory proteins (Figure 9–21).

Figure 9–19 Heterodimerization of leucine zipper proteins can alter their DNA-binding specificity. Leucine zipper homodimers bind to symmetric DNA sequences, as shown in the left-hand and center drawings. These two proteins recognize different DNA sequences, as indicated by the *red* and *blue* regions in the DNA. The two different monomers can combine to form a heterodimer, which now recognizes a hybrid DNA sequence, composed from one *red* and one *blue* region.

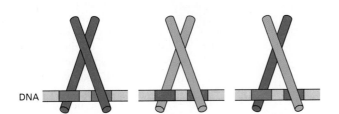

DNA

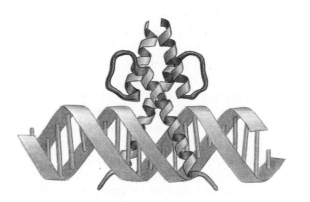

Figure 9–20 **A helix-loop-helix dimer bound to DNA.** The two monomers are held together in a four-helix bundle: each monomer contributes two α helices connected by a flexible loop of protein (*red*). A specific DNA sequence is bound by the two α helices that project from the four-helix bundle. (Adapted from Ferre-D'Amare et al., *Nature* 363:38–45, 1993. © 1993 Macmillan Magazines Ltd.)

It Is Not Yet Possible to Predict the DNA Sequence Recognized by a Gene Regulatory Protein [17]

The various DNA-binding motifs that we have discussed provide structural frameworks from which specific amino acid side chains extend to contact specific base pairs in the DNA. It is reasonable to ask, therefore, whether there is a simple amino acid–base pair recognition code: is a G-C base pair, for example, always contacted by a particular amino acid side chain? The answer appears to be no. We discussed in Chapter 3 how protein surfaces of virtually any shape and chemistry can be made from just 20 different amino acids, and a gene regulatory protein utilizes different combinations of its side chains to create a surface that is precisely complementary to that of the DNA sequence that it recognizes. It seems, therefore, that the same base pair can be recognized in many ways. Nevertheless, molecular biologists may soon understand protein-DNA recognition well enough to be able to design proteins that will recognize any specified DNA sequence.

A Gel-Mobility Shift Assay Allows Sequence-specific DNA-binding Proteins to Be Detected Readily [18]

Genetic analyses, which provided a route to the gene regulatory proteins of bacteria, yeast, and *Drosophila,* are usually not possible in vertebrates. Therefore, the isolation of vertebrate gene regulatory proteins had to await the development of different approaches. Many of these approaches rely on the detection in a cell extract of a DNA-binding protein that specifically recognizes a DNA sequence known to be important in controlling the expression of a particular gene. The most common way to detect sequence-specific DNA-binding proteins is to use a technique that is based on the effect of a bound protein on the migration of DNA molecules in an electric field.

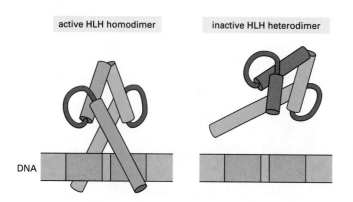

active HLH homodimer

inactive HLH heterodimer

DNA

Figure 9–21 **Inhibitory regulation by truncated HLH proteins.** The HLH motif is responsible for both dimerization and DNA binding. On the left, an HLH homodimer recognizes a symmetric DNA sequence. On the right, the binding of a full-length HLH protein to a truncated HLH protein that lacks the DNA-binding α helix generates a heterodimer that is unable to bind DNA tightly. If present in excess, the truncated protein molecule blocks the homodimerization of the full-length HLH protein and thereby prevents it from binding to DNA.

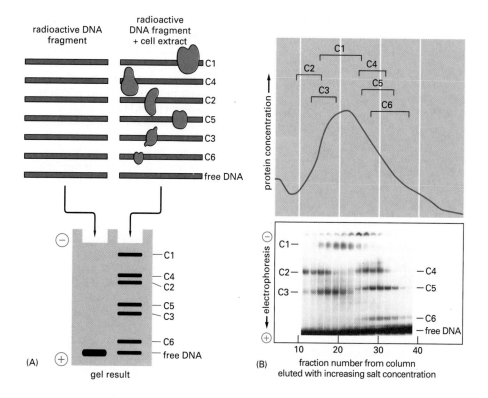

Figure 9–22 A gel-mobility shift assay. The principle of the assay is shown schematically in (A). In this example an extract of an antibody-producing cell line is mixed with a radioactive DNA fragment containing about 160 nucleotides of a regulatory DNA sequence from a gene encoding the light chain of the antibody made by the cell line. The effect of the proteins in the extract on the mobility of the DNA fragment is analyzed by polyacrylamide-gel electrophoresis followed by autoradiography. The free DNA fragments run rapidly to the bottom of the gel, while those fragments bound to proteins are retarded; the finding of six retarded bands suggests that the extract contains six different sequence-specific DNA-binding proteins (indicated as C1–C6) that bind to this DNA sequence. (For simplicity, any DNA fragments with more than one protein bound have been omitted from the figure.) In (B) the extract was fractionated by a standard chromatographic technique (*top*), and each fraction was mixed with the radioactive DNA fragment, applied to one lane of a polyacrylamide gel, and analyzed as in (A). (B, modified from C. Scheidereit, A. Heguy, and R.G. Roeder, *Cell* 51:783–793, 1987. © Cell Press.)

A DNA molecule is highly negatively charged and will therefore move rapidly toward a positive electrode when it is subjected to an electric field. When analyzed by polyacrylamide-gel electrophoresis, DNA molecules are separated according to their size because smaller molecules are able to penetrate the fine gel meshwork more easily than large ones. Protein molecules bound to a DNA molecule will cause it to move more slowly through the gel; in general, the larger the bound protein, the greater the retardation of the DNA molecule. This phenomenon provides the basis for the *gel-mobility shift assay*, which allows even trace amounts of a sequence-specific DNA-binding protein to be readily detected. In this assay a short DNA fragment of specific length and sequence (produced either by DNA cloning or by chemical synthesis) is radioactively labeled and mixed with a cell extract; the mixture is then loaded onto a polyacrylamide gel and subjected to electrophoresis. If the DNA fragment corresponds to a chromosomal region where, for example, several sequence-specific proteins bind, autoradiography will reveal a series of DNA bands, each retarded to a different extent and representing a distinct DNA-protein complex. The proteins responsible for each band on the gel can then be separated from one another by subsequent fractionations of the cell extract (Figure 9–22).

DNA Affinity Chromatography Facilitates the Purification of Sequence-specific DNA-binding Proteins [19]

A particularly powerful purification method called **DNA affinity chromatography** can be used once the DNA sequence that a gene regulatory protein recognizes has been determined. A double-stranded oligonucleotide of the correct sequence is synthesized by chemical methods and linked to an insoluble porous matrix such as agarose; the matrix with the oligonucleotide attached is then used to construct a column that selectively binds proteins that recognize the particular DNA sequence (Figure 9–23). Purifications as great as 10,000-fold can be achieved by this means with relatively little effort.

Although most proteins that bind to a specific DNA sequence are present in a few thousand copies per higher eucaryotic cell (and generally represent only about one part in 50,000 of the total cell protein), enough pure protein can usually

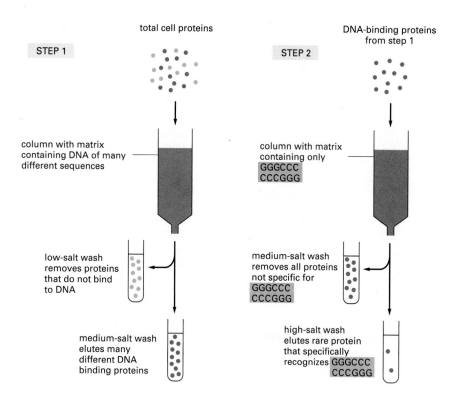

STEP 1

total cell proteins

STEP 2

DNA-binding proteins from step 1

column with matrix containing DNA of many different sequences

column with matrix containing only GGGCCC CCCGGG

low-salt wash removes proteins that do not bind to DNA

medium-salt wash removes all proteins not specific for GGGCCC CCCGGG

medium-salt wash elutes many different DNA binding proteins

high-salt wash elutes rare protein that specifically recognizes GGGCCC CCCGGG

Figure 9–23 DNA affinity chromatography. In the first step all the proteins that can bind DNA are separated from the remainder of the cellular proteins on a column containing a huge number of different DNA sequences. Most sequence-specific DNA-binding proteins have a weak (nonspecific) affinity for bulk DNA and are therefore retained on the column. This affinity is due largely to ionic attractions, and the proteins can be washed off the DNA by a solution that contains a moderate concentration of salt. In the second step the mixture of DNA-binding proteins is passed through a column that contains only DNA of a particular sequence. Typically, all the DNA-binding proteins will stick to the column, the great majority by nonspecific interactions. These are again eluted by solutions of moderate salt concentration, leaving on the column only those proteins (typically one or only a few) that bind specifically and therefore very tightly to the particular DNA sequence. These remaining proteins can be eluted from the column by solutions containing a very high concentration of salt.

be isolated by affinity chromatography to obtain a partial amino acid sequence. This sequence can then be used to synthesize an oligonucleotide probe that can in turn be used to identify the corresponding cDNA clone because it will specifically hybridize with the sequence that codes for the protein (discussed in Chapter 7). The clone provides the complete amino acid sequence of the protein as well as the means to produce the protein in unlimited amounts.

In some cases a second method, which can be even more powerful than DNA affinity chromatography, can be used more directly to obtain a cDNA clone that encodes a sequence-specific DNA-binding protein. This method begins with a cDNA library cloned in an appropriately designed expression vector (discussed in Chapter 7). An individual colony of bacteria (if the expression vector is a plasmid) or bacteriophage plaque (if the expression vector is a virus) will produce large amounts of the protein that is encoded by the cDNA it contains. To find the rare colony that produces the protein of interest, an oligonucleotide containing the desired protein's recognition sequence is radioactively labeled and used to probe paper-blot replicas of the culture that carry aliquots of thousands of individual colonies (see Figure 7–26). Those few colonies that produce proteins that specifically bind the radiolabeled oligonucleotide are selectively grown and tested further to find the one that produces the desired protein.

Because these powerful methods have only recently been developed, only a fraction of the many thousands of sequence-specific DNA-binding proteins thought to be present in higher eucaryotic cells have been isolated so far.

Summary

Gene regulatory proteins recognize short stretches of double-helical DNA of defined sequence and thereby determine which of the thousands of genes in a cell will be transcribed. Hundreds of gene regulatory proteins have been identified in a wide variety of organisms. Although each of these proteins has unique features, most bind to DNA as homodimers or heterodimers and recognize DNA through one of a small number of structural motifs, including the helix-turn-helix motif, the homeodomain motif, zinc finger motifs, the leucine zipper motif, and the helix-loop-helix motif. The precise amino acid sequence that is folded into the motif determines the particular DNA sequence that is recognized. Several powerful techniques are available that make use of the DNA-sequence specificity of gene regulatory proteins to identify and isolate these proteins and the genes that encode them.

How Genetic Switches Work [20]

In the previous section we described the basic components of genetic switches—gene regulatory proteins and the specific DNA sequences that these proteins recognize. In this section we discuss how these components operate to turn genes on and off in response to a variety of signals.

Only 40 years ago the idea that genes could be switched on and off was revolutionary. This concept was a major advance, and it came originally from the study of how bacteria adapt to changes in the composition of their growth medium. Parallel studies on the lambda bacteriophage led to many of the same conclusions and helped to establish the basis of our understanding of how gene expression is regulated. The same principles apply to eucaryotic cells, although the enormous complexity of gene regulation in higher organisms, combined with the complication that the DNA in these organisms is packaged into chromatin, creates special challenges and some novel opportunities for control, as we shall see. We begin, however, with the simplest example—an on-off switch in bacteria that responds to a single signal.

The Tryptophan Repressor Is a Simple Switch That Turns Genes On and Off in Bacteria [21]

The chromosome of the bacterium *E. coli,* a single-celled organism, consists of a single circular DNA molecule of about 5×10^6 nucleotide pairs. This DNA is, in principle, sufficient to encode about 4000 proteins, although only a fraction of these are made at any one time. *E. coli* regulates the expression of many of its genes according to the food sources that are available in the environment. Five *E. coli* genes code for enzymes that manufacture the amino acid tryptophan, and these are arranged in a cluster on the chromosome and are transcribed from a single *promoter* as one long mRNA molecule (Figure 9–24). As described in Chapter 6, the promoter is the specific DNA sequence that directs RNA polymerase to bind to DNA, to open the DNA double helix, and to begin synthesizing an RNA molecule. When, however, tryptophan is present in the growth medium and enters the cell (when the bacterium is in the gut of a mammal that has just eaten a meal of protein, for example), these enzymes are no longer needed and their production is shut off.

We now understand the molecular basis for this switch in considerable detail. Within the promoter that directs transcription of the tryptophan biosynthetic genes lies an **operator.** This operator is simply a short region of regulatory DNA of defined nucleotide sequence that is recognized by a helix-turn-helix gene regulatory protein called the **tryptophan repressor** (see Figure 9–11). The promoter and operator are arranged so that occupancy of the operator by the tryptophan

Figure 9–24 The clustered genes in *E. coli* that code for enzymes that manufacture the amino acid tryptophan. These five genes are transcribed as a single mRNA molecule, a feature that allows their expression to be controlled coordinately. Clusters of genes transcribed as a single mRNA molecule are common in bacteria. Each such cluster is called an *operon.*

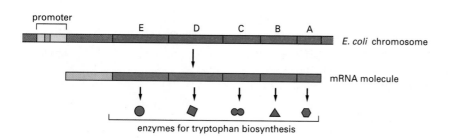

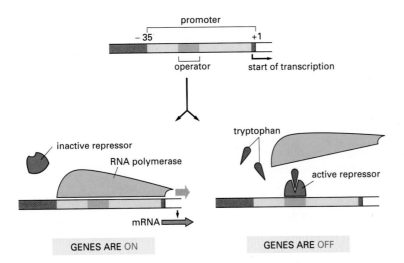

GENES ARE ON

GENES ARE OFF

Figure 9–25 Switching the tryptophan genes on and off. If the level of tryptophan inside the cell is low, RNA polymerase binds to the promoter and transcribes the five genes of the tryptophan (*trp*) operon. If the level of tryptophan is high, however, the tryptophan repressor is activated to bind to the operator, where it blocks the binding of RNA polymerase to the promoter. Whenever the level of intracellular tryptophan drops, the repressor releases its tryptophan and becomes inactive, allowing the polymerase to begin transcribing these genes.

repressor blocks access to the promoter by RNA polymerase, thereby preventing expression of the tryptophan-producing enzymes (Figure 9–25). This block is regulated in an ingenious way: the repressor protein can bind to its operator DNA only if the repressor has also bound two molecules of the amino acid tryptophan. As shown in Figure 9–26, tryptophan binding tilts the helix-turn-helix motif of the repressor so that it is presented properly to the DNA major groove; without tryptophan, the motif swings inward and the protein is unable to bind to the operator. Thus the tryptophan repressor is a simple device that switches production of the tryptophan biosynthetic enzymes on and off according to the availability of free tryptophan. Because the active, DNA-binding form of the protein serves to turn genes *off,* this mode of gene regulation is called **negative control,** and the gene regulatory proteins that function in this way are called *transcriptional repressors* or *gene repressor proteins.*

Transcriptional Activators Turn Genes On [22]

We saw in Chapter 6 that purified *E. coli* RNA polymerase can bind to a promoter and initiate DNA transcription. Some bacterial promoters, however, are only marginally functional on their own, either because they are recognized poorly by RNA polymerase or because the polymerase has difficulty opening the DNA helix as it starts transcription. In either case these poorly functioning promoters can be rescued by gene regulatory proteins that bind to a nearby site on the DNA, contacting the RNA polymerase in a way that dramatically increases the

Figure 9–26 The binding of tryptophan to the tryptophan repressor protein changes the conformation of the repressor. The conformational change enables this gene regulatory protein to bind tightly to a specific DNA sequence (the operator), thereby blocking transcription of the genes encoding the enzymes required to produce tryptophan (the *trp* operon). The three-dimensional structure of this bacterial helix-turn-helix protein, as determined by x-ray diffraction with and without tryptophan bound, is illustrated. Tryptophan binding increases the distance between the two recognition helices in the homodimer, allowing the repressor to fit snugly on the operator. (Adapted from R. Zhang et al., *Nature* 327:591–597, 1987. © 1987 Macmillan Magazines Ltd.)

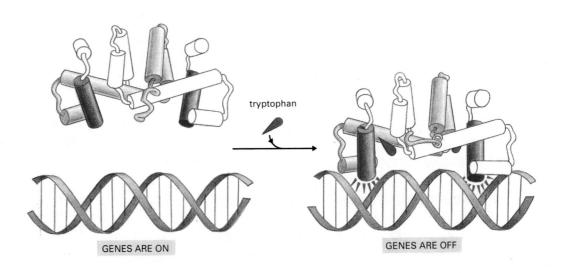

GENES ARE ON

tryptophan

GENES ARE OFF

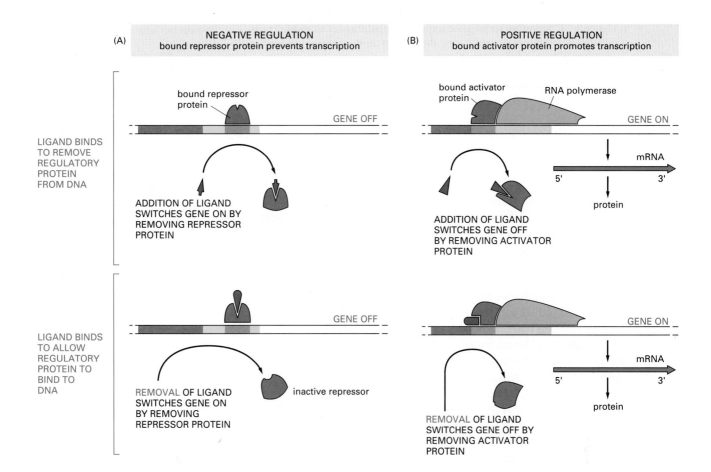

NEGATIVE REGULATION
bound repressor protein prevents transcription

(B) **POSITIVE REGULATION**
bound activator protein promotes transcription

LIGAND BINDS TO REMOVE REGULATORY PROTEIN FROM DNA

bound repressor protein

GENE OFF

ADDITION OF LIGAND SWITCHES GENE ON BY REMOVING REPRESSOR PROTEIN

bound activator protein

RNA polymerase

GENE ON

mRNA

5'   3'

protein

ADDITION OF LIGAND SWITCHES GENE OFF BY REMOVING ACTIVATOR PROTEIN

LIGAND BINDS TO ALLOW REGULATORY PROTEIN TO BIND TO DNA

GENE OFF

REMOVAL OF LIGAND SWITCHES GENE ON BY REMOVING REPRESSOR PROTEIN

inactive repressor

GENE ON

mRNA

5'   3'

protein

REMOVAL OF LIGAND SWITCHES GENE OFF BY REMOVING ACTIVATOR PROTEIN

probability that a transcript will be initiated. Because the active, DNA-binding form of such a protein turns genes *on,* this mode of gene regulation is called **positive control,** and the gene regulatory proteins that function in this manner are known as *transcriptional activators* or *gene activator proteins.*

As in negative control by a transcriptional repressor, a transcriptional activator can operate as a simple on-off genetic switch. The bacterial activator protein *CAP (catabolite activator protein),* for example, activates genes that enable *E. coli* to use alternative carbon sources when glucose, its preferred carbon source, is not available. Falling levels of glucose induce an increase in the intracellular signaling molecule cyclic AMP, which binds to the CAP protein, enabling it to bind to its specific DNA sequence near target promoters and thereby turn on the appropriate genes. In this way the expression of a target gene is switched on or off, depending on whether cyclic AMP levels in the cell are high or low, respectively. A summary of the different ways that positive and negative control can be used to regulate genes is shown in Figure 9–27.

In many respects transcriptional activators and transcriptional repressors are similar in design. The tryptophan repressor and the transcriptional activator CAP, for example, both use a helix-turn-helix motif and both require a small cofactor in order to bind DNA. In fact, some bacterial proteins (including CAP and the bacteriophage lambda repressor) are known to act as either activators or repressors, depending on the exact placement of the DNA sequence they recognize with respect to the promoter: if the binding site for the protein overlaps the promoter, the polymerase cannot bind and the protein acts as a repressor (Figure 9–28).

Figure 9–27 Summary of the mechanisms by which specific gene regulatory proteins control gene transcription in procaryotes. (A) Negative regulation; (B) positive regulation. Note that the addition of an inducing ligand can turn on a gene either by removing a gene repressor protein from the DNA (*upper left panel*) or by causing a gene activator protein to bind (*lower right panel*). Likewise, the addition of an inhibitory ligand can turn off a gene either by removing a gene activator protein from the DNA (*upper right panel*) or by causing a gene repressor protein to bind (*lower left panel*).

A Transcriptional Activator and a Transcriptional Repressor Control the *lac* Operon [23]

More complicated types of genetic switches can be constructed by combining positive and negative controls. The *lac operon* in *E. coli,* for example, unlike the

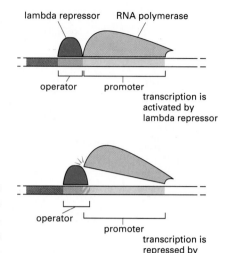

Figure 9–28 Some bacterial gene regulatory proteins act as either transcriptional activators or repressors, depending on the precise placement of their binding sites in DNA. An example is the bacteriophage lambda repressor. The protein acts as a transcriptional activator by providing a favorable contact for RNA polymerase (*top*). At the bottom of the figure the operator is located one base pair closer to the promoter, and, instead of helping polymerase, the repressor now competes with it for binding to the DNA. The lambda repressor recognizes its operator by a helix-turn-helix motif, as shown in Figure 9–11.

trp operon, is under both negative and positive transcriptional controls by the lac repressor protein and CAP, respectively. The *lac* operon codes for proteins required to transport the disaccharide lactose into the cell and to break it down. CAP, as we have seen, enables bacteria to use alternative carbon sources such as lactose in the absence of glucose. It would be wasteful, however, for CAP to induce expression of the *lac* operon if lactose is not present, and the lac repressor ensures that the *lac* operon is shut off in the absence of lactose. This arrangement enables the *lac* operon to respond to and integrate two different signals, so that it is expressed only when two conditions are met: lactose must be present and glucose must be absent. Any of the other three possible signal combinations maintain the gene in the off state (Figure 9–29).

The simple logic of this genetic switch first attracted the attention of biologists over 50 years ago. As explained above, the molecular basis of the switch was uncovered by a combination of genetics and biochemistry, providing the first insight into how gene expression is controlled. Although the same basic strategies are used to control gene expression in higher organisms, the genetic switches that are used are much more complex.

Regulation of Transcription in Eucaryotic Cells Is Complex [24]

The two-signal switching mechanism that regulates the *lac* operon is elegant and simple. However, it is difficult to imagine how additional complexity could be

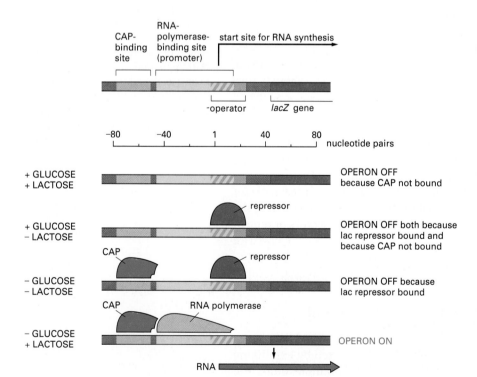

Figure 9–29 Dual control of the *lac* operon. Glucose and lactose levels control the initiation of transcription of the *lac* operon through their effects on the lac repressor protein and CAP. Lactose addition increases the concentration of allolactose, which binds to the repressor protein and removes it from the DNA. Glucose addition decreases the concentration of cyclic AMP; because cyclic AMP no longer binds to CAP, this gene activator protein dissociates from the DNA, turning off the operon. As shown in Figure 9–8, CAP is known to induce a bend in the DNA when it binds; for simplicity, the bend is not shown here. *LacZ*, the first gene of the *lac* operon, encodes the enzyme β-galactosidase, which breaks down lactose to galactose and glucose.

added so that a much larger number of signals could regulate transcription from the operon: one would quickly run out of room to insert new regulatory DNA sequences. How have eucaryotes overcome these limitations to create more complex genetic switches?

The regulation of transcription in eucaryotes differs in two important ways from that typically found in bacteria. First, eucaryotic RNA polymerases cannot initiate transcription on their own. They require a set of proteins called *general transcription factors,* which must be assembled at the promoter before transcription can begin. This assembly process provides multiple steps at which the rate of transcription initiation can be speeded up or slowed down in response to regulatory signals, and many eucaryotic gene regulatory proteins operate by influencing these steps. Second, most gene regulatory proteins in eucaryotes can act even when they are bound to DNA thousands of nucleotide pairs away from the promoter that they influence, which means that a single promoter can be controlled by an almost unlimited number of regulatory sequences scattered along the DNA. We consider these two features of eucaryotic gene regulation in turn.

Eucaryotic RNA Polymerase Requires General Transcription Factors [25]

The initial finding that purified eucaryotic RNA polymerase enzymes could not initiate transcription *in vitro* led to the discovery and purification of additional proteins, the **general transcription factors,** required for this process. These proteins were not simply missing subunits of the polymerase; they had to assemble into a complex on the DNA at the promoter in order to recruit the RNA polymerase to this site.

Figure 9–30 shows how the general transcription factors are thought to assemble at promoters utilized by *RNA polymerase II (Pol II),* the polymerase that transcribes the vast majority of eucaryotic genes. The assembly process starts with the binding of TFIID to the *TATA sequence,* a short double-helical DNA sequence primarily composed of T and A nucleotides. The TATA sequence is a component of nearly all promoters utilized by Pol II and is typically located 25 nucleotides upstream from the transcription start site. TFIID is composed of many subunits; that responsible for recognizing the TATA sequence is called TBP (TATA-binding protein) (Figure 9–31). Once TFIID is bound to this DNA site, the other general transcription factors, along with RNA Pol II, are added in turn.

After Pol II has been tethered to the promoter, it must be released from the complex of general transcription factors to begin transcription. A key step in the initiation of transcription is carried out by TFIIH, one subunit of which is a protein kinase that phosphorylates Pol II. For at least some promoters, this phosphorylation is thought to disengage the polymerase and allow transcription to begin.

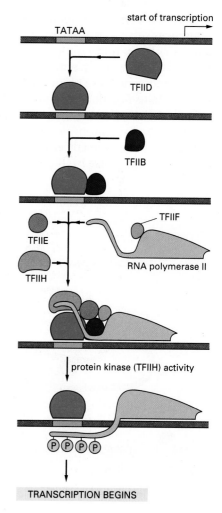

Figure 9–30 **Assembly of the general transcription factors required for the initiation of transcription by RNA polymerase II.** In the first step, TFIID binds specifically to a TATA sequence. Next, TFIIB enters the complex, followed by RNA polymerase II (Pol II) escorted by TFIIF. TFIIE and TFIIH then assemble into the complex. In the presence of ATP, TFIIH phosphorylates Pol II, which releases the polymerase so that it can initiate transcription. The site of phosphorylation is a long polypeptide tail that extends from the largest subunit of Pol II. In mammals it is composed of 52 repeats of the amino acid sequence YSPTSPS (see Figure 8–42), and it is the serine (S) and threonine (T) side chains in the repeat that are phosphorylated. The general transcription factors have been highly conserved in evolution: some of those isolated from human cells, for example, can be replaced in biochemical experiments by the equivalent factors from yeast.

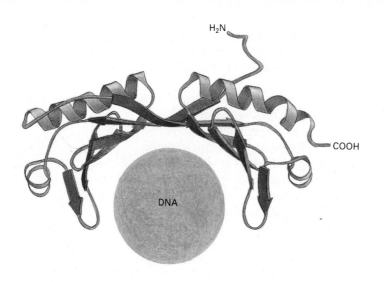

Figure 9–31 Three-dimensional structure of TBP. The structure of this subunit of TFIID resembles a saddle that fits over the DNA double helix. Although the protein is organized with twofold symmetry, it is a single polypeptide chain. It seems likely that this protein was once a symmetric dimer and that the gene encoding the original monomer duplicated and fused during evolution. TBP recognizes DNA using a motif that is distinct from those discussed earlier in the chapter. No other protein is known to bind DNA in this way, reflecting, perhaps, TBP's unique role in the cell: it participates in the initiation of transcription of all genes. Although not shown in the figure, the DNA is severely deformed when TBP is bound. This deformation—two kinks separated by partially unwound DNA—may provide a landmark that helps to attract the other general transcription factors. (Adapted from D.B. Nikolov et al., *Nature* 360:40–45, 1992. ©1992 Macmillan Magazines Ltd.)

The other two RNA polymerases found in eucaryotes, RNA polymerase I (Pol I) and RNA polymerase III (Pol III), also require a set of general transcription factors. Although TBP is required for all three polymerases, the other factors are different from those that assemble at Pol II promoters, and the binding of TBP at Pol I and Pol III promoters does not depend on a TATA sequence in the DNA.

Enhancers Control Genes at a Distance [26]

It was surprising to many biologists when, in 1979, it was discovered that DNA sequences thousands of nucleotide pairs away from a eucaryotic promoter could activate transcription from the promoter. It is now known that such **enhancer** sequences serve as specific binding sites for gene regulatory proteins that activate or *enhance* transcription and that this sort of "action-at-a-distance" is the rule rather than the exception for gene regulatory proteins in eucaryotic cells. This phenomenon also occurs, although less commonly, in procaryotes. How do such proteins function over these long distances? Many models have been proposed, but it seems that the simplest of these is correct. The DNA between the enhancer and the promoter loops out to allow the proteins bound to the enhancer to interact directly either with one of the general transcription factors or with RNA polymerase itself (Figure 9–32). The DNA thus acts as a tether, caus-

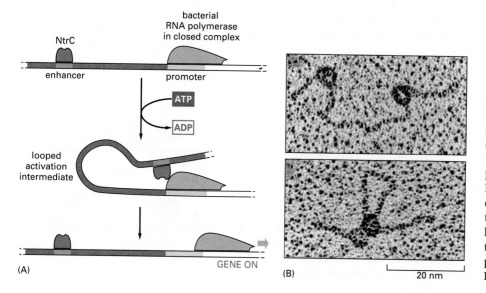

(A)

(B)

20 nm

Figure 9–32 Gene activation at a distance. (A) NtrC is a bacterial gene regulatory protein that activates transcription by facilitating the transition between closed and open RNA polymerase complexes (discussed in Chapter 8). Although not usually the case for procaryotic RNA polymerases, the transition stimulated by NtrC requires the energy produced by ATP hydrolysis. (B) The interaction of NtrC and RNA polymerase, with the intervening DNA looped out, can be seen in the electron microscope. Although transcriptional activation by DNA looping is unusual in bacteria, it is typical of eucaryotic gene regulatory proteins. (B, courtesy of Harrison Echols.)

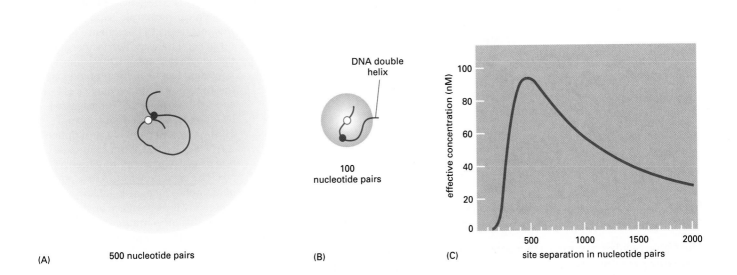

(A) 500 nucleotide pairs

(B)

DNA double helix

100 nucleotide pairs

(C)

ing a protein bound to an enhancer even thousands of nucleotide pairs away to collide repeatedly with proteins bound to the promoter. This is the same effect as would be obtained by increasing the protein's local concentration at the promoter (Figure 9–33).

A Eucaryotic Gene Control Region Consists of a Promoter Plus Regulatory DNA Sequences [27]

Because eucaryotic gene regulatory proteins can control transcription when bound to DNA far away from the promoter, the DNA sequences that control the expression of a gene can be spread over long stretches of DNA. Here we use the term **gene control region** to refer to the DNA sequences required to initiate gene transcription plus those required to regulate the rate at which initiation occurs. Thus a eucaryotic gene control region consists of the **promoter,** where the general transcription factors and the polymerase assemble, plus all of the **regulatory sequences** to which gene regulatory proteins bind to control the rate of these assembly processes at the promoter (Figure 9–34).

In higher eucaryotes it is not unusual to find the regulatory sequences of a gene dotted over distances as great as 50,000 nucleotide pairs, although much of this DNA serves as "spacer" sequence and is not recognized by gene regulatory proteins. In this chapter we use the term **gene** to refer only to the DNA that is transcribed into RNA (see Figure 9–34), although the classical view of a gene would include the gene control region as well. The different definitions arise from the different ways in which genes were historically identified, and modern discoveries have complicated even the narrowly defined sense of the word—a point we return to later in this chapter.

Although most gene regulatory proteins bind to enhancer sequences and activate gene transcription, some function as negative regulators, as we see below. In contrast to the small number of general transcription factors, which are abundant proteins and assemble on the promoters of all genes transcribed by Pol II, there are thousands of different gene regulatory proteins. These vary from gene to gene, and each is usually present in very small amounts in a cell. Most of them recognize their specific DNA sequences using one of the DNA-binding motifs discussed previously. These proteins allow individual genes of an organism to be turned on or off specifically. In a higher eucaryote different selections of gene regulatory proteins are present in different cell types.

Figure 9–33 Binding of two proteins to separate sites on the DNA double helix can greatly increase their probability of interacting. (A) The tethering of one protein to the other via an intervening DNA loop of 500 nucleotide pairs increases their frequency of collision. The intensity of *blue* coloring reflects the probability that the *red* protein will be located at each position in space relative to the *white* protein. (B) The flexibility of DNA is such that an average sequence makes a smoothly graded 90° bend (a curved turn) about once every 200 nucleotide pairs. Thus, when two proteins are tethered by only 100 nucleotide pairs, their contact is relatively restricted. In such cases the protein interaction is facilitated when the two protein-binding sites are separated by a multiple of about 10 nucleotide pairs, which places both proteins on the same side of the DNA helix (which has 10 nucleotides per turn) and thus on the inside of the DNA loop, where they can best reach each other. (C) Theoretical effective concentration of the *red* protein at the site where the *white* protein is bound as a function of their separation. (C, courtesy of Gregory Bellomy, modified from M.C. Mossing and M.T. Record, *Science* 233:889–892, 1986. © 1986 the AAAS.)

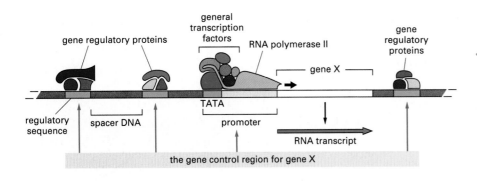

gene regulatory proteins

general transcription factors

RNA polymerase II

gene regulatory proteins

gene X

TATA

regulatory sequence

spacer DNA

promoter

RNA transcript

the gene control region for gene X

Figure 9–34 The gene control region of a typical eucaryotic gene. The _promoter_ is the DNA sequence where the general transcription factors and the polymerase assemble. The most important feature of the promoter is the TATA box, a short sequence of T-A and A-T base pairs that is recognized by the general transcription factor TFIID. The start point of transcription is typically located about 25 nucleotide pairs downstream from the TATA box. The _regulatory sequences_ serve as binding sites for gene regulatory proteins, whose presence on the DNA affects the rate of transcription initiation. These sequences can be located adjacent to the promoter, far upstream of it, or even downstream of the gene. DNA looping is thought to allow gene regulatory proteins bound at any of these positions to interact with the proteins that assemble at the promoter. Whereas the general transcription factors that assemble at the promoter are similar for all polymerase II transcribed genes, the gene regulatory proteins and the locations of their binding sites relative to the promoter are different for each gene.

Many Gene Activator Proteins Accelerate the Assembly of General Transcription Factors [28]

Most gene regulatory proteins that activate gene transcription—that is, most **gene activator proteins**—have a modular design consisting of at least two distinct domains. One domain usually contains one of the structural motifs discussed previously that recognizes a specific regulatory DNA sequence. In the simplest cases another domain contacts the transcription machinery and accelerates the rate of transcription initiation. This type of modular design was first revealed by experiments in which genetic engineering techniques were used to create a hybrid protein containing the _activation domain_ of one protein fused to the _DNA-binding domain_ of a different protein (Figure 9–35).

In one class of gene activator proteins the activation domain contains a cluster of negatively charged (acidic) amino acids on its surface. These _acidic activators_ work by accelerating the assembly of the general transcription factors at the promoter. In principle, any one of the assembly steps shown in Figure 9–30 could be the rate-limiting step of transcription initiation. At some promoters it is the entry of TFIIB into the complex that appears to be the limiting step, and

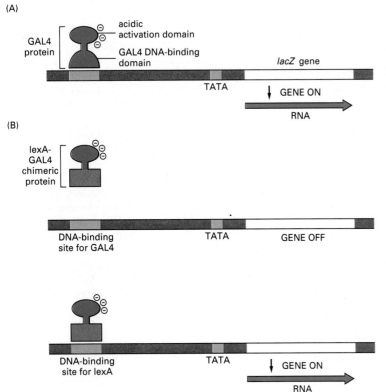

(A)

GAL4 protein

acidic activation domain

GAL4 DNA-binding domain

lacZ gene

TATA

GENE ON

RNA

(B)

lexA-GAL4 chimeric protein

DNA-binding site for GAL4

TATA

GENE OFF

DNA-binding site for lexA

TATA

GENE ON

RNA

Figure 9–35 The modular structure of a gene activator protein. Outline of a domain-swap experiment that reveals the presence of independent DNA-binding and transcription-activating domains in the yeast gene activator protein GAL4. A functional activator can be reconstituted from the carboxyl-terminal portion of the GAL4 protein if it is attached to the DNA-binding domain of a bacterial gene regulatory protein (the lexA protein) by gene fusion techniques. When the resulting bacterial-yeast hybrid protein is produced in yeast cells, it will activate transcription from yeast genes provided that the specific DNA-binding site for the bacterial protein has been inserted next to them. (A) The normal activation of gene transcription produced by the GAL4 protein. (B) The chimeric gene regulatory protein requires the lexA-protein DNA-binding site for its activity.

GAL4 is normally responsible for activating the transcription of yeast genes that code for the enzymes that convert galactose to glucose. For the experiments shown here, the control region for one of these genes was fused to the _E. coli lacZ_ gene, which codes for the enzyme β-galactosidase (see Figure 9–29). β-galactosidase is very simple to detect biochemically and thus provides a convenient way to monitor the expression level specified by a gene control region; _lacZ_ thus serves as a _reporter gene_ (see p. 321).

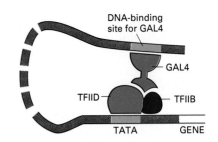

Figure 9–36 **A model for the action of acidic activators.** The GAL4 gene activator protein, bound to DNA in the rough vicinity of the promoter, facilitates the addition of TFIIB to the nascent complex of general transcription factors. DNA-bound activator proteins typically increase the rate of transcription by up to 1000-fold, which is consistent with a relatively weak and nonspecific interaction between the activator and the general transcription factors (a 1000-fold change in affinity corresponds to a change in ΔG of ~4 kcal/mole, which could be accounted for by only a few weak, noncovalent bonds).

acidic activators are thought to overcome this block by helping TFIIB assemble into the complex (Figure 9–36). Other activator proteins seem to work by loading TFIID onto the DNA, whereas yet others may activate transcription by a different mechanism.

In principle, the protein assembly process at a promoter could have more than one slow step, and a maximal level of transcription could be dependent on several gene activator proteins bound upstream, each speeding up a different step. It is not difficult to see how multiple gene regulatory proteins, each binding to a different regulatory sequence, could control the transcription of a eucaryotic gene. But whatever its precise mechanism of action, a gene regulatory protein must be bound to DNA, either directly or indirectly, to influence transcription of its target promoter.

Gene Repressor Proteins Can Inhibit Transcription in Various Ways [29]

Eucaryotic gene activator proteins work in a way that is similar in principle to that described earlier for bacterial gene activators: they catalyze polymerase action by contacting the transcription machinery. Not all eucaryotic gene regulatory proteins activate transcription, however. As mentioned previously, many act as **gene repressor proteins** to suppress transcription. Unlike bacterial repressors,

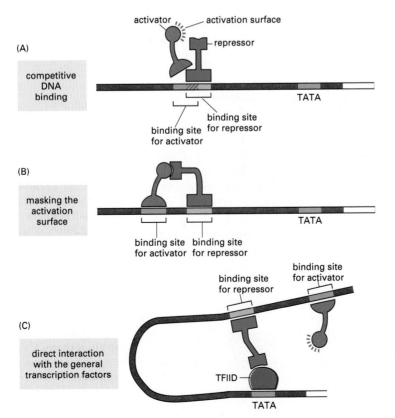

Figure 9–37 **Three ways in which eucaryotic gene repressor proteins can operate.** In the first (A) gene activator proteins and gene repressor proteins compete for binding to the same regulatory DNA sequence. In the second (B) both proteins can bind DNA, but the repressor complexes with the activation domain of the activator protein and thereby prevents it from contacting the transcription machinery. In the third (C) the repressor interacts with an early stage of the assembling complex of general transcription factors, blocking further assembly. A fourth mechanism of negative control—inactivation of a transcriptional activator by heterodimerization—was illustrated in Figure 9–21.

most do not act by directly competing with the polymerase for access to the DNA. Although their precise mechanisms of action are not yet understood, it appears that different repressors work in different ways, three of which are described in Figure 9–37.

In addition to gene repressor proteins that act on individual genes, eucaryotic cells utilize another mechanism to shut off the expression of genes in large regions of chromosomes. This mechanism depends on altering chromatin structure, as we discuss later.

Eucaryotic Gene Regulatory Proteins Often Assemble into Small Complexes on DNA [30]

Although some eucaryotic gene regulatory proteins may work individually, most act as part of a complex composed of several polypeptides, each with a distinct function. The complex often assembles only in the presence of the appropriate DNA sequence. In some well-studied cases, for example, two gene regulatory proteins with a weak affinity for each other cooperate to bind to a DNA sequence, neither protein being able to bind to the DNA site on its own. Once bound to DNA, the protein dimer creates a distinct surface that is recognized by a third protein that carries an activator domain that stimulates transcription (Figure 9–38). This example illustrates an important general point: protein-protein interactions that are too weak to cause proteins to assemble in solution can cause the proteins to assemble on DNA; in this way the DNA sequence acts as a nucleation site for the assembly of a protein complex.

An individual gene regulatory protein can often participate in more than one type of regulatory complex. A protein might function, for example, in one case as part of a complex that activates transcription and in another case as part of a complex that represses transcription (see Figure 9–38). Thus individual eucaryotic gene regulatory proteins are not necessarily dedicated activators or repressors; instead, they function as regulatory units that are used to generate complexes whose function depends on the final assembly of all of the individual components.

We saw earlier how the formation of gene regulatory heterodimers in solution provides a mechanism for the combinatorial control of gene expression. The assembly of small complexes of gene regulatory proteins on DNA provides a second mechanism for combinatorial control (see Figure 9–38).

Complex Genetic Switches That Regulate *Drosophila* Development Are Built Up from Smaller Modules [31]

Given that gene regulatory proteins can be positioned at multiple sites along long stretches of DNA, that these proteins can assemble into complexes at each site, and that the complexes can influence in different ways the ordered assembly of the general transcription factors at the promoter, there would seem to be almost limitless possibilities for creating elaborate switches for the control of eucaryotic gene transcription.

A particularly striking example of such a complex, multicomponent genetic switch is that controlling the transcription of the *Drosophila even-skipped (eve)*

Figure 9–38 **Eucaryotic gene regulatory proteins often assemble into small complexes on DNA.** Five gene regulatory proteins are shown in (A). The nature and function of the complex they form depends on the specific DNA sequence that nucleates their assembly. In (B) one complex that assembles activates gene transcription, while another represses transcription. Note that the *green* protein is shared by both the activating and the repressing complexes.

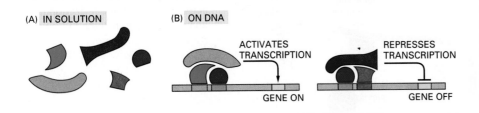

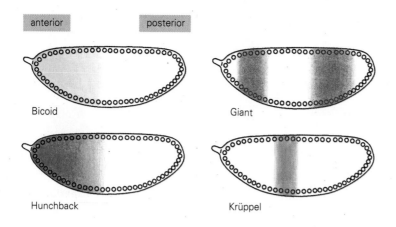

anterior   posterior

Bicoid

Giant

Hunchback

Krüppel

Figure 9–39 **The nonuniform distribution of four gene regulatory proteins in an early *Drosophila* embryo.** At this stage the embryo is a syncytium, with multiple nuclei in a common cytoplasm. Although it is not clear in these drawings, all of these proteins are concentrated in the nuclei.

gene, whose expression plays an important role in the development of the *Drosophila* embryo. If this gene is inactivated by mutation, many parts of the embryo fail to form, and the embryo dies early in development. As discussed in Chapter 21, at the earliest stage of development where *eve* is expressed, the embryo is a single giant cell containing multiple nuclei in a common cytoplasm. The cytoplasm is not uniform, however: it contains a mixture of gene regulatory proteins that are distributed unevenly along the length of the embryo, thus providing *positional information* that distinguishes one part of the embryo from another (Figure 9–39). (The way these differences are set up is discussed in Chapter 21.) Although the nuclei are initially identical, they rapidly begin to express different genes because they are exposed to different gene regulatory proteins. The nuclei near the anterior end of the developing embryo, for example, are exposed to a set of gene regulatory proteins that is distinct from the set that influences nuclei at the posterior end of the embryo.

The regulatory DNA sequences of the *eve* gene are designed to read the concentrations of gene regulatory proteins at each position along the length of the embryo and to interpret this information in such a way that the *eve* gene is expressed in seven stripes, each initially five to six nuclei wide and positioned precisely along the anterior-posterior axis of the embryo (Figure 9–40). How is this remarkable feat of information processing carried out? Although the molecular details are not yet understood, several general principles have emerged from studies of *eve* and other *Drosophila* genes that are similarly regulated.

The regulatory region of the *eve* gene is very large (approximately 20,000 nucleotide pairs). It is formed from a series of relatively simple regulatory modules, each of which contains multiple regulatory sequences and is responsible for specifying a particular stripe of *eve* expression along the embryo. This modular organization of the *eve* gene control region is revealed by experiments in which a particular regulatory module (say, that specifying stripe 2) is removed from its normal setting upstream of the *eve* gene, placed in front of a reporter gene, and reintroduced into the *Drosophila* genome (Figure 9–41A). When developing embryos derived from flies carrying this genetic construct are examined, the reporter gene is found to be expressed in precisely the position of stripe 2 (Figure 9–41B). Similar experiments reveal the existence of other regulatory modules, each of which specifies one of the other six stripes.

The *Drosophila eve* Gene Is Regulated by Combinatorial Controls [32]

A detailed study of the stripe 2 regulatory module has provided insights into how it reads and interprets positional information. It contains recognition sequences for two gene regulatory proteins (Bicoid and Hunchback) that activate *eve* tran-

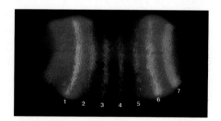

Figure 9–40 **The seven stripes of the protein encoded by the *even-skipped* (eve) gene in a developing *Drosophila* embryo.** Two and one-half hours after fertilization, the egg was fixed and stained with antibodies that recognize the Eve protein (*green*) and antibodies that recognize the Giant protein (*red*). Where Eve and Giant proteins are both present, the staining appears *yellow*. At this point in development, the egg contains approximately 4000 nuclei. The Eve and Giant proteins are both located in the nuclei, and the *eve* stripes are about four nuclei wide. The staining pattern of the Giant protein is also shown in Figure 9–39. (Courtesy of Michael Levine.)

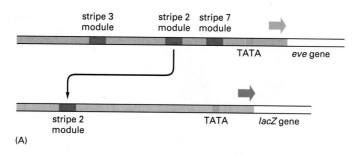

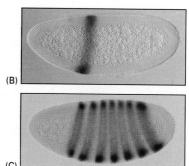

(B)

(C)

Figure 9–41 **Experiment demonstrating the modular construction of the *eve* gene regulatory region.** (A) A 480-nucleotide-pair piece of the *eve* regulatory region was removed and inserted upstream of a test promoter that directs the synthesis of the enzyme β-galactosidase (the product of the *E. coli lacZ* gene). (B) When this artificial construct was reintroduced into the genome of *Drosophila* embryos, the embryos expressed β-galactosidase (detectable by histochemical staining) precisely in the position of the second of the seven *eve* stripes (C). (B and C, courtesy of Stephen Small and Michael Levine.)

scription and two (Krüppel and Giant) that repress it (Figure 9–42). (The gene regulatory proteins of *Drosophila* often have colorful names reflecting the phenotype that results if the gene encoding the protein is inactivated by mutation.) The relative concentrations of these four proteins determine whether protein complexes form at the stripe 2 module that turn on transcription of the *eve* gene. Figure 9–43 shows the distributions of the four gene regulatory proteins across the region of a *Drosophila* embryo where stripe 2 forms. Although the precise details are not known, it seems likely that either one of the two repressor proteins, when bound to the DNA, will turn off the stripe 2 module, whereas *both* Bicoid and Hunchback must bind for maximal activation of the module. This simple regulatory unit thereby integrates these four positional signals so as to turn on the stripe 2 module (and therefore the expression of the *eve* gene) only in those nuclei that are located where the levels of both Bicoid and Hunchback are high and both Krüppel and Giant are absent. This combination of activators and repressors occurs only in one region of the early embryo; everywhere else, therefore, the stripe 2 module is off (and therefore silent).

We have previously discussed two mechanisms of combinatorial control of gene expression—heterodimerization of gene regulatory proteins in solution (see Figure 9–19) and the assembly of combinations of gene regulatory proteins into small complexes on DNA (see Figure 9–38). It is likely that both mechanisms participate in the complex regulation of *eve* expression. In addition, the regulation of stripe 2 just described illustrates a third type of combinatorial control. Because the individual regulatory sequences in the *eve* stripe 2 module are strung out along the DNA, many sets of gene regulatory proteins can be bound simultaneously and influence the promoter of a gene. The promoter integrates the transcriptional cues provided by all of the bound proteins (Figure 9–44).

The regulation of *eve* expression is an extreme example of combinatorial control. Seven combinations of gene regulatory proteins—one combination for each stripe—activate *eve* expression, while many other combinations (all those found in the interstripe regions) keep the stripe elements silent. The other stripe regulatory modules are thought to be constructed along lines similar to those described for stripe 2, being designed to read positional information provided by other combinations of gene regulatory proteins. The entire gene control region

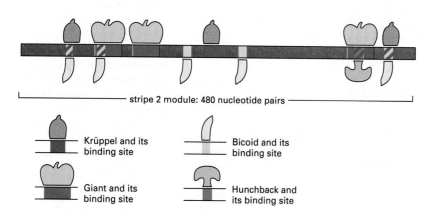

stripe 2 module: 480 nucleotide pairs

Krüppel and its binding site

Giant and its binding site

Bicoid and its binding site

Hunchback and its binding site

Figure 9–42 **Close-up view of the *eve* stripe 2 unit.** The segment of the *eve* gene control region identified in the previous figure contains regulatory sequences, each of which binds one or another of four gene regulatory proteins. It is known from genetic experiments that these four regulatory proteins are responsible for the proper expression of stripe 2 of *eve*. Flies that are deficient in the two gene activators Bicoid and Hunchback, for example, fail to express stripe 2 of *eve* efficiently. In flies deficient in either of the two gene repressors, Giant and Krüppel, stripe 2 expands and covers an abnormally broad region of the embryo. The DNA-binding sites for these gene regulatory proteins were determined by cloning the genes encoding the proteins, overexpressing the proteins in *E. coli*, purifying them, and performing DNA-footprinting experiments as described in Chapter 7. The top diagram indicates that, in some cases, the binding sites for the gene regulatory proteins overlap and the proteins can compete for binding to the DNA. For example, the binding of Krüppel and Bicoid to the site at the far right is thought to be mutually exclusive.

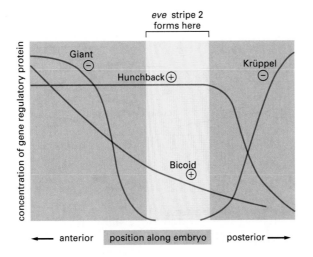

eve stripe 2 forms here

Giant ⊖

Hunchback ⊕

Krüppel ⊖

Bicoid ⊕

← anterior   position along embryo   posterior →

Figure 9–43 Distribution of the gene regulatory proteins responsible for ensuring that *eve* is expressed in stripe 2. The distributions of these proteins were visualized by staining a developing *Drosophila* embryo with antibodies directed against each of the four proteins (Figure 9–39). The expression of *eve* in stripe 2 occurs only at the position where the two activators (Bicoid and Hunchback) are present and the two repressors (Giant and Krüppel) are absent. In fly embryos that lack Krüppel, for example, stripe 2 expands posteriorly. Likewise, stripe 2 expands posteriorly if the DNA-binding sites for Krüppel in the stripe 2 module (see Figure 9–41) are inactivated by mutation and this regulatory region is reintroduced into the genome.

The *eve* gene itself encodes a gene regulatory protein, which, after its pattern of expression is set up in seven stripes, in turn regulates the expression of other *Drosophila* genes. As development proceeds, the embryo is thus subdivided into finer and finer regions that eventually give rise to the different body parts of the adult fly, as discussed in Chapter 21.

is strung out over 20,000 nucleotide pairs of DNA and binds more than 20 different proteins. A large and complex control region is thereby built from a series of smaller modules, each of which consists of a unique arrangement of short DNA sequences recognized by specific gene regulatory proteins. An important requirement of this strategy is the absence of cross-talk between the modules: the state of one module should not affect that of the others. How this molecular insulation is achieved is unknown. In this way, however, a single gene can respond to an enormous number of combinatorial inputs.

Complex Mammalian Gene Control Regions Are Also Constructed from Simple Regulatory Modules [33]

It has been estimated that several percent of the coding capacity of a mammalian genome is devoted to the synthesis of proteins that serve as regulators of gene transcription. This reflects the exceedingly complex controls that regulate the expression of mammalian genes. It is not unusual, for example, to find a gene with a control region that is 50,000 nucleotide pairs in length, in which many modules, each containing a number of regulatory sequences that bind gene regulatory proteins, are interspersed with long stretches of spacer DNA.

One of the best-understood examples of a complex mammalian regulatory region is found in the human β-globin gene, which is expressed exclusively in red blood cells and at a specific time in their development. A complex array of gene regulatory proteins controls the expression of the gene, some acting as activators and others as repressors (Figure 9–45). The concentrations (or activities) of many of these gene regulatory proteins are thought to change during development, and

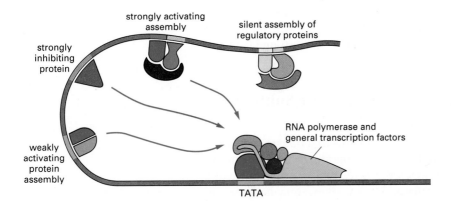

strongly activating assembly

silent assembly of regulatory proteins

strongly inhibiting protein

RNA polymerase and general transcription factors

weakly activating protein assembly

TATA

Figure 9–44 Integration at a promoter. Multiple sets of gene regulatory proteins can work together to influence a promoter, as they do in the *eve* stripe 2 module illustrated previously in Figure 9–42. It is not yet understood in detail how the integration of multiple inputs is achieved.

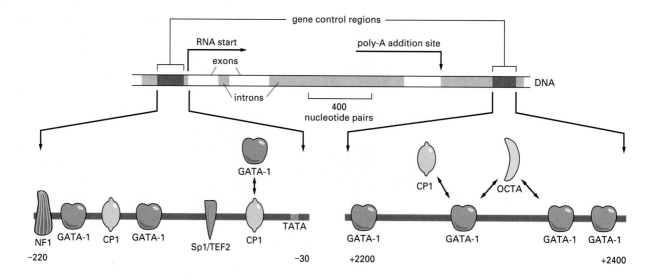

only a particular combination of all the proteins triggers transcription of the gene. We see later that the β-globin gene is also subject to a second, higher layer of control that involves global changes in chromatin structure.

The Activity of a Gene Regulatory Protein Can Itself Be Regulated [34]

The strategies for regulating the *eve* gene and the human β-globin gene are similar in that the gene control regions respond to a bewildering array of gene regulatory proteins. *Drosophila* is unusual, however, in the way that the spatial distribution of gene regulatory proteins in the cytoplasm controls gene expression. As discussed previously, the early *Drosophila* embryo is a single giant cell that contains thousands of nuclei in a common cytoplasm, and the gene regulatory proteins themselves are distributed in complex spatial patterns so that different nuclei are exposed to different concentrations of the proteins. These gene regulatory proteins enter the nuclei directly to activate or repress transcription of their target genes. In most embryos of other organisms, individual nuclei are in separate cells, and extracellular positional information must either pass across the plasma membrane or, more usually, generate signals in the cytosol in order to influence the genome.

The mechanisms by which extracellular signals communicate their message across the plasma membrane to gene regulatory proteins inside the cell are discussed in Chapter 15. Here we need deal only with the final steps in the intracellular signaling cascades activated by extracellular signals—the steps in which the activity of gene regulatory proteins is altered. In many cases the gene regulatory protein is present in the cell in an inactive form and a signal alters the protein so as to activate it. The protein may be activated by phosphorylation catalyzed by a protein kinase, for example, or it may be released from a tight complex with a second protein that otherwise holds the gene regulatory protein in the cytosol, preventing it from entering the nucleus. These and some other ways of controlling the activity of gene regulatory proteins are illustrated in Figure 9–46.

Bacteria Use Interchangeable RNA Polymerase Subunits to Help Regulate Gene Transcription [35]

We have seen the importance of gene regulatory proteins that bind to regulatory sequences in DNA and signal to the transcription apparatus whether or not to start the synthesis of an RNA chain. Although this is the main way of controlling transcriptional initiation in both eucaryotes and procaryotes, some bacteria and

Figure 9–45 Model for the control of the human β-globin gene. The diagram shows some of the gene regulatory proteins thought to control expression of the gene during red blood cell development (see Figure 9–52). Some of the gene regulatory proteins shown, such as CP1, are found in many types of cells, while others, such as GATA-1, are present in only a few types of cells including red blood cells and therefore are thought to contribute to the cell-type specificity of β-globin gene expression. As indicated by the double-headed arrows, several of the binding sites for GATA-1 overlap those of other gene regulatory proteins; it is thought that occupancy of these sites by GATA-1 excludes binding of other proteins. (Adapted from B. Emerson, In Gene Expression: General and Cell-Type-Specific [M. Karin, ed.], pp. 116–161. Boston: Birkhauser, 1993.)

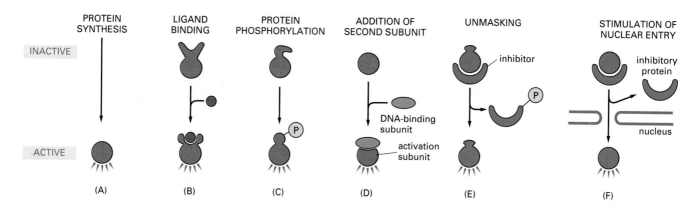

PROTEIN SYNTHESIS | LIGAND BINDING | PROTEIN PHOSPHORYLATION | ADDITION OF SECOND SUBUNIT | UNMASKING | STIMULATION OF NUCLEAR ENTRY

INACTIVE

inhibitor

inhibitory protein

ACTIVE

DNA-binding subunit

activation subunit

P

nucleus

(A)   (B)   (C)   (D)   (E)   (F)

their viruses use an additional strategy based on interchangeable subunits of RNA polymerase. As described in Chapter 8, a *sigma (σ) subunit* is required for the bacterial RNA polymerase to recognize a promoter. Some bacteria make several different sigma subunits, each of which can interact with the RNA polymerase core and direct it to different specific promoters. This scheme permits one large set of genes to be turned off and a new set to be turned on simply by replacing one sigma subunit with another. The strategy is efficient because it bypasses the need to deal with the genes one by one, and it is often used by bacterial viruses to activate several sets of genes rapidly and sequentially (Figure 9–47).

In a sense, eucaryotes employ an analogous strategy through the use of three distinct RNA polymerases (I, II, and III) that share some of their subunits. Procaryotes, in contrast, use only one type of core RNA polymerase molecule but modify it with different sigma subunits.

Gene Switches Have Gradually Evolved

We have seen that the control regions of eucaryotic genes are often spread out over long stretches of DNA, whereas those of procaryotic genes are typically closely packed around the start point of transcription. Several bacterial gene regulatory proteins, however, recognize DNA sequences that are located many nucleotide pairs away from the promoter. The example of DNA looping in *E. coli* shown previously in Figure 9–32 resembles the way that eucaryotic gene regulatory proteins act at a distance. In fact, this case provided one of the first examples of DNA looping in gene regulation and greatly influenced later studies of eucaryotic gene regulatory proteins.

Figure 9–46 Some ways in which the activity of gene regulatory proteins is regulated in eucaryotic cells. (A) The protein is synthesized only when needed and is rapidly degraded by proteolysis so that it does not accumulate. (B) Activation by ligand binding. (C) Activation by phosphorylation. (D) Formation of a complex between a DNA-binding protein and a separate protein with a transcription-activating domain. (E) Unmasking of an activation domain by the phosphorylation of an inhibitor protein. (F) Stimulation of nuclear entry by removal of an inhibitory protein that otherwise keeps the regulatory protein from entering the nucleus.

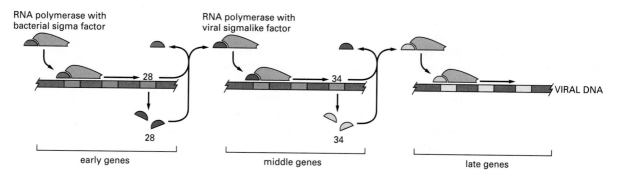

RNA polymerase with bacterial sigma factor

RNA polymerase with viral sigmalike factor

28

34

VIRAL DNA

28

34

early genes

middle genes

late genes

Figure 9–47 Interchangeable RNA polymerase subunits as a strategy to control gene expression in a bacterial virus. The bacterial virus SPO1, which infects the bacterium *B. subtilis*, uses the bacterial polymerase to transcribe its early genes. One of the early genes, called 28, encodes a sigmalike factor that binds to RNA polymerase and displaces the bacterial sigma factor. This new form of polymerase specifically initiates transcription of the SPO1 "middle" genes. One of the middle genes encodes a second sigmalike factor that displaces the 28 product and directs RNA polymerase to transcribe the "late" genes. This last set of genes produces the proteins that package the virus chromosome into a virus coat and lyse the cell. Thus, by this strategy, sets of virus genes are expressed in a particular order, allowing for rapid, yet temporally controlled, viral replication.

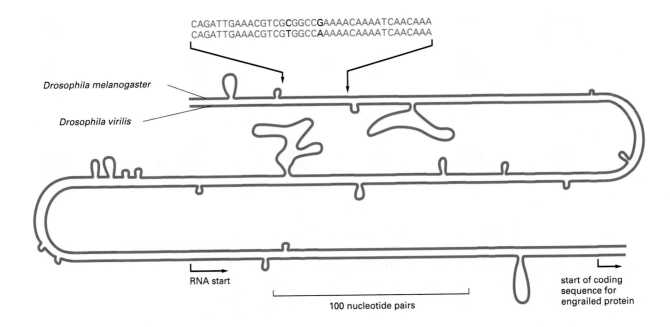

CAGATTGAAACGTCGCGGCCGAAAACAAAATCAACAAA
CAGATTGAAACGTCGTGGCCAAAAACAAAATCAACAAA

Drosophila melanogaster

Drosophila virilis

RNA start

100 nucleotide pairs

start of coding
sequence for
engrailed protein

It seems likely that the close-packed arrangement of bacterial genetic switches developed from more extended forms of switches in response to the evolutionary pressure on bacteria to maintain a small genome size. (The same argument has been used to explain the lack of introns in bacteria, as discussed in Chapter 8.) This compression comes at a price, however, as it is difficult to imagine how the compact switches could be easily altered to incorporate new levels of control. The extended form of eucaryotic control regions, in contrast, with discrete regulatory modules separated by long stretches of spacer DNA, would be expected to facilitate reshuffling of modules during evolution, both to create new regulatory circuits and to modify old ones. Unraveling the history of how gene control regions evolved presents a fascinating challenge, and many clues can be found in present-day DNA sequences (Figure 9–48).

Figure 9–48 A comparison of part of the control region upstream from the *engrailed* gene in two species of *Drosophila*. A DNA sequence comparison between *Drosophila melanogaster* and *Drosophila virilis* is shown, with regions of 90% sequence conservation shown in *red*. One example of an actual sequence match is illustrated in detail at the top. The conserved sequences presumably mark the sites where important gene regulatory proteins bind, whereas loops indicate places where insertions or deletions of nucleotides have occurred since these two species evolved from a common ancestor about 60 million years ago. (Courtesy of Judith A. Kassis and Patrick H. O'Farrell.)

Summary

The transcription of individual genes is switched on and off in cells by gene regulatory proteins. In procaryotes these proteins usually bind to specific DNA sequences close to the RNA polymerase start site and, depending on the nature of the regulatory protein and the precise location of its binding site relative to the start site, either activate or repress transcription of the gene. The flexibility of the DNA helix, however, also allows proteins bound at distant sites to affect the RNA polymerase at the promoter by the looping out of the intervening DNA. Such action at a distance is extremely common in eucaryotic cells, where gene regulatory proteins bound to sequences thousands of nucleotide pairs from the promoter can control gene expression.

Although procaryotic RNA polymerases can initiate transcription on their own, eucaryotic polymerases require the prior assembly of general transcription factors at the promoter. These factors assemble in a particular order, beginning with the binding of TFIID to the TATA box, a DNA sequence found just upstream of most eucaryotic RNA polymerase start sites. The ordered assembly of general transcription factors provides several steps at which the initiation of transcription can be regulated, and many eucaryotic gene regulatory proteins are thought to work by facilitating (positive control) or hindering (negative control) the assembly process.

Whereas the transcription of a typical procaryotic gene is controlled by only one or two gene regulatory proteins, the regulation of higher eucaryotic genes is much more complex, commensurate with the larger genome size and the large variety of cell types. The control region of the Drosophila eve gene, for example, encompasses 20,000 nucleotide pairs of DNA and has binding sites for over 20 gene regulatory proteins.

Some of these proteins are transcriptional activators, while others are transcriptional repressors. These proteins bind to regulatory sequences organized in a series of regulatory modules strung together along the DNA.

Chromatin Structure and the Control of Gene Expression [36]

As discussed in Chapter 8, the genomes of eucaryotes are highly compacted to allow the very long DNA molecules to fit inside the cell and to be managed easily. The first level of compaction is the wrapping of DNA around histones to form nucleosomes. In a second level of compaction nucleosomes are packed into 30-nm filaments. Finally, an even higher order of packing (still poorly understood) is observed in heterochromatin, which is confined to selected regions of the genome that show an unusually condensed interphase structure.

How do gene regulatory proteins and the general transcription factors gain access to DNA that is packed into these compact protein-DNA structures, and how does the packing affect the control of gene expression? We see in this section that two general principles have emerged from studies of chromatin structure and its influence on gene expression. First, nucleosomes do not usually present a serious obstruction to either gene regulatory proteins or RNA polymerases. Enhancers can still function despite them, histones that block a promoter can be displaced, and, once transcription has begun, Pol II can transcribe through the nucleosomes without dislodging them. Even bacterial polymerases, which do not encounter nucleosomes *in vivo*, can transcribe through them, suggesting that the nucleosome is built to be traversed easily (Figure 9–49). The second general principle is that some forms of higher-order DNA packaging render the DNA inaccessible both to gene regulatory proteins and to the general transcription factors. Higher-order DNA packaging thus plays a crucial part in the control of gene expression in eucaryotes, serving to silence large sections of the genome—in some cases reversibly, in other cases not.

Transcription Can Be Activated on DNA That Is Packaged into Nucleosomes [37]

In the previous section we described a simple model for how transcription of a eucaryotic gene is activated by a gene regulatory protein (see Figure 9–36). How

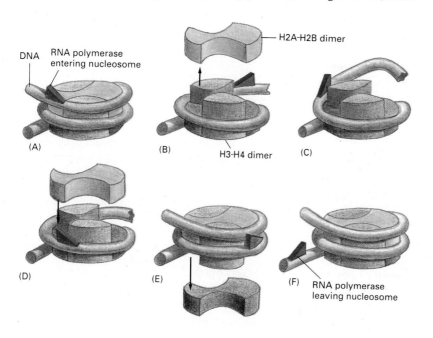

Figure 9–49 **A tentative model to account for the ability of RNA polymerases to transcribe through nucleosomes without causing their displacement.** The polymerase first displaces an H2A-H2B dimer (see Figure 8–10), allowing the polymerase to enter the nucleosome. In the next step the polymerase pulls the DNA away from the H3-H4 dimer it next encounters and continues transcribing. The displaced H2A-H2B dimer is recaptured by the nucleosome, and the second, symmetrically disposed H2A-H2B dimer is now displaced, allowing the process to repeat and permitting the polymerase to exit from the nucleosome. (After K.E. van Holde et al., *J. Biol. Chem.* 267:2837–2840, 1992.)

must this model be modified to take account of the presence of nucleosomes? To begin with the first step, how do nucleosomes affect the binding of a gene activator protein to its regulatory DNA sequence? In some cases the regulatory sequences reside in short nucleosome-free regions, and so no problem arises. It is uncertain how such nucleosome-free regions are maintained, but, as discussed in Chapter 8, some stretches of DNA are too stiff to accommodate the tight folding necessary for nucleosome formation. In other cases, however, the regulatory sequence is packaged into a nucleosome and yet at least some gene activator proteins can still recognize and bind to it. Once bound, the regulatory proteins appear to destabilize the nucleosome, which is then at least partially disassembled. Which types of gene regulatory proteins can achieve this feat and how they accomplish it remain unknown.

The general transcription factors, in contrast, seem unable to assemble onto a promoter that is packaged into a nucleosome. In fact, such packaging may have evolved in part to ensure that leaky, or basal, transcription initiation (that is, initiation without a gene activator protein bound upstream) does not occur. The binding of a gene activator protein thousands of nucleotide pairs away from a nucleosome-packaged promoter, however, can apparently displace a nucleosome from a promoter and thereby allow the assembly of the general transcription factors. The displacement either could be due to a separate activity of the gene activator protein or could be an indirect consequence of the activator contacting the general transcription factors to facilitate their assembly on the DNA (Figure 9–50).

Some Forms of Chromatin Silence Transcription [38]

Although transcription can occur on DNA that is packaged into nucleosomes, the DNA in some special forms of chromatin appears to be inaccessible to gene ac-

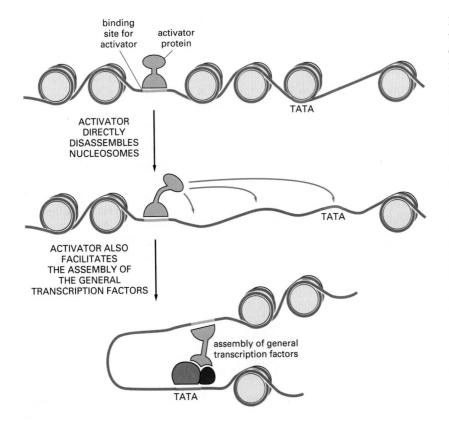

binding site for activator

activator protein

TATA

ACTIVATOR DIRECTLY DISASSEMBLES NUCLEOSOMES

TATA

ACTIVATOR ALSO FACILITATES THE ASSEMBLY OF THE GENERAL TRANSCRIPTION FACTORS

assembly of general transcription factors

TATA

Figure 9–50 **One model to explain the displacement of nucleosomes during the initiation of transcription in eucaryotes.** A bound gene activator protein possesses a separate activity that directly removes nucleosomes from the promoter, exposing the promoter to the general transcription factors and enabling them to assemble. In a different model (not shown) nucleosome displacement occurs as an indirect consequence of the activator protein promoting the assembly of the general transcription factors; the activator protein, for example, could permit the general factors to begin to assemble in the presence of a nucleosome, and, once partially assembled, these factors would destabilize the nucleosome. The process underlying nucleosome displacement is poorly understood. The histones may simply leave the DNA or, according to an alternative model, they may disassemble but remain bound to DNA (see Figure 8–40).

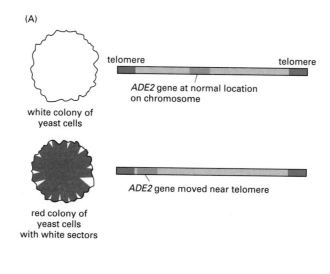

(A)

telomere   telomere

ADE2 gene at normal location
on chromosome

white colony of
yeast cells

ADE2 gene moved near telomere

red colony of
yeast cells
with white sectors

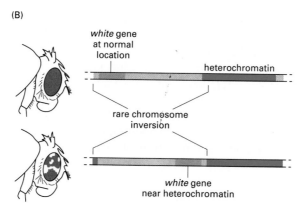

(B)

white gene
at normal
location

heterochromatin

rare chromosome
inversion

white gene
near heterochromatin

Figure 9–51 Position effects on gene expression. (A) The yeast *ADE2* gene at its normal chromosomal location is expressed in all cells. When moved near the end of a yeast chromosome, the gene is silenced in most but not all cells of the population. The absence of the *ADE2* gene product results in a block in the adenine biosynthetic pathway, which leads to the accumulation of a red pigment. The founder cell for the sectored colony shown had its *ADE2* gene shut off, so most of the colony is *red*. Normally, yeast colonies are white. The *white* sectors at the edges of the *red* colony are clones of cells where the *ADE2* gene has spontaneously become active. The finding of such sectoring indicates that the active and inactive states of *ADE2* expression are heritable when this gene is near the telomere, a topic discussed in more detail in the next section.

(B) Position effects can also be observed for the *Drosophila white* gene. Wild-type flies with a normal *white* gene have red eyes. If the *white* gene is inactivated by mutation, the eyes become white (hence the name of the gene). In flies with a chromosomal inversion that moves the *white* gene near a heterochromatic region, the eyes are mottled, with red and white patches. The white patches represent cells where the *white* gene is silenced and red patches represent cells that express the *white* gene. The difference is thought to arise from variations in how far along the chromosome the heterochromatin spreads early in eye development. As in the case of yeast *ADE2* gene, once established, the state of *white* expression is heritable, producing patches of many cells that express *white* as well as patches of cells where *white* is silenced. (After L.L. Sandell and V.A. Zakian, *Trends Cell Biol.* 2:10–14, 1992.)

tivator proteins. These *inactive* forms of chromatin, including the especially highly condensed form called *heterochromatin* (discussed in Chapter 8), are assumed to contain special proteins that make the DNA unusually inaccessible.

An observation in the yeast *S. cerevisiae* illustrates how some types of chromatin can shut off gene transcription. The *ADE2* gene, whose expression is particularly easy to monitor, is expressed when present at its normal chromosomal location. When this gene is experimentally relocated to the end of a chromosome, however, its transcription is turned off, even though the cell contains all of the proteins required to transcribe the gene. The DNA near the ends of yeast chromosomes (the *telomeres*) is packaged into an especially inaccessible form of chromatin, and it is this packaging that is thought to be responsible for maintaining the translocated *ADE2* gene in an inactive state, a process called *silencing*. The silencing of genes located near chromosome ends extends for approximately 10,000 nucleotide pairs and applies to many genes in addition to *ADE2*; the silencing seems to weaken gradually with distance from the telomere. The mechanism of silencing is not known, but it seems likely to involve a cooperative assembly of proteins on the DNA that, once established, is heritable following DNA replication (Figure 9–51A).

The silencing of the *ADE2* gene is an example of a **position effect,** in which the activity of a gene is dependent on its position in the genome. Position effects were first recognized in *Drosophila* (Figure 9–51B), but they have now been observed in a number of other organisms and are thought to reflect the different states of chromatin present at different locations in the genome and the tendency of these states to spread to encompass nearby genes. We revisit this topic later in the chapter when we analyze mechanisms of cell memory.

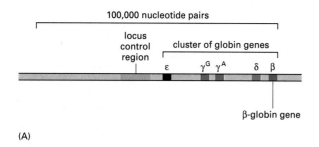

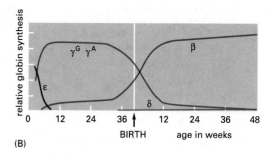

(A)

(B)

An Initial Decondensation Step May Be Required Before Mammalian Globin Genes Can Be Transcribed [39]

Another example of how highly condensed chromatin can prevent gene expression comes from studies of chick and human β-globin gene clusters. The five genes of the cluster, spread over 50,000 nucleotide pairs of DNA, are transcribed exclusively in erythroid cells (that is, cells of the red blood cell lineage). Moreover, each gene is turned on at a different stage of development (Figure 9–52) and in different organs: the ε-globin gene is expressed in the embryonic yolk sac, γ in the yolk sac and the fetal liver, and δ and β primarily in the adult bone marrow. We previously described a series of gene regulatory proteins that are necessary to turn on the human β-globin gene at the appropriate time and place (see Figure 9–45), and each of the other globin genes has a similar set of regulatory proteins, many of which are shared among these genes. In addition to the individual regulation of each of the globin genes, however, the entire cluster appears to be subject to an on-off control that involves global changes in chromatin structure.

Some of the first evidence for such changes came from studies of the sensitivity of the globin genes in isolated nuclei to digestion by the nuclease enzyme DNaseI. In cells where the globin genes are not expressed, the DNA in these genes is resistant to DNaseI, indicating that they are tightly packaged into chromatin. In erythroid cells, by contrast, the entire gene cluster is sensitive to DNaseI, indicating that the chromatin has changed to make the DNA more accessible to the enzyme. The DNA is still folded into nucleosomes, but the higher-order packing of the chromatin has loosened. This change in DNA packing occurs even before the individual globin genes are transcribed, suggesting that the genes are regulated in two steps. In the first step the chromatin of the entire globin locus is decondensed, which is presumed to allow some of the gene regulatory proteins access to the DNA. In the second step the remaining gene regulatory proteins assemble on the DNA and direct the expression of individual genes (Figure 9–53).

The extensive change in chromatin structure that occurs in the first step is thought to require a region of DNA (called the *locus control region*, or *LCR*) that lies far upstream from the gene clusters (see Figure 9–52). The importance of the LCR can be seen in patients with a certain type of thalassemia, a severe genetic form of anemia. In these patients the β-globin locus is found to have undergone deletions that remove all or part of the LCR, and although the β-globin gene and its nearby regulatory regions are intact, the gene remains silent in erythroid cells. Moreover, the β-globin gene in the erythroid cells remains DNaseI resistant, indicating that it fails to undergo the normal chromatin decondensation step during erythroid cell development.

Subsequent experiments in transgenic mice have confirmed the profound effects of the LCR on the expression of globin genes. When, for example, the human β-globin gene plus its local regulatory sequences (the region shown in Figure 9–45) is inserted into different positions in the mouse genome, it is expressed at low levels that depend on the site of insertion. This behavior is typical for mammalian genes, and it indicates that local position effects influence the expression of the gene (Table 9–2). When the LCR is included with the gene,

Figure 9–52 The cluster of β-like globin genes in humans. (A) The large chromosomal region shown spans 100,000 nucleotide pairs and contains the five globin genes and a locus control region (discussed in the text). (B) Changes in the expression of the β-like globin genes at various stages of human development. Each of the globin chains encoded by these genes combines with an α-globin chain to form the hemoglobin in red blood cells. (A, after F. Grosveld, G.B. van Assendelft, D.R. Greaves, and G. Kollias, *Cell* 51:975–985, 1987. © Cell Press.)

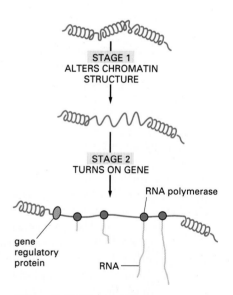

Figure 9–53 The two stages postulated to be involved in some gene activations, such as that of the human globin gene cluster. In stage 1 the structure of a large local region of chromatin is modified to decondense it in preparation for transcription. In stage 2 gene regulatory proteins (represented by a single protein in this simplified figure) bind to specific sites on the altered chromatin to induce RNA synthesis (transcription).

Table 9-2 The Expression of a Gene Transferred to a Mouse Generally Shows Chromosome Position Effects, with Different Levels of Gene Activity in Independently Derived Transgenic Animals

	Percent of Total mRNA in				Gene Copies per Cell
	Yolk Sac	**Liver**	**Gut**	**Brain**	
Endogenous gene	**20**	**5**	**0.1**	**0**	**2**
Transgenic animal 1	3.4	1	0.1	0	4
Transgenic animal 2	4.8	30	1.3	0	4
Transgenic animal 3	4.4	13	4.7	0	4
Transgenic animal 4	0.4	0.4	0	0	12

In these experiments, carried out with the mouse alpha-fetoprotein gene, the DNA fragment injected into the fertilized mouse egg included 14,000 nucleotide pairs of upstream (5'-flanking) sequence, where three enhancers that affect the expression of this gene are located. Hybridization was used to compare the level of mRNA produced by the injected gene to that normally produced by the endogenous mouse alpha-fetoprotein gene in the indicated fetal tissues. (Data from R.E. Hammer et al., *Science* 235:53–58, 1987.)

however, β-globin is expressed at high levels in erythroid cells regardless of the site of insertion, indicating that the LCR can override these position effects.

Although several proteins that specifically bind to the LCR have been identified, the mechanism that alters the chromatin structure of the entire β-globin locus is not known. Some ideas for how such changes may be brought about are discussed in the next section.

The Mechanisms That Form Active Chromatin Are Not Understood

The hypothetical model for globin activation outlined in Figure 9–53 implies that eucaryotes may contain sequence-specific DNA-binding proteins that function to decondense the chromatin in a local chromosomal domain that extends for tens of thousands of DNA nucleotide pairs. Alternatively, the observed differences in the chromatin structure of active genes could be an automatic consequence of the assembly of transcription factors or RNA polymerase (or both) onto a promoter rather than being a prerequisite for these events. We saw earlier that the assembly of the general transcription factors at promoters appears to be accompanied by changes in nucleosome distribution at the assembly site; perhaps this small perturbation can spread for long distances by some unknown propagation mechanism.

Whether or not eucaryotes turn out to have proteins that are specifically designed to decondense domains of chromatin, it is worth speculating on how proteins might accomplish this task. At present we can only guess at the mechanism. Three possibilities are outlined in Figure 9–54. These very different types of models indicate how far we are from understanding the transition from inactive to active chromatin.

Superhelical Tension in DNA Allows Action at a Distance [40]

One of the three models outlined in Figure 9–54 invokes topological changes in a closed loop of DNA double helix that can lead to the formation of DNA supercoils, a conformation that DNA adopts in response to *superhelical tension*. DNA supercoiling is most readily studied in small circular DNA molecules, such as the chromosomes of some viruses and plasmids. The same considerations apply, however, to any region of DNA bracketed by two ends that are unable to rotate freely—as, for example, in a loop of chromatin that is tightly clamped at its base.

Chromatin Structure and the Control of Gene Expression

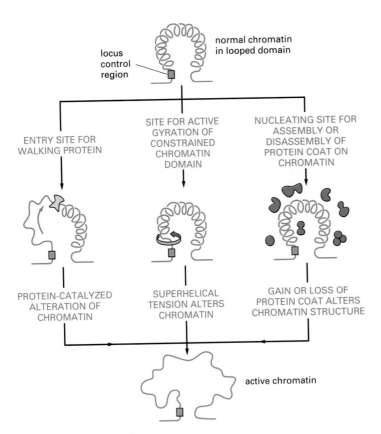

Figure 9–54 **Three models proposed to explain the long-range influence of a locus control region (LCR) on gene activity.** The loop of chromatin shown usually is postulated to represent an entire chromosomal looped domain (discussed in Chapter 8), which can contain 100,000 or more nucleotide pairs of DNA. Each model proposes a different function for the LCR site. The actual mechanism that causes long-range effects is unknown, and other mechanisms are also conceivable.

A simple way of visualizing the topological constraints that cause DNA supercoiling is illustrated in Figure 9–55. In a DNA double helix with two fixed ends, one DNA supercoil forms to compensate for each 10 nucleotide pairs that are opened (unwound) in the helix; the formation of this supercoil is energetically favorable because it restores the normal helical twist to the base-paired regions that remain.

In bacteria such as *E. coli* a special type II DNA topoisomerase, called *DNA gyrase,* uses the energy of ATP hydrolysis to pump supercoils continuously into the DNA, thereby maintaining the DNA under constant tension. These supercoils are *negative supercoils,* having the opposite handedness from the *positive supercoils,* shown in Figure 9–55, that form when a region of DNA helix opens. Therefore, rather than creating a positive supercoil, a negative supercoil is removed from bacterial DNA when a region of helix opens. Because superhelical tension in the DNA is reduced during this event, the opening of the DNA helix in *E. coli* is energetically favored compared to helix opening in DNA that is not supercoiled.

The eucaryotic type II DNA topoisomerases remove superhelical tension rather than generate it (discussed in Chapter 8). Consequently, most of the DNA

Figure 9–55 **Superhelical tension in DNA causes DNA supercoiling.** (A) For a DNA molecule with one free end (or a nick in one strand that serves as a swivel), the DNA double helix rotates by one turn for every 10 nucleotide pairs opened. (B) If rotation is prevented, superhelical tension is introduced into the DNA by helix opening. One way to accommodate this tension would be to increase the helical twist from 10 to 11 nucleotide pairs per turn in the double helix that remains in this example; the DNA helix, however, resists such a deformation in a springlike fashion, preferring to relieve the superhelical tension by bending into supercoiled loops. As a result, one DNA supercoil forms in the DNA double helix for every 10 nucleotide pairs opened. The supercoil formed in this case is a *positive supercoil* (see Figure 9–56).

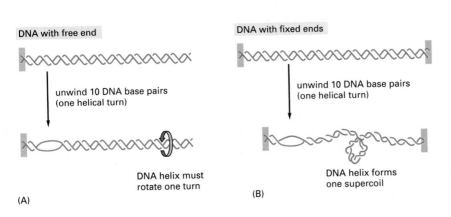

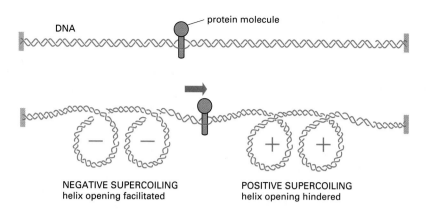

DNA

protein molecule

NEGATIVE SUPERCOILING
helix opening facilitated

POSITIVE SUPERCOILING
helix opening hindered

Figure 9–56 Supercoiling of a DNA segment induced by a protein tracking through the DNA double helix. The two ends of the DNA shown here are unable to rotate freely relative to each other, and the protein molecule is assumed also to be prevented from rotating freely as it moves. Under these conditions the movement of the protein will cause an excess of helical turns to accumulate in the DNA helix ahead of the protein and a deficit of helical turns to arise in the DNA behind the protein, as shown. Experimental evidence suggests that a moving RNA polymerase molecule causes supercoiling in this way; the positive superhelical tension ahead of it makes the DNA helix more difficult to open, but this tension should facilitate the unwrapping of the DNA in nucleosomes, as the release of DNA from the histone core helps to relax positive superhelical tension.

in eucaryotic cells is not under tension. But helix unwinding is associated with the initiation step of transcription. Moreover, a moving RNA polymerase molecule (as well as other proteins that unwind DNA) will tend to generate positive superhelical tension in the DNA in front of it and negative superhelical tension behind it (Figure 9–56). Through such topological effects an event that occurs at a single DNA site can produce forces that are felt throughout an entire domain of chromatin. It is not yet known, however, whether an effect of this kind is involved in creating large-scale changes in chromatin structure.

Summary

The genomes of eucaryotic organisms are packaged into chromatin. Some forms of chromatin are so highly compacted that the packaged genes are transcriptionally silent. Although the structural details of this type of packaging are not well understood, it is thought to be a device utilized by cells to silence large regions of their genomes. In some cases, this silencing can be reversed and the packaged genes activated, but the way this happens is also not understood.

In less compacted forms of chromatin the DNA is still packaged in nucleosomes. Nucleosomes positioned at the start point of transcription block the assembly of the general transcription factors. These nucleosomes appear to be displaced by an unknown mechanism when transcription is activated by gene regulatory proteins.

The Molecular Genetic Mechanisms That Create Specialized Cell Types [41]

Although unicellular organisms must be able to switch genes on and off, multicellular organisms require special gene switching mechanisms for generating and maintaining their different types of cells. In particular, once a cell in a multicellular organism becomes committed to differentiate into a specific cell type, the choice of fate is generally maintained through many subsequent cell generations, which means that the changes in gene expression involved in the choice must be remembered. This phenomenon of *cell memory* is a prerequisite for the creation of organized tissues and for the maintenance of stably differentiated cell types. In contrast, the simplest changes in gene expression in both eucaryotes and procaryotes are only transient; the tryptophan repressor, for example, switches off the tryptophan genes in bacteria only in the presence of tryptophan; as soon as tryptophan is removed from the medium, the genes are switched back on, and the descendants of the cell will have no memory that their ancestors had been exposed to tryptophan. Even in procaryotes, however, some changes in gene expression can be stably inherited.

In this section we examine how gene regulatory devices can be combined to form switches that, once thrown, are remembered by subsequent cell generations. We begin by considering some of the best-understood genetic mechanisms of cell differentiation, which operate in bacterial and yeast cells.

DNA Rearrangements Mediate Phase Variation in Bacteria [42]

We have seen that cell differentiation in higher eucaryotes usually occurs without detectable changes in DNA sequence. In some procaryotes, in contrast, a stably inherited pattern of gene regulation is achieved by DNA rearrangements that activate or inactivate specific genes. Since changes in DNA sequence are faithfully copied during subsequent DNA replications, an altered state of gene activity will be inherited by all the progeny of the cell in which the rearrangement occurred. Some of these rearrangements are reversible and produce an alternating pattern of gene activity that can be detected by observations over long time periods and many generations.

A well-studied example of this differentiation mechanism, known as **phase variation,** occurs in *Salmonella* bacteria. Although this mode of differentiation has no known counterpart in higher eucaryotes, it can nevertheless have considerable impact on them because it is an important means whereby disease-causing bacteria evade detection by the immune system. The switch in *Salmonella* gene expression is brought about by the occasional inversion of a specific 1000-nucleotide-pair piece of DNA and affects the expression of the cell-surface protein flagellin, for which the bacterium has two different genes. The inversion is catalyzed by a site-specific recombination enzyme and changes the orientation of a promoter that is within the 1000 nucleotide pairs. With the promoter in one orientation, the bacteria synthesize one type of flagellin; with the promoter in the other orientation, they synthesize the other (Figure 9–57). Because inversions occur only rarely, whole clones of bacteria will grow up with one type of flagellin or the other.

Phase variation almost certainly evolved because it protects the bacterial population against the immune response of its vertebrate host. If the host makes antibodies against one type of flagellin, a few bacteria whose flagellin has been altered by gene inversion will still be able to survive and multiply.

Bacteria isolated from the wild very often exhibit phase variation for one or more phenotypic traits. These "instabilities" are usually lost with time from standard laboratory strains of bacteria, and only some of the underlying mechanisms

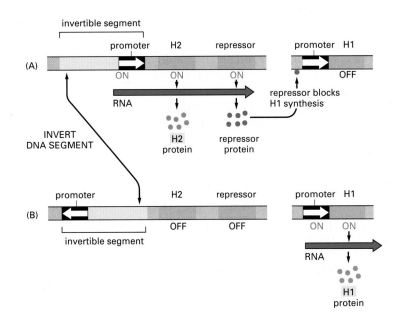

Figure 9–57 Switching gene expression by DNA inversion in bacteria. Alternating transcription of two flagellin genes in a *Salmonella* bacterium is caused by a simple site-specific recombination event that inverts a small DNA segment containing a promoter that in one orientation (A) activates transcription of the H2 flagellin gene as well as a repressor protein that blocks the expression of the H1 flagellin gene. When the promoter is inverted, it no longer turns on H2 or the repressor, and the H1 gene, which is thereby released from repression, is expressed instead (B). The recombination mechanism is activated only rarely (about once every 10^5 cell divisions). Therefore, the production of one or other flagellin tends to be faithfully inherited in each clone of cells.

have been studied. Not all involve DNA inversion. A bacterium that causes a common sexually transmitted human disease (*Neisseria gonorrhoeae*), for example, avoids immune attack by means of an inherited variation in its surface properties that is generated by *gene conversion* (discussed in Chapter 6) rather than by gene inversion. This mechanism is dependent on the recA recombination protein, and it transfers DNA sequences to an expressed gene from a set of silent "gene cassettes"; it has the advantage of creating more than 100 variants of the major bacterial surface protein.

Several Gene Regulatory Proteins Determine Cell Type Identity in Yeasts [43]

Because they are so easy to grow and to manipulate genetically, yeasts have been analyzed in great detail as model organisms for studying the mechanisms of gene control in eucaryotic cells. The common baker's yeast, *Saccharomyces cerevisiae*, has been especially illuminating because of its ability to differentiate into three cell types, even though the control mechanism differs in some basic ways from that used generally by animal and plant cells. *S. cerevisiae* is a single-celled eucaryote that exists in either a haploid or a diploid state. Diploid cells form by a process known as **mating,** in which two haploid cells fuse. In order for two haploid cells to mate, they must differ in *mating type* (sex). In *S. cerevisiae* there are two mating types, α and **a**, which are specialized for mating with each other. Each produces a specific diffusible signaling molecule (mating factor) and a receptor protein that jointly enable the cell to recognize and fuse with its opposite cell type. The resulting diploid cells, called **a**/α, are distinct from either parent: they are unable to mate but can form spores (sporulate) when they run out of food, giving rise to haploid cells by meiosis.

The mechanisms by which these three cell types are established and maintained illustrate several of the strategies we have discussed for changing the pattern of gene expression. The mating type of the haploid cell is determined by a single locus, the **mating-type (*MAT*) locus,** which in an **a**-type cell encodes a single gene regulatory protein, **a**1, and in an α cell encodes two gene regulatory proteins, α1 and α2. The **a**1 protein has no effect in the **a**-type haploid cell that produces it but becomes important later in the diploid cell that results from mating; meanwhile, the **a**-type haploid cell produces the proteins specific to its mating type by default. In contrast, the α2 protein acts in the α cell as a transcriptional repressor that turns off the **a**-specific genes, while the α1 protein acts as a transcriptional activator that turns on the α-specific genes. Once cells of the two mating types have fused, the combination of the **a**1 and α2 regulatory proteins generates a completely new pattern of gene expression, unlike that of either parent cell. The mechanism by which the mating-type-specific genes are expressed in different patterns in the three cell types is illustrated in Figure 9–58. It was among the first examples of combinatorial gene control to be identified and remains one of the best understood at the molecular level.

Although in most laboratory strains of *S. cerevisiae* the **a** and α cell types are stably maintained through many cell divisions, some strains isolated from the wild can switch repeatedly between the **a** and α cell types by a mechanism of gene rearrangement whose effects are reminiscent of phase variation in *N. gonorrhoeae*, although the exact mechanism seems to be peculiar to yeast. On either side of the *MAT* locus in the yeast chromosome, there is a silent locus encoding the mating-type gene regulatory proteins: the silent locus on one side encodes α1 and α2; the silent locus on the other side encodes **a**1. Every other cell division, the active gene in the *MAT* locus is excised and replaced by a newly synthesized copy of the silent locus determining the opposite mating type. Because the change involves the removal of one gene from the active "slot" and its replacement by another, this mechanism is called the *cassette mechanism*. The change is reversible because, although the original gene at the *MAT* locus is dis-

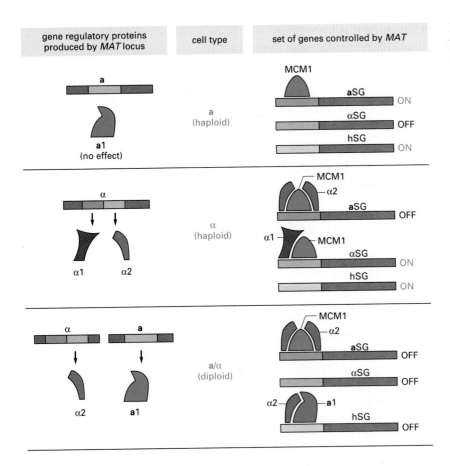

gene regulatory proteins produced by *MAT* locus	cell type	set of genes controlled by *MAT*

a
a1
(no effect)
a
(haploid)
MCM1
aSG — ON
αSG — OFF
hSG — ON

α
α1   α2
α
(haploid)
MCM1 — α2
aSG — OFF
α1 — MCM1
αSG — ON
hSG — ON

α   **a**
α2   **a**1
a/α
(diploid)
MCM1 — α2
aSG — OFF
αSG — OFF
α2 — **a**1
hSG — OFF

Figure 9–58 Control of cell type in yeasts. Yeast cell type is determined by three gene regulatory proteins (α1, α2, and **a**1) produced by the *MAT* locus. Different sets of genes are transcribed in haploid cells of type **a**, in haploid cells of type α, and in diploid cells (type **a**/α). The haploid cells express a set of haploid-specific genes (hSG) and either a set of α-specific genes (αSG) or a set of **a**-specific genes (**a**SG). The diploid cells express none of these genes. The α1, α2, and **a**1 proteins control many target genes in each type of cell by binding, in various combinations, to specific regulatory sequences upstream of these genes. Note that the α1 protein is a gene activator protein, whereas the α2 protein is a gene repressor protein, and both work in combination with a ubiquitous gene regulatory protein called MCM1. In the diploid cell type α2 and **a**1 form a heterodimer that turns off a different set of genes (including the gene encoding the α1 activator protein) from that turned off by the α2 and MCM1 proteins. This relatively simple system of gene regulatory proteins is an example of combinatorial control of gene expression (see Figure 9–38). The **a**1 and α2 proteins both recognize their DNA-binding sites by using the homeodomain motif (see Figure 9–13).

carded, a silent copy remains in the genome. New DNA copies made from the silent genes function as disposable cassettes that will be inserted in alternation into the *MAT* locus, which serves as the "playing head" (Figure 9–59).

Genetic tests suggest that the silent cassettes are maintained in a transcriptionally inactive form by the same mechanism that is responsible for silencing genes located at the ends of the yeast chromosomes (see Figure 9–51A): the DNA at a silent locus appears to be packaged into inactive chromatin.

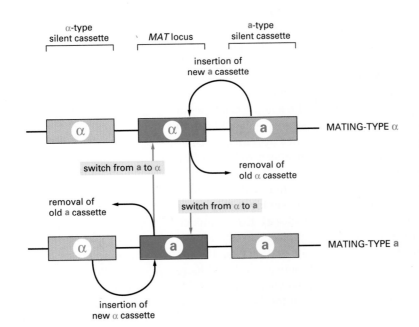

Figure 9–59 Cassette model of yeast mating-type switching. Cassette switching occurs by a gene-conversion process that involves a specialized enzyme that makes a double-stranded cut at a specific DNA sequence in the *MAT* locus. The DNA near the cut is then excised and replaced by a copy of the silent cassette of opposite mating type.

Two Proteins That Repress Each Other's Synthesis Determine the Heritable State of Bacteriophage Lambda [44]

The observation that a whole vertebrate or plant can be specified by the genetic information present in a single somatic cell nucleus (see Figure 9–1) eliminates the possibility that *irreversible* change in DNA sequence is the major mechanism in the differentiation of higher eucaryotic cells (although such changes are a crucial part of lymphocyte differentiation—discussed in Chapter 23). *Reversible* DNA sequence changes, resembling those just described for *Salmonella* and yeasts, in principle could still be responsible for some of the inherited changes in gene expression observed in higher organisms, but there is currently no evidence that such mechanisms are used.

Other mechanisms that we have touched upon in this chapter, however, are also capable of producing an inherited pattern of gene regulation with discrete stable states. None of these mechanisms is fully understood in any vertebrate system, but studies on bacteriophage lambda have provided insight at the molecular level into a switch that can flip-flop between two stable self-maintaining states, which could be a model for some similar switches that operate in the development of higher eucaryotes.

We mentioned earlier that this bacterial virus can in favorable conditions become integrated into the *E. coli* cell DNA, to be replicated automatically each time the bacterium divides instead of multiplying in the cytoplasm and killing its host. The switch between these two states is mediated by proteins encoded by the bacteriophage genome. The genome contains a total of about 50 genes, which are transcribed in very different patterns in the two states. A virus destined to integrate, for example, must produce the lambda *integrase* protein, which is needed to insert the lambda DNA into the bacterial chromosome, but must repress production of the viral proteins responsible for virus multiplication. Once one transcriptional pattern or the other has been established, it is stably maintained.

We cannot discuss the details of this complex gene regulatory system here, but we outline a few of its general features. At the heart of the system are two gene regulatory proteins synthesized by the virus: the **lambda repressor protein** (*cI protein*), which we have already encountered, and the **cro protein.** These proteins repress each other's synthesis, an arrangement giving rise to just two stable states. In state 1 (the *prophage state*) the lambda repressor occupies the operator, blocking the synthesis of cro and also activating its own synthesis. In state 2 (the *lytic state*) the cro protein occupies a different site in the operator, blocking the synthesis of repressor but allowing its own synthesis (Figure 9–60). In the prophage state most of the DNA of the stably integrated bacteriophage is not transcribed; in the lytic state this DNA is extensively transcribed, replicated, packaged into new bacteriophage, and released by host cell lysis.

When the host bacteria are growing well, an infecting virus tends to adopt state 1, allowing the DNA of the virus to multiply along with the host chromosome. When the host cell is damaged, an integrated virus converts from state 1 to state 2 in order to multiply in the cell cytoplasm and make a quick exit. This conversion is signaled by bacterial regulatory proteins that inactivate the repres-

Figure 9–60 A simplified version of the regulatory system that determines the mode of growth of bacteriophage lambda in the *E. coli* host cell. In stable state 1 (the prophage state) the bacteriophage synthesizes a repressor protein, which activates its own synthesis and turns off the synthesis of several other bacteriophage proteins, including the cro protein. In stable state 2 (the lytic state) the bacteriophage synthesizes the cro protein, which turns off the synthesis of the repressor protein, so that many bacteriophage proteins are made and the viral DNA replicates freely in the *E. coli* cell, eventually producing many new bacteriophage particles and killing the cell. This example shows how two gene regulatory proteins can be combined in a circuit to produce two heritable states. As shown in Figure 9–11, both the lambda repressor and the cro protein recognize the operator through a helix-turn-helix motif.

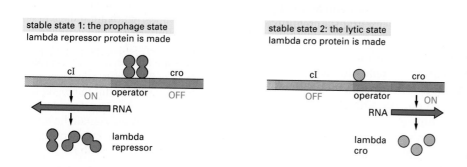

stable state 1: the prophage state
lambda repressor protein is made

cI　　　　　　　　cro

ON　operator　OFF

RNA

lambda repressor

stable state 2: the lytic state
lambda cro protein is made

cI　　　　　　　　cro

OFF　operator　ON

RNA

lambda cro

The Molecular Genetic Mechanisms That Create Specialized Cell Types

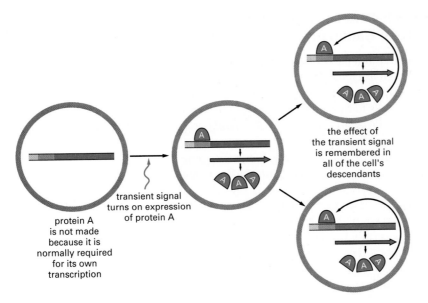

Figure 9–61 Schematic diagram showing how a positive feedback loop can create cell memory. Protein A is a gene regulatory protein that activates its own transcription. All of the descendants of the original cell will therefore "remember" that the progenitor cell had experienced a transient signal that initiated the production of the protein.

the effect of the transient signal is remembered in all of the cell's descendants

transient signal turns on expression of protein A

protein A is not made because it is normally required for its own transcription

sor protein. In the absence of such interference, however, the lambda repressor both turns off production of the cro protein and turns on its own synthesis, and this *positive feedback loop* helps to maintain the prophage state. Positive feedback loops are a feature of many cell memory circuits (Figure 9–61).

Bacteriophage lambda illustrates an important general principle: a sophisticated pattern of inherited behavior can be achieved with only a few gene regulatory proteins that reciprocally affect one another's synthesis and activities. We know that variations of this simple strategy are used by eucaryotic cells to establish and maintain heritable patterns of gene transcription. Several gene regulatory proteins that are involved in establishing the *Drosophila* body plan (discussed in Chapter 21), for example, stimulate their own transcription, thereby creating a positive feedback loop that promotes their continued synthesis; at the same time these proteins repress the transcription of genes encoding other important gene regulatory proteins.

Expression of a Critical Gene Regulatory Protein Can Trigger Expression of a Whole Battery of Downstream Genes [45]

In general, a combination of multiple gene regulatory proteins, rather than a single protein, determines where and when a gene is transcribed in eucaryotes. But even if control is combinatorial, a single gene regulatory protein can be decisive in switching a cell from one developmental pathway or state of differentiation to another. A striking example comes from experiments on muscle cell differentiation *in vitro*.

A mammalian skeletal muscle cell is typically extremely large and contains many nuclei. It is formed by the fusion of many muscle precursor cells called *myoblasts*. The mature muscle cell is distinguished from other cells by a large number of characteristic proteins, including specific types of actin, myosin, tropomyosin, and troponin (all part of the contractile apparatus), creatine phosphokinase (for the specialized metabolism of muscle cells), and acetylcholine receptors (to make the membrane sensitive to nerve stimulation). In proliferating myoblasts these muscle-specific proteins and their mRNAs are absent or are present in very low concentrations. As myoblasts begin to fuse with one another, the corresponding genes are all switched on coordinately as part of a general transformation of the pattern of gene expression.

This entire program of muscle differentiation can be triggered in cultured skin fibroblasts and certain other cell types by introducing any one of a family

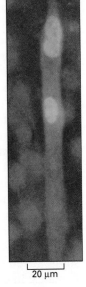

Figure 9–62 The effect of expressing the MyoD protein in fibroblasts. As shown in this immunofluorescence micrograph, skin fibroblasts from a chick embryo have been converted to muscle cells by the experimentally induced expression of the *myoD* gene. The fibroblasts were grown in culture and transfected three days earlier with a recombinant DNA plasmid containing the *myoD* coding sequence. Although only a few percent of the fibroblasts take up the DNA and produce the MyoD protein, these cells have fused to form elongated myotubes, which are stained here with an antibody that detects a muscle-specific protein. The stained cells are intermixed with a confluent layer of fibroblasts, whose nuclei are barely visible in this micrograph. Control cultures transfected with another plasmid contain no muscle cells. (Courtesy of Stephen Tapscott and Harold Weintraub.)

20 µm

of helix-loop-helix proteins—the so-called *myogenic proteins* (MyoD, Myf5, or myogenin, for example)—normally expressed only in muscle cells (Figure 9–62). Binding sites for these regulatory proteins can be detected in the regulatory DNA sequences adjacent to many muscle-specific genes. From studies in transgenic mice, it seems likely that MyoD and Myf5 act by turning on myogenin: if the myogenin gene is eliminated by targeted gene disruption, muscle cells fail to differentiate.

It is probable that the fibroblasts and other cell types that are converted to muscle cells by myogenic proteins have already accumulated a number of gene regulatory proteins that can cooperate with the myogenic proteins to switch on muscle-specific genes. In this view it is a specific *combination* of gene regulatory proteins, rather than a single protein, that determines muscle differentiation. This idea is consistent with the finding that some cell types fail to be converted to muscle by myogenin or its relatives; these cells presumably have not accumulated the other gene regulatory proteins required.

As we see next, combinatorial gene control has important implications for both the evolution and the development of multicellular organisms.

Combinatorial Gene Control Is the Norm in Eucaryotes [46]

We have already discussed how multiple gene regulatory proteins can act in combination to regulate the expression of an individual gene. But, as the example of the myogenic proteins shows, combinatorial gene control means more than this: not only does each gene have many gene regulatory proteins to control it, but each regulatory protein contributes to the control of many genes. Moreover,

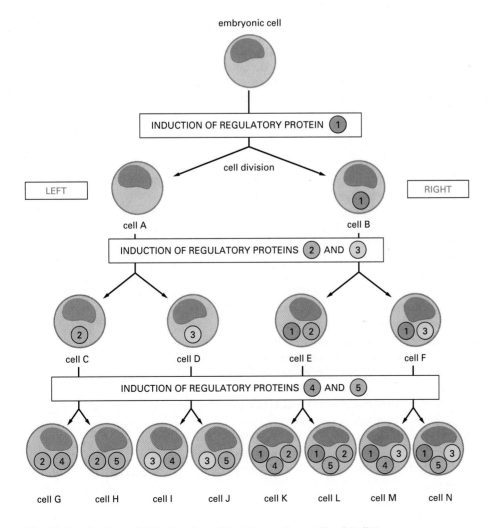

Figure 9–63 **The importance of combinatorial gene control for development.** A highly schematic scheme illustrating how combinations of a few gene regulatory proteins can generate many cell types during development. In this simple scheme a "decision" to make one of a pair of different gene regulatory proteins (shown as numbered circles) is made after each cell division. Sensing its relative position in the embryo, the daughter cell toward the left side of the embryo is always induced to synthesize the even-numbered protein of each pair, while the daughter cell toward the right side of the embryo is induced to synthesize the odd-numbered protein. The production of each gene regulatory protein is assumed to be self-perpetuating (thereby contributing to cell memory). Therefore, the cells in the enlarging clone contain an increasing number of regulatory proteins. Note that, in this purely hypothetical example, eight cell types (G through N) have been created with 5 different gene regulatory proteins. With continuation of such a scheme, more than 10,000 cell types could have been specified by only 25 different gene regulatory proteins.

The Molecular Genetic Mechanisms That Create Specialized Cell Types although some gene regulatory proteins, like MyoD or myogenin, are specific to a single cell type, more typically production of a given gene regulatory protein is itself switched on in a variety of cell types, at several sites in the body, and at several times in development. This point is illustrated schematically in Figure 9–63, which shows how combinatorial gene control makes it possible to generate a great deal of biological complexity with relatively few gene regulatory proteins.

With combinatorial control, a given gene regulatory protein does not necessarily have a single, simply definable function as commander of a particular battery of genes or specifier of a particular cell type. Instead, it may serve many purposes that overlap with those of other gene regulatory proteins. These proteins can be likened to the words of a language: they are used with different meanings in a variety of contexts and rarely alone; it is the well-chosen combination that conveys the information that specifies a gene regulatory event.

A consequence of combinatorial gene control is that the effect of adding a new gene regulatory protein to a cell will depend on the cell's past history, since this history will determine which gene regulatory proteins are already present. Thus during development a cell can accumulate a series of gene regulatory proteins that need not initially alter gene expression. When the final member of the requisite combination of gene regulatory proteins is added, however, the regulatory message is completed, leading to large changes in gene expression. Such a scheme, as we have seen, could explain how the addition of a single regulatory protein to a fibroblast can produce the dramatic transformation of the fibroblast into a muscle cell. It also can account for the important difference, discussed in Chapter 21, between the process of *cell determination*, where a cell becomes committed to a particular developmental fate, and the process of *cell differentiation*, where a committed cell expresses its specialized character. It is an essential feature of this scheme that once a gene regulatory protein has been made, it may act to maintain its own expression, thereby contributing to *cell memory*, which we discuss further below.

Combinatorial gene control also has an important consequence for evolution. Because gene regulatory proteins are not dedicated to a particular circuit or to a particular target gene, a subtle change in one gene regulatory protein can affect the expression pattern of many genes and thereby cause a substantial change in cell behaviors.

An Inactive X Chromosome Is Inherited [47]

We saw earlier how gene regulatory proteins can produce heritable patterns of gene expression in both procaryotic and eucaryotic cells. One possible mechanism, based on positive feedback in the control of gene expression, was illustrated in Figure 9–61. An additional mechanism operates only in eucaryotes, where long-range patterns of chromatin structure can be stably inherited. Perhaps the most dramatic example known is the inactivation of one of the two X chromosomes in female mammalian cells.

The X and Y chromosomes are the *sex chromosomes* of mammals: female cells contain two X chromosomes, while male cells contain one X and one Y chromosome. Presumably because a double dose of X-chromosome products would be lethal, the female cells have evolved a mechanism for permanently inactivating one of the two X chromosomes in each cell. In mice this occurs between the third and the sixth day of development, when, at random, one or other of the two X chromosomes in each cell becomes highly condensed into heterochromatin. This chromosome is seen in the light microscope during interphase as a distinct structure known as a *Barr body*, located near the nuclear membrane, and it replicates late in S phase. Most of its DNA is not transcribed. Because the inactive state of this X chromosome is faithfully inherited, every female is a mosaic composed of a mixture of clonal groups of cells in which only the paternally inherited X chro-

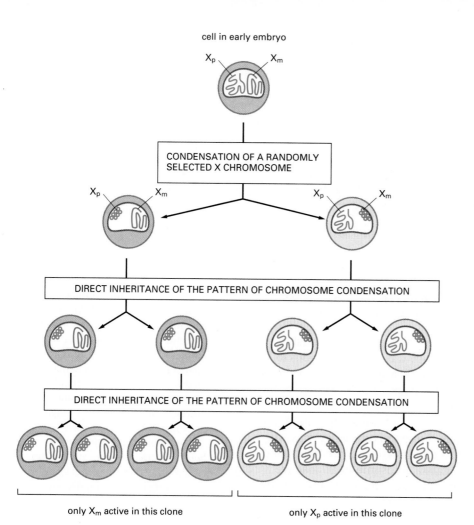

cell in early embryo

X_p X_m

CONDENSATION OF A RANDOMLY
SELECTED X CHROMOSOME

X_p X_m X_p X_m

DIRECT INHERITANCE OF THE PATTERN OF CHROMOSOME CONDENSATION

DIRECT INHERITANCE OF THE PATTERN OF CHROMOSOME CONDENSATION

only X_m active in this clone   only X_p active in this clone

Figure 9–64 X inactivation. The clonal inheritance of a condensed inactive X chromosome that occurs in female mammals.

mosome (X_p) is active and a roughly equal number of clonal groups of cells in which only the maternally inherited X chromosome (X_m) is active. In general, the cells expressing X_p and those expressing X_m are distributed in small clusters in the adult animal, reflecting the tendency of sister cells to remain close neighbors during the later stages of embryonic development and growth (Figure 9–64).

The process that forms the condensed chromatin (the heterochromatin) in an X chromosome tends to spread continuously along the chromosome. This can be seen in studies with mutant animals in which one of the X chromosomes has become joined to a portion of an *autosome* (a nonsex chromosome). In such hybrid chromosomes regions of the autosome adjacent to an inactivated X chromosome are often condensed into heterochromatin, causing the genes they contain to be inactivated in a heritable way. This suggests that X-chromosome inactivation occurs by a cooperative process that can be thought of as a chromatin "crystallization" event that spreads linearly along the DNA from a nucleation site on the X chromosome. In fact, a unique *inactivation center* has been located genetically on the X chromosome: broken fragments of X chromosome do not undergo inactivation unless they include this center.

Once the condensed chromatin structure is established on an X chromosome, some unknown process causes the structure to be faithfully inherited during all subsequent replications of the DNA. The change is not absolutely permanent, however, as the condensed X chromosome is reactivated in the formation of germ cells in the female.

The Molecular Genetic Mechanisms That Create Specialized Cell Types

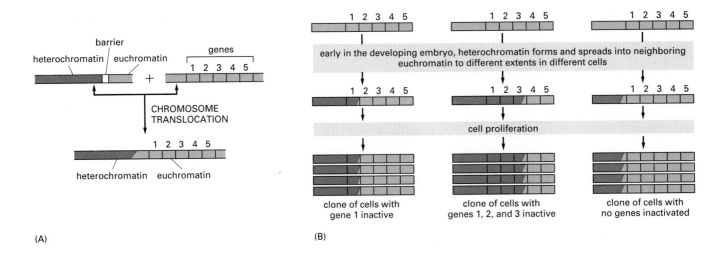

(A)

(B)

Drosophila and Yeast Genes Can Also Be Inactivated by Heritable Features of Chromatin Structure [48]

Earlier we discussed two examples of *position effects* on gene expression—one in *Drosophila* and one in yeast—that seem in many ways to be analogous to X-chromosome inactivation (see Figure 9–51). In both cases a specifically condensed form of chromatin prevents the expression of genes, and in both cases the condensed state of chromatin is heritable.

In flies with chromosomal rearrangements, breaking and rejoining events that place the middle of a region of heterochromatin next to a region of normal chromatin (*euchromatin*) tend to inactivate the nearby euchromatic genes. The situation is analogous to fusing a mammalian autosome to an inactive X chromosome, as just described, and the inactivation events are similarly patterned: the zone of inactivation spreads from the chromosome breakpoint to involve one or more adjacent genes. Moreover, while the extent of the spreading effect is different in different cells, the inactivated zone established in an embryonic cell is stably inherited by all of the cell's progeny (Figure 9–65). The example of position effect in yeast described previously in Figure 9–51 also shares some of the features of X-chromosome inactivation, including the spreading effect and the heritability of the condensed chromatin state.

It has not yet been proved that X-chromosome inactivation and the position effects in flies and yeast all occur by related mechanisms. Nevertheless, the parallels are striking. The recent identification and cloning of several *Drosophila* and yeast genes required for the position effects have provided an experimental entry point for exploring the molecular mechanisms involved. Figure 9–66 shows one hypothetical scheme that could, in principle, account for both the spreading effect and the heritable nature of the condensed chromatin state.

Regardless of its molecular basis, the packing of selected regions of the genome into condensed chromatin is a type of genetic regulatory mechanism that is not available to bacteria. The crucial feature of this uniquely eucaryotic form of gene regulation is the storing of the stable memory of gene states in an inherited chromatin structure rather than in a stable feedback loop of self-regulating gene regulatory proteins that can diffuse from place to place in the nucleus. Whether mechanisms of this type operate only to inactivate large regions of chromosomes or whether they can also operate at the level of one or a few genes is not known.

Figure 9–65 Position-effect variegation in *Drosophila*. (A) Heterochromatin (*red*) is normally prevented from spreading into adjacent regions of euchromatin (*green*) by special barrier sequences of unknown nature. In flies that inherit certain chromosomal translocations, however, this barrier is no longer present. (B) During the early development of such flies, the heterochromatin now spreads into neighboring chromosomal DNA, proceeding for different distances in different cells. The spreading soon stops, but the established pattern of heterochromatin is inherited, so that large clones of progeny cells are produced that have the same neighboring genes condensed into heterochromatin and thereby inactivated (hence the "variegated" appearance of some of these flies; see Figure 9–51B). This phenomenon shares many features with X-chromosome inactivation in mammals.

The Pattern of DNA Methylation Can Be Inherited When Vertebrate Cells Divide [49]

The nucleotides in DNA can be covalently modified, and in vertebrate cells the methylation of cytosine seems to provide an important mechanism for distin-

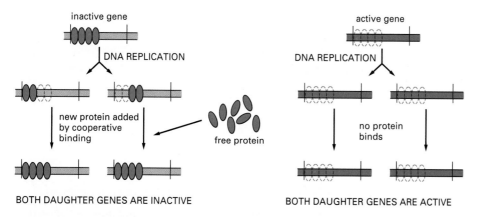

inactive gene

DNA REPLICATION

new protein added by cooperative binding

free protein

BOTH DAUGHTER GENES ARE INACTIVE

active gene

DNA REPLICATION

no protein binds

BOTH DAUGHTER GENES ARE ACTIVE

Figure 9–66 A general scheme that permits the direct inheritance of states of gene expression during DNA replication. In this hypothetical model, portions of a cooperatively bound cluster of chromosomal proteins are transferred directly from the parental DNA helix (*top left*) to both daughter helices. The inherited cluster then causes each of the daughter DNA helices to bind additional copies of the same proteins. Because the binding is cooperative, DNA synthesized from an identical parental DNA helix that lacks the bound proteins (*top right*) will remain free of them. If the bound proteins turn off gene transcription, then the inactive gene state will be directly inherited, as illustrated. If the cooperative protein binding requires specific DNA sequences, these events will be limited to specific gene control regions; if the binding can be propagated all along the chromosome, however, it could account for the spreading effect associated with the heritable chromatin states discussed in the text.

guishing genes that are active from those that are not. The covalently modified *5-methylcytosine (5-methyl C)* has the same relation to cytosine that thymine has to uracil and likewise has no effect on base-pairing (Figure 9–67). The methylation in vertebrate DNA is restricted to cytosine (C) nucleotides in the sequence CG, which is base-paired to exactly the same sequence (in opposite orientation) on the other strand of the DNA helix. Consequently, a simple mechanism permits the existing pattern of **DNA methylation** to be inherited directly by the daughter DNA strands. An enzyme called *maintenance methylase* acts preferentially on those CG sequences that are base-paired with a CG sequence that is *already* methylated. As a result, the pattern of DNA methylation on the parental DNA strand will act as a template for the methylation of the daughter DNA strand, causing this pattern to be inherited directly following DNA replication (Figure 9–68).

Bacteria produce enzymes that are useful for studying methylation in vertebrate cells. They use the methylation of either an A or a C at a specific site to protect themselves from the action of their own restriction nucleases. The restriction nuclease HpaII, for example, cuts the sequence CCGG but fails to cleave it if the central C is methylated. Thus the susceptibility of a DNA molecule to cleavage by HpaII can be used to detect whether CG sequences at specific DNA sites are methylated. The inheritance of methylation patterns can be studied in vertebrate cells in culture by first using bacterial methylating enzymes to introduce methyl groups on cytosines and then using bacterial restriction nucleases to follow the inheritance of these groups. The enzyme used to introduce 5-methyl C bases into specific CG sequences is the HpaII-methylase that normally protects the bacterium against its own HpaII restriction nuclease. If this enzyme is used to methylate the central C in the sequence CCGG on a cloned DNA molecule that is introduced into cultured vertebrate cells, the maintenance methylase can be shown to work as expected: each individual methylated CG is generally retained through many cell divisions, whereas unmethylated CG sequences remain unmethylated.

The maintenance methylase explains the automatic inheritance of 5-methyl C nucleotides, but since it normally does not methylate fully unmethylated DNA, it leaves unanswered the question of how the methyl group is first added in a vertebrate organism. If a fully unmethylated DNA molecule is injected into a fertilized mouse egg, methyl groups will be added to nearly every CG site (an important exception will be described below). This is presumed to reflect the presence of a novel *establishment methylase* activity in the egg. As we shall see, *de novo* methylation can also occur during the differentiation of specialized cell types, although it is not known how it occurs.

Figure 9–67 Formation of 5-methylcytosine occurs by methylation of a cytosine base in the DNA double helix. In vertebrates this event is confined to selected cytosine (C) nucleotides located in the sequence CG.

cytosine

5-methylcytosine

methylation

The Molecular Genetic Mechanisms That Create Specialized Cell Types

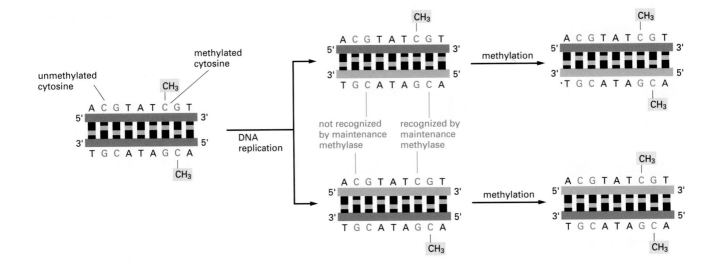

DNA Methylation Reinforces Developmental Decisions in Vertebrate Cells [50]

Although DNA methylation was once proposed to play a dominant part in generating different mammalian cell types, it is now viewed as having a more subtle role. In some invertebrates, including *Drosophila*, DNA methylation does not occur, yet the control of gene expression and the diversification of cell types appear to be similar in *Drosophila* and vertebrates. Several observations are consistent with the idea that DNA methylation in vertebrates is associated with gene inactivation but that it usually only reinforces decisions that are first brought about by other mechanisms. Thus tests with the HpaII restriction nuclease indicate that in general the DNA of inactive genes is more heavily methylated than that of active genes. When an inactive gene that contains methylated DNA is turned on during the course of normal development, however, it generally loses most of its methyl groups only after the gene has been activated. Conversely, the female X chromosome, discussed above, is first condensed and inactivated and only later acquires an increased level of methylation on some of its genes.

What, then, does methylation do, and why is it useful to the organism? There are at least two important clues. First, the DNA corresponding to a muscle-specific actin gene can be prepared in both its fully methylated and its fully unmethylated form. When these two versions of the gene are introduced into cultured muscle cells, both are transcribed at the same high rate. When, however, they are introduced into fibroblasts, which normally do not transcribe the gene, the unmethylated gene is transcribed at a low rate, whereas neither the exogenously added methylated gene nor the endogenous gene of the fibroblast (which is also methylated) is transcribed at all. Second, biochemical experiments have identified a vertebrate protein that binds tightly to DNA that contains clustered 5-methyl C nucleotides. The binding of this protein is thought to package the methylated DNA in a way that makes it unusually resistant to the transcriptional activation machinery. These two observations suggest that DNA methylation is used in vertebrates mainly to ensure that once a gene is turned off, it stays completely off (Figure 9–69).

Experiments designed to test whether a DNA sequence that is transcribed at high levels in one vertebrate cell type is transcribed at all in another have demonstrated that rates of gene transcription can differ between two cell types by a factor of more than 10^6. Thus unexpressed vertebrate genes are much less "leaky" in terms of transcription than are unexpressed genes in bacteria, in which the

Figure 9–68 How DNA methylation patterns are faithfully inherited. In vertebrate DNAs a large fraction of the cytosine nucleotides in the sequence CG are methylated (see Figure 9–67). Because of the existence of a methyl-directed methylating enzyme (the maintenance methylase), once a pattern of DNA methylation is established, each site of methylation is inherited in the progeny DNA, as shown. This means that changes in DNA methylation patterns will be perpetuated in all of the progeny of a cell.

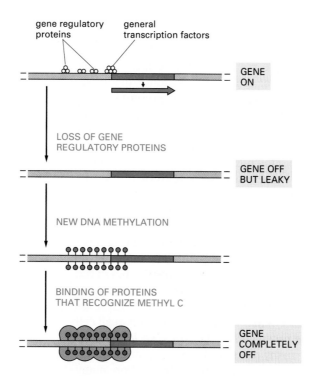

gene regulatory proteins   general transcription factors

GENE ON

LOSS OF GENE REGULATORY PROTEINS

GENE OFF BUT LEAKY

NEW DNA METHYLATION

BINDING OF PROTEINS THAT RECOGNIZE METHYL C

GENE COMPLETELY OFF

Figure 9–69 How DNA methylation may help turn off genes. The binding of gene regulatory proteins and general transcription factors near an active promoter prevents DNA methylation by some unknown mechanism. If most of these sequence-specific DNA-binding proteins dissociate, however, as generally occurs when a gene is turned off, the DNA becomes methylated, which enables other proteins to bind, and these shut down the gene completely.

largest known differences in transcription rates between expressed and unexpressed gene states are about 1000-fold. DNA methylation of unexpressed vertebrate genes may account for at least part of this difference. In addition, as we discuss next, DNA methylation is required for at least one special type of cellular memory.

Genomic Imprinting Requires DNA Methylation [51]

Mammalian cells are diploid, containing one set of genes inherited from the father and one set from the mother. In a few cases the expression of a gene has been found to depend on whether it is inherited from the mother or the father. This phenomenon is called **genomic imprinting.** Although not originally discovered in this way, genomic imprinting has been dramatically illustrated in experiments in transgenic mice. It is possible, for example, to make transgenic mice in which one of the two normal copies of the gene coding for *insulinlike growth factor-2 (IGF-2)* has been inactivated by mutation. These heterozygous mice develop normally if it is the maternally derived *Igf-2* gene that is defective, whereas if the paternally derived *Igf-2* gene is defective, they are stunted, growing to less than half the size of normal mice. Further analysis of these and normal mice provided an explanation. In both the transgenic and the wild-type mice only the paternally derived *Igf-2* gene is transcribed, while the maternally derived gene is silent; the maternally derived gene in this case is said to be *imprinted.*

Although the mechanism of imprinting is uncertain, it seems very likely that DNA methylation is involved. Thus, in transgenic mice defective in the maintenance methylase, the imprinting of the maternal *Igf-2* gene does not occur, implying that the mechanism that distinguishes between the paternal and maternal copies of the *Igf-2* gene requires DNA methylation. Interestingly, the mice lacking the maintenance methylase die as young embryos. This could result from defective imprinting, but it is also conceivable that a failure to reinforce developmental decisions by methylation is the primary defect, leading to leaky transcription of the many thousands of genes that are normally turned off in each vertebrate cell.

CG-rich Islands Are Associated with About 40,000 Genes in Mammals [52]

Because of the way DNA repair enzymes work, methylated C nucleotides in the genome tend to be eliminated in the course of evolution. Accidental deamination of an unmethylated C gives rise to U, which is not normally present in DNA and thus is recognized easily by the DNA repair enzyme uracil DNA glycosylase, excised, and then replaced with a C (discussed in Chapter 6). But accidental deamination of a 5-methyl C cannot be repaired in this way, for the deamination product is a T and so indistinguishable from the other, nonmutant T nucleotides in the DNA. Although a special repair system exists to remove these mutant Ts (see p. 250), many of the deaminations escape detection, so that those C nucleotides in the genome that are methylated tend to mutate to T over evolutionary time.

During the course of evolution, more than three out of every four CGs have been lost in this way, leaving vertebrates with a remarkable deficiency of this dinucleotide. The CG sequences that remain are very unevenly distributed in the genome; they are present at 10 to 20 times their average density in selected regions, called **CG islands,** that are 1000 to 2000 nucleotide pairs long. These islands, with some important exceptions, seem to remain unmethylated in all cell types. They are thought to surround the promoters of the so-called *housekeeping genes*—those genes that code for the many proteins that are essential for cell viability and are therefore expressed in most cells (Figure 9–70). In addition, many *tissue-specific genes*, which code for proteins needed only in selected types of cells, are also associated with CG islands.

The distribution of CG islands can be explained if we assume that CG methylation was adopted in vertebrates as a way of hindering the initiation of transcription in inactive segments of the genome (Figure 9–71). In the germ line of vertebrates—the cell lineage giving rise to eggs and sperm—most of the genome is inactive and methylated. Over long periods of evolutionary time, the methylated CG sequences in these inactive regions have presumably been lost through accidental deamination events that were not correctly repaired. The CG sequences in the regions surrounding the promoters of many genes, however, including all housekeeping genes, are kept demethylated in cells of the germ line, and so they can be readily repaired after spontaneous deamination events. Such regions are preserved as CG islands.

The mammalian genome (about 3×10^9 nucleotide pairs) contains an estimated 40,000 CG islands. Most of the islands mark the 5′ ends of a transcription unit and thus, presumably, a gene. It is possible to clone specifically the DNA surrounding the CG islands, and this technique provides a convenient way of finding new genes.

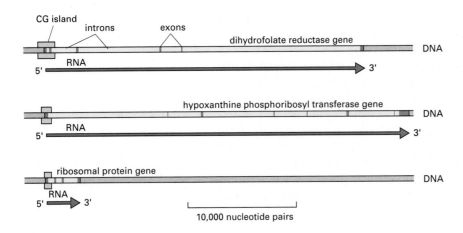

Figure 9–70 The CG islands surrounding the promoter in three mammalian housekeeping genes. The *yellow* boxes show the extent of each island. Note also that, as for most genes in mammals, the exons (*dark red*) are very short relative to the introns (*light red*). (Adapted from A.P. Bird, *Trends Genet.* 3:342–347, 1987.)

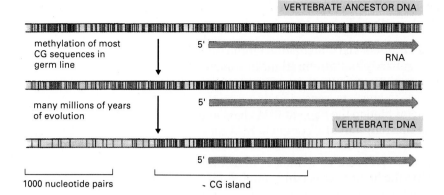

methylation of most CG sequences in germ line

5'

RNA

many millions of years of evolution

5'

VERTEBRATE DNA

5'

1000 nucleotide pairs

CG island

Figure 9–71 A mechanism to explain both the marked deficiency of CG sequences and the presence of CG islands in vertebrate genomes. A *black line* marks the location of an unmethylated CG dinucleotide in the DNA sequence, while a *red line* marks the location of a methylated CG dinucleotide.

Summary

The many types of cells in animals and plants are created largely through mechanisms that cause different genes to be transcribed in different cells. Since many specialized animal cells can maintain their unique character when grown in culture, the gene regulatory mechanisms involved in creating them must be stable once established and heritable when the cell divides, endowing the cell with a memory of its developmental history. Procaryotes and yeasts provide unusually accessible model systems in which to study gene regulatory mechanisms, some of which may be relevant to the creation of specialized cell types in higher eucaryotes. One such mechanism involves a competitive interaction between two (or more) gene regulatory proteins, each of which inhibits the synthesis of the other; this can create a flip-flop switch that switches a cell between two alternative patterns of gene expression. Direct or indirect positive feedback loops, which enable gene regulatory proteins to perpetuate their own synthesis, provide a general mechanism for cell memory.

In eucaryotes gene transcription is generally controlled by combinations of gene regulatory proteins. It is thought that each type of cell in a higher eucaryotic organism contains a specific combination of gene regulatory proteins that ensures the expression of only those genes appropriate to that type of cell. A given gene regulatory protein may be expressed in a variety of circumstances and typically is involved in the regulation of many genes.

In addition to diffusible gene regulatory proteins, inherited states of chromatin condensation are also utilized by eucaryotic cells to regulate gene expression. In vertebrates DNA methylation also plays a part, mainly as a device to reinforce decisions about gene expression that are made initially by other mechanisms.

Posttranscriptional Controls

Although controls on the initiation of gene transcription are the predominant form of regulation for most genes, other controls can act later in the pathway from RNA to protein to modulate the amount of gene product that is made. Although these **posttranscriptional controls,** which operate after RNA polymerase has bound to the gene's promoter and begun RNA synthesis, are less common than *transcriptional control,* for many genes they are crucial. It seems that every step in gene expression that could be controlled in principle is likely to be regulated under some circumstances for some genes.

We consider the varieties of posttranscriptional regulation in temporal order, according to the sequence of events that might be experienced by an RNA molecule after its transcription has begun (Figure 9–72).

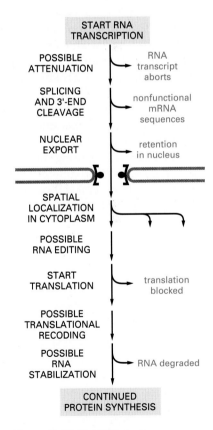

START RNA TRANSCRIPTION

POSSIBLE ATTENUATION — RNA transcript aborts

SPLICING AND 3'-END CLEAVAGE — nonfunctional mRNA sequences

NUCLEAR EXPORT — retention in nucleus

SPATIAL LOCALIZATION IN CYTOPLASM

POSSIBLE RNA EDITING

START TRANSLATION — translation blocked

POSSIBLE TRANSLATIONAL RECODING

POSSIBLE RNA STABILIZATION — RNA degraded

CONTINUED PROTEIN SYNTHESIS

Figure 9–72 Possible posttranscriptional controls on gene expression. Only a few of these controls are likely to be used for any one gene.

Transcription Attenuation Causes the Premature Termination of Some RNA Molecules [53]

In bacteria the expression of certain genes is inhibited by premature termination of transcription, a phenomenon called **transcription attenuation.** In some of these cases the nascent RNA chain adopts a structure that causes it to interact with the RNA polymerase in such a way as to abort its transcription. When the gene product is required, regulatory proteins bind to the nascent RNA chain and interfere with attenuation, allowing the transcription of a complete RNA molecule.

In eucaryotes transcription attenuation can occur by a number of distinct mechanisms. In both adenovirus and HIV (the human AIDS virus), for example, the proteins that assemble at the promoter seem to determine whether or not the polymerase will be able to pass through specific sites of attenuation downstream. These proteins can differ from one cell type to the next, and the cell can control the degree of attenuation for particular genes.

Alternative RNA Splicing Can Produce Different Forms of a Protein from the Same Gene [54]

As discussed in Chapter 8, many genes are first transcribed as long mRNA precursors that are then shortened by a series of processing steps to produce the mature mRNA molecule. One of these steps is *RNA splicing,* in which the intron sequences are removed from the mRNA precursor. Often a cell can splice the primary transcript in different ways and thereby make different polypeptide chains from the same gene—a process called **alternative RNA splicing** (Figure 9–73). A substantial proportion of higher eucaryotic genes produce multiple proteins in this way. When different splicing possibilities exist at several positions in the transcript, a single gene can produce dozens of different proteins. Usually, however, the splice alternatives are more limited, and only a few kinds of proteins are synthesized from each transcription unit.

In some cases alternative RNA splicing occurs because there is an "intron sequence ambiguity": the standard spliceosome mechanism for removing intron sequences (discussed in Chapter 8) is unable to distinguish cleanly between two or more alternative pairings of 5' and 3' splice sites, so that different choices are made haphazardly on different occasions. Where such *constitutive alternative splicing* occurs, several versions of the protein encoded by the gene are made in all cells in which the gene is expressed.

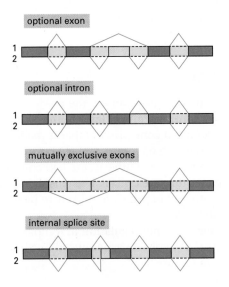

optional exon

optional intron

mutually exclusive exons

internal splice site

Figure 9–73 **Four patterns of alternative RNA splicing.** In each case a single type of RNA transcript is spliced in two alternative ways to produce two distinct mRNAs (1 and 2). The *dark blue boxes* mark exon sequences that are retained in both mRNAs. The *light blue boxes* mark possible exon sequences that are included in only one of the mRNAs; these boxes are joined by *red lines* to indicate where intron sequences (*yellow*) are removed. (Adapted with permission from A. Andreadis, M.E. Gallego, and B. Nadal-Ginard, *Annu. Rev. Cell Biol.* 3:207–242, 1987. © 1987 Annual Reviews, Inc.)

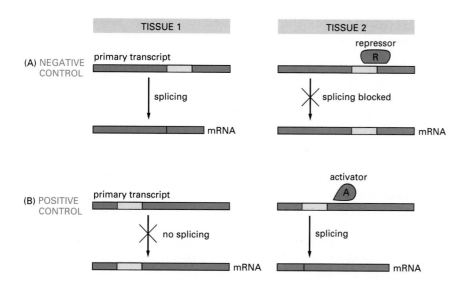

Figure 9–74 **Negative and positive control of alternative RNA splicing.** (A) Negative control, in which a repressor protein binds to the primary RNA transcript in tissue 2, thereby preventing the splicing machinery from removing an intron sequence. (B) Positive control, in which the splicing machinery is unable to remove a particular intron sequence without assistance from an activator protein.

In many cases, however, alternative RNA splicing is *regulated* rather than constitutive. In the simplest examples regulated splicing is used to switch from the production of a nonfunctional protein to the production of a functional one. The transposase that catalyzes the transposition of the *Drosophila* P element, for example, is produced in a functional form in germ cells and a nonfunctional form in somatic cells of the fly, allowing the P element to spread throughout the genome of the fly without causing damage in somatic cells. The difference in transposon activity has been traced to the presence of an intron sequence in the transposase RNA that is removed only in germ cells.

RNA splicing can be regulated either negatively, by a regulatory molecule that prevents the splicing machinery from gaining access to a particular splice site on the RNA, or positively, by a regulatory molecule that directs the splicing machinery to an otherwise overlooked splice site (Figure 9–74). In the case of the *Drosophila* transposase, the key splicing event is blocked in somatic cells by negative regulation.

In addition to switching from the production of a functional protein to the production of a nonfunctional one, the regulation of RNA splicing can generate different versions of a protein in different cell types, according to the needs of the cell. The tyrosine protein kinase encoded by the *src* proto-oncogene, for example, is produced in a specialized form in nerve cells by this mechanism (Figure 9–75). Cell-type-specific forms of many other proteins are produced in the same way.

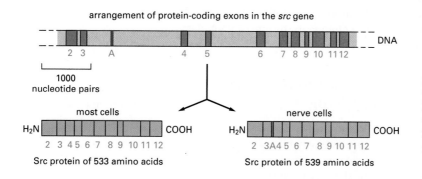

arrangement of protein-coding exons in the *src* gene

Figure 9–75 **Regulated alternative RNA splicing produces cell-type-specific forms of a gene product.** Here two slightly different tyrosine protein kinases are produced from the *src* gene because exon sequence A is included only in nerve cells. The neural form of the Src protein contains an extra site for phosphorylation and is also thought to have a higher specific activity. Only the protein-coding exons (*colored*) are shown in this diagram (exon 1, which forms the 5' leader on the mRNA, is not shown). (After J.B. Levy et al., *Mol. Cell Biol.* 7:4142–4145, 1987.)

Sex Determination in *Drosophila* Depends on a Regulated Series of RNA Splicing Events [55]

In *Drosophila* the primary signal for determining whether the fly develops as a male or female is the X chromosome/autosome ratio. Individuals with an X chromosome/autosome ratio of 1 (normally two X chromosomes and two sets of autosomes) develop as females, while those with a ratio of 0.5 (normally one X chromosome and two sets of autosomes) develop as males. This ratio is somehow assessed early in development and is remembered by each cell thereafter. Three crucial gene products are involved in transmitting information about this ratio to the many other genes that specify male and female characteristics (Figure 9–76). As explained in Figure 9–77, sex determination in *Drosophila* depends on a cascade of regulated RNA splicing events that involves these three gene products.

Drosophila sex determination provides the best-understood example of a regulatory cascade based on RNA splicing. It is not clear why the fly should use this strategy. Other organisms (the nematode, for example) use an entirely different scheme for sex determination—one based on transcriptional and translational controls. Moreover, the *Drosophila* male-determination pathway requires that a number of nonfunctional RNA molecules be continually produced, which seems unnecessarily wasteful. One speculation is that this RNA-splicing cascade is an ancient control device, left over from a stage of evolution where RNA was the predominant biological molecule and controls of gene expression had to be based almost entirely on RNA-RNA interactions.

A Change in the Site of RNA Transcript Cleavage and Poly-A Addition Can Change the Carboxyl Terminus of a Protein [56]

In eucaryotes the 3' end of an mRNA molecule is not determined by the termination of RNA synthesis by the RNA polymerase as it is in bacteria. Instead, it is determined by an RNA cleavage reaction that is catalyzed by additional factors while the transcript is elongating (see Figure 8–49). A cell can control the site of this cleavage so as to change the carboxyl terminus of the resultant protein (which is encoded by the 3' end of the mRNA).

A well-studied example is the switch from the synthesis of membrane-bound to secreted antibody molecules that occurs during the development of B lymphocytes. Early in the life history of a B lymphocyte, the antibody it produces is anchored in the plasma membrane, where it serves as a receptor for antigen. Antigen stimulation causes these cells to multiply and to start secreting their antibody. The secreted form of the antibody is identical to the membrane-bound form except at the extreme carboxyl terminus. In this part of the protein the membrane-bound form has a long string of hydrophobic amino acids that traverses the lipid bilayer of the membrane, whereas the secreted form has a much shorter string of hydrophilic amino acids. The switch from membrane-bound to secreted antibody therefore requires a different nucleotide sequence at the 3' end of the mRNA; this difference is generated through a change in the length of the primary RNA transcript caused by a change in the site of RNA cleavage, as described in Figure 9–78.

The Definition of a Gene Has Had to Be Modified Since the Discovery of Alternative RNA Splicing [57]

The discovery that eucaryotic genes usually contain introns and that their coding sequences can be put together in more than one way raised new questions about the definition of a gene. A gene was first clearly defined in molecular terms in the early 1940s from work on the biochemical genetics of the fungus *Neurospora*. Until then, a **gene** had been defined operationally as a region of the ge-

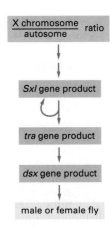

Figure 9–76 Sex determination in *Drosophila*. The gene products shown act in a sequential cascade to determine the sex of the fly according to the X chromosome/autosome ratio. The genes are called *sex-lethal (Sxl)*, *transformer (tra)*, and *doublesex (dsx)* because of the phenotypes that result when the gene is inactivated by mutation. The function of these gene products is to transmit the information about the X chromosome/autosome ratio to the many other genes that are involved in creating the sex-related phenotypes. These other genes function as two alternative sets: those that specify female features and those that specify male features (see Figure 9–77).

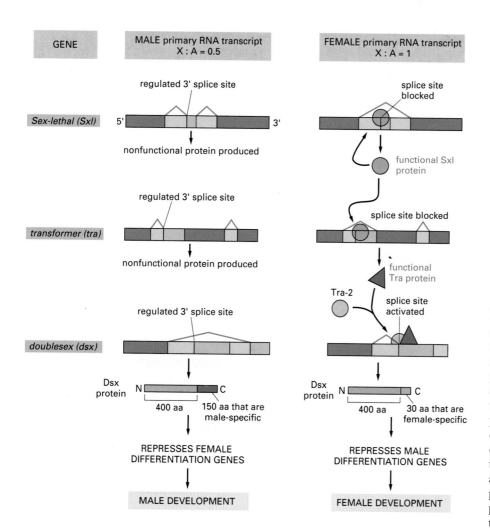

| GENE | MALE primary RNA transcript X : A = 0.5 | FEMALE primary RNA transcript X : A = 1 |

Sex-lethal (Sxl)

regulated 3' splice site

5' ─── 3'

nonfunctional protein produced

splice site blocked

functional Sxl protein

transformer (tra)

regulated 3' splice site

nonfunctional protein produced

splice site blocked

functional Tra protein

Tra-2

splice site activated

doublesex (dsx)

regulated 3' splice site

Dsx protein N ─── C

400 aa   150 aa that are male-specific

Dsx protein N ─── C

400 aa   30 aa that are female-specific

REPRESSES FEMALE DIFFERENTIATION GENES

MALE DEVELOPMENT

REPRESSES MALE DIFFERENTIATION GENES

FEMALE DEVELOPMENT

Figure 9–77 The cascade of changes in gene expression that determines the sex of a fly depends on alternative RNA splicing. An X chromosome/autosome ratio of 0.5 results in male development. Male is the "default" pathway in which the *Sxl* and *tra* genes are both transcribed, but the RNAs are spliced constitutively to produce only nonfunctional RNA molecules, and the *dsx* transcript is spliced to produce a protein that turns off the genes that specify female characteristics. An X chromosome/autosome ratio of 1 triggers the female differentiation pathway in the embryo by transiently activating a promoter within the *Sxl* gene that causes a functional Sxl protein to be synthesized. Sxl is a splicing regulatory protein with two sites of action: (1) it binds to a constitutively produced *Sxl* RNA transcript, causing a female-specific splice that continues the production of a functional Sxl protein, and (2) it binds to the constitutively produced *tra* RNA and causes an alternative splice of this transcript, which now produces an active Tra regulatory protein. The Tra protein acts with the constitutively produced Tra-2 protein to produce the female-specific spliced form of the *dsx* transcript; this encodes the female form of the Dsx protein, which turns off the genes that specify male features. The components in this pathway were all initially identified through the study of *Drosophila* mutants that are altered in their sexual development. The *dsx* gene, for example, derives its name (*doublesex*) from the observation that a fly lacking this gene product expresses both male- and female-specific features. Note that, whereas both the Sxl and Tra proteins bind to specific RNA sites, Sxl is a repressor that acts negatively to block a splice, whereas the Tra proteins are activators that act positively to induce a splice (see Figure 9–74).

nome that segregates as a single unit during meiosis and gives rise to a definable phenotypic trait, such as a red or a white eye in *Drosophila* or a round or wrinkled seed in peas. The work on *Neurospora* showed that most genes correspond to a region of the genome that directs the synthesis of a single enzyme. This led to the hypothesis that one gene encodes one polypeptide chain. The hypothesis proved fruitful for subsequent research; and, as more was learned about the mechanism of gene expression in the 1960s, a gene became identified as that stretch of DNA that was transcribed into the RNA coding for a single polypeptide chain (or a single structural RNA such as a tRNA or an rRNA molecule). The discovery of split genes in the late 1970s could be readily accommodated by the original definition of a gene, provided that a single polypeptide chain was specified by the RNA transcribed from any one DNA sequence. But it is now clear that many DNA sequences in higher eucaryotic cells produce two or more distinct proteins by means of alternative RNA splicing. How then is a gene to be defined?

In those relatively rare cases in which two very different eucaryotic proteins are produced from a single transcription unit, the two proteins are considered to be produced by distinct genes that overlap on the chromosome. It seems unnecessarily complex, however, to consider most of the protein variants produced by alternative RNA splicing as being derived from overlapping genes. A more sensible alternative is to modify the original definition to include as a gene any DNA sequence that is transcribed as a single unit and encodes one set of closely related polypeptide chains (*protein isoforms*). This definition of a gene also accommodates those DNA sequences that encode protein variants produced by posttranscriptional processes other than RNA splicing, such as ribosomal frameshifting and RNA editing, which we discuss later.

Posttranscriptional Controls

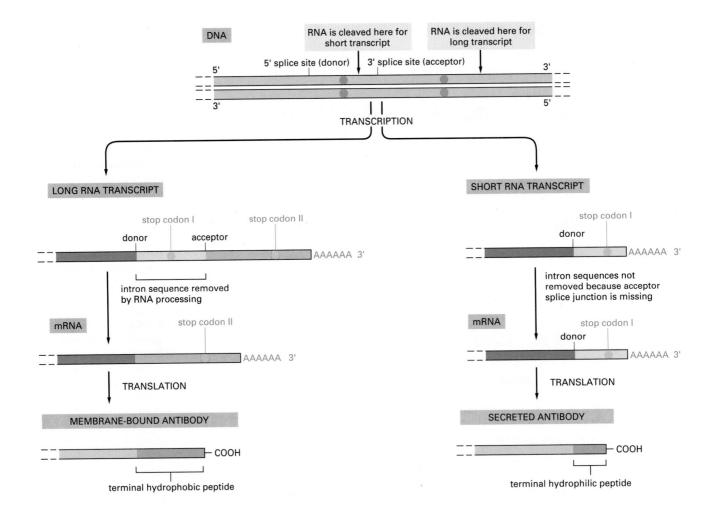

RNA Transport from the Nucleus Can Be Regulated [58]

An average primary RNA transcript seems to be at most 10 times longer than the mature mRNA molecule generated from it by RNA splicing. Yet it has been estimated that only about one-twentieth of the total mass of RNA made ever leaves the nucleus. It seems, therefore, that a substantial fraction of the primary transcripts (perhaps half) may be completely degraded in the nucleus without ever generating an RNA molecule that reaches the cytoplasm. The discarded RNAs may consist of sequences that are never made into an mRNA molecule; on the other hand, some may represent potential mRNA molecules that are functional in some cell types but in others fail to get delivered to the cytoplasm. This might be either because they are selectively targeted for intranuclear degradation or because their exit from the nucleus is selectively blocked.

Although there is very little solid evidence for either form of control, each of them remains a real possibility. In particular, RNA export through the nuclear pores is an active process, and for most RNAs it requires a specific nucleotide cap at the 5' end of the RNA molecule and a poly-A tail at the 3' end (discussed in Chapter 8). A requirement of this type makes sense, since it keeps junk RNA fragments (such as the intron sequences removed by RNA splicing) out of the cytosol. But having the proper types of ends is not enough for transport: each mRNA precursor molecule remains tethered to sites inside the nucleus until all of the spliceosome components have dissociated from it (see Figure 8–54). Therefore, any mechanism that prevents the completion of RNA splicing on particular RNA molecules could, in principle, block the exit of those RNAs from the nucleus.

Figure 9–78 Regulation of the site of RNA cleavage and poly-A addition determines whether an antibody molecule is secreted or remains membrane-bound. In unstimulated B lymphocytes (*left*) a long RNA transcript is produced, and the intron sequence near its 3' end is removed by RNA splicing to give rise to an mRNA molecule that codes for a membrane-bound antibody molecule. In contrast, after antigen stimulation (*right*) the primary RNA transcript is cleaved upstream from the splice site in front of the last exon sequence. As a result, some of the intron sequence that is removed from the long transcript remains as coding sequence in the short transcript. These are the sequences that encode the hydrophilic carboxyl-terminal portion of the secreted antibody molecule.

Some mRNAs Are Localized to Specific Regions of the Cytoplasm [59]

Once a newly made eucaryotic mRNA molecule has passed through a nuclear pore and entered the cytosol, it encounters ribosomes that translate it into a polypeptide chain. If the mRNA encodes a protein that is destined to be secreted or expressed on the cell surface, it will be directed to the endoplasmic reticulum (ER) by a signal sequence at the protein's amino terminus; the signal sequence will be recognized as soon as it emerges from the ribosome by components of the cell's protein-sorting apparatus. This apparatus then directs the entire complex of ribosome, mRNA, and nascent protein to the membrane of the ER, where the remainder of the polypeptide chain is synthesized, as discussed in Chapter 12. In other cases the entire protein is synthesized by free ribosomes in the cytosol, and signals in the completed polypeptide chain may then direct the protein to other sites in the cell.

In still other cases, however, mRNAs are directed to specific intracellular locations by signals in the mRNA sequence itself, before the sequence has been translated into an amino acid sequence. The signal is typically located in the 3′ *untranslated region (UTR)* of the mRNA molecule—a region that extends from the stop codon, which terminates protein synthesis, to the start of the poly-A tail (see Figure 9–78). A striking example is seen in the *Drosophila* egg, where the mRNA encoding the bicoid gene regulatory protein is attached to the cortical cytoskeleton at the anterior tip of the developing egg. When the translation of this mRNA is triggered by fertilization, a gradient of the bicoid protein is generated that plays a crucial part in directing the development of the anterior part of the embryo (shown in Figure 9–39 and discussed in more detail in Chapter 21). Some mRNAs in somatic cells are localized in a similar way. The mRNA that encodes actin, for example, is localized to the actin-filament-rich cell cortex in mammalian fibroblasts by means of a 3′ UTR signal, presumably because it is advantageous for the cell to position its mRNAs close to the sites where the protein produced from the mRNA is required. This form of posttranscriptional gene regulation, where mRNA is specifically localized to one part of the cell, has been recognized only recently, and it is still unclear how many mRNAs are localized in this way. The role of the UTR in localizing mRNAs to a particular region of the cytoplasm is illustrated in Figure 9–79.

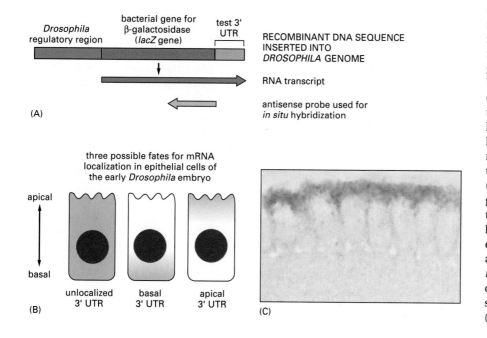

(A)

(B)

(C)

Figure 9–79 The importance of the UTR in localizing mRNAs to specific regions of the cytoplasm. *Drosophila* can be transfected with a recombinant DNA molecule coding for an mRNA in which a bacterial reporter sequence (encoding β-galactosidase) is linked to a chosen 3′ UTR sequence. According to the choice of 3′ UTR, the mRNA may be unlocalized in the embryonic cells, localized at their basal ends, or localized at their apical ends. (A) The recombinant DNA molecule used to test the effects of different 3′ UTRs. (B) *In situ* hybridization (for β-galactosidase RNA sequences) shows that the 3′ UTR determines the localization of the mRNA in the embryonic cells. (C) Photograph of an apically localized mRNA detected by *in situ* hybridization. The cells containing this mRNA are arranged in stripes along the axis of the embryo. (C, courtesy of David Ish-Horowitz.)

RNA Editing Can Change the Meaning of the RNA Message [60]

The molecular mechanisms used by cells are a continual source of surprises. An example is the phenomenon of *trans* RNA splicing, which occurs in all transcripts in the trypanosomes (the parasitic protozoa responsible for sleeping sickness). All mRNAs in trypanosomes possess a common 5′ capped leader sequence that is transcribed separately and added to the 5′ ends of RNA transcripts by splicing of two initially unconnected RNA molecules. *Trans* splicing is also used to add a 5′ leader to several mRNAs in nematodes and to combine the separate RNA transcripts that form the coding sequence of some chloroplast and mitochondrial proteins in plant cells. Cutting and pasting between RNA transcripts could speed up the evolution of new proteins, and the few cases where exons are known to be joined in this way are suspected to be evolutionary remnants of a much more extensive process that dominated ancient cells.

Another startling discovery is the process of **RNA editing,** whereby the nucleotide sequences of RNA transcripts are altered. In this process, discovered in RNA transcripts that code for proteins in the mitochondria of trypanosomes, one or more U nucleotides are inserted (or, less frequently, removed) from selected regions of a transcript, causing major modifications in both the original reading frame and the sequence, thereby changing the meaning of the message. For some genes the editing is so extensive that over half of the nucleotides in the mature mRNA are U nucleotides that were inserted during the editing process. The information that specifies exactly how the initial RNA transcript is to be altered is contained in a set of 40- to 80-nucleotide-long RNA molecules that are separately transcribed. These so-called *guide RNAs* have a 5′ end that is complementary in sequence to one end of the region of the transcript to be edited; this is followed by a sequence that specifies the set of nucleotides to be inserted into the transcript and then a continuous run of U nucleotides. The editing mechanism is surprisingly complex, with the U nucleotides at the 3′ end of the guide RNA being transferred directly into the transcript, as illustrated in Figure 9–80.

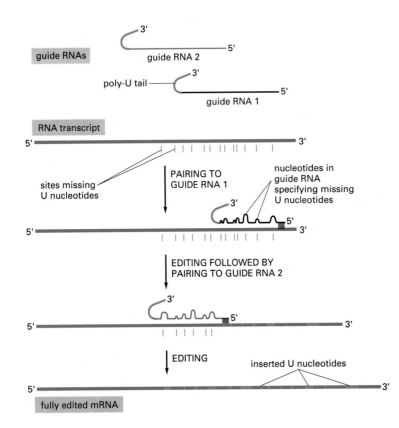

Figure 9–80 RNA editing in the mitochondria of trypanosomes. Guide RNAs contain at their 3′ end a stretch of poly U, which donates U nucleotides to sites on the RNA transcript that mispair with the guide RNA; thus the poly-U tail gets shorter as editing proceeds (not shown). Editing generally starts near the 3′ end and progresses toward the 5′ end of the RNA transcript, as shown, because the "anchor sequence" at the 5′ end of most guide RNAs can pair only with edited sequences.

Extensive editing of mRNA sequences has also been found in the mitochondria of many plants, with nearly every mRNA being edited to some extent. In this case, however, bases are changed from C to U in the RNA, without nucleotide insertions or deletions. Often many of the Cs in an mRNA are affected by editing, changing 10% or more of the amino acids that the mRNA encodes.

We can only speculate as to why the mitochondria of trypanosomes and plants make use of such extensive RNA editing. The suggestions that seem most reasonable are based on the premise that mitochondria contain a primitive genetic system. There is evidence that editing is regulated to produce different mRNAs under different conditions, so that RNA editing can be viewed as a primitive way to change the expression of genes. Trypanosomes are extremely ancient single-celled eucaryotes, which diverged very early on from the lineage leading to plants, animals, and yeasts (see Figure 1–16). Perhaps, therefore, the extreme version of RNA editing found in their mitochondria is a holdover from very ancient cells, where most catalyses were carried out by RNA molecules rather than by proteins.

RNA editing of a much more limited kind occurs in mammals. The first case discovered involved the *apolipoprotein-B* gene, where RNA editing produces two types of transcripts: in one of these a DNA-encoded C is changed to a U, creating a stop codon that causes a truncated version of this large protein to be made in a tissue-specific manner. In another case a nucleotide change in the middle of an mRNA molecule changes a single amino acid in a transmitter-gated ion channel in the brain, significantly altering the channel's permeability to Ca^{2+}. For apolipoprotein-B the editing is catalyzed in a very straightforward way: a protein binds to a specific sequence in the mRNA and then catalyzes the deamination of a C to a U. It is not known whether the other cases of mammalian and plant RNA editing are protein-mediated in this way or whether, instead, they make use of short RNA templates, as in trypanosomes.

We now turn to controls that operate on the translation of mRNAs into proteins.

Procaryotic and Eucaryotic Cells Use Different Strategies to Specify the Translation Start Site on an mRNA Molecule [61]

In bacterial mRNAs a conserved stretch of six nucleotides, the *Shine-Dalgarno sequence*, is always found a few nucleotides upstream of the initiating AUG codon. This sequence forms base pairs with the 16S RNA in the small ribosomal subunit and thereby correctly positions the initiating AUG codon in the ribosome. This interaction makes a major contribution to the efficiency of initiation and provides the bacterial cell with a simple way to regulate protein synthesis. Many translational control mechanisms in procaryotes involve blocking the Shine-Dalgarno sequence, either by covering it with a bound protein or by incorporating it into a base-paired region in the mRNA molecule.

Eucaryotic mRNAs do not contain a Shine-Dalgarno sequence. Instead, the selection of an AUG codon as a translation start site is largely determined by its proximity to the cap at the 5′ end of the mRNA molecule, which is the site at which the small ribosomal subunit binds to the mRNA and begins scanning for an initiating AUG codon (discussed in Chapter 6). The nucleotides immediately surrounding the start site in eucaryotic mRNAs also influence the efficiency of AUG recognition during the scanning process. If this recognition site is poor enough, scanning ribosomal subunits will ignore the first AUG codon in the mRNA and skip to the second or third AUG codon instead. This phenomenon, known as "leaky scanning," is a strategy frequently used to produce two or more proteins, differing in their amino termini, from the same mRNA. It allows some genes to produce the same protein with and without a signal sequence attached at its amino terminus, for example, so that the protein is directed to two different compartments in the cell.

Another important difference between eucaryotic and procaryotic translation is that eucaryotic ribosomes dissociate rapidly from the mRNA when translation terminates. Thus reinitiation at an internal AUG codon after translation of a preceding open reading frame is much less efficient in eucaryotes than in procaryotes. Together, these differences—scanning from the 5' cap and a limited ability to reinitiate at internal AUG codons—explain why the vast majority of eucaryotic mRNAs encode only a single protein and why the first AUG codon from the 5' end is usually the functional start site for translation.

A few eucaryotic cell and viral mRNAs initiate translation by an alternative mechanism that involves internal initiation rather than scanning. These mRNAs contain complex nucleotide sequences, called *internal ribosome entry sites,* where ribosomes bind in a cap-independent fashion and start translation at the next AUG codon downstream. The details of this mechanism are not known.

The Phosphorylation of an Initiation Factor Regulates Protein Synthesis [62]

Eucaryotic cells decrease their overall rate of protein synthesis in response to a variety of situations, including the deprivation of growth factors, infection by viruses, heat shock, and entry into M phase of the cell cycle. Much of this regulation is thought to involve the initiation factor **eIF-2,** which is phosphorylated by specific protein kinases to decrease the overall rate of protein synthesis.

The normal function of eIF-2 is outlined in Figure 9–81. This protein forms a complex with GTP and mediates the binding of the methionyl initiator tRNA to the small ribosomal subunit, which then binds to the 5' cap of the mRNA and begins scanning along the mRNA. After an AUG codon is recognized, the bound GTP is hydrolyzed to GDP by the eIF-2 protein, causing a conformational change in the protein and releasing it from the small ribosomal subunit. The large ribosomal subunit then joins the small one to form a complete ribosome that begins protein synthesis.

Because eIF-2 binds very tightly to GDP, a guanine nucleotide releasing protein (see Figure 15–50), designated eIF-2B, is required to cause GDP release so that a new GTP molecule can bind and eIF-2 can be reused (Figure 9–82A). The reuse of eIF-2 is inhibited when it is phosphorylated because phosphorylated eIF-2 binds to eIF-2B unusually tightly, preventing the completion of nucleotide exchange. There is more eIF-2 than eIF-2B in cells, and even a fraction of phosphorylated eIF-2 can trap nearly all of the available eIF-2B, thereby preventing the reuse of even the nonphosphorylated eIF-2 and greatly slowing protein synthesis (Figure 9–82B).

When the activity of a general translation factor, such as eIF-2, is reduced by phosphorylation, one might expect that the translation of all mRNAs would be reduced equally. Contrary to this expectation, however, the phosphorylation of

Figure 9–81 The role of eIF-2 in the initiation of protein synthesis.

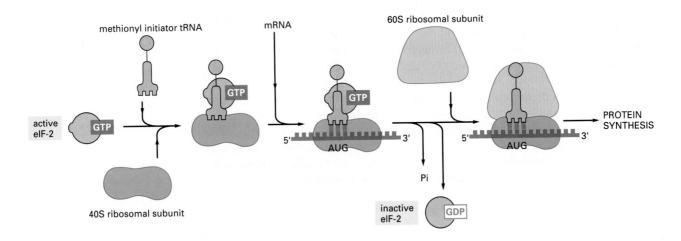

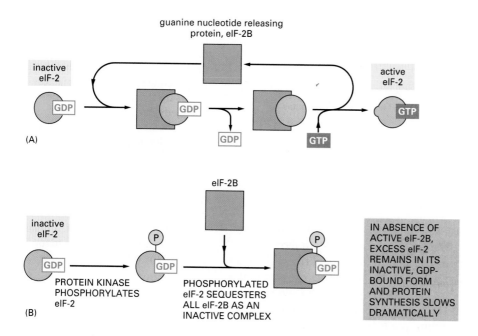

(A)

guanine nucleotide releasing
protein, eIF-2B

inactive
eIF-2

GDP

GDP

GDP

GTP

GTP

active
eIF-2

GTP

(B)

eIF-2B

inactive
eIF-2

GDP

P

GDP

PROTEIN KINASE
PHOSPHORYLATES
eIF-2

PHOSPHORYLATED
eIF-2 SEQUESTERS
ALL eIF-2B AS AN
INACTIVE COMPLEX

P

GDP

IN ABSENCE OF
ACTIVE eIF-2B,
EXCESS eIF-2
REMAINS IN ITS
INACTIVE, GDP-
BOUND FORM
AND PROTEIN
SYNTHESIS SLOWS
DRAMATICALLY

Figure 9–82 **The eIF-2 cycle.** (A) The recycling of used eIF-2 by a guanine nucleotide releasing protein (eIF-2B). (B) eIF-2 phosphorylation controls protein synthesis rates by tying up eIF-2B.

eIF-2 can have selective effects, even enhancing the translation of specific mRNAs. This can enable yeast cells, for example, to adapt to starvation for specific nutrients by shutting down the synthesis of all proteins except those that are required for synthesis of the missing nutrients. The details have been worked out for a specific yeast mRNA that encodes a protein called GCN4, a gene regulatory protein that is required for the activation of many genes encoding proteins that are important for amino acid synthesis. The GCN4 protein is produced by a specific activation of the translation of its mRNA following amino acid starvation that is induced when eIF-2 becomes phosphorylated. By a complex mechanism depending on competition between correct and incorrect ("decoy") sites of initiation of translation near the 5′ end of the GCN4 mRNA, the reduction of eIF-2 activity actually leads to an increase in the synthesis of the GCN4 protein.

Regulation of the level of eIF-2 is also important in mammalian cells as part of the mechanism by which they can be induced to enter a nonproliferating, resting state (called G_0) in which the rate of protein synthesis is reduced to about one-fifth the rate in proliferating cells (discussed in Chapter 17).

Proteins That Bind to the 5′ Leader Region of mRNAs Mediate Negative Translational Control [63]

The translation of some mRNA molecules is blocked by specific *translation repressor proteins* that bind near the 5′ end of the mRNAs, where translation would otherwise begin. This type of mechanism is called **negative translation control** (Figure 9–83). It was first discovered in bacteria, where it enables excess ribosomal proteins to repress the translation of their own mRNAs—a form of negative feedback regulation.

Figure 9–83 **Negative translational control.** This form of control is mediated by a sequence-specific RNA-binding protein that acts as a translation repressor. Binding of the protein to an mRNA molecule decreases the translation of the mRNA. Several cases of this type of translational control are known; the illustration is modeled on the mechanism that causes more ferritin (an iron storage protein) to be synthesized when the free iron concentration in the cytosol rises; the iron-sensitive translation repressor protein is called aconitase (see also Figure 9–86).

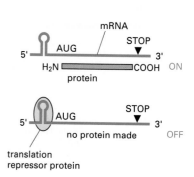

mRNA

5′

AUG

STOP

3′

H2N

COOH

ON

protein

5′

AUG

STOP

3′

no protein made

OFF

translation
repressor protein

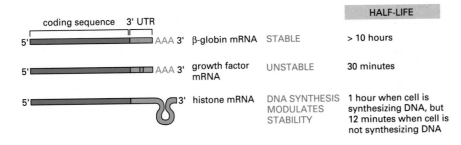

		HALF-LIFE
β-globin mRNA	STABLE	> 10 hours
growth factor mRNA	UNSTABLE	30 minutes
histone mRNA	DNA SYNTHESIS MODULATES STABILITY	1 hour when cell is synthesizing DNA, but 12 minutes when cell is not synthesizing DNA

Figure 9–84 RNA stability. Three normal mRNAs with very different half-lives. The continuous rapid degradation of the mRNA molecules that encode various growth factors allows their concentration to be changed rapidly in response to extracellular signals. A signal for rapid degradation in the 3′ untranslated region (UTR) determines the half-life of the growth factor RNA (see Figure 9–85). Histones are needed mainly to form the new chromatin produced during DNA synthesis; a large change in the stability of their mRNAs helps to confine histone synthesis to the S phase of the cell cycle.

In eucaryotic cells a particularly well-studied form of negative translational control allows the synthesis of the intracellular iron storage protein *ferritin* to be increased rapidly if the level of soluble iron atoms in the cytosol rises. The iron regulation depends on a sequence of about 30 nucleotides in the 5′ leader of the ferritin mRNA molecule. This *iron-response element* folds into a stem-loop structure that binds a translation repressor protein called *aconitase*, which blocks the translation of any RNA sequence downstream (see Figure 9–83). Aconitase is an iron-binding protein, and exposure of the cell to iron causes it to dissociate from the ferritin mRNA, releasing the block to translation and increasing the production of ferritin by as much as 100-fold.

Gene Expression Can Be Controlled by a Change in mRNA Stability [64]

Most mRNAs in a bacterial cell are very unstable, having a half-life of about 3 minutes. Because bacterial mRNAs are both rapidly synthesized and rapidly degraded, a bacterium can adapt quickly to environmental changes.

The mRNAs in eucaryotic cells are more stable. Some, such as that encoding β-globin, have a half-life of more than 10 hours. Others, however, have a half-life of 30 minutes or less. The unstable mRNAs often code for regulatory proteins, such as growth factors and gene regulatory proteins, whose production levels change rapidly in cells (Figure 9–84). Many of these RNAs are unstable because they contain specific sequences that stimulate their degradation. A long sequence rich in A and U nucleotides in the 3′ untranslated region (UTR) of several mRNAs, for example, can, if transferred to other stable mRNAs by recombinant DNA techniques, cause them to be unstable. This AU-rich sequence appears to accelerate mRNA degradation by stimulating the removal of the poly-A tail found at the 3′ end of almost all eucaryotic mRNAs. Other unstable mRNAs contain recognition sites in their 3′ UTR for specific endonucleases that cleave the mRNA (Figure 9–85).

TWO MECHANISMS OF EUCARYOTIC mRNA DECAY

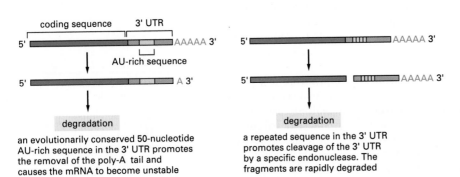

an evolutionarily conserved 50-nucleotide AU-rich sequence in the 3′ UTR promotes the removal of the poly-A tail and causes the mRNA to become unstable

a repeated sequence in the 3′ UTR promotes cleavage of the 3′ UTR by a specific endonuclease. The fragments are rapidly degraded

Figure 9–85 Control of RNA degradation. Special sequences in the 3′ untranslated region (UTR) of unstable mRNAs are responsible for their unusually rapid degradation. As indicated, AU-rich sequences found in the 3′ UTR of many short-lived mRNAs cause a rapid removal of the poly-A tail, which in turn makes the RNA unstable. Other mRNAs contain sequences in their 3′ UTR that serve as sites for specific endonucleolytic cleavage.

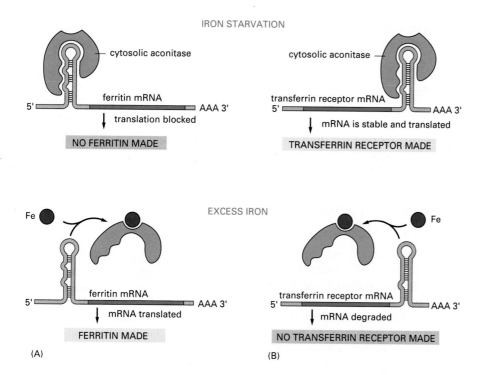

IRON STARVATION

cytosolic aconitase

ferritin mRNA

5'————————————————AAA 3'

↓ translation blocked

NO FERRITIN MADE

cytosolic aconitase

transferrin receptor mRNA

5'————————————————AAA 3'

↓ mRNA is stable and translated

TRANSFERRIN RECEPTOR MADE

EXCESS IRON

Fe ●   ●

ferritin mRNA

5'————————————————AAA 3'

↓ mRNA translated

FERRITIN MADE

(A)

●   ● Fe

transferrin receptor mRNA

5'————————————————AAA 3'

↓ mRNA degraded

NO TRANSFERRIN RECEPTOR MADE

(B)

Figure 9–86 Two posttranslational controls mediated by iron. In response to an increase in iron concentration in the cytosol, a cell increases its synthesis of ferritin in order to bind the extra iron (A) and decreases its synthesis of transferrin receptors in order to import less iron (B). Both responses are mediated by the same iron-responsive regulatory protein, aconitase, which recognizes common features in a stem-and-loop structure in the mRNAs encoding ferritin and transferrin receptor. Aconitase dissociates from the mRNA when it binds iron. Because the transferrin receptor and ferritin are regulated by different types of mechanisms, their levels respond oppositely to iron concentrations even though they are regulated by the same iron-responsive regulatory protein. (Adapted from M.W. Hentze et al., *Science* 238:1570–1573, 1987; J.L. Casey et al., *Science* 240:924–928, 1988.)

The stability of an mRNA can be changed in response to extracellular signals. Steroid hormones, for example, affect a cell not only by increasing the transcription of specific genes, but also by increasing the stability of several of the mRNAs encoded by these genes. Conversely, the addition of iron to cells decreases the stability of the mRNA that encodes the receptor protein that binds the iron-transporting protein transferrin, causing less of this receptor to be made. Interestingly, the stability of the transferrin receptor mRNA seems to be modulated by the iron-sensitive RNA-binding protein aconitase, which, as we discussed above, also controls ferritin mRNA translation. Here aconitase binds to the 3' UTR of the transferrin receptor mRNA and causes an increase in receptor production, presumably by inhibiting the function of sequences that otherwise cause rapid degradation of the mRNA. On the addition of iron, aconitase is released from the mRNA, decreasing mRNA stability (Figure 9–86).

Selective mRNA Degradation Is Coupled to Translation [65]

The control of mRNA stability in eucaryotic cells is best understood for the mRNAs that encode histones. These mRNAs have a half-life of about 1 hour during the DNA synthesis (S) phase of the cell cycle, when new histones are needed, but become unstable and are degraded within minutes when DNA synthesis stops. If DNA synthesis during S phase is inhibited with a drug, histone mRNAs immediately become unstable, perhaps because the accumulation of free histones in the absence of new DNA for them to bind increases the degradation rate of their mRNAs.

The regulation of histone mRNA stability depends on a short 3' stem-and-loop structure that replaces the poly-A tail present at the 3' end of other mRNAs (see Figure 9–84). A special cleavage reaction, which requires base-pairing to a small RNA in a ribonucleoprotein particle, creates this 3' end after the histone mRNA is synthesized by RNA polymerase II. If the 3' end is transferred to other mRNAs by recombinant DNA methods, they also become unstable when DNA synthesis stops. Thus, as for other types of mRNAs, the degradation rate of histone mRNA is strongly influenced by signals near the 3' end, where mRNA degradation is thought to begin.

If a stop codon is inserted into the middle of a coding sequence in a histone mRNA, the mRNA is no longer rapidly degraded. It has therefore been suggested that the nuclease responsible for degrading mRNAs is bound to the ribosome and that most of the histone mRNA has to be translated before the nuclease can begin digesting the 3′ end of the mRNA. This hypothesis would explain why most unstable mRNAs are selectively stabilized when cells are treated with the protein synthesis inhibitor cycloheximide. Coupling the degradation of mRNA to translation may be a mechanism for ensuring that newly synthesized mRNA molecules in eucaryotic cells are not destroyed before they have been translated at least once.

Cytoplasmic Control of Poly-A Length Can Affect Translation in Addition to mRNA Stability [66]

The initial polyadenylation of an RNA molecule (discussed in Chapter 8) occurs in the nucleus apparently automatically for nearly all eucaryotic mRNA precursors (the exception being the histone mRNAs discussed above). Once in the cytosol, the 200-nucleotide-long poly-A tails on most mRNAs gradually shorten over the course of days. Tails shorter than about 30 As, however, are not observed, suggesting that this is the minimum tail length required for mRNA stability. In addition, the poly-A tail length of some mRNAs is specifically controlled—either by selective poly-A addition or by selective poly-A removal (Figure 9–87A).

Maturing oocytes and eggs provide a striking example of the control of gene expression by means of poly-A addition to specific mRNAs. Many of the normal mRNA degradation pathways seem to be disabled in these giant cells, so that the cells can build up large stores of mRNAs in preparation for fertilization. Many mRNAs are stored with only 10 to 30 As at their 3′ end, and in this form they are not translated. At specific times during oocyte maturation and postfertilization, when the proteins encoded by these mRNAs are required, poly A is added to selected mRNAs, greatly stimulating the initiation of their translation. A model that explains how poly A added at the 3′ end can affect the initiation of translation near the 5′ end of the mRNA is illustrated in Figure 9–87B.

Figure 9–87 Control of the poly-A tail length affects both mRNA stability and mRNA translation. (A) Most translated mRNAs have poly-A tails that exceed a minimum length of about 30 As. The tails on selected mRNAs can be either elongated or rapidly cleaved in the cytosol, and this will have an effect on the translation of these mRNAs. (B) A model proposed to explain the observed stimulation of translation by an increase in poly-A tail length. The large ribosomal subunits, on finishing a protein chain, may be directly recycled from near the 3′ end of an mRNA molecule back to the 5′ end to start a new protein by special poly-A-binding proteins (*red*).

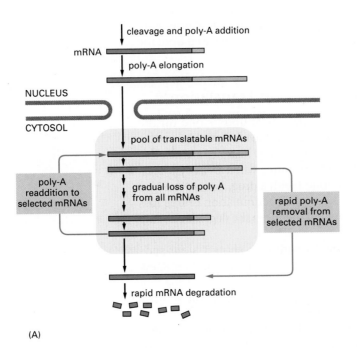

(A)

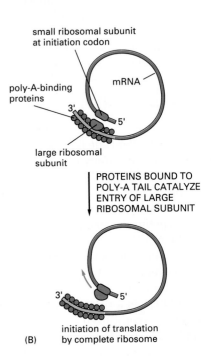

(B)

A Few mRNAs Contain a Recoding Signal That Interrupts the Normal Course of Translation [67]

The translational controls thus far discussed affect the rate at which new protein chains are initiated on an mRNA molecule. Usually, the completion of the synthesis of a protein is automatic once this synthesis has begun. In special cases, however, a process called **translational recoding** can alter the final protein that is made.

The most frequently observed form of recoding is *translational frameshifting*. This type of recoding is commonly used by retroviruses, where it allows more than one protein to be synthesized from a single mRNA. These viruses commonly make both the capsid proteins (*Gag proteins*) and the viral reverse transcriptase and integrase (*Pol proteins*) from the same RNA transcript. The virus needs many more copies of the Gag proteins than it does of the Pol proteins, and many viruses achieve this quantitative adjustment by having the *gag* and *pol* genes in different reading frames, with a stop codon at the end of the *gag* coding sequence that can be eliminated by a rare translational frameshift. The frameshift occurs at a particular codon in the mRNA and requires a specific *recoding signal*, thought to be a structural feature of the RNA sequence downstream of this site (Figure 9–88).

Similar principles are used more rarely in other forms of recoding at specific mRNA sites. Thus recoding signals have been discovered in some mRNAs that can cause a +1 frameshift, rather than the –1 frameshift shown in Figure 9–88 that operates in the viral RNA. In other cases a specific nucleotide sequence around a stop codon causes the stop codon to be leaky, so that additional amino acids are often inserted at the end of the polypeptide chain. More surprising is the recently discovered mechanism that inserts the modified amino acid *selenocysteine* into a protein chain. Selenocysteine, which is essential for the function of some enzymes, contains a selenium atom in place of the sulfur atom of cysteine and is attached to a special tRNA molecule. The tRNA has an affinity for a recoding signal present in those mRNA molecules that code for proteins utilizing selenocysteine. Selenocysteine is incorporated at special UGA codons, which in most other settings would serve as a stop codon to terminate protein synthesis.

RNA-catalyzed Reactions in Cells Are Likely to Be of Extremely Ancient Origin [68]

All of the posttranscriptional control mechanisms discussed in this section depend on a particular RNA molecule being specifically recognized for special treatment, such as splicing, editing, or degradation. In some cases this recognition is

Figure 9–88 Translational frameshifting is necessary to produce the reverse transcriptase and integrase of a retrovirus. The viral reverse transcriptase and integrase are produced by cleavage of the large Gag-Pol fusion protein, whereas the viral capsid proteins are produced by cleavage of the more abundant Gag protein. Both the Gag and fusion proteins start identically, but the Gag protein terminates at an in-frame stop codon (not shown); the indicated frameshift eliminates this stop codon, allowing the synthesis of the longer fusion protein. The frameshift occurs because features in the local RNA structure (including the RNA loop shown) cause the tRNA^{Leu} attached to the carboxyl terminus of the growing polypeptide chain occasionally to slip backward by one nucleotide on the ribosome, so that it pairs with a UUU codon instead of the UUA codon that had specified its incorporation; the next codon (AGG) in the new reading frame specifies an arginine rather than a glycine. The sequence shown is from the AIDS virus, HIV. (Adapted from T. Jacks et al., *Nature* 331:280–283, 1988.)

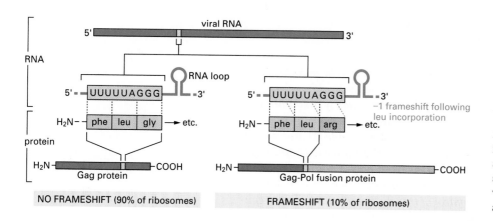

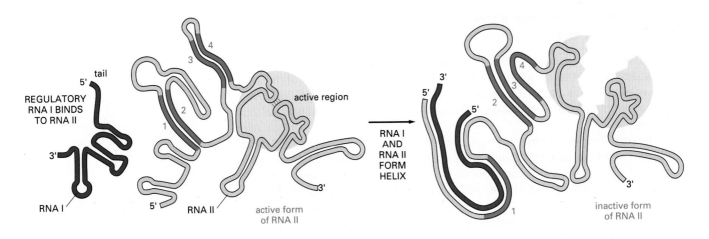

Figure 9–89 **Antisense RNA strategy for regulating plasmid numbers in bacteria.** A regulatory interaction between two RNA molecules maintains a constant plasmid copy number in the ColE1 family of bacterial DNA plasmids. RNA I (about 100 nucleotides long) is a regulatory RNA that inhibits the activity of RNA II (about 500 nucleotides long), which normally helps initiate plasmid DNA replication. The concentration of RNA I increases in proportion to the number of plasmid DNA molecules in the cell, so that as plasmid numbers increase, plasmid replication is inhibited. RNA I is complementary in sequence to the 5′ end of RNA II. In RNA II sequence 2 is complementary to both sequence 1 and sequence 3, and it is displaced from one to the other by the binding of RNA I; RNA I thereby alters the conformation of sequence 4, inactivating RNA II. (After H. Masukata and J. Tomizawa, *Cell* 44:125–136, 1986.)

accomplished by specialized RNA-binding proteins. In other cases, however, the recognition of specific RNA sequences is carried out by other RNA molecules, which use complementary RNA-RNA base-pairing as part of their recognition mechanism. RNA-RNA pairings, for example, are known to play a central part in translation, in RNA splicing, in several other forms of RNA processing, and in the RNA editing that occurs in trypanosomes. In attempting to dissect posttranscriptional mechanisms, we have largely entered an RNA world.

RNA molecules also have other regulatory roles in cells. The *antisense RNA* strategy for experimentally manipulating cells so that they fail to express a particular gene (see p. 326) mimics a normal mechanism that is known to regulate the expression of a few selected genes in bacteria and may be used much more widely than is now realized. A well-understood example of this kind of mechanism provides a feedback control on the initiation of DNA replication for a large family of bacterial DNA plasmids. The control system limits the number of copies of the plasmid made in the cell, thereby preventing the plasmid from killing its host cell by overreplicating (Figure 9–89).

Studies of RNA-catalyzed reactions are of special interest from an evolutionary perspective. As discussed in Chapter 1, the first cells are thought to have lacked DNA and may have contained very few, if any, proteins. Many of the RNA-catalyzed reactions in present-day cells seem to represent molecular fossils—descendants of the complex network of RNA-mediated reactions that are presumed to have dominated cell metabolism more than 3.5 billion years ago. Recombinant DNA technology has allowed large amounts of pure RNAs of any sequence to be produced *in vitro* with purified RNA polymerases (see Figure 7–36), making it possible to study the detailed chemistry of RNA-catalyzed reactions. From an understanding of many such reactions, biologists hope to be able to trace the path by which a living cell first evolved.

Summary

Many steps in the pathway from RNA to protein are regulated by cells to control gene expression. Most genes are thought to be regulated at multiple levels, although control of the initiation of transcription (transcriptional control) usually predominates. Some genes, however, are transcribed at a constant level and turned on and off solely by posttranscriptional regulatory processes. These processes include (1) attenuation of the RNA transcript by its premature termination, (2) alternative RNA splice-site selection, (3) control of 3′-end formation by cleavage and poly-A addition, (4) control of transport from the nucleus to the cytosol, (5) localization of mRNAs to particular parts of the cell, (6) RNA editing, (7) control of translational initiation, (8) regulated mRNA degradation, and (9) translational recoding. Most of these control processes require the recognition of specific sequences or structures in the RNA molecule being regulated. This recognition can be accomplished by either a regulatory protein or a regulatory RNA molecule.

References

General

Branden, C.; Tooze, J. Introduction to Protein Structure. New York: Garland, 1991.

Darnell, J.; Lodish, H.; Baltimore, D. Molecular Cell Biology. New York: Scientific American Books, 1986.

Lewin, B. Genes, 4th ed. New York: Wiley, 1990.

Schleif, R. Genetics and Molecular Biology. Reading, MA: Addison-Wesley, 1986.

Stent, G.S.; Calendar, R. Molecular Genetics: An Introductory Narrative, 2nd ed. San Francisco: W.H. Freeman, 1978.

Watson, J.D.; Hopkins, N.H.; Roberts, J.W.; Steitz, J.A.; Weiner, A.M. Molecular Biology of the Gene, 4th ed. Menlo Park, CA: Benjamin-Cummings, 1987.

Cited

1. Gurdon, J.B. The developmental capacity of nuclei taken from intestinal epithelium cells of feeding tadpoles. *J. Embryol. Exp. Morphol.* 10:622–640, 1962.

   Gurdon, J.B. The generation of diversity and pattern in animal development. *Cell* 68:185–199, 1992.

   Nomura, K.; Komamine, A. Identification and isolation of single cells that produce somatic embryos at a high frequency in a carrot suspension culture. *Plant Physiol.* 79:988–991, 1985.

   Steward, F.C.; Mapes, M.O.; Mears, K. Growth and organized development of cultured cells. *Am. J. Bot.* 45:705–713, 1958.

   Wareing, P.F.; Phillips, I.D.J. The Control of Growth and Differentiation in Plants, 3rd ed. Oxford, UK: Pergamon Press, 1986.

2. Garrels, J.I. Changes in protein synthesis during myogenesis in a clonal cell line. *Dev. Biol.* 73:134–152, 1979.

3. Lucas, P.C.; Granner, D.K. Hormone response domains in gene transcription. *Annu. Rev. Biochem.* 61:1131–1173, 1992.

   Miesfeld, R.L. The structure and function of steroid receptor proteins. *Crit. Rev. Biochem. Mol. Biol.* 24:101–117, 1989.

   Pilkis, S.J.; Granner, D.K. Molecular physiology of the regulation of hepatic gluconeogenisis and glycolysis. *Annu. Rev. Physiol.* 54:885–909, 1992.

   Yamamoto, K. Steroid receptor regulated transcription of specific genes and gene networks. *Annu. Rev. Genet.* 19:209–252, 1985.

4. Darnell, J.E., Jr. Variety in the level of gene control in eucaryotic cells. *Nature* 297:365–371, 1982.

   Derman, E., et al. Transcriptional control in the production of liver-specific mRNA's. *Cell* 23:731–739. 1981.

5. Ptashne, M. A Genetic Switch, 2nd ed. Cambridge, MA: Cell Press and Blackwell Press, 1992.

6. Gilbert, W.; Müller-Hill, B. The lac operator is DNA. *Proc. Natl. Acad. Sci. USA* 58:2415–2421, 1967.

   Jacob, F.; Monod, J. Genetic regulatory mechanisms in the synthesis of proteins. *J. Mol. Biol.* 3:318–356, 1961.

   Kaiser, A.D. Mutations in a temperate bacteriophage affecting its ability to lysogenize, *E. coli. Virology* 3:42–61, 1957.

   Ptashne, M. Specific binding of the λ phage repressor to λ DNA. *Nature* 214:232–234, 1967.

7. Seeman, N.C.; Rosenberg, J.M.; Rich, A. Sequence-specific recognition of double helical nucleic acids by proteins. *Proc. Natl. Acad. Sci. USA* 73:804–808, 1976.

8. Crothers, D.M.; Steitz, T.A. Transcriptional activation by *Escherichia coli* CAP protein. In Transcriptional Regulation (S.L. McKnight, K.R. Yamamoto, eds.). Cold Spring Harbor, NY: Cold Spring Harbor Laboratory Press, 1992.

   Dickerson, R.E. DNA structure from A to Z. *Methods Enzymol.* 211:67–111, 1992.

   Hagerman, P.J. Sequence-directed curvature of DNA. *Annu. Rev. Biochem.* 59:755–781, 1990.

   Harrison, S.C. Molecular characteristics of the regulatory switch in phages 434 and λ. In Transcriptional Regulation (S.L. McKnight, K.R. Yamamoto, eds.). Cold Spring Harbor, NY: Cold Spring Harbor Laboratory Press, 1992.

   van der Vliet, P.C.; Verrijzer, C.P. Bending of DNA by transcription factors. *Bioessays* 15:25–32, 1993.

9. Miller, J.H.; Reznikoff, W.S., eds. The Operon. Cold Spring Harbor, NY: Cold Spring Harbor Laboratory Press, 1978.

   Mitchell, P.J.; Tijan, R. Transcriptional regulation in mammalian cells by sequence-specific DNA binding proteins. *Science* 245:371–378, 1989.

   Verdier, J.-M. Regulatory DNA-binding proteins in yeast: an overview. *Yeast* 6:271–297, 1990.

10. Harrison, S.C. A structural taxonomy of DNA-binding domains. *Nature* 353:715–719, 1991.

    Pabo, C.O.; Sauer, R.T. Transcription factors: structural families and principles of DNA recognition. *Annu. Rev. Biochem.* 61:1053–1095, 1992.

    Steitz, T.A. Structural studies of protein–nucleic acid interaction: the sources of sequence-specific binding. *Q. Rev. Biophys.* 23:205–280, 1990.

    Wright, P.E. Solution structures of DNA-binding domains of eukaryotic transcription factors. In Transcriptional Regulation (S.L. McKnight, K.R. Yamamoto, eds.). Cold Spring Harbor, NY: Cold Spring Harbor Laboratory Press, 1992.

11. Harrison, S.C.; Aggarwal, A.K. DNA recognition by proteins with the heliz-turn-helix motif. *Annu. Rev. Biochem.* 59:933–969, 1990.

    Laughon, A; Scott, M.P. Sequence of a *Drosophila* segmentation gene: protein structure homology with DNA-binding proteins. *Nature* 310:25–31, 1984.

    Sauer, R.T.; Yocum, R.R.; Doolittle, R.F.; Lewis, M.; Pabo, C.O. Homology among DNA-binding proteins suggests use of a conserved super-secondary structure. *Nature* 298:447–451, 1982.

12. Scott, M.P.; Tamkun, J.W.; Hartzell, G.W. The structure and function of the homeodomain. *Biochim. Biophys. Acta* 989:25–48, 1989.

    Wüthrich, K.; Gehring, W. Transcriptional regulation by homeodomain proteins: structural, functional and genetic aspects. In Transcriptional Regulation (S.L. McKnight, K.R. Yamamoto, eds.). Cold Spring Harbor, NY: Cold Spring Harbor Laboratory Press, 1992.

13. Coleman, J.E. Zinc proteins: enzymes, storage proteins, transcription factors, and replication proteins. *Annu. Rev. Biochem.* 61:897–946, 1992.

    Miller, J.; McLachlan, A.D.; Klug, A. Repetitive zinc-binding domains in the protein transcription factor

IIIA from xenopus oocytes. *EMBO J.* 4:1609–1614, 1985.

Rhodes, D.; Klug, A. Zinc fingers. *Sci. Am.* 268(2):56–59, 62–65, 1993.

14. Knight, K.L.; Sauer, R.T. Biochemical and genetic analysis of operator contacts made by residues within the beta-sheet DNA binding motif of Mnt repressor. *EMBO J.* 11:215–223, 1992.

Schwabe, J.W.R. DNA between the sheets. *Curr. Biol.* 2:661–663, 1992.

15. Alber, T. Protein-DNA interactions: how GCN4 binds DNA. *Curr. Biol.* 3:182–184, 1993.

Lamb, P; McKnight, S.L. Diversity and specificity in transcriptional regulation: the benefits of heterotypic dimerization. *Trends Biochem. Sci.* 16:417–422, 1991.

Landschulz, W.H.; Johnson, P.F.; McKnight, S.L. The leucine zipper: a hypothetical structure common to a new class of DNA-binding proteins. *Science* 240:1759–1764, 1988.

McKnight, S.L. Molecular zippers in gene regulation. *Sci. Am.* 264(4):54–64, 1991.

O'Shea, E.; Rutkowski, R.; Kim, P.S. Evidence that the leucine zipper is a coiled coil. *Science* 243:538–542, 1989.

16. Benezra, R.; Davis, R.L.; Lockshon, D.; Turner, D.L.; Weintraub, H. The protein Id: a negative regulator of helix-loop-helix DNA-binding proteins. *Cell* 61:49–59, 1990.

Garrell, J.; Campuzano, S. The helix-loop-helix domain: a common motif for bristles, muscles and sex. *Bioessays* 13:493–498, 1991.

Lassar, A.B.; Davis, R.L.; Wright, W.E.; et al. Functional activity of myogenic HLH proteins requires hetero-oligomerization with E12/E47-like proteins *in vivo*. *Cell* 66:305–315, 1991.

Murre, C.; McCaw, P.S.; Baltimore, D. A new DNA binding and dimerization motif in immunoglobin enhancer binding, daughterless, MyoD, and myc proteins. *Cell* 56:777–783, 1989.

Tapscott, S.J.; Weintraub, H. MyoD and the regulation of myogenesis by helix-loop-helix proteins. *J. Clin. Invest.* 87:1133–1138, 1991.

17. Pabo, C.O.; Sauer, R.T. Transcription factors: structural families and principles of DNA recognition. *Annu. Rev. Biochem.* 61:1053–1095, 1992.

18. Fried, M.; Crothers, D.M. Equilibria and kinetic of lac repressor-operator interactions by polyacrylamide gel electrophoresis. *Nucleic Acids Res.* 9:6505–6525, 1987.

Shaffer, C.D.; Wallrath, L.L.; Elgin, S.C.R. Regulating genes by packaging domains: bits of heterochromatin in euchromatin? *Trends Genet.* 9:35–37, 1993.

19. Kadonga, J.T. Purification of sequence-specific binding proteins by DNA affinity chromatography. *Methods Enzymol.* 208:10–23, 1991.

Kadonaga, J.T.; Tjian R. Affinity purification of sequence-specific DNA binding proteins. *Proc. Nat. Acad. Sci.U.S.A.* 83:5889–5893, 1986.

Rosenfeld, P.J.; Kelly, T.J. Purification of nuclear factor I by DNA recognition site affinity chromatography. *J. Biol. Chem.* 261:1986.

Vinson, C.R.; LaMarco, K.L.; Johnsonn, P.F.; Landschulz, W.H.; McKnight, S.L. *In situ* detection of sequence-specific DNA binding activity specified by a recombinant bacteriophage. *Genet. Dev.* 2:801–806, 1988.

20. Beckwith, J. The operon: an historical account. In *Escherichia coli* and *Salmonella typhimurium:* Cellular and Molecular Biology (F.C. Neidhart, J.L. Ingraham, K.B. Low, et al., eds.), Vol. 2, pp. 1439–1443. Washington, DC: American Society for Microbiology, 1987.

Jacob, F.; Monod, J. Genetic regulatory mechanisms in the synthesis of proteins. *J. Mol. Biol.* 3:318–356, 1961.

Ptashne, M. A Genetic Switch, 2d ed. Cambridge, MA: Cell Press and Blackwell Press, 1992.

Yanofsky, C. Transcriptional regulation: elegance in design and discovery. In Transcriptional Regulation (S.L. McKnight, K.R. Yamamoto, eds.). Cold Spring Harbor, NY: Cold Spring Harbor Laboratory Press, 1992.

21. Sigler, P.B. The molecular mechanism of *trp* repression. In Transcriptional Regulation (S.L. McKnight, K.R. Yamamoto, eds.). Cold Spring Harbor, NY: Cold Spring Harbor Laboratory Press, 1992.

Yanofsky, C.; Crawford, I.P. The tryptophan operon. In *Escherichia coli* and *Salmonella typhimurium:* Cellular and Molecular Biology (F.C. Neidhart, J.L. Ingraham, K.B. Low, et al., eds.), Vol. 2, pp. 1231–1240. Washington, DC: American Society for Microbiology, 1987.

22. Bushman, F.D. Activators, deactivators and deactivated activators. *Curr. Biol.* 2:673–675, 1992.

Englesberg, E.; Irr, J.; Power, J.; Lee, N. Positive control of enzyme synthesis by gene C in the L-arabinose system. *J. Bacteriol.* 90:946–957, 1965.

Hoopes, B.C.; McClure, W.R. Strategies in regulation of transcription initiation. In *Escherichia coli* and *Salmonella typhimurium:* Cellular and Molecular Biology (F.C. Neidhart, J.L. Ingraham, K.B. Low, et al., eds.), Vol. 2, pp. 1231–1240. Washington, DC: American Society for Microbiology, 1987.

Johnson, A.D.; Poteete, A.R.; Lauer, G.; et al. λ repressor and Cro-components of an efficient molecular switch. *Nature* 294:217–223, 1981.

Ptashne, M.; Jeffrey, A.; Johnson, A.D.; et al. How the λ repressor and Cro work. *Cell* 19:1–11, 1980.

23. Beckwith, J. The lactose operon. In *Escherichia coli* and *Salmonella typhimurium:* Cellular and Molecular Biology (F.C. Neidhart, J.L. Ingraham, K.B. Low, et al., eds.), Vol. 2, pp. 1444–1452. Washington, DC: American Society for Microbiology, 1987.

Gralla, J.D. *lac* repressor. In Transcriptional Regulation (S.L. McKnight, K.R. Yamamoto, eds.). Cold Spring Harbor, NY: Cold Spring Harbor Laboratory Press, 1992.

24. Beardsley, T. Smart genes. *Sci. Am.* 265(2):86–95, 1991.

Buratowski, S.; Sharp, P.A. Initiation of transcription by RNA polymerase II. In Transcriptional Regulation (S.L. McKnight, K.R. Yamamoto, eds.). Cold Spring Harbor, NY: Cold Spring Harbor Laboratory Press, 1992.

Johnson, P.F.; McKnight, S.L. Eukaryotic transcriptional regulatory proteins. *Annu. Rev. Biochem.* 58:799–839, 1989.

Zawel, L.; Reinberg, D. Initiation of transcription by RNA polymerase II: a multi-step process. *Prog. Nucleic Acid Res. Mol. Biol.* 44:68–108, 1993.

25. Geiduschek, E.P.; Kassavetis, G.A. RNA polymerase III transcription complexes. In Transcriptional Regulation (S.L. McKnight, K.R. Yamamoto, eds.). Cold Spring Harbor, NY: Cold Spring Harbor Laboratory Press, 1992.

Gill, G. Complexes with a common core. *Curr. Biol.* 2:565–567, 1992.

Hernandez, N. TBP, a universal eukaryotic transcription factor? *Genes Dev.* 7:1291–1308, 1993.

Hisatake, K., et al. The p250 subunit of native TATA box-binding factor TFIID is the cell-cycle regulatory protein CCG1. *Nature* 362:179–181, 1993.

Peterson, M.G.; Tjian, R. Transcription: the tell-tail trigger. *Nature* 358:620–621, 1992.

Reeder, R.H. Regulation of transcription by RNA polymerase I. In Transcriptional Regulation (S.L. McKnight, K.R. Yamamoto, eds.). Cold Spring Harbor, NY: Cold Spring Harbor Laboratory Press, 1992.

Roeder, R.G. The complexities of eukaryotic transcription initiation: regulation of preinitiation complex assembly. *Trends Biochem. Sci.* 16:402–408, 1991.

Ruppert, S.; Wang, E.H.; Tjian, R. Cloning and expression of human TAF$_{II}$250: a TBP-associated factor implicated in cell-cycle regulation. *Nature* 362:175–179, 1993.

White, R.J.; Jackson, S.P. The TATA-binding protein: a central role in transcription by RNA polymerases I, II and III. *Trends Genet.* 8:284–288, 1992.

26. Drapkin, R.; Merino, A.; Reinberg, D. Regulation of RNA polymerase II transcription. *Curr. Opin. Cell Biol.* 5:469–476, 1993.

Hochschild, A. Protein-protein interactions and DNA loop formation. In DNA Topology and Its Biological Effects (N.R. Cozzarelli, J.C. Wang, eds.). Cold Spring Harbor, NY: Cold Spring Harbor Laboratory Press, 1990.

Kustu, S.; North, A.K.; Weiss, D.S. Prokaryotic transcriptional enhancers and enhancer-binding proteins. *Trends Biochem. Sci.* 16:397–402, 1991.

Müller, H.-P.; Schaffner, W. Transcriptional enhancers can act in trans. *Trends Genet.* 6:300–304, 1990.

North, A.K.; Klose, K.E.; Stedman, K.M.; Kustu, S. Prokaryotic enhancer-binding proteins reflect eukaryote-like modularity: the puzzle of nitrogen regulatory protein C. *J. Bacteriol.* 175:4267–4273, 1991.

Schleif, R. DNA looping. *Annu. Rev. Biochem.* 61:199–223, 1992.

27. Adhya, S. Multipartite genetic control elements: communication by DNA looping. *Annu. Rev. Genet.* 23:227–250, 1989.

Johnson, P.F.; McKnight, S.L. Eukaryotic transcriptional regulatory proteins. *Annu. Rev. Biochem.* 58:799–839, 1989.

Mitchell, P.J.; Tjian, R. Transcriptional regulation in mammalian cells by sequence-specific DNA-binding proteins. *Science* 245:371–378, 1989.

Serfling, E.; Jasin, M.; Schaffner, W. Enhancers and eukaryotic gene transcription. *Trends Genet.* 1:224–230, 1985.

28. Brent, R.; Ptashne, M. A eukaryotic transcriptional activator bearing the DNA specificity of a prokaryotic repressor. *Cell* 43:729–736, 1985.

Greenblat, J. Protein-protein interactions as critical determinants of regulated initiation and termination of transcription. In Transcriptional Regulation (S.L. McKnight, K.R. Yamamoto, eds.). Cold Spring Harbor, NY: Cold Spring Harbor Laboratory Press, 1992.

Lin, Y-S.; Green, M.R. Mechanism of action of an acidic transcriptional activator *in vitro*. *Cell* 64:971–981, 1991.

Ptashne, M.; Gann, A.A.F. Activators and targets. *Nature* 346:329–331, 1990.

29. Goodbourn, S. Negative regulation of transcriptional initiation in eukaryotes. *Biochim. Biophys. Acta* 1032:53–77, 1990.

Herschbach, B.M.; Johnson, A.D. Transcripted repression in eukaryotes. *Annu. Rev. Cell Biol.* 9:479–511, 1993.

Jackson, M.E. Negative regulation of eukaryotic transcription. *J. Cell Sci.* 100:1–7, 1991.

30. Guarente, L. Mechanism and regulation of transcriptional activation in eukaryotes: conserved features from yeasts to humans. In Transcriptional Regulation (S.L. McKnight, K.R. Yamamoto, eds.). Cold Spring Harbor, NY: Cold Spring Harbor Laboratory Press, 1992.

Herr, W. Oct–1 and Oct–2: differential transcriptional regulation by proteins that bind to the same DNA sequence. In Transcriptional Regulation (S.L. McKnight, K.R. Yamamoto, eds.). Cold Spring Harbor, NY: Cold Spring Harbor Laboratory Press, 1992.

Johnson, A. A combinatorial regulatory circuit in budding yeast. In Transcriptional Regulation (S.L. McKnight, K.R. Yamamoto, eds.). Cold Spring Harbor, NY: Cold Spring Harbor Laboratory Press, 1992.

Martin, K.J.; Green, M.R. Transcriptional activation by viral immediate-early proteins: variations on a common theme. In Transcriptional Regulation (S.L. McKnight, K.R. Yamamoto, eds.). Cold Spring Harbor, NY: Cold Spring Harbor Laboratory Press, 1992.

Thompson, C.C.; McKnight, S.I. Anatomy of an enhancer. *Trends Genet.* 8:232–236, 1992.

Yamamoto, K.R.; Pearch, D.; Thomas, J.; Miner, J.N. Combinatorial regulation at a mammalian composite response element. In Transcriptional Regulation (S.L. McKnight, K.R. Yamamoto, eds.). Cold Spring Harbor, NY: Cold Spring Harbor Laboratory Press, 1992.

31. Pankratz, M.J.; Jackle, H. Making stripes in the *Drosophila* embryo. *Trends Genet.* 6:287–292, 1990.

St. Johnston, D.; Nusslein-Volhard, C. The origin of pattern and polarity in the *Drosophila* embryo. *Cell* 68:201–219, 1992.

Scott, M.P.; O'Farrell, P.H. Spatial programming of gene expression in early *Drosophila* embryogenesis. *Annu. Rev. Cell Biol.* 2:49–80, 1986.

Small, S.; Levine, M. The initiation of pair-rule stripes in the *Drosophila* blastoderm. *Curr. Opin. Genet. Dev.* 1:255–260, 1991.

32. Jackle H.; Sauer, F. Transcriptional cascades in *Drosophila*. *Curr. Opin. Cell Biol.* 5:505–512, 1993.

Small, S.; Kraut, R.; Hoey, T.; Warrior, R.; Levine, M. Transcriptional regulation of a pair-rule stripe in *Drosophila*. *Genes Dev.* 5:827–839, 1991.

33. Crossley, M.; Orkin, S.H. Regulation of the beta-globin locus. *Curr. Opin. Genet. Dev.* 3:232–237, 1993.

Evans, T.; Felsenfeld, G.; Reitman, M. Control of globin gene transcription. *Annu. Rev. Cell Biol.* 6:95–124, 1990.

Higgs, D.R.; Wood, W.G. Understanding erythroid differentiation. *Curr. Biol.* 3:548–550, 1993.

Minie, M.; Clark, D.; Trainor, C., et al. Developmental regulation of globin gene expression. *J. Cell Sci.* 16(Suppl.):15–20, 1992.

34. Berk, A.J. Regulation of eukaryotic transcription factors by post-translational modification. *Biochim. Biophys. Acta* 1009:103–109, 1989.

Hunter, T.; Karin, M. The regulation of transcription by phosphorylation. *Cell* 70:375–387, 1992.

35. Stragier, P.; Losick, R. Cascades of sigma factors revisited. *Mol. Microbiol.* 4:1801–1806, 1990.

36. Grunstein, M. Histones as regulators of genes. *Sci. Am.* 267(4):68–74B, 1992.

Kornberg, R.D.; Lorch, Y. Chromatin structure and transcription. *Annu. Rev. Cell Biol.* 8:563–587, 1992.

Svaren, J.; Hörz, W. Histones, nucleosomes and transcription. *Curr. Opin. Genet. Dev.* 3:219–225, 1993.

Winston, F.; Carlson, M. Yeast SNF/SWI transcriptional activators and the SPT/SIN chromatin connection. *Trends Genet.* 8:387–391, 1992.

Wolffe, A. Chromatin: Structure and Function. San Diego, CA: Academic Press, 1992.

37. Croston, G.E.; Kadonaga, J.T. Role of chromatin structure in the regulation of transcription by RNA polymerase II. *Curr. Opin. Cell Biol.* 5:417–423, 1993.

Fedor, M.J. Chromatin structure and gene expression. *Curr. Opin. Cell Biol.* 4:436–443, 1992.

Grunstein, M. Nucleosomes: regulators of transcription. *Trends Genet.* 6:395–400, 1990.

Kornberg, R.D.; Lorch, Y. Irresistible force meets immovable object: transcription and the nucleosome. *Cell* 67:833–836, 1991.

Struhl, K. Gene expression: chromatin and transcription factors: who's on first? *Curr. Biol.* 3:220–221, 1993.

Workman, J.L.; Taylor, I.C.A.; Kingston, R.E. Activation domains of stably bound GAL4 derivatives alleviate repression of promoters by nucleosomes. *Cell* 64:533–544, 1991.

38. Aparicio, O.M.; Billington, B.L.; Gottschling, D.E. Modifiers of position effect are shared between telomeric and silent mating-type loci in *S. cerevisiae*. *Cell* 66:1279–1287, 1991.

Gottschling, D.E.; Aparicio, O.M.; Billington, B.L.; Zakian, V.A. Position effect at *S. cerevisiae* telomeres: reversible repression of Pol II transcription. *Cell* 63:751–762, 1990.

Henikoff, S. Position effect and related phenomena. *Curr. Opin. Genet. Dev.* 2:907–912, 1992.

Wilson, C.; Bellen, H.J.; Gehring, W.J. Position effects on eukaryotic gene expression. *Annu. Rev. Cell Biol.* 6:679–714, 1990.

39. Dillon, N.; Grosveld, F. Transcriptional regulation of multigene loci: multilevel control. *Trends Genet.* 9:134–137, 1993.

Evans, T.; Felsenfeld, G.; Reitman, M. Control of globin gene transcription. *Annu. Rev. Cell Biol.* 6:95–124, 1990.

Grosveld, F.; Antoniou, M.; Berry, M.; et al. The regulation of human globin gene switching. *Philos. Trans. R. Soc. Lond. (Biol.)* 339:183–191, 1993.

Minie, M.; Clark, D.; Trainor, C.; et al. Developmental regulation of globin gene expression. *J. Cell Sci.* 16 (Suppl.):15–20, 1992.

40. Ausio, J. Structure and dynamics of transcriptionally active chromatin. *J. Cell Sci.* 102:1–5, 1992.

Freeman, L.A.; Garrard, W.T. DNA supercoiling in chromatin structure and gene expression. *Crit. Rev. Eukar. Gene Express.* 2:165–209, 1992.

Hsieh, T.S. DNA topoisomerases. *Curr. Opin. Cell Biol.* 4:396–400, 1992.

Wang, J.C. DNA topoisomerases: why so many? *J. Biol. Chem.* 266:6659–6662, 1991.

Svaren, J.; Chalkley, R. The structure and assembly of active chromatin. *Trends Genet.* 6:52–56, 1990.

41. Gene expression and differentiation. *Curr. Opin. Genet. Dev.* 2:197–303, 1992.

42. Berg, D.E.; Howe, M.M., eds. Mobile DNA. Washington, DC: American Society for Microbiology, 1989.

Johnson, R.C. Mechanism of site-specific DNA inversion in bacteria. *Curr. Opin. Genet. Dev.* 1:404–411, 1991.

Meyer, T.F.; van Putten, J.P. Genetic mechanisms and biological implications of phase variation in pathogenic neisseriae. *Clin. Microbiol. Rev.* 2 (Suppl.):S139–S145, 1989.

Pillus, L. An acquired state: epigenetic mechanisms in transcription. *Curr. Opin. Cell Biol.* 4:453–458, 1992.

Prescott, D.M. DNA gains, losses, and rearrangements in eukaryotes. *Dev. Biol.* 6:13–29, 1989.

Robertson, B.D.; Meyer, T.F. Genetic variation in pathogenic bacteria. *Trends Genet.* 8:422–427, 1992.

van de Putte, P.; Goosen, N. DNA inversions in phages and bacteria. *Trends Genet.* 8:457–462, 1992.

43. Dolan, J.W.; Fields, S. Cell-type-specific transcription in yeast. *Biochim. Biophys. Acta* 1088:155–169, 1991.

Herskowitz, I. A regulatory hierarchy for cell specialization in yeast. *Nature* 342:749–757, 1989.

Johnson, A. A combinatorial regulatory circuit in budding yeast. In Transcriptional Regulation (S.L. McKnight, K.R. Yamamoto, eds.). Cold Spring Harbor, NY: Cold Spring Harbor Laboratory Press, 1992.

44. Ptashne, M. A Genetic Switch, 2d ed. Cambridge, MA: Cell Press and Blackwell Press, 1992.

Ptashne, M.; Jeffrey, A.; Johnson, A.D.; et al. How the λ repressor and Cro work. *Cell* 19:1–11, 1980.

Scott, M.P.; O'Farrell, P.H. Spatial programming of gene expression in early *Drosophila* embryogenesis. *Annu. Rev. Cell Biol.* 2:49–80, 1986.

45. Braun, T.; Rudnicki, M.A.; Arnold, H.-H.; Jaenisch, R. Targeted inactivation of the muscle regulatory gene Myf–5 results in abnormal rib development and perinatal death. *Cell* 71:369–382, 1992.

Hasty, P.; Bradley, A.; Morris, J.H.; et al. Muscle deficiency and neonatal death in mice with a targeted mutation in the myogenin gene. *Nature* 364:501–506, 1993.

Miller, J.B.; Everitt, E.A.; Smith, T.H.; Block, N.E.; Dominov, J.A. Cellular and molecular diversity in skeletal muscle development: news from *in vitro* and *in vivo*. *Bioessays* 15:191–196, 1993.

Nabeshima, Y.; Hanaoka, K.; Hayasaka, M.; et al. Myogenin gene disruption results in perinatal lethality because of severe muscle defect. *Nature* 364:532–535, 1993.

Weintraub, H.; Davis, R.; Tapscott, S.; et al. The myoD gene family: nodal point during specification of the muscle cell lineage. *Science* 251:761–766, 1991.

46. Buckingham, M. Making muscle in mammals. *Trends Genet.* 8:144–149, 1992.

Garcia-Bellido, A. Homeotic and atavic mutations in insects. *Am. Zool.* 17:613–629, 1977.

Giere, A. Molecular models and combinatorial principles in cell differentiation and morphogenesis. *Cold Spring Harbor Symp. Quant. Biol.* 38:951–961, 1974.

Johnson, P.F.; McKnight, S.L. Eukaryotic transcriptional regulatory proteins. *Annu. Rev. Biochem.* 58:799–839, 1989.

Sprague, G.F., Jr. Combinatorial associations of regulatory proteins and the control of cell type in yeast. *Adv. Genet.* 27:33–62, 1990.

47. Ballabio, A.; Willard, H.F. Mammalian X-chromosome inactivation and the XIST gene. *Curr. Opin. Genet. Dev.* 2:439–447, 1992.

Lyon, M.F. X-inactivation: controlling the X chromosome. *Curr. Biol.* 3:242–244, 1993.

Lyon, M.F. Epigenetic inheritance in mammals. *Trends Genet.* 9:123–128, 1993.

Lyon, M.F. Some milestones in the history of X-chromosome inactivation. *Annu. Rev. Genet.* 26:17–28, 1992.

Riggs, A.D.; Pfeiffer, G.P. X-chromosome inactivation and cell memory. *Trends Genet.* 8:169–174, 1992.

48. Eissenberg, J.C. Position effects variegation in *Drosophila:* towards a genetics of chromatin assembly. *Bioessays* 11:14–17, 1989.

Henikoff, S. Position-effect variegation after 60 years. *Trends Genet.* 6:422–426, 1990.

49. Barras, F.; Marinus, M.G. The great GATC: DNA methylation in *E. coli. Trends Genet.* 5:139–143, 1989.

Holliday, R. Epigenetic inheritance based on DNA methylation. *Exs* 64:452–468, 1993.

50. Bird, A. The essentials of DNA methylation. *Cell* 70:5–8, 1992.

Cedar, H. DNA methylation and gene activity. *Cell* 53:3–4, 1988.

Graessmann, M.; Graessmann, A. DNA methylation, chromatin structure and the regulation of gene expression. *Exs* 64:404–424, 1993.

Tate, P.H.; Bird, A.P. Effects of DNA methylation on DNA-binding proteins and gene expression. *Curr. Opin. Genet. Dev.* 3:226–231, 1993.

51. Bird, A.P. Genomic imprinting: imprints on islands. *Curr. Biol.* 3:275–277, 1993.

Haig, D.; Graham, C. Genomic imprinting and the strange case of the insulin-like growth factor II receptor. *Cell* 64:1045–1046, 1991.

Li, E.; Bestor, T.H.; Jaenisch, R. Targeted mutation of the DNA methyltransferase gene results in embryonic lethality. *Cell* 69:915–926, 1992.

Sasaki, H.; Allen, N.D.; Surani, M.A. DNA methylation and genomic imprinting in mammals. *Exs* 64:469–486, 1993.

52. Antequera, F.; Bird, A. CpG islands. *Exs* 64:169–185, 1993.

53. Jones, K.A. HIV *trans*-activation and transcriptional control mechanisms. *New Biol.* 1:127–135, 1989.

Landrick, R.; Yanofsky, C. Transcription attenuation. In *Escherichia coli* and *Salmonella typhimurium:* Cellular and Molecular Biology (F.C. Neidhart, J.L. Ingraham, K.B. Low, et al., eds.), Vol. 2, pp. 1276–1301. Washington, DC: American Society for Microbiology, 1987.

Yanofsky, C. Operon-specific control by transcription attenuation. *Trends Genet.* 3:356–360, 1987.

54. Black, D.L. Activation of *c-src* neuron-specific splicing by an unusual RNA element *in vivo* and *in vitro. Cell* 69:795–807, 1992.

Green, M.R. Biochemical mechanisms of constitutive and regulated pre-mRNA splicing. *Annu. Rev. Cell Biol.* 7:559–599, 1991.

Rio, D.C. RNA binding proteins, splice site selection, and alternative pre-mRNA splicing. *Gene Express.* 2:1–5, 1992.

55. Baker, B.S. Sex in flies: the splice of life. *Nature* 340:521–524, 1989.

56. Peterson, M.L. Balanced efficiencies of splicing and cleavage-polyadenylation are required for *mu-s* and *mu-m* mRNA regulation. *Gene Express.* 2:319–327, 1992.

57. Beadle, G. Genes and the chemistry of the organism. *Am. Sci.* 34:31–53, 1946.

58. Izaurralde, E.; Mattaj, I.W. Transport of RNA between nucleus and cytoplasm. *Semin. Cell Biol.* 3:279–288, 1992.

Maquat, L.E. Nuclear mRNA export. *Curr. Opin. Cell Biol.* 3:1004–1012, 1991.

59. Davis, I.; Ish-Horowicz, D. Apical localization of pair-rule transcripts requires 3′ sequences and limits protein diffusion in the *Drosophila* blastoderm embryo. *Cell* 67:927–940, 1991.

Kislauskis, E.H.; Singer, R.H. Determinants of mRNA localization. *Curr. Opin. Cell Biol.* 4:975–978, 1992.

60. Agabian, N. *Trans* splicing of nuclear pre-mRNAs. *Cell* 61:1157–1160, 1990.

Chan, L. RNA editing: exploring one mode with apolipoprotein B mRNA. *Bioessays* 15:33–41, 1993.

Sloof, P.; Benne, R. RNA editing in trypanosome mitochondria: guidelines for models. *FEBS Lett.* 325:146–151, 1993.

Sollner-Webb, B. RNA editing. *Curr. Opin. Cell Biol.* 3:1056–1061, 1991.

61. Fouillot, N.; Tlouzeau, S.; Rossignol, J.-M.; Jean-Jean, O. Translation of the hepatitis B virus P gene by ribosomal scanning as an alternative to internal initiation. *J. Virol.* 67:4886–4895, 1993.

Lindahl, L.; Hinnebusch, A. Diversity of mechanisms in the regulation of translation in prokaryotes and lower eukaryotes. *Curr. Opin. Genet. Dev.* 2:720–726, 1992.

Oh, S.K.; Sarnow, P. Gene regulation: translational initiation by internal ribosome binding. *Curr. Opin. Genet. Dev.* 3:295–300, 1993.

62. Hinnebusch, A.G. Involvement of an initiation factor and protein phosphorylation in translational control of GCN4 mRNA. *Trends Biochem. Sci.* 15:148–152, 1990.

Rhoads, R.E. Regulation of eukaryotic protein synthesis by initiation factors. *J. Biol. Chem.* 268:3017–3020, 1993.

Sarre, T.F. The phosphorylation of eukaryotic initiation factor 2: a principle of translational control in mammalian cells. *Biosystems* 22:311–325, 1989.

63. Melefors, O.; Hentze, M.W. Translational regulation by mRNA/protein interactions in eukaryotic cells: ferritin and beyond. *Bioessays* 15:85–90, 1993.

64. Sachs, A.B. Messenger RNA degradation in eukaryotes. *Cell* 74:413–421, 1993.

   Theil, E.C. Regulation of ferritin and transferrin receptor mRNAs. *J. Biol. Chem.* 265:4771–4774, 1990.

65. Marzluff, W.F. Histone 3′ ends: essential and regulatory functions. *Gene Express.* 2:93–97, 1992.

66. Wickens, M. In the beginning is the end: regulation of poly(A) addition and removal during early development. *Trends Biochem. Sci.* 15:320–324, 1990.

67. Bock, A.; Forchhammer, K.; Heider, J.; Baron, C. Selenoprotein synthesis: an expansion of the genetic code. *Trends Biochem. Sci.* 16:463–467, 1991.

Hatfield, D.; Oroszlan, S. The where, what and how of ribosomal frameshifting in retroviral protein synthesis. *Trends Biochem. Sci.* 15:186–190, 1990.

Weiss, R.B. Ribosomal frameshifting, jumping and readthrough. *Curr. Opin. Cell Biol.* 3:1051–1055, 1991.

68. Eguchi, Y.; Itoh, T.; Tomizawa, J. Antisense RNA. *Annu. Rev. Biochem.* 60:631–652, 1991.

Noller, H.F.; Hoffarth, V.; Zimniak, L. Unusual resistance of peptidyl transferase to protein extraction procedures. *Science* 256:1416–1419, 1992.

Weiner, A.M. mRNA splicing and autocatalytic introns: distant cousins or the products of chemical determinism? *Cell* 72:161–164, 1993.

Internal Organization of the Cell

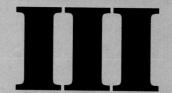

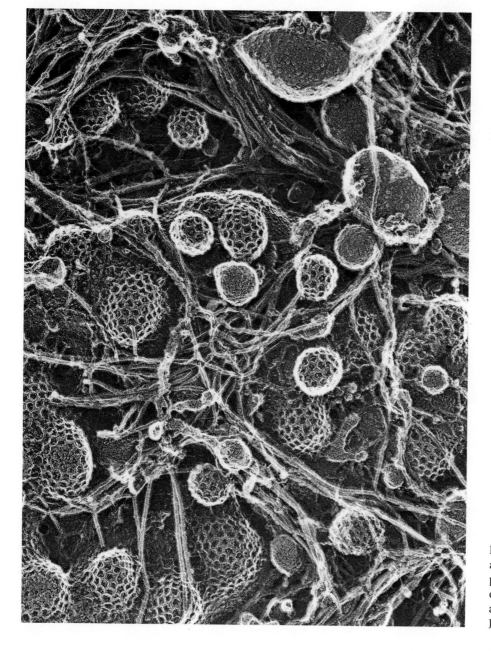

Electron micrograph of coated pits and vesicles on the inner face of the plasma membrane of a mouse liver cell. Numerous keratin filaments are also seen. (From N. Hirokawa and J. Heuser, *Cell* 30:395–406, 1982.)

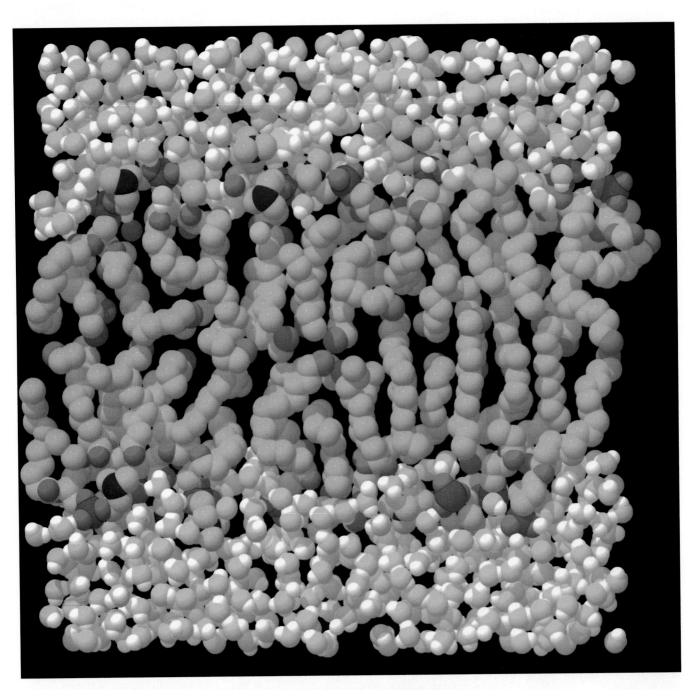

A computer simulation of a 0.8-nm slab of a phospholipid bilayer in water. Phosphorus atoms are shown in *green*, nitrogen atoms in *dark blue*, lipid oxygen atoms in *red*, terminal-chain groups in *magenta*, and other carbon atoms in *gray*; the carbon hydrogen atoms are omitted. The water molecules are shown in *yellow* (oxygen) and *white* (hydrogen). (From R.M. Venable, Y. Zhang, B.J. Hardy, and R.W. Pastor, *Science* 262, 223–228, 1993.)

Membrane Structure

- **The Lipid Bilayer**
- **Membrane Proteins**

Cell membranes are crucial to the life of the cell. The **plasma membrane** encloses the cell, defines its boundaries, and maintains the essential differences between the cytosol and the extracellular environment. Inside the cell the membranes of the endoplasmic reticulum, Golgi apparatus, mitochondria, and other membrane-bounded organelles in eucaryotic cells maintain the characteristic differences between the contents of each organelle and the cytosol. Ion gradients across membranes, established by the activities of specialized membrane proteins, can be used to synthesize ATP, to drive the transmembrane movement of selected solutes, or, in nerve and muscle cells, to produce and transmit electrical signals. In all cells the plasma membrane also contains proteins that act as sensors of external signals, allowing the cell to change its behavior in response to environmental cues; these protein sensors, or *receptors*, transfer information rather than ions or molecules across the membrane.

Despite their differing functions, all biological membranes have a common general structure: each is a very thin film of lipid and protein molecules, held together mainly by noncovalent interactions. Cell membranes are dynamic, fluid structures, and most of their molecules are able to move about in the plane of the membrane. The lipid molecules are arranged as a continuous double layer about 5 nm thick (Figure 10–1). This *lipid bilayer* provides the basic structure of

Figure 10–1 Three views of a cell membrane. (A) An electron micrograph of a plasma membrane (of a human red blood cell) seen in cross-section. (B and C) Schematic drawings showing two-dimensional and three-dimensional views of a cell membrane. (A, courtesy of Daniel S. Friend.)

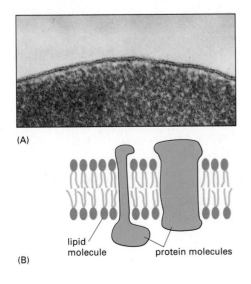

(A)

(B)

lipid molecule

protein molecules

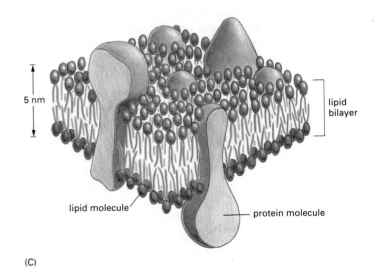

5 nm

lipid bilayer

lipid molecule

protein molecule

(C)

the membrane and serves as a relatively impermeable barrier to the passage of most water-soluble molecules. Protein molecules "dissolved" in the lipid bilayer mediate most of the other functions of the membrane, transporting specific molecules across it, for example, or catalyzing membrane-associated reactions, such as ATP synthesis. In the plasma membrane some proteins serve as structural links that connect the membrane to the cytoskeleton and/or to either the extracellular matrix or an adjacent cell, while others serve as receptors to detect and transduce chemical signals in the cell's environment. As would be expected, cell membranes are asymmetrical structures: the lipid and protein compositions of the outside and inside faces differ from one another in ways that reflect the different functions performed at the two surfaces of the membrane.

In this chapter we consider the structure and organization of the two main constituents of biological membranes—the lipids and the proteins. Although we shall focus mainly on the plasma membrane, most of the concepts discussed are applicable to the various internal membranes in cells as well. The functions of cell membranes are considered in later chapters. Their role in ATP synthesis, for example, is discussed in Chapter 14; their role in the transmembrane transport of small molecules, in Chapter 11; and their roles in cell signaling and cell adhesion in Chapters 15 and 19, respectively. In Chapters 12 and 13 we discuss the internal membranes of the cell and the protein traffic through and between them.

The Lipid Bilayer [1]

The **lipid bilayer** has been firmly established as the universal basis for cell-membrane structure. It is easily seen by ordinary electron microscopy, although specialized techniques, such as x-ray diffraction and freeze-fracture electron microscopy, are needed to reveal the details of its organization. The bilayer structure is attributable to the special properties of the lipid molecules, which cause them to assemble spontaneously into bilayers even in simple artificial conditions.

Membrane Lipids Are Amphipathic Molecules, Most of Which Spontaneously Form Bilayers [1]

Lipid molecules are insoluble in water but dissolve readily in organic solvents. They constitute about 50% of the mass of most animal cell membranes, nearly all of the remainder being protein. There are approximately 5×10^6 lipid molecules in a 1 μm × 1 μm area of lipid bilayer, or about 10^9 lipid molecules in the plasma membrane of a small animal cell. All of the lipid molecules in cell membranes are *amphipathic* (or amphiphilic)—that is, they have a *hydrophilic* ("water-loving," or *polar*) end and a *hydrophobic* ("water-hating," or *nonpolar*) end. The most abundant are the **phospholipids.** These have a polar *head group* and two hydrophobic *hydrocarbon tails* (Figure 10–2). The tails are usually fatty acids, and they can differ in length (they normally contain between 14 and 24 carbon atoms). One tail usually has one or more *cis*-double bonds (that is, it is *unsaturated*), while the other tail does not (that is, it is *saturated*). As indicated in Figure 10–2, each double bond creates a small kink in the tail. Differences in the length and saturation of the fatty acid tails are important because they influence the ability of phospholipid molecules to pack against one another, and for this reason they affect the fluidity of the membrane (discussed below).

It is the shape and amphipathic nature of the lipid molecules that cause them to form bilayers spontaneously in aqueous solution. When lipid molecules are surrounded on all sides by water, they tend to aggregate so that their hydrophobic tails are buried in the interior and their hydrophilic heads are exposed to water. Depending on their shape, they can do this in either of two ways: they can form

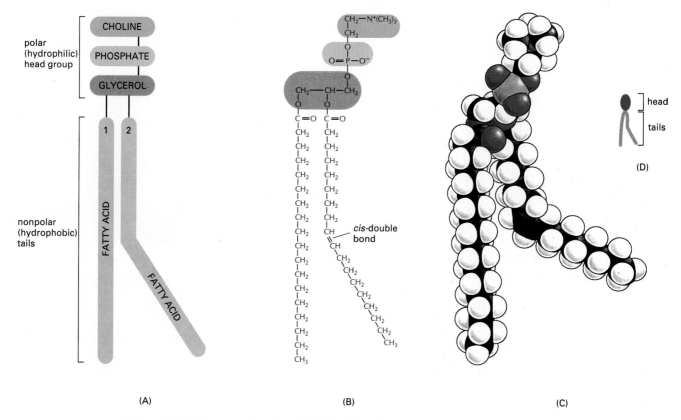

(A)   (B)   (C)

Figure 10–2 **The parts of a phospholipid molecule.**
Phosphatidylcholine, represented schematically (A), in formula
(B), as a space-filling model (C), and as a symbol (D). The kink
due to the *cis*-double bond is exaggerated in these drawings for
emphasis.

spherical *micelles,* with the tails inward, or they can form bimolecular sheets, or
bilayers, with the hydrophobic tails sandwiched between the hydrophilic head
groups (Figure 10–3).

Because of their cylindrical shape, membrane phospholipid molecules spontaneously form bilayers in aqueous environments. Moreover, these lipid bilayers tend to close on themselves to form sealed compartments, thereby eliminating free edges where the hydrophobic tails would be in contact with water. For
the same reason compartments formed by lipid bilayers tend to reseal when they
are torn.

A lipid bilayer has other characteristics besides its self-sealing properties that
make it an ideal structure for cell membranes. One of the most important of these
is its fluidity, which is crucial to many membrane functions.

The Lipid Bilayer Is a Two-dimensional Fluid [2]

It was only in the early 1970s that it was first recognized that individual lipid
molecules are able to diffuse freely within lipid bilayers. The initial demonstration came from studies of synthetic lipid bilayers. Two types of synthetic bilayers have been very useful in experimental studies: (1) bilayers made in the form
of spherical vesicles, called **liposomes,** which can vary in size from about 25 nm
to 1 μm in diameter depending on how they are produced (Figure 10–4); and (2)
planar bilayers, called **black membranes,** formed across a hole in a partition
between two aqueous compartments (Figure 10–5).

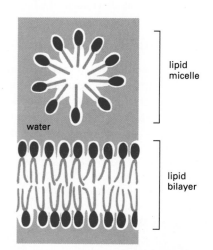

Figure 10–3 **A lipid micelle and a
lipid bilayer seen in cross-section.**
Lipid molecules form such structures
spontaneously in water. The shape of
the lipid molecule determines which
of these structures is formed. Wedge-
shaped lipid molecules (*above*) form
micelles, whereas cylinder-shaped
phospholipid molecules (*below*) form
bilayers.

The Lipid Bilayer

Various techniques have been used to measure the motion of individual lipid molecules and their different parts. One can construct a lipid molecule, for example, whose polar head group carries a "spin label," such as a nitroxyl group (>N–O); this contains an unpaired electron whose spin creates a paramagnetic signal that can be detected by electron spin resonance (ESR) spectroscopy. (The principles of this technique are similar to those of nuclear magnetic resonance, which is discussed in Chapter 4.) The motion and orientation of a spin-labeled lipid in a bilayer can be deduced from the ESR spectrum. Such studies show that phospholipid molecules in synthetic bilayers very rarely migrate from the monolayer on one side to that on the other. This process, called "flip-flop," occurs less than once a month for any individual molecule. On the other hand, lipid molecules readily exchange places with their neighbors *within* a monolayer (~10^7 times a second). This gives rise to a rapid lateral diffusion, with a diffusion coefficient (D) of about 10^{-8} cm²/sec, which means that an average lipid molecule diffuses the length of a large bacterial cell (~2 μm) in about 1 second. These studies have also shown that individual lipid molecules rotate very rapidly about their long axis and that their hydrocarbon chains are flexible (Figure 10–6).

Similar studies have been carried out with labeled lipid molecules in isolated biological membranes and in relatively simple whole cells such as mycoplasmas, bacteria, and nonnucleated red blood cells. The results are generally the same as for synthetic bilayers, and they demonstrate that the lipid component of a biological membrane is a two-dimensional liquid in which the constituent molecules are free to move laterally; as in synthetic bilayers, individual phospholipid molecules are normally confined to their own monolayer. This confinement creates a problem for their synthesis. Phospholipid molecules are synthesized in only one monolayer of a membrane, mainly being produced in the cytosolic monolayer of the endoplasmic reticulum (ER) membrane; if none of these newly made molecules could migrate to the other half of the lipid bilayer, new bilayer could not be made. The problem is solved by a special class of ER-membrane-bound enzymes called *phospholipid translocators*, which catalyze the rapid flip-flop of specific phospholipids from the monolayer where they are made to the opposite monolayer, as discussed in Chapter 12.

The Fluidity of a Lipid Bilayer Depends on Its Composition [3]

The precise fluidity of cell membranes is biologically important. Certain membrane transport processes and enzyme activities, for example, can be shown to cease when the bilayer viscosity is experimentally increased beyond a threshold level. The fluidity of a lipid bilayer depends on both its composition and temperature, as is readily demonstrated in studies of synthetic bilayers. A synthetic bi-

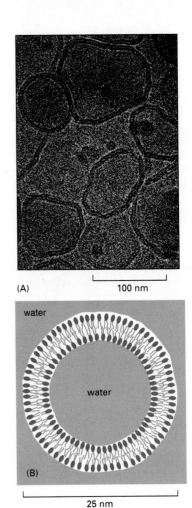

(A)

100 nm

water

water

(B)

25 nm

Figure 10–4 **Liposomes.** (A) An electron micrograph of unfixed, unstained phospholipid vesicles (liposomes) in water. The bilayer structure of the vesicles is readily apparent. (B) A drawing of a small spherical liposome seen in cross-section. Liposomes are commonly used as model membranes in experimental studies. (A, courtesy of Jean Lepault.)

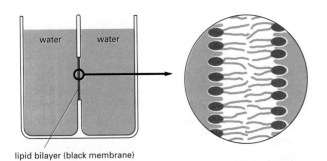

water   water

lipid bilayer (black membrane)

Figure 10–5 **A cross-sectional view of a synthetic lipid bilayer, called a black membrane.** This planar bilayer is formed across a small hole in a partition separating two aqueous compartments. Black membranes are used to measure the permeability properties of synthetic membranes.

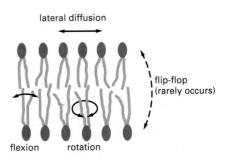

lateral diffusion

flip-flop (rarely occurs)

flexion   rotation

Figure 10–6 **Phospholipid mobility.** The types of movement possible for phospholipid molecules in a lipid bilayer.

layer made from a single type of phospholipid changes from a liquid state to a rigid crystalline (or gel) state at a characteristic freezing point. This change of state is called a *phase transition,* and the temperature at which it occurs is lower (that is, the membrane becomes more difficult to freeze) if the hydrocarbon chains are short or have double bonds. A shorter chain length reduces the tendency of the hydrocarbon tails to interact with one another, and *cis*-double bonds produce kinks in the hydrocarbon chains that make them more difficult to pack together, so that the membrane remains fluid at lower temperatures (Figure 10–7). Bacteria, yeast, and other organisms whose temperatures fluctuate with that of their environment adjust the fatty acid composition of their membrane lipids so as to maintain a relatively constant fluidity; as the temperature falls, for instance, fatty acids with more *cis*-double bonds are synthesized, so that the decrease in bilayer fluidity that would otherwise result from the drop in temperature is avoided.

The lipid bilayer of many cell membranes is not composed exclusively of phospholipids, however; it often also contains *cholesterol* and *glycolipids.* Eucaryotic plasma membranes contain especially large amounts of **cholesterol** (Figure 10–8)—up to one molecule for every phospholipid molecule. The cholesterol molecules enhance the permeability-barrier properties of the lipid bilayer. They orient themselves in the bilayer with their hydroxyl groups close to the polar head groups of the phospholipid molecules; their rigid, platelike steroid rings interact with—and partly immobilize—those regions of the hydrocarbon chains that are closest to the polar head groups (Figure 10–9). By decreasing the mobility of the first few CH_2 groups of the hydrocarbon chains of the phospholipid molecules, cholesterol makes the lipid bilayer less deformable in this region and thereby decreases the permeability of the bilayer to small water-soluble molecules. Although cholesterol tends to make lipid bilayers less fluid, at the high concentrations found in most eucaryotic plasma membranes it also prevents the hydrocarbon chains from coming together and crystallizing. In this way it inhibits possible phase transitions.

The lipid compositions of several biological membranes are compared in Table 10–1. Note that bacterial plasma membranes are often composed of one main type of phospholipid and contain no cholesterol; the mechanical stability of these membranes is enhanced by the overlying cell wall (see Figure 11–14). The plasma membranes of most eucaryotic cells, by contrast, are more varied, not only in containing large amounts of cholesterol, but also in containing a mixture of different phospholipids. Four major phospholipids predominate in the plasma membrane of many mammalian cells: *phosphatidylcholine, sphingomyelin, phosphatidylserine,* and *phosphatidylethanolamine.* The structures of these molecules are shown in Figure 10–10. Note that only phosphatidylserine carries a net negative charge, the importance of which we shall see later; the other three

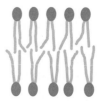

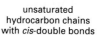

unsaturated hydrocarbon chains with *cis*-double bonds

saturated straight hydrocarbon chains

Figure 10–7 Influence of *cis*-double bonds in hydrocarbon chains. The double bonds make it more difficult to pack the chains together and therefore make the lipid bilayer more difficult to freeze.

Figure 10–8 The structure of cholesterol. Cholesterol is represented by a formula in (A), by a schematic drawing in (B), and as a space-filling model in (C).

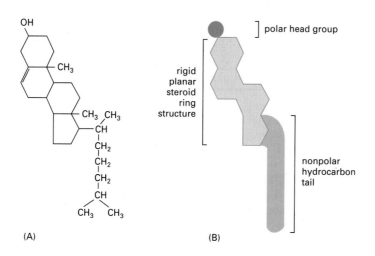
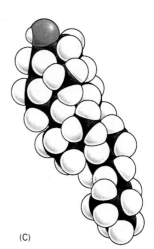

are electrically neutral at physiological pH, carrying one positive and one negative charge. Together these four phospholipids constitute more than half the mass of lipid in most membranes (see Table 10–1). Other phospholipids, such as the *inositol phospholipids*, are present in smaller quantities but are functionally very important. The inositol phospholipids have a crucial role in cell signaling, as discussed in Chapter 15.

One might wonder why eucaryotic membranes contain such a variety of phospholipids, with head groups that differ in size, shape, and charge. One can begin to understand why if one thinks of the membrane lipids as constituting a two-dimensional solvent for the proteins in the membrane, just as water constitutes a three-dimensional solvent for proteins in an aqueous solution. As we shall see, some membrane proteins can function only in the presence of specific phospholipid head groups, just as many enzymes in aqueous solution require a particular ion for activity.

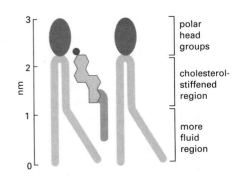

Figure 10–9 Cholesterol in a lipid bilayer. Schematic drawing of a cholesterol molecule interacting with two phospholipid molecules in one leaflet of a lipid bilayer.

The Lipid Bilayer Is Asymmetrical [4]

The lipid compositions of the two halves of the lipid bilayer in those membranes that have been analyzed are strikingly different. In the human red blood cell membrane, for example, almost all of the lipid molecules that have choline—$(CH_3)_3N^+CH_2CH_2OH$—in their head group (that is, phosphatidylcholine and sphingomyelin) are in the outer half of the lipid bilayer, whereas almost all of the phospholipid molecules that contain a terminal primary amino group (phosphatidylethanolamine and phosphatidylserine) are in the inner half (Figure 10–11). Because the negatively charged phosphatidylserine is located in the inner monolayer, there is a significant difference in charge between the two halves of the bilayer.

Most of the membranes in a eucaryotic cell, including the plasma membrane, are synthesized in the endoplasmic reticulum (ER), and it is here that the phospholipid asymmetry is generated by the phospholipid translocators in the ER that move specific phospholipid molecules from one monolayer to the other (discussed in Chapter 12). This lipid asymmetry can be functionally important. The enzyme *protein kinase C*, for example, is activated in response to various extracellular signals; it binds to the cytoplasmic face of the plasma membrane, where

Table 10–1 Approximate Lipid Compositions of Different Cell Membranes

Lipid	Percentage of Total Lipid by Weight					
	Liver Plasma Membrane	Erythrocyte Plasma Membrane	Myelin	Mitochondrion (inner and outer membranes)	Endoplasmic Reticulum	*E. coli*
Cholesterol	17	23	22	3	6	0
Phosphatidyl-ethanolamine	7	18	15	35	17	70
Phosphatidylserine	4	7	9	2	5	trace
Phosphatidyl-choline	24	17	10	39	40	0
Sphingomyelin	19	18	8	0	5	0
Glycolipids	7	3	28	trace	trace	0
Others	22	13	8	21	27	30

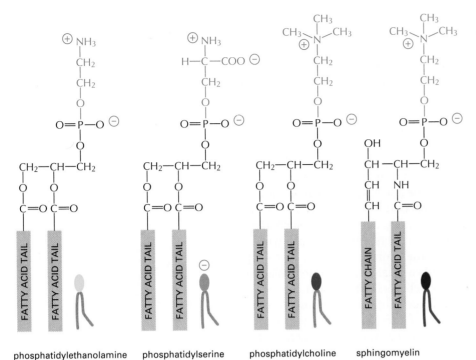

Figure 10–10 **Four major phospholipids in mammalian plasma membranes.** Note that different head groups are represented by different symbols in this figure and the next. All of the lipid molecules shown are derived from glycerol except for sphingomyelin, which is derived from serine.

phosphatidylethanolamine   phosphatidylserine   phosphatidylcholine   sphingomyelin

phosphatidylserine is concentrated, and requires this negatively charged phospholipid in order to act. Similarly, specific inositol phospholipids are concentrated in the cytoplasmic half of the plasma membrane of the eucaryotic cell. These minor phospholipids are cleaved into two fragments by specific enzymes that are activated by extracellular signals. Both fragments then act inside the cell as diffusible mediators to help relay the signal into the cell interior, as we discuss in Chapter 15.

Glycolipids Are Found on the Surface of All Plasma Membranes [5]

The lipid molecules that show the most striking and consistent asymmetry in distribution in cell membranes are the sugar-containing lipid molecules called **glycolipids.** These intriguing molecules are found exclusively in the noncytoplasmic half of the lipid bilayer, where they are thought to self-associate into microaggregates by forming hydrogen bonds with one another. In the plasma membrane their sugar groups are exposed at the cell surface (see Figure 10–11), suggesting some role in interactions of the cell with its surroundings. The asymmetric distribution of glycolipids in the bilayer results from the addition of sugar groups to the lipid molecules in the lumen of the Golgi apparatus, which is topologically equivalent to the exterior of the cell (discussed in Chapter 13).

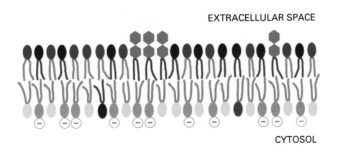

EXTRACELLULAR SPACE

CYTOSOL

Figure 10–11 **The asymmetrical distribution of phospholipids and glycolipids in the lipid bilayer of human red blood cells.** The symbols used for the phospholipids are those introduced in Figure 10–10. In addition, glycolipids are drawn with hexagonal polar head groups (*blue*). Cholesterol (not shown) is thought to be distributed about equally in both monolayers.

Glycolipids probably occur in all animal cell plasma membranes, generally constituting about 5% of the lipid molecules in the outer monolayer. They are also found in some intracellular membranes. The most complex of the glycolipids, the **gangliosides,** contain oligosaccharides with one or more sialic acid residues, which give gangliosides a net negative charge (Figure 10–12). Gangliosides are most abundant in the plasma membrane of nerve cells, where they constitute 5–10% of the total lipid mass, although they are also found in much smaller quantities in most cell types. So far more than 40 different gangliosides have been identified.

There are only hints as to what the functions of glycolipids might be. A possible clue comes from their localization: in the plasma membrane of epithelial cells, for example, glycolipids are confined to the apical surface, where they may help to protect the membrane from the harsh conditions (such as low pH and degradative enzymes) frequently found there. Charged glycolipids, such as gangliosides, may be important for their electrical effects: their presence will alter the electrical field across the membrane and the concentrations of ions—especially Ca^{2+}—at its external surface. Glycolipids may also play a role in electrical insulation, since in the myelin membrane, which electrically insulates nerve cell axons, the noncytoplasmic half of the bilayer is filled with them. They are also thought to function in cell-recognition processes. The ganglioside G_{M1} (see Figure 10–12), for example, acts as a cell-surface receptor for the bacterial toxin that causes the debilitating diarrhea of cholera. Cholera toxin binds to and enters only those cells that have G_{M1} on their surface, including intestinal epithelial cells. Its entry into a cell leads to a prolonged increase in the concentration of intracellular cyclic AMP (discussed in Chapter 15), which in turn causes a large efflux of Na^+ and water into the intestine. Although binding bacterial toxins cannot be the *normal* function of gangliosides, such observations suggest that these glycolipids might also serve as receptors for normal extracellular molecules. This suggestion is supported by increasing evidence that glycolipids can help cells to bind to the extracellular matrix, as well as to other cells, as we discuss later.

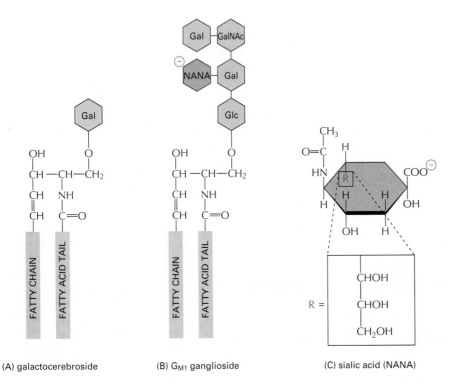

(A) galactocerebroside   (B) G_{M1} ganglioside   (C) sialic acid (NANA)

Figure 10–12 Glycolipid molecules. Galactocerebroside (A) is called a *neutral glycolipid* because the sugar that forms its head group is uncharged. A ganglioside (B) always contains one or more negatively charged sialic acid residues (also called *N*-acetylneuraminic acid, or NANA), whose structure is shown in (C). Whereas in bacteria and plants almost all glycolipids are derived from glycerol, as are most phospholipids, in animal cells they are almost always produced from sphingosine, an amino alcohol derived from *serine*, as is the case for the phospholipid sphingomyelin (see Figure 10–10). Gal = galactose; Glc = glucose, GalNAc = *N*-acetylgalactos-amine; these three sugars are uncharged.

Summary

Biological membranes consist of a continuous double layer of lipid molecules in which various membrane proteins are embedded. This lipid bilayer is fluid, with individual lipid molecules able to diffuse rapidly within their own monolayer. Most types of lipid molecules, however, very rarely flip-flop spontaneously from one monolayer to the other. Membrane lipid molecules are amphipathic, and some of them (the phospholipids) assemble spontaneously into bilayers when placed in water; the bilayers form sealed compartments that reseal if torn. There are three major classes of membrane lipid molecules—phospholipids, cholesterol, and glycolipids—and the lipid compositions of the inner and outer monolayers are different, reflecting the different functions of the two faces of a cell membrane. Different mixtures of lipids are found in the membranes of cells of different types, as well as in the various membranes of a single eucaryotic cell. Some membrane-bound proteins require specific lipid head groups in order to function, which is thought to explain, in part at least, why eucaryotic membranes contain so many kinds of lipid molecules.

Membrane Proteins [6]

Although the basic structure of biological membranes is provided by the lipid bilayer, most of the specific functions are carried out by proteins. Accordingly, the amounts and types of proteins in a membrane are highly variable: in the myelin membrane, which serves mainly as electrical insulation for nerve cell axons, less than 25% of the membrane mass is protein, whereas in the membranes involved in energy transduction (such as the internal membranes of mitochondria and chloroplasts), approximately 75% is protein. The usual plasma membrane is somewhere in between, with about 50% of its mass as protein. Because lipid molecules are small in comparison to protein molecules, there are always many more lipid molecules than protein molecules in membranes—about 50 lipid molecules for each protein molecule in a membrane that is 50% protein by mass. Like membrane lipids, membrane proteins often have oligosaccharide chains attached to them. Thus the surface that the cell presents to the exterior consists largely of carbohydrate, which forms a *glycocalyx* or *cell coat* (discussed later).

Membrane Proteins Can Be Associated with the Lipid Bilayer in Various Ways [7]

Different membrane proteins are associated with the membranes in different ways, as illustrated in Figure 10–13. Many membrane proteins extend through the lipid bilayer, with part of their mass on either side (see examples 1 and 2 in Figure 10–13). Like their lipid neighbors, these **transmembrane proteins** are amphipathic, having regions that are hydrophobic and regions that are hydrophilic. Their hydrophobic regions pass through the membrane and interact with the hydrophobic tails of the lipid molecules in the interior of the bilayer. Their hydrophilic regions are exposed to water on one or the other side of the membrane. The hydrophobicity of some of these membrane proteins is increased by the covalent attachment of a fatty acid chain that is inserted in the cytoplasmic leaflet of the lipid bilayer (see example 1 in Figure 10–13). Other membrane proteins are located entirely in the cytosol and are associated with the bilayer only by means of one or more covalently attached fatty acid chains or other types of lipid chains called *prenyl groups* (see example 3 in Figure 10–13 and Figure 10–14). Yet other membrane proteins are entirely exposed at the external cell surface, being attached to the bilayer only by a covalent linkage (via a specific oligosaccharide) to phosphatidylinositol in the outer lipid monolayer of the plasma membrane (see example 4 in Figure 10–13). The lipid-linked proteins in example

3 are made as soluble proteins in the cytosol and are subsequently directed to the membranes by the covalent attachment of a lipid group (Figure 10–14). The proteins in example 4, however, are made as single-pass transmembrane proteins in the ER; while still in the ER, the transmembrane segment of the protein is cleaved off and a **glycosylphosphatidylinositol (GPI) anchor** is added, leaving the protein bound to the noncytoplasmic surface of the membrane solely by this anchor (discussed in Chapter 12). Proteins bound to the membrane by a GPI anchor can be readily distinguished by the use of the enzyme phosphatidylinositol-specific phospholipase C, which specifically cuts these proteins free from their anchors and thereby releases them from the membrane.

Some proteins that do not extend into the hydrophobic interior of the lipid bilayer at all are bound to one or the other face of the membrane by noncovalent interactions with other membrane proteins (see examples 5 and 6 in Figure 10–13). Many of these can be released from the membrane by relatively gentle extraction procedures, such as exposure to solutions of very high or low ionic strength or extreme pH, which interfere with protein-protein interactions but leave the lipid bilayer intact; these proteins are referred to operationally as **peripheral membrane proteins.** By contrast, transmembrane proteins, many proteins held in the bilayer by lipid groups, and some other tightly bound proteins cannot be released in these ways and therefore are called **integral membrane proteins.**

How a membrane-bound protein is associated with the lipid bilayer usually reflects the function of the protein. Thus only transmembrane proteins can function on both sides of the bilayer or transport molecules across it. Some cell-surface receptors, for example, are transmembrane proteins that bind signaling molecules in the extracellular space and generate different intracellular signals on the opposite side of the plasma membrane. Proteins that function on only one side of the lipid bilayer, by contrast, are often associated exclusively with either the lipid monolayer or a protein domain on that side. A number of proteins involved in intracellular signaling, for example, are bound to the cytosolic half of the plasma membrane by one or more covalently attached lipid groups.

In Most Transmembrane Proteins the Polypeptide Chain Is Thought to Cross the Lipid Bilayer in an α-helical Conformation [6, 8]

A transmembrane protein always has a unique orientation in the membrane. This reflects both the asymmetrical manner in which it is synthesized and inserted into the lipid bilayer in the ER and the different functions of its cytoplasmic and

Figure 10–13 Six ways in which membrane proteins associate with the lipid bilayer. Most transmembrane proteins are thought to extend across the bilayer as a single α helix (1) or as multiple α helices (2); some of these "single-pass" and "multipass" proteins have a covalently attached fatty acid chain inserted in the cytoplasmic monolayer (1). Other membrane proteins are attached to the bilayer solely by a covalently attached lipid—either a fatty acid chain or prenyl group—in the cytoplasmic monolayer (3) or, less often, via an oligosaccharide, to a minor phospholipid, phosphatidylinositol, in the noncytoplasmic monolayer (4). Finally, many proteins are attached to the membrane only by noncovalent interactions with other membrane proteins (5) and (6). How the structure in (3) is formed is illustrated in Figure 10–14. The details of how membrane proteins become associated with the lipid bilayer in these different ways are discussed in Chapter 12.

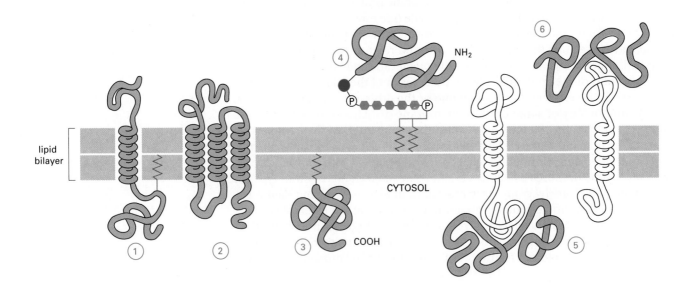

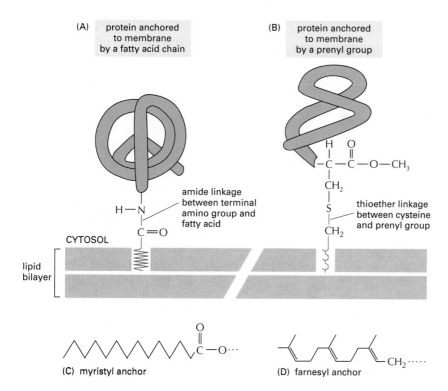

(A) protein anchored to membrane by a fatty acid chain

(B) protein anchored to membrane by a prenyl group

amide linkage between terminal amino group and fatty acid

thioether linkage between cysteine and prenyl group

CYTOSOL

lipid bilayer

(C) myristyl anchor

(D) farnesyl anchor

Figure 10–14 The covalent attachment of either of two types of lipid groups can help localize a water-soluble protein to a membrane after its synthesis in the cytosol. (A) A fatty acid chain (either myristic or palmitic acid) is attached via an amide linkage to an amino-terminal glycine. (B) A prenyl group (either farnesyl or a longer geranylgeranyl group—both related to cholesterol) is attached via a thioether linkage to a cysteine residue that is four residues from the carboxyl terminus. Following this prenylation, the terminal three amino acids are cleaved off and the new carboxyl terminus is methylated before insertion into the membrane. The structures of two lipid anchors are shown underneath: (C) a myristyl anchor (a 14-carbon saturated fatty acid chain), and (D) a farnesyl anchor (a 15-carbon unsaturated hydrocarbon chain).

Figure 10–15 A segment of a transmembrane polypeptide chain crossing the lipid bilayer as an α helix. Only the α-carbon backbone of the polypeptide chain is shown, with the hydrophobic amino acids in *green* and *yellow*. The polypeptide segment shown is part of the bacterial photosynthetic reaction center illustrated in Figure 10–33, the structure of which was determined by x-ray diffraction. (Based on data from J. Deisenhofer et al., *Nature* 318:618–624, 1985, and H. Michel et al., *EMBO J.* 5:1149–1158, 1986.)

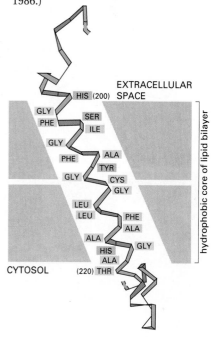

noncytoplasmic domains. These domains are separated by the membrane-spanning segments of the polypeptide chain, which contact the hydrophobic environment of the lipid bilayer and are composed largely of amino acid residues with nonpolar side chains. Because the peptide bonds themselves are polar and because water is absent, all peptide bonds in the bilayer are driven to form hydrogen bonds with one another. The hydrogen bonding between peptide bonds is maximized if the polypeptide chain forms a regular α helix as it crosses the bilayer, and this is how the great majority of the membrane-spanning segments of polypeptide chains are thought to traverse the bilayer (Figure 10–15): in **single-pass transmembrane proteins,** the polypeptide crosses only once (see example 1 in Figure 10–13), whereas in **multipass transmembrane proteins,** the polypeptide chain crosses multiple times (see example 2 in Figure 10–13). An alternative way for the peptide bonds in the lipid bilayer to satisfy their hydrogen-bonding requirements is for multiple transmembrane strands of polypeptide chain to be arranged as a β sheet in the form of a closed barrel (a so-called *β barrel*). This form of multipass transmembrane structure is seen in *porin* proteins, which we discuss later. The strong drive to maximize hydrogen bonding in the absence of water also means that a polypeptide chain that enters the bilayer is likely to pass entirely through it before changing direction, since chain bending requires a loss of regular hydrogen-bonding interactions. Probably for this reason, there is still no established example of a membrane protein in which the polypeptide chain extends only partway across the lipid bilayer.

Because transmembrane proteins are notoriously difficult to crystallize, only a few have been studied in their entirety by x-ray crystallography (discussed later); the folded three-dimensional structures of almost all of the others are uncertain. DNA cloning and sequencing techniques, however, have revealed the amino acid sequences of large numbers of transmembrane proteins, and it is often possible to predict from an analysis of the protein's sequence which parts of the polypeptide chain extend across the lipid bilayer as an α helix. Segments containing about 20–30 amino acid residues with a high degree of hydrophobicity are long enough to span a membrane as an α helix, and they can often be identified by means of a *hydropathy plot* (Figure 10–16). Because ten or fewer residues are sufficient to traverse a lipid bilayer as an extended β strand, the same strategy is unlikely to identify the membrane-spanning segments of a β barrel.

Membrane Proteins

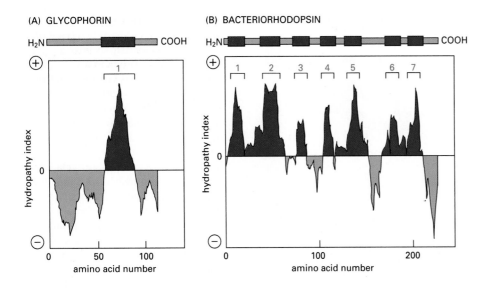

(A) GLYCOPHORIN

H_2N ▭▬▭ COOH

(B) BACTERIORHODOPSIN

H_2N ▭▬▭▬▭▬▭ COOH

Figure 10–16 **Localization of potential α-helical membrane-spanning segments in a polypeptide chain through the use of hydropathy plots.** The free energy needed to transfer successive segments of a polypeptide chain from a nonpolar solvent to water is calculated from the amino acid composition of each segment using data on model compounds. This calculation is made for segments of a fixed size (usually around 10 amino acid residues), beginning with each successive amino acid in the chain. The "hydropathy index" of the segment is plotted on the Y axis as a function of its location in the chain. A positive value indicates that free energy is required for transfer to water (that is, the segment is hydrophobic), and the value assigned is an index of the amount of energy needed. Peaks in the hydropathy index appear at the positions of hydrophobic segments in the amino acid sequence. Two examples of membrane proteins that we discuss later in this chapter are shown: (A) glycophorin has a single membrane-spanning α helix and one corresponding peak in the hydropathy plot; (B) bacteriorhodopsin has seven membrane-spanning α helices and seven corresponding peaks in the hydropathy plot. (Adapted from D. Eisenberg, *Annu. Rev. Biochem.* 53:595–624, 1984.)

The great majority of transmembrane proteins are glycosylated. As in the case of glycolipids, the sugar residues are added in the lumen of the endoplasmic reticulum and Golgi apparatus (discussed in Chapters 12 and 13), and for this reason the oligosaccharide chains are always present on the noncytoplasmic side of the membrane. A further asymmetry arises as a result of the reducing environment of the cytosol, which prevents the formation of intrachain (and interchain) disulfide (S—S) bonds between cysteine residues on the cytosolic side of membranes. These bonds do form on the noncytosolic side, where they can play an important part in stabilizing either the folded structure of the polypeptide chain (Figure 10–17) or its association with other polypeptide chains.

Membrane Proteins Can Be Solubilized and Purified in Detergents [9]

In general, transmembrane proteins (and some other tightly bound membrane proteins) can be solubilized only by agents that disrupt hydrophobic associations and destroy the lipid bilayer. The most useful among these for the membrane

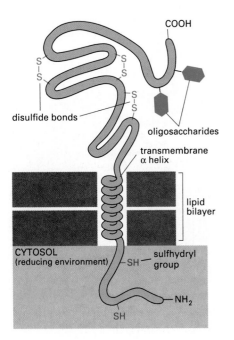

Figure 10–17 **A typical single-pass transmembrane protein.** Note that the polypeptide chain traverses the lipid bilayer as a right-handed α helix and that the oligosaccharide chains and disulfide bonds are all on the noncytosolic surface of the membrane. Disulfide bonds do not form between the sulfhydryl groups in the cytoplasmic domain of the protein because the reducing environment in the cytosol maintains these groups in their reduced (–SH) form.

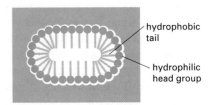

Figure 10–18 **A detergent micelle in water, shown in cross-section.** Because they have both polar and nonpolar ends, detergent molecules are amphipathic.

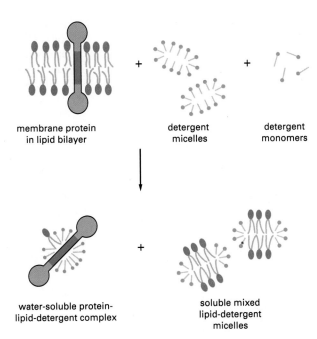

Figure 10–19 **Solubilizing membrane proteins with a mild detergent.** The detergent disrupts the lipid bilayer and brings the proteins into solution as protein-lipid-detergent complexes. The phospholipids in the membrane are also solubilized by the detergent.

membrane protein in lipid bilayer

detergent micelles

detergent monomers

water-soluble protein-lipid-detergent complex

soluble mixed lipid-detergent micelles

biochemist are **detergents,** which are small amphipathic molecules that tend to form micelles in water (Figure 10–18). When mixed with membranes, the hydrophobic ends of detergents bind to the hydrophobic regions of the membrane proteins, thereby displacing the lipid molecules. Since the other end of the detergent molecule is polar, this binding tends to bring the membrane proteins into solution as detergent-protein complexes (although some tightly bound lipid molecules may also remain) (Figure 10–19). The polar (hydrophilic) ends of detergents can be either charged (ionic), as in the case of *sodium dodecyl sulfate (SDS)*, or uncharged (nonionic), as in the case of the *Triton* detergents. The structures of these two commonly used detergents are illustrated in Figure 10–20.

With strong ionic detergents, such as SDS, even the most hydrophobic membrane proteins can be solubilized. This allows them to be analyzed by *SDS polyacrylamide-gel electrophoresis* (discussed in Chapter 4), a procedure that has revolutionized the study of membrane proteins. Such strong detergents unfold (denature) proteins by binding to their internal "hydrophobic cores," thereby rendering the proteins inactive and unusable for functional studies. Nonetheless, proteins can be readily purified in their SDS-denatured form, and in some cases the purified protein can be renatured, with recovery of functional activity, by removing the detergent.

Many hydrophobic membrane proteins can be solubilized and then purified in an active, if not entirely normal, form by the use of mild detergents, such as Triton X-100, that bind to the membrane-spanning segments of the protein. In this way functionally active membrane protein systems can be reconstituted from purified components, providing a powerful means to analyze their activities (Figure 10–21).

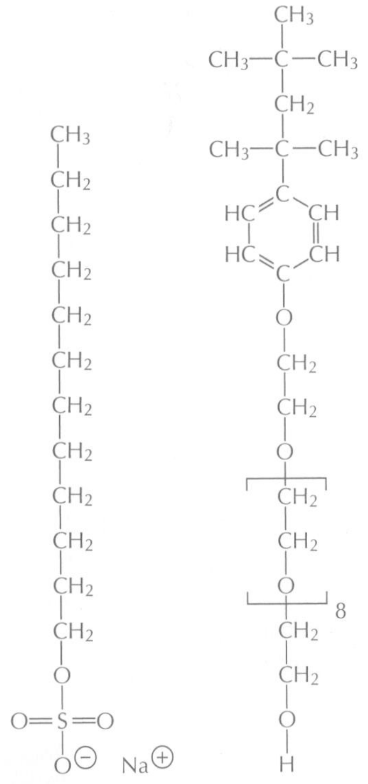

The Cytoplasmic Side of Membrane Proteins Can Be Readily Studied in Red Blood Cell Ghosts [10]

More is known about the plasma membrane of the human red blood cell (Figure 10–22) than about any other eucaryotic membrane. There are a number of reasons for this. Red blood cells are available in large numbers (from blood banks, for example) relatively uncontaminated by other cell types. Since they have no nucleus or internal organelles, the plasma membrane is their only membrane, and it can be isolated without contamination by internal membranes (thus avoiding a serious problem encountered in plasma membrane preparations from other

sodium dodecyl sulfate (SDS)

Triton X-100

Figure 10–20 **The structures of two commonly used detergents.** Sodium dodecyl sulfate (SDS) is an anionic detergent, and Triton X-100 is a nonionic detergent. The hydrophobic portion of each detergent is shown in *green*, and the hydrophilic portion is shown in *blue*. Note that the bracketed portion of Triton X-100 is repeated about eight times.

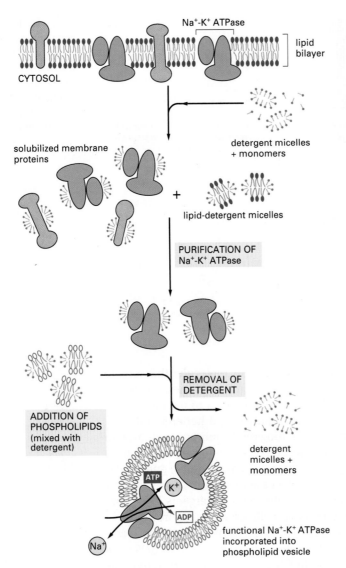

Na+-K+ ATPase

lipid bilayer

CYTOSOL

solubilized membrane proteins

detergent micelles + monomers

+

lipid-detergent micelles

PURIFICATION OF Na+-K+ ATPase

ADDITION OF PHOSPHOLIPIDS (mixed with detergent)

REMOVAL OF DETERGENT

detergent micelles + monomers

ATP

K+

ADP

Na+

functional Na+-K+ ATPase incorporated into phospholipid vesicle

Figure 10–21 The use of mild detergents for solubilizing, purifying, and reconstituting functional membrane protein systems. In this example functional Na+-K+ ATPase molecules are purified and incorporated into phospholipid vesicles. The Na+-K+ ATPase is an ion pump that is present in the plasma membrane of most animal cells; it uses the energy of ATP hydrolysis to pump Na+ out of the cell and K+ in, as discussed in Chapter 11.

cell types in which the plasma membrane typically constitutes less than 5% of the cell's membrane). It is easy to prepare empty red blood cell membranes, or "ghosts," by putting the cells in a medium with a lower salt concentration than the cell interior. Water then flows into the red cells, causing them to swell and burst (lyse) and release their hemoglobin (the major nonmembrane protein). Membrane ghosts can be studied while they are still leaky (in which case any reagent can interact with molecules on both faces of the membrane), or they can

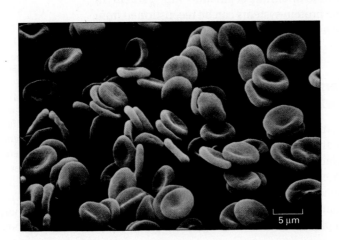

5 µm

Figure 10–22 A scanning electron micrograph of human red blood cells. The cells have a biconcave shape and lack nuclei. (Courtesy of Bernadette Chailley.)

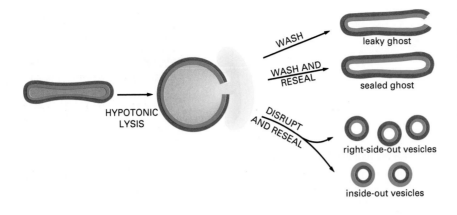

Figure 10–23 The preparation of sealed and unsealed red blood cell ghosts and of right-side-out and inside-out vesicles. As indicated, the red cells tend to rupture in only one place, giving rise to ghosts with a single hole in them. The smaller vesicles are produced by mechanically disrupting the ghosts; the orientation of the membrane in these vesicles can be either right-side-out or inside-out, depending on the ionic conditions used during the disruption procedure.

be allowed to reseal so that water-soluble reagents cannot reach the internal face. Moreover, since sealed *inside-out* vesicles can also be prepared from red blood cell ghosts (Figure 10–23), the external side and internal (cytoplasmic) side of the membrane can be studied separately. The use of sealed and unsealed red cell ghosts led to the first demonstration that some membrane proteins extend across the lipid bilayer (discussed below) and that the lipid compositions of the two halves of the bilayer are different. Like most of the basic principles initially demonstrated in red blood cell membranes, these findings were later extended to the membranes of nucleated cells.

The "sidedness" of a membrane protein can be determined in several ways. One is to use a covalent labeling reagent (for example, one carrying a radioactive or fluorescent marker) that is water soluble and therefore cannot penetrate the lipid bilayer; such a marker will attach covalently only to the portion of the protein on the exposed side of the membrane. The membranes are then solubilized with detergent and the proteins separated by SDS polyacrylamide-gel electrophoresis. The labeled proteins can be detected either by their radioactivity (by autoradiography of the gel) or by their fluorescence (by exposing the gel to ultraviolet light). By using such *vectorial labeling*, it is possible to determine how a particular protein, detected as a band on a gel, is oriented in the membrane: for example, if it is labeled from both the external side (when intact cells or sealed ghosts are labeled) and the internal (cytoplasmic) side (when sealed inside-out vesicles are labeled), then it must be a transmembrane protein. An alternative approach is to expose either the external or internal surface to membrane-impermeant proteolytic enzymes: if a protein is partially digested from both surfaces, it must be a transmembrane protein. In addition, labeled antibodies that bind only to one part of a protein can be used to determine if that part of a transmembrane protein is exposed on one side of the membrane or the other.

When the plasma membrane proteins of the human red blood cell are studied by SDS polyacrylamide-gel electrophoresis, approximately 15 major protein bands are detected, varying in molecular weight from 15,000 to 250,000. Three of these proteins—*spectrin*, *glycophorin*, and *band 3*—account for more than 60% (by weight) of the total membrane protein (Figure 10–24). Each of these proteins is arranged in the membrane in a different manner. We shall, therefore, use them as examples of three major ways that proteins are associated with membranes, not only in red blood cells, but in other cells as well.

Spectrin Is a Cytoskeletal Protein Noncovalently Associated with the Cytoplasmic Side of the Red Blood Cell Membrane [11]

Most of the protein molecules associated with the human red blood cell membrane are peripheral membrane proteins associated with the cytoplasmic side of the lipid bilayer. The most abundant of these proteins is **spectrin,** a long, thin,

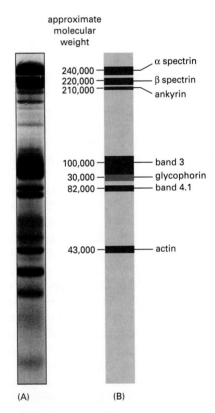

approximate molecular weight

240,000	α spectrin
220,000	β spectrin
210,000	ankyrin
100,000	band 3
30,000	glycophorin
82,000	band 4.1
43,000	actin

(A)　　　　(B)

Figure 10–24 SDS polyacrylamide-gel electrophoresis pattern of the proteins in the human red blood cell membrane. The gel in (A) is stained with Coomassie blue. The positions of some of the major proteins in the gel are indicated in the drawing in (B); glycophorin is shown in *red* to distinguish it from band 3. Other bands in the gel are omitted from the drawing. The large amount of carbohydrate in glycophorin molecules slows their migration so that they run almost as slowly as the much larger band 3 molecules. (A, courtesy of Ted Steck.)

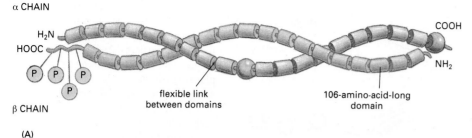

α CHAIN

H₂N

HOOC

P P P

P

β CHAIN

(A)

COOH

NH₂

flexible link
between domains

106-amino-acid-long
domain

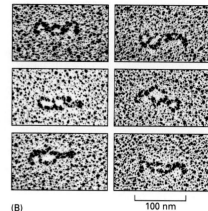

(B)

100 nm

Figure 10–25 Spectrin molecules from human red blood cells. The protein is shown schematically in (A) and in electron micrographs in (B). Each spectrin heterodimer consists of two antiparallel, loosely intertwined, flexible polypeptide chains called α and β; these are attached noncovalently to each other at multiple points, including both ends. The phosphorylated "head" end, where two dimers associate to form a tetramer, is on the left. Both the α and β chains are composed largely of repeating domains 106 amino acids long. In (B) the spectrin molecules have been shadowed with platinum. (A, adapted from D.W. Speicher and V.T. Marchesi, *Nature* 311:177–180, 1984; B, courtesy of D.M. Shotton, with permission from D.M. Shotton, B.E. Burke, and D. Branton, *J. Mol. Biol.* 131:303–329, 1979, © Academic Press Inc. [London] Ltd.)

flexible rod about 100 nm in length that constitutes about 25% of the membrane-associated protein mass (about 2.5×10^5 copies per cell). It is the principal component of the protein meshwork (the *cytoskeleton*) that underlies the red blood cell membrane, maintaining the structural integrity and biconcave shape of this membrane (see Figure 10–22): if the cytoskeleton is dissociated from red blood cell ghosts in low-ionic-strength solutions, the membrane fragments into small vesicles.

Spectrin is a heterodimer formed from two large, structurally similar subunits (Figure 10–25). The heterodimers self-associate head-to-head to form 200-nm-long tetramers. The tail ends of four or five tetramers are linked together by binding to short actin filaments and to other cytoskeletal proteins (including the *band 4.1 protein*) in a "junctional complex." The final result is a deformable, netlike meshwork that underlies the entire cytoplasmic surface of the membrane (Figure 10–26). It is this spectrin-based cytoskeleton that enables the red cell to withstand the stress on its membrane as it is forced through narrow capillaries. Mice and humans with genetic abnormalities of spectrin are anemic and have red cells that are spherical (instead of concave) and abnormally fragile; the severity of the anemia increases with the degree of the spectrin deficiency.

The protein mainly responsible for attaching the spectrin cytoskeleton to the red cell plasma membrane was identified by monitoring the binding of radiolabeled spectrin to red cell membranes from which spectrin and various other peripheral proteins had been removed. These experiments showed that the binding of spectrin depends on a large intracellular attachment protein called **ankyrin,** which attaches both to spectrin and to the cytoplasmic domain of the transmembrane protein band 3 (see Figure 10–26). By connecting some of the band 3 molecules to spectrin, ankyrin links the spectrin network to the membrane; it also greatly reduces the rate of diffusion of these band 3 molecules in the lipid bilayer. The spectrin-based cytoskeleton is also attached to the membrane by a second mechanism, which depends on the band 4.1 protein mentioned above. This protein, which binds to spectrin and actin, also binds to the cytoplasmic domain of both band 3 and *glycophorin*, the other major transmembrane protein in red blood cells.

An analogous but much more elaborate and complicated cytoskeletal network exists beneath the plasma membrane of nucleated cells. This network, which constitutes the cortical region (or *cortex*) of the cytoplasm, is rich in

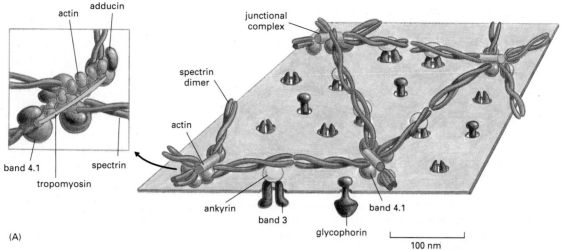

(A)

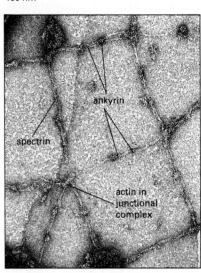

(B)

Figure 10–26 **The spectrin-based cytoskeleton on the cytoplasmic side of the human red blood cell membrane.** The structure is shown schematically in (A) and in an electron micrograph in (B). The arrangement shown in (A) has been deduced mainly from studies on the interactions of purified proteins *in vitro*. Spectrin dimers associate head-to-head to form tetramers that are linked together into a netlike meshwork by junctional complexes composed of short actin filaments (containing 13 actin monomers), tropomyosin, which probably determines the length of the actin filaments, band 4.1, and adducin (enlarged in the box on the *left*). The cytoskeleton is linked to the membrane by the indirect binding of spectrin tetramers to some band 3 proteins via ankyrin molecules, as well as by the binding of band 4.1 proteins to both band 3 and glycophorin (not shown). The electron micrograph in (B) shows the cytoskeleton on the cytoplasmic side of a red blood cell membrane after fixation and negative staining. The spectrin meshwork has been purposely stretched out to allow the details of its structure to be seen; in the normal cell the meshwork shown would occupy only about one-tenth of this area. (B, courtesy of T. Byers and D. Branton, *Proc. Natl. Acad. Sci. USA* 82:6153–6157, 1985.)

actin filaments, which are thought to be attached to the plasma membrane in numerous ways. Proteins that are structurally homologous to spectrin, ankyrin, and band 4.1 are present in the cortex of nucleated cells, but their organization and functions are less well understood than they are in red blood cells. The cortical cytoskeleton in nucleated cells and its interactions with the plasma membrane are discussed in Chapter 16.

Glycophorin Extends Through the Red Blood Cell Lipid Bilayer as a Single α Helix [10]

Glycophorin is one of the two major proteins exposed on the outer surface of the human red blood cell and was the first membrane protein for which the complete amino acid sequence was determined. Like the model transmembrane protein shown in Figure 10–17, glycophorin is a small, single-pass transmembrane glycoprotein (131 amino acid residues) with most of its mass on the external surface of the membrane, where its hydrophilic amino-terminal end is located. This part of the protein carries all of the carbohydrate (about 100 sugar residues in 16 separate oligosaccharide side chains), which accounts for 60% of the molecule's mass. In fact, the great majority of the total red blood cell surface carbohydrate (including more than 90% of the sialic acid and, therefore, most of the negative charge of the surface) is carried by glycophorin molecules. The hydrophilic carboxyl-terminal tail of glycophorin is exposed to the cytosol, while a hydrophobic α-helical segment 23 amino acid residues long spans the lipid bilayer (see Figure 10–16A).

Despite there being nearly a million glycophorin molecules per cell, their function remains unknown. Indeed, individuals whose red blood cells lack a major subset of these molecules appear to be perfectly healthy. Although glycophorin itself is found only in red blood cells, its structure is representative of a common class of membrane proteins that traverse the lipid bilayer as a single α helix. Many cell-surface receptors, for example, belong to this class.

Band 3 of the Red Blood Cell Is a Multipass Membrane Protein That Catalyzes the Coupled Transport of Anions [12]

Unlike glycophorin, the **band 3 protein** is known to play an important part in the function of red blood cells. It derives its name from its position relative to the other membrane proteins after electrophoresis in SDS polyacrylamide gels (see Figure 10–24). Like glycophorin, band 3 is a transmembrane protein, but it is a multipass membrane protein, traversing the membrane in a highly folded conformation: the polypeptide chain (about 930 amino acids long) is thought to extend across the bilayer up to 14 times. Each red blood cell contains about 10^6 band 3 polypeptide chains, which are arranged as dimers in the membrane.

The main function of red blood cells is to carry O_2 from the lungs to the tissues and to help carry CO_2 from the tissues to the lungs. The band 3 protein is crucial for the second of these functions. CO_2 is only sparingly soluble in water and so it is carried in the blood plasma as bicarbonate (HCO_3^-), which is formed and broken down inside red blood cells by an enzyme that catalyzes the reaction $H_2O + CO_2 \leftrightarrow HCO_3^- + H^+$. The band 3 protein acts as an *anion transporter*, which allows HCO_3^- to cross the membrane in exchange for Cl^-. By making the red cell membrane freely permeable to HCO_3^-, this transporter increases the amount of CO_2 that the blood can deliver to the lungs.

Band 3 proteins can be seen as distinct *intramembrane particles* by the technique of **freeze-fracture electron microscopy,** in which cells are frozen in liquid nitrogen and the resulting block of ice is fractured with a knife. The fracture plane tends to pass through the hydrophobic middle of membrane lipid bilayers, separating them into their two monolayers (Figure 10–27). The exposed *fracture faces* are then shadowed with platinum, and the resulting platinum replica is examined with an electron microscope. When examined in this way, human red blood cell membranes are studded with intramembrane particles that are relatively homogeneous in size (7.5 nm in diameter) and randomly distributed (Figure 10–

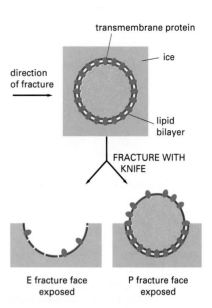

Figure 10–27 Freeze-fracture electron microscopy. The drawing shows how the technique provides images of the hydrophobic interior of the cytoplasmic (or protoplasmic) half of the bilayer (called the P face) and the external half of the bilayer (called the E face). After the fracturing process shown here, the exposed fracture faces are shadowed with platinum and carbon, the organic material is digested away, and the resulting platinum replica is examined in the electron microscope (see also Figure 4–27).

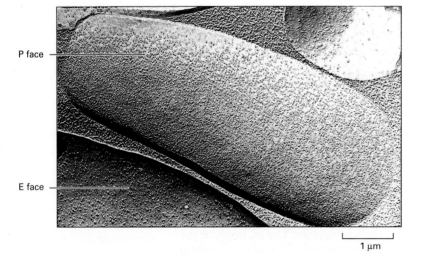

Figure 10–28 Freeze-fracture electron micrograph of human red blood cells. Note that the density of intramembrane particles on the protoplasmic (P) face is higher than on the external (E) face. (Courtesy of L. Engstrom and D. Branton.)

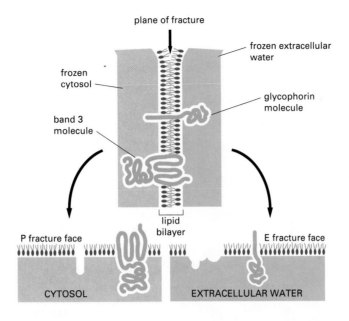

plane of fracture

frozen extracellular water

frozen cytosol

glycophorin molecule

band 3 molecule

lipid bilayer

P fracture face

E fracture face

CYTOSOL

EXTRACELLULAR WATER

Figure 10–29 **Probable fates of band 3 and glycophorin molecules in the human red blood cell membrane during freeze-fracture.** When the lipid bilayer is split, either the inside or outside half of each trans-membrane protein is pulled out of the frozen monolayer with which it is associated; the protein tends to remain with the monolayer that contains the main bulk of the protein. For this reason band 3 molecules usually remain with the inner (P) fracture face; since they have sufficient mass above the fracture plane, they are readily seen as intramembrane particles. Glycophorin molecules usually remain with the outer (E) fracture face, but it is thought that their cytoplasmic tails have insufficient mass to be seen.

28). The particles are thought to be principally band 3 molecules: when synthetic lipid bilayers are reconstituted with purified band 3 protein molecules, typical 7.5-nm intramembrane particles are observed when the bilayers are fractured. Figure 10–29 illustrates why band 3 molecules are seen in freeze-fracture electron microscopy of red blood cell membranes but glycophorin molecules probably are not.

In Chapter 11 we consider how a multipass transmembrane protein such as band 3 could, in principle, mediate the passive transport of polar molecules across the nonpolar lipid bilayer. But a detailed understanding of how a membrane transport protein actually works requires precise information about its three-dimensional structure in the bilayer. The first plasma membrane transport protein for which such detail became known was *bacteriorhodopsin,* a protein that serves as a light-activated proton (H⁺) pump in the plasma membrane of certain bacteria. The structure of bacteriorhodopsin is similar to that of many other membrane proteins, and it merits a brief digression here.

Bacteriorhodopsin Is a Proton Pump That Traverses the Lipid Bilayer as Seven α Helices [13]

The "purple membrane" of the bacterium *Halobacterium halobium* is a specialized patch in the plasma membrane that contains a single species of protein molecule, **bacteriorhodopsin** (Figure 10–30). Each bacteriorhodopsin molecule contains a single light-absorbing group, or chromophore (called *retinal*), which gives the protein its purple color; retinal is related to vitamin A and is identical to the chromophore found in rhodopsin of the photoreceptor cells of the vertebrate eye (discussed in Chapter 15). Retinal is covalently linked to a lysine side chain of the protein; when activated by a single photon of light, the excited chromophore immediately changes its shape and causes a series of small conformational changes in the protein that results in the transfer of one H⁺ from the inside to the outside of the cell (see Figure 14–36). In bright light each bacteriorhodopsin molecule can pump several hundred protons per second. The light-driven proton transfer establishes an H⁺ gradient across the plasma membrane, which in turn drives the production of ATP by a second protein in the cell's plasma membrane. Thus bacteriorhodopsin is part of a solar energy transducer that provides energy to the bacterial cell.

To understand the function of a multipass transmembrane protein in molecular detail, it is necessary to locate each of its atoms precisely, which gener-

Figure 10–30 **Schematic drawing of the bacterium *Halobacterium halobium* showing the patches of purple membrane that contain bacteriorhodopsin molecules.** These bacteria, which live in saltwater pools where they are exposed to a large amount of sunlight, have evolved a variety of light-activated proteins, including bacteriorhodopsin, which is a light-activated proton pump in the plasma membrane.

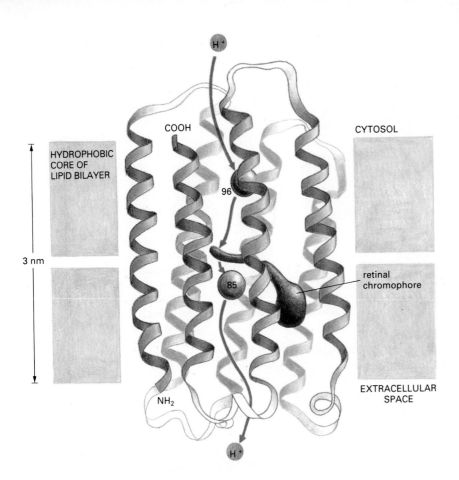

Figure 10–31 The three-dimensional structure of a bacteriorhodopsin molecule. The polypeptide chain crosses the lipid bilayer as seven α helices. The location of the chromophore and the probable pathway taken by protons during the light-activated pumping cycle are shown. When activated by a photon, the chromophore is thought to pass an H+ to the side chain of aspartic acid 85 (*pink sphere marked 85*). Subsequently, three other H+ transfers are thought to complete the cycle—from aspartic acid 85 to the extracellular space, from aspartic acid 96 (*pink sphere marked 96*) to the chromophore, and from the cytosol to aspartic acid 96 (see Figure 14–36). (Adapted from R. Henderson et al. *J. Mol. Biol.* 213:899–929, 1990.)

ally requires x-ray diffraction studies of large three-dimensional crystals of the protein. But because of their amphipathic nature, these proteins are extremely difficult to crystallize. The numerous bacteriorhodopsin molecules in the purple membrane, however, are arranged as a planar two-dimensional crystal, which has made it possible to determine their three-dimensional structure and orientation in the membrane to a resolution of about 0.3 nm by an alternative approach, which uses a combination of electron microscopy and electron diffraction analysis. This procedure (referred to as *electron crystallography*) is analogous to the study of three-dimensional crystals of soluble proteins by x-ray diffraction analysis, although less structural detail has so far been obtained. As illustrated in Figure 10–31, these studies have shown that each bacteriorhodopsin molecule is folded into seven closely packed α helices (each containing about 25 amino acids), which pass roughly at right angles through the lipid bilayer.

Bacteriorhodopsin is a member of a very large superfamily of membrane proteins with similar structures but different functions. The light receptor protein *rhodopsin* in rod cells of the vertebrate retina, for example, and many cell-surface receptor proteins that bind extracellular signaling molecules are also folded into seven transmembrane α helices. These proteins function as signal transducers rather than as transporters: each responds to an extracellular signal by activating another protein inside the cell, which generates chemical signals in the cytosol, as we discuss in Chapter 15.

Porins Are Pore-forming Transmembrane Proteins That Cross the Lipid Bilayer as a β Barrel [14]

As discussed earlier, some multipass transmembrane proteins have their transmembrane segments arranged as a closed β sheet (a *β barrel*) rather than as

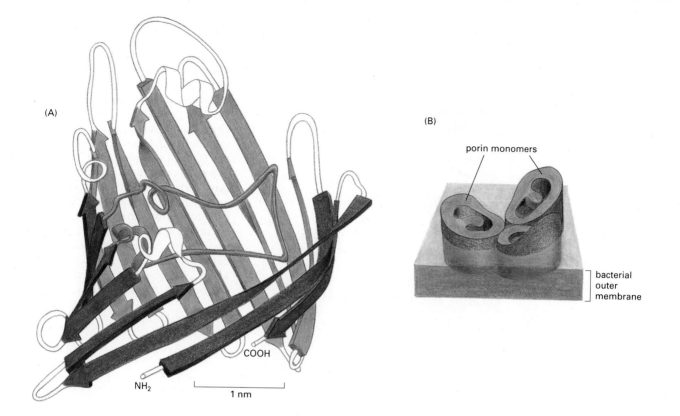

(A)

(B)

porin monomers

bacterial
outer
membrane

COOH

NH₂

1 nm

Figure 10–32 **The three-dimensional structure of a porin trimer of *Rhodobacter capsulatus* determined by x-ray crystallography.** (A) Each monomer consists of a 16-stranded antiparallel β barrel that forms a transmembrane water-filled channel. (B) The monomers tightly associate to form trimers, which have three separate channels for the diffusion of small solutes through the bacterial outer membrane. A long loop of polypeptide chain (shown in *red*), which connects two β strands, protrudes into the lumen of each channel, narrowing it to a cross-section of 0.6 × 1 nm. (Adapted from M.S. Weiss et al., *FEBS Lett.* 280: 379–382, 1991.)

α helices. The best studied examples of such proteins are the **porins**, which are found in the outer membrane of many bacteria. They are among the few transmembrane proteins whose complete atomic structure has been solved by x-ray crystallography.

Many bacteria, including *E. coli*, have an *outer membrane* surrounding their plasma membrane (see Figure 11–14). The outer membrane is penetrated by various pore-forming porin proteins, which allow selected hydrophilic solutes of up to 600 daltons to diffuse across the outer lipid bilayer. The porins (and the structurally related pore-forming proteins in the outer membrane of mitochondria and chloroplasts) have a β sheet instead of an α helix as the major transmembrane motif.

The atomic structure of a porin protein isolated from a photosynthetic bacterium was determined by x-ray crystallography in 1990. It consists of a trimer in which each monomer forms a tubular β barrel, which traverses the lipid bilayer and has a water-filled pore at its center. The barrel is formed from a 16-stranded antiparallel β sheet, which is sufficiently curved to roll up into a cylindrical structure (Figure 10–32). Polar side chains line the aqueous channel on the inside, while nonpolar side chains project from the outside of the barrel to interact with the hydrophobic core of the lipid bilayer.

Membrane Proteins Often Function as Large Complexes [15]

By far the most complex transmembrane protein structure that has been studied by x-ray crystallography to date is a bacterial *photosynthetic reaction center*, whose atomic structure was solved in 1985. It was the first transmembrane protein to be crystallized and analyzed by x-ray diffraction. The results of this analysis were of general importance to membrane biology because they showed for the first time how multiple polypeptides can associate in a membrane to form a complex protein machine (Figure 10–33). In Chapter 14 we discuss how such photosynthetic complexes function to capture light energy and use it to pump H⁺ across

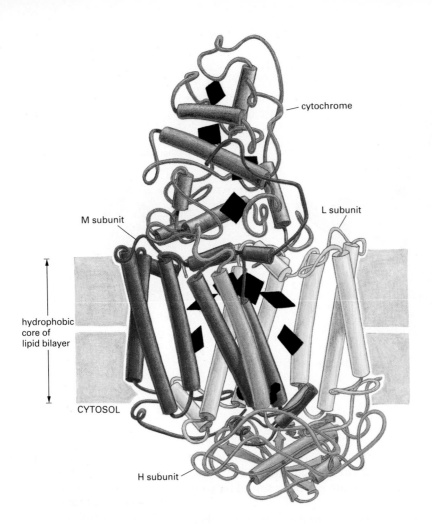

cytochrome

L subunit

M subunit

hydrophobic
core of
lipid bilayer

CYTOSOL

H subunit

Figure 10–33 **The three-dimensional structure of the photosynthetic reaction center of the bacterium *Rhodopseudomonas viridis.*** The structure was determined by x-ray diffraction analysis of crystals of this transmembrane protein complex. The complex consists of four subunits, L, M, H, and a cytochrome. The L and M subunits form the core of the reaction center, and each contains five α helices that span the lipid bilayer. The locations of the various electron carrier coenzymes are shown in *black.* (Adapted from a drawing by J. Richardson based on data from J. Deisenhofer, O. Epp, K. Miki, R. Huber, and H. Michel, *Nature* 318:618–624, 1985.)

the membrane. Membrane proteins are often arranged in large complexes, not only for harvesting various forms of energy, but also for transducing extracellular signals into intracellular ones (discussed in Chapter 15).

Many Membrane Proteins Diffuse in the Plane of the Membrane [16]

Like membrane lipids, membrane proteins do not tumble (*flip-flop*) across the bilayer, but they do rotate about an axis perpendicular to the plane of the bilayer (*rotational diffusion*). In addition, many membrane proteins are able to move laterally within the membrane (*lateral diffusion*). The first direct evidence that some plasma membrane proteins are mobile in the plane of the membrane was provided in 1970 by an experiment in which mouse cells were artificially fused with human cells to produce hybrid cells (*heterocaryons*). Two differently labeled antibodies were used to distinguish selected mouse and human plasma membrane proteins. Although at first the mouse and human proteins were confined to their own halves of the newly formed heterocaryon, the two sets of proteins diffused and mixed over the entire cell surface within half an hour or so (Figure 10–34). Further evidence for membrane protein mobility was soon provided by the discovery of the processes called *patching* and *capping*. When ligands, such as antibodies, that have more than one binding site (so-called *multivalent* ligands) bind to specific proteins on the surface of cells, the proteins tend to become aggregated, through cross-linking, into large clusters (or "patches"), indicating that the protein molecules are able to diffuse laterally in the lipid bilayer. Once such clusters have formed on the surface of a cell capable of locomotion, such as a white blood cell, they are actively moved to one pole of the cell to form a "cap" (Figure 10–35).

The lateral diffusion rates of membrane proteins can be measured using the technique of *fluorescence recovery after photobleaching (FRAP)*. The method usually involves marking the cell-surface protein of interest with a specific fluorescent ligand, such as a fluorescent antibody. (It is important that monovalent fragments of antibodies, which have only one antigen-binding site, be used to avoid cross-linking neighboring molecules.) The fluorescent ligand is then bleached in a small area by a laser beam, and the time taken for adjacent membrane proteins carrying unbleached fluorescent ligand to diffuse into the bleached area is measured (Figure 10–36). From such measurements diffusion coefficients can be calculated for the particular cell-surface protein that was marked. The values of the diffusion coefficients for different membrane proteins in different cells are highly variable, but they are typically about one-tenth or one-hundredth of the corresponding values for the phospholipid molecules in the same membrane.

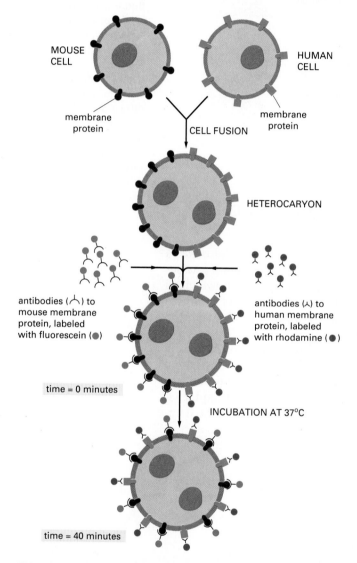

Figure 10–34 Experiment demonstrating the mixing of plasma membrane proteins on mouse-human hybrid cells. The mouse and human proteins are initially confined to their own halves of the newly formed heterocaryon plasma membrane, but they intermix with time. The two antibodies used to visualize the proteins can be distinguished in a fluorescence microscope because fluorescein is green whereas rhodamine is red. (Based on observations of L.D. Frye and M. Edidin, *J. Cell Sci.* 7:319–335, 1970, by permission of The Company of Biologists.)

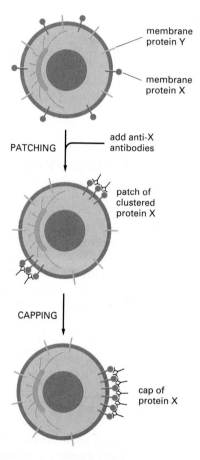

Figure 10–35 Antibody-induced patching and capping of a cell-surface protein on a white blood cell. The bivalent antibodies cross-link the protein molecules to which they bind. This causes them to cluster into large patches, which are actively swept to the tail end of the cell to form a "cap." The centrosome, which governs the head-tail polarity of the cell, is shown in *orange*.

Membrane Proteins

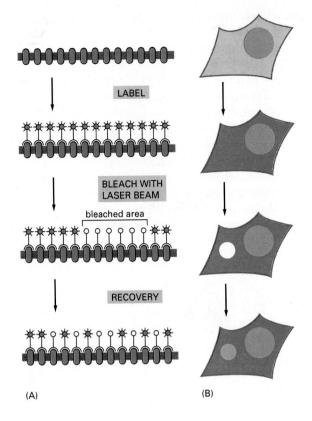

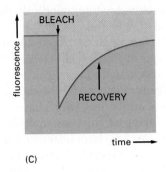

LABEL

BLEACH WITH
LASER BEAM

bleached area

RECOVERY

(A)

(B)

(C)

Figure 10–36 Measuring the rate of lateral diffusion of a plasma membrane protein by the FRAP technique. A specific protein is labeled on the cell surface with a fluorescent monovalent antibody that binds only to that protein (for simplicity, no other proteins are shown). After the antibodies are bleached in a small area using a laser beam, the fluorescence intensity recovers as the bleached molecules diffuse away and unbleached molecules diffuse into the irradiated area (shown in side view in A and top view in B). (C) A graph showing the rate of recovery. The greater the diffusion coefficient of the membrane protein, the faster the recovery.

Cells Can Confine Proteins and Lipids to Specific Domains Within a Membrane [17]

The recognition that biological membranes are two-dimensional fluids was a major advance in understanding membrane structure and function. It has become clear, however, that the picture of a membrane as a lipid sea in which all proteins float freely is greatly oversimplified. Many cells have ways of confining membrane proteins to specific domains in a continuous lipid bilayer. In epithelial cells, such as those that line the gut or the tubules of the kidney, for example, certain plasma membrane enzymes and transport proteins are confined to the apical surface of the cells, whereas others are confined to the basal and lateral surfaces (Figure 10–37). This asymmetric distribution of membrane proteins is often essential for the function of the epithelium, as we discuss in Chapter 11. The lipid compositions of these two membrane domains are also different, dem-

Figure 10–37 Diagram of an epithelial cell showing how a plasma membrane protein is restricted to a particular domain of the membrane. Protein A (in the apical membrane) and protein B (in the basal and lateral membranes) can diffuse laterally in their own domains but are prevented from entering the other domain, at least partly by the specialized cell junction called a *tight junction*. Lipid molecules in the outer (noncytoplasmic) monolayer of the plasma membrane are likewise unable to diffuse between the two domains; lipids in the inner (cytoplasmic) monolayer, however, are able to do so (not shown).

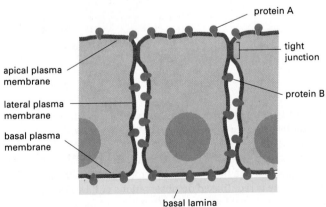

Figure 10–38 **Three domains in the plasma membrane of guinea pig sperm defined with monoclonal antibodies.** A guinea pig sperm is shown schematically in (A), while each of the three pairs of micrographs shown in (B), (C), and (D) shows cell-surface immunofluorescence staining with a different monoclonal antibody (on the *right*) next to a phase-contrast micrograph (on the *left*) of the same cell. The antibody shown in (B) labels only the anterior head, that in (C) only the posterior head, whereas that in (D) labels only the tail. (Courtesy of Selena Carroll and Diana Myles.)

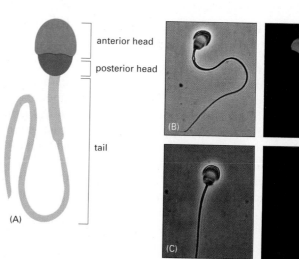

onstrating that epithelial cells can prevent the diffusion of lipid as well as protein molecules between the domains. Experiments with labeled lipids, however, suggest that only lipid molecules in the outer monolayer of the membrane are confined in this way. The separation of both protein and lipid molecules is thought to be maintained, at least in part, by the barriers set up by a specific type of intercellular junction (called a *tight junction,* discussed in Chapter 19). Clearly, the membrane proteins that form these intercellular junctions cannot be allowed to diffuse laterally in the interacting membranes.

A cell can also create membrane domains without using intercellular junctions. The mammalian spermatozoon, for instance, is a single cell that consists of several structurally and functionally distinct parts covered by a continuous plasma membrane. When a sperm cell is examined by immunofluorescence microscopy using a variety of antibodies that react with cell-surface antigens, the plasma membrane is found to consist of at least three distinct domains (Figure 10–38). In some cases, the antigens are able to diffuse within the confines of their own domain; it is not known how they are prevented from leaving it.

In the two examples just considered, the diffusion of protein and lipid molecules is confined to specialized domains within a continuous plasma membrane. Cells also have more drastic ways of immobilizing certain membrane proteins. One is exemplified by the purple membrane of *Halobacterium.* There the bacteriorhodopsin molecules assemble into large two-dimensional crystals in which the individual protein molecules are relatively fixed in relationship to one another; large aggregates of this kind diffuse very slowly. A more common way of restricting the lateral mobility of specific membrane proteins is to tether them to macromolecular assemblies either inside or outside the cell. We have seen that some red blood cell membrane proteins are anchored to the cytoskeleton inside; in other cell types plasma membrane proteins can be anchored to the cytoskeleton, or to the extracellular matrix, or to both. The four known ways of immobilizing specific membrane proteins are summarized in Figure 10–39.

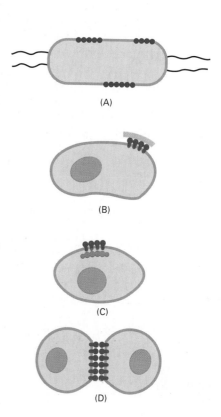

Figure 10–39 **Four ways in which the lateral mobility of specific plasma membrane proteins can be restricted.** The proteins can self-assemble into large aggregates (such as bacteriorhodopsin in the purple membrane of *Halobacterium*) (A); they can be tethered by interactions with assemblies of macromolecules outside (B) or inside (C) the cell; or they can interact with proteins on the surface of another cell (D).

Membrane Proteins

The Cell Surface Is Coated with Sugar Residues [18]

Plasma membrane proteins, as a rule, do not protrude naked from the exterior of the cell but are decorated, clothed, or hidden by carbohydrates, which are present on the surface of all eucaryotic cells. These carbohydrates occur both as oligosaccharide chains covalently bound to membrane proteins (glycoproteins) and lipids (glycolipids) and as polysaccharide chains of *integral membrane proteoglycan* molecules. Proteoglycans, which consist of long polysaccharide chains linked covalently to a protein core, are found mainly outside the cell as part of the extracellular matrix (discussed in Chapter 19); but in the case of integral membrane proteoglycans, the protein core either extends across the lipid bilayer or is attached to the bilayer by a glycosylphosphotidylinositol (GPI) anchor.

The term *cell coat,* or *glycocalyx,* is often used to describe the carbohydrate-rich zone on the cell surface. This zone can be visualized by a variety of stains, such as ruthenium red (Figure 10–40), as well as by its affinity for carbohydrate-binding proteins called **lectins,** which can be labeled with a fluorescent dye or some other visible marker. Although most of the carbohydrate is attached to intrinsic plasma membrane molecules, the glycocalyx usually also contains both glycoproteins and proteoglycans that have been secreted into the extracellular space and then adsorbed onto the cell surface (Figure 10–41). Many of these adsorbed macromolecules are components of the extracellular matrix, so that where the plasma membrane ends and the extracellular matrix begins is largely a matter of semantics.

The oligosaccharide side chains of glycoproteins and glycolipids are enormously diverse in their arrangement of sugars. Although they usually contain fewer than 15 sugar residues, they are often branched, and the sugars can be bonded together by a variety of covalent linkages, unlike the amino acid residues in a polypeptide chain, which are all linked by identical peptide bonds. Even three sugar residues can be put together to form hundreds of different trisaccharides. In principle, both the diversity and the exposed position of these oligosaccharides on the cell surface make them especially well suited to function in specific cell-recognition processes, but for many years there was little evidence for this suspected function. It seemed that the role of the cell coat might be merely to protect against mechanical and chemical damage and to keep foreign objects and other cells at a distance, preventing undesirable protein-protein interactions. Indeed, this probably is an important part of its function. Recently, however, plasma-membrane-bound lectins have been identified that recognize specific oligosaccharides on cell-surface glycolipids and glycoproteins to mediate a variety of transient cell-cell adhesion processes, including those occurring in sperm-egg interactions, blood clotting, lymphocyte recirculation, and inflammatory responses.

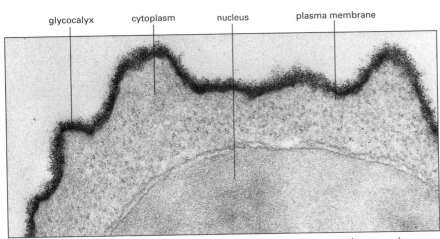

glycocalyx   cytoplasm   nucleus   plasma membrane

Figure 10–40 The cell coat, or glycocalyx. Electron micrograph of the surface of a lymphocyte stained with ruthenium red to show the cell coat. (Courtesy of A.M. Glauert and G.M.W. Cook.)

200 nm

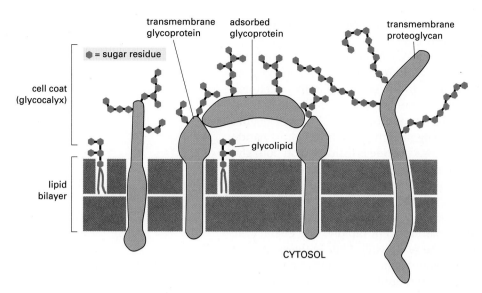

transmembrane glycoprotein

adsorbed glycoprotein

transmembrane proteoglycan

⬡ = sugar residue

cell coat (glycocalyx)

glycolipid

lipid bilayer

CYTOSOL

Figure 10–41 Simplified diagram of the cell coat (glycocalyx). The cell coat is made up of the oligosaccharide side chains of glycolipids and integral membrane glycoproteins and the polysaccharide chains on integral membrane proteoglycans. In addition, adsorbed glycoproteins and adsorbed proteoglycans (not shown) contribute to the glycocalyx in many cells. Note that all of the carbohydrate is on the noncytoplasmic surface of the membrane.

Selectins Are Cell-Surface Carbohydrate-binding Proteins That Mediate Transient Cell-Cell Adhesions in the Bloodstream [19]

One of the best understood examples of protein-carbohydrate recognition occurs in inflammatory responses, when white blood cells (of the class called *neutrophils*) are recruited from the blood into a site of inflammation in a tissue, usually to help combat a local infection. Initially, the neutrophils adhere weakly to the endothelial cells lining the blood vessels at the site, and then they adhere

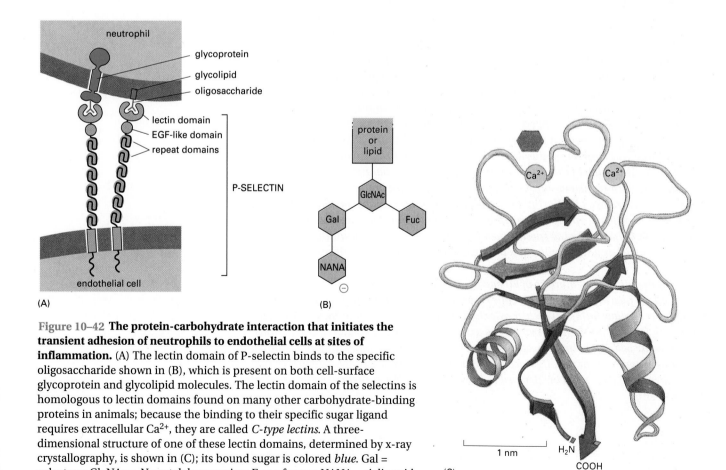

neutrophil

glycoprotein

glycolipid

oligosaccharide

lectin domain
EGF-like domain
repeat domains

P-SELECTIN

endothelial cell

(A)

protein or lipid

GlcNAc

Gal

Fuc

NANA

⊖

(B)

Ca²⁺   Ca²⁺

1 nm

H₂N

COOH

(C)

Figure 10–42 The protein-carbohydrate interaction that initiates the transient adhesion of neutrophils to endothelial cells at sites of inflammation. (A) The lectin domain of P-selectin binds to the specific oligosaccharide shown in (B), which is present on both cell-surface glycoprotein and glycolipid molecules. The lectin domain of the selectins is homologous to lectin domains found on many other carbohydrate-binding proteins in animals; because the binding to their specific sugar ligand requires extracellular Ca^{2+}, they are called *C-type lectins*. A three-dimensional structure of one of these lectin domains, determined by x-ray crystallography, is shown in (C); its bound sugar is colored *blue*. Gal = galactose; GlcNAc = *N*-acetylglucosamine; Fuc = fucose; NANA = sialic acid.

Membrane Proteins

503

more strongly and migrate out of the blood vessels by crawling between adjacent endothelial cells. It is the initial adhesion process that involves protein-carbohydrate recognition. Local chemical mediators released by cells at the site of inflammation signal the endothelial cells in the region to express a transmembrane glycoprotein called *P-selectin*, which belongs to the **selectin** family of cell-cell adhesion molecules. The selectins contain a carbohydrate-binding lectin domain at the end of an extended protein "stalk" that extends from the cell surface (Figure 10–42A). The lectin domain of P-selectin recognizes the specific oligosaccharide as shown in Figure 10–42B. Because this oligosaccharide is expressed on both glycolipid and glycoprotein molecules on the surface of neutrophils, the neutrophils stick specifically to the endothelial cells lining blood vessels at the inflamed site.

The binding of each lectin domain to its specific oligosaccharide is of relatively low affinity, and it seems that both the association and the subsequent dissociation of the domain from the oligosaccharide occur rapidly. This enables the selectins to bind the passing blood cells to the vessel wall, while at the same time allowing the attached cells to roll along the endothelium at the inflamed site, propelled by the flow of blood. The rolling continues until another cell-cell adhesion mechanism, mediated by a different class of transmembrane proteins, called *integrins* (discussed in Chapter 19), is activated and strengthens the adhesion, allowing the neutrophils to stop rolling and crawl out of the blood vessel into the tissue. A similar sequence of selectin-mediated and integrin-mediated adhesion events occurs when lymphocytes migrate out of the bloodstream into a lymph node (see Figure 23–9).

Various selectins are expressed on the surface of white blood cells, platelets, and endothelial cells, where they function in a wide range of transient cell-cell interactions in the bloodstream.

Summary

Whereas the lipid bilayer determines the basic structure of biological membranes, proteins are responsible for most membrane functions, serving as specific receptors, enzymes, transport proteins, and so on. Many membrane proteins extend across the lipid bilayer: in some of these transmembrane proteins the polypeptide chain crosses the bilayer as a single α helix (single-pass proteins); in others, including those responsible for the transmembrane transport of ions and other small water-soluble molecules, the polypeptide chain crosses the bilayer multiple times, either as a series of α helices or as a β sheet in the form of a closed barrel (multipass proteins). Other membrane-associated proteins do not span the bilayer but instead are attached to one or the other side of the membrane. Many of these are bound by noncovalent interactions with transmembrane proteins, but others are bound via covalently attached lipid groups. Like the lipid molecules in the bilayer, many membrane proteins are able to diffuse rapidly in the plane of the membrane. On the other hand, cells have ways of immobilizing specific membrane proteins and of confining both membrane protein and lipid molecules to particular domains in a continuous lipid bilayer.

In the plasma membrane of all eucaryotic cells most of the proteins exposed on the cell surface and some of the lipid molecules in the outer lipid monolayer have oligosaccharide chains covalently attached to them. Some plasma membranes also contain integral proteoglycan molecules with surface-exposed polysaccharide chains. This sugar coating helps to protect the cell surface from mechanical and chemical damage, and some of the oligosaccharide chains are recognized by cell-surface carbohydrate-binding proteins (lectins) that mediate specific, transient, cell-cell adhesion events.

References

General

Bretscher, M.S. The molecules of the cell membrane. *Sci. Am.* 253(4):100–109, 1985.

Gennis, R.B. Biomembranes: Molecular Structure and Function. New York: Springer-Verlag, 1989.

Jain, M.K. Introduction to Biological Membranes, 2nd ed. New York: Wiley, 1988.

Singer, S.J.; Nicolson, G.L. The fluid mosaic model of the structure of cell membranes. *Science* 175:720–731, 1972.

Cited

1. Ceve, G.; Marsh, D. Phospholipid Bilayers: Physical Principles and Models. New York: Wiley, 1987.

   Quinn, P.J.; Cherry, R.J., eds. Structural and Dynamic Properties of Lipids and Membranes. London: Portland Press, 1992.

   Tanford, C. The Hydrophobic Effect: Formation of Micelles and Biological Membranes. New York: Wiley, 1980.

   Vance, D.E.; Vance, J.E., eds. Biochemistry of Lipids, Lipoproteins and Membranes, New Comprehensive Biochemistry, Vol. 20. New York: Elsevier, 1991.

   van Meer, G. Lipid traffic in animal cells. *Annu. Rev. Cell Biol.* 5:247–275, 1989.

2. Bangham, A.D. Models of cell membranes. In Cell Membranes: Biochemistry, Cell Biology and Pathology (G. Weissmann, R. Claiborne, eds.), pp. 24–34. New York: Hospital Practice, 1975.

   Devaux, P.F. Lipid transmembrane asymmetry and flip-flop in biological membranes and in lipid bilayers. *Curr. Opin. Struct. Biol.* 3:489–494, 1993.

   Edidin, M. Rotational and lateral diffusion of membrane proteins and lipids: phenomena and function. *Curr. Top. Membr. Transp.* 29:91–127, 1987.

   Kornberg, R.D.; McConnell, H.M. Lateral diffusion of phospholipids in a vesicle membrane. *Proc. Natl. Acad. Sci. USA* 68:2564–2568, 1971.

3. Bretscher, M.S.; Munro, S. Cholesterol and the Golgi apparatus. *Science* 261:1280–1281, 1993.

   Chapman, D.; Benga, G. Biomembrane fluidity—studies of model and natural membranes. In Biological Membranes (D. Chapman, ed.), Vol. 5, pp. 1–56. London: Academic Press, 1984.

   de Kruiff, B., et al. Lipid polymorphism and membrane function. In The Enzymes of Biological Membranes, Vol 1, Membrane Structure and Dynamics, 2nd ed. (A.N. Martonosi, ed.), pp. 131–204. New York: Plenum, 1985.

   Kimelberg, H.K. The influence of membrane fluidity on the activity of membrane-bound enzymes. In Dynamic Aspects of Cell Surface Organization. Cell Surface Reviews (G. Poste, G.L. Nicolson, eds.), Vol. 3, pp. 205–293. Amsterdam: Elsevier, 1977.

   Yeagle, P.L. Cholesterol and the cell membrane. *Biochim. Biophys. Acta* 822:267–287, 1985.

4. Bretscher, M. Membrane structure: some general principles. *Science* 181:622–629, 1973.

   Devaux, P.F. Protein involvement in transmembrane lipid asymmetry. *Annu. Rev. Biophys. Biomol. Struct.* 21:417–439, 1992.

   Newton, A.C. Interaction of proteins with lipid headgroups: lessons from protein kinase C. *Annu. Rev. Biophys. Biomol. Struct.* 22:1–25, 1993.

   Rothman, J.; Lenard, J. Membrane asymmetry. *Science* 195:743–753, 1977.

5. Hakomori, S. Glycosphingolipids. *Sci. Am.* 254(5):44–53, 1986.

   Weigandt, H. The gangliosides. *Adv. Neurochem.* 4:149–223, 1982.

6. Branden, C.; Tooze, J. Introduction to Protein Structure, pp. 202–214. New York: Garland, 1991.

   Unwin, N.; Henderson, R. The structure of proteins in biological membranes. *Sci. Am.* 250(2):78–94, 1984.

7. Chow, M.; Der, C.J.; Buss, J.E. Structure and biological effects of lipid modifications on proteins. *Curr. Opin. Cell Biol.* 4:629–636, 1992.

   Cross, G.A. Glycolipid anchoring of plasma membrane proteins. *Annu. Rev. Cell Biol.* 6:1–39, 1990.

   Englund, P.T. The structure and biosynthesis of glycosyl phosphatidylinositol protein anchors. *Annu. Rev. Biochem.* 62:121–138, 1993.

   Jennings, M.L. Topography of membrane proteins. *Annu. Rev. Biochem.* 58:999–1027, 1989.

   Singer, S.J. The structure and insertion of integral proteins in membranes. *Annu. Rev. Cell Biol.* 6:247–296, 1990.

8. Eisenberg, D. Three-dimensional structure of membrane and surface proteins. *Annu. Rev. Biochem.* 53:595–623, 1984.

   Engelman, D.M.; Steitz, T.A.; Goldman, A. Identifying nonpolar transbilayer helices in amino acid sequences of membrane proteins. *Annu. Rev. Biophys. Biophys. Chem.* 15:321–353, 1986.

   Fasman, G.D.; Gilbert, W.A. The prediction of transmembrane protein sequences and their conformation: an evaluation. *Trends Biochem. Sci.* 15:89–92, 1990.

   Kyte, J.; Doolittle, R.F. A simple method for displaying the hydropathic character of a protein. *J. Mol. Biol.* 157:105–132, 1982.

   Popot, J.-L. Integral membrane protein structure: transmembrane α helices as autonomous folding domains. *Curr. Opin. Struct. Biol.* 3:512–540, 1993.

9. Helenius, A.; Simons, K. Solubilization of membranes by detergents. *Biochim. Biophys. Acta* 415:29–79, 1975.

   Montal, M. Functional reconstitution of membrane proteins in planar lipid bilayer membranes. In Techniques of the Analysis of Membrane Proteins (C.I. Ragan, R.J. Cherry, eds.), pp. 97–128. London: Chapman & Hall, 1986.

   Racker, E. Reconstitution of Transporters, Receptors and Pathological States. Orlando, FL: Academic Press, 1985.

   Silvius, J.R. Solubilization and functional reconstitution of biomembrane components. *Annu. Rev. Biophys. Biomol. Struct.* 21:323–348, 1992.

10. Agre, P.; Parker, J.C., eds. Red Blood Cell Membranes: Structure, Function, Clinical Implications. New York: Marcel Dekker, 1989.

    Bretscher, M. Membrane structure: some general principles. *Science* 181:622–629, 1973.

    Marchesi, V.T.; Furthmayr, H.; Tomita, M. The red cell membrane. *Annu. Rev. Biochem.* 45:667–698, 1976.

    Steck, T.L. The organization of proteins in the human red blood cell membrane. *J. Cell Biol.* 62:1–19, 1974.

11. Bennett, V. ; Lambert, S. The spectrin skeleton: f~
    cells to brain. *J. Clin. Invest.* 87:1483–1489,

    Bennett, V.; Gilligan, D.M. The spectrin-bas
    brane structure and micron-scale organiz~~~~~~ ~~ ~~~
    plasma membrane. *Annu. Rev. Cell Biol.* 9:27–66,
    1993.

    Branton, D.; Cohen, C.M.; Tyler, J. Interaction of
    cytoskeletal proteins on the human erythrocyte
    membrane. *Cell* 24:24–32, 1981.

    Davies, K.A.; Lux, S.E. Heredity disorders of the red cell
    membrane skeleton. *Trends Genet.* 5:222–227, 1989.

    Pumplin, D.W.; Bloch, R.J. The membrane skeleton.
    Trends Cell Biol. 3:113–117, 1993.

12. Alper, S.L. The band 3-related anion exchanger (AE)
    gene family. *Annu. Rev. Physiol.* 53:549–564, 1991.

    Jennings, M.L. Structure and function of the red blood
    cell anion transport protein. *Annu. Rev. Biophys.
    Chem.* 18:397–430, 1989.

    Kopito, R.R. Molecular biology of the anion exchanger
    gene family. *Int. Rev. Cytol.* 123:177–199, 1990.

    Reithmeier, R.A.F. The erythrocyte anion transporter
    (band 3). *Curr. Opin. Struct. Biol.* 3:515–523, 1993.

13. Henderson, R.; Baldwin, J.M.; Ceska, T.A.; Zemlin, F.;
    Beckmann, E.; Downing, K.H. Model for the struc-
    ture of bacteriorhodopsin based on high-resolution
    electron cryo-microscopy. *J. Mol. Biol.* 213:899–929,
    1990.

    Henderson, R.; Unwin, P.N.T. Three-dimensional
    model of purple membrane obtained by electron
    microscopy. *Nature* 257:28–32, 1975.

    Mathies, R.A.; Lin, S.W.; Ames, J.B.; Pollard, W.T. From
    femtoseconds to biology: mechanism of bacterio-
    rhodopsin's light-driven proton pump. *Annu. Rev.
    Biophys. Biophys. Chem.* 20:491–518, 1991.

    Stoeckenius, W. The rhodopsin-like pigments of
    halobacteria: light-energy and signal transducers in
    an archaebacterium. *Trends Biochem. Sci.* 10:483–
    486, 1985.

14. Cowan, S.W. Bacterial porins: lessons from three high-
    resolution structures. *Curr. Opin. Struct. Biol.* 3:501–
    507, 1993.

    Schirmer, T.; Rosenbusch, J.P. Prokaryotic and eukary-
    otic porins. *Curr. Opin. Struct. Biol.* 1:539–545, 1991.

    Schulz, G.E. Bacterial porins: structure and function.
    Curr. Opin. Cell Biol. 5:701–707, 1993.

    Weiss, M.S.; Wacker, T.; Nestel, U.; et al. The structure
    of porin from *rhodobacter capsulatus* at 0.6 nm reso-
    lution. *FEBS Lett.* 256:143–146, 1989.

15. Deisenhofer, J.; Epp, O.; Miki, K.; Huber, R.; Michel, H.
    The structure of the protein subunits in the photo-
    synthetic reaction centre of *Rhodopseudomonas
    viridis* at 3 Å resolution. *Nature* 318:618–624, 1985.

    Deisenhofer, J.; Michel, H. High-resolution structures of
    photosynthetic reaction centers. *Annu. Rev. Biophys.
    Biophys. Chem.* 20:247–266, 1991.

    Rees, D.C.; Komiya, H.; Yeates, T.O.; Allen, J.P.; Feher,
    G. The bacterial photosynthetic reaction center as a

    ~~~~~~~~~~~~~~~~~~~~~~~~~~~~~
    lin studied by electron microscopy. *Nature New Biol.*
    241:257–259, 1973.

    Edidin, M. Molecular associations and membrane do-
    mains. *Curr. Top. Membr. Transp.* 36:81–96, 1990.

    Frye, L.D.; Edidin, M. The rapid intermixing of cell sur-
    face antigens after formation of mouse-human het-
    erokaryons. *J. Cell Sci.* 7:319–335, 1970.

    Jacobson, K.; Ishihara, A.; Inman, R. Lateral diffusion of
    proteins in membranes. *Annu. Rev. Physiol.* 49:163–
    175, 1987.

    Sheetz, M.P. Glycoprotein mobility and dynamic do-
    mains in fluid plasma membranes. *Annu. Rev.
    Biophys. Biomol. Struct.* 22:417–431, 1993.

17. Gumbiner, B.; Louvard, D. Localized barriers in the
    plasma membrane: a common way to form do-
    mains. *Trends Biochem. Sci.* 10:435–438, 1985.

    Jacobson, K.; Vaz, W.L.C. Domains in biological mem-
    branes. *Comm. Mol. Cell Biophys.* 8:1–114, 1992.

    Myles, D.G.; Primakoff, P. Sperm surface domains. In
    Hybridoma Technology in the Biosciences and
    Medicine (T.A. Springer, ed.), pp. 239–250. New York:
    Plenum, 1985.

    Simons, K.; van Meer, G. Lipid sorting in epithelial cells.
    Biochemistry 27:6197–6202, 1988.

    Edidin, M. Patches, posts and fences: proteins and
    plasma membrane domains. *Trends Cell Biol.* 2:376–
    380, 1992.

18. Drickamer, K.; Taylor, M.E. Biology of animal lectins.
    Annu. Rev. Cell Biol. 9:237–264, 1993.

    Dwek, R.A., et al. Analysis of glycoprotein-associated
    oligosaccharides. *Annu. Rev. Biochem.* 62:65–100,
    1993.

    Parelch, R.B. Effects of glycosylation on protein func-
    tion. *Curr. Opin. Struct. Biol.* 1:750–754, 1991.

    Paulson, J.C. Glycoproteins: what are the sugar chains
    for? *Trends Biochem. Sci.* 14:272–276, 1989.

    Sharon, N.; Lis, H. Carbohydrates in cell recognition.
    Sci. Am. 268(1):82–89, 1993.

19. Bevilacqua, M.P.; Nelson, R.M. Selectins. *J. Clin. Invest.*
    91:379–387, 1993.

    Cummings, R.D.; Smith, D.F. The selectin family of car-
    bohydrate-binding proteins: structure and impor-
    tance of carbohydrate ligands for cell adhesion.
    BioEssays 14:849–856, 1992.

    Lasky, L.A. Selectins: interpreters of cell-specific carbo-
    hydrate information during inflammation. *Science*
    258:964–969, 1992.

    Lawrence, M.B.; Springer, T.A. Leukocytes roll on a
    selectin at physiologic flow rates: distinction from
    and prerequisite for adhesion through integrins. *Cell*
    65:859–873, 1991.

Membrane Transport of Small Molecules and the Ionic Basis of Membrane Excitability

- **Principles of Membrane Transport**
- **Carrier Proteins and Active Membrane Transport**
- **Ion Channels and Electrical Properties of Membranes**

Because of its hydrophobic interior, the lipid bilayer of cell membranes serves as a barrier to the passage of most polar molecules. This barrier function is crucially important as it allows the cell to maintain concentrations of solutes in its cytosol that are different from those in the extracellular fluid and in each of the intracellular membrane-bounded compartments. To make use of this barrier, however, cells have had to evolve ways of transferring specific water-soluble molecules across their membranes in order to ingest essential nutrients, excrete metabolic waste products, and regulate intracellular ion concentrations. Transport of inorganic ions and small water-soluble organic molecules across the lipid bilayer is achieved by specialized transmembrane proteins, each of which is responsible for the transfer of a specific ion or molecule or a group of closely related ions or molecules. Cells can also transfer macromolecules and even large particles across their membranes, but the mechanisms involved in most of these cases are different from those used for transferring small molecules, and they are discussed in Chapters 12 and 13. The importance of membrane transport is indicated by the fact that almost 20% of the genes identified so far in *E. coli* are associated with such transport processes.

We begin this chapter by considering some general principles that will guide our discussion of how small water-soluble molecules traverse cell membranes. We then consider, in turn, the two main classes of membrane proteins that mediate the transfer: *carrier proteins,* which have moving parts to shift specific molecules across the membrane, and *channel proteins,* which form a narrow hydrophilic pore, allowing the passive movement of small inorganic ions. Carrier proteins can be coupled to a source of energy to catalyze active transport, and a combination of selective passive permeability and active transport creates large differences in the composition of the cytosol compared with either the extracellular fluid (Table 11–1) or the fluid within membrane-bounded organelles. In particular, by generating ionic concentration differences across the lipid bilayer, cell membranes are able to store potential energy in the form of electrochemical gradients, which are used to drive various transport processes, to convey electrical signals in electrically excitable cells, and (in mitochondria, chloroplasts, and bacteria) to make most of the cell's ATP. We focus the discussion mainly on transport across the plasma membrane, but similar mechanisms operate across the other membranes of the eucaryotic cell, as discussed in later chapters. In the last part of this chapter we concentrate mainly on the functions of ion channels in nerve cells, for it is in these cells that channel proteins perform at their highest level of sophistication, enabling networks of nerve cells to carry out all of the astonishing feats that the human brain is capable of.

Table 11–1 Comparison of Ion Concentrations Inside and Outside a Typical Mammalian Cell

Component	Intracellular Concentration (mM)	Extracellular Concentration (mM)
Cations		
Na$^+$	5–15	145
K$^+$	140	5
Mg^{2+}	0.5	1–2
Ca^{2+}	10^{-4}	1–2
H$^+$	7×10^{-5} ($10^{-7.2}$ M or pH 7.2)	4×10^{-5} ($10^{-7.4}$ M or pH 7.4)
Anions*		
Cl$^-$	5–15	110

*The cell must contain equal quantities of + and – charges (that is, be electrically neutral). Thus, in addition to Cl$^-$, the cell contains many other anions not listed in this table; in fact, most cellular constituents are negatively charged (HCO$_3^-$, PO$_4^{3-}$, proteins, nucleic acids, metabolites carrying phosphate and carboxyl groups, etc.). The concentrations of Ca^{2+} and Mg^{2+} given are for the free ions. There is a total of about 20 mM Mg^{2+} and 1–2 mM Ca^{2+} in cells, but this is mostly bound to proteins and other substances and, in the case of Ca^{2+}, stored within various organelles.

Principles of Membrane Transport [1]

We begin this section by describing the permeability properties of protein-free, synthetic lipid bilayers. We then introduce some of the terms used to describe the various forms of membrane transport and some strategies for characterizing the proteins and processes involved.

Protein-free Lipid Bilayers Are Highly Impermeable to Ions [2]

Given enough time, virtually any molecule will diffuse across a protein-free lipid bilayer down its concentration gradient. The rate at which it does so, however, varies enormously, depending partly on the size of the molecule and mostly on its relative solubility in oil. In general, the smaller the molecule and the more soluble it is in oil (that is, the more hydrophobic, or nonpolar, it is), the more rapidly it will diffuse across a bilayer. *Small nonpolar* molecules, such as O$_2$ (32 daltons) and CO$_2$ (44 daltons), readily dissolve in lipid bilayers and therefore rapidly diffuse across them. *Uncharged polar* molecules also diffuse rapidly across a bilayer if they are small enough. Water (18 daltons), ethanol (46 daltons), and urea (60 daltons), for example, cross rapidly; glycerol (92 daltons) diffuses less rapidly; and glucose (180 daltons), hardly at all (Figure 11–1).

By contrast, lipid bilayers are highly impermeable to *charged* molecules (ions), no matter how small: the charge and high degree of hydration of such molecules prevents them from entering the hydrocarbon phase of the bilayer. Thus synthetic bilayers are 10^9 times more permeable to water than to even such small ions as Na$^+$ or K$^+$ (Figure 11–2).

There Are Two Main Classes of Membrane Transport Proteins—Carriers and Channels [1]

Like synthetic lipid bilayers, cell membranes allow water and nonpolar molecules to permeate by simple diffusion. Cell membranes, however, also have to allow passage of various polar molecules, such as ions, sugars, amino acids, nucle-

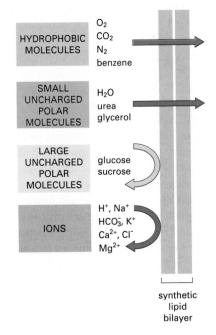

Figure 11–1 The relative permeability of a synthetic lipid bilayer to different classes of molecules. The smaller the molecule and, more important, the fewer hydrogen bonds it makes with water, the more rapidly the molecule diffuses across the bilayer.

otides, and many cell metabolites that pass across synthetic lipid bilayers only very slowly. Special membrane proteins are responsible for transferring such solutes across cell membranes. These proteins, referred to as **membrane transport proteins,** occur in many forms and in all types of biological membranes. Each protein transports a particular class of molecule (such as ions, sugars, or amino acids) and often only certain molecular species of the class. The specificity of transport proteins was first indicated in the mid-1950s by studies in which single gene mutations were found to abolish the ability of bacteria to transport specific sugars across their plasma membrane. Similar mutations have now been discovered in humans suffering from a variety of inherited diseases that affect the transport of a specific solute in the kidney or intestine or both. Individuals with the inherited disease *cystinuria,* for example, are unable to transport certain amino acids (including cystine, the disulfide-linked dimer of cysteine) from either the urine or the intestine into the blood; the resulting accumulation of cystine in the urine leads to the formation of cystine stones in the kidneys.

All membrane transport proteins that have been studied in detail have been found to be multipass transmembrane proteins—that is, their polypeptide chains traverse the lipid bilayer multiple times. By forming a continuous protein pathway across the membrane, these proteins are thought to enable the specific hydrophilic solutes to cross the membrane without coming into direct contact with the hydrophobic interior of the lipid bilayer.

There are two major classes of membrane transport proteins: *carrier proteins* and *channel proteins*. **Carrier proteins** (also called *carriers, permeases,* or *transporters*) bind the specific solute to be transported and undergo a series of conformational changes in order to transfer the bound solute across the membrane. **Channel proteins,** on the other hand, need not bind the solute. Instead, they form hydrophilic pores that extend across the lipid bilayer; when these pores are open, they allow specific solutes (usually inorganic ions of appropriate size and charge) to pass through them and thereby cross the membrane (Figure 11–3). Not surprisingly, transport through channel proteins occurs at a very much faster rate than transport mediated by carrier proteins.

Active Transport Is Mediated by Carrier Proteins Coupled to an Energy Source [1, 3]

All channel proteins and many carrier proteins allow solutes to cross the membrane only passively ("downhill")—a process called **passive transport** (or **facilitated diffusion**). If the transported molecule is uncharged, it is simply the difference in its concentration on the two sides of the membrane (its *concentration gradient*) that drives passive transport and determines its direction. If the solute carries a net charge, however, both its concentration gradient and the electrical potential difference across the membrane (the *membrane potential*) influence its transport. The concentration gradient and the electrical gradient can be combined to calculate a net driving force, or **electrochemical gradient,** for each charged solute. We discuss this in more detail in Chapter 14. In fact, almost all plasma membranes have an electrical potential difference (voltage gradient) across them, with the inside usually negative with respect to the outside. This potential difference favors the entry of positively charged ions into the cell but opposes the entry of negatively charged ions.

Cells also require transport proteins that will actively pump certain solutes across the membrane against their electrochemical gradient ("uphill"); this process, known as **active transport,** is always mediated by carrier proteins. In active transport the pumping activity of the carrier protein is directional because it is tightly coupled to a source of metabolic energy, such as ATP hydrolysis or an ion gradient, as discussed later. Thus transport by carrier proteins can be either active or passive, whereas transport by channel proteins is always passive (Figure 11–4).

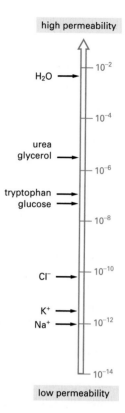

Figure 11–2 Permeability coefficients (cm/sec) for the passage of various molecules through synthetic lipid bilayers. The rate of flow of a solute across the bilayer is directly proportional to the difference in its concentration on the two sides of the membrane. Multiplying this concentration difference (in mol/cm^3) by the permeability coefficient (cm/sec) gives the flow of solute in moles per second per square centimeter of membrane. A concentration difference of tryptophan of 10^{-4} mol/cm^3 ($10^{-4}/10^{-3}$ L = 0.1 M), for example, would cause a flow of 10^{-4} mol/cm$^3 \times 10^{-7}$ cm/sec = 10^{-11} mol/sec through 1 cm^2 of membrane, or 6×10^4 molecules/sec through 1 μm^2 of membrane.

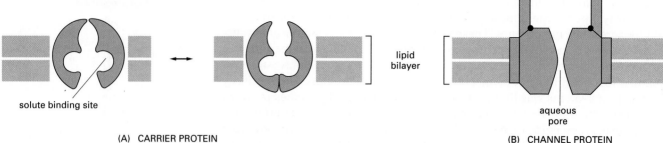

(A) CARRIER PROTEIN

(B) CHANNEL PROTEIN

Recombinant DNA Technology Has Revolutionized the Study of Membrane Transport Proteins [4]

Because of the difficulty in obtaining three-dimensional crystals of multipass membrane proteins for x-ray crystallographic studies, the detailed three-dimensional structures of the membrane transport proteins we shall discuss are not known. We therefore do not understand the molecular mechanisms they use to transport specific solutes across the lipid bilayer. But important insights have been gained by other methods, especially by the use of recombinant DNA technology.

Once the DNA encoding a transport protein has been cloned and sequenced, the amino acid sequence of the polypeptide can be deduced and the number of transmembrane α helices can be estimated by analysis of a hydropathy plot (discussed in Chapter 10). Antibodies made against synthetic peptides corresponding to specific segments of the polypeptide chain can be used to help determine which segments are exposed on one or the other side of the membrane. The DNA sequence encoding specific parts of the protein can be altered by site-specific mutagenesis, and the corresponding mutant mRNA can be injected into cultured mammalian cells or *Xenopus* oocytes, where it will direct the synthesis of a mutant protein whose transport function can be readily assessed. In this way functionally important amino acid residues and protein segments can be identified. One must interpret the results of such studies with caution, however, as a small change in one part of a protein can sometimes have large effects on the protein's overall folded conformation and, therefore, on its function, even when the altered part is not directly involved in that function. Nonetheless, this strategy is proving to be invaluable, as we shall see.

Recombinant DNA technology has also contributed to our understanding of membrane transport proteins in a second way. Once the DNA encoding a protein has been isolated, it is often a relatively simple matter to use this DNA as a probe to isolate related DNA sequences encoding homologous proteins. Such studies have revealed that membrane transport is mediated by a surprisingly small number of protein families, whose members have related structures and,

Figure 11–3 A schematic view of the two classes of membrane transport proteins. A *carrier protein* is thought to alternate between two conformations, so that the solute binding site is sequentially accessible on one side of the bilayer and then on the other. In contrast, a *channel protein* is thought to form a water-filled pore across the bilayer through which specific ions can diffuse.

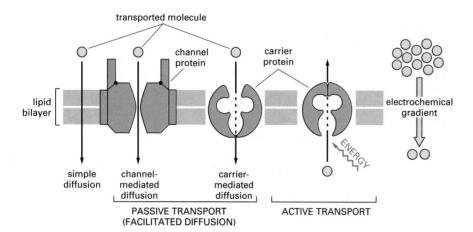

Figure 11–4 Comparison of passive transport down an electrochemical gradient with active transport against an electrochemical gradient. Whereas simple diffusion and passive transport by membrane transport proteins (facilitated diffusion) occur spontaneously, active transport requires an input of metabolic energy. Only carrier proteins can carry out active transport, but both carrier proteins and channel proteins can mediate facilitated diffusion.

presumably, related mechanisms of action and a common evolutionary origin. A family, however, can contain a very large number of different proteins, and, even for a specific family member, there are often many variants (called *isoforms*) produced either from different genes or from differently processed RNA transcripts from a single gene. In some cases the isoforms differ in their transport activity, time of expression in development, tissue distribution, location in the cell, or any combination of these properties. In other cases the significance of the heterogeneity is unclear.

Before discussing in detail the different classes of membrane transport proteins and the insights gained by these techniques, we pause briefly to consider another class of molecules that can selectively increase the permeability of lipid bilayers. These molecules provide a simple illustration of some of the principles discussed earlier.

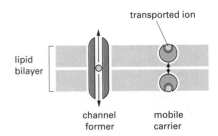

Figure 11–5 A mobile ion carrier and a channel-forming ionophore. In both cases net ion flow occurs only down an electrochemical gradient.

Ionophores Can Be Used as Tools to Increase the Permeability of Membranes to Specific Ions [5]

Ionophores are small hydrophobic molecules that dissolve in lipid bilayers and increase their permeability to specific inorganic ions. Most are synthesized by microorganisms (presumably as biological weapons against competitors or prey). They are widely used by cell biologists as tools to increase the ion permeability of membranes in studies on synthetic bilayers, cells, or cell organelles. There are two classes of ionophores—**mobile ion carriers** and **channel formers** (Figure 11–5). Both types operate by shielding the charge of the transported ion so that it can penetrate the hydrophobic interior of the lipid bilayer. Since ionophores are not coupled to energy sources, they permit net movement of ions only down their electrochemical gradients.

Valinomycin is an example of a mobile ion carrier. It is a ring-shaped polymer that transports K^+ down its electrochemical gradient by picking up K^+ on one side of the membrane, diffusing across the bilayer, and releasing K^+ on the other side. The ionophore *A23187* is another example of a mobile ion carrier, but it transports divalent cations such as Ca^{2+} and Mg^{2+}. It normally acts as an ion-exchange shuttle, carrying two H^+ out of the cell for every divalent cation it carries in. When cells are exposed to A23187, Ca^{2+} enters the cytosol from the extracellular fluid down a steep electrochemical gradient. Accordingly, this ionophore is widely used in cell biology to increase the concentration of free Ca^{2+} in the cytosol, thereby mimicking certain cell-signaling mechanisms (discussed in Chapter 15).

Gramicidin A is an example of a channel-forming ionophore. As a linear peptide of only 15 amino acid residues, all with hydrophobic side chains, it is the simplest and best characterized ion channel. Two gramicidin molecules are thought to come together end to end across the bilayer to form a transmembrane

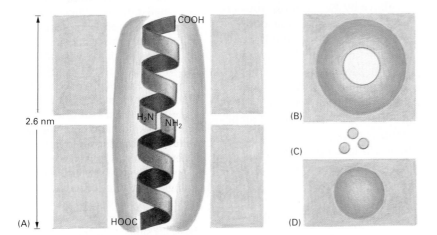

Figure 11–6 The structure of a gramicidin channel. The channel is formed by the association of two identical peptides at their amino-terminal ends. Each chain is folded into a β helix, which resembles a rolled-up β pleated sheet. (A) is a side view and (B) a top view. The peptide backbones that line the channel are shown in *blue* and *dark green*, while the *light green* represents the protruding hydrophobic side chains. The lipid bilayer is shown in *gray*. (C) shows the size of unhydrated K^+ ions, while (D) shows a membrane-spanning α helix in top view for comparison with the β helix. Note that the α helix does not contain a pore and, therefore, cannot, on its own, form a channel. (After S. Weinstein, B.A. Wallace, E.R. Blout, J.S. Morrow, and W. Veatch, *Proc. Natl. Acad. Sci.* 76:4230, 1979.)

channel (Figure 11–6), which selectively allows monovalent cations to flow down their electrochemical gradients. These dimers are unstable and are constantly forming and dissociating, so that the average open time for a channel is about 1 second. With a large electrochemical gradient, gramicidin A can transport about 20,000 cations per open channel each millisecond, which is 1000 times more ions than can be transported by a single mobile carrier molecule in the same time. Gramicidin is made by certain bacteria, perhaps to kill other microorganisms by collapsing the H^+, Na^+, and K^+ gradients that are essential for cell survival, and it has been useful as an antibiotic.

Summary

Lipid bilayers are highly impermeable to most polar molecules. In order to transport small water-soluble molecules into or out of cells or intracellular membrane-bounded compartments, cell membranes contain various transport proteins, each of which is responsible for transferring a particular solute or class of solutes across the membrane. There are two classes of membrane transport proteins—carriers and channels; both form continuous protein pathways across the lipid bilayer. Whereas transport by carrier proteins can be either active or passive, transport by channel proteins is always passive. Ionophores, which are small hydrophobic molecules made by microorganisms, can be used as tools to increase the permeability of cell membranes to specific inorganic ions in studies on cells or organelles.

Carrier Proteins and Active Membrane Transport [1,6]

The process by which a carrier protein transfers a solute molecule across the lipid bilayer resembles an enzyme-substrate reaction, and the carriers involved behave like specialized membrane-bound enzymes. Each type of carrier protein has one or more specific binding sites for its solute (substrate). When the carrier is saturated (that is, when all these binding sites are occupied), the rate of transport is maximal. This rate, referred to as V_{max}, is characteristic of the specific carrier. In addition, each carrier protein has a characteristic binding constant for its solute, K_M, equal to the concentration of solute when the transport rate is half its maximum value (Figure 11–7). As with enzymes, the binding of solute can be blocked specifically by competitive inhibitors (which compete for the same binding site and may or may not be transported by the carrier) or by noncompetitive inhibitors (which bind elsewhere and specifically alter the structure of the carrier). In contrast with ordinary enzyme-substrate reactions, however, the transported solute is usually not covalently modified by the carrier protein.

Some carrier proteins simply transport a single solute from one side of the membrane to the other at a rate determined as above by V_{max} and K_M; they are called **uniporters.** Others, with more complex kinetics, function as **coupled transporters,** in which the transfer of one solute depends on the simultaneous or sequential transfer of a second solute, either in the same direction (**symport**) or in the opposite direction (**antiport**) (Figure 11–8). Most animal cells, for example, take up glucose from the extracellular fluid, where its concentration is high relative to that in the cytosol, by *passive* transport through glucose carriers that operate as uniporters. There are a variety of these glucose carriers, all belonging to the same family of homologous proteins with 12 putative transmembrane α helices. By contrast, intestinal and kidney cells take up glucose from the lumen of the intestine and kidney tubules, respectively, where the concentration of the sugar is low. These cells actively transport glucose across their plasma membrane by symport with Na^+. As discussed in Chapter 10, the band 3 protein of the human red blood cell is an anion carrier that operates as an antiporter to exchange Cl^- for HCO_3^-.

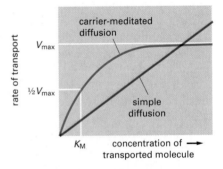

Figure 11–7 **Kinetics of simple diffusion compared to carrier-mediated diffusion.** Whereas the rate of the former is always proportional to the solute concentration, the rate of the latter reaches a maximum (V_{max}) when the carrier protein is saturated. The solute concentration when transport is at half its maximal value approximates the binding constant (K_M) of the carrier for the solute and is analogous to the K_M of an enzyme for its substrate. The graph applies to a carrier transporting a single solute; the kinetics of coupled transport of two or more solutes (see text) are more complex but show basically similar phenomena.

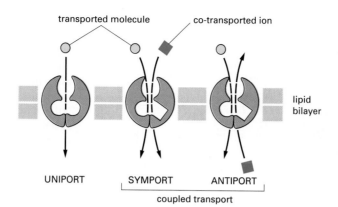

Figure 11–8 **Three types of carrier-mediated transport.** The schematic diagram shows carrier proteins functioning as uniports, symports, and antiports.

Although the molecular details are unknown, carrier proteins are thought to transfer the solute across the lipid bilayer by undergoing reversible conformational changes that alternately expose the solute binding site first on one side of the membrane and then on the other. A schematic model of how such a carrier protein might operate is shown in Figure 11–9. Because carriers are now known to be multipass transmembrane proteins, it is highly unlikely that they ever tumble in the membrane or shuttle back and forth across the lipid bilayer as was once believed.

As we discuss below, it requires only a relatively minor modification of the model shown in Figure 11–9 to link the carrier protein to a source of energy (such as ATP hydrolysis [see Figure 11–11] or an ion gradient) in order to pump a solute uphill against its electrochemical gradient. In fact, comparison of some bacterial carrier proteins with mammalian ones supports the idea that there need be little difference in molecular design between carrier proteins that mediate active transport and those that operate passively. Some carriers that in bacteria use the energy stored in the H^+ gradient across the bacterial plasma membrane to drive the active uptake of various sugars are structurally similar to the passive glucose carriers of animal cells. This suggests an evolutionary relationship between these carrier proteins; and given the importance of sugars as an energy source, it would not be surprising if this superfamily of sugar carriers were an ancient one.

We begin our discussion of active transport by considering a carrier protein that plays a crucial part in generating and maintaining the Na^+ and K^+ gradients across the plasma membrane of animal cells.

The Plasma Membrane Na⁺-K⁺ Pump Is an ATPase [7]

The concentration of K^+ is typically 10 to 20 times higher inside cells than outside, whereas the reverse is true of Na^+ (see Table 11–1, p. 508). These concentration differences are maintained by a **Na⁺-K⁺ pump** that is found in the plasma membrane of virtually all animal cells. The pump operates as an antiporter, actively pumping Na^+ out of the cell against its steep electrochemical gradient and

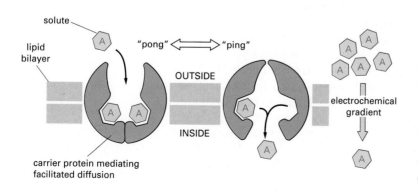

Figure 11–9 **A hypothetical model showing how a conformational change in a carrier protein could mediate the facilitated diffusion of a solute.** The carrier protein shown can exist in two conformational states: in state "pong" the binding sites for solute A are exposed on the outside of the bilayer; in state "ping" the same sites are exposed on the other side of the bilayer. The transition between the two states is proposed to occur randomly and to be completely reversible. Therefore, if the concentration of A is higher on the outside of the bilayer, more A will bind to the carrier protein in the pong conformation than in the ping conformation, and there will be a net transport of A down its electrochemical gradient.

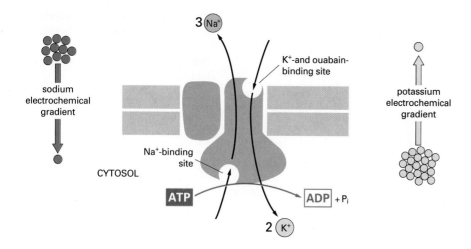

Figure 11–10 **The Na⁺-K⁺ ATPase.**
This carrier protein actively pumps
Na⁺ out of and K⁺ into a cell against
their electrochemical gradients. For
every molecule of ATP hydrolyzed
inside the cell, three Na⁺ are pumped
out and two K⁺ are pumped in. The
specific pump inhibitor ouabain and
K⁺ compete for the same site on the
external side of the ATPase.

pumping K⁺ in. As explained below, the Na⁺ gradient produced by the pump regulates cell volume through its osmotic effects and is also exploited to drive transport of sugars and amino acids into the cell. Almost one-third of the energy requirement of a typical animal cell is consumed in fueling this pump; in electrically active nerve cells, which, as we shall see, are repeatedly gaining small amounts of Na⁺ and losing small amounts of K⁺ during the propagation of nerve impulses, this figure approaches two-thirds of the cell's energy requirement.

A major advance in understanding the Na⁺-K⁺ pump came with the discovery in 1957 of an enzyme that hydrolyzes ATP to ADP and phosphate and requires Na⁺ and K⁺ for maximal activity. An important clue linking this **Na⁺-K⁺ ATPase** with the Na⁺-K⁺ pump was the observation that a known inhibitor of the pump, *ouabain*, also inhibits the ATPase. But the crucial evidence that ATP hydrolysis provides the energy for driving the pump came from studies of resealed red blood cell ghosts, in which the concentrations of ions, ATP, and drugs on either side of the membrane could be varied and the effects on ion transport and ATP hydrolysis observed. It was found that (1) the transport of Na⁺ and K⁺ is tightly coupled to ATP hydrolysis, so that one cannot occur without the other; (2) ion transport and ATP hydrolysis can occur only when Na⁺ and ATP are present inside the ghosts and K⁺ is present on the outside; (3) ouabain is inhibitory only when present outside the ghosts, where it competes for the K⁺-binding site; and (4) for every molecule of ATP hydrolyzed (100 ATP molecules can be hydrolyzed by each ATPase molecule each second), three Na⁺ ions are pumped out and two K⁺ ions are pumped in (Figure 11–10).

Although these experiments provided compelling evidence that ATP supplies the energy for pumping Na⁺ and K⁺ ions across the plasma membrane, they did not explain how ATP hydrolysis is coupled to ion transport. A partial explanation was provided by the finding that, during the pumping cycle, the terminal phosphate group of the ATP is transferred to an aspartic acid residue of the ATPase and is subsequently removed, as explained in Figure 11–11.

The Na⁺-K⁺ pump in red blood cell ghosts can be driven in reverse to produce ATP: when the Na⁺ and K⁺ gradients are experimentally increased to such an extent that the energy stored in their electrochemical gradients is greater than the chemical energy of ATP hydrolysis, these ions move down their electrochemical gradients and ATP is synthesized from ADP and phosphate by the Na⁺-K⁺ ATPase. Thus the phosphorylated form of the ATPase (step 2 in Figure 11–11) can relax either by donating its phosphate to ADP (step 2 to step 1) or by changing its conformation (step 2 to step 3). Whether the overall change in free energy is used to synthesize ATP or to pump Na⁺ out of the ghost depends on the relative concentrations of ATP, ADP, and phosphate and on the electrochemical gradients for Na⁺ and K⁺.

The Na⁺-K⁺ ATPase has been purified and found to consist of a large, multipass, transmembrane catalytic subunit (about 1000 amino acids long) and

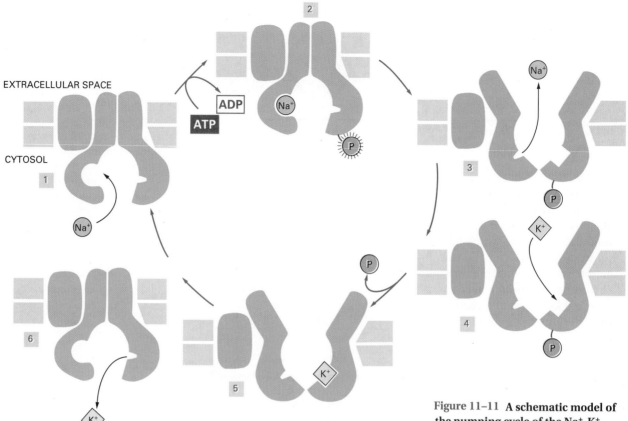

EXTRACELLULAR SPACE

CYTOSOL

ADP

ATP

Figure 11–11 A schematic model of the pumping cycle of the Na⁺-K⁺ ATPase. The binding of Na⁺ (1) and the subsequent phosphorylation by ATP of the cytoplasmic face of the ATPase (2) induce the protein to undergo a conformational change that transfers the Na⁺ across the membrane and releases it on the outside (3). Then the binding of K⁺ on the extracellular surface (4) and the subsequent dephosphorylation (5) return the protein to its original conformation, which transfers the K⁺ across the membrane and releases it into the cytosol (6). These changes in conformation are analogous to the ping ⇌ pong transitions shown in Figure 11–9 except that here the Na⁺-dependent phosphorylation and the K⁺-dependent dephosphorylation of the protein cause the conformational transitions to occur in an orderly manner, enabling the protein to do useful work. Although for simplicity only one Na⁺- and one K⁺-binding site are shown, in the real pump there are thought to be three Na⁺- and two K⁺-binding sites. Moreover, although the ATPase is shown as alternating between two conformational states, there is evidence that it goes through a more complex series of conformational changes during the actual pumping cycle.

an associated smaller, single-pass glycoprotein. The former has binding sites for Na⁺ and ATP on its cytoplasmic surface and a binding site for K⁺ on its external surface, and is reversibly phosphorylated and dephosphorylated during the pumping cycle. The function of the glycoprotein is uncertain, except that it is required for the intracellular transport of the catalytic subunit to the plasma membrane. A functional Na⁺-K⁺ pump can be reconstituted from the purified complex: the ATPase is solubilized in detergent, purified, and mixed with appropriate phospholipids. When the detergent is removed, membrane vesicles are formed that pump Na⁺ and K⁺ in opposite directions in the presence of ATP (see Figure 10–22).

The Na⁺-K⁺ ATPase Is Required to Maintain Osmotic Balance and Stabilize Cell Volume [8]

Since the Na⁺-K⁺ ATPase drives three positively charged ions out of the cell for every two it pumps in, it is *electrogenic;* that is, it drives a net current across the membrane, tending to create an electrical potential, with the inside negative relative to the outside. This effect of the pump, however, seldom contributes more than 10% to the membrane potential. The remaining 90%, as we shall see later, depends on the pump only indirectly.

On the other hand, the Na⁺-K⁺ ATPase does have a direct role in regulating cell volume: it controls the solute concentration inside the cell, thereby regulating the osmotic forces that can make a cell swell or shrink (Figure 11–12). As explained in Panel 11–1, cells contain a high concentration of solutes, including numerous negatively charged organic molecules that are confined inside the cell (the so-called *fixed anions*) and their accompanying cations that are required for charge balance, and this creates a large osmotic gradient that tends to "pull" water into the cell. For animal cells this effect is counteracted by an opposite osmotic gradient due to a high concentration of inorganic ions—chiefly Na⁺ and Cl⁻—in the extracellular fluid. The Na⁺-K⁺ ATPase maintains osmotic balance by

Carrier Proteins and Active Membrane Transport

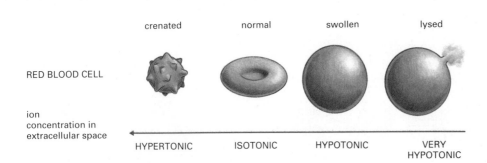

crenated   normal   swollen   lysed

RED BLOOD CELL

ion concentration in extracellular space

HYPERTONIC   ISOTONIC   HYPOTONIC   VERY HYPOTONIC

Figure 11–12 Response of a human red blood cell to changes in osmolarity (also called *tonicity*) of the extracellular fluid. Because the plasma membrane is freely permeable to water, water will move into or out of cells down its concentration gradient, a process called *osmosis*. If cells are placed in a *hypotonic solution* (i.e., a solution having a low solute concentration and therefore a high water concentration), there will be a net movement of water into the cells, causing them to swell and burst (lyse). Conversely, if cells are placed in a *hypertonic solution*, they will shrink.

pumping out the Na^+ that leaks in down its steep electrochemical gradient; the Cl^- is kept out by the membrane potential.

The importance of the Na^+-K^+ ATPase in controlling cell volume is indicated by the observation that many animal cells swell, and sometimes burst, if they are treated with ouabain, which inhibits the Na^+-K^+ ATPase. There are, of course, other ways for a cell to cope with its osmotic problems. Plant cells and many bacteria are prevented from bursting by the semirigid cell wall that surrounds their plasma membrane; in amoebae the excess water that flows in osmotically is collected in contractile vacuoles, which periodically discharge their contents to the exterior (see Panel 11–1). But for most animal cells, the Na^+-K^+ ATPase is crucial.

Some Ca^{2+} Pumps Are Also Membrane-bound ATPases [9]

Eucaryotic cells maintain very low concentrations of free Ca^{2+} in their cytosol (~10^{-7} M) in the face of very much higher extracellular Ca^{2+} concentrations (~10^{-3} M). Even a small influx of Ca^{2+} significantly increases the concentration of free Ca^{2+} in the cytosol, and the flow of Ca^{2+} down its steep concentration gradient in response to extracellular signals is one means of transmitting these signals rapidly across the plasma membrane. The maintenance of a steep Ca^{2+} gradient is therefore important to the cell. The Ca^{2+} gradient is in part maintained by Ca^{2+} pumps in the plasma membrane that actively transport Ca^{2+} out of the cell. One of these is an ATPase, while the other is an antiporter that is driven by the Na^+ electrochemical gradient.

The best-understood Ca^{2+} pump is a membrane-bound ATPase in the *sarcoplasmic reticulum* of muscle cells. The sarcoplasmic reticulum—a specialized type of endoplasmic reticulum—forms a network of tubular sacs in the cytoplasm of muscle cells and serves as an intracellular store of Ca^{2+}. (When an action potential depolarizes the muscle cell membrane, Ca^{2+} is released from the sarcoplasmic reticulum into the cytosol, stimulating the muscle to contract, as discussed in Chapter 16.) The Ca^{2+} pump, which accounts for about 90% of the membrane protein of the organelle, is responsible for pumping Ca^{2+} from the cytosol into the sarcoplasmic reticulum. (The endoplasmic reticulum of nonmuscle cells contains a similar Ca^{2+} ATPase, but in smaller quantities, so that it is harder to purify.)

The Ca^{2+} ATPase can be analyzed biochemically by the same methods as the Na^+-K^+ ATPase and is found to function in a closely similar way. DNA sequencing studies show, in fact, that the Na^+-K^+ ATPase and Ca^{2+} ATPases are homologous proteins. In each case the large catalytic subunit exists in multiple isoforms, is thought to have about 10 putative membrane-spanning α helices, and is phosphorylated and dephosphorylated during the pumping cycle.

Membrane-bound Enzymes That Synthesize ATP Are Transport ATPases Working in Reverse [10]

The plasma membrane of bacteria, the inner membrane of mitochondria, and the thylakoid membrane of chloroplasts all contain an enzyme that is analogous

SOURCES OF INTRACELLULAR OSMOLARITY

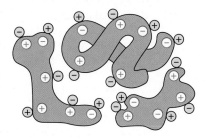

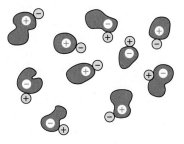

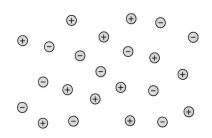

Macromolecules themselves contribute very little to the osmolarity of the cell interior since, despite their large size, each one counts only as a single molecule and there are relatively few of them compared to the number of small molecules in the cell. However, most biological macromolecules are highly charged, and they attract many inorganic ions of opposite charge. Because of their large numbers, these counterions make a major contribution to intracellular osmolarity.

As the result of active transport and metabolic processes, the cell contains a high concentration of small organic molecules, such as sugars, amino acids, and nucleotides, to which its plasma membrane is impermeable. Because most of these metabolites are charged, they also attract counterions. Both the small metabolites and their counterions make a further major contribution to intracellular osmolarity.

The osmolarity of the extracellular fluid is usually due mainly to small inorganic ions. These leak slowly across the plasma membrane into the cell. If they were not pumped out, and if there were no other molecules inside the cell that interacted with them so as to influence their distribution, they would eventually come to equilibrium with equal concentrations inside and outside the cell. However, the presence of charged macromolecules and metabolites in the cell that attract these ions gives rise to the Donnan effect: it causes the total concentration of inorganic ions (and therefore their contribution to the osmolarity) to be greater inside than outside the cell at equilibrium.

THE PROBLEM

Because of the above factors, a cell that does nothing to control its osmolarity will have a higher concentration of solutes inside than outside. As a result, water will be higher in concentration outside the cell than inside. This difference in water concentration across the plasma membrane will cause water to move continuously into the cell by osmosis, causing it to rupture.

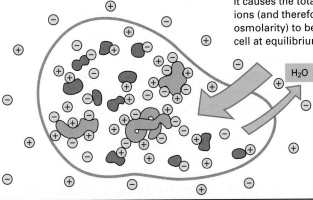

THE SOLUTION

Animal cells and bacteria control their intracellular osmolarity by actively pumping out inorganic ions, such as Na^+, so that their cytoplasm contains a lower total concentration of inorganic ions than the extracellular fluid, thereby compensating for their excess of organic solutes.

Plant cells are prevented from swelling by their rigid walls and so can tolerate an osmotic difference across their plasma membranes: an internal turgor pressure is built up, which at equilibrium forces out as much water as enters.

Many protozoa avoid becoming swollen with water, despite an osmotic difference across the plasma membrane, by periodically extruding water from special contractile vacuoles.

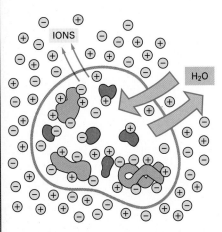

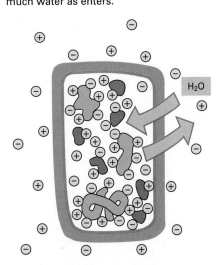

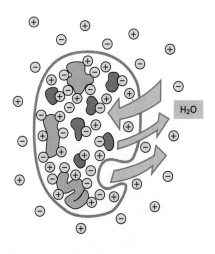

Panel 11–1 Intracellular water balance: the problem and its solution.

to the transport ATPases discussed above, but it normally works in reverse. Instead of ATP hydrolysis driving ion transport, H^+ gradients across these membranes drive the synthesis of ATP from ADP and phosphate. The H^+ gradients are generated during the electron-transport steps of oxidative phosphorylation (in aerobic bacteria and mitochondria) or photosynthesis (in chloroplasts) or by the light-activated H^+ pump (bacteriorhodopsin) in *Halobacterium*. The enzyme that normally synthesizes ATP, called *ATP synthase*, can, like the transport ATPases, work in either direction, depending on the conditions: it can hydrolyze ATP and pump H^+ across the membrane, or it can synthesize ATP when H^+ flows through the enzyme in the reverse direction. ATP synthase is responsible for producing nearly all of the ATP in most cells and is discussed in detail in Chapter 14.

Active Transport Can Be Driven by Ion Gradients [11]

Many active transport systems are driven by the energy stored in ion gradients rather than by ATP hydrolysis. The free energy released during the movement of an inorganic ion down an electrochemical gradient is used as the driving force to pump other solutes uphill, against their electrochemical gradient. Thus all of these proteins function as coupled transporters—some as symporters, others as antiporters. In the plasma membrane of animal cells Na^+ is the usual co-transported ion whose electrochemical gradient provides the driving force for the active transport of a second molecule. The Na^+ that enters the cell during transport is subsequently pumped out by the Na^+-K^+ ATPase, which, by maintaining the Na^+ gradient, indirectly drives the transport. (For this reason ion-driven carriers are said to mediate *secondary active transport*, whereas transport ATPases are said to mediate *primary active transport*.) Intestinal and kidney epithelial cells, for instance, contain a variety of symport systems that are driven by the Na^+ gradient across the plasma membrane; each system is specific for importing a small group of related sugars or amino acids into the cell. In these systems the solute and Na^+ bind to different sites on a carrier protein; because the Na^+ tends to move into the cell down its electrochemical gradient, the sugar or amino acid is, in a sense, "dragged" into the cell with it. The greater the electrochemical gradient for Na^+, the greater the rate of solute entry; conversely, if the Na^+ concentration in the extracellular fluid is reduced, solute transport decreases.

In bacteria and yeasts, as well as in many membrane-bounded organelles of animal cells, most active transport systems driven by ion gradients depend on H^+ rather than Na^+ gradients, reflecting the predominance of H^+ ATPases and the virtual absence of Na^+-K^+ ATPases in these membranes. The active transport of many sugars and amino acids into bacterial cells, for example, is driven by the electrochemical H^+ gradient across the plasma membrane.

Na^+-driven Carrier Proteins in the Plasma Membrane Regulate Cytosolic pH [12]

The structure and function of most macromolecules are greatly influenced by pH, and most proteins operate optimally at a particular pH. Lysosomal enzymes, for example, function best at the low pH (~5) found in lysosomes, whereas cytosolic enzymes function best at the close to neutral pH (~7.2) found in the cytosol. It is crucial, therefore, that cells be able to control the pH of their intracellular compartments.

Most cells have one or more types of Na^+-driven antiporters in their plasma membrane that regulate intracellular (cytosolic) pH (pH_i), keeping it at about 7.2. These proteins use the energy stored in the Na^+ gradient to reduce acidity by getting rid of excess H^+, which either leaks in or is produced in the cell by acid-forming reactions. Two mechanisms are used: either H^+ is directly transported out of the cell or HCO_3^- is brought into the cell to neutralize H^+ in the cytosol. One of these antiporters, which uses the first mechanism, is a *Na^+-H^+ exchanger*, which couples an influx of Na^+ to an efflux of H^+. Another, which uses a combination of the two mechanisms, is a *Na^+-driven Cl^--HCO_3^- exchanger* that couples an influx of Na^+ and HCO_3^- to an efflux of Cl^- and H^+ (so that $NaHCO_3$ goes in and HCl comes out). The Na^+-driven Cl^--HCO_3^- exchanger is twice as effective as the Na^+-H^+ exchanger, in the sense that it pumps out one H^+ and neutralizes another for each Na^+ that enters the cell. If HCO_3^- is available, as is usually the case, this antiporter is the most important carrier protein regulating pH_i. Both exchangers are regulated by pH_i and increase their activity as pH_i falls.

In some cells a third Na^+-dependent transporter plays a part in pH_i regulation. This *Na^+-HCO_3^- symporter* transports one Na^+ into the cell together with two or more HCO_3^- ions. In contrast with the other two transporters just described, this symporter, therefore, is *electrogenic* and has the net effect of carrying a negative charge into the cell. The lower the voltage inside the cell, the harder it is for the symporter to operate. Consequently, the pH_i in cells with this symporter—notably glial cells in the nervous system—is sensitive to changes in the membrane potential. This sensitivity is thought to allow these cells to help regulate extracellular pH locally in the brain in response to changes in electrical activity.

A Na^+-independent *Cl^--HCO_3^- exchanger*, similar to the band 3 protein in the membrane of red blood cells discussed in Chapter 10, also plays an important part in pH_i regulation in some nucleated cells. Like the Na^+-dependent transporters, the Cl^--HCO_3^- exchanger is regulated by pH_i, but the movement of HCO_3^- in this case is normally *out* of the cell, down its electrochemical gradient. The rate of HCO_3^- efflux and Cl^- influx increases as pH_i rises, thereby decreasing pH_i whenever the cytosol becomes too alkaline.

As discussed in Chapter 13, the low pH in lysosomes, as well as in endosomes and secretory vesicles, is maintained by *H^+ ATPases*, which use the energy of ATP hydrolysis to pump H^+ into these organelles from the cytosol.

An Asymmetrical Distribution of Carrier Proteins in Epithelial Cells Underlies the Transcellular Transport of Solutes [13]

In epithelial cells, such as those involved in absorbing nutrients from the gut, carrier proteins are distributed asymmetrically in the plasma membrane and thereby contribute to the **transcellular transport** of absorbed solutes. As shown in Figure 11–13, Na^+-linked symporters located in the apical (absorptive) domain of the plasma membrane actively transport nutrients into the cell, building up substantial concentration gradients, while Na^+-independent transport proteins in the basal and lateral (basolateral) domain allow nutrients to leave the cell passively down these concentration gradients. The Na^+-K^+ ATPase that maintains the Na^+ gradient across the plasma membrane of these cells is located in the basolateral domain. Related mechanisms are thought to be used by kidney and intestinal epithelial cells to pump water from one extracellular space to another.

In many of these epithelial cells the plasma membrane area is greatly increased by the formation of thousands of **microvilli,** which extend as thin, fingerlike projections from the apical surface of each cell (see Figure 11–13). Such microvilli can increase the total absorptive area of a cell by as much as 25-fold, thereby greatly increasing its transport capabilities.

Some Bacterial Transport ATPases Are Homologous to Eucaryotic Transport ATPases Involved in Drug Resistance and Cystic Fibrosis: The ABC Transporter Superfamily [14]

The last type of carrier protein that we discuss is a family of transport ATPases that are of great clinical importance, even though their normal functions in eucaryotic cells are only just beginning to be discovered. The first of these proteins to be characterized were found in bacteria. We have already mentioned that the

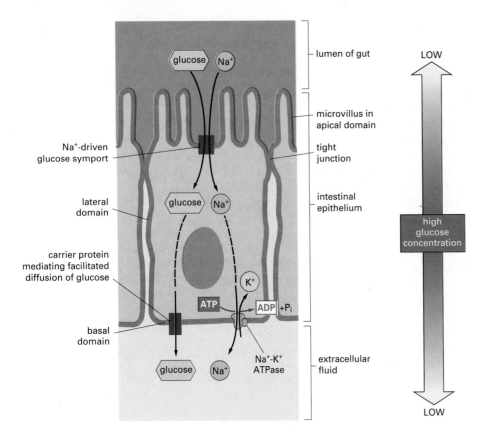

Figure 11–13 **The transcellular transport of glucose across an intestinal epithelial cell depends on the asymmetrical distribution of transport proteins in the cell's plasma membrane.** The process shown results in the transport of glucose from the gut lumen to the extracellular fluid (from where it passes into the blood). Glucose is pumped into the cell through the apical domain of the membrane by a Na^+-powered glucose symport, and glucose passes out of the cell (down its concentration gradient) by facilitated diffusion mediated by a different glucose carrier protein in the basal and lateral membrane domains. The Na^+ gradient driving the glucose symport is maintained by the Na^+-K^+ ATPase in the basal and lateral plasma membrane domains, which keeps the internal concentration of Na^+ low.

Adjacent cells are connected by impermeable junctions (called *tight junctions*). The junctions have a dual function in the transport process illustrated: they prevent solutes from crossing the epithelium between cells, allowing a concentration gradient of glucose to be maintained across the cell sheet, and they also serve as diffusion barriers within the plasma membrane, which help confine the various carrier proteins to their respective membrane domains (see Figure 10–37).

plasma membranes of all bacteria contain carrier proteins that use the H^+ gradient across the membrane to pump a variety of nutrients into the cell. Many also have transport ATPases that use the energy of ATP hydrolysis to import certain sugars, amino acids, and small peptides. In bacteria such as *E. coli*, which have double membranes (Figure 11–14), the transport ATPases are located in the inner membrane, and an auxiliary mechanism exists to capture the nutrients and deliver them to the transporters (Figure 11–15).

The transport ATPases in the bacterial plasma membrane belong to the largest and most diverse family of transport proteins known. It is called the **ABC transporter superfamily** because each member contains a highly conserved ATP-binding cassette (Figure 11–16). Over 50 ABC transporters have been described, and although each one is usually specific for a particular substrate or class of substrates, the variety of substrates transported by this superfamily is great and includes amino acids, sugars, inorganic ions, polysaccharides, peptides, and even proteins. Whereas most family members have been described in procaryotes, an increasing number are being discovered in eucaryotes. The first of these to be identified were discovered because of their ability to pump hydrophobic drugs out of eucaryotic cells. One of these is the **multidrug resistance (MDR) protein,** whose overexpression in human cancer cells can make these cells simultaneously resistant to a variety of chemically unrelated cytotoxic drugs that are widely used in cancer chemotherapy. Treatment with any one of these drugs can result in the selection of cells that overexpress the MDR transport protein; the transporter pumps the drugs out of the cell, thereby reducing their toxicity and conferring resistance to a wide variety of therapeutic agents. A related and equally sinister phenomenon occurs in the protist *Plasmodium falciparum*, which causes malaria. More than 200 million people are infected with this parasite, which remains a major cause of human death, killing more than a million people a year. The control of malaria is hampered by the development of resistance to the

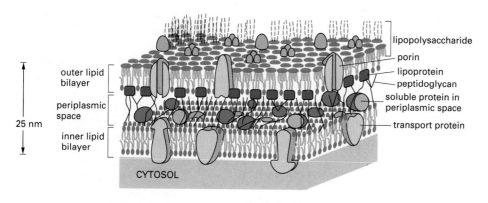

Figure 11–14 Schematic view of a small section of the double membrane of an *E. coli* bacterium. The inner membrane is the cell's plasma membrane. Between the inner and outer lipid bilayer membranes there is a highly porous, rigid peptidoglycan composed of protein and polysaccharide that constitutes the bacterial cell wall; it is attached to lipoprotein molecules in the outer membrane and fills the *periplasmic space* (only a little of the peptidoglycan is shown). This space also contains a variety of soluble protein molecules. The dashed green threads at the top represent the polysaccharide chains of the special lipopolysaccharide molecules that form the external monolayer of the outer membrane; for clarity, only a few of these chains are shown.

Bacteria with double membranes are called *gram negative* because they do not retain the dark blue dye used in the gram staining procedure. Bacteria with single membranes (but thicker cell walls), such as staphylococci and streptococci, retain the blue dye and therefore are called *gram positive*; their single membrane is analogous to the inner (plasma) membrane of gram-negative bacteria.

antimalarial drug *chloroquine*, and resistant *P. falciparum* have been shown to have amplified a gene that encodes an ABC transporter that pumps out the chloroquine.

The number of known members of the ABC superfamily of transporters in eucaryotic cells is growing rapidly, and the normal functions of some of them are becoming apparent. In yeasts an ABC transporter is responsible for exporting a mating pheromone (which is a peptide 13 amino acids long) across the yeast cell plasma membrane. In most vertebrate cells an ABC transporter in the endoplasmic reticulum (ER) membrane actively transports a wide variety of peptides, produced by protein degradation, from the cytosol into the ER. This is the first step in a pathway of great importance in the surveillance of cells by the immune system (discussed in Chapter 23). The transported protein fragments, having entered the ER, are eventually carried to the cell surface, where they are displayed for scrutiny by cytotoxic T lymphocytes, which will kill the cell if the fragments appear foreign (as they will if they derive from a virus inside the cell). Yet another member of the ABC family has been discovered through studies of the common genetic disease *cystic fibrosis*. This disease is caused by a mutation in a gene encoding an ABC transporter that functions as a Cl⁻ channel in the plasma membrane of epithelial cells. The channel is unusual in that it requires both ATP hydrolysis and cyclic-AMP-dependent phosphorylation in order to open. As

Figure 11–15 The auxiliary transport system associated with transport ATPases in bacteria with double membranes. The solute diffuses through channel-forming proteins (called *porins*) in the outer membrane and binds to a *periplasmic substrate-binding protein*. As a result, the substrate-binding protein undergoes a conformational change that enables it to bind to a transport ATPase in the plasma membrane, which then picks up the solute and actively transfers it across the bilayer in a reaction driven by ATP hydrolysis. The peptidoglycan is omitted for simplicity; its porous structure allows the substrate-binding proteins and water-soluble solutes to move through it by simple diffusion.

Carrier Proteins and Active Membrane Transport

521

Table 11–2 Some Carrier Protein Families

Family*	Representative Members
Sugar transporters	passive glucose transporters in mammalian cells some H$^+$-driven sugar transporters in bacteria
Cation-transporting ATPases	Na$^+$-K$^+$ ATPases Ca^{2+} ATPases
ABC transporters	multidrug resistance (MDR) ATPase in mammalian cells periplasmic substrate-binding-protein-dependent ATPases in bacteria chloroquine-resistance ATPase in *P. falciparum* mating pheromone exporter in yeast peptide pump in vertebrate ER membrane cystic fibrosis transmembrane regulator (CFTR) protein
Anion (Cl$^-$-HCO$_3^-$) antiporters	band 3 in red blood cells anion exchangers in other cells
Cation antiporters	Na$^+$-H$^+$ exchanger
Cation/anion antiporters	Na-dependent Cl$^-$-HCO$_3^-$ exchanger
Na$^+$-driven symporters	Na$^+$-glucose symporter in intestinal cells Na$^+$-proline symporter in bacteria Na$^+$-HCO$_3^-$ symporter in glial cells

*The members of a family are similar in amino acid sequence and, therefore, are thought to have evolved from a common ancestral protein.

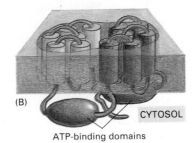

ATP-binding domains

Figure 11–16 A schematic drawing of a typical ABC transporter. (A) Topology diagram. (B) Hypothetical arrangement of the polypeptide chain in the membrane. The transporter consists of four domains: two highly hydrophobic domains, each with six putative membrane-spanning segments that somehow form the translocation pathway, and two ATP-binding catalytic domains (or cassettes). In some cases the two halves of the transporter are formed by a single polypeptide (*as shown*), whereas in other cases they are formed by two separate polypeptides.

there is evidence that the MDR protein may also function as a Cl$^-$ channel in some cells (in this case, regulated by cell volume rather than by cyclic AMP), it seems clear that at least some ABC transporters can function as both carriers and ion channels. How ABC transporters can function in these two different ways and transfer such diverse types of molecules across a membrane is a mystery.

Some of the families of structurally related carrier proteins that we have discussed are summarized in Table 11–2.

Summary

Carrier proteins bind specific solutes and transfer them across the lipid bilayer by undergoing conformational changes that expose the solute binding site sequentially on one side of the membrane and then on the other. Some carrier proteins simply transport a single solute "downhill," whereas others can act as pumps to transport a solute "uphill" against its electrochemical gradient, using energy provided by ATP hydrolysis or by a "downhill" flow of another solute (such as Na$^+$) to drive the requisite series of conformational changes. DNA cloning and sequencing studies show that carrier proteins belong to a small number of families, each of which comprises proteins of similar amino acid sequence that are thought to have evolved from a common ancestral protein and to operate by a similar mechanism. The family of cation-transporting ATPases, which includes the ubiquitous Na$^+$-K$^+$ pump, is an important example; each of these ATPases contains a large catalytic subunit that is sequentially phosphorylated and dephosphorylated during the pumping cycle. The superfamily of ABC transporters is especially important clinically: it includes proteins that are responsible for cystic fibrosis, as well as for drug resistance in cancer cells and in malaria-causing parasites.

Ion Channels and Electrical Properties of Membranes [15]

Unlike carrier proteins, **channel proteins** form hydrophilic pores across membranes. One class of channel proteins found in virtually all animal phyla forms *gap junctions* between two adjacent cells; each plasma membrane contributes equally to the formation of the channel, which connects the cytoplasm of the two cells. These channels are discussed in Chapter 19 and will not be considered further here. Both gap junctions and *porins*, the channel-forming proteins of the outer membranes of bacteria, mitochondria, and chloroplasts (discussed in Chapter 10), have relatively large and permissive pores, which would be disastrous if they directly connected the inside of a cell to an extracellular space. In contrast, most channel proteins in the plasma membrane of animal and plant cells connect the cytosol to the cell exterior and necessarily have narrow, highly selective pores. These proteins are concerned specifically with inorganic ion transport and so are referred to as **ion channels.** For transport efficiency, channels have an advantage over carriers in that more than 1 million ions can pass through one channel each second, which is a rate 1000 times greater than the fastest rate of transport mediated by any known carrier protein. On the other hand, channels cannot be coupled to an energy source to carry out active transport, so the transport they mediate is always passive ("downhill"). Thus the function of ion channels is to allow specific inorganic ions, mainly Na^+, K^+, Ca^{2+}, or Cl^-, to diffuse rapidly down their electrochemical gradients across the lipid bilayer, although this does not mean that the transport through ion channels cannot be regulated. Indeed, we shall see that the ability to control ion fluxes in this way is essential for many cell functions. Nerve cells, in particular, have made a specialty of using ion channels, and we shall consider how they utilize a diversity of such channels for receiving, conducting, and transmitting signals.

Ion Channels Are Ion Selective and Fluctuate Between Open and Closed States [15]

Two important properties distinguish ion channels from simple aqueous pores. First, they show *ion selectivity,* permitting some inorganic ions to pass but not others. This suggests that their pores must be narrow enough in places to force permeating ions into intimate contact with the walls of the channel so that only ions of appropriate size and charge can pass. It is thought that the permeating ions have to shed most of their associated water molecules in order to pass, in single file, through the narrowest part of the channel; this limits their rate of passage. Thus, as ion concentrations are increased, the flux of ions through a channel increases proportionally but then levels off (saturates) at a maximum rate.

The second important distinction between ion channels and simple aqueous pores is that ion channels are not continuously open. Instead, they have "gates," which open briefly and then close again, as shown schematically in Figure 11–17. In most cases the gates open in response to a specific stimulus. The main types of stimuli that are known to cause ion channels to open are a change in the voltage across the membrane (*voltage-gated channels*), a mechanical stress (*mechanically gated channels*), or the binding of a ligand (*ligand-gated channels*). The ligand can be either an extracellular mediator—specifically, a *neurotransmitter* (*transmitter-gated channels*)—or an intracellular mediator, such as an ion (*ion-gated channels*), or a nucleotide (*nucleotide-gated channels*) (Figure 11–18). The activity of many ion channels is regulated in addition by protein phosphorylation and dephosphorylation; this type of channel regulation is discussed, together with nucleotide-gated ion channels, in Chapter 15.

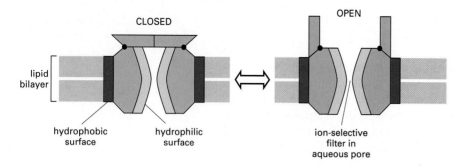

CLOSED

OPEN

lipid bilayer

hydrophobic surface

hydrophilic surface

ion-selective filter in aqueous pore

Figure 11–17 Schematic drawing of a typical ion channel, which fluctuates between closed and open conformations. A transmembrane protein complex, seen in cross-section, forms a hydrophilic pore across the lipid bilayer only when the gate is open. Polar amino acid side chains are thought to line the wall of the pore, while hydrophobic side chains interact with the lipid bilayer. The pore narrows to atomic dimensions in one region (the "ion-selective filter"), where the ion selectivity of the channel is largely determined.

More than 100 types of ion channels have been described thus far, and new ones are still being discovered. They are responsible for the electrical excitability of muscle cells, and they mediate most forms of electrical signaling in the nervous system. A single nerve cell might typically contain 10 kinds of ion channels or more, located in different domains of its plasma membrane. But ion channels are not restricted to electrically excitable cells. They are present in all animal cells and are found in plant cells and microorganisms: they propagate the leaf-closing response of the mimosa plant, for example, and allow the single-celled paramecium to reverse direction after a collision.

Perhaps the most common ion channels are those that are permeable mainly to K^+. These channels are found in the plasma membrane of almost all animal cells. An important subset of K^+ channels are open even in an unstimulated or "resting" cell and are hence sometimes called *K^+ leak channels*. Although this term covers a variety of different K^+ channels depending on the cell type, they serve a common function: by making the plasma membrane much more permeable to K^+ than to other ions, they play a critical part in maintaining the *membrane potential*—the voltage difference that is present across all plasma membranes.

The Membrane Potential in Animal Cells Depends Mainly on K^+ Leak Channels and the K^+ Gradient Across the Plasma Membrane [15, 16]

A **membrane potential** arises when there is a difference in the electrical charge on the two sides of a membrane, due to a slight excess of positive ions over negative on one side and a slight deficit on the other. Such charge differences can result both from active electrogenic pumping (see p. 515) and from passive ion diffusion. We shall see in Chapter 14 that most of the membrane potential of the mitochondrion is generated by electrogenic H^+ pumps in the mitochondrial inner membrane. Electrogenic pumps also generate most of the electrical potential across the plasma membrane in plants and fungi. In typical animal cells, however, passive ion movements make the largest contribution to the electrical potential across the plasma membrane.

As explained earlier, the Na^+-K^+ ATPase helps to maintain osmotic balance across the animal cell membrane by keeping the intracellular concentration of Na^+ low. Because there is little Na^+ inside the cell, other cations have to be plentiful there to balance the charge carried by the cell's fixed anions—the negatively charged organic molecules that are confined inside the cell. This balancing role is performed largely by K^+, which is actively pumped into the cell by the Na^+-K^+ ATPase and can also move freely in or out through the **K^+ leak channels** in the plasma membrane. Because of the presence of these channels, K^+ comes almost to an equilibrium, where an electrical force exerted by an excess of negative charges attracting K^+ into the cell balances the tendency of K^+ to leak out down its concentration gradient. The membrane potential is the manifestation of this electrical force, and its equilibrium value can be calculated from the steepness

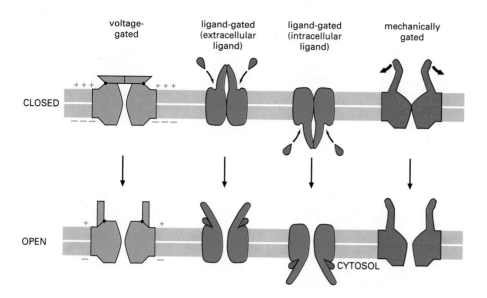

Figure 11–18 Gated ion channels. Schematic drawing of the different ways in which ion channels are gated.

voltage-gated

ligand-gated (extracellular ligand)

ligand-gated (intracellular ligand)

mechanically gated

CLOSED

OPEN

CYTOSOL

of the K^+ concentration gradient. The following argument may help to make this clear.

Suppose that initially there is no voltage gradient across the plasma membrane (the membrane potential is zero), but the concentration of K^+ is high inside the cell and low outside. K^+ will tend to leave the cell through the K^+ leak channels, driven by its concentration gradient. As K^+ moves out, it will leave behind unbalanced negative charge, thereby creating an electrical field, or membrane potential, which will tend to oppose the further efflux of K^+. The net efflux of K^+ will halt when the membrane potential reaches a value where this electrical driving force on K^+ exactly balances the effect of its concentration gradient—that is, when the electrochemical gradient for K^+ is zero. Although Cl^- ions also equilibrate across the membrane, because their charge is negative, the membrane potential keeps most of these ions out of the cell. The equilibrium condition, in which there is no net flow of ions across the plasma membrane, defines the **resting membrane potential** for this idealized cell. A simple but very important formula, the **Nernst equation,** expresses the equilibrium condition quantitatively and, as explained in Panel 11–2, makes it possible to calculate the theoretical resting membrane potential if the ratio of internal and external ion concentrations is known. As the plasma membrane of a real cell is not exclusively permeable to K^+ and Cl^-, however, the actual resting membrane potential is usually not exactly equal to that predicted by the Nernst equation for K^+ or Cl^-.

The Resting Potential Decays Only Slowly When the Na^+-K^+ Pump Is Stopped [15, 16]

The number of ions that must move across the plasma membrane to set up the membrane potential is minute. Thus one can think of the membrane potential as arising from movements of charge that leave ion *concentrations* practically unaffected and result in only a very slight discrepancy in the number of positive and negative ions on the two sides of the membrane (Figure 11–19). Moreover, these movements of charge are generally rapid, taking only a few milliseconds or less.

It is illuminating to consider what happens to the membrane potential in a real cell if the Na^+-K^+ ATPase is suddenly inactivated. First, there is an immediate slight drop in the membrane potential. This is because the pump is electrogenic and, when active, makes a small direct contribution to the membrane potential by pumping out three Na^+ for every two K^+ that it pumps in. Switching off the pump, however, does not abolish the major component of the resting

THE NERNST EQUATION AND ION FLOW

The flow of any ion through a membrane channel protein is driven by the electrochemical gradient for that ion. This gradient represents the combination of two influences: the voltage gradient and the concentration gradient of the ion across the membrane. When these two influences just balance each other the electrochemical gradient for the ion is zero and there is no *net* flow of the ion through the channel. The voltage gradient (membrane potential) at which this equilibrium is reached is called the equilibrium potential for the ion. It can be calculated from an equation that will be derived below, called the Nernst equation.

The Nernst equation is

$$V = \frac{RT}{zF} \ln \frac{C_o}{C_i}$$

where

V = the equilibrium potential in volts (internal potential minus external potential)

C_o and C_i = outside and inside concentrations of the ion, respectively

R = the gas constant (2 cal mol^{-1} $^\circ$K^{-1})

T = the absolute temperature ($^\circ$K)

F = Faraday's constant (2.3 x 10^4 cal V^{-1} mol^{-1})

z = the valence (charge) of the ion

ln = logarithm to the base e

The Nernst equation is derived as follows:

A molecule in solution (a solute) tends to move from a region of high concentration to a region of low concentration simply due to the pressure of numbers. Consequently, movement down a concentration gradient is accompanied by a favorable free-energy change ($\Delta G < 0$), whereas movement up a concentration gradient is accompanied by an unfavorable free-energy change ($\Delta G > 0$). (Free energy is introduced and discussed in Panel 14–1, pp. 668–669.) The free-energy change per mole of solute moved across the plasma membrane (ΔG_{conc}) is equal to $-RT \ln C_o / C_i$. If the solute is an ion, moving it into a cell across a membrane whose inside is at a voltage V relative to the outside will cause an additional free-energy change (per mole of solute moved) of $\Delta G_{volt} = zFV$. At the point where the concentration and voltage gradients just balance, $\Delta G_{conc} + \Delta G_{volt} = 0$ and the ion distribution is at equilibrium across the membrane. Thus,

$$zFV - RT \ln \frac{C_o}{C_i} = 0$$

and, therefore,

$$V = \frac{RT}{zF} \ln \frac{C_o}{C_i} = 2.3 \frac{RT}{zF} \log_{10} \frac{C_o}{C_i}$$

For a univalent ion,

$$2.3 \frac{RT}{F} = 58 \text{ mV at } 20^\circ C \quad \text{and} \quad 61.5 \text{ mV at } 37^\circ C$$

Thus, for such an ion at 37°C, V = + 61.5 mV for C_o / C_i = 10, whereas V = 0 for C_o / C_i = 1.

The K$^+$ equilibrium potential (V_K), for example, is 61.5 $\log_{10}$([K$^+$]$_o$ / [K$^+$]$_i$) millivolts (–89 mV for a typical cell where [K$^+$]$_o$ = 5 mM and [K$^+$]$_i$ = 140 mM). At V_K, there is no net flow of K$^+$ across the membrane. Similarly, when the membrane potential has a value of 61.5 $\log_{10}$([Na$^+$]$_o$ /[Na$^+$]$_i$), the Na$^+$ equilibrium potential (V_{Na}), there is no net flow of Na$^+$.

For any particular membrane potential, V_M, the net force tending to drive a particular type of ion out of the cell, is proportional to the difference between V_M and the equilibrium potential for the ion: hence, for K$^+$ it is $V_M - V_K$ and for Na$^+$ it is $V_M - V_{Na}$.

The number of ions that go to form the layer of charge adjacent to the membrane is minute compared with the total number inside the cell. For example, the movement of 6000 Na$^+$ ions across 1 µm^2 of membrane will carry sufficient charge to shift the membrane potential by about 100 mV. Because there are about 3 x 10^7 Na$^+$ ions in 1 µm^3 of bulk cytoplasm, such a movement of charge will generally have a negligible effect on the ion concentration gradients across the membrane.

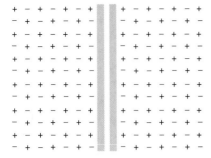

exact balance of charges on each side of the membrane; membrane potential = 0

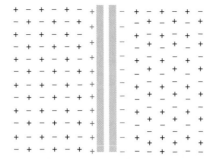

a few of the positive ions (*red*) cross the membrane from right to left, leaving their negative counterions (*red*) behind; this sets up a nonzero membrane potential

Figure 11–19 A small flow of ions carries sufficient charge to cause a large change in the membrane potential. The ions that give rise to the membrane potential lie in a thin (<1 nm) surface layer close to the membrane, held there by their electrical attraction to their oppositely charged counterparts (counterions) on the other side of the membrane. For a typical cell 1 microcoulomb of charge (6×10^{12} univalent ions) per square centimeter of membrane, transferred from one side of the membrane to the other, will change the membrane potential by roughly 1 V. This means, for example, that in a spherical cell of diameter 10 μm, the number of K^+ ions that have to flow out to alter the membrane potential by 100 mV is only about 1/100,000 of the total number of K^+ ions in the cytosol.

potential, which is generated by the K^+ equilibrium mechanism outlined above. This component persists as long as the Na^+ concentration inside the cell stays low and the K^+ ion concentration high—typically for many minutes. The plasma membrane, however, is somewhat permeable to all small ions, including Na^+. Therefore, without the Na^+-K^+ ATPase, the ion gradients set up by pumping will eventually run down, and the membrane potential established by diffusion through the K^+ leak channels will fall as well. As Na^+ enters, the osmotic balance is upset, and water seeps into the cell (see Panel 11–1, p. 517). But if the cell does not burst, it eventually comes to a new resting state where Na^+, K^+, and Cl^- are all at equilibrium across the membrane. The membrane potential in this state is much less than it was in the normal cell with an active Na^+-K^+ pump.

The potential difference across the plasma membrane of an animal cell at rest varies between –20 mV and –200 mV, depending on the organism and cell type. Although the K^+ gradient always has a major influence on this potential, the gradients of other ions (and the disequilibrating effects of ion pumps) also have a significant effect: the more permeable the membrane for a given ion, the more strongly the membrane potential tends to be driven toward the equilibrium value for that ion. Consequently, almost any change of a membrane's permeability to ions causes a change in the membrane potential. This is the key principle relating the electrical excitability of cells to the activities of ion channels.

Nerve cells, or *neurons*, have made a profession of electrical excitability, and most of what we know about the topic has come from studies of these remarkable cells. To put the following account of electrical excitability in context, therefore, we must digress to review briefly how a typical neuron is organized.

The Function of a Nerve Cell Depends on Its Elongated Structure [15]

The fundamental task of the **neuron** is to receive, conduct, and transmit signals. To perform these functions, neurons in general are extremely elongated: a single nerve cell in a human being, extending, say, from the spinal cord to a muscle in the foot, may be a meter long. Every neuron consists of a *cell body* (containing the nucleus) with a number of long, thin processes radiating outward from it. Usually there is one long **axon,** to conduct signals away from the cell body toward distant targets, and several shorter branching **dendrites,** which extend from the cell body like antennae and provide an enlarged surface area to receive signals from the axons of other nerve cells (Figure 11–20). Signals are also received on the cell body itself. The axon commonly divides at its far end into many branches and so can pass on its message to many target cells simultaneously. Likewise, the extent of branching of the dendrites can be very great—in some cases sufficient to receive as many as 100,000 inputs on a single neuron.

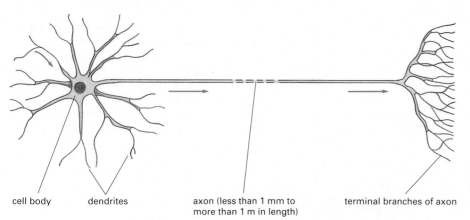

Figure 11–20 **Schematic diagram of a typical vertebrate neuron.** The arrows indicate the direction in which signals are conveyed. The single axon conducts signals away from the cell body, while the multiple dendrites receive signals from the axons of other neurons. The nerve terminals end on the dendrites or cell body of other neurons or on other cell types, such as muscle or gland cells.

cell body dendrites axon (less than 1 mm to more than 1 m in length) terminal branches of axon

Despite the varied significance of the signals carried by different classes of neurons, the form of the signal is always the same, consisting of changes in the electrical potential across the neuron's plasma membrane. Communication occurs because an electrical disturbance produced in one part of the cell spreads to other parts. Such a disturbance becomes weaker with increasing distance from its source unless energy is expended to amplify it as it travels. Over short distances this attenuation is unimportant; in fact, many small neurons conduct their signals passively, without amplification. For long-distance communication, however, passive spread is inadequate. Thus larger neurons employ an active signaling mechanism, which is one of their most striking features: an electrical stimulus that exceeds a certain threshold strength triggers an explosion of electrical activity that is propagated rapidly along the neuron's plasma membrane and is sustained by automatic amplification all along the way. This traveling wave of electrical excitation, known as an **action potential,** or *nerve impulse,* can carry a message without attenuation from one end of a neuron to the other at speeds as great as 100 meters/second or more. Action potentials are the direct consequence of the properties of voltage-gated cation channels, as we shall now see.

Voltage-gated Cation Channels Are Responsible for the Generation of Action Potentials in Electrically Excitable Cells [15, 17]

The plasma membrane of all electrically excitable cells—not only neurons but also muscle, endocrine, and egg cells—contains **voltage-gated cation channels,** which are responsible for generating the action potentials. An action potential is triggered by a *depolarization* of the plasma membrane—that is, by a shift in the membrane potential to a less negative value. (We shall see later how this may be caused by the action of a neurotransmitter.) In nerve and skeletal muscle cells

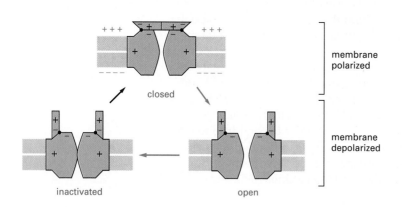

Figure 11–21 **The voltage-gated Na$^+$ channel can adopt at least three conformations (states).** Internal forces, represented here by attractions between charges on different parts of the channel, stabilize each state against small disturbances, but a sufficiently violent collision with other molecules can cause the channel to flip from one of these states to another. The state of lowest energy depends on the membrane potential because the different conformations have different charge distributions. When the membrane is at rest (highly polarized), the closed conformation has the lowest free energy and is therefore most stable; when the membrane is depolarized, the energy of the *open* conformation is lower and so the channel has a high probability of opening. But the free energy of the *inactivated* conformation is lower still, and so, after a randomly variable period spent in the open state, the channel becomes inactivated. Thus the open conformation corresponds to a metastable state that can exist only transiently. The *red arrows* indicate the sequence that follows a sudden depolarization, while the *black arrow* indicates the return to the original conformation as the lowest energy state after the membrane is repolarized.

closed

membrane polarized

membrane depolarized

inactivated

open

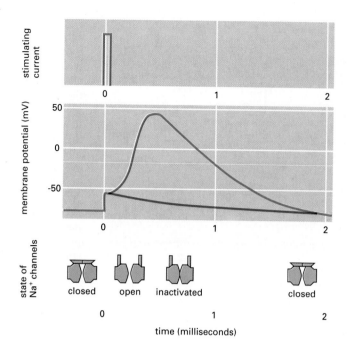

Figure 11–22 **An action potential.**
The action potential is triggered by a brief pulse of current (shown in the *upper graph*), which partially depolarizes the membrane, as shown in the plot of membrane potential versus time (*middle graph*). The *green curve* shows how the membrane potential would have simply relaxed back to the resting value after the initial depolarizing stimulus if there had been no voltage-gated ion channels in the membrane; this relatively slow return of the membrane potential to its initial value of –70 mV in the absence of open Na^+ channels is automatic because of the efflux of K^+ through K^+ channels, which drives the membrane back toward the K^+ equilibrium potential. The *red curve* shows the course of the action potential that is caused by the opening and subsequent inactivation of voltage-gated Na^+ channels, whose state is shown at the bottom. The membrane cannot fire a second action potential until the Na^+ channels have returned to the closed conformation (see Figure 11–21); until then the membrane is refractory to stimulation.

a stimulus that causes sufficient depolarization promptly causes **voltage-gated Na^+ channels** to open, allowing a small amount of Na^+ to enter the cell down its electrochemical gradient. The influx of positive charge depolarizes the membrane further, thereby opening more Na^+ channels, which admit more Na^+ ions, causing still further depolarization. This process continues in a self-amplifying fashion until, within a fraction of a second, the electrical potential in the local region of membrane has shifted from its resting value of about –70 mV almost as far as the Na^+ equilibrium potential of about +50 mV (see Panel 11–2, p. 526). At this point, when the net electrochemical driving force for the flow of Na^+ is almost zero, the cell would come to a new resting state, with all of its Na^+ channels permanently open, if the open conformation of the channel were stable.

The cell is saved from such a permanent electrical spasm because the Na^+ channels have an automatic inactivating mechanism, which causes the channels to reclose rapidly even though the membrane is still depolarized. The Na^+ channels remain in this *inactivated* state, unable to reopen, until a few milliseconds after the membrane potential returns to its initial negative value. A schematic illustration of these three distinct states of the voltage-gated Na^+ channel—closed, open, and inactivated—is shown in Figure 11–21. How they contribute to the rise and fall of the action potential is shown in Figure 11–22.

The description just given of an action potential concerns only a small patch of plasma membrane. The self-amplifying depolarization of the patch, however, is sufficient to depolarize neighboring regions of membrane, which then go through the same cycle. In this way the action potential spreads as a traveling wave from the initial site of depolarization to involve the entire plasma membrane, as shown in Figure 11–23.

In addition to the inactivation of Na^+ channels, in many nerve cells a second mechanism operates to help bring the activated plasma membrane more rapidly back toward its original negative potential, ready to transmit a second impulse. **Voltage-gated K^+ channels** open, so that the transient influx of Na^+ is rapidly overwhelmed by an efflux of K^+, which quickly drives the membrane back toward the K^+ equilibrium potential, even before the inactivation of the Na^+ channels is complete. These K^+ channels respond to changes in membrane potential in much the same way as the Na^+ channels do, but with slightly slower kinetics; for this reason they are sometimes called *delayed K^+ channels*.

Ion Channels and Electrical Properties of Membranes

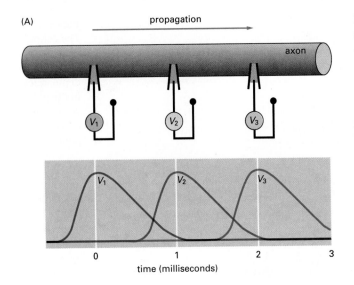

(A)

propagation

axon

V_1 V_2 V_3

V_1 V_2 V_3

0   1   2   3
time (milliseconds)

Figure 11–23 The propagation of an action potential along an axon. (A) shows the voltages that would be recorded from a set of intracellular electrodes placed at intervals along the axon. (B) shows the changes in the Na$^+$ channels and the current flows (*brown lines*) that give rise to the traveling disturbance of the membrane potential. The region of the axon with a depolarized membrane is shaded in *red*. Note that an action potential can only travel away from the site of depolarization because Na$^+$-channel inactivation prevents the depolarization from spreading backward. (See also Figure 11–22.)

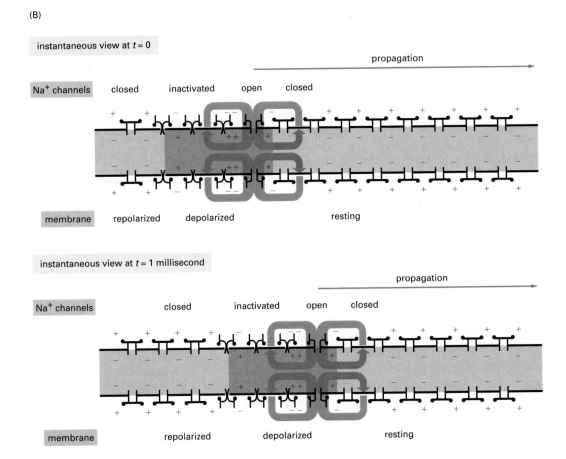

(B)

instantaneous view at $t = 0$

propagation

Na$^+$ channels   closed   inactivated   open   closed

membrane   repolarized   depolarized   resting

instantaneous view at $t = 1$ millisecond

propagation

Na$^+$ channels   closed   inactivated   open   closed

membrane   repolarized   depolarized   resting

The electrochemical mechanism of the action potential was first established by a famous series of experiments carried out in the 1940s and 1950s. Because the techniques for studying electrical events in small cells had not yet been developed, the experiments exploited the giant neurons in the squid. Despite the many technical advances made since then, the logic of the original analysis continues to serve as a model for present-day work. Panel 11–3 outlines some of the key original experiments.

1. Action potentials are recorded with an intracellular electrode

The squid giant axon is about 0.5–1 mm in diameter and several centimeters long. An electrode in the form of a glass capillary tube containing a conducting solution can be thrust down the axis of the axon so that its tip lies deep in the cytoplasm. With its help, one can measure the voltage difference between the inside and the outside of the axon—that is, the membrane potential—as an action potential sweeps past the electrode. The action potential is triggered by a brief electrical stimulus to one end of the axon. It does not matter which end, because the excitation can travel in either direction; and it does not matter how big the stimulus is, as long as it exceeds a certain threshold: the action potential is all or none.

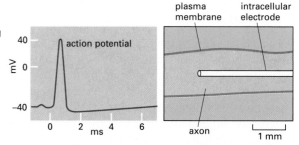

2. Action potentials depend only on the neuronal plasma membrane and on gradients of Na^+ and K^+ across it

The three most plentiful ions, both inside and outside the axon, are Na^+, K^+, and Cl^-. As in other cells, the Na^+-K^+ pump maintains a concentration gradient: the concentration of Na^+ is about 9 times lower inside the axon than outside, while the concentration of K^+ is about 20 times higher inside than outside. Which ions are important for the action potential?

The squid giant axon is so large and robust that it is possible to extrude the cytoplasm from it, like toothpaste from a tube, and then to perfuse it internally with pure artificial solutions of Na^+, K^+, and Cl^- or SO_4^{2-}. Remarkably, if (and only if) the concentrations of Na^+ and K^+ inside and outside approximate those found naturally, the axon will still propagate action potentials of the normal form. The important part of the cell for electrical signaling, therefore, must be the plasma membrane; the important ions are Na^+ and K^+; and a sufficient source of free energy to power the action potential must be provided by their concentration gradients across the membrane, because all other sources of metabolic energy have presumably been removed by the perfusion.

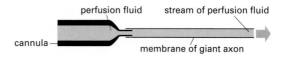

3. At rest, the membrane is chiefly permeable to K^+; during the action potential, it becomes transiently permeable to Na^+

At rest the membrane potential is close to the equilibrium potential for K^+. When the external concentration of K^+ is changed, the resting potential changes roughly in accordance with the Nernst equation for K^+ (see Panel 11–2). At rest, therefore, the membrane is chiefly permeable to K^+: K^+ leak channels provide the main ion pathway through the membrane.

If the external concentration of Na^+ is varied, there is no effect on the resting potential. However, the height of the peak of the action potential varies roughly in accordance with the Nernst equation for Na^+. During the action potential, therefore, the membrane appears to be chiefly permeable to Na^+: Na^+ channels have opened. In the aftermath of the action potential, the membrane potential reverts to a negative value that depends on the external concentration of K^+ and is even closer to the K^+ equilibrium potential than the resting potential is: the membrane has lost most of its permeability to Na^+ and has become even more permeable to K^+ than before—that is, Na^+ channels have closed, and additional K^+ channels have opened.

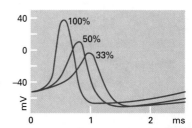

The form of the action potential when the external medium contains 100%, 50%, or 33% of the normal concentration of Na^+.

4. Voltage clamping reveals how the membrane potential controls opening and closing of ion channels

The membrane potential can be held constant ("voltage clamped") throughout the axon by passing a suitable current through a bare metal wire inserted along the axis of the axon while monitoring the membrane potential with another intracellular electrode. When the membrane is abruptly shifted from the resting potential and held in a depolarized state (A), Na^+ channels rapidly open until the Na^+ permeability of the membrane is much greater than the K^+ permeability; they then close again spontaneously, even though the membrane potential is clamped and unchanging. K^+ channels also open but with a delay, so that the K^+ permeability increases as the Na^+ permeability falls (B). If the experiment is now very promptly repeated, by returning the membrane briefly to the resting potential and then quickly depolarizing it again, the response is different: prolonged depolarization has caused the Na^+ channels to enter an inactivated state, so that the second depolarization fails to cause a rise and fall similar to the first. Recovery from this state requires a relatively long time—about 10 milliseconds—spent at the repolarized (resting) membrane potential.

In a normal unclamped axon, an inrush of Na^+ through the opened Na^+ channels produces the spike of the action potential; inactivation of Na^+ channels and opening of K^+ channels bring the membrane rapidly back down to the resting potential.

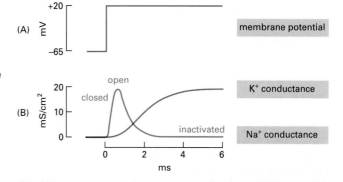

Panel 11–3 Some classical experiments on the squid giant axon.

Myelination Increases the Speed and Efficiency of Action Potential Propagation in Nerve Cells [15, 18]

The axons of many vertebrate neurons are insulated by a **myelin sheath,** which greatly increases the rate at which an axon can conduct an action potential. The importance of myelination is dramatically demonstrated by the demyelinating disease *multiple sclerosis,* in which myelin sheaths in some regions of the central nervous system are destroyed by an unknown mechanism; where this happens, the propagation of nerve impulses is greatly slowed, often with devastating neurological consequences.

Myelin is formed by specialized supporting, or *glial,* cells—*Schwann cells* in peripheral nerves and *oligodendrocytes* in the central nervous system. These glial cells wrap layer upon layer of their own plasma membrane in a tight spiral around the axon (Figure 11–24), thereby insulating the axonal membrane so that almost no current leaks across it. The sheath is interrupted at regularly spaced *nodes of Ranvier,* where almost all the Na+ channels in the axon are concentrated. Because the ensheathed portions of the axon membrane have excellent cable properties (i.e., they behave electrically much like well-designed undersea telegraph cables), a depolarization of the membrane at one node almost immediately spreads passively to the next node. Thus an action potential propagates along a myelinated axon by jumping from node to node, a process called *saltatory conduction.* This type of conduction has two main advantages: action potentials travel faster, and metabolic energy is conserved because the active excitation is confined to the small regions of axonal plasma membrane at nodes of Ranvier.

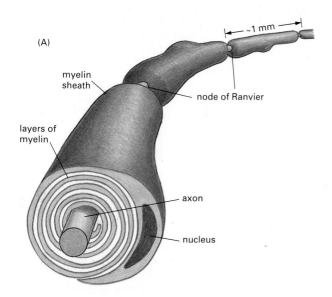

(A)

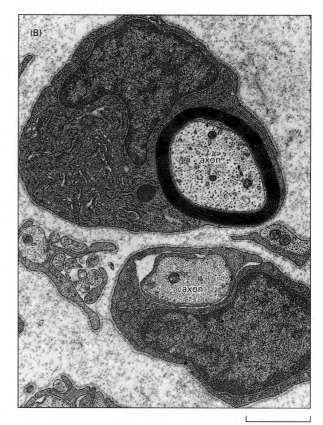

(B)

Figure 11–24 **Myelination.** (A) Schematic diagram of a myelinated axon from a peripheral nerve. Each Schwann cell wraps its plasma membrane concentrically around the axon to form a segment of myelin sheath about 1 mm long. For clarity, the layers of myelin are not shown compacted together as tightly as they are in reality (see part B). (B) Electron micrograph of a section from a nerve in the leg of a young rat. Two Schwann cells can be seen: one (*below*) is just beginning to myelinate its axon; the other has formed an almost mature myelin sheath. (B, from C. Raine, in Myelin [P. Morell, ed.]. New York: Plenum, 1976.)

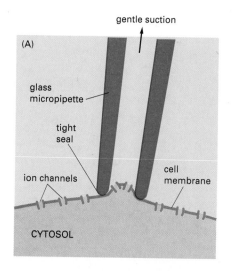

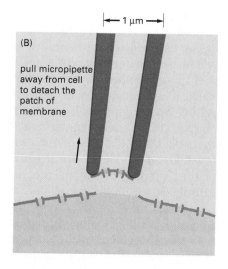

(A) gentle suction

glass micropipette

tight seal

ion channels

cell membrane

CYTOSOL

(B) pull micropipette away from cell to detach the patch of membrane

|← 1 μm →|

Figure 11–25 **The technique of patch-clamp recording.** Because of the extremely tight seal between the micropipette and the membrane, current can enter or leave the micropipette only by passing through the channels in the *patch* of membrane covering its tip. The term *clamp* is used because an electronic device is employed to maintain, or "clamp," the membrane potential at a set value while recording the ionic current through individual channels. Recordings of the current through these channels can be made with the patch still attached to the rest of the cell, as in (A), or detached, as in (B). The advantage of the detached patch is that it is easy to alter the composition of the solution on either side of the membrane to test the effect of various solutes on channel behavior. A detached patch can also be produced with the opposite orientation, so that the cytoplasmic surface of the membrane faces the inside of the pipette. (See also Figures 4–54 and 4–55.)

Patch-Clamp Recording Indicates That Individual Na$^+$ Channels Open in an All-or-Nothing Fashion [15, 19]

Neuron and skeletal muscle cell plasma membranes contain many thousands of voltage-gated Na$^+$ channels, and the current crossing the membrane is the sum of the currents flowing through all of these. This aggregate current can be recorded with an intracellular microelectrode, as shown in Figure 11–23. Remarkably, however, it is also possible to record current flowing through individual channels. This is achieved by means of **patch-clamp recording,** a method that has revolutionized the study of ion channels by allowing transport through a single molecule of channel protein to be studied in a small patch of membrane covering the mouth of a micropipette (Figure 11–25). With this simple but powerful technique, the detailed properties of ion channels can be studied in all sorts of cell types, and this has led to the discovery that even cells that are not electrically excitable usually have a variety of gated ion channels in their plasma membrane. Many of these cells, such as yeasts, are too small to be investigated by the traditional electrophysiologist's method of impalement with an intracellular microelectrode.

Patch-clamp recording indicates that individual voltage-gated Na$^+$ channels open in an all-or-nothing fashion: the times of its opening and closing are random, but when open, the channel always has the same very large conductance, allowing more than 8000 ions to pass per millisecond. Therefore, the aggregate current crossing the membrane of an entire cell does not indicate the *degree* to

Figure 11–26 **Patch-clamp measurements for a single voltage-gated Na$^+$ channel.** A tiny patch of plasma membrane was detached from an embryonic rat muscle cell as in Figure 11–25. The membrane was depolarized by an abrupt shift of potential, as indicated in (A). The three current records shown in (B) are from three experiments performed on the same patch of membrane. Each major current step in (B) represents the opening and closing of a single channel. Comparison of the three records shows that, whereas the times of channel opening and closing vary greatly, the rate at which current flows through an open channel is practically constant. The minor fluctuations in the current records arise largely from electrical noise in the recording apparatus. The sum of the currents measured in 144 repetitions of the same experiment is shown in (C). This aggregate current is equivalent to the usual Na$^+$ current that would be observed flowing through a relatively large region of membrane containing 144 channels. Comparison of (B) and (C) reveals that the time course of the aggregate current reflects the probability that any individual channel will be in the open state; this probability decreases with time as the channels in the depolarized membrane adopt their inactivated conformation. (Data from J. Patlak and R. Horn, *J. Gen. Physiol.* 79:333–351, 1982, by copyright permission of the Rockefeller University Press.)

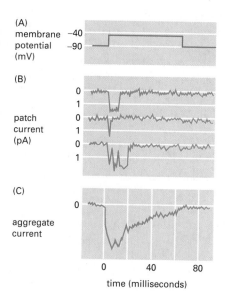

(A) membrane potential (mV)   −40   −90

(B) patch current (pA)   0 1   0 1   0 1

(C) aggregate current   0

time (milliseconds)   0   40   80

which a typical individual channel is open but rather the *total number* of channels in its membrane that are open at any one time (Figure 11–26).

The phenomenon of voltage gating can be understood in terms of simple physical principles. The interior of the resting neuron or muscle cell is at an electric potential about 50–100 mV more negative than the external medium. Although this potential difference seems small, it exists across a plasma membrane only about 5 nm thick, so that the resulting voltage *gradient* is about 100,000 V/cm. Proteins in the membrane are thus subjected to a very large electrical field. These proteins, like all others, have a number of charged groups on their surface, as well as polarized bonds between their various atoms. The electrical field therefore exerts forces on the molecular structure. For many membrane proteins the effects of changes in the membrane electrical field are probably insignificant, but voltage-gated ion channels can adopt a number of alternative conformations whose stabilities depend on the strength of the field. Each conformation can "flip" to another conformation if given a sufficient jolt by the random thermal movements of the surroundings, and it is the relative stability of the closed, open, and inactivated conformations against flipping that is altered by changes in the membrane potential (see Figure 11–21).

Voltage-gated Cation Channels Are Evolutionarily and Structurally Related [20]

Na$^+$ channels are not the only kind of voltage-gated cation channel that can generate an action potential: the action potentials in some muscle, egg, and endocrine cells, for example, depend on *voltage-gated Ca^{2+} channels* rather than on Na$^+$ channels. Moreover, voltage-gated Na$^+$, K$^+$, or Ca^{2+} channels of unknown function are found in some cell types that are not normally electrically active.

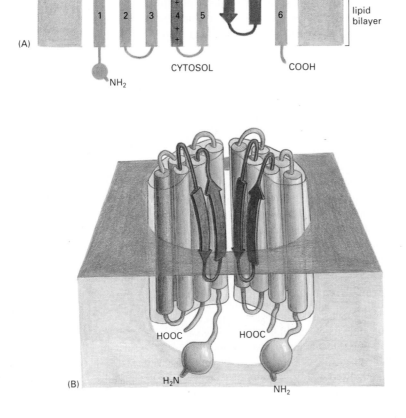

Figure 11–27 **A model for the structure of a voltage-gated K$^+$ channel.** (A) A topology diagram showing the major functional domains of the polypeptide chain of one subunit, with the six putative transmembrane α helices labeled 1 to 6. Four such subunits, each having about 600 amino acids, are thought to assemble to form a transmembrane pore; only two are shown in (B). In voltage-gated Na$^+$ and Ca^{2+} channels the four subunits are domains of a single very large polypeptide chain, but otherwise the overall structure is thought to be similar. The 20 amino acid segment (shown in *red*), contained in the region linking helices 5 and 6, is thought to extend across the membrane as two antiparallel β strands to line the pore as shown. The fourth α helix (*blue*) has positively charged residues at every third position, which is thought to allow this helix to serve as a *voltage sensor*. In at least some K$^+$ channels, the amino-terminal domain is involved in rapid channel inactivation, as illustrated in Figure 11–28.

There is a surprising amount of structural and functional diversity within each of these three classes of voltage-gated cation channels, which is generated both by multiple genes and by alternative splicing of RNA transcripts produced from the same gene. Nonetheless, the amino acid sequences of the known voltage-gated Na^+, K^+, and Ca^{2+} channels show striking similarities, suggesting that they all belong to a large superfamily of evolutionarily and structurally related proteins.

Each type of voltage-gated cation channel is composed of either four homologous protein domains (in the case of Na^+ and Ca^{2+} channels) or four identical subunits (in the case of K^+ channels). These four domains or subunits are thought to be arranged like staves of a barrel surrounding a central pore, as illustrated for a voltage-gated K^+ channel in Figure 11–27. The K^+ channels are especially convenient for studying the relationship between the structure and function of voltage-gated cation channels because they are formed from relatively small identical subunits rather than from a single large polypeptide chain. The amino acid sequence suggests that each subunit of a K^+ channel contains six membrane-spanning α helices, but, unexpectedly, none of these seems to line the ion-conducting pore. Studies of K^+ channel proteins that have been modified by the use of recombinant DNA techniques suggest that a 20 amino acid segment of polypeptide chain extends across the membrane as an antiparallel β sheet to line the pore: when this segment is exchanged between two K^+ channels with differing permeability properties, the permeability characteristics are found to depend solely on this segment.

A similar approach has been used to identify two other important functional regions of voltage-gated K^+ channel proteins. The amino-terminal 19 amino acid residues of at least one such protein are involved in rapid *channel inactivation*. If this region is altered, the kinetics of channel inactivation are changed, and if the region is entirely removed, inactivation is abolished. Amazingly, in the latter case, inactivation can be restored by exposing the cytoplasmic face of the plasma membrane to a small synthetic peptide corresponding to the missing amino terminus. These findings suggest that the amino terminus of each K^+ channel subunit acts like a tethered ball that occludes the cytoplasmic end of the pore soon after it opens (Figure 11–28); a similar mechanism is thought to operate in the rapid inactivation of voltage-gated Na^+ channels, although a different segment of the protein seems to be involved. Finally, one of the transmembrane α helices that is highly conserved in all known voltage-gated cation channels contains regularly spaced, positively charged amino acid residues (see Figure 11–27A). This helix has been implicated as the *voltage sensor* in these channels: if any of these charged residues are changed, the response of the channel to shifts in membrane potential is altered.

The same type of analysis, combining recombinant DNA technology and electrophysiological techniques, has been used to characterize another class of ion channels. These channels open in response to the binding of a specific *neurotransmitter* rather than to a change in membrane potential.

Figure 11–28 The "ball-and-chain" model of rapid inactivation for a voltage-gated K^+ channel. When the membrane is depolarized, the channel opens and begins to conduct ions. The open channel is then susceptible to occlusion (inactivation) by the amino-terminal 19 amino acid "ball," which is linked to the channel proper by a segment of unfolded polypeptide chain that serves as the "chain." For simplicity, only two balls are shown; in fact there are four, one from each subunit. A similar mechanism, using a different segment of the polypeptide chain, is thought to operate in Na^+ channel inactivation.

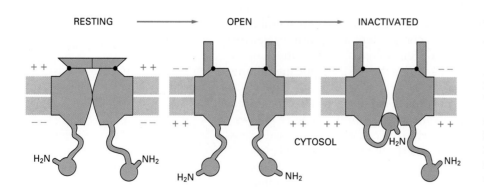

RESTING   OPEN   INACTIVATED

CYTOSOL

Ion Channels and Electrical Properties of Membranes

Transmitter-gated Ion Channels Convert Chemical Signals into Electrical Ones at Chemical Synapses [15]

Neuronal signals are transmitted from cell to cell at specialized sites of contact known as **synapses.** The usual mechanism of transmission is indirect. The cells are electrically isolated from one another, the *presynaptic cell* being separated from the *postsynaptic cell* by a narrow *synaptic cleft.* A change of electrical potential in the presynaptic cell triggers it to release a small signaling molecule known as a **neurotransmitter,** which is stored in membrane-bounded *synaptic vesicles* and is released by exocytosis. The neurotransmitter rapidly diffuses across the synaptic cleft and provokes an electrical change in the postsynaptic cell by binding to *transmitter-gated ion channels* (Figure 11–29). After the neurotransmitter has been secreted, it is rapidly removed, either by specific enzymes in the synaptic cleft or by re-uptake—either by the nerve terminal that released it or by surrounding glial cells. Re-uptake is mediated by a variety of Na^+-dependent neurotransmitter carrier proteins. Rapid removal ensures both spatial and temporal precision of signaling at a synapse: it prevents the neurotransmitter from influencing neighboring cells and clears the synaptic cleft before the next pulse of neurotransmitter is released, so that the timing of repeated, rapid signaling events can be accurately communicated to the postsynaptic cell. As we shall see, signaling via such *chemical synapses* is far more versatile and adaptable than direct electrical coupling via gap junctions at *electrical synapses* (discussed in Chapter 19), which are also used by neurons but to a much lesser extent.

Transmitter-gated ion channels are specialized for rapidly converting extracellular chemical signals into electrical signals at chemical synapses. The channels are concentrated in the plasma membrane of the postsynaptic cell in the region of the synapse and open transiently in response to the binding of neurotransmitter molecules, thereby producing a brief permeability change in the membrane (see Figure 11–29). Unlike the voltage-gated channels responsible for action potentials, transmitter-gated channels are relatively insensitive to the membrane potential and, therefore, cannot by themselves produce a self-amplifying excitation. Instead, they produce local permeability changes, and hence changes of membrane potential, that are graded according to how much neurotransmitter is released at the synapse and how long it persists there. An action potential can be triggered from this site only if the local membrane potential increases enough to open a sufficient number of nearby voltage-gated cation channels that are present in the same target cell membrane.

Chemical Synapses Can Be Excitatory or Inhibitory [15, 21]

Transmitter-gated ion channels differ from one another in several important ways. First, as receptors, they have a highly selective binding site for the neurotransmitter that is released from the presynaptic nerve terminal. Second, as channels, they are selective as to the type of ions that they let pass across the plasma membrane; this determines the nature of the postsynaptic response. **Excitatory neurotransmitters,** such as *acetylcholine, glutamate,* and *serotonin,* open cation channels, causing an influx of Na^+ that depolarizes the postsynaptic membrane toward the threshold potential for firing an action potential. **Inhibitory neurotransmitters,** such as *γ-aminobutyric acid (GABA)* and *glycine,* by contrast, open Cl^- channels, and this suppresses firing by keeping the postsynaptic membrane polarized.

We have already discussed how the opening of cation channels depolarizes a membrane. The effect of opening Cl^- channels can be understood as follows. The concentration of Cl^- is much higher outside the cell than inside (see Table 11–1, p. 508). For this reason opening Cl^- channels will tend to hyperpolarize the membrane by letting more negatively charged chloride ions into the cell, unless the membrane potential is already so negative that it is sufficient to counter the steep Cl^- gradient. (In fact, for many neurons, the equilibrium potential for Cl^-

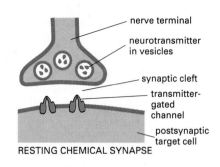

RESTING CHEMICAL SYNAPSE

nerve terminal

neurotransmitter in vesicles

synaptic cleft

transmitter-gated channel

postsynaptic target cell

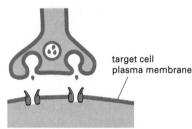

ACTIVE CHEMICAL SYNAPSE

target cell plasma membrane

Figure 11–29 A chemical synapse. When an action potential reaches the nerve terminal, it stimulates the terminal to release its neurotransmitter; the neurotransmitter is contained in synaptic vesicles and is released to the cell exterior when the vesicles fuse with the plasma membrane of the nerve terminal. The released neurotransmitter binds to and opens the transmitter-gated ion channels concentrated in the plasma membrane of the target cell at the synapse. The resulting ion flows alter the membrane potential of the target cell, thereby transmitting a signal from the excited nerve.

is close to the resting potential—or even more negative.) In either case the opening of Cl⁻ channels makes it more difficult to depolarize the membrane and hence to excite the cell. The importance of the inhibitory neurotransmitters is demonstrated by the effects of toxins that block their action: strychnine, for example, by binding to glycine receptors and blocking the action of glycine, causes muscle spasms, convulsions, and death.

Not all chemical signaling in the nervous system, however, operates through ligand-gated ion channels. Many of the signaling molecules that are secreted by nerve terminals, including a large variety of *neuropeptides*, bind to receptors that regulate ion channels only indirectly. These so-called *G-protein-linked receptors* and *enzyme-linked receptors* are discussed in detail in Chapter 15. Whereas signaling mediated by excitatory and inhibitory neurotransmitters binding to transmitter-gated ion channels is generally immediate, simple, and brief, signaling mediated by ligands binding to G-protein-linked receptors and enzyme-linked receptors tends to be far slower and more complex and longer lasting in its consequences.

The Acetylcholine Receptors at the Neuromuscular Junction Are Transmitter-gated Cation Channels [22]

The best-studied example of a transmitter-gated ion channel is the acetylcholine receptor of skeletal muscle cells. This channel is opened transiently by acetylcholine released from the nerve terminal at a **neuromuscular junction**—the specialized chemical synapse between a motor neuron and a skeletal muscle cell (Figure 11–30). This synapse has been intensively investigated because it is readily accessible to electrophysiological study, unlike most of the synapses in the central nervous system.

The acetylcholine receptor has a special place in the history of ion channels. It was the first ion channel to be purified, the first to have its complete amino acid sequence determined, the first to be functionally reconstituted in synthetic lipid bilayers, and the first for which the electrical signal of a single open channel was recorded. Its gene was also the first channel protein gene to be cloned and sequenced. There were at least two reasons for the rapid progress in purifying and characterizing this receptor. First, there is an unusually rich source of the receptor in the electric organs of electric fish and rays (these organs are modified muscles designed to deliver a large electric shock to prey). Second, there are neurotoxins (such as *α-bungarotoxin*) in the venom of certain snakes that bind with high affinity ($K_a = 10^9$ liters/mole) and specificity to the receptor and therefore can be used to purify it by affinity chromatography. Fluorescent or radiolabeled α-bungarotoxin can also be used to localize and count acetylcholine receptors. In this way it has been shown that the receptors are densely packed in the muscle cell plasma membrane at a neuromuscular junction (about 20,000 such receptors/μm²), with relatively few receptors elsewhere in the same membrane.

The acetylcholine receptor of skeletal muscle is composed of five transmembrane polypeptides, two of one kind and three others, encoded by four separate genes. The four genes are strikingly similar in sequence, implying that they evolved from a single ancestral gene. The two identical polypeptides in the pentamer each have binding sites for acetylcholine. When two acetylcholine molecules bind to the pentameric complex, they induce a conformational change that opens the channel. The channel remains open for about 1 millisecond and then closes; like that of the voltage-gated Na⁺ channel, the open form of the acetylcholine receptor channel is short-lived and quickly flips to a closed state of lower free energy (Figure 11–31). Subsequently, the acetylcholine molecules dissociate from the receptor and are hydrolyzed by a specific enzyme (acetylcholinesterase) located in the neuromuscular junction. Once freed of its bound neurotransmitter, the acetylcholine receptor reverts to its initial resting state.

The general shape of the acetylcholine receptor and the likely arrangement of its subunits have been determined by a combination of electron microscopy

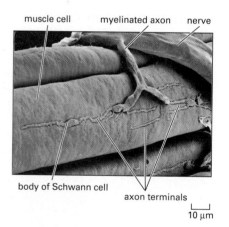

muscle cell   myelinated axon   nerve

body of Schwann cell

axon terminals

10 μm

Figure 11–30 Low-magnification scanning electron micrograph of a neuromuscular junction in a frog. The termination of a single axon on a skeletal muscle cell is shown. (From J. Desaki and Y. Uehara, *J. Neurocytol.* 10:101–110, 1981, by permission of Chapman & Hall.)

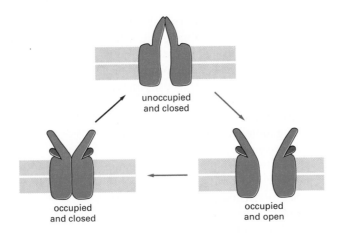

Figure 11–31 **Three conformations of the acetylcholine receptor.** The binding of two acetylcholine molecules opens this transmitter-gated ion channel. But even with acetylcholine bound, the receptor is thought to stay in the open conformation only briefly, before the channel recloses. The acetylcholine then dissociates from the receptor, which enables the receptor to return to its original conformation.

unoccupied and closed

occupied and closed

occupied and open

and low-angle x-ray diffraction of two-dimensional crystals: the five subunits are arranged in a ring to form a water-filled transmembrane channel that consists of a narrow pore through the lipid bilayer bounded by wide cylindrical entrances (Figure 11–32). Clusters of negatively charged amino acid residues at either end of the pore help to exclude negative ions and to encourage any positive ion of diameter less than 0.65 nm to pass through. The normal traffic consists chiefly of Na^+ and K^+, together with some Ca^{2+}. Thus, unlike voltage-gated cation channels, there is little selectivity among cations, and the relative contributions of the different cations to the current through the channel depend chiefly on their concentrations and on the electrochemical driving forces. When the muscle cell membrane is at its resting potential, the net driving force for K^+ is near zero, since the voltage gradient nearly balances the K^+ concentration gradient across the membrane (see Panel 11–2, p. 526). For Na^+, on the other hand, the voltage gradient and the concentration gradient both act in the same direction to drive the ion into the cell. (The same is true for Ca^{2+}, but the extracellular concentration of Ca^{2+} is so much lower than that of Na^+ that Ca^{2+} makes only a small contribution to the total inward current.) Therefore, the opening of the acetylcholine receptor channels leads to a large net influx of Na^+ (peak rate of about 30,000 ions per channel each millisecond). This influx causes a membrane depolarization that signals the muscle to contract, as discussed below.

Transmitter-gated Ion Channels Are Major Targets for Psychoactive Drugs [23]

The ion channels that open directly in response to the neurotransmitters acetylcholine, serotonin, GABA, and glycine contain subunits that are structurally similar, suggesting that they are evolutionarily related and probably form transmem-

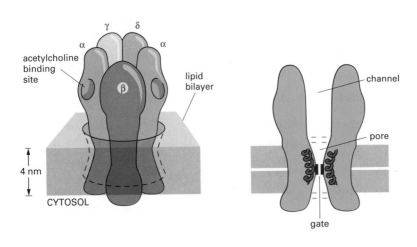

Figure 11–32 **A model for the structure of the acetylcholine receptor.** Five homologous subunits (α, α, β, γ, δ) combine to form a transmembrane aqueous pore. The pore is lined by a ring of five transmembrane α helices, one contributed by each subunit. The ring of α helices is probably surrounded by a continuous rim of transmembrane β sheet, made up of the other transmembrane segments of the five subunits. In its closed conformation the pore is thought to be occluded by the hydrophobic side chains of five leucine residues, one from each α helix, which form a gate near the middle of the lipid bilayer. The negatively charged side chains at either end of the pore ensure that only positively charged ions pass through the channel. Both of the α subunits contain an acetylcholine binding site; when acetylcholine binds to both sites, the channel undergoes a conformational change that opens the gate, possibly by causing the leucine residues to move outward. (Adapted from N. Unwin, *Cell/Neuron* 72/10[Suppl.] 31–41, 1993. © Cell Press.)

acetylcholine binding site

lipid bilayer

4 nm

CYTOSOL

channel

pore

gate

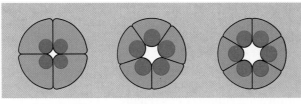

Figure 11–33 **Three classes of channel proteins.** The postulated relationship between the number of protein subunits and pore diameter. (Adapted from B. Hille, Ionic Channels of Excitable Membranes, 2nd ed. Sunderland, MA: Sinauer, 1992.)

voltage-gated cation channels

transmitter-gated ion channels

gap junctions

brane pores in the same way, even though their neurotransmitter-binding specificities and ion selectivities are distinct. Glutamate-gated ion channels are constructed from a distinct family of subunits but, nonetheless, seem to have a similar overall structure. In each case the channel is formed by homologous polypeptide subunits, which probably form a pentamer resembling the acetylcholine receptor (see Figure 11–32). Comparison of Figures 11–27 and 11–32 emphasizes the contrast with voltage-gated cation channels, which are constructed from a ring of four subunits (or domains). Perhaps as a result, the transmitter-gated channels have wider pores, which are correspondingly less stringent in their ion selectivity. Gap junctions, which are constructed from a ring of six subunits, have even wider pores, which permit the passage of small organic molecules in addition to inorganic ions (discussed in Chapter 19). This possible relationship between subunit number and pore width is illustrated in Figure 11–33.

For each class of transmitter-gated ion channels, alternative forms of each type of subunit exist, either encoded by distinct genes or generated by alternative RNA splicing of the same gene product. These combine in different combinations to form an extremely diverse set of distinct channel subtypes, with different ligand affinities, different channel conductances, different rates of opening and closing, and different sensitivities to drugs and toxins. Vertebrate neurons, for example, have acetylcholine-gated ion channels that differ from those of muscle cells in that they are formed usually from two subunits of one type and three of another; but there are at least seven genes coding for different versions of the first type of subunit and at least three coding for different versions of the second, with further diversity due to alternative RNA splicing. Subsets of acetylcholine-sensitive neurons serving different functions in the brain are characterized by different combinations of these subunits. This, in principle and already to some extent in practice, makes it possible to design drugs targeted against narrowly defined groups of neurons or synapses, thereby influencing particular brain functions specifically. Indeed, transmitter-gated ion channels have for a long time been important targets for drugs. A surgeon, for example, can make muscles relax for the duration of an operation by blocking the acetylcholine receptors on skeletal muscle cells with *curare*, a drug from a plant that was originally used by South American Indians to poison arrows. Most drugs used in the treatment of insomnia, anxiety, depression, and schizophrenia exert their effects at chemical synapses, and many of these act by binding to transmitter-gated channels: both barbiturates and tranquilizers such as Valium and Librium, for example, bind to GABA receptors, potentiating the inhibitory action of GABA by allowing lower concentrations of this neurotransmitter to open Cl^- channels. The new molecular biology of ion channels, by revealing both their diversity and the details of their structure, holds out the hope of designing a new generation of psychoactive drugs that will act still more selectively to alleviate the miseries of mental illness.

Ion channels are the basic molecular components from which neuronal devices for signaling and computation are built. To provide a glimpse of how sophisticated the functions of these devices can be, we consider several examples that demonstrate how groups of ion channels work together in synaptic communication between electrically excitable cells.

Ion Channels and Electrical Properties of Membranes

Neuromuscular Transmission Involves the Sequential Activation of Five Different Sets of Ion Channels [15]

The importance of ion channels to electrically excitable cells can be illustrated by following the process whereby a nerve impulse stimulates a muscle cell to contract. This apparently simple response requires the sequential activation of five different sets of ion channels—all within a few milliseconds (Figure 11–34).

1. The process is initiated when the nerve impulse reaches the nerve terminal and depolarizes the plasma membrane of the terminal. The depolarization transiently opens voltage-gated Ca^{2+} channels in this membrane. As the Ca^{2+} concentration outside cells is more than 1000 times greater than the free Ca^{2+} concentration inside, Ca^{2+} flows into the nerve terminal. The increase in Ca^{2+} concentration in the cytosol of the nerve terminal triggers the localized release of acetylcholine into the synaptic cleft.

2. The released acetylcholine binds to acetylcholine receptors in the muscle cell plasma membrane, transiently opening the cation channels associated with them. The resulting influx of Na^+ causes a localized membrane depolarization.

3. The local depolarization of the muscle cell plasma membrane opens voltage-gated Na^+ channels in this membrane, allowing more Na^+ to enter, which further depolarizes the membrane. This, in turn, opens neighboring voltage-gated Na^+ channels and results in a self-propagating depolarization (an action potential) that spreads to involve the entire plasma membrane (see Figure 11–23).

4. The generalized depolarization of the muscle cell plasma membrane activates voltage-gated Ca^{2+} channels in specialized regions (the transverse [T] tubules—discussed in Chapter 16) of this membrane. This, in turn, causes *Ca^{2+} release channels* in an adjacent region of the sarcoplasmic reticulum membrane to open transiently and release the Ca^{2+} stored in the sarcoplasmic reticulum into the cytosol. It is the sudden increase in the cytosolic Ca^{2+} concentration that causes the myofibrils in the muscle cell to contract. It is not certain how the activation of the voltage-gated Ca^{2+} channels in the T-tubule membrane leads to the opening of the Ca^{2+} release channels in the sarcoplasmic reticulum membrane. The two membranes are closely apposed, however, with the two types of channels joined together in a specialized structure (see Figure 16–92). It is possible, therefore, that a voltage-induced change in the conformation of the plasma membrane Ca^{2+} channel directly opens the Ca^{2+} release channel in the sarcoplasmic reticulum through a mechanical coupling (discussed in Chapter 16).

Figure 11–34 The system of ion channels at a neuromuscular junction. These gated ion channels are essential for the stimulation of muscle contraction by a nerve impulse. The various channels are numbered in the sequence in which they are activated, as described in the text.

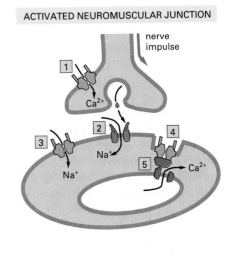

RESTING NEUROMUSCULAR JUNCTION

VOLTAGE-GATED Ca^{2+} CHANNEL

nerve terminal

acetylcholine

ACETYLCHOLINE-GATED CATION CHANNEL

VOLTAGE-GATED Ca^{2+} CHANNEL

VOLTAGE-GATED Na^+ CHANNEL

sarcoplasmic reticulum

muscle plasma membrane

GATED Ca^{2+} RELEASE CHANNEL

ACTIVATED NEUROMUSCULAR JUNCTION

nerve impulse

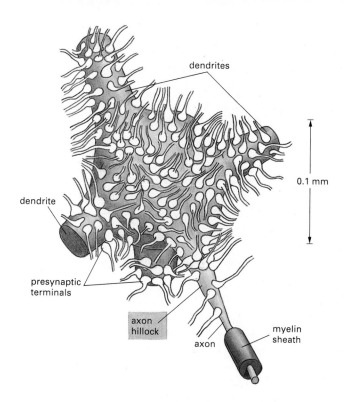

Figure 11–35 **A motor neuron cell body in the spinal cord.** Many thousands of nerve terminals synapse on the cell body and dendrites. These deliver signals from other parts of the organism to control the firing of action potentials along the single axon of this large cell.

dendrites

0.1 mm

dendrite

presynaptic terminals

axon hillock

axon

myelin sheath

While the activation of muscle contraction by a motor neuron is complex, an even more sophisticated interplay of ion channels is required for a neuron to integrate a large number of input signals at synapses and compute an appropriate output, as we now discuss.

The Grand Postsynaptic Potential in a Neuron Represents a Spatial and Temporal Summation of Many Small Postsynaptic Potentials [15, 24]

In the central nervous system a single neuron can receive inputs from thousands of other neurons. Several thousand nerve terminals, for example, make synapses on an average motor neuron in the spinal cord; its cell body and dendrites are almost completely covered with them (Figure 11–35). Some of these synapses transmit signals from the brain or spinal cord; others bring sensory information from muscles or from the skin. The motor neuron must combine the information received from all these sources and react either by firing action potentials along its axon or by remaining quiet.

Of the many synapses on a neuron, some will tend to excite it, others to inhibit it. Neurotransmitter released at an excitatory synapse causes a small depolarization in the postsynaptic membrane called an *excitatory postsynaptic potential (excitatory PSP),* while neurotransmitter released at an inhibitory synapse generally causes a small hyperpolarization called an *inhibitory PSP.* Because the membrane of the dendrites and cell body of most neurons contains few voltage-gated Na^+ channels, an individual excitatory PSP generally does not trigger an action potential. Instead, each incoming signal is reflected in a local PSP of graded magnitude, which decreases with distance from the site of the synapse. If signals arrive simultaneously at several synapses in the same region of the dendritic tree, the total PSP in that neighborhood will be roughly the sum of the individual PSPs, with inhibitory PSPs making a negative contribution to the total. The PSPs from each neighborhood spread passively and converge on the cell body. Because the cell body is small compared with the dendritic tree, its membrane potential will be roughly uniform and will be a composite of the effects of all the signals impinging on the cell, weighted according to the distances of the synapses from the

Ion Channels and Electrical Properties of Membranes

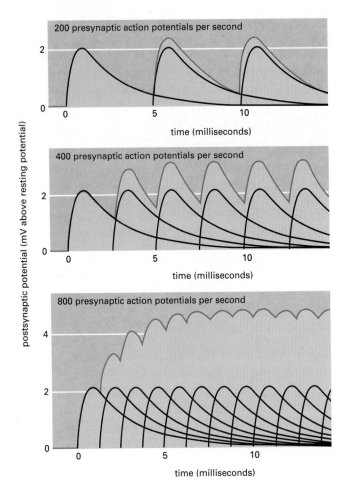

Figure 11–36 **The principle of temporal summation.** Each presynaptic action potential arriving at a synapse produces a small postsynaptic potential, or PSP (*black lines*). When successive action potentials arrive at the same synapse, each PSP produced adds to the tail of the preceding one to produce a larger combined PSP (*green lines*). The greater the frequency of incoming action potentials, the greater the size of the combined PSP.

cell body. The **grand postsynaptic potential (grand PSP)** of the cell body is thus said to represent a **spatial summation** of all the stimuli being received. If excitatory inputs predominate, the grand PSP will be a depolarization; if inhibitory inputs predominate, it will usually be a hyperpolarization.

While spatial summation combines the effects of signals received at different sites on the membrane, **temporal summation** combines the effects of signals received at different times. If an action potential arrives at a synapse and triggers neurotransmitter release before a previous PSP at the synapse has decayed completely, the second PSP adds to the remaining tail of the first. If many action potentials arrive in quick succession, each PSP adds to the tail of the preceding PSP, building up to a large sustained average PSP whose magnitude reflects the rate of firing of the presynaptic neuron (Figure 11–36). This is the essence of temporal summation: it translates the *frequency* of incoming signals into the *magnitude* of a net PSP.

Temporal and spatial summation together provide the means by which the rates of firing of many presynaptic neurons jointly control the membrane potential (the grand PSP) in the body of a single postsynaptic cell. The final step in the neuronal computation made by the postsynaptic cell is the generation of an output, usually in the form of action potentials, to relay a signal to other cells. The output signal reflects the magnitude of the grand PSP in the cell body. While the grand PSP is a continuously graded variable, however, action potentials are always all-or-nothing and uniform in size. The only variable in signaling by action potentials is the time interval between one action potential and the next. For long-distance transmission the magnitude of the grand PSP is therefore translated, or *encoded,* into the *frequency* of firing of action potentials (Figure 11–37). This encoding is achieved by a special set of gated ion channels that are present

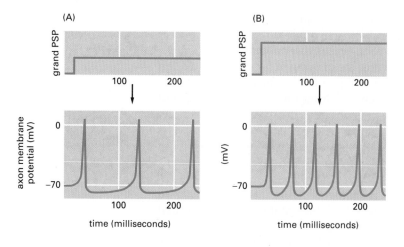

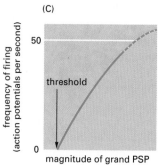

Figure 11–37 **The encoding of the grand PSP in the form of the frequency of firing of action potentials by an axon.** A comparison of (A) and (B) shows how the firing frequency of an axon increases with an increase in the grand PSP, while (C) summarizes the general relationship.

at high density at the base of the axon, adjacent to the cell body, in a region known as the *axon hillock* (see Figure 11–35).

Neuronal Computation Requires a Combination of At Least Three Kinds of K⁺ Channels [15, 25]

We have seen that the intensity of stimulation received by a neuron is encoded for long-distance transmission as the frequency of action potentials the neuron fires: the stronger the stimulation, the higher the frequency of action potentials. Action potentials are initiated at the **axon hillock,** a unique region of each neuron where voltage-gated Na⁺ channels are plentiful. But to perform its special function of encoding, the membrane of the axon hillock also contains at least four other classes of ion channels—three selective for K⁺ and one selective for Ca²⁺. The three varieties of K⁺ channels have different properties; we shall refer to them as the *delayed,* the *early,* and the *Ca²⁺-activated K⁺ channels.* To understand the need for multiple types of channels, consider first what would happen if the only voltage-gated ion channels present in the nerve cell were the Na⁺ channels. Below a certain threshold level of synaptic stimulation, the depolarization of the axon hillock membrane would be insufficient to trigger an action potential. With gradually increasing stimulation, the threshold would be crossed; the Na⁺ channels would open; and an action potential would fire. The action potential would be terminated in the usual way by inactivation of the Na⁺ channels. Before another action potential could fire, these channels would have to recover from their inactivation. But that would require a return of the membrane voltage to a very negative value, which would not occur as long as the strong depolarizing stimulus (from PSPs) was maintained. An additional channel type is needed, therefore, to repolarize the membrane after each action potential to prepare the cell to fire again. This task is performed by the **delayed K⁺ channels,** which we discussed previously in relation to the propagation of the action potential (see p. 529). They are voltage-gated, but because of their slower kinetics, they open only during the falling phase of the action potential, when the Na⁺ channels are inactive. Their opening permits an efflux of K⁺ that drives the membrane back toward the K⁺ equilibrium potential, which is so negative that the Na⁺ channels rapidly recover from their inactivated state. Repolarization of the membrane also causes the delayed K⁺ channels to close. The axon hillock is now reset so that the depolarizing stimulus from synaptic inputs can fire another action potential. In this way, sustained stimulation of the dendrites and cell body leads to repetitive firing of the axon.

Repetitive firing in itself, however, is not enough. The frequency of the firing has to reflect the intensity of the stimulation, and a simple system of Na⁺ channels and delayed K⁺ channels is inadequate for this purpose. Below a certain threshold level of steady stimulation, the cell will not fire at all; above that threshold it will abruptly begin to fire at a relatively rapid rate. The **early K⁺ channels** solve the problem. These too are voltage-gated and open when the membrane is depolarized, but their specific voltage sensitivity and kinetics of inactivation are such that they act to reduce the rate of firing at levels of stimulation that are only just above the threshold required for firing. Thus they remove the discontinuity in the relationship between the firing rate and the intensity of stimulation. The result is a firing rate that is proportional to the strength of the depolarizing stimulus over a very broad range (see Figure 11–37).

The process of encoding is usually further modulated by the two other types of ion channels in the axon hillock that were mentioned at the outset—*voltage-gated Ca^{2+} channels* and *Ca^{2+}-activated K^+ channels*. They act together to decrease the response of the cell to an unchanging, prolonged stimulation—a process called **adaptation.** The Ca^{2+} channels are similar to the Ca^{2+} channels that mediate release of neurotransmitter from presynaptic axon terminals; they open when an action potential fires, transiently allowing Ca^{2+} into the axon hillock. The **Ca^{2+}-activated K^+ channel** is both structurally and functionally different from any of the channel types described earlier. It opens in response to a raised concentration of Ca^{2+} at the cytoplasmic face of the nerve cell membrane. Suppose that a strong depolarizing stimulus is applied for a long time, triggering a long train of action potentials. Each action potential permits a brief influx of Ca^{2+} through the voltage-gated Ca^{2+} channels, so that the intracellular Ca^{2+} concentration gradually builds up to a level high enough to open the Ca^{2+}-activated K^+ channels. Because the resulting increased permeability of the membrane to K^+ makes the membrane harder to depolarize, it increases the delay between one action potential and the next. In this way a neuron that is stimulated continuously for a prolonged period becomes gradually less responsive to the constant stimulus. Such adaptation, which can also occur by other mechanisms, allows a neuron, and indeed the nervous system generally, to react sensitively to *change,* even against a high background level of steady stimulation. It is one of the strategies that help us, for example, to feel a light touch on the shoulder and yet ignore the constant pressure of our clothing. We discuss adaptation in more detail in Chapter 15.

Other neurons do different computations, reacting to their synaptic inputs in myriad ways, reflecting the different assortments of members of the various ion channel families that they have in their membranes. There are, for example, at least five known types of voltage-gated Ca^{2+} channels in the vertebrate nervous system and at least four known types of voltage-gated K^+ channels. The multiplicity of genes evidently allows for a host of different types of neurons, whose electrical behavior is specifically tuned to the particular tasks they must perform.

One of the crucial properties of the nervous system is its ability to learn and remember, which seems to depend largely on long-term changes in specific synapses. We end this chapter by considering a remarkable type of ion channel that has a special role in some forms of learning and memory. It is located at many synapses in the central nervous system, where it is gated by both voltage and the excitatory neurotransmitter glutamate. It is also the site of action of the psychoactive drug phencyclidine, or angel dust.

Long-term Potentiation in the Mammalian Hippocampus Depends on Ca²⁺ Entry Through NMDA-Receptor Channels [26]

Practically all animals can learn, but mammals seem to learn exceptionally well (or so we like to think). In a mammal's brain the *hippocampus*, a part of the cerebral cortex, plays a special role in learning: when it is destroyed on both sides of the brain, the ability to form new memories is largely lost, although previous long-established memories remain. Correspondingly, some synapses in the hippocampus show dramatic functional alterations with repeated use: whereas oc-

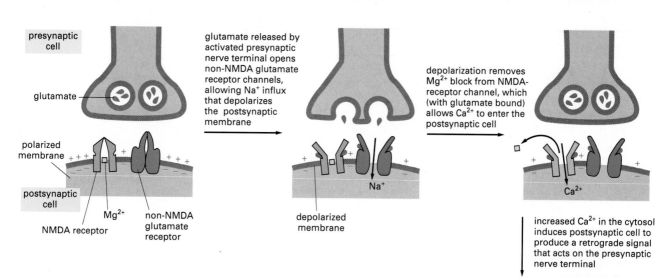

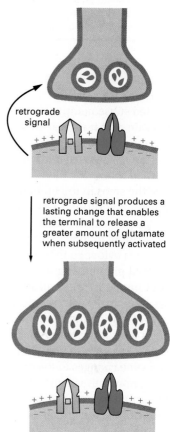

glutamate released by activated presynaptic nerve terminal opens non-NMDA glutamate receptor channels, allowing Na⁺ influx that depolarizes the postsynaptic membrane

depolarization removes Mg^{2+} block from NMDA-receptor channel, which (with glutamate bound) allows Ca^{2+} to enter the postsynaptic cell

increased Ca^{2+} in the cytosol induces postsynaptic cell to produce a retrograde signal that acts on the presynaptic nerve terminal

retrograde signal produces a lasting change that enables the terminal to release a greater amount of glutamate when subsequently activated

Figure 11–38 The signaling events in long-term potentiation.

casional single action potentials in the presynaptic cells leave no lasting trace, a short burst of repetitive firing causes **long-term potentiation (LTP),** such that subsequent single action potentials in the presynaptic cells evoke a greatly enhanced response in the postsynaptic cells. The effect lasts hours, days, or weeks, according to the number and intensity of the bursts of repetitive firing. Only the synapses that were activated show the potentiation; synapses that have remained quiet on the same postsynaptic cell are not affected. But if, while the cell is receiving a burst of repetitive stimulation via one set of synapses, a single action potential is delivered at *another* synapse on its surface, that latter synapse also will undergo long-term potentiation, even though a single action potential delivered there at another time would leave no such lasting trace.

The underlying rule in the hippocampus seems to be that *long-term potentiation occurs on any occasion where a presynaptic cell fires (once or more) at a time when the postsynaptic membrane is strongly depolarized* (either through recent repetitive firing of the same presynaptic cell or by other means). There is good evidence that this rule reflects the behavior of a particular class of ion channels in the postsynaptic membrane. Glutamate is the main excitatory neurotransmitter in the mammalian central nervous system, and in the hippocampus, as elsewhere, most of the depolarizing current responsible for excitatory PSPs is carried by glutamate-gated ion channels that operate in the standard way. But the current has in addition a second and more intriguing component, which is mediated by a separate subclass of glutamate-gated ion channels, known as **NMDA receptors** because they are selectively activated by the artificial glutamate analog N-methyl-D-aspartate. The NMDA-receptor channels are doubly gated, opening only when two conditions are satisfied simultaneously: glutamate must be bound to the receptor, and the membrane must be strongly depolarized. The second condition is required to release Mg^{2+} that normally blocks the resting channel, and it means that NMDA receptors are normally only activated when conventional glutamate-gated ion channels are activated as well and depolarize the membrane. The NMDA receptors are critical for long-term potentiation. When they are selectively blocked with a specific inhibitor, long-term potentiation does not occur, even though ordinary synaptic transmission continues. An animal treated with this inhibitor shows specific deficits in its learning abilities but behaves almost normally otherwise.

How do the NMDA receptors mediate such a remarkable effect? The answer is that these channels, when open, are highly permeable to Ca^{2+}, which acts as an intracellular mediator in the postsynaptic cell, triggering a cascade of changes that are responsible for long-term potentiation. Thus long-term potentiation is prevented when Ca^{2+} levels are held artificially low in the postsynaptic cell by injecting the Ca^{2+} chelator EGTA into it and can be induced by transiently raising extracellular Ca^{2+} levels artificially high.

Table 11–3 Some Ion Channel Families

Family*	Representative Subfamilies	
Voltage-gated cation channels	voltage-gated Na⁺ channels voltage-gated K⁺ channels (including delayed and early) voltage-gated Ca²⁺ channels	
Transmitter-gated ion channels	acetylcholine-gated cation channels serotonin-gated cation channels glutamate-gated cation channels**	} excitatory
	GABA-gated Cl⁻ channels glycine-gated Cl⁻ channels	} inhibitory

*The members of a family are similar in amino acid sequence and, therefore, are thought to have derived from a common ancestor; within subfamilies the resemblances are usually even closer.
**These channels are formed by a distinct family of subunits but are thought to have a similar overall structure to the other transmitter-gated ion channels.

The long-term changes, although initiated in the postsynaptic cell, affect the presynaptic cell as well so that it releases more glutamate than normal when it is activated subsequently. The nature of the lasting change in the presynaptic cell is uncertain, but it is clear that some message must pass retrogradely from the postsynaptic cell to the presynaptic cell when long-term potentiation is induced. The nature of the retrograde signal is also unknown, although both nitric oxide and carbon monoxide have been suggested as candidates. A tentative model of some of the steps in the induction of long-term potentiation is presented in Figure 11–38. In addition to the long-lasting changes in the presynaptic cell illustrated in Figure 11–38, there are also long-lasting changes in the postsynaptic cell that contribute to long-term potentiation.

There is evidence that NMDA receptors play an important part in learning and related phenomena in other parts of the brain as well as in the hippocampus. In Chapter 21 we shall see, moreover, that NMDA receptors have a crucial role in adjusting the anatomical pattern of synaptic connections in the light of experience during the development of the nervous system.

Thus neurotransmitters released at synapses, besides relaying transient electrical signals, can also alter concentrations of intracellular mediators that bring about lasting changes in the efficacy of synaptic transmission. It is still uncertain, however, how these changes endure for weeks, months, or a lifetime in the face of the normal turnover of cell constituents.

Some of the ion channel families that we have discussed are summarized in Table 11–3.

Summary

Channel proteins form aqueous pores across the lipid bilayer and allow inorganic ions of appropriate size and charge to cross the membrane down their electrochemical gradients at rates that are about 1000 times greater than those achieved by any known carrier. These ion channels are "gated" and usually open transiently in response to a specific perturbation in the membrane, such as a change in membrane potential (voltage-gated channels) or the binding of a neurotransmitter (transmitter-gated channels).

K⁺-selective leak channels play an important part in determining the resting membrane potential across the plasma membrane in most animal cells. Voltage-gated cation channels are responsible for the generation of self-amplifying action potentials in electrically excitable cells such as neurons and skeletal muscle cells.

Transmitter-gated ion channels convert chemical signals to electrical signals at chemical synapses: excitatory neurotransmitters, such as acetylcholine and glutamate, open transmitter-gated cation channels and thereby depolarize the postsynaptic membrane toward the threshold potential for firing an action potential; inhibitory neurotransmitters, such as GABA and glycine, open transmitter-gated Cl⁻ channels and thereby suppress firing by keeping the postsynaptic membrane polarized. A subclass of glutamate-gated ion channels, called NMDA-receptor channels, are highly permeable to Ca^{2+}, which can trigger the long-term changes in synapses that are thought to be involved for some forms of learning and memory.

Ion channels work together in complex ways to control the behavior of electrically excitable cells. A typical neuron, for example, receives thousands of excitatory and inhibitory inputs, which combine by spatial and temporal summation to produce a grand postsynaptic potential (PSP) in the cell body. The magnitude of the grand PSP is translated into the rate of firing of action potentials by a mixture of cation channels in the membrane of the axon hillock.

References

Cited

1. Hille, B. Ionic Channels of Excitable Membranes, 2nd ed. Sunderland, MA: Sinauer, 1992.

   Martonosi, A.N., ed. The Enzymes of Biological Membranes, Vol. 3, Membrane Transport, 2nd ed. New York: Plenum Press, 1985.

   Stein, W.D. Channels, Carriers and Pumps: An Introduction to Membrane Transport. San Diego, CA: Academic Press, 1990.

   Tosteson, D.C., ed. Membrane Transport: People and Ideas. Bethesda, MD: American Physiology Society, 1989.

2. Anderson, O.S. Permeability properties of unmodified lipid bilayer membranes. In Membrane Transport in Biology (G. Giebisch, D.C. Tosteson, H.H. Ussing, eds.), Vol. 1, pp. 369–446. New York: Springer-Verlag, 1978.

   Finkelstein, A. Water movement through membrane channels. *Curr. Top. Membr. Transp.* 21:295–308, 1984.

   Walter, A.; Gutknecht, J. Permeability of small nonelectrolytes through lipid bilayer membranes. *J. Membr. Biol.* 90:207–217, 1986.

3. Stein, W.D., ed. Ion Pumps: Structure, Function and Regulation. New York: Liss, 1988.

   Tanford, C. Mechanism of free energy coupling in active transport. *Annu. Rev. Biochem.* 52:379–409, 1983.

4. Griffith, J.K., et al. Membrane transport proteins: implications of sequence comparisons. *Curr. Opin. Cell Biol.* 4:684–695, 1992.

   Kaback, H.R. Molecular biology of active transport: from membrane to molecule to mechanism. *Harvey Lect.* 83:77–105, 1989.

   Lodish, H.F. Anion-exchange and glucose transport proteins: structure, function and distribution. *Harvey Lect.* 82:19–46, 1988.

   Numa, S. A molecular view of neurotransmitter receptors and ionic channels. *Harvey Lect.* 83:121–165, 1989.

   Seeburg, P.H. The molecular biology of mammalian glutamate receptor channels. *Trends Neurosci.* 16:359–365, 1993.

5. Dobler, M. Ionophores and Their Structures. New York: Wiley-Interscience, 1981.

   Pressman, B.C. Biological applications of ionophores. *Annu. Rev. Biochem.* 45:501–530, 1976.

   Wallace, B.A. Gramicidin channels and pores. *Annu. Rev. Biophys. Biophys. Chem.* 19:127–157, 1990.

6. Henderson, P.J.F. The 12-transmembrane helix transporters. *Curr. Opin. Cell Biol.* 5:708–721, 1993.

   Läuger, P. Electrogenic Ion Pumps. Sunderland, MA: Sinauer, 1991.

   Stein, W.D. Transport and Diffusion Across Cell Membranes. Orlando, FL: Academic Press, 1986.

7. Glynn, I.M.; Ellory, C., eds. The Sodium Pump. Cambridge, UK: Company of Biologists, 1985.

   Horisberger, J.D.; Lemas, V.; Kraehenbühl, J.-P.; Rossier, B.C. Structure-function relationship of Na,K-ATPase. *Annu. Rev. Physiol.* 53:565–584, 1991.

   Mercer, R.W. Structure of the Na,K-ATPase. *Int. Rev. Cytol.* 137C:139–168, 1993.

   Shull, G.E.; Schwartz, A.; Lingrel, J.B. Amino acid sequence of the catalytic subunit of the (Na^+-K^+) ATPase deduced from a complementary DNA. *Nature* 316:619–695, 1985.

8. Glynn, I.M. The Na^+-K^+ transporting adenosine triphosphatase. In The Enzymes of Biological Membranes, 2nd ed. (A. Martonosi, ed.), Vol. 3, pp. 34–114. New York: Plenum Press, 1985.

   Sweadner, K.J.; Goldin, S.M. Active transport of sodium and potassium ions: mechanism, function and regulation. *New Engl. J. Med.* 302:777–783, 1980.

9. Carafoli, E. Calcium pump of the plasma membrane. *Physiol. Rev.* 71:129–153, 1991.

   Jencks, W.P. How does a calcium pump pump calcium? *J. Biol. Chem.* 264:18855–18858, 1989.

   MacLennan, D.H.; Brandl, C.J.; Korczak, B.; Green, N.M. Amino acid sequence of a Ca^{2+}, Mg^{2+}-dependent ATPase from rabbit muscle sarcoplasmic reticulum, deduced from its complementary DNA sequence. *Nature* 316:696–700, 1985.

   Sachs, G.; Munson, K. Mammalian phosphorylating ion-motive ATPases. *Curr. Opin. Cell Biol.* 3:685–694, 1991.

Schatzmann, H.J. The calcium pump of the surface membrane and of the sarcoplasmic reticulum. *Annu. Rev. Physiol.* 51:473–486, 1989.

10. Hinkle, P.C.; McCarty, R.E. How cells make ATP. *Sci. Am.* 238(3):104–123, 1978.

Nicholls, D.G. An Introduction to the Chemiosmotic Theory, 2nd ed. San Diego, CA: Academic Press, 1987.

Senior, A.E. ATP synthesis by oxidative phosphorylation. *Physiol. Rev.* 68:177–231, 1988.

11. Kinne, R.; Hannafin, J.A.; Konig, B. Role of the NaCl-KCl cotransport system in active chloride absorption and secretion. *Ann. N.Y. Acad. Sci.* 456:198–206, 1985.

Scott, D.M. Sodium cotransport systems: cellular, molecular and regulatory aspects. *Bioessays* 7:71–78, 1987.

Semenza, G.; Kessler, M.; Schmidt, U.; Venter, J.C.; Fraser, C.M. The small-intestinal sodium glucose cotransporter(s). *Ann. N.Y. Acad. Sci.* 456:83–96, 1985.

Wright, E.M.; Hager, K.M.; Turk, E. Sodium cotransport proteins. *Curr. Opin. Cell Biol.* 4:696–702, 1992.

Wright, J.K.; Seckler, R.; Overath, P. Molecular aspects of sugar:ion transport. *Annu. Rev. Biochem.* 55:225–248, 1986.

12. Boron, W.F. Intracellular pH regulation in epithelial cells. *Annu. Rev. Physiol.* 48:377–388, 1986.

Boron, W.F. Intracellular pH Regulation. In Membrane Transport Processes in Organized Systems (T.E. Andreoli, J.F. Hoffman, D.D. Fanestil, S.G. Schultz, eds.), pp. 39–51. New York: Plenum Press, 1987.

Chesler, M.; Kaila, K. Modulation of pH by neuronal activity. *Trends Neurosci.* 15:396–402, 1992.

Rudnick, G. ATP-driven H^+-pumping into intracellular organelles. *Annu. Rev. Physiol.* 48:403–413, 1986.

Thomas, R.C. Cell growth factors bicarbonate and pH_i response. *Nature* 337:601, 1989.

13. Almers, W.; Stirling, C. Distribution of transport proteins over animal cell membranes. *J. Membr. Biol.* 77:169–186, 1984.

Handler, J.S. Overview of epithelial polarity. *Annu. Rev. Physiol.* 51:729–740, 1989.

14. Ames, G.F.L. Bacterial periplasmic transport systems: structure, mechanism and evolution. *Annu. Rev. Biochem.* 55:397–425, 1986.

Gottesman, M.M.; Pastan, I. Biochemistry of multidrug resistance mediated by the multidrug transporter. *Annu. Rev. Biochem.* 62:385–427, 1993.

Higgins, C.F. ABC transporters: from microorganisms to man. *Annu. Rev. Cell Biol.* 8:67–113, 1992.

Kartner, N.; Ling, V. Multidrug resistance in cancer. *Sci. Am.* 261(3):44–51, 1989.

Welsh, M.J.; Anderson, M.P.; Rich, D.P.; et al. Cystic fibrosis transmembrane conductance regulator: a chloride channel with novel regulation. *Neuron* 8:821–829, 1992.

15. Hall, Z.W. An Introduction to Molecular Neurobiology, pp. 33–178. Sunderland, MA: Sinauer, 1992.

Hille, B. Ionic Channels of Excitable Membranes, 2nd ed. Sunderland, MA: Sinauer, 1992.

Jessell, T.M.; Kandel, E.R. Synaptic transmission: a bidirectional and self-modifiable form of cell-cell communication. *Cell* 72 (Suppl. 1):1–30, 1993.

Kandel, E.R.; Schwartz, J.H.; Jessell, T.M. Principles of Neural Science, 3rd ed., pp. 34–224. New York: Elsevier, 1991.

Nicholls, J.G.; Martin, A.R.; Wallace, B.G. From Neuron to Brain, 3rd ed. Sunderland, MA: Sinauer, 1992.

Unwin, N. The structure of ion channels in membranes of excitable cells. *Neuron* 3:665–676, 1989.

16. Baker, P.F.; Hodgkin, A.L.; Shaw, T.L. The effects of changes in internal ionic concentration on the electrical properties of perfused giant axons. *J. Physiol.* 164:355–374, 1962.

Hodgkin, A.L.; Keynes, R.D. Active transport of cations in giant axons from *Sepia* and *Loligo*. *J. Physiol.* 128:26–60, 1955.

17. Hodgkin, A.L.; Huxley, A.F. Currents carried by sodium and potassium ions through the membrane of the giant axon of *Loligo*. *J. Physiol.* 116:449–472, 1952.

Hodgkin, A.L.; Huxley, A.F. A quantitative description of membrane current and its application to conduction and excitation in nerve. *J. Physiol.* 117:500–544, 1952.

Katz, B. Nerve, Muscle and Synapse. New York: McGraw-Hill, 1966.

18. Huxley, A.F.; Stämpfli, R. Evidence for saltatory conduction in peripheral myelinated nerve fibres. *J. Physiol.* 108:315–339, 1949.

Morell, P., ed. Myelin, 2nd ed. New York: Plenum Press, 1984.

Morell, P.; Norton, W.T. Myelin. *Sci. Am.* 242(5):88–118, 1980.

19. Neher, E.; Sakmann, B. The patch clamp technique. *Sci. Am.* 266(3):28–35, 1992.

Sakmann, B. Elementary steps in synaptic transmission revealed by currents through single ion channels. *Science* 256:503–512, 1992.

20. Armstrong, C.M. Voltage-dependent ion channels and their gating. *Physiol. Rev.* (Suppl. on Forty Years of Membrane Current in Nerve) 72:S5–S13, 1992.

Catterall, W.A. Cellular and molecular biology of voltage-gated sodium channels. *Physiol. Rev.* (Suppl. on Forty Years of Membrane Current in Nerve) 72:S15–S48, 1992.

Hoshi, T.; Zagotta, W.N.; Aldrich, R.W. Biophysical and molecular mechanisms of *Shaker* potassium channel inactivation. *Science* 250:533–538, 1990.

Jan, L.Y.; Jan, Y.N. Structural elements involved in specific K^+ channel functions. *Annu. Rev. Physiol.* 54:537–555, 1992.

Perney, T.M.; Kaczmarek, L.K. The molecular biology of K^+ channels. *Curr. Opin. Cell Biol.* 3:663–670, 1991.

21. Hökfelt, T. Neuropeptides in perspective: the last ten years. *Neuron* 7:867–879, 1991.

22. Changeux, J.-P.; Galzi, J.-L.; Devillers-Thiery, A.; Bertrand, D. The functional architecture of the acetylcholine nicotinic receptor explored by affinity labelling and site-directed mutagenesis. *Q. Rev. Biophys.* 25:395–432, 1992.

Karlin, A. Explorations of the nicotinic acetylcholine receptor. *Harvey Lect.* 85:71–107, 1991.

Lester, H.A. The permeation pathway of neurotransmitter-gated ion channels. *Annu. Rev. Biophys. Biomol. Struct.* 21:267–292, 1992.

Unwin, N. Neurotransmitter action: opening of ligand-gated ion channels. *Cell* 72 (Suppl.):31–41, 1993.

Unwin, N. Nicotinic acetylcholine receptor at 9 Å resolution. *J. Mol. Biol.* 229:1101–1124, 1993.

23. Betz, H. Ligand-gated ion channels in the brain: the amino acid receptor superfamily. *Neuron* 5:383–392, 1990.

Sargent, P.B. The diversity of neuronal nicotinic acetylcholine receptors. *Annu. Rev. Neurosci.* 16:403–443, 1993.

Snyder, S.H. Drugs and the Brain. New York: W.H. Freeman/Scientific American Books, 1987.

Tallman, J.F.; Gallager, D.W. The GABAergic system: a locus of benzodiazepine action. *Annu. Rev. Neurosci.* 8:21–44, 1985.

24. Barrett, J.N. Motoneuron dendrites: role in synaptic integration. *Fed. Proc.* 34:1398–1407, 1975.

Fuortes, M.G.F.; Frank, K.; Becker, M.C. Steps in the production of motoneuron spikes. *J. Gen. Physiol.* 40:735–752, 1957.

25. Baxter, D.A.; Byrne, J.H. Ionic conductance mechanisms contributing to the electrophysiological properties of neurons. *Curr. Opin. Neurobiol.* 1:105–112, 1991.

Connor, J.A.; Stevens, C.F. Prediction of repetitive firing behavior from voltage clamp data on an isolated neurone soma. *J. Physiol.* 213:31–53, 1971.

Meech, R.W. Calcium-dependent potassium activation in nervous tissues. *Annu. Rev. Biophys. Bioeng.* 7:1–18, 1978.

Rogawski, M.A. The A-current: how ubiquitous a feature of excitable cells is it? *Trends Neurosci.* 8:214–219, 1985.

Tsien, R.W., et al. Multiple types of neuronal calcium channels and their selective modulation. *Trends Neurosci.* 11:431–438, 1988.

26. Bekkers, J.M.; Stevens, C.F. Computational implications of NMDA receptor channels. *Cold Spring Harb. Symp. Quant. Biol.* 55:131–135, 1990.

Bliss, T.V.; Collingridge, G.L. A synaptic model of memory: long-term potentiation in the hippocampus. *Nature* 361:31–39, 1993.

Daw, N.W.; Stein, P.S.; Fox, K. The role of NMDA receptors in information processing. *Annu. Rev. Neurosci.* 16:207–222, 1993.

Madison, D.V.; Malenka, R.C.; Nicoll, R.A. Mechanisms underlying long-term potentiation of synaptic transmission. *Annu. Rev. Neurosci.* 14:379–397, 1991.

Stevens, C.F. Quantal release of neurotransmitter and long-term potentiation. *Cell* 72 (Suppl.):55–63, 1993.

References

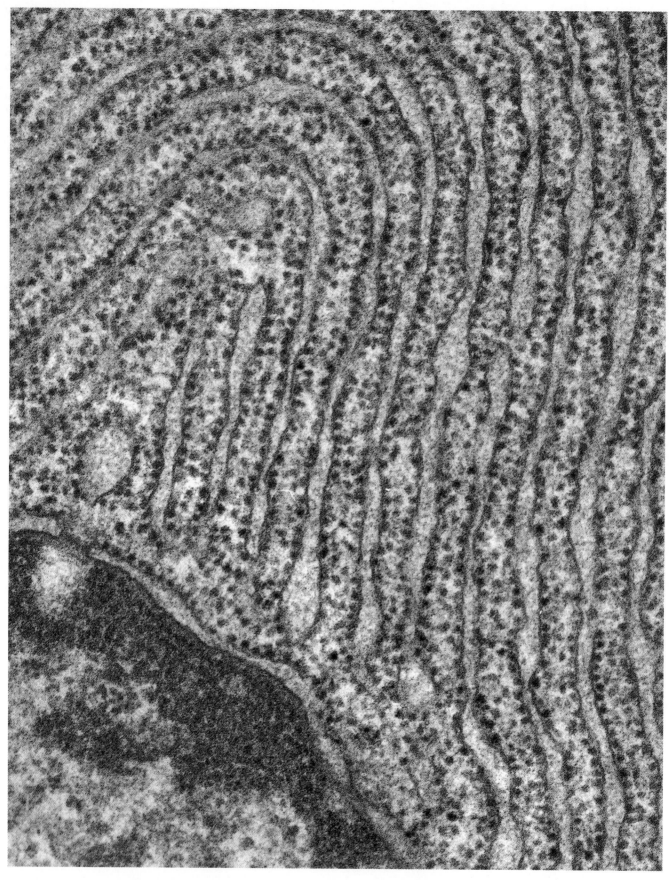

Thin section of an exocrine cell from a dog's pancreas. At the bottom left is a portion of the nucleus and its nuclear envelope. The cytosol is filled with closely packed sheets of endoplasmic reticulum studded with ribosomes. (Courtesy of Lelio Orci.)

Intracellular Compartments and Protein Sorting

Unlike a bacterium, which generally consists of a single intracellular compartment surrounded by a plasma membrane, a eucaryotic cell is elaborately subdivided into functionally distinct, membrane-bounded compartments. Each compartment, or **organelle,** contains its own characteristic set of enzymes and other specialized molecules, and complex distribution systems transport specific products from one compartment to another. To understand the eucaryotic cell, it is essential to know what occurs in each of these compartments, how molecules move between them, and how the compartments themselves are created and maintained.

Proteins play a central part in the compartmentalization of a eucaryotic cell. They catalyze the reactions that occur in each organelle and selectively transport small molecules into and out of its interior, or *lumen.* Proteins also serve as organelle-specific surface markers that direct new deliveries of proteins and lipids to the appropriate organelle. A mammalian cell contains about 10 billion (10^{10}) protein molecules of perhaps 10,000 kinds, and the synthesis of almost all of them begins in the cytosol. Each newly synthesized protein is then delivered specifically to the cell compartment that requires it. We shall make the intracellular transport of proteins the central theme of this chapter as well as of the next. By tracing the protein traffic from one compartment to another, one can begin to make sense of the otherwise bewildering maze of intracellular membranes.

The Compartmentalization of Higher Cells

In this introductory section we give a brief overview of the compartments of the cell and of the relationships between them. In doing so, we organize the organelles conceptually into a small number of discrete families, discussing how proteins are directed to specific organelles and how they cross organelle membranes.

All Eucaryotic Cells Have the Same Basic Set of Membrane-bounded Organelles [1]

Many vital biochemical processes take place in or on membrane surfaces. Lipid metabolism, for example, is catalyzed mostly by membrane-bound enzymes, and oxidative phosphorylation and photosynthesis both require a membrane in or-

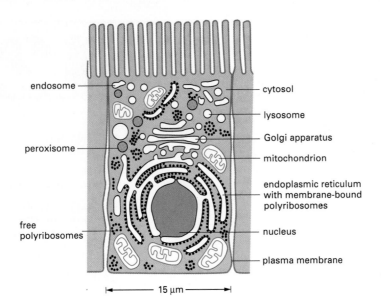

Figure 12–1 The major intracellular compartments of an animal cell. The cytosol (*gray*), endoplasmic reticulum, Golgi apparatus, nucleus, mitochondrion, endosome, lysosome, and peroxisome are distinct compartments isolated from the rest of the cell by at least one selectively permeable membrane.

der to couple the transport of H+ to the synthesis of ATP. Intracellular membrane systems, however, do more for the cell than just provide increased membrane area: they create enclosed compartments that are separate from the cytosol, thus providing the cell with functionally specialized aqueous spaces. Because the lipid bilayer of organelle membranes is impermeable to most hydrophilic molecules, the membrane of each organelle must contain transport proteins that are responsible for the import and export of specific metabolites. Each organelle membrane must also have a mechanism for importing, and incorporating into the organelle, the specific proteins that make the organelle unique.

The major intracellular compartments common to eucaryotic cells are illustrated in Figure 12–1. The *nucleus* contains the main genome and is the principal site of DNA and RNA synthesis. The surrounding *cytoplasm* consists of the *cytosol* and the cytoplasmic organelles suspended in it. The cytosol constitutes a little more than half the total volume of the cell and is the site of protein synthesis and of most of the cell's intermediary metabolism—that is, the many reactions by which some small molecules are degraded and others are synthesized to provide the building blocks of macromolecules (discussed in Chapter 2).

About half the total area of membrane in a cell encloses the labyrinthine spaces of the *endoplasmic reticulum (ER)*. The ER has many ribosomes bound to its cytosolic surface; these are engaged in the synthesis of integral membrane proteins and soluble proteins, most of which are destined for secretion or for other organelles. We shall see that this reflects an important difference between how proteins are directed to the ER and how they are directed to other cytoplasmic organelles: whereas proteins are translocated into other organelles only after their synthesis is complete, they are translocated into the ER during their synthesis, and hence the ribosomes on which they are made are tethered to the ER membrane. The ER also produces the lipid for the rest of the cell and functions as a store for Ca^{2+} ions. The *Golgi apparatus* consists of organized stacks of disclike compartments called Golgi *cisternae;* it receives lipids and proteins from the ER and dispatches them to a variety of destinations, usually covalently modifying them en route.

Mitochondria and (in plants) *chloroplasts* generate most of the ATP used to drive cellular reactions that require an input of free energy. *Lysosomes* contain digestive enzymes that degrade defunct intracellular organelles, as well as macromolecules and particles taken in from outside the cell by endocytosis. On their way to lysosomes, endocytosed material must first pass through a series of compartments called *endosomes. Peroxisomes* (also known as *microbodies*) are small vesicular compartments that contain enzymes utilized in a variety of oxidative

Intracellular Compartment	Percent of Total Cell Volume	Approximate Number per Cell*
Cytosol	54	1
Mitochondria	22	1700
Rough ER cisternae	9	1
Smooth ER cisternae plus Golgi cisternae	6	
Nucleus	6	1
Peroxisomes	1	400
Lysosomes	1	300
Endosomes	1	200

*All the cisternae of the rough and smooth endoplasmic reticulum are thought to be joined to form a single large compartment. The Golgi apparatus, in contrast, is organized into a number of discrete sets of stacked cisternae in each cell, and the extent of interconnection between these sets has not been clearly established.

reactions. In general, each membrane-bounded organelle carries out the same set of basic functions in all cell types but varies in abundance and can have additional properties that differ from cell type to cell type according to the specialized functions of differentiated cells.

On average, the membrane-bounded compartments together occupy nearly half the volume of a cell (Table 12–1), and a large amount of intracellular membrane is required to make them all. In the two mammalian cells analyzed in Table 12–2, for example, the endoplasmic reticulum has a total membrane surface area that is, respectively, 25 times and 12 times that of the plasma membrane. In terms of its area and mass the plasma membrane is only a minor membrane in most eucaryotic cells (Figure 12–2).

Table 12–2 Relative Amounts of Membrane Types in Two Types of Eucaryotic Cells

Membrane Type	Percent of Total Cell Membrane	
	Liver Hepatocyte*	Pancreatic Exocrine Cell*
Plasma membrane	2	5
Rough ER membrane	35	60
Smooth ER membrane	16	<1
Golgi apparatus membrane	7	10
Mitochondria		
Outer membrane	7	4
Inner membrane	32	17
Nucleus		
Inner membrane	0.2	0.7
Secretory vesicle membrane	not determined	3
Lysosome membrane	0.4	not determined
Peroxisome membrane	0.4	not determined
Endosome membrane	0.4	not determined

*These two cells are of very different sizes, since the average hepatocyte has a volume of about 5000 μm^3 compared with about 1000 μm^3 for the pancreatic exocrine cell. Total cell membrane areas are estimated at about 110,000 μm^2 and 13,000 μm^2, respectively.

The Compartmentalization of Higher Cells

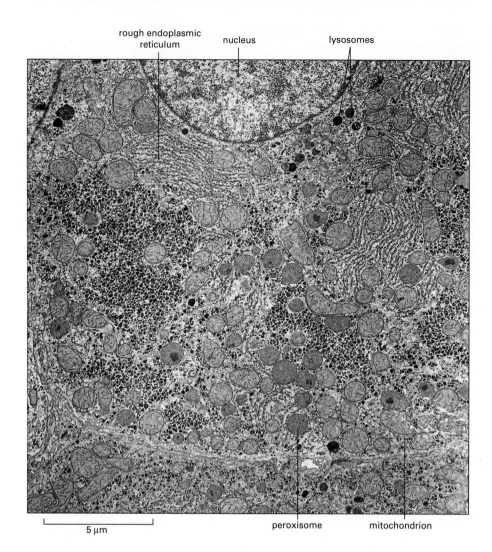

rough endoplasmic reticulum   nucleus   lysosomes

5 μm

peroxisome   mitochondrion

Figure 12–2 Electron micrograph of part of a liver cell seen in cross-section. Examples of most of the major intracellular compartments are indicated. (Courtesy of Daniel S. Friend.)

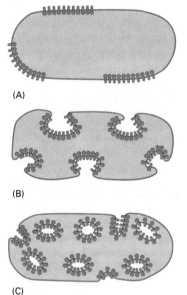

(A)

(B)

(C)

Figure 12–3 Organization of specialized membranes in bacteria. (A) Membrane patches on the cell surface consisting of clusters of specialized membrane proteins. (B) Invaginated patches of plasma membrane that increase the amount of membrane available for a specialized function such as photosynthesis. (C) Internalization of the specialized invaginated membrane to form vesicles, whose interior surface is topologically equivalent to the exterior surface of the cell. Membrane-bounded vesicles of this type are present in some types of photosynthetic bacteria; their topological relationship to the cell surface is similar to that of the ER, Golgi apparatus, endosomes, and lysosomes in eucaryotic cells.

Membrane-bounded organelles are not randomly distributed in the cytosol; instead they often have characteristic positions. In most cells, for example, the Golgi apparatus is located close to the nucleus, whereas the network of ER tubules extends from the nucleus throughout the entire cytosol. These characteristic distributions seem to depend on interactions of the organelles with the cytoskeleton: the localization of both the ER and the Golgi apparatus, for example, is dependent on an intact microtubule array; if the microtubules are experimentally depolymerized with a drug, the Golgi apparatus fragments and disperses throughout the cell and the ER network collapses toward the cell center, or *centrosome*, from which the microtubule array emanates (discussed in Chapter 16).

The Topological Relationships of Membrane-bounded Organelles Can Be Interpreted in Terms of Their Evolutionary Origins [2]

To understand the relationships between the compartments of the cell, it is helpful to consider how they might have evolved. The precursors of the first eucaryotic cells are thought to have been simple organisms that resembled bacteria, which generally have a plasma membrane but no internal membranes. The plasma membrane in such cells therefore provides all membrane-dependent functions, including the pumping of ions, ATP synthesis, protein secretion, and lipid synthesis. Typical present-day eucaryotic cells are 10 to 30 times larger in linear dimension and 1000 to 10,000 times greater in volume than a typical bacterium such as *E. coli*. The profusion of internal membranes can be seen in part as an

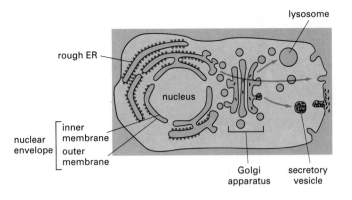

Figure 12–4 Topological relationships between compartments in a eucaryotic cell. Topologically equivalent spaces are shown in *red*. In principle, cycles of vesicle budding and fusion permit any lumen to communicate with any other and with the cell exterior. The *blue arrows* indicate the outward direction of vesicle traffic from the ER to Golgi apparatus to plasma membrane (or lysosomes), and the *black dots* represent protein molecules that are secreted by the cell. Some organelles, most notably mitochondria and (in plant cells) chloroplasts, however, do not take part in this vesicular communication and so are isolated from the traffic between organelles shown here.

adaptation to this increase in size: the eucaryotic cell has a much smaller ratio of surface to volume, and its area of plasma membrane is presumably too small to sustain the many vital functions for which membranes are required.

The evolution of internal membranes evidently has gone hand in hand with specialization of membrane function. Some present-day bacteria have specialized patches of plasma membrane in which a selected set of membrane proteins coalesce to carry out a group of related functions (Figure 12–3A). These specialized membrane patches, such as the "purple membrane" containing bacteriorhodopsin in *Halobacterium* (discussed in Chapter 10) and the chromatophores in photosynthetic bacteria (discussed in Chapter 14), represent primitive organelles. In some photosynthetic bacteria the patches have become elaborated into extensive invaginations of the plasma membrane (Figure 12–3B); in others the invaginations seem to have pinched off completely, forming sealed membrane-bounded vesicles specialized for photosynthesis (Figure 12–3C).

A eucaryotic organelle that originated by the type of pathway illustrated in Figure 12–3 might be expected to have an interior that is topologically equivalent to the exterior of the cell. We shall see that this is the case for the ER, Golgi apparatus, endosome, and lysosome—as well as for the many vesicular intermediates (*transport vesicles*) in the secretory and endocytic pathways. We can therefore think of all of these organelles as members of the same family, and, as we discuss in detail in the next chapter, their interiors communicate extensively with one another and with the outside of the cell via transport vesicles that bud off from one organelle and fuse with another (Figure 12–4). Ribosomes are found attached to the cytosolic side of the plasma membrane in bacteria, and so the evolutionary origin of the ER membrane from the plasma membrane may explain why ribosomes are attached to the ER membrane in eucaryotic cells.

This evolutionary scheme also offers a reasonable explanation for the architecture of the cell nucleus with its double membrane. In bacteria the single chromosome is attached at special sites to the inside of the plasma membrane. It is possible therefore that the double-layered nuclear envelope originated as a deep invagination of the plasma membrane, as shown in Figure 12–5A. This scheme would explain why the nuclear compartment is topologically equivalent to the cytosol. In fact, in higher eucaryotic cells the nuclear envelope breaks down during mitosis, allowing the nuclear contents to disperse in the cytosol, a situation that never occurs for the contents of any other membrane-bounded organelle. As the scheme in Figure 12–5A also predicts, the space between the two nuclear membranes is topologically equivalent to the exterior of the cell and is continuous with the lumen of the ER.

As discussed in Chapter 14, mitochondria and plastids (of which chloroplasts are one form) differ from the other membrane-bounded organelles in that they contain their own genomes. The nature of these genomes and the close resemblance of the proteins in these organelles to those in some present-day bacteria strongly suggest that mitochondria and plastids evolved from bacteria that were engulfed by other cells with which they initially lived in symbiosis (discussed in Chapters 1 and 14). According to the hypothetical scheme shown in Figure 12–5B, the inner membrane of mitochondria and plastids corresponds to the original

The Compartmentalization of Higher Cells

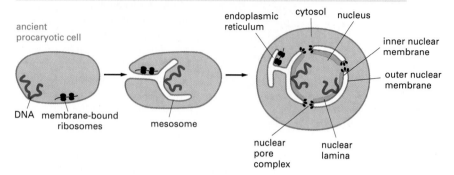

(A) PROPOSED EVOLUTIONARY PATHWAY FOR NUCLEUS AND ENDOPLASMIC RETICULUM

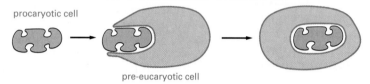

(B) PROPOSED EVOLUTIONARY PATHWAY FOR MITOCHONDRIA

Figure 12–5 **Hypotheses for the evolutionary origins of some membrane-bounded organelles.** The origins of mitochondria, chloroplasts, ER, and the cell nucleus could explain the topological relationships of these intracellular compartments in eucaryotic cells. (A) A possible pathway for the evolution of the cell nucleus and the ER. In some bacteria the single DNA molecule is attached to an invagination of the plasma membrane, called a *mesosome*. Such an invagination in a very ancient procaryotic cell could have spread to form an envelope around the DNA while still allowing access of the DNA to the cell cytosol (as is required for DNA to direct protein synthesis). This envelope is presumed to have eventually pinched off completely from the plasma membrane, producing a nuclear compartment surrounded by a double membrane. As illustrated, the nuclear envelope is organized by a fibrous shell called the *nuclear lamina* and is penetrated by communicating channels called *nuclear pore complexes*. Because it is surrounded by two membranes that are in continuity where they are penetrated by these pores, the nuclear compartment is topologically equivalent to the cytosol. The lumen of the ER is continuous with the space between the inner and outer nuclear membranes and topologically equivalent to the extracellular space. (B) Mitochondria (and chloroplasts) are thought to have originated when a bacterium was engulfed by a larger pre-eucaryotic cell. They retain their autonomy. This may explain why the lumens of these organelles remain isolated from the vesicular traffic that interconnects the lumens of many other intracellular compartments.

plasma membrane of the bacterium, while the lumen of these organelles evolved from the bacterial cytosol. As might be expected from such origins, these two organelles remain isolated from the extensive vesicular traffic that connects the interiors of most of the other membrane-bounded organelles to one another and to the outside of the cell.

This evolutionary scheme groups the intracellular compartments in eucaryotic cells into five distinct families: (1) the nucleus and the cytosol, which communicate through the nuclear pores and are thus topologically continuous (although functionally distinct); (2) all organelles that function in the secretory and endocytic pathways—including the ER, Golgi apparatus, endosomes, lysosomes, and numerous classes of transport vesicles; (3) the mitochondria; (4) the plastids (in plants only); and (5) the peroxisomes (whose evolutionary origins are discussed later).

Proteins Can Move Between Compartments in Different Ways [3]

All proteins begin being synthesized on ribosomes in the cytosol, except for the few that are synthesized on the ribosomes of mitochondria and plastids. Their subsequent fate depends on their amino acid sequence, which can contain **sorting signals** that direct their delivery to locations outside the cytosol. Most proteins do not have a sorting signal and consequently remain in the cytosol as permanent residents. Many others, however, have specific sorting signals that direct their transport from the cytosol into the nucleus, the ER, mitochondria, plastids (in plants), or peroxisomes; sorting signals can also direct the transport of proteins from the ER to other destinations in the cell.

To understand the general principles by which sorting signals operate, it is important to distinguish three fundamentally different ways by which proteins move from one compartment to another. (1) The protein traffic between the cytosol and nucleus occurs between topologically equivalent spaces, which are in continuity through the nuclear pore complexes. This process is called **gated transport** because the nuclear pore complexes function as selective gates that can actively transport specific macromolecules and macromolecular assemblies, although they also allow free diffusion of smaller molecules. (2) In **transmembrane transport** membrane-bound *protein translocators* directly transport specific proteins across a membrane from the cytosol into a space that is topologically distinct. The transported protein molecule usually must unfold in order to snake

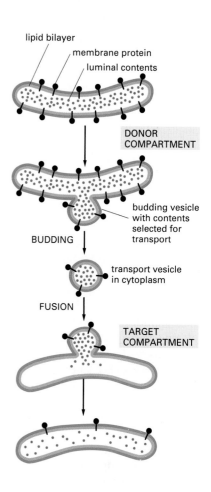

Figure 12–6 **The "sidedness" of membranes is preserved during vesicular transport.** Note that the original orientation of both proteins and lipids in the donor-compartment membrane is preserved in the target-compartment membrane and that soluble molecules are transferred from lumen to lumen.

lipid bilayer
membrane protein
luminal contents

DONOR COMPARTMENT

budding vesicle with contents selected for transport

BUDDING

transport vesicle in cytoplasm

FUSION

TARGET COMPARTMENT

through the membrane. The initial transport of selected proteins from the cytosol into the ER lumen or into mitochondria, for example, occurs in this way. (3) In **vesicular transport,** *transport vesicles* ferry proteins from one compartment to another. The vesicles become loaded with a cargo of molecules derived from the lumen of one compartment as they pinch off from its membrane; they discharge their cargo into a second compartment by fusing with its membrane. The transfer of soluble proteins from the ER to the Golgi apparatus, for example, occurs in this way. Because the transported proteins do not cross a membrane, they move only between compartments that are topologically equivalent (Figure 12–6). We discuss vesicular transport in more detail in Chapter 13. The three ways in which proteins are transported between different compartments are summarized in Figure 12–7.

Each of the three modes of protein transfer is usually selectively guided by sorting signals in the transported protein that are recognized by complementary receptor proteins in the target organelle. If a large protein is to be imported into the nucleus, for example, it must possess a sorting signal that is recognized by receptor proteins associated with the nuclear pore complex. If a protein is to be transferred directly across a membrane, it must possess a sorting signal that is recognized by the translocator in the membrane to be crossed. Likewise, if a protein is to be incorporated into certain types of transport vesicles or to be retained in certain organelles, its sorting signal must be recognized by a complementary receptor in the appropriate membrane.

Signal Peptides and Signal Patches Direct Proteins to the Correct Cellular Address [4]

There are at least two types of sorting signals on proteins. One type resides in a continuous stretch of amino acid sequence, typically 15 to 60 residues long. This **signal peptide** is often (but not always) removed from the finished protein by a specialized **signal peptidase** once the sorting process has been completed. The other type consists of a specific three-dimensional arrangement of atoms on the

Figure 12–7 **A simplified "road map" of protein traffic.** Proteins can move from one compartment to another by gated transport (*red*), transmembrane transport (*blue*), or vesicular transport (*green*). The signals that direct a given protein's movement through the system, and thereby determine its eventual location in the cell, are contained in its amino acid sequence. The journey begins with the synthesis of a protein on a ribosome and terminates when the final destination is reached. At each intermediate station (*boxes*) a decision is made as to whether the protein is to be retained or transported further. In principle, a signal could be required either for retention in or for exit from each of the compartments shown, with the alternative fate being the *default pathway* (one that requires no signal). The vesicular transport of proteins from the ER through the Golgi apparatus to the cell surface, for example, appears not to require any specific sorting signals; specific sorting signals therefore are required to retain in the ER and the Golgi apparatus those specialized proteins that are resident there.

We shall use this figure repeatedly as a guide throughout this chapter and the next, highlighting the particular pathway being discussed.

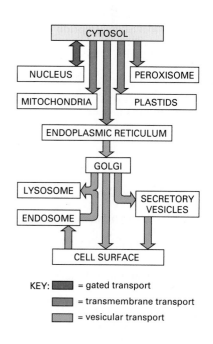

CYTOSOL

NUCLEUS | PEROXISOME

MITOCHONDRIA | PLASTIDS

ENDOPLASMIC RETICULUM

GOLGI

LYSOSOME

SECRETORY VESICLES

ENDOSOME

CELL SURFACE

KEY: = gated transport
= transmembrane transport
= vesicular transport

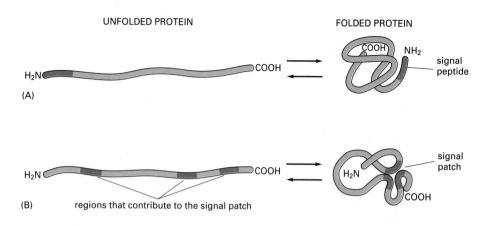

UNFOLDED PROTEIN

H₂N ━━━━━━━━━━━━━━━ COOH

(A)

FOLDED PROTEIN

COOH   NH₂
                signal
                peptide

H₂N ━━━━━━━━━━━━━━━ COOH

(B)   regions that contribute to the signal patch

H₂N

                signal
                patch

COOH

Figure 12–8 **Two ways that a sorting signal can be built into a protein.** (A) The signal resides in a single discrete stretch of amino acid sequence, called a *signal peptide,* that is exposed in the folded protein. Signal peptides often occur at the end of the polypeptide chain (as shown), but they can also be located elsewhere. (B) A *signal patch* can be formed by the juxtaposition of amino acids from regions that are physically separated before the protein folds (as shown); alternatively, separate patches on the surface of the folded protein that are spaced a fixed distance apart could form the signal. In either case the transport signal depends on the three-dimensional conformation of the protein, which makes it difficult to locate the signal precisely.

protein's surface that forms when the protein folds up. The amino acid residues that comprise this **signal patch** can be distant from one another in the linear amino acid sequence, and they generally remain in the finished protein (Figure 12–8). Signal peptides are used to direct proteins from the cytosol into the ER, mitochondria, chloroplasts, peroxisomes, and nucleus, and they are also used to retain soluble proteins in the ER. Signal patches identify certain enzymes that are to be marked with specific sugar residues that then direct them from the Golgi apparatus into lysosomes; signal patches are also used in other sorting steps that have been less well characterized.

Different types of signal peptides are used to specify different destinations in the cell. Proteins destined for initial transfer to the ER usually have a signal peptide at their amino terminus, which characteristically includes a sequence composed of about 5 to 10 hydrophobic amino acids. Most of these proteins will in turn pass from the ER to the Golgi apparatus, but those with a specific sequence of four amino acids at their carboxyl terminus are retained as permanent ER residents. Proteins destined for mitochondria have signal peptides of yet another type, in which positively charged amino acids alternate with hydrophobic ones. Proteins destined for peroxisomes usually have a specific signal sequence of three amino acids at their carboxyl terminus. Many proteins destined for the nucleus carry a signal peptide formed from a cluster of positively charged amino

Table 12–3 **Some Typical Signal Peptides**

Function of Signal Peptide	Example of Signal Peptide
Import into ER	⁺H₃N-Met-Met-Ser-Phe-Val-Ser- Leu-Leu-Leu-Val Gly-Ile-Leu-Phe-Trp-Ala -Thr-Glu-Ala-Glu- Gln-Leu-Thr-Lys-Cys-Glu-Val-Phe-Gln-
Retain in lumen of ER	-Lys-Asp-Glu-Leu-COO⁻
Import into mitochondria	⁺H₃N-Met-Leu-Ser-Leu-Arg-Gln-Ser-Ile-Arg-Phe- Phe-Lys-Pro-Ala-Thr-Arg-Thr-Leu-Cys-Ser- Ser-Arg-Tyr-Leu-Leu-
Import into nucleus	-Pro-Pro-Lys-Lys-Lys-Arg-Lys-Val-
Import into peroxisomes	-Ser-Lys-Leu-
Attach to membranes via the covalent linkage of a myristic acid to the amino terminus	⁺H₃N-Gly-Ser-Ser-Lys-Ser-Lys-Pro-Lys-

Positively charged amino acids are shown in *red* and negatively charged amino acids in *green.* An extended block of hydrophobic amino acids is enclosed in a *yellow* box. H₃N⁺ indicates the amino terminus of a protein; COO⁻ indicates the carboxyl terminus.

A transfection approach for defining signal sequences

One way to show that a signal sequence is required and sufficient to target a protein to a specific intracellular compartment is to create a fusion protein in which the signal sequence is attached by genetic engineering techniques to a protein that is normally resident in the cytosol. After the cDNA encoding this protein is transfected into cells, the location of the fusion protein is determined by immunostaining or by cell fractionation.

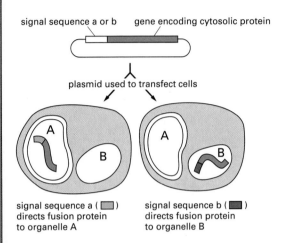

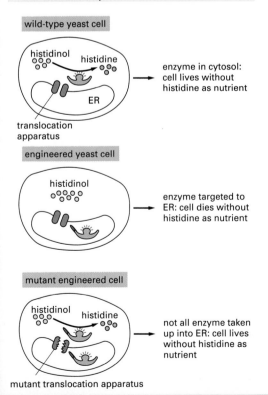

By altering the signal sequence using site-directed mutagenesis, one can determine which structural features are important for its function.

A biochemical approach for studying the mechanism of protein translocation

In this approach a labeled protein containing a specific signal sequence is transported into isolated organelles *in vitro*. The labeled protein is usually produced by cell-free translation of a purified mRNA encoding the protein; radioactive amino acids are used to label the newly synthesized protein so that it can be distinguished from the many other proteins that are present in the *in vitro* translation system.

Three methods are commonly used to test if the labeled protein has been translocated into the organelle:

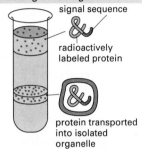

1. The labeled protein co-fractionates with the organelle during centrifugation.

signal sequence

radioactively labeled protein

protein transported into isolated organelle

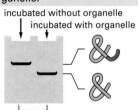

2. The signal sequence is removed by a specific protease that is present inside the organelle.

incubated without organelle
incubated with organelle

radioactive proteins on SDS gel

3. The protein is protected from digestion when proteases are added to the incubation medium but is susceptible if a detergent is first added to disrupt the organelle membrane.

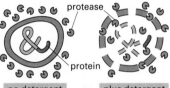

protease

protein

no detergent   plus detergent

By exploiting such *in vitro* assays, one can determine what components (proteins, ATP, GTP, etc.) are required for the translocation process.

Genetic approaches for studying the mechanism of protein translocation

wild-type yeast cell

histidinol → histidine

enzyme in cytosol: cell lives without histidine as nutrient

translocation apparatus

ER

engineered yeast cell

histidinol

enzyme targeted to ER: cell dies without histidine as nutrient

mutant engineered cell

histidinol → histidine

not all enzyme taken up into ER: cell lives without histidine as nutrient

mutant translocation apparatus

Yeast cells with mutations in genes that encode components of the translocation machinery have been useful for studying protein translocation. Because mutant cells that cannot translocate proteins across their membranes will die, the trick is to design a strategy that allows weak mutations that cause only a partial defect in protein translocation to be isolated.

One way uses genetic engineering to design special yeast cells. The enzyme histidinol dehydrogenase, for example, normally resides in the cytosol, where it is required to produce the essential amino acid histidine from its precursor histidinol. A yeast strain is constructed in which the histidinol dehydrogenase gene is replaced by a re-engineered gene encoding a fusion protein with an added signal sequence that misdirects the enzyme into the endoplasmic reticulum (ER). When such cells are grown without histidine, they die because all of the histidinol dehydrogenase is sequestered in the ER, where it is of no use. Cells with a mutation that partially inactivates the mechanism for translocating proteins from the cytosol to the ER, however, will survive because enough of the dehydrogenase will be retained in the cytosol to produce histidine. Often one obtains a cell in which the mutant protein still functions partially at normal temperature but is completely inactive at higher temperature. A cell carrying such a temperature-sensitive mutation dies at higher temperature, whether or not histidine is present, as it cannot transport any protein into the ER. This allows the normal gene that was disabled by the mutation to be identified by transfecting the mutant cells with a yeast plasmid vector into which random yeast genomic DNA fragments have been cloned: the specific DNA fragment that rescues the mutant cells when they are grown at high temperature should encode the wild-type version of the mutant gene.

acids, which is commonly found at internal sites of the polypeptide chain. Some typical signal peptides are listed in Table 12–3.

The importance of each of these signal peptides for protein targeting has been shown by experiments in which the peptide is transferred from one protein to another by genetic engineering techniques: placing the amino-terminal ER signal peptide at the beginning of a cytosolic protein, for example, redirects the protein to the ER. Even though their amino acid sequences can vary greatly, the signal peptides of all proteins having the same destination are functionally interchangeable: physical properties, such as hydrophobicity, often appear to be more important in the signal-recognition process than the exact amino acid sequence.

Signal patches are far more difficult to analyze than signal peptides, and so less is known about their structure. Because they result from a complex three-dimensional protein-folding pattern, they cannot be easily transferred experimentally from one protein to another.

The main ways of studying how proteins are directed from the cytosol to a specific compartment and how they are translocated across membranes are illustrated in Panel 12–1 (p. 559).

Cells Cannot Construct Their Membrane-bounded Organelles *de Novo:* They Require Information in the Organelle Itself [5]

When a cell reproduces by division, it has to duplicate its membrane-bounded organelles. In general, cells do this by enlarging the existing organelles by incorporating new molecules into them; the enlarged organelles then divide and are distributed to the two daughter cells. Thus each daughter cell inherits from its mother a complete set of specialized cell membranes. This inheritance is essential because a cell could not make such membranes *de novo.* If the ER were completely removed from a cell, for example, how could the cell reconstruct it? The membrane proteins that define the ER and carry out many of its functions are themselves products of the ER. A new ER could not be made without an existing ER or, at the very least, a membrane that contains the translocators required to import specific proteins into the ER (and lacks the translocators required to import the proteins that function in other organelles).

Thus it seems that the information required to construct a membrane-bounded organelle does not reside exclusively in the DNA that specifies the organelle's proteins. *Epigenetic* information in the form of at least one distinct protein that preexists in the organelle membrane is also required, and this information is passed from parent cell to progeny cell in the form of the organelle itself. Presumably, such information is essential for the propagation of the cell's compartmental organization, just as the information in DNA is essential for the propagation of its nucleotide and amino acid sequences.

Summary

Eucaryotic cells contain intracellular membranes that enclose nearly half the cell's total volume in separate intracellular compartments called organelles. The main types of membrane-bounded organelles that are present in all eucaryotic cells are the endoplasmic reticulum, Golgi apparatus, nucleus, mitochondria, lysosomes, endosomes, and peroxisomes; plant cells also contain plastids, such as chloroplasts. Each organelle contains a distinct set of proteins that mediates its unique functions.

Each newly synthesized organelle protein finds its way from the ribosome where it is made to the organelle where it functions by following a specific pathway, guided by signals in its amino acid sequence that function as signal peptides or signal patches. The signal peptides and patches are recognized by complementary receptor proteins in the target organelle. Proteins that function in the cytosol do not contain signal peptides or signal patches and therefore remain in the cytosol after they are synthesized.

The Transport of Molecules into and out of the Nucleus [6]

The **nuclear envelope** encloses the DNA and defines the nuclear compartment. It is formed from two concentric membranes that, as we have seen, are continuous with the endoplasmic reticulum. Although the inner and outer nuclear membranes are continuous, the two membranes maintain distinct protein compositions. The **inner nuclear membrane** contains specific proteins that act as binding sites for the feltlike *nuclear lamina* that supports it. The inner membrane is surrounded by the **outer nuclear membrane,** which closely resembles the membrane of the rough endoplasmic reticulum (Figure 12–9). Like the membrane of the rough ER, the outer nuclear membrane is studded with ribosomes engaged in protein synthesis. The proteins made on these ribosomes are transported into the space between the inner and outer nuclear membranes (the *perinuclear space*), which is continuous with the ER lumen (see Figure 12–9).

Bidirectional traffic occurs continuously between the cytosol and the nucleus. The many proteins that function in the nucleus—including histones, DNA and RNA polymerases, gene regulatory proteins, and RNA-processing proteins—are selectively imported into the *nuclear compartment* from the cytosol where they are made. At the same time, tRNAs and mRNAs are synthesized in the nuclear compartment and then exported to the cytosol. Like the import process, the export process is selective; mRNAs, for example, are exported only after they have been properly modified by RNA-processing reactions in the nucleus. In some cases the transport process is complex: ribosomal proteins, for instance, are made in the cytosol, imported into the nucleus—where they assemble with newly made ribosomal RNA into particles—and then exported again to the cytosol as part of a ribosomal subunit; each of these steps involves selective transport across the nuclear envelope.

Nuclear Pores Perforate the Nuclear Envelope [7]

The nuclear envelope in all eucaryotes, from yeasts to humans, is perforated by *nuclear pores*. Each pore is formed by a large, elaborate structure known as the **nuclear pore complex,** which has an estimated molecular mass of about 125 million and is thought to be composed of more than 100 different proteins, arranged with a striking octagonal symmetry (Figures 12–10 and 12–11).

Each pore complex contains one or more open aqueous channels through which water-soluble molecules that are smaller than a certain size can passively

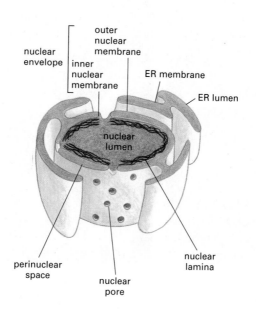

Figure 12–9 The nuclear envelope. The double-membrane envelope is penetrated by nuclear pores and is continuous with the endoplasmic reticulum. The ribosomes that are bound to the cytosolic surface of the ER membrane and outer nuclear membrane are not shown.

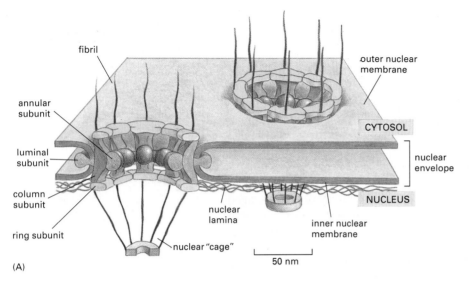

(A)

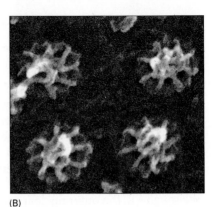

(B)

Figure 12–10 The arrangement of the nuclear pore complexes in the nuclear envelope. (A) A sketch showing a small region of the nuclear envelope. In cross-section the nuclear pore complex appears composed of three parts: (1) a column component which forms the bulk of the pore wall; (2) an annular component, which extends "spokes" toward the center of the pore; and (3) a luminal component, which is formed by a large transmembrane glycoprotein that is thought to help anchor the complex to the nuclear membrane. In addition, fibrils protrude from both the cytosolic and nuclear sides of the complex. On the nuclear side the fibrils converge to form cagelike structures, which are shown in a scanning electron micrograph of the nuclear side of the nuclear envelope of an oocyte in (B). (B, from M.W. Goldberg and T.D. Allen, *J. Cell Biol.* 119:1429–1440, 1992, by copyright permission of the Rockefeller University Press.)

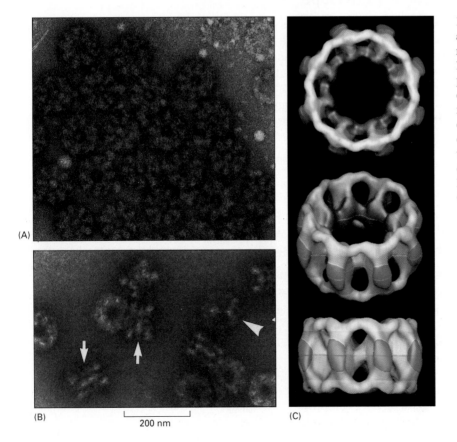

(A)

(B)

200 nm

(C)

Figure 12–11 Electron micrograph and computer reconstruction of nuclear pore complexes. (A) and (B) Negatively stained views of nuclear pore complexes released from the envelope by detergent. In (B) some nuclear pore complexes can be seen on their side. (C) Three-dimensional computer reconstructions showing top, tilted, and side views of pore complexes. (From J.E. Hinshaw and R. Milligan, *Cell* 69:1133–1141, 1992. © Cell Press.)

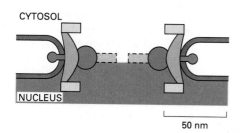

size of proteins
that enter nucleus
by free diffusion

size of proteins
that enter nucleus
by active transport

Figure 12–12 Possible paths for free diffusion through the nuclear pore complex. The drawing shows a hypothetical diaphragm inserted into the pore to restrict the size of the open channel to 9 nm, which is the pore size estimated from diffusion measurements. Nine nanometers is a much smaller diameter than that of the central opening apparent on the images of the nuclear pore complex derived from electron micrographs (or from that measured during active transport when the pore dilates to allow transport of particles of up to 26 nm in diameter). Thus it is likely that some pore components are lost during the preparation of specimens for electron microscopy and that these normally restrict free diffusion through the central opening. Such components may form a diaphragm (or plug) that opens and closes to allow passage of large objects during active transport, which is mediated by a sorting signal (discussed below). Although plugs can be seen in some preparations, it is not clear whether they are components of the pore complex or material that is being transported through it. Three-dimensional computer reconstructions suggest that the channels permitting free diffusion might not be located at the center of the pore complex but near its rim between the column components (see Figure 12–10A); this would mean that passive diffusion and active transport take place through different parts of the pore complex.

diffuse. The effective size of these channels has been determined by injecting labeled molecules (that are not nuclear components) into the cytosol and then measuring their rate of diffusion into the nucleus. Small molecules (5000 daltons or less) diffuse in so fast that the nuclear envelope can be considered to be freely permeable to them. A protein of 17,000 daltons takes 2 minutes to equilibrate between the cytosol and nucleus, while a protein of 44,000 daltons takes 30 minutes. A globular protein larger than about 60,000 daltons seems hardly able to enter the nucleus at all. A quantitative analysis of such data suggests that the nuclear pore complex contains a pathway for free diffusion equivalent to a water-filled cylindrical channel about 9 nm in diameter and 15 nm long; such a channel would occupy only a small fraction of the total pore volume (Figure 12–12).

Because many cellular proteins are too large to pass by diffusion through the nuclear pores, the nuclear envelope allows the nuclear compartment and the cytosol to maintain different complements of proteins. Mature cytosolic ribosomes, for example, are about 30 nm in diameter and thus cannot diffuse through the 9-nm channels; their exclusion from the nucleus ensures that all protein synthesis is confined to the cytosol. But how does the nucleus export newly made ribosomal subunits or import large molecules, such as DNA and RNA polymerases, which have subunit molecular weights of 100,000 to 200,000? As we discuss next, these and many other protein and RNA molecules bind to specific receptor proteins located in the pore complexes and are then actively transported across the nuclear envelope through the complexes.

Nuclear Localization Signals Direct Nuclear Proteins to the Nucleus [8]

In general, the more active the nucleus is in transcription, the greater the number of pore complexes its envelope contains. The nuclear envelope of a typical mammalian cell contains 3000 to 4000 pore complexes. If the cell is synthesizing DNA, it needs to import about 10^6 histone molecules from the cytosol every 3 minutes in order to package newly made DNA into chromatin, which means that, on average, each pore complex needs to transport about 100 histone molecules per minute. If the cell is growing rapidly, each pore complex also needs to transport about 6 newly assembled large and small ribosomal subunits per minute from the nucleus, where they are produced, to the cytosol, where they are used. And that is only a very small part of the total traffic that passes through the nuclear pores.

When proteins are experimentally extracted from the nucleus and microinjected back into the cytosol, even the very large ones efficiently reaccumulate in the nucleus. The selectivity of this nuclear protein import resides in **nuclear localization signals,** which are present only in nuclear proteins. The signals have been precisely defined in many nuclear proteins using recombinant DNA technology. They can be located almost anywhere in the amino acid sequence and generally consist of a short sequence (typically from four to eight amino acids) that varies for different nuclear proteins but is rich in the positively charged amino acids lysine and arginine and usually contains proline. In many nuclear proteins this sequence is split into two blocks of two to four amino acids each,

The Transport of Molecules into and out of the Nucleus

(A) LOCALIZATION OF T-ANTIGEN CONTAINING
WILD-TYPE NUCLEAR IMPORT SIGNAL

Pro — Pro —Lys—Lys—Lys—Arg—Lys— Val —

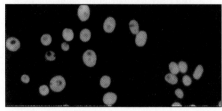

(B) LOCALIZATION OF T-ANTIGEN CONTAINING
A MUTATED NUCLEAR IMPORT SIGNAL

Pro — Pro —Lys—Thr—Lys—Arg—Lys— Val —

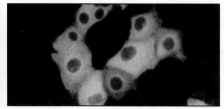

Figure 12–13 The function of a nuclear localization signal. Immunofluorescence micrographs showing the cellular location of SV40 virus T-antigen containing or lacking a short peptide that serves as a nuclear localization signal. The wild-type T-antigen protein contains the lysine-rich sequence indicated and is imported to its site of action in the nucleus, as indicated by immunofluorescence staining with antibody against the T-antigen (A). T-antigen with an altered nuclear localization signal (a threonine replacing a lysine) remains in the cytosol (B). (From D. Kalderon, B. Roberts, W. Richardson, and A. Smith, *Cell* 39:499–509, 1984. © Cell Press.)

with the blocks separated from each other by about ten amino acids. The signals are thought to form loops on the protein surface.

Nuclear localization signals were first identified in the large viral protein called *T-antigen*, which is encoded by the SV40 virus and is needed for viral DNA replication in the host cell nucleus. The T-antigen normally accumulates in the nucleus shortly after being synthesized in the cytosol. A mutation in a single amino acid, however, prevents nuclear import (Figure 12–13). On the assumption that this mutation is in a nuclear localization signal sequence, short lengths of the DNA encoding this region of the normal T-antigen were fused to a gene coding for a cytosolic protein. The shortest sequence that caused the resulting fusion protein to be imported into the nucleus encoded a stretch of eight contiguous amino acids, which is normally located in an internal region of the T-antigen polypeptide chain (see Figure 12–13). Further experiments showed that the signal sequence could function even when it was linked as a short peptide to selected lysine side chains on the surface of a cytosolic protein, suggesting that the precise location of a nuclear localization signal within the amino acid sequence of a nuclear protein is not important. In fact, many nuclear proteins contain more than one nuclear localization signal.

Macromolecules Are Actively Transported into and out of the Nucleus Through Nuclear Pores [9]

The active transport of nuclear proteins through nuclear pore complexes can be directly visualized by coating gold particles with a nuclear protein, injecting the particles into the cytosol, and then following their fate by electron microscopy (Figures 12–14 and 12–15).

The initial interaction of a nuclear protein with the nuclear pore complex requires one or more cytosolic proteins that bind to the nuclear localization signals and help direct the nuclear protein to the pore complex, where it appears to bind to the fibrils that project from the rim of the complex. The nuclear protein then moves to the center of the pore complex, where it is actively transported across the nuclear envelope by a process that requires ATP hydrolysis (Figure 12–16). Studies with various sizes of gold beads indicate that the opening can dilate up to about 26 nm in diameter during the transport process: a poorly defined structure in the center of the nuclear pore complex appears to function like a close-fitting diaphragm that opens just the right amount when activated by a signal on an appropriate large protein (see Figure 12–12). The molecular basis of this mechanism, and how it operates to pump macromolecules both into and out of the nucleus, is a mystery.

It seems likely that the export of new ribosomal subunits and messenger RNA molecules through the nuclear pores also depends on a selective transport system. If 20-nm-diameter gold spheres, similar to those used in the experiments shown in Figure 12–15, are coated with small RNA molecules (tRNA or 5S RNA) and then injected into the nucleus of a frog oocyte, they are rapidly transported through the nuclear pores into the cytosol. If they are injected into the cytosol

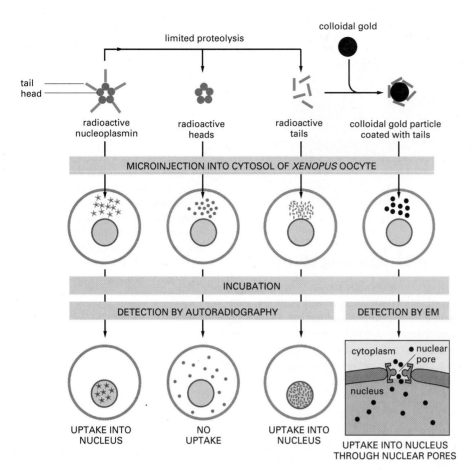

limited proteolysis

colloidal gold

tail
head

radioactive
nucleoplasmin

radioactive
heads

radioactive
tails

colloidal gold particle
coated with tails

MICROINJECTION INTO CYTOSOL OF *XENOPUS* OOCYTE

INCUBATION

DETECTION BY AUTORADIOGRAPHY

DETECTION BY EM

UPTAKE INTO
NUCLEUS

NO
UPTAKE

UPTAKE INTO
NUCLEUS

cytoplasm

nuclear
pore

nucleus

UPTAKE INTO NUCLEUS
THROUGH NUCLEAR PORES

Figure 12–14 The uptake of nuclear proteins via nuclear pores.
Nucleoplasmin is a large nuclear protein with distinct head and tail domains. The heads can be cleaved from the tails by limited proteolysis. When injected into the cytosol of a frog oocyte, intact nucleoplasmin molecules rapidly accumulate in the nucleus even though they are too large to diffuse passively through the pore complex. The signal for this nuclear import resides in the tail domains, since injected tails are taken up by the nucleus but heads are not. The role of nuclear pores in this signal-directed import is demonstrated by electron microscopy using nucleoplasmin tails coupled to spheres of colloidal gold, which are easily visualized because of their high electron density. The attached nucleoplasmin tails direct the entry of the gold particles into the nucleus via nuclear pores (see Figure 12–15).

of the oocyte, on the other hand, they remain there. Thus it seems that, in addition to receptors that recognize nuclear protein import signals, the pore contains one or more receptors that recognize RNA molecules (or the proteins bound to them) destined for the cytosol. Using differently sized gold particles, one set coated with RNA and injected into the nucleus and the other set coated with nuclear protein import signals, it can be shown that a single pore complex allows traffic in both directions.

The mechanism of macromolecular transport across nuclear pores is fundamentally different from the transport mechanisms involved in the transfer of proteins across the membranes of other organelles in that it occurs through a large, regulated aqueous pore rather than through a protein transporter that spans one or more lipid bilayers. It is thought that a nuclear protein is transported through the pores while it is in a fully folded conformation, just as a newly formed ribosomal subunit is transported as an assembled particle; by contrast, proteins have to be unfolded during their transport into other organelles, as we discuss later.

Figure 12–15 Visualizing the specific import of a protein through nuclear pores. This electron micrograph shows colloidal gold spheres coated with nucleoplasmin (see Figure 12–14) entering the nucleus by means of nuclear pores (indicated by *red brackets*). The same result is obtained when the gold spheres are coated with the tail regions of nucleoplasmin molecules. These gold particles are much larger in diameter than the diffusion channel in the pore complex, implying that a pore has been induced to widen to permit their passage. Because the gold particles line up in the cytosol before they contact and enter the pore complex, it has been suggested that the fibrils that extend into the cytosol from the pore complex (see Figure 12–10A) guide the particles to their destination (From C. Feldherr, E. Kallenbach, and N. Schultz, *J. Cell Biol.* 99:2216–2222, 1984, by copyright permission of the Rockefeller University Press.)

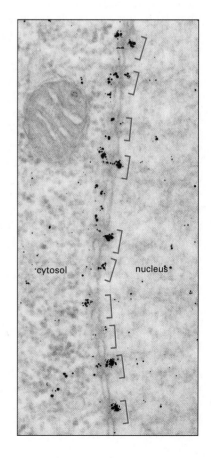

cytosol

nucleus

The Transport of Molecules into and out of the Nucleus

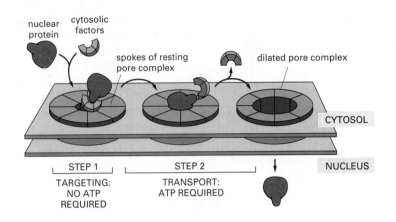

nuclear   cytosolic
protein   factors

spokes of resting
pore complex   dilated pore complex

CYTOSOL

STEP 1   STEP 2   NUCLEUS

TARGETING: TRANSPORT:
NO ATP   ATP REQUIRED
REQUIRED

Figure 12–16 **Highly schematic view of the mechanism of active transport through nuclear pores.** The proteins and structures involved in the active transport process are not known. A diverse set of related cytosolic proteins, however, is required for the initial binding of nuclear proteins to the complex. These proteins, called *nucleoporins,* contain a simple sugar (*N*-acetylglucosamine) that aided their identification through the use of lectins and specific antibodies. The fibrils that project from the pore complex and are thought to help guide nuclear proteins to the center of the pore are not shown.

The Nuclear Envelope Is Disassembled During Mitosis [10]

The **nuclear lamina** is a meshwork of interconnected protein subunits called **nuclear lamins.** These are a special class of intermediate filament proteins (discussed in Chapter 16) that polymerize into a two-dimensional lattice (Figure 12–17). The nuclear lamina is thought to give shape and stability to the nuclear envelope, to which it is anchored by attachment to both the nuclear pore complexes and the inner nuclear membrane. As the chromatin is also thought to interact directly with the nuclear lamina, the lamina provides a structural link between the DNA and the nuclear envelope.

When a nucleus disassembles during mitosis, the nuclear lamina depolymerizes, at least partly as a consequence of the phosphorylation of the nuclear lamins at the onset of mitosis. At the same time the nuclear pore complexes disassemble into their various components. Depolymerization of the nuclear lamina is probably a prerequisite for the nuclear envelope to break up into membrane vesicles, which, together with the nuclear contents, disperse throughout the cytosol. Reassembly of the lamina occurs when the nuclear lamins are dephosphorylated and, as a result, repolymerize on the surface of the chromosomes; the re-assembled lamina then binds the vesicles of nuclear envelope membrane, which fuse with one another to re-form an envelope around each chromosome or group of chromosomes. During this process the nuclear pore complexes also re-assemble. The enveloped chromosomes then come together, and their membranes fuse to form a single nuclear envelope, which actively reimports all those proteins that contain nuclear localization signals (Figure 12–18). Because the new nuclear envelope is so closely applied to the surface of the chromosomes, it excludes all of the proteins in the cell except those bound to the mitotic chromosomes. Thus large proteins are kept out of the interphase nucleus unless they contain nuclear localization signals.

Nuclear localization signals are not cleaved off after transport into the nucleus. This is presumably because nuclear proteins need to be imported repeatedly, once after every cell division. In contrast, once a protein molecule has been imported into any of the other membrane-bounded organelles, it is passed on from generation to generation within that compartment and need never be translocated again; the signal peptide on these molecules is often removed following protein translocation.

Transport Between Nucleus and Cytosol Can Be Regulated by Preventing Access to the Transport Machinery [11]

As discussed in Chapter 9, the activity of some gene regulatory proteins is controlled by keeping them out of the nuclear compartment until they are needed there. The nuclear localization signal of some of these proteins can be inactivated by phosphorylation. Others are bound to inhibitory cytosolic proteins that either

1 µm

Figure 12–17 **The nuclear lamina.** Electron micrograph of a portion of the nuclear lamina in a *Xenopus* oocyte prepared by freeze drying and metal shadowing. The lamina is formed by a regular lattice of specialized intermediate filaments. (Courtesy of Ueli Aebi.)

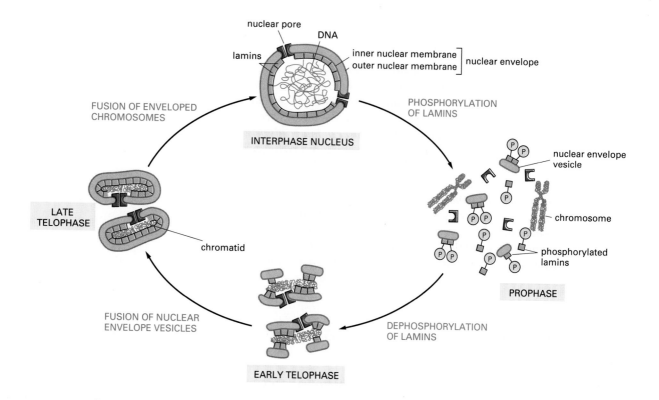

Figure 12–18 **Breakdown and re-formation of the nuclear envelope during mitosis.** The phosphorylation of the lamins is thought to help trigger the disassembly of the nuclear lamina, which in turn causes the nuclear envelope to break up into vesicles. Dephosphorylation of the lamins is thought to help reverse the process.

anchor them in the cytosol—presumably through interactions with the cytoskeleton or specific organelles—or mask their nuclear localization signals. When the cell receives an appropriate stimulus, the protein is released from its cytosolic anchor or mask and is transported into the nucleus (Figure 12–19).

Export of RNA from the nucleus may be controlled in a similar way. Like active import into the nucleus, export also requires a signal: in the case of most messenger RNA molecules, this is provided by a unique modification, the cap structure, at the 5′ end of the RNA (discussed in Chapter 8). Incompletely processed pre-messenger RNAs include this cap but are anchored to the nuclear transcription and splicing machinery, which releases an RNA molecule only after its processing is completed: genetic studies in yeast have shown that mutations that prevent the pre-messenger RNA from properly engaging with the splicing machinery lead to the export of the unspliced RNA. Other RNAs, like transfer RNA or ribosomal RNA, which lack a 5′ cap, must first be assembled with proteins and are then exported as part of these complexes. Presumably, nuclear export signals are contained in the protein subunits of these complexes, and these signals become activated after proper assembly with the RNA components, but the nature of these signals is not known.

Figure 12–19 **The nuclear import of the glucocorticoid receptor.** The glucocorticoid receptor is a gene regulatory protein that, in the non-hormone-treated cell, is bound in the cytosol to the chaperone protein hsp90. When activated by the binding of the appropriate steroid hormone, it is released from hsp90 and is directed into the nucleus by a nuclear localization signal; once in the nucleus, it binds to specific DNA sequences and regulates the transcription of a discrete set of genes (discussed in Chapters 9 and 15).

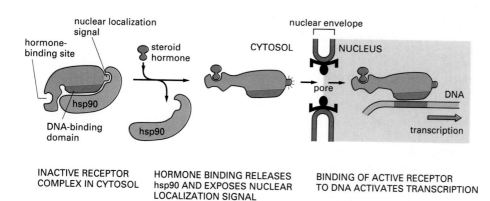

The Transport of Molecules into and out of the Nucleus

Summary

The nuclear envelope consists of an inner and an outer nuclear membrane. The outer membrane is continuous with the ER membrane, and the space between it and the inner membrane is continuous with the ER lumen. RNA molecules, which are made in the nucleus, and ribosomal subunits, which are assembled there, are exported to the cytosol, while all of the proteins that function in the nucleus are synthesized in the cytosol and are then imported. The extensive traffic of materials between nucleus and cytosol occurs through nuclear pores that provide a direct passageway across the nuclear envelope.

Proteins containing nuclear localization signals are actively transported inward through the pores, while RNA molecules and newly made ribosomal subunits are actively transported outward through the pores. Because the nuclear localization signals are not removed, nuclear proteins can be imported repeatedly, as is required each time the nucleus reassembles following mitosis. The transport of nuclear proteins and RNA molecules through the pores can be regulated by denying these molecules access to the transport machinery in the nuclear pore complexes.

The Transport of Proteins into Mitochondria and Chloroplasts [12]

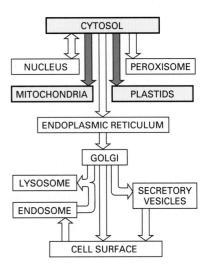

As discussed in Chapter 14, mitochondria and chloroplasts are double-membrane-bounded organelles that specialize in the synthesis of ATP, using energy derived from electron transport and oxidative phosphorylation in mitochondria and from photosynthetic phosphorylation in chloroplasts. Although both organelles contain their own DNA, ribosomes, and other machinery for protein synthesis, most of their proteins are encoded in the cell nucleus and imported from the cytosol. Moreover, each imported protein must reach the particular organelle subcompartment in which it functions. For mitochondria there are two subcompartments: the internal **matrix space** and the **intermembrane space.** These compartments are formed by the two distinct mitochondrial membranes: the **inner membrane,** which encloses the matrix space, and the **outer membrane,** which is in contact with the cytosol (Figure 12–20A). Chloroplasts have the same two subcompartments plus an additional subcompartment, the *thylakoid space*, which is surrounded by the *thylakoid membrane* (Figure 12–20B). Each of the subcompartments contains a distinct set of proteins. The growth of mitochondria and chloroplasts by the import of proteins from the cytosol is therefore a major feat, requiring that proteins be translocated across a number of membranes in succession and end up in the appropriate place.

The relatively few proteins encoded by the genomes of these organelles are located mostly in the inner membrane in mitochondria and in the thylakoid membrane in chloroplasts. These organelle-encoded polypeptides generally form subunits of protein complexes whose other subunits are encoded by nuclear genes and are imported from the cytosol. The formation of such hybrid protein complexes requires a balanced synthesis of the two types of subunits; how protein synthesis is coordinated on different types of ribosomes located two membranes apart is still largely a mystery.

Translocation into the Mitochondrial Matrix Depends on a Matrix Targeting Signal [13]

Proteins imported into the mitochondrial matrix are usually taken up from the cytosol within a minute or two of their release from polyribosomes. These **mitochondrial precursor proteins** almost always have a signal peptide (20–80 residues long) at their amino terminus that is rapidly removed after import by a protease (the *signal peptidase*) in the mitochondrial matrix. The signal peptide

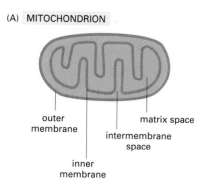

(A) MITOCHONDRION

outer membrane

inner membrane

intermembrane space

matrix space

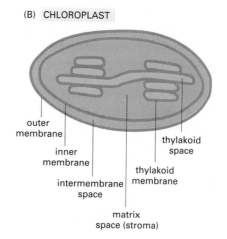

(B) CHLOROPLAST

outer membrane

inner membrane

intermembrane space

thylakoid membrane

matrix space (stroma)

thylakoid space

Figure 12–20 The subcompartments of mitochondria and chloroplasts. The topology of the chloroplast can be derived from that of the mitochondrion in a simple way: pinching off the invaginations of the inner mitochondrial membrane to create vesicles would generate a compartment that is topologically equivalent to the thylakoid vesicle in chloroplasts. Thylakoid vesicles may have evolved in this way.

can be remarkably simple. Molecular genetic experiments in which the signal peptide is progressively reduced in length have shown that, for one mitochondrial protein, only 12 amino acids at the amino terminus are needed to signal mitochondrial import. These 12 residues can be attached to any cytosolic protein and will direct the protein into the mitochondrial matrix. Physical studies of full-length signal peptides suggest that they can form amphipathic α-helical structures in which positively charged residues are clustered on one side of the helix while uncharged hydrophobic residues are clustered on the opposite side (Figure 12–21). This configuration is thought to be recognized by specific receptor proteins on the mitochondrial surface.

Translocation into the Mitochondrial Matrix Requires Both the Electrochemical Gradient Across the Inner Membrane and ATP Hydrolysis [14]

Almost everything we know about the molecular mechanism of protein import into mitochondria has been learned from analysis of cell-free, reconstituted transport systems. Mitochondria are first purified by differential centrifugation of homogenized cells and are then incubated with radiolabeled mitochondrial precursor proteins. The precursor proteins are generally taken up rapidly and efficiently into such mitochondria during a brief *in vitro* incubation. By changing the conditions in these experiments *in vitro*, it is possible to establish the biochemical requirements for protein transport into the mitochondria.

Vectorial movement and transport require energy. In most biological systems the energy is supplied by ATP hydrolysis. In the case of mitochondrial import, however, an electrochemical gradient across the inner mitochondrial membrane is required in addition to ATP hydrolysis. This gradient is maintained by the pumping of H+ from the matrix to the intermembrane space, driven by electron transport processes in the inner membrane. The mitochondrial outer membrane,

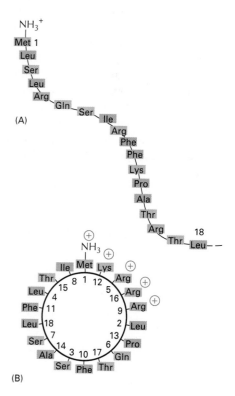

Figure 12–21 A signal peptide for mitochondrial protein import. Cytochrome oxidase is a large multiprotein complex located in the mitochondrial inner membrane, where it functions as the terminal enzyme in the electron-transport chain (discussed in Chapter 14). (A) The first 12 amino acids of the precursor to subunit IV of this enzyme serve as a signal peptide for the import of the subunit into the mitochondrion. (B) When the full-length signal peptide is folded as an α helix with 3.6 residues per turn and viewed from the top, the positively charged residues (*red*) are seen to be clustered on one face of the helix while the nonpolar residues (*green*) are clustered on the opposite face. Mitochondrial signal peptide sequences always have the potential to form such an amphipathic helix, which is thought to be recognized by specific receptor proteins on the mitochondrial surface.

The Transport of Proteins into Mitochondria and Chloroplasts

like that of gram-negative bacteria (see Figure 11–14), contains large amounts of a pore-forming protein called *porin* and is thus freely permeable to inorganic ions and metabolites (but not to most proteins), so that no gradient can be maintained across it. The energy in the electrochemical gradient across the inner membrane is tapped to help drive most of the cell's ATP synthesis, but it is also used to drive the translocation of proteins bearing a mitochondrial import signal peptide: when ionophores that collapse the mitochondrial membrane potential are added, import is blocked. It is still uncertain how the electrochemical gradient contributes to protein translocation. The role of ATP hydrolysis is much better understood, as we see below.

Mitochondrial Proteins Are Imported into the Matrix in a Two-Stage Process at Contact Sites That Join the Inner and Outer Membranes [15]

As a first step in mitochondrial import, the mitochondrial precursor proteins have to bind to receptor proteins that reside in the mitochondrial outer membrane and recognize the mitochondrial signal peptides. The next step is the translocation process itself.

A protein could reach the mitochondrial matrix by crossing the two membranes one at a time, or it could pass through both at once. To distinguish these possibilities, a cell-free import system is cooled to a low temperature, arresting the proteins at an intermediate step in the translocation process. The proteins that accumulate at this step have already had their amino-terminal signal peptide removed by the matrix signal peptidase, indicating that their amino terminus must be in the matrix space; yet the bulk of the protein can still be attacked from outside the mitochondria by externally added proteolytic enzymes (Figure 12–22). This result demonstrates that the precursor proteins can pass through both mitochondrial membranes at once to enter the matrix. It is thought that there are two protein translocators, one in the outer membrane and one in the inner membrane, whose functions are usually coupled to allow translocation across both membranes at the same time. Electron microscopists have noted numerous **contact sites** at which the inner and outer mitochondrial membranes appear to be joined, and it seems likely that translocation occurs at or near these sites.

Although precursor proteins are transported through both membranes at once, it is clear from the experiments described in Figure 12–22 that the import process occurs in two distinct stages, only the second of which is arrested at a low temperature. Of these, it is only the initial penetration, which is not affected by low temperatures, that requires the membrane potential: when cooled mitochondria containing partly translocated intermediates are warmed up, import is rapidly completed (see Figure 12–22) even if the potential across the inner membrane is collapsed. This second stage of the transport process, however, requires

Figure 12–22 Proteins transiently span both the inner and outer mitochondrial membranes during their translocation into the matrix. When isolated mitochondria are incubated with a precursor protein at 5°C, the precursor is only partially translocated. The amino-terminal signal peptide (*red*) is cleaved off in the matrix; most of the polypeptide chain remains outside the mitochondria, where it is accessible to proteolytic enzymes. Upon warming to 25°C, the translocation is completed. Once inside the mitochondrion, the polypeptide chain is protected from externally added proteolytic enzymes unless detergents are added to disrupt the mitochondrial membranes, which allows the imported proteins to be digested.

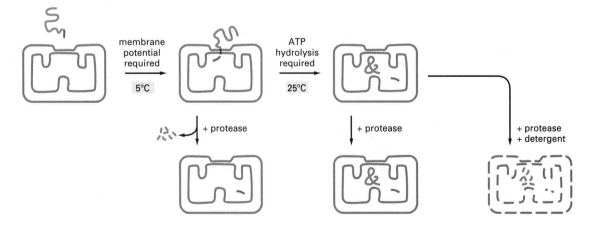

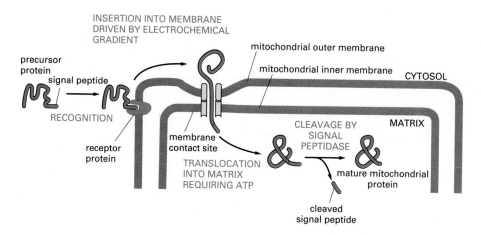

INSERTION INTO MEMBRANE
DRIVEN BY ELECTROCHEMICAL
GRADIENT

precursor protein

signal peptide

RECOGNITION

receptor protein

membrane contact site

mitochondrial outer membrane

mitochondrial inner membrane

CYTOSOL

CLEAVAGE BY SIGNAL PEPTIDASE

MATRIX

TRANSLOCATION INTO MATRIX REQUIRING ATP

mature mitochondrial protein

cleaved signal peptide

Figure 12–23 Protein import by mitochondria. The amino-terminal signal peptide of the precursor protein is recognized by receptors that reside in the outer membrane. The protein is thought to be translocated across both mitochondrial membranes at or near special contact sites, driven first by the electrochemical gradient across the inner membrane and then by ATP hydrolysis. The signal peptide is cleaved off by a signal peptidase in the matrix to form the mature protein; the free signal peptide is rapidly degraded.

ATP. Thus the first stage, which involves the insertion of the signal peptide and adjoining sequences into both mitochondrial membranes, is driven by the electrochemical gradient, and the second stage, in which the remainder of the polypeptide chain moves into the matrix, requires both ATP hydrolysis and a physiological temperature (Figure 12–23).

Proteins Are Imported into the Mitochondrial Matrix in an Unfolded State [16]

Transport of mitochondrial precursor proteins across the two mitochondrial membranes at a contact site is guided by members of the chaperone family of proteins, which are discussed in Chapter 5. It is difficult to envisage how a folded, water-soluble protein could straddle two (or even one) lipid bilayer while retaining its native three-dimensional conformation. It is now known that cytosolic chaperone proteins (called *chaperonins*) belonging to the hsp70 family, as well as helping to ensure the correct folding of cytosolic proteins, play an essential part in protein import into both mitochondria and the ER by binding the precursor in its unfolded state during translocation. As discussed in Chapter 5, the release of newly synthesized polypeptides from the hsp70 family of chaperone proteins requires ATP hydrolysis, and this partly accounts for the ATP dependence of the later stages of mitochondrial import.

The essential role of the chaperone proteins in translocation across internal cellular membranes was first indicated by genetic studies in yeasts. When the genes encoding certain members of the hsp70 family of chaperone proteins are inactivated, mitochondrial precursor proteins fail to be imported into mitochondria and accumulate in the cytosol instead. It is thought that newly synthesized precursor proteins, as they are released from polyribosomes in the cytosol, bind to hsp70 proteins, which prevent the precursor proteins from aggregating or folding up spontaneously before they bind to the protein translocator in the target membrane. The energy liberated by the hydrolysis of ATP is used to release the bound hsp70 proteins as the translocated protein is passed across the membrane. Experimentally, the requirement for hsp70 and ATP in the cytosol can be bypassed if the precursor proteins are artificially unfolded, for example, by a denaturation step in a concentrated solution of urea.

Sequential Binding of the Imported Protein to Mitochondrial hsp70 and hsp60 Drives Its Translocation and Assists Protein Folding [17]

Imported proteins are not only delivered to the mitochondrion by chaperone proteins: once they are extruded into the interior, they are received by closely related hsp70 proteins in the matrix space. *Mitochondrial hsp70* is crucial to the

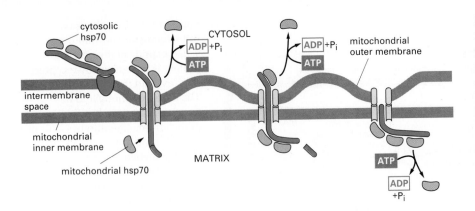

intermembrane space

cytosolic hsp70

CYTOSOL

ADP +P_i

ATP

ADP +P_i

ATP

mitochondrial outer membrane

mitochondrial inner membrane

MATRIX

ATP

ADP +P_i

mitochondrial hsp70

Figure 12–24 Protein import into the mitochondrial matrix requires hsp70 proteins on both sides of the mitochondrial double membrane. After the initial insertion of the signal peptide and of adjacent portions of the polypeptide chain, the unfolded chain slides in a channel that spans both membranes. Bound *cytosolic hsp70* is released from the protein in a step that depends on ATP hydrolysis; concomitantly, *mitochondrial hsp70* binds to regions of the polypeptide chain as they become exposed in the matrix, thereby pulling the protein into the interior of the mitochondrion.

import process, as mitochondria containing mutant forms of the protein fail to import precursor proteins. Like its cytosolic cousin, mitochondrial hsp70 has a high affinity for unfolded polypeptide chains, and it binds tightly to an imported protein as soon as it emerges from the translocator. The hsp70 then releases the protein in an ATP-dependent step. This energy-driven cycle of binding and subsequent release could provide the driving force for protein import after the protein has initially inserted into the translocator (Figure 12–24): the sequential binding of multiple mitochondrial hsp70 proteins may pull the unfolded protein through a transmembrane channel into the matrix.

After the initial interaction with mitochondrial hsp70, many imported proteins are passed on to another chaperone protein, *mitochondrial hsp60*. As discussed in Chapter 5, hsp60 attaches to the unfolded polypeptide chain and facilitates its folding in an ATP-consuming reaction. Much of our current understanding of the function of hsp60 in facilitating protein folding is derived from studies on import of proteins into mitochondria.

Protein Transport into the Mitochondrial Inner Membrane and the Intermembrane Space Requires Two Signals [18]

Many of the functions of mitochondria require proteins that either are integrated into the mitochondrial inner membrane or operate in the intermembrane space. These proteins are transported from the cytosol by the same mechanism that transports proteins into the matrix, but they are then prevented by various means from ending their journey in the matrix. In many cases the precursor proteins are first transferred all the way into the matrix, as was illustrated in Figure 12–23. A very hydrophobic amino acid sequence, however, is strategically placed after the amino-terminal signal peptide that initiates import. Once the amino-terminal signal is cleaved by the matrix signal peptidase, the hydrophobic sequence can function as a new amino-terminal signal peptide to translocate the protein back again from the matrix into or across the inner membrane. Presumably, the final step in this pathway involves a mechanism that is also used to direct proteins encoded in the mitochondrion to the inner membrane (Figure 12–25A), and we see later that a similar mechanism is used for translocating proteins into or across the ER membrane and the procaryotic plasma membrane, where it has been more extensively investigated.

An alternative route to the inner membrane may avoid the excursion into the matrix space altogether. The translocator in the inner membrane binds to the hydrophobic sequence that follows the amino-terminal signal peptide that initiates import and prevents further translocation across the inner membrane. The two translocators in the outer and inner membranes become uncoupled, which causes the remainder of the protein to be pulled into the intermembrane space (Figure 12–25B). Different proteins may use one or the other of these two pathways to the inner membrane or intermembrane space. It is not clear which one is more commonly used.

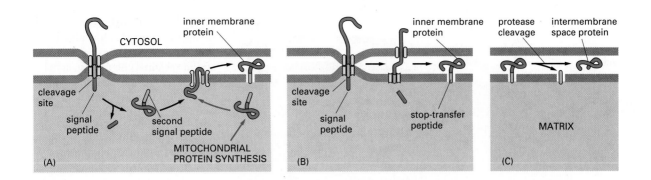

After proteins destined for the intermembrane space have been inserted via their hydrophobic signal peptides into the inner membrane, some are cleaved by a signal peptidase in the intermembrane space to release the mature polypeptide chain as a soluble protein (Figure 12–25C). Many of these proteins ultimately become attached as peripheral membrane proteins to the outer surface of the inner membrane, where they form subunits of protein complexes that also contain transmembrane proteins.

Two Signal Peptides Are Required to Direct Proteins to the Thylakoid Membrane in Chloroplasts [19]

Protein transport into chloroplasts resembles transport into mitochondria in many respects: both occur posttranslationally, both require energy, and both utilize amphipathic amino-terminal signal peptides that are removed after use. There is at least one important difference, however: mitochondria exploit the electrochemical gradient across their inner membrane to help drive the transport, whereas chloroplasts, which have an electrochemical gradient across their thylakoid but not their inner membrane, appear to employ only ATP hydrolysis to power import across their double-membrane outer envelope.

Although the signal peptides for import into chloroplasts resemble those for import into mitochondria, mitochondria and chloroplasts are both present in the same plant cells, and proteins must choose appropriately between them. In plants, for example, a bacterial enzyme is directed specifically to mitochondria if it is experimentally joined to an amino-terminal signal sequence of a mitochondrial protein; the same enzyme joined to an amino-terminal signal sequence of a chloroplast protein ends up in chloroplasts. The different signal sequences, therefore, can be distinguished, presumably by the import receptors on each organelle.

Chloroplasts have an extra membrane-bounded compartment, the **thylakoid.** Many chloroplast proteins, including protein subunits of the photosynthetic system and of the ATP synthase, are embedded in the membrane of the thylakoid compartment. Like the precursors of some mitochondrial proteins, these proteins are transported from the cytosol to their final destination in two steps. First, they pass across the double membrane at contact sites into the matrix space of the chloroplast (termed the **stroma**), and then they are translocated into the thylakoid membrane (or across this membrane into the thylakoid space). The precursors of these proteins have a hydrophobic thylakoid signal peptide following the amino-terminal chloroplast signal peptide. After the amino-terminal signal peptide has been used to import the protein into the stroma, it is removed by a stromal signal peptidase (analogous to the matrix signal peptidase in mitochondria). This cleavage unmasks the thylakoid signal peptide, which then initiates transport across the thylakoid membrane (Figure 12–26). As with mitochondria, the second step is the pathway used to insert chloroplast-encoded proteins into the thylakoid membrane; the protein translocator required presumably originated in the chloroplast's bacterial ancestor.

Figure 12–25 Import of proteins from the cytosol to the mitochondrial intermembrane space or inner membrane. (A) A pathway that requires two signal peptides and two translocation events is thought to be used to move some proteins from the cytosol to the inner membrane. The protein is first imported into the matrix space as in Figure 12–23. Cleavage of the signal peptide (*red*) used for the initial translocation, however, unmasks an adjacent hydrophobic signal peptide (*orange*) at the new amino terminus. This signal causes the protein to be integrated into the inner membrane by the same pathway that is used to insert proteins encoded by the mitochondrial genome into this membrane. (B) In an alternative mechanism, the hydrophobic sequence that follows the matrix targeting signal binds to the translocator and stops the translocation across the inner membrane. The remainder of the protein is then pulled into the intermembrane space and the hydrophobic sequence is released into the inner membrane. This mechanism is called the stop-transfer pathway and is discussed in detail later. (C) Some soluble proteins of the intermembrane space may also use the pathways shown in (A) and (B) before they are released into the intermembrane space by a second signal peptidase (with its active site in the intermembrane space), which removes the hydrophobic signal peptide.

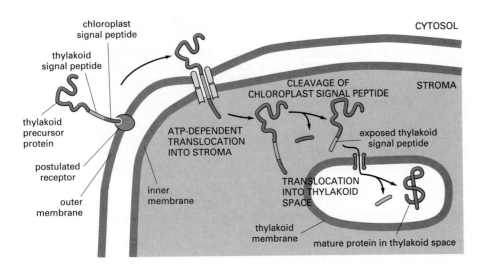

CYTOSOL

chloroplast signal peptide

thylakoid signal peptide

thylakoid precursor protein

postulated receptor

outer membrane

inner membrane

ATP-DEPENDENT TRANSLOCATION INTO STROMA

CLEAVAGE OF CHLOROPLAST SIGNAL PEPTIDE

STROMA

exposed thylakoid signal peptide

TRANSLOCATION INTO THYLAKOID SPACE

thylakoid membrane

mature protein in thylakoid space

Figure 12–26 Translocation into the thylakoid space of chloroplasts. The precursor polypeptide contains an amino-terminal chloroplast signal peptide (*red*) followed immediately by a thylakoid signal peptide (*orange*). The chloroplast signal peptide initiates translocation into the stroma through a membrane contact site by a mechanism similar to that used for translocation into the mitochondrial matrix. The signal peptide is then cleaved off, unmasking the thylakoid signal peptide, which initiates translocation across the thylakoid membrane.

Summary

Although mitochondria and chloroplasts have their own genetic systems, they produce only a small proportion of their own proteins. Instead, the two organelles import most of their proteins from the cytosol using similar mechanisms. The transport processes involved have been most extensively studied in mitochondria, especially in yeasts. A protein is translocated into the mitochondrial matrix space by passing through sites of adhesion between the outer and inner membranes called contact sites. Translocation into mitochondria is driven by both ATP hydrolysis and the electrochemical gradient across the inner membrane, whereas translocation into chloroplasts is driven by ATP hydrolysis alone. The transported protein crosses the membranes of the mitochondrion or chloroplast in an unfolded state. Chaperone proteins of the cytosolic hsp70 family maintain the precursor proteins in an unfolded, translocation-competent state. Mitochondrial hsp70 in the matrix binds to the incoming polypeptide chain and is thought to pull the protein chain into the matrix. Once the protein is in the matrix, another stress protein, hsp60, helps the translocated protein fold up. Only proteins that contain a specific signal peptide are translocated into mitochondria or chloroplasts. The signal peptide is usually located at the amino terminus and is cleaved off after import. Transport across or into the inner membrane can occur as a second step if a hydrophobic signal peptide is also present in the imported protein; this second signal peptide is unmasked when the first signal peptide is removed. In the case of chloroplasts, import from the stroma into the thylakoid likewise requires a second signal peptide.

Peroxisomes [20]

Peroxisomes differ from mitochondria and chloroplasts in many ways. Most notably, they are surrounded by only a single membrane, and they do not contain DNA or ribosomes. In spite of these differences, peroxisomes are thought to acquire their proteins by a similar process of selective import from the cytosol. Because peroxisomes have no genome, however, *all* of their proteins must be imported. Peroxisomes thus resemble the ER in being self-replicating membrane-bounded organelles that exist without genomes of their own.

Because we do not discuss peroxisomes elsewhere, we shall digress to consider some of the functions of this diverse family of organelles before discussing their biosynthesis. Peroxisomes are found in all eucaryotic cells. They contain oxidative enzymes, such as *catalase* and *urate oxidase*, at such high concentrations that in some cells the peroxisomes stand out in electron micrographs because of the presence of a crystalloid core, largely composed of urate oxidase (Figure 12–27).

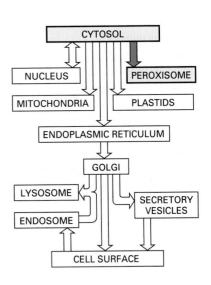

CYTOSOL

NUCLEUS

PEROXISOME

MITOCHONDRIA

PLASTIDS

ENDOPLASMIC RETICULUM

GOLGI

LYSOSOME

SECRETORY VESICLES

ENDOSOME

CELL SURFACE

Like the mitochondrion, the peroxisome is a major site of oxygen utilization. One hypothesis is that the peroxisome is a vestige of an ancient organelle that carried out all of the oxygen metabolism in the primitive ancestors of eucaryotic cells. When the oxygen produced by photosynthetic bacteria first began to accumulate in the atmosphere, it would have been highly toxic to most cells. Peroxisomes might have served to lower the intracellular concentration of oxygen while also exploiting its chemical reactivity to carry out useful oxidative reactions. According to this view, the later development of mitochondria rendered the peroxisome largely obsolete because many of the same reactions—which had formerly been carried out in peroxisomes without producing energy—were now coupled to ATP formation by means of oxidative phosphorylation. The oxidative reactions carried out by peroxisomes in present-day cells would therefore be those that have important functions not taken over by mitochondria.

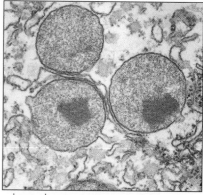

200 nm

Figure 12–27 **Peroxisomes.** Electron micrograph of three peroxisomes in a rat liver cell. The paracrystalline electron-dense inclusions are the enzyme urate oxidase. (Courtesy of Daniel S. Friend.)

Peroxisomes Use Molecular Oxygen and Hydrogen Peroxide to Carry Out Oxidative Reactions [21]

Peroxisomes are so called because they usually contain one or more enzymes that use molecular oxygen to remove hydrogen atoms from specific organic substrates (designated here as R) in an oxidative reaction that produces *hydrogen peroxide* (H_2O_2):

$$RH_2 + O_2 \rightarrow R + H_2O_2$$

Catalase utilizes the H_2O_2 generated by other enzymes in the organelle to oxidize a variety of other substrates—including phenols, formic acid, formaldehyde, and alcohol—by the "peroxidative" reaction: $H_2O_2 + R'H_2 \rightarrow R' + 2H_2O$. This type of oxidative reaction is particularly important in liver and kidney cells, whose peroxisomes detoxify various toxic molecules that enter the bloodstream. About a quarter of the ethanol we drink is oxidized to acetaldehyde in this way. In addition, when excess H_2O_2 accumulates in the cell, catalase converts it to H_2O ($2H_2O_2 \rightarrow 2H_2O + O_2$).

A major function of the oxidative reactions carried out in peroxisomes is the breakdown of fatty acid molecules. In a process called *β oxidation,* the alkyl chains of fatty acids are shortened sequentially by blocks of two carbon atoms at a time that are converted to acetyl CoA and exported from the peroxisomes to the cytosol for reuse in biosynthetic reactions. β oxidation in mammalian cells occurs both in mitochondria and peroxisomes; in yeast and plant cells, however, this essential reaction is exclusively found in peroxisomes.

Peroxisomes are unusually diverse organelles and even in the different cells of a single organism may contain very different sets of enzymes. They also can adapt remarkably to changing conditions. Yeast cells grown on sugar, for example, have small peroxisomes. But when some yeasts are grown on methanol, they develop large peroxisomes that oxidize methanol; and when grown on fatty acids, they develop large peroxisomes that break down fatty acids to acetyl CoA by β oxidation.

Peroxisomes also have very important roles in plants. Two very different types have been studied extensively. One type is present in leaves, where it catalyzes the oxidation of a side product of the crucial reaction that fixes CO_2 in carbohydrate (Figure 12–28A). This process is called *photorespiration* because it uses up O_2 and liberates CO_2. The other type of peroxisome is present in germinating seeds, where it plays an essential role in converting the fatty acids stored in seed lipids into the sugars needed for the growth of the young plant. Because this conversion of fats to sugars is accomplished by a series of reactions known as the *glyoxylate cycle,* these peroxisomes are also called *glyoxysomes* (Figure 12–28B). In the glyoxylate cycle two molecules of acetyl CoA produced by fatty acid breakdown in the peroxisome are used to make succinic acid, which leaves the peroxisome and is converted into glucose. The glyoxylate cycle does not occur in animal cells, and animals are thus unable to convert the fatty acids in fats into carbohydrates.

Peroxisomes

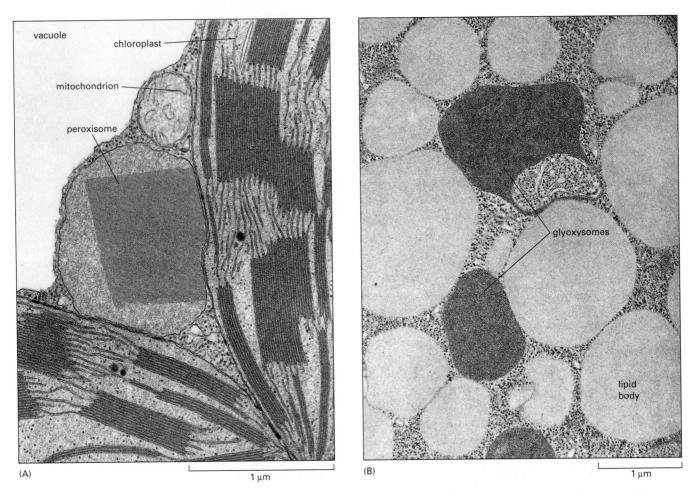

(A) vacuole
chloroplast
mitochondrion
peroxisome

(B) glyoxysomes
lipid body

1 µm

1 µm

Figure 12–28 Electron micrographs of two types of peroxisomes found in plant cells. (A) A leaf peroxisome with a paracrystalline core in a tobacco leaf mesophyll cell. Its close association with chloroplasts is thought to facilitate the exchange of materials between these organelles during photorespiration. (B) Peroxisomes in a fat-storing cotyledon cell of a tomato seed 4 days after germination. Here the peroxisomes (*glyoxysomes*) are associated with the lipid bodies where fat is stored, reflecting their central role in fat mobilization and gluconeogenesis during seed germination. (A, courtesy of P.J. Gruber and E.H. Newcomb; B, courtesy of S.E. Frederick and E.H. Newcomb.)

A Short Signal Sequence Directs the Import of Proteins into Peroxisomes [22]

A specific sequence of three amino acids located near the carboxyl terminus of many peroxisomal proteins functions as an import signal (see Table 12–3); if this sequence is experimentally attached to a cytosolic protein, the protein is imported into peroxisomes. The importance of this import process and of peroxisomes is dramatically demonstrated by the inherited human disease *Zellweger syndrome*, in which a defect in importing proteins into peroxisomes leads to a severe peroxisomal deficiency. These individuals, whose cells contain "empty" peroxisomes, have severe abnormalities in their brain, liver, and kidneys, and they die soon after birth. One form of this disease has been shown to be due to a mutation in the gene encoding a peroxisomal integral membrane protein called peroxisome assembly factor-1.

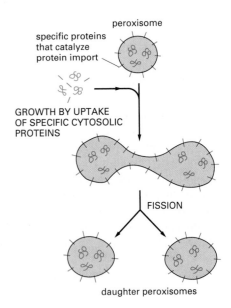

peroxisome

specific proteins that catalyze protein import

GROWTH BY UPTAKE OF SPECIFIC CYTOSOLIC PROTEINS

FISSION

daughter peroxisomes

Figure 12–29 A model for how peroxisomes are assembled. The peroxisome membrane contains specific import receptor proteins. All peroxisomal proteins, including new copies of the import receptor, are synthesized by cytosolic ribosomes and then imported into the organelle. Thus peroxisomes form only from preexisting peroxisomes by a process of growth and fission. Presumably, the lipids required to make new peroxisome membrane are also imported. We discuss later how lipids made in the ER can be transported through the cytosol to other organelles.

Peroxisomes presumably have at least one unique protein exposed on their cytosolic surface to act as a receptor that recognizes the signal on the proteins to be imported. At one time it was thought that the membrane of the peroxisome forms by budding from the ER, while the content is imported from the cytosol. There is now evidence, however, suggesting that new peroxisomes arise only from preexisting ones, by organelle growth and fission, as described elsewhere for mitochondria and plastids and for the ER itself (Figure 12–29).

Summary

Peroxisomes are specialized for carrying out oxidative reactions using molecular oxygen. They generate hydrogen peroxide, which they also use for oxidative purposes—destroying the excess by means of the catalase they contain. Like mitochondria and plastids, peroxisomes are self-replicating organelles. Because they contain no DNA or ribosomes, they have to import all of their proteins from the cytosol. A specific three amino acid sequence near the carboxyl terminus of many of these proteins functions as a peroxisomal import signal.

The Endoplasmic Reticulum [23]

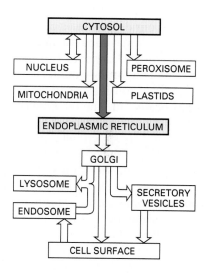

All eucaryotic cells have an **endoplasmic reticulum (ER).** Its membrane typically constitutes more than half of the total membrane of an average animal cell (see Table 12–2). It is organized into a netlike labyrinth of branching tubules and flattened sacs extending throughout the cytosol (Figure 12–30). The tubules and sacs are all thought to interconnect, so that the ER membrane forms a continuous sheet enclosing a single internal space. This highly convoluted space is called the **ER lumen** or the *ER cisternal space,* and it often occupies more than 10% of the total cell volume (see Table 12–1). The ER membrane separates the ER lumen from the cytosol, and it mediates the selective transfer of molecules between these two compartments.

The ER plays a central part in lipid and protein biosynthesis. Its membrane is the site of production of all the transmembrane proteins and lipids for most of the cell's organelles, including the ER itself, the Golgi apparatus, lysosomes, endosomes, secretory vesicles, and the plasma membrane. The ER membrane also makes a major contribution to mitochondrial and peroxisomal membranes by producing most of their lipids. In addition, almost all of the proteins that will be secreted to the cell exterior—as well as those destined for the lumen of the ER, Golgi apparatus, or lysosomes—are initially delivered to the ER lumen.

Membrane-bound Ribosomes Define the Rough ER [24]

The ER captures selected proteins from the cytosol as they are being synthesized. These proteins are of two types: (1) transmembrane proteins, which are only partly translocated across the ER membrane and become embedded in it, and (2) water-soluble proteins, which are fully translocated across the ER membrane and are released into the ER lumen. Some of the transmembrane proteins will remain in the ER, but many are destined to reside in the plasma membrane or the membrane of another organelle; the water-soluble proteins are destined either for the lumen of an organelle or for secretion. All of these proteins, regardless of their subsequent fate, are directed to the ER membrane by the same kind of signal peptide and are translocated across it by the same mechanism.

In mammalian cells the import of proteins into the ER begins before the polypeptide chain is completely synthesized—that is, it occurs **co-translationally.** This distinguishes the process from the import of proteins into mitochondria, chloroplasts, nuclei, and peroxisomes, which is **posttranslational** and requires different signal peptides. Since one end of the protein is usually translocated into the ER as the rest of the polypeptide chain is being made, the protein is never released into the cytosol and therefore is never in danger of folding up before

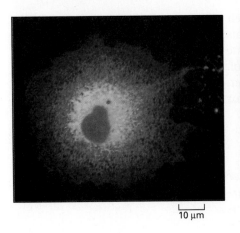
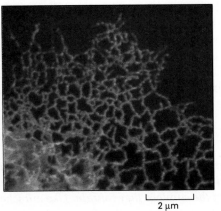

L___ 10 µm ___L   L___ 2 µm ___L

Figure 12–30 **The endoplasmic reticulum.** Fluorescence micrograph of a cultured mammalian cell stained with an antibody that binds to a protein retained in the ER. The ER extends as a network throughout the entire cytosol, so that all regions of the cytosol are close to some portion of the ER membrane. (Courtesy of Hugh Pelham.)

reaching the translocator in the membrane. In contrast to the posttranslational import of proteins into the mitochondria and chloroplasts, cytosolic chaperonins are therefore not required to keep the protein unfolded. The ribosome that is synthesizing the protein is directly attached to the ER membrane. These membrane-bound ribosomes coat the surface of the ER, creating regions termed **rough endoplasmic reticulum** (Figure 12–31).

There are, therefore, two spatially separate populations of ribosomes in the cytosol. **Membrane-bound ribosomes,** attached to the cytosolic side of the ER membrane, are engaged in the synthesis of proteins that are being concurrently translocated into the ER. **Free ribosomes,** unattached to any membrane, make all other proteins encoded by the nuclear genome. Membrane-bound and free ribosomes are structurally and functionally identical. They differ only in the proteins they are making at any given time. When a ribosome happens to be making a protein with an **ER signal peptide,** the signal directs the ribosome to the ER membrane. Since many ribosomes can bind to a single mRNA molecule, a **polyribosome** is usually formed, which becomes attached to the ER membrane via the signal peptides on multiple growing polypeptide chains (Figure 12–32). The individual ribosomes associated with such an mRNA molecule can return to the cytosol when they finish translation near the 3′ end of the mRNA molecule. The mRNA itself, however, tends to remain attached to the ER membrane by a changing population of ribosomes that are also held at the membrane by a ribosome receptor that helps to bind it there. In contrast, if an mRNA molecule encodes a protein that lacks an ER signal peptide, the polyribosome that forms remains free in the cytosol and its protein product is discharged there. Therefore, only those mRNA molecules that encode proteins with an ER signal peptide bind to rough ER membranes; those mRNA molecules that encode all other proteins remain free in the cytosol. The individual ribosomal subunits are thought to move

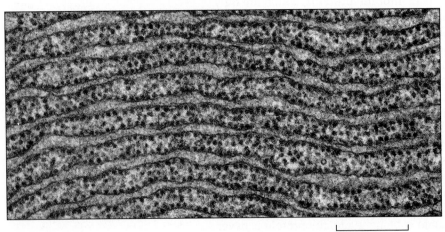

L___ 0.2 µm ___L

Figure 12–31 **The rough ER.** Electron micrograph of the rough ER, which receives its name from the many ribosomes on its cytosolic surface. (Courtesy of L. Orci.)

randomly between these two segregated populations of mRNA molecules (Figure 12–33).

Smooth ER Is Abundant in Some Specialized Cells [25]

Regions of ER that lack bound ribosomes are called **smooth endoplasmic reticulum**, or **smooth ER**. In the great majority of cells such regions are scanty, and there is only a small region of the ER that is partly smooth and partly rough. This region is said to consist of **transitional elements** because it is from here that transport vesicles carrying newly synthesized proteins and lipids bud off for transport to the Golgi apparatus. In certain specialized cells, however, the smooth ER is abundant and has additional functions. In particular, it is usually prominent in cells that specialize in lipid metabolism: cells that synthesize steroid hormones from cholesterol, for example, have an expanded smooth ER compartment to accommodate the enzymes needed to make cholesterol and to modify it to form the hormones (Figure 12–34).

The main cell type in the liver, the *hepatocyte*, provides another example. It is the principal site of the production of lipoprotein particles; these particles carry lipids via the bloodstream to other sites in the body. The enzymes that synthesize the lipid components of lipoproteins are located in the membrane of the smooth ER, which also contains enzymes that catalyze a series of reactions to detoxify both lipid-soluble drugs and various harmful compounds produced by metabolism. The most extensively studied of the *detoxification reactions* are catalyzed by the *cytochrome P450* family of enzymes, which catalyze a series of reactions whereby water-insoluble drugs or metabolites that would otherwise accumulate to toxic levels in cell membranes are rendered sufficiently water-soluble to leave the cell and be excreted in the urine. Because the rough ER alone cannot house enough of these and other necessary enzymes, a major portion of the membrane in a hepatocyte normally consists of smooth ER (Figure 12–35 and see Table 12–2).

When large quantities of certain compounds, such as the drug phenobarbital, enter the circulation, detoxification enzymes are synthesized in the liver in unusually large amounts, and the smooth ER doubles in surface area within a few days. Once the drug disappears, the excess smooth ER membrane is specifically and rapidly removed by a lysosome-dependent process called *autophagocytosis* (discussed in Chapter 13). How these dramatic changes are regulated is not known.

Another function of the ER in most eucaryotic cells is to sequester Ca^{2+} from the cytosol. The release of Ca^{2+} into the cytosol from the ER, and its subsequent reuptake, mediate many rapid responses to extracellular signals, as discussed in Chapter 15. The storage of Ca^{2+} in the ER lumen is facilitated by the high concen-

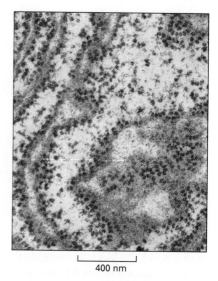

400 nm

Figure 12–32 Polyribosomes. Thin-section electron micrograph of polyribosomes attached to the ER membrane. The plane of section in some places cuts through the ER roughly parallel to the membrane, giving a face-on view of the rosettelike pattern of the polyribosomes. (Courtesy of George Palade.)

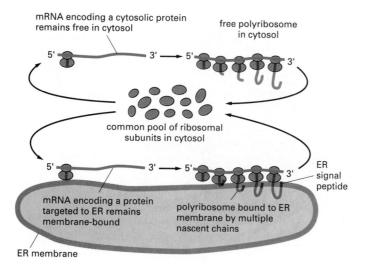

Figure 12–33 Free and membrane-bound ribosomes. A common pool of ribosomes is used to synthesize both the proteins that stay in the cytosol and those that are transported into the ER. It is the ER signal peptide on a newly formed polypeptide chain that directs the engaged ribosome to the ER membrane. The mRNA molecule may remain permanently bound to the ER as part of a polyribosome, while the ribosomes that move along it are recycled; at the end of each round of protein synthesis, the ribosomal subunits are released and rejoin the common pool in the cytosol.

The Endoplasmic Reticulum

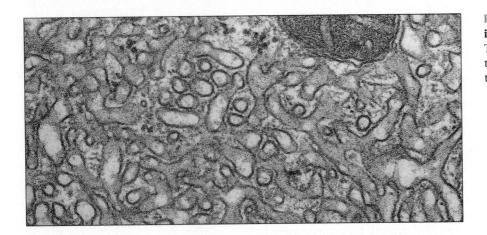

Figure 12–34 **Abundant smooth ER in a steroid-hormone-secreting cell.** This electron micrograph is of a testosterone-secreting Leydig cell in the human testis.

trations of Ca^{2+}-binding proteins there. In some cell types, and perhaps in most, specific regions of the ER are specialized for Ca^{2+} storage. Muscle cells, for example, have an abundant specialized smooth ER, called the *sarcoplasmic reticulum*, which sequesters Ca^{2+} from the cytosol by means of a Ca^{2+}-ATPase that pumps in Ca^{2+}; the release and reuptake of Ca^{2+} by the sarcoplasmic reticulum mediates the contraction and relaxation of the myofibrils during each round of muscle contraction (discussed in Chapter 16).

We shall now return to the two major roles of the ER: the synthesis and modification of proteins and the synthesis of lipids.

Rough and Smooth Regions of ER Can Be Separated by Centrifugation [26]

In order to study the functions and biochemistry of the ER, it is necessary to isolate the ER membrane. At first sight this seems a hopeless task since the ER is intricately interleaved with other components of the cytosol. Fortunately, when tissues or cells are disrupted by homogenization, the ER is fragmented and reseals into many small (~100 nm in diameter) closed vesicles called **microsomes,** which are relatively easy to purify. Microsomes derived from rough ER are studded with ribosomes and are called *rough microsomes*. The ribosomes are always found on the *outside* surface, so that the interior of the microsome is biochemically equivalent to the luminal space of the ER (Figure 12–36). Because they can be readily purified in functional form, rough microsomes are especially useful for studying the many processes carried out by the rough ER. To the biochemist they represent small authentic versions of the rough ER, still capable of protein synthesis, protein glycosylation, and lipid synthesis.

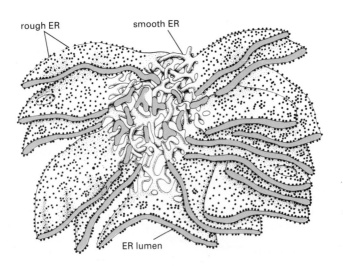

rough ER   smooth ER

ER lumen

Figure 12–35 **Three-dimensional reconstruction of a region of the smooth and rough ER in a liver cell.** The rough ER forms oriented stacks of flattened cisternae, each having a luminal space 20 to 30 nm wide. The smooth ER membrane is connected to these cisternae and forms a fine network of tubules 30 to 60 nm in diameter. (After R.V. Krstić, Ultrastructure of the Mammalian Cell. New York: Springer-Verlag, 1979.)

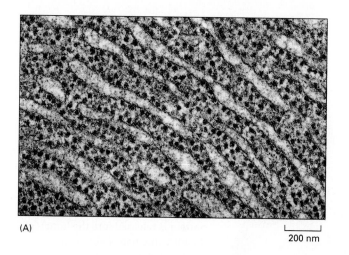

(A)

200 nm

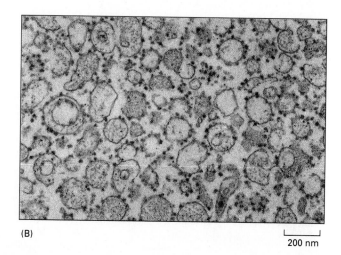

(B)

200 nm

Many vesicles of a size similar to that of rough microsomes but lacking attached ribosomes are also found in these homogenates. Such *smooth microsomes* are derived in part from smooth portions of the ER and in part from vesiculated fragments of plasma membrane, Golgi apparatus, endosomes, and mitochondria (the ratio depending on the tissue). Thus, whereas rough microsomes can be equated with rough portions of ER, the origins of smooth microsomes cannot be so easily assigned. The microsomes of the liver are an exception. Because of the unusually large quantities of smooth ER in the hepatocyte, most of the smooth microsomes in liver homogenates are derived from smooth ER.

The ribosomes attached to them make rough microsomes more dense than smooth microsomes. As a result, the rough and smooth microsomes can be separated from each other by equilibrium centrifugation (Figure 12–37). When the separated rough and smooth microsomes of liver are compared with respect to such properties as enzyme activity or polypeptide composition, they are very similar, although not identical: apparently most of the components of the ER membrane can diffuse freely between the rough and smooth regions, as would be expected for a continuous fluid membrane. The rough microsomes, however, contain more than 20 proteins that are not present in smooth microsomes, showing that some restraining mechanism must exist for a subset of ER membrane proteins. Some of the proteins in this subset help bind ribosomes to the rough ER, while others presumably produce the flattened shape of this part of the ER (see Figure 12–35). It is not clear whether these membrane proteins are retained by forming large two-dimensional aggregates in the lipid bilayer or whether they are instead held in place by interactions with a network of structural proteins on one or the other face of the ER membrane.

Figure 12–36 Electron micrographs of microsomes. When cells are disrupted by homogenization, the cisternae of rough ER (A) break up into small closed vesicles called *rough microsomes* (B). Similarly, the smooth ER breaks up into small vesicles that lack ribosomes and are called *smooth microsomes.* (A, courtesy of Daniel S. Friend; B, courtesy of George Palade.)

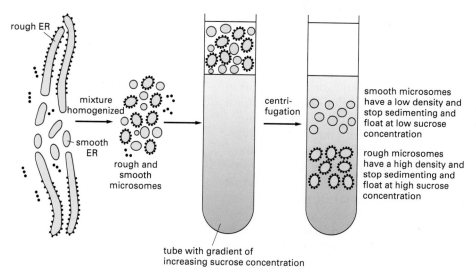

Figure 12–37 The isolation procedure used to purify rough and smooth microsomes from the ER. When sedimented to equilibrium through a gradient of sucrose, the two types of microsomes separate from each other on the basis of their different densities.

rough ER

mixture homogenized

smooth ER

rough and smooth microsomes

centrifugation

tube with gradient of increasing sucrose concentration

smooth microsomes have a low density and stop sedimenting and float at low sucrose concentration

rough microsomes have a high density and stop sedimenting and float at high sucrose concentration

The Endoplasmic Reticulum

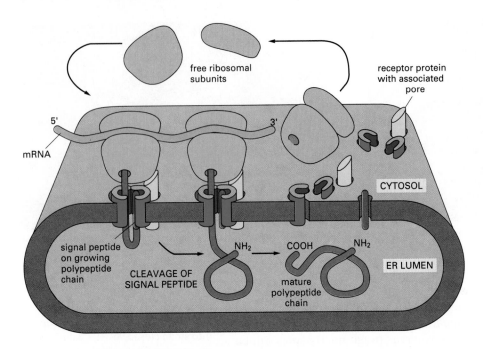

Figure 12–38 **The original signal hypothesis.** A simplified view of protein translocation across the ER membrane, as originally proposed. When the signal peptide emerges from the ribosome, it directs the ribosome to a receptor protein on the ER membrane. As it is synthesized, the polypeptide is postulated to be translocated across the ER membrane through a protein pore associated with the receptor. The signal peptide is clipped off during translation by a signal peptidase, and the mature protein is released into the lumen of the ER immediately after being synthesized. We now know that the hypothesis is correct in outline but that additional components besides those shown in this figure are required. The signal peptidase, for example, is a complex of five different membrane-bound polypeptide chains, with one complex apparently associated with every translocation pore.

Signal Peptides Were First Discovered in Proteins Imported into the Rough ER [27]

Signal peptides (and the signal peptide strategy of protein import) were first discovered in the early 1970s in secreted proteins that are translocated across the ER membrane as a first step toward their eventual discharge from the cell. In the key experiment the mRNA encoding a secreted protein was translated by ribosomes *in vitro*. When microsomes were omitted from this cell-free system, the protein synthesized was slightly larger than the normal secreted protein, the extra length being due to the presence of an amino-terminal *leader peptide*. In the presence of microsomes derived from the rough endoplasmic reticulum, however, a protein of the correct size was produced. These results were explained by the **signal hypothesis,** which postulated that the leader serves as a *signal peptide* that directs the secreted protein to the ER membrane and is then cleaved off by a signal peptidase in the ER membrane before the polypeptide chain is completed (Figure 12–38).

According to the signal hypothesis, the secreted protein should be extruded into the lumen of the microsome during its *in vitro* synthesis. This can be demonstrated with protease treatment: a newly synthesized protein made in the absence of microsomes is degraded when protease is added to the medium, whereas the same protein made in the presence of microsomes remains intact because it is protected by the microsomal membrane. When proteins without ER signal peptides are similarly synthesized *in vitro,* they are not imported into microsomes and therefore are degraded by protease treatment.

The signal hypothesis has been thoroughly tested by genetic and biochemical experiments and is found to apply to both plant and animal cells, as well as to protein translocation across the bacterial plasma membrane and, as we have seen, the membranes of mitochondria, chloroplasts, and peroxisomes. Amino-terminal ER signal peptides guide not only secreted proteins but also the precursors of all proteins made in the ER, including soluble proteins and membrane proteins. The signaling function of these peptides has been demonstrated directly by using recombinant DNA techniques to attach signal sequences to proteins that do not normally have them; the resulting fusion proteins are directed to the ER.

Cell-free systems in which proteins are imported into microsomes have provided powerful assay procedures for identifying, purifying, and studying the various components of the molecular machinery responsible for the ER import process.

A Signal-Recognition Particle (SRP) Directs ER Signal Peptides to a Specific Receptor in the Rough ER Membrane [28]

The ER signal peptide is guided to the ER membrane by at least two components: a **signal-recognition particle (SRP),** which cycles between the ER membrane and the cytosol and binds to the signal peptide, and an *SRP receptor,* also known as a *docking protein,* in the ER membrane. The SRP was discovered when it was found that washing microsomes with salt eliminated their ability to import secreted proteins. Import could be restored by adding back the supernatant containing the salt extract. The "translocation factor" in the salt extract was then purified and found to be a complex particle consisting of six different polypeptide chains bound to a single small RNA molecule (Figure 12–39). SRP and SRP receptor are present in all eucaryotic cells and probably in procaryotic cells as well.

The SRP binds to the ER signal peptide as soon as the peptide emerges from the ribosome. This causes a pause in protein synthesis, which presumably gives the ribosome enough time to bind to the ER membrane before the synthesis of the polypeptide chain is completed, thereby ensuring that the protein is not released into the cytosol. This may provide a safety mechanism as many secreted proteins and lysosomal proteins are hydrolases that could wreak havoc in the cytosol. Cells that secrete large amounts of hydrolases take the added precaution of having high concentrations of hydrolase inhibitors in their cytosol.

Once formed, the SRP ribosome complex binds to the **SRP receptor,** which is an integral membrane protein exposed on the cytosolic surface of the rough ER membrane. The SRP is then released, and a poorly characterized translocation apparatus transfers the growing polypeptide chain across the membrane (Figure 12–40). Because one of the SRP proteins and both chains of the SRP receptor contain GTP-binding domains, it is thought that conformational changes that occur during cycles of GTP binding and hydrolysis (discussed in Chapter 5) ensure that SRP release occurs only after the ribosome has become properly engaged with the translocation apparatus in the ER membrane.

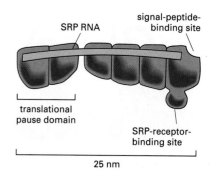

Figure 12–39 A highly schematic drawing of a signal-recognition particle (SRP). An SRP is an elongated complex containing six protein subunits and one RNA molecule (SRP RNA). One end of the SRP binds to an ER signal peptide on a growing polypeptide chain, while the other end binds to the ribosome itself and stops translation. The RNA in the particle may mediate an interaction with ribosomal RNA. (Adapted from V. Siegel and P. Walter, *Nature* 320:82–84, 1986.)

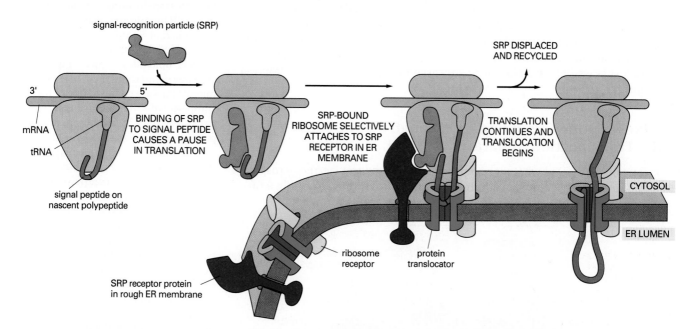

Figure 12–40 How ER signal peptides and SRP direct ribosomes to the ER membrane. The SRP and the SRP receptor are thought to act in concert. The SRP binds to the exposed ER signal peptide and to the ribosome, thereby inducing a pause in translation. The SRP receptor in the ER membrane, which is composed of two different polypeptide chains, binds the SRP ribosome complex. In a poorly understood reaction that involves multiple GTP-binding proteins, the SRP is released, leaving the ribosome on the ER membrane. A multisubunit protein translocation apparatus in the ER membrane then inserts the polypeptide chain into the membrane and transfers it across the lipid bilayer.

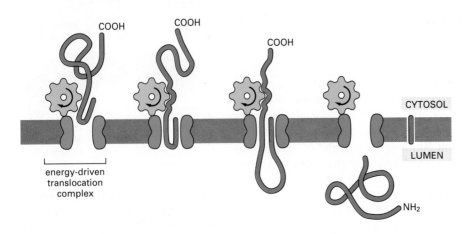

Figure 12–41 **Translocation of a protein across the bacterial plasma membrane.** In this schematic model, after a receptor in the translocation complex has bound the amino-terminal signal peptide, an energy-driven protein translocator threads the protein through the membrane, unfolding the polypeptide chain in the process. The energy is provided both by ATP hydrolysis and an electrochemical gradient across the membrane.

Translocation Across the ER Membrane Does Not Always Require Ongoing Polypeptide Chain Elongation [29]

As we have seen, translocation of proteins into mitochondria, chloroplasts, and peroxisomes occurs *posttranslationally,* after the protein is completed and released into the cytosol, whereas translocation across the ER membrane usually occurs during translation (*co-translationally*). This explains why ribosomes are bound to the ER but usually not to other organelles. A ribosome attached to the rough ER may utilize the energy of protein synthesis to force its growing polypeptide chain through a channel formed by a translocator in the ER membrane. Recent studies *in vitro*, however, have shown that a small minority of protein precursors can be imported into the ER after their synthesis has been completed, thus demonstrating that translocation does not always require ongoing translation. Posttranslational protein translocation may occur even more commonly across the ER membrane in yeast cells and across the bacterial plasma membrane (which is thought to be evolutionarily related to the ER; see Figure 12–5). In both cases the translocation requires ATP hydrolysis, and in bacteria an electrochemical gradient is needed as well. A translocation apparatus has been reconstituted from purified bacterial components, one of which is an ATPase that is thought to help thread the protein through the membrane (Figure 12–41). Since the proteins that use a posttranslational translocation pathway are first released into the cytosol, they are prevented from folding up by binding to cytosolic chaperone proteins, just as we have seen for the posttranslational import into mitochondria and chloroplasts.

The Polypeptide Chain Passes Through an Aqueous Pore in the Translocation Apparatus [30]

It has long been debated whether polypeptide chains are transferred across the ER membrane in direct contact with the lipid bilayer or through a pore in a pro-

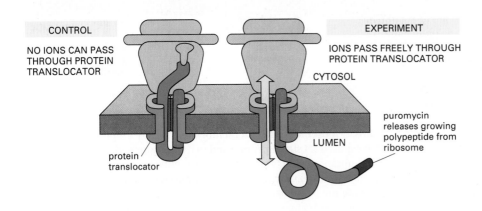

Figure 12–42 **The demonstration of protein-translocating aqueous pores in the ER membrane.** The experimental set-up is similar to that shown in Figure 10–5, with an artificial lipid bilayer separating two aqueous compartments. When rough microsomes are added to one of the compartments, they occasionally fuse with the lipid bilayer, incorporating a portion of ER membrane (with its bound ribosomes) into the bilayer. When the drug puromycin (*dark blue*) is added to the same compartment, it couples covalently to the carboxyl terminus of the growing polypeptide chain and releases it from the ribosome; pores of uniform size can now be detected as discrete increases in the electrical conductance across the membrane (the ion flow responsible for the increased electrical conductance is indicated by the *yellow arrow*). If the ribosomes are removed from the membrane with a high-salt wash, pores are no longer detected, indicating that ribosome binding is required to open (or assemble) the pore (not shown).

tein translocator. There is now strong evidence for a protein translocator. Normally, the pore in the translocator is plugged with the growing polypeptide chain that is in transit across the membrane; when the nascent chains are experimentally released from the ribosomes with the drug puromycin, however, the pores can be detected by the ion currents that flow through them. Although the pore is large enough to allow the passage of an unfolded polypeptide chain, it closes when the ribosome is removed from the membrane (Figure 12–42). Thus the pore seems to be a dynamic structure, opening when a ribosome with a growing polypeptide chain attaches to the membrane and closing when the ribosome detaches after the synthesis of the protein is completed.

The ER Signal Peptide Is Removed from Most Soluble Proteins After Translocation [31]

We have seen that in chloroplasts and mitochondria the signal peptides are cleaved from the precursor proteins once they have crossed the membrane. Similarly, amino-terminal ER signal peptides are removed by a signal peptidase on the luminal side of the ER membrane. The peptide by itself, however, is not sufficient to signal cleavage by the peptidase; this requires an adjacent cleavage site that is specifically recognized by the peptidase. We shall see below that ER signal peptides that are contained within the polypeptide chain rather than at the amino terminus do not have these recognition sites and are never cleaved; instead, they can serve to retain transmembrane proteins in the lipid bilayer after the translocation process has been completed.

The amino-terminal ER signal peptide of a soluble protein itself has two signaling functions: in addition to directing the protein to the ER membrane, it is thought to serve as a **start-transfer signal,** which remains bound to the translocation apparatus while the rest of the protein is threaded continuously through the membrane as a large loop. Once the carboxyl terminus of the protein has passed through the membrane, the signal peptide is released from the translocator pore, cleaved off by the signal peptidase, and rapidly degraded to amino acids by other proteases in the ER while the protein is released into the ER lumen (Figure 12–43).

Figure 12–43 **The translocation of a soluble protein across the ER membrane.** In this hypothetical model the protein translocator in the membrane is postulated to exist in two alternative states—active or inactive. On binding an ER signal peptide (which acts as a start-transfer signal), the translocator adopts an active state and begins to transfer the polypeptide chain across the lipid bilayer as a loop. In this state it forms an aqueous pore across the membrane that can be detected electrophysiologically if the polypeptide chain is released (see Figure 12–42). After the protein has been completely translocated, the translocator reverts to an inactive conformation, which can no longer conduct ions across the membrane but is open to the lipid bilayer, allowing the hydrophobic signal peptide to diffuse out into the bilayer, where it is rapidly degraded. In this and the following two figures the ribosomes have been omitted for clarity.

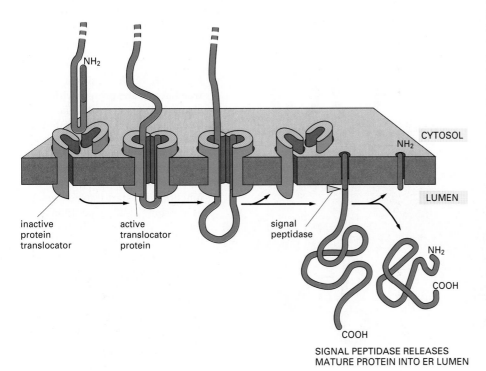

inactive protein translocator

active translocator protein

signal peptidase

CYTOSOL

LUMEN

NH₂

NH₂

NH₂

COOH

COOH

SIGNAL PEPTIDASE RELEASES MATURE PROTEIN INTO ER LUMEN

In Single-Pass Transmembrane Proteins a Single Internal ER Signal Peptide Remains in the Lipid Bilayer as a Membrane-spanning α Helix [32]

The translocation process for proteins destined to remain in the membrane is more complex than it is for soluble proteins, as some parts of the polypeptide chain are translocated across the lipid bilayer whereas others are not. Nevertheless, all modes of insertion of membrane proteins can be considered as variants of the sequence of events just described for transferring a soluble protein into the lumen of the ER. We begin by describing the three ways in which **single-pass transmembrane proteins** (see Figure 10–13) become inserted into the ER.

In the simplest case an amino-terminal signal peptide initiates translocation, just as for a soluble protein, but an additional hydrophobic segment in the polypeptide chain stops the transfer process before the entire polypeptide chain is translocated. This **stop-transfer peptide** anchors the protein in the membrane after the ER signal (start-transfer) peptide is released from the translocator and is cleaved off (Figure 12–44). The stop-transfer peptide forms a single α-helical membrane-spanning segment, with the amino terminus of the protein on the luminal side of the membrane and the carboxyl terminus on the cytosolic side.

In the other two cases the signal peptide is internal, rather than at the amino-terminal end of the protein. Like the amino-terminal ER signal peptides, the internal signal peptide is recognized by SRP, which brings the ribosome making the protein to the ER membrane and serves as a start-transfer signal that initiates the translocation of the protein. After release from the translocator, the internal start-transfer peptide remains in the lipid bilayer as a single membrane-spanning α helix. Internal start-transfer peptides, however, can bind to the translocation apparatus in either of two orientations, and the orientation of the inserted start-transfer peptide, in turn, determines which protein segment (the one preceding or the one following the start-transfer peptide) is moved across the membrane into the ER lumen. In one case the resulting membrane protein has its carboxyl terminus on the luminal side (Figure 12–45A), while in the other it has its amino terminus on the luminal side (Figure 12–45B). The orientation of the start-transfer peptide depends on the distribution of nearby charged amino acids, as described in the figure legend.

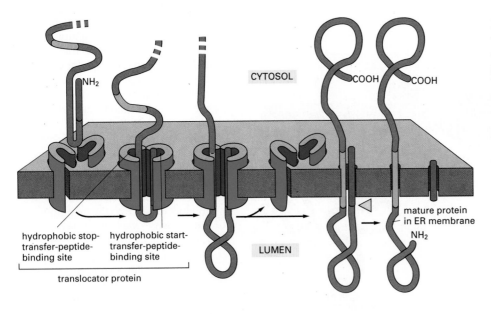

hydrophobic stop-transfer-peptide-binding site

hydrophobic start-transfer-peptide-binding site

translocator protein

CYTOSOL

LUMEN

NH₂

COOH

COOH

mature protein in ER membrane

NH₂

Figure 12–44 How a single-pass transmembrane protein with a cleaved ER signal peptide is integrated into the ER membrane. In this hypothetical model the co-translational translocation process is initiated by an amino-terminal ER signal peptide (*red*) that functions as a start-transfer signal as in Figure 12–43. In addition to the start-transfer peptide, however, the protein also contains a stop-transfer peptide (*orange*). When the stop-transfer peptide enters the translocator and interacts with a binding site, the translocator flips into its inactive state and discharges the protein laterally into the lipid bilayer.

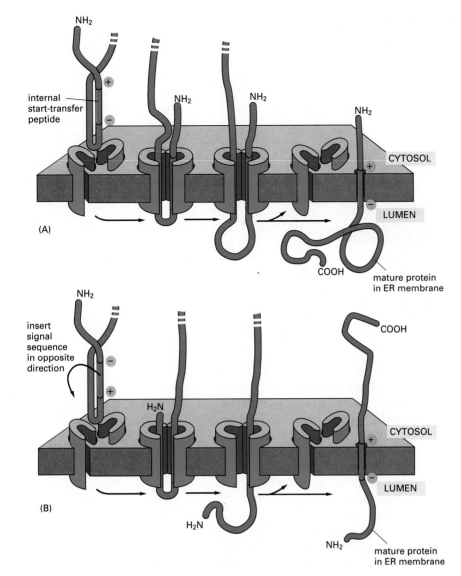

Figure 12–45 **How a single-pass membrane protein with an internal signal peptide is integrated into the ER membrane.** In this hypothetical model an internal ER signal peptide that functions as a start-transfer signal will bind to the translocator in such a way that its more positively charged end remains in the cytosol. If there are more positively charged amino acids immediately preceding the hydrophobic core of the start transfer peptide than there are following it on its carboxyl-terminal end, the start-transfer peptide will be inserted into the translocator in the orientation shown in (A), and the arm of the inserted loop carboxyl-terminal to the start-transfer sequence will be passed across the membrane. If, however, there are more positively charged amino acids immediately following the hydrophobic core of the start-transfer peptide than there are preceding it on its amino-terminal end, the start-transfer peptide will be inserted into the translocator in the orientation shown in (B), and the arm of the inserted loop amino-terminal to the start-transfer peptide will be passed across the membrane. Because translocation cannot start before a start-transfer sequence appears outside the ribosome, translocation of the amino-terminal portion of the protein shown in (B) can occur only after this portion has been fully synthesized. Note that there are two ways to insert a single-pass membrane-spanning protein whose amino terminus is located in the ER lumen: that shown in Figure 12–44 and that shown in (B) here.

Combinations of Start- and Stop-Transfer Signals Determine the Topology of Multipass Transmembrane Proteins [33]

In **multipass transmembrane proteins** the polypeptide chain passes back and forth repeatedly across the lipid bilayer (see Figure 10–13). It is thought that an internal signal peptide serves as a start-transfer signal in these proteins to initiate translocation, which continues until a stop-transfer peptide is reached. In double-pass transmembrane proteins, for example, the polypeptide is released into the bilayer at this point (Figure 12–46). In more complex multipass proteins, in which many hydrophobic α helices span the bilayer, a second start-transfer peptide reinitiates translocation further down the polypeptide chain until the next stop-transfer peptide causes polypeptide release, and so on for subsequent start-transfer and stop-transfer peptides (Figure 12–47).

Whether a given hydrophobic signal sequence will function as a start-transfer or stop-transfer peptide must depend on its location in a polypeptide chain, since its function can be switched by changing its location in the protein using recombinant DNA techniques. Thus the distinction between start-transfer and stop-transfer peptides results mostly from their relative order in the growing polypeptide chain. It seems that the SRP begins scanning an unfolded polypeptide chain for hydrophobic segments at its amino terminus and proceeds toward the car-

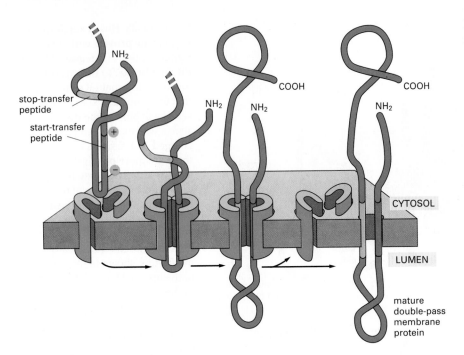

Figure 12–46 How a double-pass membrane protein with an internal signal sequence is integrated into the ER membrane. In this hypothetical model an internal ER signal peptide acts as a start-transfer signal (as in Figure 12–45) and initiates the transfer of the carboxyl terminal arm of the polypeptide chain. When a stop-transfer peptide enters the translocator, it discharges the protein laterally into the membrane.

boxyl terminus, in the direction that the protein is synthesized. By recognizing the first appropriate hydrophobic segment to emerge from the ribosome, the SRP sets the "reading frame": if translocation is initiated, the next appropriate hydrophobic segment will be recognized as a stop-transfer peptide, causing the region of the polypeptide chain in between to be threaded across the membrane. A similar scanning process continues until all of the hydrophobic regions in the protein have been inserted into the membrane.

Because membrane proteins are always inserted from the cytosolic side of the ER in this programmed manner, all copies of the same polypeptide chain will have the same orientation in the lipid bilayer. This generates an asymmetrical ER membrane in which the protein domains exposed on one side are different from those domains exposed on the other. This asymmetry is maintained during the many membrane budding and fusion events that transport the proteins made in the ER to other cell membranes (discussed in Chapter 13). Thus the way in which a newly synthesized protein is inserted into the ER membrane determines the orientation of the protein in the other membranes as well.

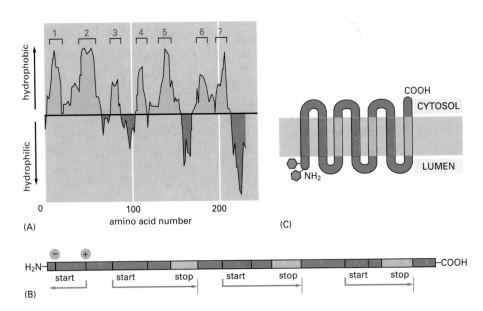

Figure 12–47 The insertion of the multipass membrane protein rhodopsin into the ER membrane. Rhodopsin is the light-sensitive protein in rod photoreceptor cells in the mammalian retina. (A) A hydrophobicity plot identifies seven short hydrophobic regions in rhodopsin. (B) The most amino-terminal region serves as a start-transfer peptide that causes the preceding amino-terminal portion of the protein to be passed across the ER membrane. Subsequent hydrophobic peptides will function in alternation as start-transfer and stop-transfer peptides. (C) The final integrated rhodopsin has its amino terminus located in the ER lumen and its carboxyl terminus located in the cytosol. The *blue hexagons* represent covalently attached oligosaccharides.

When proteins are dissociated from a membrane and reconstituted in artificial lipid vesicles, a random mixture of right-side-out and inside-out protein orientations usually results. Thus the protein asymmetry observed in cell membranes seems not to be an inherent property of the protein but to result solely from the process by which proteins are inserted into the ER membrane from the cytosol.

Translocated Polypeptide Chains Fold and Assemble in the Lumen of the Rough ER [34]

Many of the proteins in the lumen of the ER are in transit, *en route* to other destinations; others, however, are normally resident there and are present at high concentrations. These **ER resident proteins** contain an **ER retention signal** of four amino acids at their carboxyl terminus that is responsible for retaining the protein in the ER (see Table 12–3). Some of these proteins function as catalysts that help the many proteins that are translocated into the ER to fold and assemble correctly. One such ER resident protein is *protein disulfide isomerase (PDI)*, which catalyzes the oxidation of free sulfhydryl (SH) groups to form disulfide (S—S) bonds. Almost all cysteine residues in protein domains exposed to either the extracellular space or the lumen of organelles in the secretory and endocytic pathways are disulfide bonded; disulfide bonds do not form, however, in domains exposed to the cytosol because of the reducing environment there.

Another ER resident protein is a chaperone protein known as **binding protein (BiP)**, which is structurally related to the hsp70 proteins and, like them, recognizes incorrectly folded proteins, as well as protein subunits that have not yet assembled into their final oligomeric complexes. BiP, like other chaperone proteins, is thought to bind to exposed amino acid sequences that would normally be buried in the interior of correctly folded or assembled polypeptide chains. The bound BiP both prevents the proteins from aggregating and helps to keep them in the ER (and thus out of the Golgi apparatus and later parts of the secretory pathway); it may also help them to fold normally. Like the hsp70 family of proteins, which bind unfolded proteins in the cytosol and facilitate their import into mitochondria and chloroplasts, BiP hydrolyzes ATP to provide the energy for its role in protein folding.

As we have seen earlier for mitochondrial hsp70 (see Figure 12–24), the binding of BiP to an unfolded protein chain emerging in the ER lumen may help pull the protein into the ER. This pulling process may be particularly important for proteins that enter the ER posttranslationally because in this case there is no ribosome attached to the membrane to help push the nascent protein through the translocator during the protein's synthesis.

Most Proteins Synthesized in the Rough ER Are Glycosylated by the Addition of a Common *N*-linked Oligosaccharide [35]

The covalent addition of sugars to proteins is one of the major biosynthetic functions of the ER. Most of the soluble and membrane-bound proteins that are made in the ER, including those destined for transport to the Golgi apparatus, lysosomes, plasma membrane, or extracellular space are **glycoproteins.** In contrast, very few proteins in the cytosol are glycosylated, and those that are carry a much simpler sugar modification in which a single *N*-acetylglucosamine group is added to a serine or threonine residue of the protein.

An important advance in understanding the process of **protein glycosylation** was the discovery that a preformed oligosaccharide (composed of *N*-acetylglucosamine, mannose, and glucose and containing a total of 14 sugar residues) is transferred *en bloc* to proteins in the ER. Because this oligosaccharide is transferred to the side-chain NH₂ group of an asparagine amino acid in the protein, it is said to be *N-linked* or *asparagine-linked* (Figure 12–48). The transfer is cata-

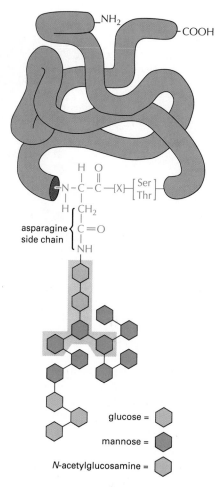

Figure 12–48 The asparagine-linked (*N*-linked) oligosaccharide that is added to most proteins in the rough ER membrane. The five sugar residues in the *gray box* form the "core region" of this oligosaccharide. For many glycoproteins only the core sugars survive the extensive oligosaccharide trimming process that takes place in the Golgi apparatus. Only asparagines in the sequences *Asn-X-Ser* or *Asn-X-Thr* (where X is any amino acid except proline) become glycosylated. These two sequences occur much less frequently in glycoproteins than in nonglycosylated cytosolic proteins; evidently there has been selective pressure against these sequences during protein evolution, presumably because glycosylation at too many sites would interfere with protein folding.

The Endoplasmic Reticulum

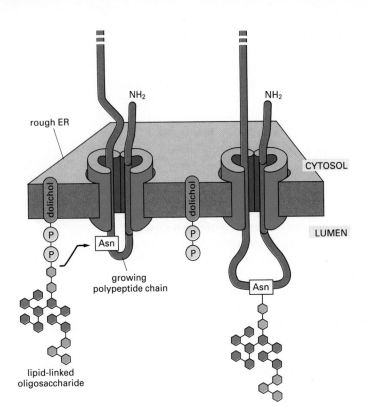

rough ER

NH₂

NH₂

CYTOSOL

dolichol

dolichol

P

P

Asn

growing
polypeptide chain

LUMEN

P

P

Asn

lipid-linked
oligosaccharide

Figure 12–49 Protein glycosylation in the rough ER. Almost as soon as a polypeptide chain enters the ER lumen, it is glycosylated on target asparagine amino acids. The oligosaccharide shown in Figure 12–48 is transferred to the asparagine as an intact unit in a reaction catalyzed by a membrane-bound *oligosaccharyl transferase* enzyme. There is one copy of this enzyme associated with each protein translocator in the ER membrane.

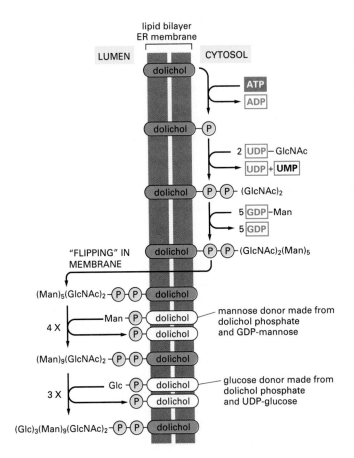

lipid bilayer
ER membrane

LUMEN

CYTOSOL

dolichol

ATP
ADP

dolichol — P

2 UDP — GlcNAc
UDP + UMP

dolichol — P—P — (GlcNAc)₂

5 GDP—Man
5 GDP

dolichol — P—P — (GlcNAc)₂(Man)₅

"FLIPPING" IN
MEMBRANE

(Man)₅(GlcNAc)₂ — P—P — dolichol

4 X

Man — P — dolichol

P — dolichol

mannose donor made from
dolichol phosphate
and GDP-mannose

(Man)₉(GlcNAc)₂ — P—P — dolichol

3 X

Glc — P — dolichol

P — dolichol

glucose donor made from
dolichol phosphate
and UDP-glucose

(Glc)₃(Man)₉(GlcNAc)₂ — P—P — dolichol

Figure 12–50 Synthesis of the lipid-linked precursor oligosaccharide in the rough ER membrane. The oligosaccharide is assembled sugar by sugar onto the carrier lipid dolichol (a polyisoprenoid—see Panel 2–4). Dolichol is long and very hydrophobic: its 22 five-carbon units can span the thickness of a lipid bilayer more than three times, so that the attached oligosaccharide is firmly anchored in the membrane. The first sugar group is linked to dolichol by a pyrophosphate bridge. This high-energy bond activates the oligosaccharide for its transfer from the lipid to an asparagine side chain of a nascent polypeptide on the luminal side of the rough ER. The synthesis of the oligosaccharide starts on the cytosolic side of the ER membrane and continues on the luminal face after the $(Man)_5(GlcNAc)_2$ lipid intermediate is flipped across the bilayer. All of the subsequent glycosyl transfer reactions on the luminal side of the ER involve transfers from dolichol-P-glucose and dolichol-P-mannose; these activated, lipid-linked monosaccharides are synthesized from dolichol phosphate and UDP-glucose or GDP-mannose (as appropriate) on the cytosolic side of the ER and are then thought to be flipped across the ER membrane. GlcNAc = *N*-acetylglucosamine; Man = mannose; Glc = glucose.

lyzed by a membrane-bound enzyme, an *oligosaccharyl transferase*, which has its active site exposed on the luminal side of the ER membrane; this explains why cytosolic proteins are not glycosylated in this way. The precursor oligosaccharide is held in the ER membrane by a special lipid molecule called **dolichol**, and it is transferred to the target asparagine in a single enzymatic step immediately after that amino acid emerges in the ER lumen during protein translocation (Figure 12–49). Since most proteins are co-translationally imported into the ER, *N*-linked oligosaccharides are almost always added during protein synthesis.

The lipid-linked precursor oligosaccharide is linked to the dolichol by a high-energy pyrophosphate bond, which provides the activation energy that drives the glycosylation reaction illustrated in Figure 12–49. The entire oligosaccharide is built up sugar by sugar on this membrane-bound lipid molecule prior to its transfer to a protein. The sugars are first activated in the cytosol by the formation of *nucleotide-sugar intermediates*, which then donate their sugar (directly or indirectly) to the lipid in an orderly sequence. Partway through this process, the lipid-linked oligosaccharide is flipped from the cytosolic to the luminal side of the ER membrane (Figure 12–50).

All of the diversity of the *N*-linked oligosaccharide structures on mature glycoproteins results from later modification of the original precursor structure. While still in the ER, three glucose residues (see Figure 12–48) and one mannose residue are quickly removed from the oligosaccharides of most glycoproteins. This oligosaccharide "trimming" or "processing" continues in the Golgi apparatus and is discussed in Chapter 13.

The *N*-linked oligosaccharides are by far the most common ones found on glycoproteins. Less frequently, oligosaccharides are linked to the hydroxyl group on the side chain of a serine, threonine, or hydroxylysine amino acid. These *O-linked oligosaccharides* are formed in the Golgi apparatus by pathways that are not yet fully understood (discussed in Chapter 13).

Some Membrane Proteins Exchange a Carboxyl-Terminal Transmembrane Tail for a Covalently Attached Glycosylphosphatidylinositol (GPI) Anchor After Entry into the ER [36]

As discussed in Chapter 10, several cytosolic enzymes catalyze the covalent addition of a single fatty acid chain or prenyl group to selected proteins to help direct these proteins to cell membranes. A related process is catalyzed by enzymes in the rough ER: the carboxyl terminus of some membrane proteins destined for the plasma membrane is covalently attached to a sugar residue of a glycolipid. This linkage forms in the lumen of the ER by the mechanism illustrated in Figure 12–51, and it adds a **glycosylphosphatidylinositol (GPI) anchor,** which

Figure 12–51 The attachment of a glycosylphosphatidylinositol anchor. Immediately after the completion of protein synthesis, the precursor protein remains anchored in the ER membrane by a hydrophobic carboxyl-terminal sequence of 15 to 20 amino acids, with the rest of the protein in the ER lumen. Within less than a minute, an enzyme in the ER cuts the protein free from its membrane-bound carboxyl terminus and simultaneously attaches the new carboxyl terminus to an amino group on a preassembled glycosylphosphatidylinositol intermediate. The signal that specifies this modification is contained within the hydrophobic carboxyl-terminal sequence and a few amino acids adjacent to it on the luminal side of the ER membrane; if this signal is added to other proteins, they too become modified in this way. Because of the covalently linked lipid anchor, the protein remains membrane-bound with all of its amino acids exposed initially on the luminal side of the ER and eventually on the cell exterior.

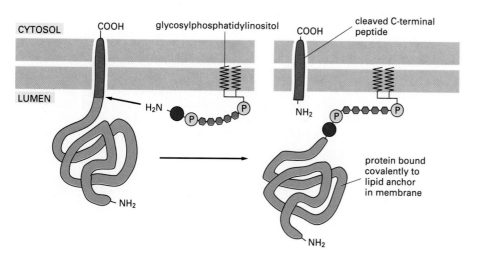

contains two fatty acids, to the protein. At the same time the transmembrane segment of the protein is cleaved off. An increasing number of plasma membrane proteins have been shown to be modified in this way. Since these proteins are attached to the exterior of the plasma membrane only by their GPI anchors, in principle they can be released from cells in soluble form in response to signals that activate a specific phospholipase in the plasma membrane. Trypanosome parasites, for example, use this mechanism to shed their coat of GPI-anchored surface proteins if attacked by the immune system.

Most Membrane Lipid Bilayers Are Assembled in the ER [37]

The ER membrane produces nearly all of the lipids required for the elaboration of new cell membranes, including both phospholipids and cholesterol. The major phospholipid made is *phosphatidylcholine* (also called lecithin), which can be formed in three steps from choline, two fatty acids, and glycerol phosphate. Each step is catalyzed by enzymes in the ER membrane that have their active sites facing the cytosol, where all of the required metabolites are found. Thus phospholipid synthesis occurs exclusively in the cytosolic half of the ER bilayer. In the first step acyl transferases successively add two fatty acids to glycerol phosphate to produce phosphatidic acid, a compound sufficiently water-insoluble to remain in the lipid bilayer after it has been synthesized. It is this step that enlarges the lipid bilayer; the later steps determine the head group of a newly formed lipid molecule, and therefore the chemical nature of the bilayer, but do not result in net membrane growth (Figure 12–52). The two other major membrane phospholipids—phosphatidylethanolamine (PE) and phosphatidylserine (PS)—as well as the minor phospholipid phosphatidylinositol (PI), are all synthesized in this way.

Figure 12–52 The synthesis of phosphatidylcholine. This phospholipid is synthesized from fatty acyl-coenzyme A (fatty acyl CoA), glycerol 3-phosphate, and cytidine-bisphosphocholine (CDP-choline).

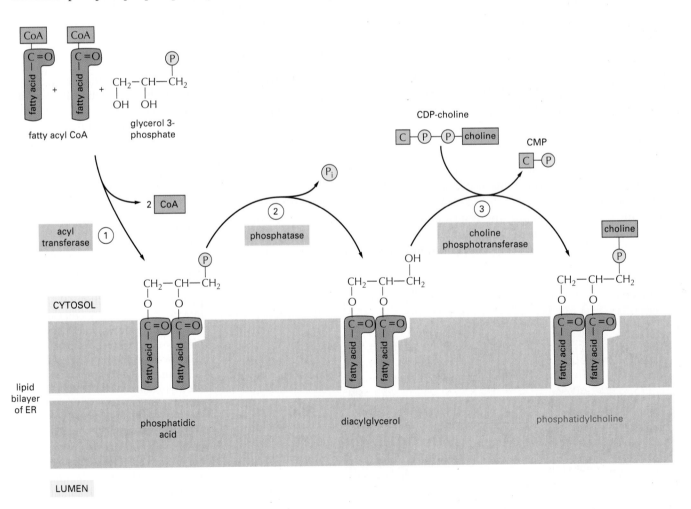

As phospholipid synthesis takes place in the cytosolic half of the ER bilayer, there needs to be a mechanism that transfers some of the newly formed phospholipid molecules to the other half of the bilayer. In synthetic lipid bilayers, lipids do not "flip-flop" in this way. In the ER, however, phospholipids equilibrate across the membrane within minutes, which is almost 100,000 times faster than can be accounted for by spontaneous "flip-flop." This rapid transbilayer movement is thought to be mediated by **phospholipid translocators** that are head-group-specific. In particular, the ER membrane seems to contain a translocator (a *"flippase"*) that transfers choline-containing phospholipids—but not ethanolamine-, serine-, or inositol-containing phospholipids—between cytosolic and luminal faces. This means that phosphatidylcholine reaches the luminal face much more readily than the other phospholipids. In this way the translocator is responsible for the asymmetric distribution of the lipids in the bilayer (Figure 12–53).

The ER also produces cholesterol and ceramide. *Ceramide* is made by condensing the amino acid serine with a fatty acid to form the amino alcohol *sphingosine;* a second fatty acid is then added to form ceramide. The ceramide is exported to the Golgi apparatus, where it serves as the precursor for the synthesis of two types of lipids: oligosaccharide chains are added to form *glycosphingolipids* (glycolipids), and phosphocholine head groups are transferred from phosphatidylcholine to other ceramide molecules to form *sphingomyelin.* Thus both glycolipids and sphingomyelin are produced relatively late in the process of membrane synthesis. Because they are produced by enzymes exposed to the Golgi lumen, they are found exclusively in the noncytosolic half of the lipid bilayers that contain them.

Phospholipid Exchange Proteins Help Transport Phospholipids from the ER to Mitochondria and Peroxisomes [38]

As discussed in Chapter 13, the plasma membrane and the membranes of the Golgi apparatus, lysosomes, and endosomes all form part of a membrane system that communicates with the ER by means of transport vesicles that transfer both proteins and lipids. Mitochondria, plastids, and peroxisomes do not belong to this system, and they require different mechanisms for the import of proteins and lipids for growth. We have already seen that most (for mitochondria and plastids) or all (for peroxisomes) of the proteins in these organelles are imported from the cytosol. Although mitochondria modify some of the lipids they import, they do not synthesize lipids *de novo;* instead, their lipids have to be imported from the ER, either directly, or indirectly by way of other cellular membranes. In either case, special mechanisms are required for the transfer.

Water-soluble carrier proteins—called **phospholipid exchange proteins** (or *phospholipid transfer proteins*)—have been shown in *in vitro* experiments to have the ability to transfer individual phospholipid molecules between membranes. Each exchange protein recognizes only specific types of phospholipids. It functions by "extracting" a molecule of the appropriate phospholipid from a membrane and diffusing away with the lipid buried within its binding site. When it encounters another membrane, the exchange protein tends to discharge the bound phospholipid molecule into the new lipid bilayer (Figure 12–54). It has been proposed that phosphatidylserine is imported into mitochondria in this way and then decarboxylated to yield phosphatidylethanolamine, while phosphatidylcholine is imported intact.

Exchange proteins act to distribute phospholipids at random among all membranes present. In principle, such a random exchange process can result in a net transport of lipids from a lipid-rich to a lipid-poor membrane, allowing phosphatidylcholine and phosphatidylserine molecules, for example, to be transferred from the ER, where they are synthesized, to a mitochondrial or peroxisomal

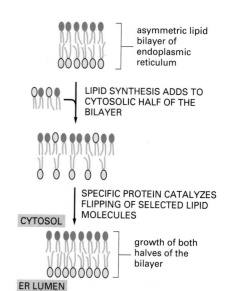

Figure 12–53 The role of phospholipid translocators in lipid bilayer synthesis. Since new lipid molecules are added only to the cytosolic half of the bilayer and lipid molecules do not flip spontaneously from one monolayer to the other, membrane-bound phospholipid translocator proteins ("flippases") are required to transfer selected lipid molecules from the cytosolic half to the luminal half so that the membrane grows as a bilayer. Because the flippase in the ER membrane preferentially recognizes and transfers choline-containing head groups, an asymmetric bilayer is generated, with the luminal monolayer (which produces the outer half of the plasma membrane bilayer) highly enriched for phosphatidylcholine.

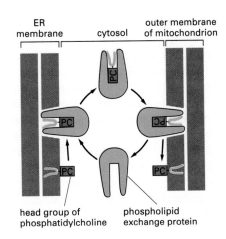

Figure 12–54 Phospholipid exchange proteins. Because phospholipids are insoluble in water, their passage between membranes requires a carrier protein. Phospholipid exchange proteins are water-soluble proteins that carry a single molecule of phospholipid at a time; they can pick up a lipid molecule from one membrane and release it at another and thereby redistribute phospholipids between membrane-bounded compartments. The transfer of phosphatidylcholine (PC) from ER to mitochondria can occur without the input of additional energy because the concentration of PC is high in the ER membrane (where it is made) and low in the mitochondrial outer membrane. One would predict that there must be a flippase in the outer mitochondrial membrane to equilibrate the lipids between the two leaflets of the bilayer, and there must be a mechanism to transfer lipids between the outer and inner mitochondrial membrane. These postulated pathways, however, remain to be discovered.

membrane. It might be that mitochondria and peroxisomes are the only "lipid-poor" organelles in the cytosol and that such an exchange process is sufficient, although other, more specific mechanisms probably also exist for transporting phospholipids to these organelles.

Summary

The extensive ER network serves as a factory for the production of almost all of the cell's lipids. In addition, a major portion of the cell's protein synthesis occurs on the cytosolic surface of the ER: all proteins destined for secretion and all proteins destined for the ER itself, the Golgi apparatus, the lysosomes, the endosomes, and the plasma membrane are first imported into the ER from the cytosol. In the ER lumen, the proteins fold and oligomerize, disulfide bonds are formed, and N-linked oligosaccharides are added.

Only proteins that carry a special hydrophobic signal peptide are imported into the ER. The ER signal peptide is recognized by a signal recognition particle (SRP), which binds both the growing polypeptide chain and the ribosome and directs them to a receptor protein on the cytosolic surface of the rough ER membrane. This binding to the membrane initiates the translocation process that threads a loop of polypeptide chain across the ER membrane through a hydrophilic pore in a protein translocator.

Soluble proteins destined for the ER lumen, for secretion, or for transfer to the lumen of other organelles pass completely into the ER lumen. Transmembrane proteins destined for the ER or for other cell membranes are translocated across the ER membrane but are not released into the lumen; instead, they remain anchored in the lipid bilayer by one or more membrane-spanning α-helical regions in their polypeptide chain. These hydrophobic portions of the protein can act either as start-transfer or stop-transfer signals during the translocation process. When a polypeptide contains multiple alternating start-transfer and stop-transfer signals, it will pass back and forth across the bilayer multiple times.

The asymmetry of lipid synthesis, protein insertion, and glycosylation in the ER establishes the polarity of the membranes of all of the other organelles that the ER supplies with lipids and membrane proteins.

References

General

Blobel, G. Control of intracellular protein traffic. *Methods Enzymol.* 96:663–682, 1983.

Neupert, W.; Lill, R. New Comprehensive Biochemistry: Membrane Biogenesis and Protein Targeting, Vol. 22. Amsterdam: Elsevier, 1992.

Osborne, M.A.; Silver, P.A. Nucleocytoplasmic transport in the yeast *Saccharomyces cerevisiae. Annu. Rev. Biochem.* 62:219–254, 1993.

Palade, G. Intracellular aspects of the process of protein synthesis. *Science* 189:347–358, 1975.

Warren, G. Membrane partitioning during cell division. *Annu. Rev. Biochem.* 62:323–348, 1993.

Cited

1. Bolender, R.P. Stereological analysis of the guinea pig pancreas. *J. Cell Biol.* 61:269–287, 1974.

   Fulton, A.B. How crowded is the cytoplasm? *Cell* 30:345–347, 1982.

   Kelly, R.B. Associations between microtubules and intracellular organelles. *Curr. Opin. Cell Biol.* 2:105–108, 1990.

   Weibel, E.R.; Staubli, W.; Gnagi, H.R.; Hess, F.A. Correlated morphometric and biochemical studies on the liver cell. *J. Cell Biol.* 42:68–91, 1969.

2. Blobel, G. Intracellular protein topogenesis. *Proc. Natl. Acad. Sci. USA* 77:1496–1500, 1980.

   Morden, C.W.; Delwiche, C.F.; Kuhsel, M.; Palmer, J.D. Gene phylogenies and the endosymbiotic origin of plastids. *Biosystems* 28:75–90, 1992.

   Schwarz, R.M.; Dayhoff, M.O. Origins of prokaryotes, eukaryotes, mitochondria, and chloroplasts. *Science* 199:395–403, 1978.

   Voelker, D.R. Organelle biogenesis and intracellular lipid transport in eukaryotes. *Microbiol. Rev.* 55:543–560, 1991.

3. Bradshaw, R.A. Protein translocation and turnover in eukaryotic cells. *Trends Biochem. Sci.* 14:276–279, 1989.

   Sabatini, D.D.; Kreibich, G.; Morimoto, T.; Adesnik, M. Mechanisms for the incorporation of proteins in membranes and organelles. *J. Cell Biol.* 92:1–22, 1982.

4. Blobel, G.; Walter, P.; Chang, C.N.; et al. Translocation of proteins across membranes: the signal hypothesis and beyond. *Symp. Soc. Exp. Biol.* 33:9–36, 1979.

   Deshaies, R.J.; Schekman, R. A yeast mutant defective at an early stage in import of secretory protein precursors into the endoplasmic reticulum. *J. Cell Biol.* 105:633–645, 1987.

   Garoff, H. Using recombinant DNA techniques to study protein targeting in the eukaryotic cell. *Annu. Rev. Cell Biol.* 1:403–445, 1985.

   Oliver, D.B.; Beckwith, J. *E. coli* mutant pleiotropically defective in the export of secreted proteins. *Cell* 25:765–772, 1981.

   von Heijne, G. Protein targeting signals. *Curr. Opin. Cell Biol.* 2:604–608, 1990.

5. Jones, H.D.; Schliwa, M.; Drubin, D.G. Video microscopy of organelle inheritance and motility in budding yeast. *Cell Motil. Cytoskeleton* 25:129–142, 1993.

   Warren, G. Membrane partitioning during cell division. *Annu. Rev. Biochem.* 62:323–348, 1993.

6. Dingwall, C.; Laskey, R. The nuclear membrane. *Science* 258:942–947, 1992.

   Newport, J.W.; Forbes, D.J. The nucleus: structure, function and dynamics. *Annu. Rev. Biochem.* 56:535–565, 1987.

   Nigg, E.A.; Baeuerle, P.A.; Luhrmann, R. Nuclear import-export: in search of signals and mechanisms. *Cell* 66:15–22, 1991.

   Silver, P.A. How proteins enter the nucleus. *Cell* 64:489–497, 1991.

7. Akey, C.W.; Radermacher, M. Architecture of the *Xenopus* nuclear pore complex revealed by three-dimensional cryo-electron microscopy. *J. Cell Biol.* 122:1–19, 1993.

   Forbes, D.J. Structure and function of the nuclear pore complex. *Annu. Rev. Cell Biol.* 8:495–527, 1992.

   Gerace, L. Molecular trafficking across the nuclear pore complex. *Curr. Opin. Cell Biol.* 4:637–645, 1992.

   Hinshaw, J.E.; Carragher, B.O.; Milligan, R.A. Architecture and design of the nuclear pore complex. *Cell* 69:1133–1141, 1992.

   Pante, N.; Aebi, U. The nuclear pore complex. *J. Cell Biol.* 122:977–984, 1993.

8. Dingwall, C.; Laskey, R.A. Nuclear targeting sequences—a consensus? *Trends Biochem. Sci.* 16:478–481, 1991.

   Goldfarb, D.S.; Gariépy, J.; Schoolnik, G.; Kornberg, R.D. Synthetic peptides as nuclear localization signals. *Nature* 322:641–644, 1986.

   Kalderon, D.; Roberts, B.L.; Richardson, W.D.; Smith, A.E. A short amino acid sequence able to specify nuclear location. *Cell* 39:499–509, 1984.

   Yamasake, L.; Lanford, R.E. Nuclear transport: a guide to import receptors. *Trends Cell Biol.* 2:123–127, 1992.

9. Adam, S.A.; Gerace, L. Cytosolic proteins that specifically bind nuclear localization signals are receptors for nuclear import. *Cell* 66:837–847, 1991.

   Dingwall, C. Transport across the nuclear envelope: enigmas and explanations. *Bioessays* 13:213–218, 1991.

   Feldherr, C.M.; Akin, D. EM visualization of nucleocytoplasmic transport processes. *Electron Microsc. Rev.* 3:73–86, 1990.

   Michaud, N.; Goldfarb, D.S. Multiple pathways in nuclear transport: the import of U2 snRNP occurs by a novel kinetic pathway. *J. Cell Biol.* 112:215–223, 1991.

   Newmeyer, D.D. The nuclear pore complex and nucleocytoplasmic transport. *Curr. Opin. Cell Biol.* 5:395–407, 1993.

10. Chaudhary, N.; Courvalin, J. Stepwise reassembly of the nuclear envelope at the end of mitosis. *J. Cell Biol.* 122:295–306, 1993.

    Glass, J.R.; Gerace, L. Lamins A and C bind and assemble at the surface of mitotic chromosomes. *J. Cell Biol.* 111:1047–1057, 1990.

    Peter, M.; Heitlinger, E.; Haner, M.; Aebi, U.; Nigg, E.A. Disassembly of *in vitro* formed lamin head-to-tail polymers by *cdc2* kinase. *EMBO J.* 10:1535–1554, 1991.

Weise, C.; Wilson, K.L. Nuclear membrane dynamics. *Curr. Opin. Cell Biol.* 5:387–394, 1993.

11. Dingwall, C. If the cap fits. . . . *Curr. Biol.* 1:65–66, 1991.

Eckner, R.; Ellmeier, W.; Bimstiel, M.L. Mature mRNA 3′ end formation stimulates RNA export from the nucleus. *EMBO J.* 10:3513–3522, 1991.

Izaurralde, E.; Mattaj, I.W. Transport of RNA between nucleus and cytoplasm. *Semin. Cell Biol.* 3:279–288, 1992.

Miller, M.W.; Hanover, J.A. Regulation of macromolecular traffic mediated by the nuclear pore complex. *Cell Biol. Int. Rep.* 16:791–798, 1992.

Moll, T.; Tebb, G.; Surana, U.; Robitsch, H.; Nasmyth, K. The role of phosphorylation and the *cdc28* protein kinase in cell cycle-regulated nuclear import of the *S. cerevisiae* transcription factor SWI5. *Cell* 66:743–758, 1991.

12. Archer, E.K.; Keegstra, K. Current views on chloroplast protein import and hypotheses on the origin of the transport mechanism. *J. Bioenerg. Biomem.* 22:789–810, 1990.

Attardi, G.; Schatz, G. Biogenesis of mitochondria. *Annu. Rev. Cell Biol.* 4:289–333, 1988.

Glick, B.; Schatz, G. Import of proteins into mitochondria. *Annu. Rev. Genet.* 25:21–44, 1991.

Pfanner, N.; Rassow, J.; van der Klei, I.J.; Neupert, W. A dynamic model of the mitochondrial protein import machinery. *Cell* 68:999–1002, 1992.

Smeekens, S.; Weisbeek, P.; Robinson, C. Protein transport into and within chloroplasts. *Trends Biochem. Sci.* 15:73–76, 1990.

13. Roise, D.; Schatz, G. Mitochondrial presequences. *J. Biol. Chem.* 263:4509–4511, 1988.

Vestweber, D.; Schatz, G. DNA-protein conjugates can enter mitochondria via the protein import pathway. *Nature* 338:170–172, 1987.

14. Beasley, E.M.; Wachter, C.; Schatz, G. Putting energy into mitochondrial protein import. *Curr. Opin. Cell Biol.* 4:646–651, 1992.

Weinhues, U.; Neupert, W. Protein translocation across mitochondrial membranes. *Bioessays* 14:17–23, 1992.

15. Pfanner, N.; Rassow, J.; Wienhues, U.; et al. Contact sites between inner and outer membranes: structure and role in protein translocation into the mitochondria. *Biochim. Biophys. Acta* 1018:239–242, 1990.

Pon, L.; Moll, T.; Vestweber, D.; Marshallsay, B.; Schatz, G. Protein import into mitochondria: ATP-dependent protein translocation activity in a submitochondrial fraction enriched in membrane contact sites and specific proteins. *J. Cell Biol.* 109:2603–2316, 1989.

Schleyer, M.; Neupert, W. Transport of proteins into mitochondria: translocational intermediates spanning contact sites between outer and inner membranes. *Cell* 43:339–350, 1985.

16. Deshaies, R.J.; Koch, B.D.; Werner-Washburne, M.; Craig, E.A.; Schekman, R. A subfamily of stress proteins facilitates translocation of secretory and mitochondrial precursor polypeptides. *Nature* 332:800–805, 1988.

Eilers, M.; Schatz, G. Protein unfolding and the energetics of protein translocation across biological membranes. *Cell* 52:481–483, 1988.

Wienhues, U.; Becker, K.; Schleyer, M.; et al. Protein folding causes an arrest of preprotein translocation into mitochondria *in vivo*. *J. Cell Biol.* 115:1601–1609, 1991.

17. Hendrick, J.P.; Hartl, F.U. Molecular chaperone functions of heat-shock proteins. *Annu. Rev. Biochem.* 62:349–384, 1993.

Kelley, W.L.; Georgopoulos, C. Chaperones and protein folding. *Curr. Opin. Cell Biol.* 4:984–991, 1992.

Koll, H.; Guiard, B.; Rassow, J.; et al. Antifolding activity of *hsp60* couples protein import into the mitochondrial matrix with export to the intermembrane space. *Cell* 68:1163–1175, 1992.

Scherer P.E.; Krieg, U.C.; Hwang, S.T.; Vestweber, D.; Schatz, G. A precursor protein partly translocated into yeast mitochondria is bound to a 70 kd mitochondrial stress protein. *EMBO J.* 9:4315–4322, 1990.

18. Glick, B.S.; Beasley, E.M.; Schatz, G. Protein sorting in mitochondria. *Trends Biochem. Sci.* 17:453–459, 1992.

Glick, B.S.; Brandt, A.; Cunningham, K.; et al. Cytochromes c1 and b2 are sorted to the intermembrane space of mitochondria by a stop-transfer mechanism. *Cell* 69:809–822, 1992.

Hartl, F.U.; Ostermann, J.; Guiard, B.; Neupert, W. Successive translocation into and out of the mitochondrial matrix: targeting of proteins to the intermembrane space by a bipartite signal peptide. *Cell* 51:1027–1037, 1987.

19. Knight, J.S.; Madueno, F.; Gray, J.C. Import and sorting of protein by chloroplasts. *Biochem. Soc. Trans.* 21:31–36, 1993.

Robinson, C.; Klosgen, R.B.; Herrmann, R.G; Shackleton, J.B. Protein translocation across the thylakoid membrane—a tale of two mechanisms. *FEBS Lett.* 325:67–69, 1993.

Smeekens, S.; Weisbeek, P.; Robinson, D. Protein transport into and within chloroplasts. *Trends Biochem. Sci.* 15:73–76, 1990.

Theg, S.M.; Geske, F.S. Biophysical characterization of a transit peptide directing chloroplast protein import. *Biochemistry* 31:5053–5060, 1992.

20. deDuve, C. Microbodies in the living cell. *Sci. Am.* 248(5):74–84, 1983.

Lazarow, P.B. Genetic approaches to studying peroxisome biogenesis. *Trends Cell Biol.* 3:89–93, 1993.

Lazarow, P.B.; Fujiki, Y. Biogenesis of peroxisomes. *Annu. Rev. Cell Biol.* 1:489–530, 1985.

21. Taiz, L.; Zeiger, E. Plant Physiology, pp. 229–233; 288–290. Redwood City, CA: Benjamin/Cummings, 1991.

Titus, D.E.; Becker, W.M. Investigation of the glyoxysome-peroxisome transition in germinating cucumber cotyledons using double-label immunoelectron microscopy. *J. Cell Biol.* 101:1288–1299, 1985.

Tolbert, N.E.; Essner, E. Microbodies: peroxisomes and glyoxysomes. *J. Cell Biol.* 91:271s–283s, 1981.

22. Borst, P. Peroxisome biogenesis revisited. *Biochim. Biophys. Acta* 1008:1–13, 1989.

Lazarow, P.B. Peroxisome biogenesis. *Curr. Opin. Cell Biol.* 1:630–634, 1989.

Shimozawa, N.; Tsukamoto, T.; Suzuki, Y.; et al. A human gene responsible for Zellweger syndrome that affects peroxisome assembly. *Science* 255:1132–1134, 1992.

Subramani, S. Targeting of proteins into the peroxisomal matrix. *J. Memb. Biol.* 125:99–106, 1992.

Valle, D.; Gartner, J. Human genetics: penetrating the peroxisome. *Nature* 361:682–683, 1993.

23. DePierre, J.W.; Dallner, G. Structural aspects of the membrane of the endoplasmic reticulum. *Biochim. Biophys. Acta* 415:411–472, 1975.

Lee, C.; BoChen, L. Dynamic behavior of endoplasmic reticulum in living cells. *Cell* 54:37–46, 1988.

Preuss, D.; Mulholland, J.; Kaiser, C.; et al. Structure of the yeast endoplasmic reticulum: localization of ER proteins using immunofluorescence and immuno-electron microscopy. *Yeast* 7:891–911, 1991.

Vertel, B.M.; Walters, L.M.; Mills, D. Subcompartments of the endoplasmic reticulum. *Semin. Cell Biol.* 3:325–341, 1992.

24. Adelman, M.R.; Sabatini, D.D.; Blobel, G. Ribosome-membrane interaction: nondestructive disassembly of rat liver rough microsomes into ribosomal and membranous components. *J. Cell Biol.* 56:206–229, 1973.

Borgese, N.; Mok, W.; Kreibich, G.; Sabatini, D.D. Ribosomal-membrane interaction: *in vitro* binding of ribosomes to microsomal membranes. *J. Mol. Biol.* 88:559–580, 1974.

Garcia, P.D.; Walter, P. Full-length pre-pro-α-factor can be translocated across the mammalian microsomal membrane only if translation has not terminated. *J. Cell Biol.* 106:1043–1048, 1988.

Nunnari, J.; Walter, P. Protein targeting to and translocation across the membrane of the endoplasmic reticulum. *Curr. Opin. Cell Biol.* 4:573–580, 1992.

Rapoport, T.A. Transport of proteins across the endoplasmic reticulum membrane. *Science* 258:931–936, 1992.

25. Jones, A.L.; Fawcett, D.W. Hypertrophy of the agranular endoplasmic reticulum in hamster liver induced by phenobarbital. *J. Histochem. Cytochem.* 14:215–232, 1966.

Mori, H.; Christensen, A.K. Morphometric analysis of Leydig cells in the normal rat testis. *J. Cell Biol.* 84:340–354, 1980.

Villa, A.; Podini, P.; Panzeri, M.; et al. The endoplasmic-sarcoplasmic reticulum of smooth muscle: immuno-cytochemistry of *vas deferens* fibers reveals specialized subcompartments differently equipped for the control of Ca²⁺ homeostasis. *J. Cell Biol.* 121:1041–1051, 1993.

26. Dallner, G. Isolation of rough and smooth microsomes—general. *Meth. Enzym.* 31:191–201, 1974.

deDuve, C. Tissue fractionation—past and present. *J. Cell Biol.* 50:20d–55d, 1971.

27. Blobel, G.; Dobberstein, B. Transfer of proteins across membranes. *J. Cell Biol.* 67:835–851, 1975.

Kaiser, C.A.; Preuss, D.; Grisafi, P.; Botstein, D. Many random sequences functionally replace the secretion signal sequence of yeast invertase. *Science* 235:312–317, 1987.

Milstein, C.; Brownlee, G.; Harrison, T.; Mathews, M.B. A possible precursor of immunoglobulin light chains. *Nat. New Biol.* 239:117–120, 1972.

Simon, K.; Perara, E.; Lingappa, V. Translocation of globin fusion proteins across the endoplasmic reticulum membrane in *Xenopus laevis* oocytes. *J. Cell Biol.* 104:1165–1172, 1987.

von Heijne, G. Signal sequences: the limits of variation. *J. Mol. Biol.* 184:99–105, 1985.

28. Gilmore, R. The protein translocation apparatus of the rough endoplasmic reticulum, its associated proteins, and the mechanism of translocation. *Curr. Opin. Cell Biol.* 3:580–584, 1991.

Meyer, D.I.; Krause, E.; Dobberstein, B. Secretory protein translocation across membranes—the role of the "docking protein." *Nature* 297:647–650, 1982.

Siegel, V.; Walter, P. Each of the activities of signal recognition particle (SRP) is contained within a distinct domain: analysis of biochemical mutants of SRP. *Cell* 52:39–49, 1988.

Simon, S. Translocation of proteins across the endoplasmic reticulum. *Curr. Opin. Cell Biol.* 5:581–588, 1993.

Walter, P.; Lingappa, V.R. Mechanism of protein translocation across the endoplasmic reticulum membrane. *Annu. Rev. Cell Biol.* 2:499–516, 1986.

29. Hann, B.C.; Walter, P. The signal recognition particle in *S. cerevisiae*. *Cell* 67:131–134, 1991.

Sanders, S.L.; Schekman, R. Polypeptide translocation across the endoplasmic reticulum membrane. *J. Biol. Chem.* 267:13791–13794, 1992.

Wickner, W.; Driessen, A.J.; Hartl, F.U. The enzymology of protein translocation across the *Escherichia coli* plasma membrane. *Annu. Rev. Biochem.* 60:101–124, 1991.

Zimmermann, R.; Meyer, D.I. 1986: a year of new insights into how proteins cross membranes. *Trends Biochem. Sci.* 11:512–515, 1986.

30. Crowley, K.S.; Reinhart, G.D.; Johnson, A.E. The signal sequence moves through a ribosomal tunnel into a noncytoplasmic aqueous environment at the ER membrane early in translocation. *Cell* 73:1101–1115, 1993.

Deshaies, R.J.; Sanders, S.L.; Feldheim, D.A.; Schekman, R. Assembly of yeast Sec proteins involved in translocation into the endoplasmic reticulum into a membrane-bound multisubunit complex. *Nature* 349:806–808, 1991.

Gorlich, D.; Prehn, S.; Hartmann, E.; Kalies, K.U.; Rapoport, T.A. A mammalian homolog of SEC61p and SECYp is associated with ribosomes and nascent polypeptides during translocation. *Cell* 71:489–503, 1992.

Nicchitta, C.; Migliaccio, G.; Blobel, G. Reconstitution of secretory protein translocation from detergent-solubilized rough microsomes. *Methods Cell Biol.* 34:263–285, 1991.

Simon, S.M.; Blobel, G. A protein-conducting channel in the endoplasmic reticulum. *Cell* 65:371–380, 1991.

31. Blobel, G.; Dobberstein, B. Transfer of proteins across membranes. *J. Cell Biol.* 67:835–851, 1975.

Dalbey, R.E.; von Heijne, G. Signal peptidases in prokaryotes and eukaryotes—a new protease family. *Trends Biochem. Sci.* 17:474–478, 1992.

32. High, S.; Andersen, S.S.; Gorlich, D.; et al. Sec61p is adjacent to nascent type I and type II signal-anchor proteins during their membrane insertion. *J. Cell Biol.* 121:743–750, 1993.

References

High, S.; Dobberstein, B. Mechanisms that determine the transmembrane disposition of proteins. *Curr. Opin. Cell Biol.* 4:581–586, 1992.

Singer, S.J. The structure and insertion of integral proteins in membranes. *Annu. Rev. Cell Biol.* 6:247–296, 1990.

Thrift, R.N.; Andrew, D.W.; Walter, P.; Johnson, A.E. A nascent membrane protein is located adjacent to ER membrane proteins throughout its integration and translation. *J. Cell Biol.* 112:809–821, 1991.

von Heijne, G. Transcending the impenetrable: how proteins come to terms with membranes. *Biochim. Biophys. Acta* 974:307–333, 1988.

33. Engelman, D.M.; Steitz, T.A.; Goldman, A. Identifying nonpolar transbilayer helices in amino acid sequences of membrane proteins. *Annu. Rev. Biophys. Biophys. Chem.* 15:321–353, 1986.

Hartmann, E.; Rapoport, T.A.; Lodish, H.F. Predicting the orientation of eukaryotic membrane-spanning proteins. *Proc. Natl. Acad. Sci. USA* 86:5786–5790, 1989.

Kyte, J.; Doolittle, R.F. A simple method for displaying the hydropathic character of a protein. *J. Mol. Biol.* 157:105–132, 1982.

Wessels, H.P.; Spiess, M. Insertion of a multispanning membrane protein occurs sequentially and requires only one signal sequence. *Cell* 55:61–70, 1988.

34. Doms, R.W.; Lamb, R.A.; Rose, J.K.; Helenius, A. Folding and assembly of viral membrane proteins. *Virology* 193:545–562, 1993.

Gaut, J.R.; Hendershot, L.M. The modification and assembly of proteins in the endoplasmic reticulum. *Curr. Biol.* 5:589–595, 1993.

Gething, M.J.; Sambrook, J. Transport and assembly processes in the endoplasmic reticulum. *Semin. Cell Biol.* 1:65–72, 1990.

Helenius, A.; Marquardt, T.; Braakman, I. The endoplasmic reticulum as a protein-folding compartment. *Trends Cell Biol.* 2:227–231, 1992.

Marquardt, T.; Helenius, A. Misfolding and aggregation of newly synthesized proteins in the endoplasmic reticulum. *J. Cell Biol.* 117:505–513, 1992.

35. Abeijon, C.; Hirschberg, C.B. Topography of glycosylation reactions in the endoplasmic reticulum. *Trends Biochem. Sci.* 17:32–66, 1992.

Hart, G.W. Glycosylation. *Curr. Opin. Cell Biol.* 4:1017–1023, 1992.

Kelleher, D.J.; Kreibich, G.; Gilmore, R. Oligosaccharyltransferase activity is associated with a protein complex composed of ribophorins I and II and a 48 kd protein. *Cell* 69:55–65, 1992.

Kornfeld, R.; Kornfeld, S. Assembly of asparagine-linked oligosaccharides. *Annu. Rev. Biochem.* 54:631–664, 1985.

Snider, M. A function for ribophorins. *Curr. Biol.* 2:43–45, 1992.

36. Brown, D.A. Interactions between GPI-anchored proteins and membrane lipids. *Trends Cell Biol.* 2:338–343, 1992.

Ferguson, M.A. Colworth Medal Lecture. Glycosylphosphatidylinositol membrane anchors: the tale of a tail. *Biochem. Soc. Trans.* 20:243–256, 1992.

Moran, P.; Caras, I. Proteins containing an uncleaved signal for glycophosphatidylinositol membrane anchor attachment are retained in a post-ER compartment. *J. Cell Biol.* 119:763–772, 1992.

37. Bishop, W.R.; Bell, R.M. Assembly of phospholipids into cellular membranes: biosynthesis, transmembrane movement, and intracellular translocation. *Annu. Rev. Cell Biol.* 4:579–610, 1988.

Davidowicz, E.A. Dynamics of membrane lipid metabolism and turnover. *Annu. Rev. Biochem.* 56:43–61, 1987.

van Meer, G. Transport and sorting of membrane lipids. *Curr. Opin. Cell Biol.* 5:661–674, 1993.

38. Dawidowicz, E.A. Lipid exchange: transmembrane movement, spontaneous movement, and protein-mediated transfer of lipids and cholesterol. *Curr. Top. Memb. Transp.* 29:175–202, 1987.

Dowhan, W. Phospholipid-transfer proteins. *Curr. Opin. Cell Biol.* 3:621–625, 1991.

Pagano, R.E. Lipid traffic in eukaryotic cells: mechanisms for intracellular transport and organelle-specific enrichment of lipids. *Curr. Opin. Cell Biol.* 2:652–663, 1990.

Vesicular Traffic in the Secretory and Endocytic Pathways

13

- Transport from the ER Through the Golgi Apparatus
- Transport from the *Trans* Golgi Network to Lysosomes
- Transport from the Plasma Membrane via Endosomes: Endocytosis
- Transport from the *Trans* Golgi Network to the Cell Surface: Exocytosis
- The Molecular Mechanisms of Vesicular Transport and the Maintenance of Compartmental Diversity

Every cell must communicate with its environment. In a procaryotic cell all of this communication takes place across the plasma membrane: digestive enzymes, for example, are secreted to the cell exterior, and the small metabolites generated by digestion are then taken up by transport proteins in the plasma membrane. Eucaryotic cells, by contrast, have evolved an elaborate internal membrane system that allows them to take up macromolecules by a process called *endocytosis* and deliver them to digestive enzymes that are stored intracellularly in lysosomes; as a consequence, metabolites generated by digestion are delivered from the lysosomes directly to the cytosol as they are produced. Besides providing for regulated digestion of macromolecules by the *endocytic pathway,* the internal membrane system provides a means whereby eucaryotic cells can regulate the delivery of newly synthesized proteins and carbohydrates to the exterior. Because each molecule that travels along this *biosynthetic-secretory pathway* passes through multiple compartments, the cell can modify the molecule in a series of controlled steps, store it until needed, and then deliver it to a specific cell-surface domain by a process called *exocytosis.* The endocytic and biosynthetic-secretory pathways are shown in color in Figure 13–1.

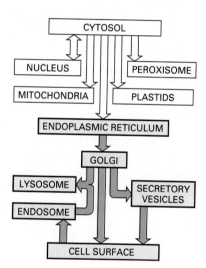

Figure 13–1 The secretory and endocytic pathways. In this "road map" of biosynthetic protein traffic, which was introduced in Chapter 12, both the secretory and endocytic pathways are colored.

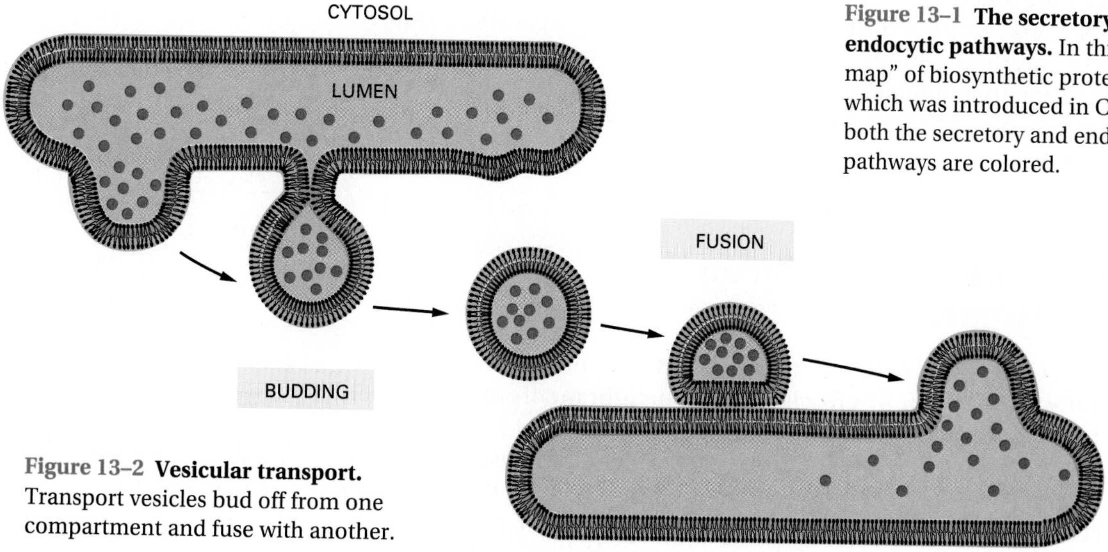

Figure 13–2 Vesicular transport. Transport vesicles bud off from one compartment and fuse with another.

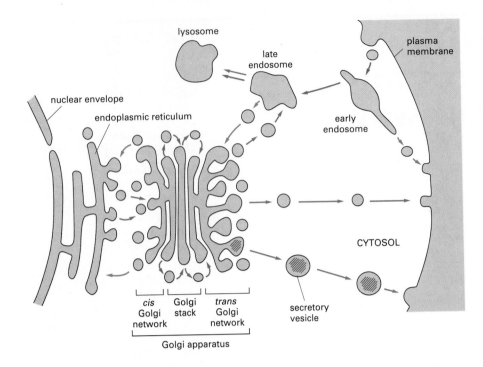

lysosome

late endosome

plasma membrane

nuclear envelope

endoplasmic reticulum

early endosome

CYTOSOL

cis Golgi network   Golgi stack   trans Golgi network

secretory vesicle

Golgi apparatus

Figure 13–3 The intracellular compartments of the eucaryotic cell involved in the biosynthetic-secretory and endocytic pathways. Each compartment encloses a space that is topologically equivalent to the outside of the cell, and they all communicate with one another by means of transport vesicles. In the biosynthetic-secretory pathway (*red arrows*) protein molecules are transported from the ER to the plasma membrane or (via late endosomes) to lysosomes. In the endocytic pathway (*green arrows*) molecules are ingested in vesicles derived from the plasma membrane and delivered to early endosomes and then (via late endosomes) to lysosomes. Many endocytosed molecules are retrieved from early endosomes and returned to the cell surface for reuse; similarly, some molecules are retrieved from the late endosome and returned to the Golgi apparatus, and some are retrieved from the Golgi apparatus and returned to the ER. All of these retrieval pathways are shown with *blue arrows*.

The lumen of each compartment along the biosynthetic-secretory and endocytic pathways is topologically equivalent to the exterior of the cell, and the compartments are all in constant communication with one another, at least partly by means of numerous transport vesicles, which continually bud off from one membrane and fuse with another (Figure 13–2). The traffic is highly organized: the biosynthetic-secretory pathway leads outward from the ER toward the Golgi apparatus and cell surface, with a side route leading to lysosomes, while the endocytic pathway leads inward toward endosomes and lysosomes from the plasma membrane (Figure 13–3).

To perform its function, each transport vesicle that buds from a compartment must take up only the appropriate proteins and must fuse only with the appropriate target membrane. A vesicle carrying cargo from the Golgi apparatus to the plasma membrane, for example, must exclude proteins that are to stay in the Golgi apparatus, and it must fuse only with the plasma membrane and not with any other organelle. While participating in this constant flow of membrane components, each organelle must maintain its own distinct identity. In this chapter we consider the function of the Golgi apparatus, lysosomes, secretory vesicles, and endosomes, and we trace the pathways by which these organelles are interconnected. In the final section we consider the molecular mechanisms of budding and fusion that underlie all vesicular transport, and we discuss the fundamental problem of how, in the face of this transport, the differences between the compartments are maintained.

Transport from the ER Through the Golgi Apparatus [1]

As discussed in Chapter 12, newly synthesized proteins enter the biosynthetic-secretory pathway in the ER by crossing the ER membrane from the cytosol. Subsequent transport, from the ER to the Golgi apparatus and from the Golgi apparatus to the cell surface and elsewhere, is mediated by transport vesicles, which transfer proteins from membrane to membrane or from lumen to lumen (or to extracellular space) by cycles of vesicle budding and fusion (see Figure 12–7). The pathway from the ER via the Golgi apparatus to the cell surface is often

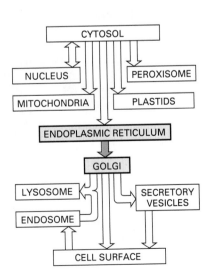

CYTOSOL

NUCLEUS | PEROXISOME

MITOCHONDRIA | PLASTIDS

ENDOPLASMIC RETICULUM

GOLGI

LYSOSOME | SECRETORY VESICLES

ENDOSOME

CELL SURFACE

referred to as the *default pathway* because proteins seem not to require special signals to follow it: any protein that enters the ER (and folds and assembles properly) will automatically be transported through the Golgi apparatus to the cell surface unless it contains signals that either detain it in an earlier compartment en route or divert it, via the Golgi apparatus, to lysosomes or secretory vesicles.

In this section we focus mainly on the **Golgi apparatus** (also called the *Golgi complex*), which is a major site of carbohydrate synthesis as well as a sorting and dispatching station for the products of the ER. Many of the cell's polysaccharides are made in the Golgi apparatus, including the pectin and hemicellulose of the plant cell wall and most of the glycosaminoglycans of the extracellular matrix in animals (discussed in Chapter 19). But the Golgi apparatus also lies on the exit route from the ER, and a large proportion of the carbohydrates it makes are attached as oligosaccharide side chains to the proteins and lipids that the ER sends to it. Certain oligosaccharide groups serve as tags to direct specific proteins into vesicles that will transport them to lysosomes; other proteins and lipids, once they have acquired their appropriate oligosaccharides in the Golgi apparatus, are dispatched in transport vesicles to other destinations.

The Golgi Apparatus Consists of an Ordered Series of Compartments [2]

The Golgi apparatus is usually located near the cell nucleus, and in animal cells it is often close to the *centrosome,* or *cell center.* It consists of a collection of flattened, membrane-bounded *cisternae* and thus resembles a stack of plates. Each of these **Golgi stacks** usually consists of four to six cisternae (Figure 13–4). The number of Golgi stacks per cell varies greatly depending on the cell type: some animal cells contain one large stack, while certain plant cells contain hundreds of small ones.

Swarms of small vesicles are associated with the Golgi stacks, clustered on the side abutting the ER and along the dilated rims of each cisterna (see Figure 13–4). These *Golgi vesicles* are thought to transport proteins and lipids both to and from the Golgi apparatus and between the Golgi cisternae. During their passage through the Golgi apparatus, the transported molecules undergo an ordered series of covalent modifications.

Each Golgi stack has two distinct faces: a ***cis* face** (or entry face) and a ***trans* face** (or exit face). Both the *cis* and *trans* faces are closely connected to special compartments, which are composed of a network of interconnected tubular and

Figure 13–4 **The Golgi apparatus.** (A) Three-dimensional reconstruction from electron micrographs of the Golgi apparatus in a secretory animal cell. (B) Electron micrograph of a Golgi apparatus in a plant cell (the green alga *Chlamydomonas*) seen in cross-section. Two adjacent Golgi stacks are shown. In plant cells the Golgi apparatus is generally more distinct and more clearly separated from other intracellular membranes than in animal cells. (A, redrawn from A. Rambourg and Y. Clermont, *Eur. J. Cell Biol.* 51:189–200, 1990; B, courtesy of George Palade.)

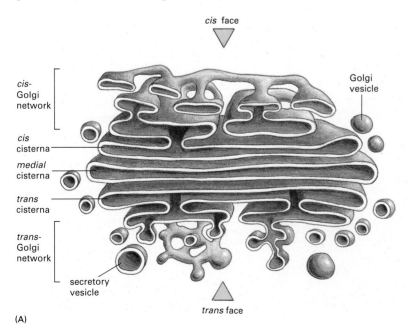

cis face

cis-Golgi network

cis cisterna

medial cisterna

trans cisterna

trans-Golgi network

secretory vesicle

Golgi vesicle

trans face

(A)

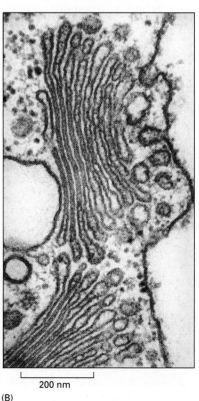

200 nm

(B)

secretion of mucus through
apical surface

Figure 13–5 A goblet cell of the small intestine. This cell is specialized for secreting mucus, a mixture of glycoproteins and proteoglycans synthesized in the ER and Golgi apparatus. The Golgi apparatus in these cells is highly polarized, which facilitates the discharge of mucus by exocytosis at the apical surface. (After R.V. Krstić, Illustrated Encyclopedia of Human Histology. New York: Springer-Verlag, 1984.)

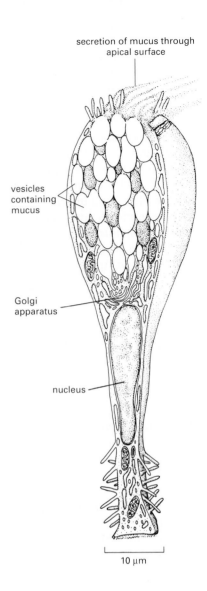

secretion of mucus through apical surface

vesicles containing mucus

Golgi apparatus

nucleus

10 µm

cisternal structures. These are the *cis* **Golgi network** (also called the *intermediary* or *salvage compartment*) and the *trans* **Golgi network,** respectively. Proteins and lipids enter the *cis* Golgi network in transport vesicles from the ER and exit from the *trans* Golgi network in transport vesicles destined for the cell surface or another compartment. Both networks are thought to be important for protein sorting: proteins entering the *cis* Golgi network can either move onward in the Golgi apparatus or be returned to the ER; proteins exiting the *trans* Golgi network are sorted according to whether they are destined for lysosomes, secretory vesicles, or the cell surface.

The Golgi apparatus is especially prominent in cells that are specialized for secretion, such as the goblet cells of the intestinal epithelium, which secrete large amounts of polysaccharide-rich mucus into the gut. In such cells unusually large vesicles are found on the *trans* side of the Golgi apparatus, which faces the plasma membrane domain where secretion occurs (Figure 13–5).

ER-Resident Proteins Are Selectively Retrieved from the *Cis* Golgi Network [3]

Vesicles destined for the Golgi apparatus bud from a specialized region of the ER called the **transitional elements,** whose membrane lacks bound ribosomes and is often located between the rough ER and the Golgi apparatus (Figure 13–6). Vesicles budding from the transitional elements of the ER are thought to be nonselective. They will transport any protein in the ER to the Golgi apparatus, although it remains possible that there are signals which accelerate the process. There is one strict requirement, however, for the exit of a protein from the ER: it must be correctly folded and assembled. Proteins that are misfolded or incompletely assembled into their protein complexes are retained in the ER, either bound to the special binding protein BiP (discussed in Chapter 12) or in aggregates that cannot be packaged, and are eventually degraded within the ER. Thus exit from the ER can be regarded as a quality checkpoint: unless folding and subunit assembly are successfully completed, the protein is discarded. In fact, the ER seems to be one of the main sites in the cell where proteins are degraded (the other being lysosomes, as we discuss later, and the cytosol, as we discuss in Chapter 5).

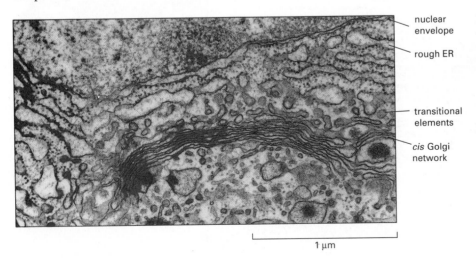

nuclear envelope

rough ER

transitional elements

cis Golgi network

1 µm

Figure 13–6 Electron micrograph of transitional elements and *cis* Golgi network. Transport vesicles bud from the transitional elements of the ER, which are nearly free of ribosomes and fuse with the *cis* Golgi network, thereby transferring newly made proteins and lipids from the ER to the Golgi apparatus. (Courtesy of Brij J. Gupta.)

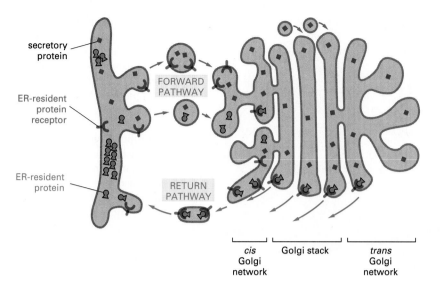

secretory
protein

ER-resident
protein
receptor

ER-resident
protein

FORWARD
PATHWAY

RETURN
PATHWAY

cis
Golgi
network

Golgi stack

trans
Golgi
network

Correctly folded proteins do not need a special signal to be transported out of the ER, but those, such as BiP, that are resident in the ER lumen do need such a signal to be retained there. Retention of soluble ER-resident proteins is mediated by a short, four–amino acid sorting signal, identified as KDEL (Lys-Asp-Glu-Leu) or a similar sequence (see Table 12–3). If this *ER retention signal* is removed from BiP, for example, by genetic engineering, the protein is secreted from the cell; and if the signal is transferred to a protein that is normally secreted, the protein is now retained in the ER. The retention signal works not by anchoring resident proteins in the lumen of the ER but by the selective retrieval of ER-resident proteins after they have escaped in transport vesicles and been delivered to the *cis* Golgi network. In the *cis* Golgi network a specific membrane-bound receptor protein binds to the ER retention signal and packages any proteins displaying the signal into special transport vesicles that return the proteins to the ER. Thus for these resident proteins the ER is like an open prison: there is nothing to stop them leaving, but if they leave, they are brought back (Figure 13–7).

Golgi Proteins Return to the ER When Cells Are Treated with the Drug Brefeldin A [4]

The continuous retrieval of ER-resident proteins from the *cis* Golgi network means that transport between these two organelles occurs in both directions. As mentioned in the legend to Figure 13–7, receptors for the ER retention signal are

Figure 13–7 The mechanism used to retain resident proteins in the ER. ER-resident proteins that escape to the *cis* Golgi network are returned to the ER by vesicular transport. A membrane receptor in the *cis* Golgi network captures the proteins and carries them in transport vesicles back to the ER. The ionic conditions in the ER dissociate the ER proteins from the receptor, and the receptor is then returned to the *cis* Golgi network for reuse. Receptors for the ER retention signal are also found in the *cis, medial,* and *trans* Golgi cisternae. Thus the retrieval of ER proteins begins in the *cis* Golgi network, but the return pathway operates from the later Golgi cisternae as well. The retention is aided by interactions between ER-resident proteins in the ER lumen. These interactions retard the exit of ER proteins relative to proteins that are destined for secretion (secretory proteins). Experiments in which the ER retention signal is removed from BiP show that BiP leaves the ER and is eventually secreted from cells but that its exit from the ER occurs much more slowly than that of *bona fide* secretory proteins, indicating that it is held there by weak interactions with other proteins.

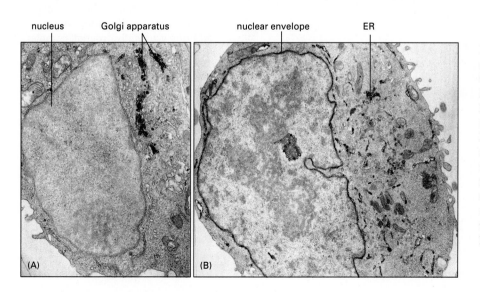

nucleus   Golgi apparatus   nuclear envelope   ER

(A)   (B)

Figure 13–8 Electron micrographs showing the effect of brefeldin A treatment on the Golgi apparatus. A histochemical stain shows the location of a Golgi enzyme (a mannosidase) before (A) and two hours after (B) brefeldin A treatment of cultured fibroblasts. Note that after the treatment the enzyme relocates from the *cis* and *medial* Golgi cisternae to the ER and to the nuclear envelope, which is continuous with the ER. (Courtesy of Jennifer Lippincott-Schwartz and Lydia Yuan.)

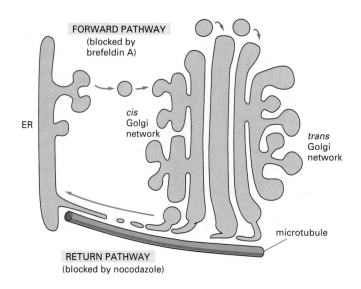

FORWARD PATHWAY
(blocked by brefeldin A)

ER

cis
Golgi
network

trans
Golgi
network

microtubule

RETURN PATHWAY
(blocked by nocodazole)

Figure 13–9 The postulated return pathway from the Golgi apparatus to the ER. Whereas the forward pathway requires transport vesicles and occurs independently of microtubules, the return pathway is thought to involve membrane tubes that are pulled from the Golgi apparatus to the ER along microtubules (which are disrupted by drugs such as nocodazole). As we discuss later, brefeldin A prevents the assembly of the coats that are required for budding of the transport vesicles, and this may block the forward vesicular transport steps while leaving the backward membrane-tube-dependent transport process intact.

also found in the later Golgi compartments, suggesting that a return pathway from these compartments to the ER exists. The importance of the return pathway from Golgi to ER is dramatically illustrated by studies using the drug **brefeldin A,** which blocks protein secretion by disrupting the Golgi apparatus. In brefeldin-A-treated cells the Golgi apparatus largely disappears and the Golgi proteins end up in the ER, where they intermix with ER proteins. When the drug is removed, the normal Golgi apparatus reforms and the Golgi proteins return to their proper Golgi compartments (Figure 13–8).

To explain these observations, it has been proposed that brefeldin A blocks the forward transport from the ER through the Golgi apparatus without affecting the return transport from the Golgi to the ER (Figure 13–9). In this way the drug would cause the Golgi apparatus to empty into the ER via the return pathway, and when the drug is removed, forward traffic would resume and deliver the Golgi proteins back to their proper compartment.

Oligosaccharide Chains Are Processed in the Golgi Apparatus [5]

As described in Chapter 12, a single species of **N-linked oligosaccharide** is attached *en bloc* to many proteins in the ER, and this oligosaccharide is then trimmed while the protein is still in the ER. Further modifications and additions occur in the Golgi apparatus, depending on the protein. The outcome is that two

Figure 13–10 The two main classes of asparagine-linked (*N*-linked) oligosaccharides found in mature glycoproteins. A *complex oligosaccharide* is shown in (B) and a *high-mannose oligosaccharide* in (C). Each complex oligosaccharide consists of a *core region* (shown in color in A), derived from the original *N*-linked oligosaccharide added in the ER and typically containing two *N*-acetylglucosamines (GlcNAc) and three mannoses (Man), together with a *terminal region* that contains a variable number of trisaccharide units (*N*-acetylglucosamine–galactose–sialic acid) linked to the core mannoses. Frequently the terminal region is truncated and contains only GlcNAc and galactose (Gal) or just GlcNAc. In addition, a fucose residue may be added, usually to the core GlcNAc attached to the asparagine (Asn). Thus, although the steps of processing and subsequent sugar addition are rigidly ordered, complex oligosaccharides can be heterogeneous: while the complex oligosaccharide shown has three terminal branches, for example, two and four branches are also common, depending on the glycoprotein and the cell in which it is made. Hybrid oligosaccharides with one Man branch and one GlcNAc and Gal branch are also found. The indicated three amino acids constitute the sequence recognized by the oligosaccharyl transferase enzyme that adds the initial oligosaccharide to the protein. Ser = serine; Thr = threonine; X = any amino acid.

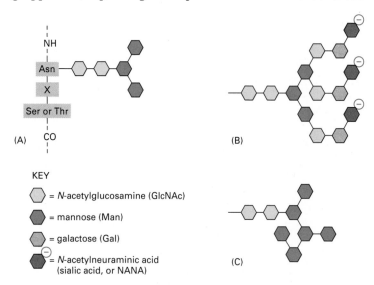

NH

Asn

X

Ser or Thr

(A) CO

(B)

(C)

KEY

◯ = *N*-acetylglucosamine (GlcNAc)

⬢ = mannose (Man)

⬡ = galactose (Gal)

⬢⊖ = *N*-acetylneuraminic acid (sialic acid, or NANA)

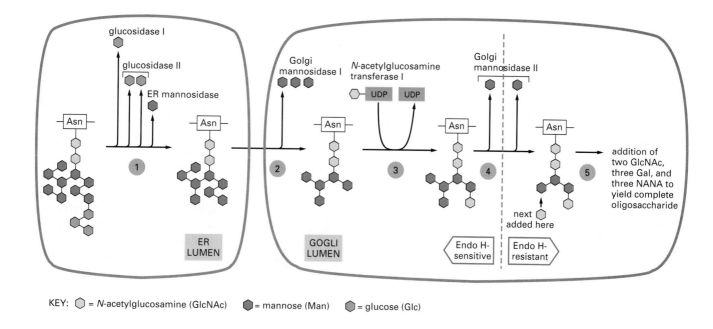

KEY: ◯ = N-acetylglucosamine (GlcNAc)   ⬡ = mannose (Man)   ⬡ = glucose (Glc)

broad classes of N-linked oligosaccharides, the *complex oligosaccharides* and the *high-mannose oligosaccharides,* are found attached to mammalian glycoproteins. Sometimes both types are attached (in different places) to the same polypeptide chain. **High-mannose oligosaccharides** have no new sugars added to them in the Golgi apparatus. They contain just two N-acetylglucosamines and many mannose residues, often approaching the number originally present in the lipid-linked oligosaccharide precursor added in the ER. **Complex oligosaccharides,** by contrast, can contain more than the original two N-acetylglucosamines as well as a variable number of galactose and sialic acid residues and, in some cases, fucose. Sialic acid is of special importance because it is the only sugar in glycoproteins that bears a net negative charge (Figure 13–10).

The complex oligosaccharides are generated by a combination of trimming the original oligosaccharide added in the ER and the addition of further sugars. Whether a given oligosaccharide remains high-mannose or is processed is determined largely by its configuration on the protein. If the oligosaccharide is accessible to the processing enzymes in the Golgi apparatus, it is likely to be converted to a complex form; if it is inaccessible, it is likely to remain in a high-mannose form. The processing that generates complex oligosaccharide chains follows the highly ordered pathway shown in Figures 13–11 and 13–12.

The Golgi Cisternae Are Organized as a Series of Processing Compartments [6]

Proteins exported from the ER enter the first of the Golgi processing compartments (the *cis* **compartment**), which is thought to be continuous with the *cis* Golgi network; they then move to the next compartment (the *medial* **compartment,** consisting of the central cisternae of the stack) and finally to the *trans* **compartment,** where glycosylation is completed. The lumen of the *trans* compartment is thought to be continuous with the *trans* **Golgi network,** where proteins are segregated into different transport vesicles and dispatched to their final destinations—the plasma membrane, lysosomes, or secretory vesicles.

These oligosaccharide processing pathways occur in a correspondingly organized sequence in the Golgi stack, with each cisterna containing its own set of processing enzymes. Proteins are modified in successive stages as they move from cisterna to cisterna across the stack, so that the stack forms a multistage processing unit. This compartmentalization might seem unnecessary, since each oligosaccharide processing enzyme can accept a glycoprotein as a substrate only

Figure 13–11 Oligosaccharide processing in the ER and the Golgi apparatus. The processing pathway is highly ordered so that each step shown is dependent on the previous one. Processing begins in the ER with the removal of the glucoses from the oligosaccharide initially transferred to the protein. Then a mannosidase in the ER membrane removes a specific mannose. The remaining steps occur in the Golgi stack, where Golgi mannosidase I first removes three more mannoses and N-acetylglucosamine transferase I then adds an N-acetylglucosamine, which enables mannosidase II to remove two additional mannoses. This yields the final core of three mannoses that is present in a complex oligosaccharide. At this stage the bond between the two N-acetylglucosamines in the core becomes resistant to attack by a highly specific endoglycosidase (*Endo H*). Since all later structures in the pathway are also Endo H-resistant, treatment with this enzyme is widely used to distinguish complex from high-mannose oligosaccharides. Finally, as shown in Figure 13–10, additional N-acetylglucosamines, galactoses, and sialic acids are added. Some oligosaccharides will escape processing in the Golgi apparatus, whereas others will follow the pathway shown to varying extents: the extent of processing depends on the protein and on the location of the asparagine residue in the protein to which the oligosaccharide is attached.

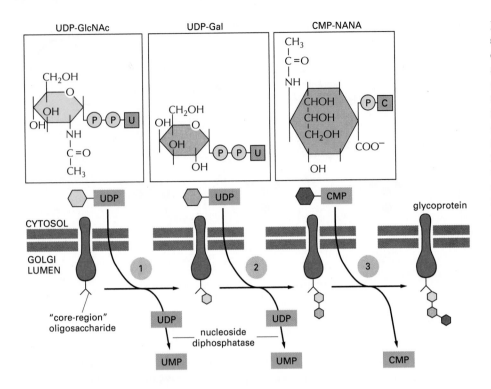

after it has been properly processed by the preceding enzyme. Nonetheless, it is clear that processing occurs in a spatial as well as a biochemical sequence: enzymes catalyzing early processing steps are localized in cisternae toward the *cis* face of the Golgi stack, whereas enzymes catalyzing later processing steps are localized in cisternae toward the *trans* face.

The transport of proteins between the different Golgi cisternae is thought to be mediated by transport vesicles, which bud from one cisterna and fuse with the next. Like the transport vesicles that shuttle from the ER to the Golgi apparatus, the vesicles shuttling between the Golgi cisternae are also thought to be nonselective for their cargo: any soluble or membrane protein that is not otherwise attached as a permanent resident to Golgi membranes can enter the transport vesicles and be moved forward in the biosynthetic-secretory pathway from the *cis* to the *medial* to the *trans* Golgi cisternae. Indeed, much of what we know about the molecular mechanism of vesicular transport was originally described using *in vitro* systems designed to measure protein transport between the Golgi cisternae (see Panel 13–1, pp. 638–639). Electron microscopists have seen small membrane tubules that seem to interconnect some Golgi stacks, and it is possible that some transfer of material from one cisterna to the next may also occur through these structures.

The functional differences among the *cis, medial,* and *trans* subdivisions of the Golgi apparatus were discovered by localizing the enzymes involved in processing *N*-linked oligosaccharides in distinct regions of the organelle, both by physical fractionation of the organelle and by labeling the enzymes in electron-microscope sections with antibodies. The removal of mannose residues and the addition of *N*-acetylglucosamine, for example, were shown to occur in the *medial* compartment, while the addition of galactose and sialic acid was found to occur in the *trans* compartment and the *trans* Golgi network (Figure 13–13). The functional compartmentalization of the Golgi apparatus is summarized in Figure 13–14.

Proteoglycans Are Assembled in the Golgi Apparatus [7]

It is not only the *N*-linked oligosaccharide chains on proteins that are altered as the proteins pass through the Golgi cisternae en route from the ER to their final

Figure 13–13 Histochemical stains demonstrate that the Golgi apparatus is biochemically compartmentalized. A series of electron micrographs shows the Golgi apparatus unstained (A), stained with osmium (B), which is preferentially reduced by the cisternae of the *cis* compartment, and stained to reveal the location of a specific enzyme (C and D). The enzyme nucleoside diphosphatase (see Figure 13–12) is found in the *trans* Golgi cisternae (C), while the enzyme acid phosphatase is found in the *trans* Golgi network (D). (Courtesy of Daniel S. Friend.)

destinations; many proteins are also modified in other ways. Some have sugars added to the OH groups of selected serine or threonine side chains, for example. This **O-linked glycosylation,** like the extension of *N*-linked oligosaccharide chains, is catalyzed by a series of glycosyl transferase enzymes that use the sugar nucleotides in the lumen of the Golgi apparatus to add sugar residues to a protein one at a time. Usually, *N*-acetylgalactosamine is added first, followed by a variable number of additional sugar residues, ranging from just a few to 10 or more.

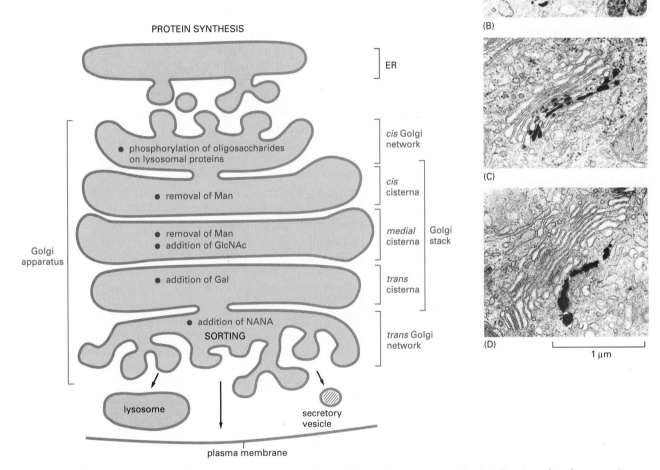

Figure 13–14 The functional compartmentalization of the Golgi apparatus. The localization of each processing step shown was determined by a combination of techniques, including biochemical subfractionation of the Golgi apparatus membranes and electron microscopy after staining with antibodies specific to some of the processing enzymes. The locations of many other processing reactions have not been determined. Although only three distinguishable cisternal compartments have so far been demonstrated, each of these sometimes consists of a group of two or more cisternae in sequence, and it is possible that there are finer subdivisions still to be discovered. Alternatively, it may be that there are only three functionally distinct compartments and that the extra cisternae represent multiple copies of one of the three functional units. It is not clear, however, whether each processing enzyme is completely restricted to a particular cisterna or whether its distribution is graded across the stack—such that early acting enzymes are present mostly in the *cis* Golgi cisternae while later acting enzymes are mostly in the *trans* Golgi cisternae.

Transport from the ER Through the Golgi Apparatus

The Golgi apparatus confers the heaviest glycosylation of all on *proteoglycan core proteins*, which it modifies to produce **proteoglycans.** As discussed in Chapter 19, this process involves the polymerization of one or more *glycosaminoglycan chains* (long unbranched polymers composed of repeating disaccharide units) via a xylose link onto serines on the core protein. Many proteoglycans are secreted and become components of the extracellular matrix while others remain anchored to the plasma membrane. Still others form a major component of slimy materials such as the mucus that is secreted to form a protective coating over many epithelia.

The sugars incorporated into glycosaminoglycans are heavily sulfated in the Golgi apparatus immediately after these polymers are made, and this helps to give proteoglycans their large negative charge. Some tyrosine residues in proteins also become sulfated at this stage. In both cases the sulfation depends on a sulfate donor (3'-phosphoadenosine-5'-phosphosulfate, or PAPS) that is transported from the cytosol into the lumen of the *trans* Golgi network.

The Carbohydrate in Cell Membranes Faces the Side of the Membrane That Is Topologically Equivalent to the Outside of the Cell [8]

Because all oligosaccharide chains are added on the luminal side of the ER and Golgi apparatus, the distribution of carbohydrate on membrane proteins and lipids is asymmetrical. As with the asymmetry of the lipid bilayer itself, the asymmetric orientation of these glycosylated molecules is maintained during their transport to the plasma membrane, secretory vesicles, or lysosomes. As a result, the oligosaccharides of all of the glycoproteins and glycolipids in the corresponding intracellular membranes face the lumen, while those in the plasma membrane face the outside of the cell (Figure 13–15).

What Is the Purpose of Glycosylation? [9]

There is an important difference between the construction of an oligosaccharide and the synthesis of other macromolecules such as DNA, RNA, and protein. Whereas nucleic acids and proteins are copied from a template in a repeated series of identical steps using the same enzyme(s), complex carbohydrates require

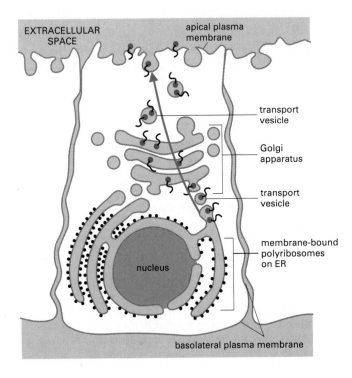

EXTRACELLULAR SPACE
apical plasma membrane
transport vesicle
Golgi apparatus
transport vesicle
membrane-bound polyribosomes on ER
nucleus
basolateral plasma membrane

Figure 13–15 The sugars on membrane glycoproteins and glycolipids are oriented away from the cytosol. The orientation of a transmembrane protein in the ER membrane is preserved when the protein is transported to other membranes. The *red ball* on the end of each glycoprotein molecule represents an *N*-linked oligosaccharide that is added to protein in the ER lumen. Note that these sugar residues are confined to the lumen of each of the internal organelles and become exposed to the extracellular space after a transport vesicle fuses with the plasma membrane. The same is true of the sugar residues on the glycolipids and *O*-linked oligosaccharides produced in the Golgi apparatus.

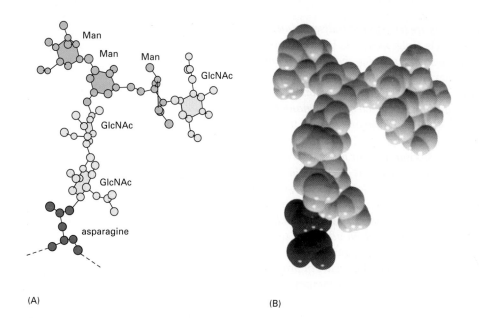

(A)

(B)

Figure 13–16 **The three-dimensional structure of a small *N*-linked oligosaccharide.** The structure was determined by x-ray crystallographic analysis of a glycoprotein. This oligosaccharide contains only 6 sugar residues, whereas there are 14 sugar residues in the *N*-linked oligosaccharide that is initially transferred to proteins in the ER (see Figure 12–48). (A) Backbone model showing all atoms except hydrogens; (B) space-filling model, with the asparagine shown with dark atoms. (Courtesy of Richard Feldmann.)

a different enzyme at each step, each product being recognized as the exclusive substrate for the next enzyme in the series. Given the complicated pathways that have evolved to synthesize them, it seems likely that the oligosaccharides on glycolipids and glycoproteins have important functions, but for the most part these functions are not known.

N-linked glycosylation, for example, is prevalent in all eucaryotes, including yeasts, but is absent from procaryotes. Because one or more *N*-linked oligosaccharides are present on most proteins transported through the ER and Golgi apparatus—a pathway that is unique to eucaryotic cells—it was once thought that their function was to aid this transport process. Drugs that block steps in glycosylation, however, do not generally interfere with transport (with the important exception of transport to lysosomes, which is discussed below), and mutant cells in culture that are blocked in various glycosylation steps in the Golgi apparatus, nevertheless, are viable and transport proteins normally. Although some proteins do not fold correctly without their normal oligosaccharide and therefore precipitate in the ER and fail to be transported, most proteins retain their normal activities in the absence of glycosylation.

Because chains of sugars have limited flexibility, even a small *N*-linked oligosaccharide protrudes from the surface of a glycoprotein (Figure 13–16) and can thus limit the approach of other macromolecules to the surface of the glycoprotein. In this way, for example, the presence of oligosaccharides tends to make a glycoprotein relatively resistant to protease digestion. It may be that the oligosaccharides on cell-surface proteins originally provided an ancestral eucaryotic cell with a protective coat that, unlike the rigid bacterial cell wall, allowed the cell freedom to change shape and move. But these sugar chains have since become modified to serve other purposes as well: the oligosaccharides attached to the cell-surface proteins called selectins, for example, function in cell-cell adhesion processes, as discussed in Chapter 10.

Summary

The Golgi apparatus receives newly synthesized proteins and lipids from the ER and distributes them to the plasma membrane, lysosomes, and secretory vesicles. It is a polarized structure containing one or more stacks of disc-shaped cisternae, which are organized as a series of at least three biochemically and functionally distinct compartments, termed cis, medial, *and* trans *cisternae. Both the* cis *and* trans *cisternae are connected to sorting stations, called the* cis *Golgi network and the* trans *Golgi network, respectively. Correctly folded proteins are transferred indiscriminately from the lumen and membrane of the ER to the* cis *Golgi network, but the resident ER proteins are returned. Proteins destined for secretory vesicles, the plasma membrane, and lysosomes move through the Golgi stack in the* cis-*to-*trans *direction, passing from one cisterna to the next; they finally reach the* trans *Golgi network, from which each type of protein departs for its specific destination. Each of these many transport steps is mediated by transport vesicles, which bud off from one membrane and then fuse with another.*

The Golgi apparatus, unlike the ER, contains many sugar nucleotides, which are used by a variety of glycosyl transferase enzymes to carry out glycosylation reactions on lipid and protein molecules as they pass through the Golgi apparatus. N-linked oligosaccharides, for example, which are added to proteins in the ER, are often initially trimmed by removal of mannoses, and then additional sugars—including N-acetylglucosamine, galactose, and sialic acid—are added. In addition, the Golgi is the site where O-linked glycosylation occurs and where glycosaminoglycan chains are added to core proteins to form proteoglycans. Sulfation of the sugars in proteoglycans and of selected tyrosines on proteins also occurs in a late Golgi compartment.

Transport from the *Trans* Golgi Network to Lysosomes

All of the proteins that pass through the Golgi apparatus, except those that are retained there as permanent residents, are sorted in the *trans* Golgi network according to their final destination. The mechanism of sorting is especially well understood for those proteins destined for the lumen of lysosomes, and in this section we consider this selective transport process. We begin with a brief account of lysosomal structure and function.

Lysosomes Are the Principal Sites of Intracellular Digestion [10]

Lysosomes are membranous bags of hydrolytic enzymes used for the controlled intracellular digestion of macromolecules. They contain about 40 types of hydrolytic enzymes, including proteases, nucleases, glycosidases, lipases, phospholipases, phosphatases, and sulfatases. All are **acid hydrolases.** For optimal activity they require an acid environment, and the lysosome provides this by maintaining a pH of about 5 in its interior. In this way the contents of the cytosol are doubly protected against attack by the cell's own digestive system. The membrane of the lysosome normally keeps the digestive enzymes out of the cy-

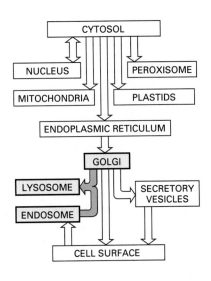

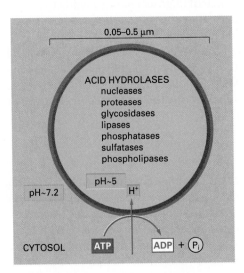

Figure 13–17 **Lysosomes.** The acid hydrolases are hydrolytic enzymes that are active under acidic conditions. The lumen is maintained at an acidic pH by an H⁺ ATPase in the membrane that pumps H⁺ into the lysosome.

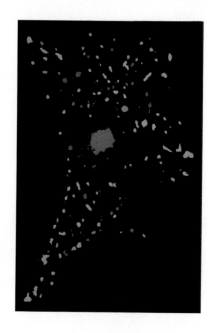

Figure 13–18 The low pH in lysosomes and endosomes. Proteins labeled with a pH-sensitive fluorescent probe (fluorescein) and then endocytosed by cells can be used to measure the pH in endosomes and lysosomes. The different colors reflect the pH that the fluorescent probe encounters in these organelles. The pH in lysosomes (*red*) is about 5, while the pH in various types of endosomes (*blue* and *green*) ranges from 5.5 to 6.5. This method was originally developed in the 1890s by Metchnikoff, who fed litmus particles to phagocytic cells and observed that the color of the particles changed from blue to red after ingestion. (Courtesy of Fred Maxfield and Kenneth Dunn.)

tosol, but even if they should leak out, they can do little damage at the cytosolic pH of about 7.2.

Like all other intracellular organelles, the lysosome not only contains a unique collection of enzymes, but also has a unique surrounding membrane. Transport proteins in this membrane allow the final products of the digestion of macromolecules, such as amino acids, sugars, and nucleotides, to be transported to the cytosol, from where they can be either excreted or reutilized by the cell. An H^+ pump in the lysosomal membrane utilizes the energy of ATP hydrolysis to pump H^+ into the lysosome, thereby maintaining the lumen at its acidic pH (Figure 13–17). Most of the lysosomal membrane proteins are unusually highly glycosylated, which is thought to help protect them from the lysosomal proteases in the lumen.

As we discuss later, endocytosed materials are initially delivered to organelles called *endosomes* before being delivered to lysosomes. Endosomes also have H^+ pumps that keep their lumen at a low pH, although not as low as that of lysosomes (Figure 13–18). We shall see that these pH differences are often used to load and unload cargo molecules from their receptors during vesicular transport along the endocytic pathway.

Lysosomes Are Heterogeneous [11]

Lysosomes were initially discovered by biochemical fractionations of cell extracts; only later were they seen clearly in the electron microscope. They are extraordinarily diverse in shape and size but can be identified as members of a single family of organelles by histochemistry, using the precipitate produced by the action of an acid hydrolase on its substrate to show which organelles contain the enzyme (Figure 13–19). By this criterion, lysosomes are found in all eucaryotic cells.

The heterogeneity of lysosomal morphology contrasts with the relatively uniform structures of most other cellular organelles. The diversity reflects the wide variety of digestive functions mediated by acid hydrolases, including the breakdown of intra- and extracellular debris, the destruction of phagocytosed microorganisms, and the production of nutrients for the cell. For this reason lysosomes are sometimes viewed as a heterogeneous collection of distinct organelles whose common feature is a high content of hydrolytic enzymes. It is especially hard to apply a narrower definition than this in plant cells, as we see next.

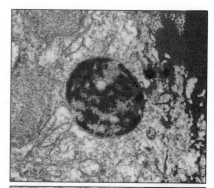

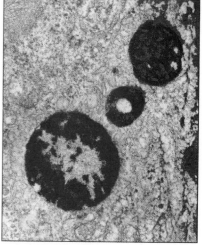

Figure 13–19 Histochemical visualization of lysosomes. Electron micrographs of two sections of a cell stained to reveal the location of acid phosphatase, a marker enzyme for lysosomes. The larger membrane-bounded organelles, containing dense precipitates of lead phosphate, are lysosomes, whose diverse morphology reflects variations in the amount and nature of the material they are digesting. The precipitates are produced when tissue fixed with glutaraldehyde (to fix the enzyme in place) is incubated with a phosphatase substrate in the presence of lead ions. Two small vesicles thought to be carrying acid hydrolases from the Golgi apparatus are indicated by *red arrows* in the top panel. (Courtesy of Daniel S. Friend.)

200 nm

Transport from the *Trans* Golgi Network to Lysosomes

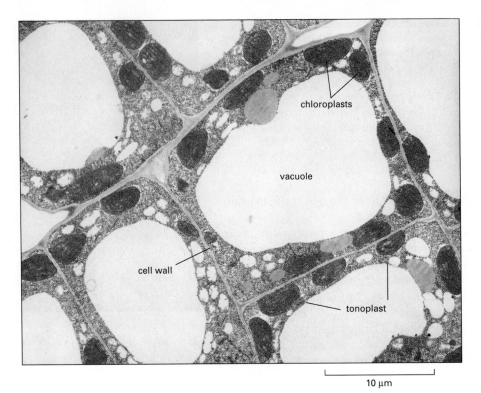

chloroplasts

vacuole

cell wall

tonoplast

10 μm

Figure 13–20 **The plant cell vacuole.**
This electron micrograph of cells in a young tobacco leaf shows that the cytosol is confined by the enormous vacuole to a thin layer, containing chloroplasts, pressed against the cell wall. The membrane of the vacuole is called the tonoplast. (Courtesy of J. Burgess.)

Plant and Fungal Vacuoles Are Remarkably Versatile Lysosomes [12]

Most plant and fungal cells (including yeasts) contain one or several very large, fluid-filled vesicles called **vacuoles.** They typically occupy more than 30% of the cell volume and as much as 90% in some cell types (Figure 13–20). Vacuoles are related to lysosomes of animal cells, containing a variety of hydrolytic enzymes, but their functions are remarkably diverse. The plant vacuole can act as a storage organelle for nutrients and for waste products, as a degradative compartment, as an economical way of increasing cell size (Figure 13–21), and as a controller of *turgor pressure* (the osmotic pressure that pushes outward on the cell wall and keeps the plant from wilting). Different vacuoles with distinct functions (for example, digestion and storage) are often present in the same cell.

The vacuole is important as a homeostatic device, enabling plant cells to withstand wide variations in their environment. When the pH in the environment

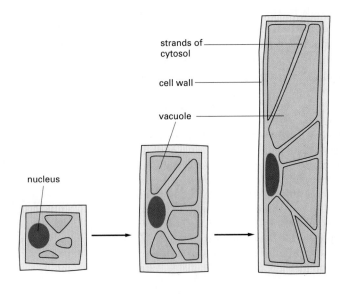

strands of cytosol

cell wall

vacuole

nucleus

Figure 13–21 **The role of the vacuole in controlling the size of plant cells.**
A large increase in cell volume can be achieved without increasing the volume of the cytosol. Localized weakening of the cell wall orients a turgor-driven cell enlargement that accompanies the uptake of water into an expanding vacuole (see Figure 19–67). The cytosol is eventually confined to a thin peripheral layer that is connected to the nuclear region by strands of cytosol, which are stabilized by bundles of actin filaments (not shown).

drops, for example, the flux of H^+ into the cytosol is balanced, at least in part, by increased transport of H^+ into the vacuole so as to keep the pH in the cytosol constant. Similarly, many plant cells maintain an almost constant turgor pressure in the face of large changes in the tonicity of the fluid in their immediate environment. They do so by changing the osmotic pressure of the cytosol and vacuole—in part by the controlled breakdown and resynthesis of polymers such as polyphosphate in the vacuole and in part by altering rates of transport of sugars, amino acids, and other metabolites across the plasma membrane and the vacuolar membrane. The turgor pressure controls these fluxes by regulating the activities of the distinct sets of transporters in each lipid bilayer.

Substances stored in plant vacuoles in different species range from rubber to opium to the flavoring of garlic. Often, the stored products have a metabolic function. Proteins, for example, can be preserved for years in the vacuoles of the storage cells of many seeds, such as those of peas and beans. When the seeds germinate, the proteins are hydrolyzed and the mobilized amino acids provide a food supply for the developing embryo. Anthocyanin pigments that are stored in vacuoles color the petals of many flowers to attract pollinating insects, while noxious molecules that are released from vacuoles when a plant is eaten or damaged provide a defense against predators.

Materials Are Delivered to Lysosomes by Multiple Pathways [13]

Lysosomes in general are meeting places in which several streams of intracellular traffic converge. Digestive enzymes are delivered to them by a route that leads outward from the ER via the Golgi apparatus, while substances to be digested are fed in by at least three paths, according to their source.

Of the three paths to degradation in lysosomes, the best studied is that followed by macromolecules taken up from the external medium by endocytosis. In brief (for the details will be discussed later), the endocytosed molecules are initially delivered into small, irregularly shaped intracellular vesicles called *early endosomes*. From these, some of the ingested molecules are selectively retrieved and recycled to the plasma membrane, while others pass on into *late endosomes*. Here, by fusion of two streams of transport vesicles, the materials coming in for digestion first meet the lysosomal hydrolases coming out from the Golgi apparatus. The interior of the late endosomes is mildly acidic (pH ~6), and it is thought to be the site where the hydrolytic digestion of the endocytosed molecules begins. Mature lysosomes form from the late endosomes, although it is not known precisely how this occurs. During the conversion process some distinct endosomal membrane proteins are lost, and there is a further decrease in internal pH.

A second pathway to degradation in lysosomes is used in all cell types for disposal of obsolete parts of the cell itself—a process called **autophagy.** In a liver cell, for example, an average mitochondrion has a lifetime of about 10 days, and electron microscopic images of normal cells reveal lysosomes containing (and presumably digesting) mitochondria as well as other organelles. The process seems to begin with the enclosure of an organelle by membranes derived from the ER, creating an *autophagosome*, which then fuses with a lysosome (or a late endosome). The process is highly regulated, and selected cell components can somehow be marked for destruction during cell remodeling: the smooth ER that proliferates in a liver cell in response to the drug phenobarbital (discussed in Chapter 12), for example, is selectively removed by autophagy when the drug is withdrawn.

As we discuss later, the third pathway that provides materials to lysosomes for degradation occurs mainly in cells specialized for the *phagocytosis* of large particles and microorganisms. Such professional phagocytes (macrophages and neutrophils in vertebrates) engulf objects to form a *phagosome*, which is then converted to a lysosome in the manner described for the autophagosome. The three pathways are summarized in Figure 13–22.

Transport from the *Trans* Golgi Network to Lysosomes

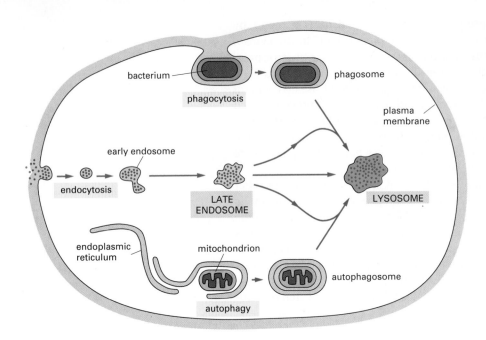

Figure 13–22 Three pathways to degradation in lysosomes. Each pathway leads to the intracellular digestion of materials derived from a different source. The compartments resulting from the three pathways can sometimes be distinguished morphologically—hence the terms "autophagolysosome," "phago-lysosome," and so on. Such lysosomes, however, may differ only because of the different materials they are digesting.

Some Cytosolic Proteins Are Directly Transported into Lysosomes for Degradation [14]

There may be yet a fourth route for proteins to enter a lysosome for degradation: some proteins contain certain signals on their surface [called KFERQ sequences, KFERQ standing for lysine (K), phenylalanine (F), glutamate (E), arginine (R), and glutamine (Q)] that cause the proteins bearing them to be selectively delivered to lysosomes for degradation. It is possible that the KFERQ sequences attach these proteins to cytosolic organelles that are on the way to being autophago-cytosed, thereby dragging the proteins into the lysosome indirectly. Alternatively, there may be a specific transporter in the lysosomal membrane that recognizes these signals and transfers the proteins directly across the lysosomal membrane.

There are precedents for nonconventional mechanisms for moving proteins directly across membranes. A number of proteins that are secreted from cells, such as basic fibroblast growth factor or interleukin-1, for example, arrive at the cell surface without ever entering the classical secretory pathway through the ER and Golgi apparatus. In most cases it is not known which membrane the protein crosses or how its transmembrane transport is catalyzed. In the case of a small yeast peptide, the pheromone a-factor, the transport is known to be mediated directly across the plasma membrane by an ATP-driven peptide pump that belongs to the protein family of the ABC transporters (discussed in Chapter 11). Thus it is possible that similar pumps provide "private" transport systems, each specialized for the transfer of a small, specific subset of proteins across a particular membrane.

Lysosomal Enzymes Are Sorted from Other Proteins in the *Trans* Golgi Network by a Membrane-bound Receptor Protein That Recognizes Mannose 6-Phosphate [15]

We now consider more closely the system that delivers the other half of the traffic into lysosomes—the specialized lysosomal hydrolases and membrane proteins. Both classes of proteins are synthesized in the rough ER and transported through the Golgi apparatus. Transport vesicles that deliver these proteins to late endosomes—which later form lysosomes—bud from the *trans* Golgi network, incorporating lysosomal proteins while excluding the many other proteins being packaged into different transport vesicles for delivery elsewhere.

How are lysosomal proteins recognized and selected with the required accuracy? For the lysosomal hydrolases the answer is known. They carry a unique marker in the form of **mannose 6-phosphate (M6P)** groups, which are added exclusively to the *N*-linked oligosaccharides of these soluble lysosomal enzymes, probably while they are in the lumen of the *cis* Golgi network. The M6P groups are recognized by **M6P receptor proteins,** which are transmembrane proteins present in the *trans* Golgi network. These receptor proteins bind the lysosomal hydrolases and help package them into specific transport vesicles that bud from the *trans* Golgi network and subsequently fuse with a late endosome, delivering their contents to the lumen of this organelle. As we discuss later, the M6P receptor proteins are selected for packaging into specific transport vesicles in the *trans* Golgi network by special coat proteins that assemble on the cytosolic surface of the membrane and help the vesicles bud from the membrane of the *trans* Golgi network.

The Mannose 6-Phosphate Receptor Shuttles Back and Forth Between Specific Membranes [16]

The M6P receptor protein binds its specific oligosaccharide at pH 7 in the *trans* Golgi network and releases it at pH 6, which is the pH in the interior of late endosomes. Thus in the late endosomes the lysosomal hydrolases dissociate from the M6P receptor and can begin to digest the endocytosed material delivered from early endosomes. Having released their bound enzymes, the M6P receptors are retrieved into transport vesicles that bud from late endosomes and return to the membrane of the *trans* Golgi network for reuse (Figure 13–23). It is not clear whether transport back to the Golgi apparatus requires a specific signal peptide in the cytoplasmic tail of the M6P receptor or whether it will occur by default. This process of *membrane recycling* from late endosome back to the Golgi apparatus resembles the recycling that occurs between other subcompartments of the secretory and endocytic pathways that we discuss later.

The sorting of lysosomal hydrolases from other proteins is presently the best-understood example of the many sorting processes mediated by transport vesicles in a eucaryotic cell. Although an oligosaccharide marker is not likely to be used elsewhere, the general strategy is probably typical of other vesicle-mediated sorting processes. Cargo molecules are recognized and picked up by

Figure 13–23 The transport of newly synthesized lysosomal hydrolases to lysosomes. The precursors of lysosomal hydrolases are covalently modified by the addition of mannose 6-phosphate groups (M6P) in the *cis* Golgi network. They then become segregated from all other types of proteins in the *trans* Golgi network because a specific class of transport vesicles (called clathrin-coated vesicles) budding from the *trans* Golgi network concentrates mannose 6-phosphate-specific receptors, which bind the modified lysosomal hydrolases. These vesicles subsequently fuse with late endosomes. At the low pH of the late endosome the hydrolases dissociate from the receptors, which are recycled to the Golgi apparatus for further rounds of transport. In late endosomes the phosphate is removed from the mannose on the hydrolases, further ensuring that the hydrolases do not return to the Golgi apparatus with the receptor.

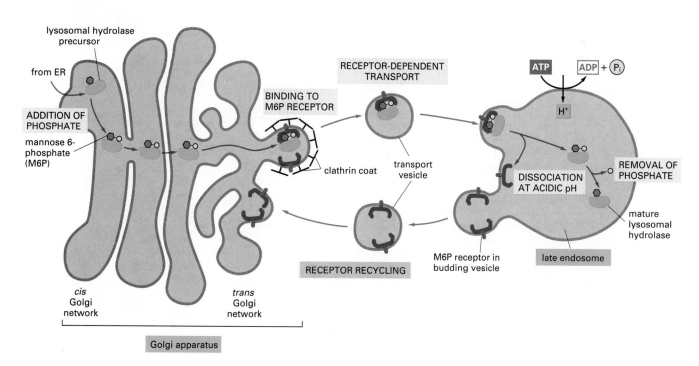

membrane-bound *cargo receptors* during the budding of specific clathrin-coated vesicles. These loaded vesicles move off and fuse with a specific target membrane, the cargo molecules are released in the target compartment, and the empty receptors are recycled back to their original compartment.

Not all of the cargo that is tagged for delivery to lysosomes gets to its proper destination. It seems that some of the lysosomal hydrolase molecules escape the normal packaging process in the *trans* Golgi network and instead are transported via the default pathway to the cell surface, where they are secreted into the extracellular fluid. Some M6P receptors, however, also take a detour to the plasma membrane, where they help undo the error in lysosomal hydrolase routing by recapturing the escaped enzymes and returning them by *receptor-mediated endocytosis* to lysosomes via early and late endosomes.

This *scavenger pathway* was originally discovered through studies of cells from humans who are genetically defective in a specific lysosomal hydrolase. An example is *Hurler's disease,* in which the enzyme required for breakdown of glycosaminoglycans is missing. In these mutant individuals the lysosomes accumulate massive quantities of the particular molecules that cannot be digested. For this reason these diseases are called *lysosomal storage diseases.* The same cellular abnormality is seen when cells from the mutant individuals are grown in culture. If the mutant cells are co-cultured with cells from a normal individual, however, the abnormality is no longer seen. The mutant cells are rescued because they can scavenge the lysosomal hydrolase that they lack from the culture medium, into which the normal cells have discharged it. We see below that studies of these diseases played a crucial part in the discovery of the lysosomal hydrolase-sorting mechanism.

A Signal Patch in the Polypeptide Chain Provides the Cue for Tagging a Lysosomal Enzyme with Mannose 6-Phosphate [17]

The sorting system that segregates lysosomal hydrolases and dispatches them to late endosomes works because M6P groups are added to only the appropriate glycoproteins in the Golgi apparatus. This requires specific recognition of the hydrolases by the Golgi enzyme responsible for adding M6P. Since all glycoproteins leave the ER with identical *N*-linked oligosaccharide chains, the signal for adding the M6P units to oligosaccharides must reside somewhere in the polypeptide chain of each hydrolase.

Two enzymes act sequentially to catalyze the addition of M6P groups to lysosomal hydrolases (Figure 13–24). The first is a phosphotransferase with a *recognition site* that specifically binds the hydrolase and a separate *catalytic site* for the phosphotransferase reaction; the signal recognized by the recognition site is a conformation-dependent *signal patch* in the hydrolase rather than a signal peptide (see Figure 12–8). Once the hydrolase is bound, the phosphotransferase adds GlcNAc-phosphate to one or two of the mannose residues on each oligosaccharide chain (Figure 13–25). A second enzyme then cleaves off the GlcNAc residue, creating the mannose 6-phosphate marker (see Figure 13–24).

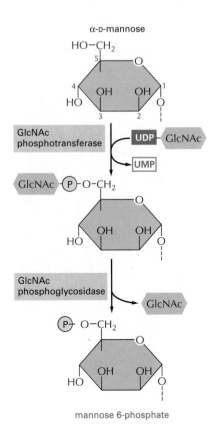

α-D-mannose

mannose 6-phosphate

Figure 13–24 Synthesis of the mannose 6-phosphate marker on a lysosomal hydrolase. The synthesis occurs in two steps. First, GlcNAc phosphotransferase transfers GlcNAc-P to the 6 position of several mannoses on the *N*-linked oligosaccharides of the lysosomal hydrolase. Second, a phospho-glycosidase cleaves off the GlcNAc, creating the mannose 6-phosphate marker. The first enzyme is specifically activated by a signal patch present on lysosomal hydrolases (see Figure 13–25), while the phospho-glycosidase is a nonspecific enzyme. This modification of selected mannose residues in the *cis* Golgi network protects these mannoses from removal by the mannosidases that will later be encountered in the *medial* Golgi compartment.

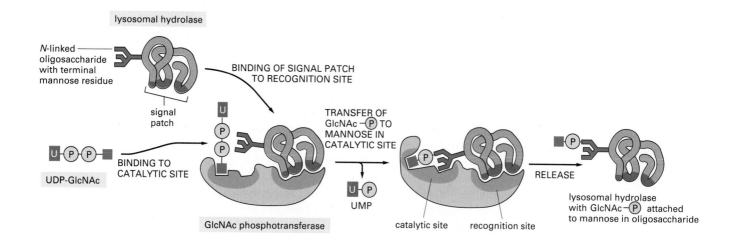

Since most lysosomal hydrolases have multiple oligosaccharides, they acquire many M6P residues, providing a strong and easily recognized signal for the M6P receptor. While a lysosomal hydrolase typically binds to the recognition site of the phosphotransferase with an affinity constant (K_a) of about 10^5 liters/mole, the multiply phosphorylated hydrolase binds to the M6P receptor with a K_a of about 10^9 liters/mole, a 10,000-fold increase in affinity.

Defects in the GlcNAc Phosphotransferase Cause a Lysosomal Storage Disease in Humans [18]

Lysosomal storage diseases are caused by genetic defects that affect one or more of the lysosomal hydrolases and result in accumulation of their undigested substrates in lysosomes, with severe pathological consequences. Most often, there is a mutation in a structural gene that codes for an individual lysosomal hydrolase; this is the case in Hurler's disease, mentioned above. The most dramatic form of lysosomal storage disease, however, is a very rare disorder called inclusion-cell disease (*I-cell disease*). In this disease almost all of the hydrolytic enzymes are missing from the lysosomes of fibroblasts, and their undigested substrates accumulate in lysosomes, which consequently form large "inclusions" in the patients' cells. I-cell disease is due to a single gene defect, and like most genetic enzyme deficiencies, it is recessive—that is, it is seen only in individuals in whom both copies of the gene are defective.

In these individuals all the hydrolases missing from lysosomes are found in the blood; because they fail to be sorted properly in the Golgi apparatus, the hydrolases are secreted rather than transported to lysosomes. The missorting has been traced to a defective or missing GlcNAc-phosphotransferase. Because lysosomal enzymes are not phosphorylated in the *cis* Golgi network, they are not segregated by M6P receptors into the appropriate transport vesicles in the *trans* Golgi network and instead are carried to the cell surface and secreted by the default pathway. This was, in fact, the first evidence for such a default pathway; and it was through a biochemical comparison of normal lysosomal hydrolases with those from patients with I-cell disease that mannose 6-phosphate was discovered to be the lysosomal sorting signal and the whole lysosomal hydrolase-sorting pathway was elucidated.

In I-cell disease the lysosomes in some cell types, such as hepatocytes, contain a normal complement of lysosomal enzymes, implying that there is another pathway for directing hydrolases to lysosomes that is used by some cell types but not others. The nature of this M6P-independent pathway is unknown. Similarly, the lysosomal membrane proteins are sorted from the *trans* Golgi network to late endosomes by an M6P-independent pathway in all cells. It is unclear why cells should need more than one sorting pathway to construct a lysosome, although

Figure 13–25 The recognition of a lysosomal hydrolase. The GlcNAc phosphotransferase enzyme that recognizes lysosomal hydrolases in the Golgi apparatus has separate catalytic and recognition sites. The catalytic site binds both high-mannose *N*-linked oligosaccharides and UDP-GlcNAc. The recognition site binds to a signal patch that is present only on the surface of lysosomal hydrolases.

it is perhaps not surprising that different mechanisms should operate for soluble and membrane-bound proteins.

Summary

Lysosomes are specialized for intracellular digestion. They contain unique membrane proteins and a wide variety of hydrolytic enzymes that operate best at pH 5, the internal pH of lysosomes, which is maintained by an ATP-driven H^+ pump in the lysosomal membrane. Newly synthesized lysosomal proteins are transferred into the lumen of the ER, transported through the Golgi apparatus, and then carried from the trans Golgi network to late endosomes by means of transport vesicles.

The lysosomal hydrolases contain N-linked oligosaccharides that are covalently modified in a unique way in the cis Golgi network so that their mannose residues are phosphorylated. These mannose 6-phosphate (M6P) groups are recognized by an M6P receptor protein in the trans Golgi network that segregates the hydrolases and helps to package them into budding transport vesicles, which deliver their contents to late endosomes, and thereby to lysosomes. These transport vesicles act as shuttles that move the M6P receptor back and forth between the trans Golgi network and late endosomes. The low pH in the late endosome dissociates the lysosomal hydrolases from this receptor, making the transport of the hydrolases unidirectional.

Transport from the Plasma Membrane via Endosomes: Endocytosis [19]

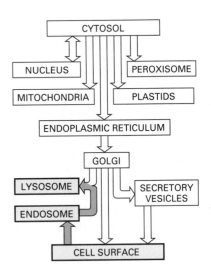

The routes that lead inward to lysosomes from the cell surface start with the process of **endocytosis,** by which cells take up macromolecules, particulate substances, and, in specialized cases, even other cells.

Material to be ingested is progressively enclosed by a small portion of the plasma membrane, which first invaginates and then pinches off to form an intracellular vesicle containing the ingested substance or particle. Two main types of endocytosis are distinguished on the basis of the size of the endocytic vesicles formed: *pinocytosis* ("cellular drinking"), which involves the ingestion of fluid and solutes via small vesicles (≤ 150 nm in diameter), and *phagocytosis* ("cellular eating"), which involves the ingestion of large particles, such as microorganisms or cell debris, via large vesicles called *phagosomes,* generally > 250 nm in diameter. Although most eucaryotic cells are continually ingesting fluid and solutes by pinocytosis, large particles are ingested mainly by specialized phagocytic cells.

Specialized Phagocytic Cells Can Ingest Large Particles [20]

Phagocytosis is a special form of endocytosis in which large particles such as microorganisms and cell debris are ingested via large endocytic vesicles called *phagosomes.* In protozoa phagocytosis is a form of feeding: large particles taken up into phagosomes end up in lysosomes, and the products of the subsequent digestive processes pass into the cytosol to be utilized as food. Few cells in multicellular organisms are able to ingest large particles efficiently, however, and in the gut of animals, for example, large particles of food are broken down extracellularly before import into cells. Phagocytosis is important in most animals for purposes other than nutrition, and it is mainly carried out by specialized cells that are "professional" phagocytes. In mammals there are two classes of white blood cells that act as professional phagocytes—**macrophages** (which are widely distributed in tissues as well as in blood) and **neutrophils.** These two types of cells develop from a common precursor cell (discussed in Chapter 22), and they defend us against infection by ingesting invading microorganisms. Macrophages also play an important part in scavenging senescent and damaged cells and cellular debris. In quantitative terms the latter function is far more important: macrophages phagocytose more than 10^{11} senescent red blood cells in each of us every day, for example.

Whereas the endocytic vesicles involved in pinocytosis are small and uniform, **phagosomes** have diameters that are determined by the size of the ingested particle, and they can be almost as large as the phagocytic cell itself (Figure 13–26). The phagosomes fuse with lysosomes, and the ingested material is degraded; indigestible substances will remain in lysosomes, forming *residual bodies*. Some of the internalized plasma membrane components are retrieved from the phagosome by transport vesicles and returned to the plasma membrane.

In order to be phagocytosed, particles must first bind to the surface of the phagocyte. Not all particles that bind are ingested, however. Phagocytes have a variety of specialized surface receptors that are functionally linked to the phagocytic machinery of the cell. Unlike pinocytosis, which is a constitutive process that occurs continuously, phagocytosis is a triggered process that requires that activated receptors transmit signals to the cell interior to initiate the response. The best-characterized triggers are antibodies, which protect us by binding to the surface of infectious microorganisms to form a coat in which the tail region of each antibody molecule (called the Fc region) is exposed on the exterior. This antibody coat is then recognized by specific *Fc receptors* on the surface of macrophages and neutrophils (see Figure 23–20). The binding of antibody-coated particles to these receptors induces the phagocytic cell to extend pseudopods that engulf the particle and fuse at their tips to form a phagosome (Figure 13–27).

Several other classes of receptors that promote phagocytosis have been characterized—those that recognize *complement* (a class of molecules that circulate in the blood and collaborate with antibodies in targeting undesirable cells for destruction, discussed in Chapter 23), for example, and those that directly recognize oligosaccharides on the surface of certain microorganisms. Which macrophage receptors recognize senescent or damaged cells is uncertain, although cell adhesion proteins of the integrin family are thought to be involved in some cases (discussed in Chapter 22).

Pinocytic Vesicles Form from Coated Pits in the Plasma Membrane [21]

Virtually all eucaryotic cells continually ingest bits of their plasma membrane in the form of small pinocytic (endocytic) vesicles that are later returned to the cell

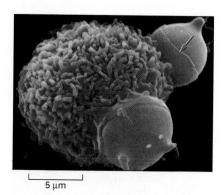

Figure 13–26 Phagocytosis by a macrophage. Scanning electron micrograph of a mouse macrophage phagocytosing two chemically altered red blood cells. The *red arrows* point to edges of thin processes (pseudopods) of the macrophage that are extending as collars to engulf the red cells. (Courtesy of Jean Paul Revel.)

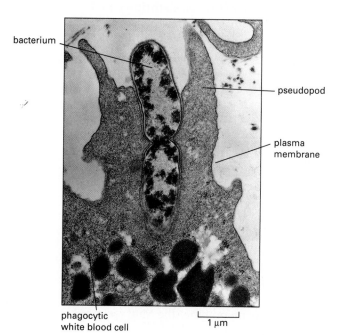

bacterium

pseudopod

plasma membrane

phagocytic white blood cell

1 μm

Figure 13–27 Phagocytosis by a neutrophil. Electron micrograph of a neutrophil phagocytosing a bacterium, which is in the process of dividing. (Courtesy of Dorothy F. Bainton.)

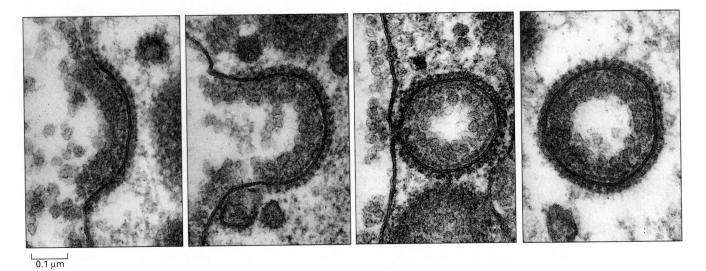

0.1 μm

surface. The rate at which plasma membrane is internalized in this process of **pinocytosis** varies from cell type to cell type, but it is usually surprisingly large. A macrophage, for example, ingests 25% of its own volume of fluid each hour. This means that it must ingest 3% of its plasma membrane each minute, or 100% in about half an hour. Fibroblasts endocytose at a somewhat lower rate, whereas some amoebae ingest their plasma membrane even more rapidly. Since a cell's surface area and volume remain unchanged during this process, it is clear that as much membrane as is being removed by endocytosis is being added to the cell surface by *exocytosis*—the converse process, discussed later. In this sense endocytosis and exocytosis are linked processes that can be considered to constitute an **endocytic-exocytic cycle.**

The endocytic part of the cycle usually begins at specialized regions of the plasma membrane called **clathrin-coated pits,** which typically occupy about 2% of the total plasma membrane area. In electron micrographs of plasma membranes studied by the rapid-freeze, deep-etch technique, these pits appear as invaginations of the plasma membrane coated on their inner (cytosolic) surface with a densely packed material. These coats are made of the protein *clathrin*, which, with other proteins, forms a characteristic basket or cage, which we discuss later. The lifetime of clathrin-coated pits is short: within a minute or so of being formed, they invaginate into the cell and pinch off to form **clathrin-coated vesicles** (Figure 13–28). It has been estimated that about 2500 clathrin-coated vesicles leave the plasma membrane of a cultured fibroblast every minute. These coated vesicles are even more transient than the coated pits: within seconds of being formed, they shed their coat and are able to fuse with early endosomes. Since extracellular fluid is trapped in clathrin-coated pits as they invaginate to form coated vesicles, substances dissolved in the extracellular fluid are internalized—a process called **fluid-phase endocytosis**.

Figure 13–28 The formation of clathrin-coated vesicles from the plasma membrane. Electron micrographs illustrating the probable sequence of events in the formation of a clathrin-coated vesicle from a clathrin-coated pit. The clathrin-coated pits and vesicles shown are larger than those seen in normal-sized cells. They are involved in taking up lipoprotein particles into a very large hen oocyte to form yolk. The lipoprotein particles bound to their membrane-bound receptors can be seen as a dense, fuzzy layer on the extracellular surface of the plasma membrane. (Courtesy of M.M. Perry and A.B. Gilbert, *J. Cell Sci.* 39:257–272, 1979, by permission of The Company of Biologists.)

Clathrin-coated Pits Can Serve as a Concentrating Device for Internalizing Specific Extracellular Macromolecules [22]

In most animal cells, clathrin-coated pits and vesicles provide an efficient pathway for taking up specific macromolecules from the extracellular fluid, a process called **receptor-mediated endocytosis.** The macromolecules bind to complementary cell-surface receptors (which are transmembrane proteins), accumulate in coated pits, and enter the cell as receptor-macromolecule complexes in clathrin-coated vesicles. The process is very similar to the packaging of lysosomal hydrolases in the Golgi apparatus. There too, as we have seen, the molecules to be transported bind to specific receptors in the membrane (the M6P receptors) and so become captured in membrane vesicles that detach from their original compartment and are released into the cytosol. Moreover, the budding of vesicles loaded with lysosomal enzymes from the Golgi apparatus also involves the formation of a clathrin coat (see Figure 13–23). As we discuss later, for both the plasma membrane and the Golgi membrane, the assembly of the coat on the cytosolic surface is thought to drive the membrane to invaginate there.

Receptor-mediated endocytosis provides a selective concentrating mechanism that increases the efficiency of internalization of particular ligands more than 1000-fold, so that even minor components of the extracellular fluid can be specifically taken up in large amounts without taking in a correspondingly large volume of extracellular fluid. A particularly well-understood and physiologically important example is the process whereby mammalian cells take up cholesterol.

Cells Import Cholesterol by Receptor-mediated Endocytosis [23]

Many animal cells take up cholesterol through receptor-mediated endocytosis and in this way acquire most of the cholesterol they require to make new membrane. If the uptake is blocked, cholesterol accumulates in the blood and can contribute to the formation in blood vessel walls of atherosclerotic plaques—the deposits of lipid and fibrous tissue that cause strokes and heart attacks by blocking blood flow. In fact, it was through a study of humans with a strong genetic predisposition for atherosclerosis that the mechanism of receptor-mediated endocytosis was first clearly revealed.

Most cholesterol is transported in the blood bound to protein in the form of particles known as **low-density lipoproteins, or LDL** (Figure 13–29). When a cell needs cholesterol for membrane synthesis, it makes transmembrane receptor proteins for LDL and inserts them into its plasma membrane. Once in the plasma membrane, the LDL receptors diffuse until they associate with clathrin-coated pits that are in the process of forming (Figure 13–30A). Since coated pits constantly pinch off to form coated vesicles, any LDL particles bound to LDL receptors in the coated pits are rapidly internalized in coated vesicles. After shedding their clathrin coats, these vesicles deliver their contents to early endosomes, which are located near the cell periphery. Once in the endosomal compartment, the LDL moves inward and is delivered via late endosomes to lysosomes, where the cholesteryl esters in the LDL particles are hydrolyzed to free cholesterol, which thereby becomes available to the cell for new membrane synthesis. If too much free cholesterol accumulates in a cell, it shuts off both the cell's own cholesterol synthesis and the synthesis of LDL receptor proteins, so that the cell ceases either to make or to take up cholesterol.

This regulated pathway for the uptake of cholesterol is disrupted in individuals who inherit defective genes encoding LDL receptor proteins and whose cells, consequently, are deficient in the capacity to take up LDL from the blood. The resulting high levels of blood cholesterol predispose these individuals to develop atherosclerosis prematurely, and most die at an early age of heart attacks resulting from coronary artery disease. In some cases the receptor is lacking altogether; in others the receptors are defective—either in the extracellular binding site for LDL or in the intracellular binding site that attaches the receptor to the coat of a clathrin-coated pit (see Figure 13–30B). In the latter case normal numbers of LDL-binding receptor proteins are present, but they fail to become localized in the clathrin-coated regions of the plasma membrane; although LDL binds to the surface of these mutant cells, it is not internalized, directly demonstrating the importance of clathrin-coated pits in the receptor-mediated endocytosis of cholesterol.

More than 25 different receptors are known to participate in receptor-mediated endocytosis of different types of molecules, and they all apparently utilize the same clathrin-coated-pit pathway. Many of these receptors, like the LDL

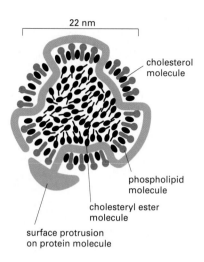

Figure 13–29 A low-density lipoprotein (LDL) particle. Each spherical particle has a mass of 3×10^6 daltons. It contains a core of about 1500 cholesterol molecules esterified to long-chain fatty acids that is surrounded by a lipid monolayer composed of about 800 phospholipid and 500 unesterified cholesterol molecules. A single molecule of a 500,000-dalton protein organizes the particle and mediates the specific binding of LDL to cell-surface receptor proteins.

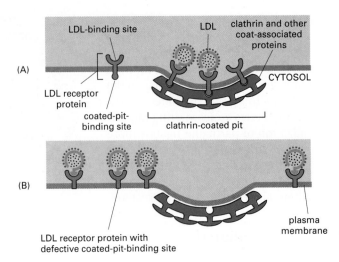

Figure 13–30 Normal and mutant LDL receptors. (A) LDL receptor proteins binding to a coated pit in the plasma membrane of a normal cell. The human LDL receptor is a single-pass transmembrane glycoprotein composed of about 840 amino acid residues, only 50 of which are on the cytoplasmic side of the membrane. (B) A mutant cell in which the LDL receptor proteins are abnormal and lack the site in the cytoplasmic domain that enables them to bind to coated pits. Such cells bind LDL but cannot ingest it. In most human populations 1 in 500 individuals inherits one defective LDL receptor gene and, as a result, is likely to die prematurely from a heart attack caused by atherosclerosis.

receptor, enter coated pits irrespective of whether they have bound their specific ligands; others enter only with a specific ligand bound, suggesting that a ligand-induced conformational change is required for them to bind to the pits. Since most plasma membrane proteins fail to accumulate in clathrin-coated pits, the pits must function as molecular filters, collecting certain plasma membrane proteins (receptors) and excluding others. Electron microscopic studies of cultured cells exposed to different ligands (labeled to make them visible in the electron microscope) have demonstrated that many kinds of receptors cluster in the same coated pit. The plasma membrane of one clathrin-coated pit can probably accommodate about 1000 receptors of assorted varieties. Although all of the receptor-ligand complexes that utilize this endocytic pathway apparently are delivered to the same endosomal compartment, the subsequent fates of the endocytosed molecules vary, as we now discuss.

Endocytosed Materials Often End Up in Lysosomes [24]

The **endosomal compartment** can be made visible in the electron microscope by adding a readily detectable tracer molecule, such as the enzyme peroxidase, to the extracellular medium and leaving the cells for varying lengths of time to take it up by endocytosis. The distribution of the molecule after its uptake reveals the endosomal compartment as a complex set of heterogeneous membrane-bounded tubes and vesicles extending from the periphery of the cell to the perinuclear region, where it is often close to the Golgi apparatus, although they are clearly distinct. Two sets of endosomes can be readily distinguished in such labeling experiments: the tracer molecule appears in **early endosomes,** just beneath the plasma membrane, within a minute or so and in **late endosomes,** close to the Golgi apparatus and near the nucleus, after 5 to 15 minutes (Figure 13–31).

As mentioned earlier, the interior of the endosomal compartment is kept acidic (pH ~6) by ATP-driven H^+ pumps in the endosomal membrane that pump H^+ into the lumen from the cytosol; in general, late endosomes are more acidic than early endosomes. This acidic environment plays a crucial part in the function of these organelles. A similar or identical *vacuolar H^+ ATPase* is thought to acidify all endocytic and exocytic organelles, including phagosomes, lysosomes, selected compartments of the Golgi apparatus, and many transport and secretory vesicles.

We have already seen how endocytosed materials that reach the late endosomes become mixed with newly synthesized acid hydrolases and end up in lysosomes. Many molecules, however, are specifically diverted from this journey to destruction and are recycled instead from the early endosomes back to the

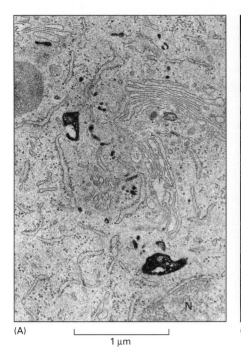

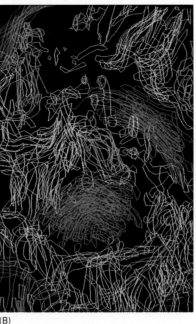

(A)

1 μm

(B)

Figure 13–31 The relationship of late endosomes to other membrane-bounded compartments. (A) Baby hamster kidney (BHK) cells in culture were incubated in a solution containing the enzyme peroxidase for 15 minutes, which was long enough for the peroxidase to be taken up by fluid-phase endocytosis and delivered to late endosomes but not long enough for it to be delivered to lysosomes. After the cells were fixed and exposed to a peroxidase substrate, the product of the enzymatic reaction was made electron dense by fixation with osmium tetroxide. (B) Serial reconstructions of late endosomes (*red*), ER (*yellow*), and Golgi apparatus (*blue*) prepared from electron micrographs, one of which is shown in (A). The reconstruction was drawn from 18 serial thin sections. The nucleus is indicated by N in (A) and is shown in *green* in (B). (A, from M. Marsh, G. Griffiths, G. Dean, I. Mellman, and A. Helenius, *Proc. Natl. Acad. Sci. USA* 83:2899–2903, 1986; B, courtesy of Mark Marsh.)

plasma membrane via transport vesicles. Only those molecules that are not retrieved from endosomes are degraded.

Specific Proteins Are Removed from Early Endosomes and Returned to the Plasma Membrane [25]

The early endosomal compartment acts as the main sorting station in the endocytic pathway, just as the *trans* Golgi network serves this function in the biosynthetic-secretory pathway. In the acidic environment of the early endosome, many internalized receptor proteins change their conformation and release their ligand, just as the M6P receptors unload their cargo of acid hydrolases in the even more acidic late endosomes. Those endocytosed ligands that dissociate from their receptors in the early endosome are usually doomed to destruction in lysosomes, along with the other non-membrane-bound contents of the endosome. Some other endocytosed ligands, however, remain bound to, and thereby share the fate of, their receptors.

The fates of the receptor proteins—and of any ligands remaining bound to them—vary according to the specific type of receptor. (1) Most receptors return to the same plasma membrane domain from which they came; (2) some receptors progress to lysosomes, where they are degraded; and (3) some receptors proceed to a different domain of the plasma membrane, thereby mediating a process called *transcytosis* (Figure 13–32).

Figure 13–32 Possible fates for transmembrane receptor proteins that have been endocytosed. Three pathways from the endosomal compartment in an epithelial cell are shown. Receptors that are not specifically retrieved from endosomes follow the pathway from the endosomal compartment to lysosomes, where they are degraded. Retrieved receptors are returned either to the same plasma membrane domain from which they came (*recycling*) or to a different domain of the plasma membrane (*transcytosis*). If the ligand that is endocytosed with its receptor stays bound to the receptor in the acidic environment of the endosome, it will follow the same pathway as the receptor; otherwise it will be delivered to lysosomes.

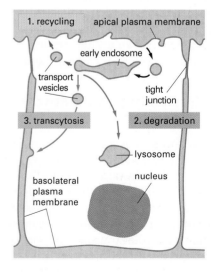

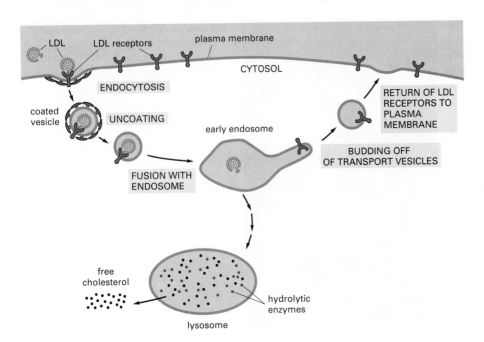

Figure 13–33 Receptor-mediated endocytosis of LDL. Note that the LDL dissociates from its receptors in the acidic environment of the endosome. After a number of steps (see Figure 13–34) the LDL ends up in lysosomes, where it is degraded to release free cholesterol. In contrast, the LDL receptor proteins are returned to the plasma membrane via transport vesicles that bud off from the tubular region of the endosome, as shown. For simplicity, only one LDL receptor is shown entering the cell and returning to the plasma membrane. Whether it is occupied or not, an LDL receptor typically makes one round trip into the cell and back to the plasma membrane every 10 minutes, making a total of several hundred trips in its 20-hour life-span.

The LDL receptor follows the first pathway. It dissociates from its ligand LDL in the endosome and is recycled to the plasma membrane for reuse, while the discharged LDL is carried to lysosomes (Figure 13–33). A similar but more complex recycling occurs following the endocytosis of **transferrin,** a protein that carries iron in the blood. Cell-surface transferrin receptors deliver transferrin with its bound iron to early endosomes by receptor-mediated endocytosis. The low pH in the endosome induces transferrin to release its bound iron, but the iron-free transferrin itself (called apotransferrin) remains bound to its receptor and is recycled back to the plasma membrane as a receptor-apotransferrin complex. When it has returned to the neutral pH of the extracellular fluid, the apotransferrin dissociates from the receptor and is thereby freed to pick up more iron and begin the cycle again. Thus the transferrin protein shuttles back and forth between the extracellular fluid and the endosomal compartment, avoiding lysosomes and delivering the iron that cells need to grow.

The second pathway that endocytosed receptors can follow from endosomes is taken by the receptor that binds *epidermal growth factor (EGF)*, a small protein that stimulates epidermal and various other cells to divide. Unlike LDL receptors, these receptors accumulate in coated pits only after binding EGF. Moreover, most of them do not recycle but end up in lysosomes, where they are degraded along with the ingested EGF. EGF binding therefore leads to a decrease in the concentration of EGF receptors on the cell surface—a process called *receptor down-regulation*. As a result, the concentration of signaling ligand in the extracellular fluid regulates the number of its complementary receptor molecules on the target-cell surface (discussed in Chapter 15).

The Relationship Between Early and Late Endosomes Is Uncertain [26]

It is unclear how endocytosed molecules move from one endosomal compartment to another so as to end up in lysosomes. One view is that early endosomes slowly move inward to become late endosomes, which, as a result of fusion with hydrolase-bearing transport vesicles from the *trans* Golgi network, continuous membrane retrieval, and increasing acidification, are converted to lysosomes. Another view is that early and late endosomes are separate stationary compartments and that transport between them occurs via an intermediate transport compartment—either by a dynamic network of tubes or by the pinching off of pieces of the early endosome that are transported to the cell interior, where they

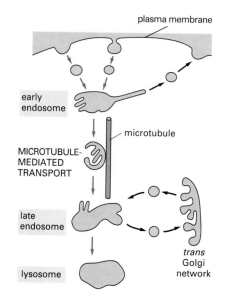

Figure 13–34 The endocytic pathway from the plasma membrane to lysosomes. Transport from early to the late endosome is mediated by large *endosomal carrier vesicles,* which contain large amounts of invaginated membrane and are therefore called *multivesicular bodies.* It is uncertain whether they should be regarded as middle-aged endosomes moving toward the cell interior as they mature or as distinct transport compartments. The movement occurs along microtubules and can be experimentally blocked with microtubule-depolymerizing drugs. Eventually, the late endosome is thought to convert into a lysosome. Transport vesicles recycle material between the early endosome and the cell surface, while a different set of transport vesicles recycle material between the late endosome and the *trans* Golgi network.

eventually fuse with late endosomes (Figure 13–34). Early and late endosomes do, in fact, differ in their protein composition: in particular, they are associated with different rab proteins, which play an important part in directing vesicular transport, as we discuss later (see Table 13–1, p. 644).

Macromolecules Can Be Transferred Across Epithelial Cell Sheets by Transcytosis [27]

Some receptors on the surface of polarized epithelial cells transfer specific macromolecules from one extracellular space to another by a process called **transcytosis.** These receptors follow the third pathway from endosomes (see Figure 13–32). A newborn rat, for example, obtains antibodies from its mother's milk (which help protect it against infection) by transporting them across the epithelium of its gut. The lumen of the gut is acidic, and at this low pH the antibodies in the milk bind to specific receptors on the apical (absorptive) surface of the gut epithelial cells and are internalized via clathrin-coated pits and vesicles and are delivered to early endosomes. The receptor-antibody complexes remain intact and are retrieved in transport vesicles that bud from the early endosome and subsequently fuse with the basolateral domain of the plasma membrane. On exposure to the neutral pH of the extracellular fluid that bathes the basolateral surface of the cells, the antibodies dissociate from their receptors and eventually enter the newborn's bloodstream. The secretion of these antibodies into the mother rat's milk also occurs by transcytosis, but in the reverse direction, from blood to milk. Other mammals, including humans, also transport antibodies into milk in this way, but the antibodies remain in the infant's gut and do not, as in the rat, enter the bloodstream.

The variety of pathways that different receptors follow from endosomes implies that, in addition to binding sites for their ligands and binding sites for coated pits, many receptors also possess sorting signals that guide them into the appropriate type of transport vesicle leaving the endosome and thereby to the appropriate target membrane in the cell.

Epithelial Cells Have Two Distinct Early Endosomal Compartments But a Common Late Endosomal Compartment [28]

In polarized epithelial cells, endocytosis occurs from both the basolateral and the apical domains of the plasma membrane. Material endocytosed from either domain first enters an early endosomal compartment that is unique to that domain. This arrangement allows endocytosed receptors to be recycled back to their original membrane domain, unless they contain signals that mark them for transcytosis to the other domain. Molecules endocytosed from either domain that are not retrieved from the early endosomes are transported to a common late endosomal compartment near the cell center and are eventually degraded in lysosomes (Figure 13–35).

Transport from the Plasma Membrane via Endosomes: Endocytosis

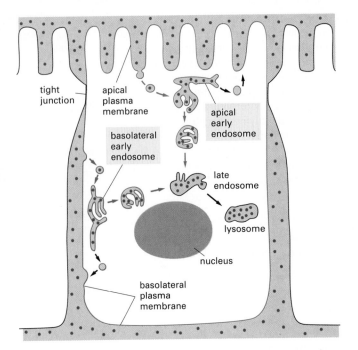

Figure 13–35 Two distinct early endosomal compartments in an epithelial cell. The basolateral and the apical domain of the plasma membrane communicate with distinct early endosomal compartments, although endocytosed molecules from both domains that do not contain signals for recycling or transcytosis meet in a common late endosomal compartment before being digested in lysosomes.

Labels in figure: tight junction; apical plasma membrane; basolateral early endosome; apical early endosome; late endosome; lysosome; nucleus; basolateral plasma membrane

Summary

Cells ingest macromolecules by endocytosis, in which localized regions of the plasma membrane invaginate and pinch off to form endocytic vesicles; many of the endocytosed particles and molecules end up in lysosomes, where they are degraded. Endocytosis occurs both constitutively and as a triggered response to extracellular signals.

Endocytosis is so extensive in many cells that a large fraction of the plasma membrane is internalized every hour. The plasma membrane components (proteins and lipids) are continually returned to the cell surface in a large-scale endocytic-exocytic cycle that is largely mediated by clathrin-coated pits and vesicles. Many cell-surface receptors that bind specific extracellular macromolecules become localized in clathrin-coated pits and consequently are internalized in clathrin-coated vesicles—a process called receptor-mediated endocytosis. The coated endocytic vesicles rapidly shed their clathrin coats and fuse with early endosomes. Most ligands dissociate from their receptors in the acidic environment of the endosome and eventually end up in lysosomes, while most receptors are recycled via transport vesicles back to the cell surface for reuse. But receptor-ligand complexes can follow other pathways from the endosomal compartment. In some cases both the receptor and the ligand end up being degraded in lysosomes, causing "receptor down-regulation." In other cases both are transferred to a different plasma membrane domain, and the ligand is consequently released by exocytosis at a surface of the cell different from that where it originated—a process called transcytosis.

Transport from the *Trans* Golgi Network to the Cell Surface: Exocytosis [29]

Having considered the cell's internal digestive system and the various types of incoming membrane traffic that converge on lysosomes, we now return to the Golgi apparatus and examine the secretory pathways that lead out to the cell exterior. Transport vesicles destined for the plasma membrane normally leave the *trans* Golgi network in a steady stream. The membrane proteins and the lipids in these vesicles provide new components for the cell's plasma membrane, while the soluble proteins inside the vesicles are secreted to the extracellular space. The fusion of the vesicles with the plasma membrane is called **exocytosis.** In this way,

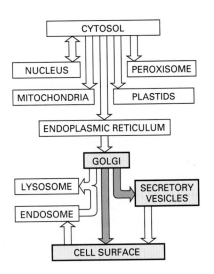

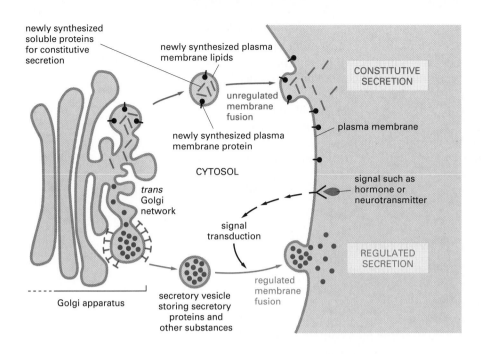

newly synthesized
soluble proteins
for constitutive
secretion

newly synthesized plasma
membrane lipids

CONSTITUTIVE
SECRETION

unregulated
membrane
fusion

newly synthesized plasma
membrane protein

plasma membrane

CYTOSOL

trans
Golgi
network

signal such as
hormone or
neurotransmitter

signal
transduction

REGULATED
SECRETION

regulated
membrane
fusion

Golgi apparatus

secretory vesicle
storing secretory
proteins and
other substances

Figure 13–36 The regulated and constitutive secretory pathways. The two pathways diverge in the *trans* Golgi network. Many soluble proteins are continually secreted from the cell by the *constitutive secretory pathway* (also called the *default pathway*), which operates in all cells. This pathway also supplies the plasma membrane with newly synthesized lipids and proteins. Specialized secretory cells also have a *regulated secretory pathway,* by which selected proteins in the *trans* Golgi network are diverted into secretory vesicles, where the proteins are concentrated and stored until an extracellular signal stimulates their secretion. The regulated secretion of small molecules, such as histamine, occurs by a similar pathway: these molecules are actively transported from the cytosol into preformed secretory vesicles. There they are often complexed to specific macromolecules (proteoglycans in the case of histamine), so that they can be stored at high concentration without generating an excessively high osmotic pressure.

for example, cells produce and secrete most of the proteoglycans and glycoproteins of the *extracellular matrix,* which is discussed in Chapter 19.

All cells require this **constitutive secretory pathway.** Specialized secretory cells, however, have a second secretory pathway in which soluble proteins and other substances are initially stored in *secretory vesicles* for later release. This is the **regulated secretory pathway,** which is found mainly in cells that are specialized for secreting products such as hormones, neurotransmitters, or digestive enzymes rapidly on demand (Figure 13–36). In this section we consider the role of the Golgi apparatus in the two secretory pathways and compare the two mechanisms of secretion.

Many Proteins and Lipids Seem to Be Carried Automatically from the ER and Golgi Apparatus to the Cell Surface [30]

In a cell capable of regulated secretion, at least three classes of proteins must be separated before they leave the *trans* Golgi network—those destined for lysosomes (via late endosomes), those destined for secretory vesicles, and those destined for immediate delivery to the cell surface. We have already noted that proteins destined for lysosomes are tagged for packaging into specific departing

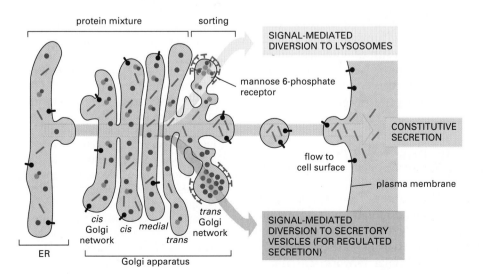

protein mixture

sorting

SIGNAL-MEDIATED
DIVERSION TO LYSOSOMES

mannose 6-phosphate
receptor

CONSTITUTIVE
SECRETION

flow to
cell surface

plasma membrane

cis
Golgi
network

cis

medial

trans

trans
Golgi
network

ER

Golgi apparatus

SIGNAL-MEDIATED
DIVERSION TO SECRETORY
VESICLES (FOR REGULATED
SECRETION)

Figure 13–37 The best-understood pathways of protein sorting in the *trans* Golgi network. Proteins with the mannose 6-phosphate marker are diverted to lysosomes (via late endosomes) in clathrin-coated transport vesicles (see Figure 13–23). Proteins with signals directing them to secretory vesicles are concentrated in large clathrin-coated vesicles that rapidly lose their coats to become secretory vesicles—a pathway that is present only in specialized secretory cells. In unpolarized cells proteins with no special features are thought to be delivered to the cell surface by default via the constitutive secretory pathway. In polarized cells, however, secreted and plasma membrane proteins are selectively directed to either the apical or the basolateral plasma membrane domain, so that at least one of these two pathways must be mediated by a specific signal, as we discuss later.

vesicles (by M6P in the case of lysosomal hydrolases), and analogous signals are thought to direct proteins packaged into secretory vesicles. Most other proteins are transported directly to the cell surface by a nonselective **default pathway** (proteins that must be directed selectively to one cell-surface domain or another are exceptions) (Figure 13–37). Thus, in an unpolarized cell such as a white blood cell or a fibroblast, it seems that any protein in the lumen of the ER will automatically be carried through the Golgi apparatus to the cell surface by the constitutive secretory pathway unless it is either specifically retained as a resident of the ER or Golgi apparatus or selected for the pathways that lead to regulated secretion or to lysosomes. In polarized cells, where different products have to be delivered to different domains of the cell surface, we shall see that the options are slightly more complex.

Secretory Vesicles Bud from the *Trans* Golgi Network [31]

Cells that are specialized for secreting some of their products rapidly on demand concentrate and store these products in **secretory vesicles** (frequently called *secretory granules* or *dense core vesicles* because they have dense cores when viewed in the electron microscope). Secretory vesicles form by clathrin-coated budding from the *trans* Golgi network, and they release their contents to the cell exterior by exocytosis in response to extracellular signals. The secreted product can be either a small molecule (such as histamine) or a protein (such as a hormone or digestive enzyme).

Proteins destined for secretory vesicles (often called *secretory proteins*) are packaged into appropriate vesicles in the *trans* Golgi network by a mechanism that is believed to involve the selective aggregation of the secretory proteins, which can be detected in the electron microscope as clumps of electron-dense material in the lumen of the *trans* Golgi network. The "sorting signal" that directs secretory proteins into such aggregates is not known, but it is thought to be a signal patch that is common to proteins of this class. When a gene encoding a secretory protein is transferred to a secretory cell that normally does not make the protein, the foreign protein is appropriately packaged into secretory vesicles.

It is unclear how the aggregates of secretory proteins are segregated into secretory vesicles. Secretory vesicles have unique proteins in their membrane, some of which might serve as receptors for aggregated material in the *trans* Golgi network. The aggregates are much too big, however, for each molecule of the secreted protein to be bound by its own cargo receptor, as proposed for transport of the lysosomal enzymes. The uptake of the aggregates into secretory vesicles may therefore more closely resemble the uptake of particles by phagocytosis at the cell surface, which can similarly be mediated by clathrin-coated membranes.

After the immature secretory vesicles bud from the *trans* Golgi network, their clathrin coat is removed and their contents become greatly condensed—as much as 200-fold relative to their concentration in the Golgi lumen (Figure 13–38). The condensation occurs suddenly, probably as the result of acidification of the vesicle lumen induced by ATP-driven H^+ pumps in the vesicle membrane. Because the mature vesicles are so densely filled, the secretory cell can disgorge large amounts of material promptly by exocytosis when triggered to do so (Figure 13–39).

Proteins Are Often Proteolytically Processed During the Formation of Secretory Vesicles [32]

Condensation is not the only process to which secretory proteins are subject as the secretory vesicles mature. Many polypeptide hormones and neuropeptides, as well as many secreted hydrolytic enzymes, are synthesized as inactive protein precursors from which the active molecules have to be liberated by proteolysis. These cleavages are thought to begin in the *trans* Golgi network, and they con-

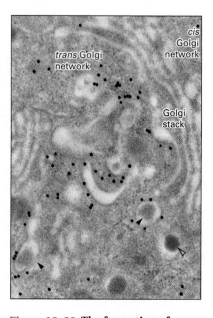

Figure 13–38 The formation of secretory vesicles. This electron micrograph shows secretory vesicles forming from the *trans* Golgi network in an insulin-secreting β cell of the pancreas. An antibody conjugated to gold spheres (*black dots*) has been used to locate clathrin molecules. The immature secretory vesicles (*black arrowheads*), which contain insulin precursor protein (proinsulin), are coated with clathrin. The clathrin coat is rapidly shed once the vesicle has formed and is not part of the mature secretory vesicle, which has a highly condensed core (*open arrowhead*). (Courtesy of Lelio Orci.)

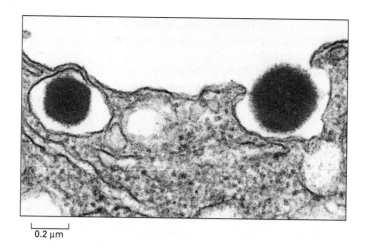

Figure 13–39 **Exocytosis of secretory vesicles.** The electron micrograph shows the release of insulin from a secretory vesicle of a pancreatic β cell. (Courtesy of Lelio Orci, from L. Orci, J-D. Vassali, and A. Perrelet, *Sci. Am.* 256:85–94, 1988.)

0.2 μm

tinue in the secretory vesicles and sometimes in the extracellular fluid after secretion has occurred. Many secreted polypeptides have, for example, an amino-terminal *pro-piece* that is cleaved off shortly before secretion to yield the mature protein. These proteins are thus synthesized as *pre-pro-proteins*, the *pre-piece* consisting of the ER signal peptide that is cleaved off earlier in the rough ER. In other cases peptide signaling molecules are made as *polyproteins* that contain multiple copies of the same amino acid sequence. In still more complex cases a variety of peptide signaling molecules are synthesized as parts of a single polyprotein that acts as a precursor for multiple end products, which are individually cleaved from the initial polypeptide chain; the same polyprotein may be processed in various ways to produce different peptides in different cell types (Figure 13–40).

Why is proteolytic processing so common in the secretory pathway? Some of the peptides produced in this way, such as the *enkephalins* (five amino acid neuropeptides with morphinelike activity), are undoubtedly too short in their mature forms to be co-translationally transported into the ER lumen or to include the necessary signals for packaging into secretory vesicles. For secreted hydrolytic enzymes, or any protein whose activity could be harmful inside the cell that makes it, delaying activation of the protein until it reaches a secretory vesicle or until after it has been secreted has the clear advantage of preventing it from acting prematurely inside the cell in which it is synthesized.

Secretory Vesicles Wait Near the Plasma Membrane Until Signaled to Release Their Contents [33]

Once loaded, a secretory vesicle has to get to the site of secretion, where it must wait until the cell receives the signal to secrete. In some cells the site of secretion is far removed from the Golgi apparatus. Nerve cells provide the most extreme example. Secretory proteins such as peptide neurotransmitters that are to be released from the end of the axon are made and packaged into vesicles in the cell body, where the ribosomes, ER, and Golgi apparatus are located. They must

Figure 13–40 **Alternative processing pathways of the prohormone pro-opiocortin.** The initial cleavages are made by membrane-bound proteases that cut next to pairs of positively charged amino acid residues (Lys-Arg, Lys-Lys, Arg-Lys, or Arg-Arg pairs), and trimming reactions then produce the final secreted products. Different cell types contain different processing enzymes, so that the same prohormone precursor can be used to produce different peptide hormones. In the anterior lobe of the pituitary gland, for example, only corticotropin (ACTH) and β-lipotropin are produced from pro-opiocortin, whereas in the intermediate lobe of the pituitary, mainly α-MSH, γ-lipotropin, β-MSH, and β-endorphin are produced.

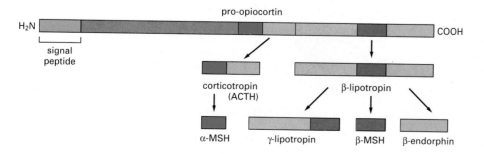

then travel out along the axon—a distance of anything up to a meter or more—to reach the axon terminal. As discussed in Chapter 16, the vesicles use motor proteins attached to their surface to propel themselves along axonal microtubules, whose uniform orientation guides this traffic down the axon in the proper direction. Microtubules may have a similar role in guiding secretory vesicles to the appropriate surface of polarized epithelial cells.

The last step along the regulated secretory pathway is the triggered release of the product by exocytosis. The signal to secrete is often a chemical messenger, such as a hormone, that binds to receptors on the cell surface. The resulting activation of the receptors generates intracellular signals, often including a transient increase in the concentration of free Ca^{2+} in the cytosol. In the case of a nerve axon the signal for exocytosis is usually an electrical excitation—an action potential—that has itself been triggered by a chemical transmitter binding to receptors elsewhere on the cell surface. The action potential causes an influx of Ca^{2+} into the axon terminal through voltage-gated Ca^{2+} channels. By an unknown mechanism, the sudden flush of Ca^{2+}, or some other intracellular signal in the secretory cell, triggers exocytosis, causing the secretory vesicles to fuse with the plasma membrane and release their contents to the extracellular space.

Regulated Exocytosis Is a Localized Response of the Plasma Membrane and Its Underlying Cytoplasm [34]

Histamine is a small molecule secreted by **mast cells** by the regulated pathway in response to specific ligands that bind to receptors on the mast-cell surface. It is responsible for many of the unpleasant symptoms, such as itching and sneezing, that accompany allergic reactions. When mast cells are incubated in fluid containing a soluble stimulant, exocytosis occurs all over the cell surface (Figure 13–41). Yet this is not a generalized response of the whole cell. This has been demonstrated by artificially attaching the stimulating ligand to a solid bead so that it can interact only with a localized region of the mast cell surface; now exocytosis is restricted to the region where the cell contacts the bead (Figure 13–42). Clearly, individual segments of the plasma membrane can function independently. As a result, the mast cell, unlike a nerve cell, does not respond as a whole when it is triggered; the activation of receptors, the resulting intracellular signals, and the subsequent exocytosis are all localized in the particular region of the cell that has been excited.

Secretory-Vesicle Membrane Components Are Recycled [35]

When a secretory vesicle fuses with the plasma membrane, its contents are discharged from the cell by exocytosis and its membrane becomes part of the

Figure 13–41 **Electron micrographs of exocytosis in rat mast cells.** The cell in (A) has not been stimulated. The cell in (B) has been activated to secrete its stored histamine by a soluble extracellular ligand. Histamine-containing secretory vesicles are dark, while those that have released their histamine are light. The material remaining in the spent vesicles consists of a network of proteoglycans to which the stored histamine was bound. Once a secretory vesicle has fused with the plasma membrane, the secretory vesicle membrane often serves as a target to which other secretory vesicles fuse. Thus the cell in (B) contains several large cavities lined by the fused membranes of many spent secretory vesicles, which are now in continuity with the plasma membrane. This continuity is not always apparent in one plane of section through the cell. (From D. Lawson, C. Fewtrell, B. Gomperts, and M. Raff, *J. Exp. Med.* 142:391–402, 1975, by copyright permission of the Rockefeller University Press.)

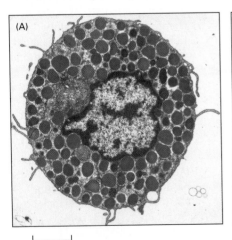
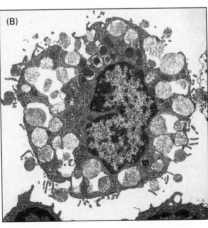

5 μm

Figure 13–42 Exocytosis as a localized response. Electron micrograph of a mast cell that has been activated to secrete histamine by a stimulant coupled to a large solid bead. Exocytosis has occurred only in the region of the cell that is in contact with the bead. (From D. Lawson, C. Fewtrell, and M. Raff, *J. Cell Biol.* 79:394–400, 1978, by copyright permission of the Rockefeller University Press.)

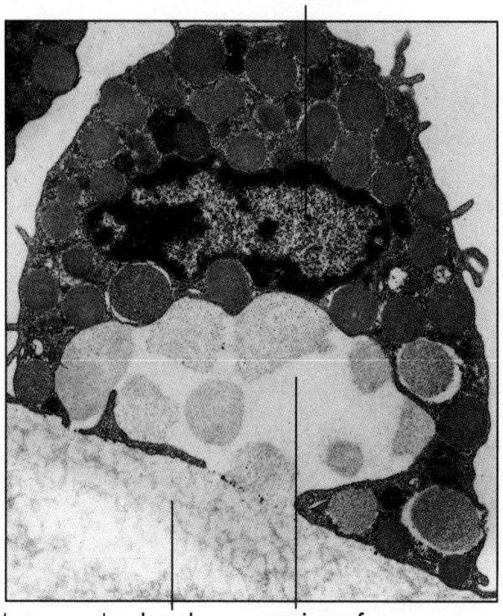

plasma membrane. Although this should greatly increase the surface area of the plasma membrane, it does so only transiently because membrane components are removed from the surface by endocytosis almost as fast as they are added by exocytosis. This removal returns the proteins of the secretory vesicle membrane to the *trans* Golgi network (probably via endosomes), where they can be used again. Such recycling maintains a steady-state distribution of membrane components among the various cellular compartments. The amount of secretory vesicle membrane that is temporarily added to the plasma membrane can be enormous: in a pancreatic acinar cell discharging digestive enzymes, about 900 μm^2 of vesicle membrane is inserted into the apical plasma membrane (whose area is only 30 μm^2) when the cell is stimulated to secrete.

Synaptic Vesicles Form from Endosomes [36]

Nerve cells (and some endocrine cells) contain two types of secretory vesicles. As we have just discussed, these cells package proteins and peptides in dense-core secretory vesicles in the standard way for release by the regulated secretory pathway. In addition, however, they make use of another specialized class of tiny (~50-nm diameter) secretory vesicles that are generated in a different way. These **synaptic vesicles** store the small neurotransmitter molecules, such as acetylcholine, glutamate, and γ-aminobutyric acid (GABA), that serve for rapid signaling from cell to cell at chemical synapses (and for local signaling in some endocrine tissues). The vesicles are triggered to release their contents within a fraction of a millisecond when an action potential arrives at a nerve terminal, and some neurons fire more than 1000 times per second, releasing synaptic vesicles each time. This demands very rapid replenishment of the vesicles that are believed to be generated not from the Golgi membrane but by local recycling from the plasma membrane in the following way. It is thought that the membrane components of the synaptic vesicles are initially delivered to the plasma membrane by the constitutive secretory pathway and then retrieved by endocytosis and delivered to endosomes, from which they are reassembled and bud off to form synaptic vesicles. The membrane components of the vesicles include carrier proteins specialized for the uptake of neurotransmitter from the cytosol, where it is synthesized. Once filled with neurotransmitter, the vesicles return to the plasma membrane, where they wait until the cell is stimulated. After they release their contents, their membrane components are retrieved in the same way and used again (Figure 13–43). The whole cycle, from endocytosis to exocytosis, can be observed by adding a tracer molecule, such as peroxidase, to the external medium and following its fate as it is first taken up into endosomes and then returned to the cell surface in synaptic vesicles.

Polarized Cells Direct Proteins from the *Trans* Golgi Network to the Appropriate Domain of the Plasma Membrane [37]

Most cells in tissues are *polarized* and have two (and sometimes more) distinct plasma membrane domains to which different types of secretory vesicles must be targeted. This raises the general problem of how the delivery of membrane from the Golgi apparatus is organized so as to maintain differences between one cell-surface domain and another. A typical epithelial cell, for example, has an

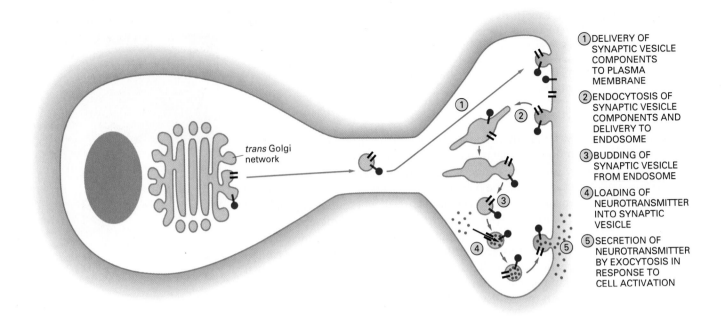

trans Golgi
network

① DELIVERY OF
SYNAPTIC VESICLE
COMPONENTS
TO PLASMA
MEMBRANE

② ENDOCYTOSIS OF
SYNAPTIC VESICLE
COMPONENTS AND
DELIVERY TO
ENDOSOME

③ BUDDING OF
SYNAPTIC VESICLE
FROM ENDOSOME

④ LOADING OF
NEUROTRANSMITTER
INTO SYNAPTIC
VESICLE

⑤ SECRETION OF
NEUROTRANSMITTER
BY EXOCYTOSIS IN
RESPONSE TO
CELL ACTIVATION

Figure 13–43 The formation of synaptic vesicles. These tiny uniform vesicles are found only in nerve cells and in some endocrine cells, where they store and secrete small neurotransmitters.

apical domain, which faces the lumen and often has specialized features such as cilia or a brush border of microvilli, and a *basolateral domain*, which covers the rest of the cell. The two domains are demarcated by a ring of *tight junctions* (see Figure 23–35). These specialized cell-cell junctions (discussed in Chapter 19) prevent proteins, and lipids in the outer leaflet of the lipid bilayer, from diffusing between the apical and basolateral regions, so that not only the protein but also the lipid composition of the two membrane domains is different. In particular, the apical membrane domain is greatly enriched in glycolipids, which are thought to help protect this exposed surface from damage by, for example, the digestive enzymes and the low pH encountered in sites such as the lumen of the gut. Plasma membrane proteins that are linked to the lipid bilayer by a glycosylphosphatidylinositol (GPI) anchor also are found exclusively in the apical plasma membrane. If a sequence providing for attachment of a GPI anchor is added (by genetic manipulation) to a protein that would normally be delivered to the basolateral surface, the protein is delivered to the apical surface instead. GPI-linked proteins seem to associate with glycolipids and may be targeted to the same region of the cell surface as a result of this association, but the nature of the targeting mechanism is not known.

In principle, differences between plasma membrane domains need not depend on targeted delivery of the appropriate membrane components. Instead, the membrane components, for example, could be delivered to all regions of the cell surface indiscriminately but then be selectively stabilized in some locations and selectively eliminated in others. This strategy of random delivery and selective retention or removal seems to be used in certain cases; there are many clear examples, however, where deliveries are specifically directed. Thus epithelial cells often secrete one set of products—such as digestive enzymes or mucus, in the case of cells in the lining of the gut—at their apical surface and another set of products—such as laminin and other basal lamina components—at their basolateral surface. Such cells must have ways of directing vesicles carrying different cargoes, wrapped in different types of membrane, to different plasma membrane domains. By examining polarized cells in culture, it has been found that proteins from the ER destined for different domains travel together until they reach the *trans* Golgi network. Here they are separated and dispatched in secretory or transport vesicles to the appropriate plasma membrane domain. In some cases both basolateral and apical proteins have distinct sorting signals that direct them to the appropriate domain—either directly or indirectly via endosomes (Figure 13–44); in other cases only proteins destined for one of the two membrane

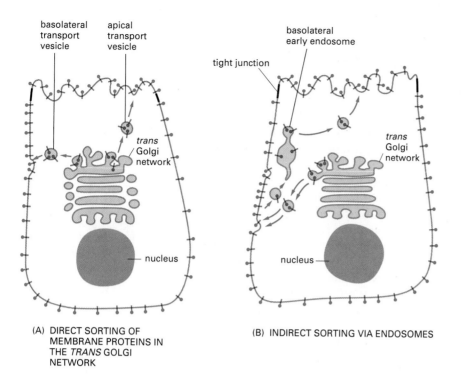

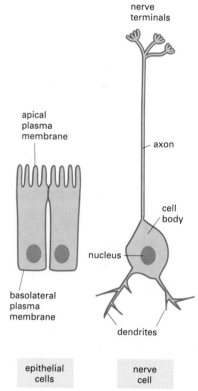

Figure 13–44 **Sorting of plasma membrane proteins in a polarized epithelial cell.** Newly synthesized proteins can reach their proper plasma membrane domain by either a direct (A) or an indirect (B) pathway. In the indirect pathway a protein is retrieved from the inappropriate plasma membrane domain by endocytosis and then transported to the correct domain via early endosomes—that is, by transcytosis.

(A) DIRECT SORTING OF MEMBRANE PROTEINS IN THE *TRANS* GOLGI NETWORK

(B) INDIRECT SORTING VIA ENDOSOMES

domains have a sorting signal, while the other domain is reached by a default pathway that requires no signal.

A nerve cell is an extreme example of a polarized cell: the plasma membrane of its axon terminals is specialized for signaling to other cells, and the plasma membrane of its cell body and dendrites is specialized to receive signals from other nerve cells. Both of these plasma membrane domains are not only functionally distinct but also have a distinct membrane protein composition. Studies of protein traffic in nerve cells in culture indicate that, so far as vesicular transport from the *trans* Golgi network to the cell surface is concerned, the plasma membrane of the nerve cell body and dendrites is equivalent to the basolateral membrane of a polarized epithelial cell, while the plasma membrane of the axon and nerve terminals is equivalent to the apical membrane of such a cell (Figure 13–45). Thus a protein that is targeted to a specific domain in the epithelial cell is usually found to be targeted to the corresponding domain in the nerve cell.

Summary

Proteins can be secreted from cells by exocytosis in either a constitutive or a regulated fashion. In the regulated pathways molecules are stored in secretory vesicles or synaptic vesicles, which do not fuse with the plasma membrane to release their contents until an extracellular signal is received. A selective condensation of the proteins directed to secretory vesicles accompanies their packaging into these vesicles in the trans *Golgi network. Synaptic vesicles are confined to nerve cells and some endocrine cells; they form from endosomes and are responsible for the regulated secretion of small neurotransmitters. Whereas the regulated pathways operate only in specialized secretory cells, a constitutive secretory pathway operates in all cells, mediated by continual vesicular transport from the* trans *Golgi network to the plasma membrane.*

Proteins made in the ER are automatically delivered to the trans *Golgi network and then to the plasma membrane by the constitutive, or default, pathway unless they are diverted into other pathways or retained by specific sorting signals. In polarized cells, however, the transport pathways from the* trans *Golgi network to the plasma membrane must operate selectively to ensure that different sets of membrane proteins, secreted proteins, and lipids are delivered to the appropriate plasma membrane domains.*

Figure 13–45 **Comparison of two types of polarized cells.** In terms of the mechanisms used to direct proteins to them, the plasma membrane of the nerve cell body and dendrites seems to be equivalent to the basolateral plasma membrane domain of a polarized epithelial cell, whereas the plasma membrane of the axon and nerve terminals seems to be equivalent to the apical membrane of an epithelial cell.

Transport from the *Trans* Golgi Network to the Cell Surface: Exocytosis

The Molecular Mechanisms of Vesicular Transport and the Maintenance of Compartmental Diversity [38]

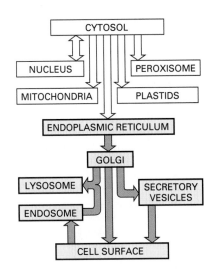

We come now to the most fundamental question in vesicular traffic. We have seen that the cell contains 10 or more chemically distinct membrane-bounded compartments and that vesicular transport mediates a continual exchange of components among them (see Figure 13–3). In the presence of this massive exchange, how does each compartment maintain its specialized character?

To answer this question, we must first consider what defines the character of a compartment. Above all, it seems to be the nature of the enclosing membrane: markers displayed on the cytosolic surface of the membrane guide the targeting of vesicles, ensuring that they fuse only with the correct compartment and so dictating the pattern of traffic between one compartment and another.

Given the presence of distinct membrane markers for each compartment, the problem is to explain how specific membrane components are kept at high concentration in one compartment and at low concentration in another. The answer depends, most fundamentally, on the mechanisms of transport vesicle formation and fusion, by which patches of membrane, enriched or depleted in specific components, are transferred from one compartment to another. In this section, therefore, we examine what is known about these processes at a molecular level.

We have already seen that the creation of a transport vesicle involves the assembly of a special coat on the cytosolic face of the budding membrane. Such coats serve as devices to suck membrane that is enriched in certain membrane proteins and depleted in others out of one compartment, so that specific proteins can be delivered to another compartment. We shall consider how coats form and what they are made of. We also discuss how transport proteins dock at the appropriate target membrane and how they then fuse with that membrane to deliver their contents to the target organelle. We shall see how a combination of genetics and biochemistry has uncovered a variety of GTP-binding proteins that help control vesicular transport. By coupling GTP hydrolysis to other catalytic events, they help give vesicular transport its directionality by linking vesicle budding and fusion to the expenditure of free energy, and they guarantee its fidelity by monitoring the accuracy with which a transport vesicle recognizes its specific target membrane. The basic genetic and biochemical strategies that have been used to study the molecular machinery involved in vesicular transport are outlined in Panel 13–1, pages 638–639.

Before discussing the details of the machinery, however, it is helpful to strip the fundamental problem down to its barest essentials, to see the basic general principles that must apply to any unidirectional vesicular transport process.

Maintenance of Differences Between Compartments Requires an Input of Free Energy [38]

Suppose there are two membrane-bounded compartments connected by transport vesicles that shuttle between them and that the difference between the two compartments lies solely in the concentration of a single type of membrane-bound protein, P. If the system is left simply to drift toward equilibrium through the traffic of transport vesicles between the compartments, the concentrations of P will equalize and the difference between the compartments will disappear. The difference can be maintained by using free energy to transfer P molecules actively in one direction, against their concentration gradient. Protein P might be sequestered into the budding membrane of a forming transport vesicle in the low-concentration compartment, for example, but be kept out of the vesicles budding from the high-concentration compartment by a conformational change driven directly or indirectly by ATP or GTP hydrolysis on the membrane forming the bud (Figure 13–46). Although this example is much simpler than any

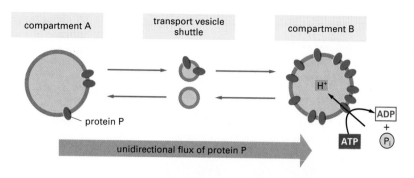

Figure 13–46 Chemical energy is used to give unidirectionality to vesicular transport. In this hypothetical example protein P is an ATP-driven H+ pump that is present in low concentration in compartment A and in high concentration in compartment B. Because of the high concentration of P in compartment B, the lumen of this organelle will be at a much lower pH than that of compartment A. If P undergoes a pH-dependent conformational change that allows it to enter budding vesicles at the higher pH of compartment A but prevents it from doing so at the lower pH of compartment B, then a unidirectional flux of P will occur. As long as the pH difference between the two compartments is maintained through the continuous use of free energy in the form of ATP hydrolysis to drive the H+ pump, the concentration gradient of P between the two compartments will be self-sustaining. As discussed in Chapter 12, most membranes are never created *de novo* but grow by expansion of existing membrane. Thus, although this simple model fails to address how the gradient of P between the two compartments was initially established, it does provide an example of how a cell could use energy to maintain the character of its compartments.

known system used by cells, it serves to illustrate why there must be an input of free energy for transport vesicles to mediate selective directional transport between any two membrane-bounded compartments.

Selective directional vesicular transport is of central importance in the organization of the eucaryotic cell. We begin our discussion of the molecular mechanisms that underlie it by considering how transport vesicles are formed.

There Is More Than One Type of Coated Vesicle [39]

Most transport vesicles form from specialized *coated regions* of membranes and so bud off as **coated vesicles** with a distinctive cage of proteins covering the surface facing the cytosol. Before the vesicle can fuse with a target membrane, this coat has to be discarded in order to let the two membranes interact directly.

There are two well-characterized types of coated vesicles—clathrin-coated and coatomer-coated (Figure 13–47). *Clathrin-coated vesicles*, as we saw earlier, mediate selective transport of transmembrane receptors, such as the M6P receptor from the *trans* Golgi network or the LDL receptor from the plasma membrane, together with any soluble molecules that these receptors may have bound and trapped in the vesicle lumen. *Coatomer-coated vesicles*, by contrast, mediate nonselective vesicular transport from the ER and Golgi cisternae.

There may be a third type of coated vesicle. The plasma membrane of most cells has morphologically and biochemically distinct invaginations called *caveolae* (Figure 13–48); although their function is uncertain, one possibility is that they bud off to form calveolin-coated vesicles. If so, it is unclear what they transport or what their destination is, and we shall not discuss them further.

Coated vesicles appear to mediate directional transfer of specific types of membrane. This transfer is usually balanced by a counterflow of membrane in the opposite direction, either in the form of vesicles of less well-characterized

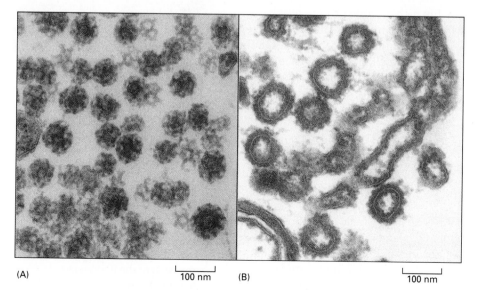

(A) 100 nm (B) 100 nm

Figure 13–47 Comparison of clathrin-coated and coatomer-coated vesicles. (A) Electron micrograph of clathrin-coated vesicles. (B) Electron micrograph of Golgi cisternae from a cell-free system in which coatomer-coated vesicles bud in the test tube. Note that the clathrin-coated vesicles have a more obviously regular structure. (Courtesy of Lelio Orci, from L. Orci, B. Glick, and J. Rothman, *Cell* 46:171–184, 1986. © Cell Press.)

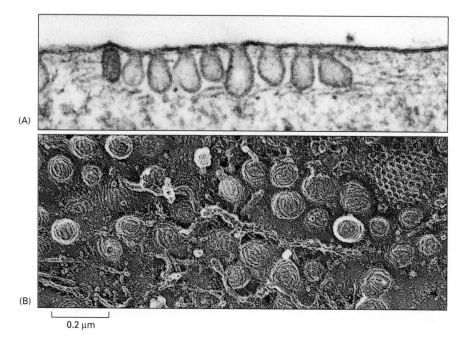

(A)

(B)

0.2 μm

Figure 13–48 **Caveolae on the plasma membrane of a human fibroblast.** (A) Electron micrograph of a fibroblast in cross-section showing caveolae as deep indentations in the plasma membrane. (B) Deep-etch electron micrograph showing numerous caveolae at the cytoplasmic side of the plasma membrane. Their coat appears to be made of concentrically arranged threads that contain the transmembrane protein *caveolin*. Note that caveolae differ in both size and structure from clathrin-coated pits, one of which is seen at the top right of (B). (Courtesy of R.G.W. Anderson, from K.G. Rothberg et al., *Cell* 68:673–682, 1992. © Cell Press.)

types or by means of elongated sacs or tubes of membrane that are dragged along microtubules (see Figure 13–9).

The Assembly of a Clathrin Coat Drives Bud Formation [40]

The major protein component of **clathrin-coated vesicles** is **clathrin** itself, a protein complex that has been highly conserved in evolution. It consists of three large and three small polypeptide chains that together form a three-legged structure called a *triskelion*. Clathrin triskelions assemble into a basketlike convex framework of hexagons and pentagons to form coated pits on the cytoplasmic surface of membranes (Figure 13–49). Under appropriate conditions, isolated triskelions will spontaneously reassemble into typical polyhedral cages in a test tube, even in the absence of the membrane vesicles that these baskets normally enclose (Figure 13–50).

The formation of a clathrin-coated bud is believed to be driven by forces generated by the assembly of the coat proteins on the cytosolic surface of the membrane (Figure 13–51). It is not known what initiates the assembly process at a particular region of membrane or how the coated bud pinches off to form a coated vesicle. Once the vesicle pinches off, the coat is lost very rapidly. The mechanism of shedding is also uncertain, but a chaperone protein of the hsp70 family has been shown to act *in vitro* as an *uncoating ATPase* that uses the energy of ATP hydrolysis to remove the coat from clathrin-coated vesicles. Some additional control mechanism must operate in the cell, however, to prevent the clathrin coat from being removed from a coated bud before it has had time to form a vesicle, especially since the coated bud persists much longer than the coat on the vesicle. One possibility is that uncoating is controlled by Ca^{2+}, which can bind to clathrin light chains and destabilize clathrin coats. Ca^{2+} pumps in the plasma membrane pump Ca^{2+} out of the cell and thereby keep Ca^{2+} concentrations extremely low at the cytosolic face of the membrane, allowing coated pits to persist; but once coated vesicles form and migrate away from the membrane, they encounter a higher concentration of Ca^{2+}, which may be the trigger for uncoating.

While clathrin-coated pinocytic vesicles are usually small and uniform in size, clathrin is also involved in the formation of much larger vesicles, including secretory vesicles that contain large protein aggregates and phagosomes that contain large particles. In these cases clathrin forms patches rather than complete coats on the forming vesicles. The assembly of the clathrin patches is thought to help

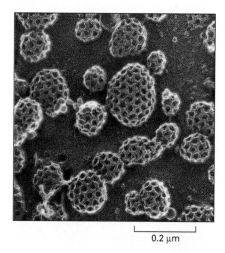

0.2 μm

Figure 13–49 **Clathrin-coated pits and vesicles.** This rapid-freeze, deep-etch electron micrograph shows numerous clathrin-coated pits and vesicles on the inner surface of the plasma membrane of cultured fibroblasts. The cells were rapidly frozen in liquid helium, fractured, and deep-etched to expose the cytoplasmic surface of the plasma membrane. (From J. Heuser, *J. Cell Biol.* 84:560–583, 1980, by copyright permission of the Rockefeller University Press.)

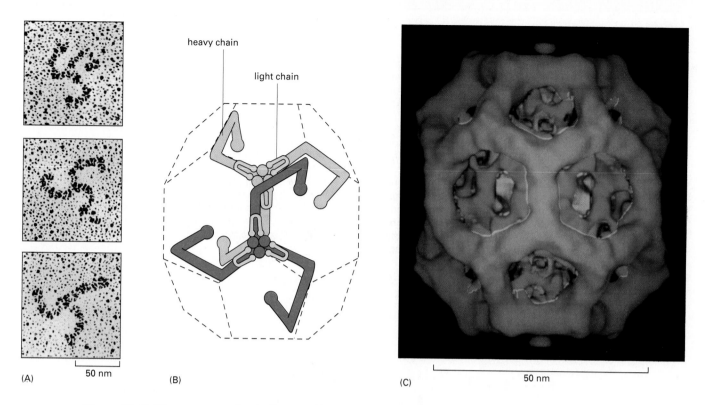

heavy chain

light chain

(A)   50 nm   (B)   (C)   50 nm

Figure 13–50 The structure of a clathrin coat. (A) Electron micrographs of clathrin triskelions shadowed with platinum. Although this feature cannot be seen in these micrographs, each triskelion is composed of 3 clathrin heavy chains and 3 clathrin light chains. (B) A schematic drawing of the probable arrangement of triskelions on the cytosolic surface of a clathrin-coated vesicle. Two triskelions are shown, with the heavy chains of one in *red* and of the other in *gray;* the light chains are shown in *yellow.* The overlapping arrangement of the flexible triskelion arms provides both mechanical strength and flexibility. Note that the end of each leg of the triskelion turns inward, so that its amino-terminal domain forms an intermediate shell. (C) A three-dimensional reconstruction of a clathrin coat composed of 36 triskelions organized in a network of 12 pentagons and 6 hexagons. The outer, *red* polygonal shell represents the overlapping legs of the clathrin triskelions; the intermediate, *green* shell, the amino-terminal domains of the triskelions; and the inner, *blue* shell, the adaptor proteins that we discuss later. Although the coat shown is too small to enclose a membrane vesicle, the clathrin coats on vesicles are constructed in a similar way from 12 pentagons plus a larger number of hexagons. (A, from E. Ungewickell and D. Branton, *Nature* 289:420–422, 1981, © 1981 Macmillan Journals Ltd.; B, from I.S. Nathke et al., *Cell* 68:899–910, 1992. © Cell Press; C, from G.P.A. Vigers, R.A. Crowther, and B.M.F. Pearse, *EMBO J.* 5:2079–2085, 1986.)

Figure 13–51 The assembly and disassembly of a clathrin coat. The assembly of the coat is thought to introduce curvature into the membrane, which leads in turn to the formation of uniformly sized coated buds. The pinching off of the bud to form a vesicle involves the more complex process of membrane fusion, which we discuss later. Although coats consist of multiple protein components, only clathrin is shown in this simplified schematic drawing. Whereas the coat of clathrin-coated vesicles is rapidly removed shortly after the vesicle forms, we shall see later that coatomer coats are removed after the vesicle docks on its target membrane.

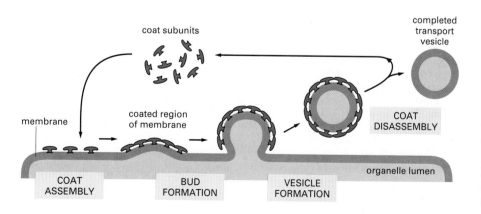

coat subunits

completed transport vesicle

membrane

coated region of membrane

COAT DISASSEMBLY

organelle lumen

COAT ASSEMBLY   BUD FORMATION   VESICLE FORMATION

CELL-FREE SYSTEMS FOR STUDYING THE COMPONENTS AND MECHANISM OF VESICULAR TRANSPORT

Vesicular transport can be reconstituted in cell-free systems. This was first achieved for the Golgi stack. When Golgi stacks are isolated from cells and incubated with cytosol and with ATP as a source of energy, transport vesicles bud from their rims and appear to transport proteins between cisternae. By following the progressive processing of the oligosaccharides on a glycoprotein as it moves from one Golgi compartment to the next, it is possible to follow the process of vesicular transport.

To follow the transport, two distinct populations of Golgi stacks are incubated together. The "donor" population is isolated from mutant cells that lack the enzyme N-acetylglucosamine (GlcNAc) transferase I and that have been infected with a virus; because of the mutation, the major viral glycoprotein fails to be modified with GlcNAc in the Golgi apparatus of the mutant cells. The "acceptor" Golgi stacks are isolated from uninfected wild-type cells and thus contain a good copy of GlcNAc transferase I, but lack the viral glycoprotein. In the mixture of Golgi stacks the viral glycoprotein acquires GlcNAc, indicating that it must have been transported between the Golgi stacks—presumably by vesicles that bud from the *cis* compartment of the donor Golgi and fuse with the *medial* compartment of the acceptor Golgi. This transport-dependent glycosylation is monitored by measuring the transfer of ^{3}H-GlcNAc from UDP-^{3}H-GlcNAc to the viral glycoprotein. Transport occurs only when ATP and cytosol are added. By fractionating the cytosol, a number of specific cytosolic proteins have been identified that are required for the budding and fusion of transport vesicles.

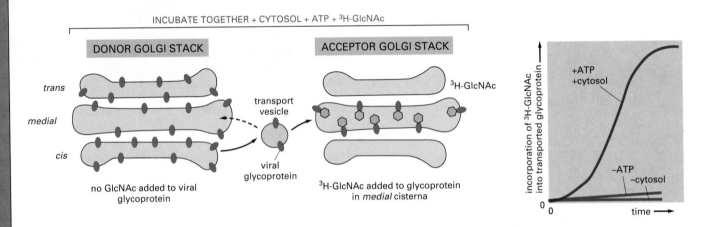

Similar cell-free systems have been used to study transport from the *medial* to the *trans* Golgi network, from the *trans* Golgi network to the plasma membrane, from endosomes to lysosomes, and from the *trans* Golgi network to late endosomes.

GENETIC APPROACHES FOR STUDYING VESICULAR TRANSPORT

Genetic studies of mutant yeast cells defective for secretion at high temperature have identified more than 25 genes that are involved in the secretory pathway. Many of the mutant genes encode *temperature-sensitive* proteins. these function normally at 25°C, but when the mutant cells are shifted to 35°C, some of them fail to transport proteins from the ER to the Golgi apparatus, others from one Golgi cisternae to another, and still others from the Golgi apparatus to the vacuole (the yeast lysosome) or to the plasma membrane.

Once a protein required for secretion has been identified in this way, one can identify genes that encode proteins that interact with it by making use of a phenomenon called *multicopy suppression*. A temperature-sensitive mutant protein at high temperature often has too low an affinity for the proteins it normally interacts with to bind to them. If the interacting proteins are produced at much higher concentration than normal, however, sufficient binding occurs to cure the defect. For this reason yeast cells with a temperature-sensitive mutation in a gene involved in vesicular transport are often transfected with a yeast plasmid vector into which random yeast genomic DNA fragments have been cloned. Because this plasmid is maintained in cells at high copy number, those that carry intact genes will overproduce the normal gene product, allowing rare cells to survive at the high temperature. The relevant DNA fragments, which presumably encode proteins that interact with the original mutant protein, can then be isolated from the surviving cell clones.

The genetic and biochemical approaches complement each other, and many of the proteins involved in vesicular transport have been identified independently by biochemical studies of mammalian cell-free systems and by genetic studies in yeast.

SEMI-INTACT CELL SYSTEMS FOR STUDYING VESICULAR TRANSPORT

Vesicular transport can also be studied in cells whose plasma membrane has been permeabilized to allow small molecules and macromolecules to leave and enter the cell freely. Permeabilization is achieved by physical rupture or treatment with bacterial toxins that punch large holes in the plasma membrane. Such semi-intact cells are particularly useful for studying the transport from extended membrane systems that become extensively fragmented during conventional homogenization procedures, such as the ER and the *trans* Golgi network.

Semi-intact cells have been used to isolate transport vesicles that mediate transport from the *trans* Golgi network to the apical plasma membrane. Epithelial cells are infected with a virus that buds from the apical plasma membrane. The cells are grown at a reduced temperature (20°C) so that membrane traffic from the *trans* Golgi network to the cell surface is blocked, trapping the major viral protein in the *trans* Golgi network (A). When the cells are permeabilized (by laying a piece of nitrocellulose paper on top of them and then ripping it off again to rupture the plasma membrane), the cytosol leaks out of the cells, leaving the internal membrane systems behind (B). The temperature block is then released, and fresh cytosol and ATP are added, causing transport vesicles that contain the viral glycoprotein to bud off from the *trans* Golgi network and leave the semi-intact cells through the holes in the plasma membrane (C). The vesicles are collected and purified using an antibody that recognizes the cytosolic tail of the transmembrane viral glycoprotein, allowing the proteins that make up the transport vesicles to be identified (D).

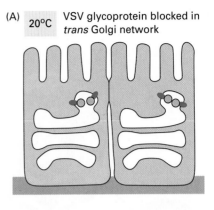

(A) 20°C VSV glycoprotein blocked in *trans* Golgi network

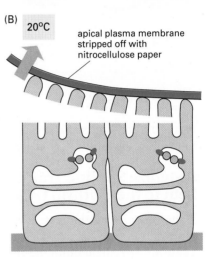

(B) 20°C apical plasma membrane stripped off with nitrocellulose paper

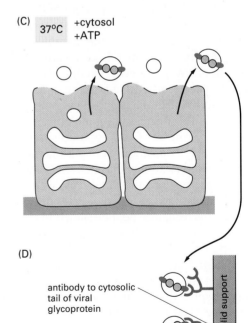

(C) 37°C +cytosol +ATP

(D) antibody to cytosolic tail of viral glycoprotein

solid support

apical transport vesicles are purified by binding to an antibody column

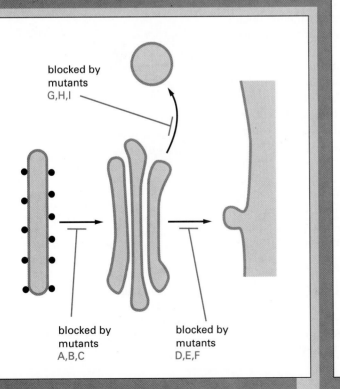

blocked by mutants G,H,I

blocked by mutants A,B,C

blocked by mutants D,E,F

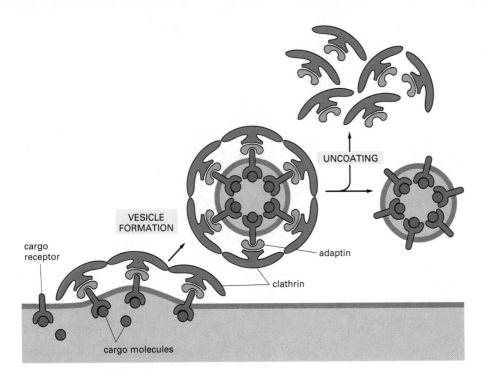

bend the membrane, but the large size of the cargo prevents the membrane from bending enough to allow the formation of a complete coat.

Adaptins Recognize Specific Transmembrane Proteins and Link Them to the Clathrin Cage [41]

The assembly of the coat on clathrin-coated vesicles is thought to serve at least two functions: it provides the mechanical force to pull the membrane into a bud, and it helps to capture specific membrane receptors and their bound cargo molecules. A second major coat protein in these vesicles, a multisubunit complex called **adaptin,** plays a role in both of these functions. Adaptins are required both to bind the clathrin coat to the membrane and to trap various transmembrane receptor proteins, which in turn capture specific cargo molecules inside the vesicle. In this way a selected set of cargo molecules, bound to their specific *cargo receptors,* is incorporated into the lumen of each newly formed clathrin-coated transport vesicle (Figure 13–52).

Clathrin-coated vesicles are not all alike. We have seen, for example, that some, in transit from the Golgi apparatus to late endosomes, are rich in M6P receptors; others, in transit from the plasma membrane to early endosomes, are rich in receptors for extracellular materials such as LDL. Although the cage of clathrin itself seems to be the same in each case, the adaptins are different and mediate the capture of the different types of cargo receptors.

The adaptins recognize peptide signals in the cytoplasmic tail of cargo receptors. A characteristic stretch of four amino acid residues, which are thought to form a sharp turn in the polypeptide chain, forms an essential part of the *endocytosis signal* shared by those cell-surface receptors that function in receptor-mediated endocytosis from the plasma membrane (Figure 13–53). By contrast, a stretch of phosphorylated amino acids at the carboxyl terminus of M6P receptors is recognized by adaptins in the *trans* Golgi network.

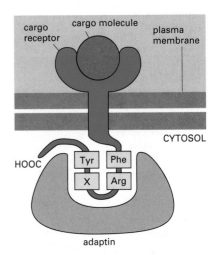

Figure 13–53 **The peptide signal for endocytosis.** The various cell-surface receptor proteins that are endocytosed in clathrin-coated vesicles are thought to share this signal, which is recognized by the adaptins that function in receptor-mediated endocytosis from the plasma membrane. The amino acids shown form an essential part of the signal.

Coatomer-coated Vesicles Mediate Nonselective Vesicular Transport [42]

Coatomer-coated vesicles are thought to mediate the nonselective vesicular transport of the default pathway, which includes transport from the endoplas-

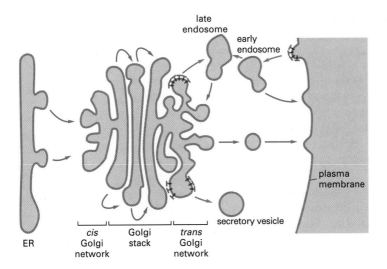

Figure 13–54 Selected and nonselected vesicular transport in nonpolarized cells. Nonselected (constitutive) transport (*blue arrows*) is postulated to be mediated by coatomer-coated vesicles, while various forms of selected (signal-mediated) transport (*red arrows*) are postulated to be carried out by clathrin-coated vesicles. In polarized cells an additional signaled pathway from the *trans* Golgi network is required.

mic reticulum to the Golgi apparatus, from one Golgi cisterna to another, and from the *trans* Golgi network to the plasma membrane (Figure 13–54). None of these transport steps requires that the forming vesicle capture a specific cargo in its lumen.

The coat of these vesicles consists in part of a large protein complex called **coatomer,** comprising seven individual coat-protein subunits (called COPs). At least one of these shows sequence homology to adaptins of clathrin-coated vesicles, but there are important differences in the way that coatomer and clathrin coats behave. In contrast to clathrin coats, coatomer coats do not self-assemble but require ATP to drive their formation, and instead of disassembling as soon as the vesicle has pinched off from the donor membrane, the coatomer coat is retained until the vesicle docks with its target membrane.

Both the assembly and disassembly of the coatomer coat are thought to depend on a protein called **ARF,** which may also play a role in the assembly of clathrin coats. This is one of the many GTP-binding proteins that are key components in the control of vesicular transport. Before discussing the particulars of ARF, we pause to review some general properties of regulatory GTP-binding proteins.

Vesicular Transport Depends on Regulatory GTP-binding Proteins [43]

As discussed in Chapter 5, cells contain large families of regulatory GTP-binding proteins. These proteins act as molecular switches that can flip between two conformational states—an active, charged state with GTP bound and an inactive, discharged state with GDP bound—and they function as regulators of many complex cellular processes. GTP-binding proteins operate in a cycle that typically depends on two auxiliary components: a *guanine-nucleotide-releasing protein (GNRP)* to catalyze exchange of GDP for GTP and a *GTPase-activating protein (GAP)* to trigger the hydrolysis of the bound GTP.

Many regulatory GTP-binding proteins have a covalently attached lipid group that helps them bind to membranes, and they are involved in a great variety of membrane-dependent transactions in the cell. Two structurally distinct classes are recognized: the *monomeric GTP-binding proteins* (also called *monomeric GTPases*), consisting of a single polypeptide chain, and the *trimeric GTP-binding proteins* (also called *G proteins*), consisting of three different subunits. Although studies with inhibitors indicate that both classes have essential roles in vesicular transport, the roles of the monomeric GTPases are better understood, and so we shall focus our discussion on them.

ARF Seems to Signal the Assembly and Disassembly of the Coatomer Coat [44]

ARF is a monomeric GTPase with a fatty acid tail, and it is thought to play a crucial role in both the assembly and disassembly of coatomer coats. It is found in high concentration in the cytosol in its discharged, GDP-bound state. It seems that the donor membrane from which a coatomer-coated vesicle is to bud contains a specific guanine-nucleotide-releasing protein that causes ARF to release its GDP and bind GTP in its place (as GTP is present in much higher concentration in the cytosol than GDP). The binding of GTP is thought to cause the ARF to expose its fatty acid tail, which inserts into the lipid bilayer of the donor membrane. The tightly bound ARF now recruits coatomer subunits, which bind to it. The assembly of the coatomer coat, which consists of both the GTP-charged ARF and the coatomer proteins, pulls the membrane into a bud, which then pinches off as a coated vesicle (Figure 13–55).

When the coatomer-coated vesicle docks with its target membrane, a specific GTPase-activating protein in the target membrane triggers the ARF to hydrolyze its bound GTP to GDP. This is thought to lead to a conformational change in ARF so that its fatty acid chain pops out of the membrane, causing the vesicle's coat to disassemble and allowing membrane fusion to proceed, as discussed later. Thus ARF can be viewed as a protein that senses the circumstances and gives the appropriate signal, either for coat assembly and vesicle budding or for coat disassembly and vesicle docking, as the case may be. Most important, given a guanine-nucleotide-releasing protein in the donor membrane and a GTPase-activating protein in the target membrane, the direction of transport is defined: because of its cycle of GTP hydrolysis and GDP/GTP exchange, ARF facilitates transfer in one direction only.

Organelle Marker Proteins Called SNAREs Help Guide Vesicular Transport [45]

Transport vesicles, whether or not they are selective in the way they pick up cargo from the donor compartment, have to be highly selective as to the target membrane with which they fuse. This suggests that all types of transport vesicles in the cell should display surface markers that identify them according to their origin and cargo and that are recognized by complementary receptors in the proper target membrane. Although the mechanism of this recognition is not known for certain, an attractive hypothesis is that it involves proteins called **SNAREs** (for

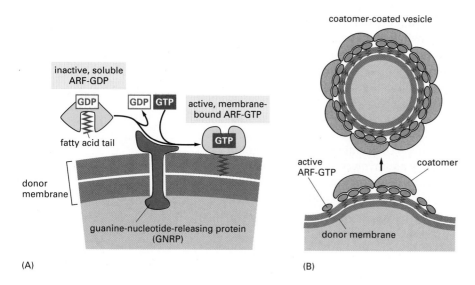

(A)

(B)

Figure 13–55 A current model of coatomer-coated vesicle formation. (A) Inactive, soluble ARF-GDP binds to a guanine-nucleotide-releasing protein in the donor membrane, causing the ARF to release its GDP and bind GTP. A GTP-triggered conformational change in ARF exposes its fatty acid chain, which inserts into the donor membrane. (B) Membrane-bound, active ARF-GTP recruits coatomer subunits to the membrane. This causes the membrane to form a bud. A subsequent membrane-fusion event pinches off and releases the coated vesicle. The drug brefeldin A blocks coatomer-coat assembly by inhibiting the exchange reaction of GDP to GTP. This blocks coatomer-coated vesicular traffic from the ER through the Golgi apparatus, causing the Golgi apparatus to empty into the ER, as explained on page 604.

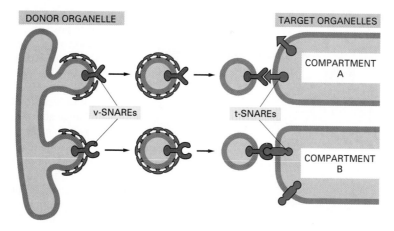

COMPARTMENT A

COMPARTMENT B

v-SNAREs

t-SNAREs

Figure 13–56 The postulated role of SNAREs in guiding vesicular transport. Complementary sets of vesicle-SNAREs (v-SNAREs) and target-membrane SNAREs (t-SNAREs) determine the selectivity of transport-vesicle docking. v-SNAREs, which are co-packaged with the coat proteins during the budding of transport vesicles from the donor membrane, bind to complementary t-SNAREs in the target membrane.

reasons that are discussed below), which exist as complementary sets—v-SNAREs on the vesicle membrane and t-SNAREs on the target membrane (Figure 13–56). SNAREs are best characterized in nerve cells, where they are thought to mediate the docking of synaptic vesicles at the nerve terminal plasma membrane in preparation for exocytosis: a v-SNARE is located on the synaptic vesicle and a complementary t-SNARE on the cytoplasmic face of the plasma membrane.

Rab Proteins Are Thought to Ensure the Specificity of Vesicle Docking [46]

Because there are many different membrane systems in the cell, the docking process must be highly selective. A vesicle is likely to inspect many potential target membranes before its v-SNARE finds a complementary t-SNARE. According to one view, this crucial recognition step is controlled by members of a family of monomeric GTPases called **Rab proteins,** which check that the fit between a v-SNARE and a t-SNARE is correct. In this view Rab proteins become attached to the surface of budding coated vesicles in the donor membrane. When a vesicle encounters the correct target membrane, the binding of v-SNARE to t-SNARE causes the vesicle to remain bound for long enough to allow the Rab protein to hydrolyze its bound GTP, which locks the vesicle onto the target membrane, readying it for subsequent fusion (Figure 13–57).

Eucaryotic cells contain many types of Rab proteins, each associated with a particular membrane-bounded organelle involved in the secretory or endocytic pathways. Each of these organelles has at least one Rab protein on its cytosolic surface (Table 13–1). The first Rab protein (called Sec4) was discovered in yeast by selecting for mutations (called *SEC* mutations) that interfere with secretion. It was subsequently shown to be a component of secretory vesicles and is required for their docking at the plasma membrane; mutations that disrupt it prevent secretory vesicles from discharging their contents to the exterior. The amino acid sequences of Rab proteins are most dissimilar near their carboxyl-terminal tails, and tail-swapping experiments using genetic engineering techniques indicate that it is the tail that determines the intracellular location of each family member, presumably by enabling it to bind to a complementary guanine-nucleotide-releasing factor on the surface of a particular organelle (see Figure 13–57). Before SNAREs became candidates for the organelle marker proteins that guide vesicular transport, Rab proteins were thought to play this role because of their remarkable organelle-specific distribution. It is now known, however, that at least some Rab proteins are functionally interchangeable as long as they are experimentally engineered to become localized to the new organelle. Thus they cannot be the sole explanation for the selectivity of vesicular transport.

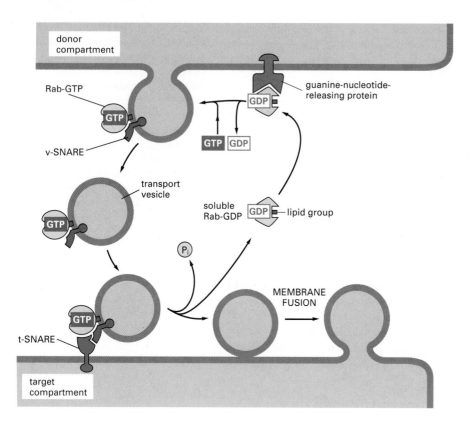

Figure 13–57 Postulated role of Rab proteins in ensuring specificity in the docking of transport vesicles. The guanine-nucleotide-releasing protein in the donor membrane recognizes a specific Rab protein and induces it to exchange GDP for GTP. This exchange alters the conformation of the Rab protein, exposing its covalently attached lipid group, which helps anchor the protein in the membrane. The Rab-GTP remains bound to the surface of the transport vesicle after it pinches off from the donor membrane. v-SNARE on the vesicle surface binds to t-SNARE in the target membrane, docking the vesicle. The Rab protein now hydrolyzes its bound GTP, locking the vesicle onto the target membrane and releasing Rab-GDP into the cytosol, from where it can be reused in a new round of transport. The vesicle then fuses with the target membrane. Note that the vesicle coats have been omitted from the drawings for clarity.

Vesicle Fusion Is Catalyzed by a "Membrane-Fusion Machine" [47]

Once a transport vesicle has recognized its target membrane and docked there, the vesicle has to unload its cargo by membrane fusion. Membrane fusion does not always follow immediately, however. As we have seen, in regulated exocytosis fusion does not occur until it is triggered by an extracellular signal.

Docking and fusion are two distinct and separable processes. It is possible, for example, to prevent fusion while permitting docking by keeping the cytosolic concentration of Ca^{2+} very low. This results in an accumulation of vesicles attached to but not fused with their target membrane. Docking requires only that the two membranes come close enough for proteins protruding from the lipid bilayers to interact and adhere. Fusion requires a much closer approach, bringing the lipid bilayers to within 1.5 nm of each other so that they can join. For this

Table 13–1 Subcellular Locations of Some Rab Proteins

Protein*	Organelle
Rab1 (YPT1)	ER and Golgi complex
Rab2	transitional ER, *cis* Golgi network
Rab3A	secretory vesicles
Rab4	early endosomes
Rab5	early endosomes, plasma membrane
Rab6	*medial* and *trans* Golgi cisternae
Rab7	late endosomes
Rab9	late endosomes, *trans* Golgi network
Sec4	secretory vesicles

*All of these proteins are found in mammalian cells except for Sec4 and YPT1, which are yeast proteins.

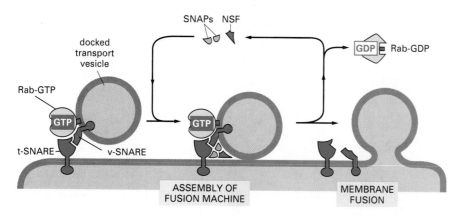

docked
transport
vesicle

SNAPs NSF

GDP Rab-GDP

Rab-GTP

GTP

GTP

t-SNARE

v-SNARE

ASSEMBLY OF
FUSION MACHINE

MEMBRANE
FUSION

Figure 13–58 A current model of protein-mediated vesicle fusion. A complex membrane-fusion machine catalyzes the fusion of a transport vesicle with its target membrane. Only two of the protein components of the fusion complex have been characterized: NSF (*N*-ethylmaleimide-*s*ensitive *f*usion protein) and SNAPs (*s*oluble *N*SF *a*ttachment *p*roteins). (NEM is a chemical that modifies free SH groups exposed on protein surfaces and thereby inactivates proteins whose exposed SH groups are required for activity.) SNAREs were first identified as SNAP receptors (hence their name): they bind to both v-SNAREs and t-SNAREs. The binding of the SNAPs allows NSF to bind. This complex, with the help of acyl CoA and as yet unidentified proteins, catalyzes the fusion of the two lipid bilayers. NSF is an ATPase that hydrolyzes ATP to release the complex once it has done its job (not shown).

close approach water must be displaced from the hydrophilic surface of the membrane—a process that is energetically highly unfavorable. It seems likely that all membrane fusions in cells are catalyzed by specialized fusion proteins that provide a way to cross this energy barrier. The mechanism is still poorly understood. In the case of coatomer-coated transport vesicles at least, fusion with the target membrane requires ATP, GTP, acyl CoA, and several protein components. Two known essential protein components, called NSF and SNAPs (for reasons explained in the legend to Figure 13–58), cycle between the membranes to be fused and the cytosol. The SNAPs bind to both v-SNARE on the vesicle membrane and t-SNARE on the target membrane to initiate the assembly of the fusion apparatus, which catalyzes the fusion of the two lipid bilayers at the vesicle-target-membrane interface (Figure 13–58).

The Best-characterized Membrane-Fusion Protein Is Made by a Virus [48]

Membrane fusion is important in other processes besides vesicular transport, and in particular the simpler membrane fusions that are catalyzed by viral fusion proteins are understood in some detail. Viral fusion proteins play a crucial part in permitting the entry of enveloped viruses (which have a lipid-bilayer-based membrane coat) into the cells that they infect (discussed in Chapter 6). Viruses such as the influenza virus, for example, enter the cell by receptor-mediated endocytosis and are delivered to endosomes. The low pH in endosomes activates a fusion protein in the viral envelope that catalyzes the fusion of the viral and endosomal membranes, thereby allowing the viral nucleic acid to escape into the cytosol, where it can replicate (Figure 13–59).

The genes encoding several viral fusion proteins have been cloned and used to transfect eucaryotic cells in culture. These transfected cells express the viral proteins on their surface, and under appropriate conditions they fuse to form giant multinucleated cells. In the best-studied case, that of the influenza virus, the three-dimensional structure of the fusion protein has been determined by x-ray crystallography. It has been shown that low pH induces a large conformational change in the fusion protein, exposing a previously buried hydrophobic region on the surface of the protein that can interact with the lipid bilayer of a target membrane. A cluster of such hydrophobic regions on closely spaced fusion-protein molecules is thought to bring the two lipid bilayers into close apposition and to destabilize them so that the bilayers fuse (Figure 13–60).

Recently, a mammalian fusion protein has been identified that resembles viral fusion proteins, and it is thought to mediate the fusion of the plasma membranes of sperm and egg that occurs at fertilization (discussed in Chapter 20). As all of these examples emphasize, under normal circumstances membranes do not fuse easily. Membrane fusion requires special proteins and is subject to highly selective controls—a constraint that is crucial both for maintaining the identify

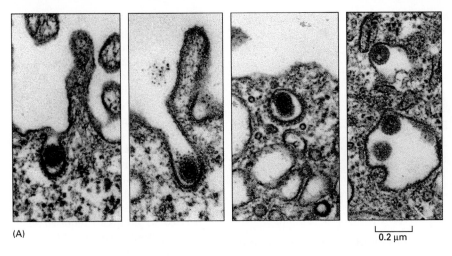

(A)

0.2 μm

Figure 13–59 The entry of fowl plague virus into cells. (A) Electron micrographs showing how the virus is endocytosed in a clathrin-coated vesicle, is delivered to an endosome, and then escapes by fusing with the endosomal membrane. (B) Schematic drawing showing how fusion proteins on the surface of the virus mediate its escape from the endosome. (A, Courtesy of Karl Matlin and Hubert Reggio, from K.S. Matlin et al., *J. Cell Biol.* 91:601–613, 1981, by copyright permission of the Rockefeller University Press.)

of the cell itself and for maintaining the individuality of each of the intracellular compartments.

Summary

The differences between the membranous compartments of a cell are maintained by an input of free energy, driving directed, selective transport of particular membrane components from one compartment to another. Transport vesicles bud from specialized coated regions of the donor membrane. The assembly of the coat helps to drive the formation of the vesicle. There are two well-characterized types of coated vesicles: clathrin-coated vesicles mediate selective vesicular transport from the plasma membrane and the trans *Golgi network, while coatomer-coated vesicles mediate nonselective vesicular transport from the ER and Golgi cisternae. Adaptins provide a*

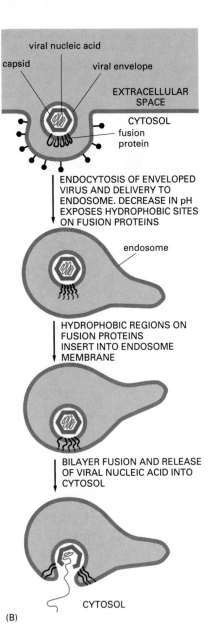

(B)

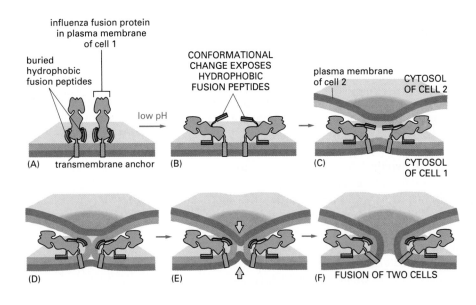

Figure 13–60 A model for how a membrane-fusion protein catalyzes lipid bilayer fusion. A cell that expresses the influenza fusion protein on its surface rapidly fuses with neighboring cells after exposure to low pH. The fusion process proceeds through an intermediate (D and E) in which only the outer leaflets of the membranes are fused, while the inner two leaflets are still separate. Indeed, mutant forms of the fusion protein have been obtained that allow the reaction to proceed only to this intermediate state.

molecular link between clathrin coats and specific membrane receptors and thereby mediate the selective uptake of cargo molecules into clathrin-coated vesicles. Coated vesicles have to lose their coat to fuse with their appropriate target membrane in the cell: clathrin coats are lost soon after the vesicle pinches off from the donor membrane, whereas coatomer coats are lost after the vesicle has docked on the target membrane.

Several classes of monomeric GTPases, including ARF and the Rab proteins, help regulate various steps in vesicular transport, including vesicle budding, docking, and fusion. ARF, Rab, and v-SNARE proteins are incorporated during budding into the transport vesicles and help ensure that the vesicles deliver their contents only to the appropriate membrane-bounded compartment: ARF is thought to mediate coatomer (and probably clathrin) coat assembly and coatomer coat disassembly, while Rab proteins are thought to help ensure the specificity of vesicle docking by locking the vesicle onto the target membrane only when complementary vesicle and target membrane SNAREs interact. Membrane fusion is then catalyzed by a number of cytosolic proteins, including SNAPs and NSF, that assemble into a fusion complex at the docking site.

References

General

Gal, S.; Raikhel, N.V. Protein sorting in the endomembrane system of plant cells. *Curr. Opin. Cell Biol.* 5:636–640, 1993.

Gruenberg, J.; Clague, M.J. Regulation of intracellular membrane transport. *Curr. Opin. Cell Biol.* 4:593–599, 1992.

Pryer, N.K.; Wuesthube, L.J.; Schekman, R. Vesicle-mediated protein sorting. *Annu. Rev. Biochem.* 61:471–516, 1992.

Rothman, J.E. The reconstitution of intracellular protein transport in cell-free systems. *Harvey Lect.* 86:65–85, 1992.

Cited

1. Balch, W.E. Molecular dissection of early stages of the eukaryotic secretory pathway. *Curr. Opin. Cell Biol.* 2:634–641, 1990.

   Hauri, H.-P.; Schweizer, A. The endoplasmic reticulum-Golgi intermediate compartment. *Curr. Opin. Cell Biol.* 4:600–608, 1992.

   Saraste, J.; Kuismanen, E. Pathways of protein sorting and membrane traffic between the rough endoplasmic reticulum and the Golgi complex. *Semin. Cell Biol.* 3:343–355, 1992.

   Schwaninger, R.; Plutner, H.; Bokoch, G.M.; Balch, W.E. Multiple GTP-binding proteins regulate vesicular transport from the ER to Golgi membranes. *J. Cell Biol.* 119:1077–1096, 1992.

2. Driouich, A.; Faye, L.; Staehelin, L.A. The plant Golgi apparatus: a factory for complex polysaccharides and glycoproteins. *Trends Biochem. Sci.* 18:210–214, 1993.

   Griffing, L.R. Comparisons of Golgi structure and dynamics in plant and animal cells. *J. Electron Microsc. Tech.* 17:179–199, 1991.

   Mollenhauer, H.H.; Morre, D.J. Perspectives on Golgi apparatus form and function. *J. Electron Microsc. Tech.* 17:2–14, 1991.

   Rambourg, A.; Clermont, Y. Three-dimensional electron microscopy: structure of the Golgi apparatus. *Eur. J. Cell Biol.* 51:189–200, 1990.

3. Hauri, H.-P.; Schweizer, A. The endoplasmic reticulum-Golgi intermediate compartment. *Curr. Opin. Cell Biol.* 4:600–608, 1992.

   Lippincott-Schwartz, J. Bidirectional membrane traffic between the endoplasmic reticulum and Golgi apparatus. *Trends Cell Biol.* 3:81–88, 1993.

   Pelham, H.R. Recycling of proteins between the endoplasmic reticulum and Golgi complex. *Curr. Opin. Cell Biol.* 3:585–591, 1991.

4. Doms, R.W.; Russ, G.; Yewdell, J.W. Brefeldin A redistributes resident and itinerant Golgi proteins to the endoplasmic reticulum. *J. Cell Biol.* 109:61–72, 1989.

   Graham, T.R.; Scott, P.A.; Emr, S.D. Brefeldin A reversibly blocks early but not late protein transport steps in the yeast secretory pathway. *EMBO J.* 12:869–877, 1993.

   Lippincott-Schwartz, J.; Yuan, L.; Tipper, C.; et al. Brefeldin A's effects on endosomes, lysosomes, and the TGN suggest a general mechanism for regulating organelle structure and membrane traffic. *Cell* 67:601–616, 1991.

   Pelham, H.R. Multiple targets for Brefeldin A. *Cell* 67:449–451, 1991.

   Takizawa, P.A.; Yucel, J.K.; Veit, B.; et al. Complete vesiculation of Golgi membranes and inhibition of protein transport by a novel sea sponge metabolite, ilimaquinone. *Cell* 73:1079–1090, 1993.

5. Balch, W.E.; Dunphy, W.G.; Braell, W.A.; Rothman, J.E. Reconstitution of the transport of proteins between successive compartments of the Golgi measured by the coupled incorporation of *N*-acetylglucosamine. *Cell* 39:405–416, 1984.

   Kornfeld, R.; Kornfeld, S. Assembly of asparagine-linked oligosaccharides. *Annu. Rev. Biochem.* 54:631–634, 1985.

   Schachter, H.; Roseman, S. Mammalian glycosyltransferases: their role in the synthesis and function of complex carbohydrates and glycolipids. In The Biochemistry of Glycoproteins and Proteoglycans (W.J. Lennarz, ed.), Chap. 3. New York: Plenum Press, 1980.

6. Armstrong, J. Latest episodes in the Golgi serial. *Curr. Biol.* 2:335–337, 1992.

Graham, T.R.; Emr, S.D. Compartmental organization of Golgi-specific protein modification and vacuolar protein sorting events defined in a yeast *sec18* (NSF) mutant. *J. Cell Biol.* 114:207–218, 1991.

Machamer, C. Targeting and retention of Golgi membrane proteins. *Curr. Opin. Cell Biol.* 5:606–612, 1993.

Mellman, I.; Simons, K. The Golgi complex: *in vitro veritas? Cell* 68:829–840, 1992.

Rothman, J.E.; Orci, L. Movement of proteins through the Golgi stack: a molecular dissection of vesicular transport. *FASEB J.* 4:1460–1468, 1990.

7. Esko, J.D. Genetic analysis of proteoglycan structure, function and metabolism. *Curr. Opin. Cell Biol.* 3:805–816, 1991.

Jentoft, N. Why are proteins O-glycosylated? *Trends Biochem. Sci.* 15:291–294, 1990.

Ruoslahti, G. Structure and biology of proteoglycans. *Annu. Rev. Cell Biol.* 4:229–255, 1988.

8. Hirschberg, C.B.; Snider, M.D. Topography of glycosylation in the rough endoplasmic reticulum and Golgi apparatus. *Annu. Rev. Biochem.* 56:63–87, 1987.

9. Varki, A. Biological roles of oligosaccharides: all of the theories are correct. *Glycobiology* 3:97–130, 1993.

West, C.M. Current ideas on significance of protein glycosylation. *Mol. Cell Biochem.* 72:3–20, 1986.

10. Bainton, D. The discovery of lysosomes. *J. Cell Biol.* 91:66s–76s, 1981.

Kornfeld, S.; Mellman, I. The biogenesis of lysosomes. *Annu. Rev. Cell Biol.* 5:483–525, 1989.

11. Holzman, E. Lysosomes: A Survey. New York:Springer-Verlag, 1976.

12. Boller, T.; Wiemken, A. Dynamics of vacuolar compartmentation. *Annu. Rev. Plant Physiol.* 37:137–164, 1986.

Klionsky, D.J.; Herman, P.K.; Emr, S.D. The fungal vacuole: composition, function, and biogenesis. *Microbiol. Rev.* 54:266–292, 1990.

Matile, P. Biochemistry and function of vacuoles. *Annu. Rev. Plant Physiol.* 29:193–213, 1978.

Vitale, A.; Chrispeels, M.J. Sorting of proteins to the vacuoles of plant cells. *Bioessays* 14:151–160, 1992.

13. Dunn, W.A., Jr. Studies on the mechanisms of autophagy: formation of the autophagic vacuole. [Two articles.] *J. Cell Biol.* 110:1923–1933, 1935–1945, 1990.

Helenius, A.; Mellman, I.; Wall, D.; Hubbard, A. Endosomes. *Trends Biochem. Sci.* 8:245–250, 1983.

Lawrence, B.P.; Brown, W.J. Autophagic vacuoles rapidly fuse with pre-existing lysosomes in cultured hepatocytes. *J. Cell Science* 102:515–526, 1992.

Noda, T.; Farquhar, M.G. A non-autophagic pathway for diversion of ER secretory proteins to lysosomes. *J. Cell Biol.* 119:85–97, 1992.

Takeshige, K.; Baba, M.; Tsuboi, S.; Noda, T.; Ohsumi, Y. Autophagy in yeast demonstrated with protease-deficient mutants and conditions for its induction. *J. Cell Biol.* 119:301–311, 1992.

14. Dice, J.F. Selective degradation of cytosolic proteins by lysosomes. *Ann. N.Y. Acad. Sci.* 674:658–664, 1992.

Klionsky, D.J.; Cueva, R.; Yaver, D.S. Aminopeptidase I of *Saccharomyces cerevisiae* is localized to the vacu-ole independent of the secretory pathway. *J. Cell Biol.* 119:287–299, 1992.

15. Dahms, N.M.; Lobel, P.; Kornfeld, S. Mannose 6-phosphate receptors and lysosomal enzyme targeting. *J. Biol. Chem.* 264:12115–12118, 1989.

von Figura, K. Molecular recognition and targeting of lysosomal proteins. *Curr. Opin. Cell Biol.* 3:642–646, 1991.

16. Brown, W.J.; Goodhouse, J.; Farquhar, M.G. Mannose-6-phosphate receptors for lysosomal enzymes cycle between the Golgi complex and endosomes. *J. Cell Biol.* 103:1235–1247, 1986.

Duncan, J.R.; Kornfeld, S. Intracellular movement of two mannose–6-phosphate receptors: return to the Golgi apparatus. *J. Cell Biol.* 106:617–628, 1988.

Johnson, K.F.; Kornfeld, S. The cytoplasmic tail of the mannose 6-phosphate/insulin-like growth factor-II receptor has two signals for lysosomal enzyme sorting in the Golgi. *J. Cell Biol.* 119:249–257, 1992.

17. Baranski, T.J.; Cantor, A.B.; Kornfeld, S. Lysosomal enzyme phosphorylation: protein recognition determinants in both lobes of procathepsin D mediate its interaction with UDP-GlcNAc:lysosomal enzyme *N*-acetylglucosamine-1-phosphotransferase. *J. Biol. Chem.* 267:23342–23348, 1992.

Baranski, T.J.; Faust, P.L.; Kornfeld, S. Generation of a lysosomal enzyme targeting signal in the secretory protein pepsinogen. *Cell* 63:281–291, 1990.

von Figura, K. Molecular recognition and targeting of lysosomal proteins. *Curr. Opin. Cell Biol.* 3:642–646, 1991.

18. Amara, J.; Cheng, S.H.; Smith, A. Intracellular protein trafficking defects in human disease. *Trends Cell Biol.* 2:145–149, 1992.

Kornfeld, S. Trafficking of lysosomal enzymes in normal and disease states. *J. Clin. Invest.* 77:1–6, 1986.

Neufeld, E.F.; Lim, T.W.; Shapiro, L.J. Inherited disorders of lysosomal metabolism. *Annu. Rev. Biochem.* 44:357–376, 1975.

19. Watts, C.; Marsh, M. Endocytosis: what goes in and how? *J. Cell Sci.* 103:1–8, 1992.

20. Aggeler, J.; Werb, Z. Initial events during phagocytosis by macrophages viewed from outside and inside the cell: membrane-particle interactions and clathrin. *J. Cell Biol.* 94:613–623, 1982.

Frank, M.M.; Fries, L.F. The role of complement in inflammation and phagocytosis. *Immun. Today* 12:322–326, 1991.

Griffin, F.M., Jr.; Silverstein, S.C. Segmental response of the macrophage plasma membrane to a phagocytic stimulus. *J. Exp. Med.* 139:323–336, 1974.

Mellman, I. Endocytosis and antigen processing. *Semin. Immunol.* 2:229–237, 1990.

21. Brodsky, F.M. Living with clathrin: its role in intracellular membrane traffic. *Science* 242:1396–1402, 1988.

Robinson, M.S. Coated vesicles and protein sorting. *J. Cell Sci.* 87:203–204, 1987.

22. Brodsky, F.M.; Hill, B.L.; Acton, S.L.; et al. Clathrin light chains: arrays of protein motifs that regulate coated-vesicle dynamics. *Trends Biochem. Sci.* 16:208–213, 1991.

Morris, S.; Ahle, S.; Ungewickell, E. Clathrin-coated vesicles. *Curr. Opin. Cell Biol.* 1:684–690, 1989.

Smythe, E.; Warren, G. The mechanism of receptor-mediated endocytosis. *Eur. J. Biochem.* 202:689–699, 1991.

23. Brown, M.S.; Goldstein, J.L. Koch's postulates for cholesterol. *Cell* 71:187–188, 1992.

Brown, M.S.; Goldstein, J.L. A receptor-mediated pathway for cholesterol homeostasis. *Science* 232:34–47, 1986.

24. Helenius, A.; Mellman, I.; Wall, D.; Hubbard, A. Endosomes. *Trends Biochem. Sci.* 8:245–250, 1983.

McCoy, K.L. Contribution of endosomal acidification to antigen processing. *Semin. Immunol.* 2:239–246, 1990.

Mellman, I.; Fuchs, R.; Helenius, A. Acidification of the endocytic and exocytic pathways. *Annu. Rev. Biochem.* 55:663–700, 1986.

Robinson, D.G.; Depta, H. Coated vesicles. *Annu. Rev. Plant Physiol. Plant Mol. Biol.* 39:53–99, 1988.

Schmid, S.L. Toward a biochemical definition of the endosomal compartment. Studies using free flow electrophoresis. *Subcell. Biochem.* 19:1–28, 1993.

25. Bomsel, M.; Mostov, K. Sorting of plasma membrane proteins in epithelial cells. *Curr. Opin. Cell Biol.* 3:647–653, 1991.

Hansen, S.H.; Sandvig, K.; van Deurs, B. Molecules internalized by clathrin-independent endocytosis are delivered to endosomes containing transferrin receptors. *J. Cell Biol.* 123:89–97, 1993.

Mayor, S.; Presley, J.F.; Maxfield, F.R. Sorting of membrane components from endosomes and subsequent recycling to the cell surface occurs by a bulk flow process. *J. Cell Biol.* 121:1257–1269, 1993.

Sorkin, A.; Waters, C.M. Endocytosis of growth factor receptors. *Bioessays* 15:375–382, 1993.

26. Dunn, K.W.; Maxfield, F.R. Delivery of ligands from sorting endosomes to late endosomes occurs by maturation of sorting endosomes. *J. Cell Biol.* 117:301–310, 1992.

Hopkins, C.R. Selective membrane protein trafficking: vectorial flow and filter. *Trends Biochem. Sci.* 17:27–32, 1992.

Nelson, W.J. Cytoskeleton functions in membrane traffic in polarized epithelial cells. *Semin. Cell Biol.* 2:375–385, 1991.

Parton, R.G.; Schrotz, P.; Bucci, C.; Gruenberg, J. Plasticity of early endosomes. *J. Cell Sci.* 103:335–348, 1992.

Rodman, J.S.; Mercer, R.W.; Stahl, P.D. Endocytosis and transcytosis. *Curr. Opin. Cell Biol.* 2:664–672, 1990.

27. Apodaca, G.; Bomsel, M.; Arden, J.; et al. The polymeric immunoglobulin receptor. A model protein to study transcytosis. *J. Clin. Invest.* 87:1877–1882, 1991.

Mellman, I.; Fuchs, R.; Helenius, A. Acidification of the endocytic and exocytic pathways. *Annu. Rev. Biochem.* 55:663–700, 1986.

Mostov, K. The polymeric immunoglobulin receptor. *Semin. Cell Biol.* 2:411–418, 1991.

28. Bomsel, M.; Prydz, K.; Parton, R.G.; Gruenberg, J.; Simons, K. Endocytosis in filter-grown Madin-Darby canine kidney (MDCK) cells. *J. Cell Biol.* 109:3243–58, 1989.

Fujita, M.; Reinhart, F.; Neutra, M. Convergence of apical and basolateral endocytic pathways at apical late endosomes in absorptive cells of suckling rat ileum *in vivo. J. Cell Sci.* 97:385–394, 1990.

Hauri, H.P.; Matter, K. Protein traffic in intestinal epithelial cells. *Semin. Cell Biol.* 2:355–364, 1991.

Matlin, K.S. W(h)ither default? Sorting and polarization in epithelial cells. *Curr. Opin. Cell Biol.* 4:623–628, 1992.

Simionescu, M.; Simionsecu, N. Endothelial transport of macromolecules: transcytosis and endocytosis. A look from cell biology. *Cell Biol. Rev.* 25:1–78, 1991.

29. Hong, W., Tang, B.L. Protein trafficking along the exocytotic pathway. *Bioessays* 15:231–238, 1993.

Kelly, R.B. Secretory granule and synaptic vesicle formation. *Curr. Opin. Cell Biol.* 3:654–660, 1991.

30. Burgess, T.L.; Kelly, R.B. Constitutive and regulated secretion of proteins. *Annu. Rev. Cell Biol.* 3:243–294, 1987.

Fuller, S.D.; Bravo, R.; Simons, K. An enzymatic assay reveals that proteins destined for the apical or basolateral domains of an epithelial cell line share the same late Golgi compartments. *EMBO J.* 4:297–307, 1985.

Stoller, T.J.; Shields, D. The propeptide of preprosomatostatin mediates intracellular transport and secretion of alpha-globin from mammalian cells. *J. Cell Biol.* 108:1647–1655, 1989.

van Meer, G. Transport and sorting of membrane lipids. *Curr. Opin. Cell Biol.* 5:661–673, 1993.

31. Davidson, H.W.; McGowan, C.H.; Balch, W.E. Evidence for the regulation of exocytic transport by protein phosphorylation. *J. Cell Biol.* 116:1343–1355, 1992.

Orci, L.; Ravazzola, M.; Amherdt, M.; et al. The trans-most cisternae of the Golgi complex: a compartment for sorting of secretory and plasma membrane proteins. *Cell* 51:1039–1051, 1987.

Tooze, J. Secretory granule formation. The morphologist's view. *Cell Biophys.* 19:117–130, 1991.

32. Bruzzone, R. The molecular basis of enzyme secretion. *Gastroenterology.* 99:1157–1176, 1990.

Douglass, J.; Civelli, O.; Herbert, E. Polyprotein gene expression: generation of diversity of neuroendocrine peptides. *Annu. Rev. Biochem.* 53:665–715, 1984.

Neurath, H. Proteolytic processing and regulation. *Enzyme* 45:239–243, 1991.

33. Burgoyne, R.D.; Morgan, A. Regulated exocytosis. *Biochem. J.* 293:305–316, 1993.

34. Arvan, P.; Castle, D. Protein sorting and secretion granule formation in regulated secretory cells. *Trends Cell Biol.* 2:327–331, 1992.

Lawson, D.; Fewtrell, C.; Raff, M. Localized mast cell degranulation induced by concanavalin A-sepharose beads: implications for the Ca^{2+} hypothesis of stimulus-secretion coupling. *J. Cell Biol.* 79:394–400, 1978.

Thomas, P.; Almers, W. Exocytosis and its control at the synapse. *Curr. Opin. Neurobiol.* 2:308–311, 1992.

35. Heuser, J. The role of coated vesicles in recycling of synaptic vesicle membrane. *Cell Biol. Int. Rep.* 13:1063–1076, 1989.

Meldolesi, J. Dynamics of cytoplasmic membranes in guinea pig pancreatic acinar cells. I. Synthesis and turnover of membrane proteins. *J. Cell Biol.* 61:1–13, 1974.

References

von Grafenstein, H.; Roberts, C.S.; Baker, P.F. Kinetic analysis of the triggered exocytosis/endocytosis secretory cycle in cultured bovine adrenal medullary cells. *J. Cell Biol.* 103:2343–2352, 1986.

36. Bauerfeind, R.; Huttner, W. Biogenesis of constitutive secretory vesicles, secretory granules, and synaptic vesicles. *Curr. Opin. Cell Biol.* 5:628–635, 1993.

Kelly, R.B. Secretory granule and synaptic vesicle formation. *Curr. Opin. Cell Biol.* 3:654–660, 1991.

Zimmermann, H.; Volknandt, W.; Henkel, A.; et al. The synaptic vesicle membrane: origin, axonal distribution, protein components, exocytosis and recycling. *Cell Biol. Int. Rep.* 13:993–1006, 1989.

37. Hubbard, A.L. Targeting of membrane and secretory proteins to the apical domain in epithelial cells. *Semin. Cell Biol.* 2:365–374, 1991.

Hunziker, W.; Mellman, I. Relationships between sorting in the exocytic and endocytic pathways of MDCK cells. *Semin. Cell Biol.* 2:397–410, 1991.

Lisanti, M.P.; Rodriguez-Boulan, E. Polarized sorting of GPI-linked proteins in epithelia and membrane microdomains. *Cell Biol. Int. Rep.* 15:1023–1049, 1991.

Mostov, K.; Apodaca, G.; Aroeti, B.; Okamoto, C. Plasma membrane protein sorting in polarized epithelial cells. *J. Cell Biol.* 116:577–583, 1992.

Simons, K;. Wandinger-Ness, A. Polarized sorting in epithelia. *Cell* 62:207–210, 1990.

38. Bennett, M.K.; Scheller, R.H. The molecular machinery for secretion is conserved from yeast to neurons. *PNAS* 90:2559–2563, 1993.

Franzusoff, A. Beauty and the yeast: compartmental organization of the secretory pathway. *Semin. Cell Biol.* 3:309–324, 1992.

Kreis, T.E. Regulation of vesicular and tubular membrane traffic of the Golgi complex by coat proteins. *Curr. Opin. Cell Biol.* 4:609–615, 1992.

Nakano, A.; Muramatsu, M. A novel GTP-binding protein, Sar1p, is involved in transport from the endoplasmic reticulum to the Golgi apparatus. *J. Cell Biol.* 109:2677–2691, 1989.

Rothman, J.E. The reconstitution of intracellular protein transport in cell-free systems. *Harvey Lect.* 86:65–85, 1992.

Schekman, R. Genetic and biochemical analysis of vesicular traffic in yeast. *Curr. Opin. Cell Biol.* 4:587–592, 1992.

Wandinger-Ness, A.; Bennett, M.K.; Antony, C.; Simons, K. Distinct transport vesicles mediate the delivery of plasma membrane proteins to the apical and basolateral domains of MDCK cells. *J. Cell Biol.* 111:987–1000, 1990.

Warren, G. Intracellular transport: vesicular consumption. *Nature* 345:382–383, 1990.

Waters, M.G.; Griff, I.C.; Rothman, J.E. Proteins involved in vesicular transport and membrane fusion. *Curr. Opin. Cell Biol.* 3:615–620, 1991.

39. Anderson, R.G.W. Plasmalemma caveolae and GPI-anchored membrane proteins. *Curr. Opin. Cell Biol.* 5:647–652, 1993.

Dupree, P.; Parton, R.G.; Raposo, G.; Kurzchalia, T.V.; Simons, K. Caveolae and sorting in the *trans*-Golgi network of epithelial cells. *EMBO J.* 12:1597–1605, 1993.

Kreis, T.E. Regulation of vesicular and tubular membrane traffic of the Golgi complex by coat proteins. *Curr. Opin. Cell Biol.* 4:609–615, 1992.

Orci, L.; Glick, B.S.; Rothman, J.E. A new type of coated vesicular carrier that appears not to contain clathrin: its possible role in protein transport within the Golgi stack. *Cell* 46:171–184, 1986.

Sargiacomo, M.; Sudol, M.; Tang, Z.-L.; Lisanti, M.P. Signal transducing molecules and glycosyl-phosphatidylinositol-linked proteins form a caveolin-rich insoluble complex in MDCK cells. *J. Cell Biol.* 122:789–808, 1993.

Schmid, S. Biochemical requirements for the formation of clathrin and COP-coated transport vesicles. *Curr. Opin. Cell Biol.* 5:621–627, 1993.

40. Anderson, R. Dissecting clathrin-coated pits. *Trends Cell Biol.* 2:177–179, 1992.

DeLuca-Flaherty, C.; McKay, D.B.; Parham, P.; Hill, B.L. Uncoating protein (hsc70) binds a conformationally labile domain of clathrin light chain LCa to stimulate ATP hydrolysis. *Cell* 62:875–887, 1990.

Gao, B.; Biosca, J.; Craig, E.A.; Greene, L.E.; Eisenberg, E. Uncoating of coated vesicles by yeast hsp70 proteins. *J. Biol. Chem.* 266:19565–19571, 1991.

Nathke, I.S.; Heuser, J.;. Lupas, A.; et al. Folding and trimerization of clathrin subunits at the triskelion hub. *Cell* 68:899–910, 1992.

Pearse, B.M.; Robinson, M.S. Clathrin, adaptors, and sorting. *Annu. Rev. Cell Biol.* 6:151–171, 1990.

41. Chang, M.P.; Mallet, W.G.; Mostov, K.E.; Brodsky, F.M. Adaptor self-aggregation, adaptor-receptor recognition and binding of α-adaptin subunits to the plasma membrane contribute to recruitment of adaptor (AP2) components of clathrin-coated pits. *EMBO J.* 12:2169–2180, 1993.

Pearse, B.M. Characterization of coated-vesicle adaptors: their reassembly with clathrin and with recycling receptors. *Methods Cell Biol.* 31:229–246, 1989.

Sosa, M.A.; Schmidt, B.; von Figura, K.; Hille-Rehfeld, A. *In vitro* binding of plasma membrane-coated vesicle adaptors to the cytoplasmic domain of lysosomal acid phosphatase. *J. Biol. Chem.* 268:12537–12543, 1993.

Trowbridge, I.S. Endocytosis and signals for internalization. *Curr. Opin. Cell Biol.* 3:634–641, 1991.

42. Kreis, T.E. Regulation of vesicular and tubular membrane traffic of the Golgi complex by coat proteins. *Curr. Opin. Cell Biol.* 4:609–615, 1992.

Pepperkok, R.; Scheel, J.; Horstmann, H.; et al. Beta-COP is essential for biosynthetic membrane transport from the endoplasmic reticulum to the Golgi complex *in vivo*. *Cell* 74:71–82, 1993.

Pfeffer, S.R.; Rothman, J.E. Biosynthetic protein transport and sorting by the endoplasmic reticulum and Golgi. *Annu. Rev. Biochem.* 56:829–852, 1987.

43. Balch, W.E. From G Minor to G Major. *Curr. Biol.* 2:157–160, 1992.

Bourne, H.R.; Sanders, D.A.; McCormick, F. The GTPase superfamily: a conserved switch for diverse cell functions. *Nature* 348:125–132, 1990.

Goud, B.; McCaffrey, M. Small GTP-binding proteins and their role in transport. *Curr. Opin. Cell Biol.* 3:626–633, 1991.

Melancon, P. Vesicle traffic: "G Whizz." *Curr. Biol.* 3:230–233, 1993.

44. Magee, T.; Newman, C. The role of lipid anchors for small G proteins in membrane trafficking. *Trends Cell Biol.* 2:318–323, 1992.

Orci, L.; Palmer, D.J.; Amherdt, M.; Rothman, J.E. Coated vesicle assembly in the Golgi requires only coatomer and ARF proteins from the cytosol. *Nature* 364:732–734, 1993.

Serafini, T.; Orci, L.; Amherdt, M.; et al. ADP-ribosylation factor is a subunit of the coat of Golgi-derived COP-coated vesicles: a novel role for a GTP-binding protein. *Cell* 67:239–253, 1991.

45. Bennett, M.K.; Carciaarraras, J.E.; Elferink, L.A.; et al. The syntaxin family of vesicular transport receptors. *Cell* 74:863–873, 1993.

Sollner, T.; Whiteheart, S.W.; Brunner, M.; et al. SNAP receptors implicated in vesicle targeting and fusion. *Nature* 362:318–324, 1993.

46. Goud, B. Small GTP-binding proteins as compartmental markers. *Semin. Cell Biol.* 3:301–307, 1992.

Lombardi, D.; Soldati, T.; Riederer, M.A.; et al. Rab9 functions in transport between late endosomes and the *trans* Golgi network. *EMBO J.* 12:677–682, 1993.

Marsh, M.; Cutler, D. Membrane traffic: taking the Rabs off endocytosis. *Curr. Biol.* 3:30–33, 1993.

Walworth, N.C.; Brennwald, P.; Kabcenell, A.K.; Garrett, M.; Novick, P. Hydrolysis of GTP by Sec4 protein plays an important role in vesiclular transport and is stimulated by a GTPase-activating protein in *Saccharomyces cerevisiae. Mol. Cell Biol.* 12:2017–2028, 1992.

Zerial, M.; Stenmark, H. Rab GTPases in vesicular transport. *Curr. Opin. Cell Biol.* 5:613–620, 1993.

47. Sztul, E.S.; Melancon, P.; Howell, K.E. Targeting and fusion in vesicular transport. *Trends Cell Biol.* 2:381–386, 1992.

Waters, M.G.; Griff, I.C.; Rothman, J.E. Proteins involved in vesicular transport and membrane fusion. *Curr. Opin. Cell Biol.* 3:615–620, 1991.

Wilson, D.W.; Whiteheart, S.W.; Wiedmann, M.; Brunner, M.; Rothman, J.E. A multisubunit particle implicated in membrane fusion. *J. Cell Biol.* 117:531–538, 1992.

Zimmerberg, J.; Vogel, S.S.; Chernomordik, L.V. Mechanisms of membrane fusion. *Annu. Rev. Biophys. Biomol. Struct.* 22:433–466, 1993.

48. Blumenthal, R.; Schoch, C.; Puri, A.; Clague, M.J. A dissection of steps leading to viral envelope protein-mediated membrane fusion. *Ann. N.Y. Acad. Sci.* 635:285–296, 1991.

Doms, R.W. Protein conformational changes in virus-cell fusion. *Methods Enzymol.* 221:61–72, 1993.

Stegmann, T.; Doms, R.W.; Helenius, A. Protein-mediated membrane fusion. *Annu. Rev. Biophys. Biophys. Chem.* 18:187–211, 1989.

White, J.M. Membrane fusion. *Science* 258:917–924, 1992.

Wiley, D.C.; Skehel, J.J. The structure and function of the hemagglutinin membrane glycoprotein of influenza virus. *Annu. Rev. Biochem.* 56:365–394, 1987.

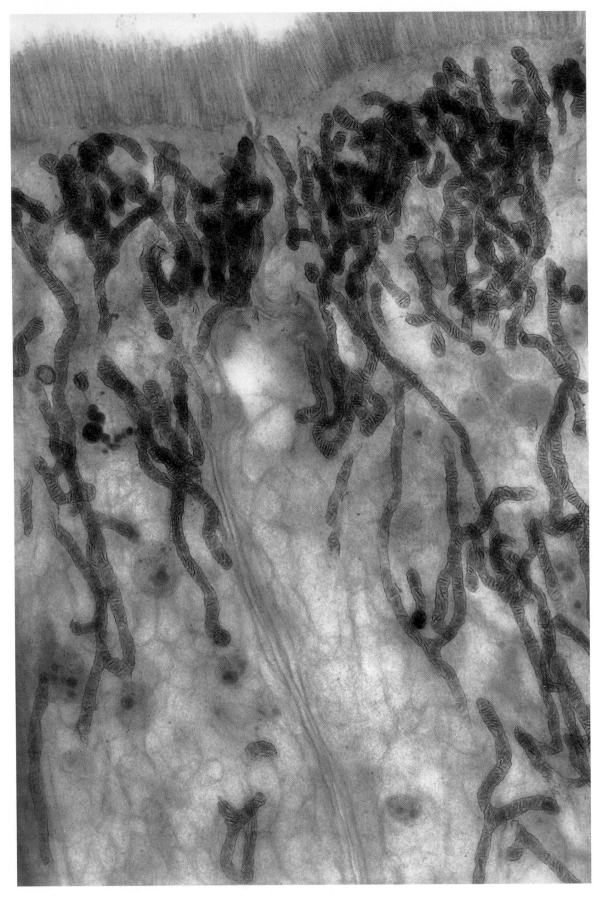

Snakelike mitochondria of a snail epithelial cell, as visualized in a high-voltage electron microscope. (Courtesy of Pierre Favard.)

Energy Conversion: Mitochondria and Chloroplasts

14

- The Mitochondrion

- The Respiratory Chain and ATP Synthase

- Chloroplasts and Photosynthesis

- The Evolution of Electron-Transport Chains

- The Genomes of Mitochondria and Chloroplasts

Mitochondria, which are present in virtually all eucaryotic cells, and **plastids** (most notably **chloroplasts**), which occur only in plants, are membrane-bounded organelles that convert energy to forms that can be used to drive cellular reactions. Consistent with their importance in metabolism, they generally occupy a major fraction of the total cell volume. In electron micrographs the most striking morphological feature of mitochondria and chloroplasts is the large amount of internal membrane they contain. As we shall see, this membrane has a crucial role in the function of these *energy-converting organelles* by providing a framework for electron-transport processes.

Although mitochondria convert energy derived from chemical fuels whereas chloroplasts convert energy derived from sunlight, the two types of organelles are organized similarly; moreover, both produce large amounts of ATP by the same mechanism. This striking conclusion emerged from painstaking studies carried out over the past 30 years.

The common pathway by which mitochondria, chloroplasts, and even bacteria harness energy for biological purposes operates by a process known as *chemiosmotic coupling*. The energy from the oxidation of foodstuffs or from sunlight is used to drive membrane-bound *proton pumps (H^+ pumps)* that transfer H^+ from one side of the membrane to the other. These pumps generate an *electrochemical proton gradient* across the membrane, which is used to drive various energy-requiring reactions when the protons flow back "downhill" through membrane-embedded protein machines (Figure 14–1): Foremost among these machines is the enzyme *ATP synthase*, which uses the energy of the H^+ flow to synthesize ATP from ADP and P_i. Other proteins couple the H^+ flow to the transport of specific metabolites into and out of the organelles. In bacteria the electrochemical proton gradient itself is as important a store of directly usable energy as is the ATP it generates: the gradient not only drives many transport processes, it also drives the rapid rotation of the bacterial flagellum, which allows the bacterium to swim.

How does the energy derived from food or light drive the H^+ pumps that are at the heart of the chemiosmotic mechanism? The answer lies in the reactions in which *electrons* are transferred from one compound to another. In the mitochondrion, for example, electrons released from a carbohydrate food molecule in the course of its degradation to CO_2 are transferred by a circuitous route to O_2, reducing the O_2 to form water. The free energy released as the electrons flow down this path from a high-energy state to a low-energy state is used to drive the

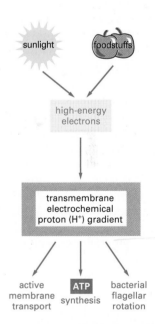

Figure 14–1 Chemiosmotic coupling. Energy from sunlight or the oxidation of foodstuffs is first used to create an *electrochemical proton gradient* across a membrane. This gradient serves as a versatile energy store and is used to drive a variety of reactions in mitochondria, chloroplasts, and bacteria.

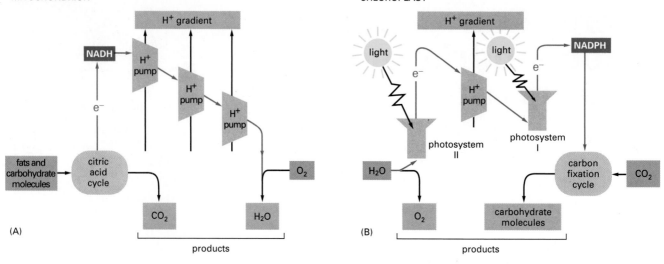

(A)   (B)

Figure 14–2 **The mitochondrion and chloroplast as electrical energy-conversion devices.** Inputs are *light green*, products are *blue*, and the path of electron flow is indicated by *red arrows*. Note that the electron-motive force generated by the two chloroplast photosystems enables the chloroplast (B) to drive electron transfer from H_2O to carbohydrate, which is *opposite* to the direction of electron transfer in the mitochondrion (A).

H⁺ pumps as part of an elaborate electron-transport process that takes place in the major mitochondrial membrane. The mechanism is analogous to an electric cell driving a current through a set of electric motors. But in biological systems electrons are carried between one site and another not by conducting wires but by diffusible molecules that can pick up electrons at one location and deliver them to another. One of the most important of these *electron carriers* is NAD⁺, which can take up two electrons (plus a H⁺) to become NADH, which is a water-soluble small molecule that ferries electrons from the site where food molecules are degraded to the first of a series of electron carriers embedded in the mitochondrial membrane. These carriers diffuse in the plane of the membrane and ferry electrons from one H⁺ pump to another. The third H⁺ pump in the series catalyzes the final transfer of the electrons to O_2 (Figure 14–2A). The entire set of proteins and small molecules involved in this orderly sequence of electron transfers within the membrane is called an **electron-transport chain**.

Although the chloroplast can be described in similar terms, and several of its main components are very similar to those of the mitochondrion, the chloroplast membrane contains some crucial components not found in the mitochondrial membrane. Foremost among these are the *photosystems*, where light energy is captured and harnessed to drive the transfer of electrons, much as man-made photocells in solar panels absorb light energy and use it to drive an electric current. The electron-motive force generated by the chloroplast photosystems drives electron transfer in the direction opposite to that in mitochondria: electrons are taken *from* water to produce O_2, and they are donated (via NADPH) *to* CO_2 to synthesize carbohydrate. Thus the chloroplast generates O_2 and carbohydrate, while the mitochondrion consumes them (Figure 14–2B).

It is generally believed that the energy-converting organelles of eucaryotes evolved from procaryotes that were engulfed by primitive eucaryotic cells and developed a symbiotic relationship with them about 1.5×10^9 years ago. This would explain why mitochondria and chloroplasts contain their own DNA, which codes for some of their proteins. Since their initial uptake by a host cell, these organelles have lost much of their own genomes and have become heavily dependent on proteins that are encoded by genes in the nucleus, synthesized in the cytosol, and then imported into the organelle. Conversely, the host cells have become dependent on these organelles for much of the ATP they need to carry out biosyntheses, ion pumping, and movement—as well as requiring selected biosynthetic reactions that occur inside these organelles.

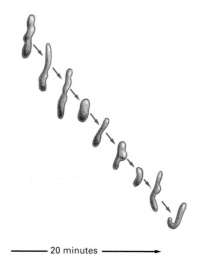

—— 20 minutes ——→

Figure 14–3 **Mitochondrial plasticity.** Rapid changes of shape are observed when a mitochondrion is visualized in a living cell.

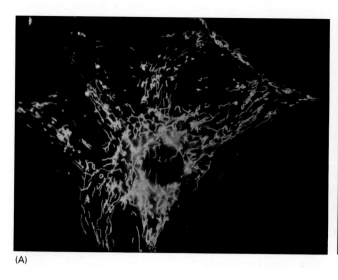
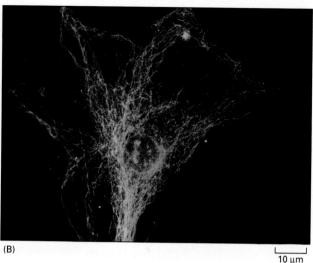

(A)   (B)

10 μm

The Mitochondrion [1]

Mitochondria occupy a substantial portion of the cytoplasmic volume of eucaryotic cells, and they have been essential for the evolution of complex animals. Without mitochondria present-day animal cells would be dependent on anaerobic glycolysis for all of their ATP. However, when glucose is converted to pyruvate by glycolysis, only a very small fraction of the total free energy potentially available from the glucose is released. In mitochondria the metabolism of sugars is completed: the pyruvate is imported into the mitochondrion and oxidized by molecular oxygen (O_2) to CO_2 and H_2O. The energy released is harnessed so efficiently that about 30 molecules of ATP are produced for each molecule of glucose oxidized. By contrast, only 2 molecules of ATP are produced by glycolysis alone.

Mitochondria are usually depicted as stiff, elongated cylinders with a diameter of 0.5 to 1 μm, resembling bacteria. Time-lapse microcinematography of living cells, however, shows that mitochondria are remarkably mobile and plastic organelles, constantly changing their shape (Figure 14–3) and even fusing with one another and then separating again. As they move about in the cytoplasm, they often appear to be associated with microtubules (Figure 14–4), which may determine the unique orientation and distribution of mitochondria in different types of cells. Thus the mitochondria in some cells form long moving filaments or chains, while in others they remain fixed in one position where they provide

Figure 14–4 Relationship between mitochondria and microtubules. (A) Light micrograph of chains of elongated mitochondria in a living mammalian cell in culture. The cell was stained with a vital fluorescent dye (rhodamine 123) that specifically labels mitochondria. (B) Immunofluorescence micrograph of the same cell stained (after fixation) with fluorescent antibodies that bind to microtubules. Note that the mitochondria tend to be aligned along microtubules. (Courtesy of Lan Bo Chen.)

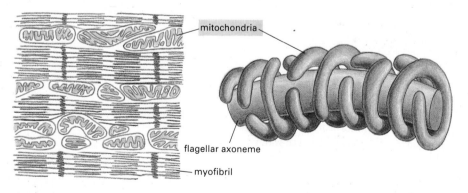

mitochondria

flagellar axoneme

myofibril

CARDIAC MUSCLE   SPERM TAIL

Figure 14–5 Localization of mitochondria near sites of high ATP utilization in cardiac muscle and a sperm tail. During the development of the flagellum of the sperm tail, microtubules wind helically around the axoneme, where they are thought to help localize the mitochondria in the tail; these microtubules then disappear.

where much ATP is needed

ATP directly to a site of unusually high ATP consumption—packed between adjacent myofibrils in a cardiac muscle cell, for example, or wrapped tightly around the flagellum in a sperm (Figure 14–5).

Although mitochondria are large enough to be seen in the light microscope and were first identified in the nineteenth century, real progress in understanding their function depended on procedures developed in 1948 for isolating intact mitochondria. For technical reasons many biochemical studies have been carried out with mitochondria purified from liver; each liver cell contains 1000 to 2000 mitochondria, which in total occupy roughly a fifth of the cell volume.

The Mitochondrion Contains an Outer Membrane and an Inner Membrane That Create Two Internal Compartments [2]

Each mitochondrion is bounded by two highly specialized membranes that play a crucial part in its activities. Together they create two separate mitochondrial compartments: the internal **matrix space** and a much narrower **intermembrane space.** If purified mitochondria are gently disrupted and then fractionated into separate components (Figure 14–6), the biochemical composition of each of the two membranes and of the spaces enclosed by them can be determined. As described in Figure 14–7, each contains a unique collection of proteins.

The **outer membrane** contains many copies of a transport protein called *porin* (see Chapter 10), which forms large aqueous channels through the lipid bilayer. This membrane thus resembles a sieve that is permeable to all molecules of 5000 daltons or less, including small proteins. Such molecules can enter the intermembrane space, but most of them cannot pass the impermeable inner membrane. Thus, while the intermembrane space is chemically equivalent to the cytosol with respect to the small molecules it contains, the matrix space contains a highly selected set of small molecules.

As we explain in detail later, the major working part of the mitochondrion is the matrix space and the **inner membrane** that surrounds it. The inner membrane is highly specialized. It contains a high proportion of the "double" phospholipid *cardiolipin,* which contains four fatty acids and may help make the membrane especially impermeable to ions. It also contains a variety of transport proteins that make it selectively permeable to those small molecules that are metabolized or required by the many mitochondrial enzymes concentrated in the matrix space. The matrix enzymes include those that metabolize pyruvate and fatty acids to produce acetyl CoA and those that oxidize acetyl CoA in the citric acid cycle. The principal end products of this oxidation are CO_2, which is released from the cell as waste, and NADH, which is the main source of electrons for transport along the **respiratory chain**—the name given to the electron-transport chain in mitochondria. The enzymes of the respiratory chain are embedded in the inner mitochondrial membrane, and they are essential to the process of <u>oxidative phosphorylation</u>, which generates most of the animal cell's ATP.

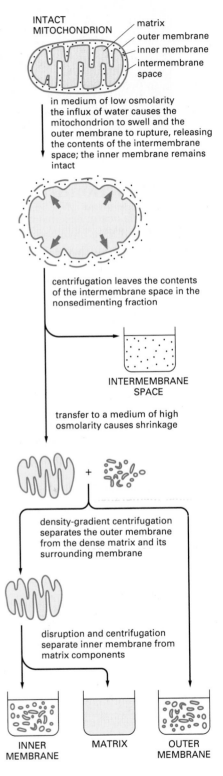

INTACT MITOCHONDRION — matrix, outer membrane, inner membrane, intermembrane space

in medium of low osmolarity the influx of water causes the mitochondrion to swell and the outer membrane to rupture, releasing the contents of the intermembrane space; the inner membrane remains intact

centrifugation leaves the contents of the intermembrane space in the nonsedimenting fraction

INTERMEMBRANE SPACE

transfer to a medium of high osmolarity causes shrinkage

density-gradient centrifugation separates the outer membrane from the dense matrix and its surrounding membrane

disruption and centrifugation separate inner membrane from matrix components

INNER MEMBRANE   MATRIX   OUTER MEMBRANE

Figure 14–6 Fractionation of purified mitochondria into separate components. These techniques have made it possible to study the different proteins in each mitochondrial compartment. The method shown, which allows the processing of large numbers of mitochondria at the same time, takes advantage of the fact that in media of low osmotic strength water flows into mitochondria and greatly expands the matrix space (*yellow*). While the cristae of the inner membrane allow it to unfold to accommodate the expansion, the outer membrane—which has no folds to begin with— breaks, releasing a structure composed of only the inner membrane and the matrix.

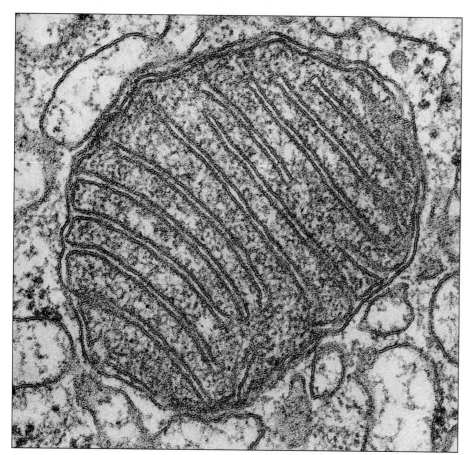

100 nm

Matrix. The matrix contains a highly concentrated mixture of hundreds of enzymes, including those required for the oxidation of pyruvate and fatty acids and for the citric acid cycle. The matrix also contains several identical copies of the mitochondrial DNA genome, special mitochondrial ribosomes, tRNAs, and various enzymes required for expression of the mitochondrial genes.

Inner membrane. The inner membrane is folded into numerous cristae, which greatly increases its total surface area. It contains proteins with three types of functions: (1) those that carry out the oxidation reactions of the respiratory chain, (2) an enzyme complex called *ATP synthase* that makes ATP in the matrix, and (3) specific transport proteins that regulate the passage of metabolites into and out of the matrix. Since an electrochemical gradient that drives the ATP synthase is established across this membrane by the respiratory chain, it is important that the membrane be impermeable to most small ions.

Outer membrane. Because it contains a large channel-forming protein (called porin), the outer membrane is permeable to all molecules of 5000 daltons or less. Other proteins in this membrane include enzymes involved in mitochondrial lipid synthesis and enzymes that convert lipid substrates into forms that are subsequently metabolized in the matrix.

Intermembrane space. This space contains several enzymes that use the ATP passing out of the matrix to phosphorylate other nucleotides.

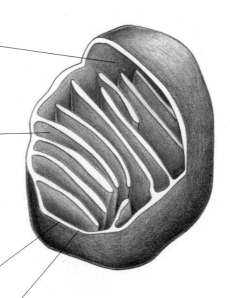

The inner membrane is usually highly convoluted, forming a series of infoldings, known as **cristae,** in the matrix space. These convolutions greatly increase the area of the inner membrane, so that in a liver cell, for example, it constitutes about a third of the total cell membrane. The number of cristae is three times greater in the mitochondrion of a cardiac muscle cell than in the mitochondrion of a liver cell, presumably because of the greater demand for ATP in heart cells. There are also substantial differences in the mitochondrial enzymes of different cell types. In this chapter we shall largely ignore the differences, however, and focus instead on the enzymes and properties that are common to all mitochondria.

Mitochondrial Oxidation Begins When Large Amounts of Acetyl CoA Are Produced in the Matrix Space from Fatty Acids and Pyruvate [3]

Oxidative metabolism in mitochondria is fueled not only by the pyruvate produced from sugars by glycolysis in the cytosol but also by fatty acids. Pyruvate and fatty acids are selectively transported from the cytosol into the mitochondrial matrix, where they are broken down into the two-carbon acetyl group on **acetyl coenzyme A (acetyl CoA)** (Figure 14–8); the acetyl group is then fed into the citric acid cycle for further degradation, and the process ends with the passage of acetyl-derived high-energy electrons along the respiratory chain.

To ensure a continuous supply of fuel for oxidative metabolism, animal cells store fatty acids in the form of fats and glucose in the form of glycogen. Quantitatively, fat is a far more important storage form than glycogen, in part because its oxidation releases more than six times as much energy as the oxidation of an equal mass of glycogen in its hydrated form. An average adult human stores enough glycogen for only about a day of normal activities but enough fat to last for nearly a month. If our main fuel reservoir had to be carried as glycogen instead of fat, body weight would need to be increased by an average of about 60 pounds.

Most of our fat is stored in adipose tissue, from which it is released into the bloodstream for other cells to utilize as needed. The need arises after a period of not eating; even a normal overnight fast results in the mobilization of fat, so that

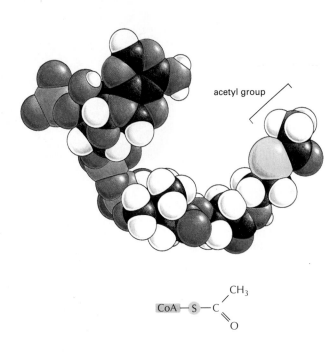

Figure 14–8 Acetyl coenzyme A (acetyl CoA). This central intermediate is produced during the breakdown of foodstuffs in the mitochondrion. A space-filling model is shown above a common abbreviation (see also Figure 2–20). The sulfur atom (S) forms a thioester linkage to acetate. Because this is a "high-energy" linkage, which releases a large amount of free energy when it is hydrolyzed, the acetate group can be readily transferred to other molecules, such as oxaloacetate (see Figure 14–14).

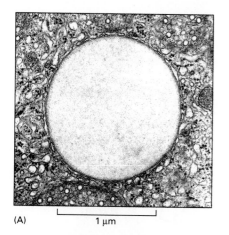

(A)

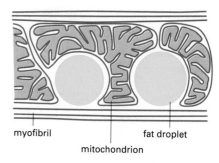

(B)

Figure 14–9 **Fat.** (A) Electron micrograph of a lipid droplet in the cytoplasm; the droplet contains triacylglycerols, the main form of stored fat. (B) The structure of triacylglycerol, with its glycerol portion in *green*. (A, courtesy of Daniel S. Friend.)

in the morning most of the acetyl CoA entering the citric acid cycle is derived from fatty acids rather than from glucose. After a meal, however, most of the acetyl CoA entering the citric acid cycle comes from glucose derived from food, and any excess glucose is used to replenish depleted glycogen stores or to synthesize fats. (While animal cells readily convert sugars to fats, they cannot convert fatty acids to sugars.)

A fat molecule is composed of three molecules of fatty acid held in ester linkage to glycerol. Such **triacylglycerols** (*triglycerides*) have no charge and are virtually insoluble in water, coalescing into droplets in the cytosol (Figure 14–9). A single very large fat droplet accounts for most of the volume of *adipocytes* (fat cells), the large cells specialized for fat storage in adipose tissue. Much smaller fat droplets are common in cells that rely on the breakdown of fatty acids for their energy supply, such as cardiac muscle cells; these droplets are often closely associated with mitochondria (Figure 14–10). In all cells, enzymes in the outer and inner mitochondrial membranes mediate the movement of fatty acids derived from fat molecules into the mitochondrial matrix. In the matrix each fatty acid molecule (as *fatty acyl CoA*) is broken down completely by a cycle of reactions

Figure 14–10 Fat droplets in a cardiac muscle cell. The droplets are surrounded by mitochondria that oxidize the fatty acids derived from their triacylglycerols.

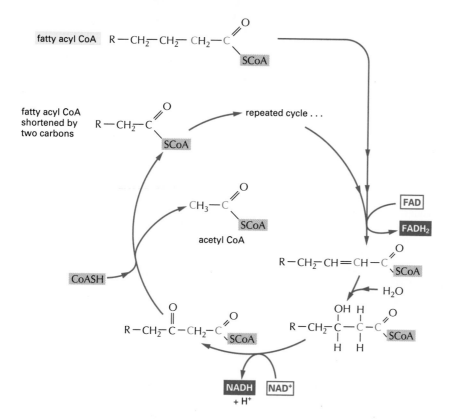

Figure 14–11 The fatty acid oxidation cycle. The cycle is catalyzed by a series of four enzymes in the mitochondrial matrix. Each turn of the cycle shortens the fatty acid chain by two carbons (shown in *red*), as indicated, and generates one molecule of acetyl CoA and one molecule each of NADH and $FADH_2$. The NADH is freely soluble in the matrix. The $FADH_2$, in contrast, remains tightly bound to the enzyme *fatty acyl-CoA dehydrogenase*; its two electrons will be rapidly transferred to the respiratory chain in the mitochondrial inner membrane, regenerating FAD.

The Mitochondrion

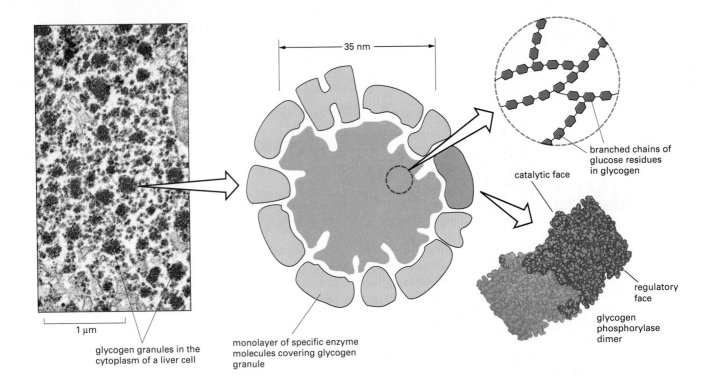

35 nm

branched chains of glucose residues in glycogen

catalytic face

regulatory face

glycogen phosphorylase dimer

1 μm

glycogen granules in the cytoplasm of a liver cell

monolayer of specific enzyme molecules covering glycogen granule

that trims two carbons at a time from its carboxyl end, generating one molecule of acetyl CoA in each turn of the cycle (Figure 14–11). The acetyl CoA produced is fed into the citric acid cycle to be oxidized further.

Glycogen is a large, branched polymer of glucose that is contained in granules in the cytoplasm (Figure 14–12); its synthesis and degradation are highly regulated according to need. When the need arises, cells break down glycogen to release glucose 1-phosphate, which is then subjected to *glycolysis*. The reactions of glycolysis convert the six-carbon glucose molecule (and related sugars) to two three-carbon pyruvate molecules, which still retain most of the energy that can be derived from the complete oxidation of sugars. This energy is harvested only after the pyruvate is transported from the cytosol into the mitochondrial matrix, where it encounters a giant multienzyme complex, the *pyruvate dehydrogenase complex*. This complex—containing multiple copies of three enzymes, five coenzymes, and two regulatory proteins—rapidly converts pyruvate to acetyl CoA, releasing CO_2 as a by-product (Figure 14–13). This acetyl CoA joins the acetyl CoA produced from fatty acids to fuel the citric acid cycle.

Figure 14–12 Electron micrograph and schematic drawing of a glycogen granule. Glycogen is the major storage form of carbohydrate in vertebrate cells. It is a polymer of glucose, and each glycogen granule is a single, highly branched molecule. The synthesis and degradation of glycogen are catalyzed by enzymes bound to the granule surface, including the synthetic enzyme *glycogen synthase* and the degradative enzyme *glycogen phosphorylase*. (Courtesy of Robert Fletterick and Daniel S. Friend.)

The Citric Acid Cycle Oxidizes the Acetyl Group on Acetyl CoA to Generate NADH and FADH$_2$ for the Respiratory Chain [4]

In the nineteenth century biologists noticed that in the absence of air (anaerobic conditions) cells produce lactic acid (or ethanol), while in its presence (aero-

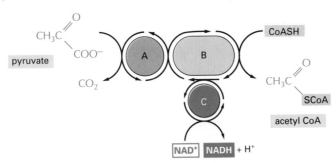

CH₃C
O
COO⁻
pyruvate
A
CO₂
B
CoASH
CH₃C
O
SCoA
acetyl CoA
C
NAD⁺ NADH + H⁺

Figure 14–13 The reactions carried out by the pyruvate dehydrogenase complex. The complex converts pyruvate to acetyl CoA in the mitochondrial matrix; NADH is also produced in this reaction. A, B, and C are the three enzymes *pyruvate decarboxylase, lipoamide reductase-transacetylase,* and *dihydrolipoyl dehydrogenase,* whose activities are coupled as shown. The structure of the complex, which is larger than a ribosome, is shown in Figure 2–41; the complex also contains a protein kinase and a protein phosphatase that regulate its activity, turning it off whenever ATP levels are high.

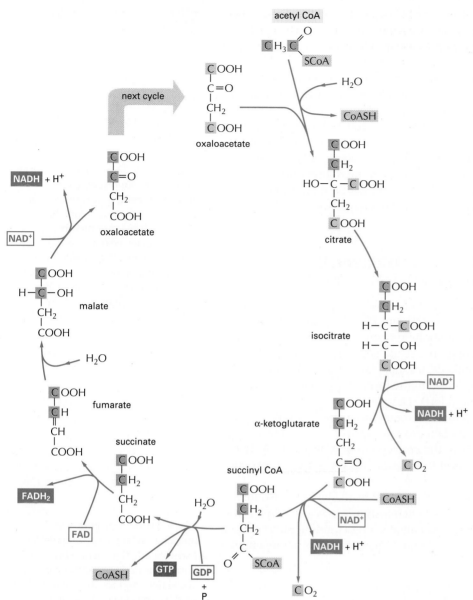

Figure 14–14 The citric acid cycle.
The intermediates are shown as their free acids, although the carboxyl groups are actually ionized. Each of the indicated steps is catalyzed by a different enzyme located in the mitochondrial matrix. The two carbons from acetyl CoA that enter this turn of the cycle (shadowed in *red*) will be converted to CO_2 in subsequent turns of the cycle: it is the two carbons shadowed in *blue* that are converted to CO_2 in this cycle. Three molecules of NADH are formed. The GTP molecule produced can be converted to ATP by the exchange reaction GTP + ADP → GDP + ATP. The molecule of $FADH_2$ formed remains protein-bound as part of the *succinate dehydrogenase complex* in the mitochondrial inner membrane; this complex feeds the electrons acquired by $FADH_2$ directly to ubiquinone (see below).

bic conditions) they use O_2 to produce CO_2 and H_2O. Efforts to define the pathways of aerobic metabolism eventually focused on the oxidation of pyruvate and led in 1937 to the discovery of the **citric acid cycle,** also known as the *tricarboxylic acid cycle* or the *Krebs cycle.* The citric acid cycle accounts for about two-thirds of the total oxidation of carbon compounds in most cells, and its end products are CO_2 and high-energy electrons, which pass via NADH and $FADH_2$ to the respiratory chain. CO_2 is released as a waste product, while the high-energy electrons move along the respiratory chain, eventually combining with O_2 to produce H_2O.

The citric acid cycle begins when the acetyl CoA formed from fatty acids or pyruvate reacts with the four-carbon compound *oxaloacetate* to produce the six-carbon *citric acid* for which the cycle is named. Then, as a result of seven sequential enzyme-mediated reactions, two carbon atoms are removed as CO_2 and oxaloacetate is regenerated. Each such turn of the cycle produces two CO_2 molecules from two carbon atoms that entered in *previous* cycles (Figure 14–14). But the net result, insofar as the acetyl group on acetyl CoA is concerned, is

$$CH_3COOH \text{ (as acetyl CoA)} + 2H_2O + 3NAD^+ + \text{protein-bound FAD} \rightarrow$$
$$2CO_2 + 3H^+ + 3NADH + \text{protein-bound } FADH_2$$

This reaction also produces one molecule of ATP (via GTP) by the direct transfer of a phosphate from a sugar-phosphate intermediate to GDP; a very similar *substrate-level phosphorylation* reaction occurs in glycolysis, as explained in Chapter 2.

The most important contribution of the citric acid cycle to metabolism is the extraction of high-energy electrons during the oxidation of the two acetyl carbon atoms to CO_2. These electrons, which are transiently held by NADH and $FADH_2$, are quickly passed to the respiratory chain in the mitochondrial inner membrane. $FADH_2$, which is part of the succinate dehydrogenase complex in the inner membrane, passes its electrons directly to the respiratory chain. The NADH, in contrast, forms a soluble pool of reducing equivalents in the mitochondrial matrix and passes on its electrons after a random collision with a membrane-bound dehydrogenase enzyme. We now consider how the energy stored in these electrons is used to synthesize ATP.

A Chemiosmotic Process Converts Oxidation Energy into ATP on the Inner Mitochondrial Membrane [5]

Although the citric acid cycle constitutes part of aerobic metabolism, none of the reactions leading to the production of NADH and $FADH_2$ makes direct use of molecular oxygen; only in the final catabolic reactions that take place on the mitochondrial inner membrane is oxygen directly consumed. Nearly all of the energy available from burning carbohydrates, fats, and other foodstuffs in the earlier stages of oxidation is initially saved in the form of high-energy electrons removed from substrates by NAD^+ and FAD. These electrons, carried by NADH and $FADH_2$, are then combined with molecular oxygen by means of the respiratory chain. Because the large amount of energy released is harnessed by the enzymes in the inner membrane to drive the conversion of ADP + P_i to ATP, the term **oxidative phosphorylation** is used to describe this last series of reactions (Figure 14–15).

As previously mentioned, the generation of ATP by oxidative phosphorylation via the respiratory chain depends on a chemiosmotic process. When it was first proposed in 1961, this mechanism explained a long-standing puzzle in cell biology. Nonetheless, the idea was so novel that it was some years before enough

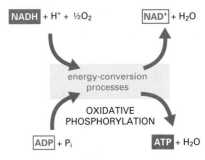

Figure 14–15 The major net energy conversion catalyzed by the mitochondrion. *In this process of oxidative phosphorylation, the mitochondrial inner membrane serves as a device that converts one form of chemical bond energy to another, changing a major part of the energy of NADH (and $FADH_2$) oxidation into phosphate-bond energy in ATP.*

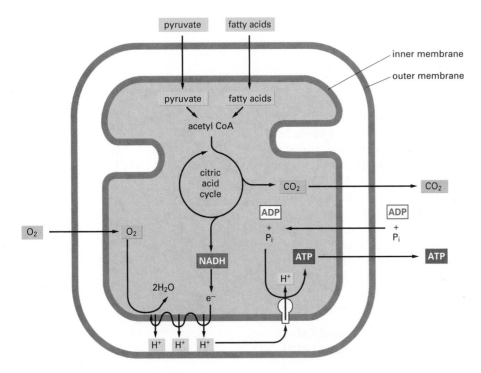

Figure 14–16 A summary of mitochondrial energy metabolism. Pyruvate and fatty acids enter the mitochondrion, are broken down to acetyl CoA, and are then metabolized by the citric acid cycle, which produces NADH (and $FADH_2$, which is not shown). In the process of oxidative phosphorylation, high-energy electrons from NADH (and $FADH_2$) are then passed to oxygen by means of the respiratory chain in the inner membrane, producing ATP by a chemiosmotic mechanism.

NADH generated by glycolysis in the cytosol also passes electrons to the respiratory chain (not shown). Since NADH cannot pass across the mitochondrial inner membrane, the electron transfer from cytosolic NADH must be accomplished indirectly by means of one of several "shuttle" systems that transport another reduced compound into the mitochondrion; after being oxidized, this compound is returned to the cytosol, where it is reduced by NADH again.

Table 14–1 Chemiosmotic Coupling

The chemiosmotic hypothesis, as proposed in the early 1960s, consisted of four independent postulates. In terms of mitochondrial function they were as follows:

1. The mitochondrial respiratory chain in the inner membrane is proton translocating; it pumps H^+ out of the matrix space when electrons are transported along the chain.

2. The mitochondrial ATP synthase also translocates protons across the inner membrane. Being reversible, it can use the energy of ATP hydrolysis to pump H^+ across the membrane, but if a large enough electrochemical proton gradient is present, protons flow in the reverse direction through the complex and drive ATP synthesis.

3. The mitochondrial inner membrane is equipped with a set of carrier proteins that mediate the entry and exit of essential metabolites and selected inorganic ions.

4. The mitochondrial inner membrane is otherwise impermeable to H^+, OH^-, and generally to anions and cations.

supporting evidence accumulated to make it generally accepted. It was originally believed that the energy for ATP synthesis via the respiratory chain was supplied by the same process that operates during substrate-level phosphorylations: that is, the energy of oxidation was thought to generate a high-energy bond between a phosphate group and some intermediate compound, and the conversion of ADP to ATP was thought to be driven by the energy released when this bond was broken. Despite intensive efforts, however, the expected intermediates could not be detected.

A summary of our present view of mitochondrial energy metabolism is presented in Figure 14–16. According to the *chemiosmotic hypothesis,* the high-energy chemical intermediates are replaced by a link between chemical processes ("chemi") and transport processes ("osmotic"—from the Greek *osmos,* push)—hence **chemiosmotic coupling** (Table 14–1). As the high-energy electrons from the hydrogens on NADH and $FADH_2$ are transported down the respiratory chain in the mitochondrial inner membrane, the energy released as they pass from one carrier molecule to the next is used to pump protons (H^+) across the inner membrane from the mitochondrial matrix into the intermembrane space. This creates an *electrochemical proton gradient* across the mitochondrial inner membrane, and the backflow of H^+ down this gradient is in turn used to drive the membrane-bound enzyme *ATP synthase,* which catalyzes the conversion of ADP + P_i to ATP, completing the process of oxidative phosphorylation.

In the remainder of this section we briefly outline the type of reactions that make oxidative phosphorylation possible, saving the details of the respiratory chain for later.

Electrons Are Transferred from NADH to Oxygen Through Three Large Respiratory Enzyme Complexes [6]

Although the mechanism by which energy is harvested by the respiratory chain differs from that in other catabolic reactions, the principle is the same. The energetically favorable reaction $H_2 + \frac{1}{2}O_2 \rightarrow H_2O$ is made to occur in many small steps, so that most of the energy released can be converted into a storage form instead of being lost to the environment as heat. As in the formation of ATP and NADH in glycolysis or the citric acid cycle, this involves employing an indirect pathway for the reaction. The respiratory chain is unique in that the hydrogen atoms are first separated into protons and electrons. The electrons pass through a series of electron carriers in the mitochondrial inner membrane. At several steps along the way, protons and electrons are transiently recombined. But only when the electrons reach the end of this electron-transport chain are the protons re-

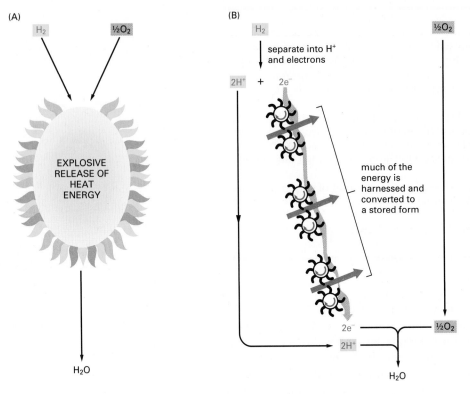

(A)

H₂   ½O₂

EXPLOSIVE RELEASE OF HEAT ENERGY

H₂O

(B)

H₂

separate into H⁺ and electrons

2H⁺   +   2e⁻

much of the energy is harnessed and converted to a stored form

½O₂

2e⁻

½O₂

2H⁺

H₂O

Figure 14–17 Comparison of biological oxidations with combustion. Highly schematic illustration showing how most of the energy that would be released as heat if hydrogen were burned (A) is instead harnessed and stored in a form useful to the cell by means of the electron-transport chain in the mitochondrial inner membrane (B). The rest of the oxidation energy is released as heat by the mitochondrion. In reality, the protons and electrons shown are removed from hydrogen atoms that are covalently linked to NADH or FADH₂ molecules (see Figure 14–18).

Figure 14–18 The biological oxidation of an alcohol to an aldehyde. The components of two complete hydrogen atoms are lost from the alcohol: a hydride ion is transferred to NAD⁺, and a proton escapes to the aqueous solution. Only the nicotinamide ring portion of the NAD⁺ and NADH molecules is shown here (see Figure 2–24). The steps illustrated occur on a protein surface, being catalyzed by specific chemical groups on the enzyme alcohol dehydrogenase (not shown). (Modified with permission from P.F. Cook, N.J. Oppenheimer, and W.W. Cleland, *Biochemistry* 20:1817–1825, 1981. © 1981 American Chemical Society.)

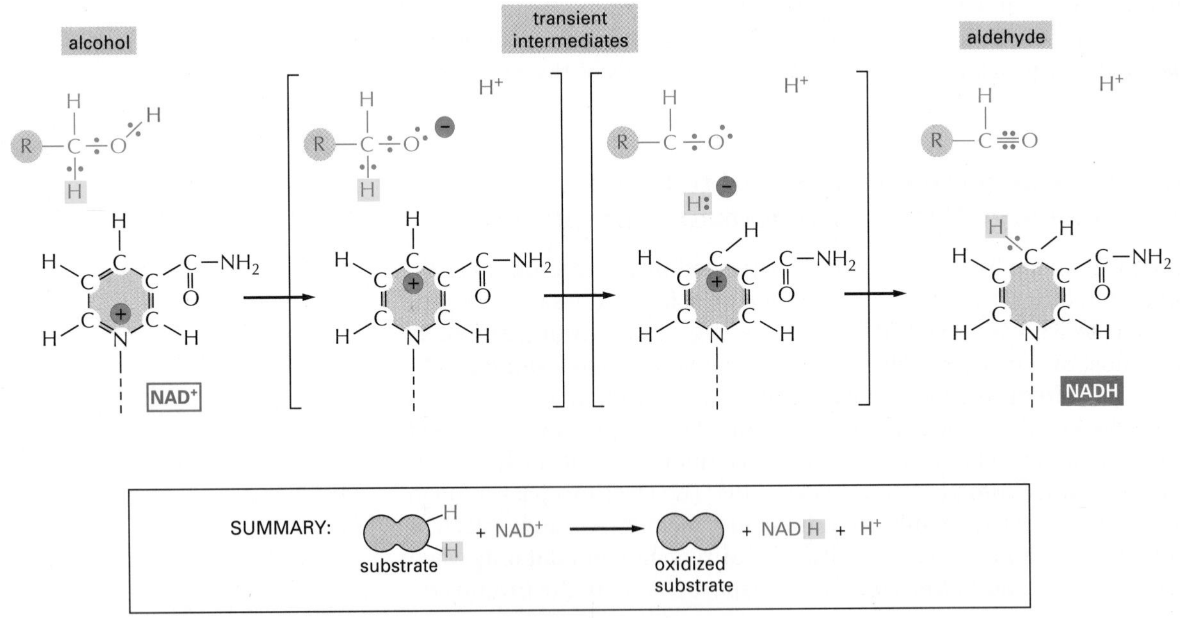

turned permanently, when they are used to neutralize the negative charges created by the final addition of the electrons to the oxygen molecule (Figure 14–17).

We shall outline the oxidation process starting from NADH, the major collector of reactive electrons derived from the oxidation of food molecules. Each hydrogen atom consists of one electron (e^-) and one proton (H⁺). The mechanism by which electrons are acquired by NADH was discussed in Chapter 2 and is shown in greater detail in Figure 14–18. As this example makes clear, each molecule of NADH carries a *hydride ion* (a hydrogen atom plus an extra electron, which we can denote as H:⁻, illustrating each of its two electrons as a dot), rather than a single hydrogen atom. Because protons are freely available in aqueous solutions, however, carrying the hydride ion on NADH is equivalent to carrying two hydrogen atoms, or a hydrogen molecule (H:⁻ + H⁺ → H₂).

alcohol   transient intermediates   aldehyde

SUMMARY: substrate + NAD⁺ → oxidized substrate + NAD H + H⁺

The process of electron transport begins when the hydride ion is removed from NADH to regenerate NAD^+ and is converted into a proton and two electrons $(H:^- \rightarrow H^+ + 2e^-)$. The two electrons are passed to the first of the more than 15 different electron carriers in the respiratory chain. The electrons start with very high energy and gradually lose it as they pass along the chain. For the most part, the electrons pass from one metal atom to another, each metal atom being tightly bound to a protein molecule, which alters the electron affinity of the metal atom. The various types of electron carriers in the respiratory chain will be discussed in detail later. Most important, the many proteins involved are grouped into three large *respiratory enzyme complexes*, each containing transmembrane proteins that hold the complex firmly in the mitochondrial inner membrane. Each complex in the chain has a greater affinity for electrons than its predecessor, and electrons pass sequentially from one complex to another until they are finally transferred to oxygen, which has the greatest affinity of all for electrons.

Energy Released by the Passage of Electrons Along the Respiratory Chain Is Stored as an Electrochemical Proton Gradient Across the Inner Membrane [7]

Oxidative phosphorylation is made possible by the close association of the electron carriers with protein molecules. The proteins guide the electrons along the respiratory chain so that the electrons move sequentially from one enzyme complex to another—with no short circuits. Most important, the transfer of electrons is coupled to oriented H^+ uptake and release and to allosteric changes in selected protein molecules. The net result is that the energetically favorable flow of electrons pumps H^+ across the inner membrane, from the matrix space to the intermembrane space. This movement of H^+ has two major consequences. (1) It generates a pH gradient across the inner mitochondrial membrane, with the pH higher in the matrix than in the cytosol, where the pH is generally close to 7. (Since small molecules equilibrate freely across the outer membrane of the mitochondrion, the pH in the intermembrane space is the same as in the cytosol.) (2) It generates a voltage gradient (membrane potential) across the inner mitochondrial membrane, with the inside negative and the outside positive (as a result of the net outflow of positive ions).

The pH gradient (ΔpH) drives H^+ back into the matrix and OH^- out of the matrix and thus reinforces the effect of the membrane potential (ΔV), which acts to attract any positive ion into the matrix and to push any negative ion out. Together, the ΔpH and the ΔV are said to constitute an **electrochemical proton gradient** (Figure 14–19).

The electrochemical proton gradient exerts a **proton-motive force,** which can be measured in units of millivolts (mV). Since each ΔpH of 1 pH unit has an effect equivalent to a membrane potential of about 60 mV, the total proton-

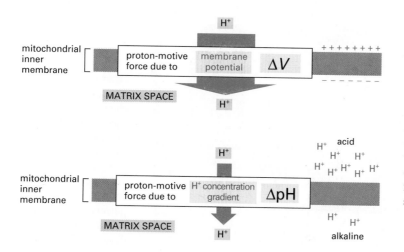

Figure 14–19 **The two components of the electrochemical proton gradient.** The total proton-motive force across the mitochondrial inner membrane consists of a large force due to the membrane potential (traditionally designated Δψ by experts, but designated ΔV in this text) and a smaller force due to the H^+ concentration gradient (ΔpH). Both forces act to drive H^+ into the matrix space.

motive force equals $\Delta V - 60(\Delta pH)$. In a typical cell the proton-motive force across the inner membrane of a respiring mitochondrion is about 200 mV and is made up of a membrane potential of about 140 mV and a pH gradient of about –1 pH unit.

The Energy Stored in the Electrochemical Proton Gradient Is Used to Produce ATP and to Transport Metabolites and Inorganic Ions into the Matrix Space [8]

The mitochondrial inner membrane contains an unusually high proportion of protein, being approximately 70% protein and 30% phospholipid by weight. Many of the proteins belong to the electron-transport chain, which establishes the electrochemical proton gradient across the membrane. Another major component is the enzyme *ATP synthase*, which catalyzes the synthesis of ATP. This is a large protein complex through which H^+ flows down its electrochemical gradient into the matrix. Like a turbine, ATP synthase converts one form of energy to another, synthesizing ATP from ADP and P_i in the mitochondrial matrix in a reaction that is coupled to the inward flow of H^+ (Figure 14–20).

ATP synthesis is not the only process that is driven by the electrochemical H^+ gradient. The enzymes in the mitochondrial matrix, where the citric acid cycle and other metabolic reactions take place, must be supplied with high concentrations of substrates, and ATP synthase must be supplied with ADP and phosphate. Thus many charged substrates must be transported across the inner membrane. This is achieved by various membrane carrier proteins, many of which actively transport specific molecules against their electrochemical gradients, a process that requires an input of energy. As discussed in Chapter 11, the energy often comes from *co-transporting* another molecule down its electrochemical gradient. The transport of ADP into the matrix space, for example, is mediated by an ADP-ATP antiport system: for each ADP molecule that moves in, an ATP molecule moves out in a process driven by the voltage gradient (the net outward movement of one negative charge is favorable). The transport of phosphate into the matrix space is mediated by a carrier protein that couples the inward movement of phosphate to the inward flow of H^+ down its electrochemical gradient so that the phosphate is dragged in. Pyruvate is transported into the matrix in the same way (Figure 14–21). The electrochemical H^+ gradient is also used to import Ca^{2+}, which is thought to be important in regulating the activity of selected mitochondrial enzymes; the import of Ca^{2+} into mitochondria may also be important for removing Ca^{2+} from the cytosol when cytosolic Ca^{2+} levels become dangerously high.

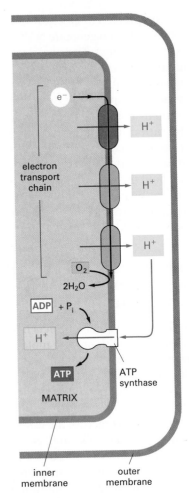

Figure 14–20 The general mechanism of oxidative phosphorylation. As a high-energy electron is passed along the electron-transport chain, some of the energy released is used to drive three respiratory enzyme complexes that pump H^+ out of the matrix space. The resulting electrochemical proton gradient across the inner membrane drives H^+ back through the ATP synthase, a transmembrane protein complex that uses the energy of the H^+ flow to synthesize ATP from ADP and P_i in the matrix.

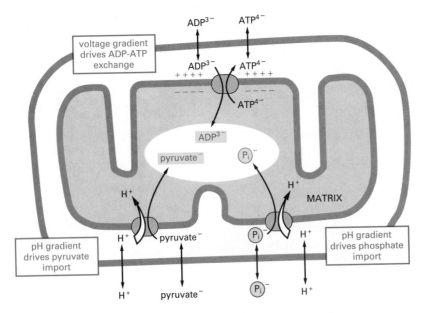

Figure 14–21 Some of the active transport processes driven by the electrochemical proton gradient across the mitochondrial inner membrane. The charge on each of the transported molecules is indicated for comparison with the membrane potential, which is negative inside, as shown. The outer membrane is freely permeable to all of these compounds. Membrane transport mechanisms are discussed in Chapter 11.

The more energy from the electrochemical proton gradient is used to transport molecules and ions into the mitochondrion, the less there is to drive the ATP synthase. If isolated mitochondria are incubated in a high concentration of Ca^{2+}, for example, they cease ATP production completely; all the energy in their electrochemical proton gradient is diverted to pumping Ca^{2+} into the matrix. Similarly, in certain specialized cells the electrochemical proton gradient is short-circuited so that the mitochondria produce heat instead of ATP, as we discuss later. In general, the use of the energy stored in the electrochemical proton gradient is regulated by cells so that it is directed toward those activities that are most needed at the time.

The Rapid Conversion of ADP to ATP in Mitochondria Maintains a High Ratio of ATP to ADP in Cells [8]

Because of the antiporter in the inner membrane that pumps ADP into the matrix space in exchange for ATP (see Figure 14–21), ADP molecules produced by ATP hydrolysis in the cytosol rapidly enter mitochondria for recharging, while the ATP molecules formed in the mitochondrial matrix by oxidative phosphorylation are rapidly pumped into the cytosol where they are needed. A typical ATP molecule in the human body shuttles into and out of a mitochondrion for recharging (as ADP) thousands of times a day, keeping the concentration of ATP in a cell about 10 times higher than that of ADP.

As discussed in Chapter 2, biosynthetic enzymes in cells guide their substrates along specific reaction paths, often driving energetically unfavorable reactions by coupling them to the energetically favorable hydrolysis of ATP (see Figure 2–29). The ATP pool is thereby used to drive cellular processes in much the same way that a battery can be used to drive electric engines: if the activity of the mitochondria is halted, ATP levels fall and the cell's battery runs down, so that, eventually, energetically unfavorable reactions can no longer be driven by ATP hydrolysis.

It might seem that this state would not be reached until the concentration of ATP is zero, but in fact it is reached much sooner than that, at a concentration of ATP that depends on the concentrations of ADP and P_i. To explain why, we must consider some elementary principles of thermodynamics.

The Difference Between $\Delta G°$ and ΔG: A Large Negative Value of ΔG Is Required for ATP Hydrolysis to Be Useful to the Cell [9]

The second law of thermodynamics states that chemical reactions proceed spontaneously in the direction that corresponds to an increase in the disorder of the *universe*. In Chapter 2 we noted that reactions that release energy to their surroundings as heat (such as the hydrolysis of ATP) tend to increase the disorder of the universe by increasing random molecular motions. For this reason reactions go in the direction that converts *free energy* (energy that is available to do work) into heat. Thus the reaction A $\rightleftharpoons$ B will go in the direction A $\rightarrow$ B when the associated free-energy change, ΔG, is negative, just as a tensed spring left to itself will relax and lose its stored energy to its surroundings as heat. For a chemical reaction, however, ΔG depends not only on the energy stored in each individual molecule but also on the concentrations of the molecules in the reaction mixture. This is because, for a reversible reaction A $\rightleftharpoons$ B, a large excess of B over A will tend to drive the reaction in the direction B $\rightarrow$ A; that is, there will be more molecules making the transition B $\rightarrow$ A than there are making the transition A $\rightarrow$ B. Just how much of a concentration difference is needed to compensate for a given amount of heat release is not obvious; it depends on *entropy* changes, which can be calculated as outlined in Panel 14–1, pages 668–669.

The ΔG for a given reaction can be written as the sum of two parts: the first, called the **standard free-energy change, $\Delta G°$**, depends on the intrinsic charac-

THE IMPORTANCE OF FREE ENERGY FOR CELLS

Life is possible because of the complex network of interacting chemical reactions occurring in every cell. In viewing the metabolic pathways that comprise this network, one might suspect that the cell has had the ability to evolve an enzyme to carry out any reaction that it needs. But this is not so. Although enzymes are powerful catalysts, they can speed up only those reactions that are thermodynamically possible; other reactions proceed in cells only because they are *coupled* to very favorable reactions that drive them. The question of whether a reaction can occur spontaneously, or instead needs to be coupled to another reaction, is central to cell biology. The answer is obtained by reference to a quantity called the *free energy:* the total change in free energy during a set of reactions determines whether or not the entire reaction sequence can occur. In this panel we shall explain some of the fundamental ideas—derived from a special branch of chemistry and physics called *thermodynamics*—that are required for understanding what free energy is and why it is so important to cells.

ENERGY RELEASED BY CHANGES IN CHEMICAL BONDING IS CONVERTED INTO HEAT

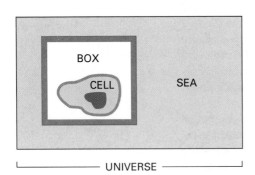

An *enclosed system* is defined as a collection of molecules that does not exchange matter with the rest of the universe (for example, the "cell in a box" shown above). Any such system will contain molecules with a total energy E. This energy will be distributed in a variety of ways: some as the translational energy of the molecules, some as their vibrational and rotational energies, but most as the bonding energies between the individual atoms that make up the molecules. Suppose that a reaction occurs in the system. The first law of thermodynamics places a constraint on what types of reactions are possible: it states that "in any process, the total energy of the universe remains constant." For example, suppose that reaction A → B occurs somewhere in the box and releases a great deal of chemical bond energy. This energy will initially increase the intensity of molecular motions (translational, vibrational, and rotational) in the system, which is equivalent to raising its temperature. However, these increased motions will soon be transferred out of the system by a series of molecular collisions that heat up first the walls of the box and then the outside world (represented by the sea in our example). In the end, the system returns to its initial temperature, by which time all the chemical bond energy released in the box has been converted into heat energy and transferred out of the box to the surroundings. According to the first law, the change in the energy in the box (ΔE_{box}, which we shall denote as ΔE) must be equal and opposite to the amount of heat energy transferred, which we shall designate as h: that is, $\Delta E = -h$. Thus, the energy in the box (E) decreases when heat leaves the system.

E also can change during a reaction due to work being done on the outside world. For example, suppose that there is a small increase in the volume (ΔV) of the box during a reaction. Since the walls of the box must push against the constant pressure (P) in the surroundings in order to expand, this does work on the outside world and requires energy. The energy used is $P(\Delta V)$, which according to the first law must decrease the energy in the box (E) by the same amount. In most reactions chemical bond energy is converted into both work and heat. *Enthalpy* (H) is a composite function that includes both of these ($H = E + PV$). To be rigorous, it is the change in enthalpy (ΔH) in an enclosed system and not the change in energy that is equal to the heat transferred to the outside world during a reaction. Reactions in which H decreases release heat to the surroundings and are said to be "exothermic," while reactions in which H increases absorb heat from the surroundings and are said to be "endothermic." Thus, $-h = \Delta H$. However, the volume change is negligible in most biological reactions, so to a good approximation

$$-h = \Delta H \cong \Delta E$$

THE SECOND LAW OF THERMODYNAMICS

Consider a container in which 1000 coins are all lying heads up. If the container is shaken vigorously, subjecting the coins to the types of random motions that all molecules experience due to their frequent collisions with other molecules, one will end up with about half the coins oriented heads down. The reason for this reorientation is that there is only a single way in which the original orderly state of the coins can be reinstated (every coin must lie heads up), whereas there are many different ways (about 10^{298}) to achieve a disorderly state in which there is an equal mixture of heads and tails; in fact, there are more ways to achieve a 50-50 state than to achieve any other state. Each state has a probability of occurrence that is proportional to the number of ways it can be realized. The second law of thermodynamics states that "systems will change spontaneously from states of lower probability to states of higher probability." Since states of lower probability are more "ordered" than states of high probability, the second law can be restated: "the universe constantly changes so as to become more disordered."

THE ENTROPY, S

The second law (but not the first law) allows one to predict the *direction* of a particular reaction. But to make it useful for this purpose, one needs a convenient measure of the probability or, equivalently, the degree of disorder of a state. The entropy (S) is such a measure. It is a logarithmic function of the probability such that the *change in entropy* (ΔS) that occurs when the reaction A $\rightarrow$ B converts one mole of A into one mole of B is

$$\Delta S = R \ln p_B/p_A$$

where p_A and p_B are the probabilities of the two states A and B, R is the gas constant (2 cal deg^{-1} mole^{-1}), and ΔS is measured in entropy units (eu). In our initial example of 1000 coins, the relative probability of all heads (state A) versus half heads and half tails (state B) is equal to the ratio of the number of different ways that the two results can be obtained. One can calculate that $p_A = 1$ and $p_B = 1000!/(500! \times 500!) = 10^{298}$. Therefore, the entropy change for the reorientation of the coins when their container is vigorously shaken and an equal mixture of heads and tails is obtained is $R \ln (10^{298})$, or about 1370 eu per mole of such containers (6×10^{23} containers). We see that, because ΔS defined above is positive for the transition from state A to state B ($p_B/p_A > 1$), reactions with a large *increase* in S (that is, for which $\Delta S > 0$) are favored and will occur spontaneously.

As discussed in Chapter 2, heat energy causes the random commotion of molecules. Because the transfer of heat from an enclosed system to its surroundings increases the number of different arrangements that the molecules in the outside world can have, it increases their entropy. It can be shown that the release of a fixed quantity of heat energy has a greater disordering effect at low temperature than at high temperature and that the value of ΔS for the surroundings, as defined above (ΔS_{sea}), is precisely equal to the amount of heat transferred to the surroundings from the system (h) divided by the absolute temperature (T):

$$\Delta S_{sea} = h/T$$

THE GIBBS FREE ENERGY, G

When dealing with an enclosed biological system, one would like to have a simple way of predicting whether a given reaction will or will not occur spontaneously in the system. We have seen that the crucial question is whether the entropy change for the universe is positive or negative when that reaction occurs. In our idealized system, the cell in a box, there are two separate components to the entropy change of the universe—the entropy change for the system enclosed in the box and the entropy change for the surrounding "sea"—and both must be added together before any prediction can be made. For example, it is possible for a reaction to absorb heat and thereby decrease the entropy of the sea ($\Delta S_{sea} < 0$) and at the same time to cause such a large degree of disordering inside the box ($\Delta S_{box} > 0$) that the total $\Delta S_{universe} = \Delta S_{sea} + \Delta S_{box}$ is greater than 0. In this case the reaction will occur spontaneously, even though the sea gives up heat to the box during the reaction. An example of such a reaction is the dissolving of sodium chloride in a beaker containing water (the "box"), which is a spontaneous process even through the temperature of the water drops as the salt goes into solution.

Chemists have found it useful to define a number of new "composite functions" that describe *combinations* of physical properties of a system. The properties that can be combined include the temperature (T), pressure (P), volume (V), energy (E), and entropy (S). The enthalpy (H) is one such composite function. But by far the most useful composite function for biologists is the *Gibbs free energy, G*. It serves as an accounting device that allows one to deduce the entropy change of the universe resulting from a chemical reaction in the box, while avoiding any separate consideration of the entropy change in the sea. The definition of G is

$$G = H - TS$$

where, for a box of volume V, H is the enthalpy described above (E + PV), T is the absolute temperature, and S is the entropy. Each of these quantities applies to the inside of the box only. The change in free energy during a reaction in the box (the G of the products minus the G of the starting materials) is denoted as ΔG and, as we shall now demonstrate, it is a direct measure of the amount of disorder that is created in the universe when the reaction occurs.

At constant temperature the change in free energy (ΔG) during a reaction equals $\Delta H - T\Delta S$. Remembering that $\Delta H = -h$, the heat absorbed from the sea, we have

$$-\Delta G = -\Delta H + T\Delta S$$
$$-\Delta G = h + T\Delta S, \text{ so } -\Delta G/T = h/T + \Delta S$$

But h/T is equal to the entropy change of the sea (ΔS_{sea}), and the ΔS in the above equation is ΔS_{box}. Therefore

$$-\Delta G/T = \Delta S_{sea} + \Delta S_{box} = \Delta S_{universe}$$

We conclude that **the free-energy change is a direct measure of the entropy change of the universe**. A reaction will proceed in the direction that causes the change in the free energy (ΔG) to be less than zero, because in this case there will be a positive entropy change in the universe when the reaction occurs.

For a complex set of coupled reactions involving many different molecules, the total free-energy change can be computed simply by adding up the free energies of all the different molecular species after the reaction and comparing this value to the sum of free energies before the reaction; for common substances the required free-energy values can be found from published tables. In this way one can predict the direction of a reaction and thereby readily check the feasibility of any proposed mechanism. Thus, for example, from the observed values for the magnitude of the electrochemical proton gradient across the inner mitochondrial membrane and the ΔG for ATP hydrolysis inside the mitochondrion, one can be certain that ATP synthase requires the passage of more than one proton for each molecule of ATP that it synthesizes.

The value of ΔG for a reaction is a direct measure of how far the reaction is from equilibrium. The large negative value for ATP hydrolysis in a cell merely reflects the fact that cells keep the ATP hydrolysis reaction as much as 10 orders of magnitude away from equilibrium. If a reaction reaches equilibrium, $\Delta G = 0$, the reaction then proceeds at precisely equal rates in the forward and backward direction. For ATP hydrolysis, equilibrium is reached when the vast majority of the ATP has been hydrolyzed, as occurs in a dead cell.

ters of the reacting molecules; the second depends on their concentrations. For the simple reaction A → B,

$$\Delta G = \Delta G^\circ + RT \ln \frac{[B]}{[A]}$$

where [A] and [B] denote the concentrations of A and B, and ln is the natural logarithm. ΔG° therefore equals the value of ΔG when the molar concentrations of A and B are equal (ln 1 = 0). Chemical equilibrium is reached when the concentration effect is just balanced by the effect of ΔG°, so that there is no net change of free energy to drive the reaction in either direction; then $\Delta G = 0$, and so the concentrations of A and B are such that

$$-RT \ln \frac{[B]}{[A]} = \Delta G^\circ$$

which means that there is chemical equilibrium when

$$\frac{[B]}{[A]} = e^{-\Delta G^\circ / RT}$$

When ATP is hydrolyzed to ADP and P_i under the conditions that normally exist in a cell, the free-energy change is roughly –11 to –13 kcal/mole. This extremely favorable ΔG depends on having a high concentration of ATP in the cell compared to the concentration of ADP and P_i. When ATP, ADP, and P_i are all present at the same concentration of 1 mole/liter (so-called "standard conditions"), the ΔG for ATP hydrolysis is the *standard free-energy change* (ΔG°), which is only –7.3 kcal/mole. At much lower concentrations of ATP relative to ADP and P_i, ΔG will become zero. At this point the rate at which ADP and P_i will join to form ATP will be equal to the rate at which ATP hydrolyzes to form ADP and P_i. In other words, when $\Delta G = 0$, the reaction is at *equilibrium* (Figure 14–22).

It is ΔG, not ΔG°, that indicates how far a reaction is from equilibrium and determines if it can be used to drive other reactions. Because the efficient conversion of ADP to ATP in mitochondria maintains such a high concentration of ATP relative to ADP and P_i, the ATP-hydrolysis reaction in cells is kept very far from equilibrium and ΔG is correspondingly very negative. Without this disequilibrium ATP hydrolysis could not be used to direct the reactions of the cell, and many biosynthetic reactions would run backward rather than forward.

Cellular Respiration Is Remarkably Efficient [10]

By means of oxidative phosphorylation, each pair of electrons in NADH is thought to provide energy for the formation of about 2.5 molecules of ATP. The pair of electrons in $FADH_2$, being at a lower energy, generates only about 1.5 ATP molecules. In all, about 10 molecules of ATP can be formed from each molecule of acetyl CoA that enters the citric acid cycle, which means that about 20 ATP molecules are produced from 1 molecule of glucose and 84 ATP molecules from 1 molecule of palmitate, a 16-carbon fatty acid. If one includes the energy-yielding reactions that occur before acetyl CoA is formed, the complete oxidation of 1 molecule of glucose gives a net yield of about 30 ATPs, while the complete oxidation of 1 molecule of palmitate gives a net yield of about 110 ATPs. These numbers are approximate maximal values. As previously discussed, the actual amount of ATP made in the mitochondrion depends on what fraction of the electrochemical gradient energy is used for purposes other than ATP synthesis.

When the free-energy changes for burning fats and carbohydrates directly into CO_2 and H_2O are compared to the total amount of energy generated and stored in the phosphate bonds of ATP during the corresponding biological oxidations, it is seen that the efficiency with which oxidation energy is converted into ATP bond energy is often greater than 40%. This is considerably better than the efficiency of most nonbiological energy-conversion devices. If cells worked with

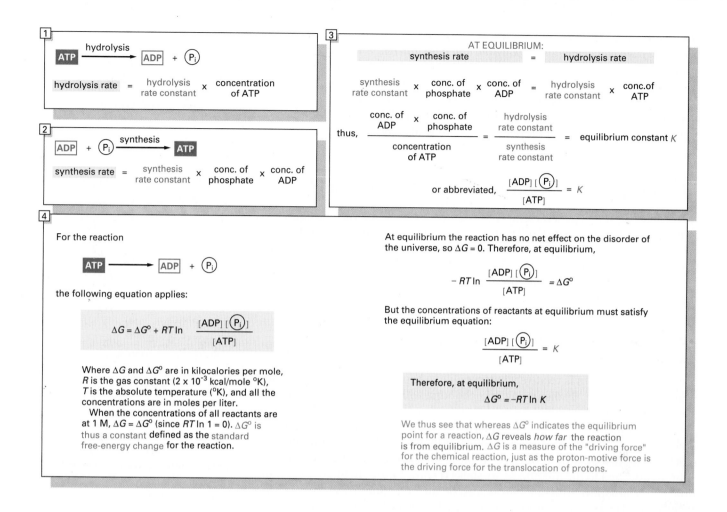

1

$$ATP \xrightarrow{\text{hydrolysis}} ADP + P_i$$

hydrolysis rate = $\begin{array}{c}\text{hydrolysis} \\ \text{rate constant}\end{array}$ × $\begin{array}{c}\text{concentration} \\ \text{of ATP}\end{array}$

2

$$ADP + P_i \xrightarrow{\text{synthesis}} ATP$$

synthesis rate = $\begin{array}{c}\text{synthesis} \\ \text{rate constant}\end{array}$ × $\begin{array}{c}\text{conc. of} \\ \text{phosphate}\end{array}$ × $\begin{array}{c}\text{conc. of} \\ \text{ADP}\end{array}$

3

AT EQUILIBRIUM:

synthesis rate = hydrolysis rate

$\begin{array}{c}\text{synthesis} \\ \text{rate constant}\end{array}$ × $\begin{array}{c}\text{conc. of} \\ \text{phosphate}\end{array}$ × $\begin{array}{c}\text{conc. of} \\ \text{ADP}\end{array}$ = $\begin{array}{c}\text{hydrolysis} \\ \text{rate constant}\end{array}$ × $\begin{array}{c}\text{conc.of} \\ \text{ATP}\end{array}$

thus, $\dfrac{\begin{array}{c}\text{conc. of} \\ \text{ADP}\end{array} \times \begin{array}{c}\text{conc. of} \\ \text{phosphate}\end{array}}{\begin{array}{c}\text{concentration} \\ \text{of ATP}\end{array}}$ = $\dfrac{\begin{array}{c}\text{hydrolysis} \\ \text{rate constant}\end{array}}{\begin{array}{c}\text{synthesis} \\ \text{rate constant}\end{array}}$ = equilibrium constant K

or abbreviated, $\dfrac{[ADP]\,[P_i]}{[ATP]} = K$

4

For the reaction

$$ATP \longrightarrow ADP + P_i$$

the following equation applies:

$$\Delta G = \Delta G^{\circ} + RT \ln \frac{[ADP]\,[P_i]}{[ATP]}$$

Where ΔG and ΔG° are in kilocalories per mole, R is the gas constant (2×10^{-3} kcal/mole °K), T is the absolute temperature (°K), and all the concentrations are in moles per liter.
   When the concentrations of all reactants are at 1 M, $\Delta G = \Delta G^{\circ}$ (since $RT \ln 1 = 0$). ΔG° is thus a constant defined as the standard free-energy change for the reaction.

At equilibrium the reaction has no net effect on the disorder of the universe, so $\Delta G = 0$. Therefore, at equilibrium,

$$-RT \ln \frac{[ADP]\,[P_i]}{[ATP]} = \Delta G^{\circ}$$

But the concentrations of reactants at equilibrium must satisfy the equilibrium equation:

$$\frac{[ADP]\,[P_i]}{[ATP]} = K$$

Therefore, at equilibrium,

$$\Delta G^{\circ} = -RT \ln K$$

We thus see that whereas ΔG° indicates the equilibrium point for a reaction, ΔG reveals *how far* the reaction is from equilibrium. ΔG is a measure of the "driving force" for the chemical reaction, just as the proton-motive force is the driving force for the translocation of protons.

the efficiency of an electric motor or a gasoline engine (10–20%), an organism would have to eat voraciously in order to maintain itself. Moreover, since wasted energy is liberated as heat, large organisms would need more efficient mechanisms for giving up heat to the environment.

   Students sometimes wonder why the chemical interconversions in cells follow such complex pathways. The oxidation of sugars to CO_2 plus H_2O could certainly be accomplished more directly, eliminating the citric acid cycle and many of the steps in the respiratory chain. Although this would have made respiration easier to learn, it would have been a disaster for the cell. Oxidation produces huge amounts of free energy, which can be utilized efficiently only in small bits. The complex oxidative pathways involve many intermediates, each differing only slightly from its predecessor. The energy released is thereby parceled out into small packets that can be efficiently converted to high-energy bonds in useful molecules such as ATP and NADH by means of coupled reactions (see Figure 2–17).

Figure 14–22 The basic relationship between free-energy changes and equilibrium, as illustrated by the ATP hydrolysis reaction. The rate constants in boxes (1) and (2) are determined from experiments in which product accumulation is measured as a function of time. The equilibrium constant shown here, K, is in units of moles per liter. (See Panel 14–1, pp. 668–669, for a discussion of free energy and Figure 3–9 for a definition of the equilibrium constant.)

Summary

The mitochondrion carries out most cellular oxidations and produces the bulk of the animal cell's ATP. The mitochondrial matrix space contains a large variety of enzymes, including those that convert pyruvate and fatty acids to acetyl CoA and those that oxidize this acetyl CoA to CO_2 through the citric acid cycle. Large amounts of NADH (and $FADH_2$) are produced by these oxidation reactions. The energy available from combining oxygen with the reactive electrons carried by NADH and $FADH_2$ is harnessed by an electron-transport chain in the mitochondrial inner membrane called the respiratory chain. The respiratory chain pumps H^+ out of the matrix to

create a transmembrane electrochemical proton (H⁺) gradient, which includes con-
tributions from both a membrane potential and a pH difference. The transmembrane
gradient in turn is used both to synthesize ATP and to drive the active transport of
selected metabolites across the mitochondrial inner membrane. The combination of
these reactions is responsible for an efficient ATP-ADP exchange between the mito-
chondrion and the cytosol that keeps the cell's ATP pool highly charged, so that ATP
can be used to drive many of the cell's energy-requiring reactions.

The Respiratory Chain and ATP Synthase [11]

Having considered in general terms how mitochondria function, let us now look in more detail at the respiratory chain—the electron-transport chain that is so crucial to all oxidative metabolism. Most of the elements of the chain are intrinsic components of the inner mitochondrial membrane, and they provide some of the clearest examples of the many complicated interactions that can occur among the individual proteins located in a biological membrane.

Functional Inside-out Particles Can Be Isolated from Mitochondria [12]

The respiratory chain is relatively inaccessible to experimental manipulation in intact mitochondria. By disrupting mitochondria with ultrasound, however, it is possible to isolate functional *submitochondrial particles,* which consist of bro-ken cristae that have resealed into small closed vesicles about 100 nm in diam-eter (Figure 14–23). When these submitochondrial particles are examined in an electron microscope, their outside surfaces are seen to be studded with tiny spheres attached to the membrane by stalks (Figure 14–24). In intact mitochon-dria these lollipoplike structures are located on the *inner* (matrix) side of the inner membrane. Thus the submitochondrial particles are inside-out vesicles of inner membrane, with what was previously their matrix-facing surface exposed to the surrounding medium. As a result, they can readily be provided with the mem-brane-impermeable metabolites that would normally be present in the matrix space. When NADH, ADP, and inorganic phosphate are added, such preparations transport electrons from NADH to O_2 and couple this oxidation to ATP synthe-sis, catalyzing the reaction ADP + P_i → ATP. This cell-free system provides an assay that makes it possible to purify the many proteins responsible for oxida-tive phosphorylation in a functional form.

ATP Synthase Can Be Purified and Added Back to Membranes [13]

The first experiments to show that the various membrane proteins that catalyze oxidative phosphorylation can be separated without destroying their activity were performed in 1960. The tiny protein spheres studding the surface of submito-chondrial particles were stripped from the particles, leaving the stem of the lol-lipop and the other inner membrane proteins still in the particle membrane. The stripped particles could still oxidize NADH in the presence of oxygen, but they could no longer synthesize ATP. On the other hand, the purified spheres on their own acted as ATPases, hydrolyzing ATP to ADP and P_i. When the purified spheres (referred to as *F_1ATPase*) were added back to stripped submitochondrial particles, however, the reconstituted particles once again made ATP from ADP and P_i.

Subsequent work showed that the F_1ATPase is part of a larger transmem-brane complex (~500,000 daltons) containing at least nine different polypeptide chains (Figure 14–25), which is now known as **ATP synthase** (also called *F_0F_1ATPase*). ATP synthase constitutes about 15% of the total inner membrane protein, and very similar enzyme complexes are present in both chloroplast and

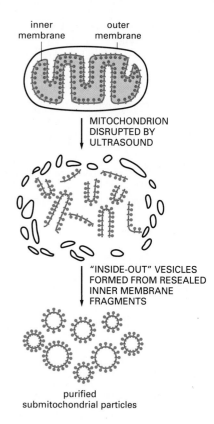

inner membrane   outer membrane

MITOCHONDRION DISRUPTED BY ULTRASOUND

"INSIDE-OUT" VESICLES FORMED FROM RESEALED INNER MEMBRANE FRAGMENTS

purified submitochondrial particles

Figure 14–23 Preparation of submitochondrial particles from purified mitochondria. The particles are pieces of broken-off cristae that form closed vesicles.

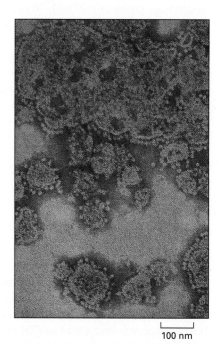

100 nm

Figure 14–24 Electron micrograph of submitochondrial particles. This preparation has been negatively stained. (Courtesy of Efraim Racker.)

bacterial membranes. The transmembrane portion of the protein complex acts as a H^+ carrier, and the F_1ATPase portion (the lollipop head) normally synthesizes ATP when protons pass through it down their electrochemical gradient. When separated from the H^+ carrier, however, the F_1ATPase goes into reverse and catalyzes only ATP hydrolysis.

One of the most convincing demonstrations of the function of ATP synthase came from an experiment performed in 1974. By that time methods had been developed for transferring detergent-solubilized integral membrane proteins into lipid vesicles (liposomes) formed from purified phospholipids. It thus became possible to form a hybrid membrane that contained both a complete purified mitochondrial ATP synthase and bacteriorhodopsin (a bacterial light-driven H^+ pump, discussed in Chapter 10) but none of the proteins of the mitochondrial respiratory chain. When these vesicles were exposed to light, the H^+ pumped into the vesicle lumen by the bacteriorhodopsin flowed back out through the ATP synthase, causing ATP to be made in the medium outside (Figure 14–26). Because a direct interaction between a bacterial H^+ pump and a mammalian ATP synthase seems highly unlikely, this experiment strongly suggests that in mitochondria the proton translocation driven by electron transport and the ATP synthesis are separate events.

ATP Synthase Can Function in Reverse to Hydrolyze ATP and Pump H^+ [13]

ATP synthase can either use the energy of ATP hydrolysis to pump H^+ across the inner mitochondrial membrane or it can harness the flow of H^+ down an electrochemical proton gradient to make ATP (Figure 14–27). It thus acts as a *reversible coupling device,* interconverting electrochemical-proton-gradient and chemical-bond energies. Its direction of action depends on the balance between the steepness of the electrochemical proton gradient and the local ΔG for ATP hydrolysis.

The enzyme complex is called ATP synthase because it is normally driven by the large electrochemical proton gradient maintained by the respiratory chain (see Figure 14–20) to make most of the cell's ATP. The exact number of protons needed to make each ATP molecule is not known with certainty. To facilitate the calculations to be described below, however, we shall assume that one molecule of ATP is made by the ATP synthase for every three protons driven through it.

Figure 14–25 **ATP synthase.** As indicated, the F_1ATPase portion is formed from multiple subunits (*Greek letters*), as is the transmembrane H^+ carrier.

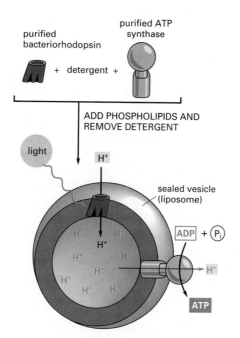

Figure 14–26 **An experiment demonstrating that the ATP synthase is driven by proton flow.** By combining a light-driven bacterial proton pump (bacteriorhodopsin), an ATP synthase purified from ox heart mitochondria, and phospholipids, vesicles were produced that synthesized ATP in response to light.

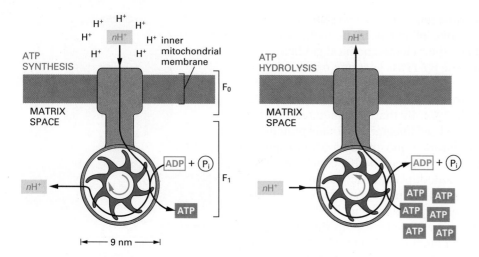

Figure 14–27 ATP synthase is a reversible coupling device that interconverts the energies of the electrochemical proton gradient and chemical bonds. The ATP synthase can either synthesize ATP by harnessing the proton-motive force (*left*) or pump protons against their electrochemical gradient by hydrolyzing ATP (*right*). As explained in the text, the direction of operation at any given instant depends on the net free-energy change for the coupled processes of H$^+$ translocation across the membrane and the synthesis of ATP from ADP and P$_i$.

We have previously shown how the free-energy change (ΔG) for ATP hydrolysis depends on the concentrations of the three reactants ATP, ADP, and P$_i$ (Figure 14–22); the ΔG for ATP synthesis is the negative of this value. The ΔG for proton translocation across the membrane is proportional to the proton-motive force. The conversion factor between them is the faraday. Thus, $\Delta G_{H^+} = -0.023$ (proton-motive force), where ΔG_{H^+} is in kilocalories per mole (kcal/mole) and the proton-motive force is in millivolts (mV). For an electrochemical H$^+$ gradient of 200 mV, $\Delta G_{H^+} = -4.6$ kcal/mole.

Whether the ATP synthase works in its ATP-synthesizing or its ATP-hydrolyzing direction at any instant depends on the exact balance between the favorable free-energy change for moving the three protons across the membrane into the matrix space (ΔG_{3H^+}, which is less than zero) and the unfavorable free-energy change for ATP *synthesis* in the matrix ($\Delta G_{\text{ATP synthesis}}$, which is greater than zero). As previously discussed, the value of $\Delta G_{\text{ATP synthesis}}$ depends on the exact concentrations of the three reactants ATP, ADP, and P$_i$ in the mitochondrial matrix space (see Figure 14–22). The value of ΔG_{3H^+}, on the other hand, is proportional to the value of the proton-motive force across the inner mitochondrial membrane. The following example will help to explain how the balance between these two free-energy changes affects the ATP synthase.

As explained in the legend to Figure 14–27, a single H$^+$ moving into the matrix down an electrochemical gradient of 200 mV liberates 4.6 kcal/mole of free energy, while the movement of three protons liberates three times this much free energy ($\Delta G_{3H^+} = -13.8$ kcal/mole). Thus, if the proton-motive force remains constant at 200 mV, the ATP synthase will synthesize ATP until a ratio of ATP to ADP and P$_i$ is reached where $\Delta G_{\text{ATP synthesis}}$ is just equal to +13.8 kcal/mole (here $\Delta G_{\text{ATP synthesis}} + \Delta G_{3H^+} = 0$). At this point there will be no further net ATP synthesis or hydrolysis by the ATP synthase.

Suppose that a large amount of ATP is suddenly hydrolyzed by energy-requiring reactions in the cytosol—causing the ATP:ADP ratio in the matrix to fall. Now the value of $\Delta G_{\text{ATP synthesis}}$ will decrease (see Figure 14–22), and ATP synthase will begin to synthesize ATP again to restore the original ATP:ADP ratio. Alternatively, if the proton-motive force drops suddenly and is then maintained at a constant 160 mV, ΔG_{3H^+} will change to −11.0 kcal/mole. As a result, ATP synthase will start hydrolyzing some of the ATP in the matrix until a new balance of ATP to ADP and P$_i$ is reached (where $\Delta G_{\text{ATP synthesis}} = +11.0$ kcal/mole) and so on.

In many bacteria ATP synthase is routinely reversed in a transition between aerobic and anaerobic metabolism, as we shall see later. The reversibility of the ATP synthase is a property shared by other membrane proteins that couple ion movement to ATP synthesis or hydrolysis. Both the Na$^+$-K$^+$ pump and the Ca^{2+} pump described in Chapter 11, for example, hydrolyze ATP and use the energy released to pump specific ions across a membrane. If either of these pumps is exposed to an abnormally steep gradient of the ions it transports, however, it will act in reverse—synthesizing ATP from ADP and P$_i$ instead of hydrolyzing it. Thus, like ATP synthase, such pumps are able to convert the electrochemical energy stored in a transmembrane ion gradient directly into phosphate bond energy in ATP.

The Respiratory Chain Pumps H⁺ Across the Inner Mitochondrial Membrane [14]

The respiratory chain embedded in the inner mitochondrial membrane normally generates the electrochemical proton gradient that drives ATP synthesis. The ability of the respiratory chain to translocate H⁺ outward from the matrix space can be demonstrated experimentally under special conditions. A suspension of isolated mitochondria, for example, can be provided with a suitable substrate for oxidation, and the H⁺ flow through ATP synthase can be blocked. In the absence of air the injection of a small amount of oxygen into such a preparation causes a brief burst of respiration, which lasts for 1 to 2 seconds before all the oxygen is consumed. During this respiratory burst a sudden acidification of the medium resulting from the extrusion of H⁺ from the matrix space can be measured with a sensitive pH electrode.

In a similar experiment carried out with a suspension of submitochondrial particles, the medium becomes more basic when oxygen is injected, since H⁺ is pumped *into* each vesicle because of its inside-out orientation.

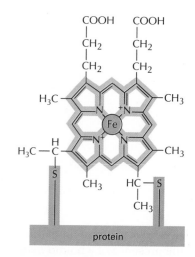

Figure 14–28 **The structure of the heme group attached covalently to cytochrome c.** The porphyrin ring is shown in *blue*. There are five different cytochromes in the respiratory chain. Because the hemes in different cytochromes have slightly different structures and are held by their respective proteins in different ways, each of the cytochromes has a different affinity for an electron.

Spectroscopic Methods Have Been Used to Identify Many Electron Carriers in the Respiratory Chain [15]

Many of the electron carriers in the respiratory chain absorb visible light and change color when they are oxidized or reduced. In general, each has an absorption spectrum and reactivity that is distinct enough to allow its behavior to be traced spectroscopically even in crude mixtures. It was therefore possible to purify these components long before their exact functions were known. Thus the *cytochromes* were discovered in 1925 as compounds that undergo rapid oxidation and reduction in living organisms as disparate as bacteria, yeasts, and insects. By observing cells and tissues with a spectroscope, three types of cytochromes were identified by their distinctive absorption spectra and designated cytochromes a, b, and c. This nomenclature has survived even though cells are now known to contain several cytochromes of each type and the classification into types is not functionally important.

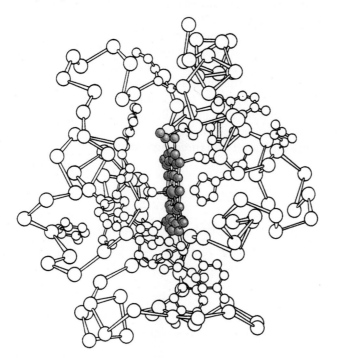

Figure 14–29 **The three-dimensional structure of cytochrome c, an electron carrier in the electron-transport chain.** This small protein contains just over 100 amino acids and is held loosely on the membrane by ionic interactions (see Figure 14–33). The iron atom (*orange*) on the bound heme (*blue*) can carry a single electron (see also Figure 3–59).

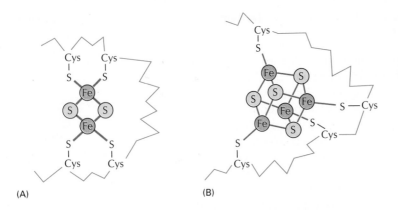

Figure 14–30 **The structures of two types of iron-sulfur centers.** (A) A center of the 2Fe2S type. (B) A center of the 4Fe4S type. Although they contain multiple iron atoms, each iron-sulfur center can carry only one electron at a time. There are more than six different iron-sulfur centers in the respiratory chain.

The **cytochromes** constitute a family of colored proteins that are related by the presence of a bound *heme group* whose iron atom changes from the ferric (Fe III) to the ferrous (Fe II) state whenever it accepts an electron. The heme group consists of a *porphyrin* ring with a tightly bound iron atom held by four nitrogen atoms at the corners of a square (Figure 14–28). A related porphyrin ring is responsible for the red color of blood and the green color of leaves, being bound to iron in hemoglobin and to magnesium in chlorophyll. The best understood of the many proteins in the respiratory chain is *cytochrome c,* whose three-dimensional structure has been determined by x-ray crystallography (Figure 14–29).

Iron-sulfur proteins are a second major family of electron carriers. In these proteins either two or four iron atoms are bound to an equal number of sulfur atoms and to cysteine side chains, forming an **iron-sulfur center** on the protein (Figure 14–30). There are more iron-sulfur centers than cytochromes in the respiratory chain, but their spectroscopic detection requires electron spin resonance (ESR) spectroscopy, and they are less well characterized.

The simplest of the electron carriers is a small hydrophobic molecule dissolved in the lipid bilayer known as *ubiquinone,* or *coenzyme Q.* A **quinone (Q)** can pick up or donate either one or two electrons, and it temporarily picks up a proton from the medium along with each electron that it carries (Figure 14–31).

In addition to six different hemes linked to cytochromes, more than six iron-sulfur centers, and ubiquinone, there are also two copper atoms and a flavin serving as electron carriers tightly bound to respiratory-chain proteins in the pathway from NADH to oxygen. The pathway involves about 40 different proteins in all. The order of the individual electron carriers in the chain has been determined by sophisticated spectroscopic measurements (Figure 14–32), and many of the proteins were initially isolated and characterized as individual polypeptides. A major advance in understanding the respiratory chain, however, was the later realization that most of the proteins are organized into three large enzyme complexes.

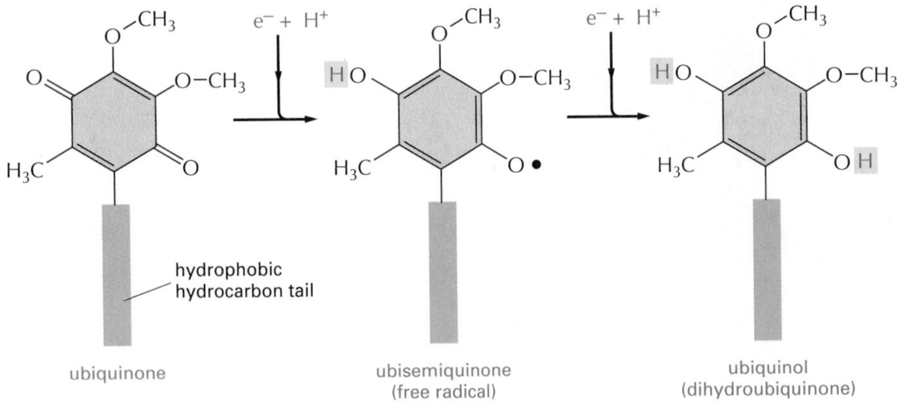

Figure 14–31 **Quinones.** Each of these electron carriers in the respiratory chain picks up one H⁺ from the aqueous environment for every electron it accepts, and it can carry either one or two electrons as part of a hydrogen atom (*yellow*). When it donates its electrons to the next carrier in the chain, these protons are released. In mitochondria the quinone is ubiquinone (coenzyme Q), shown here; the long hydrophobic tail, which confines ubiquinone to the membrane, consists of 6 to 10 five-carbon isoprene units, depending on the organism. The corresponding electron carrier in plants is plastoquinone, which is almost identical. For simplicity, both ubiquinone and plastoquinone will normally be referred to as *quinone* and abbreviated as Q.

$e^- + H^+$ $e^- + H^+$

ubiquinone

hydrophobic
hydrocarbon tail

ubisemiquinone
(free radical)

ubiquinol
(dihydroubiquinone)

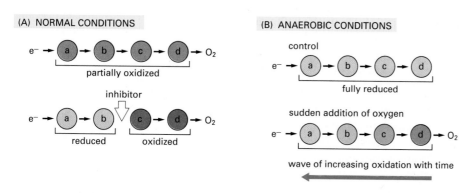

(A) NORMAL CONDITIONS

e^- → a → b → c → d → O_2
partially oxidized

inhibitor

e^- → a → b ⇨ c → d → O_2
reduced   oxidized

(B) ANAEROBIC CONDITIONS

control

e^- → a → b → c → d
fully reduced

sudden addition of oxygen

e^- → a → b → c → d → O_2

wave of increasing oxidation with time

Figure 14–32 **The general methods used to determine the path of electrons along an electron-transport chain.** The extent of oxidation of electron carriers a, b, c, and d is continuously monitored by following their distinct spectra, which differ in their oxidized and reduced states. In this schematic an increased degree of oxidation is indicated by a *darker red.* (A) Under normal conditions, where oxygen is abundant, all carriers are in a partially oxidized state. Addition of a specific inhibitor causes the downstream carriers to become more oxidized (*red*) and the upstream carriers to become more reduced. (B) In the absence of oxygen all carriers are in their fully reduced state (*gray*). The sudden addition of oxygen converts each carrier to its partially oxidized form with a delay that is greatest for the most upstream carriers.

The Respiratory Chain Contains Three Large Enzyme Complexes Embedded in the Inner Membrane [16]

Membrane proteins are difficult to purify as intact complexes because they are insoluble in most aqueous solutions, and some of the detergents required to solubilize them can destroy normal protein-protein interactions. In the early 1960s, however, it was found that relatively mild ionic detergents, such as deoxycholate, will solubilize selected components of the mitochondrial inner membrane in their native form. This permitted the identification and purification of the three major membrane-bound **respiratory enzyme complexes** in the pathway from NADH to oxygen (Figure 14–33). As we shall see, each of these complexes acts as an electron-transport-driven H+ pump; they were initially characterized, however, in terms of the electron carriers that they interact with and contain.

1. The **NADH dehydrogenase complex** is the largest of the respiratory enzyme complexes, with a mass of about 800,000 daltons and more than 22 polypeptide chains. It accepts electrons from NADH and passes them through a flavin and at least five iron-sulfur centers to ubiquinone, which transfers its electrons to a second respiratory enzyme complex, the *b-c₁ complex.*

2. The **cytochrome b-c₁ complex** contains at least 8 different polypeptide chains and is thought to function as a dimer of about 500,000 daltons. Each monomer contains three hemes bound to cytochromes and an iron-sulfur protein. The complex accepts electrons from ubiquinone and passes them on to cytochrome c, which carries its electron to the *cytochrome oxidase complex.*

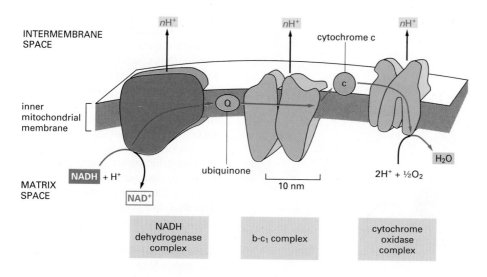

INTERMEMBRANE SPACE

nH^+ nH^+ nH^+

cytochrome c

c

inner mitochondrial membrane

Q

H_2O

NADH + H⁺

NAD⁺

ubiquinone

MATRIX SPACE

10 nm

$2H^+ + \frac{1}{2}O_2$

NADH dehydrogenase complex   b-c₁ complex   cytochrome oxidase complex

Figure 14–33 **The path of electrons through the three respiratory enzyme complexes.** The size and shape of each complex is shown, as determined from images of two-dimensional crystals (crystalline sheets) viewed in the electron microscope at various tilt angles. During the transfer of two electrons from NADH to oxygen (*red lines*) ubiquinone and cytochrome c serve as carriers between the complexes.

3. The **cytochrome oxidase complex** (cytochrome aa₃) is the best characterized of the three complexes. It is isolated as a dimer of about 300,000 daltons; each monomer contains at least 9 different polypeptide chains, including two cytochromes and two copper atoms. The complex accepts electrons from cytochrome c and passes them to oxygen.

The cytochromes, iron-sulfur centers, and copper atoms can carry only one electron at a time. Yet each NADH donates two electrons, and each O_2 molecule must receive four electrons to produce water. There are several electron-collecting and electron-dispersing points along the electron-transport chain where these changes in electron number are accommodated.

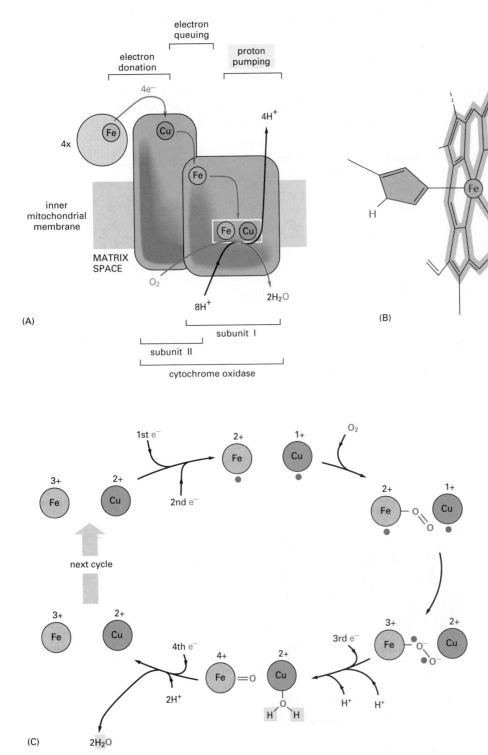

Figure 14–34 The reaction of O_2 with electrons in cytochrome oxidase. (A) The arrangement of electron carriers in cytochrome oxidase. Subunit I, which has 12 membrane-spanning α helices, contains two heme-linked iron atoms; one of these serves as an electron queuing point that feeds electrons into the bimetallic center (*boxed*), which is formed by the other iron and a closely opposed copper atom. Note that four protons are pumped out of the matrix for each O_2 molecule that reacts and that this requires a total of four electrons. (B) An enlarged view of the bimetallic iron-copper center with O_2 bound. (C) An outline of the pathway used for oxygen reduction at the bimetallic center, giving some idea of the complexity of the reactions involved. Electrons are shown as *red dots* until they become incorporated into hydrogen atoms (*yellow*). (Based on G.T. Babcock and M. Wikström, *Nature* 356:301–309, 1992. © 1992 Macmillan Magazines Ltd.)

An Iron-Copper Center in Cytochrome Oxidase Catalyzes Efficient O_2 Reduction [17]

Because oxygen has a high affinity for electrons, it releases a large amount of free energy when it is reduced to form water. Thus the evolution of cellular respiration, in which O_2 is converted to water, enabled organisms to harness much more energy than can be derived from anaerobic metabolism. This is presumably why all higher organisms respire. For biological systems to use O_2 in this way, however, requires a very sophisticated chemistry. We can tolerate O_2 in the air we breathe because it has trouble picking up its first electron, which allows its initial reaction in cells to be controlled closely by enzymatic catalysis. But once a molecule of O_2 has picked up one electron to form a superoxide radical (O_2^-), it becomes dangerously reactive and will rapidly take up an additional three electrons wherever it can find them. The cell can use O_2 for respiration only because cytochrome oxidase holds onto oxygen at a special bimetallic center (Figure 14–34B), where it remains clamped between a heme-linked iron atom and a copper atom until it has picked up a total of four electrons; only then can the two oxygen atoms of the oxygen molecule be safely released as two molecules of water (Figure 14–34C).

Although cytochrome oxidase contains many protein subunits, most of these appear to have a subsidiary role, helping to regulate either the activity or the assembly of the three subunits that form the core of the enzyme. One of the core subunits contains the bimetallic center where oxygen is bound, and it is responsible for pumping the four protons that are transferred across the inner mitochondrial membrane for each O_2 molecule that is reduced to water (see Figure 14–34).

The cytochrome oxidase reaction is estimated to account for 90% of the total oxygen uptake in most cells. Cyanide and azide are toxic to cells because they bind tightly to this complex and thereby block all electron transport.

Electron Transfers Are Mediated by Random Collisions Between Diffusing Donors and Acceptors in the Mitochondrial Inner Membrane [18]

The two components that carry electrons between the three major enzyme complexes of the respiratory chain—ubiquinone and cytochrome c—diffuse rapidly in the plane of the inner mitochondrial membrane. The expected rate of random collisions between these mobile carriers and the enzyme complexes can account for the observed rates of electron transfer (each complex donates and receives an electron about once every 5 to 20 milliseconds). Thus there is no need to postulate a structurally ordered chain of electron-transfer proteins in the lipid bilayer; indeed, the three enzyme complexes appear to exist as independent entities in the plane of the inner membrane, and the ordered transfer of electrons is due entirely to the specificity of the functional interactions among the components of the chain.

This view is supported by the observation that the various components of the respiratory chain are present in different amounts. For each molecule of NADH dehydrogenase complex in heart mitochondria, for example, it is estimated that there are 3 molecules of b-c_1 complex, 7 molecules of cytochrome oxidase complex, 9 molecules of cytochrome c, and 50 molecules of ubiquinone; very different ratios are found in the mitochondria of some other cells. These components form a chain in the sense that each interacts specifically only with the carrier adjacent to it in the sequence shown in Figure 14–33, and there is a net flow of electrons from NADH dehydrogenase to cytochrome oxidase because each of the enzyme complexes in the sequence has a higher affinity for electrons than its predecessor. The affinity of a molecule for electrons is its *redox potential*. The changes in redox potential from one electron carrier to the next are exploited to pump proteins out of the mitochondrial matrix, as we now discuss.

A Large Drop in Redox Potential Across Each of the Three Respiratory Enzyme Complexes Provides the Energy for H⁺ Pumping [19]

Pairs of compounds such as H_2O and $\frac{1}{2}O_2$, or NADH and NAD^+, are called **conjugate redox pairs,** since one compound is converted to the other by adding one or more electrons plus one or more protons—the protons being readily available in any aqueous solution. Thus, for example,

$$\frac{1}{2}O_2 + 2e^- + 2H^+ \rightarrow H_2O$$

Many readers will know that a 50-50 (equimolar) mixture of the members of a *conjugate acid-base pair* acts as a *buffer,* maintaining a defined "H⁺ pressure," or pH, which is a measure of the dissociation constant of the acid. In exactly the same way a 50-50 mixture of the members of a conjugate redox pair maintains a defined "electron pressure," or **redox** (reduction-oxidation) **potential, *E*,** that is a measure of the electron carrier's affinity for electrons.

By placing electrodes in contact with solutions that contain the appropriate conjugate redox pairs, one can measure the redox potential of each of the various electron carriers that participate in biological oxidation-reduction reactions. For biological systems each redox potential is determined at pH 7.0, where $[H^+] = 10^{-7}$ M. Those pairs of compounds that have the most negative redox potentials have the weakest affinity for electrons and therefore contain carriers with the strongest tendency to donate electrons, whereas pairs that have the most positive redox potentials have the strongest affinity for electrons and contain carriers with the strongest tendency to accept electrons. Thus a 50-50 mixture of NADH and NAD^+ has a redox potential of –320 mV, indicating that NADH has a strong tendency to donate electrons; a 50-50 mixture of H_2O and $\frac{1}{2}O_2$ has a redox potential of +820 mV, indicating that O_2 has a strong tendency to accept electrons.

Redox potentials can be readily determined for all the electron carriers in the respiratory chain that can be distinguished by their spectra, and they can be shown to increase as one passes along the chain of electron carriers. As most cytochromes have higher redox potentials than iron-sulfur centers, they generally serve as electron carriers near the O_2 end of the respiratory chain, whereas the iron-sulfur proteins serve as carriers near the NADH end.

An outline of the redox potentials measured along the respiratory chain is shown in Figure 14–35. The potentials drop in three large steps, one across each major enzyme complex. The change in redox potential between any two electron

Figure 14–35 The redox potential (denoted E_0' or E_h) increases as electrons flow down the respiratory chain to oxygen. The standard free-energy change, $\Delta G°$ (in kilocalories per mole), for the transfer of the two electrons donated by an NADH molecule can be obtained from the right-hand ordinate ($\Delta G° = -n(0.023)\Delta E_0'$, where *n* is the number of electrons transferred across a redox potential change of $\Delta E_0'$ mV). Electrons flow through an enzyme complex by passing in sequence to the four or more electron carriers in each complex. As indicated, part of the favorable free-energy change is harnessed by each enzyme complex to pump H⁺ across the mitochondrial inner membrane. Although the number of H⁺ pumped per electron (*n*) is uncertain, it is estimated that the NADH dehydrogenase and b-c_1 complexes each pump two H⁺ per electron, whereas the cytochrome oxidase complex pumps one.

The two electrons transported from $FADH_2$, generated by fatty acid oxidation (see Figure 14–11) and by the citric acid cycle (see Figure 14–14), are passed directly to ubiquinone, and they therefore cause less H⁺ pumping than the two electrons transported from NADH (not shown).

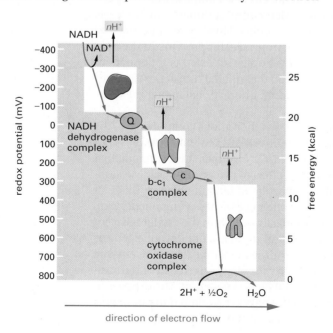

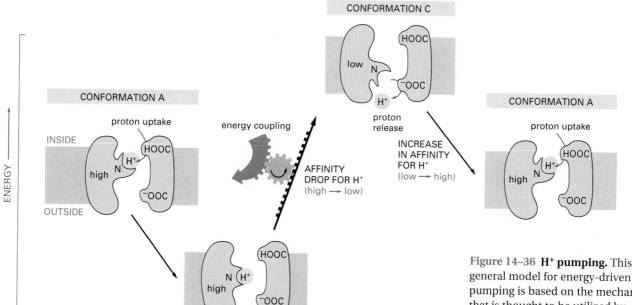

Figure 14–36 **H⁺ pumping.** This general model for energy-driven H⁺ pumping is based on the mechanism that is thought to be utilized by bacteriorhodopsin. The transmembrane protein shown is driven through a cycle of three conformations, denoted here as A, B, and C. In conformation C the protein has a low affinity for H⁺, causing it to release an H⁺ on the outside of the lipid bilayer; in conformation A the protein has a high affinity for H⁺, causing it to pick up an H⁺ on the inside of the lipid bilayer. As indicated, the transition from conformation B to conformation C is energetically unfavorable but is driven by being coupled to an energetically favorable reaction occurring elsewhere on the protein (*blue arrow*). The other conformational changes lead to states of lower energy and proceed spontaneously. The cycle A → B → C → A therefore goes only one way, causing H⁺ to be pumped from the inside to the outside. For bacteriorhodopsin the energy for the transition B → C is provided by light, whereas in the mitochondria this energy is provided by electron transport.

carriers is directly proportional to the free energy released by an electron transfer between them (see Figure 14–35). Each complex acts as an energy-conversion device, harnessing this free-energy change to pump H⁺ across the inner membrane, thereby creating an electrochemical proton gradient as electrons pass through. This conversion can be demonstrated by incorporating each purified complex separately into liposomes: when an appropriate electron donor and acceptor is added so that electrons can pass through the complex, H⁺ is translocated across the liposome membrane.

The Mechanism of H⁺ Pumping Is Best Understood in Bacteriorhodopsin [20]

Because some respiratory enzyme complexes pump one H⁺ per electron across the inner mitochondrial membrane whereas others pump two, the molecular mechanism by which electron transport is coupled to H⁺ pumping is presumably different for the three different enzyme complexes. The details of the actual mechanisms are not known. In the case of the b-c₁ complex, the quinones clearly play a part. As mentioned previously, a quinone picks up a H⁺ from the aqueous medium along with each electron it carries and liberates it when it releases the electron (see Figure 14–31). Since ubiquinone is freely mobile in the lipid bilayer, it could accept electrons near the inside surface of the membrane and donate them to the b-c₁ complex near the outside surface, thereby transferring one H⁺ across the bilayer for every electron transported. Two protons are pumped per electron in the b-c₁ complex, however, and there is evidence for a so-called *Q-cycle*, in which ubiquinone is recycled through the complex in an ordered way that makes this two-for-one transfer possible.

Allosteric changes in protein conformations driven by electron transport can also pump H⁺, just as H⁺ is pumped when ATP is hydrolyzed by the ATP synthase running in reverse. For both the NADH dehydrogenase complex and the cytochrome oxidase complex, it seems likely that electron transport drives orderly allosteric changes in protein conformation that cause a portion of the protein to pump H⁺ across the inner mitochondrial membrane. This type of proton pumping is best understood for bacteriorhodopsin, a light-driven H⁺ pump found in the plasma membrane of certain highly specialized bacteria (see Figure 10–32).

A general mechanism for H$^+$ pumping based on structural and functional studies of this protein is presented in Figure 14–36.

H$^+$ Ionophores Dissipate the H$^+$ Gradient and Thereby Uncouple Electron Transport from ATP Synthesis [21]

Since the 1940s several substances, such as 2,4-dinitrophenol, have been known to act as *uncoupling agents,* uncoupling electron transport from ATP synthesis. The addition of these low-molecular-weight organic compounds to cells stops ATP synthesis by mitochondria without blocking their uptake of oxygen. In the presence of an uncoupling agent electron transport and H$^+$ pumping continue at a rapid rate, but no H$^+$ gradient is generated. The explanation for this effect is both simple and elegant: uncoupling agents are lipid-soluble weak acids that act as H$^+$ carriers (H$^+$ ionophores) and provide a pathway in addition to the ATP synthase for the flow of H$^+$ across the inner mitochondrial membrane. As a result of this "short-circuiting," the proton-motive force is dissipated completely, and ATP can no longer be made.

Respiratory Control Normally Restrains Electron Flow Through the Chain [22]

When an uncoupler such as dinitrophenol is added to cells, mitochondria increase their oxygen uptake substantially because of an increased rate of electron transport. This increase reflects the existence of **respiratory control.** The control is thought to act via a direct inhibitory influence of the electrochemical proton gradient on the rate of electron transport. When the gradient is collapsed by an uncoupler, electron transport is free to run unchecked at the maximal rate. As the gradient increases, electron transport becomes more difficult and the process slows. Moreover, if an artificially large electrochemical proton gradient is experimentally created across the inner membrane, normal electron transport stops completely and a *reverse electron flow* can be detected in some sections of the respiratory chain. This observation suggests that respiratory control reflects a simple balance between the free-energy change for electron-transport-linked proton pumping and the free-energy change for electron transport—that is, the magnitude of the electrochemical proton gradient affects both the rate and the direction of electron transport, just as it affects the directionality of the ATP synthase (see Figure 14–27).

Respiratory control is just one part of an elaborate interlocking system of feedback controls that coordinates the rates of glycolysis, fatty acid breakdown, the citric acid cycle, and electron transport. The rates of all of these processes are adjusted to the ATP:ADP ratio, increasing whenever increased utilization of ATP causes the ratio to fall. The ATP synthase in the inner mitochondrial membrane, for example, works faster as the concentrations of its substrates ADP and P$_i$ increase. As it speeds up, the enzyme lets more H$^+$ flow into the matrix and thereby dissipates the electrochemical proton gradient more rapidly. The falling gradient, in turn, enhances the rate of electron transport.

Similar controls, including feedback inhibition of several key enzymes by ATP (see Figure 14–13, for example), act to adjust the rates of NADH production to the rate of NADH utilization by the respiratory chain and so on. As a result of these many control mechanisms, the body oxidizes fats and sugars 5 to 10 times more rapidly during a period of strenuous exercise than during a period of rest.

Natural Uncouplers Convert the Mitochondria in Brown Fat into Heat-generating Machines [23]

In some specialized fat cells mitochondrial respiration is normally uncoupled from ATP synthesis. In these cells, known as brown fat cells, most of the energy of oxidation is dissipated as heat rather than being converted into ATP. The inner membranes of the large mitochondria in these cells contain a special transport protein that allows protons to move down their electrochemical gradient without activating ATP synthase. As a result, the cells oxidize their fat stores at a rapid rate and produce more heat than ATP. Tissues containing brown fat thereby serve as "heating pads" that revive hibernating animals and protect sensitive areas of newborn human babies from the cold.

All Bacteria Use Chemiosmotic Mechanisms to Harness Energy [24]

Bacteria use enormously diverse energy sources. Some, like animal cells, are aerobic and synthesize ATP from sugars that they oxidize to CO_2 and H_2O by glycolysis and the citric acid cycle through a respiratory chain in their plasma membrane similar to that in the mitochondrial inner membrane. Others are strict anaerobes, deriving their energy either from glycolysis alone (by fermentation) or, in addition, from an electron-transport chain that employs a molecule other than oxygen as the final electron acceptor. The alternative electron acceptor can be a nitrogen compound (nitrate or nitrite), a sulfur compound (sulfate or sulfite), or a carbon compound (fumarate or carbonate), for example. The electrons are transferred to these acceptors by a series of electron carriers in the plasma membrane that are comparable to those in mitochondrial respiratory chains.

Despite this diversity, the plasma membrane of the vast majority of bacteria contains an ATP synthase that is very similar to that in mitochondria (and chloroplasts). In aerobic bacteria the electron-transport chain pumps H^+ out of the cell and thereby establishes a proton-motive force that drives the ATP synthase to make ATP. In anaerobic bacteria that lack an electron-transport chain, the ATP synthase works in reverse, using the ATP produced by glycolysis to pump H^+ and establish a proton-motive force across the bacterial plasma membrane.

Thus most bacteria, including the strict anaerobes, maintain a proton-motive force across their plasma membrane. It can be harnessed to drive a flagellar motor that enables the bacterium to swim and is used to pump Na^+ out of the bacterium via a Na^+-H^+ antiporter that takes the place of the Na^+-K^+ ATPase of eucaryotic cells. It is also used for the active transport of nutrients, such as most amino acids and many sugars, into bacteria: each nutrient is dragged into the cell along with one or more H^+ through a specific symporter (Figure 14–37). In animal cells, by contrast, most inward transport across the plasma membrane is driven by the Na^+ gradient established by the Na^+-K^+ ATPase.

Some unusual bacteria have adapted to live in a very alkaline environment and yet must maintain their cytoplasm at a physiological pH. For these cells any attempt to generate an electrochemical H^+ gradient would be opposed by a large H^+ concentration gradient in the wrong direction (H^+ higher inside than outside). Presumably for this reason, at least some of these bacteria substitute Na^+ for H^+ in all of their chemiosmotic mechanisms. The respiratory chain pumps Na^+ out of the cell, the transport systems and flagellar motor are driven by an inward flux of Na^+, and a Na^+-driven ATP synthase synthesizes ATP. The existence of such bacteria demonstrates that the principle of chemiosmosis is more fundamental than the proton-motive force on which it is normally based.

Summary

The respiratory chain in the inner mitochondrial membrane contains three major enzyme complexes through which electrons pass on their way from NADH to O_2. Each of these can be purified, inserted into synthetic lipid vesicles, and then shown to pump H^+ when electrons are transported through it. In the native membrane the mobile electron carriers ubiquinone and cytochrome c complete the electron-transport chain

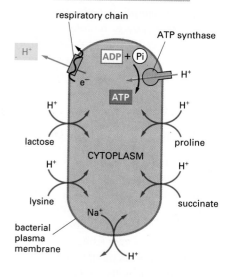

(A) AEROBIC CONDITIONS

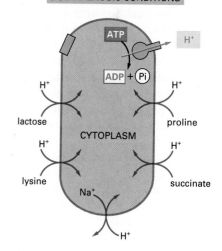

(B) ANAEROBIC CONDITIONS

Figure 14–37 H^+-driven transport in bacteria. A proton-motive force generated across the plasma membrane pumps nutrients into the cell and expels sodium. In (A) the electrochemical proton gradient is generated in an aerobic bacterium by a respiratory chain and is then used by ATP synthase to make ATP and to transport some nutrients into the cell. In (B) the same bacterium growing under anaerobic conditions can derive its ATP from glycolysis. Part of this ATP is hydrolyzed by ATP synthase to establish the transmembrane proton-motive force that drives transport processes.

by shuttling between the enzyme complexes. The path of electron flow is NADH → NADH dehydrogenase complex → ubiquinone → b-c₁ complex → cytochrome c → cytochrome oxidase complex → molecular oxygen (O_2).

The respiratory enzyme complexes couple the energetically favorable transport of electrons to the pumping of H^+ out of the matrix. The resulting electrochemical proton gradient is harnessed to make ATP by another transmembrane protein complex, ATP synthase, through which H^+ flows back into the matrix. The ATP synthase is a reversible coupling device that normally converts a backflow of H^+ into ATP phosphate-bond energy by catalyzing the reaction $ADP + P_i → ATP$, but it can also work in the opposite direction and hydrolyze ATP to pump H^+ if the electrochemical proton gradient is reduced. Its universal presence in mitochondria, chloroplasts, and bacteria testifies to the central importance of chemiosmotic mechanisms in cells.

Chloroplasts and Photosynthesis [25]

All animals and most microorganisms rely on the continual uptake of large amounts of organic compounds from their environment. These compounds provide both the carbon skeletons for biosynthesis and the metabolic energy that drives all cellular processes. It is believed that the first organisms on primitive earth had access to an abundance of organic compounds produced by geochemical processes (see Chapter 1) but that most of these original compounds were used up billions of years ago. Since that time virtually all of the organic materials required by living cells have been produced by *photosynthetic organisms,* including many types of photosynthetic bacteria. The most advanced photosynthetic bacteria are the cyanobacteria, which have minimal nutrient requirements. They use electrons from water and the energy of sunlight to convert atmospheric CO_2 into organic compounds. In the course of splitting water [in the reaction $nH_2O + nCO_2 \xrightarrow{light} (CH_2O)_n + nO_2$], they liberate into the atmosphere the oxygen required for oxidative phosphorylation. As we shall see, it is thought that the evolution of cyanobacteria from more primitive photosynthetic bacteria first made possible the development of aerobic life forms.

In plants, which developed later, photosynthesis occurs in a specialized intracellular organelle—the **chloroplast**. Chloroplasts carry out photosynthesis during the daylight hours. The products of photosynthesis are used directly by the photosynthetic cells for biosynthesis and are also converted to a low-molecular-weight sugar (usually sucrose) that is exported to meet the metabolic needs of the many nonphotosynthetic cells of the plant. Alternatively, the products can be stored as an osmotically inert polysaccharide (usually starch) that is kept available as a source of sugar for future use.

Biochemical evidence suggests that chloroplasts are descendants of oxygen-producing photosynthetic bacteria that were endocytosed and lived in symbiosis with primitive eucaryotic cells. Mitochondria are also generally believed to be descended from endocytosed bacteria. The many differences between chloroplasts and mitochondria are thought to reflect their different bacterial ancestors as well as their subsequent evolutionary divergence. Nevertheless, the fundamental mechanisms involved in light-driven ATP synthesis in chloroplasts and in respiration-driven ATP synthesis in mitochondria are very similar.

The Chloroplast Is One Member of a Family of Organelles That Is Unique to Plants—the Plastids [26]

Chloroplasts are the most prominent member of the **plastid** family of organelles. Plastids are present in all living plant cells, each cell type having its own characteristic complement. All plastids share certain features. Most notably, all plastids in a particular plant species contain multiple copies of the same relatively

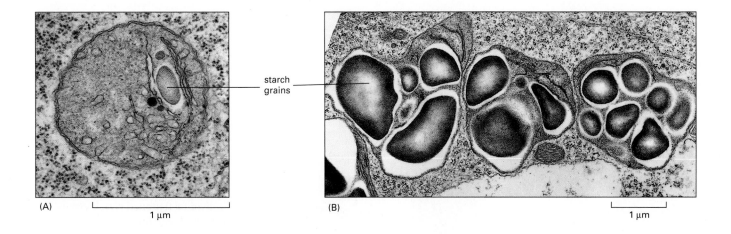

starch grains

(A)

1 μm

(B)

1 μm

small genome (see Table 14–3, p. 706) and are enclosed by an envelope composed of two concentric membranes.

All plastids develop from *proplastids,* which are relatively small organelles present in the immature cells of plant meristems (Figure 14–38A). Proplastids develop according to the requirements of each differentiated cell, and which type is present is determined in large part by the nuclear genome. If a leaf is grown in darkness, its proplastids enlarge and develop into *etioplasts,* which have a semicrystalline array of internal membranes that contain a yellow chlorophyll precursor instead of chlorophyll. When exposed to light, the etioplasts rapidly develop into chloroplasts by converting this precursor to chlorophyll and by synthesizing new membrane, pigments, photosynthetic enzymes, and components of the electron-transport chain.

Leucoplasts are plastids that occur in many epidermal and internal tissues that do not become green and photosynthetic. They are little more than enlarged proplastids. A common form of leucoplast is the *amyloplast* (Figure 14–38B), which accumulates starch in storage tissues. In some plants, such as potatoes, the amyloplasts can grow to be as large as an average animal cell.

It is important to realize that plastids are not just sites for photosynthesis and the deposition of storage materials. Plants have exploited their plastids in the cellular compartmentalization of intermediary metabolism. Plastids produce more than the energy and reducing power (as ATP and NADPH) that is used for the plant's biosynthetic reactions. Purine and pyrimidine, most amino acid, and all of the fatty acid synthesis of plants takes place in the plastids, whereas in animal cells these compounds are produced in the cytosol.

Figure 14–38 Plastid diversity. (A) A *proplastid* from a root tip cell of a bean plant. Note the double membrane; the inner membrane gives rise to the relatively sparse internal membranes. (B) Three *amyloplasts* (a form of leucoplast), or starch-storing plastids, in a root tip cell of soybean. (From B. Gunning and M. Steer, Ultrastructure and the Biology of Plant Cells. London: Arnold, 1975.)

Chloroplasts Resemble Mitochondria But Have an Extra Compartment [27]

Chloroplasts carry out their energy interconversions by chemiosmotic mechanisms in much the same way that mitochondria do, and they are organized on the same principles (Figures 14–39 and 14–40). They have a highly permeable outer membrane, a much less permeable inner membrane, in which special carrier proteins are embedded, and a narrow intermembrane space. The inner membrane surrounds a large space called the **stroma,** which is analogous to the mitochondrial matrix and contains various enzymes, ribosomes, RNA, and DNA.

There is, however, an important difference between the organization of mitochondria and that of chloroplasts. The inner membrane of the chloroplast is not folded into cristae and does not contain an electron-transport chain. Instead, the electron-transport chain as well as the photosynthetic light-absorbing system and an ATP synthase are all contained in a third distinct membrane that

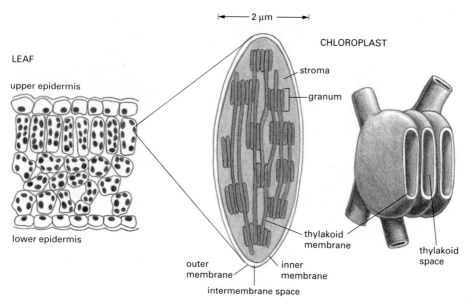

LEAF

upper epidermis

lower epidermis

CHLOROPLAST

2 μm

stroma

granum

thylakoid
membrane

outer
membrane

inner
membrane

intermembrane space

thylakoid
space

Figure 14–39 The chloroplast. This photosynthetic organelle contains three distinct membranes (the outer membrane, the inner membrane, and the thylakoid membrane) that define three separate internal compartments (the intermembrane space, the stroma, and the thylakoid space). The thylakoid membrane contains all of the energy-generating systems of the chloroplast. In electron micrographs this membrane appears to be broken up into separate units that enclose individual flattened vesicles (see Figure 14–40), but these are probably joined into a single, highly folded membrane in each chloroplast. As indicated, the individual thylakoids are interconnected, and they tend to stack to form aggregates called grana.

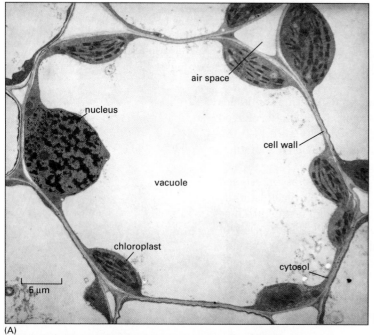

air space

nucleus

cell wall

vacuole

chloroplast

cytosol

5 μm

(A)

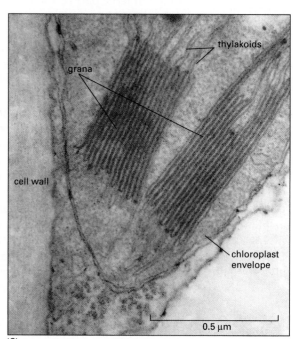

thylakoids

grana

cell wall

chloroplast
envelope

0.5 μm

(C)

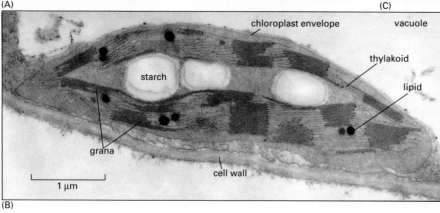

chloroplast envelope

vacuole

thylakoid

starch

lipid

grana

cell wall

1 μm

(B)

Figure 14–40 Electron micrographs of chloroplasts. (A) A wheat leaf cell in which a thin rim of cytoplasm containing chloroplasts surrounds a large vacuole. (B) A thin section of a single chloroplast, showing the starch granules and lipid droplets that have accumulated in the stroma as a result of the biosyntheses occurring there. (C) A high-magnification view of a granum, showing its stacked thylakoid membrane. (Courtesy of K. Plaskitt.)

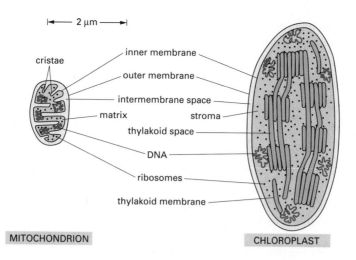

Figure 14–41 Comparison of a mitochondrion and a chloroplast. The chloroplast is generally much larger and contains a thylakoid membrane and thylakoid space. The mitochondrial inner membrane is folded into cristae.

forms a set of flattened disclike sacs, the **thylakoids** (see Figure 14–39). The lumen of each thylakoid is thought to be connected with the lumen of other thylakoids, thereby defining a third internal compartment called the *thylakoid space,* which is separated from the stroma by the *thylakoid membrane.*

The structural similarities and differences between mitochondria and chloroplasts are illustrated in Figure 14–41. Superficially, the chloroplast resembles a greatly enlarged mitochondrion in which the cristae have been converted into a series of interconnected submitochondrial particles in the matrix space. The knobbed end of the chloroplast ATP synthase, where ATP is made, protrudes from the thylakoid membrane into the stroma, just as it protrudes into the matrix from the membrane of each mitochondrial crista.

Two Unique Reactions in Chloroplasts: The Light-driven Production of ATP and NADPH and the Conversion of CO_2 to Carbohydrate [25]

The many reactions that occur during photosynthesis can be grouped into two broad categories. (1) In the **photosynthetic electron-transfer reactions** (also called the "light reactions") energy derived from sunlight energizes an electron in *chlorophyll,* enabling the electron to move along an electron-transport chain in the thylakoid membrane in much the same way that an electron moves along the respiratory chain in mitochondria. The chlorophyll obtains its electrons from water, with the liberation of O_2. During the electron-transport process H^+ is pumped across the thylakoid membrane, and the resulting proton-motive force drives the synthesis of ATP in the stroma. As the final step in this series of reactions, high-energy electrons are loaded (together with H^+) onto $NADP^+$, converting it to NADPH. All of these reactions are confined to the chloroplast. (2) In the **carbon-fixation reactions** (also called the "dark reactions") the ATP and NADPH produced by the photosynthetic electron-transfer reactions serve as the source of energy and reducing power, respectively, to drive the conversion of CO_2 to carbohydrate. The carbon-fixation reactions, which begin in the chloroplast stroma and continue in the cytosol, produce sucrose in the leaves of the plant; from there it is exported to other tissues as a source of both organic molecules and energy for growth.

Thus the formation of oxygen (which requires light energy directly) and the conversion of carbon dioxide to carbohydrate (which requires light energy only indirectly) are separate processes (Figure 14–42). We shall see, however, that elaborate feedback mechanisms interconnect the two in order to balance biosynthesis. Changes in the cell's ATP and NADPH requirements, for example, regulate the production of these two molecules in the thylakoid membrane, and sev-

Chloroplasts and Photosynthesis

eral of the chloroplast enzymes required for carbon fixation are inactivated in the dark and reactivated by light-stimulated electron-transport processes.

Carbon Fixation Is Catalyzed by Ribulose Bisphosphate Carboxylase [28]

We have seen earlier in this chapter how cells produce ATP by using the large amount of free energy released when carbohydrates are oxidized to CO_2 and H_2O. Clearly, therefore, the reverse reaction, in which CO_2 and H_2O combine to make carbohydrate, must be a very unfavorable one and must be coupled to other, very favorable reactions to drive it.

The central reaction of carbon fixation, in which an atom of inorganic carbon is converted to organic carbon, is illustrated in Figure 14–43: CO_2 from the atmosphere combines with the five-carbon compound ribulose 1,5-bisphosphate plus water to give two molecules of the three-carbon compound 3-phosphoglycerate. This "carbon-fixing" reaction, which was discovered in 1948, is catalyzed in the chloroplast stroma by a large enzyme called *ribulose bisphosphate carboxylase* (~500,000 daltons). Since each copy of the complex works sluggishly (processing only about 3 molecules of substrate per second compared to 1000 molecules per second for a typical enzyme), many copies are needed. Ribulose bisphosphate carboxylase often constitutes more than 50% of the total chloroplast protein and is thought to be the most abundant protein on earth.

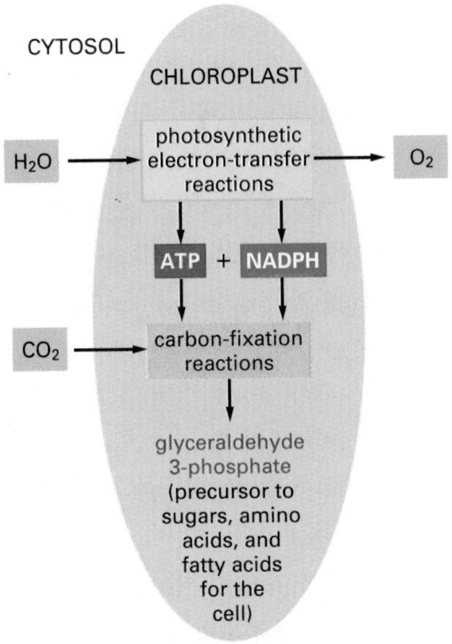

Three Molecules of ATP and Two Molecules of NADPH Are Consumed for Each CO_2 Molecule That Is Fixed in the Carbon-Fixation Cycle [29]

The actual reaction in which CO_2 is fixed is energetically favorable because of the reactivity of the energy-rich compound *ribulose 1,5-bisphosphate*, to which each molecule of CO_2 is added (see Figure 14–43). But to produce a supply of ribulose 1,5-bisphosphate requires a series of reactions that use up large amounts of NADPH and ATP. The elaborate pathway by which this compound is regenerated was worked out in one of the most successful early applications of radioisotopes. As outlined in Figure 14–44, 3 molecules of CO_2 are fixed by ribulose bisphosphate carboxylase to produce 6 molecules of 3-phosphoglycerate (containing $6 \times 3 = 18$ carbon atoms in all: 3 from the CO_2 and 15 from ribulose 1,5-bisphosphate). The 18 carbon atoms then undergo a cycle of reactions that regenerates the 3 molecules of ribulose 1,5-bisphosphate used in the initial carbon-fixation step (containing $3 \times 5 = 15$ carbon atoms). This leaves 1 molecule of *glyceraldehyde 3-phosphate* (3 carbon atoms) as the net gain. In this **carbon-fixation cycle** (or Calvin-Benson cycle), 3 molecules of ATP and 2 molecules of NADPH are consumed for each CO_2 molecule converted into carbohydrate.

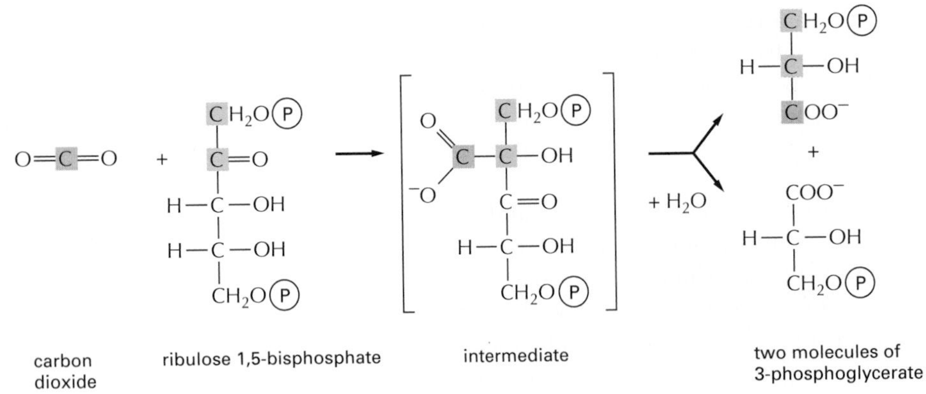

Figure 14–42 Photosynthesis in a chloroplast. Water is oxidized and oxygen is released in the photosynthetic electron-transfer reactions, while carbon dioxide is assimilated (fixed) to produce carbohydrate in the carbon-fixation reactions.

carbon dioxide   ribulose 1,5-bisphosphate   intermediate   two molecules of 3-phosphoglycerate

Figure 14–43 The initial reaction in carbon fixation. This reaction, in which carbon dioxide is converted into organic carbon, is catalyzed in the chloroplast stroma by the abundant enzyme *ribulose bisphosphate carboxylase*. The product, 3-phosphoglycerate, is also an important intermediate in glycolysis: the two carbon atoms shaded in *blue* are used to produce *phosphoglycolate* when the enzyme adds oxygen instead of CO_2 (see below).

The net equation is

$$3CO_2 + 9ATP + 6NADPH + water \rightarrow$$
$$\text{glyceraldehyde 3-phosphate} + 8P_i + 9ADP + 6NADP^+$$

Thus both *phosphate-bond energy* (as ATP) and *reducing power* (as NADPH) are required for the formation of organic molecules from CO_2 and H_2O. We return to this important point later.

The glyceraldehyde 3-phosphate produced in chloroplasts by the carbon-fixation cycle is a three-carbon sugar that serves as a central intermediate in glycolysis. Much of it is exported to the cytosol, where it can be converted into fructose 6-phosphate and glucose 1-phosphate by reversal of several reactions in glycolysis (see Figure 2–21). Glucose 1-phosphate is then converted to the sugar nucleotide UDP-glucose, and this combines with fructose 6-phosphate to form sucrose phosphate, the immediate precursor of the disaccharide **sucrose.** Sucrose is the major form in which sugar is transported between plant cells: just as glucose is transported in the blood of animals, sucrose is exported from the leaves via vascular bundles, providing the carbohydrate required by the rest of the plant.

Most of the glyceraldehyde 3-phosphate that remains in the chloroplast is converted to *starch* in the stroma. Like glycogen in animal cells, **starch** is a large polymer of glucose that serves as a carbohydrate reserve. The production of starch is regulated so that it is produced and stored as large grains in the chloroplast stroma (see Figure 14–38B) during periods of excess photosynthetic capacity. This occurs through reactions in the stroma that are the reverse of those in glycolysis: they convert glyceraldehyde 3-phosphate to glucose 1-phosphate, which is then used to produce the sugar nucleotide ADP-glucose, the immediate precursor of starch. At night the starch is broken down to help support the metabolic needs of the plant.

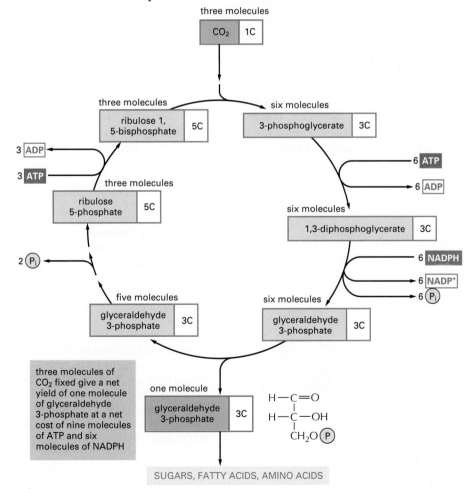

Figure 14–44 **The carbon-fixation cycle, which forms organic molecules from CO_2 and H_2O.** The number of carbon atoms in each type of molecule is indicated in the *white box*. There are many intermediates between glyceraldehyde 3-phosphate and ribulose 5-phosphate, but they have been omitted here for clarity. The entry of water into the cycle is also not shown.

Chloroplasts and Photosynthesis

Carbon Fixation in Some Plants Is Compartmentalized to Facilitate Growth at Low CO_2 Concentrations [30]

Although ribulose bisphosphate carboxylase preferentially adds CO_2 to ribulose 1,5-bisphosphate, it can use O_2 in addition to CO_2, and if the concentration of CO_2 is low, it will add O_2 instead. This is the first step in a pathway called *photorespiration,* whose ultimate effect is to use up O_2 and liberate CO_2 without the production of useful energy stores. In many plants about one-third of the CO_2 fixed is lost again as CO_2 because of photorespiration.

Photorespiration can be a serious liability for plants in hot, dry conditions, where they close their stomata (the gas exchange pores in their leaves) to avoid excessive water loss. This causes the CO_2 levels in the leaf to fall precipitously and thereby favors photorespiration. A special adaptation, however, occurs in the leaves of many plants, such as corn and sugar cane, that live in hot, dry environments. In these plants the carbon-fixation cycle shown in Figure 14–44 occurs only in the chloroplasts of specialized *bundle-sheath cells,* which contain all of the plant's ribulose bisphosphate carboxylase. These cells are protected from the air and are surrounded by a specialized layer of *mesophyll cells* that "pump" CO_2 into the bundle-sheath cells, supplying the ribulose bisphosphate carboxylase with a high concentration of CO_2, which greatly reduces photorespiration.

The CO_2 pump is produced by a reaction cycle that begins in the cytosol of the mesophyll cells with a CO_2-fixation step catalyzed by an enzyme that binds carbon dioxide (as bicarbonate) and combines it with an activated three-carbon molecule to produce a four-carbon molecule. The four-carbon molecule diffuses into the bundle-sheath cells, where it is broken down to release the CO_2 and generate a molecule with three carbons. The pumping cycle is completed when this three-carbon molecule is returned to the mesophyll cells and converted to its original activated form. Because the CO_2 is initially captured by converting it into a compound containing four carbons, the CO_2-pumping plants are called C_4 *plants.* All other plants are called C_3 *plants* (Figure 14–45) because they capture CO_2 directly into the three-carbon compound 3-phosphoglycerate.

As for any vectorial transport process, pumping CO_2 into the bundle-sheath cells in C_4 plants costs energy. In hot, dry environments, however, this cost may be much less than the energy lost by photorespiration in C_3 plants, and so C_4 plants have a potential advantage. Moreover, because C_4 plants can carry out photosynthesis at a lower concentration of CO_2 inside the leaf, they need to open their stomata less and therefore can fix about twice as much net carbon as C_3 plants per unit of water lost.

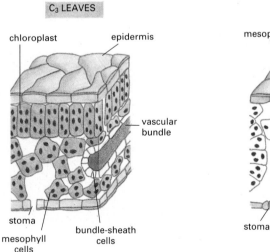

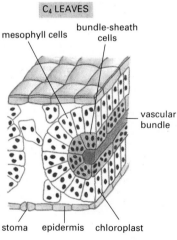

Figure 14–45 A comparison of the anatomy of the leaf in a C_3 plant and a C_4 plant. The cells with *green* cytosol in the leaf interior contain chloroplasts that carry out the normal carbon-fixation cycle. In C_4 plants the mesophyll cells are specialized for CO_2 pumping rather than for carbon fixation, and they thereby create a high $CO_2{:}O_2$ ratio in the bundle-sheath cells, which are the only cells in these plants where the carbon-fixation cycle occurs. The vascular bundles carry the sucrose made in the leaf to other tissues.

Photosynthesis Depends on the Photochemistry of Chlorophyll Molecules [31]

Having discussed the carbon-fixation reactions, we now return to the question of how the photosynthetic electron-transfer reactions in the chloroplast generate the ATP and the NADPH needed to drive the production of carbohydrates from CO_2 and H_2O (see Figure 14–42). The required energy is derived from sunlight absorbed by **chlorophyll** molecules (Figure 14–46). The process of energy conversion begins when a chlorophyll molecule is excited by a quantum of light (a photon) and an electron is moved from one molecular orbital to another of higher energy. Such an excited molecule is unstable and will tend to return to its original, unexcited state in one of three ways: (1) by converting the extra energy into heat (molecular motions) or to some combination of heat and light of a longer wavelength (fluorescence), which is what happens when light energy is absorbed by an isolated chlorophyll molecule in solution; (2) by transferring the energy—but not the electron—directly to a neighboring chlorophyll molecule by a process called *resonance energy transfer;* or (3) by transferring the high-energy electron to another nearby molecule (an *electron acceptor*) and then returning to its original state by taking up a low-energy electron from some other molecule (an *electron donor,* Figure 14–47). The last two mechanisms are exploited in the process of photosynthesis.

A Photosystem Contains a Reaction Center Plus an Antenna Complex [32]

Multiprotein complexes called **photosystems** catalyze the conversion of the light energy captured in excited chlorophyll molecules to useful forms. A photosystem consists of two closely linked components: a *photochemical reaction center* consisting of a complex of proteins and chlorophyll molecules that enable light energy to be converted into chemical energy and an *antenna complex* consisting of pigment molecules that capture light energy and feed it to the reaction center.

The **antenna complex** is important for capturing light. In chloroplasts it consists of a cluster of several hundred chlorophyll molecules linked together by proteins that hold them tightly on the thylakoid membrane. Depending on the plant, varying amounts of accessory pigments called *carotenoids,* which can help collect light of other wavelengths, are also located in each complex. When a chlorophyll molecule in the antenna complex is excited, the energy is rapidly transferred from one molecule to another by resonance energy transfer until it reaches

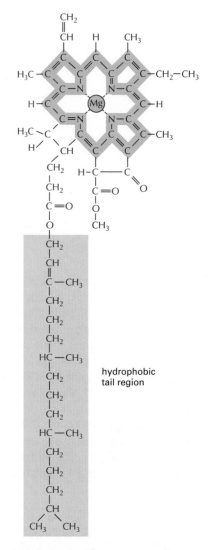

Figure 14–46 The structure of chlorophyll. A magnesium atom is held in a porphyrin ring, which is related to the porphyrin ring that binds iron in heme (compare with Figure 14–28). Electrons are delocalized over the bonds shown in color.

hydrophobic tail region

Figure 14–47 Three ways for an excited chlorophyll molecule to return to its original, unexcited state. The light energy absorbed by an isolated chlorophyll molecule is completely released as light and heat by process 1. In photosynthesis, by contrast, chlorophylls undergo process 2 in the antenna complex and process 3 in the reaction center, as described in the text.

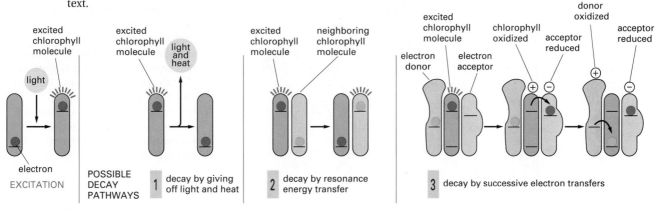

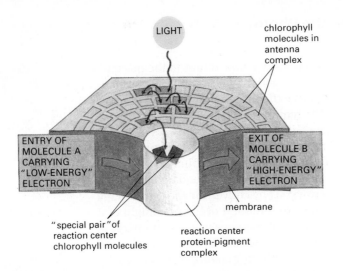

LIGHT

chlorophyll molecules in antenna complex

ENTRY OF MOLECULE A CARRYING "LOW-ENERGY" ELECTRON

EXIT OF MOLECULE B CARRYING "HIGH-ENERGY" ELECTRON

membrane

"special pair" of reaction center chlorophyll molecules

reaction center protein-pigment complex

Figure 14–48 **The reaction center and antenna in a photosystem.** Molecules A (electron donor) and B (electron acceptor) differ according to the photosystem.

a special pair of chlorophyll molecules in the photochemical reaction center. Each antenna complex thereby acts as a "funnel," collecting light energy and directing it to a specific site where it can be used effectively (Figure 14–48).

The **photochemical reaction center** is a transmembrane protein-pigment complex that lies at the heart of photosynthesis. It is thought to have evolved more than 3 billion years ago in primitive photosynthetic bacteria. The special pair of chlorophyll molecules in the reaction center acts as an irreversible trap for excitation quanta because its excited electron is immediately passed to a chain of electron acceptors that are precisely positioned as neighbors in the same protein complex (Figure 14–49). By moving the high-energy electron rapidly away from the chlorophylls, the reaction center transfers it to an environment where it is much more stable. The electron is thereby suitably positioned for subsequent photochemical reactions, which require more time to complete.

In a Reaction Center, Light Energy Captured by Chlorophyll Creates a Strong Electron Donor from a Weak One [33]

The electron transfers involved in the photochemical reactions just outlined have been analyzed extensively by rapid spectroscopic methods, especially in the photosystem of purple bacteria, which is simpler than the evolutionarily related photosystem in chloroplasts. The bacterial reaction center is a large protein-pigment complex that can be solubilized with detergent and purified in active form. In 1985 its complete three-dimensional structure was determined by x-ray crys-

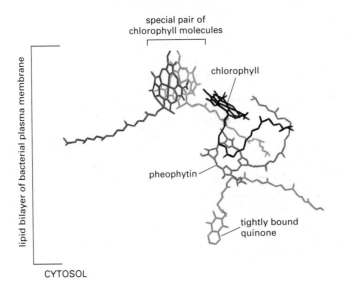

special pair of chlorophyll molecules

chlorophyll

pheophytin

tightly bound quinone

lipid bilayer of bacterial plasma membrane

CYTOSOL

Figure 14–49 **The arrangement of the electron carriers in a bacterial photochemical reaction center as determined by x-ray crystallography.** The pigment molecules shown are held in the interior of a transmembrane protein and are surrounded by the lipid bilayer, as indicated. An electron in the special pair is excited by resonance from an antenna complex chlorophyll (process 2 in Figure 14–47), and the excited electron is then transferred stepwise from the special pair to the quinone (see Figure 14–50).

Figure 14–50 The electron transfers that occur in the photochemical reaction center of a purple bacterium. A similar set of reactions is believed to occur in the evolutionarily related photosystem II in plants. At the top right is a schematic diagram showing the molecules that carry electrons, which are those in Figure 14–49, plus an exchangeable quinone (Q_B) and a freely mobile quinone (Q) dissolved in the lipid bilayer. Electron carriers 1 through 5 are each bound in a specific position on a 596-amino-acid transmembrane protein formed from two separate subunits (see Figure 10–33). Following excitation by a photon of light, a high-energy electron passes from pigment molecule to pigment molecule, very rapidly creating a charge separation as shown in the sequence in steps A through D below, where the pigment molecule carrying high-energy electrons is colored *green*. Step E occurs more slowly. Once released into the bilayer, the quinone with two electrons loses its charge by picking up two protons (see Figure 14–31).

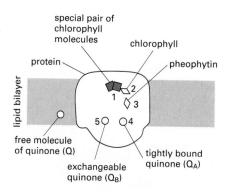

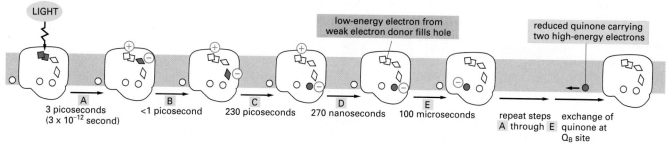

tallography (see Figure 10–33 and Figure 14–49). This structure, combined with kinetic data, provides the best picture we have of the initial electron-transfer reactions that underlie photosynthesis.

The sequence of transfers that take place in the reaction center of purple bacteria is shown in Figure 14–50. As outlined previously (Figure 14–48), in a reaction center, light causes a net electron transfer from a weak electron donor to a molecule that is a strong electron donor in its reduced form. In this way the excitation energy that would otherwise be released as fluorescence or heat or both is used instead to raise the energy of an electron and thereby create a strong electron donor where none had been before. In this bacterium the weak electron donor is a cytochrome (*orange box*), and the strong electron donor is a quinone (*yellow box*). In the chloroplasts of higher plants, as we discuss later, water, rather than cytochrome, serves as the initial electron donor, which is why oxygen is released by photosynthesis in plants.

In Plants and Cyanobacteria Noncyclic Photophosphorylation Produces Both NADPH and ATP [31, 34]

Photosynthesis in plants and cyanobacteria produces both ATP and NADPH directly by a two-step process called **noncyclic photophosphorylation.** Because two photosystems are used in series to energize an electron, the electron can be transferred all the way from water to NADPH. As the high-energy electrons pass through the coupled photosystems to generate NADPH, some of their energy is siphoned off for ATP synthesis.

In the first of the two photosystems—called *photosystem II* for historical reasons—the oxygens of two water molecules bind to a cluster of manganese atoms in a poorly understood water-splitting enzyme that enables electrons to be removed one at a time to fill the holes created by light in chlorophyll molecules in the reaction center. As soon as four electrons have been removed from the two water molecules (requiring four quanta of light), O_2 is released; photosystem II thus catalyzes the reaction $2H_2O + 4 \text{ photons} \rightarrow 4H^+ + 4e^- + O_2$.

The core of the reaction center in photosystem II is homologous to the bacterial reaction center just described, and it likewise produces strong electron donors in the form of reduced quinone molecules in the membrane. The quinones pass their electrons to a H^+ pump called the b_6-f *complex,* which closely resembles the b-c_1 complex in the respiratory chain of mitochondria and a re-

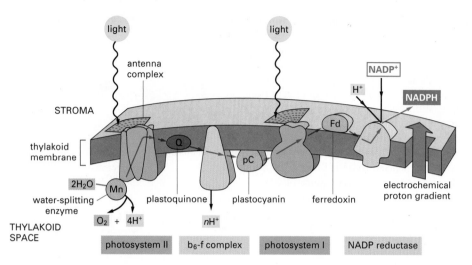

light   light

Figure 14–51 Electron flow during photosynthesis in the thylakoid membrane. The mobile electron carriers in the chain are plastoquinone (which closely resembles the ubiquinone of mitochondria), plastocyanin (a small copper-containing protein), and ferredoxin (a small protein containing an iron-sulfur center). The b_6-f *complex* closely resembles the b-c_1 complex of mitochondria and the b-c complex of bacteria (see Figure 14–61): all three complexes accept electrons from quinones and pump H^+ across the membrane. Note that the H^+ released by water oxidation and the H^+ taken up during NADPH formation also contribute to the generation of the electrochemical H^+ gradient, which drives ATP synthesis by an ATP synthase present in this same membrane (not shown).

lated complex in bacteria. As in mitochondria, the complex pumps H^+ into the thylakoid space across the thylakoid membrane (or out of the cytosol across the plasma membrane in cyanobacteria), and the resulting electrochemical gradient drives the synthesis of ATP by an ATP synthase (Figures 14–51 and 14–52). The final electron acceptor in this electron-transport chain is the second photosystem (*photosystem I*), which accepts an electron into the hole created by light in the chlorophyll molecule in its reaction center. Each electron that enters photosystem I is boosted to a very high energy level that allows it to be passed to the iron-sulfur center in ferredoxin and then to $NADP^+$ to generate NADPH; this last step also involves the uptake of a H^+ from the medium (Figure 14–52).

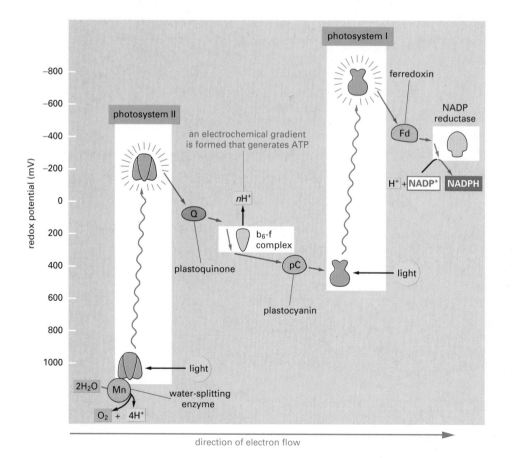

Figure 14–52 Changes in redox potential during photosynthesis. The redox potential for each molecule is indicated by its position along the vertical axis. Photosystem II closely resembles the reaction center in purple bacteria. Photosystem I is different: it passes electrons from its excited chlorophyll through a series of tightly bound iron-sulfur centers. The net electron flow through the two photosystems in series is from water to $NADP^+$, and it produces NADPH as well as ATP, which is synthesized by an ATP synthase (not shown) that harnesses the electrochemical proton gradient produced by the three sites of H^+ activity that are highlighted in Figure 14–51. This *Z scheme* for ATP production is called *noncyclic photophosphorylation* to distinguish it from a cyclic scheme that utilizes only photosystem I (see the text).

The scheme for photosynthesis shown in Figure 14–52 is known as the *Z scheme*. By means of its two electron-energizing steps, one catalyzed by each photosystem, an electron is passed from water, which normally holds on to its electrons very tightly (redox potential = +820 mV), to NADPH, which normally holds on to its electrons more loosely (redox potential = –320 mV). There is not enough energy in a single quantum of visible light to energize an electron all the way from the bottom of photosystem II to the top of photosystem I, which is probably the energy change required to pass an electron efficiently from water to NADP$^+$. The use of two separate photosystems in series also means that there is enough energy left over to enable the electron-transport chain that links the two photosystems to pump H$^+$ across the thylakoid membrane (or the plasma membrane of cyanobacteria), which allows ATP synthase to harness some of the light-derived energy for ATP production.

Chloroplasts Can Make ATP by Cyclic Photophosphorylation Without Making NADPH [31, 35]

In the noncyclic photophosphorylation scheme just discussed, high-energy electrons leaving photosystem II are harnessed to generate ATP and are passed on to photosystem I to drive the production of NADPH. This produces slightly more than one molecule of ATP for every pair of electrons that passes from H$_2$O to NADP$^+$ to generate a molecule of NADPH. But one and a half molecules of ATP per NADPH are needed for carbon fixation (see Figure 14–44). To produce extra ATP, the chloroplasts in some species of plants can switch photosystem I into a cyclic mode so that it produces ATP instead of NADPH. In this process, called *cyclic photophosphorylation*, the high-energy electrons from photosystem I are transferred back to the b$_6$-f complex rather than being passed on to NADP$^+$, and the electron is then recycled to photosystem I at a low energy. The only net result, besides the conversion of some light energy to heat, is that H$^+$ is pumped across the thylakoid membrane by the b$_6$-f complex to increase the electrochemical proton gradient that drives the ATP synthase.

In summary, cyclic photophosphorylation involves only photosystem I, and it produces ATP without the formation of either NADPH or O$_2$. Thus the relative activities of cyclic and noncyclic electron flows can determine how much light energy is converted into reducing power (NADPH) and how much into high-energy phosphate bonds (ATP).

The Electrochemical Proton Gradient Is Similar in Mitochondria and Chloroplasts [36]

The presence of the thylakoid space separates a chloroplast into three rather than the two internal compartments of a mitochondrion. The net effect of H$^+$ translocation in the two organelles, however, is similar. As illustrated in Figure 14–53, in chloroplasts H$^+$ is pumped out of the stroma (pH 8) into the thylakoid space (pH about 5), creating a gradient of 3 to 3.5 pH units. This represents a proton-motive force of about 200 mV across the thylakoid membrane (nearly all of which is contributed by the pH gradient rather than by a membrane potential), which drives ATP synthesis by the ATP synthase embedded in this membrane.

Like the stroma, the mitochondrial matrix has a pH of about 8, but this is created by pumping H$^+$ out of the mitochondrion into the cytosol (pH about 7) rather than into an interior space in the organelle. Thus the pH gradient is relatively small, and most of the proton-motive force across the mitochondrial inner membrane, which is about the same as that across the chloroplast thylakoid membrane, is caused by the resulting membrane potential. For both mitochondria and chloroplasts, however, the catalytic site of the ATP synthase is at a pH of about 8 and is located in a large organelle compartment (matrix or stroma) packed full of soluble enzymes. Consequently, it is here that all of the organelle's ATP is made (Figure 14–53).

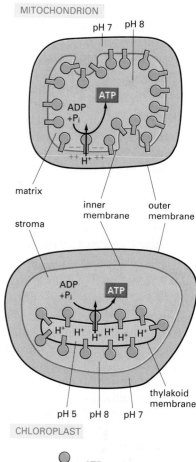

= ATP synthase

Figure 14–53 Comparison of the flow of H$^+$ and the orientation of ATP synthase in mitochondria and chloroplasts. Those compartments with a similar pH have been colored similarly. The proton-motive force across the thylakoid membrane consists almost entirely of the pH gradient; a high permeability of this membrane to Mg^{2+} and Cl$^-$ ions allows the flow of these ions to dissipate most of the membrane potential. Mitochondria presumably need a large membrane potential because they could not tolerate having their matrix at pH 10, as would be required to generate their proton-motive force without one.

Although there are many similarities between mitochondria and chloroplasts, the structure of chloroplasts makes their electron- and proton-transport processes easier to study: by breaking both the inner and outer membranes of a chloroplast, isolated thylakoid discs can be obtained intact. These thylakoids resemble submitochondrial particles in that they have a membrane whose electron-transport chain has its utilization sites for $NADP^+$, ADP, and phosphate all freely accessible to the outside. But isolated thylakoids retain their undisturbed native structure and are much more active than isolated submitochondrial particles. For this reason several of the experiments that first demonstrated the central role of chemiosmotic mechanisms were carried out with chloroplasts rather than with mitochondria.

Like the Mitochondrial Inner Membrane, the Chloroplast Inner Membrane Contains Carrier Proteins That Facilitate Metabolite Exchange with the Cytosol [37]

If chloroplasts are isolated in a way that leaves their inner membrane intact, this membrane can be shown to have a selective permeability, reflecting the presence of specific carrier proteins. Most notably, much of the glyceraldehyde 3-phosphate produced by CO_2 fixation in the chloroplast stroma is transported out of the chloroplast by an efficient antiport system that exchanges three-carbon sugar-phosphates for inorganic phosphate.

Glyceraldehyde 3-phosphate normally provides the cytosol with an abundant source of carbohydrate, which is used by the cell as the starting point for many other biosyntheses—including the production of sucrose for export. But this is not all it provides. Once the glyceraldehyde 3-phosphate reaches the cytosol, it is readily converted (by part of the glycolytic pathway) to 3-phosphoglycerate, generating one molecule of ATP and one of NADH. (A very similar two-step reaction working in reverse forms glyceraldehyde 3-phosphate in the carbon-fixation cycle—see Figure 14–44.) As a result, the export of glyceraldehyde 3-phosphate from the chloroplast provides not only the main source of fixed carbon to the rest of the cell, but also the reducing power and ATP needed for metabolism outside the chloroplast.

Chloroplasts Carry Out Other Biosyntheses

The chloroplast carries out many biosyntheses in addition to photosynthesis. All of the cell's fatty acids and a number of amino acids, for example, are made by enzymes in the chloroplast stroma. Similarly, the reducing power of light-activated electrons drives the reduction of nitrite (NO_2^-) to ammonia (NH_3) in the chloroplast; this ammonia provides the plant with nitrogen for the synthesis of amino acids and nucleotides. The metabolic importance of the chloroplast for plants and algae therefore extends far beyond its role in photosynthesis.

Summary

Chloroplasts and photosynthetic bacteria obtain high-energy electrons by means of photosystems that capture the electrons excited when sunlight is absorbed by chlorophyll molecules. Photosystems are composed of an antenna complex attached to a photochemical reaction center, which is a precisely ordered complex of proteins and pigments in which the photochemistry of photosynthesis occurs. By far the best-understood photochemical reaction center is that of the purple photosynthetic bacteria, for which the complete three-dimensional structure is known. Whereas these bacteria contain only a single photosystem, there are two photosystems in chloroplasts and cyanobacteria. The two photosystems are normally linked in series and transfer electrons from water to $NADP^+$ to form NADPH, with the concomitant production of a transmembrane electrochemical proton gradient; molecular oxygen (O_2) is generated as a by-product.

Compared to mitochondria, chloroplasts have an additional internal membrane (the thylakoid membrane) and internal space (the thylakoid space). All electron-transport processes occur in the thylakoid membrane: to make ATP, H^+ is pumped into the thylakoid space and a backflow of H^+ through an ATP synthase then produces the ATP in the chloroplast stroma. This ATP is used in conjunction with the NADPH made by photosynthesis to drive a large number of biosynthetic reactions in the chloroplast stroma, including the all-important carbon-fixation cycle, which creates carbohydrate from CO_2. Along with other chloroplast products, this carbohydrate is exported to the cell cytosol, where—as glyceraldehyde 3-phosphate—it provides organic carbon, ATP, and reducing power to the rest of the cell.

The Evolution of Electron-Transport Chains [38]

Much of the structure, function, and evolution of cells and organisms can be related to their need for energy. We have seen that the fundamental mechanisms for harnessing energy from such disparate sources as light and the oxidation of glucose are the same. Apparently, an effective method for synthesizing ATP arose early in evolution and has since been conserved with only small variations. How did the crucial individual components—ATP synthase, redox-driven H^+ pumps, and photosystems—first arise? Hypotheses about events occurring on an evolutionary time scale are difficult to test. But clues abound, both in the many different primitive electron-transport chains that survive in some present-day bacteria and in geological evidence concerning the environment of the earth billions of years ago.

The Earliest Cells Probably Produced ATP by Fermentation [39]

As explained in Chapter 1, the first living cells are thought to have arisen more than 3.5×10^9 years ago, when the earth was not more than about 10^9 years old. Because the environment lacked oxygen but was rich in geochemically produced

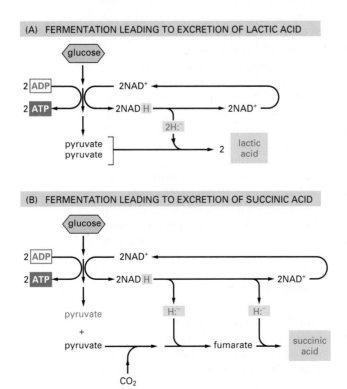

Figure 14–54 Two types of fermentation processes. The end products are highlighted by *green boxes*. In both cases two molecules of NAD^+ are used for each molecule of glucose that undergoes glycolysis, and these are regenerated by the transfer of hydride ions from NADH. In (A) the hydride ions are transferred to pyruvate to produce two molecules of lactic acid, which is excreted. In (B) the two hydride ions are successively transferred from two NADH molecules to compounds derived from pyruvate that produce succinic acid; for each molecule of succinic acid excreted, a molecule of pyruvate (*red*) is saved for biosyntheses inside the cell. In both (A) and (B) an organic acid must be excreted to re-form NAD^+ and thereby enable glycolysis to continue in the absence of oxygen.

organic molecules, the earliest metabolic pathways for producing ATP presumably resembled present-day forms of fermentation.

In the process of **fermentation,** ATP is made by a phosphorylation event that harnesses the energy released when a hydrogen-rich organic molecule, such as glucose, is partly oxidized (see Figure 2–14). Without O_2 to serve as the final electron acceptor, the electrons lost from the oxidized organic molecules must be transferred (via NADH or NADPH) to a different organic molecule (or to a different part of the same molecule), which thereby becomes more reduced. At the end of the fermentation process one (or more) of the organic molecules produced is excreted into the medium as a metabolic waste product; others, such as pyruvate, are retained by the cell for biosynthesis.

The excreted end products are different in different organisms, but they tend to be organic acids (carbon compounds that carry a COOH group). Among the most important of such products in bacterial cells are lactic acid (which also accumulates in anaerobic mammalian glycolysis) and formic, acetic, propionic, butyric, and succinic acids. Two fermentation pathways of present-day bacteria are illustrated in Figure 14–54.

The Evolution of Energy-conserving Electron-transport Chains Enabled Anaerobic Bacteria to Use Non-fermentable Organic Compounds as a Source of Energy [40]

The early fermentation processes would have provided not only the ATP but also the reducing power (as NADH or NADPH) required for essential biosyntheses, and many of the major metabolic pathways probably evolved while fermentation was the only mode of energy production. With time, however, the metabolic activities of these procaryotic organisms must have changed the local environment, forcing organisms to evolve new biochemical pathways. The accumulation of waste products of fermentation, for example, might have resulted in the following series of changes:

Stage 1. The continuous excretion of organic acids lowered the pH of the environment, favoring the evolution of proteins that function as transmembrane H^+ pumps that could pump H^+ out of the cell to protect it from the dangerous effects of intracellular acidification. One of these pumps may have used the energy available from ATP hydrolysis and could have been the ancestor of the present-day ATP synthase.

Stage 2. At the same time that nonfermentable organic acids were accumulating in the environment and favoring the evolution of an ATP-consuming H^+ pump, the supply of geochemically generated fermentable nutrients, which provided the energy for the pumps and for all other cellular processes, was dwindling. This favored bacteria that could excrete H^+ without hydrolyzing ATP, allowing the ATP to be conserved for other cellular activities. Selective pressures of this kind might have led to the first membrane-bound proteins that could use electron transport between molecules of different redox potential as the energy source for transporting H^+ across the plasma membrane. Some of these proteins would have found their electron donors and electron acceptors among the nonfermentable organic acids that had accumulated. Many such electron-transport proteins can be found in present-day bacteria: some bacteria that grow on formic acid, for example, pump H^+ by using the small amount of redox energy derived from the transfer of electrons from formic acid to fumarate (Figure 14–55). Others have similar electron-transport components devoted solely to the oxidation and reduction of inorganic substrates (see Figure 14–57, for example).

Stage 3. Eventually some bacteria developed H^+-pumping electron-transport systems that were efficient enough to harness more redox energy than they needed just to maintain their internal pH. Now, bacteria that car-

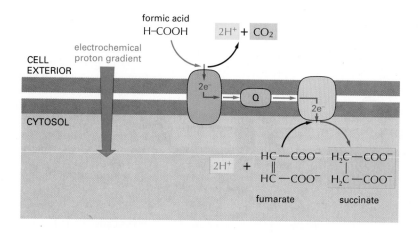

CELL EXTERIOR

electrochemical proton gradient

formic acid
H—COOH

$2H^+$ + CO_2

$2e^-$

Q

$2e^-$

CYTOSOL

$2H^+$ +

HC—COO⁻
‖
HC—COO⁻
fumarate

H_2C—COO⁻
|
H_2C—COO⁻
succinate

Figure 14–55 The oxidation of formic acid in some present-day bacteria. In such anaerobic bacteria, including *E. coli*, the oxidation is mediated by an energy-conserving electron-transport chain in the plasma membrane. As indicated, the starting materials are formic acid and fumarate, and the products are succinate and CO_2. Note that H^+ is consumed inside the cell and generated outside the cell, which is equivalent to pumping H^+ to the cell exterior. Thus this membrane-bound electron-transport system can generate an electrochemical proton gradient across the plasma membrane. The redox potential of the formic-acid–CO_2 pair is –420 mV, while that of the fumarate-succinate pair is +30 mV.

ried both types of H^+ pumps were at an advantage. In these cells a large electrochemical proton gradient generated by excessive H^+ pumping allowed protons to leak back into the cell through the ATP-driven H^+ pumps, thereby running them in reverse, so that they functioned as ATP synthases to make ATP. Because such bacteria required much less of the increasingly scarce supply of fermentable nutrients, they proliferated at the expense of their neighbors.

These three hypothetical stages in the evolution of oxidative phosphorylation mechanisms are summarized in Figure 14–56.

By Providing an Inexhaustible Source of Reducing Power, Photosynthetic Bacteria Overcame a Major Obstacle in the Evolution of Cells [41]

The evolutionary steps just outlined would have solved the problem of maintaining both a neutral intracellular pH and an abundant store of energy, but they would not have solved another problem that was equally serious. The depletion of organic nutrients from the environment meant that organisms had to find some alternative source of carbon to make the sugars that served as the precursors for so many other cellular molecules. Although the CO_2 in the atmosphere provided an abundant potential carbon source, to convert it into an organic molecule such as a carbohydrate requires that the fixed CO_2 be reduced by a strong electron donor, such as NADH or NADPH, which can provide the high-energy electrons needed to generate each (CH_2O) unit from CO_2 (see Figure 14–44). Early in cellular evolution, strong reducing agents (electron donors) would have been plentiful as products of fermentation. But as the supply of fermentable nutrients dwindled and a membrane-bound ATP synthase began to produce most of the ATP, the plentiful supply of NADH and other reducing agents would have disappeared. It thus became imperative for cells to evolve a new way of generating strong reducing agents.

Presumably, the main reducing agents still available were the organic acids produced by the anaerobic metabolism of carbohydrates, inorganic molecules such as hydrogen sulfide (H_2S) generated geochemically, and water. But the reducing power of all of these molecules is far too weak to be useful for CO_2 fixation. An early supply of strong electron donors could have been generated by using the electrochemical proton gradient across the plasma membrane to drive a *reverse electron flow*. This would have required the evolution of membrane-bound enzyme complexes resembling a NADH dehydrogenase, and mechanisms of this kind survive in the anaerobic metabolism of some present-day bacteria (Figure 14–57). The major evolutionary breakthrough in energy metabolism, however, was almost certainly the development of photochemical reaction centers that could use the energy of sunlight to produce molecules such as NADH.

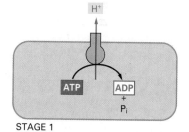

H^+

ATP

ADP + P_i

STAGE 1

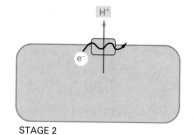

H^+

e^-

STAGE 2

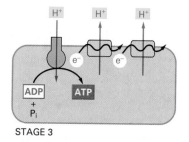

H^+ H^+ H^+

e^- e^-

ADP + P_i

ATP

STAGE 3

Figure 14–56 The evolution of oxidative phosphorylation mechanisms. One possible sequence is shown.

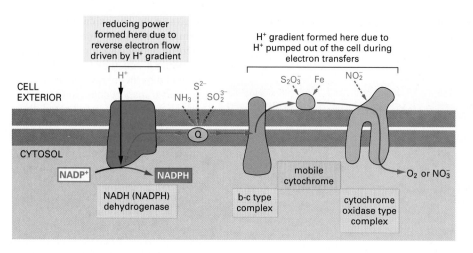

reducing power formed here due to reverse electron flow driven by H^+ gradient

H^+ gradient formed here due to H^+ pumped out of the cell during electron transfers

CELL EXTERIOR

H^+

NH_3 S^{2-} SO_2^{3-}

$S_2O_3^-$ Fe NO_2^-

Q

CYTOSOL

$NADP^+$ NADPH

NADH (NADPH) dehydrogenase

b-c type complex

mobile cytochrome

cytochrome oxidase type complex

O_2 or NO_3^-

Figure 14–57 Some of the electron-transport pathways in present-day bacteria. These pathways generate all the cell's ATP and reducing power from the oxidation of inorganic molecules, such as iron, ammonia, nitrite, and sulfur compounds. As indicated, some species can grow anaerobically by substituting nitrate for oxygen as the terminal electron acceptor. Most use the carbon-fixation cycle and synthesize their organic molecules entirely from carbon dioxide. The forward electron flows cause H^+ to be pumped out of the cell, and the resulting H^+ gradient drives the production of ATP by an ATP synthase (not shown). The NADPH required for carbon fixation is produced by a reverse electron flow that is also driven by the H^+ gradient, as indicated.

It is thought that this occurred early in the process of cellular evolution—probably more than 3×10^9 years ago, in the ancestors of the green sulfur bacteria. Present-day green sulfur bacteria use light energy to transfer hydrogen atoms (as an electron plus a proton) from H_2S to NADPH, thereby creating the strong reducing power required for carbon fixation (Figure 14–58). Because the electrons removed from H_2S are at a much more negative redox potential than those of H_2O (–230 mV compared to +820 mV for H_2O), one quantum of light absorbed by the single photosystem in these bacteria is sufficient to achieve a high enough redox potential to generate NADPH via a relatively simple photosynthetic electron-transport chain.

The More Complex Photosynthetic Electron-Transport Chains of Cyanobacteria Produced Atmospheric Oxygen and Permitted New Life Forms [42]

The next step, which is thought to have occurred with the development of the cyanobacteria at least 3×10^9 years ago, was the evolution of organisms capable of using water as the electron source for CO_2 reduction. This entailed the evolution of a water-splitting enzyme and also required the addition of a second photosystem, acting in series with the first, to bridge the enormous gap in redox potential between H_2O and NADPH. Present-day structural homologies between photosystems suggest that this change involved the cooperation of a photosystem derived from green bacteria (photosystem I) with a photosystem derived

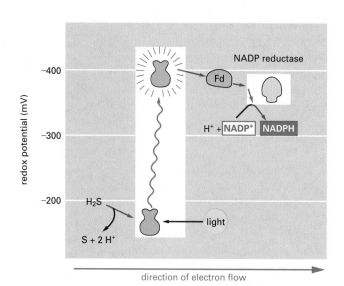

redox potential (mV)

–400

–300

–200

NADP reductase

Fd

H^+ + $NADP^+$ NADPH

H_2S

$S + 2 H^+$

light

direction of electron flow

Figure 14–58 The general flow of electrons in a relatively primitive form of photosynthesis observed in present-day green sulfur bacteria. The photosystem in green bacteria resembles photosystem I in plants and cyanobacteria in using a series of iron-sulfur centers as primary electron acceptors that eventually donate their high-energy electrons to ferredoxin (Fd).

from purple bacteria (photosystem II). The biological consequences of this evolutionary step were far-reaching. For the first time there were organisms that made only very minimal chemical demands on their environment and therefore could spread and evolve in ways denied the earlier photosynthetic bacteria, which needed H_2S or organic acids as a source of electrons. Consequently, large amounts of biologically synthesized, reduced organic materials accumulated. Moreover, oxygen entered the atmosphere for the first time.

Oxygen is highly toxic because the oxidation reactions it brings about can randomly alter biological molecules. Many present-day anaerobic bacteria, for example, are rapidly killed when exposed to air. Thus organisms on the primitive earth would have had to evolve protective measures against the rising O_2 levels in the environment. Late evolutionary arrivals, such as ourselves, have numerous detoxifying mechanisms that protect our cells from the ill effects of oxygen.

The increase in atmospheric O_2 was very slow at first and would have allowed a gradual evolution of protective devices. The early seas contained large amounts of ferrous iron (Fe[II]), and nearly all the O_2 produced by early photosynthetic bacteria was utilized in converting Fe(II) to Fe(III). This conversion caused the precipitation of huge amounts of ferric oxides, and the extensive banded iron formations beginning about 2.7×10^9 years ago help to date the rise of the cyanobacteria. By about 2×10^9 years ago the supply of ferrous iron was exhausted and the deposition of further iron precipitates ceased. The geological evidence suggests that O_2 levels in the atmosphere then began to rise, reaching current levels between 0.5 and 1.5×10^9 years ago (Figure 14–59).

The availability of O_2 made possible the development of bacteria that relied on aerobic metabolism to make their ATP. As explained previously, these organisms could harness the large amount of energy released by breaking down carbohydrates and other reduced organic molecules all the way to CO_2 and H_2O. Components of preexisting electron-transport complexes were modified to produce a cytochrome oxidase, so that the electrons obtained from organic or inorganic substrates could be transported to O_2 as the terminal electron acceptor. Many present-day purple photosynthetic bacteria can switch between photosynthesis and respiration, depending on the availability of light and O_2, by surprisingly minor reorganizations of their electron-transport chains.

As organic materials accumulated on earth as a result of photosynthesis, some photosynthetic bacteria (including the precursors of *E. coli*) lost their ability to survive on light energy alone and came to rely entirely on respiration. It is believed that mitochondria first arose some 1.5×10^9 years ago, when a primitive eucaryotic cell endocytosed such a respiration-dependent bacterium. Plants are believed to have evolved when a descendant of this early aerobic eucaryotic

Figure 14–59 The relationship between changes in atmospheric O_2 levels and some of the major stages that are believed to have occurred during the evolution of living organisms on earth. As indicated, geological evidence suggests that there was more than a billion-year delay between the rise of cyanobacteria (thought to be the first organisms to release O_2) and the time that high O_2 levels began to accumulate in the atmosphere. This delay was probably due largely to the rich supply of dissolved ferrous iron in the oceans, which reacted with the released O_2 to form enormous iron oxide deposits.

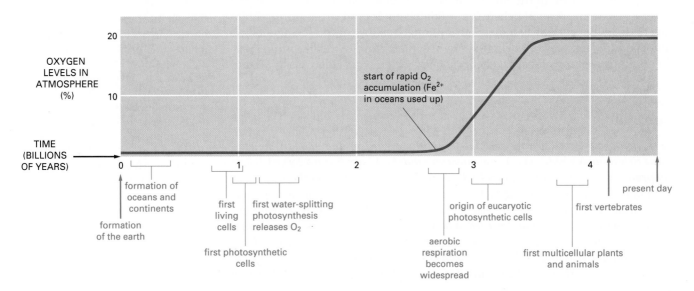

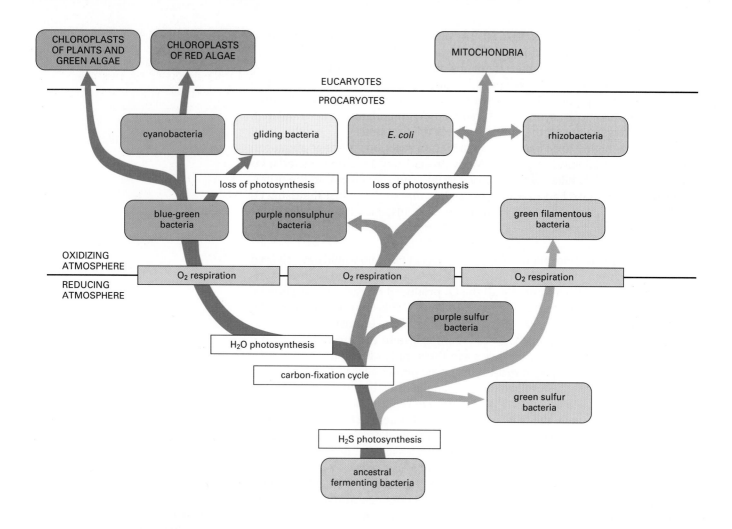

The following labels appear in the figure:

CHLOROPLASTS OF PLANTS AND GREEN ALGAE

CHLOROPLASTS OF RED ALGAE

MITOCHONDRIA

EUCARYOTES

PROCARYOTES

cyanobacteria

gliding bacteria

E. coli

rhizobacteria

loss of photosynthesis

loss of photosynthesis

blue-green bacteria

purple nonsulphur bacteria

green filamentous bacteria

OXIDIZING ATMOSPHERE

REDUCING ATMOSPHERE

O_2 respiration

O_2 respiration

O_2 respiration

purple sulfur bacteria

H_2O photosynthesis

carbon-fixation cycle

green sulfur bacteria

H_2S photosynthesis

ancestral fermenting bacteria

cell endocytosed a photosynthetic bacterium that became the precursor of chloroplasts; present-day chloroplasts are so different in different types of algae, however, that chloroplasts probably evolved separately in these different lineages. Figure 14–60 outlines some of the suspected evolutionary pathways just discussed.

Evolution is always conservative, taking parts of the old and building upon them to create something new. Thus parts of the electron-transport chains that were derived to service anaerobic bacteria 3 to 4 billion years ago probably survive, in altered form, in the mitochondria and chloroplasts of today's higher eucaryotes. As one example, there is a striking homology in structure and function between the enzyme complexes that pump H+ in the central segment of the mitochondrial respiratory chain (the b-c$_1$ complex) and the corresponding segments of the electron-transport chains of both bacteria and chloroplasts (Figure 14–61).

Summary

Early cells are believed to have been bacteriumlike organisms living in an environment rich in highly reduced organic molecules that had been formed by geochemical processes over the course of hundreds of millions of years. They probably derived most of their ATP by converting these reduced organic molecules to a variety of organic acids, which were then released as waste products. By acidifying the environment, these fermentations may have led to the evolution of the first membrane-bound H+ pumps, which could maintain a neutral pH in the cell interior. The properties of present-day bacteria suggest that an electron-transport-driven H+ pump and an ATP-driven H+ pump first arose in this anaerobic environment. Reversal of the ATP-driven

Figure 14–60 A phylogenetic tree of the probable evolution of mitochondria and chloroplasts and their bacterial ancestors. Oxygen respiration is thought to have begun developing about 2×10^9 years ago. As indicated, it seems to have evolved independently in the green, purple, and blue-green (cyanobacterial) lines of photosynthetic bacteria. It is thought that an aerobic purple bacterium that had lost its ability to photosynthesize gave rise to the mitochondrion, while several different blue-green bacteria gave rise to chloroplasts. Nucleotide sequence analyses suggest that mitochondria arose from bacteria that resembled the rhizobacteria, agrobacteria, and rickettsias—three closely related species known to form intimate associations with present-day eucaryotic cells.

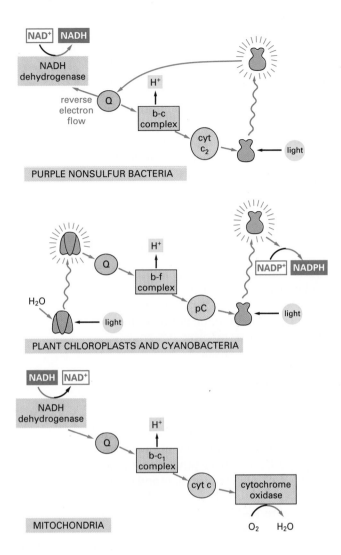

Bacteria, chloroplasts, and mitochondria all contain a membrane-bound enzyme complex that closely resembles the cytochrome b-c$_1$ complex of mitochondria. These complexes all accept electrons from a quinone carrier (designated as Q) and pump H$^+$ across their respective membranes. Moreover, in reconstituted *in vitro* systems the different complexes can substitute for one another, and the amino acid sequences of their protein components reveal that they are evolutionarily related.

pump would have allowed it to function as an ATP synthase. As more effective electron-transport chains developed, the energy released by redox reactions between inorganic molecules and accumulated nonfermentable compounds produced a large electrochemical proton gradient, which could be harnessed by the ATP-driven pump for ATP production.

Because preformed organic molecules were replenished only very slowly by geochemical processes, the proliferation of bacteria that used them as the source of both carbon and reducing power could not go on forever. The depletion of fermentable organic nutrients presumably led to the evolution of bacteria that could use CO_2 to make carbohydrates. By combining parts of the electron-transport chains that had developed earlier, light energy was harvested by a single photosystem in photosynthetic bacteria to generate the NADPH required for carbon fixation. The subsequent appearance of the more complex photosynthetic electron-transport chains of the cyanobacteria allowed H_2O to be used as the electron donor for NADPH formation, rather than the much less abundant electron donors required by other photosynthetic bacteria. Life could then proliferate over large areas of the earth, so that reduced organic molecules accumulated again. About 2 billion years ago, the O_2 released by photosynthesis in cyanobacteria began to accumulate in the atmosphere. Once both organic molecules and O_2 were abundant, electron-transport chains became adapted for the transport of electrons from NADH to O_2, and efficient aerobic metabolism developed in many bacteria. Exactly the same aerobic mechanisms operate in the mitochondria of eucaryotes, and there is considerable evidence that both mitochondria and chloroplasts evolved from aerobic bacteria that were endocytosed by primitive eucaryotic cells.

The Genomes of Mitochondria and Chloroplasts

Cells must generate new cytoplasmic organelles if they are to grow and divide. They must also replenish organelles that are degraded as part of the continual process of organelle turnover in nonproliferating cells. Organelle biosynthesis requires the ordered synthesis of the requisite proteins and lipids and the delivery of each component to the correct organelle subcompartment. In Chapter 12 we discussed how selected proteins and lipids are imported into mitochondria and chloroplasts from elsewhere in the cell. Here we describe the contributions that these energy-converting organelles make to their own biogenesis.

The Biosynthesis of Mitochondria and Chloroplasts Involves the Contribution of Two Separate Genetic Systems [43]

While most of the proteins in mitochondria and chloroplasts are encoded by nuclear DNA and imported into the organelle from the cytosol after they are synthesized on cytosolic ribosomes, some are encoded by organelle DNA and synthesized on ribosomes within the organelle. The protein traffic between the cytosol and these organelles seems to be unidirectional, as no protein is known to be exported from mitochondria or chloroplasts to the cytosol.

The contributions from the two genetic systems to the construction of mitochondria and chloroplasts are closely coordinated in the cell. Isolated organelles in a test tube continue to make organelle DNA, RNA, and proteins for brief periods, however, thereby providing one means of determining which proteins are encoded in organelle DNA and which in nuclear DNA. Another approach uses specific inhibitors on intact cells. The drug *cycloheximide*, for example, inhibits cytosolic protein synthesis but does not inhibit organelle protein synthesis. Conversely, various antibiotics (such as chloramphenicol, tetracycline, and erythromycin) inhibit protein synthesis in mitochondria and chloroplasts but have little effect on cytosolic protein synthesis (Figure 14–62). These inhibitors are widely used in studies of the functions of these organelles.

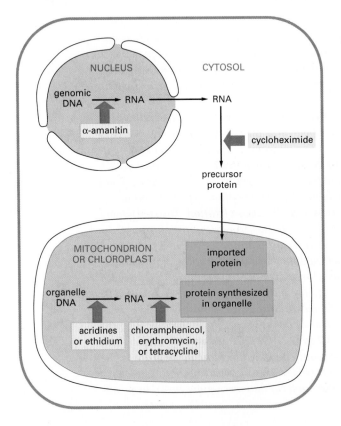

Figure 14–62 An overview of the biosynthesis of mitochondrial and chloroplast proteins. Each *red arrow* indicates the site of action of an inhibitor that is specific for either organelle or cytosolic protein synthesis.

Organelle Growth and Division Maintain the Number of Mitochondria and Chloroplasts in a Cell [44]

Mitochondria and chloroplasts are never made *de novo*. They always arise by the growth and division of existing mitochondria and chloroplasts. Observations of living cells indicate that mitochondria not only divide but also fuse with one another. On average, however, each organelle must double in mass and then divide in half once in each cell generation. Electron microscopic studies suggest that organelle division begins by an inward furrowing of the inner membrane, as occurs in cell division in many bacteria (Figures 14–63 and 14–64), implying that it is a controlled process rather than a chance pinching in two.

In most cells individual energy-converting organelles divide throughout interphase, out of phase with one another and with the division of the cell. Similarly, the replication of organelle DNA is not limited to the S phase, when nuclear DNA replicates, but occurs throughout the cell cycle. Individual organelle DNA molecules seem to be selected at random for replication, so that in a given cell cycle some may replicate more than once and others not at all. Nonetheless, under constant conditions the process is regulated to ensure that the total number of organelle DNA molecules doubles in every cell cycle, so that each cell type maintains a constant amount of organelle DNA.

The number of organelles per cell can be regulated according to need; a large increase in mitochondria (as much as five- to tenfold), for example, is observed if a resting skeletal muscle is repeatedly stimulated to contract for a prolonged period. Moreover, in special circumstances, organelle division is precisely controlled by the cell: thus, in some algae that contain only one or a few chloroplasts, the organelle divides just prior to cytokinesis in a plane that is identical to the future plane of cell division.

The Genomes of Chloroplasts and Mitochondria Are Usually Circular DNA Molecules [45]

Organelle DNA molecules are relatively small and simple, and, except for the mitochondrial genomes of some algae and protozoans, they are circular. The chloroplast genome (which is identical to the genomes of the other plastids in a plant) has a similar size in all organisms examined, but the mitochondrial genome is very much larger in plants than in animals (Table 14–2).

Many organelle DNA molecules are about the same size as typical viral DNAs. In mammals, for example, the mitochondrial genome is a DNA circle of about

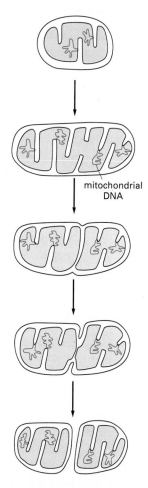

Figure 14–63 Diagram of a dividing mitochondrion. The pathway shown has been postulated from static views of dividing mitochondria like that in Figure 14–64.

mitochondrial DNA

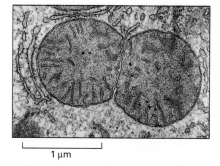

Figure 14–64 Electron micrograph of a dividing mitochondrion in a liver cell. (Courtesy of Daniel S. Friend.)

1 µm

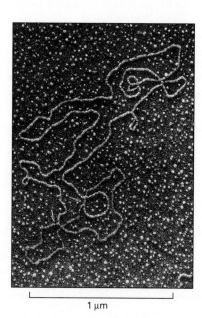

1 µm

Figure 14–65 Electron micrograph of an animal mitochondrial DNA molecule caught during the process of DNA replication. The circular DNA genome has replicated only between the two points marked by arrows (*yellow strands*). (Courtesy of David Clayton.)

Table 14-2 The Size of Organelle Genomes*

Type of DNA	Size (thousands of nucleotide pairs)
Chloroplast DNA	
Higher plants	120–200
Chlamydomonas (green alga)	180
Mitochondrial DNA	
Animals (including flatworms, insects, and mammals)	16–19
Higher plants	150–2500
Fungi	
Schizosaccharomyces pombe (fission yeast)	17
Aspergillus nidulans	32
Neurospora crassa	60
Saccharomyces cerevisiae (budding yeast)	78
Chlamydomonas (green alga)	16 (linear molecule)
Protozoa	
Trypanosoma brucei	22
Paramecium	40 (linear molecule)

*These genomes are circular DNA molecules unless indicated otherwise.

16,500 base pairs (less than 10^{-5} times the size of the nuclear genome). It is nearly the same size in animals as diverse as *Drosophila* and sea urchins (Figure 14–65). Plants, however, contain a circular mitochondrial genome that is 10 to 150 times larger, depending on the plant. The largest of these are about half the size of typical bacterial genomes, which are also circular DNA molecules.

All mitochondria and chloroplasts contain multiple copies of the organelle DNA molecule (Table 14–3). The molecules are usually distributed in several clusters in the matrix of the mitochondrion and in the stroma of the chloroplast, where they are thought to be attached to the inner membrane. Although it is not known how the DNA is packaged, the genome structure is likely to resemble that in bacteria rather than eucaryotic chromatin. As in bacteria, for example, there are no histones.

In mammalian cells mitochondrial DNA makes up less than 1% of the total cellular DNA. In other cells, however, such as the leaves of higher plants or the

Table 14-3 Relative Amounts of Organelle DNA in Some Cells and Tissues

Organism	Tissue or Cell Type	DNA Molecules per Organelle	Organelles per Cell	Organelle DNA as Percent of Total Cellular DNA
Mitochondrial DNA				
Rat	liver	5–10	1000	1
Yeast*	vegetative	2–50	1–50	15
Frog	egg	5–10	10^7	99
Chloroplast DNA				
Chlamydomonas	vegetative	80	1	7
Maize	leaves	20–40	20–40	15

*The large variation in the number and size of mitochondria per cell in yeast is due to mitochondrial fusion and fragmentation.

very large egg cells of amphibia, a much larger fraction of the cellular DNA may be present in the energy-converting organelles (see Table 14–3), and a larger fraction of RNA and protein synthesis takes place there.

Mitochondria and Chloroplasts Contain Complete Genetic Systems [46]

Despite the small number of proteins encoded in their genomes, mitochondria and plastids carry out their own DNA replication, DNA transcription, and protein synthesis. These processes take place in the matrix in mitochondria and in the stroma in chloroplasts. Although the proteins that mediate these genetic processes are unique to the organelle, most of them are encoded in the nuclear genome. This is all the more surprising because the protein-synthesis machinery of the organelles resembles that of bacteria rather than that of eucaryotes. The resemblance is particularly close in the case of chloroplasts:

1. Chloroplast ribosomes are very similar to *E. coli* ribosomes, both in their sensitivity to various antibiotics (such as chloramphenicol, streptomycin, erythromycin, and tetracycline) and in their structure. Not only are the nucleotide sequences of the ribosomal RNAs of chloroplasts and *E. coli* strikingly similar, but chloroplast ribosomes are able to use bacterial tRNAs in protein synthesis. In all these respects, chloroplast ribosomes differ from those found in the cytosol of the same plant cell.
2. Protein synthesis in chloroplasts starts with *N*-formylmethionine, as in bacteria, and not with methionine, as in the cytosol of eucaryotic cells.
3. Unlike nuclear DNA, chloroplast DNA can be transcribed by the RNA polymerase enzyme from *E. coli* to produce chloroplast mRNAs, and these mRNAs are efficiently translated by an *E. coli* protein-synthesizing system.

Although mitochondrial genetic systems are much less similar to those of present-day bacteria than are the genetic systems of chloroplasts, their ribosomes are also sensitive to antibacterial antibiotics, and protein synthesis in mitochondria also starts with *N*-formylmethionine.

The Chloroplast Genome of Higher Plants Contains About 120 Genes [47]

The best-studied chloroplast genomes are those of green algae and higher plants, whose chloroplasts are very similar circular DNA molecules. The complete nucleotide sequences have been determined for the chloroplasts of tobacco and liverwort. The results indicate that these two distantly related higher plants contain nearly identical chloroplast genes. In addition to four ribosomal RNAs, these genomes encode about 20 chloroplast ribosomal proteins, selected subunits of the chloroplast RNA polymerase, several proteins that are part of photosystems I and II, subunits of the ATP synthase, portions of enzyme complexes in the electron-transport chain, one of the two subunits of ribulose bisphosphate carboxylase, and 30 tRNAs (Figure 14–66). In addition, the DNA sequences present seem to encode at least 40 proteins whose functions are unknown. Paradoxically, all of the known proteins encoded in the chloroplast are part of larger protein complexes that also contain one or more subunits encoded in the nucleus. Possible reasons will be discussed later.

The similarities between the genomes of chloroplasts and bacteria are striking. The basic regulatory sequences, such as transcription promoters and terminators, are virtually identical in the two cases. Protein sequences encoded in chloroplasts are clearly recognizable as bacterial, and several clusters of genes with related functions (for example, those encoding ribosomal proteins) are organized in the same way in the genomes of chloroplasts, *E. coli*, and cyanobacteria.

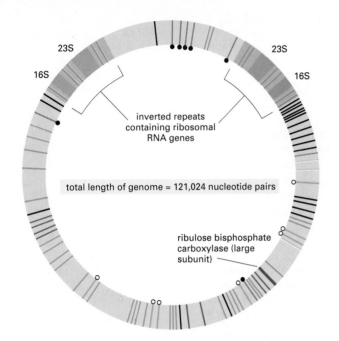

23S

23S

16S

16S

inverted repeats
containing ribosomal
RNA genes

total length of genome = 121,024 nucleotide pairs

ribulose bisphosphate
carboxylase (large
subunit)

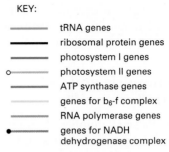

Figure 14–66 The organization of the liverwort chloroplast genome. The complete nucleotide sequence of this genome has been determined. The organization of the chloroplast genome is very similar in all higher plants, although the size varies from species to species depending on how much of the DNA surrounding the genes encoding the chloroplast's 16S and 23S ribosomal RNAs is present in two copies.

Detailed comparisons of large numbers of homologous nucleotide sequences should help to clarify the exact evolutionary pathway from bacteria to chloroplasts, but several conclusions can already be drawn. (1) Chloroplasts in higher plants arose from photosynthetic bacteria. (2) The chloroplast genome has been stably maintained for at least several hundred million years, the estimated time of divergence of liverwort and tobacco. (3) Many of the genes of the original bacterium are now present in the nuclear genome, where they have been transferred and stably maintained. In higher plants, for example, two-thirds of the 60 or so chloroplast ribosomal proteins are encoded in the cell nucleus, although the genes have a clear bacterial ancestry and the chloroplast ribosomes retain their original bacterial properties.

Mitochondrial Genomes Have Several Surprising Features [48]

The chloroplast genome was not the first organelle genome to be sequenced completely. The relatively small size of the human mitochondrial genome made it a particularly attractive target for molecular geneticists equipped with newly devised DNA-sequencing techniques, and in 1981 the complete sequence of its 16,569 nucleotides was published. By comparing this sequence with known mitochondrial tRNA sequences and with the partial amino acid sequences available for proteins encoded by the mitochondrial DNA, it has been possible to locate

Figure 14–67 The organization of the human mitochondrial genome. The genome contains 2 rRNA genes, 22 tRNA genes, and 13 protein-coding sequences. The DNAs of several other animal mitochondrial genomes have also been completely sequenced and have the same genes and gene organization.

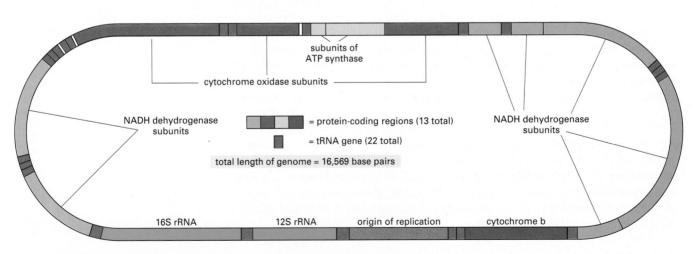

subunits of
ATP synthase

cytochrome oxidase subunits

NADH dehydrogenase
subunits

NADH dehydrogenase
subunits

= protein-coding regions (13 total)

= tRNA gene (22 total)

total length of genome = 16,569 base pairs

16S rRNA   12S rRNA   origin of replication   cytochrome b

Table 14-4 Some Differences Between the "Universal" Code and Mitochondrial Genetic Codes*

Codon	"Universal" Code	Mitochondrial Codes			
		Mammals	*Drosophila*	Yeasts	Plants
UGA	STOP	*Trp*	*Trp*	*Trp*	STOP
AUA	Ile	*Met*	*Met*	*Met*	Ile
CUA	Leu	Leu	Leu	*Thr*	Leu
AGA AGG	Arg	*STOP*	*Ser*	Arg	Arg

*Italics and color shading indicate that the code differs from the "universal" code.

all of the human mitochondrial genes on the circular DNA molecule (Figure 14–67).

Compared to nuclear, chloroplast, and bacterial genomes, the human mitochondrial genome has several surprising features. (1) Unlike other genomes, nearly every nucleotide appears to be part of a coding sequence, either for a protein or for one of the rRNAs or tRNAs. Since these coding sequences run directly into each other, there is very little room left for regulatory DNA sequences. (2) Whereas 30 or more tRNAs specify amino acids in the cytosol and in chloroplasts, only 22 tRNAs are required for mitochondrial protein synthesis. The normal codon-anticodon pairing rules are relaxed in mitochondria, so that many tRNA molecules recognize any one of the four nucleotides in the third (wobble) position. Such "2 out of 3" pairing allows one tRNA to pair with any one of four codons and permits protein synthesis with fewer tRNA molecules. (3) Perhaps most surprising, comparison of mitochondrial gene sequences and the amino acid sequences of the corresponding proteins indicates that the genetic code is different, so that 4 of the 64 codons have different "meanings" from those of the same codons in other genomes (Table 14–4).

The observation that the genetic code is nearly the same in all organisms provides strong evidence that all cells have evolved from a common ancestor. How, then, does one explain the few differences in the genetic code in mitochondria? A hint comes from the recent finding that the mitochondrial genetic code is different in different organisms. Thus UGA, which is a stop codon elsewhere, is read as tryptophan in mitochondria of mammals, fungi, and protozoans but as *stop* in plant mitochondria. Similarly, the codon AGG normally codes for arginine, but it codes for *stop* in the mitochondria of mammals and for serine in *Drosophila* (see Table 14–4). Such variation suggests that a random drift can occur in the genetic code in mitochondria. Presumably, the unusually small number of proteins encoded by the mitochondrial genome makes an occasional change in the meaning of a rare codon tolerable, whereas such a change in a large genome would alter the function of many proteins and thereby destroy the cell.

Animal Mitochondria Contain the Simplest Genetic Systems Known [49]

Comparisons of DNA sequences in different organisms reveal that the rate of nucleotide substitution during evolution has been 10 times greater in mitochondrial genomes than in nuclear genomes, which presumably is due to a reduced fidelity of mitochondrial DNA replication, DNA repair, or both. Because only about 16,500 DNA nucleotides need to be replicated and expressed as RNAs and proteins in animal cell mitochondria, the error rate *per nucleotide* copied by DNA replication, maintained by DNA repair, transcribed by RNA polymerase, or translated into protein by mitochondrial ribosomes can be relatively high without damaging one of the relatively few gene products. This could explain why the mechanisms that carry out these processes are relatively simple compared to those used for the same purpose elsewhere in cells. The presence of only 22 tRNAs and the unusually small size of the rRNAs (less than two-thirds the size of the *E. coli* rRNAs), for example, would be expected to reduce the fidelity of protein synthesis in mitochondria, although this has not yet been tested adequately.

The relatively high rate of evolution of mitochondrial genes makes mitochondrial DNA sequence comparisons especially useful for estimating the dates of relatively recent evolutionary events, such as the steps in primate development.

Why Are Plant Mitochondrial Genomes So Large? [50]

Mitochondrial genomes are much larger in plant than in animal cells, and they vary remarkably in their DNA content, ranging from about 150,000 to about 2.5×10^6 nucleotide pairs. Yet these genomes seem to encode only a few more proteins than do animal mitochondrial genomes. The paradox is compounded by the observation that in one family of plants, the cucurbits, mitochondrial genomes vary in size by as much as sevenfold. The green alga *Chlamydomonas* has a linear mitochondrial genome of only 16,000 nucleotide pairs, the same size as in animals.

Although very little sequence information is available for higher plant mitochondrial DNA molecules, almost all of the 70,000 nucleotide pairs in the large mitochondrial genome of the yeast *Saccharomyces cerevisiae* have been sequenced, and only about one-third of them code for protein. This finding raises the possibility that much of the extra DNA in yeast mitochondria, and possibly in plant mitochondria as well, is "junk DNA" of little consequence to the organism.

Some Organelle Genes Contain Introns [51]

The processing of precursor RNAs plays an important role in the two mitochondrial systems studied in most detail—human and yeast. In human cells both strands of the mitochondrial DNA are transcribed at the same rate from a single promoter region on each strand, producing two different giant RNA molecules, each containing a full-length copy of one DNA strand. Transcription is, therefore, completely symmetric. The transcripts made on one strand—called the *heavy strand (H strand)* because of its density in CsCl—are extensively processed by nuclease cleavage to yield the two rRNAs, most of the tRNAs, and about 10 poly-A-containing RNAs. In contrast, the *light strand (L strand)* transcript is processed to produce only eight tRNAs and one small poly-A-containing RNA; the remaining 90% of this transcript apparently contains no useful information (being complementary to coding sequences synthesized on the other strand) and is degraded. The poly-A-containing RNAs are the mitochondrial mRNAs: although they lack a cap structure at their 5' end, they carry a poly-A tail at their 3' end that is added posttranscriptionally by a mitochondrial poly-A polymerase.

Unlike human mitochondrial genes, some plant and fungal (including yeast) mitochondrial genes contain *introns*, which must be removed by RNA splicing. Introns have also been found in about 20 plant chloroplast genes. Many of the introns in organelle genes consist of related nucleotide sequences that are capable of splicing themselves out of the RNA transcripts by RNA-mediated catalysis (see p. 109), although these self-splicing reactions are generally aided by proteins. The presence of introns in organelle genes is surprising, as introns are not common in the genes of the bacteria whose ancestors are thought to have given rise to mitochondria and plant chloroplasts.

In yeasts the same mitochondrial gene may have an intron in one strain but not in another. Such "optional introns" seem to be able to move in and out of genomes like transposable elements. On the other hand, introns in other yeast mitochondrial genes have been found in a corresponding position in the mitochondria of *Aspergillus* and *Neurospora*, implying that they were inherited from a common ancestor of these three fungi. It seems likely that the intron sequences themselves are of ancient origin and that, while they have been lost from many bacteria, they have been preferentially retained in those organelle genomes where RNA splicing is regulated to help control gene expression.

Mitochondrial Genes Can Be Distinguished from Nuclear Genes by Their Non-Mendelian (Cytoplasmic) Inheritance [52]

Most experiments on the mechanisms of mitochondrial biogenesis have been performed with *Saccharomyces cerevisiae* (baker's yeast). There are several reasons for this. First, when grown on glucose, this yeast has an ability to live by glycolysis alone and can therefore survive without functional mitochondria, which are required for oxidative phosphorylation. This makes it possible to grow cells with mutations in mitochondrial or nuclear DNA that drastically interfere with mitochondrial biogenesis; such mutations are lethal in many other eucaryotes. Second, yeasts are simple unicellular eucaryotes that are easy to grow and characterize biochemically. Finally, these yeast cells normally reproduce asexually by budding (asymmetrical mitosis), but they can also reproduce sexually. During sexual reproduction two haploid cells mate and fuse to form a diploid zygote, which can either grow mitotically or divide by meiosis to produce new haploid

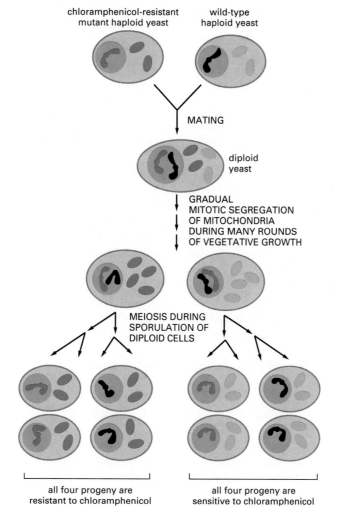

Figure 14–68 The difference in the pattern of inheritance between mitochondrial and nuclear genes of yeast. For each nuclear gene two of the four cells that result from meiosis inherit the gene from one of the original haploid parent cells and the remaining two cells inherit the gene from the other (*Mendelian inheritance*). In contrast, because of the gradual mitotic segregation of mitochondria during vegetative growth (see text), it is possible for all four of the cells that result from meiosis to inherit their mitochondrial genes from only one of the two original haploid cells (*non-Mendelian, or cytoplasmic, inheritance*). In this example the mitochondrial gene is one that (in its mutant form) makes protein synthesis in the mitochondrion resistant to chloramphenicol, a protein synthesis inhibitor that acts specifically on energy-converting organelles and bacteria. Yeast cells that contain the mutant gene can be detected by their ability to grow in the presence of chloramphenicol on a substrate, such as glycerol, that cannot be used for glycolysis. With glycolysis blocked, ATP must be provided by functional mitochondria, and therefore only cells that carry chloramphenicol-resistant mitochondria will grow.

cells. The ability to control the alternation between asexual and sexual reproduction in the laboratory greatly facilitates genetic analyses. Because mutations in mitochondrial genes are not inherited according to the Mendelian rules that govern the inheritance of nuclear genes, genetic studies reveal which of the genes involved in mitochondrial function are located in the nucleus and which in the mitochondria.

An example of **non-Mendelian (cytoplasmic) inheritance** of mitochondrial genes in a haploid yeast cell is illustrated in Figure 14–68. In this example we follow the inheritance of a mutant gene that makes mitochondrial protein synthesis resistant to chloramphenicol. When a chloramphenicol-resistant haploid cell mates with a chloramphenicol-sensitive wild-type haploid cell, the resulting diploid zygote will contain a mixture of mutant and wild-type mitochondria. But when the zygote undergoes mitosis to produce a diploid daughter, the mutant and wild-type mitochondria will be distributed at random between the mother and the daughter cell, so that each daughter is likely to inherit more mutant or more wild-type mitochondria. With successive mitotic divisions, either the mutant or the wild-type mitochondria will gradually be diluted out of some daughters by the same random process, leaving mitochondria of only one type. Thereafter, all of the progeny from that daughter will have mitochondria that are genetically identical. Thus this random process, called *mitotic segregation,* will eventually produce diploid yeast cells with only a single type of mitochondrial DNA. When such diploid cells undergo meiosis to form four haploid daughter cells, each of the four daughters receives the same mitochondrial genes. This type of inheritance is called *non-Mendelian,* or *cytoplasmic,* to contrast it with the Mendelian inheritance of nuclear genes (see Figure 14–68). When it occurs, it demonstrates that the gene in question is located outside the nuclear chromosomes and therefore probably in the yeast mitochondria.

Organelle Genes Are Maternally Inherited in Many Organisms [53]

The consequences of cytoplasmic inheritance are more profound for some organisms, including ourselves, than they are for yeasts. In yeasts, when two haploid cells mate, they are equal in size and contribute equal amounts of mitochondrial DNA to the zygote (see Figure 14–68). Mitochondrial inheritance in yeasts is therefore *biparental:* both parents contribute equally to the mitochondrial gene pool of the progeny (although, as we have just seen, after several generations of vegetative growth the *individual* progeny often contain mitochondria from only one parent). In higher animals, by contrast, the egg cell always contributes much more cytoplasm to the zygote than does the sperm. One would expect mitochondrial inheritance in higher animals, therefore, to be nearly *uniparental* (or more precisely, *maternal*). Such *maternal inheritance* has been demonstrated in laboratory animals. When animals carrying type A mitochondrial DNA are crossed with animals carrying type B, the progeny contain only the maternal type of mitochondrial DNA. Similarly, by following the distribution of variant mitochondrial DNA sequences in large families, human mitochondrial DNA has been shown to be maternally inherited.

In about two-thirds of higher plants the chloroplasts from the male parent (contained in pollen grains) do not enter the zygote, so that chloroplast as well as mitochondrial DNA is maternally inherited. In other plants the pollen chloroplasts enter the zygote, making chloroplast inheritance biparental. In such plants defective chloroplasts are a cause of *variegation:* a mixture of normal and defective chloroplasts in a zygote may sort out by mitotic segregation during plant growth and development, thereby producing alternating green and white patches in leaves. The green patches contain normal chloroplasts, while the white patches contain defective chloroplasts.

Petite Mutants in Yeasts Demonstrate the Overwhelming Importance of the Cell Nucleus for Mitochondrial Biogenesis [54]

Genetic studies of yeasts have played a crucial part in the analysis of mitochondrial biogenesis. A striking example is provided by studies of yeast mutants that contain large deletions in their mitochondrial DNA, so that all mitochondrial protein synthesis is abolished. Not surprisingly, these mutants cannot make respiring mitochondria. Some of these mutants lack mitochondrial DNA altogether. Because they form unusually small colonies when grown in media with low glucose, all mutants with such defective mitochondria are called *cytoplasmic petite mutants.*

Although petite mutants cannot synthesize proteins in their mitochondria and therefore cannot make mitochondria that produce ATP, they nevertheless contain mitochondria. These mitochondria have a normal outer membrane and an inner membrane with poorly developed cristae (Figure 14–69), and they contain virtually all of the mitochondrial proteins that are specified by nuclear genes and imported from the cytosol—including DNA and RNA polymerases, all of the citric acid cycle enzymes, and most inner membrane proteins—demonstrating the overwhelming importance of the nucleus in mitochondrial biogenesis. The petite mutants also show that an organelle that divides by fission can replicate indefinitely in the cytoplasm of proliferating eucaryotic cells even in the complete absence of its own genome. Many biologists believe that peroxisomes normally replicate in this way (see Figure 12–29).

For chloroplasts the nearest equivalent to yeast mitochondrial petite mutants are mutants of unicellular algae such as *Euglena.* Cells in which no chloroplast protein synthesis occurs still contain chloroplasts and are perfectly viable if oxidizable substrates are provided. If the development of mature chloroplasts is blocked in higher plants, however, either by raising the plants in the dark or because chloroplast DNA is defective or absent, the plants die as soon as their food stores run out.

Mitochondria and Chloroplasts Contain Tissue-specific Proteins [55]

Mitochondria can have specialized functions in particular types of cells. The *urea cycle,* for example, is the central metabolic pathway in mammals for disposing of cellular breakdown products that contain nitrogen. These products are excreted in the urine as urea. Nuclear-encoded enzymes in the mitochondrial matrix carry out several steps in the cycle. Urea synthesis occurs in only a few tissues, such as the liver, and the required enzymes are synthesized and imported into mitochondria only in these tissues. In addition, the respiratory enzyme complexes in the mitochondrial inner membrane of mammals contain several tissue-specific, nuclear-encoded subunits that are thought to act as regulators of electron transport. Thus some humans with a genetic muscle disease have a defective subunit of cytochrome oxidase; since the subunit is specific to skeletal muscle cells, their other cells, including their heart muscle cells, function normally, allowing the individuals to survive. As would be expected, tissue-specific differences are also found among the nuclear-encoded proteins in chloroplasts.

Mitochondria Import Most of Their Lipids; Chloroplasts Make Most of Theirs [56]

The biosynthesis of new mitochondria and chloroplasts requires lipids in addition to nucleic acids and proteins. Chloroplasts tend to make the lipids they require. In spinach leaves, for example, all cellular fatty acid synthesis takes place

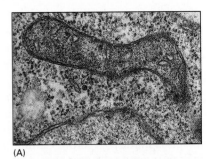

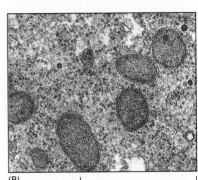

(A)

(B)

1 μm

Figure 14–69 Electron micrographs of yeast cells showing the structure of normal mitochondria (A) and mitochondria in a petite mutant (B). In petite mutants all of the mitochondrion-encoded gene products are missing, and so the organelle is constructed entirely from nucleus-encoded proteins. (Courtesy of Barbara Stevens.)

in the chloroplast, although desaturation of the fatty acids occurs elsewhere. The major glycolipids of the chloroplast are also synthesized locally.

Mitochondria, on the other hand, import most of their lipids. In animal cells the phospholipids phosphatidylcholine and phosphatidylserine are synthesized in the endoplasmic reticulum and then transferred to the outer membrane of mitochondria. In addition to decarboxylating imported phosphatidylserine to phosphatidylethanolamine, the main reaction of lipid biosynthesis catalyzed by the mitochondria themselves is the conversion of imported lipids to cardiolipin (bisphosphatidylglycerol). Cardiolipin is a "double" phospholipid that contains four fatty-acid tails; it is found mainly in the mitochondrial inner membrane, where it constitutes about 20% of the total lipid.

We have discussed the important question of how specific cytosolic proteins are imported into mitochondria and chloroplasts in detail in Chapter 12.

Both Mitochondria and Chloroplasts Probably Evolved from Endosymbiotic Bacteria [57]

As discussed in Chapter 1, the procaryotic character of the organelle genetic systems, especially striking in chloroplasts, suggests that mitochondria and chloroplasts evolved from bacteria that were endocytosed more than a billion years ago. According to this **endosymbiont hypothesis,** eucaryotic cells started out as anaerobic organisms without mitochondria or chloroplasts and then established a stable endosymbiotic relation with a bacterium, whose oxidative phosphorylation system they subverted for their own use (Figure 14–70). The endocytic event that led to the development of mitochondria is presumed to have occurred when oxygen entered the atmosphere in substantial amounts, about 1.5×10^9 years ago, before animals and plants separated (see Figure 14–59). Plant and algal

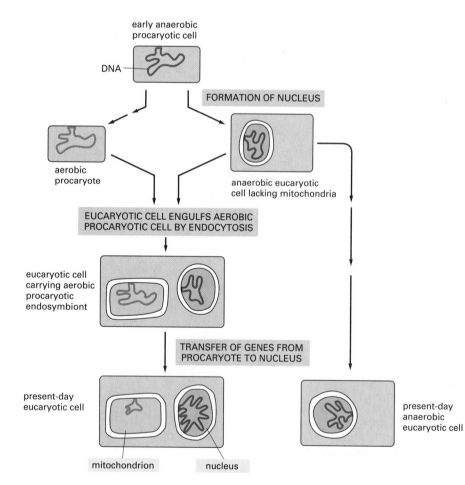

Figure 14–70 A suggested evolutionary pathway for the origin of mitochondria. *Microsporidia* and *Giardia* are two present-day anaerobic single-celled eucaryotes (protozoa) without mitochondria. Because they have an rRNA sequence that suggests a great deal of evolutionary distance from all other known eucaryotes, it has been postulated that their ancestors were also anaerobic and resembled the eucaryote that first engulfed the precursors of mitochondria.

chloroplasts seem to have been derived later from an endocytic event involving an oxygen-evolving photosynthetic bacterium. In order to explain the different pigments and properties of the chloroplasts found in present-day higher plants and algae, it is usually assumed that at least three separate events of this kind occurred.

Since most of the genes encoding present-day mitochondrial and chloroplast proteins are in the cell nucleus, it seems that an extensive transfer of genes from organelle to nuclear DNA has occurred during eucaryote evolution. This would explain why some of the nuclear genes encoding mitochondrial proteins resemble bacterial genes: the amino acid sequence of the chicken mitochondrial enzyme *superoxide dismutase*, for example, resembles the corresponding bacterial enzyme much more than it resembles the superoxide dismutase found in the cytosol of the same eucaryotic cells. Further evidence that such DNA transfers have occurred during evolution comes from the discovery of some noncoding DNA sequences in nuclear DNA that seem to be of recent mitochondrial origin; they have apparently integrated into the nuclear genome as "junk DNA."

What type of bacterium gave rise to the mitochondrion? Protein and nucleotide sequence analyses have provided evidence for the evolutionary tree shown previously in Figure 14–60. It appears that mitochondria are descendants of a particular type of purple photosynthetic bacterium that had previously lost its ability to carry out photosynthesis and was left with only a respiratory chain. It is not clear that all mitochondria have originated from a single endosymbiotic event, however. While the mitochondria from protozoans have distinctly procaryotic features, for example, some of them are sufficiently different from plant and animal mitochondria to suggest a separate origin.

Why Do Mitochondria and Chloroplasts Have Their Own Genetic Systems? [58]

Why do mitochondria and chloroplasts require their own separate genetic systems when other organelles that share the same cytoplasm, such as peroxisomes and lysosomes, do not? The question is not trivial because maintaining a separate genetic system is costly: more than 90 proteins—including many ribosomal proteins, aminoacyl-tRNA synthases, DNA and RNA polymerases, and RNA-processing and -modifying enzymes—must be encoded by nuclear genes specifically for this purpose (Figure 14–71). The amino acid sequences of most of these proteins in mitochondria and chloroplasts differ from those of their counterparts in the nucleus and cytosol, and there is reason to think that these organelles have relatively few proteins in common with the rest of the cell. This means that the nucleus must provide at least 90 genes just to maintain each organelle genetic system. The reason for such a costly arrangement is not clear, and the hope that the nucleotide sequences of mitochondrial and chloroplast genomes would provide the answer has proved unfounded. We cannot think of compelling reasons why the proteins made in mitochondria and chloroplasts should be made there rather than in the cytosol.

At one time it was suggested that some proteins have to be made in the organelle because they are too hydrophobic to get to their site in the membrane from the cytosol. More recent studies, however, make this explanation implausible. In many cases even highly hydrophobic subunits are synthesized in the cytosol. Moreover, although the individual protein subunits in the various mitochondrial enzyme complexes are highly conserved in evolution, their site of synthesis is not. The diversity in the location of the genes coding for the subunits of functionally equivalent proteins in different organisms is difficult to explain by any hypothesis that postulates a specific evolutionary advantage of present-day mitochondrial or chloroplast genetic systems.

Perhaps the organelle genetic systems are an evolutionary dead end. In terms of the endosymbiont hypothesis, this would mean that the process whereby the

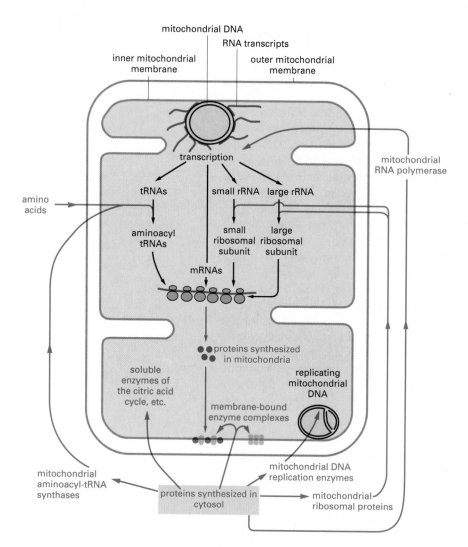

inner mitochondrial membrane

mitochondrial DNA

RNA transcripts

outer mitochondrial membrane

mitochondrial RNA polymerase

transcription

tRNAs

small rRNA

large rRNA

amino acids

aminoacyl tRNAs

small ribosomal subunit

large ribosomal subunit

mRNAs

proteins synthesized in mitochondria

soluble enzymes of the citric acid cycle, etc.

membrane-bound enzyme complexes

replicating mitochondrial DNA

mitochondrial aminoacyl-tRNA synthases

proteins synthesized in cytosol

mitochondrial DNA replication enzymes

mitochondrial ribosomal proteins

Figure 14–71 The origins of mitochondrial RNAs and proteins. The proteins imported from the cytosol play a major part in creating the genetic system of the mitochondrion in addition to contributing most of the organelle protein. The mitochondrion itself contributes only mRNAs, rRNAs, and tRNAs to its genetic system. Not indicated in this diagram are the additional nucleus-encoded proteins that regulate the expression of individual mitochondrial genes at posttranscriptional levels.

endosymbionts transferred most of their genes to the nucleus stopped before it was complete. Further transfers may have been ruled out, in the case of mitochondria, by recent alterations in the mitochondrial genetic code that made the remaining mitochondrial genes nonfunctional if they were transferred to the nucleus.

Summary

Mitochondria and chloroplasts grow and divide in two in a coordinated process that requires the contribution of two separate genetic systems—that of the organelle and that of the cell nucleus. Most of the proteins in these organelles are encoded by nuclear DNA, synthesized in the cytosol, and then imported individually into the organelle. Some organelle proteins and RNAs are encoded by the organelle DNA and are synthesized in the organelle itself. The human mitochondrial genome contains about 16,500 nucleotides and encodes 2 ribosomal RNAs, 22 transfer RNAs, and 13 different polypeptide chains. Chloroplast genomes are about 10 times larger and contain about 120 genes. But partially functional organelles will form in normal numbers even in mutants that lack a functional organelle genome, demonstrating the overwhelming importance of the nucleus for the biogenesis of both organelles.

The ribosomes of chloroplasts closely resemble bacterial ribosomes, while mitochondrial ribosomes show both similarities and differences that make their origin more difficult to trace. Protein similarities, however, suggest that both organelles originated when a primitive eucaryotic cell entered into a stable endosymbiotic re-

lationship with a bacterium: a purple bacterium is thought to have given rise to the mitochondrion, and (later) a relative of a cyanobacterium is thought to have given rise to the plant chloroplast. Although many of the genes of these ancient bacteria still function to make organelle proteins, most of them have become integrated into the nuclear genome, where they encode bacterial-like enzymes that are synthesized on cytosolic ribosomes and then imported into the organelle.

References

General

Ernster, L., ed. Bioenergetics. New York: Elsevier, 1984.

Harold, F.M. The Vital Force: A Study of Bioenergetics. New York: W.H. Freeman, 1986.

Lehninger, A.L. Principles of Biochemistry, Chaps. 16, 17, and 23. New York: Worth, 1982.

Nicholls, D.G. Bioenergetics: An Introduction to the Chemiosmotic Theory. New York: Academic Press, 1982.

Stryer, L. Biochemistry, 3rd ed., Chaps. 16, 17, and 22. New York: W.H. Freeman, 1988.

Cited

1. Bereiter-Hahn, J. Behavior of mitochrondria in the living cell. *Int. Rev. Cytol.* 122:1–63, 1990.

    Ernster, L.; Schatz, G. Mitochondria: a historical review. *J. Cell Biol.* 91:227s–255s, 1981.

    Fawcett, D.W. The Cell, 2nd ed., pp. 410–485. Philadelphia: Saunders, 1981.

    Tzagoloff, A. Mitochondria. New York: Plenum Press, 1982.

2. DePierre, J.W.; Ernster, L. Enzyme topology of intracellular membranes. *Annu. Rev. Biochem.* 46:201–262, 1977.

    Krstic, R.V. Ultrastructure of the Mammalian Cell, pp. 28–57. New York: Springer-Verlag, 1979.

    Pollak, J.K.; Sutton, R. The differentiation of animal mitochondria during development. *Trends Biochem. Sci.* 5:23–27, 1980.

    Srere, P.A. The structure of the mitochondrial inner membrane-matrix compartment. *Trends Biochem. Sci.* 7:375–378, 1982.

3. Geddes, R. Glycogen: a metabolic viewpoint. *Biosci. Rep.* 6:415–428, 1986.

    McGilvery, R.W. Biochemistry: A Functional Approach, 3rd ed. Philadelphia: Saunders, 1983.

    Newsholme, E.A.; Start, C. Regulation in Metabolism. New York: Wiley, 1973.

4. Baldwin, J.E.; Krebs, H. The evolution of metabolic cycles. *Nature* 291:381–382, 1981.

    Krebs, H.A. The history of the tricarboxylic acid cycle. *Perspect. Biol. Med.* 14:154–170, 1970.

5. Mitchell, P. Coupling of phosphorylation to electron and hydrogen transfer by a chemi-osmotic type of mechanism. *Nature* 191:144–148, 1961.

    Racker, E. From Pasteur to Mitchell: a hundred years of bioenergetics. *Fed. Proc.* 39:210–215, 1980.

6. Hatefi, Y. The mitochondrial electron transport and oxidative phosphorylation system. *Annu. Rev. Biochem.* 54:1015–1070, 1985.

7. Nicholls, D.G. Bioenergetics: An Introduction to the Chemiosmotic Theory, Chap. 3. New York: Academic Press, 1982.

Wood, W.B.; Wilson, J.H.; Benbow, R.M.; Hood, L.E. Biochemistry: A Problems Approach, 2nd ed. Menlo Park, CA: Benjamin-Cummings, 1981. (See problems, Chapters 9, 12, and 14.)

8. Durand, R.; Briand, Y.; Touraille, S.; Alziari, S. Molecular approaches to phosphate transport in mitochondria. *Trends Biochem. Sci.* 6:211–214, 1981.

    Hinkle, P.C.; McCarty, R.E. How cells make ATP. *Sci. Am.* 238(3):104–123, 1978.

    Klingenberg, M. The ADP, ATP shuttle of the mitochondrion. *Trends Biochem. Sci.* 4:249–252, 1979.

    LaNoue, K.F.; Schoolwerth, A.C. Metabolite transport in mitochondria. *Annu. Rev. Biochem.* 48:871–922, 1979.

9. Eisenberg, D.; Crothers, D. Physical Chemistry with Applications to the Life Sciences, Chaps. 4 and 5. Menlo Park, CA: Benjamin-Cummings, 1979.

    Kauzmann, W. Thermodynamics and Statistics: With Applications to Gases. New York: W.A. Benjamin, 1967.

10. Hinkle, P.C.; Kumar, M.A.; Resetar, A.; Harris, D.L. Mechanistic stoichiometry of mitochondrial oxidative phosphorylation. *Biochemistry* 30:3576–3582, 1991.

11. Hatefi, Y. The mitochondrial electron transport and oxidative phosphorylation system. *Annu. Rev. Biochem.* 54:1015–1069, 1985.

12. Racker, E. A New Look at Mechanisms in Bioenergetics. New York: Academic Press, 1976. (A personal account of the concepts and history.)

13. Bianchet, M.; Ysern, X.; Hullihen, J.; Pedersen, P.L.; Amzel, L.M. Mitochondrial ATP synthase. Quaternary structure of the F_1 moiety at 3.6 Å determined by x-ray diffraction analysis. *J. Biol. Chem.* 266:21197–21201, 1991.

    Futai, M.; Noumi, T.; Maeda, M. ATP synthase (H^+-ATPase): results by combined biochemical and molecular biological approaches. *Annu. Rev. Biochem.* 58:111–136, 1989.

    Pederson, P.L.; Carafoli, E. Ion motive ATPases. II. Energy coupling and work output. *Trends Biochem. Sci* 12:186–189, 1987.

    Racker, E.; Stoeckenius, W. Reconstitution of purple membrane vesicles catalyzing light-driven proton uptake and adenosine triphosphate formation. *J. Biol. Chem.* 249:662–663, 1974.

14. Fillingame, R.H. The proton-translocating pumps of oxidative phosphorylation. *Annu. Rev. Biochem.* 49:1079–1113, 1980.

    Malmstrom, B.G. The mechanism of proton translocation in respiration and photosynthesis. *FEBS Lett.* 250:9–21, 1989.

15. Chance, B.; Williams, G.R. A method for the localization of sites for oxidative phosphorylation. *Nature* 176:250–254, 1955.

Dickerson, R.E. The structure and history of an ancient protein. *Sci. Am.* 226(4):58–72, 1972. (The conformation and evolution of cytochrome c.)

Keilin, D. The History of Cell Respiration and Cytochromes. Cambridge, UK: Cambridge University Press, 1966.

Spiro, T.G., ed. Iron-Sulfur Proteins. New York: Wiley-Interscience, 1982.

16. Capaldi, R.A. Structure and function of cytochrome c oxidase. *Annu. Rev. Biochem.* 59:569–596, 1990.

Casey, R.P. Membrane reconstitution of the energy-conserving enzymes of oxidative phosphorylation. *Biochim. Biophys. Acta* 768:319–347, 1984.

Hofhaus, G.; Weiss, H.; Leonard, K. Electron microscopic analysis of the peripheral and membrane parts of mitochondrial NADH dehydrogenase (complex I). *J. Mol. Biol.* 221:1027–1043, 1991.

Weiss, H.; Linke, P.; Haiker, H.; Leonard, K. Structure and function of the mitochondrial ubiquinol: cytochrome c reductase and NADH: ubiquinone reductase. *Biochem. Soc. Trans.* 15:100–102, 1987.

17. Babcock, G.T.; Wikström, M. Oxygen activation and the conservation of energy in cell respiration. *Nature* 356:301–308, 1992.

18. Hackenbrock, C.R. Lateral diffusion and electron transfer in the mitochondrial inner membrane. *Trends Biochem. Sci.* 6:151–154, 1981.

19. Dutton, P.L; Wilson, D.F. Redox potentiometry in mitochondrial and photosynthetic bioenergetics. *Biochim. Biophys. Acta* 346:165–212, 1974.

Hamamoto, T.; Carrasco, N.; Matsushita, K.; Kabak, H.R.; Montal, M. Direct measurement of the electrogenic activity of O-type cytochrome oxidase from *E. coli* reconstituted into planar lipid bilayers. *Proc. Natl. Acad. Sci. USA* 82:2570–2573, 1985.

Lehninger, A.L. Bioenergetics: The Molecular Basis of Biological Energy Transformations, 2nd ed. Menlo Park, CA: Benjamin-Cummings, 1971.

20. Henderson, R.; Baldwin, J.M.; Ceska, T.A.; Zemlin, F.; Beckmann, E.; Downing, K.H. Model for the structure of bacteriorhodopsin based on high-resolution electron cryo-microscopy. *J. Mol. Biol.* 213:899–929, 1990.

Mathies, R.A.; Lin, S.W.; Ames, J.B.; Pollard, W.T. From femtoseconds to biology: mechanism of bacteriorhodopsin's light-driven proton pump. *Annu. Rev. Biophys. Biophys. Chem.* 20:491–518, 1991.

Prince, R.C. The proton pump of cytochrome oxidase. *Trends Biochem. Sci.* 13:159–160, 1988.

Trumpower, B.L. The proton motive Q cycle. *J. Biol. Chem.* 265:11409–11412, 1990.

21. Hanstein, W.G. Uncoupling of oxidative phosphorylation. *Trends Biochem. Sci.* 1:65–67, 1976.

22. Brand, M.D.; Murphy, M.P. Control of electron flux through the respiratory chain in mitochondria and cells. *Biol. Rev. Camb. Phil. Soc.* 62:141–193, 1987.

Racker, E. A New Look at Mechanisms in Bioenergetics. New York: Academic Press, 1976.

23. Klingenberg, M. Mechanism and evolution of the uncoupling protein of brown adipose tissue. *Trends Biochem. Sci.* 15:108–112, 1990.

Nicholls, D.G.; Rial, E. Brown fat mitochondria. *Trends Biochem. Sci.* 9:489–491, 1984.

24. Gottschalk, G. Bacterial Metabolism, 2nd ed. New York: Springer-Verlag, 1986.

MacNab, R.M. The bacterial flagellar motor. *Trends Biochem. Sci.* 9:185–189, 1984.

Neidhardt, F.C., et al, eds. *Escherichia coli* and *Salmonella typhimurium*: Cellular and Molecular Biology. Washington, DC: American Society for Microbiology, 1987.

Skulachev, V.P. Sodium bioenergetics. *Trends Biochem. Sci.* 9:483–485, 1984.

Thauer, R.; Jungermann, K.; Decker, K. Energy conservation in chemotrophic anaerobic bacteria. *Bacteriol. Rev.* 41:100–180, 1977.

25. Bogorad, L. Chloroplasts. *J. Cell Biol.* 91:256s–270s, 1981. (A historical review.)

Clayton, R.K. Photosynthesis: Physical Mechanisms and Chemical Patterns. Cambridge, UK: Cambridge University Press, 1980. (Excellent general treatment.)

Haliwell, B. Chloroplast Metabolism—The Structure and Function of Chloroplasts in Green Leaf Cells. Oxford, UK: Clarendon Press, 1981.

Hoober, J.K. Chloroplasts. New York: Plenum Press, 1984.

26. Anderson, J.M. Photoregulation of the composition, function, and structure of thylakoid membranes. *Annu. Rev. Plant Physiol.* 37:93–136, 1986.

Gruissem, W. Chloroplast gene expression: how plants turn their plastids on. *Cell* 56:161–170, 1989.

Thomson, W.W.; Whatley, J.M. Development of non-green plastids. *Annu. Rev. Plant Physiol.* 31:375–394, 1980.

27. Cramer, W.A.; Widger, W.R.; Herrmann, R.G.; Trebst, A. Topography and function of thylakoid membrane proteins. *Trends Biochem. Sci.* 10:125–129, 1985.

Miller, K.R. The photosynthetic membrane. *Sci. Am.* 241(4)102–113, 1979.

28. Akazawa, T.; Takabe, T.; Kobayashi, H. Molecular evolution of ribulose–1,5-bisphosphate carboxylase/oxygenase (RuBisCO). *Trends Biochem. Sci.* 9:380–383, 1984.

Lorimer, G.H. The carboxylation and oxygenation of ribulose–1, 5-bisphosphate: the primary events in photosynthesis and photorespiration. *Annu. Rev. Plant Physiol.* 32:349–383, 1981.

Lundqvist, T.; Schneider, G. Crystal structure of activated ribulose–1,5-bisphosphate carboxylase complexed with its substrate, ribulose–1,5-bisphosphate. *J. Biol. Chem.* 266:12604–12611, 1991.

29. Bassham, J.A. The path of carbon in photosynthesis. *Sci. Am.* 206(6):88–100, 1962.

Preiss, J. Starch, sucrose biosynthesis and the partition of carbon in plants are regulated by orthophosphate and triose-phosphates. *Trends Biochem. Sci.* 9:24–27, 1984.

30. Bjorkman, O.; Berry, J. High-efficiency photosynthesis. *Sci. Am.* 229(4):80–93, 1973. (C^4 plants.)

Edwards, G.; Walker, D. C^3, C^4 Mechanisms, and Cellular and Environmental Regulation of Photosynthesis. Berkeley: University of California Press, 1983.

Heber, U.; Krause, G.H. What is the physiological role of photorespiration? *Trends Biochem. Sci.* 5:32–34, 1980.

Langdale, J.A.; Nelson, T. Spatial regulation of photosynthetic development in C^4 plants. *Trends Genet.* 7:191–196, 1991.

31. Clayton, R.K. Photosynthesis: Physical Mechanisms and Chemical Patterns. Cambridge, UK: Cambridge University Press, 1980.

    Parson, W.W. Photosynthesis and other reactions involving light. In Biochemistry (G. Zubay, ed.), 2nd ed., pp. 564–597. New York: Macmillan, 1988; 3rd ed., Carmel, IN: Brown & Benchmark, 1993.

32. Barber, J. Photosynthethic reaction centres: a common link. *Trends Biochem. Sci.* 12:321–326, 1987.

    Govindjee; Govindjee, R. The absorption of light in photosynthesis. *Sci. Am.* 231(6):68–82, 1974.

    Zuber, H. Structure of light-harvesting antenna complexes of photosynthetic bacteria, cyanobacteria, and red algae. *Trends Biochem. Sci.* 11:414–419, 1986.

33. Boxer, S.G. Mechanisms of long-distance electron transfer in proteins: lessons from photosynthetic reaction centers. *Annu. Rev. Biophys. Biophys. Chem.* 19:267–299, 1990.

    Deisenhofer, J.; Epp, O.; Miki, K.; Huber, R.; Michel, H. Structure of the protein subunits in the photosynthetic reaction centre of *Rhodopseudomonas virdis* at 3Å resolution. *Nature* 318:618–624, 1985.

    Deisenhofer, J.; Michel, H. Structures of bacterial photosynthetic reaction centers. *Annu. Rev. Cell Biol.* 7:1–23, 1991.

34. Blankenship, R.E.; Prince, R.C. Excited-state redox potentials and the Z scheme of photosynthesis. *Trends Biochem. Sci.* 10:382–383, 1985.

    Govindjee; Coleman, W.J. How plants make oxygen. *Sci. Am.* 262(2):50–58, 1990.

    Krauss, N.; Hinrichs, W.; Witt, I.; et al. Three-dimensional structure of system I of photosynthesis at 6 Å resolution. *Nature* 361:326–331, 1993.

35. Anderson, J.M. Photoregulation of the composition, function, and structure of thylakoid membranes. *Annu. Rev. Plant Physiol.* 37:93–136, 1986.

36. Hinkle, P.C.; McCarty, R.E. How cells make ATP. *Sci. Am.* 238(3):104–123, 1978.

    Jagendorf, A.T. Acid-base transitions and phosphorylation by chloroplasts. *Fed. Proc.* 26:1361–1369, 1967.

37. Douce, R.; Joyard, J. Biochemistry and function of the plastid envelope. *Annu. Rev. Cell Biol.* 6:173–216, 1990.

    Flügge, U.I.; Heldt, H.W. The phosphate-triose phosphate-phosphoglycerate translocator of the chloroplast. *Trends Biochem. Sci.* 9:530–533, 1984.

38. Wilson, T.H.; Lin, E.C.C. Evolution of membrane bioenergetics. *J. Supramol. Struct.* 13:421–446, 1980.

    Woese, C.R. Bacterial evolution. *Microbiol. Rev.* 51:221–271, 1987.

39. Gest, H. The evolution of biological energy-transducing systems. *FEMS Microbiol. Lett.* 7:73–77, 1980.

    Gottschalk, G. Bacterial Metabolism, 2nd ed. New York: Springer-Verlag, 1986. (Chapter 8 covers fermentations.)

    Miller, S.M.; Orgel, L.E. The Origins of Life on the Earth. Englewood Cliffs, NJ: Prentice-Hall, 1974.

40. Cross, R.L.; Taiz, L. Gene duplication as a means for altering H$^+$/ATP ratios during the evolution of F$_0$F$_1$ ATPases and synthases. *FEBS Lett.* 259:227–229, 1990.

Knowles, C.J., ed. Diversity of Bacterial Respiratory Systems, Vol. 1. Boca Raton, FL: CRC Press, 1980.

Nelson, N.; Taiz, L. The evolution of H$^+$-ATPases. *Trends Biochem. Sci.* 14:113–116, 1989.

41. Clayton, R.K.; Sistrom, W.R., eds. The Photosynthetic Bacteria. New York: Plenum Press, 1978.

    Deamer, D.W., ed. Light Transducing Membranes: Structure, Function and Evolution. New York: Academic Press, 1978.

    Gromet-Elhanan, Z. Electrochemical gradients and energy coupling in photosynthetic bacteria. *Trends Biochem. Sci.* 2:274–277, 1977.

    Olson, J.M.; Pierson, B.K. Evolution of reaction centers in photosynthetic prokaryotes. *Int. Rev. Cytol.* 108:209–248, 1987.

42. Dickerson, R.E. Cytochrome c and the evolution of energy metabolism. *Sci. Am.* 242(3):136–153, 1980.

    Gabellini, N. Organization and structure of the genes for the cytochrome b/c$_1$ complex in purple photosynthetic bacteria. A phylogenetic study describing the homology of the b/c$_1$ subunits between prokaryotes, miotochondria, and chloroplasts. *J. Bioenerg. Biomembr.* 20:59–83, 1988.

    Lockhart, P.J.; Penny, D.; Hendy, M.D.; et al. Controversy on chloroplast origins. *FEBS Lett.* 301:127–131, 1992.

    Schopf, J.W.; Hayes, J.M.; Walter, M.R. Evolution of earth's earliest ecosystems: recent progress and unsolved problems. In Earth's Earliest Biosphere: Its Origin and Evolution (J.W. Schopf, ed.), pp. 361–384. Princeton, NJ: Princeton University Press, 1983.

43. Attardi, G.; Schatz, G. Biogenesis of mitochondria. *Annu. Rev. Cell Biol.* 4:289–333, 1988.

    Ellis, R.J., ed. Chloroplast Biogenesis. Cambridge, UK: Cambridge University Press, 1984.

44. Clayton, D.A. Replication and transcription of vertebrate mitochondrial DNA. *Annu. Rev. Cell Biol.* 7:453–478, 1991.

    Posakony, J.W.; England, J.M.; Attardi, G. Mitochondrial growth and division during the cell cycle in HeLa cells. *J. Cell Biol.* 74:468–491, 1977.

45. Borst, P.; Grivell, L.A.; Groot, G.S.P. Organelle DNA. *Trends Biochem. Sci.* 9:128–130, 1984.

    Gray, M.W. Origin and evolution of mitochondrial DNA. *Annu. Rev. Cell Biol.* 5:25–50, 1989.

    Sugiura, M. The chloroplast chromosomes in land plants. *Annu. Rev. Cell Biol.* 5:51–70, 1989.

46. Grivell, L.A. Mitochondrial DNA. *Sci. Am.* 248(3):60–73, 1983.

    Hoober, J.K. Chloroplasts, New York: Plenum Press, 1984.

    Rochaix, J.D. Molecular genetics of chloroplasts and mitochondria in the unicellular green alga Chlamydomonas. *FEMS Microbiol. Rev.* 46:13–34, 1987.

47. Clegg, M.T. Chloroplast gene sequences and the study of plant evolution. *Proc. Natl. Acad. Sci. USA.* 90:363–367, 1993.

    Ohyama, K., et al. Chloroplast gene organization deduced from complete sequence of liverwort. *Marchantia polymorpha* chloroplast DNA. *Nature* 322:572–574, 1986.

Shinozaki, K., et al. The complete nucleotide sequence of the tobacco chloroplast genome: its gene organization and expression. *EMBO J.* 5:2034–2049, 1986.

48. Anderson, S., et al. Sequence and organization of the human mitochondrial genome. *Nature* 290:457–465, 1981.

    Bibb, M.J.; Van Etten, R.A.; Wright, C.T.; Walberg, M.W.; Clayton, D.A. Sequence and gene organization of mouse mitochondrial DNA. *Cell* 26:167–180, 1981.

    Breitenberger, C.A.; RajBhandary, U.L. Some highlights of mitchondrial research based on analysis of *Neurospora crassa* mitchondrial DNA. *Trends Biochem. Sci.* 10:478–482, 1985.

    Fox, T.D. Natural variation in the genetic code. *Annu. Rev. Genet.* 21:67–91, 1987.

49. Attardi, G. Animal mitochondrial DNA: an extreme example of genetic economy. *Int. Rev. Cytol.* 93:93–145, 1985.

    Wilson, A. The molecular basis of evolution. *Sci. Am.* 253(4):164–173, 1985.

50. Levings, C.S. Molecular biology of plant mitochondria. *Cell* 56:171–179, 1989.

    Mulligan, R.M.; Walbot, V. Gene expression and recombination in plant mitochondrial genomes. *Trends Genet.* 2:263–266, 1986.

    Newton, K.J. Plant mitochondrial genomes: organization, expression and variation. *Annu. Rev. Plant Physiol. Plant Mol. Biol.* 39:503–532, 1988.

51. Clayton, D.A. Transcription of the mammalian mitochondrial genome. *Annu. Rev. Biochem.* 53:573–594, 1984.

    Grivell, L.A.; Schweyen, R.J. RNA splicing in yeast mitochondria: taking out the twists. *Trends Genet.* 5:39–41, 1989.

    Gruissem, W.; Barken, A.; Deng, S.; Stern, D. Transcriptional and post-transcriptional control of plastid mRNA in higher plants. *Trends Genet.* 4:258–262, 1988.

    Mullet, J.E. Chloroplast development and gene expression. *Annu. Rev. Plant Physiol. Plant Mol. Biol.* 39:475–502, 1988.

    Rochaix, J.-D. Post-transcriptional steps in the expression of chloroplast genes. *Annu. Rev. Cell Biol.* 8:1–28, 1992.

52. Birky, C.W., Jr. Transmission genetics of mitochondria and chloroplasts. *Annu. Rev. Genet.* 12:471–512, 1978.

53. Giles, R.E.; Blanc, H.; Cann, H.M.; Wallace, D.C. Maternal inheritance of human mitochondrial DNA. *Proc. Natl. Acad. Sci. USA* 77:6715–6719, 1980.

54. Attardi, G.; Schatz, G. Biogenesis of mitochondria. *Annu. Rev. Cell Biol.* 4:289–333, 1988.

    Bernardi, G. The petite mutation in yeast. *Trends Biochem. Sci.* 4:197–201, 1979.

    Montisano, D.F.; James, T.W. Mitochondrial morphology in yeast with and without mitochondrial DNA. *J. Ultrastruct. Res.* 67:288–296, 1979.

55. Capaldi, R.A. Mitochondrial myopathies and respiratory chain proteins. *Trends Biochem. Sci.* 13:144–148, 1988.

    DiMauro, S., et al. Mitochondrial myopathies. *J. Inherit. Metab. Dis.* 10(Suppl. 1):113–128, 1987.

    Wallace, D.C. Diseases of the mitochondrial DNA. *Annu. Rev. Biochem.* 61:1175–1212, 1992.

56. Bishop, W.R.; Bell, R.M. Assembly of phospholipids into cellular membranes: biosynthesis, transmembrane movement, and intracellular translocation. *Annu. Rev. Cell Biol.* 4:579–611, 1988.

57. Butow, B.A.; Doeherty, R.; Parikh, V.S. A path from mitochondria to the yeast nucleus. *Philos. Trans. R. Soc. Lond. (Biol.)* 319:127–133, 1988.

    Cavalier-Smith, T. The origin of eukaryotic and archaebacterial cells. *Ann. N.Y. Acad. Sci.* 503:17–54, 1987.

    Gellissen, G.; Michaelis, G. Gene transfer: mitochondria to nucleus. *Ann. N.Y. Acad. Sci.* 503:391–401, 1987.

    Margulis, I. Symbiosis in Cell Evolution. New York: W.H. Freeman, 1981.

    Schwartz, R.M.; Dayhoff, M.O. Origins of prokaryotes, eukaryotes, mitochondria, and chloroplasts. *Science* 199:395–403, 1978.

    Whatley, J.M.; John, P.; Whatley, F.R. From extracellular to intracellular: the establishment of mitochondria and chloroplasts. *Proc. R. Soc. Lond. (Biol.)* 204:165–187, 1979.

58. von Heijne, G. Why mitochondria need a genome. *FEBS Lett.* 198:1–4, 1986.

Cell Signaling

15

- **General Principles of Cell Signaling**

- **Signaling via G-Protein-linked Cell-Surface Receptors**

- **Signaling via Enzyme-linked Cell-Surface Receptors**

- **Target-Cell Adaptation**

- **The Logic of Intracellular Signaling: Lessons from Computer-based "Neural Networks"**

The fossil record suggests that sophisticated unicellular organisms resembling present-day bacteria were present on earth 3.5 billion years ago but that it apparently required more than another 2.5 billion years for the first multicellular organisms to appear (see Figure 1–17). Why was multicellularity so slow to evolve? Although the answer cannot be known, it seems likely to be related to the need in a multicellular organism for elaborate signaling mechanisms that enable its cells to communicate with one another so as to coordinate their behavior for the benefit of the organism as a whole. Intercellular signals, interpreted by complex machinery in the responding cell, allow each cell to determine its position and specialized role in the body and ensure, for example, that each cell divides only when its neighbors dictate that it should do so. The importance of such "social controls" on cell division becomes apparent when the controls fail, resulting in cancer, which usually kills the multicellular organism.

As more and more powerful techniques become available to study cells and the mechanisms they use to communicate with one another, the intricacy of the signaling processes used by higher eucaryotes is slowly coming into focus. An animal cell contains an elaborate system of proteins that enables the cell to respond to signals from other cells. The system includes cell-surface and intracellular receptor proteins, protein kinases, protein phosphatases, GTP-binding proteins, and the many intracellular proteins with which these signaling proteins interact. In this chapter we first discuss the general principles of intercellular signaling. In the subsequent two sections we consider in turn the two main families of cell-surface receptor proteins and how they generate intracellular signals. We then examine how cells continuously adapt in order to respond sensitively to small changes in the concentration of an extracellular signaling molecule. Finally, we consider an analogy with computer-based neural networks, which provides insights into how complex intracellular signaling networks operate.

General Principles of Cell Signaling [1]

Mechanisms enabling one cell to influence the behavior of another almost certainly existed in the world of unicellular organisms long before multicellular organisms appeared on earth. Evidence comes from studies of some present-day unicellular eucaryotes such as yeasts. Although these cells normally lead independent lives, they can communicate and influence one another's proliferation

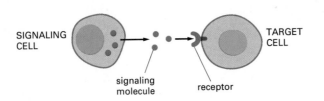

SIGNALING BY SECRETED MOLECULES

SIGNALING CELL

TARGET CELL

signaling molecule

receptor

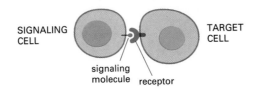

SIGNALING BY PLASMA-MEMBRANE-BOUND MOLECULES

SIGNALING CELL

TARGET CELL

signaling molecule   receptor

Figure 15–1 Intercellular signaling in animals. Two ways that animal cells communicate with one another are illustrated.

in preparation for sexual mating. In the budding yeast *Saccharomyces cerevisiae*, for example, when a haploid individual is ready to mate, it secretes a peptide *mating factor* that signals cells of opposite mating types to stop proliferating and prepare to conjugate; the subsequent fusion of two haploid cells of the opposite mating type produces a diploid cell, which can then undergo meiosis and sporulate to generate haploid cells with new assortments of genes.

Studies of yeast mutants that are unable to mate have identified many proteins that are required in the signaling process. These proteins form a signaling network that includes cell-surface receptors, GTP-binding proteins, and protein kinases, each of which has close relatives among the proteins involved in signaling in animal cells. Through gene duplication and divergence, however, the signaling systems in animals have become much more elaborate than those in yeasts.

Extracellular Signaling Molecules Are Recognized by Specific Receptors on or in Target Cells [2]

Whereas yeast cells communicate with one another for mating by secreting several kinds of small peptides, cells in higher animals communicate by means of hundreds of kinds of signaling molecules, including proteins, small peptides, amino acids, nucleotides, steroids, retinoids, fatty acid derivatives, and even dissolved gases such as nitric oxide and carbon monoxide. Most of these signaling molecules are secreted from the *signaling cell* by exocytosis (discussed in Chapter 13). Others are released by diffusion through the plasma membrane, while some remain tightly bound to the cell surface and influence only cells that contact the signaling cell (Figure 15–1).

Regardless of the nature of the signal, the *target cell* responds by means of a specific protein called a **receptor.** It specifically binds the signaling molecule and then initiates a response in the target cell. Many of the extracellular signaling molecules act at very low concentrations (typically $\leq 10^{-8}$ M), and the receptors that recognize them usually bind them with high affinity (affinity constant $K_a \geq 10^8$ liters/mole; see Figure 3–9). In most cases the receptors are transmembrane proteins on the target-cell surface; when they bind an extracellular signaling molecule (a *ligand*), they become activated so as to generate a cascade of intracellular signals that alter the behavior of the cell. In some cases, however, the receptors are inside the target cell and the signaling ligand has to enter the cell to activate them: these signaling molecules therefore must be sufficiently small and hydrophobic to diffuse across the plasma membrane (Figure 15–2).

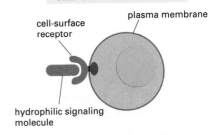

CELL-SURFACE RECEPTORS

cell-surface receptor

plasma membrane

hydrophilic signaling molecule

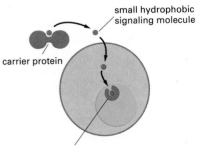

INTRACELLULAR RECEPTORS

small hydrophobic signaling molecule

carrier protein

intracellular receptor

Figure 15–2 Extracellular signaling molecules bind to either cell-surface receptors or intracellular receptors. Most signaling molecules are hydrophilic and are therefore unable to cross the plasma membrane directly; instead, they bind to cell-surface receptors, which in turn generate one or more signals inside the target cell. Some small signaling molecules, by contrast, diffuse across the plasma membrane and bind to receptors inside the target cell—either in the cytosol or in the nucleus (as shown). Many of these small signaling molecules are hydrophobic and nearly insoluble in aqueous solutions; they are therefore transported in the bloodstream and other extracellular fluids bound to carrier proteins, from which they dissociate before entering the target cell.

In this chapter we concentrate mainly on the communication between animal cells that is mediated by secreted chemical signals. This emphasis reflects the state of current knowledge: secreted molecules are very much easier to study than those that are membrane-bound, and we know much more about how they work. Contact-dependent signaling via membrane-bound molecules, although harder to study and less well understood, nonetheless, is crucially important, especially during development and in immune responses; its molecular basis can be very similar to that for signaling at a distance, as we see later.

Secreted Molecules Mediate Three Forms of Signaling: Paracrine, Synaptic, and Endocrine [2]

Signaling molecules that a cell secretes may be carried far afield to act on distant targets, or they may act as **local mediators,** affecting only cells in the immediate environment of the signaling cell. This latter process is called **paracrine signaling** (Figure 15–3A). For paracrine signals to be delivered only to their proper targets, the secreted signaling molecules must not be allowed to diffuse too far; for this reason they are often rapidly taken up by neighboring target cells, destroyed by extracellular enzymes, or immobilized by the extracellular matrix.

For a large, complex multicellular organism, short-range signaling is not sufficient on its own to coordinate the behavior of the organism's cells. Sets of specialized cells have evolved with a specific role in signaling between widely separate parts of the body. The most sophisticated of these are nerve cells, or neurons, which typically extend long processes (axons) that contact target cells far away. When activated by signals from the environment or from other nerve cells, a neuron sends electrical impulses (action potentials) along its axon; when an impulse reaches the nerve terminals at the end of the axon, it stimulates the terminals to secrete a chemical signal called a **neurotransmitter.** The nerve terminals contact their target cell at specialized cell junctions called *chemical synapses,* which are designed to ensure that the neurotransmitter is delivered to the postsynaptic target cell rapidly and specifically (Figure 15–3B). This **synaptic signaling** process is discussed in detail in Chapter 11 and will not be considered further here.

The other specialized signaling cells that control the behavior of the organism as a whole are **endocrine cells.** They secrete their signaling molecules, called **hormones,** into the bloodstream (of an animal) or the sap (of a plant), which carries the signal to target cells distributed widely throughout the body (Figure 15–3C). The distinctive ways that endocrine cells and nerve cells coordinate cell behavior in animals are contrasted in Figure 15–4.

Because endocrine signaling relies on diffusion and blood flow, it is relatively slow. Nerve cells, by contrast, can achieve much greater speed and precision.

Figure 15–3 Three forms of signaling mediated by secreted molecules. Many of the same types of signaling molecules are used in paracrine, synaptic, and endocrine signaling. The crucial differences lie in the speed and selectivity with which the signals are delivered to their targets.

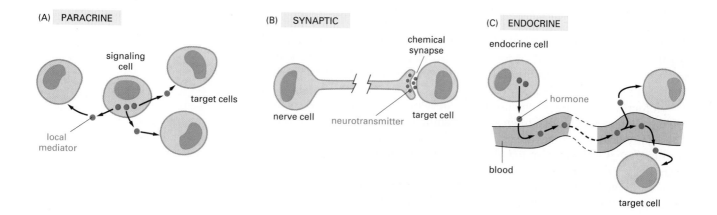

(A) PARACRINE

signaling cell

target cells

local mediator

(B) SYNAPTIC

chemical synapse

nerve cell   neurotransmitter   target cell

(C) ENDOCRINE

endocrine cell

hormone

blood

target cell

Figure 15–4 The contrast between endocrine and synaptic signaling.
Endocrine cells and nerve cells work together to coordinate the diverse activities of the billions of cells in a higher animal. Endocrine cells secrete many different hormones into the blood to signal specific target cells. The target cells have receptors for binding specific hormones and thereby "pull" the appropriate hormones from the extracellular fluid. In synaptic signaling, by contrast, the specificity arises from the contacts between nerve processes and the specific target cells they signal: usually only a target cell that is in synaptic contact with a nerve cell is exposed to the neurotransmitter released from the nerve terminal (although some neurotransmitters act in a paracrine mode as local mediators that influence multiple target cells in the area). Whereas different endocrine cells must use different hormones in order to communicate specifically with their target cells, many nerve cells can use the same neurotransmitter and still communicate in a specific manner.

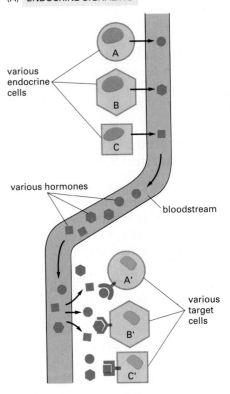

(A) ENDOCRINE SIGNALING

various endocrine cells

various hormones

bloodstream

various target cells

They can transmit information over long distances by electrical impulses that travel at rates of up to 100 meters per second; once released from a nerve terminal, a neurotransmitter has to diffuse less than 100 nm to the target cell, a process that takes less than a millisecond. Another difference between endocrine and synaptic signaling is that whereas hormones are greatly diluted in the bloodstream and interstitial fluid and therefore must be able to act at very low concentrations (typically $< 10^{-8}$ M), neurotransmitters are diluted much less and can achieve high local concentrations. The concentration of acetylcholine in the synaptic cleft of an active neuromuscular junction, for example, is about 5×10^{-4} M. Correspondingly, neurotransmitter receptors have a relatively low affinity for their ligand, which means that the neurotransmitter can dissociate rapidly from the receptor to terminate a response. (Neurotransmitters are quickly removed from the synaptic cleft either by specific hydrolytic enzymes or by specific membrane transport proteins that pump the neurotransmitter back into either the nerve terminal or neighboring glial cells.)

Autocrine Signaling Can Coordinate Decisions by Groups of Identical Cells [3]

All of the forms of signaling discussed so far allow one cell type to influence another. By the same mechanisms, however, cells can send signals to other cells of the same type, and it follows from this that they can also send signals to themselves. In such **autocrine signaling** a cell secretes signaling molecules that can bind back to its own receptors. During development, for example, once a cell has been directed into a particular path of differentiation, it may begin to secrete autocrine signals that reinforce this developmental decision.

Because autocrine signaling is most effective when carried out simultaneously by neighboring cells of the same type, it may be used to encourage groups of identical cells to make the same developmental decisions (Figure 15–5). Thus autocrine signaling is thought to be one possible mechanism underlying the "community effect" observed in early development, where a group of identical cells can respond to a differentiation-inducing signal but a single isolated cell of the same type cannot.

Autocrine signaling is not confined to development, however. **Eicosanoids** are signaling molecules that often act in an autocrine mode in mature mammals. These fatty-acid derivatives are made by cells in all mammalian tissues. They are continuously synthesized in the plasma membrane and released to the cell exterior, where they are rapidly degraded by enzymes in extracellular fluid. Made from precursors (mainly *arachidonic acid*) that are cleaved from membrane phospholipids by phospholipases (Figure 15–6), they have a wide variety of biological activities, influencing the contraction of smooth muscle and the aggrega-

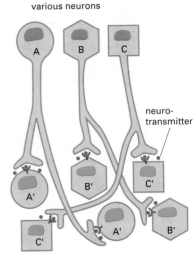

(B) SYNAPTIC SIGNALING

various neurons

neuro-transmitter

various target cells

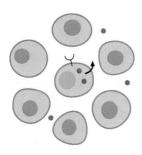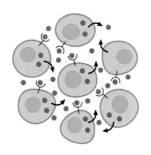

A SINGLE SIGNALING CELL
RECEIVES WEAK AUTOCRINE
SIGNAL

IN A GROUP OF IDENTICAL SIGNALING
CELLS, EACH CELL RECEIVES A STRONG
AUTOCRINE SIGNAL

Figure 15–5 Autocrine signaling.
A group of identical cells produces a
higher concentration of a secreted
signal than does a single cell.

Figure 15–6 The synthesis of an eicosanoid. Eicosanoids are continuously synthesized in membranes from 20-carbon fatty acid chains that contain at least three double bonds, as shown for the synthesis of prostaglandin PGE$_2$ in (A). The subscript refers to the two carbon-carbon double bonds outside the ring of PGE$_2$. There are four major classes of eicosanoids—*prostaglandins, prostacyclins, thromboxanes,* and *leukotrienes*—and they are all made mainly from arachidonic acid. The synthesis of all but the leukotrienes involves the enzyme *cyclooxygenase;* the synthesis of leukotrienes involves the enzyme *lipoxygenase* (B). These synthetic pathways are targets for a large number of therapeutic drugs, since eicosanoids play an important part in pain, fever, and inflammation. Corticosteroid hormones such as cortisone, for example, which inhibit the activity of the phospholipase in the first step of the eicosanoid synthesis pathway shown, are widely used clinically to treat noninfectious inflammatory diseases, such as some forms of arthritis. Nonsteroid anti-inflammatory drugs such as aspirin and ibuprofen, by contrast, block the first oxidation step, which is catalyzed by cyclooxygenase. Certain prostaglandins that are produced in large amounts in the uterus at the time of childbirth to stimulate the contraction of the uterine smooth muscle cells are widely used as pharmacological agents to induce abortion.

tion of platelets, for example, and participating in pain and inflammatory responses. When cells are activated by tissue damage or by some types of chemical signals, the rate of eicosanoid synthesis is increased; the resulting increase in the local level of eicosanoid influences both the cells that make it and their immediate neighbors.

Gap Junctions Allow Signaling Information to Be Shared by Neighboring Cells [4]

Another way to coordinate the activities of neighboring cells is through **gap junctions.** These are specialized cell-cell junctions that can form between closely apposed plasma membranes, directly connecting the cytoplasms of the joined cells via narrow water-filled channels (see Figure 19–15). The channels allow the exchange of small intracellular signaling molecules (*intracellular mediators*), such as Ca^{2+} and cyclic AMP, but not of macromolecules, such as proteins or nucleic acids. Thus cells connected by gap junctions can communicate with each other directly without having to deal with the barrier presented by the intervening plasma membranes (Figure 15–7).

As discussed in Chapter 19, the pattern of gap-junction connections in a tissue can be revealed either electrically, with intracellular electrodes, or visually, after the microinjection of water-soluble dyes. Studies of this kind indicate that

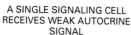
membrane
phospholipid

phospholipase A$_2$

arachidonic acid
(20 carbons)
extended conformation

COOH

arachidonic acid,
folded conformation

COOH

OXIDATION STEPS

prostaglandin (PGE$_2$)

COOH

OH   OH

(A)

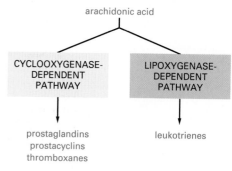

arachidonic acid

CYCLOOXYGENASE-
DEPENDENT
PATHWAY

LIPOXYGENASE-
DEPENDENT
PATHWAY

prostaglandins
prostacyclins
thromboxanes

leukotrienes

(B)

the cells in a developing embryo make and break gap-junction connections in specific and interesting patterns, suggesting that these junctions play an important part in the signaling processes that occur between these cells. One suspects that, like autocrine signaling described above, gap-junction communication helps adjacent cells of a similar type to coordinate their behavior. It is not known, however, which particular small molecules are important as carriers of signals through gap junctions; nor has the precise function of gap-junction communication in animal development been defined.

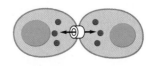

Figure 15–7 Signaling via gap junctions. Cells connected by gap junctions share small molecules, including small intracellular signaling molecules, and therefore can respond to extracellular signals in a coordinated way.

Each Cell Is Programmed to Respond to Specific Combinations of Signaling Molecules [5]

Any given cell in a multicellular organism is exposed to many—perhaps hundreds—of different signals from its environment. These signals can be soluble, or bound to the extracellular matrix, or bound to the surface of a neighboring cell, and they can act in many millions of possible combinations. The cell must respond to this babel selectively, according to its own specific character, acquired through progressive cell specialization in the course of development. Thus a cell may be programmed to respond to one set of signals by differentiating, to another set by proliferating, and to yet another by carrying out some specialized function.

Most cells in higher animals, moreover, are programmed to depend on a specific set of signals simply for survival: when deprived of the appropriate signals (in a culture dish, for example), a cell will activate a suicide program and kill itself—a process called *programmed cell death*, which is discussed further in Chapter 21 (Figure 15–8). Different types of cells require different sets of survival signals and so are restricted to different environments in the body.

Because signaling molecules generally act in combinations, an animal can control the behavior of its cells in highly specific ways using a limited diversity of such molecules: hundreds of such signals can be used in millions of combinations.

Different Cells Can Respond Differently to the Same Chemical Signal [6]

The specific way a cell reacts to its environment varies, first, according to the set of receptor proteins that the cell possesses through which it is tuned to detect a particular subset of the available signals and, second, according to the intracellular machinery by which the cell integrates and interprets the information that it receives. Thus a single signaling molecule often has different effects on different target cells. The neurotransmitter acetylcholine, for example, stimulates the con-

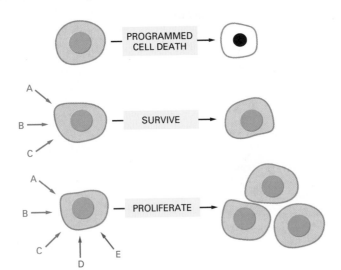

Figure 15–8 Combinatorial signaling. Each cell type displays a set of receptors that enables it to respond to a corresponding set of signaling molecules produced by other cells. These signaling molecules work in combinations to regulate the behavior of the cell. As shown here, many cells require multiple signals (*green arrows*) to survive and additional signals (*red arrows*) to proliferate; if deprived of all signals, these cells undergo programmed cell death.

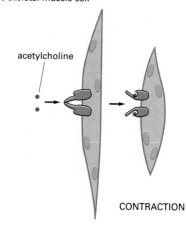

Figure 15–9 The same signaling molecule can induce different responses in different target cells. In some cases this is because the signaling molecule binds to different receptor proteins, as illustrated in (A) and (B). In other cases the signaling molecule binds to identical receptor proteins that activate different response pathways in different cells, as illustrated in (B) and (C). In all of the cases shown the signaling molecule is *acetylcholine* (D).

traction of skeletal muscle cells but decreases the rate and force of contraction in heart muscle cells. This is because the acetylcholine receptor proteins on skeletal muscle cells are different from those on heart muscle cells. But receptor differences are not always the explanation for the different effects. In many cases the same signaling molecule binds to identical receptor proteins and yet produces very different responses in different types of target cells, reflecting differences in the internal machinery to which the receptors are coupled (Figure 15–9).

The Concentration of a Molecule Can Be Adjusted Quickly Only If the Lifetime of the Molecule Is Short [6]

It is natural to think of signaling systems in terms of the changes produced when a signal is delivered. But it is just as important to consider what happens when a signal is withdrawn. During development transient signals often produce lasting effects: they can trigger a change that persists indefinitely, through cell memory mechanisms such as those discussed in Chapters 9 and 21. But in most cases, especially in adult tissues, when a signal ceases, the response fades. The signal acts on a system of molecules that is undergoing continual turnover, and when the signal is shut off, the replacement of the old molecules by new ones wipes out the traces of its action. It follows that the speed of reaction to shutting off the signal depends on the rate of turnover of the molecules that the signal affects. It may not be as obvious that this turnover rate also determines the promptness of the response when the signal is turned on.

Consider, for example, two intracellular molecules X and Y, both of which are normally maintained at a concentration of 1000 molecules per cell. Molecule X has a slow turnover rate: it is synthesized and degraded at a rate of 10 molecules per second, so that each molecule has an average lifetime in the cell of 100 seconds. Molecule Y turns over 10 times as quickly: it is synthesized and degraded at a rate of 100 molecules per second, with each molecule having an average lifetime of 10 seconds. If a signal acting on the cell boosts the rates of synthesis of both X and Y tenfold without any change in the molecular lifetimes, at the end of 1 second the concentration of Y will have increased by nearly 900 molecules per cell ($10 \times 100 - 100$) while the concentration of X will have increased by only 90 molecules per cell. In fact, after its synthesis rate has been either increased or decreased abruptly, the time required for a molecule to shift halfway from its old to its new equilibrium concentration is equal to its normal half-life—that is, it is equal to the time that would be required for its concentration to fall by half if all synthesis were stopped (Figure 15–10).

The same principles apply to proteins as well as to small molecules and to molecules in the extracellular space as well as to those in cells. Many intracellular proteins that are rapidly degraded have short half-lives, some surviving less than 10 minutes; in most cases these are proteins with key regulatory roles, whose concentrations are rapidly regulated in the cell by changes in their rates of synthesis. Likewise, any covalent modifications of proteins that occur as part of a rapid signaling process—most commonly the addition of a phosphate group to an amino acid side chain—must be continuously removed at a rapid rate to make such signaling possible. We discuss some of these molecular events in detail later, for the case of signaling pathways that operate via cell-surface receptors. But the principles apply generally, as the next example illustrates.

(A) skeletal muscle cell

acetylcholine

CONTRACTION

(B) heart muscle cell

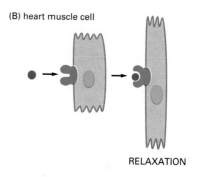

RELAXATION

(C) secretory cell

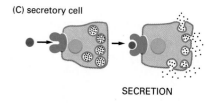

SECRETION

(D) acetylcholine

$$H_3C - \overset{\overset{\displaystyle O}{\|}}{C} - O - CH_2 - CH_2 - \overset{\overset{\displaystyle CH_3}{|}}{\underset{\underset{\displaystyle CH_3}{|}}{N^+}} - CH_3$$

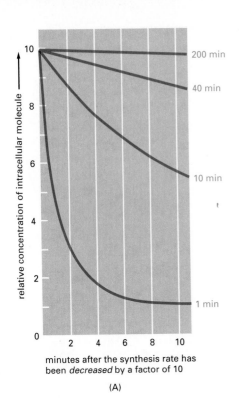

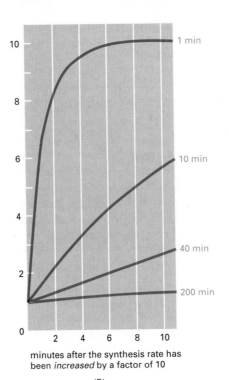

Figure 15–10 **The importance of rapid turnover.** The figure shows the predicted relative rates of change in the intracellular concentrations of molecules with differing turnover times when their rates of synthesis are either decreased (A) or increased (B) suddenly by a factor of 10. In both cases the concentrations of those molecules that are normally being rapidly degraded in the cell (*red lines*) change quickly, whereas the concentrations of those that are normally being slowly degraded (*green lines*) change proportionally more slowly. The numbers (in *blue*) on the right-hand side are the half-lives assumed for each of the different molecules.

Nitric Oxide Gas Signals by Binding Directly to an Enzyme Inside the Target Cell [7]

Although most extracellular signals are mediated by hydrophilic molecules that bind to receptors on the surface of the target cell, some signaling molecules are hydrophobic enough and/or small enough to pass readily across the target-cell plasma membrane; once inside, they directly regulate the activity of specific intracellular proteins. A remarkable example is the gas **nitric oxide (NO),** which only recently has been recognized to act as a signaling molecule in vertebrates. When acetylcholine is released by autonomic nerves in the walls of a blood vessel, for example, it causes smooth muscle cells in the vessel wall to relax. The acetylcholine acts indirectly by inducing the endothelial cells to make and release NO, which then signals the smooth muscle cells to relax. This effect of NO on blood vessels provides an explanation for the mechanism of action of nitroglycerine, which has been used for almost 100 years to treat patients with angina (pain due to inadequate blood flow to heart muscle). The nitroglycerine is converted to NO, which relaxes blood vessels, thereby reducing the workload on the heart and, as a consequence, the oxygen requirement of the heart muscle. NO is also produced as a local mediator by activated macrophages and neutrophils to help them kill invading microorganisms. In addition, it is used by many types of nerve cells to signal neighboring cells: NO released by autonomic nerves in the penis, for example, causes the local blood vessel dilation that is responsible for penile erection.

NO is made by the enzyme *NO synthase* by the deamination of the amino acid arginine. Because it diffuses readily across membranes, the NO diffuses out of the cell where it is produced and passes directly into neighboring cells. It acts only locally because it has a short half-life—about 5–10 seconds—in the extracellular space before it is converted to nitrates and nitrites by oxygen and water. In many target cells, such as endothelial cells, NO reacts with iron in the active site of the enzyme *guanylyl cyclase,* stimulating it to produce the intracellular mediator *cyclic GMP,* which we discuss later. The effects of NO can be rapid, occurring within seconds, because the rate of turnover of cyclic GMP is high: rapid production from GTP by guanylyl cyclase is balanced by rapid degradation

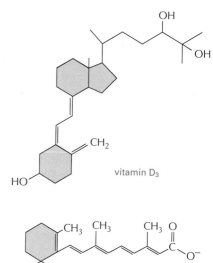

to GMP by a phosphodiesterase. There is recent evidence that *carbon monoxide (CO)* is also used as an intercellular signal and can act in the same way as NO, by stimulating guanylyl cyclase.

Gases such as NO and CO are not the only signaling molecules that can pass directly across the target-cell plasma membrane. A group of small, hydrophobic, nongaseous hormones and local mediators also enter target cells in this way, but instead of binding to enzymes, they bind to intracellular receptor proteins that directly regulate gene transcription.

Figure 15–11 Some signaling molecules that bind to intracellular receptors. Note that all of them are small and hydrophobic. The active, hydroxylated form of vitamin D_3 is shown.

Steroid Hormones, Thyroid Hormones, Retinoids, and Vitamin D Bind to Intracellular Receptors That Are Ligand-activated Gene Regulatory Proteins [8]

Steroid hormones, thyroid hormones, retinoids, and *vitamin D* are small hydrophobic molecules that differ greatly from one another in both chemical structure (Figure 15–11) and function. Nonetheless, they all act by a similar mechanism. They diffuse directly across the plasma membrane of target cells and bind to intracellular receptor proteins. Ligand binding activates the receptors, which then directly regulate the transcription of specific genes. These receptors are structurally related and constitute the **intracellular receptor superfamily** (or *steroid-hormone receptor superfamily*) (Figure 15–12).

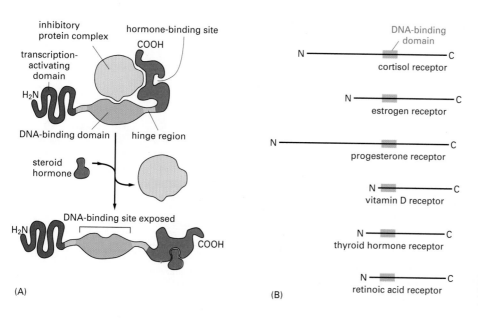

(A)

(B)

Figure 15–12 The intracellular receptor superfamily. (A) A model of an intracellular receptor protein. In its inactive state the receptor is bound to an inhibitory protein complex that contains a heat-shock protein called Hsp90 (discussed in Chapter 5). The binding of ligand to the receptor causes the inhibitory complex to dissociate, thereby activating the receptor by exposing its DNA-binding site. The model shown is based on the receptor for cortisol, but all of the receptors in this superfamily have a related structure, as shown in (B), where the short DNA-binding domain in each receptor is shown in *green*. Domain-swap experiments suggest that many of the hormone-binding, transcription-activating, and DNA-binding domains in these receptors can function as interchangeable modules. It is thought that all of the intracellular receptor proteins bind to DNA as either homodimers or heterodimers.

Steroid hormones, including cortisol, the steroid sex hormones, vitamin D (in vertebrates), and the moulting hormone ecdysone (in insects), are all made from cholesterol. *Cortisol* is produced in the cortex of the adrenal gland and influences the metabolism of many cell types. The steroid sex hormones are made in the testis and ovary and are responsible for the secondary sex characteristics that distinguish males from females. **Vitamin D** is synthesized in the skin in response to sunlight; after it is converted to an active form in the liver or kidneys, it functions to regulate Ca^{2+} metabolism, promoting Ca^{2+} uptake in the gut and reducing its excretion in the kidney. The **thyroid hormones,** which are made from the amino acid tyrosine, act to increase metabolism in a wide variety of cell types, while the **retinoids,** such as retinoic acid, which are made from vitamin A, play important roles as local mediators in vertebrate development. Although all of these signaling molecules are relatively insoluble in water, they are made soluble for transport in the bloodstream and other extracellular fluids by binding to specific carrier proteins, from which they dissociate before entering a target cell (see Figure 15–2).

Besides the fundamental difference in the way they signal their target cells, most water-insoluble signaling molecules differ from water-soluble ones in the length of time that they persist in the bloodstream or tissue fluids. Most water-soluble hormones are removed and/or broken down within minutes of entering the blood, and local mediators and neurotransmitters are removed from the extracellular space even faster—within seconds or milliseconds. Steroid hormones, by contrast, persist in the blood for hours and thyroid hormones for days. Consequently, water-soluble signaling molecules usually mediate responses of short duration, whereas the water-insoluble ones tend to mediate longer-lasting responses.

The intracellular receptors for the steroid and thyroid hormones, retinoids, and vitamin D all bind to specific DNA sequences adjacent to the genes that the ligand regulates. Some, such as cortisol receptors, are located primarily in the cytosol and bind to DNA only following ligand binding (see Figure 15–12); others, such as retinoid receptors, are located primarily in the nucleus and bind to DNA even in the absence of ligand. In either case, ligand binding alters the conformation of the receptor protein, which then activates (or occasionally suppresses) gene transcription. In many cases the response takes place in two steps: the direct induction of the transcription of a small number of specific genes

Figure 15–13 Early primary response (A) and delayed secondary response (B) that result from the activation of an intracellular receptor protein. The response to a steroid hormone is illustrated, but the same principles apply for all ligands that activate this family of receptor proteins. Some of the primary-response proteins turn on secondary-response genes, whereas others turn off the primary-response genes. The actual number of primary- and secondary-response genes is greater than shown. As expected, drugs that inhibit protein synthesis suppress the transcription of secondary-response genes but not primary-response genes.

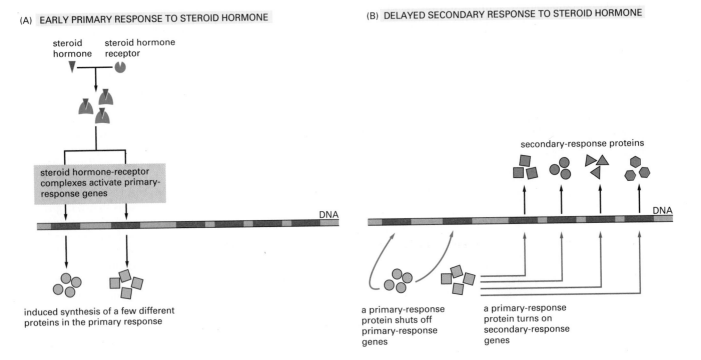

(A) EARLY PRIMARY RESPONSE TO STEROID HORMONE

steroid hormone   steroid hormone receptor

steroid hormone-receptor complexes activate primary-response genes

DNA

induced synthesis of a few different proteins in the primary response

(B) DELAYED SECONDARY RESPONSE TO STEROID HORMONE

secondary-response proteins

DNA

a primary-response protein shuts off primary-response genes

a primary-response protein turns on secondary-response genes within about 30 minutes is known as the *primary response;* the products of these genes in turn activate other genes and produce a delayed, *secondary response.* Thus a simple hormonal trigger can cause a very complex change in the pattern of gene expression (Figure 15–13).

The responses to steroid and thyroid hormones, vitamin D, and retinoids, like responses to extracellular signals in general, are determined as much by the nature of the target cell as by the nature of the signaling molecule. Even when different types of cells have the identical intracellular receptor, the set of genes that the receptor regulates is different. This is because more than one type of gene regulatory protein generally must bind to a eucaryotic gene in order to activate its transcription. An intracellular receptor can activate a gene, therefore, only if the right combination of other gene regulatory proteins is also present, and some of these are cell-type specific. Thus thyroid hormone, vitamin D, and each steroid hormone and retinoid induces a characteristic set of responses in an animal because (1) only certain types of cells have receptors for it and (2) each of these cell types contains a different combination of other cell-type-specific gene regulatory proteins that collaborate with the activated receptor to influence the transcription of specific sets of genes. The molecular details of how intracellular receptors and other gene regulatory proteins control specific gene transcription are discussed in Chapter 9.

There Are Three Known Classes of Cell-Surface Receptor Proteins: Ion-Channel-linked, G-Protein-linked, and Enzyme-linked [9]

Recombinant DNA techniques have revolutionized the study of the receptors and intracellular proteins involved in cell signaling. Because these proteins often constitute less than 0.01% of the total mass of protein in the cell, it has been extremely difficult to purify them. Cloning the DNA sequences that encode the proteins has greatly accelerated the process of characterization, and most of the signaling proteins discussed in this chapter have been characterized in this way. A major contribution of these DNA-cloning and -sequencing studies has been to reveal that the bewildering diversity of known receptor proteins can be reduced to a much smaller number of large families. The intracellular receptors that we have just discussed constitute one such family. We now consider the family groups that can be identified within the other, larger, class of signal receptors—those located on the cell surface.

All water-soluble signaling molecules (including neurotransmitters, protein hormones, and protein growth factors), as well as some lipid-soluble ones, bind to specific receptor proteins on the surface of the target cells they influence. These cell-surface receptor proteins act as signal transducers: they bind the signaling ligand with high affinity and convert this extracellular event into one or more intracellular signals that alter the behavior of the target cell.

Most cell-surface receptor proteins belong to one of three classes, defined by the transduction mechanism used. **Ion-channel-linked receptors,** also known as *transmitter-gated ion channels,* are involved in rapid synaptic signaling between electrically excitable cells. This type of signaling is mediated by a small number of neurotransmitters that transiently open or close the ion channel formed by the protein to which they bind, briefly changing the ion permeability of the plasma membrane and thereby the excitability of the postsynaptic cell (Figure 15–14A). The ion-channel-linked receptors belong to a family of homologous, multipass transmembrane proteins. They are discussed in Chapter 11 and will not be considered further here.

G-protein-linked receptors act indirectly to regulate the activity of a separate plasma-membrane-bound target protein, which can be an enzyme or an ion channel. The interaction between the receptor and the target protein is mediated by a third protein, called a *trimeric GTP-binding regulatory protein (G protein)* (Figure 15–14B). The activation of the target protein either alters the concentra-

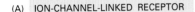

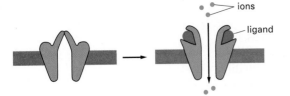

(A) ION-CHANNEL-LINKED RECEPTOR

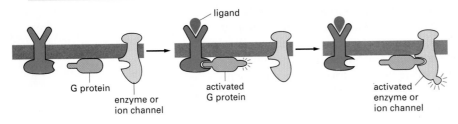

ions
ligand

(B) G-PROTEIN-LINKED RECEPTOR

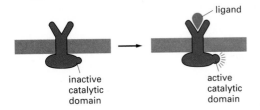

ligand

G protein

enzyme or
ion channel

activated
G protein

activated
enzyme or
ion channel

(C) ENZYME-LINKED RECEPTOR

ligand

inactive
catalytic
domain

active
catalytic
domain

Figure 15–14 Three classes of cell-surface receptors. Although many enzyme-linked receptors have intrinsic enzyme activity as shown in (C), many others rely on associated enzymes (not shown).

tion of one or more intracellular mediators (if the target protein is an enzyme) or alters the ion permeability of the plasma membrane (if the target protein is an ion channel). The intracellular mediators act in turn to alter the behavior of yet other proteins in the cell. All of the G-protein-linked receptors belong to a large superfamily of homologous, seven-pass transmembrane proteins.

Enzyme-linked receptors, when activated, either function directly as enzymes or are associated with enzymes (Figure 15–14C). Most are single-pass transmembrane proteins, with their ligand-binding site outside the cell and their catalytic site inside. Compared with the other two classes, enzyme-linked recep-

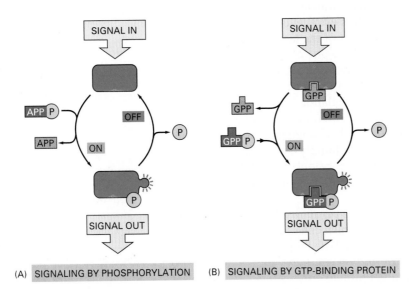

SIGNAL IN

APP P

APP

OFF

ON

P

P

SIGNAL OUT

(A) SIGNALING BY PHOSPHORYLATION

SIGNAL IN

GPP

GPP

GPP P

GPP

OFF

ON

P

GPP P

SIGNAL OUT

(B) SIGNALING BY GTP-BINDING PROTEIN

Figure 15–15 Two major intracellular signaling mechanisms share common features. In both cases a signaling protein is activated by the addition of a phosphate group and inactivated by the removal of the phosphate. In (A) the phosphate is added covalently to the signaling protein by a protein kinase; in (B) a signaling protein is induced to exchange its bound GDP for GTP. To emphasize the similarity in the two mechanisms, ATP is shown as APP(P), ADP as APP, GTP as GPP(P), and GDP as GPP.

tors are heterogeneous, although the great majority are protein kinases, or are associated with protein kinases, that phosphorylate specific sets of proteins in the target cell.

Activated Cell-Surface Receptors Trigger Phosphate-Group Additions to a Network of Intracellular Proteins [9, 10]

Much of the remainder of this chapter is concerned with how G-protein-linked receptors and enzyme-linked receptors operate. Signals received at the surface of a cell by both of these classes of receptors are often relayed to the nucleus, where they alter the expression of specific genes and thereby alter the behavior of the cell. Elaborate sets of intracellular signaling proteins form the relay systems. The majority of these proteins are of one of two kinds: proteins that become phosphorylated by protein kinases, and proteins that are induced to bind GTP when the signal arrives. In both cases the proteins gain one or more phosphates in their activated state and lose the phosphates when the signal decays (Figure 15–15). These proteins in turn generally cause the phosphorylation of downstream proteins as part of a *phosphorylation cascade*.

The phosphorylation cascades are mediated by two main types of protein kinases: *serine/threonine kinases*, which phosphorylate proteins on serines and (less often) threonines, and *tyrosine kinases*, which phosphorylate proteins on tyrosines. An occasional kinase can do both. It is estimated that about 1% of our genes encode protein kinases and that a single mammalian cell may contain more than 100 distinct kinds of these enzymes, most of which are serine/threonine kinases. Although fewer than 0.1% of the phosphorylated proteins in cells contain phosphotyrosine, we shall see that this small minority plays a crucial part in signaling by most enzyme-linked receptors.

As discussed previously, complex cell behaviors, such as survival or proliferation, are generally stimulated by specific combinations of signals rather than by a single signal acting alone (see Figure 15–8). The cell has to integrate the information coming from separate signals so as to make a proper response—to live or die, or to proliferate or stay quiescent. The integration seems to depend on interactions between the various protein phosphorylation cascades that are activated by different extracellular signals. In particular, some of the signaling proteins in the cascades function as integrating devices, equivalent to microprocessors in a computer: in response to multiple signal inputs, they produce an output that is calibrated to cause the desired biological effect. Two examples of how such integrating proteins could operate are illustrated in Figure 15–16.

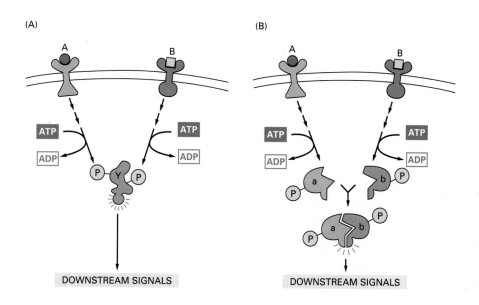

Figure 15–16 **Signal integration.** In (A) signals A and B activate different cascades of protein phosphorylations, each of which leads to the phosphorylation of protein Y but at different sites on the protein. Protein Y is activated only when both of these sites are phosphorylated, and therefore it is active only when signals A and B are simultaneously present. In (B) signals A and B lead to the phosphorylation of two proteins, a and b, which then bind to each other to create the active protein ab. In both of the examples illustrated the proteins themselves are phosphorylated: an equivalent form of control can also occur, however, by the exchange of GTP for GDP on a GTP-binding protein (see Figure 15–15).

The complexity of such signal-response systems, with multiple interacting relay chains of signaling proteins, is daunting. But recombinant DNA technology, combined with classical genetic analyses in *Drosophila*, the nematode *C. elegans*, and yeasts, as well as more conventional biochemical and pharmacological methods, is rapidly uncovering the intricate details of these mechanisms by which activated receptor proteins change the behavior of the cell.

Summary

Each cell in a multicellular animal is programmed during development to respond to a specific set of signals that act in various combinations to regulate the behavior of the cell and to determine whether the cell lives or dies and whether it proliferates or stays quiescent. Most of these signals mediate paracrine signaling, in which local mediators are rapidly taken up, destroyed, or immobilized, so that they act only on neighboring cells. In addition, centralized control is exerted both by endocrine signaling, in which hormones secreted by endocrine cells are carried in the blood to target cells throughout the body, and by synaptic signaling, in which neurotransmitters secreted by nerve cells act locally on the postsynaptic cells that their axons contact.

Cell signaling requires both extracellular signaling molecules and a complementary set of receptor proteins in each cell that enable it to bind and respond to them in a programmed and characteristic way. Some small hydrophobic signaling molecules, including the steroid and thyroid hormones and the retinoids, diffuse across the plasma membrane of the target cell and activate intracellular receptor proteins, which directly regulate the transcription of specific genes. Some dissolved gases, such as nitric oxide and carbon monoxide, act as local mediators by diffusing across the plasma membrane of the target cell and activating an intracellular enzyme—usually guanylyl cyclase, which produces cyclic GMP in the target cell. But most extracellular signaling molecules are hydrophilic and are able to activate receptor proteins only on the surface of the target cell; these receptors act as signal transducers, converting the extracellular binding event into intracellular signals that alter the behavior of the target cell. There are three main families of cell-surface receptors, each of which transduces extracellular signals in a different way. Ion-channel-linked receptors are transmitter-gated ion channels that open or close briefly in response to the binding of a neurotransmitter. G-protein-linked receptors indirectly activate or inactivate plasma-membrane-bound enzymes or ion channels via trimeric GTP-binding proteins (G proteins). Enzyme-linked receptors either act directly as enzymes or are associated with enzymes; the enzymes are usually protein kinases that phosphorylate specific proteins in the target cell. Through cascades of highly regulated protein phosphorylations, elaborate sets of interacting proteins relay most signals from the cell surface to the nucleus, thereby altering the cell's pattern of gene expression and, as a consequence, its behavior. Cross-talk between different signaling cascades enables a cell to integrate information from the multiple signals that it receives.

Signaling via G-Protein-linked Cell-Surface Receptors [11]

G-protein-linked receptors are the largest family of cell-surface receptors. More than 100 members have already been defined in mammals. Many of these have been identified by *homology cloning*, in which low stringency hybridization with existing cDNA probes is used to detect related DNA sequences (see Figure 7–17). Other family members have been found by *expression cloning*, using their ligand-binding or cell-activation properties to identify them. In one form of this approach, a library of cDNA molecules prepared from cells or tissues that express the receptor are copied into RNA molecules, which are then injected into *Xenopus* oocytes. The oocytes translate the RNA molecules into proteins. These proteins are inserted into the plasma membrane, where their ligand-binding or cell-activation properties allow them to be detected.

G-protein-linked receptors mediate the cellular responses to an enormous diversity of signaling molecules, including hormones, neurotransmitters, and local mediators, which are as varied in structure as they are in function: the list includes proteins and small peptides, as well as amino acid and fatty acid derivatives. The same ligand can activate many different family members. At least 9 distinct G-protein-linked receptors are activated by adrenaline, for example, another 5 or more by acetylcholine, and at least 15 by serotonin.

Despite the chemical and functional diversity of the signaling molecules that bind to them, all of the G-protein-linked receptors whose amino acid sequences are known from DNA sequencing studies have a similar structure and are almost certainly evolutionarily related. They consist of a single polypeptide chain that threads back and forth across the lipid bilayer seven times (Figure 15–17). As we discuss later, this superfamily of seven-pass transmembrane receptor proteins includes *rhodopsin,* the light-activated protein in the vertebrate eye, as well as olfactory receptors in the vertebrate nose. Other family members are found in unicellular organisms: the receptors in yeasts that recognize the yeast mating factors are an example. This ancient structural motif is also shared by bacteriorhodopsin, a bacterial light-activated H^+ pump discussed in Chapter 10, although, unlike the other family members, bacteriorhodopsin is not a receptor and does not act via a G protein. Taken together, these findings suggest that the G-protein-linked receptors that mediate cell-cell signaling in multicellular organisms may have evolved from sensory receptors possessed by their unicellular ancestors. The members of this receptor family have conserved not only their amino acid sequence but also their functional relationship to G proteins by means of which they broadcast into the interior of the cell the message that an extracellular ligand is present. It is the intracellular sequence of events beginning with the activation of G proteins that mainly concern us in this section.

Trimeric G Proteins Relay the Intracellular Signal from G-Protein-linked Receptors [11, 12]

The **trimeric GTP-binding proteins (G proteins)** that functionally couple these receptors to their target enzymes or ion channels in the plasma membrane are structurally distinct from the single-chain GTP-binding proteins (called *monomeric GTP-binding proteins* or *monomeric GTPases*) that help relay intracellular signals and regulate vesicular traffic and many other processes in eucaryotic cells. The monomeric GTPases are discussed later in this chapter as well as in other chapters. Both classes of GTP-binding proteins, however, are GTPases and function as molecular switches that can flip between two states: active, when GTP is bound, and inactive, when GDP is bound. "Active" in this context usually means that the molecule acts as a signal to trigger other events in the cell. When an extracellular ligand binds to a G-protein-linked receptor, the receptor changes its conformation and switches on the trimeric G proteins that associate with it by causing them to eject their GDP and replace it with GTP. The switch is turned off when the G protein hydrolyzes its own bound GTP, converting it back to GDP. But before that occurs, the active protein has an opportunity to diffuse away from the receptor and deliver its message for a prolonged period to its downstream target.

Most G-protein-linked receptors activate a chain of events that alters the concentration of one or more small intracellular signaling molecules. These small molecules, often referred to as **intracellular mediators** (also called *intracellular messengers* or *second messengers*), in turn pass the signal on by altering the behavior of selected cellular proteins. Two of the most widely used intracellular mediators are *cyclic AMP (cAMP)* and *Ca^{2+}*: changes in their concentrations are stimulated by distinct pathways in most animal cells, and most G-protein-linked receptors regulate one or the other of them, as outlined in Figure 15–18.

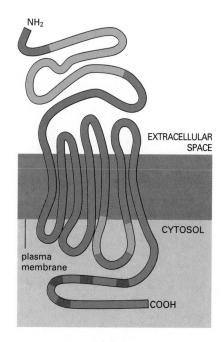

Figure 15–17 A schematic drawing of a G-protein-linked receptor. Receptors that bind protein ligands have a large extracellular ligand-binding domain formed by the part of the polypeptide chain shown in *light green.* Receptors for small ligands such as adrenaline have small extracellular domains, and the ligand-binding site is usually deep within the plane of the membrane, formed by amino acids from several of the transmembrane segments. The parts of the intracellular domains that are mainly responsible for binding to trimeric G proteins are shown in *orange,* while those that become phosphorylated during receptor desensitization (discussed later) are shown in *red.*

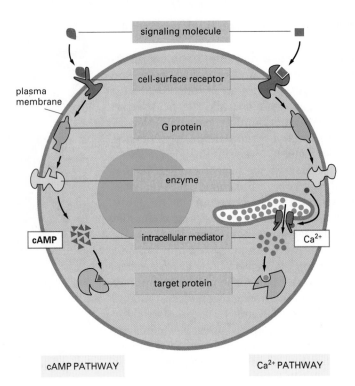

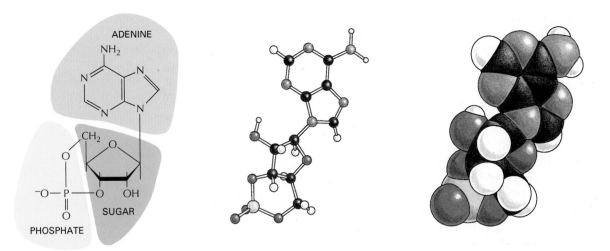

Figure 15–18 Two major pathways by which G-protein-linked cell-surface receptors generate small intracellular mediators. In both cases the binding of an extracellular ligand alters the conformation of the cytoplasmic domain of the receptor, causing it to bind to a G protein that activates (or inactivates) a plasma membrane enzyme. In the cyclic AMP (cAMP) pathway the enzyme directly produces cyclic AMP. In the Ca^{2+} pathway the enzyme produces a soluble mediator (inositol trisphosphate, discussed later) that releases Ca^{2+} from the endoplasmic reticulum. Like other small intracellular mediators, both cyclic AMP and Ca^{2+} relay the signal by acting as allosteric effectors: they bind to specific proteins in the cell, altering their conformation and thereby their activity.

Some Receptors Increase Intracellular Cyclic AMP by Activating Adenylyl Cyclase via a Stimulatory G Protein (G_s) [13]

Cyclic AMP (Figure 15–19) was first identified as an intracellular mediator of hormone action in 1959 and has since been found to act as an intracellular signaling molecule in all procaryotic and animal cells that have been studied. For cyclic AMP to function as an intracellular mediator, its intracellular concentration (normally $\leq 10^{-7}$ M) must be able to change up or down in response to extracellular signals: upon hormonal stimulation, cyclic AMP levels can change fivefold in seconds. As explained earlier (see Figure 15–10), such responsiveness requires that rapid synthesis of the molecule be balanced by rapid breakdown or removal. Cyclic AMP is synthesized from ATP by a plasma-membrane-bound enzyme **adenylyl cyclase**, and it is rapidly and continuously destroyed by one or more **cyclic AMP phosphodiesterases**, which hydrolyze cyclic AMP to adenosine 5′-monophosphate (5′-AMP) (Figure 15–20).

Figure 15–19 Cyclic AMP. It is shown as a formula, a ball-and-stick model, and a space-filling model. (C, H, N, O, and P indicate carbon, hydrogen, nitrogen, oxygen, and phosphorus atoms, respectively.)

Many extracellular signaling molecules work by controlling cyclic AMP levels, and they do so by altering the activity of adenylyl cyclase (Figure 15–21) rather than the activity of phosphodiesterase. Just as the same steroid hormone produces different effects in different target cells, so different target cells respond very differently to external signals that change intracellular cyclic AMP levels (Table 15–1). All ligands that activate adenylyl cyclase in a given type of target cell, however, usually produce the same effect: at least four hormones activate adenylyl cyclase in fat cells, for example, and all of them stimulate the breakdown of triglyceride (the storage form of fat) to fatty acids (see Table 15–1). The different receptors for these hormones activate a common pool of adenylyl cyclase molecules, to which they are coupled by a trimeric G protein. Because this G protein is involved in enzyme *activation*, it is called **stimulatory G protein (G_s).** Individuals who are genetically deficient in G_s show decreased responses to certain hormones and, consequently, have metabolic abnormalities, abnormal bone development, and are mentally retarded.

The best-studied examples of receptors coupled to the activation of adenylyl cyclase are the **β-adrenergic receptors,** which mediate some of the actions of *adrenaline* and *noradrenaline* (Figure 15–22, and see Table 15–1). An adrenaline-activated adenylyl cyclase system can be reconstituted in synthetic phospholipid vesicles using purified β-adrenergic receptors, G_s, and adenylyl cyclase molecules, indicating that no other proteins are required for the activation process. But precisely how does G_s mediate the coupling? The answer depends on the trimeric structure of the G protein, as we now discuss.

Trimeric G Proteins Are Thought to Disassemble When Activated [11, 12, 14]

A trimeric G protein is composed of three different polypeptide chains, called α, β, and γ. The G_s *α chain* (α_s) binds and hydrolyzes GTP and activates adenylyl cyclase. The G_s *β chain* and *γ chain* form a tight complex (βγ), which anchors G_s to the cytoplasmic face of the plasma membrane, at least partly by a lipid chain (a prenyl group) that is covalently attached to the γ subunit. In its inactive form G_s exists as a trimer with GDP bound to α_s. When stimulated by binding to a ligand-activated receptor, α_s exchanges its GDP for GTP. This is thought to cause α_s to dissociate from βγ, allowing α_s to bind instead to an adenylyl cyclase molecule, which it activates to produce cyclic AMP.

If cells are to be able to respond rapidly to changes in the concentration of an extracellular signaling molecule, the activation of adenylyl cyclase must be reversed quickly once the signaling ligand dissociates from its receptor. This ability to respond rapidly to change is assured because the lifetime of the active form of α_s is short: the GTPase activity of α_s is stimulated when α_s binds to adenylyl cyclase, so that the bound GTP is hydrolyzed to GDP, rendering both α_s

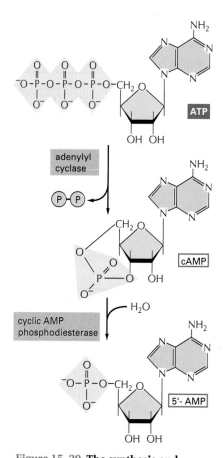

Figure 15–20 The synthesis and degradation of cyclic AMP (cAMP). A pyrophosphatase makes the synthesis of cyclic AMP an irreversible reaction by hydrolyzing the released pyrophosphate ⓟ–ⓟ (not shown).

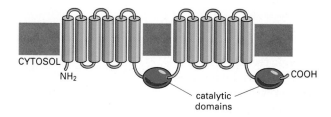

Figure 15–21 Adenylyl cyclase. In vertebrates the enzyme usually contains about 1100 amino acid residues and is thought to have two clusters of six transmembrane segments separating two similar cytoplasmic catalytic domains. There are at least six types of this form of adenylyl cyclase in mammals (types I–VI). All of them are stimulated by G_s, but type I, which is found mainly in the brain, is also stimulated by complexes of Ca^{2+} bound to the Ca^{2+}-binding protein calmodulin (discussed later).

Table 15–1 Some Hormone-induced Cellular Responses Mediated by Cyclic AMP

Target Tissue	Hormone	Major Response
Thyroid gland	thyroid-stimulating hormone (TSH)	thyroid hormone synthesis and secretion
Adrenal cortex	adrenocorticotropic hormone (ACTH)	cortisol secretion
Ovary	luteinizing hormone (LH)	progesterone secretion
Muscle	adrenaline	glycogen breakdown
Bone	parathormone	bone resorption
Heart	adrenaline	increase in heart rate and force of contraction
Liver	glucagon	glycogen breakdown
Kidney	vasopressin	water resorption
Fat	adrenaline, ACTH, glucagon, TSH	triglyceride breakdown

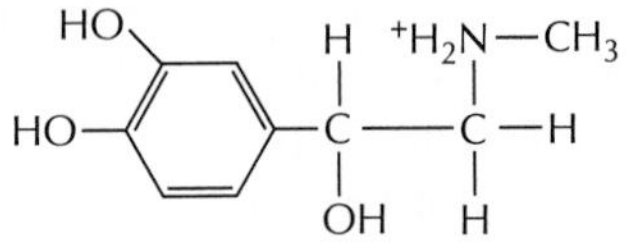

adrenaline

Figure 15–22 Adrenaline. This hormone (also called epinephrine) is made from tyrosine and is secreted by the adrenal gland when a mammal is stressed.

and the adenylyl cyclase inactive. The α_s then reassociates with $\beta\gamma$ to re-form an inactive G_s molecule (Figure 15–23).

The importance of the GTPase activity of α_s in shutting off the response can be readily demonstrated in a test tube. If cells are broken open and exposed to an analogue of GTP (GTPγS) in which the terminal phosphate cannot be hydrolyzed, cyclic AMP production after hormone treatment is greatly prolonged. A similar phenomenon is seen in patients suffering from *cholera,* where the bacterial toxin responsible for the symptoms of the disease inhibits the self-inactivating mechanism of α_s. **Cholera toxin** is an enzyme that catalyzes the transfer of ADP ribose from intracellular NAD^+ to α_s. The ADP ribosylation alters the α_s so that it can no longer hydrolyze its bound GTP. An adenylyl cyclase molecule activated by such an altered α_s subunit thus remains in the active state indefinitely. The resulting prolonged elevation in cyclic AMP levels within intestinal epithelial cells causes a large efflux of Na^+ and water into the gut, which is responsible for the severe diarrhea that is characteristic of cholera.

Some Receptors Decrease Cyclic AMP by Inhibiting Adenylyl Cyclase via an Inhibitory Trimeric G Protein (G_i) [11, 12, 15]

The same signaling molecule can either increase or decrease the intracellular concentration of cyclic AMP depending on the type of receptor to which it binds. When adrenaline binds to *β-adrenergic receptors,* for example, it activates adenylyl cyclase, whereas when it binds to *α_2-adrenergic receptors,* it inhibits the enzyme. The difference reflects the type of G proteins that couple these receptors to the cyclase. While the β-adrenergic receptors are functionally coupled to adenylyl cyclase by G_s, the α_2-adrenergic receptors are coupled to this enzyme by an **inhibitory G protein (G_i).** G_i can contain the same $\beta\gamma$ complex as G_s, but it has a different α subunit (α_i). When activated, α_2-adrenergic receptors bind to G_i, causing α_i to bind GTP and dissociate from the $\beta\gamma$ complex. Both the released α_i and $\beta\gamma$ are thought to contribute to the inhibition of adenylyl cyclase. α_i inhibits the cyclase, probably indirectly, whereas $\beta\gamma$ may inhibit cyclic AMP synthesis in two ways—directly, by binding to the cyclase itself, and indirectly, by binding to any free α_s subunits in the same cell, thereby preventing them from activating cyclase molecules. We see later that G_i also acts to open K^+ channels in the plasma membrane, and it seems likely that this function is more important than the inhibition of adenylyl cyclase.

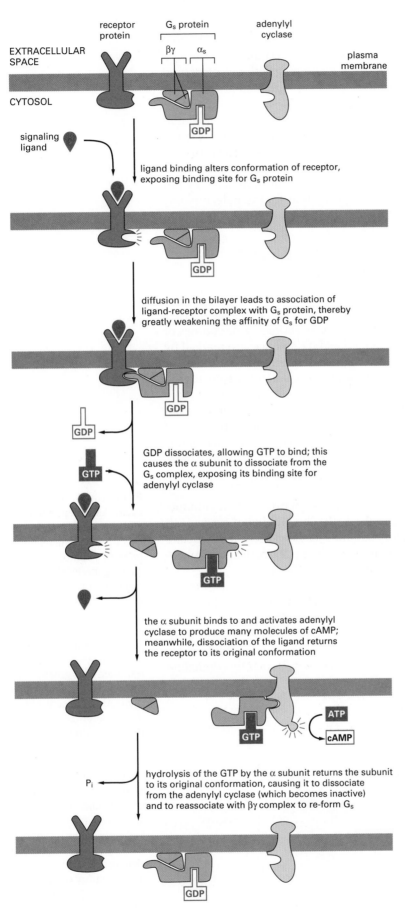

receptor protein

Gs protein

βγ   αs

adenylyl cyclase

EXTRACELLULAR SPACE

plasma membrane

CYTOSOL

GDP

signaling ligand

ligand binding alters conformation of receptor, exposing binding site for Gs protein

GDP

diffusion in the bilayer leads to association of ligand-receptor complex with Gs protein, thereby greatly weakening the affinity of Gs for GDP

GDP

GDP

GTP

GDP dissociates, allowing GTP to bind; this causes the α subunit to dissociate from the Gs complex, exposing its binding site for adenylyl cyclase

GTP

the α subunit binds to and activates adenylyl cyclase to produce many molecules of cAMP; meanwhile, dissociation of the ligand returns the receptor to its original conformation

ATP

cAMP

GTP

Pi

hydrolysis of the GTP by the α subunit returns the subunit to its original conformation, causing it to dissociate from the adenylyl cyclase (which becomes inactive) and to reassociate with βγ complex to re-form Gs

GDP

Figure 15–23 A current model of how Gs couples receptor activation to adenylyl cyclase activation. As long as the extracellular signaling ligand remains bound, the receptor protein can continue to activate molecules of Gs protein, thereby amplifying the response. More important, an αs can remain active and continue to stimulate a cyclase molecule for many seconds after the signaling ligand dissociates from the receptor, providing even greater amplification.

Whereas cholera toxin catalyzes the ADP ribosylation of α_s and thereby inactivates the GTPase activity of α_s, **pertussis toxin,** made by the bacterium that causes pertussis (whooping cough), catalyzes the ADP ribosylation of α_i. The ADP ribosylation of α_i prevents the G_i complex from interacting with receptors, and so the complex remains bound to GDP and is unable to inhibit adenylyl cyclase or open K[+] channels.

The trimeric G proteins are remarkably versatile intracellular signaling molecules. In the examples considered so far, either the α subunit or both the α and the $\beta\gamma$ subunits are the active components. But in other cases receptors are coupled to their target proteins only by the released $\beta\gamma$ complex. Moreover, $\beta\gamma$ complexes can also act as conditional regulators of effector proteins: they can enhance the activation of some forms of adenylyl cyclase, for example, but only if the cyclase has already been activated by α_s.

Cyclic-AMP-dependent Protein Kinase (A-Kinase) Mediates the Effects of Cyclic AMP [16]

Cyclic AMP exerts its effects in animal cells mainly by activating the enzyme **cyclic-AMP-dependent protein kinase (A-kinase),** which catalyzes the transfer of the terminal phosphate group from ATP to specific serines or threonines of selected proteins. The amino acids phosphorylated by A-kinase are marked by the presence of two or more basic amino acids on their amino-terminal side. Covalent phosphorylation of the appropriate amino acids in turn regulates the activity of the target protein.

A-kinase is found in all animal cells and is thought to account for all of the effects of cyclic AMP in most of these cells. (The only other known function of cyclic AMP in animals is to regulate a special class of ion channels in smell-responsive olfactory neurons, as we discuss later.) The substrates for A-kinase differ in different cell types, explaining why the effects of cyclic AMP vary depending on the target cell.

In the inactive state A-kinase consists of a complex of two catalytic subunits and two regulatory subunits that bind cyclic AMP. The binding of cyclic AMP alters the conformation of the regulatory subunits, causing them to dissociate from the complex. The released catalytic subunits are thereby activated to phosphorylate specific substrate protein molecules (Figure 15–24).

Cyclic-AMP-mediated protein phosphorylation was first demonstrated in studies of glycogen metabolism in skeletal muscle cells. Glycogen is the major storage form of glucose, and both its synthesis and degradation in skeletal muscle cells are regulated by adrenaline. When an animal is frightened or otherwise stressed, for example, the adrenal gland secretes adrenaline into the blood, "alerting" various tissues in the body. Among other effects, the circulating adrenaline

Figure 15–24 The activation of cyclic-AMP-dependent protein kinase (A-kinase). The binding of cyclic AMP to the regulatory subunits induces a conformational change, causing these subunits to dissociate from the complex, thereby activating the catalytic subunits. Each regulatory subunit has two cyclic-AMP-binding sites, and the release of the catalytic subunits requires the binding of more than two cyclic AMP molecules to the tetramer. This greatly sharpens the response of the kinase to changes in cyclic AMP concentration, as we discuss later. There are at least two types of A-kinase in most mammalian cells: type I is mainly in the cytosol, whereas type II is bound via its regulatory subunit to the plasma membrane, nuclear membrane, and microtubules. In both cases, however, once the catalytic subunits are freed and active, they can migrate into the nucleus (where they can phosphorylate gene regulatory proteins), while the regulatory subunits remain in the cytoplasm. The three-dimensional structure of the protein kinase domain of the A-kinase catalytic subunit is shown in Figure 5–12.

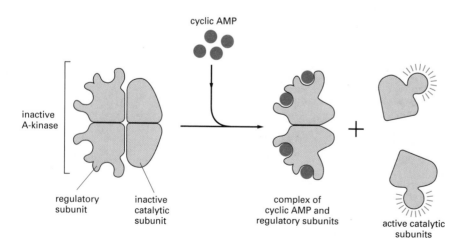

cyclic AMP

inactive
A-kinase

regulatory
subunit

inactive
catalytic
subunit

complex of
cyclic AMP and
regulatory subunits

active catalytic
subunits

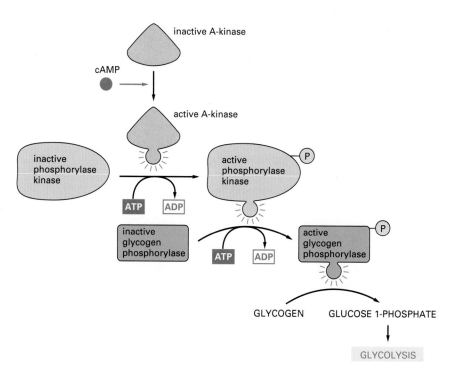

Figure 15–25 The stimulation of glycogen breakdown by cyclic AMP in skeletal muscle cells. The binding of cyclic AMP to A-kinase activates this enzyme to phosphorylate and thereby activate phosphorylase kinase, which in turn phosphorylates and activates glycogen phosphorylase, the enzyme that breaks down glycogen. The A-kinase also directly and indirectly increases the phosphorylation of glycogen synthase, which inhibits the enzyme, thereby shutting off glycogen synthesis (not shown).

induces muscle cells to break down glycogen to glucose 1-phosphate and at the same time to stop synthesizing glycogen. The glucose is then oxidized by glycolysis to provide ATP for sustained muscle contraction. In this way adrenaline prepares the muscle cells for anticipated strenuous activity. Adrenaline acts by binding to β-adrenergic receptors on the muscle cell surface, thereby causing an increase in the level of cyclic AMP in the cytosol. The cyclic AMP activates A-kinase, which phosphorylates two other enzymes. The first, *phosphorylase kinase*, which was the first protein kinase to be discovered (in 1956), phosphorylates in turn the enzyme *glycogen phosphorylase*, thereby activating the phosphorylase to release glucose residues from the glycogen molecule (Figure 15–25). The second enzyme phosphorylated by activated A-kinase is *glycogen synthase*, which performs the final step in glycogen synthesis from glucose. This phosphorylation inhibits the enzyme's activity, thereby shutting off glycogen synthesis. By means of this cascade of interactions, an increase in cyclic AMP levels both stimulates glycogen breakdown and inhibits glycogen synthesis, thus maximizing the amount of glucose available to the cell.

In some animal cells an increase in cyclic AMP activates the transcription of specific genes. In cells that secrete the peptide hormone *somatostatin*, for example, cyclic AMP turns on the gene that encodes this hormone. The regulatory region of the somatostatin gene contains a short DNA sequence, called the *cyclic AMP response element (CRE)*, that is also found in the regulatory region of other genes that are activated by cyclic AMP. This sequence is recognized by a specific gene regulatory protein called **CRE-binding (CREB) protein.** When CREB is phosphorylated by A-kinase on a single serine residue, it is activated to turn on the transcription of these genes; the phosphorylation stimulates the transcriptional activity of CREB without affecting its DNA-binding properties. If this serine residue is mutated, CREB is inactivated and no longer stimulates gene transcription in response to a rise in cyclic AMP levels.

Serine/Threonine Protein Phosphatases Rapidly Reverse the Effects of A-Kinase [17]

Since it is usually important that the effects of cyclic AMP are transient, cells must be able to dephosphorylate the proteins that have been phosphorylated by A-kinase. In general, the dephosphorylation of phosphorylated serines and threonines is catalyzed by four groups of **serine/threonine phosphoprotein phosphatases**—protein phosphatases I, IIA, IIB, and IIC. Except for protein phosphatase-IIC (which is a minor phosphatase, unrelated to the others), all of these phosphatases are composed of a homologous catalytic subunit complexed with one or more regulatory subunits. *Protein phosphatase-I* plays an important role in the response to cyclic AMP, as we discuss below. *Protein phosphatase-IIA* has a broad specificity and seems to be the main phosphatase responsible for reversing many of the phosphorylations catalyzed by serine/threonine kinases; it plays an important part in regulating the cell cycle. *Protein phosphatase-IIB*, also called *calcineurin*, is activated by Ca^{2+} and is especially abundant in the brain.

The activity of any protein regulated by phosphorylation depends on the balance at any instant between the activities of the kinases that phosphorylate it and the phosphatases that are constantly dephosphorylating it. Protein phosphatase-I is responsible for dephosphorylating many of the proteins phosphorylated by A-kinase. It inactivates CREB, for example, by removing its activating phosphate, thereby turning off the transcriptional response caused by a rise in cyclic AMP. In skeletal muscle cells it dephosphorylates each of the three key enzymes in the glycogen pathway that, as mentioned earlier, are phosphorylated in response to adrenaline by A-kinase and switch the cells from synthesizing glycogen to degrading it. Protein phosphatase-I tends to counteract these phosphorylations, but its activity is suppressed in adrenaline-stimulated muscle cells by yet another target of A-kinase, which is a specific *phosphatase inhibitor protein*. When this inhibitor protein is phosphorylated by A-kinase, it binds to protein phosphatase-I and inactivates it (Figure 15–26). By simultaneously activating phosphorylase kinase and inhibiting the opposing action of protein phosphatase-I, the A-kinase causes a much larger change in glycogen metabolism than could be obtained by its action on any one of these enzymes alone.

Having discussed how trimeric G proteins couple receptors to adenylyl cyclase to alter the levels of cyclic AMP in cells, we now consider how G proteins couple receptors to another crucial enzyme—*phospholipase C*. The activation of this enzyme leads to an increase in the concentration of Ca^{2+} in the cytosol, and Ca^{2+} is even more widely used as an intracellular mediator than cyclic AMP.

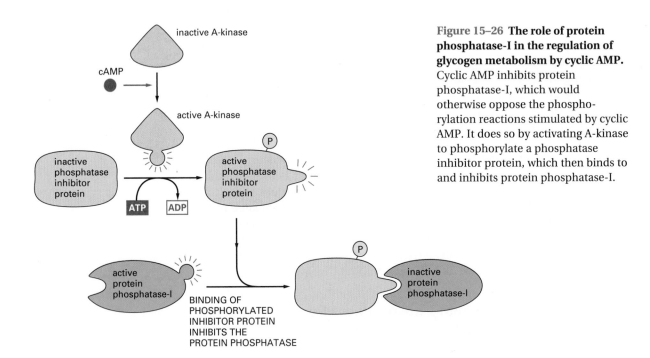

Figure 15–26 The role of protein phosphatase-I in the regulation of glycogen metabolism by cyclic AMP. Cyclic AMP inhibits protein phosphatase-I, which would otherwise oppose the phosphorylation reactions stimulated by cyclic AMP. It does so by activating A-kinase to phosphorylate a phosphatase inhibitor protein, which then binds to and inhibits protein phosphatase-I.

To Use Ca²⁺ as an Intracellular Signal, Cells Must Keep Resting Cytosolic Ca²⁺ Levels Low [18]

The concentration of free Ca^{2+} in the cytosol of any cell is extremely low ($\leq 10^{-7}$ M), whereas its concentration in the extracellular fluid ($\sim 10^{-3}$ M) and in the endoplasmic reticulum (ER) is high. Thus there is a large gradient tending to drive Ca^{2+} into the cytosol across both the plasma membrane and the ER membrane. When a signal transiently opens Ca^{2+} channels in either of these membranes, Ca^{2+} rushes into the cytosol, dramatically increasing the local Ca^{2+} concentration and triggering Ca^{2+}-responsive proteins in the cell.

For this signaling mechanism to work, the resting concentration of Ca^{2+} in the cytosol must be kept low, and this is achieved in several ways. All eucaryotic cells have a Ca^{2+}-ATPase in their plasma membrane that uses the energy of ATP hydrolysis to pump Ca^{2+} out of the cytosol. Cells such as muscle and nerve cells, which make extensive use of Ca^{2+} signaling, have an additional Ca^{2+} pump in their plasma membrane that couples the efflux of Ca^{2+} to the influx of Na^{+}. This Na^{+}-Ca^{2+} exchanger has a relatively low affinity for Ca^{2+} and therefore begins to operate efficiently only when cytosolic Ca^{2+} levels rise to about 10 times their normal level, as occurs after repeated muscle or nerve cell stimulation. A Ca^{2+} pump in the ER membrane also plays an important part in keeping the cytosolic Ca^{2+} concentration low: this Ca^{2+}-ATPase enables the ER to take up large amounts of Ca^{2+} from the cytosol against a steep concentration gradient, even when Ca^{2+} levels in the cytosol are low.

Normally, the concentration of free Ca^{2+} in the cytosol varies from about 10^{-7} M, when the cell is at rest, to about 5×10^{-6} M, when the cell is activated by an extracellular signal. But when a cell is damaged and cannot pump Ca^{2+} out of the cytosol efficiently, the Ca^{2+} concentration can rise beyond that to dangerously high levels ($> 10^{-5}$ M). In these circumstances a low-affinity, high-capacity Ca^{2+} pump in the inner mitochondrial membrane comes into action and uses the electrochemical gradient generated across this membrane during the electron-transfer steps of oxidative phosphorylation to take up Ca^{2+} from the cytosol. These mechanisms are summarized in Figure 15–27.

Figure 15–27 Controls on cytosolic Ca²⁺. The schematic drawing shows the main ways in which cells maintain a very low concentration of free Ca^{2+} in the cytosol in the face of high concentrations of Ca^{2+} in the extracellular fluid. Ca^{2+} is actively pumped out of the cytosol to the cell exterior (A) and into the ER (B). In addition, various molecules in the cell bind free Ca^{2+} tightly. Mitochondria can also pump Ca^{2+} out of the cytosol, but they do so efficiently only when Ca^{2+} levels are extremely high—usually as a result of cell damage.

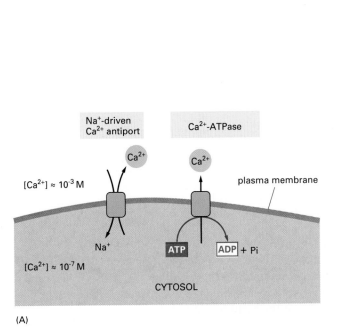

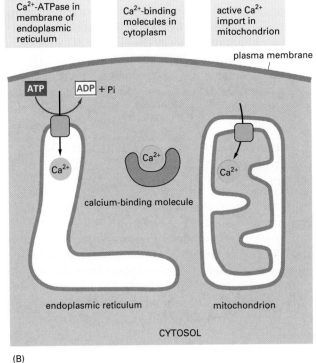

(A)

(B)

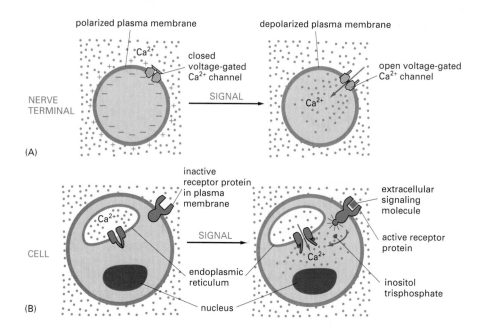

Figure 15–28 **Two common pathways by which Ca^{2+} can enter the cytosol in response to extracellular signals.** In (A) Ca^{2+} enters a nerve terminal from the extracellular fluid through voltage-gated Ca^{2+} channels when the nerve terminal membrane is depolarized by an action potential. In (B) the binding of an extracellular signaling molecule to a cell-surface receptor generates inositol trisphosphate, which stimulates the release of Ca^{2+} from the ER.

Ca^{2+} Functions as a Ubiquitous Intracellular Messenger [19]

The first direct evidence that Ca^{2+} functions as an intracellular mediator came from an experiment done in 1947 showing that the intracellular injection of a small amount of Ca^{2+} causes a skeletal muscle cell to contract. In recent years it has become clear that Ca^{2+} also acts as an intracellular messenger in a wide variety of other cellular responses, including secretion and cell proliferation. Two pathways of Ca^{2+} signaling have been well defined, one used mainly by electrically active (excitable) cells and the other used by almost all eucaryotic cells. The first of these pathways has been particularly well studied in nerve cells, in which depolarization of the plasma membrane causes an influx of Ca^{2+} into the nerve terminal, initiating the secretion of neurotransmitter; the Ca^{2+} enters through voltage-gated Ca^{2+} channels that open when the plasma membrane of the nerve terminal is depolarized by an invading action potential (see Figure 11–34). In the second, ubiquitous pathway the binding of extracellular signaling molecules to cell-surface receptors causes the release of Ca^{2+} from the ER. The events at the cell surface are coupled to the opening of Ca^{2+} channels in the ER through yet another intracellular messenger molecule, *inositol trisphosphate* (Figure 15–28), as we discuss next.

Some G-Protein-linked Receptors Activate the Inositol Phospholipid Signaling Pathway by Activating Phospholipase C-β [20]

A role for inositol phospholipids (*phosphoinositides*) in signal transduction was first suggested in 1953, when it was found that some extracellular signaling molecules stimulate the incorporation of radioactive phosphate into **phosphatidylinositol (PI),** a minor phospholipid in cell membranes. It was later shown that this incorporation results from the breakdown and subsequent resynthesis of inositol phospholipids. The inositol phospholipids found to be most important in signal transduction were two phosphorylated derivatives of PI, *PI phosphate (PIP)* and *PI bisphosphate (PIP$_2$)*, which are thought to be located mainly in the inner half of the plasma membrane lipid bilayer (Figure 15–29). Although PIP$_2$ is less plentiful in animal cell membranes than PI, it is the hydrolysis of PIP$_2$ that matters most.

The chain of events leading to PIP$_2$ breakdown begins with the binding of a signaling molecule to a G-protein-linked receptor in the plasma membrane.

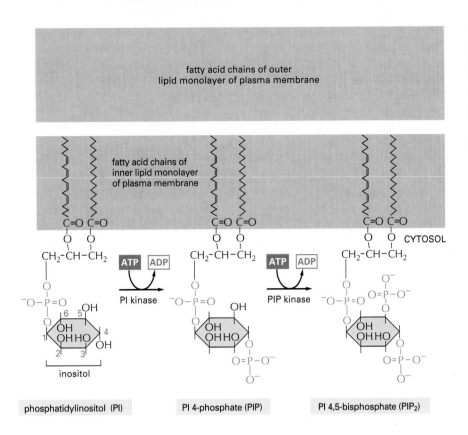

fatty acid chains of outer
lipid monolayer of plasma membrane

fatty acid chains of
inner lipid monolayer
of plasma membrane

CYTOSOL

PI kinase

PIP kinase

inositol

phosphatidylinositol (PI)

PI 4-phosphate (PIP)

PI 4,5-bisphosphate (PIP$_2$)

Figure 15–29 Inositol phospholipids (phosphoinositides). The polyphosphoinositides (PIP and PIP$_2$) are produced by the phosphorylation of phosphatidylinositol (PI). Although all three inositol phospholipids may be broken down in the signaling response, it is the breakdown of PIP$_2$ that is most critical, even though it is the least abundant, constituting less than 10% of the total inositol lipids and less than 1% of the total phospholipids.

More than 25 different cell-surface receptors have been shown to utilize this transduction pathway; several examples of responses mediated in this way are given in Table 15–2. Although the details of the activation process are not as well understood as they are in the cyclic AMP pathway, the same type of multistep mechanism is thought to operate in the plasma membrane. An activated receptor stimulates a trimeric G protein called **G$_q$,** which in turn activates a *phospho-inositide-specific phospholipase C* called **phospholipase C-β.** In less than a second, this enzyme cleaves PIP$_2$ to generate two products: *inositol trisphosphate* and *diacylglycerol* (Figure 15–30). At this step the signaling pathway splits into two branches. Since both molecules play crucial parts in signaling the cell, we consider them in turn.

Inositol Trisphosphate (IP$_3$) Couples Receptor Activation to Ca^{2+} Release from the ER [21]

The **inositol trisphosphate (IP$_3$)** produced by PIP$_2$ hydrolysis is a small water-soluble molecule that leaves the plasma membrane and diffuses rapidly through the cytosol. There it releases Ca^{2+} from the ER by binding to *IP$_3$-gated Ca^{2+}-release channels* in the ER membrane. The channels are structurally similar to the

Table 15–2 Some Cellular Responses Mediated by G-Protein-linked Receptors Coupled to the Inositol-Phospholipid Signaling Pathway

Target Tissue	Signaling Molecule	Major Response
Liver	vasopressin	glycogen breakdown
Pancreas	acetylcholine	amylase secretion
Smooth muscle	acetylcholine	contraction
Mast cells	antigen	histamine secretion
Blood platelets	thrombin	aggregation

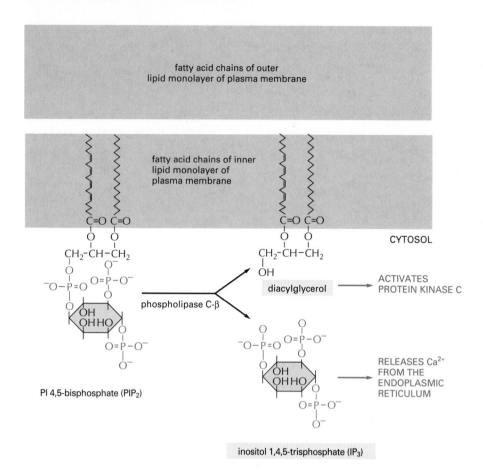

fatty acid chains of outer
lipid monolayer of plasma membrane

fatty acid chains of inner
lipid monolayer of
plasma membrane

CYTOSOL

PI 4,5-bisphosphate (PIP$_2$)

phospholipase C-β

diacylglycerol → ACTIVATES PROTEIN KINASE C

RELEASES Ca^{2+} FROM THE ENDOPLASMIC RETICULUM

inositol 1,4,5-trisphosphate (IP$_3$)

Figure 15–30 The hydrolysis of PIP$_2$. Two intracellular mediators are produced when PIP$_2$ is hydrolyzed: *inositol trisphosphate (IP$_3$)*, which diffuses through the cytosol and releases Ca^{2+} from the ER, and *diacylglycerol*, which remains in the membrane and helps activate the enzyme protein kinase C (see below). There are at least three classes of phospholipase C—β, γ, and δ—and it is the β class that is activated by G-protein-linked receptors. We shall see later that the γ class is activated by a second class of receptors, called *receptor tyrosine kinases*, that activate the inositol-phospholipid signaling pathway without an intermediary G protein.

Ca^{2+}-release channels (*ryanodine receptors*) in the sarcoplasmic reticulum of muscle cells, which release the Ca^{2+} that triggers muscle contraction (see Figure 16–92). Both types of channels are regulated by positive feedback, in which the released Ca^{2+} can bind back to the channels to increase the Ca^{2+} release, which tends to make the release occur in a sudden, all-or-none fashion. In many cells, including muscle cells, both types of Ca^{2+}-release channels are present.

Two mechanisms operate to terminate the initial Ca^{2+} response: (1) IP$_3$ is rapidly dephosphorylated (and thereby inactivated) by specific phosphatases, and (2) Ca^{2+} that enters the cytosol is rapidly pumped out, mainly out of the cell.

Not all of the IP$_3$ is dephosphorylated, however: some is instead phosphorylated to form inositol 1,3,4,5-tetrakisphosphate (IP$_4$), which may mediate slower and more prolonged responses in the cell or promote the refilling of the intracellular Ca^{2+} stores from the extracellular fluid, or both. The enzyme that catalyzes the production of IP$_4$ is activated by the increase in cytosolic Ca^{2+} induced by IP$_3$, providing a form of negative feedback regulation on IP$_3$ levels.

Ca^{2+} Oscillations Often Prolong the Initial IP$_3$-induced Ca^{2+} Response [22]

When Ca^{2+}-sensitive fluorescent indicators, such as aequorin or fura-2 (discussed in Chapter 4), are used to monitor cytosolic Ca^{2+} in individual cells in which the inositol phospholipid signaling pathway has been activated, the initial Ca^{2+} signal is often seen to propagate as a wave through the cytosol from a localized region of the cell. Moreover, the initial transient increase in Ca^{2+} is often followed by a series of Ca^{2+} "spikes," each lasting seconds or minutes; these **Ca^{2+} oscillations** can persist for as long as receptors are activated on the cell surface (Figure 15–31).

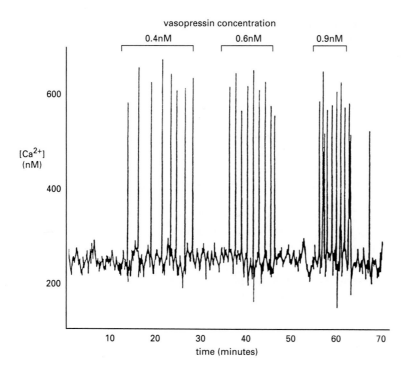

vasopressin concentration

Figure 15–31 Vasopressin-induced Ca²⁺ oscillations in a liver cell. The cell was loaded with the Ca^{2+}-sensitive protein aequorin and then exposed to increasing concentrations of vasopressin. Note that the frequency of the Ca^{2+} spikes increases with increasing concentration of vasopressin but that the amplitude of the spikes is not affected. (Adapted from N.M. Woods, K.S.R. Cuthbertson, and P.H. Cobbold, *Nature* 319:600–602, 1986. © 1986 Macmillan Magazines Ltd.)

The mechanisms responsible for the propagation of Ca^{2+} waves and for generating the oscillations are uncertain, although a number of models have been proposed. In most models both the propagation and oscillations depend on positive feedback, whereby Ca^{2+} activates its own release, thereby producing an all-or-none Ca^{2+} spike. The models differ mainly in whether Ca^{2+} acts directly on the Ca^{2+}-release channels in the ER to stimulate its own release or whether it acts indirectly, by increasing the activity of phospholipase C, thereby generating surges of IP_3, which in turn induce surges of Ca^{2+} release.

The biological significance of the Ca^{2+} oscillations is also uncertain. Their frequency often depends on the concentration of the extracellular signaling ligand (see Figure 15–31) and might, in principle, be translated into a frequency-dependent cellular response. In hormone-secreting pituitary cells, for example, stimulation by an extracellular signaling molecule induces repeated Ca^{2+} spikes, each of which is associated with a burst of hormone secretion. It has been suggested that this arrangement might maximize secretory output while avoiding the toxic effects of a sustained rise in cytosolic Ca^{2+}.

Diacylglycerol Activates Protein Kinase C (C-Kinase) [23]

At the same time that the IP_3 produced by hydrolysis of PIP_2 is increasing the concentration of Ca^{2+} in the cytosol, the other cleavage product of PIP_2—**diacylglycerol**—is exerting different effects. Diacylglycerol has two potential signaling roles. First, it can be further cleaved to release arachidonic acid, which either can act as a messenger in its own right or be used in the synthesis of eicosanoids (see Figure 15–6). Second, and more important, it activates a crucial serine/threonine protein kinase that phosphorylates selected proteins in the target cell.

The enzyme activated by diacylglycerol is called **protein kinase C (C-kinase, or PKC)** because it is Ca^{2+}-dependent. The initial rise in cytosolic Ca^{2+} induced by IP_3 is thought to alter the C-kinase so that it translocates from the cytosol to the cytoplasmic face of the plasma membrane. There it is activated by the combination of Ca^{2+}, diacylglycerol, and the negatively charged membrane phospholipid phosphatidylserine. Of the eight or more distinct isoforms of C-kinase in mammals, at least four are activated by diacylglycerol.

Because the diacylglycerol produced initially by the cleavage of PIP₂ is rapidly metabolized, it cannot sustain the activity of C-kinase, as would be required for long-term responses such as cell proliferation or differentiation. Prolonged activation of C-kinase depends on a second wave of diacylglycerol production, catalyzed by phospholipases that cleave the major membrane phospholipid phosphatidylcholine. It is uncertain how these later-acting phospholipases become activated.

When activated, C-kinase phosphorylates specific serine or threonine residues on target proteins that vary depending on the cell type. The highest concentrations of C-kinase are found in the brain, where (among other things) it phosphorylates ion channels in nerve cells, thereby changing their properties and altering the excitability of the nerve cell plasma membrane.

In many cells the activation of C-kinase increases the transcription of specific genes. At least two pathways are known. In one, C-kinase activates a protein kinase cascade that leads to the phosphorylation and activation of a DNA-bound gene regulatory protein; in another, C-kinase activation leads to the phosphorylation of an inhibitor protein, thereby releasing a cytoplasmic gene regulatory protein so that it can migrate into the nucleus and stimulate the transcription of specific genes (Figure 15–32).

The two branches of the inositol phospholipid signaling pathway are summarized in Figure 15–33. As indicated in the figure, each branch of the pathway can be mimicked by the addition of specific pharmacological agents to intact cells. The effects of IP₃ can be mimicked by using a *Ca²⁺ ionophore,* such as A23187 or ionomycin, which allows Ca²⁺ to move into the cytosol from the extracellular fluid (discussed in Chapter 11). The effects of diacylglycerol can be mimicked by *phorbol esters,* plant products that bind to C-kinase and activate it di-

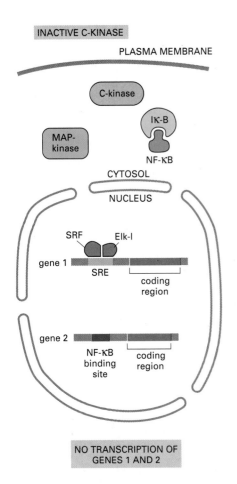

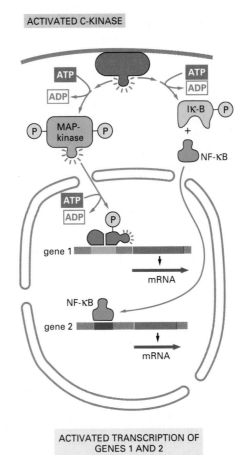

Figure 15–32 **Two intracellular pathways by which activated C-kinase can activate the transcription of specific genes.** In one (*red arrows*) C-kinase activates a phosphorylation cascade that leads to the phosphorylation of a pivotal protein kinase called *MAP-kinase* (discussed later), which in turn phosphorylates and activates the gene regulatory protein Elk-1. Elk-1 is bound to a short DNA sequence (called *serum response element, SRE*) in association with another DNA-binding protein (called *serum response factor, SRF*). In the other pathway (*green arrows*) C-kinase activation leads to the phosphorylation of Iκ-B, which releases the gene regulatory protein NF-κB so that it can migrate into the nucleus and activate the transcription of specific genes.

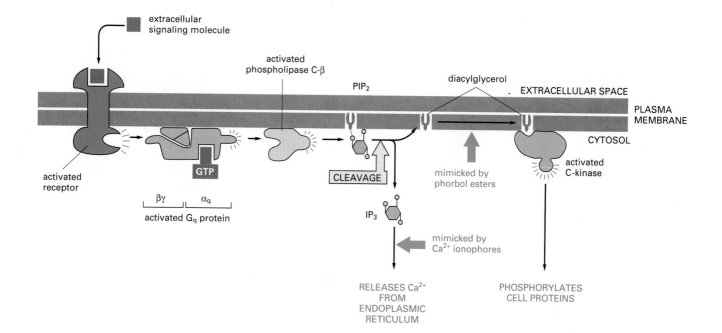

Figure 15–33 **The two branches of the inositol phospholipid pathway.** The activated receptor binds to a specific trimeric G protein (G_q), causing the α subunit to dissociate and activate phospholipase C-β, which cleaves PIP_2 to generate IP_3 and diacylglycerol. The diacylglycerol (together with bound Ca^{2+} and phosphatidylserine—not shown) activates C-kinase. Both phospholipase C-β and C-kinase are water-soluble enzymes that translocate from the cytosol to the inner face of the plasma membrane in the process of being activated. The effects of IP_3 can be mimicked experimentally in intact cells by treatment with Ca^{2+} ionophores, while the effects of diacylglycerol can be mimicked by treatment with phorbol esters, which bind to C-kinase and activate it.

rectly. Using these reagents, it has been shown that the two branches of the pathway often collaborate in producing a full cellular response. A number of cell types, for example, can be stimulated to proliferate in culture when treated with both a Ca^{2+} ionophore and a C-kinase activator but not when they are treated with either reagent alone.

Calmodulin Is a Ubiquitous Intracellular Ca^{2+} Receptor [24]

Since the free Ca^{2+} concentration in the cytosol is usually $\leq 10^{-7}$ M and generally does not rise above 6×10^{-6} M even when the cell is activated by an influx of Ca^{2+}, any structure in the cell that is to serve as a direct target for Ca^{2+}-dependent regulation must have an affinity constant (K_a) for Ca^{2+} of around 10^6 liters/mole. Moreover, since the concentration of free Mg^{2+} in the cytosol is relatively constant at about 10^{-3} M, these Ca^{2+}-binding sites must have a selectivity for Ca^{2+} over Mg^{2+} of at least 1000-fold. Several specific Ca^{2+}-binding proteins fulfill these criteria.

The first such protein to be discovered was *troponin C* in skeletal muscle cells; its role in muscle contraction is discussed in Chapter 16. A closely related Ca^{2+}-binding protein, known as **calmodulin,** is found in all eucaryotic cells that have been examined. A typical animal cell contains more than 10^7 molecules of calmodulin, which can constitute as much as 1% of the total protein mass of the cell. Calmodulin functions as a multipurpose intracellular Ca^{2+} receptor, mediating many Ca^{2+}-regulated processes. It is a highly conserved, single polypeptide chain of about 150 amino acids, with four high-affinity Ca^{2+}-binding sites (Figure 15–34A), and it undergoes a conformational change when it binds Ca^{2+}.

The allosteric activation of calmodulin by Ca^{2+} is analogous to the allosteric activation of A-kinase by cyclic AMP, except that Ca^{2+}/calmodulin has no enzyme activity itself but acts by binding to other proteins. In some cases calmodulin serves as a permanent regulatory subunit of an enzyme complex, but in most cases the binding of Ca^{2+} enables calmodulin to bind to various target proteins in the cell and thereby alter their activity. When Ca^{2+}/calmodulin binds to its target protein, it can undergo a further and more dramatic change in conformation (Figure 15–34B).

Among the targets regulated by Ca^{2+}/calmodulin are various enzymes and membrane transport proteins. In many cells, for example, Ca^{2+}/calmodulin binds to and activates the plasma membrane Ca^{2+}-ATPase that pumps Ca^{2+} out of the

Signaling via G-Protein-linked Cell-Surface Receptors

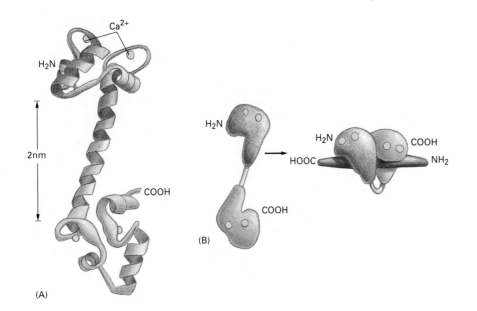

Figure 15–34 The structure of Ca²⁺/calmodulin based on x-ray diffraction and NMR studies. (A) The molecule has a "dumbbell" shape, with two globular ends connected by a long, exposed α helix. Each end has two Ca^{2+}-binding domains, each with a loop of 12 amino acid residues in which aspartic acid and glutamic acid side chains form ionic bonds with Ca^{2+}. The two Ca^{2+}-binding sites in the carboxyl-terminal part of the molecule have a tenfold higher affinity for Ca^{2+} than those in the amino-terminal part. In solution the molecule is flexible, displaying a range of forms, from extended (as shown) to more compact. (B) The structural changes in Ca^{2+}/calmodulin that occurs when it binds to a target protein (in this example a peptide that consists of the Ca^{2+}/calmodulin-binding domain of a Ca^{2+}/calmodulin-dependent protein kinase [myosin light-chain kinase, discussed below]). Note that the Ca^{2+}/calmodulin has "jack-knifed" to surround the peptide. (A, based on x-ray crystallographic data from Y.S. Babu et al., *Nature* 315:37–40, 1985. © 1985 Macmillan Magazines Ltd.; B, based on x-ray crystallographic data from W.E. Meador, A.R. Means, and F.A. Quiocho, *Science* 257:1251–1255, 1992, and on NMR data from M. Ikura et al., *Science* 256:632–638, 1992. © 1992 the AAAS.)

cell. Thus, if the concentration of Ca^{2+} in the cytosol rises, the pump is activated, which helps return the cytosolic Ca^{2+} level to normal. Most effects of Ca^{2+}/calmodulin, however, are more indirect and are mediated by *Ca²⁺/calmodulin-dependent protein kinases.*

Ca²⁺/Calmodulin-dependent Protein Kinases (CaM-Kinases) Mediate Most of the Actions of Ca²⁺ in Animal Cells [25]

Most of the effects of Ca^{2+} in cells are mediated by protein phosphorylations catalyzed by a family of **Ca²⁺/calmodulin-dependent protein kinases (CaM-kinases).** These kinases phosphorylate serines or threonines in proteins, and, as in the case of cyclic AMP, the response of a target cell to an increase in free Ca^{2+} concentration in the cytosol depends on which CaM-kinase-regulated target proteins are present in the cell. The first CaM-kinases to be discovered—*myosin light-chain kinase,* which activates smooth muscle contraction, and *phosphorylase kinase,* which activates glycogen breakdown—have narrow substrate specificities. More recently, however, a number of CaM-kinases have been identified that have much broader specificities, and these seem to be responsible for mediating many of the actions of Ca^{2+} in animal cells.

The best-studied example of such a *multifunctional CaM-kinase* is **CaM-kinase II,** which is found in all animal cells but is especially enriched in the nervous system. It constitutes up to 2% of the total protein mass in some regions of the brain, where it is highly concentrated in synapses. When neurons that use *catecholamines* (dopamine, noradrenaline, or adrenaline) as their neurotransmitter are activated, for example, the influx of Ca^{2+} through voltage-gated Ca^{2+} channels in the plasma membrane stimulates the cells to secrete their neurotransmitter. The Ca^{2+} influx also activates CaM-kinase II to phosphorylate, and thereby activate, *tyrosine hydroxylase,* which is the rate-limiting enzyme in catecholamine synthesis. In this way, both the secretion and resynthesis of the neurotransmitter are stimulated when the cell is activated.

CaM-kinase II has a remarkable property: it can function as a molecular memory device, switching to an active state when exposed to Ca^{2+}/calmodulin and then remaining active even after the Ca^{2+} is withdrawn. This is because the kinase phosphorylates itself (a process called *autophosphorylation*) as well as other cell proteins when it is activated by Ca^{2+}/calmodulin. In its autophosphorylated state the enzyme remains active in the absence of Ca^{2+}, thereby prolonging the duration of the kinase activity beyond the duration of the

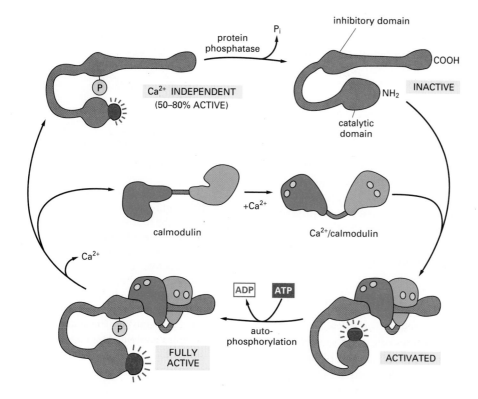

protein
phosphatase

P_i

inhibitory domain

COOH

P

Ca²⁺ INDEPENDENT
(50–80% ACTIVE)

INACTIVE

NH₂

catalytic
domain

calmodulin

+Ca²⁺

Ca²⁺/calmodulin

Ca²⁺

ADP   ATP

auto-
phosphorylation

P

FULLY
ACTIVE

ACTIVATED

Figure 15–35 The activation of CaM-kinase II. The enzyme is a protein complex of about 12 subunits. The subunits are of four homologous kinds (α, β, γ, and δ), which are expressed in different proportions in different cell types. Only the α subunit, which is expressed only in the brain, is shown (in *gray*). In the absence of Ca²⁺/calmodulin the enzyme is inactive as the result of an interaction between the inhibitory domain and the catalytic domain. The binding of Ca²⁺/calmodulin alters the conformation of the protein, allowing the catalytic domain to phosphorylate the inhibitory domain of neighboring subunits in the complex, as well as other proteins in the cell (not shown). The autophosphorylation of the enzyme complex (by mutual phosphorylation of its subunits) prolongs the activity of the enzyme in two ways: (1) it traps the bound Ca²⁺/calmodulin so that it does not dissociate from the enzyme complex until cytosolic Ca²⁺ levels return to basal values for at least 10 seconds (not shown); (2) it converts the enzyme to a Ca²⁺-independent form so that the kinase remains active even after the Ca²⁺/calmodulin dissociates from it. Activity continues until the autophosphorylation process is overwhelmed by a protein phosphatase.

initial activating Ca²⁺ signal; the activity is maintained until phosphatases overwhelm the autophosphorylating activity of the enzyme and shut it off (Figure 15–35).

Because of these properties, CaM-kinase II activation can serve as a memory trace of a prior Ca²⁺ pulse, and it seems to play an important part in some types of memory and learning in the vertebrate nervous system. Mutant mice that are missing the brain-specific subunit illustrated in Figure 15–35 have specific defects in their ability to remember the location of an object—that is, in spatial learning.

The Cyclic AMP and Ca²⁺ Pathways Interact [26]

The cyclic AMP and Ca²⁺ intracellular signaling pathways interact at several levels in the hierarchy of control. First, cytosolic Ca²⁺ and cyclic AMP levels can influence each other. For example, some forms of the enzymes that break down and make cyclic AMP—cyclic AMP phosphodiesterase and adenylyl cyclase, respectively—are regulated by Ca²⁺-calmodulin complexes. Conversely, A-kinase can phosphorylate some Ca²⁺ channels and pumps and alter their activity; A-kinase phosphorylates the IP₃ receptor in the ER, for example, which can either inhibit or promote IP₃-induced Ca²⁺ release, depending on the cell type. Second, the enzymes directly regulated by Ca²⁺ and cyclic AMP can influence each other. Some CaM-kinases are phosphorylated by A-kinase, for example. Third, these enzymes can have interacting effects on shared downstream target molecules. Thus A-kinase and CaM-kinases frequently phosphorylate different sites on the same proteins, which are thereby regulated by both cyclic AMP and Ca²⁺; the CREB gene regulatory protein, which we discussed earlier (see p. 741), is one example.

As an example of how Ca²⁺ and cyclic AMP pathways can interact, consider the *phosphorylase kinase* of skeletal muscle, whose role in glycogen degradation we have already discussed. This kinase phosphorylates glycogen phosphorylase, causing it to break down glycogen (see Figure 15–25). The kinase is a multisubunit enzyme, but only one of its four subunits actually catalyzes the phosphorylation reaction: the other three subunits are regulatory and enable the enzyme complex to be activated both by cyclic AMP and by Ca^{2+}. The four subunits are designated α, β, γ, and δ. The γ subunit carries the catalytic activity; the δ subunit is calmodulin and is largely responsible for the Ca^{2+} dependence of the enzyme; the α and β subunits are targets for cyclic AMP-mediated regulation, both being phosphorylated by the A-kinase (Figure 15–36).

The same Ca^{2+} signal that initiates muscle contraction also ensures that there is adequate glucose to power the contraction. The large influx of Ca^{2+} into the cytosol discussed in Chapter 16 alters the conformation of the calmodulin subunit of phosphorylase kinase, increasing kinase activity and thereby increasing the rate of the glycogen breakdown catalyzed by glycogen phosphorylase several hundredfold within seconds. In addition, the Ca^{2+} influx activates CaM-kinases that phosphorylate and inhibit glycogen synthase, thereby shutting off glycogen synthesis. By contrast, the adrenaline-induced A-kinase phosphorylations previously discussed adjust muscle cell metabolism in anticipation of an increased energy demand, allowing the enzyme to be activated when fewer calcium ions are bound to calmodulin, thereby making it more sensitive to Ca^{2+}.

Some Trimeric G Proteins Directly Regulate Ion Channels [27]

Trimeric G proteins do not act exclusively by regulating the activity of enzymes and altering the concentration of cyclic nucleotides or Ca^{2+} in the cytosol. In some cases they directly activate or inactivate ion channels in the plasma membrane of the target cell, thereby altering the ion permeability, and hence the excitability, of the membrane. Acetylcholine released by the vagus nerve, for example, reduces both the rate and strength of heart muscle cell contraction. The effect is mediated by a special class of acetylcholine receptors that activate the inhibitory G protein, G_i, discussed previously. (These receptors, which can be activated by the fungal alkaloid muscarine, are called *muscarinic acetylcholine receptors* to distinguish them from the very different *nicotinic acetylcholine receptors,* which are ion-channel-linked receptors on skeletal muscle cells and nerve cells that can be activated by nicotine.) Once activated, the α subunit of G_i inhibits adenylyl cyclase (as described previously), while the βγ complex directly opens K^+ channels in the muscle cell plasma membrane. The opening of these K^+ channels makes it harder to depolarize the cell, which contributes to the inhibitory effect of acetylcholine on the heart.

Other trimeric G proteins regulate the activity of ion channels less directly, either by regulating channel phosphorylation (by A-kinase, C-kinase, or CaM-kinase, for example) or by causing the production or destruction of cyclic nucleotides that directly activate or inactivate ion channels. Such *cyclic-nucleotide-gated ion channels* play a crucial role in both smell (olfaction) and vision.

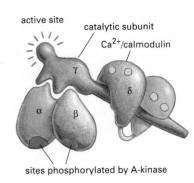

Figure 15–36 Phosphorylase kinase. This highly schematized drawing shows the four subunits of the enzyme from mammalian muscle. The γ subunit has the catalytic activity of the active enzyme; the α and β subunits and the δ subunit (calmodulin) mediate the regulation of the enzyme by cyclic AMP and Ca^{2+}, respectively. The actual enzyme complex contains four copies of each subunit.

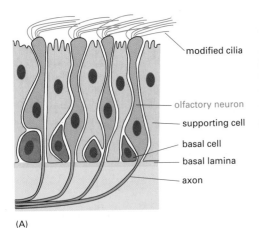

(A)

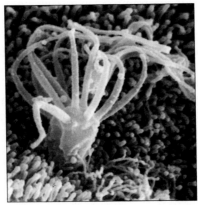

(B)

Figure 15–37 Olfactory receptor neurons. (A) Schematic drawing of olfactory epithelium in the nose. The olfactory receptor neurons possess modified cilia, which project from the surface of the epithelium and contain the olfactory receptors as well as the signal transduction machinery. The axon, which extends from the opposite end of the receptor neuron, conveys electrical signals to the brain when the cell is activated by an odorant. The basal cells act as stem cells, producing new receptor neurons throughout life. (B) A scanning electron micrograph of the cilia on the surface of an olfactory neuron. (B, from E.E. Morrison and R.M. Costanzo, *J. Comp. Neurol.* 297:1–13, 1990. © 1990 Wiley-Liss, Inc.)

Smell and Vision Depend on G-Protein-linked Receptors and Cyclic-Nucleotide-gated Ion Channels [28]

Humans can distinguish more than 10,000 different smells (*odorants*), which are detected by specialized olfactory receptor neurons in the lining of the nose. These cells recognize odorants by means of specific G-protein-linked **olfactory receptors,** which are displayed on the surface of the modified cilia that extend from each cell (Figure 15–37). Many of these receptors act through cyclic AMP: when stimulated by odorant binding, they activate an olfactory-specific trimeric G protein (G_{olf}), which in turn activates adenylyl cyclase; the resulting increase in cyclic AMP opens *cyclic-AMP-gated cation channels,* which allows an influx of Na^+ that depolarizes the cell and initiates a nerve impulse that travels along the axon to the brain. Other olfactory receptors act via the inositol phospholipid pathway and IP_3-gated Ca^{2+} channels in the plasma membrane, but less is known about this transduction mechanism.

It is thought that there are hundreds of different olfactory receptors, each encoded by a different gene and each recognizing different odorants but all belonging to the G-protein-linked receptor superfamily. Although it is known that each olfactory cell responds to a specific set of odorants, it is not yet clear if each cell contains only one type of receptor that recognizes a set of odorants or whether each cell contains a set of receptors, each specific for a single odorant. G-protein-linked receptors also seem to mediate some forms of taste, but less is known about them.

Cyclic-nucleotide-gated ion channels are also involved in signal transduction in vertebrate vision, but here the crucial cyclic nucleotide is **cyclic GMP** (Figure 15–38) rather than cyclic AMP. Like cyclic AMP, the concentration of cyclic GMP in cells is controlled by rapid synthesis (by *guanylyl cyclase*) and rapid degradation (by *cyclic GMP phosphodiesterase*).

In visual transduction, receptor activation is caused by light, and it leads to a fall rather than a rise in the level of the cyclic nucleotide. The pathway has been especially well studied in **rod photoreceptors (rods)** in the vertebrate retina. Rods are responsible for monochromatic vision in dim light, whereas *cone photoreceptors (cones)* are responsible for color vision in bright light. A rod photoreceptor is a highly specialized cell with an outer and an inner segment, a cell body, and a synaptic region where the rod passes a chemical signal to a retinal nerve cell, which relays the signal along the visual pathway (Figure 15–39). The phototransduction apparatus is in the outer segment, which contains a stack of *discs,* each formed by a closed sac of membrane in which photosensitive **rhodopsin** molecules are embedded. The plasma membrane surrounding the outer segment contains cyclic-GMP-gated Na^+ channels. These Na^+ channels are kept open in the dark by cyclic GMP molecules bound to the channels. Paradoxically, light causes a hyperpolarization (which inhibits synaptic signaling) rather than a depolarization of the plasma membrane (which could stimulate synaptic signaling), because the activation by light of rhodopsin molecules in the disc membrane leads to the *closure* of the Na^+ channels in the surrounding plasma membrane (Figure 15–40).

Rhodopsin, as we noted earlier, is a seven-pass transmembrane molecule homologous to other members of the G-protein-linked receptor family, and, like its cousins, it acts through a trimeric G protein. The activating extracellular signal, however, is not a molecule but a photon of light. Each rhodopsin molecule contains a covalently attached chromophore, 11-*cis* retinal, which isomerizes almost instantaneously to all-*trans* retinal when it absorbs a single photon. The isomerization alters the shape of the retinal, forcing a slower conformational change in

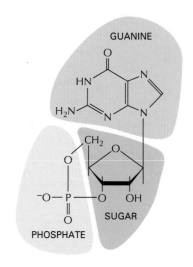

Figure 15–38 Cyclic GMP.

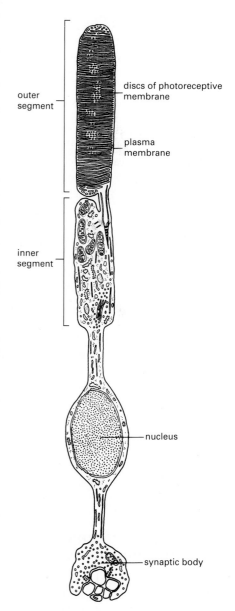

Figure 15–39 Drawing of a rod photoreceptor cell. There are about 1000 discs in the outer segment, and the disc membranes are not connected to the plasma membrane.

Signaling via G-Protein-linked Cell-Surface Receptors

753

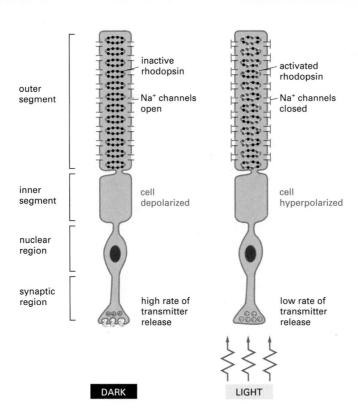

Figure 15–40 **The response of a rod photoreceptor cell to light.** Photons are absorbed by rhodopsin molecules in the outer-segment discs. This leads to the closure of Na$^+$ channels in the plasma membrane, which hyperpolarizes the membrane and reduces the rate of neurotransmitter release from the synaptic region. Because the neurotransmitter acts to inhibit many of the postsynaptic retinal neurons, illumination serves to free the neurons from inhibition and thus, in effect, excites them.

the protein (opsin). The activated protein then binds to the trimeric G-protein *transducin* (G$_t$), causing the α subunit (α_t) to dissociate and activate **cyclic GMP phosphodiesterase,** which hydrolyzes cyclic GMP, so that cyclic GMP levels in the cytosol drop. As a consequence, cyclic GMP dissociates from the plasma membrane Na$^+$ channels, allowing them to close. In this way the signal passes from the disc membrane to the plasma membrane, and a light signal is converted into an electrical one.

The Na$^+$ channels are also permeable to Ca^{2+}, so that when they close, the normal influx of Ca^{2+} is inhibited, causing the Ca^{2+} concentration in the cytosol to fall; the fall in Ca^{2+} stimulates guanylyl cyclase to replenish the cyclic GMP, rapidly returning the cell toward the state it was in before the light was switched on. The activation of guanylyl cyclase by the fall in Ca^{2+} is mediated by a Ca^{2+}-sensitive protein called *recoverin,* which, in contrast to calmodulin, is inactive when Ca^{2+} is bound to it and active when it is Ca^{2+}-free; it stimulates the cyclase when Ca^{2+} levels are low following a light response. This Ca^{2+}-dependent mechanism is of crucial importance in two ways. First, it allows the photoreceptor to revert quickly to its resting, dark state in the aftermath of a flash of light, making it possible to perceive the shortness of the flash. Second, it helps to enable the photoreceptor to *adapt,* stepping down the response when it is exposed to light continuously. Adaptation, as we explain later, means that the receptor cell can function as a sensitive detector of *changes* in stimulus intensity over an enormously wide range of baseline levels of stimulation.

The various trimeric G proteins that we have discussed in this chapter are summarized in Table 15–3.

Extracellular Signals Are Greatly Amplified by the Use of Intracellular Mediators and Enzymatic Cascades [29]

Despite the differences in molecular details, all of the signaling systems that are triggered by G-protein-linked receptors share certain features and are governed by similar general principles. Most of them depend on complex cascades, or relay chains, of intracellular mediators. By contrast with the more direct signaling

Table 15–3 The Major Families of Trimeric G proteins*

Family	Some Family Members	α Subunits	Functions	Modified by Bacterial Toxin
I	G_s	α_s	activates adenylyl cyclase; activates Ca^{2+} channels	cholera activates
	G_{olf}	α_{olf}	activates adenylyl cyclase in olfactory sensory neurons	cholera activates
II	G_i	α_i	inhibits adenylyl cyclase; activates K^+ channels	pertussis inhibits
	G_o	α_o	activates K^+ channels; inactivates Ca^{2+} channels; activates phospholipase C-β	pertussis inhibits
	G_t (transducin)	α_t	activates cyclic GMP phospho-diesterase in vertebrate rod photoreceptors	cholera activates and pertussis inhibits
III	G_q	α_q	activates phospholipase C-β	no effect

*Families are determined by amino acid sequence relatedness of the α subunits. Only selected examples are shown. About 20 α subunits and at least 4 β subunits and 7 γ subunits have been described in mammals.

pathways used by intracellular receptors discussed earlier and by ion-channel-linked receptors discussed in Chapter 11, catalytic cascades of intracellular mediators provide numerous opportunities for amplifying the responses to extracellular signals. In the visual transduction cascade just described, for example, a single activated rhodopsin molecule catalyzes the activation of hundreds of molecules of transducin at a rate of about 1000 transducin molecules per second. Each activated transducin molecule activates a molecule of cyclic GMP phosphodiesterase, each of which hydrolyzes about 4000 molecules of cyclic GMP per second. This catalytic cascade lasts for about 1 second and results in the hydrolysis of more than 10^5 cyclic GMP molecules for a single quantum of light absorbed, which transiently closes hundreds of Na^{2+} channels in the plasma membrane (Figure 15–41).

Similarly, when an extracellular signaling molecule binds to a receptor that indirectly activates adenylyl cyclase via G_s, each receptor protein may activate many molecules of G_s protein, each of which can activate a cyclase molecule. As each G_s molecule activated persists in its active form for seconds before it hydrolyzes its bound GTP to shut itself off, it can keep its bound cyclase molecule active for seconds, so that the cyclase can catalyze the conversion of a large number of ATP molecules to cyclic AMP molecules (Figure 15–42). The same type of amplification operates in the inositol-phospholipid pathway. As a result, a nanomolar (10^{-9} M) concentration of an extracellular signal often induces micromolar (10^{-6} M) concentrations of an intracellular second messenger such as cyclic AMP or Ca^{2+}. Since these messengers themselves function as allosteric effectors to activate specific enzymes or ion channels, a single extracellular signaling molecule can cause many thousands of molecules to be altered within the target cell. Moreover, each protein in the relay chain of signals can be a separate target for regulation, as, for example, in the glycogen breakdown cascade in skeletal muscle cells.

Any such amplifying cascade of stimulatory signals requires that there should be counterbalancing mechanisms at every step of the cascade to restore the system to its resting state when stimulation ceases. Cells therefore have efficient mechanisms for rapidly degrading (and resynthesizing) cyclic nucleotides and for buffering and removing cytosolic Ca^{2+}, as well as for inactivating the responding enzymes and transport proteins once they have been activated. This is not only essential for turning a response off, it is also important for defining the resting

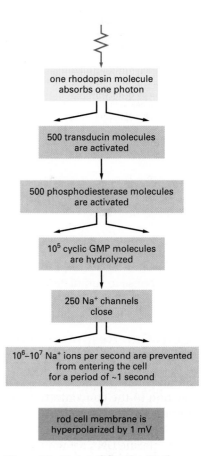

Figure 15–41 Amplification in the light-induced catalytic cascade in vertebrate rods. The divergent arrows indicate the steps where amplification occurs.

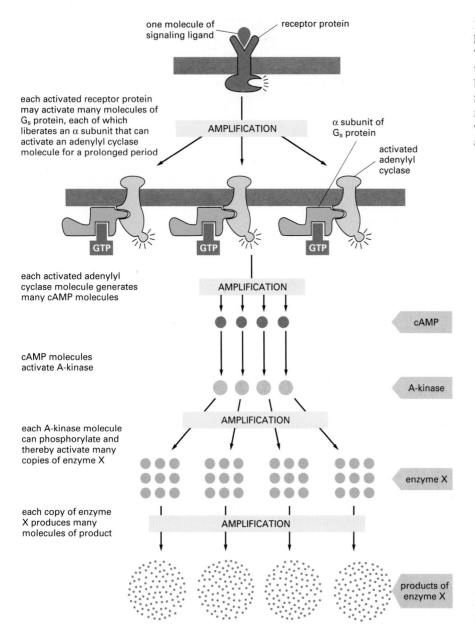

one molecule of
signaling ligand

receptor protein

each activated receptor protein
may activate many molecules of
G_s protein, each of which
liberates an α subunit that can
activate an adenylyl cyclase
molecule for a prolonged period

AMPLIFICATION

α subunit of
G_s protein

activated
adenylyl
cyclase

GTP   GTP   GTP

each activated adenylyl
cyclase molecule generates
many cAMP molecules

AMPLIFICATION

cAMP

cAMP molecules
activate A-kinase

A-kinase

each A-kinase molecule
can phosphorylate and
thereby activate many
copies of enzyme X

AMPLIFICATION

enzyme X

each copy of enzyme
X produces many
molecules of product

AMPLIFICATION

products of
enzyme X

Figure 15–42 Amplification in a ligand-induced catalytic cascade. The first amplification step requires that the signaling ligand remain bound to the receptor long enough for the complex to activate many G_s molecules; in many cases the ligand will dissociate too quickly for this amplification to occur.

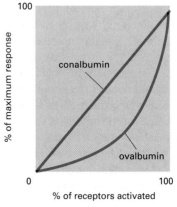

Figure 15–43 Response of chick oviduct cells to the steroid sex hormone estradiol. When activated, estradiol receptors turn on the transcription of several genes. Dose-response curves for two of these genes, one coding for the egg protein *conalbumin* and the other coding for the egg protein *ovalbumin,* are shown. The linear response curve for conalbumin indicates that each activated receptor molecule that binds to the conalbumin gene increases the activity of the gene by the same amount. In contrast, the lag followed by the steep increase in the response curve for ovalbumin suggests that more than one activated receptor (in this case two receptors) must bind simultaneously to the ovalbumin gene in order to initiate its transcription. (Adapted from E.R. Mulvihill and R.D. Palmiter, *J. Biol. Chem.* 252:2060–2068, 1977.)

state from which a response takes off. As we saw earlier (see p. 727), in general the response to stimulation can be rapid only if the inactivating mechanisms also are rapid.

Cells Can Respond Suddenly to a Gradually Increasing Concentration of an Extracellular Signal [30]

Some cellular responses to signaling ligands are smoothly graded in simple proportion to the concentration of the ligand. The primary responses to steroid hormones (see Figure 15–13) often follow this pattern, presumably because each intracellular hormone receptor protein binds a single molecule of hormone and each specific DNA recognition sequence in a steroid-hormone-responsive gene acts independently. As the concentration of hormone increases, the concentration of hormone-receptor complexes increases proportionally, as does the number of complexes bound to specific recognition sequences in the responsive genes; the cellular response is therefore a gradual and linear one.

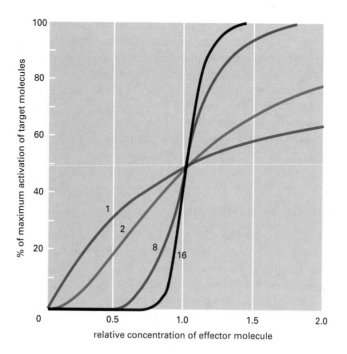

Figure 15–44 **Activation curves as a function of signal-molecule concentration.** The curves show how the sharpness of the response increases as the number of effector molecules that must bind simultaneously to activate a target macromolecule increases. The curves shown are those expected if the activation requires the simultaneous binding of 1, 2, 8, or 16 effector molecules, respectively.

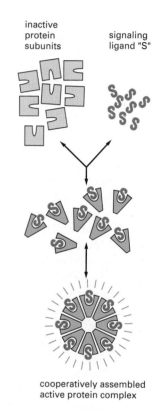

Figure 15–45 **One type of signaling mechanism expected to show a steep thresholdlike response.** Here the simultaneous binding of eight molecules of a signaling ligand to a set of eight subunits is required to form an active protein complex: the ability of the subunits to assemble into the active complex depends on an allosteric conformational change that the subunits undergo when they bind their ligand. The binding of the ligand in the formation of such a complex is generally a cooperative process, causing a steep response as the ligand concentration is changed, as explained in Chapter 5. At low ligand concentrations the number of active complexes will increase roughly in proportion to the eighth power of the ligand concentration.

Other responses to signaling ligands, however, begin more abruptly as the concentration of ligand increases. Some may even occur in a nearly all-or-none manner, being undetectable below a threshold concentration of ligand and then reaching a maximum as soon as this concentration is exceeded. What might be the molecular basis for such steep or even switchlike responses to graded signals?

One mechanism for steepening the response is to require that more than one intracellular effector molecule or complex bind to some target macromolecule in order to induce a response. In some steroid-hormone-induced responses, for example, it appears that more than one hormone-receptor complex must bind simultaneously to specific regulatory sequences in the DNA in order to activate a particular gene. As a result, as the hormone concentration rises, gene activation begins more abruptly than it would if only one bound complex were sufficient for activation (Figure 15–43). A similar mechanism operates in the activation of A-kinase and calmodulin, as discussed earlier. Two or more Ca^{2+} ions, for example, must bind before calmodulin adopts its activating conformation; as a result, a fiftyfold increase in activation occurs when the free intracellular Ca^{2+} concentration increases only tenfold. Such responses become sharper as the number of cooperating molecules increases, and if the number is large enough, responses approaching the all-or-none type can be achieved (Figures 15–44 and 15–45).

Responses are also sharpened when a ligand activates one enzyme and at the same time inhibits another that catalyzes the opposite reaction. We have already discussed one example of this common type of regulation in the stimulation of glycogen breakdown in skeletal muscle cells, where a rise in the intracellular cyclic AMP level both activates phosphorylase kinase and inhibits the opposing action of phosphoprotein phosphatase.

The above mechanisms can produce responses that are very steep but, nevertheless, always smoothly graded according to the concentration of the signaling ligand. Another mechanism, however, can produce true all-or-none responses, such that raising the signal above a critical threshold level trips a sudden switch in the responding system. All-or-none threshold responses of this type generally depend on *positive feedback*. Thus, by positive feedback nerve and muscle cells generate all-or-none *action potentials* in response to neurotransmitters (discussed in Chapter 11). The activation of acetylcholine receptors at a neuromuscular junction, for example, opens cation channels in the muscle cell plasma

Signaling via G-Protein-linked Cell-Surface Receptors

membrane. The result is a net influx of Na$^+$ that locally depolarizes the membrane. This causes voltage-gated Na$^+$ channels to open in the same membrane region, producing a further influx of Na$^+$, which further depolarizes the membrane and thereby opens more Na$^+$ channels. If the initial depolarization exceeds a certain threshold, this positive feedback has an explosive "runaway" effect, producing an action potential that propagates to involve the entire muscle membrane. As discussed earlier, a similar phenomenon occurs when Ca^{2+} is released from the ER or sarcoplasmic reticulum by Ca^{2+}-release channels: the released Ca^{2+} can bind back to the channels, increasing Ca^{2+} release, thereby producing an all-or-none Ca^{2+} spike.

The Effect of Some Signals Can Be Remembered by the Cell [31]

An accelerating positive feedback mechanism can operate through signaling proteins that are enzymes rather than ion channels. Suppose, for example, that a particular signaling ligand activates an enzyme located downstream in the signal relay pathway and that two or more molecules of the product of the enzymatic reaction bind back to the same enzyme to activate it further (Figure 15–46). The consequence will be a very low rate of synthesis of the enzyme product in the absence of the ligand, increasing slowly with the concentration of ligand until, at some threshold level of ligand, enough of the product is being synthesized to activate the enzyme in a self-accelerating, runaway fashion; the concentration of the enzyme product then suddenly increases to a much higher level. In this way the cell can translate a gradual change in the concentration of a signaling ligand into a switchlike change in the level of a particular enzyme product, creating an all-or-none response by the cell.

This type of mechanism has an important property that makes it unsuitable for some purposes and uniquely valuable for others. If such a system has been switched on by raising the concentration of signaling ligand above threshold, it will generally remain switched on even when the signal disappears: instead of faithfully reflecting the current level of signal, the response system displays a memory. We have already discussed one example—CaM-kinase II, which is activated by Ca^{2+}/calmodulin to phosphorylate itself and other proteins. The autophosphorylation keeps the kinase active long after Ca^{2+} levels return to normal and Ca^{2+}/calmodulin has dissociated from the enzyme (see Figure 15–35). A self-activating memory mechanism can also operate further downstream in a signaling pathway, at the level of gene transcription. The signals that trigger muscle cell determination, for example, turn on a series of muscle-specific gene regulatory proteins that stimulate the transcription of their own genes as well as genes producing many other muscle cell proteins (see p. 445).

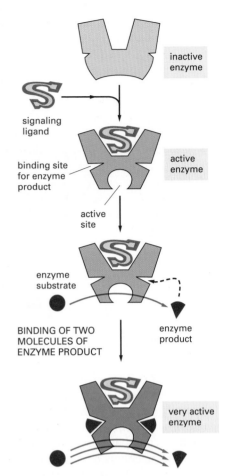

Figure 15–46 An accelerating positive feedback mechanism. The initial binding of the signaling ligand activates the enzyme to generate a product that binds back to the enzyme, further increasing the enzyme's activity.

Summary

G-protein-linked receptors indirectly activate or inactivate plasma-membrane-bound enzymes or ion channels via trimeric GTP-binding regulatory proteins (G proteins) that shut themselves off by hydrolyzing their bound GTP. Some G-protein-linked receptors activate or inactivate adenylyl cyclase, thereby altering the intracellular concentration of the intracellular mediator cyclic AMP. Others activate a phosphoinositide-specific phospholipase C (phospholipase C-β), which hydrolyzes phosphatidylinositol bisphosphate (PIP$_2$) to generate two intracellular mediators— inositol trisphosphate (IP$_3$), which releases Ca^{2+} from the ER and thereby increases the concentration of Ca^{2+} in the cytosol, and diacylglycerol, which remains in the plasma membrane and activates C-kinase. A rise in cyclic AMP or Ca^{2+} levels affects cells by stimulating A-kinase and CaM-kinases, respectively. C-kinase, A-kinase, and CaM-kinases phosphorylate specific target proteins on serine or threonine residues and thereby alter the activity of the proteins. Each type of cell has characteristic sets of target proteins that are regulated in these ways, enabling the cell to make its own distinctive response to these intracellular mediators. Through the intracellular signaling cascades activated by G-protein-linked receptors, the responses to extracellular signals can be greatly amplified.

The various responses mediated by these receptors are rapidly turned off when the extracellular signaling ligand is removed. This is because the G proteins self-inactivate by hydrolyzing their bound GTP, IP$_3$ is rapidly dephosphorylated by a phosphatase (or phosphorylated by a kinase), diacylglycerol is rapidly broken down, cyclic nucleotides are hydrolyzed by phosphodiesterases, Ca^{2+} is rapidly pumped out of the cytosol, and proteins are dephosphorylated by protein phosphatases. The continuous rapid turnover of these intracellular mediators makes possible rapid increases in their concentrations when cells respond to extracellular signals. In addition, cells make use of both cooperativity and positive feedback to sharpen their responses.

Signaling via Enzyme-linked Cell-Surface Receptors

Like G-protein-linked receptors, enzyme-linked receptors are transmembrane proteins with their ligand-binding domain on the outer surface of the plasma membrane. Instead of having a cytosolic domain that associates with a trimeric G protein, however, their cytosolic domains either have an intrinsic enzyme activity or associate directly with an enzyme. Whereas a G-protein-linked receptor protein has seven transmembrane segments, each subunit of a catalytic receptor usually has only one.

There are five known classes of enzyme-linked receptors: (1) *receptor guanylyl cyclases*, which catalyze the production of cyclic GMP in the cytosol; (2) *receptor tyrosine kinases*, which phosphorylate specific tyrosine residues on a small set of intracellular signaling proteins; (3) *tyrosine-kinase-associated receptors*, which associate with proteins that have tyrosine kinase activity; (4) *receptor tyrosine phosphatases*, which remove phosphate groups from tyrosine residues of specific intracellular signaling proteins; and (5) *receptor serine/threonine kinases*, which phosphorylate specific serine or threonine residues on some intracellular proteins.

We begin our discussion with the receptor guanylyl cyclases.

Receptor Guanylyl Cyclases Generate Cyclic GMP Directly [31]

Atrial natriuretic peptides (ANPs) are a family of closely related peptide hormones secreted by muscle cells in the atrium of the heart when blood pressure rises. ANPs stimulate the kidney to excrete Na$^+$ and water and induce the smooth muscle cells in the walls of blood vessels to relax. Both of these effects tend to lower blood pressure. The ANP receptor that mediates these responses is present on kidney cells and the smooth muscle cells of blood vessels; it is a single-pass transmembrane protein that has an extracellular binding site for ANPs and an intracellular guanylyl cyclase catalytic domain. The binding of ANPs activates the cyclase to produce cyclic GMP, which in turn binds to and activates a *cyclic GMP-dependent protein kinase (G-kinase)*, which phosphorylates specific proteins on serine or threonine residues. Thus **receptor guanylyl cyclases** use cyclic GMP as an intracellular mediator in the same way that some G-protein-linked receptors use cyclic AMP, except that the linkage between ligand binding and cyclase activity is direct rather than via a trimeric G protein.

While relatively few known receptors belong to the guanylyl cyclase family, many known receptors belong to the *tyrosine kinase* family, which we consider next.

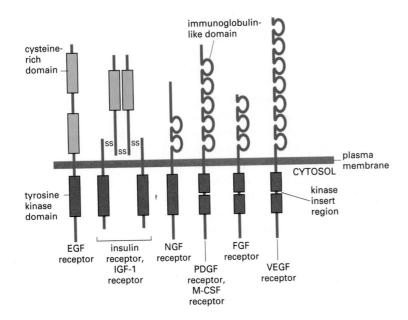

Figure 15–47 Six subfamilies of receptor tyrosine kinases. Only one or two members of each subfamily are indicated. Note that the tyrosine kinase domain is interrupted by a "kinase insert region" in some of the subfamilies. The functional significance of the cysteine-rich and immunoglobulinlike domains is unknown.

The Receptors for Most Growth Factors Are Transmembrane Tyrosine-specific Protein Kinases [32]

The first receptor protein recognized to be a **tyrosine-specific protein kinase** (in 1982) was the receptor for *epidermal growth factor (EGF)*. EGF is a small protein (53 amino acids) that stimulates the proliferation of epidermal cells and a variety of other cell types. Its receptor is a single-pass transmembrane protein of about 1200 amino acids, with a large glycosylated extracellular portion that binds EGF. An intracellular tyrosine kinase domain is activated when EGF binds to the receptor. Once activated, the receptors transfer a phosphate group from ATP to selected tyrosine side chains, both on the receptor proteins themselves and on specific cellular proteins. Many other receptors for growth and differentiation factors are also **receptor tyrosine kinases.** These include the receptors for *platelet-derived growth factor (PDGF), fibroblast growth factors (FGFs), hepatocyte growth factor (HGF), insulin, insulinlike growth factor-1 (IGF-1), nerve growth factor (NGF), vascular endothelial growth factor (VEGF),* and *macrophage colony stimulating factor (M-CSF),* all of which are proteins. As shown in Figure 15–47, the family of receptor tyrosine kinases can be divided into a number of structural subfamilies; in each case the receptors phosphorylate themselves to initiate the intracellular signaling cascade.

How does the binding of a specific protein to the extracellular portion of a receptor tyrosine kinase activate the catalytic domain on the other side of the plasma membrane? It is difficult to imagine how a conformational change could propagate across the lipid bilayer through a single transmembrane α helix. The puzzle was solved when it was demonstrated that ligand binding causes the EGF receptor to assemble into dimers, which enables the two cytoplasmic domains to cross-phosphorylate each other on multiple tyrosine residues. This cross-phosphorylation is referred to as *autophosphorylation* because it occurs within the receptor dimer. In the case of PDGF receptors the ligand is a dimer that crosslinks two receptors together (Figure 15–48). EGF, by contrast, is a monomer that is thought to induce a conformational change in the extracellular domain of its receptors to induce receptor dimerization. It is thought that receptor dimerization is a general mechanism for activating enzyme-linked receptors with a single transmembrane domain.

Receptor dimerization can be exploited experimentally to inactivate specific receptors in order to determine their importance for a particular cell response. The strategy involves transfecting cells with DNA that encodes a mutant form of a receptor tyrosine kinase that dimerizes normally but has an inactive kinase

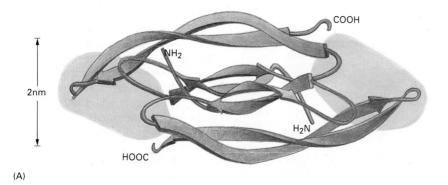

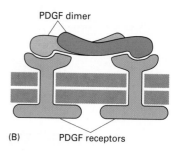

(A)

(B) PDGF receptors

domain. When coexpressed at a high level with normal receptors, the mutant receptor acts in a *dominant-negative* way (see Figure 7–44), disabling the normal receptors by forming an inactive dimer with them.

The signaling pathways triggered by EGF, PDGF, and FGF receptors have been analyzed in the greatest detail. In each case the autophosphorylated tyrosines serve as high-affinity binding sites for a number of intracellular signaling proteins in the target cell. Each of these proteins binds to a different phosphorylated site on the activated receptor, recognizing surrounding features of the polypeptide chain in addition to the phosphotyrosine. Once bound, many of these proteins become phosphorylated themselves on tyrosines and are thereby activated. In this way tyrosine autophosphorylation is thought to serve as a switch to trigger the transient assembly of an intracellular signaling complex, which serves to relay the signal into the cell interior. Different receptor tyrosine kinases bind different combinations of these signaling proteins and therefore activate different responses.

The receptor for insulin and IGF-1 acts in a slightly different way. First, because the receptors are tetramers to start with (see Figure 15–47), ligand binding is thought to induce an allosteric interaction of the two receptor halves rather than receptor dimerization. Second, insulin binding causes the receptor to phosphorylate its catalytic domains, which activates them to phosphorylate a separate protein (called *insulin receptor substrate-1*, or *IRS-1*) on multiple tyrosines. The phosphotyrosines on IRS-1 then serve as high-affinity binding sites for the docking and activation of intracellular signaling proteins, many of which also bind directly to other activated receptor tyrosine kinases.

Phosphorylated Tyrosine Residues Are Recognized by Proteins with SH2 Domains [33]

A whole menagerie of intracellular signaling proteins have been found to bind to the phosphotyrosines on activated receptor tyrosine kinases. Some of these proteins—a *GTPase-activating protein (GAP)*, *phospholipase C-γ (PLC-γ)*, and the *Src-like nonreceptor protein tyrosine kinases*—have functions that are more or less understood, as we see in the following pages. Phospholipase C-γ, for example, functions in the same way as phospholipase C-β to activate the inositol phospholipid signaling pathway, discussed earlier in connection with signaling via G proteins. Other proteins that bind to activated receptor tyrosine kinases are more of a mystery. The enzyme *phosphatidylinositol 3′-kinase (PI3-kinase)*, for example, is thought to be important in regulating cell proliferation; its immediate action is to phosphorylate the inositol ring of phosphotidylinositol at the 3 position (see Figure 15–29), but the functions of the special phosphoinositides produced in this way are unknown.

Although the intracellular signaling proteins that bind to phosphotyrosine residues on activated receptor tyrosine kinases (and to IRS-1) have varied structures and functions, they usually share two highly conserved noncatalytic domains, called *SH2* and *SH3* for *Src homology regions 2* and *3* because they were

Signaling via Enzyme-linked Cell-Surface Receptors

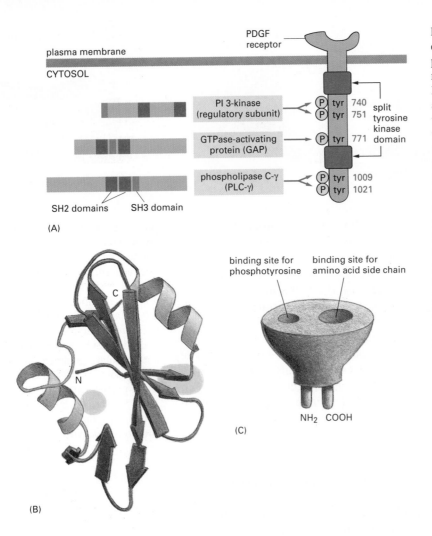

Figure 15–49 Binding of SH2-containing intracellular signaling proteins to an activated PDGF receptor. (A) Schematic drawing of a PDGF receptor, showing five tyrosine autophosphorylation sites, three in the kinase insert region and two on the carboxyl-terminal tail, to which the three signaling proteins shown on the left bind as indicated. (There are two additional tyrosine autophosphorylation sites on the receptor that are not shown, located between the membrane-spanning domain and the start of the kinase domain, which serve as a binding site for Src-like nonreceptor tyrosine kinases.) The numbers on the right indicate the position of the tyrosines in the polypeptide chain. These binding sites have been identified by using recombinant DNA technology to mutate specific tyrosines in the receptor: mutation of tyrosines 1009 and 1021, for example, prevents the binding and activation of PLC-γ, so that receptor activation no longer stimulates the inositol phospholipid signaling pathway. The locations of the SH2 (*red*) and SH3 (*blue*) domains in the three signaling proteins are indicated. (B) The three-dimensional structure of an SH2 domain as determined by x-ray crystallography. The binding pocket for phosphotyrosine is shown in *yellow* on the right, while a pocket for binding a specific amino acid side chain (isoleucine in this case) is shown in *yellow* on the left. (C) The SH2 domain is a compact module, which can be inserted almost anywhere in a protein without disturbing the protein's folding or function. Because each domain has distinct sites for recognizing phosphotyrosine and for recognizing a particular amino acid side chain, different SH2 domains recognize phosphotyrosine in the context of different flanking amino acid sequences. (B, based on data from G. Waksman et al., *Cell* 72:1–20, 1993. © Cell Press.)

first found in the Src protein (famous for its role in cancer research, as discussed in Chapter 24). The **SH2 domains** recognize phosphorylated tyrosines and enable proteins that contain them to bind to the activated receptor tyrosine kinases as well as to other intracellular signaling proteins that have been transiently phosphorylated on tyrosines (Figure 15–49). The function of the SH3 domain is less clear, but it is thought to bind to other proteins in the cell.

A protein-binding function for the SH3 domain is suggested by studies of a class of small "adaptor" proteins that consist only of SH2 and SH3 domains. These *small SH adaptor proteins* have no intrinsic catalytic function and serve to couple tyrosine-phosphorylated proteins such as activated receptor tyrosine kinases to other proteins that do not have their own SH2 or SH3 domains. One such SH adaptor protein, called *Sem-5,* was discovered through genetic studies in the nematode worm *C. elegans,* as discussed in Chapter 21. Mutations in the *sem-5* gene block the signaling pathways from several receptor tyrosine kinases and have profound effects on the development of the worm. The signaling pathways are blocked equally effectively by mutations that disrupt the single SH2 domain or either of the two SH3 domains in the molecule, which suggests that both SH3 domains are required to bind a downstream signaling component (see Figure 15–53). In fact, homologues of the Sem-5 protein seem to be present in most animal cells, and there is both genetic and biochemical evidence that they couple activated receptor tyrosine kinases to the important downstream signaling protein Ras, to which we now turn.

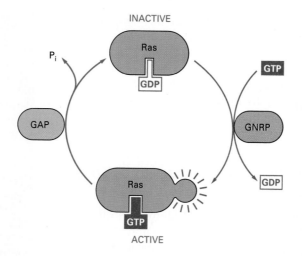

INACTIVE

Ras
GDP

P_i

GAP

GTP

GNRP

GDP

Ras
GTP

ACTIVE

Figure 15–50 **The regulation of Ras activity.** GTPase-activating proteins (GAPs) inactivate Ras by stimulating it to hydrolyze its bound GTP. Guanine nucleotide releasing proteins (GNRPs) activate Ras by stimulating it to give up its GDP; as the concentration of GTP in the cytosol is 10 times greater than the concentration of GDP, Ras will tend to bind GTP once GDP is ejected. Two Ras-regulating GAPs (*Ras GAPs*) have been characterized in mammalian cells so far—*p120*[GAP] and *neurofibromin* (so called because it is encoded by the gene that is mutated in the common human genetic disease neurofibromatosis, which is associated with tumors of nerves). Although both Ras GAPs are expressed ubiquitously, one or the other seems to predominate (depending on the cell type) in maintaining most of the Ras protein (~95%) in an inactive GDP-bound state.

The Ras Proteins Provide a Crucial Link in the Intracellular Signaling Cascades Activated by Receptor Tyrosine Kinases [34]

The *Ras proteins* belong to the large **Ras superfamily of monomeric GTPases,** which also contains two other subfamilies: (1) the *Rho* and *Rac* proteins, involved in relaying signals from cell-surface receptors to the actin cytoskeleton (discussed in Chapter 16), and (2) the *Rab family,* involved in regulating the traffic of intracellular transport vesicles (discussed in Chapter 13). Like almost all of these monomeric GTPases, the Ras proteins contain a covalently attached prenyl group that helps anchor the protein to a membrane, in this case to the cytoplasmic face of the plasma membrane where the protein functions.

The **Ras proteins** help relay signals from receptor tyrosine kinases to the nucleus to stimulate cell proliferation or differentiation. If Ras function is inhibited by the microinjection of either neutralizing anti-Ras antibodies or dominant-negative mutant forms of Ras, the cell proliferation or differentiation responses normally induced by activated receptor tyrosine kinases do not occur; conversely, if a hyperactive mutant Ras protein is introduced into a cell, the effect on proliferation or differentiation is usually the same as that induced by the binding of ligand to the cell-surface receptors. In fact, Ras proteins were first discovered as the hyperactive products of mutant *ras* genes, which promote cancer by disrupting the normal controls on cell proliferation and differentiation; about 30% of human cancers have such mutations in a *ras* gene.

Like the other monomeric GTPases and the trimeric G proteins, Ras proteins function as switches, cycling between two distinct conformational states—active when GTP is bound (see Figure 15–18) and inactive when GDP is bound. Ras hydrolyzes GTP at least 100 times more slowly than the α subunit of the stimulatory trimeric G protein G_s discussed earlier, and because GTP is present in the cytosol at 10 times higher concentration than GDP, this means that in the absence of other influences, once GTP is bound, Ras would remain constantly active. In the cell, however, two classes of signaling proteins regulate Ras activity by influencing the transition between the active and inactive states. **GTPase-activating proteins (GAPs)** increase the rate of hydrolysis of bound GTP by Ras, thereby inactivating it. These negative regulators are counteracted by **guanine nucleotide releasing proteins (GNRPs),** which promote the exchange of bound nucleotide by stimulating the loss of GDP and the subsequent uptake of GTP from the cytosol; they therefore tend to activate Ras (Figure 15–50). In principle, therefore, receptor tyrosine kinases could activate Ras either by activating a GNRP or by inhibiting a GAP. Activated receptor tyrosine kinases bind GAPs directly, as we noted earlier, and bind GNRPs only indirectly, as we discuss below. Nonetheless, it is the indirect coupling to GNRPs that is usually responsible for driving Ras into its active, GTP-bound state.

Ras proteins and the proteins that regulate Ras activity have been highly conserved in evolution, and genetic analyses in *Drosophila* and *C. elegans* provided the first clues as to how receptor tyrosine kinases activate Ras. Genetic studies of photoreceptor cell development in the *Drosophila* eye have been particularly informative.

An SH Adaptor Protein Couples Receptor Tyrosine Kinases to Ras: Evidence from the Developing *Drosophila* Eye [35]

The *Drosophila* compound eye consists of about 800 identical units called *ommatidia*, each composed of 8 photoreceptor cells (R1–R8) and 12 accessory cells (Figure 15–51). The eye develops from a simple epithelial sheet, and the cells that make up each ommatidium are recruited from the sheet in a fixed sequence by a series of cell-cell interactions. Beginning with the development of R8, each differentiating cell induces its uncommitted immediate neighbor to adopt a specific fate and assemble into the developing ommatidium (Figure 15–52).

The development of the R7 photoreceptor, which is required for the detection of ultraviolet light, has been studied most intensively, beginning in 1976 with the description of a mutant fly called *sevenless (sev)*, in which the only observed defect is that R7 fails to develop. Such mutants are easy to select on the basis of their blindness to ultraviolet light. The *sev* gene was eventually isolated and shown to encode a receptor tyrosine kinase that is expressed on R7 precursor cells. Further genetic analysis of mutants in which R7 development is blocked but the Sev protein itself is not affected led to the identification of the gene *bride-of-sevenless (boss)*, which encodes the ligand for the Sev receptor protein. Boss is a seven-pass transmembrane protein that is expressed exclusively on the surface of the adjacent R8 cell, and when it binds to and activates Sev, it induces the R7 precursor cell to differentiate into an R7 photoreceptor. The Sev protein is also expressed on several other precursor cells in the developing ommatidium. But none of these cells contact R8; therefore, the Sev protein is not activated, and the cells do not differentiate into R7 photoreceptors. Although one suspects that multicellular organisms make widespread use of cell-surface-bound signaling ligands like Boss, such molecules have been hard to identify and characterize by standard biochemical techniques. The identification of Boss illustrates the power of the genetic approach.

The components of the intracellular signaling pathway activated by Sev in the R7 precursor cell proved more difficult to identify than the receptor or its ligand because they are used by cells in a variety of developing organs besides the eye and mutations that inactivate them are lethal. Some of these mutations, however, are lethal only when both copies of the gene are affected and so can be maintained in heterozygous animals, carrying one normal and one mutant copy of the gene. By isolating such mutants, as well as by using other genetic strategies, several genes encoding some of the intracellular signaling proteins were identified. One encodes a Ras protein. Whereas flies in which both copies of the *ras* gene are inactivated by mutation die, flies with only one inactivated copy survive but lack R7. Conversely, if one of the *ras* genes is rendered overactive by mutation, R7 develops even in mutants in which both *sev* and *boss* are inactive.

Figure 15–51 Scanning electron micrograph of a compound eye of *Drosophila*. The eye is composed of about 800 identical units (ommatidia), each having a separate lens that focuses light onto eight photoreceptor cells at its base. (Courtesy of Ernst Hafen.)

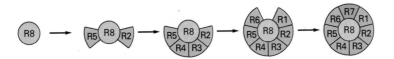

Figure 15–52 The assembly of photoreceptor cells in a developing *Drosophila* ommatidium. Schematic drawing of the sequential recruitment of photoreceptors, beginning with R8 and ending with R7, which is the last photoreceptor cell to develop.

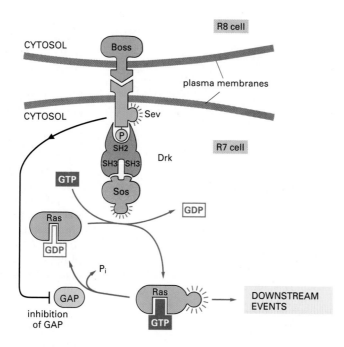

CYTOSOL

R8 cell

Boss

plasma membranes

CYTOSOL

Sev

P

SH2

SH3 SH3

Drk

R7 cell

GTP

Sos

GDP

Ras

GDP

P_i

GAP

inhibition
of GAP

Ras

GTP

DOWNSTREAM
EVENTS

Figure 15–53 Early cell-signaling events in R7 development. The activation of the Sev receptor tyrosine kinase on the surface of the R7 precursor cell by the Boss protein on the surface of R8 is coupled to the activation of the guanine nucleotide releasing protein Sos by the small SH adaptor protein Drk. Drk recognizes a specific autophosphorylated tyrosine on the Sev protein by means of an SH2 domain and interacts with Sos by means of two SH3 domains. Sos stimulates the inactive Ras protein to give up its bound GDP and take up GTP, which activates Ras to relay the signal downstream. (Although not shown here, Ras is bound to the cytosolic face of the plasma membrane.) It is thought that activated Sev also represses GAP, which otherwise would counteract the function of Sos by stimulating Ras to hydrolyze its bound GTP and become inactive. The coupling of receptor tyrosine kinases to Ras seems to occur by the same mechanism in mammalian cells.

Thus activation of Ras seems to be necessary and sufficient to induce R7 differentiation. A second gene, *son-of-sevenless (sos)*, encodes a guanine nucleotide releasing protein (GNRP), which is required for the Sev receptor tyrosine kinase to activate Ras. A third gene encodes a Sem-5-like protein called *Drk* (**d**own-stream of **r**eceptor **k**inases), which couples the Sev receptor to the Sos protein; the SH2 domain of Drk binds to activated Sev, while the two SH3 domains are thought to bind to Sos. A fourth gene encodes a GTPase-activating protein (GAP). If this gene is inactivated, R7 develops even if *sev* has been inactivated, presumably because Ras is hyperactive in the absence of inhibition by GAP (Figure 15–53). In this signaling system, therefore, and in most others that have been studied, the activation of Ras by receptor tyrosine kinases depends on the activation of a GNRP rather than on the inactivation of a GAP.

Once activated, Ras relays the signal downstream by activating a serine/threonine phosphorylation cascade that is highly conserved in eucaryotic cells from yeasts to humans. A crucial component in this cascade is a novel type of protein kinase called *MAP-kinase,* which we consider next.

Ras Activates a Serine/Threonine Phosphorylation Cascade That Activates MAP-Kinase [36]

The tyrosine phosphorylations and the activation of Ras, which are stimulated by receptor tyrosine kinases at the cytoplasmic surface of the plasma membrane, are very short-lived: the phosphorylations are quickly reversed by tyrosine-specific protein phosphatases (discussed later), and activated Ras inactivates itself by hydrolyzing its bound GTP to GDP. To stimulate cells to proliferate or differentiate, these short-lived signaling events need to be converted into longer-lasting ones that can sustain the signal and relay it downstream to the nucleus. The relay systems involve multiple, interacting cascades of serine/threonine phosphorylations, which are much longer-lived than tyrosine phosphorylations. Many serine/threonine kinases are involved in these cascades, but one family, which contains at least five members, seems to play an especially important role—the **mitogen-activated protein (MAP) kinases** (also called *extracellular-signal-regulated kinases [ERKs]*). These kinases are turned on by a wide range of extracellular proliferation- and differentiation-inducing signals, some of which activate receptor tyrosine kinases, while others activate G-protein-linked receptors.

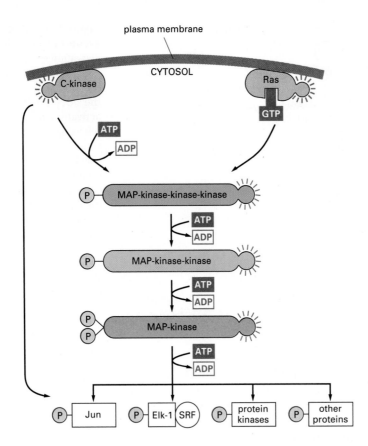

plasma membrane

CYTOSOL

C-kinase

Ras

GTP

ATP
ADP

(P) MAP-kinase-kinase-kinase

ATP
ADP

(P) MAP-kinase-kinase

ATP
ADP

(P)(P) MAP-kinase

ATP
ADP

(P) Jun   (P) Elk-1 (SRF)   (P) protein kinases   (P) other proteins

Figure 15–54 The serine/threonine phosphorylation cascade activated by Ras and C-kinase. In the pathway activated by receptor tyrosine kinases via Ras, the MAP-kinase-kinase-kinase is often a serine/threonine kinase called *Raf*, which is thought to be activated by the binding of activated Ras. In the pathway activated by G-protein-linked receptors via C-kinase, the MAP-kinase-kinase-kinase can either be Raf or a different serine/threonine kinase. A similar serine/threonine phosphorylation cascade involving structurally and functionally related proteins operates in yeasts and in all animals that have been studied, where it integrates and amplifies signals from different extracellular stimuli. Receptor tyrosine kinases may also activate a more direct signaling pathway to the nucleus by directly phosphorylating, and thereby activating, gene regulatory proteins that contain SH2 domains.

An unusual feature of the MAP-kinases is that their full activation requires phosphorylation of both a threonine and a tyrosine, which are separated in the protein by a single amino acid. The protein kinase that catalyzes both of these phosphorylations is called *MAP-kinase-kinase*. The requirement for both tyrosine and threonine phosphorylation ensures that MAP-kinases are kept inactive unless specifically activated by MAP-kinase-kinase, whose only known substrates are MAP-kinases. MAP-kinase-kinase is itself activated by serine/threonine phosphorylation catalyzed by a MAP-kinase-kinase-kinase, which is thought to be activated by the binding of activated Ras (Figure 15–54).

Once a MAP-kinase is activated, it relays signals downstream by phosphorylating various proteins in the cell, including other protein kinases and gene regulatory proteins. Often, transcription of a set of *immediate early genes* is activated within minutes of a cell's being stimulated by a growth factor. One protein complex that plays an important part in this transcriptional activation is the complex discussed earlier that is formed by the serum response factor (SRF) and Elk-1. The complex is constitutively bound to a specific DNA sequence (the serum response element) found in the regulatory region of a gene called *fos* and some other immediate early genes. When activated, MAP-kinases migrate from the cytosol into the nucleus and phosphorylate Elk-1, thereby activating it to turn on the transcription of the *fos* gene (see Figure 15–32). In addition, MAP-kinases may phosphorylate the Jun protein, which combines with the newly made Fos protein to form an active gene regulatory protein called AP-1. The AP-1 protein then turns on additional genes, although its exact role in stimulating cell proliferation remains to be defined.

C-kinase also can phosphorylate Jun, and, as illustrated in Figure 15–54, it can activate MAP-kinase-kinase-kinase, so that both Jun and MAP-kinase-kinase-kinase are examples of integration points where several signaling pathways converge.

Tyrosine-Kinase-associated Receptors Depend on Nonreceptor Tyrosine Kinases for Their Activity [37]

Many of the cell-surface receptor proteins that have been isolated and characterized do not fall into any of the main families of receptors that we have described so far: they are neither ion-channel-linked nor G-protein-linked, and they lack an obvious catalytic domain. This large and heterogeneous assortment includes receptors for most of the local mediators (called *cytokines*) that regulate proliferation and differentiation in the hemopoietic system as well as receptors for some hormones (growth hormone and prolactin, for example) and the antigen-specific receptors on T and B lymphocytes. Many receptors in this category work through associated tyrosine kinases, which phosphorylate various target proteins when the receptor binds its ligand. The kinases involved with these **tyrosine-kinase-associated receptors** are mostly members of the well-characterized *Src family* of nonreceptor protein tyrosine kinases mentioned earlier, or of the recently described *Janus family* of nonreceptor protein tyrosine kinases. It is thought that these receptors function in much the same way as receptor tyrosine kinases except that their kinase domain is encoded by a separate gene and is noncovalently associated with the receptor polypeptide chain. As with receptor tyrosine kinases, their activation is thought often to involve ligand-induced receptor dimerization (Figure 15–55).

There are at least eight members of the **Src family** of nonreceptor protein tyrosine kinases in mammals: *Src, Yes, Fgr, Fyn, Lck, Lyn, Hck,* and *Blk*. They contain SH2 and SH3 domains and are all located on the cytoplasmic side of the plasma membrane, held there partly by their interaction with transmembrane receptor proteins and partly by covalently attached lipid chains. Different family members are associated with different receptors and phosphorylate overlapping but distinct sets of target proteins. Lyn, Fyn, and Lck, for example, are each associated with different sets of receptors in lymphocytes. In each case the Src-family tyrosine kinase is activated when an extracellular ligand binds to the appropriate receptor protein.

The same Src-family kinase can associate both with receptors that do not have intrinsic tyrosine kinase activity and with receptors that do. In various nonlymphocyte cells, for example, Fyn binds via its SH2 domain to activated PDGF receptors, which phosphorylate the Fyn protein on tyrosine residues and thereby activate its kinase activity. It is not surprising, therefore, that receptor tyrosine kinases and Src-family-tyrosine-kinase-associated receptors activate some of the same signaling pathways.

Much less is known about the **Janus family** of nonreceptor protein tyrosine kinases, which includes JAK1, JAK2, and Tyk2. They are involved in signaling from a number of tyrosine-kinase-associated receptors, including those for growth hormone, prolactin, and various cytokines that act on hemopoietic cells (discussed in Chapter 22).

For many tyrosine-kinase-associated receptors, ligand binding causes the assembly of two or more different transmembrane receptor subunits. A well-studied example is the receptor for *interleukin-2 (IL-2)*, a local mediator secreted by T lymphocytes that stimulates lymphocyte proliferation (discussed in Chapter 23). The **IL-2 receptor** is composed of three polypeptide chains (α, β, and γ), which are thought to assemble after ligand binding to form a functional receptor complex, as illustrated in Figure 15–56.

Some Receptors Are Protein Tyrosine Phosphatases [38]

As mentioned previously, the tyrosine residues that are phosphorylated by protein tyrosine kinases are very rapidly dephosphorylated by **protein tyrosine phosphatases.** These enzymes are structurally unrelated to the serine/threonine pro-

Figure 15–55 The three-dimensional structure of human growth hormone bound to its receptor. The hormone (*red*) has cross-linked two identical receptors (one shown in *green* and the other in *blue*) to form a receptor homodimer. (It was entirely unexpected that a monomeric ligand such as growth hormone would cross-link its receptors, as it requires that the two identical receptors recognize different parts of the hormone.) Ligand-induced dimerization is thought to bring together the cytoplasmic domains of the two single-pass, transmembrane receptor proteins. This in turn activates a nonreceptor tyrosine kinase (not shown). The structures shown were determined by x-ray crystallographic studies of complexes formed between the hormone and the extracellular receptor domain produced by recombinant DNA technology. (From A.M. deVos, M. Ultech, and A.A. Kossiakoff, *Science* 255:306–312, 1992. © 1992 the AAAS.)

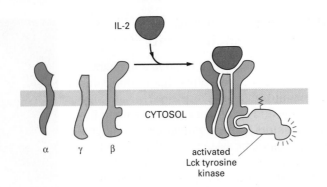

Figure 15–56 **The ligand-induced assembly of an IL-2 receptor.** The low-affinity binding of IL-2 to the α chain is thought to trigger the assembly of the heterotrimeric high-affinity receptor, which then binds and activates the Lck tyrosine kinase through an interaction with the cytoplasmic tail of the β subunit.

tein phosphatases discussed earlier, and they remove phosphate groups only from selected phosphotyrosines on particular types of proteins. These phosphatases are found in both soluble and membrane-bound forms and come in many more varieties than do serine/threonine phosphatases. Their high specific activity ensures that tyrosine phosphorylations are very short-lived and that the level of tyrosine phosphorylation in resting cells is very low. Protein tyrosine phosphatases do not simply continuously reverse the effects of protein tyrosine kinases, however; they can be regulated to play specific roles in cell signaling, as well as in controlling the cell cycle (discussed in Chapter 17).

An important example of a regulated protein tyrosine phosphatase is the *CD45 protein,* which is found on the surface of white blood cells and plays an essential part in the activation of both T and B lymphocytes by foreign antigens. CD45 is a single-pass transmembrane glycoprotein with its tyrosine phosphatase domain exposed on the cytoplasmic side of the plasma membrane. When cross-linked by extracellular antibodies (its normal ligand is unknown), its catalytic domain is activated to remove phosphate groups from tyrosine residues on specific target proteins in the cell. One such target protein is thought to be the Lck tyrosine kinase mentioned earlier. When dephosphorylated by CD45, Lck is apparently activated to phosphorylate other proteins in the cell.

Cancer-promoting Oncogenes Have Helped Identify Many Components in the Receptor Tyrosine Kinase Signaling Pathways [39]

Cells in higher animals normally divide only when they are stimulated by growth factors, which are produced by other cells and usually act by binding to receptor tyrosine kinases. Cancer cells proliferate excessively mainly because, as a result of accumulated mutations, they are able to divide without stimulation from other cells and therefore are no longer subject to the normal "social" controls on cell proliferation (discussed in Chapter 17). Not surprisingly, many of these mutations affect genes that code for proteins involved in the signaling pathways utilized by receptor tyrosine kinases. Indeed, a number of the genes that encode the signaling proteins discussed in this section, including Ras, Src (and the other members of the Src family), Raf, Fos, and Jun, were first identified as mutant forms in cancer cells or in cancer-promoting tumor viruses. As discussed in Chapter 24, the mutant genes were called *oncogenes* before their origin from normal genes was understood; the normal genes are therefore sometimes referred to as *proto-oncogenes.*

In principle, one might expect that any mutation that results in the production of an abnormally active protein anywhere along the signaling pathways that lead from growth factor to the nucleus could promote cancer by encouraging the cell to proliferate in the absence of the appropriate extracellular signals. The evidence to date supports such a view. The *sis* oncogene, for example, encodes a functionally active form of PDGF that is expressed inappropriately; cells that carry the *sis* oncogene and also express PDGF receptors continuously stimulate themselves to proliferate. The *erbB* oncogene encodes a truncated form of the

EGF receptor that has an intracellular tyrosine kinase domain that is continuously active; cells expressing this oncogene behave as though they are constantly being signaled to proliferate by a growth factor. Cells behave in a similar way if they are infected with a tumor virus carrying the *v-src* oncogene, which encodes a constitutively active form of the Src tyrosine kinase, or if they express a *ras* oncogene, which encodes an abnormal form of a Ras protein that cannot shut itself off because it has lost the ability to hydrolyze GTP. Similarly, cells proliferate abnormally if they express an oncogene that encodes a constitutively active form of a growth-factor-activated gene regulatory protein, such as Jun or Fos. Studies of such oncogenes have not only illuminated the molecular mechanisms underlying cancer, they have also uncovered many previously unknown proteins in the signaling pathways activated by growth factors.

Some hormones that stimulate their target cells to proliferate bind to G-protein-linked receptors rather than to receptor tyrosine kinases. *Growth-hormone-releasing factor (GHF)*, for example, stimulates growth-hormone-secreting cells in the pituitary gland to proliferate; the GHF binds to GHF receptors that activate adenylyl cyclase via the stimulatory G protein G_s. The resulting rise in cyclic AMP stimulates the pituitary cells to divide. As might be expected, mutations in the α_s gene that inactivate the GTPase activity of the α subunit of G_s (and thereby render the protein constitutively active) produce an oncogene that is frequently found in human pituitary tumors.

Proteins in the TGF-β Superfamily Activate Receptors That Are Serine/Threonine Protein Kinases [40]

Transforming growth factor-βs (TGF-βs) constitute a family of local mediators that regulate the proliferation and functions of most vertebrate cell types. The five members of the family (TGF-β1 to β5) are proteins with similar structures and functions. Their effects on cells are varied. Depending on the cell type, they can suppress proliferation, stimulate the synthesis of extracellular matrix, stimulate bone formation, and attract cells by chemotaxis. The TGF-βs are synthesized as large precursors and are secreted as inactive complexes that are later activated by proteolytic processing.

A number of other extracellular signaling proteins are structurally related to the TGF-βs. These include the *activins*, which play an important part in mesoderm induction in vertebrate development (discussed in Chapter 21), and the *bone morphogenetic proteins*, which stimulate bone formation. Together these proteins constitute the *TGF-β superfamily*.

Recently, the cDNAs encoding several receptors for members of this superfamily have been cloned and sequenced. The receptors were found to be single-pass transmembrane proteins with a serine/threonine kinase domain on the cytosolic side of the plasma membrane. These are the first **receptor serine/threonine kinases** to be identified, and little is known about the signaling pathways that they activate.

Some of the tyrosine-specific and serine/threonine-specific protein kinases that we have discussed in this chapter are reviewed in Figure 15–57.

The Notch Transmembrane Receptor Mediates Lateral Inhibition by an Unknown Mechanism [41]

The class of enzyme-linked cell-surface receptors is large and varied. It includes, in addition to those we have already considered, the *integrins*, which bind to components of the extracellular matrix. As discussed in Chapter 19, these transmembrane proteins not only attach cells to the matrix through focal contacts but also generate intracellular signals at such attachment sites, presumably by triggering the assembly of an intracellular signaling complex (see Figure 17–42).

The signaling mechanisms used by some receptor proteins are still so poorly understood that it is uncertain how to classify the receptors. An important ex-

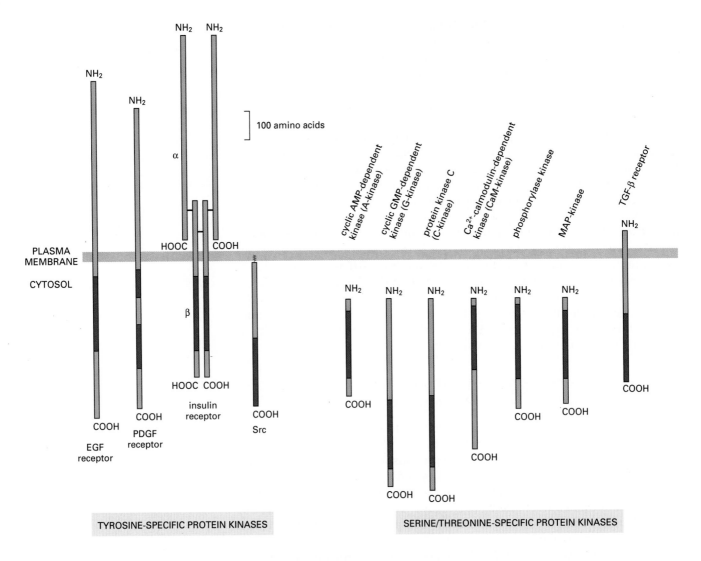

PLASMA MEMBRANE

CYTOSOL

100 amino acids

α

β

HOOC COOH

insulin receptor

COOH
Src

COOH
EGF receptor

COOH
PDGF receptor

HOOC COOH

TYROSINE-SPECIFIC PROTEIN KINASES

cyclic AMP-dependent kinase (A-kinase)

cyclic GMP-dependent kinase (G-kinase)

protein kinase C (C-kinase)

Ca²⁺-calmodulin-dependent kinase (CaM-kinase)

phosphorylase kinase

MAP-kinase

TGF-β receptor

SERINE/THREONINE-SPECIFIC PROTEIN KINASES

ample is the *Drosophila* protein **Notch.** As explained in Chapter 21, this single-pass transmembrane protein plays a crucial part in the cell-cell interactions that control the fine-grained pattern of cell diversification during fly development. Typically, a cell that lacks Notch is unresponsive to *lateral inhibition*—the inhibitory signals from its immediate neighbors that would normally cause it to differentiate in a way different from them. In a normal fly embryo, for example, a nerve cell precursor inhibits its neighbors from also becoming nerve cells, and they instead become epidermal cells; in Notch mutants the inhibition fails, and all of these cells become nerve cells.

Like the Sev protein discussed earlier, Notch is activated by binding to proteins displayed on the surface of adjacent cells rather than by soluble signaling molecules. Although the sequence of Notch is known and several of the proteins downstream from it have been identified by genetic analysis, it is still unclear how Notch relays its signal into the interior of the cell. Notch-like proteins also play an important role in vertebrate development, but here too the signaling mechanisms involved remain a mystery.

Figure 15–57 Some of the protein kinases discussed in this chapter. The size and location of their catalytic domains (*dark green*) are shown. In each case the catalytic domain is about 250 amino acid residues long. These domains are all similar in amino acid sequence, suggesting that they have all evolved from a common primordial kinase. Note that all of the tyrosine kinases shown are bound to the plasma membrane, whereas most of the serine/threonine kinases are in the cytosol.

Summary

There are five known classes of enzyme-linked receptors: (1) transmembrane guanylyl cyclases, which generate cyclic GMP directly; (2) receptor tyrosine phosphatases, which remove phosphate from phosphotyrosine side chains of specific proteins; (3) transmembrane receptor serine/threonine kinases, which add a phosphate group to serine and threonine side chains on target proteins; (4) receptor tyrosine kinases; and

(5) tyrosine-kinase-associated receptors. The last two types of receptors are by far the most numerous, and they are thought to work in a similar way: ligand binding usually induces the receptors to dimerize, which activates the kinase activity of either the receptor or its associated nonreceptor tyrosine kinase. When activated, receptor tyrosine kinases usually cross-phosphorylate themselves on multiple tyrosine residues, which then serve as docking sites for a small set of intracellular signaling proteins, which bind via their SH2 domains to specific phosphotyrosine residues. In this way a multiprotein signaling complex is activated from which the signal spreads to the cell interior.

The Ras proteins serve as crucial links in the intracellular relay system. They are monomeric GTPases that behave as molecular switches; they are activated by guanine nucleotide releasing proteins and inactivated by GAPs. When Ras proteins are activated, they initiate a cascade of serine/threonine phosphorylations that converge on MAP-kinases, which help relay the signal to the nucleus. Many of the genes that encode the proteins in the intracellular signaling cascades that are activated by receptor tyrosine kinases were first identified as oncogenes in cancer cells or tumor viruses, since their inappropriate activation causes a cell to proliferate excessively.

Target-Cell Adaptation

In responding to almost any type of stimulus, cells and organisms typically can detect the same percent change in a signal over a very wide range of stimulus intensities. At the cellular level this requires that target cells undergo a process of **adaptation** or **desensitization,** whereby, when they are exposed to a stimulus for a prolonged period, their response to it decreases. In this way a cell reversibly adjusts its sensitivity to the stimulus. In the case of chemical signaling, adaptation enables cells to respond to *changes* in the concentration of a signaling ligand (rather than to the absolute concentration of the ligand) over a very wide range of absolute concentrations. The general principle is simple: adaptation is achieved through a negative feedback that operates with a delay. The negative feedback means that a strong response modifies the machinery for making that response and so turns itself off; but thanks to the delay, a sudden change in the stimulus is able to make itself felt strongly for a short period before the negative feedback has time to act.

Adaptation to chemical signals can occur in various ways. In some cases it results from a gradual decrease in the number of specific cell-surface receptor proteins, which generally takes hours. In other cases it results from a rapid inactivation of such receptors, which can occur in minutes. In still other cases it is due to change in the proteins involved in transducing the signal following receptor activation, which usually occurs with an intermediate time course.

Slow Adaptation Depends on Receptor Down-Regulation [42]

After a protein hormone or growth factor binds to its receptor on the surface of a target cell, it is usually ingested by receptor-mediated endocytosis and delivered to endosomes (discussed in Chapter 13). Most receptors discharge their ligand in the acidic environment of endosomes and recycle back to the plasma membrane for reuse, while the ligand is delivered to lysosomes and is degraded. This process, therefore, represents a major pathway for the breakdown of many signaling proteins. Although many receptor molecules are retrieved from the endosome and recycled, a proportion of them fail to release their ligand and end up in lysosomes, where they are degraded along with the ligand. Thus, with continuous exposure to high concentrations of ligand, the number of cell-surface receptors gradually decreases, with a concomitant decrease in the sensitivity of the target cell to the ligand. By this type of mechanism, known as **receptor down-regulation,** a cell can slowly (over hours) adjust its sensitivity to the concentration of a stimulating ligand.

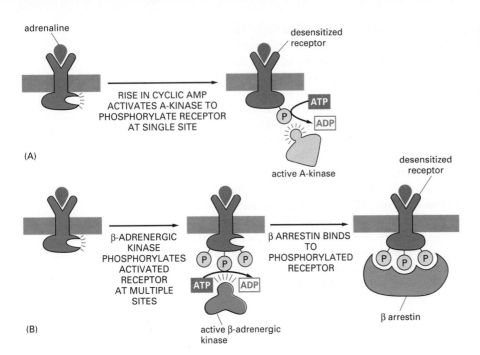

(A)

adrenaline

RISE IN CYCLIC AMP
ACTIVATES A-KINASE TO
PHOSPHORYLATE RECEPTOR
AT SINGLE SITE

desensitized
receptor

ATP

ADP

active A-kinase

(B)

β-ADRENERGIC
KINASE
PHOSPHORYLATES
ACTIVATED
RECEPTOR
AT MULTIPLE
SITES

ATP   ADP

active β-adrenergic
kinase

β ARRESTIN BINDS
TO
PHOSPHORYLATED
RECEPTOR

desensitized
receptor

β arrestin

Figure 15–58 Two mechanisms for the rapid desensitization of the β₂-adrenergic receptor. Both depend on receptor phosphorylation. Both the phosphorylation in (A) and the binding of arrestin in (B) inhibit the ability of the activated receptor to interact with G$_s$ and therefore decrease the response to adrenaline.

Rapid Adaptation Often Involves Receptor Phosphorylation [43]

Target-cell adaptation frequently involves a rapid ligand-induced phosphorylation of receptors, in addition to the slower down-regulation of the number of receptor molecules on the target cell. The best-understood example is the β₂-adrenergic receptor, which activates adenylyl cyclase via the stimulatory G protein G$_s$. When cells are exposed to a high concentration of adrenaline, they can desensitize within minutes by two pathways that depend on β₂-adrenergic receptor phosphorylation. In one, the rise in cyclic AMP caused by adrenaline binding activates A-kinase, which phosphorylates the β₂ receptor on a serine residue, thereby interfering with the receptor's ability to activate G$_s$. In the other, the activated β₂ receptor becomes a substrate for another, more specific protein kinase (called *β-adrenergic kinase*) that phosphorylates the carboxyl-terminal cytoplasmic tail of the activated receptor on multiple serine and threonine residues; this phosphorylated tail binds an inhibitory protein called *β arrestin*, which blocks the receptor's ability to activate G$_s$ (Figure 15–58). In vertebrate photoreceptor cells, rhodopsin, which, as we have seen, is structurally related to β-adrenergic receptors, is inactivated by a closely similar arrestin-based mechanism after it has been activated by the switching on of light. These cells have exceptionally rapid and sophisticated powers of adaptation, involving several mechanisms in addition to that based on arrestin; one of these was discussed earlier, on page 754.

The A-kinase-dependent mechanism that desensitizes the β₂-adrenergic receptor operates whenever cyclic AMP levels rise in the cell. Hence, the activation of any type of receptor in the target cell that activates adenylyl cyclase can desensitize the β₂ receptor—an example of *heterologous desensitization*, where one ligand desensitizes target cells to another. The β-arrestin-dependent mechanism, by contrast, operates only when the β₂ receptor itself is activated by ligand binding—an example of *homologous desensitization*, where a ligand desensitizes target cells only to itself.

Some Forms of Adaptation Are Due to Downstream Changes [44]

Although most known mechanisms of adaptation involve changes in receptor proteins, adaptation can, in principle, result from a change in any of the components in the signaling pathway. There are several cases in which target-cell adaptation has been shown to involve a change in a trimeric G protein. This occurs, for example, in the response of yeast cells to mating pheromones.

Changes downstream from G proteins can also contribute to target-cell adaptation, as in the photoreceptor (see p. 754). In morphine addicts, for example, opiate-sensitive neurons in the brain become desensitized to morphine so that the addicts require much higher doses than normal individuals to relieve pain or to feel euphoric (Figure 15–59). The adapted cells, however, usually have normal levels of functional cell-surface morphine (opiate) receptors. The mechanism of adaptation has been studied both in rats and in morphine-sensitive neural cell lines in culture. Morphine receptors activate the inhibitory G protein G_i, which inhibits adenylyl cyclase and thereby causes a decrease in intracellular cyclic AMP levels. This in turn decreases the activity of A-kinase and thereby the phosphorylation of several types of ion channels, which decreases the electrical firing of the neurons. Cells maintained for a long time in the presence of a high concentration of morphine adapt by a compensatory increase in their expression of the A-kinase and adenylyl cyclase genes, with the net effect that both adenylyl cyclase activity and intracellular cyclic AMP levels return to normal even though morphine is still bound to cell-surface receptors. Because the adapted cells have increased levels of adenylyl cyclase and A-kinase, however, when morphine is withdrawn, there is a marked increase in adenylyl cyclase and A-kinase activity, which causes cyclic AMP concentrations to rise to abnormally high levels. This increases the firing of the neurons and gives rise to the extremely unpleasant withdrawal symptoms (anxiety, sweating, tremors, hallucinations, etc.) experienced by morphine addicts who go "cold turkey."

Adaptation Plays a Crucial Role in Bacterial Chemotaxis [45]

Many of the mechanisms involved in chemical signaling between cells in multicellular animals have evolved from mechanisms used by unicellular organisms to respond to chemical changes in their environment. In fact, some of the same intracellular mediators, such as cyclic nucleotides, are used by both types of organisms. Among the best-studied reactions of unicellular organisms to extracellular signals are chemotactic responses, in which cell movement is oriented toward or away from a source of some chemical in the environment. We conclude this section with an account of bacterial chemotaxis, which, largely through the power of genetic analysis, provides a particularly clear and elegant illustration of the crucial role of adaptation in the response to chemical signals. The chemotaxis of eucaryotic cells is discussed in Chapter 16.

Motile bacteria will swim toward higher concentrations of nutrients (*attractants*), such as sugars, amino acids, and small peptides, and away from higher concentrations of various noxious chemicals (*repellents*) (Figure 15–60). This relatively simple but highly adaptive chemotactic behavior has been most studied in *E. coli* and *Salmonella typhimurium*. We concentrate here chiefly on chemo-

Figure 15–59 The structure of morphine. Why do some of our cells have receptors for a drug that comes from poppy seeds? Pharmacologists long suspected that morphine may mimic some endogenous signaling molecule that regulates pain perception and mood. In 1975 two pentapeptides with morphinelike activity, called **enkephalins,** were isolated from pig brain, and soon thereafter larger polypeptides with similar activity, called **endorphins,** were isolated from the pituitary gland and other tissues. All of these so-called *endogenous opiates* contain a common sequence of four amino acids and bind to the same cell-surface receptors as morphine (and related narcotics). Unlike morphine, however, they are rapidly degraded after release and so do not accumulate in quantities large enough to induce the tolerance seen in morphine addicts.

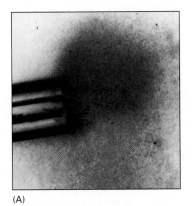
(A)

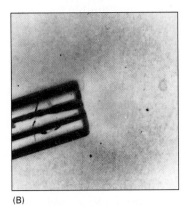
(B)

Figure 15–60 Bacterial chemotaxis. The photographs show *Salmonella typhimurium* bacteria being attracted to a small glass capillary tube containing the amino acid serine (A) and repelled from a capillary tube containing phenol (B). The pictures were taken 5 minutes after the capillary tubes had been introduced into the culture dishes containing the bacteria. This capillary tube assay is a simple method of demonstrating bacterial chemotaxis. (From B.A. Rubik and D.E. Koshland, *Proc. Natl. Acad. Sci. USA* 75:2820–2824, 1978.)

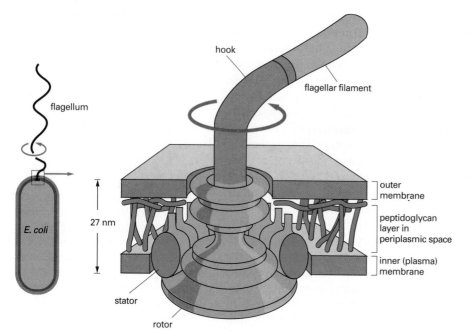

hook

flagellum

flagellar filament

outer
membrane

peptidoglycan
layer in
periplasmic space

inner (plasma)
membrane

27 nm

E. coli

stator

rotor

Figure 15–61 Schematic drawing of the bacterial flagellar motor. The flagellum is linked to a flexible hook. The hook is attached to a series of protein rings (shown in *red*), which are embedded in the outer and inner (plasma) membranes and rotate with the flagellum at about 150 revolutions per second. The rotation is thought to be driven by a flow of protons through an outer ring of proteins (the *stator*), which also contains the proteins responsible for switching the direction of rotation. (Based on data from T. Kubori et al., *J. Mol. Biol.* 226:433–446, 1992, and N.R. Francis et al., *Proc. Natl. Acad. Sci. USA* 89:6304–6308, 1992.)

taxis toward attractants; chemotaxis away from repellents depends on essentially the same mechanisms operating in reverse.

Bacteria swim by means of flagella that are completely different from the flagella of eucaryotic cells. The bacterial flagellum consists of a helical tube formed from a single type of protein subunit, called *flagellin*. Each flagellum is attached by a short flexible hook at its base to a small protein disc embedded in the bacterial membrane. Incredible though it may seem, this disc is part of a tiny "motor" that uses the energy stored in the transmembrane H^+ gradient to rotate rapidly and turn the helical flagellum (Figure 15–61).

Because the flagella on the bacterial surface have an intrinsic "handedness," different directions of rotation have different effects on movement. Counterclockwise rotation allows all the flagella to draw together into a coherent bundle so that the bacterium swims uniformly in one direction. Clockwise rotation causes them to fly apart, so that the bacterium tumbles chaotically without moving forward (Figure 15–62). In the absence of any environmental stimulus, the direction of rotation of the disc reverses every few seconds, producing a characteristic pattern of movement in which smooth swimming in a straight line is interrupted by abrupt, random changes in direction caused by tumbling (Figure 15–63A).

The normal swimming behavior of bacteria is modified by chemotactic attractants or repellents, which bind to specific receptor proteins and affect the frequency of tumbling by increasing or decreasing the time that elapses between successive changes in direction of flagellar rotation. When bacteria are swimming in a favorable direction (toward a higher concentration of an attractant or away from a higher concentration of a repellent), they tumble less frequently than when they are swimming in an unfavorable direction (or when no gradient is present). Since the periods of smooth swimming are longer when a bacterium is traveling in a favorable direction, it will gradually progress in that direction—toward an attractant (Figure 15–63B) or away from a repellent.

In its natural environment a bacterium detects a spatial gradient of attractants or repellents in the medium by swimming at a constant velocity and comparing the concentration of chemicals over *time*. (It does not monitor changes in concentration by using a *spatial* separation of receptors over its length; this would be extremely difficult given the very small size of a bacterium.) Changes over time can be produced artificially in the laboratory by the sudden addition or removal of a chemical to the culture medium. When an attractant is added in this way, tumbling is suppressed within a few tenths of a second, as expected. But

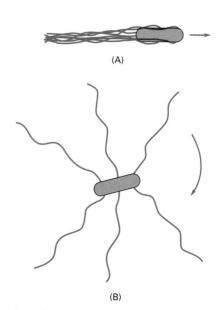

(A)

(B)

Figure 15–62 Positions of the flagella on *E. coli* during swimming. When the flagella rotate counterclockwise (A), they are drawn together into a single bundle, which acts as a propeller to produce smooth swimming. When the flagella rotate clockwise (B), they fly apart and produce tumbling.

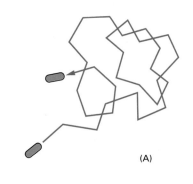

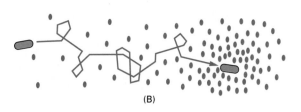

(A)

(B)

Figure 15–63 The tracks of a swimming bacterium. In the absence of a chemotactic signal (A), periods of smooth swimming (runs) are interrupted by brief tumbles that randomly change the direction of swimming. Thus runs and tumbles occur in alternating sequence, each run constituting a step in a three-dimensional random walk. In the presence of a chemotactic attractant (B), tumbling is partially suppressed whenever the bacterium happens to be swimming toward a higher concentration of the attractant, so that it gradually moves in the direction of the attractant—a biased random walk.

after some time, even in the continuing presence of the attractant, tumbling frequency returns to normal. The bacteria remain in this adapted state as long as there is no increase or decrease in the concentration of the attractant; addition of more attractant will briefly suppress tumbling, whereas removal of the attractant will briefly enhance tumbling until the bacteria again adapt to the new level. Adaptation is a crucial part of the chemotactic response in that it enables bacteria to respond to *changes* in concentration rather than to steady-state levels of an attractant and to respond to these changes over an astonishingly wide range of attractant concentrations (from less than 10^{-10} M to over 10^{-3} M for some attractants).

Bacterial Chemotaxis Is Mediated by a Family of Four Homologous Transmembrane Receptors and a Phosphorylation Relay System [46]

The unraveling of the molecular mechanisms responsible for bacterial chemotaxis has depended largely on the isolation and analysis of mutants with defective chemotactic behavior. In this way it has been shown that chemotaxis to a number of chemicals depends on a small family of closely related transmembrane receptor proteins that are responsible for transmitting chemotactic signals across the plasma membrane. These **chemotaxis receptors** are methylated during adaptation (see below) and so are also called *methyl-accepting chemotaxis proteins (MCPs)*. As we shall see, receptor activity is stimulated by an increase in repellent concentration and decreased by an increase in attractant concentration: a single receptor is affected by both sorts of molecules, with opposite consequences.

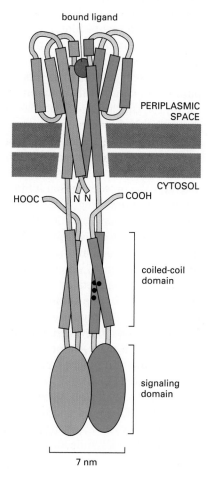

Figure 15–64 A model of the homodimeric structure of the aspartate chemotaxis receptor protein. The three-dimensional structure of the extracellular domain has been obtained by x-ray diffraction. The intracellular coiled-coil domains are predicted from amino acid sequence analysis. They contain the methylation sites (shown as *black dots*), of which there are four on each of the two polypeptide chains (the sites on one of the chains are out of view). The binding of the ligand in the periplasmic space is thought to induce a conformational change in the receptor that is propagated through the membrane by a scissorlike movement of the whole molecule. (Based on M.V. Milburn et al., *Science* 254:1342–1347, 1991. © 1991 the AAAS; and J.B. Stock et al., *J. Biol. Chem.* 267:19753–19756, 1992, ASBMB publisher.)

Target-Cell Adaptation

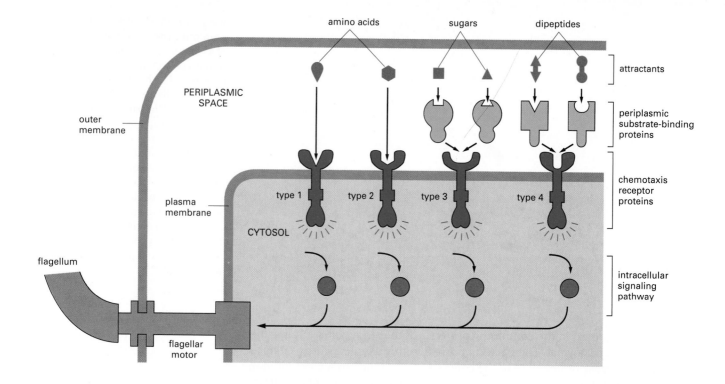

There are four types of plasma membrane chemotaxis receptors, each concerned with the response to a small group of chemicals. Type 1 and 2 receptors mediate responses to serine and aspartate, respectively, by directly binding these amino acids and transducing the binding event into an intracellular signal. A model of the structure of one of these receptors is shown in Figure 15–64. Type 3 and 4 receptors mediate responses to sugars and dipeptides, respectively, in a slightly less direct fashion (Figure 15–65).

Genetic studies indicate that four cytoplasmic proteins—*CheA, CheW, CheY,* and *CheZ*—are involved in the intracellular signaling pathway that couples the chemotactic receptors to the flagellar motor. CheY acts at the effector end of the pathway to control the direction of flagellar rotation. When activated, it binds to the motor, causing it to rotate clockwise and thereby inducing tumbling; mutants that lack this protein swim constantly without tumbling. CheA is a histidine protein kinase. When bound to both an activated chemotactic receptor and CheW, it phosphorylates itself on a histidine residue and almost immediately transfers the phosphate to an aspartic acid residue on CheY. The phosphorylation of CheY activates the protein so that it binds to the flagellar motor and causes clockwise rotation and tumbling. CheZ rapidly inactivates phosphorylated CheY by stimulating its dephosphorylation (Figure 15–66).

The binding of a repellent to a chemotactic receptor increases the activity of the receptor, which in turn increases the activity of CheA and thereby the phosphorylation of CheY, which causes tumbling. These phosphorylations occur rapidly: the time required for the tumbling response after adding a repellent is about 200 milliseconds. The binding of an attractant has the opposite effect. It decreases the activity of the receptor, which decreases the activity of CheA, so that CheY remains dephosphorylated, the motor continues to rotate counterclockwise, and the bacterium swims smoothly.

The function of CheY in bacterial chemotaxis is analogous to the function of Ras proteins in animal cell signaling. Like Ras, CheY functions as an on/off switch: it is on when phosphorylated and off when dephosphorylated, just as Ras is on with GTP bound and off with GDP bound. CheY is activated by CheA and inactivated by CheZ, just as Ras is activated by GNRPs and inactivated by GAPs (see Figure 15–50). Indeed, the three-dimensional structures of CheY and Ras are similar.

Figure 15–65 The different types of chemotaxis receptors. Chemical attractants bind to type 1 or type 2 receptors in the plasma membrane or to binding proteins in the periplasmic space (between the inner and outer bacterial membranes) that then bind to type 3 or type 4 receptors. The binding of an attractant decreases the activity of the chemotaxis receptor, shutting off the intracellular signaling cascade and causing the flagellar motor to continue to rotate counterclockwise, thereby suppressing tumbling and causing continuous smooth swimming. The attractants diffuse into the periplasmic space from outside the cell through large channels in the outer membrane (not shown).

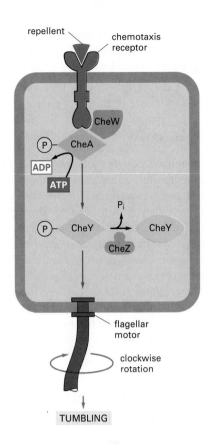

Figure 15–66 The phosphorylation relay system that enables the chemotaxis receptors to control the flagellar motor. The binding of a repellent increases the activity of the receptor, which binds CheW and CheA, thereby stimulating CheA to phosphorylate itself. CheA quickly transfers its covalently bound, high-energy phosphate directly to CheY to generate CheY-phosphate, which binds to the flagellar motor and causes it to rotate clockwise, resulting in tumbling. The binding of an attractant has the opposite effect. It decreases the activity of the receptor and therefore decreases the phosphorylation of CheA and CheY, which results in counterclockwise flagellar rotation and smooth swimming. CheZ accelerates the dephosphorylation of CheY-phosphate, thereby inactivating it. Each of the phosphorylated intermediates decays in about 10 seconds, enabling the bacterium to respond very quickly to changes in its environment (see Figure 15–10).

Receptor Methylation Is Responsible for Adaptation in Bacterial Chemotaxis [46]

Adaptation in bacterial chemotaxis results from the covalent methylation of the chemotaxis receptor proteins. When methylation is blocked by mutation, adaptation is markedly inhibited, and exposure of the mutant bacteria to an attractant results in the suppression of tumbling for days instead of for a minute or so. Binding of a chemoattractant to a chemotaxis receptor, therefore, has two separable consequences. (1) It rapidly decreases the activity of the receptor, thereby decreasing the activity of CheA and CheY and causing the flagellar motor to continue to rotate counterclockwise; this results in a suppression of tumbling. (2) It causes adaptation because, while the attractant is bound, the receptor is methylated by an enzyme in the cytoplasm, which increases the activity of the receptor over a period of a few minutes (Figure 15–67).

Receptor methylation is catalyzed by an enzyme (*methyl transferase*) that acts on the receptor protein. As many as eight methyl groups can be transferred to a single receptor, the extent of methylation increasing at higher concentrations of attractant (where each receptor spends a larger proportion of its time with ligand bound). When the attractant is removed, the receptor is demethylated by a demethylating enzyme (methylesterase). Although the level of methylation changes during a chemotactic response, it remains constant once a bacterium is adapted because a balance is reached between the rates of methylation and demethylation. The methylesterase that removes methyl groups from the chemotactic receptors is also regulated by CheA-mediated phosphorylation, and this provides another form of negative feedback regulation that makes a further contribution to adaptation.

A variety of other regulatory interactions and feedback loops are probably still to be discovered in bacterial chemotaxis. Nonetheless, all of the genes and proteins involved in this highly adaptive behavior may now have been identified, and in most cases the protein sequences are known and the proteins are available in

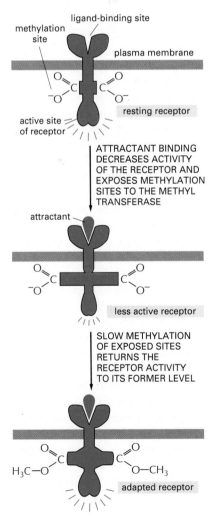

Figure 15–67 The sequential activation and adaptation (via methylation) of a chemotaxis receptor. Note that the activity of the receptor, and therefore the tumbling frequency of the bacterium, is the same in the resting and adapted states. The receptor is shown with two methylation sites for simplicity; in fact, there are eight on each receptor. As the concentration of attractant increases, the fraction of time that the receptor is occupied by the attractant increases. A higher level of attractant will thereby initially cause a greater change in the conformation of the receptor than a low level, pushing the receptor more toward its fully inactivated state. An increase in methylation ensues, however, so that within minutes the conformational strain on the receptor is exactly reversed, and the activity of the receptor increases to its previous level. The receptor has now adapted.

large quantities. It therefore seems likely that bacterial chemotaxis will be the first cell-signaling system to be understood completely in molecular terms. But even when all the molecules and their interactions have been defined, it may be difficult to comprehend how the signaling system operates as an integrated network, as we discuss next.

Summary

By adapting to high concentrations of a signaling ligand in a time-dependent, reversible manner, cells can adjust their sensitivity to the level of a stimulus and thereby respond to changes in a ligand's concentration over an enormously large range rather than to the absolute concentration of the ligand. Adaptation occurs in various ways: (1) ligand binding can induce the internalization of receptors, some of which are then degraded in lysosomes—a process called receptor down-regulation; (2) activated receptors can be reversibly inactivated by being phosphorylated or methylated; (3) G proteins can be reversibly inactivated; and (4) proteins downstream of G proteins in the signaling pathway can be up-regulated or down-regulated. At a molecular level the best-understood example of adaptation occurs in bacterial chemotaxis, in which the reversible methylation of key signal-transducing proteins in the plasma membrane helps the cell to swim toward an optimal environment.

The Logic of Intracellular Signaling: Lessons from Computer-based "Neural Networks"

Each cell in a complex multicellular animal is bombarded with chemical signals that are made by other cells—signals that regulate its metabolism, signals that alter or maintain its differentiated state, signals that determine whether it should divide, and signals that dictate whether it will live or die. In general, these signals bind to cell-surface receptors that activate several intracellular signaling pathways, so that the intracellular signals generated from different receptors will interact with one another in complex ways.

How does a cell integrate all of this information so as to behave in a way that is optimal for the animal as a whole? The task of understanding how a cell manages this feat seems overwhelming. Even in the relatively simple case of bacterial chemotaxis, where there are relatively few components, all of which are probably known, the complexities of the interactions are still too great to be easily visualized and completely understood at present. We still cannot reliably predict, for example, the behavior of a mutant bacterium in which the level of a membrane receptor or a cytoplasmic signaling protein has been altered, especially if the chemotactic stimulus includes more than one attractant or repellent or if the stimulus changes rapidly with time. How then can we hope to understand the much more complex networks of intracellular signaling pathways in animal cells, where hundreds of components, many still undiscovered, are involved?

As more quantitative data become available it is likely that computer-based simulations will play an increasingly important role in our attempts to understand how these signaling pathways operate. By modeling the pathways on a computer, one can display and manipulate the network of interacting components in ways that are not possible in cells.

Computer-based analysis is useful in another way—one that does not depend on detailed quantitative knowledge of the reactions involved. The intracellular signaling pathways, seen as a whole, form a highly interconnected network in which signals are processed along multiple parallel routes that interact with one another. One can therefore learn about the behavior of signaling networks by comparing them to other highly interconnected networks. Computer-based networks, often called **neural networks,** were originally developed to understand how nerve cells relay and process information in the brain, but they have properties that are also relevant to intracellular signaling.

Computer-based Neural Networks Can Be Trained [47]

One of the simplest types of computer-based neural networks is composed of three layers of interconnected *units*—an *input layer*, a *hidden layer*, and an *output layer*. Each of the many units in the network acts as a model nerve cell (neuron), with its individual output controlled by multiple inputs. The connections between units are analogous to synapses and have modifiable *connection weights* that control the strength with which one unit influences another. The activity of the network as a whole depends on the values of these connection weights as well as on the mathematical rules by which each unit sums its inputs to generate an output. A pattern of inputs received by the input units is transformed according to these weights and rules into a different pattern of activity in the output units via connections through the hidden units (Figure 19–68).

The most remarkable and useful feature of computer-based neural networks is that they can be *trained* to recognize specific patterns of inputs and respond to each pattern with a specific pattern of output activity. The training is achieved by presenting the network with training examples in the form of a series of inputs together with the desired pattern of outputs. The input, for example, might be a series of letters presented in any orientation on a screen and the desired output simply the correct identification of each letter. Or the input could be the amino acid sequences of a number of polypeptide chains and the output the types of secondary structures that the polypeptide chains form. Whatever the task, as each training example is presented, the output of the network is compared to the desired output and an "error" score is assigned based on how closely the actual and desired outputs match. After processing all examples in the training set, an overall performance measure is calculated that characterizes how well the network performed on the entire training set.

The goal of training is to change the weights in the network so that its performance will improve on subsequent presentations of the training set. A number of training algorithms have been devised for making these weight changes. One of the conceptually simplest methods, and perhaps the most relevant to cell signaling (as discussed below), is one in which the weights are changed randomly and those changes that result in an improved performance are preferentially maintained. Whatever algorithm is used, the training process is repeated over and over until the actual output of the network is close to the desired one. The final weights in the network are not predetermined by the training algorithm but instead emerge autonomously as a result of repeated presentation of the training examples. Once the network has "learned" the desired task, it can often recognize and give the correct output for novel input patterns that were not part of the original training set.

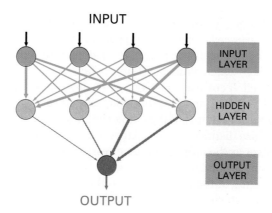

Figure 15–68 A simple neural network. The activity of each neural unit (shown as a circle) is determined by the unit's inputs. The output of each unit is usually a nonlinear function of the unit's inputs. Each connection between units has a particular strength, or "weight," which is indicated by differences in thickness of the connecting arrows.

Cell Signaling Networks Can Be Viewed as Neural Networks Trained by Evolution [48]

Although neural networks were originally used to model systems of interconnected neurons, there is no reason why the units in such a network have to represent neurons. They could be enzymes or other types of intracellular signaling molecules, for example. Consider the protein kinases involved in cell signaling. These enzymes receive inputs (in the form of phosphorylations or simple protein binding) from components of a number of intracellular signaling pathways, and these inputs collectively regulate the catalytic activity of the kinase. Each kinase in turn has an output (the phosphorylation of other proteins) that regulates the activity of specific target proteins, which can be either downstream components in the same signaling pathway or components of a parallel signaling pathway. In terms of a network, the kinases in this example function just as neurons do: they integrate various inputs and respond with an appropriate output.

In principle, at least, a highly interconnected network of intracellular signaling reactions could be trained to recognize certain input patterns of signals and to respond with specific output patterns in a way similar to that just described for a computer-based neural network. In this scheme the training would occur during evolution, by mutation and natural selection, with random mutations in the genes that encode signaling proteins serving the same function as random changes in connection weights made on a computer. A mutation that changed the activity of a signaling protein kinase, for example, might increase the weight of one or more connections in the network, while a change in the binding specificity of an SH adaptor protein could add new connections. Mutations that improved the performance of a signaling network, by enabling it, for example, to recognize a new combination of growth factors or to discriminate between two extracellular signals that were previously indistinguishable to a cell, could give the organism a selective advantage and hence be retained for further improvement. In this way an increasingly complex network of signaling reactions would evolve.

Signaling Networks Enable Cells to Respond to Complex Patterns of Extracellular Signals [47, 49]

When trained neural networks are analyzed to see how they recognize complex patterns of inputs, it is found that individual units in the hidden layer (see Figure 15–68) have become strongly connected to meaningful sets of units in the input layer, so that the hidden units come to represent signification features of the input pattern applied to the network. In a network trained to recognize letters presented in any orientation on a screen, for example, specific hidden units might come to recognize curved lines or pairs of lines at right angles, while in a network trained to pronounce written words, specific hidden units might come to represent vowel sounds or consonants.

In a similar way specific intracellular signaling molecules in a cell might come to recognize a particular combination of extracellular signals and help to translate this combination of signals into a particular cellular response. Consider the hypothetical network shown in Figure 15–69, which almost certainly is far simpler than any signaling network found in a cell. It consists of six types of cell-surface receptors that are functionally coupled to three cytosolic protein kinases. Receptors I, II, and III each recognize distinct components of the extracellular matrix, while receptors A, B, and C each recognize distinct growth factors. Kinase 3 lies downstream of kinases 1 and 2 and serves as the output of the network, perhaps stimulating the cell to proliferate by phosphorylating a set of gene regulatory proteins. Because of the different strengths of the connections in the network, some of which are excitatory and some of which are inhibitory, each of the three

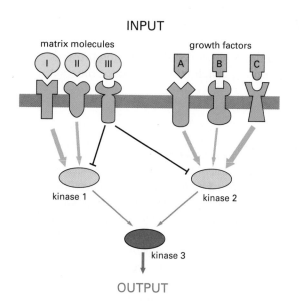

INPUT

matrix molecules | growth factors

I II III | A B C

kinase 1 | kinase 2

kinase 3

OUTPUT

Figure 15–69 A simple hypothetical signaling network. The network consists of six receptors and three cytosolic protein kinases. Each receptor activates (*green arrows*) or inhibits (*black lines*) kinase 1 or 2 or both by an unspecified mechanism. Because signals converge onto kinase 3 (the output kinase), this network will be maximally active only when specific combinations of extracellular stimuli are present. Although this network is far simpler than any likely to be found in a living cell, it could form part of a more complex signaling pathway.

kinases will be optimally stimulated by different combinations of extracellular signals. Kinase 1 is most active when the cell encounters matrix components I and II in the absence of III, while kinase 2 is most active when the cell encounters growth factors A, B, and C in the absence of the matrix component III. For kinase 3 the requirements for optimal activity are more complex, as it senses the extracellular environment only through kinases 1 and 2. It will be maximally active and stimulate the cell to proliferate when kinases 1 and 2 are simultaneously active—that is, when the cell encounters growth factors A, B, and C and matrix molecules I and II in the absence of III, which is a surprisingly complex pattern considering the simplicity of the network. According to this view, kinase 3 has "learned" through evolution to associate this particular combination of extracellular stimuli with the need of the cell to proliferate. In the same way some of the important signaling molecules in a real cell will have come to recognize and respond to relevant combinations of features in the cell's environment.

Signaling Networks Are Robust

An important consequence of the highly interconnected architecture of a neural network is that once it has been trained to perform a task, its performance is not easily destroyed by modifying or removing individual units of the network. If a specific unit responds optimally when it receives inputs from six other units, it will also respond, although less well, to a subset of five of these inputs. If random changes in the strengths of connections are introduced into a neural network that has been trained to recognize letters, for example, the ability to do this task is not abolished but only somewhat degraded.

In a similar way a cellular response that depends on highly interconnected intracellular signaling pathways will not be easily disrupted by removing or changing a single signaling element in one of these pathways. We should not be too surprised, therefore, to find that a eucaryotic cell can function nearly normally when a protein kinase is inactivated by mutation, even though that protein kinase has been highly conserved during evolution. This stability of signaling networks is probably important to perfectly normal cells, as cells do not contain precisely determined numbers of intracellular proteins; moreover, the concentrations of important metabolites can fluctuate with the metabolic state of the cell. The extensive cross-talk between signaling pathways in animal cells may have evolved, in part, to allow the pathways to function normally in the face of such fluctuations.

The neural-network-like properties of highly interconnected systems of proteins can help us to understand other aspects of cell biology besides cell signaling. The same considerations apply to the complex networks of interacting proteins that form the cell's cytoskeleton, for example, which is the subject of the next chapter.

Summary

Computer modeling can help to illuminate the complex behaviors of the interacting signaling cascades that are found in cells. In particular, computer-based neural networks have a number of properties that are likely to be shared by intracellular signaling networks. Just as neural networks can be trained to respond appropriately to specific patterns of inputs, so these signaling networks, by a Darwinian process of random change and selection, may have evolved the capacity to respond appropriately to complex combinations of extracellular signals. The highly interactive architecture of neural networks is mimicked by the networks of intracellular signaling proteins. Both networks can in principle function as pattern-recognition devices, which respond optimally to selected combinations of input stimuli. Networks of this kind are also relatively resistant to noise fluctuations or to damage, and eliminating one component does not totally disable the network.

References

General

Barritt, G.J. Communication Within Animal Cells. Oxford, UK: Oxford Science Publications, 1992.

Baulieu, E.-E.; Kelly, P.A. Hormones: From Molecules to Disease. London: Chapman and Hall, 1990.

Hardie, D.G. Biochemical Messengers: Hormones, Neurotransmitters and Growth Factors. London: Chapman and Hall, 1990.

Molecular Biology of Signal Transduction. *Cold Spring Harb. Symp. Quant. Biol.*, Vol. 53, 1988.

Morgan, N.G. Cell Signalling. Milton Keynes, UK: Open University Press, 1989.

Cited

1. Errede, B.; Levin, D.E. A conserved kinase cascade for MAP kinase activation in yeast. *Curr. Opin. Cell Biol.* 5:254–260, 1993.

   Kurjan, J. Pheromone response in yeast. *Annu. Rev. Biochem.* 61:1097–1129, 1992.

   Marsh, I.; Neimsen, A.M.; Herskowitz, I. Signal transduction during pheromone response in yeast. *Annu. Rev. Cell Biol.* 7:699–728, 1991.

2. Hardie, D.G. Biochemical Messengers: Hormones, Neurotransmitters and Growth Factors. London: Chapman and Hall, 1990.

   Kahn, C.R. Membrane receptors for hormones and neurotransmitters. *J. Cell Biol.* 70:261–286, 1976.

   Levitski, A. Receptors: A Quantitative Approach. Menlo Park, CA: Benjamin-Cummings, 1984.

   Snyder, S.H. The molecular basis of communication between cells. *Sci. Am.* 253(4):132–140, 1985.

3. Gurdon, J.; Tiller, E.; Roberts, J.; Kato, K. A community effect in muscle development. *Curr. Biol.* 3:1–11, 1993.

   Smith, W.L.; Borgeat, P. The eicosanoids: prostaglandins, thromboxanes, leukotrienes, and hydroxyeicosaenoic acids. In Biochemistry of Lipids and Membranes (D.E. Vance, J.E. Vance, eds.), pp. 325–360. Menlo Park, CA: Benjamin-Cummings, 1985.

   Weissmann, G. Aspirin. *Sci. Am.* 264(1):84–90, 1991.

4. Caveney, S. The role of gap junctions in development. *Annu. Rev. Physiol.* 47:319–335, 1985.

   Warner, A.E. The role of gap junctions in amphibian development. *J. Embryol. Exp. Morphol. Suppl.* 89:365–380, 1985.

5. Raff, M.C. Social controls on cell survival and cell death. *Nature* 356:397–400, 1992.

6. Schimke, R.T. On the roles of synthesis and degradation in regulation of enzyme levels in mammalian tissues. *Curr. Top. Cell Regul.* 1:77–124, 1969.

7. Bredt, D.S.; Snyder, S.H. Nitric oxide, a novel neuronal messenger. *Neuron* 8:3–11, 1992.

   Lowenstein, C.J.; Snyder, S.H. Nitric oxide, a novel biological messenger. *Cell* 70:705–707, 1992.

   Moncada, S.; Palmer, R.M.; Higgs, E.A. Nitric oxide: physiology, pathophysiology and pharmacology. *Pharmacol. Rev.* 43:109–142, 1991.

   Stevens, C.F. Just say NO. *Curr. Biol.* 2:108–109, 1992.

8. Andres, A.J.; Thummel, C.S. Hormones, puffs and flies: the molecular control of metamorphosis by ecdysone. *Trends Genet.* 8:132–138, 1992.

   Ashburner, M.; Chihara, C.; Meltzer, P.; Richards, G. Temporal control of puffing activity in polytene chromosomes. *Cold Spring Harb. Symp. Quant. Biol.* 38:655–662, 1974.

   Evans, R.M. The steroid and thyroid hormone receptor superfamily. *Science* 240:889–895, 1988.

   Parker, M.G., ed. Nuclear Hormone Receptors: Molecular Mechanisms, Cellular Functions and Clinical Abnormalities. London: Academic Press, 1991.

   Yamamoto, K.R. Steroid receptor regulated transcription of specific genes and gene networks. *Annu. Rev. Genet.* 19:209–252, 1985.

9. Nishizuka, Y., ed. Signal transduction: crosstalk. *Trends Biochem. Sci.* 17:367–443, 1992.

10. Bourne, H.R.; Nicoll, R. Molecular machines integrate coincident synaptic signals. *Cell/Neuron* 72/10:65–75, 1993.

    Cohen, P. Signal integration at the level of protein kinases, protein phosphatases and their substrates. *Trends Biochem. Sci.* 17:408–413, 1992.

    Pelech, S.L. Networking with protein kinases. *Curr. Biol.* 3:513–515, 1993.

    Posada, J.; Cooper, J.A. Molecular signal integration. Interplay between serine, threonine, and tyrosine phosphorylation. *Mol. Biol. Cell* 3:583–592, 1992.

11. Birnbaumer, L. G proteins in signal transduction. *Annu. Rev. Pharmacol. Toxicol.* 30:675–705, 1990.

    Dohlman, H.G.; Thorner, J.; Caron, M.G.; Lefkowitz, R.J. Model systems for the study of seven-transmembrane-segment receptors. *Annu. Rev. Biochem.* 60:653–688, 1991.

    Houslay, M.D.; Milligan, G., eds. G-Proteins as Mediators of Cellular Signalling Processes. Chichester, UK: Wiley, 1990.

    Linder, M.E.; Gilman, A.G. G proteins. *Sci. Am.* 267(1):36–43, 1992.

12. Bourne, H.R.; Sanders, D.A.; McCormick, F. The GTPase superfamily: conserved structure and molecular mechanism. *Nature* 349:117–127, 1991.

    Hepler, J.R.; Gilman, A.G. G proteins. *Trends Biochem. Sci.* 17:383–387, 1992.

13. Feder, D., et al. Reconstitution of beta$_1$-adrenoceptor-dependent adenylate cyclase from purified components. *EMBO J.* 5:1509–1514, 1986.

    Gilman, A.G. G proteins: transducers of receptor-generated signals. *Annu. Rev. Biochem.* 56:615–649, 1987.

    Pastan, I. Cyclic AMP. *Sci. Am.* 227(2):97–105, 1972.

    Schramm, M.; Selinger, Z. Message transmission: receptor controlled adenylate cyclase system. *Science* 225:1350–1356, 1984.

    Sutherland, E.W. Studies on the mechanism of hormone action. *Science* 177:401–408, 1972.

    Tang, W.-J.; Gilman, A.G. Adenylyl cyclases. *Cell* 70:869–872, 1992.

14. Lai, C.-Y. The chemistry and biology of cholera toxin. *CRC Crit. Rev. Biochem.* 9:171–206, 1980.

15. Birnbaumer, L. Receptor-to-effector signaling through G proteins: roles for beta gamma dimers as well as alpha subunits. *Cell* 71:1069–1072, 1992.

    Iniguez-Lluhi, J.; Kleuss, C.; Gilman, A.G. The importance of G-protein βγ subunits. *Trends Cell Biol.* 3:230–235, 1993.

16. Brindle, P.K.; Montminy, M.R. The CREB family of transcription activators. *Curr. Opin. Gen. Dev.* 2:199–204, 1992.

    Cohen, P. Protein phosphorylation and the control of glycogen metabolism in skeletal muscle. *Philos. Trans. R. Soc. Lond. (Biol.)* 302:13–25, 1983.

    Krebs, E.G. Role of the cyclic AMP-dependent protein kinase in signal transduction. *JAMA* 262:1815–1818, 1989.

    Pilkis, S.J.; El-Maghrabi, M.R.; Claus, T.H. Hormonal regulation of hepatic gluconeogenesis and glycolysis. *Annu. Rev. Biochem.* 57:755–784, 1988.

    Taylor, S.S.; Buechler, J.A.; Yonemoto, W. cAMP-dependent protein kinase: framework for a diverse family of regulatory enzymes. *Annu. Rev. Biochem.* 59:971–1005, 1990.

17. Cohen, P. Structure and regulation of protein phosphatases. *Annu. Rev. Biochem.* 58:453–508, 1989.

18. Carafoli, E. Intracellular calcium homeostasis. *Annu. Rev. Biochem.* 56:395–433, 1987.

    Evered, D.; Whelan, J., eds. Calcium and the Cell, Ciba Foundation Symposium 122. Chichester, UK: Wiley, 1986.

    Koch, G.L.E. The endoplasmic reticulum and calcium storage. *Bioessays* 12:527–531, 1990.

19. Augustine, G.J.; Charlton, M.P.; Smith, S.J. Calcium action in synaptic transmitter release. *Annu. Rev. Neurosci.* 10:633–693, 1987.

    Heilbrunn, L.V.; Wiercenski, F.J. The action of various cations on muscle protoplasm. *J. Cell. Comp. Physiol.* 29:15–32, 1947.

20. Bansal, V.S.; Majerus, P.W. Phosphatidylinositol-derived precursors and signals. *Annu. Rev. Cell Biol.* 6:41–67, 1990.

    Harden, T.K. G-protein-regulated phospholipase C: identification of component proteins. *Adv. Second Messenger Phosphoprotein Res.* 26:11–34, 1992.

    Majerus, P.W. Inositol phosphate biochemistry. *Annu. Rev. Biochem.* 61:225–250, 1992.

    Michell, R.H. Inositol lipids in cellular signalling mechanisms. *Trends Biochem. Sci.* 17:274–276, 1992.

    Sekar, M.C.; Hokin, L.E. The role of phosphoinositides in signal transduction. *J. Membr. Biol.* 89:193–210, 1986.

21. Berridge, M.J. Inositol trisphosphate and calcium signalling. *Nature* 361:315–325, 1993.

    Ferris, C.D.; Snyder, S.H. Inositol 1,4,5-trisphosphate-activated calcium channels. *Annu. Rev. Physiol.* 54:469–488, 1992.

    Meldolesi, J. Multifarious IP$_3$ receptors. *Curr. Biol.* 2:393–394, 1992.

    Taylor, C.W.; Marshall, I.C.B. Calcium and inositol 1,4,5-trisphosphate receptors: a complex relationship. *Trends Biochem. Sci.* 17:403–407, 1992.

22. Berridge, M.J. Inositol triphosphate and calcium oscillations. *Adv. Second Messenger Phosphoprotein Res.* 26:211–223, 1992.

    Cobbold, P.H.; Cuthbertson, K.S. Calcium oscillations: phenomena, mechanisms and significance. *Semin. Cell Biol.* 1:311–321, 1990.

    Meyer, T.; Stryer, L. Calcium spiking. *Annu. Rev. Biophys. Biophys. Chem.* 20:153–174, 1991.

    Tsien, R.W.; Tsien, R.Y. Calcium channels, stores, and oscillations. *Annu. Rev. Cell Biol.* 6:715–760, 1990.

    Tsunoda, Y. Oscillatory Ca^{2+} signaling and its cellular function. *New Biol.* 3:3–17, 1991.

23. Asaoka, Y.; Nakamura, S.-I.; Yoshida, K.; Nishizuka, Y. Protein kinase C, calcium and phospholipid degradation. *Trends Biochem. Sci.* 17:414–417, 1992.

    Hunter, T.; Karin, M. The regulation of transcription by phosphorylation. *Cell* 70:375–387, 1992.

    Liou, H.-C.; Baltimore, D. Regulation of the NF-κB/rel transcription factor and IκB inhibitor system. *Curr. Opin. Cell Biol.* 5:477–487, 1993.

    Nishizuka, Y. Intracellular signaling by hydrolysis of phospholipids and activation of protein kinase C. *Science* 258:607–614, 1992.

References

Sternweis, P.C.; Smrcka, A.V. Regulation of phospholipase C by G proteins. *Trends Biochem. Sci.* 17:502–506, 1992.

24. Gerday, C.; Bolis, L.; Gilles, R., eds. Calcium and Calcium Binding Proteins. Berlin: Springer-Verlag, 1988.

Head, J.F. A better grip on calmodulin. *Curr. Biol.* 2:609–611, 1992.

O'Neil, K.T.; DeGrado, W.F. How calmodulin binds its targets: sequence independent recognition of amphipathic α-helices. *Trends Biochem. Sci.* 15:59–64, 1990.

25. Hanson, P.I.; Schulman, H. Neuronal Ca^{2+}/calmodulin-dependent protein kinases. *Annu. Rev. Biochem.* 61:559–601, 1992.

Morris, R.G.M.; Kennedy, M.B. The pierian spring. *Curr. Biol.* 2:511–514, 1992.

Schulman, H. The multifunctional Ca^{2+}/calmodulin-dependent protein kinases. *Curr. Opin. Cell Biol.* 5:247–253, 1993.

26. Cohen, P. Protein phosphorylation and hormone action. *Proc. R. Soc. Lond. (Biol.)* 234:115–144, 1988.

27. Brown, A.M.; Birnbaumer, L. Ionic channels and their regulation by G protein subunits. *Annu. Rev. Physiol.* 52:197–213, 1990.

Hille, B. G protein-coupled mechanisms and nervous signaling. *Neuron* 9:187–195, 1992.

28. Buck, L.B. The olfactory multigene family. *Curr. Opin. Neurobiol.* 2:282–288, 1992.

Kaupp, U.B.; Koch, K.W. Role of cGMP and Ca^{2+} in vertebrate photoreceptor excitation and adaptation. *Annu. Rev. Physiol.* 54:153–176, 1992.

Lagnado, L.; Baylor, D. Signal flow in visual transduction. *Neuron* 8:995–1002, 1992.

Reed, R.R. How does the nose know? *Cell* 60:1–2, 1990.

Simon, M.I.; Strathmann, M.P.; Gautam, N. Diversity of G proteins in signal transduction. *Science* 252:802–808, 1991.

Stryer, L. Visual excitation and recovery. *J. Biol. Chem.* 266:10711–10714, 1991.

29. Lamb, T.D.; Pugh, E.N., Jr. G-protein cascades: gain and kinetics. *Trends Neurosci.* 15:291–298, 1992.

30. Lewis, J.; Slack, J.; Wolpert, L. Thresholds in development. *J. Theor. Biol.* 65:579–590, 1977.

Mulvihill, E.R.; Palmiter, R.D. Relationship of nuclear estrogen receptor levels to induction of ovalbumin and conalbumin mRNA in chick oviduct. *J. Biol. Chem.* 252:2060–2068, 1977.

31. Maack, T. Receptors of atrial natriuretic factor. *Annu. Rev. Physiol.* 54:11–27, 1992.

Miller, S.G.; Kennedy, M.B. Regulation of brain type II Ca^{2+}/calmodulin-dependent protein kinase by autophosphorylation: a Ca^{2+}-triggered molecular switch. *Cell* 44:861–870, 1986.

Mohun, T. Muscle differentiation. *Curr. Opin. Cell Biol.* 4:923–928, 1992.

Rosenzweig, A.; Seidman, C.E. Atrial natriuretic factor and related peptide hormones. *Annu. Rev. Biochem.* 60:229–256, 1991.

Yuen, P.S.T.; Garbers, D.L. Guanylyl cyclase-linked receptors. *Annu. Rev. Neurosci.* 15:193–225, 1992.

32. Carpenter, G. Receptors for epidermal growth factor and other polypeptide mitogens. *Annu. Rev. Biochem.* 56:881–914, 1987.

Fantl, W.J.; Johnson, D.E.; Williams, L.T. Signalling by receptor tyrosine kinases. *Annu. Rev. Biochem.* 62:453–481, 1993.

Schlessinger, J.; Ullrich, A. Growth factor signaling by receptor tyrosine kinases. *Neuron* 9:383–391, 1992.

Ullrich, A.; Schlessinger, J. Signal transduction by receptors with tyrosine kinase activity. *Cell* 61:203–212, 1990.

33. Clark, S.G.; Stern, M.J.; Horvitz, H.R. *C. elegans* cell-signalling gene sem-5 encodes a protein with SH2 and SH3 domains. *Nature* 356:340–344, 1992.

Koch, C.A.; Anderson, D.; Moran, M.F.; Ellis, C.; Pawson, T. SH2 and SH3 domains: elements that control interactions of cytoplasmic signaling proteins. *Science* 252:668–674, 1991.

Mayer, B.J.; Baltimore, D. Signalling through SH2 and SH3 domains. *Trends Cell Biol.* 3:8–13, 1993.

Pawson, T.; Schlessinger, J. SH2 and SH3 domains. *Curr. Biol.* 3:434–442, 1993.

34. Bollag, G.; McCormick, F. Regulators and effectors of *ras* proteins. *Annu. Rev. Cell Biol.* 7:601–632, 1991.

Downward, J. Ras regulation: putting back the GTP. *Curr. Biol.* 2:329–331, 1992.

Hall, A. Ras-related proteins. *Curr. Opin. Cell Biol.* 5:265–268, 1993.

Lowy, D.R.; Willumsen, B.M. Function and regulation of *Ras. Annu. Rev. Biochem.* 62:851–891, 1993.

35. Greenwald, I.; Rubin, G.M. Making a difference: the role of cell-cell interactions in establishing separate identities for equivalent cells. *Cell* 68:271–281, 1992.

Olivier, J.P., et al. A *Drosophila* SH2-SH3 adaptor protein implicated in coupling the sevenless tyrosine kinase to an activator of Ras guanine nucleotide exchange, Sos. *Cell* 73:179–191, 1993.

Ready, D.F. A multifaceted approach to neural development. *Trends Neurosci.* 12:102–110, 1989.

Simon, M.A.; Dodson, G.S.; Rubin, G.M. An SH3-SH2-SH3 protein is required for p21^{Ras1} activation and binds to sevenless and Sos proteins *in vitro*. *Cell* 73:169–177, 1993.

Tomlinson, A. Cellular interactions in the developing *Drosophila* eye. *Development* 104:183–193, 1988.

Warne, P.H.; Viciana, P.R.; Downward, J. Direct interaction of Ras and the amino-terminal region of Raf-1 *in vitro*. *Nature* 364:352–355, 1993.

36. Hill, C.S., et al. Functional analysis of a growth factor–responsive transcription factor complex. *Cell* 73:395–406, 1993.

Nishida, E.; Gotoh, Y. The MAP kinase cascade is essential for diverse signal transduction pathways. *Trends Biochem. Sci.* 18:128–130, 1993.

Pelech, S.L. Networking with protein kinases. *Curr. Biol.* 3:513–515, 1993.

Ruderman, J.V. MAP kinase and the activation of quiescent cells. *Curr. Opin. Cell Biol.* 5:207–213, 1993.

Thomas, G. MAP kinase by any other name smells just as sweet. *Cell* 68:3–6, 1992.

37. Argetsinger, L.S., et al. Identification of JAK2 as a growth hormone receptor-associated tyrosine kinase. *Cell* 74:237–244, 1993.

Miyajima, A.; Hara, T.; Kitamura, T. Common subunits of cytokine receptors and the functional redundancy of cytokines. *Trends Biochem. Sci.* 17:378–382, 1992.

Mustelin, T.; Burn, P. Regulation of *src* family tyrosine kinases in lymphocytes. *Trends Biochem. Sci.* 18:215–220, 1993.

Schreurs, J.; Gorman, D.M.; Miyajima, A. Cytokine receptors: a new superfamily of receptors. *Int. Rev. Cytol.* 137B:121–155, 1993.

Stahl, N.; Yancopoulos, G.D. The alphas, betas, and kinases of cytokine receptor complexes. *Cell* 74:587–590, 1993.

38. Charbonneau, H.; Tonks, N.K. 1002 protein phosphatases? *Annu. Rev. Cell Biol.* 8:463–493, 1992.

Koretzky, G.A. Role of the CD45 tyrosine phosphatase in signal transduction in the immune system. *FASEB J.* 7:420–426, 1993.

Walton, K.M.; Dixon, J.E. Protein tyrosine phosphatases. *Annu. Rev. Biochem.* 62:101–120, 1993.

39. Bishop, J.M. Molecular themes in oncogenes. *Cell* 64:235–248, 1991.

Gupta, S.K.; Gallego, C.; Johnson, G.L. Mitogenic pathways regulated by G protein oncogenes. *Mol. Biol. Cell* 3:123–128, 1992.

Kahn, P.; Graf, T., eds. Oncogenes and Growth Control. Berlin: Springer, 1986.

40. Lin, H.Y.; Lodish, H.F. Receptors for the TGF-β superfamily: multiple polypeptides and serine/threonine kinases. *Trends Cell Biol.* 3:14–19, 1993.

Massague, J. The transforming growth factor-β family. *Annu. Rev. Cell Biol.* 6:597–641, 1990.

Massague, J. Receptors for the TGF-β family. *Cell* 69:1067–1070, 1992.

Taylor, S.S., et al. Structural framework for the protein kinase family. *Annu. Rev. Cell Biol.* 8:429–462, 1992.

41. Artavanis-Tsakonas, S.; Delidakis, C.; Fehon, R.G. The *Notch* locus and the cell biology of neuroblast segregation. *Annu. Rev. Cell Biol.* 7:427–452, 1991.

Burridge, K.; Petch, L.A.; Romer, L.H. Signals from focal adhesions. *Curr. Biol.* 2:537–539, 1992.

42. Soderquist, A.M.; Carpenter, G. Biosynthesis and metabolic degradation of receptors for epidermal growth factor. *J. Membr. Biol.* 90:97–105, 1986.

43. Hausdorff, W.P.; Caron, M.G.; Lefkowitz, R.J. Turning off the signal: desensitization of β-adrenergic receptor function. *FASEB J.* 4:2881–2889, 1990.

Lefkowitz, R.J. G-protein-coupled receptor kinases. *Cell* 74:409–412, 1993.

Palczewski, K.; Benovic, J.L. G-protein-coupled receptor kinases. *Trends Biochem. Sci.* 16:387–391, 1991.

44. Cole, G.M.; Reed, S.I. Pheromone-induced phosphorylation of a G protein β subunit in *S. cerevisiae* is associated with an adaptive response to mating pheromone. *Cell* 64:703–716, 1991.

Nestler, E.J. Molecular mechanisms of drug addiction. *J. Neurosci.* 12:2439–2450, 1992.

45. Adler, J. The sensing of chemicals by bacteria. *Sci. Am.* 234(4):40–47, 1976.

Berg, H. How bacteria swim. *Sci. Am.* 233(2):36–44, 1975.

Koshland, D.E., Jr. Biochemistry of sensing and adaptation in a simple bacterial system. *Annu. Rev. Biochem.* 50:765–782, 1981.

46. Bourret, R.B.; Borkovich, K.A.; Simon, M.I. Signal transduction pathways involving protein phosphorylation in prokaryotes. *Annu. Rev. Biochem.* 60:401–441, 1991.

Hazelbauer, G.L. Bacterial chemoreceptors. *Curr. Opin. Struct. Biol.* 2:505–510, 1992.

Parkinson, J.S. Signal transduction schemes of bacteria. *Cell* 73:857–871, 1993.

Stock, J.B.; Lukat, G.S.; Stock, A.M. Bacterial chemotaxis and the molecular logic of intracellular signal transduction networks. *Annu. Rev. Biophys. Biophys. Chem.* 20:109–136, 1991.

Stoddard, B.L.; Bui, J.D.; Koshland, D.E., Jr. Structure and dynamics of transmembrane signaling by the *Escherichia coli* aspartate receptor. *Biochemistry* 31:11978–11983, 1992.

47. Hinton, G.E. How neural networks learn from experience. *Sci. Am.* 267(3):144–151, 1992.

Hopfield, J.J. Neural networks and physical systems with emergent collective computational abilities. *Proc. Natl. Acad. Sci. USA* 79:2554–2558, 1982.

Sejnowski, T.J.; Rosenberg, C.R. Parallel networks that learn to pronounce English text. *Complex Systems* 1:145–168, 1987.

48. Bray, D. Intracellular signalling as a parallel distributed process. *J. Theor. Biol.* 143:215–231, 1990.

Pelech, S.L. Networking with protein kinases. *Curr. Biol.* 3:513–515, 1993.

49. Gatmaitan, Z., et al. Regulation of growth and differentiation of a rat hepatoma cell line by the synergistic interactions of hormones and collagenous substrata. *J. Cell Biol.* 97:1179–1190, 1983.

Nishizuka, Y. Signal transduction crosstalk. *Trends Biochem. Sci.* 17:367–374, 1992.

Rozengurt, E.; Mendoza, S.A. Synergistic signals in mitogenesis: role of ion fluxes, cyclic nucleotides and protein kinase in Swiss 3T3 cells. *J. Cell Sci. Suppl.* 3:229–242, 1985.

References

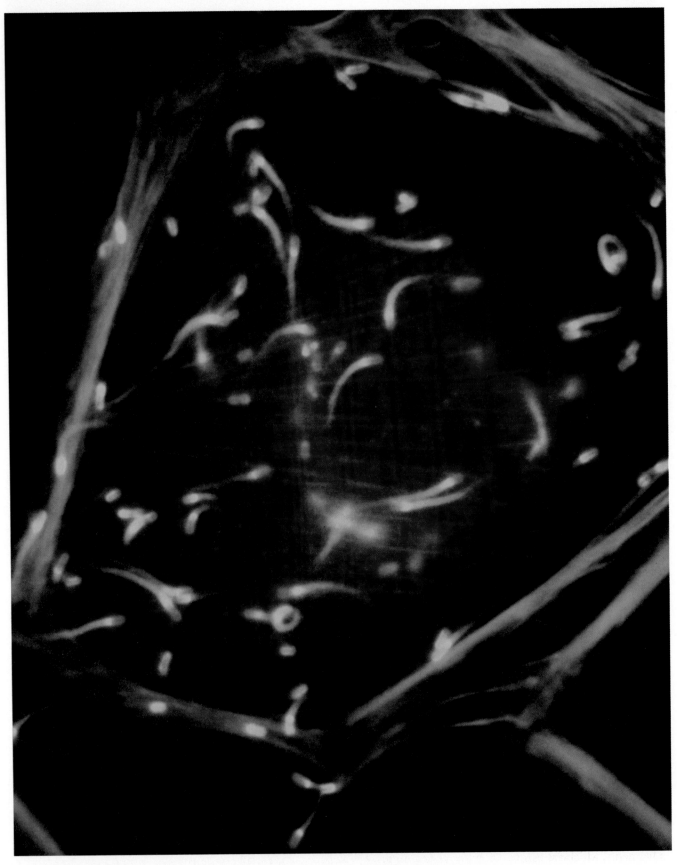

Fluorescence micrograph of the bacterium *Listeria monocytogenes*. The bacteria (*red*) move and spread by inducing the formation of actin filaments (*green*) in the cytosol of the host cells. Regions where the *red* and *green* fluorescence overlap appear *yellow*. (Courtesy of Tim Mitchison and Julie Theriot.)

The Cytoskeleton

The ability of eucaryotic cells to adopt a variety of shapes and to carry out coordinated and directed movements depends on a complex network of protein filaments that extends throughout the cytoplasm (Figure 16–1). This network is called the **cytoskeleton**, although, unlike a skeleton made of bone, it is a highly dynamic structure that reorganizes continuously as the cell changes shape, divides, and responds to its environment. In fact, the cytoskeleton might equally well be called the "cytomusculature" because it is directly responsible for such movements as the crawling of cells on a substratum, muscle contraction, and the many changes in shape of a developing vertebrate embryo; it also provides the machinery for intracellular movements, such as the transport of organelles from one place to another in the cytoplasm and the segregation of chromosomes at mitosis. The cytoskeleton is apparently absent from bacteria, and it may have been a crucial factor in the evolution of eucaryotic cells.

The diverse activities of the cytoskeleton depend on three types of protein filaments—*actin filaments, microtubules,* and *intermediate filaments*. Each type of filament is formed from a different protein subunit: actin for actin filaments, tubulin for microtubules, and a family of related fibrous proteins, such as vimentin or lamin, for intermediate filaments. Actin and tubulin have been especially highly conserved throughout the evolution of eucaryotes; their protein filaments bind a large variety of accessory proteins, which enable the same filament to participate in distinct functions in different regions of a cell. Some of these accessory proteins link filaments to one another or to other cell components, such as the plasma membrane. Others control where and when actin filaments and microtubules are assembled in the cell by regulating the rate and extent of their polymerization. Yet others are motor proteins, which hydrolyze ATP to produce force and directed movement along the filament.

We begin this chapter by introducing the three main types of cytoskeletal filaments and by illustrating some of the general principles by which they function. After this overview we consider each type of filament in turn: first, intermediate filaments, whose ropelike structure seems to have the relatively simple function of providing cells with mechanical strength; second, microtubules, which are thought to be the primary organizers of the cytoskeleton; finally, actin filaments, which are essential for many movements of the cell, especially those of its surface.

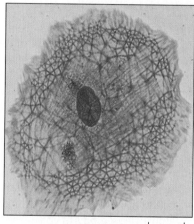

10 μm

Figure 16–1 The cytoskeleton. A cell in culture has been fixed and stained with Coomassie blue, a general stain for proteins. Note the variety of filamentous structures that extend throughout the cell. (Courtesy of Colin Smith.)

The Nature of the Cytoskeleton

A eucaryotic cell contains a billion or so protein molecules, which constitute about 60% of its dry mass. There are thought to be about 10,000 different types of protein in an individual vertebrate cell, and most of them are highly organized spatially. This organization is present at multiple levels. In all cells proteins are arranged into functional complexes, most consisting of perhaps 5 to 10 proteins but others as large or larger than ribosomes. A further level of organization involves the confinement of functionally related proteins within the same membrane or aqueous compartment of a membrane-bounded organelle, such as the nucleus, mitochondria, or Golgi apparatus. An even higher level of organization is created and maintained by the cytoskeleton. It enables the living cell, like a city, to have many specialized services concentrated in different areas but extensively interconnected by paths of communication. In this section we review some of the basic strategies that enable the cytoskeleton to control the spatial location of protein complexes and organelles, as well as to provide communication paths between them.

The Cytoplasm of a Eucaryotic Cell Is Spatially Organized by Actin Filaments, Microtubules, and Intermediate Filaments [1]

How can a eucaryotic cell, with a diameter of 10 μm or more, be spatially organized by cytoskeletal protein molecules that are typically 2000 times smaller in linear dimensions? The answer lies in *polymerization*. For each of the three major types of cytoskeletal protein, thousands of identical protein molecules assemble into linear filaments that can be long enough, if necessary, to stretch from one side of the cell to the other. Such filaments connect protein complexes and organelles in different regions of the cell and serve as tracks for transport between them. In addition, they provide mechanical support, which is especially important for animal cells, since they do not have rigid external walls. The cytoskeleton forms an internal framework for the large volume of cytoplasm, supporting it like a framework of girders supporting a building.

It is easy to see how filaments arose in evolution: any protein with an appropriately oriented pair of complementary self-binding sites on its surface can form a long helical filament (see p. 124). Each of the three principal types of protein filaments that make up the cytoskeleton is a helical polymer that has a different arrangement in the cell and a distinct function (Figure 16–2). By themselves, however, the three types of filaments could provide neither shape nor strength to the cell. Their functions depend on a large retinue of accessory proteins that link the filaments to one another and to other cell components. Accessory proteins are also essential for the controlled assembly of the protein filaments in particular locations, and they provide the motors that either move organelles along the filaments or move the filaments themselves.

Dynamic Microtubules Emanate from the Centrosome [2]

Microtubules are polar structures: one end (the *plus end*) is capable of rapid growth, while the other end (the *minus end*) tends to lose subunits if not stabilized. In most cells, the minus ends of microtubules are stabilized by embedding them in a structure called the **centrosome**, and the rapidly growing ends are then free to add tubulin molecules (Figure 16–3). The centrosome generally lies next to the nucleus, near the center of the cell.

At any one time, several hundred microtubules are growing outward from a centrosome, with some extending for many microns, so that their plus end is at the edge of the cell. Each of these microtubules is a highly dynamic structure that can shorten as well as lengthen: after growing outward for many minutes by add-

ACTIN FILAMENTS

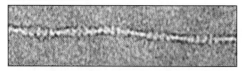

25 nm

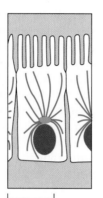

25 μm

Actin filaments (also known as *microfilaments*) are two-stranded helical polymers of the protein actin. They appear as flexible structures, with a diameter of 5–9 nm, that are organized into a variety of linear bundles, two-dimensional networks, and three-dimensional gels. Although actin filaments are dispersed throughout the cell, they are most highly concentrated in the *cortex*, just beneath the plasma membrane.

MICROTUBULES

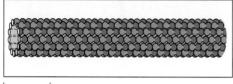

25 nm

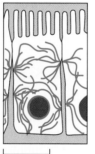

25 μm

Microtubules are long, hollow cylinders made of the protein tubulin. With an outer diameter of 25 nm, they are much more rigid than actin filaments. Microtubules are long and straight and typically have one end attached to a single microtubule organizing center (MTOC) called a *centrosome*, as shown here.

INTERMEDIATE FILAMENTS

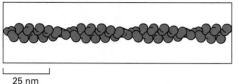

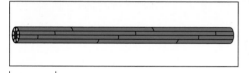

25 nm

25 μm

Intermediate filaments are ropelike fibers with a diameter of around 10 nm; they are made of intermediate filament proteins, which constitute a large and heterogeneous family. One type of intermediate filament forms a meshwork called the nuclear lamina just beneath the inner nuclear membrane. Other types extend across the cytoplasm, giving cells mechanical strength and carrying the mechanical stresses in an epithelial tissue by spanning the cytoplasm from one cell-cell junction to another.

---

ing subunits, its plus end may undergo a sudden transition that causes it to lose subunits, so that the microtubule shrinks rapidly inward and may disappear. The microtubule network that emanates starlike from the centrosome is constantly sending out new microtubules to replace the old ones that have depolymerized (Figure 16–4).

The Microtubule Network Can Find the Center of the Cell [3]

What determines how the cytoplasmic array of microtubules is normally positioned in a cell? Important clues have been provided by experiments on cultured pigment cells isolated from fish scales: large flat cells containing many pigment granules. The granules, which can be dark brown, yellow, red, or iridescent, depending on the species of fish, are attached to microtubules and can either aggregate in the center of the cell or disperse throughout the cytoplasm. The movement of the pigment granules occurs along the microtubules and can be controlled by the fish to change its skin color. In a cultured pigment cell, the

Figure 16–2 The three types of protein filaments that form the cytoskeleton. Each type of filament is shown in an electron micrograph and as a schematic diagram showing how it is built from subunits. The distribution of each filament in one type of epithelial cell is also shown schematically. The colors used here for each type of filament are used in this way throughout the chapter. (Micrographs of actin filaments, microtubules, and intermediate filaments courtesy of Roger Craig, Richard Wade, and Roy Quinlan, respectively.)

The Nature of the Cytoskeleton

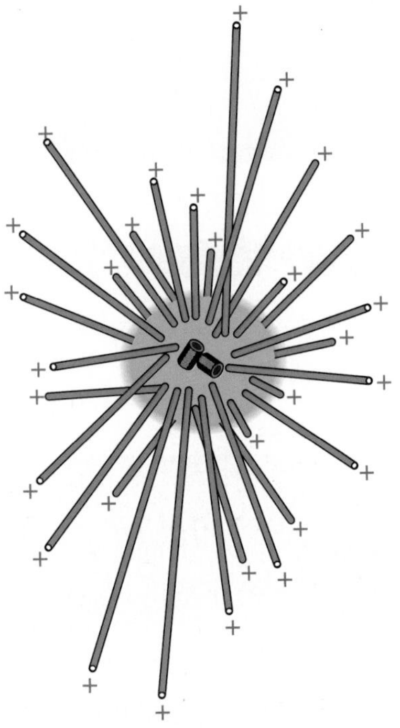

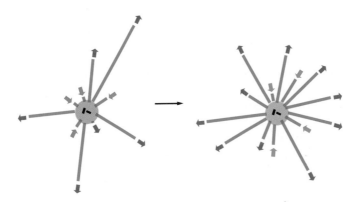

Figure 16–3 **A centrosome with attached microtubules.** As indicated, the slow-growing minus end of each microtubule is embedded in the *centrosome* matrix (*light green*) that surrounds a pair of structures called *centrioles.* By nucleating the growth of new microtubules, this matrix helps to determine the number of microtubules in a cell.

movement can be conveniently controlled by applying hormones or other reagents that change the concentration of cyclic AMP in the cytosol: raising cyclic AMP causes the granules to disperse, whereas lowering it causes the granules to aggregate. The pigment granules therefore provide a useful marker for the arrangement of microtubules in the cell (Figure 16–5).

If one part of a fish pigment cell is cut off with a needle, the cell fragment can survive for long periods even though it lacks a nucleus. The same operation, performed when the pigment granules are dispersed, causes some granules to be trapped in the cell fragment. If the pigment granules in the fragment are induced to aggregate by hormonal treatment immediately after the surgery, they move toward the site of the cut. But if they are induced to aggregate 4 hours after the surgery, they do not move to the cut site but instead move to the exact center of the cell fragment. Further investigation shows that this change results from a major rearrangement of the microtubules within the fragment, so that their minus ends are now at the center of the fragment, just as they were at the center of the intact cell. In effect, the isolated cell fragment has become a minicell with respect to its microtubule organization, the microtubules having reorganized around a new microtubule organizing center (Figure 16–6).

This simple experiment suggests that the cytoplasmic array of microtubules emanating from the centrosome can act as a surveying device that is able to find the center of the cell. This is a useful starting point if the array is to be able to organize the cell interior. But it is only a starting point; as we see later in this introductory section, a cell can position the array by specifically moving its centrosome to a location displaced from the cell center.

Motor Proteins Use the Microtubule Network as a Scaffold to Position Membrane-bounded Organelles [4]

As we have just seen in the case of fish pigment cells, cytoskeletal filaments serve not only as structural supports but also as lines of transport. If a living vertebrate cell is observed in a light microscope, its cytoplasm is seen to be in continual motion. Over the course of minutes, mitochondria and smaller membrane-bounded organelles change their positions by periodic *saltatory movements,* which are much more sustained and directional than the continual small Brownian movements caused by random thermal motions. These and other intracellular movements in eucaryotic cells are generated by **motor proteins,** which bind to either an actin filament or a microtubule and use the energy derived from repeated cycles of ATP hydrolysis to move steadily along it (see p. 208). Dozens of differ-

Figure 16–4 **Growth and shrinkage in a microtubule array.** The array of microtubules anchored in a centrosome is continually changing, as new microtubules grow (*red arrows*) and old microtubules shrink (*blue arrows*).

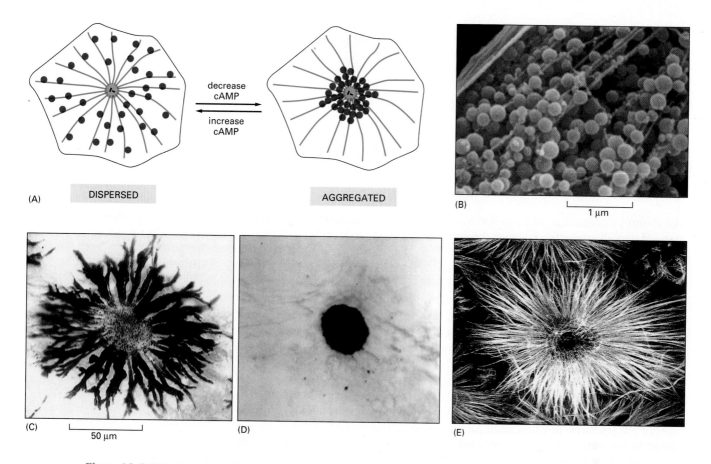

Figure 16–5 **Fish pigment cells.** These giant cells, which are responsible for changes in skin coloration in several species of fish, contain large pigment granules (*brown*), which can change their location in the cell in response to a neuronal or hormonal stimulus. (A) Schematic view of a pigment cell, showing the dispersal and aggregation of pigment granules, which occur along microtubules. (B) Scanning electron micrograph of a pigment cell following a brief exposure to detergent. The plasma membrane and soluble contents of the cytoplasm have been removed, exposing the array of microtubules and associated pigment granules. (C and D) Bright-field images of the same cell in a scale of an African cichlid fish, showing its pigment granules either dispersed throughout the cytoplasm or aggregated in the center of the cell. (E) An immunofluorescence picture of another cell from the same fish stained with antibodies to tubulin, showing large bundles of parallel microtubules extending from the centrosome to the periphery of the cell. (B, from M.A. McNiven and K.R. Porter, *J. Cell Biol.* 103:1547–1555, 1986, by copyright permission of the Rockefeller University Press; C, D, and E, courtesy of Leah Haimo.)

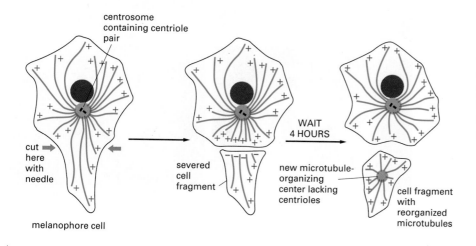

Figure 16–6 **An experiment showing that a microtubule array can find the center of a cell.** After the arm of a fish pigment cell is cut off with a needle, the microtubules in the detached cell fragment reorient with their minus ends near the center of the fragment.

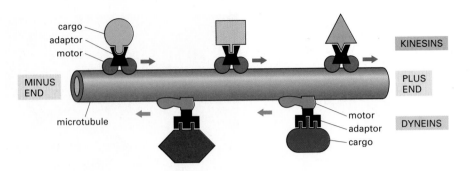

Figure 16–7 The motor proteins that move along microtubules. Kinesins move toward the plus end of a microtubule, whereas dyneins move toward the minus end. As indicated, both types of microtubule motor proteins exist in many forms, each of which is thought to transport a different cargo.

ent motor proteins have now been identified. They differ in the type of filament they bind to, the direction in which they move along the filament, and the "cargo" they carry.

The first motor protein to be discovered was *myosin,* a protein that moves along actin filaments and is especially abundant in skeletal muscle, where it forms a major part of the contractile apparatus. Other types of myosins were subsequently found in nonmuscle cells. All myosins have similar motor domains (the part of the protein that generates movement), but they differ markedly in the domains that are responsible for attaching the myosin molecule to other components of the cell.

The motor proteins that move along microtubules are distinct from the myosins and belong to one of two families: the *kinesins,* which generally move toward the plus end of a microtubule (away from the centrosome), and the *dyneins,* which move toward the minus end (toward the centrosome). As with the myosins, each type of microtubule-dependent motor protein carries a distinct cargo with it as it moves (Figure 16–7).

Microtubule-dependent motor proteins play an important part in positioning membrane-bounded organelles within a eucaryotic cell. The membrane tubules of the endoplasmic reticulum (ER), for example, align with microtubules and extend almost to the edge of the cell, whereas the Golgi apparatus is located near the centrosome. When cells are treated with a drug that depolymerizes microtubules, both of these organelles change their location: the ER collapses to the

Figure 16–8 The placement of organelles by microtubules. (A) Schematic diagram of a cell showing the typical arrangement of microtubules (*green*), endoplasmic reticulum (*blue*), and Golgi apparatus (*yellow*). The nucleus is shown in *brown* and the centrosome in *light green.* (B) Cell stained with antibodies to endoplasmic reticulum (*upper panel*) or to microtubules (*lower panel*). Motor proteins pull the endoplasmic reticulum along microtubules, stretching it like a net from its attachments to the nuclear envelope. (C) Cell stained with antibodies to the Golgi apparatus (*upper panel*) or to microtubules (*lower panel*). In this case motor proteins move the Golgi apparatus inward to its position near the centrosome. (B, courtesy of Mark Terasaki and Lan Bo Chen; C, courtesy of Viki Allan and Thomas Kreis.)

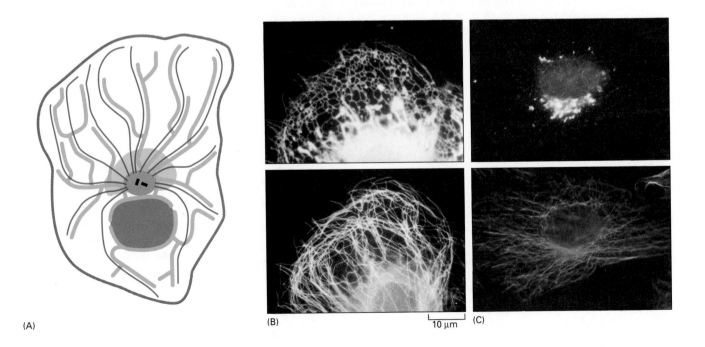

(A) (B) 10 μm (C)

center of the cell, while the Golgi apparatus fragments into small vesicles that disperse throughout the cytoplasm. When the drug is removed, the organelles return to their original positions, dragged by motor proteins moving along the re-formed microtubules. Thus the normal position of each of these organelles is thought to be determined by a receptor protein on the cytosolic surface of its membrane that binds a specific microtubule-dependent motor—a kinesin for the ER and a dynein for the Golgi apparatus (Figure 16–8).

The Actin Cortex Can Generate and Maintain Cell Polarity [5]

In general, microtubules in the cytoplasm function as individuals, whereas actin filaments work in networks or bundles. Actin filaments lying just beneath the plasma membrane, for example, are cross-linked into a network by various actin-binding proteins to form the **cell cortex**. As we discuss later, the network is highly dynamic and functions with various myosins to control cell-surface movements. The location and orientation of the cortical actin filaments are controlled by nucleation sites in the plasma membrane, and different regions of the membrane direct the formation of distinct actin-filament-based structures.

Localized extracellular signals that impinge on a portion of the cell surface can induce a local restructuring of the actin cortex beneath the corresponding part of the plasma membrane. In a reciprocal way the organization of the actin cortex can have a major influence on the behavior of the overlying plasma membrane. Mechanisms based on cortical actin filaments, for example, can push the plasma membrane outward to form long, thin *microspikes* or sheetlike *lamellipodia,* or they can pull the plasma membrane inward to divide the cell in two (Figure 16–9).

In extreme cases the actin cortex can integrate movements of an animal cell over its entire surface and maintain cell polarity independently of the microtubule array. This is illustrated by experiments on a nonpigmented type of cell isolated from fish scales. These epidermal cells, known as *keratocytes,* migrate unusually rapidly in culture, traveling at speeds of 30 μm/minute or more. Immunostaining with antibodies indicates that intermediate filaments and microtubules are present only in the trailing region around the cell nucleus, whereas the flattened leading edge of the cells is rich in actin filaments (Figure 16–10). Furthermore, cells that are treated with a drug that depolymerizes microtubules migrate just as rapidly as untreated cells, whereas the migration is immediately halted by agents that interfere with actin filaments. Evidently, actin filaments (acting with other proteins) are able to move a keratocyte over a surface and also maintain this cell's distinctive shape and polarity; the details of the mechanism involved, however, are unclear.

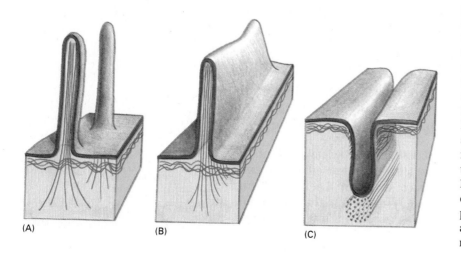

(A) (B) (C)

Figure 16–9 Actin filaments often shape the plasma membrane of animal cells. Three examples of plasma membrane changes caused by the cortical network of actin filaments. (A) Thin, spiky protrusions such as microspikes form on the surface of cells by the assembly of supporting bundles of actin filaments anchored in the cell cortex. (B) Sheetlike extensions, called lamellipodia, also form on the surface, in this case supported by a flattened web of actin filaments rather than discrete bundles. (C) Invaginations of the cell surface, as occur during cell division, are produced by a contractile bundle of actin filaments associated with the motor protein myosin.

The Nature of the Cytoskeleton

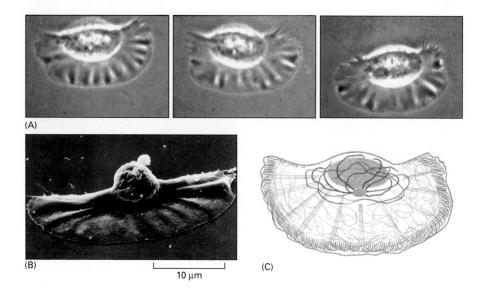

(A)

(B)   └─────────┘   (C)
                        10 μm

Figure 16–10 **Migratory cells from fish epidermis.** (A) Light micrographs of a keratocyte in culture taken at 15-second intervals. The cell shown is migrating at about 15 μm/second. (B) Keratocyte seen by scanning electron microscopy, showing its highly flattened leading edge, with the body of the cell, containing the nucleus, trailing at the rear. (C) Distribution of cytoskeletal filaments in this unusual type of cell. Actin filaments (*red*) fill the flattened leading margin of the cell and are responsible for its migration. Microtubules (*green*) and intermediate filaments (*blue*) are restricted to the region close to the nucleus. (Micrographs courtesy of Juliet Lee.)

Actin Filaments and Microtubules Usually Act Together to Polarize the Cell [6]

In a living cell the three major types of cytoskeletal filaments are connected to one another and their functions are coordinated. The distribution of intermediate filaments in an epithelial cell in culture, for example, is radically altered if the microtubules are depolymerized by drug treatment: the intermediate filaments, which are normally arrayed throughout the cytoplasm, pull back to a region close to the nucleus. There are also many situations in which microtubules and actin filaments act in a coordinated way to polarize the whole cell. We discuss just one example: the killing of specific target cells by cytotoxic T lymphocytes.

Cytotoxic T cells kill other cells that carry foreign antigens on their surface. This is an important part of a vertebrate's immune response to infection, as discussed in Chapter 23. When receptors on the surface of the T cell recognize antigen on the surface of a target cell, the receptors signal to the underlying cortex of the T cell, altering the cytoskeleton in several ways. First, proteins associated with actin filaments in the T cell reorganize under the zone of contact between the two cells. The centrosome then reorients, moving with its microtubules to the zone of T-cell-target contact (Figure 16–11A). The microtubules, in turn, position the Golgi apparatus right under the contact zone, focusing the killing machinery—which is associated with secretion from the Golgi—on the target cell.

In this example, as in many others, a cell becomes polarized in the following general way. First, the plasma membrane senses some difference on one side of the cell that generates a transmembrane signal. The actin cortex is then reorganized in a local area beneath the affected membrane, which in turn moves the centrosome to that part of the cell, presumably by pulling on its microtubules. The centrosome in turn positions the internal membrane systems in a polarized way. The net result is a cell with a strong directional focus (Figure 16–11B).

The Functions of the Cytoskeleton Are Difficult to Study

Although the main subunits of the three classes of cytoskeletal polymers, as well as many of the hundreds of accessory proteins that associate with them, have been isolated and their amino acid sequences determined, it has been frustratingly difficult to establish how these proteins function in the cell. Besides the complexity that stems from the large number of proteins involved, two general

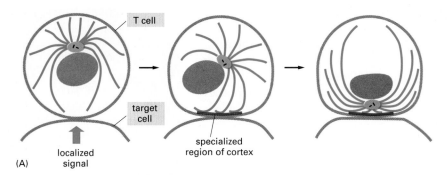

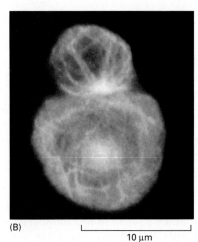

(A)   localized
      signal

(B) |———————————|
         10 µm

T cell

target
cell

specialized
region of cortex

Figure 16–11 The polarization of a cytotoxic T cell after target-cell recognition. (A) Changes in the cytoskeleton of a cytotoxic T cell after it makes contact with a target cell. (B) Immunofluorescence micrograph in which both the T cell (*top*) and its target cell (*bottom*) have been stained with an antibody against microtubules. The centrosome and the microtubules radiating from it in the T cell are oriented toward the point of cell-cell contact. In contrast, the microtubule array in the target cell is not polarized. (B, reproduced from B. Geiger, D. Rosen, and G. Berke, *J. Cell Biol.* 95:137–143, 1982, by copyright permission of the Rockefeller University Press.)

features make the cytoskeleton especially difficult to understand. First, the function of the cytoskeleton depends on complex assemblies of proteins, which bind in cooperative groups to the cytoskeletal filaments. It is relatively straightforward to examine the effect on a filament of a single accessory protein but very much more difficult to analyze the effects of a mixture of many different proteins. This problem is not unique to the cytoskeleton, but it is especially acute here. Secondly, the functions of the cytoskeleton are much more difficult to analyze than the functions of many other large protein complexes. The processes of RNA and DNA synthesis, for example, which involve the formation of new polymers held together by covalent bonds, can be readily analyzed *in vitro*, in part because the products of the *in vitro* reactions can easily be measured and compared with the corresponding products made in a cell. The cytoskeleton, in contrast, exerts forces and generates movements without any major chemical change. This makes it especially difficult to assay the function of a cytoskeletal system that has been reconstituted *in vitro* from purified components.

Summary

The cytoplasm of eucaryotic cells is spatially organized by a network of protein filaments known as the cytoskeleton. This network contains three principal types of filaments: microtubules, actin filaments, and intermediate filaments. Microtubules are stiff structures that usually have one end anchored in the centrosome and the other free in the cytoplasm. In many cells microtubules are highly dynamic structures that alternately grow and shrink by the addition and loss of tubulin subunits. Motor proteins move in one direction or the other along microtubules, carrying specific membrane-bounded organelles to desired locations in the cell. Actin filaments are also dynamic structures, but they normally exist in bundles or networks rather than as single filaments. A layer called the cortex is formed just beneath the plasma membrane from actin filaments and a variety of actin-binding proteins. This actin-rich layer controls the shape and surface movements of most animal cells. Intermediate filaments are relatively tough, ropelike structures that provide mechanical stability to cells and tissues. The three types of filaments are connected to one another, and their functions are coordinated.

Intermediate Filaments [7]

Intermediate filaments are tough and durable protein fibers found in the cytoplasm of most, but not all, animal cells. They are called "intermediate" because in electron micrographs their apparent diameter (8–10 nm) is between that of the thin actin filaments and the thick myosin filaments of muscle cells, where they were first described (they are also intermediate in diameter between actin filaments and microtubules). In most animal cells an extensive network of intermediate filaments surrounds the nucleus and extends out to the cell periphery, where they interact with the plasma membrane (Figure 16–12). In addition, a tightly woven basketwork of intermediate filaments—the *nuclear lamina*—underlies the nuclear envelope.

Intermediate filaments are particularly prominent in the cytoplasm of cells that are subject to mechanical stress. They are present in large numbers, for example, in epithelia, where they are linked from cell to cell at specialized junctions, along the length of nerve cell axons, and in all kinds of muscle cells. When cells are treated with concentrated salt solutions and nonionic detergents, the intermediate filaments remain behind while most of the rest of the cytoskeleton is lost. In fact, the term "cytoskeleton" was originally coined to describe this unusually stable and insoluble fiber system.

Intermediate Filaments Are Polymers of Fibrous Proteins [8]

Unlike actin and tubulin, which are globular proteins, the many types of intermediate filament protein monomers are all highly elongated fibrous molecules that have an amino-terminal *head*, a carboxyl-terminal *tail*, and a central *rod domain* (Figure 16–13). The central rod domain consists of an extended α-helical region containing long tandem repeats of a distinctive amino acid sequence motif called the *heptad repeat*. As discussed in Chapter 3, this seven amino acid motif promotes the formation of coiled-coil dimers between two parallel α helices (see Figure 3–48). Long stretches of heptad repeats are also found in many other elongated cytoskeletal proteins with coiled-coil dimeric structures, including tropomyosin and the tail of myosin, which we discuss later.

In the next stage of assembly, two of the coiled-coil dimers associate in an antiparallel manner to form a tetrameric subunit (Figure 16–14). Soluble tetramers are found in small amounts in cells, suggesting that they are the fundamental subunit from which intermediate filaments assemble. The antiparallel arrangement of dimers implies that the tetramer, and hence the intermediate filament that it forms, is a nonpolarized structure—that is, it is the same at both ends and symmetrical along its length. This distinguishes intermediate filaments from microtubules and actin filaments, which are polarized and whose functions depend on this polarity. The final stages of intermediate filament assembly are less well characterized, but it seems that tetramers add to an elongating intermedi-

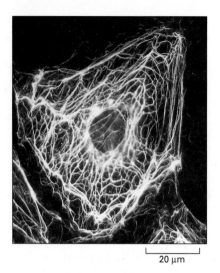

20 μm

Figure 16–12 The intermediate filaments in the cytoplasm of a tissue culture cell. Rat kangaroo epithelial cells (Ptk2 cells) in interphase were labeled with antibodies to one class of intermediate filaments (called keratin filaments) and examined by fluorescence microscopy. (Courtesy of Mary Osborn.)

Figure 16–13 The domain organization of intermediate filament protein monomers. Most intermediate filament proteins share a similar rod domain that is usually about 310 amino acids long and forms an extended α helix. The amino-terminal and carboxyl-terminal domains are non-α-helical and vary greatly in size and sequence in different intermediate filaments.

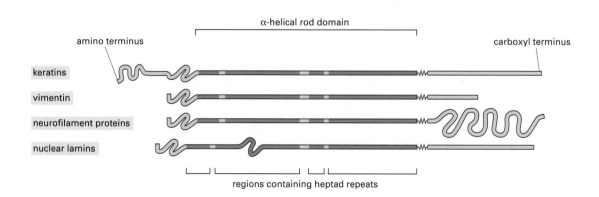

ate filament in a simple binding reaction in which they align along the axis of the filament and pack together in a helical pattern (see Figure 16–14).

The central rod domain, which is structurally similar in all intermediate filament proteins, mediates the lateral interactions that form the assembled filament. The globular head and tail domains, by contrast, can vary greatly in both size and amino acid sequence without affecting the basic axial structure of the filament; they often project from the surface of the filament and mediate its interactions with other components. This structural design means that intermediate filaments can be made from proteins of a surprisingly wide range of sizes (from about 40,000 to about 200,000 daltons).

In most cells, almost all intermediate filament protein molecules are in the fully polymerized state, with very little free tetramer. Nonetheless, a cell can regulate the assembly of its intermediate filaments and determine their number, length, and position. One mechanism of control involves the phosphorylation of specific serine residues in the amino-terminal head domain of intermediate filament proteins. In the most dramatic example, phosphorylation of the protein subunits that form the nuclear lamina causes them to disassemble completely at mitosis; when mitosis finishes, the specific serines are dephosphorylated and the nuclear lamina re-forms (see Figure 12–18). Cytoplasmic intermediate fila-

Figure 16–14 A current model of intermediate filament construction. The monomer shown in (A) pairs with an identical monomer to form a dimer (B) in which the conserved central rod domains are aligned in parallel and wound together into a coiled-coil. Two dimers then line up side by side to form an antiparallel tetramer of four polypeptide chains (C). Within each tetramer the dimers are staggered with respect to one another, thereby allowing it to associate with another tetramer, as shown in (D). In the final 10-nm intermediate filament, tetramers are packed together in a ropelike array (E). An electron micrograph of the final filament is shown upper left. (Diagram based on data from Murray Stewart; micrograph courtesy of Roy Quinlan.)

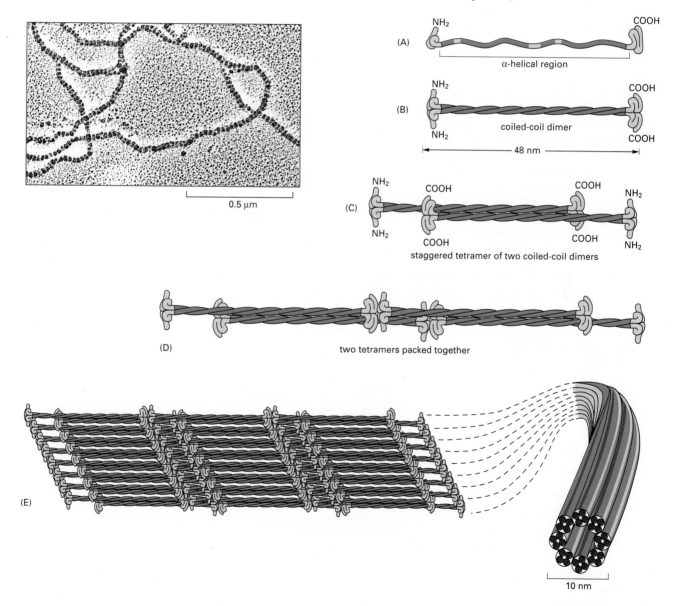

ments can also undergo a radical reorganization during mitosis, as well as in response to some extracellular signals. Although these changes are usually accompanied by an increase in subunit phosphorylation, other factors may also help mediate them.

Epithelial Cells Contain a Highly Diverse Family of Keratin Filaments [9]

The cytoplasmic intermediate filaments in vertebrate cells can be grouped into three classes: (1) *keratin filaments,* (2) *vimentin* and *vimentin-related filaments,* and (3) *neurofilaments,* each formed by polymerization of their corresponding subunit proteins (Table 16–1). By far the most diverse family of these subunits is the **keratins** (also called *cytokeratins*), which form **keratin filaments,** primarily in epithelial cells. There are over 20 distinct keratins in human epithelia. At least 8 more keratins, called *hard keratins,* are specific to hair and nails. (The keratins of epithelial cells, hair, and nails are sometimes referred to as *α-keratins* to distinguish them from the evolutionarily distinct β-keratins found in bird feathers, which have an entirely different structure and are not discussed in this chapter.)

Based on their amino acid sequence, the keratins can be subdivided into two types: the *type I (acidic) keratins* and the *type II (neutral/basic) keratins.* In reassembly experiments it is found that heterodimers of type I and type II keratins can form intermediate filaments but homodimers cannot, which explains why keratin filaments are always heteropolymers formed from equal numbers of type I and type II keratin polypeptides.

A single epithelial cell can make a variety of keratins, all of which copolymerize into a single keratin filament system. The simplest epithelia, such as those found in early embryos and in some adult tissues such as the liver, contain only a single type I and a single type II keratin. Epithelia in other locations, such as the tongue, bladder, and sweat glands, contain six or more keratins—the particular blend depending on the cell's location in the organ. The diversity is most pronounced in skin, where distinct sets of keratins are expressed by the cells in the different layers of the epidermis (see Figure 22–19). There are also keratins characteristic of actively proliferating epithelial cells. This heterogeneity of keratins is clinically useful: in the diagnosis of epithelial cancers (*carcinomas*), the particular set of keratins expressed can be used to determine the epithelial

Table 16–1 Major Types of Intermediate Filament Proteins in Vertebrate Cells

Type of IF	Component Polypeptides (mass in daltons)	Cellular Location
Nuclear lamins	lamins A, B, and C (65,000–75,000)	nuclear lamina of eucaryotic cells
Vimentinlike proteins	vimentin (54,000)	many cells of mesenchymal origin, often expressed transiently during development
	desmin (53,000)	muscle
	glial fibrillary acidic protein (50,000)	glial cells (astrocytes and Schwann cells)
	peripherin (66,000)	neurons
Keratins	type I (acidic) (40,000–70,000) type II (neutral/basic) (40,000–70,000)	epithelial cells and their derivatives (e.g., hair and nails)
Neuronal intermediate filaments	neurofilament proteins NF-L, NF-M, and NF-H (60,000–130,000)	neurons

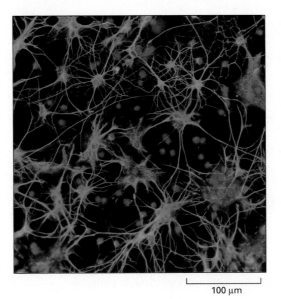

Figure 16–15 **An immuno-fluorescence micrograph of glial filaments in cultured astrocytes.** The bundles of intermediate filaments (*green*) are stained with antibodies to glial fibrillary acidic protein. Nuclei are stained with a *blue* DNA-binding dye. (Courtesy of Nancy L. Kedersha.)

100 μm

tissue in which the tumor originated and thus help to decide the type of treatment that is likely to be most effective.

Many Nonepithelial Cells Contain Their Own Distinctive Cytoplasmic Intermediate Filaments [10]

Unlike keratins, *vimentin* and the *vimentin-related proteins* can form intermediate filaments that are polymers of a single protein species. *Vimentin* itself is the most widely distributed of the cytoplasmic intermediate filament proteins, being present in many cells of mesodermal origin, including fibroblasts, endothelial cells, and white blood cells; in addition, many cells express it transiently during development. *Desmin* is found mainly in muscle cells: it is distributed throughout the cytoplasm of smooth muscle cells, and it links together adjacent myofibrils (ordered bundles of filamentous actin and myosin, discussed later) in skeletal and heart muscle cells. *Glial fibrillary acidic protein* forms *glial filaments* in astrocytes in the central nervous system and in some Schwann cells in peripheral nerves (Figure 16–15). All of these proteins co-polymerize readily with one another, and co-polymers of vimentin and a vimentin-related protein are found in a number of adult cell types. By contrast, none of these proteins co-polymerize with keratins: when keratins and vimentin-related proteins are expressed in the same cell, they form separate filament systems.

Nerve cells contain a variety of unique intermediate filaments, which are expressed in different regions of the nervous system or at specific stages of development. By far the most abundant are the *neurofilaments*, which extend along the length of an axon and form its primary cytoskeletal component, especially in mature nerve cells. In mammals, three *neurofilament proteins* have long been recognized: termed *NF-L, NF-M,* and *NF-H,* for low, middle, and high molecular weight, respectively, all three are usually found in each neurofilament. NF-M and NF-H have especially long carboxyl-terminal tails, which are thought to project from the neurofilament axis and contribute to the regular side-to-side spacing of neuro-filaments in an axon (Figure 16–16).

If a cell in culture is stained with an antibody to a cytoplasmic intermediate filament protein, a delicate network of threadlike filaments is usually seen surrounding the nucleus and extending through the cytoplasm to the plasma membrane (see Figure 16–12). In epithelial cells, keratin filaments are attached to specialized cell junctions—both to *desmosomes,* which bond neighboring cells together, and to *hemidesmosomes,* which anchor cells to the underlying basal lamina (discussed in Chapter 19). Because the keratin filaments in each cell are connected via desmosomes to those of its neighbors, they form a continuous

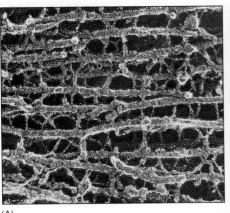
(A)

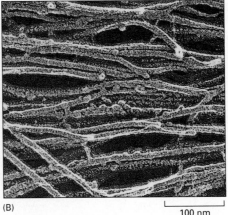
(B)
⊢——————⊣
100 nm

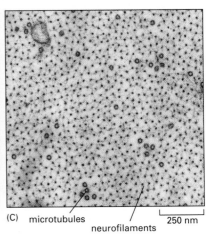
(C)   microtubules
      ‖   ‖   ⊢————⊣
           neurofilaments   250 nm

network throughout the entire epithelium (Figure 16–17). Similarly, desmin filaments are often anchored to specialized cell junctions in muscle cells.

The Nuclear Lamina Is Constructed from a Special Class of Intermediate Filament Proteins—the Lamins [11]

The **nuclear lamina** is a meshwork of intermediate filaments that lines the inside surface of the inner nuclear membrane in eucaryotic cells (Figure 16–18). It is typically 10–20 nm thick and is interrupted in the region of nuclear pores to provide a passageway for macromolecules entering and leaving the nucleus. In mammalian cells the nuclear lamina is composed of **lamins,** which are homologous to other intermediate filament proteins but differ from them in at least four ways: (1) Their central rod domain is somewhat longer (see Figure 16–13). (2) They contain a nuclear transport signal that directs them from the cytosol, where they are made, into the nucleus. (3) They assemble into a two-dimensional, sheetlike lattice, which is thought to require their association with other proteins. (4) The meshwork they form is unusually dynamic and rapidly disassembles at the start of mitosis and reassembles at the end of mitosis; as already mentioned, the disassembly and reassembly are mediated by the phosphorylation and dephosphorylation of several serine residues on the lamins.

Unlike microtubules and actin filaments, which are a defining characteristic of eucaryotic cells, cytoplasmic intermediate filaments have been described

Figure 16–16 Electron micrographs of two types of intermediate filaments in cells of the nervous system. (A) Freeze-etch image of neurofilaments in a nerve cell axon, showing the extensive cross-linking through protein cross-bridges—an arrangement believed to provide great tensile strength in this long cell process. The cross-links are formed by the long, nonhelical extensions at the carboxyl terminus of the largest neurofilament protein. (B) Freeze-etch image of glial filaments in glial cells illustrating that these filaments are smooth and have few cross-bridges. (C) Conventional electron micrograph of a cross-section of an axon showing the regular side-to-side spacing of the neurofilaments, which greatly outnumber the microtubules. (A and B, courtesy of Nobutaka Hirokawa; C, courtesy of John Hopkins.)

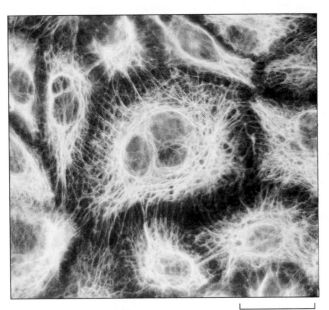
⊢————⊣
20 μm

Figure 16–17 Keratin filaments join cells together in cell sheets. Immunofluorescence micrograph of the network of keratin filaments in a sheet of epithelial cells in culture. The filaments in each cell are indirectly connected to those of its neighbors by desmosomes. (Courtesy of Michael Klymkowsky.)

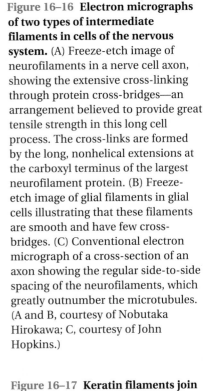

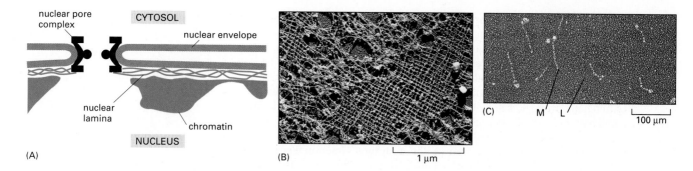

(A) diagram labels: nuclear pore complex, CYTOSOL, nuclear envelope, nuclear lamina, chromatin, NUCLEUS

(B) 1 μm

(C) M   L   100 μm

Figure 16–18 The nuclear lamina. (A) Schematic drawing showing the nuclear lamina in cross-section in the region of a nuclear pore. The lamina is associated with both the chromatin and the inner nuclear membrane. (B) Electron micrograph of a portion of the nuclear lamina in a frog oocyte prepared by freeze-drying and metal shadowing. The lamina is formed from a square lattice of intermediate filaments composed of nuclear lamins (not always as highly organized as that shown here). (C) Electron micrograph of metal-shadowed isolated lamin dimers (marked L). They have an overall form similar to muscle myosin (marked M), with a rodlike tail and two globular heads, but they are much smaller molecules. The globular heads are formed from the two large carboxyl-terminal domains. (B and C, courtesy of Ueli Aebi.)

only in multicellular animals, and even in these organisms they are not required in every cell type. The specialized glial cells that make myelin in the vertebrate central nervous system, for example, do not contain intermediate filaments. Moreover, intermediate filaments can be disrupted in muscle cells, fibroblasts, and epithelial cells in culture without detectable effects on cell behavior.

It seems likely that the first type of intermediate filament protein to appear in evolution was a nuclear lamin and that the various kinds of cytoplasmic intermediate filaments are later adaptations of this primitive form. The intermediate filament proteins in invertebrates, for example, more closely resemble lamins than vertebrate cytoplasmic intermediate filament proteins.

Intermediate Filaments Provide Mechanical Stability to Animal Cells [12]

There is increasing evidence that a major function of cytoplasmic intermediate filaments is to resist mechanical stress. In the human genetic disease *epidermolysis bullosa simplex,* mutations in keratin genes that are normally expressed in the basal cell layer of the epidermis disrupt the keratin filament network in these cells, making them very sensitive to mechanical injury: a gentle squeeze can cause the mutant basal cells to rupture, and the skin in affected individuals is blistered. A similar condition can be produced in transgenic mice

Figure 16–19 Blistering of the skin caused by a mutant keratin gene. A mutant gene encoding a truncated keratin protein (lacking both the amino- and carboxyl-terminal domains) was expressed in a transgenic mouse. The defective protein assembles with the normal keratins and thereby disrupts the keratin filament network in the basal cells. Light micrographs of normal (A) and mutant (B) skin show that the blistering results from the rupturing of cells in the basal layer of the mutant epidermis. The sketch in (C) of three cells observed by electron microscopy in the basal layer of the mutant epidermis shows that the cells rupture between the nucleus and the hemidesmosomes, which connect the keratin filaments to the underlying basal lamina. (From P.A. Coulombe, M.E. Hutton, R. Vassar, and E. Fuchs, *J. Cell Biol.* 115:1661–1674, 1991, by copyright permission of the Rockefeller University Press.)

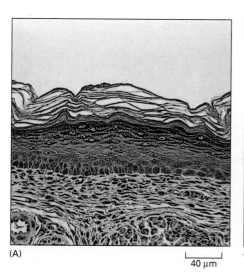

(A)   40 μm

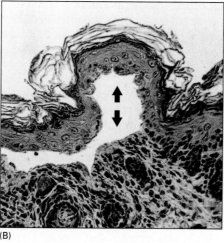

(B)

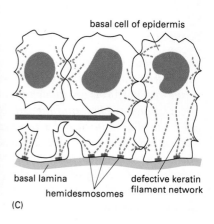

(C) labels: basal cell of epidermis, basal lamina, hemidesmosomes, defective keratin filament network

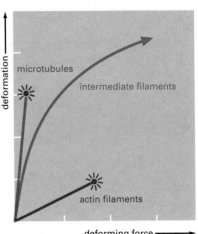

Figure 16–20 Mechanical properties of actin, tubulin, and vimentin polymers. Networks composed of either microtubules or actin filaments or vimentin filaments, all at equal concentration, were exposed to a shear force in a viscometer and the resulting degree of stretch measured. The results show that microtubule networks are easily deformed but that they rupture (indicated by *red starburst*) and begin to flow without limit when stretched beyond 50% of their original length. Actin filament networks are much more rigid, but they also rupture easily. Vimentin networks, by contrast, are easily deformed, but unlike microtubule networks, they withstand large stresses and strains without rupture. Vimentin filaments are therefore well suited to maintain cell integrity. (Adapted from P. Jamney et al., *J. Cell Biol.* 113:155–160, 1991.)

that express mutant keratins of this type (Figure 16–19). In both humans and mice the epidermis can be so weakened that individuals carrying the mutation can die from mechanical trauma. Cytoplasmic intermediate filaments are thought to strengthen nonepithelial cells in a similar way.

The structure of intermediate filaments is ideally suited for such a mechanical function. Because the fibrous subunits associate side by side in overlapping arrays, the filaments can withstand very much larger stretching forces than microtubules or actin filaments (Figure 16–20). In the skin, keratin filaments in the outermost layers of the epidermis become covalently cross-linked to one another and to associated proteins, and as the cells die, the cross-linked keratins persist as a major part of the protective outer layer of the animal. Specialized epithelial cells at particular locations in the skin provide regional variation by generating surface appendages rich in keratin, such as hairs and nails.

But if intermediate filaments function simply to provide tensile strength to cells and tissues, why are there so many different types? And what is the function of the head and tail domains of the proteins, which show such large variations in sequence? Detailed answers to these questions cannot be given at present, but it is clear that the way that intermediate filaments are linked to other cellular components varies greatly among cell types. The desmin filaments that tie the edges of the myofibrils together in skeletal muscle cells are likely to have binding sites for specific myofibril-associated proteins. Neurofilaments in axons are linked side by side by their carboxyl-terminal tail domains to provide a continuous rope of filaments that can be a meter or more in length. Some keratins are specialized to form the tough, protective outer layer of the skin, while others specifically strengthen epithelia undergoing shape changes during morphogenesis. These different functional requirements must be accommodated by the variable regions of the different intermediate filament proteins, which project from the surface of the intermediate filaments and determine their ability to associate with one another and with other components in the cell. In a sense, therefore, the variable regions of intermediate filament proteins serve functions similar to those of the accessory proteins of actin filaments and microtubules. The difference is that the variable regions are an integral part of the intermediate filament subunit, rather than being a separate protein.

Summary

Intermediate filaments are strong, ropelike polymers of fibrous polypeptides that resist stretch and play a structural or tension-bearing role in the cell. A variety of tissue-specific forms are known that differ in the type of polypeptide they contain: these include the keratin filaments of epithelial cells, the neurofilaments of nerve cells, the glial filaments of astrocytes and Schwann cells, the desmin filaments of muscle cells, and the vimentin filaments of fibroblasts and many other cell types. Nuclear lamins, which form the fibrous lamina that underlies the nuclear envelope, are a separate family of intermediate filament proteins.

The monomers of the different types of intermediate filaments differ in amino acid sequence and have very different molecular weights. But they all contain a homologous central rod domain that forms an extended coiled-coil structure when the protein dimerizes. Two coiled-coil dimers associate with each other to form a symmetrical tetramer, which in turn assembles in large overlapping arrays to form the nonpolarized intermediate filament. The rod domains of the subunits form the structural core of the intermediate filament, whereas the domains at either end can project outward. One function of the variable terminal domains may be to allow each type of filament to associate with specific other components in the cell, so as to position the filaments appropriately for a particular cell type.

Microtubules [13]

Microtubules, as we have seen, are long, stiff polymers that extend throughout the cytoplasm and govern the location of membrane-bounded organelles and other cell components. In this section we discuss the assembly of these remarkable structures from tubulin molecules and explain how their polymerization and depolymerization are controlled by the nucleotide GTP. We then examine some ways in which selected microtubules are stabilized in the cell by their association with specific accessory proteins. Finally, we discuss the importance of microtubule-dependent motors that transport membrane vesicles and various protein complexes along microtubules.

Microtubules Are Hollow Tubes Formed from Tubulin [14]

Microtubules are formed from molecules of **tubulin**, each of which is a heterodimer consisting of two closely related and tightly linked globular polypeptides called *α-tubulin* and *β-tubulin*. Although tubulin is present in virtually all eukaryotic cells, the most abundant source for biochemical studies is the vertebrate brain. Extraction procedures yield 10 to 20% of the total soluble protein in brain as tubulin, reflecting the unusually high density of microtubules in the elongated processes of nerve cells.

Tubulin molecules themselves are diverse. In mammals there are at least six forms of α-tubulin and a similar number of forms of β-tubulin, each encoded by a different gene. The different forms of tubulin are very similar, and they will generally co-polymerize into mixed microtubules in the test tube, although they can have distinct locations in the cell and perform subtly different functions. The microtubules in six specialized touch-sensitive neurons in the nematode *Caenorhabditis elegans*, for example, contain a specific form of β-tubulin, and mutations in the gene for this protein result in the specific loss of touch-sensitivity with no apparent defect in other cell functions.

A microtubule can be regarded as a cylindrical structure in which the tubulin heterodimers are packed around a central core, which appears empty in electron micrographs. More accurately, perhaps, one can view the structure as being built from 13 linear protofilaments, each composed of alternating α- and β-tubulin subunits and bundled in parallel to form a cylinder (Figure 16–21). Since the 13

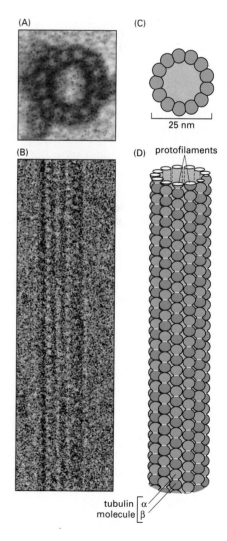

Figure 16–21 Microtubules. (A) Electron micrograph of a microtubule seen in cross-section, with its ring of 13 distinct subunits, each of which corresponds to a separate tubulin molecule (an α/β heterodimer). (B) Cryoelectron micrograph of a microtubule assembled *in vitro*. (C and D) Schematic diagrams of a microtubule, showing how the tubulin molecules pack together to form the cylindrical wall. (C) The 13 molecules in cross-section. (D) A side view of a short section of a microtubule, with the tubulin molecules aligned into long parallel rows, or *protofilaments*. Each of the 13 protofilaments is composed of a series of tubulin molecules, each an α/β heterodimer. Note that a microtubule is a polar structure, with a different end of the tubulin molecule (α or β) facing each end of the microtubule. (A, courtesy of Richard Linck; B, courtesy of Richard Wade; D, drawn from data supplied by Joe Howard.)

Microtubules

protofilaments are aligned in parallel with the same polarity, the microtubule itself is a polar structure, and it is possible to distinguish a *plus* (fast-growing) and a *minus* (slow-growing) end.

Microtubules Are Highly Labile Structures That Are Sensitive to Specific Antimitotic Drugs [15]

Many of the microtubule arrays in cells are labile and depend on this lability for their function. One of the most striking examples is the mitotic spindle, which forms after the cytoplasmic microtubules disassemble at the onset of mitosis. The mitotic spindle is the target of a variety of specific *antimitotic drugs* that act by interfering with the exchange of tubulin subunits between the microtubules and the free tubulin pool. One of these is *colchicine* (Figure 16–22), an alkaloid extracted from the meadow saffron that has been used medicinally in the treatment of gout since ancient Egyptian times. Each molecule of colchicine binds tightly to one tubulin molecule and prevents its polymerization, but it cannot bind to tubulin once the tubulin has polymerized into a microtubule. The exposure of a dividing cell to colchicine, or to the closely related drug *colcemid*, causes the rapid disappearance of the mitotic spindle, indicating that a chemical equilibrium is maintained through continual exchange of subunits between the spindle microtubules and the pool of free tubulin. Because the temporary disruption of spindle microtubules preferentially kills many abnormally dividing cells, antimitotic drugs, such as vinblastine and vincristine (whose effects are similar to those of colcemid), are widely used in the treatment of cancer.

The drug *taxol* (Figure 16–22), extracted from the bark of yew trees, has the opposite effect. It binds tightly to microtubules and stabilizes them, and when added to cells, it causes much of the free tubulin to assemble into microtubules. The stabilization of microtubules by taxol arrests dividing cells in mitosis, indicating that microtubules must be able not only to polymerize but also to depolymerize during mitosis. Taxol is also widely used as an anticancer drug.

Elongation of a Microtubule Is Rapid, Whereas the Nucleation of a New Microtubule Is Slow [16]

Microtubule polymerization and depolymerization are complex and interesting processes with important biological roles. Most of what we know about the dynamic behavior of microtubules has come from studying the polymerization of purified tubulin molecules *in vitro*. Pure tubulin will polymerize into microtubules at 37°C in a test tube as long as Mg^{2+} and GTP are present. If the polymerization is followed either by light-scattering measurements or by microscopy, it shows an initial lag phase, after which microtubules form rapidly until a plateau level of polymerization is reached. The lag phase occurs because it is much easier to add subunits to an existing microtubule, a process called *elongation*, than to start a new microtubule *de novo*, a process called *nucleation*.

Figure 16–22 Chemical structures of colchicine and taxol. A third drug, colcemid, is a close relative of colchicine in which the group shown in *yellow* is replaced by —CH_3. Its binding to tubulin, unlike that of colchicine, is readily reversible.

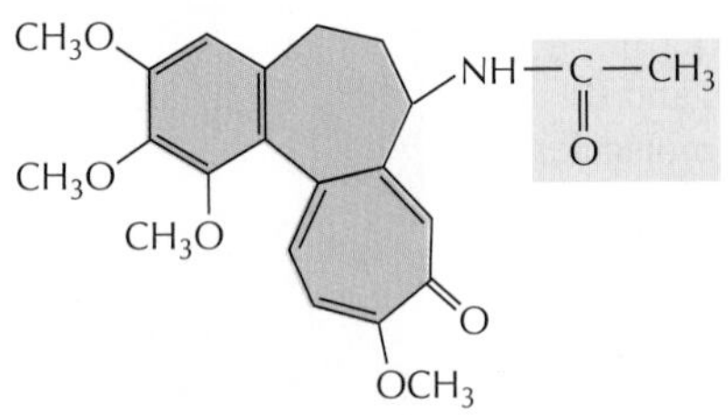

colchicine

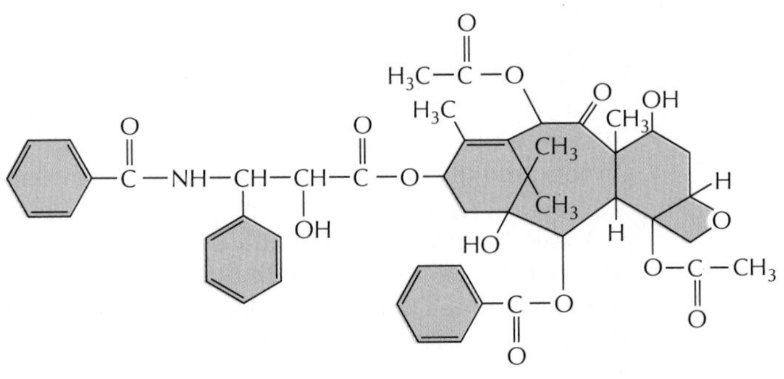

taxol

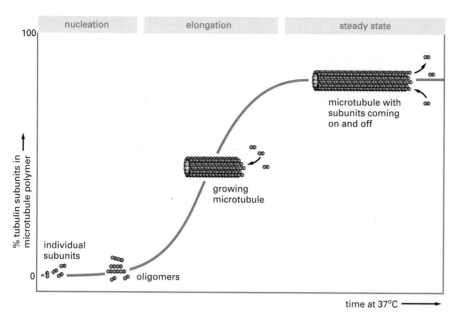

Figure 16–23 Polymerization of pure tubulin. A mixture of tubulin, buffer, and GTP is warmed to 37°C at time zero. The amount of microtubule polymer, measured by light-scattering, follows a sigmoidal curve. During the lag phase individual tubulin molecules associate to form metastable aggregates, some of which go on to nucleate microtubules. The lag phase reflects a kinetic barrier to this nucleation process. During the rapid elongation phase, subunits add to the free ends of existing micro-tubules. During the plateau phase, polymerization and depolymerization are balanced because the amount of free tubulin has dropped to the point where a *critical concentration* has been reached. For simplicity, subunits are shown coming on and off the microtubule at only one end.

During the rapid polymerization phase, the high concentration of free tubulin causes microtubules to polymerize faster than they depolymerize (see below). When the plateau of polymerization is reached, however, not all of the tubulin will have polymerized because subunits are dissociating (depolymerizing) from the ends of microtubules as well as adding to them. The rate of polymerization drops with time because this rate is proportional to the concentration of free tubulin; the final concentration of free tubulin at the plateau, where the polymerization and depolymerization rates are exactly balanced, is called the *critical concentration* (Figure 16–23).

We saw at the beginning of the chapter that the microtubules in a cell usually grow from a specific nucleating site (in most cases, the centrosome); because of a kinetic barrier to nucleation in solution, tubulin polymerization occurs only at this site. As in the test tube, not all the tubulin in the cell becomes polymerized. A typical fibroblast cell contains approximately 20 micromolar tubulin (2mg/ml), of which 50% is in microtubules and 50% is free.

The Two Ends of a Microtubule Are Different and Grow at Different Rates [17]

The structural polarity of a microtubule, which reflects the regular orientation of its tubulin subunits, makes the two ends of the polymer different in ways that have a profound effect on its rate of growth. If purified tubulin molecules are allowed to polymerize for a short time at the ends of fragments of stable microtubules and the mixture is then examined in the electron microscope, one end can be seen to elongate at three times the rate of the other (Figure 16–24). The fast-growing end is thereby defined as the **plus end** and the other as the **minus end.**

It is possible to detect the polarity of microtubules in cross-section by adding free tubulin molecules to existing microtubules: under special conditions the tubulin monomers, instead of adding to the ends of the microtubules, add to the sides, forming curved protofilament sheets. In cross-section the sheets resemble hooks and, depending on the orientation of the microtubule, will appear to point either clockwise or counterclockwise (Figure 16–25). In this way it has been shown that the plus ends of the microtubules in a cell extend away from micro-tubule-nucleating sites such as the centrosome, the poles of a mitotic spindle, or the basal body of a cilium (Figure 16–26).

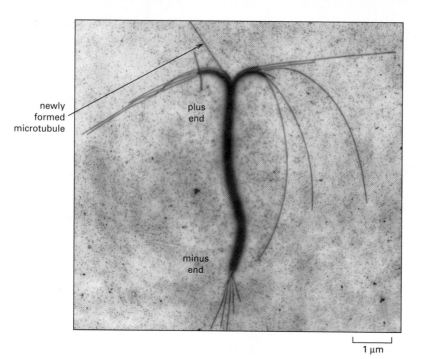

Figure 16–24 **Electron micrograph showing preferential polymerization of tubulin onto the plus ends of microtubules.** A stable bundle of microtubules obtained from the core of a cilium (discussed later) was incubated with tubulin subunits under polymerizing conditions. Microtubules grow fastest from the plus end of the microtubule bundle (the end above the bundle in this figure). (Courtesy of Gary Borisy.)

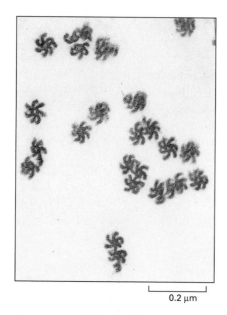

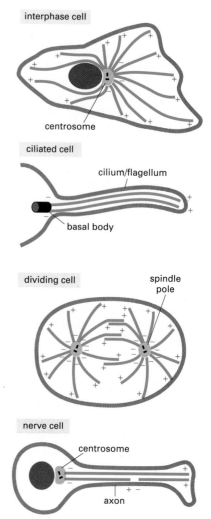

Figure 16–25 **Microtubule polarity as revealed by the hook-decoration method.** All the microtubules in this electron micrograph (seen in cross-section) have the same orientation. The hooks formed by the added tubulin curve clockwise, which indicates that the microtubules are being viewed as though looking along each filament from its plus end toward its minus end. Microtubule polarity can also be determined by decoration with dynein molecules (not shown). (Courtesy of Ursula Euteneuer.)

Figure 16–26 **The orientation of microtubules in cells.** The minus ends of microtubules are generally embedded in a microtubule-organizing center, while the plus ends are often located near the plasma membrane.

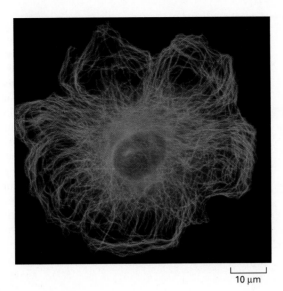

Figure 16–27 **The interphase array of microtubules in a cultured fibroblast.** The microtubules (*green*) are stained with an antibody to tubulin; the cell nucleus (*blue*) is stained with a fluorescent DNA-binding dye. (Courtesy of Nancy L. Kedersha.)

10 μm

Centrosomes Are the Primary Site of Nucleation of Microtubules in Animal Cells [18]

The microtubules in the cytoplasm of an interphase cell in culture can be visualized by staining the cell with fluorescent anti-tubulin antibodies after the cells have been fixed. The microtubules are seen in greatest density around the nucleus and radiate out into the cell periphery in fine lacelike threads (Figure 16–27). The origin of the microtubules is seen most clearly if they are first depolymerized with colcemid and then allowed to repolymerize after the drug is washed out. The new microtubules grow out from the centrosome to form a small starlike structure called an **aster** and then elongate toward the cell periphery until the original microtubule distribution is reestablished (Figure 16–28). If the microtubules in cultured cells are decorated with tubulin hooks to determine their polarity, they are all seen to have their plus ends facing away from the *centrosome,* indicating that this organizing center has the capacity to nucleate microtubule polymerization with a specific polarity.

The **centrosome** is the major microtubule-organizing center in almost all animal cells. In interphase it is typically located to one side of the nucleus, close to the outer surface of the nuclear envelope. Embedded in the centrosome is a pair of cylindrical structures arranged at right angles to each other in an L-shaped configuration. These are *centrioles,* and we discuss their structure later. The centrosome duplicates and splits into two equal parts during interphase, each half containing a duplicated centriole pair. These two daughter centrosomes move to opposite sides of the nucleus when mitosis begins, and they form the two poles of the mitotic spindle (see Figure 18–5).

Surrounding each centriole pair, in both interphase and metaphase, is a region of the cytoplasm that stains darkly when viewed by electron microscopy and appears in the best micrographs to be made of a network of small fibers (Figure 16–29A). This is the **pericentriolar material,** or **centrosome matrix,** and it is the

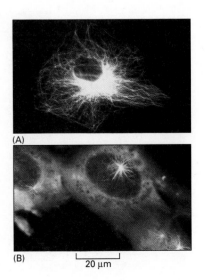

(A)

(B)   20 μm

Figure 16–28 **Microtubules growing out from the centrosome after the removal of colcemid.** Immunofluorescence micrographs showing the arrangement of microtubules in cultured cells as revealed by staining with anti-tubulin antibodies. A normal tissue-culture cell is shown in (A). The cells shown in (B) were treated with colcemid for 1 hour to depolymerize their microtubules and were then allowed to recover; microtubules appear first in a starlike aster and then elongate toward the periphery of the cell. (A, courtesy of Eric Karsenti and Marc Kirschner; B, from M. Osborn and K. Weber, *Proc. Natl. Acad. Sci. USA* 73:867–871, 1976.)

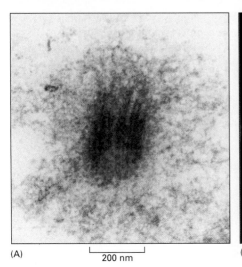

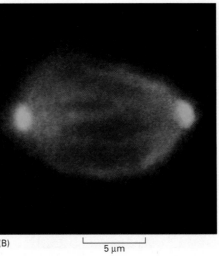

(A) |⎯⎯ 200 nm ⎯⎯| (B) |⎯⎯ 5 μm ⎯⎯|

Figure 16–29 The centrosome matrix. (A) Electron micrograph of a centrosome in a purified preparation. The matrix surrounds a barrel-shaped centriole, and it appears as a fibrous material that contains fine granules. (B) Light micrograph of a dividing human cell in culture stained with an antibody to β-tubulin (*green*) and with an antibody to γ-tubulin (*yellow*), a protein that is located in the centrosome in cells from a wide variety of organisms. The superimposition of the red and green staining causes the γ-tubulin-containing regions at the spindle poles to be yellow. (A, courtesy of Stephen Fuller; B, courtesy of M. Katherine Jung and Berl R. Oakley.)

part of the centrosome that nucleates microtubule polymerization. The protein composition of the centrosome matrix is only partly known, as is the mechanism by which it nucleates microtubules. However, it contains a number of centrosome-specific proteins, including a special minor form of tubulin, called *γ-tubulin* (Figure 16–29B), which may interact with the normal α/β tubulin dimer to help nucleate microtubules.

Not all microtubule-organizing centers contain centrioles. In mitotic cells of higher plants, for example, the microtubules terminate in poorly defined regions of electron density that are completely devoid of centrioles. Similarly, centrioles are not present in the meiotic spindle of mouse oocytes, although they appear later in the developing embryo. In fungi and diatoms the microtubule-organizing center is a plaque called the *spindle pole body*, which is embedded in the nuclear envelope. Despite these morphological differences (Figure 16–30), all of the organizing centers contain a matrix that nucleates microtubule polymerization, and they usually contain γ-tubulin and other centrosome-specific proteins. Thus the molecular mechanism of microtubule nucleation is likely to be highly conserved.

Microtubules Depolymerize and Repolymerize Continually in Animal Cells [19]

In a cell such as a cultured fibroblast the entire microtubule array is turning over rapidly. The half-life of an individual microtubule is about 10 minutes, while the

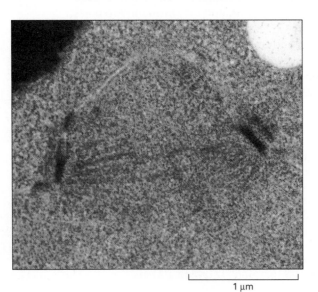

|⎯⎯ 1 μm ⎯⎯|

Figure 16–30 A microtubule-organizing center in a fungal cell. Electron micrograph of the spindle pole body in yeast. (Courtesy of John Kilmartin.)

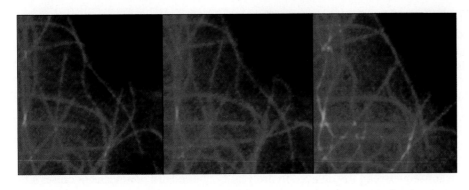

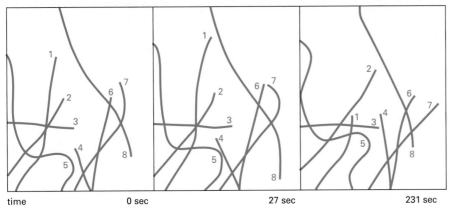

| time | 0 sec | 27 sec | 231 sec |

Figure 16–31 Microtubule dynamics in a living cell. A fibroblast was injected with tubulin that had been covalently linked to rhodamine, so that approximately 1 tubulin subunit in 10 in the cell was labeled with a fluorescent dye. The fluorescence at an edge of the cell was then observed using an extremely sensitive electronic imaging device. Below are tracings of the micrographs that show selected microtubules more clearly. Note, for example, that microtubule #1 first grows and then shrinks rapidly, whereas microtubule #4 grows continuously. (From P.J. Sammak and G.G. Borisy, *Nature* 332:724–736, 1988. © 1988 Macmillan Journals Ltd.)

average lifetime of a tubulin molecule, between its synthesis and proteolytic degradation, is more than 20 hours. Thus each tubulin molecule will participate in the formation and dismantling of many microtubules in its lifetime, a process that can be investigated by direct observation of living cells. One way is to inject tubulin that has been covalently linked to a fluorescent dye and then follow the behavior of microtubules that incorporate the tagged tubulin using fluorescence microscopy. Alternatively, in certain very flat cells one can visualize microtubules directly, without labeling them, using video-enhanced differential-interference-contrast microscopy (see Figure 4–12). When microtubules in a cell are watched over time by either method, a remarkable phenomenon is observed. Individual microtubules grow toward the cell periphery at a constant rate for some period and then suddenly shrink rapidly back toward the centrosome. They may shrink partially and then recommence growing, or they may disappear completely, to be replaced by a different microtubule (Figure 16–31). These fluctuations in length occur over many micrometers and involve the polymerization and then depolymerization of tens of thousands of tubulin subunits. Transitions between prolonged periods of polymerization and depolymerization are also seen when pure microtubules are studied in a test tube (Figure 16–32). This behavior, called *dynamic instability*, plays a major role in positioning microtubules in the cell, as we discuss below.

GTP Hydrolysis Can Explain the Dynamic Instability of Individual Microtubules [20]

The **dynamic instability** of microtubules requires an input of energy to shift the chemical balance between polymerization and depolymerization—energy that comes from the hydrolysis of GTP. GTP binds to the β-tubulin subunit of the heterodimeric tubulin molecule, and when a tubulin molecule adds to the end of a microtubule, this GTP molecule is hydrolyzed to GDP. (The α-tubulin subunit also carries GTP, but this cannot be exchanged for free GTP and is not hydrolyzed, so we can consider it a fixed part of the tubulin protein structure.)

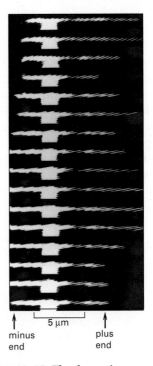

minus end   5 μm   plus end

Figure 16–32 The dynamic instability of microtubule growth. Fluctuations in length of a single microtubule in a solution of pure tubulin as seen by video-enhanced dark-field microscopy. Images of the same microtubule were recorded at. intervals of 1 to 2 minutes and displayed in sequential order on a monitor screen. The two ends go through cycles of elongation and shortening independently, with the plus end showing the greatest fluctuations. (From T. Horio and H. Hotani, *Nature* 321:605–607, 1986. © 1986 Macmillan Journals Ltd.)

The role of GTP hydrolysis in microtubule polymerization has been examined using analogues of GTP that cannot be hydrolyzed. Tubulin molecules containing such nonhydrolyzable GTP analogues form microtubules normally, indicating that, while the binding of this nucleotide is required for microtubule polymerization, its hydrolysis is not. These microtubules, however, are abnormally stable and do not depolymerize like normal microtubules when the tubulin concentration in the surrounding fluid is lowered or when they are treated with colchicine. Thus the normal role of GTP hydrolysis is apparently to allow microtubules to depolymerize by weakening the bonds between tubulin subunits in the microtubule.

Dynamic instability is thought to be a consequence of the delayed hydrolysis of GTP after tubulin assembly. When a microtubule grows rapidly, tubulin molecules add to a polymer end faster than the GTP they carry can be hydrolyzed. This results in the presence of a *GTP cap* on the end of the microtubule, and because tubulin molecules carrying GTP bind to one another with higher affinity than tubulin molecules carrying GDP, the GTP cap will encourage a growing microtubule to continue growing. Conversely, once a microtubule has lost its GTP cap—for example, if the instantaneous rate of polymerization slows down—it will start to shrink and then tend to go on shrinking.

A model for the structural changes that accompany dynamic instability is shown schematically in Figure 16–33. Some general principles that apply to the polymerization of both actin filaments and microtubules are discussed in Panel 16–1, pages 824–825.

Cells can modify the dynamic instability of their microtubules for specific purposes. In each M phase of the cell cycle, for example, the rapidity with which microtubules form and break down is greatly increased, so that the chromosomes can readily capture growing microtubules and a mitotic spindle can rapidly assemble (discussed in Chapter 18). Conversely, when a cell differentiates and takes on a defined morphology, the dynamic instability of its microtubules is often suppressed by proteins that bind to the microtubules and stabilize them against depolymerization. The ability to stabilize microtubules in a particular configuration provides an important mechanism by which a cell can organize its cytoplasm.

Figure 16–33 GTP hydrolysis after polymerization destabilizes microtubules. Analysis of the growth and shrinkage of microtubules *in vitro* suggests the following model for dynamic instability. (A) Addition of tubulin heterodimers carrying GTP to the end of a protofilament causes it to grow in a linear conformation that can readily pack into the cylindrical wall of the microtubule, thereby becoming stabilized. Hydrolysis of GTP after assembly changes the conformation of the subunits and tends to force the protofilament into a curved shape that is less able to pack into the microtubule wall. (B) In an intact microtubule, protofilaments made from GDP-containing subunits are forced into a linear conformation by the many lateral bonds within the microtubule wall, especially in the stable cap of GTP-containing subunits. Loss of the GTP cap, however, allows the GDP-containing protofilaments to relax to their more curved conformation. This leads to progressive disruption of the microtubule and the eventual disassembly of protofilaments into free tubulin dimers.

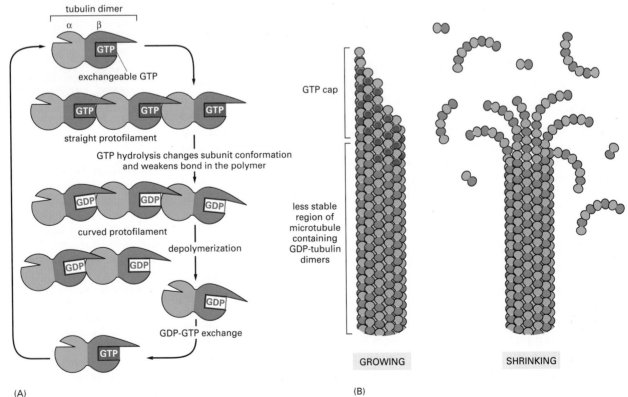

(A)

(B)

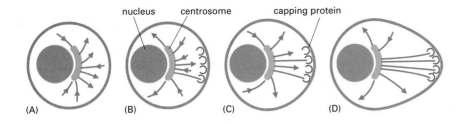

nucleus   centrosome   capping protein

(A)   (B)   (C)   (D)

Figure 16–34 The selective stabilization of microtubules can polarize a cell. A newly formed microtubule will persist only if both of its ends are protected from depolymerizing. In cells the minus ends of microtubules are generally protected by the organizing centers from which these filaments grow. The plus ends are initially free but can be stabilized by other proteins. Here, for example, a nonpolarized cell is depicted in (A) with new microtubules growing and shrinking from a centrosome in all directions randomly. The array of microtubules then encounters hypothetical structures in a specific region of the cell cortex that can cap (stabilize) the free plus end of the microtubules (B). The selective stabilization of those microtubules that happen by chance to encounter these structures will lead to a rapid redistribution of the arrays and convert the cell to a polarized form (C and D).

The Dynamic Instability of Microtubules Provides an Organizing Principle for Cell Morphogenesis [21]

Cytoplasmic microtubules in animal cells tend to radiate out in all directions from the centrosome, where their minus ends are anchored. Most animal cells are polarized, however, and the assembly and disassembly of tubulin molecules are spatially controlled so that microtubules extending toward specific regions of the cell predominate. It is not known for certain how this is achieved, but it seems likely that the mechanisms depend on the dynamic instability of microtubules.

We have seen that individual microtubules *in vitro* tend to exist in one of two states—steady growth or rapid, "catastrophic" disassembly—and that microtubules in a cell can also exist in these two states. The inherent instability of microtubules helps to explain how they can become organized in specific directions in a cell—toward the leading edge of a crawling cell, for example. The array of microtubules radiating from the centrosome is continually changing as new microtubules grow and replace others that have depolymerized. A microtubule that grows from a centrosome can be stabilized if its plus end is somehow stabilized, or *capped,* so as to prevent its depolymerization. If capped by a structure in a particular region of the cell, it will establish a relatively stable link between that structure and the centrosome. Microtubules originating in the centrosome can thus be selectively stabilized by events elsewhere in the cell. Cell polarity is thought to be determined in this way by unknown structures or factors localized in particular regions of the cell cortex that "capture" the plus ends of microtubules (Figure 16–34).

In many cells the initial stabilization of microtubules at their plus ends is consolidated to produce a more permanent polarization of the cell, as we now discuss.

Microtubules Undergo a Slow "Maturation" Revealed by Posttranslational Modifications of Their Tubulin [22]

Tubulin subunits can be covalently modified after they polymerize. Two such modifications are especially interesting in that they provide a form of molecular clock, which can be used to tell how long it has been since a given microtubule polymerized. These modifications are the acetylation of α-tubulin on a particular lysine and the removal of the tyrosine residue from the carboxyl terminus of α-tubulin. *Acetylation* and *detyrosination* are both relatively slow enzymatic reactions that occur only on microtubules and not on free tubulin molecules; moreover, they are rapidly reversed as soon as a tubulin molecule depolymerizes. Thus the longer the time that has elapsed since a particular microtubule polymerized, the higher will be the fraction of its subunits that are acetylated and detyrosinated. Complete modification takes several hours, so that in fibroblasts, where microtubules turn over rapidly, relatively few of them are modified. In nerve axons, by contrast, the majority of microtubules are stable and most are modified.

Acetylation and detyrosination can be detected by specific antibodies, and they provide a useful indication of the stability of microtubules in cells in which it is difficult to study microtubule dynamics directly. The role of these modifications is unknown, but it is thought that they provide sites for the binding of specific microtubule-associated proteins that further stabilize mature microtubules.

Microtubule-associated Proteins (MAPs) Bind to Microtubules and Modify Their Properties [23]

Whereas the posttranslational modification of tubulin marks certain microtubules as "mature" and may promote their stability, the most far-reaching and versatile modifications of microtubules are those conferred by the binding of other proteins. These **microtubule-associated proteins,** or **MAPs,** serve both to stabilize microtubules against disassembly and to mediate their interaction with other cell components. As one might expect from the diverse functions of microtubules, there are many kinds of MAPs; some are widely distributed in most cells, whereas others are found only in specific cell types.

Two major classes of MAPs can be isolated from brain in association with microtubules: *HMW proteins* (high-molecular-weight proteins), which have molecular weights of 200,000 to 300,000 or more and include *MAP-1* and *MAP-2;* and *tau proteins*, which have molecular weights of 55,000 to 62,000. Proteins in both classes have two domains, only one of which binds to microtubules; the other is thought to help link the microtubule to other cell components (Figure 16–35). Because the microtubule-binding domain binds to several unpolymerized tubulin molecules simultaneously, these MAPs speed up the nucleation step of tubulin polymerization *in vitro*. More important, they inhibit the dissociation of tubulin from the microtubule ends and thus stabilize the microtubules once they have formed. Staining with antibodies to MAP-2 and tau shows that both proteins bind along the entire length of cytoplasmic microtubules.

Many other MAPs have been isolated. Some act as structural components and provide permanent links to other cell components, including other parts of the cytoskeleton. Others are microtubule motors, which use the energy of ATP hydrolysis to move along microtubules, as we discuss below.

MAPs Help Create Functionally Differentiated Cytoplasm [24]

Many cell types specifically stabilize microtubules in specialized regions of cytoplasm. An especially well-studied example is provided by nerve cells, which extend two kinds of processes—*axons* and *dendrites*. Axons, which are uniform in diameter and can be many centimeters long, are responsible for propagating electrical signals away from the cell body, whereas dendrites, which taper away from the cell body and rarely exceed 500 μm in length, are responsible for receiving electrical information from other neurons and relaying it to the cell body. Most nerve cells form several dendrites but only a single axon (see Figure 11–20).

Axons and dendrites are both packed with microtubules, although with different arrangements. In axons microtubules are very long and are all oriented with their plus ends away from the cell body. In dendrites the microtubules are

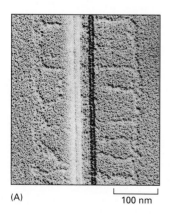

(A)

100 nm

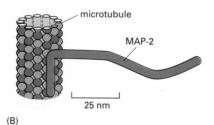

microtubule

MAP-2

25 nm

(B)

Figure 16–35 **A microtubule-associated protein.** (A) Electron micrograph showing the regularly spaced side arms formed on a microtubule by a large microtubule-associated protein (known as MAP-2) isolated from vertebrate brain. Portions of the protein project away from the microtubule, as shown schematically in (B). (Electron micrograph courtesy of William Voter and Harold Erickson.)

shorter and their polarity is mixed: some have their plus ends pointing away from the cell body, while others have their plus ends pointing toward the cell body. When the distribution of MAPs in cultured neurons is studied with specific antibodies, certain forms of the tau protein are found to be present only in axons; MAP-2, on the other hand, is present in both dendrites and the cell body but completely excluded from axons (Figure 16–36). Axons and dendrites are different in many other ways as well: mRNAs, ribosomes, and some kinds of ion channels, for example, are present in dendrites and the cell body but are excluded from axons, while certain cell-adhesion molecules and the Na⁺ channels involved in the generation of action potentials are selectively localized to axons. Thus both the cytoplasm and the plasma membrane of a nerve cell are divided into axonal and dendritic *compartments*. These compartments within a single cell differ from membrane-bounded compartments such as the endoplasmic reticulum or mitochondria, since they are not separated from each other by a membrane; instead, the difference seems to be one of structural organization and the types of proteins present.

The generation of axons and dendrites during the differentiation of nerve cells is discussed in Chapter 21. Although it is unclear how the cytoplasm and plasma membrane of a nerve cell become compartmentalized, MAPs may be essential for this process. When the production of tau protein is inhibited in cultured neurons by treatment with specific antisense oligonucleotides, the formation of axons is suppressed, whereas the formation of dendrites is unaffected. Conversely, when nonneuronal cells are genetically manipulated so that they express tau protein (which is normally expressed only in nerve cells), they form long axonlike processes, which contain bundles of microtubules arranged with their plus ends pointing away from the cell body, just as in nerve cells.

Because different components of the cell move along microtubules in different directions, one can postulate that an initial difference in microtubule polarity is created by a different distribution of MAPs, which will in turn lead to further differences between dendrites and axons. Secretory vesicles, for example, move toward the plus end of microtubules and therefore will be carried down the axon to the nerve terminals where they function; conversely, if ribosomes and mRNAs move toward the minus end of microtubules, they could be excluded from axons.

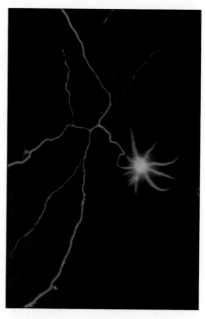

Figure 16–36 An example of the cytoplasmic compartmentalization of nerve cells. This micrograph shows the distribution of tau protein (*green*) and MAP-2 (*orange*) in a hippocampal neuron in culture. Whereas tau is confined to the axon, MAP-2 is confined to the cell body and dendrites. The antibody used to detect tau binds only to dephosphorylated tau, which is confined to the axon; other data show that phosphorylated tau is present in dendrites. (Courtesy of James W. Mandell and Gary A. Banker.)

Kinesin and Dynein Direct Organelle Movement Along Microtubules [25]

Important advances in cell biology have often followed the introduction of a new experimental technique, and it was the improved ability to see small faint objects by video-enhanced light microscopy that led to the discovery of the microtubule motors responsible for organelle transport. Once it became possible to visualize single microtubules in an unfixed specimen, investigators could follow the movement of organelles and other particles along these microtubules *in vitro*. Alternatively, they could observe and measure the gliding movement of individual microtubules over glass surfaces coated with cell extracts.

Such *in vitro* motility assays were used to identify and isolate two classes of microtubule-dependent motor proteins—the *kinesins* and the *cytoplasmic dyneins*. **Cytoplasmic dyneins** are involved in organelle transport and mitosis and are closely related to *ciliary dynein*, the motor protein in cilia and flagella (discussed later). **Kinesins** are more diverse than the dyneins, and different family members are involved in organelle transport, in mitosis, in meiosis, and in the transport of synaptic vesicles along axons. Both the cytoplasmic dyneins and the kinesins are composed of two heavy chains plus several light chains. Each heavy chain contains a conserved, globular, ATP-binding head and a tail composed of a string of rodlike domains. The two head domains are ATPase motors that bind to microtubules, while the tails generally bind to specific cell components and thereby specify the type of cargo that the protein transports (Figure 16–37).

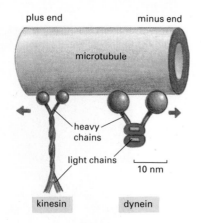

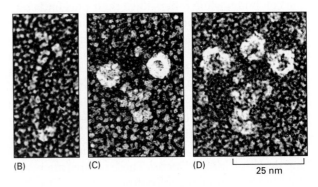

(A) plus end ... minus end
microtubule
heavy chains
light chains
10 nm
kinesin   dynein

(B) (C) (D) 25 nm

The Rate and Direction of Movement Along a Microtubule Are Specified by the Head Domain of Motor Proteins [26]

Most known motor proteins move in only one direction along microtubules—either toward the plus end or toward the minus end. This directionality can be analyzed *in vitro* by allowing polystyrene beads coated with the motor protein to move along microtubules that have been polymerized on centrosomes. Because the microtubules in such arrays have their plus ends outermost, the direction of movement can be readily determined with a light microscope. Whereas polystyrene beads coated with crude extracts of cytoplasm move in both directions, beads coated with kinesin isolated from axons move only outward toward the plus end of the microtubules. Beads coated with cytoplasmic dyneins, by contrast, move toward the minus ends of the microtubules, which are embedded in the centrosome.

Studies of intact nerve axons have confirmed the results obtained in *in vitro* experiments: organelle movement away from the cell body is driven mainly by kinesin, whereas organelle movement back from the nerve terminal toward the cell body is driven by cytoplasmic dynein (Figure 16–38). Since all proteins are made in the nerve cell body, cytoplasmic dynein must be carried first in a nonfunctional state to the nerve terminal before it can begin to work to transport organelles back to the cell body.

Surprisingly, not all kinesins move organelles toward the plus end of microtubules. A *Drosophila* kinesin called *Ncd*, for example, which is required for normal meiosis, differs from axonal kinesin in both the direction and the rate at which it moves along microtubules: whereas axonal kinesin walks toward the plus end at approximately 2 μm/second, the Ncd protein walks toward the *minus* end at about 0.1 μm/second.

The mechanism by which these motor proteins convert the energy of ATP hydrolysis into vectorial movement is not known. Finding out how two closely related head domains can move in opposite directions along a microtubule will require detailed structural studies and is likely to illuminate the energy transduction process itself.

Figure 16–37 Microtubule motor proteins. Kinesins and cytoplasmic dyneins are microtubule motor proteins that generally move in opposite directions along a microtubule (A). These proteins (drawn here to scale) are complexes composed of two identical heavy chains plus several smaller light chains. Each heavy chain forms a globular head region that attaches the protein to microtubules in an ATP-dependent fashion. (B and C) Freeze-etch electron micrographs of a kinesin molecule (B) and a molecule of cytoplasmic dynein (C). Whereas both kinesin and cytoplasmic dynein are two-headed molecules, ciliary dynein (D) has three heads (see Figure 16–44). (Freeze-etch electron micrographs prepared by John Heuser.)

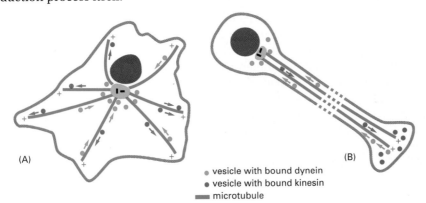

(A)   (B)
● vesicle with bound dynein
● vesicle with bound kinesin
▬ microtubule

Figure 16–38 Vesicle transport in two directions. Kinesin and cytoplasmic dynein carry their cargo in opposite directions along microtubules, as illustrated in a fibroblast (A) and in the axon of a neuron (B).

Summary

Microtubules are stiff polymers of tubulin molecules. They assemble by addition of GTP-containing tubulin molecules to the free end of the microtubule, with one end (the plus end) growing faster than the other. Hydrolysis of the bound GTP takes place after assembly and weakens the bonds that hold the microtubule together. Slowly growing microtubules are especially unstable and liable to catastrophic disassembly, but they can be stabilized in cells by association with other structures that cap their two ends. Microtubule-organizing centers such as centrosomes protect the minus ends of microtubules and continually nucleate the formation of new microtubules, which grow out in random directions. Any microtubule that happens to encounter a structure that stabilizes its free plus end will be selectively retained, while other microtubules will depolymerize. It is thought that this selective process largely determines the position of the microtubule arrays in a cell.

The tubulin subunits in microtubules that have been selectively stabilized are modified by acetylation and detyrosination. These alterations are thought to label the microtubule as "mature" and provide sites for the binding of specific microtubule-associated proteins (MAPs), which further stabilize the microtubule against disassembly. Microtubule motor proteins constitute an important class of MAPs that use the energy of ATP hydrolysis to move unidirectionally along a microtubule, carrying specific cargo. In general, dyneins move cargo toward the minus ends of microtubules, while most kinesins move cargo toward the plus ends. Such motor proteins are largely responsible for the spatial organization and directed movements of organelles in the cytoplasm.

Cilia and Centrioles [27]

Ciliary beating is an extensively studied form of cellular movement. **Cilia** are tiny hairlike appendages about 0.25 μm in diameter with a bundle of microtubules at their core; they extend from the surface of many kinds of cells and are found in most animal species, many protozoa, and some lower plants. The primary function of cilia is to move fluid over the surface of the cell or to propel single cells through a fluid. Protozoa, for example, use cilia both to collect food particles and for locomotion. On the epithelial cells lining the human respiratory tract, huge numbers of cilia ($10^9/cm^2$ or more) sweep layers of mucus, together with trapped particles of dust and dead cells, up toward the mouth, where they are swallowed and eliminated. Cilia also help to sweep eggs along the oviduct, and a related structure, the flagellum, propels sperm.

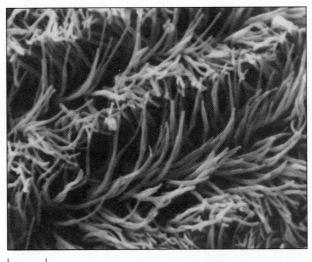

Figure 16–39 **Cilia.** Scanning electron micrograph of a field of cilia in the gut of a marine worm. (From J.S. Mellor and J.S. Hyams, *Micron* 9:91–94, 1978. © 1978, by permission of Pergamon Press Ltd.)

2 μm

Figure 16–40 **The contrasting motions of beating cilia and flagella.** (A) The beat of a cilium such as that on an epithelial cell from the human respiratory tract resembles the breast stroke in swimming. A fast *power stroke* (stages 1 and 2), in which fluid is driven over the surface of the cell, is followed by a slow *recovery stroke* (stages 3, 4, and 5). Each cycle typically requires 0.1 to 0.2 second and generates a force perpendicular to the axis of the axoneme. For comparison, the wavelike movements of the flagellum of a sperm cell from a tunicate are shown in (B). The cell was photographed on moving film with stroboscopic illumination at 400 flashes per second. Note that waves of constant amplitude move continuously from the base to the tip of a flagellum. The cell is thereby pushed forward, a distinctly different effect from that caused by a cilium. (B, courtesy of C.J. Brokaw.)

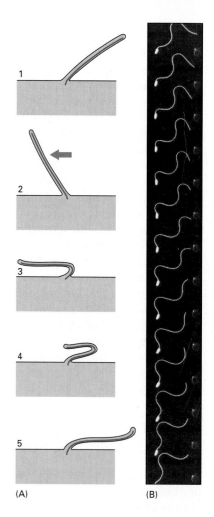

Cilia Move by the Bending of an Axoneme—a Complex Bundle of Microtubules [27]

Fields of cilia bend in coordinated unidirectional waves (Figure 16–39). Each cilium moves with a whiplike motion: a forward active stroke, in which the cilium is fully extended and beating against the surrounding liquid, is followed by a recovery phase, in which the cilium returns to its original position with an unrolling movement that minimizes viscous drag (Figure 16–40A). The cycles of adjacent cilia are almost but not quite in synchrony, creating the wavelike patterns that can be seen in fields of beating cilia under the microscope.

The simple **flagella** of sperm and of many protozoa are much like cilia in their internal structure, but they are usually very much longer. Instead of making whiplike movements, they propagate quasi-sinusoidal waves (Figure 16–40B). Nevertheless, the molecular basis for their movement is the same as that in cilia. It should be noted that the flagella of bacteria (described in Chapter 15) are completely different from the cilia and flagella of eucaryotic cells.

The movement of a cilium or a flagellum is produced by the bending of its core, which is called the **axoneme.** The axoneme is composed entirely of microtubules and their associated proteins. The microtubules are modified and arranged in a pattern whose curious and distinctive appearance was one of the most striking revelations of early electron microscopy: nine special doublet

(A)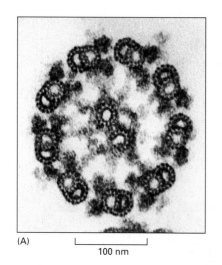

100 nm

(B)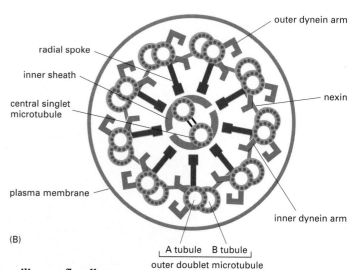

Figure 16–41 **The arrangement of microtubules in a cilium or flagellum.** (A) Electron micrograph of the flagellum of a green algal cell (*Chlamydomonas*) shown in cross-section, illustrating the distinctive "9 + 2" arrangement of microtubules. (B) Diagram of the parts. The various projections from the microtubules link them together and occur at regular intervals along the length of the axoneme. (A, courtesy of Lewis Tilney.)

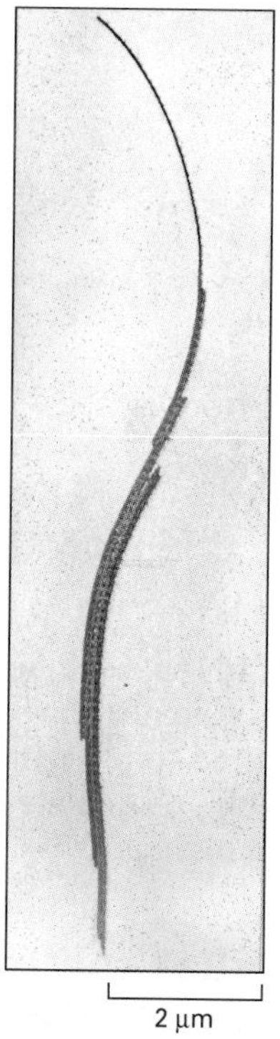

Figure 16–42 Microtubule sliding in an axoneme. Electron micrograph of an isolated axoneme (from a cilium of *Tetrahymena*) that has been briefly exposed to the proteolytic enzyme trypsin to loosen the protein ties that normally hold it together. Following treatment with ATP, the individual microtubule doublets slide against each other, as shown schematically in Figure 16–43A. Because there are nine microtubule doublets in the axoneme, the original structure can increase as much as ninefold in length. (From F.D. Warner and D.R. Mitchell, *J. Cell Biol.* 89:35–44, 1981, by copyright permission of the Rockefeller University Press.)

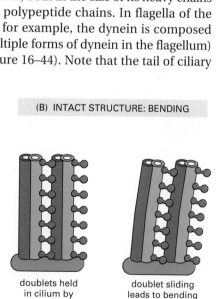

2 μm

microtubules are arranged in a ring around a pair of single microtubules (Figure 16–41). This "9 + 2" array is characteristic of almost all forms of cilia and eucaryotic flagella—from those of protozoa to those found in humans. The microtubules extend continuously for the length of the axoneme, which is usually about 10 μm long but may be as long as 200 μm in some cells.

While each member of the pair of single microtubules (the *central pair*) is a complete microtubule, each of the outer doublets is composed of one complete and one partial microtubule fused together so that they share a common tubule wall. In transverse sections each complete microtubule appears to be formed from a ring of 13 subunits, while the incomplete tubule of the outer doublet is formed from only 11.

Dynein Drives the Movements of Cilia and Flagella [28]

The microtubules of an axoneme are associated with numerous proteins, which project at regular positions along the length of the microtubules. Some serve as cross-links that hold the bundle of microtubules together. Others generate the force that drives the bending motion, while still others form a mechanically activated relay system that controls the motion to produce the desired waveform. The most important of these accessory proteins is **ciliary dynein,** whose heads interact with adjacent microtubules to generate a sliding force between the microtubules. Because of the multiple links that hold adjacent microtubule doublets together, what would be a sliding movement between free microtubules (Figure 16–42) is converted to a bending motion in the cilium (Figure 16–43).

Like cytoplasmic dynein, ciliary dynein has a motor domain, which hydrolyzes ATP to move along a microtubule toward its minus end, and a tail region that carries a cargo, which in this case is an adjacent microtubule. Ciliary dynein is considerably larger than cytoplasmic dynein, both in the size of its heavy chains and in the number and complexity of its polypeptide chains. In flagella of the unicellular green algae *Chlamydomonas,* for example, the dynein is composed of either 2 or 3 heavy chains (there are multiple forms of dynein in the flagellum) and 10 or more smaller polypeptides (Figure 16–44). Note that the tail of ciliary

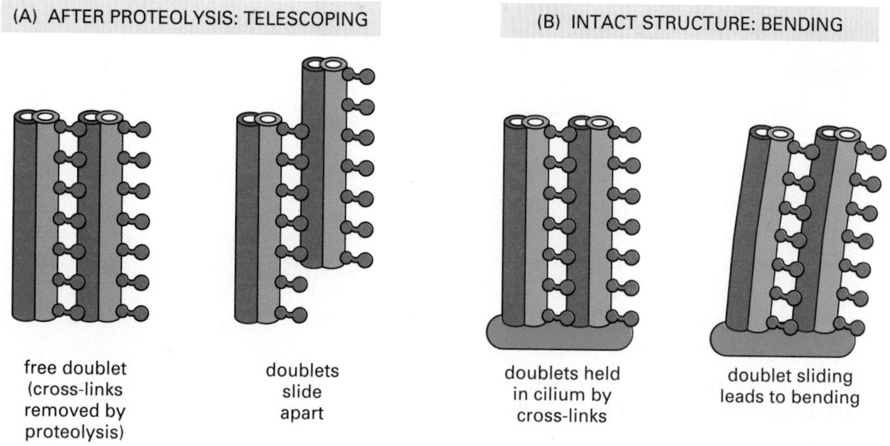

(A) AFTER PROTEOLYSIS: TELESCOPING

(B) INTACT STRUCTURE: BENDING

free doublet
(cross-links
removed by
proteolysis)

doublets
slide
apart

doublets held
in cilium by
cross-links

doublet sliding
leads to bending

Figure 16–43 The bending of an axoneme. (A) The sliding of outer microtubule doublets against each other causes the axoneme to elongate if the proteins that link the doublets together are removed by proteolysis. (B) If the doublets are tied to each other at one end, the axoneme bends.

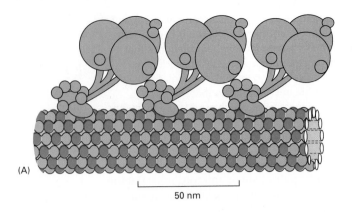

(A)

50 nm

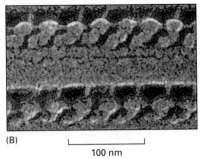

(B)

100 nm

Figure 16–44 Ciliary dynein. Ciliary dynein is a large protein assembly (nearly 2 million daltons) composed of 9 to 12 polypeptide chains, the largest of which is the heavy chain of 512,000 daltons. (A) The heavy chains are believed to form the major portion of the globular head and stem domains, and many of the smaller chains are clustered around the base of the stem. The base of the molecule binds tightly to an A microtubule in an ATP-independent manner, while the large globular heads have an ATP-dependent-binding site for a B microtubule (see Figure 16–41). When the heads hydrolyze their bound ATP, they move toward the minus end of this second microtubule, thereby producing a sliding force between the adjacent microtubule doublets in a cilium or flagellum (see Figure 16–43). The three-headed form of ciliary dynein, formed from three heavy chains, is illustrated here. (B) Freeze-etch electron micrograph of a cilium showing the dynein arms projecting at regular intervals from the doublet microtubules. (B, courtesy of John Heuser.)

dynein binds only to the A tubule and not to the B tubule, which has a slightly different structure. The resulting asymmetry in the arrangement of the dynein molecules is required to prevent a fruitless tug-of-war between neighboring microtubules, which presumably explains why each of the nine outer microtubules is an A-B doublet.

Flagella and Cilia Grow from Basal Bodies That Are Closely Related to Centrioles [29]

If the two flagella of the green alga *Chlamydomonas* are sheared from the cell, they rapidly re-form by elongating from structures called *basal bodies*. The basal bodies have the same structure as the centrioles that are found embedded in the center of animal centrosomes. Indeed, in some organisms, basal bodies and centrioles seem to be functionally interconvertible: during each mitosis in *Chlamydomonas,* for example, the flagella are resorbed and the basal bodies move into the cell interior and become embedded in the spindle poles.

Centrioles and basal bodies are cylindrical structures about 0.2 µm wide and 0.4 µm long. Nine groups of three microtubules, fused into triplets, form the wall of the centriole, each triplet being tilted inward like the blades of a turbine (Figure 16–45). Adjacent triplets are linked at intervals along their length, while faint protein spokes can often be seen in electron micrographs to radiate out to each triplet from a central core, forming a pattern like a cartwheel (see Figure 16–45A).

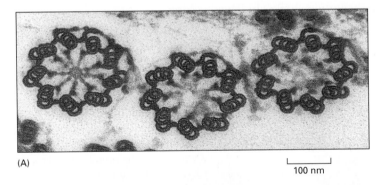

(A)

100 nm

Figure 16–45 Basal bodies. (A) Electron micrograph of a cross-section through three basal bodies in the cortex of a protozoan. (B) Diagram of a basal body viewed from the side. Each basal body forms the lower portion of a ciliary axoneme, and it is composed of nine sets of *triplet* microtubules, each triplet containing one complete microtubule (the A tubule) fused to two incomplete microtubules (the B and C tubules). Other proteins [shown in *red* in (B)] form links that hold the cylindrical array of microtubules together. The structure of a centriole is essentially the same. (A, courtesy of D.T. Woodrow and R.W. Linck.)

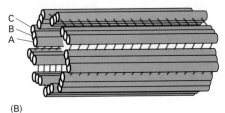

C
B
A

(B)

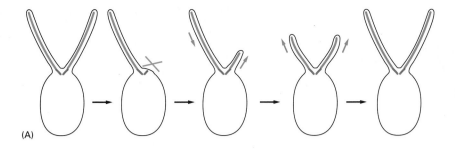

(A)

During the formation or regeneration of a cilium, each doublet microtubule of the axoneme grows from two of the microtubules in the triplet microtubules of the basal body so that the ninefold symmetry of the basal body microtubules is preserved in the ciliary axoneme. Autoradiographic evidence suggests that the addition of tubulin and other proteins of the axoneme takes place at the distal tip of the structure, at the plus end of the microtubules. How the central pair of single microtubules forms in the axoneme is not known; there is no central pair in basal bodies or centrioles.

It is not known how the length of flagella and cilia is determined. The length is constant for a given species of cell, and it is not limited by the availability of components or the kinetics of elongation. If one of the two flagella is removed in *Chlamydomonas,* for example, the remaining flagellum begins to shrink while the lost flagellum simultaneously regenerates. Once the shrinking flagellum and the regrowing flagellum reach the same length, they then both grow out together to reach their final characteristic length. This experiment suggests that flagellar length is constantly monitored in some way (Figure 16–46).

Centrioles Usually Arise by the Duplication of Preexisting Centrioles [30]

The otherwise continuous increase in cell mass throughout the animal cell cycle is punctuated by two discrete duplication events: the replication of DNA and the doubling of the centrosome, which usually has a centriole pair at its center. The two centrioles of the pair are positioned at right angles to each other (Figure 16–47). In cultured fibroblasts centriole doubling begins at around the time that DNA synthesis begins: first the two members of a pair separate, and then a daughter centriole is formed perpendicular to each original centriole (see Figure 18–4). An immature centriole contains a ninefold symmetric array of *single* microtubules; each microtubule then presumably acts as a template for the assembly of the triplet microtubule of mature centrioles.

Figure 16–46 Flagellar length in *Chlamydomonas* is monitored by an active process. (A) When one flagellum is physically detached (*blue cross*), it starts to grow back by polymerization off the basal body (*red*). At the same time the remaining flagellum begins to shrink. When both are half their normal length, they grow out together. Growth stops when both flagella reach the final, accurately specified length. (B) Color photo of *Chlamydomonas,* where the *orange* color results from the auto-fluorescence of chlorophyll and the *green* from the binding of a fluorescent antibody to a plasma membrane glycoprotein. (B, courtesy of Robert A. Bloodgood.)

(B)

10 μm

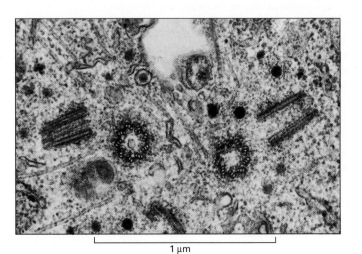

Figure 16–47 An electron micrograph showing a newly replicated pair of centrioles. One centriole of each pair has been cut in cross-section and the other in longitudinal section, indicating that the two members of each pair are aligned at right angles to each other. (From M. McGill, D.P. Highfield, T.M. Monahan, and B.R. Brinkley, *J. Ultrastruct. Res.* 57:43–53, 1976.)

1 μm

Cilia and Centrioles

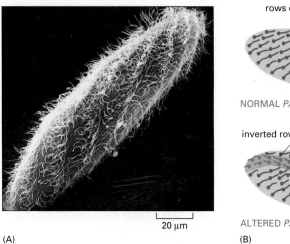

(A)

20 μm

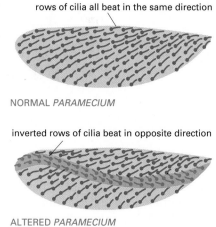

rows of cilia all beat in the same direction

NORMAL *PARAMECIUM*

inverted rows of cilia beat in opposite direction

ALTERED *PARAMECIUM*

(B)

Figure 16–48 Cortical inheritance of pattern in a ciliated protozoan. (A) Scanning electron micrograph of a *Paramecium*, which swims by synchronously beating its cilia. (B) Schematic diagram of the rows of cilia on the surface of a normal *Paramecium* and on a *Paramecium* in which rows of cilia have been inverted so that they beat in the opposite direction. Such altered patterns are propagated indefinitely as the *Paramecium* divides, even though the information in the DNA is unchanged. (A, courtesy of Sidney Tamm.)

The two centrioles of a pair are not identical: the daughter centriole not only has a distinct orientation but differs also in detailed morphology and function. In many vertebrate cells, for example, one of the two centrioles is distinguished by its ability to nucleate a so-called *primary cilium*—an isolated nonmotile cilium that has no known function.

Parent/daughter differences also exist in basal bodies and can lead to asymmetries in the cytoskeleton. In ciliated protozoa, basal body replication is coordinated with cell division and the stereospecificity of the duplication process is thought to be important for maintaining the orientation of cilia on the cell surface. This was clearly demonstrated in a classic experiment performed in the 1960s on *Paramecium,* a large protozoan whose surface is covered with rows of motile cilia. Normally, all of the rows are aligned with the same polarity through the coordinated replication of basal bodies, which consistently produce daughter basal bodies with the same orientation relative to the cell surface. The array of cilia growing from these basal bodies enables the cell to swim with great efficiency. By grafting experiments, however, it is possible to disturb this pattern and produce some inverted rows of cilia that beat in the direction opposite to that of their neighbors (Figure 16–48). Once established, such altered patterns are passed on from parent to daughter *Paramecium* for more than 100 generations. This form of heredity has nothing to do with DNA: the modified cells inherit a particular pattern of ciliary rows through the stereospecific replication of their basal bodies.

Summary

The axoneme of a cilium and a eucaryotic flagellum contains a cylindrical bundle of nine outer doublet microtubules. Dynein side arms extend between adjacent microtubule doublets and hydrolyze ATP to generate a sliding force between the doublets. Accessory proteins bundle the ring of microtubule doublets together and convert the sliding force into the bending movement that underlies ciliary beating. The complex structure of the ciliary axoneme forms by the self-assembly of its component proteins and is nucleated by a centriole (basal body), which serves as a template for the distinct 9 + 2 pattern of microtubules that forms the core axoneme. The centriole duplicates in a highly controlled process in which a daughter centriole is nucleated from the side of a mother centriole and grows at right angles to it. Oriented replication of basal bodies underlies the heritable pattern of beating cilia on the surface of ciliated protozoa.

Actin Filaments [31]

All eucaryotic species contain **actin.** This cytoskeletal protein is the most abundant protein in many eucaryotic cells, often constituting 5% or more of the total cell protein. Vertebrate skeletal muscle cells are the usual source of actin for experiments done *in vitro,* as about 20% of their mass is actin. If dry powdered muscle is treated with a very dilute salt solution, the *actin filaments* dissociate into their actin subunits. Each actin molecule is a single polypeptide 375 amino acids long that has a molecule of ATP tightly associated with it.

Actin filaments can form both stable and labile structures in cells. Stable actin filaments form the core of microvilli and are a crucial component of the contractile apparatus of muscle cells. Many cell movements, however, depend on labile structures constructed from actin filaments. In this section we focus on the question of how the cell controls the assembly of dynamic actin filaments from pools of soluble actin subunits in the cytosol.

Actin Filaments Are Thin and Flexible [32]

Actin filaments appear in electron micrographs as threads about 8 nm wide. They consist of a tight helix of uniformly oriented actin molecules (also known as globular actin, or *G actin*) (Figure 16–49). Like a microtubule, an actin filament is a polar structure, with two structurally different ends—a relatively inert and slow-growing *minus end* and a faster-growing *plus end.* Because of the oriented "arrowhead" appearance of the complex formed between actin filaments and the motor protein myosin, which we describe later, the minus end is also referred to as the "pointed end" and the plus end as the "barbed end." The three-dimensional structure of the actin molecule has been solved by x-ray diffraction analysis, and this information has been used to deduce the structure of an actin filament at the level of individual amino acids (Figure 16–50).

Some lower eucaryotes, such as yeasts, have only one actin gene, encoding a single protein. All higher eucaryotes, however, have several isoforms encoded by a family of actin genes. At least six types of actin are present in mammalian tissues; these fall into three classes, depending on their isoelectric point. Alpha actins are found in various types of muscle, whereas β and γ actins are the principal constituents of nonmuscle cells. Although there are subtle differences in the properties of different forms of actin, the amino acid sequences have been highly conserved in evolution, and all assemble into filaments that are essentially identical in most tests performed *in vitro.*

The total length of all of the actin filaments in a cell is at least 30 times greater than the total length of the microtubules, reflecting a fundamental difference in the way these two cytoskeletal polymers are organized and function in cells. Actin filaments are thinner and more flexible, and usually much shorter, than microtubules. We shall see that actin filaments rarely occur in isolation in the cell but rather in cross-linked aggregates and bundles, which are much stronger than the individual filaments.

Actin and Tubulin Polymerize by Similar Mechanisms [33]

Polymerization of pure actin *in vitro* requires ATP as well as both monovalent and divalent cations, which are usually K^+ and Mg^{2+}. The reaction is often studied either by observing the change in the light emission from a fluorescent probe that has been covalently attached to the actin or by monitoring the large increase in viscosity caused by the polymerization. When K^+ and Mg^{2+} are added to monomeric actin in the presence of ATP, there is initially a lag phase, as new filaments

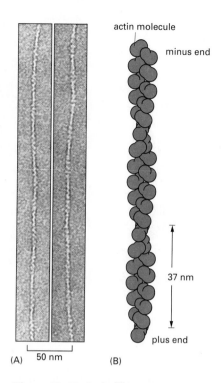

Figure 16–49 Actin filaments.
(A) Electron micrographs of negatively stained actin filaments.
(B) The helical arrangement of actin molecules in an actin filament. (A, courtesy of Roger Craig.)

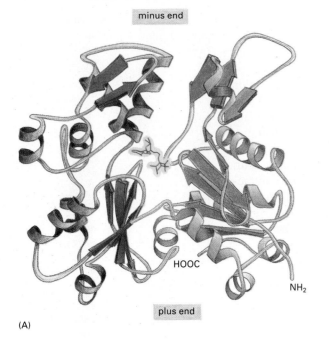

minus end

plus end

(A)

HOOC

NH₂

(B)

ATP

(C)

ADP
ADP
ADP
ADP
ADP
ADP

minus end

plus end

(D)

are nucleated, and then a rapid polymerization phase, as the short filaments elongate. The lag in polymerization with pure actin is due to the same kinetic barrier to nucleation that we discussed for tubulin polymerization (see Figure 16–23). For actin the rate of nucleation is proportional to the cube of the actin concentration, suggesting that the nucleating structure for the spontaneous polymerization of pure actin is a trimer of actin molecules. By contrast, the rate at which each filament elongates is proportional, as for microtubules, to the concentration of the free subunit, indicating that the filament elongates by the addition of one actin molecule at a time.

The polymerization rate is different at the two ends of the actin filament, and this difference is greater than for microtubules: the plus (or barbed) end of actin filaments polymerizes at up to 10 times the rate of the minus (or pointed) end. The *critical concentration* for actin polymerization—that is, the free actin monomer concentration at which the proportion of actin in polymer stops increasing—is around 0.2 micromolar (about 8 μg/ml). This concentration is very much lower than the concentration of unpolymerized actin in a cell, and the cell has evolved special mechanisms to prevent most of its monomeric actin from assembling into filaments, as we discuss later.

Shortly after polymerization, the terminal phosphate of the ATP bound to the actin molecule is hydrolyzed, leaving the resulting ADP trapped in the polymer. The hydrolysis of ATP during actin polymerization is analogous to the GTP hydrolysis that accompanies microtubule assembly, but in the case of actin we can understand the conformational changes involved because the three-dimensional structure of actin is known. The actin molecule is clam-shaped and binds ATP in the crevice between its two halves; like a clam shell, it can open and close. When actin polymerizes, the shell is clamped shut by interactions between amino acids on both lips of the shell and the back side of the next subunit in the polymer. It is thought that ATP hydrolysis is triggered by the closing of the clam shell as each actin molecule is incorporated into the filament, leaving ADP trapped inside (Figure 16–51).

ATP Hydrolysis Is Required for the Dynamic Behavior of Actin Filaments [34]

The role of ATP hydrolysis in actin polymerization is similar to the role of GTP hydrolysis in tubulin polymerization, as explained in Panel 16–1 (pp. 824–825).

Figure 16–50 The structure of actin. (A) The three-dimensional structure of an actin molecule, deduced by x-ray diffraction analysis. A single molecule of ATP (*yellow*) is tightly bound in a crevice between the two domains of the protein. (B) A schematic drawing of an actin molecule that emphasizes its two domains and the binding site for ATP that lies between them. (C) Schematic drawing of the actin filament showing how the actin molecules interact with each other to form a helical polymer. Note that as the actin molecules assemble into the polymer, they hydrolyze their tightly bound ATP molecules (see Figure 16–51). (D) The structure of the actin molecule fitted onto the image of an actin filament obtained by electron microscopy. Each ball in the model represents a single amino acid; those that interact with myosin (discussed later) are shown in *green*. The difference in structure of the plus and minus ends of the actin filament is apparent. (A, adapted from W. Kabsch et al., *Nature* 347:37–44, 1990. © 1990 Macmillan Magazines Ltd.; D, from K.C. Holmes et al., *Nature* 347:44–49, 1990. © 1990 Macmillan Magazines Ltd.)

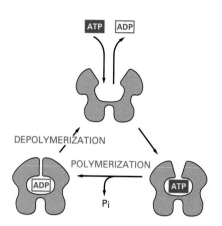

Figure 16–51 The trapping of ADP in an actin filament. An actin molecule has a structure that is related to that of the ubiquitous enzyme hexokinase (see Figure 5–2), with two domains that are hinged around an ATP-binding site. The bound ATP is hydrolyzed to ADP immediately after the molecule becomes incorporated into an actin filament. In order for the ADP to be replaced by ATP, the hinge would have to open. But in the actin filament the two domains in each actin molecule are held together by interactions with neighboring subunits, thereby keeping the hinge closed and trapping the ADP in the actin filament until the filament depolymerizes.

In neither case is hydrolysis required to form the filament; instead, it serves to weaken the bonds in the polymer and thereby promote depolymerization. There are, however, important differences in the behavior of the bound nucleotide in the subunits of these two polymers. An especially interesting difference is that ATP-ADP exchange (the replacement of bound ADP by ATP) is relatively slow for free actin (half-time of minutes), while GTP-GDP exchange is very rapid for free tubulin (half-time of seconds); thus, when actin molecules are released by disassembly of a filament, there is a relatively long delay before they can be re-used in filament assembly. In principle, this property of actin allows the cell to maintain a high cytosolic concentration of unpolymerized actin molecules in the form of ADP actin; furthermore, the ADP-actin monomer in a cell can be stabilized by binding to another protein, and this could provide a way to regulate actin polymerization.

The effect of ATP hydrolysis on actin is subtle, and there are still many questions about its precise consequences for the cell. Actin filaments, unlike microtubules, do not seem to show drastic dynamic instability *in vitro*. Instead, they can engage in an interesting dynamic behavior called *treadmilling*, which occurs when actin molecules are added continually to the plus end of the filament and are lost continually from the minus end, with no net change in filament length (see Panel 16-1, pp. 824–825). Treadmilling, like dynamic instability, is a nonequilibrium behavior that requires an input of energy, which is provided by the ATP hydrolysis that accompanies polymerization. This phenomenon is thought to contribute to the rapid exchange of the subunits of actin filaments that takes place in cells.

It is remarkable that actin and tubulin have both evolved nucleoside triphosphate hydrolysis for the same basic reason—to enable them, having polymerized, to depolymerize readily. Actin and tubulin are completely unrelated in amino acid sequence: actin is distantly related in structure to the glycolytic enzyme hexokinase, whereas tubulin is distantly related to a large family of GTPases that includes the heterotrimeric G proteins and monomeric GTPases such as Ras. (Both types of structures are discussed in detail in Chapter 5.) The convergent evolution of the capacity for nucleotide hydrolysis in actin and tubulin demonstrates just how important it is to microtubule and actin filament function: the dynamic assembly and disassembly of these cytoskeletal polymers that hydrolysis makes possible lies at the heart of cytoplasmic organization.

The Functions of Actin Filaments Are Inhibited by Both Polymer-stabilizing and Polymer-destabilizing Drugs [35]

Drugs that stabilize or destabilize actin filaments provide important tools to investigate their dynamic behavior in cells. The *cytochalasins* are fungal products that prevent actin from polymerizing by binding to the plus end of actin filaments. The *phalloidins* are toxins isolated from the *Amanita* mushroom that bind tightly all along the side of actin filaments and stabilize them against depolymerization. (One remedy for *Amanita* mushroom poisoning is to eat a large quantity of raw meat: the high concentration of actin filaments in the muscle tissue binds the phalloidin and thereby reduces its toxicity.) Both of these drugs cause

Actin Filaments

ON RATES AND OFF RATES

A linear polymer of protein molecules, such as an actin filament or a microtubule, assembles (polymerizes) and disassembles (depolymerizes) by the addition and removal of subunits at the ends of the polymer. The rate of addition of monomers is given by the rate constant k_{on}, which has units of M^{-1} sec^{-1}. The rate of loss is given by k_{off} (units of sec^{-1}).

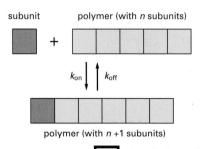

subunit   polymer (with *n* subunits)

k_{on} | k_{off}

polymer (with *n* +1 subunits)

NUCLEATION

A helical polymer is stabilized by multiple contacts between adjacent subunits. In the case of actin, two actin molecules bind relatively weakly to each other, but addition of a third actin monomer to form a trimer makes the entire group more stable.

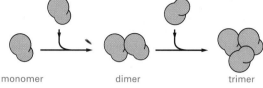

monomer   dimer   trimer

Further addition can take place onto this trimer, which therefore acts as a nucleation site for polymerization. For tubulin the nucleation site is larger and has a more complicated structure (possibly a ring of 13 or more molecules)—but the principle is the same.

The assembly of a nucleation site is relatively slow, and this explains the lag phase seen during polymerization. The lag phase can be reduced or abolished entirely if premade nucleation sites, such as fragments of already polymerized microtubules or actin filaments, are added to serve this role.

THE CRITICAL CONCENTRATION

The number of monomers that add to the polymer (actin filament or microtubule) per second will be proportional to the concentration of the free subunit ($k_{on}C$), but the subunits will leave the polymer end at a constant rate (k_{off}) that does not depend on C. As the polymer grows, subunits are used up and C is observed to drop until it reaches a constant value, called the critical concentration (C_c). At this concentration the rate of subunit addition equals the rate of subunit loss.

At this equilibrium phase,

$$k_{on} C = k_{off}$$

so that

$$C_c = \frac{k_{off}}{k_{on}} = \frac{1}{K}$$

(where K is the equilibrium constant for subunit addition, see Figure 3–9).

TIME COURSE OF POLYMERIZATION

The assembly of a protein into a long helical polymer such as a cytoskeletal filament or a bacterial flagellum typically shows the following time course:

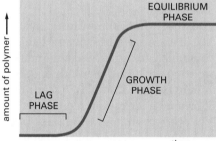

The lag phase is due to a kinetic barrier to nucleation.

The growth phase occurs as monomers add to the exposed ends of the growing polymer.

The equilibrium phase is reached when the growth of the polymer due to monomer addition is precisely balanced by the shrinkage of the polymer due to monomer loss.

PLUS AND MINUS ENDS

The two ends of an actin filament or microtubule polymerize at different rates. The fast-growing end is called the plus end, whereas the slow-growing end is called the minus end. The difference in the rates of growth at the two ends is made possible by changes in the conformation of each subunit as it enters the polymer.

free subunit   subunit in polymer

This conformational change affects the rates at which subunits add to the two ends.

Even though k_{on} and k_{off} will have different values for the plus and minus ends of the polymer, their ratio k_{off}/k_{on}—and hence C_c—must be the same at both ends. This is because exactly the same subunit interactions are broken when a subunit is lost at either end, and the final state of the subunit after dissociation is

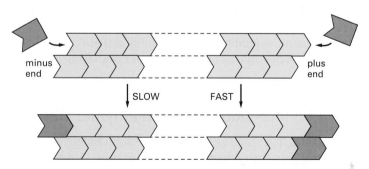

minus end   plus end

SLOW   FAST

identical. Therefore the ΔG for subunit loss, which determines the equilibrium constant for its association with the end (see Table 3–3, p. 97), is identical at both ends: if the plus end grows four times faster than the minus end, it must also shrink four times faster. Thus, for C > C_c both ends grow; for C < C_c both ends shrink.

TREADMILLING

One consequence of the nucleotide hydrolysis that accompanies polymer formation is to change the critical concentration at the two ends of the polymer. Since k^D_{off} and k^T_{on} refer to different reactions, their ratio k^D_{off}/k^T_{on} need not be the same at both ends of the polymer, so that:

$$C_c \text{ (minus end)} > C_c \text{ (plus end)}$$

Thus, if both ends of a polymer are exposed, polymerization will proceed until the concentration of free monomer reaches a value that is above C_c for the plus end but below C_c for the minus end. At this steady state, subunits will undergo a net assembly at the plus end and a net disassembly at the minus end at an identical rate. The polymer will maintain a constant length, even though there is a net flux of subunits through the polymer, known as treadmilling.

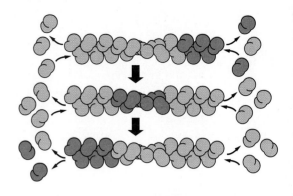

NUCLEOTIDE HYDROLYSIS

Each actin molecule carries a tightly bound ATP molecule that is hydrolyzed to a tightly bound ADP molecule soon after its assembly into polymer. Similarly, each tubulin molecule carries a tightly bound GTP that is converted to a GDP molecule soon after the molecule assembles into the polymer.

free monomer   subunit in polymer

(T = ATP or GTP)
(D = ADP or GDP)

Hydrolysis of the bound nucleotide reduces the binding affinity of the subunit for neighboring subunits and makes it more likely to dissociate from each end of the filament (see Figure 16–33 for a possible mechanism). It is usually the T form that adds to the filament and the D form that leaves.

Considering events at the plus end only:

k^T_{on}

k^D_{off}

As before, the polymer will grow until $C = C_c$. For illustrative purposes, we can ignore k^D_{on} and k^T_{off} since they are usually very small, so that polymer growth ceases when

$$k^T_{on} C = k^D_{off} \quad \text{or} \quad C_c = \frac{k^D_{off}}{k^T_{on}}$$

This is a steady state and not a true equilibrium, because the ATP or GTP that is hydrolyzed must be replenished by a nucleotide exchange reaction of the free subunit (D → T).

DYNAMIC INSTABILITY and TREADMILLING

DYNAMIC INSTABILITY and TREADMILLING are two behaviors observed in cytoskeletal polymers. Both are associated with nucleoside triphosphate hydrolysis. Dynamic instability is believed to predominate in microtubules, whereas treadmilling may predominate in actin filaments.

ATP CAPS / GTP CAPS

The rate of addition of subunits to a growing actin filament or microtubule can be faster than the rate at which their bound nucleotide is hydrolyzed. Under such conditions subunits at the end form a "cap" of subunits containing the nucleoside triphosphate—an ATP cap on an actin filament or a GTP cap on a microtubule.

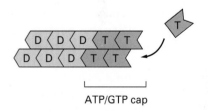

ATP/GTP cap

DYNAMIC INSTABILITY

Microtubules depolymerize about 100 times faster from an end containing GDP tubulin than from one containing GTP tubulin. A GTP cap favors growth, but if it is lost, then depolymerization ensues.

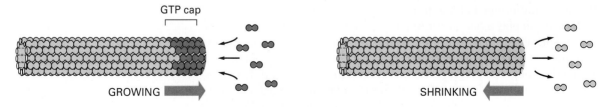

GTP cap

GROWING   SHRINKING

Individual microtubules can therefore alternate between periods of slow growth and rapid disassembly, a phenomenon called dynamic instability.

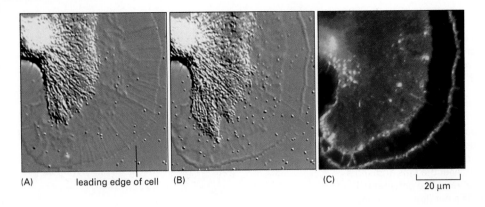

(A)   leading edge of cell   (B)   (C)   20 μm

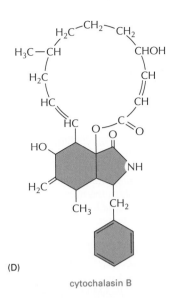

(D)   cytochalasin B

dramatic changes in the actin cytoskeleton. We saw earlier for microtubules that both polymer-destabilizing drugs such as colchicine and polymer-stabilizing drugs such as taxol are toxic to cells, and the same is true for drugs affecting the stability of actin filaments, indicating that the function of actin filaments also depends on a dynamic equilibrium between the filaments and actin monomer.

Cytochalasin has found its greatest use in studying cell locomotion. In particular, the leading edge of a moving cell contains actin filaments that are continually polymerizing and are therefore very sensitive to cytochalasin. In most moving cells cytochalasin causes the leading edge rapidly to retract. If the plasma membrane of the leading edge is very firmly attached to the substratum, however, cytochalasin causes the actin filaments to retract but leaves the membrane behind, stuck to the substratum (Figure 16–52).

Phalloidin is widely used, as a fluorescent derivative, to stain actin filaments in fixed cells, and it also has a profound effect on living cells. When it is microinjected into a living fibroblast, for example, it drives all of the actin monomer into filaments at random positions in the cytoplasm, causing a drastic blebbing and contraction that often destroys the cell.

Figure 16–52 **The effect of cytochalasin on the leading edge of the growth cone of a nerve cell in culture.** A living growth cone is viewed by Nomarski differential-interference-contrast microscopy both before (A) and after (B) treatment with cytochalasin. The cell in (B) has then been stained with rhodamine phalloidin to reveal the actin filaments (C). Note how the region behind the leading edge of the cytochalasin-treated growth cone is devoid of actin filaments. The chemical structure of cytochalasin B is shown in (D). (A, B, and C, courtesy of Paul Forscher.)

The Actin Molecule Binds to Small Proteins That Help to Control Its Polymerization [36]

In a fibroblast cell approximately 50% of the actin is in filaments and 50% is in monomer. The monomer concentration is typically 50–200 micromolar (2–8 mg/ml) in a variety of cell types; this is surprisingly high, given the low critical concentration of pure actin (less than 1 micromolar), and it reflects the presence of special proteins that bind to the actin molecule and inhibit its addition to the ends of actin filaments. The most abundant of these actin-monomer-binding proteins in many cells is **thymosin,** an unusually small protein with a molecular weight of about 5000. In the cells in which it has been most carefully studied (blood platelets and neutrophils), it is present in concentrations that are sufficient to sequester all of the monomeric actin. It is not clear how this protein inhibits actin polymerization: it could sterically block polymerization by covering a site where one monomer binds to another, or it could trap ADP on actin by inhibiting ADP-ATP exchange, thereby making the actin molecule unlikely to polymerize (Figure 16–53).

Another actin-monomer-binding protein is **profilin,** which is present in all cells and is thought to play a part in controlling actin polymerization in response to extracellular stimuli. Profilin, which in many cells is largely associated with the plasma membrane, accelerates the exchange of ATP for ADP when bound to actin monomers and is thought to play a part in promoting the regulated polymerization of actin during cell movement, although this is still controversial. A mutant yeast cell that is deficient in profilin has a deficit of actin filaments, which supports a role for this molecule in stimulating the polymerization of actin.

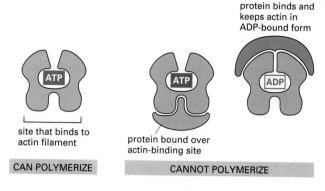

protein binds and keeps actin in ADP-bound form

site that binds to actin filament

CAN POLYMERIZE

protein bound over actin-binding site

CANNOT POLYMERIZE

Figure 16–53 Two possible mechanisms by which an actin-monomer-binding protein could inhibit actin polymerization. It is thought that thymosin inhibits actin polymerization in one of these ways.

In addition to thymosin and profilin, cells contain other abundant proteins that are able to bind actin monomers, and some of these, such as *actin-depolymerizing factor (ADF)*, inhibit the assembly of actin into filaments. Evidently cells have a variety of mechanisms, the details of which are not yet understood, by which they hold stocks of actin monomer in reserve in order to assemble actin filaments only when and where they are needed.

Many Cells Extend Dynamic Actin-containing Microspikes and Lamellipodia from Their Leading Edge [37]

Dynamic surface extensions containing actin filaments are a common feature of animal cells, especially when the cells are moving or changing shape. The large, free-living cells of *Amoeba proteus,* for example, produce *pseudopodia*—stubby distensions of the actin cortex—with which they walk over surfaces. Many cells in vertebrate tissues are also capable of independent migration over surfaces, especially when put into tissue culture. The leading edge of a crawling fibroblast regularly extends a thin, sheetlike process known as a **lamellipodium,** which contains a dense meshwork of actin filaments. Many cells also extend thin, stiff protrusions called *microspikes,* which are about 0.1 μm wide and 5 to 10 μm long and contain a loose bundle of about 20 actin filaments oriented with their plus ends pointing outward (see Figure 16–9). The growing tip (growth cone) of a developing nerve cell axon extends even longer microspikes, called *filopodia,* which can be up to 50 μm long.

Figure 16–54 Actin filaments at the leading edge of a fibroblast in culture. (A) Whole-mount electron micrograph of the leading edge of a cultured cell that has been extracted with nonionic detergent to remove the plasma membrane and most of the soluble proteins. Note the oriented network of actin filaments in the lamellipodium, in which a microspike is embedded. A schematic view of the actin filaments in the lamellipodium is shown in (B). (A, from J.V. Small, *J. Cell Biol.* 91:695–705, 1981, by copyright permission of the Rockefeller University Press.)

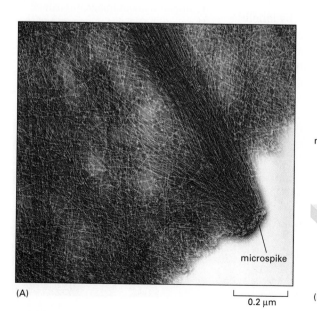

(A)

microspike

0.2 μm

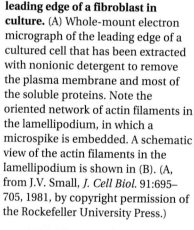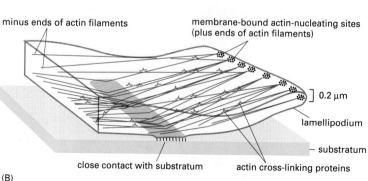

minus ends of actin filaments

membrane-bound actin-nucleating sites (plus ends of actin filaments)

0.2 μm

lamellipodium

substratum

close contact with substratum

actin cross-linking proteins

(B)

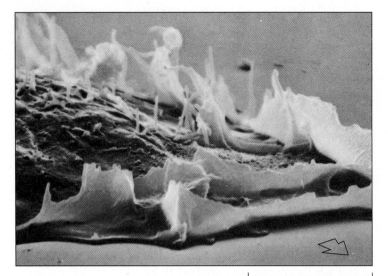

Figure 16–55 Lamellipodia and microspikes at the leading edge of a human fibroblast migrating in culture. The arrow in this scanning electron micrograph shows the direction of cell movement. As the cell moves forward, lamellipodia and microspikes that fail to attach to the tissue culture dish sweep backward over its dorsal surface—a movement known as *ruffling*. (Courtesy of Julian Heath.)

5 μm

A lamellipodium can be viewed as a two-dimensional version of a microspike; indeed, short microspikes often project from the edges of a lamellipodium. When carefully fixed and stained for examination in an electron microscope, the actin filaments in the lamellipodium of a moving cell appear to be more organized than they are in other regions of the cell cortex. Many of the filaments project outward in an orderly array, with their plus ends inserted into the leading edge of the plasma membrane (Figure 16–54). The lamellipodium behaves as a structural unit; if it fails to adhere to the substratum, it is usually swept rapidly backward over the top of the cell as a "ruffle" (Figure 16–55).

Both lamellipodia and microspikes are motile structures that can form and retract with great speed. As we discuss next, it is thought that microspikes and lamellipodia are generated by local actin polymerization at the plasma membrane and that such actin polymerization can rapidly push out the plasma membrane without tearing it.

The Leading Edge of Motile Cells Nucleates Actin Polymerization [38]

When the behavior of actin filaments at the leading edge is studied by labeling a small patch of actin and following its movement, it is seen that actin is continually moving back toward the cell body at a speed of about 1 μm/minute, suggesting that actin is continuously polymerizing near the tip of the leading edge and continuously depolymerizing at more internal sites (Figure 16–56). This highly

Figure 16–56 Actin filament dynamics in the lamellipodium of a cultured fibroblast. Actin molecules labeled with the fluorescent dye rhodamine were microinjected into the cell, where they became incorporated into actin filaments. A small spot on the actin filaments at the leading edge of the cell was bleached with a laser beam. The cell was then photographed at intervals using a fluorescence microscope equipped with an image intensifier. The rapid backward movement of the bleached spot suggests that actin polymerizes continuously at the tip of the leading edge and depolymerizes at its base. (Courtesy of Y.L. Wang.)

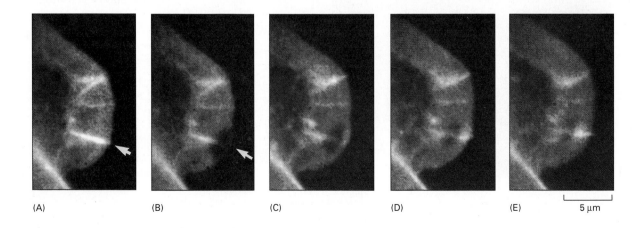

(A)   (B)   (C)   (D)   (E)   5 μm

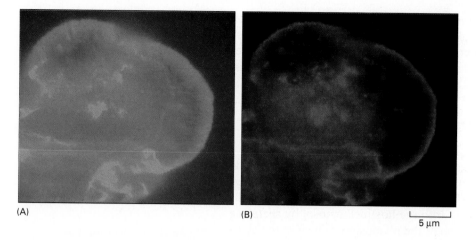

(A)   (B)   5 µm

Figure 16–57 The tip of the leading edge nucleates actin filaments. Fibroblast cells in culture were gently permeabilized using a nonionic detergent and were then incubated with rhodamine-labeled actin molecules (*red*). After 5 minutes the cells are fixed and stained with fluorescein-labeled phalloidin (*green*). (A) All of the actin filaments, most of which were formed prior to lysis, are shown in *green*. (B) The location of the newly formed actin filaments (*red*) polymerized from the added rhodamine-actin show that the leading edge is the predominant site of actin filament nucleation in the cell. (From M.H. Symons and T.J. Mitchison, *J. Cell Biol.* 114:503–513, 1991, by copyright permission of the Rockefeller University Press.)

dynamic behavior of actin filaments at the leading edge is thought to be crucial for such processes as directed cell locomotion and chemotaxis. It gives the impression that the leading edge is propelling itself forward by pushing actin filaments backward.

The leading edge of a cell seems to organize actin filaments much as a centrosome organizes microtubules but with one crucial difference: it not only nucleates the growth of new filaments, but also seems to be the site at which monomers are added subsequently to enable the filaments to elongate. This role can be demonstrated by gently lysing a fibroblast and then adding rhodamine-tagged actin monomers, which are seen to polymerize preferentially at the tip of the leading edge (Figure 16–57). Moreover, if the actin filaments in a cell are decorated to reveal their polarity, the fast-growing plus end of each actin filament is found to be attached to the membrane at the leading edge.

There are many unanswered questions about the mechanism by which the leading edge nucleates actin filament polymerization. Does the leading edge hold on to the plus end of a filament that it nucleates, for example, or does it nucleate a new filament and then quickly release it? Because of the continuous backward movement of actin (see Figure 16–56), any model that postulates that the leading edge holds onto actin filament ends would require that the filaments in the lamellipodium undergo continuous treadmilling by insertion of actin monomers at the site where the filaments are held by the membrane. In an alternative model, individual actin filaments are released and move away from the membrane (presumably as a cross-linked meshwork) soon after they form (Figure 16–58).

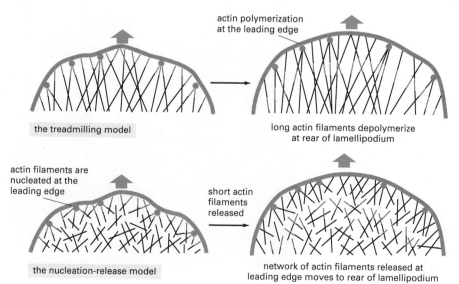

actin polymerization at the leading edge

the treadmilling model

long actin filaments depolymerize at rear of lamellipodium

actin filaments are nucleated at the leading edge

short actin filaments released

the nucleation-release model

network of actin filaments released at leading edge moves to rear of lamellipodium

Figure 16–58 Two models that could explain the backward flux of actin in a lamellipodium. The *blue arrows* indicate the direction of cell movement. Although the differences between the two models are emphasized here, both processes could occur simultaneously in the cell. (Adapted from J.A. Theriot and T.J. Mitchison, *Nature* 352:126–131, 1991. Reprinted with permission from *Nature.* © 1991 Macmillan Magazines Ltd.)

The rapid assembly of actin filaments at the leading edge of a moving cell requires that actin monomers be released from the actin-monomer-binding proteins that normally restrain their polymerization into filaments. We discuss below how signals in the cell's environment may regulate the release of actin monomers for polymerization at the tip of the leading edge.

Some Pathogenic Bacteria Use Actin to Move Within and Between Cells [39]

Listeria monocytogenes, a bacterium that causes a severe form of food poisoning, has provided unexpected insights into the mechanism by which the local polymerization of actin is controlled in cells. This pathogenic bacterium enters cells by being phagocytosed; it then escapes into the host cell cytosol by secreting enzymes that break down the membrane of the phagosome. Once in the cytosol, the bacteria not only grow and divide, but they also spread to adjoining cells by mobilizing the actin-based motility system of the host cell. By nucleating actin filaments at one region of its surface, an individual bacterium moves through the cytosol at rates of 10 μm/minute or more, laying down a tail of actin filaments behind it. When it collides with the plasma membrane of the host cell, it keeps moving outward, inducing the formation of a long, thin microspike with a bacterium at its tip. This projection is often engulfed by a neighboring cell, allowing the bacterium to enter its cytoplasm without exposure to the extracellular environment, thereby avoiding recognition by antibodies produced by the host (Figure 16–59).

This form of movement suggests that the bacterium may be using actin to propel itself forward in the same way that the plasma membrane of a eucaryotic cell uses actin to propel itself forward during the formation of a normal microspike or lamellipodium.

Figure 16–59 The actin-based movement of a bacterium within and between mammalian cells. (A) The bacterium *Listeria monocytogenes* spreads from cell to cell by inducing the assembly of actin filaments in the host cell cytosol. (B) Fluorescence micrograph of the bacterium moving in a cell that has been stained to reveal both bacteria and actin filaments. Note the cometlike tail of actin filaments (*green*) behind each moving bacterium (*red*). Regions of overlap of *red* and *green* fluorescence appear *yellow*. (B, courtesy of Tim Mitchison and Julie Theriot.)

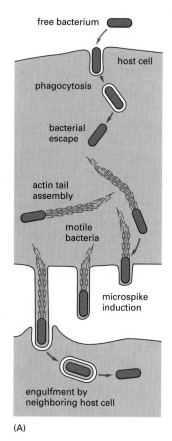

(A)

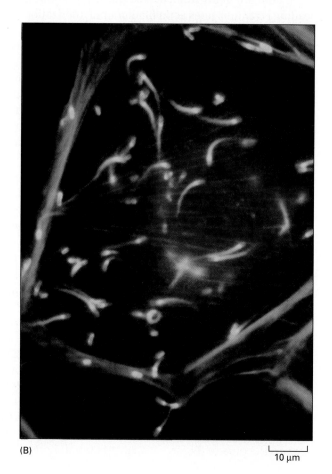

(B)

10 μm

If the actin filaments in the tail behind a *Listeria* bacterium migrating in the cytosol are marked with a fluorescent tag and observed by fluorescence microscopy, they are found to be stationary. The filaments form at the rear of the bacterium and are left behind like a rocket trail as the bacterium advances, depolymerizing again within a minute or so as they encounter depolymerizing factors in the cytosol. Assembly is induced by a specific protein on the surface of the bacterium that acts indirectly by sequestering host-cell proteins, including profilin. Since bacterium-induced movement can be reproduced in a concentrated cell-free extract, details of the mechanism should emerge from biochemical studies. These details should help us to understand how actin nucleation and polymerization occur in the microspikes and lamellipodia of a normal, uninfected cell and how these processes power the forward movement of the cell.

Polymerization of Actin in the Cell Cortex Is Controlled by Cell-Surface Receptors [40]

The production of movement is of little use unless it is properly directed according to the environment. As discussed earlier, the dynamic cortical meshwork of actin filaments rearranges rapidly in response to signals from outside the cell that impinge on the plasma membrane. The actin cytoskeleton can therefore be considered to be an integral part of the cell's signal-transduction systems, discussed in Chapter 15: when certain growth factors are added to the medium bathing quiescent cells in culture, for example, they immediately cause actin-containing lamellipodia to form and move over the cell surface.

The response of the actin cortex to external signals conveying spatial information can be highly localized. We considered one example earlier when we discussed the polarization of a cytotoxic T cell that is induced by contact with the target cell it subsequently kills (see Figure 16–11). A signal-induced polarization of the actin cortex also occurs in animal cells that are capable of *chemotaxis*, which is defined as movement in a direction controlled by a gradient of a diffusible chemical sensed by the cell. One well-studied example is the chemotactic movement of certain white blood cells (*neutrophils*) toward a source of bacterial infection. Neutrophils have receptor proteins on their surface that enable them to detect the very low concentrations of the N-formylated peptides derived from bacterial proteins (only procaryotes begin protein synthesis with N-formyl methionine). The neutrophils can be guided to their targets by a difference of only 1% in the concentration of these diffusible peptides on one side of the cell versus the other.

Another example of chemotaxis is provided by the cellular slime mold ***Dictyostelium discoideum.*** These eucaryotes live on the forest floor as independent motile cells called *amoebae,* which feed on bacteria and yeast and, under optimal conditions, divide every few hours. When their food supply is exhausted, the amoebae stop dividing and gather together to form tiny (1–2 mm), multicellular, wormlike structures, which crawl about as glistening slugs and leave trails of slime behind them (Figure 16–60). As the slug migrates, the cells begin to differentiate, initiating a process that ends with the production of a tiny plantlike structure consisting of a stalk and a *fruiting body* some 30 hours after the beginning of aggregation (Figure 16–61). The fruiting body contains large numbers of spores, which can survive for long periods of time even in extremely hostile environments. Only when conditions are favorable do the spores germinate to produce the free-living amoebae that start the cycle again.

The *Dictyostelium* amoebae aggregate by chemotaxis, migrating toward a source of cyclic AMP, which is secreted by the starved amoebae. Like neutrophils, the amoebae reorient their leading edge in order to migrate up a shallow chemoattractant gradient. And when they are exposed to a local source of cyclic AMP leaking from a micropipette, they extend actin-containing processes directly toward the pipette (Figure 16–62). This experiment shows that eucaryotic chemo-

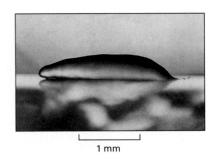

Figure 16–60 Light micrograph of a migrating slug of the cellular slime mold *Dictyostelium discoideum.* (Courtesy of David Francis.)

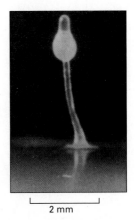

Figure 16–61 Light micrograph of a *Dictyostelium discoideum* fruiting body. (Courtesy of John Bonner.)

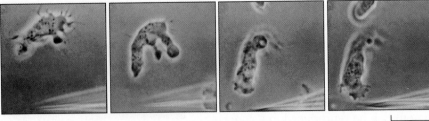

50 μm

Figure 16–62 A chemotactic response in a *Dictyostelium* amoeba. The amoeba has receptors for cyclic AMP in its plasma membrane that enable it to crawl toward an extracellular source of cyclic AMP. In this experiment cyclic AMP was released from the tip of the micropipette seen at the bottom of the micrographs; the response illustrated occupied less than a minute. (Courtesy of Günther Gerisch.)

taxis involves detecting a spatial gradient of attractant concentration directly, in contrast to bacterial chemotaxis, which uses a time-dependent variation in concentration to detect gradients, as discussed in Chapter 15.

The cytoskeletal reaction of *Dictyostelium* amoebae to cyclic AMP can be examined by making lysates of these cells very shortly after bulk stimulation with cyclic AMP in solution. As shown in Figure 16–63, a dramatic burst of actin polymerization occurs 5–10 seconds after adding cyclic AMP, which corresponds to the time required for flattening of the cells on the substratum. Between 20 and 40 seconds after the pulse of cyclic AMP, actin depolymerizes and the cells round up. Then there is a more prolonged burst of actin polymerization as actin-binding proteins are recruited into the cytoskeleton from soluble pools; during this latter period the cells that respond to cyclic AMP begin to extend lamellipodia and other actin-rich processes.

Heterotrimeric G Proteins and Small GTPases Relay Signals from the Cell Surface to the Actin Cortex [41]

How does cyclic AMP binding to its receptor in *Dictyostelium* amoebae trigger massive actin polymerization? The receptor is known to activate a heterotrimeric G protein. The cytoplasm contains a reservoir of actin monomers, which, as we saw earlier, are stabilized by actin-monomer-binding proteins. Stimulation of actin polymerization requires that these actin molecules be made available in a form that can polymerize and also that nucleation sites for actin filaments be provided to overcome the kinetic barrier to nucleation. The actin-monomer-binding protein profilin binds tightly to the inositol phospholipids in the plasma membrane that generate intracellular signals in response to extracellular ligands

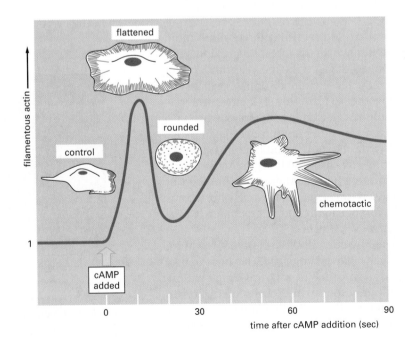

Figure 16–63 The effect of cyclic AMP on the actin cortex of a *Dictyostelium* amoeba. The graph (*green line*) shows the relative amounts of filamentous actin associated with the cytoskeleton at different times following the sudden addition of cAMP.

(see Figure 15–30). According to one hypothesis, activation of this signaling pathway (which occurs via a heterotrimeric G protein) could release profilin from the plasma membrane into the cytosol. Profilin can catalyze ATP-ADP exchange on actin *in vitro*, and so when it is released from the plasma membrane, it may rapidly convert inactive ADP actin to active ATP actin to induce the local formation of actin filaments.

G proteins have also been implicated in the signaling processes that activate the actin cortex during the chemotactic response of neutrophils and the activation of blood platelets. There is evidence that two Ras-related small GTPases known as **Rho** and **Rac** act downstream; these proteins have been shown to have distinct effects on the actin cytoskeleton in fibroblasts. Microinjection of Rac protein into cultured cells causes a dramatic increase in the formation of lamellipodia within 5 minutes. Moreover, a dominant-negative mutant form of Rac inhibits the formation of lamellipodia normally induced by various growth factors, indicating that this response to growth factors depends on Rac. Microinjection of Rho protein leads to the appearance of large bundles of actin filaments known as *stress fibers* and to the enhancement of *focal contacts*, where the cell is attached to the substratum externally and stress fibers are anchored internally (as we discuss later). Rho is also thought to be needed to assemble the contractile ring during cell division. Thus Rac and Rho not only control the polymerization of actin into filaments but also govern the organization of these filaments into specific types of structures.

Mechanisms of Cell Polarization Can Be Analyzed in Yeast Cells [42]

Further clues to how cells may orient the activities of their cytoskeleton have come from the behavior of yeast cells. The ease of genetic analysis in yeasts has made them an important source of fundamental information about biological mechanisms that are common to all eucaryotic cells. In particular, studies on the interactions between yeast cells during mating have begun to identify mechanisms by which eucaryotic cells become structurally polarized. In the budding yeast *Saccharomyces cerevisiae*, cells of two mating types, a and α, secrete hormones, known as a-factor and α-factor, respectively. These hormones act by binding to cell-surface receptors that belong to the large G-protein-linked receptors discussed in Chapter 15. One consequence of the binding of α-factor to its receptors on an a-cell is to cause the cell to become polarized so that it adopts a shape known as a "shmoo" (Figure 16–64). If an α-factor gradient is present, the shmoo tip is directed toward the highest concentration of this signaling molecule.

During this polarization response the yeast cell undergoes cytoskeletal reorganizations that parallel those of an animal cell that is becoming polarized. Actin filaments congregate at the pointed shmoo tip, where they are thought to direct the local secretion of cell-wall components—possibly by directing the transport vesicles carrying these components to the shmoo tip. At the same time, the microtubule organizing center (in this case the *spindle pole body*, see Figure 17–24) moves to the side of the nucleus that is closest to the shmoo tip, and

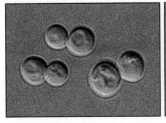
(A)

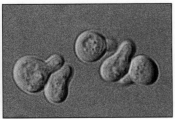
(B)

(C)

Figure 16–64 Morphological polarization of yeast cells. Cells of *Saccharomyces cerevisiae* are usually spherical (A), but they become polarized when treated with mating factor (B). The polarized cells are called "shmoos," after Al Capp's famous cartoon character (C). (A and B, courtesy of Michael Snyder; C,© 1948 Capp Enterprises, Inc., all rights reserved.)

microtubules extend from it toward the tip. By screening for mutant cells that fail to form a shmoo during mating, many of the genes involved in yeast-cell polarization are being identified. It is likely that some of the proteins that these genes encode will also be involved in polarizing an animal cell.

Summary

Actin is a highly conserved cytoskeletal protein that is present at high concentrations in nearly all eucaryotic cells. Purified actin exists as a monomer in low ionic strength solutions and spontaneously assembles into actin filaments on addition of salt provided ATP is present. As with tubulin, the polymerization of actin is a dynamic process that is regulated by the hydrolysis of a tightly bound nucleotide (ATP in this case). In cells, approximately half of the actin is kept in a monomeric form through its binding to small proteins such as thymosin. In the cortex of animal cells, actin molecules continually polymerize and depolymerize to generate cell-surface protrusions such as lamellipodia and microspikes. Polymerization can be regulated by extracellular signals binding to cell-surface receptors that act through heterotrimeric G proteins and the small GTPases Rac and Rho.

Actin-binding Proteins [43]

Actin is involved in a remarkably wide range of structures, from stiff and relatively permanent extensions of the cell surface to the dynamic three-dimensional networks at the leading edge of a migrating cell. Very different structures based on actin coexist in every living cell. In every case the fundamental structure of the actin filament is the same. It is the length of these filaments, their stability, and the number and geometry of their attachments (both to one another and to other components of the cell) that varies in different cytoskeletal assemblies. These properties in turn depend on a large retinue of **actin-binding proteins,** which bind to actin filaments and modulate their properties and functions.

In this section we describe some of the most important actin-binding proteins and the structures they form. Many of these are found at the perimeter of the cell in the actin-rich layer just beneath the plasma membrane called the *cell cortex.* This layer gives an animal cell mechanical strength and enables it to perform a variety of surface movements, such as phagocytosis, cytokinesis (cell division), and cell locomotion.

A Simple Membrane-attached Cytoskeleton Provides Mechanical Support to the Plasma Membrane of Erythrocytes [44]

As noted in Chapter 10, the proteins **spectrin** and **ankyrin** were first discovered as prominent components of the membrane-associated cytoskeleton of mammalian red blood cells (erythrocytes). These unusual cells have lost their nucleus and internal membranes, and so the plasma membrane is the only membrane. It is supported by a two-dimensional network of spectrin tetramers that are connected at their ends by very short actin filaments. The spectrin is linked to the cytoplasmic tail of an abundant transmembrane carrier protein (band 3) by means of ankyrin bridges (see Figure 10–26). Close relatives of spectrin (also called *fodrin*) and of ankyrin are found in the cortex of many vertebrate cells. Thus the detailed arrangement of proteins in the erythrocyte cortex provides a simplified model for the actin-based cytoskeletal network that supports the plasma membrane in all other animal cells.

The actin filaments in the erythrocyte cortex are very short, acting only as cross-linking elements between spectrin tetramers. Those in a more typical cell cortex, by contrast, are much longer and thus project into the cytoplasm, where they form the basis of a three-dimensional actin filament network. It is uncertain whether ankyrinlike molecules anchor these more typical cortical arrays to the plasma membrane, although in some epithelial cells the transmembrane Na$^+$/K$^+$ ATPase (discussed in Chapter 11) is thought to link the plasma membrane to the cortical actin filament network through such molecules.

The cortical actin filament network generally determines the shape and mechanical properties of the plasma membrane. Many types of membrane attachments are needed for actin filaments to perform their various functions in the cortex; coupling to transmembrane proteins through ankyrin is only one. More dynamic attachments also exist, but the proteins that mediate them are just beginning to be characterized.

Cross-linking Proteins with Different Properties Organize Particular Actin Assemblies [43]

The cortical actin filaments in animal cells are organized into three general types of arrays (Figure 16–65). In *parallel bundles,* as found in microspikes and filopodia, the filaments are oriented with the same polarity and are often closely spaced (10–20 nm apart). In *contractile bundles,* as found in stress fibers and in the contractile ring that divides cells in two during mitosis, filaments are arranged with opposite polarities; they are more loosely spaced (30–60 nm apart) and contain the motor protein myosin-II (discussed later). In the *gel-like networks* of the cell cortex the filaments are arranged in a relatively loose, open array with many orthogonal interconnections. How are these different arrangements of the same actin filament generated and maintained within a single cell? While we do not know the complete answer, *actin filament cross-linking proteins* are clearly of central importance.

Actin filament cross-linking proteins can be divided into two classes—*bundling proteins* and *gel-forming proteins*—according to their effect on pure actin filaments *in vitro.* Bundling proteins cross-link actin filaments into a parallel array and are important for forming both the tight parallel arrays and the looser contractile bundles of actin filaments described above. Gel-forming proteins, by contrast, cross-link actin filaments at crosswise intersections, creating loose gels.

Fimbrin and *α-actinin* are widely distributed bundling proteins. **Fimbrin** is enriched in the parallel filament bundles at the leading edge of cells, particularly in microspikes and filopodia, and it is thought to be responsible for the tight association of actin filaments in these arrays. The second actin-bundling protein, **α-actinin**, is concentrated in stress fibers, where it is thought to be partly respon-

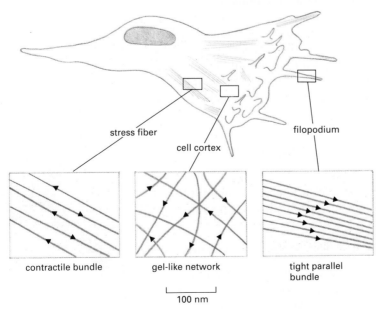

stress fiber

cell cortex

filopodium

contractile bundle

gel-like network

tight parallel bundle

100 nm

Figure 16–65 Three types of cortical arrays of actin filaments. A crawling cell is shown with three areas enlarged to show the arrangement of actin filaments drawn to scale. Arrowheads point toward the plus end of the filaments.

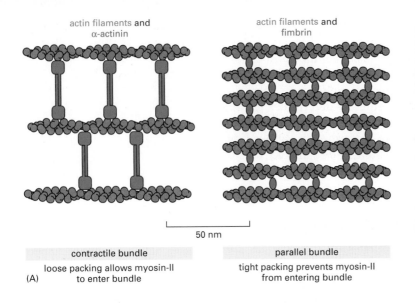

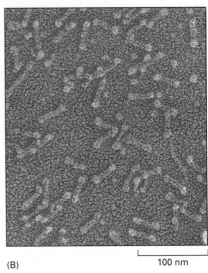

actin filaments and
α-actinin

actin filaments and
fimbrin

50 nm

contractile bundle
loose packing allows myosin-II
to enter bundle

(A)

parallel bundle
tight packing prevents myosin-II
from entering bundle

(B)

100 nm

sible for the relatively loose cross-linking of actin filaments in these contractile bundles; it also helps to form the anchorage for the ends of stress fibers where they terminate on the plasma membrane at *focal contacts*. As explained later, myosin is the motor protein in stress fibers and other contractile arrays that is responsible for their contractility. It seems likely that the very close packing of actin filaments caused by fimbrin excludes myosin, whereas the looser packing caused by α-actinin allows myosin molecules to enter; likewise, the very different spacing causes each of the two bundling proteins to exclude the other (Figure 16–66).

Filamin is a widely distributed gel-forming protein. Although it is not present in stress fibers or the leading edge, it is enriched elsewhere in the cortex. Filamin is a homodimer that promotes the formation of a loose and highly viscous network by clamping together two actin filaments that cross each other (Figure 16–67). It is an abundant protein in many animal cells, reflecting the prevalence of the loose-network type of actin organization.

Actin-binding Proteins with Different Properties Are Built Up from Similar Modules [45]

Fimbrin, α-actinin, filamin, and spectrin each contain two actin-filament-binding domains, which is not surprising given that each needs to cross-link two filaments. Unexpectedly, however, in all of these proteins the actin-binding domains have a similar structure. The length and flexibility of the spacer sequences that separate the two actin-binding sites differ in the four proteins, and these differences determine the different properties of the four cross-linkers. Evidently, these proteins have diverged from a common ancestral actin-binding protein by adding different spacer sequences (Figure 16–68).

Gelsolin Fragments Actin Filaments in Response to Ca²⁺ Activation [46]

Extracts prepared from many types of animal cells form a gel in the presence of ATP when they are warmed to 37°C. Although this *gelation* depends on both actin

Figure 16–66 The formation of two types of actin filament bundles. (A) α-actinin, which is a homodimer, cross-links actin filaments into loose bundles, which allow the motor protein myosin-II (not shown) to participate in the assembly. Fimbrin cross-links actin filaments into tight bundles, which exclude this motor protein. Fimbrin and α-actinin tend to exclude each other because of the very different spacing of the actin filament bundles that they form. (B) Electron micrograph of purified α-actinin molecules. (B, courtesy of John Heuser.)

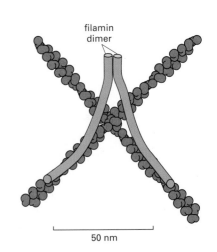

filamin
dimer

50 nm

Figure 16–67 Filamin cross-links actin filaments into a three-dimensional network with the physical properties of a gel. Each filamin homodimer is about 160 nm long when fully extended and forms a flexible, high-angle link between two adjacent actin filaments. Filamin can constitute 1% of the cell protein, or about one molecule per 50 actin monomers.

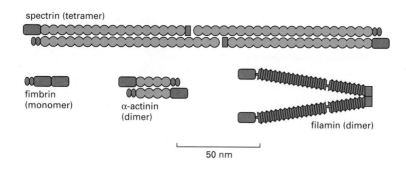

Figure 16–68 The modular structures of four actin-binding proteins. Each of the proteins shown has two actin-binding sites (*red*) that are related in sequence. Fimbrin has two directly adjacent actin-binding sites, so that it holds its two actin filaments very close together (14 nm apart), aligned with the same polarity (see Figure 16–66). The two actin-binding sites in α-actinin are more widely separated and are linked by a somewhat flexible spacer 30 nm long, so that it forms actin filament bundles with a greater separation between the filaments (40 nm apart) than does fimbrin. Filamin has two actin-binding sites that are very widely spaced, with a V-shaped linkage between them, so that it cross-links actin filaments into a network with the filaments oriented almost at right angles to one another (see Figure 16–67). Spectrin is a tetramer of two α and two β subunits, and the tetramer has two actin-binding sites spaced about 200 nm apart. The spacer regions of these various proteins are built in a modular fashion from repeating units that include α-helical motifs (*light green*), β-sheet motifs (*dark green*), and Ca^{2+}-binding domains (*blue ovals*).

filaments and a cross-linking protein such as filamin, the gels exhibit more complex behavior than simple mixtures of actin filaments and filamin. If the Ca^{2+} concentration is raised above 10^{-7} M, for example, the semisolid actin gel begins to liquefy—a process known as *solation*—and regions of the solating gel show vigorous local streaming when examined under a microscope. Clearly, there must be components besides actin and filamin in the extracts to account for this behavior. These components are likely to be involved in the *cytoplasmic streaming* observed in some large cells, where vigorous flowing movements are required to maintain an even distribution of metabolites and other cytoplasmic components. These movements seem to be associated with sudden local changes in the cytoplasm from a solid gel-like consistency to a more fluid state.

A number of proteins have been isolated from cell extracts that, when added to a gel formed from purified actin filaments and filamin, cause it to change to a more fluid state in the presence of Ca^{2+}. The best characterized of these is **gelsolin**, which, when activated by the binding of Ca^{2+}, severs an actin filament and forms a cap on the newly exposed plus end of the filament, thus breaking up the cross-linked network of actin filaments. Similar proteins are found in the cortex of many types of vertebrate cells; these *severing proteins* are activated by concentrations of Ca^{2+} (about 10^{-6} M) that occur only transiently in the cytosol.

One of the postulated functions of severing proteins is to help loosen or liquefy the cell cortex locally to allow membrane fusion events. When a phagocytic white blood cell engulfs a microorganism, for example, the resulting phagosome is initially coated on its cytoplasmic side with a thick network of actin filaments originating from the cortex. In order for this phagosome to fuse with lysosomes, these actin filaments must be depolymerized to allow intimate contact between the phagosome and lysosome membranes. This removal of actin can be prevented by artificially reducing the Ca^{2+} ion concentration, and it is thought that removal may depend on a local rise in Ca^{2+} through the action of gelsolin (or a similar protein). Gelsolin is also thought to be required for a cell to crawl along a substratum, although its exact role in this process is not clear.

While a mixture of purified actin filaments, filamin, and gelsolin is capable of undergoing Ca^{2+}-dependent gel-to-sol transitions, it will not contract or show the streaming movements displayed by the cruder actin-rich gels obtained from cells. These activities require another type of actin-binding protein—the motor protein myosin. If myosin is selectively removed from the crude actin-rich gels, contractions and streaming no longer occur, suggesting that an interaction between actin and myosin generates the force for cytoplasmic streaming.

Multiple Types of Myosin Are Found in Eucaryotic Cells [47]

Time-lapse cinematography reveals the cortex of cells to be continually moving. In the previous section we emphasized the importance of actin filament polymerization and depolymerization in these movements, but, as with microtubules, motor proteins are also important. All of the actin filament motor proteins identified to date belong to the *myosin* family. **Myosins** were originally isolated on the basis of their ability to hydrolyze ATP to ADP and P_i when stimulated by binding to actin filaments, and this remains a useful biochemical criterion for their

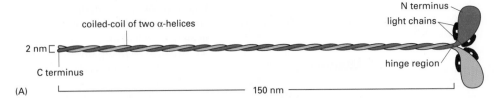

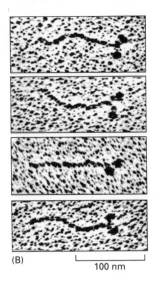

Figure 16–69 Myosin-II. (A) A myosin-II molecule is composed of two
heavy chains (each about 2000 amino acids long) and four light chains. The
light chains are of two types (one containing about 190 and the other about
170 amino acids), and one molecule of each type is present on each myosin
head (see Figure 5–23). Dimerization occurs by the two α helices wrapping
around each other to form an α-helical coiled-coil, driven by the association of regularly spaced hydrophobic amino acids (see Figure 3–48). The
coiled-coil arrangement makes an extended rod in solution, and this part of
the molecule is termed the rod domain, or the tail. This type of structural
motif is found in many other cytoskeletal proteins, enabling them to form
an extended structure. (B) The two globular heads and the tail can be
clearly seen in electron micrographs of myosin molecules shadowed with
platinum. (B, courtesy of David Shotton.)

identification. It is also possible to observe the motor activity of myosins directly
by adsorbing them onto a glass coverslip: when fluorescent actin filaments are
added together with ATP, the filaments can be observed with a fluorescence
microscope to glide over the myosin-coated glass surface. Novel myosins have
also been identified by DNA sequencing even before being characterized biochemically or functionally.

Myosin, along with actin, was first discovered in skeletal muscle, and much
of what we know about the interaction of these two proteins was learned there.
Muscle myosin belongs to the **myosin-II** subfamily of myosins, all of which have
two heads and a long, rodlike tail: each head has both ATPase and motor activity. A myosin-II protein is composed of two identical heavy chains, each of which
is complexed to a pair of light chains. The amino-terminal portion of the heavy
chain forms the motor-domain head, while the carboxyl-terminal half of the
heavy chain forms an extended α helix. Two heavy chains associate by twisting
their α-helical tail domains together into a coiled-coil to form a stable dimer that
has two heads and a single rodlike tail (Figure 16–69).

A major role of the rodlike tail of myosin-II is to allow the molecules to polymerize into bipolar filaments. This polymerization is crucial for the function of
myosin-II, which is to move groups of oppositely oriented actin filaments past
each other, as seen most clearly in muscle contraction. Myosin-II is relatively
abundant in the cell cortex; in fibroblasts, for example, there is roughly one
myosin-II molecule per 100 actin molecules. Myosin-II filaments in the contractile ring are responsible for driving membrane furrowing during cell division, as
discussed in Chapter 18, and they are thought to generate tension in stress fibers
as well as much of the cortical tension that keeps the cell surface taut. Their role
in muscle contraction is described at the end of the chapter.

In addition to myosin-II, which is generally the most abundant myosin in the
cell, nonmuscle cells contain various smaller myosins, the best characterized of
which is called **myosin-I** (Figure 16–70). Myosin-I is thought to be more like the
original, more primitive myosin from which myosin-II evolved. A single cell can
contain multiple smaller myosins, each encoded by a different gene and performing a distinct function; the cellular slime mold *Dictyostelium*, for example, has
at least nine. The common feature of all myosins is a conserved motor domain
(motor head); the other domains vary from myosin to myosin and determine the
specific role of the molecule in the cell. Thus myosin tails may have a membrane-
binding site and/or a site that binds to a second actin filament independently of

Figure 16–70 **Two myosin family members.** On the left, myosin-I and myosin-II are drawn to scale and aligned with respect to their conserved ATP-binding and actin-binding sites. The relative shapes of the folded proteins are shown on the right.

the head domain. Depending on its tail, a myosin molecule can move a vesicle along an actin filament, attach an actin filament to the plasma membrane, or cause two actin filaments to align closely and then slide past each other (Figure 16–71).

All known myosins hydrolyze ATP to move along actin filaments from the minus end toward the plus end. Given the importance of oppositely directed motor proteins that move along microtubules (see Figure 16–37), it would not be surprising to discover an additional class of motor proteins that move toward the minus end of an actin filament.

There Are Transient Musclelike Assemblies in Nonmuscle Cells [48]

In higher eucaryotic cells, organized contractile bundles of actin filaments and myosin-II filaments often form transiently to perform a specific function and then disassemble. Most notably, cell division in animal cells is made possible by a beltlike bundle of actin filaments and myosin-II filaments known as the **contractile ring.** This ring appears beneath the plasma membrane during the M phase of the cell-division cycle; forces generated by it pull inward on the plasma membrane and thereby constrict the middle of the cell, leading to the eventual separation of the two daughter cells by a process known as cytokinesis (Figure 16–72). The contractile ring must be assembled from actin, myosin, and other proteins at the start of cell division, a process that can be monitored by staining dividing cells with fluorescent anti-myosin antibodies. In sea urchin eggs that are about to divide, for example, myosin-II molecules are at first distributed evenly beneath the plasma membrane and then move to the equatorial region as the contractile ring forms. Once cell division is complete, the myosin-II molecules disperse. It is not known how this process is controlled, but it seems likely that Ca^{2+} is involved, as Ca^{2+}-dependent phosphorylation of myosin-II both increases its inter-

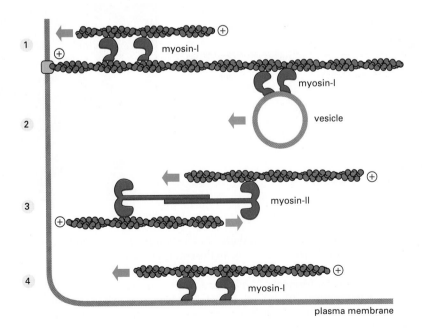

Figure 16–71 **Possible roles of myosin-I and myosin-II in a typical eucaryotic cell.** The short tail of a myosin-I molecule contains sites that bind either to other actin filaments or to membranes. This allows the head domain to move one actin filament relative to another (1), a vesicle relative to an actin filament (2), or an actin filament and membrane relative to each other (4). In addition, small antiparallel assemblies of myosin-II molecules can slide actin filaments over each other, thus mediating local contractions in an actin filament bundle (3). In all four cases the head group "walks" toward the plus end of the actin filament it contacts.

Actin-binding Proteins

Figure 16–72 Musclelike contractile assemblies in nonmuscle cells. Each assembly contains myosin-II filaments in addition to actin filaments.

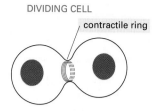

DIVIDING CELL

contractile ring

action with actin and promotes its assembly into short bipolar filaments (Figure 16–73).

Stress fibers, which are prominent components of the cytoskeleton of fibroblast cells in culture (see Figure 16–72), are a second example of a temporary contractile bundle of actin filaments and myosin-II. Although smaller and less highly organized, they resemble the tiny *myofibrils* in muscle (discussed later) in their structure and function. At one end they insert into the plasma membrane at special sites called *focal contacts,* where the external face of the cell is closely attached to the extracellular matrix (Figure 16–74); at the other end they insert into a second focal contact or into a meshwork of intermediate filaments that surrounds the cell nucleus. Stress fibers form in response to tension generated across a cell and are disassembled at mitosis when the cell rounds up and loses its attachments to the substratum. They also disappear rapidly if tension is released by suddenly detaching one end of the stress fiber from the focal contact by means of a laser beam. Stress fibers within fibroblasts in tissues are thought to allow the cells to exert tension on the matrix of collagen surrounding them—an essential process in both wound healing and morphogenesis (see Figure 19–48). In epithelia, actin filament bundles spanning the cytoplasm from one cell-cell junction to another can appear and disappear in a similar way; such filament bundles, linked end to end via the cell-cell junctions, can form cables that transmit and generate tension along lines of particular stress in the multicellular sheet.

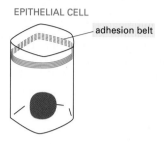

EPITHELIAL CELL

adhesion belt

Not all contractile assemblies of actin filaments and myosin in nonmuscle cells are transitory. Those associated with the intercellular anchoring junctions called *adhesion belts,* for example, are often more lasting. Adhesion belts (discussed in Chapter 19) are found near the apical surface of epithelial cells (see Figure 16–72). Among other functions, they are thought to play an important part in the folding of epithelial cell sheets during embryogenesis.

CULTURED FIBROBLAST

stress fiber

The mechanism of contraction of all of these cytoskeletal bundles is based on the ATP-driven sliding of interdigitated actin and myosin filaments, and it is thought to require a particular type of ordered assembly, which will be explained later when we discuss muscle.

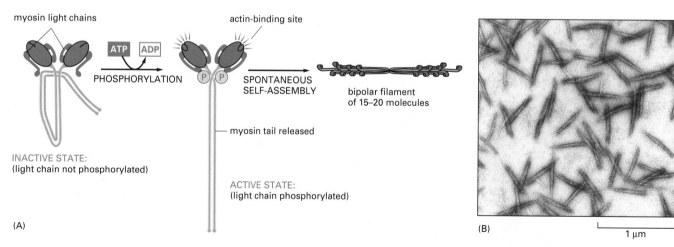

Figure 16–73 The controlled assembly of myosin-II into filaments. (A) The controlled phosphorylation of one of the two light chains has at least two effects *in vitro:* it causes a change in the conformation of the myosin head, exposing its actin-binding site, and it releases the myosin tail from a "sticky patch" on the myosin head, thereby allowing the myosin molecules to assemble into short bipolar filaments. The enzyme responsible for this phosphorylation (myosin light-chain kinase) is described later in connection with smooth muscle (see Figure 16–98). (B) Negatively stained short filaments of myosin-II that have been induced to assemble by the phosphorylation of their light chains. (B, courtesy of John Kendrick-Jones.)

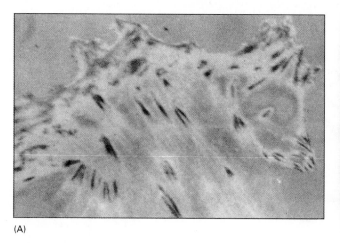

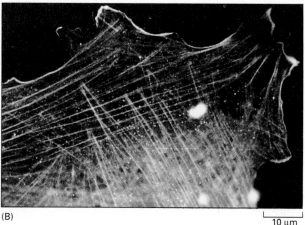

10 μm

Focal Contacts Allow Actin Filaments to Pull Against the Substratum [49]

To pull on the extracellular matrix or on another cell, a stress fiber must be strongly anchored in the plasma membrane at the appropriate site. Attachments between actin filaments inside the cell and extracellular matrix on the outside of the cell are mediated by transmembrane linker glycoproteins in the plasma membrane. Those formed by cultured fibroblasts with the extracellular matrix are the best characterized. When fibroblasts grow on a culture dish, most of their cell surface is separated from the substratum by a gap of more than 50 nm; but at **focal contacts** (*adhesion plaques*), this gap is reduced to 10 to 15 nm. Here the plasma membrane is attached to components of the extracellular matrix that have become adsorbed to the culture dish. Staining with anti-actin antibodies clearly shows these regions to be the sites where the ends of stress fibers attach to the plasma membrane (see Figure 16–74).

The main transmembrane linker proteins of focal contacts are members of the *integrin* family, whose external domain binds to an extracellular matrix component while the cytoplasmic domain is linked to actin filaments in stress fibers. The linkage is indirect and is mediated by multiple *attachment proteins* (Figure 16–75). The cytoplasmic domain of the integrin binds to the protein *talin*, which in turn binds to *vinculin*, a protein found also in other actin-containing cell junctions, such as *adherens junctions* (discussed in Chapter 19). Vinculin associates with α-actinin and is thereby linked to an actin filament. Although the exact topology of protein interactions in the focal contact has not been established, a possible arrangement is shown in Figure 16–75B.

Besides their role as anchors for the cell, focal contacts can also relay signals from the extracellular matrix to the cytoskeleton. Several protein kinases, including the tyrosine kinase encoded by the *src* gene, are localized to focal contacts, and there are indications that their activity changes with the type of substratum on which the cell rests. These kinases can phosphorylate various target proteins, including components of the cytoskeleton, and hence regulate the survival, growth, morphology, movement, and differentiation of cells in response to the extracellular matrix in their environment.

Microvilli Illustrate How Bundles of Cross-linked Actin Filaments Can Stabilize Local Extensions of the Plasma Membrane [50]

Microvilli are fingerlike extensions found on the surface of many animal cells. They are especially abundant on those epithelial cells that require a very large surface area to function efficiently. A single absorptive epithelial cell in the human small intestine, for example, has several thousand microvilli on its apical

Figure 16–74 The relation between focal contacts and stress fibers in cultured fibroblasts. Focal contacts are best seen in living cells by reflection-interference microscopy (A). In this technique, light is reflected from the lower surface of a cell attached to a glass slide, and the focal contacts appear as dark patches. (B) Immunofluorescence staining of the same cell (after fixation) with antibodies to actin shows that most of the cell's actin filament bundles (or stress fibers) terminate at or close to a focal contact. (Courtesy of Grenham Ireland.)

Actin-binding Proteins

841

(A)

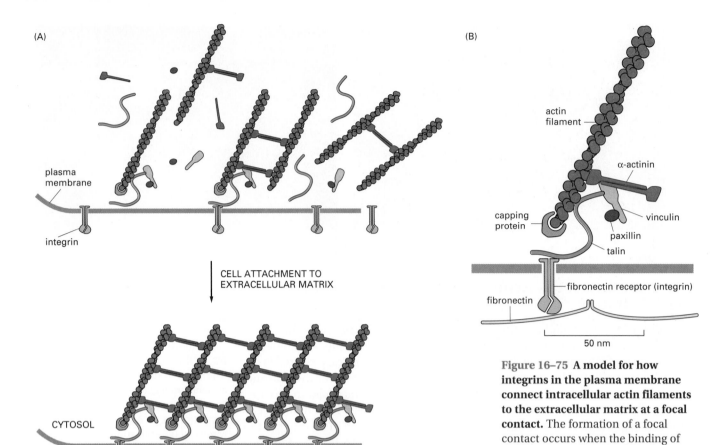

plasma
membrane

integrin

CELL ATTACHMENT TO
EXTRACELLULAR MATRIX

CYTOSOL

EXTRACELLULAR MATRIX

fibronectin

(B)

actin
filament

α-actinin

capping
protein

vinculin

paxillin

talin

fibronectin receptor (integrin)

fibronectin

50 nm

Figure 16–75 A model for how integrins in the plasma membrane connect intracellular actin filaments to the extracellular matrix at a focal contact. The formation of a focal contact occurs when the binding of matrix glycoproteins (such as fibronectin) on the outside of the cell causes the integrin molecules to cluster at the contact site, as illustrated schematically in (A). A possible arrangement of some of the intracellular attachment proteins that mediate the linkage between an integrin and actin filaments is shown in (B).

surface. Each is about 0.08 μm wide and 1 μm long, making the cell's absorptive surface area 20 times greater than it would be without them. The plasma membrane that covers these microvilli is highly specialized, bearing a thick extracellular coat of polysaccharide and digestive enzymes. The cytoskeleton of the microvillus has been studied in detail—a task that is made easier by its highly ordered structure, compared with the less specialized regions of cell cortex.

At the core of each intestinal microvillus is a rigid bundle of 20 to 30 parallel actin filaments that extend from the tip of the microvillus down into the cell cortex. The actin filaments in the bundle are all oriented with their plus ends pointing away from the cell body and are held together at regular intervals by actin-bundling proteins. Although fimbrin, the bundling protein in microspikes and filopodia, helps to bundle actin filaments into microvilli, the most important bundling protein is **villin,** which is found only in microvilli (Figure 16–76). Villin, like fimbrin, cross-links actin filaments into tight parallel bundles, but it has a different actin-binding sequence. When villin is introduced into cultured fibroblasts, which do not normally contain villin and have only a few small microvilli,

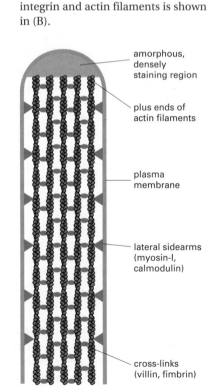

amorphous,
densely
staining region

plus ends of
actin filaments

plasma
membrane

lateral sidearms
(myosin-I,
calmodulin)

cross-links
(villin, fimbrin)

Figure 16–76 A microvillus. A bundle of parallel actin filaments held together by the actin-bundling proteins villin and fimbrin forms the core of a microvillus. Lateral arms (composed of myosin-I and the Ca^{2+}-binding protein calmodulin) connect the sides of the actin filament bundle to the overlying plasma membrane. The plus ends of the actin filaments are all at the tip of the microvillus, where they are embedded in an amorphous, densely staining substance of unknown composition.

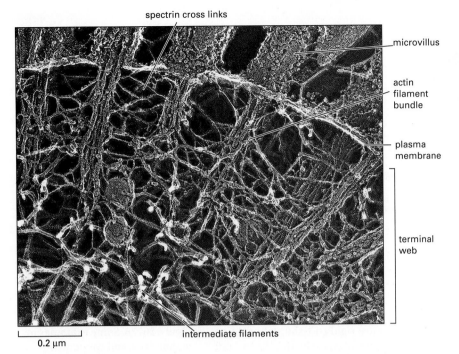

spectrin cross links

microvillus

actin filament bundle

plasma membrane

terminal web

intermediate filaments

0.2 μm

Figure 16–77 Freeze-etch electron micrograph of an intestinal epithelial cell, showing the terminal web beneath the apical plasma membrane. Bundles of actin filaments forming the core of microvilli extend into the terminal web, where they are linked together by a complex set of cytoskeletal proteins that includes spectrin and myosin-II. Beneath the terminal web is a layer of intermediate filaments. (From N. Hirokawa and J.E. Heuser, *J. Cell Biol.* 91:399–409, 1981, by copyright permission of the Rockefeller University Press.)

the existing microvilli become greatly elongated and stabilized, and new ones may also be induced, suggesting that villin is mainly responsible for the formation of the long microvilli in epithelial cells.

At the base of the microvillus the actin filament bundle is anchored into a specialized region of cortex at the apical end of the intestinal epithelial cell. This cortex, known as the *terminal web,* contains a dense network of spectrin molecules overlying a layer of intermediate filaments (Figure 16–77). The spectrin is thought to provide rigidity and stability to the cortex in this region, and the anchoring of the actin filaments to the terminal web is thought to stiffen the microvilli, keeping their actin bundles projecting outward at a right angle to the apical cell surface.

The actin filament bundle is attached to the overlying plasma membrane of the microvillus by lateral bridges that can be seen in electron micrographs. The bridges are composed of a form of myosin-I that has several molecules of calmodulin (discussed in Chapter 15) bound to its tail region. The myosin is oriented with this tail region embedded in the membrane and its active ATP-binding head contacting the actin filaments. It is a mystery why a motor protein is used to link actin filaments to the membrane in microvilli. If the myosin-I in microvilli is motile, it should move toward the plus end of the actin filaments at the microvillus tip. This has lead to speculation that the myosin-I helps to pull the membrane up over the microvillus core, forming vesicles at its tip that are then released into the lumen of the intestine, where the digestive enzymes they carry continue their action.

The Behavior of the Cell Cortex Depends on a Balance of Cooperative and Competitive Interactions Among a Large Set of Actin-binding Proteins

The preceding examples show that the same actin filament can interact with different sets of actin-binding proteins at different locations in the cortex and that the actin-binding proteins can be segregated to different parts of the cell. What prevents the various sets of actin-binding proteins from mixing in the cytoplasm? It seems likely that both cooperative and competitive interactions among these proteins are important. One class of actin-binding proteins not yet discussed, for example, binds along the length of the actin filaments. The most widespread

Actin-binding Proteins

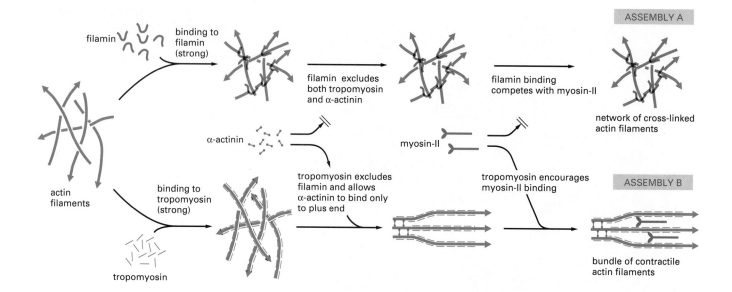

filamin

binding to filamin (strong)

ASSEMBLY A

filamin excludes both tropomyosin and α-actinin

filamin binding competes with myosin-II

network of cross-linked actin filaments

α-actinin

actin filaments

myosin-II

binding to tropomyosin (strong)

tropomyosin excludes filamin and allows α-actinin to bind only to plus end

tropomyosin encourages myosin-II binding

ASSEMBLY B

tropomyosin

bundle of contractile actin filaments

members of this class are the **tropomyosins,** which are rigid rod-shaped proteins named for similarities in their x-ray diffraction pattern to myosin-II. Like the tail of myosin-II, tropomyosin is a dimer of two identical α-helical chains that wind around each other in a coiled-coil. By binding along the length of an actin filament, the tropomyosin stabilizes and stiffens the filament. It also inhibits the binding of filamin to actin filaments, which probably explains why tropomyosin and filamin tend to be differentially distributed in cells. By contrast, tropomyosin binding to an actin filament increases the binding of myosin-II to the filament—an example of a *cooperative* interaction (Figure 16–78).

We can now begin to see how stress fibers and cortical networks of actin filaments can coexist in a common cytoplasm. At one site in a cell—perhaps nucleated at a forming focal adhesion under the influence of activated Rho protein—tropomyosin, myosin-II, and α-actinin associate with actin filaments and exclude filamin; the contractile activity of myosin-II then promotes further organizational changes to produce a stress fiber. At another site in the cell tropomyosin-deficient actin filaments bind filamin, producing a loose network that provides few sites where α-actinin can bind to two filaments at once, so that it is excluded; bending of the filaments in the loose meshwork may also discourage tropomyosin-binding, since this molecule prefers a straight filament. While this picture is partly speculative, it illustrates the basic pathway by which a combination of co-operative and competitive interactions can give rise to spatially differentiated actin filament arrays in a common cytoplasm. It is not known how the postulated local differences that initiate the formation of these assemblages are established; nor is it known how many distinct types of actin filament arrays can coexist in the same cell—there are certainly more than the two we have just mentioned.

Some of the actin-binding proteins discussed in this section are summarized in Figure 16–79.

The Migration of Animal Cells Can Be Divided into Three Distinct Actin-dependent Subprocesses [51]

The crawling movements of animal cells are among the most difficult to explain at the molecular level. Different parts of the cell change at the same time, and there is not a single, easily identifiable locomotory organelle (analogous to a flagellum, for example). Although actin forms the basis of animal cell migration, it undergoes many different transformations as the cell moves forward, assembling into lamellipodia and microspikes, associating with focal contacts, forming stress fibers, and so on. A complete account would have to give a molecular explana-

Figure 16–78 Some examples of competitive and cooperative interactions between actin-binding proteins. The arrowhead at the end of each actin filament indicates the minus end. Tropomyosin and filamin both bind strongly to actin filaments, but their binding is competitive. Because tropomyosin binds cooperatively to actin filaments, either tropomyosin or filamin will predominate over large regions of the actin filament network. Other actin-binding proteins, such as α-actinin or myosin-II, will be excluded from specific sites by a competitive interaction; thus, for example, α-actinin binds all along pure actin filaments *in vitro*, but it binds relatively weakly to actin filaments in cells, where it is largely confined to sites near the plus ends because of competition with other proteins. Alternatively, binding can be enhanced through cooperative interaction; thus tropomyosin appears to enhance the binding of myosin-II to actin filaments. Multiple interactions of these types between the many different types of actin-binding proteins are thought to be responsible for the complex variety of actin networks found in all eucaryotic cells (see Figure 16–79 for key to symbols).

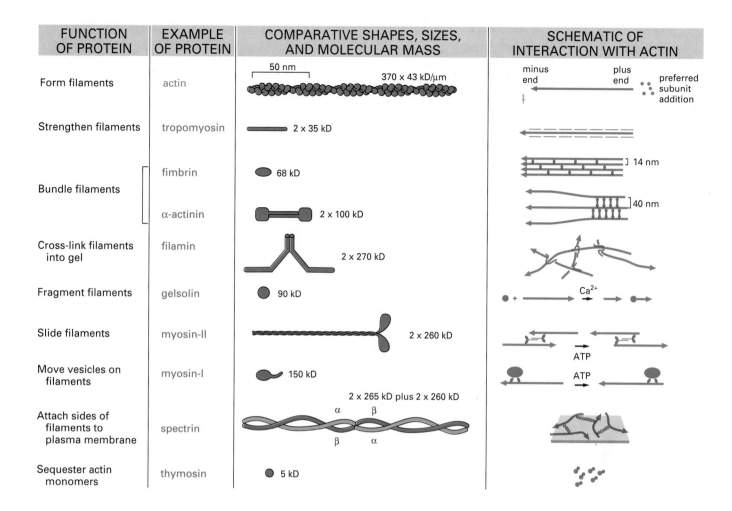

FUNCTION OF PROTEIN	EXAMPLE OF PROTEIN	COMPARATIVE SHAPES, SIZES, AND MOLECULAR MASS	SCHEMATIC OF INTERACTION WITH ACTIN
Form filaments	actin	50 nm — 370 x 43 kD/μm	minus end — plus end; preferred subunit addition
Strengthen filaments	tropomyosin	2 x 35 kD	
Bundle filaments	fimbrin	68 kD	14 nm
	α-actinin	2 x 100 kD	40 nm
Cross-link filaments into gel	filamin	2 x 270 kD	
Fragment filaments	gelsolin	90 kD	Ca²⁺
Slide filaments	myosin-II	2 x 260 kD	ATP
Move vesicles on filaments	myosin-I	150 kD	ATP
Attach sides of filaments to plasma membrane	spectrin	2 x 265 kD plus 2 x 260 kD; α β β α	
Sequester actin monomers	thymosin	5 kD	

Figure 16–79 **Some of the major classes of actin-binding proteins found in most vertebrate cells.** Actin is shown in *red*, while the actin-binding proteins are shown in *green*. The molecular mass of each protein is given in kilodaltons (kD).

tion for these transformations, explain how they are coordinated in time and space, and also account for important biophysical parameters such as the development of tension in the cortex and the formation of strong adhesions between the cell and its substratum.

In broad terms, three distinct processes can be identified in the crawling movements of animal cells: *protrusion*, in which lamellipodia and microspikes (or filopodia) are extended from the front of the cell; *attachment*, where the actin cytoskeleton makes a connection with the substratum; and *traction*, where the body of the cell moves forward.

Protrusion is a function of the leading edge of the cell. Actin-rich lamellipodia and microspikes (or filopodia) extend forward over the substratum, a process that is accompanied by actin polymerization, as described previously. It seems likely that the protrusion is driven by actin polymerization at the leading edge (see Figure 16–58), although this is still debated. Myosin-I motors attached to the plasma membrane could also drive the cell forward by actively walking along actin filaments. Yet another possibility, which has been suggested to apply in particular to the locomotion of giant amoebae, is that protrusions are squeezed out of the front of the cell by hydrostatic pressure generated by the contraction of the cortex elsewhere in the cell.

The attachment of cortical actin filaments to the substratum was discussed earlier when we described focal contacts, although these are specialized attachment structures present in fibroblasts in culture and associated with the ends of stress fibers. Rapidly motile cells—such as *Dictyostelium* amoebae and white blood cells—make more diffuse contacts with the substratum. It is thought, however, that similar principles apply to these contacts: transmembrane receptors for extracellular matrix proteins link the plasma membrane to the substratum, and actin filaments in the cytoplasm interact with the cytoplasmic domains of

Actin-binding Proteins

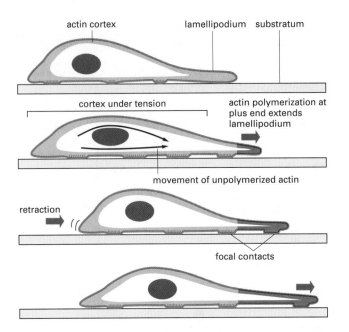

actin cortex   lamellipodium   substratum

cortex under tension

actin polymerization at
plus end extends
lamellipodium

movement of unpolymerized actin

retraction

focal contacts

Figure 16–80 One model of how forces generated in the actin-rich cortex might move a cell forward. The actin-dependent extension and firm attachment of a lamellipodium at the leading edge stretches the actin cortex. The cortical tension then draws the body of the cell forward to relax some of the tension. New focal contacts are made and old ones are disassembled as the cell crawls forward. The same cycle can be repeated over and over again, moving the cell forward in a stepwise fashion. The newly polymerized cortical actin is shown in *red*.

these receptors through actin-binding proteins. The details of these important interactions are uncertain, but it is clear that the cell contacts with the substratum must be continually made and broken as the cell moves forward.

Traction is perhaps the most mysterious part of cell locomotion. In many cases it is thought that the force for cell locomotion is generated near the front of the cell and that the nucleus and bulk cytoplasm are dragged forward passively. The force generation can be viewed in different ways. The leading part of the cell might actively contract like a muscle fiber and thus pull on the back of the cell. In another view polymerization of actin filaments at the front of the cell extends the actin cortex forward, and the rear of the cell is then carried forward by the contractile force of the resulting cortical tension (Figure 16–80).

The Mechanism of Cell Locomotion Can Be Dissected Genetically [52]

One of the most powerful ways to analyze the mechanism of a complex cellular process is to examine the effect of mutations that result in the deletion, overexpression, or modification of specific proteins. In the case of eucaryotic cell locomotion, the amoeboid cells of the slime mold *Dictyostelium* are particularly suitable for genetic analysis. These cells have a shape and a manner of moving that closely resemble those of the cells of higher organisms. But because they are haploid, they are readily manipulated by reverse genetic methods. Thus it has been possible to delete a number of actin-binding proteins from these cells and examine the consequences for cell locomotion.

The role of myosin-II, for example, has been tested genetically by two methods. In one, a defective form of the myosin-II gene is substituted for the existing gene by homologous recombination (see Figure 7–47), leading to a mutant strain with no myosin-II. The second strategy is to use anti-sense RNA (see Figure 7–43) to inactivate myosin-II mRNA, which has essentially the same effect. In a normal crawling *Dictyostelium* myosin-II is concentrated near the rear of the cell and myosin-I is concentrated in the leading edge (Figure 16–81). In the mutants the myosin-I is unchanged but myosin-II is gone.

Remarkably, *Dictyostelium* cells without myosin-II can still move over the substratum and respond chemotactically to a source of cyclic AMP, although both processes are somewhat impaired. Thus myosin-II is not absolutely essential for cell locomotion. Although protrusive activity at the leading edge of such mutant cells is quite normal, movement of the cell body forward is somewhat impaired, suggesting that myosin-II plays a role in generating traction. Nevertheless, myosin-I and/or actin polymerization must be able to drive the cell forward at a reasonable rate without the help of myosin-II.

Not surprisingly, the mutant cells are unable to form a contractile ring following mitosis and therefore develop into multinucleated giant cells. These cells eventually divide by using cell locomotion to tear themselves in two. It is interesting to speculate that such locomotion-dependent cytokinesis may represent a primitive cell division mechanism and that myosin-II might have evolved from myosin-I through natural selection for a more efficient cytokinetic apparatus.

Summary

The varied forms and functions of actin in eucaryotic cells depend on a versatile repertoire of actin-binding proteins that cross-link actin filaments into loose gels, bind them into stiff bundles, attach them to the plasma membrane, or forcibly move them relative to one another. Tropomyosin, for example, binds along the length of actin filaments, making them more rigid and altering their affinity for other proteins. Filamin cross-links actin filaments into a loose gel. Fimbrin and a-actinin form bundles of parallel actin filaments. Gelsolin mediates Ca²⁺-dependent fragmentation of actin filaments, thereby causing a rapid solation of actin gels. Various forms of myosin use the energy of ATP hydrolysis to move along actin filaments, either carrying membrane-bounded organelles from one location in the cell to another or moving adjacent actin filaments against each other. Sets of actin-binding proteins are thought to act cooperatively in generating the movements of the cell surface, including cytokinesis, phagocytosis, and cell locomotion. These movements are difficult to analyze because of the many components involved, but genetic approaches, in which genes encoding specific actin-binding proteins are mutated, can show the function of individual proteins in each process.

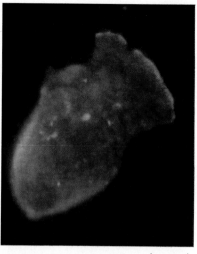

5 μm

Figure 16–81 The locations of myosin-I and myosin-II in a normal crawling *Dictyostelium* amoeba. The two forms of myosin were stained with specific antibodies, each coupled to a different fluorescent dye, and examined in a fluorescence microscope. Myosin-II (*red*) shows the highest accumulation in the posterior cortex, whereas myosin-I (*green*) is mainly restricted to the leading edge of lamellipodia at the front of the cell. Some myosin is also seen in phagocytic vesicles in the cytoplasm. (Courtesy of Yoshio Fukui.)

Muscle [53]

Many of the proteins that associate with actin filaments in eucaryotic cells were first discovered in muscle. Muscle contraction is the most familiar and the best understood of all the kinds of movement of which animals are capable. In vertebrates, for example, running, walking, swimming, and flying all depend on the ability of *skeletal muscle* to contract rapidly on its scaffolding of bone, while involuntary movements such as heart pumping and gut peristalsis depend on the contraction of *cardiac* and *smooth muscle,* respectively.

Although muscle is the best-understood example of actin-based motility, it was a relatively late development in evolution, and it is highly specialized compared with more typical animal cells. In particular, the actin- and myosin-based contractile units of muscle cells, the *myofibrils,* are not labile like the actin- and myosin-based structures of nonmuscle cells.

Myofibrils Are Composed of Repeating Assemblies of Thick and Thin Filaments [54]

The long thin *muscle fibers* of skeletal muscle are huge single cells formed during development by the fusion of many separate cells (discussed in Chapter 22). The nuclei of the contributing cells are retained in this large cell and lie just beneath the plasma membrane. But the bulk of the cytoplasm (about two-thirds of its dry mass) is made up of **myofibrils,** which are the contractile elements of the muscle cell. They are cylindrical structures 1 to 2 μm in diameter and are often as long as the muscle cell itself (Figure 16–82).

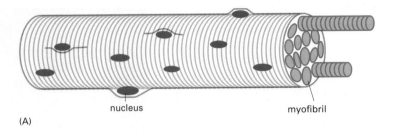

(A)

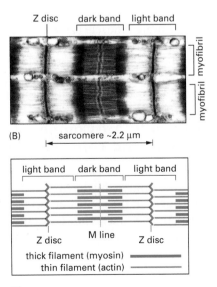

(B)

Figure 16–82 Skeletal muscle cells (also called muscle fibers). (A) In an adult human these huge multinucleated cells are typically 50 µm in diameter, and they can be several centimeters long. (B) Fluorescence micrograph of rat muscle showing the peripherally located nuclei (*blue*). (B, courtesy of Nancy L. Kedersha.)

Each myofibril consists of a chain of tiny contractile units, or **sarcomeres,** each about 2.2 µm long, which give the vertebrate myofibril its striated appearance. At high magnification a series of broad light and dark bands can be seen in each sarcomere; a dense line in the center of each light band separates one sarcomere from the next and is known as the Z line, or *Z disc* (Figure 16–83).

Each sarcomere comprises a miniature, precisely arranged assembly of parallel and partly overlapping filaments. *Thin filaments* composed of actin with

Figure 16–83 Skeletal muscle myofibrils. (A) Low-magnification electron micrograph of a longitudinal section through a skeletal muscle cell of a rabbit, showing the regular pattern of cross-striations. The cell contains many myofibrils aligned in parallel (see Figure 16–82). (B) Detail of the skeletal muscle cell shown in (A), showing portions of two adjacent myofibrils and the definition of a sarcomere. (C) Schematic diagram of a single sarcomere, showing the origin of the dark and light bands seen in the electron micrographs. Z discs, at either end of the sarcomere, are attachment sites for thin filaments (actin filaments); the M line, or midline, is the location of specific proteins that link adjacent thick filaments (myosin-II filaments) to each other. The *broad green bands,* which mark the location of the thick filaments, are sometimes referred to as *A bands* because they appear anisotropic in polarized light (that is, their refractive index changes with the plane of polarization). The *light red bands,* which contain only thin filaments and therefore have a lower density of protein, are relatively isotropic in polarized light and are sometimes called *I bands.* (A and B, courtesy of Roger Craig.)

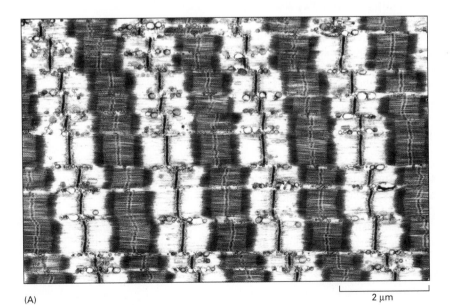

(A)

(B) sarcomere ~2.2 µm

(C)

light band | dark band | light band
Z disc | M line | Z disc
thick filament (myosin)
thin filament (actin)

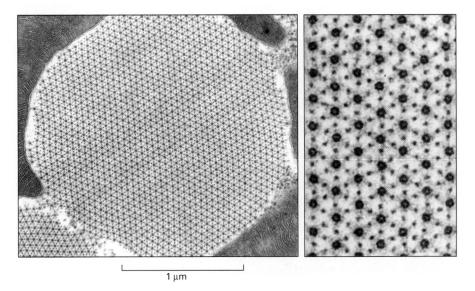

Figure 16–84 **Electron micrographs of an insect flight muscle viewed in cross-section.** The myosin and actin filaments are packed together with almost crystalline regularity. Unlike their vertebrate counterparts, these myosin filaments have a hollow center, as seen in the enlargement on the right. A longitudinal section of this muscle is shown in Figure 16–86. The geometry of the hexagonal lattice is slightly different in vertebrate muscle. (From J. Auber, *J. de Microsc.* 8:197–232, 1969.)

associated proteins are attached to the Z discs at either end of the sarcomere. They extend in toward the middle of the sarcomere, where they overlap with *thick filaments,* which are polymers of specific muscle isoforms of myosin-II (Figure 16–83C). When this region of overlap is examined in cross-section by electron microscopy, the myosin filaments are seen to be arranged in a regular hexagonal lattice, with the actin filaments placed regularly between them (Figure 16–84).

Contraction Occurs as the Myosin and Actin Filaments Slide Past Each Other

Sarcomere shortening is caused by the myosin filaments sliding past the actin filaments with no change in the length of either type of filament (Figure 16–85). This *sliding filament model,* first proposed in 1954, was crucial to understanding the contractile mechanism.

The ultrastructural basis for the force-generating interaction is visible at very high magnification in electron micrographs. The myosin filaments are seen to possess numerous tiny side arms, or *cross-bridges,* that extend about 13 nm to make contact with adjacent actin filaments (Figure 16–86). These cross-bridges are myosin-II heads, and when a muscle contracts, the myosin and actin filaments are pulled past each other by the cross-bridges acting cyclically, like banks of tiny oars.

As stated previously, the globular head, or motor domain, of the myosin-II molecule both binds to actin filaments and hydrolyzes ATP. Isolated myosin-II heads, which can be prepared by papain digestion, retain both the ATPase activity and the actin-filament-binding properties of the intact myosin-II molecule and therefore can be used to analyze the interaction between actin and myosin.

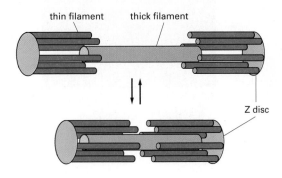

thin filament   thick filament

Z disc

Figure 16–85 **The sliding filament model of muscle contraction.** The actin and myosin filaments slide past one another without shortening.

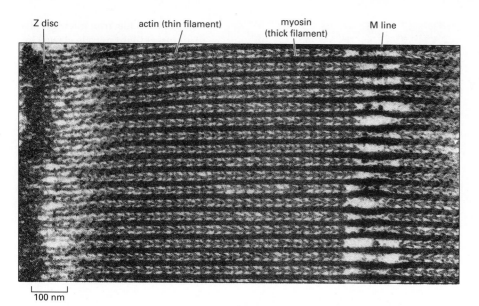

Z disc   actin (thin filament)   myosin (thick filament)   M line

100 nm

Figure 16–86 Electron micrograph of a longitudinal section of an insect flight muscle. This very thin section shows clearly the alternating myosin and actin filaments and the cross-bridges that link the two. Note that insect flight muscle has an unusually high degree of overlap between the myosin and actin filaments. (Courtesy of Mary C. Reedy.)

Each actin molecule in an actin filament is capable of binding one myosin-II head to form a complex that reveals the structural polarity of the actin filament. With negative staining, such complexes can be seen in the electron microscope to have a regular and distinctive form: each myosin head forms a lateral projection, and the superimposed image of many such projections gives the appearance of arrowheads along the actin filament. The *pointed end* created by these myosin arrows corresponds to the slow-growing minus end of the actin filament described earlier (see p. 821). The other, *barbed end* corresponds to the fast-growing plus end (Figure 16–87).

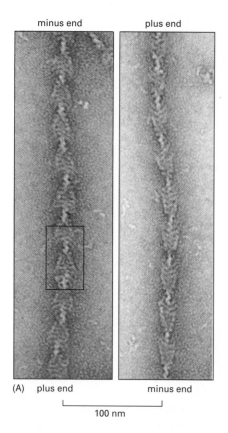

minus end   plus end

(A)   plus end   minus end

100 nm

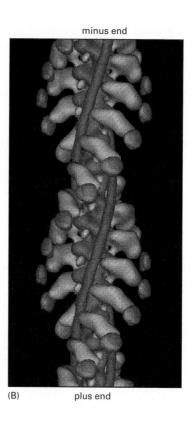

minus end

(B)   plus end

Figure 16–87 Actin filaments decorated with isolated myosin-II heads. (A) In the electron microscope the helical arrangement of the bound myosin heads, which are tilted in one direction, gives the appearance of arrowheads and indicates the polarity of the actin filament. The pointed end corresponds to the *minus end*, the barbed end to the *plus end.* (B) A three-dimensional reconstruction from electron micrographs of a similar decorated actin filament. The region shown corresponds to the boxed area in (A). The actin filament is shown in *red,* the myosin heads are *yellow,* the myosin light chains are *gray,* and the position of tropomyosin is shown in *purple.* (A, courtesy of Roger Craig; B, courtesy of Ron Milligan.)

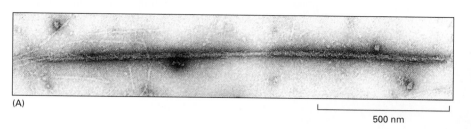

(A)

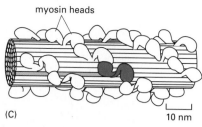

myosin heads

10 nm

(C)

bare zone   myosin heads

(B)

Figure 16–88 The myosin-II thick filament. (A) Electron micrograph of a myosin-II thick filament isolated from frog muscle. Note the central bare zone. (B) Schematic diagram, not drawn to scale. The myosin-II molecules aggregate together by means of their tail regions, with their heads projecting to the outside. The bare zone in the center of the filament consists entirely of myosin-II tails. (C) A small section of a myosin-II filament as reconstructed from electron micrographs. An individual myosin molecule is highlighted in *green*. (A, courtesy of Murray Stewart; C, based on R.A. Crowther, R. Padron, and R. Craig, *J. Mol. Biol.* 184:429–439, 1985.)

As shown in Figure 16–88, myosin heads face in opposite directions on either side of the bare central region of a myosin-II filament. Since the heads must interact with actin filaments in the region of overlap, the actin filaments on either side of the sarcomere should be of opposite polarity. This has been demonstrated by using myosin-II heads to decorate the actin filaments attached to isolated Z discs. All of the myosin arrowheads are found to point away from the Z disc. Therefore, the plus end of each actin filament is embedded in the Z disc, while the minus end points toward the myosin filaments (Figure 16–89).

A Myosin Head "Walks" Toward the Plus End of an Actin Filament [55]

Muscle contraction is driven by the interaction between myosin-II heads and adjacent actin filaments, during which the myosin head hydrolyzes ATP. The ATP hydrolysis and subsequent dissociation of the tightly bound products (ADP and P_i) produce an ordered series of allosteric changes in the conformation of myosin. As a result, part of the energy released is coupled to the production of movement. A major advance in understanding these concerted changes in protein structure, and hence in understanding how ATP hydrolysis is coupled to directed movement of the myosin molecule, came with the determination of the three-dimensional structure of the myosin head by x-ray diffraction analysis (Figure 16–90). In conjunction with a wealth of other data, this structure suggests that unidirectional movement is generated by the sequence of events illustrated in Figure 16–91.

Figure 16–89 The myosin and actin filaments of a sarcomere overlap with the same relative polarity on either side of the midline.

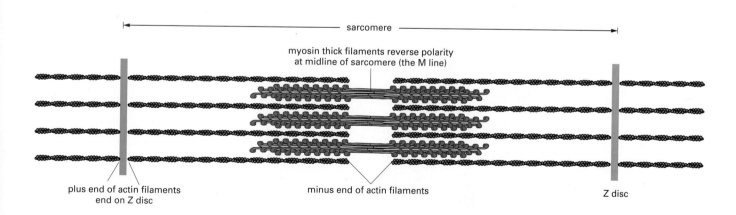

sarcomere

myosin thick filaments reverse polarity at midline of sarcomere (the M line)

plus end of actin filaments end on Z disc

minus end of actin filaments

Z disc

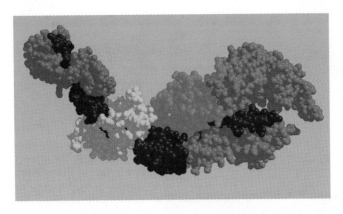

Figure 16–90 Space-filling model of the head of muscle myosin. The model is oriented so that the actin-binding surface is located at the lower right-hand corner. Three domains of the myosin heavy chain are colored *green*, *red*, and *blue*, respectively, whereas the two light chains are shown in *yellow* and *purple*. (From I. Rayment et al., *Science*, 261:50–58, 1993. © 1993 the AAAS.)

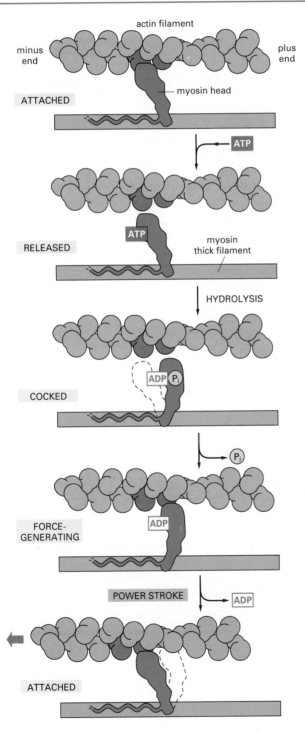

ATTACHED—At the start of the cycle shown in this figure, a myosin head lacking a bound nucleotide is locked tightly onto an actin filament in a *rigor* configuration (so named because it is responsible for *rigor mortis*, the rigidity of death). In an actively contracting muscle this state is very short-lived, being rapidly terminated by the binding of a molecule of ATP.

RELEASED—A molecule of ATP binds to the large cleft on the "back" of the head (that is, on the side farthest from the actin filament) and immediately causes a slight change in the conformation of the domains that make up the actin-binding site (see Figure 16–90). This reduces the affinity of the head for actin and allows it to move along the filament. (The space drawn here between the head and actin emphasizes this change, although in reality the head probably remains very close to the actin.)

COCKED—The cleft closes like a clam shell around the ATP molecule, triggering a large shape change that causes the head to be displaced along the filament by a distance of about 5 nm. Hydrolysis of ATP occurs, but the ADP and P_i produced remain tightly bound to the protein.

FORCE-GENERATING—The weak binding of the myosin head to a new site on the actin filament causes release of the inorganic phosphate produced by ATP hydrolysis, concomitantly with the tight binding of the head to actin. This release triggers the power stroke—the force-generating change in shape during which the head regains its original conformation. In the course of the power stroke, the head loses its bound ADP, thereby returning to the start of a new cycle.

ATTACHED—At the end of the cycle, the myosin head is again locked tightly to the actin filament in a rigor configuration. Note that the head has moved to a new position on the actin filament.

Figure 16–91 The cycle of changes by which a myosin molecule walks along an actin filament. (Based on I. Rayment et al., *Science* 261:50–58, 1993. © 1993 the AAAS.)

Because each turn of the cycle illustrated in Figure 16–91 results in the hydrolysis and release of one ATP molecule, the series of conformational changes just described is driven by a large favorable change in free energy, making it unidirectional. Each individual myosin head, therefore, "walks" in a single direction along an adjacent actin filament, always moving toward the filament's plus end (see Figure 16–89). As it undergoes its cyclical change in conformation, the myosin head pulls against the actin filament, causing this filament to slide against the myosin filament. Once an individual myosin head has detached from the actin filament, it is carried along by the action of other myosin heads in the same myosin filament, so that a snapshot of an entire myosin filament in a contracting muscle would show some of the myosin heads attached to actin filaments and others unattached. (A certain amount of springlike elasticity in the myosin molecule is essential to allow this to happen.) Each myosin filament has about 300 myosin heads (294 in frog muscle), and each head cycles about 5 times per second in the course of a rapid contraction—sliding the myosin and actin filaments past one another at rates of up to 15 µm/second.

Muscle Contraction Is Initiated by a Sudden Rise in Cytosolic Ca²⁺ [56]

The force-generating molecular interaction just described takes place only when a signal passes to the skeletal muscle from its motor nerve. The signal from the nerve triggers an action potential in the muscle cell plasma membrane, and this electrical excitation spreads rapidly into a series of membranous folds, the *transverse tubules*, or *T tubules*, that extend inward from the plasma membrane around each myofibril. The signal is then relayed across a small gap to the **sarcoplasmic reticulum,** an adjacent sheath of anastomosing flattened vesicles that surrounds each myofibril like a net stocking (Figure 16–92A).

In the junctional region, large *Ca²⁺ release channels* extend like pillars from the sarcoplasmic reticulum membrane to make contact with the T-tubule membrane on the other side (Figure 16–92C). When voltage-sensitive proteins in the

Figure 16–92 T tubules and the sarcoplasmic reticulum. (A) Drawing of the two systems of membranes that relay the signal to contract from the muscle cell plasma membrane to all of the myofibrils in the cell. (B) Electron micrograph showing two T tubules. Note the position of the large Ca²⁺ release channels in the sarcoplasmic reticulum membrane; they look like square-shaped "feet" that connect to the adjacent T-tubule membrane. (C) Schematic diagram showing how a Ca²⁺ release channel in the sarcoplasmic reticulum membrane is thought to be opened by a voltage-sensitive transmembrane protein in the adjacent T-tubule membrane. (B, courtesy of Clara Franzini-Armstrong.)

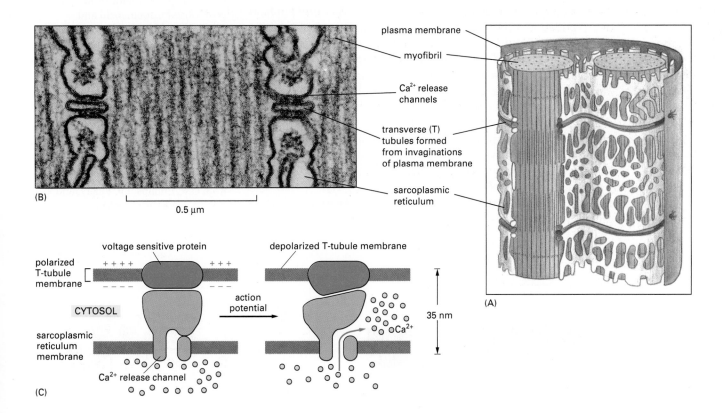

T-tubule membrane are activated by the incoming action potential, they trigger some of the Ca^{2+} release channels to open, probably by direct mechanical coupling. Ca^{2+} ions then escape from the sarcoplasmic reticulum (where they are stored in high concentration) into the cleft of the junction, causing more of the Ca^{2+} release channels to open, thereby amplifying the response. Ca^{2+} ions flooding into the cytosol then initiate the contraction of each myofibril.

Because the signal from the muscle-cell plasma membrane is passed within milliseconds (via the T tubules and sarcoplasmic reticulum) to every sarcomere in the cell, all of the myofibrils in the cell contract at the same time. The increase in Ca^{2+} concentration in the cytosol is transient because the Ca^{2+} is rapidly pumped back into the sarcoplasmic reticulum by an abundant Ca^{2+}-ATPase in its membrane (discussed in Chapter 11). Typically, the cytosolic Ca^{2+} concentration is restored to resting levels within 30 milliseconds, causing the myofibrils to relax.

Troponin and Tropomyosin Mediate the Ca^{2+} Regulation of Skeletal Muscle Contraction [57]

The Ca^{2+} dependence of vertebrate skeletal muscle contraction, and hence its dependence on motor commands transmitted via nerves, is due entirely to a set of specialized accessory proteins closely associated with actin filaments. If myosin is mixed with pure actin filaments in a test tube, the ATPase activity of myosin is stimulated whether or not Ca^{2+} is present; in a normal myofibril, on the other hand, where the actin filaments are associated with accessory proteins, the stimulation of myosin ATPase activity depends on Ca^{2+}.

One of these accessory proteins is a muscle form of *tropomyosin,* the rod-shaped molecule introduced earlier that binds in the groove of the actin helix (see Figure 16–78). The other major accessory protein involved in Ca^{2+} regulation in vertebrate skeletal muscle is *troponin,* a complex of three polypeptides—troponins T, I, and C (named for their *T*ropomyosin-binding, *I*nhibitory, and *C*alcium-binding activities). The troponin complex has an elongated shape, with subunits C and I forming a globular head region and T forming a long tail. The tail of *troponin T* binds to tropomyosin and is thought to be responsible for positioning the complex on the thin filament (Figure 16–93A). *Troponin I* binds to actin, and when it is added to troponin T and tropomyosin, the complex inhibits the interaction of actin and myosin, even in the presence of Ca^{2+}.

The further addition of *troponin C* completes the troponin complex and makes its effects sensitive to Ca^{2+}. Troponin C binds up to four molecules of Ca^{2+}, and with Ca^{2+} bound, it relieves the inhibition of myosin binding to actin produced by the other two troponin components. Troponin C is closely related to calmodulin, which mediates Ca^{2+}-signaled responses in all cells, including the activation of smooth muscle myosin. Troponin C may therefore be regarded as a specialized form of calmodulin that has evolved permanent binding sites for troponin I and troponin T, thereby ensuring that the myofibril responds extremely rapidly to an increase in Ca^{2+} concentration.

There is only one molecule of the troponin complex for every seven actin monomers in an actin filament (see Figure 16–93A). Structural studies suggest

Figure 16–93 The control of skeletal muscle contraction by troponin. (A) A muscle thin filament showing the positions of tropomyosin and troponin along the actin filament. Each tropomyosin molecule has seven evenly spaced regions of homologous sequence, each of which is thought to bind to an actin monomer as shown. (B) A thin filament shown end-on, illustrating how Ca^{2+} binding to troponin is thought to relieve the tropomyosin blockage of the interaction of the myosin head with actin. (A, adapted from G.N. Phillips, J.P. Fillers, and C. Cohen, *J. Mol. Biol.* 192:111–131, 1986.)

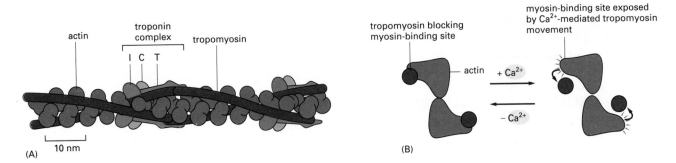

that in a resting muscle the binding of troponin I to actin moves the tropomyosin molecules to a position on the actin filaments that in an actively contracting muscle is occupied by the myosin heads and thus inhibits the interaction of actin and myosin. When the level of Ca^{2+} is raised, troponin C causes the troponin I to release its hold on actin, thereby allowing the tropomyosin molecules to shift their position slightly so that the myosin heads can bind to the actin filament (Figure 16–93B).

Other Accessory Proteins Maintain the Architecture of the Myofibril and Provide It with Elasticity [58]

The remarkable speed and power of muscle contraction depend on the filaments of actin and myosin in each myofibril being held at the optimal distance from one another and in correct alignment. More than a dozen structural proteins contribute to the precise architecture of the myofibril: the order in which they assemble, and the controls over this process, are important topics of contemporary research.

Actin filaments are anchored by their plus ends to the Z disc, where they are held in a square lattice arrangement by other proteins. One of the most important structural proteins in this region is α-actinin, the actin cross-linking protein discussed earlier that is abundant in most animal cells and is concentrated in the Z-disc region of the myofibril (Figure 16–94). Myosin filaments are also held in a regular lattice—in this case a hexagonal one—through associated proteins that bind midway along the bipolar thick filaments.

Skeletal muscle cells contain two extraordinarily large proteins, called *titin* and *nebulin,* which form a network of fibers associated with the actin and myosin filaments. Titin, which has a molecular weight of 3×10^6, is the largest polypeptide yet described. Stringlike titin molecules extend from the thick filaments to the Z disc; they are thought to act like springs to keep the myosin thick filaments centered in the sarcomere (Figure 16–95). By contrast, nebulin, which is also large, is closely associated with the actin thin filaments and consists almost entirely of a repeating, 35-amino-acid actin-binding motif. The number of these motifs, and hence the total length of the nebulin molecule, is that needed to extend from one end of the actin filament to the other. Nebulin therefore could act as a "molecular ruler" to regulate the assembly of actin and the length of the actin filaments during muscle development (see Figure 3–52).

The myofibrils are bound to one another side by side by a system of desmin intermediate filaments, and the entire array is then anchored to the plasma membrane of the muscle cell by various proteins, including a flexible, elongated actin-binding protein called *dystrophin.* This protein, which is either absent or defective in patients with muscular dystrophy, has a close structural resemblance to spectrin, and it may link specific muscle membrane proteins to actin filaments in the myofibril.

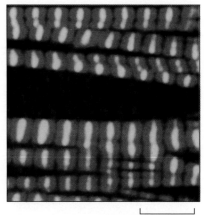

5 μm

Figure 16–94 Location of α-actinin in muscle. This confocal immunofluorescence image shows a group of myofibrils from a cultured heart muscle cell. Actin is stained *red* with rhodamine-labeled phalloidin, and α-actinin is stained *green* with a fluorescein-labeled antibody, but because actin and α-actinin are co-localized in the Z disc, this region actually appears *yellow.* (From M.H. Lu et al., *J. Cell Biol.* 117:1017–1022, 1992, by copyright permission of the Rockefeller University Press.)

Figure 16–95 Location of titin and nebulin in a skeletal muscle sarcomere. Each giant titin molecule extends from the Z disc to the M line—a distance of over 1 μm. Part of each titin molecule is closely associated with myosin molecules in the thick filament; the rest of the molecule is elastic and changes length as the muscle contracts and relaxes. Each nebulin molecule extends from the Z disc along the length of one thin actin filament and could thereby determine thin filament length.

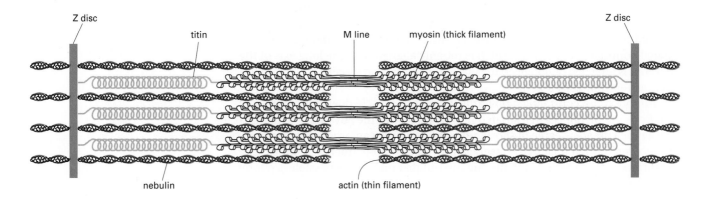

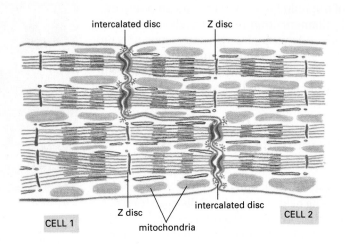

intercalated disc   Z disc

Z disc

intercalated disc

mitochondria

CELL 1

CELL 2

Figure 16–96 The structure of heart muscle. Schematic diagram of heart muscle showing two cells joined end to end by specialized junctions known as intercalated discs. Actin filaments from sarcomeres in adjacent cells insert into the dense material associated with the plasma membrane in the region of each intercalated disc as though they were Z discs. Thus the myofibrils continue across the muscle, ignoring cell boundaries.

The Same Contractile Machinery, in Modified Form, Is Found in Heart Muscle and Smooth Muscle [59]

Thus far we have described only one of the three major types of muscle present in vertebrates—skeletal muscle. The others are *heart (cardiac) muscle,* which contracts about 3 billion times in the course of an average human life-span, and *smooth muscle,* which produces the slower and longer-lasting contractions characteristic of organs such as the intestines. All three types of muscle cells, together with another class of contractile cells known as *myoepithelial cells* (see Figure 22–36E), contract by an actin and myosin-II sliding filament mechanism.

Like skeletal muscle, **heart muscle** is striated, reflecting a very similar organization of actin filaments and myosin filaments. It is also triggered to contract by a similar mechanism: an action potential triggers the sarcoplasmic reticulum to release Ca^{2+}, which activates contraction by means of a troponin-tropomyosin complex. Heart muscle cells, however, are not syncytial but are cells with a single nucleus. They are joined end to end by special structures called *intercalated discs* (Figure 16–96). The intercalated discs serve at least three functions. (1) They attach one cell to the next by means of desmosomes (discussed in Chapter 19). (2) They connect the actin filaments of the myofibrils of adjacent cells (performing a function analogous to that of the Z discs inside the cells). (3) They contain gap junctions, which allow an action potential to spread rapidly from one cell to the next, synchronizing the contractions of the heart muscle cells.

The most "primitive" muscle, in the sense of being most like nonmuscle cells, has no striations and is therefore called **smooth muscle.** It forms the contractile portion of the stomach, intestine, and uterus, the walls of arteries, and many other structures in which slow and sustained contractions are needed. It is composed of sheets of highly elongated spindle-shaped cells, each with a single nucleus. The cells contain both myosin-II and actin filaments, but these are not arranged in the strictly ordered pattern found in skeletal and cardiac muscle and do not form distinct myofibrils. Instead, the filaments form a more loosely arranged contractile apparatus, which is roughly aligned with the long axis of the cell—but is attached obliquely to the plasma membrane at disclike junctions connecting adjacent cells together.

Although the contractile apparatus in smooth muscle does not contract as rapidly as the myofibrils in a striated muscle cell, it has the advantage of permitting a much greater degree of shortening and therefore can produce large movements even though it lacks the leverage provided by attachments to bones. The organization of the actin filaments and myosin that makes this possible is poorly understood; one model is presented in Figure 16–97.

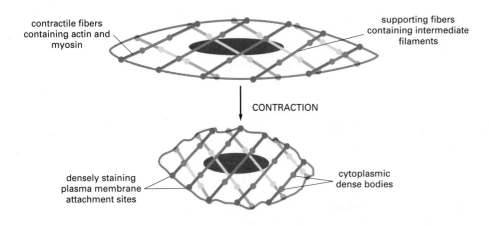

contractile fibers containing actin and myosin

supporting fibers containing intermediate filaments

CONTRACTION

densely staining plasma membrane attachment sites

cytoplasmic dense bodies

Figure 16–97 **A model for the contractile apparatus in a smooth muscle cell.** In this hypothetical view, bundles of contractile filaments containing actin and myosin (*red*) are anchored at one end to sites in the plasma membrane and at the other end, through cytoplasmic "dense bodies," to noncontractile bundles of intermediate filaments (*blue*). The contractile actin-myosin bundles are oriented obliquely to the long axis of the cell (which is generally much more elongated than shown), and their contraction greatly shortens the cell. Only a few of the many bundles are shown.

The Activation of Myosin in Many Cells Depends on Myosin Light-Chain Phosphorylation [60]

The highly specialized contractile mechanisms that we have described in muscle cells evolved from the simpler force-generating mechanisms found in all eucaryotic cells. Not surprisingly, the myosin-II in nonmuscle cells most closely resembles the myosin-II in smooth muscle cells, the least specialized type of muscle. Contraction in smooth muscle cells is triggered by a rise in cytosolic Ca^{2+}, but unlike the mechanism in skeletal and heart muscle, contraction is initiated mainly by phosphorylation of one of the two myosin-II light chains, which in turn controls the interaction of myosin with actin. A similar mechanism regulates nonmuscle myosin-II activity.

The two light chains on each head of the myosin-II molecule (see Figure 5–23) are different, and one of them is phosphorylated during nonmuscle and smooth muscle contraction. When this light chain is phosphorylated, the myosin head can interact with an actin filament and thereby cause contraction; when it is dephosphorylated, the myosin head tends to dissociate from actin and becomes inactive. In both smooth muscle and nonmuscle cells the phosphorylation is catalyzed by the enzyme *myosin light-chain kinase,* whose action requires the binding of a Ca^{2+}/calmodulin complex. As a result, contraction is controlled by the level of cytosolic Ca^{2+}, as in cardiac and skeletal muscle (Figure 16–98).

Light-chain phosphorylation can also influence the state of aggregation of myosin-II molecules in the cell, as already mentioned in connection with motility in nonmuscle cells (see Figure 16–73). The phosphorylation of myosin-II occurs relatively slowly, so that even though it has assembled into a contractile bundle with actin, maximum contraction often requires nearly a second (compared with the few milliseconds required for a striated muscle cell). But rapid activation of contraction is not important in smooth muscle or nonmuscle cells: myosin-IIs in such cells hydrolyze ATP about 10 times more slowly than skeletal muscle myosin, producing a slow cross-bridge cycle and a slow contraction.

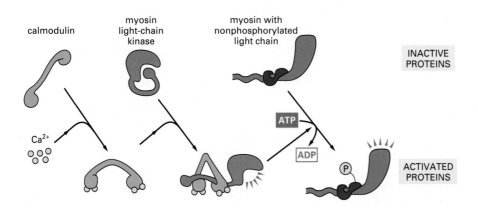

calmodulin

myosin light-chain kinase

myosin with nonphosphorylated light chain

INACTIVE PROTEINS

Ca^{2+}

ATP

ADP

P

ACTIVATED PROTEINS

Figure 16–98 **The regulation of smooth muscle contraction by Ca^{2+}.** The contraction is activated in the presence of Ca^{2+} by *myosin light-chain kinase,* which catalyzes the phosphorylation of a particular site on one of the two types of myosin light chains. Nonmuscle myosin molecules are regulated by the same mechanism (see Figure 16–73).

Muscle

Summary

Muscle contraction is produced by the sliding of actin filaments against myosin filaments. The head regions of myosin molecules, which project from myosin filaments, engage in an ATP-driven cycle in which they attach to adjacent actin filaments, undergo a conformational change that pulls the myosin filament against the actin filament, and then detach. This cycle is facilitated by special accessory proteins in muscle that hold the actin and myosin filaments in parallel overlapping arrays with the correct orientation and spacing for sliding to occur. Two other accessory proteins—troponin and tropomyosin—allow the contraction of skeletal and cardiac muscle to be regulated by Ca^{2+}.

In smooth muscle cells, and in most nonmuscle cells, actin and myosin produce contraction in fundamentally the same way as in skeletal and cardiac muscle. The contractile units are smaller, however, and less highly ordered in such cells; both their activity and their state of assembly are controlled by Ca^{2+}-regulated phosphorylation of a myosin light chain.

References

General

Amos, L.A.; Amos, W.B. Molecules of the Cytoskeleton. New York: Guilford Press, 1991.

Bershadsky, A.D.; Vasiliev, J.M. Cytoskeleton. New York: Plenum Press, 1988.

Bray, D. Cell Movements. New York: Garland Publishing, 1992.

Kreis, T.E.; Vale, R.D., eds. Guidebook to Cytoskeletal and Motor Proteins. Oxford, UK: Oxford University Press, 1993.

Cited

1. Cell motility. *Trends Cell Biol.* 3:363–412, 1993.
   Elson, E.L. Cellular mechanics as an indicator of cytoskeletal structure and function. *Annu. Rev. Biophys. Biophys. Chem.* 17:397–430, 1988.
   Ingber, D.E. Cellular tensegrity: defining new rules of biological design that govern the cytoskeleton. *J. Cell Sci.* 104:613–627, 1993.

2. Gelfand, V.I.; Bershadsky, A.D. Microtubule dynamics: mechanism, regulation, and function. *Annu. Rev. Cell Biol.* 7:93–116, 1991.
   Karsenti, E.; Maro, B. Centrosomes and the spatial distribution of microtubules in animal cells. *Trends Biochem. Sci.* 11:460–463, 1986.

3. McNiven, M.A.; Porter, K.R. Organization of microtubules in centrosome-free cytoplasm. *J. Cell Biol.* 106:1593–1605, 1988.

4. Corthésy-Theulaz, I.; Pauloin, A.; Pfeffer, S.R. Cytoplasmic dynein participates in the centrosomal localization of the Golgi complex. *J. Cell Biol.* 118:1333–1345, 1992.
   Kreis, T.E. Role of microtubules in the organisation of the Golgi apparatus. *Cell Motil. Cytoskeleton* 15:67–70, 1990.
   Lee, C.; Chen, L.B. Dynamic behavior of endoplasmic reticulum in living cells. *Cell* 54:37–46, 1988.
   Terasaki, M. Recent progress on structural interactions of the endoplasmic reticulum. *Cell Motil. Cytoskeleton* 15:71–75, 1990.
   Vale, R. Intracellular transport using microtubule-based motors. *Annu. Rev. Cell Biol.* 3:347–378, 1987.

5. Euteneuer, U.; Schliwa, M. Persistent, directional motility of cells and cytoplasmic fragments in the absence of miocrotubules. *Nature* 310:58–61, 1984.

Lee, J.; Ishihara, A.; Theriot, J.A.; Jacobson, K. Principles of locomotion for simple-shaped cells. *Nature* 362:167–171, 1993.

6. Euteneuer, U.; Schliwa, M. Mechanism of centrosome positioning during the wound response in BSC-1 cells. *J. Cell Biol.* 116:1157–1166, 1992.
   Gyoeva, F.K.; Gelfand, V.I. Coalignment of vimentin intermediate filaments with microtubules depends on kinesin. *Nature.* 353:445–448, 1991.
   Kupfer, A.; Singer, S.J. Cell biology of cytotoxic and helper T cell functions: immunofluorescence microscopic studies of single cells and cell couples. *Annu. Rev. Immunol.* 7:309–337, 1989.

7. Albers, K.; Fuchs, E. The molecular biology of intermediate filament proteins. *Int. Rev. Cytol.* 134:243–279, 1992.
   Cary, R.B.; Klymkowsky, M.W. Finding filament function. *Curr. Biol.* 2:43–45, 1992.
   Steinert, P.M.; Roop, D.R. Molecular and cellular biology of intermediate filaments. *Annu. Rev. Biochem.* 57:593–625, 1988.

8. Chou, Y.-H.; Bischoff, J.R.; Beach, D.; Goldman, R.D. Intermediate filament reorganization during mitosis is mediated by p34[cdc2] phosphorylation of vimentin. *Cell* 62:1063–1071, 1990.
   Okabe, S.; Miyasaka, H.; Hirokawa, N. Dynamics of the neuronal intermediate filaments. *J. Cell Biol.* 121:375–386, 1993.
   Stewart, M. Intermediate filament structure and assembly. *Curr. Opin. Cell Biol.* 5:3–11, 1993.

9. Coulombe, P.A. The cellular and molecular biology of keratins: beginning a new era. *Curr. Opin. Cell Biol.* 5:17–29, 1993.
   Osborn, M.; Weber, K. Tumor diagnosis by intermediate filament typing: a novel tool for surgical pathology. *Lab. Invest.* 48:372–394, 1983.

10. Cleveland, D.W.; Monteiro, M.J.; Wong, P.C.; et al. Involvement of neurofilaments in the radial growth of axons. *J. Cell Sci.* Suppl. 15:85–95, 1991.
    Garrod, D.R. Desmosomes and hemidesmosomes. *Curr. Opin. Cell Biol.* 5:30–40, 1993.

11. Nigg, E.A. Assembly and cell cycle dynamics of the nuclear lamina. *Semin. Cell Biol.* 3:245–253, 1992.

12. Fuchs, E.; Coulombe, P.A. Of mice and men: genetic skin diseases of keratin. *Cell.* 69:899–902, 1992.

Janmey, P.A. Mechanical properties of cytoskeletal polymers. *Curr. Opin. Cell Biol.* 3:4–11, 1991.

13. Dustin, P. Microtubules, 2nd ed. New York: Springer-Verlag, 1984.

Hyams, J.F.; Lloyd, C.W. Microtubules. New York: Wiley-Liss, 1993.

14. Amos, L.A.; Baker, T.S. The three-dimensional structure of tubulin protofilaments. *Nature* 279:607–612, 1979.

Luduena, R.F. Are tubulin isotypes functionally significant? *Mol. Biol. Cell* 4:445–457, 1993.

Sullivan, K.F. Structure and utilization of tubulin isotypes. *Annu. Rev. Cell Biol.* 4:687–716, 1988.

Wade, R.H.; Chrétien, D. Cryoelectron microscopy of microtubules. *J. Struct. Biol.* 110:1–27, 1993.

15. DeBrabander, M. Microtubule dynamics during the cell cycle: the effects of taxol and nocodazole on the microtubule system of Pt K2 cells at different stages of the mitotic cycle. *Int. Rev. Cytol.* 101:215–274, 1986.

Inoué, S. Cell division and the mitotic spindle. *J. Cell Biol.* 91:131s–147s, 1981.

Salmon, E.D.; McKeel, M.; Hays, T. Rapid rate of tubulin dissociation from microtubules in the mitotic spindle *in vivo* measured by blocking polymerization with colchicine. *J. Cell Biol.* 99:1066–1075, 1984.

16. Mandelkow, E.-M.; Mandlekow, E. Microtubule oscillations. *Cell Motil. Cytoskeleton.* 22:235–244, 1992.

17. Bergen, L.G.; Borisy, G.G. Head-to-tail polymerization of microtubules *in vitro*. Electron microscope analysis of seeded assembly. *J. Cell Biol.* 84:141–150, 1980.

McIntosh, J.R.; Euteneuer, U. Tubulin hooks as probes for microtubule polarity: an analysis of the method and an evaluation of data on microtubule polarity in the mitotic spindle. *J Cell Biol.* 98:525–533, 1984.

Mitchison, T.J. Localization of an exchangeable GTP binding site at the plus end of microtubules. *Science* 261:1044–1047, 1993.

18. Glover, D.M.; Gonzalez, C.; Raff, J.W. The centrosome. *Sci. Am.* 268(6):62–68, 1993.

Joshi, H.C.; Palacios, M.J.; McNamara, L.; Cleveland, D.W. γ-tubulin is a centrosomal protein required for cell cycle-dependent microtubule nucleation. *Nature* 356:80–83, 1992.

Mazia, D. Centrosomes and mitotic poles. *Exp. Cell Res.* 153:1–15, 1984.

19. Sammak, P.J.; Borisy, G.G. Direct observation of microtubule dynamics in living cells. *Nature* 332:724–726, 1988.

Tanaka, E.M.; Kirschner, M.W. Microtubule behavior in the growth cones of living neurons during axon elongation. *J. Cell Biol.* 115:345–363, 1991.

20. Erickson, H.P.; O'Brien, E.T. Microtubule dynamic instability and GTP hydrolysis. *Annu. Rev. Biophys. Biomol. Struct.* 21:145–166, 1992.

Mandelkow, E.-M.; Mandelkow, E.; Milligan, R.A. Microtubule dynamics and microtubule caps—a time-resolved cryo-electron microscopy study. *J. Cell Biol.* 114:977–991, 1991.

Mitchison, T.J.; Kirschner, M.W. Dynamic instability of microtubule growth. *Nature* 312:237–242, 1984.

21. Gelfand, V.I.; Bershadsky, A.D. Microtubule dynamics: mechanism, regulation, and function. *Annu. Rev. Cell Biol.* 7:93–116, 1991.

Kirschner, M.W.; Mitchison, T.J. Beyond self-assembly: from microtubules to morphogenesis. *Cell* 45:329–342, 1986.

22. Greer, K.; Rosenbaum, J.L. Post-translational modifications of tubulin. In Cell Movement, Vol. 2: Kinesin, Dynein, and Microtubule Dynamics (F.D. Warner, J.R. McIntosh, eds.), 47–66. New York: Wiley-Liss, 1989.

MacRae, T.H. Towards an understanding of microtubule function and cell organization: an overview. *Biochem. Cell Biol.* 70:835–841, 1992.

23. Drechsel, D.N.; Hyman, A.A.; Cobb, M.H.; Kirschner, M.W. Modulation of the dynamic instability of tubulin assembly by the microtubule-associated protein tau. *Mol. Biol. Cell.* 3:1141–1154, 1992.

Lee, G. Non-motor microtubule-associated proteins. *Curr. Opin. Cell Biol.* 5:88–94, 1993.

Olmsted, J.B. Microtubule-associated proteins. *Annu. Rev. Cell Biol.* 2:421–457, 1986.

24. Chen, J.; Kanai, Y.; Cowan, N.J.; Hirokawa, N. Projection domains of MAP2 and tau determine spacings between microtubules in dendrties and axons. *Nature* 360:674–677, 1992.

Knops, J.; Kosik, K.S.; Lee, G.; et al. Overexpression of tau in a nonneuronal cell induces long cellular processes. *J. Cell Biol.* 114:725–733, 1991.

25. Skoufias, D.A.; Scholey, J.M. Cytoplasmic microtubule-based motor proteins. *Curr. Opin. Cell Biol.* 5:95–104, 1993.

Vale, R.D.; Reese, T.S.; Sheetz, M.P. Identification of a novel force-generating protein, kinesin, involved in microtubule-based motility. *Cell* 42:39–50, 1985.

Vallee, R.B.; Wall, J.S.; Paschal, B.M.; Shpetner, H.S. Microtubule-associated protein 1C from brain is a two-headed cytosolic dynein. *Nature* 332:561–563, 1988.

26. Bloom, G.S. Motor proteins for cytoplasmic microtubules. *Curr. Opin. Cell Biol.* 4:66–73, 1992.

Endow, S.A.; Titus, M.A. Genetic approaches to molecular motors. *Annu. Rev. Cell Biol.* 8:29–66, 1992.

Gelfand, V.I.; Scholey, J.M. Every motion has its motor. *Nature* 359:480–482, 1992.

27. Blair, D.F.; Dutcher, S.K. Flagella in prokaryotes and lower eukaryotes. *Curr. Opin. Gen. Dev.* 2:756–767, 1992.

Fawcett, D.W.; Porter, K.R. A study of the fine structure of ciliated epithelia. *J. Morphol.* 94:221–281, 1954.

Gibbons, I.R. Cilia and flagella of eukaryotes. *J. Cell Biol.* 91:107s–124s, 1981.

28. Burgess, S.A.; Dover, S.D.; Woolley, D.M. Architecture of the outer arm dynein ATPase in an avian sperm flagellum, with further evidence for the B-link. *J. Cell Sci.* 98:17–26, 1991.

Smith, E.F.; Sale, W.S. Regulation of dynein-driven microtubule sliding by the radial spokes in flagella. *Science* 257:1557–1559, 1992.

Warner, F.D.; Satir, P.; Gibbons, I.R., eds. Cell Movement, Vol. 1. The Dynein ATPases. New York: Wiley-Liss, 1989.

Witman, G.B. Axonemal dyneins. *Curr. Opin. Cell Biol.* 4:74–79, 1992.

29. Huang, B. *Chlamydomonas reinhardtii*: a model system for genetic analysis of flagellar structure and motility. *Int. Rev. Cytol.* 99:181–215, 1986.

References

Johnson, K.A.; Rosenbaum, J.L. Polarity of flagellar assembly in *Chlamydomonas. J. Cell Biol.* 119:1605–1611, 1992.

Ringo, D.L. Flagellar motion and the fine structure of flagellar apparatus in *Chlamydomonas reinhardtii. J. Cell Biol.* 33:543–571, 1967.

30. Beisson, J.; Sonnenborn, T.M. Cytoplasmic inheritance of the organization of the cell cortex in *Paramecium aurelia. Proc. Natl. Acad. Sci. USA* 53:275–282, 1965.

Holmes, J.A.; Dutcher, S.K. Cellular asymmetry in *Chlamydomonas reinhardtii. J. Cell Sci.* 94:273–285, 1989.

Palazzo, R.E.; Vaisberg, E.; Cole, R.W.; Rieder, C.L. Centriole duplication in lysates of *Spisula solidissima* oocytes. *Science* 256:219–221, 1992.

31. Kabsch, W.; Vandekerckhove, J. Structure and function of actin. *Annu. Rev. Biophys. Biomol. Struct.* 21:49–76, 1992.

32. Bonder, E.M.; Fishkind, D.J.; Mooseker, M.S. Direct measurement of critical concentrations and assembly rate constants at the two ends of an actin filament. *Cell* 34:491–501, 1983.

Janmey, P.A. Mechanical properties of cytoskeletal polymers. *Curr. Opin. Cell Biol.* 3:4–11, 1991.

Kabsch, W.; Mannherz, H.G.; Suck, D.; Pai, E.F.; Holmes, K.C. Atomic structure of the actin:DNase I complex. *Nature* 347:37–44, 1990.

33. Carlier, M.-F. Actin: protein structure and filament dynamics. *J. Biol. Chem.* 266:1–4, 1991.

Mitchison, T.J. Compare and contrast actin filaments and microtubules. *Mol. Biol. Cell* 3:1309–1315, 1992.

Oosawa, F.; Asakura, S. Thermodynamics of the Polymerization of Protein. London: Academic Press, 1975.

34. Carlier, M.-F. Role of nucleotide hydrolysis in the dynamics of actin filaments and microtubules. *Int. Rev. Cytol.* 115:139–170, 1989.

Wegner, A. Head to tail polymerization of actin. *J. Mol. Biol.* 108:139–150, 1976.

35. Cooper, J.A. Effects of cytochalasin and phalloidin on actin. *J. Cell Biol.* 105:1473–1478, 1987.

Forscher, P.; Smith, S.J. Actions of cytochalasins on the organization of actin filaments and microtubules in a neuronal growth cone. *J. Cell Biol.* 107:1505–1516, 1988.

Wehland, J.; Weber, K. Actin rearrangement in living cells revealed by microinjection of a fluorescent phalloidin derivative. *Eur. J. Cell Biol.* 24:176–183, 1981.

36. Cassimeris, L.; Safer, D.; Nachmias, V.T.; Zigmond, S.H. Thymosin β4 sequesters the majority of G-actin in resting human polymorphonuclear leukocytes. *J. Cell Biol.* 119:1261–1270, 1992.

Fechheimer, M.; Zigmond, S.H. Focusing on unpolymerized actin. *J. Cell Biol.* 123:1–5, 1993.

Goldschmidt-Clermont, P.J.; Furman, M.I.; Wachsstock, D.; et al. The control of actin nucleotide exchange by thymosin beta 4 and profilin. A potential regulatory mechanism for actin polymerization in cells. *Mol. Biol. Cell* 3:1015–1024, 1992.

37. Abercrombie, M. The crawling movement of metazoan cells. *Proc. R. Soc. Lond. (Biol.)* 207:129–147, 1980.

Fukui, Y. Toward a new concept of cell motility: cytoskeletal dynamics in amoeboid movements and cell division. *Int. Rev. Cytol.* 144:85–127, 1993.

Small, J.V. Microfilament-based motility in non-muscle cells. *Curr. Opin. Cell Biol.* 1:75–79, 1989.

38. Cao, L.; Fishkind, D.J.; Wang, Y-L. Localization and dynamics of nonfilamentous actin in cultured cells. *J. Cell Biol.* 123:173–181, 1993.

Condeelis, J. The formation of cell protrusions. *Annu. Rev. Cell Biol.* 9:441–444, 1993.

Theriot, J.A.; Mitchison, T.J. The nucleation-release model of actin filament dynamics in cell motility. *Trends Cell Biol.* 2:219–222, 1992.

39. Kocks, C.; Hellio, R.; Gounon, P.; Ohayon, H.; Cossart, P. Polarized distribution of *Listeria monocytogenes* surface protein ActA at the site of directional actin assembly. *J. Cell Sci.* 105:699–710, 1993.

Theriot, J.A.; Mitchison, T.J.; Tilney, L.G.; Portnoy, D.A. The rate of actin-based motility of intracellular *Listeria monocytogenes* equals the rate of actin polymerization. *Nature* 357:257–260, 1992.

Tilney, L.G.; Portnoy, D.A. Actin filaments and the growth, movement, and spread of the intracellular bacterial parasite, *Listeria monocytogenes. J. Cell Biol.* 109:1597–1608, 1989.

40. Devreotes, P.N.; Zigmond, S.H. Chemotaxis in eukaryotic cells: a focus on leukocytes and *Dictyostelium. Annu. Rev. Cell Biol.* 4:649–686, 1988.

Sauterer, R.A.; Eddy, R.J.; Hall, A.L.; Condeelis, J.S. Purification and characterization of aginactin, a newly identified agonist-regulated actin-capping protein from *Dictyostelium* amoebae. *J. Biol. Chem.* 266:24533–24539, 1991.

41. Ridley, A.J.; Hall, A. The small GTP-binding protein rho regulates the assembly of focal adhesions and actin stress fibers in response to growth factors. *Cell* 70:389–399, 1992.

Zigmond, S.H., ed. Sensory Adaptation and Motor Activation in Chemotaxis. *Semin. in Cell Biol.*, Vol. 1(2), 1990.

42. Chenevert, J.; Corrado, K.; Bender, A.; Pringle, J.; Herskowitz, I. A yeast gene (*BEM1*) necessary for cell polarization whose product contains two SH domains. *Nature* 356:77–79, 1992.

Flescher, E.G.; Madden, K.; Snyder, M. Components required for cytokinesis are important for bud site selection in yeast. *J. Cell Biol.* 122:373–386, 1993.

43. Bretscher, A. Microfilament structure and function in the cortical cytoskeleton. *Annu. Rev. Cell Biol.* 7:337–374, 1991.

Hartwig, J.H.; Kwiatkowski, D.J. Actin-binding proteins. *Curr. Opin. Cell Biol.* 3:87–97, 1991.

44. Bennett, V.; Gilligan, D.M. The spectrin-based membrane structure and micron-scale organization of the plasma membrane. *Annu. Rev. Cell Biol.* 9:27–66, 1993.

Carraway, K.L.; Carraway, C.A.C. Membrane-cytoskeleton interactions in animal cells. *Biochim. Biophys. Acta* 988:147–171, 1989.

45. Matsudaira, P. Modular organization of actin crosslinking proteins. *Trends Biochem. Sci.* 16:87–92, 1991.

Vandekerckhove, J.; Vancompernolle, K. Structural relationships of actin-binding proteins. *Curr. Opin. Cell Biol.* 4:36–42, 1992.

46. McLaughlin, P.J.; Gooch, J.T.; Mannherz, H.-G.; Weeds, A.G. Structure of gelsolin segment 1-actin complex and the mechanism of filament severing. *Nature* 364:685–692, 1993.

    Matsudaira, P.; Janmey, P. Pieces in the actin-severing protein puzzle. *Cell* 54:139–140, 1988.

47. Cheney, R.E.; Riley, M.A.; Mooseker, M.S. Phylogenetic analysis of the myosin superfamily. *Cell Motil. Cytoskeleton* 24:215–223, 1993.

    Korn, E.D.; Hammer, J.A., 3d. Myosin I. *Curr. Opin. Cell Biol.* 2:57–61, 1990.

    Pollard, T.D.; Doberstein, S.K.; Zot, H.G. Myosin-I. *Annu. Rev. Physiol.* 53:653–681, 1991.

    Titus, M.A. Myosins. *Curr. Opin. Cell Biol.* 5:77–81, 1993.

48. Langanger, G.; Moeremans, M.; Daneels, G.; et al. The molecular organization of myosin in stress fibers of cultured cells. *J. Cell Biol.* 102:200–209, 1986.

    Strome, S. Determination of cleavage planes. *Cell* 72:3–6, 1993.

49. Burridge, K.; Fath, K.; Kelly, T.; Nuckolls, G.; Turner, C. Focal adhesions: transmembrane junctions between the extracellular matrix and the cytoskeleton. *Annu. Rev. Cell Biol.* 4:487–525, 1988.

    Burridge, K.; Turner, C.E.; Romer, L.H. Tyrosine phosphorylation of paxillin and pp125[FAK] accompanies cell adhesion to extracellular matrix: a role in cytoskeletal assembly. *J. Cell Biol.* 119:893–903, 1992.

50. Mooseker, M.S. Organization, chemistry, and assembly of the cytoskeketal apparatus of the intestinal brush border. *Annu. Rev. Cell Biol.* 1:209–241, 1985.

51. Heath, J.P.; Holifield, B.F. Cell locomotion: new research tests old ideas on membrane and cytoskeletal flow. *Cell Motil. Cytoskeleton* 18:245–257, 1991.

    Lackie, J.M.Cell Movement and Cell Behavior. London: Allen and Unwin, 1986.

    Stossel, T.P. On the crawling of animal cells. *Science* 260:1086–1094, 1993.

    Trinkaus, J.P. Cells into Organs: The Forces That Shape the Embryo, 2nd ed. Englewood Cliffs, NJ: Prentice-Hall, 1984.

52. Brown, S.S. Phenotypes of cytoskeletal mutants. *Curr. Opin. Cell Biol.* 5:129–134, 1993.

    DeLozanne, A.; Spudich, J.A. Disruption of the *Dictyostelium* myosin heavy chain gene by homologous recombination. *Science* 236:1086–1091, 1987.

    Gerisch, G.; Noegel, A.A.; Schleicher, M. Genetic alteration of proteins in actin-based motility systems. *Annu. Rev. Physiol.* 53:607–628, 1991.

    Knecht, D.A.; Loomis, W.F. Antisense RNA inactivation of myosin heavy chain gene depression in *Dictyosteilum discoideum*. *Science* 236:1081–1086, 1987.

53. Cooke, R. The mechanism of muscle contraction. *CRC Crit. Rev. Biochem.* 21:53–118, 1986.

    Huxley, A.F. Reflections on Muscle. Princeton, NJ: Princeton University Press, 1980.

    Huxley, H.E. The mechanism of muscular contraction. *Science* 164:1356–1366, 1969.

    Squire, J.M. Muscle: Design, Diversity and Disease. Menlo Park, CA: Benjamin-Cummings, 1986.

54. Amos, L.A. Structure of muscle filaments studied by electron microscopy. *Annu. Rev. Biophys. Biophys. Chem.* 14:291–313, 1985.

55. Burton, K. Myosin step size: estimates from motility assays and shortening muscle. *J. Muscle Res. Cell Motil.* 13:590–607, 1992.

    Holmes, K.C.; Popp, D.; Gebhard, W.; Kabsch, W. Atomic model of the actin filament. *Nature* 347:44–49, 1990.

    Irving, M. Muscle mechanics and probes of the crossbridge cycle. In Fibrous Protein Structure (J.M. Squire; P.J. Vibert, eds.). San Diego, CA: Academic Press, 1987.

    Pollard, T.D. The myosin crossbridge problem. *Cell* 48:909–910, 1987.

    Rayment, I.; Rypniewski, W.R.; Schmidt-Base, K.; et al. Three-dimensional structure of myosin sub-fragment-1: a molecular motor. *Science* 261:50–58, 1993.

56. Katz, B. Nerve, Muscle and Synapse. New York: McGraw-Hill, 1966.

    Lai, F.A.; Erickson, H.P.; Rousseau, E.; Liu, Q.-Y.; Meissner, G. Purification and reconstitution of the calcium release channel from skeletal muscle. *Nature* 331:315–319, 1988.

    Saito, A.; Inui, M.; Radermacher, M.; Frank, J.; Fleischer, S. Ultrastructure of the calcium release channel of sarcoplasmic reticulum. *J. Cell Biol.* 107:211–219, 1988.

57. Murray, J.M.; Weber, A. The cooperative action of muscle proteins. *Sci. Am.* 230(2):59–71, 1974.

    Phillips, G.N.; Fillers, J.P.; Cohen, C. Tropomyosin crystal structure and muscle regulation. *J. Mol. Biol.* 192:111–131, 1986.

    Zot, A.S.; Potter, J.D. Structural aspects of troponin-tropomyosin regulation of skeletal muscle contraction. *Annu. Rev. Biophys. Biophys. Chem.* 16:535–559, 1987.

58. Ervasti, J.M.; Campbell, K.P. Dystrophin and the membrane skeleton. *Curr. Opin. Cell Biol.* 5:82–87, 1993.

    Trinick, J. Molecular rulers in muscle? *Curr. Biol.* 2:75–77, 1992.

    Wang, K. Sarcomere-associated cytoskeletal lattices in striated muscle. Review and hypothesis. In Cell and Muscle Motility (J.W. Shay, ed., Vol. 6), New York: Plenum Press, 1985.

59. Fawcett, D.W. A Textbook of Histology, 11th ed. Philadelphia: Saunders, 1986.

    Korn, E.D.; Hammer, J.A. Myosins of nonmuscle cells. *Annu. Rev. Biophys. Biophys. Chem.* 17:23–45, 1988.

    Sellers, J.R.; Adelstein, R.S. Regulation of contractile activity. In The Enzymes (P. Boyer; E.G. Krebs, eds.), San Diego, CA: Academic Press, 1987.

60. Citi, S.; Kendrick-Jones, J. Regulation of non-muscle myosin structure and function. *Bioessays* 7:155–159, 1987.

    Fishkind, D.J.; Cao, L.G.; Wang, Y.-L. Microinjection of the catalytic fragment of myosin light chain kinase into dividing cells: effects on mitosis and cytokinesis. *J. Cell Biol.* 114:967–975, 1991.

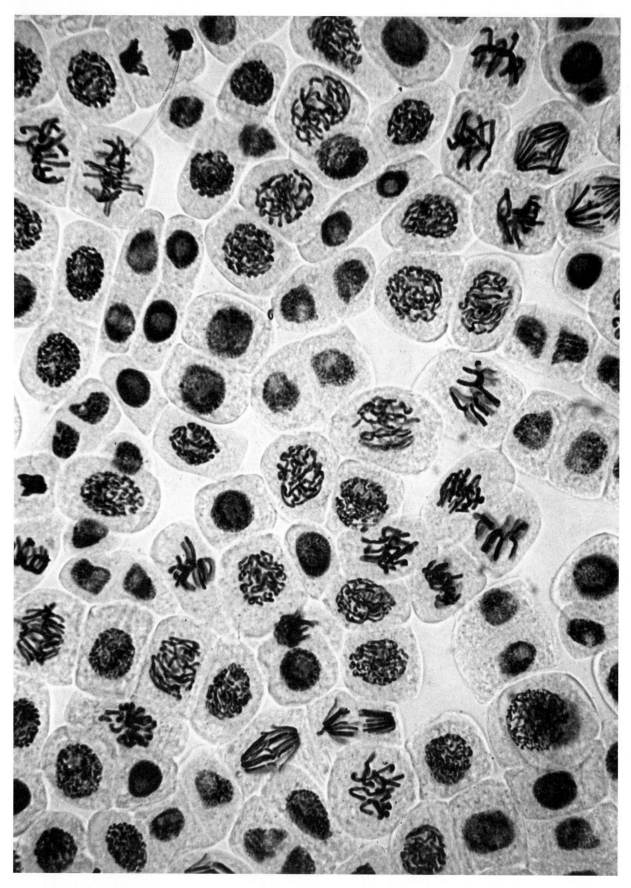

Cells from a root tip of a plant in various stages of the cell cycle.
(Courtesy of John McLeish.)

The Cell-Division Cycle

17

Cells reproduce by duplicating their contents and then dividing in two. This **cell-division cycle** is the fundamental means by which all living things are propagated. In unicellular species, such as bacteria and yeasts, each cell division produces an additional organism. In multicellular species many rounds of cell division are required to make a new individual, and cell division is needed in the adult body, too, to replace cells that are lost by wear and tear or by programmed cell death. Thus an adult human must manufacture many millions of new cells each second simply to maintain the status quo, and if all cell division is halted—for example, by a large dose of ionizing radiation—the individual will die within a few days.

The details of the cell cycle may vary, but certain requirements are universal. First and foremost, to produce a pair of genetically identical daughter cells, the DNA must be faithfully replicated, and the replicated chromosomes must be segregated into two separate cells (Figure 17–1). The cell cycle comprises, at a minimum, the set of processes that a cell must perform to accomplish these tasks. The vast majority of cells also double their mass and duplicate all their cytoplasmic organelles in each cell cycle. Thus a complex set of cytoplasmic and nuclear processes have to be coordinated with one another during the cell cycle. The central problem is to explain how this coordination is achieved.

Our understanding of the cell cycle has undergone a revolution in recent years. In the past the cell cycle was monitored by observing the events of chromosome segregation with a light microscope and by following DNA replication by measuring the incorporation of radioactive precursors into DNA. The focus of attention, therefore, was on the chromosomes, and there seemed to be large differences between the cell cycles of different organisms and different types of cells. Recent experiments have provided a new and simpler perspective, revealing a **cell-cycle control system** that coordinates the cycle as a whole. The proteins of this control system first appeared over a billion years ago and have been so well conserved in evolution that many of them function perfectly when transferred from a human cell to a yeast cell. We can therefore study the control system in a variety of eucaryotic organisms and use the findings from all of them to assemble a unified picture of how cells grow and divide.

This chapter is concerned with how the processes of the cell cycle are controlled and coordinated with one another. Other chapters discuss in detail some of the individual processes themselves: Chapters 6 and 8 deal with the machinery of DNA replication, and Chapter 18 describes the cytoskeletal apparatus that segregates the chromosomes and divides the cell in two. Here, after briefly out-

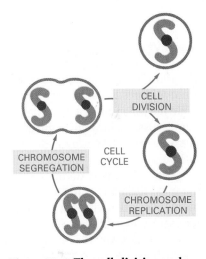

Figure 17–1 The cell-division cycle.

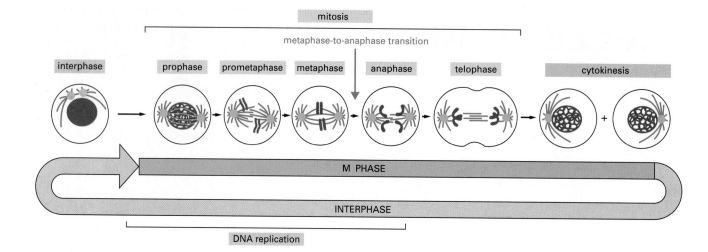

mitosis

metaphase-to-anaphase transition

interphase   prophase   prometaphase   metaphase   anaphase   telophase   cytokinesis

M PHASE

INTERPHASE

DNA replication

lining the events of the cell cycle, we discuss experiments performed on early animal embryos and yeasts that have provided the main insights into the basic control system. We then consider how extracellular signals act on the control system to regulate cell division in a multicellular organism.

The General Strategy of the Cell Cycle

The duration of the cell cycle varies greatly from one cell type to another. Fly embryos have the shortest known cell cycles, each lasting as little as 8 minutes, while the cell cycle of a mammalian liver cell can last longer than a year. We begin our discussion, however, with a more typical example and describe the sequence of events in a fairly rapidly dividing mammalian cell, with a cycle time of about 24 hours.

The cell cycle is traditionally divided into several distinct phases, of which the most dramatic is **mitosis,** the process of nuclear division, leading up to the moment of cell division itself. As discussed in detail in Chapter 18, in mitosis the nuclear envelope breaks down, the contents of the nucleus condense into visible chromosomes, and the cell's microtubules reorganize to form the *mitotic spindle* that will eventually separate the chromosomes. As mitosis proceeds, the cell seems to pause briefly in a state called *metaphase*, in which the chromosomes, already duplicated, are aligned on the mitotic spindle, poised for segregation. The separation of the duplicated chromosomes marks the beginning of *anaphase*, during which the chromosomes move to the poles of the spindle, where they decondense and re-form intact nuclei. The cell is then pinched in two by a process called **cytokinesis,** which is traditionally viewed as the end of the mitotic phase, or **M phase,** of the cell cycle (Figure 17–2).

In most cells the whole of M phase takes only about an hour, which is only a small fraction of the total cycle time. The much longer period that elapses between one M phase and the next is known as **interphase.** Under the microscope this appears, deceptively, as an uneventful interlude in which the cell simply grows in size. But other techniques reveal that interphase is actually a busy time for the proliferating cell, during which elaborate preparations for cell division are occurring in a closely ordered sequence. In particular, it is during interphase that the DNA in the nucleus is replicated.

Replication of the Nuclear DNA Occurs During a Specific Part of Interphase—the S Phase [1]

Replication of the nuclear DNA usually occupies only a portion of interphase, called the **S phase** of the cell cycle (S = synthesis). The interval between the

Figure 17–2 The events of cell division as seen under a microscope. The easily visible processes of nuclear division (mitosis) and cell fission (cytokinesis), which are together called the M phase, typically occupy only a small fraction of the cell cycle. The other, much longer part of the cycle is known as *interphase*. During M phase an abrupt change in the biochemical state of the cell occurs at the transition from metaphase to anaphase; a cell can pause in metaphase before this transition point, but once the point is passed, the cell will carry on smoothly to the end of mitosis and through cytokinesis into interphase.

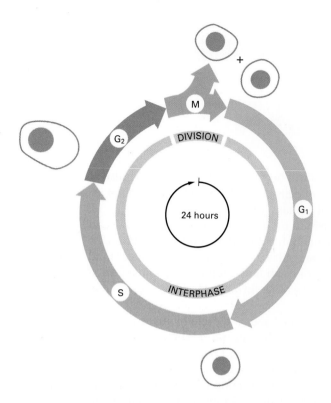

Figure 17–3 **The four successive phases of a standard eucaryotic cell cycle.** During interphase the cell grows continuously; during M phase it divides. DNA replication is confined to the part of interphase known as S phase. G_1 phase is the gap between M phase and S phase; G_2 is the gap between S phase and M phase.

completion of mitosis and the beginning of DNA synthesis is called the **G_1 phase** (G = gap), and the interval between the end of DNA synthesis and the beginning of mitosis is called the **G_2 phase.** G_1 and G_2 provide additional time for growth: if interphase lasted only long enough for DNA replication, the cell would not have time to double its mass before it divided. During G_1 the cell monitors its environment and its own size and, when the time is ripe, takes a decisive step that commits it to DNA replication and completion of a division cycle. The G_2 phase provides a safety gap, allowing the cell to ensure that DNA replication is complete before it plunges into mitosis. G_1, S, G_2, and M are the traditional subdivisions of the **standard cell cycle** (Figure 17–3). We shall see that most, but not all, cell cycles conform to this standard scheme.

Because cells require time to grow before they divide, the standard cell cycle is generally quite long—12 hours or more for fast growing tissues in a mammal, for example. Although the lengths of all phases of the cycle are variable to some extent, by far the greatest variation, in most of the commonly studied types of cells, occurs in the duration of G_1. Cells in G_1, if they have not yet committed themselves to DNA replication, can pause in their progress around the cycle and enter a specialized resting state, often called G_0 (G zero), where they can remain for days, weeks, or even years before resuming proliferation.

The shortest eucaryotic division cycles of all—shorter even than those of many bacteria—are the *early embryonic cell cycles* that occur in certain animal embryos immediately after fertilization, serving to subdivide a giant egg cell into many smaller cells as quickly as possible. In these cycles no growth occurs, the G_1 and G_2 phases are drastically curtailed, and the time from one division to the next is between 8 and 60 minutes, spent half in S phase, half in M (Figure 17–4). We shall have more to say about these early embryonic cycles later.

How can one tell where a cell is in the cycle? Cells in S phase can be recognized by supplying them with labeled molecules of thymidine—a compound that cells use exclusively for the synthesis of DNA. The label can be radioactive, usually in the form of ³H-thymidine, or chemical, usually in the form of bromodeoxyuridine (BrdU), an artificial thymidine analog. Cell nuclei that have incorporated the labeled compound are recognized by autoradiography (Figure 17–5) or by staining with anti-BrdU antibody, respectively. Typically, in a population of growing cells that are all proliferating rapidly but asynchronously, about 30%

The General Strategy of the Cell Cycle

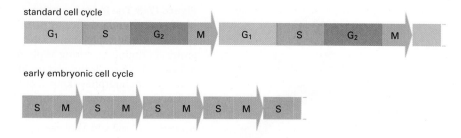

standard cell cycle

| G₁ | S | G₂ | M | G₁ | S | G₂ | M |

early embryonic cell cycle

| S | M | S | M | S | M | S | M | S |

Figure 17–4 The standard cell cycle compared with the early embryonic cell cycle. In the early embryonic cycle no growth occurs, so that each of the two daughter cells of each division is half the size of the parent cell. The cycle time is extraordinarily short, and S phases and M phases alternate without any intervening G_1 or G_2 phases.

will be in S phase at any instant and so will become labeled by a brief pulse of the DNA precursor. From the fraction of cells that are labeled (the *labeling index*), one can estimate the duration of S phase as a fraction of the whole cycle. Similarly, from the fraction of cells seen in mitosis (the *mitotic index*), one can estimate the duration of M phase as a fraction of the whole cycle. In addition, by giving a pulse of ^{3}H-thymidine or BrdU and allowing the cells to continue around the cycle for measured lengths of time, one can find out how long it takes an S-phase cell to progress through G_2 into M phase, through M phase into G_1, and finally through G_1 back to S phase.

Alternatively, one can assess where a cell is in the cycle by measuring the DNA content, which doubles during S phase. This approach is greatly facilitated by the use of a *fluorescence-activated cell sorter* (Figure 17–6), which allows large numbers of cells to be analyzed automatically. One can go on to discover the lengths of the G_1, S, and G_2 + M phases by following a population of cells that are selected to be all in one particular phase and using DNA measurements to monitor the subsequent progress of these cells through the cycle.

Discrete Cell-Cycle Events Occur Against a Background of Continuous Growth [2]

In conditions that favor growth, the total protein content of a typical cell increases more or less continuously throughout the cycle. Likewise, RNA synthesis continues at a steady rate, except during M phase, when the chromosomes are apparently too condensed to allow transcription. When the pattern of synthesis of individual proteins is analyzed, the vast majority are seen to be synthesized throughout the cycle. For most of the constituents of the cell, therefore, growth is a steady, continuous process, interrupted only briefly at M phase, when the nucleus and then the cell divide into two.

DNA synthesis and the visible events of mitosis are, however, not the only discrete processes occurring against this background of continuous growth. The centrosome, for example, has to be duplicated in preparation for mitosis, so as to form the two poles of the mitotic spindle (see Figure 17–2). And production of a few key proteins—although only a very few—is switched on at a high rate at

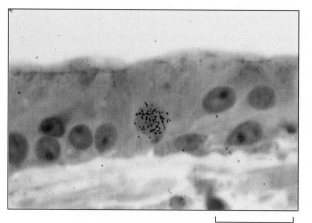

Figure 17–5 Labeling of S-phase cells by autoradiography. The tissue has been exposed for a short period to ^{3}H-thymidine. Silver grains (*black dots*) in the photographic emulsion over a nucleus indicate that the cell incorporated ^{3}H-thymidine into its DNA, and thus was in S phase, sometime during the labeling period. In this specimen, showing sensory epithelium from the inner ear, the presence of an S-phase cell is evidence of cell proliferation occurring in response to damage. (Courtesy of Mark Warchol and Jeffrey Corwin.)

20 µm

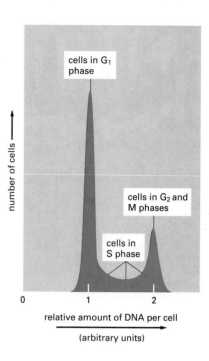

Figure 17–6 Analysis of DNA content with a fluorescence-activated cell sorter. The graph shows typical results obtained for a growing cell population when the DNA content of its individual cells is determined. The fluorescence-activated cell sorter (see Figure 4–31) is used here simply to make measurements on the individual cells, rather than to sort them. The cells are stained with a dye that becomes fluorescent when it binds to DNA, so that the amount of fluorescence is directly proportional to the amount of DNA in each cell. The cells fall into three categories: those that have an unreplicated complement of DNA (one arbitrary unit) and are therefore in G_1 phase, those that have a fully replicated complement of DNA (two arbitrary units) and are in G_2 or M phase, and those that have an intermediate amount of DNA and are in S phase. The distribution of cells in the case illustrated indicates that there are greater numbers of cells in G_1 than in G_2 + M, implying that G_1 is longer than G_2 + M in this population.

a specific stage of the cycle. Histones, for example, which are required for the formation of new chromatin, are made at a high rate only in S phase, and the same is true for some of the enzymes that manufacture deoxyribonucleotides and replicate DNA.

The turning on and off of genes and the starting and stopping of processes such as DNA synthesis and mitosis are the overt consequences of a much less easily observed series of sudden transitions in the state of the cell-cycle control system, whose broad principles we discuss next.

A Central Control System Triggers the Essential Processes of the Cell Cycle [3]

From the point of view of its control, the cell cycle operates like an automatic clothes washing machine. The function of the washing machine is to take in water and detergent, wash the clothes, rinse them, and spin them dry. These essential processes of the washing cycle are analogous to the essential processes of DNA replication, mitosis, and so on in the cell cycle (Figure 17–7). In both cases a central controller triggers each process in turn in a specific sequence. Although the controller could in principle operate as a simple clock that allots a fixed time for each process, usually, in both the washing machine and the cell cycle, it is itself regulated at certain critical points of the cycle by feedback from the processes

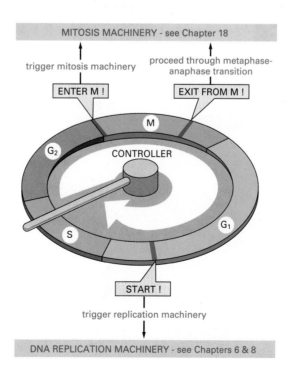

Figure 17–7 The control of the cell cycle. The essential processes, such as DNA replication and mitosis and cytokinesis, are triggered by a central cell-cycle control system. By analogy with a washing machine, the control system is drawn as an indicator that rotates clockwise, triggering essential processes when it reaches specific points on the outer dial.

The General Strategy of the Cell Cycle

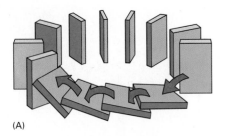 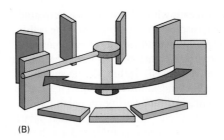

(A)   (B)

Figure 17–8 Two alternative conceptions of cell-cycle control. The *green slabs* represent essential processes of the cell cycle, such as DNA replication, chromosome condensation, spindle formation, and so on. In (A) the performance of one process is the trigger for the next, as in a chain of falling dominoes. In (B) the processes are not directly coupled in this way but are caused to occur in succession by a control device that operates independently of them. The mechanism shown in (B) corresponds more closely to the real cell cycle, where a cyclic control device, which we shall call the *cell-cycle control system,* triggers the *downstream processes* of the cycle.

that are being performed. In the washtub, sensors monitor the water level, for example, and send signals back to the controller to prevent the next process from beginning before the previous one has finished. Without such feedback a delay or interruption in any of the processes can cause a disaster.

This distinction between the control system and the machinery performing the essential processes of the cell cycle was not generally recognized until recently. Instead, it was thought that each of the major essential processes might somehow directly trigger the next process, as in a chain of falling dominoes (Figure 17–8). The turning point in our understanding came with the identification of key components of the central cell-cycle control system and the recognition that these were distinct from the molecules that perform the essential processes of DNA replication, chromosome segregation, and so forth.

The **cell-cycle control system** is a cyclically operating biochemical device constructed from a set of interacting proteins that induce and coordinate the essential *downstream processes* that duplicate and divide the cell's contents ("downstream" in this context meaning simply that they occupy a subordinate position in the hierarchy of cell-cycle control). In the standard cell cycle the control system is regulated by brakes that can stop the cycle at specific *checkpoints.* Here, feedback signals conveying information about the downstream processes can delay progress of the control system itself, so as to prevent it from triggering the next downstream process before the previous one has finished.

The brakes are important also in another way: they allow the cell-cycle control system to be regulated by signals from the environment. These environmen-

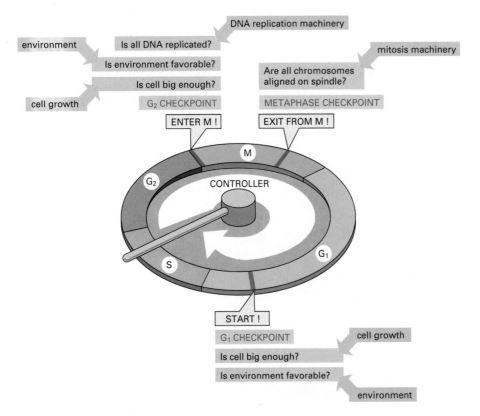

Figure 17–9 Checkpoints and inputs of regulatory information to the cell-cycle control system. Feedback from downstream processes and signals from the environment can prevent the control system from passing through certain specific checkpoints. The most prominent checkpoints are where the control system activates the triggers shown in *yellow* boxes.

tal controls generally act on the control system at one or other of two major checkpoints in its cycle—one in G_1, just before entry into S phase; the other in G_2, at the entry to mitosis. In higher eucaryotic cells signals that arrest the cycle usually act at the G_1 control point. This checkpoint is called *Start* in yeast, and in mammalian cells we shall call it simply the G_1 checkpoint (Figure 17–9). When circumstances forbid cell division, it is at this point in the cycle that many cells halt. In a continuously cycling cell the G_1 checkpoint is the point where the cell-cycle control system triggers a process that will initiate S phase, and the G_2 checkpoint is where it triggers a process that will initiate M phase.

The Cell-Cycle Control System Is a Protein-Kinase-based Machine [3]

For historical reasons most of what we know about the mechanism of the cell-cycle control system has been learned from studies on the G_2 checkpoint at the entry to mitosis. In our initial description of this control system, consequently, we focus on the mechanisms that drive the cell past the G_2 checkpoint into M phase. It is thought likely that a similar mechanism operates at the G_1 checkpoint, although the precise components are different.

The cell-cycle control system is based on two key families of proteins. The first is the family of **cyclin-dependent protein kinases (Cdk** for short)**,** which induce downstream processes by phosphorylating selected proteins on serines and threonines. The second is a family of specialized activating proteins, called *cyclins,* that bind to Cdk molecules and control their ability to phosphorylate appropriate target proteins (Figure 17–10). The cyclic assembly, activation, and disassembly of cyclin-Cdk complexes are the pivotal events driving the cell cycle. Cyclins are so called because they undergo a cycle of synthesis and degradation in each division cycle of the cell. There are two main classes of cyclins: **mitotic cyclins,** which bind to Cdk molecules during G_2 and are required for entry into mitosis, and **G_1 cyclins,** which bind to Cdk molecules during G_1 and are required for entry into S phase (Figure 17–11). In yeast cells, which have played an essential part in research on the cell cycle, the same member of the Cdk family provides

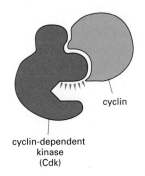

Figure 17–10 Two key components of the cell-cycle control system. A complex of cyclin with Cdk acts as a protein kinase to trigger downstream processes. Without cyclin, Cdk is inactive.

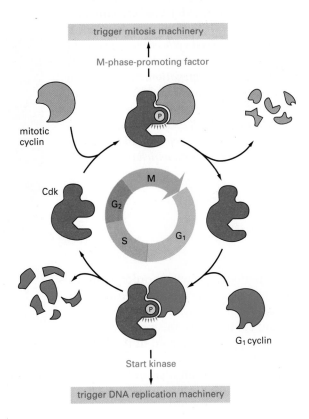

Figure 17–11 The core of the cell-cycle control system. Cdk is thought to associate successively with different cyclins to trigger the different downstream processes of the cycle. Cdk activity is terminated by cyclin degradation.

the kinase activity at both checkpoints; in mammalian cells there are at least two different Cdk proteins, one for each checkpoint.

In outline, the events that drive the cell into mitosis are as follows: Mitotic cyclin accumulates gradually during G_2 and binds to Cdk to form a complex known as *M-phase-promoting factor (MPF)*. This complex is at first inactive, but through the action of other enzymes that phosphorylate and dephosphorylate it, it is converted to an active form. The ultimate activation of MPF is almost explosive. This is believed to be due to a positive feedback mechanism whereby active MPF increases the activity of the enzymes that activate MPF: thus the concentration of active MPF builds up at an accelerating pace until a critical flashpoint is reached, whereupon a flood of active MPF triggers the downstream events that propel the cell into mitosis. MPF is inactivated equally suddenly by the degradation of mitotic cyclin at the metaphase-anaphase boundary, enabling the cell to exit from mitosis.

Each step of Cdk activation or inactivation marks a *cell-cycle transition* and presumably has an effect on the cell-cycle control system itself, initiating reactions that will eventually lead it to trigger the next downstream process. The mechanism operating at the G_1 checkpoint is much less well understood than that at the G_2 checkpoint, but the principles are believed to be similar: just as the assembly of MPF ultimately triggers the events of mitosis, so the assembly of a related complex comprising a Cdk protein and G_1 cyclin is thought to drive the cell past the G_1 checkpoint, triggering the events that lead to DNA replication. The downstream events induced by the activation of Cdk at the G_1 and G_2 checkpoints are completely different, even though in yeast the same Cdk protein serves for both. The particular proteins that are phosphorylated by activated Cdk protein are therefore thought to depend on the cyclin component of the complex. We now turn to the evidence on which this view of the cell cycle is based.

Summary

In each division cycle a cell must replicate its DNA. Most cells also grow and duplicate all of their contents. During M phase the replicated chromosomes are segregated into separate nuclei (by mitosis) and the cell splits into two (by cytokinesis). The other, much longer part of the cycle is known as interphase. This period of continuous cell growth includes S phase, when DNA replication occurs, and two gaps, the G_1 and G_2 phases, between S phase and M phase. The sequence of cell-cycle events is governed by a cell-cycle control system, which cyclically triggers the essential processes of cell reproduction, such as DNA replication and chromosome segregation. At the heart of this system is a set of protein complexes formed from two basic types of components: protein kinase subunits (called Cdk proteins) and activating proteins called cyclins. At least two such protein complexes regulate the normal cell cycle—one at a late G_1 checkpoint, just before S phase, and the other late in G_2, just before M Phase. These protein complexes exert control through their kinase activities, which are abruptly switched on or off at particular points in the cycle.

The Early Embryonic Cell Cycle and the Role of MPF [4]

In a standard cycle the cell has to undergo a period of growth in order to duplicate all of its DNA, as well as all of the components of its cytoplasm, before it can divide. This takes time—a variable amount of time if the environment is variable. Thus the standard cell cycle is prolonged, and control mechanisms are required to guarantee that all the necessary preparations are completed in the correct sequence before division occurs.

The early embryos of many animal species undergo cell-division cycles that are not standard: they occur without growth and at extraordinary speed and lack

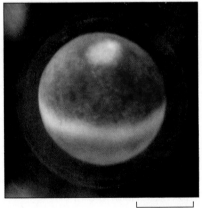

0.5 mm

Figure 17–12 A mature *Xenopus* egg, ready for fertilization. The pale spot near the top shows the site of the nucleus, which has displaced the brown pigment in the surface layer of the egg cytoplasm and whose envelope has broken down during the process of egg maturation. (Courtesy of Tony Mills.)

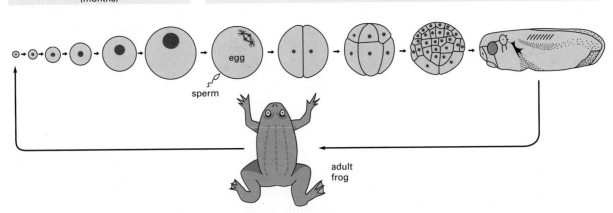

oocyte grows without dividing (months)

fertilized egg divides without growing (hours)

egg

sperm

adult frog

most of the usual checks and controls. These **early embryonic cell cycles** reveal the workings of the cell-cycle control system stripped down and simplified to the bare minimum needed to achieve the most fundamental requirement—the duplication of the genome and its segregation into two daughter cells. Although these cell cycles are exceptional, they illuminate the mechanisms of the division cycle in all eucaryotic cells. In this section we see how studies of the early embryonic cycles of the frog *Xenopus* led to the identification of MPF as a central component of the cell-cycle control system; we discuss how the simplified control system performs its repeated cycles of activation and inactivation of MPF; and we see how each such cycle of the control system drives precisely one round of chromosome replication and segregation.

Growth of the *Xenopus* Oocyte Is Balanced by Cleavage of the Egg [5]

The egg of *Xenopus*, like that of many other species, is a giant spherical cell (Figure 17–12), just over a millimeter in diameter. It thus has more than 100,000 times more cytoplasm than an average cell in the body but only a single nucleus. The cytoplasm is filled with a stockpile of the materials that will be required for the construction of the tadpole. These materials have accumulated during a long period of growth of the immature egg, known as the *oocyte*, while it is in the ovary of the mother. In this way the early embryo is relieved of the need to find nourishment and can develop quickly into a free-living organism that can fend for itself (Figure 17–13).

In preparation for its later sexual fusion with a sperm, the growing oocyte has also embarked on the process of *meiosis*, by which it eventually reduces its chromosome number to half the normal value. The reduction is achieved through a single round of DNA replication followed by two specialized cell divisions in which no further DNA replication occurs (as discussed in Chapter 20). Oocyte growth is spread over a long period, during which progress through meiosis is arrested at a stage traditionally called meiotic prophase but better described as the G_2 phase of the first meiotic division cycle: this point of arrest, just before entry into M phase, corresponds to the G_2 checkpoint in a standard cell cycle.

To produce a mature egg, hormones act on the oocyte, releasing the G_2 arrest and causing the oocyte to progress through meiosis until it comes to a halt at an unusual second point of arrest, in M phase of its second meiotic division (Figure 17–14). In this state the egg travels down the oviduct and gets fertilized as it is laid.

Fertilization triggers an astonishingly rapid sequence of cell divisions in which the single giant cell *cleaves*, without growing, to generate an embryo consisting of thousands of smaller cells (see Figure 17–13). In this process almost the only macromolecules synthesized are DNA—required to produce the thousands

Figure 17–13 Oocyte growth and egg cleavage in *Xenopus*. The oocyte grows for many months in the ovary of the mother frog, without dividing, and finally matures into an egg. Upon fertilization, the egg cleaves very rapidly—initially at a rate of one division cycle every 30 minutes—forming a multicellular tadpole within a day or two. The cells get progressively smaller with each division as no growth can occur until the tadpole begins feeding. The drawings in the top row are all on the same scale (but the frog below is not).

The Early Embryonic Cell Cycle and the Role of MPF

871

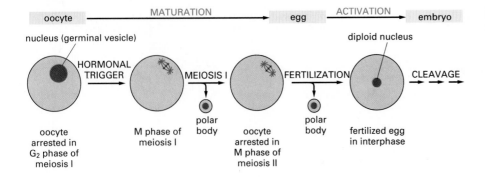

Figure 17–14 *Xenopus* egg
maturation and activation.
Hormones acting on the fully grown
oocyte drive it from its state of G_2
arrest into M phase of meiosis: it
completes the first meiotic division
and becomes arrested in metaphase
of the second meiotic division. In this
state, now called a *mature egg*, it is
laid. The two meiotic divisions follow
one another in quick succession,
without an intervening S phase; the
chromosomes remain condensed
throughout. Fertilization releases the
metaphase arrest, so that the egg
completes its second meiotic division
and enters interphase of the first
embryonic cell cycle. Each of the two
meiotic divisions generates one large
cell—the future egg—and one tiny
cell, called a polar body, which
eventually degenerates.

of nuclei—and a small amount of protein. After a first division that takes about
90 minutes, the next 11 cleavage divisions occur, more or less synchronously, at
30-minute intervals, producing about 4096 (2^{12}) cells within about 7 hours. Each
cycle consists of a 15-minute M phase and a 15-minute interphase that is largely
occupied by DNA synthesis. There are no detectable G_1 or G_2 phases.

A Cytoplasmic Regulator, MPF, Controls Entry into Mitosis [6, 7]

Two crucial experiments established the existence of a cytoplasmic control
mechanism that operates in all dividing cells to initiate mitosis. The first was
based on a *Xenopus* oocyte assay technique that has been crucially important in
identifying the components of the cell-cycle control system.

The *Xenopus* oocyte, as we have seen, is arrested in meiotic G_2, whereas the
egg is arrested in meiotic M phase. The *Xenopus* oocyte and egg therefore provide abundant sources of cytoplasm from these defined stages of the cell cycle.
Moreover, because they are so big, it is easy to inject substances into their cytoplasm. When M-phase cytoplasm from a mature unfertilized egg is injected into
a G_2-phase oocyte, the recipient oocyte is driven into M phase and completes its
maturation (Figure 17–15). The activity identified in the egg cytoplasm in this way
was initially called **maturation-promoting factor** because it induces the maturation of an immature oocyte into a mature egg.

The other crucial experiment was performed with mammalian cells in culture. Mammalian cells are generally not large enough for cytoplasmic injections
to be easy, but a logically equivalent test can be performed by fusing a mitotic
cell with an interphase cell, so that the nucleus of the interphase cell is exposed
to any active components that may be present in the cytoplasm of the mitotic
cell. In such experiments the interphase cell is driven directly into mitosis,
whether it has replicated its DNA or not (Figure 17–16).

It was subsequently shown that the activity that drives the oocyte to maturation and the interphase cell into mitosis is the same: maturation-promoting
factor is identical to M-phase-promoting factor—two names for one substance,
MPF.

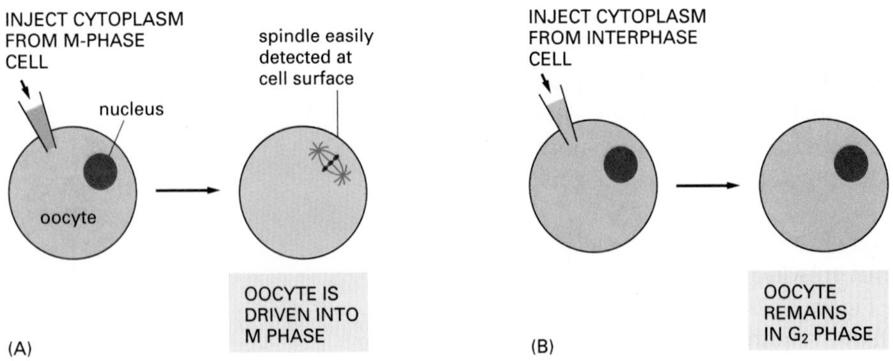

INJECT CYTOPLASM
FROM M-PHASE
CELL

nucleus

spindle easily
detected at
cell surface

oocyte

OOCYTE IS
DRIVEN INTO
M PHASE

INJECT CYTOPLASM
FROM INTERPHASE
CELL

OOCYTE
REMAINS
IN G_2 PHASE

(A)

(B)

Figure 17–15 **Assaying for MPF by
injection into a *Xenopus* oocyte.** MPF
can be detected because it drives the
oocyte into M phase. The large
nucleus (or "germinal vesicle") of the
oocyte breaks down as the mitotic
spindle forms.

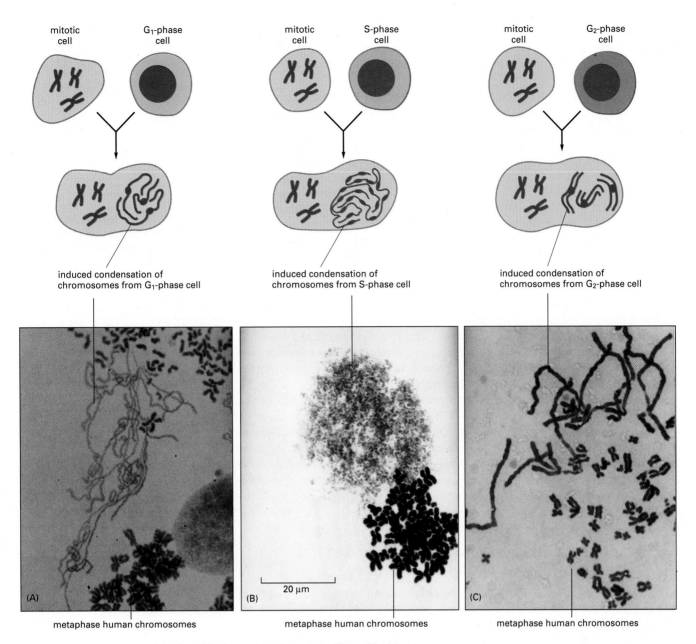

mitotic cell G_1-phase cell

mitotic cell   S-phase cell

mitotic cell G_2-phase cell

induced condensation of chromosomes from G_1-phase cell

induced condensation of chromosomes from S-phase cell

induced condensation of chromosomes from G_2-phase cell

(A)

(B)   20 μm

(C)

metaphase human chromosomes

metaphase human chromosomes

metaphase human chromosomes

Figure 17–16 Results of fusing a mitotic mammalian cell with an interphase mammalian cell. The cells are induced to fuse by adding an appropriate agent to the culture medium (see p. 160). The interphase nucleus is driven directly into a mitotic state, with chromosomes condensed regardless of their state of replication. The photographs show fusions of mitotic human cells with interphase marsupial (PtK) cells. In (A) the PtK cell was in G_1 phase; consequently, its prematurely condensed chromosomes are still single chromatids. In (B) the PtK cell was in S phase, and its chromatin adopts a "pulverized" appearance. In (C) the PtK cell was in G_2 phase, and now the chromatids, although very long compared to the normal (human) metaphase chromosomes, are double. (From K. Sperling and P. Rao, *Humangenetik* 23:235–258, 1974.)

Oscillations in MPF Activity Control the Cell-Division Cycle [8]

Xenopus eggs and oocytes provide a good source of material for both purifying MPF and assaying its activity. Once purified, MPF was found to be a protein kinase that phosphorylates proteins on serine and threonine residues and to consist of two essential subunits—a cyclin-dependent kinase (Cdk) called Cdc2 and a mitotic cyclin (Figure 17–17). Even before it was purified and biochemically characterized, however, it was possible to demonstrate the crucial role of MPF in the cell cycle.

Experiments using the oocyte assay to test samples of cytoplasm from early embryonic cell cycles as well as from maturing oocytes and eggs showed that MPF activity is high in every mitosis and low in every interphase, peaking every 30 minutes in a cleaving egg. Moreover, MPF activity was found to be a universal feature of the eucaryotic cell cycle: in every species, from yeasts to mammals, cells that are in mitosis contain MPF activity that can be detected by the oocyte injection assay.

It was quickly established that the surge of MPF activity that occurs every 30 minutes in the cleaving *Xenopus* embryo is generated by a cytoplasmic oscillator that operates even in the absence of a nucleus. By constricting an activated egg before it has completed its first division, one can split it into two separate parts—one containing a nucleus, the other not. As expected, the nucleated part continues with the normal program of rapid cleavages. Remarkably, the nonnucleated part also goes through a series of oscillations, seen as repeated cycles of contraction and stiffening of the cortical cytoplasm. These recurrent spasms occur in almost perfect synchrony with the cleavage divisions of the nucleated half-egg (Figure 17–18). A starfish egg whose nucleus has been removed goes even further: provided it retains a centrosome, it will cleave repeatedly to form smaller and smaller cells, all without nuclei.

If samples of cytoplasm are taken at intervals from the nonnucleated *Xenopus* cell and assayed by injection into oocytes, it can be shown that the visible oscillations reflect oscillations of MPF activity. In these giant egg cells the oscillations that drive the division cycles evidently run independently of any signals from the relatively tiny amount of nuclear DNA that is normally present. We shall see later that this is not the case in standard cell cycles, where strict control mechanisms operate to ensure that chromosome replication and cell division are correctly coupled.

Cyclin Accumulation and Destruction Control the Activation and Inactivation of MPF [9]

Although division cycles in the cleaving embryo can occur in the absence of DNA, they cannot occur in the absence of protein synthesis: blocking protein synthesis in early interphase prevents both the activation of MPF and the next mitosis. The explanation for this observation became clear with the identification of **cyclin.**

The synthesis of proteins in fertilized sea urchin eggs (whose cell cycle is similar to that of *Xenopus* embryos) was examined by fertilizing a batch of eggs in sea water containing a radioactive amino acid (^{35}S-methionine), removing samples periodically, and running them on a gel. Once they had been separated on the gel, the freshly made proteins could be visualized by virtue of their radio-

Figure 17–17 The two key subunits of MPF. The Cdk subunit is named Cdc2 after the gene in fission yeast that encodes it.

Figure 17–18 A *Xenopus* egg split into two parts—one nucleated, the other not. A loop of fine human hair is tightened around the egg so as to cut it in two. The half containing the nucleus goes on to divide; the other half, lacking a nucleus, does not divide. Time-lapse cinematography shows, however, that the nonnucleated half periodically changes its height through changes in the stiffness of the of the cell cortex, oscillating in close synchrony with the divisions of the nucleated half. (From K. Hara, P. Tydeman, and M. Kirschner, *Proc. Natl. Acad. Sci. USA* 77:462–466, 1980.)

45 minutes   90 minutes   116 minutes   1 mm

time postfertilization ⟶

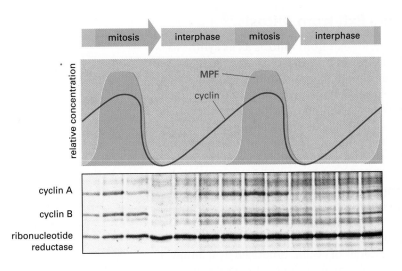

Figure 17–19 **Rise and fall in levels of MPF and cyclin during the early embryonic cell cycle.** The cyclin measurements have been made chiefly in the eggs of marine invertebrates, where cyclin accounts for 5% of the protein synthesized during a brief pulse with radioactive amino acids. The gels below the graph show the amounts of labeled cyclin of two varieties, A and B, at different stages of the cycle of a clam egg. The bottom line in the gel shows the synthesis of a house-keeping enzyme that serves as a standard of comparison. (Adapted from T. Hunt, F.C. Luca, and J.V. Ruderman, *J. Cell Biol.* 116:707–724, 1992.)

activity. These experiments revealed that whereas most proteins accumulate continuously after fertilization, one class of proteins shows a periodic pattern: they accumulate steadily during each interphase until the metaphase-anaphase transition, and then are suddenly destroyed (Figure 17–19). This pattern of events earned the proteins the name cyclin and led to the hypothesis that one or more cyclins had to build up to a threshold concentration to activate MPF and that destruction of cyclin was coupled to inactivation of MPF and exit from mitosis.

The development of cell-free extracts that would undergo the cell cycle *in vitro* was crucial in elucidating the role of cyclin. These extracts are prepared by centrifuging frog eggs so as to break them open and collect their cytoplasm. Sperm nuclei added to such a cytoplasmic extract can swell, replicate their DNA, and pass through mitosis, just as they would in a fertilized egg: they thus serve as indicators of the cell-cycle phase of the extract (Figure 17–20). If all of the mRNA in the extract is then destroyed, the cell cycle arrests in interphase, just as it does when protein synthesis is inhibited in the embryo. Remarkably, the addition of purified cyclin mRNA alone restores the ability of the extract to activate MPF and induce mitosis, indicating that it is only the lack of cyclin synthesis in the mRNA-depleted extract that arrests the cycle. The simplest hypothesis would be that the activation of MPF is normally triggered as soon as the cyclin concentration reaches a threshold value. In reality, as already discussed, the timing of MPF activation is not governed exclusively by cyclin but depends as well on the regulation of the Cdk subunit of MPF by other proteins. These other controls are hard to investigate in the *Xenopus* embryo, but they can be deciphered by the powerful tools of yeast genetics, as we discuss later.

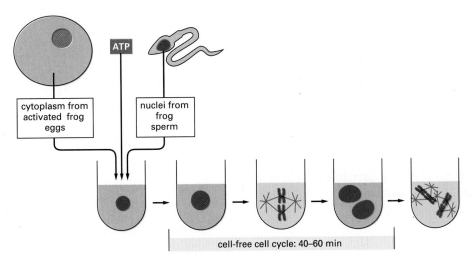

Figure 17–20 **Cycling in a cell-free system.** A large batch of activated frog eggs are broken open by gentle centrifugation, which also separates their cytoplasm from other components. The undiluted cytoplasm is collected, and sperm nuclei are added to it, together with ATP. The sperm nuclei decondense and then go through repeated cycles of mitosis and DNA replication, indicating that the cell-cycle control system is operating in this cell-free cytoplasmic extract.

The Early Embryonic Cell Cycle and the Role of MPF

Degradation of Cyclin Triggers Exit from Mitosis [10]

The destruction of cyclin is as important for exit from mitosis as its synthesis is for entry. Normally, cyclin is suddenly destroyed by proteolysis at the metaphase-anaphase transition. This process requires a signal sequence in the cyclin polypeptide chain that targets it for degradation (by providing a site for attachment of ubiquitin—see p. 218), and it is dependent on the activation of MPF. The role of cyclin degradation has been tested by constructing a truncated form of cyclin that is still able to stimulate the activation of MPF but cannot be degraded because it lacks the appropriate signal sequence. When mRNA for the indestructible cyclin is added to the frog cell-cycle extract, the extract enters mitosis but cannot escape from it. Thus the activation of MPF that induces mitosis requires cyclin synthesis, and the inactivation of MPF that leads to the next interphase requires cyclin degradation.

Cyclin is an essential component of MPF and is found in all eucaryotic cells. As mentioned earlier, there are many varieties of cyclin, produced by a family of related genes. The main mitotic cyclin is known as *cyclin B*. Standard cell cycles depend also on several other cyclins: the G_1 *cyclins*, in particular, appear to play a key role in the activation of the protein kinase that drives cells out of G_1 and commits them to embark on DNA replication. For this too the chief evidence comes from genetic studies in yeasts and is discussed later.

MPF Can Act Autocatalytically to Stimulate Its Own Activation [6]

Xenopus oocyte experiments have suggested an explanation for one further feature of MPF—its sudden, all-or-none activation at the onset of mitosis. When a small amount of active MPF is injected into an arrested oocyte, the oocyte is triggered to generate additional active MPF, implying that production of active MPF is an autocatalytic process. The operation of the cell-cycle control system in the early embryo thus can be outlined as follows: The gradual accumulation of cyclin acts like a slow-burning fuse, which eventually ignites an autocatalytic explosion of MPF activity that triggers the early events of mitosis and initiates the destruction of cyclin. The destruction of cyclin terminates the MPF activity, and a new round of cyclin accumulation begins.

Active MPF Induces the Downstream Events of Mitosis [11]

MPF has to bring about many radical changes in the cell to drive it into mitosis. The chromosomes must condense, the nuclear envelope must break down, and the cytoskeleton must be reorganized to form a mitotic spindle. MPF induces all of these essential events through its protein kinase activity. It brings about some changes directly by phosphorylating key architectural components of the cell. Other events may be induced indirectly—for example, by phosphorylations that activate other protein kinases that act in a cascade to alter the state of the cell.

The breakdown of the nucleus requires the disassembly of the nuclear lamina—the underlying shell of polymerized lamin filaments (discussed in Chapter 12). MPF catalyzes this process directly, forcing the lamin molecules to disassemble by phosphorylating them on key serine residues (see Figure 12–18). In cells containing genetically engineered mutant lamins that lack these MPF phosphorylation sites, the nuclear lamina fails to break down at mitosis, although the membranous nuclear envelope breaks up into vesicles and the other events of mitosis take place almost normally.

Another of the molecules that MPF can phosphorylate directly is histone H1, which plays a part in the packaging of DNA into nucleosomes. It is possible, but not proven, that its phosphorylation may help to induce chromosome condensation.

MPF changes the behavior of microtubules in mitosis by phosphorylating microtubule-associated proteins. In interphase the centrosome nucleates long microtubules that extend throughout the cytoplasm. At mitosis this cytoplasmic array of microtubules disassembles and the centrosomes nucleate a larger number of shorter, less stable microtubules, which interact with one another to form the mitotic spindle (discussed in Chapter 18). This transformation reflects a chemical change in the centrosome, the microtubules, or both, and it can be reproduced *in vitro* by adding MPF to a cell-free system containing centrosomes and tubulin and other components of interphase cytoplasm. Both of the components of MPF—mitotic cyclin and the Cdc2 kinase—can be found bound strongly to centrosomes in the living cell: the mitotic cyclin is thought to recognize components of the centrosome and recruit the Cdc2 kinase to the site.

Although the pathways are not known in detail, it is clear that MPF directly or indirectly induces the phosphorylations of many proteins. To escape from mitosis, cells have to reverse these phosphorylations, and it has been shown in flies and fission yeasts that mutations that inactivate protein phosphatase I—one of the major general-purpose phosphatases in the cell—will prevent or greatly delay the downstream events that normally follow the inactivation of MPF, such as reconstruction of the nuclear envelope.

The Cell-Cycle Control System Allows Time for One Round of DNA Replication in Each Interphase [12]

We have now seen in outline how components of the cell-cycle control system are periodically activated and inactivated to drive the cell into and out of mitosis and how the activated components trigger the downstream processes. The downstream processes themselves take time, however. How does the control system ensure that it allows enough time for each of these processes to go to completion before it triggers the next?

Of the processes that must be given adequate time for completion, one of the most critical is DNA replication. If the control system activates the M-phase trigger before DNA replication is complete, the cell enters a suicidal mitosis with its chromosomes only partially replicated. We shall see later that in standard cell cycles a feedback signal from incompletely replicated DNA protects the cell from this type of disaster by arresting the progress of the control system.

For the first few embryonic cycles of the developing frog, however, no such feedback operates. Here, the time required for each downstream process is predictable and invariant, thanks to the stockpiles of nutrients in the egg and the protected environment of the embryo; consequently, it is sufficient for the control system to go through its cycle at a suitable rate, with appropriate preset delays between the activation of one trigger and the next. The DNA polymerase and other components required for DNA replication are permanently at the ready and can complete a round of DNA replication faster than the control system can complete its cycle. In fact, the control system in the early embryo is oblivious to the progress of DNA replication, so that if replication is artificially halted by an inhibitor of DNA synthesis, the cell is driven on regardless into a disastrous mitosis. During the early cleavages of the embryo, it seems that the quantity of DNA is simply too tiny in relation to the quantity of cytoplasm for DNA-dependent signals to make themselves felt by the cytoplasmic control system. Over the first 12 division cycles, as the egg is subdivided and new nuclei are formed, the ratio of nuclear DNA to cytoplasm increases by a factor of more than 4000, and toward the end of this period the feedback controls of the standard cell cycle begin to operate.

A Re-replication Block Ensures That No Segment of DNA Is Replicated More Than Once in a Cell Cycle [7]

Another type of control is required to solve a complementary problem: while it is essential in each cell cycle that all of the cell's DNA should be replicated, it is equally important that none of the cell's DNA should be replicated more than once. The cell solves this problem not by feedback regulation of the cell-cycle control system but by a self-limiting device that is built into the process of DNA replication itself: as discussed in Chapter 8, each segment of chromatin, as soon as it has been replicated, becomes altered in some way so as to prevent it from being replicated again during the current cycle.

Although the nature of this **re-replication block** is still unknown, there is clear evidence for its existence, and it appears to be fundamental to the operation of the cell cycle in all eucaryotes. A neat demonstration is provided by experiments on mammalian cells undergoing standard division cycles in culture. As we saw earlier in Figure 17–16, two cells that are in different phases of the cycle can be fused with each other. When a cell in G_1, with its DNA still unreplicated, is fused with a cell in S phase, the G_1 nucleus in the hybrid cytoplasm is induced to begin DNA synthesis immediately. The S-phase cell evidently contains inducers of DNA synthesis in its cytoplasm, and the G_1 nucleus is susceptible to them. In contrast, when a G_2 cell, which has just finished DNA synthesis, is fused with an S-phase cell, the G_2 nucleus does not resume DNA synthesis, even though DNA synthesis continues in the S-phase nucleus in the shared cytoplasm. It seems that the machinery for DNA replication is present in the hybrid cell, but the G_2 nucleus is refractory to its action (Figure 17–21). Passage from G_1 via S into G_2 has created a block to further DNA replication.

Passage Through Mitosis Removes the Re-replication Block [13]

The block to re-replication has to be removed at some stage between the end of G_2 and the beginning of the next S phase to allow the next cycle of DNA replication to begin. It is not known precisely when or how the block is lifted, but in the early embryonic cell cycle, at least, this change appears to depend on the breakdown of the nuclear envelope at mitosis. The experiments that show this also illustrate how the cell-cycle control system drives the chromosomes through cycles of replication and segregation in strict alternation.

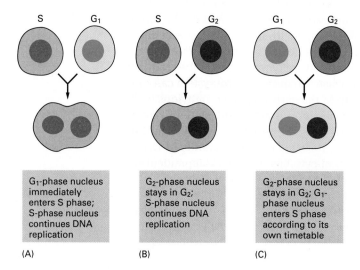

(A) G$_1$-phase nucleus immediately enters S phase; S-phase nucleus continues DNA replication

(B) G$_2$-phase nucleus stays in G$_2$; S-phase nucleus continues DNA replication

(C) G$_2$-phase nucleus stays in G$_2$; G$_1$-phase nucleus enters S phase according to its own timetable

Figure 17–21 The re-replication block: evidence from cell-fusion experiments with cultured mammalian cells. The results show that S-phase cytoplasm contains factors that drive a G_1 nucleus directly into DNA synthesis (A); but a G_2 nucleus, having already replicated its DNA, is refractory to their action (B). Note that fusion of a G_2 cell with a G_1 cell does not drive the G_1 nucleus into DNA synthesis (C); therefore, the cytoplasmic factors for DNA replication that were present in the S-phase cell disappear when the cell moves from S phase into G_2.

As discussed previously, the early embryonic cell-cycle control system can operate in cell-free frog-egg extracts, where it can be arrested in interphase by blocking protein synthesis or be driven into M phase by adding MPF. Sperm nuclei added to an interphase-arrested extract undergo exactly one round of DNA replication and then halt. This is not because there has been any change in the extract (which can still support replication of fresh nuclei added later) but because of a re-replication block imposed on the sperm nuclei. If MPF is added to the extract, the nuclei break down, and when they re-form (following inactivation of MPF), they undergo a single additional round of replication and halt again. Thus each burst of MPF activity gives the nuclei a license to undergo one round of replication. Although the nature of this license is not known, there is evidence to suggest that it depends on a factor from the cytoplasm that has to get into the nucleus to act and is able to do so only when the nuclear envelope breaks down at mitosis.

Thus the control of DNA replication in the early embryo can be summarized as follows. The trigger for mitosis is activated at fixed time intervals, and at each mitosis the DNA receives a license for replication. The DNA replication machinery has time to complete one round of replication before the next mitosis begins, and it is prevented from going beyond one round by the re-replication block. This is enough to ensure that S phases precisely replicate the chromosomes and alternate regularly with M phases, which segregate the chromosomes into separate cells.

In standard cell-division cycles the control system is more complex, but the re-replication block plays a similar role in guaranteeing that the DNA is replicated only once in each cycle. Like most "rules" in the cell cycle, however, the block to re-replication has some exceptions. In flies, for example, many of the larval cells go through multiple rounds of chromosome replication without intervening mitosis or cell division, thereby forming giant polytene chromosomes in which hundreds or thousands of copies of the genome are bundled together in parallel (see Figure 8–19). In this special case the re-replication block is somehow removed without disruption of the nuclear envelope.

To take our account of the cell-cycle control system further, and to see how cells coordinate the more complex events of the standard cell cycle, we must now turn from the early frog embryo to yeasts, where the problem can be approached genetically.

Summary

The early embryos of many animal species undergo exceptionally rapid cell cycles, through which the large egg cell becomes subdivided into many smaller cells without growing. These early embryonic cell cycles, in which S and M phases alternate in quick succession without intervening G_1 or G_2 phases, demonstrate the workings of the cell-cycle control system in its simplest form. The key component of the control system is a protein kinase, MPF, whose activation, by an explosive autocatalytic process, drives the cell into mitosis; inactivation of MPF then allows the cell to exit from mitosis and replicate its DNA. MPF is cyclically activated and inactivated in the early embryonic cycles independently of the nucleus. It consists of two major subunits— a cyclin-dependent kinase called Cdc2, and cyclin. Cdc2 must associate with cyclin to become active as MPF. Destruction of the cyclin inactivates MPF at the point of exit from mitosis, and accumulation of freshly synthesized cyclin allows reactivation of MPF in the next cycle. In these early embryos the time taken to reactivate MPF after exit from mitosis is just long enough to allow one round of DNA replication. A re-replication block is imposed on each segment of DNA as it is replicated, and the re-replication block is removed only on passage through mitosis. In this way the cell-cycle control system drives alternating rounds of DNA replication and chromosome segregation.

Yeasts and the Molecular Genetics of Cell-Cycle Control [14]

Yeasts are unicellular fungi—a large, heterogeneous group of eucaryotic organisms. They are ideal for genetic studies of eucaryotic cell biology because they reproduce almost as rapidly as bacteria and have a genome size less than 1/100th that of a mammal. Yeasts and frogs have complementary strengths and weaknesses for studies on the cell cycle. Yeasts are very well suited for identifying, cloning, and characterizing the genes involved in controlling the cycle; but the yeast cells are too tiny for microinjection studies, and their cell cycles are more complex and cannot yet be reproduced in a cell-free system *in vitro*. Yet these two very different organisms use fundamentally similar cell-cycle machinery, and in combination they have allowed the cell-cycle control system to be dissected into its component parts.

The two yeast species that we shall discuss are the **budding yeast** *Saccharomyces cerevisiae*, used by brewers and bakers, and the **fission yeast** *Schizosaccharomyces pombe*, whose second name comes from the African beer it is used to produce. Although the evolutionary lineages leading to budding and fission yeasts diverged many hundreds of millions of years ago, the two organisms have similar life cycles. Both yeasts can proliferate in either a diploid or a haploid state: the diploid cells, as an alternative to dividing in the ordinary way, can go through meiosis to form haploid cells (see Chapter 20); the haploid cells, as an alternative to dividing, can mate (*conjugate*) with one another to form diploid cells (Figure 17–22). The haploid phase makes it easy to isolate and study mutations that inactivate a gene without the complication of having a second gene copy in the cell. In both species of yeasts, food and sex play important parts in controlling the cell-division cycle, so that these organisms can be used to investigate the general question of how the division cycle is regulated by environmental factors and by cell-cell interactions.

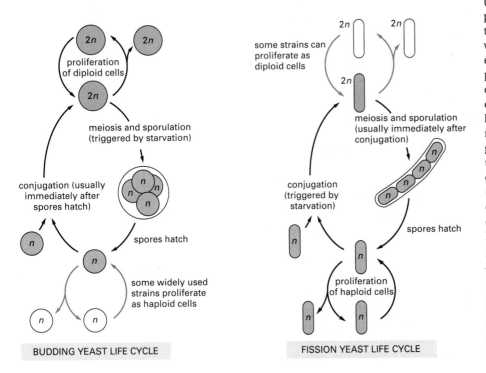

Figure 17–22 The life cycles of a budding yeast (*Saccharomyces cerevisiae*) and a fission yeast (*Schizosaccharomyces pombe*). The proportion of the life cycle spent in the diploid or haploid state varies with the species and according to the environment. When nutrients are plentiful, normal wild-type varieties of budding yeast proliferate as diploid cells, with a cell-cycle time of about 2 hours. If starved, they go through meiosis to form haploid spores, which germinate when conditions improve to become haploid cells that can either proliferate or fuse sexually (conjugate) in G_1 phase to re-form diploid cells, depending on the environment and other factors. Fission yeasts, by contrast, typically proliferate as haploid cells, which fuse in response to starvation to form diploid cells that promptly go through meiosis and sporulation to regenerate haploid cells. The most widely used laboratory strains of budding yeasts are mutants that can proliferate, like fission yeasts, as haploids.

BUDDING YEAST LIFE CYCLE

FISSION YEAST LIFE CYCLE

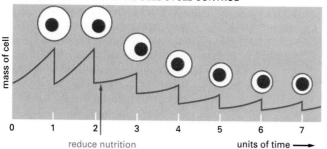

(A) WITHOUT NUTRITIONAL CELL-CYCLE CONTROL

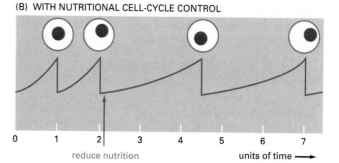

(B) WITH NUTRITIONAL CELL-CYCLE CONTROL

Cell Growth Requires a Prolonged Interphase with Cell-Cycle Checkpoints [3, 15]

The yeast cell cycle is more complex than that of the early frog embryo because a yeast, like the vast majority of cells, has to grow before it divides. Since it takes a cell longer to double its total mass than it does to replicate its DNA and segregate its chromosomes, the cell cycle of growing cells includes G_1 and G_2 interludes—G_1 between the end of M phase and the onset of DNA synthesis, and G_2 between the end of DNA synthesis and the onset of M phase. G_1 and G_2 allow for controls that couple the length of the cell cycle to the cell's growth rate.

To maintain a constant average cell size, the length of the cell cycle must exactly match the time it takes a cell to double in size. If the cycle time is shorter than the time required to double in size, the cells will get smaller in each generation; conversely, if the cycle time is longer than the size-doubling time, the cells will get bigger in each generation. Because a cell's growth rate is at the mercy of the environment, varying according to the supply of nutrients and other factors, the length of the cell cycle has to be correspondingly adjustable (Figure 17–23). To achieve this coordination, the cell-cycle control system has specific *size checkpoints* where the control system halts and waits until the cell has reached a critical size. These checkpoints occur both in G_1, allowing the system to halt before it triggers a new round of DNA replication, and in G_2, allowing the system to halt before it triggers mitosis.

Although both of these checkpoints operate in all cells, the G_1 checkpoint is more prominent in some cells and the G_2 checkpoint is more prominent in others, depending on where the stricter size criteria, or the majority of environmental controls, are applied. Thus in budding yeast the G_1 checkpoint, called Start, is the most important size checkpoint, and a cell that is large enough to pass this checkpoint will generally pass the G_2 checkpoint; the G_1 checkpoint is also the one at which most environmental controls act in the yeasts and in mammalian cells. In fission yeast, by contrast, the G_2, or *mitotic entry*, checkpoint is the more stringent size checkpoint.

Fission and Budding Yeasts Change Their Shape as They Progress Around the Cell Cycle [14]

Although yeasts serve as useful models for study of the standard cell cycle, their cycles differ from those of animal and plant cells in some respects. Like all fungi they keep their nuclear envelope intact throughout mitosis: a mitotic spindle forms *inside* the nucleus, and after chromosome segregation has been completed, the nucleus pinches in half.

Certain details of the cell cycle, moreover, are different in the two types of yeast. Fission yeasts are rod-shaped cells that grow by elongation at their ends. After mitosis the cell divides in two by laying down a septum at the center of the

Figure 17–23 Control of cell size through control of the cell cycle. The diagrams show the relationship between growth rate, cell size, and the cell-division cycle in a free-living organism such as a yeast. (A) If cell division continued at an unchanged rate when cells were starved, the daughter cells produced at each division would become progressively smaller, the size of the daughter being reduced to the small amount of material its parent cell could synthesize in one cycle time. (B) Yeast cells actually respond to poor nutritional conditions by slowing the rate of cell division: because a cell cannot proceed past a certain point in the division cycle until it has attained a certain standard size, the rate of cell division slows down and cell size remains more or less unchanged. (One unit of time in the diagram is the cycle time observed when nutrients are in excess.)

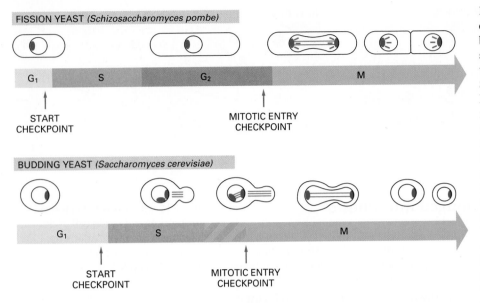

FISSION YEAST (Schizosaccharomyces pombe)

G₁   S   G₂   M

START
CHECKPOINT

MITOTIC ENTRY
CHECKPOINT

BUDDING YEAST (Saccharomyces cerevisiae)

G₁   S   M

START
CHECKPOINT

MITOTIC ENTRY
CHECKPOINT

Figure 17–24 A comparison of the cell cycles of fission yeast and budding yeast. The *fission yeast* shown in the upper panel has a typical eucaryotic cell cycle with G₁, S, G₂, and M phases. Unlike that of higher eucaryotic cells, however, the nuclear envelope of the yeast cells does not break down: the microtubules of the mitotic spindle form inside the nucleus and are attached to *spindle pole bodies* (*dark green*) at its periphery. The cell divides by forming a partition (known as the cell plate) and splitting in two. The *budding yeast* has normal G₁ and S phases. However, a microtubule-based spindle begins to form very early in the cycle, during S phase; thus there does not appear to be a normal G₂ phase. In contrast with fission yeasts, the cell divides by budding. As in fission yeasts, but in contrast with higher eucaryotic cells, the nuclear envelope remains intact during mitosis. The condensed mitotic chromosomes (*red*) are readily visible in fission yeasts, but are less easily seen in budding yeasts.

rod. Budding yeasts, by contrast, divide by forming a bud. The bud is initiated during G₁, grows steadily, and finally separates from its mother after mitosis (Figure 17–24). The presence of the bud is a signal that the cell has passed Start and embarked on a division cycle, and the size of the bud provides an indication of how far beyond that point the cell has progressed around the cycle.

Cell-Division-Cycle Mutations Halt the Cycle at Specific Points; *wee* Mutations Let the Cycle Skip Past a Size Checkpoint [16]

The foundations of our present understanding of the cell cycle in yeasts come from a systematic search for mutations in genes encoding components of the cell-cycle machinery.

To identify genes that directly control the cell cycle, the search was focused on two mutant phenotypes. In the first all of the cells in the mutant population arrest at the same specific point in the cell cycle. The affected genes in these mutants are called cell-division-cycle, or ***cdc,*** genes; each *cdc* mutant is typically deficient in a gene product required to get the cell past the specific point in the cycle at which the mutant cells arrest. The second type of mutation is called ***wee,*** from the Scottish word for small, because the mutant cells divide at a smaller size than normal. *Wee* mutants are expected to be deficient in a product that normally *inhibits* passage through a size checkpoint.

Since a mutant that cannot complete a division cycle cannot be propagated, *cdc* mutants can be selected and maintained only if their phenotype is *conditional,* that is, if the gene product fails to function only in certain specific conditions. Most conditional cell-cycle mutants are *temperature-sensitive mutants* in which the mutant protein fails to function at high temperatures but functions well enough to allow cell division at low temperatures. A temperature-sensitive *cdc* mutant strain of cells can be grown at low temperature (the *permissive condition*) and then raised to a higher temperature (the *restrictive condition*) to switch off the function of the affected gene. All of the cells will continue their cycling until they reach the point where the function of the mutant gene is required for further progress, and at this point they will all stop (Figure 17–25). In budding yeast a uniform cell-cycle arrest of this type can be detected just by looking at the cells; the presence or absence of a bud and its size provide a simple indication of the point in the cycle where the mutants are blocked. In the fission yeast, more laborious tests must be used to identify the process in the cell cycle that has failed.

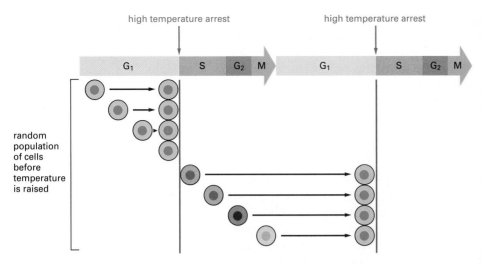

Figure 17–25 Behavior of a temperature-sensitive cell-division-cycle (*cdc*) mutant. At the permissive (low) temperature the cells divide more or less normally and are found in all phases of the cycle (the phase of the cell is indicated by its color). On warming to the restrictive (high) temperature, where the mutant gene product functions abnormally, the mutant cells continue progress through the cycle until they come to the specific step that they are unable to complete (initiation of S phase, in this example). Because the *cdc* mutants nevertheless continue growing, they become abnormally large (not shown). By contrast, non-*cdc* mutants, if deficient in processes (such as ATP production) necessary throughout the cycle for biosynthesis and growth, halt haphazardly at any stage of the cycle as soon as their biochemical reserves run out.

There are now about 70 known cell-division-cycle genes, many of which have been cloned and sequenced. Several of them encode already familiar proteins involved in the downstream processes of the cell cycle, such as DNA polymerases and enzymes that synthesize the precursors for DNA synthesis, which are required for passage through S phase. A substantial number of other *cdc* genes, however, together with one key *wee* gene, code for components and regulators of the cell-cycle control system itself.

The Subunits of MPF in Yeasts Are Homologous to Those of MPF in Animals [17]

As mentioned earlier, the most important size checkpoint in the cell cycle of *S. pombe* is the mitotic entry checkpoint located late in G_2. This is where the fission yeast cell-cycle control system normally pauses to allow time for growth when the cell is smaller than it should be: once past this checkpoint, the cell is committed to mitosis. Thus studies of the fission yeast provide a natural route toward identifying the genes encoding the components of the cell-cycle control system that drive a cell into mitosis, as well as the regulators that act on the system at this point. The mitotic entry point is also the point at which MPF is activated in the early embryonic cell cycles of *Xenopus*, and we shall now see how genetic studies on fission yeast converged with the biochemical studies in *Xenopus* and enabled the gene encoding the Cdk subunit of MPF to be identified and the mechanism of MPF activation to be clarified.

The fission yeast homolog of the protein kinase subunit of MPF is encoded by a gene called **cdc2.** Investigations of fission yeast cell-cycle mutants revealed the *cdc2* gene product as a pivotal, decisive component for driving the cell into mitosis: when it is defective, mitosis fails to occur; when it is released from normal control, mitosis occurs prematurely. Genetic tests identified three other genes, *wee*1, *cdc*25, and *cdc*13, whose products regulate the function of the *cdc2* gene product (Figure 17–26). The characterization of these regulatory gene products has led to our present understanding of the regulation of MPF, as we discuss below.

Figure 17–26 Some fission yeast cell-division-cycle mutants compared with a normal cell. In the *cdc2⁻* mutant, which has a recessive phenotype, the *cdc2* gene is inactivated, and the cell fails to pass the mitotic entry checkpoint, continues growing, and so becomes enormously large. Conversely, the dominant *cdc2* mutant, *cdc2*^{D}, in which the *cdc2* gene product is hyperactive, has a *wee* phenotype: it passes the checkpoint prematurely, at an abnormally small size. The *cdc*25 mutant behaves like a *cdc2⁻* mutant, and the *wee*1 mutant behaves like a *cdc2*^{D} mutant: *cdc*25 *and wee*1 affect the activation of the kinase encoded by *cdc2* (Photographs courtesy of Sergio Moreno.)

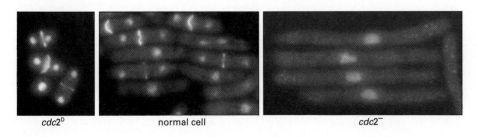

cdc2^D normal cell   *cdc2⁻*

The discovery of the relationship between MPF and the *cdc2* gene product depended on the cloning of *cdc2*. There is a simple and powerful strategy for cloning the normal counterpart of any mutant *cdc* gene. The mutant cells, which are incapable of dividing at high temperature (the restrictive condition), are transfected with plasmids containing random fragments of DNA from normal cells. Occasional mutant cells will receive a DNA fragment that includes a copy of the normal *cdc* gene that the mutant lacks, and these few cells will be able to divide under the restrictive conditions. By recovering the plasmid DNA from their progeny, one obtains the *cdc* gene. It can then be sequenced and characterized, and antibodies can be made against its protein product.

Sequencing revealed that the fission yeast *cdc2* gene encodes a protein kinase. Antibodies against this kinase recognized the cyclin-dependent kinase subunit of purified MPF from a frog. Further gene sequencing showed that the fission yeast Cdc13 protein is homologous to mitotic cyclin (cyclin B) of animals. The fundamental similarity between yeasts and vertebrates was even more dramatically emphasized by a third finding: when mutant yeast cells deficient in *cdc2* were transfected with plasmids containing fragments of human DNA, some were rescued and divided normally. The rescuing fragment, when cloned and sequenced, contained a human gene homologous to the fission yeast *cdc2* gene. From all this and other evidence, it is clear that in both yeasts and vertebrates entry into mitosis is driven by essentially the same kinase and that, to become active as MPF, this kinase has to be complexed with cyclin. For historical reasons the cyclin-dependent kinase subunit of MPF is commonly referred to as Cdc2 both in the fission yeast and in animal cells.

MPF Activity Is Regulated by Phosphorylation and Dephosphorylation [18, 19]

We now return to the regulatory proteins identified by *cdc* and *wee* mutations and examine how they control the activation of MPF. We saw earlier that cyclin by itself, although necessary, is not sufficient to activate Cdc2; however, once bound to cyclin B during interphase, Cdc2 becomes a substrate for two protein kinases. The first kinase is the Wee1 protein, which phosphorylates a tyrosine residue close to the catalytic site of Cdc2, blocking its kinase activity and preventing it from acting prematurely as MPF. The second kinase, called MO15 (identified in the frog), phosphorylates a threonine residue in another region of the Cdc2 molecule. This phosphorylation will ultimately activate MPF, but as long as the tyrosine residue is also phosphorylated, the cyclin-Cdc2 complex is inactive. In both frogs and yeasts it is the removal of the inhibitory tyrosine phosphate that finally activates MPF. In fission yeasts this removal is catalyzed by Cdc25 protein, which is a protein phosphatase (Figure 17–27). The series of conformational changes in the Cdc2 protein molecule underlying this activation mechanism is shown in Figure 5–16.

The balance of Wee1 activity and Cdc25 activity is presumably such that at first only a small amount of active MPF is generated. But this active MPF is thought to stimulate further MPF activation, probably through stimulation of the Cdc25 phosphatase activity, inhibition of the Wee1 kinase activity, or both, cre-

Figure 17–27 Genesis of MPF activity. Cdc2 becomes associated with cyclin as the level of cyclin gradually increases; this enables Cdc2 to be phosphorylated by an activating kinase on an "activating" site as well by Wee1 kinase on Cdc2's catalytic site. The latter phosphorylation inhibits Cdc2 activity until this phosphate group is removed by the Cdc25 phosphatase. Active MPF is thought to stimulate its own activation by activating Cdc25 and inhibiting Wee1, either directly or indirectly.

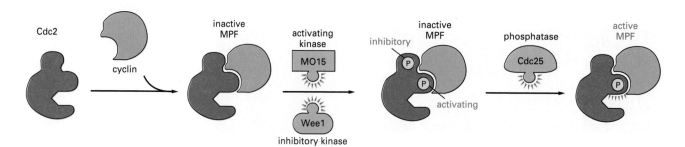

ating a positive feedback effect. Thus, during G$_2$ phase, cyclin gradually accumulates, causing a slowly increasing level of MPF activity that eventually undergoes an autocatalytic explosion and, by rising above a critical point, drives the cell irreversibly into mitosis.

The MPF-Activation Mechanism Controls Size in Fission Yeast [15, 18, 19, 20, 21]

Although the above account of MPF activation appears to apply to all eucaryotes, there are variations. The timing of MPF activation, for example, may depend in some organisms chiefly on the timing of cyclin accumulation but in others on the timing of Cdc25 activity. Moreover, the MPF-activation mechanism is exploited for different regulatory purposes in different organisms and in different cell types within an organism. In the early embryo it serves simply to set the time between one mitosis and the next. In fission yeast, by contrast, it gives both a delay between the end of S phase and the onset of mitosis and an opportunity for cell-size control.

The precise mechanism of cell-size control in fission yeast is not yet known, but one can imagine in a general way how it might work. The activation of MPF is governed by the balance of inhibition by Wee1 protein and activation by Cdc25 protein, as well as by other possible regulators. The concentrations and activities of these molecules will change in different ways as the cell gets bigger. If, for example, Wee1 protein were to become diluted relative to Cdc25 protein as a result of cell growth, growth would swing the regulatory balance in favor of MPF activation, and growth beyond a critical size would trigger an autocatalytic MPF explosion. Whether cell size is a crucial regulator of MPF activation (as in fission yeast) or not (as in most other cells) will depend on the quantitative details of the activation mechanism and the regulators impinging on it rather than on its basic mode of operation.

Although there is evidence for a size control in many other types of cells, the mechanism is even more obscure than in fission yeast. A possible clue comes from comparisons of cells that differ in ploidy (that is, in the number of copies of the genome that they contain) but are otherwise biochemically and genetically similar (see Figure 17–49). It seems to be a general rule that cell size is roughly proportional to ploidy, suggesting that the control mechanism depends on some sort of titration—direct or indirect—of the quantity of a cytoplasmic component against the quantity of DNA.

For Most Cells the Major Cell-Cycle Checkpoint Is in G$_1$ at Start [14, 15]

For budding yeast and for the standard division cycles of most of the commonly studied cell types of multicellular eucaryotes, the major checkpoint where the cell-cycle control system pauses to allow time for growth is not in G$_2$, as it is in fission yeast, but in late G$_1$. This G$_1$ checkpoint is called **Start**. If the budding yeast cell can pass Start, it will generally pass the mitotic entry checkpoint once it has completed S phase. Thus, for budding yeast, as for most animal cells, the G$_1$ checkpoint is the point of no return beyond which the cell will complete the cycle even if conditions change. Genetic studies on budding yeast have shown that passage through this checkpoint, like passage through the G$_2$ checkpoint, hinges on the activity of a cyclin-dependent protein kinase.

The Cdc2 Protein Associates with G$_1$ Cyclins to Drive a Cell Past Start [14, 18, 22]

For a cell colony to thrive in the wild it is not enough simply to pass Start: it is essential to pass it at the right moment. For budding yeast at least three conditions are crucial—cell size, the availability of nutrients, and the demands of sex. If the cell is too small, the control system pauses to give time for growth. If the cell is starved, the control system also pauses, delaying the cell's attempt to duplicate itself. And when the cell is required to mate, a peptide mating factor secreted by a neighboring yeast cell arrests the cell cycle in G_1 and prepares the cell for fusion with a haploid partner (see Figure 17–22). Thus size, food, and sex together govern a three-way choice that the cell faces as it approaches Start (Figure 17–28). Mutations affecting the response of budding yeast cells to these influences have identified the cell-cycle control components that govern progress from G_1 to S.

The search for *cdc* genes in budding yeast revealed several that are necessary for a cell to pass Start. A mutant with a defect in any of these genes comes to a halt in G_1 despite being large enough to pass Start. One of the *cdc* genes identified in this way, named *CDC28* in budding yeast, turned out to be homologous to the fission yeast *cdc2*: the two genes have similar sequences and are functionally interchangeable. This discovery exposed a remarkable link between Start (the predominant checkpoint in budding yeast) and mitotic entry (the predominant checkpoint in fission yeast). We now know that the cell cycles of both types of yeast include both types of checkpoint, and in both types of yeast the product of the same gene serves both to drive the cell into mitosis and to drive it past Start so as to initiate the replication of DNA. In this chapter, for clarity and to emphasize its universal role, we call this gene *cdc2* regardless of species.

The Cdc2 protein has distinct activities at the two different checkpoints and is associated with different cyclins. In G_2, as we have seen, it associates with mitotic cyclin to form MPF; in G_1 it associates with **G_1 cyclin** to form a complex that we shall refer to as **Start kinase.** Start kinase and MPF presumably phosphorylate different sets of target proteins, or phosphorylate them differently, or both. The specificity of Cdc2 action, therefore, appears to depend on the type of cyclin that is associated with it (Figure 17–29).

The G_1 cyclin class of proteins was discovered in budding yeast through studies of mutant cells that passed Start either prematurely or under conditions where nonmutant cells would not. Three genes identified by these mutations turned out to code for proteins distantly related to mitotic cyclins. This earned the proteins the name of "G1 cyclins" and immediately suggested that the gene products might play an activating role at Start analogous to the role of mitotic cyclins at the G_2 checkpoint (Figure 17–30).

The mutant phenotypes could be shown to result from an excess of G_1-cyclin activity. Surprisingly, however, *deletion* of any single one of the G_1-cyclin genes had practically no effect; only when all three of the identified genes were simultaneously deleted did the cells become arrested in G_1, unable to pass Start. It seems therefore that there are at least three G_1 cyclins in the normal yeast cell, that their function is required for the cell to get past Start, and that they are, to some extent at least, functionally interchangeable so that the cell cycle can still continue if one or two of the three are missing.

In a normal cycle the three G_1 cyclins are thought to collaborate to ensure that cells pass Start briskly, decisively, and irreversibly by an explosive activation

Figure 17–28 The three options open to a haploid budding yeast cell as it approaches Start. Note that a diploid cell, instead of mating, would have the option of entering meiosis and forming spores.

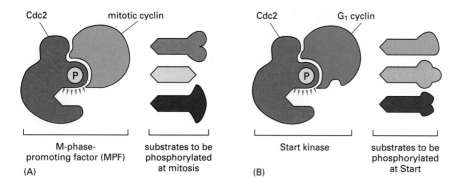

(A)   (B)

M-phase-promoting factor (MPF)   substrates to be phosphorylated at mitosis   Start kinase   substrates to be phosphorylated at Start

Cdc2   mitotic cyclin   Cdc2 G_1 cyclin

Figure 17–29 MPF and Start kinase contrasted. Both include Cdc2, but it is associated with different types of cyclin. This suggests that the cyclin determines which proteins the Cdc2 will phosphorylate, either because it binds the specific substrates (as shown) or because it directs the kinase to the appropriate location in the cell (not shown).

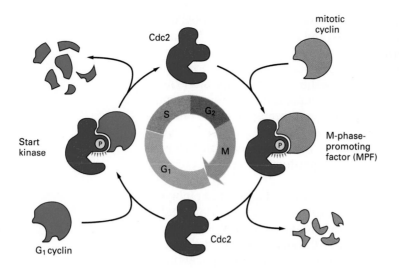

Figure 17–30 **The Cdc2 cycle in yeast.** Cdc2 is permanently present, but its state of association with cyclins changes, defining the division-cycle phase of the cell.

of Start kinase analogous to the explosive activation of MPF at the onset of mitosis. Although many uncertainties remain, the mechanism is again thought to depend on positive feedback, although by a different pathway: the Cdc2 kinase when bound to one of the G_1 cyclins is believed to form an active complex that induces transcription of the genes encoding the other two G_1 cyclins, which bind to Cdc2 in turn and further increase their own production until a threshold level of the active complex is reached.

After the cell has passed Start, the G_1 cyclins, like the G_2 cyclins at mitosis, disappear from the cell, and, so far as is known, they then play no further part until the G_1 phase of the next cycle.

The G_1 Cyclins Mediate Multiple Controls That Operate at Start [15, 22, 23]

The budding yeast cell-cycle control system appears to be regulated at Start largely through brakes and accelerators that control the production of the various G_1 cyclins.

Budding yeast cells regulate their size by making the attainment of a certain size a precondition for passing Start—a strategy that probably is also used by many other cells. Thus cells that begin G_1 when they are abnormally small take extra time to grow before passing Start, while those that begin G_1 when they are abnormally large will pass Start sooner than usual. We have already discussed how cell size may govern passage past the G_2 checkpoint in fission yeast by affecting the relative concentrations or activities of regulatory molecules. Similar principles may apply to how cell size governs passage through Start.

The mechanisms by which environmental factors regulate passage through Start are somewhat better understood. A shortage of nutrients reduces the rates of synthesis of cyclins relative to their rates of degradation and thereby reduces their concentrations in the cell. As a result, the cell may fail to attain the threshold concentrations of G_1 cyclins required to trigger the Start kinase explosion. Mating pheromones are also thought to block progress past Start by decreasing the levels of G_1 cyclins, but in this case they do so by causing the G_1 cyclins to be held inactive and degraded.

Start Kinase Triggers Production of Components Required for DNA Replication [2, 24]

Directly or indirectly, the activity of Start kinase at Start has to induce chromosome replication, just as the activity of MPF at the end of G_2 has to induce mitosis. Chromosome replication requires complex equipment—enzymes such as

DNA polymerase, ligase, and topoisomerase, as well as enzymes for nucleotide synthesis, structural proteins such as histones, and initiation factors that act at origins of replication to start the process of DNA synthesis (discussed in Chapter 6). The genes for at least some of these proteins are transcribed cyclically during each S phase, and there is strong circumstantial evidence that Start kinase activates their transcription. With the exception of the histones, however, most of these proteins appear to persist (in yeast, at least) throughout the cycle. It is not clear, therefore, whether Start kinase triggers S phase by phosphorylating regulatory components so as to allow activity of a replication machine that is already present or triggers the S phase by causing production of parts of the machine that were previously missing.

In early embryonic cell cycles the situation is much simpler. The embryo is provided with an abundant store of maternally derived RNA transcripts, and all the apparatus for chromosome replication, including, apparently, Start kinase, remains available throughout the cycle. Passage through M phase, by lifting the re-replication block, is sufficient to permit a new round of DNA replication.

Feedback Controls Ensure That Cells Complete One Cell-Cycle Process Before They Start the Next [3, 25]

Each of the major actions of the cell-cycle control system—the activation of MPF at the onset of mitosis, its inactivation at the metaphase-anaphase transition, the activation of Start kinase at Start—triggers a complex downstream process that takes time to complete. If the control system proceeds to its next action before the downstream process is completed, the consequences are likely to be fatal or mutagenic to the cell. Thus, if the cell is driven into mitosis before it has finished replicating its DNA, it will pass on broken and incomplete sets of chromosomes to its daughters; if it progresses into anaphase and starts to divide in two before all the chromosomes are aligned on the mitotic spindle, the chromosomes will not be allocated equally between the daughter cells. Such disasters are avoided in most cells by feedback controls that operate at certain checkpoints to arrest the cell-cycle control system until the requisite process is completed.

The best-studied feedback control is the one that delays mitosis until DNA replication is complete. The phenomenon is easily demonstrated in mammalian cells undergoing standard cell cycles. If these cells are treated while in S phase with an inhibitor of DNA synthesis, such as *aphidicolin* (which specifically inhibits DNA polymerase) or *hydroxyurea* (which blocks the synthesis of deoxy-ribonucleotides), the cells halt in S phase and will not progress to mitosis until the inhibitor is removed and DNA replication has been completed. If cells in which DNA synthesis is inhibited receive *caffeine* together with the inhibitor, however, the cells progress suicidally into mitosis with their DNA incompletely replicated. In the absence of an inhibitor of DNA synthesis, caffeine does no harm: the control system still follows the standard timetable, and this gives the cells time to finish DNA replication before mitosis begins (Figure 17–31). Apparently, therefore,

Figure 17–31 Incompletely replicated DNA blocks onset of mitosis. In the experiments schematized here mammalian cells in culture were treated with caffeine and hydroxyurea, either alone or in combination. Hydroxyurea blocks DNA synthesis, arresting the cells in S phase and delaying mitosis. But if caffeine is added as well as hydroxyurea, the delaying mechanism fails, and the cells proceed into mitosis according to their normal schedule with incompletely replicated DNA.

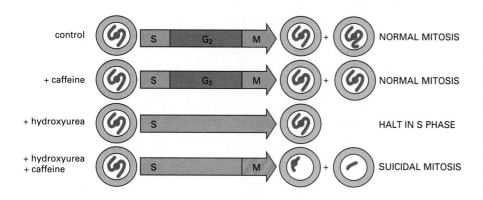

control   S   G$_2$ M   NORMAL MITOSIS

+ caffeine   S   G$_2$ M   NORMAL MITOSIS

+ hydroxyurea   S   HALT IN S PHASE

+ hydroxyurea + caffeine   S   M   SUICIDAL MITOSIS

the cells possess a feedback control mechanism that is inactivated (in an unknown way) by caffeine. This feedback control acts as a safety device. In normal circumstances the safety device does not have to be called into play, for there is sufficient time for the DNA to replicate completely before mitosis begins. Indeed, as we noted earlier, in the early cell cycles of the frog embryo there is no such safety device, and inhibiting DNA replication does not delay the entry into mitosis.

Damaged DNA Generates a Signal to Delay Mitosis [25, 26]

Several other feedback controls operate to restrain the cell-cycle control system until particular conditions are satisfied. As already mentioned, chromosomes that are not attached to the mitotic spindle generate a feedback signal that blocks the inactivation of MPF (discussed in Chapter 18). Another well-characterized feedback control operates at the mitotic entry checkpoint to prevent cells with damaged DNA from entering mitosis until the damage is repaired.

The response to DNA damage is usually studied using experimentally induced lesions in DNA, such as those created by x-rays. Many radiation-sensitive (*rad*) mutations have been isolated in budding yeast, and at least one, called *rad*9, has been shown to code for an essential component of the feedback control mechanism. Mutants lacking *rad*9 still possess the machinery for DNA repair, but they fail to delay in G_2 when they have been irradiated. As a result, they proceed into mitosis with damaged chromosomes; not surprisingly, they are killed by doses of radiation that normal cells would survive. Some of the genes involved in regulating the cell cycle in yeasts are summarized in Table 17–1.

An additional safety device operates in mammalian cells to restrain them from entering S phase with damaged DNA. The mechanism seems to depend on a protein called **p53,** which accumulates in the cell in response to DNA damage and halts the cell-cycle control system in G_1. Mutations in the p53 gene play a crucial part in the genesis of a large proportion of human cancers, apparently by disabling the feedback control and thereby increasing the frequency of cancer-promoting genetic alterations, as we discuss in Chapter 24.

Feedback Controls in the Cell Cycle Generally Depend on Inhibitory Signals [27]

A general argument suggests why feedback controls such as those we have mentioned are based on a negative signal that arrests the cell-cycle control system rather than on a positive signal that moves the control system forward when a downstream process is completed.

Consider, for example, the monitoring of the attachment of chromosomes to the mitotic spindle. A cell needs to be able to detect the attachment of the last unattached chromosome to the microtubules of the spindle: if it proceeds into anaphase and starts to segregate its chromosomes into separate daughter cells before this has occurred, one daughter will receive an incomplete chromosome set, while the other daughter will receive a surplus. In a cell with many chromosomes, if each sent a positive signal to the cell-cycle control system once it was attached, the attachment of the last one would be hard to detect because it would be signaled by only a small fractional change in the total intensity of the "go" signal. On the other hand, if each unattached chromosome sends a negative signal to inhibit progress of the cell-cycle control system, the attachment of the last chromosome will be easily detected because it will mean a change from some "stop" signal to none. A similar argument would imply that mitosis is dependent on completion of DNA replication, not because of a requirement for a positive signal from fully replicated DNA but because of an inhibitory signal from unreplicated DNA or from the replication forks that are necessarily present as long as the chromosomes are still being replicated.

Table 17–1 Some Yeast Cell-Division-Cycle Genes and Their Functions

Gene Name in Budding Yeast	Gene Name in Fission Yeast	Gene Product	Phenotype of Loss-of-Function Mutant
CDC2	pol3	catalytic subunit of DNA polymerase δ	arrest in S phase
CDC9	cdc17	DNA ligase	arrest in G_2 with imperfectly replicated DNA
CDC28	cdc2	serine/threonine protein kinase	arrest at Start or at mitotic entry (G_2) checkpoint
SWI6	cdc10	gene regulatory protein required for transcription of G_1 cyclins	failure to enter S phase
CLN1,2,3	?	G_1 cyclins	arrest at Start (if all three genes inactive)
CLB1,2,3,4	cdc13	mitotic cyclins	arrest at mitotic entry (G_2) checkpoint
WEE1	wee1	tyrosine protein kinase	premature passage past mitotic entry (G_2) checkpoint, hence small size
	cdc25	tyrosine protein phosphatase	arrest at mitotic entry (G_2) checkpoint
RAD9	?	protein of unknown function	failure to delay mitosis when DNA is damaged (loss of feedback control)
DIS2 S1	dis2	protein phosphatase I	arrest in mitosis

The table shows only a small selection of the *cdc* and related genes that have been identified. The terminology is confusing. In each species of yeast, the *cdc* genes were numbered roughly in order of their discovery—*CDC1*, *CDC2*, *CDC3*, . . . in budding yeast; *cdc*1, *cdc*2, *cdc*3, . . . in fission yeast. The numbering sequences, consequently, do not correspond, so that, for example, the counterpart of the fission yeast gene *cdc*2 is called *CDC*28 in budding yeast (both genes encode the key cyclin-dependent kinase of the cell-cycle control system). To make matters worse, there is no regular convention for naming the homologs of these genes in other organisms.

Some of the controls that operate on the cell cycle are summarized in Figure 17–32.

Summary

Yeasts are genetically tractable model organisms for the study of standard cell cycles in which cells grow as well as divide. Normal cells keep their division cycle in step with growth, so as to maintain a standard average size, by controls that operate at two size checkpoints in the cell cycle—one in G_1, called Start (most important in bud-

Figure 17–32 **Summary of feedback, size, and damage controls in the cell cycle.** The *red* T bars represent checks on progress of the cell-cycle control system arising from intracellular processes that are uncompleted or deranged.

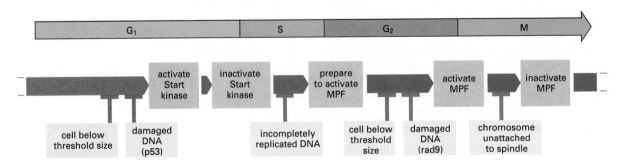

ding yeast and in cells of higher animals), the other in G_2, called the mitotic entry checkpoint (most important in fission yeast). At these points the cell-cycle control system is halted if the yeast cell has not yet reached a critical size. Genes coding for components of the cell-cycle control system can be identified through mutations that make the cell halt at a specific point in the cycle or allow it to proceed past a checkpoint and divide at an abnormally small size. One of these genes, identified in fission yeast as cdc2, *encodes a protein that is homologous to, and functionally interchangeable with, the cyclin-dependent kinase subunit of vertebrate MPF. It plays a central role throughout the yeast cell cycle: in G_2 it associates with mitotic cyclin to form MPF and drive the cell past the mitotic entry checkpoint into mitosis; in G_1 it associates with G_1 cyclin to form Start kinase and drive the cell past Start. Each of these kinase activations is thought to occur by an explosive autocatalytic mechanism involving several other regulatory components, making the cell-cycle control system responsive to multiple controls. These controls include feedback signals from incompletely replicated or damaged DNA that prevent the control system from passing the next checkpoint until replication is complete or the damage is repaired.*

Cell-Division Controls in Multicellular Animals

For unicellular organisms, where each cell division generates a new individual, natural selection favors the cells that grow and divide the fastest and survive hard times the best. Their proliferation is typically restrained only by the availability of nutrients and by the occasional demands of sex. In multicellular species, by contrast, natural selection acts not on each individual cell but on the organism as a whole. To produce and maintain the intricate organization of the body, the component cells must obey strict controls that limit their proliferation. At any instant most cells in the adult are not growing or dividing but instead are in a resting state, performing their specialized function while retired from the division cycle. Because nutrients are plentiful in the tissues of the body, the cells must refrain from proliferating in circumstances where a yeast or bacterium would proliferate readily. What accounts for this difference?

We shall see that for the cells of a multicellular animal, nutrients are not enough: in order to grow and divide, a cell must receive specific positive signals from other cells. Many of these signals are protein *growth factors,* which bind to complementary receptors in the plasma membrane to stimulate cell proliferation. These positive signals act by overriding intracellular negative controls that otherwise restrain growth and block progress of the cell-cycle control system. Thus, while a well-fed yeast cell proliferates unless it gets a negative signal (such as a mating factor) to halt, an animal cell halts unless it gets a positive signal to proliferate.

In this section we focus on mammalian cells and address four questions: (1) What is the nature of the mammalian cell-cycle control system? (2) How is proliferation in mammalian cells regulated, and how is it studied? (3) What are the extracellular signals that determine whether the cell will grow and divide? (4) How do those signals override the intracellular restraints and exert their effect on the cell-cycle control system? The last question is much the hardest. We shall therefore leave it until last and begin by examining how closely the components of the mammalian cell-cycle control system resemble those of the yeast cell-cycle control system.

The Mammalian Cell-Cycle Control System Is More Elaborate Than That of the Yeast [28]

All of the cell-cycle genes that we have discussed in yeasts, including *cdc2,* the cyclins, *cdc25,* and *wee*1, are present in mammals too. All of these have been shown to be functionally similar to their yeast counterparts, to the extent that a

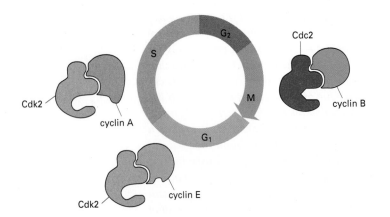

Figure 17–33 Cyclins and Cdk proteins in the standard vertebrate cell cycle. Vertebrates have many different cyclin genes and many different *cdk* genes. Their products act in different cyclin/Cdk combinations at different stages of the cycle. The diagram shows only a few of these molecules and is speculative. The roles of Cdk2 and cyclin A, in particular, are still uncertain.

mutant yeast lacking its own functional gene copy can be rescued by transfection with the mammalian gene. As we noted previously, this has provided an efficient way to clone the mammalian cell-cycle genes.

The cell-cycle control system of most multicellular organisms, however, is more complex than that of yeasts. Gene duplication and divergence apparently has generated multiple variants of the basic cell-cycle genes, and these variants, existing side by side in a single cell, are specialized to function in slightly different ways. Thus, whereas in yeasts one cyclin-dependent kinase gene is sufficient for all the steps in the cell cycle, human cells depend on at least two and probably more. Two of these—*cdc*2 and a related gene called *cdk*2—are known to be present in *Xenopus* and *Drosophila* as well; both encode kinases whose activation depends on binding to cyclins. Like yeasts, higher animals also have multiple cyclins: at least six types have been found so far, named cyclins A, B, C, D, E, and F; some of these are themselves families of closely related molecules. Induction of mitosis in the vertebrate cell depends on Cdc2 protein complexed with cyclin B; there is evidence that Cdk2 protein complexed with cyclin E may induce passage past the G_1 checkpoint (the vertebrate counterpart of Start), and cyclin A complexed to Cdk2 protein may be required subsequently to activate the DNA replication machinery (Figure 17–33). It seems, therefore, that separate mammalian Cdk proteins perform the various functions that in yeast can be carried out by a single one.

In cells that are steadily cycling, the concentrations of most of the various cyclins rise and fall at different times in the cell cycle, while the concentrations of the cyclin-dependent kinases stay roughly constant, as in yeast. The precise functions of most of the proteins of the cell-cycle control system in the mammalian cell, however, are not yet understood in detail.

The Regulation of Mammalian Cell Growth and Proliferation Is Commonly Studied in Cultured Cell Lines [29]

Mammalian cells are not easily accessible to detailed observation in the intact animal. Most studies on mammalian cell proliferation therefore use cells that are growing in culture (Figure 17–34). This gives rise to a complication, however. When cells from normal mammalian tissues are cultured in standard conditions, they usually can be propagated only for a limited number of division cycles—about 50 for typical cells derived from humans, for example; after this they cease dividing and eventually die—a process called *cell senescence,* which we discuss later. But during the propagation of some cell cultures, especially those derived from rodents, a few cells often arise that escape senescence and divide indefinitely as *cell lines.* Although these cells resemble normal cells in most respects, their immortality reflects the presence of one or more mutations that have altered their proliferative properties. Nevertheless, in spite of their slight abnormalities, cell lines are used widely for cell-cycle studies—and for cell biology generally—

10 μm

Figure 17–34 Scanning electron micrograph of mammalian cells proliferating in culture. The cells are rat fibroblasts. (Courtesy of Guenter Albrecht-Buehler.)

because they provide an unlimited source of cells of a standardized, genetically homogeneous type.

Growth Factors Stimulate the Proliferation of Mammalian Cells [30]

Mammalian cells were first cultured in blood clots, and for many decades all efforts to define the minimal requirements for cell proliferation failed; even in a medium containing all the obvious chemically defined nutrients, including glucose, amino acids, and vitamins, cells would only grow if the medium was supplemented with *serum*, the blood-derived fluid that remains after blood has clotted. Like yeast cells deprived of nutrients, mammalian cells deprived of serum stop growing and become arrested, usually between mitosis and S phase, in a quiescent state called G_0. It was eventually shown that the essential components provided by serum are certain highly specific proteins called **growth factors,** most of which are required only in very low concentrations (on the order of 10^{-9} to 10^{-11} M).

One of the first such factors to be identified was **platelet-derived growth factor,** or **PDGF,** and it is typical of many others discovered since. The path to its isolation began with the observation that cultured fibroblasts proliferate when provided with serum but not when provided with *plasma*—the liquid prepared by removing the cells from blood without allowing clotting to occur. When blood clots, platelets incorporated in the clot are triggered to release the contents of their secretory vesicles (Figure 17–35). The superior ability of serum to support proliferation suggested that platelets contain one or more growth factors. This hypothesis was confirmed by showing that extracts of platelets could serve instead of serum to support fibroblast proliferation. A crucial growth factor in the extracts was shown to be a protein, which was subsequently purified and named PDGF. In the body PDGF liberated from blood clots probably plays a major role in stimulating cell division (and other processes) during wound healing.

PDGF is only 1 of about 50 known proteins that act as growth factors. For each type of growth factor there is a specific receptor or set of receptors, which some cells express on their surface and others do not. Cells respond to a given protein growth factor only if they display the appropriate receptor protein (discussed in Chapter 15). Other classes of molecules besides proteins can also function as growth factors; steroid hormones, which act on intracellular receptor proteins, are an example. The growth factors can be divided into broad- and narrow-specificity classes. The broad-specificity factors, like PDGF and *epidermal growth factor (EGF)*, affect many classes of cells. Thus PDGF acts on a range of target cells including fibroblasts, smooth muscle cells, and neuroglial cells, while EGF acts not only on epidermal cells but also on many other cell types, both epithelial and nonepithelial. At the opposite extreme lie narrow-specificity factors such as *erythropoietin*, which induces proliferation only of red blood cell precursors.

Because they are present in such small amounts, growth factors are difficult to isolate; once the DNA encoding a growth factor has been identified and cloned, however, it can often be used as a probe to identify and isolate a whole family of related genes that encode other members of the same growth-factor family. An example is the *fibroblast growth factor (FGF)* family, which includes at least seven members. Today, when a new growth factor is isolated by a biological assay, it is frequently found to be identical, or closely related, to an already known growth factor.

In intact animals proliferation of most cell types depends on a specific *combination* of growth factors rather than a single growth factor. Thus a fairly small number of growth factor families may serve, in different combinations, to regulate selectively the proliferation of each of the many types of cells in a higher animal.

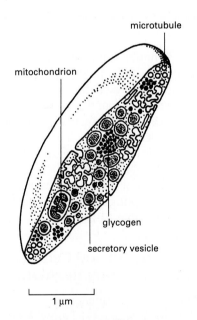

Figure 17–35 **A platelet.** Platelets are miniature cells without a nucleus that circulate in the blood and help to mediate blood clotting at sites of damage. They also release various factors that stimulate healing. The platelet in the diagram is cut open to show the secretory vesicles it contains; some of these contain platelet-derived growth factor (PDGF).

Table 17–2 Some Protein Growth Factors and Their Actions

Factor	Related Family Members	Broad or Narrow Specificity	Representative Actions
Platelet-derived growth factor (PDGF) —three subtypes		broad	stimulate proliferation of connective-tissue cells and some neuroglial cells
Epidermal growth factor (EGF)	transforming growth factor α (TGF-α); Lin-3 protein (in *C. elegans*)	broad	stimulate proliferation of many cell types; act as inductive signal in embryonic development
Insulinlike growth factor I (IGF-I)	insulinlike growth factor II (IGF-II); insulin	broad	promote cell survival; stimulate cell metabolism; collaborate with other growth factors to stimulate cell proliferation
Transforming growth factor β (TGF-β) —multiple subtypes	activins; bone morphogenetic proteins (BMPs); Decapentaplegic protein (in *Drosophila*); Vg1 protein (in *Xenopus*)	broad	potentiate or inhibit responses of most cells to other growth factors, depending on cell type; regulate differentiation of some cell types; act as inductive signals in embryonic development
Fibroblast growth factor (FGF) —multiple subtypes		broad	stimulate proliferation of many cell types; inhibit differentiation of various types of stem cells; act as inductive signals in embryonic development
Interleukin-2 (IL-2)		narrow	stimulate proliferation of activated T lymphocytes
Nerve growth factor (NGF)	brain-derived neurotrophic factor (BDNF); neurotrophin-3 (NT-3); neurotrophin-4 (NT-4)	narrow	promote survival and nerve process outgrowth of specific classes of neurons
Erythropoietin		narrow	promote proliferation, differentiation, and survival of red blood cell precursors
Interleukin-3 (IL-3)	hemopoietic colony stimulating factors (CSFs)—multiple types	narrow	stimulate proliferation and survival of various types of blood cell precursors

Although some growth factors are present in the circulation, most originate from cells in the neighborhood of the affected cell and act as local mediators. In addition to growth factors that stimulate cell division, there are factors, such as some members of the *transforming growth factor beta (TGF-β)* family, that act on some cells to stimulate cell proliferation and others to inhibit it, or stimulate at one concentration and inhibit at another. Indeed, most growth factors have a multitude of other actions besides the regulation of cell growth and division: they can control the proliferation, survival, differentiation, migration, or function of cells depending on the circumstance.

A sampling of some of the many growth factors discussed in this book is given in Table 17–2.

Cell Growth and Cell Division Can Be Independently Regulated [31]

One important function of growth factors is to regulate protein synthesis and thus the rate at which cells grow. Most factors that stimulate cell proliferation also stimulate cell growth, but the correspondence is not always exact. Some factors will make cells of a given type grow but do not get them past the G_1 checkpoint in their cycle, while other factors will get them past the G_1 checkpoint but do not make them grow. It seems that in a mammal there is not so rigid a rule coupling cell size and cell division as there is in yeasts.

Growth factors that act independently on cell growth and proliferation are important in the whole animal. Differently specialized cells vary enormously in their ratio of cytoplasm to DNA, and some cells in G_0, such as neurons, can grow very large without ever dividing (Figure 17–36). Such variation in cell size is controlled partly by external factors and partly by intracellular mechanisms that depend on the cell type. The growth of certain types of neurons, for example, depends on the *nerve growth factor (NGF)* that is secreted by the target cells the neurons innervate: the greater the amount of NGF that a neuron has access to, the larger it becomes.

Cells Can Delay Division by Entering a Specialized Nongrowing State [32]

When proliferating cells in culture are deprived of serum, they stop growing but continue to pass through the cell cycle until they reach the G_1 phase. On reaching this part of the cycle, they come to a halt in a specialized, nongrowing state—the G_0 ("G zero") resting state we referred to earlier. The G_0 state is distinct from the state of proliferating cells in any phase of their cycle. The rate of protein synthesis, for example, is drastically reduced, often to as little as 20% of its value in proliferating cells. Thus, the absence of appropriate growth factors sends cells into a sort of cell-cycle sleep, where the cell-cycle control system is disabled from progressing past the G_1 checkpoint. Depriving a cell of nutrients, such as amino acids, also stops growth and blocks passage past a G_1 checkpoint, but from the point of view of the cell such starvation is a very different experience from that of failing to receive the wakening signal that growth factors provide. In fact, it is debatable whether the same checkpoint mechanism is involved in the two situations and uncertain how closely either type of arrest in mammalian cells resembles the Start checkpoint mechanism of yeast (which is why different names, such as "restriction point" or "commitment point," have been used to refer to the G_1 checkpoint in mammalian cells). In the tissues of the body nutrients are generally plentiful but growth factors are in short supply; thus growth factors are thought to exert the critical control.

It is the ability to enter G_0 that accounts for the enormous variability of the length of the cell cycle in multicellular organisms. In the human body, for example, some cells, such as neurons and skeletal muscle cells, do not divide at all; others, such as liver cells, normally divide only once every year or two; while certain epithelial cells in the gut divide more than twice a day in order to renew the lining of the gut continually (Figure 17–37). Most cells in vertebrates fall somewhere between these extremes: they can divide if need arises but normally do so infrequently. The rate at which a cell divides varies according to external circumstances as well as the internal character of the particular cell type. Blood loss stimulates proliferation of blood cell precursors. Acute liver damage provokes surviving liver cells to proliferate with a cycle time of only a day or two until the loss is made good, and cells in the neighborhood of a wound are stimulated to divide to repair the lesion.

Almost all the variation in proliferation rates in the adult body lies in the time cells spend delayed between mitosis and the G_1 checkpoint, with slowly dividing cells remaining parked in a G_0 state for weeks or even years. By contrast, the time taken for a cell to progress from the beginning of S phase through mitosis is usually brief (typically 12 to 24 hours in mammals) and remarkably constant, irrespective of the interval from one division to the next. We must now turn to the experimental evidence that identified the G_1 checkpoint as a specific point in the cycle and consider what entry into G_0 means in molecular terms.

Serum Deprivation Prevents Passage Through the G_1 Checkpoint [32, 33]

Cells in culture can be observed under the microscope by time-lapse cinematography. In this way the time each cell takes between one division and the next is easily monitored. If the culture is filmed for at least one cell cycle before applying an experimental treatment, it is also possible to establish the time elapsed since

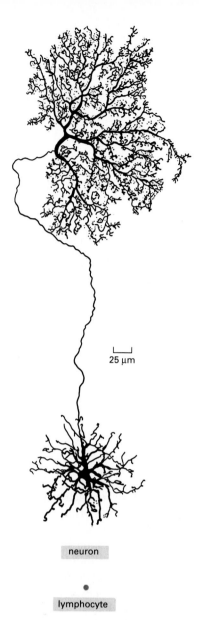

25 µm

neuron

lymphocyte

Figure 17–36 A mammalian neuron (from the retina) and a lymphocyte compared for size. Both of these cells contain the same amount of DNA. A neuron grows progressively larger during its development while remaining in a G_0 state. During this time the ratio of cytoplasm to DNA increases enormously (by a factor of more than 10^5 for some neurons). (Neuron from B.B. Boycott in Essays on the Nervous System [R. Bellairs and E.G. Gray, eds.]. Oxford, U.K.: Clarendon Press, 1974.)

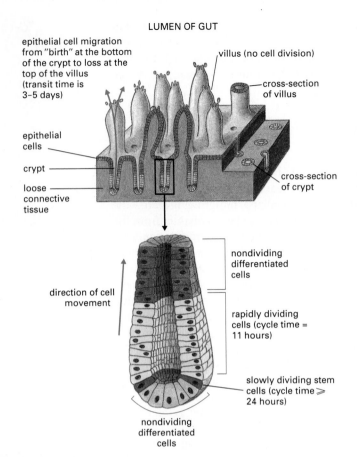

LUMEN OF GUT

epithelial cell migration from "birth" at the bottom of the crypt to loss at the top of the villus (transit time is 3–5 days)

villus (no cell division)

cross-section of villus

epithelial cells

crypt

loose connective tissue

cross-section of crypt

direction of cell movement

nondividing differentiated cells

rapidly dividing cells (cycle time = 11 hours)

slowly dividing stem cells (cycle time ≥ 24 hours)

nondividing differentiated cells

Figure 17–37 Cell division and migration in the epithelium lining the small intestine of the mouse. All cell division is confined to the bottom portion of the tube-shaped epithelial infoldings known as *crypts*. Newly generated cells move upward to form the epithelium that covers the villi, where they function in the digestion and absorption of foodstuffs from the lumen of the gut. Most epithelial cells have a very short lifetime, being shed from the tip of a villus within 5 days after emerging from the crypt. A ring of about 20 slowly dividing "immortal" cells (shown in *red*), however, remains anchored near the base of each crypt. These *stem cells* will divide to give rise to two daughter cells. On average, one daughter remains in place as an undifferentiated stem cell, while the other usually migrates upward to differentiate and join the villus epithelium. (Adapted from C.S. Potten, R. Schofield, and L.G. Lajtha, *Biochim. Biophys. Acta* 560:281–299, 1979.)

the last mitosis for each cell at the time of the treatment. In this way one can test the effects of tampering with external conditions at different stages in the division cycle. Such studies have been done mainly with fibroblast cells. A simple experiment of this kind showed that depriving the cells of serum (that is, growth factors) for just 1 hour has dramatic effects. All cells less than 3.5 hours past mitosis when serum was withdrawn took an extra 8 hours to reach mitosis after serum was added back to the medium. Cells more than 3.5 hours old, by contrast, showed no such delay but continued with the current cycle (Figure 17–38). This behavior defines the G_1 checkpoint as lying 3.5 hours after mitosis for the chosen line of cells. Cells past this point are irrevocably committed to replicate their DNA and complete the current division cycle, but cells between mitosis and the checkpoint stop at the checkpoint if appropriate growth factors are absent. The extra delay before these arrested cells could undergo mitosis after serum was returned to the medium suggests that a 1-hour serum deprivation between 0 and 3.5 hours induces them to enter an altered state—G_0—from which they require 8 hours to emerge.

The effect of serum deprivation is to depress protein synthesis and cell growth and can be mimicked with low doses of inhibitors of protein synthesis, such as cycloheximide. Experiments using either serum deprivation or cycloheximide have shown that depressing protein synthesis briefly in late G_2 can also induce entry into G_0, but in this case the cells first undergo mitosis and come to a G_0 halt when they reach the G_1 checkpoint. On the other hand, cells that are

Figure 17–38 Effect of brief serum deprivation at different points in the cell cycle. Fibroblast cells deprived for 1 hour during the interval between mitosis and the G_1 checkpoint (*yellow bar*) are delayed by approximately 8 hours in their journey to the next mitosis; cells deprived in a similar way after the G_1 checkpoint (*dark green bar*) suffer no such delay (although they may be delayed in the following cycle).

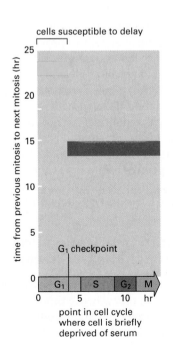

cells susceptible to delay

time from previous mitosis to next mitosis (hr)

G_1 checkpoint

G_1 | S | G_2 | M

point in cell cycle where cell is briefly deprived of serum

briefly deprived during S phase or early G_2 proceed through the G_1 checkpoint with little or no delay, presumably because they are already more than 8 hours away from the checkpoint. The machinery that responds so dramatically to a brief withdrawal of serum must therefore have the following properties: it must be needed to pass the G_1 checkpoint but not to enter mitosis, and although it can be rapidly disabled, it must require on the order of 8 hours to be regenerated once growth factors, or protein synthesis, are restored.

The Cell-Cycle Control System Can Be Rapidly Disassembled But Only Slowly Reassembled [34]

How are these phenomena related to the behavior of the cell-cycle control system? A simple interpretation would be that some molecular component of the cell-cycle control system disappears from the cell rapidly—within an hour—when serum is withdrawn but takes a long time—8 hours—to reappear when serum is restored. The disappearance and reappearance can occur at any time in the division cycle, it seems, but it is only at the G_1 checkpoint that the component is required.

An obvious speculation is that a Cdk protein itself is the critical component and that mammalian cells require this protein to pass the G_1 checkpoint, just as yeast cells require Cdc2 protein to pass Start. In fact, when quiescent G_0 cells are compared with cycling cells, it is found that they are severely depleted both in Cdk protein (or at least in one or more types of Cdk protein) and in all of the G_1 cyclins, even though the Cdk proteins and some of the G_1 cyclins (cyclins C and D) are present at a nearly constant level during all the phases of the cycle in the cycling cells. The G_0 cells, therefore, have not merely halted their cell-cycle control system: they have dismantled it.

When serum is supplied to G_0 cells, there is a lag of several hours before the concentrations of Cdk and G_1 cyclins are returned to their cycling levels, corresponding to the delay before the cells resume cycling. If serum deprivation halts cell proliferation by rapidly dismantling the cell-cycle control system rather than simply stopping it, it is not surprising that, when the environment becomes favorable again, cells must spend time slowly reassembling the control system in order to begin cycling again.

Having considered how the cell-cycle control system of the individual cell responds to growth factors, we next discuss how growth factors and other influences adjust the proliferative behavior of cells in tissues to maintain the form and function of the body.

Neighboring Cells Compete for Growth Factors [35]

Cell proliferation in the body has to be regulated so as to maintain both the numbers of cells and their spatial organization. This regulation depends on interactions of cells with one another and with the extracellular matrix. Consider an epithelial sheet in an adult mammal, for example. As cells die, new cells must be produced to take their places. Cell proliferation must be precisely controlled to balance the cell loss, so that the epithelial sheet neither grows nor shrinks. The new cells must be fitted into the structure correctly, so that the architecture of the sheet is not disrupted. In fact, in most epithelia it is only cells retaining contact with the underlying basal lamina that divide. These cells on the basal lamina are sensitive to at least two sorts of signals that govern their readiness to divide: those that carry information about the local cell population density, and those that reflect a cell's attachments to other cells and to the basal lamina. Both types of controls can be demonstrated and analyzed in the simplified conditions of cell culture, although most of the work has been done with fibroblasts and it is not clear how these findings relate to organized arrays of cells such as those in epithelia or in a three-dimensional organ.

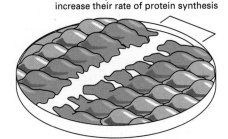
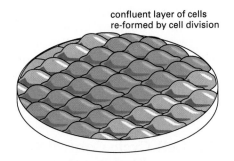

confluent layer of cells with line of cells removed by scraping

cells at margin spread and flatten, and increase their rate of protein synthesis

confluent layer of cells re-formed by cell division

Dissociated cells plated on a dish in the presence of serum will adhere to the surface, spread out, and divide until a *confluent monolayer* is formed in which each cell is attached to the dish and contacts its neighbors on all sides. At this point normal cells, unlike cancerous ("transformed") cells, stop dividing—a phenomenon known as *density-dependent inhibition of cell division*. If such a monolayer is "wounded" with a needle so as to create a cell-free strip on the dish, the cells at the edges of the strip spread into the empty space and divide (Figure 17–39). Such phenomena were originally described in terms of "contact inhibition" of cell division, but it is probably misleading to imply that cell-cell contact interactions are solely responsible. The cell population density at which cell proliferation ceases in the confluent monolayer increases with increasing concentration of growth factors in the medium. Passing a stream of fresh medium over a confluent layer of fibroblasts reduces the diffusional limitation to the supply of growth factors and induces the cells under the stream of medium to divide at densities where they would normally be inhibited from doing so (Figure 17–40). Thus density-dependent inhibition of cell proliferation seems to reflect, in part at least, the ability of a cell to deplete the medium locally of growth factors, thereby depriving its neighbors.

Calculations using the known concentrations of growth factors in serum and the rate at which cells remove the factors from the culture medium support this suggestion. PDGF, for example, is typically present in the medium at concentrations of about 10^{-10} M (about one molecule in a sphere of 3 μm diameter). A fibroblast has about 10^5 PDGF receptors, each with a very high affinity for the growth factor. Each cell therefore has enough receptors to bind all the PDGF molecules within a sphere of diameter ~150 μm. Thus it is clear that neighboring cells compete for minute quantities of growth factors. This type of competition could be important for cells in tissues as well as in culture, preventing them from proliferating beyond a certain population density, which is determined by the amount of growth factor available.

Normal Animal Cells in Culture Need Anchorage in Order to Pass Start [36]

Competition for growth factors is not the only influence on the rate of cell division observed in cell culture. The shape of a cell as it spreads and crawls out over a substratum to occupy vacant space also strongly affects its ability to divide.

Figure 17–39 Regulation of cell division in a "wounded" cell monolayer. Cells scattered on the surface of a culture dish normally proliferate until they touch one another, forming a confluent monolayer. The diagrams show the consequences of scraping away a strip of cells. The remaining cells at the margins of the vacant "wound" area flatten out and resume growth and division, which continue until the "wound" is "healed." Once the monolayer is again confluent, cell proliferation ceases almost entirely.

Figure 17–40 Effect of fresh medium on a confluent cell monolayer. Cells in a confluent monolayer do not divide (indicated in *gray*). The cells resume dividing (indicated in *green*) when exposed directly to fresh medium. Apparently, in the undisturbed confluent monolayer proliferation has halted because the medium close to the cells is depleted of growth factors, for which the cells compete.

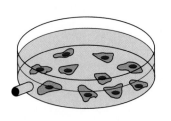

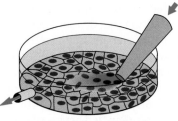
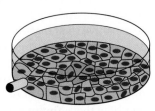

cells proliferate

confluent monolayer: cells no longer proliferate

fresh medium pumped across cells

flow of medium stimulates cell proliferation

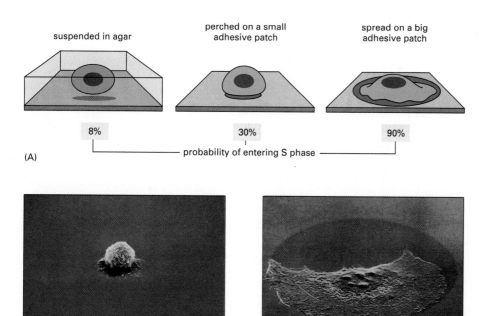

suspended in agar | perched on a small adhesive patch | spread on a big adhesive patch

8%   30%   90%

(A)

—— probability of entering S phase ——

(B)

(C)

50 µm

Figure 17–41 The dependence of cell division on cell shape and anchorage. In the experiment shown here, cells are either held in suspension or allowed to settle on patches of an adhesive material (palladium) on a nonadhesive substratum; the patch diameter, which is variable, determines the extent to which an individual cell spreads and the probability that it will divide. ³H-thymidine is added to the culture medium, and after 1 or 2 days the culture is fixed and auto-radiographed to discover the percentage of cells that have entered S phase. (A) Cells of the 3T3 cell line divide rarely when held rounded up in suspension, but adherence even to a very tiny patch—one that is too small to allow spreading—enables them to divide much more frequently. (B, C) Scanning electron micrographs showing a cell perched on a small patch as compared with a cell spread on a large patch. Note that, in contrast to fibroblasts, some cell types in the body (in particular, blood-cell precursors) will divide readily in suspension. (B and C from C. O'Neill, P. Jordan, and G. Ireland, *Cell* 44:489–496, 1986. © Cell Press.)

When normal fibroblast or epithelial cells are cultured in suspension, unattached to any solid surface and therefore rounded up, they almost never divide—a phenomenon known as *anchorage dependence* of cell division. The relationship of cell spreading to proliferation can be demonstrated by culturing cells on substrata of varying stickiness or by allowing them to settle on a nonsticky surface that is dotted with minute sticky patches on which an individual cell may adhere but beyond which it cannot spread. The frequency with which a cell divides increases as the cell becomes more spread out. Perhaps well-spread cells can capture more molecules of growth factor and take up larger quantities of nutrients because of their larger surface area. But some fibroblast cell lines, although scarcely able to proliferate at all in suspension, will divide readily once they have touched down and formed a *focal contact,* even if the site of adhesion is a tiny patch on which there is no space for a cell to spread (Figure 17–41). Focal contacts are sites of anchorage for intracellular actin filaments and extracellular matrix molecules, and these and other observations strongly hint that the control of cell division is somehow coupled to the organization of the cytoskeleton or is dependent on intracellular signals generated at sites of adhesion, or both (Figure 17–42). In fact,

Figure 17–42 Focal contacts as sites of production of intracellular signals. This fluorescence micrograph shows a fibroblast cultured on a substratum coated with the extracellular matrix molecule fibronectin. Actin filaments have been labeled so as to fluoresce green, while proteins containing phosphotyrosine have been labeled with an antibody that is tagged so as to fluoresce red. Where the two components overlap, the resulting color is yellow. The actin filaments terminate at focal contacts, where the cell adheres to the substratum. Proteins containing phosphotyrosine are also concentrated at these sites. This is thought to reflect the operation of a tyrosine-kinase intracellular signaling mechanism activated by trans-membrane integrin proteins that bind to fibronectin extracellularly and (indirectly) to actin filaments intracellularly (see p. 999). Signals generated at such adhesion sites are thought to help regulate cell proliferation, both in fibroblasts and in epithelial cells. (Courtesy of Keith Burridge.)

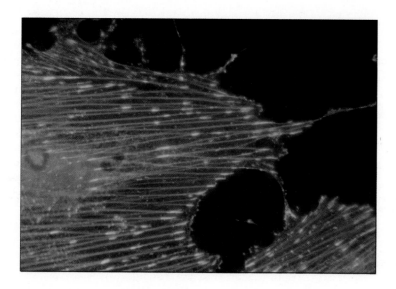

Cell-Division Controls in Multicellular Animals

for some cell types specific extracellular matrix molecules, such as laminin or fibronectin, may act as growth factors. The basal epidermal cells of the skin provide an example, discussed in Chapter 22.

Like other controls of cell proliferation, anchorage control operates at the G_1 checkpoint: cells require anchorage to pass this point but then do not require it to complete the cycle. In fact, they commonly loosen their attachments and round up as they pass through M phase. This cycle of attachment and detachment presumably allows cells to rearrange their contacts with other cells and with the extracellular matrix so as to accommodate the daughter cells produced by cell division and then bind them securely into the tissue before they are allowed to begin the next division cycle.

Although the mechanism of anchorage dependence is uncertain and some of the details of the phenomena just described may be peculiarities of fibroblasts in culture, it is likely that intracellular signals generated at adhesion sites play an important part in controlling the cell-cycle control system in many different cell types.

Studies of Cancer Cells Reveal Genes Involved in the Control of Cell Proliferation [37]

How do the various external influences we have examined affect the interior of the cell so as to regulate cell proliferation? In particular, how does the binding of a growth factor to cell-surface receptors influence the operation of the cell-cycle control system? Much of what we know on this topic has come from studies of cancer cells, where the control of cell proliferation is disrupted. These cells are mutants, and because they proliferate excessively and give rise to tumors, they bring their mutant genes to our attention.

As discussed in Chapter 24, analysis of the genetic alterations in cancer cells has revealed a large number of genes that encode proteins involved in the control of cell proliferation. These genes can be crudely classified as *proliferation genes* and *antiproliferation genes*. The products of the former help to promote cell growth and the assembly of the cell-cycle control system and to drive the cell past the G_1 checkpoint; the products of the latter help to apply the brakes that halt the control system and cause it to be dismantled. A mutation in a proliferation gene that causes its product to be overexpressed or hyperactive results in the excessive cell proliferation characteristic of cancer. The mutant gene is then classified as an **oncogene** (that is, a cancer-causing gene), and the normal proliferation gene is called a **proto-oncogene.** Conversely, a cell may be released from the normal proliferation restraints and enabled to divide as a cancer cell if an antiproliferation gene undergoes a mutation that makes it inactive. Thus the

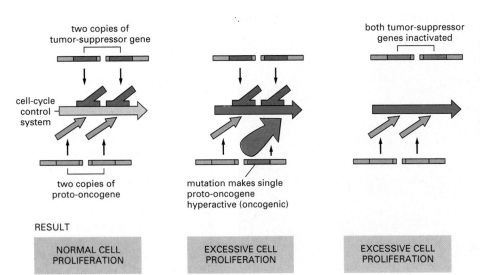

two copies of tumor-suppressor gene

both tumor-suppressor genes inactivated

cell-cycle control system

two copies of proto-oncogene

mutation makes single proto-oncogene hyperactive (oncogenic)

RESULT

NORMAL CELL PROLIFERATION

EXCESSIVE CELL PROLIFERATION

EXCESSIVE CELL PROLIFERATION

Figure 17–43 Tumor-suppressor genes versus proto-oncogenes. The product of a tumor-suppressor gene inhibits assembly and activation of the cell-cycle control system; the product of a proto-oncogene does the opposite. Unrestrained proliferation can result from mutations that either inactivate both copies of the tumor-suppressor gene or strongly overactivate one copy of the proto-oncogene.

antiproliferation genes found in normal cells are often referred to as **tumor-suppressor genes.** For an ordinary diploid cell both copies of a tumor-suppressor gene must typically be lost or inactivated to bring about the loss of growth control (that is, the mutant phenotype is recessive), whereas only one copy of a proto-oncogene need be activated to bring about a similar effect (that is, the mutant phenotype is dominant) (Figure 17–43).

Because they have been easier to isolate, the set of known proliferation genes is much larger than the set of known antiproliferation genes and probably includes genes that encode representatives of all the major classes of proteins involved in relaying the stimulatory signals from the growth factor receptors to the interior of the cell.

Growth Factors Trigger Cascades of Intracellular Signals [30, 38]

The first step in the action of a protein growth factor is its binding to a transmembrane receptor at the surface of the target cell. The intracellular portion of the receptor then catalyzes the production of molecules that act as intracellular signals, relaying the stimulus to yet other molecules. The details of the initial steps in several of these signaling cascades have been well worked out for several classes of receptors and are described in Chapter 15. The complexities arise because the cascades generally are not simple linear relay chains but instead branch to activate many interacting components that operate in parallel, forming a highly interconnected signaling network.

We saw in Chapter 15 that growth-factor receptors activate intracellular phosphorylation cascades that lead to changes in gene expression. The genes that growth factors induce fall into two classes: **early-response genes** are induced within 15 minutes of growth factor treatment, and their induction does not require protein synthesis; **delayed-response genes,** by contrast, are not induced until at least 1 hour after growth factor treatment, and their induction requires protein synthesis. It seems that the delayed-response genes are induced by the products of the early-response genes, several of which are known to be gene regulatory proteins (Figure 17–44).

Both classes of genes are silent and not transcribed in cells in G_0 but are induced to high levels when growth factors are added to the medium. If the exposure to growth factors is then maintained, the level of expression of the genes gradually falls back—for some genes apparently to zero, for certain others to a new non-zero steady value. Products of the latter class of genes are therefore present at a constant low level in steadily cycling cells (Figure 17–45). Thus the transcription of these genes indicates the presence of growth factors in the medium, and a high level of expression indicates a sudden increase in growth factor concentration. The signals we get from our own senses (of smell, for example) behave in much the same way.

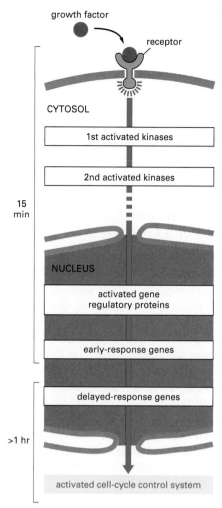

Figure 17–44 Typical signaling pathway for stimulation of cell proliferation by a growth factor. This greatly simplified diagram shows some of the major steps. It omits many of the intermediate steps in the relay system. Intracellular signaling pathways are discussed in detail in Chapter 15.

Figure 17–45 The response of Myc to a growth factor. Myc is the product of the early-response gene *myc*. The graph shows the changes in the concentration of Myc protein following a sudden increase in growth factor concentration to a new steady value, which causes the cell to exit G_0 and to proliferate. The changes in Myc concentration reflect changes in *myc* gene transcription, stimulated by exposure of the cell to the growth factor. Myc protein itself inhibits *myc* transcription, and this negative feedback is thought to explain why the level of Myc declines from its initial peak to a lower steady value.

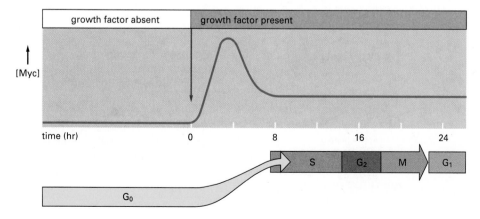

The best-studied early-response genes are the *myc, fos,* and *jun* proto-oncogenes. All three genes encode gene regulatory proteins that act as homo- or heterodimers (discussed in Chapter 9). When overexpressed or hyperactivated by mutation in certain types of cells, all of them can cause uncontrolled proliferation. There is evidence suggesting that *myc* in particular may have a critical role in the normal control of cell proliferation. Cells in which *myc* expression is specifically prevented will not divide even in the presence of growth factors. Conversely, cells in which *myc* expression is specifically switched on independently of growth factors cannot enter G_0, and if they are in G_0 when Myc protein is provided, they will leave G_0 and begin to divide even in the absence of growth factors—a behavior that ultimately causes them to undergo programmed cell death.

Cyclins and Cdk Are Induced by Growth Factor After a Long Delay [34]

The delayed-response genes do not begin to be transcribed until well after the addition of growth factor, and their transcription requires the products of the early-response genes such as *myc.* Among the products of the delayed-response genes are some of the essential components of the cell-cycle control system itself, including Cdk proteins and several cyclins, which, from the timing of their expression, are suspected to be involved with Cdk proteins in driving the cells past the G_1 checkpoint, in initiating S phase, or both.

Thus one can tentatively trace a chain of stimulatory effects that leads from the binding of growth factor to the initiation of DNA replication. These stimulatory signals are thought to act by overcoming specific inhibitory devices that ensure that the cell refrains from proliferating in the absence of a positive signal to do so. The inhibitory devices are proteins encoded by the antiproliferation genes discussed earlier, which were originally discovered as tumor-suppressor genes in human cancers. The best-understood antiproliferation gene is the *retinoblastoma* gene.

The Retinoblastoma Protein Acts to Hold Proliferation in Check [39]

The **retinoblastoma (Rb) gene** was identified originally through studies of an inherited predisposition to a rare cancer that occurs in the eyes of children, as discussed in Chapter 24. Loss of both copies of this gene leads to excessive cell proliferation in the immature retina, suggesting that the gene product normally helps keep proliferation in check. Cloning of the retinoblastoma gene made it possible to explore how the gene product exerts this effect.

The Rb protein is an abundant molecule in the nucleus of mammalian cells. It binds to many other proteins, including several important gene regulatory proteins, but its binding capacity depends on its state of phosphorylation. When Rb is dephosphorylated, it binds a set of regulatory proteins that favor cell proliferation, holding them sequestered and out of action; the phosphorylation of Rb makes it release these proteins, allowing them to act (Figure 17–46). In nor-

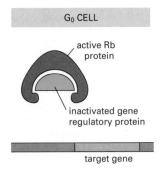
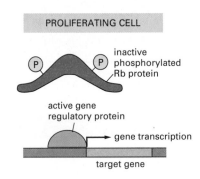

Figure 17–46 Action of the retinoblastoma (Rb) protein. Dephosphorylated Rb binds to, and holds inactive, gene regulatory proteins that stimulate transcription of target genes (such as *myc*) required for cell proliferation. Phosphorylated Rb detaches, releasing the stimulatory proteins that activate proliferation.

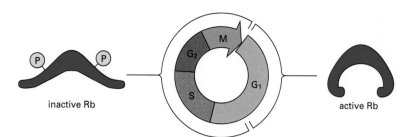

mal cells the Rb protein is permanently present, no matter whether the cells are in G_0 or cycling, but its state of phosphorylation changes. In the G_0 cell it contains little phosphate and appears to hinder the transcription of genes, such as *fos* and *myc*, that are required for proliferation. These genes are transcribed at a high level in mutant cells that lack a functional copy of the *Rb* gene and at a much lower level in these cells when a functional copy of the *Rb* gene is put back into them by transfection.

Growth factors relieve the inhibition exerted by Rb by causing the protein to become phosphorylated on multiple serines and threonines. The cells now begin to express Cdk protein, pass the G_1 checkpoint, and embark on DNA synthesis. In proliferating cells the phosphorylation of the Rb protein increases and decreases in every cycle: it rises late in G_1, remains high in S and G_2, then falls back to a dephosphorylated state as the cell goes through mitosis (Figure 17–47). *In vitro* the Rb protein is a good substrate for phosphorylation by protein kinases of the Cdk family, suggesting a possible way in which the state of Rb phosphorylation could be tightly linked to the state of the cell-cycle control system.

The dephosphorylated (active) Rb protein is thought to function in G_1 as part of the braking mechanism to inhibit passage past Start; it may also enable the cell to enter G_0 by shutting off the production of key components of the cell-cycle control system—as well as of other proteins—when the environment becomes unfavorable for proliferation. But in most cells the situation is complicated by the presence of more than one Rb-like protein, and many cell types appear to behave normally even when Rb itself is missing. (Transgenic mice that lack the *Rb* gene progress almost normally through the first half of embryonic development, but then die, showing defects only in certain specific tissues.) Improving our understanding of anti-proliferation genes such as Rb is an important task, for deficiencies in these genes play a part in a remarkably large proportion of human cancers.

The Probability of Entering G_0 Increases with the Number of Times That a Cell Divides: Cell Senescence [40]

Cell proliferation in a higher animal is not simply governed by the environment of the cell but depends on the cell's long-term history in complex ways: each differentiated cell type, at each stage of animal development, is obedient to slightly different rules, reflecting differences in its internal control machinery. Perhaps the simplest, but also the most mysterious, example of long-term effects on cell division is seen in the phenomenon of **cell senescence.**

Most normal cells in the body of a mammal or bird show a striking reluctance to continue proliferating forever, even when carefully nourished *in vitro*. This distinguishes them from germ-line cells and from established cultured cell lines, which are thought to have undergone a genetic change that makes them "immortal." Fibroblasts taken from a normal human fetus, for example, will go through only about 50 population doublings when cultured in a standard growth medium. Toward the end of this time, proliferation slows down and finally halts, and the cells enter a G_0 state from which they never recover. Similar cells taken from a 40-year-old stop dividing after about 40 doublings, while cells from an 80-year-old stop after about 30 doublings. Fibroblasts from animals with a shorter life-

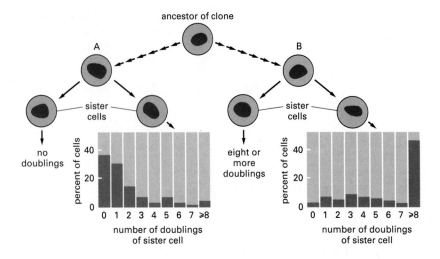

Figure 17–48 **Evidence for cell variation in a heritable ability to divide.** Individual cells in a clone vary in the number of division cycles they will undergo. Here, different pairs of sister cells from the same clone have been studied, and histograms have been drawn to show the numbers of cells that divide a given number of times. If one sister fails to divide at all, the other sister usually does likewise or divides only a few times (*left side*); and if one sister undergoes eight or more doublings, the other also usually undergoes eight or more doublings (*right side*). This shows that there are heritable differences between cells of the clone in the numbers of division cycles of which they are capable. The different heritable states are not perfectly stable, however, so sister cells sometimes behave differently. Further studies show that, as the cell population ages, the cells undergo random transitions toward states of reduced division capability. (Data from J.R. Smith and R.G. Whitney, *Science* 207:82–84, 1980.)

span cease dividing after a smaller number of division cycles in culture. This phenomenon has been called cell senescence because of the correspondence with aging of the body as a whole. Its relationship to the aging of the organism, however, is obscure.

Cell senescence is puzzling in several ways, and neither its function, if it has one, nor its mechanism is clear. Many theories have been proposed to explain the phenomenon. Some suggest, for example, that it results from an accumulation of deleterious random mutations, reflecting merely imprecision in the machinery of cell reproduction; others suggest that it is the effect of a mechanism that has evolved to protect us against cancer by limiting the growth of tumors. But there are good arguments against both of these interpretations. It is known, however, that the rate of senescence for at least some cell types is strongly dependent on the concentrations of growth factors in the medium. The process apparently reflects changes of proliferative potential that can be regulated by the cell's environment.

Short programmed sequences of cell divisions terminating in differentiation are a familiar feature of embryonic development, but it is hard to imagine how a cell could keep a long-term account of its division cycles and halt after completing 50. In fact, although senescence occurs at a predictable time for a given cell population, it is not strictly programmed at the level of the individual cell. In a clone of apparently identical normal fibroblasts monitored under standard culture conditions, some cells divide many times, others only a few times. Individual cells seem to stop dividing as a result of a random transition. This transition occurs with an increasing probability in each successive cell generation until there are no proliferating cells left in the population (Figure 17–48).

A possible interpretation lies in the behavior of the *telomeres*—the special repetitive DNA sequences required at the ends of chromosomes. As discussed in Chapter 8, when a cell divides these sequences are not replicated in the same manner as the rest of the genome but are synthesized by an enzyme, *telomerase*, that operates less exactly, creating random variation in the number of repeats of the telomeric DNA sequence. Cell senescence is closely correlated with a progressive reduction in the number of these repeats, suggesting that senescence may be caused by failure to maintain the length of telomeres, perhaps because the somatic cells (in contrast with germ-line cells) are deficient in telomerase.

Intricately Regulated Patterns of Cell Division Generate and Maintain the Body [41]

While life may end with a haphazard process of senescence, it begins with a series of division cycles that are controlled according to precise and intricate rules. This is true of multicellular organisms in general but is most strikingly illustrated by

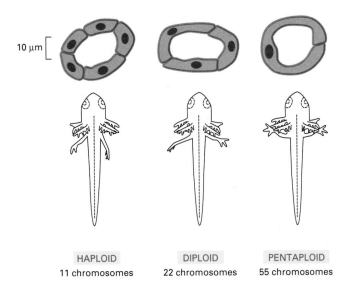

Figure 17–49 Drawings of representative sections of kidney tubules from salamander larvae of different ploidy. Pentaploid salamanders have cells that are bigger than those of haploid salamanders, but the animals and their individual organs are the same size because each tissue in the pentaploid animal contains fewer cells. This indicates that the number of cells is regulated by some mechanism based on size and distance rather than on the counting of cell divisions or of cell numbers. (After G. Fankhauser, in Analysis of Development [B.H. Willier, P.A. Weiss, and V. Hamburger, eds.], pp. 126–150. Philadelphia: Saunders, 1955.)

the nematode worm *Caenorhabditis elegans*. The fertilized egg of *C. elegans* divides to produce an adult with precisely 959 somatic cell nuclei, each one of which is generated by its own characteristic and absolutely predictable sequence of cell divisions. In general such phenomena are not merely a matter of counting out cell divisions according to a clocklike schedule. Salamanders of different ploidy, for example, are the same size but have different numbers of cells; individual cells in a pentaploid animal are five times the volume of those in a haploid animal, and in each organ the pentaploids have generated only one-fifth as many cells as their haploid cousins, so that the organs are about the same size in the two animals (Figures 17–49 and 17–50). Evidently, in this case (and, in fact, in most others) the size of organs is not controlled by counting cells or cell cycles but by some mechanism that can actually measure distances. Such mechanisms require complex positional controls in which growth factors may play an important part. As we shall see in Chapter 21, some of the genes that govern programs of cell proliferation in the embryo are beginning to be characterized. In the early *Drosophila* embryo it is beginning to be possible to explain the patterns of cell division in terms of the activities of identified components of the cell-cycle control system. But developmental biologists are far from understanding in detail how cyclin-dependent kinases, cyclins, growth factors, cell contacts, proliferation genes, and antiproliferation genes—all the nuts and bolts of cell-division control that we have discussed so far—fit together to implement the complex programs of cell division in development.

In the same way, we have as yet only a fragmentary understanding of the complex network of controls that govern cell division in the society of cells that form an adult body. When a skin wound heals in a vertebrate, for example, about a dozen cell types, ranging from fibroblasts to Schwann cells, must be regenerated in appropriate numbers and in appropriate positions to reconstruct the lost tissue. These questions of cell-division control in tissues, which will be discussed further in Chapter 22, are central to the understanding of cancer, where the social controls go wrong, and of innumerable other diseases. At the heart of the matter is the operation of the cell cycle itself—the fundamental mechanism of cell reproduction on which all life depends. The revelations that have come from

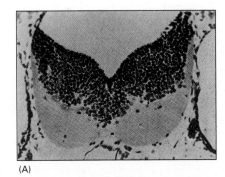
(A)

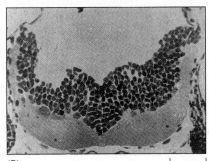
(B)

Figure 17–50 Micrographs comparing cells in the brains of haploid and tetraploid salamanders. (A) Cross-section of the hindbrain of a haploid salamander. (B) Corresponding cross-section through the hindbrain of a tetraploid salamander showing how reduced cell numbers compensate for increased cell size. (From G. Fankhauser, *Int. Rev. Cytol.* 1:165–193, 1952.)

Cell-Division Controls in Multicellular Animals

studies in frogs and yeasts are already beginning to cast new light on urgent human problems.

Summary

The cells of multicellular animals, such as mammals, appear to have the same basic cell-cycle control system as is found in yeasts, with a major checkpoint in G_1. But, unlike yeasts, they normally refrain from proliferating unless they receive specific signals from other cells to do so. If deprived of these signals, a typical mammalian cultured cell will come to a halt in a quiescent, nongrowing variant of the G_1 state called G_0, in which the production of components of the cell-cycle control system— and of many other proteins—is switched off. Important extracellular cues for cell proliferation are provided by protein growth factors. These are present at very low concentrations, and competition for them appears to limit cell population densities. Most cells, in addition to a requirement for specific growth factors, must be anchored to a substratum before they will divide.

Growth factors regulate cell proliferation through a complex network of intracellular signaling cascades, which ultimately regulate gene transcription and the assembly and activation of the cell-cycle control system. Many of the protein components of these signaling pathways have been identified through studies of cancer cells, where mutations have occurred either to activate genes whose products promote proliferation (proto-oncogenes) or to inactivate genes whose products normally restrain proliferation (tumor-suppressor genes). Among the proto-oncogenes are some gene regulatory proteins, such as Myc, that in some cell types are necessary and sufficient to induce cell proliferation. Conversely, there are tumor-suppressor genes, such as Rb, whose products block proliferation by binding to and sequestering gene regulatory proteins. In addition to these short-term controls on cell proliferation in multicellular organisms, there are long-term controls, such as those responsible for the progressive loss of proliferation potential when cells senesce or for the elaborate programs of cell division required during embryonic development. As yet, these are poorly understood.

References

General

The Cell Cycle. *Cold Spring Harbor Symp. Quant. Biol.* Vol. 56, 1991.

Cooper, S. Bacterial Growth and Division. San Diego, CA: Academic Press, 1991.

Mitchison, J.M. The Biology of the Cell Cycle. Cambridge, UK: Cambridge University Press, 1971.

Murray, A.; Hunt, T. The Cell Cycle. Oxford, UK: Oxford University Press, 1993.

Pollack, R., ed. Readings in Mammalian Cell Culture, 2nd ed. Cold Spring Harbor, NY: Cold Spring Harbor Laboratory, 1981.

Varmus, H.; Weinberg, R.A. Genes and the Biology of Cancer. New York: Scientific American Library, 1993.

Wilson, E.B. The Cell in Development and Heredity, 3rd ed. New York: Macmillan, 1928.

Cited

1. Baserga, R. The Biology of Cell Reproduction. Cambridge: Harvard University Press, 1985.

   Howard, A.; Pelc, S.R. Nuclear incorporation of P^{32} as demonstrated by autoradiographs. *Exp. Cell Res.* 2:178–187, 1951.

   Krishan, A. Rapid flow cytofluorometric analysis of mammalian cell cycle by propidium iodide staining. *J. Cell Biol.* 66:188–193, 1975.

2. Bailly, E.; Bornens, M. Centrosome and cell division. *Nature* 355:300–301, 1992.

   Creanor, J.; Mitchison, J.M. Patterns of protein synthesis during the cell cycle of the fission yeast *Schizosaccharomyces pombe. J. Cell Sci.* 58:263–285, 1982.

   Elledge, S.J.; Zhou, Z.; Allen, J.B. Ribonucleotide reductase: regulation, regulation, regulation. *Trends Biochem. Sci.* 17:119–123, 1992.

   Johnston, L.H. Periodic events in the cell cycle. *Curr. Opin. Cell Biol.* 2:274–279, 1990.

   Xu, H.; Kim, U.-J.; Schuster, T.; Grunstein, M. Identification of a new set of cell cycle-regulatory genes that regulate S-phase transcription of histone genes in *Saccharomyces cerevisiae. Mol. Cell Biol.* 12:5249–5259, 1992.

3. Hartwell, L.H.; Weinert, T.A. Checkpoints: controls that ensure the order of cell cycle events. *Science* 246:629–634, 1989.

   Murray, A.W.; Kirschner, M.W. Dominoes and clocks: the union of two views of the cell cycle. *Science* 246:614–621, 1989.

Murray, A.W.; Kirschner, M.W. What controls the cell cycle? *Sci. Am.* 264(3):56–63, 1991.

Nurse, P. Universal control mechanism regulating onset of M-phase. *Nature* 344:503–508, 1990.

O'Farrell, P.H. Cell cycle control: many ways to skin a cat. *Trends Cell Biol.* 2:159–163, 1992.

4. Kirschner M. The cell cycle then and now. *Trends Biochem. Sci.* 17:281–285, 1992.

5. Browder, L.W.; Erickson, C.A.; Jeffery, W.R. Developmental Biology, 3rd ed., Chaps. 3 and 5. Philadelphia: Saunders, 1991.

Kirschner, M.; Newport, J.; Gerhart, J. The timing of early developmental events in *Xenopus. Trends Genet.* 1:41–47, 1985.

6. Lohka, M.J.; Hayes, M.K.; Maller, J.L. Purification of maturation-promoting factor, an intracellular regulator of early mitotic events. *Proc. Natl. Acad. Sci. USA* 85:3009–3013, 1988.

Masui, Y; Markert, C.L. Cytoplasmic control of nuclear behavior during meiotic maturation of frog oocytes. *J. Exp. Zool.* 177:129–146, 1971.

Newport, J.W.; Kirschner, M.W. Regulation of the cell cycle during early *Xenopus* development. *Cell* 37:731–742, 1984.

7. Johnson, R.T.; Rao, P.N. Mammalian cell fusion: induction of premature chromosome condensation in interphase nuclei. *Nature* 226:717–722, 1970.

Rao, P.N.; Johnson, R.T. Mammalian cell fusion studies on the regulation of DNA synthesis and mitosis. *Nature* 225:159–164, 1970.

8. Gerhart, J.; Wu, M.; Kirschner, M. Cell cycle dynamics of an M-phase-specific cytoplasmic factor in *Xenopus laevis* oocytes and eggs. *J. Cell Biol.* 98:1247–1255, 1984.

Hara, K.; Tydeman, P.; Kirschner, M. A cytoplasmic clock with the same period as the division cycle in *Xenopus* eggs. *Proc. Natl. Acad. Sci. USA* 77:462–466, 1980.

Harvey, E.B. A comparison of the development of nucleate and non-nucleate eggs of *Arbacia punctulata. Biol. Bull.* 79:166–187, 1940.

Picard, A.; Harricane, M.-C.; Labbé, J.C.; Dorée, M. Germinal vesicle components are not required for the cell-cycle oscillator of the early starfish embryo. *Dev. Biol.* 128:121–128, 1988.

9. Evans, T.; Rosenthal, E.T.; Youngblom, J.; Distel, D.; Hunt, T. Cyclin: a protein specified by maternal mRNA in sea urchin eggs that is destroyed at each cleavage division. *Cell* 33:389–396, 1983.

Minshull, J.; Blow, J.J.; Hunt, T. Translation of cyclin mRNA is necessary for extracts of activated *Xenopus* eggs to enter mitosis. *Cell* 56:947–956, 1989.

Murray, A.W.; Kirschner, M.W. Cyclin synthesis drives the early embryonic cell cycle. *Nature* 339:275–280, 1989.

Murray, A.W.; Solomon, M.J.; Kirschner, M.W. The role of cyclin synthesis and degradation in the control of maturation promoting factor activity. *Nature* 339:280–286, 1989.

10. Glotzer, M.; Murray, A.W.; Kirschner, M.W. Cyclin is degraded by the ubiquitin pathway. *Nature* 349:132–138, 1991.

Sherr, C.J. Mammalian G1 cyclins. *Cell* 73:1059–1065, 1993.

11. Bailly, E.; Pines, J.; Hunter, T.; Bornens, M. Cytoplasmic accumulation of cyclin B1 in human cells: association with a detergent-resistant compartment and with the centrosome. *J. Cell Sci.* 101:529–545, 1992.

Belmont, L.D.; Hyman, A.A.; Sawin, K.E.; Mitchison, T.J. Real-time visualization of cell cycle-dependent changes in microtubule dynamics in cytoplasmic extracts. *Cell* 62:579–589, 1990.

Heald, R.; McKeon, F. Mutations of phosphorylation sites in lamin A that prevent nuclear lamina disassembly in mitosis. *Cell* 61:579–589, 1990.

Nigg, E.A. Targets of cyclin-dependent protein kinases. *Curr. Opin. Cell Biol.* 5:187–193, 1993.

Peter, M.; Nakagawa, J.; Dorée, M.; Labbé, J.C.; Nigg, E.A. *In vitro* disassembly of the nuclear lamina and M phase-specific phosphorylation of lamins by cdc2 kinase. *Cell* 61:591–602, 1990.

Verde, F.; Dogterom, M.; Stelzer, E.; Karsenti, E.; Leibler, S. Control of microtubule dynamics and length by cyclin A- and cyclin B-dependent kinases in *Xenopus* egg extracts. *J. Cell Biol.* 118:1097–1108, 1992.

Yanagida, M.; Kinoshita, N.; Stone, E.M.; Yamano, H. Protein phosphatases and cell division cycle control. *Ciba Found. Symp.* 170:130–146, 1992.

12. Dasso, M; Newport, J.W. Completion of DNA replication is monitored by a feedback system that controls the initiation of mitosis *in vitro*: studies in *Xenopus. Cell* 61:811–823, 1990.

Kimelman, D.; Kirschner, M.; Scherson, T. The events of the midblastula transition in *Xenopus* are regulated by changes in the cell cycle. *Cell* 48:399–407, 1987.

Smythe, C.; Newport, J.W. Coupling of mitosis to the completion of S phase in *Xenopus* occurs via modulation of the tyrosine kinase that phosphorylates p34^{cdc2}. *Cell* 68:787–797, 1992.

Raff, J.W.; Glover, D. Nuclear and cytoplasmic mitotic cycles continue in *Drosophila* embryos in which DNA synthesis is inhibited with aphidicolin. *J. Cell Biol.* 107:2009–2019, 1988.

13. Blow, J.J.; Laskey, R.A. A role for the nuclear envelope in controlling DNA replication within the cell cycle. *Nature* 332:546–548, 1988.

Coverly, D.; Downes, C.S.; Romanowski, P.; Laskey, R.A. Reversible effects of nuclear membrane permeabilization on DNA replication: evidence for a positive licensing factor. *J. Cell Biol.* 122:985–992, 1993.

Hennessy, K.M.; Lee, A.; Chen, E.; Botstein, D. A group of interacting yeast DNA replication genes. *Genes Dev.* 5:958–969, 1991.

Smith, A.V.; Orr-Weaver, T.L. The regulation of the cell cycle during *Drosophila* embryogenesis: the transition to polyteny. *Development* 112:997–1008, 1991.

14. Forsburg, S.L.; Nurse, P. Cell cycle regulation in the yeasts *Saccharomyces cerevisiae* and *Schizosaccharomyces pombe. Annu. Rev. Cell Biol.* 7:227–256, 1991.

Hartwell, L.H. Twenty-five years of cell cycle genetics. *Genetics* 129:975–980, 1991.

Hartwell, L.H.; Culotti, J.; Pringle, J.R.; Reid, B.J. Genetic control of cell division cycle in yeast. *Science* 183:46–51, 1974.

Pringle, J.R.; Hartwell, L.H. The *Saccharomyces cerevisiae* cell cycle. In The Molecular Biology of the Yeast *Saccharomyces*, Life Cycle and Inheritance (J.N.

Strathern, E.W. Jones, J.R. Broach, eds.), pp. 97–142. Cold Spring Harbor, NY: Cold Spring Harbor Laboratory, 1981.

Watson, J.D.; Hopkins, N.H.; Roberts, J.W.; Steitz, J.A.; Weiner, A.M. Molecular Biology of the Gene, 4th ed., pp. 550–618. Menlo Park, CA: Benjamin-Cummings, 1987.

15. Johnston, G.C.; Pringle, J.R.; Hartwell, L.H. Coordination of growth with cell division in the yeast *Saccharomyces cerevisiae. Exp. Cell Res.* 105:79–98, 1977.

Nurse, P. Genetic control of cell size at cell division in yeast. *Nature* 256:547–551, 1975.

16. Hartwell, L.H.; Culotti, J.; Reid, B. Genetic control of cell division in yeast. I. Detection of mutants. *Proc. Natl. Acad. Sci. USA.* 66:352–359, 1970.

Murray, A.; Hunt, T. The Cell Cycle. Oxford, UK: Oxford University Press, 1993.

Nurse, P.; Thuriaux, P.; Nasmyth, K. Genetic control of the cell division cycle in the fission yeast *Schizosaccharomyces pombe. Mol. Gen. Genet.* 146:167–178, 1976.

Russell, P.; Nurse, P. Negative regulation of mitosis by *wee1+*, a gene encoding a protein kinase homolog. *Cell* 49:559–567, 1987.

17. Dunphy, W.G.; Brizuela, L.; Beach, D.; Newport, J. The *Xenopus* cdc2 protein is a component of MPF, a cytoplasmic regulator of mitosis. *Cell* 54:423–431, 1988.

Gautier, J.; Norbury, C.; Lohka, M; Nurse, P; Maller, J. Purified maturation-promoting factor contains the product of a *Xenopus* homolog of the fission yeast cell cycle control gene *cdc2+. Cell* 54:433–439, 1988.

Booher, R.N.; Alfa, C.E.; Hymans, J.S.; Beach, D.H. The fission yeast cdc2/cdc13/suc1 protein kinase: regulation of catalytic activity and nuclear localization. *Cell* 58:485–497, 1989.

Lee, M.G.; Nurse, P. Complementation used to clone a human homologue of the fission yeast cell cycle control gene *cdc2. Nature* 327:31–35, 1988.

18. Nasmyth, K. Control of the yeast cell cycle by the Cdc28 protein kinase. *Curr. Opin. Cell Biol.* 5:166–179, 1993.

Solomon, M.J. Activation of the various cyclin/cdc2 protein kinases. *Curr. Opin. Cell Biol.* 5:180–186, 1993.

19. Fesquet, D., et al. The MO15 gene encodes the catalytic subunit of a protein kinase that activates cdc2 and other cyclin-dependent kinases (CDKs) through phosphorylation of Thr161 and its homologues. *EMBO J.* 12:3111–3121, 1993.

Gould, K.L.; Nurse, P. Tyrosine phosphorylation of the fission yeast cdc2+ protein kinase regulates entry into mitosis. *Nature* 342:39–45, 1989.

Kumagai, A.; Dunphy, W.G. Regulation of the cdc25 protein during the cell cycle in *Xenopus* extracts. *Cell* 70:139–151, 1992.

20. Edgar, B.A.; O'Farrell, P.H. The three postblastoderm cell cycles of *Drosophila* embryogenesis are regulated in G2 by *string. Cell* 62:469–480, 1990.

O'Farrell, P.H. Cell cycle control: many ways to skin a cat. *Trends Cell Biol.* 2:159–163, 1992.

21. Fankhauser, G. Nucleo-cytoplasmic relations in amphibian development. *Int. Rev. Cytol.* 1:165–193, 1952.

Henery, C.C.; Bard, J.B.L.; Kaufman, M.H. Tetraploidy in mice, embryonic cell number, and the grain of the developmental map. *Dev. Biol.* 152:233–241, 1992.

Prescott, D.M. Changes in nuclear volume and growth rate and prevention of cell division in *Amoeba proteus* resulting from cytoplasmic amputions. *Exp. Cell Res.* 11:94–98, 1956.

22. Beach, D.; Durkacz, B.; Nurse, P. Functionally homologous cell cycle control genes in budding and fission yeast. *Nature* 300:706–709, 1982.

Cross, F. DAF1, a mutant gene affecting size control, pheromone arrest, and cell cycle kinetics of *Saccharomyces cerevisiae. Mol. Cell Biol.* 8:4675–4684, 1988.

Dirick, L.; Nasmyth, K. Positive feedback in the activation of G1 cyclins in yeast. *Nature* 351:754–757, 1991.

Reed, S.I. The role of p34 kinases in the G1 to S-phase transition. *Annu. Rev. Cell Biol.* 8:529–561, 1992.

Richardson, H.E.; Wittenberg, C.; Cross, F.; Reed, S.I. An essential G1 function for cyclin-like proteins in yeast. *Cell* 59:1127–1133, 1989.

23. Peter, M.; Gartner, A.; Horecka, J.; Ammerer, G.; Herskowitz, I. FAR1 links the signal transduction pathway to the cell cycle machinery in yeast. *Cell* 73:747–760, 1993.

24. Roberts, J.M. Turning DNA replication on and off. *Curr. Opin. Cell Biol.* 5:201–206, 1993.

25. Broek, D.; Bartlett, R.; Crawford, K.; Nurse, P. Involvement of p34cdc2 in establishing the dependency of S phase on mitosis. *Nature* 349:388–393, 1991.

Hartwell, L.H.; Weinert, T.A. Checkpoints: controls that ensure the order of cell cycle events. *Science* 246:629–634, 1989.

Murray, A.W. Creative blocks: cell-cycle checkpoints and feedback controls. *Nature* 359:599–604, 1992.

Schlegel, R.; Pardee, A.B. Caffeine-induced uncoupling of mitosis from completion of DNA replication in mammalian cells. *Science* 232:1264–1266, 1986.

26. Enoch, T.; Nurse, P. Mutation of fission yeast cell cycle control genes abolishes dependence of mitosis on DNA replication. *Cell* 60:665–673, 1990.

Kuerbitz, S.J.; Plunkett, W.V.; Walsh, W.V.; Kastan, M.B. Wild-type p53 is a cell cycle checkpoint determinant following irradiation. *Proc. Natl. Acad. Sci. USA* 89:7491–7495, 1992.

Weinert, T.A.; Hartwell, L.H. The RAD9 gene controls the cell cycle response to DNA damage in *Saccharomyces cerevisiae. Science* 241:317–322, 1988.

Weinert, T.A.; Hartwell, L.H. Cell cycle arrest of cdc mutants and specificity of the RAD9 checkpoint. *Genetics* 134:63–80, 1993.

27. Li, R.; Murray, A.W. Feedback control of mitosis in budding yeast. *Cell* 66:519–531, 1991.

28. Elledge, S.J.; Spottswood, M.R. A new human p34 protein kinase, CDK2, identified by complementation of a cdc28 mutation in *Saccharomyces cerevisiae*, is a homolog of *Xenopus* Eg1. *EMBO J.* 10:2653–2659, 1991.

Fang, F.; Newport, J.W. Evidence that the G1-S and G2-M transitions are controlled by different cdc2 proteins in higher eukaryotes. *Cell* 66:731–742, 1991.

Koff, A., et al. Formation and activation of a cyclin E-cdk2 complex during the G1 phase of the human cell cycle. *Science* 257:1689–1694, 1992.

Lew, D.J.; Dulic, V.; Reed, S.I. Isolation of three novel human cyclins by rescue of G1 cyclin (Cln) function in yeast. *Cell* 66:1197–1206, 1991.

Rosenblatt, J.; Gu, Y.; Morgan, D.O. Human cyclin-dependent kinase 2 is activated during the S and G_2 phases of the cell cycle and associates with cyclin A. *Proc. Natl. Acad. Sci. USA* 89:2824–2828, 1992.

Sherr, C.J. Mammalian G1 cyclins. *Cell* 73:1059–1065, 1993.

Solomon, M.J. Activation of the various cyclin/cdc2 protein kinases. *Curr. Opin. Cell Biol.* 5:180–186, 1993.

29. Todaro, G.J.; Green, H. Quantitative studies of the growth of mouse embryo cells in culture and their development into established cell lines. *J. Cell Biol.* 17:299–313, 1963.

30. Cross, M.; Dexter, T.M. Growth factors in development, transformation, and tumorigenesis. *Cell* 64:271–280, 1991.

Heldin, C.-H. Structural and functional studies on platelet-derived growth factor. *EMBO J.* 11:4251–4259, 1992.

Massagué, J. The transforming growth factor-β family. *Annu. Rev. Cell Biol.* 6:597–641, 1990.

Massagué, J.; Pandiella, A. Membrane-anchored growth factors. *Annu. Rev. Biochem.* 62:515–541, 1993.

Ross, R.; Raines, E.W.; Bowen-Pope, D.F. The biology of platelet-derived growth factor. *Cell* 46:155–169, 1986.

Sporn, M.B.; Roberts, A.B., eds. Peptide Growth Factors and Their Receptors. Berlin: Springer-Verlag, 1990.

31. Loughlin, J.H; Fallon, J.H., eds. Neurotrophic Growth Factors. San Diego, CA: Academic Press, 1992.

Purves, D. Body and Brain: A Trophic Theory of Neural Connections. Cambridge: Harvard University Press, 1988.

32. Brooks, R.F. Regulation of the fibroblast cell cycle by serum. *Nature* 260:248–250, 1976.

Goss, R.J. The Physiology of Growth. New York: Academic Press, 1978.

Pardee, A.B. A restriction point for control of normal animal cell proliferation. *Proc. Natl. Acad. Sci. USA* 71:1286–1290, 1974.

Pardee, A.B. G_1 events and regulation of cell proliferation. *Science* 246:603–608, 1989.

Pledger, W.J.; Stiles, C.D.; Antoniades, H.N.; Scher, C.D. An ordered sequence of events is required before BALB/c–3T3 cells become comitted to DNA synthesis. *Proc. Natl. Acad. Sci. USA* 75:2839–2843, 1978.

33. Larsson, O.; Zetterberg, A.; Engstrom, W. Consequences of parental exposure to serum-free medium for progeny cell division. *J. Cell Sci.* 75:259–268, 1985.

Zetterberg, A. Control of mammalian cell proliferation. *Curr. Opin. Cell Biol.* 2:296–300, 1990.

Zetterberg, A.; Larsson, O. Kinetic analysis of regulatory events in G1 leading to proliferation or quiescence of swiss 3T3 cells. *Proc. Natl. Acad. Sci. USA* 82:5365–5369, 1985.

34. Furukawa, Y.; Piwnica-Worms, H.; Ernst, T.J.; Kanakura, Y.; Griffin, J.D. *cdc2* gene expression at the G_1 to S transition in human T lymphocytes. *Science* 250:805–808, 1990.

Lee, M.G.; Norbury, C.J.; Spurr, N.K.; Nurse, P. Regulated expression and phosphorylation of a possible mammalian cell-cycle control protein. *Nature* 333:676–679, 1988.

Matsushime, H.; Roussel, M.F.; Ashmun, R.A.; Sherr, C.J. Colony-stimulating factor 1 regulates novel cyclins during the G1 phase of the cell cycle. *Cell* 65:701–713, 1991.

Ninomiya-Tsuji, J.; Nomoto, S.; Yasuda, H.; Reed, S.I.; Matsumoto, K. Cloning of a human cDNA encoding a CDC2-related kinase by complementation of a budding yeast *cdc28* mutation. *Proc. Natl. Acad. Sci. USA* 88:9006–9010, 1991.

Sherr, C.J. Mammalian G1 cyclins. *Cell* 73:1059–1065, 1993.

35. Dunn, G.A.; Ireland, G.W. New evidence that growth in 3T3 cell cultures is a diffusion-limited process. *Nature* 312:63–65, 1984.

Holley, R.W.; Kiernan, J.A. "Contact inhibition" of cell division in 3T3 cells. *Proc. Natl. Acad. Sci. USA* 60:300–304, 1968.

Stoker, M.G.P. Role of diffusion boundary layer in contact inhibition of growth. *Nature* 246:200–203, 1973.

36. Folkman, J.; Moscona, A. Role of cell shape in growth control. *Nature* 273:345–349, 1978.

Han, E.K.-H.; Guadagno, T.M.; Dalton, S.L.; Assoian, R.K. A cell cycle and mutational analysis of anchorage-independent growth: cell adhesion and TGF-β1 control G1/S transit specifically. *J. Cell Biol.* 122:461–471, 1993.

O'Neill, C.; Jordan, P.; Ireland, G. Evidence for two distinct mechanisms of anchorage stimulation in freshly explanted and 3T3 mouse fibroblasts. *Cell* 44:489–496, 1986.

Symington, B.E. Fibronectin receptor modulates cyclin-dependent kinase activity. *J. Biol. Chem.* 267:25744–25747, 1992.

Turner, C.E.; Burridge, K. Transmembrane molecular assemblies in cell-extracellular matrix interactions. *Curr. Opin. Cell Biol.* 3:849–853, 1991.

37. Bishop, J.M. Molecular themes in oncogenesis. *Cell* 64:235–248, 1991.

Hartwell, L. Defects in a cell cycle checkpoint may be responsible for the genomic instability of cancer cells. *Cell* 71:543–546, 1992.

Varmus, H.; Weinberg, R.A. Genes and the Biology of Cancer. New York: Scientific American Library, 1993.

38. Almendral, J.M., et al. Complexity of the early genetic response to growth factors in mouse fibroblasts. *Mol. Cell Biol.* 8:2140–2148, 1988.

Evan, G.I.; Littlewood, T.D. The role of c-*myc* in cell growth. *Curr. Opin. Genet. Dev.* 3:44–49, 1993.

Naeve, G.S.; Sharma, A.; Lee, A.S. Temporal events regulating the early phases of the mammalian cell cycle. *Curr. Opin. Cell Biol.* 3:261–268, 1991.

Pazin, M.J.; Williams, L.T. Triggering signaling cascades by receptor tyrosine kinases. *Trends Biochem. Sci.* 17:374–378, 1992.

39. Cobrinik, D.; Dowdy, S.F.; Hinds, P.W.; Mittnacht, S.; Weinberg, R.A. The retinoblastoma protein and the regulation of cell cycling. *Trends Biochem. Sci.* 17:312–315, 1992.

Hollingsworth, R.E.; Chen, P.-L.; Lee, W.-H. Integration of cell cycle control with transcriptional regulation by the retinoblastoma protein. *Curr. Opin. Cell Biol.* 5:194–200, 1993.

40. Allsop, R.C., et al. Telomere length predicts replicative capacity of human fibroblasts. *Proc. Natl. Acad. Sci. USA* 89:10114–10118, 1992.

References

Finch, C.E. Longevity, Senescence, and the Genome. Chicago: University of Chicago Press, 1990.

Hayflick, L. The limited *in vitro* lifetime of human diploid cell strains. *Exp. Cell Res.* 37:614–636, 1965.

Lundblad, V.; Szostak, J.W. A mutant with a defect in telomere elongation leads to senescence in yeast. *Cell* 57:633–643, 1989.

Rheinwald, J.G.; Green, H. Epidermal growth factor and the multiplication of cultured human epidermal keratinocytes. *Nature* 265:421–424, 1977.

Smith, J.R.; Whitney, R.G. Intraclonal variation in proliferative potential of human diploid fibroblasts: stochastic mechanism for cellular aging. *Science* 207:82–84, 1980.

41. Glover, D.M. Mitosis in the *Drosophila* embryo—in and out of control. *Trends Genet.* 7:125–132, 1991.

Wood, W.B., ed. The Nematode *Caenorhabditis elegans*. Cold Spring Harbor, NY: Cold Spring Harbor Laboratory, 1988.

The Mechanics of Cell Division

- **An Overview of M Phase**
- **Mitosis**
- **Cytokinesis**

In this chapter we discuss the mechanical events of the **M phase** (or cell-division phase) of the cell cycle. This phase, which includes the various stages of nuclear division (*mitosis*) and cytoplasmic division (*cytokinesis*), is the culmination of the cell cycle. In a comparatively brief period, the contents of the parental cell, which were doubled by the biosynthetic activities of the preceding interphase, are segregated into two daughter cells (Figure 18–1).

At the molecular level M phase is initiated by a cascade of protein phosphorylations triggered by the activation of the mitosis-inducing protein kinase *MPF*, and it is terminated by the dephosphorylations that follow the inactivation of MPF through proteolysis of its cyclin subunits (discussed in Chapter 17). The protein phosphorylations that occur during M phase are responsible for the many morphological changes that accompany mitosis: the chromosomes condense, the nuclear envelope breaks down, the endoplasmic reticulum and Golgi apparatus fragment, the cell loosens its adhesions to other cells and the extracellular matrix, and the cytoskeleton is transformed to bring about the highly organized movements that will segregate the chromosomes and partition the cell. Because M phase involves a complete reorganization of the cell interior, the number of proteins that become phosphorylated is thought to be large, and essentially every part of the cell is affected in some way.

An Overview of M Phase [1]

With minor variations, the processes that occur in M phase to divide one cell into two follow the same sequence in all eucaryotes. They are dominated by cytoskeletal rearrangements as the cells assemble, use, and dismantle the machinery required to pull the duplicated chromosome sets apart and to split the cytoplasm into two halves.

Three Features Are Unique to M Phase: Chromosome Condensation, the Mitotic Spindle, and the Contractile Ring

The first readily visible manifestation of an impending M phase is a progressive compaction of the dispersed interphase chromatin into threadlike chromosomes. This **chromosome condensation** is required for the subsequent organized seg-

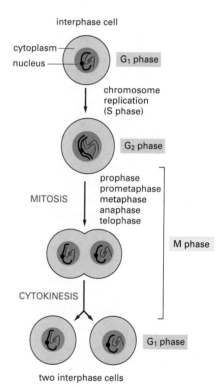

Figure 18–1 The M phase of the cell cycle. M phase starts at the end of G_2 phase and ends at the start of the next G_1 phase. It includes the five stages of nuclear division (mitosis), as well as cytoplasmic division (cytokinesis).

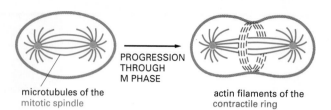

microtubules of the
mitotic spindle

PROGRESSION
THROUGH
M PHASE

actin filaments of the
contractile ring

Figure 18–2 **The cytoskeleton in M phase.** The mitotic spindle assembles first and segregates the chromosomes. The contractile ring assembles later and divides the cell in two.

regation of the chromosomes into daughter cells, and it is accompanied by extensive phosphorylation of histone H1 molecules (up to six phosphates per molecule). Since histone H1 is present at a concentration of about one molecule per nucleosome and is known to be involved in packing nucleosomes together, its phosphorylation by MPF at the onset of the M phase is thought to contribute to chromosome condensation.

Chromosome condensation is the prelude to two distinct mechanical processes: (1) *mitosis*—the segregation of the chromosomes and the formation of two nuclei in place of one—and (2) *cytokinesis*—the splitting of the cell as a whole into two. These processes are carried out by two distinct cytoskeletal structures that appear transiently in M phase. The first to form is a bipolar **mitotic spindle,** composed of microtubules and their associated proteins. The mitotic spindle aligns the replicated chromosomes in a plane that bisects the cell; each chromosome then separates into two daughter chromosomes, which are moved by the spindle to opposite spindle poles. The second cytoskeletal structure required in M phase in animal cells is a **contractile ring** of actin filaments and myosin-II that forms slightly later, just beneath the plasma membrane in a plane perpendicular to the axis of the spindle (Figure 18–2); as the ring contracts it pulls the membrane inward so as to divide the cell in two, thereby ensuring that each daughter cell receives not only one complete set of chromosomes but also half of the cytoplasmic constituents in the parental cell. The two cytoskeletal structures contain different sets of proteins and can be formed independently of each other in some specialized cells. Their formation is usually closely coordinated, however, so that cytoplasmic division (cytokinesis) occurs immediately after the end of nuclear division (mitosis). The same sequence occurs in plant cells, even though their rigid walls necessitate a different mechanism for cytokinesis, as we discuss later.

The description of M phase just given applies only to eucaryotic cells. Bacterial cells do not contain either actin filaments or microtubules. They generally have only one chromosome, whose replicated copies are segregated to daughter cells without special condensation by a mechanism that involves chromosome attachment to the bacterial plasma membrane. The need for complex mitotic machinery probably arose only with the evolution of cells that contained greatly increased amounts of DNA packaged in a number of discrete chromosomes. The primary function of this machinery is to divide the replicated chromosomes precisely between the two daughter cells: in yeast cells, where its accuracy has been determined, it makes an error in chromosome segregation only about once every 10^5 cell divisions.

Cell Division Depends on the Duplication of the Centrosome [2]

As discussed in Chapter 16 the principal *microtubule organizing center (MTOC)* in most animal cells is the **centrosome,** a cloud of poorly defined pericentriolar material (the *centrosome matrix*) associated with a pair of *centrioles* (Figure 18–3). During interphase the centrosome matrix nucleates a cytoplasmic array of microtubules, which project outward toward the cell perimeter with their minus ends attached to the centrosome. Before a eucaryotic cell divides, it must duplicate its centrosome to provide one for each of its two daughter cells. In fact,

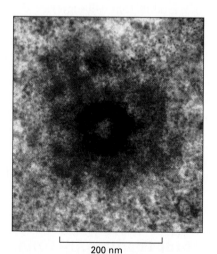

200 nm

Figure 18–3 **A centrosome.** An electron micrograph of a thick section of a centrosome in a mammalian cell in culture. (Courtesy of P. Witt and G. Borisy.)

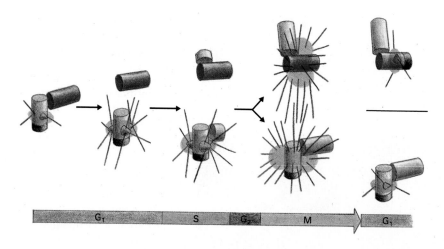

duplicated centrosomes are required to create the two daughter cells, for centrosomes form the two poles of the mitotic spindle.

Centrosomes in most animal species share a remarkable and distinctive structural feature in the form of a pair of **centrioles** (discussed in Chapter 16). The centrioles, however, which are associated with the centrosome matrix, are not required for the nucleation of microtubules: plant centrosomes lack centrioles altogether, and centrioles are also missing during the early divisions of the cleaving mouse egg; moreover, drug treatments of cultured mammalian cells can create tripolar mitotic spindles, in which one spindle pole lacks centrioles yet appears to function normally. It is therefore thought that the centrosome matrix, which contains a set of centrosome-specific proteins, is the most fundamental part of the centrosome. When present, the centrioles associated with the matrix are duplicated in a strictly ordered manner, and their behavior may help in the creation of precisely two centrosomes as the cell enters each M phase (Figure 18–4).

The process of centrosome duplication and separation is known as the *centrosome cycle*. During interphase of each cell cycle, the centrioles and other components of the centrosome are duplicated but remain together as a single complex on one side of the nucleus. As mitosis begins, this complex splits in two and

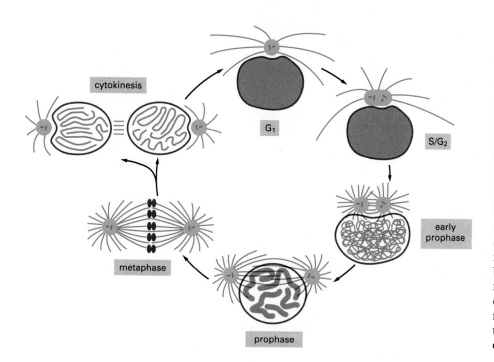

An Overview of M Phase

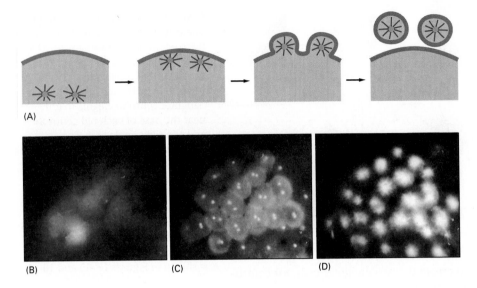

Figure 18–6 Cytoplasmic organization by centrosomes and their asters. If an early *Drosophila* embryo is injected with aphidicolin to block nuclear division, the centrosomes and the asters they nucleate detach from the nuclei in this large syncytial cell and migrate to the cell surface. At the posterior pole, where germ cells would normally form, membrane buds are organized by the detached centrosomes, resulting in the formation of germ "cells" without nuclei. (A) Schematic illustration of the budding process that is caused by the aster. (B–D) Fluorescence micrographs of germ "cells" with no nuclei: (B) is stained for DNA, showing that no nuclei are present; (C) is stained with an antibody to centrosomes; (D) is stained to show the microtubules associated with the centrosomes. (B–D, courtesy of Jordan Raff.)

each centriole pair becomes part of a separate microtubule organizing center that nucleates a radial array of microtubules called an **aster**. The two asters move to opposite sides of the nucleus to form the two poles of the mitotic spindle. As mitosis ends and the nuclear envelope re-forms around the separated chromosomes, each daughter cell receives a centrosome (the former spindle pole) in association with its chromosomes (Figure 18–5).

The centrosome cycle can operate with a surprising degree of independence from other processes of the cell cycle. Thus, if the nucleus is physically removed from a sea urchin egg, or if nuclear DNA replication is blocked by the DNA synthesis inhibitor aphidicolin, cycles of centrosome doubling and division proceed almost normally, giving first two centrosomes, then four, then eight. And in early *Drosophila* embryos similarly treated with aphidicolin, the proliferating centrosomes in the interior of the embryo dissociate from their blocked nuclei and march stepwise through the cytoplasm toward the plasma membrane; once they reach this membrane the centrosomes, through their asters, can reshape the membrane and its underlying cortex, generating cells that contain centrosomes but no nuclei (Figure 18–6). Cleaving eggs, with their stockpiles of cell components that free them from dependence on gene transcription, are admittedly exceptional in their behavior; in other cells the centrosome cycle depends on the presence of a functional cell nucleus. Yet it is clear that the division cycle of the cell as a whole depends on—and is at least in part organized by—the microtubule aster, which is in turn organized by the centrosome.

Centrosomes, and the centrioles usually associated with them, have perplexed and tantalized cell biologists for more than a hundred years. What are they made of, how are they replicated, and how did they originate in the course of evolution? These fundamental questions remain to be answered.

M Phase Is Traditionally Divided into Six Stages

The basic strategy of cell division is remarkably constant among eucaryotic organisms. The first five stages of the M phase constitute mitosis (originally defined as the period in which the chromosomes are visibly condensed); the sixth stage, overlapping with the end of mitosis, is cytokinesis. These six stages form a dynamic sequence, the complexity and beauty of which are hard to appreciate from written descriptions or from a set of static pictures.

The description of cell division is based on observations from two sources: light microscopy of living cells (often combined with microcinematography) and light and electron microscopy of fixed and stained cells. A brief summary of the various stages of cell division is given in Panel 18–1. The five stages of mitosis—

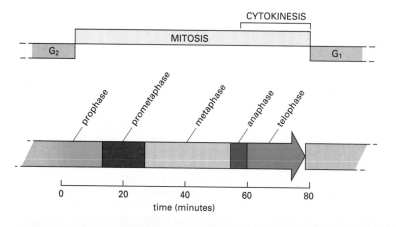

Figure 18–7 A typical time course for mitosis and cytokinesis (M phase) in a mammalian cell. The times vary for different cell types and are much shorter in embryonic cell cycles. Note that cytokinesis begins before mitosis ends. The beginning of prophase (and therefore of M phase as a whole) is defined as the point in the cell cycle at which condensed chromosomes first become visible—a somewhat arbitrary criterion, since the extent of chromosome condensation appears to increase continuously during late G₂. The beginning of prometaphase is defined as the time when the nuclear envelope breaks down.

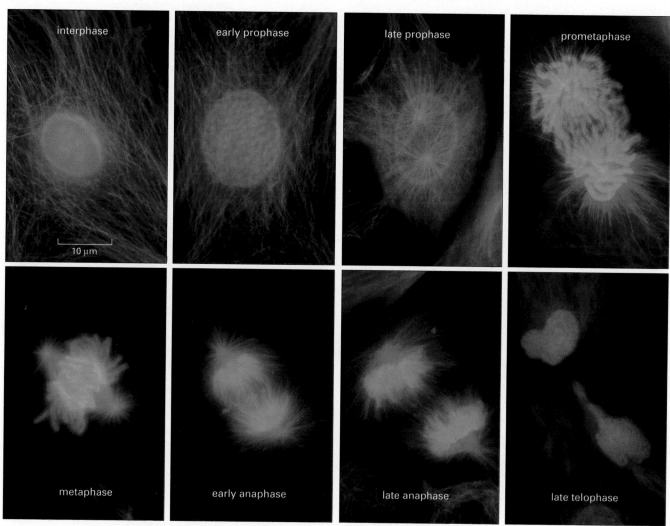

Figure 18–8 The course of mitosis in a typical animal cell. In these micrographs of cultured newt lung cells, the microtubules have been visualized by indirect immunofluorescence, while chromatin is stained with a blue fluorescent dye. During *interphase* the centrosome, consisting of matrix associated with a centriole pair, forms the focus for the interphase microtubule array. By *early prophase* the single centrosome contains two centriole pairs (not visible); at *late prophase* the centrosome divides and the resulting two asters move apart. The nuclear envelope breaks down at *prometaphase*, allowing the spindle microtubules to interact with the chromosomes. At *metaphase* the bipolar spindle structure is clear and all the chromosomes are aligned across the middle of the spindle. The paired daughter chromosomes, called *chromatids*, all separate synchronously at *early anaphase* and, under the influence of the microtubules, begin to move toward the poles. By *late anaphase* the spindle poles have moved farther apart, increasing the separation of the two groups of chromosomes. At *telophase* the daughter nuclei re-form, and by *late telophase* cytokinesis is almost complete, with the midbody persisting between the daughter cells. (Photographs courtesy of C.L. Rieder, J.C. Waters, and R.W. Cole.)

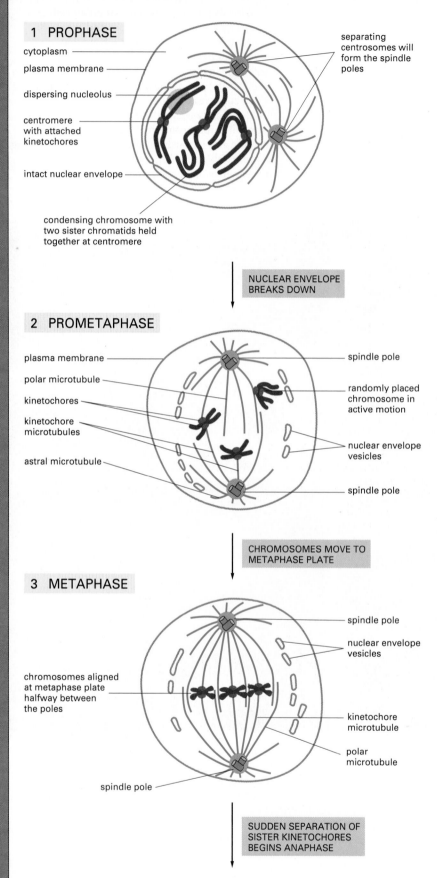

1 PROPHASE

cytoplasm

plasma membrane

dispersing nucleolus

centromere with attached kinetochores

intact nuclear envelope

separating centrosomes will form the spindle poles

condensing chromosome with two sister chromatids held together at centromere

NUCLEAR ENVELOPE BREAKS DOWN

2 PROMETAPHASE

plasma membrane

polar microtubule

kinetochores

kinetochore microtubules

astral microtubule

spindle pole

randomly placed chromosome in active motion

nuclear envelope vesicles

spindle pole

CHROMOSOMES MOVE TO METAPHASE PLATE

3 METAPHASE

chromosomes aligned at metaphase plate halfway between the poles

spindle pole

nuclear envelope vesicles

kinetochore microtubule

polar microtubule

spindle pole

SUDDEN SEPARATION OF SISTER KINETOCHORES BEGINS ANAPHASE

1 PROPHASE

As viewed in the microscope, the transition from the G_2 phase to the M phase of the cell cycle is not a sharply defined event. The chromatin, which is diffuse in interphase, slowly condenses into well-defined chromosomes. Each chromosome has duplicated during the preceding S phase and consists of two sister *chromatids;* each of these contains a specific DNA sequence known as a *centromere,* which is required for proper segregation. Toward the end of prophase, the cytoplasmic microtubules that are part of the interphase cytoskeleton disassemble and the main component of the mitotic apparatus, the *mitotic spindle,* begins to form. This is a bipolar structure composed of microtubules and associated proteins. The spindle initially assembles outside the nucleus between separating centrosomes.

2 PROMETAPHASE

Prometaphase starts abruptly with disruption of the nuclear envelope, which breaks into membrane vesicles that are indistiguishable from bits of endoplasmic reticulum. These vesicles remain visible around the spindle during mitosis. The spindle microtubules, which have been lying outside the nucleus, can now enter the nuclear region. Specialized protein complexes called *kinetochores* mature on each centromere and attach to some of the spindle microtubules, which are then called *kinetochore microtubules.* The remaining microtubules in the spindle are called *polar microtubules,* while those outside the spindle are called *astral microtubules.* The kinetochore microtubules exert tension on the chromosomes, which are thereby thrown into agitated motion.

3 METAPHASE

The kinetochore microtubules eventually align the chromosomes in one plane halfway between the spindle poles. Each chromosome is held in tension at this *metaphase plate* by the paired kinetochores and their associated microtubules, which are attached to opposite poles of the spindle.

4 ANAPHASE

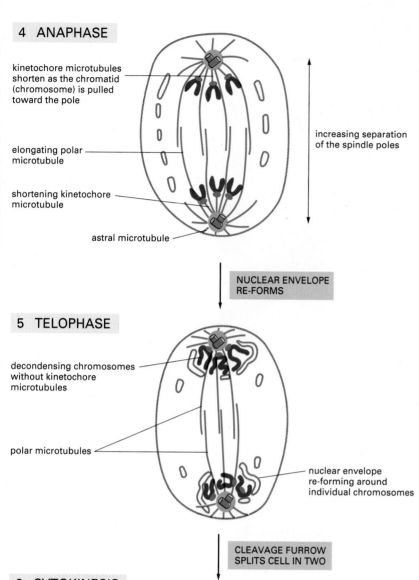

kinetochore microtubules shorten as the chromatid (chromosome) is pulled toward the pole

elongating polar microtubule

shortening kinetochore microtubule

astral microtubule

increasing separation of the spindle poles

NUCLEAR ENVELOPE RE-FORMS

5 TELOPHASE

decondensing chromosomes without kinetochore microtubules

polar microtubules

nuclear envelope re-forming around individual chromosomes

CLEAVAGE FURROW SPLITS CELL IN TWO

6 CYTOKINESIS

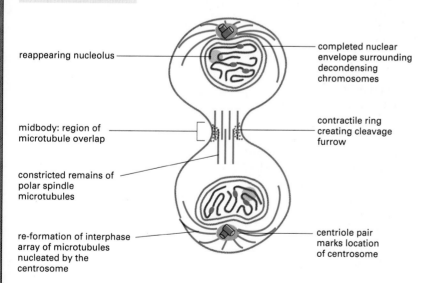

reappearing nucleolus

midbody: region of microtubule overlap

constricted remains of polar spindle microtubules

re-formation of interphase array of microtubules nucleated by the centrosome

completed nuclear envelope surrounding decondensing chromosomes

contractile ring creating cleavage furrow

centriole pair marks location of centrosome

4 ANAPHASE

Triggered by a specific signal, anaphase begins abruptly as the paired kinetochores on each chromosome separate, allowing each chromatid (now called a chromosome) to be pulled slowly toward the spindle pole it faces. All of the newly separated chromosomes move at the same speed, typically about 1 μm per minute. Two catagories of movement can be distinguished. During *anaphase A,* kinetochore microtubules shorten as the chromosomes approach the poles. During *anaphase B,* the polar microtubules elongate and the two poles of the spindle move farther apart. Anaphase typically lasts only a few minutes.

5 TELOPHASE

In telophase (*telos,* end) the separated daughter chromosomes arrive at the poles and the kinetochore microtubules disappear. The polar microtubules elongate still more, and a new nuclear envelope re-forms around each group of daughter chromosomes. The condensed chromatin expands once more, the nucleoli—which had disappeared at prophase—begin to reappear, and mitosis is at an end.

6 CYTOKINESIS

The cytoplasm divides by a process known as *cleavage,* which usually starts sometime during anaphase. The process is illustrated here as it occurs in animal cells. The membrane around the middle of the cell, perpendicular to the spindle axis and between the daughter nuclei, is drawn inward to form a *cleavage furrow,* which gradually deepens until it encounters the narrow remains of the mitotic spindle between the two nuclei. This thin bridge, or *midbody,* may persist for some time before it narrows and finally breaks at each end, leaving two separate daughter cells.

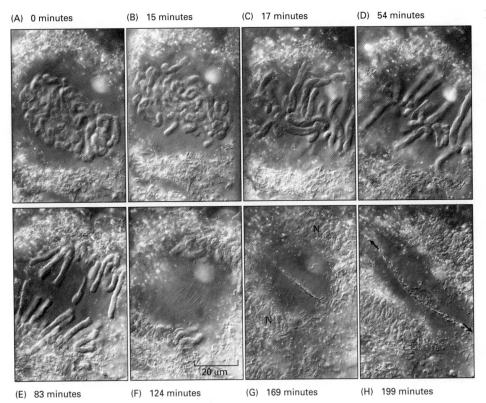

(A) 0 minutes   (B) 15 minutes   (C) 17 minutes   (D) 54 minutes

(E) 83 minutes   (F) 124 minutes   (G) 169 minutes   (H) 199 minutes

20 μm

Figure 18–9 The course of mitosis in a typical plant cell. These micrographs of a living *Haemanthus* (lily) cell were taken at the times indicated using differential-interference-contrast microscopy. The cell has unusually large chromosomes that are easy to see. At the light microscope level shown here, the major events of cell division have been known for more than 100 years. (A) *Prophase:* the chromosomes have condensed and are clearly visible in the cell nucleus. (B) and (C) *Prometaphase:* the nuclear envelope has broken down and the chromosomes are interacting with microtubules that emanate from the two spindle poles. Plants do not have centrioles, but their spindle poles contain proteins related to those found in the centrosomal matrix of animal cells. Note that only 2 minutes have elapsed between the stages shown in (B) and (C). (D) *Metaphase:* the chromosomes have lined up at the metaphase plate with their kinetochores located halfway between the two spindle poles. (E) *Anaphase:* the chromosomes have separated into their two sister chromatids, which are moving to opposite poles. (F) *Telophase:* the chromosomes are decondensing to form the two nuclei that are seen later [marked N in (G)]. (G) and (H) *Cytokinesis:* two successive stages in the formation of the *cell plate* (a new cell wall) are shown; the cell plate appears as a line whose direction of outgrowth is indicated by arrows in (H). (Courtesy of Andrew Bajer.)

prophase, prometaphase, metaphase, anaphase, and *telophase*—occur in strict sequential order, while cytokinesis begins during anaphase and continues through the end of M phase (Figure 18–7). Light micrographs of cell division in a typical animal and a typical plant cell are shown in Figures 18–8 and 18–9, respectively. Innumerable variations in all of the stages of cell division shown schematically in Panel 18–1 occur in the animal and plant kingdoms. We shall mention some of these after we have looked more closely at the general mechanisms of cell division.

Large Cytoplasmic Organelles Are Fragmented During M Phase to Ensure That They Are Faithfully Inherited [3]

The process of cell division must ensure that all the essential classes of cell components are inherited by each daughter cell. As discussed in Chapter 12 organelles like mitochondria and chloroplasts, for example, cannot assemble spontaneously from their individual components; they can arise only from the growth and fission of the corresponding preexisting organelle, and a daughter cell cannot contain any unless it has inherited one or more. Likewise, it may not be possible to make new copies of the Golgi apparatus and endoplasmic reticulum without the prior presence of at least part of the corresponding structure. How are the various membrane-bounded organelles segregated when a higher eucaryotic cell divides? Organelles present in very large numbers will be safely inherited if, on average, their numbers simply double once each cell generation. Other organelles, such as the Golgi apparatus and the endoplasmic reticulum (ER), break up into a set of smaller fragments and vesicles during mitosis, presumably because in this highly vesiculated form they can be more evenly distributed when a cell divides. The ER vesicles seem to associate with microtubules of the mitotic spindle, which may help distribute them evenly between the two daughter cells.

Summary

The process of cell division (M phase of the cell cycle) consists of nuclear division (mitosis) followed by cytoplasmic division (cytokinesis). The nuclear division is mediated by a microtubule-based mitotic spindle, which separates the chromosomes, while the cytoplasmic division is mediated by an actin-filament-based contractile ring. Mitosis is largely organized by the microtubule asters that form around each of the two centrosomes produced when the centrosome duplicates. Centrosome duplication begins during the S and G_2 phases of the cell cycle, and the duplicated centrosomes separate and move to opposite sides of the nucleus at the onset of M phase to form the two poles of the mitotic spindle. Large membrane-bounded organelles, such as the Golgi apparatus and the endoplasmic reticulum, break up into many smaller fragments during M phase, which ensures their even distribution into daughter cells during cytokinesis.

Mitosis [4]

The segregation of the chromosome is brought about by a complex machine with many moving parts. This machine, the mitotic spindle, is built mainly from microtubules, which are used both to push and to pull: pushing drives the poles of the spindle apart, while pulling draws the chromosomes toward those poles.

As we have seen, the spindle starts to form outside the nucleus while the chromosomes are condensing during prophase. When the nuclear envelope breaks down at prometaphase, the microtubules of the spindle are able to capture the chromosomes, which eventually become aligned at its midpoint, forming the so-called *metaphase plate* (see Panel 18–1). At anaphase the sister chromatids abruptly split apart and are drawn to opposite poles of the spindle, while elongation of the spindle increases the separation between the poles. The spindle continues to elongate during telophase as the chromosomes arriving at the poles are released from the spindle microtubules and the nuclear envelope re-forms around them.

The assembly of the spindle, the capture of the chromosomes and their alignment at the metaphase plate, and the subsequent elongation of the spindle with the movement of the chromosomes to the poles are all critically dependent on processes occurring at or near the ends of microtubules: these are not only sites of microtubule assembly and disassembly but are also sites of force production. We shall see that three sets of spindle microtubules can be distinguished: the **polar microtubules,** which overlap at the midline of the spindle and are respon-

Figure 18–10 The three classes of microtubules of the fully formed mitotic spindle. (A) A confocal image of a mitotic spindle at metaphase from a *Drosophila* embryo, with the microtubules fluorescently labeled. (B) A schematic diagram identifying the three classes of microtubules in a spindle. In reality, the chromosomes are much larger than shown, and multiple microtubules are attached to each kinetochore. (A, courtesy of William Theurkauf.)

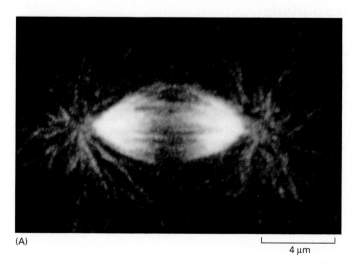

(A)

4 μm

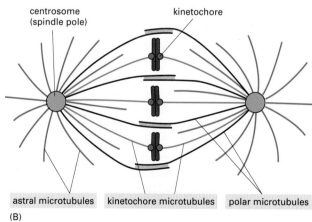

centrosome (spindle pole)   kinetochore

astral microtubules   kinetochore microtubules   polar microtubules

(B)

sible for pushing the poles of the spindle apart; the **kinetochore microtubules,** which attach to the specialized *kinetochore* that forms at the centromere of each duplicated chromosome and maneuver the chromosomes in the spindle; and the **astral microtubules,** which radiate in all directions from the centrosomes and are thought to contribute to the forces that separate the poles and position them in relation to the rest of the cell (Figure 18–10). The behavior of each of these sets of microtubules is different because of the different structures that their ends engage with and the different types of processes that consequently occur there.

Formation of the Mitotic Spindle in an M-Phase Cell Is Accompanied by Striking Changes in the Dynamic Properties of Microtubules [5]

The *interphase microtubule array* radiating from the centrosome is in a state of dynamic instability in which microtubules are continually polymerizing and depolymerizing and fresh microtubules are continually being nucleated to balance the loss of those that disappear completely by depolymerization (discussed in Chapter 16). During late prophase the rates of these processes change dramatically. The half-life of an average microtubule decreases about twentyfold (from about 5 minutes to as little as 15 seconds) (Figure 18–11), and this is thought to reflect a steep increase in the probability that a typical growing microtubule will convert to a shrinking one by a change at its plus end (see p. 810). At the same time there is a large increase in the number of microtubules radiating from the centrosome, apparently because of an alteration in the centrosome itself, boosting the rate at which fresh microtubules are nucleated there: prophase centrosomes can be shown *in vitro* to have a greatly increased capacity for nucleating the growth of microtubules. These two changes are sufficient to explain why the onset of M phase is characterized by a rapid transition from relatively few long microtubules that extend from the centrosome to the cell periphery (the interphase microtubule array) to large numbers of short microtubules that surround each centrosome (the astral microtubules). In some cells a third mechanism may operate in addition: the breakdown of the interphase microtubules may be accelerated by proteins that sever them. The molecular basis for these changes in microtubule dynamics is uncertain, but it seems likely to involve the phosphorylation by MPF of one or more proteins that interact with microtubules.

The rapidly growing and shrinking microtubules emanate in all directions from the prophase centrosomes. Subsets of these astral microtubules become selectively stabilized in different ways to form an organized mitotic spindle.

Interactions Between Oppositely Oriented Microtubules Drive Spindle Assembly [6]

During prophase, while the nuclear envelope is still intact, some of the elongating microtubules that radiate from each centrosome appear to engage with microtubules of opposite polarity growing from the other centrosome (Figure 18–12). In the region of overlap, midway between the two centrosomes, cross-links are thought to form, stabilizing the ends of these *polar microtubules* against disassembly and binding the two oppositely oriented sets together to form the basic framework of the characteristic bipolar spindle. The proteins that cross-link the polar microtubules are thought to be plus-end-directed microtubule motor molecules that not only serve to stabilize the bipolar spindle but act to push the antiparallel, overlapping microtubules past each other in a direction that tends to force the two poles apart. Thus, although the polar microtubules are stabilized against catastrophic disassembly, they are not static; indeed, we shall see later that, even in the metaphase spindle (Figure 18–13), which gives the appearance of stability, the polar microtubules continuously undergo net addition of monomers at their plus ends and net loss at their minus ends at the poles.

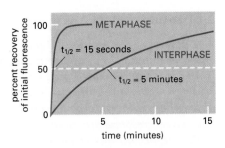

Figure 18–11 Microtubules in an M-phase cell are much more dynamic, on average, than the microtubules at interphase. Mammalian cells in culture were injected with tubulin that had been covalently linked to a fluorescent dye. After the fluorescent tubulin had become incorporated into the cell's microtubules, all of the fluorescence in a small region was bleached by an intense laser beam. The recovery of fluorescence in the bleached region of microtubules, caused by their replacement by microtubules formed from unbleached fluorescent tubulin from the soluble pool, was then monitored as a function of time. The time for 50% recovery of fluorescence ($t_{1/2}$) is thought to be equal to the time required for half of the microtubules in the region to depolymerize and reform. (Data from W.M. Saxton et al., *J. Cell Biol.* 99:2175–2187, 1984, by copyright permission of the Rockefeller University Press.)

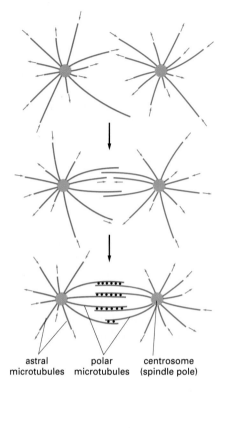

Figure 18–12 A model for how a bipolar mitotic spindle may form by the selective stabilization of interacting microtubules. New microtubules grow out in random directions from two nearby centrosomes. The microtubules are anchored to the centrosome by their minus ends. Their plus ends are "dynamically unstable" and switch suddenly from uniform growth (*outward-pointing red arrows*) to rapid shrinkage (*inward-pointing red arrows*), during which the entire microtubule often depolymerizes. When two microtubules from opposite centrosomes interact in an overlap zone, microtubule-associated proteins are thought to cross-link the microtubules together (*black beads*) in a way that caps their plus ends, stabilizing them by decreasing their probability of depolymerizing. There is evidence that the cross-linking proteins are plus-end-directed microtubule motor molecules that tend to drive the microtubules in the directions that push the poles of the spindle apart.

astral microtubules   polar microtubules   centrosome (spindle pole)

The polar microtubules presumably serve to separate the centrosomes as the spindle assembles during prophase. Later they will drive the elongation of the spindle at anaphase. In the meantime, however, during the middle stages of mitosis elongation of the spindle is held in check; while the polar microtubules exert a push, holding the spindle poles apart, the kinetochore microtubules, attaching the spindle poles to the chromosomes, exert a counterbalancing pull, as we shall now see.

Replicated Chromosomes Attach to Microtubules by Their Kinetochores [7]

At the start of M phase, each chromosome has replicated and consists of two *sister chromatids* joined together along their length, with a constriction in a unique region called the **centromere,** where the chromatin seems to be especially condensed (Figure 18–14). During late prophase specialized protein complexes known as **kinetochores** assemble on each centromere, with the two kinetochores on each chromosome (one on each sister chromatid) facing in opposite directions. Through their kinetochores the replicated chromosomes will bind to a distinct subset of the microtubules of the mitotic spindle (Figure 18–15). These **kinetochore microtubules** will serve eventually, at anaphase, to pull the two sister chromatids toward opposite poles of the spindle. But before that stage, while the sister chromatids are still linked, there is already a tension generated in the kinetochore microtubules, tugging outward on the chromosomes and inward on the spindle poles.

In most organisms the kinetochore is a large multiprotein complex that can be seen in the electron microscope as a platelike multilaminar structure (Figure

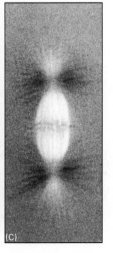

(A) 10 μm (B) (C)

Figure 18–13 A mitotic spindle. An isolated metaphase spindle is shown viewed by three types of light microscopy: (A) differential-interference-contrast microscopy, (B) phase-contrast microscopy, and (C) polarized-light microscopy. (Courtesy of E.D. Salmon and R.R. Segall, from *J. Cell Biol.* 86:355–365, 1980, by copyright permission of the Rockefeller University Press.)

Figure 18–14 The centromere.
Scanning electron micrograph of a human mitotic chromosome, consisting of two sister chromatids joined along their length. The constricted region is the centromere. (Courtesy of Terry D. Allen.)

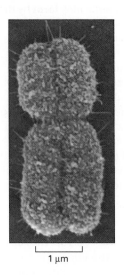

1 μm

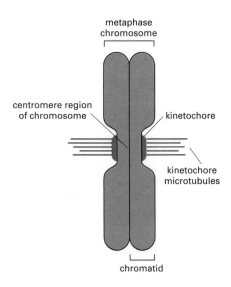

metaphase chromosome

centromere region of chromosome

kinetochore

kinetochore microtubules

chromatid

18–16). We shall see that it is not just a passive anchorage site for microtubules: it plays a central role in controlling the assembly and disassembly of the kinetochore microtubules, generating tension in them and, ultimately, driving chromosome movement.

Kinetochore Protein Complexes Assemble on Specific Centromeric DNA Sequences in Yeast Chromosomes [8]

The information that specifies the construction of a kinetochore at a specific site on a chromosome is contained in the DNA sequence at the centromere. In budding yeasts centromeric DNA can be readily identified genetically by its effect on the propagation of plasmids that contain it: a plasmid including centromeric DNA can be stably inherited in mitosis and meiosis in the same way as a chromosome, whereas a plasmid that lacks this DNA (and is present in low numbers) will segregate irregularly to daughter cells and be lost within a few cell divisions. (We discuss in Chapter 7 how this property of centromeres has been exploited in the construction of vectors for cloned DNA.) Molecular genetic experiments show that each of the 17 chromosomes of the yeast *Saccharomyces cerevisiae* contains a different centromeric sequence about 110 base pairs long (Figure 18–17). Although each sequence is unique, they all contain substantial regions of homology and can be inverted or swapped from chromosome to chromosome without loss of function.

Some of the proteins that form the yeast kinetochore have been identified by their ability to bind to centromeric DNA. Element III of the yeast centromere

Figure 18–15 Kinetochore microtubules. Schematic drawing of a metaphase chromosome showing its two sister chromatids attached to kinetochore microtubules. Each kinetochore forms a plaque on the surface of the centromere. The plus ends of the microtubules bind to the kinetochores (not shown).

Figure 18–16 The kinetochore. (A) A metaphase chromosome stained with a DNA-binding fluorescent dye. (B) A metaphase chromosome stained with human autoantibodies that react with specific kinetochore proteins, showing two kinetochores, one associated with each chromatid. (C) Electron micrograph of an anaphase chromatid with microtubules attached to its kinetochore. While most kinetochores have a trilaminar structure, the one shown (from a green alga) has an unusually complex structure with additional layers. (A and B, courtesy of Bill Brinkley; C, from J.D. Pickett-Heaps and L.C. Fowke, *Aust. J. Biol. Sci.* 23:71–92, 1970, reproduced by permission of CSIRO.)

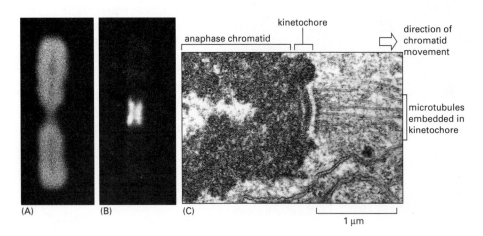

anaphase chromatid

kinetochore

direction of chromatid movement

microtubules embedded in kinetochore

(A) (B) (C)

1 μm

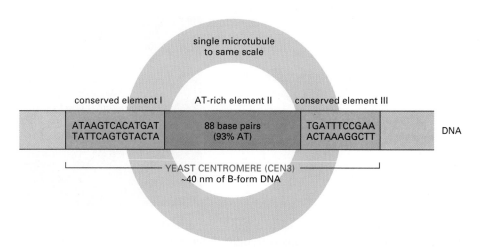

single microtubule
to same scale

conserved element I   AT-rich element II   conserved element III

| ATAAGTCACATGAT | 88 base pairs | TGATTTCCGAA | |
| TATTCAGTGTACTA | (93% AT) | ACTAAAGGCTT | DNA |

YEAST CENTROMERE (CEN3)
~40 nm of B-form DNA

Figure 18–17 The DNA sequence of a typical centromere in the yeast *Saccharomyces cerevisiae*. The DNA sequence shown is both necessary and sufficient to cause faithful chromosome segregation. It serves to assemble the kinetochore proteins that attract a single microtubule (shown in *green*) to the kinetochore.

(see Figure 18–17), for example, binds a complex of three proteins. This protein complex is thought to initiate formation of a simple kinetochore that latches onto the end of a single microtubule. One of the purified centromere-binding proteins is a minus-end-directed microtubule motor, which seems sufficient to propel the yeast kinetochore along its attached microtubule toward the mitotic pole. Other proteins bind to other sequences of the centromere and stabilize its interaction with the microtubule.

Mammalian centromeres consist of different and much longer DNA sequences, many of them repetitive, and they form much larger kinetochores, which bind 30 or 40 microtubules instead of one. It is thought that much of the repetitive sequence (called satellite DNA) in mammalian centromeres is required for the special chromatin organization at the centromere. It seems likely, however, that there are DNA sequences embedded in this structural DNA that bind protein complexes similar to those bound to the yeast centromere.

An unexpected opportunity to study the proteins of mammalian kinetochores has come from the finding that human patients suffering from certain types of *scleroderma* (a disease of unknown cause that is associated with a progressive fibrosis of connective tissue in skin and other organs) produce autoantibodies that react specifically with kinetochores. When these antibodies are used to stain dividing cells by immunofluorescence, a pattern of fluorescent spots is seen, with each spot marking the position of a kinetochore. Unexpectedly, a similar pattern is obtained if nondividing cells are stained, with the number of spots per cell corresponding to the number of chromosomes, suggesting that a kinetochore precursor is attached to each centromere even in interphase nuclei (Figure 18–18). Scleroderma autoantibodies have also made it possible to clone the genes that encode several of the many proteins associated with mammalian kineto-

Figure 18–18 Immunofluorescence staining of kinetochores in cultured cells. The marsupial cells (PtK cells) shown in (A) and (B) contain relatively few chromosomes and are in G₁ phase in (A), where there is one kinetochore stained per chromosome, and in G₂ phase in (B), where there are two kinetochores stained per chromosome. The same auto-antibodies were used to label the kinetochores as were used in Figure 18–16B. A different type of cell, which is in metaphase, is shown in (C); the kinetochores are stained with antibody against a kinetochore protein, and the DNA is stained with a *red* dye. (Courtesy of B.R. Brinkley and D. He.)

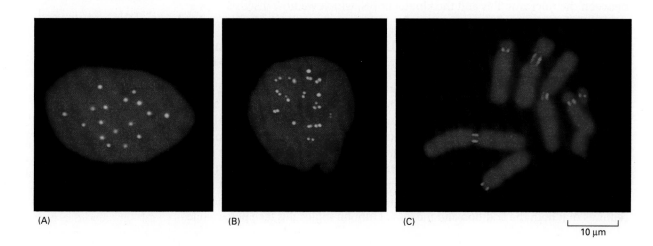

(A)   (B)   (C)

10 μm

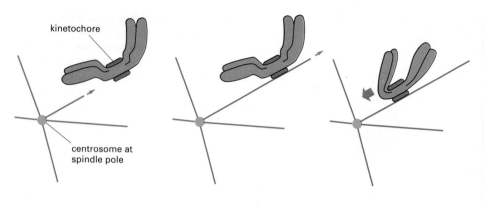

kinetochore

centrosome at
spindle pole

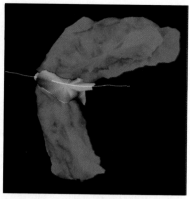

Figure 18–19 Capture of kinetochores by microtubules. The kinetochore binds to the side of a growing microtubule and slides along it toward the spindle pole. On the left, the *red arrow* indicates the direction of microtubule growth, while the *gray arrow* indicates the direction of chromosome sliding. On the right is shown a computer-aided three-dimensional reconstruction, from several thin-sections, of a prometaphase chromosome from a newt. This chromosome (*blue*) had just started to move toward the spindle pole after its kinetochore (*orange*) had attached to a single mictotubule (*white*). (From C.L. Rieder and S.P. Alexander, *J. Cell Biol.* 110:81–95, 1990, by copyright permission of the Rockefeller University Press.)

chores, so that these normally rare proteins can now be produced in large quantities by recombinant DNA technologies. In this way their interactions with one another, with DNA, and with microtubules are being characterized.

Kinetochores Capture Microtubules Nucleated by the Spindle Poles [9]

In prophase, as the polar microtubules form and the spindle lengthens, many microtubules are still growing and shrinking rapidly from the centrosome at each pole, with their plus ends constantly probing at random through the cell. Prophase ends and **prometaphase** begins when the rapid phosphorylation of the nuclear lamina triggers the breakdown of the nuclear envelope and the microtubules gain access to the chromosomes. By a process that has been likened to a fisherman casting a line, a randomly probing microtubule every so often passes close enough to a kinetochore to capture the chromosome. In newt cells, where the initial capture event can be visualized, the kinetochore can be seen first to bind to the side of the microtubule and then to slide rapidly along it toward the spindle pole (Figure 18–19). The lateral attachment to the microtubule and poleward sliding movement probably reflect the action of dynein (which is discussed in Chapter 16), a minus-end-directed microtubule motor protein, which can be shown by staining with antidynein antibodies to be present in kinetochores.

Each kinetochore quickly seizes hold of more microtubules as it is dragged toward the spindle pole, where the microtubule density is high. During this process the initial side-on interactions with microtubules are converted into end-on interactions, with the kinetochore bound to the plus end of each microtubule. Eventually, the opposite kinetochore, on the other sister chromatid, will be captured by the free end of a microtubule from the opposite spindle pole, positioning the chromatids for segregation and establishing tension in the kinetochore microtubule system. This tension is due to the special properties of the linkage between the microtubules and the kinetochores, which we now discuss.

The Plus Ends of Kinetochore Microtubules Can Add and Lose Tubulin Subunits While Attached to the Kinetochore [10]

Just as the polar microtubules continue to add subunits at one end and lose them at the other after they have established their cross-linking connections at the spindle equator, so the kinetochore microtubules are also in a state of dynamic flux. If chemically marked tubulin is injected into mitotic cells during metaphase, it is found to be incorporated continually into these microtubules near their point of attachment to the kinetochore (Figure 18–20). We shall see shortly that during anaphase the reverse reaction occurs, and tubulin molecules are lost from these microtubules at the kinetochore as it moves toward a spindle pole. This requires that both the addition and the loss of tubulin molecules take place while

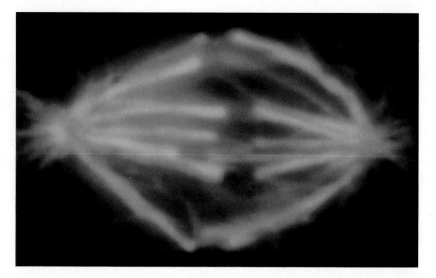

Figure 18–20 **Experiment demonstrating that metaphase kinetochore molecules add subunits at their kinetochore-attached (plus) end.** In this experiment fluoroscein-labeled tubulin was injected into a living cell, where it became incorporated into the spindle microtubules (*green*). Rhodamine-labeled tubulin was then introduced into the cell at metaphase, where it became incorporated into kinetochore microtubules at the kinetochore-attached (plus) end (*orange*). Some incorporation is also seen at the centrosome. (Courtesy of G. Borisy.)

the kinetochore maintains a firm mechanical attachment to the microtubules. This attachment remains under tension throughout, whether tubulin molecules are being added or removed. The kinetochore thus seems to act like a sliding collar, maintaining a lateral association with polymerized tubulin subunits near the end of the microtubule while allowing the addition or loss of tubulin molecules to occur at that end (see Figure 18–28, below). When tubulin molecules are added the collar remains at the end of the microtubule by slipping back; when tubulin molecules are lost the collar remains at the end by traveling forward.

Spindle Poles Repel Chromosomes [11]

While the kinetochore microtubules tend to pull the chromosomes poleward, another, more mysterious force acts in an opposite direction, repelling any large object that approaches the poles too closely. This can be seen if the arms of a chromosome are cut free from the kinetochore by laser microsurgery. While the kinetochore is drawn to the nearest pole, the arms tend to move away from it (Figure 18–21). The origin of this "astral exclusion force," or "polar wind," is not known. It may be the result of pushing by the growing ends of the free microtubules that are nucleated continually at the pole; alternatively, the freed chromosome arms may bind plus-end-directed microtubule motors and migrate along such microtubules; or some other cell components may be moving in that way and sweeping the chromosome arms along with them. The astral exclusion force has been proposed to play an important part in aligning chromosomes at the spindle equator in metaphase, as discussed below.

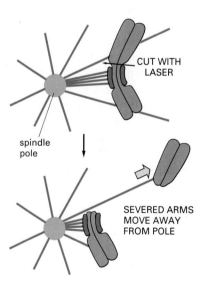

Sister Chromatids Attach by Their Kinetochores to Opposite Spindle Poles [12]

The eventual outcome of the interaction between kinetochores and microtubules is the faithful segregation of the two sister chromatids of each chromosomal pair to opposite sides of the cell. This requires that sister kinetochores bind microtubules coming from opposite poles; if sister kinetochores both moved to the same pole, the resulting cell division would produce two defective cells, one missing a chromosome and one with an extra copy. Much of the complexity of mitosis reflects this requirement for accurate segregation, and the most common mitotic error is for both sister chromatids to end up in one of the two daughter cells. This error can occur if a replicated chromosome fails to attach to the spindle, if it attaches to only one half of the spindle, or if the two sister chromatids fail to separate at anaphase.

Figure 18–21 **Demonstration of the astral exclusion force.** In this experiment prometaphase chromosomes that are temporarily attached to a single pole by kinetochore microtubules are cut in half with a laser beam. The half that is freed from the kinetochore is pushed rapidly away from the pole, whereas the half that remains attached to the kinetochore moves toward the pole, reflecting a decreased repulsion.

Since sister kinetochores face in opposite directions, they will naturally tend to attach to opposite poles, but mistakes do occur. In early prometaphase the two sister kinetochores on a chromosome can become attached to the same spindle pole. This and other incorrect configurations, which would cause a failure of the chromosome to segregate properly if they persisted, are almost always corrected, however. It seems that a balanced arrangement, in which each sister kinetochore is attached to a different spindle pole and only to that spindle pole, has the greatest stability. Why this might be is suggested by an experiment designed to test how chromosomes are attached to the spindle.

Extremely fine glass needles are used to poke and pull at chromosomes inside a living cell, in this case a grasshopper spermatocyte in meiosis, whose unusually flexible plasma membrane tolerates the manipulation. By micromanipulating chromosomes during prometaphase, it is possible to force the two kinetochores on a single chromosome to engage with the same spindle pole. This arrangement is normally unstable, but it can be made stable and will persist until anaphase if the needle is used to keep the kinetochore microtubules on the incorrectly associated chromosome under tension by gently tugging on the chromosome in the direction away from the pole to which both its kinetochores are attached. It therefore appears that kinetochore microtubules are stabilized by tension and that, normally, only chromosomes attached to *both* poles will be stably attached to the spindle.

Balanced Bipolar Forces Hold Chromosomes on the Metaphase Plate [13]

During prometaphase the chromosomes can be seen to move about as if jerked first this way and then that before they align at an equal distance from the two spindle poles, thereby forming the *metaphase plate* and defining the beginning of **metaphase.** But what brings them to this precise position? If acted on simply by equal and opposite constant pulls toward the two poles of the spindle, the chromosomes could be held in equilibrium at any position along the spindle axis. There has to be some change in the forces acting according to the distances from the spindle poles such that any chromosome displaced away from the equatorial plane experiences a net force to bring it back there.

One hypothesis proposes that the pull exerted by a kinetochore microtubule increases as the microtubule gets longer; the net force on a chromosome then will be toward whichever pole is the more distant, with zero net force only when the chromosome lies equidistant from both poles (Figure 18–22A). A plausible mechanism for generating a force proportional to microtubule length would be to have plus-end-directed microtubule motors immobilized in a matrix associ-

Figure 18–22 Two hypotheses for how chromosomes line up at the metaphase plate. In both cases the chromosomes enter the spindle randomly during prometaphase. In (A) the chromosomes eventually line up at the equator (the metaphase plate) because a *pulling* force on each kinetochore increases as it gets farther from a pole. In (B) the chromosomes end up at the equator because the astral exclusion forces *push* them there. In both cases the chromosomes are held under tension at the equator by balanced forces.

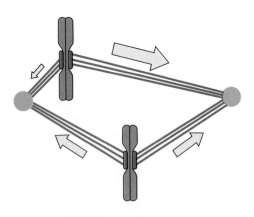

(A) "PULL" pulling force proportional to length of kinetochore microtubules

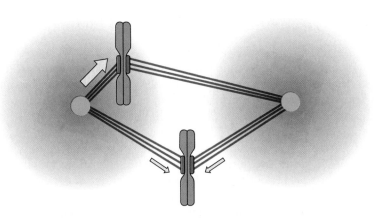

(B) "PUSH" astral exclusion force decreases with distance from pole

ated with the spindle: the longer the kinetochore microtubule, the more of these motors would be engaged, propelling the microtubule back toward its spindle pole. The position-dependent component of the force on the chromosome would be supplementary to the force developed by the activity of the motor proteins at the kinetochore itself.

Another hypothesis (Figure 18–22B) proposes that the astral exclusion force is responsible for centering each chromosome on the spindle by pushing the chromosome more strongly away from the spindle pole the more closely it approaches (as would be expected, for example, if the force is due to free microtubules growing out from the pole). In this way too, a balance would be achieved only when the chromosome is equidistant from both poles. The two hypotheses are not mutually exclusive, and it is possible that both mechanisms, and/or some other mechanism, may operate in some spindles.

The forces that bring the chromosomes to the metaphase plate continue to operate throughout metaphase, and the chromosomes can be seen to oscillate back and forth, continually adjusting their positions. If one of a pair of kinetochore attachments is severed with a laser beam during metaphase, the entire chromosome immediately moves toward the one pole to which it remains attached. Similarly, if the attachment between the two chromatids is severed, the two chromatids separate and move toward opposite poles, just as they do in anaphase. Thus the forces that will pull the chromatids poleward in anaphase are not suddenly switched on at the metaphase-anaphase transition: they are present as soon as microtubules attach to the kinetochores.

Microtubules Are Dynamic in the Metaphase Spindle [14]

The metaphase spindle is a complex and beautiful assembly suspended in a state of dynamic equilibrium and is already tensed for the actions that will begin at anaphase. All of the spindle microtubules, despite their appearance of stability, are continually exchanging subunits with the pool of soluble free tubulin: treatment with the drug colchicine, which blocks tubulin assembly into microtubules, causes the whole spindle to disassemble rapidly. The dynamic behavior of kinetochore and polar microtubules has been studied directly by allowing the microtubules to incorporate tubulin that has been covalently coupled to photoactivatable "caged" fluorescein (see p. 184). When such metaphase spindle microtubules are marked with a bar of fluorescence (by illumination with a laser beam), the marks move continuously poleward (Figure 18–23). This experiment shows that in both the kinetochore and polar microtubules the

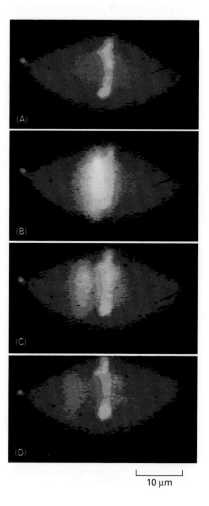

Figure 18–23 **The dynamic behavior of microtubules in the metaphase spindle studied by photoactivation of fluorescence.** A metaphase spindle formed *in vitro* from an extract of *Xenopus* eggs has incorporated three fluorescent markers: rhodamine-labeled tubulin (*red*) to mark all of the microtubules, a *blue* DNA-binding dye that labels the chromosomes, and caged-fluorescein-labeled tubulin, which is also incorporated into all of the microtubules but is invisible because it is nonfluorescent until activated by ultraviolet light. The distribution of the chromosomes and microtubules in the spindle is shown in (A). In (B) a beam of ultraviolet light is used to uncage the caged-fluorescein-labeled tubulin locally, mainly just to the left side of the metaphase plate. Over the next few minutes (after 1½ minutes in C, after 2½ minutes in D) the uncaged fluorescein-tubulin signal is seen to move toward the left spindle pole, indicating that tubulin is continuously moving poleward even though the spindle (visualized by the *red* rhodamine-tubulin fluorescence) remains largely unchanged. The caged fluorescein signal also diminishes in intensity, indicating that the individual microtubules are continually depolymerizing and being replaced. (From K.E. Sawin and T.J. Mitchison, *J. Cell Biol.* 112:941–954, 1991, by copyright permission of the Rockefeller University Press.)

10 μm

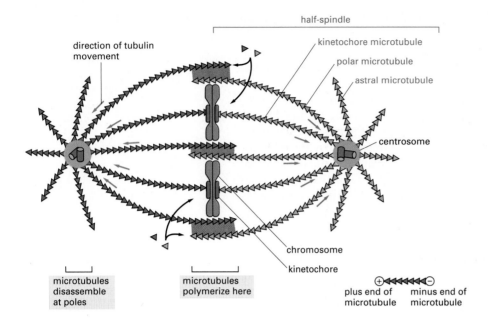

half-spindle

direction of tubulin
movement

kinetochore microtubule

polar microtubule

astral microtubule

centrosome

chromosome

kinetochore

microtubules
disassemble
at poles

microtubules
polymerize here

$\oplus$◄◄◄◄◄$\ominus$

plus end of
microtubule

minus end of
microtubule

Figure 18–24 Simplified diagram of the mitotic spindle at metaphase. The spindle is constructed from two half-spindles (*green* and *orange*), each composed of kinetochore, polar, and astral microtubules. The polarity of the microtubules is indicated by the arrowheads, which point toward the plus end. The polar microtubules emanating from opposite spindle poles have a region of overlap (*shaded gray*), where microtubule-associated proteins may cross-link them. Note that the microtubules are antiparallel in the overlap zone.

tubulin subunits are moving continually toward the poles—a process referred to as *treadmilling* (Figure 18–24, and see p. 823). The marks get dimmer with time, indicating that many individual microtubules depolymerize completely and are replaced. Since the polar and kinetochore microtubules of the metaphase spindle stay at a constant length, there must be an exact balance between the addition of tubulin subunits at the plus ends at the spindle equator and the removal of tubulin subunits at the minus ends attached to the poles (Figure 18–25A).

The continuous treadmilling of tubulin subunits toward the poles of the metaphase spindle has profound implications for how the spindle is constructed. It is thought that many of the bonds between the polar microtubules in the spindle, and between microtubules and other structures such as kinetochores and centrosomes, are made by microtubule motor proteins. Because motor proteins continually associate and dissociate from the microtubule as they go through their catalytic cycle, they are ideally designed to hold onto a microtubule whose subunits are moving. They can sustain a tension even when the microtubule is lengthening by addition of subunits, and they can sustain a compression even when the microtubule is shortening by subunit loss.

Figure 18–25 The behavior of kinetochore microtubules at metaphase and anaphase. (A) During metaphase, subunits are added to the plus end of a microtubule at the kinetochore and are removed from the minus end at the spindle pole. Thus a constant poleward flux of tubulin subunits occurs, while the microtubules maintain a constant length and remain under tension. (B) At anaphase the chromatid is released from attachment to its sister at the metaphase plate and the kinetochore moves rapidly up the microtubule, removing subunits from its plus end as it goes. Its attached chromatid is thereby carried to the spindle pole. Part of the chromatid movement is due to the simultaneous loss of tubulin subunits from the minus end of the microtubules at the pole.

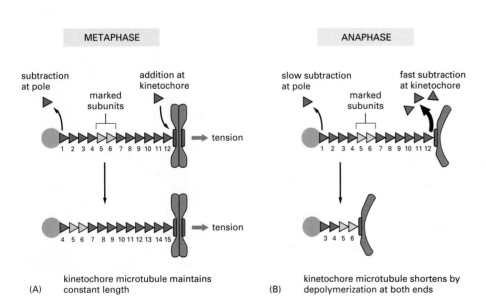

METAPHASE

subtraction
at pole

addition at
kinetochore

marked
subunits

tension

1 2 3 4 5 6 7 8 9 10 11 12

tension

4 5 6 7 8 9 10 11 12 13 14 15

(A) kinetochore microtubule maintains
constant length

ANAPHASE

slow subtraction
at pole

fast subtraction
at kinetochore

marked
subunits

1 2 3 4 5 6 7 8 9 10 11 12

3 4 5 6

(B) kinetochore microtubule shortens by
depolymerization at both ends

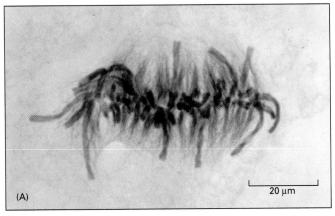 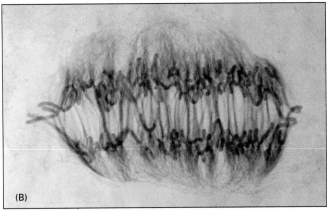

20 μm

(A)   (B)

Sister Chromatids Separate Suddenly at Anaphase [15]

As discussed in Chapter 17, **anaphase** is triggered by the degradation of cyclin and the consequent inactivation of MPF. This leads abruptly to the synchronous splitting of each chromosome into its sister chromatids, each with one kinetochore. The separation of the sister chromatids frees them from the bond that holds them at the metaphase plate, and the kinetochores can begin to move toward the spindle pole to which they are attached by their kinetochore microtubules (Figure 18–25B). It is thought that the chromatids are held together along their length by chromosomal proteins that are suddenly altered in some way at the start of anaphase, allowing the chromatids to separate and begin their journey poleward (Figure 18–26).

Anaphase Is Delayed Until All Chromosomes Are Positioned at the Metaphase Plate [16]

Metaphase occupies a substantial portion of the mitotic period (see Figure 18–7), partly because cells pause here until all of their chromosomes are lined up appropriately at the metaphase plate. Cells arrest in metaphase for hours or days, for example, if the mitotic spindle is disassembled by treatment with colchicine. (This method of cell-cycle arrest is commonly used to collect large numbers of cells in mitosis so that their condensed chromosomes can be analyzed.) The normal inactivation of MPF that signals the metaphase-to-anaphase transition is blocked so long as the spindle is disassembled in this way. It is thought that any kinetochore that is not attached to microtubules generates a diffusible signal that somehow stabilizes MPF and that this normally guarantees that all of the chromosomes are aligned before progression into anaphase is allowed. Thus, disrupting the spindle with drugs would be expected to produce a strong signal that greatly prolongs metaphase.

Two Distinct Processes Separate Sister Chromatids at Anaphase [17]

Once each chromosome has split in response to the anaphase trigger, its two chromatids move to opposite spindle poles, where they will assemble into the nucleus of a new cell. Their movement is the consequence of two independent processes mediated by the spindle. The first, referred to as **anaphase A,** consists of the poleward movement of chromatids, accompanied by the shortening of the kinetochore microtubules (see Figure 18–25B). The second, referred to as **anaphase B,** consists of the separation of the poles themselves, accompanied by the elongation of the polar microtubules (Figure 18–27). Historically, these two processes were distinguished by their differential sensitivities to drugs, and it is now clear that they occur by different mechanisms.

Figure 18–26 Chromatid separation at anaphase. In the transition from metaphase (A) to anaphase (B), sister chromatids are pulled apart by spindle microtubules—as seen in these *Haemanthus* (lily) endosperm cells stained with gold-labeled antibodies to tubulin. (Courtesy of Andrew Bajer.)

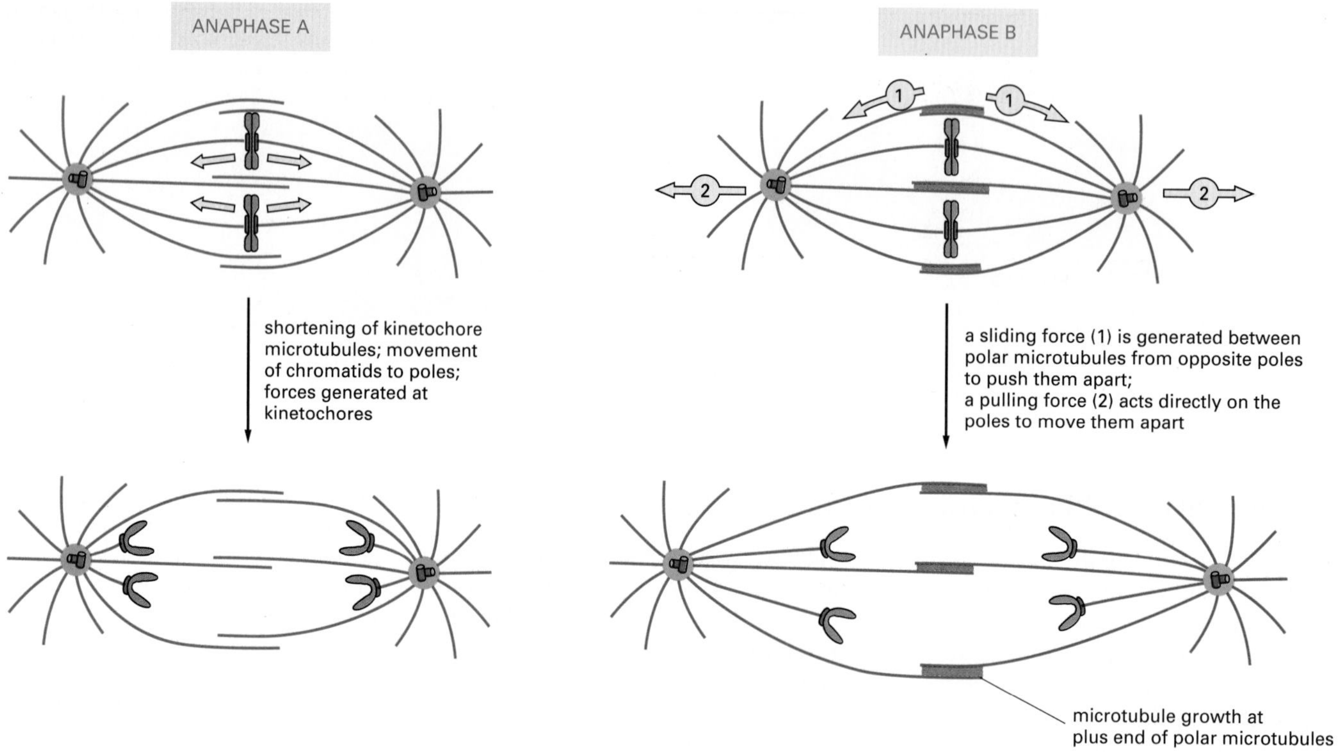

ANAPHASE A

ANAPHASE B

shortening of kinetochore microtubules; movement of chromatids to poles; forces generated at kinetochores

a sliding force (1) is generated between polar microtubules from opposite poles to push them apart;
a pulling force (2) acts directly on the poles to move them apart

microtubule growth at plus end of polar microtubules

The relative contributions of anaphase A and anaphase B to the final separation of the chromosomes vary considerably depending on the organism. In mammalian cells anaphase B begins shortly after the chromatids have begun their voyage to the poles and stops when the spindle is about 1.5–2 times its metaphase length. In some other cells, such as yeasts and certain protozoa, anaphase B predominates and the spindle elongates to 15 times its metaphase length.

Kinetochore Microtubules Disassemble During Anaphase A [18]

A surprisingly large force acts on a chromosome as it moves from the metaphase plate to the spindle pole. Measurements using the deflection of fine glass needles give an estimate of 10^{-5} dynes per chromosome, which is more than 10,000 times greater than the force required simply to move chromosomes at their observed rate through the cytoplasm. Evidently, powerful motors are responsible for moving the chromosomes, and the speed of chromosome movement must be limited by something other than viscous drag. As discussed above, the same motors may generate the tension on chromosomes at the metaphase plate.

As each chromosome moves poleward, its kinetochore microtubules depolymerize, so that they have nearly disappeared at telophase. The site of subunit loss can be determined by making marks on the kinetochore microtubules using fluorescent tubulin subunits and laser beams, as described earlier. In such experiments, kinetochore microtubules are seen to depolymerize at both ends during anaphase, with most of the depolymerization occurring at the kinetochores, where they previously polymerized. In a cultured newt cell, for example, anaphase A movement occurs at 2 microns per minute, which is equivalent to microtubules depolymerizing at about 50 subunits per second. Of this depolymerization, 60% to 80% occurs at the kinetochores, while the remainder occurs at the poles, at the same rate as observed during metaphase (see Figure 18–25). The finding that the kinetochore is the major site of microtubule depolymerization in anaphase suggests that it is also the main site where the force for chromosome movement is generated.

Figure 18–27 **The two processes that separate sister chromatids at anaphase.** In *anaphase A* the chromatids are *pulled* toward opposite poles by forces associated with shortening of their kinetochore microtubules. The force driving this movement is thought to be generated mainly at the kinetochore. In *anaphase B* the two spindle poles move apart. It is likely that the forces driving anaphase B are similar to those that cause the centrosome to split and separate into two spindle poles at prophase (see Figure 18–5). There is evidence that two separate forces are responsible for anaphase B: the elongation and sliding of the polar microtubules past one another *pushes* the two poles apart, and outward forces exerted by the astral microtubules at each spindle pole act to *pull* the poles away from each other, toward the cell surface.

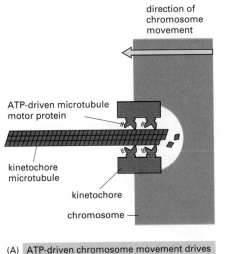

direction of
chromosome
movement

ATP-driven microtubule
motor protein

kinetochore
microtubule

kinetochore

chromosome

(A) ATP-driven chromosome movement drives
microtubule disassembly

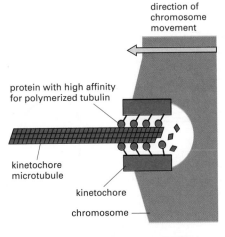

direction of
chromosome
movement

protein with high affinity
for polymerized tubulin

kinetochore
microtubule

kinetochore

chromosome

(B) microtubule disassembly drives
chromosome movement

Figure 18–28 Two alternative models of how the kinetochore may generate a poleward force on its chromosome during anaphase. (A) Microtubule motor proteins are part of the kinetochore and use the energy of ATP hydrolysis to pull the chromosome along its bound microtubules. (B) Chromosome movement is driven by microtubule disassembly: as tubulin subunits dissociate, the kinetochore is obliged to slide poleward in order to maintain its binding to the walls of the microtubule.

The mechanism by which the kinetochore, and thus the chromosome, moves toward the spindle pole during anaphase A is, however, unknown. Two possible models are presented schematically in Figure 18–28. In one, a minus-end-directed motor protein in the kinetochore hydrolyzes ATP to move along its attached microtubule, with the plus end of the microtubule depolymerizing as it becomes exposed. In the other model depolymerization of the microtubule itself causes the kinetochore to move passively because the kinetochore remains attached to the end of the microtubule as depolymerization proceeds. The recent finding of minus-end-directed motor proteins in the kinetochore has shifted the balance of opinion in favor of the first model, although microtubule depolymerization probably contributes to the movement, perhaps by acting as a "governor" that limits the rate of poleward migration.

Two Separate Forces May Contribute to Anaphase B [19]

Anaphase B increases the distance between the two spindle poles and, in contrast to anaphase A, is accompanied by the *elongation* of microtubules. The movement is thought to be driven by the same mechanism that pushes the centrosomes apart in prometaphase. As the two poles move apart, the polar microtubules between them lengthen, apparently by polymerization at their distal, plus ends.

Figure 18–29 Sliding of overlap microtubules at anaphase. Electron micrographs showing spindle elongation and the reduction in the degree of polar microtubule overlap during mitosis in a diatom. (A) Metaphase. (B) Late anaphase. (Courtesy of Jeremy D. Pickett-Heaps.)

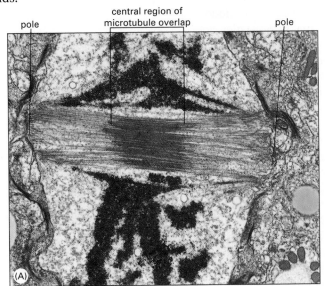

pole

central region of
microtubule overlap

pole

(A)

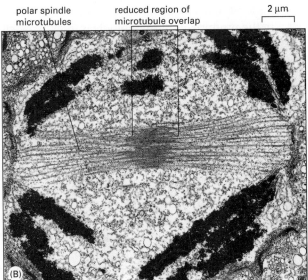

polar spindle
microtubules

reduced region of
microtubule overlap

2 μm

(B)

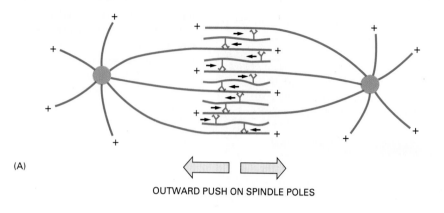

(A)

OUTWARD PUSH ON SPINDLE POLES

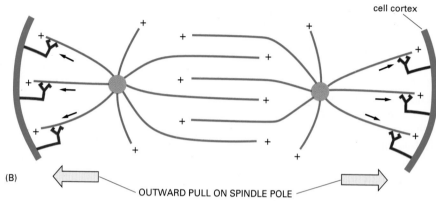

cell cortex

(B)

OUTWARD PULL ON SPINDLE POLE

Figure 18–30 Model of how microtubule motor proteins are thought to act in anaphase B. In (A) plus-end-directed motor proteins of the kinesin family cross-link adjacent, overlapping, antiparallel polar microtubules and slide the microtubules past each other, thereby *pushing* the spindle poles apart. The *black arrows* indicate the direction of microtubule sliding. In (B) minus-end-directed motor proteins bind to the cell cortex and to those astral microtubules that point away from the spindle and *pull* the spindle poles apart.

Both the extent of spindle pole separation at anaphase and the degree of overlap of the polar microtubules in the midzone vary greatly from species to species. In many diatoms, which have the distinctive feature that mitosis occurs within the nuclear envelope, the overlapping arrays of spindle microtubules are especially prominent (Figure 18–29). Painstaking reconstruction of the three-dimensional architecture of complete diatom spindles from hundreds of serial thin sections examined in the electron microscope shows that the polar microtubules from each half-spindle overlap in a central region near the spindle equator and that microtubules of opposite polarity are arranged with regular spacing between them. During anaphase these two sets of antiparallel polar microtubules appear to slide away from each other in the region of overlap.

Anaphase movements can also be studied in lysed diatom cells. In this model system the prominent mitotic spindle is freely accessible to macromolecules, so that the effects of various macromolecular probes (including specific antibodies) can be tested. Anaphase B movement in diatoms is blocked by an antibody that recognizes microtubule motor proteins of the kinesin family, suggesting that the movement is probably driven by a motor protein of this family. The motor protein is thought to reside in the overlap zone, where it serves to push the half-spindles apart (Figure 18–30A).

The astral microtubules, which elongate in all directions from the spindle poles during anaphase, are thought to play an additional role in anaphase B movement. In both fungi and animal spindles, severing the central spindle does not block the separation of the spindle poles but in fact accelerates it. This suggests that the astral microtubules that point away from the spindle generate a pulling force that assists the separation of poles during anaphase B, possibly by interacting with minus-end-directed motor proteins attached to the cell cortex or other cytoplasmic structures (Figure 18–30B). Consistent with this suggestion, the injection of antibodies against the minus-end-directed motor protein dynein causes the spindle in cultured cells to collapse. This pulling force is believed to operate also during spindle assembly in prophase, and similar forces are thought to guide the specific positioning of spindles prior to asymmetric cell cleavages, as we discuss later.

At Telophase the Nuclear Envelope Initially Re-forms Around Individual Chromosomes [20]

By the end of anaphase the chromosomes have separated into two equal groups, one at each pole of the spindle. In **telophase,** the final stage of mitosis, a nuclear envelope reassembles around each group of chromosomes to form the two daughter interphase nuclei.

At least three parts of the nuclear-envelope complex (discussed in Chapter 12) must be considered during its breakdown and reassembly in mitosis: (1) the *outer and inner nuclear membranes,* which are continuous with the endoplasmic reticulum membrane; (2) the underlying *nuclear lamina* (a thin sheetlike network of intermediate filaments formed from the nuclear lamins), which interacts with the inner nuclear membrane, chromatin, and nuclear pores; and (3) the *nuclear pores,* which are formed by large protein complexes. At prophase the phosphorylation of the lamins occurs at many sites in each polypeptide chain and causes them to disassemble, thereby disrupting the nuclear lamina. Subsequently, perhaps in response to a different signal, the nuclear envelope membranes break up into small membrane vesicles.

The sudden transition from metaphase to anaphase initiates dephosphorylation of the many proteins that were phosphorylated at prophase: although the relevant phosphatases are active throughout mitosis, it is not until MPF is switched off that the phosphatases can act unopposed. Shortly thereafter, at telophase, nuclear membrane vesicles associate with the surface of individual chromosomes and fuse to re-form the nuclear membranes, which partially enclose clusters of chromosomes before coalescing to re-form the complete nuclear envelope (see Figure 12–18). During this process the nuclear pores reassemble and the dephosphorylated lamins reassociate to form the nuclear lamina; one of the lamina proteins (lamin B) remains with the nuclear membrane fragments throughout mitosis and may help nucleate reassembly. After the nucleus re-forms, the pores pump in nuclear proteins, the chromosomes decondense, and RNA synthesis resumes, causing the nucleolus to reappear. Chromatin decondensation is thought to be triggered, in part, by the dephosphorylation of histone H1 molecules.

Both nuclear disassembly and nuclear assembly will occur in crude extracts of *Xenopus* eggs, provided that these extracts are prepared from cells at the correct stage of the cell cycle (mitotic cells for disassembly and interphase cells for assembly). In these extracts the entire process involving the lamina, nuclear pores, and nuclear membranes appears to progress normally in response to phosphorylation and dephosphorylation cycles. *In vitro* systems of this type provide an assay for the identification and purification of the proteins that catalyze nuclear envelope assembly and disassembly in the cell, including the proteins (such as MPF) that regulate these processes. While DNA must be added to these extracts for a nucleus to assemble, a complete nuclear envelope will form around purified DNA molecules from virtually any organism, including a bacterial virus. Therefore, while DNA-binding proteins must be involved, it is unlikely that specific DNA sequences are recognized during nuclear reassembly.

Surprisingly, nuclear-envelope breakdown is not a requirement for mitosis. In fact, we shall see below that in lower eucaryotes the nuclear envelope does not disassemble during mitosis. These organisms are said to have a closed, rather than an open, spindle.

Summary

Mitosis begins with prophase, which is marked by an increase in the phosphorylation of specific proteins, triggered by the activity of the mitosis-inducing protein ki-

nase MPF. One consequence of this phosphorylation is an unusually dynamic micro-tubule array nucleated on the duplicated centrosomes that form the spindle poles. One subset of the microtubules from each centrosome becomes stabilized, apparently by cross-linking to microtubules from the opposite centrosome, to form the polar microtubules, which are thought to push the poles apart. After the nuclear envelope breaks down in pro-metaphase, the kinetochores on condensed chromosomes can capture and stabilize other subsets of microtubules from the large numbers that continually grow out from each spindle pole. Kinetochore microtubules from opposite spindle poles pull in opposite directions on the two kinetochores on each duplicated chromosome, creating a tension that stabilizes the kinetochore attachment. Balanced forces on the kinetochores also bring the chromosomes to the spindle equator to form the metaphase plate. The tubulin subunits in the spindle microtubules at metaphase undergo a continuous treadmilling from the spindle equator toward the poles. At anaphase sister chromatids suddenly detach from each other and are pulled to opposite poles (the anaphase A movement). Meanwhile, the two spindle poles move apart (the anaphase B movement). In telophase, the final stage of mitosis, the nuclear envelope re-forms on the surface of each group of separated chromosomes as the proteins phosphorylated at the onset of M phase are dephosphorylated.

Cytokinesis [21]

During **cytokinesis** the cytoplasm divides by a process called *cleavage*. Although nuclear and cytoplasmic division are generally linked, they are separable events, and in some normal circumstances, nuclear division is not followed by cytokinesis. The early *Drosophila* embryo, for example, undergoes 13 rounds of nuclear division without cytoplasmic division, forming a single large cell containing 6000 nuclei arranged in a monolayer near the surface. Uninucleated cells are then generated by cytoplasmic cleavage around all of these nuclei (see Figure 21–51). In a typical cell, however, cytokinesis accompanies every mitosis, beginning in anaphase, continuing in telophase, and reaching completion as the following interphase begins.

The Mitotic Spindle Determines the Site of Cytoplasmic Cleavage During Cytokinesis [22]

Normally, the mitotic spindle determines where cleavage occurs as well as when. The first visible sign of cleavage in animal cells is a puckering and *furrowing* of the plasma membrane during anaphase (Figure 18–31). The furrowing invariably occurs in the plane of the metaphase plate, at right angles to the long axis of the mitotic spindle. This ensures that the **cleavage furrow** cuts between the two groups of separating chromatids and thus that the two daughter cells receive identical copies of the genetic material. If the spindle is moved by micromanipu-

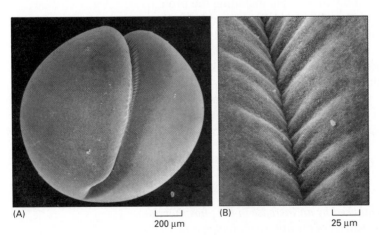

(A) 200 μm

(B) 25 μm

Figure 18–31 Scanning electron micrographs of early cleavage in a fertilized frog egg. The furrowing of the cell membrane is caused by the activity of the *contractile ring* underneath it. The cleavage furrow is unusually obvious and well defined in this giant spherical cell. (A) Low-magnification view of egg surface. (B) Surface of furrow at higher magnification. (From H.W. Beams and R.G. Kessel, *Am. Sci.* 64:279–290, 1976. Reprinted by permission of *American Scientist*, journal of Sigma Xi.)

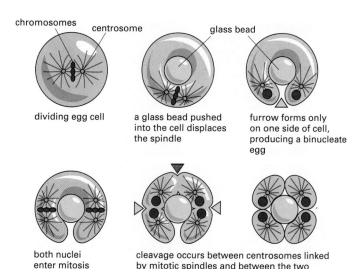

chromosomes centrosome glass bead

dividing egg cell

a glass bead pushed into the cell displaces the spindle

furrow forms only on one side of cell, producing a binucleate egg

both nuclei enter mitosis

cleavage occurs between centrosomes linked by mitotic spindles and between the two centrosomes that are simply adjacent, and four daughter cells are formed

Figure 18–32 An experiment that shows the influence of the position of microtubule asters on the subsequent plane of cleavage. If a mitotic spindle is mechanically pushed to one side of the cell, the membrane furrowing is incomplete, failing to occur on the opposite side of the cell. Subsequent cleavages occur not only in the conventional relation to each of the two subsequent mitotic spindles (*yellow arrowheads*) but also between the two adjacent asters that are not linked by a mitotic spindle (but in this abnormal cell share the same cytoplasm)(*red arrowhead*). Apparently, the contractile bundle of actin filaments that produces the cleavage furrow always forms in the region midway between two asters, which implies that the asters somehow alter the adjacent region of cell cortex.

lation early enough in anaphase, the incipient furrow disappears and a new one develops in accord with the new spindle site. Ingenious experiments with fertilized sand dollar eggs show that a cleavage furrow will form midway between the asters originating from two centrosomes even when the centrosomes are not connected by a mitotic spindle (Figure 18–32). Thus, the microtubule asters, and not the chromosomes, signal to the cortex to initiate a furrow. Later, once the furrowing process is well underway, cleavage proceeds even if the spindle and its asters are removed by suction or destroyed by colchicine.

The mechanism by which a pair of microtubule asters signals the site of cleavage at the cortex is not known, but it is a classic example of communication between the microtubule and actin filament systems of the cytoskeleton. One hypothesis is that the two sets of astral microtubules contact the cell cortex and set the future cleavage site, possibly by moving actin filaments. Another is that the cortex senses a gradient of Ca^{2+}, which is set up by Ca^{2+}-sequestering vesicles that are known to be present at the spindle poles. Whatever the signal, it is clear that the cortex is poised to detect and amplify it. The whole cortex is under tension (see Figure 16–80), and if this tension is locally increased at the site of furrow initiation or decreased at the poles, actin filaments will locally align and new filaments will be drawn in, along with myosin-II, from the surrounding cortex. Thus, the furrow may be a self-amplifying system, which self-organizes on the basis of a subtle initial cue.

The Spindle Is Specifically Repositioned to Create Asymmetric Cell Divisions [23]

Most cells divide symmetrically. The cleavage furrow forms around the equator of the parent cell, so that the two daughter cells produced are of equal size and have similar properties. This cleavage symmetry results from the prior placement of the spindle, which tends to center itself in the cytoplasm, with the spindle axis lined up along a particular axis of the cell. This position is thought to be established by pulling forces exerted on astral microtubules by minus-end-directed motor proteins in the cytoplasm (see Figure 18–30B).

There are many instances during embryonic development, however, in which cells divide asymmetrically. In these cases the furrow creates two cells that are different in size and are destined to develop along divergent paths. Divisions of this kind are often precisely defined spatially. They may bear a precise relationship to the surface of an epithelial sheet, for example, or segregate regions of cytoplasm that have a different complement of organelles. Asymmetric positioning of the furrow is always anticipated by asymmetric positioning of the mitotic

Cytokinesis

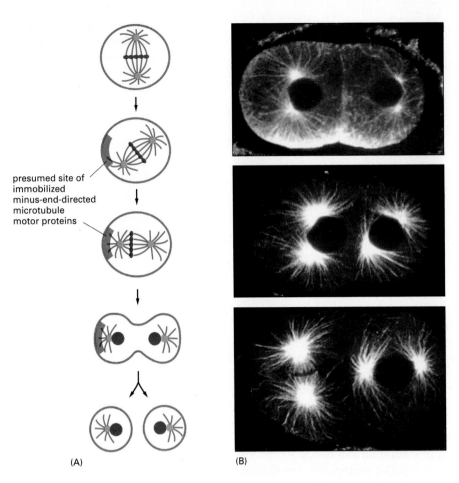

presumed site of immobilized minus-end-directed microtubule motor proteins

(A)

(B)

Figure 18–33 Spindle rotation. (A) Diagram showing a possible mechanism underlying the controlled rotation of a mitotic spindle. The *red bar* represents a specialized region of cortex toward which one spindle pole is pulled by its astral microtubules. (B) Light micrographs showing a precisely programmed rotation of a mitotic spindle in an embryo of the nematode worm *C. elegans* at the two-cell stage in preparation for cleavage to form four cells in a specific pattern. The spindle in the cell on the right rotates almost 90° clockwise. (Courtesy of Tony Hyman and John White.)

spindle. To orient the plane of division, the spindle rotates in a controlled manner to adopt a suitable position in the cell. It seems likely that these spindle movements are directed by programmed changes in local regions of the cell cortex and that the cortex then moves the spindle poles via their astral microtubules (Figure 18–33). A similar mechanism is thought to position the centrosome in a polarized cell (see p. 792).

Asymmetric cell division is particularly important in plant cells, since these cells cannot move after division, so that division planes alone determine tissue morphology. Plants use a different mechanism for establishing their cleavage plane, as we discuss later.

Actin and Myosin Generate the Forces for Cleavage [24]

Cleavage is accomplished by the contraction of a thin ring composed mainly of an overlapping array of actin filaments and bipolar myosin-II filaments. This **contractile ring** defines the cleavage furrow. It consists of circumferentially oriented filaments bound to the cytoplasmic face of the plasma membrane by uncharacterized attachment proteins (Figure 18–34). The contractile ring assembles in early anaphase. Once assembled, it develops a force large enough to bend a fine glass needle inserted into the cell. There is compelling evidence that musclelike sliding of actin and myosin filaments in the contractile ring generates this force. In lysed mitotic cells, for example, cleavage is stopped by the addition of an inactivated myosin subfragment that blocks sites on actin that normally bind myosin. Similarly, an injection of antimyosin antibodies into fertilized sea urchin eggs causes the cleavage furrow to relax without affecting nuclear division.

At each point on its circumference, the contractile ring contains a bundle of about 20 actin filaments. During a normal cell division the ring does not get thicker as the furrow invaginates, suggesting that it continuously reduces its volume by losing filaments. Thus, like other cytoskeletal structures and unlike

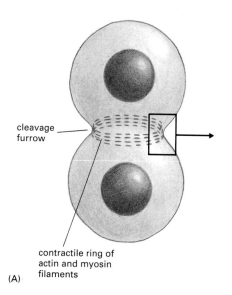

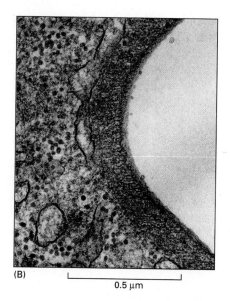

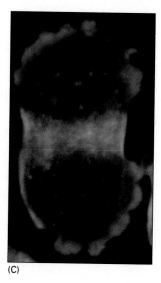

cleavage
furrow

contractile ring of
actin and myosin
filaments

(A)

(B)

0.5 μm

(C)

skeletal muscle, the furrow is dynamic. The contractile ring is finally dispensed with altogether when cleavage ends, as the plasma membrane of the cleavage furrow narrows to form the **midbody,** which remains as a tether between the two daughter cells. The midbody contains the remains of the two sets of polar microtubules packed tightly together with a dense matrix material (Figure 18–35).

The process of cytokinesis requires wholesale reorganization of actin and myosin filaments in the cortex to allow the assembly of the contractile ring. Other functions of the cortex, most notably cell adhesion, are also affected in M phase. Tissue culture cells that are spread out in interphase due to strong adhesive contacts with extracellular matrix molecules on the substratum, for example, round

Figure 18–34 The contractile ring.
(A) A schematic drawing of a cleavage furrow in a dividing cell. (B) Electron micrograph of the ingrowing edge of the cleavage furrow of a dividing animal cell. (C) A dividing slime mold amoeba stained for actin (*red*) and myosin-II (*green*). The myosin staining in the contractile ring somewhat masks the actin that is also present there. (B, from H.W. Beams and R.G. Kessel, *Am. Sci.* 64:279–290, 1976, reprinted by permission of *American Scientist*, journal of Sigma Xi; C, courtesy of Yoshio Fukui.)

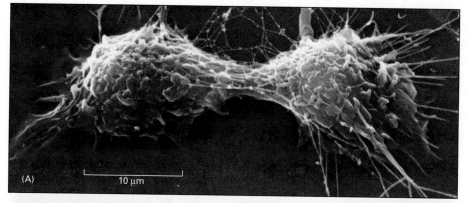

(A)

10 μm

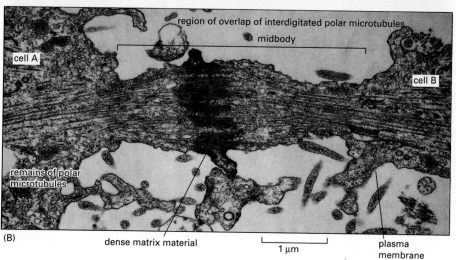

region of overlap of interdigitated polar microtubules

midbody

cell A

cell B

remains of polar microtubules

(B)

dense matrix material

1 μm

plasma membrane

Figure 18–35 Cytokinesis. (A) Scanning electron micrograph of an animal cell in culture in the process of dividing; the midbody still joins the two daughter cells. (B) Electron micrograph of the midbody of a dividing animal cell. Cleavage is virtually complete, but the daughter cells remain attached by this thin strand of cytoplasm. (A, courtesy of Guenter Albrecht-Buehler; B, courtesy of J.M. Mullins.)

Cytokinesis

up when they enter M phase; following cytokinesis the daughter cells flatten out again. It is thought that the activities of transmembrane adhesion proteins such as the integrins (discussed in Chapter 19) are in some way down-regulated during M phase or that the attachment proteins that connect these proteins to actin filaments in the cortex are modified. Like other manifestations of M phase, these changes are presumably mediated by protein phosphorylation.

In Special Cases, Selected Cell Components Can Be Segregated to One Daughter Cell Only [25]

Whereas most cell divisions produce two similar daughter cells, some produce daughter cells that are manifestly unequal. This is especially true of the initial cleavages in which a large fertilized egg is subdivided into smaller cells destined to form different parts of the body. We have already discussed how the asymmetric positioning of the spindle during cell division can produce two cells of unequal size, but the genesis of biochemically different daughter cells is a special problem.

A striking example of this process is provided by the behavior of a set of distinctive granules in the egg of the nematode worm *C. elegans*. These "P-granules" are spread evenly throughout the cytoplasm of the unfertilized egg, but they move to the posterior end of the cell just before the first cleavage and are therefore inherited by only one of the two daughter cells. The same type of segregation process is repeated in several subsequent cell divisions, so that the granules end up in only the primordial germ cells, which give rise to the eggs and sperm (Figure 18–36). It is possible that the granules play a part in controlling the distinctive fate of these cells.

The P-granules will still move to the posterior end of the cell even in mutants where the mitotic spindle is disoriented and turned at right angles to its normal position. Moreover, segregation of the granules is blocked by the drug cytochalasin D, which inhibits actin-filament polymerization, but not by colchicine, suggesting that the oriented movement of P-granules is dependent on actin filaments but not on microtubules. Although the unequal segregation of these components thus seems to be based on some asymmetric property of the actin-based cytoskeleton, the molecular mechanism for their oriented movement is unknown.

Cytokinesis Occurs by a Special Mechanism in Higher Plant Cells [26]

Most higher plant cells are enclosed by a rigid *cell wall,* and the mechanism of cytokinesis in these cells is different from the one we have just described for

Figure 18–36 Asymmetric segregation. These micrographs illustrate the controlled asymmetric segregation of a cytoplasmic component to one daughter cell during each of the first few cell divisions of a fertilized egg of the nematode *C. elegans*. Above are cells viewed by differential-interference-contrast light microscopy and stained with a *blue* DNA-specific fluorescent dye to show cell nuclei; below are the same cells stained with an antibody against P-granules. These small granules of unknown function (0.5–1 μm in diameter) are distributed randomly throughout the cytoplasm in the unfertilized egg. (Courtesy of Susan Strome.)

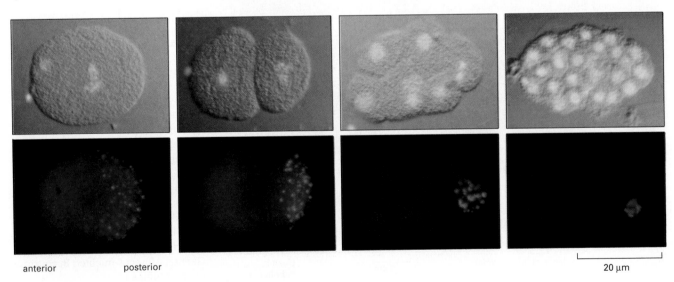

anterior   posterior   20 μm

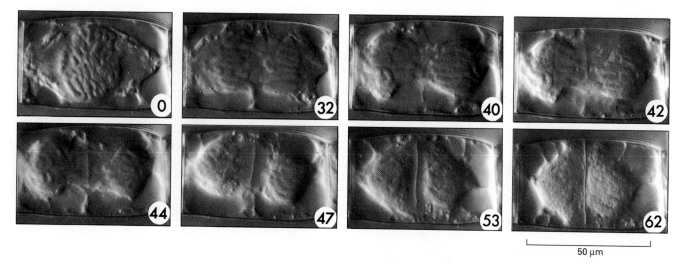

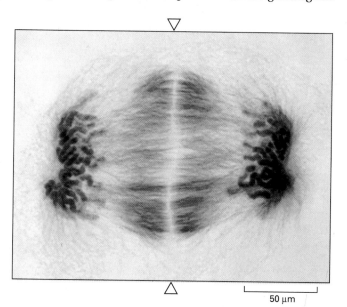

animal cells. Rather than pinching off the two daughter cells by means of a contractile ring at the cell surface, the cytoplasm of the plant cell is partitioned by the construction of a new cell wall inside the cell (Figure 18–37). This cross-wall precisely determines the positions of the two daughter cells relative to neighboring cells. It follows that the planes of cell division, together with cell enlargement, determine plant form.

The new cross-wall, or **cell plate,** starts to assemble in a plane between the two daughter nuclei and in association with the residual polar spindle microtubules, which form a cylindrical structure called the **phragmoplast.** The phragmoplast, corresponding to the microtubules in the animal cell midbody, contains two sets of microtubules that interdigitate at their growing (plus) ends (Figure 18–38). The microtubules have their plus ends embedded in an electron dense disc in the equatorial plane. As outlined in Figure 18–39, small membrane-bounded vesicles, largely derived from the Golgi apparatus and filled with cell-wall precursors, seem to contact the microtubules on each side of the phragmoplast and to be transported along them until they reach the equatorial region. Here they fuse to form a disclike membrane-bounded structure, the *early cell plate.* The polysaccharide molecules delivered by these vesicles assemble within the early cell plate to form the matrix material of the primary cell wall. This disc now has to expand laterally to reach the original parent cell wall. To make this possible the microtubules of the early phragmoplast are continuously reorganized at the periphery of the early cell plate. There they attract more vesicles, which fuse at the equator to extend the edge of the plate. This process is repeated until the growing cell

Figure 18–37 Sequential light micrographs of a dividing plant cell. The elapsed time in minutes is shown at the bottom left corner of each photograph. The vesicles that align to form the cell plate can be seen after 42 minutes. The plate then extends sideways until it reaches and fuses with the mother cell wall. (Courtesy of Peter Hepler.)

Figure 18–38 Light micrograph of cytokinesis in a plant cell in telophase. The early cell plate (between the two *arrows*) is forming in a plane perpendicular to the plane of the page. The two arrays of microtubules that contribute to the phragmoplast are stained using gold-labeled antibodies to tubulin, while the DNA in the two sets of daughter chromosomes is stained with a fluorescent dye. (Courtesy of Andrew Bajer.)

Cytokinesis

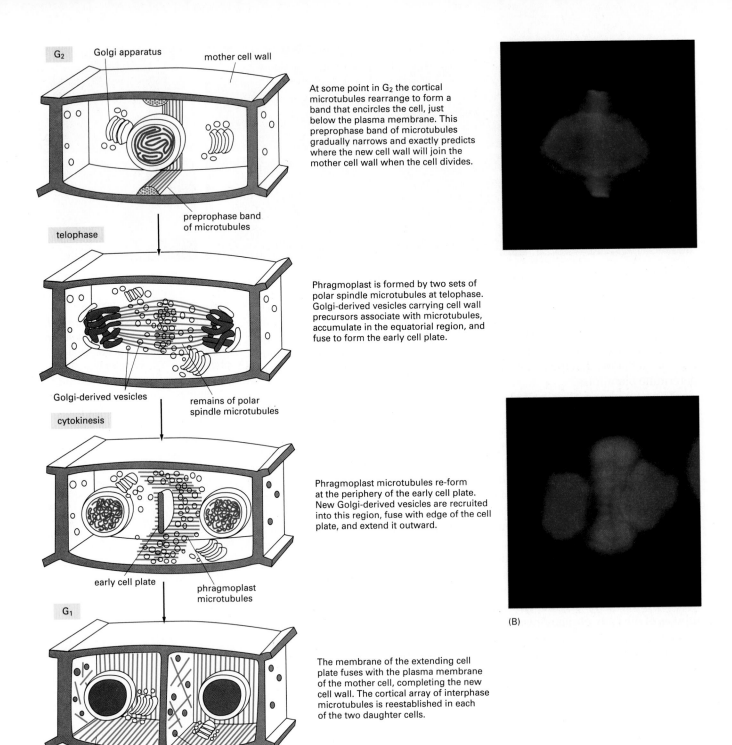

G₂

Golgi apparatus

mother cell wall

At some point in G₂ the cortical microtubules rearrange to form a band that encircles the cell, just below the plasma membrane. This preprophase band of microtubules gradually narrows and exactly predicts where the new cell wall will join the mother cell wall when the cell divides.

preprophase band of microtubules

telophase

Phragmoplast is formed by two sets of polar spindle microtubules at telophase. Golgi-derived vesicles carrying cell wall precursors associate with microtubules, accumulate in the equatorial region, and fuse to form the early cell plate.

Golgi-derived vesicles

remains of polar spindle microtubules

cytokinesis

Phragmoplast microtubules re-form at the periphery of the early cell plate. New Golgi-derived vesicles are recruited into this region, fuse with edge of the cell plate, and extend it outward.

early cell plate

phragmoplast microtubules

G₁

The membrane of the extending cell plate fuses with the plasma membrane of the mother cell, completing the new cell wall. The cortical array of interphase microtubules is reestablished in each of the two daughter cells.

cortical array of interphase microtubules

plasmodesmata

(A)

(B)

Figure 18–39 The special features of mitosis and cytokinesis in a higher plant cell. (A) Diagrams showing plant cells in G₂, telophase, cytokinesis, and early G₁, to emphasize the dynamic changes in the distribution of microtubules through the cell cycle. (B) Immunofluorescence micrographs of onion root-tip cells showing the preprophase band of microtubules in G₂ and the phragmoplast at cytokinesis. Microtubules are stained *green*, and the chromosomes are stained *blue*. (B, courtesy of Kim Findlay.)

plate reaches the plasma membrane of the parent cell. The plasma membrane and the membrane surrounding the cell plate fuse, completely separating the two new daughter cells (see Figure 18–37). Actin filaments are also abundant in the phragmoplast, aligned with the microtubules, but their functional role in cell plate formation is unclear. Sometime later, cellulose microfibrils are laid down within the cell plate to complete the new cell wall.

A Cytoskeletal Framework Determines the Plane of Plant Cell Division [27]

The mitotic spindle by itself is generally not sufficient to determine the exact position and shape of the cell plate. The plate's future site of junction with the mother cell wall seems to be defined at some point in G_2, before mitosis has begun. Thus, the first visible sign that a higher plant cell has become committed to divide in a particular plane is seen just after the interphase cortical array of microtubules disappears in preparation for mitosis. At this time a circumferential band of microtubules appears and forms a ring around the entire cell just beneath the plasma membrane. Because this array of microtubules appears in G_2 before prophase begins, it is called the **preprophase band** (see Figure 18–39). The band becomes narrower as the cell progresses to prophase, and it disappears before metaphase is reached, yet the boundary of the division plane has somehow been imprinted: when the new cell plate forms later during cytokinesis, it grows outward to fuse with the parental wall precisely at the zone that was formerly occupied by the preprophase band (see Figure 18–39). Even if the cell contents are displaced by centrifugation after the preprophase band has disappeared, the growing cell plate will tend to find its way back to the plane defined by the former preprophase band.

It is now known that the preprophase band contains numerous actin filaments in addition to microtubules. The actin filaments are not confined to the cell cortex; in vacuolated cells they also form a radial, disclike array of strands, which crosses the cell and connects to and supports the central dividing nucleus. After the microtubules in the preprophase band depolymerize, the radial actin strands remain and provide a "memory" of the predetermined division plane. During cytokinesis, as the phragmoplast grows out centrifugally like a circular ripple in a pond, the edges of the growing cell plate are connected to the site of the preprophase band by actin filaments. Thus, actin seems to have an important function in the division of walled cells even though contraction plays no obvious part. Actin is also associated with septum formation in fungal cells, which suggests that it may help guide cytokinesis in all eucaryotes.

The Elaborate M Phase of Higher Organisms Evolved Gradually from Procaryotic Fission Mechanisms [28]

In procaryotic cells, division of the DNA and of the cytoplasm are coupled in a very direct way. When DNA replicates, the two copies of the chromosome are attached to specialized regions of the plasma membrane, which are thought to be separated gradually by the growth of the membrane between them. Fission takes place between the two attachment sites, so that each daughter cell captures one chromosome (Figure 18–40). With the evolution of the eucaryotes, the genome increased in complexity and the chromosomes increased in number and in size. For these organisms a more elaborate mechanism for dividing the chromosomes between daughter cells was apparently required.

Clearly, the mitotic apparatus could not have evolved all at once. In many primitive eucaryotes, such as the dinoflagellate *Cryphthecodinium cohnii*, mitosis still depends on a membrane-attachment mechanism, with the nuclear membrane taking over the part played by the plasma membrane in procaryotes. The intermediate status of this large single-celled alga is also reflected in the biochem-

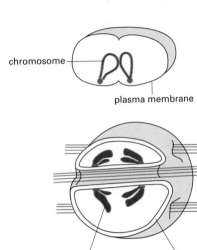

chromosome

plasma membrane

BACTERIA
daughter chromosomes attached to the plasma membrane are separated by the ingrowth of plasma membrane between them

TYPICAL DINOFLAGELLATES
several bundles of microtubules pass through tunnels in the intact nuclear envelope to establish the polarity of division; chromosomes move apart in association with the inner nuclear membrane without being attached to the microtubule bundles

chromosomes   intact nuclear envelope

HYPERMASTIGOTES AND SOME UNUSUAL DINOFLAGELLATES
a single central spindle between two centriole pairs is formed in a tunnel through the intact nuclear envelope; chromosomes are attached by their kinetochores to the nuclear membrane and interact with the spindle poles via kinetochore microtubules

centrioles

polar microtubules   kinetochore microtubules

YEASTS AND DIATOMS
nuclear envelope remains intact; polar spindle microtubules form inside the nucleus and are associated with the nuclear envelope; a single kinetochore microtubule attaches each chromosome to a pole

kinetochore microtubules

polar microtubules   centrioles

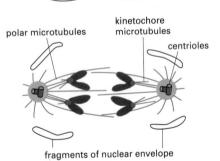

ANIMALS
the spindle begins to form outside the nucleus; at prometaphase the nuclear envelope breaks down to allow chromosomes to capture spindle microtubules, which now become kinetochore microtubules

fragments of nuclear envelope

Figure 18–40 Different chromosome separation mechanisms are used by different organisms. Some of these may have been intermediate stages in the evolution of the mitotic spindle of higher organisms. For all of the examples except bacteria, only the central nuclear region of the cell is shown.

istry of its chromosomes, which, like those of procaryotes, have relatively little associated protein. The nuclear membrane in *C. cohnii* remains intact throughout mitosis, and the spindle microtubules remain entirely outside the nucleus. Where these spindle microtubules press on the outside of the nuclear envelope, the envelope becomes indented in a series of parallel channels (see Figure 18–40). The chromosomes become attached to the inner membrane of the nuclear envelope opposite these channels, and the separation of the chromosomes is entirely mediated on the inside of this channeled nuclear membrane. Thus the extranuclear "spindle" is used to order the nuclear membrane and thereby define the plane of division. Kinetochores in these species seem to be integrated into the nuclear membrane and may therefore have evolved from some membrane component. The evolutionary origin of microtubules themselves is mysterious. They are important for chromosome segregation in even the most primitive eucaryotes, but they are also present in flagellar axonemes (see p. 816). Whether the flagellum or the spindle came first is unclear.

A somewhat more advanced, although still extranuclear, spindle is seen in *hypermastigotes,* in which the nuclear envelope again remains intact throughout mitosis. These large protozoa from the guts of insects provide a particularly clear illustration of the independence of spindle elongation and the chromosome movements that separate the chromatids, since the sister kinetochores become separated by the growth of the nuclear membrane (to which they are attached) before becoming attached to the spindle. Only when the kinetochores are near the poles of the spindle do they acquire the kinetochore fibers needed to attach them to the spindle. Because the spindle fibers remain separated from the chromosomes by the nuclear envelope, the kinetochore fibers, which are formed outside the nucleus, must somehow attach to the chromosomes through the nuclear membranes. After this attachment has occurred, the kinetochores are drawn poleward in a conventional manner (see Figure 18–40).

A further stage in the evolution of mitotic mechanisms may be represented by organisms that form spindles inside an intact nucleus. In both yeasts and diatoms the spindle is attached to chromosomes by their kinetochores and the chromosomes are segregated in a way closely similar to that described for mammalian cells—except that the entire process generally occurs within the confines of the nuclear envelope (see Figure 18–40). It is thought that the "open" mitosis of higher organisms and the "closed" mitosis of yeasts and diatoms evolved separately from a common ancestor resembling the modern hypermastigote spindle.

At present, there is no convincing explanation for why higher plants and animals have evolved a mitotic apparatus that requires the controlled and reversible dissolution of the nuclear envelope.

Summary

Cell division ends as the cytoplasmic contents are divided by the process of cytokinesis and the chromosomes decondense and resume RNA synthesis. Cytokinesis appears to be guided by organized bundles of actin filaments in eucaryotic cells as diverse as animals, plants, and fungi. In animal cells the mitotic spindle determines when and where cytokinesis occurs, with the contractile ring of actin and myosin filaments forming midway between the spindle pole asters. Whereas most cells divide symmetrically, in some cases the spindle is specifically positioned to create an asymmetric cell division: a particular cell can divide into one small cell and one large one, for example, or a specific cytoplasmic component can be moved to one side of a cell prior to cytokinesis so that it is inherited by only one of the two otherwise equal daughter cells. Cytokinesis occurs by a special mechanism in higher plant cells, where the cytoplasm is partitioned by the construction of a new cell wall, the cell plate, inside the cell. The position of the cell plate is determined by the position of a preprophase band of microtubules and actin filaments. The organization of mitosis in some protozoa and in fungi differs from that in animals and plants, suggesting how the complex process of eucaryotic cell division may have evolved.

References

General

Amos, L.A.; Amos, W.B. Molecules of the Cytoskeleton. New York: Guilford Press, 1991.

Conrad, G.W.; Schroeder, T.E., eds. Cytokinesis: mechanisms of furrow formation during cell division. *Ann. N.Y. Acad. Sci.,* Vol. 582. New York: Academy of Sciences, 1990.

Hyams, J.S.; Brinkely, B.R., eds. Mitosis: Molecules and Mechanisms. San Diego, CA: Academic Press, 1989.

Mazia, D. Mitosis and the physiology of cell division. In The Cell (J. Brachet, A.E. Mirsky, eds.), Vol. 3, pp. 77–412. London: Academic Press, 1961.

Wilson, E.B. The Cell in Development and Heredity, 3rd ed. with corrections. New York: Macmillan, 1925, 1928. Reprinted, New York: Garland, 1987.

Zimmerman, A.M.; Forer, A., eds. Mitosis/Cytokinesis. New York: Academic Press, 1981.

Cited

1. McIntosh, J.R.; McDonald, K.L. The mitotic spindle. *Sci. Am.* 261(4):48–56, 1989.

   Sawin, K.E.; Scholey, J.M. Motor proteins in cell division. *Trends Cell Biol.* 1:122–129, 1991.

2. Gard, D.L.; Hafezi, S.; Zhang, T.; Doxsey, S.J. Centrosome duplication continues in cycloheximide-

treated *Xenopus* blastulae in the absence of a detectable cell cycle. *J. Cell Biol.* 110:2033–2042, 1990.

Maniotis, A.; Schliwa, M. Microsurgical removal of centrosomes blocks cell reproduction and centriole generation in BSC-1 cells. *Cell* 67:495–504, 1991.

Mazia, D. The chromosome cycle and the centrosome cycle in the mitotic cycle. *Int. Rev. Cytol.* 100:49–92, 1987.

Raff, J.W.; Glover, D.M. Centrosomes, and not nuclei, initiate pole cell formation in *Drosophila* embryos. *Cell* 57:611–619, 1989.

3. Lucocq, J.M.; Warren, G. Fragmentation and partitioning of the Golgi apparatus during mitosis in HeLa cells. *EMBO J.* 6:3239–3246, 1987.

McConnell, S.J.; Yaffe, M.P. Intermediate filament formation by a yeast protein essential for organelle inheritance. *Science* 260:687–689, 1993.

Warren, G. Mitosis and membranes. *Nature* 342:857–858, 1989.

4. Gelfand, V.I.; Scholey, J.M. Every motion has its motor. *Nature* 359:480–482, 1992.

Karsenti, E. Mitotic spindle morphogenesis in animal cells. *Semin. Cell Biol.* 2:251–260, 1991.

McIntosh, J.R.; Koonce, M.P. Mitosis. *Science* 246:622–628, 1989.

Wadsworth, P. Mitosis: spindle assembly and chromosome motion. *Curr. Opin. Cell Biol.* 5:123–128, 1993.

5. Belmont, L.D.; Hyman, A.A.; Sawin, K.E.; Mitchison, T.J. Real-time visualization of cell cycle dependent changes in microtubule dynamics in cytoplasmic extracts. *Cell* 62:579–589, 1990.

Buendia, B.; Draetta, G.; Karsenti, E. Regulation of the microtubule nucleating activity of centrosomes in *Xenopus* egg extracts: role of cyclin A-associated protein kinase. *J. Cell Biol.* 116:1431–1442, 1992.

Gotoh, Y.; Nishida, E.; Matsuda, S.; et al. *In vitro* effects on microtubule dynamics of purified *Xenopus* M phase-activated MAP kinase. *Nature* 349:251–254, 1991.

Kuriyama, R.; Borisy, G.G. Microtubule-nucleating activity of centrosomes in Chinese hamster ovary cells is independent of the centriole cycle but coupled to the mitotic cycle. *J. Cell Biol.* 91:822–826, 1981.

Vale, R.D. Severing of stable microtubules by a mitotically activated protein in *Xenopus* egg extracts. *Cell* 64:827–839, 1991.

6. McDonald, K.L.; Edwards, M.K.; McIntosh, J.R. Cross-sectional structure of the central mitotic spindle of *Diatoma vulgare*. *J. Cell Biol.* 83:443–461, 1979.

Nislow, C.; Lombillo, V.A.; Kuriyama, R.; McIntosh, J.R. A plus-end-directed motor enzyme that moves antiparallel microtubules *in vitro* localizes to the interzone of mitotic spindles. *Nature* 359:543–547, 1992.

Saunders, W.S.; Hoyt, M.A. Kinesin-related proteins required for structural integrity of the mitotic spindle. *Cell* 70:451–458, 1992.

Sawin, K.E.; Mitchison, T.J. Mitotic spindle assembly by two different pathways *in vitro*. *J. Cell Biol.* 112:925–940, 1991.

7. Rieder, C.L. The formation, structure and composition of the mammalian kinetochore and kinetochore fiber. *Int. Rev. Cytol.* 79:1–58, 1982.

8. Bloom, K. The centromere frontier: kinetochore components, microtubule-based motility, and the CEN-value paradox. *Cell* 73:621–624, 1993.

Clarke, L.; Carbon, J. The structure and function of yeast centromeres. *Annu. Rev. Genet.* 19:29–56, 1985.

Earnshaw, W.C.; Tomkiel, J.E. Centromere and kinetochore structure. *Curr. Opin. Cell Biol.* 4:86–93, 1992.

Haaf, T.; Warburton, P.E.; Willard, H.F. Integration of human α-satellite DNA into simian chromosomes: centromere protein binding and disruption of normal chromosome segregation. *Cell* 70:681–696, 1992.

9. Euteneuer, U.; McIntosh, J.R. Structural polarity of kinetochore microtubules in PtK$_1$ cells. *J. Cell Biol.* 89:338–345, 1981.

Rieder, C.L.; Alexander, S.P. Kinetochores are transported polewards along a single astral microtubule during chromosome attachment to the spindle in newt lung cells. *J. Cell Biol.* 110:81–95, 1990.

Vallee, R. Dynein and the kinetochore. *Nature* 345:206–207, 1990.

10. Mitchison, T.J. Microtubule dynamics and kinetochore function in mitosis. *Annu. Rev. Cell Biol.* 4:527–549, 1988.

11. Rieder, C.L.; Davison, E.A.; Jensen, L.C.W.; Cassimeris, L.; Salmon, E.D. Oscillatory movements of mono-oriented chromosomes and their position relative to the spindle pole result from the ejection properties of the aster and half-spindle. *J. Cell Biol.* 103:581–591, 1986.

12. Nicklas, R.B.; Krawitz, L.E.; Ward, S.C. Odd chromosome movement and inaccurate chromosome distribution in mitosis and meiosis after treatment with protein kinase inhibitors. *J. Cell Sci.* 104:961–973, 1993.

Nicklas, R.B.; Kubai, D.F. Microtubules, chromosome movement, and reorientation after chromosomes are detached from the spindle by micromanipulation. *Chromosoma* 92:313–324, 1985.

13. Bajer, A.S.; Mole-Bajer, J. Spindle Dynamics and Chromosome Movements. New York: Academic Press, 1972.

Hays, T.S.; Salmon, E.D. Poleward force at the kinetochore in metaphase depends on the number of kinetochore microtubules. *J. Cell Biol.* 110:391–404, 1990.

McNeill, P.A.; Berns, M.W. Chromosome behavior after laser microirradiation of a single kinetochore in mitotic PtK2 cells. *J. Cell Biol.* 88:543–553, 1981.

Oestergren, G. The mechanism of coordination in bivalents and multivalents. The theory of orientation by pulling. *Hereditas* 37:85–156, 1951.

14. Inoue, S.; Ritter, H.J. Dynamics of mitotic spindle organization and function. In Molecules and Cell Movement (S. Inoue, R.E. Stephens, eds.), Society of General Physiologists Series, Vol. 30, pp. 3–30. New York: Raven Press, 1975.

Mitchison, T.J. Polewards microtubule flux in the mitotic spindle: evidence from photoactivation of fluorescence. *J. Cell Biol.* 109:637–652, 1989.

Salmon, E.D.; Leslie, R.J.; Saxton, W.M.; McIntosh, J.R. Spindle microtubule dynamics in sea urchin embryos. Analysis using fluorescence-labeled tubulin and measurements of fluorescence redistribution after laser photobleaching. *J. Cell Biol.* 99:2165–2174, 1984.

Sawin, K.E.; Endow, S.A. Meiosis, mitosis and microtubule motors. *Bioessays* 15:399–407, 1993.

15. Murray, A.W.; Szostak, J.W. Chromosome segregation in mitosis and meiosis. *Annu. Rev. Cell Biol.* 1:289–315, 1985.

Shamu, C.E.; Murray, A.W. Sister chromatid separation in frog egg extracts requires DNA topoisomerase II activity during anaphase. *J. Cell Biol.* 117:921–934, 1992.

Surana, U.; Amon, A.; Dowzer, C.; et al. Destruction of the CDC28/CLB mitotic kinase is not required for the metaphase to anaphase transition in budding yeast. *EMBO J.* 12:1969–1978, 1993.

16. Holloway, S.; Glotzer, M.; King, R.W.; Murray, A.W. Anaphase is initiated by proteolysis rather than by the inactivation of maturation-promoting factor. *Cell* 73:1393–1402, 1993.

Hoyt, M.A.; Totis, L.; Roberts, B.T. *S. cerevisiae* genes required for cell cycle arrest in response to loss of microtubule function. *Cell* 66:507–517, 1991.

Levan, A. The effect of colchicine in root mitoses in *Allium*. *Hereditas* 40:471–486, 1938.

Li, R.; Murray, A.W. Feedback control of mitosis in budding yeast. *Cell* 66:519–532, 1991.

Rieder, C.L.; Palazzo, R.E. Colcemid and the mitotic cycle. *J. Cell Sci.* 102:387–392, 1992.

Zirkle, R.E. Ultraviolet-microbeam irradiation of newt-cell cytoplasm: spindle destruction, false anaphase, and delay of true anaphase. *Radiat. Res.* 41:516–537, 1970.

17. Ris, H. The anaphase movement of chromosomes in the spermatocytes of grasshoppers. *Biol. Bull. (Woods Hole).* 96:90–106, 1949.

18. Coue, M.; Lombillio, V.A.; McIntosh, J.R. Microtubule depolymerization promotes particle and chromosome movement *in vitro*. *J. Cell Biol.* 112:1165–1175, 1991.

Gorbsky, G.J.; Sammak, P.J.; Borisy, G.G. Microtubule dynamics and chromosome motion visualized in living anaphase cells. *J. Cell Biol.* 106:1185–1192, 1988.

Mitchison, T.J.; Salmon, E.D. Poleward kinetochore fiber movement occurs during both metaphase and anaphase-A in newt lung cells. *J. Cell Biol.* 119:569–582, 1992.

Nicklas, R.B. The forces that move chromosomes in mitosis. *Annu. Rev. Biophys. Biophys. Chem.* 17:431–449, 1988.

19. Aist, J.R.; Bayles, C.J.; Tao, W.; Berns, M.W. Direct experimental evidence for the existence, structural basis and function of astral forces during anaphase B *in vivo*. *J. Cell Sci.* 100:279–288, 1991.

Hiramoto, Y.; Nakano, Y. Micromanipulation studies on the mitotic apparatus in sand dollar eggs. *Cell Motil. Cytoskeleton.* 10:172–184, 1988.

Hogan, C.J.; Stephens, L.; Shimizu, T.; Cande, W.Z. Physiological evidence for involvement of a kinesin-related protein during anaphase spindle elongation in diatom central spindles. *J. Cell Biol.* 119:1277–1286, 1992.

McIntosh, J.R.; McDonald, K.L.; Edwards, M.K.; Ross, B.M. Three-dimensional structure of the central mitotic spindle of *Diatoma vulgare*. *J. Cell Biol.* 83:428–442, 1979.

20. Gerace, L.; Blobel, G. The nuclear lamina is reversibly depolymerized during mitosis. *Cell* 19:277–287, 1980.

Gerace, L.; Burke, B. Functional organization of the nuclear envelope. *Annu. Rev. Cell Biol.* 4:335–374, 1988.

Moreno, S.; Nurse, P. Substrates for p34^{cdc2}: *in vivo veritas? Cell* 61:549–551, 1990.

Newport, J. Nuclear reconstitution *in vitro*: stages of assembly around protein-free DNA. *Cell* 48:205–217, 1987.

Nigg, E.A. Assembly and cell cycle dynamics of the nuclear lamina. *Semin. Cell Biol.* 3:245–253, 1992.

Swanson, J.A.; McNeil, P.L. Nuclear reassembly excludes large macromolecules. *Science* 238:548–550, 1987.

21. Mabuchi, I. Biochemical aspects of cytokinesis. *Int. Rev. Cytol.* 101:175–213, 1986.

Salmon, E.D. Cytokinesis in animal cells. *Curr. Opin. Cell Biol.* 1:541–547, 1989.

Satterwhite, L.L.; Pollard, T.D. Cytokinesis. *Curr. Opin. Cell Biol.* 4:43–52, 1992.

Schweisguth, F.; Vincent, A.; Lepesant, J.A. Genetic analysis of the cellularization of the *Drosophila* embryo. *Biol. Cell* 72:15–23, 1991.

22. Bray, D.; White, J.G. Cortical flow in animal cells. *Science* 239:883–888, 1988.

Fluck, R.A.; Miller, A.L.; Jaffe, L.F. Slow calcium waves accompany cytokinesis in *Medaka* fish eggs. *J. Cell Biol.* 115:1259–1265, 1991.

Rappaport, R. Establishment of the mechanism of cytokinesis in animal cells. *Int. Rev. Cytol.* 105:245–281, 1986.

23. Hyman, A.A.; White, G.J. Determination of cell division axes in the early embryogenesis of *Caenorhabditis elegans*. *J. Cell Biol.* 105:2123–2135, 1987.

Schroeder, T.E. Fourth cleavage of sea urchin blastomeres: microtubule patterns and myosin localization in equal and unequal cell divisions. *Dev. Biol.* 124:9–22, 1987.

24. Cao, L.-G.; Wang, Y.-L. Mechanism of the formation of contractile ring in dividing cultured animal cells. II. Cortical movement of microinjected actin filaments. *J. Cell Biol.* 111:1905–1911, 1990.

Fukui, Y.; DeLozanne, A.; Spudich, J.A. Structure and function of the cytoskeleton of a *Dictyostelium* myosin-defective mutant. *J. Cell Biol.* 110:367–378, 1990.

Karess, R.E.; Chang, X.-J.; Edward, K.A.; et al. The regulatory light chain of non-muscle myosin is encoded by *spaghetti-squash*, a gene required for cytokinesis in *Drosophila*. *Cell* 65:1177–1189, 1991.

Mabuchi, I.; Okuno, M. The effect of myosin antibody on the division of starfish blastomeres. *J. Cell Biol.* 74:251–263, 1977.

Schroeder, T.E. Dynamics of the contractile ring. In Molecules and Cell Movement (S. Inoue, R.E. Stephens, eds.), Society of General Physiologists Series, Vol. 30, pp. 305–334. New York: Raven Press, 1975.

25. Cowing, D.W.; Kenyon, C. Expression of the homeotic gene *mab-5* during *Caenorhabditis elegans* embryogenesis. *Development* 116:481–490, 1992.

Kemphues, K.J.; Priess, J.R.; Morton, D.G.; Cheng, N. Identification of genes required for cytoplasmic localization in early *C. elegans* embryos. *Cell* 52:311–320, 1988.

Strome, S. Generation of cell diversity during early embryogenesis in the nematode *Caenorhabditis elegans*. *Int. Rev. Cytol.* 114:81–123, 1989.

26. Lambert, A.-M. Microtubule-organizing centers in higher plants. *Curr. Opin. Cell Biol.* 5:116–122, 1993.

Lloyd, C.W., ed. The Cytoskeletal Basis of Plant Growth and Form. London: Academic Press, 1991.

References

Staiger, C.; Doonan, J. Cell division in plants. *Curr. Opin. Cell Biol.* 5:226–231, 1993.

Staiger, C.J.; Lloyd, C.W. The plant cytoskeleton. *Curr. Opin. Cell Biol.* 3:33–42, 1991.

27. Gunning, B.E.S.; Wick, S.M. Preprophase bands, phragmoplasts and spatial control of cytokinesis. *J. Cell Sci. Suppl.* 2:157–179, 1985.

Marks, J.; Hagan, I.; Hyams, J.S. Growth polarity and cytokinesis in fission yeast: the role of the cytoskeleton. *J. Cell Sci. Suppl.* 5:229–241, 1986.

Pickett-Heaps, J.D.; Northcote, D.H. Organization of microtubules and endoplasmic reticulum during mitosis and cytokinesis in wheat meristems. *J. Cell Sci.* 1:109–120, 1966.

Wick, S.M. Spatial aspects of cytokinesis in plant cells. *Curr. Opin. Cell Biol.* 3:253–260, 1991.

28. de Boer, P.A.J. Chromosome segregation and cytokinesis in bacteria. *Curr. Opin. Cell Biol.* 5:232–237, 1993.

Donachie, W.D.; Robinson, A.C. Cell division: parameter values and the process. In *Escherichia coli* and *Salmonella typhimurium:* Cellular and Molecular Biology (F.C. Neidhardt et al., eds.), pp. 1578–1593. Washington, DC: American Society for Microbiology, 1987

Heath, I.B. Variant mitoses in lower eukaryotes: indicators of the evolution of mitosis? *Int. Rev. Cytol.* 64:1–80, 1980.

Kubai, D.F. The evolution of the mitotic spindle. *Int. Rev. Cytol.* 43:167–227, 1975.

Wise, D. The diversity of mitosis: the value of evolutionary experiments. *Biochem. Cell Biol.* 66:515–529, 1988.

Cells in Their Social Context

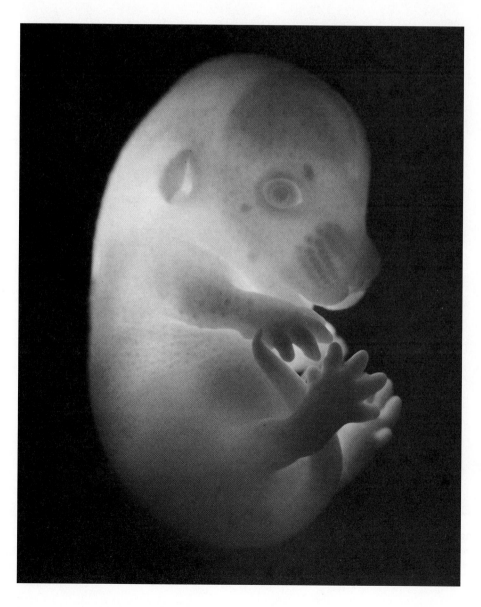

A mouse embryo at 15 days
of gestation.

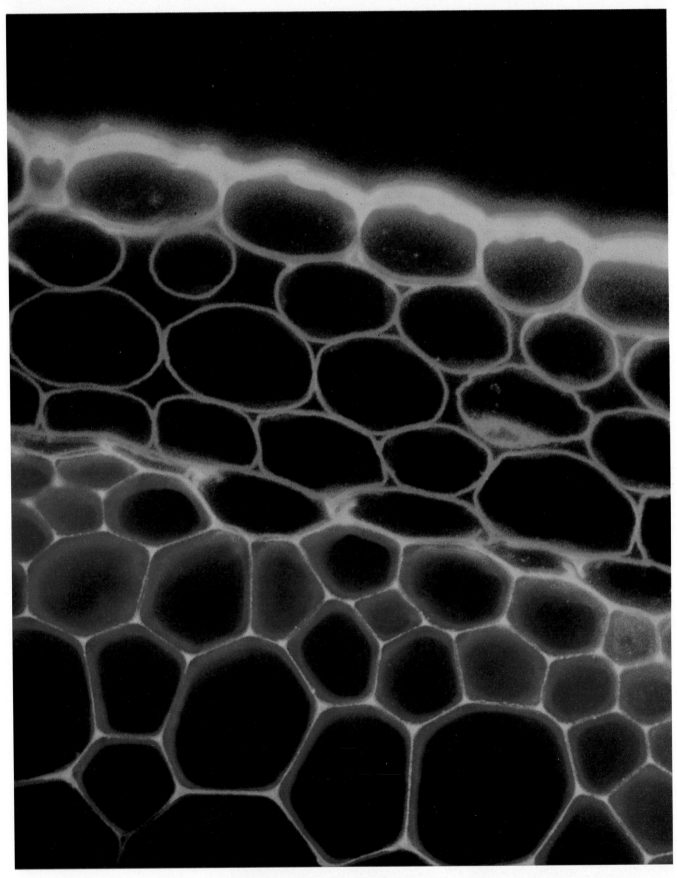

A cross-section of the stem of the flowering plant *Arabidopsis*. By using fluorescent probes for cellulose (*blue*) and pectin (*green*), the relative distribution of these two polysaccharides in the cell wall, or extracellular matrix, has been revealed. (Courtesy of Paul Linstead.)

Cell Junctions, Cell Adhesion, and the Extracellular Matrix

19

Most of the cells in multicellular organisms are organized into cooperative assemblies called *tissues*, which in turn are associated in various combinations to form larger functional units called *organs*. The cells in tissues are usually in contact with a complex network of secreted extracellular macromolecules referred to as the *extracellular matrix*. This matrix helps to hold cells and tissues together, and in animals it provides an organized lattice within which cells can migrate and interact with one another. In many cases the cells in a tissue are also held in place by direct cell-cell adhesions.

In vertebrates the major types of tissues are *nerve, muscle, blood, lymphoid, epithelial,* and *connective tissues*. Connective tissues and epithelial tissues represent two extremes in which the structural roles played by the matrix and by cell-cell adhesions are radically different (Figure 19–1). In **connective tissues** (discussed in Chapter 22) extracellular matrix is plentiful and cells are sparsely distributed within it. The matrix is rich in fibrous polymers, especially collagen, and it is the matrix—rather than the cells—that bears most of the mechanical stress to which the tissue is subjected. The cells are attached to components of the matrix, on which they may exert force, but direct attachments between one cell and another are relatively unimportant. In **epithelial tissues,** by contrast, cells are tightly bound together into sheets (called *epithelia*). Extracellular matrix is scanty and consists mainly of a thin mat called the *basal lamina*, which under-

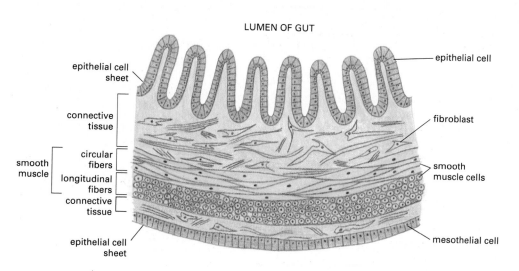

LUMEN OF GUT

epithelial cell sheet
epithelial cell
connective tissue
fibroblast
circular fibers
smooth muscle
longitudinal fibers
connective tissue
smooth muscle cells
epithelial cell sheet
mesothelial cell

Figure 19–1 Simplified drawing of a cross-section through part of the wall of the intestine. This long, tubelike organ is constructed from epithelial tissues (*red*), connective tissues (*green*), and muscle tissues (*yellow*). Each tissue is an organized assembly of cells held together by cell-cell adhesions, extracellular matrix, or both.

lies the cellular sheet; most of the volume is occupied by cells. Here the cells themselves, rather than the matrix, bear most of the mechanical stresses, by means of strong intracellular protein filaments (components of the cytoskeleton) that criss-cross the cytoplasm of each epithelial cell; to transmit mechanical stress from one cell to the next, the filaments are directly or indirectly attached to transmembrane proteins in the plasma membrane, where specialized junctions are formed between the surfaces of adjacent cells and with the underlying basal lamina.

Epithelial cell sheets line all the cavities and free surfaces of the body, and the specialized junctions between the cells enable these sheets to form barriers to the movement of water, solutes, and cells from one body compartment to another. As illustrated in Figure 19–1, epithelial sheets almost always rest on a supporting bed of connective tissue, which may attach them to other tissues (such as muscle) that do not themselves have either strictly epithelial or strictly connective-tissue organization.

In this chapter we first discuss the structure and function of specialized cell-cell and cell-matrix junctions (collectively called *cell junctions*). We then consider how animal cells recognize one another and initiate the formation of cell junctions in the process of assembling into tissues and organs. Finally, we discuss the structure and organization of the extracellular matrix in animals and of the cell wall in plants.

Cell Junctions [1]

Specialized **cell junctions** occur at many points of cell-cell and cell-matrix contact in all tissues, but they are particularly important and plentiful in epithelia. Most of these junctions are too small to be resolved by light microscopy. They can be visualized, however, using either conventional or freeze-fracture electron microscopy, both of which show that the interacting plasma membranes (and often the underlying cytoplasm and the intervening intercellular space as well) are highly specialized in these regions. Cell junctions can be classified into three functional groups: (1) **occluding junctions,** which can seal cells together in an epithelial cell sheet in a way that prevents even small molecules from leaking from one side of the sheet to the other; (2) **anchoring junctions,** which mechanically attach cells (and their cytoskeletons) to their neighbors or to the extracellular matrix; and (3) **communicating junctions,** which mediate the passage of chemical or electrical signals from one interacting cell to its partner.

Table 19–1 A Functional Classification of Cell Junctions

1. **Occluding junctions** (tight junctions)

2. **Anchoring junctions**
   - a. actin filament attachment sites
     - i. cell-cell adherens junctions (e.g., adhesion belts)
     - ii. cell-matrix adherens junctions (e.g., focal contacts)
     - iii. septate junctions (invertebrates only)
   - b. intermediate filament attachment sites
     - i. cell-cell (desmosomes)
     - ii. cell-matrix (hemidesmosomes)

3. **Communicating junctions**
   - a. gap junctions
   - b. chemical synapses
   - c. plasmodesmata (plants only)

The major kinds of intercellular junctions within each class are listed in Table 19–1. We shall discuss each of them in turn, except for chemical synapses, which are formed exclusively by nerve cells and are discussed in Chapters 11 and 15.

Tight Junctions Form a Selective Permeability Barrier Across Epithelial Cell Sheets [2]

Despite the many structural and biochemical differences among various types of epithelia, all have at least one important function in common: they serve as selective permeability barriers, separating fluids on each side that have different chemical compositions. **Tight junctions** play two distinct roles in this selective-barrier function, as we shall illustrate by considering the epithelium of the mammalian small intestine, or gut.

The epithelial cells lining the small intestine keep most of the gut contents in the inner cavity (the lumen). At the same time, however, the cells must transport selected nutrients across the cell sheet from the lumen into the extracellular fluid permeating the connective tissue on the other side (see Figure 19–1), from where the nutrients diffuse into small blood vessels. This *transcellular transport* depends on two sets of membrane-bound carrier proteins: one is confined to the *apical surface* of the epithelial cell (the surface facing the lumen) and actively transports selected molecules into the cell from the lumen of the gut; the other, which is confined to the *basolateral* (basal and lateral) *surface*, allows the same molecules to leave the cell by facilitated diffusion into the extracellular fluid on the other side. If this directional transport is to be maintained, the apical set of carrier proteins must not be allowed to migrate to the basolateral surface of the cell, and the basolateral set must not be allowed to migrate to the apical surface. Furthermore, the spaces between epithelial cells must be sealed so that the transported molecules cannot diffuse back into the gut lumen through the intercellular space (Figure 19–2).

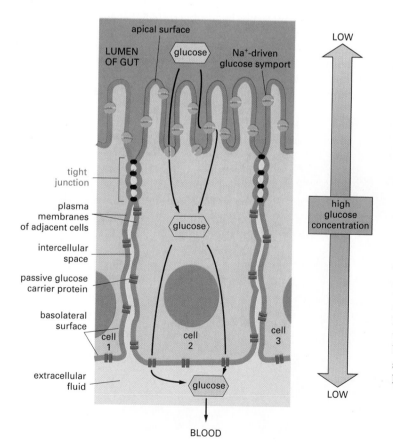

Figure 19–2 The role of tight junctions in transcellular transport. Transport proteins are confined to different regions of the plasma membrane in epithelial cells of the small intestine. This segregation permits a vectorial transfer of nutrients across the epithelial sheet from the gut lumen to the blood. In the example shown, glucose is actively transported into the cell by Na⁺-driven glucose symports at the apical surface, and it diffuses out of the cell by facilitated diffusion mediated by glucose carriers in the basolateral membrane. Tight junctions are thought to confine the transport proteins to their appropriate membrane domains by acting as diffusion barriers within the lipid bilayer of the plasma membrane; these junctions also block the backflow of glucose from the basal side of the epithelium into the gut lumen.

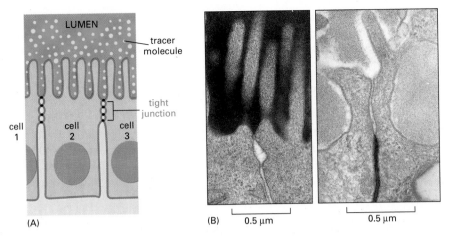

(A)

(B) 0.5 μm   0.5 μm

Figure 19–3 Tight junctions allow cell sheets to serve as barriers to solute diffusion. (A) Schematic drawing showing how a small extracellular tracer molecule added on one side of an epithelial cell sheet cannot traverse the tight junctions that seal adjacent cells together. (B) Electron micrographs of cells in an epithelium where a small, extracellular, electron-dense tracer molecule has been added to either the apical side (on the *left*) or the basolateral side (on the *right*); in both cases the tracer is stopped by the tight junction. (B, courtesy of Daniel Friend.)

The tight junctions between the epithelial cells are thought to block both these kinds of diffusion. First, they function as barriers to the diffusion of membrane proteins between apical and basolateral domains of the plasma membrane (see Figure 19–2). This undesirable diffusion of membrane constituents occurs if tight junctions are disrupted, for example, by removing the extracellular Ca^{2+} required for tight-junction integrity. Second, they seal neighboring cells together so that water-soluble molecules cannot leak between the cells: if a low-molecular-weight tracer is added to one side of an epithelial cell sheet, it will usually not pass beyond the tight junction (Figure 19–3). The seal is not absolute or invariable, however. Although all tight junctions are impermeable to macromolecules, their permeability to small molecules varies greatly in different epithelia. Tight junc-

Figure 19–4 Structure of a tight junction between epithelial cells of the small intestine. The junctions are shown schematically in (A) and in freeze-fracture (B) and conventional (C) electron micrographs. Note that the cells are oriented with their apical ends down. In (B) the plane of the micrograph is parallel to the plane of the membrane, and the tight junction appears as a beltlike band of anastomosing *sealing strands* that encircle each cell in the sheet. The sealing strands are seen as ridges of intramembrane particles on the cytoplasmic fracture face of the membrane (the P face) or as complementary grooves on the external face of the membrane (the E face) (see Figure 19–5). In (C) the junction is seen as a series of focal connections between the outer leaflets of the two interacting plasma membranes, each connection corresponding to a sealing strand in cross-section. (B and C, from N.B. Gilula, in Cell Communication [R.P. Cox, ed.], pp. 1–29. New York: Wiley, 1974. Reprinted by permission of John Wiley & Sons, Inc.)

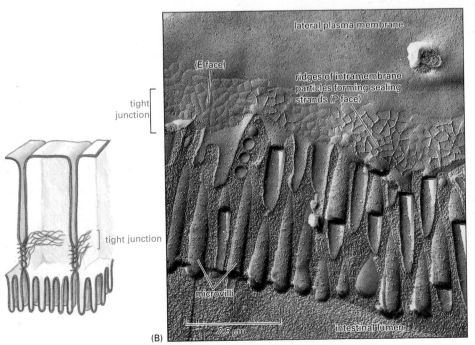

(B)

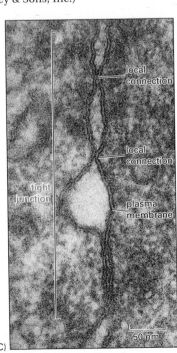

(C)

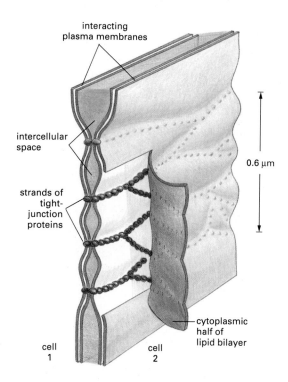

interacting
plasma membranes

intercellular
space

strands of
tight-
junction
proteins

0.6 μm

cytoplasmic
half of
lipid bilayer

cell
1

cell
2

Figure 19–5 A current model of a tight junction. It is postulated that the sealing strands that hold adjacent plasma membranes together are formed by continuous strands of transmembrane junctional proteins, which make contact across the intercellular space and create a seal. In this schematic the cytoplasmic half of one membrane has been peeled back by the artist to expose the protein strands. Two peripheral proteins associated with the cytoplasmic side of tight junctions have been characterized, but the putative transmembrane protein has not yet been identified. In freeze-fracture electron microscopy the tight-junction proteins would remain with the cytoplasmic (P face) half of the lipid bilayer to give the pattern of intramembrane particles seen in Figure 19–4B, instead of staying in the other half as shown here.

tions in the epithelium lining the small intestine, for example, are 10,000 times more leaky to inorganic ions such as Na+ than those in the epithelium lining the urinary bladder. Epithelial cells can transiently alter their tight junctions in order to permit an increased flow of solutes and water through breaches in the junctional barriers. This pathway (called *paracellular transport*) is especially important in the absorption of amino acids and monosaccharides from the lumen of the intestine (where their concentration is sometimes high enough to drive passive transport in the desired direction).

The molecular structure of tight junctions is still uncertain, but freeze-fracture electron microscopy shows them to be composed of an anastomosing network of strands that completely encircles the apical end of each cell in the epithelial sheet (Figure 19–4A and B). In conventional electron micrographs they are seen as a series of focal connections between the outer leaflets of the two interacting plasma membranes (Figure 19–4C). The ability of tight junctions to restrict the passage of ions through the spaces between cells increases logarithmically with increasing numbers of strands in the network, as if each strand acts as an independent barrier. The strands are thought to be composed of long rows of specific transmembrane proteins in each of the two interacting plasma membranes, which join directly to each other to occlude the intercellular space (Figure 19–5).

Anchoring Junctions Connect the Cytoskeleton of a Cell to Those of Its Neighbors or to the Extracellular Matrix

Anchoring junctions are widely distributed in animal tissues. They enable groups of cells, such as those in an epithelium, to function as robust structural units by connecting the cytoskeletal elements of a cell either to those of another cell or to the extracellular matrix (Figure 19–6). They are most abundant in tissues that are subjected to severe mechanical stress, such as heart muscle and skin epithelium (epidermis). They occur in three structurally and functionally different forms: (1) *adherens junctions*, (2) *desmosomes*, and (3) *hemidesmosomes*. Adherens junctions are connection sites for actin filaments; desmosomes and hemidesmosomes are connection sites for intermediate filaments (see Table 19–1, p. 950).

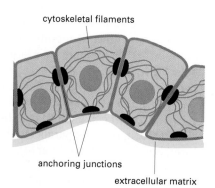

cytoskeletal filaments

anchoring junctions

extracellular matrix

Figure 19–6 Anchoring junctions in an epithelial tissue. Highly schematized drawing of how such junctions join cytoskeletal filaments from cell to cell and from cell to extracellular matrix.

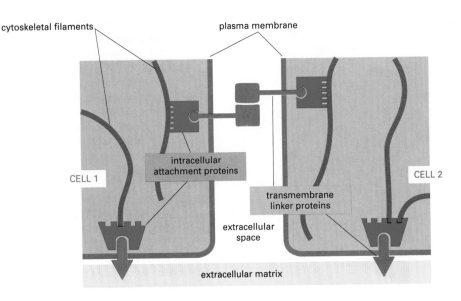

cytoskeletal filaments

plasma membrane

intracellular
attachment proteins

CELL 1

CELL 2

transmembrane
linker proteins

extracellular
space

extracellular matrix

Figure 19–7 **Construction of an anchoring junction.** Highly schematized drawing showing the two classes of proteins that constitute such a junction: intracellular attachment proteins and transmembrane linker proteins.

Before we discuss the different classes of anchoring junctions, it is worth considering briefly the general principles of their construction. As illustrated in Figure 19–7, these junctions are composed of two classes of proteins: (1) *intracellular attachment proteins*, which form a distinct *plaque* on the cytoplasmic face of the plasma membrane and connect the junctional complex to either actin filaments or intermediate filaments; and (2) *transmembrane linker proteins*, whose cytoplasmic domains bind to one or more intracellular attachment proteins, while their extracellular domains interact either with the extracellular matrix or with the extracellular domains of transmembrane linker proteins on another cell.

Much less is known about *septate junctions*, which are unique to invertebrates. They are probably best classified as anchoring junctions, for they act as connection sites for actin filaments; but it has been suggested that they can function as permeability barriers in some cases.

Adherens Junctions Connect Bundles of Actin Filaments from Cell to Cell or from Cell to Extracellular Matrix [3]

Cell-cell adherens junctions occur in various forms. In many nonepithelial tissues they take the form of small punctate or streaklike attachments that connect actin filaments in the cortical cytoplasm of adjacent cells. In epithelial sheets they often form a continuous **adhesion belt** (or *zonula adherens*) around each of the interacting cells in the sheet, located near the apex of each cell just below the tight junction. The adhesion belts in adjacent epithelial cells are directly apposed, and the interacting plasma membranes are held together by transmembrane linker proteins that are members of a large family of Ca^{2+}-dependent cell-cell adhesion molecules called *cadherins*, which we discuss later. At one time an adhesion belt was called a belt desmosome, a misleading name because the adhesion belt is chemically and functionally very different from a real desmosome.

Within each cell a contractile bundle of actin filaments lies adjacent to the adhesion belt, running parallel to the plasma membrane, to which it is attached through a set of intracellular attachment proteins that includes α-, β-, and γ-catenin (discussed later), *vinculin*, α-*actinin*, and *plakoglobin*. The actin bundles in adjacent cells are thus linked, via the cadherins and attachment proteins, into an extensive transcellular network (Figure 19–8). The contraction of this network, which depends on myosin motor proteins, is thought to help mediate a fundamental process in animal morphogenesis—the folding of epithelial cell sheets into tubes and other related structures (Figure 19–9).

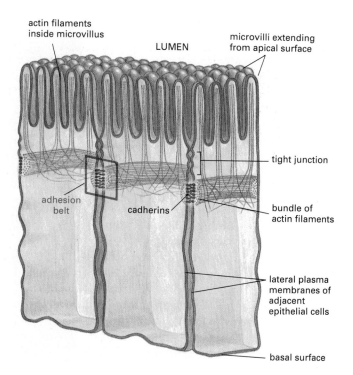

actin filaments inside microvillus

LUMEN

microvilli extending from apical surface

tight junction

adhesion belt

cadherins

bundle of actin filaments

lateral plasma membranes of adjacent epithelial cells

basal surface

Figure 19–8 Adhesion belts between epithelial cells in the small intestine. This beltlike anchoring junction encircles each of the interacting cells. Its most obvious feature is a contractile bundle of actin filaments running along the cytoplasmic surface of the junctional plasma membrane. The actin filaments are joined from cell to cell by transmembrane linker proteins (cadherins), whose extracellular domain binds to the extracellular domain of an identical cadherin molecule on the adjacent cell (see Figure 19–7).

Cell-matrix adherens junctions enable cells to get a hold on the extracellular matrix by connecting their actin filaments to the matrix. Cultured fibroblasts migrating on an artificial substratum coated with extracellular matrix molecules, for example, grip the substratum at specialized regions of the plasma membrane called **focal contacts,** or *adhesion plaques,* where bundles of actin filaments terminate. Many cells in tissues make analogous focal contacts with the surrounding extracellular matrix. The transmembrane linker proteins that mediate these adhesions and serve as links between the matrix and the actin filament bundles in these plaques are members of a large family of cell-surface matrix receptors called *integrins,* which we discuss later. The extracellular domain of the integrin at a focal contact binds to a protein component of the extracellular matrix, while its intracellular domain binds indirectly to bundles of actin filaments via a complex of attachment proteins, including *talin,* α-actinin, and vinculin (Figure 19–10).

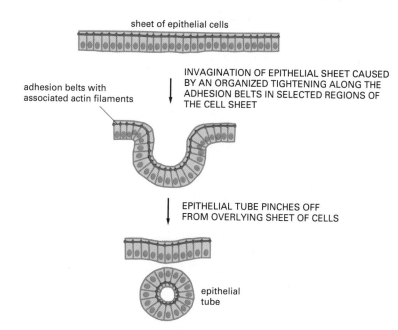

sheet of epithelial cells

adhesion belts with associated actin filaments

INVAGINATION OF EPITHELIAL SHEET CAUSED BY AN ORGANIZED TIGHTENING ALONG THE ADHESION BELTS IN SELECTED REGIONS OF THE CELL SHEET

EPITHELIAL TUBE PINCHES OFF FROM OVERLYING SHEET OF CELLS

epithelial tube

Figure 19–9 The folding of an epithelial sheet to form an epithelial tube. It is thought that the oriented contraction of the bundle of actin filaments running along adhesion belts causes the epithelial cells to narrow at their apex and that this plays an important part in the rolling up of the epithelial sheet into a tube (although cellular rearrangements are also thought to play an important part). An example is the formation of the neural tube in early vertebrate development (discussed in Chapter 21).

Cell Junctions

Septate junctions are widespread in invertebrate tissues. They share a number of features with adhesion belts, with which they sometimes coexist: (1) they form a continuous band around the apical borders of epithelial cells, (2) they are thought to help hold cells together, and (3) they serve as sites of attachment for actin filaments. They have a highly distinctive morphology, for the interacting plasma membranes are joined by poorly characterized junctional proteins that are arranged in parallel rows with a regular periodicity (Figure 19–11).

Desmosomes Connect Intermediate Filaments from Cell to Cell; Hemidesmosomes Connect Them to the Basal Lamina [4]

Desmosomes and *hemidesmosomes* act as rivets to distribute tensile or shearing forces through an epithelium and its underlying connective tissue.

Desmosomes are buttonlike points of intercellular contact that rivet cells together (Figure 19–12A). Inside the cell they serve as anchoring sites for ropelike intermediate filaments, which form a structural framework for the cytoplasm of great tensile strength (Figure 19–12B). Thus, through desmosomes, the intermediate filaments of adjacent cells are connected indirectly to form a continuous network throughout the tissue. The particular type of intermediate filaments attached to the desmosomes depends on the cell type: they are *keratin filaments* in most epithelial cells, for example, and *desmin filaments* in heart muscle cells.

The general structure of a desmosome is illustrated in Figure 19–12C. It has a dense cytoplasmic plaque composed of a complex of intracellular attachment proteins responsible for connecting the cytoskeleton to the transmembrane linker proteins, which interact through their extracellular domains to hold the adjacent plasma membranes together. As in adhesion belts, the transmembrane linker proteins belong to the cadherin family of Ca^{2+}-dependent cell-cell adhesion molecules. The importance of desmosomes in holding cells together is demonstrated by some forms of the potentially fatal skin disease *pemphigus*, in which individuals make antibodies against one of their own desmosomal cadherin proteins; these antibodies bind to and disrupt desmosomes between skin epithelial cells (keratinocytes), causing severe blistering as a result of the leakage of body fluids into the loosened epithelium. The antibodies disrupt desmosomes only in skin, suggesting that these desmosomes are biochemically different from those in other tissues.

Hemidesmosomes, or half-desmosomes, resemble desmosomes morphologically but are both functionally and chemically distinct. Instead of joining adjacent epithelial cell membranes, they connect the basal surface of epithelial cells to the underlying *basal lamina*—a specialized mat of extracellular matrix at the interface between the epithelium and connective tissue. Moreover, whereas the keratin filaments associated with desmosomes make lateral attachments to the desmosomal plaques (see Figure 19–12C), many of those associated with

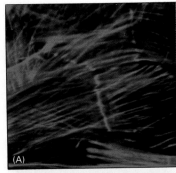

Figure 19–10 **The localization of vinculin at a focal contact.** In these immunofluorescence micrographs, cells in culture have been double-labeled with antibodies against actin (*green*) and vinculin (*red*). Note that vinculin is located at focal contacts, where bundles of actin filaments terminate at the plasma membrane. (From B. Geiger, E. Schmid, and W. Franke, *Differentiation* 23:189–205, 1983.)

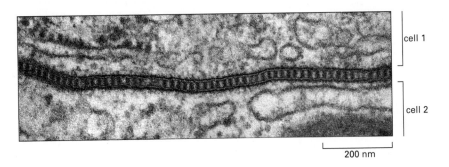

cell 1

cell 2

200 nm

Figure 19–11 **A septate junction.** Electron micrograph of a septate junction between two epithelial cells of a mollusk. The interacting plasma membranes, seen in cross-section, are connected by parallel rows of junctional proteins. The rows, which have a regular periodicity, are seen as dense bars or septa. (From N.B. Gilula, in Cell Communication [R.P. Cox, ed.], pp. 1–29. New York: Wiley, 1974. Reprinted by permission of John Wiley & Sons, Inc.)

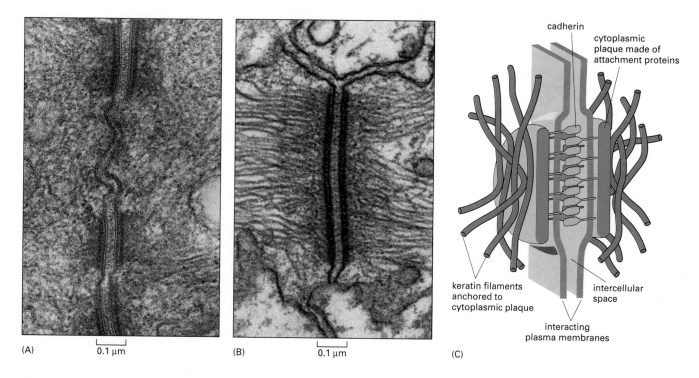

(A) 0.1 μm   (B) 0.1 μm   (C)

cadherin

cytoplasmic plaque made of attachment proteins

keratin filaments anchored to cytoplasmic plaque

intercellular space

interacting plasma membranes

Figure 19–12 Desmosomes. (A) An electron micrograph of three desmosomes between two epithelial cells in the intestine of a rat. (B) An electron micrograph of a single desmosome between two epidermal cells in a developing newt, showing clearly the attachment of intermediate filaments. (C) A schematic drawing of a desmosome. On the cytoplasmic surface of each interacting plasma membrane is a dense plaque composed of a mixture of intracellular attachment proteins (including *plakoglobin* and *desmoplakins*). Each plaque is associated with a thick network of keratin filaments, which are attached to the surface of the plaque. Transmembrane linker proteins, which belong to the cadherin family of cell-cell adhesion molecules, bind to the plaques and interact through their extracellular domains to hold the adjacent membranes together by a Ca²⁺-dependent mechanism. (A, from N.B. Gilula, in Cell Communication [R.P. Cox, ed.], pp. 1–29, New York: Wiley, 1974. Reprinted by permission of John Wiley & Sons, Inc.; B, from D.E. Kelly, *J. Cell Biol.* 28:51–59, 1966, by copyright permission of the Rockefeller University Press.)

hemidesmosomes have their ends buried in the plaque (Figure 19–13). As in focal contacts, the transmembrane linker proteins in hemidesmosomes belong to the integrin family of extracellular matrix receptors, rather than to the cadherin family of cell-cell adhesion proteins used in desmosomes. The intracellular attachment proteins in hemidesmosomes are also different from those in desmosomes.

Thus, although the terminology for the various anchoring junctions is a muddle, the molecular principles (for vertebrates at least) are simple (Table 19–2). Integrins in the plasma membrane anchor a cell to extracellular matrix molecules; cadherins in the plasma membrane anchor it to cadherins in the membrane of an adjacent cell. In both cases there is an intracellular coupling to cytoskeletal filaments, which can be either actin or intermediate filaments depending on the types of intracellular attachment proteins employed. Moreover, for all these classes of anchoring junctions, the adhesion depends on extracellular divalent cations, although the significance of this dependence is unknown.

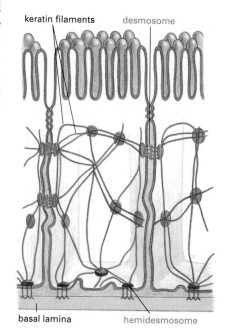

keratin filaments   desmosome

basal lamina   hemidesmosome

Figure 19–13 The distribution of desmosomes and hemidesmosomes in epithelial cells of the small intestine. The keratin filament networks of adjacent cells are indirectly connected to one another through desmosomes and to the basal lamina through hemidesmosomes.

Cell Junctions

Table 19–2 **Anchoring Junctions**

Junction	Transmembrane Linker Protein	Extracellular Ligand	Intracellular Cytoskeletal Attachment	Some Intracellular Attachment Proteins
Adherens (cell-cell)	cadherin (E-cadherin)	cadherin in neighboring cell	actin filaments	catenins, vinculin, α-actinin, plakoglobin
Desmosome	cadherin (desmogleins & desmocollins)	cadherin in neighboring cell	intermediate filaments	desmoplakins, plakoglobin
Adherens (cell-matrix)	integrin	extracellular matrix proteins	actin filaments	talin, vinculin, α-actinin
Hemidesmo-some	integrin ($\alpha_6\beta_4$, see p. 997)	extracellular matrix (basal lamina) proteins	intermediate filaments	desmoplakinlike protein

Gap Junctions Allow Small Molecules to Pass Directly from Cell to Cell [5]

Perhaps the most intriguing cell junction of all is the **gap junction.** It is one of the most widespread, being found in large numbers in most animal tissues and in practically all animal species. It appears in conventional electron micrographs as a patch where the membranes of two adjacent cells are separated by a uniform narrow gap of about 2–4 nm. This gap, however, is spanned by channel-forming protein molecules that allow inorganic ions and other small water-soluble molecules to pass directly from the cytoplasm of one cell to the cytoplasm of the other, thereby coupling the cells both electrically and metabolically. Such *cell coupling* has important functional implications, many of which are only beginning to be understood.

Cell-cell communication of this type was first demonstrated physiologically in 1958, but it took more than 10 years to show that the physiological coupling correlates with the presence of gap junctions seen in the electron microscope. The initial evidence for cell coupling came from electrophysiological studies of specific pairs of interacting nerve cells in the nerve cord of a crayfish. When a voltage gradient was applied across the junctional membrane through electrodes inserted into each of the two interacting cells, an unexpectedly large current flowed, indicating that inorganic ions (which carry current in living tissues) could pass freely from one cell interior to the other. Later experiments showed that small fluorescent dye molecules injected into one cell can likewise pass readily into adjacent cells without leaking into the extracellular space, provided that the molecules are no bigger than about 1000 daltons (Figure 19–14). This suggests a maximal functional pore size for the connecting channels of about 1.5 nm, implying that coupled cells share their small molecules (such as inorganic ions, sugars, amino acids, nucleotides, and vitamins) but not their macromolecules (proteins, nucleic acids, and polysaccharides).

The evidence that gap junctions mediate electrical and chemical coupling between cells in contact with each other comes from several sources. Gap-junction structures can almost always be found where coupling can be demonstrated by electrical or chemical criteria. Conversely, coupling between vertebrate cells is not found where there are no gap junctions. Moreover, dye and electrical coupling can be blocked by a microinjection of antibodies directed against a major gap-junction protein. More recently, molecular methods have provided direct proof: when a gap-junction protein is reconstituted in synthetic lipid bilayers or when mRNA encoding the protein is injected into either a frog oocyte or a gap-

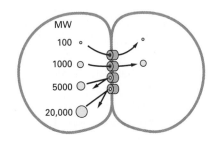

Figure 19–14 Determining the size of a gap-junction channel. When fluorescent molecules of various sizes are injected into one of two cells coupled by gap junctions, molecules smaller than about 1000 daltons can pass into the other cell but larger molecules cannot.

junction-deficient cell line, channels with the properties expected of gap-junction channels can be demonstrated electrophysiologically.

Gap-Junction Connexons Are Composed of Six Subunits [6]

Gap junctions are constructed from transmembrane proteins that form structures called *connexons*. When the connexons in the plasma membranes of two cells in contact are aligned, they form a continuous aqueous channel, which connects the two cell interiors (Figure 19–15). The connexons protrude from each cell surface, holding the interacting plasma membranes at a fixed distance from each other—hence the term gap junction, emphasizing the contrast with a tight junction, where the lipid bilayers appear to be in direct contact (compare Figures19–5 and 19–15). Each connexon is seen as an intramembrane particle in freeze-fracture electron micrographs, and each gap junction can contain a cluster of up to several hundred connexons (Figure 19–16).

A connexon is composed of a ring of six identical protein subunits called *connexins*, each of which contains four putative membrane-spanning α helices. The six subunits are thought to associate to form a connexon with a central aqueous pore that is lined by one transmembrane α helix from each subunit. The six connexins form a larger and more permeable channel than do either the five subunits of the neurotransmitter-gated ion channels or the four subunits (or domains) of the voltage-gated cation channels, which are discussed in Chapter 11 (see Figure 11–33).

Gap junctions in different tissues can have somewhat different properties. The permeability of their individual channels can vary, for example. This is now known to reflect differences in the connexins that form the junctions. In rats, for instance, there are at least 11 distinct connexins, each encoded by a separate gene and each having a distinctive, but sometimes overlapping, tissue distribution. Some cell types express more than one type of connexin, but it is unclear whether different connexin proteins ever assemble into the same connexon. Despite the differences between various connexin proteins, their basic structure and function have been highly conserved in evolution. Thus, in cell culture at least, a cell expressing one type of connexin can often form a functional gap junction with a cell expressing a different connexin, even if the two cells are from different vertebrates.

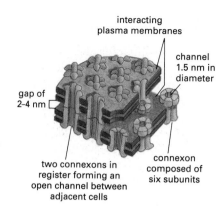

Figure 19–15 A model of a gap junction. The drawing shows the interacting plasma membranes of two adjacent cells. The apposed lipid bilayers (*red*) are penetrated by protein assemblies called *connexons* (*green*), each of which is thought to be formed by six identical protein subunits (called *connexins*). Two connexons join across the intercellular gap to form a continuous aqueous channel connecting the two cells.

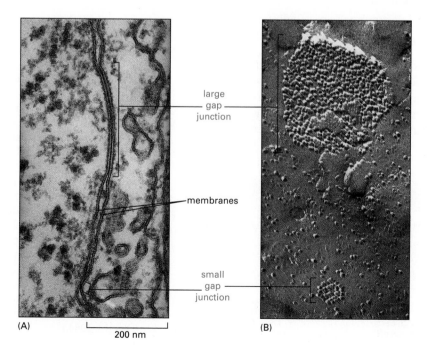

Figure 19–16 Gap junctions as seen in the electron microscope. Thin-section (A) and freeze-fracture (B) electron micrographs of a large and a small gap junction between fibroblasts in culture. In (B) each gap junction is seen as a cluster of homogeneous intramembrane particles associated exclusively with the cytoplasmic fracture face (P face) of the plasma membrane. Each intramembrane particle corresponds to a connexon, illustrated in Figure 19–15. (From N.B. Gilula, in Cell Communication [R.P. Cox, ed.], pp. 1–29. New York: Wiley, 1974. Reprinted by permission of John Wiley & Sons, Inc.)

Most Cells in Early Embryos Are Coupled by Gap Junctions [7]

In tissues containing electrically excitable cells, coupling via gap junctions serves an obvious function. Electrical coupling between nerve cells, for example, allows action potentials to spread rapidly from cell to cell without the delay that occurs at chemical synapses; this is advantageous where speed and reliability are crucial, as in certain escape responses in fish and insects. Similarly, in higher vertebrates, electrical coupling synchronizes the contractions of heart muscle cells and of smooth muscle cells responsible for the peristaltic movements of the intestine.

It is less obvious why gap junctions occur in tissues that do not contain electrically excitable cells. In principle, the sharing of small metabolites and ions provides a mechanism for coordinating the activities of individual cells in such tissues and for smoothing out random fluctuations from cell to cell. The activities of cells in an epithelial cell sheet, for example, such as the beating of cilia, might be coordinated via gap junctions. More generally, since intracellular mediators such as cyclic AMP and Ca^{2+} can pass through gap junctions, responses of coupled cells to extracellular signaling molecules might be propagated and coordinated in this way.

Cell coupling via gap junctions appears to be important in embryogenesis. In early vertebrate embryos (beginning with the late eight-cell stage in mouse embryos) most cells are electrically coupled to one another. As specific groups of cells in the embryo develop their distinct identities and begin to differentiate, however, they commonly uncouple from surrounding tissue. As the neural plate folds up and pinches off to form the neural tube, for instance (see Figure 19–9), its cells uncouple from the overlying ectoderm. Meanwhile the cells within each group remain coupled with one another and so tend to behave as a cooperative assembly, all following a similar developmental pathway in a coordinated fashion.

It is possible that the coupling of cells in embryos provides a pathway for long-range cell signaling within a developing epithelium. A small molecule, for example, could pass through gap junctions from a region of the tissue where its intracellular concentration is kept high to a region where it is kept low, thereby setting up a smooth concentration gradient. The local concentration could then provide cells with "positional information" to control their differentiation according to their location in the embryo (discussed in Chapter 21). Whether gap junctions actually serve this purpose is not known.

The Permeability of Gap Junctions Is Regulated [8]

Like conventional ion channels, individual gap-junction channels do not remain continuously open; instead, they flip between open and closed states. Moreover, the permeability of gap junctions is rapidly (within seconds) and reversibly decreased by experimental manipulations that decrease cytosolic pH or increase the cytosolic concentration of free Ca^{2+}. These observations indicate that gap-

Figure 19–17 **A proposed model for how gap-junction channels may close in response to a rise in Ca^{2+} or a fall in pH in the cytosol.** A small rotation of each subunit closes the channel. The model is based on an image analysis of electron micrographs of rapidly frozen tissue in which the structure of gap junction channels in their presumed open state was compared with their structure in a Ca^{2+}-induced closed state. It is possible that a similar mechanism operates in the opening and closing of the gated ion channels discussed in Chapter 11. (After P.N.T. Unwin and P.D. Ennis, *Nature* 307:609–613, 1984.)

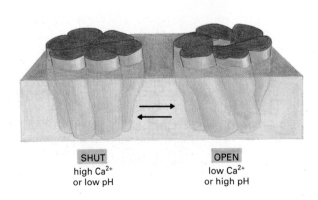

SHUT
high Ca^{2+}
or low pH

OPEN
low Ca^{2+}
or high pH

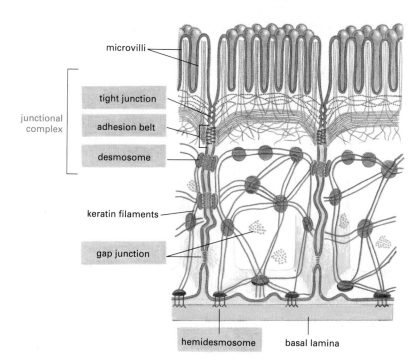

Figure 19–18 Summary of the various cell junctions found in animal cell epithelia. This drawing is based on epithelial cells of the small intestine.

junction channels are dynamic structures that, like conventional ion channels, are *gated*: they can undergo a reversible conformational change that closes the channel in response to changes in the cell. An attractive model for the type of conformational change that might be involved is shown in Figure 19–17.

The physiological role of pH regulation of gap-junction permeability is unknown. There is one case, however, where the reason for the Ca^{2+} control seems clear. When a cell is damaged, its plasma membrane can become leaky. Ions present at high concentration in the extracellular fluid, such as Ca^{2+} and Na^+, then move into the cell, and valuable metabolites leak out. If the cell were to remain coupled to its healthy neighbors, these too would suffer a dangerous disturbance of their internal chemistry. But the influx of Ca^{2+} into the sick cell causes its gap-junction channels to close immediately, effectively isolating the cell and preventing damage from spreading in this way.

Figure 19–18 summarizes the various types of junctions formed by vertebrate cells in an epithelium. In the most apical portion of the cell, the relative positions of the junctions are the same in nearly all epithelia: the tight junction occupies the most apical portion of the cell, followed by the adhesion belt and then by a special parallel row of desmosomes; together these form a structure called a *junctional complex*. Gap junctions and additional desmosomes are less regularly organized.

In Plants, Plasmodesmata Perform Many of the Same Functions as Gap Junctions [9]

The tissues of a plant are organized on different principles from those of an animal. This is because the plant cells are imprisoned within rigid *cell walls*, consisting of an extracellular matrix rich in cellulose, as we discuss later. The system of cell walls eliminates the need for anchoring junctions to hold the cells in place, but the need for direct cell-cell communication remains. Thus, in contrast to animal cells, plant cells have only one class of intercellular junctions, *plasmodesmata*, which, like gap junctions, directly connect the cytoplasms of adjacent cells.

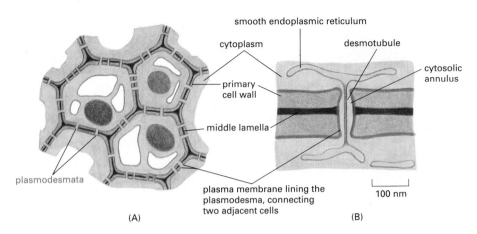

smooth endoplasmic reticulum

cytoplasm

desmotubule

cytosolic annulus

primary cell wall

middle lamella

plasmodesmata

plasma membrane lining the plasmodesma, connecting two adjacent cells

(A)

(B)

100 nm

Figure 19–19 Plasmodesmata. (A) The cytoplasmic channels of plasmodesmata pierce the plant cell wall and connect all cells in a plant together. (B) Each plasmodesma is lined with plasma membrane common to two connected cells. It usually also contains a fine tubular structure, the desmotubule, derived from smooth endoplasmic reticulum.

In plants, however, the cell wall between a typical pair of adjacent cells is at least 0.1 μm thick, so a structure very different from a gap junction is required to mediate communication across it. **Plasmodesmata** (singular, *plasmodesma*) solve the problem. With a few specialized exceptions, every living cell in a higher plant is connected to its living neighbors by plasmodesmata, which form fine cytoplasmic channels through the intervening cell walls. As shown in Figure 19–19A, the plasma membrane of one cell is continuous with that of its neighbor at each plasmodesma, and the cytoplasms of the two cells are connected by a roughly cylindrical channel with a diameter of 20 to 40 nm. Thus the cells of a plant can be viewed as forming a syncytium in which many cell nuclei share a common cytoplasm. Running through the center of the channel in most plasmodesmata is a narrower cylindrical structure, the *desmotubule*, which is continuous with elements of the smooth endoplasmic reticulum in each of the connected cells (Figures 19–19B and 19–20). Between the outside of the desmotubule and the inner face of the cylindrical channel formed by plasma membrane is an annulus of cytosol through which small molecules can pass from cell to cell. Plasmodesmata are normally created in all new cell walls as they are assembled during the cytokinesis phase of a cell division; they form around elements of smooth endoplasmic reticulum that become trapped across the developing cell plate (discussed in Chapter 18).

In spite of the radical difference of structure between plasmodesmata and gap junctions, they seem to function in remarkably similar ways. Evidence obtained by injecting tracer molecules of different sizes suggests that plasmodesmata allow the passage of molecules with a molecular weight of less than about 800, which is similar to the molecular-weight cutoff for gap junctions. As with gap junctions, transport through plasmodesmata is regulated. Dye-injection experiments, for example, show that there can be barriers to the movement of even low-molecular-weight molecules between certain cells that are connected by apparently normal plasmodesmata; the mechanisms that restrict communication in these cases are not understood. Conversely, certain plant viruses can enlarge plasmodesmata and use this route to pass from cell to cell, thereby spreading the infection. These viruses produce special proteins that bind to components of the plasmodesmata and dramatically increase the effective pore size of the channel. It is not clear, however, how these proteins work.

Summary

Many cells in tissues are linked to one another and to the extracellular matrix at specialized contact sites called cell junctions. Cell junctions fall into three functional classes: occluding junctions, anchoring junctions, and communicating junctions. Tight junctions are occluding junctions that play a critical part in maintaining the concentration differences of small hydrophilic molecules across epithelial cell sheets

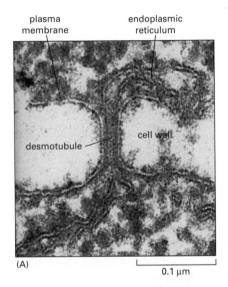

plasma membrane

endoplasmic reticulum

desmotubule

cell wall

(A)

0.1 μm

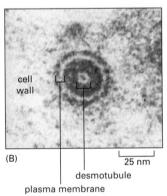

cell wall

(B)

25 nm

desmotubule

plasma membrane

Figure 19–20 Plasmodesmata as seen in the electron microscope. (A) Longitudinal section of a plasmodesma from a water fern. The plasma membrane lines the pore and is continuous from one cell to the next. Endoplasmic reticulum and its association with the central desmotubule can be seen. (B) A similar plasmodesma in cross-section. (Courtesy of R. Overall.)

by (1) sealing the plasma membranes of adjacent cells together to create a continuous, impermeable, or semipermeable barrier to diffusion across the cell sheet and (2) acting as barriers in the lipid bilayer to restrict the diffusion of membrane transport proteins between the apical and the basolateral domains of the plasma membrane in each epithelial cell.

The main types of anchoring junctions in vertebrate tissues are adherens junctions, desmosomes, and hemidesmosomes. Adherens junctions are connecting sites for bundles of actin filaments, whereas desmosomes and hemidesmosomes are connecting sites for intermediate filaments. Septate junctions also serve as connecting sites for actin filaments, but only in invertebrate tissues. Gap junctions are communicating junctions composed of clusters of channel proteins that allow molecules smaller than about 1000 daltons to pass directly from the inside of one cell to the inside of the other. Cells connected by such junctions share many of their inorganic ions and other small molecules and are therefore chemically and electrically coupled. Gap junctions are important in coordinating the activities of electrically active cells, and they are thought to play a coordinating role in other groups of cells as well. Plasmodesmata are the only intercellular junctions in plants; they function like gap junctions even though their structure is entirely different.

Cell-Cell Adhesion [10]

To form an anchoring junction, cells must first adhere. A bulky cytoskeletal apparatus must then be assembled around the molecules that directly mediate the adhesion. The result is a well-defined structure—a desmosome, a hemidesmosome, or an adherens or septate junction—that is easily identified in the electron microscope. Indeed, electron microscopy provided the basis for the original classification of cell junctions. In the early stages of development of a cell junction, however, before the cytoskeletal apparatus has assembled, and especially in embryonic tissues, the cells often adhere to one another without clearly displaying these characteristic structures: in the electron microscope one may simply see two plasma membranes separated by a small gap of a definite width. Functional tests may show, nevertheless, that the two cells are sticking to one another, and biochemical analysis can reveal the molecules responsible for the adhesion.

Thus, while cell-cell junctions and cell-cell adhesion might seem to be two names for the same phenomenon, they correspond in practice to two different experimental approaches—one through electron microscopic description, the other through functional tests and biochemistry—and two different emphases— one on mature, adult structure, the other on developmental function. It is only in recent years that these two approaches have begun to converge in a unified view of the molecular basis of cell junctions and cell adhesion. In the previous section we concentrated on the structures of mature cell junctions. In this section we turn to functional and biochemical studies of the cell-cell adhesion mechanisms that have to operate before a full-blown cell-cell anchoring junction can be constructed; later in the chapter we discuss functional and biochemical studies of cell-matrix adhesion mechanisms. We begin with a developmental question: what mechanisms ensure that an embryonic cell will attach to appropriate neighbors at the right time?

There Are Two Basic Ways in Which Animal Cells Assemble into Tissues [11]

Many simple tissues, including most epithelia, derive from precursor cells whose progeny are prevented from wandering away by being attached to the extracellular matrix or to other cells or to both (Figure 19–21). But the cells, as they accumulate, do not simply remain passively stuck together as a disorderly pile; instead, as we shall see, the tissue architecture is actively maintained by selective adhesions that the cells make and progressively adjust. Thus, if cells of different embryonic tissues are artificially mingled, they will often spontaneously sort out to restore a more normal arrangement.

Such selective adhesion is even more essential for the development of tissues that have more complex origins involving cell migration, whereby one population of cells invades another and assembles with them, and perhaps with other migrant cells, to form an orderly structure. In vertebrate embryos, for example, cells from the *neural crest* break away from the epithelial neural tube with which they are initially associated and migrate along specific paths to many other regions. There they assemble with other cells and with one another and differentiate into a variety of tissues, including those of the peripheral nervous system (Figure 19–22). Such a process requires, first, some mechanism for directing the cells to their final destination, such as the secretion of a soluble chemical that attracts migrating cells (by *chemotaxis*) or the laying down of adhesive molecules in the extracellular matrix or on cell surfaces to guide the migrating cells along the right paths (by *pathway guidance*). Once a migrating cell reaches its destination, it must recognize and join other cells of the appropriate type in order to assemble into a tissue.

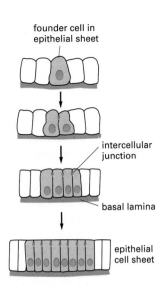

Figure 19–21 **The simplest mechanism by which cells assemble to form a tissue.** The progeny of the founder cell are retained in the epithelial sheet by the basal lamina and by cell-cell adhesion mechanisms, including the formation of intercellular junctions.

Dissociated Vertebrate Cells Can Reassemble into Organized Tissues Through Selective Cell-Cell Adhesion [12]

Unlike adult vertebrate tissues, which are difficult to dissociate, embryonic vertebrate tissues are easily dissociated by treatment with low concentrations of a proteolytic enzyme such as trypsin, sometimes combined with the removal of extracellular Ca^{2+} and Mg^{2+} with a divalent-cation chelator (such as EDTA). These reagents disrupt the protein-protein interactions (many of which are divalent-cation-dependent) that hold cells together. Remarkably, such dissociated cells often reassemble *in vitro* into structures that resemble the original tissue. Such findings suggest that tissue structure is not just a product of history; it is actively maintained and stabilized by the system of affinities that cells have for one another and for the extracellular matrix. Thus, by studying the reassembly of dissociated cells in culture, one can hope to illuminate the role of cell-cell and cell-matrix adhesion in creating and maintaining the organization of tissues in the body.

Experiments on cultured cells from the *epidermis* (the epithelium of the skin) provide an instructive example. The epidermal cells, known as *keratinocytes*, adhere tightly to one another and form a multilayered sheet that rests on a basal lamina. The keratinocytes in the basal layer are relatively undifferentiated and proliferate steadily, releasing progeny into the upper layers. There cell division halts and terminal differentiation occurs (see Figure 22–21). Given a suitable substratum, dissociated keratinocytes in culture will likewise proliferate and differentiate. If the concentration of Ca^{2+} in the culture medium is kept abnormally low, however, so that Ca^{2+}-dependent cell-cell adhesion systems cannot operate, the keratinocytes grow as a monolayer in which proliferating and differentiating cells are intermingled. If the Ca^{2+} concentration is then raised, the spatial organization of the cells is soon transformed: the monolayer is converted into a multilayered epithelium in which the proliferating cells form the basal layer ad-

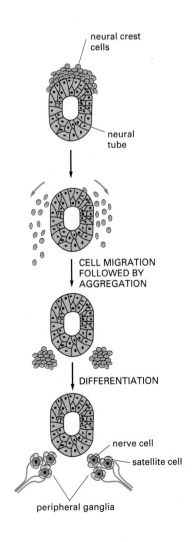

Figure 19–22 **An example of a more complex mechanism by which cells assemble to form a tissue.** Neural crest cells escape from the epithelium forming the upper surface of the neural tube and migrate away to form a variety of cell types and tissues throughout the embryo. Here they are shown assembling and differentiating to form two collections of nerve cells in the peripheral nervous system. Such a collection of nerve cells is called a *ganglion*. Other neural crest cells differentiate in the ganglion to become supporting (satellite) cells surrounding the neurons. Although it is not shown, the neural crest cells proliferate rapidly as they migrate.

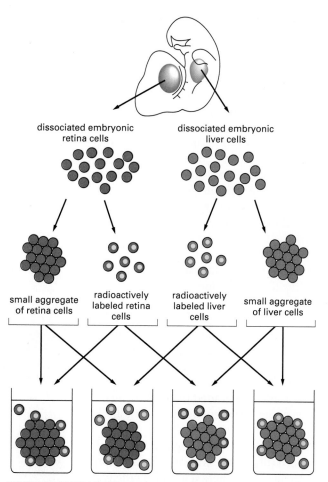

dissociated embryonic
retina cells

dissociated embryonic
liver cells

small aggregate
of retina cells

radioactively
labeled retina
cells

radioactively
labeled liver
cells

small aggregate
of liver cells

MIXING OF RADIOACTIVELY LABELED CELLS WITH CELL AGGREGATES

Figure 19–23 Organ-specific adhesion of dissociated vertebrate embryo cells determined by a radioactive cell-binding assay. The rate of cell adhesion can be measured by determining the number of radioactively labeled cells bound to the cell aggregates after various periods of time. The rate of adhesion is greater between cells of the same kind. In a commonly used modification of this assay, cells labeled with a fluorescent or radioactive marker are allowed to bind to a monolayer of unlabeled cells in culture.

herent to the substratum and the differentiating cells are segregated into the upper layers, just as in normal skin. This result suggests that the normal stratified arrangement of keratinocytes, ordered according to their state of differentiation, is maintained by Ca^{2+}-dependent cell adhesion mechanisms. One such mechanism involves integrin matrix receptors, which we discuss later; these are absent from differentiated epidermal cells but are present on basal cells, which use the integrins to adhere to the basal lamina. Others involve cadherin cell-cell adhesion molecules, which we discuss below.

A still more striking example of the same phenomenon is seen when dissociated cells from two embryonic vertebrate organs such as liver and retina are mixed together and artificially formed into a pellet: the mixed aggregates gradually sort out according to their organ of origin. Similarly, disaggregated cells are found to adhere more readily to aggregates of their own organ than to aggregates of other organs (Figure 19–23). Evidently there are cell-cell recognition systems that make cells of the same differentiated tissue preferentially adhere to one another; these adhesive preferences are presumably important in stabilizing tissue architecture.

What is the molecular basis of this selective cell-cell adhesion in vertebrates? Two distinct classes of **cell-cell adhesion molecules (CAMs)** operate in most multicellular animals, one Ca^{2+}-dependent and the other Ca^{2+}-independent, and it is the Ca^{2+}-dependent molecules that seem to be primarily responsible for the tissue-specific cell-cell adhesion seen in early vertebrate embryos. Both classes of adhesion molecules were initially identified by making antibodies against cell-surface molecules and then testing the antibodies for their ability to inhibit cell-cell adhesion in a test tube. Those rare antibodies that inhibit are then used to characterize and isolate the adhesion molecule recognized by the antibodies.

Cell-Cell Adhesion

The Cadherins Mediate Ca²⁺-dependent Cell-Cell Adhesion in Vertebrates [13]

The **cadherins** are responsible for Ca²⁺-dependent cell-cell adhesion in vertebrate tissues, as mentioned in our account of cell-cell junctions. The first three cadherins that were discovered were named according to the main tissues in which they were found: *E-cadherin* is present on many types of epithelial cells; *N-cadherin* on nerve, muscle, and lens cells; and *P-cadherin* on cells in the placenta and epidermis. All are also found transiently on various other tissues during development. In addition, new types of cadherins are continually being discovered, and at least a dozen are currently known. Virtually all vertebrate cells seem to express one or more cadherins, each encoded by a separate gene, the particular set expressed being characteristic of the cell type. Experiments *in vitro* and *in vivo* demonstrate that cadherins are the main adhesion molecules holding cells together in early embryonic tissues. *In vitro*, the removal of extracellular Ca²⁺ or treatment with anti-cadherin antibodies disrupts the tissue, and if cadherin-mediated adhesion is left intact, antibodies against other adhesion molecules are without effect; *in vivo*, mutations that inactivate the function of cadherins cause embryos to fall apart early in development.

Most cadherins are single-pass transmembrane glycoproteins composed of about 700–750 amino acid residues. The large extracellular part of the polypeptide chain is usually folded into five domains, each containing about 100 amino acid residues; four of these domains are homologous and contain presumptive Ca²⁺-binding sites (Figure 19–24). In the absence of Ca²⁺, the cadherins undergo a large conformational change and, as a result, are rapidly degraded by proteolytic enzymes. The biological significance of the striking Ca²⁺ dependence of cadherin protein function is unknown.

E-cadherin (also called *uvomorulin*) is the best-characterized cadherin. We encountered it earlier when we discussed cell junctions, since it is usually concentrated in *adhesion belts* in mature epithelial cells, where it connects the cortical actin cytoskeletons of the cells it holds together. E-cadherin is also the first cadherin expressed during mammalian development, where it helps to cause *compaction*, an important morphological change that occurs at the eight-cell stage of mouse embryo development. During compaction the loosely attached cells, called *blastomeres,* become tightly packed together and joined by intercellular junctions. Antibodies against E-cadherin block blastomere compaction, whereas antibodies that react with various other cell-surface molecules on these cells do not.

It seems likely that cadherins also play crucial roles in later stages of vertebrate development, since their appearance and disappearance correlate with major morphogenetic events in which tissues segregate from one another. As the neural tube forms and pinches off from the overlying ectoderm, for example, the neural tube cells lose E-cadherin and acquire N-cadherin, while the cells in the overlying ectoderm continue to express E-cadherin (Figure 19–25). Moreover, the neural crest cells that form the peripheral nervous system have large amounts of N-cadherin on their surface when they are associated with the neural tube, lose it while they are migrating, and then reexpress it when they aggregate to form a ganglion (see Figure 19–22). Thus three cell groups that originate from one cell layer exhibit distinct patterns of cadherin expression when separating from one another, suggesting that the switches in cadherin expression are involved in the separation process.

Cadherins Mediate Cell-Cell Adhesion by a Homophilic Mechanism [14]

How do cell-cell adhesion molecules such as the cadherins bind cells together? Three possibilities are illustrated in Figure 19–26: (1) molecules on one cell might bind to other molecules of the same kind on adjacent cells (so-called *homophilic*

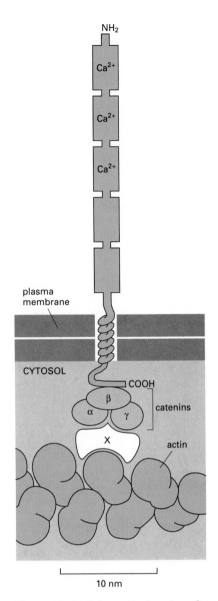

Figure 19–24 Schematic drawing of a typical cadherin molecule. The extracellular part of the protein is folded into five similar domains, three of which contain Ca²⁺-binding sites. The extracellular domain farthest from the membrane is thought to mediate cell-cell adhesion; the sequence His-Ala-Val in this domain seems to be involved, as peptides with this sequence inhibit cadherin-mediated adhesion. The cytoplasmic tail interacts with the actin cytoskeleton via a number of intracellular attachment proteins, including three catenin proteins. α-catenin is structurally related to vinculin. X represents uncharacterized attachment proteins involved in coupling cadherins to actin filaments.

binding); (2) molecules on one cell might bind to molecules of a different kind on adjacent cells (so-called *heterophilic* binding); and (3) cell-surface receptors on adjacent cells might be linked to one another by secreted multivalent linker molecules. All of these mechanisms have been found to operate in animals. Cadherins, however, usually utilize a homophilic mechanism. This has been shown by using a line of cultured fibroblasts called *L cells*, which do not express cadherins and do not adhere to one another. When L cells are transfected with DNA encoding E-cadherin, the transfected cells now adhere to one another by a Ca^{2+}-dependent mechanism and the adhesion is inhibited by anti-E-cadherin antibodies. Since the transfected cells do not bind to untransfected L cells, E-cadherin must bind cells together by the interaction of two E-cadherin molecules on different cells.

If L cells expressing different cadherins are mixed together, they sort out and aggregate separately, indicating that different cadherins preferentially bind to their own type. A similar segregation of cells occurs if L cells expressing different amounts of the same cadherin are mixed together. It seems likely, therefore, that both qualitative and quantitative differences in the expression of cadherins play a crucial part in forming tissues; differences in cadherins probably also explain most of the classical experiments demonstrating organ- and tissue-specific adhesion in a test tube.

Most cadherins, such as E-, N-, and P-cadherins, function as transmembrane linker proteins that mediate interactions between the actin cytoskeletons of the cells they join together. They are, as we have seen, the adhesion proteins around which cell-cell adherens junctions are constructed. A highly conserved cytoplasmic domain of these cadherins interacts with the actin cortex by means of at least three intracellular attachment proteins called *catenins* (see Figure 19–24). This interaction is required for cell-cell adhesion: E-cadherin molecules lacking their cytoplasmic domain are unable to hold cells together. Those cadherins that are localized in desmosomes interact with intermediate filaments rather than actin filaments; their cytoplasmic domain is different and binds to a different set of attachment proteins, which in turn bind to intermediate filaments.

Cadherins are not the only proteins that mediate Ca^{2+}-dependent cell-cell adhesion: some integrins can also bind cells together through heterophilic interactions with other cell-surface proteins, although most integrins mediate the attachment of cells to the extracellular matrix, as we discuss later. In addition, a family of cell-surface carbohydrate-binding proteins (*lectins*) called **selectins** function in a variety of transient cell-cell adhesion interactions in the bloodstream; they enable white blood cells, for example, to bind transiently to endothelial cells lining small blood vessels and thereby to migrate out of the blood into tissues at sites of inflammation. Selectins contain a highly conserved lectin domain that, in the presence of Ca^{2+}, binds to a specific oligosaccharide on another cell—another example of heterophilic cell-cell adhesion (see Figure 10–42). Since selectins have been discussed in Chapter 10, they will not be considered further here.

We discuss later how cells can regulate the adhesive activity of their integrins. In a similar way it seems likely that some cells, at least, can regulate the adhesive activity of their cadherins, although much less is known about cadherin regulation than integrin regulation. Such regulation may be important for the cellular

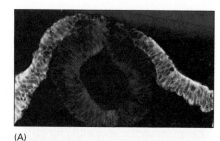
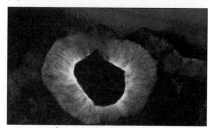

(A)

(B)   100 µm

Figure 19–25 Distribution of E- and N-cadherin in the developing nervous system. Immunofluorescence micrographs of a cross-section of a chick embryo showing the developing neural tube labeled with antibodies against E-cadherin (A) and N-cadherin (B). Note that the overlying ectoderm cells express only E-cadherin, while the cells in the neural tube have lost E-cadherin and have acquired N-cadherin. (Courtesy of Kohei Hatta and Masatoshi Takeichi.)

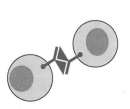
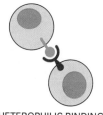
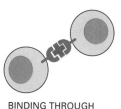

HOMOPHILIC BINDING   HETEROPHILIC BINDING   BINDING THROUGH AN EXTRACELLULAR LINKER MOLECULE

Figure 19–26 Three mechanisms by which cell-surface molecules can mediate cell-cell adhesion. Although all of these mechanisms can operate in animals, the one that depends on an extracellular linker molecule seems to be least common.

Cell-Cell Adhesion

rearrangements that occur within epithelia when these cell sheets change their shape and organization during animal development.

Ca²⁺-independent Cell-Cell Adhesion Is Mediated Mainly by Members of the Immunoglobulin Superfamily of Proteins [15]

The molecules responsible for Ca²⁺-independent cell-cell adhesion belong mainly to the large and ancient immunoglobulin (Ig) superfamily of proteins, so-called because they contain one or more Ig-like domains that are characteristic of antibody molecules (discussed in Chapter 23). The best-studied example is the **neural cell adhesion molecule (N-CAM),** which is expressed by a variety of cell types, including most nerve cells. It is the most prevalent of the Ca²⁺-independent cell-cell adhesion molecules in vertebrates, and, like cadherins, it is thought to bind cells together by a homophilic interaction (between N-CAM molecules on adjacent cells). Some Ig-like cell-cell adhesion proteins, however, use a heterophilic mechanism; some of these, called *intercellular adhesion molecules (ICAMs),* are expressed on activated endothelial cells, where they bind to integrins on the surface of white blood cells and thereby help to trap these blood cells at sites of inflammation.

There are at least 20 forms of N-CAM. Unlike the cadherins, each of which is encoded by a separate gene, the different N-CAM mRNAs are generated by alternative splicing of an RNA transcript produced from a single gene. The large extracellular part of the polypeptide chain in all forms of N-CAM is folded into five Ig-like domains. Most N-CAMs are single-pass transmembrane proteins with variable-sized intracellular domains, which are thought to be involved in cell signaling or binding to the cytoskeleton. One form does not cross the lipid bilayer and is attached to the plasma membrane by a glycosylphosphatidylinositol (GPI) anchor, while another is secreted and may become incorporated into the extracellular matrix (Figure 19–27). Further variation arises from the glycosylation of N-CAM: some forms carry a large quantity of sialic acid (in the highly unusual form of several chains, each containing hundreds of repeating sialic acid residues), while others carry very much less. By virtue of their negative charge, the long sialic acid chains hinder cell adhesion, thereby modifying the adhesive function of the N-CAM. Indeed, it is possible that N-CAM that is heavily loaded with sialic acid may, in some cases, serve to prevent adhesion rather than cause it. In some neurons, for example, the presence of these polysialic acid chains promotes nerve process outgrowth, presumably by making it easier for the growing tips of the processes to let go of the cells to which they are stuck.

There is substantial evidence that N-CAM and its Ig-like relatives play an important part in vertebrate development. When antibodies against either N-CAM or another Ig-related neural cell-cell adhesion molecule called *L1* are injected along the pathway of nerve processes growing from the retina to the brain, they disturb the normal growth pattern of the nerve processes. When used in culture, these antibodies inhibit the tendency of developing nerve cell processes to adhere to one another to form bundles (fascicles). Like N-cadherin, N-CAM is expressed in large amounts on cells of the developing neural tube, but when neural crest cells dissociate from the neural tube and migrate away, they lose N-CAM, only to reexpress it later when they reaggregate to form a neural ganglion (see Figure 19–22). As in the case of cadherins, N-CAM is also expressed transiently during critical stages in the development of many non-neural tissues.

Although cadherins and Ig family members are frequently expressed on the same cells, the adhesions mediated by the cadherins are much stronger, and they almost certainly play the major role in holding cells together, segregating cell collectives into discrete tissues, and maintaining tissue integrity. N-CAM and other members of the Ig family seem to contribute more to the regulation or fine-tuning of these adhesive interactions during development and regeneration. Thus an injection of N-cadherin mRNA into a fertilized frog egg results in the

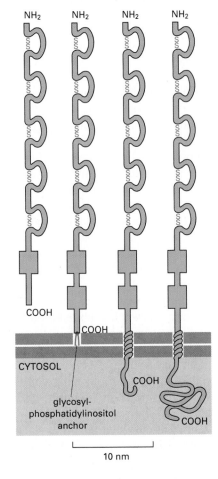

Figure 19–27 Schematic drawing of four forms of N-CAM. The extracellular part of the polypeptide chain in each case is folded into five immunoglobulinlike domains (and one or two other domains called fibronectin type III repeats for reasons that will become clear later). Disulfide bonds (shown in *red*) connect the ends of each loop forming each Ig-like domain.

overexpression of N-cadherin in places where it is not normally expressed and leads to a gross disruption of normal tissue architecture. By contrast, the same experiment performed with N-CAM mRNA leads to relatively minor disturbances in development even though N-CAM is overexpressed in many abnormal locations.

The most critical test of the requirement for a protein in a particular biological process is not to overexpress it but instead to inhibit its production by disrupting the gene. While this can now be done in some vertebrates, it is most readily done in genetically tractable invertebrates such as *Drosophila* and the nematode *C. elegans*. A number of Ig-like proteins that mediate Ca^{2+}-independent cell-cell adhesion have been defined in *Drosophila*. One of these, **fasciclin II**, is a close relative of N-CAM: like N-CAM, it has five Ig-like domains and operates by homophilic binding. It is expressed mainly on a subset of nerve cell processes and on some of the glial cells they contact during development. If both copies of the fasciclin II gene are inactivated by mutation, the gross structure of the nervous system is normal. However, at least two of the nerve cell processes that normally express fasciclin II and adhere together now fail to recognize each other and therefore do not form a bundle. This observation is consistent with the view that Ig-like cell-cell adhesion molecules play subtle but important roles in development.

Multiple Types of Cell-Surface Molecules Act in Parallel to Mediate Selective Cell-Cell and Cell-Matrix Adhesion [16]

Morphological, cell biological, and biochemical studies all indicate that even a single cell type utilizes multiple molecular mechanisms in adhering to other cells

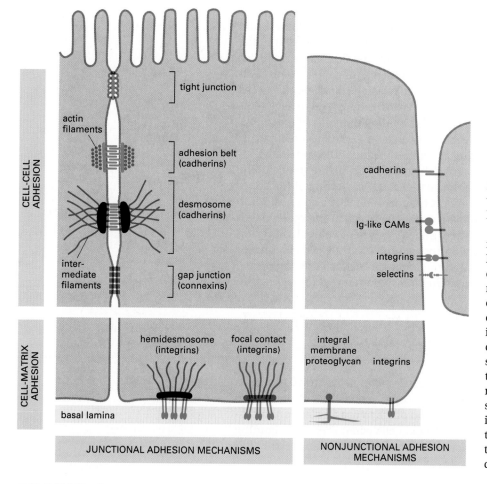

Figure 19–28 **A summary of the junctional and nonjunctional adhesive mechanisms used by animal cells in binding to one another and to the extracellular matrix.** The junctional mechanisms are shown in epithelial cells, while the nonjunctional mechanisms are shown in nonepithelial cells. A junctional interaction is operationally defined as one that can be seen as a specialized region of contact by conventional and/or freeze-fracture electron microscopy. Note that the integrins and cadherins are involved in both nonjunctional and junctional cell-cell (cadherins) and cell-matrix (integrins) contacts. The cadherins generally mediate homophilic interactions, whereas the integrins mediate heterophilic interactions (see Figure 19–26). Both the cadherins and integrins act as transmembrane linkers and depend on extracellular divalent cations to function; for this reason, most cell-cell and cell-matrix contacts are divalent-cation-dependent. The selectins and integrins can also act as heterophilic cell-cell adhesion molecules: the selectins bind to carbohydrate, while the cell-binding integrins bind to members of the immunoglobulin superfamily. The integrins and integral membrane proteoglycans that mediate nonjunctional adhesion to the extracellular matrix are discussed later.

Cell-Cell Adhesion

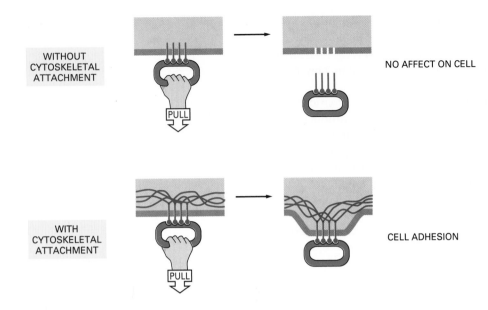

WITHOUT CYTOSKELETAL ATTACHMENT

PULL

NO AFFECT ON CELL

WITH CYTOSKELETAL ATTACHMENT

PULL

CELL ADHESION

Figure 19–29 Importance of the cytoskeleton in cell adhesion. This drawing illustrates why cell-adhesion molecules must be linked to the cytoskeleton in order to mediate robust cell-cell or cell-matrix adhesion. In reality, many adhesion proteins would probably be pulled from the cell with bits of attached membrane, and the holes left in the membrane would immediately reseal.

and to the extracellular matrix. Some of these mechanisms involve organized cell junctions; others do not (Figure 19–28). Just as each cell in a multicellular animal contains an assortment of cell-surface receptors that enables the cell to respond specifically to a complementary assortment of soluble chemical signals such as hormones and growth factors, so each cell in a tissue has a particular combination (and concentration) of cell-surface receptors (cell adhesion molecules) that enables it to bind in its own characteristic way to other cells and to the extracellular matrix. And just as receptors for soluble chemical signals generate intracellular signals that alter the cell's behavior, so too can cell adhesion molecules, although the signaling mechanisms are less well understood for these molecules.

Unlike receptors for soluble chemical signals, which bind their specific ligand with high affinity, the receptors that bind to molecules on cell surfaces or in the extracellular matrix usually do so with relatively low affinity. The latter receptors therefore rely on the enormous increase in binding strength gained through simultaneous binding of multiple receptors to multiple ligands on an opposing cell or in the adjacent matrix. One could call this the "Velcro principle." We have seen, however, that the interaction of the extracellular binding domains of these cell-surface molecules is not enough to ensure cell adhesion: at least in the case of cadherins and, as we shall see, integrins, the adhesion molecules must also attach (via attachment proteins) to the cortical cytoskeleton inside the cell. The cytoskeleton is thought to assist and stabilize the lateral clustering of the adhesion molecules so as to facilitate multipoint binding, and it is also required to enable the adhering cell to exert traction on the adjacent cell or matrix (and vice versa) (Figure 19–29). Thus the mixture of specific types of cell-cell adhesion molecules and matrix receptors present on any two cells, as well as their concentration, cytoskeletal linkages, and distribution on the cell surface, will determine the total affinity with which the two cells bind to each other and to the matrix.

Nonjunctional Contacts May Initiate Tissue-specific Cell-Cell Adhesions That Junctional Contacts Then Orient and Stabilize [16]

Which, if any, of the several types of intercellular junctions discussed earlier in this chapter are involved as cells migrate and recognize one another during the formation of tissues and organs? One way to find out is to use an electron microscope to examine the contacts between adjacent cells when they are migrating over each other in developing embryos or in adult tissues undergoing repair after injury. Such studies show that, with the exception of cells reorganizing within an epithelium, these contacts generally do not involve the formation of organized intercellular junctions. Nevertheless, the interacting plasma membranes often come close together and run parallel, separated by a space of 10–20 nm. As several known transmembrane proteins extend above the plasma membrane by 10–20 nm or more, two cell-surface proteins could readily interact directly with each other across the 10–20-nm gap to mediate the adhesion. This type of nonjunctional contact may be optimal for cell locomotion—close enough to give traction but not tight enough to immobilize the cell.

As anchoring junctions (adherens junctions, desmosomes, hemidesmosomes, and, in insects, septate junctions) are generally not seen between migrating embryonic cells, the formation of such junctions might be an important mechanism for immobilizing cells within an organized tissue once it has formed. In addition, within epithelia the formation of intercellular junctions is thought to be necessary for mechanical strength and to help polarize and orient the constituent cells. A reasonable hypothesis is that nonjunctional cell-cell adhesion proteins initiate tissue-specific cell-cell adhesions, which are then oriented and stabilized by the assembly of full-blown intercellular junctions. As many of the transmembrane proteins involved can diffuse in the plane of the plasma membrane, they can accumulate at sites of cell-cell (and cell-matrix) contact and therefore be used for junctional as well as nonjunctional adhesions. This has been demonstrated to occur for some integrins and cadherins, which help initiate cell adhesion and then later become integral parts of cell junctions.

As an increasing number of monoclonal antibodies and peptide fragments are characterized, each of which blocks a single type of cell-cell adhesion molecule or matrix receptor—and as the genes that encode these cell-surface proteins become available for manipulation in cells in culture and in experimental animals—it should be possible to inactivate the various types of cell-cell adhesion proteins and matrix receptors individually and in different combinations in order to decipher the rules of recognition and binding used in the morphogenesis of complex tissues.

Summary

Cells dissociated from various tissues of vertebrate embryos preferentially reassociate with cells from the same tissue when they are mixed together. This tissue-specific recognition process in vertebrates is mainly mediated by a family of Ca^{2+}-dependent cell-cell adhesion proteins called cadherins, which hold cells together by a homophilic interaction between transmembrane cadherin proteins on adjacent cells. In order to hold cells together, the cadherins must be attached to the cortical cytoskeleton. Most animal cells also have Ca^{2+}-independent cell-cell adhesion systems that mainly involve members of the immunoglobulin superfamily, which includes the neural cell adhesion molecule N-CAM. As even a single cell type uses multiple molecular mechanisms in adhering to other cells (and to the extracellular matrix), the specificity of cell-cell adhesion seen in embryonic development must result from the integration of a number of different adhesion systems, some of which are associated with specialized cell junctions while others are not.

The Extracellular Matrix of Animals [17]

Tissues are not made up solely of cells. A substantial part of their volume is *extracellular space*, which is largely filled by an intricate network of macromolecules constituting the **extracellular matrix** (Figure 19–30). This matrix is composed of a variety of versatile proteins and polysaccharides that are secreted locally and assembled into an organized meshwork in close association with the surface of the cell that produced them. Whereas we discussed cell junctions chiefly in the context of epithelial tissues, our account of extracellular matrix concentrates chiefly on **connective tissues** (Figure 19–31). In these tissues the matrix is frequently more plentiful than the cells that it surrounds, and it determines the tissue's physical properties. Connective tissues form the architectural framework of the vertebrate body, but the amounts found in different organs vary greatly: from skin and bone, in which they are the major component, to brain and spinal cord, in which they are only minor constituents.

Variations in the relative amounts of the different types of matrix macromolecules and the way they are organized in the extracellular matrix give rise to an amazing diversity of forms, each adapted to the functional requirements of the particular tissue. The matrix can become calcified to form the rock-hard structures of bone or teeth, or it can form the transparent matrix of the cornea, or it can adopt the ropelike organization that gives tendons their enormous tensile strength. At the interface between an epithelium and connective tissue, the matrix forms a *basal lamina,* a thin but tough mat that plays an important part in controlling cell behavior. We shall focus on the extracellular matrix of vertebrates, but other organisms make many unique and interesting related materials, as in the cell walls of bacteria, the cuticles of worms and insects, the shells of mollusks, and, as we discuss later, the cell walls of plants.

Until recently the vertebrate extracellular matrix was thought to serve mainly as a relatively inert scaffolding to stabilize the physical structure of tissues. But now it is clear that the matrix plays a far more active and complex role in regulating the behavior of the cells that contact it—influencing their development, migration, proliferation, shape, and function. The extracellular matrix has a correspondingly complex molecular composition. Although our understanding of its organization is still fragmentary, there has been rapid progress in characterizing many of its major components.

The Extracellular Matrix Is Made and Oriented by the Cells Within It [17]

The macromolecules that constitute the extracellular matrix are mainly produced locally by cells in the matrix. As we discuss later, these cells also help to pattern the matrix, in that the orientation of their cytoskeleton influences the orientation of the matrix they produce. In most connective tissues the matrix macromolecules are secreted largely by cells called **fibroblasts** (Figure 19–32). In some specialized connective tissues such as cartilage and bone, however, they are secreted by cells of the fibroblast family that have more specific names: *chondroblasts,* for example, form cartilage, and *osteoblasts* form bone. The two main classes of extracellular macromolecules that make up the matrix are (1) polysaccharide chains of the class called *glycosaminoglycans (GAGs),* which are

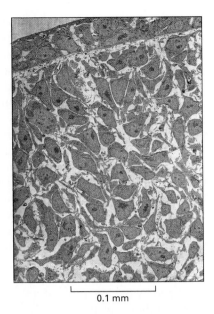

0.1 mm

Figure 19–30 **Cells surrounded by spaces filled with extracellular matrix.** The particular cells shown in this low-power electron micrograph are those in an embryonic chick limb bud. The cells have not yet acquired their specialized characteristics. (Courtesy of Cheryll Tickle.)

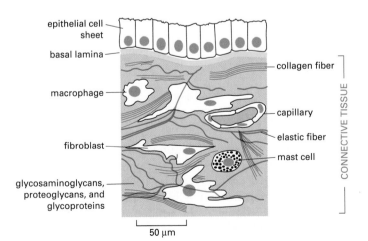

50 μm

Figure 19–31 **The connective tissue underlying an epithelial cell sheet.** It consists largely of extracellular matrix that is secreted by the fibroblasts.

usually found covalently linked to protein in the form of *proteoglycans,* and (2) fibrous proteins of two functional types: mainly structural (for example, *collagen* and *elastin*) and mainly adhesive (for example, *fibronectin* and *laminin*). We shall see (in Figure 19–57) that the members of both classes come in a great variety of shapes and sizes. Glycosaminoglycan and proteoglycan molecules in connective tissue form a highly hydrated, gel-like "ground substance" in which the fibrous proteins are embedded; the polysaccharide gel resists compressive forces on the matrix, and the collagen fibers provide tensile strength. The aqueous phase of the polysaccharide gel permits the rapid diffusion of nutrients, metabolites, and hormones between the blood and the tissue cells; the collagen fibers both strengthen and help to organize the matrix, and rubberlike elastin fibers give it resilience. The adhesive proteins help cells attach to the appropriate part of the extracellular matrix: fibronectin, for example, promotes the attachment of fibroblasts and various other cells to the matrix in connective tissues, while laminin promotes the attachment of epithelial cells to the basal lamina.

Glycosaminoglycan (GAG) Chains Occupy Large Amounts of Space and Form Hydrated Gels [18]

Glycosaminoglycans (GAGs) are unbranched polysaccharide chains composed of repeating disaccharide units. They are called GAGs because one of the two sugar residues in the repeating disaccharide is always an amino sugar (*N*-acetylglucosamine or *N*-acetylgalactosamine), which in most cases is sulfated. The second sugar is usually a uronic acid (glucuronic or iduronic). Because there are sulfate or carboxyl groups on most of their sugar residues, GAGs are highly negatively charged (Figure 19–33). Four main groups of GAGs have been distinguished by their sugar residues, the type of linkage between these residues, and the number and location of sulfate groups: (1) *hyaluronan,* (2) *chondroitin sulfate* and *dermatan sulfate,* (3) *heparan sulfate* and *heparin,* and (4) *keratan sulfate.*

Polysaccharide chains are too inflexible to fold up into the compact globular structures that polypeptide chains typically form. Moreover, they are strongly hydrophilic. Thus GAGs tend to adopt highly extended conformations that occupy a huge volume relative to their mass (Figure 19–34), and they form gels even at very low concentrations. Their high density of negative charges attracts a cloud of cations, such as Na^+, that are osmotically active, causing large amounts of water to be sucked into the matrix. This creates a swelling pressure, or turgor, that enables the matrix to withstand compressive forces (in contrast to collagen fibrils, which resist stretching forces). The cartilage matrix that lines the knee joint, for example, can support pressures of hundreds of atmospheres by this mechanism.

10 µm

Figure 19–32 Scanning electron micrograph of fibroblasts in connective tissue. The tissue is from the cornea of a rat. The extracellular matrix surrounding the fibroblasts is composed largely of collagen fibrils (there are no elastic fibers in the cornea). The glycoproteins, glycosaminoglycans, and proteoglycans, which normally form a hydrated gel filling the interstices of the fibrous network, have been removed by enzyme and acid treatment. (From T. Nishida et al. *Invest. Ophthalmol. Vis. Sci.* 29:1887–1890, 1988.)

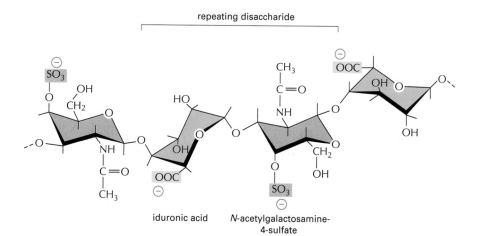

repeating disaccharide

iduronic acid

N-acetylgalactosamine-4-sulfate

Figure 19–33 The repeating disaccharide sequence of a dermatan sulfate glycosaminoglycan (GAG) chain. These chains are typically 70 to 200 sugar residues long. There is a high density of negative charges along the chain resulting from the presence of both carboxyl and sulfate groups.

The amount of GAGs in connective tissue is usually less than 10% by weight of the amount of the fibrous proteins. Because they form porous hydrated gels, however, the GAG chains fill most of the extracellular space, providing mechanical support to tissues while still allowing the rapid diffusion of water-soluble molecules and the migration of cells. The importance of GAGs is illustrated by a rare human genetic disease in which there is a severe deficiency in the synthesis of the dermatan sulfate disaccharide shown in Figure 19–33. The affected individuals are dwarves, have a prematurely aged appearance, and have generalized defects in their skin, joints, muscles, and bones.

It should be emphasized, however, that in invertebrates and in plants other types of polysaccharides often dominate the structure of the extracellular matrix. Thus in higher plants, as we discuss later, cellulose (polyglucose) chains are packed tightly together in ribbonlike crystalline arrays to form the microfibrillar component of the cell wall. In insects, crustaceans, and other arthropods, *chitin* (poly-*N*-acetylglucosamine) similarly forms the main component of the exoskeleton. Together, cellulose and chitin are the most abundant biopolymers on earth.

Hyaluronan Is Thought to Facilitate Cell Migration During Tissue Morphogenesis and Repair [19]

Hyaluronan (also called *hyaluronic acid* or *hyaluronate*) is the simplest of the GAGs. It consists of a regular repeating sequence of up to 25,000 nonsulfated disaccharide units (Figure 19–35). It is found in variable amounts in all tissues and fluids in adult animals and is especially abundant in early embryos. Because of its simplicity, hyaluronan is thought to represent the earliest evolutionary form of GAG, but it is not typical of the majority of GAGs. All of the others (1) contain sulfated sugars, (2) tend to contain a number of different disaccharide units arranged in more complex sequences, (3) have much shorter chains, consisting of fewer than 300 sugar residues, and (4) are covalently linked to protein to form proteoglycans. Moreover, whereas other GAGs are synthesized inside the cell and released by exocytosis, hyaluronan is spun out directly from the cell surface by an enzyme complex that is embedded in the plasma membrane.

Hyaluronan is thought to play a part in resisting compressive forces in tissues and joints. It also has an important role as a space filler during embryonic development, where it can be used to force a change in the shape of a structure. Like styrofoam, it can be quickly and cheaply produced: a small quantity expands with water to occupy a large volume (see Figure 19–34). Hyaluronan synthesized from the basal side of an epithelial sheet, for example, often serves to create a cell-free space into which cells subsequently migrate; this occurs in the formation of the heart, the cornea, and several other organs. When cell migration ends, the excess hyaluronan is generally degraded by the enzyme *hyaluronidase*. Hyaluronan is also produced in large quantities during wound repair, and it is an important constituent of joint fluid, where it serves as a lubricant.

globular protein (MW 50,000)

glycogen (MW ~ 400,000)

spectrin (MW 460,000)

collagen (MW 290,000)

hyaluronan (MW 8 x 10⁶)

300 nm

Figure 19–34 The relative dimensions and volumes occupied by various macromolecules. Several proteins, a glycogen granule, and a single hydrated molecule of hyaluronan are shown.

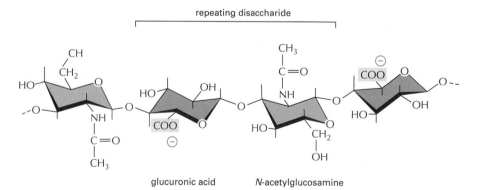

repeating disaccharide

glucuronic acid

N-acetylglucosamine

Figure 19–35 The repeating disaccharide sequence in hyaluronan, a relatively simple GAG. It consists of a single long chain of up to 25,000 sugar residues. Note the absence of sulfate groups.

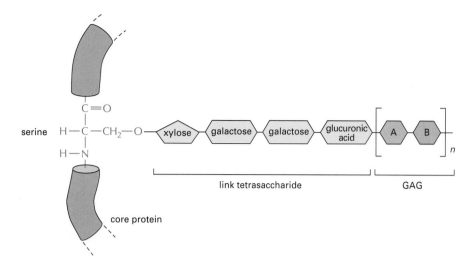

Figure 19–36 The linkage between a GAG chain and its core protein in a proteoglycan molecule. A specific link tetrasaccharide is first assembled on a serine residue. In most cases it is not clear how the serine residue is selected, but it seems to be a specific local conformation of the polypeptide chain, rather than a specific linear sequence of amino acids, that is recognized. The rest of the GAG chain, consisting mainly of a repeating disaccharide unit, is then synthesized, with one sugar residue being added at a time. In *chondroitin sulfate* the disaccharide is composed of D-glucuronic acid and *N*-acetyl-D-galactosamine; in *heparan sulfate* it is D-glucosamine (or L-iduronic acid) and *N*-acetyl-D-glucosamine; in *keratan sulfate* it is D-galactose and *N*-acetyl-D-glucosamine.

Many of the functions of hyaluronan depend on specific hyaluronan-binding proteins and proteoglycans, some of which are constituents of the extracellular matrix, while others are integral components of the surface of cells. A number of these molecules (sometimes referred to as *hyaladherins*) have been shown to have homologous hyaluronan-binding domains containing a characteristic cluster of positively charged amino acid residues.

Proteoglycans Are Composed of GAG Chains Covalently Linked to a Core Protein [20]

Except for hyaluronan, all GAGs are found covalently attached to protein in the form of **proteoglycans,** which are made by most animal cells. As is the case for almost all glycoproteins, the polypeptide chain, or *core protein,* of a proteoglycan is made on membrane-bound ribosomes and threaded into the lumen of the endoplasmic reticulum. The polysaccharide chains are assembled on the core protein mainly in the Golgi apparatus: first a special *link tetrasaccharide* is attached to a serine residue on the core protein to serve as a primer for polysaccharide growth; then one sugar residue at a time is added by specific glycosyl transferases (Figure 19–36). While still in the Golgi apparatus, many of the polymerized sugar residues are covalently modified by a sequential and coordinated series of sulfation reactions and epimerization reactions. The epimerizations alter the configuration of the substituents around individual carbon atoms in the sugar molecule; the sulfations greatly increase the negative charge of the proteoglycans.

The proteoglycans are usually easily distinguished from other glycoproteins by the nature, quantity, and arrangement of their sugar side chains: by definition, at least one of the sugar side chains of a proteoglycan must be a GAG. Glycoproteins contain from 1% to 60% carbohydrate by weight in the form of numerous relatively short, branched oligosaccharide chains. The core protein in a proteoglycan is usually a glycoprotein, but it can contain as much as 95% carbohydrate by weight, mostly in the form of long unbranched GAG chains, each typically about 80 sugar residues long. Proteoglycans can thus be much larger than glycoproteins. The *aggrecan* proteoglycan, for example, which is a major component of cartilage, has a mass of about 3×10^6 daltons; it has over 100 GAG chains, approximately 1 for every 20 amino acid residues. On the other hand, many proteoglycans are much smaller and have only 1 to 10 GAG chains; an example is *decorin*, which is secreted by fibroblasts and has a single GAG chain (Figure 19–37).

In principle, proteoglycans have the potential for almost limitless heterogeneity. Core proteins range in molecular weight from 10,000 to more than 600,000 daltons and vary greatly in the number and types of their attached GAG chains. Moreover, the underlying repeating pattern of disaccharides in each GAG can be

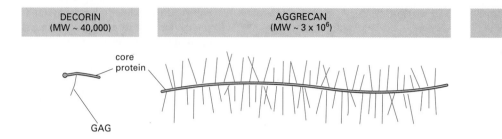

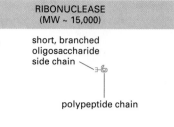

| DECORIN (MW ~ 40,000) | AGGRECAN (MW ~ 3 x 10⁶) | RIBONUCLEASE (MW ~ 15,000) |

core protein

GAG

short, branched oligosaccharide side chain

polypeptide chain

100 nm

modified by a complex pattern of sulfate groups. The heterogeneity of these GAGs makes it difficult to identify and classify proteoglycans in terms of their sugars. The sequences of many core proteins have been determined with the aid of recombinant DNA techniques, and they too are extremely diverse. Although a few small families have been recognized, there is no common structural feature that clearly distinguishes proteoglycan core proteins from other proteins, and many have one or more domains that are homologous to domains found in other proteins of the extracellular matrix or plasma membrane. Thus it is probably best to regard proteoglycans as a diverse group of highly glycosylated glycoproteins whose functions are mediated by both their core proteins and GAG chains.

Proteoglycans Can Regulate the Activities of Secreted Signaling Molecules [21]

Given the structural diversity of proteoglycan molecules, it would be surprising if their function in the extracellular matrix were limited to providing hydrated space around and between cells. Their GAG chains, for example, can form gels of varying pore size and charge density, and they could therefore serve as selective sieves to regulate the traffic of molecules and cells according to their size, charge, or both. There is evidence that a heparan sulfate proteoglycan called *perlecan* has this role in the basal lamina of the kidney glomerulus, which filters molecules passing into the urine from the bloodstream (discussed below).

Proteoglycans are thought to play a major part in chemical signaling between cells. They bind various secreted signaling molecules, such as certain protein growth factors, in a test tube, and it is likely that they do so in tissues. Such binding can enhance or inhibit the activity of the growth factor. *Fibroblast growth factor (FGF)*, for example, which stimulates a variety of cell types to proliferate, binds to heparan sulfate chains of proteoglycans both *in vitro* and in tissues; for some cells, this binding seems to be a required step for FGF to activate its cell-surface receptor (which is a transmembrane tyrosine kinase, discussed in Chapter 15). Whereas in most cases the signaling molecules bind to the GAG chains of the proteoglycan, this is not always so: the ubiquitous growth regulatory protein *transforming growth factor β (TGF-β)* binds to the core proteins of several matrix proteoglycans, including decorin; binding to decorin inhibits the activity of the TGF-β.

Proteoglycans also bind and regulate the activities of other types of secreted proteins, such as proteolytic enzymes (proteases) and protease inhibitors. Binding to a proteoglycan could control the activity of a protein in any of the following ways: (1) it could immobilize the protein close to the site where it is produced, thereby restricting its range of action; (2) it could sterically block the activity of the protein; (3) it could provide a reservoir of the protein for delayed release; (4) it could protect the protein from proteolytic degradation, thereby prolonging its action; and (5) it could alter or concentrate the protein for more effective presentation to cell-surface receptors. Proteoglycans are thought to act in all these ways to help regulate the activities of secreted proteins.

Figure 19-37 Examples of a large (aggrecan) and a small (decorin) proteoglycan found in the extracellular matrix. They are compared to a typical secreted glycoprotein molecule (pancreatic ribonuclease B). All are drawn to scale. The core proteins of both aggrecan and decorin contain oligosaccharide chains as well as the GAG chains, but these are not shown. Aggrecan typically consists of about 100 chondroitin sulfate chains and about 30 keratan sulfate chains linked to a serine-rich core protein of almost 3000 amino acids. Decorin "decorates" the surface of collagen fibrils, hence its name.

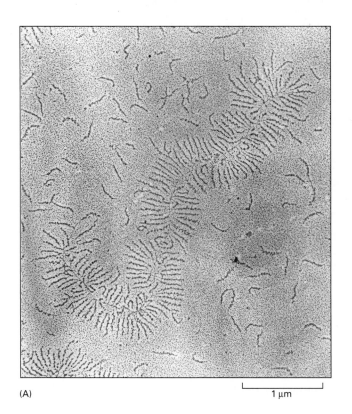

(A)

1 µm

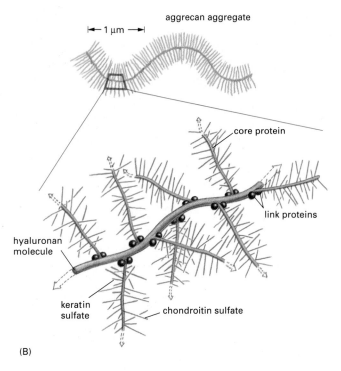

(B)

GAG Chains May Be Highly Organized in the Extracellular Matrix [20, 22]

GAGs and proteoglycans associate to form huge polymeric complexes in the extracellular matrix. Molecules of *aggrecan*, for example, the major proteoglycan in cartilage, shown in Figure 19–37, assemble with hyaluronan in the extracellular space into aggregates that are as big as a bacterium (Figure 19–38).

Moreover, besides associating with one another, GAGs and proteoglycans associate with fibrous matrix proteins such as collagen and with protein meshworks such as the basal lamina, creating extremely complex structures. The arrangement of proteoglycan molecules in living tissues is generally hard to determine. As they are highly water soluble, they may be washed out of the extracellular matrix when tissue sections are exposed to aqueous solutions during fixation; and changes of pH, ionic, or osmotic conditions can drastically alter their conformation. Thus specialized methods have to be used to visualize them *in vivo* (Figure 19–39).

Cell-Surface Proteoglycans Act as Co-Receptors

Not all proteoglycans are secreted components of the extracellular matrix. Some, such as *serglycin*, are constituents of intracellular secretory vesicles, where they help to package and store secretory molecules. Others are integral components of plasma membranes and have their core protein either inserted across the lipid bilayer or attached to the lipid bilayer by a glycosylphosphatidylinositol (GPI) anchor.

Among the best-characterized plasma membrane proteoglycans are the *syndecans*, which have a membrane-spanning core protein. The extracellular domain of this transmembrane proteoglycan carries a variable number of chondroitin sulfate and heparan sulfate GAG chains, while its intracellular domain is thought to interact with the actin cytoskeleton in the cell cortex. Syndecans are found on the surface of many types of cells, including fibroblasts and epithelial cells, where they serve along with integrins as receptors for collagen, fibronectin, and other matrix proteins to which the syndecans bind. As discussed above,

Figure 19–38 An aggrecan aggregate from fetal bovine cartilage. (A) Electron micrograph of an aggrecan aggregate shadowed with platinum. Many free aggrecan molecules are also seen. (B) Schematic drawing of the giant aggrecan aggregate shown in (A). It consists of about 100 aggrecan monomers (each like the one shown in Figure 19–37) noncovalently bound to a single hyaluronan chain through two link proteins that bind to both the core protein of the proteoglycan and to the hyaluronan chain, thereby stabilizing the aggregate; the link proteins are members of the hyaladherin family of hyaluronan-binding proteins discussed previously. The molecular weight of such a complex can be 10^8 or more, and it occupies a volume equivalent to that of a bacterium, which is about 2×10^{-12} cm^3. (A, courtesy of Lawrence Rosenberg.)

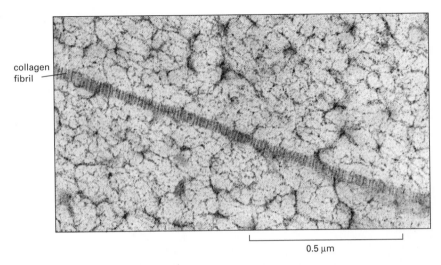

collagen fibril

0.5 μm

Figure 19–39 Electron micrograph of proteoglycans in the extracellular matrix of rat cartilage. The tissue was rapidly frozen at –196°C and fixed and stained while still frozen (a process called *freeze substitution*) to prevent the GAG chains from collapsing. The proteoglycan molecules are seen to form a fine filamentous network in which a single striated collagen fibril is embedded. The more darkly stained parts of the proteoglycan molecules are the core proteins; the faintly stained threads are the GAG chains. (Reproduced from E.B. Hunziker and R.K. Schenk, *J. Cell Biol.* 98:277–282, 1984, by copyright permission of the Rockefeller University Press.)

syndecans also bind fibroblast growth factor (FGF) and present it to FGF receptor proteins on the same cell. Similarly, another plasma membrane proteoglycan, called *betaglycan,* binds transforming growth factor β (TGF-β) and presents it to TGF-β receptors.

Thus plasma membrane proteoglycans act as *co-receptors* that collaborate with conventional cell-surface receptor proteins, both in binding cells to the extracellular matrix and in initiating the response of cells to some growth factors. The proteoglycans that are discussed in this chapter are summarized in Table 19–3.

Collagens Are the Major Proteins of the Extracellular Matrix [23]

The **collagens** are a family of highly characteristic fibrous proteins found in all multicellular animals. They are secreted by connective tissue cells, as well as by a variety of other cell types. As a major component of skin and bone, they are the most abundant proteins in mammals, constituting 25% of the total protein mass in these animals. The characteristic feature of a typical collagen molecule is its long, stiff, triple-stranded helical structure, in which three collagen polypeptide

Table 19–3 Some Common Proteoglycans

Proteoglycan	Approximate Molecular Weight of Core Protein	Type of GAG Chains	Number of GAG Chains	Location	Functions
Aggrecan	210,000	chondroitin sulfate + keratan sulfate	~130	cartilage	mechanical support; forms large aggregates with hyaluronan
Betaglycan	36,000	chondroitin sulfate/ dermatan sulfate	1	cell surface and matrix	binds TGF-β
Decorin	40,000	chondroitin sulfate/ dermatan sulfate	1	widespread in connective tissues	binds to type I collagen fibrils and TGF-β
Perlecan	600,000	heparan sulfate	2–15	basal laminae	structural and filtering function in basal lamina
Serglycin	20,000	chondroitin sulfate/ dermatan sulfate	10–15	secretory vesicles in white blood cells	helps to package and store secretory molecules
Syndecan-1	32,000	chondroitin sulfate + heparan sulfate	1–3	fibroblast and epithelial cell surface	cell adhesion; binds FGF

Figure 19–40 **The structure of a typical collagen molecule.** (A) A model of part of a single collagen α chain in which each amino acid is represented by a sphere. The chain contains about 1000 amino acid residues and is arranged as a left-handed helix with three amino acid residues per turn and with glycine as every third residue. Therefore an α chain is composed of a series of triplet Gly-X-Y sequences in which X and Y can be any amino acid (although X is commonly proline and Y is commonly hydroxyproline). (B) A model of a part of a collagen molecule in which three α chains, each shown in a different color, are wrapped around one another to form a triple-stranded helical rod. Glycine is the only amino acid small enough to occupy the crowded interior of the triple helix. Only a short length of the molecule is shown; the entire molecule is 300 nm long. (From model by B.L. Trus.)

1.5 nm

chains, called *α chains*, are wound around one another in a ropelike superhelix. Collagens are extremely rich in proline and glycine, both of which are important in the formation of the triple-stranded helix. Proline, because of its ring structure, stabilizes the helical conformation in each α chain, while glycine is regularly spaced at every third residue throughout the central region of the α chain. Being the smallest amino acid (having only a hydrogen atom as a side chain), glycine allows the three helical α chains to pack tightly together to form the final collagen superhelix (Figure 19–40).

So far, about 25 distinct collagen α chains have been identified, each encoded by a separate gene. Different combinations of these genes are expressed in different tissues. Although in principle more than 10,000 types of triple-stranded collagen molecules could be assembled from various combinations of the 25 or so α chains, only about 15 types of collagen molecules have been found. The main types of collagen found in connective tissues are types I, II, III, V, and XI—type I being the principal collagen of skin and bone and by far the most common. These are the **fibrillar collagens** and have the ropelike structure we have described for a typical collagen molecule. After being secreted into the extracellular space, these collagen molecules assemble into ordered polymers called *collagen fibrils*, which are thin (10–300 nm in diameter) structures, many hundreds of micrometers long in mature tissues and clearly visible in electron micrographs (Figure 19–41, and see Figure 19–39). The collagen fibrils often aggregate into larger, cablelike bundles, which can be seen in the light microscope as *collagen fibers* several micrometers in diameter. Types IX and XII are called *fibril-associated collagens* as they decorate the surface of collagen fibrils; they are thought to link these fibrils to one another and to other components in the extracellular matrix. Types IV and VII are *network-forming collagens*: type IV molecules assemble into a feltlike sheet or meshwork that constitutes a major part of mature basal laminae, while type VII molecules form dimers that assemble into specialized structures called *anchoring fibrils*, which help attach the basal lamina of multilayered epithelia to the underlying connective tissue and therefore are especially abundant in the skin. The collagen types that we discuss are listed in Table 19–4.

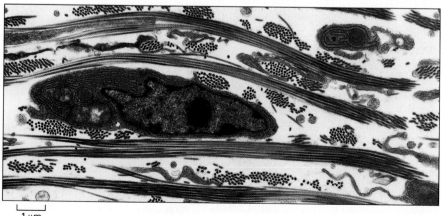

1 μm

Figure 19–41 **Electron micrograph of fibroblasts surrounded by collagen fibrils in the connective tissue of embryonic chick skin.** The fibrils, which are organized into bundles that run approximately at right angles to one another, are produced by the fibroblasts. These cells contain abundant endoplasmic reticulum, where secreted proteins such as collagen are synthesized. (From C. Ploetz, E.I. Zycband, and D.E. Birk, *J. Struct. Biol.* 106:73–81, 1991.)

The Extracellular Matrix of Animals

Table 19–4 Some Types of Collagen and Their Properties

	Type	Molecular Formula	Polymerized Form	Tissue Distribution
FIBRIL-FORMING (FIBRILLAR)	I	$[\alpha1(I)]_2\alpha2(I)$	fibril	bone, skin, tendon, ligaments, cornea, internal organs (accounts for 90% of body collagen)
	II	$[\alpha1(II)]_3$	fibril	cartilage, intervertebral disc, notochord, vitreous humor of the eye
	III	$[\alpha1(III)]_3$	fibril	skin, blood vessels, internal organs
	V	$[\alpha1(V)]_2\alpha2(V)$	fibril (with type I)	as for type I
	XI	$\alpha1(XI)\alpha2(XI)\alpha3(XI)$	fibril (with type II)	as for type II
FIBRIL-ASSOCIATED	IX	$\alpha1(IX)\alpha2(IX)\alpha3(IX)$ with type II fibrils	lateral association	cartilage
	XII	$[\alpha1(XII)]_3$ with some type I fibrils	lateral association	tendon, ligaments, some other tissues
NETWORK-FORMING	IV	$[\alpha1(IV)]_2\alpha2(IV)$	sheetlike network	basal laminae
	VII	$[\alpha1(VII)]_3$	anchoring fibrils	beneath stratified squamous epithelia

Note that types I, IV, V, and XI are each composed of 2 or 3 types of α chain, whereas types II, III, VII, and XII are composed of only 1 type of α chain each. Only 9 types of collagen are shown, but about 15 types of collagen and about 25 types of α chain have been defined so far.

Many proteins that contain a repeated pattern of amino acids have evolved by duplications of DNA sequences. The fibrillar collagens apparently arose in this way. Thus the genes that encode the α chains of most of these collagens are very large (up to 44 kilobases in length) and contain about 50 exons. Most of the exons are 54, or multiples of 54, nucleotides long, suggesting that these collagens arose by multiple duplications of a primordial gene containing 54 nucleotides and encoding exactly 6 Gly-X-Y repeats (see Figure 19–40).

Collagens Are Secreted with a Nonhelical Extension at Each End [23, 24]

The individual collagen polypeptide chains are synthesized on membrane-bound ribosomes and injected into the lumen of the endoplasmic reticulum (ER) as larger precursors, called *pro-α chains*. These precursors not only have the short amino-terminal signal peptide required to direct the nascent polypeptide to the ER, they also have additional amino acids, called *propeptides,* at both their amino- and carboxyl-terminal ends. In the lumen of the ER selected proline and lysine residues are hydroxylated to form hydroxyproline and hydroxylysine, respectively, and some of the hydroxylysine residues are glycosylated. Each pro-α chain then combines with two others to form a hydrogen-bonded, triple-stranded helical molecule known as *procollagen*. The secreted forms of fibrillar collagens (but not the other types of collagen) are converted to *collagen molecules* in the extracellular space by the removal of the propeptides (see Figure 19–43).

Hydroxylysine and *hydroxyproline* residues (Figure 19–42) are infrequently found in other animal proteins, although hydroxyproline is abundant in some proteins found in the plant cell wall. In collagen the hydroxyl groups of these amino acids are thought to form interchain hydrogen bonds that help stabilize the triple-stranded helix, and conditions that prevent proline hydroxylation, such as a deficiency of ascorbic acid (vitamin C), have serious consequences. In scurvy, the disease caused by a dietary deficiency of vitamin C that was common in sailors until the last century, the defective pro-α chains that are synthesized fail to form a stable triple helix and are immediately degraded within the cell. Conse-

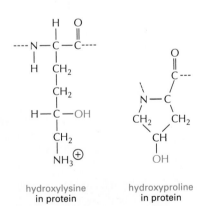

hydroxylysine in protein   hydroxyproline in protein

Figure 19–42 Hydroxylysine and hydroxyproline residues. These modified amino acids are common in collagen; they are formed by enzymes that act after the lysine and proline are incorporated into procollagen molecules.

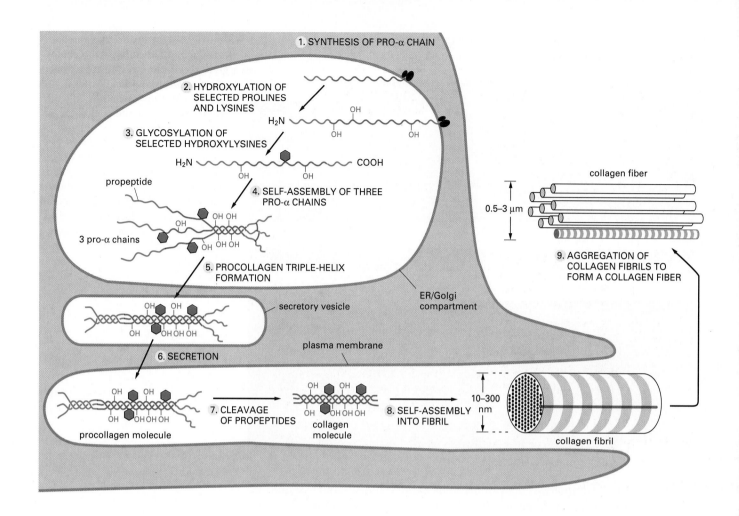

1. SYNTHESIS OF PRO-α CHAIN

2. HYDROXYLATION OF SELECTED PROLINES AND LYSINES

3. GLYCOSYLATION OF SELECTED HYDROXYLYSINES

propeptide

3 pro-α chains

4. SELF-ASSEMBLY OF THREE PRO-α CHAINS

5. PROCOLLAGEN TRIPLE-HELIX FORMATION

ER/Golgi compartment

secretory vesicle

plasma membrane

6. SECRETION

procollagen molecule

7. CLEAVAGE OF PROPEPTIDES

collagen molecule

8. SELF-ASSEMBLY INTO FIBRIL

10–300 nm

collagen fibril

collagen fiber

0.5–3 μm

9. AGGREGATION OF COLLAGEN FIBRILS TO FORM A COLLAGEN FIBER

quently, with the gradual loss of the preexisting normal collagen in the matrix, blood vessels become extremely fragile and teeth become loose in their sockets. This implies that in these particular tissues degradation and replacement of collagen is relatively rapid. In many other adult tissues, however, the turnover of collagen (and other extracellular matrix macromolecules) is thought to be very slow: in bone, to take an extreme example, collagen molecules persist for about 10 years before they are degraded and replaced. By contrast, most cellular proteins have half-lives of hours or days.

After Secretion Fibrillar Procollagen Molecules Are Cleaved to Collagen Molecules, Which Assemble into Fibrils [23, 24, 25]

After secretion the propeptides of the fibrillar procollagen molecules are removed by specific proteolytic enzymes outside the cell. This converts the procollagen molecules to collagen molecules, which assemble in the extracellular space to form much larger **collagen fibrils**. The propeptides have at least two functions: (1) they guide the intracellular formation of the triple-stranded collagen molecules, and (2) because they are removed only after secretion, they prevent the intracellular formation of large collagen fibrils, which could be catastrophic for the cell. The process of fibril formation is driven, in part, by the tendency of the collagen molecules, which are more than 1000-fold less soluble than procollagen molecules, to self-assemble. The fibrils begin to form close to the cell surface, often in deep infoldings of the plasma membrane formed by the tandem fusion of secretory vesicles with the cell surface. The underlying cortical cytoskeleton can therefore influence the sites, rates, and orientation of fibril assembly.

Figure 19–43 The intracellular and extracellular events involved in the formation of a collagen fibril. Note that collagen fibrils are shown assembling in the extracellular space contained within a large infolding in the plasma membrane. As one example of how the collagen fibrils can form ordered arrays in the extracellular space, they are shown further assembling into large collagen fibers, which are visible in the light microscope. The covalent cross-links that stabilize the extracellular assemblies are not shown.

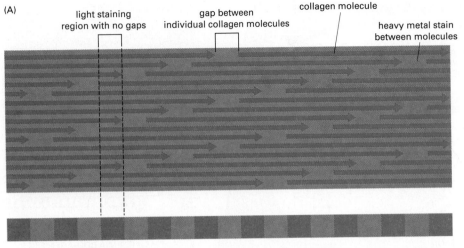

(A)

light staining region with no gaps | gap between individual collagen molecules | collagen molecule | heavy metal stain between molecules

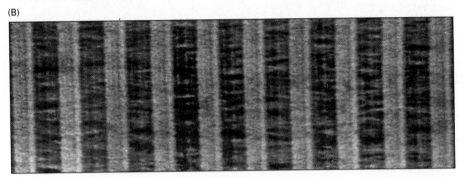

(B)

Figure 19–44 How the staggered arrangement of collagen molecules gives rise to the striated appearance of a negatively stained fibril. (A) Since the negative stain fills only the space between the molecules, the stain in the gaps between the individual molecules in each row accounts for the dark staining bands. An electron micrograph of a portion of a negatively stained fibril is shown below (B). The staggered arrangement of the collagen molecules maximizes the tensile strength of the aggregate. (B, courtesy of Robert Horne.)

Figure 19–43 summarizes the various steps in the synthesis and assembly of collagen fibrils. Given the large number of enzymatic steps involved in forming a collagen fibril, it is not surprising that there are many human genetic diseases that affect fibril formation. Mutations affecting type I collagen cause *osteogenesis imperfecta*, characterized by weak bones that easily fracture. Mutations affecting type II collagen cause *chondrodysplasias*, characterized by abnormal cartilage, which leads to bone and joint deformities. Mutations affecting type III collagen cause Ehlers-Danlos syndrome, characterized by fragile skin and blood vessels and hypermobile joints.

When viewed in an electron microscope, collagen fibrils have characteristic cross-striations every 67 nm, reflecting the regularly staggered packing of the individual collagen molecules in the fibril (Figure 19–44). After the fibrils form in the extracellular space, they are greatly strengthened by the formation of covalent cross-links between lysine residues of the constituent collagen molecules (Figure 19–45). The types of covalent bonds involved are found only in collagen and elastin. If cross-linking is inhibited, the tensile strength of the fibrils is drastically reduced; collagenous tissues become fragile, and structures such as skin, tendons, and blood vessels tend to tear. The extent and type of cross-linking varies from tissue to tissue: collagen is especially highly cross-linked in the Achilles tendon, for example, where tensile strength is crucial.

Figure 19–45 The covalent intramolecular and intermolecular cross-links formed between modified lysine side chains within a collagen fibril. The cross-links are formed in several steps. First, certain lysine and hydroxylysine residues are deaminated by the extracellular enzyme lysyl oxidase to yield highly reactive aldehyde groups. The aldehydes then react spontaneously to form covalent bonds with each other or with other lysine or hydroxylysine residues. Most of the cross-links form between the short nonhelical segments at each end of the collagen molecules.

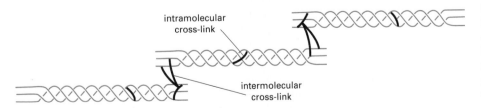

intramolecular cross-link

intermolecular cross-link

Fibril-associated Collagens Help Organize the Fibrils [26]

In contrast to GAGs, which resist compressive forces, collagen fibrils form structures that resist tensile forces. The fibrils come in a variety of diameters and are organized in different ways in different tissues. In mammalian skin, for example, they are woven in a wickerwork pattern so that they resist tensile stress in multiple directions. In tendons they are organized in parallel bundles aligned along the major axis of tension. And in mature bone and in the cornea they are arranged in orderly plywoodlike layers, with the fibrils in each layer lying parallel to each other but nearly at right angles to the fibrils in the layers on either side. The same arrangement occurs in tadpole skin, which serves to illustrate this organization (Figure 19–46).

The connective tissue cells themselves must determine the size and arrangement of the collagen fibrils. The cells can express one or more of the genes for the different types of fibrillar procollagen molecules. But even fibrils composed of the same mixture of fibrillar collagen molecules have different arrangements in different tissues. How is this achieved? Part of the answer may be that cells can regulate the disposition of the collagen molecules after secretion by guiding collagen fibril formation in close association with the plasma membrane (see Figure 19–43). In addition, as the spatial organization of collagen fibrils at least partly reflects their interactions with other molecules in the matrix, cells can influence this organization by secreting, along with their fibrillar collagens, different kinds and amounts of other matrix macromolecules. The **fibril-associated collagens,** such as *type IX* and *XII* collagen molecules, are thought to be especially important in this regard. They differ from the fibrillar collagens in several ways. (1) Their triple-stranded helical structure is interrupted by one or two short nonhelical domains, which makes the molecules more flexible than fibrillar collagen molecules. (2) They are not cleaved after secretion and so retain their propeptides. (3) They do not aggregate with one another to form fibrils in the extracellular space. Instead, they bind in a periodic manner to the surface of fibrils formed by the fibrillar collagens: type IX molecules bind to type-II-collagen-containing fibrils in cartilage, the cornea, and the vitreous of the eye (Figure 19–47), whereas type XII molecules bind to type-I-collagen-containing fibrils in tendons and various other tissues. The fibril-associated collagens are thought to mediate interactions of collagen fibrils with one another and with other matrix macromolecules. In this way they play a part in determining the organization of the fibrils in the matrix.

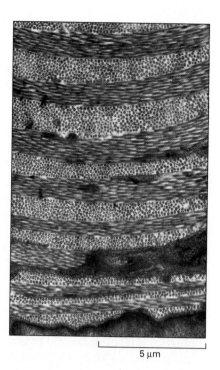

Figure 19–46 **Electron micrograph of a cross-section of tadpole skin.** Note the plywoodlike arrangement of collagen fibrils, in which successive layers of fibrils are laid down nearly at right angles to each other. This arrangement is also found in mature bone and in the cornea. (Courtesy of Jerome Gross.)

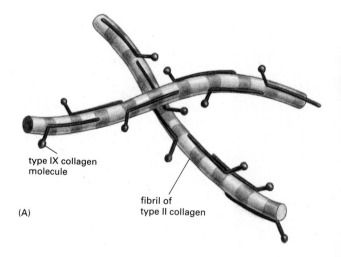

(A)

type IX collagen molecule

fibril of type II collagen

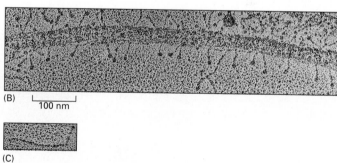

(B) └ 100 nm ┘

(C)

Figure 19–47 **Type IX collagen.** (A) Schematic drawing of type IX collagen molecules binding in a periodic pattern to the surface of a type-II-collagen-containing fibril. (B) Electron micrograph of a rotary-shadowed type-II-collagen-containing fibril in cartilage sheathed in type IX collagen molecules; an individual type IX collagen molecule is shown in (C). (B and C, from L. Vaughan et al., *J. Cell Biol.* 106:991–997, 1988, by copyright permission of the Rockefeller University Press.)

Cells Help Organize the Collagen Fibrils They Secrete by Exerting Tension on the Matrix [27]

There is yet another way that collagen-secreting cells determine the spatial organization of the collagen matrix they produce. Fibroblasts work on the collagen they have secreted, crawling over it and tugging on it—helping to compact it into sheets and draw it out into cables. This mechanical role of fibroblasts in shaping collagen matrices has been demonstrated dramatically in culture. When fibroblasts are mixed with a meshwork of randomly oriented collagen fibrils that form a gel in a culture dish, the fibroblasts tug on the meshwork, drawing in collagen from their surroundings and causing the gel to contract to a small fraction of its initial volume; by similar activities, a cluster of fibroblasts will surround itself with a capsule of densely packed and circumferentially oriented collagen fibers.

If two small pieces of embryonic tissue containing fibroblasts are placed far apart on a collagen gel, the intervening collagen becomes organized into a compact band of aligned fibers that connect the two explants (Figure 19–48). The fibroblasts subsequently migrate out from the explants along the aligned collagen fibers. Thus the fibroblasts influence the alignment of the collagen fibers, and the collagen fibers in turn affect the distribution of the fibroblasts. Fibroblasts presumably play a similar role in generating long-range order in the extracellular matrix inside the body—in helping to create tendons and ligaments, for example, and the tough, dense layers of connective tissue that ensheathe and bind together most organs.

Elastin Gives Tissues Their Elasticity [28]

Many vertebrate tissues, such as skin, blood vessels, and lungs, need to be both strong and elastic in order to function. A network of **elastic fibers** in the extracellular matrix of these tissues gives them the required resilience so that they can recoil after transient stretch. Elastic fibers are at least five times more extensible than a rubber band of the same cross-sectional area. Long, inelastic collagen fibrils are interwoven with the elastic fibers to limit the extent of stretching and prevent the tissue from tearing.

The main component of elastic fibers is **elastin,** a highly hydrophobic protein (about 750 amino acid residues long), which, like collagen, is unusually rich in proline and glycine but, unlike collagen, is not glycosylated and contains little hydroxyproline and no hydroxylysine. Elastin molecules are secreted into the

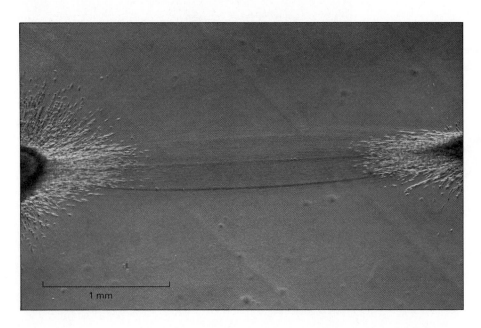

Figure 19–48 The shaping of the extracellular matrix by cells. This micrograph shows a region between two pieces of embryonic chick heart (rich in fibroblasts as well as heart muscle cells) that has grown in culture on a collagen gel for four days. A dense tract of aligned collagen fibers has formed between the explants, presumably as a result of the fibroblasts in the explants tugging on the collagen. (From D. Stopak and A.K. Harris, *Dev. Biol.* 90:383–398, 1982.)

1 mm

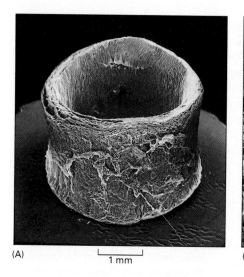 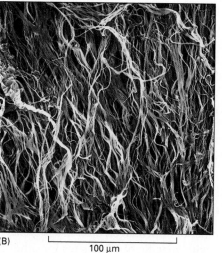

(A)

1 mm

(B)

100 μm

Figure 19–49 A network of elastic fibers. These scanning electron micrographs show a low-power view of a segment of a dog's aorta (A) and a high-power view of the dense network of longitudinally oriented elastic fibers in the outer layer of the same blood vessel (B). All of the other components have been digested away with enzymes and formic acid. (From K.S. Haas, S.J. Phillips, A.J. Comerota, and J.W. White, *Anat. Rec.* 230:86–96, 1991.)

extracellular space and assemble into elastic fibers close to the plasma membrane, generally in cell-surface infoldings. After secretion the elastin molecules become highly cross-linked to one another to generate an extensive network of fibers and sheets (Figure 19–49). The cross-links are formed between lysine residues by a mechanism similar to the one that operates in cross-linking collagen molecules.

The elastin protein is composed largely of two types of short segments that alternate along the polypeptide chain—hydrophobic segments, which are responsible for the elastic properties of the molecule, and alanine- and lysine-rich α-helical segments, which form cross-links between adjacent molecules. Each segment is encoded by a separate exon. There is still controversy, however, about the conformation of elastin molecules in elastic fibers and about how the structure of these fibers accounts for their rubberlike properties. In one view the elastin polypeptide chain, like the polymer chains in ordinary rubber, adopts a loose "random coil" conformation, and it is the random-coil structure of the component molecules cross-linked into the elastic fiber network that allows the network to stretch and recoil like a rubber band (Figure 19–50).

Elastic fibers are not composed solely of elastin, however. The elastin core is covered with a sheath of *microfibrils*, each microfibril having a diameter of about 10 nm. While elastic fibers always contain microfibrils, the same microfibrils can be found in extracellular matrices that do not contain elastin. Mi-

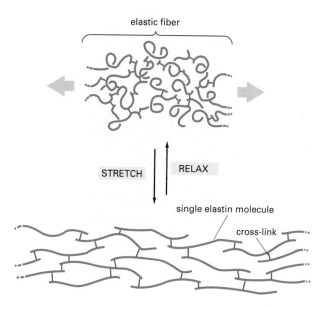

elastic fiber

STRETCH | RELAX

single elastin molecule

cross-link

Figure 19–50 Stretching a network of elastin molecules. The molecules are joined together by covalent bonds (indicated in *red*) to generate a cross-linked network. In the model shown each elastin molecule in the network can expand and contract as a random coil, so that the entire assembly can stretch and recoil like a rubber band.

The Extracellular Matrix of Animals

crofibrils are composed of a number of distinct glycoproteins, including the large glycoprotein *fibrillin*, which seems to be essential for the integrity of elastic fibers. Mutations in the fibrillin gene result in *Marfan's syndrome*, a relatively common human genetic disease that affects connective tissues that are rich in elastic fibers; in the most severely affected individuals, the aorta (whose wall is normally full of elastin—see Figure 19–49) is prone to rupture. Microfibrils are thought to play an important part in the assembly of elastic fibers. They appear before elastin in developing tissues and seem to form a scaffold on which the secreted elastin molecules are deposited. As the elastin is deposited, the microfibrils become displaced to the periphery of the growing fiber.

Fibronectin Is an Extracellular Adhesive Protein That Helps Cells Attach to the Matrix [29]

The extracellular matrix contains a number of noncollagen *adhesive proteins* that typically have multiple domains, each with specific binding sites for other matrix macromolecules and for receptors on the surface of cells. These proteins thus contribute to both organizing the matrix and helping cells attach to it. The first of them to be well characterized was **fibronectin,** a large glycoprotein found in all vertebrates. Fibronectin is a dimer composed of two very large subunits joined by a pair of disulfide bonds near their carboxyl termini. Each subunit is folded into a series of functionally distinct rodlike domains separated by regions of flexible polypeptide chain (Figure 19–51A and B). The domains in turn consist of smaller modules, each of which is serially repeated and usually encoded by a separate exon, suggesting that the fibronectin gene, like the collagen genes, evolved by multiple exon duplications. The main type of module, called the **type III fibronectin repeat,** is about 90 amino acid residues in length, and it occurs at least 15 times in each subunit (Figure 19–51C). It is also found in some other matrix proteins, as well as in some plasma membrane and cytoplasmic proteins.

One way to analyze a complex multifunctional protein molecule like fibronectin is to chop it into pieces and determine the function of its individual domains. The protein is treated with low concentrations of a proteolytic enzyme, which cuts the polypeptide chain in the connecting regions between the rodlike domains, leaving the domains themselves intact, so that their binding activity can

Figure 19–51 The structure of a fibronectin dimer. As shown schematically in (A), the two polypeptide chains are similar but generally not identical (being made from the same gene but from differently spliced mRNAs). They are joined by two disulfide bonds near the carboxyl terminus. Each chain is almost 2500 amino acid residues long and is folded into five or six rodlike domains connected by flexible polypeptide segments. Individual domains are specialized for binding to a particular molecule or to a cell, as indicated for three of the domains. For simplicity, not all of the known binding sites are shown (there are other cell-binding sites, for example). (B) Electron micrographs of individual molecules shadowed with platinum; *arrows* mark the carboxyl termini. (C) The three-dimensional structure of a type III fibronectin repeat, as determined by nuclear magnetic resonance studies. It is the main type of repeating module in fibronectin and is also found in many other proteins. The Arg-Gly-Asp (RGD) sequence shown is part of the major cell-binding site (shown in *blue* in [A]) that we discuss in the text. (B, from J. Engel et al., *J. Mol. Biol.* 150:97–120, 1981. Academic Press Inc. [London] Ltd.; C, adapted from A.L. Main, T.S. Harvey, M. Baron, J. Boyd, and I.D. Campbell, *Cell* 71:671–678, 1992. © Cell Press.)

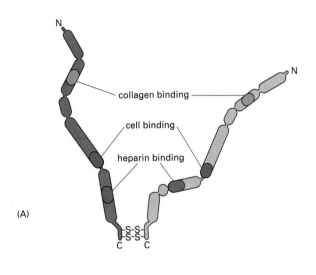

(A)

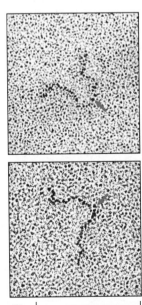

(B) 100 nm

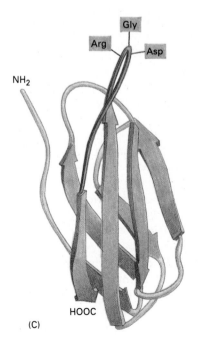

(C)

be tested. In this way it was shown that one domain binds to collagen, another to heparin, another to specific receptors on the surface of various types of cells, and so on (see Figure 19–51). Once a domain with cell-binding activity had been isolated, for example, its amino acid sequence could be determined and synthetic peptides corresponding to different segments of the domain prepared. These peptides were used to localize the main region responsible for cell binding and then to identify a specific tripeptide sequence (Arg-Gly-Asp, or RGD), which is found in one of the type III repeats (see Figure 19–51C), as a central feature of the binding site. Even very short peptides containing this **RGD sequence** will compete with fibronectin for the binding site on cells and so will inhibit the attachment of the cells to a fibronectin matrix. If these peptides are coupled to a solid surface, they cause cells to adhere to it. The RGD sequence is not confined to fibronectin. It is found in a number of extracellular matrix proteins, and it is recognized by several members of the integrin family of cell-surface matrix receptors that bind these proteins (discussed below). Each receptor, however, specifically recognizes its own small set of matrix molecules, indicating that tight receptor binding requires more than just the RGD sequence.

Multiple Forms of Fibronectin Are Produced by Alternative RNA Splicing [29, 30]

There are multiple forms (isoforms) of fibronectin, including one called *plasma fibronectin*, which is soluble and circulates in the blood and other body fluids, where it is thought to enhance blood clotting, wound healing, and phagocytosis. All of the other forms assemble on the surface of cells and are deposited in the extracellular matrix as highly insoluble *fibronectin filaments*. In these cell-surface and matrix forms, fibronectin dimers are cross-linked to one another by additional disulfide bonds; unlike fibrillar collagen molecules, which can be made to self-assemble into fibrils in a test tube, fibronectin molecules assemble into filaments only on the surface of certain cells, suggesting that additional proteins are needed for filament formation.

All forms of fibronectin are encoded by a single large gene that is about 50 kilobases long and contains about 50 exons of similar size. Transcription produces a single large RNA molecule that can be alternatively spliced in three regions, depending on the cell type and stage of development. In humans about 20 different messenger RNAs are produced, each encoding a somewhat different fibronectin subunit. Plasma fibronectin, for example, which is secreted mainly by liver cells, lacks two of the type III repeats that are found in cell- and matrix-associated forms of fibronectin. In some cases alternative splicing adds or deletes a cell-type-specific cell binding site: one such site is used by lymphocytes to adhere to fibronectin.

Alternative splicing presumably allows a cell to produce the type of fibronectin that is most suitable for the needs of the tissue. The pattern of fibronectin RNA splicing in the early embryo is different from that seen later in development; but if adult skin is injured, the pattern of fibronectin RNA splicing in the base of the wound switches back to the pattern seen in early development. These observations suggest that the forms of fibronectin produced in the early embryo and in wound healing are especially appropriate for promoting the cell migrations and proliferation required for tissue development and repair.

The crucial importance of fibronectin in animal development has been dramatically demonstrated by gene "knockout" experiments. Mice with both copies of their fibronectin gene inactivated by mutation, for example, die early in embryogenesis. The mutant mice have multiple morphological defects, including abnormalities in the formation of the notochord, somites, heart, blood vessels, neural tube, and extraembryonic membranes.

Glycoproteins in the Matrix Help Define Cell Migration Pathways [31]

Fibronectin is important not only for cell adhesion to the matrix but also for guiding cell migrations in vertebrate embryos. Large amounts of fibronectin, for example, are found along the pathway followed by migrating prospective meso-dermal cells during amphibian gastrulation (discussed in Chapter 21). The migration of these cells can be inhibited by injecting into the developing amphibian embryo various ligands that disrupt the ability of the cells to bind to fibronectin: antibodies against fibronectin, peptides containing the RGD cell-binding tripeptide but lacking the matrix-binding domains of fibronectin, and antibodies against an integrin that serves as a fibronectin receptor on these cells all inhibit the migration. Fibronectin presumably promotes cell migration by helping cells attach to the matrix. The effect must be delicately balanced so that the migrating cells can grip the matrix without becoming immobilized on it.

Many types of adhesive molecules in the matrix are believed to play a part in guiding morphogenetic cell movements, and new ones are continually being discovered. **Tenascin,** for example, is a large glycoprotein complex of six identical or similar disulfide-linked polypeptide chains, which radiate from a center like the spokes of a wheel (see Figure 19–57). As in fibronectin, each of the polypeptide chains is composed of several types of short amino acid sequences that are repeated many times; a fibronectin type III repeat, for instance, occurs eight or more times in each chain. Each polypeptide chain is folded into a number of functionally distinct domains, one of which binds the cell-surface transmembrane proteoglycan syndecan, while another binds fibronectin. Tenascin has a much more restricted distribution than fibronectin and is most abundant in the extracellular matrix of embryonic tissues. Unlike fibronectin, it can either promote or inhibit cell adhesion, depending on the cell type; the adhesive and anti-adhesive functions are thought to be mediated by different protein domains. There is increasing evidence that anti-adhesive interactions, like adhesive ones, play an important part in guiding cell migration, as we discuss in Chapter 21.

Type IV Collagen Molecules Assemble into a Sheetlike Meshwork to Help Form Basal Laminae [32]

Our discussion of extracellular matrix thus far has focused on the volume-filling material between cells. But in certain places, especially beneath epithelia, extracellular matrix can also be organized as a thin tough sheet—a *basal lamina*. The construction of basal laminae depends on some specialized types of extracellular matrix molecules, including a specialized variety of collagen, to which we now turn.

Type IV collagen molecules have a more flexible structure than the fibrillar collagens: their triple-stranded helix is interrupted in 26 regions, allowing multiple bends. Like the fibril-associated collagens, they are not cleaved after secretion but retain the terminal regions that hinder side-to-side packing into long fibrils. Instead, they interact via their uncleaved terminal domains to assemble extracellularly into a flexible, sheetlike, multilayered network. Electron microscopic studies of preparations of assembling type IV collagen molecules suggest that these molecules associate by their carboxyl termini to form head-to-head dimers, which then form an extended lattice via amino-terminal associations with three other molecules and the further lateral associations shown in Figure 19–52. Disulfide and other covalent cross-links between the collagen molecules stabilize these associations. The resulting meshwork forms an insoluble scaffolding to which other components of the basal lamina bind via their specific associations with type IV collagen molecules.

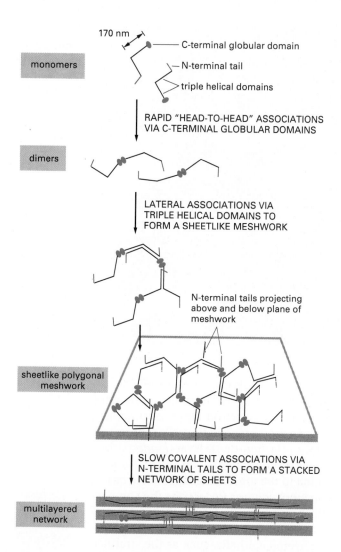

monomers

170 nm

C-terminal globular domain

N-terminal tail

triple helical domains

RAPID "HEAD-TO-HEAD" ASSOCIATIONS VIA C-TERMINAL GLOBULAR DOMAINS

dimers

LATERAL ASSOCIATIONS VIA TRIPLE HELICAL DOMAINS TO FORM A SHEETLIKE MESHWORK

N-terminal tails projecting above and below plane of meshwork

sheetlike polygonal meshwork

SLOW COVALENT ASSOCIATIONS VIA N-TERMINAL TAILS TO FORM A STACKED NETWORK OF SHEETS

multilayered network

Figure 19–52 How type IV collagen molecules are thought to assemble into a multilayered network. The model is based on electron micrographs of rotary-shadowed preparations of these molecules assembling *in vitro*. (Based on P.D. Yurchenco, E.C. Tsilibary, A.S. Charonis, and H. Furthmayr, *J. Histochem. Cytochem.* 34:93–102, 1986.)

Basal Laminae Are Composed Mainly of Type IV Collagen, Heparan Sulfate Proteoglycan, Laminin, and Entactin [33]

Basal laminae are flexible thin (40–120 nm thick) mats of specialized extracellular matrix that underlie all epithelial cell sheets and tubes; they also surround individual muscle cells, fat cells, and Schwann cells (which wrap around peripheral nerve cell axons to form myelin). The basal lamina thus separates these cells and cell sheets from the underlying or surrounding connective tissue. In other locations, such as the kidney glomerulus and lung alveolus, a basal lamina lies between two cell sheets and functions as a highly selective filter (Figure 19–53). Basal laminae serve more than simple structural and filtering roles, however. They are able to determine cell polarity, influence cell metabolism, organize the proteins in adjacent plasma membranes, induce cell differentiation, and serve as specific highways for cell migration.

Figure 19–53 Three ways in which basal laminae (*yellow lines*) are organized. They surround certain cells (such as muscle cells), underlie epithelial cell sheets, and are interposed between two cell sheets (as in the kidney glomerulus). Note that in the kidney glomerulus both cell sheets have gaps in them, so that the basal lamina serves as the permeability barrier determining which molecules will pass into the urine from the blood.

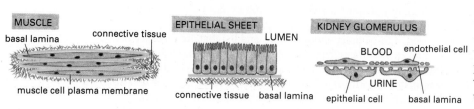

MUSCLE

basal lamina

connective tissue

muscle cell plasma membrane

EPITHELIAL SHEET

LUMEN

connective tissue   basal lamina

KIDNEY GLOMERULUS

BLOOD

endothelial cell

URINE

epithelial cell   basal lamina

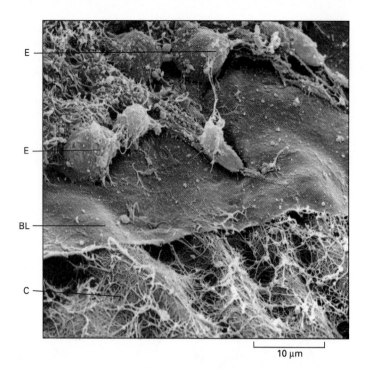

Figure 19–54 **Scanning electron micrograph of a basal lamina in the cornea of a chick embryo.** Some of the epithelial cells (E) have been removed to expose the upper surface of the matlike basal lamina (BL). A network of collagen fibrils (C) in the underlying connective tissue interacts with the lower face of the lamina. (Courtesy of Robert Trelstad.)

10 µm

The basal lamina is largely synthesized by the cells that rest on it (Figure 19–54). As seen in the electron microscope after conventional fixation and staining, most basal laminae consist of two distinct layers: an electron-lucent layer (*lamina lucida* or *rara*) adjacent to the basal plasma membrane of the cells that rest on the lamina—typically epithelial cells—and an electron-dense layer (*lamina densa*) just below. In some cases a third layer containing collagen fibrils (*lamina fibroreticularis*) connects the basal lamina to the underlying connective tissue. Some cell biologists use the term *basement membrane* to describe the composite of all three layers, which is usually thick enough to be seen in the light microscope; others use the terms basal lamina and basement membrane interchangeably. In the basal lamina of some multilayered epithelia, such as the stratified squamous epithelium that forms the epidermis of the skin, the lamina densa is

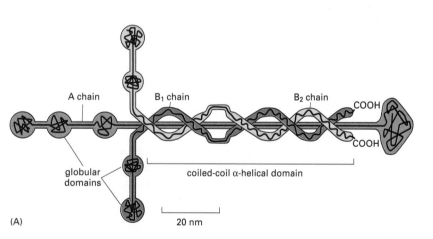

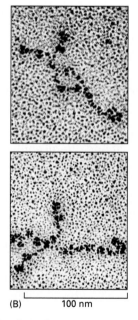

Figure 19–55 **The structure of laminin.** A schematic drawing of a laminin molecule is shown in (A), and electron micrographs of laminin molecules shadowed with platinum are shown in (B). This multidomain glycoprotein is composed of three polypeptides (A, B_1, and B_2) that are disulfide bonded into an asymmetric crosslike structure. Each of the polypeptide chains is more than 1500 amino acid residues long. Three types of A chains, three types of B_1 chains, and two types of B_2 chains have been identified, which in principle can associate to form 18 different laminin isoforms. Several such isoforms have been found, each with a characteristic tissue distribution. There are also several isoforms of type IV collegen, each with a distinctive tissue distribution. Thus basal laminae are chemically diverse, which is not surprising in view of their functional diversity. (B, from J. Engel et al., *J. Mol. Biol.* 150:97–120, 1981. © Academic Press Inc. [London] Ltd.)

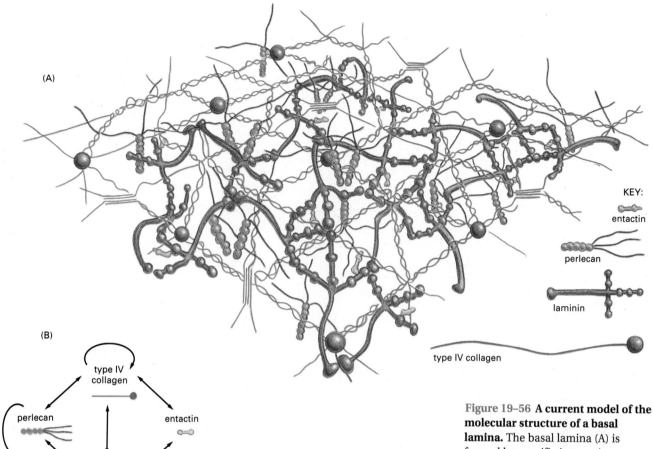

(A)

(B)

KEY:

entactin

perlecan

laminin

type IV collagen

type IV collagen

perlecan

entactin

laminin

Figure 19–56 A current model of the molecular structure of a basal lamina. The basal lamina (A) is formed by specific interactions between the proteins type IV collagen, laminin, and entactin plus the proteoglycan perlecan (B). Arrows in (B) connect molecules that can bind directly to each other. (Based on P.D. Yurchenco and J.C. Schittny, *FASEB J.* 4:1577–1590, 1990.)

tethered to the underlying connective tissue by specialized *anchoring fibrils* made of *type VII* collagen molecules. In one type of skin-blistering disease these connections are either absent or destroyed, and the epidermis and its basal lamina become detached from the underlying connective tissue.

Although its precise composition varies from tissue to tissue and even from region to region in the same lamina, most mature basal laminae contain type IV collagen (see Figure 19–52), the large heparan sulfate proteoglycan *perlecan*, and the glycoproteins *laminin* and *entactin*. **Laminin** is one of the first extracellular matrix proteins synthesized in a developing embryo, and early in development basal laminae contain little or no type IV collagen and consist mainly of a laminin network. Laminin is a large (~850,000 daltons) flexible complex of three very long polypeptide chains arranged in the shape of an asymmetric cross and held together by disulfide bonds (Figure 19–55). Like many other proteins in the extracellular matrix, it consists of a number of functional domains: one binds to type IV collagen, one to heparan sulfate, one to entactin, and two or more to laminin receptor proteins on the surface of cells. Like type IV collagen, laminin molecules can self-assemble *in vitro* into a feltlike sheet, largely through interactions between the ends of the laminin arms. A single dumbbell-shaped *entactin* molecule binds tightly to each laminin molecule where the short arms meet the long one; as entactin also binds to type IV collagen, it is thought to act as an additional bridge between the type IV collagen and laminin networks in basal laminae (Figure 19–56).

The shapes and sizes of some of the extracellular matrix molecules discussed in this chapter are compared in Figure 19–57.

The Extracellular Matrix of Animals

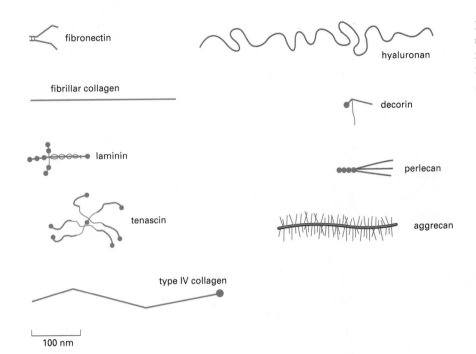

Figure 19–57 The comparative shapes and sizes of some of the major extracellular matrix macromolecules. Protein is shown in *green*, glycosaminoglycan in *red*.

fibronectin

hyaluronan

fibrillar collagen

decorin

laminin

perlecan

tenascin

aggrecan

type IV collagen

100 nm

Basal Laminae Perform Diverse and Complex Functions [34]

In the kidney glomerulus an unusually thick basal lamina acts as a molecular filter, preventing the passage of macromolecules from the blood into the urine as urine is formed (see Figure 19–53). The heparan sulfate proteoglycan seems to be important for this function: when the GAG chains are removed by specific enzymes, the filtering properties of the lamina are destroyed. The basal lamina can also act as a selective barrier to the movement of cells. The lamina beneath an epithelium, for example, usually prevents fibroblasts in the underlying connective tissue from making contact with the epithelial cells. It does not, however, stop macrophages, lymphocytes, or nerve processes from passing through it. The basal lamina plays an important part in tissue regeneration after injury. When tissues such as muscles, nerves, and epithelia are damaged, the basal lamina survives and provides a scaffolding along which regenerating cells can migrate. In this way the original tissue architecture is readily reconstructed. In some cases, as in the skin or cornea, the basal lamina becomes chemically altered following injury—for example, by the addition of fibronectin, which promotes the cell migration required for wound repair.

A particularly striking example of the instructive role of the basal lamina in regeneration comes from studies on the *neuromuscular junction*, the site where a nerve cell transmits its stimulus to a skeletal muscle cell. At the neuromuscular junction the nerve terminals of a motor neuron form a *synapse* with a muscle cell. The basal lamina that surrounds the muscle cell separates the nerve and muscle cell plasma membranes at the synapse. The synaptic region of the basal lamina has a distinctive chemical character: special isoforms of type IV collagen and laminin are found there, for example. This *junctional basal lamina* plays a central role in reconstructing a synapse after nerve or muscle injury. The evidence comes mainly from experiments in frogs. If a frog muscle and its motor nerve are destroyed, the basal lamina around each muscle cell remains and the sites of the old neuromuscular junctions are still recognizable. If the motor nerve but not the muscle is allowed to regenerate, the nerve axons regularly seek out the original synaptic sites on the empty basal laminae and differentiate there to form normal-looking nerve terminals. Thus the junctional basal lamina by itself can guide the regeneration of motor nerve terminals.

Similar experiments show that the basal lamina also controls the localization of the acetylcholine receptors that cluster in the muscle cell plasma membrane

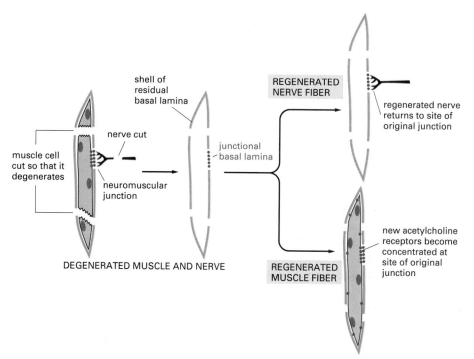

shell of
residual
basal lamina

REGENERATED
NERVE FIBER

regenerated nerve
returns to site of
original junction

nerve cut

muscle cell
cut so that it
degenerates

junctional
basal lamina

neuromuscular
junction

DEGENERATED MUSCLE AND NERVE

new acetylcholine
receptors become
concentrated at
site of original
junction

REGENERATED
MUSCLE FIBER

Figure 19–58 Regeneration experiments indicating the special character of the junctional basal lamina at a neuromuscular junction. When the nerve, but not the muscle, is allowed to regenerate after both the nerve and muscle have been damaged (*upper* part of figure), the junctional lamina directs the regenerating nerve to the original synaptic site. When the muscle, but not the nerve, is allowed to regenerate (*lower* part of figure), the junctional lamina causes newly made acetylcholine receptors to accumulate at the original synaptic site (the muscle regenerates from satellite cells located between the basal lamina and the original muscle cell—not shown, but see p. 1178). These experiments show that the junctional basal lamina controls the localization of other components of the synapse—on both sides of the lamina.

at a neuromuscular junction (discussed in Chapter 11). If the muscle and nerve are both destroyed but now the muscle is allowed to regenerate while the nerve is prevented from doing so, the acetylcholine receptors synthesized by the regenerated muscle localize predominantly in the region of the old junctions, even though the nerve is absent (Figure 19–58).

Thus the junctional basal lamina apparently coordinates the local spatial organization of the components in each of the two cells that form a neuromuscular junction. Extracts prepared from junctional basal lamina contain a novel matrix protein called *agrin*, which, when added to cultured muscle cells, initiates the assembly of synaptic structures in their plasma membrane. Motor neurons have been shown to make agrin, and it is thought that they deposit it (and other specialized macromolecules) in the basal lamina at the developing neuromuscular junction and that these matrix-localized molecules help assemble and stabilize the synaptic connection. Agrin is also made by many other types of neurons, raising the possibility that it directs the assembly of receptors and other postsynaptic macromolecules in synapses throughout the nervous system.

It is likely that basal laminae also play a sophisticated part in guiding cell migrations during embryonic development. Thus, in the nematode worm *Caenorhabditis elegans* mutation of a gene coding for a lamininlike protein selectively disrupts the paths taken by certain of the mesoderm cells and nerve axons that migrate over the basal lamina underlying the epidermis. Remarkably, only migrations along the dorsoventral axis of the embryo are affected; migrations along the anteroposterior axis of the same basal lamina occur normally. Such findings, and those on the regeneration of the frog neuromuscular junction, suggest that we still have much to learn about the chemistry and functional specializations of basal laminae.

The Degradation of Extracellular Matrix Components Is Tightly Controlled [35]

The regulated turnover of extracellular matrix macromolecules is critical to a variety of important biological processes. Rapid degradation occurs, for example, when the uterus involutes following childbirth or when the tadpole tail is resorbed during metamorphosis. A more localized degradation of matrix components is required when cells migrate through a basal lamina, as when white blood cells migrate across the vascular basal lamina into tissues in response to infection or injury, or when cancer cells migrate from their site of origin to distant organs via the bloodstream or lymphatic vessels, a process known as *metastasis*. Even in the seemingly static extracellular matrix of adult animals there is a slow continual turnover due to degradation and resynthesis.

In each of these cases matrix components are degraded by extracellular proteolytic enzymes that are secreted locally by cells. Most of these proteases belong to one of two general classes: many are **metalloproteases,** which depend on bound Ca^{2+} or Zn^{2+} for activity, while the others are **serine proteases,** which have a highly reactive serine residue in their active site. Together, metalloproteases and serine proteases cooperate to degrade matrix proteins such as collagen, laminin, and fibronectin. Some of the metalloproteases, such as the *collagenases*, are highly specific, cleaving particular proteins at a small number of sites, which are often positioned in such a fashion that the structural integrity of the matrix is destroyed by relatively limited proteolysis; in this way cell migration can be greatly facilitated by a relatively small amount of proteolysis.

An important serine protease involved in matrix degradation is *urokinase-type plasminogen activator (U-PA)*. This acts as the specific trigger in a proteolytic cascade: its immediate target is *plasminogen*, an inactive serine protease precursor that is abundant in the bloodstream and accumulates at sites of tissue remodeling such as wounds, tumors, and sites of inflammation. U-PA cleaves a single bond in plasminogen to yield the active protease *plasmin*. In contrast to U-PA, plasmin has a broad specificity, cleaving a variety of proteins, including fibrin (a component of blood clots), fibronectin, and laminin.

Several mechanisms operate to ensure that the degradation of matrix components is tightly controlled. First, like plasminogen, many proteases are secreted as inactive precursors that can be activated locally. Second, the action of proteases is confined to specific areas by various secreted protease inhibitors, such as the *tissue inhibitors of metalloproteases (TIMPs)* and the serine protease inhibitors known as *serpins*. These inhibitors are specific for particular proteases and bind tightly to the activated enzyme to block its activity. An attractive idea is that inhibitors are secreted by cells at the margins of areas of active degradation in order to protect uninvolved matrix; they may also protect cell-surface proteins that are required for cell adhesion or migration. Third, many cells have receptors on their surface that bind proteases such as U-PA, thereby confining the enzyme

Figure 19–59 Importance of cell-surface-receptor-bound protease. In (A) human prostate cancer cells make and secrete the serine protease U-PA, which binds to cell-surface U-PA receptor proteins. In (B) the same cells have been transfected with DNA that encodes an excess of an inactive form of U-PA, which binds to the U-PA receptors but has no protease activity; by occupying most of the U-PA receptors, the inactive U-PA prevents the active protease from binding to the cell surface. Both types of cells secrete active U-PA, grow rapidly, and produce tumors when injected into experimental animals. But the cells in (A) metastasize widely, whereas the cells in (B) do not. In order to metastasize, tumor cells have to crawl through basal laminae and other extracellular matrices on the way into and out of the bloodstream. This experiment therefore suggests that proteases must be cell-surface bound to mediate migration through the matrix.

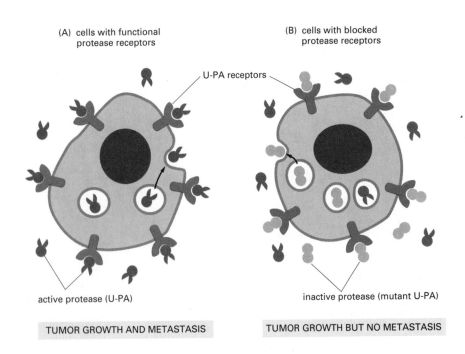

(A) cells with functional protease receptors

(B) cells with blocked protease receptors

U-PA receptors

active protease (U-PA)

inactive protease (mutant U-PA)

TUMOR GROWTH AND METASTASIS

TUMOR GROWTH BUT NO METASTASIS

to where it is needed: receptor-bound U-PA is found on nerve growth cones and at the leading edge of migrating white blood cells, for example, where it may serve to clear a pathway for their migration, and it seems to be required for some types of cancer cells to metastasize (Figure 19–59).

Summary

Cells in connective tissues are embedded in an intricate extracellular matrix that not only binds the cells together, but also influences their development, polarity, and behavior. The matrix contains various protein fibers interwoven in a hydrated gel composed of a network of glycosaminoglycan (GAG) chains.

The GAGs are a heterogeneous group of negatively charged polysaccharide chains, which (except for hyaluronan) are covalently linked to protein to form proteoglycan molecules. They occupy a large volume and form hydrated gels in the extracellular space. Proteoglycans are also found on the surface of cells, where they function as co-receptors to help cells bind to the matrix and respond to growth factors.

The fiber-forming proteins can be divided roughly into two functional types: mainly structural (collagens and elastin) and mainly adhesive (such as fibronectin and laminin). The fibrillar collagens (types I, II, III, V, and XI) are ropelike, triple-stranded helical molecules that aggregate into long fibrils in the extracellular space; these in turn can assemble into a variety of highly ordered arrays. Fibril-associated collagen molecules, such as types IX and XII, decorate the surface of collagen fibrils and influence the interactions of the fibrils with one another and with other matrix components. Type IV collagen molecules assemble into a sheetlike meshwork that is a crucial component of all mature basal laminae, which also contain the proteins laminin and entactin, as well as the heparan sulfate proteoglycan perlecan. Elastin molecules form an extensive cross-linked network of fibers and sheets that can stretch and recoil, imparting elasticity to the matrix. Fibronectin and laminin are examples of large, multidomain, adhesive glycoproteins in the matrix; fibronectin is widely distributed in connective tissues, whereas laminin is found mainly in basal laminae. By means of their multiple binding domains, such proteins help organize the extracellular matrix and help cells adhere to it.

Extracellular Matrix Receptors on Animal Cells: The Integrins

To understand how the extracellular matrix interacts with cells, one has to identify the cell-surface molecules (*matrix receptors*) that bind the matrix components as well as the extracellular matrix components themselves. Because of the multiple interactions among matrix macromolecules in the extracellular space, it is largely a matter of semantics where the plasma membrane components end and the extracellular matrix begins. The ultimate link to the cell, however, requires a transmembrane protein that ties the matrix to the cell's cortical cytoskeleton. Although we have seen that some proteoglycans with transmembrane core proteins function as co-receptors for matrix components, the principal receptors on animal cells for binding most extracellular matrix proteins, including collagen, fibronectin, and laminin, are the **integrins,** a large family of homologous transmembrane linker proteins.

Integrins differ from cell-surface receptors for hormones and for other soluble signaling molecules in that they bind their ligand with relatively low affinity ($K_a = 10^6$–10^9 liters/mole) and are usually present at about 10- to 100-fold higher concentration on the cell surface. This arrangement makes sense, as binding simultaneously but weakly to large numbers of matrix molecules allows cells to explore their environment without losing all attachment to it. If the binding were too tight, cells would presumably become irreversibly glued to the matrix and be unable to move—a problem that does not arise if attachment depends on multiple weak adhesions. This is an example of the "Velcro principle" mentioned earlier.

Integrins Are Transmembrane Heterodimers [36]

Integrins are crucially important receptor proteins because they are the main way that cells both bind to and respond to the extracellular matrix. They are composed of two noncovalently associated transmembrane glycoprotein subunits called α and β, both of which contribute to the binding of the matrix protein (Figure 19–60). Whereas some integrins seem to bind only one matrix macromolecule such as fibronectin or laminin, others bind more than one: an integrin that is present on fibroblasts, for example, binds collagen, fibronectin, and laminin. One subfamily of integrins recognizes the RGD sequence present in these and other matrix proteins, while other integrins recognize various other sequences or domains. Since the same integrin molecule in different cell types can have different ligand-binding activities, it seems that additional cell-type-specific factors can interact with integrins to modulate their binding activity.

The binding of integrins to their ligands depends on extracellular divalent cations (Ca^{2+} or Mg^{2+}, depending on the integrin), reflecting the presence of three or four divalent-cation-binding domains in the large extracellular part of the α chain. This property can be used to purify integrins: detergent-solubilized plasma membrane proteins are passed over an affinity column that contains an extracellular matrix protein or an RGD-containing peptide, and the bound integrins are then eluted from the column by washing in a divalent-cation-free solution. The type of divalent cation can influence both the affinity and specificity of the binding of an integrin to its ligands, but, as in the case of the cadherins, the physiological significance of this cation regulation is unknown.

Many matrix proteins in vertebrates are recognized by multiple integrins: for example, at least 8 integrins bind fibronectin, and at least 5 bind laminin. About 20 integrin heterodimers, made from 9 types of β subunits and 14 types of α subunits, have been defined, and new ones are still being discovered. This diversity is further increased by the alternative splicing of some integrin RNAs. The β_1 chains, which form dimers with at least 9 distinct α chains, are found on almost all vertebrate cells; $\alpha_5\beta_1$, for example, is a fibronectin receptor, and $\alpha_6\beta_1$ is a laminin receptor on many types of cells. β_2 chains, by contrast, which form dimers with 3 types of α chains, are expressed exclusively on the surface of white blood cells, and they play an essential role in enabling these cells to fight infection. One of these β_2 integrins ($\alpha_L\beta_2$) is called *LFA-1* (for lymphocyte function associated); another ($\alpha_M\beta_2$) is called *Mac-1* because it is found mainly on macrophages. The β_2 integrins mainly mediate cell-cell, rather than cell-matrix, interactions by binding to specific ligands on another cell, such as an endothelial cell lining a blood vessel; the ligands, sometimes referred to as *counterreceptors*, are members of the immunoglobulin superfamily of cell adhesion molecules discussed earlier. The β_2 integrins enable white blood cells, for example, to attach firmly to and cross the endothelial lining of blood vessels at sites of infection. Humans with the genetic disease called *leucocyte adhesion deficiency* are unable to synthesize the β_2 subunit; as a consequence, their white blood cells lack the entire family of β_2 receptors, and they suffer repeated bacterial infections. β_3 integrins are found on a variety of cells, including blood platelets, and they bind several matrix proteins, including *fibrinogen*; platelets interact with fibrinogen during blood clotting, and humans with *Glanzmann's disease*, who are genetically deficient in β_3 integrins, bleed excessively.

Two integrins that share a common β subunit have been described in *Drosophila*. If both copies of the *Drosophila* gene encoding this β subunit are mutated, the flies die as embryos; they develop normally until the first muscle contractions begin, at which point the muscles tear away from their extracellular matrix attachment sites.

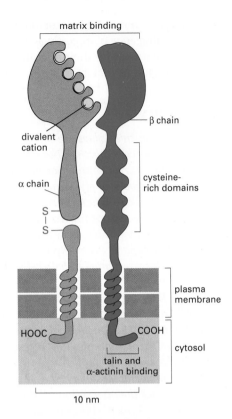

Figure 19–60 The subunit structure of an integrin cell-surface matrix receptor. Electron micrographs of isolated receptors suggest that the molecule has approximately the shape shown, with the globular head projecting more than 20 nm from the lipid bilayer. By binding to a matrix protein outside the cell and to the actin cytoskeleton (via the attachment proteins talin and α-actinin) inside the cell, the protein serves as a transmembrane linker. The α and β chains are both glycosylated (not shown) and are held together by noncovalent bonds. In the fibronectin receptor shown, the α chain is made initially as a single 140,000-dalton polypeptide chain, which is then cleaved into one small transmembrane chain and one large extracellular chain that remain held together by a disulfide bond; this extracellular chain is folded into four divalent-cation-binding domains. The extracellular part of the β chain contains a repeating cysteine-rich region, where intrachain disulfide bonding occurs; the β chain has a mass of about 100,000 daltons.

Integrins Must Interact with the Cytoskeleton in Order to Bind Cells to the Extracellular Matrix [37]

Integrins function as transmembrane linkers (or "integrators") mediating the interactions between the cytoskeleton and the extracellular matrix that are required for cells to grip the matrix. Most integrins connect to bundles of actin filaments. (The integrin found in hemidesmosomes—$\alpha_6\beta_4$—is an exception in that it connects to intermediate filaments.) Following the binding of a typical integrin to its ligand in the matrix, the cytoplasmic tail of the β chain binds to both *talin* and α-actinin and thereby initiates the assembly of a complex of intracellular attachment proteins that link the integrin to actin filaments in the cell cortex; this is thought to be how *focal contacts* form between cells and the extracellular matrix, as discussed earlier. If the cytoplasmic domain of the β chain is deleted using recombinant DNA techniques, the mutant integrins still bind to their ligands but no longer mediate robust cell adhesion or cluster at focal contacts. It seems that integrins must interact with the cytoskeleton in order to bind cells to the matrix, just as cadherins must interact with the cytoskeleton in order to hold cells together. As discussed earlier, a transmembrane attachment to the cytoskeleton appears to be an important general requirement for both cell-matrix and cell-cell adhesions: without such internal anchorage the attachment site is liable to be ripped out of the cell (see Figure 19–29). The cytoskeletal attachments may also help to cluster the integrins together to give a strong aggregate bond (the Velcro principle again).

As we shall discuss next, the interactions that integrins mediate between the extracellular matrix and the cytoskeleton operate in both directions and play an important part in orienting both the cells and the matrix in a tissue.

Integrins Enable the Cytoskeleton and Extracellular Matrix to Communicate Across the Plasma Membrane [37, 38]

The matrix can influence the organization of a cell's cytoskeleton. This can be vividly demonstrated with transformed (cancerlike) fibroblasts in culture (discussed in Chapter 24). Transformed cells often make less fibronectin than normal cultured cells and behave differently. They adhere poorly to the substratum, for example, and fail to flatten out or develop the organized intracellular actin filament bundles known as *stress fibers*. This may contribute to the tendency of cancer cells to break away from the primary tumor and spread to other parts of the body. In some cases the fibronectin deficiency seems to be at least partly responsible for this abnormal morphology: if the cells are grown on a matrix of organized fibronectin filaments, they flatten out and assemble intracellular stress fibers that are aligned with the extracellular fibronectin filaments.

This interaction between the extracellular matrix and the cytoskeleton is reciprocal in that intracellular actin filaments can influence the orientation of secreted fibronectin molecules. Extracellular fibronectin filaments, for example, assemble on or near the surface of cultured fibroblasts in alignment with adjacent intracellular stress fibers (Figure 19–61). If these cells are treated with the drug cytochalasin, which disrupts actin filaments, the fibronectin filaments dissociate from the cell surface (just as they do during mitosis when a cell rounds up). These reciprocal interactions between extracellular fibronectin and intracellular actin filaments across the fibroblast plasma membrane are mediated mainly by integrins.

Since the cytoskeletons of cells can exert forces that orient the matrix macromolecules that the cells secrete, and the matrix macromolecules can in turn organize the cytoskeletons of cells that contact them, the extracellular matrix can in principle propagate order from cell to cell (Figure 19–62), creating large-scale oriented structures, as we saw earlier (see p. 984). The integrins serve as adapters in this ordering process, mediating the interactions between cells and the matrix around them.

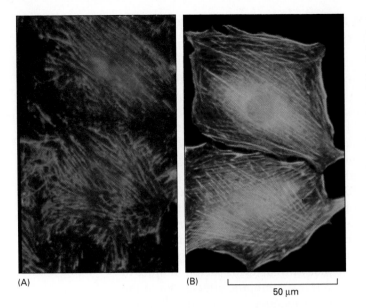

(A)　(B)

50 μm

Figure 19–61 **Coalignment of extracellular fibronectin filaments and intracellular actin filament bundles.** The fibronectin is visualized in two rat fibroblasts in culture by the binding of rhodamine-coupled anti-fibronectin antibodies (A). The actin is visualized by the binding of fluorescein-coupled anti-actin antibodies (B). (From R.O. Hynes and A.T. Destree, *Cell* 15:875–886, 1978. © Cell Press.)

Cells Can Regulate the Activity of Their Integrins [39]

Whereas the matrix-binding integrins of many cells in tissues are constantly in an adhesive-competent state, the integrins on blood cells often have to be activated before they can mediate cell adhesion. Such regulated adhesion presumably allows blood cells to circulate unimpeded until they are activated by an appropriate stimulus; because the integrins do not need to be synthesized *de novo*, the response can be rapid. Platelets, for example, can be activated by contact with a damaged blood vessel or any one of a number of soluble signaling molecules. The stimulus triggers intracellular signaling pathways, which in turn rapidly and permanently activate a β_3 integrin in the platelet membrane, altering its conformation so that its extracellular domain becomes able to bind the blood-clotting protein fibrinogen with high affinity, thereby promoting platelet aggregation and blood clot formation (Figure 19–63A). Similarly, the weak binding of T lymphocytes, either to their specific antigen on the surface of an antigen-presenting cell or to a virus-infected cell (discussed in Chapter 23), triggers intracellular signaling pathways in the T cells, which leads to the rapid but transient activation of an integrin (*LFA-1*) on the T cells. The activated integrin enables the T lymphocytes to adhere strongly to the target cell, so that they remain in contact long enough to become stimulated; the integrin then returns to an inactive state to allow the T lymphocytes to disengage. The mechanisms by which intracellular signaling events activate the extracellular binding site of an integrin on a blood cell are largely unknown.

Other intracellular events can inactivate integrins. The phosphorylation of a serine residue on the cytoplasmic tail of a β_1 integrin during mitosis in cultured cells, for example, impairs the ability of the integrin to bind fibronectin, which may explain why these cells round up and detach from the substratum during mitosis. Similarly, in some cancer cells the phosphorylation of a tyrosine residue on the cytoplasmic tail of a fibronectin-binding integrin reduces the ability of the integrin to bind to talin, and this is thought to contribute to the relatively poor adhesion of these cells to fibronectin (Figure 19–63B).

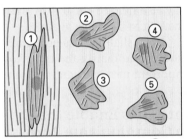

orientation of cytoskeleton in cell ① orients the assembly of secreted extra-cellular matrix molecules in the vicinity

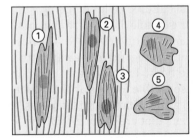

the oriented extracellular matrix reaches cells ② and ③ and orients the cytoskeleton of those cells

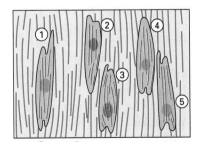

cells ② and ③ now secrete an oriented matrix in their vicinity; in this way the ordering of cytoskeletons is propagated to cells ④ and ⑤

Figure 19–62 **How the extracellular matrix could propagate order from cell to cell within a tissue.** For simplicity, the figure represents a hypothetical scheme in which one cell influences the orientation of its neighboring cells. It is more likely, however, that the cells would mutually affect one another's orientation.

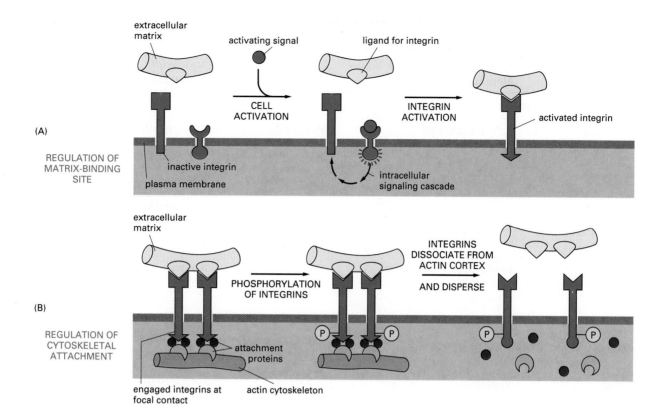

Figure 19–63 **Cells can regulate the activity of their integrins.** In (A) cell activation leads to a change in the extracellular binding site of the integrin so that it can now mediate cell adhesion. In (B) the tyrosine phosphorylation of the cytoplasmic tail of the integrins impairs their ability to bind to the actin cytoskeleton. As integrins must bind to the cytoskeleton to mediate robust cell-matrix adhesion, the phosphorylation causes the integrins to relax their grip on the extracellular matrix.

Integrins Can Activate Intracellular Signaling Cascades [40]

Extracellular matrix macromolecules have striking effects on the behavior of cells in culture, influencing their shape, polarity, movement, metabolism, development, and differentiated functions. Many of these effects involve changes in gene expression, and almost all of them are mediated by integrins. We have already discussed how integrins function as transmembrane linkers that connect extracellular matrix molecules to actin filaments in the cell cortex and thereby regulate the shape, orientation, and movement of cells. But there is increasing evidence that the clustering of integrins at the sites of contact with the matrix (or another cell) can also activate several intracellular signaling pathways, including the inositol phospholipid pathway; in addition, several intracellular proteins, including a tyrosine kinase located in focal contacts, become phosphorylated on tyrosine residues. Although the molecular mechanisms are not known, it seems likely that clustered integrins generate intracellular signals by initiating the assembly of a signaling complex at the cytoplasmic face of the plasma membrane, in much the same way that growth factor receptor tyrosine kinases operate (discussed in Chapter 15). Signaling by both integrins and growth factor receptors frequently seems to be required for an optimal cellular response: many cells in culture, for example, will not proliferate in response to growth factors unless the cells are attached via integrins to extracellular matrix molecules. The challenge is to determine how these signaling cascades interact to influence complex cell behaviors such as gene expression and cell proliferation.

The cell adhesion molecules discussed in this chapter are summarized in Table 19–5.

Summary

Integrins are the principal receptors used by animal cells to bind to the extracellular matrix. They are heterodimers that function as transmembrane linkers that mediate bidirectional interactions between the extracellular matrix and the actin

Table 19–5 Cell Adhesion Molecule Families

	Some Family Members	Ca²⁺- or Mg²⁺-dependence	Homophilic or Heterophilic	Cytoskeleton Associations	Cell Junction Associations
CELL-CELL ADHESION					
Cadherins	E, N, P cadherins	yes	homophilic	actin filaments (via catenins)	adhesion belts
	desmosomal cadherins	yes	homophilic	intermediate filaments (via desmoplakins, plakoglobin and other proteins)	desmosomes
Ig family members	N-CAM, L1	no	homophilic or heterophilic	unknown	no
Selectins (blood cells + endo-thelial cells only)	P-selectin (see p. 504)	yes	heterophilic	unknown	no
Integrins on blood cells	LFA-1 ($a_L b_2$), Mac-1 ($a_M b_2$)	yes	heterophilic	actin filaments	no
CELL-MATRIX ADHESION					
Integrins	many types	yes	heterophilic	actin filaments (via talin, vinculin, and other proteins)	focal contacts
	$a_6 b_4$	yes	heterophilic	intermediate filaments	hemidesmosomes
Transmembrane Proteoglycans	syndecans	no	heterophilic	actin filaments	no

cytoskeleton. They also function as signal transducers, activating various intracellular signaling pathways when activated by matrix binding. A cell can regulate the adhesive activity of its integrins by altering either their matrix-binding site or their attachment to actin filaments.

The Plant Cell Wall

The plant cell wall is an elaborate extracellular matrix that encloses each cell in a plant. It was the thick cell walls of cork, visible in a primitive microscope, that in 1663 enabled Robert Hooke to distinguish cells clearly and to name them as such. The walls of neighboring plant cells, cemented together to form the intact plant (Figure 19–64), are generally thicker, stronger, and, most important of all, more rigid than the extracellular matrix produced by animal cells. In evolving relatively rigid walls, which can be up to many micrometers in thickness, early plant cells forfeited the ability to crawl about and adopted a sedentary life-style that has persisted in all present-day plants.

The Composition of the Cell Wall Depends on the Cell Type [41]

Most newly formed cells in a multicellular plant are produced in special regions called *meristems*, as explained in Chapter 21. These new cells are generally small in comparison to their final size, and to accommodate subsequent cell growth, their walls, called **primary cell walls,** are thin and only semirigid. Once growth stops and the wall no longer needs to be able to expand, either the primary wall

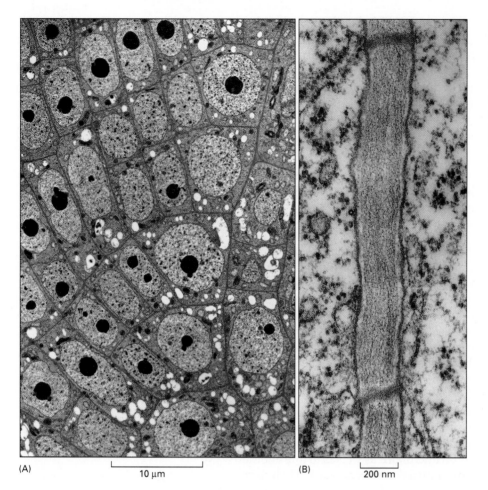

Figure 19–64 **Plant cell walls.** (A) Electron micrograph of the root tip of a rush, showing the organized pattern of cells that results from an ordered sequence of cell divisions in cells with rigid cell walls. (B) Section of a typical cell wall separating two adjacent plant cells. The two dark transverse bands correspond to plasmodesmata that span the wall. (A, courtesy of Brian Gunning; B, courtesy of Jeremy Burgess.)

(A)   (B)

10 µm   200 nm

is simply retained or, far more commonly, a rigid, **secondary cell wall** is produced, either by thickening the primary wall or by depositing new layers with a different composition underneath the old ones. In addition to a structural or "skeletal" role, the cell wall also protects the underlying cell and functions in the transport of fluid within the plant. When plant cells become specialized, they generally produce specially adapted types of walls, according to which the different types of cells in a plant can be recognized and classified.

Although the primary cell walls of higher plants vary greatly in both composition and organization, like all extracellular matrices they are constructed according to a common principle: they derive their tensile strength from long fibers and their resistance to compression from the matrix of protein and polysaccharide in which the fibers are embedded. In the cell walls of higher plants the fibers are generally made from the polysaccharide *cellulose*, the most abundant organic macromolecule on earth. The rest of the matrix is composed predominantly of two other types of polysaccharide, *hemicellulose* and *pectin*, together with structural proteins. All of these molecules are held together by a combination of covalent and noncovalent bonds to form a highly complex structure whose composition depends on the cell type.

The Tensile Strength of the Cell Wall Allows Plant Cells to Develop Turgor Pressure [42]

The aqueous extracellular environment of a plant cell consists of the fluid contained in the walls that surround the cell. Although the fluid in the plant cell wall contains more solutes than does the water in the plant's external milieu (for example, soil), it is still hypotonic in comparison to the cell interior. This osmotic imbalance causes the cell to develop a large internal hydrostatic pressure, or

turgor pressure, that pushes outward on the cell wall, just as an inner tube pushes outward on a tire. The turgor pressure increases just to the point where the cell is in osmotic equilibrium, with no net influx of water despite the salt imbalance (see Panel 11–1, p. 517). This pressure is vital to plants because it is the main driving force for cell expansion during growth, and it provides much of the mechanical rigidity of living plant tissues. Compare the wilted leaf of a dehydrated plant, for example, with the turgid leaf of a well-watered one. It is the mechanical strength of the cell wall that allows plant cells to sustain this internal pressure.

The Cell Wall Is Built from Cellulose Microfibrils Interwoven with a Network of Polysaccharides and Proteins [43]

The tensile strength of the primary cell wall is provided by cellulose. A cellulose molecule consists of a linear chain of at least 500 glucose residues that are covalently linked to one another to form a ribbonlike structure, which is stabilized by hydrogen bonds within the chain. In addition, intermolecular hydrogen bonds between adjacent cellulose molecules cause them to adhere strongly to one another in overlapping parallel arrays, forming a bundle of 60 to 70 cellulose chains, all of which have the same polarity. These highly ordered crystalline aggregates, many micrometers long, are called **cellulose microfibrils.** Sets of microfibrils are arranged in layers, or lamellae, with each microfibril about 20–40 nm from its neighbors and connected to them by long *hemicellulose* molecules that are bound by hydrogen bonds to the surface of the microfibrils. The primary cell wall consists of several such lamellae arranged in a plywoodlike network (Figure 19–65).

Hemicelluloses are a heterogeneous group of branched polysaccharides that bind tightly to the surface of each cellulose microfibril as well as to one another and thereby help to cross-link microfibrils into a complex network. Their function is analogous to that of the fibril-associated collagens discussed earlier. There are many classes of hemicelluloses, but they all have a long linear backbone composed of one type of sugar, from which short side chains of other sugars protrude. Both the backbone sugar and the side-chain sugars vary according to the plant species and its stage of development. It is the sugar molecules in the backbone that form hydrogen bonds with cellulose microfibrils.

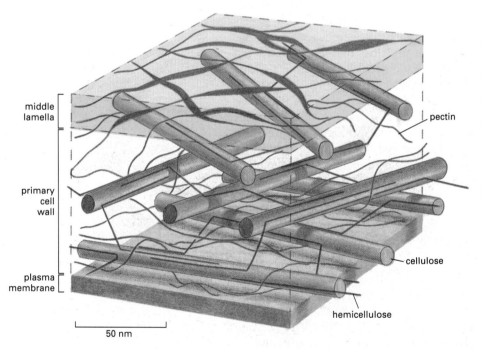

middle
lamella

primary
cell
wall

plasma
membrane

pectin

cellulose

hemicellulose

50 nm

Figure 19–65 Scale model of a portion of a primary cell wall showing the two major polysaccharide networks. The orthogonally arranged layers of cellulose microfibrils (*green*) are cross-linked into a network by H-bonded hemicellulose (*red*). This network is coextensive with a network of pectin polysaccharides (*blue*). The cellulose and hemicellulose network provides tensile strength, while the pectin network resists compression. Cellulose, hemicellulose, and pectin are typically present in roughly equal quantities in a primary cell wall. The middle lamella is pectin rich and cements adjacent cells together.

Coextensive with this network of cellulose microfibrils and hemicelluloses is another cross-linked polysaccharide network based on *pectins* (see Figure 19–65). These are a heterogeneous group of branched polysaccharides that contain many negatively charged galacturonic acid residues. Because of their negative charge, pectins are highly hydrated and accompanied by a cloud of cations, resembling the glycosaminoglycans of animal cells in the large amount of space they occupy. When Ca^{2+} is added to a solution of pectin molecules, it cross-links them to produce a semirigid gel (it is pectin that is added to fruit juice to make jelly). Certain pectins are particularly abundant in the *middle lamella*, the specialized central region of the wall that cements together the walls of adjacent cells, and such Ca^{2+} cross-links are thought to help hold cell wall components together. Thus many plant tissues, if treated with a Ca^{2+} chelating agent, dissociate into their constituent cells. Although covalent bonds also play a part in linking the different plant cell-wall components together, very little is known about their nature.

In addition to the two polysaccharide-based networks that are present in all plant primary cell walls, there is a variable contribution from structural proteins. One class of proteins contains high levels of hydroxyproline, like collagen. These proteins are thought to strengthen the wall, and they are produced in greatly increased amounts as a local response to attack by microorganisms. During normal differentiation cells use structural proteins to modify local regions of their walls, as required to create the wide range of functionally specialized secondary walls characteristic of mature cell types (see Panel 1–1, pp. 18–19).

In order for a plant cell to grow or change its shape, the cell wall has to stretch or deform. But because of their crystalline structure, individual cellulose microfibrils are unable to stretch. Thus stretching or deformation of the cell wall must involve either the sliding of microfibrils past one another, or the separation of adjacent microfibrils, or both. As we discuss next, the direction in which the growing cell enlarges depends on the orientation of the strain-resisting cellulose microfibrils in the primary wall, which in turn depends on the orientation of microtubules in the underlying cell cortex at the time the wall was deposited.

Microtubules Orient Cell-Wall Deposition [44]

The final shape of a growing plant cell, and hence the final form of the plant, is determined by controlled cell expansion. Expansion occurs in response to turgor pressure in a direction that depends on the arrangement of the cellulose microfibrils in the wall. Cells anticipate their future morphology, therefore, by controlling the orientation of microfibrils that they deposit in the wall. Unlike most other matrix macromolecules, which are made in the endoplasmic reticulum and Golgi apparatus and secreted, cellulose, like hyaluronan, is spun out from the surface of the cell by a plasma-membrane-bound enzyme complex (*cellulose synthase*), which uses sugar nucleotide precursors supplied from the cytosol. As they are being synthesized, the nascent cellulose chains spontaneously assemble into microfibrils that form on the extracellular surface of the plasma membrane—forming a layer, or lamella, in which all the microfibrils have more or less the same alignment (Figure 19–65). Each new lamella forms internally to the previous one, so that the wall consists of concentrically arranged lamellae, with the oldest on the outside. The most recently laid down microfibrils in elongating cells commonly lie perpendicular to the axis of cell elongation (Figure 19–66); although the orientation of the microfibrils in the outer lamellae that were laid down earlier might be different, it is the orientation of these inner lamellae that have a dominant influence on the direction of cell expansion (Figure 19–67).

An important clue to understanding how this orientation is brought about was provided by the discovery that most cytoplasmic microtubules are arranged in the cortex of the plant cell with the same orientation as the cellulose microfibrils that are currently being deposited in that region. These cortical microtubules, forming what is called the *cortical array*, lie close to the cytoplasmic face

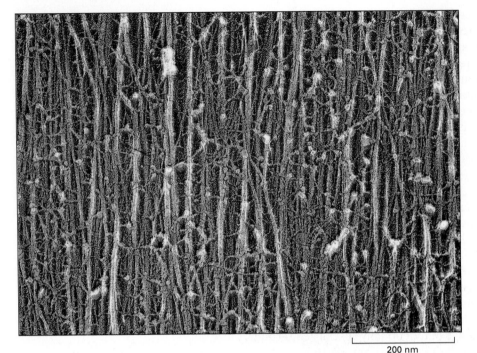

200 nm

of the plasma membrane, held there by poorly characterized proteins (Figure 19–68). The congruent orientation of the cortical array of microtubules (lying just inside the plasma membrane) and cellulose microfibrils (lying just outside) is seen in many types and shapes of plant cells and is present during both primary and secondary cell-wall deposition.

If the entire system of cortical microtubules is disassembled by treating a plant tissue with a microtubule-depolymerizing drug, the consequences for subsequent cellulose deposition are not as straightforward as might be expected. The drug treatment has no effect on the production of new cellulose microfibrils, and in some cases cells can continue to deposit new microfibrils in the preexisting orientation. Any developmental change in the microfibril pattern that would normally occur between successive lamellae, however, is invariably blocked. It seems that a preexisting orientation of microfibrils can be propagated even in the absence of microtubules, but any change in the deposition of cellulose microfibrils requires that intact microtubules be present to determine the new orientation.

These observations are consistent with the following model. The cellulose-synthesizing complexes embedded in the plasma membrane are thought to spin out long cellulose molecules. As the synthesis of cellulose molecules and their self-assembly into microfibrils proceeds, the distal end of each microfibril presumably forms indirect cross-links to the previous layer of wall material. At the growing, proximal end the synthesizing complexes would therefore need to move along the membrane in the direction of synthesis. Since the growing cellulose microfibrils are very stiff, each layer of microfibrils would tend to be spun out from the membrane in the same orientation as the previously laid down layer, with the cellulose synthase complex following along the preexisting tracks of oriented microfibrils outside the cell. Oriented microtubules inside the cell, however, can change this predetermined direction in which the synthase complexes move: they can create boundaries in the plasma membrane that act like the banks of a canal to constrain movement of the synthase complexes to a parallel axis (Figure 19–69). In this view, cellulose synthesis can occur independently of microtubules but is constrained spatially when cortical microtubules are present to define membrane domains within which the enzyme complex can move.

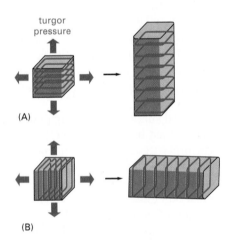

turgor pressure

(A)

(B)

Figure 19–67 How the orientation of cellulose microfibrils within the cell wall influences the direction in which the cell elongates. The cells in (A) and (B) start off with identical shapes (shown here as cubes) but with different orientations of cellulose microfibrils in their walls. Although turgor pressure is uniform in all directions, cell-wall weakening causes each cell to elongate in a direction perpendicular to the orientation of the microfibrils, which have great tensile strength. The final shape of an organ, such as a shoot, is determined by the direction in which its cells expand.

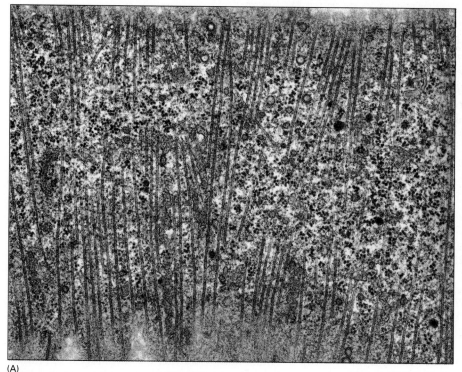

(A)

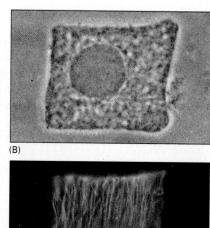

(B)

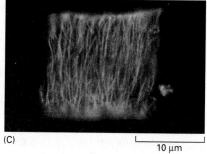

(C)

10 µm

Plant cells change their direction of elongation, and thus their future plane of cell growth and division, by a sudden change in the orientation of their entire cortical array of microtubules. Inasmuch as plant cells cannot move (being constrained by their walls), the entire morphology of a multicellular plant depends on the coordinated, highly patterned control of these cortical microtubule orientations during plant development. It is not known how the organization of these microtubules is controlled, although it has been shown that they can rapidly reorient in response to extracellular stimuli, including low-molecular-weight plant growth factors such as *ethylene* and *gibberellic acid*.

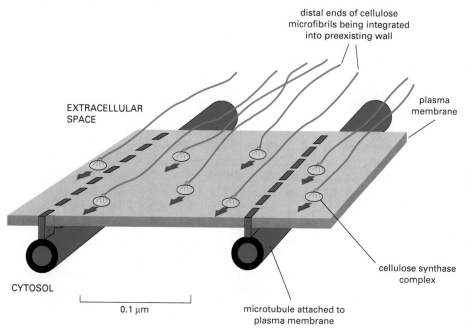

distal ends of cellulose microfibrils being integrated into preexisting wall

EXTRACELLULAR SPACE

plasma membrane

CYTOSOL

0.1 µm

cellulose synthase complex

microtubule attached to plasma membrane

Figure 19–69 One model of how the orientation of newly deposited cellulose microfibrils might be determined by the orientation of cortical microtubules. The large cellulose synthase complexes are integral membrane proteins that continuously synthesize cellulose microfibrils on the outer face of the plasma membrane. The distal ends of the stiff microfibrils become integrated into the texture of the wall, and their elongation at the proximal end pushes the synthase complex along in the plane of the membrane. Because the cortical array of microtubules is attached to the plasma membrane in a way that confines this complex to defined membrane channels, the microtubule orientation determines the axis along which the microfibrils are laid down.

Summary

Plant cells are surrounded by a rigid extracellular matrix in the form of a cell wall, which is responsible for many of the unique features of a plant's life-style. The cell wall is composed of tough cellulose microfibrils embedded in a highly cross-linked matrix of polysaccharides (mainly pectins and hemicellulose) and glycoproteins. A cortical array of microtubules can determine the orientation of newly deposited cellulose microfibrils, which in turn determines the manner in which the cell expands and therefore the cell's final shape and cell-division patterns.

References

Cited

1. Bock, G.; Clark, S., eds. Junctional Complexes of Epithelial Cells. Ciba Symposium 125. New York: Wiley, 1987.

   Farquhar, M.G.; Palade, G.E. Junctional complexes in various epithelia. *J. Cell Biol.* 17:375–412, 1963.

   Gilula, N.B. Junctions between cells. In Cell Communication (R.P. Cox, ed.), pp. 1–29. New York: Wiley, 1974.

   Goodenough, D.A.; Revel, J.P. A fine structural analysis of intercellular junctions in the mouse liver. *J. Cell Biol.* 45:272–290, 1970.

   Staehelin, L.A.; Hull, B.E. Junctions between living cells. *Sci. Am.* 238(5):141–152, 1978.

2. Anderson, J.M.; Balda, M.S.; Fanning, A.S. The structure and regulation of tight junctions. *Curr. Opin. Cell Biol.* 5:772–778, 1993.

   Diamond, J.M. The epithelial junction: bridge, gate and fence. *Physiologist* 20:10–18, 1977.

   Handler, J.S. Overview of epithelial polarity. *Annu. Rev. Physiol.* 51:729–740, 1989.

   Madara, J.L. Tight junction dynamics: is paracellular transport regulated? *Cell* 53:497–498, 1988.

   Madara, J.L.; Dharmsathaphorn, K. Occluding junction structure-function relationships in cultured epithelial monolayer. *J. Cell Biol.* 101:2124–2133, 1985.

   Schneeberger, E.E.; Lynch, R.D. Structure, function and regulation of cellular tight junctions. *Am. J. Physiol.* 262:647–661, 1992.

   Simons, K.; Fuller, S.D. Cell surface polarity in epithelia. *Annu. Rev. Cell Biol.* 1:243–288, 1985.

   van Meer, G.; Gumbiner, B.; Simons, K. The tight junction does not allow lipid molecules to diffuse from one epithelial cell to the next. *Nature* 322:639–641, 1986.

3. Burridge, K.; Fath, K.; Kelly, T.; Nuckolls, G.; Turner, C. Focal adhesions: transmembrane junctions between the extracellular matrix and the cytoskeleton. *Annu. Rev. Cell Biol.* 4:487–526, 1988.

   Geiger, B.; Ginsberg, D. The cytoplasmic domain of adherens-type junctions. *Cell Motil. Cytoskel.* 20:1–6, 1991.

   Geiger, B.; Volk, T.; Volberg, T. Molecular heterogeneity of adherens junctions. *J. Cell Biol.* 101:1523–1531, 1985.

   Noirot-Timothee, C.; Noirot, C. Septate and scalariform junctions in arthropods. *Int. Rev. Cytol.* 63:97–140, 1980.

   Turner, C.E.; Burridge, K. Transmembrane molecular assemblies in cell-extracellular matrix interactions. *Curr. Opin. Cell Biol.* 3:849–853, 1991.

4. Buxton, R.S.; Magee, A.I. Structure and interactions of desmosomal and other cadherins. *Semin. Cell Biol.* 3:157–167, 1992.

   Buxton, R.S., et al. Nomenclature of the desmosomal cadherins. *J. Cell Biol.* 121:481–483, 1993.

   Legan, P.K.; Collins, J.E.; Garrod, D.R. The molecular biology of desmosomes and hemidesmosomes: what's in a name? *Bioessays* 14:385–393, 1992.

   Schwarz, M.A.; Owaribe, K.; Kartenbeck, J.; Franke, W.W. Desmosomes and hemidesmosomes: constitutive molecular components. *Annu. Rev. Cell Biol.* 6:461–491, 1990.

5. Bennett, M.V.L., et al. Gap junctions: new tools, new answers, new questions. *Neuron* 6:305–320, 1991.

   Beyer, E.C. Gap junctions. *Int. Rev. Cytol.* 137:1–38, 1993.

   Furshpan, E.J.; Potter, D.D. Low-resistance junctions between cells in embryos and tissue culture. *Curr. Top. Dev. Biol.* 3:95–127, 1968.

6. Kumar, N.M.; Gilula, N.B. Molecular biology and genetics of gap junction channels. *Semin. Cell Biol.* 3:3–16, 1992.

   Stauffer, K.A.; Unwin, N. Structure of gap junction channels. *Semin. Cell Biol.* 3:17–20, 1992.

7. Warner, A. Gap junctions in development—a perspective. *Semin. Cell Biol.* 3:81–91, 1992.

8. Bennett, M.V.L.; Verselis, V.K. Biophysics of gap junctions. *Semin. Cell Biol.* 3:29–47, 1992.

   Sgaz, J.C.; Berthoud, V.M.; Moreno, A.P.; Spray, D.C. Gap junctions: multiplicity of controls in differentiated and undifferentiated cells and possible functional implications. *Adv. Second Messenger Phosphoprotein Res.* 27:163–198, 1993.

9. Deom, C.M.; Lapidot, M.; Beachy, R.N. Plant virus movement proteins. *Cell* 69:221–224, 1992.

   Gunning, B.E.S.; Robards, A.W., eds. Intercellular Communication in Plants: Studies on Plasmodesmata. New York: Springer-Verlag, 1976.

   Lucas, W.J.; Wolf, S. Plasmodesmata: the intercellular organelles of green plants. *Trends Cell Biol.* 3:308–315, 1993.

10. Piggot, R. The Adhesion Molecule Facts Book. San Diego, CA: Academic Press, 1993.

11. Bronner-Fraser, M. Environmental influences on neural crest cell migration. *J. Neurobiol.* 24:233–247, 1993.

    Le Douarin, N.; Smith, J. Development of the peripheral nervous system from the neural crest. *Annu. Rev. Cell Biol.* 4:375–404, 1988.

12. Gerisch, G. Univalent antibody fragments as tools for the analysis of cell interactions in *Dictyostelium. Curr. Top. Dev. Biol.* 14:243–270, 1980.

    Hennings, H.; Holbrook, K.A. Calcium regulation of cell-cell contact and differentiation of epidermal cells in culture. An ultrastructural study. *Exp. Cell Res.* 143:127–142, 1983.

    Moscona, A.A.; Hausman, R.E. Biological and biochemical studies on embryonic cell-cell recognition. In Cell and Tissue Interactions, Society of General Physiologists Series (J.W. Lash, M.M. Burger, eds.), Vol. 32, pp. 173–185. New York: Raven Press, 1977.

    Roth, S.; Weston, J. The measurement of intercellular adhesion. *Proc. Natl. Acad. Sci. USA* 58:974–980, 1967.

13. Geiger, B.; Ayalon, O. Cadherins. *Annu. Rev. Cell Biol.* 8:307–332, 1992.

    Takeichi, M. Cadherins: a molecular family important in selective cell-cell adhesion. *Annu. Rev. Biochem.* 59:237–252, 1990.

14. Hynes, R.O. Specificity of cell adhesion in development: the cadherin superfamily. *Curr. Opin. Genet. Dev.* 2:621–624, 1992.

    Kemler, R. Classical cadherins. *Semin. Cell Biol.* 3:149–155, 1992.

    Stappert, J.; Kemler, R. Intracellular associations of adhesion molecules. *Curr. Opin. Neurobiol.* 3:60–66, 1993.

    Takeichi, M. Cadherin cell adhesion receptors as a morphogenetic regulator. *Science* 251:1451–1455, 1991.

    Tsukita, S.; Tsukita, S.; Nagatuchi, A.; Yonemura, S. Molecular linkage between cadherins and actin filaments in cell-cell adherens junctions. *Curr. Opin. Cell Biol.* 4:834–839, 1992.

15. Edelman, G.M.; Crossin, K.L. Cell adhesion molecules: implications for a molecular histology. *Annu. Rev. Biochem.* 60:155–190, 1991.

    Grenningloh, G.; Rehm, E.J.; Goodman, C.S. Genetic analysis of growth cone guidance in *Drosophila:* fasciclin II functions as a neuronal recognition molecule. *Cell* 67:45–57, 1991.

    Kintner, C. Molecular bases of early neural development in *Xenopus* embryos. *Annu. Rev. Neurosci.* 15:251–284, 1992.

    Walsh, F.S.; Doherty, P. Factors regulating the expression and function of calcium-independent cell adhesion molecules. *Curr. Opin. Cell Biol.* 5:791–796, 1993.

16. Ekblom, P.; Vestweber, D.; Kemler, R. Cell-matrix interactions and cell adhesion during development. *Annu. Rev. Cell Biol.* 2:27–48, 1986.

    Hortsch, M.; Goodman, C.S. Cell and substrate adhesion molecules in *Drosophila. Annu. Rev. Cell Biol.* 7:505–557, 1991.

    Hynes, R.O.; Lander, A.D. Contact and adhesive specificities in the associations, migrations, and targeting of cells and axons. *Cell* 68:303–322, 1992.

    Schweighoffer, T.; Shaw, S. Adhesion cascades: diversity through combinatorial strategies. *Curr. Opin. Cell Biol.* 4:824–829, 1992.

17. Birk, D.E.; Silver, F.H.; Trelstad, R.L. Matrix assembly. In Cell Biology of Extracellular Matrix (E.D. Hay, ed.), 2nd ed., pp. 221–254. New York: Plenum Press, 1991.

    Hay, E.D., ed. Cell Biology of Extracellular Matrix, 2nd ed. New York: Plenum Press, 1991.

    Kreis, T.; Vale, R. Guidebook to the Extracellular Matrix and Adhesion Proteins. Oxford, UK: Oxford University Press, 1993.

    Piez, K.A.; Reddi, A.H., eds. Extracellular Matrix Biochemistry. New York: Elsevier, 1984.

    Reichardt, L.F.; Tomaselli, K.J. Extracellular matrix molecules and their receptors: functions in neural development. *Annu. Rev. Neurosci.* 14:531–570, 1991.

18. Jackson, R.L.; Busch, S.J.; Cardin, A.D. Glycosaminoglycans: molecular properties, protein interactions and role in physiological processes. *Physiol. Rev.* 71:481–539, 1991.

    Quentin, E.; Gladen, A.; Rodén, L.; Kresse, H. A genetic defect in the biosynthesis of dermatan sulfate proteoglycan: galactosyltransferase I deficiency in fibroblasts from a patient with a progeroid syndrome. *Proc. Natl. Acad. Sci. USA* 87:1342–1346, 1990.

19. Prehm, P. Hyaluronate is synthesized at plasma membranes. *Biochem. J.* 220:597–600, 1984.

    Toole, B. Proteoglycans and hyaluronan in morphogenesis and differentiation. In Cell Biology of Extracellular Matrix (E.D. Hay, ed.), 2nd ed., pp. 305–341. New York: Plenum Press, 1991.

20. Hardingham, T.E.; Fosang, A.J. Proteoglycans: many forms and many functions. *FASEB J.* 6:861–870, 1992.

    Hascall, V.C.; Heinegård, D.K.; Wight, T.N. Proteoglycans: metabolism and pathology. In Cell Biology of Extracellular Matrix (E.D. Hay, ed.), 2nd ed., pp. 149–176. New York: Plenum Press, 1991.

    Kjellén, L.; Lindahl, U. Proteoglycans: structures and interactions. *Annu. Rev. Biochem.* 60:443–475, 1991.

    Ruoslahti, E. Structure and biology of proteoglycans. *Annu. Rev. Cell Biol.* 4:229–255, 1988.

21. Flaumenhaft, R.; Rifkin, D.B. The extracellular regulation of growth factor action. *Mol. Biol. Cell* 3:1057–1065, 1992.

    Massagné, J. A helping hand from proteoglycans. *Curr. Biol.* 2:117–119, 1991.

    Ruoslahti, E.; Yamaguchi, Y. Proteoglycans as modulators of growth factor activities. *Cell* 64:867–869, 1991.

    Wight, T.N.; Kinsella, M.G.; Qwarnström, E.E. The role of proteoglycans in cell adhesion, migration and proliferation. *Curr. Opin. Cell Biol.* 4:793–801, 1992.

22. Bernfield, M., et al. Biology of syndecans: a family of transmembrane heparan sulfate proteoglycans. *Annu. Rev. Cell Biol.* 8:365–393, 1992.

    Scott, J.E. Supramolecular organization of extracellular matrix of glycosaminoglycans *in vitro* and in the tissues. *FASEB J.* 6:2639–2645, 1992.

23. Burgeson, R.E. New collagens, new concepts. *Annu. Rev. Cell Biol.* 4:551–577, 1988.

    Kucharz, E.J. The Collagens: Biochemistry and Pathophysiology. New York: Springer-Verlag, 1992.

    Linsenmayer, T.F. Collagen. In Cell Biology of Extracellular Matrix (E.D. Hay, ed.), 2nd ed., pp. 7–44. New York: Plenum Press, 1991.

    Mayne, R.; Brewton, R.G. New members of the collagen superfamily. *Curr. Opin. Cell Biol.* 5:883–890, 1993.

    Van der Rest, M.; Garrone, R. Collagen family of proteins. *FASEB J.* 5:2814–2823, 1991.

24. Olsen, B.R. Collagen biosynthesis. In Cell Biology of Extracellular Matrix (E.D. Hay, ed.), 2nd ed., pp. 177–220. New York: Plenum Press, 1991.

25. Eyre, D.R.; Paz, M.A.; Gallop, P.M. Cross-linking in collagen and elastin. *Annu. Rev. Biochem.* 53:717–748, 1984.

Ploetz, C.; Zycband, E.I.; Birk, D.E. Collagen fibril assembly and deposition in the developing dermis: segmental deposition in extracellular compartments. *J. Struct. Biol.* 106:73–81, 1991.

Prockop, D.J. Mutations in collagen genes as a cause of connective-tissue diseases. *N. Engl. J. Med.* 326:540–546, 1992.

26. Linsenmayer, T.F. Collagen. In Cell Biology of Extracellular Matrix (E.D. Hay, ed.), 2nd ed., pp. 7–44. New York: Plenum Press, 1991.

Van der Rest, M.; Mayne, R. Type IX collagen. In Structure and Function of Collagen Types (R. Mayne, R.E. Burgeson, eds.), pp. 195–219. New York: Academic Press, 1987.

27. Stopak, D.; Harris, A.K. Connective tissue morphogenesis by fibroblast traction I. Tissue culture observations. *Dev. Biol.* 90:383–398, 1982.

28. Cleary, E.G.; Gibson, M.A. Elastin-associated microfibrils and microfibrillar proteins. *Int. Rev. Connect. Tissue Res.* 10:97–209, 1983.

Gosline, J.M.; Rosenbloom, J. Elastin. In Extracellular Matrix Biochemistry (K.A. Piez, A.H. Reddi, eds.), New York: Elsevier, 1984.

Indik, Z., et al. Structure of the elastin gene and alternative splicing of elastin mRNA: implications for human disease. *Am. J. Med. Genet.* 34:81–90, 1989.

Mecham, R.P.; Heuser, J.E. The elastic fiber. In Cell Biology of Extracellular Matrix (E.D. Hay, ed.), 2nd ed., pp. 79–110. New York: Plenum Press, 1991.

Ramirez, F.; Pereira, L.; Zhang, H.; Lee, B. The fibrillin-marfan syndrome connection. *Bioessays* 15:589–594, 1993.

29. Hynes, R.O. Fibronectins. *Sci. Am.* 254(6):42–51, 1986.

Hynes, R.O. Fibronectins. New York: Springer-Verlag, 1989.

Mosher, D.R., ed. Fibronectin. New York: Academic Press, 1988.

Ruoslahti, E.; Pierschbacher, M.D. New perspectives in cell adhesion: RGD and integrins. *Science* 238:491–497, 1987.

D'Souza, S.E.; Ginsberg, M.H.; Plow, E.F. Arginyl-glycyl-aspartic acid (RGD): a cell adhesion motif. *Trends Biochem. Sci.* 16:246–250, 1991.

30. George,E.L.;Georges-Lebouesse,E.H.;Patel-King,R.S.; Rayburn, H.; Hynes, R.O. Defects in mesoderm, neural tube and vascular development in mouse embryos lacking fibronectin. *Development* 119:1079–1091.

Schwarzbauer, J.E. Alternative splicing of fibronectin: three variants, three functions. *Bioessays* 13:527–533, 1991.

31. Chiquet-Ehrismann, R. Anti-adhesive molecules of the extracellular matrix. *Curr. Opin. Cell Biol.* 3:800–804, 1991.

Erickson, H.P. Tenascin-C, tenascin-R and tenascin-X: a family of talented proteins in search of functions. *Curr. Opin. Cell Biol.* 5:869–876, 1993.

Yamada, K.M. Fibronectin and other cell interactive glycoproteins. In Cell Biology of Extracellular Matrix (E.D. Hay, ed.), 2nd ed., pp. 111–148. New York: Plenum Press, 1991.

32. Yurchenco, P.D.; Furthmayr, H. Self-assembly of basement membrane collagen. *Biochemistry* 23:1839–1850, 1984.

Yurchenco, P.D.; Ruben, G.C. Basement membrane structure *in situ*: evidence for lateral associations in the type IV collagen network. *J. Cell Biol.* 105:2559–2568, 1987.

33. Burgeson, R.E., et al. Type VII collagen. In Structure and Function of Collagen Types (R. Mayne, R.E. Burgeson, eds.), pp. 145–172. New York: Academic Press, 1990.

Inoue, S. Ultrastructure of basement membranes. *Int. Rev. Cytol.* 117:57–98, 1989.

Martin, G.R.; Timpl, R. Laminin and other basement membrane components. *Annu. Rev. Cell Biol.* 3:57–85, 1987.

Rohrbach, D.H.; Timpl, R., eds. Molecular and Cellular Aspects of Basement Membranes. San Diego, CA: Academic Press, 1993.

Tryggvason, K. The laminin family. *Curr. Opin. Cell Biol.* 5:877–882, 1993.

Yurchenco, P.D.; Schittny, J.C. Molecular architecture of basement membranes. *FASEB J.* 4:1577–1590, 1990.

34. Farquhar, M.G. The glomerular basement membrane: a selective macromolecular filter. In Cell Biology of Extracellular Matrix (E.D. Hay, ed.), 2nd ed., pp. 365–418. New York: Plenum Press, 1991.

McMahan, U.J., et al. Agrin isoforms and their role in synaptogenesis. *Curr. Opin. Cell Biol.* 4:869–874, 1992.

Sanes, J.R.; Engvall, E.; Butkowski, R.; Hunter, D.D. Molecular heterogeneity of basal laminae: isoforms of lamin and collagen IV at the neuromuscular junction and elsewhere. *J. Cell Biol.* 111:1685–1699, 1990.

Wadsworth, W.G.; Hedgecock, E.M. Guidance of neuroblast migrations and axonal projections in *Caenorhabditis elegans. Curr. Opin. Neurobiol.* 2:36–41, 1992.

35. Birkedal-Hansen, H.; Werb, Z.; Welgus, H.G. Matrix Metalloproteinases and Inhibitors. Stuttgart, Ger.: Gustav, Fisher, Verlag, 1992.

Chen, W.-T. Membrane proteases: roles in tissue remodeling and tumour invasion. *Curr. Opin. Cell Biol.* 4:802–809, 1992.

Matrisian, L.M. The matrix-degrading metalloproteinases. *Bioessays* 14:455–463, 1992.

Stetler-Stevenson, W.G.; Aznavoorian, S.; Liotta, L.A. Tumor cell interactions with the extracellular matrix during invasion and metastasis. *Annu. Rev. Cell Biol.* 9:541–574, 1993.

36. Anderson, D.C.; Springer, T.A. Leukocyte adhesion deficiency: an inherited defect in the Mac-1, LFA-1 and P150,95 glycoprotein. *Annu. Rev. Med.* 38:175–194, 1987.

Brown, N.H. Integrins hold *Drosophila* together. *Bioessays* 15:383–390, 1993.

Buck, C.A.; Horwitz, A.F. Cell surface receptors for extracellular matrix molecules. *Annu. Rev. Cell Biol.* 3:179–205, 1987.

Hynes, R.O. Integrins: a family of cell surface receptors. *Cell* 48:549–554, 1987.

Ruoslahti, E. Integrins are receptors for extracellular matrix. In Cell Biology of Extracellular Matrix (E.D. Hay, ed.), 2nd ed., pp. 343–364. New York: Plenum Press, 1991.

37. Sastry, S.K.; Horwitz, A.F. Integrin cytoplasmic domains: mediators of cytoskeletal linkages and extra- and intercellular initiated transmembrane signaling. *Curr. Opin. Cell Biol.* 5:819–831, 1993.

Solowska, J., et al. Expression of normal and mutant avian integrin subunits in rodent cells. *J. Cell Biol.* 109:853–861, 1989.

Turner, C.E.; Burridge, K. Transmembrane molecular assemblies in cell-extracellular matrix interactions. *Curr. Opin. Cell Biol.* 5:849–853, 1991.

38. Burridge, K.; Fath, K.; Kelly, T.; Nuckolls, G.; Turner, C. Focal adhesions: transmembrane junctions between the extracellular matrix and the cytoskeleton. *Annu. Rev. Cell Biol.* 4:487–526, 1988.

39. Ginsberg, M.H.; Du, X.; Plow, E.F. Inside-out integrin signalling. *Curr. Opin. Cell Biol.* 4:766–771, 1992.

Hynes, R.O. Integrins: versatility, modulation, and signaling in cell adhesion. *Cell* 69:11–25, 1992.

Phillips, D.R.; Charo, I.F.; Scarborough, R.M. GPIIb-IIIa: the responsive integrin. *Cell* 65:359–362, 1992.

40. Damsky, C.H.; Werb, Z. Signal transduction by integrin receptors for extracellular matrix: cooperative processing of extracellular information. *Curr. Opin. Cell Biol.* 5:772–781, 1992.

Juliano, R.L.; Haskill, S. Signal transduction from the extracellular matrix. *J. Cell Biol.* 120:577–585, 1993.

Schwartz, M.A. Transmembrane signalling by integrins. *Trends Cell Biol.* 2:304–308, 1992.

41. Bacic, A.; Harris, P.J.; Stone, B.A. Structure and function of plant cell walls. In Biochemistry of Plants: A Comprehensive Treatise (J. Preiss, ed.), Vol. 14: Carbohydrates, pp. 298–371. San Diego, CA: Academic Press, 1988.

Fry, S.C. The Growing Plant Cell Wall: Chemical and Metabolic Analysis. New York: Wiley, 1988.

Tanner, W.; Loewus, F.A., eds. Encyclopedia of Plant Physiology, New Series, Vol. 13B: Plant Carbohydrates II, Extracellular Carbohydrates. Heidelberg: Springer-Verlag, 1982.

42. Street, H.E.; Öpik, H. The Physiology of Flowering Plants, 3rd ed. London: Edward Arnold, 1984.

43. Bolwell, G.P. Dynamic aspects of the plant extracellular matrix. *Int. Rev. Cytol.* 146:261–324, 1993.

McNeil, M.; Darvill, A.G.; Fry, S.C.; Albersheim, P. Structure and function of the primary cell walls of plants. *Annu. Rev. Biochem.* 53:625–663, 1984.

Levy, S.; Staehelin, L.A. Synthesis, assembly and function of plant cell wall macromolecules. *Curr. Opin. Cell Biol.* 5:856–862, 1992.

Roberts, K. Structures at the plant cell surface. *Curr. Opin. Cell Biol.* 2:920–928, 1990.

Varner, J.E.; Lin, L.-S. Plant cell wall architecture. *Cell* 56:231–239, 1989.

44. Brown, R.M. Cellulose microfibril assembly and orientation: recent developments. *J. Cell Sci.* (Suppl. 2):13–32, 1985.

Delmer, D.P. Cellulose biosynthesis. *Annu. Rev. Plant Physiol.* 38:259–290, 1987.

Herth, W. Plant cell wall formation. In Botanical Microscopy (A.W. Robards, ed.), pp. 285–310. New York: Oxford University Press, 1985.

Lloyd, C.W., ed. The Cytoskeleton in Plant Growth and Development. New York: Academic Press, 1982.

References

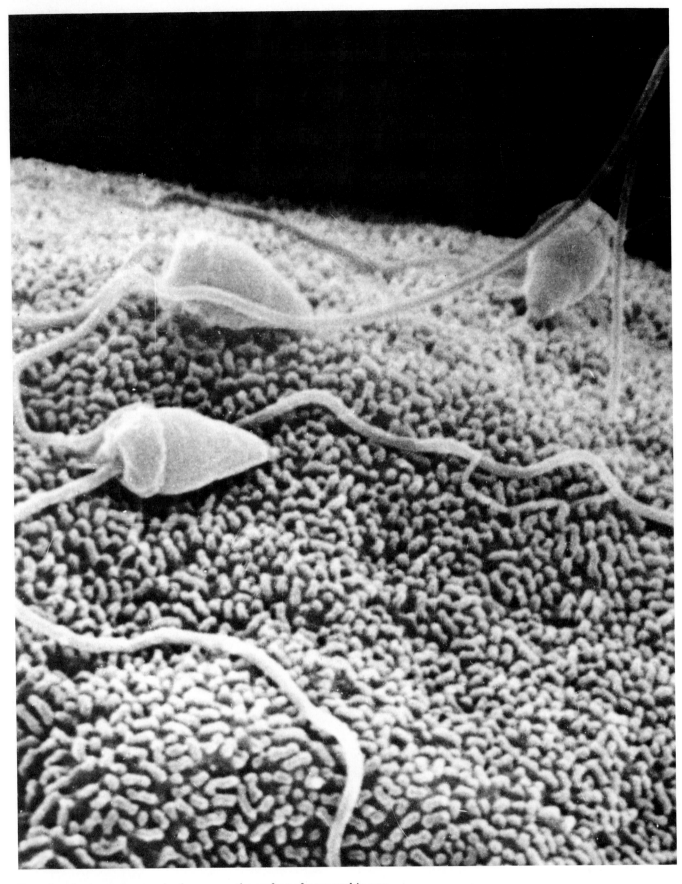

Scanning electron micrograph of sperm on the surface of a sea urchin egg.
(Courtesy of Brian Dale.)

Germm Cells and Fertilization

Sex is not absolutely necessary. Single-celled organisms can reproduce by simple mitotic division, and many plants propagate vegetatively, by forming multicellular offshoots that later detach from the parent. Likewise, in the animal kingdom, a solitary multicellular *Hydra* can produce offspring by budding (Figure 20–1). Sea anemones and marine worms can split into two half-organisms, each of which then regenerates its missing half. There are even species of lizards that consist only of females and reproduce without mating. While such **asexual reproduction** is simple and direct, it gives rise to offspring that are genetically identical to the parent organism. **Sexual reproduction,** on the other hand, involves the mixing of genomes from two individuals to produce offspring that differ genetically from one another and from both their parents. This mode of reproduction apparently has great advantages, as the vast majority of plants and animals have adopted it. Even many procaryotes and other organisms that normally reproduce asexually engage in occasional bouts of sexual reproduction, thereby creating new combinations of genes. This chapter is concerned with the cellular machinery of sexual reproduction. Before discussing in detail how the machinery works, however, we shall pause to consider why it exists and what benefits it brings.

The Benefits of Sex

The sexual reproductive cycle involves an alternation of **haploid** generations of cells, each carrying a single set of chromosomes, with **diploid** generations of cells, each carrying a double set of chromosomes. The mixing of genomes is achieved by fusion of two haploid cells to form a diploid cell. Later, new haploid cells are generated when a descendant of this diploid cell divides by the process of *meiosis* (Figure 20–2). During meiosis the chromosomes of the double chromosome set exchange DNA by *genetic recombination* before being shared out, in fresh combinations, into single chromosome sets. In this way each cell of the new haploid generation receives a novel assortment of genes, with some genes on each chromosome originating from one ancestral cell of the previous haploid generation and some from the other. Thus, through cycles of haploidy, cell fusion, diploidy, and meiosis, old combinations of genes are broken up and new combinations are created.

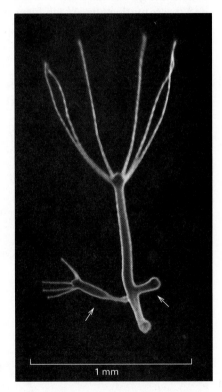

1 mm

Figure 20–1 Photograph of a *Hydra* from which two new organisms are budding (*arrows*). The offspring, which are genetically identical to their parent, will eventually detach and live independently. (Courtesy of Amata Hornbruch.)

In Multicellular Animals the Diploid Phase Is Complex and Long, the Haploid Simple and Fleeting

Cells proliferate by mitotic division. In most organisms that reproduce sexually, this proliferation occurs during the diploid phase. Some primitive organisms, such as fission yeasts, are exceptional in that the haploid cells proliferate mitotically and the diploid cells, once formed, proceed directly to meiosis. A less extreme exception occurs in plants, where mitotic cell divisions occur in both the haploid and the diploid phases. In all but the most primitive plants, however, the haploid phase is very brief and simple, while the diploid phase is extended into a long period of development and proliferation. For almost all multicellular animals, including vertebrates, practically the whole of the life cycle is spent in the diploid state: the haploid cells exist only briefly, do not divide at all, and are highly specialized for sexual fusion (Figure 20–3).

Haploid cells that are specialized for sexual fusion are called **gametes.** Typically, two types of gametes are formed: one is large and nonmotile and is referred to as the *egg* (or *ovum*); the other is small and motile and is referred to as the *sperm* (or *spermatozoon*) (Figure 20–4). During the diploid phase that follows fusion of gametes, the cells proliferate and diversify to form a complex multicellular organism. In most animals a useful distinction can be drawn between the cells of the **germ line,** from which the next generation of gametes will be derived, and the **somatic cells,** which form the rest of the body and ultimately leave no progeny. In a sense, the somatic cells exist only to help the cells of the germ line (the **germ cells**) survive and propagate.

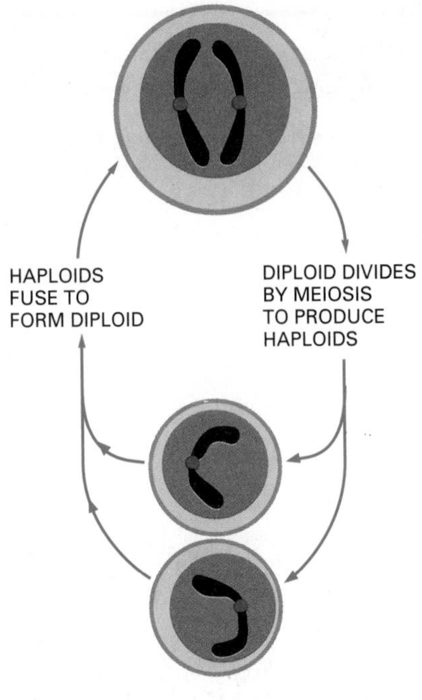

Figure 20–2 The sexual life cycle. It involves an alternation of haploid and diploid generations of cells.

Sexual Reproduction Gives a Competitive Advantage to Organisms in an Unpredictably Variable Environment [1]

The machinery of sexual reproduction is elaborate, and the resources spent on it are large. What benefits does it bring, and why did it evolve? Through genetic

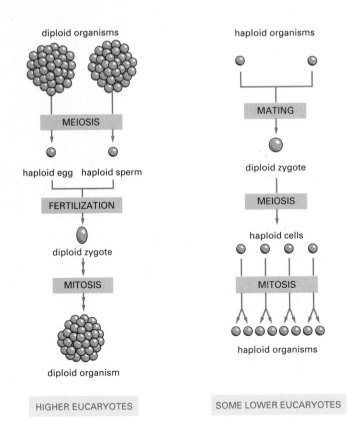

Figure 20–3 **Haploid and diploid cells in the life cycle of higher and some lower eucaryotes.** Cells in higher eucaryotic organisms proliferate in the diploid phase to form a multicellular organism; only the gametes are haploid. In some lower eucaryotes, by contrast, the haploid cells proliferate, and the only diploid cell is the *zygote*, which exists transiently following mating. The haploid cells are shown in *red* and the diploid cells in *blue.*

HIGHER EUCARYOTES   SOME LOWER EUCARYOTES

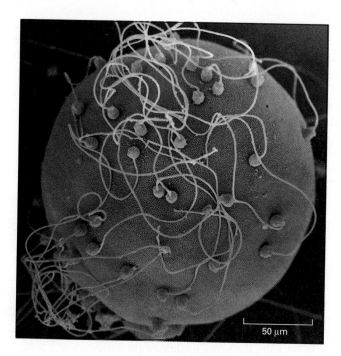

Figure 20–4 **Scanning electron micrograph of a clam egg with sperm bound to its surface.** Although many sperm are bound to the egg, only one will fertilize it, as we discuss later. (Courtesy of David Epel.)

50 μm

recombination sexual individuals beget unpredictably dissimilar offspring, whose haphazard genotypes are at least as likely to represent a change for the worse as a change for the better. Why, then, should sexual individuals have a competitive advantage over individuals that breed true, by an asexual process? This problem continues to perplex population geneticists, but the general conclusion seems to be that the reshuffling of genes in sexual reproduction helps a species survive in an unpredictably variable environment. If a parent produces many offspring with a wide variety of gene combinations, there is a better chance that at least one of the offspring will have the assortment of features necessary for survival.

Many other ideas have been proposed to explain the competitive advantages of sexual reproduction. One of these suggests how one of the first steps in the evolution of sex might have occurred. Evolution depends to a large extent on competition among individuals carrying alternative *alleles*, or variants, created by mutation of particular genes. Suppose that two individuals in a population each undergo a beneficial mutation affecting a different *genetic locus* and therefore a different function. In a strictly asexual species each of these individuals will give rise to a clone of mutant progeny, and the two clones will compete until one or the other triumphs: one of the two beneficial mutations will spread through the population, while the other will eventually be lost. But suppose that one of the original mutants has evolved a genetically determined mechanism that enables it occasionally to incorporate genes from other cells. During the period of competition acquisition of genes from a cell of the competing clone is likely to create a cell that carries both beneficial mutations. Such a cell will be the most successful of all, and its success will ensure the propagation of the trait that enabled it to incorporate genes from other cells. This rudimentary sexual capability will thus be favored by natural selection.

Whatever the origins of sex may be, it is striking that practically all complex present-day organisms have evolved largely through generations of sexual, rather than asexual, reproduction. Asexual organisms, although plentiful, seem mostly to have remained simple and primitive.

We shall now examine the detailed cellular mechanisms of sex, beginning with the events of *meiosis*, in which genetic recombination occurs and diploid cells of the germ line divide to produce haploid *gametes*. Then we shall consider the gametes themselves and, finally, the process of *fertilization*, in which the gametes fuse to form a new diploid organism.

The Benefits of Sex

Summary

Sexual reproduction involves a cyclic alternation of diploid and haploid states: diploid cells divide by meiosis to form haploid cells, and the haploid cells from two individuals fuse in pairs at fertilization to form new diploid cells. In the process, genomes are mixed and recombined to produce individuals with novel assortments of genes. Most of the life cycle of higher plants and animals is spent in the diploid phase; the haploid phase is very brief. Sexual reproduction has probably been favored by evolution because the random recombination of genetic information improves the chances of producing at least some offspring that will survive in an unpredictably variable environment.

Meiosis [2]

The realization that germ cells are haploid, and must therefore be produced by a special type of cell division, came from an observation that was also among the first to suggest that chromosomes carry genetic information. In 1883 it was discovered that, whereas the fertilized egg of a particular worm contains four chromosomes, the nucleus of the egg and that of the sperm each contain only two chromosomes. The chromosome theory of heredity therefore explained the long-standing paradox that maternal and paternal contributions to the character of the progeny seem often to be equal, despite the enormous difference in size between the egg and sperm (see Figure 20–4).

The finding also implied that germ cells must be formed by a special kind of nuclear division in which the chromosome complement is precisely halved. This type of division is called **meiosis,** from the Greek, meaning diminution. (There is no connection with the term mitosis, which is from the Greek *mitos*, meaning a thread, and refers to the threadlike appearance of the chromosomes as they condense during nuclear division—a process that occurs in both ordinary and meiotic divisions.) The behavior of the chromosomes during meiosis turned out to be considerably more complex than expected. Consequently, it was not until the early 1930s, as a result of painstaking cytological and genetic studies, that the essential features of meiosis were established.

Meiosis Involves Two Nuclear Divisions Rather Than One

With the exception of the chromosomes that determine sex (the *sex chromosomes*), a diploid nucleus contains two closely similar versions of each of the other chromosomes (the *autosomes*), one from the male parent (paternal chromosome) and one from the female parent (maternal chromosome). The two versions are called **homologues,** and in most cells they maintain a completely separate existence as independent chromosomes. When each chromosome is duplicated by DNA replication, the twin copies of the fully replicated chromosome at first remain closely associated and are called **sister chromatids.** In an ordinary cell division (described in Chapter 18) the sister chromatids line up on the spindle during mitosis with their kinetochore fibers pointing toward opposite poles. The sister chromatids then separate from each other at anaphase to become individual chromosomes. In this manner each daughter cell formed by ordinary cell division inherits one copy of each paternal chromosome and one copy of each maternal chromosome.

In contrast, a haploid gamete produced by the divisions of a diploid cell during meiosis must contain half the original number of chromosomes—only one chromosome in place of each homologous pair of chromosomes—so that the gamete is endowed with either the maternal or the paternal copy of each gene but not both. This requirement makes an extra demand on the machinery for cell division. The mechanism that has evolved to accomplish the additional sorting requires that homologues recognize each other and become physically paired

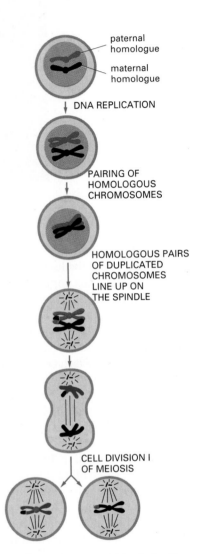

paternal homologue

maternal homologue

DNA REPLICATION

PAIRING OF HOMOLOGOUS CHROMOSOMES

HOMOLOGOUS PAIRS OF DUPLICATED CHROMOSOMES LINE UP ON THE SPINDLE

CELL DIVISION I OF MEIOSIS

Figure 20–5 Events through the first cell division of meiosis. For clarity, only one pair of homologous chromosomes is shown. The pairing of homologous chromosomes (homologues) is unique to meiosis. Each chromosome has been duplicated and exists as attached sister chromatids before the pairing occurs. As shown by the formation of chromosomes that are part *red* and part *black*, the chromosome pairing in meiosis involves *crossing-over* (genetic recombination) between homologous chromosomes, as explained in the text.

before they line up on the spindle. This pairing of the maternal and the paternal copy of each chromosome is unique to meiosis. How the correct chromosomes recognize each other is still unclear, as will be discussed later.

Given a mechanism for pairing the maternal and paternal homologues and for their subsequent separation on the spindle, cells could, in principle, carry out meiosis by a simple modification of a single mitotic cell cycle in which chromosome duplication (S phase) was omitted: if the unduplicated homologues paired before M phase, the ensuing cell division would then produce two haploid cells directly. For unknown reasons, the actual meiotic process is more complex. Before the homologues pair, each one replicates to produce two sister chromatids as in an ordinary cell division. It is only after DNA replication has been completed that the special features of meiosis become evident. Rather than separating, the sister chromatids behave as a unit, as if chromosome duplication had not oc-

Figure 20–6 Comparison of meiosis and normal cell division. As in the previous figure, only one pair of homologous chromosomes is shown. In meiosis, following DNA replication, two nuclear (and cell) divisions are required to produce the haploid gametes. Each diploid cell that enters meiosis therefore produces four haploid cells, whereas each diploid cell that divides by mitosis produces two diploid cells.

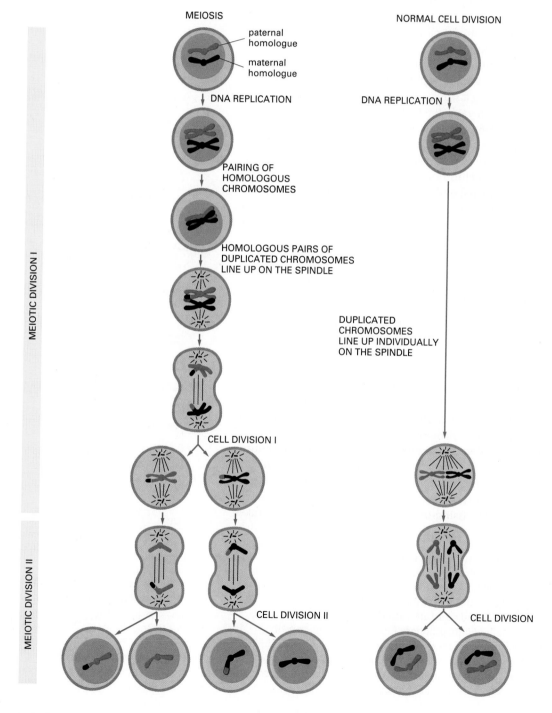

curred: each duplicated homologue pairs with its partner, forming a structure called a *bivalent,* which contains four chromatids. The pairing, as we shall see, allows genetic recombination to occur, whereby a fragment of a maternal chromatid may be exchanged for a corresponding fragment of a homologous paternal chromatid. The bivalents line up on the spindle, and at anaphase the two duplicated homologues (each consisting of two sister chromatids) separate and move to opposite poles. Because the joined sister chromatids behave as a unit, each daughter cell inherits two copies of one of the two homologues when the meiotic cell divides; these two copies are identical except where genetic recombination has occurred (Figure 20–5). The two progeny of this division (**division I of meiosis**) therefore contain a diploid amount of DNA but differ from normal diploid cells in two ways: (1) both of the two DNA copies of each chromosome derive from only one of the two homologous chromosomes in the original cell (except where there has been genetic recombination), and (2) these two copies are inherited as closely associated sister chromatids, as if they were a single chromosome (see Figure 20–5).

Formation of the actual gamete nuclei can now proceed simply through a second cell division, **division II of meiosis,** without further DNA replication. The chromosomes align on a second spindle and the sister chromatids separate, as in normal mitosis, to produce cells with a haploid DNA content. Meiosis thus consists of two cell divisions following a single phase of DNA replication, so that four haploid cells are produced from each cell that enters meiosis. Meiosis and mitosis are compared in Figure 20–6.

Occasionally, the meiotic process occurs abnormally and homologues fail to separate—a phenomenon known as **nondisjunction.** In this case some of the haploid cells that are produced lack a chromosome, while others have more than one copy. Such gametes form abnormal embryos, most of which die. Some survive, however: *Down's syndrome* in humans, for example, is caused by an extra copy of chromosome 21 resulting from nondisjunction during meiotic division I or II.

Genetic Reassortment Is Enhanced by Crossing-over Between Homologous Nonsister Chromatids [3]

Unless they are identical twins, which develop from a single zygote, no two offspring of the same parents are genetically the same. This is because, long before the two gametes fuse, two kinds of randomizing genetic reassortment have occurred during meiosis.

One kind of reassortment is a consequence of the random distribution of the maternal and paternal homologues between the daughter cells at meiotic division I, as a result of which each gamete acquires a different mixture of maternal and paternal chromosomes. From this process alone, one individual could, in principle, produce 2^n genetically different gametes, where n is the haploid number of chromosomes (Figure 20–7A). In humans, for example, each individual can produce at least $2^{23} = 8.4 \times 10^6$ genetically different gametes. But the actual number of variants is very much greater than this because a second type of reassortment, called **chromosomal crossing-over,** occurs during meiosis. It takes

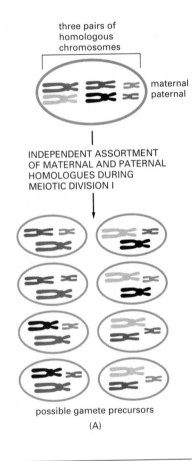

three pairs of homologous chromosomes

maternal
paternal

INDEPENDENT ASSORTMENT OF MATERNAL AND PATERNAL HOMOLOGUES DURING MEIOTIC DIVISION I

possible gamete precursors

(A)

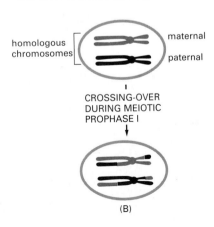

homologous chromosomes

maternal
paternal

CROSSING-OVER DURING MEIOTIC PROPHASE I

(B)

Figure 20–7 **Two major contributions to the reassortment of genetic material that occurs during meiosis.** (A) The independent assortment of the maternal and paternal homologues during the first meiotic division produces 2^n different haploid gametes for an organism with n chromosomes. Here $n = 3$, and there are 8 different possible gametes. (B) Crossing-over during meiotic prophase I exchanges segments of homologous chromosomes and thereby reassorts genes in individual chromosomes. Because of the many small differences in DNA sequence that always exist between any two homologues, both mechanisms increase the genetic variability of organisms that reproduce sexually.

place during the long prophase of meiotic division I, in which parts of homologous chromosomes are exchanged. On average, between two and three crossover events occur on each pair of human chromosomes during meiotic division I. This process scrambles the genetic constitution of each of the chromosomes in gametes, as illustrated in Figure 20–7B.

Chromosomal crossing-over involves breaking the DNA double helix in a maternal chromatid and in a homologous paternal chromatid so as to exchange fragments between the two nonsister chromatids in a reciprocal fashion by a process known as **general genetic recombination.** The molecular details of this process are discussed in Chapter 6. The consequences of each crossover event can be observed cytologically at the latest stages of prophase of meiotic division I, when the chromosomes are highly condensed. At this stage the sister chromatids are tightly apposed along their entire length. The two duplicated homologues (maternal and paternal) are seen to be physically connected at specific points. Each connection, called a **chiasma** (plural **chiasmata**), corresponds to a crossover between two nonsister chromatids (Figure 20–8).

At this stage of meiosis, each pair of duplicated homologues, or *bivalent,* is held together by at least one chiasma. Many bivalents contain more than one chiasma, indicating that multiple crossovers can occur between homologues.

Meiotic Chromosome Pairing Culminates in the Formation of the Synaptonemal Complex [4]

Elaborate morphological changes occur in the chromosomes as they pair (*synapse*) and then begin to unpair (*desynapse*) during the first meiotic prophase. This prophase is traditionally divided into five sequential stages—*leptotene, zygotene, pachytene, diplotene,* and *diakinesis*—defined by these morphological changes. The most striking event is the initiation of intimate chromosome synapsis at **zygotene,** when a complex structure called the *synaptonemal complex* begins to develop between the two sets of sister chromatids in each bivalent. **Pachytene** is said to begin as soon as synapsis is complete, and it generally persists for days, until desynapsis begins the **diplotene** stage, in which the chiasmata are first seen (Figure 20–9).

Genetic recombination requires a close apposition between the recombining chromosomes. The **synaptonemal complex,** which forms just before pachytene and dissolves just afterward, keeps the homologous chromosomes in a bivalent together and closely aligned, and it has been suggested that it may play a part in the recombination process. It consists of a long ladderlike protein core, on opposite sides of which the two homologues are aligned to form a long linear chromosome pair. The sister chromatids in each homologue are kept tightly packed together, and their DNA extends from the same side of the protein ladder in a series of loops (Figure 20–10).

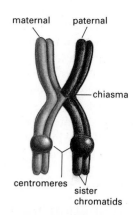

Figure 20–8 Paired homologous chromosomes during the transition to metaphase of meiotic division I. A single crossover event has occurred earlier in prophase to create one chiasma. Note that the four chromatids are arranged as two distinct pairs of sister chromatids and that the two chromatids in each pair are tightly aligned along their entire lengths as well as joined at their centromeres. The entire unit of four chromatids is referred to as a *bivalent.*

Figure 20–9 Time course of chromosome synapsis and desynapsis during meiotic prophase I. A single bivalent is shown. The *pachytene stage* is defined as the period during which a fully formed synaptonemal complex exists. In gametes of female animals the subsequent *diplotene stage* is an enormously prolonged period of cell growth during which the chromosomes are decondensed and very active in transcription. This ends with *diakinesis*—the stage of transition to metaphase—in which the chromosomes recondense and transcription halts. In male gametes diplotene and diakinesis are briefer and less distinct.

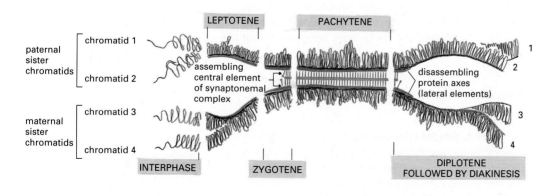

It is not known how homologous chromosomes become aligned. It is unlikely that continuous connections all along the interacting chromosomes are involved, since the chromatin of one homologue is well separated from the chromatin of its partner in the synaptonemal complex. It has been proposed that the initial interaction between homologous chromosomes is mediated by complementary DNA base-pair interactions at discrete sites along the chromosomes. This recognition may occur at zygotene or even earlier, when the chromosomes are not very condensed; following chromosome condensation, the formation of the synaptonemal complex would then pack the remaining portions of the chromosomes together.

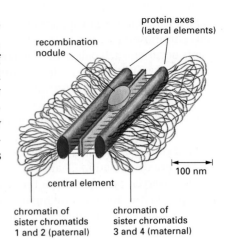

Figure 20–10 **A typical synaptonemal complex, showing the lateral and central elements of the complex.** A recombination nodule is also shown. Only a short section of the long ladderlike complex is shown. A similar synaptonemal complex is present in organisms as diverse as yeast and human, but very little is known about the protein molecules that form it.

Recombination Nodules Are Thought to Mediate Chromatid Exchanges [5]

Although the synaptonemal complex may provide the structural framework for recombination events, it probably is not the engine that brings them about. The active recombination process is thought to be mediated instead by **recombination nodules,** which are very large protein-containing assemblies with a diameter of about 90 nm. (For comparison, a large globular protein molecule of molecular weight 400,000 has a diameter of about 10 nm.) Recombination nodules sit at intervals on the synaptonemal complex, placed like basketballs on a ladder between the two homologous chromosomes (see Figure 20–10). They are thought to mark the site of a large multienzyme "recombination machine," which brings local regions of DNA on the maternal and paternal chromatids together across the 100-nm-wide synaptonemal complex.

The evidence that the recombination nodule serves this function is indirect: (1) The total number of nodules is about equal to the total number of chiasmata seen later in prophase. (2) The nodules are distributed along the synaptonemal complex in the same way that crossover events are distributed. Like the crossover events themselves, for example, the nodules are absent from those regions of the synaptonemal complex that hold heterochromatin together. Moreover, both genetic and cytological measurements indicate that the occurrence of one crossover event prevents a second crossover event occurring at any nearby chromosomal site; similarly, the nodules tend not to occur very near one another. (3) Some *Drosophila* mutations cause an abnormal distribution of crossover events along the chromosomes, as well as a greatly diminished recombination frequency. In these mutants correspondingly fewer recombination nodules are found, with a changed distribution that parallels the changed crossover distribution. This correlation strongly suggests that a recombination nodule determines the site of each crossover event. (4) Genetic recombination is thought to involve a limited amount of DNA synthesis at the site of each crossover event (discussed in Chapter 6). Electron microscopic autoradiography shows that radioactive DNA precursors are preferentially incorporated into pachytene DNA at or near recombination nodules.

Because there are about as many recombination nodules as crossover events, it seems that recombination nodules are extremely efficient in causing the chromatids on opposite homologues to recombine. Little is known, however, about their structure or mechanism of action.

Chiasmata Play an Important Part in Chromosome Segregation in Meiosis

In addition to reassorting genes, chromosomal crossing-over is crucial in most organisms for the correct segregation of the two homologues to separate daughter nuclei. This is because the chiasma created by each crossover event plays a role analogous to that of the centromere in an ordinary mitotic division, holding the maternal and paternal homologues together on the spindle until anaphase I. In mutant organisms that have a reduced frequency of meiotic chromosome crossing-over, some of the chromosome pairs lack chiasmata. These pairs fail to seg-regate normally, and a high proportion of the resulting gametes contain too many or too few chromosomes—an example of nondisjunction.

There are at least two major differences in the way chromosomes separate in meiotic division I and in normal mitosis (see Figure 20–6). (1) During normal mitosis (and in meiotic division II, which resembles a normal mitosis) the sister chromatids are held together only at the centromere; the kinetochores (protein complexes associated with the centromeres, discussed in Chapter 18) on each sister chromatid have attached kinetochore fibers pointing in opposite directions, so that the chromatids are drawn into different daughter cells at anaphase. At metaphase I of meiosis, by contrast, the kinetochores on both sister chroma-tids appear to have fused so that their attached kinetochore fibers all point in the same direction and the arms of the sister chromatids are closely apposed;

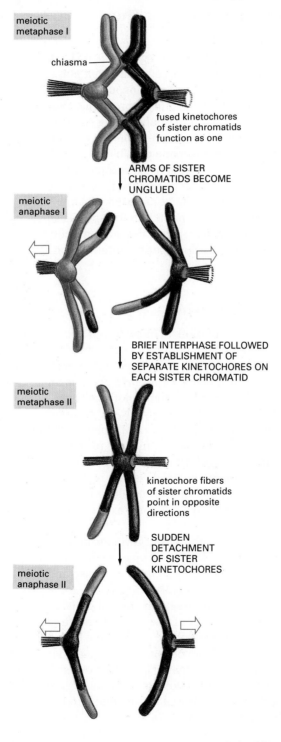

Figure 20–11 Comparison of the mechanisms of chromosome align-ment (at metaphase) and separation (at anaphase) in meiotic division I and meiotic division II. The mechan-isms used in meiotic division II are the same as those used in normal mitosis (discussed in Chapter 18).

Meiosis

moreover, the homologous maternal and paternal chromosomes are held together at the chiasmata. (2) During normal mitosis (and meiotic division II) the movement of chromatids to the poles is triggered by a mechanism that detaches the two sister kinetochores from each other (thus beginning anaphase), allowing the sister chromatids to segregate into different daughter cells. In anaphase I of meiosis, however, movement to the poles is initiated by the disruption of the poorly understood forces keeping the arms of sister chromatids together and by the simultaneous dissolution of the chiasmata linking the maternal maternal and paternal chromosomes; consequently, the sister chromatids remain paired, but the maternal and paternal homologues segregate into different daughter cells. The difference between the way chromosomes separate in meiotic divisions I and II are illustrated in Figure 20–11.

Pairing of the Sex Chromosomes Ensures That They Also Segregate [6]

We have explained how homologous chromosomes pair during meiotic division I so that they segregate accurately between the daughter cells. But what about the **sex chromosomes,** which in male mammals are not homologous? Females have two X chromosomes, which pair and segregate like other homologues. But males have one X and one Y chromosome, which must pair during the first metaphase of meiosis if the sperm are to contain either one Y or one X chromosome and not both or neither. The necessary pairing is made possible by a small region of homology between the X and the Y at one end of these chromosomes. In this region the two chromosomes pair and cross over during the first meiotic prophase. The chiasma corresponding to this small amount of genetic recombination is sufficient to keep the X and Y chromosomes paired on the spindle so that only two types of sperm are normally produced: sperm containing one Y chromosome, which will give rise to male embryos, and sperm containing one X chromosome, which will give rise to female embryos.

Meiotic Division II Resembles a Normal Mitosis

After the long prophase I (which can occupy 90% or more of meiosis) has ended, two successive cell divisions, without an intervening period of DNA synthesis, bring meiosis to an end (see Figure 20–6). The entire first meiotic cell cycle, which ends with an initial meiotic cell division, is called *meiotic division I,* and it is far more complex and requires much more time than the second meiotic cell cycle, called *meiotic division II* (Figure 20–12). Even the preparatory DNA replication during the first cell cycle tends to take much longer than a normal S phase, and cells can then spend days, months, or even years in the first meiotic prophase, depending on the species and on the gamete being formed. (Although it is traditionally called prophase, this prolonged phase of meiotic division I resembles the G_2 phase of an ordinary cell division in that the nuclear envelope remains intact and disappears only when the spindle fibers begin to form as prophase I gives way to metaphase I.)

After the end of meiotic division I, nuclear membranes re-form around the two daughter nuclei and a brief interphase begins. During this period the chromosomes may decondense somewhat, but usually they soon recondense and prophase II begins. As there is no DNA synthesis during this interval, in some organisms the chromosomes seem to pass almost directly from one division phase into the next. In all organisms prophase II is brief: the nuclear envelope breaks down as the new spindle forms, after which metaphase II, anaphase II, and telophase II usually follow in quick succession. As in an ordinary mitosis, a separate set of kinetochore fibers forms on each sister chromatid, and these two sets of fibers extend in opposite directions. Moreover, the two sister chromatids are kept together on the metaphase plate until they are released by the sudden sepa-

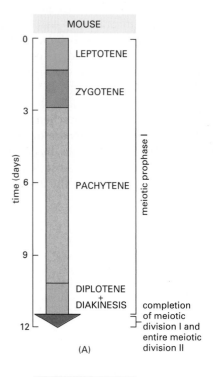

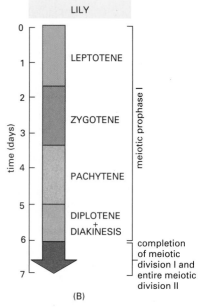

Figure 20–12 Comparison of times required for each of the stages of meiosis. Approximate times for both a male mammal (mouse) and the male tissue of a plant (lily) are shown. Times differ for male and female gametes (egg and sperm) of the same species, as well as for the same gametes of different species. Meiosis in a human male, for example, lasts for 24 days, compared with 12 days in the mouse. Meiotic prophase I, however, is always much longer than all the other meiotic stages combined.

ration of their kinetochores at anaphase (see Figure 20–11). Thus division II, unlike division I, closely resembles a normal mitosis. The difference is that one copy of each chromosome is present instead of two homologues. After nuclear envelopes have formed around the four haploid nuclei produced at telophase II, meiosis is complete (see Figure 20–6). The principles of meiosis are the same in plants and animals and in males and females. But the production of gametes involves more than just meiosis, and the other processes required vary widely among organisms and are very different for eggs and sperm. We shall focus our discussion of gametogenesis mainly on vertebrates. As we shall see, by the end of meiosis a vertebrate egg is fully mature (and in some cases even fertilized), whereas a sperm that has completed meiosis has only just begun its differentiation.

Summary

The formation of both eggs and sperm begins in a similar way, with meiosis. In this process two successive cell divisions following one round of DNA replication give rise to four haploid cells from a single diploid cell. Meiosis is dominated by prophase of meiotic division I, which can occupy 90% or more of the total meiotic period. Each chromosome as it enters this prophase consists of two tightly joined sister chromatids. Chromosomal crossover events occur during this prolonged prophase I, when homologous chromosomes are aligned in register. Each crossover event is thought to be mediated by a recombination nodule, and it results in the formation of a chiasma, which persists until anaphase I. In the first meiotic cell division one member of each chromosome pair, still composed of linked sister chromatids, is distributed to each daughter cell. A second cell division, without DNA replication, then rapidly ensues in which each sister chromatid is segregated into a separate haploid cell.

Eggs

In all vertebrate embryos certain cells are singled out early in development as progenitors of the gametes. These **primordial germ cells** migrate to the developing gonads, called the *genital ridges,* which will form the ovaries in females and the testes in males. After a period of mitotic proliferation, these cells undergo meiosis and differentiate into mature gametes—either eggs or sperm. Later, the fusion of egg and sperm after mating initiates embryogenesis, with the subsequent production in the embryo of new primordial germ cells, which begins the cycle again.

An Egg Is the Only Cell in a Higher Animal That Is Able to Develop into a New Individual

In one respect at least, eggs are the most remarkable of animal cells: once activated, they can give rise to a complete new individual within a matter of days or weeks. No other cell in a higher animal has this capacity. Activation is usually the consequence of *fertilization*—fusion of a sperm with the egg. The sperm itself, however, is not strictly required. An egg can be activated artificially by a variety of nonspecific chemical or physical treatments; a frog egg, for example, can be activated by pricking it with a needle. Indeed, some organisms, including even a few vertebrates such as some lizards, normally reproduce from eggs that become activated in the absence of sperm—that is, **parthenogenetically.**

Although an egg can give rise to every cell type in the adult organism (that is, it is *totipotent*), it is itself a highly specialized cell, uniquely equipped for the single function of generating a new individual. We shall now briefly consider some of its specialized features before discussing how it develops to the point at which it is ready for fertilization.

An Egg Is Highly Specialized for Independent Development, with Large Nutrient Reserves and an Elaborate Coat [7]

The eggs of most animals are giant cells, containing stockpiles of all the materials needed for initial development of the embryo to carry it through to the stage where the new individual can begin feeding. Before this stage, the single giant cell cleaves into many smaller cells, but no net growth occurs. Mammals are an exception in that the embryo can start to grow early by taking up nutrients from the mother; thus a mammalian egg, though still a large cell, does not have to be as large as the egg of a frog or a bird, for example. In general, eggs are typically spherical or ovoid, with a diameter of about 100 μm in humans and sea urchins (whose feeding larvae are tiny), 1 mm to 2 mm in frogs and fishes, and many centimeters in birds and reptiles (Figures 20–13). A typical somatic cell, by contrast, has a diameter of only about 10 or 20 μm (Figure 20–14).

The egg cytoplasm contains nutritional reserves in the form of **yolk,** which is rich in lipids, proteins, and polysaccharides and is usually contained within discrete structures called *yolk granules.* In some species each yolk granule is membrane-bounded, whereas in others it is not. In eggs that develop into large animals outside the mother's body, yolk can account for more than 95% of the volume of the cell, whereas in mammals, whose embryos are largely nourished by their mothers, there is little if any.

The **egg coat** is another peculiarity of eggs. It is a specialized form of extracellular matrix consisting largely of glycoprotein molecules, some secreted by the egg and others by surrounding cells. In many species the major coat is a layer immediately surrounding the egg plasma membrane; in nonmammalian eggs, such as those of sea urchins or chickens, it is called the *vitelline layer,* whereas in mammalian eggs it is called the *zona pellucida* (Figure 20–15). This layer protects the egg from mechanical damage, and in many eggs it also acts as a species-specific barrier to sperm, admitting only those of the same or closely related species (discussed below). Nonmammalian eggs often have additional layers overlying the vitelline layer that are secreted by surrounding cells. As frog eggs, for example, pass from the ovary through the oviduct (the tube that conveys them to the outside), they acquire several layers of gelatinous coating secreted by epithelial cells lining the oviduct. Similarly, the "white" (albumin) and shell of chicken eggs are added (after fertilization) as the eggs pass along the oviduct. The vitelline layer of insect eggs is covered by a thick, tough layer called the *chorion,* which is secreted by the *follicle cells* that surround each egg in the ovary.

Many eggs (including those of mammals) contain specialized secretory vesicles just under the plasma membrane in the outer region, or *cortex,* of the egg cytoplasm. When the egg is activated by a sperm, these **cortical granules** release their contents by exocytosis; the contents of the granules act to alter the egg coat so as to prevent more than one sperm from fusing with the egg (discussed below).

Eggs Develop in Stages [8]

A developing egg is called an **oocyte.** Its differentiation into a mature egg (or *ovum*) involves a series of changes whose timing is geared to the steps of meiosis in which the germ cells go through their two final, highly specialized divisions. Oocytes have evolved special mechanisms for arresting progress through meiosis: they remain suspended in prophase I for prolonged periods while the oocyte grows in size, and in many cases they later arrest in metaphase II while awaiting fertilization.

While the details of oocyte development (**oogenesis**) vary in different species, the general stages are similar, as outlined in Figure 20–16. Primordial germ cells migrate to the forming gonad to become *oogonia,* which proliferate by ordinary

human egg

chicken egg

frog egg

Figure 20–13 The actual sizes of three eggs. The human egg is 0.1 mm in diameter.

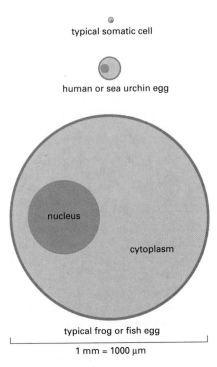
typical somatic cell

human or sea urchin egg

nucleus

cytoplasm

typical frog or fish egg

1 mm = 1000 μm

Figure 20–14 The relative sizes of various eggs. They are compared to a typical somatic cell.

cell division cycles for a period before differentiating into *primary oocytes*. At this stage the first meiotic division begins: the DNA replicates so that each chromosome consists of two chromatids, the homologous chromosomes pair along their long axes, and crossing-over occurs between the chromatids of these paired chromosomes. After these events the cell remains arrested in prophase of division I of meiosis (in a state equivalent, as we previously pointed out, to a G_2 phase of an ordinary division cycle) for a period lasting from a few days to many years, depending on the species. During this long period (or, in some cases, at the onset of sexual maturity) the primary oocytes synthesize a coat and cortical granules and, in the case of large nonmammalian oocytes, they accumulate ribosomes, yolk, glycogen, lipid, and the mRNA that will later direct the synthesis of proteins required for early embryonic growth and the unfolding of the developmental program. In many oocytes the intensive biosynthetic activities are reflected in the structure of the chromosomes, which decondense and form lateral loops, taking on a characteristic "lampbrush" appearance, signifying that they are very busily engaged in RNA synthesis (discussed in Chapter 8).

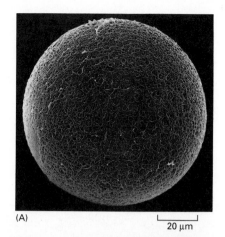

(A)   20 μm

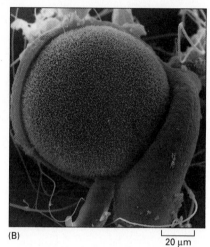

(B)   20 μm

Figure 20–15 The zona pellucida. (A) Scanning electron micrograph of a hamster egg showing the zona pellucida. In (B) the zona (to which many sperm are attached) has been peeled back to reveal the underlying plasma membrane of the egg, which contains numerous microvilli. (From D.M. Phillips, *J. Ultrastruct. Res.* 72:1–12, 1980.)

The next phase of oocyte development is called *oocyte maturation* and usually does not occur until sexual maturity, when it is stimulated by hormones. Under these hormonal influences the cell resumes its progress through division I of meiosis: the chromosomes recondense, the nuclear envelope breaks down (this is generally taken to mark the beginning of maturation), and the replicated homologous chromosomes segregate at anaphase I into two daughter nuclei, each containing half the original number of chromosomes. To end division I, the cytoplasm divides asymmetrically to produce two cells that differ greatly in size: one is a small *polar body*, and the other is a large **secondary oocyte,** the precursor of the egg. At this stage each of the chromosomes is still composed of two sister chromatids. These chromatids do not separate until division II of meiosis, when they are partitioned into separate cells by a process that is identical to a normal mitosis, as previously described. After this final chromosome separation at anaphase II, the cytoplasm of the large secondary oocyte again divides asymmetrically to produce the mature **egg** (or **ovum**) and a second small polar body, each with a haploid number of single chromosomes (see Figure 20–16). Because of these two asymmetrical divisions of their cytoplasm, oocytes maintain their large size despite undergoing the two meiotic divisions. Both of the polar bodies are small, and they eventually degenerate.

In most vertebrates oocyte maturation proceeds to metaphase of meiosis II and then arrests until fertilization. At **ovulation** the arrested secondary oocyte is released from the ovary, and if fertilization occurs, the oocyte is stimulated to complete meiosis.

Oocytes Grow to Their Large Size Through Special Mechanisms [8, 9]

A somatic cell with a diameter of 10 to 20 μm typically takes about 24 hours to double its mass in preparation for cell division. At this rate of biosynthesis such a cell would take a very long time to reach the hundredfold greater mass of a mammalian egg with a diameter of 100 μm or the 10^5-fold greater mass of an insect egg with a diameter of 1000 μm. Yet some insects live only a few days and manage to produce eggs with diameters even greater than 1000 μm. It is clear that eggs must have special mechanisms for achieving their large size.

One simple strategy for rapid growth is to have extra gene copies in the cell. Thus the oocyte delays completion of the first meiotic division so as to grow while it contains the diploid chromosome set in duplicate. In this way it has twice as much DNA available for RNA synthesis as does an average somatic cell in the G_1 phase of the cell cycle. Some oocytes go to even greater lengths to accumulate extra DNA: they produce many extra copies of certain genes. We discuss in Chap-

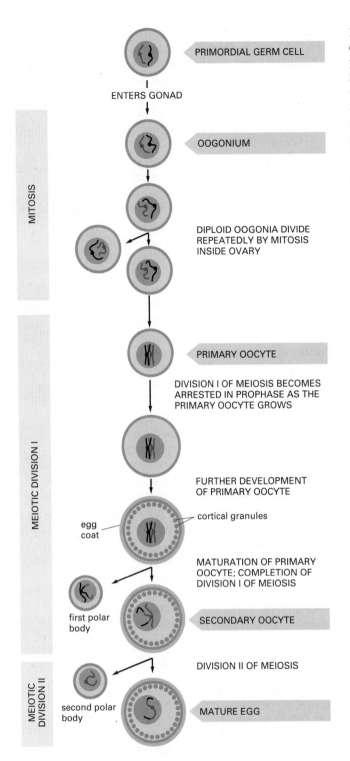

MITOSIS

MEIOTIC DIVISION I

MEIOTIC DIVISION II

PRIMORDIAL GERM CELL

ENTERS GONAD

OOGONIUM

DIPLOID OOGONIA DIVIDE
REPEATEDLY BY MITOSIS
INSIDE OVARY

PRIMARY OOCYTE

DIVISION I OF MEIOSIS BECOMES
ARRESTED IN PROPHASE AS THE
PRIMARY OOCYTE GROWS

FURTHER DEVELOPMENT
OF PRIMARY OOCYTE

egg
coat

cortical granules

MATURATION OF PRIMARY
OOCYTE; COMPLETION OF
DIVISION I OF MEIOSIS

first polar
body

SECONDARY OOCYTE

DIVISION II OF MEIOSIS

second polar
body

MATURE EGG

Figure 20–16 The stages of oogenesis. *Oogonia* develop from primordial germ cells that migrate into the developing gonad early in embryogenesis. After a number of mitotic divisions, oogonia begin meiotic division I, after which they are called *primary oocytes.* In mammals primary oocytes are formed very early (between 3 and 8 months of gestation in the human embryo) and remain arrested in prophase of meiotic division I until the female becomes sexually mature. At this point a small number periodically mature under the influence of hormones, completing meiotic division I to become *secondary oocytes,* which eventually undergo meiotic division II to become mature eggs (*ova*). The stage at which the egg or oocyte is released from the ovary and is fertilized varies from species to species. In most vertebrates oocyte maturation is arrested at metaphase of meiosis II and the secondary oocyte completes meiosis II only after fertilization. All of the polar bodies eventually degenerate. In most animals the developing oocyte is surrounded by specialized accessory cells that help isolate and nourish it (not shown).

ter 8 how the somatic cells of most organisms require 100 to 500 copies of the ribosomal RNA genes in order to produce enough ribosomes for protein synthesis. Eggs require even greater numbers of ribosomes to support protein synthesis during early embryogenesis, and in the oocytes of many animals the ribosomal RNA genes are specifically amplified; some amphibian eggs, for example, contain 1 or 2 million copies of these genes.

Oocytes may also depend partly on the synthetic activities of other cells for their growth. Yolk, for example, is usually synthesized outside the ovary and imported into the oocyte. In birds, amphibians, and insects yolk proteins are made by liver cells (or their equivalents), which secrete these proteins into the blood. Within the ovaries oocytes take up the yolk proteins from the extracellular fluid by receptor-mediated endocytosis (see Figure 13–28). Nutritive help can also come from neighboring accessory cells in the ovary. These can be of two types. In some invertebrates some of the progeny of the oogonia become **nurse cells** instead of becoming oocytes. These cells usually are connected to the oocyte by cytoplasmic bridges through which macromolecules can pass directly into the oocyte cytoplasm (Figure 20–17). For the insect oocyte the nurse cells manufacture many of the products—ribosomes, mRNA, protein, and so on—that a vertebrate oocyte has to manufacture for itself.

The other accessory cells in the ovary that help nourish developing oocytes are ordinary somatic cells called **follicle cells,** which are found in both invertebrates and vertebrates. They are arranged as an epithelial layer around the oocyte (Figure 20–18, and see Figure 20–17), to which they are connected only by gap junctions, which permit the exchange of small molecules but not macromolecules. While these cells are unable to provide the oocyte with preformed macromolecules through these communicating junctions, they may help to supply the smaller precursor molecules from which macromolecules are made. In addition, follicle cells frequently secrete macromolecules that contribute to the egg coat, or are taken up by receptor-mediated endocytosis into the growing oocyte, or act on egg cell-surface receptors to control the spatial patterning and axial asymmetries of the egg (discussed in Chapter 21).

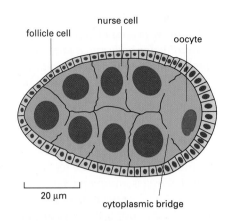

Figure 20–17 Nurse cells and follicle cells associated with a *Drosophila* oocyte. The nurse cells and the oocyte arise from a common oogonium, each of which gives rise to one oocyte and 15 nurse cells (only 7 of which are seen in this plane of section). These cells remain joined by cytoplasmic bridges, which result from incomplete cell division. The follicle cells develop inde-pendently, from mesodermal cells.

Summary

Eggs develop in stages from primordial germ cells that migrate into the developing gonad very early in development to become oogonia. After mitotic proliferation oogonia become primary oocytes that begin meiotic division I and then arrest at prophase for days or years, depending on the species. During this prophase-I arrest period, primary oocytes grow, synthesize a coat, and accumulate ribosomes, mRNAs,

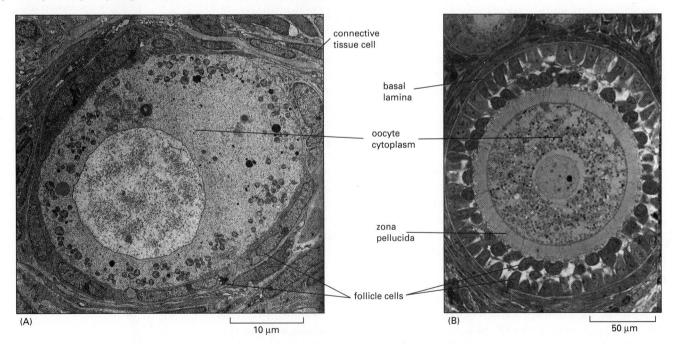

Figure 20–18 Electron micrographs of developing primary oocytes in the rabbit ovary. (A) An early stage of primary oocyte development. Neither a zona pellucida nor cortical granules have developed, and the oocyte is surrounded by a single layer of flattened follicle cells. (B) A more mature primary oocyte, which is shown at a sixfold lower magnification because it is much larger than the oocyte in (A). This oocyte has acquired a thick zona pellucida and is surrounded by several layers of follicle cells and a basal lamina, which isolate the oocyte from the other cells in the ovary. The primary oocyte together with its surrounding follicle cells is called a *primary follicle.* The follicle cells are connected to one another and to the oocyte by gap junctions. (Copyright 1979. Urban & Schwarzenberg, Baltimore–Munich. Reproduced with permission from The Cellular Basis of Mammalian Reproduction, edited by Jonathan Van Blerkom and Pietro Motta. All rights reserved.)

Eggs

and proteins, often enlisting the help of other cells, including surrounding accessory cells. In the process of maturation primary oocytes complete meiotic division I to form a small polar body and a large secondary oocyte, which proceeds into metaphase of meiotic division II. There, in many species, the oocyte is arrested until stimulated by fertilization to complete meiosis and begin embryonic development.

Sperm

In most species there are just two types of gametes, and they are radically different. The egg is among the largest cells in an organism, while the **sperm** (**spermatozoon**, plural **spermatozoa**) is often the smallest. The egg and the sperm are optimized in opposite ways for the propagation of the genes they carry. The egg is nonmotile and aids the survival of the maternal genes by providing large stocks of raw materials for growth and development, as well as providing an effective protective wrapping. The sperm, by contrast, is optimized to propagate the paternal genes by exploiting this maternal investment: it is usually highly motile and streamlined for speed and efficiency in the task of fertilization. Competition between sperm is fierce, and the vast majority fail in their mission: of the billions of sperm released during the reproductive life of a human male, only a few ever manage to fertilize an egg.

Sperm Are Highly Adapted for Delivering Their DNA to an Egg [10]

Typical sperm are "stripped-down" cells, equipped with a strong flagellum to propel them through an aqueous medium but unencumbered by cytoplasmic organelles such as ribosomes, endoplasmic reticulum, or Golgi apparatus, which are unnecessary for the task of delivering the DNA to the egg. On the other hand, sperm contain many mitochondria strategically placed where they can most efficiently power the flagellum. Sperm usually consist of two morphologically and functionally distinct regions enclosed by a single plasma membrane: the *tail*, which propels the sperm to the egg and helps it burrow through the egg coat, and the *head*, which contains a condensed haploid nucleus (Figure 20–19). The DNA in the nucleus is extremely tightly packed, so that its volume is minimized for transport. The chromosomes of many sperm have dispensed with the histones of somatic cells and are packed instead with simple, highly positively charged proteins called *protamines*.

In the head of most animal sperm, closely apposed to the anterior end of the nuclear envelope, is a specialized secretory vesicle called the **acrosomal vesicle** (see Figure 20–19). This contains hydrolytic enzymes that help the sperm to penetrate the egg's outer coat. When a sperm contacts an egg, the contents of the vesicle are released by exocytosis in the so-called *acrosomal reaction;* in some sperm this reaction also exposes or releases specific proteins that help bind the sperm tightly to the egg coat.

The motile tail of a sperm is a long flagellum whose central axoneme emanates from a basal body situated just posterior to the nucleus. As described in Chapter 16, the axoneme consists of two central singlet microtubules surrounded by nine evenly spaced microtubule doublets. The flagellum of some sperm (including those of mammals) differs from other flagella in that the usual 9 + 2 pattern of the axoneme is further surrounded by nine *outer dense fibers* composed mainly of keratin (Figure 20–20). These dense fibers are stiff and noncontractile, and it is not known what part they play in the active bending of the flagellum, which is caused by the sliding of adjacent microtubule doublets past one another. Flagellar movement is driven by dynein motor proteins, which use the energy of ATP hydrolysis to slide the microtubules, as discussed in Chapter 16; the ATP is

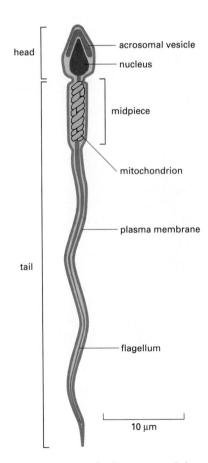

Figure 20–19 A human sperm. It is shown in longitudinal section.

generated by highly specialized mitochondria in the anterior part of the sperm tail (called the *midpiece*), where the ATP is needed (see Figures 20–19 and 20–20).

Sperm Are Produced Continuously in Many Mammals [11]

In mammals there are major differences in the way eggs are produced (oogenesis) and the way sperm are produced (**spermatogenesis**). In human females, for

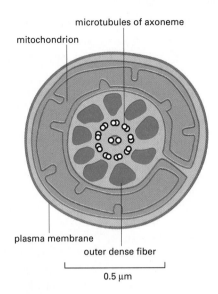

Figure 20–20 Drawing of the midpiece of a mammalian sperm as seen in cross-section in an electron microscope. The core of the flagellum is composed of an axoneme surrounded by nine dense fibers. The axoneme consists of two singlet microtubules surrounded by nine microtubule doublets. The mitochondrion (shown in *green*) is well placed for providing the ATP required for flagellar movement; its unusual spiral structure (see Figure 20–19) results from the fusion of individual mitochondria during spermatid differentiation.

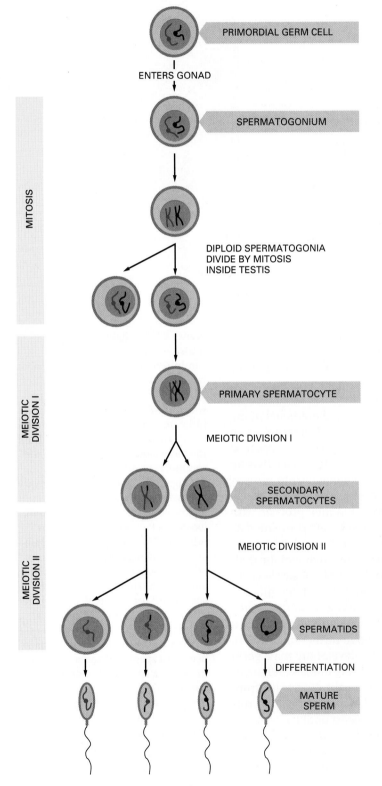

Figure 20–21 The stages of spermatogenesis. *Spermatogonia* develop from primordial germ cells that migrate into the testis early in embryogenesis. When the animal becomes sexually mature, the spermatogonia begin to proliferate rapidly, generating some progeny that retain the capacity to continue dividing indefinitely (as stem-cell spermatogonia) and other progeny (maturing spermatogonia) that will, after a limited number of further normal division cycles, embark on meiosis to become *primary spermatocytes*. These continue through meiotic division I to become *secondary spermatocytes*. After they complete meiotic division II, the secondary spermatocytes produce haploid *spermatids* that differentiate into mature sperm (*spermatozoa*). Spermatogenesis differs from oogenesis (see Figure 20–16) in several ways: (1) new cells enter meiosis continually from the time of puberty, (2) each cell that begins meiosis gives rise to four mature gametes rather than one, and (3) mature sperm form by an elaborate process of cell differentiation that begins after meiosis is complete.

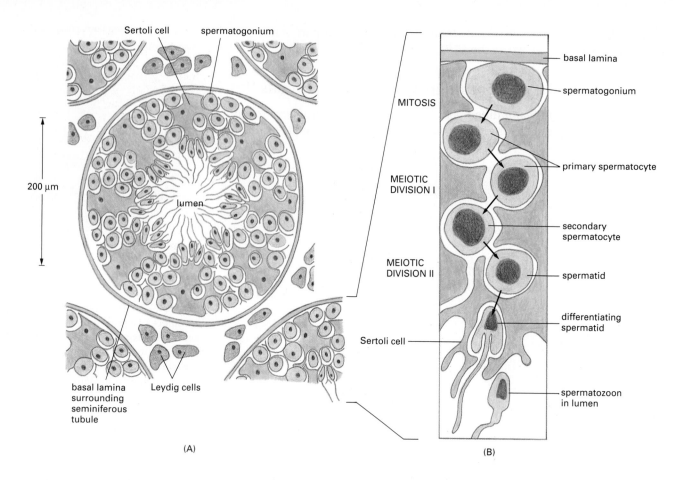

Sertoli cell   spermatogonium

MITOSIS

MEIOTIC
DIVISION I

MEIOTIC
DIVISION II

200 μm

lumen

basal lamina
surrounding
seminiferous
tubule

Leydig cells

Sertoli cell

(A)

basal lamina

spermatogonium

primary spermatocyte

secondary
spermatocyte

spermatid

differentiating
spermatid

spermatozoon
in lumen

(B)

example, oogonia proliferate only in the fetus, enter meiosis before birth, and become arrested as oocytes in the first meiotic prophase, in which state they may remain for up to 50 years. Individual oocytes mature from this strictly limited stock and are ovulated at intervals, generally one at a time, beginning at puberty. In human males, on the other hand, meiosis and spermatogenesis do not begin in the testes until puberty and then go on continuously in the epithelial lining of very long, tightly coiled tubes, called *seminiferous tubules.* Immature germ cells, called *spermatogonia,* are located around the outer edge of these tubes next to the basal lamina, where they proliferate continuously by ordinary cell division cycles. Some of the daughter cells stop proliferating and differentiate into *primary spermatocytes.* These cells enter the first meiotic prophase, in which their paired homologous chromosomes participate in crossing-over, and then proceed with division I of meiosis to produce two *secondary spermatocytes,* each containing 22 duplicated autosomal chromosomes and either a duplicated X or a duplicated Y chromosome. The two secondary spermatocytes proceed through meiotic division II to produce four *spermatids,* each with a haploid number of single chromosomes. These haploid spermatids then undergo morphological differentiation into sperm (Figure 20–21), which escape into the lumen of the seminiferous tubule (Figure 20–22). The sperm subsequently pass into the *epididymis,* a coiled tube overlying the testis, where they are stored and undergo further maturation.

An intriguing feature of spermatogenesis is that the developing male germ cells fail to complete cytoplasmic division (cytokinesis) during mitosis and meiosis. Consequently, large clones of differentiating daughter cells descended from one maturing spermatogonium remain connected by cytoplasmic bridges, forming a *syncytium* (Figure 20–23). The cytoplasmic bridges persist until the very end of sperm differentiation, when individual sperm are released into the tubule lumen. This accounts for the observation that mature sperm arise synchronously in any given area of a seminiferous tubule. But what is the function of the syncytial arrangement?

Figure 20–22 Highly simplified drawing of a cross-section of a seminiferous tubule in a mammalian testis. (A) All of the stages of spermatogenesis shown take place while the developing gametes are in intimate association with *Sertoli cells,* which are large cells that extend from the basal lamina to the lumen of the seminiferous tubule; they are analogous to follicle cells in the ovary. Spermatogenesis depends on testosterone secreted by *Leydig cells,* located between the seminiferous tubules. (B) Dividing spermatogonia are found along the basal lamina. Some of these cells stop dividing and enter meiosis to become primary spermatocytes. Eventually sperm are released into the lumen. In man it takes about 24 days for a spermatocyte to complete meiosis to become a spermatid and another 5 weeks for a spermatid to develop into a sperm. Sperm undergo further maturation and become motile in the epididymis and are only then fully mature sperm.

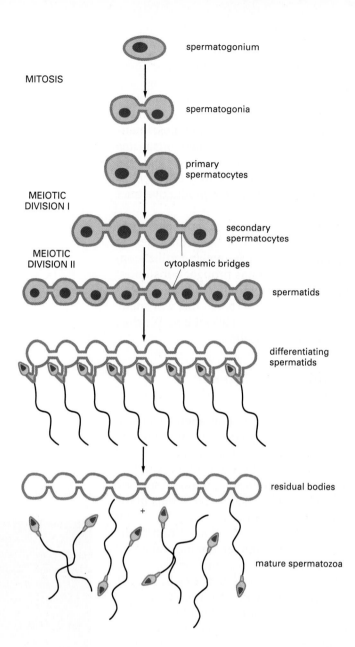

MITOSIS

MEIOTIC
DIVISION I

MEIOTIC
DIVISION II

spermatogonium

spermatogonia

primary
spermatocytes

secondary
spermatocytes

cytoplasmic bridges

spermatids

differentiating
spermatids

residual bodies

mature spermatozoa

Figure 20–23 Cytoplasmic bridges in developing sperm cells and their precursors. The progeny of a single maturing spermatogonium remain connected to one another by cytoplasmic bridges throughout their differentiation into mature sperm. For the sake of simplicity, only two connected maturing spermatogonia are shown entering meiosis, eventually to form eight connected haploid spermatids. In fact, the number of connected cells that go through two meiotic divisions and differentiate together is very much larger than shown here.

Unlike oocytes, sperm undergo most of their differentiation after their nuclei have completed meiosis to become haploid. The presence of cytoplasmic bridges between them, however, means that each developing haploid sperm shares a common cytoplasm with its neighbors, so that it can be supplied with all the products of a complete diploid genome. Thus the diploid genome directs sperm differentiation just as it directs egg differentiation.

Summary

A sperm is usually a small, compact cell, highly specialized for the task of fertilizing an egg. Whereas in human females the total pool of oocytes is produced before birth, in males new germ cells enter meiosis continually from the time of sexual maturation, each primary spermatocyte giving rise to four mature sperm. Sperm differentiation occurs after meiosis, when the nuclei are haploid. Because the maturing spermatogonia and spermatocytes fail to complete cytokinesis, however, the progeny of a single spermatogonium develop as a large syncytium. This allows sperm differentiation to be directed by the products of both parental chromosomes.

Fertilization [12]

Once released, egg and sperm alike are destined to die within minutes or hours unless they find each other and fuse in the process of **fertilization.** Through fertilization the egg and sperm are saved: the egg is activated to begin its developmental program, and the nuclei of the two gametes come together to form the genome of a new organism. The mechanism of fertilization has been most intensively studied in marine invertebrates, especially sea urchins. In these organisms fertilization occurs in sea water, into which huge numbers of both sperm and eggs are released. Such *external fertilization* is more accessible to study than the *internal fertilization* of mammals, which occurs in the female reproductive tract following mating.

In the late 1950s, however, it became possible to fertilize mammalian eggs (more accurately, secondary oocytes—see Figure 20–16) *in vitro*, opening the way to an analysis of the cellular and molecular events in mammalian fertilization. Progress in understanding mammalian fertilization has brought substantial medical benefit: mammalian eggs that have been fertilized *in vitro* can develop into normal individuals when transplanted into the uterus; in this way many previously infertile women have been able to produce normal children. We shall focus our brief discussion, then, on mammalian fertilization.

Binding to the Zona Pellucida Induces the Sperm to Undergo an Acrosomal Reaction [13]

Of the 300 million human sperm ejaculated during coitus, only about 200 reach the site of fertilization in the oviduct. Once there, the sperm must first migrate through the shell of follicular cells that surrounds the ovulated egg and then bind to and traverse the egg coat—the *zona pellucida*. Finally, it must bind and fuse with the egg plasma membrane. To become competent to accomplish these tasks, ejaculated mammalian sperm must normally be modified by secretions in the female reproductive tract, a process called **capacitation,** which requires about 5–6 hours in humans. Capacitation seems to involve both an alteration in the lipid and glycoprotein composition of the sperm plasma membrane and an increase in sperm metabolism and motility; its mechanism is unclear.

Once a capacitated sperm has penetrated the layer of follicle cells, it binds to the **zona pellucida** (see Figure 20–15). The zona pellucida usually acts as a barrier to fertilization across species, and removing it often removes this barrier. Human sperm, for example, will fertilize hamster eggs from which the zona pellucida has been removed with specific enzymes; not surprisingly, such hybrid zygotes do not develop. Zona-free hamster eggs, however, are used in fertility clinics to assess the fertilizing capacity of human sperm *in vitro* (Figure 20–24).

The zona pellucida of mammalian eggs is composed of only three glycoproteins. Two of them, *ZP2* and *ZP3*, assemble into filaments, while the other, *ZP1*, cross-links the filaments into a three-dimensional network. **ZP3** acts as a sperm receptor: the species-specific binding of sperm to the zona pellucida is mediated by a molecule (which may be the enzyme *galactosyl transferase*) on the surface of the sperm head that binds to *O*-linked oligosaccharides on ZP3 in the zona. On binding, the sperm is induced to undergo the **acrosomal reaction**, in which the contents of the acrosome are released by exocytosis (Figure 20–25). In the mouse, at least, the trigger for the acrosomal reaction is ZP3 in the zona, which activates a complex intracellular signaling mechanism that induces an influx of Ca^{2+} into the sperm cytosol, which is thought to initiate exocytosis.

The acrosomal reaction releases proteases and hyaluronidase, which are essential for the penetration of the sperm through the zona pellucida, and it exposes other proteins on the sperm surface that bind to ZP2 and thereby help the sperm maintain its tight binding to the zona while boring through it. In addition, the acrosomal reaction exposes a protein in the sperm plasma membrane

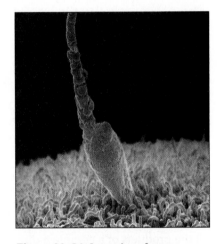

Figure 20–24 Scanning electron micrograph of a human sperm contacting a hamster egg. The zona pellucida of the egg has been removed, exposing the plasma membrane, which contains numerous microvilli. The ability of an individual's sperm to penetrate hamster eggs is used as an assay of male fertility; penetration of more than 10–25% of the eggs is considered to be normal. (Courtesy of David M. Phillips.)

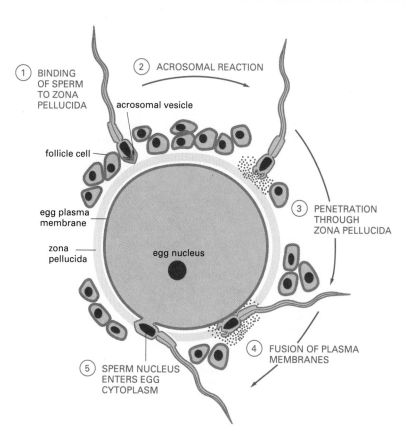

Figure 20–25 The acrosomal reaction that occurs when a mammalian sperm fertilizes an egg. In mice a single glycoprotein in the zona pellucida, ZP3, is thought to be responsible for both binding the sperm and inducing the acrosomal reaction. Note that a mammalian sperm interacts tangentially with the egg plasma membrane so that fusion occurs at the equator, rather than at the tip, of the sperm head. In mice the zona pellucida is 7 μm in diameter and sperm cross it at a rate of about 1 μm/min.

that mediates the binding and fusion of this membrane with that of the egg, as we see below.

The Egg Cortical Reaction Helps to Ensure That Only One Sperm Fertilizes the Egg [14]

Although many sperm can bind to an egg, normally only one fuses with the egg plasma membrane and injects its nucleus and other organelles into the egg cytoplasm. If more than one sperm fuses—a condition called *polyspermy*—multipolar or extra mitotic spindles are formed, resulting in faulty segregation of chromosomes during cell division; nondiploid cells are produced, and development quickly stops. Two mechanisms operate to ensure that only one sperm fertilizes the egg. A rapid depolarization of the egg plasma membrane, which is caused by the fusion of the first sperm, is thought to prevent further sperm from fusing and thereby acts as a fast *primary block to polyspermy*. But the membrane potential returns to normal soon after fertilization, so that a second mechanism is required to ensure a longer-term, *secondary block to polyspermy*. This is provided by the egg **cortical reaction.**

When the sperm fuses with the egg plasma membrane, it activates the inositol phospholipid cell-signaling pathway (discussed in Chapter 15) in the egg. This, in turn, causes a local increase in cytosolic Ca^{2+}, which spreads through the cell in a wave. The rise in Ca^{2+} in the cytosol is thought to activate the egg and initiate the cortical reaction, in which the cortical granules release their contents by exocytosis. If the cytosolic concentration of Ca^{2+} is increased artificially—either directly by an injection of Ca^{2+} or indirectly by the use of Ca^{2+}-carrying ionophores (discussed in Chapter 11)—the eggs of all animals so far tested, including mammals, are activated. Conversely, preventing the increase in Ca^{2+} by injecting the Ca^{2+} chelator EGTA inhibits activation of the egg in response to fertilization. The enzymes released by the cortical reaction change the structure of the zona pellucida, which becomes "hardened," so that sperm no longer bind to it, thereby providing a slow, secondary block to polyspermy. Among the changes

Fertilization

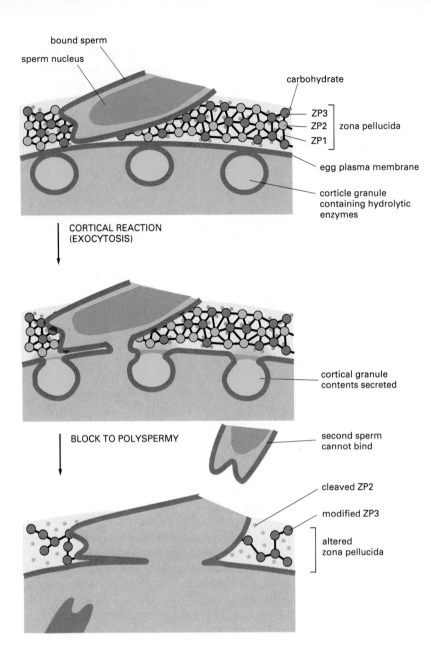

bound sperm
sperm nucleus
carbohydrate
ZP3
ZP2 } zona pellucida
ZP1
egg plasma membrane
corticle granule containing hydrolytic enzymes

CORTICAL REACTION (EXOCYTOSIS)

cortical granule contents secreted

BLOCK TO POLYSPERMY

second sperm cannot bind

cleaved ZP2

modified ZP3

altered zona pellucida

Figure 20–26 Schematic drawing of how the cortical reaction in a mouse egg is thought to prevent additional sperm from entering the egg. The released contents of the cortical granules both remove carbohydrate from ZP3 so it no longer can bind to the sperm plasma membrane and partly cleave ZP2, hardening the zona pellucida. Together these changes provide a block to polyspermy.

that occur in the zona is the proteolytic cleavage of ZP2 and the hydrolysis of sugar groups on ZP3 (Figure 20–26).

A Transmembrane Fusion Protein in the Sperm Plasma Membrane Catalyzes Sperm-Egg Fusion [15]

After a sperm penetrates the extracellular coat of the egg, it interacts with the egg plasma membrane overlying the tips of microvilli on the egg surface (see Figure 20–24). Neighboring microvilli then rapidly elongate and cluster around the sperm to ensure that it is held firmly so that it can fuse with the egg. After fusion, the entire sperm is drawn head-first into the egg as the microvilli are resorbed. In hamsters a single transmembrane protein called **PH-30,** which becomes exposed on the sperm surface during the acrosomal reaction, is thought to mediate both the binding of the sperm to the egg plasma membrane and the fusion of the two plasma membranes.

The protein is composed of two glycosylated transmembrane subunits called α and β, which are held together by noncovalent bonds (Figure 20–27). The extracellular domain of the α subunit contains a hydrophobic region of about 20 amino acid residues that resembles the fusogenic regions of viral fusion proteins, which mediate the fusion of enveloped viruses with the cells that they infect (discussed in Chapter 13). It has long been suspected that the various membrane fusions that occur within and between eucaryotic cells are catalyzed by fusion proteins resembling those present in enveloped viruses; PH-30 is the first such cellular protein to be defined.

The extracellular amino-terminal domain of the β subunit of PH-30 resembles a domain found in some proteins that bind to *integrins*, the cell-surface receptors that help animal cells to adhere to the extracellular matrix (discussed in Chapter 19). This and other indirect evidence suggest that the PH-30 β subunit binds to an integrin in the egg plasma membrane and thereby helps the sperm adhere to the surface of the egg in preparation for fusion.

As the cell biology of mammalian fertilization becomes better understood and the molecules that mediate the various steps in the process are defined, new strategies for contraception become possible. One approach currently being investigated, for example, is to immunize males or females with molecules that are required for reproduction in the hope that the antibodies produced will inhibit the activities of these molecules. In addition to the various hormones and hormone receptors involved in reproduction, ZP3 and PH-30 might be appropriate target molecules. An alternative approach would be to administer oligosaccharides or peptides corresponding to ligands, such as the postulated integrin-binding domain of PH-30, that operate in fertilization. Small molecules of this type might block fertilization by competing with the normal ligand.

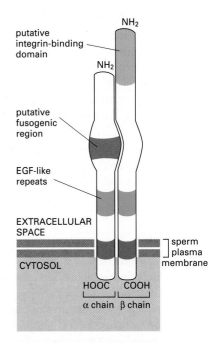

Figure 20–27 **The PH-30 protein in the hamster sperm plasma membrane.** The α and β subunits, which are both glycosylated (not shown), are noncovalently associated. Amino acid sequence similarity between the two subunits, such as the EGF-like repeat in the same location, suggests that the subunits evolved from a common progenitor protein.

The Sperm Provides a Centriole for the Zygote [16]

Once fertilized, the egg is called a **zygote.** Fertilization is not complete, however, until the two haploid nuclei (called *pronuclei*) have come together and combined their chromosomes into a single diploid nucleus. In fertilized mammalian eggs the two pronuclei do not fuse directly as they do in many other species: they approach each other but remain distinct until after the membrane of each pronucleus breaks down in preparation for the first mitotic division (Figure 20–28).

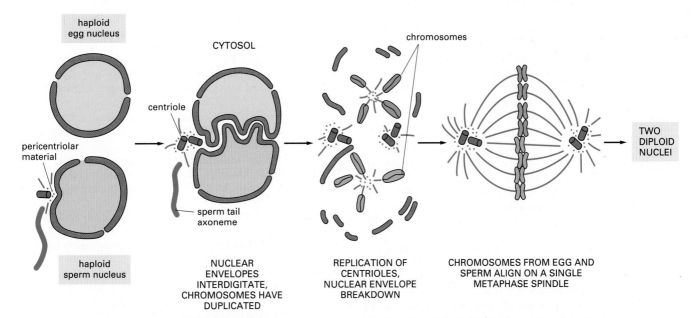

Figure 20–28 **The coming together of the sperm and egg pronuclei following mammalian fertilization.** The pronuclei migrate toward the center of the egg. When they come together, their nuclear envelopes interdigitate. The centrioles replicate, the nuclear envelopes break down, and the chromosomes of both gametes are eventually integrated into a single mitotic spindle, which mediates the first cleavage division of the zygote. (Adapted from drawings and electron micrographs provided by Daniel Szöllösi.)

In most animals, including humans, the sperm contributes more than DNA to the zygote: it also donates a centriole—an organelle that is curiously lacking in the unfertilized eggs of these animals; the egg has a centrosome, but this does not contain a centriole. The sperm centriole enters the egg along with the sperm nucleus and tail, and in some species it replicates and helps organize the assembly of the first mitotic spindle in the zygote (see Figure 20–28). This explains why multipolar or extra mitotic spindles form in cases of polyspermy, where several sperm contribute centrioles to the egg.

Fertilization marks the beginning of one of the most remarkable phenomena in all of biology—the process of embryogenesis, in which the zygote develops into a new individual. This is the subject of the next chapter.

Summary

Mammalian fertilization begins when the head of a sperm binds in a species-specific manner to the zona pellucida surrounding the egg. This induces the acrosomal reaction in the sperm, which releases the contents of its acrosomal vesicle, including enzymes that help the sperm digest its way through the zona to the egg plasma membrane in order to fuse with it. Fusion is catalyzed by a transmembrane protein located in the sperm plasma membrane. This event activates the egg to undergo the cortical reaction, in which cortical granules release their contents, including enzymes that alter the zona pellucida and thereby prevent the fusion of additional sperm. Development of the zygote begins after sperm and egg haploid pronuclei have come together, pooling their chromosomes to form a single diploid nucleus.

References

General

Austin, C.R.; Short, R.V., eds. Reproduction in Mammals: I. Germs Cells and Fertilization, 2nd ed. Cambridge, UK: Cambridge University Press, 1982.

Gilbert, S.F. Developmental Biology, 3rd ed., pp. 34–60. Sunderland, MA: Sinauer, 1991.

Knobil, E.; Neill, J.D., eds. The Physiology of Reproduction, Vols. 1 and 2. New York: Raven Press, 1988.

Metz, C.B.; Monroy, A., eds. Biology of Fertilization, Vols. 1–3. New York: Academic Press, 1985.

Wassarman, P.M., ed. Elements of Mammalian Fertilization. Boca Raton, FL: CRC Press, 1990.

Cited

1. Maynard Smith, J. Evolution of Sex. Cambridge, UK: Cambridge University Press, 1978.

   Williams, G.C. Sex and Evolution. Princeton, NJ: Princeton University Press, 1975.

2. Evans, C.W.; Dickinson, H.G., eds. Controlling Events in Meiosis. Symposia of the Society for Experimental Biology, Vol. 38. Cambridge, UK: Company of Biologists, 1984.

   Moens, P.B. Meiosis. New York: Academic Press, 1987.

3. John, B.; Lewis, K.R. The Meiotic Mechanism. Oxford Biology Readers (J.J. Head, ed.). Oxford, UK: Oxford University Press, 1976.

   Jones, G.H. The control of chiasma distribution. In Controlling Events in Meiosis (C.W. Evans, H.G. Dickinson, eds.). Symposia of the Society for Experimental Biology, Vol. 38, pp. 293–320. Cambridge, UK: Company of Biologists, 1984.

Orr-Weaver, T.L.; Szostak, J.W. Fungal recombination. *Microbiol. Rev.* 49:33–58, 1985.

4. Moses, M.J. Synaptonemal complex. *Annu. Rev. Genet.* 2:363–412, 1968.

   Roeder, G.S. Chromosome synapsis and genetic recombination: their roles in meiotic chromosome segregation. *Trends Genet.* 6:385–389, 1990.

   von Wettstein, D.; Rasmussen, S.W.; Holm, P.B. The synaptonemal complex in genetic segregation. *Annu. Rev. Genet.* 18:331–413, 1984.

5. Carpenter, A.T.C. Recombination nodules and synaptonemal complex in recombination-defective females of *Drosophila melanogaster*. *Chromosoma* 75:259–263, 1979.

   Carpenter, A.T.C. Gene conversion, recombination nodules, and the initiation of meiotic synapsis. *Bioessays* 6:232–236, 1987.

6. Buckle, V.; Mondello, C.; Darling, S.; Craig, I.W.; Goodfellow, P.N. Homologous expressed genes in the human sex chromosome pairing region. *Nature* 317:739–741, 1985.

   Chandley, A.C. Meiosis in man. *Trends Genet.* 4:79–84, 1988.

   Solari, A.J. The behavior of the XY pair in mammals. *Int. Rev. Cytol.* 38:273–317, 1974.

7. Austin, C.R. The egg. In Reproduction in Mammals: I. Germ Cells and Fertilization (C.R. Austin, R.V. Short, eds.), 2nd ed., pp. 46–62. Cambridge, UK: Cambridge University Press, 1982.

   Dietl, J., ed. The Mammalian Egg Coat: Structure and Function. Berlin: Springer-Verlag, 1989.

   Gilbert, S.F. Developmental Biology, pp. 37–41. Sunderland, MA: Sinauer, 1991.

8. Baker, T.G. Oogenesis and ovulation. In Reproduction in Mammals: I. Germs Cells and Fertilization (C.R. Austin, R.V. Short, eds.), 2nd ed., pp. 17–45. Cambridge, UK: Cambridge University Press, 1982.

   Browder, L.W. Oogenisis. New York: Plenum Press, 1985.

   Metz, C.B.; Monroy, A., eds. Model Systems and Oogenesis. Biology of Fertilization, Vol. 1. Orlando, FL: Academic Press, 1985.

9. Gilbert, S.F. Developmental Biology, pp. 812–815. Sunderland, MA: Sinauer, 1991.

   Spradling, A. Developmental genetics of oogenesis. In *Drosophila* Development (M. Bate, A. Martinez-Arias, eds.), pp. 1–69. Cold Spring Harbor, NY: Cold Spring Harbor Laboratory Press, 1993.

10. Bellvé, A.R.; O'Brien, D.A. The mammalian spermatozoon: structure and temporal assembly. In Mechanism and Control of Animal Fertilization (J.F. Hartmann, ed.), pp. 56–137. New York: Academic Press, 1983.

    Fawcett, D.W.; Bedford, J.M., eds. The Spermatozoon. Baltimore: Urban & Schwarzenberg, 1979.

11. Clermont, Y. Kinetics of spermatogenesis in mammals: seminiferous epithelium cycle and spermatogonial renewal. *Physiol. Rev.* 52:198–236, 1972.

    Metz, C.B.; Monroy, A., eds. Biology of the Sperm. Biology of Fertilization, Vol. 2. Orlando, FL: Academic Press, 1985.

    Setchell, B.P. Spermatogenesis and spermatozoa. In Reproduction in Mammals: I. Germ Cells and Fertilization (C.R. Austin, R.V. Short, eds.), 2nd ed., pp. 63–101. Cambridge, UK: Cambridge University Press, 1982.

12. Dunbar, B.S.; O'Rand, M.G., eds. A Comparative Overview of Mammalian Fertilization. New York: Plenum Press, 1991.

    Garbers, D.L. Molecular basis of fertilization. *Annu. Rev. Biochem.* 58:719–742, 1989.

    Longo, F.J. Fertilization. London: Chapman and Hall, 1987.

    Metz, C.B.; Monroy, A., eds. The Fertilization Response of the Egg. Biology of Fertilization, Vol. 3. Orlando, FL: Academic Press, 1985.

    Wassarman, P.M. Early events in mammalian fertilization. *Annu. Rev. Cell Biol.* 3:109–142, 1987.

    Yanagimachi, R. Mammalian fertilization. In The Physiology of Reproduction (E. Knobil, J.D. Neill, eds.), Vol. 1, pp. 135–185. New York: Raven Press, 1988.

13. Florman, H.M.; Babcock, D.F. Progress toward understanding the molecular basis of capacitation. In Elements of Mammalian Fertilization (P.M. Wassarman, ed.), Vol. 1: Basic Concepts, pp. 105–132. Boca Raton, FL: CRC Press, 1990.

    Kopf, G.S.; Gerton, G.L. The mammalian sperm acrosome and acrosome reaction. In Elements of Mammalian Fertilization (P.M. Wassarman, ed.), Vol. 1: Basic Concepts, pp. 153–204. Boca Raton, FL: CRC Press, 1990.

    Wassarman, P.M.; Mortillo, S. Structure of the mouse egg extracellular coat, the zona pellucida. *Int. Rev. Cytol.* 130:85–110, 1991.

14. Ducibella, T. Mammalian egg cortical granules and the cortical reaction. In Elements of Mammalian Fertilization (P.M. Wassarman, ed.), Vol. 1: Basic Concepts, pp. 205–232. Boca Raton, FL: CRC Press, 1990.

    Jaffe, L.A.; Cross, N.L. Electrical regulation of sperm-egg fusion. *Annu. Rev. Physiol.* 48:191–200, 1986.

    Wassarman, P.M. Mouse gamete adhesion molecules. *Biol. Reprod.* 46:186–191, 1992.

15. Blobel, C.P., et al. A potential fusion peptide and an integrin ligand domain in a protein active in sperm-egg fusion. *Nature* 356:248–252, 1992.

16. Schatten, H., et al. Behavior of centrosomes during fertilization and cell division in mouse oocytes and in sea urchin eggs. *Proc. Natl. Acad. Sci. USA* 83:105–109, 1986.

    Sathananthan, A.H., et al. Centrioles in the beginning of human development. *Proc. Natl. Acad. Sci. USA* 88:4806–4810, 1991.

    Szöllösi, D., et al. Sperm penetration into immature mouse oocytes and nuclear changes during maturation: an EM study. *Biol. Cell* 69:53–64, 1990.

The fly *Drosophila melanogaster.* (Courtesy of E.B. Lewis.)

Cellular Mechanisms of Development

A multicellular animal or plant is an ordered clone of cells, all containing the same genome but specialized in different ways. Although the final structure may be enormously complex, it is generated by a limited repertoire of cell activities. Cells grow, divide, and die. They form mechanical attachments and generate forces for movement. They differentiate by switching on or off the production of specific sets of proteins. They produce molecular signals to influence neighboring cells, and they respond to signals that neighboring cells deliver to them. The genome, repeated identically in every cell, defines the rules according to which these various possible cell activities are called into play. Through its operation in each cell individually, it guides the whole intricate multicellular process of development by which an adult organism is generated from a fertilized egg.

In this chapter, rather than follow any one organism in detail, we illustrate the general principles of development by reference to the species that display each principle best. We discuss first how cell movements and cell divisions shape the animal embryo and how differences between cells arise in a spatially ordered fashion. We then consider how cell memory serves to perpetuate the spatial pattern of differences and allows new details to be filled in as an animal grows. In the central part of the chapter we examine the underlying genetic control mechanisms, taking the nematode worm *Caenorhabditis elegans* and the fly *Drosophila melanogaster* as examples. We shall see that molecular genetics has revealed remarkable similarities in the development of the most diverse types of animals. Because worms and flies are our cousins, what we learn from them provides a key to the development of mammals also.

The last two sections of the chapter can be read as separate modules: we review the development of flowering plants and ask how far it obeys the same principles as animal development, and then we discuss the special mechanisms by which the nervous system develops its astonishing circuitry.

Morphogenetic Movements and the Shaping of the Body Plan [1]

We begin by considering how the geometrical structure of the early vertebrate embryo is formed. The focus will be on the question of how cells move into the correct positions. Later sections will consider how cells adopt the correct differentiated characters. It is traditional to distinguish three phases in the development of a vertebrate—and indeed of many other types of animal. In the first phase the fertilized egg *cleaves* to form many smaller cells, and these become organized into an epithelium and perform a complex series of movements, called *gastrulation* and *neurulation,* that create the basic body plan, with a rudimentary gut cavity and a neural tube. In the second phase the rudiments of the various organs, such as limbs, eyes, heart, and so on, are formed—a process called *organogenesis.* In the third phase the tiny structures that have been generated in this way proceed to grow to their adult size. These phases are not sharply distinct but overlap considerably in time. To follow the course of events from the fertilized egg to the beginning of organogenesis, we take as our chief example the frog *Xenopus laevis* (Figure 21–1), whose early development has been particularly well studied. As in other amphibians, the entire process from fertilization onward takes place outside the mother, and the developing embryo is robust and easy to manipulate experimentally.

The Polarity of the Amphibian Embryo Depends on the Polarity of the Egg [2]

The *Xenopus* egg is a large cell, just over a millimeter in diameter, enclosed in a transparent extracellular capsule, or jelly coat. Most of the cell's volume is occupied by yolk platelets, which are membrane-bounded aggregates chiefly of lipid and protein. The yolk is concentrated toward the lower end of the egg, called the **vegetal pole**; the upper end is called the **animal pole**. The animal and vegetal regions contain different selections of mRNA molecules as well as different quantities of yolk and other cell components, and they have different fates. Roughly speaking, the vegetal end of the egg is destined to form internal tissues (in particular, the gut), and the animal end, external ones (such as the skin). Fertilization initiates a complex series of movements that will eventually tuck vegetal regions into the interior to form the gut and in the process will establish the three principal axes of the body: *anteroposterior,* from head to tail; *dorsoventral,* from back to belly; and *mediolateral,* from the median plane outward to the left or to the right.

The animal-vegetal asymmetry of the unfertilized egg is sufficient to define only one of the eventual body axes—the anteroposterior—but fertilization triggers a distortion of the egg contents that creates an additional asymmetry defining a dorsoventral difference: the outer, actin-rich cortex of the egg cytoplasm abruptly rotates relative to the central core of the egg, so that the animal pole of the cortex is slightly shifted to the future ventral side (Figure 21–2). The direction of the rotation is determined by the point of sperm entry—perhaps through an effect of the centrosome that the sperm brings into the egg. Because pigment granules in the egg are displaced by the rotation, a band of slightly diminished pigmentation, called the *gray crescent,* becomes visible in some amphibian species opposite the sperm entry point. In the neighborhood of the gray crescent the cortex of the vegetal hemisphere has become juxtaposed with core cytoplasm of

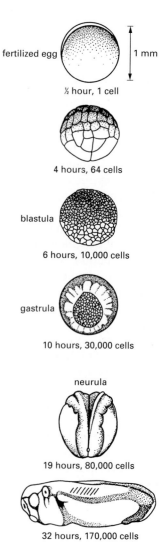

fertilized egg   1 mm

½ hour, 1 cell

4 hours, 64 cells

blastula

6 hours, 10,000 cells

gastrula

10 hours, 30,000 cells

neurula

19 hours, 80,000 cells

32 hours, 170,000 cells

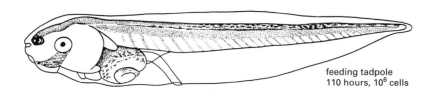

feeding tadpole
110 hours, 10^6 cells

Figure 21–1 Synopsis of the development of *Xenopus laevis* from newly fertilized egg to feeding tadpole. The adult frog is shown in the photograph at the top. The developmental stages are viewed from the side, except for the 10-hour and 19-hour embryos, which are viewed from below and from above, respectively. All stages except the adult are shown at the same scale. (Photograph courtesy of Jonathan Slack; drawings after P.D. Nieuwkoop and J. Faber, Normal Table of *Xenopus laevis* [Daudin]. Amsterdam: North-Holland, 1956.)

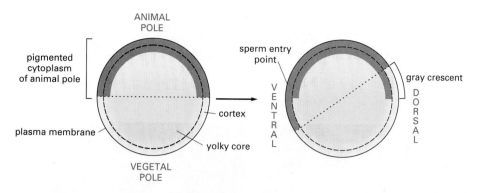

ANIMAL
POLE

pigmented
cytoplasm
of animal pole

sperm entry
point

gray crescent

plasma membrane

cortex

V
E
N
T
R
A
L

D
O
R
S
A
L

yolky core

VEGETAL
POLE

Figure 21–2 The first morphogenetic movement following fertilization of a frog's egg. The egg cortex (a layer a few micrometers deep) rotates through about 30° relative to the core of the egg in a direction determined by the site of sperm entry. In species where the cytoplasm capping the animal pole is appropriately pigmented, the rotation creates a visible gray crescent opposite the site of sperm entry.

the animal hemisphere, creating a special region that is crucial in organizing the dorsoventral axis of the body, as we discuss later.

The sperm entry point corresponds, roughly speaking, to the future belly; the opposite side will form the back and dorsal structures, including the spinal cord. Treatments that block the rotation allow cleavage to occur normally but produce an embryo with a central gut and no dorsoventral asymmetry.

Cleavage Produces Many Cells from One [3]

The cortical rotation is completed in about an hour after fertilization and sets the scene for *cleavage,* in which the single large egg cell subdivides by repeated mitosis into many smaller cells, or **blastomeres,** without any change in total mass (Figure 21–3). To survive, the embryo must quickly reach a stage where it can begin to feed, swim, and escape from predators, and these first cell divisions are extraordinarily rapid, with a cycle time of about 30 minutes. The very high rate of DNA replication and mitosis seems to preclude gene transcription (although protein synthesis occurs), and the cleaving embryo is almost entirely dependent on reserves of RNA, protein, membrane, and other materials that accumulated in the egg while it developed as an oocyte in the mother. The only crucial biosynthesis obviously required is that of DNA, and unusually rapid DNA replication is made possible by the use of an exceptionally large number of replication origins, closely spaced in the chromosomal DNA.

After about 12 cycles of cleavage (7 hours), the cell division rate slows down abruptly, and transcription of the embryo's genome begins. This change, known as the *mid-blastula transition,* seems to be triggered by attainment of a critical ratio of DNA to cytoplasm: the transition can be hastened or delayed by artificially increasing or decreasing the amount of DNA in the egg.

The Blastula Consists of an Epithelium Surrounding a Cavity [4]

From the outset the cells of the embryo are not only bound together mechanically, they are also coupled by *gap junctions* through which ions and other small molecules can pass, conveying messages that may help to coordinate the behavior of the cells. Meanwhile, in the outermost regions of the embryo, *tight junctions* between the blastomeres create a seal, isolating the interior of the embryo from the external medium. At about the 16-cell stage, Na^+ begins to be pumped across the cell membranes into the spaces between cells in the interior of the embryo, and water follows because of the resulting osmotic pressure gradient. As a result, the intercellular crevices deep inside the embryo enlarge to form a single cavity, the **blastocoel,** and the embryo is now termed a **blastula** (Figure 21–4). The cells that form the exterior of the blastula have become organized as an epithelial sheet, which will be crucial in coordinating their subsequent behavior.

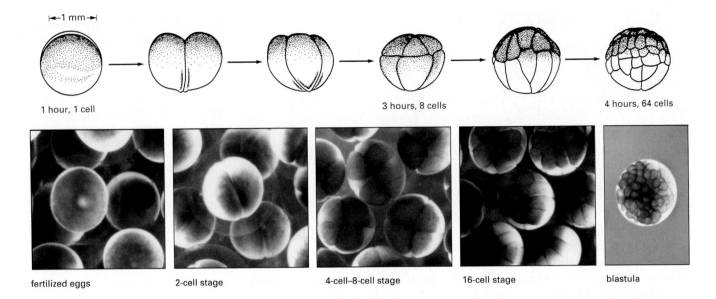

1 hour, 1 cell

3 hours, 8 cells

4 hours, 64 cells

fertilized eggs   2-cell stage   4-cell–8-cell stage   16-cell stage   blastula

Gastrulation Transforms a Hollow Ball of Cells into a Three-layered Structure with a Primitive Gut [5, 6]

Once the cells of the blastula have become arranged into an epithelial sheet, the stage is set for the coordinated movements of **gastrulation**. This dramatic process transforms the simple hollow ball of cells into a multilayered structure with a central gut tube and bilateral symmetry: by a complicated invagination, many of the cells on the outside of the embryo are moved inside it. Subsequent development depends on the interactions of the inner, outer, and middle layers of cells thus formed.

Gastrulation—the formation of a gut by tucking cells from the exterior of the early embryo into the interior—is a fundamental step in the development of practically every type of animal. The transparent embryo of the sea urchin provides one of the clearest and simplest illustrations of the process. Figure 21–5 shows the sequence of events, starting with a simple hollow blastula. Briefly, cells at the vegetal pole invaginate, forming a hollow tube that eventually makes contact with the epithelium near the opposite end of the embryo to form the mouth. Meanwhile, cells escape from the invaginating epithelium at certain sites and move into the body cavity to form embryonic connective tissue, or *mesenchyme.*

In the three-layered structure created by gastrulation, the innermost layer, the tube of the primitive gut, is the *endoderm;* the outermost layer, the epithe-

Figure 21–3 The stages of cleavage in *Xenopus.* The drawings show a series of side views. The photographs show views from above. The cleavage divisions rapidly subdivide the egg into many smaller cells. All the cells divide synchronously for the first 12 cleavages, but the divisions are asymmetric, so that the lower, vegetal cells, encumbered with yolk, are fewer and larger.

The asymmetries of the egg and the detailed patterns of cleavage vary from one animal species to another. In mammals, whose small, symmetrical eggs contain little yolk, the first three cleavages divide the cell evenly into eight equal blastomeres. At the other extreme, exemplified by the very yolky bird egg, cleavage does not cut all the way through the yolk, and all the nuclei remain clustered at the animal pole; the embryo then develops from a cap of cells on top of the yolk. (Photographs courtesy of Jonathan Slack.)

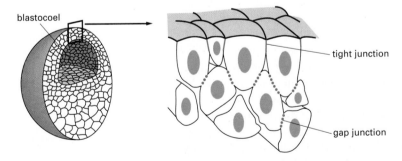

Figure 21–4 The blastula. At this stage the cells are arranged to form an epithelium surrounding a fluid-filled cavity, the blastocoel. The cells are electrically coupled via gap junctions, and tight junctions close to the outer surface create a seal that isolates the interior of the embryo from the external medium. Note that in *Xenopus* the wall of the blastocoel is several cells thick, and only the outermost cells are tightly bound together as an epithelium.

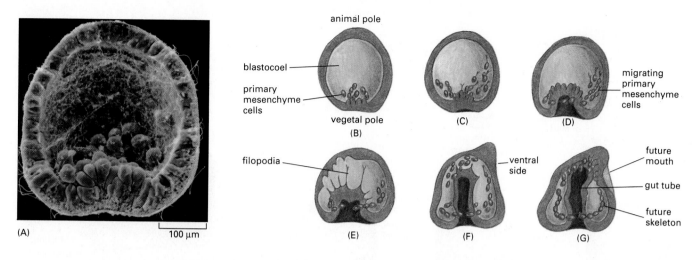

Figure 21–5 Gastrulation in a sea urchin. The starting point for sea-urchin gastrulation is a very simple blastula: a sheet of about 1000 cells, one cell thick, surrounding a spherical cavity. (A) Scanning electron micrograph showing the initial intucking of the epithelium at the vegetal pole. (B) A first group of mesenchyme cells break loose from the epithelium at the vegetal pole of the blastula. (C) These cells then crawl over the inner face of the wall of the blastula. (D) Meanwhile the epithelium at the vegetal pole is continuing to tuck inward. (E and F) The invaginating epithelium extends into a long gut tube: the invaginating cells actively change their packing, without much altering their average shape, so as to convert the initial squat dome-shaped invagination into a long narrow gut tube. This type of tissue movement, in which a sheet of cells elongates along one dimension while narrowing along another, provides an important means of remodeling during animal development and is called *convergent extension.* At the same time certain cells in the rounded tip of the invaginating sheet extend long filopodia into the blastocoel cavity; these contact the walls of the cavity, adhere there, and contract, thereby helping to steer the invagination movement. (G) The end of the gut tube makes contact with the wall of the blastula at the site of the future mouth opening. Here the epithelia will fuse and a hole will form. (A, from R.D. Burke, R.L. Myers, T.L. Sexton, and C. Jackson, *Dev. Biol.* 146:542–557, 1991; B–G, after L. Wolpert and T. Gustafson, *Endeavour* 26:85–90, 1967.)

lium that has remained external, is the *ectoderm;* and between the two, the looser layer of tissue composed of mesenchyme cells is the *mesoderm.* These are the three primary **germ layers** common to higher animals. The organization of the embryo into the three layers corresponds roughly to the organization of the adult—gut on the inside, epidermis on the outside, and connective tissue and muscle in between. Very crudely, these three layers of adult tissues may be said to derive from the endoderm, the ectoderm, and the mesoderm, respectively, although there are exceptions.

In *Xenopus* the geometry of gastrulation is more complex than in the sea urchin. But it is important to grasp the basic principles, for it is through the movements of gastrulation that the main axes of the vertebrate body are created. The details of the process are described in Figure 21–6. A central part is played by the tissue near the site of the gray crescent, to one side of the vegetal pole. Here, gastrulation starts with a short indentation that gradually extends to form the **blastopore**—a line of invagination that eventually curves around to encircle the vegetal pole. The site where the invagination starts defines the *dorsal lip* of the blastopore; this tissue plays a leading part in the ensuing complex series of movements and gives rise to the dorsal structures of the main body axis. As in the sea urchin, the end result of the whole process is a three-layered structure: an outermost sheet of ectoderm, an innermost tube of endoderm forming the rudiment of the gut, and between them a layer of mesoderm. Again, the mouth develops as a hole formed at an anterior site where endoderm and ectoderm come into direct contact without intervening mesoderm.

The transformation that is brought about by gastrulation can be summarized by plotting on the surface of the embryo at the beginning of gastrulation a *fate map* showing which regions are destined to give rise to which parts of the adult body; such a map is shown in Figure 21–6B.

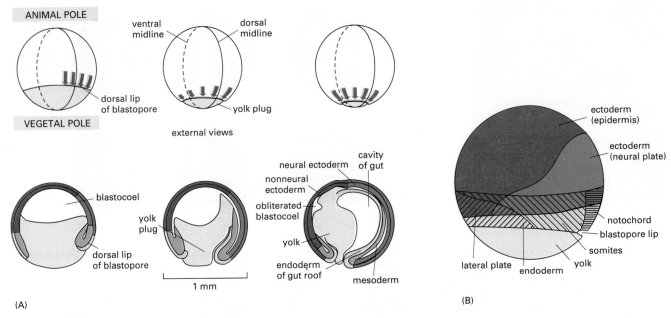

Figure 21–6 Gastrulation in *Xenopus*. (A) The external views (*above*) show the embryo as a semitransparent object, seen from the side; the cross-sections (*below*) are cut in the median plane (the plane of the dorsal and ventral midlines). The directions of cell movement are indicated by *red arrows*. Gastrulation begins when a short indentation, the beginning of the blastopore, becomes visible in the exterior of the blastula. This indentation gradually extends, curving around to form a complete circle surrounding a plug of very yolky cells (destined to be enclosed in the gut and digested). Sheets of cells meanwhile turn in around the lip of the blastopore and move deep into the interior of the embryo. At the same time the external epithelium in the region of the animal pole actively spreads to take the place of the cell sheets that have turned inward. Eventually, the epithelium of the animal hemisphere extends in this way to cover the whole external surface of the embryo, and, as gastrulation reaches completion, the blastopore circle shrinks almost to a point. (B) A fate map for the early *Xenopus* embryo (viewed from the side) as it begins gastrulation, showing the origins of the cells that will come to form the three germ layers as a result of the movements of gastrulation. The various parts of the mesoderm (lateral plate, somites, and notochord) derive from deep-lying cells that segregate from the epithelium in the cross-hatched region; the other cells, including the more superficial cells in the cross-hatched region, will give rise to ectoderm (*blue* and *red, above*) or endoderm (*yellow, below*). Roughly speaking, the first cells to turn into the interior, or *involute,* will move forward inside the embryo to form the most anterior endodermal and mesodermal structures, while the last to involute will form the most posterior structures. (After R.E. Keller, *J. Exp. Zool.* 216:81–101, 1981.)

Gastrulation Movements Are Organized Around the Dorsal Lip of the Blastopore [6, 7]

The dorsal lip of the blastopore plays a central role not just in a geometrical sense, but also as the source of a controlling influence. If the dorsal lip of the blastopore is excised from a normal embryo at the beginning of gastrulation and grafted into another embryo but in a different position, the host embryo initiates gastrulation both at the site of its own dorsal lip and at the site of the graft (Figure 21–7). The movements of gastrulation at the second site entail the formation of a second whole set of body structures, and a double embryo (Siamese twins) results.

By carrying out such grafts between species with differently pigmented cells, so that host tissue can be distinguished from implanted tissue, it has been shown that the grafted blastopore lip recruits host epithelium into its own system of invaginating endoderm and mesoderm. Evidently, the dorsal lip of the blastopore is the source of some signal (or signals) coordinating both the movements of gastrulation and, directly or indirectly, the pattern of specialization of the tissues in its neighborhood. Because of this crucial role in organizing the formation of the main body axis, the dorsal lip of the blastopore is known as the *Organizer* (or *Spemann's Organizer,* after its co-discoverer). It is the oldest and most famous

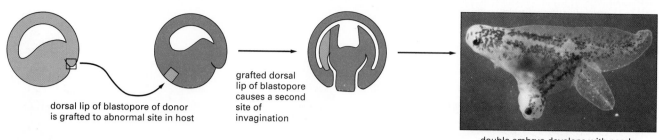

dorsal lip of blastopore of donor
is grafted to abnormal site in host

grafted dorsal
lip of blastopore
causes a second
site of
invagination

double embryo develops with nearly
all its tissues of host origin

Figure 21–7 The role of the Organizer. Diagram of an experiment showing that the dorsal lip of the blastopore (Spemann's Organizer) initiates and controls the movements of gastrulation and thereby, if transplanted, organizes the formation of a second set of body structures. The photograph shows a two-headed, two-tailed axolotl tadpole resulting from such an operation; the results are similar for *Xenopus*. (Photo courtesy of Jonathan Slack.)

example of an *embryonic signaling center*—a function we discuss later when we consider how cell diversification is controlled.

Active Changes of Cell Packing Provide a Driving Force for Gastrulation [1, 6, 8]

Gastrulation begins with changes in the shape of the cells at the site of the blastopore. In the amphibian these are called *bottle cells:* they have broad bodies and narrow necks that anchor them to the surface of the epithelium (Figure 21–8), and they may help to force the epithelium to curve and so to tuck inward, producing the initial indentation seen from outside. Once this first tuck has formed, cells can continue to pass into the interior as a sheet to form the gut and mesoderm. Just as in the sea urchin, the movement seems to be driven by a combination of mechanisms but mainly by active repacking of the cells, especially those in the dorsal part of the *marginal zone* neighboring the blastopore lip (see Figure 21–8). Here **convergent extension** occurs. Small square fragments of dorsal marginal-zone tissue isolated in culture will spontaneously narrow and elongate through a rearrangement of the cells, just as they would in the embryo in the process of converging toward the dorsal midline, turning inward around the blastopore lip, and elongating to form the main axis of the body. A current view of the cellular mechanism underlying convergent extension is illustrated in Figure 21–9.

The Three Germ Layers Formed by Gastrulation Have Different Fates [9, 10, 11, 12]

The **endoderm** forms a tube, the primordium of the digestive tract, from the mouth to the anus. It gives rise not only to the pharynx, esophagus, stomach, and intestines, but also to many associated glands. The salivary glands, the liver, the pancreas, the trachea, and the lungs, for example, all develop from extensions of the wall of the originally simple digestive tract and grow to become systems of branching tubes that open into the gut or pharynx. While the endoderm forms

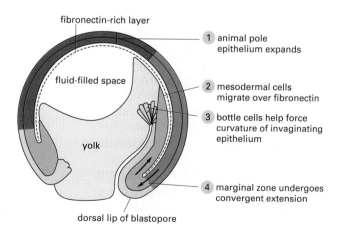

fibronectin-rich layer

fluid-filled space

yolk

dorsal lip of blastopore

1 animal pole epithelium expands

2 mesodermal cells migrate over fibronectin

3 bottle cells help force curvature of invaginating epithelium

4 marginal zone undergoes convergent extension

Figure 21–8 Cell movements in gastrulation. A section through a gastrulating *Xenopus* embryo, cut in the same plane as in Figure 21–6, indicating the four main types of movement that gastrulation involves. The animal pole epithelium expands by cell rearrangement, becoming thinner as it spreads. Migration of mesodermal cells over fibronectin-rich matrix lining the roof of the blastocoel may help to pull the invaginated tissues forward. But the main driving force for gastrulation in *Xenopus* is convergent extension in the marginal zone. (After R.E. Keller, *J. Exp. Zool.* 216:81–101, 1981.)

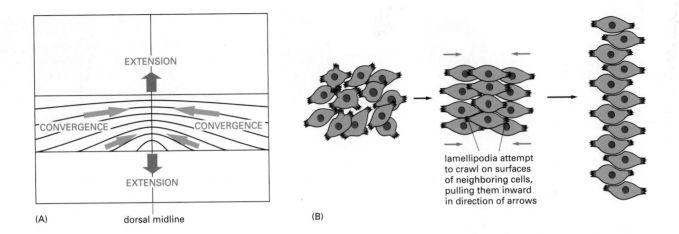

(A)   dorsal midline   (B)

the epithelial components of these structures—the lining of the gut and the secretory cells of the pancreas, for example—the supporting muscular and fibrous elements arise from the **mesoderm**.

The differentiation of the mesoderm is guided by the Organizer at the dorsal lip of the blastopore, which is thought to be a source of signaling molecules that regulate choices between alternative mesodermal fates. Signals from the differently specialized groups of mesoderm cells in turn control the basic pattern of specializations of the endoderm and ectoderm and in particular initiate formation of the nervous system, as we shall see. The mesodermal layer is divided in the postgastrulation embryo into separate parts on the left and right of the body. Defining the central axis of the vertebrate body, and effecting this separation, is the very early specialization of the mesoderm known as the **notochord**. This is a slender rod of cells, about 80 μm in diameter, with ectoderm above it, endoderm below it, and mesoderm on either side (see Figure 21–12). It derives from the cells of the Organizer itself. As these pass around the dorsal lip of the blastopore and move into the interior of the embryo, they form a column of tissue that elongates dramatically by convergent extension. The cells of the notochord also become swollen with vacuoles, so that the rod elongates still further and stretches out the embryo. In the most primitive chordates, which have no vertebrae, the notochord persists as a primitive substitute for a vertebral column. In vertebrates it serves as a core around which other mesodermal cells gather to form the vertebrae. Thus the notochord is the precursor of the vertebral column, both in an evolutionary and in a developmental sense.

In general, the mesoderm gives rise to the muscles and to the connective tissues of the body—at first to the loose, space-filling, three-dimensional mesh of cells known as *mesenchyme* (see Figure 19–30) and ultimately to cartilage, bone, and fibrous tissue, including the dermis (the inner layer of the skin). In addition, the tubules of the urogenital system form from it and so does the vascular system, including the heart, the blood vessels, and the cells of the blood. These specialized mesodermal tissues derive from cells at different distances from the dorsal lip of the blastopore, with notochord having the most dorsal origin and blood cells the most ventral.

At the end of gastrulation the sheet of **ectoderm** covers the embryo and thus eventually forms the epidermis (the outer layer of the skin). It also gives rise to the entire nervous system. In a process known as **neurulation**, a broad central region of the ectoderm thickens, rolls up into a tube, and pinches off from the rest of the cell sheet (Figure 21–10). The tube thus created from the ectoderm is called the **neural tube**; it will form the brain and the spinal cord. The mechanics of neurulation depend, like gastrulation, on changes of cell packing and cell shape, and Figure 21–11 shows how the cytoskeleton can be organized to bring about cell shape changes that can make an epithelium roll up into a tube.

Figure 21–9 Convergent extension and its cellular basis. (A) The pattern of convergent extension in the marginal zone of a gastrula as viewed from the dorsal aspect. *Blue arrows* represent convergence toward the dorsal midline, *red arrows* represent extension of the anteroposterior axis. The simplified diagram does not attempt to show the accompanying movement of involution, whereby the cells are tucking into the interior of the embryo. (B) Schematic diagram of the cell behavior that underlies convergent extension. The cells form lamellipodia, with which they attempt to crawl over one another. Alignment of the lamellipodial movements along a common axis leads to convergent extension. The process is presumably cooperative because cells that are already aligned exert forces that tend to align their neighbors in the same way. (B, after J. Shih and R. Keller, *Development* 116:901–914, 1992.)

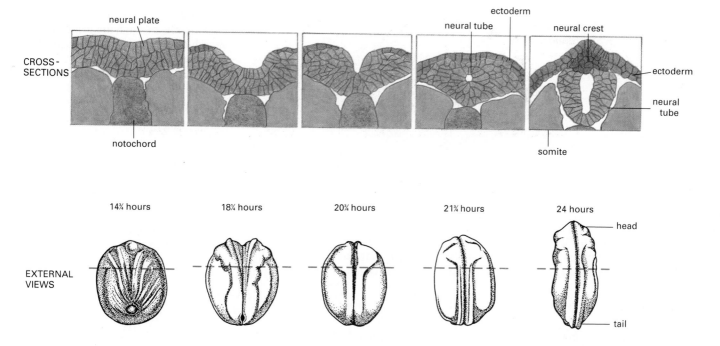

CROSS-SECTIONS

neural plate

notochord

ectoderm

neural tube

neural crest

ectoderm

neural tube

somite

EXTERNAL VIEWS

14¾ hours · 18¼ hours · 20¾ hours · 21¾ hours · 24 hours

head

tail

Neurulation is induced by an interaction with the underlying notochord and the mesoderm adjacent to it. If a piece of such dorsal mesoderm is taken from the area just beneath the future neural tube of one gastrulating amphibian embryo and implanted directly beneath the ectoderm of another gastrulating embryo in, say, the belly region, the ectoderm in that region will thicken and roll up to form a piece of misplaced neural tube.

Along the line where the neural tube pinches off from the future epidermis, a number of ectodermal cells break loose from the epithelium and migrate as individuals out through the mesoderm. These are the cells of the **neural crest**; they will form almost all of the peripheral nervous system (including most of the sensory and all of the sympathetic ganglia and the Schwann cells that make the myelin sheaths of peripheral nerves) as well as the pigment cells of the skin. In the head many of the neural crest cells will differentiate into cartilage, bone, and other connective tissues, which elsewhere in the body arise from the mesoderm. This is one of several instances that run counter to the general scheme in which the three germ layers give rise to cells in three corresponding concentric layers of the adult body.

The sense organs, by which light, sound, smell, and so forth impinge on the nervous system, also have ectodermal origins: some derive from the neural tube, some from the neural crest, and some from the exterior layer of ectoderm (see Figure 21–102). The retina, for example, originates as an outgrowth of the brain and so is derived from cells of the neural tube, while the olfactory cells of the nose differentiate directly from the ectodermal epithelium lining the nasal cavity.

The Mesoderm on Either Side of the Body Axis Breaks Up into Somites from Which Muscle Cells Derive [13]

On either side of the newly formed neural tube lies a broad expanse of mesoderm (Figure 21–12). The thicker, more medial and dorsal part of this mesoderm gives rise to the muscular and skeletal tissues of the central body axis. It consists at first of a single continuous slab of tissue on each side of the body. To form the repetitive series of vertebrae and segmental muscles, this slab soon breaks up into separate blocks, or **somites** (Figure 21–13). The somites form one after another, starting in the head and ending at the tail (Figure 21–13). Segmentation is accompanied by changes in the connections between the mesoderm cells, but the

Figure 21–10 **Neural tube formation in *Xenopus*.** The external views are from the dorsal aspect. The cross-sections are cut in a plane indicated by the broken lines. (After T.E. Schroeder, *J. Embryol. Exp. Morphol.* 23:427–462, 1970. © Company of Biologists Ltd.)

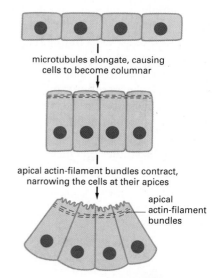

microtubules elongate, causing cells to become columnar

apical actin-filament bundles contract, narrowing the cells at their apices

apical actin-filament bundles

Figure 21–11 **The bending of an epithelium through cell shape changes mediated by microtubules and actin filaments.** The diagram is based on observations of neurulation in newts and salamanders, where the epithelium is only one cell layer thick. As the apical ends of the cells become narrower, their upper-surface membrane becomes puckered.

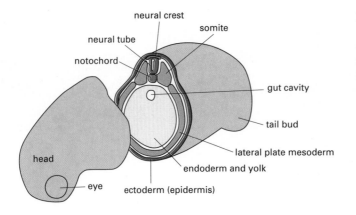

Figure 21–12 **A cross-section (schematic) through the trunk of an amphibian embryo after the neural tube has closed.** (After T. Mohun, R. Tilly, R. Mohun, and J.M.W. Slack, *Cell* 22:9–15, 1980. © Cell Press.)

mechanism that controls the regular spacing of the clefts that separate one somite from the next remains a mystery (although it is known that the physical process of somite formation is foreshadowed by a segmental pattern of expression of certain genes).

Each somite corresponds to one unit in the final sequence of articulated elements. The bulk of the somite forms the skeletal muscles of the segment, while a subset of its cells go to form the corresponding vertebrae and other connective tissues such as dermis. The somites are also the source of almost all skeletal muscle cells elsewhere in the body: these derive from precursors that migrate away from the somites before differentiating overtly.

Changing Patterns of Cell Adhesion Molecules Regulate Morphogenetic Movements [14]

The tissue movements in the embryo go hand in hand with changes in the chemical characters of the cells. By switching on production of a cytoskeletal protein, for example, a cell may alter its shape or the way it moves. By changing the set of adhesion molecules it displays on its surface, it may break old attachments and make new ones. Cells in one region may develop surface properties that make them cohere with one another and become segregated from a neighboring group of cells whose surface chemistry is different.

Classical experiments on early amphibian embryos showed that the effects of selective cell-cell adhesion can be so powerful that they bring about an approximate reconstruction of the normal structure even after the cells have been artificially dissociated into a random mixture (Figure 21–14). As discussed in Chapter 19, studies on chick and mouse embryos suggest that this behavior de-

Figure 21–13 **Somite formation in *Xenopus*.** (A) Photograph of embryos at three successive stages, seen in side view and stained with a muscle-specific antibody to show the progress of somite formation. (B) Explanatory drawings. A side view of the embryo is shown at the top; the broken line indicates the plane of the horizontal section shown below. The bottom drawing is a schematic high-magnification view of the mesoderm cells in the process of rearranging to form somites. In *Xenopus* the future somite cells are initially all oriented at right angles to the body axis and then rotate in groups during somite formation. The main part of each somite will form muscle and is called the *myotome;* the inner part facing the notochord is the source of the cells that form the vertebrae and ribs and is called the *sclerotome;* the outer, dorsal part (in higher vertebrates, though not in *Xenopus*) will contribute to the dermis (the connective tissue of the skin) and is called the *dermatome.*

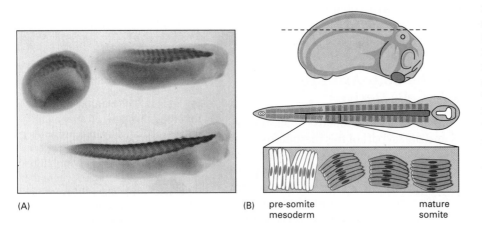

(A)

(B) pre-somite mesoderm   mature somite

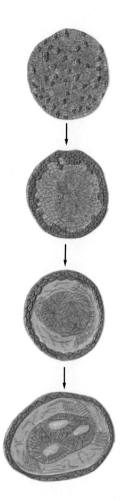

Figure 21–14 (*left*) **Sorting out.** Cells from different parts of an early amphibian embryo will sort out according to their origins. In the classical experiment shown here mesoderm cells, neural plate cells, and epidermal cells have been disaggregated and then reaggregated in a random mixture. They sort out into an arrangement reminiscent of a normal embryo, with a "neural tube" internally, epidermis externally, and mesoderm in between. (Modified from P.L. Townes and J. Holtfreter, *J. Exp. Zool.* 128:53–120, 1955.)

Figure 21–15 (*right*) **Cadherins in the early embryo.** The changing patterns of expression of three cadherins at successive stages in the early chick or mouse embryo, as seen in cross-sections through the developing neural tube and somites. Cells expressing the same type of cadherins tend to stick to each other and to segregate from other cells. The pattern of cadherins thus helps to regulate the pattern of morphogenetic movements involved in formation of the neural tube, notochord, somites, neural crest, and sclerotomes. (After M. Takeichi, *Trends Genet.* 8:213–217, 1987.)

ectoderm (neural plate)

mesoderm

endoderm

neural crest

epidermis

neural tube

somite

notochord

sclerotome

KEY
cadherin type

E   N   E+N   E+P   N+P

pends, at least in part, on a family of homologous Ca^{2+}-dependent cell-cell adhesion glycoproteins—the *cadherins*. These molecules and other, Ca^{2+}-independent cell-cell-adhesion molecules such as N-CAM are differentially expressed in the various tissues of the early embryo, and antibodies against them interfere with the normal selective adhesion between cells of a similar type.

Changes in the patterns of expression of the various cadherins correlate closely with the changing patterns of association among cells during gastrulation, neurulation, and somite formation (Figure 21–15); these transformations of the early embryo may be regulated and driven in part by the cadherin pattern. In particular, cadherins appear to have a major role in controlling the formation and dissolution of epithelial sheets and clusters of cells. They not only glue one cell to another, but also provide anchorage for intracellular actin filaments at the sites of cell-cell adhesion (discussed in Chapter 19): in this way they help to regulate the pattern of stresses and movements in the developing tissue according to the pattern of adhesions.

Besides sticking to one another, cells can stick to components of the extracellular matrix such as fibronectin and laminin. These adhesions are typically mediated by *integrins*, which, like cadherins, serve as transmembrane linkers between sites of attachment on the outside of the cell and actin filaments inside. Cell-matrix interactions of this sort are important for the movements of certain special classes of cells that lose adhesions to their neighbors and migrate as individuals through the embryo by crawling through the spaces between other cells. As a result of such invasions, to be discussed next, most tissues in the adult vertebrate body include admixtures of cells derived from widely separate parts of the early embryo.

Morphogenetic Movements and the Shaping of the Body Plan

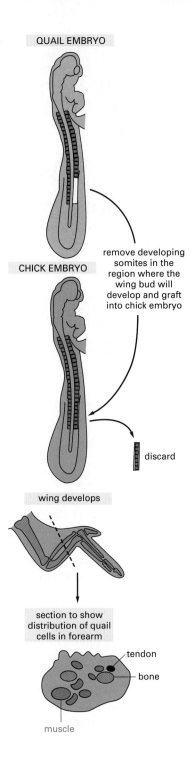

QUAIL EMBRYO

CHICK EMBRYO

remove developing somites in the region where the wing bud will develop and graft into chick embryo

discard

wing develops

section to show distribution of quail cells in forearm

tendon

bone

muscle

Figure 21–16 Migratory origin of limb muscle cells. If quail somite cells are substituted for the somite cells of a chick embryo at 2 days of incubation and the wing of the chick is sectioned a week later, it is found that the muscle cells in the chick wing derive from the transplanted quail somites.

Embryonic Tissues Are Invaded in a Strictly Controlled Fashion by Migratory Cells [11, 15, 16]

We have already mentioned two classes of migratory cells—those of the neural crest and those that leave the somites to give rise to skeletal muscle. Other important migrants are the precursors of the blood cells, of the germ cells, and of many groups of neurons within the central nervous system.

Cell migrations can be traced by marking the cells at the beginning of their journey, using either a nontoxic dye or, better, a heritable genetic label. Much of our knowledge has come from studies in which cells are grafted from quail embryos into chick embryos. Although the quail is similar in most respects to the chick, its cells can be distinguished in histological sections by a large, strongly staining mass of heterochromatin associated with the nucleolus. This nucleolar marker makes it possible to identify grafted cells that have migrated from the site where they were implanted. For example, if quail somite tissue is substituted for the somite tissue of a very young chick embryo before the limb buds appear, all the muscle cells in the limbs that subsequently develop have a quail origin (Figure 21–16). Evidently the future muscle cells migrate from the somites into the prospective wing region and remain there, inconspicuously mixed with the connective-tissue cells of the limb bud, until the time comes for them to differentiate.

In a similar way one can trace the dispersal of cells from the neural crest. These migrate along certain specific pathways through the embryo (Figure 21–17) and settle in precisely defined locations. As a migrant cell travels through the embryo, it repeatedly extends projections that probe its immediate surroundings (Figure 21–18), testing for subtle cues to which it is particularly sensitive by virtue of its specific assortment of cell-surface receptor proteins. Inside the cell these receptor proteins are connected to the cytoskeleton, which moves the cell along.

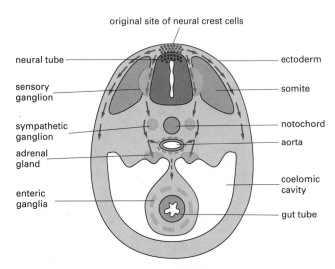

original site of neural crest cells

neural tube

ectoderm

sensory ganglion

somite

sympathetic ganglion

notochord

aorta

adrenal gland

enteric ganglia

coelomic cavity

gut tube

Figure 21–17 The main pathways of neural crest cell migration. A chick embryo is shown in a schematic cross-section through the middle part of the trunk. The cells that take the pathway just beneath the ectoderm will form pigment cells of the skin; those that take the deep pathway via the somites will form sensory ganglia, sympathetic ganglia, and parts of the adrenal gland. The enteric ganglia, in the wall of the gut, are formed from neural crest cells that migrate along the length of the body, originating from either the neck region or the sacral region. (See also Figure 19–22.)

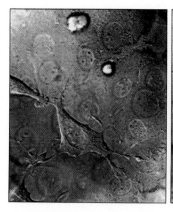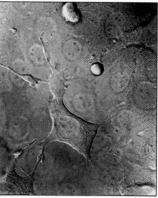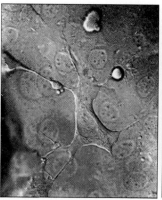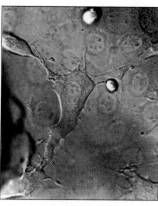

Some extracellular matrix materials, such as fibronectin, provide adhesive sites that help the cell to advance; others, such as chondroitin sulfate proteoglycan, inhibit locomotion and repel immigration. The nonmigrant cells along the pathway may likewise have inviting or repellent surfaces, or may even extend filopodia that touch the migrant cell and affect its behavior. An incessant tug-of-war between opposing tentative attachments made by the migrant cell leads to a net movement in the most favored direction until the cell finds a site where it can form a lasting attachment. Other factors such as chemotaxis and interactions among the migratory cells may also play an important part.

Yet another means of controlling the distribution of migrant cells is through regulation of their survival and proliferation. Germ cells, blood cell precursors, and pigment cells derived from the neural crest all appear to be governed in this respect by the same basic control mechanism. This involves a transmembrane receptor, called the *Kit* protein, in the membrane of the migrant cells and a ligand, called the *Steel* factor, produced by the cells of the tissue through which the cells migrate and/or in which they come to settle. Individuals with mutations in the genes for either of these proteins are deficient in their pigmentation, their supply of blood cells, and their production of germ cells (Figure 21–19). The Steel factor appears to be required in a membrane-bound form in order to activate Kit correctly and enable all these cell types to survive and proliferate.

Figure 21–18 **A neural crest cell migrating.** This series of photographs of a living zebra fish embryo, viewed by interference contrast optics, shows a neural crest cell putting out tentative processes in several directions and withdrawing them before finally setting off in a ventral direction (downward in the final photograph). The photographs are taken at intervals of about 5 minutes. (Courtesy of Suresh Jesuthasan.)

The Vertebrate Body Plan Is First Formed in Miniature and Then Maintained as the Embryo Grows [9]

The embryo at the stage when the somites are forming and neural crest cells are setting off on their migrations is typically a few millimeters long and consists of about 10^5 cells. While we have been speaking thus far mainly of *Xenopus,* the scale and general form are much the same for a fish, a salamander, a chick, or a human (see Figure 1–36). Later these species of embryo will grow to be very different

Figure 21–19 **Effect of mutations in the *kit* gene.** Both the baby and the mouse are heterozygous for a loss-of-function mutation that leaves them with only half the normal quantity of *kit* gene product. In both cases pigmentation is defective because pigment cells depend on the *kit* product as a receptor for a survival factor. (Courtesy of R.A. Fleischman, from *Proc. Natl. Acad. Sci. USA* 88:10885–10889, 1991. © 1991 Macmillan Magazines Ltd.)

in size and shape, but at this early stage they all share the basic vertebrate body plan. The central nervous system is represented by the neural tube, with an enlargement at one end for the brain; the gut and its derivatives, by a tube of endoderm; the segments of the trunk, by the somites; the other connective tissues, including the vascular system, by the more peripheral unsegmented mesoderm; and the epidermal layer of the skin, by the ectoderm. During subsequent development all of these components will enlarge, by a factor of as much as a hundred or more in length or a million or more in volume and cell number. But the same basic organization of the body will be preserved.

Summary

The eggs of most animals are large cells, containing stores of nutrients and other cell components specified by the maternal genome. In amphibians the first major movement after fertilization is a rotation of the cortex of the egg relative to its core. The asymmetry created by this rotation, together with the original asymmetry in the distribution of the contents of the egg before fertilization, defines the future anteroposterior and dorsoventral axes of the body. During the subsequent cleavage divisions the egg subdivides into many smaller cells, but no growth occurs.

A cavity soon develops in the interior of the embryo, while the surrounding cells become organized into an epithelial sheet. Part of the epithelium then invaginates, transforming the embryo into a three-layered structure with an internal epithelial tube of endoderm, an external epithelial covering of ectoderm, and a middle layer of mesodermal cells that have broken loose from the original epithelial sheet. In this process of gastrulation the epithelial cells actively change their packing, and this is thought to provide a major driving force for the movements.

The endoderm will form the lining of the gut and its derivatives, the ectoderm will form the epidermis and the nervous system, and the mesoderm will form muscles, connective tissues, vascular system, and urogenital tract. The development of all these structures depends on interactions between the three germ layers and involves further cell movements. The dorsal mesoderm, for example, induces the overlying ectoderm to thicken, roll up, and pinch off to form the neural tube and neural crest. In the middle of the dorsal mesoderm a rod of specialized cells called the notochord elongates to form the central axis of the embryo. The long slabs of mesoderm on either side of the notochord become segmented into somites, from which the vertebrae and skeletal muscles will be derived. At several sites migrant cells, such as those of the neural crest, break loose from their original neighbors and migrate through the embryo to colonize new sites. Specific cell-adhesion molecules, such as cadherins and integrins, help to guide the migrations and control the selective cohesion of cells in epithelia.

Cell Diversification in the Early Animal Embryo [17, 18]

A fertilized egg may develop into a daisy or an oak tree, a sea urchin or a human being. The outcome is governed by the genome: the linear sequence of A, G, C, and T nucleotides in the DNA of the organism must direct the production of a variety of chemically different cell types arranged in a precise pattern in space. Developmental biology aims to explain how. The whole discussion of this problem, in this and subsequent sections, rests on one fundamentally important fact: the cells in the body inherit the same genome from the egg. No matter how different they may appear—in muscle, bone, or nerve, in root, stem, or leaf—they all contain the same set of genetic instructions.

One of the earliest and most powerful demonstrations of this principle came from experiments on nuclear transplantation using amphibian eggs (Figure 21–20). A typical amphibian egg is so large that, using a fine glass pipette, one can readily inject into it a nucleus taken from another cell. The nucleus of the egg itself is destroyed beforehand by ultraviolet irradiation. The egg is activated to begin development by the act of pricking with the fine pipette used to inject the transplanted nucleus. Thus one can test whether the nucleus from a differentiated somatic cell contains a complete genome equivalent to that of a normal fertilized egg and equally serviceable for development. The answer is yes: a complete swimming tadpole can be produced, for example, from an egg whose own nucleus has been replaced by a nucleus derived from a keratinocyte cell from an adult frog's skin or by a nucleus from a frog red blood cell. These experiments admittedly have limitations. They have been successful only with nuclei from a limited range of differentiated cell types and in only a few species. But there is now an overwhelming body of evidence pointing to the same conclusion. With just a few exceptions (see Figure 23–37), the genome remains intact during development. Genes can be switched on or off, and the cells of the body differ not because they contain different genes but because they *express* different genes. In Chapter 9 we examine the intracellular mechanisms for regulating gene expression. In this chapter we have to consider not only how the differences between cells originate, but how they are coordinated in space and time within a multicellular organism. The present section discusses how the first steps of cell diversification are coordinated in early embryos, taking frog and mouse as examples.

Initial Differences Among *Xenopus* Blastomeres Arise from the Spatial Segregation of Determinants in the Egg [2, 17, 19]

In most animal and plant species the egg itself is chemically asymmetrical, with certain components concentrated in specific regions of the cytoplasm or mem-

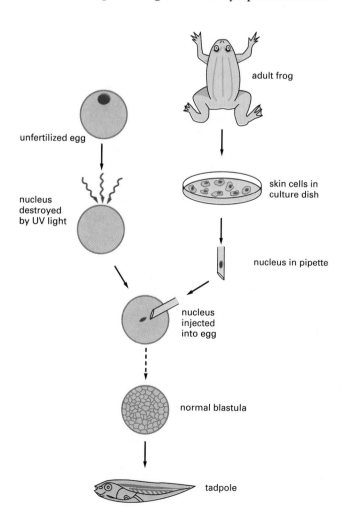

unfertilized egg

adult frog

nucleus destroyed by UV light

skin cells in culture dish

nucleus in pipette

nucleus injected into egg

normal blastula

tadpole

Figure 21–20 Nuclear transplantation. Diagram of an experiment showing that the nucleus of a differentiated cell from the skin of an adult frog contains all the genetic material necessary to control the formation of an entire tadpole. The broken arrow in the lower part of the figure is to indicate that, to give the transplanted genome time to adjust to an embryonic environment, a further transfer step is required in which one of the nuclei is taken from the early embryo that begins to develop and is put back into a second enucleated egg. (Modified from J.B. Gurdon, Gene Expression During Cell Differentiation. Oxford, UK: Oxford University Press, 1973.)

brane. As a result, there are differences from the outset between the cells that form by cleavage because they receive different portions of the localized materials. The importance of such *localized determinants* in the egg varies from species to species. It is traditional to distinguish two theoretical extremes: in *mosaic* development the whole future pattern of the body is delineated by localized determinants in the egg, and subsequent cell-cell interactions count for nothing; in *regulative* development localized determinants in the egg count for nothing, and the body pattern is generated entirely by subsequent cell-cell interactions. In reality, most higher animals and plants lie between these extremes. None, so far as is known, is truly mosaic—regulative interactions always play an important part; mammalian eggs, as we shall see, appear to be entirely regulative. *Xenopus* represents a typical intermediate case.

The asymmetries of the *Xenopus* egg are manifest in several ways—in the eccentric location of the nucleus, in the distribution of yolk and pigment granules, in the cytoskeleton, and, perhaps most significantly, in the distribution of certain specific mRNAs. The egg asymmetries endow the early blastomeres with different characters according to whether they are animal or vegetal, dorsal or ventral. Treatments such as centrifugation or ultraviolet irradiation that displace the contents of the uncleaved egg or prevent the cortical rotation that usually follows fertilization lead to drastic disturbances of the embryonic body plan, and equally drastic disturbances result if the early blastomeres are artificially rearranged.

Inductive Interactions Generate New Types of Cells in a Progressively More Detailed Pattern [20]

The initial differences between the early blastomeres define only the crude beginnings of the pattern of the embryo. To generate the full range of cell types, the blastomeres must interact with one another. If the early *Xenopus* embryo is placed in a medium devoid of Ca^{2+} and Mg^{2+}, the blastomeres lose their cohesiveness and can then be separated and allowed to develop on their own; some go on to develop features characteristic of ectoderm, while others develop features characteristic of endoderm, but none of them switches on expression of genes characteristic of mesoderm, such as the muscle-specific actin gene. But when cells from the animal pole of a blastula are placed next to vegetal cells, some of the animal pole cells are diverted from the ectodermal pathway of development into the mesodermal pathway (Figure 21–21). The switching of cells from one pathway into another by the influence of an adjacent group of cells is called

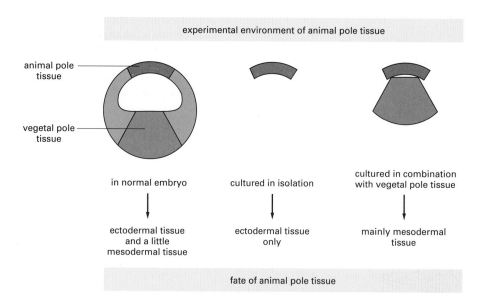

Figure 21–21 **Mesoderm induction in** ***Xenopus.*** Cells from the animal pole of a blastula, normally destined to form only ectoderm, will form mesodermal tissues if they are cultured in conjunction with cells from the vegetal pole. In normal development an inductive interaction of this sort presumably occurs at an earlier stage; the equatorial region of the blastula is already capable of forming mesodermal tissues when it is cultured in isolation.

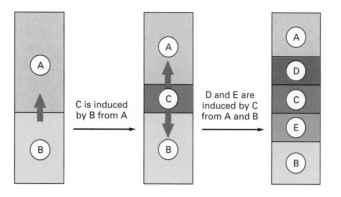

Figure 21–22 **Patterning by sequential induction.** A series of inductive interactions can generate many kinds of cells, starting from only a few.

induction. During normal development, inductive interactions may occur between cells that have been adjacent from the outset—as in mesoderm induction—or between cells that are brought together through morphogenetic movements such as gastrulation. By a series of successive inductions, it is possible to generate many different kinds of cells from interactions between a few kinds (Figure 21–22).

As we emphasized earlier, asymmetries in the *Xenopus* egg define not only the animal-vegetal axis, and thereby the partitioning of the embryo into ectoderm, mesoderm, and endoderm, but also the dorsoventral and anteroposterior axes of the body. For the organization of the dorsoventral axis, an inductive mechanism again seems to operate. Grafting experiments indicate that, while all vegetal blastomeres can induce mesoderm, they do not all do so in the same way: the *dorsal* vegetal blastomeres are unique in that they induce the cells above them to take on the special character of Spemann's Organizer. The Organizer in its turn, as we saw earlier, produces a signal that induces an array of specializations in the mesoderm next to it. Later still, the pattern created in the mesoderm will induce patterns of local specialization in the ectoderm and endoderm that it contacts.

Thus there seem to be at least three inductive signals at work in the earliest stages of *Xenopus* development: from ventral vegetal blastomeres, from dorsal vegetal blastomeres, and from the Organizer (Figure 21–23). What are these signals in chemical terms? Members of at least four families of secreted signaling proteins seem to be involved. Although their precise roles in normal development are still not clear, all are thought to be present in the early *Xenopus* embryo, and all have dramatic inductive effects when supplied artificially. For at least two of the four, artificial blockade of function produces embryos with major parts of the body missing (Figure 21–24).

Such observations do not explain, however, how the localization of the inductive signals that pass between blastomeres in the *Xenopus* embryo is governed by the pattern of asymmetries in the uncleaved egg. In the case of the *Vg1 protein*, which is a member of the TGF-β superfamily of secreted signaling factors, one can glimpse how this may come about. A store of maternal mRNA coding for the protein is localized in the vegetal part of the egg before fertilization. It is thought that the protein is produced in precursor form in the vegetal regions, and

Figure 21–23 **The three-signal model for mesoderm induction in the early *Xenopus* embryo.** At least three signals, acting as shown, seem to be needed to explain the results of grafting experiments. Each "signal" may actually be a complex combination of signaling molecules. (After J. Slack, From Egg to Embryo, 2nd ed. Cambridge, UK: Cambridge University Press, 1991.)

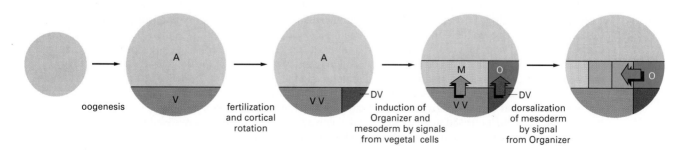

oogenesis → fertilization and cortical rotation → induction of Organizer and mesoderm by signals from vegetal cells → dorsalization of mesoderm by signal from Organizer

Cell Diversification in the Early Animal Embryo

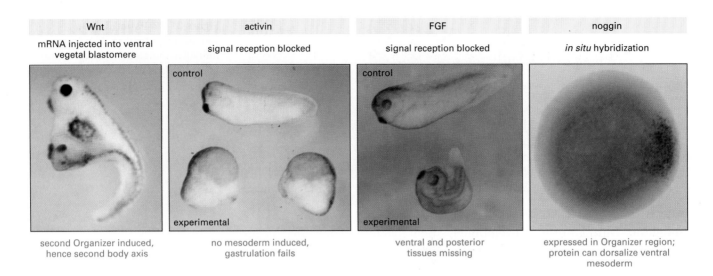

Wnt	activin	FGF	noggin
mRNA injected into ventral vegetal blastomere	signal reception blocked	signal reception blocked	*in situ* hybridization
second Organizer induced, hence second body axis	no mesoderm induced, gastrulation fails	ventral and posterior tissues missing	expressed in Organizer region; protein can dorsalize ventral mesoderm

Figure 21–24 Some signaling molecules involved in mesoderm induction in *Xenopus*. The result of one representative experiment is shown for each of four classes of factors. Although all four classes of factors can have powerful effects on mesoderm induction, their exact roles in relation to the three-signal model (Figure 21–23) are not yet certain. The Wnt, activin, and FGF (fibroblast growth factor) families of factors are well known as cell-cell signaling molecules in other contexts; activin (like Vg1—see Figure 21–25) belongs to the TGF-β superfamily of growth factors. Reception of activin or FGF signals can be blocked by injecting mRNA coding for a defective form of the corresponding receptor protein, which lacks the intracellular domain and interferes with the function of the normal receptor. (Photographs from S. Sokol et al., *Cell* 67:741–752, 1992. © Cell Press; A. Hemmati-Brivanlou and D.A. Melton, *Nature* 359:609–614, 1992. © 1992 Macmillan Magazines Ltd.; E. Amaya, T.J. Musci, and M.W. Kirschner, *Cell* 66:257–270, 1991. © Cell Press; and W.C. Smith and R.M. Harland, *Cell* 70:829–840, 1992. © Cell Press).

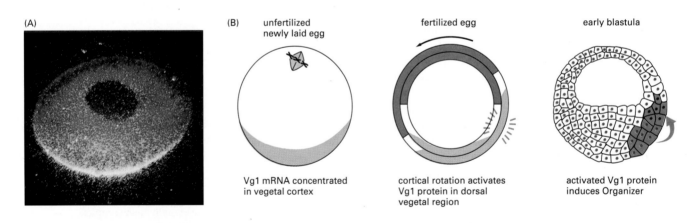

(A)

(B) unfertilized newly laid egg — Vg1 mRNA concentrated in vegetal cortex

fertilized egg — cortical rotation activates Vg1 protein in dorsal vegetal region

early blastula — activated Vg1 protein induces Organizer

Figure 21–25 Localization of Vg1 and its suspected role as an inducer in the *Xenopus* embryo. (A) *In situ* hybridization with a probe for Vg1 mRNA, showing its localization in the vegetal cortical region of the oocyte (the future egg). (B) Diagrams illustrating a hypothesis as to how Vg1 acts. Vg1 mRNA is synthesized in the oocyte and becomes localized, by unknown mechanisms, in the vegetal cortical regions of the cell. In the same way as for other TGF-β superfamily members, the active form of Vg1 protein is a fragment cleaved from the full-length precursor. The control of the activating cleavage step is not understood. When mRNA coding for full-length Vg1 is injected into an early embryo, very little of the active fragment is produced and no effect on embryo patterning is seen. But if the mRNA is modified to code for a precursor that is readily cleaved to produce the Vg1 active fragment, the effects are dramatic: an entire body axis can be induced, in a way that suggests that the Vg1 fragment is mimicking the signal that normally comes from dorsal vegetal blastomeres and induces development of the Organizer. According to one proposal, Vg1 acts as this signal in normal development, and the production of the active Vg1 fragment is localized to dorsal vegetal blastomeres by a two-step process. First, the mRNA is delivered to the vegetal end of the egg; then the cortical rotation that follows fertilization creates special conditions in the dorsal part of the vegetal cortex, such that the precursor protein is cleaved there to produce the active fragment. This then is released from the dorsal vegetal blastomeres to induce formation of an Organizer. (A, courtesy of Douglas Melton.)

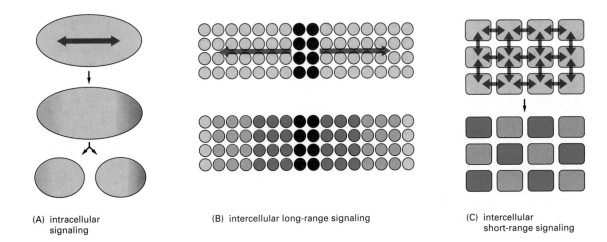

(A) intracellular
signaling

(B) intercellular long-range signaling

(C) intercellular
short-range signaling

that it may be activated in the dorsal vegetal region and released from dorsal veg-etal blastomeres to induce the Organizer (Figure 21–25).

A Simple Morphogen Gradient Can Organize a Complex Pattern of Cell Responses [21]

There are many ways in which signals passing from one part of an embryo to another can control pattern formation (Figure 21–26). The Organizer exemplifies one strategy of particular interest: a small patch of tissue in a specific region acquires a specialized character and becomes the source of a signal that spreads into neighboring tissue and controls its behavior. The signal, for example, may take the form of a diffusible molecule secreted from the signaling center. Suppose that this substance is slowly degraded as it diffuses through the neighboring tissue. The steady-state concentration then will be high near the source and decrease gradually with increasing distance, so that a concentration gradient is established (Figure 21–27). Cells at different distances from the source will be exposed to different concentrations and may become different as a result. A substance such as this, whose concentration is read by cells to discover their position relative to a certain landmark or beacon, is termed a **morphogen**. Morphogen gradients are thought to be a common way of providing cells with positional information or controlling their pattern of differentiation, although there are still only a few cases where a morphogen has been identified chemically.

How do cells respond to a morphogen gradient? The concentration of a diffusible morphogen should be smoothly graded, but many of the important specializations in development are discrete: there is no graded series of mature kinds of cells intermediate between cartilage and muscle, or bone and nerve, for example. In theory, sharp distinctions can arise in a population of initially uniform cells through a *threshold* in their response to a smoothly graded signal. If there is a positive feedback in each responding cell that amplifies the effect of a small increment in the signal, cells exposed to only slightly different intensities of the signal can be launched on radically different courses of development according to whether their exposure is above or below a certain threshold level. If there are several thresholds of response to one signal, a single morphogen can control the pattern of several different cell choices. It has been shown, for example, that

Figure 21–26 **Three kinds of signaling for three styles of pattern formation.** (A) Intracellular signals can organize cytoplasmic determinants in the egg, which are inherited by different blastomeres when the egg divides. (B) Long-range diffusible signals from a signaling center can direct the global pattern of cell specialization in the surrounding tissue. (C) Short-range, cell-cell contact interactions can create a fine-grained mosaic of cells in different states; they often play a crucial part in deciding the final step of differentiation in intricate tissues such as the retina and other sensory epithelia.

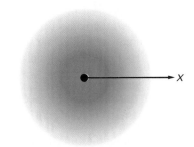

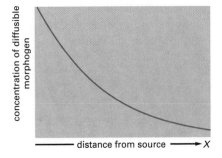

Figure 21–27 **A morphogen gradient.** If a substance is produced at a point source and is degraded as it diffuses from that point, a concentration gradient results with a maximum at the source. The substance can serve as a morphogen, whose local concentration controls the behavior of cells according to their distance from the source.

when cells from the animal pole of an early *Xenopus* embryo are exposed to the signaling molecule activin (see Figure 21–24), they will develop as epidermis if the activin concentration is low, as muscle if it is a little higher, and as notochord if it is a little higher still. The normal role of activin in the intact *Xenopus* embryo, however, is uncertain, and the nature of the signals emanating from Spemann's Organizer is still unclear.

Cells Can React Differently to a Signal According to the Time When They Receive It: The Role of an Intracellular Clock [22]

As development proceeds, embryonic cells generally change their character even if their environment is unchanged. If cells taken from the animal pole of a *Xenopus* blastula, for example, are kept in isolation *in vitro*, they will spontaneously differentiate into epidermis at roughly the normal time. In this sense, the cells behave as though governed by some sort of intracellular clock. Because cells are spontaneously changing their internal state, they may respond differently to an inductive signal according to the time when they receive it. If a fragment of animal pole epithelium is taken from an early gastrula and grafted over the eye rudiment of a later embryo, for example, it will be induced to differentiate (inappropriately) into a piece of tissue resembling neural tube; if it is allowed to age for a few hours *in vitro* before grafting into the same environment, it will be induced to differentiate (appropriately) into a lens; if it is cultured *in vitro* for a longer period still, it loses competence to respond to the inductive influence from the eye rudiment in either of these ways.

There is an important general lesson here: cellular diversity and spatial patterning can arise from a simple *unchanging* inductive signal acting on a succession of otherwise identical cells at different times (Figure 21–28). We have seen, for example, that the parts of the central body axis are formed sequentially during gastrulation, with anterior parts involuting around the blastopore lip first and posterior parts last. According to one theory, the difference in the age at which the cells pass the dorsal lip and are acted on by Spemann's Organizer could be the source of the differences of cell character between the anterior and posterior parts of the mesoderm and endoderm and therefore of the body as a whole.

Thus the general strategy of pattern formation can be summarized as follows: (1) patterns begin from simple asymmetries, (2) details are filled in sequentially through inductive cell-cell interactions, and the pattern of cell diversification that results depends both on (3) the positional signals between cells and on (4) intracellular programs that change a cell's response to these signals with time.

In different species these four basic elements may be combined in different ways. We now consider the special case of the early mammalian embryo, which has some remarkable regulative properties.

In Mammals the Protected Uterine Environment Permits an Unusual Style of Early Development [9, 23]

The mammalian embryo does many things differently from other animals. Developing in the protected environment of the uterus, it does not have the same need as the embryos of most other species to complete the early stages of development rapidly. Moreover, the development of a placenta quickly provides nutrition from the mother, so that the egg does not have to contain large stores of raw materials such as yolk. The egg of a mouse has a diameter of only about 80 μm and therefore a volume about 2000 times smaller than that of a typical amphibian egg. Its cleavage divisions occur no more quickly than the divisions of many ordinary somatic cells, and gene transcription has already begun by the two-cell stage. Furthermore, while the later stages of mammalian development are fundamentally similar to those of other vertebrates such as *Xenopus*, mam-

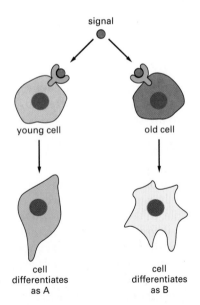

Figure 21–28 The significance of timing. An unchanging signal acting on otherwise similar cells at different ages can evoke different responses. Spatial patterns can be produced in this way by allowing an unchanging signal to act at different times on different members of an array of initially similar cells.

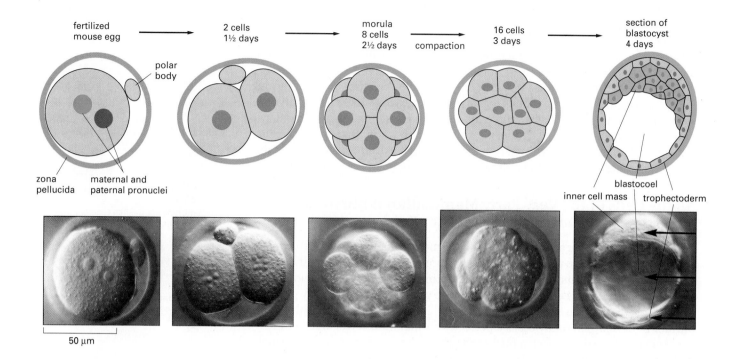

fertilized mouse egg → 2 cells 1½ days → morula 8 cells 2½ days → compaction → 16 cells 3 days → section of blastocyst 4 days

polar body

zona pellucida   maternal and paternal pronuclei

blastocoel
inner cell mass   trophectoderm

50 μm

mals begin by taking a large developmental detour to generate a complicated set of structures—notably the amniotic sac and the placenta—that enclose and protect the embryo proper and provide for the exchange of metabolites with the mother. These structures, like the rest of the body, derive from the fertilized egg but are called *extraembryonic* because they are discarded at birth and form no part of the adult.

The early stages of mouse development are summarized in Figure 21–29. The egg is surrounded initially by a transparent cell coat, the *zona pellucida*. Upon fertilization, the egg cleaves within this coat to form a mulberry-shaped cluster of cells called the *morula*. Sometime between the 8-cell and 16-cell stages, the surface of the morula becomes smoother and more nearly spherical as the cells change their cohesiveness and become compacted together (Figure 21–30), with tight junctions forming between the outer cells and sealing off the interior of the morula from the external medium. Soon after, the internal intercellular spaces enlarge to create a central fluid-filled cavity—the blastocoel. At this stage the morula is said to have become a *blastocyst*. The cells of the blastocyst form a spherical shell enclosing the blastocoel, with one pole distinguished by a thicker accumulation of cells. As shown in Figure 21–29, the entire outer cell layer is the *trophectoderm*; the cluster of cells inside the trophectoderm at the thicker pole is called the *inner cell mass*.

Figure 21–29 The early stages of mouse development. (Photographs courtesy of Patricia Calarco, from G. Martin, *Science* 209:768–776, 1980. Copyright 1980 the AAAS.)

Figure 21–30 Scanning electron micrographs of the early mouse embryo. The zona pellucida has been removed. (A) Two-cell stage. (B) Four-cell stage (a polar body is visible in addition to the four blastomeres—see Figure 20–16). (C) Eight-to-sixteen-cell morula—compaction occurring. (D) Blastocyst. (Courtesy of Patricia Calarco; D, from P. Calarco and C.J. Epstein, *Dev. Biol.* 32:208–213, 1973.)

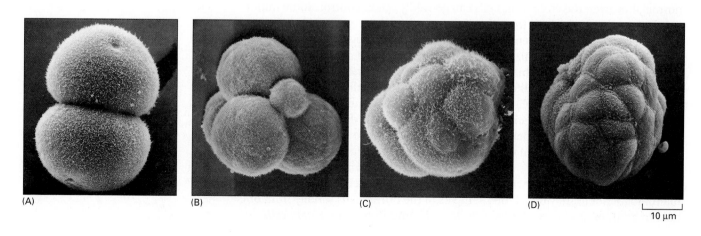

(A)   (B)   (C)   (D)

10 μm

Cell Diversification in the Early Animal Embryo

The whole of the embryo proper is derived from the inner cell mass. The trophectoderm is the precursor of the placenta and is the earliest component of the system of extraembryonic structures. Once the zona pellucida has been shed, the cells of the trophectoderm come into close contact with the wall of the uterus, in which the embryo becomes implanted. Meanwhile the inner cell mass grows and begins to differentiate. Part of it gives rise to some further extraembryonic structures, such as the yolk sac, while the rest of it goes on to form the embryo proper by processes of gastrulation, neurulation, and so on, that are largely homologous to those seen in other vertebrates, although extreme distortions of the geometry sometimes make the homology hard to discern.

All the Cells of the Very Early Mammalian Embryo Have the Same Developmental Potential [24]

Up to the eight-cell stage, each cell of the early mammalian embryo can form any part of the later embryo or adult. If the early embryo is split in two, a pair of identical twins can be produced—two complete normal individuals from a single cell. Similarly, if one of the cells in a two-cell mouse embryo is destroyed by pricking it with a needle and the resulting "half-embryo" is placed in the uterus of a foster mother to develop, in many cases a perfectly normal mouse will emerge.

Conversely, two eight-cell mouse embryos can be combined to form a single giant morula, which then develops into a mouse of normal size (Figure 21–31). Such creatures, formed from aggregates of genetically different groups of cells, are called **chimeras**. Chimeras can also be made by injecting cells from an early embryo of one genotype into a blastocyst of another genotype. The injected cells become incorporated into the inner cell mass of the host blastocyst, and a chimeric animal develops. It is even possible to make a chimera by injecting a single cell in this way; thus one can assay the developmental capabilities of the single cell. One of the major conclusions derived from these studies is that the cells of the very early mammalian embryo (up to the eight-cell stage) are initially identical and unrestricted in their capabilities: they are all *totipotent*. Localized determinants apparently have no part to play in the mammalian egg, and the pattern of cell diversification in the embryo is generated later, entirely through interactions of the cells with one another and with their environment.

Mammalian Embryonic Stem Cells Show How Environmental Cues Can Control the Pace as well as the Pathway of Development [25]

Mammalian early development is highly *regulative*. The fate of each cell is governed by interactions with its neighbors. The mouse experiments just described illustrate this well. The cells in a half-embryo or in a chimeric double embryo must adjust their behavior so as to generate an animal that is normal in both pattern and size. When the circumstances of development are more grossly abnormal, however, the embryonic cells can go wildly out of control. Some important lessons can be learned from these phenomena.

If a normal early mouse embryo is grafted into the kidney or testis of an adult, it rapidly becomes disorganized, and the normal controls on cell proliferation break down. The result is a bizarre growth known as a *teratoma*, which consists of a disorganized mass of cells containing many varieties of differentiated tissue—skin, bone, glandular epithelium, and so on—mixed with undifferentiated stem cells that continue to divide and generate yet more of these differentiated tissues. Teratomas with similar properties can also arise spontaneously from germ cells in the gonads as the result of various developmental accidents.

It is possible to derive transplantable cancers from teratomas. Such *teratocarcinomas* will grow without limit until they kill their host. They can be maintained indefinitely by grafting samples of the tumor cells serially from one host to another, and they always include some undifferentiated *stem cells,* to-

8-cell-stage mouse embryo whose parents are white mice

8-cell-stage mouse embryo whose parents are black mice

zona pellucida of each egg is removed by treatment with protease

embryos are pushed together and fuse when incubated at 37°C

development of fused embryos continues *in vitro* to blastocyst stage

blastocyst transferred to pseudopregnant mouse, which acts as a foster mother

the baby mouse has four parents (but its foster mother is not one of them)

Figure 21–31 A procedure for creating a chimeric mouse. Two morulae of different genotypes are combined.

gether with a variety of differentiated cell types to which the stem cells give rise. The teratocarcinoma stem cells can also be maintained in culture as permanent cell lines.

One might think that teratocarcinoma stem cells originate, as in other cancers, through mutations in genes responsible for the normal controls of cell behavior (discussed in Chapter 24). The following observations, however, suggest that this is not the case. Stem cells with very similar properties can be derived by placing a normal inner cell mass in culture and dispersing the cells as soon as they proliferate. Once dispersed, some of the cells, if kept in suitable culture conditions, will continue dividing indefinitely without altering their character. The resulting *embryonic stem (ES)* cell lines are similar to teratocarcinoma-derived cell lines, but they can be generated at such high frequency from normal embryos that it is unlikely that they arise by mutation. Instead, it appears that separating the cells from their normal neighbors and placing them in the appropriate culture medium has arrested the normal program of change of cell character with time and so enabled the cells to carry on dividing indefinitely without differentiating. The presence in the medium of a protein growth factor known as *leukemia inhibitory factor (LIF)* seems to be critical for this suspension of developmental progress. With a slightly more complex cocktail of growth factors, embryonic germ cells can be induced to behave in the same way in culture.

The state in which the ES, teratocarcinoma, or germ-cell-derived stem cells are arrested seems to be equivalent to that of normal inner-cell-mass cells. This can be shown by taking the cells from their culture dish and injecting them into the blastocoel cavity of a normal blastocyst (Figure 21–32). The injected cells become incorporated in the inner cell mass of the blastocyst and can contribute to the formation of an apparently normal chimeric mouse. Descendants of the injected stem cells can be found in practically any of the tissues of this mouse, where they differentiate in a well-behaved manner appropriate to their location and can even form viable germ cells. This capability of ES cells forms the basis for a widely used technique that allows mice to be generated with a genetically engineered mutation in any chosen gene whose DNA has been cloned. To produce such "gene-knockout" mice, mutant ES cells are made by selecting for a DNA insertion that replaces the chosen gene by an artificially altered version; the mutant ES cells are then used to produce chimeric mice that carry the mutation in their germ cells (see p. 329).

The extraordinarily adaptable behavior of ES cells shows that environmental cues not only guide choices between different pathways of differentiation, but in certain cases, they can also stop or start the developmental clock—the processes that drive a cell to progress from an embryonic to an adult state.

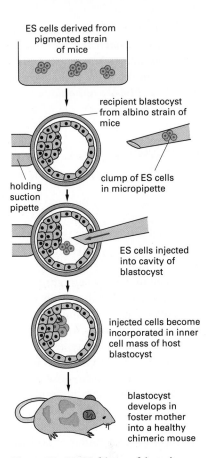

ES cells derived from pigmented strain of mice

recipient blastocyst from albino strain of mice

clump of ES cells in micropipette

holding suction pipette

ES cells injected into cavity of blastocyst

injected cells become incorporated in inner cell mass of host blastocyst

blastocyst develops in foster mother into a healthy chimeric mouse

Figure 21–32 **Making a chimeric mouse with ES or teratocarcinoma stem cells.** The experiment shows that the stem cells can combine with the cells of a normal blastocyst to form a healthy chimeric mouse.

Summary

In the course of embryonic development many types of cells are generated from the fertilized egg. The genomes of the differentiated cells remain the same; it is the pattern of gene expression that changes. Some of the differences between cells in the early embryo generally originate from the unequal distribution of cytoplasmic determinants localized in the egg before cleavage, but most of them arise later from local differences in the environments of the cells in the embryo. In Xenopus, for example, the animal and vegetal cells of the early embryo inherit different cytoplasmic determinants from the egg, and an influence from the vegetal cells then induces some of the animal cells to develop as mesoderm instead of ectoderm. This mesoderm induction seems to be mediated by families of growth factor proteins that also help regulate growth and differentiation in the mature organism.

Mammalian eggs are exceptional in that they are essentially symmetrical. Thus all the cells in an early mammalian embryo are initially alike and become different only through their interactions with one another. Through cell-cell interactions cells from two different early mouse embryos can adjust their fates and collaborate to form a single chimeric mouse. Early mouse embryo cells removed from the normal influ-

ences of their neighbors can proliferate inappropriately to give rise to terato-carcinomas, from which embryonic stem cells can be obtained. But when implanted into a normal early embryo, such cells revert to normal behavior, and their progeny differentiate according to their environment and can contribute to the formation of a healthy chimeric animal.

Cell Memory, Cell Determination, and the Concept of Positional Values

Cells must not only become different, they must also *remain* different after the original cues responsible for cell diversification have disappeared. Despite the continual turnover and resynthesis of almost all cell components, most cell types in the adult body have at least some distinctive features that are stably and heritably maintained even when the environment is changed. Thus, when a pigment cell divides, its daughters remain pigment cells; when a keratinocyte from the skin divides, its daughters remain keratinocytes; even though a fibroblast may be convertible into some other sort of connective-tissue cell such as a cartilage cell, it never changes into a neuron or a liver cell; and so on. Such durable differences between cell types are ultimately due to the different influences that the cells have been subjected to in the embryo, but the differences are maintained because the cells somehow remember the effects of those past influences and pass them on to their descendants. As we discuss in this section, cell memory—and the types of information, especially positional information, that cells retain as a consequence—are central elements of the patterning mechanisms that make a complex multicellular organism possible.

Cells Often Become Determined for a Future Specialized Role Long Before They Differentiate Overtly [15, 26]

Cell memory is most obvious in the persistence and stability of the differentiated states of cells in the adult body (discussed in Chapter 22). But the final character of a cell has usually been decided by a complex sequence of cues delivered to its progenitors during development and is often fixed long before differentiation becomes manifest. Through a series of decisions taken before, during, and just after gastrulation, for example, certain cells in the somites of a vertebrate become specialized at a very early stage as precursors of skeletal muscle cells; they then migrate from the somites into various other regions including those where the limbs will form (see Figure 21–16). These muscle cell precursors lack the large quantities of specialized contractile proteins found in mature muscle cells; indeed, they look superficially just like the other cells of the limb rudiment. But after several days they begin manufacturing large quantities of specialized muscle proteins, whereas the other limb cells with which they are mingled differentiate into various types of connective-tissue cells. Thus the developmental choice between muscle and connective tissue has been made by each cell long before it is expressed in overt differentiation, and it is meanwhile recorded in each cell as a molecular change that has no obvious effect on the cell's outward appearance.

A cell that has made a developmental choice in the above sense is said to be *determined*. Since the concept is a basic part of the language of developmental biology, it is useful to have a formal definition: a cell is **determined** if it has undergone a self-perpetuating change of internal character that distinguishes it and its progeny from other cells in the embryo and commits them to a specialized course of development. The term **differentiation** is generally reserved for *overt* cell specialization, that is, for a specialization of cell character that is grossly apparent. Usually, a cell becomes determined before it differentiates, although in some cases the two processes occur simultaneously. Indeed, it is possible for

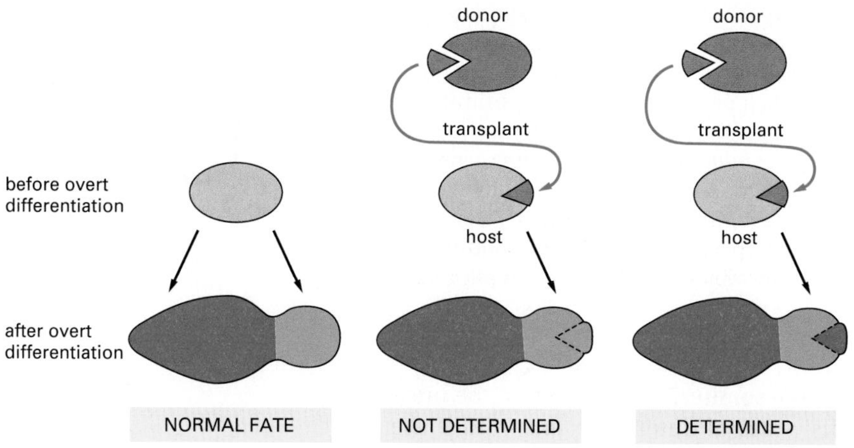

Figure 21–33 **The standard test of determination.**

before overt differentiation

after overt differentiation

donor

transplant

host

donor

transplant

host

NORMAL FATE

NOT DETERMINED

DETERMINED

differentiation to occur without determination, if the overt specialization of cell character is reversible.

The Time of Cell Determination Can Be Discovered by Transplantation Experiments [27]

To prove that a cell or group of cells is determined, one must show that it has a distinctive character that is maintained even when its circumstances are altered by experimental manipulation. The standard technique is to transplant the cells to a test environment (Figure 21–33).

A simple example of such an experiment comes from studies on amphibian embryos. As noted earlier, one can plot a fate map for a blastula or an early gastrula, showing which of its parts will normally develop into what. The cells in one region, for example, are fated to become epidermis if development proceeds normally, while those in another region are fated to form brain. To establish when these two groups of cells become *determined* to follow their particular modes of differentiation, a block of cells is cut from the prospective epidermal region and put in the position of prospective brain, and vice versa. If the cells are transplanted at the early gastrula stage, they show no memory of their origins and differentiate in the fashion appropriate to their new locations. If, however, the same experiment is done at a somewhat later stage, in the late gastrula, the prospective brain cells transplanted to an epidermal site will differentiate as misplaced neural tissue, and the prospective epidermal cells transplanted to a brain site will differentiate there as misplaced epidermis. This shows that both groups of cells have become determined sometime between the early and late gastrula stages.

Cell Determination and Differentiation Reflect the Expression of Regulatory Genes [28]

The phenomenon of determination raises three molecular questions: what molecule or molecules define a cell's state of determination; what is the memory mechanism that maintains that state; and how is determination coupled to differentiation? In general, the character of a cell is governed by the combination of gene regulatory proteins that it contains. These control its pattern of gene expression. In the well-studied case of muscle, as discussed in Chapter 9, a critical part is played by the *MyoD family* of closely related *myogenic proteins* (MyoD, Myf5, MRF4, and myogenin). In suitable circumstances these can activate the expression of muscle-specific genes such as muscle actin and muscle myosin, and introduction of a MyoD family member into fibroblasts and various other cell types can convert them into muscle precursor cells. In normal development genes coding for proteins of the MyoD family begin to be switched on very early

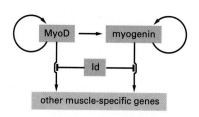

Figure 21–34 Genetic control circuitry for muscle cell determination. In this simplified diagram only two representative members of the MyoD family of genes are shown—*myoD* itself and *myogenin*. Mutual activation and self-activation of these genes by their own products create positive feedback that tends to make expression of the genes self-sustaining. Id is a helix-loop-helix protein encoded by the *inhibitor-of-DNA-binding* gene; by dimerizing with other helix-loop-helix proteins, and in particular by competing with MyoD family members for the requisite partners, it is thought to hinder expression of muscle-specific genes. The full control system for muscle differentiation is, however, certainly more complicated than this diagram suggests.

in the muscle precursor cells as they leave the somites, suggesting that the presence of these proteins defines the cells' state of determination. And if the myogenin gene is deleted by targeted genetic recombination, for example, muscle cells fail to develop.

The set of genes subject to activation by MyoD family members includes at least some of the genes of that family themselves. For this reason, expression of one member of the family generally leads to expression of others as well. In addition, at least some of these regulatory proteins act back directly on their own gene, so as to maintain expression of the gene once it has been turned on. The *positive feedback* resulting from mutual activation and self-activation provides a possible mechanism for cell memory, as discussed in Chapter 9.

This still leaves a problem. The muscle precursor cells do not start to manufacture large quantities of muscle-specific proteins until days, weeks, or even years after leaving the somites. How can they remain undifferentiated for so long after they have become determined? The mechanism is thought to depend on other proteins that interact with MyoD family members and regulate their action. As discussed in Chapter 9, MyoD and its relatives belong to the helix-loop-helix superfamily, whose members dimerize with one another in order to bind to DNA and activate gene expression. The efficacy in gene activation depends on the choice of partner for dimerization. By regulating the availability of appropriate dimerization partners for a protein of the MyoD family, the cell can apparently switch from a determined state, where the protein is able to maintain production of MyoD family members only, to a differentiated state, where the protein activates the full panoply of muscle-specific genes (Figure 21–34).

The State of Determination May Be Governed by the Cytoplasm or Be Intrinsic to the Chromosomes [29]

Cell memory, as manifested in the phenomenon of determination, presents one of the most challenging problems in molecular biology. In Chapter 9 we discuss some of the molecular mechanisms by which certain patterns of gene expression can become self-sustaining. In the context of cell determination three broad categories of cell memory can be distinguished, which may be called *cytoplasmic, autocrine,* and *nuclear memory,* respectively. The mechanism that has just been outlined for the myogenic proteins is an example of cytoplasmic memory. Here, components encoded by the set of active genes are present in the cytoplasm and act back on the genome, directly or indirectly, to maintain the selective expression of that specific set of genes. An implication of this mechanism is that if a nucleus is taken from one type of differentiated cell and injected into the cytoplasm of another type, the pattern of gene expression should alter to match the character of the host cytoplasm. The nuclear transplantation experiments on amphibian eggs that we discussed earlier provide an example of this sort of behavior.

The autocrine memory mechanism is a variant of the cytoplasmic. It depends again on the synthesis of products that stimulate their own production, but with the special feature that these products are secreted into the extracellular medium and act back on the cell's exterior to keep the cell in the state where it produces them. This mechanism has an important side effect: since neighboring cells share the same extracellular environment, they will tend to behave cooperatively, adopting the same state because they are exposed to the substances that they themselves produce, and an individual cell transplanted into a new environment will tend to switch its character to match that of the cells that surround it on all sides. Thus a group of cells may behave as determined, even though an individual cell in isolation does not. Such "community effects" in cell determination seem to be common and have been especially well documented in the early *Xenopus* embryo.

In contrast with cytoplasmic and autocrine memory, nuclear memory depends on self-sustaining changes that are intrinsic to the chromosomes—changes that define the selection of genes to be expressed and yet leave the DNA sequence unaltered. X-chromosome inactivation (see p. 446) and genomic imprinting (see p. 451) are well-established examples. Nuclear memory is based on inherited modifications in the chromatin or the DNA; unlike cytoplasmic memory, it allows two identical genes to coexist in different states in a single cell, one being expressed and the other not, even though both are exposed to the same intracellular environment.

Our ignorance is still profound concerning cell memory, and it is not yet possible in most cases even to classify the memory mechanism as cytoplasmic or nuclear.

Cells in Developing Tissues Remember Their Positional Values [30]

In an animal embryo positional signals and interactions operate over small distances, on the order of a millimeter or less, and through cell memory these influences leave their mark on cell character. As the body grows, further influences act locally in each of its parts, creating new distinctions within each class of cells and embroidering progressively finer levels of detail on the original basic body plan.

Thus before cells become committed to a particular mode of differentiation, they usually become *regionally specified:* they acquire distinct biochemical address labels, or **positional values**, that reflect their location in the body. The positional value of a cell will guide its behavior in subsequent steps of pattern formation—the way it responds to later positional signals, the ways in which it interacts with its neighbors, and the range of modes of differentiation ultimately open to it and its progeny. The cues that control the choice of positional value are said to provide the cell with *positional information.*

The existence and nature of remembered positional values is dramatically demonstrated by grafting experiments that have been carried out between the developing leg and wing of the chick embryo. The leg and the wing of the adult both consist of muscle, bone, skin, and so on—almost exactly the same range of differentiated tissues. The difference between the two limbs lies not in the types of tissues, but in the way in which those tissues are arranged in space. So how does the difference come about? At first sight it might seem simplest to explain the difference in terms of the presence of a different spatial distribution of signals in the developing forelimb and hindlimb, which directly tells cells which differentiated state to adopt. A simple grafting experiment shows that this view is profoundly wrong.

In the chick embryo the leg and the wing originate at about the same time in the form of small tongue-shaped buds projecting from the flank (Figure 21–35). The cells in the two pairs of limb buds appear similar and uniformly undifferentiated at first (see Figure 19–30). A small block of undifferentiated tissue at the base of the leg bud, from the region that would normally give rise to part of the thigh, can be cut out and grafted into the tip of the wing bud. Remarkably, the graft forms not the appropriate part of the wing tip, nor a misplaced piece

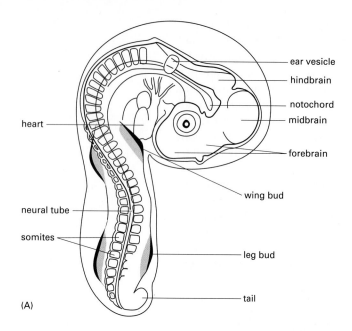

(A)

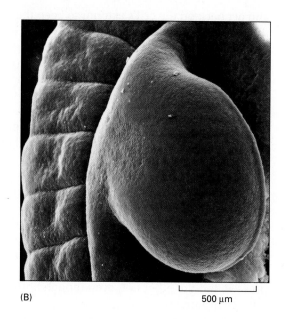

(B)   500 μm

of thigh tissue, but a toe (Figure 21–36). This experiment shows that the early leg-bud cells are already determined as leg but are not yet irrevocably committed to form a particular part of the leg: they can still respond to cues in the wing bud so that they form structures appropriate to the tip of the limb rather than the base. The signaling system that controls the differences between the parts of the limb is apparently the same for leg and wing. The difference between the two limbs results from a difference in the internal states of their cells at the outset of limb development. Even though the cells look the same and are destined to give rise to the same range of differentiated cell types, they are *nonequivalent*, with different positional values. In this way the final specification of how a limb cell should behave is built up combinatorially: first it is supplied with information as to whether it is to be leg or wing; then signals within the growing limb bud specify more fine-grained components of positional value, reflecting the precise position within the limb.

One of the most remarkable revelations of modern molecular genetics has been that almost all animals seem to use the same highly conserved molecular machinery to record positional values along the head-to-tail axis of the body, and some of these same gene products also operate to specify positional values in the limbs of vertebrates. We shall postpone the discussion of these master regulators of the body pattern until we have introduced the fruit fly, *Drosophila*, where the machinery was first discovered and characterized.

The Pattern of Positional Values Controls Cell Proliferation and Is Regulated by Intercalation [31]

A crucial aspect of pattern formation is the regulation of cell proliferation, through which the parts of the pattern attain their appropriate sizes. In many cases growth and the pattern of positional values both depend in a closely coupled way on continuing cell-cell interactions. A simple rule has been deduced from studies of the regeneration that occurs in various organisms when fragments of tissue with different positional values are juxtaposed and allowed time to grow and adjust. The principles appear to be general, but they are perhaps most clearly illustrated by studies on the leg of the cockroach (Figure 21–37).

Figure 21–35 Chick limb development. (A) A chick embryo after 3 days of incubation, illustrating the positions of the early limb buds. (B) Scanning electron micrograph showing a dorsal view of the wing bud and adjacent somites 1 day later; the bud has grown to become a tongue-shaped projection about 1 mm long, 1 mm broad, and 0.5 mm thick. (A, after W.H. Freeman and B. Bracegirdle, An Atlas of Embryology. London: Heinemann, 1967; B, courtesy of Paul Martin.)

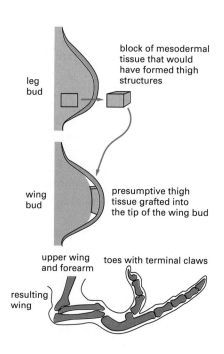

Figure 21–36 Prospective thigh tissue grafted into the tip of a chick wing bud forms toes. (After J.W. Saunders et al., *Dev. Biol.* 1:281–301, 1959.)

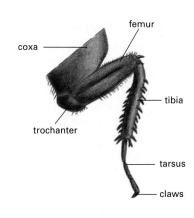

Figure 21–37 **The cockroach leg.** With each successive molt the leg grows bigger (by cell proliferation) but does not change its basic structure. The leg is covered by a cuticle that is secreted by a sheet of epidermal cells and replaced at each molt. The pattern of the cuticle reflects the pattern of positional values in the underlying epidermal sheet.

Cockroaches belong to the class of insects in which there is no radical metamorphosis from larva to adult but a gradual progression through a series of juvenile forms separated by molts, in which the old coat of cuticle is shed and a larger one is laid down. The juvenile cockroach has well-differentiated limbs, but the differentiated cells—unlike those in human limbs—are still able to respond to the cues that governed the development of the limb pattern, and they can regenerate that pattern if it is disturbed. Thus the workings of the pattern-formation system can be tested by operations done long after the period of embryonic development.

If two cockroach legs are amputated through one of their middle segments—through the tibia, say—but at different levels, the distal fragment of the one can be grafted onto the proximal stump of the other in such a way that the composite leg heals with the middle part of the tibia missing. Yet the leg that emerges after the animal has molted appears normal: the missing middle part of the pattern has regenerated (Figure 21–38A). More surprising is the result of a variant of this operation. The tibia of one cockroach leg is cut through near the proximal end and that of another leg near the distal end. The large detached portion of the first leg is then stuck onto the large remaining stump of the second leg to give an excessively long leg with a middle part present in duplicate (Figure 21–38B). The animal is left to molt. The leg that results, far from being more nearly normal, is now even longer because a third middle part of a tibia has developed between the two already present. As shown in Figure 21–38B, the bristles on this freshly formed region point in the direction opposite to that of the bristles on the rest of the tibia.

Many different operations of this type can be performed. All of them point to the existence in the insect epidermis of a system of positional values that

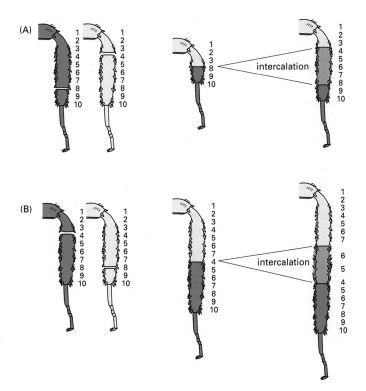

Figure 21–38 **Intercalary regeneration.** When mismatched portions of the cockroach tibia are grafted together, new tissue (*green*) is intercalated (by cell proliferation) to fill in the gap in the pattern of positional values (numbered from 1 to 10). In case (A) intercalation restores the missing part. In case (B) intercalation generates a third middle part of a tibia between the two middle parts already present. The bristles indicate the polarity of the intercalated tissue. In both cases continuity is restored in the final pattern of positional values.

Cell Memory, Cell Determination, and the Concept of Positional Values

makes the cells at different positions along the limb axis nonequivalent, and that is intimately coupled to the control of cell proliferation. It is convenient to describe the positional value by a number that goes from a maximum at one end of the limb segment to a minimum at the other. In the operations described above, cells with widely different positional values are brought together. As a result, new cells are formed by proliferation of the cells in the neighborhood of the junction. These new cells acquire positional values interpolated between those of the two sets of cells that were brought into confrontation (Figure 21–38). This behavior is summed up in the **rule of intercalation**: *discontinuities of positional value provoke local cell proliferation, and the newly formed cells take on intermediate positional values so as to restore continuity in the pattern.* Cell proliferation ceases only when cells with all the missing positional values have been intercalated in the initial gap and have become spread out to the normal spatial separation from one another. This process as a whole is called **intercalary regeneration**.

The rule of intercalation, with the corollary that cell proliferation continues until a certain spacing of positional values has been attained, is a powerful organizing principle in those systems to which it applies. Beginning with a pattern specified approximately and in miniature—for example, by a morphogen gradient—it can bring about the construction of a complete accurate pattern of positional values and regulate the growth of each part of the pattern to a standard size: all that is necessary is that the initial pattern should be qualitatively—that is, topologically—correct. The same rule appears to govern many processes of organogenesis and regeneration not only in insects but also in crustaceans and amphibians. Even in creatures such as mammals, where lost structures generally do not regenerate in the adult, the rule of intercalation may help to regulate growth and pattern formation during embryonic development. Unfortunately, the molecular mechanisms that underlie this crucial form of growth control are unknown.

Summary

Embryonic cells must not only become different, they must also remain different even after the influence that initiated cell diversification has disappeared. This requires cell memory, which enables cells to become determined for a particular specialized role long before they differentiate overtly. The mechanisms of cell memory may be cytoplasmic, involving molecules in the cytoplasm that act back on the nucleus to maintain their own synthesis, autocrine, involving secreted molecules that act back on the cell, or nuclear, involving processes of chromatin or DNA modification. In some cases the state of determination has been related to the expression of specific regulatory genes, such as the myogenic genes for muscle cells.

The different kinds of cells in an embryo are produced in a regular spatial pattern. The formation of this pattern usually begins with asymmetries in the egg and continues by means of cell-cell interactions in the embryo. The spatial signals that coordinate pattern formation supply cells with positional information, and a cell's remembered record of this information is called its positional value. Cells in the early forelimb and hindlimb rudiments of a vertebrate embryo, for example, acquire different positional values, making forelimb and hindlimb cells nonequivalent in their intrinsic character, long before the detailed pattern of cell differentiation has been determined.

In many animals the pattern of positional values is closely coupled to the control of cell proliferation according to a simple rule of intercalation. According to the rule, discontinuities of positional value provoke local cell proliferation, and the newly formed cells take on intermediate positional values that restore continuity in the pattern. This mechanism is likely to operate in normal embryonic development to correct inaccuracies in the initial specification of positional information.

The Nematode Worm: Developmental Control Genes and the Rules of Cell Behavior [32]

For cells, as for computers, memory makes complex programs of behavior possible, and many cells together, each one stepping through its complex developmental program, can generate a very complex adult body. Some of the steps that a cell takes in the course of development are autonomous, while others are affected by signals from other cells. Thus the cells of the embryo can be likened to an array of little computers, or *automata,* operating in parallel and exchanging information with one another. The rules that determine cell behavior are encoded in the cell's genes. Each cell contains the same genome and therefore behaves according to the same rules, but it can exist in a variety of states; the rules direct development along various alternative paths according to a combination of the past information the cell has remembered and the present environmental signals it receives. Computer modeling shows that even a very simple set of rules for the individual automata (cells) in such a system can lead to the production of astonishingly complex patterns; one cannot deduce the rules simply by observing the normal development of the pattern. The challenge, therefore, is to decipher the underlying cellular rules of development by experimentation and to find out how they are specified by the genes.

In this enterprise the nematode worm *Caenorhabditis elegans* offers some exceptional advantages, and it has become one of the foremost model systems in developmental genetics. We use it here to illustrate some general principles. A detailed discussion of the developmental genetics of pattern formation, however, is reserved for the next section, on *Drosophila,* where more years of research and a much larger army of research workers have provided a fuller picture.

Caenorhabditis elegans Is Anatomically and Genetically Simple [33]

As an adult, *C. elegans* is about 1 mm long and consists of only about 1000 somatic cells and 1000–2000 germ cells (exactly 959 somatic cell nuclei plus about 2000 germ cells are counted in one sex; exactly 1031 somatic cell nuclei plus about 1000 germ cells in the other) (Figure 21–39). The anatomy has been reconstructed, cell by cell, by electron microscopy of serial sections. The body plan of this simple worm is fundamentally the same as that of most higher animals in that it has a roughly bilaterally symmetrical, elongate body composed of the same basic tissues (nerve, muscle, gut, skin) organized in the same basic way (mouth and brain at the anterior end, anus at the posterior). The outer body wall is composed of two layers: the protective hypodermis, or "skin," and the underlying muscular layer. A simple tube of endodermal cells forms the intestine. A second

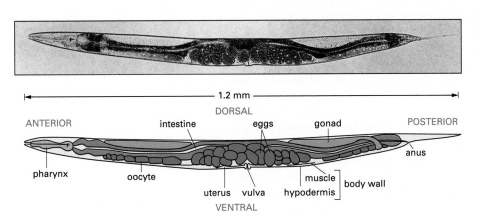

Figure 21–39 *Caenorhabditis elegans.* A side view of an adult hermaphrodite is shown. Note that the tissue called hypodermis in the nematode corresponds to the epidermis of other animals. (From J.E. Sulston and H.R. Horvitz, *Dev. Biol.* 56:110–156, 1977.)

tube, located between the intestine and the body wall, constitutes the gonad; its wall is composed of somatic cells, with the germ cells inside it. *C. elegans* has two sexes—a hermaphrodite and a male. The hermaphrodite can be viewed most simply as a female that produces a limited number of sperm: she can reproduce either by self-fertilization, using her own sperm, or by cross-fertilization after transfer of male sperm by mating. Self-fertilization allows a single heterozygous worm to produce homozygous progeny, a special feature that helps to make *C. elegans* an exceptionally convenient organism for genetic studies.

The relative simplicity of *C. elegans* anatomy is reflected in a similar simplicity of its genome. The animal has six homologous pairs of chromosomes, estimated to carry a total of 3000 "essential" genes (that is, genes in which mutations are lethal or have an easily observable effect on the phenotype) and four or five times that number of nonessential genes. The haploid genome consists of approximately 10^8 nucleotide pairs of DNA, which is about 20 times more than *E. coli*, about the same as *Drosophila*, and 30 times less than humans. Currently, more than 900 essential genes have been identified by mutation. These include genes that influence visible features such as the shape or behavior of the worm, genes that code for known proteins such as myosin, and genes that control the course of development. Nearly the entire genome has been mapped as a large set of overlapping DNA segments, represented by a library of ordered genomic clones (see p. 314), and a systematic effort has begun to determine the complete DNA sequence of the organism.

Nematode Development Is Almost Perfectly Invariant [34]

C. elegans begins life as a single cell, the fertilized egg, which gives rise, through repeated cell divisions, to 558 cells that form a small worm inside the egg shell. After hatching, further divisions result in the growth and sexual maturation of the worm as it passes through four successive larval stages separated by molts. After the final molt to the adult stage, the hermaphrodite worm begins to produce its own eggs. The entire developmental sequence, from egg to egg, takes only about three days.

Because C. *elegans* is small and transparent, its individual cells can be followed as they divide, migrate, differentiate, and die in the living embryo, and their pedigree can be traced from egg to adult organism. By this simple technique of direct observation, the behavior and lineage of all of the cells from the single-cell egg to the adult animal have been described. This has made possible a detailed **lineage analysis** that would be very difficult in larger animals, where individual cells at early stages usually must be specially marked if they and their progeny are to be identified later. Moreover, in larger animals the details of cell lineage show many random variations, even between genetically identical individuals. In the nematode, by contrast, the somatic structures develop by an invariant, predictable cell lineage, and each of the many cell divisions is precisely timed. This means that a given precursor cell follows the same pattern of cell divisions in every individual, and with very few exceptions the fate of each descendant cell can be predicted from its position in the lineage tree (Figure 21–40).

The full description of cell lineage in *C. elegans* leads to an immediate answer to a fundamental question. The nematode, like most animals, is formed from a relatively large number of cells that can be classified into a much smaller number of differentiated cell types. Given the importance of cell ancestry, one might be tempted to guess that all the cells of a given type are descendants of a single "founder cell" committed exclusively to that developmental pathway. Lineage analysis shows, however, that this is not generally true, either for nematodes or for other animals. Thus in *C. elegans* (with a few exceptions such as the intestinal cells and the germ-line cells) each class of differentiated cells—hypodermal, neuronal, muscular, gonadal—is derived from several founder cells originating in separate branches of the lineage tree (see Figure 21–40). Thus cells of similar character need not be close relatives. Conversely (but rarely), cells of very differ-

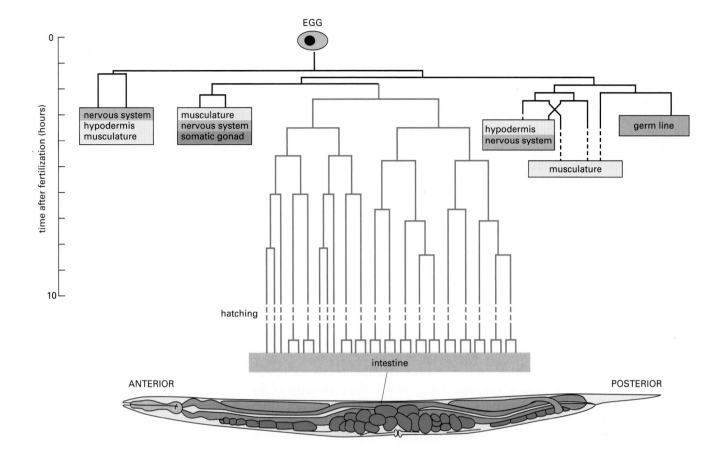

ANTERIOR   POSTERIOR

ent character may be closely related by lineage; for example, some of the neurons in *C. elegans* are sisters of muscle cells.

The problem, then, is to understand the rules that operate in each branch of the lineage tree to generate a specific array of cell types, each in appropriate numbers.

Developmental Control Genes Define the Rules of Cell Behavior That Generate the Body Plan [35]

To explain how the genome specifies the developmental rules, one has to be able to identify the genes that control the cells' developmental choices. Mutations in such genes will disturb development, but they are not the only mutations that do so. Some mutations, for example, will cut short all cell lineages and cause premature death of the embryo simply because they disrupt "housekeeping" genes that every cell needs in order to survive and proliferate. Other mutations will affect genes for proteins that particular types of differentiated cells require in order to carry out their specialized function; the body plan will then be essentially normal, but certain cell types, though still identifiable, will malfunction. Mutations in genes that are involved specifically in controlling developmental choices, by contrast, will disturb the body plan: they typically give rise to cells of the normal differentiated types arranged in an abnormal pattern or in abnormal numbers as a result of specific alterations in the lineage tree. **Developmental control genes** identified in this way can be classified according to the parts of the lineage tree that are affected and, hence, if we know the rules of cell behavior that generate that part of the lineage tree, according to the rules of cell behavior for which they are responsible.

To illustrate the principles of genetic analysis of a developmental mechanism, we discuss one example of a cell-cell interaction in the nematode—the induction of the vulva.

Figure 21–40 The lineage tree for the cells that form the intestine of *C. elegans*. The egg (*top*) is drawn to the same scale as the adult (*bottom*). Note that although the intestinal cells form a single clone (as do the germ-line cells), the cells of most other tissues do not.

Induction of the Vulva Depends on a Large Set of Developmental Control Genes [36]

The vulva—the egg-laying orifice in a hermaphrodite—is a ventral opening in the hypodermis (skin) formed by 22 cells that arise by specific lineages from three precursor cells in the hypodermis. A single nondividing cell in the gonad, called the *anchor cell,* attaches, or "anchors," the developing vulva to the overlying gonad (the uterus) to create a passageway through which the eggs can pass to the outside world. Microsurgical experiments show that the anchor cell is responsible for inducing the three nearest hypodermal cells to form a vulva (Figure 21–41). If the anchor cell is destroyed by focusing a laser beam on it, these cells, instead of forming a vulva, give rise to ordinary hypodermal cells. And if the anchor cell is shifted relative to the hypodermal cells, there is a corresponding shift in the site at which the vulva develops: flanking the three cells that normally give rise to the vulva lie three others that are also capable of doing so if exposed to the anchor-cell signal. Thus the anchor cell induces vulval differentiation in *C. elegans* just as the vegetal blastomeres induce mesodermal differentiation in the early *Xenopus* embryo. Only the anchor cell is necessary for this induction: if all the gonadal cells except the anchor cell are destroyed, the vulva still develops normally.

To identify genes involved in a given step of development, one searches for mutations that disrupt the process by screening the progeny of a large population of animals that have been exposed to mutagens. In this way many mutants are found that have a "vulvaless" phenotype, where none of the hypodermal cells behave as though they have received the anchor-cell signal. Another large group of mutants have a converse "multivulva" phenotype, in which all six hypodermal cells capable of responding to the anchor-cell signal behave as though they have

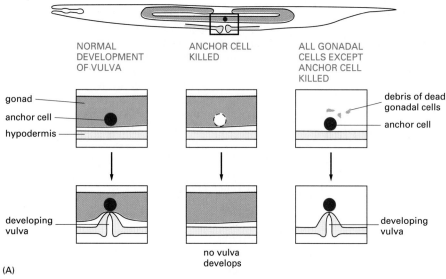

Figure 21–41 **Induction of the vulva.** (A) Experiments showing that an inductive influence from the anchor cell is required for normal development of the vulva. (B) Magnified view of the cells of the ventral hypodermis adjacent to the gonad, in the neighborhood of the anchor cell, with the normal lineage diagrams of their progeny sketched below. All six of these cells (and no others) are capable of responding to the vulva-inducing influence, but only three of them are normally exposed to it.

actually received it, so that the worm forms several vulvalike structures instead of one. Individual mutations giving a similar phenotype are then tested in pairs to see whether they affect the same or different genes, as explained in Panel 21–1, pages 1072–1073. Once a set of relevant genes has been identified in this way, still more components of the system usually can be discovered by searching for mutations in other genes that will suppress the ill effects of mutations in an already identified gene. Such *extragenic suppressor mutations* can be rare, and it is only in genetically favorable organisms such as *C. elegans* that one can easily find them; but when found, they often identify genes whose protein products interact directly with those of the already identified gene (because the alteration in the shape of the one protein molecule, for example, can be compensated for by a complementary alteration in the shape of its partner). More than 30 distinct identified genes have been implicated in the control of vulval development.

Genetic and Microsurgical Tests Reveal the Logic of Developmental Control; Gene Cloning and Sequencing Help to Reveal Its Biochemistry [37]

We focus here on just five of the vulval control genes, called *lin-3, let-23, sem-5, let-60,* and *lin-45.* Impairment of the function of any one of them by mutation has the same consequence—a vulvaless phenotype. Conversely, a genetic change causing excess of any of the gene products or excessive or unregulated activity—in other words, a gain of function—can have an opposite, multivulva effect. Each of the five genes therefore is needed for induction of the vulva, in a way that suggests they might all be links in a single chain of cause and effect; all of them, that is, might belong to a single *genetic pathway.* We saw in Chapter 15 how the signaling pathway that controls specialization of a particular cell type in the *Drosophila* eye has been defined by genetic analysis. A similar kind of analysis has been used to determine the order in which the vulval control genes act, as explained in Figure 21–42. The five genes do indeed appear to lie in a single genetic pathway, with *lin-3* the most upstream, then *let-23, sem-5, let-60,* and lastly *lin-45.* Thus, for example, a gain-of-function mutation in *lin-3* has no effect on the phenotype in an animal that also carries a loss-of-function mutation of *let-23;* the double mutant is vulvaless because the upstream component can do nothing when the downstream component upon which it should operate is missing.

The next problem is to relate the gene actions to specific cells in the embryo. Is *lin-3,* for example, needed in the anchor cells that produce the inductive signal or in the hypodermal cells that respond to it? For *lin-3* a simple answer has come from molecular genetics: the gene has been cloned and has been shown to be expressed in the anchor cell and nowhere else in the neighborhood (Figure 21–43). The other four genes, by contrast, appear to function in the hypodermal cells, and a gain-of-function mutation in one of them can cause a multivulva phenotype even when the anchor cell has been destroyed.

Figure 21–42 **How genes can be ordered in a genetic pathway by tests with double mutants.** A gene A is said to lie *upstream* from a gene B if its product normally acts by regulating the activity of B (or of the product of B) and *downstream* if the relationship is the other way around. If the upstream gene affects the phenotype *only* by regulating the downstream gene activity, the two genes are said to be links in a single *genetic pathway.* In this case mutations of either gene will result in a similar range of phenotypes. To discover the ordering of the genes in the pathway, one uses double mutants to see which of two genes is the more direct determinant of the phenotype. The approach depends on finding specific mutations of A and B that have opposite effects on the phenotype when taken singly. In the example shown a (dominant) gain-of-function mutation in gene B, making it active independently of regulation by upstream genes, is combined with a (recessive) loss-of-function mutation in A or C: the (A,B) double mutant has a gain-of-function phenotype, implying that A lies upstream from B, while the (B,C) double mutant has a loss-of-function phenotype, implying that C lies downstream from B.

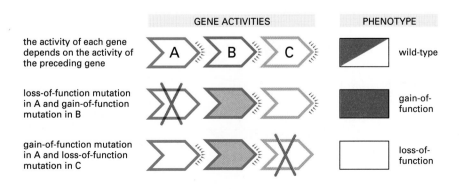

the activity of each gene depends on the activity of the preceding gene

loss-of-function mutation in A and gain-of-function mutation in B

gain-of-function mutation in A and loss-of-function mutation in C

GENE ACTIVITIES

A B C

PHENOTYPE

wild-type

gain-of-function

loss-of-function

GENES AND PHENOTYPES

Gene: a functional unit of inheritance, usually corresponding to the segment of DNA coding for a single protein.

Genome: an organism's set of genes.

locus: the site of the gene in the genome

alleles: alternative forms of a gene

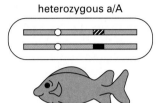

Wild-type: the normal, naturally occurring type

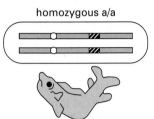

Mutant: differing from the wild-type because of a genetic change (a mutation)

GENOTYPE: the specific set of alleles forming the genome of an individual

PHENOTYPE: the visible character of the individual

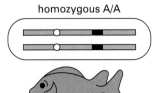

homozygous A/A   heterozygous a/A   homozygous a/a

allele A is dominant (relative to a); allele a is recessive (relative to A)

In the example above, the phenotype of the heterozygote is the same as that of one of the homozygotes; in cases where it is different from both, the two alleles are said to be co-dominant.

CHROMOSOMES

a chromosome at the beginning of the cell cycle, in G_1 phase; the single long bar represents one long double helix of DNA

centromere

short "p" arm   long "q" arm

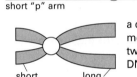

a chromosome at the end of the cell cycle, in metaphase; it is duplicated and condensed, consisting of two identical sister chromatids (each containing one DNA double helix) joined at the centromere.

short "p" arm   long "q" arm

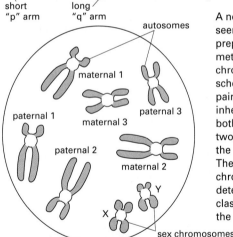

autosomes

maternal 1

paternal 1   maternal 3   paternal 3

paternal 2   maternal 2

Y

X

sex chromosomes

A normal diploid chromosome set, as seen in a metaphase spread, prepared by bursting open a cell at metaphase and staining the scattered chromosomes. In the example shown schematically here, there are three pairs of autosomes (chromosomes inherited symmetrically from both parents, regardless of sex) and two sex chromosomes—an X from the mother and a Y from the father. The numbers and types of sex chromosomes and their role in sex determination are variable from one class of organisms to another, as is the number of pairs of autosomes.

THE HAPLOID-DIPLOID CYCLE OF SEXUAL REPRODUCTION

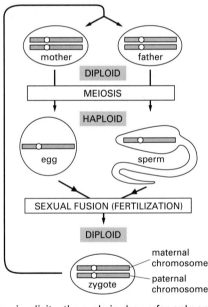

mother   father

DIPLOID

MEIOSIS

HAPLOID

egg   sperm

SEXUAL FUSION (FERTILIZATION)

DIPLOID

maternal chromosome

paternal chromosome

zygote

For simplicity, the cycle is shown for only one chromosome/chromosome pair.

MEIOSIS AND GENETIC RECOMBINATION

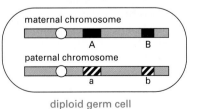

maternal chromosome

A   B

paternal chromosome

a   b

diploid germ cell

genotype $\dfrac{AB}{ab}$

MEIOSIS AND RECOMBINATION

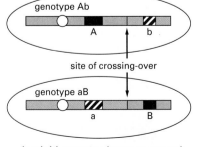

genotype Ab

A   b

site of crossing-over

genotype aB

a   B

haploid gametes (eggs or sperm)

The bigger the distance between two loci, the bigger the chance that they will be recombined by crossing-over occurring at a site between them. If they get recombined in x% of gametes, they are said to be separated by a genetic map distance of x map units (or x centimorgans).

TYPES OF MUTATIONS

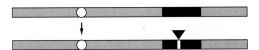

point mutation: maps to a single site in the genome, corresponding to a single nucleotide pair or a very small part of a single gene

inversion: inverts a segment of a chromosome

deletion: deletes a segment of a chromosome

translocation: breaks off a segment from one chromosome and attaches it to another

lethal mutation: causes the developing organism to die prematurely.

conditional mutation: produces its phenotypic effect only under certain conditions, called the *restrictive* conditions; under other conditions—the *permissive* conditions—the effect is not seen. For a *temperature-sensitive* mutation, the restrictive condition typically is high temperature, while the permissive condition is low temperature.

loss-of-function mutation: reduces or abolishes the activity of the gene. These are the commonest class of mutations. Loss-of-function mutations are usually *recessive*—the organism can usually function normally so long as it retains at least one normal copy of the affected gene.

null mutation: completely abolishes the activity of the gene.

gain-of-function mutation: increases the activity of the gene or makes it active in inappropriate circumstances; these mutations are usually *dominant.*

dominant negative mutation: dominant-acting mutation that blocks gene activity, causing a loss-of-function phenotype even in the presence of a normal copy of the gene. The phenomenon occurs when the mutant gene product interferes with the function of the normal gene product.

suppressor mutation: suppresses the phenotypic effect of another mutation, so that the double mutant appears normal. An *intragenic* suppressor mutation lies within the gene affected by the first mutation; an *extragenic* suppressor mutation lies in a second gene—often one whose product interacts directly with the product of the first.

TWO GENES OR ONE?

Given two mutations that produce the same phenotype, how can we tell whether they are mutations in the same gene? If the mutations are recessive (as they most often are), the answer can be found by a complementation test.

In the simplest type of complementation test, an individual who is homozygous for one mutation is mated with an individual who is homozygous for the other. The phenotype of the offspring gives the answer to the question.

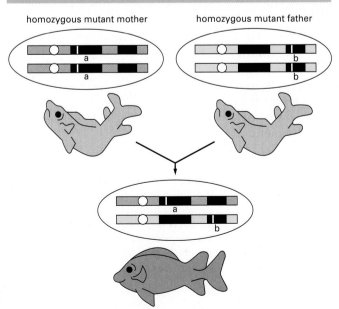

COMPLEMENTATION:
MUTATIONS IN TWO DIFFERENT GENES

homozygous mutant mother   homozygous mutant father

hybrid offspring shows normal phenotype:
one normal copy of each gene is present

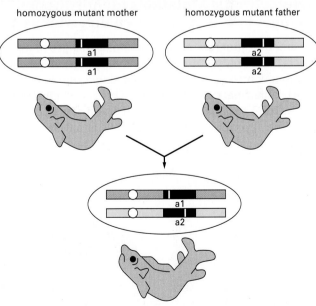

NONCOMPLEMENTATION:
TWO INDEPENDENT MUTATIONS IN THE SAME GENE

homozygous mutant mother   homozygous mutant father

hybrid offspring shows mutant phenotype:
no normal copies of the mutated gene are present

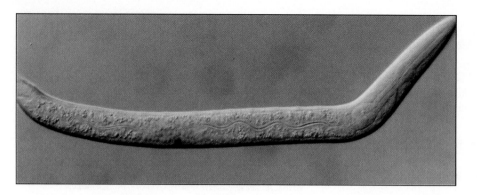

Figure 21–43 **Expression of *lin-3* in the anchor cell.** A *C. elegans* embryo has been transfected with an artificial reporter gene consisting of the control region of *lin-3* coupled to the gene for the enzyme β-galactosidase, whose presence is easily detected by a histochemical reaction that gives a blue reaction product. Only the anchor cell is stained blue, implying that it is only in the anchor cell that the *lin-3* gene is normally switched on. (Courtesy of Russell Hill.)

To complete the picture, we have to relate the genetically defined pathway to protein molecules and biochemistry. The five genes *lin-3, let-23, sem-5, let-60,* and *lin-45* have all been cloned and sequenced, and in each case the sequence indicates the probable function: *lin-3* codes for a protein similar to a secreted signaling molecule well known in vertebrates—epidermal growth factor (EGF); *let-23* codes for a receptor tyrosine kinase homologous to the members of the vertebrate EGF receptor family; *sem-5*, as we saw in Chapter 15, codes for a protein containing the SH2 and SH3 domains, found in many proteins that directly bind to such receptors and mediate their effects on other intracellular components; and *let-60* and *lin-45* are respectively homologous to the vertebrate *ras* and *raf* genes, whose products relay signals intracellularly from such receptors into the cell interior, as discussed in Chapter 15. Presumably, therefore, the Lin-3 protein is the signal molecule secreted by the anchor cell, the Let-23 protein is the transmembrane receptor in the hypodermal cells to which it binds, and the Sem-5, Let-60, and Lin-45 proteins are links in the intracellular signaling chain through which binding of the ligand to the receptor exerts its ultimate effects on gene expression and cell determination (Figure 21–44). In fact, the genetic analysis of this system in the developing nematode worm provides one of the clearest accounts we have of the organization of a signaling pathway that appears to have been conserved throughout most of the animal kingdom. A very similar pathway, as we saw in Chapter 15, emerges from analysis of the *sevenless* mutant in *Drosophila* (see p. 764).

Heterochronic Mutations Identify Genes That Specify Changes in the Rules of Cell Behavior as Time Goes By [38]

As computer programmers know, small changes in a program can have drastic effects on the output produced when a program is executed. Likewise, a mutation in a control gene that alters a single rule of cell behavior can result in a grossly abnormal cell lineage tree. This is well illustrated by *heterochronic mutations* in *C. elegans,* which cause certain sets of cells to behave in a way that would be appropriate for normal cells at a different stage in development. A

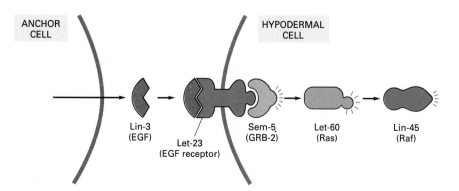

Figure 21–44 **The pathway for vulval induction in *C. elegans*.** The diagram shows the functions of the gene products that have been identified. The names of the homologous vertebrate proteins are indicated in parentheses.

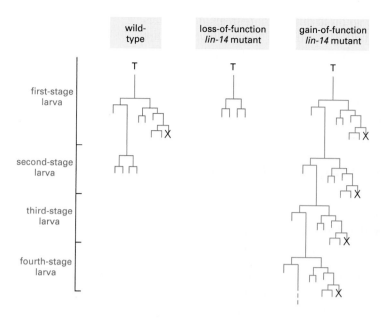

| wild-type | loss-of-function *lin-14* mutant | gain-of-function *lin-14* mutant |

first-stage larva

second-stage larva

third-stage larva

fourth-stage larva

Figure 21–45 Heterochronic mutations in the *lin-14* gene of *C. elegans.* The effects on only one of the many affected lineages are shown. The loss-of-function (recessive) mutation in *lin-14* causes premature occurrence of the pattern of cell division and differentiation characteristic of a late larva; the gain-of-function (dominant) mutation has the opposite effect. The cross denotes a programmed cell death. *Green* lines represent cells that contain Lin-14 protein, *red* lines those that do not. In normal development disappearance of Lin-14 is triggered by the beginning of larval feeding. (After V. Ambros and H.R. Horvitz, *Science* 226:409–416, 1984. © the AAAS; and P. Arasu, B. Wightman, and G. Ruvkun, *Growth Dev.Aging.* 5:1825–1833, 1991.)

daughter cell may behave like its parent or grandparent, for example, and the offspring of the daughter may behave again in the same way, and so on, with the result that a portion of the lineage pattern is reiterated indefinitely.

Figure 21–45 shows lineage diagrams for a set of mutations in a gene called *lin-14*, illustrating this phenomenon: instead of progressing through the normal series of cell divisions characteristic of the first, second, third, and fourth larval stages and then halting, many of the cells in gain-of-function *lin-14* mutants repeatedly go through the patterns of cell divisions characteristic of the first larval stage, continuing through as many as five or six molt cycles and persisting in the manufacture of an immature type of cuticle. Loss-of-function mutations in the *lin-14* gene have the reverse effect, causing cells to adopt mature states precociously, skipping intermediate stages, so that the animal reaches its final state prematurely and with an abnormally small number of cells.

The *lin-14* gene has been cloned, and the protein it encodes has been found to be concentrated in cell nuclei. In a normal individual the protein is present in most of the somatic cells of the late embryo and early first larval stage, but its concentration then declines to near zero by the second larval stage. Those *lin-14* mutants that enter an adult state precociously are found to have a reduced level of the Lin-14 protein, whereas those mutants that carry on with repeated first larval stage cycles are found to express the Lin-14 protein for an abnormally long time (because of a mutation in a regulatory portion of the gene). Thus the effect of the Lin-14 protein is to keep the cells in an immature state, and normal maturation depends on its disappearance. This gene product is presumably only one of many whose changing concentrations in cells specify changes in the rules of cell behavior as development proceeds.

The Tempo of Development Is Not Controlled by the Cell-Division Cycle [39]

The example we have just discussed brings us to a fundamental general problem in development. The genome has to define a set of rules for cell division as well as for cell specialization, and the two processes have to be coordinated. How is the division cycle regulated in development, and how is it coordinated with cell specialization?

One suggestion is that changes of internal state might be locked to passage through the division cycle: the cell would click to the next state as it went through mitosis, so to speak. This seems a tempting idea, especially when one pictures development in terms of lineage diagrams, but the evidence is largely against it.

The Nematode Worm: Developmental Control Genes and the Rules of Cell Behavior

Cells in developing embryos frequently go on to differentiate in an almost normal way even when cell division is artificially prevented. Necessarily, there are some abnormalities, if only because a single undivided cell cannot differentiate normally in two ways at once. But in most cases that have been studied, it seems clear that cell divisions are not the ticks of a clock that sets the tempo of development. Rather, the cell changes its chemical state with time regardless of cell division, and this changing state controls both the decision to divide and the decision as to when and how to specialize.

Cells Die Tidily as a Part of the Program of Development [40]

A *C. elegans* hermaphrodite generates 1030 somatic cell nuclei in the course of its development, but 131 of the cells die. These **programmed cell deaths** occur in an absolutely predictable pattern, and they create no mess. Whereas cells that die from damage or poisoning typically swell and burst, spilling their contents over their neighbors, these normal cell deaths occur by a process known as **apoptosis**, in which the cell nucleus becomes condensed, the cell itself shrivels, and the shrunken corpse is rapidly engulfed and digested by neighboring cells (Figure 21–46). Programmed cell death is a regular feature of normal animal development and is probably the fate of a substantial fraction of the cells produced in most animals.

Because apoptosis occurs quickly and leaves no trace, the deaths easily go unnoticed. Yet cell death may be as important as cell division in generating an individual with the right cell types in the right numbers and places. In vertebrates, for example, it regulates the numbers of neurons (as we discuss later), eliminates undesirable types of lymphocytes (discussed in Chapter 23), disposes of cells that have finished their job (as when a tadpole loses its tail at metamorphosis), and helps to sculpt the shapes of developing organs (creating the gaps between digits by doing away with the cells that lie between the digit rudiments in the limb bud, for example).

Normal cell deaths are thought to be suicides in which the cell activates a death program and kills itself. The best evidence that animal cells have an intrinsic death program comes from genetic studies in *C. elegans*, where two genes, called *ced-3* and *ced-4* (*ced* stands for "cell death abnormal"), have been identified that are required for the 131 normal cell deaths to occur. If either gene is inactivated by mutation, the cells that are normally fated to die survive instead, differentiating as recognizable cell types such as neurons. Conversely, overexpression or misplaced expression of *ced-3* and *ced-4* (as a result of loss-of-function mutations that inactivate another gene, *ced-9*, which normally represses the death program) causes many cells to die that would normally survive.

The amino acid sequences of these three Ced proteins are known. The Ced-4 protein is novel and is thought to act upstream of Ced-3, which is a protease. Ced-9 is 23% identical in amino acid sequence to a mammalian protein called Bcl-2 (the product of the proto-oncogene *bcl-2*), which acts like Ced-9 to suppress programmed cell death in many types of mammalian cells. Remarkably, when the human *bcl-2* gene is transferred to *C. elegans*, it acts to inhibit normal cell death in the worm and is even able to rescue *ced-9* mutants that otherwise die early in development. These important findings indicate that both the mechanism of programmed cell death and its regulation have been highly conserved in evolution from worms to humans, confirming that the ability to commit suicide in this way is a fundamental property of animal cells.

Summary

Two things make the nematode Caenorhabditis elegans *an attractive organism for investigating the genetic basis of development: first, genetic analysis is easy because the generation time is short and the genome small; second, the normal course of development is extraordinarily reproducible and has been chronicled in detail, so*

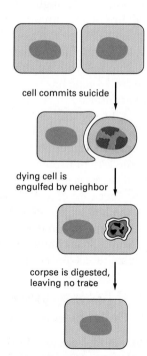

Figure 21–46 Apoptotic cell death in *C. elegans*. Death depends on expression of the *ced-3* and *ced-4* genes in the dying cell itself, whereas the subsequent engulfment and disposal of the remains depend on expression of other genes in the neighboring cells.

cell commits suicide

dying cell is
engulfed by neighbor

corpse is digested,
leaving no trace

that a cell at any given position in the body has the same lineage in every individual, and this lineage is fully known. As in other organisms, development depends on an interplay of cell-cell interactions and cell-autonomous processes. Cell-destruction experiments show, for example, that the development of the vulva depends on an inductive signal, and the genes required for this induction can be identified through mutations that disrupt vulval development. Molecular genetic analysis reveals the individual functions of these genes and shows that several of them code for components of a signaling pathway that operates in vertebrates too. Lineage analysis of mutants leads to the discovery of many other important classes of genes, including genes whose products serve to specify changes in the rules of cell behavior with time during development and genes that are responsible for programmed cell death—an invariable feature of development in all animals.

Drosophila and the Molecular Genetics of Pattern Formation. I. Genesis of the Body Plan [41]

The structure of an organism is controlled by its genes: classical genetics is based on this proposition. Yet for almost a century, and even long after the role of DNA in inheritance had become clear, the mechanisms of the genetic control of body structure remained an intractable mystery. In recent years this chasm in our understanding has begun to be filled. In the previous section we used the nematode worm to illustrate some of the general principles of how developmental control genes orchestrate the events of development. But it is the fly *Drosophila melanogaster* (Figure 21–47), more than any other organism, that has really transformed our understanding of how genes govern the patterning of the body. Decades of genetic study, culminating in massive systematic searches, have yielded a large catalogue of developmental control genes in the fly whose specific function is to define the spatial pattern of cell types and body parts. It has become possible not only to identify the key genes, but also to watch them at work: by *in situ* hybridization using DNA or RNA probes, one can observe directly how the internal states of the cells in the embryo are defined by the sets of regulatory genes that they express. By analyzing mutants, transgenic animals, and animals that are a patchwork of mutant and nonmutant cells, one can go on to discover how each gene operates as part of a system to specify the organization of the body. Moreover, the fly has provided a crucial key to our own development; for the genes controlling the pattern of the body in *Drosophila* turn out to have close counterparts in higher animals, including ourselves.

Figure 21–47 *Drosophila melanogaster.* Dorsal view of a normal adult fly. (A) Photograph. (B) Labeled drawing. (Photograph courtesy of E.B. Lewis.)

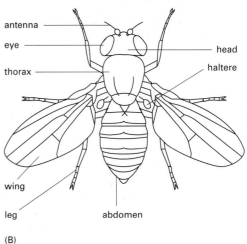

Our account of *Drosophila* developmental genetics is divided into two sections. The first deals with events in the early embryo and describes how the basic body plan is created, with a head rudiment at one end, a posterior rudiment at the other, and in between them an ordered series of segments—the basic modular units from which all insects are constructed. The second section deals with later events and discusses the genetic apparatus that endows cells with positional values that make the cells of one segment different from those of the next; these processes ensure that, for example, the head will develop antennae and the thorax legs—and not, as happens in some mutants we shall encounter, the other way around.

The Insect Body Is Constructed by Modulation of a Fundamental Pattern of Repeating Units [41, 42]

The timetable of *Drosophila* development, from egg to adult, is summarized in Figure 21–48. The period of *embryonic development* begins at fertilization and takes about a day, at the end of which the embryo hatches out of the egg shell to become a *larva*. The larva then passes through three stages, or *instars*, separated by molts in which it sheds its old coat of cuticle and lays down a larger one. At the end of the third instar it pupates. Inside the *pupa* a radical remodeling of the body takes place, and eventually, about nine days after fertilization, an adult fly, or *imago*, emerges.

The fly consists of a head, three thoracic segments (numbered T1 to T3), and eight or nine abdominal segments (numbered A1 to A9). Each segment, although

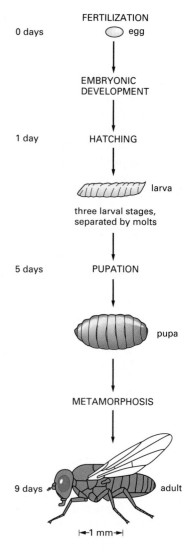

Figure 21–48 Synopsis of *Drosophila* development from egg to adult fly.

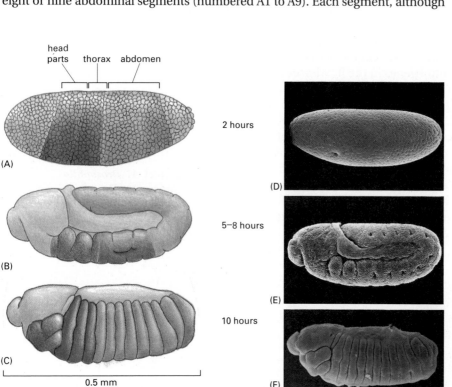

Figure 21–49 The origins of the *Drosophila* body segments during embryonic development. The embryos are seen in side view in drawings (A–C) and corresponding scanning electron micrographs (D–F). (A and D) At 2 hours the embryo is at the *syncytial blastoderm* stage (see Figure 21–51) and no segmentation is visible, although a fate map can be drawn showing the future segmented regions (*color* in A). (B and E) At 5–8 hours the embryo is at the *extended germ band* stage: gastrulation has occurred, segmentation has begun to be visible, and the segmented axis of the body has lengthened, curving back on itself at the tail end so as to fit into the egg shell. (C and F) At 10 hours the body axis has contracted and become straight again, and all the segments are clearly defined. The head structures, visible externally at this stage, will subsequently become tucked into the interior of the larva, to emerge again only when the larva goes through pupation to become an adult. (D and E, courtesy of Rudi Turner and Anthony Mahowald; F, courtesy of Jane Petschek.)

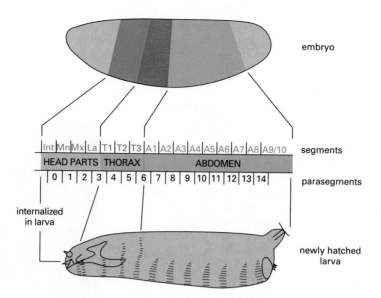

Figure 21–50 **The segments of the *Drosophila* larva and their correspondence with regions of the blastoderm.** Note that the ends of the blastoderm correspond to nonsegmental structures that form largely internal parts of the larva, as do the segmental rudiments of the adult head parts. Segmentation in *Drosophila* can be described in terms of either segments or parasegments: the relationship is shown in the middle part of the figure. Parasegments often correspond more simply to patterns of gene expression. The exact number of abdominal segments is debatable: eight are clearly defined, and a ninth is probably present.

different from the others, is built according to a similar plan. Segment T1, for example, carries a pair of legs, T2 carries a pair of legs plus a pair of wings, and T3 carries a pair of legs plus a pair of halteres—small knob-shaped balancers important in flight, evolved from the second pair of wings that more primitive insects possess. The quasi-repetitive segmentation develops in the early embryo during the first few hours after fertilization, but it is more obvious in the larva, where the segments look more similar than in the adult. In the embryo it can be seen that the rudiments of the head, or at least the future adult mouth parts, are likewise segmental (Figure 21–49). At the two ends of the animal, however, there are highly specialized terminal structures that are not segmentally derived.

It is partly a matter of convention where one draws the boundary between one segmental unit and the next. In discussing patterns of gene expression, we shall see that it is convenient to speak in terms of a total of 14 parasegments (numbered P1 to P14) that are half a segment out of register with traditionally defined segments (Figure 21–50).

Drosophila Begins Its Development as a Syncytium [41, 43]

The egg of *Drosophila* is about 400 μm long and about 160 μm in diameter, with a clearly defined polarity. Like the eggs of other insects, it begins its development in an unusual way: a series of nuclear divisions without cell division creates a syncytium. The early nuclear divisions are synchronous and extremely rapid, occurring about every 8 minutes. The first nine divisions generate a cloud of nuclei, most of which migrate from the middle of the egg toward the surface, where they form a monolayer called the syncytial blastoderm. After another four rounds of nuclear division, plasma membranes grow inward from the egg surface to enclose each nucleus, thereby converting the syncytial blastoderm into a *cellular blastoderm* consisting of about 6000 separate cells (Figure 21–51). A small subset of nuclei populating the extreme posterior end of the egg are segregated into cells a few cycles earlier; these *pole cells* are the primordial germ cells that will give rise to eggs or sperm.

As in a cleaving amphibian egg, the very rapid cycles of DNA replication seem to hinder transcription, so that up to the cellular blastoderm stage development depends largely—although not exclusively—on stocks of maternal mRNA and protein that accumulated in the egg before fertilization. After cellularization, cell division continues in a more conventional way, asynchronously and at a slower rate, and the rate of transcription increases dramatically.

The cellular blastoderm corresponds to the hollow blastula of an amphibian or a sea urchin, even though its interior is filled with yolk rather than being a

Drosophila I. Genesis of the Body Plan

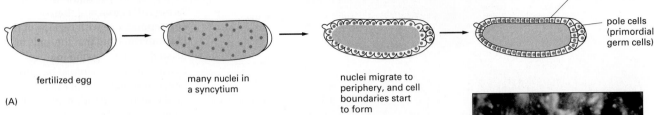

fertilized egg

many nuclei in
a syncytium

nuclei migrate to
periphery, and cell
boundaries start
to form

somatic cells

pole cells
(primordial
germ cells)

(A)

Figure 21–51 **Development of the *Drosophila* egg from fertilization to the
cellular blastoderm stage.** (A) Schematic drawings. (B) Surface view and
(C) optical section photographs of blastoderm nuclei undergoing mitosis at
the transition from the syncytial to the cellular blastoderm stage. Actin is
stained *green*, tubulin *orange*. (A, after H.A. Schneiderman, in Insect
Development [P.A. Lawrence, ed.], pp. 3–34. Oxford, UK: Blackwell, 1976; B
and C, courtesy of W. Theurkauf.)

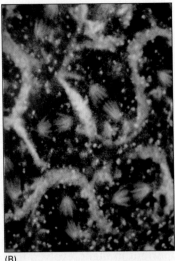

(B)

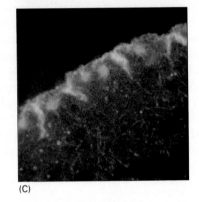

(C)

fluid-filled cavity. Gastrulation follows as soon as cellularization is complete, and
although the geometry of this process is very different in the insect, the general
outcome is similar. Through coordinated cell movements, endodermal cells are
invaginated into the interior to form the gut extending along the axis of the embryo. Mesoderm surrounds the gut rudiment and occupies the space between it
and an enveloping layer of ectoderm on the exterior.

By marking and following the cells through their complex gastrulation movements, one can draw a fate map for the monolayer of cells on the surface of the
blastoderm (Figure 21–52). The fate map is especially simple for a cross-section
through the middle of the embryo, with prospective mesoderm ventrally and
ectoderm on each side above it. As in a vertebrate, the cords of nerve cells that
run the length of the body derive from part of the ectoderm: a subset of the cells
in this *neurogenic ectoderm* will detach from their neighbors, escape from the
epithelial sheet, and move into the interior of the embryo as neuronal precursors.
For mesoderm, ectoderm, and nerve cord, the position of the cells along the
anteroposterior axis is roughly preserved during gastrulation because their movements are in the transverse plane. The gut, however, is formed by invagination
of two groups of cells from the opposite extremities of the embryo; these two
invaginations meet in the middle to form eventually a continuous gut tube.

As gastrulation nears completion, a series of indentations and bulges appear
in the surface of the embryo, marking the subdivision of the body into parasegments along its anteroposterior axis (see Figure 21–49). More subtle tests show
that the main features of this segmental pattern are already established at the
cellular blastoderm stage, before gastrulation begins.

Two Orthogonal Systems Define the Ground Plan of the Embryo [44]

Two coordinates are needed to define each position in the blastoderm, and, correspondingly, one can distinguish two sets of **egg-polarity genes** that act independently at the outset of development to specify the two main axes of the embryo—the dorsoventral and the anteroposterior. These genes define the spatial
coordinates of the embryo by setting up morphogen gradients in the egg.

The egg-polarity genes were found by exhaustive searches for mutants in
which the polarity of the embryo is disrupted. In this way 12 *dorsoventral egg-
polarity genes* were discovered. All but one of these have the same loss-of-function mutant phenotype, in which the embryo is *dorsalized*—that is, all its cells
take on a dorsal character, so that the normal ventral structures fail to form. The
remaining gene has the opposite loss-of-function phenotype—the embryo is
ventralized. We shall see that all these genes are components of a single system
that sets up a dorsoventral morphogen gradient in the early embryo.

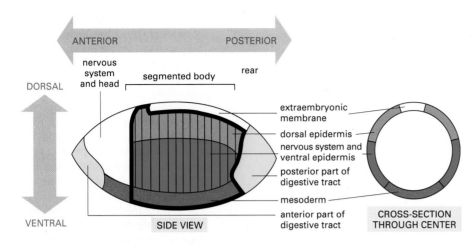

Figure 21–52 Fate map of a *Drosophila* embryo at the cellular blastoderm stage. The embryo is shown in side view and in cross-section, displaying the relationship between the dorsoventral subdivision into future major tissue types and the anteroposterior pattern of future segments. A heavy line encloses the region that will form segmental structures. During gastrulation the cells along the ventral midline invaginate to form mesoderm, while the cells fated to form the gut invaginate near each end of the embryo. Thus, with respect to their role in gut formation, the opposite ends of the embryo, although far apart in space, are close in function and in final fate. (After V. Hartenstein, G.M. Technau, and J.A. Campos-Ortega, *Wilhelm Roux' Arch. Dev. Biol.* 194:213–216, 1985.)

The anteroposterior set of genes, by contrast, can be subdivided according to their mutant phenotypes into three subsystems, responsible for specifying different parts of the anteroposterior axis (Figure 21–53). The *anterior group* (4 genes) governs the anterior part of the axis. The *posterior group* (11 genes) governs the posterior part of the axis. Lastly, the *terminal group* (6 known genes)

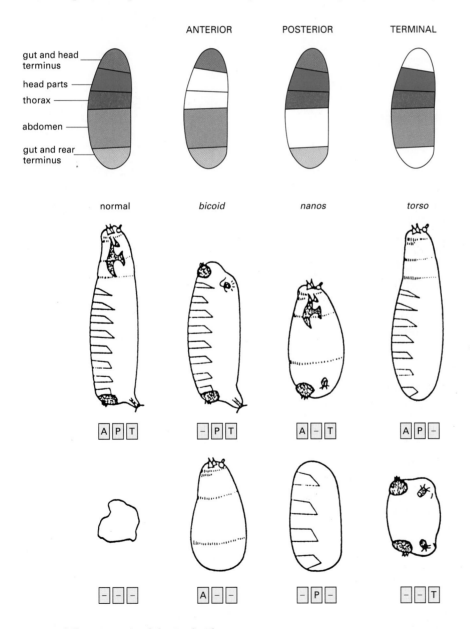

Figure 21–53 The domains of the anterior, posterior, and terminal systems of egg-polarity genes. The upper diagrams show the fates of the different regions of the egg/early embryo and indicate in *white* the parts that fail to develop if the anterior, posterior, or terminal system is defective. The middle row shows schematically the appearance of a normal larva and of mutant larvae that are defective in a gene of the anterior system (for example, *bicoid*), of the posterior system (for example, *nanos*), or of the terminal system (for example, *torso*). The bottom row of drawings shows the appearances of larvae in which none or only one of the three gene systems is functional. The lettering beneath each larva specifies which systems are intact (A P T for a normal larva, –P T for a larva where the anterior system is defective but the posterior and terminal systems are intact, and so on). Inactivation of a particular gene system causes loss of the corresponding set of body structures; the body parts that form correspond to the gene systems that remain functional. Note that larvae with a defect in the anterior system can still form terminal structures at their anterior end, but these are of the type normally found at the rear end of the body rather than the front of the head. (Slightly modified from D. St. Johnston and C. Nüsslein-Volhard, *Cell* 68:201–219, 1992.© Cell Press.)

***Drosophila* I. Genesis of the Body Plan**

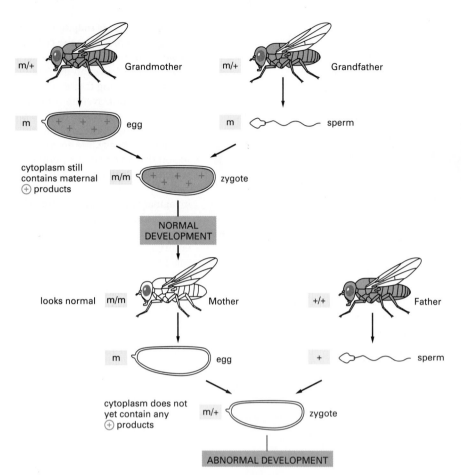

Figure 21–54 Inheritance of a recessive maternal-effect mutation. The pattern of inheritance is traced, starting with heterozygous (m/+) grandparents, for a mutation (m) that is recessive to the normal gene (+). The genotype of each animal or cell is shown to the left of it. *Red* color denotes presence of the normal (+) gene product. The gene product acts only at the beginning of development, and the appearance of the mature animal reflects the set of maternally specified components present in the egg. Note that the sperm makes no significant contribution of these gene products to the egg. The pattern of inheritance of a dominant maternal-effect mutation is different but can be worked out in a similar way.

governs the two extreme ends of the embryo, comprising the specialized nonsegmental terminal structures and in particular the pair of regions—one anterior, one posterior—from which the gut is derived. Like the dorsoventral system, each of these three subsystems sets up a morphogen gradient—one in the anterior half of the embryo, one in the posterior half (although this is somewhat controversial), and one operating symmetrically at both of the extreme ends of the embryo. Loss-of-function mutations that inactivate a particular subsystem cause a loss of the corresponding anterior, posterior, or terminal structures.

The four primary spatial signals—anterior, posterior, terminal, and ventral—organize the subsequent patterning of the embryo by governing the expression of other sets of genes, which serve to interpret, refine, and record the positional information that the primary signals supply.

The Patterning of the Embryo Begins with Influences from the Cells Surrounding the Egg [44, 45]

The egg-polarity genes are transcribed from the maternal genome during oogenesis, and their products act very soon after fertilization or in some cases even before. Thus the phenotype of the embryo is determined by the alleles present in the mother (and in her oocytes) rather than by the combination of maternal and paternal genes possessed by the embryo itself. Genes acting in this way are called **maternal-effect genes**. They are discovered by looking for the appropriate mutant phenotypes in the embryos produced from eggs laid by mothers who themselves appear normal but who carry a genetic mutation that makes their eggs abnormal (Figure 21–54). Most often, the maternal-effect mutation is recessive, and the mothers who make the defective eggs are homozygous for the mutant gene.

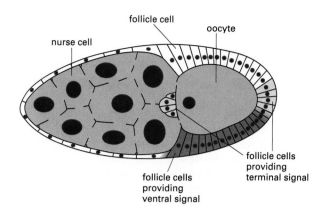

Figure 21–55 A *Drosophila* oocyte in its follicle. The oocyte is derived from a germ cell that divides four times to give a family of 16 cells that remain in communication with one another via cytoplasmic bridges. One member of the family group becomes the oocyte, while the others become nurse cells, which make many of the components required by the oocyte and pass them into it via the cytoplasmic bridges. The follicle cells that partially surround the oocyte have a separate ancestry; they are the sources of terminal and ventral egg-polarizing signals.

Once a gene has been identified, its site of action can be investigated by creating and analyzing *genetic mosaics*—flies containing marked clonal patches of cells in which the gene of interest is missing or mutated. (We explain later how this astonishing trick of genetic microsurgery is performed.) In the case of the egg-polarity genes it can be shown in this way that, while most are required in the oocyte lineage itself, a few crucial ones are required instead in the follicle cells that surround the oocyte in the ovary. The genes required in the follicle cells supply cues that act on the outside of the egg to localize the sources of the dorsoventral and terminal morphogen gradients that will develop inside it (Figure 21–55). In addition, localized products supplied to the growing oocyte by the giant nurse cells connected to it at one end serve to define the anteroposterior polarity of the egg.

To see how the patterns are set up inside the egg, we focus first on the **dorsoventral system**.

The Dorsoventral Axis Is Specified Inside the Embryo by a Gene Regulatory Protein with a Graded Intranuclear Concentration [44, 45, 46]

The role of the follicle cells in establishing the dorsoventral gradient in the *Drosophila* egg is to provide a localized signaling molecule that binds to a receptor on the outside of the egg and thereby controls the distribution of a gene regulatory protein inside the egg. The system can be analyzed genetically in much the same way as described earlier for the system mediating vulval induction in the nematode worm. Seven of the genes in the dorsoventral system are concerned with producing the localized extracellular signal; one, called *Toll*, encodes the transmembrane receptor for the signaling molecule, and the products of the remaining three act inside the embryo, downstream from *Toll*. The final maternal-effect gene in the signaling pathway codes for a gene regulatory protein and is called *dorsal*. The extracellular signaling molecule produced by the follicle cells is generated in active form only at the ventral surface of the egg and forms a gradient that is reflected in a graded activation of the Toll protein and ultimately in a graded concentration of the Dorsal protein in the nuclei of the embryo.

The Dorsal protein belongs to the same family as the NF-κB gene regulatory protein of vertebrates (see Figure 15–32) and is thought to act in a similar way. In the newly laid egg both the *dorsal* mRNA (detected by *in situ* hybridization) and the protein it encodes (detected with antibodies) are distributed uniformly in the cytoplasm. After the nuclei have migrated to the surface of the embryo to form the blastoderm, however, a remarkable redistribution of the Dorsal protein occurs: dorsally the protein remains in the cytoplasm, but ventrally it is concentrated in the nuclei, and between these two extremes there is a smooth gradient of nuclear localization (Figure 21–56). The partitioning of Dorsal protein between nucleus and cytoplasm appears to be governed, in part at least, by the product

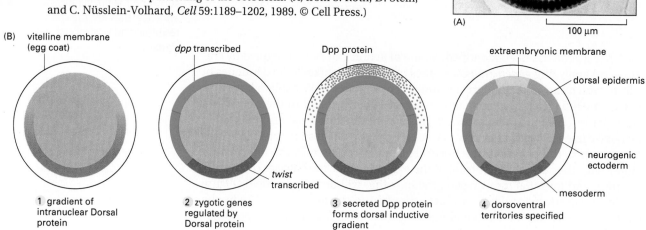

Figure 21–56 The gradient of the Dorsal protein and its interpretation.
(A) The concentration gradient of Dorsal protein in the nuclei of the
blastoderm, as revealed by an antibody. (B) The interpretation of the Dorsal
gradient by genes that demarcate the different dorsoventral territories; for
simplicity, only two representative genes are shown. Subsequent processes
will further subdivide these territories. The *decapentaplegic (dpp)* gene in
particular codes for a secreted factor that will act as a local morphogen to
control the detailed patterning of the ectoderm. (A, from S. Roth, D. Stein,
and C. Nüsslein-Volhard, *Cell* 59:1189–1202, 1989. © Cell Press.)

(A) 100 µm

(B) vitelline membrane (egg coat)

dpp transcribed

Dpp protein

extraembryonic membrane

dorsal epidermis

neurogenic ectoderm

mesoderm

twist transcribed

1 gradient of intranuclear Dorsal protein

2 zygotic genes regulated by Dorsal protein

3 secreted Dpp protein forms dorsal inductive gradient

4 dorsoventral territories specified

of a gene called *cactus*. The Cactus protein is homologous to the I-κB protein that
inhibits NF-κB in vertebrate cells by preventing it from migrating into the nucleus
(see Figure 15–32). By analogy, the Cactus protein is thought to bind to the Dorsal
protein, trapping it in the cytoplasm; the signal transmitted by the Toll protein
is thought to lead to the phosphorylation of the Dorsal protein, causing it to dissociate from the Cactus protein so that it can enter nuclei.

Once inside a nucleus the Dorsal protein turns on or off the expression of
different sets of genes depending on its concentration. In this way the gradient
of nuclear localization of the protein creates a dorsoventral series of territories—
distinctive bands of cells that run the length of the embryo. Most ventrally, where
the concentration of Dorsal protein is highest, it switches on, for example, expression of a gene called *twist*, specific for mesoderm. Most dorsally, where the concentration of Dorsal protein is lowest, a gene called *decapentaplegic (dpp)* is
permitted to switch on, specifying dorsal structures. And in an intermediate region, where the concentration of Dorsal protein is high enough to repress *dpp*
but too low to activate *twist*, the cells are specified to become neurogenic ectoderm (see Figure 21–56).

Products of the genes directly regulated by the Dorsal protein generate in
turn more local signals that define finer subdivisions of the dorsoventral axis. In
particular, *dpp* codes for a secreted protein of the TGF-β superfamily that is
thought to form a local morphogen gradient in the dorsal part of the embryo. The
action of this protein is reminiscent of the action of activin, also a TGF-β
family member, in early *Xenopus* development. From experiments with injected
dpp mRNA, it seems that the highest concentrations of Dpp protein cause
development of the most dorsal tissue of all—extraembryonic membrane—
intermediate concentrations cause development of dorsal ectoderm, and very
low concentrations allow development of neurogenic ectoderm.

Like the dorsoventral system, the **terminal system** depends on a transmembrane receptor that detects localized signals provided by follicle cells to gener-

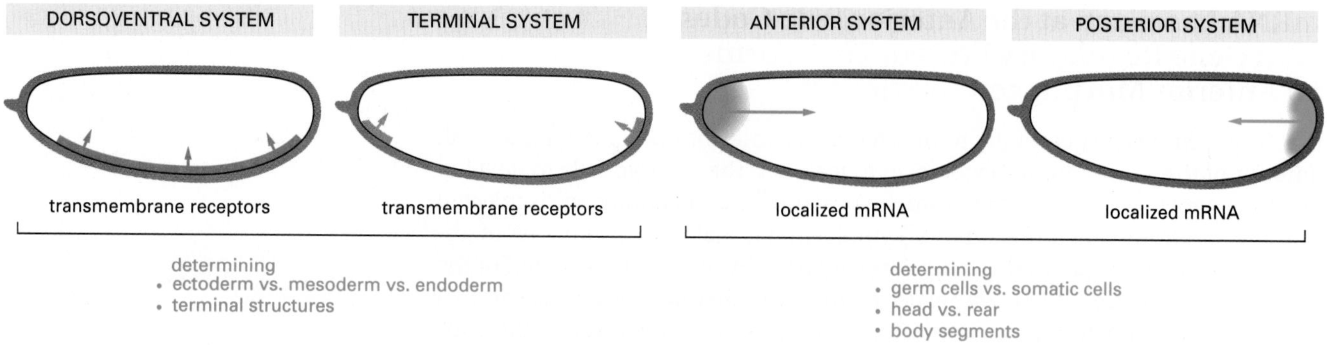

DORSOVENTRAL SYSTEM	TERMINAL SYSTEM	ANTERIOR SYSTEM	POSTERIOR SYSTEM
transmembrane receptors	transmembrane receptors	localized mRNA	localized mRNA

determining
- ectoderm vs. mesoderm vs. endoderm
- terminal structures

determining
- germ cells vs. somatic cells
- head vs. rear
- body segments

ate gradients of gene regulatory proteins inside the embryo (Figure 21–57). These gradients serve to specify gut endoderm, as well as some specialized terminal structures, and so can be viewed, with the dorsoventral system, as part of the apparatus for defining the three basic germ layers of the insect. The dorsoventral and terminal systems in the fly, therefore, employing secreted molecules that act as inductive signals, are comparable with the inductive mechanisms for specifying germ layers in the early *Xenopus* embryo.

The anterior and posterior systems of egg-polarity genes, by contrast, set up gradients that depend instead on localized accumulations of specific mRNAs inside the egg (see Figure 21–57). These gradients govern the differences between head and rear and specify the series of body segments along the head-to-rear axis, as we shall see in detail for the anterior system. First, we pause briefly to discuss a special role of the **posterior system**: the specification of germ cells.

The Posterior System Specifies Germ Cells as well as Posterior Body Segments [47]

In practically all animals that have been studied, the *primordial germ cells*—the precursors of the next generation of gametes—are singled out at a very early stage of development from the *somatic cells*—those that will form all the other tissues of the body (see Figure 21–51). In many species the egg contains localized cytoplasmic components—visible as *polar granules* in *C. elegans* and *Drosophila*, or as *germ plasm* in *Xenopus*—that are segregated into the primordial germ cells during egg cleavage and are suspected to include or to be associated with the determinants of germ-cell character. These components are generally concentrated at the posterior or vegetal end of the egg, and the cells that inherit them migrate from that site to colonize the gonads.

In *Drosophila* maternal-effect genes required for the formation of germ cells can be identified through the discovery of mutants that produce offspring in which germ cells are lacking. These abnormal offspring are found to lack posterior body segments also, indicating that the genes belong to the posterior system of egg-polarity genes. The products of many of these genes turn out to be localized at the posterior pole—among them, presumably, the determinants of germ-cell character. The morphogen gradient that organizes the posterior body segments depends on the machinery that creates and localizes the germ cell determinants. A key gene in this system is called *oskar*. Normally, *oskar* mRNA and protein are localized at the posterior pole of the egg. In their absence no germ cells develop there, and if *oskar* mRNA is artificially misdirected to the anterior end of the egg, germ cells will form there instead. The localized *oskar* mRNA, moreover, can be shown to control the localization of other components—products of other posterior-group genes involved in the development of posterior body segments as well as germ cells. By their localization at the posterior pole of the egg, these products can become specifically incorporated in the cells that form there, determining their fate as germ cells.

Drosophila I. Genesis of the Body Plan

Figure 21–57 **The organization of the four egg-polarity gradient systems.**

mRNA Localized at the Anterior Pole Codes for a Gene Regulatory Protein That Forms an Anterior Morphogen Gradient [44, 48]

If a *Drosophila* egg is carefully punctured at its anterior end, allowing a small amount of the most anterior cytoplasm to leak out, the embryo fails to develop head segments. And if cytoplasm from the posterior end of another egg is injected into the site from which the anterior cytoplasm has leaked, a second set of abdominal segments will develop, with reversed polarity, in the anterior half of the recipient egg (Figure 21–58). This experiment shows that the segmental patterning of the anteroposterior axis is controlled by substances localized at the ends of the egg. These substances have been identified by the genetic approach, starting with a search for mutations that mimic the effects of losing anterior or posterior cytoplasm. Most notably, mothers that are homozygous for a mutation in the egg-polarity gene *bicoid* produce embryos that lack head and thoracic structures and have abdominal structures extended over an abnormally large fraction of the body length. Such a mutant embryo can be rescued from abnormal development, however, if cytoplasm from the anterior end of a normal egg is injected into its anterior end. Thus the normal *bicoid* gene is required to make some product at the anterior end of the egg that can act as the source of a long-range influence controlling the pattern of development of the anterior parts.

In situ hybridization studies show that *bicoid* mRNA is originally synthesized in the ovary by the nurse cells connected with the oocyte (see Figure 21–55). As the *bicoid* mRNA passes through the cytoplasmic bridges into the oocyte, it becomes anchored by part of its 3′ untranslated tail to a component of the cytoplasm—presumably a part of the cytoskeleton—at the oocyte's anterior end. Translation of this mRNA begins only when the egg is laid, giving rise to a concentration gradient of Bicoid protein with its high point at the anterior end of the embryo. The concentration gradient can be altered genetically by constructing mutants that contain multiple copies of the normal *bicoid* gene: as the gene dosage increases in the mother, so does the protein concentration increase in the egg. The segments of the resultant embryo are correspondingly shifted toward the posterior pole, as though their locations were determined by positional information derived from the local concentration of the Bicoid protein (Figure 21–59). This protein therefore fits exactly the definition of a morphogen. Like Dorsal, the Bicoid protein binds to DNA and functions by regulating the expression of other genes.

Figure 21–58 Localized determinants at the ends of the *Drosophila* egg control its anteroposterior polarity. A little anterior cytoplasm is allowed to leak out of the anterior end of the egg and is replaced by an injection of posterior cytoplasm. The resulting double-posterior larva (*photograph on right*) is compared with a normal control (*photograph on left*); the substitution of cytoplasm at one end of the egg has had a long-range effect, converting all the more anterior segments into a mirror-image duplicate of the last three abdominal segments. The larvae are shown in dark-field illumination. (From H.G. Frohnhöfer, R. Lehmann, and C. Nüsslein-Volhard, *J. Embryol. Exp. Morphol.* 97[Suppl]:169–179, 1986, by permission of the Company of Biologists Ltd.)

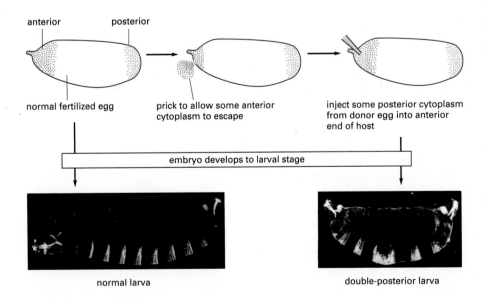

anterior   posterior

normal fertilized egg

prick to allow some anterior cytoplasm to escape

inject some posterior cytoplasm from donor egg into anterior end of host

embryo develops to larval stage

normal larva

double-posterior larva

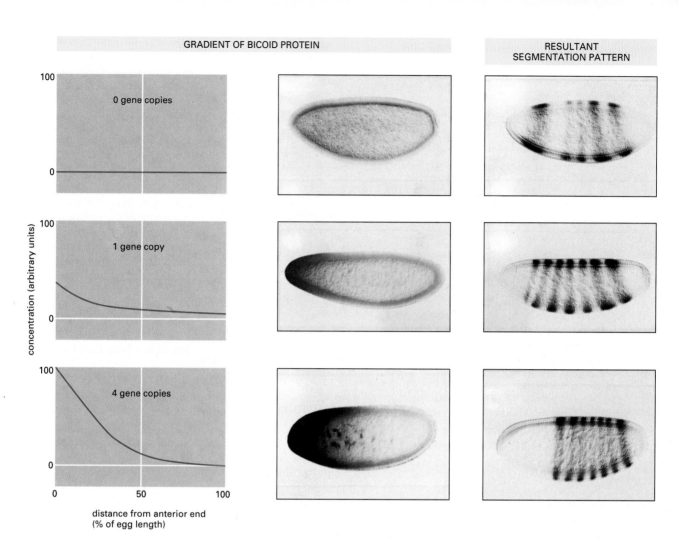

Figure 21–59 The gradient of Bicoid protein in the *Drosophila* egg and its effects on the pattern of segments. The gradient is revealed by staining with an antibody against the Bicoid protein; the segment pattern is revealed by an antibody against the product of a pair-rule gene, *even-skipped* (discussed later). Three embryos are compared, containing zero, one, and four copies, respectively, of the normal *bicoid* gene. With zero dosage of *bicoid*, segments with an anterior character do not form; with increasing gene dosage they form progressively farther from the anterior end of the egg, as expected if their position is determined by the local concentration of the Bicoid protein. Measurements of this concentration, as indicated by the intensity of staining, are shown in the graphs. Despite the considerable differences of position and spacing of the segment rudiments in the embryos with one and four doses of the gene, both embryos will develop into normally proportioned larvae and adults. A mechanism that may be responsible for this regulation is discussed on page 1064. (Slightly adapted from W. Driever and C. Nüsslein-Volhard, *Cell* 54:83–104, 1988. © Cell Press.)

Three Classes of Segmentation Genes Subdivide the Embryo [49]

Graded global cues are thus provided inside the egg by the products of the egg-polarity genes. For the anterior system the cues derive from the *bicoid* mRNA that is localized at the anterior end of the egg before fertilization, and they take the form of an anteroposterior gradient of the Bicoid gene regulatory protein. The gradient guides the creation of a series of discrete body segments. This process depends on a collection of **segmentation genes**, about 25 of which have been characterized. Mutations in any one of these genes will alter the number of segments or their basic internal organization without affecting the global polarity of the egg. The segmentation genes act at later stages than the egg-polarity genes, when the embryo is transcribing its own genome instead of relying on stored maternal mRNA. Because the embryonic gene transcripts, rather than maternal

Drosophila I. Genesis of the Body Plan

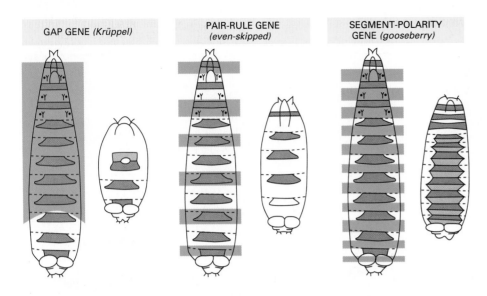

GAP GENE (Krüppel)	PAIR-RULE GENE (even-skipped)	SEGMENT-POLARITY GENE (gooseberry)

Figure 21–60 Examples of the phenotypes of mutations affecting the three types of segmentation genes. In each case the areas shaded in *green* on the normal larva (*left*) are deleted in the mutant or are replaced by mirror-image duplicates of the unaffected regions. By convention, dominant mutations are written with an initial capital letter and recessive mutations are written with a lower-case letter. Several of the patterning mutations of *Drosophila* are classed as dominant because they have a perceptible effect on the phenotype of the heterozygote, even though the characteristic major, lethal effects are recessive—that is, visible only in the homozygote. (Modified from C. Nüsslein-Volhard and E. Wieschaus, *Nature* 287:795–801, 1980. © 1980 Macmillan Magazines Ltd.)

transcripts, determine the phenotype, these genes are classed as **zygotic-effect genes** rather than maternal-effect genes.

The segmentation genes fall into three groups according to their mutant phenotypes and the stages at which they act (Figure 21–60). First come a set of at least six **gap genes**, whose products mark out the coarsest subdivisions of the embryo. Mutations in a gap gene eliminate one or more groups of adjacent segments, and mutations in different gap genes cause different but partially overlapping defects. In the mutant *Krüppel*, for example, the larva lacks eight segments, from T1 to A5 inclusive.

The next segmentation genes to act are a set of eight **pair-rule genes**. Mutations in these cause a series of deletions affecting alternate segments, leaving the embryo with only half as many segments as usual. While all the pair-rule mutants display this two-segment periodicity, they differ in the precise positioning of the deletions relative to the segmental or parasegmental borders. The pair-rule mutant *even-skipped*, for example, which is discussed in Chapter 9, lacks the whole of each even-numbered parasegment, while the pair-rule mutant *fushi tarazu (ftz)* lacks the whole of each odd-numbered parasegment, and the pair-rule mutant *hairy* lacks a series of regions that are of similar width but out of register with the parasegmental units.

Finally, there are at least 10 **segment-polarity genes**. Mutations in these genes produce larvae with a normal number of segments but with a part of each segment deleted and replaced by a mirror-image duplicate of all or part of the rest of the segment. In *gooseberry* mutants, for example, the posterior half of each segment (that is, the anterior half of each parasegment) is replaced by an approximate mirror image of the adjacent anterior half-segment (see Figure 21–60).

We see later that, in parallel with the segmentation process, a further set of genes, the *homeotic selector genes*, serve to define and preserve the differences between one parasegment and the next.

The phenotypes of the various segmentation mutants suggest that the segmentation genes form a coordinated system that subdivides the embryo progressively into smaller and smaller domains along the anteroposterior axis distinguished by different patterns of gene expression. Again, molecular genetics provides the tools to investigate how this system works.

The Localized Expression of Segmentation Genes Is Regulated by a Hierarchy of Positional Signals [50, 51]

Most of the segmentation genes have been cloned, and cDNA sequencing reveals that about three-quarters of them, including all of the gap genes, code for gene regulatory proteins. Their actions on one another and on other genes can there-

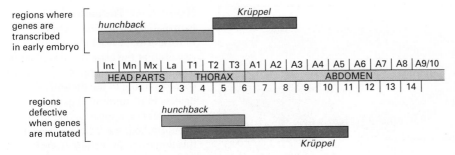
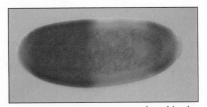
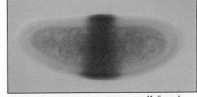

regions where genes are transcribed in early embryo

Krüppel

hunchback

Int	Mn	Mx	La	T1	T2	T3	A1	A2	A3	A4	A5	A6	A7	A8	A9/10
HEAD PARTS				THORAX			ABDOMEN								
	1	2	3	4	5	6	7	8	9	10	11	12	13	14	

regions defective when genes are mutated

hunchback

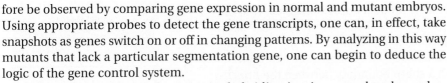

Krüppel

(A)

hunchback

(B)   *Krüppel*

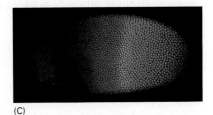

(C)

Figure 21–61 The spatial domains of the gap genes *hunchback* and *Krüppel*. Both genes code for gene regulatory proteins of the zinc-finger class. (A) Diagram of the main, anterior domains of *hunchback* and *Krüppel* showing how the defect caused by an absence of functional *hunchback* or *Krüppel* product extends outside the region where the gene transcripts are normally found. (B) The normal distribution of *hunchback* and *Krüppel* transcripts as seen by *in situ* hybridization at the blastoderm stage. (C) The normal distribution of Krüppel protein (*red*) and Hunchback protein (*green*) as demonstrated with fluorescent antibodies. A region of overlap, where both proteins are present, appears *yellow;* more sensitive staining would reveal more extensive overlap. The proteins spread outside their respective gene transcription domains and are thought to act as local morphogens helping to regulate expression of other genes (including gap genes and pair-rule genes). (A, adapted from M. Hülskamp and D. Tautz, *BioEssays* 13:261–268, 1991; B, courtesy of Diethard Tautz; C, courtesy of Jim Langeland, Steve Paddock, Sean Carroll, and the Howard Hughes Medical Institute.)

fore be observed by comparing gene expression in normal and mutant embryos. Using appropriate probes to detect the gene transcripts, one can, in effect, take snapshots as genes switch on or off in changing patterns. By analyzing in this way mutants that lack a particular segmentation gene, one can begin to deduce the logic of the gene control system.

We have already seen how *in situ* hybridization in normal embryos has helped to show that the *bicoid* gene transcripts are the source of a positional signal: the transcripts are localized at one end of the egg, even though the effects of a mutation in the gene are spread over a large part of the embryo. In a similar way it can be shown that the gap genes in their turn generate (directly or indirectly) positional signals that help to control the pattern of development in neighborhoods extending beyond their own expression domain. Mutants that are defective in the gap gene *Krüppel* or *hunchback,* for example, show abnormalities within the region where the gene transcripts are detected in a normal embryo and also for several segments beyond (Figure 21–61). As with the Bicoid protein, it is thought that the gene regulatory proteins encoded by gap genes such as *Krüppel* and *hunchback* spread out as diffusible morphogens from the sites where the genes are transcribed.

The next finer level of spatial patterning is marked out by the pair-rule genes. Some of these, too, may code for proteins that spread by diffusion to exert effects on cells neighboring the site of gene transcription; others, by contrast, appear to affect the development only of those regions in which they are transcribed. Transcripts of the normal *ftz* gene, for example, occur in seven circumferential "zebra stripes" at the blastoderm stage (Figure 21–62), each of the stripes being roughly four cells wide, matching in width and location the rudiments of the even-numbered parasegments that would be missing in a *ftz* mutant.

Taken together, these observations imply that the products of the egg-polarity genes provide global positional signals that cause particular gap genes to be expressed in particular regions, and the products of the gap genes then provide a second tier of positional signals that act more locally to regulate finer details of patterning by influencing the expression of yet other genes, including the pair-

Drosophila I. Genesis of the Body Plan

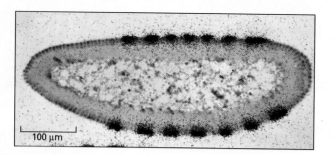

100 μm

Figure 21–62 The pattern of *ftz* gene expression in the *Drosophila* blastoderm. *In situ* hybridization reveals that the gene is transcribed in a pattern of seven stripes corresponding to the pattern of defects in *ftz* mutants. The bands of *ftz* expression appear as black patches of autoradiographic silver grains in this longitudinal section. (Courtesy of Philip Ingham.)

rule genes. In this way the global gradients produced by the egg-polarity genes organize the creation of a fine-grained pattern through a process of sequential subdivision using a hierarchy of sequential positional controls. This is a reliable strategy: because the global positional signals do not have to specify fine details, the individual nuclei that respond to them do not have to react with extreme precision to small differences of signal concentration (Figure 21–63).

The Product of One Segmentation Gene Controls the Expression of Another to Create a Detailed Pattern [41, 50, 51, 52]

The hierarchy of control relationships between the successive tiers of segmentation genes can be demonstrated by observing the expression pattern of one such gene when another is inactivated by mutation. In a mutant embryo that lacks the normal *Krüppel* product, for example, the usual *ftz* stripes fail to develop in just that region of the blastoderm corresponding to the defect in the *Krüppel* mutant. Thus the *Krüppel* product, directly or indirectly, regulates *ftz* gene expression. In a *ftz* mutant, by contrast, the distribution of the normal *Krüppel* product is not disturbed, indicating that the *ftz* product does not regulate *Krüppel* gene expression.

There are also interactions between genes in the same tier of the regulatory hierarchy. The gap genes *Krüppel* and *hunchback*, for example, are expressed in adjacent regions of the blastoderm, with a sharp boundary between the *hunch-*

Figure 21–63 Two strategies for using signal concentration gradients to specify a fine-grained pattern of cells in different states. In (A) there is only one signal gradient, and cells select their states by responding accurately to small changes of signal concentration. In (B) the initial signal gradient controls establishment of a small number of more local signals, which control establishment of other still more narrowly local signals, and so on. Because there are multiple local signals, the cells do not have to respond very precisely to any single signal in order to create the correct spatial array of cell states. Case B corresponds more closely to the strategy of the real embryo.

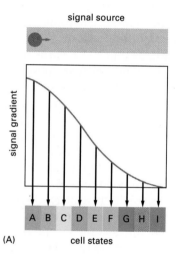

(A)

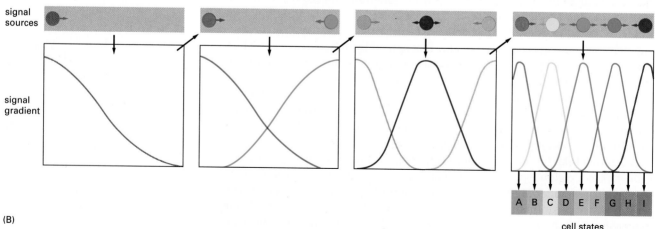

(B)

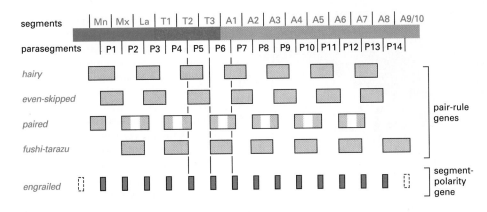

| segments | | Mn | Mx | La | T1 | T2 | T3 | A1 | A2 | A3 | A4 | A5 | A6 | A7 | A8 | A9/10 |

| parasegments | P1 | P2 | P3 | P4 | P5 | P6 | P7 | P8 | P9 | P10 | P11 | P12 | P13 | P14 |

hairy

even-skipped

paired

fushi-tarazu

pair-rule genes

engrailed

segment-polarity gene

Figure 21–64 How pair-rule genes define segments in the *Drosophila* blastoderm. The diagram shows the pattern of transcription of four of the eight known pair-rule genes and of one of the segment-polarity genes, *engrailed*. Although each pair-rule gene by itself defines only a simple alternation with a repeat distance of two segments, the whole set of pair-rule genes in combination, by their pattern of adjacency and overlap, potentially defines a much finer subdivision of the blastoderm into stripes only one cell wide, such as those in which the *engrailed* gene is expressed. (After M. Akam, *Development* 101:1–22, 1987, by permission of the Company of Biologists Ltd.)

back territory anteriorly and the *Krüppel* territory posteriorly (see Figure 21–61). A repression of *Krüppel* gene expression by the Hunchback gene regulatory protein helps to establish this boundary, ensuring that the expression domains of the two genes are properly correlated. Interactions of this sort also guide the regular periodic pattern of expression of the pair-rule genes, setting up an exactly reproducible arrangement of mutual exclusions and overlaps that repeats itself reliably in every double-segment unit in the blastoderm of every normal embryo (Figures 21–64 and 21–65). In this way different bands of cells around the blastoderm are distinguished by different combinations of pair-rule gene expression, down to the finest possible level of detail—the width of a single cell, which corresponds to about a quarter of the width of a prospective segment or parasegment.

This whole elaborate patterning process depends on the long stretches of DNA sequence that control the expression of each of the segmentation genes. These regulatory regions bind multiple copies of the gene regulatory proteins produced by a subset of other segmentation genes, and the gene is turned on or off according to the combination of proteins bound. In Chapter 9 (see p. 426) we focus on one particular segmentation gene and discuss how the decision whether to transcribe the gene is made on the basis of all these inputs.

Egg-Polarity, Gap, and Pair-Rule Genes Create a Transient Pattern That Is Remembered by Other Genes [41, 50, 52]

Within the first few hours after fertilization, the gap genes and the pair-rule genes are activated one after another. Their mRNA products appear first in patterns that only approximate the final picture; then, within a short time—through a series of interactive adjustments—the fuzzy initial distribution of gene products resolves itself into a regular, crisply defined system of stripes (see Figure 21–65). But this system itself is unstable and transient. As the embryo proceeds through gastrulation and beyond, the regular segmental pattern of gap and pair-rule gene products disintegrates. Their actions, however, have stamped a permanent set of labels (positional values) on the cells of the blastoderm. These positional labels are recorded in the persistent activation of certain of the segment-polarity genes and of the homeotic selector genes, which serve to maintain the segmental organization of the larva and adult.

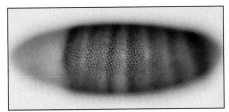

3 hours after fertilization

3½ hours after fertilization

Figure 21–65 The formation of *ftz* and *eve* stripes in the *Drosophila* blastoderm. Genes *ftz* and *eve* are both pair-rule genes. Their expression patterns (shown in *brown* for *ftz* and in *gray* for *eve*) are at first blurred but rapidly resolve into sharply defined stripes. (From P.A. Lawrence, The Making of a Fly. Oxford, UK: Blackwell, 1992.)

Drosophila I. Genesis of the Body Plan

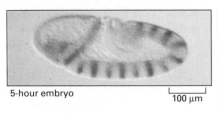

5-hour embryo   100 μm

adult   500 μm

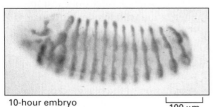

10-hour embryo   100 μm

Figure 21–66 The pattern of expression of *engrailed*, a segment-polarity gene. The *engrailed* pattern is shown in a 5-hour embryo (at the extended germ-band stage), a 10-hour embryo, and an adult (whose wings have been removed in this preparation). The pattern is revealed by an antibody (*brown*) against the Engrailed protein (for the 5- and 10-hour embryos) or (for the adult) by constructing a strain of *Drosophila* containing the control sequences of the *engrailed* gene coupled to the coding sequence of the enzyme β-galactosidase, whose presence is easily detected histochemically through the *blue* product of a reaction that it catalyzes. Note that the *engrailed* pattern, once established, is preserved throughout the animal's life. (Courtesy of Tom Kornberg and Cory Hama.)

Segment-Polarity Genes Label the Basic Subdivisions of Every Parasegment [53]

Segment-polarity genes are expressed in a pattern that repeats itself from one parasegment to the next. The gene *engrailed* provides a good example (Figure 21–66). Its RNA transcripts are seen in the cellular blastoderm in a series of 14 bands, each approximately one cell wide, corresponding to the anteriormost portions of the future parasegments. These bands appear in a fixed relationship to the bands of expression of the pair-rule genes (see Figure 21–64). Again, the pattern is governed in a combinatorial fashion by the products of the previous set of genes in the hierarchy and is refined and elaborated by interactions among the segment-polarity genes themselves. Through expression of different segment-polarity genes in different bands of cells, each future parasegment is already subdivided at the cellular blastoderm stage into at least three distinct regions. The chemical distinctions will persist, maintained by continued transcription of at least some of the segment-polarity genes, after the pair-rule gene products have largely disappeared (see Figure 21–66). Some of the segment-polarity genes thus expressed—including, in particular, one called *wingless*—encode secreted proteins that act also during subsequent development as spatial signals within the parasegment to regulate the details of its internal patterning and growth.

Besides regulating the segment-polarity genes, the products of pair-rule genes collaborate with the products of gap genes (and perhaps egg-polarity genes) to cause the precisely localized activation of a further set of spatial labels—the homeotic selector genes, which permanently distinguish one parasegment from another. In the next section we examine these selector genes in detail and consider their role in cell memory.

Summary

Like other insects, Drosophila *is constructed from a series of repeating modular units called segments, with specialized nonsegmental structures at each end of the body. Each major subdivision of each segment is distinguished by the expression of a particular selection of control genes that defines its "address." The pattern originates with asymmetry in the egg: positional information is supplied by four gradients set up by the products of four groups of maternal-effect genes called egg-polarity genes. The four groups of genes control four distinctions fundamental to the body plan of animals: dorsal versus ventral, endoderm versus mesoderm and ectoderm, germ cells versus somatic cells, and head versus rear. The egg-polarity genes operate by setting up graded distributions of gene regulatory proteins in the egg and early embryo, but the gradients are set up differently for the different egg axes.*

The dorsoventral polarity is defined by a localized signal from the follicle cells that surround the egg. The signal molecule binds to transmembrane receptors in the ventral surface of the egg, leading ultimately to a graded intranuclear concentration of the gene regulatory protein Dorsal along the dorsoventral axis of the early embryo. The Dorsal protein regulates expression of other genes, including dpp, *whose product acts in turn as a morphogen to specify finer subdivisions of the dorsoventral axis, like the early inductive signals that operate in* Xenopus.

In the case of the anterior group of egg-polarity genes, the gradient arises from a localized deposit of mRNA, the product of the bicoid *gene, at the anterior end of the egg. Because the insect egg develops initially as a syncytium, the Bicoid protein translated from this mRNA is able to diffuse in the cytosol along the length of the embryo, guiding the global organization of its anterior half. The Bicoid concentration gradient initiates the orderly expression of gap genes, pair-rule genes, segment-polarity genes, and homeotic selector genes. These, through a hierarchy of interactions, become expressed in some regions of the embryo and not others, progressively subdividing the body into a regular series of segmental and subsegmental units.*

Drosophila and the Molecular Genetics of Pattern Formation. II. Homeotic Selector Genes and the Patterning of Body Parts [41, 50]

The first glimpses of the system of genes for pattern formation came over 70 years ago, with the discovery of the first of a set of mutations in *Drosophila* that cause bizarre disturbances of the organization of the adult fly. In the mutation *Antennapedia*, for example, legs sprout from the head in place of antennae (Figure 21–67), while in the mutation *bithorax*, portions of an extra pair of wings appear where normally there should be the much smaller appendages called halteres. These mutations transform parts of the body into structures appropriate to other positions and are called *homeotic*. A whole set of **homeotic selector genes** determines the anteroposterior character of the segments of the fly. In this section we follow *Drosophila* development through to the final steps in the formation of the adult fly to see how the homeotic selector genes do their job. At the end of the section we see that the same genes have a central role in patterning the body parts of other animals, including ourselves.

The Homeotic Selector Genes of the Bithorax Complex and the Antennapedia Complex Specify the Differences Among Parasegments [54, 55]

The homeotic selector genes of interest to us here all lie in one or the other of two tight gene clusters known as the **bithorax complex** and the **Antennapedia complex**. Each complex contains several genes with analogous functions: those in the bithorax complex control the differences among the abdominal and thoracic segments of the body, while those in the Antennapedia complex control the differences among thoracic and head segments. In some other insects the corresponding groups of genes all lie in a single complex, called the *HOM complex;* the Antennapedia and bithorax complexes are thus thought to be the two halves of a single HOM complex that has become split in the course of the fly's evolution. Each homeotic selector gene has a characteristic domain of action, defined as the region of the body that is transformed as a result of mutation in that gene. Typically, this domain has sharp boundaries that are roughly half a segment out of register with the conventional segment boundaries, indicating that the domain is a parasegment or a block of parasegments (see Figure 21–50).

Many of the mutations of homeotic selector genes have a recessive lethal phenotype and allow the embryo to survive only to around the time of hatching. Observations of embryos or very early larvae therefore give the clearest and in some respects most complete picture of the role of the homeotic selector genes.

Figure 21–67 A homeotic mutation. The fly shown here is an *Antennapedia* mutant. Its antennae are converted into leg structures by a mutation in the *Antennapedia* gene that causes it to be expressed in the head. Compare with the normal fly shown in Figure 21–47. (Courtesy of Matthew Scott.)

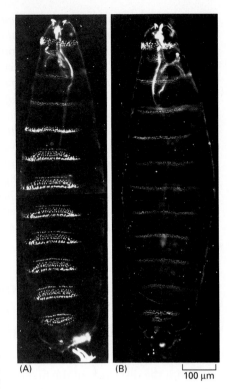

Figure 21–68 The effect of deleting most of the genes of the bithorax complex. (A) A normal *Drosophila* larva shown in dark-field illumination; (B) the mutant larva with the bithorax complex largely deleted. In the mutant the parasegments posterior to P5 all have the appearance of P5. (Courtesy of Gary Struhl; A, from *Nature* 293:36–41. © 1981 Macmillan Journals Ltd.)

(A)   (B)   100 µm

Larvae that are deficient in all the genes of the bithorax complex have a particularly simple structure: the head and anterior thorax are normal as far as the P4 parasegment, but all of the remaining 10 parasegments are converted to the character of P4. Partial deletions of the bithorax complex cause transformations that are less extensive (Figure 21–68). These observations, and analogous findings for the Antennapedia complex, illustrate the essential role of the homeotic selector genes in defining the differences among the parasegments: when the genes are missing, the distinctions between one parasegment and another are not made.

Homeotic Selector Genes Encode a System of Molecular Address Labels [50, 56]

Like the segmentation genes, the homeotic selector genes are first activated in the blastoderm. Since all of the DNA in the Antennapedia and bithorax complexes has been cloned, nucleic acid probes are available to map the spatial pattern of transcription of each of the homeotic selector genes by *in situ* hybridization. The conclusions from these studies are striking: to a first approximation each homeotic selector gene is normally expressed in just those regions that develop abnormally, as though misplaced, when the gene is mutated or absent.

The products of the selector genes can thus be viewed as molecular address labels possessed by the cells of each parasegment. If the address labels are changed, the parasegment behaves as though it were located somewhere else. Because the segmentation genes help to control the activation of the homeotic selector genes, the pattern of homeotic selector gene expression is in exact register with the parasegmental boundaries defined by the pair-rule and segment-polarity gene products. In this way the combination of a particular homeotic selector gene product (or set of such products) with a particular set of segmentation gene products reliably defines a unique address carried only by the cells in one subdivision of one segment.

Although the pattern of expression of the homeotic selector genes undergoes complex adjustments as development proceeds, these genes continue to play a crucial part throughout the subsequent development of the fly. They somehow equip cells with a memory of their positional value.

The Control Regions of the Homeotic Selector Genes Act as Memory Chips for Positional Information [54, 57, 58, 59]

The products of the homeotic selector genes, as discussed in Chapter 9, are gene regulatory proteins, all homologous to one another and all containing a highly conserved *homeobox sequence*, which codes for a DNA-binding *homeodomain* (60 amino acids long) in the corresponding proteins. Although many other genes also contain a homeobox, the particular type of homeobox sequence found in the homeotic selector genes is characteristic.

There are eight homeotic selector genes in the Antennapedia and bithorax complexes (which, for convenience, we shall refer to collectively as the HOM complex). Their coding sequences are interspersed amid a much larger quantity—a total of about 650,000 nucleotide pairs—of regulatory DNA. This DNA includes binding sites for the products of egg-polarity and segmentation genes—genes such as *bicoid, hunchback,* and *even-skipped*. The regulatory DNA in the

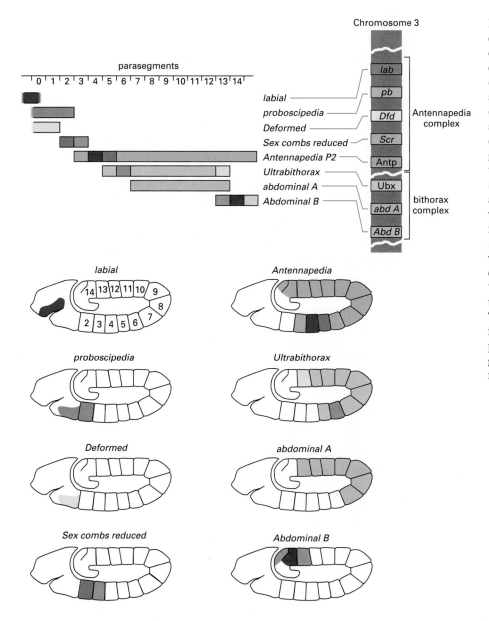

Figure 21–69 The patterns of expression compared to the chromosomal locations of the genes of the HOM complex. The sequence of genes in each of the two subdivisions of the chromosomal complex corresponds to the spatial sequence in which the genes are expressed. Note that most of the genes are expressed at a high level throughout one parasegment (*dark color*) and at a lower level in some adjacent parasegments (*medium color* where the presence of the transcripts is necessary for a normal phenotype, *light color* where it is not). In regions where the expression domains overlap, it is usually the most "posterior" of the locally active genes that determines the local phenotype. The drawings in the lower part of the figure represent the gene expression patterns in embryos at the extended germ band stage, about 5 hours after fertilization.

HOM complex acts as an interpreter of the multiple items of positional information supplied to it by all these factors, and, in response to them, it makes a decision to transcribe or not to transcribe a particular set of homeotic selector genes. There are, however, some deep mysteries about how the HOM control system is organized and how it operates.

One remarkable feature is that the sequence in which the genes are ordered along the chromosome in both the Antennapedia and the bithorax complexes corresponds almost exactly to the order in which they are expressed along the axis of the body (Figure 21–69). It is as though the genes are activated serially by some process that spreads farther and farther along the chromosome in proportion to some intracellular indicator of distance along the body axis. It is not clear whether this ordering is merely an accident of evolution or truly reflects involvement of some activation mechanism that propagates along the chromosome, although we shall see later that it is a feature of the HOM complex that has been highly conserved in the course of evolution.

There is a further puzzle. The HOM complex serves to make each parasegment different from the next, but the number of homeotic selector genes is smaller than the number of parasegments. The bithorax complex, for example, contains just three genes, but it is responsible for the differences between 10 parasegments (see Figure 21–69). Moreover, there are many mutations, mapping to different sites in the complex, that alter the anteroposterior character of only a single parasegment or even of a part of a parasegment. Most of these mutations lie in noncoding *control regions* and are also ordered along the chromosome in a sequence that matches in detail the anatomical ordering of the regions they affect. This suggests that the differences between body regions are defined not simply by the presence of different homeotic selector gene products but, more subtly, by persistent differences of some sort in the states of the control regions associated with those genes. A control region, in this view, is to be pictured not as a simple on-off switch but as something more like a computer microchip: it receives inputs (in the form of gene regulatory factors and other molecules that bind to it), it produces an output (in the form of a directive to transcribe or not to transcribe the homeotic selector gene), and it can store a memory trace (a record of positional information) that affects the way the output is computed from the inputs. The positional value of a cell thus will not necessarily be reflected in a certain fixed level of expression of the homeotic selector gene but rather in a particular way of regulating that gene in response to changing conditions.

All this remains speculative as long as we have no answer to a third and most fundamental question about the HOM complex: what mechanism maintains the memory trace? As discussed earlier (see p. 1062), one possibility is that the mechanism involves positive feedback, where the product of a gene, once it is made, stimulates its own transcription. At least some of the homeotic selector genes seem to have this property. The gene *Deformed* (in the Antennapedia complex), for example, has multiple binding sites for the Deformed protein in its upstream control region, and in some cells these are sufficient for it to keep itself activated once activity has been triggered. Such self-stimulatory effects, however, are not sufficient by themselves to maintain the memory trace in most cells. A whole additional set of genes, called the *Polycomb group*, have been found to be required to keep silent those homeotic selector genes that should not be expressed: if any of the Polycomb-group genes are inactivated by mutations, the homeotic selector genes are initially switched on in a normal pattern but then become activated indiscriminately all over the embryo (Figure 21–70A). The Polycomb protein is bound to the chromatin of the genes it controls (Figure 21–70B). Moreover, related genes appear to be involved elsewhere in the control of

Figure 21–70 Action of genes of the Polycomb group. (A) Photograph of a mutant embryo defective for the gene *extra sex combs (esc)* and derived from a mother also lacking this gene. The gene belongs to the Polycomb group. Essentially all segments have been transformed to resemble the most posterior abdominal segment (compare with Figure 21–68). In the mutant the pattern of expression of the homeotic selector genes, which is roughly normal initially, is unstable in such a way that all these genes soon become switched on all along the body axis. (B) The normal pattern of binding of Polycomb protein to *Drosophila* giant chromosomes, visualized with an antibody against Polycomb. The protein is bound to the Antennapedia complex (ANT-C) and the bithorax complex (BX-C) as well as about 60 other sites. (A, from G. Struhl, *Nature* 293:36–41, 1981. © 1981 Macmillan Journals Ltd.; B, courtesy of B. Zink and R. Paro, from R. Paro, *Trends Genet.* 6:416–421, 1990.)

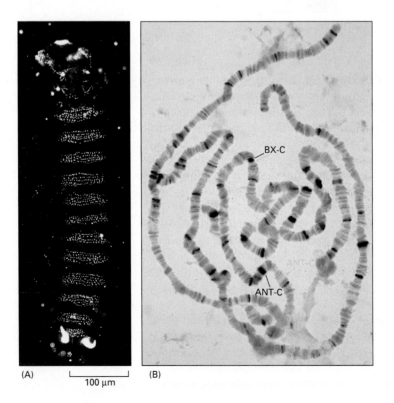

(A)

100 µm

(B)

BX-C

ANT-C

chromatin structure, suggesting that the memory of positional value may be carried by some persistent local modification of the chromatin in the HOM gene complex.

The Adult Fly Develops from a Set of Imaginal Discs That Carry Remembered Positional Information [60]

The basic pattern of expression of the homeotic selector genes is established in the *Drosophila* embryo and determines the structure not only of the larva, but also, much later, that of the adult fly. To appreciate fully the role of these genes as carriers of a positional memory, it is necessary to have some idea of the curious way in which the adult, or *imago*, finally develops.

The adult fly is formed largely from groups of cells, called *imaginal cells,* that are set aside, apparently undifferentiated, in each segment of the larva. The imaginal cells for most of the adult body originate from the embryonic epidermis—the epithelium that covers the body. They remain connected with the epidermis of the larva, and they will form mainly the epidermal structures of the adult fly. The imaginal cells for the head, thorax, and genitalia are organized into **imaginal discs**; other clusters of imaginal cells will form the abdomen. There are also groups of imaginal cells in the viscera of the larva to give rise to the internal organs of the fly. Detailed studies have focused chiefly on the imaginal discs. There are 19 of these, arranged as 9 pairs on either side of the larva plus 1 disc in the midline (Figure 21–71). The discs are pouches of epithelium, shaped like crumpled and flattened balloons, that evaginate (turn inside out), extend, and differentiate at metamorphosis. The eyes and antennae develop from one pair of discs, the wings and part of the thorax from another, the first pair of legs from another, and so on.

The cells of one imaginal disc look like those of another, and when they differentiate, they will give rise to generally similar sets of specialized cell types. But grafting experiments show that they are in fact already regionally determined and nonequivalent. If one imaginal disc is transplanted into the position of another in the larva and the larva is then left to go through metamorphosis, the grafted disc is found to differentiate autonomously into the structure appropriate to its origin, regardless of its new site. This implies that the imaginal disc cells are governed by a memory of their original position. By an ingenious grafting procedure that lets the imaginal disc cells proliferate for an extended period before differentiating, it can be shown that this cell memory is stably heritable (with rare lapses) through an indefinitely large number of cell generations (Figure 21–72).

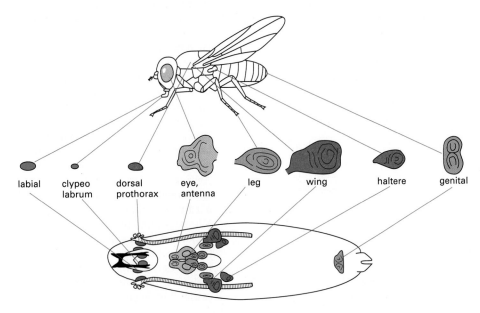

labial   clypeo   dorsal   eye,   leg   wing   haltere   genital
          labrum   prothorax   antenna

Figure 21–71 The imaginal discs in the *Drosophila* larva and the adult structures they give rise to. (After J.W. Fristrom et al., in Problems in Biology: RNA in Development [E.W. Hanley, ed.], p. 382. Salt Lake City: University of Utah Press, 1969.)

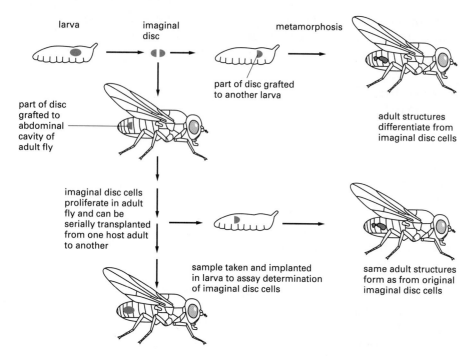

larva   imaginal disc   metamorphosis

part of disc grafted to another larva

adult structures differentiate from imaginal disc cells

part of disc grafted to abdominal cavity of adult fly

imaginal disc cells proliferate in adult fly and can be serially transplanted from one host adult to another

sample taken and implanted in larva to assay determination of imaginal disc cells

same adult structures form as from original imaginal disc cells

Figure 21–72 Experiments to test the state of determination of imaginal disc cells. The method of assay is to implant the cells in a larva that is about to undergo metamorphosis; the cells then differentiate to form recognizable adult structures, which lie, however, inside the body of the host fly after metamorphosis and are not integrated with it. The disc cells can either be assayed immediately or be implanted in the abdomen of adult flies, which serve as a natural culture chamber. Hormonal conditions in the adult allow the imaginal disc cells that have thus bypassed metamorphosis to continue to proliferate for an indefinite period, without differentiating, before the assay for cell determination is done. In both cases the cells generally differentiate to form the structures appropriate to the disc from which they derived originally.

The homeotic selector genes are essential components of the memory mechanism. If they are eliminated from imaginal disc cells at any stage in the long period leading up to differentiation at metamorphosis, the cells will differentiate into incorrect structures, as though they belonged to a different segment of the body. This can be demonstrated by the very powerful technique of *x-ray-induced mitotic recombination*—in effect, a form of genetic surgery on individual cells by means of which mutant clones of cells of a specified genotype can be generated at a chosen time in development, as we now explain.

Homeotic Selector Genes Are Essential for the Memory of Positional Information in Imaginal Disc Cells [61]

A short pulse of x-irradiation, as a side effect of the damage it does to DNA, can provoke crossing over between homologous chromosomes in a dividing cell—an event that would normally occur only at meiosis. As explained in Figure 21–73, if the cell is heterozygous for a gene in the crossed-over chromosomal region, the

Figure 21–73 Mitotic recombination (B) compared with normal mitosis (A). The diagrams follow the fate of a single pair of homologous chromosomes, one from the father (*shaded*), the other from the mother (*unshaded*). These chromosomes contain a locus for a pigmentation gene (or other marker gene) with a wild-type allele *A* (small *white square* on paternal chromosome) and a recessive mutant allele *a* (small *red square* on maternal chromosome) such that a homozygous *A/A* or heterozygous *A/a* cell has a normal appearance (shown as *white*) and a homozygous *a/a* cell has an altered appearance (shown as *orange*). Recombination by exchange of DNA between the maternal and paternal chromosomes can give rise to a pair of daughter cells, one homozygous *A/A* and therefore still normal in appearance, the other homozygous *a/a* and therefore visibly different. Mitotic recombination is a rare accidental event and occurs without the specialized apparatus that facilitates recombination during meiosis. A pulse of x-irradiation causes it to occur more frequently.

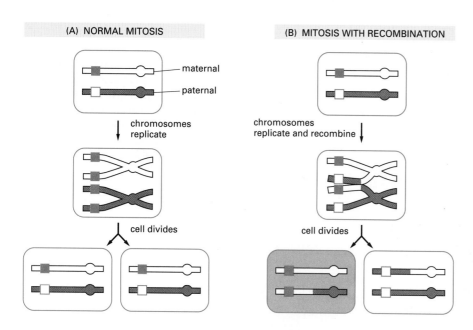

(A) NORMAL MITOSIS

maternal
paternal

chromosomes replicate

cell divides

(B) MITOSIS WITH RECOMBINATION

chromosomes replicate and recombine

cell divides

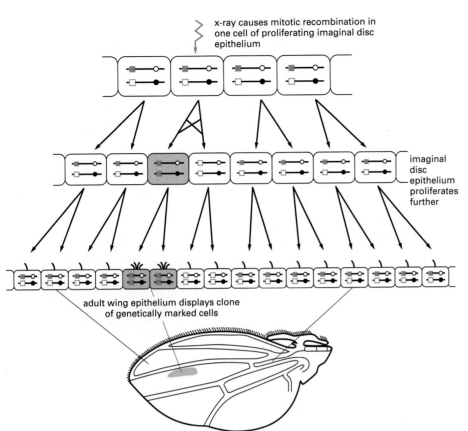

Figure 21–74 **How mitotic recombination is used to produce a clone of genetically marked mutant cells in the *Drosophila* wing.** The earlier the stage at which recombination occurs, the larger the eventual clone will be.

x-ray causes mitotic recombination in one cell of proliferating imaginal disc epithelium

imaginal disc epithelium proliferates further

adult wing epithelium displays clone of genetically marked cells

process can result in a pair of daughter cells that are homozygous, the one receiving two copies of the maternal allele of the gene, the other receiving two copies of the paternal allele. The occurrence of the cross-over can be detected if the animal is chosen to be also heterozygous for a mutation in a marker gene—a pigmentation gene, for example—that lies near the gene of interest and so undergoes crossing over in company with it. In this way marked homozygous mutant clones of cells can be created to order (Figure 21–74).

The major effects of mutations in homeotic selector genes are generally recessive: only the homozygous mutant organism shows the homeotic transformation. By exploiting mitotic recombination, one can create a clonal patch of marked homozygous homeotic mutant cells in an imaginal disc and examine their behavior in a heterozygous, phenotypically normal background. The finding is that the marked cells, and only the marked cells, show the homeotic transformation (provided that they lie in the normal domain of action of the homeotic selector gene), and this applies whether the recombination event was provoked early in development or late. A 2-day larva heterozygous for a mutation that destroys the function of the *Ultrabithorax (Ubx)* gene (in the bithorax complex), for example, can be x-irradiated to produce isolated clones of homozygous cells in its imaginal discs that contain no functional *Ubx* gene. These clones, if they lie in the haltere disc, will give rise to patches of wing-type tissue in the haltere. These and other observations indicate that each cell's memory of positional information depends on the continued activity of the normal homeotic selector gene. This memory, furthermore, is expressed in a cell-autonomous fashion—each cell maintains its state independently, depending on its own history and genome, regardless of its neighbors.

The Homeotic Selector Genes and Segment-Polarity Genes Define Compartments of the Body [53, 62]

The remembered distinctions specified by the homeotic selector genes are discrete: there is an abrupt difference of gene expression between cells in adjacent

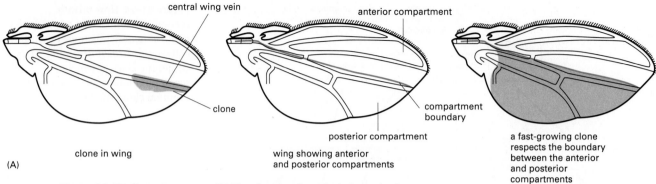

central wing vein

anterior compartment

clone

compartment boundary

posterior compartment

clone in wing

wing showing anterior and posterior compartments

a fast-growing clone respects the boundary between the anterior and posterior compartments

(A)

Figure 21–75 Compartments. (A) The shapes of marked clones in the *Drosophila* wing reveal the existence of a compartment boundary. The border of each marked clone is straight where it abuts the boundary. Even when a marked clone has been genetically altered so that it grows more rapidly than the rest of the wing and is therefore very large, it respects the boundary in the same way (*last drawing*). Note that the compartment boundary does not coincide with the central wing vein. (B) The pattern of expression of the *engrailed* gene in the wing, revealed by the same technique as in Figure 21–66. The compartment boundary coincides with the boundary of *engrailed* gene expression. (A, after F.H.C. Crick and P.A. Lawrence, *Science* 189:340–347, 1975. © 1975 the AAAS; B, courtesy of Cory Hama and Tom Kornberg.)

(B)

500 μm

parasegments. The same is true for at least some of the segment-polarity genes, such as *engrailed* (see Figure 21–66), whose differential expression corresponds to an abrupt difference between cells in the posterior part of a parasegment and cells in its anterior part. Thus, through the differential expression of these two classes of genes, the body is subdivided into a series of discrete regions comprising cells in different states of determination. At the frontier between one such region and the next, the cells appear to be prevented from mixing, as though selective cohesion between cells with the same molecular address label keeps them segregated from cells with a different label (Figure 21–75). Thus, for example, when a clone of genetically marked but otherwise normal cells is created in the wing by mitotic recombination, the clone is observed to be confined strictly to one side or the other of a precisely specified boundary at the frontier between the two parasegments from which the wing is constructed. A subdivision of the body defined in this way—in the wing or any other organ—is called a **compartment** (Figure 21–75).

By definition, a compartment boundary is a frontier where two populations of cells in different states of determination are prohibited from mixing. Because the state of determination is not normally reversible, each compartment has to be a self-sufficient unit. It cannot recruit cells from the adjacent compartment or transfer surplus cells into it. It can and does, however, regulate its internal organization and its size in obedience to the rule of intercalation, discussed earlier, by adjustments that do not violate this constraint. Thus, in the regulation of pattern and growth, each compartment seems to behave as a more or less independent module during normal development (although during regeneration after a drastic disturbance cells sometimes do switch their character and their compartmental allegiance).

Some of the morphogenetic signals operating in the imaginal disc to control these processes have been identified. They appear to include products of the *dpp* and *wingless* genes, which, as we saw, are both active in patterning the early embryo also (see Figure 21–56). But we do not yet know in molecular genetic terms how these signaling systems are organized or how they collaborate with the homeotic selector genes to give each compartment its characteristic internal pattern and make it stop growing when it has reached its proper size.

Thus, in following the genetic pathways of pattern formation to later and later stages and finer and finer levels of detail, we come to a point where the chain of cause and effect becomes obscure. At the very last stage in the process, however, as cells prepare for terminal differentiation, the trail can be picked up again, and we can trace the genetic mechanisms that control some of the most minute details of patterning of the fly's body surface as displayed in its array of *sensory bristles*.

Localized Expression of Specific Gene Regulatory Proteins Foreshadows the Production of Sensory Bristles [63]

Flies have many bristles on their body—some big, some small. The big ones are landmark structures on the surface of the fly: they are relatively few and far between and occupy exactly predictable positions. The small ones are more closely spaced and occur in fields covering precisely defined regions of the body surface. The bristles are miniature sense organs—components of the peripheral nervous system. Some respond to chemical stimuli, others to mechanical stimuli, but they are all constructed in a similar way. The structure is seen at its simplest and most stereotyped in the *mechanosensory bristles*. Each of these, whether big or small, consists of exactly four cells: a shaft cell, a socket cell, a glial sheath cell, and a neuron (Figure 21–76). Movement of the shaft of the bristle excites the neuron, which sends a signal to the central nervous system.

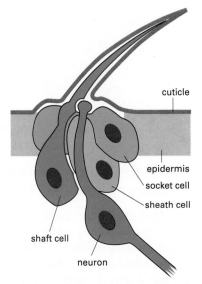

Figure 21–76 The basic structure of a mechanosensory bristle. The four cells of the bristle are shown diagrammatically.

The cells of the bristle of the adult fly derive from the imaginal disc epithelium, and all four of them are granddaughters of a single *sensory mother cell* that becomes distinct from the neighboring prospective epidermal cells during the last larval instar (Figure 21–77). To account for the pattern of bristle differentiation, we have to explain first how the genesis of sensory mother cells is controlled and then how the four granddaughters of each such cell become different from one another.

Two genes, called *achaete* and *scute,* are crucial in initiating the formation of bristles in the imaginal disc epithelium. These genes have similar and overlapping functions and code for closely related gene regulatory proteins of the helix-loop-helix class (discussed in Chapter 9). They belong to a group of closely linked homologous genes, all located in the *achaete-scute complex. In situ* hybridization shows that *achaete* and *scute* are expressed in the imaginal disc in precisely the regions where bristles will form. Mutations that eliminate the expression of these genes at some of their usual sites block development of bristles at just those sites, and mutations that cause expression in additional, abnormal sites cause bristles to develop there. But expression of *achaete* and *scute* is transient, and only a minority of the cells initially expressing the genes go on to become sensory mother cells; the others become ordinary epidermis. The state that is specified by expression of *achaete* and *scute* is called *proneural*. The proneural cells are primed to take the neurosensory pathway of differentiation, but which of them will actually do so depends on competitive interactions among them.

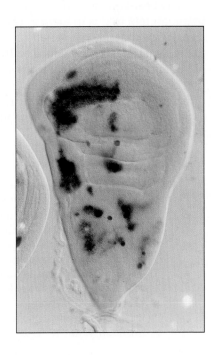

Figure 21–77 Sensory mother cells in the wing imaginal disc. The sensory mother cells (*bluish* here) are easily revealed in this special strain of *Drosophila,* which contains an artificial *lacZ* reporter gene that, by chance, has inserted itself in the genome next to a control region that causes it to be expressed selectively in sensory mother cells. Animals such as this provide a way to detect and track down specific control regions in the genome—the so-called *enhancer-trap* technique. The *purple* stain shows the expression pattern of the *scute* gene; this foreshadows the production of sensory mother cells and fades as the sensory mother cells successively develop. (From P. Cubas, J.-F. de Celis, S. Campuzano, and J. Modolell, *Genes Dev.* 5:996–1008, 1991.)

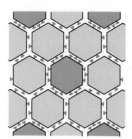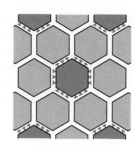

Figure 21–78 **Lateral inhibition.** At first, all cells in the patch are equivalent; each one has a tendency to differentiate as a sensory mother cell, and each sends an inhibitory signal to its neighbors to discourage them from differentiating in that way. This creates a competitive situation. As soon as an individual cell gains any advantage in the competition, that advantage becomes magnified. The winning cell, as it becomes more strongly committed to differentiating as a sensory mother, also inhibits its neighbors more strongly, and they, conversely, as they lose their capacity to differentiate as sensory mothers, also lose their capacity to inhibit other cells from doing so. Lateral inhibition thus makes adjacent cells follow different fates; it is the opposite of the community effect discussed on page 1063.

Lateral Inhibition Regulates the Fine-grained Pattern of Differentiated Cell Types [63, 64]

Proneural cells, expressing *achaete* or *scute* or both genes together, occur in groups in the imaginal disc epithelium—a small, isolated cluster of fewer than 30 cells for a big bristle, a broad, continuous patch of hundreds or thousands of cells for a field of small bristles. In the former case just one member of the cluster becomes a sensory mother cell; in the latter case many cells scattered throughout the proneural region do so. The sensory mother cells are almost always separated from one another by a certain minimum number of epidermal cells. Experiments with genetic mosaics show that a cell that becomes committed to the sensory-mother-cell pathway of differentiation sends a signal to its neighbors not to do the same thing: it exerts a **lateral inhibition** (Figure 21–78). If a cell that would normally become a sensory mother is genetically disabled from doing so, a neighboring proneural cell, freed from lateral inhibition, will become a sensory mother instead.

The genes responsible for lateral inhibition were first identified as such through studies of mutant embryos. In the embryo, both the achaete-scute complex and the genes for lateral inhibition govern development of the central and peripheral nervous system in just the same way that they later govern development of the sense organs of the peripheral nervous system in imaginal discs. In both situations mutations abolishing lateral inhibition have a simple and striking effect: neural cells are produced in vast excess at the expense of epidermal cells (Figure 21–79). Genes are generally named according to their mutant phenotype; hence, the genes responsible for lateral inhibition are called, confusingly, **neurogenic genes**. They form a genetic system with at least seven members.

The best-known neurogenic gene is called *Notch*. It codes for a transmembrane protein that is thought to serve as the receptor for the lateral-inhibition signal. Experiments with genetic mosaics show that cells lacking *Notch* are blind to the signal and follow a neural pathway of differentiation. Another related transmembrane protein, encoded by the neurogenic gene *Delta*, appears to be a ligand that binds to *Notch* and activates it; lateral inhibition, it seems, is transmitted via direct cell-to-cell contact. Downstream from *Notch* the products of other neurogenic genes act intracellularly to interpret the signal and suppress neural differentiation.

Figure 21–79 **The result of switching off lateral inhibition.** The photograph shows part of the thorax of a fly containing a mutant patch (created by x-ray-induced mitotic recombination) in which the neurogenic gene *Delta* has been partially inactivated. The reduction of lateral inhibition has caused almost all the cells in the mutant patch (*in the center of the picture*) to develop as sensory mother cells, producing a great excess of sensory bristles there. Mutant patches of cells carrying more extreme mutations, causing a total loss of lateral inhibition, form no visible bristles because all of the progeny of the sensory mother cells develop as neurons instead of diversifying to form both neurons and the external parts of the bristle structure. (Courtesy of Patricia Simpson.)

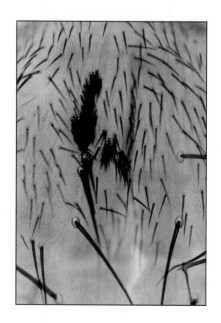

The same lateral inhibition mechanism dependent on *Notch* can be shown to operate twice in the formation of bristles—first, to force the neighbors of sensory mother cells to follow a different pathway and become epidermal and, second, to make the four granddaughters of the sensory mother cell follow different pathways of differentiation so as to form the four components of the bristle. At both stages the default pathway is the neural pathway, and lateral inhibition mediates a competitive interaction that forces adjacent cells to differentiate in contrasting ways.

The same set of neurogenic genes in *Drosophila* not only mediates lateral inhibition repeatedly during development of the nervous system but also is required for the detailed patterning of many other tissues of the fly. Indeed, lateral inhibition is a key strategy in the control of multicellular patterns of differentiation throughout the animal world and almost certainly in plants also; the types of spacing patterns that it can generate are ubiquitous, from the stomata on a leaf to the photoreceptors in the eye. As homologues of the neurogenic genes are found in vertebrates, it may be that the same conserved molecular mechanisms operate in at least some of these cases. In the final part of this section we consider how far *Drosophila* does actually provide a universal model for the molecular genetics of pattern formation.

The Developmental Control Genes of *Drosophila* Have Homologues in Vertebrates [65]

The theory of evolution tells us that all animals are our cousins. It is easy enough to see the family resemblances between a human being and a mouse, or even a fish, and to chart the homologies between the parts of their bodies and the parts of our own. But when we compare ourselves with flies or worms, from which we are separated by about 600 million years, the correspondences are far from clear. True, one can recognize some familiar cell types—neurons, striated muscle cells, and spermatozoa, for example. With a little less confidence, one can see similarities in the body plan, with its central gut tube and its head at one end. But how deep do these similarities go? The fossil record gives us no clear answer, but molecular genetics has begun to supply one.

Comparisons of gene sequences show that an astonishingly large proportion of the genes in an animal such as a fly have unmistakable homologues in vertebrates, and vice versa. Such homologies have been recognized for a majority of the developmental control genes we have mentioned in this chapter. But are these control genes used in the same combinations and for homologous purposes, so that the genetic *system* governing development is conserved? When we compare a human being with a fly, there seem at first sight to be fundamental differences, in development as well as final structure. Vertebrate eggs do not, for example, develop through a syncytial stage as insects do, and their initial multicellular patterning therefore cannot be controlled by morphogen gradients such as that of Bicoid in *Drosophila*, set up by intracellular diffusion of a protein through a cytoplasm that is shared by many nuclei. And yet, when we turn to slightly later stages, we encounter a remarkable pattern of anatomical correspondences. These could never have been discerned without the help of molecular genetics, which reveals in very different animals similar positional markers expressed in body parts that we might not otherwise judge to have anything in common. The HOM gene complex has been central to this new appreciation of our relation to flies and worms.

Mammals Have Four Homologous HOM Complexes [59, 66]

Because the homeodomain of the homeotic selector genes has been highly conserved in evolution, it has been relatively easy to discover homologues of the *Drosophila* genes in other classes of animals. They have been found in almost

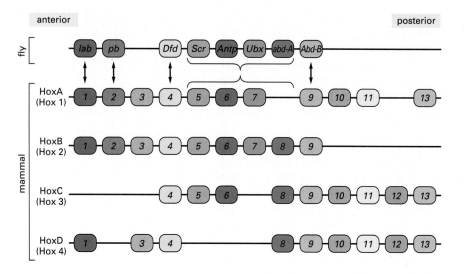

anterior   posterior

Figure 21–80 The HOM complex of an insect and the Hox complexes of a mammal compared. The genes of the Antennapedia and bithorax complexes of *Drosophila* are shown in their chromosomal order in the top line; the corresponding genes of the four mammalian (mouse or human) Hox complexes are shown below, also in chromosomal order. Genes with the most anterior expression domains are to the left, those with the most posterior expression domains to the right. The five complexes are aligned so that genes with the most closely corresponding sequences lie in the same column. The complexes are thought to have evolved as follows: first, in some common ancestor of worms, flies, and vertebrates, a single primordial homeotic selector gene underwent repeated duplication to form a series of such genes in tandem—a HOM complex. In the *Drosophila* sublineage this single complex became split into separate Antennapedia and bithorax complexes. Meanwhile, in the lineage leading to the mammals the whole complex was repeatedly duplicated to give the four Hox complexes. Thus *labial (lab)* in *Drosophila* is identifiable by its sequence as the counterpart of *Hoxa-1, Hoxb-1,* and *Hoxd-1; proboscipedia (pb)* is the counterpart of *Hoxa-2* and *Hoxb-2;* and so on. The parallelism is not perfect because apparently some individual genes have been duplicated and others lost since the complexes diverged. (Based on M.P. Scott, *Cell* 71:551–553, 1992. © Cell Press.)

every sort of creature—in *Hydra,* in nematodes and earthworms, in beetles and mollusks and sea urchins, in fish, frogs, birds, and mammals. Remarkably, in those cases that have been investigated adequately, these genes seem to be grouped in complexes similar to the insect HOM complex. In the mouse there are four such complexes—called the HoxA, HoxB, HoxC, and HoxD complexes—each on a different chromosome. Individual genes in each complex can be recognized by their homeobox sequences, as counterparts of specific members of the *Drosophila* set. It appears that each of the four mammalian Hox complexes is, roughly speaking, the equivalent of a complete insect HOM complex (that is, an Antennapedia complex plus a bithorax complex) (Figure 21–80). The ordering of the genes within each Hox complex is essentially the same as in the insect HOM complex, suggesting that all four vertebrate complexes originated by duplications of a single primordial complex and have preserved its basic organization. Most tellingly, when the expression patterns of the Hox genes are examined in the vertebrate embryo by *in situ* hybridization, it turns out that the members of each complex are expressed in a head-to-tail series along the axis of the body, just as they are in *Drosophila* (Figure 21–81). The pattern is most clearly seen in the neural tube. With minor exceptions this anatomical ordering matches the chromosomal ordering of the genes in each complex, and corresponding genes in the four different Hox complexes have almost identical anteroposterior domains of expression.

The gene expression domains define a detailed system of correspondences between insect body regions and vertebrate body regions. As shown in Figure 21–82, the parasegments of the fly correspond to a similarly labeled series of segments in the anterior part of the vertebrate embryo. These are most clearly demarcated in the hindbrain, where they are called *rhombomeres.* In the tissues lateral to the hindbrain the segmentation is seen in the series of *branchial arches,* prominent in all vertebrate embryos—the precursors of the system of gills in fish and of the jaws and structures of the neck in mammals; each pair of rhombomeres in the hindbrain corresponds to one branchial arch (see Figure 21–82). In the hindbrain, as in *Drosophila,* the boundaries of the expression domains of the Hox genes are aligned with the boundaries of the anatomical segments. And as in *Drosophila* compartments, the cells of one rhombomere do not mix with those of the next rhombomere.

It is not yet clear, however, how similar in detail the mechanisms that set up the hindbrain and branchial arch segmentation of a vertebrate are to those that generate the parasegments of an insect. Although, for example, vertebrates have homologues of the *engrailed* and *wingless* genes, these are not expressed in a repetitive segmental fashion in the hindbrain.

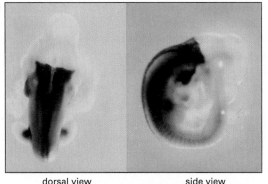

Hoxb-2

dorsal view   side view

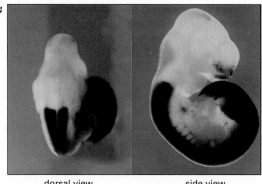

Hoxb-4

dorsal view   side view

Hox Genes Specify Positional Values in Vertebrates as in Insects [67]

Despite uncertainties over mechanisms of segmentation, there can be little doubt that our head-to-tail axis is homologous to that of an insect and that essentially the same sets of genes mark out the anteroposterior positional values of our cells. The Hox genes appear to have not only similar expression patterns to the insect HOM genes but also similar controlling functions. Because the vertebrate has four Hox gene complexes acting more or less in parallel along its body axis, in place of the insect's single HOM complex, it is not enough to eliminate or misexpress a single Hox gene to produce a full-blown homeotic transformation of one region into the character of another. Nevertheless, genetically engineered mice with alterations in single Hox genes do show localized abnormalities that can be interpreted as incomplete homeotic transformations.

This illustrates one of the fundamental difficulties in analyzing the genetics of developmental control systems in vertebrates. The vertebrate genome is very big, and it owes its size, in large measure, to gene duplications in the course of evolution. Thus it contains multiple variant copies of genes that are represented singly in a fly or a nematode: the four Hox complexes corresponding to the single HOM complex are typical in this respect. The multiple versions of a gene have overlapping and partially interchangeable functions, and this partial redundancy makes it very difficult to identify the basic role of any single gene, just as it is hard, by removing or inserting a single screw, to demonstrate the function of the multiple screws that hold a door on its hinges. Herein lies the cardinal importance of the insights that simpler model organisms such as *Drosophila* and *Caenorhabditis elegans* have to offer.

Figure 21–81 Expression domains of Hox genes in a mouse. The photographs show whole embryos displaying the expression domains of genes of the HoxB complex (*blue* stain). The expression domains can be revealed by *in situ* hybridization or, as in these examples, by constructing transgenic mice containing the control sequence of a Hox gene coupled to the coding sequence of β-galactosidase, whose presence is detected histochemically. Each gene is expressed in a long expanse of tissue with a sharply defined anterior limit. The earlier the position of the gene in its chromosomal complex, the more anterior the anatomical limit of its expression. Thus, with minor exceptions, the anatomical domains of the successive genes form a nested set, ordered according to the ordering of the genes in the chromosomal complex. (Courtesy of Robb Krumlauf.)

Figure 21–82 Correspondences between insect and vertebrate body regions as defined by HOM/Hox gene expression. A *Drosophila* embryo is shown at the extended germ band stage, with its parasegments colored according to the HOM genes that they express. The color code is as in Figure 21–80, and the same color code is used for the pattern of HoxB gene expression in the neural tube of a vertebrate embryo. For simplicity, the expression in other tissues of the vertebrate is not shown. Both in the fly and in the vertebrate, in regions where the expression domains of two or more HOM/Hox genes overlap, the coloring corresponds to the most "posterior" of the genes expressed. Where several genes have the same boundary to their expression domain, their common territory is shown striped. Note that just as the expression domains in the fly are related to parasegments, so the expression domains in the vertebrate are related to the rhombomeres (segments in the hindbrain). Each pair of rhombomeres is associated with a branchial arch (a modified gill rudiment), to which it sends innervation. The pattern of Hox gene expression in the branchial arches (not shown) matches that in the associated rhombomeres.

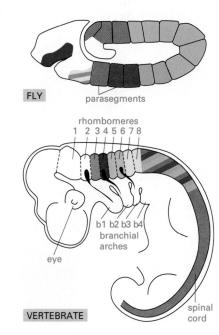

FLY

parasegments

rhombomeres
1 2 3 4 5 6 7 8

b1 b2 b3 b4
branchial
arches

eye

spinal
cord

VERTEBRATE

This is not to say that an individual gene in a vertebrate set is superfluous; for as evolution proceeds, the duplicated genes diverge and begin to take on new and more specialized functions that distinguish them from one another. Old components can be adapted to organize the development of new types of structures in addition to the old. The limbs of higher vertebrates provide a beautiful example.

Subsets of Hox Genes Are Expressed in Order Along Two Orthogonal Axes in the Vertebrate Limb Bud [68]

Earlier in this chapter we used the developing limb buds of the chick embryo to show that cells in different regions are distinguished from one another by a property that we called their positional value. This remembered characteristic of the cells controls whether they will form the structures appropriate to leg or wing, upper arm or forearm, thumb or little finger. Molecular genetics has revealed what "positional value" means in molecular terms in the limb bud.

We have seen that along the main body axis, both in flies and in vertebrates, positional values are defined by the state of expression of HOM/Hox genes. *In situ* hybridization shows that the same is true in the limb buds of a mouse or chick embryo—but with a twist. Instead of finding the corresponding genes of all four Hox complexes expressed in similar, overlapping patterns, as in the hindbrain, one finds a subset of members of the HoxD complex expressed in a series of domains ordered along one limb axis (very roughly, the anteroposterior) and a subset of members of the HoxA complex expressed in series along a different axis (more or less proximodistal) (Figure 21–83). To test whether these genes actually control limb patterning, a retrovirus has been used as an expression vector in the chick embryo to introduce a particular Hox gene into the limb bud cells and force expression of the gene in an inappropriate site. When cells in the region from which the first toe will develop are thus caused to express the Hox gene characteristic of the second toe (*Hoxd-11*), their behavior is transformed,

Figure 21–83 Hox gene expression patterns in vertebrate limb buds. In (A) the pattern of expression of the posteriorly expressed members of the HoxD complex in a 12½-day mouse embryo is shown schematically. In (B) the expression patterns of chicken HoxD (ChoxD) and chicken HoxA (ChoxA) genes in the forelimb bud of a 4-day chick embryo are compared. The HoxD genes, in both chick and mouse, mark out an anteroposterior pattern of domains; the HoxA genes mark out a proximodistal pattern. (A, after D. Duboule, *BioEssays* 14:375–384, 1992. © ICSU Press; B, after Y. Yokouchi, H. Sasaki, and A. Kuroiwa, *Nature* 353:443–445, 1991. © 1991 Macmillan Magazines Ltd.)

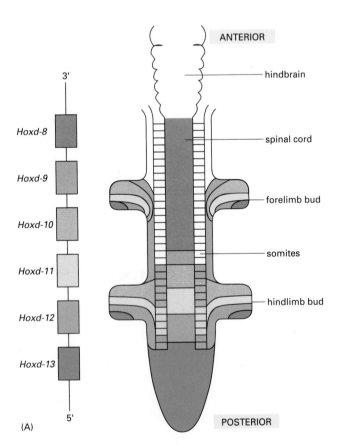

and at the site of the first toe a duplicate of the second toe develops. Evidently, when vertebrates evolved limbs, they co-opted the different sets of Hox genes in different ways to control limb patterning as well as the patterning of the main body axis.

A central problem now for vertebrate embryology is to find out how the Hox genes themselves are regulated. Several studies show that retinoic acid can control Hox gene expression both in the limb bud and along the main body axis, but how this control is exerted and what part it plays in normal development are as yet open questions.

The HOM/Hox genes provide at present the most spectacular example of conserved developmental control machinery. But the flood of genetic homologies discovered through gene sequencing in the past few years gives every reason to expect that many further developmental parallels between vertebrates and invertebrates, no less profound, will soon become apparent.

Classical and molecular genetic studies of small, tractable organisms such as flies and worms give us a key to unlock the mysteries of development in the animal world as a whole. But can we take the generalization a step further still, to the world of plants, or does plant development rest on an entirely different set of principles and mechanisms? This is the question that we tackle in the next section.

Summary

Homeotic selector genes specify the differences between body segments along the head-to-rear axis: they provide the cells with a record of their positional value. Mutations in homeotic selector genes can convert one body segment to the character of another, and deletion of the genes en masse *results in a larva whose body segments are all alike. Similar transformations are seen in the external structures of the adult fly, which are derived from the imaginal discs of the larva. Transplantation experiments show that the cells in the discs retain a long-term memory of their positional value, and this memory depends on the continued presence of the homeotic selector genes.*

The homeotic selector genes all code for DNA-binding proteins containing a characteristic highly conserved homeobox sequence. They are grouped in two clusters in the genome, thought to be the separated parts of a single ancestral gene cluster called the HOM complex. The chromosomal ordering of the genes in each part of the complex matches the spatial ordering of their expression domains in the body. The molecular mechanism of the memory phenomenon is unknown, but it is thought to depend on self-perpetuating changes in the state of the control regions in the HOM complex.

The expression patterns of the HOM genes and segment-polarity genes jointly subdivide the body into compartments whose cells do not mix. Subsequent processes generate a fine-grained pattern of cell differentiation inside each compartment. Lateral inhibition, mediated by the so-called neurogenic genes, plays a key part in this final stage of cell diversification, causing cells that are in contact with one another to differentiate in different ways and so helping to organize the creation of minutely specialized sets of cells forming structures such as sensory bristles.

A large proportion of the developmental control genes identified in flies and worms have homologues in other types of animals, including vertebrates. In some cases the corresponding genes have been shown to have corresponding developmental functions, implying that fundamental mechanisms of animal development have been conserved even where the outward appearance of the body has evolved out of all recognition. Practically all animals appear to have HOM gene complexes organized in a similar way to those of insects: in mammals there are four such complexes, called Hox complexes, and their products are thought to specify positional values that control the anteroposterior pattern of parts in the region of the hindbrain and trunk. The Hox complexes have also acquired new functions as specifiers of positional information in the more recently evolved parts of the vertebrate body, in particular in the limbs.

Plant Development [69]

Plants and animals are separated by about a billion years of evolutionary history. They have evolved their multicellular organization independently but using the same initial tool kit—the set of genes inherited from their common unicellular eucaryotic ancestor. Most of the contrasts in their developmental strategies spring from two basic peculiarities of plants. First, they get their energy from sunlight, not by ingesting other organisms. This dictates a different body plan. Second, their cells are encased in semirigid cell walls that are cemented together, preventing them from moving as animal cells do. This dictates a different set of mechanisms for shaping the body and different developmental mechanisms to cope with a changeable environment.

Animal development is largely buffered against environmental changes, and the embryo generates the same genetically determined body structure unaffected by external conditions. The development of most plants, by contrast, is dramatically influenced by the environment: because they cannot match themselves to their environment by moving to another place, plants adapt instead by altering the course of their development. Their strategy is opportunistic. A given type of organ—a leaf, a flower, or a root, say—can be produced from the fertilized egg by many different paths according to environmental cues. A begonia leaf pegged to the ground may sprout a root; the root may throw up a shoot; the shoot, given sunlight, may grow leaves and flowers.

The mature plant is typically made of many copies of a small set of standardized *modules,* as described in Figure 21–84. The positions and times at which those modules are generated are strongly influenced by the environment, causing the overall structure of the plant to vary. The choices between alternative modules and their organization into a whole plant depend on external cues and long-range hormonal signals that play a much smaller part in the control of animal development.

But although the global structure of a plant—its pattern of roots or branches, its numbers of leaves or flowers—can be highly variable, its detailed organization on a small scale is not. A leaf, a flower, or indeed an early plant embryo, is as precisely structured as any organ of an animal. The internal organization of a plant module raises essentially the same problems in the genetic control of pattern formation as does animal development, and they are solved in analogous ways. In this section we focus on the cellular mechanisms of development in flowering plants. We examine both the contrasts and the similarities with animals.

Embryonic Development Starts by Establishing a Root-Shoot Axis and Then Halts Inside the Seed [70]

Flowering plants, despite their staggering variety, are of relatively recent origin. The earliest known fossil examples are 125 million years old, as against 350 million years for vertebrate animals. This helps to explain why certain features of their form and development are remarkably constant. Their basic strategy of sexual reproduction is briefly summarized in Panel 21–2, page 1109. The fertilized egg, or *zygote,* of a higher plant begins by dividing asymmetrically to establish the polarity of the future embryo. One product of this division is a small cell with dense cytoplasm, which will become the embryo proper. The other is a large vacuolated cell that divides further and forms a structure called the *suspensor,* which in some ways is comparable to the umbilical cord in mammals. The suspensor attaches the embryo to the adjacent nutritive tissue and provides a pathway for the transport of nutrients.

During the next step in development the diploid embryo cell proliferates to form a ball of cells that quickly acquires a polarized structure. This comprises two key groups of proliferating cells—one at the suspensor end of the embryo that will generate a *root* and one at the opposite pole that will generate a *shoot* (Figure 21–85). The main root-shoot axis established in this way is analogous to the

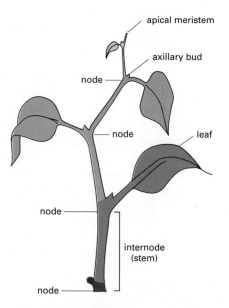

Figure 21–84 A simple example of the modular construction of plants. Each module (shown in different shades of *green*) consists of a stem, a leaf, and a bud containing a potential growth center, or *meristem.* The bud forms at the branch point, or *node,* where the leaf diverges from the stem. Modules arise sequentially from the continuous activity of the apical meristem.

THE FLOWER

Flowers, which contain the reproductive cells of higher plants, arise from vegetative shoot apical meristems (see Figures 21–84 and 21–88). They terminate further vegetative growth from that meristem. Environmental factors, often the rhythms of day length and temperature, trigger the switch from vegetative to floral development. The germ cells thus arise late in plant development from somatic cells rather than from a germ cell line, as in animals.

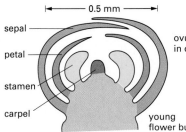

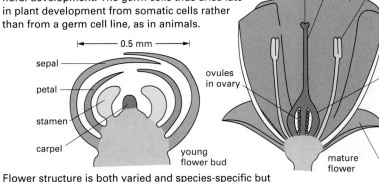

Flower structure is both varied and species-specific but generally comprises four concentrically arranged sets of structures that may each be regarded as modified leaves.

Petal: distinctive leaflike structures, usually brightly colored, facilitate pollination via, for example, attracted insects.

Stamen: an organ containing cells that undergo meiosis and form haploid pollen grains, each of which contains two male sperm cells. Pollen transferred to a stigma germinates, and the pollen tube delivers the two nonmotile sperm to the ovary.

Carpel: an organ containing one or more ovaries, each of which contains ovules. Each ovule houses cells that undergo meiosis and form an embryo sac containing the female egg cell. At fertilization, one sperm cell fuses with the egg cell and will form the future diploid embryo, while the other fuses with two cells in the embryo sac to form the triploid endosperm tissue.

Sepals: leaflike structures that form a protective covering during early flower development.

THE SEED

A seed contains a dormant embryo, a food store, and a seed coat. At the end of its development a seed's water content can drop from 90% to 5%. The seed is usually protected in a *fruit* whose tissues are of maternal origin.

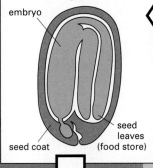

THE EMBRYO

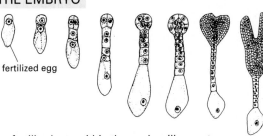

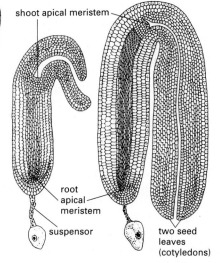

The fertilized egg within the ovule will grow to form an embryo using nutrients transported from the endosperm by the suspensor. A complex series of cell divisions, illustrated here for the common weed called shepherd's purse, produces an embryo with a root apical meristem, a shoot apical meristem, and either one (monocots) or two (dicots) seed leaves, or cotyledons.

Development is arrested at this stage, and the ovule, containing the embryo, now becomes a seed, adapted for dispersal and survival.

GERMINATION

For the embryo to resume its growth the seed must germinate, a process dependent upon both internal factors (dormancy) and environmental factors including water, temperature, and oxygen. The food reserves for the early phase of germination may either be the endosperm (maize) or the cotyledons (pea and bean).

The primary root usually emerges first from the seed to ensure an early water supply for the seedling. The cotyledon(s) may appear above the ground, as in the garden bean shown here, or they may remain in the soil, as in peas. In both cases the cotyledons eventually wither away.

The apical meristem can now show its capacity for continuous growth, producing a typical pattern of nodes, internodes, and buds (see Figure 21–84).

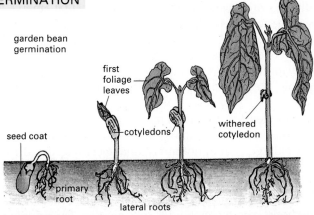

Panel 21–2 Features of early development in flowering plants.

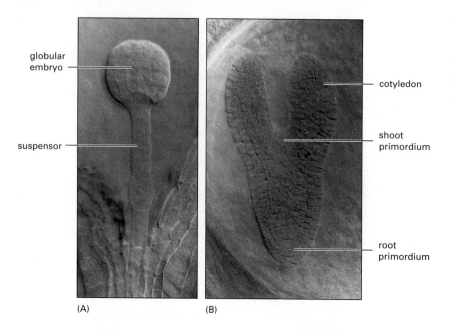

globular
embryo

suspensor

(A)   (B)

cotyledon

shoot
primordium

root
primordium

Figure 21–85 Two stages of embryogenesis in a plant, *Arabidopsis thaliana.* (From G. Jürgens, U. Mayer, R.A. Torres-Ruiz, T. Berleth, and S. Miséra, *Development [Suppl.]* 1:27–38, 1991.)

head-to-tail axis of an animal. At the same time it begins to be possible to distinguish the future *epidermal cells,* forming the outermost layer of the embryo, the future *ground tissue cells,* occupying most of the interior, and the future *vascular tissue cells,* forming the central core. These three sets of cells can be compared to the three germ layers of an animal embryo. Slightly later in development, the rudiment of the shoot begins to produce the embryonic seed leaves, or *cotyledons*—one in the case of monocots and two in the case of dicots. Soon after this stage, development usually halts and the embryo becomes packaged in a *seed,* specialized for dispersal and for survival in harsh conditions. The embryo in a seed is stabilized by dehydration, and it can remain dormant for a very long time—even hundreds of years. When rehydrated, the seeds germinate and embryonic development resumes.

The Repetitive Modules of a Plant Are Generated Sequentially by Meristems [71]

Roughly speaking, the embryo of an insect or a vertebrate animal is a rudimentary miniature scale model of the later organism, and the details of body structure are filled in progressively as it enlarges. The plant embryo grows into an adult in a quite different way: the parts of the adult plant are created sequentially by groups of cells that proliferate to lay down additional structures at the plant's periphery. These all-important groups of cells are called **apical meristems** (see Figure 21–84). Each meristem consists of a self-renewing population of stem cells. As these divide, they leave behind a trail of progeny that emerge from the meristem region, enlarge, and finally differentiate. Although the shoot and root apical meristems generate all the basic varieties of cells that are needed to build leaves, roots, and stems, many cells outside the apical meristems also retain a capacity for further proliferation. In this way trees and other perennial plants, for example, are able to increase the girth of their stems and roots as the years go by.

The rudiments of the apical meristems of root and shoot are already determined in the embryo. As soon as the seed coat ruptures during germination, a dramatic enlargement of nonmeristematic cells occurs, driving the emergence first of a root, to establish an immediate foothold in the soil, and then of a shoot (Figure 21–86). This is followed by rapid and continual cell divisions in the apical meristems: in the apical meristem of a maize root, for example, cells divide every 12 hours, producing 5×10^5 cells per day. The rapidly growing roots and

shoot apical meristem
(hidden)

cotyledon
(seed leaf)

root apical meristem

Figure 21–86 A seedling of *Arabidopsis.* The brown objects to the right of the young seedling are the two halves of the discarded seed coat. (Courtesy of Catherine Duckett.)

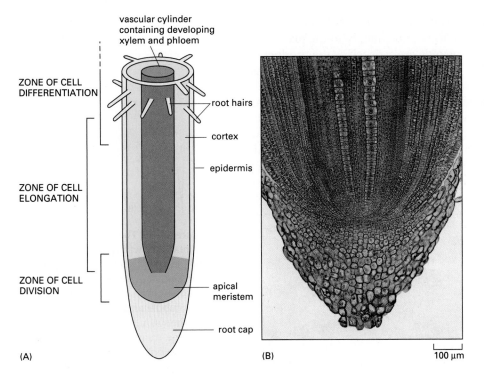

ZONE OF CELL DIFFERENTIATION

ZONE OF CELL ELONGATION

ZONE OF CELL DIVISION

vascular cylinder containing developing xylem and phloem

root hairs

cortex

epidermis

apical meristem

root cap

(A)

(B)

100 μm

Figure 21–87 **A growing root tip.** (A) The organization of the final 2 mm of a growing root tip. The approximate zones in which cells can be found dividing, elongating, and differentiating are indicated. (B) The apical meristem and root cap of a corn root tip, showing the orderly files of cells produced. (B, from P.H. Raven, R.F. Evert, and S.E. Eichhorn, Biology of Plants, 4th ed. New York: Worth, 1986.)

shoots probe the environment—the roots increasing the plant's capacity for taking up water and minerals from the soil, the shoots increasing its capacity for photosynthesis (see Panel 21–2, p. 1109).

The Shaping of Each New Structure Depends on Oriented Cell Division and Expansion [72]

Plant cells, imprisoned within their cell walls, cannot crawl about and cannot be shuffled as the plant grows, but they can divide, and they can swell, stretch, and bend. The morphogenesis of a developing plant therefore depends on orderly cell divisions followed by strictly oriented cell expansion. Most cells produced in the root-tip meristem, for example, go through three distinct phases of development—division, growth (elongation), and differentiation. These three steps, which overlap in both space and time, give rise to the characteristic architecture of a root tip. Although the process of cell differentiation often begins while a cell is still enlarging, it is comparatively easy to distinguish in a root tip a zone of cell division, a zone of oriented cell elongation (which accounts for the growth in length of the root), and a zone of cell differentiation (Figure 21–87).

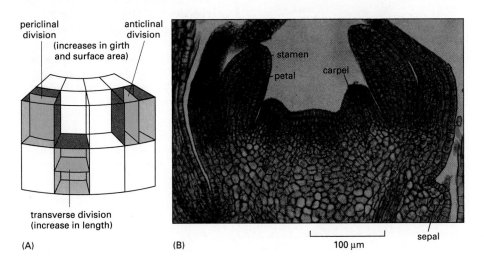

periclinal division

anticlinal division

(increases in girth and surface area)

transverse division (increase in length)

(A)

stamen

petal

carpel

sepal

(B)

100 μm

Figure 21–88 **The relationship between division plane, cell expansion, and morphogenesis.** (A) Three planes of cell division found in a typical plant organ. Variations in the relative proportion of each, combined with oriented cell expansion, can account for the morphogenetic patterns found in plants. (B) A longitudinal section of a young flower bud of a periwinkle. The small domes of cells destined to become the different floral parts have arisen by a combination of new planes of cell division and directional cell expansion determined by the reinforcing hoops of cellulose in the cell wall. (From N.H. Boke, Am. J. Bot. 36:535–547, 1949.)

Because it affects the direction of cell elongation, the exact plane in which cells divide is crucial to plant morphogenesis, and changes in the plane of division are often associated with morphogenetic events such as the production of a leaf or petal primordium (Figure 21–88). The special intracellular mechanisms controlling the plane of cell division in plants are discussed in Chapter 18.

In the phase of controlled expansion that generally follows cell division, the daughter cells may often increase in volume by a factor of 50 or more. This expansion is driven by an osmotically based turgor pressure that presses outward on the plant cell wall, and its direction is determined by the orientation of the cellulose fibrils in the cell wall, which constrain expansion along one axis (see Figure 19–68). The orientation of the cellulose in turn is apparently controlled by the orientation of arrays of microtubules just inside the plasma membrane, which are thought to guide cellulose deposition (discussed in Chapter 19). These orientations can be rapidly changed by plant growth regulators, such as *ethylene* and *gibberellic acid* (Figure 21–89), but the molecular mechanisms underlying these dramatic cytoskeletal rearrangements are still unknown.

Each Plant Module Grows from a Microscopic Set of Primordia in a Meristem [73]

The apical meristems are self-perpetuating: they carry on with their functions indefinitely, as long as the plant survives, and they are responsible for its continuous growth and development. But apical meristems also give rise to a second type of outgrowth, whose development is strictly limited and culminates in the formation of a structure such as a leaf or a flower, with a determinate size and shape and a short lifespan. Thus, as a vegetative shoot elongates, its apical meristem lays down behind itself an orderly sequence of *nodes,* where leaves have grown out, and *internodes* (segments of stem). In this way the continuous activity of the meristem produces an ever increasing number of similar **modules**, each consisting of a stem, a leaf, and a bud (see Figure 21–84). The modules are connected to one another by supportive and transport tissue, and successive modules are precisely located relative to each other, giving rise to a repetitively patterned structure. This iterative mode of development is characteristic of plants and is seen in many other structures besides the stem-leaf system (Figure 21–90).

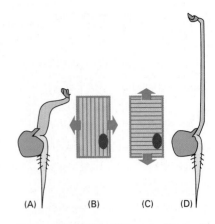

Figure 21–89 The different effects of the plant growth regulators ethylene and gibberellic acid. These regulators exert rapid and opposing effects on the orientation of the cortical microtubule array in cells of young pea shoots. A typical cell in an ethylene-treated plant (B) shows a net longitudinal orientation of microtubules, while a typical cell in a gibberellic-acid-treated plant (C) shows a net transverse orientation. New cellulose microfibrils are deposited parallel to the microtubules. Since this influences the direction of cell expansion, gibberellic acid and ethylene encourage growth in opposing directions: ethylene-treated seedlings will develop short, fat shoots (A), while gibberellic-acid-treated seedlings will develop long, thin shoots (D).

Figure 21–90 Repetitive patterning in plants. Accurate placing of successive modules from a single apical meristem produces these elaborate but regular patterns in leaves (A), flowers (B), and fruits (C). (A, from John Sibthorp, Flora Graeca. London: R. Taylor, 1806–1840; B, from Pierre Joseph Redouté, Les Liliacées. Paris: chez l'Auteur, 1807; C, from Christopher Jacob Trew, Uitgezochte planten. Amsterdam: Jan Christiaan Sepp, 1771—all courtesy of the John Innes Foundation.)

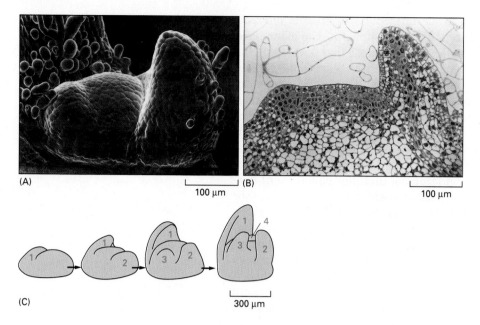

(A)

100 µm

(B)

100 µm

(C)

300 µm

Figure 21–91 A shoot apex from a young tobacco plant. (A) A scanning electron micrograph shows the shoot apex with two sequentially emerging leaf primordia, seen here as lateral swellings on either side of the domed apical meristem. (B) A thin section of a similar apex shows that the youngest leaf primordium arises from a small group of cells (about 100) in the outer four or five layers of cells. (C) A very schematic drawing showing that the sequential appearance of leaf primordia takes place over a small distance and very early in shoot development. Growth of the apex will eventually form internodes that will separate the leaves in order along the stem (see Figure 21–84). (A and B, from R.S. Poethig and I.M. Sussex, *Planta* 165:158–169, 1985.)

Although the final module is large, its organization, like that of an animal embryo, is mapped out at first on a microscopic scale. At the apex of the shoot, within a space of a millimeter or less, one finds a small, low central dome surrounded by a set of distinctive swellings in various stages of enlargement (Figure 21–91). The central dome is the apical meristem itself; each of the surrounding swellings is the primordium of a leaf. This small region, therefore, contains the already distinct rudiments of several entire modules. Through a well-defined program of cell proliferation and cell enlargement, each leaf primordium and its adjacent cells will grow to form a leaf, a node, and an internode. Meanwhile, the apical meristem itself will give rise to new leaf primordia, so as to generate more and more modules in a potentially unending succession. The serial organization of the modules of the plant is thus controlled by events at the shoot apex. Local signals within this tiny region determine the pattern of primordia—the position of one leaf rudiment relative to the next, the spacing between them, and their location relative to the apical meristem itself.

Almost nothing is known of the mechanisms that mediate these central patterning processes in the plant kingdom. All the strategies that we discussed for animal pattern formation, such as those based on local morphogens, timing mechanisms, and lateral inhibition, are possibilities here too. Detailed studies of the fate and lineage of cells in the shoot apex and the root apex are beginning to provide some of the essential background information, however, and some of the key developmental control genes are beginning to be identified. The gene regulatory protein encoded by the gene *Knotted*, for example, is expressed in the central part of the meristem, and overexpression in tobacco causes leaf cells to behave as meristem, generating new organs from the leaf itself.

Long-range Hormonal Signals Coordinate Developmental Events in Separate Parts of the Plant [74]

If a stem is to branch, new meristems must be created, and it is through control of this process that the environment exerts an important part of its influence over the form of a plant. At each node, in the acute angle, or *axil*, between the leaf branch and the stem, a bud is formed. This contains a nest of cells, derived from the apical meristem, that have kept a meristematic character (and express *Knotted*). They have the capacity to become the apical meristem of a new branch, but they also have the alternative option of remaining quiescent. The plant's pattern of branching is regulated through this choice, which the environment helps to dictate. Separate parts of the plant experience different environments and react to them individually by changes in their mode of development. The plant, however, must continue to function as a whole. This demands that developmental choices and events in one part of the plant should affect developmental choices elsewhere. There must be long-range signals to bring about such coordination.

As gardeners know, for example, by pinching off the tip of a branch one can stimulate side growth: removal of the apical meristem relieves the quiescent axillary meristems of an inhibition and allows them to form new twigs. In this case the long-range signal from the apical meristem, or at least a key component of the system of signals, has been identified. It is an *auxin*, a member of one of five known classes of **plant growth regulators** (sometimes called *plant hormones*), all of which have powerful influences on plant development. The four other known classes are the **gibberellins**, the **cytokinins**, **abscisic acid**, and the gas **ethylene**. As shown in Figure 21–92, all are small molecules that readily penetrate cell walls. They are all synthesized by most plant cells and can either act locally or be transported to influence target cells at a distance. Auxin, for example, is transported from cell to cell at a rate of about 1 cm per hour from the tip of a shoot toward its base. Each growth regulator has multiple effects, and these are modulated by the other growth regulators as well as by environmental cues and nutritional status. Thus auxin alone can promote root formation, but in conjunction with gibberellin it can promote stem elongation, with cytokinin it can suppress lateral shoot outgrowth, and with ethylene it can stimulate lateral root growth. The receptors that recognize these growth regulators are only now being characterized, and their mechanisms of action remain unknown.

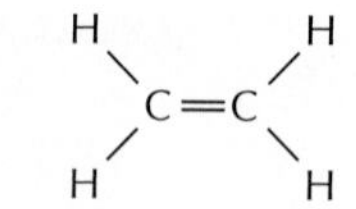

Arabidopsis Serves as a Model Organism for Plant Molecular Genetics [75]

By screening systematically for mutations affecting the pattern of the plant embryo, it has been possible to begin to identify the genes that govern plant development and to start to work out how they function. This approach requires a plant that is, like *Drosophila* or *Caenorhabditis elegans*, small, quick to reproduce, and convenient for genetics. The role of "model plant" has fallen on a small weed, the common wall cress ***Arabidopsis thaliana*** (Figure 21–93), which can be grown indoors in test tubes in large numbers and produces thousands of offspring per plant after 8 to 10 weeks. *Arabidopsis* also has the advantage for molecular analysis of having one of the smallest plant genomes known (7×10^7 nucleotide pairs), comparable to yeast (2×10^7 nucleotide pairs), *C. elegans* (10^8 nucleotide pairs), and *Drosophila* (10^8 nucleotide pairs). Cell culture and genetic transformation methods have been established, large numbers of interesting mutants have been isolated, and an ordered collection of genomic DNA clones is now available. *Arabidopsis* has, in common with *C. elegans*, one significant advantage over *Drosophila* for genetics: like many flowering plants, it can reproduce as a hermaphrodite because a single flower produces both eggs and the pollen that can fertilize them. Therefore, when a flower that is heterozygous for a recessive lethal mutation is self-fertilized, one-fourth of its seeds will display the homozygous embryonic phenotype (Figure 21–94).

By using mutagens to create tens of thousands of mutant plants and inspecting their progeny in this way, a total of about 50 distinct genes governing embryonic pattern formation in *Arabidopsis* have thus far been identified. As in *Drosophila*, the patterning genes can be grouped according to their homozygous mutant phenotypes (Figure 21–95). Some are required for formation of the seedling root, some for the seedling stem, and some for the seedling apex with its cotyledons. Another class is required for formation of the three major tissue types—epidermis, ground tissue, and vascular tissue—and yet another class for the organized changes of cell shape that give the embryo and seedling their elongated form. Given this catalogue of key genes, it should soon be possible to take the next step and discover how they function. But from the range of mutant

ethylene

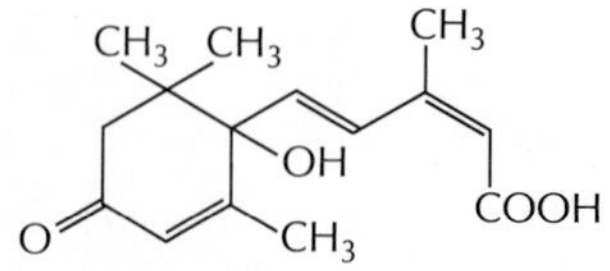

abscisic acid (ABA)

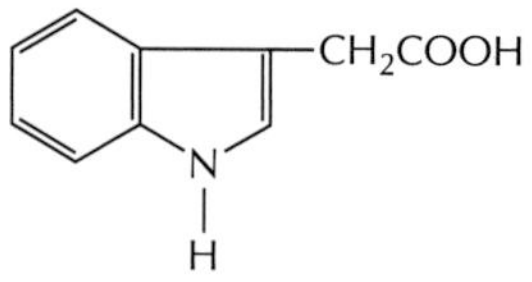

indole-3-acetic acid (IAA) [an auxin]

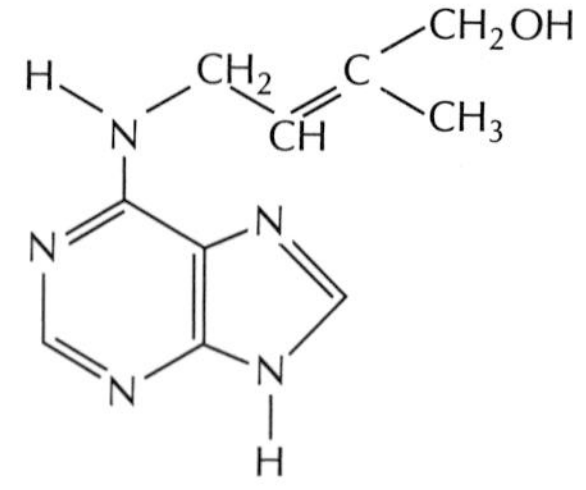

zeatin [a cytokinin]

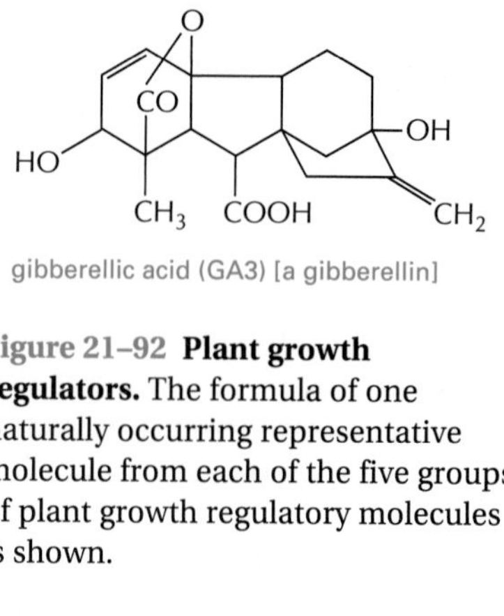

gibberellic acid (GA3) [a gibberellin]

Figure 21–92 Plant growth regulators. The formula of one naturally occurring representative molecule from each of the five groups of plant growth regulatory molecules is shown.

Figure 21–93 *Arabidopsis thaliana.* This small plant is a member of the mustard (or crucifer) family. It is a weed of no economic use but of great value for genetic studies of plant development. (Courtesy of Chris Sommerville.)

15 mm

phenotypes, it already seems likely that the initial patterning of the plant embryo will be largely explicable within the same conceptual framework that we have presented for animals. As we now see, the same can be said for the later developmental processes by which a flower is made.

Homeotic Selector Genes Specify the Parts of a Flower [76]

Meristems face other developmental choices besides that between quiescence and growth, and these also are frequently regulated by the environment. The most important is the decision to form a flower (Figure 21–96).

The switch from meristematic growth to flower formation is typically triggered by light. By poorly understood mechanisms based on light absorption by specific proteins known as *phytochromes,* the cells in the meristem are able to alter their pattern of gene expression in response to a change in day length and thereby undergo the change of state that initiates flower development. By this switch in its state the apical meristem abandons its chances of continuing vegetative growth and gambles its future on the production of gametes. Its cells embark on a strictly finite program of growth and differentiation: by a modification of the ordinary mechanisms for generating leaves, a series of whorls of specialized appendages are formed in a precise order—typically sepals first, then

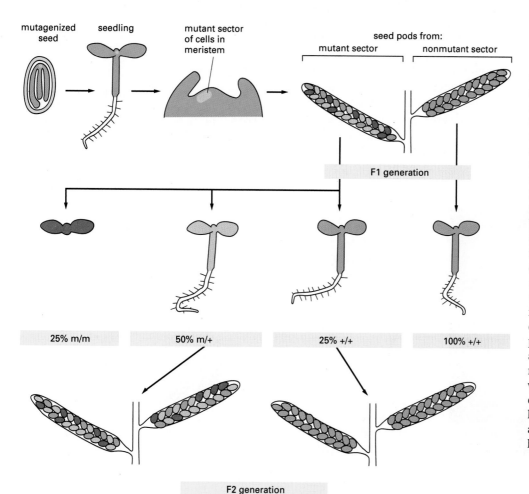

Figure 21–94 **Production of mutants in *Arabidopsis.*** A seed, containing a multicellular embryo, is treated with a chemical mutagen and left to grow into a plant. In general, this plant will be a mosaic of clones of cells carrying different induced mutations. An individual flower produced by this plant will usually be composed of cells belonging to the same clone, all carrying the same mutation, *m,* in heterozygous form (*m/+*). Self-fertilization of individual flowers by their own pollen results in seed pods, each of which contains a family of embryos of whose members half, on average, will be heterozygous (m/+), one quarter will be homozygous mutant (*m/m*), and one quarter will be homozygous wild-type (+/+).

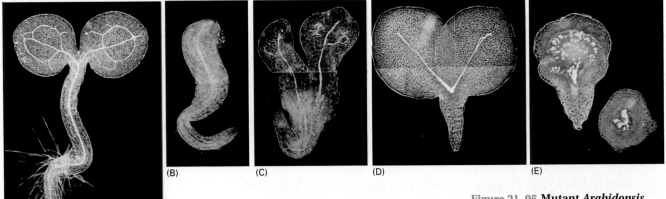

(B)   (C)   (D)   (E)

(A)

Figure 21–95 Mutant *Arabidopsis* seedlings. A normal seedling (A) compared with four types of mutant (B–E) defective in different parts of their apico-basal pattern: (B) has structures missing at its apex, (C) has an apex and a root but lacks a stem between them, (D) lacks a root, and (E) forms stem tissues but is defective at both ends. The seedlings have been "cleared" so as to show the vascular tissue inside them (pale strands). (From U. Mayer et al., *Nature* 353:402–407, 1991. © 1991 Macmillan Magazines Ltd.)

petals, then stamens carrying anthers containing pollen, and lastly carpels containing eggs (see Panel 21–2, p. 1109). By the end of this process the meristem has disappeared, but among its progeny it has created germ cells.

The series of modified leaves forming a flower can be compared to the series of body segments forming a fly. In plants, as in flies, one can find homeotic mutations that convert one part of the pattern to the character of another. The mutant phenotypes can be grouped into three classes (Figure 21–97) in which different but overlapping sets of organs are altered. The first class, exemplified by the *apetala2* mutant of *Arabidopsis*, has its two outermost whorls transformed: the sepals are converted into carpels and the petals into stamens. The second class, exemplified by *apetala3*, has its two middle whorls transformed: the petals are converted into sepals and the stamens into carpels. The third class, exemplified by *agamous*, has its two innermost whorls transformed, with a more drastic consequence: the stamens are converted into petals, the carpels are missing, and in their place the central cells of the flower behave as a floral meristem, which begins the developmental performance all over again, generating another abnormal set of sepals and petals nested inside the first and, potentially, another

Figure 21–96 The structure of an *Arabidopsis* flower. (A) Photograph. (B) Drawings. (C) Schematic cross-sectional view. The basic plan, as shown in (C), is common to most flowering dicotyledonous plants. (A, courtesy of Leslie Sieburth.)

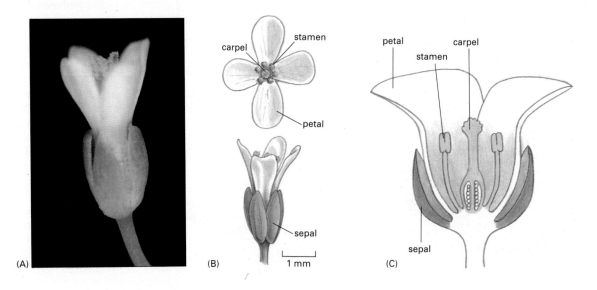

(A)

(B)

carpel
stamen
petal

sepal

1 mm

(C)

petal
stamen
carpel

sepal

Figure 21–97 *Arabidopsis* **flowers showing homeotic mutations.** (A) In *agamous*, stamens are converted into petals and carpels into floral meristem; (B) In *apetala3*, petals are converted into sepals and stamens into carpels; (C) In *apetala2*, sepals are converted into carpels and petals into stamens. Another gene, *pistillata*, has a mutant phenotype similar to *apetala3;* thus three functional classes of homeotic selector genes can be identified. (D) In a triple mutant where these three functions are defective, all the organs of the flower are converted into leaves. (A–C, courtesy of Leslie Sieburth; D, courtesy of Mark Running.)

nested inside that, and so on, indefinitely. These phenotypes identify three classes of homeotic selector genes, which, like the homeotic selector genes of *Drosophila*, all code for gene regulatory proteins. These define the differences of cell state that give the different parts of a normal flower their different characters. *In situ* hybridization confirms that the genes are expressed in the patterns expected on this interpretation (Figure 21–98). In a triple mutant where all three genetic functions are absent, one obtains in place of a flower an indefinite succession of tightly nested leaves. Leaves therefore represent a "ground state" in which none of these homeotic selector genes are expressed, while the other types of organ result from expressing the genes in different combinations.

Similar studies have been carried out in the snapdragon *Antirrhinum majus*, and a similar set of phenotypes and genes have been identified. Gene sequencing reveals that, despite the large evolutionary distance between *Antirrhinum* and *Arabidopsis*, the corresponding homeotic phenotypes arise from mutations in homologous genes: plants, no less than animals, have conserved their homeotic selector gene systems. Again, the set of these genes appears to have arisen through gene duplication: several of them, required in different organs of the flower, have clearly homologous sequences. These are not of the homeobox class but are related to another family of gene regulatory proteins (the so-called MADS family) found in yeast and in vertebrates.

Investigation of the molecular genetics of plant development has only just begun. So far, almost nothing is known, for example, about the genetic systems responsible for local cell-cell communication and positional signaling in plant pattern formation. Yet it is clear already that plants and animals, despite their differences, have independently found very similar solutions to many of the fundamental problems of multicellular development.

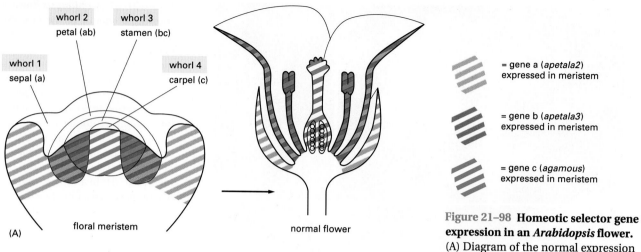

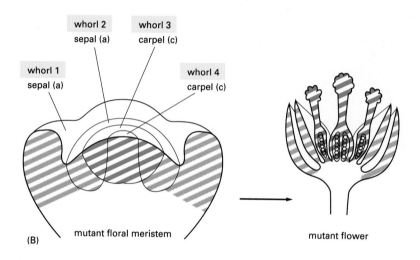

Figure 21–98 **Homeotic selector gene expression in an *Arabidopsis* flower.** (A) Diagram of the normal expression patterns of the three genes whose mutant phenotypes are illustrated in Figure 21–97. All three genes code for gene regulatory proteins. The colored shading on the flower indicates which organ develops from each whorl of the meristem, and does not imply that the homeotic selector genes are still expressed at this stage. (B) The patterns in a mutant where the *apetala3* gene is defective. Because the character of the organs in each whorl is defined by the set of homeotic selector genes that they express, the stamens and petals are converted into sepals and carpels. The consequence of a deficiency of a gene of class a, such as *apetala2*, is slightly more complex: the absence of this class a gene product allows the class c gene to be expressed in the outer two whorls as well as the inner two, causing these outer whorls to develop as carpels and stamens, respectively. Deficiency of a class c gene prevents the central region from undergoing terminal differentiation as a carpel and causes it instead to continue growth as a meristem, generating more and more sepals and petals.

Summary

The development of a flowering plant, like that of an animal, begins with division of a fertilized egg to form an embryo with a polarized organization: the apical part of the embryo will form the shoot, the basal part, the root, and the middle part, the stem. At first, cell division occurs throughout the body of the embryo. As the embryo grows, however, addition of new cells becomes restricted to small regions known as meristems. Apical meristems, at shoot tips and root tips, will persist throughout the life of the plant, enabling it to grow by sequentially adding new body parts at its periphery. Typically, the shoot generates a repetitive series of modules, each consisting of a segment of stem, a leaf, and an axillary bud. An axillary bud is a potential new meristem, capable of giving rise to a side branch, and the environment can control the development of the plant by regulating bud activation. Environmental cues can also cause the apical meristem to switch from a leaf-forming to a flower-forming mode. Long-range signaling mediated by plant hormones coordinates such developmental events occurring in separate parts of the plant.

The internal organization of each plant module, however, is controlled through strictly local pattern formation mechanisms analogous to those that govern animal development. These operate in the neighborhood of the apical meristem, where the relative positions of the rudiments of leaves and other organs are initially mapped out on a microscopic scale. The pattern of modified leaves—sepals, petals, stamens, and carpels—in a flower is set up similarly. The genetic basis of pattern formation in plants can be analyzed in the same way as in animals. The small weed Arabidopsis thaliana *is widely used as a "model plant" for such studies. Genes governing the or-*

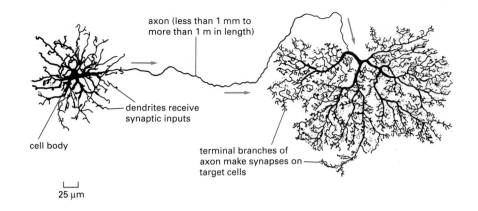

axon (less than 1 mm to
more than 1 m in length)

dendrites receive
synaptic inputs

cell body

terminal branches of
axon make synapses on
target cells

⊢——⊣
25 μm

Figure 21–99 A typical neuron of a vertebrate. The arrows indicate the direction in which signals are conveyed. The neuron shown is from the retina of a monkey. The longest and largest neurons in a human extend for about 1 million μm and have an axon diameter of 15 μm. (Drawing of neuron from B.B. Boycott in Essays on the Nervous System [R. Bellairs and E.G. Gray, eds.]. Oxford, UK: Clarendon Press, 1974.)

ganization of the embryo, analogous to the egg-polarity and segmentation genes of **Drosophila,** *can be identified. And the sequence of parts in a flower is controlled by homeotic selector genes closely analogous (although not homologous) to those of animals.*

Neural Development [77]

Nerve cells, or **neurons**, are among the most ancient of all specialized animal cell types, as important to jellyfish and sea anemones as they are to worms, flies, and people. Their structure is like that of no other class of cells, and the development of the nervous system poses problems that have no parallel in other tissues. A neuron is extraordinary above all for its enormously extended shape, with a long **axon** and **dendrites** connecting it through **synapses** to other cells (Figure 21–99). The central challenge of neural development is to explain how the axons and dendrites grow out, find their right partners, and synapse with them selectively to create a functional network (Figure 21–100).

Most of the components of a typical nervous system—the various classes of neurons, sensory cells, and muscles—originate in widely separate locations in the embryo and are initially unconnected. Thus, in the first phase of neural development (Figure 21–101), the different parts develop according to their own local programs, following principles of cell diversification common to other tissues of the body, as already discussed. The next phase involves a type of morphogenesis unique to the nervous system: a provisional but orderly set of connections is set up between the separate parts of the system through the outgrowth of axons and dendrites along specific routes, so that the parts can begin to interact. In the third and final phase, which continues into adult life, the connections are adjusted and refined through interactions among the far-flung components in a way that depends on the electrical signals that pass between them.

Stocks of Neurons Are Generated at the Outset of Neural Development and Are Not Subsequently Replenished [78]

The nervous system develops from the ectoderm in all animals. In vertebrates, on which we concentrate here, it derives chiefly, as we saw earlier in this chapter, from two sets of cells—those of the *neural tube* (an invagination of the

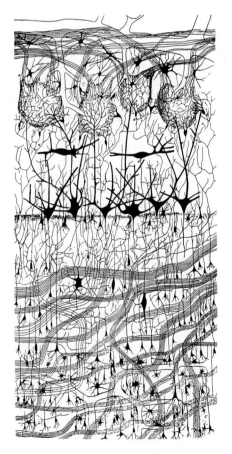

Figure 21–100 The complex organization of nerve cell connections. This semischematic drawing depicts a section through a small part of a mammalian brain—the olfactory bulb of a dog, stained by the Golgi technique. The black objects are neurons; the thin lines are axons and dendrites, through which the various sets of neurons are interconnected according to precise rules. (From C. Golgi, *Riv. sper. freniat. Reggio-Emilia* 1:405–425, 1875; reproduced in M. Jacobson, Developmental Neurobiology, 3rd ed. New York: Plenum, 1992.)

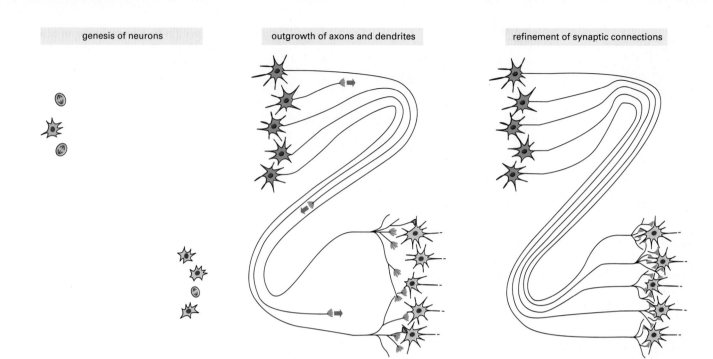

genesis of neurons · outgrowth of axons and dendrites · refinement of synaptic connections

Figure 21–101 **The three phases of neural development.**

ectoderm) and those of the *neural crest* (a population of cells that break loose from the neural ectoderm and migrate to other regions of the embryo). The neural tube forms the central nervous system (the spinal cord and brain, including the retina of the eye), while the neural crest gives rise to most of the neurons and supporting cells of the peripheral nervous system (Figure 21–102).

The **neural tube**, with which we shall be mainly concerned, consists initially of a single-layered epithelium (Figure 21–103). This will generate both the neurons and the associated supporting, or **glial**, cells of the central nervous system. In the process it becomes transformed into a thicker and more complex structure with many layers of cells of various types.

Because differentiated neurons do not divide, each one can be assigned a "birthday," defined as the time of the final mitosis that generated it from a dividing neuronal precursor cell. In both higher vertebrates and invertebrates, the birthdays of the neurons of a given type generally all occur within a strictly limited period of development, after which no further neurons of that type are produced. Each region of the developing neural tube has its own program of cell divisions, and neurons with different birthdays are generally destined for different functions. Since neural stem cells usually do not persist once the production of nerve cells is complete, nerve cell numbers thereafter can only be regulated downward, through cell death, as we shall see.

The Time and Place of a Neuron's Birth Determine the Connections It Will Form [79]

Before sending out its axon and dendrites, the immature neuron or its precursor commonly migrates from its birthplace and settles in some other location. In

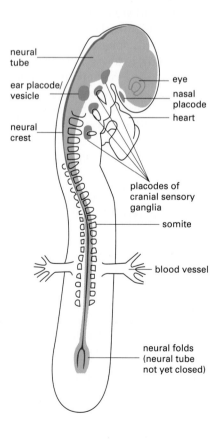

Figure 21–102 **Diagram of an early (2½-day) chick embryo, showing the origins of the nervous system.** The neural tube (*light green*) has already closed, except at the tail end, and lies internally, beneath the ectoderm, of which it was originally a part (see Figure 21–10). The neural crest (*red*) lies dorsally beneath the ectoderm, in or above the roof of the neural tube. In addition, thickenings, or *placodes* (*dark green*), in the ectoderm of the head give rise to some of the sensory transducer cells and neurons of that region, including those of the ear and the nose. The cells of the retina of the eye, by contrast, originate as part of the neural tube.

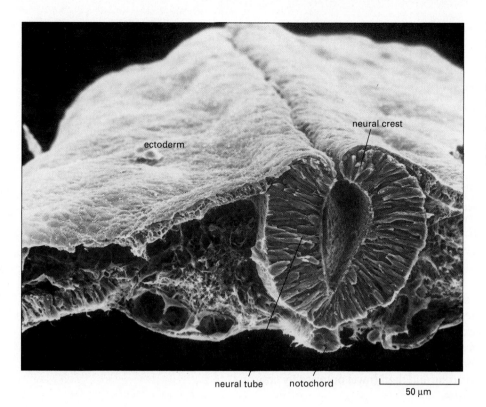

Figure 21–103 Formation of the neural tube. The scanning electron micrograph shows a cross-section through the trunk of a 2-day chick embryo. The neural tube is about to close and pinch off from the ectoderm; at this stage it consists (in the chick) of an epithelium that is only one cell thick. (Courtesy of Jean-Paul Revel.)

ectoderm

neural crest

neural tube   notochord

50 μm

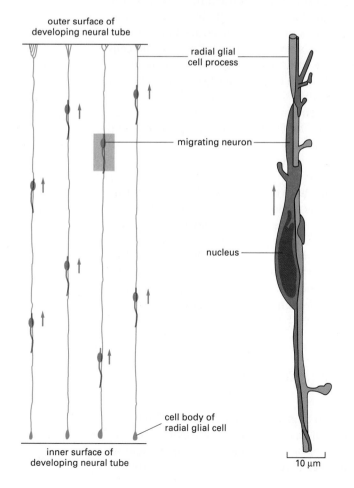

outer surface of developing neural tube

radial glial cell process

migrating neuron

nucleus

cell body of radial glial cell

inner surface of developing neural tube

10 μm

Figure 21–104 Migration of immature neurons along radial glial cells. The diagrams are based on reconstructions from sections of the cerebral cortex of a monkey (part of the neural tube). The neurons are born close to the inner, luminal surface of the neural tube and migrate outward. The radial glial cells can be considered as persisting cells of the original columnar epithelium of the neural tube that become extraordinarily stretched as the wall of the tube thickens. (After P. Rakić, *J. Comp. Neurol.* 145:61–84, 1972.)

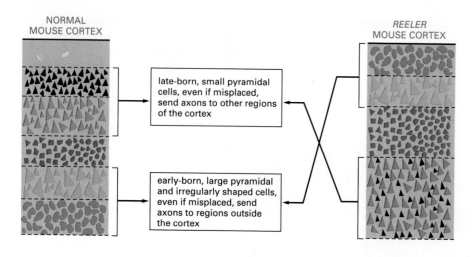

NORMAL MOUSE CORTEX

REELER MOUSE CORTEX

late-born, small pyramidal cells, even if misplaced, send axons to other regions of the cortex

early-born, large pyramidal and irregularly shaped cells, even if misplaced, send axons to regions outside the cortex

Figure 21–105 Comparison of the layering of neurons in the cortex of normal and *reeler* mice. In the *reeler* mutant an abnormality of cell migration causes an approximate inversion of the normal relationship between neuronal birthday and position. The misplaced neurons nevertheless differentiate according to their birthdays and make the connections appropriate to their birthdays.

the central nervous system glial cells often provide a pathway for the migration. The neural tube of a vertebrate embryo, for example, contains a scaffolding of *radial glial cells.* Each of these cells extends from the inner to the outer surface of the tube, a distance that may be as much as 2 cm in the cerebral cortex of the developing brain of a primate. Prospective neurons go through their final cell division close to the lumen of the neural tube and then travel outward by crawling along the radial glial cells (Figure 21–104).

Successive cohorts of migrant cells, born at different times, settle in different positions. In the cerebral cortex, for example, the neurons become arranged in layers according to their birthdays as a result of a migration in which the cells that are born later migrate outward past those born earlier. By transplanting cells between young and old embryos, it can be shown that these different choices of destination are already specified before the cells set off on their migration; they reflect differences in the intrinsic characters of the cells produced at different times—differences that will also dictate the synaptic connections that the cells later form. Thus, in the cerebral cortex the early-born cells (in inner layers) will send their axons to regions outside the cortex, while the late-born cells (in outer layers) will send their axons to regions within the cortex. This relationship between birthday and axonal connections is maintained even in a mutant mouse in which the migrations are abnormal and the final positions of the early- and late-born cells are inverted, confirming that the connections reflect the intrinsic character, rather than the final location, of the neurons (Figure 21–105).

No less important than the time of birth of a neuron is the place of its birth. Cells in different regions of the neural tube have different positional values that govern the connections they will form. These position-dependent differences are evident in the pattern of expression of the *Hox* genes, as we have already seen, and of a large number of other genes that code for gene regulatory proteins and other regulatory molecules. The mechanisms that create the molecular differences between prospective neurons are poorly understood, but they seem, where known, to be similar in principle to the mechanisms of pattern formation discussed earlier. The question we have to confront now, however, is a different one: how do the newborn nerve cells, equipped with their specific markers, proceed to set up an orderly pattern of connections?

Each Axon or Dendrite Extends by Means of a Growth Cone at Its Tip [77, 80]

As a rule the axon and the dendrites begin to grow out from the nerve cell body soon after the cell body has reached its final location. The sequence of events was originally observed in intact embryonic tissue by the method of Golgi staining (Figure 21–106). This technique, and other methods developed subsequently,

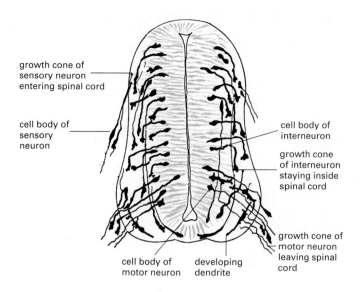

growth cone of sensory neuron entering spinal cord

cell body of sensory neuron

cell body of interneuron

growth cone of interneuron staying inside spinal cord

growth cone of motor neuron leaving spinal cord

cell body of motor neuron

developing dendrite

Figure 21–106 **Growth cones in the developing spinal cord of a 3-day chick embryo.** The drawing shows a cross-section stained by the Golgi technique. Most of the neurons, apparently, have as yet only one elongated process—the future axon. The growth cones of the interneurons remain inside the spinal cord, those of the motor neurons emerge from it (to make their way toward muscles), and those of the sensory neurons grow into it from outside (where their cell bodies lie). Many of the cells in the more central regions of the embryonic spinal cord are still proliferating and have not yet begun to differentiate as neurons or glial cells. (From S. Ramón y Cajal, Histologie du Système Nerveux de l'Homme et des Vertébrés. Paris: Maloine, 1909–1911; reprinted, Madrid: C.S.I.C., 1972.)

reveal an irregular, spiky enlargement at the tip of each developing nerve cell process. This structure, which is called the **growth cone**, appears to be crawling through the surrounding tissue. It comprises both the engine that produces the movement and the steering apparatus that directs the tip of each process along the proper path.

Much of what we know about the properties of growth cones has come from studies in tissue or cell culture. One can watch as a neuron begins to put out its processes, all at first alike, until one of the growth cones puts on a sudden turn of speed, identifying its process as the axon, with its own axon-specific set of proteins (Figure 21–107). The contrast between axon and dendrite established at this stage will cause the two types of process to grow out for different distances, to follow different paths, and to play different parts in synapse formation.

For an isolated neuron in culture the distinction between axon and dendrite is not always easy to see, and it is convenient to refer to both types of process as *neurites.* The growth cone at the end of a typical rapidly growing neurite moves forward at a speed of about 1 mm per day. It consists of a broad, flat expansion, like the palm of a hand, with many long *microspikes* or *filopodia* extending from it like fingers (Figure 21–108). These are continually active: some are retracting back into the growth cone while others are elongating, waving about, and touching down and adhering to the substratum. The "webs" or "veils" between the filopodia form *lamellipodia* with a typical ruffling membrane. All these features, as well as the configuration of the cytoskeleton internally, suggest that the growth cone is crawling forward in much the same way as the leading edge of a cell such as a neutrophil or fibroblast, as discussed in Chapter 16.

Figure 21–107 **Formation of axon and dendrites in culture.** A young neuron has been isolated from the brain of a mammal and put to develop in culture, where it sends out processes. One of these processes, the future axon, has begun to grow out faster than the rest (the future dendrites) and has bifurcated. (A) A phase-contrast picture; (B) the pattern of staining with fluorescent phalloidin, which binds to filamentous actin. Actin is concentrated in the growth cones at the tips of the processes that are actively extending and at some other sites of lamellipodial activity. (Courtesy of Kimberly Goslin, from Z.W. Hall, An Introduction to Molecular Neurobiology. Sunderland, MA: Sinauer, 1992.)

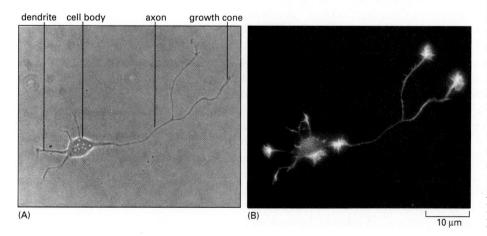

dendrite   cell body   axon   growth cone

(A)

(B)

10 μm

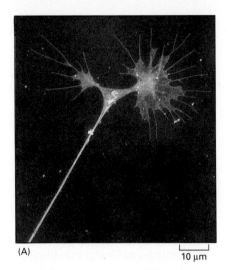

(A)

10 μm

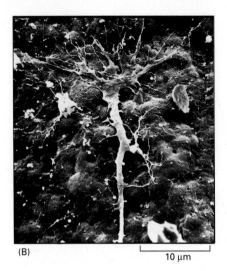

(B)

10 μm

Figure 21–108 **Neural growth cones.**
(A) Scanning electron micrograph of growth cones at the end of a neurite put out by a chick sympathetic neuron in culture. Here a previously single growth cone has recently divided in two. Note the many filopodia and the taut appearance of the neurite, due to tension generated by the forward movement of the growth cones, which are often the only firm points of attachment to the substratum. (B) Scanning electron micrograph of the growth cone of a sensory neuron *in vivo* crawling over the inner surface of the epidermis of a *Xenopus* tadpole. (A, from D. Bray, in Cell Behaviour [R. Bellairs, A. Curtis, and G. Dunn, eds.]. Cambridge, UK: Cambridge University Press, 1982; B, from A. Roberts, *Brain Res.* 118:526–530, 1976.)

With its filopodia and lamellipodia the growth cone explores the regions that lie ahead and on either side. When such a protrusion contacts an unfavorable surface, it withdraws; when it contacts a more favorable surface, it persists longer, steering the growth cone as a whole to move in that direction. In this way the growth cone can be guided by subtle variations in the surface properties of the substrata over which it moves.

The Growth Cone Pilots the Developing Neurite Along a Precisely Defined Path *in Vivo* [81–84]

In living animals growth cones generally travel toward their targets along predictable routes, exploiting a multitude of different cues to find their way. Most often, they take routes that have been pioneered by other neurites, which they follow by contact guidance. As a result, nerve fibers in a mature animal are usually found grouped together in tight parallel bundles (called *fascicles* or *fiber tracts*). Such crawling of growth cones along axons is thought to be mediated by homophilic cell-cell-adhesion molecules—membrane glycoproteins that help a cell displaying them to stick to any other cell that displays them also. As discussed in Chapter 19, two of the most important classes of such molecules are those that belong to the *immunoglobulin superfamily*, such as *N-CAM*, and those of the Ca²⁺-dependent *cadherin* family, such as *N-cadherin*. Members of both families are generally present on the surfaces of growth cones, of axons, and of various other cell types that growth cones crawl over, including glial cells in the central nervous system and muscle cells in the periphery of the body. Growth cones also migrate over components of the extracellular matrix, especially laminin, which they bind to by means of cell-surface matrix receptors of the *integrin* family (discussed in Chapter 19).

In some cases one can demonstrate the importance of a given cell-cell or cell-matrix adhesion molecule by blocking its function with an antibody and observing a disturbance of axon outgrowth. But usually a growth cone employs several adhesion systems to migrate, and antibodies against any single one of them have little effect; only when multiple antibodies are applied, so as to block all of them together, is the growth cone severely hindered in its navigation. In principle, different combinations of adhesion molecules allow for great variety in the surface properties of growth cones and for subtle and complex pathway selection according to the combinations of molecules on the surfaces of cells along the way.

It is still uncertain how far different combinations of adhesion proteins such as N-CAM, N-cadherin, and integrins in the growth cone membrane are sufficient to explain why some growth cones take one route while others take another or how a set of axons, on reaching their target region, are able to form synapses there in an orderly array. Adhesion molecules are certainly not the only influences at work. The contacts a growth cone makes with cell surfaces and matrix can give rise to intracellular signals that can, for example, actively inhibit forward movement. Substances that diffuse through the extracellular medium can also give rise to gradients that provide guidance. In the developing spinal cord, for example, there is a group of neurons whose axons travel ventrally, toward the *floor plate* of the neural tube, to cross by that route to the other side of the tube. When these neurons are placed in culture a short distance from an explanted fragment of floor plate, their axons will again orient their outgrowth toward it, implying that the specialized cells in the floor plate secrete molecules that have a chemotactic guiding effect.

Target Tissues Release Neurotrophic Factors That Control Nerve Cell Growth and Survival [82, 85]

Most types of neurons in the vertebrate central and peripheral nervous system are produced in excess; up to 50% or more of them then die soon after they reach their target, even though they appear perfectly normal and healthy up to the time of their death. About half of all the motor neurons that send axons to skeletal muscle, for example, die within a few days after making contact with their target muscle cells. This large-scale death of neurons is thought to reflect the outcome of a competition. Each type of target cell releases a limited amount of a specific **neurotrophic factor** that the neurons innervating that target require to survive: the neurons apparently compete to take up the factor, and those that do not get enough die by programmed cell death. This seemingly wasteful process provides a simple and elegant means of adjusting the number of neurons of each type to the number of target cells that they innervate.

The first neurotrophic factor to be identified, and still the best characterized, is known simply as **nerve growth factor**, or **NGF**. It was discovered by accident in the course of experiments in which foreign tissues and tumors were transplanted into chick embryos. Transplants of one particular tumor became exceptionally densely innervated and caused a striking enlargement of certain groups of peripheral neurons in the vicinity of the graft. Just two classes of neurons were affected: *sensory neurons* and *sympathetic neurons* (a subclass of peripheral neurons that control contractions of smooth muscle and secretion from exocrine glands). The cause of this phenomenon was traced to a specific protein, NGF, and it was shown that if anti-NGF antibodies are administered to mice while the nervous system is still developing, most sympathetic neurons and some sensory neurons die. Sympathetic neurons and some sensory neurons also die in culture in the absence of NGF; if NGF is present, they survive and send out neurites (Figure 21–109). Some classes of neurons in the central nervous system are dependent on NGF in a similar way.

NGF is produced by the tissues that are innervated by NGF-dependent neurons. Experimental manipulations confirm that the larger the quantity of target tissue, the larger the number of surviving neurons, and this effect can be shown to be mediated by NGF because it can be mimicked by direct manipulation of NGF concentrations. Later in life, after the phase of cell death is over, NGF has a continuing role in regulating the density of innervation by controlling the extent of local sprouting of axon branches. This mechanism is important in restoring innervation in tissues such as skin and smooth muscle after an injury. NGF acts in the intact animal just as it does in a culture dish (see Figure 21–109), both to sustain cell survival and as a local stimulus for growth cone activity, thus adjusting the supply of innervation according to the requirements of the target.

NGF is only one of a family of homologous neurotrophic factors (called *neurotrophins*) that are responsible for this type of regulation in different parts of the vertebrate nervous system. They bind to a complementary family of transmembrane receptor proteins (named after a proto-oncogene called *trk* that codes

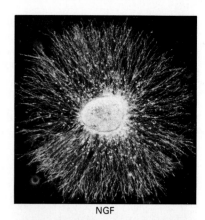

NGF

control

Figure 21–109 NGF effects on neurite outgrowth. Dark-field photomicrographs of a sympathetic ganglion cultured for 48 hours with (*above*) or without (*below*) NGF. Neurites grow out from the sympathetic neurons only if NGF is present in the medium. Each culture also contains Schwann (glial) cells that have migrated out of the ganglion; these are not affected by NGF. Neuronal survival and maintenance of growth cones for neurite extension represent two distinct effects of NGF. The effect on growth cones is local, direct, rapid, and independent of communication with the cell body; when NGF is removed, the deprived growth cones halt their movements within a minute or two. The effect of NGF on cell survival is less immediate and is associated with uptake of NGF by endocytosis and its intracellular transport back to the cell body. (Courtesy of Naomi Kleitman.)

Figure 21–110 Connections between eye and brain in a *Xenopus* tadpole. In this specimen a tracer molecule has been injected into one eye (dark object at *left*), taken up by the neurons there, and carried along their axons, revealing the paths they take to the optic tectum in the brain. (Courtesy of Jeremy Taylor.)

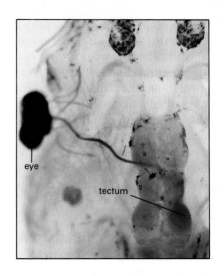

for one of them), which belong to the tyrosine-kinase class of receptors discussed in Chapter 15. It is hoped that the neurotrophic factors will prove useful in the treatment of neurological diseases, such as Alzheimer's disease and motor neuron disease (Lou Gehrig's disease), in which neurons degenerate and die inappropriately.

We now return to the problem of the spatial patterning of nerve connections.

The Positional Values of Neurons Guide the Formation of Orderly Neural Maps: The Doctrine of Neuronal Specificity [86]

The inputs from sense organs are generally *mapped* or *projected* in an orderly way onto the sensory regions in the central nervous system, and the outputs from the motor regions of the central nervous system are mapped in an orderly way onto the muscles. Thus, similar nerve cells in different regions of the vertebrate retina send their axons to synapse with neurons in correspondingly different regions of the *optic tectum* in the midbrain (Figure 21–110), and similar motor neurons at different locations in the spinal cord send their axons to different muscles.

In principle, the growth cones could be simply channeled to different destinations as a direct consequence of their different starting positions, like drivers on a multilane highway where it is forbidden to change lanes. This possibility was tested in the visual system by a famous experiment in the 1940s. If the optic nerve of a frog is cut, it will regenerate. The retinal axons grow back to the optic tectum, restoring normal vision. If, in addition, the eye is rotated in its socket at the time of cutting of the nerve, so as to put originally ventral retinal cells in the position of dorsal retinal cells, vision is still restored, but with an awkward flaw: the animal behaves as though it sees the world upside down. This is because the misplaced retinal cells make the connections appropriate to their *original*, not their actual, positions (Figure 21–111). The cells are evidently endowed with positional values, carrying a record of their original position, so that cells on opposite sides of the retina are intrinsically different. As in the cortex of

Figure 21–111 The regeneration of connections between eye and brain in an amphibian after one eye has been rotated. The axons from each part of the rotated retina regenerate so as to reconnect with the part of the tectum appropriate to the *original* positions of the retinal bodies. Thus, for example, light falling on the ventral part of the rotated retina is perceived as though it were falling on the dorsal part, and the animal sees the world upside down; if food is dangled above it, it makes a lunge downward, and so on.

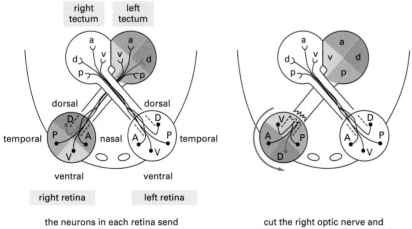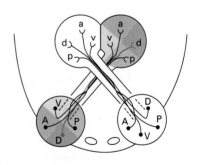

the neurons in each retina send their axons to the opposite tectum, setting up an orderly map (A–a, V–v, etc.)

cut the right optic nerve and rotate the right eye; the severed ends of the axons from the retina degenerate

the neurons in each part of the retina, even though now misplaced, regenerate connections with the same part of the tectum that they were connected with before the *reeler* mouse (see Figure 21–105), it is the intrinsic character, rather than the position, that decides the choice of target site. Such nonequivalence among neurons is referred to as **neuronal specificity**.

Axons from Opposite Sides of the Retina Respond Differently to a Gradient of Repulsive Molecules in the Tectum [87]

On reaching the tectum, the retinal axons must choose, according to their individual character, which region of tectum to innervate. Axons from the nasal retina (the side closest to the nose), for example, project to the posterior tectum, and axons from the temporal retina (the side farthest from the nose) project to the anterior tectum. This choice is governed by differences in the intrinsic characters of the cells in different parts of the tectum. Thus the neuronal map depends on a correspondence between two systems of positional markers, one in the retina and the other in the tectum.

Experiments *in vitro* with tissues from the chick embryo give some insight into the nature of the tectal markers and the way in which the retinal axons respond to them. Fragments of retina are placed in culture and allowed to send out axons over a substratum that is carpeted with membrane vesicles prepared from tectal cells (Figure 21–112). The carpet is laid out in stripes, with bands of anterior tectal membrane alternating with bands of posterior tectal membrane. Axons from nasal retina, depending on details of the preparation, either show no preference and grow indiscriminately in all of the bands or show a preference, appropriately, for posterior tectal membrane. Axons from temporal retina consistently grow only along the bands of anterior tectal membrane, in accordance with their normal destiny. Surprisingly, this is not because the anterior tectal membrane is particularly adhesive or attractive to them but because the posterior tectal membrane is particularly repellent: filopodia that touch it withdraw and collapse. In fact, the growth cones of the temporal axons (but not those of nasal axons) will collapse and retract if a suspension of posterior tectal membranes is dripped onto them. No such collapse occurs in response to anterior tectal membrane.

The peculiar effects of the posterior tectal membrane on the temporal retinal cells have been traced to a specific inhibitory glycoprotein that is distributed in a gradient from posterior to anterior in the tectum. In other parts of the nervous system other surface molecules can be shown to have analogous functions as growth cone repellents. These crude systems of markers are adequate to define the anteroposterior *orientation* of the map in the frog optic tectum. Other mechanisms of an entirely different sort, however, are required to make the map precise.

Figure 21–112 Selectivity of retinal axons growing over tectal membranes. The culture substratum has been coated with alternating stripes of membrane prepared either from posterior tectum (P) or from anterior tectum (A); the anterior tectal stripes are made visible by staining them with a fluorescent marker in the vertical strips at the sides of the picture. Axons of neurons from the temporal half of the retina (growing in from the left) follow the stripes of anterior tectal membrane but avoid the posterior tectal membrane, while axons of neurons from the nasal half of the retina (growing in from the right) do the converse. Thus anterior tectum differs from posterior tectum and nasal retina from temporal retina, and the differences guide selective axon outgrowth. These experiments have been done with cells from the chick embryo. (From Y. von Boxberg, S. Diess, and U. Schwarz, *Neuron* 10:345–357, 1993.)

P
A
P
A
P
A
P
A
P

temporal   nasal

Diffuse Patterns of Synaptic Connections Are Sharpened by Activity-dependent Synapse Elimination [88, 89]

In a normal animal the retinotectal map is initially fuzzy and imprecise. Studies in frogs and fish show that each retinal axon at first branches widely in the tectum and makes a profusion of synapses, distributed over a large area of tectum that overlaps with the territories innervated by other axons. These territories are subsequently trimmed back by elimination of synapses and retraction of axon branches. This refinement of the map through synapse elimination is governed by two competition rules that jointly create spatial order: (1) axons from separate regions of retina, which tend to be excited at different times, compete to dominate the available tectal territory, but (2) axons from neighboring sites in the retina, which tend to be excited at the same time, innervate neighboring territories in the tectum because they collaborate to retain synapses on shared tectal cells (Figure 21–113). The mechanism underlying both these rules depends on electrical activity and signaling at the synapses that are formed. If all action potentials are blocked by a toxin that binds to voltage-gated Na+ channels, synapse elimination is inhibited and the map remains fuzzy.

This phenomenon of **activity-dependent synapse elimination** is encountered in almost every part of the developing vertebrate nervous system. Synapses are first formed in abundance and distributed over a broad target field; then the system of connections is pruned back by competitive processes that depend on electrical activity and synaptic signaling. The elimination of synapses in this way is distinct from the elimination of surplus neurons by cell death, and it occurs after the period of normal neuronal death is over.

The cellular mechanisms of synapse elimination are beginning to be clarified by experiments on the innervation of skeletal muscle in vertebrate embryos, where typically each muscle cell at first receives synapses from several neurons but in the end is left innervated by only one. Co-cultures of motor neurons with muscle cells can be used to analyze the mechanism *in vitro*. One can identify a muscle cell that is innervated by a single neuron and then directly excite the muscle cell repeatedly with puffs of acetylcholine delivered through a micropipette close to its surface. The synapse made on the muscle cell by the neuron is found to be permanently weakened by this treatment unless the neuron itself is stimulated electrically so that it fires in synchrony with the acetylcholine puffs delivered to the muscle cell, in which case the synapse remains strong (Figure 21–

Figure 21–113 Sharpening of the retinotectal map by synapse elimination. At first the map is fuzzy because each retinal axon branches widely to innervate a broad region of tectum overlapping the regions innervated by other retinal axons. The map is then refined by synapse elimination. Where axons from separate parts of the retina synapse on the same tectal cell, competition occurs, eliminating the connections made by one of the axons. But axons from cells that are close neighbors in the retina cooperate, maintaining their synapses on shared tectal cells. Thus each retinal axon ends up innervating a small tectal territory, adjacent to and partly overlapping the territory innervated by axons from neighboring sites in the retina.

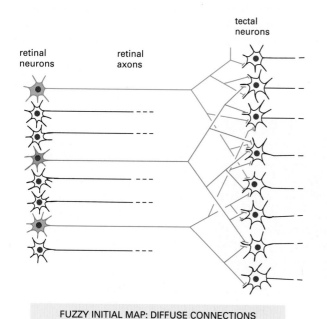

FUZZY INITIAL MAP: DIFFUSE CONNECTIONS

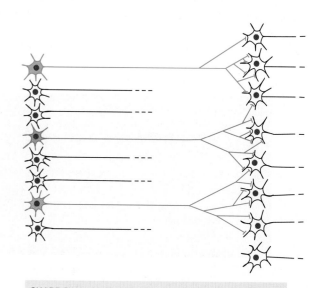

SHARP FINAL MAP: DIFFUSE CONNECTIONS ELIMINATED

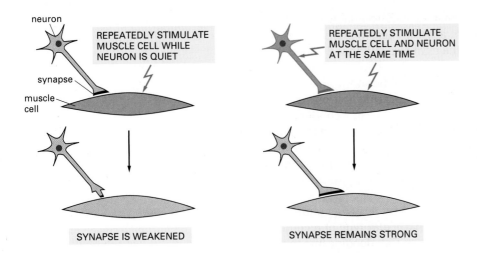

neuron

REPEATEDLY STIMULATE MUSCLE CELL WHILE NEURON IS QUIET

REPEATEDLY STIMULATE MUSCLE CELL AND NEURON AT THE SAME TIME

synapse

muscle cell

SYNAPSE IS WEAKENED

SYNAPSE REMAINS STRONG

Figure 21–114 Synapse elimination and its dependence on the pattern of excitation. In the experiment illustrated schematically here, a neuron and a muscle cell from an embryo have been allowed to form a synapse *in vitro*. The muscle cell is then stimulated with puffs of acetylcholine (mimicking neural stimulation) either alone or in synchrony with electrical excitation of the neuron. The results illustrate a general principle: each excitation of a target cell tends to cause the rejection of any synapse where the presynaptic axon terminal has just been quiet but to maintain synapses where the presynaptic axon terminal has just been active.

114). Weakening, or *repression*, of the synapse reflects a change on its presynaptic side, which causes the axon terminal to release less neurotransmitter when the neuron fires. It can be shown that this synaptic repression depends on the entry of Ca^{2+} into the muscle cell through the cation channels associated with the acetylcholine receptors. Somehow, a sudden rise in intracellular Ca^{2+} causes the postsynaptic cell to send a rebuff to any axon terminals synapsing on its surface in that neighborhood, but the axon terminals are immune to this rebuff if they themselves have just been active.

These and many other findings suggest a simple interpretation of the competition rules for synapse elimination in the retinotectal system. Axons from different parts of the retina fire at different times and so compete. Each time one of them fires, the synapse(s) made by the other on a shared tectal target cell are weakened, until one of the axons is left in sole command of that cell. Axons from neighboring retinal cells, on the other hand, tend to fire in synchrony with one another: they therefore do not compete but instead maintain synapses on shared tectal cells, creating a precisely ordered map in which neighboring cells of the retina project to neighboring sites in the tectum (see Figure 21–113).

Experience Molds the Pattern of Synaptic Connections in the Brain [89, 90]

The same "firing rule" relating synapse maintenance to neural activity helps to organize our developing brains in the light of experience. In the brain of a mammal axons relaying inputs from the two eyes are brought together in the visual region of the cerebral cortex, where they form two overlapping maps of the external visual field, one as perceived through the right eye, the other as perceived through the left. The organization and development of the cortical projections from the two eyes have been studied in great detail, both by anatomical tracing and by physiological tests in which single cortical cells are monitored to find out what kinds of visual stimulus will excite them. These studies reveal an extraordinary sensitivity to experience early in life: if, during a certain *critical period*, one eye is kept covered so as to deprive it of visual stimulation, while the other eye is allowed normal stimulation, the deprived eye loses its synaptic connections to the cortex and becomes almost entirely, and irreversibly, blind. In accordance with the firing rule, a competition has occurred in which synapses in the visual cortex made by inactive axons are eliminated while synapses made by active axons are consolidated. In this way cortical territory is allocated to axons that carry information and is not wasted on those that are silent.

But the firing rule also operates in more subtle ways to establish the nerve connections that enable us to see. For example, the ability to see depth—stereo vision—depends on the presence in the visual cortex of cells that receive inputs

Neural Development

from both eyes at once, conveying information about the same part of the visual field as seen from two slightly different angles. These binocularly driven cells allow us to compare the view through the right eye with that through the left so as to derive information about the relative distances of objects from us. If, however, the two eyes are prevented during the critical period from ever seeing the same scene at the same time—for example, by covering first one eye and then the other on alternate days or simply as a consequence of a childhood squint—almost no binocularly driven cells are retained in the cortex, and the capacity for stereo perception is irretrievably lost. Evidently, in accordance with the firing rule, the inputs from each eye to a binocularly driven neuron are maintained only if the two inputs are frequently triggered to fire in synchrony, as occurs when the two eyes look together at the same scene.

We saw in Chapter 15 that synaptic changes underlying memory in many parts of the brain hinge on the behavior of a particular type of receptor for the neurotransmitter glutamate—the **NMDA receptor**. Ca^{2+} flooding into the postsynaptic cell through the channels opened by this receptor triggers lasting changes in the strengths of the synapses on that cell, just as Ca^{2+} entering a muscle cell via acetylcholine-receptor channels during development affects the synapses made on it by motor neurons. The changes that are induced by the NMDA-dependent mechanism in the adult brain obey rules closely akin to the developmental firing rule. In fact, the refinement and remodeling of synaptic connections that we have just described in the developing visual systems of mammals and amphibians can be blocked by an inhibitor of the NMDA receptor. Both memory and the developmental adjustments, therefore, may depend on essentially the same machinery. The molecular basis of this device through which experience molds our brains is one of the central challenges that the nervous system presents to cell biology.

Summary

The development of the nervous system proceeds in three phases: first, nerve cells are generated through cell division; then, having ceased dividing, they send out axons and dendrites to form profuse synapses with other, remote cells so that communication can begin; last, the system of synaptic connections is refined and remodeled according to the pattern of electrical activity in the neural network.

Axons and dendrites grow out by means of growth cones at their tips, following specific pathways delineated by cells and extracellular matrix along the way. The guidance depends on many different classes of adhesion molecules and intercellular signals as well as on factors that inhibit and repel growth cones. Growth cones from different, nonequivalent neurons respond differently to these cues, and in this way neural maps are set up—orderly projections of one array of neurons onto another. After the growth cones have reached their targets, two major sorts of adjustment occur. First, many of the innervating neurons die as a result of a competition for survival factors such as NGF (nerve growth factor) secreted by the target tissue. This cell death adjusts the quantity of innervation according to the size of the target. Second, individual synapses are pruned away in some places, reinforced in others, so as to create a more precisely ordered pattern of connections. This process depends on electrical activity: synapses that are frequently active are reinforced, and different neurons contacting the same target cell tend to maintain their synapses on the shared target only if they are both frequently active at the same time. In this way the structure of the brain can be adjusted to reflect the connections between events in the external world. The underlying molecular mechanism may be similar to that responsible for the formation of memories in adult life.

References

General

Browder, L.W.; Erickson, C.A.; Jeffery, W.R. Developmental Biology, 3rd ed. Philadelphia: Saunders, 1991.

Gilbert, S.F. Developmental Biology, 3rd ed. Sunderland, MA: Sinauer, 1991.

Larsen, W.J. Human Embryology. New York: Churchill Livingstone, 1993.

Slack, J.M.W. From Egg to Embryo: Regional Specification in Early Development, 2nd ed. Cambridge, UK: Cambridge University Press, 1991.

Spemann, H. Embryonic Development and Induction. New Haven: Yale University Press, 1938. Reprinted, New York: Garland, 1988.

Wilkins, A.S. Genetic Analysis of Animal Development, 2nd ed. New York: Wiley-Liss, 1993.

Wolpert, L. The Triumph of the Embryo. Oxford, UK: Oxford University Press, 1991.

Cited

1. Bard, J.B.L. Morphogenesis: The Cellular and Molecular Processes of Developmental Anatomy. Cambridge, UK: Cambridge University Press, 1990.

   Bray, D. Cell Movements. New York: Garland, 1992.

   Slack, J.M.W., ed. Early Amphibian Development. *J. Embryol. Exp. Morphol. Suppl.* 89, 1985.

   Trinkaus, J.P. Cells into Organs: The Forces That Shape the Embryo, 2nd ed. Englewood Cliffs, NJ: Prentice-Hall, 1984.

2. Gerhart, J., et al. Cortical rotation of the *Xenopus* egg: consequences for the anteroposterior pattern of embryonic dorsal development. *Development* 107 (Suppl.): 37–51, 1989.

   Vincent, J.P.; Oster, G.F.; Gerhart, J.C. Kinematics of gray crescent formation in *Xenopus* eggs: the displacement of subcortical cytoplasm relative to the egg surface. *Dev. Biol.* 113:484–500, 1986.

3. Gilbert, S.F. Developmental Biology, 3rd ed., pp. 75–114. Sunderland, MA: Sinauer, 1991.

   Kirschner, M.; Newport, J.; Gerhart, J. The timing of early developmental events in *Xenopus*. *Trends Genet.* 1:41–47, 1985.

   Wilson, E.B. The Cell in Development and Heredity, 3rd ed. with corrections. New York: Macmillan, 1925, 1928. Reprinted, New York: Garland, 1987.

4. Furshpan, E.J.; Potter, D.D. Low-resistance junctions between cells in embryos and tissue culture. *Curr. Top. Dev. Biol.* 3:95–128, 1968.

   Guthrie, S.C.; Gilula, N.B. Gap junctional communication and development. *Trends Neurosci.* 12:12–16, 1989.

   Kalt, M.R. The relationship between cleavage and blastocoel formation in *Xenopus laevis*. II. Electron microscopic observations. *J. Embryol. Exp. Morphol.* 26:51–66, 1971.

5. Gustafson, T.; Wolpert, L. Cellular movement and contact in sea urchin morphogenesis. *Biol. Rev.* 42:442–498, 1967.

   Hardin, J.D.; Cheng, L.Y. The mechanisms and mechanics of archenteron elongation during sea urchin gastrulation. *Dev. Biol.* 115:490–501, 1986.

   Stern, C.D.; Ingham, P.W., eds. Gastrulation. *Development Suppl.* 1992.

6. Fristrom, D. The cellular basis of epithelial morphogenesis. *Tissue Cell* 20:645–690, 1988.

   Gerhart, J.; Keller, R. Region-specific cell activities in amphibian gastrulation. *Annu. Rev. Cell Biol.* 2:201–229, 1986.

7. Spemann, H.; Mangold, H. Induction of embryonic primordia by implantation of organizers from a different species. *Roux's Archiv.* 100:599–638, 1924. (English translation in Foundations of Experimental Embryology, 2nd ed. [B.H. Willier, J.M. Oppenheimer, eds.] New York: Hafner, 1974.)

8. Shi, J.; Keller, R. Cell motility driving mediolateral intercalation in explants of *Xenopus laevis*. *Development* 116:901–914, 1992.

   Spemann, H. Embryonic Development and Induction. New Haven: Yale University Press, 1938. Reprinted, New York: Garland, 1988.

9. Balinsky, B.I. Introduction to Embryology, 5th ed. Philadelphia: Saunders, 1981.

   Beddington, R.S.P.; Smith, J.C. Control of vertebrate gastrulation: inducing signals and responding genes. *Curr. Opin. Genet. Dev.* 3:655–661, 1993.

   Langman, J. Medical Embryology, 5th ed. Baltimore, Williams & Wilkins, 1985.

   Romer, A.S.; Parsons, T.S. The Vertebrate Body, 6th ed. Philadelphia: Saunders, 1986.

10. Adams, D.S.; Keller, R.; Koehl, M.A. The mechanics of notochord elongation, straightening and stiffening in the embryo of *Xenopus laevis*. *Development* 110:115–130, 1990.

    Yasuo, H.; Satoh, N. Function of vertebrate T gene. *Nature* 364:582–583, 1993.

11. Anderson, D.J. The neural crest lineage problem: neuropoiesis? *Neuron* 3:1–12, 1989.

    Le Douarin, N. The Neural Crest. Cambridge, UK: Cambridge University Press, 1982.

    Marusich, M.F.; Weston, J.A. Development of the neural crest. *Curr. Opin. Genet. Dev.* 1:221–229, 1991.

12. Morriss-Kay, G.; Tuckett, F. The role of microfilaments in cranial neurulation in rat embryos: effects of short-term exposure to cytochalasin D. *J. Embryol. Exp. Morphol.* 88:333–348, 1985.

    Odell, G.M.; Oster, G.; Alberch, P.; Burnside, B. The mechanical basis of morphogenesis. I. Epithelial folding and invagination. *Dev. Biol.* 85:446–462, 1981.

    Ruiz i Altaba, A.; Jessell, T.M. Midline cells and the organization of the vertebrate neuraxis. *Curr. Opin. Genet. Dev.* 3:633–640, 1993.

13. Keynes, R.J.; Stern, C.D. Mechanisms of vertebrate segmentation. *Development* 103:413–429, 1988.

14. Hynes, R.O.; Lander, A.D. Contact and adhesive specificities in the associations, migrations, and targeting of cells and axons. *Cell* 68:303–322, 1992.

    Nose, A.; Nagafuchi, A.; Takeichi, M. Expressed recombinant cadherins mediate cell sorting in model systems. *Cell* 54:993–1001, 1988.

    Takeichi, M. Cadherin cell adhesion receptors as a morphogenetic regulator. *Science* 251:1451–1455, 1991.

    Townes, P.L.; Holtfreter, J. Directed movements and selective adhesion of embryonic amphibian cells. *J. Exp. Zool.* 128:53–120, 1955.

15. Chevallier, A.; Kieny, M.; Mauger, A. Limb-somite relationship: origin of the limb musculature. *J. Embryol. Exp. Morphol.* 41:245–258, 1977.

    Christ, B.; Jacob, H.J.; Jacob, M. Experimental analysis of the origin of the wing musculature in avian embryos. *Anat. Embryol.* 150:171–186, 1977.

16. Bronner-Fraser, M. Mechanisms of neural crest migration. *Bioessays* 15:221–230, 1993.

    Bronner-Fraser, M. An antibody to a receptor for fibronectin and laminin perturbs cranial neural crest development *in vivo. Dev. Biol.* 117:528–536, 1986.

    Erickson, C.A. Cell migration in the embryo and adult organism. *Curr. Opin. Cell Biol.* 2:67–74, 1990.

    Morrison-Graham, K.; Takahashi, Y. Steel factor and c-kit receptor: from mutants to a growth-factor system. *Bioessays* 15:77–83, 1993.

17. Gurdon, J.B. The generation of diversity and pattern in animal development. *Cell* 68:185–199, 1992.

18. DiBerardino, M.A.; Orr, N.H.; McKinnell, R.G. Feeding tadpoles cloned from *Rana* erythrocyte nuclei. *Proc. Natl. Acad. Sci. USA* 83:8231–8234, 1986.

    Gurdon, J.B. The Control of Gene Expression in Animal Development. Cambridge: Harvard University Press, 1974.

    Gurdon, J.B. Transplanted nuclei and cell differentiation. *Sci. Am.* 219(6):24–35, 1968.

    McKinnell, R.G. Cloning-Nuclear Transplantation in Amphibia. Minneapolis: University of Minnesota Press, 1978.

19. Davidson, E.H. Gene Activity in Early Development, 3rd ed., pp. 411–524. Orlando, FL: Academic Press, 1986.

    Horvitz, H.R.; Herskowitz, I. Mechanisms of asymmetric cell division: two Bs or not two Bs, that is the question. *Cell* 68:237–255, 1992.

    Rebagliati, M.R.; Weeks, D.L.; Harvey, R.P.; Melton, D.A. Identification and cloning of localized maternal RNAs from *Xenopus* eggs. *Cell* 42:769–777, 1985.

    Wilson, E.B. The Cell in Development and Heredity, 3rd ed., pp. 1035–1121. New York: Macmillan, 1925, 1928. Reprinted, New York: Garland, 1987.

20. Amaya, E.; Musci, T.J.; Kirschner, M.W. Expression of a dominant negative mutant of the FGF receptor disrupts mesoderm formation in *Xenopus* embryos. *Cell* 66:257–270, 1991.

    Cho, K.W.Y.; Blumberg, B.; Steinbeisser, H.; De Robertis, E.M. Molecular nature of Spemann's organizer: the role of the *Xenopus* homeobox gene *goosecoid. Cell* 67:1111–1120, 1991.

    Dale, L.; Howes, G.; Price, B.M.J.; Smith, J.C. Bone morphogenetic protein 4: a ventralizing factor in *Xenopus* development. *Development* 115:573–585, 1992.

    Hemmati-Brivanlou, A.; Melton, D.A. A truncated activin receptor inhibits mesoderm induction and formation of axial structures in *Xenopus* embryos. *Nature* 359:609–614, 1992.

    Slack, J.M.W. From Egg to Embryo: Regional Specification in Early Development, 2nd ed. Cambridge, UK: Cambridge University Press, 1991.

    Smith, W.C.; Harland, R.M. Injected Xwnt-8 RNA acts early in *Xenopus* embryos to promote formation of a vegetal dorsalizing center. *Cell* 67:753–765, 1991.

    Smith, W.C.; Knecht, A.K.; Wu, M.; Harland, R.M. Secreted *noggin* protein mimics the Spemann organizer in dorsalizing *Xenopus* mesoderm. *Nature* 361:547–549, 1993.

    Thomsen, G.H.; Melton, D.A. Processed Vg1 protein is an axial mesoderm inducer in *Xenopus. Cell* 74:433–441, 1993.

21. Green, J.B.A.; Smith, J.C. Graded changes in the dose of a *Xenopus* activin A homologue elicit stepwise transitions in embryonic cell fate. *Nature* 347:391–394, 1990.

    Lewis, J.; Slack, J.M.W.; Wolpert, L. Thresholds in development. *J. Theor. Biol.* 65:579–590, 1977.

    Wolpert, L. Positional information and pattern formation. *Curr. Top. Dev. Biol.* 6:183–224, 1971.

22. Servetnick, M.; Grainger, R.M. Changes in neural and lens competence in *Xenopus* ectoderm: evidence for an autonomous developmental timer. *Development* 112:177–188, 1991.

23. Austin, C.R.; Short, R.V., eds. Embryonic and Fetal Development, 2nd ed. Reproduction in Mammals, Ser., Book 2. Cambridge, UK: Cambridge University Press, 1982.

    Fleming, T.P.; Johnson, M.H. From egg to epithelium. *Annu. Rev. Cell Biol.* 4:459–485, 1988.

    Kaufman, M.H. The Atlas of Mouse Development. London: Academic Press, 1992.

24. Gardner, R.L. Clonal analysis of early mammalian development. *Philos. Trans. R. Soc. Lond. (Biol.)* 312:163–178, 1985.

    Kelly, S.J. Studies of the developmental potential of 4- and 8-cell stage mouse blastomeres. *J. Exp. Zool.* 200:365–376, 1977.

    McLaren, A. Mammalian Chimeras. Cambridge, UK: Cambridge University Press, 1976.

    Tarkowski, A.K. Experiments on the development of isolated blastomeres of mouse eggs. *Nature* 184:1286–1287, 1959.

25. Capecchi, M.R. The new mouse genetics: altering the genome by gene targeting. *Trends Genet.* 5:70–76, 1989.

    Illmensee, K.; Stevens, L.C. Teratomas and chimeras. *Sci. Am.* 240(4):120–132, 1979.

    Papaioannou, V.E.; Gardner, R.L.; McBurney, M.W.; Babinet, C.; Evans, M.J. Participation of cultured teratocarcinoma cells in mouse embryogenesis. *J. Embryol. Exp. Morphol.* 44:93–104, 1978.

    Robertson, E.J. Pluripotential stem cell lines as a route into the mouse germ line. *Trends Genet.* 2:9–13, 1986.

    Williams, R.L., et al. Myaloid leukemia inhibitory factor maintains the developmental potential of embryonic stem cells. *Nature* 336:685–687, 1988.

26. Weiss, P.A. Principles of Development, pp. 289–437. New York: Holt, 1939.

27. Spemann, H. Über die Determination der ersten Organanlagen des Amphibienembryo I–VI. *Arch. Entw. Mech. Org.* 32:448–555, 1918.

28. Buckingham, M. Making muscle in mammals. *Trends Genet.* 8:144–149, 1992.

    Weintraub, H., et al. The *myoD* gene family: nodal point during specification of the muscle cell lineage. *Science* 251:761–766, 1991.

29. Gurdon, J.B. A community effect in animal development. *Nature* 336:772–774, 1988.

    Lyon, M.F. Epigenetic inheritance in mammals. *Trends Genet.* 9:123–128, 1993.

Paro, R. Mechanisms of heritable gene repression during development of *Drosophila. Curr. Opin. Cell Biol.* 5:999–1005, 1993.

Riggs, A.D.; Pfeiffer, G.P. X-chromosome inactivation and cell memory. *Trends Genet.* 8:169–174, 1992.

30. Held, L.I., Jr. Models for Embryonic Periodicity. Farmington, CT: Karger, 1992.

Lewis, J.H.; Wolpert, L. The principle of non-equivalence in development. *J. Theor. Biol.* 62:479–490, 1976.

Saunders, J.W., Jr.; Gasseling, M.T.; Cairns, J.M. The differentiation of prospective thigh mesoderm grafted beneath the apical ectodermal ridge of the wing bud in the chick embryo. *Dev. Biol.* 1:281–301, 1959.

Theories of biological pattern formation. *Philos. Trans. R. Soc. Lond. (Biol.)* 295:425–617, 1981.

Wolpert, L. Pattern formation in biological development. *Sci. Am.* 239(4):154–164, 1978.

31. Bohn, H. Tissue interactions in the regenerating cockroach leg. In Insect Development (P.A. Lawrence, ed.), Royal Entomological Society of London Symposium No. 8, pp. 170–185. Oxford, UK: Blackwell, 1976.

Bryant, P.J. The polar coordinate model goes molecular. *Science* 259:471–472, 1993.

Bryant, P.J.; Bryant, S.V.; French, V. Biological regeneration and pattern formation. *Sci. Am.* 237(1):66–81, 1977.

Bryant, P.J.; Simpson, P. Intrinsic and extrinsic control of growth in developing organs. *Q. Rev. Biol.* 59:387–415, 1984.

Lewis, J. Simpler rules for epimorphic regeneration: the polar-coordinate model without polar coordinates. *J. Theor. Biol.* 88:371–392, 1981.

32. Wolfram, S. Cellular automata as models of complexity. *Nature* 311:419–424, 1984.

33. Sulston, J., et al. The *C. elegans* genome sequencing project: a beginning. *Nature* 356:37–41, 1992.

Wood, W.B., et al. The nematode *Caenorhabditis elegans.* Cold Spring Harbor, NY: Cold Spring Harbor Laboratory, 1988.

34. Kenyon, C. Cell lineage and the control of *Caenorhabditis elegans* development. *Philos. Trans. R. Soc. Lond. (Biol.)* 312:21–38, 1985.

Sulston, J.E.; Horvitz, H.R. Post-embryonic cell lineage of the nematode, *Caenorhabditis elegans. Dev. Biol.* 56:110–156, 1977.

Sulston, J.E.; Schierenberg, E.; White, J.G.; Thompson, J.N. The embryonic cell lineage of the nematode *Caenorhabditis elegans. Dev. Biol.* 100:64–119, 1983.

35. Schnabel, R. Early determinative events in *Caenorhabditis elegans. Curr. Opin. Genet. Dev.* 1:179–184, 1991.

Sternberg, P.W.; Horvitz, H.R. The genetic control of cell lineage during nematode development. *Annu. Rev. Genet.* 18:489–524, 1984.

36. Ferguson, E.L.; Sternberg, P.W.; Horvitz, H.R. A genetic pathway for the specification of the vulval cell lineages of *Caenorhabditis elegans. Nature* 326:259–267, 1987.

Sulston, J.E.; White, J.G. Regulation and cell autonomy during postembryonic development of *Caenorhabditis elegans. Dev. Biol.* 78:577–597, 1980.

37. Han, M.; Golden, A.; Han, Y.; Sternberg, P.W. *C. elegans lin-45 raf* gene participates in *let-60 ras*-stimulated vulval differentiaion. *Nature* 363:133–140, 1993.

Hill, R.J.; Sternberg, P.W. The gene *lin-3* encodes an inductive signal for vulval development in *C. elegans. Nature* 358:470–476, 1992.

Sternberg, P.W.; Horvitz, H.R. Signal transduction during *C. elegans* vulval induction. *Trends Genet.* 7:366–371, 1991.

38. Ambros, V.; Horvitz, H.R. The lin-14 locus of *Caenorhabditis elegans* controls the time of expression of specific postembryonic developmental events. *Genes Dev.* 1:398–414, 1987.

Chalfie, M.; Horvitz, H.R.; Sulston, J.E. Mutations that lead to reiterations in the cell lineages of *C. elegans. Cell* 24:59–69, 1981.

Ruvkun, G.; Wightman, B.; Bürglin, T.; Arasu, P. Dominant gain-of-function mutations that lead to misregulation of the *C. elegans* heterochronic gene *lin-14*, and the evolutionary implications of dominant mutations in pattern-formation genes. *Development (Suppl. 1)* 1991: 47–54.

39. Cooke, J. Properties of the primary organisation field in the embryo of *Xenopus laevis.* IV. Pattern formation and regulation following early inhibition of mitosis. *J. Embryol. Exp. Morphol.* 30:49–62, 1973.

Edgar, L.G.; McGhee, J.D. DNA synthesis and the control of embryonic gene expression in *C. elegans. Cell* 53:589–599, 1988.

Harris, W.A.; Hartenstein, V. Neuronal determination without cell division in *Xenopus* embryos. *Neuron* 6:499–515, 1991.

Rollins, M.B.; Andrews, M.T. Morphogenesis and regulated gene activity are independent of DNA replication in *Xenopus* embryos. *Development* 112:559–569, 1991.

Satoh, N. Towards a molecular understanding of differentiation mechanisms in Ascidian embryos. *Bioessays* 7:51–56, 1987.

40. Ellis, R.E.; Yuan, J.V.; Horvitz, H.R. Mechanisms and functions of cell death. *Annu. Rev. Cell Biol.* 7:663–698, 1991.

Vaux, D.L.; Weisman, I.L.; Kim, S.K. Prevention of programmed cell death in *Caenorhabditis elegans* by human *bcl-2. Science* 258:1955–1957, 1992.

41. Lawrence, P.A. The Making of a Fly: The Genetics of Animal Design. Oxford, UK: Blackwell Scientific, 1992.

42. Martínez-Arias, A.; Lawrence, P.A. Parasegments and compartments in the *Drosophila* embryo. *Nature* 313:639–642, 1985.

43. Campos-Ortega, J.A.; Hartenstein, V. The Embryonic Development of *Drosophila melanogaster.* Berlin: Springer, 1985.

Foe, V.E.; Alberts, B.M. Studies of nuclear and cytoplasmic behavior during the five mitotic cycles that precede gastrulation in *Drosophila* embryogenesis. *J. Cell Sci.* 61:31–70, 1983.

Leptin, M.; Grunewald, B. Cell shape changes during gastrulation in *Drosophila. Development* 110:73–84, 1990.

Technau, G.M. A single cell approach to problems of cell lineage and commitment during embryogenesis of *Drosophila melanogaster. Development* 100:1–12, 1987.

44. St. Johnston, D.; Nüsslein-Volhard, C. The origin of pattern and polarity in the *Drosophila* embryo. *Cell* 68:201–219, 1992.

45. Anderson, K.V.; Schneider, D.S.; Morisato, D.; Ferguson, E.L. Extracellular morphogens in *Drosophila* dorsal-ventral patterning. *Cold Spring Harbor Symp. Quant. Biol.* 58, 1993.

46. Ferguson, E.L.; Anderson, K.V. *Decapentaplegic* acts as a morphogen to organize dorsal-ventral pattern in the *Drosophila* embryo. *Cell* 7:451–461, 1992.

 Govind, S.; Steward, R. Coming to grips with cactus. *Curr. Biol.* 3:351–354, 1993.

 Klingler, M.; Erdelyi, M.; Szabad, J.; Nüsslein-Volhard, C. Function of *torso* in determining the terminal anlagen of the *Drosophila* embryo. *Nature* 335:275–277, 1988.

 Roth, S.; Stein, D.; Nüsslein-Volhard, C. A gradient of nuclear localization of the *dorsal* protein determines dorsoventral pattern in the *Drosophila* embryo. *Cell* 59:1189–1202, 1989.

 Rushlow, C.A.; Han, K.; Manley, J.L.; Levine, M. The graded distribution of the *dorsal* morphogen is initiated by selective nuclear transport in *Drosophila*. *Cell* 59:1165–1177, 1989.

47. Ephrussi, A.; Lehmann, R. Induction of germ cell formation by *oskar*. *Nature* 358:387–392, 1992.

 Gilbert, S.F. Developmental Biology, 3rd ed., pp. 273–281. Sunderland, MA: Sinauer, 1991.

 Wilson, J.E.; Macdonald, P.M. Formation of germ cells in *Drosophila*. *Curr. Opin. Genet. Dev.* 3:562–565, 1993.

48. Driever, W.; Nüsselin-Volhard, C. A gradient of *bicoid* protein in *Drosophila* embryos. *Cell* 54:83–93, 1988.

 Driever, W.; Nüsslein-Volhard, C. The *bicoid* protein determines position in the *Drosophila* embryo in a concentration-dependent manner. *Cell* 54:95–104, 1988.

 Lawrence, P.A. Background to *bicoid*. *Cell* 54:1–2, 1988.

 Nüsslein-Volhard, C.; Frohnhöfer, H.G.; Lehmann, R. Determination of anteroposterior polarity in *Drosophila*. *Science* 238:1675–1681, 1987.

 Sander, K. Embryonic pattern formation in insects: basic concepts and their experimental foundations. In Pattern Formation: A Primer in Developmental Biology (G. Malacinski, S.V. Bryant, eds.), pp. 235–268. New York: Macmillan, 1984.

49. Nüsslein-Volhard, C.; Wieschaus, E. Mutations affecting segment number and polarity in *Drosophila*. *Nature* 287:795–801, 1980.

50. Akam, M. The molecular basis for metameric pattern in the *Drosophila* embryo. *Development* 101:1–22, 1987.

 Ingham, P.W. The molecular genetics of embryonic pattern formation in *Drosophila*. *Nature* 335:25–34, 1988.

 Scott, M.P.; O'Farrell, P.H. Spatial programming of gene expression in early *Drosophila* embryogenesis. *Annu. Rev. Cell Biol.* 2:49–80, 1986.

51. Hafen, E.; Kuroiwa, A.; Gehring, W.J. Spatial distribution of transcripts from the segmentation gene *fushi tarazu* during *Drosophila* embryonic development. *Cell* 37:833–841, 1984.

 Hoch, M.; Jäckle, H. Transcriptional regulation and spatial patterning in *Drosophila*. *Curr. Opin. Genet. Dev.* 3:566–573, 1993.

 Hülskamp, M.; Tautz, D. Gap genes and gradients—the logic behind the gaps. *Bioessays* 13:261–268, 1991.

52. Small, S.; Levine, M. The initiation of pair-rule stripes in the *Drosophila* blastoderm. *Curr. Opin. Genet. Dev.* 1:255–260, 1991.

53. Ingham, P.W.; Martínez-Arias, A. Boundaries and fields in early embryos. *Cell* 68:221–235, 1992.

 Kornberg, T.B.; Tabata, T. Segmentation of the *Drosophila* embryo. *Curr. Opin. Genet. Dev.* 3:585–593, 1993.

 Struhl, G.; Basler, K. Organizing activity of wingless protein in *Drosophila*. *Cell* 72:527–540, 1993.

54. Lewis, E.B. A gene complex controlling segmentation in *Drosophila*. *Nature* 276:565–570, 1978.

55. Beeman, R.W.; Stuart, J.J.; Brown, S.J.; Denell, R.E. Structure and function of the homeotic gene complex (HOM-C) in the beetle, *Tribolium castaneum*. *Bioessays* 14:439–444, 1993.

 Morata, G.; Lawrence, P.A. Homoeotic genes, compartments and cell determination in *Drosophila*. *Nature* 265:211–216, 1977.

 Wakimoto, B.T.; Kaufman, T.C. Analysis of larval segmentation in lethal genotypes associated with the *Antennapedia* gene complex in *Drosophila melanogaster*. *Dev. Biol.* 81:51–64, 1981.

56. Akam, M.E. Segments, lineage boundaries and the domains of expression of homeotic genes. *Philos. Trans. R. Soc. Lond. (Biol.)* 312:179–187, 1985.

 Harding, K.; Wedeen, C.; McGinnis, W.; Levine, M. Spatially regulated expression of homeotic genes in *Drosophila*. *Science* 229:1236–1242, 1985.

57. Akam, M. The molecular basis for metameric pattern in the *Drosophila* embryo. *Development* 101:1–22, 1987.

 McGinnis, W., et al. A conserved DNA sequence in homeotic genes of the *Drosophila* Antennapedia and Bithorax complexes. *Nature* 308:428–433, 1984.

 Scott, M.P.; Weiner, A.J. Structural relationships among genes that control development: sequence homology between the *Antennapedia, Ultrabithorax,* and *fushi tarazu* loci of *Drosophila*. *Proc. Natl. Acad. Sci. USA* 81:4115–4119, 1984.

58. Bender, W., et al. Molecular genetics of the bithorax complex in *Drosophila melanogaster*. *Science* 221:23–29, 1983.

 Bergson, C.; McGinnis, W. An autoregulatory enhancer element of the *Drosophila* homeotic gene *Deformed*. *EMBO J.* 9:4287–4297, 1990.

 Paro, R. Mechanisms of heritable gene repression during development of *Drosophila*. *Curr. Opin. Cell Biol.* 5:999–1005, 1993.

 Peifer, M.; Karch, F.; Bender, W. The bithorax complex: control of segmental identity. *Genes Dev.* 1:891–898, 1987.

59. McGinnis, W.; Krumlauf, R. Homeobox genes and axial patterning. *Cell* 68:283–302, 1992.

60. Ashburner, M.; Wright, T.F., eds. The Genetics and Biology of *Drosophila*, Vol. 2C. London: Academic Press, 1978.

 Gehring, W.; Nöthiger, R. The imaginal discs of *Drosophila*. In Developmental Systems: Insects (S. Counce, C.H. Waddington, eds.), Vol. 2, pp. 211–290. New York: Academic Press, 1973.

 Hadorn, E. Transdetermination in cells. *Sci. Am.* 219(5):110–120, 1968.

61. Nöthiger, R. Clonal analysis in imaginal discs. In Insect Development (P.A. Lawrence, ed.), Royal Entomological Society of London Symposium No. 8, pp. 109–117. Oxford, UK: Blackwell, 1976.

Stern, C. Genetic Mosaics and Other Essays. Cambridge: Harvard University Press, 1968.

Struhl, G. Genes controlling segment specification in the *Drosophila* thorax. *Proc. Natl. Acad. Sci. USA* 79:7380–7384, 1982.

62. Couso, J.P.; Bate, M.; Martinez-Arias, A. A *wingless*-dependent polar coordinate system in *Drosophila* imaginal discs. *Science* 259:484–489, 1993.

Garcia-Bellido, A.; Lawrence, P.A.; Morata, G. Compartments in animal development. *Sci. Am.* 241(1):102–111, 1979.

Kornberg, T.; Sidén, I.; O'Farrell, P.; Simon, M. The *engrailed* locus of *Drosophila: in situ* localization of transcripts reveals compartment-specific expression. *Cell* 40:45–53, 1985.

Posakony, L.G.; Raftery, L.A.; Gelbart, W.M. Wing formation in *Drosophila melanogaster* requires *decapentaplegic* gene function along the anterior-posterior compartment boundary. *Mech. Dev.* 33:69–82, 1991.

Simpson, P.; Morata, G. Differential mitotic rates and patterns of growth in compartments in the *Drosophila* wing. *Dev. Biol.* 85:299–308, 1981.

63. Campuzano, S.; Modolell, J. Patterning of the *Drosophila* nervous system: the *achaete-scute* complex. *Trends Genet.* 8:202–208, 1992.

Ghysen, A.; Dambly-Chaudière, C. Genesis of the *Drosophila* peripheral nervous system. *Trends. Genet.* 5:251–255, 1989.

Ghysen, A.; Dambly-Chaudière, C.; Jan, L.Y.; Jan, Y.N. Cell interactions and gene interactions in peripheral neurogenesis. *Genes Dev.* 7:723–733, 1993.

64. Hartenstein, V.; Posakony, J.W. A dual function of the *Notch* gene in *Drosophila* sensillum development. *Dev. Biol.* 142:13–30, 1990.

Heitzler, P.; Simpson, P. The choice of cell fate in the epidermis of *Drosophila*. *Cell* 64:1083–1092, 1991.

Sternberg, P.W. Falling off the knife edge. *Curr. Biol.* 3:763–765, 1993.

65. McGinnis, W.; Garber, R.L.; Wirz, J.; Kuroiwa, A.; Gehring, W.J. A homologous protein-coding sequence in *Drosophila* homeotic genes and its conservation in other metazoans. *Cell* 37:403–408, 1984.

Müller, M.M.; Carrasco, A.E.; De Robertis, E.M. A homeobox-containing gene expressed during oogenesis in *Xenopus*. *Cell* 39:157–162, 1984.

66. Bürglin, T.R.; Ruvkun, G. The *Caenorhabditis elegans* homeobox gene cluster. *Curr. Opin. Genet. Dev.* 3:615–620, 1993.

De Robertis, E.M.; Oliver, G.; Wright, C.V.E. Homeobox genes and the vertebrate body plan. *Sci. Am.* 263(1):26–33, 1990.

Lumsden, A. The cellular basis of segmentation in the developing hindbrain. *Trends Neurosci.* 13:329–335, 1990.

McMahon, A.P. The *Wnt* family of developmental regulators. *Trends Genet.* 8:236–242, 1992.

Wilkinson, D.G.; Krumlauf, R. Molecular approaches to the segmentation of the hindbrain. *Trends Neurosci.* 13:335–339, 1990.

67. Holland, P. Homeobox genes in vertebrate evolution. *Bioessays* 14:267–273, 1992.

Krumlauf, R. Mouse *Hox* genetic functions. *Curr. Opin. Genet. Dev.* 3:621–625, 1993.

68. Izpisúa-Belmonte, J.-C.; Tickle, C.; Dollé, P.; Wolpert, L.; Duboule, D. Expression of the homeobox *Hox-4* genes and the specification of position in chick wing development. *Nature* 350:585–589, 1991.

Morgan, B.A.; Izpisúa-Belmonte, J.-C.; Duboule, D.; Tabin, C.J. Targeted misexpression of *Hox-4.6* in the avian limb bud causes apparent homeotic transformations. *Nature* 358:236–239, 1992.

Morgan, B.A.; Tabin, C.J. The role of homeobox genes in limb development. *Curr. Opin. Genet. Dev.* 3:668–674, 1993.

Yokouchi, Y.; Sasaki, H.; Kuroiwa, A. Homeobox gene expression correlated with the bifurcation process of limb cartilage development. *Nature* 353:443–445, 1991.

69. Cutter, E.G. Plant Anatomy, 2nd ed., Part 1, Cells and Tissues; Part 2, Organs. London: Arnold, 1978.

Esau, K. Anatomy of Seed Plants, 2nd ed. New York: Wiley, 1977.

Walbot, V. On the life strategies of plants and animals. *Trends Genet.* 1:165–169, 1985.

70. Chasan, R.; Walbot, V. Mechanisms of plant reproduction: questions and approaches. *Plant Cell* 5:1139–1146, 1993.

Johri, B.M. Embryology of Angiosperms. Berlin: Springer-Verlag, 1984.

West, M.A.L.; Harada, J.J. Embryogenesis in higher plants: an overview. *Plant Cell* 5:1161–1169, 1993.

71. Harper, J.L., ed. Growth and Form of Modular Organisms. London: Royal Society, 1986. (There are several relevant papers in this collection.)

Meinke, D.W. (guest ed.). Plant developmental genetics. *Seminars Dev. Biol.* 4:1–89, 1993.

72. Gunning, B.E.S. Microtubules and cytomorphogenesis in a developing organ: the root primordium of *Azolla pinnata*. In Cytomorphogenesis in Plants (O. Kiermayer, ed.), pp. 301–325. New York: Springer, 1981.

Lloyd, C.W., ed. The Cytoskeletal Basis of Plant Growth and Form. New York: Academic Press, 1991.

73. Becraft, P.W.; Freeling, M. Cell interactions in plants. *Curr. Opin. Genet. Dev.* 2:571–575, 1992.

Lyndon, R.F. Plant Development: The Cellular Basis. London: Unwin Hyman, 1990.

McDaniel, C.N.; Poethig, R.S. Cell lineage patterns in the shoot apical meristem of the germinating corn embryo. *Planta* 175:13–22, 1988.

Sinha, N.R.; Williams, R.E.; Hake, S. Overexpression of a maize homeobox gene, *Knotted-1*, causes a switch from determinate to indeterminate cell fates. *Gene Dev.* 7:787–795, 1993.

Steeves, T.A.; Sussex, I.M. Patterns in Plant Development, 2nd ed. New York: Cambridge University Press, 1989.

74. Rothenberg, M.; Ecker, J.R. Mutant analysis as an experimental approach towards understanding plant hormone action. *Seminars Dev. Biol.* 4:3–13, 1993.

75. Finkelstein, R.; Estelle, M.; Martinez-Zapater, J.; Somerville, C. *Arabidopsis* as a tool for the identification of genes involved in plant development. In Temporal and Spatial Regulation of Plant Genes

(D.P.S. Verma and R.B. Goldberg, eds.), pp. 1–25. New York: Springer-Verlag, 1988.

Jürgens, G. Pattern formation in the flowering plant embryo. *Curr. Opin. Genet. Dev.* 2:567–570, 1992.

Koncz, C.; Chua, N.-H.; Schell, J., eds. Methods in *Arabidopsis* Research. River Edge, NJ: World Scientific Publishing, 1992.

Mayer, U.; Torres Ruiz, R.A.; Berleth, T.; Miséra, S.; Jürgens, G. Mutations affecting body organization in the *Arabidopsis* embryo. *Nature* 353:402–407, 1991.

Meyerowitz, E.M. *Arabidopsis*, a useful weed. *Cell* 56:263–269, 1989.

76. Coen, E.S.; Carpenter, R. The metamorphosis of flowers. *Plant Cell* 5:1175–1181, 1993.

Coen, E.S.; Meyerowitz, E.M. The war of the whorls: genetic interactions controlling flower development. *Nature* 353:31–37, 1991.

Schwarz-Sommer, Z.; Huijser, P.; Nacken, W.; Saedler, H.; Sommer, H. Genetic control of flower development by homeotic genes in *Antirrhinum majus*. *Science* 250:931–936, 1990.

77. The Brain. *Cold Spring Harbor Symp. Quant. Biol.* 55, 1990.

Brown, M.C.; Hopkins, W.G.; Keynes, R.J. Essentials of Neural Development. Cambridge, UK: Cambridge University Press, 1990.

Hall, Z.W. An Introduction to Molecular Neurobiology. Sunderland, MA: Sinauer, 1992.

Jacobson, M. Developmental Neurobiology, 3rd ed. New York: Plenum Press, 1991.

Nicholls, J.G.; Martin, A.R.; Wallace, B.G. From Neuron to Brain, 3rd ed. Sunderland, MA: Sinauer, 1992.

Patterson, P.H.; Purves, D. Readings in Developmental Neurobiology. Cold Spring Harbor, NY: Cold Spring Harbor Laboratory, 1982. (An anthology.)

Purves, D.; Lichtman, J.W. Principles of Neural Development. Sunderland, MA: Sinauer, 1985.

78. Williams, R.W.; Herrup, K. The control of neuron number. *Annu. Rev. Neurosci.* 11:423–453, 1988.

79. Caviness, V.S. Neocortical histogenesis in normal and reeler mice: a developmental study based on [3H]thymidine autoradiography. *Dev. Brain Res.* 4:293–302, 1982.

Hatten, M.E. The role of migration in central nervous system neuronal development. *Curr. Opin. Neurobiol.* 3:38–44, 1993.

McConnel, S.K. The generation of neuronal diversity in the central nervous system. *Annu. Rev. Neurosci.* 14:269–300, 1991.

Rakic, P. Mode of cell migration to the superficial layers of the fetal monkey neocortex. *J. Comp. Neurol.* 145:61–84, 1972.

80. Goslin, K.; Banker, G. Experimental observations on the development of polarity by hippocampal neurons in culture. *J. Cell Biol.* 108:1507–1516, 1989.

Harrison, R.G. The outgrowth of the nerve fiber as a mode of protoplasmic movement. *J. Exp. Zool.* 9:787–846, 1910.

Mitchison, T.; Kirschner, M. Cytoskeletal dynamics and nerve growth. *Neuron* 1:761–772, 1988.

81. Chang, S.; Rathjen, F.G.; Raper, J.A. Extension of neurites on axons is impaired by antibodies against specific neural cell adhesion molecules. *J. Cell Biol.* 104:355–362, 1987.

Dodd, J.; Jessell, T.M. Axon guidance and the patterning of neuronal projections in vertebrates. *Science* 242:692–699, 1988.

Neugebauer, K.M.; Tomaselli, K.J.; Lilien, J.; Reichardt, L.F. N-cadherin, NCAM and integrins promote retinal neurite outgrowth on astrocytes *in vitro*. *J. Cell Biol.* 107:1177–1187, 1988.

Rutishauser, U. Adhesion molecules of the nervous system. *Curr. Opin. Cell Biol.* 3:709–715, 1993.

82. Campenot, R.B. Local control of neurite development by nerve growth factor. *Proc. Natl. Acad. Sci. USA* 74:4516–4519, 1977.

83. Luo, Y.; Raible, D.; Raper, J.A. Collapsin: a protein in brain that induces the collapse and paralysis of neuronal growth cones. *Cell* 75:217–227, 1993.

Schwab, M.E.; Kapfhammer, J.P.; Bandtlow, C.E. Inhibitors of neurite outgrowth. *Annu. Rev. Neurosci.* 16:565–595, 1993.

84. Patel, N.; Poo, M.-M. Orientation of neurite growth by extracellular electric fields. *J. Neurosci.* 2:483–496, 1982.

Tessier-Lavigne, M.; Placzek, M. Target attraction: are developing axons guided by chemotropism? *Trends Neurosci.* 15:303–310, 1991.

Wilson, S.W. Clues from clueless. (Axonal guidance mutants in *Drosophila*.) *Curr. Biol.* 3:536–539, 1993.

85. Chao, M. Neurotrophin receptors: a window into neuronal differentiation. *Neuron* 9:583–593, 1992.

Ebendal, T.; Olson, L.; Seiger, A.; Hedlund, K.-O. Nerve growth factors in the rat iris. *Nature* 286:25–28, 1980.

Greene, L.A. The importance of both early and delayed responses in the biological actions of nerve growth factor. *Trends Neurosci.* 7:91–94, 1984.

Levi-Montalcini, R.; Calissano, P. The nerve growth factor. *Sci. Am.* 240(6):68–77, 1979.

Purves, D. Body and Brain: A Trophic Theory of Neural Connections. Cambridge: Harvard University Press, 1988.

Thoenen, H. The changing scene of neurotrophic factors. *Trends Neurosci.* 14:165–170, 1991.

86. Sanes, J.R. Topographic maps and molecular gradients. *Curr. Opin. Neurobiol.* 3:67–74, 1993.

Sperry, R.W. Chemoaffinity in the orderly growth of nerve fiber patterns and connections. *Proc. Natl. Acad. Sci. USA* 50:703–710, 1963.

87. Cox, E.C.; Muller, B.; Bonhoeffer, F. Axonal guidance in the chick visual system: posterior tectal membranes induce collapse of growth cones from temporal retina. *Neuron* 4:31–37, 1990.

Stahl, B.; von Boxberg, Y.; Müller, B.; et al. Directional cues for retinal axons. *Cold Spring Harbor Symp.* 55:351–357, 1990.

Tessier-Lavigne, M. Down the slippery slope. *Curr. Biol.* 2:353–355, 1992.

88. Brown, M.C.; Jansen, J.K.S.; Van Essen, D. Polyneuronal innervation of skeletal muscle in new-born rats and its elimination during maturation. *J. Physiol.* 261:387–422, 1976.

Jennings, C.G.B.; Burden, S.J. Development of the neuromuscular synapse. *Curr. Opin. Neurobiol.* 3:75–81, 1976.

Lo, Y.-J.; Poo, M.-M. Activity-dependent synaptic competition *in vitro*: heterosynaptic suppression of developing synapses. *Science* 254:1019–1022, 1991.

Shatz, C.J. The developing brain. *Sci. Am.* 267(3):60–67, 1992.

89. Constantine-Paton, M.; Cline, H.T.; Debski, E. Patterned activity, synaptic convergence, and the NMDA receptor in developing visual pathways. *Annu. Rev. Neurosci.* 13:129–154, 1990.

Goodman, C.S.; Shatz, C.J. Developmental mechanisms that generate precise patterns of neuronal connectivity. *Cell* 72/*Neuron* 10 (Suppl.):77–98, 1993.

Hockfield, S.; Kalb, R.G. Activity-dependent structural changes during neuronal development. *Curr. Opin. Neurobiol.* 3:87–92, 1993.

90. Barlow, H. Visual experience and cortical development. *Nature* 258:199–204, 1975.

Cline, H.T.; Debski, E.A.; Constantine-Paton, M. N-methyl-D-aspartate receptor antagonist desegregates eye-specific stripes. *Proc. Natl. Acad. Sci. USA* 84:4342–4345, 1987.

Hubel, D.H.; Wiesel, T.N. Binocular interaction in striate cortex of kittens reared with artificial squint. *J. Neurophysiol.* 28:1041–1059, 1965.

Hubel, D.H.; Wiesel, T.N.; Le Vay, S. Plasticity of ocular dominance columns in monkey striate cortex. *Philos. Trans. R. Soc. Lond. (Biol.)* 278:377–409, 1977.

Stryker, M.P.; Harris, W.A. Binocular impulse blockade prevents the formation of ocular dominance columns in cat visual cortex. *J. Neurosci.* 6:2117–2133, 1986.

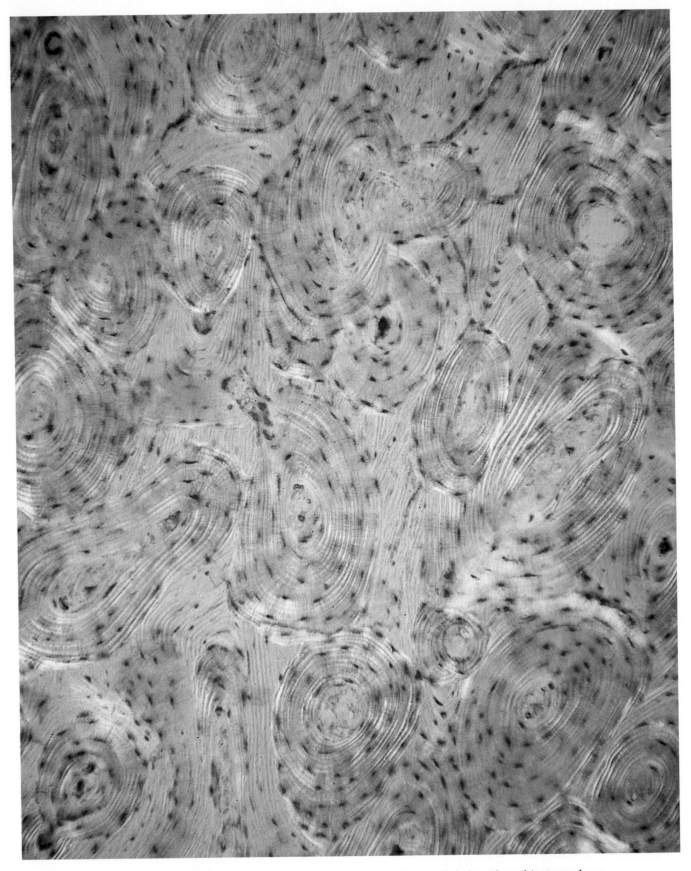

The consequences of tissue turnover, as seen in a section of compact bone. The small dark spidery objects are the spaces occupied by mature bone cells (osteocytes) embedded in the bone matrix. The alternating light and dark bands are layers of matrix containing oriented collagen (made visible with the help of polarized light). The pattern results from a process of continual erosion and reconstruction, in which old bone is tunneled away by osteoclasts (macrophagelike cells) and new bone is deposited in layers by osteoblasts (osteocyte precursors).

Differentiated Cells and the Maintenance of Tissues

22

In the space of a few days or weeks, a single fertilized egg gives rise to a complex multicellular organism consisting of differentiated cells arranged in a precise pattern. As a rule, the pattern of the body of an animal is set up in this way on a small scale and then grows. During embryonic development the different cell types become determined, each in its proper place. In the subsequent period of growth the cells proliferate, but with certain exceptions, their specialized characters remain more or less fixed. The organism may continue to become bigger throughout life, as do most crustaceans and fish, or it may stop growing when it reaches a certain size, as do birds and mammals. But even when growth stops, cell proliferation in many species continues. Thus in our own adult tissues new cells are continually produced. The adult body of a vertebrate can be likened to a stable ecosystem in which one generation of individuals (cells in this case) succeeds another but the organization of the system as a whole remains unchanged. It is still not known how the precise balance between cell proliferation and cell death is achieved.

This chapter discusses how cells are born, live, and die in multicellular tissues and how the organization of these tissues is maintained. We concentrate on higher vertebrates, and in considering the problems of tissue maintenance and renewal, we shall try to convey something of the remarkable variety of structure, function, and life history to be found among their specialized cell types.

Maintenance of the Differentiated State [1]

Although the tissues of the body differ in many ways, they all have certain basic requirements, usually provided for by a mixture of cell types, as illustrated for the skin in Figure 22–1. They all need mechanical strength, which is often provided by a supporting framework of *extracellular matrix,* mainly secreted by *fibroblasts.* In addition, almost all tissues need a blood supply to bring nutrients and remove waste products, and so they are pervaded by blood vessels lined with *endothelial cells.* Likewise, most tissues are innervated by *nerve cell* axons, which are ensheathed by *Schwann cells. Macrophages* are usually present to dispose of dying cells and to remove unwanted extracellular matrix, as are *lymphocytes* and other white blood cells to combat infection. *Melanocytes* may be present to provide a protective or decorative pigmentation. Most of these cell types, ancillary to the specialized function of the tissue, originate outside it and invade the tis-

(A)

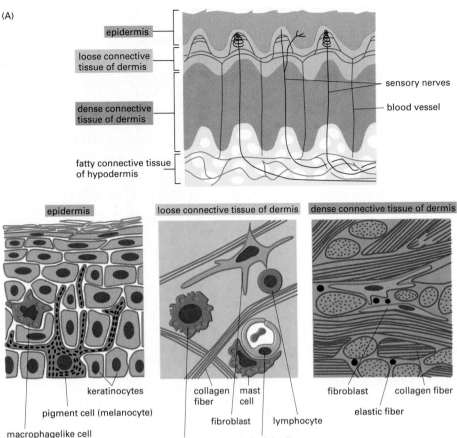

epidermis

loose connective
tissue of dermis

dense connective
tissue of dermis

fatty connective tissue
of hypodermis

sensory nerves

blood vessel

epidermis

keratinocytes

pigment cell (melanocyte)

macrophagelike cell
(Langerhans cell)

loose connective tissue of dermis

collagen
fiber

mast
cell

fibroblast

lymphocyte

macrophage

endothelial cell
forming capillary

dense connective tissue of dermis

fibroblast

collagen fiber

elastic fiber

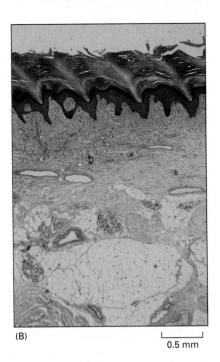

(B)

0.5 mm

Figure 22–1 Mammalian skin. (A) Schematic diagrams showing the cellular architecture of thick skin. (B) Photograph of a cross-section through the sole of a human foot, stained with hematoxylin and eosin. The skin can be viewed as a large organ composed of two main tissues: epithelial tissue (the *epidermis*), which lies outermost, and connective tissue, which consists of the tough dermis (from which leather is made) and the underlying fatty *hypodermis*. Each tissue is composed of a variety of cell types. The dermis and hypodermis are richly supplied with blood vessels and nerves. Some nerve fibers extend also into the epidermis.

sue either early in the course of its development (endothelial cells, nerve cell axons, Schwann cells, and melanocytes) or continually during life (macrophages and other white blood cells). This complex supporting apparatus is required to maintain the principal specialized cells of the tissue: the contractile cells of the muscle, the secretory cells of the gland, or the blood-forming cells of the bone marrow, for example.

Almost every tissue is therefore an intricate mixture of many cell types that must remain different from one another while coexisting in the same environment. Moreover, the organization of the mixture must be preserved even though, in almost all adult tissues, cells are continually dying and being replaced. The retention of tissue form and function is made possible largely through two fundamental properties of cells. Because of *cell memory* (see Chapter 21), specialized cells autonomously maintain their distinctive character and pass it on to their progeny. At the same time each type of specialized cell continually senses its environment and adjusts its proliferation and properties to suit the circumstances; in fact, the very survival of most cells depends on signals from other cells. The intracellular mechanisms thought to be responsible for cell memory are discussed in Chapter 9, while the ways in which cells respond to environmental signals are considered in Chapter 15. In this preliminary section on the behavior of cells in tissues, we briefly review some of the evidence for the stability and heritability of the differentiated state and consider to what extent this state can be modified by environmental influences.

Most Differentiated Cells Remember Their Essential Character Even in a Novel Environment [2]

Cell culture experiments demonstrate that even when cells are removed from their usual environment, they and their progeny generally remain true to their

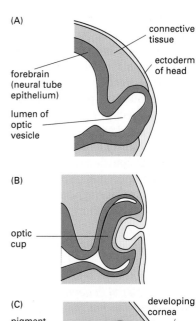

Figure 22–2 **The development of the vertebrate eye.** The retina develops from the *optic vesicle,* an epithelial outpocketing of the forebrain region of the neural tube. (A) The neural epithelium makes contact with the ectoderm covering the exterior of the head. (B) This contact induces the ectoderm to invaginate to form a lens. At the same time the outer part of the optic vesicle invaginates, reducing the vesicle lumen to an interface between two layers that together form a cuplike structure. (C) The layer of the optic cup closest to the lens differentiates into the *neural retina,* which contains the photoreceptor cells and the neurons that relay visual stimuli to the brain (see Figure 22–6). The other layer differentiates into the retinal *pigment epithelium.* Its cells are heavily loaded with melanin granules and thus form a dark enclosure for the photoreceptive system (serving to reduce the amount of scattered light, much as a coat of black paint does inside a camera).

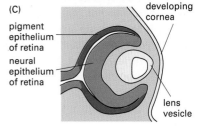

original instructions. Consider, for example, the epithelial cells that form the pigmented layer of the retina (Figure 22–2). Because they display their specialized character by manufacturing dark brown granules of melanin, it is easy to monitor their state of differentiation. When these cells are isolated from the retina of a chick embryo and grown in culture, they proliferate to form clones. Single cells taken from these clones breed true, giving subclones of similar pigment epithelial cells. The differentiated state can be maintained in this way through more than 50 cell generations.

The behavior of the cells is not, however, independent of their environment. In certain media or in conditions of extreme crowding, they may survive but synthesize little or no pigment. But even when they fail to express their differentiated character, they remain *determined* as pigment cells: when they are returned to more favorable culture conditions, they synthesize pigment once again. There are one or two known exceptions to this rule. In some vertebrate species, under certain conditions, retinal pigment cells will *transdifferentiate* into lens cells or into cells of the neural retina, but no manipulation of the conditions has been found to cause them to differentiate instead into blood cells, for example, or into liver cells or heart cells. Similarly, most types of specialized cells, including blood cells, liver cells, and heart cells, maintain their essential character in culture.

In the body, just as in culture, most specialized cells behave as though their basic character has been irreversibly determined by their developmental history. Epidermal cells, for example, remain epidermal cells even in the most alien surroundings: if a suspension of dissociated epidermal cells is prepared from the tail skin of a rat and injected beneath the capsule of the kidney, the cells grow there to form cysts lined with unmistakable epidermis, resembling that on the surface of the body.

The Differentiated State Can Be Modulated by a Cell's Environment [1, 3]

Although radical transformations are largely forbidden, the character of many differentiated cells can be strongly influenced by the environment. The possible adjustments can be classified mostly as *modulations* of the differentiated state—that is, reversible changes between closely related cell phenotypes. Liver cells, for example, adjust their synthesis of specific enzymes (through changes in specific mRNA levels) according to the ambient concentrations of the steroid hormone hydrocortisone, and the production of milk proteins by mammary gland cells can be switched on or off by changes in the extracellular matrix. Fibroblasts and their relatives—the family of *connective-tissue cells*—are a special case. These cells are exceptionally adaptable and can undergo various interconversions: fibroblasts, for example, can apparently change reversibly into cartilage cells. Such transformations are important in the healing of wounds and bone fractures and in other pathological processes; they are discussed later in this chapter. Even these conversions of one differentiated cell type into another, however, are narrowly restricted: the converted cell remains a member of the family of connective-tissue cells. Important but restricted changes of differentiated state occur also in many normal adult tissues where new differentiated cells are generated from *stem cells*—precursors that do not themselves display the mature differentiated character but are specialized to divide and to yield progeny that will. Distinct types of stem cells are committed to the production of distinct types of differentiated cells and are not interconvertible.

The majority of adult tissues, therefore, are composed of a number of distinct, irreversibly determined cell lineages. The numbers and spatial relationships of these components have to be maintained throughout life by mechanisms that do not require one type of differentiated cell to transform into another but depend on complex interactions between the different cell types.

Summary

Most differentiated cells in adult tissues will maintain their specialized character even when placed in a novel environment. Although states of differentiation are generally stable and not interconvertible, even highly specialized cells can alter their properties to a limited extent in response to environmental cues. In many adult tissues, moreover, new differentiated cells are continually generated from stem cells that appear undifferentiated. Especially striking cell transformations occur within the family of connective-tissue cells that includes fibroblasts and cartilage cells.

Tissues with Permanent Cells [4]

Not all the populations of differentiated cells in the body are subject to cell turnover. Some cell types, having been generated in appropriate numbers in the embryo, are retained throughout adult life; they seem never to divide, and they cannot be replaced if they are lost. Almost all nerve cells are permanent in this sense. So are a few other types of cells, including—in mammals—the muscle cells of the heart, the auditory hair cells of the ear (Figure 22–3), and the lens cells of the eye.

While all these cells have extremely long life-spans and necessarily live in protected environments, they are dissimilar in other respects, and it is difficult to give a general reason why they should be permanent and irreplaceable. For

Figure 22–3 **Auditory hair cells.** (A) Diagrammatic cross-section of the auditory apparatus (the organ of Corti) in the inner ear of a mammal, showing the auditory hair cells held in an elaborate structure of supporting cells and overlaid by a mass of extracellular matrix (called the tectorial membrane). (B) Scanning electron micrograph showing the apical surface of some of the outer auditory hair cells, with their characteristic organ-pipe arrays of giant microvilli (called *stereocilia*). The auditory hair cells function as transducers, generating an electrical signal in response to sound vibrations that rock the organ of Corti and so cause the stereocilia to tilt. In mammals the auditory hair cells produced in the embryo have to last a lifetime: if they are destroyed by disease or by excessively loud noise, they are not regenerated and permanent deafness results. (B, from R.G. Kessel and R.H. Kardon, Tissues and Organs: A Text-Atlas of Scanning Electron Microscopy. San Francisco: Freeman, 1979. Copyright © 1979 W.H. Freeman and Company.)

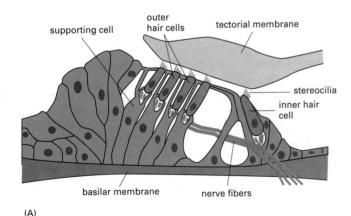

(A)

(B)

5 µm

heart muscle cells and auditory hair cells it is difficult to give any reason at all. In the case of nerve cells it seems likely that cell turnover in the adult would be disadvantageous as a rule, since it would be difficult to reestablish in the adult the precise and complex patterns of nerve connections that are set up under very different circumstances during development. Moreover, any memories recorded in the form of slight modifications of the structure or interconnections of individual nerve cells would presumably be obliterated. In the lens of the eye, on the other hand, the permanence of the cells appears to be simply an inevitable consequence of the way the tissue grows.

The Cells at the Center of the Lens of the Adult Eye Are Remnants of the Embryo [5]

Very little of the adult body consists of the same molecules that were laid down in the embryo. The **lens** of the eye is an exception: it is one of the few structures containing cells that are not only preserved but are preserved without turnover of their contents.

The lens is formed from the ectoderm at the site where the developing optic vesicle makes contact with it: the ectoderm here thickens, invaginates, and finally pinches off as a *lens vesicle* (see Figure 22–2). The lens thus originates as a spherical shell of cells formed from an epithelium, one cell layer thick, surrounding a central cavity. The cells at the rear of the lens vesicle (those facing the retina) soon undergo a striking transformation. They synthesize and become filled with *crystallins,* the characteristic proteins of the lens. In the process they elongate enormously, differentiating into *lens fibers* (Figure 22–4). Eventually, their nuclei disintegrate and protein synthesis ceases. In this way the part of the lens vesicle epithelium facing the retina is expanded into a thick refractile body consisting of many long, lifeless cells packed side by side (Figure 22–5). The central cavity of the vesicle is obliterated, and the front part of the epithelium of the lens vesicle—the part facing the external world—remains as a thin sheet of low cuboidal cells. Growth of the lens depends on the proliferation of these cells at

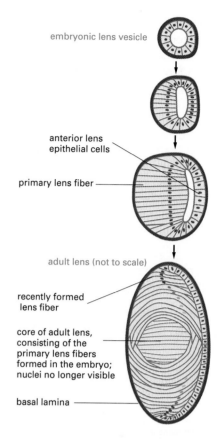

Figure 22–4 **The development of the lens of a human eye.** Proliferation occurs only in the anterior lens epithelial cells, which move posteriorly and differentiate into lens fibers.

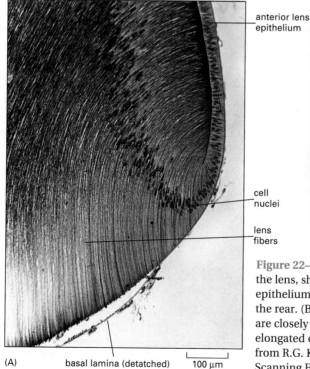

Figure 22–5 **The structure of the mature lens.** (A) Light micrograph of part of the lens, showing the junction between the thin sheet of anterior lens epithelium that covers the front of the lens and the differentiated lens fibers to the rear. (B) Scanning electron micrograph of part of the lens. The lens fibers are closely stacked, like planks in a lumberyard. Each one is a single, lifeless, elongated cell that can be up to 12 mm long. (A, courtesy of Peter Gould; B, from R.G. Kessel and R.H. Kardon, Tissues and Organs: A Text-Atlas of Scanning Electron Microscopy. San Francisco: Freeman, 1979. Copyright © 1979 W.H. Freeman and Company.)

the front, pushing some of the cells from this region around the rim of the lens and toward the back (see Figures 22–4 and 22–5A). As cells move to the rear, they stop dividing, step up their rate of synthesis of crystallins, and differentiate into lens fibers. Additional lens fibers continue to be recruited in this way throughout life, although at an ever decreasing rate.

The types of crystallins filling the earliest generations of lens fibers are different from those of the later generations, just as the hemoglobins of fetal red blood cells are different from those of adult red blood cells. But whereas old red blood cells are discarded, old lens fibers are not. Thus at the core of the adult lens lie fibers that were laid down in the embryo and are still packed with the distinctive types of crystallins manufactured in that earlier period. Differences of refractive index between the early embryonic types of crystallins and those that are laid down later help to free the lens of the eye from the optical aberrations that bedevil simple lenses made out of homogeneous media such as glass.

Most Permanent Cells Renew Their Parts: The Photoreceptor Cells of the Retina [6]

There are few cells as immutable as lens fibers. As a rule, even those cells that persist throughout life without dividing undergo renewal of their component

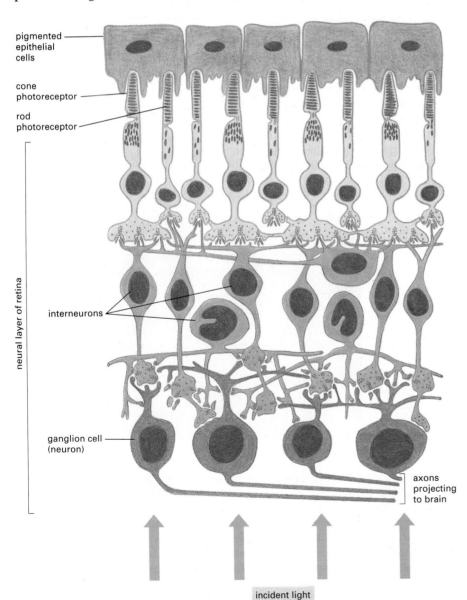

pigmented epithelial cells

cone photoreceptor

rod photoreceptor

neural layer of retina

interneurons

ganglion cell (neuron)

axons projecting to brain

incident light

Figure 22–6 The structure of the retina. The stimulation of the photoreceptors by light is relayed via the interneurons to the ganglion cells, which convey the signal to the brain. The spaces between neurons and between photoreceptors in the neural retina are occupied by a population of specialized supporting cells, which are not shown here. (Modified from J.E. Dowling and B.B. Boycott, *Proc. R. Soc. Lond. (Biol.)* 166:80–111, 1966.)

parts. Thus, while they do not divide, heart muscle cells, auditory hair cells, and nerve cells are metabolically active and capable not only of synthesizing new RNA and protein, but also of altering their size and structure during adult life. Heart muscle cells, for example, replace the bulk of their protein molecules in the course of a week or two, and they will adjust the balance of protein synthesis and degradation so as to grow bigger if the load on the heart is increased—for example, by a sustained increase in blood pressure. Nerve cells also replace their protein molecules continuously; moreover, many nerve cells can regenerate axons and dendrites that have been cut off.

The turnover of cell components is dramatically illustrated in the highly specialized neural cells that form the **photoreceptors** of the retina. The neural retina (see Figure 22–2) consists of several cell layers organized in a way that seems perverse. The neurons that transmit signals from the eye to the brain (called *retinal ganglion cells*) lie closest to the external world, so that the light, focused by the lens, must pass through them to reach the photoreceptor cells. The photoreceptors, which are classified as **rods** or **cones,** according to their shape, lie with their photoreceptive ends, or *outer segments,* partly buried in the pigment epithelium (Figure 22–6). Rods and cones contain different photosensitive complexes of protein with visual pigment: rods are especially sensitive at low light levels, while cones (of which there are three types, each with different

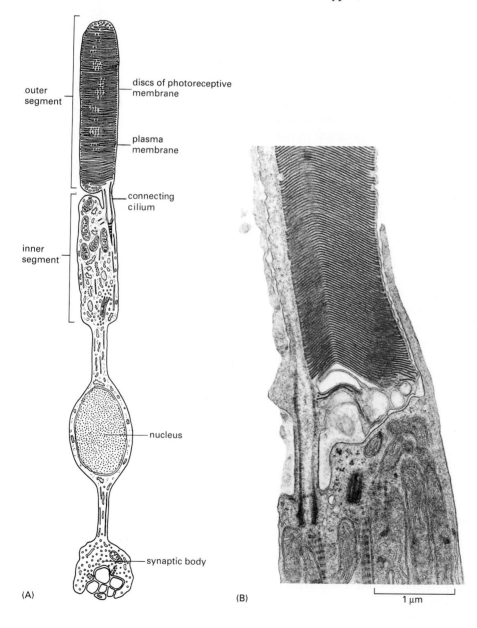

outer segment

discs of photoreceptive membrane

plasma membrane

connecting cilium

inner segment

nucleus

synaptic body

(A)

(B)

1 µm

Figure 22–7 A rod photoreceptor. (A) Schematic drawing. The actual number of photoreceptive discs in the outer segment is about 1000. (B) Electron micrograph of part of a rod photoreceptor, showing the base of the outer segment and the modified cilium that connects it to the inner segment. (A, from T.L. Lentz, Cell Fine Structure. Philadelphia: Saunders, 1971; B, from M.J. Hogan, J.A. Alvarado, and J.E. Weddell, Histology of the Human Eye: An Atlas and Textbook. Philadelphia: Saunders, 1971.)

Tissues with Permanent Cells

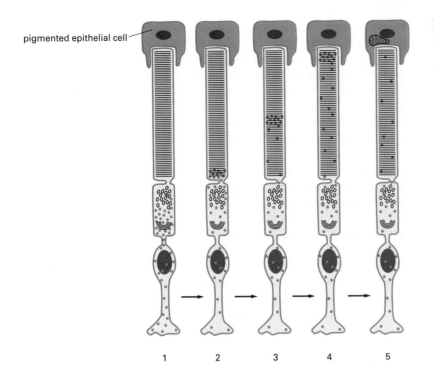

pigmented epithelial cell

Figure 22–8 Turnover of membrane protein in a rod cell. Following a pulse of ^{3}H-leucine, the passage of radiolabeled proteins through the cell is followed by autoradiography. *Red dots* indicate sites of radioactivity. The method reveals only the ^{3}H-leucine that has been incorporated into proteins; the rest is washed out during the preparation of the tissue. The incorporated leucine is first seen concentrated in the neighborhood of the Golgi apparatus (1). From there it passes to the base of the outer segment into a newly synthesized disc of photoreceptive membrane (2). New discs are formed at a rate of three or four per hour (in a mammal), displacing the older discs toward the pigment epithelium (3–5).

1   2   3   4   5

spectral responses) detect color and fine detail. The outer segment of a photo-receptor appears to be a modified cilium with a characteristic ciliumlike arrangement of microtubules in the region where the outer segment is connected to the rest of the cell (Figure 22–7). The remainder of the outer segment is almost entirely filled with a dense stack of membranes in which the photosensitive complexes are embedded; light absorbed here produces an electrical response, as discussed in Chapter 15. At their opposite ends the photoreceptors form synapses on interneurons, which relay the signal to the retinal ganglion cells (see Figure 22–6).

The photoreceptors are permanent cells that do not divide. But the photosensitive protein molecules are not permanent. There is a steady turnover, which can be demonstrated by showing that injected radioactive amino acids are incorporated into these molecules. In rods (although not, curiously, in cones) this turnover is organized in an orderly production line, which can be analyzed by following the passage of a cohort of radiolabeled protein molecules through the cell after a short pulse of radioactive amino acid (Figure 22–8). The radiolabeled proteins can be followed from the Golgi apparatus in the inner segment of the cell to the base of the stack of membranes in the outer segment. From here they are gradually displaced toward the tip as new material is fed into the base of the stack. Finally (after about 10 days in the rat), on reaching the tip of the outer segment, the labeled proteins and the layers of membrane in which they are embedded are phagocytosed (chewed off and digested) by the cells of the pigment epithelium.

Summary

Some cells in mammals—including nerve cells, heart muscle cells, sensory receptor cells for light and sound, and lens fibers—persist throughout life without dividing and without being replaced. In mature lens fibers the cell nuclei have degenerated and protein synthesis has stopped, so that the core of the adult lens consists of lens proteins laid down early in embryonic life. In most other permanent cells biosynthetic activity continues, and there is a steady turnover of cell components. In the rod cells of the retina, for example, new layers of photoreceptive membrane are synthesized close to the nucleus and are steadily displaced outward until they are eventually engulfed and digested by cells of the pigment epithelium.

Renewal by Simple Duplication [7]

Most of the differentiated cell populations in a vertebrate are not permanent: the cells are continually dying and being replaced. New differentiated cells can be produced during adult life in either of two ways: (1) they can form by the *simple duplication* of existing differentiated cells, which divide to give pairs of daughter cells of the same type; or (2) they can be generated from relatively undifferentiated *stem cells* by a process that involves a change of cell phenotype, as will be explained in detail later in this chapter.

Rates of renewal vary from one tissue to another. The turnover time may be as short as a week or less, as in the epithelial lining of the small intestine (which is renewed by means of stem cells), or as long as a year or more, as in the pancreas (which is renewed by simple duplication). Many tissues whose normal rates of renewal are very slow can be stimulated to produce new cells at higher rates when the need arises.

In this section we discuss two examples of cell populations that are renewed by simple duplication—liver cells and endothelial cells.

The Liver Functions as an Interface Between the Digestive Tract and the Blood [7, 8]

Digestion is a complex process. The cells that line the digestive tract secrete into the lumen of the gut a variety of substances, such as hydrochloric acid and digestive enzymes, to break down food molecules into simpler nutrients. The cells absorb these nutrients from the gut lumen, process them, and then release them into the blood for utilization by other cells of the body. All of these activities are adjusted according to the composition of the food consumed and the levels of metabolites in the circulation. The complex set of tasks is performed by a division of labor: some of the cells are specialized for the secretion of hydrochloric acid, others for the secretion of enzymes, others for absorption of nutrients, others for the production of peptide hormones, such as gastrin, that regulate digestive and metabolic activities, and so on (Figure 22–9). Some of these cell types lie closely intermingled in the wall of the gut; others are segregated in large glands that communicate with the gut and originate in the embryo as outgrowths of the gut epithelium.

The liver is the largest of these glands. It develops at a site where a major vein runs close to the wall of the primitive gut tube, and the adult organ retains a singularly close relationship with the blood. The cells in the liver that derive from the primitive gut epithelium—the **hepatocytes**—are arranged in folded sheets, facing blood-filled spaces called *sinusoids* (Figure 22–10A). The blood is separated from the surface of the hepatocytes by a single layer of flattened endothelial cells that covers the sides of each hepatocyte sheet (Figure 22–10B). This structure facilitates the chief functions of the liver, which center on the exchange of metabolites between hepatocytes and the blood.

The liver is the main site at which nutrients that have been absorbed from the gut and then transferred to the blood are processed for use by other cells of the body. It receives a major part of its blood supply directly from the intestinal tract (via the portal vein). Hepatocytes are responsible for the synthesis, degradation, and storage of a vast number of substances; they play a central part in the carbohydrate and lipid metabolism of the body as a whole; and they secrete most of the protein found in blood plasma. At the same time the hepatocytes remain connected with the lumen of the gut via a system of minute channels (or *canaliculi*) and larger ducts (see Figure 22–10B) and secrete into the gut by this route both waste products of their metabolism and an emulsifying agent, *bile*, which helps in the absorption of fats. In contrast to the rest of the digestive tract, there seems to be remarkably little division of labor within the population of hepatocytes: each hepatocyte appears to be able to perform the same broad range of metabolic and secretory tasks.

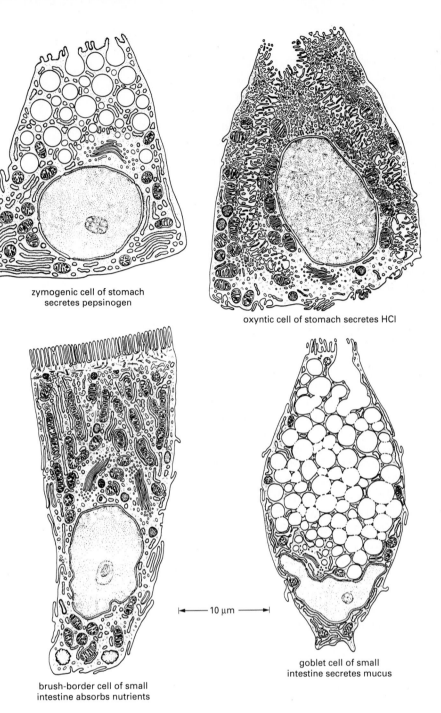

zymogenic cell of stomach
secretes pepsinogen

oxyntic cell of stomach secretes HCl

brush-border cell of small
intestine absorbs nutrients

goblet cell of small
intestine secretes mucus

|← 10 μm →|

Figure 22–9 **Some of the specialized cell types found in the epithelial lining of the gut.** Neighboring positions in the epithelial sheet are often occupied by cells of dissimilar types (see Figure 22–16B). (After T.L. Lentz, Cell Fine Structure. Philadelphia: Saunders, 1971.)

Hepatocytes have a life-style different from the cells lining the lumen of the gut. The latter, exposed to the abrasive and corrosive contents of the gut, cannot live for long and must be rapidly replaced by a continual supply of new cells (see Figure 22–17). Hepatocytes, removed from direct contact with the contents of the gut, live much longer and are normally renewed at a slow rate.

Liver Cell Loss Stimulates Liver Cell Proliferation [9]

Even in a slowly renewing tissue, a small but persistent imbalance between the rate of cell production and the rate of cell death will lead to disaster. If 2% of the hepatocytes in a human divided each week but only 1% died, the liver would grow to exceed the weight of the rest of the body within 8 years. Homeostatic mechanisms must operate to adjust the rate of cell proliferation and/or the rate of cell death in order to keep the organ at its standard size.

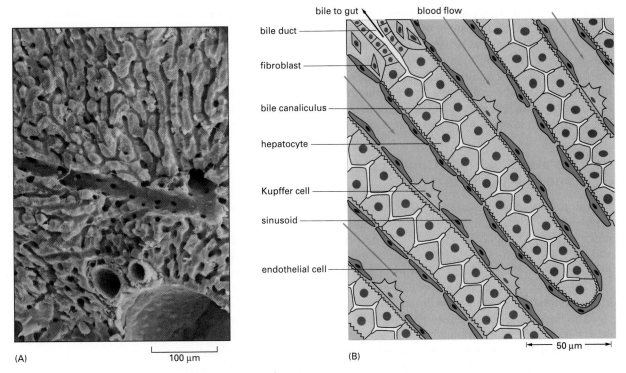

Figure 22–10 The structure of the liver. (A) Scanning electron micrograph of a portion of the liver, showing the irregular sheets of hepatocytes and the many small channels, or sinusoids, for the flow of blood. The larger channels are vessels that distribute and collect the blood that flows through the sinusoids. (B) The fine structure of the liver (highly schematized). The hepatocytes are separated from the bloodstream by a single thin sheet of endothelial cells with interspersed macrophagelike *Kupffer cells*. Small holes in the endothelial sheet allow exchange of molecules and small particles between the hepatocytes and the bloodstream while protecting the hepatocytes from buffeting by direct contact with the circulating blood cells. Besides exchanging materials with the blood, the hepatocytes form a system of minute bile canaliculi into which they secrete bile, which is ultimately discharged into the gut via bile ducts. The real structure is less regular than this diagram suggests. (A, from R.G. Kessel and R.H. Kardon, Tissues and Organs: A Text-Atlas of Scanning Electron Microscopy. San Francisco: Freeman, 1979. Copyright © 1979 W.H. Freeman and Company.)

Direct evidence for homeostatic control of liver cell proliferation comes from experiments in which large numbers of hepatocytes are removed surgically or are intentionally killed by poisoning with carbon tetrachloride. Within a day or so after either sort of damage, a surge of cell division occurs among the surviving hepatocytes, and the lost tissue is quickly replaced. If two-thirds of a rat's liver is removed, for example, a liver of nearly normal size can regenerate from the remainder within about 2 weeks. In cases of this kind a signal for liver regeneration can be demonstrated in the circulation: if the circulations of two rats are connected surgically and two-thirds of the liver of one of them is excised, cell division is stimulated in the unmutilated liver of the other. One of the signals responsible for the increased cell proliferation has been identified as a protein called *hepatocyte growth factor*. It stimulates hepatocytes to divide in culture, and its concentration in the bloodstream rises steeply (by poorly understood mechanisms) in response to liver damage. The same factor affects several other cell types in a variety of ways and is also known as *scatter factor* because it causes some kinds of epithelial cells to become motile so that they dissociate and migrate away from one another. It is not clear why it is specifically the liver that is stimulated to grow after liver damage.

The balance between cell births and cell deaths in the adult liver (and other organs too) does not depend exclusively on the regulation of cell proliferation: cell survival controls seem also to play a part. If an adult rat is treated with the drug phenobarbital, for example, hepatocytes are stimulated to divide, causing the liver to enlarge. When the phenobarbital treatment is stopped, hepatocyte cell

Renewal by Simple Duplication

death greatly increases until the liver returns to its original size, usually within a week or so. The mechanism of this type of cell survival control is unknown, but it has been suggested that hepatocytes, like most vertebrate cells, depend on signals from other cells for their survival and that the normal level of these signals can support only a certain standard number of hepatocytes. If the number of hepatocytes rises above this (as a result of phenobarbital treatment, for example), hepatocyte death will automatically increase to bring their number back down. It is not known how the appropriate levels of survival factors are maintained.

Regeneration Requires Coordinated Growth of Tissue Components [10]

Like all organs, the liver comprises a mixture of cell types. Besides the hepatocytes and the endothelial cells that line its sinusoids, it contains both specialized macrophages (*Kupffer cells*), which engulf particulate matter in the bloodstream and dispose of worn-out red blood cells, and a small number of fibroblasts, which provide a tenuous supporting framework of connective tissue (see Figure 22–10B). All of these cell types are capable of division. For optimal regeneration their proliferation must be properly coordinated.

The importance of balanced regeneration of cell types is demonstrated by what happens when an imbalance occurs. If hepatocytes, for example, are poisoned repeatedly with carbon tetrachloride or with alcohol at such frequent intervals that they cannot recover fully between attacks, the fibroblasts take advantage of the situation and the liver becomes irreversibly clogged with connective tissue, leaving little space for the hepatocytes to grow even after the toxic agents are withdrawn. This condition, called *cirrhosis,* is common in chronic alcoholics. In a similar way the regeneration of severely damaged skeletal muscle is often seriously hindered by the overgrowth of its connective tissue so that scar tissue replaces the contractile muscle fibers. These imbalances, however, require unusual tissue damage; in ordinary circumstances of tissue renewal, poorly understood mechanisms regulate cell proliferation and cell survival so as to ensure that the proper mixture of cell types is maintained.

Endothelial Cells Line All Blood Vessels [11]

By contrast with the above examples of ill-coordinated behavior of fibroblasts, the **endothelial cells** that form the lining of blood vessels have a remarkable capacity to adjust their number and arrangement to suit local requirements. Almost all tissues depend on a blood supply, and the blood supply depends on endothelial cells. They create an adaptable life-support system spreading into almost every region of the body. If it were not for endothelial cells extending and remodeling the network of blood vessels, tissue growth and repair would be impossible.

The largest blood vessels are arteries and veins, which have a thick, tough wall of connective tissue and smooth muscle (Figure 22–11A). The wall is lined by an exceedingly thin single layer of endothelial cells, separated from the surrounding outer layers by a basal lamina. The amounts of connective tissue and smooth muscle in the vessel wall vary according to the vessel's diameter and function, but the endothelial lining is always present (Figure 22–11B). In the finest branches of the vascular tree—the capillaries and sinusoids—the walls consist of nothing but endothelial cells and a basal lamina (Figure 22–12). Thus endothelial cells line the entire vascular system, from the heart to the smallest capillary, and control the passage of materials—and the transit of white blood cells—into and out of the bloodstream. A study of the embryo reveals, moreover, that arteries and veins develop from small vessels constructed solely of endothelial cells and a basal lamina: connective tissue and smooth muscle are added later where required, under the influence of signals from the endothelial cells.

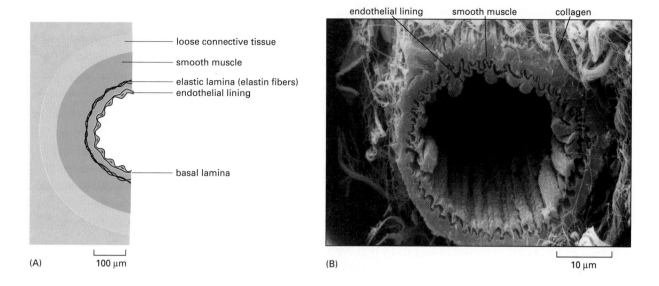

endothelial lining   smooth muscle   collagen

(A)   100 μm

(B)   10 μm

New Endothelial Cells Are Generated by Simple Duplication of Existing Endothelial Cells [12]

Throughout the vascular system of the adult, endothelial cells retain a capacity for cell division and movement. If, for example, a part of the wall of the aorta is damaged and denuded of endothelial cells, neighboring endothelial cells proliferate and migrate in to cover the exposed surface. Newly formed endothelial cells will even cover the inner surface of plastic tubing used by surgeons to replace parts of damaged blood vessels.

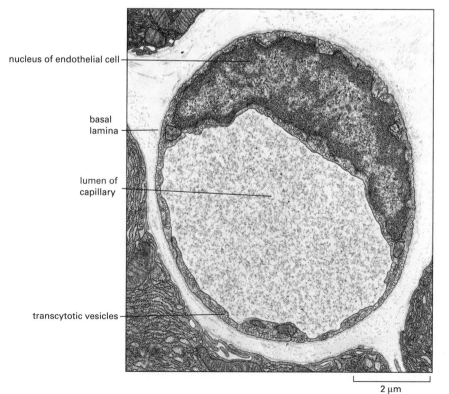

nucleus of endothelial cell

basal lamina

lumen of capillary

transcytotic vesicles

2 μm

Figure 22–11 A small artery in cross-section. (A) Schematic diagram of a part of the wall. The endothelial cells, although inconspicuous, are the fundamental component. Compare with the capillary in Figure 22–12. (B) Scanning electron micrograph of a cross-section through an arteriole (a very small artery), showing the inner lining of endothelial cells and the surrounding layer of smooth muscle and collagenous connective tissue. A slight contraction of the smooth muscle has thrown the endothelial lining of the vessel into folds. In fixation the endothelial lining has shrunk away from the muscular wall, leaving a small gap. (B, from R.G. Kessel and R.H. Kardon, Tissues and Organs: A Text-Atlas of Scanning Electron Microscopy. San Francisco: Freeman, 1979. Copyright © 1979 W.H. Freeman and Company.)

Figure 22–12 Electron micrograph of a small capillary in cross-section. The wall is formed by a single endothelial cell surrounded by a basal lamina. Note the small "transcytotic" vesicles, which according to one theory provide transport of large molecules in and out of this type of capillary: materials are taken up into the vesicles by endocytosis at the luminal surface of the cell and discharged by exocytosis at the external surface, or vice versa. (From R.P. Bolender, *J. Cell Biol.* 61:269–287, 1974, by copyright permission of the Rockefeller University Press.)

The proliferation of endothelial cells can be demonstrated by using ^{3}H-thymidine to label cells synthesizing DNA. In normal vessels the proportion of endothelial cells that become labeled is especially high at branch points in arteries, where turbulence and the resulting wear on the endothelial cells presumably stimulate cell turnover. On the whole, however, endothelial cells turn over very slowly, with a cell lifetime of months or even years.

Endothelial cells not only repair the lining of established blood vessels, they also create new blood vessels. They must do this in embryonic tissues to keep pace with growth, in normal adult tissues to support recurrent cycles of remodeling and reconstruction, and in damaged adult tissues to support repair.

New Capillaries Form by Sprouting [13, 14]

New vessels always originate as capillaries, which sprout from existing small vessels. This process of **angiogenesis** occurs in response to specific signals. The process can be readily observed in rabbits by making a small hole in the ear and fixing glass coverslips on either side to create a thin transparent viewing chamber into which the cells that surround the wound can grow. Angiogenesis can also be conveniently observed in naturally transparent structures such as the cornea of the eye. Irritants applied to the cornea induce the growth of new blood vessels from the rim of the tissue surrounding the cornea, which has a rich blood supply, in toward the center of the cornea, which normally has none. Thus the cornea becomes vascularized through an invasion of endothelial cells into the tough collagen-packed corneal tissue.

Observations such as these reveal that endothelial cells that will form a new capillary grow out from the side of a capillary or small venule by extending long processes or pseudopodia (Figure 22–13). The cells at first form a solid sprout, which then hollows out to form a tube. This process continues until the sprout encounters another capillary, with which it connects, allowing blood to circulate. Experiments in culture show that endothelial cells in a medium containing suitable growth factors will spontaneously form capillary tubes even if they are isolated from all other types of cells. The first sign of tube formation in culture is the appearance in a cell of an elongated vacuole that is at first completely encompassed by cytoplasm (Figure 22–14A). Contiguous cells develop similar vacuoles, and eventually the cells arrange their vacuoles end to end so that the vacuoles become continuous from cell to cell, forming a capillary channel (Figure 22–14B). The process is strongly dependent on the nature of the extracellular matrix in the environment of the cells: formation of capillary tubes is promoted by basal lamina components, such as laminin, which the endothelial cells themselves can secrete. The capillary tubes that develop in a pure culture of endothelial cells do not contain blood, and nothing travels through them, indicating that blood flow and pressure are not required for the formation of a capillary network.

Figure 22–13 Angiogenesis. A new blood capillary forms by the sprouting of an endothelial cell from the wall of an existing small vessel. This schematic diagram is based on observations of cells in the transparent tail of a living tadpole. (After C.C. Speidel, *Am. J. Anat.* 52:1–79, 1933.)

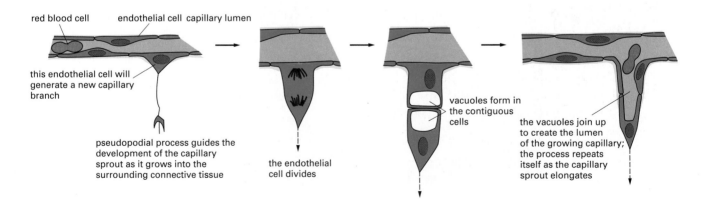

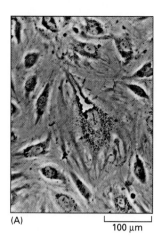

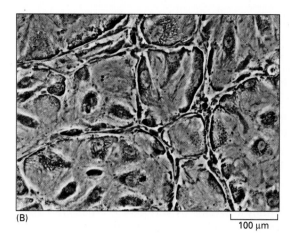

(A) 100 µm   (B) 100 µm

Figure 22–14 **Capillary formation *in vitro*.** Endothelial cells in culture spontaneously develop internal vacuoles that join up, giving rise to a network of capillary tubes. Photographs (A) and (B) show successive stages in the process; the arrow in (A) indicates a vacuole forming initially in a single endothelial cell. The cultures are set up from small patches of two to four endothelial cells taken from short segments of capillary. These cells will settle on the surface of a collagen-coated culture dish and form a small flattened colony that enlarges gradually as the cells proliferate. The colony spreads across the dish, and eventually, after about 20 days, capillary tubes begin to form in the central regions. Once tube formation has started, branches soon appear, and after 5 to 10 more days an extensive network of tubes is visible, as seen in (B). (From J. Folkman and C. Haudenschild, *Nature* 288:551–556, 1980. © Macmillan Journals Ltd.)

Angiogenesis Is Controlled by Growth Factors Released by the Surrounding Tissues [14]

In living animals endothelial cells form new capillaries wherever there is a need for them. It is thought that when cells in tissues are deprived of oxygen, they release angiogenic factors that induce new capillary growth. Probably for this reason, nearly all cells in a vertebrate are located within 50 µm of a capillary. Similarly, after wounding a burst of capillary growth is stimulated in the neighborhood of the damaged tissue (Figure 22–15). Local irritants or infections also cause a proliferation of new capillaries, most of which regress and disappear when the inflammation subsides.

Angiogenesis is also important in tumor growth. The growth of a solid tumor is limited by its blood supply: if it were not invaded by capillaries, a tumor would be dependent on the diffusion of nutrients from its surroundings and could not enlarge beyond a diameter of a few millimeters. To grow further, a tumor must induce the formation of a capillary network that invades the tumor mass. A small sample of such a tumor implanted in the cornea will cause blood vessels to grow quickly toward the implant from the vascular margin of the cornea, and the growth rate of the tumor increases abruptly as soon as the vessels reach it.

In all of these cases the invading endothelial cells must respond to a signal produced by the tissue that requires a blood supply. The response of the endothelial cells includes at least four components. First, the cells must breach the basal lamina that surrounds an existing blood vessel; endothelial cells during angiogenesis have been shown to produce *proteases*, which enable them to digest their way through the basal lamina of the parent capillary or venule. Second,

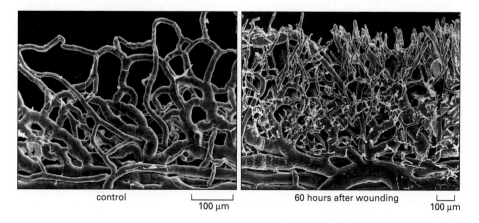

control 100 µm   60 hours after wounding 100 µm

Figure 22–15 **New capillary formation in response to wounding.** Scanning electron micrographs of casts of the system of blood vessels surrounding the margin of the cornea show the reaction to wounding. The casts are made by injecting a resin into the vessels and letting the resin set; this reveals the shape of the lumen, as opposed to the shape of the cells. Sixty hours after wounding many new capillaries have begun to sprout toward the site of injury, which is just above the top of the picture. Their oriented outgrowth reflects a chemotactic response of the endothelial cells to an angiogenic factor released at the wound. (Courtesy of Peter C. Burger.)

the endothelial cells must move toward the source of the signal. Third, they must proliferate. Fourth, they must form tubes. In certain circumstances some of the components of this complex response can be elicited in the absence of the others. But there are also identified growth factors that can evoke all four components of the angiogenic response together. Foremost among these factors is a protein known as *vascular endothelial growth factor* (VEGF—a distant relative of platelet-derived growth factor [PDGF]). This acts selectively on endothelial cells to stimulate angiogenesis in many different circumstances, and it seems to be the agent by which some tumors acquire their rich blood supply. Other growth factors, including some members of the *fibroblast growth factor* family, also stimulate angiogenesis but at the same time influence other cell types besides endothelial cells. Angiogenic factors such as these are released during tissue repair, inflammation, and tissue growth; they are made by various cell types, including macrophages, mast cells, and fat cells. A number of natural inhibitors have also been identified that can block the formation of new blood vessels. Thus angiogenesis, like the control of cell proliferation in general, seems to be regulated by complex combinations of signals rather than by one signal alone.

Summary

Most populations of differentiated cells in vertebrates are subject to turnover through cell death and cell division. In some cases, such as that of hepatocytes in the liver, the fully differentiated cells simply divide to produce daughter cells of the same type. Both the proliferation and the survival of hepatocytes are controlled to maintain appropriate total cell numbers. If a large part of the liver is destroyed, the remaining hepatocytes increase their division rate to restore the loss; and if hepatocyte proliferation is transiently increased by drug treatment, the increase in cell numbers is soon compensated for by an increase in cell death, returning cell numbers to normal. Such control mechanisms normally keep the numbers of cells of each type in a tissue in appropriate balance. In response to unusual damage, however, repair may be unbalanced, as when the fibroblasts in a repeatedly damaged liver grow too rapidly in relation to the hepatocytes and replace them with connective tissue.

Endothelial cells form a single cell layer that lines all blood vessels and regulates exchanges between the bloodstream and the surrounding tissues. New blood vessels develop from the walls of existing small vessels by the outgrowth of endothelial cells, which have the capacity to form hollow capillary tubes even when isolated in culture. In the living animal anoxic, damaged, or growing tissues stimulate angiogenesis by releasing angiogenic growth factors. These factors attract nearby endothelial cells and stimulate them to secrete proteases, to proliferate, and to form new capillaries.

Renewal by Stem Cells: Epidermis [7, 15]

We turn now from cell populations that are renewed by simple duplication to those that are renewed by means of **stem cells.** These populations vary widely—not only in cell character and rate of turnover, but also in the geometry of cell replacement. In the lining of the small intestine, for example, cells are arranged as a single-layered epithelium. This epithelium covers the surfaces of the *villi* that project into the lumen of the gut, and it lines the *crypts* that descend into the underlying connective tissue (Figure 22–16). The stem cells lie in a protected position in the depths of the crypts. The differentiated cells generated from them are carried upward by a sliding movement in the plane of the epithelial sheet until they reach the exposed surfaces of the villi; at the tips of the villi the cells die and are shed into the lumen of the gut. A contrasting example is found in the epithelium that forms the outer covering of the skin, called the *epidermis.* The epidermis is a many-layered epithelium, and the differentiating cells travel outward from their site of origin in a direction perpendicular to the plane of the cell sheet. In

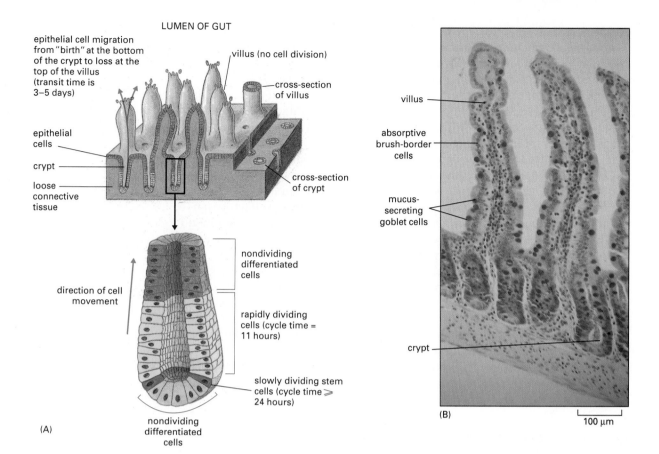

LUMEN OF GUT

epithelial cell migration from "birth" at the bottom of the crypt to loss at the top of the villus (transit time is 3–5 days)

villus (no cell division)

cross-section of villus

epithelial cells

crypt

loose connective tissue

cross-section of crypt

nondividing differentiated cells

direction of cell movement

rapidly dividing cells (cycle time = 11 hours)

slowly dividing stem cells (cycle time ≥ 24 hours)

(A)

nondividing differentiated cells

villus

absorptive brush-border cells

mucus-secreting goblet cells

crypt

(B)

100 μm

Figure 22–16 **Renewal of the gut lining.** (A) The pattern of cell turnover and the proliferation of stem cells in the epithelium that forms the lining of the small intestine. The nondividing differentiated cells at the base of the crypts also have a finite lifetime, terminated by programmed cell death, and are continually replaced by progeny of the stem cells. (B) Photograph of a section of part of the lining of the small intestine, showing the villi and crypts. Note how mucus-secreting goblet cells (stained *red*) are interspersed among the absorptive brush-border cells in the epithelium of the villi. See Figure 22–9 for the structure of these cells.

the case of blood cells the spatial pattern of production is complex and appears chaotic. Before going into such details, however, we must pause to consider what a stem cell is.

Stem Cells Can Divide Without Limit and Give Rise to Differentiated Progeny [16]

The defining properties of a stem cell are as follows:

1. It is not itself terminally differentiated (that is, it is not at the end of a pathway of differentiation).
2. It can divide without limit (or at least for the lifetime of the animal).
3. When it divides, each daughter has a choice: it can either remain a stem cell, or it can embark on a course leading irreversibly to terminal differentiation (Figure 22–17).

Stem cells are required wherever there is a recurring need to replace differentiated cells that cannot themselves divide. In several tissues the terminal state of cell differentiation is obviously incompatible with cell division. The cell nucleus may be digested, for example, as in the outermost layers of the skin, or it may be extruded, as in the mammalian red blood cell. Alternatively, the cytoplasm may be heavily encumbered with structures, such as the myofibrils of striated muscle cells, that would hinder mitosis and cytokinesis. In other terminally differentiated cells the chemistry of differentiation may be incompatible with cell division in some more subtle way. In any such case, renewal must depend on stem cells.

The job of the stem cell is not to carry out the differentiated function but rather to produce cells that will. Consequently, stem cells often have a nondescript appearance, making them hard to identify. But that is not to say that stem cells are all alike. Although not terminally differentiated, they are nevertheless *determined* (see p. 1060): the muscle satellite cell, as a source of skeletal muscle; the epidermal stem cell, as a source of keratinized epidermal cells; the spermatogonium, as a source of spermatozoa; the basal cell of olfactory epithelium, as a source of olfactory neurons (Figure 22–18); and so on. Those stem cells that give rise to only one type of differentiated cell are called *unipotent*, and those that give rise to several cell types are called *pluripotent*.

Tissues that form from stem cells raise many important questions. We need to consider what factors determine whether a stem cell divides or stays quiescent, what decides whether a given daughter cell remains a stem cell or differentiates, and in what ways the differentiation of a daughter cell is regulated after it has become committed to differentiate. On the opposite side of the balance sheet, we have to consider how cells die and are disposed of and how their survival is controlled. We begin our discussion with the epidermis, for its simple spatial organization makes it relatively easy to study the natural history of its stem cells and the fate of their progeny.

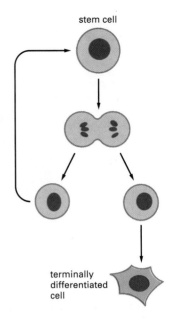

Figure 22–17 **The definition of a stem cell.** Each daughter produced when a stem cell divides can either remain a stem cell or go on to become terminally differentiated.

Epidermal Stem Cells Lie in the Basal Layer [17, 18]

The epidermal layer of the skin and the epithelial lining of the digestive tract are the two tissues that suffer the most direct and damaging encounters with the external world. In both, mature differentiated cells are rapidly lost from the most exposed positions and are replaced by the proliferation of less differentiated cells in more sheltered niches.

The **epidermis** is a multilayered epithelium composed largely of *keratinocytes* (so called because their characteristic differentiated activity is the synthesis of intermediate filament proteins called *keratins*) (Figure 22–19). These cells change their appearance from one layer to the next. Those in the innermost layer, attached to an underlying basal lamina, are termed *basal cells,* and it is normally only these that undergo mitosis. Above the basal cells are several layers of larger *prickle cells* (Figure 22–20), whose numerous desmosomes—each a site of anchorage for thick tufts of keratin filaments—are just visible in the light microscope as tiny prickles around the cell surface (hence the name). Beyond the prickle cells lies the thin granular cell layer (see Figure 22–19). This marks the boundary between the inner, metabolically active strata and the outermost layer, consisting

Figure 22–18 **A schematic cross-section of olfactory epithelium.** In this epithelium, which is specialized for sensing smells, three cell types can be distinguished—supporting cells, basal cells, and olfactory neurons. Autoradiographic experiments show that the basal cells are the stem cells for production of the olfactory neurons, which constitute one of the very few exceptions to the rule that neurons are permanent cells. Each olfactory neuron survives for about a month (in a mammal) before it is replaced. Six to eight modified cilia project from the globular head of the olfactory neuron and are believed to contain the smell receptors. The axon extending from the other end of the neuron conveys the message to the brain. A new axon must grow out and make appropriate connections in the brain whenever a basal cell differentiates into an olfactory neuron.

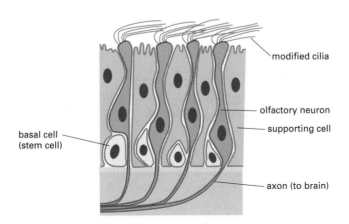

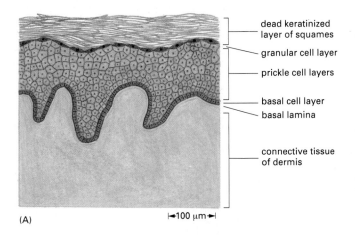

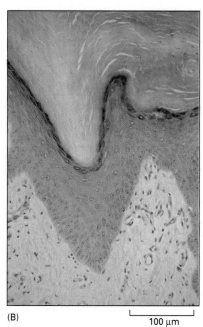

(A)

(B)

|←100 μm→|

100 μm

dead keratinized
layer of squames
granular cell layer
prickle cell layers
basal cell layer
basal lamina
connective tissue
of dermis

of dead cells whose intracellular organelles have disappeared. These outermost cells are reduced to flattened scales, or *squames,* filled with densely packed keratin. The plasma membranes of both the squames and the outer granular cells are reinforced on their cytoplasmic surface by a thin (12-nm), tough, cross-linked layer containing an intracellular protein called *involucrin.* The squames themselves are normally so compressed and thin that their boundaries are hard to make out in the light microscope, but soaking in sodium hydroxide makes them swell slightly, and with suitable staining a remarkably ordered geometric arrangement can often be seen in regions where the skin is thin. The squames are found to be stacked in hexagonal columns that interlock neatly at their edges (Figure 22–21), a typical column being 10–20 cells high and resting on about 10 basal cells.

Figure 22–19 Cross-section of mammalian epidermis. (A) Schematic diagram. (B) Photomicrograph of a section through the sole of the foot (hematoxylin and Van Gieson stain). The *granular cells* between the prickle cells and the flattened squames are in the penultimate stages of keratinization; they appear granular because they contain darkly staining aggregates of a material called *keratohyalin,* which is thought to be involved in the intracellular compaction and cross-linking of the keratin. Keratohyalin consists mainly of a protein known as *filaggrin.* In addition to the cells destined for keratinization, the deep layers of the epidermis include small numbers of cells of different character (not shown here)—including macrophagelike *Langerhans* cells, derived from bone marrow; *melanocytes,* derived from the neural crest; and *Merkel cells,* which are associated with nerve endings in the epidermis. See also Figure 22–1.

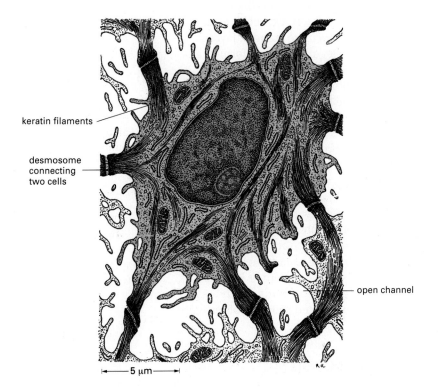

keratin filaments

desmosome
connecting
two cells

open channel

|←5 μm→|

Figure 22–20 A prickle cell. Drawing from an electron micrograph of a section of the epidermis, showing the bundles of keratin filaments that traverse the cytoplasm and are inserted at the desmosome junctions that bind the prickle cell (*red*) to its neighbors. Note that between adjacent cells there are open channels that allow nutrients to diffuse freely through the metabolically active layers of the epidermis. Further out, at the level of the granular cells, there is a waterproof barrier that is thought to be created by a sealant material that the granular cells secrete from vesicles called *membrane-coating granules.* (From R.V. Krstić, Ultrastructure of the Mammalian Cell: An Atlas. Berlin: Springer, 1979.)

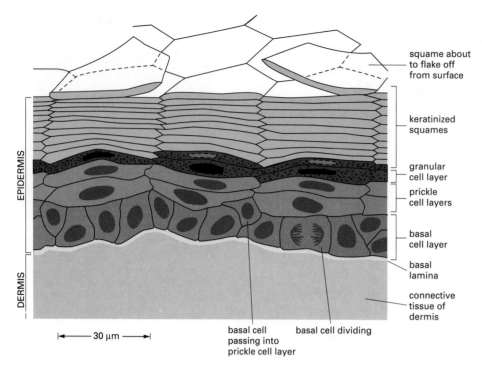

squame about to flake off from surface

keratinized squames

granular cell layer

prickle cell layers

basal cell layer

basal lamina

connective tissue of dermis

EPIDERMIS

DERMIS

⊢— 30 μm —⊣

basal cell passing into prickle cell layer

basal cell dividing

Figure 22–21 The columnar organization of squames in the epidermal layer of thin skin. The structure is revealed by swelling the keratinized squames in a solution containing sodium hydroxide. This type of organization occurs only where the epidermis is thin. Some studies suggest that each such column is a "proliferative unit," corresponding to a single stem cell among the 10–12 basal cells on which the column rests.

Differentiating Epidermal Cells Synthesize a Sequence of Different Keratins as They Mature [18]

Having described the static picture, let us now set it in motion. While some basal cells are dividing, adding to the population in the basal layer, others (their sisters or cousins) are slipping out of the basal cell layer into the prickle cell layer, taking the first step on their outward journey. When they reach the granular layer, the cells start to lose their nuclei and cytoplasmic organelles and are transformed into the keratinized squames of the keratinized layer. These finally flake off from the surface of the skin (and become a main constituent of household dust). The period from the time a cell is born in the basal layer of the human skin to the time it is shed from the surface varies from 2 to 4 weeks, depending on the region of the body.

The accompanying molecular transformations can be studied by analyzing either thin slices of epidermis cut parallel to the surface or successive layers of cells stripped off by repeatedly applying and removing strips of adhesive tape. The keratin molecules, for example, which are plentiful in all layers of the epidermis, can be extracted and characterized. They are of many types (discussed in Chapter 16), encoded by a large family of homologous genes, with the variety further increased through alternative splicing of the gene transcripts. As the new keratinocyte at the base of the column is transformed into the squame at the top (see Figure 22–21), it expresses a succession of different selections from its keratin gene repertoire. During this process other characteristic proteins, such as involucrin, also begin to be synthesized as part of a coordinated program of terminal cell differentiation.

Epidermal Stem Cells Are a Subset of Basal Cells [19]

If each patch of epidermis is maintained indefinitely by proliferation of its basal cells, there must be among these basal cells at least one whose line of descendants will not die out in the lifetime of the animal. We shall call such a cell an *immortal stem cell* (Figure 22–22). In principle, the division of an immortal stem cell could generate two initially similar daughters whose different fates would be governed by subsequent circumstances. At the opposite extreme, the stem cell division could be always asymmetric: one and only one of the daughters would inherit a special character required for immortality, while the other would be somewhat altered already at the time of its birth in a way that forced it to differentiate and ultimately to die. In the latter case there could never be any increase in the existing number of immortal stem cells, and this is contradicted by the facts. If a patch of epidermis is destroyed, the damage is repaired by surrounding healthy epidermal cells that migrate and proliferate to cover the denuded area. In this process a new self-renewing patch of epidermis is established, implying that additional immortal stem cells have been generated to make up for the loss.

Thus the fate of the daughters of a stem cell must be governed at least partly by the circumstances. One possible determining factor might be contact with the basal lamina or with the exposed connective tissue at a wound, with a loss of contact triggering the start of terminal differentiation, and maintenance of contact tending to preserve stem cell potential. Studies *in vitro* indicate that this is not the only determinant of basal cell fate, however.

Basal keratinocytes can be dissociated from intact epidermis and will proliferate in a culture dish, giving rise to new basal cells and to terminally differentiated cells. Even within a population of cultured basal keratinocytes that all appear undifferentiated, there is great variation in the ability to proliferate. When cells are taken singly and tested for their ability to found new colonies, some appear unable to divide at all, others go through only a few division cycles and then halt, and still others can divide enough times to form large colonies. The basal cells differ also in their expression of extracellular matrix receptors of the integrin family (discussed in Chapter 19): the cells that have more of these receptors, and so are better able to bind to basal laminal components, are the ones with the greater proliferative potential. This suggests that not all basal cells are alike *in vivo* and that mere contact with the basal lamina is not enough to keep them as stem cells. Rather, it appears that stem cells are a small subset—about 10%—of the basal cell population and are programmed to generate a certain proportion of progeny that become committed to terminal differentiation even before they have left the basal layer. In fact, if the keratinocytes are cultured in a Ca^{2+}-deficient medium, which keeps them as a monolayer and therefore all in a basal position, some of them will actually embark on terminal differentiation despite their location, as indicated by the synthesis of involucrin; these differentiating cells emerge from the basal layer as soon as the Ca^{2+} concentration is raised.

Nevertheless, contact with extracellular matrix has a critical influence on the choice of cell fate, which is evidently not programmed rigidly. If the cells are held in suspension, instead of being allowed to settle and attach to the bottom of the culture dish, they all stop dividing and differentiate. Some of the cells will refrain from differentiating even in suspension, however, if the medium includes fibronectin (a minor component of basal lamina and a major component of the extracellular matrix that keratinocytes migrate onto during wound healing). The cells that show this response to fibronectin are those that possess appropriate integrins. In normal conditions possession of such receptors presumably holds the cells bound to the basal lamina, keeping open their option to remain as stem cells; loss or inactivation of the receptors leads to ejection from the basal layer, confirming the decision to differentiate; and ejection from the basal layer through other causes leads to loss of the receptors, forcing the cell to differentiate prematurely.

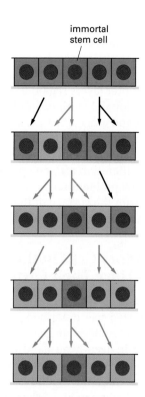

Figure 22–22 **An immortal stem cell.** Each self-renewing patch of epidermis must contain in each cell generation at least one "immortal" stem cell, whose descendants will still be present in the patch in the distant future. The arrows indicate lines of descent. An immortal stem cell is shown here occupying the same position in each cell generation. Other basal cells might be born chemically different in a way that commits them to leave the basal layer and differentiate; or they too might be stem cells, equivalent to the immortal stem cell in character and mortal only in the sense that their progeny happen subsequently to be jostled out of the basal layer and shed from the skin.

Basal Cell Proliferation Is Regulated According to the Thickness of the Epidermis [20]

Whatever the influence of the basal lamina may be, additional controls must operate to regulate the rate of production and the rate of sloughing of epidermal cells. If the outer layers of the epidermis are stripped away, for example, the division rate of the basal cells increases. After a transient overshoot, normal thickness is restored, and the division rate in the basal layer declines to normal. It is as though the removal of the outer differentiated layers releases the cells in the proliferative basal layer from an inhibitory influence, which is restored as soon as the outer layers regain their full thickness.

Although keratinocytes in culture are known to respond to a variety of hormones and growth factors, including epidermal growth factor (EGF), the molecular mechanisms that regulate their proliferation in the body remain an unsolved problem of great clinical importance. The consequences of faulty control of basal cell proliferation are seen in *psoriasis*. In this common skin disorder the rate of basal cell proliferation is greatly increased—the epidermis thickens, and cells are shed from the surface of the skin within as little as a week after emerging from the basal layer, before they have had time to keratinize fully.

Secretory Cells in the Epidermis Are Secluded in Glands That Have Their Own Population Kinetics [21]

In certain specialized regions of the body surface other types of cells besides the keratinized cells described above develop from the embryonic epidermis. In particular, secretions such as sweat, tears, saliva, and milk are produced by cells segregated in deep-lying glands that originate as ingrowths of the epidermis but have patterns of renewal quite different from those of keratinizing regions.

The mammary gland is of special interest because of the hormonal control of its cell division and differentiation. Milk production must be switched on when a baby is born and switched off when the baby is weaned. A "resting" mammary gland consists of branching systems of ducts embedded in connective tissue; these ducts are lined, in their secretory portions, by a single layer of relatively inactive epithelial cells that serve as stem cells. As a first step toward large-scale milk production, the hormones that circulate during pregnancy cause the duct cells to proliferate and the terminal portions of the ducts to grow and branch, forming little dilated outpocketings, or *alveoli*, containing secretory cells (Figure 22–23). Milk secretion begins only when these cells are stimulated by the different combination of hormones circulating in the mother after the birth of the baby. Later, when suckling stops, the secretory cells die and most of the alveoli disappear; macrophages rapidly clear away the dead cells, and the gland reverts to its resting state. Degradation of the basal lamina seems to play a critical part in this process of *involution*.

Cell division in the mammary gland is regulated not only by hormones but also by local signals passing between cells within the epithelium and between the epithelial cells and the connective tissue, or *stroma*, in which the epithelial cells are embedded. Mutations in genes involved in these local controls promote the development of cancer, as we discuss in Chapter 24, and it is through studies of breast cancer that several of these control mechanisms have come to light.

Summary

Many tissues, especially those with a rapid turnover—such as the lining of the gut, the epidermal layer of the skin, and the blood-forming tissues—are renewed by means of stem cells. Stem cells, by definition, are not terminally differentiated and have the ability to divide throughout the lifetime of the organism, yielding some progeny that differentiate and others that remain stem cells. In the skin the stem cells of the epidermis lie in the basal layer, attached to the basal lamina. The progeny of the stem cells differentiate on leaving this layer and, as they move outward, synthesize a succession of different types of keratin until, eventually, their nuclei degenerate, producing an outer layer of dead keratinized cells that are continually shed from the surface. Only a minority of basal cells are stem cells. The fate of the daughters of a stem cell is controlled in part by interactions with the basal lamina and in part by other

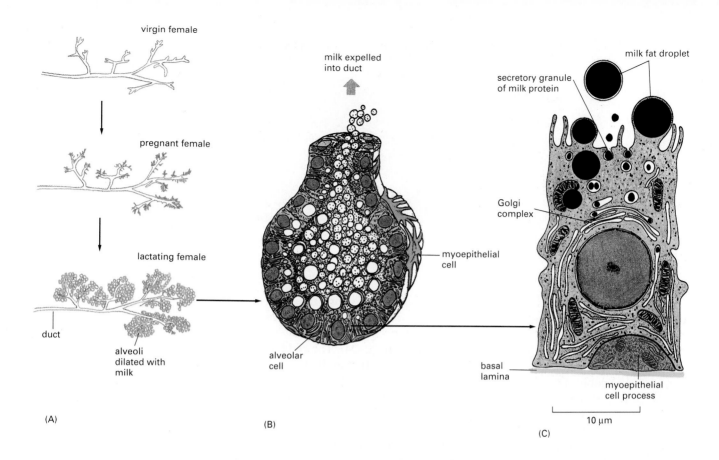

Figure 22–23 The mammary gland. (A) Schematic diagram of the growth of alveoli from the ducts of the mammary gland during pregnancy and lactation. Only a small part of the gland is shown. The "resting" gland contains a small amount of inactive glandular tissue embedded in a large amount of fatty connective tissue. During pregnancy an enormous proliferation of the glandular tissue takes place at the expense of the fatty connective tissue, with the secretory portions of the gland developing preferentially to create alveoli. (B) One of the milk-secreting alveoli with a basket of *myoepithelial cells* (*green*) embracing it. The myoepithelial cells contract and expel milk from the alveolus in response to the hormone oxytocin, which is released as a reflex response to the stimulus of suckling. (C) A single type of secretory alveolar cell produces both the milk proteins and the milk fat. The proteins are secreted in the normal way by exocytosis, while the fat is released as droplets surrounded by plasma membrane detached from the cell. (B, after R. Krstić, Die Gewebe des Menschen und der Säugetiere. Berlin: Springer-Verlag, 1978; C, from D.W. Fawcett, A Textbook of Histology, 11th ed. Philadelphia: Saunders, 1986.)

poorly understood factors. These factors allow two stem cells to be generated from one during repair processes, and they regulate the rate of basal cell proliferation according to the thickness of the epidermis. Glands connected to the epidermis, such as the mammary glands, have their own stem cells and their own distinct patterns of cell renewal.

Renewal by Pluripotent Stem Cells: Blood Cell Formation [22, 23]

The blood contains many types of cells with very different functions, ranging from the transport of oxygen to the production of antibodies. Some of these cells function entirely within the vascular system, while others use the vascular system only as a means of transport and perform their function elsewhere. All blood cells, however, have certain similarities in their life history. They all have limited life-spans and are produced throughout the life of the animal. Most remarkably, they are all generated ultimately from a common stem cell in the bone marrow. This

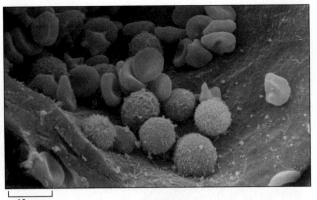

Figure 22–24 **Scanning electron micrograph of mammalian blood cells in a small blood vessel.** The larger, more spherical cells with a rough surface are white blood cells; the smaller, smoother, flattened cells are red blood cells. (From R.G. Kessel and R.H. Kardon, Tissues and Organs: A Text-Atlas of Scanning Electron Microscopy. San Francisco: Freeman, 1979. Copyright © 1979 W.H. Freeman and Company.)

10 μm

hemopoietic (or blood-forming) *stem cell* is thus pluripotent, giving rise to all of the types of terminally differentiated blood cells as well as some other types of cells, such as bone osteoclasts, which we discuss later.

Blood cells can be classified as red or white (Figure 22–24). The **red blood cells,** or **erythrocytes,** remain within the blood vessels and transport O_2 and CO_2 bound to hemoglobin. The **white blood cells,** or **leucocytes,** combat infection and in some cases phagocytose and digest debris. Leucocytes, unlike erythrocytes, must make their way across the walls of small blood vessels and migrate into tissues to perform their tasks. In addition, the blood contains large numbers of *platelets,* which are not entire cells but small detached cell fragments or "minicells" derived from the cortical cytoplasm of large cells called *megakaryocytes.* Platelets adhere specifically to the endothelial cell lining of damaged blood vessels, where they help repair breaches and aid in the process of blood clotting.

There Are Three Main Categories of White Blood Cells: Granulocytes, Monocytes, and Lymphocytes [22, 23]

All red blood cells are similar to one another, as are all platelets, but there are many distinct types of white blood cells. They are traditionally grouped into three major categories, called granulocytes, monocytes, and lymphocytes, on the basis of their appearance in the light microscope.

The **granulocytes** all contain numerous lysosomes and secretory vesicles (or granules) and are subdivided into three classes on the basis of the morphology and staining properties of these organelles (Figure 22–25). The differences in staining reflect major differences of chemistry and function. *Neutrophils* (also called *polymorphonuclear leucocytes* because of their multilobed nucleus) are the most common type of granulocyte; they phagocytose and destroy small organisms—especially bacteria. *Basophils* secrete histamine (and, in some species, serotonin) to help mediate inflammatory reactions; they are closely related in function to *mast cells,* which reside in connective tissues but are also generated from the hemopoietic stem cells. *Eosinophils* help destroy parasites and modulate allergic inflammatory responses.

Once they leave the bloodstream, **monocytes** (see Figure 22–25D) mature into *macrophages,* which together with neutrophils are the main "professional phagocytes" in the body. As discussed in Chapter 13, both types of phagocytic cells contain specialized lysosomes that fuse with newly formed phagocytic vesicles (phagosomes), exposing phagocytosed microorganisms to a barrage of enzymatically produced, highly reactive molecules of superoxide (O_2^-) and hypochlorite (HOCl, the active ingredient in bleach), as well as to a concentrated

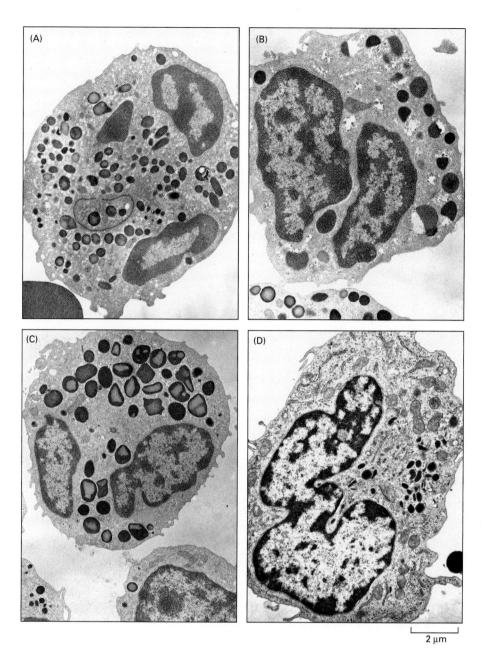

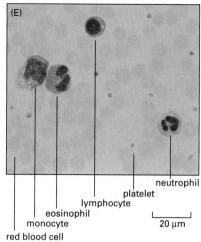

Figure 22–25 White blood cells.
(A–D) Electron micrographs showing, respectively, a neutrophil, a basophil, an eosinophil, and a monocyte. Electron micrographs of lymphocytes are shown in Figure 23–4. Each of the cell types shown here has a different function, which is reflected in the distinctive types of secretory granules and lysosomes it contains. There is only one nucleus per cell, but it has an irregular lobed shape, and in (B), (C), and (D) the connections between the lobes are out of the plane of section. (E) Light micrograph of a blood smear stained with the Romanowsky stain, which colors the white blood cells strongly. (A–D, courtesy of Dorothy Bainton; E, courtesy of David Mason.)

mixture of lysosomal hydrolases. Macrophages, however, are much larger and longer lived than neutrophils. They are responsible for removing senescent, dead, and damaged cells in many tissues, and they are unique in being able to ingest large microorganisms such as protozoa.

There are two main classes of **lymphocytes,** both involved in immune responses: *B lymphocytes* make antibodies, while *T lymphocytes* kill virus-infected cells and regulate the activities of other white blood cells. In addition, there are lymphocytelike cells called *natural killer (NK)* cells, which kill some types of tumor cells and some virus-infected cells. The production of lymphocytes is a specialized topic that is discussed in detail in Chapter 23. Here we shall concentrate mainly on the development of the other blood cells, often referred to collectively as **myeloid cells.**

The various types of blood cells and their functions are summarized in Table 22–1.

Table 22–1 Blood Cells

Type of Cell	Main Functions	Typical Concentration in Human Blood (cells/liter)
Red blood cells (erythrocytes)	transport O_2 and CO_2	5×10^{12}
White blood cells (leucocytes)		
Granulocytes		
Neutrophils (polymorphonuclear leucocytes)	phagocytose and destroy invading bacteria	5×10^9
Eosinophils	destroy larger parasites and modulate allergic inflammatory responses	2×10^8
Basophils	release histamine (and in some species serotonin) in certain immune reactions	4×10^7
Monocytes	become tissue macrophages, which phagocytose and digest invading microorganisms and foreign bodies as well as damaged and senescent cells	4×10^8
Lymphocytes		
B cells	make antibodies	2×10^9
T cells	kill virus-infected cells and regulate activities of other leucocytes	1×10^9
Natural killer (NK) cells	kill virus-infected cells and some tumor cells	1×10^8
Platelets (cell fragments, arising from *megakaryocytes* in bone marrow)	initiate blood clotting	3×10^{11}

Humans contain about 5 liters of blood, accounting for 7% of body weight. Red blood cells constitute about 45% of this volume and white cells about 1%, the rest being the liquid *blood plasma*.

The Production of Each Type of Blood Cell in the Bone Marrow Is Individually Controlled [22, 24]

Most white blood cells function in tissues other than the blood. The blood simply transports them to where they are needed. A local infection or injury in any tissue rapidly attracts white blood cells into the affected region as part of the **inflammatory response,** which helps fight the infection or heal the wound. The inflammatory response is complex and is mediated by a variety of signaling molecules produced locally by mast cells, nerve endings, platelets, and white blood cells, as well as by the activation of complement (discussed in Chapter 23). Some of these signaling molecules act on nearby capillaries, causing the endothelial cells to adhere less tightly to one another but making their surfaces adhesive to passing white blood cells. The white blood cells are thus caught like flies on flypaper and then can escape from the vessel by squeezing between the endothelial cells and crawling across the basal lamina with the aid of digestive enzymes; the initial binding to endothelial cells is mediated by *selectins* (discussed in Chapter 10), and the stronger binding required for the white blood cells to crawl out of the blood vessel is mediated by *integrins* (discussed in Chapter 19). Other molecules act as chemoattractants for specific types of white blood cells, causing these cells to become polarized and crawl toward the source of the attractant. As a result, large numbers of white blood cells enter the affected tissue (Figure 22–26).

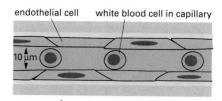

endothelial cell   white blood cell in capillary

10 μm

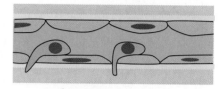

EXPOSURE TO MEDIATORS
OF INFLAMMATION RELEASED
FROM DAMAGED TISSUE

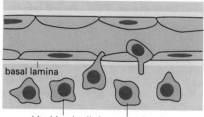

CHEMOTAXIS TOWARD
ATTRACTANTS RELEASED
FROM DAMAGED TISSUE

basal lamina

white blood cells in connective tissue

Figure 22–26 Migration of white blood cells out of the bloodstream in an inflammatory response. The response is initiated by a variety of signaling molecules produced locally by cells (mainly in the connective tissue) or by complement activation. Some of these mediators act on capillary endothelial cells, causing them to loosen their attachments to their neighbors so that the capillaries become more permeable; the endothelial cells are also stimulated to express *selectins*—cell-surface molecules that recognize specific carbohydrates that are present on the surface of leucocytes in the blood and cause them to stick to the endothelium. Other mediators act as chemoattractants, causing the bound leucocytes to crawl between the capillary endothelial cells into the tissue.

Other signaling molecules produced in the course of an inflammatory response escape into the blood and stimulate the bone marrow to produce more leucocytes and release them into the bloodstream. The bone marrow is the key target for such regulation because, with the exception of lymphocytes and some macrophages, most types of blood cells in adult mammals are generated only in the bone marrow. The regulation tends to be cell-type-specific: some bacterial infections, for example, cause a selective increase in neutrophils, while infections with some protozoa and other parasites cause a selective increase in eosinophils. (For this reason, physicians routinely use differential white blood cell counts to aid in the diagnosis of infectious and other inflammatory diseases.)

In other circumstances erythrocyte production is selectively increased—for example, if one goes to live at high altitude, where oxygen is scarce. Thus blood cell formation (**hemopoiesis**) necessarily involves complex controls in which the production of each type of blood cell is regulated individually to meet changing needs. It is a problem of great medical importance to understand how these controls operate, and much progress has been made in this area in recent years.

In intact animals hemopoiesis is more difficult to analyze than is cell turnover in a tissue such as the epidermal layer of the skin. In epidermis there is a simple, regular spatial organization that makes it easy to follow the process of renewal and to locate the stem cells. This is not true of the hemopoietic tissues. On the other hand, the hemopoietic cells have a nomadic life-style that makes them more accessible to experimental study in other ways. Dispersed hemopoietic cells can be easily transferred, without damage, from one animal to another, and the proliferation and differentiation of individual cells and their progeny can be observed and analyzed in culture. Because of this, more is known about the molecules that control blood cell production than about those that control cell production in other mammalian tissues.

Bone Marrow Contains Hemopoietic Stem Cells [22, 25]

The different types of blood cells and their immediate precursors can be recognized in the bone marrow by their distinctive appearances (Figure 22–27). They are intermingled with one another, as well as with fat cells and other *stromal cells* (connective-tissue cells) that produce a delicate supporting meshwork of collagen fibers and other extracellular-matrix components. In addition, the whole tissue is richly supplied with thin-walled blood vessels (called *blood sinuses*) into which the new blood cells are discharged. **Megakaryocytes** are also present; these, unlike other blood cells, remain in the bone marrow when mature and are one of its most striking features, being extraordinarily large (diameter up to 60 μm), with a highly polyploid nucleus. They normally lie close beside blood sinuses, and they extend processes through holes in the endothelial lining of these vessels; platelets pinch off from the processes and are swept away into the blood (Figure 22–28).

Because of the complex arrangement of the cells in bone marrow, it is difficult to identify any but the immediate precursors of the mature blood cells. The

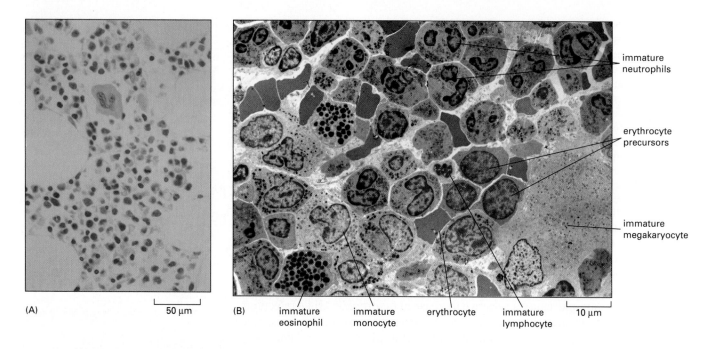

Figure 22–27 **Bone marrow.** (A) Light micrograph of a stained section. The large empty spaces correspond to fat cells, whose fatty contents have been dissolved away during specimen preparation. The giant cell with a lobed nucleus is a megakaryocyte. (B) Low-magnification electron micrograph. This tissue is the main source of new blood cells (except for T lymphocytes, which are produced in the thymus). Note that the immature blood cells of a particular type tend to cluster in "family groups." (A, courtesy of David Mason; B, from J.A.G. Rhodin, Histology: A Text and Atlas. New York: Oxford University Press, 1974.)

Figure 22–28 **Megakaryocytes.** (A) Schematic drawing of a megakaryocyte among other cells in the bone marrow. Its enormous size results from its having a highly polyploid nucleus. One megakaryocyte produces about 10,000 platelets, which split off from long processes that extend through holes in the walls of an adjacent blood sinus. (B) Scanning electron micrograph of the interior of a blood sinus in the bone marrow, showing the megakaryocyte processes. (B, from R.G. Kessel and R.H. Kardon, Tissues and Organs: A Text-Atlas of Scanning Electron Microscopy. San Francisco: Freeman, 1979. Copyright © 1979 W.H. Freeman and Company.)

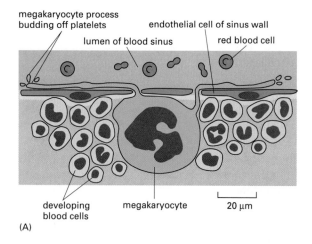

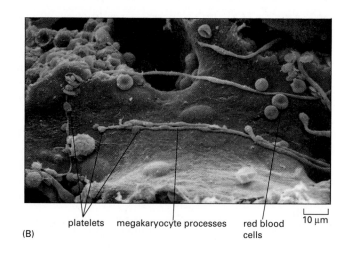

corresponding cells at still earlier stages of development, before any overt differentiation has begun, are confusingly similar in appearance, and there is no visible feature by which the ultimate stem cells can be recognized. To identify and characterize the stem cells, one needs a functional test, which involves tracing the progeny of single cells. As we shall see, this can be done *in vitro* simply by examining the colonies that isolated cells produce in culture. The hemopoietic system, however, can also be manipulated so that such clones of cells can be recognized *in vivo* in the intact animal.

If an animal is exposed to a large dose of x-irradiation, most of the hemopoietic cells are destroyed and the animal dies within a few days as a result of its inability to manufacture new blood cells. The animal can be saved, however, by a transfusion of cells taken from the bone marrow of a healthy, immunologically compatible donor. Among these cells there are small numbers (about 1 cell in 10,000) that can colonize the irradiated host and permanently reequip it with hemopoietic tissue. One of the tissues where colonies develop is the spleen, which in a normal mouse is an important additional site of hemopoiesis. When the spleen of an irradiated mouse is examined a week or two after the transfusion of cells from a healthy donor, a number of distinct nodules are seen in it, each of which is found to contain a colony of myeloid cells (Figure 22–29); after 2 weeks some colonies may contain more than a million cells. The discreteness of the nodules suggests that each might be a clone of cells descended from a single founder cell, like a bacterial colony on a culture plate; and with the help of genetic markers, it can be established that this is indeed the case.

The founder of such a colony is called a **colony-forming cell,** or **CFC** (also known as a colony-forming unit, CFU). The colony-forming cells are heterogeneous. Some give rise to only one type of myeloid cell, while others give rise to mixtures. Some go through many division cycles and form large colonies, while others divide less and form small colonies. Most of the colonies die out after generating a restricted number of terminally differentiated blood cells. A few of the colonies, however, are capable of extensive self-renewal and produce new colony-forming cells in addition to terminally differentiated blood cells. The founders of such self-renewing colonies are assumed to be the hemopoietic stem cells in the transfused bone marrow.

A Pluripotent Stem Cell Gives Rise to All Classes of Blood Cells [26]

All the types of myeloid cells can often be found together in one spleen colony, derived from a single stem cell. The hemopoietic stem cell, therefore, is *pluripotent:* it can give rise to many cell types. Although the spleen colonies do not seem to contain lymphocytes, another approach shows that these cells also derive from the same stem cell that gives rise to all of the myeloid cells. The demonstration employs genetic markers that make it possible to identify the members of a clone even after they have been released into the bloodstream. Although several types of clonal markers have been used for this, a specially engineered retrovirus (a *retroviral vector* carrying a marker gene) serves the purpose particularly well. The marker virus, like other retroviruses, can insert its own genome into the chromosomes of the cell it infects, but the genes that would enable it to generate new infectious virus particles have been removed. The marker, therefore, is confined to the progeny of the cells that were originally infected, and the progeny of one such cell can be distinguished from the progeny of another because the chromosomal sites of insertion of the virus are different. To analyze hemopoietic cell lineages, bone marrow cells are first infected with the retroviral vector *in vitro* and then are transferred into a lethally irradiated recipient; DNA probes can then be used to trace the progeny of individual infected cells in the various hemopoietic and lymphoid tissues of the host.

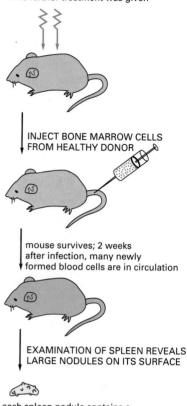

x-irradiation halts blood cell production; mouse would die if no further treatment was given

INJECT BONE MARROW CELLS FROM HEALTHY DONOR

mouse survives; 2 weeks after infection, many newly formed blood cells are in circulation

EXAMINATION OF SPLEEN REVEALS LARGE NODULES ON ITS SURFACE

each spleen nodule contains a clone of hemopoietic cells, descended from one of the injected bone marrow cells

Figure 22–29 The spleen colony assay. The spleen of a heavily irradiated animal becomes seeded with bone marrow cells transfused from a healthy donor. This assay, developed in 1961, revolutionized the study of hemopoiesis by allowing individual myeloid precursor cells to be analyzed for the first time.

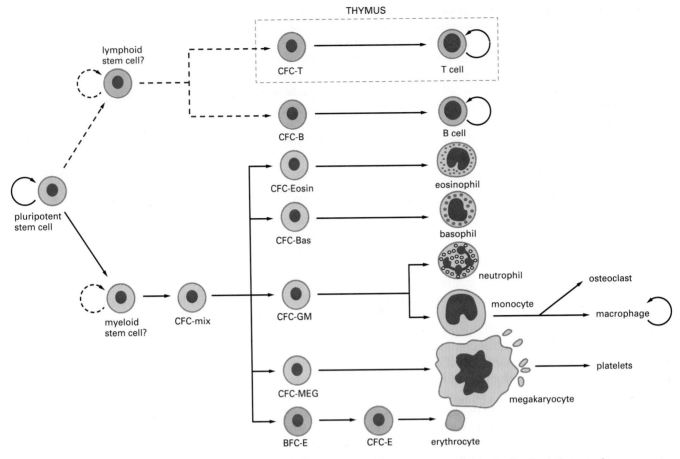

Figure 22–30 A tentative scheme of hemopoiesis. The pluripotent stem cell normally divides infrequently to generate either more pluripotent stem cells (self-renewal) or *committed progenitor cells* (labeled CFC = colony-forming cells), which are irreversibly determined to produce only one or a few types of blood cells. The progenitor cells are stimulated to proliferate by specific growth factors but progressively lose their capacity for division and develop into terminally differentiated blood cells, which usually live for only a few days or weeks.

In adult mammals all of the cells shown develop mainly in the bone marrow—except for T lymphocytes, which develop in the thymus, and macrophages and osteoclasts, which develop from blood monocytes. The most controversial part of the scheme is where the precursors for T and B lymphocytes fit into the scheme. The dashed lines reflect this uncertainty. The pluripotent stem cells also give rise to various types of tissue cells not shown in this scheme, such as NK cells, mast cells, and a variety of classes of antigen-presenting cells (discussed in Chapter 23), but the pathways by which these cells develop are uncertain.

These experiments not only confirm that all classes of blood cells—both myeloid and lymphoid—derive from a common stem cell (Figure 22–30), but they also make it possible to follow the pedigrees of the blood cells over long periods of time. After many months, when the hemopoietic system has had time to stabilize fully following the transfusion, practically all of the blood cells in the irradiated host mouse are found to be descendants of a remarkably small number—sometimes as few as a single one—of the original transfected cells. A single pluripotent stem cell evidently has the capacity to generate an indefinitely large clone of progeny, among them, presumably, many daughter stem cells with a similar capacity, as well as cells that are terminally differentiated.

The Number of Specialized Blood Cells Is Amplified by Divisions of Committed Progenitor Cells [22, 27]

Once a cell has differentiated as an erythrocyte or a granulocyte or some other type of blood cell, there seems to be no going back: the state of differentiation is not reversible. Therefore, at some stage in their development, some of the progeny of the pluripotent stem cell must become irreversibly committed or determined for a particular line of differentiation. It is clear from simple microscopic examination of the bone marrow that this commitment occurs well before the final division in which the mature differentiated cell is formed: one can recognize specialized precursor cells that are still proliferating but already show signs of having begun differentiation. It thus appears that commitment to a particular line of differentiation is followed by a series of cell divisions that amplify the number of cells of a given specialized type.

The hemopoietic system, therefore, can be viewed as a hierarchy of cells. **Pluripotent stem cells** give rise to **committed progenitor cells,** which are irreversibly determined as ancestors of only one or a few blood cell types. The committed progenitors divide rapidly but only a limited number of times. At the end of this series of *amplification divisions,* they develop into **terminally differentiated cells,** which usually divide no further and die after several days or weeks. Cells may also die at any of the earlier steps in the pathway. Studies in culture provide a way to find out how these cellular events—proliferation, differentiation, and death—are regulated.

The Factors That Regulate Hemopoiesis Can Be Analyzed in Culture [28]

Hemopoietic cells will survive, proliferate, and differentiate in culture if, and only if, they are provided with specific growth factors or accompanied by cells that produce these factors; if deprived of such factors, the cells die. Long-term proliferation of pluripotent stem cells can be achieved, for example, by culturing dispersed bone-marrow hemopoietic cells on top of a layer of bone-marrow stromal cells, presumably mimicking the environment in intact bone marrow; such cultures can generate all the types of myeloid cells. Alternatively, dispersed bone-marrow hemopoietic cells can be cultured in a semisolid matrix of dilute agar or methylcellulose, and factors derived from other cells can be added artificially to the medium. Because cells in the semisolid matrix cannot migrate, the progeny of each isolated precursor cell remain together as an easily distinguishable colony. A single committed neutrophil progenitor, for example, may be seen to give rise to a clone of thousands of neutrophils. Such culture systems, developed in the mid-1960s, provide a way to assay for the factors that support hemopoiesis and hence to purify them and explore their actions. These substances are found to be glycoproteins and are usually called **colony-stimulating factors,** or **CSFs.** Of the growing number of CSFs that have been defined and purified, some circulate in the blood and act as hormones, while others act in the bone marrow either as secreted local mediators or as membrane-bound signals that act through cell-cell contact. The best understood of the CSFs that act as hormones is the glycoprotein *erythropoietin,* which is produced in the kidney and regulates *erythropoiesis* (the formation of red blood cells).

Erythropoiesis Depends on the Hormone Erythropoietin [29]

The erythrocyte is by far the most common type of cell in the blood (see Table 22–1). When mature, it is packed full of hemoglobin and contains practically none of the usual cell organelles. In an erythrocyte of an adult mammal, even the nucleus, endoplasmic reticulum, mitochondria, and ribosomes are absent, having been extruded from the cell in the course of its development (Figure 22–31). The erythrocyte therefore cannot grow or divide; the only possible way of making more erythrocytes is by means of stem cells. Furthermore, erythrocytes have a limited life-span—about 120 days in humans or 55 days in mice. Worn-out erythrocytes are phagocytosed and digested by macrophages in the liver and spleen, which remove more than 10^{11} senescent erythrocytes in each of us each day.

A lack of oxygen or a shortage of erythrocytes stimulates cells in the kidney to synthesize and secrete increased amounts of **erythropoietin** into the blood-

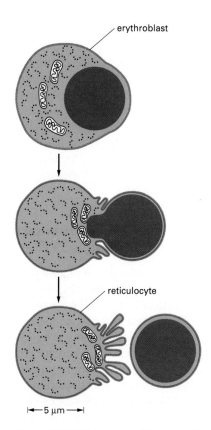

Figure 22–31 Schematic diagram of a developing red blood cell (erythroblast). The cell is shown extruding its nucleus to become an immature erythrocyte (*reticulocyte*), which then leaves the bone marrow and passes into the bloodstream. The reticulocyte will lose its mitochondria and ribosomes within a day or two to become a mature erythrocyte. Erythrocyte clones develop in the bone marrow on the surface of a macrophage, which phagocytoses and digests the nuclei discarded by the erythroblasts.

stream. The erythropoietin in turn stimulates the production of more erythrocytes. Since a change in the rate of release of new erythrocytes into the bloodstream is observed as early as 1 or 2 days after an increase in erythropoietin levels in the bloodstream, the hormone must act on cells that are very close precursors of the mature erythrocytes.

The cells that respond to erythropoietin can be identified by culturing bone marrow cells in a semisolid matrix in the presence of erythropoietin. In a few days colonies of about 60 erythrocytes appear, each founded by a single committed erythroid progenitor cell. This cell is known as an **erythrocyte colony-forming cell,** or **CFC-E,** and it gives rise to mature erythrocytes after about six division cycles or less. The CFC-Es do not yet contain hemoglobin, and they are derived from an earlier type of progenitor cell whose proliferation does not depend on erythropoietin. CFC-Es themselves depend on erythropoietin for their survival as well as for proliferation: if erythropoietin is removed from the cultures, the cells rapidly undergo programmed cell death.

A second CSF, called **interleukin 3 (IL-3),** promotes the survival and proliferation of the earlier erythroid progenitor cells. In its presence much larger erythroid colonies, each comprising up to 5000 erythrocytes, develop from cultured bone marrow cells in a process requiring a week or 10 days. These colonies derive from erythroid progenitor cells called **erythrocyte burst-forming cells,** or **BFC-Es.** The BFC-E is distinct from the pluripotent stem cell in that it has a limited capacity to proliferate and gives rise to colonies that contain erythrocytes only, even under culture conditions that enable other progenitor cells to give rise to other classes of differentiated blood cells. It is distinct from the CFC-E in that it is insensitive to erythropoietin, and its progeny must go through as many as 12 divisions before they become mature erythrocytes (for which erythropoietin must be present). The cell also differs in size from the CFC-E and can be separated from it by sedimentation. Thus the BFC-E is thought to be a progenitor cell committed to erythrocyte differentiation and an early ancestor of the CFC-E (Figure 22–32).

Multiple CSFs Influence the Production of Neutrophils and Macrophages [28, 30]

The two professional phagocytic cells, neutrophils and macrophages, develop from a common progenitor cell called the **granulocyte/macrophage** (or **GM) progenitor cell.** Like the other granulocytes (eosinophils and basophils), neutrophils circulate in the blood for only a few hours before migrating out of capillaries into the connective tissues or other specific sites, where they survive for only a few days and then die and are phagocytosed by macrophages. Macrophages, by contrast, can persist for months or perhaps even years outside the bloodstream, where they can be activated by local signals to resume proliferation.

At least seven distinct CSFs that stimulate neutrophil and macrophage colony formation in culture have been defined, and some or all of these are thought to act in different combinations to regulate the selective production of these cells *in vivo*. These CSFs are synthesized by various cell types—including endothelial cells, fibroblasts, macrophages, and lymphocytes—and their concentration in the blood typically increases rapidly in response to bacterial infection in a tissue, thereby increasing the number of phagocytic cells released from the bone marrow into the bloodstream. IL-3 is one of the least specific of the factors, acting on pluripotent stem cells as well as on most classes of committed progenitor cells, including GM-progenitor cells. Various other factors act more selectively on committed GM-progenitor cells and their differentiated progeny (Table 22–2), although in many cases they act on certain other branches of the hemopoietic family tree as well.

All of these CSFs, like erythropoietin, are glycoproteins that act at low concentrations (~10^{-12} M) by binding to specific cell-surface receptors, as discussed

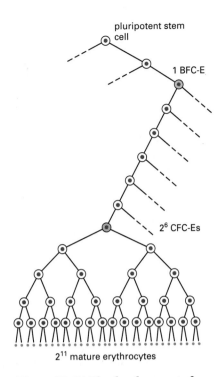

Figure 22–32 The development of red blood cells. The drawing shows the relationship between the BFC-E, the CFC-E, and the mature erythrocyte. BFC-Es and CFC-Es are both committed erythroid progenitor cells. BFC-Es respond to the factor IL-3 but not to erythropoietin, whereas CFC-Es respond to erythropoietin. The series of cell divisions that occur in this lineage under the influence of erythropoietin provides a powerful means of controlling the production of erythrocytes without upsetting the production of other types of blood cells.

Table 22–2 Some Colony-stimulating Factors (CSFs) That Influence Blood Cell Formation

Factor	Size (in mouse)	Target Cells	Producing Cells	Receptors
Erythropoietin	51,000 daltons	CFC-E	kidney cells	cytokine family
Interleukin 3 (IL-3)	25,000 daltons	pluripotent stem cell, most progenitor cells, many terminally differentiated cells	T lymphocytes, epidermal cells	cytokine family
Granulocyte/ macrophage CSF (GM-CSF)	23,000 daltons	GM progenitor cells	T lymphocytes, endothelial cells, fibroblasts	cytokine family
Granulocyte CSF (G-CSF)	25,000 daltons	GM progenitor cells and neutrophils	macrophages, fibroblasts	cytokine family
Macrophage CSF (M-CSF)	70,000 daltons (dimer)	GM progenitor cells and macrophages	fibroblasts, macrophages, endothelial cells	receptor tyrosine kinase family
Steel factor (stem cell factor)	40–50,000 daltons (dimer)	hemopoietic stem cell	stromal cells in bone marrow and many other cells	receptor tyrosine kinase family

in Chapter 15. A few of these receptors are transmembrane tyrosine kinases. The others belong to another large receptor family (sometimes called the *cytokine receptor* family), whose members are usually composed of two or more subunits, one of which is frequently shared among several receptor types (Figure 22–33). The CSFs not only operate on the precursor cells to promote the production of differentiated progeny, they also activate the specialized functions (such as phagocytosis and target-cell killing) of the terminally differentiated cells. Proteins produced artificially from the cloned genes for these factors (sometimes referred to as *recombinant* factors because they are made using recombinant DNA technology) are strong stimulators of hemopoiesis in experimental animals. They are now being used in human patients to stimulate the regeneration of hemopoietic tissue and to boost resistance to infection—an impressive demonstration of how basic cell biological research and animal experiments can lead to better medical treatment.

Factors that promote the development of the other classes of myeloid cells, such as megakaryocytes and eosinophils, have also been identified. Again, there are many of these factors, and they have overlapping actions when tested in laboratory assay systems. It is not easy to discover precisely what their individual roles are in natural circumstances. Perhaps the most direct test of the normal function

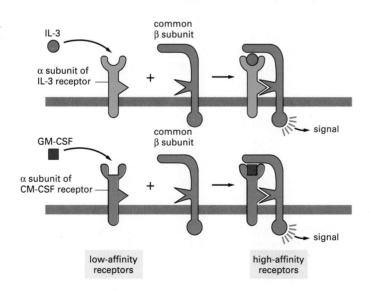

Figure 22–33 **Sharing of subunits among CSF receptors.** Human IL-3 receptors and GM-CSF receptors have different α subunits and a common β subunit. Their ligands bind to the free α subunit with low affinity, and this triggers the assembly of the heterodimer that binds the ligand with high affinity.

of a CSF is to inactivate the CSF or its receptor in a living animal and study the consequences. This has now been done for several CSFs. Anti-G-CSF antibodies, which neutralize the activity of G-CSF—a CSF that promotes neutrophil production *in vitro*—have been shown to cause a marked decrease in neutrophils when injected into healthy dogs, establishing that G-CSF is required for the normal production of neutrophils. Genetic approaches can be even more powerful. Mice with a mutation in the gene that encodes M-CSF, for example, are deficient in macrophages, as well as in osteoclasts, which also develop from monocytes. Because osteoclasts are required for bone resorption (as we discuss later), these mice produce an excessive amount of bone, which encroaches on the bone marrow and produces abnormally thickened bones and decreased blood cell formation—a condition called *osteopetrosis*.

Hemopoietic Stem Cells Depend on Contact with Cells Expressing the Steel Factor [31]

CSFs that act on the pluripotent stem cells are the most intriguing of all. IL-3, as we have seen, seems to be in this class. Another such factor of fundamental importance came to light through the analysis of mouse mutants that show a curious combination of defects: a shortage of red blood cells (anemia), of germ cells (sterility), and of pigment cells (white spotting of the skin). As discussed in Chapter 21, this syndrome results from mutations in either of two genes: one, called *c-kit*, codes for a receptor tyrosine kinase; the other, called *Steel*, codes for its ligand. The cell types affected by the mutations all derive from migratory precursors, and it seems that these precursors in each case must express the receptor (Kit) and be provided with the ligand (Steel) by their environment if progeny cells are to be produced in normal numbers.

Like IL-3, the Steel factor acts on several of the committed blood-cell lineages, including the erythroid lineage, as well as on the pluripotent stem cells. But it has little effect on its own. It mainly potentiates the effects of other CSFs, greatly increasing the number and size of clonal blood-cell colonies of all kinds in culture. It is an unusual CSF in another way too. It is made in both a membrane-bound and a secreted form, generated by alternative splicing of the mRNA, and it seems to be the membrane-bound form that is most important: mutant mice that make the secreted form of the Steel factor but not the membrane-bound form show severe defects. This implies that normal hemopoiesis requires direct cell-cell contact between the hemopoietic stem cell and a stromal cell that expresses *Steel* and that only this contact enables the Steel factor to activate the Kit receptor protein efficiently. Kit may thus behave as a coreceptor (discussed in Chapter 23), which has to be activated at the same time as receptors for factors such as IL-3 in order to stimulate hemopoiesis. This could help explain why hemopoiesis occurs only in a few special environments, such as that provided by the stromal cells of the bone marrow, while other tissues escape invasion and colonization even though there are always some hemopoietic stem cells circulating in the bloodstream.

The Behavior of a Hemopoietic Cell Depends Partly on Chance [28, 32]

Up to this point we have glossed over a central question. The CSFs are defined as factors that promote the production of colonies of differentiated blood cells. But what effect precisely does a CSF have on an individual hemopoietic cell? The factor might control the rate of cell division or the number of division cycles that the progenitor cell goes through before differentiating; it might act late in the hemopoietic lineage to facilitate differentiation; it might act early to influence commitment; or it might simply increase the probability of cell survival (Figure 22–34). By monitoring the fate of isolated individual hemopoietic cells in culture,

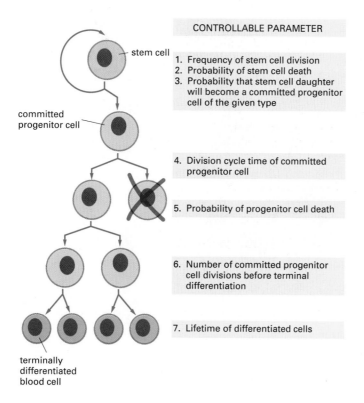

Figure 22–34 **Some of the parameters through which the production of blood cells of a specific type might be regulated.** Studies *in vitro* suggest that colony-stimulating factors (CSFs) can affect all of these aspects of hemopoiesis.

it has been possible to show that a single CSF, such as GM-CSF, can exert all these different effects. Nevertheless, it is still not clear which actions are most important *in vivo*. The behavior of the pluripotent stem cells remains especially elusive: these crucial cells are few and far between—less than 1 in 1000 of the cells in the bone marrow—and are difficult to identify unambiguously.

Studies *in vitro* indicate, moreover, that there is a large element of chance in the way a hemopoietic cell behaves. The CSFs seem to act by regulating probabilities, not by dictating directly what the cell shall do. In hemopoietic cell cultures, even if the cells have been selected to be as homogeneous a population as possible, there is a remarkable variability in the sizes and often in the characters of the colonies that develop. And if two sister cells are taken immediately after a cell division and cultured apart under identical conditions, they will frequently give rise to colonies that contain different types of blood cells or the same types of blood cells in different numbers. Thus both the programming of cell division and the process of commitment to a particular path of differentiation seem to involve random events at the level of the individual cell, even though the behavior of the multicellular system as a whole is regulated in a reliable way.

Regulation of Cell Survival Is as Important as Regulation of Cell Proliferation [33]

While such observations show that CSFs are not strictly required to instruct the hemopoietic cells how to differentiate or how many times to divide, CSFs *are* required to keep the cells alive: the default behavior of the cells in the absence of CSFs is suicide. In principle, the CSFs could regulate the numbers of the various types of blood cells entirely through selective control of cell survival in this way, and there is increasing evidence that the control of cell survival plays a central part in the normal regulation of the numbers of blood cells and, as discussed earlier for hepatocytes, of many other cell types too. In many tissues, it seems, cells are programmed to kill themselves if they do not receive specific signals for survival. We have already discussed the importance and the mechanism of *programmed cell death* during development (see p. 1076); it is no less important in

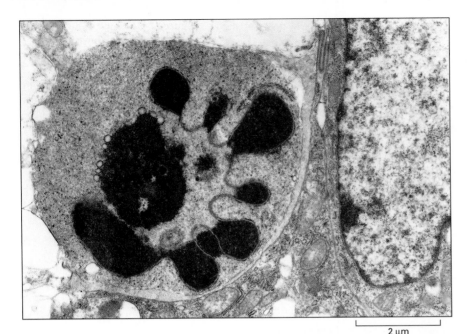

Figure 22–35 Cells dying by apoptosis. The electron micrograph shows an apoptotic cell in the mammary gland. Apoptotic cell death is a normal occurrence here, balancing the proliferation of mammary epithelial cells that occurs in each menstrual cycle. Note the disintegrating nuclear envelope and the dark clumps of condensed chromatin. For comparison, part of a normal cell is visible to one side of the picture. (Courtesy of David Ferguson.)

2 μm

the turnover and renewal of cell populations in the adult body (Figure 22–35). The genes that regulate it have been highly conserved in evolution, to the extent that at least one of them, called *bcl-2*, coding for an intracellular inhibitor of the cell death program in mammalian cells, can perform the same function in cells of a nematode worm. Too little cell death can be as dangerous to the health of the multicellular organism as too much proliferation, and mutations that inhibit cell death by causing overexpression of *bcl-2* have been implicated in the development of cancer, as discussed in Chapter 24.

The amount of programmed cell death in the vertebrate hemopoietic system is enormous: billions of neutrophils die in this way each day in an adult human, for example. Although the mechanism of programmed cell death remains a mystery, the dying cells usually undergo a characteristic morphological change called **apoptosis,** in which the cell and its nucleus shrink and condense and frequently fragment. By contrast, cells that die accidentally, as a result of acute injury, usually swell and burst—a process called *cell necrosis.* Whereas cells that die by necrosis spill their cytosolic contents into the extracellular space and elicit an inflammatory response, cells that die by apoptosis disappear in a way that is more efficient for the organism: they are so rapidly phagocytosed by macrophages (or other neighboring cells) that there is no leakage of cytosolic components and no inflammatory response. Once inside the macrophage, the apoptotic cell is quickly disassembled and its chemical building blocks reused.

To activate this disposal mechanism, apoptotic cells change their surface chemistry so that macrophages can recognize them. The recognition mechanism varies depending on the tissue and the type of blood cell. In some cases a lectin on the macrophage surface seems to recognize altered sugar groups on the apoptotic cell surface. In others an integrin (discussed in Chapter 19) on the macrophage surface recognizes an extracellular matrix protein called *thrombospondin*, which is secreted by the macrophage and seems to act as a bridge between it and the apoptotic cell; the mechanism by which thrombospondin binds to apoptotic cells is unknown. In still other cases the macrophage is thought to recognize phosphatidylserine, a negatively charged phospholipid that is normally confined to the cytosolic leaflet of the plasma membrane lipid bilayer (see Figure 10–11) but apparently relocates to the extracellular leaflet in some apoptotic blood cells. No matter which of these recognition systems is used, macrophages react to the apoptotic cells in a specific way:

they engulf and digest them, but they do not secrete inflammation-inducing signals as they do when they phagocytose and digest necrotic cells. This is a second reason why cell necrosis is associated with inflammation, whereas apoptosis is not.

Although biologists have paid much more attention to the control of cell proliferation than to the control of cell survival, it is becoming increasingly clear that both kinds of controls can serve to regulate cell numbers. Both depend on specific signals produced by other cells, ensuring that a cell divides only when more cells are required and that a cell survives only when and where it is needed. The challenge is to define all of the signals that regulate the survival and proliferation of each cell type, to determine how their levels are controlled to balance cell proliferation and cell death according to the varying needs of the organism, and to understand how an individual cell integrates these diverse extracellular signals and decides whether to live or die and whether to divide or remain quiescent.

Summary

The many types of blood cells all derive from a common pluripotent stem cell. In the adult the stem cells are found mainly in bone marrow, where they normally divide infrequently to produce more stem cells (self-renewal) and various committed progenitor cells, each able to give rise to only one or a few types of blood cells. The committed progenitor cells divide profusely under the influence of various protein signaling molecules (called colony-stimulating factors, or CSFs) and then differentiate into mature blood cells, which usually die after several days or weeks. Studies of hemopoiesis have been greatly aided by in vitro assays in which stem cells or committed progenitor cells form clonal colonies when cultured in a semisolid matrix. The progeny of stem cells appear to make their choices among alternative developmental pathways in a partly random manner. Cell death, controlled by the availability of CSFs, also plays a central part in regulating the numbers of mature differentiated blood cells; it depends on activation of an intracellular suicide program and is thought to help regulate cell numbers in many other tissues and in other kinds of animals.

Genesis, Modulation, and Regeneration of Skeletal Muscle [34]

The term "muscle" covers a multitude of cell types, all specialized for contraction but in other respects dissimilar. As noted in Chapter 16, a contractile system involving actin and myosin is a basic feature of animal cells in general, but muscle cells have developed this apparatus to a high degree. Mammals possess four main categories of cells specialized for contraction: *skeletal muscle cells, heart* (or *cardiac) muscle cells, smooth muscle cells,* and *myoepithelial cells* (Figure 22–36). These differ in function, structure, and development. Although all of them appear to generate contractile forces by means of organized filament systems based on actin and myosin, the actin and myosin molecules employed are somewhat different in amino acid sequence, are differently arranged in the cell, and are associated with different sets of proteins to control contraction.

Skeletal muscle cells, whose contractile apparatus is discussed in detail in Chapter 16, are responsible for practically all movements that are under voluntary control. These cells can be very large (2 or 3 cm long and 100 μm in diameter in an adult human) and are often referred to as *muscle fibers* because of their highly elongated shape. Each one is a syncytium, containing many nuclei within a common cytoplasm. The other types of muscle cells are more conventional, having only a single nucleus. **Heart muscle cells** resemble skeletal muscle cells in that their actin and myosin filaments are aligned in very orderly arrays to form a series of contractile units called *sarcomeres*, so that the cells have a striated appearance. **Smooth muscle cells** are so called because they, in contrast, do not appear striated. The functions of smooth muscle vary greatly, from propelling food along the digestive tract to erecting hairs in response to cold or fear. **Myoepithelial cells** also have no striations, but unlike all other muscle cells they lie in epithelia and are derived from the ectoderm. They form the dilator muscle of the iris and serve to expel saliva, sweat, and milk from the corresponding glands (see Figure 22–36E). The four main categories of muscle cells can be further divided into distinctive subtypes, each with its own characteristic features.

The mechanisms of muscle contraction are discussed in Chapter 16; here, we consider how muscle tissue is generated and maintained. We focus on the skeletal muscle cell, which has a curious mode of development, a striking ability to modulate its differentiated character, and an unusual strategy for repair.

New Skeletal Muscle Cells Form by the Fusion of Myoblasts [2, 35]

The previous chapter described how certain cells, originating from the somites of a vertebrate embryo at a very early stage, become determined as *myoblasts*

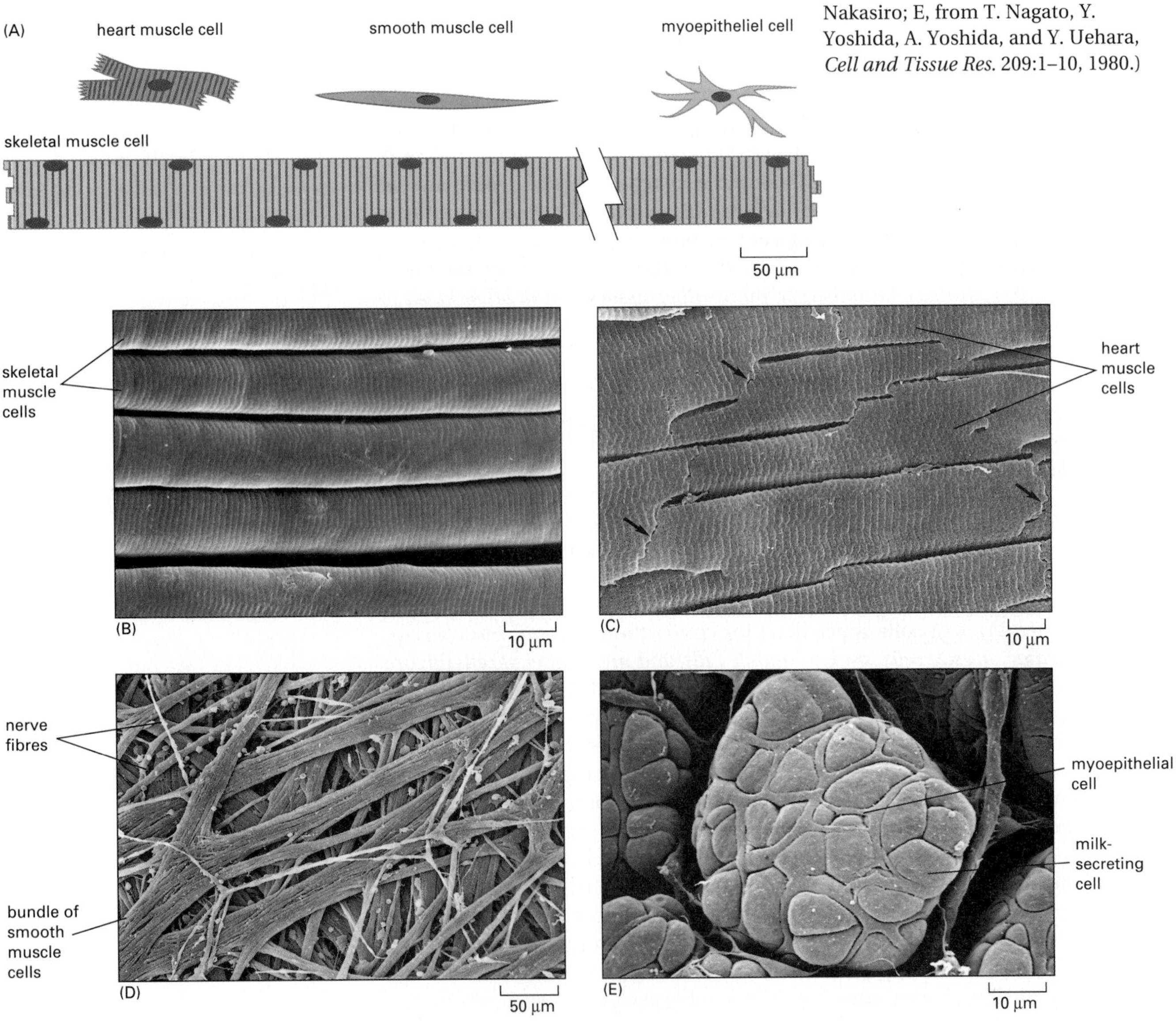

(A)

heart muscle cell   smooth muscle cell   myoepitheliel cell

skeletal muscle cell

50 μm

Figure 22–36 The four classes of muscle cells of a mammal. (A) Schematic drawings (to scale). (B–E) Scanning electron micrographs, showing (B) skeletal muscle from the neck of a hamster, (C) heart muscle from a rat, (D) smooth muscle from the urinary bladder of a guinea pig, and (E) myoepithelial cells in a secretory alveolus from a lactating rat mammary gland. The arrows in (C) point to intercalated discs—end-to-end junctions between the heart muscle cells; skeletal muscle cells in long muscles are joined end to end in a similar way. Note that the smooth muscle is shown at a lower magnification than the others. (B, courtesy of Junzo Desaki; C, from T. Fujiwara, in Cardiac Muscle in Handbook of Microscopic Anatomy [E.D. Canal, ed.]. Berlin: Springer Verlag, 1986; D, courtesy of Satoshi Nakasiro; E, from T. Nagato, Y. Yoshida, A. Yoshida, and Y. Uehara, *Cell and Tissue Res.* 209:1–10, 1980.)

skeletal muscle cells

(B)   10 μm

heart muscle cells

(C)   10 μm

nerve fibres

bundle of smooth muscle cells

(D)   50 μm

myoepithelial cell

milk-secreting cell

(E)   10 μm

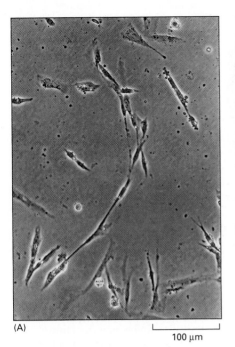
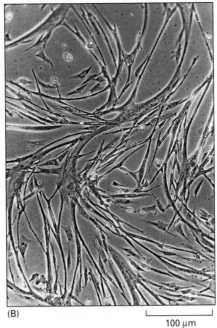
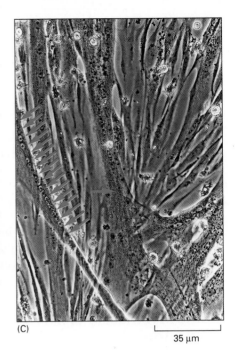

(A) 100 µm (B) 100 µm (C) 35 µm

Figure 22–37 Myoblast fusion in culture. The phase-contrast micrographs show how the cells will proliferate, line up, and fuse to form multinucleate muscle cells. (C) is at higher magnification, showing the cross-striations that are just beginning to be visible as the contractile apparatus develops (*red arrow*) and the accumulations of many nuclei within a single cell (*green arrows*). (Courtesy of Rosalind Zalin.)

(that is, as precursors of skeletal muscle cells) and migrate into the adjacent embryonic connective tissue, or mesenchyme. As discussed in Chapter 9, the commitment to be a myoblast (rather than, say, a fibroblast) depends on the activation of one or more *myogenic genes,* which encode gene regulatory proteins of the helix-loop-helix family. After a period of proliferation the myoblasts fuse with one another to form multinucleate skeletal muscle cells (Figure 22–37). As they fuse, they undergo a dramatic switch of phenotype that depends on the coordinated activation of a whole battery of muscle-specific genes. Once fusion has occurred, the nuclei never again replicate their DNA. Fusion involves specific cell-cell adhesion molecules that mediate recognition between myoblasts.

Myoblasts that have been kept proliferating in culture for as long as 2 years still retain the ability to differentiate and will fuse to form muscle cells in response to a suitable change in culture conditions. *Fibroblast growth factor (FGF)* in the medium seems to be crucial for keeping the myoblasts proliferating and preventing their differentiation: if FGF is removed, the cells rapidly stop dividing, fuse, and differentiate. The system of controls is complex, however, and in order to differentiate, myoblasts must attach to the extracellular matrix. Moreover, the process of fusion is cooperative: fusing myoblasts secrete factors that encourage other myoblasts to fuse. In the intact animal the myoblasts and muscle fibers are held in the meshes of a connective-tissue framework formed by fibroblasts. This framework guides muscle development and controls the arrangement and orientation of the muscle cells.

Muscle Cells Can Vary Their Properties by Changing the Protein Isoforms That They Contain [36]

Once formed, a skeletal muscle cell is generally retained for the entire lifetime of the animal. Over this period it grows, matures, and modulates its character according to functional requirements. The genome contains multiple variant copies of the genes encoding many of the characteristic proteins of the skeletal muscle cell, and the RNA transcripts of many of these genes can be spliced in several ways. As a result, a wealth of protein variants (*isoforms*) can be produced for the components of the contractile apparatus. As the muscle cell matures, different selections of isoforms are produced, adapted to the changing demands

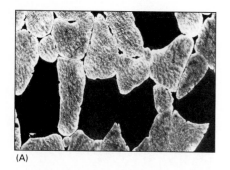

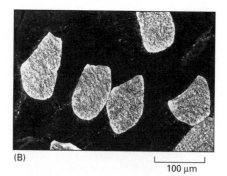

(A)

(B)

100 μm

Figure 22–38 Fast and slow muscle fibers. Two consecutive cross-sections of the same piece of adult chicken muscle have been stained with two fluorescent antibodies, each specific for a different isoform of myosin-II. In (A) cells specialized to produce fast-twitch contractions are stained with antibodies against "fast" myosin; in (B) cells specialized to produce slow, sustained contractions are stained with antibodies against "slow" myosin. The fast-twitch cells are known as *white muscle cells* because they contain relatively little of the colored oxygen-binding protein myoglobin; the slow muscle cells are called *red muscle cells* because they contain much more of it. The cells can adjust their fast or slow character through changes of gene expression according to the pattern of nerve stimulation they receive. (From G. Gauthier et al., *J. Cell Biol.* 92:471–484, 1982, by copyright permission of the Rockefeller University Press.)

for speed, strength, and endurance in the fetus, the newborn, and the adult. Within a single adult muscle, several distinct types of skeletal muscle cells, each with different sets of protein isoforms and different functional properties, can be found side by side (Figure 22–38).

Some Myoblasts Persist as Quiescent Stem Cells in the Adult [37]

A muscle can grow in three ways: its differentiated muscle cells can increase in number, in length, or in girth. Because skeletal muscle cells are unable to divide, more of them can be made only by the fusion of myoblasts. The adult number of multinucleated skeletal muscle cells is in fact attained early—before birth in humans. The subsequent enormous increase in muscle bulk is achieved by cell enlargement. Growth in length depends on recruitment of more myoblasts into the existing multinucleate cells, mainly by fusion at their ends, which increases the number of nuclei in each cell. In contrast, growth in girth, such as occurs in the muscles of weightlifters, depends on an increase in the size and numbers of the contractile myofibrils that each muscle cell contains rather than on changes in the numbers of muscle cells or of their nuclei.

In the adult, nevertheless, a few myoblasts persist as small, flattened, and inactive cells lying in close contact with the mature muscle cell and contained within its sheath of basal lamina. If the muscle is damaged or if it is treated artificially with FGF, these so-called *satellite cells* are activated to proliferate (Figure 22–39), and their progeny can fuse to form new muscle cells. Satellite cells are thus the stem cells of adult skeletal muscle, normally held in reserve in a quiescent state but available when needed as a self-renewing source of terminally differentiated cells.

Although this muscle repair mechanism operates well in small animals such as mice, it is less efficient in humans. In muscular dystrophy, for example, differentiated skeletal muscle cells die because of a genetic defect in the cytoskeletal dystrophin protein (see p. 855). As a result, satellite cells proliferate to form new muscle cells; but this regenerative response is unable to keep pace with the damage, and the muscle cells are eventually replaced by connective tissue, blocking any further possibility of regeneration.

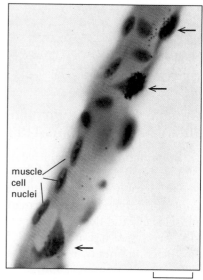

muscle cell nuclei

20 μm

Figure 22–39 Autoradiograph of a single multinucleate muscle cell with associated satellite cells. The fiber has been isolated from an adult rat and transferred into culture medium containing ^{3}H-thymidine plus an extract from damaged muscle that stimulates the satellite cells to divide. The dividing satellite cells (*arrows*) have become radioactively labeled (silver grains visible as *black dots*); the muscle cell nuclei are unable to proliferate and remain unlabeled. (From R. Bischoff, *Dev. Biol.* 115:140–147, 1986.)

Summary

Skeletal muscle cells are one of the four main categories of vertebrate cells specialized for contraction, and they are responsible for voluntary movement. Each skeletal muscle cell is a syncytium and develops by the fusion of many myoblasts. Myoblasts are stimulated to proliferate by growth factors such as FGF, but once they fuse, they can no longer divide. Myoblast fusion is generally coupled with the onset of muscle cell differentiation, in which many genes encoding muscle-specific proteins are switched on coordinately. Some myoblasts persist (in a quiescent state) as satellite cells in adult muscle; when a muscle is damaged, these cells are reactivated to proliferate and to fuse to replace the muscle cells that have been lost.

Fibroblasts and Their Transformations: The Connective-Tissue Cell Family [38]

Many of the differentiated cells in the adult body can be grouped into families whose members are closely related by origin and by character. An important example is the family of **connective-tissue cells,** whose members are not only related but are also to an unusual extent interconvertible. The family includes *fibroblasts, cartilage cells,* and *bone cells,* all of which are specialized for secretion of collagenous extracellular matrix and are jointly responsible for the architectural framework of the body, as well as *fat cells* and *smooth muscle cells,* which appear to have a common origin with them. These cell types and the interconversions that are thought to occur between them are illustrated in Figure 22–40. Connective-tissue cells play a central part in the support and repair of almost every tissue and organ, and the adaptability of their differentiated character is an important feature of the responses to many types of damage.

Fibroblasts Change Their Character in Response to Signals in the Extracellular Matrix [38, 39]

Fibroblasts appear to be the least specialized cells in the connective-tissue family. They are dispersed in connective tissue throughout the body, where they secrete a nonrigid extracellular matrix that is rich in type I and/or type III collagen, as discussed in Chapter 19. When a tissue is injured, the fibroblasts nearby migrate into the wound, proliferate, and produce large amounts of collagenous matrix, which helps to isolate and repair the damaged tissue. Their ability to thrive in the face of injury, together with their solitary life-style, may explain why fibroblasts are the easiest of cells to grow in culture—a feature that has made them a favorite subject for cell biological studies (Figure 22–41).

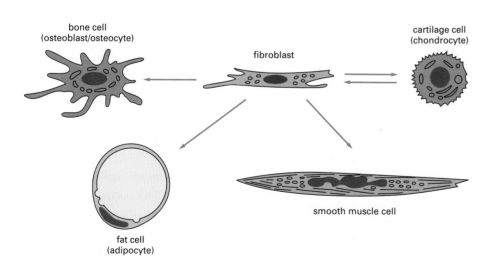

Figure 22–40 **The family of connective-tissue cells.** Arrows show the interconversions that appear to occur within the family. For simplicity, the fibroblast is shown as a single cell type, but in fact it is uncertain how many types of fibroblasts exist and whether the differentiation potential of different types may be restricted in different ways.

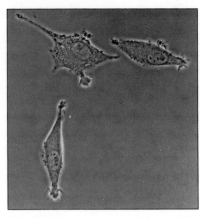

(A)

Figure 22–41 **The fibroblast.** (A) Phase-contrast micrograph of fibroblasts in culture. (B) Drawings of a living fibroblastlike cell in the transparent tail of a tadpole, showing the changes in its shape and position on successive days. Note that while fibroblasts flatten out in culture, they can have more complex, process-bearing morphologies in tissues. (A, courtesy of Daniel Zicha; B, redrawn from E. Clark, *Am. J. Anat.* 13:351–379, 1912.)

As indicated in Figure 22–40, fibroblasts also seem to be the most versatile of connective-tissue cells, displaying a remarkable capacity to differentiate into other members of the family. There are some important uncertainties about their interconversions, however. There is good evidence that fibroblasts in different parts of the body are intrinsically different, and it is far from proven that all fibroblasts in a given region are equivalent. In the absence of firm evidence to the contrary, it is simplest to suppose that they are indeed equivalent, but it is conceivable that connective tissue may contain a mixture of distinct fibroblast lineages, some capable of transformation into chondrocytes, others capable of transformation into fat cells, and so on, rather than just one type of fibroblast with multiple developmental capabilities. It is possible also that "mature" fibroblasts incapable of transformation may exist side by side with "immature" fibroblasts (often called *mesenchymal cells*) that can develop into a variety of mature cell types.

Despite these uncertainties, there is clear evidence, from studies both *in vivo* and *in vitro,* that connective-tissue cells can undergo radical changes of character. Thus, if a preparation of bone matrix, made by grinding bone into a fine powder and dissolving away the hard mineral component, is implanted in the dermal layer of the skin, some of the cells there (probably dermal fibroblasts) become transformed into cartilage cells and, a little later, others into bone cells, thereby creating a small lump of bone, complete with a marrow cavity. These experiments suggest that components in the extracellular matrix can dramatically influence connective-tissue cell differentiation. We shall see that similar cell transformations are important in the natural repair of broken bones. In fact, bone matrix has been found to contain trapped within it high concentrations of several growth factors that can affect the behavior of connective-tissue cells, including, in particular, transforming growth factor β (TGF-β) and a set of distinct *bone morphogenetic proteins (BMPs)* that belong to the TGF-β superfamily. These factors are powerful regulators of growth, differentiation, and matrix synthesis by connective-tissue cells, exerting a variety of actions depending on the target cell type and the combination of other factors and matrix components that are present. When injected into a living animal, they can induce formation of cartilage, of bone, or of fibrous matrix, according to the site and circumstances of injection.

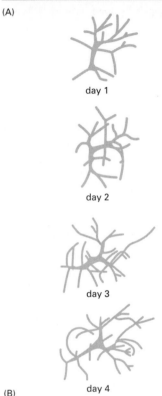

day 1

day 2

day 3

(B)

day 4

The Extracellular Matrix May Influence Connective-Tissue Cell Differentiation by Affecting Cell Shape and Attachment [40]

The extracellular matrix may influence the differentiated state of connective-tissue cells through physical as well as chemical effects. This has been shown in studies on cultured cartilage cells, or *chondrocytes.* Under appropriate culture conditions these cells will proliferate and maintain their differentiated character, continuing for many cell generations to synthesize large quantities of highly distinctive cartilage matrix, with which they surround themselves. However, under conditions where the cells are kept at relatively low density and remain as a monolayer on the culture dish, a transformation occurs. The cells lose the rounded shape that is typical of chondrocytes, flatten down on the substratum, and stop making cartilage matrix. In particular, they stop producing type II collagen—the type characteristic of cartilage—and instead start producing type I collagen—the type characteristic of fibroblasts. By the end of a month in culture, almost all the cartilage cells have switched their collagen gene expression and taken on the appearance of fibroblasts. The biochemical change must occur abruptly, since very few cells are observed to make both types of collagen simultaneously.

Several lines of evidence suggest that the biochemical change is induced at least in part by the change of cell shape and attachments. Cartilage cells that have made the transition to a fibroblastlike character, for example, can be gently detached from the culture dish and transferred to a dish of agarose. By forming a gel around them, the agarose holds the cells suspended without any attachment to a substratum, forcing them to adopt a rounded shape. In these circumstances the cells promptly revert to the character of chondrocytes and start making type II collagen again. Cell shape and anchorage may control gene expression through intracellular signals generated at focal contacts, as we saw in Chapter 16.

For most types of cells, and especially for a connective-tissue cell, the opportunities for anchorage and attachment depend on the surrounding matrix, which is usually made by the cell itself. Thus a cell can create an environment that then acts back on the cell to reinforce its differentiated state. Furthermore, the extracellular matrix that a cell secretes forms part of the environment for its neighbors as well as for the cell itself and thus tends to make neighboring cells differentiate in the same way. A group of chondrocytes forming a nodule of cartilage, for example, either in the developing body or in a culture dish, can be seen to enlarge by the conversion of neighboring fibroblasts into chondrocytes.

Different Signaling Molecules Act Sequentially to Regulate Production of Fat Cells [41]

Fat cells, or **adipocytes,** are also thought to develop from fibroblastlike cells, both during normal mammalian development and in various pathological circumstances—for example, in muscular dystrophy, where the muscle cells die and are gradually replaced by fatty connective tissue. Fat cell differentiation begins with the production of specific enzymes, followed by the accumulation of fat droplets, which then coalesce and enlarge until the cell is hugely distended, with only a thin rim of cytoplasm around the mass of lipid (Figure 22–42).

The process can be studied in culture (using fibroblast cell lines such as mouse 3T3 cells) so that the factors that influence it can be analyzed. It was initially found that the development of fat cells in culture required the presence of fetal calf serum, a common additive to culture media. The crucial factor in the serum that triggers fat cell differentiation was later identified as *growth hormone*—a protein normally secreted into the bloodstream by the pituitary gland. There is evidence that growth hormone stimulates chondrocyte as well as fat cell differentiation and that it acts in this way *in vivo* as well as *in vitro*. But growth hormone is not the only secreted signaling molecule that regulates fat cell development. Fat cell precursors that have been stimulated by growth hormone become sensitive to *IGF-1 (insulinlike growth factor-1)*, which stimulates the proliferation of the differentiating fat cells.

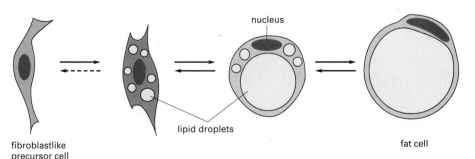

nucleus

lipid droplets

fibroblastlike precursor cell

fat cell

Figure 22–42 Development of a fat cell. A fibroblastlike precursor cell is converted into a mature fat cell by the accumulation and coalescence of lipid droplets. The process is at least partially reversible, as indicated by the arrows. The cells in the early and intermediate stages can divide, but the mature fat cell cannot.

The differentiation of fat cells, like that of chondrocytes, is also influenced by factors that affect cell shape and anchorage. The differentiation of 3T3 cells into fat cells is inhibited if the cells are allowed to flatten onto a culture dish coated with fibronectin, to which they adhere strongly. This inhibition is reversed, however, by treatment with the drug cytochalasin, which disrupts actin filaments and causes the cells to round up.

All of these experiments on connective-tissue cells illustrate a recurrent theme: differentiation is regulated by a combination of soluble signals and contacts with the extracellular matrix. The effects of each of these factors depend on the character of the responding cell, and this in turn depends on the cell's developmental history.

Bone Is Continually Remodeled by the Cells Within It [42]

Bone is a very dense, specialized form of connective tissue. Like reinforced concrete, bone matrix is predominantly a mixture of tough fibers (type I collagen fibrils), which resist pulling forces, and solid particles (calcium phosphate as hydroxyapatite crystals), which resist compression. The volume occupied by the collagen is nearly equal to that occupied by the calcium phosphate. The collagen fibrils in adult bone are arranged in regular plywoodlike layers, with the fibrils in each layer lying parallel to one another but at right angles to the fibrils in the layers on either side.

For all its rigidity, bone is by no means a permanent and immutable tissue. Throughout its hard extracellular matrix are channels and cavities occupied by living cells, which account for about 15% of the weight of compact bone. These cells are engaged in an unceasing process of remodeling: one class of cells demolishes old bone matrix while another deposits new bone matrix. This mechanism provides for continuous turnover and replacement of the matrix in the interior of the bone.

Bone can grow only by *apposition*—that is, by the laying down of additional matrix and cells on the free surfaces of the hard tissue. This process must occur in the embryo in coordination with the growth of other tissues in such a way that the pattern of the body can be scaled up without its proportions being radically disturbed. For most of the skeleton, and in particular for the long bones of the limbs and trunk, the coordinated growth is achieved by a complex strategy. A set of minute "scale models" of the bones are first formed out of cartilage. Each scale model grows, and as new cartilage is formed, the older cartilage is replaced by bone. Cartilage growth and erosion and bone deposition are so ingeniously coordinated during development that the adult bone, though it may be half a meter long, is almost the same shape as the initial cartilaginous model, which was no more than a few millimeters long.

Osteoblasts Secrete Bone Matrix, While Osteoclasts Erode It [40, 42, 43]

Cartilage is a simple tissue, consisting of cells of a single type—chondrocytes—embedded in a more or less uniform matrix. The cartilage matrix is deformable, and the tissue grows by expanding as the chondrocytes divide and secrete more matrix (Figure 22–43). Bone is more complex. The bone matrix is secreted by

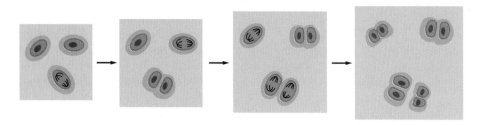

Figure 22–43 **The growth of cartilage.** The tissue expands as the chondrocytes divide and make more matrix. The freshly synthesized matrix with which each cell surrounds itself is shaded *dark green*. Cartilage may also grow by recruiting fibroblasts from the surrounding tissue and converting them into chondrocytes.

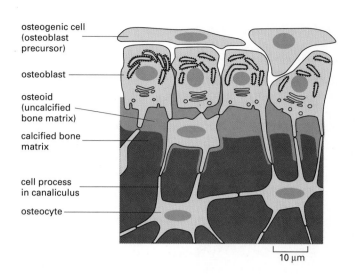

osteogenic cell (osteoblast precursor)

osteoblast

osteoid (uncalcified bone matrix)

calcified bone matrix

cell process in canaliculus

osteocyte

10 μm

Figure 22–44 Deposition of bone matrix by osteoblasts. Osteoblasts lining the surface of bone secrete the organic matrix of bone (osteoid) and are converted into osteocytes as they become embedded in this matrix. The matrix calcifies soon after it has been deposited. The osteoblasts themselves are thought to derive from osteogenic stem cells that are closely related to fibroblasts.

osteoblasts that lie at the surface of the existing matrix and deposit fresh layers of bone onto it. Some of the osteoblasts remain free at the surface, while others gradually become embedded in their own secretion. This freshly formed material (consisting chiefly of type I collagen) is called *osteoid*. It is rapidly converted into hard bone matrix by the deposition of calcium phosphate crystals in it. Once imprisoned in hard matrix, the original bone-forming cell, now called an **osteocyte,** has no opportunity to divide, although it continues to secrete further matrix in small quantities around itself. The osteocyte, like the chondrocyte, occupies a small cavity, or *lacuna*, in the matrix, but unlike the chondrocyte it is not isolated from its fellows. Tiny channels, or *canaliculi,* radiate from each lacuna and contain cell processes from the resident osteocyte, enabling it to form gap junctions with adjacent osteocytes (Figure 22–44). Although the networks of osteocytes do not themselves secrete or erode substantial quantities of matrix, they probably play a part in controlling the activities of the cells that do.

While bone matrix is deposited by osteoblasts, it is eroded by **osteoclasts** (Figure 22–45). These large multinucleated cells originate, like macrophages, from hemopoietic stem cells in the bone marrow. The precursor cells are released as

lysosomes

multiple nuclei

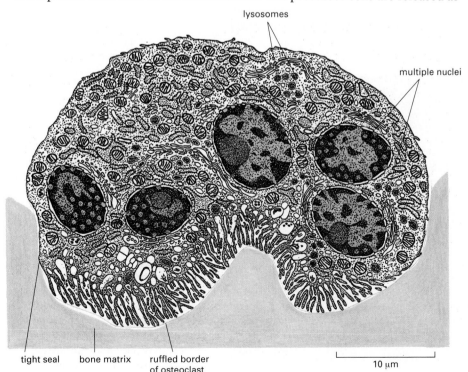

tight seal   bone matrix   ruffled border of osteoclast

10 μm

Figure 22–45 An osteoclast shown in cross-section. This giant, multinucleated cell erodes bone matrix. The "ruffled border" is a site of secretion of acids (to dissolve the bone minerals) and hydrolases (to digest the organic components of the matrix). Osteoclasts vary in shape, are motile, and often send out processes to resorb bone at multiple sites. They develop from monocytes and can be viewed as specialized macrophages. (From R.V. Krstić, Ultrastructure of the Mammalian Cell: An Atlas. Berlin: Springer, 1979.)

Fibroblasts and Their Transformations: The Connective-Tissue Cell Family

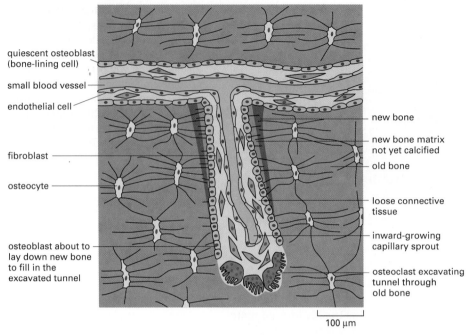

quiescent osteoblast
(bone-lining cell)

small blood vessel

endothelial cell

fibroblast

osteocyte

osteoblast about to
lay down new bone
to fill in the
excavated tunnel

new bone

new bone matrix
not yet calcified

old bone

loose connective
tissue

inward-growing
capillary sprout

osteoclast excavating
tunnel through
old bone

100 μm

Figure 22–46 The remodeling of compact bone. Osteoclasts acting together in a small group excavate a tunnel through the old bone, advancing at a rate of about 50 μm per day. Osteoblasts enter the tunnel behind them, line its walls, and begin to form new bone, depositing layers of matrix at a rate of 1 or 2 μm per day. At the same time a capillary sprouts down the center of the tunnel. The tunnel will eventually become filled with concentric layers of new bone, with only a narrow central canal remaining. Each such canal, besides providing a route of access for osteoclasts and osteoblasts, contains one or more blood vessels bringing the nutrients the bone cells must have to survive. Typically, about 5–10% of the bone in a healthy adult mammal is replaced in this way each year. (After Z.F.G. Jaworski, B. Duck, and G. Sekaly, *J. Anat.* 133:397–405, 1981.)

monocytes into the bloodstream and collect at sites of bone resorption, where they fuse to form the multinucleated osteoclasts, which cling to surfaces of the bone matrix and eat it away. Osteoclasts are capable of tunneling deep into the substance of compact bone, forming cavities that are then invaded by other cells. A blood capillary grows down the center of such a tunnel, and the walls of the tunnel become lined with a layer of osteoblasts (Figure 22–46). To produce the plywoodlike structure of compact bone, these osteoblasts lay down concentric layers of new bone, which gradually fill the cavity, leaving only a narrow canal surrounding the new blood vessel. Many of the osteoblasts become trapped in the bone matrix and survive as concentric rings of osteocytes. At the same time as some tunnels are filling up with bone, others are being bored by osteoclasts, cutting through older concentric systems. The consequences of this perpetual remodeling are beautifully displayed in the layered patterns of matrix observed in compact bone (Figure 22–47).

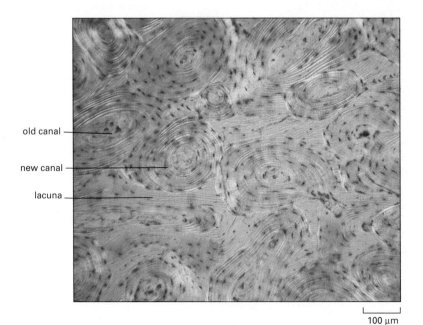

old canal

new canal

lacuna

100 μm

Figure 22–47 Transverse section through a compact outer portion of a long bone. The micrograph shows the outlines of tunnels formed by osteoclasts and then filled in by osteoblasts during successive rounds of bone remodeling. The section has been prepared by grinding; the hard matrix has been preserved but not the cells. Lacunae and canaliculi that were occupied by osteocytes are clearly visible, however. The alternating bright and dark concentric rings correspond to an alternating orientation of the collagen fibers in the successive layers of bone matrix laid down by the osteoblasts that lined the wall of the canal during life. (This pattern is revealed here by viewing the specimen between partly crossed Polaroid filters.) Note how older systems of concentric layers of bone have been partly cut through and replaced by newer systems.

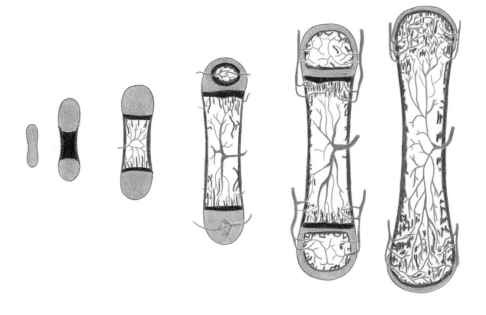

Figure 22–48 **The development of a long bone.** Long bones, such as the femur or the humerus, develop from a miniature cartilage model. Uncalcified cartilage is shown in *green*, calcified cartilage in *black*, bone in *brown*, and blood vessels in *red*. The cartilage is not converted to bone but is gradually replaced by it through the action of osteoclasts and osteoblasts, which invade the cartilage in association with blood vessels. Osteoclasts erode cartilage and bone matrix, while osteoblasts secrete bone matrix. The process of ossification begins in the embryo and is not completed until the end of puberty. The resulting bone consists of a thick-walled hollow cylinder of compact bone enclosing a large central cavity occupied by the bone marrow. Note that not all bones develop in this way. The *membrane bones* of the skull, for example, are formed directly as bony plates, not from a prior cartilage model. (Adapted from D.W. Fawcett, A Textbook of Histology, 11th ed. Philadelphia: Saunders, 1986.)

There are many unsolved problems about these processes. Bones, for example, have a remarkable ability to adapt to the load imposed on them by remodeling their structure, and this implies that the deposition and erosion of the matrix are somehow controlled by local mechanical stresses. We do not understand the mechanisms that determine whether matrix will be deposited by osteoblasts or eroded by osteoclasts at a given bone surface, but it seems likely that an important part is played by growth factors that are made by the bone cells, trapped in the matrix, and released, perhaps, when the matrix is degraded or suitably stressed.

During Development, Cartilage Is Eroded by Osteoclasts to Make Way for Bone [44]

The replacement of cartilage by bone in the course of development is also thought to depend on the activities of osteoclasts. As the cartilage matures, its cells in certain regions become greatly enlarged at the expense of the surrounding matrix, and the matrix itself becomes mineralized, like bone, by deposition of calcium phosphate crystals. The swollen chondrocytes die, leaving large empty cavities. Osteoclasts and blood vessels invade the cavities and erode the residual cartilage matrix, while osteoblasts following in their wake begin to deposit bone matrix. The only surviving remnant of cartilage in the adult long bone is a thin layer that forms a smooth covering on the bone surfaces at joints, where one bone articulates with another (Figure 22–48).

Some cells capable of forming new cartilage persist, however, in the connective tissue that surrounds a bone. If the bone is broken, the cells in the neighborhood of the fracture will carry out a repair by a rough-and-ready recapitulation of the original embryonic process, in which cartilage is first laid down to bridge the gap and is then replaced by bone.

The Structure of the Body Is Stabilized by Its Connective-Tissue Framework and by the Selective Cohesion of Cells [45]

A bone, like the body as a whole, is a dynamic system, maintaining its structure through a balance between the opposed activities of a variety of specialized cells. Any dynamic system poses a problem of stability, and this leads us to a general question about the maintenance of body structure. We have seen how cells in various types of tissues maintain their differentiated state, how new cells are

Fibroblasts and Their Transformations: The Connective-Tissue Cell Family 1185

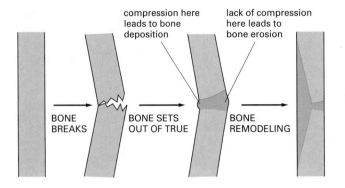

compression here leads to bone deposition

lack of compression here leads to bone erosion

BONE BREAKS

BONE SETS OUT OF TRUE

BONE REMODELING

Figure 22–49 Remodeling of a long bone in the leg after a fracture that has healed out of true. The deformity in the recently healed bone exposes it to abnormal stresses. Where the compressive forces are increased, the rate of bone deposition is increased relative to the rate of erosion; where the forces are decreased, the rate of deposition is decreased relative to the rate of erosion. In this way the bone is gradually remodeled back to its normal form.

produced in a controlled fashion to replace those that are lost, and how the extracellular matrix is remodeled and renewed. But why do the different types of cells not become progressively jumbled and misplaced? Why does the whole structure not sag, warp, or otherwise change its proportions as new parts are substituted for old?

To some extent, of course, the body does sag and warp with the passage of time—that is a part of aging. But it does so remarkably little. The skeleton, despite constant remodeling, provides a rigid framework whose dimensions scarcely change. This is partly because the parts of a bone are renewed not all at once but little by little, rather like a building whose bricks are replaced one at a time. Besides such conservatism in the mode of renewal, active homeostatic mechanisms are at work. Thus small departures of a bone from its normal shape set up altered patterns of stresses, which regulate bone remodeling in such a way as to restore the bone to its normal shape (Figure 22–49).

The growth and renewal of many of the soft parts of the body are also homeostatically controlled so that each component is adjusted to fit its niche. The epidermis spreads to keep the surface of the body covered; if it is damaged, the cells grow back to cover the lesion, halting their migration when that end is achieved; connective tissue grows to just the extent necessary to fill the gap created by a wound; and so on. In all this, something more than mere control of cell numbers is required. The various types of differentiated cells must be maintained not only in the correct relative quantities but also in the correct relative positions. Tissue turnover necessarily involves cell movements. Somehow those movements must be limited; the cells must be subject to territorial restraints.

These restraints are of various kinds. Glands and other masses of specialized cells are often contained within tough capsules of connective tissue. Many types of cells die if they find themselves outside their normal environment, deprived of specific growth factors on which their survival depends. Perhaps the most important strategy for keeping the different cells in their places is the strategy of selective cell-cell adhesion: cells of the same type tend to stick together, either in solid masses, such as smooth muscle, or in epithelial sheets, such as the lining of the gut. As described in Chapter 19, this mechanism, for example, enables dissociated epidermal cells to reassociate spontaneously to form a correctly structured epithelium. And on a larger scale, stable epithelial sheets of cells serve to divide the body into compartments, thereby keeping other cells properly segregated and confined to their correct territories.

Clearly, the checks and balances that preserve the structure of the body and the organization of its cells in the face of continual turnover and renewal are intricate and subtle. The importance of these controls is all too clearly evident when they fail, as we see when we come to the topic of cancer in the final chapter of this book.

Summary

The family of connective-tissue cells includes fibroblasts, cartilage cells, bone cells, fat cells, and smooth muscle cells. Fibroblasts seem to be able to transform into any of the other members of the family—in some cases reversibly—although it is not clear whether this is a property of a single type of fibroblast that is pluripotent or of a mixture of distinct types of fibroblasts with more restricted potentials. These transformations of connective-tissue cell type are regulated by the composition of the surrounding extracellular matrix, by cell shape, and by hormones and growth factors.

Cartilage and bone both consist of cells embedded in a solid matrix. Cartilage has a deformable matrix and can grow by swelling, whereas bone is rigid and can grow only by accretion at its surfaces. Bone is, nonetheless, subject to perpetual remodeling through the combined action of osteoclasts, which erode matrix, and osteoblasts, which secrete it. Some osteoblasts become trapped in the matrix as osteocytes and play a part in regulating the turnover of bone matrix. Most long bones develop from miniature cartilage "models," which, as they grow, serve as templates for the deposition of bone by the combined action of osteoblasts and osteoclasts. Similarly, in the repair of a bone fracture in the adult, the gap is first bridged by cartilage, which is later replaced by bone. Although bone, like most other tissues, is subject to continual turnover, this dynamic process is regulated so that the global structure is preserved. In this way, and through other mechanisms such as selective cell-cell adhesion, the organization of the body is stably maintained even though most of its components are continually being replaced.

Appendix

Cells of the Adult Human Body: A Catalogue

How many distinct cell types are there in an adult human being? In other words, how many normal adult ways are there of expressing the human genome? A large textbook of histology will mention about 200 cell types that qualify for individual names. These traditional names are not, like the names of colors, labels for parts of a continuum that has been subdivided arbitrarily: they represent, for the most part, discrete and distinctly different categories. Within a given category there is often some variation—the skeletal muscle fibers that move the eyeball are small, while those that move the leg are big; auditory hair cells in different parts of the ear may be tuned to different frequencies of sound; and so on. But there is no continuum of adult cell types intermediate in character between, say, the muscle cell and the auditory hair cell.

The traditional histological classification is based on the shape and structure of the cell as seen in the microscope and on its chemical nature as assessed very crudely from its affinities for various stains. Subtler methods reveal new subdivisions within the traditional classification. Thus modern immunology has shown that the old category of "lymphocyte" includes more than 10 quite distinct cell types. Similarly, pharmacological and physiological tests reveal that there are many varieties of smooth muscle cell—those in the wall of the uterus, for example, are highly sensitive to estrogen, and in the later stages of pregnancy to oxytocin, while those in the wall of the gut are not. Another major type of diversity is revealed by embryological experiments of the sort discussed in Chapter 21. These show that, in many cases, apparently similar cells from different regions of the body are nonequivalent, that is, they are inherently different in their developmental capacities and in their effects on other cells. Thus, within categories such as "fibroblast" there are probably many distinct cell types, different chemically in ways that are not easy to perceive directly.

For these reasons any classification of the cell types in the body must be somewhat arbitrary with respect to the fineness of its subdivisions. Here, we list only the adult human cell types that a histology text would recognize to be different, grouped into families roughly according to function. We have not attempted to subdivide the class of neurons of the central nervous system. Also, where a single cell type such as the keratinocyte is conventionally given a succession of different names as it matures, we give only two entries—one for the differentiating cell and one for the stem cell. With these serious provisos, the 210 varieties of cells in the catalogue represent a more or less exhaustive list of the distinctive ways in which a given mammalian genome can be expressed in the phenotype of a normal cell of the adult body.

Keratinizing Epithelial Cells

keratinocyte of epidermis (= differentiating epidermal cell)

basal cell of epidermis (stem cell)

keratinocyte of fingernails and toenails

basal cell of nail bed (stem cell)

hair shaft cells
  medullary
  cortical
  cuticular

hair-root sheath cells
  cuticular
  of Huxley's layer
  of Henle's layer
  external

hair matrix cell (stem cell)

Cells of Wet Stratified Barrier Epithelia

surface epithelial cell of stratified squamous epithelium of cornea, tongue, oral cavity, esophagus, anal canal, distal urethra, vagina

basal cell of these epithelia (stem cell)

cell of urinary epithelium (lining bladder and urinary ducts)

Epithelial Cells Specialized for Exocrine Secretion

cells of salivary gland
  mucous cell (secretion rich in polysaccharide)
  serous cell (secretion rich in glycoprotein enzymes)

cell of von Ebner's gland in tongue (secretion to wash over taste buds)

cell of mammary gland, secreting milk

cell of lacrimal gland, secreting tears

cell of ceruminous gland of ear, secreting wax

cell of eccrine sweat gland, secreting glycoproteins (dark cell)

cell of eccrine sweat gland, secreting small molecules (clear cell)

cell of apocrine sweat gland (odoriferous secretion, sex-hormone sensitive)

cell of gland of Moll in eyelid (specialized sweat gland)

cell of sebaceous gland, secreting lipid-rich sebum

cell of Bowman's gland in nose (secretion to wash over olfactory epithelium)

cell of Brunner's gland in duodenum, secreting alkaline solution of mucus and enzymes

cell of seminal vesicle, secreting components of seminal fluid, including fructose (as fuel for swimming sperm)

cell of prostate gland, secreting other components of seminal fluid

cell of bulbourethral gland, secreting mucus

cell of Bartholin's gland, secreting vaginal lubricant

cell of gland of Littré, secreting mucus

cell of endometrium of uterus, secreting mainly carbohydrates

isolated goblet cell of respiratory and digestive tracts, secreting mucus

mucous cell of lining of stomach

zymogenic cell of gastric gland, secreting pepsinogen

oxyntic cell of gastric gland, secreting HCl

acinar cell of pancreas, secreting digestive enzymes and bicarbonate

Paneth cell of small intestine, secreting lysozyme

type II pneumocyte of lung, secreting surfactant

Clara cell of lung (function unknown)

Cells Specialized for Secretion of Hormones

cells of anterior pituitary, secreting
  growth hormone
  follicle-stimulating hormone
  luteinizing hormone
  prolactin
  adrenocorticotropic hormone
  thyroid-stimulating hormone

cell of intermediate pituitary, secreting
  melanocyte-stimulating hormone

cells of posterior pituitary, secreting
  oxytocin
  vasopressin

cells of gut and respiratory tract, secreting
  serotonin
  endorphin
  somatostatin
  gastrin
  secretin
  cholecystokinin
  insulin
  glucagon
  bombesin

cells of thyroid gland, secreting
  thyroid hormone
  calcitonin

cells of parathyroid gland, secreting
  parathyroid hormone
  oxyphil cell (function unknown)

cells of adrenal gland, secreting
  epinephrine
  norepinephrine
  steroid hormones
  mineralocorticoids
  glucocorticoids

cells of gonads, secreting
  testosterone (Leydig cell of testis)
  estrogen (theca interna cell of ovarian follicle)
  progesterone (corpus luteum cell of ruptured ovarian follicle)

cells of juxtaglomerular apparatus of kidney
  juxtaglomerular cell (secreting renin)
  macula densa cell
  peripolar cell
  mesangial cell
  } (uncertain but probably related in function; possibly involved in secretion of erythropoietin)

Epithelial Absorptive Cells in Gut, Exocrine Glands, and Urogenital Tract

brush border cell of intestine (with microvilli)

striated duct cell of exocrine glands

gall bladder epithelial cell

brush border cell of proximal tubule of kidney

distal tubule cell of kidney

nonciliated cell of ductulus efferens

epididymal principal cell

epididymal basal cell

Cells Specialized for Metabolism and Storage

hepatocyte (liver cell)

fat cells
  white fat
  brown fat
  lipocyte of liver

Epithelial Cells Serving Primarily a Barrier Function, Lining the Lung, Gut, Exocrine Glands, and Urogenital Tract

type I pneumocyte (lining air space of lung)

pancreatic duct cell (centroacinar cell)

nonstriated duct cell of sweat gland, salivary gland, mammary gland, etc. (various)

parietal cell of kidney glomerulus

podocyte of kidney glomerulus

cell of thin segment of loop of Henle (in kidney)

collecting duct cell (in kidney)

duct cell of seminal vesicle, prostate gland, etc. (various)

Epithelial Cells Lining Closed Internal Body Cavities

vascular endothelial cells of blood vessels and lymphatics

fenestrated

continuous

splenic

synovial cell (lining joint cavities, secreting largely hyaluronic acid)

serosal cell (lining peritoneal, pleural, and pericardial cavities)

squamous cell lining perilymphatic space of ear

cells lining endolymphatic space of ear

squamous cell

columnar cells of endolymphatic sac

with microvilli

without microvilli

"dark" cell

vestibular membrane cell

stria vascularis basal cell

stria vascularis marginal cell

cell of Claudius

cell of Boettcher

choroid plexus cell (secreting cerebrospinal fluid)

squamous cell of pia-arachnoid

cells of ciliary epithelium of eye

pigmented

nonpigmented

corneal "endothelial" cell

Ciliated Cells with Propulsive Function

of respiratory tract

of oviduct and of endometrium of uterus (in female)

of rete testis and ductulus efferens (in male)

of central nervous system (ependymal cell lining brain cavities)

Cells Specialized for Secretion of Extracellular Matrix

epithelial

ameloblast (secreting enamel of tooth)

planum semilunatum cell of vestibular apparatus of ear (secreting proteoglycan)

interdental cell of organ of Corti (secreting tectorial "membrane" covering hair cells of organ of Corti)

nonepithelial (connective tissue)

fibroblasts (various—of loose connective tissue, of cornea, of tendon, of reticular tissue of bone marrow, etc.)

pericyte of blood capillary

nucleus pulposus cell of intervertebral disc

cementoblast/cementocyte (secreting bonelike cementum of root of tooth)

odontoblast/odontocyte (secreting dentin of tooth)

chondrocytes

of hyaline cartilage

of fibrocartilage

of elastic cartilage

osteoblast/osteocyte

osteoprogenitor cell (stem cell of osteoblasts)

hyalocyte of vitreous body of eye

stellate cell of perilymphatic space of ear

Contractile Cells

skeletal muscle cells

red (slow)

white (fast)

intermediate

muscle spindle—nuclear bag

muscle spindle—nuclear chain

satellite cell (stem cell)

heart muscle cells

ordinary

nodal

Purkinje fiber

smooth muscle cells (various)

myoepithelial cells

of iris

of exocrine glands

Cells of Blood and Immune System

red blood cell

megakaryocyte

macrophages and related cells

monocyte

connective-tissue macrophage (various)

Langerhans cell (in epidermis)

osteoclast (in bone)

dendritic cell (in lymphoid tissues)

microglial cell (in central nervous system)

neutrophil

eosinophil

basophil

mast cell

T lymphocyte

helper T cell

suppressor T cell

killer T cell

B lymphocyte

IgM

IgG

IgA

IgE

killer cell

stem cells and committed progenitors for the blood and immune system (various)

Sensory Transducers

photoreceptors

rod

cones

blue sensitive

green sensitive

red sensitive

hearing

inner hair cell of organ of Corti

outer hair cell of organ of Corti

acceleration and gravity

type I hair cell of vestibular apparatus of ear

type II hair cell of vestibular apparatus of ear

taste

type II taste bud cell

smell

olfactory neuron

basal cell of olfactory epithelium (stem cell for olfactory neurons)

blood pH

carotid body cell

type I

type II

touch

Merkel cell of epidermis

primary sensory neurons specialized for touch (various)

temperature

primary sensory neurons specialized for temperature

cold sensitive

heat sensitive

pain

primary sensory neurons specialized for pain (various)

configurations and forces in musculoskeletal system

proprioceptive primary sensory neurons (various)

Autonomic Neurons

cholinergic (various)

adrenergic (various)

peptidergic (various)

Supporting Cells of Sense Organs and of Peripheral Neurons

supporting cells of organ of Corti

inner pillar cell

outer pillar cell

inner phalangeal cell

outer phalangeal cell

border cell

Hensen cell

supporting cell of vestibular apparatus

supporting cell of taste bud (type I taste bud cell)

supporting cell of olfactory epithelium

Schwann cell

satellite cell (encapsulating peripheral nerve cell bodies)

enteric glial cell

Neurons and Glial Cells of Central Nervous System

neurons (huge variety of types—still poorly classified)

glial cells

astrocyte (various)

oligodendrocyte

Lens Cells

anterior lens epithelial cell

lens fiber (crystallin-containing cell)

Pigment Cells

melanocyte

retinal pigmented epithelial cell

Germ Cells

oogonium/oocyte

spermatocyte

spermatogonium (stem cell for spermatocyte)

Nurse Cells

ovarian follicle cell

Sertoli cell (in testis)

thymus epithelial cell

References

General

Burkitt, H.G.; Young, B.; Heath, J.W. Wheater's Functional Histology, 3rd ed. Edinburgh: Churchill Livingstone, 1993.

Clark, W.E. Le Gros. The Tissues of the Body, 6th ed. Oxford, UK: Clarendon Press, 1971.

Cormack, D.H. Essential Histology. Philadelphia: Lippincott, 1993.

Fawcett, D.W. (Bloom and Fawcett) A Textbook of Histology, 11th ed. Philadelphia: Saunders, 1986.

Goss, R.J. The Physiology of Growth. New York: Academic Press, 1978.

Krstić, R.V. Illustrated Encyclopedia of Human Histology. Berlin: Springer-Verlag, 1984.

Weiss, L., ed. Cell and Tissue Biology: A Textbook of Histology, 6th ed. Baltimore: Urban and Schwartzenberg, 1988.

Cited

1. Clark, W.E. Le Gros. The Tissues of the Body, 6th ed. Oxford, UK: Clarendon Press, 1971.

   Montagna, W. The skin. *Sci. Am.* 212(2):56–66, 1965.

2. Cahn, R.D.; Cahn, M.B. Heritability of cellular differentiation: clonal growth and expression of differentiation in retinal pigment cells *in vitro*. *Proc. Natl. Acad. Sci. USA* 55:106–114, 1966.

   Coon, H.G. Clonal stability and phenotypic expression of chick cartilage cells *in vitro*. *Proc. Natl. Acad. Sci. USA* 55:66–73, 1966.

   Eguchi, G.; Kodama, R. Transdifferentiation. *Curr. Opin. Cell Biol.* 5:1023–1028, 1993.

   Watt, F.M. Cell culture models of differentiation. *FASEB J.* 5:287–294, 1991.

   Yaffe, D. Retention of differentiation potentialities during prolonged cultivation of myogenic cells. *Proc. Natl. Acad. Sci. USA* 61:477–483, 1968.

3. Anderson, J.E. The effect of steroid hormones on gene transcription. In Biological Regulation and Development (R.F. Goldberger, K. Yamamoto, eds.), Vol. 3B, pp. 169–212. New York: Plenum Press, 1983.

   Hay, E.D. Extracellular matrix alters epithelial differentiation. *Curr. Opin. Cell Biol.* 5:1029–1035, 1993.

   Okada, T.S.; Kondoh, H., eds. Commitment and Instability in Cell Differentiation. *Curr. Top. Dev. Biol.* 20, 1986.

4. Goss, R.J. The Physiology of Growth. New York: Academic Press, 1978.

   Richardson, G. Hair-cell regeneration: keep the noise down. *Curr. Biol.* 3:759–762, 1993.

5. Clayton, R.M. Divergence and convergence in lens cell differentiation: regulation of the formation and specific content of lens fibre cells. In Stem Cells and Tissue Homeostasis (B. Lord, C. Potten, R. Cole, eds.), pp. 115–138. Cambridge, UK: Cambridge University Press, 1978.

   Goss, R.J. The Physiology of Growth, pp. 210–225. New York: Academic Press, 1978.

   Maisel, H., ed. The Ocular Lens: Structure, Function, and Pathology. New York: Dekker, 1985.

   Wistow, G.J.; Piatigorsky, J. Lens crystallins: the evolution and expression of proteins for a highly specialized tissue. *Annu. Rev. Biochem.* 57:479–504, 1988.

6. Fawcett, D.W. (Bloom and Fawcett) A Textbook of Histology, 11th ed. Philadelphia: Saunders, 1986.

   Gevers, W. Protein metabolism in the heart. *J. Mol. Cell. Cardiol.* 16:3–32, 1984.

   Young, R.W. Visual cells. *Sci. Am.* 223(4):80–91, 1970.

7. Leblond, C.P. The life history of cells in renewing systems. *Am. J. Anat.* 160:114–158, 1981.

   Wright, N.A.; Alison, M.R. Biology of Epithelial Cell Populations, Vols. 1–3. Oxford, UK: Oxford University Press, 1984.

8. Fawcett, D.W. (Bloom and Fawcett) A Textbook of Histology, 11th ed., pp. 679–715. Philadelphia: Saunders, 1986.

   Louvard, D.; Kedinger, M.; Hauri, H.P. The differentiating intestinal epithelial cell: establishment and maintenance of functions through interactions between cellular structures. *Annu. Rev. Cell Biol.* 8:157–195, 1992.

   Moog, F. The lining of the small intestine. *Sci. Am.* 245(5):154–176, 1981.

9. Bursch, W.; Taper, H.S.; Lauer, B.; Schulte-Hermann, R. Quantitative histological and histochemical studies on the occurrence and stages of controlled cell death (apoptosis) during regression of rat liver hyperplasia. *Virchows Arch. B Cell. Pathol.* 50:153–166, 1985

   Fausto, N. Hepatocyte differentiation and liver progenitor cells. *Curr. Opin. Cell Biol.* 2:1036–1042, 1990.

   Furlong, R.A. The biology of hepatocyte growth factor/scatter factor. *Bioessays* 14:613–617, 1992.

   Goss, R.J. The Physiology of Growth, pp. 251–266. New York: Academic Press, 1978.

   Nakamura, T., et al. Molecular cloning and expression of human hepatocyte growth factor. *Nature* 342:440–443, 1989.

10. McGee, J.O'D.; Isaacson, P.G.; Wright, N.A., eds. Oxford Textbook of Pathology, Vol. 1, pp. 365–389. Oxford, UK: Oxford University Press, 1992.

    Robbins, S.L.; Cotran, R.S.; Kumar, V. Pathologic Basis of Disease, 4th ed. Philadelphia: Saunders, 1989.

11. Campbell, J.H.; Campbell, G.R. Endothelial cell influences on vascular smooth muscle phenotype. *Annu. Rev. Physiol.* 48:295–306, 1986.

    Development of the Vascular System. *Ciba Found. Symp.* 100. London: Pitman, 1983.

    Fawcett, D.W. (Bloom and Fawcett) A Textbook of Histology, 11th ed., pp. 367–405. Philadelphia: Saunders, 1986.

    Ryan, U.S., ed. Endothelial Cells, Vols. 1–3. Boca Raton, FL: CRC Press, 1988.

12. Goss, R.J. The Physiology of Growth, pp. 120–137. New York: Academic Press, 1978.

    Hobson, B.; Denekamp, J. Endothelial proliferation in tumours and normal tissues: continuous labelling studies. *Br. J. Cancer* 49:405–413, 1984.

13. Folkman, J. What is the evidence that tumors are angiogenesis dependent? *J. Natl. Cancer Inst.* 82:4–6, 1990.

    Folkman, J. The vascularization of tumors. *Sci. Am.* 234(5):58–73, 1976.

    Folkman, J.; Haudenschild, C. Angiogenesis *in vitro*. *Nature* 288:551–556, 1980.

    Grant, D.S., et al. Two different laminin domains mediate the differentiation of human endothelial cells into capillary-like structures *in vitro. Cell* 58:933–943, 1989.

Madri, J.A.; Pratt, B.M. Endothelial cell-matrix interactions: *in vitro* models of angiogenesis. *J. Histochem. Cytochem.* 34:85–91, 1986.

14. Kalebic, T.; Garbisa, S.; Glaser, B.; Liotta, L.A. Basement membrane collagen: degradation by migrating endothelial cells. *Science* 221:281–283, 1983.

Klagsbrun, M.; D'Amore, P.A. Regulators of angiogenesis. *Annu. Rev. Physiol.* 53:217–239, 1991.

Klagsbrun, M.; Soker, S. VEGF/VPF: the angiogenesis factor found? *Curr. Biol.* 3:699–702, 1993.

Shweiki, D.; Itin, A.; Soffer, D.; Keshet, E. Vascular endothelial growth factor induced by hypoxia may mediate hypoxia-initiated angiogenesis. *Nature* 359:843–845, 1992.

15. Cairnie, A.B.; Lala, P.K.; Osmond, D.G., eds. Stem Cells of Renewing Cell Populations. New York: Academic Press, 1976.

Cheng, H.; Leblond, C.P. Origin, differentiation, and renewal of the four main epithelial cell types in the mouse small intestine. V. Unitarian theory of the origin of the four epithelial cell types. *Am. J. Anat.* 141:537–562, 1974.

16. Graziadei, P.P.C.; Monti Graziadei, G.A. Continuous nerve cell renewal in the olfactory system. In Handbook of Sensory Physiology, Vol. IX: Development of Sensory Systems (M. Jacobson, ed.), pp. 55–82. New York: Springer-Verlag, 1978.

17. Bereiter-Hahn, J.; Matoltsy, A.G.; Richards, K.S., eds. Biology of the Integument. Vol. 2: Vertebrates. New York: Springer, 1986.

MacKenzie, I.C. Ordered structure of the stratum corneum of mammalian skin. *Nature* 222:881–882, 1969.

Sengel, P. Morphogenesis of skin. Cambridge, UK: Cambridge University Press, 1976.

Stenn, K.S. The skin. In Cell and Tissue Biology: A Textbook of Histology (L.Weiss, ed.), 6th ed., pp. 539–572. Baltimore: Urban and Schwartzenberg, 1988.

18. Fuchs, E. Epidermal differentiation. *Curr. Opin. Cell Biol.* 2:1028–1035, 1990.

Fuchs, E.; Green, H. Changes in keratin gene expression during terminal differentiation of the keratinocyte. *Cell* 19:1033–1042, 1980.

Green, H. The keratinocyte as differentiated cell type. *Harvey Lect.* 74:101–139, 1979.

19. Barrandon, Y.; Green, H. Three clonal types of keratinocyte with different capacities for multiplication. *Proc. Natl. Acad. Sci. USA* 84:2302–2306, 1987.

Green, H. Cultured cells for the treatment of disease. *Sci. Am.* 265(5):96–102, 1991.

Green, H. Terminal differentiation of cultured human epidermal cells. *Cell* 11:405–415, 1977.

Jones, P.H.; Watt, F.M. Separation of human epidermal stem cells from transit amplifying cells on the basis of differences in integrin function and expression. *Cell* 73:713–724, 1993.

Martin, P.; Hopkinson-Woolley, J.; McCluskey, J. Growth factors and cutaneous wound repair. *Prog. Growth Factor Res.* 4:25–44, 1992.

Watt, F.M. Selective migration of terminally differentiating cells from the basal layer of cultured human epidermis. *J. Cell Biol.* 98:16–21, 1984.

20. Barrandon, Y.; Green, H. Cell migration is essential for sustained growth of keratinocyte colonies: the roles of transforming growth factor-alpha and epidermal growth factor. *Cell* 50:1131–1137, 1987.

Elder, J.T., et al. Overexpression of transforming growth factor alpha in psoriatic epidermis. *Science* 243:811–814, 1989.

Read, J.; Watt, F.M. A model for *in vitro* studies of epidermal homeostasis: proliferation and involucrin synthesis by cultured human keratinocytes during recovery after stripping off the suprabasal layers. *J. Invest. Dermatol.* 90:739–743, 1988.

21. Fawcett, D.W. (Bloom and Fawcett) A Textbook of Histology, 11th ed., pp. 568–576, 901–912. Philadelphia: Saunders, 1986.

Neville, M.C.: Neifert, M.R. Lactation: Physiology, Nutrition, and Breast-Feeding. New York: Plenum Press, 1983.

Patton, S. Milk. *Sci Am.* 221(1):58–68, 1969.

Richards, R.C.; Benson, G.K. Ultrastructural changes accompanying involution of the mammary gland in the albino rat. *J. Endocrinol.* 51:127–135, 1971.

Talhouk, R.S.; Bissell, M.J.; Werb, Z. Coordinated expression of extracellular matrix-degrading proteinases and their inhibitors regulates mammary epithelial function during involution. *J. Cell Biol.* 118:1271–1282, 1992.

22. Dexter, T.M.; Spooncer, E. Growth and differentiation in the hemopoietic system. *Annu. Rev. Cell Biol.* 3:423–441, 1987.

Golde, D.W. The stem cell. *Sci. Am.* 265(6):86–93, 1991.

Weiss, L., ed. Cell and Tissue Biology: A Textbook of Histology, 6th ed., pp. 423–478. Baltimore: Urban and Schwartzenberg, 1988.

Wintrobe, M.M. Blood, Pure and Eloquent. New York: McGraw-Hill, 1980.

23. McGee, J.O'D.; Isaacson, P.G.; Wright, N.A., eds. Oxford Textbook of Pathology, Vol. 1, pp. 236–258 and 321–347. Oxford, UK: Oxford University Press, 1992.

Zucker, M.B. The functioning of blood platelets. *Sci. Am.* 242(6):86–103, 1980.

24. McEver, R.P. Leukocyte-endothelial cell interactions. *Curr. Opin. Cell Biol.* 4:840–849, 1992.

Robbins, S.L.; Cotran, R.S.; Kumar, V. Pathologic Basis of Disease, 3rd ed. Philadelphia: Saunders, 1984.

Taussig, M.J. Processes in Pathology and Microbiology, 2nd ed. Oxford, UK: Blackwell, 1984.

25. Magli, M.C.; Iscove, N.N.; Odartchenko, N. Transient nature of haematopoietic spleen colonies. *Nature* 295:527–529, 1982.

Till, J.E.; McCulloch, E.A. A direct measurement of the radiation sensitivity of normal mouse bone marrow cells. *Radiat. Res.* 14:213–222, 1961.

26. Jordan, C.T.; Lemischka, I.R. Clonal and systemic analysis of long-term hematopoiesis in the mouse. *Genes Dev.* 4:220–232, 1990.

Wu, A.M.; Till, J.E.; Siminovitch, L.; McCulloch, E.A. Cytological evidence for a relationship between normal hematopoietic colony-forming cells and cells of the lymphoid system. *J. Exp. Med.* 127:455–462, 1968.

27. Till, J.E.; McCulloch, E.A. Hemopoietic stem cell differentiation. *Biochim. Biophys. Acta* 605:431–459, 1980.

28. Metcalf, D. The Hemopoietic Colony-Stimulating Factors. Amsterdam: Elsevier, 1984.

Metcalf, D. Clonal analysis of proliferation and differentiation of paired daughter cells: action of granulocyte-macrophage colony-stimulating factor on granulocyte-macrophage precursors. *Proc. Natl. Acad. Sci. USA* 77:5327–5330, 1980.

Ogawa, M. Review: differentiation and proliferation of haematopoietic stem cells. *Blood* 81:2844–2854, 1993.

29. Goldwasser, E. Erythropoietin and the differentiation of red blood cells. *Fed. Proc.* 34:2285–2292, 1975.

Heath, D.S.; Axelrad, A.A.; McLeod, D.L.; Shreeve, M.M. Separation of the erythropoietin-responsive progenitors BFU-E and CFU-E in mouse bone marrow by unit gravity sedimentation. *Blood* 47:777–792, 1976.

Ihle, J.N., et al. Biologic properties of homogeneous interleukin 3. *J. Immunol.* 131:282–287, 1983.

Suda, J., et al. Purified interleukin-3 and erythropoietin support the terminal differentiation of hemopoietic progenitors in serum-free culture. *Blood* 67:1002–1006, 1986.

30. Metcalf, D. The hemopoietic regulators—an embarrassment of riches. *Bioessays* 14:799–805, 1992.

Metcalf, D. The Florey Lecture, 1991. The colony-stimulating factors: discovery to clinical use. *Phil. Trans. Roy. Soc. Lond. (Biol.)* 333:147–173, 1991.

Stahl, N.; Yancopoulos, G.D. The alphas, betas, and kinases of cytokine receptor complexes. *Cell* 74:587–590, 1993.

Yoshida, H., et al. The murine mutation *osteopetrosis* is in the coding region of the macrophage colony stimulating factor gene. *Nature* 345:442–444, 1990.

31. Flanagan, J.G.; Chan, D.C.; Leder, P. Transmembrane form of the *kit* ligand growth factor is determined by alternative splicing and is missing in the Sl^d mutant. *Cell* 64:1025–1035, 1991.

Morrison-Graham, K.; Takahashi, Y. Steel factor and c-kit receptor: from mutants to a growth factor system. *Bioessays* 15:77–83, 1993.

32. Spangrude, G.J.; Heimfeld, S.; Weissman, I.L. Purification and characterization of mouse hematopoietic stem cells. *Science* 241:58–62, 1988.

Suda, T.; Suda, J.; Ogawa, M. Disparate differentiation in mouse hemopoietic colonies derived from paired progenitors. *Proc. Natl. Acad. Sci. USA* 81:2520–2524, 1984.

Whetton, A.D.; Dexter, T.M. Influence of growth factors and substrates on differentiation of haemopoietic stem cells. *Curr. Opin. Cell Biol.* 5:1044–1049, 1993.

33. Raff, M.C. Social controls on cell survival and cell death. *Nature* 356:397–400, 1992.

Savill, J.; Hogg, N.; Ren, Y.; Haslett, C. Thrombospondin cooperates with CD36 and the vitronectin receptor in macrophage recognition of neutrophils undergoing apoptosis. *J. Clin. Invest.* 90:1513–1522, 1992.

Vaux, D.L.; Weisman, I.L.; Kim, S.K. Prevention of programmed cell death in *Caenorhabditis elegans* by human *bcl-2*. *Science* 258:1955–1957, 1992.

Williams, G.T.; Smith, C.A.; Spooncer, E.; Dexter, T.M.; Taylor, D.R. Haemopoietic colony stimulating factors promote cell survival by suppressing apoptosis. *Nature* 343:76–79, 1990.

Wyllie, A.H.; Duvall, E. Cell death. In Oxford Textbook of Pathology (McGee, J.O'D.; Isaacson, P.G.; Wright, N.A., eds.), Vol. 1, pp. 141–157. Oxford, UK: Oxford University Press, 1992.

34. Buckingham, M. Making muscle in mammals. *Trends Genet.* 8:144–148, 1992.

Cormack, D. Ham's Histology, 9th ed., pp. 388–420. Philadelphia: Lippincott, 1987.

Emerson, C.P., Jr. Skeletal myogenesis: genetics and embryology to the fore. *Curr. Opin. Genet. Devel.* 3:265–274, 1993.

35. Devlin, R.B.; Emerson, C.P., Jr. Coordinate regulation of contractile protein synthesis during myoblast differentiation. *Cell* 13:599–611, 1978.

Konigsberg, I.R. Diffusion-mediated control of myoblast fusion. *Dev. Biol.* 26:133–152, 1971.

Olsen, E.N. Interplay between proliferation and differentiation within the myogenic lineage. *Dev. Biol.* 154:261–272, 1992.

Rosen, G.D., et al. Roles for the integrin VLA-4 and its counter receptor VCAM-1 in myogenesis. *Cell* 69:1107–1119, 1992.

Weintraub, H., et al. The *myoD* gene family: nodal point during specification of the muscle cell lineage. *Science* 251:761–766, 1991.

36. Buckingham, M.E. The control of muscle gene expression: a review of molecular studies on the production and processing of primary transcripts. *Brit. Med. Bull.* 45:608–629, 1989.

Lømo, T.; Westgaard, R.H.; Dahl, H.A. Contractile properties of muscle: control by pattern of muscle activity in the rat. *Proc. R. Soc. Lond. (Biol.)* 187:99–103, 1974.

Plasticity of the Neuromuscular System. *Ciba Found. Symp.* 138, 1988.

Stockdale, F.E. Myogenic cell lineages. *Dev. Biol.* 154:284–298, 1992.

37. Bischoff, R. Proliferation of muscle satellite cells on intact myofibers in culture. *Dev. Biol.* 115:129–139, 1986.

Goldspink, G. Development of muscle. In Differentiation and Growth of Cells in Vertebrate Tissues (G. Goldspink, ed.), pp. 69–99. London: Chapman and Hall, 1974.

Moss, F.P.; Leblond, C.P. Satellite cells as the source of nuclei in muscles of growing rats. *Anat. Rec.* 170:421–435, 1971.

Tinsley, J.M., et al. Dystrophin and related proteins. *Curr. Opin. Genet. Devel.* 3:484–490, 1993.

38. Fawcett, D.W. (Bloom and Fawcett) A Textbook of Histology, 11th ed., pp. 136–187. Philadelphia: Saunders, 1986.

Gabbiani, G.; Rungger-Brändle, E. The fibroblast. In Tissue Repair and Regeneration (L.E. Glynn, ed.), pp. 1–50. Handbook of Inflammation, Vol. 3. Amsterdam: Elsevier, 1981.

39. Clark, R.A.F.; Henson, P.M., eds. The Molecular and Cellular Biology of Wound Repair. New York: Plenum Press, 1988.

Conrad, G.W.; Hart, G.W.; Chen, Y. Differences *in vitro* between fibroblast-like cells from cornea, heart, and skin of embryonic chicks. *J. Cell Sci.* 26:119–137, 1977.

Reddi, A.H.; Gay, R.; Gay, S.; Miller, E.J. Transitions in collagen types during matrix-induced cartilage, bone, and bone marrow formation. *Proc. Natl. Acad. Sci. USA* 74:5589–5592, 1977.

Rosen, V.; Thies, R.S. The BMP proteins in bone formation and repair. *Trends Genet.* 8:97–102, 1992.

Schor, S.L.; Schor, A.M. Clonal heterogeneity in fibroblast phenotype: implications for the control of epithelial-mesenchymal interactions. *Bioessays* 7:200–204, 1987.

Wozney, J.M., et al. Novel regulators of bone formation: molecular clones and activities. *Science* 242:1528–1534, 1988.

40. Benya, P.D.; Schaffer, J.D. Dedifferentiated chondrocytes reexpress the differentiated collagen phenotype when cultured in agarose gels. *Cell* 30:215–224, 1982.

Caplan, A.I. Cartilage. *Sci. Am.* 251(4):84–94, 1984.

von der Mark, K.; Gauss, V.; von der Mark, H.; Müller, P. Relationship between cell shape and type of collagen synthesized as chondrocytes lose their cartilage phenotype in culture. *Nature* 267:531–532, 1977.

Zanetti, N.C.; Solursh, M. Induction of chondrogenesis in limb mesenchymal cultures by disruption of the actin cytoskeleton. *J. Cell Biol.* 99:115–123, 1984.

41. Johnson, P.R.; Greenwood, M.R.C. The adipose tissue. In Cell and Tissue Biology: A Textbook of Histology (L.Weiss, ed.), 6th ed., pp. 189–209. Baltimore: Urban and Schwartzenberg, 1988.

Spiegelman, B.M.; Ginty, C.A. Fibronectin modulation of cell shape and lipogenic gene expression in 3T3 adipocytes. *Cell* 35:657–666, 1983.

Sugihara, H.; Yonemitsu, N.; Miyabara, S.; Yun, K. Primary cultures of unilocular fat cells: characteristics of growth *in vitro* and changes in differentiation properties. *Differentiation* 31:42–49, 1986.

Sul, H.S. Adipocyte differentiation and gene expression. *Curr. Opin. Cell Biol.* 1:1116–1121, 1989.

Zezulak, K.M.; Green, H. The generation of insulin-like growth factor-1-sensitive cells by growth hormone action. *Science* 233:551–553, 1986.

42. Cell and Molecular Biology of Vertebrate Hard Tissues. *Ciba Found. Symp.* 136, 1988.

Fawcett, D.W. (Bloom and Fawcett) A Textbook of Histology, 11th ed., pp. 188–238. Philadelphia: Saunders, 1986.

Jee, W.S.S. The skeletal tissues. In Cell and Tissue Biology: A Textbook of Histology (L.Weiss, ed.), 6th ed., pp. 211–254. Baltimore: Urban and Schwartzenberg, 1988.

Seyedin, S.M.; Rosen, D.M. Matrix proteins of the skeleton. *Curr. Opin. Cell Biol.* 2:914–919, 1990.

43. Marcus, R. Normal and abnormal bone remodeling in man. *Annu. Rev. Med.* 38:129–141, 1987.

Osdoby, P.; Krukowski, M.; Oursler, M.J.; Salino-Hugg, T. The origin, development, and regulation of osteoclasts. *Bioessays* 7:30–34, 1987.

Vaughan, J. The Physiology of Bone, 3rd ed. Oxford, UK: Clarendon Press, 1981.

44. Cormack, D. Ham's Histology, 9th ed., pp. 312–320. Philadelphia: Lippincott, 1987.

45. Currey, J. The Mechanical Adaptations of Bone. Princeton, NJ: Princeton University Press, 1984.

Goss, R.J. The Physiology of Growth. New York: Academic Press, 1978.

Rogers, S.L. The Aging Skeleton. Springfield, IL: Thomas, 1982.

Sinclair, D. Human Growth After Birth, 4th ed. Oxford, UK: Oxford University Press, 1985.

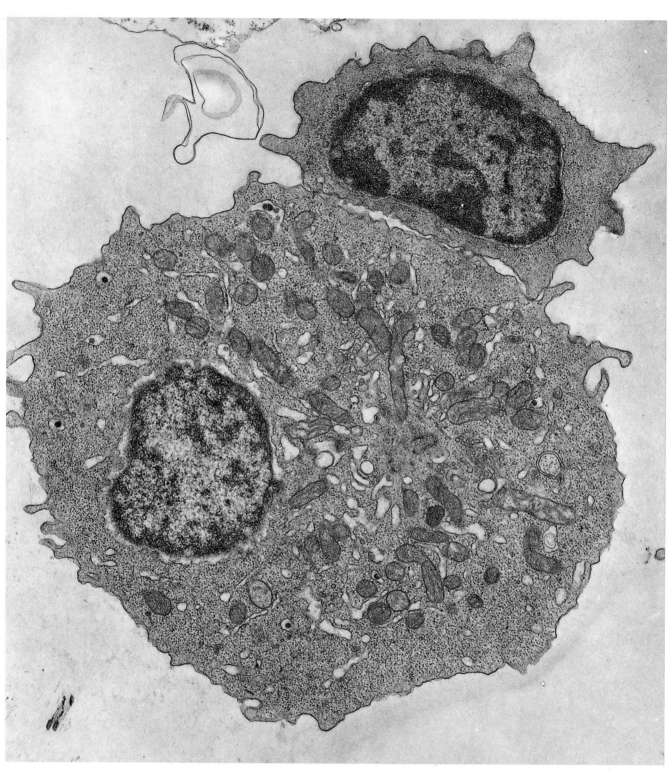

A small cytotoxic T cell preparing to kill a large tumor cell. (From D. Zagury, J. Bernard,
N. Thierness, M. Feldman, and G. Berke, *Eur. J. Immunol.* 5:818–822, 1975.)

The Immune System

Our immune system saves us from certain death by infection. Any vertebrate born with a severely defective immune system will soon die unless extraordinary measures are taken to isolate it from a host of infectious agents—bacterial, viral, fungal, and parasitic. Whereas all vertebrates have an immune system, invertebrates have more primitive defense systems, which often rely chiefly on phagocytic cells. Such cells (mainly macrophages and neutrophils) also play an important role in defending vertebrates against infection, but they are only one part of a much more complex and sophisticated defense strategy.

Immunology, the study of the immune system, grew out of the common observation that people who recover from certain infections are thereafter "immune" to the disease; that is, they rarely develop the same disease again. Immunity is highly specific: an individual who recovers from measles is protected against the measles virus but not against other common viruses, such as mumps or chicken pox. Such specificity is a fundamental characteristic of immune responses.

Many of the responses of the immune system initiate the destruction and elimination of invading organisms and any toxic molecules produced by them. Because these immune reactions are destructive, it is essential that they be made only in response to molecules that are foreign to the host and not to those of the host itself. This ability to distinguish *foreign* molecules from *self* molecules is another fundamental feature of the immune system. Occasionally, it fails to make this distinction and reacts destructively against the host's own molecules; such *autoimmune diseases* can be fatal.

Although the immune system evolved to protect vertebrates from infection by microorganisms and larger parasites, most of what we know about immunity has come from studies of the responses of laboratory animals to injections of noninfectious substances, such as foreign proteins and polysaccharides. Almost any macromolecule, as long as it is foreign to the recipient, can induce an immune response; any substance capable of eliciting an immune response is referred to as an **antigen** (*anti*body *gen*erator). Remarkably, the immune system can distinguish between antigens that are very similar—such as between two proteins that differ in only a single amino acid or between two optical isomers of the same molecule.

There are two broad classes of immune responses: (1) antibody responses and (2) cell-mediated immune responses. **Antibody responses** involve the production of antibodies, which are proteins called *immunoglobulins*. The antibodies circulate in the bloodstream and permeate the other body fluids, where they bind specifically to the foreign antigen that induced them. Binding by antibody inactivates viruses and bacterial toxins (such as tetanus or botulinum toxin) by blocking their ability to bind to receptors on host cells. Antibody binding also marks invading microorganisms for destruction, either by making it easier for a phagocytic cell to ingest them or by activating a system of blood proteins, collectively called *complement,* that kills the invaders.

Cell-mediated immune responses, the second class of immune responses, involve the production of specialized cells that react with foreign antigens on the surface of other host cells. The reacting cell, for example, can kill a virus-infected host cell that has viral antigens on its surface, thereby eliminating the infected cell before the virus has replicated. In other cases the reacting cell secretes chemical signals that activate macrophages to destroy the invading microorganisms.

The main challenge in immunology has been to understand how the immune system specifically recognizes and reacts aggressively to a virtually unlimited number of different foreign macromolecules but avoids reacting against the tens of thousands of different self macromolecules made by host cells. We begin our discussion of the immune system by considering the cells that are principally responsible for the two types of immunity. We then consider the functional and structural features of antibodies that enable them to recognize and destroy extracellular antigens. After discussing how antibody diversity is generated, we consider the special features of cell-mediated immune responses, which are crucial in the defense against intracellular microorganisms.

The Cellular Basis of Immunity

The Human Immune System Is Composed of Trillions of Lymphocytes [1]

The cells responsible for immune specificity belong to a class of white blood cells known as **lymphocytes.** They are found in large numbers in the blood and the lymph (the colorless fluid in the lymphatic vessels that connect the lymph nodes in the body) and in specialized **lymphoid organs,** such as the thymus, lymph nodes, spleen, and appendix (Figure 23–1).

There are ~2×10^{12} lymphocytes in the human body, which makes the immune system comparable in cell mass to the liver or brain. Although lymphocytes

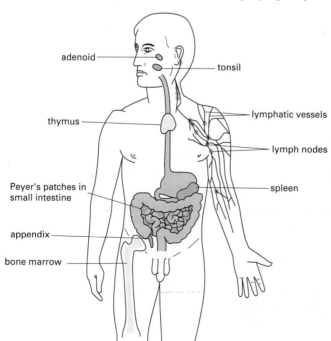

adenoid

tonsil

thymus

lymphatic vessels

lymph nodes

Peyer's patches in small intestine

spleen

appendix

bone marrow

Figure 23–1 Human lymphoid organs. Lymphocytes develop in the thymus and bone marrow (*yellow*), which are therefore referred to as *primary* (or *central*) *lymphoid organs.* The newly formed lymphocytes migrate from these primary organs to *secondary* (or *peripheral*) *lymphoid organs* (*blue*), where they can react with antigen. Only some of the secondary lymphoid organs are shown; many lymphocytes, for example, are found in the skin and lungs.

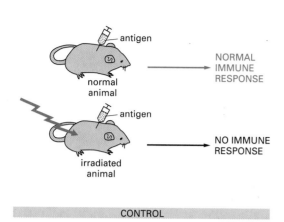

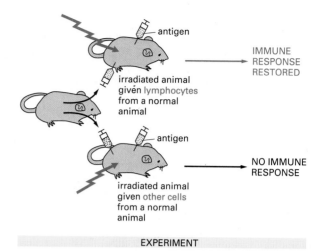

NORMAL IMMUNE RESPONSE

NO IMMUNE RESPONSE

CONTROL

IMMUNE RESPONSE RESTORED

NO IMMUNE RESPONSE

EXPERIMENT

have long been recognized as a major cellular component of the blood, their central role in immunity was demonstrated only in the late 1950s. The proof came from experiments in which mice or rats were heavily irradiated in order to kill most of their white blood cells, including lymphocytes. Since such animals are unable to make immune responses, it is possible to transfer various types of cells into them and determine which ones reverse the deficiency. Only lymphocytes restored the immune responses of irradiated animals (Figure 23–2). Since both antibody and cell-mediated responses were restored, these experiments established that lymphocytes are responsible for both types of immune response. At the time these experiments were done, lymphocytes were among the least understood of vertebrate cells; now they are among the cells we understand best.

B Lymphocytes Make Humoral Antibody Responses; T Lymphocytes Make Cell-mediated Immune Responses [2]

During the 1960s it was discovered that the two major classes of immune responses are mediated by different classes of lymphocytes: **T cells,** which develop in the *thymus,* are responsible for cell-mediated immunity; **B cells,** which in mammals develop in the adult *bone marrow* or the fetal liver, produce antibodies. This dichotomy of the lymphoid system was initially demonstrated in animals with experimentally induced immunodeficiencies. It was found that removing the thymus from a newborn animal markedly impairs cell-mediated immune responses but has much less effect on antibody responses. In birds it was also possible to demonstrate the converse effect because B cells develop in a discrete gut-associated lymphoid organ, the *bursa of Fabricius,* that is unique to birds. Removing the bursa of Fabricius at the time of hatching impaired the bird's ability to make antibodies but had little effect on cell-mediated immunity. In addition, studies of children born with impaired immunity showed that some of these children could not make antibodies but had normal cell-mediated immunity, while others had the reverse deficiency; and those with selectively impaired cell-mediated responses almost always had thymus abnormalities.

One of the puzzling features of these studies on immunodeficient animals was that individuals deficient in T cells (because their thymus was removed at birth or was abnormal) not only were unable to make cell-mediated immune responses, but also had somewhat impaired antibody responses. We now know the explanation. There are two main classes of T cells—*helper T cells* and *cytotoxic T cells.* Helper T cells enhance the responses of other white blood cells, and some of these T cells help B cells make antibody responses. Cytotoxic T cells, by contrast, kill infected cells; because they are involved *directly* in defense against infection, unlike helper T cells, they (together with B cells) are sometimes referred to as *effector cells.*

Figure 23–2 The classic experiment showing that lymphocytes are responsible for recognizing and responding to foreign antigens. An important feature of all such cell-transfer experiments is that cells are transferred between animals of the same *inbred strain.* Members of an inbred strain are genetically identical. If lymphocytes are transferred to a genetically different animal that has been irradiated, they react against the "foreign" antigens of the host and can kill the animal.

The Cellular Basis of Immunity

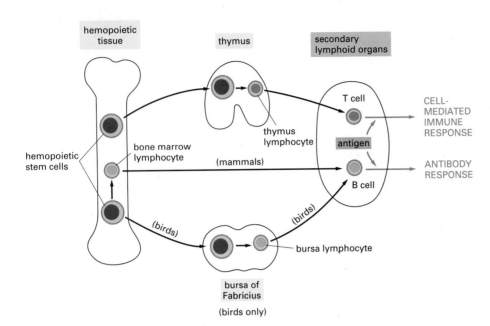

Figure 23–3 **The development of T and B cells.** The primary lymphoid organs, where lymphocytes develop from precursor cells, are labeled in *yellow boxes.*

Lymphocytes Develop in Primary Lymphoid Organs and React with Foreign Antigens in Secondary Lymphoid Organs [3]

Lymphocytes develop from *pluripotent hemopoietic stem cells,* which give rise to all of the blood cells, including red blood cells, white blood cells, and platelets. These stem cells, which are discussed in Chapter 22, are located primarily in *hemopoietic* tissues—the liver in fetuses and the bone marrow in adults. T cells develop in the thymus from precursor cells that migrate in from the hemopoietic tissues via the blood. In mammals B cells develop from stem cells in the hemopoietic tissues themselves; in birds, however, B cells develop in the bursa of Fabricius from precursor cells that migrate in from the hemopoietic tissues via the blood. Because they are sites where lymphocytes develop from precursor cells, the thymus, the hemopoietic tissues, and the bursa of Fabricius are referred to as **primary (central) lymphoid organs** (see Figure 23–1).

As we discuss later, most lymphocytes die soon after they develop in a primary lymphoid organ. Others, however, mature and migrate via the blood to the **secondary (peripheral) lymphoid organs**—mainly, the lymph nodes, spleen, and epithelium-associated lymphoid tissues found in the gastrointestinal tract, respiratory tract, and skin (see Figure 23–1). It is chiefly in the secondary lymphoid organs that T cells and B cells react with foreign antigens (Figure 23–3).

Because most of the migration of lymphocytes out of the thymus and bursa occurs early in development, removing these organs from *adult* animals has relatively little effect on immune responses, which is why their role in immunity remained undiscovered for so long. In contrast, the bone marrow in mammals continues to generate large numbers of new B cells (~5 × 10^7/day in a mouse) throughout life.

Cell-Surface Markers Make It Possible to Distinguish and Separate T and B Cells [2, 4]

T and B cells become morphologically distinguishable only after they have been stimulated by antigen. Unstimulated ("resting") T and B cells look very similar, even in an electron microscope: both are small, only marginally bigger than red blood cells, and are largely filled by the nucleus (Figure 23–4A). Both are activated by antigen to proliferate and mature further. Activated B cells develop into antibody-secreting cells, the most mature of which are *plasma cells,* which are filled

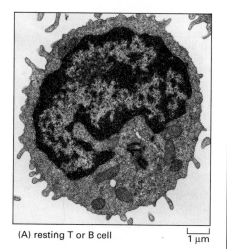

(A) resting T or B cell   1 μm

(B) active B cell (plasma cell)   1 μm

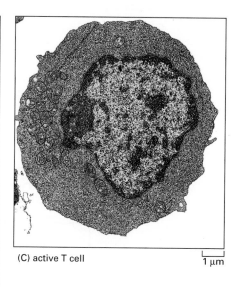

(C) active T cell   1 μm

with an extensive rough endoplasmic reticulum (Figure 23–4B). In contrast, activated T cells contain very little endoplasmic reticulum and do not secrete antibodies, although they do secrete a variety of mediators called *lymphokines*, *interleukins*, or *cytokines* (Figure 23–4C).

Since both T and B cells occur in all secondary lymphoid organs, it has been necessary to find ways to distinguish and separate the two cell types and their various subtypes in order to study their individual properties. Fortunately, there are many differences in the plasma membrane proteins of the different types of lymphocytes that can serve as distinguishing markers. Antibodies that react with the *Thy-1 protein*, for example, which is found on T but not B cells in mice, are widely used to remove or purify T cells from a mixed population of mouse lymphocytes. Similarly, antibodies against the *CD4* and *CD8 proteins*, which we discuss later, are widely used to distinguish and separate helper T cells and cytotoxic T cells, respectively, in both mice and humans.

The Immune System Works by Clonal Selection [5]

The most remarkable feature of the immune system is that it can respond to millions of different foreign antigens in a highly specific way—for example, by making antibodies that react specifically with the antigen that induced their production. How can the immune system produce such a diversity of specific antibodies? The answer began to emerge in the 1950s with the formulation of the **clonal selection theory.** According to this theory, each animal first randomly generates a vast diversity of lymphocytes, and then those cells that react against the foreign antigens that the animal actually encounters are specifically selected for action. The theory is based on the proposition that during development each lymphocyte becomes committed to react with a particular antigen before ever being exposed to it. Each cell expresses this commitment in the form of cell-surface receptor proteins that specifically fit the antigen. The binding of antigen to the receptors activates the cell, causing it to both proliferate and mature. A foreign antigen, therefore, selectively stimulates those cells that express complementary antigen-specific receptors and are thus already committed to respond to it. This is what makes immune responses antigen-specific.

The term "clonal" in clonal selection derives from the postulate that the immune system is composed of millions of different families, or *clones*, of cells, each consisting of T or B cells descended from a common ancestor. Each ances-

Figure 23–4 Electron micrographs of resting and activated lymphocytes. The resting lymphocyte in (A) could be a T cell or a B cell, for these cells are difficult to distinguish morphologically until they have been activated. The activated B cell (a plasma cell) in (B) is filled with an extensive rough endoplasmic reticulum (ER), which is distended with antibody molecules. The activated T cell in (C) has relatively little rough ER but is filled with free ribosomes. Note that the three cells are shown at the same magnification. (A, courtesy of Dorothy Zucker-Franklin; B, courtesy of Carlo Grossi; A and B, from D. Zucker-Franklin et al., Atlas of Blood Cells: Function and Pathology, 2nd ed. Milan, Italy: Edi. Ermes, 1988; C, courtesy of Stefanello de Petris.)

The Cellular Basis of Immunity

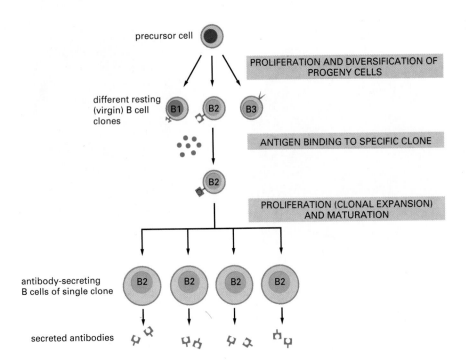

precursor cell

PROLIFERATION AND DIVERSIFICATION OF
PROGENY CELLS

different resting
(virgin) B cell
clones

B1   B2   B3

ANTIGEN BINDING TO SPECIFIC CLONE

B2

PROLIFERATION (CLONAL EXPANSION)
AND MATURATION

antibody-secreting
B cells of single clone

B2   B2   B2   B2

secreted antibodies

Figure 23–5 The clonal selection theory. An antigen activates only those lymphocyte clones that are already committed to respond to it. A cell committed to respond to a particular antigen displays cell-surface receptors that specifically recognize the antigen, and all cells within a clone display the same receptor. The immune system is thought to consist of millions of different lymphocyte clones, hundreds of which may be activated by a particular antigen (see below). Although only B cells are shown, T cells operate in a similar way.

tral cell is already committed to make one particular antigen-specific receptor protein, and all cells in a clone have the same antigen specificity (Figure 23–5). Thus, according to the clonal selection theory, the immune system functions on the "ready-made" rather than the "made-to-measure" principle.

There is compelling evidence to support the main tenets of the clonal selection theory. When lymphocytes from an animal that has not been immunized, for example, are incubated in a test tube with any of a number of radioactively labeled antigens—say, A, B, C, and D—only a very small proportion (<0.01%) bind each antigen, suggesting that only a few cells can respond to A, B, C, or D. This interpretation is confirmed by making antigen A so highly radioactive that any cell that binds it is lethally irradiated; the remaining population of lymphocytes is then no longer able to produce an immune response to A but can still respond normally to antigen B, C, or D. The same effect can be achieved by constructing an affinity column of glass beads coated with antigen A and then passing the lymphocytes through the column. The cells with receptors for A stick to the beads, while other cells pass through; as a result, the cells that emerge from the column no longer respond to A but do respond normally to other antigens (Figure 23–6).

These two experiments indicate, first, that lymphocytes are committed to respond to a particular antigen before they have been exposed to it and, second, that the committed lymphocytes have receptors on their surface that specifically bind the antigen. Two major predictions of the clonal selection theory are therefore confirmed. Although almost all of the experiments of this kind have involved B cells and antibody responses, other experiments indicate that T-cell-mediated responses also operate by clonal selection.

It is now known that the antigen-specific receptors on both T and B cells are encoded by genes that are assembled from series of gene segments by a unique form of genetic recombination that occurs early in the cell's development, before it has encountered antigen. We shall see later how the assembly process generates the enormous diversity of receptors that enables the immune system to respond to an almost unlimited diversity of antigens.

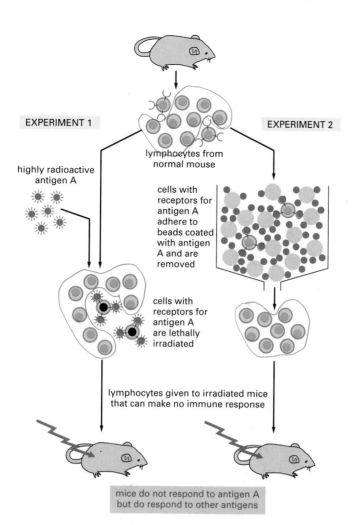

EXPERIMENT 1

EXPERIMENT 2

lymphocytes from
normal mouse

highly radioactive
antigen A

cells with
receptors for
antigen A
adhere to
beads coated
with antigen
A and are
removed

cells with
receptors for
antigen A
are lethally
irradiated

lymphocytes given to irradiated mice
that can make no immune response

mice do not respond to antigen A
but do respond to other antigens

Figure 23–6 **Two types of experiments that support the clonal selection theory.** For simplicity, cell-surface receptors are shown only on those lymphocytes committed to respond to antigen A; in fact, however, all T and B cells have antigen-specific receptors on their surface. The experiments shown have been carried out mainly with B cells since T cells recognize an antigen only when it is bound to the surface of a host cell, as we discuss later.

Most Antigens Stimulate Many Different Lymphocyte Clones [6]

Most macromolecules, including virtually all proteins and many polysaccharides, can serve as antigens. Those parts of an antigen that combine with the antigen-binding site on an antibody molecule or on a lymphocyte receptor are called **antigenic determinants** (or *epitopes*). Most antigens have a variety of antigenic determinants that stimulate the production of antibodies or T cell responses. Some determinants are more antigenic than others, so that the reaction to them may dominate the overall response. Such determinants are said to be *immunodominant*.

As one might expect of a system that works by clonal selection, even a single antigenic determinant, in general, will activate many clones, each of which produces an antigen-binding site with its own characteristic affinity for the determinant. Even a relatively simple structure like the *dinitrophenyl (DNP)* group shown in Figure 23–7, for example, can be "looked at" in many ways. When it is coupled to a protein, it usually stimulates the production of hundreds of species of anti-DNP antibodies, each made by a different B cell clone. Such responses are said to be *polyclonal*. When only a few clones respond, the response is said to be *oligoclonal;* and when the total response is made by a single B or T cell clone, it is said to be *monoclonal*. The responses to most antigens are polyclonal.

Even an antigen that activates many clones will stimulate only a tiny fraction of the total lymphocyte population. To ensure that these few lymphocytes are

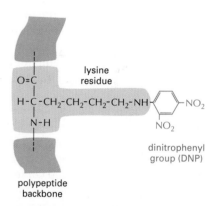

O=C

lysine
residue

H–C–CH₂–CH₂–CH₂–CH₂–NH

NO₂

NO₂

N–H

dinitrophenyl
group (DNP)

polypeptide
backbone

Figure 23–7 **The dinitrophenyl (DNP) group.** Although it is too small to induce an immune response on its own, when it is coupled covalently to a lysine side chain on a protein, as illustrated, DNP stimulates the production of many different species of antibodies that all bind specifically to it.

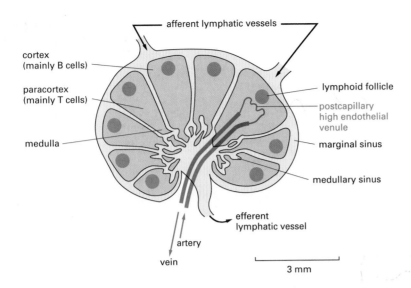

afferent lymphatic vessels

cortex (mainly B cells)

paracortex (mainly T cells)

medulla

lymphoid follicle

postcapillary high endothelial venule

marginal sinus

medullary sinus

efferent lymphatic vessel

artery

vein

3 mm

Figure 23–8 A simplified drawing of a human lymph node. B cells are located primarily in the *cortex*, where they are clustered in structures called *lymphoid follicles*. T cells are found mainly in the *paracortex*. Both types of lymphocytes enter the lymph node from the blood via specialized small veins called *postcapillary high endothelial venules* and then migrate to their respective areas. Eventually, both T cells and B cells migrate to the medullary sinuses and leave the node via the efferent lymphatic vessel. This vessel ultimately empties into the bloodstream, allowing the lymphocytes to begin another cycle of circulation through a secondary lymphoid organ.

Foreign antigens that enter the lymph node are displayed on the surface of specialized *antigen-presenting cells:* one type (*interdigitating dendritic cells*) presents antigen to T cells in the paracortical area; another type (*follicular dendritic cells*) is thought to be involved in activating memory B cells (discussed below) in the activated center (called a *germinal center*) of lymphoid follicles.

exposed to the antigen, antigens are generally collected by specialized *antigen-presenting cells* in secondary lymphoid organs, through which T and B cells continuously recirculate. Antigens that enter through the gut are trapped by gut-associated lymphoid tissues; those that enter through the skin or respiratory tract are retained locally and/or are transported via the lymph to local lymph nodes; and those that enter the blood are filtered out in the spleen.

Most Lymphocytes Continuously Recirculate [7]

The majority of T and B cells continuously recirculate between the blood and the secondary lymphoid tissues. In a lymph node, for example, lymphocytes leave the bloodstream, squeezing out between specialized endothelial cells; after percolating through the node, they accumulate in small lymphatic vessels that leave the node and connect with other lymphatic vessels, which then pass through other lymph nodes downstream. Passing into larger and larger vessels, the lymphocytes eventually enter the main lymphatic vessel (the *thoracic duct*), which carries them back into the blood. This continuous recirculation not only ensures that the appropriate lymphocytes will come into contact with antigen, it also ensures that appropriate lymphocytes encounter one another: we shall see that interactions between specific lymphocytes are a crucial part of most immune responses.

Lymphocyte recirculation depends on specific interactions between the lymphocyte cell surface and the surface of specialized endothelial cells lining small veins in the secondary lymphoid organs; because their endothelial cells are unusually tall, they are called *postcapillary high endothelial venules* (Figure 23–8). Many cell types in the blood come into contact with these high endothelial cells, but only lymphocytes adhere and then migrate out of the bloodstream. Different subpopulations of lymphocytes migrate through different lymphoid tissues: whereas most lymphocytes migrate into lymph nodes, for example, some migrate preferentially into Peyer's patches in the small intestine and constitute, in effect, a gut-specific subsystem of lymphocytes specialized for responding to antigens that enter the body from the intestine.

These migrations are guided by various **homing receptors** on lymphocytes and by the ligands for these receptors (often called *counterreceptors*) on endothelial cells. Both receptors and counterreceptors have been identified by monoclonal antibodies that bind to the surface of either lymphocytes or the specialized high endothelial cells and inhibit the ability of the lymphocytes both to bind to the endothelial cells in tissue sections of secondary lymphoid organs and to

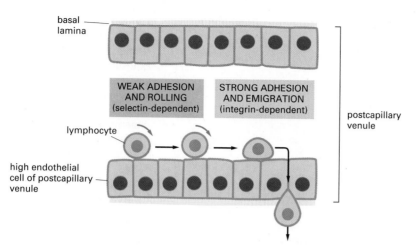

Figure 23–9 **Migration of a lymphocyte out of the bloodstream into a lymph node.** A circulating lymphocyte adheres weakly to the surface of the specialized high endothelial cells in a postcapillary venule in a lymph node. This initial adhesion, mediated by E-selectin on the lymphocyte surface, is sufficiently weak that it enables the lymphocyte to roll along the surface of the endothelial cells. The lymphocyte rapidly activates a stronger adhesion system, mediated by an integrin, that enables the cell to stop rolling and migrate out of the venule between the endothelial cells.

recirculate *in vivo.* Lymphocyte migration into lymph nodes, for example, depends on a cell adhesion protein called *E-selectin,* which belongs to the selectin family of cell-surface lectins discussed in Chapter 10. This homing receptor, which is present on most lymphocytes, binds to specific sugar groups on a highly glycosylated, mucinlike counterreceptor that is expressed exclusively on the surface of high endothelial cells lining postcapillary venules in lymph nodes (see Figure 23–8). E-selectin binding causes the lymphocytes to adhere weakly to the endothelial cells and to roll slowly along their surface. The rolling continues until another, much stronger adhesion system is activated. This strong adhesion, which is mediated by a member of the *integrin* family of cell adhesion molecules on the lymphocyte surface (discussed in Chapter 19), allows the lymphocytes to stop rolling and crawl out of the blood vessel into the lymph node (Figure 23–9).

Other homing receptors on lymphocytes are thought to be responsible for the subsequent segregation of T and B cells into distinct areas in the lymph node (see Figure 23–8). Once they are activated by antigen, most lymphocytes lose many of their original homing receptors and acquire new ones: instead of migrating through lymphoid organs they migrate through nonlymphoid tissues to sites of inflammation. The migration of activated lymphocytes and other white blood cells into sites of inflammation is largely mediated by other combinations of selectins and integrins (discussed in Chapter 10).

Immunological Memory Is Due to Clonal Expansion and Lymphocyte Maturation [8]

The immune system, like the nervous system, can remember. This is why we develop lifelong immunity to many common viral diseases after our initial exposure to the virus, and it is why immunization works. The same phenomenon can be demonstrated in experimental animals. If an animal is injected once with antigen A, its immune response (either antibody or cell-mediated) will appear after a lag period of several days, rise rapidly and exponentially, and then, more gradually, fall again. This is the characteristic course of a **primary immune response,** occurring on an animal's first exposure to an antigen. If some weeks or months or even years are allowed to pass and the animal is reinjected with antigen A, it will usually produce a **secondary immune response** that is very different from the primary response: the lag period is shorter and the response is greater. These differences indicate that the animal has "remembered" its first exposure to antigen A. If the animal is given a different antigen (for example, antigen B) instead of a second injection of antigen A, the response is typical of a primary, and not a secondary, immune response. Therefore, the secondary response reflects antigen-specific **immunological memory** for antigen A (Figure 23–10).

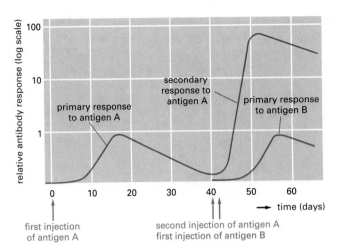

Figure 23–10 **Primary and secondary antibody responses.** The secondary response induced by a second exposure to antigen A is faster and greater than the primary response and is specific for A, indicating that the immune system has specifically remembered encountering antigen A before. Evidence for the same type of immunological memory is obtained if T-cell-mediated responses rather than B cell antibody responses are measured.

The clonal selection theory provides a useful conceptual framework for understanding the cellular basis of immunological memory. In an adult animal the T and B cells in the secondary lymphoid organs are a mixture of cells in at least three stages of maturation, which can be designated *virgin* (or *naïve*) *cells, memory cells,* and *activated cells*. When **virgin cells** encounter antigen for the first time, some of them are stimulated to multiply and become **activated cells,** which we define as cells that are actively engaged in making a response (activated T cells carry out cell-mediated responses, while activated B cells secrete antibody). Other virgin cells are stimulated to multiply and mature instead into **memory cells**—cells that are not themselves making a response but are readily induced to become activated cells by a later encounter with the same antigen (Figure 23–11). Whereas virgin cells and memory cells can live for months or even years, activated cells die by programmed cell death within days.

Memory cells respond much more readily to antigen than do virgin cells. We shall see later that one reason for the increased responsiveness of memory B cells is that their receptors have a higher affinity for antigen. In contrast, memory T cells seem to respond to antigen more readily than virgin T cells—not because they have higher-affinity receptors for antigen, but because they adhere more strongly to other cells and transduce extracellular signals more efficiently. Thus immunological memory is generated during the primary response in part because the proliferation of antigen-triggered virgin cells creates many memory cells—a process known as *clonal expansion*—and in part because virgin cells differentiate into memory cells that are able to respond more readily to antigen than do virgin cells. Antigens can persist in lymphoid tissues for a very long time following a primary response, and it is thought that continual stimulation by antigen contributes to the long-term maintenance of memory.

Figure 23–11 **A model for the cellular basis of immunological memory.** When virgin T or B cells are stimulated by their specific antigen, they proliferate and mature; some become activated to make a response, while others become memory cells. During a subsequent exposure to the same antigen, the memory cells respond more readily than did the virgin cells: they proliferate and give rise to activated cells and to more memory cells. In the model shown an individual virgin cell can give rise to either a memory cell or an activated cell, depending on the conditions. In an alternative model (not shown) the virgin cells that mature into memory cells are different from those that mature into activated cells. It is not known which of these models is correct.

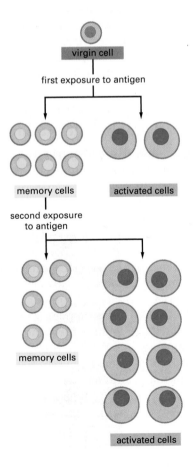

The Failure to Respond to Self Antigens Is Due to Acquired Immunological Tolerance [9]

How is the immune system able to distinguish foreign molecules from self molecules? One possibility might be that an animal inherits genes that encode receptors for foreign antigens but not self antigens, so that its immune system is genetically constituted to respond only to foreign antigens. Alternatively, the immune system could be inherently capable of responding to both foreign and self antigens but could "learn" during development not to respond to self antigens. The latter explanation has been shown to be correct. The first evidence for this was an observation made in 1945. Normally, when tissues are transplanted from one individual to another, they are recognized as foreign by the immune system of the recipient and are destroyed. Dizygotic cattle twins, however, which develop from two fertilized ova and are therefore nonidentical, sometimes exchange blood cells *in utero* as a result of the spontaneous fusion of their placentas; such twins were shown to accept skin grafts from each other. These findings were later reproduced experimentally—in chicks, by allowing the blood vessels of two embryos to fuse, and in mice, by introducing cells from one strain of mouse into a neonatal mouse of another strain, where they survived for most of the recipient animal's life. In both cases, when the animals matured, grafts from the joined or donor animal were accepted (Figure 23–12), while "third-party" grafts from a different animal were rejected. Thus the continuous presence of nonself antigens from before the time the immune system matures leads to a long-lasting unresponsiveness to the specific nonself antigens. The resulting state of antigen-specific immunological unresponsiveness is known as **acquired immunological tolerance.**

There is strong evidence that the unresponsiveness of an animal's immune system to its own macromolecules (*natural immunological tolerance*) is acquired in the same way and is not inborn. Normal mice, for example, cannot make an immune response against their own blood complement protein C5, but mutant mice that lack the gene encoding C5 (but are otherwise genetically identical to the normal mice) can make an immune response to this protein. Maintaining self-tolerance requires the constant presence of the self antigens. If an antigen such as C5 is removed, an animal regains the ability to respond to it within weeks or months. Thus it is clear that the immune system is genetically capable of responding to self but learns not to do so.

The learning process that leads to self-tolerance can involve either killing the self-reactive lymphocytes (*clonal deletion*) or functionally inactivating the cells but leaving them alive (*clonal anergy*). As we discuss later, many self-reactive lymphocytes are eliminated or inactivated when they first encounter antigen in the primary lymphoid organs. It seems that newly formed lymphocytes in these organs are not activated by binding antigen but instead are either killed or inactivated by it.

Figure 23–12 Immunological tolerance. The skin graft seen here, transplanted from an adult brown mouse to an adult white mouse, has survived for many weeks only because the latter was made immunologically tolerant by injecting cells from the brown mouse into it at the time of birth. (Courtesy of Leslie Brent, from I. Roitt, Essential Immunology, 6th ed. Oxford, U.K.: Blackwell Scientific, 1988.)

Tolerance to self antigens sometimes breaks down, causing T or B cells (or both) to react against their own tissue antigens. *Myasthenia gravis* is an example of such an **autoimmune disease.** Affected individuals make antibodies against the acetylcholine receptors on their own skeletal muscle cells; the antibodies interfere with the normal functioning of the receptors so that such patients become weak and can die because they cannot breathe.

Summary

The immune system evolved to defend vertebrates against infection. It is composed of millions of lymphocyte clones. The lymphocytes in each clone share a unique cell-surface receptor that enables them to bind a particular antigenic determinant consisting of a specific arrangement of atoms on a part of a molecule. There are two classes of lymphocytes: B cells, which are produced in the bone marrow and make antibodies, and T cells, which are produced in the thymus and make cell-mediated immune responses.

Beginning early in lymphocyte development, many lymphocytes that would react against antigenic determinants on self macromolecules are eliminated or inactivated; as a result, the immune system normally reacts only to foreign antigens. The binding of a foreign antigen to a lymphocyte initiates a response by the cell that helps to eliminate the antigen. As part of the response, some lymphocytes proliferate and mature into memory cells that are able to respond more readily to antigen than do virgin cells. Thus the next time the same antigen is encountered, the immune response to it is much faster and stronger.

The Functional Properties of Antibodies [10]

Vertebrates rapidly die of infection if they are unable to make antibodies. Antibodies defend us against infection by inactivating viruses and bacterial toxins and by recruiting the complement system and various types of white blood cells to kill extracellular microorganisms and larger parasites. Synthesized exclusively by B cells, antibodies are produced in millions of forms, each with a different amino acid sequence and a different binding site for antigen. Collectively called **immunoglobulins** (abbreviated as **Ig**), they are among the most abundant protein components in the blood, constituting about 20% of the total plasma protein by weight. In this section we describe the five classes of antibodies found in higher vertebrates, each of which mediates a characteristic biological response following antigen binding.

The Antigen-specific Receptors on B Cells Are Antibody Molecules [11]

As predicted by the clonal selection theory, all antibody molecules made by an individual B cell have the same antigen-binding site. The first antibodies made by a newly formed B cell are not secreted. Instead, they are inserted into the plasma membrane, where they serve as receptors for antigen. Each B cell has approximately 10^5 such antibody molecules in its plasma membrane. Each of these antibody molecules is noncovalently associated with an invariant set of transmembrane polypeptide chains that are involved in passing signals to the cell interior when the extracellular binding site of the antigen is occupied by antigen. These invariant polypeptides are thought to couple the antigen receptors on B cells to one or more members of the Src family of tyrosine protein kinases (including the Lyn kinase), thereby activating a phosphorylation cascade when antigen is bound (see Figure 23–54B).

Each B cell produces a single species of antibody, with a unique antigen-binding site. When a virgin or a memory B cell is activated by antigen (with the aid of helper T cells), it proliferates and matures to become an antibody-secreting cell. The activated cells make and secrete large amounts of soluble (rather than membrane-bound) antibody, which has the same unique antigen-binding site as the cell-surface antibody that served earlier as the antigen receptor (Figure 23–13). Activated B cells can begin secreting antibody while they are still small lymphocytes, but the end stage of their maturation pathway is a large plasma cell (see Figure 23–4B), which secretes antibodies at the rapid rate of about 2000 molecules per second. Plasma cells seem to have committed so much of their protein-synthesizing machinery to making antibody that they are incapable of further growth and division, and most die after several days.

B Cells Can Be Stimulated to Secrete Antibodies in a Culture Dish [12]

Two advances in the 1960s revolutionized research on B cells. The first was the development of the **hemolytic plaque assay,** which made it possible to identify and count individual activated B cells secreting antibody against a specific antigen. In the simplest form of this assay, lymphocytes (commonly from the spleen) are taken from animals that have been immunized with sheep red blood cells (SRBCs). They are then embedded in agar together with an excess of SRBCs so that the dish contains a "lawn" of immobilized SRBCs with occasional lymphocytes in it. Under these conditions the cells are unable to move, but any anti-SRBC antibody secreted by a B cell will diffuse outward and coat all SRBCs in the vicinity of the secreting cell. Once the SRBCs are coated with antibody, they can be killed by adding complement. In this way the presence of each antibody-secreting cell is indicated by the presence of a clear spot, or *plaque,* in the opaque layer of SRBCs. The same assay can be used to count cells making antibody to other antigens, such as proteins or polysaccharides, if these antigens are chemically coupled to the surface of the SRBC.

The second important advance was the demonstration that B cells can be induced to secrete antibodies by exposing them to antigen in a test tube or cell culture dish, where cell-cell interactions can be manipulated and the environment controlled. This led to the discovery that both helper T cells and specialized *antigen-presenting cells* are required for most antigens to stimulate virgin B cells to secrete antibodies, as we discuss later.

Antibodies Have Two Identical Antigen-binding Sites [13]

The simplest antibodies are Y-shaped molecules with two identical **antigen-binding sites,** one at the tip of each arm of the Y (Figure 23–14). Because of their two antigen-binding sites, they are said to be *bivalent.* As long as an antigen has three or more antigenic determinants, bivalent antibody molecules can cross-link it into a large lattice (Figure 23–15), which can be rapidly phagocytosed and degraded by macrophages. The efficiency of antigen binding and cross-linking is greatly increased by a flexible *hinge region* in antibodies, which allows the distance between the two antigen-binding sites to vary (Figure 23–16).

The protective effect of antibodies is not due simply to their ability to bind antigen. They engage in a variety of activities that are mediated by the tail of the Y-shaped molecule. This part of the molecule determines what will happen to the antigen once it is bound to the antibody. We shall see that antibodies with the same antigen-binding sites can have any one of several different tail regions, each of which confers on the antibody different functional properties, such as the ability to activate complement or to bind to phagocytic cells.

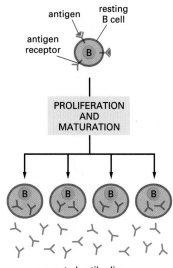

Figure 23–13 **B cell activation.** When resting B cells are activated by antigen to proliferate and mature into antibody-secreting cells, they produce and secrete antibodies with a unique antigen-binding site, which is the same as that of their original membrane-bound antibodies that served as antigen receptors.

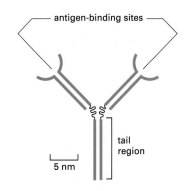

Figure 23–14 **A simple representation of an antibody molecule.** Note that its two antigen-binding sites are identical.

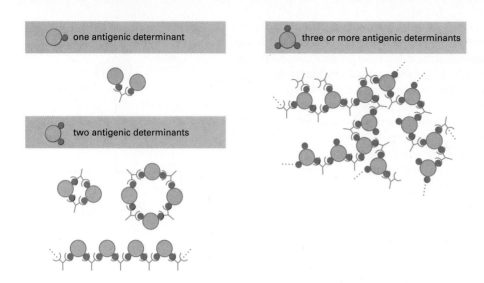

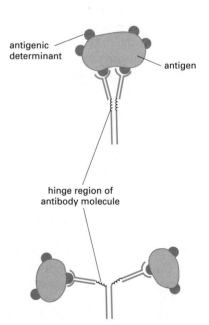

Figure 23–15 **Antibody-antigen interactions.** Because antibodies have two identical antigen-binding sites, they can cross-link antigens. The types of antibody-antigen complexes that form depend on the number of antigenic determinants on the antigen. Here a single species of antibody (a monoclonal antibody) is shown binding to antigens containing one, two, or three copies of a single type of antigenic determinant. Antigens with two antigenic determinants can form small cyclic complexes or linear chains with antibody, while antigens with three or more antigenic determinants can form large three-dimensional lattices that readily precipitate out of solution. Most antigens have many different antigenic determinants (see Figure 23–25A) and the different antibodies that recognize these different determinants can cooperate in cross-linking the antigen (not shown).

An Antibody Molecule Is Composed of Two Identical Light Chains and Two Identical Heavy Chains [13]

The basic structural unit of an antibody molecule consists of four polypeptide chains, two identical **light (L) chains** (each containing about 220 amino acids) and two identical **heavy (H) chains** (each usually containing about 440 amino acids). The four chains are held together by a combination of noncovalent and covalent (disulfide) bonds. The molecule is composed of two identical halves, each with the same antigen-binding site, and both light and heavy chains usually cooperate to form the antigen-binding surface (Figure 23–17).

The proteolytic enzymes papain and pepsin split antibody molecules into different characteristic fragments. *Papain* produces two separate and identical **Fab** (*f*ragment *a*ntigen *b*inding) **fragments,** each with one antigen-binding site, and one **Fc fragment** (so called because it readily crystallizes). Pepsin, on the other hand, produces one **F(ab')₂ fragment,** so called because it consists of two covalently linked F(ab') fragments (each slightly larger than a Fab fragment); the rest of the molecule is broken down into smaller fragments (Figure 23–18). Because F(ab')₂ fragments are bivalent, they can still cross-link antigens and form precipitates, unlike the univalent Fab fragments. Neither of these fragments has the other biological properties of intact antibody molecules because they lack the tail (Fc) region that is responsible for these properties.

There Are Five Classes of Heavy Chains, Each with Different Biological Properties [10, 14]

In higher vertebrates there are five *classes* of antibodies, IgA, IgD, IgE, IgG, and IgM, each with its own class of heavy chain—α, δ, ε, γ, and μ, respectively. IgA molecules have α chains, IgG molecules have γ chains, and so on. In addition, there are a number of subclasses of IgG and IgA immunoglobulins; for example, there are four human IgG subclasses (IgG1, IgG2, IgG3, and IgG4) having γ₁, γ₂, γ₃, and γ₄ heavy chains, respectively. The various heavy chains impart a distinctive conformation to the hinge and tail regions of antibodies and give each class (and subclass) characteristic properties of its own.

IgM, which has a μ heavy chain, is always the first class of antibody produced by a developing B cell, although many B cells eventually switch to making other classes of antibody (discussed below). The immediate precursor of a B cell, called a **pre-B cell,** initially makes μ chains, which associate with non-light-chain polypeptides (often referred to as surrogate light chains) and insert into the plasma membrane. As the synthesis of bona fide light chains increases, these

Figure 23–16 **The hinge region of an antibody molecule.** Because of its flexibility, the hinge region improves the efficiency of antigen binding and cross-linking.

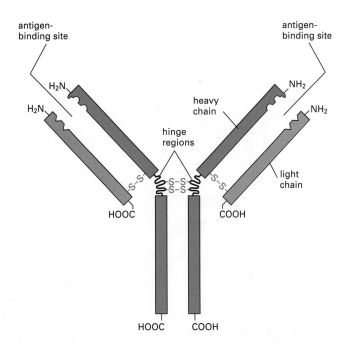

Figure 23–17 **A schematic drawing of a typical antibody molecule.** It is composed of two identical heavy chains and two identical light chains. Note that the antigen-binding sites are formed by a complex of the amino-terminal regions of both light and heavy chains, but the tail and hinge regions are formed by the heavy chains alone.

combine with the μ chains, displacing the surrogate light chains, to form a four-chain IgM molecule (with two μ chains and two light chains), which inserts into the plasma membrane. The cell now has cell-surface receptors with which it can bind antigen, and at this point it is called a *virgin B cell*. Many virgin B cells soon start to produce cell-surface **IgD** molecules as well, with the same antigen-binding site as the IgM molecules.

IgM is not only the first class of antibody to appear on the surface of a developing B cell, it is also the major class secreted into the blood in the early stages

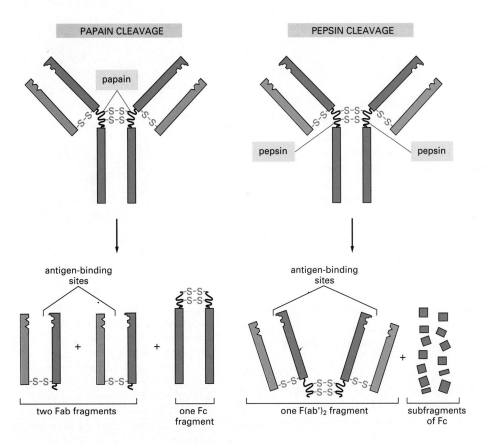

Figure 23–18 **Fab and F(ab')₂ antibody fragments.** These fragments are produced when antibody molecules are cleaved with the proteolytic enzymes papain and pepsin, respectively.

The Functional Properties of Antibodies

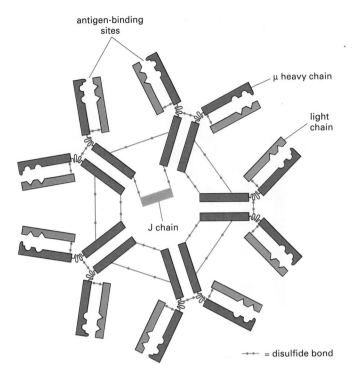

antigen-binding sites

μ heavy chain

light chain

J chain

= disulfide bond

Figure 23–19 A pentameric IgM molecule. The five subunits are held together by disulfide bonds. A single J chain, which has a structure similar to that of a single Ig domain (discussed later), is disulfide-bonded between two μ heavy chains. The J chain is required for the polymerization process. The addition of each successive four-chain IgM subunit requires a J chain, which is then discarded, except for the last one, which is retained.

of a *primary* antibody response. In its secreted form IgM is a pentamer composed of five four-chain units and thus has a total of 10 antigen-binding sites. Each pentamer contains one copy of another polypeptide chain, called a *J (joining) chain*. The J chain is produced by IgM-secreting cells and is covalently inserted between two adjacent tail (Fc) regions (Figure 23–19).

The binding of antigen to the Fab regions of the secreted pentameric IgM molecule induces the Fc regions to bind to and thereby activate the first component of the *complement system*. As we discuss later, when the antigen is on the surface of an invading microorganism, the resulting activation of complement unleashes a biochemical attack that kills the microorganism. Unlike IgM, IgD molecules are rarely secreted by an activated B cell, and their functions (other than as receptors for antigen) are unknown.

The major class of immunoglobulin in the blood is **IgG,** which is produced in large quantities during *secondary* immune responses. Besides activating the complement system, the Fc region of an IgG molecule binds to specific receptors on macrophages and neutrophils. Largely by means of such **Fc receptors,** these phagocytic cells bind, ingest, and destroy infecting microorganisms that have become coated with the IgG antibodies produced in response to the infection (Figure 23–20). Some white blood cells that express Fc receptors can also kill IgG-coated foreign eucaryotic cells without phagocytosing them.

IgG molecules are the only antibodies that can pass from mother to fetus via the placenta. Cells of the placenta that are in contact with maternal blood have Fc receptors that bind IgG molecules and mediate their passage to the fetus. The antibodies are first taken up from the maternal blood by receptor-mediated endocytosis and then transported across the cell in vesicles and released by exocytosis into the fetal blood (a process called *transcytosis,* discussed in Chapter 13). Because other classes of antibodies do not bind to these receptors, they cannot pass across the placenta. IgG is also secreted into the mother's milk and is taken up from the gut of the neonate into the blood.

IgA is the principal class of antibody in secretions (saliva, tears, milk, and respiratory and intestinal secretions) (Figure 23–21). It is transported through secretory epithelial cells from the extracellular fluid into the secreted fluid by another type of Fc receptor that is unique to secretory epithelia (Figure 23–22).

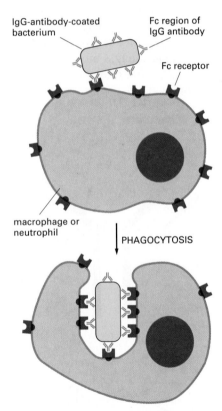

IgG-antibody-coated bacterium

Fc region of IgG antibody

Fc receptor

macrophage or neutrophil

PHAGOCYTOSIS

Figure 23–20 Antibody-activated phagocytosis. An IgG-antibody-coated bacterium is efficiently phagocytosed by a macrophage or neutrophil, which has cell-surface receptors able to bind the Fc region of IgG molecules. The binding of the antibody-covered bacterium to these Fc receptors activates the phagocytic process.

The Fc region of **IgE** molecules binds with unusually high affinity ($K_a \approx 10^{10}$ liters/mole) to yet another class of Fc receptors. These receptors are located on the surface of *mast cells* in tissues and on *basophils* in the blood, and the IgE molecules bound to them in turn serve as receptors for antigen. Antigen binding triggers the cells to secrete a variety of biologically active amines, especially *histamine* (Figure 23–23). These amines cause dilation and increased permeability of blood vessels and are largely responsible for the clinical manifestations of such *allergic* reactions as hay fever, asthma, and hives. In normal circumstances the blood vessel changes are thought to help white blood cells, antibodies, and complement components to enter sites of inflammation. Mast cells also secrete factors that attract and activate a special class of white blood cells called *eosinophils*, which can kill various types of parasites, especially if the parasites are coated with IgE or IgA antibodies.

The properties of the various classes of antibodies in humans are summarized in Table 23–1.

Antibodies Can Have Either κ or λ Light Chains, but Not Both

In addition to the five classes of heavy chains, higher vertebrates have two types of light chains, κ and λ, either of which may be associated with any of the heavy chains. An individual antibody molecule always consists of identical light chains and identical heavy chains; therefore, its two antigen-binding sites are always identical. This symmetry is crucial for the cross-linking function of secreted antibodies. An Ig molecule, consequently, may have either κ or λ light chains, but not both. No difference in the biological function of these two types of light chain has yet been identified.

The Strength of an Antibody-Antigen Interaction Depends on Both the Number of Antigen-binding Sites Occupied and the Affinity of Each Binding Site [10, 15]

The binding of an antigen to antibody, like the binding of a substrate to an enzyme, is reversible. It is mediated by the sum of many relatively weak noncovalent forces, including hydrophobic and hydrogen bonds, van der Waals forces, and ionic interactions. These weak forces are effective only when the antigen molecule is close enough to allow some of its atoms to fit into complementary re-

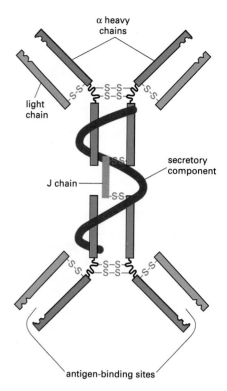

Figure 23–21 A highly schematized diagram of a dimeric IgA molecule found in secretions. In addition to the two IgA monomers, there is a single J chain and an additional polypeptide chain called the *secretory component*, which is thought to protect the IgA molecules from being digested by proteolytic enzymes in the secretions.

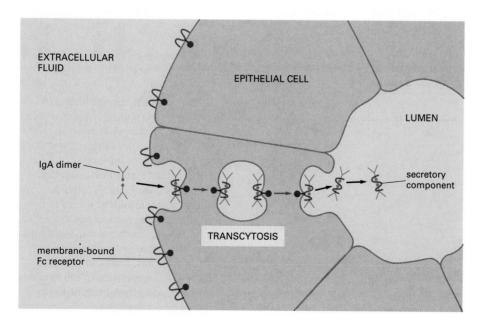

Figure 23–22 The mechanism of transport of a dimeric IgA molecule across an epithelial cell. The IgA molecule, as a J-chain-containing dimer, binds to a specialized transmembrane Fc receptor protein on the nonluminal surface of the secretory epithelial cell. The receptor-IgA complexes are ingested by receptor-mediated endocytosis, transferred across the epithelial cell cytoplasm in vesicles, and secreted into the lumen on the opposite side of the cell by exocytosis. When exposed to the lumen, the part of the Fc receptor protein that is bound to the IgA dimer (the *secretory component*) is cleaved from its transmembrane tail, thereby releasing the antibody in the form shown in Figure 23–21.

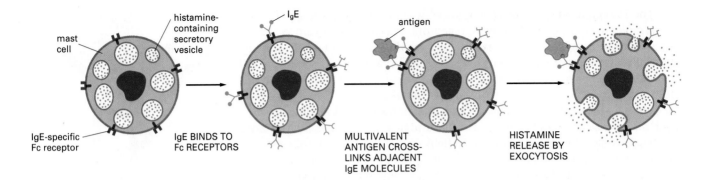

histamine-containing secretory vesicle

IgE

mast cell

antigen

IgE-specific Fc receptor

IgE BINDS TO Fc RECEPTORS

MULTIVALENT ANTIGEN CROSS-LINKS ADJACENT IgE MOLECULES

HISTAMINE RELEASE BY EXOCYTOSIS

cesses on the surface of the antibody. The complementary regions of a four-chain antibody unit are its two identical antigen-binding sites; the corresponding region on the antigen is an *antigenic determinant* (Figure 23–24). Most antigenic macromolecules have many different antigenic determinants; if two or more of them are identical (as in a polymer with a repeating structure), the antigen is said to be *multivalent* (Figure 23–25).

The reversible binding reaction between an antigen with a single antigenic determinant (denoted Ag) and a single antigen-binding site (denoted Ab) can be expressed as

$$Ag + Ab \rightleftharpoons AgAb$$

The equilibrium point depends both on the concentrations of Ab and Ag and on the strength of their interaction. Clearly, a larger fraction of Ab will become associated with Ag as the concentration of Ag is increased. The strength of the interaction is generally expressed as the **affinity constant (K_a)** (see Figure 3–9), where

$$K_a = \frac{[AgAb]}{[Ag][Ab]}$$

(the square brackets indicate the concentration of each component at equilibrium).

The affinity constant, sometimes called the association constant, can be determined by measuring the concentration of free Ag required to fill half of the antigen-binding sites on the antibody. When half the sites are filled, [AgAb] = [Ab] and $K_a = 1/[Ag]$. Thus the reciprocal of the antigen concentration that produces half-maximal binding is equal to the affinity constant of the antibody for the antigen. Common values range from as low as 5×10^4 to as high as 10^{11} liters/mole. The affinity constant at which an immunoglobulin molecule ceases to be considered an antibody for a particular antigen is somewhat arbitrary, but it is

Figure 23–23 **The role of IgE in histamine secretion by mast cells.** A mast cell (or a basophil) binds IgE molecules after they are secreted by activated B cells; the soluble IgE antibodies bind to Fc receptor proteins on the mast cell surface that specifically recognize the Fc region of these antibodies. The passively acquired IgE molecules on the mast cell serve as cell-surface receptors for antigen. Thus, unlike B cells, each mast cell (and basophil) has a set of cell-surface antibodies with a wide variety of antigen-binding sites. When an antigen molecule binds to these membrane-bound IgE antibodies so as to cross-link them to their neighbors, it activates the mast cell to release its histamine by exocytosis.

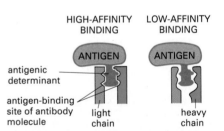

HIGH-AFFINITY BINDING

LOW-AFFINITY BINDING

ANTIGEN

ANTIGEN

antigenic determinant

antigen-binding site of antibody molecule

light chain

heavy chain

Figure 23–24 **Antigen binding to antibody.** In this highly schematized diagram, an antigenic determinant on a macromolecule is shown interacting with the antigen-binding site of two different antibody molecules, one of high and one of low affinity. The antigenic determinant is held in the binding site by various weak noncovalent forces, and the site with the better fit to the antigen has a greater affinity. Note that both the light and heavy chains of the antibody molecule usually contribute to the antigen-binding site.

Table 23–1 **Properties of the Major Classes of Antibody in Humans**

Properties	Class of Antibody				
	IgM	IgD	IgG	IgA	IgE
Heavy chains	μ	δ	γ	α	ε
Light chains	κ or λ	κ or λ	κ or λ	κ or λ	κ or λ
Number of four-chain units	5	1	1	1 or 2	1
Percent of total Ig in blood	10	<1	75	15	<1
Activates complement	++++	–	++	–	–
Crosses placenta	–	–	+	–	–
Binds to macrophages and neutrophils	–	–	+	–	–
Binds to mast cells and basophils	–	–	–	–	+

unlikely that an antibody with a K_a below 10^4 would be biologically effective; moreover, B cells with receptors that have such a low affinity for an antigen are unlikely to be activated by the antigen.

The **affinity** of an antibody for an antigenic determinant describes the strength of binding of a single copy of the antigenic determinant to a single antigen-binding site, and it is independent of the number of sites. When, however, an antigen carrying multiple copies of the same antigenic determinant combines with a multivalent antibody, the binding strength is greatly increased because all of the antigen-antibody bonds must be broken simultaneously before the antigen and antibody can dissociate. Thus a typical IgG molecule can bind at least 50–100 times more strongly to a multivalent antigen if both antigen-binding sites are engaged than if only one site is engaged. The total binding strength of a multivalent antibody with a multivalent antigen is referred to as the **avidity** of the interaction.

If the affinity of the antigen-binding sites in an IgG and an IgM molecule is the same, the IgM molecule (with 10 binding sites) will have a very much greater avidity for a multivalent antigen than an IgG molecule (which has two sites). This difference in avidity, often 10^4-fold or more, is important because antibodies produced early in an immune response usually have much lower affinities than those produced later. (The increase in the average affinity of antibodies produced with time after immunization, called *affinity maturation,* is discussed later.) Because of its high total avidity, IgM—the major Ig class produced early in immune responses—can function effectively even when each of its binding sites has only a low affinity.

Antibodies Recruit Complement to Help Fight Bacterial Infections [16]

Complement, so called because it *complements* and amplifies the action of antibody, is one of the principal means by which antibodies defend vertebrates against most bacterial infections. Individuals with a deficiency in one of the central complement components (called C3) are subject to repeated bacterial infections, just as are individuals deficient in antibodies themselves.

The complement system consists of about 20 interacting soluble proteins that are made mainly by the liver and circulate in the blood and extracellular fluid. Most are inactive until they are triggered by an immune response or, more directly, by an invading microorganism itself. The ultimate consequence of comple-

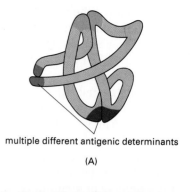

multiple different antigenic determinants

(A)

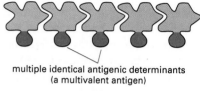

multiple identical antigenic determinants
(a multivalent antigen)

(B)

Figure 23–25 Molecules with multiple antigenic determinants. (A) A globular protein with a number of different antigenic determinants. Different regions of a polypeptide chain usually come together in the folded structure to form each antigenic determinant on the surface of the protein. (B) A polymeric structure with many *identical* antigenic determinants; such a molecule is called a *multivalent antigen.*

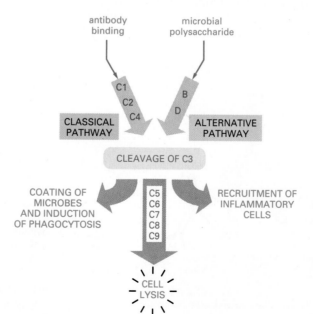

Figure 23–26 The principal stages in complement activation by the classical and alternative pathways. In both pathways the reactions of complement activation usually take place on the surface of an invading microbe, such as a bacterium. C1–C9 and factors B and D are the reacting components of the complement system; various other components (not shown) regulate the system. The *early components* are shown within *gray arrows,* while the *late components* are shown within a *brown arrow.*

The Functional Properties of Antibodies

ment activation is the assembly of the so-called *late complement components* into large protein complexes, called *membrane attack complexes,* that form holes in the membrane of a microorganism and thereby destroy the microorganism.

Because one of its main functions is to attack the membrane of microbial cells, the activation of complement is focused on the microbial cell membrane, where it is triggered either by antibody bound to the microorganism or by microbial envelope polysaccharides, both of which activate the *early complement components*. There are two sets of early components belonging to two distinct pathways of complement activation, the **classical pathway** and the **alternative pathway.** The early components of both pathways act locally to activate **C3,** which is the pivotal component of complement, whose cleavage leads not only to the assembly of membrane attack complexes but also to the recruitment of various white blood cells (Figure 23–26).

The early components and C3 are proenzymes that are activated sequentially by limited proteolytic cleavage: the cleavage of each proenzyme in the sequence activates the component to generate a serine protease, which cleaves the next proenzyme in the sequence, and so on. Since each activated enzyme cleaves many molecules of the next proenzyme in the chain, the activation of the early components consists of an amplifying *proteolytic cascade*. Thus each molecule activated at the beginning of the sequence leads to the production of many active components, including many membrane attack complexes.

Many of these cleavages liberate a small peptide fragment and thereby expose a membrane-binding site on the larger fragment, which binds tightly to the target cell membrane and helps to carry out the next reaction in the sequence, eventually leading to the formation of membrane attack complexes. In this way complement activation is confined largely to the particular cell surface where it began. The larger fragment of C3 is called C3b. It binds covalently to the surface of a target cell. There it not only acts as a protease to catalyze the subsequent steps in the complement cascade, but also is recognized by specific receptor proteins on macrophages and neutrophils that enhance the ability of these cells to phagocytose the target cell. The smaller fragment of C3 (called C3a) acts independently as a diffusible signal that promotes an inflammatory response by encouraging white blood cells to migrate into the site of infection.

The classical pathway is usually activated by clusters of IgG or IgM antibodies bound to antigens on the surface of a microorganism. The first step in this pathway is illustrated in Figure 23–27. The alternative pathway, by contrast, is activated by polysaccharides in the cell envelopes of microorganisms even in the absence of antibody, although activation of the classical pathway also activates the alternative pathway through a positive feedback loop. The alternative pathway therefore provides a first line of defense against infection before an immune response can be mounted, and it also amplifies the effects of the classical pathway once an immune response has begun.

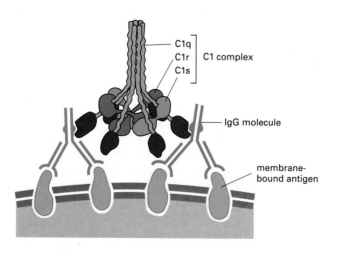

Figure 23–27 The complex structure of C1. The binding of two or more IgG molecules (or one pentameric IgM molecule—not shown) to the surface of a microorganism enables their Fc regions to bind the first component of the classical pathway, C1, which is a large complex composed of three subcomponents—C1q, C1r, and C1s. When C1q binds to antibody-antigen complexes, it activates C1r, which becomes proteolytic, cleaving C1s to begin the proteolytic cascade. Note that C1q is composed of six identical subunits, each with a globular head (which binds to antibody) and a collagenlike tail.

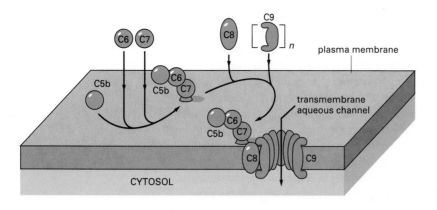

plasma membrane

transmembrane
aqueous channel

CYTOSOL

Figure 23–28 Assembly of the late complement components to form a membrane attack complex. When C3b is produced by either the classical or alternative pathway, it is immobilized on a membrane, where it causes the cleavage of a complement protein called C5 to produce C5a (not shown) and C5b. C5b remains loosely bound to C3b (not shown) and rapidly assembles with C6 and C7 to form C567, which then binds firmly via C7 to the membrane, as illustrated. This complex adds one molecule of C8 to form C5678. The binding of a molecule of C9 to C5678 induces a conformational change in the C9 that exposes a hydrophobic region and causes the C9 to insert into the lipid bilayer of the target cell next to C8. This starts a chain reaction in which the altered C9 binds a second molecule of C9, which undergoes a conformational change and inserts into the bilayer, where it can bind another molecule of C9, and so on. In this way a large transmembrane channel is formed by a chain of C9 molecules.

Membrane-immobilized C3b, produced by either the classical or alternative pathway, triggers a further cascade of reactions that leads to the assembly of **membrane attack complexes** from the late components (Figure 23–28). These complexes form in the membrane near the site of C3 activation and have a characteristic appearance in negatively stained electron micrographs, where they are seen to form aqueous pores through the membrane (Figure 23–29). For this reason, and because they perturb the structure of the lipid bilayer in their vicinity, they make the membrane leaky. Small molecules leak into and out of the cell around and through the complexes while macromolecules remain inside, so that the cell's normal mechanism for controlling water balance is disrupted. Water is therefore drawn into the cell by osmosis, causing it to swell and burst. The process is so efficient that a very small number of membrane attack complexes (perhaps even one) can lyse a red blood cell. Even an enveloped virus, which does not have a large osmotic pressure gradient across its membrane and is therefore not susceptible to such osmotic lysis, can be destroyed by these complexes, presumably because they disorganize the viral membrane.

The self-amplifying destructive properties of the complement cascade make it essential that key activated components be rapidly inactivated after they are generated to ensure that the attack does not spread to nearby host cells. Deactivation is achieved in at least two ways. First, specific inhibitor proteins in the blood terminate the cascade by either binding or cleaving certain components once they have been activated by proteolytic cleavage. Second, many of the activated components in the cascade are unstable; unless they bind immediately to either an appropriate component in the chain or a nearby membrane, they rapidly become inactive.

Summary

A typical antibody molecule is a Y-shaped protein with two identical antigen-binding sites at the tips of the Y (the Fab regions) and binding sites for complement components and/or various cell-surface receptors on the tail of the Y (the Fc region). Antibodies defend vertebrates against infection by inactivating viruses and bacterial toxins and by recruiting the complement system and various cells to kill and ingest invading microorganisms.

Each B cell clone makes antibody molecules with a unique antigen-binding site. Initially, the molecules are inserted into the plasma membrane, where they serve as receptors for antigen. Antigen binding to these receptors activates the B cells (usually with the aid of helper T cells) to multiply and mature either into memory cells or into antibody-secreting cells, which secrete antibodies with the same unique antigen-binding site as the membrane-bound antibodies.

Each antibody molecule is composed of two identical heavy chains and two identical light chains. Typically, parts of both the heavy and light chains form the antigen-binding sites. There are five classes of antibodies (IgA, IgD, IgE, IgG, and IgM), each with a distinctive heavy chain (α, δ, λ, γ, and μ, respectively). The heavy chains

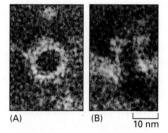

(A)   (B)   10 nm

Figure 23–29 Electron micrographs of negatively stained complement lesions in the plasma membrane of a red blood cell. The lesion in (A) is seen *en face*, while that in (B) is seen from the side as an apparent transmembrane channel. The negative stain fills the individual channels, which therefore look black. (From R. Dourmashkin, *Immunology* 35:205–212, 1978.)

also form the Fc region of the antibody, which determines what other proteins will bind to the antibody and therefore what biological properties the antibody class has. Either type of light chain (κ or λ) can be associated with any class of heavy chain, but the type of light chain does not seem to influence the properties of the antibody.

The complement system cooperates with antibodies to defend vertebrates against infection. The early components are proenzymes that circulate in the blood and are sequentially activated in an amplifying series of limited proteolytic reactions. The most important complement component is the C3 protein, which is activated by proteolytic cleavage and binds to the membrane of a microbial cell, where it helps to initiate the local assembly of the late complement components and to induce the phagocytosis of the microbial cell. The late components form large membrane attack complexes in the microbial cell membrane and thereby kill the invading microorganism.

The Fine Structure of Antibodies

Because antibodies exist in so many forms, in an unimmunized individual any one form will constitute a minute fraction of the Ig molecules in the blood. This fact presented immunochemists with a uniquely difficult problem in protein chemistry: how to obtain enough of any one antibody molecule to determine its amino acid sequence and three-dimensional structure.

The problem was solved by the discovery that the cells of a type of cancer known as **multiple myeloma** (because multiple tumors develop in the bone marrow, or myeloid tissues) secrete large amounts of a single species of antibody into the patient's blood. The antibody is homogeneous, or monoclonal, because cancer usually begins with the uncontrolled growth of a single cell, and in multiple myeloma the single cell is an antibody-secreting plasma cell. The antibody, which accumulates in the blood, is known as a **myeloma protein.**

The detailed structure of antibodies was initially determined by studying myeloma proteins from patients or from mice in which similar tumors had been purposely induced. Later it became possible to immortalize single antibody-secreting B cells by fusing them with non-antibody-secreting myeloma cells. The resultant *hybridomas* provide a ready source of monoclonal antibodies, which can be produced in unlimited amounts against any desired antigen, as discussed in Chapter 4. Today, homogeneous antibodies can also be produced in unlimited quantities by recombinant DNA technology.

Light and Heavy Chains Consist of Constant and Variable Regions [10, 13]

Comparison of the amino acid sequences of different myeloma proteins reveals a striking feature with important genetic implications. Both light and heavy chains have a variable sequence at their amino-terminal ends but a constant sequence at their carboxyl-terminal ends. When the amino acid sequences of many different myeloma κ chains are compared, for example, the carboxyl-terminal halves are the same or show only minor differences, whereas the amino-terminal halves are all different. Thus light chains have a **constant region** about

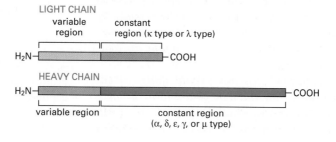

Figure 23–30 **Constant and variable regions of immunoglobulin chains.** Both light and heavy chains of an Ig molecule have distinct constant and variable regions.

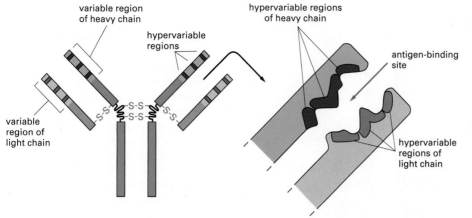

Figure 23–31 Antibody hypervariable regions. Highly schematized drawing of how the three hypervariable regions in each light and heavy chain together form the antigen-binding site of an antibody molecule. The hypervariable regions are sometimes called *complementarity-determining regions.* The actual three-dimensional structure of an antigen-binding site is shown in Figure 23–35.

110 amino acids long and a **variable region** of the same size. The variable region of the heavy chains (at their amino terminus) is also about 110 amino acids long, but the heavy-chain constant region is about 330 or 440 amino acids long, depending on the class (Figure 23–30).

It is the amino-terminal ends of the light and heavy chains that come together to form the antigen-binding site (see Figure 23–17), and the variability of their amino acid sequences provides the structural basis for the diversity of antigen-binding sites. The existence of variable and constant regions raises important questions about the genetic mechanisms that produce antibody molecules, and we consider these later. Before it became possible to investigate these genetic questions directly, other important features of antibody molecules emerged from structural studies on myeloma proteins.

The Light and Heavy Chains Each Contain Three Hypervariable Regions That Together Form the Antigen-binding Site [17]

Scrutiny of the amino acid sequences of a variety of Ig chains shows that the variability in the variable regions of both light and heavy chains is for the most part restricted to three small **hypervariable regions** in each chain. The remaining parts of the variable region, known as *framework regions,* are relatively constant. These findings led to the prediction that only the 5 to 10 amino acids in each hypervariable region form the antigen-binding site (Figure 23–31). This prediction has since been confirmed by x-ray diffraction studies of antibody molecules (see below). In agreement with the size of the antigen-binding site of an antibody molecule, the antigenic determinant that is specifically recognized by an antibody is generally comparably small: it can consist of fewer than 25 amino acid residues on the surface of a globular protein (see Figure 23–35), for example, and can be as small as a dinitrophenyl group (see Figure 23–7).

The Light and Heavy Chains Are Folded into Repeating Similar Domains [10, 18]

Both light and heavy chains are made up of repeating segments—each about 110 amino acids long and each containing one intrachain disulfide bond—that fold independently to form compact functional units, or **domains.** As shown in Figure 23–32, a light chain consists of one variable (V_L) and one constant (C_L) domain, while most heavy chains consist of a variable domain (V_H) and three constant domains (C_H1, C_H2, and C_H3). (The μ and ε chains each have one variable and four constant domains.) The variable domains are responsible for antigen binding, while the constant domains of the heavy chains (excluding C_H1) form the Fc region that determines the other biological properties of the antibody.

The Fine Structure of Antibodies

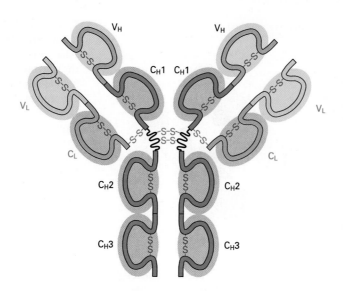

Figure 23–32 **Immunoglobulin domains.** The light and heavy chains in an Ig molecule are each folded into repeating domains that are similar to one another. The variable domains (shaded in *blue*) of the light and heavy chains (V_L and V_H) make up the antigen-binding sites, while the constant domains of the heavy chains (mainly C_H2 and C_H3) determine the other biological properties of the molecule. The heavy chains of IgM and IgE antibodies have an extra constant domain (C_H4). Hydrophobic interactions between domains on adjacent Ig chains play an important part in holding the chains together in the Ig molecule: C_L binds to C_H1, for example, and the C_H3 domains bind to each other.

The similarity between their domains suggests that Ig chains arose during evolution by a series of gene duplications, beginning with a primordial gene coding for a single 110 amino acid domain of unknown function. This hypothesis is supported by the finding that each domain of the constant region of a heavy chain is encoded by a separate coding sequence (exon) (Figure 23–33).

X-ray Diffraction Studies Have Revealed the Structure of Ig Domains and Antigen-binding Sites in Three Dimensions [19]

Even when the complete amino acid sequence of a large protein is known, it is not yet possible to deduce its three-dimensional structure; x-ray diffraction studies of protein crystals are generally needed. A number of fragments of both myeloma proteins and antibodies, as well as an intact IgG molecule, have been crystallized, and x-ray studies of their structures have confirmed the predictions of the immunochemists. More important, these studies have revealed the way in which millions of different antigen-binding sites are constructed on a common structural theme.

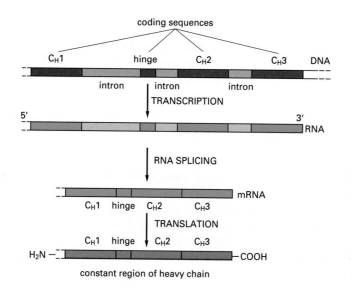

Figure 23–33 **The organization of the DNA sequences that encode the constant region of an Ig heavy chain.** The coding sequences (exons) for each domain and for the hinge region are separated by noncoding sequences (introns). The intron sequences are removed by splicing the primary RNA transcripts to form mRNA. The presence of introns in the DNA is thought to have facilitated accidental duplications of DNA segments that gave rise to the antibody genes during evolution (discussed in Chapter 8). The DNA and RNA sequences that encode the variable region of the heavy chain are not shown.

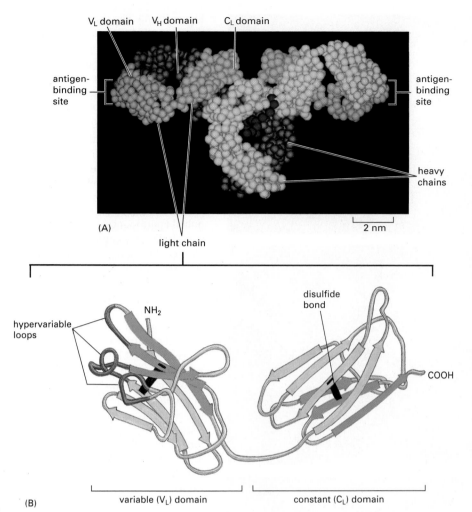

(A)

2 nm

light chain

(B)

Figure 23–34 The folded structure of an IgG antibody molecule, based on x-ray crystallography studies. (A) Each amino acid residue in the protein is shown as a small sphere. One heavy chain is shown in *light blue,* the other in *dark blue,* with the light-chain domains in *yellow.* All antibody molecules are glycosylated: the oligosaccharide chain attached to a C_H2 domain is shown in *red.* (B) The path of the polypeptide chain for an entire light chain. Both the variable and constant domains consist of two β sheets—one composed of three strands (*green*) and one composed of four strands (*yellow*). The sheets are joined by a disulfide bond (*black*). Note that all the hypervariable regions (*red*) form loops at the far end of the variable domain, where they come together to form part of the antigen-binding site. (A, after E.W. Silverton, M.A. Navia, and D.R. Davies, *Proc. Natl. Acad. Sci. USA* 74:5140, 1977; B, after M. Schiffer, R.L. Girling, K.R. Ely, and A.B. Edmundson, *Biochemistry* 12:4620, 1973. Copyright 1973 American Chemical Society.)

As illustrated in Figure 23–34, each Ig domain has a very similar three-dimensional structure based on what is called the **immunoglobulin fold.** Each domain is roughly a cylinder ($4 \times 2.5 \times 2.5$ nm) composed of a "sandwich" of two extended protein layers: one layer contains three strands of polypeptide chain and the other contains four. In each layer the adjacent strands are antiparallel and form a β sheet. The two layers are roughly parallel and are connected by a single intrachain disulfide bond. We shall see later that many other proteins on the surface of lymphocytes and other cells, many of which function as cell-cell adhesion molecules (discussed in Chapter 19), contain similar domains and hence are members of a very large *immunoglobulin (Ig) superfamily* of proteins.

The variable domains of Ig molecules are unique in that each has its particular set of three hypervariable regions, which are arranged in three *hypervariable loops* (see Figure 23–34B). The hypervariable loops of both the light and heavy variable domains are clustered together to form the antigen-binding site, as had been predicted. An important principle to emerge from these studies is that the variable region of an antibody molecule consists of a highly conserved rigid framework, with hypervariable loops attached at one end. Therefore, an enormous diversity of antigen-binding sites can be generated by changing only the length and amino acid sequence of the hypervariable loops without disturbing the overall three-dimensional structure necessary for antibody function.

X-ray analysis of crystals of antibody fragments bound to an antigenic determinant has revealed exactly how the hypervariable loops of the light and heavy

The Fine Structure of Antibodies

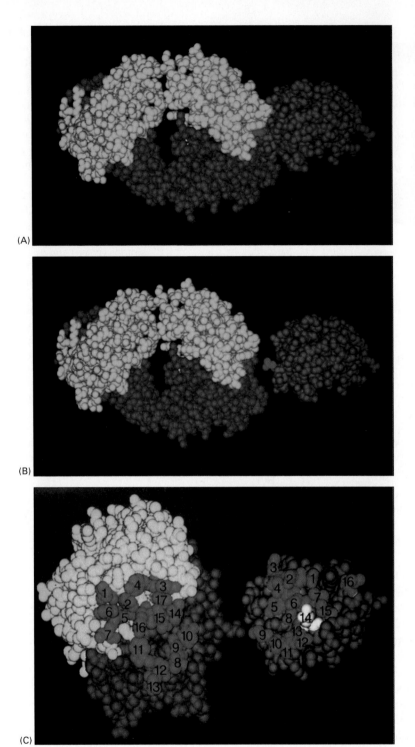

(A)

(B)

(C)

Figure 23–35 The three-dimensional structure of an antigen-antibody complex as determined by x-ray diffraction analysis. The protein antigen, which is the enzyme lysozyme, is shown in *green*. The antigen-binding site of the Fab fragment of the antibody is formed by both the light chain (*yellow*) and the heavy chain (*blue*). About 20 amino acid residues in the binding site contact a similar number of residues on the lysozyme surface. In (B) the antigen and antibody have been pulled apart to reveal their complementary contacting surfaces. The bit of the antigen that protrudes from the complementary surface (shown in *red*) is a glutamine residue. In (C) the separated molecules have been rotated about 90 degrees around the vertical axis from (A) to show the interacting surfaces; the amino acid side chains that interact are shown in *red*, with the protruding glutamine now in *pink*. In several other antibody molecules that have been studied in this way, the antigen-binding site (for a small antigenic determinant) is formed by a much deeper cleft. (From A. Amit, R. Mariuzza, S. Phillips, and R. Poljak, S*cience* 233:747–753, 1986. Copyright 1986 by the AAAS.)

variable domains cooperate to form an antigen-binding surface in particular cases (Figure 23–35). The dimensions and shape of each different site vary depending on the conformation of the polypeptide chain in the hypervariable loops, which in turn is determined by the sequence of the amino acid side chains in the loops. The shapes of binding sites vary greatly—from clefts, to grooves, to flatter undulating surfaces, and even to protrusions—depending on the antibody. Smaller ligands tend to bind to deeper pockets, whereas larger ones tend to bind to flatter surfaces. In addition, the binding site can alter its shape following antigen binding to better fit the ligand. Thus the general principles of antibody structure are now clear.

Summary

Each immunoglobulin light and heavy chain consists of a variable region of about 110 amino acids at its amino-terminal end, followed by a constant region, which is the same size as the variable region in the light chain and three or four times larger in the heavy chain. Each chain is composed of repeating, similarly folded domains: a light chain has one variable-region (V_L) and one constant-region (C_L) domain, while a heavy chain has one variable-region (V_H) and three or four constant-region (C_H) domains. The amino acid sequence variation in the variable regions of both light and heavy chains is for the most part confined to several small hypervariable regions; they form protruding surface loops that come together to form the antigen-binding site.

The Generation of Antibody Diversity

It is estimated that even in the absence of antigen stimulation a human makes at least 10^{15} different antibody molecules—its *preimmune antibody repertoire.* The antigen-binding sites of many antibodies can cross-react with a variety of related but different antigenic determinants, and the preimmune repertoire is apparently large enough to ensure that there will be an antigen-binding site to fit almost any potential antigenic determinant, albeit with low affinity.

Antibodies are proteins, and proteins are encoded by genes. Antibody diversity therefore poses a special genetic problem: how can an animal make more antibodies than there are genes in its genome? (The human genome, for example, is thought to contain fewer than 10^5 genes.) This problem is not quite as formidable as it might first appear. Because the variable regions of both the light and heavy chains contribute to an antigen-binding site, an animal with 1000 genes encoding light chains and 1000 genes encoding heavy chains could combine their products in 1000×1000 different ways to make 10^6 different antigen-binding sites (assuming that any light chain can combine with any heavy chain to make an antigen-binding site). Nonetheless, the mammalian immune system has evolved unique genetic mechanisms that enable it to generate an almost unlimited number of different light and heavy chains in a remarkably economical way by joining separate *gene segments* together before they are transcribed. Birds and fish use very different strategies for diversifying antibodies, but we shall confine our discussion to the mechanisms used by mammals.

Antibody Genes Are Assembled from Separate Gene Segments During B Cell Development [20]

The first direct evidence that DNA is rearranged during B cell development came from experiments done in 1976 in which DNA from early mouse embryos, which do not make antibodies, was compared with the DNA of a mouse myeloma cell line, which does. The experiments showed that the specific variable (V)-region and constant (C)-region coding sequences used by the myeloma cells were present on the same DNA restriction fragment in the myeloma cells but on two different restriction fragments in the embryos, demonstrating that the DNA sequences encoding an antibody molecule are rearranged at some stage in the differentiation of B cells (Figure 23–36).

It is now known that for each type of Ig chain—κ light chains, λ light chains, and heavy chains—there is a separate pool of **gene segments** from which a single polypeptide chain is eventually synthesized. Each pool is on a different chromosome and usually contains a large number of gene segments encoding the V region of an Ig chain and a smaller number of gene segments encoding the C region. During B cell development a complete coding sequence for each of the two Ig chains to be synthesized is assembled by site-specific genetic recombination (discussed in Chapter 6), bringing together the entire coding sequence for a V

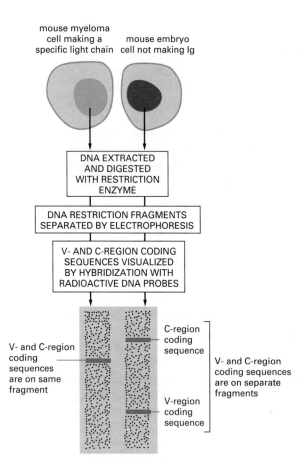

Figure 23–36 The experiment that first directly demonstrated that DNA is rearranged during B cell development. The two radioactive DNA probes used were specific for the DNA sequences encoding the C region and the V region of the myeloma light chain.

Labels in figure:

mouse myeloma cell making a specific light chain

mouse embryo cell not making Ig

DNA EXTRACTED AND DIGESTED WITH RESTRICTION ENZYME

DNA RESTRICTION FRAGMENTS SEPARATED BY ELECTROPHORESIS

V- AND C-REGION CODING SEQUENCES VISUALIZED BY HYBRIDIZATION WITH RADIOACTIVE DNA PROBES

V- and C-region coding sequences are on same fragment

C-region coding sequence

V-region coding sequence

V- and C-region coding sequences are on separate fragments

region and the coding sequence for a C region. In addition to bringing together the separate gene segments of the antibody gene, these rearrangements also activate transcription from the gene promoter through changes in the relative positions of the enhancers and silencers acting on the promoter. Thus a complete Ig chain can be synthesized only after a DNA rearrangement has occurred. As we shall see, the process of joining gene segments contributes to the diversity of antigen-binding sites in several ways.

Each V Region Is Encoded by More Than One Gene Segment [21]

When genomic DNA sequences encoding V and C regions were analyzed, it was found that a single *C gene segment* encodes the C region of an Ig chain, but two or more gene segments are combined to encode each V region. Each light-chain V region is encoded by a DNA sequence assembled from two gene segments— a long *V gene segment* and a short *joining,* or *J gene segment* (not to be confused with the protein *J chain* (see Figure 23–19), which is encoded elsewhere in the genome. Figure 23–37 illustrates the genetic mechanisms involved in producing an intact light-chain polypeptide from separate *V, J,* and *C* gene segments.

Each heavy-chain V region is encoded by a DNA sequence assembled from three gene segments—a *V* segment, a *J* segment, and a *diversity segment,* or *D gene segment.* Figure 23–38 shows the organization of the gene segments used in making heavy chains.

The large number of inherited *V, J,* and *D* gene segments available for encoding Ig chains makes a substantial contribution on its own to antibody diversity, but the combinatorial joining of these segments (called **combinatorial diversification**) greatly increases this contribution. Any of the 300 or so *V* segments in the mouse κ light-chain gene-segment pool, for example, can be joined to any of the 4 *J* segments (see Figure 23–37), so that at least 1200 (300 × 4) different κ-

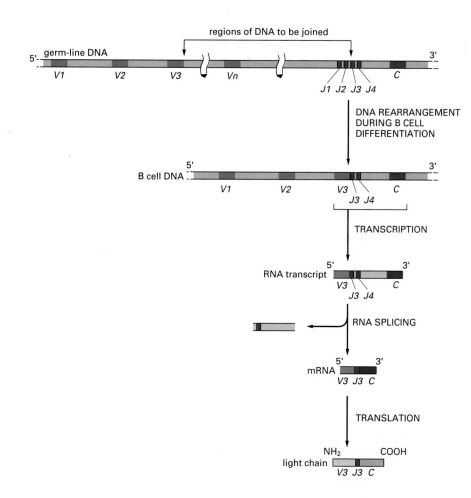

Figure 23–37 **The V-J joining process involved in making a κ light chain in the mouse.** In the "germ-line" DNA (where the immunoglobulin genes are not being expressed and are therefore not rearranged), the cluster of four *J* gene segments is separated from the *C* gene segment by a short intron and from the 300 or so *V* gene segments by thousands of nucleotide pairs. During B cell development the chosen *V* gene segment (*V3* in this case) is moved to lie precisely next to one of the *J* gene segments (*J3* in this case). The "extra" *J* gene segment (*J4*) and the intron sequence are transcribed (along with the joined *V3*, *J3*, and *C* gene segments) and then removed by RNA splicing to generate mRNA molecules in which the *V3*, *J3*, and *C* sequences are contiguous. These mRNAs are then translated into κ light chains. A *J* gene segment encodes the carboxyl-terminal 15 or so amino acids of the V region, and the V-J segment junction coincides with the third hypervariable region of the light chain.

chain V regions can be encoded by this pool. Similarly, any of the 500 or so *V* segments in the mouse heavy-chain pool can be joined to any of the 4 *J* segments and any of at least 12 *D* segments to encode at least 24,000 (500 × 4 × 12) different heavy-chain V regions. These are only rough estimates, since the exact numbers of *V* gene segments in these pools are not known.

The combinatorial diversification resulting from the assembly of different combinations of inherited *V*, *J*, and *D* gene segments, just discussed, is an important mechanism for diversifying the antigen-binding sites of antibodies. By this mechanism alone, it is estimated that a mouse could produce at least 1000 different V_L regions and on the order of 25,000 different V_H regions. These could then be combined to make 25×10^6 different antigen-binding sites. In addition, the joining mechanism itself, as we discuss next, greatly increases this number of possibilities (probably more than 10^8-fold), making it much greater than the total number of B cells in a mouse (about 5×10^8).

Imprecise Joining of Gene Segments Greatly Increases the Diversity of V Regions [21, 22]

The mechanism by which gene segments that may be hundreds of thousands of nucleotide pairs apart are joined to form a functional V_L- or V_H-region coding sequence is not known in detail. Conserved DNA sequences flank each gene segment and serve as recognition sites for a site-specific recombination system, ensuring that only appropriate gene segments recombine. Thus, for example, a

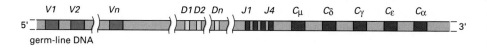

Figure 23–38 **The heavy-chain gene-segment pool in the mouse.** There are thought to be between 100 and 1000 *V* segments, at least 12 *D* segments, 4 *J* segments, and an ordered cluster of *C* segments, each encoding a different class of heavy chain. The *D* segment encodes amino acids in the third hypervariable region of the V region, as does part of the *J* segment. The figure is not drawn to scale: about 200,000 nucleotide pairs separate the *J1* and C_α gene segments, for example. Moreover, many details are omitted: for instance, there are 4 *Cγ* gene segments ($C_{γ1}$, $C_{γ2a}$, $C_{γ2b}$, and $C_{γ3}$); each *C* gene segment is composed of multiple exons (see Figure 23–33); and the V_H gene segments are clustered on the chromosome in groups of homologous families. The genetic mechanisms involved in producing a heavy chain are the same as those shown in Figure 23–37 for light chains except that two DNA rearrangement steps are required instead of one: first a *D* segment joins to a *J* segment, and then a *V* segment joins to the rearranged *DJ* segments.

V segment will always join to a *J* or *D* segment and not to another *V* segment. Two closely linked genes called *rag-1* and *rag-2* (*rag* = recombination activating genes) appear to encode the lymphocyte-specific proteins of the *V(D)J recombination system*. Thus, if a fibroblast is transfected with both of these genes, it is now able to rearrange experimentally introduced Ig gene segments just as a developing B cell normally does. Moreover, transgenic mice that are deficient in either gene are unable to initiate *V(D)J* rearrangements and consequently do not have functional B or T cells. (T cells use the same recombination system to assemble the genes that encode their antigen-specific receptors.)

In most cases of site-specific recombination, DNA joining is precise. But during the joining of antibody (and T cell receptor) gene segments, a variable number of nucleotides are often lost from the ends of the recombining gene segments, and, in the case of the heavy chain, one or more randomly chosen nucleotides may also be inserted. This random loss and gain of nucleotides at joining sites (called **junctional diversification**) enormously increases the diversity of V-region coding sequences created by recombination, specifically in the third hypervariable region. The increased diversification in this case comes at a price, since in many cases it will result in a shift in the reading frame so that a nonfunctional gene will be produced. Such "nonproductive" joining is thought to occur commonly in developing B cells.

Antigen-driven Somatic Hypermutation Fine-tunes Antibody Responses [23]

As mentioned earlier, with the passage of time after immunization there is usually a progressive increase in the affinity of the antibodies produced against the immunizing antigen. This phenomenon, known as **affinity maturation,** is unique to antibodies (it does not occur in T cell receptors) and is due to the accumulation of point mutations specifically in both heavy- and light-chain V-region coding sequences. These mutations occur long after the coding regions have been assembled, when B cells are stimulated by antigen and helper T cells to generate memory cells in the activated center (so-called *germinal center*) of a lymphoid follicle in secondary lymphoid organs (see Figure 23–8). The point mutations occur at the rate of about one mutation per V-region coding sequence per cell generation, which is about a million times greater than the spontaneous mutation rate in other genes; hence, the process is called **somatic hypermutation.** The mechanism that allows the nucleotide changes to be targeted to the DNA of a precisely specified part of the genome in this way is not known.

Only a small minority of these point mutations will result in antigen receptors that have an increased affinity for the antigen. The few B cells expressing these high-affinity receptors, however, will be preferentially stimulated by antigen to survive and proliferate, while the other B cells will undergo programmed cell death. Thus, as a result of repeated cycles of somatic hypermutation followed by antigen-driven selection, antibodies of increasingly higher affinity are produced during the course of an immune response, providing progressively better protection against harmful antigens.

Antibody Gene-Segment Joining Is Regulated to Ensure That B Cells Are Monospecific [24]

As the clonal selection theory predicts, B cells are *monospecific*. That is, all antibodies produced by one B cell have the same antigen-binding sites. This ensures that the antigen-binding sites on any one antibody molecule are identical and, therefore, that secreted antibodies can form large lattices of cross-linked antigens, thereby promoting antigen elimination (see Figure 23–15). It also ensures that an activated B cell secretes antibodies with the same specificity as that of the membrane-bound antibody on the B cell that was originally stimulated.

The requirement of monospecificity means that there must be some mechanism for ensuring that when Ig genes are activated during B cell development, only one type of V_L region and one type of V_H region are made by each B cell. Since B cells, like other somatic cells, are diploid, each cell has six gene-segment pools encoding antibodies: two heavy-chain pools (one from each parent) and four light-chain pools (one κ and one λ from each parent). If DNA rearrangements occurred independently in each heavy-chain pool and each light-chain pool, a single cell could make up to eight different antibodies, each with a different antigen-binding site. In fact, however, each B cell uses only two of the six gene-segment pools: one of the four light-chain pools and one of the two heavy-chain pools. Thus each B cell must choose not only between its κ and λ light-chain pools, but also between its maternal and paternal light-chain and heavy-chain pools (Figure 23–39). The latter choice is called **allelic exclusion** and seems to occur only in genes that encode antibodies (and T cell receptors). For other proteins that are encoded by autosomal genes (except for those encoded by genes that are subject to genomic imprinting—discussed in Chapter 9), both maternal and paternal genes in a cell appear to be expressed about equally.

The mechanism of allelic exclusion and κ versus λ light-chain choice during B cell development is uncertain. One possibility is that B cells are monospecific simply because the chance of a successful rearrangement occurring in more than one gene pool for each Ig chain is very low. This is unlikely to be the only mechanism, however, as there is evidence for some kind of *negative feedback regulation* on the V(D)J recombination system, whereby a functional rearrangement in one gene-segment pool suppresses rearrangements in the remaining pools that encode the same type of polypeptide chain. In B cell clones isolated from transgenic mice expressing a rearranged μ-chain gene, for example, the rearrangement of endogenous heavy-chain genes is usually suppressed, but only if the μ chain encoded by the transgene is inserted into the plasma membrane. Similar results have been obtained for light chains. It therefore seems that the product of an assembled heavy- or light-chain gene must be expressed on the cell surface for the feedback suppression to operate, suggesting that extracellular signals participate in the regulation process.

The assembly of V-region coding sequences in a developing B cell proceeds in an orderly sequence, one segment at a time, usually beginning with the heavy-chain pool. In this pool, D segments first join to J_H segments on both parental chromosomes. Then V_H to DJ_H joining occurs on one of these chromosomes. If this rearrangement produces a functional gene, the resulting production of complete μ chains (always the first heavy chains made) leads to their expression on the cell surface in association with a surrogate light chain. These cell-surface receptors enable the B cell to receive signals from their neighbors (called *stromal cells*), and these signals shut down all further rearrangements of V_H-region-encoding gene segments and may speed up the rate of V_L rearrangements. In mice, at least, V_L rearrangement usually occurs first in a κ gene-segment pool, and only if that fails does it occur in the other κ pool or in the λ pools. If, at any point, "in-phase" V_L-to-J_L joining leads to the production of light chains, these combine with preexisting μ chains to form IgM antibody molecules, which insert into the plasma membrane. The IgM cell-surface receptors are thought to enable the newly formed B cell to receive extracellular signals that shut down further V(D)J recombination by turning off the expression of the *rag-1* and *rag-2* genes. If a developing B cell fails to assemble both a functional V_H-region and a functional V_L-region coding sequence, it is unable to make antibody molecules and dies.

Although no biological differences between κ and λ light chains have been discovered, there is an obvious advantage in having two separate pools of gene segments that encode light chains: it increases the chance that a pre-B cell that has successfully assembled a V_H-region coding sequence will go on to assemble successfully a V_L-region coding sequence to become a B cell.

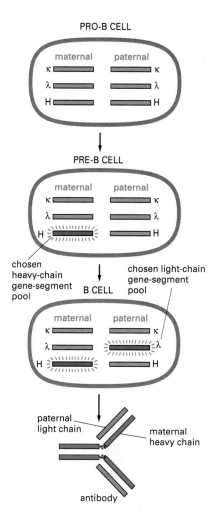

Figure 23–39 B cell development. This drawing shows the choices in Ig gene activation that a developing B cell must make in order to produce antibodies with only one type of antigen-binding site. The choice between maternal and paternal gene-segment pools is thought to be random.

When Stimulated by Antigen, B Cells Switch from Making a Membrane-bound Antibody to Making a Secreted Form of the Same Antibody [25]

We now turn from the genetic mechanisms that determine the antigen-binding site of an antibody to those that determine its biological properties—the mechanisms that determine what form of heavy-chain constant region is synthesized. The choice of the particular gene segments that encode the antigen-binding site is usually a commitment for the life of a B cell and its progeny, but the type of C_H region that is made changes during B cell development. The changes are of two types: changes from a membrane-bound form to a secreted form of the same C_H region and changes in the class of the C_H region made.

All classes of antibody can be made in a membrane-bound form as well as in a soluble, secreted form. The membrane-bound form serves as an antigen receptor on the B cell surface, while the soluble form is made only after the cell is stimulated by antigen to become an antibody-secreting cell. The sole difference between the two forms resides in the carboxyl terminus of the heavy chain: the heavy chains of membrane-bound Ig molecules have a hydrophobic carboxyl terminus, which anchors them in the lipid bilayer of the B cell plasma membrane; those of secreted Ig molecules have instead a hydrophilic carboxyl terminus, which allows them to escape from the cell. The switch in the character of the Ig molecules made occurs because the activation of B cells by antigen (and helper T cells) induces a change in the way that the Ig RNA transcripts are processed in the nucleus (see Figure 9–78).

B Cells Can Switch the Class of Antibody They Make [26]

During B cell development many B cells switch from making one class of antibody to making another—a process called **class switching.** All B cells begin their antibody-synthesizing lives by making IgM molecules and inserting them into the plasma membrane as receptors for antigen. Before they have interacted with antigen, many B cells then switch and make both IgM and IgD molecules as membrane-bound antigen receptors. Upon stimulation by antigen, some of these cells are activated to secrete IgM antibodies, which dominate the primary antibody response. Other antigen-stimulated cells switch to making IgG, IgE, or IgA antibodies; memory cells express one of these three classes of molecules on their surface, while activated B cells secrete them. The IgG, IgE, and IgA molecules are collectively referred to as *secondary* classes of antibodies because they are thought to be produced only after antigen stimulation and because they dominate secondary antibody responses. As we saw earlier, these different classes of antibodies are each specialized to attack microorganisms in different ways and in different sites.

Since the class of an antibody is determined by the constant region of its heavy chain, the fact that B cells can switch the class of antibody they make without changing the antigen-binding site implies that the same assembled V_H-region coding sequence (which specifies the antigen-binding part of the heavy chain) can sequentially associate with different C_H gene segments. This has important functional implications. It means that in an individual animal a particular antigen-binding site that has been selected by environmental antigens can be distributed among the various classes of immunoglobulin and thereby acquire the different biological properties characteristic of each class.

Class switching occurs by at least two distinct molecular mechanisms. When virgin B cells change from making membrane-bound IgM alone to the simultaneous production of membrane-bound IgM and IgD, the switch is thought to be due to a change in RNA processing. The cells produce large primary RNA tran-

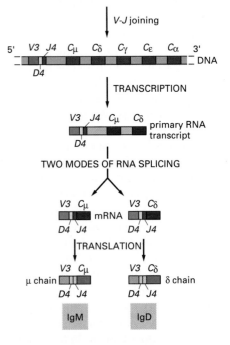

Figure 23–40 Simultaneous synthesis of IgM and IgD. B cells that simultaneously make plasma-membrane-bound IgM and IgD molecules that have the same antigen-binding sites produce long RNA transcripts that contain both C_μ and C_δ sequences. These transcripts are spliced in two ways to produce mRNA molecules that have the same V_H-region coding sequence (*V3D4J4*) joined to either a C_μ or a C_δ sequence.

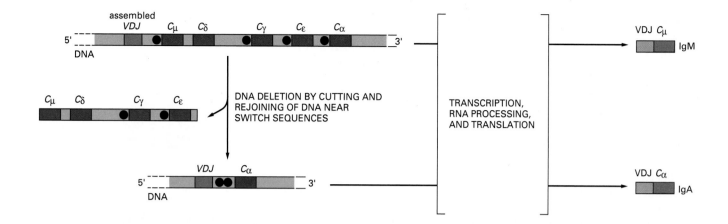

scripts that contain the assembled V_H-region coding sequence along with both the C_μ and C_δ sequences; IgM and IgD molecules are then produced by differential splicing of these RNA transcripts (Figure 23–40).

By contrast, terminal maturation to an activated B cell that secretes one of the secondary classes of antibody is accompanied by an irreversible change at the DNA level—a process called *class switch recombination*. It entails deletion of all the C_H gene segments upstream (that is, on the 5′ side as measured on the coding strand) of the particular C_H segment the cell is destined to express (Figure 23–41). Evidence that this step in class switching involves DNA deletion comes from experiments on myeloma cells: myeloma cells that secrete IgG lack the DNA coding for C_μ and C_δ regions, and those that secrete IgA lack the DNA coding for all of the other classes of heavy-chain C regions.

Figure 23–41 An example of the DNA rearrangement that occurs in class switch recombination. When a B cell making IgM antibodies from an assembled *VDJ* DNA sequence is stimulated by antigen to mature into an IgA-antibody-secreting cell, it deletes the DNA between the *VDJ* sequence and the C_α gene segment. Specific DNA sequences (*switch sequences*, shown as *black spheres*) located upstream of each C_H gene segment (except C_δ) recombine with one another to delete the intervening DNA. Class switch recombination is thought to be mediated by a *switch recombinase*, which is directed to the appropriate switch sequences when these become accessible under the influence of extracellular signals (lymphokines) secreted by helper T cells, as we discuss later.

Summary

Antibodies are produced from three pools of gene segments, encoding κ light chains, λ light chains, and heavy chains, respectively. In each pool, separate gene segments that code for different parts of the variable regions of light and heavy chains are brought together by site-specific recombination during B cell differentiation. The light-chain pools contain one or more constant (C) gene segments and sets of variable (V) and joining (J) gene segments. The heavy-chain pool contains a set of C gene segments and sets of V, diversity (D), and J gene segments. To make an antibody molecule, a V_L gene segment is recombined with a J_L gene segment to produce a DNA sequence coding for the V region of the light chain, and a V_H gene segment is recombined with a D and a J_H gene segment to produce a DNA sequence coding for the V region of the heavy chain. Each of the assembled gene segments is then co-transcribed with the appropriate C-region sequence to produce an mRNA molecule that codes for the complete polypeptide chain. By variously combining inherited gene segments that code for V_L and V_H regions, mammals can make thousands of different light chains and thousands of different heavy chains. Since the antigen-binding site is formed where V_L and V_H come together in the final antibody, the heavy and light chains can pair to form antibodies with millions of different antigen-binding sites. This number is enormously increased by the loss and gain of nucleotides at the site of gene-segment joining, as well as by somatic mutations that occur with very high frequency in the assembled V-region coding sequences following antigen stimulation.

All B cells initially make IgM antibodies. Later some make antibodies of other classes but with the same antigen-binding site as the original IgM antibodies. Such class switching allows the same antigen-binding sites to be distributed among antibodies with varied biological properties.

T Cell Receptors and Subclasses

The diverse responses of T cells are collectively called *cell-mediated immune reactions*. Like antibody responses, they are exquisitely antigen-specific and are at least as important in defending vertebrates against infection.

T cells differ from B cells, however, in several important ways. First, they act only at short range, interacting directly with another cell in the body, which they either kill or signal in some way (we shall refer to such cells as *target cells*); B cells, by contrast, secrete antibodies that can act far away. Second, T cells are specialized to recognize foreign antigen only when it is displayed on the surface of a target cell. For this reason the form of antigen recognized by T cells is different from that recognized by B cells: whereas B cells recognize intact antigen, T cells recognize peptide fragments of protein antigens that have been partially degraded inside the target cell and then carried to the cell surface and displayed there. In this way T cells are able to detect the presence of microorganisms that proliferate inside cells, as well as foreign extracellular antigens that cells have ingested.

There are two main classes of T cells—cytotoxic T cells and helper T cells. *Cytotoxic T cells* directly kill cells that are infected with a virus or some other intracellular microorganism. *Helper T cells*, by contrast, help stimulate the responses of other cells: they help activate macrophages and B cells, for example.

T Cell Receptors Are Antibodylike Heterodimers [27]

Because T cell responses depend on direct contact with a target cell, the antigen receptors made by T cells, unlike antibodies made by B cells, exist only in membrane-bound form and are not secreted. For this reason T cell receptors were difficult to isolate, and it was not until 1983 that they were first identified biochemically. On both cytotoxic and helper T cells, the receptors are composed of two disulfide-linked polypeptide chains (called α and β), each of which contains two Ig-like domains and shares with antibodies the distinctive property of a variable amino-terminal region and a constant carboxyl-terminal region (Figure 23–42).

The gene pools that encode the α and β chains are located on different chromosomes and contain, like antibody gene pools, separate *V, D, J*, and *C* gene segments, which are brought together by site-specific recombination during T cell development in the thymus. With one exception, all the mechanisms used by B cells to generate antibody diversity are also used by T cells to generate T cell receptor diversity; in particular, the same *V(D)J* recombination system is used, requiring the proteins encoded by the *rag-1* and *rag-2* genes discussed earlier. The mechanism that does not operate in T cell receptor diversification is antigen-driven somatic hypermutation, and so the affinity of the receptors remains low ($K_a \approx 10^4$ liters/mole), even late in an immune response. We discuss later how antigen-nonspecific cell-cell adhesion mechanisms greatly strengthen the binding of a T cell to its target cell, helping to compensate for the low affinity of the T cell receptors.

A small minority of T cells, instead of making α and β chains, make a different type of receptor heterodimer, composed of γ and δ chains. These cells arise early in development and are found mainly in epithelia (in the skin and gut, for example), where they may provide a first line of defense against microorganisms that attempt to penetrate these cell sheets.

As is the case for antigen receptors on B cells, the T cell receptors are tightly associated in the plasma membrane with a number of invariant proteins that are involved in passing the signal from an antigen-activated receptor to the cell interior. We discuss these proteins in more detail later.

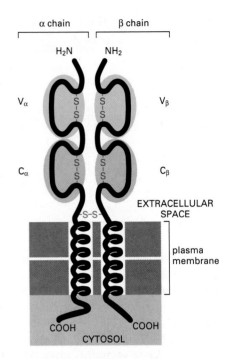

Figure 23–42 A T cell receptor heterodimer. The receptor is composed of an α and a β polypeptide chain. Each chain is about 280 amino acids long and has a large extracellular part that is folded into two Ig-like domains—one variable (V) and one constant (C). It is thought that an antigen-binding site formed by a V_α and V_β domain (shaded in *blue*) is similar in its overall dimensions and geometry to the antigen-binding site of an antibody molecule. Unlike antibodies, however, which have two binding sites for antigen, T cell receptors have only one. The α/β heterodimer shown is noncovalently associated with a large set of invariant proteins (not shown), which help activate the T cell when the T cell receptors bind to antigen. A typical T cell has about 20,000 such receptor complexes on its surface.

Different T Cell Responses Are Mediated by Distinct Classes of T Cells [28]

The two major classes of T cells have very different functions. **Cytotoxic T cells** kill cells harboring harmful microbes, while **helper T cells** help activate the responses of other white blood cells, mainly by secreting a variety of local mediators, collectively called *lymphokines, interleukins,* or *cytokines.* Thus cytotoxic T cells provide protection against pathogenic microorganisms, such as viruses and some intracellular bacteria, that multiply in the host cytoplasm, where they are sheltered from attack by antibodies. The most efficient way of preventing such microorganisms from spreading to other cells is to kill the infected cell before the microorganisms can proliferate. Helper T cells, by contrast, are crucial for stimulating responses to extracellular microorganisms and their toxic products. There are two types of helper T cells: T_H1 *cells,* which activate macrophages to destroy microorganisms that they have ingested, and T_H2 *cells,* which stimulate B cells to proliferate and secrete antibodies.

Both cytotoxic T cells and helper T cells recognize antigen in the form of peptide fragments that are generated by the degradation of foreign protein antigens inside the target cell, and both, therefore, depend on the presence in the target cell of special proteins that bind these fragments, carry them to the cell surface, and present them there to the T cells. These special proteins are called *MHC molecules* because they are encoded by a complex of genes called the *major histocompatibility complex (MHC).* There are two structurally and functionally distinct classes of MHC molecules: *class I MHC molecules,* which present foreign peptides to cytotoxic cells, and *class II MHC molecules,* which present foreign peptides to helper cells. Before we examine the different mechanisms by which protein antigens are processed for display to the two types of T cells, we must look more closely at the MHC molecules themselves, which play such an important part in T cell immunity.

Summary

There are at least two functionally distinct subclasses of T cells: cytotoxic T cells directly kill infected cells, especially those infected with a virus, while helper T cells help activate both B cells to make antibody responses and macrophages to ingest and destroy invading microorganisms. Both types of T cells express cell-surface, antibodylike receptors, which are encoded by genes that are assembled from multiple gene segments during T cell development in the thymus. These receptors recognize fragments of foreign proteins that are displayed on the surface of host cells in association with MHC molecules.

MHC Molecules and Antigen Presentation to T Cells [29]

MHC molecules were recognized long before their normal function was understood. They were initially defined as the main target antigens in **transplantation reactions.** When organ grafts are exchanged between adult individuals, either of the same species (*allografts*) or of different species (*xenografts*), they are usually rejected. In the 1950s experiments involving skin grafting between different strains of mice demonstrated that *graft rejection* is an immune response to the foreign antigens on the surface of the grafted cells. It was later shown that these reactions are mediated mainly by T cells and that they are directed against genetically "foreign" versions of cell-surface proteins called *histocompatibility* *molecules* (histo = tissue). The MHC family of proteins encoded by the clustered genes of the **major histocompatibility complex (MHC)** are by far the most important of these. MHC molecules are expressed on the cells of all higher vertebrates. They were first demonstrated in mice and called *H-2 antigens* (histocompatibility-2 antigens). In humans they are called *HLA antigens* (human-leucocyte-associated antigens) because they were first demonstrated on leucocytes (white blood cells).

Three remarkable properties of MHC molecules baffled immunologists for a long time. First, MHC molecules are overwhelmingly the preferred target antigens for T-cell-mediated transplantation reactions. Second, an unusually large fraction of T cells are able to recognize foreign MHC molecules: whereas fewer than 0.001% of an individual's T cells respond to a typical viral antigen, for example, more than 0.1% of them respond to a single foreign MHC antigen. Third, many of the loci that code for MHC molecules are the most *polymorphic* known in higher vertebrates; that is, within a species there is an extraordinarily large number of *alleles* (alternative forms of the same gene) at each locus (in some cases as many as 100), each allele being present at a relatively high frequency in the population. For this reason, and because each individual has five or more loci encoding MHC molecules (see below), it is very rare for two individuals to have an identical set of MHC proteins, making it very difficult to match donor and recipient for organ transplantation in humans (except in the case of genetically identical twins).

A vertebrate does not need to be protected against invasion by foreign vertebrate cells. So the apparent obsession of its T cells with foreign MHC molecules and the extreme polymorphism of these molecules were a great puzzle to immunologists. The puzzle was solved only after it was discovered that MHC molecules serve to focus T cells on those host cells that have foreign antigen on their surface and that the T cells respond to foreign MHC molecules in the same way as to self MHC molecules that have foreign antigen bound to them.

There Are Two Principal Classes of MHC Molecules [29]

Class I and class II MHC proteins have very similar overall structures. They are both transmembrane heterodimers whose extracellular amino-terminal domains bind antigen for presentation to T cells.

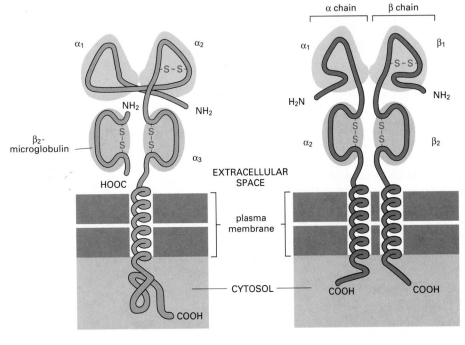

(A) CLASS I MHC PROTEIN

(B) CLASS II MHC PROTEIN

Figure 23–43 Class I and class II MHC proteins. (A) The α chain of the class I molecule has three extracellular domains, α_1, α_2, and α_3, encoded by separate exons. It is noncovalently associated with a smaller polypeptide chain, β_2-microglobulin, which is not encoded within the MHC. The α_3 domain and β_2-microglobulin are Ig-like. While β_2-microglobulin is invariant, the α chain is extremely polymorphic, mainly in the α_1 and α_2 domains. (B) In class II MHC molecules both chains are polymorphic (β more than α), mainly in the α_1 and β_1 domains; the α_2 and β_2 domains are Ig-like. Thus there are striking similarities between class I and class II MHC proteins. In both, the two outermost domains (shaded in *blue*) interact to form a groove that binds foreign antigen and presents it to T cells. All of the chains are glycosylated except for β_2-microglobulin (not shown).

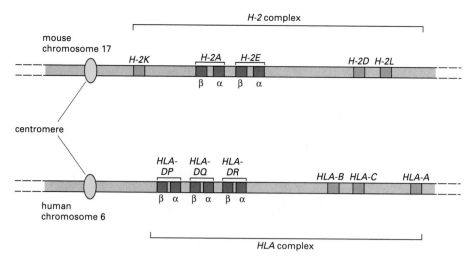

Figure 23–44 The *H-2* and *HLA* gene complexes. This simplified schematic drawing shows the location of the genetic loci that encode the transmembrane subunits of class I (*light green*) and class II (*dark green*) MHC proteins. There are three types of class I proteins (H-2K, H-2D, and H-2L in mouse, and HLA-A, HLA-B, and HLA-C in human). There are two class II MHC loci in the mouse—*H-2A* and *H-2E*—and more than three in humans, of which only three are shown—*HLA-DP*, *HLA-DQ*, and *HLA-DR*. Each class II locus encodes at least one α chain and at least one β chain, but some encode more than one α or β chain (not shown).

Each **class I MHC gene** encodes a single transmembrane polypeptide chain (called α), most of which is folded into three extracellular globular domains (α_1, α_2, α_3). Each α *chain* is noncovalently associated with a small extracellular protein called *β_2-microglobulin*, which does not span the membrane and is encoded by a gene that does not lie in the MHC gene cluster (Figure 23–43A). β_2-microglobulin and the α_3 domain, which are closest to the membrane, are both homologous to an Ig domain. The two amino-terminal domains of the α chain, which are farthest from the membrane, bind antigen and contain the polymorphic (variable) amino acids that are recognized by T cells in transplantation reactions.

Like class I MHC molecules, **class II MHC molecules** are heterodimers with two conserved Ig-like domains close to the membrane and two antigen-binding polymorphic (variable) amino-terminal domains farthest from the membrane. In these molecules, however, both chains (α and β) are encoded within the MHC, and both span the membrane (Figure 23–43B). The presence of Ig-like domains in class I and class II proteins suggests that MHC molecules and antibodies have a common evolutionary history. The locations of the genes that encode class I and class II MHC proteins in mice and humans are shown in Figure 23–44.

X-ray Diffraction Studies Reveal the Antigen-binding Site of MHC Proteins as well as the Bound Peptide [30]

Any individual has only a small number of types of MHC molecules, which together must be able to present peptide fragments from almost any foreign protein to T cells. Thus each MHC molecule has to be able to bind a very large number of different peptides. The structural basis for this versatility has emerged from the x-ray crystallographic analysis of MHC molecules.

As shown in Figure 23–45A, a class I MHC protein has a single peptide-binding site located at one end of the molecule. This site consists of a deep groove between two long α helices derived from the nearly identical α_1 and α_2 domains; the base of the groove is formed by eight β strands derived from the same two domains. The groove is large enough to accommodate an extended peptide of about 10 amino acid residues. In fact, when a class I MHC protein was first analyzed by x-ray crystallography, this groove was found to contain a small density, suspected to be bound peptide that had co-crystallized with the MHC protein (Figure 23–45B). This finding implicated the groove as the antigen-binding site and suggested that once a peptide binds to this site, it dissociates very slowly. This conclusion is supported by the observation that cells exposed for a short period to fragments of a viral protein can remain targets for specific cytotoxic T cells for days. In acidified solutions, however, the bound peptides can be eluted from isolated class I MHC molecules, and they are indeed found to be

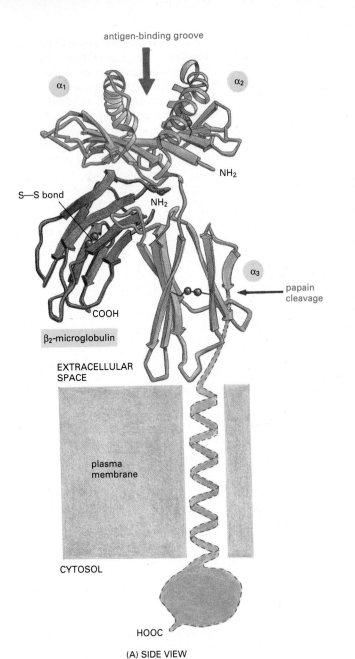

antigen-binding groove

α_1

α_2

NH$_2$

S—S bond

NH$_2$

α_3

papain cleavage

COOH

β$_2$-microglobulin

EXTRACELLULAR SPACE

plasma membrane

CYTOSOL

HOOC

(A) SIDE VIEW

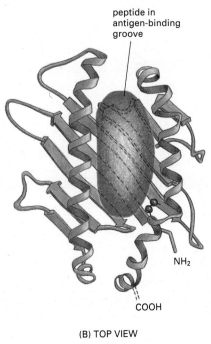

peptide in antigen-binding groove

NH$_2$

COOH

(B) TOP VIEW

Figure 23–45 (A) The structure of a human class I MHC protein as determined by x-ray diffraction analysis of crystals of the extracellular part of the molecule. The extracellular part of the protein was cleaved from the transmembrane segment by the proteolytic enzyme papain prior to crystallization. Each of the two domains closest to the plasma membrane (α_3 and β$_2$-microglobulin) resembles a typical immunoglobulin domain (see Figure 23–34B), while the two domains farthest from the membrane (α_1 and α_2) are very similar to each other and together form a peptide-binding groove at the top of the molecule. Class II MHC molecules are thought to have a very similar structure. (B) The peptide-binding groove viewed from above, containing the small peptides that co-purified with the MHC protein. This is the part of the MHC molecule that interacts with the T cell receptor. (After P.J. Bjorkman, M.A. Saper, B. Samraoui, W.S. Bennett, J.L. Strominger, and D.C. Wiley, *Nature* 329:506–512, 1987.)

8 to 10 amino acid residues long. Almost all of the polymorphic amino acids in the MHC protein (those that vary between allelic forms of this type of molecule) are located inside the groove, where they would be expected to bind antigen, or on its edges, where they would be accessible for recognition by the T cell receptor.

Different peptides have been found to bind in the groove of a class I MHC protein through a combination of invariant and variable contacts. In each case the peptide is seen as an extended chain of 8 to 10 amino acids, with the invariant peptide backbone of the terminal amino acids at each end of the peptide anchored in highly conserved pockets located at each end of the groove (Figure 23–46). Other parts of the peptide bind to "specificity pockets" formed by polymorphic portions of the MHC protein, while the side chains of some residues point outward, in a position to be recognized by receptors on cytotoxic T cells. Because the conserved pockets at the ends of the binding groove recognize features of the peptide backbone that are common to many peptides, and not the amino acid side chains, which vary, a single class I MHC protein can bind a large variety of peptides of diverse sequence. At the same time the differing specificity pockets along the groove ensure that each allelic form of MHC molecule binds and pre-

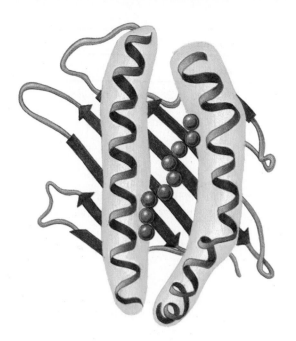

Figure 23–46 **Highly schematized drawing of a peptide in the binding groove of a class I MHC molecule.** In this top view the peptide backbone is shown as a string of *red* balls, each of which represents one of the nine amino acid residues. The backbones of the "anchor residues" at the ends of the peptide bind to conserved pockets at the ends of the groove.

sents its own characteristic set of peptides. Thus the several types of class I MHC molecules in an individual can present a broad range of foreign peptides to the cytotoxic T cells, but in each individual they do so in slightly different ways.

Class II MHC molecules have a three-dimensional structure that is very similar to that of class I molecules, but their antigen-binding grooves accommodate longer and much more heterogeneous peptides, ranging in size from 15 to 24 amino acid residues. Thus, although an individual probably makes fewer than 20 types of class II molecules, each with its own unique antigen-binding site, together these molecules seem to be able to bind and present an apparently unlimited variety of foreign peptides to helper T cells, which play a crucial part in almost all immune responses.

Class I and Class II MHC Molecules Have Different Functions [29]

Class I MHC molecules are expressed on virtually all nucleated cells, presumably because cytotoxic T cells must be able to focus on any cell in the body that happens to become infected with an intracellular microbe such as a virus. Class II molecules, by contrast, are normally confined to specialized cells, such as B cells,

Figure 23–47 **Cytotoxic and helper T cells recognize different MHC molecules.** Cytotoxic T cells recognize foreign antigens in association with class I MHC proteins on the surface of any infected host cell, whereas helper T cells recognize foreign antigens in association with class II MHC proteins on the surface of an antigen-presenting cell, such as a macrophage or a B cell. The foreign antigen bound to a class I MHC molecule is synthesized within the target cell, while the foreign antigen bound to a class II MHC molecule has been taken up by the cell by endocytosis and processed before it is presented on the cell surface (not shown). In transplantation reactions as well, helper T cells react against foreign class II MHC proteins, whereas cytotoxic T cells react against foreign class I MHC proteins.

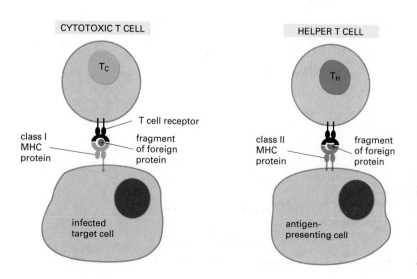

Table 23-2 Properties of Class I and Class II MHC Molecules

	Class I	Class II
Genetic loci	*H-2K, H-2D, H-2L* in mice; *HLA-A, HLA-B, HLA-C* in humans	*H-2A* and *H-2E* clusters in mice; *DP, DQ, DR,* and several others in humans
Chain structure	α chain + β$_2$-microglobulin	α chain + β chain
Cell distribution	most nucleated cells	antigen-presenting cells (including B cells), thymus epithelial cells, some others
Involved in presenting antigen to	cytotoxic T cells	helper T cells
Source of peptide fragments	proteins made in cytosol	endocytosed plasma membrane and extracellular proteins
Polymorphic domains	α$_1$ + α$_2$	α$_1$ + β$_1$

macrophages, and other antigen-presenting cells, that take up foreign antigens from the extracellular fluid and interact with helper T cells (Figure 23–47). The principal features of the two classes of MHC proteins are summarized in Table 23–2.

It is important that cytotoxic T cells focus their attack on cells that *make* foreign antigens, while helper T cells focus their help on cells that take up foreign antigens from the extracellular fluid. The former type of target cell is a menace, but the latter type is essential for the body's immune defenses. The immune system must be able to dispose of extracellular foreign antigens, as many bacteria multiply outside cells, and some secrete protein toxins, such as tetanus toxin and botulinum toxin, that can be lethal. Helper T cells help eliminate such pathogens both by helping B cells make antibodies against microbes and against their toxins and by activating macrophages to destroy ingested microbes.

To ensure that there is no misdirection of cytotoxic and helper functions, each of the two major classes of T cells, in addition to the antigen receptor that recognizes a peptide-MHC complex, also expresses a *co-receptor* that recognizes a nonpolymorphic part of the appropriate class of MHC molecule, as we now discuss.

CD4 and CD8 Proteins Act as MHC-binding Co-Receptors on Helper and Cytotoxic T Cells, Respectively [31]

The affinity of T cell receptors for peptide-MHC complexes on a target cell is usually too low to mediate a functional interaction between the two cells. *Accessory receptors* are normally required to help stabilize the interaction by increasing the overall strength of the cell-cell adhesion; when they also have a direct role in activating the T cell by generating their own intracellular signals, they are called *co-receptors*. Unlike T cell receptors or MHC molecules, the accessory receptors do not bind antigen and are invariant and nonpolymorphic.

The most important and best understood of the co-receptors on T cells are the CD4 and CD8 proteins, both of which are single-pass transmembrane proteins with extracellular Ig-like domains. Like T cell receptors, they recognize MHC proteins, but, unlike T cell receptors, they bind to nonvariable parts of the protein, far away from the peptide-binding groove. **CD4** is expressed on helper T cells and binds to class II MHC molecules, whereas **CD8** is expressed on cytotoxic T cells and binds to class I MHC molecules (Figure 23–48). Thus CD4 and CD8 contribute to T cell recognition by helping to focus the cell on particular MHC molecules, and thus on particular types of target cells. The cytoplasmic tail of these transmembrane proteins is associated with a member of the Src family of tyrosine-specific protein kinases called the *Lck* protein, which phosphorylates various cellular proteins on tyrosine residues and thereby participates in the activation of the T cell.

The CD4 and CD8 proteins not only are required to increase the strength of cell-cell adhesion and to help activate the T cell, they are also needed for T cell development: if the genes that encode CD4 or CD8 are inactivated in a mouse by targeted genetic recombination, then either helper T cells or cytotoxic T cells, respectively, do not develop. Ironically, CD4 also functions as a receptor for the AIDS virus (HIV), allowing the virus to infect helper T cells.

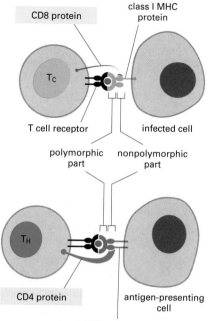

Figure 23–48 CD4 and CD8 co-receptors on the surface of T cells. Note that these proteins bind to the nonvariable part of the same MHC molecule that the T cell receptor has engaged, so that they are brought together with T cell receptors during the cell activation process. Antibodies against CD4 and CD8 are widely used to distinguish helper and cytotoxic T cells, respectively.

Summary

Class I and class II MHC molecules are the most polymorphic proteins known—that is, they show the greatest genetic variability from one individual to another—and they play a crucial role in presenting foreign protein antigens to cytotoxic and helper T cells, respectively. Whereas class I molecules are expressed on almost all vertebrate cells, class II molecules are restricted to a few cell types that interact with helper T cells, such as B lymphocytes and macrophages. Both classes of MHC molecules have Ig-like domains and a single peptide-binding groove, which binds small peptide fragments derived from foreign proteins. Each MHC molecule can bind a large and characteristic set of peptides, which are produced intracellularly by protein degradation. After they form inside the target cell, the peptide-MHC complexes are transported to the cell surface, where they are recognized by T cell receptors. In addition to their antigen-specific receptors that recognize peptide-MHC complexes on the surface of target cells, T cells express CD4 or CD8 co-receptors, which recognize nonpolymorphic regions of MHC molecules on the target cell: helper cells express CD4, which recognizes class II MHC molecules, while cytotoxic T cells express CD8, which recognizes class I MHC molecules.

Cytotoxic T Cells

As previously discussed, cytotoxic T cells defend us against microorganisms such as viruses that grow inside cells, away from the reach of antibodies. Cytotoxic T cells, unlike antibodies, can recognize such infected cells because class I MHC molecules continually ferry fragments of the microbe's proteins to the cell surface, where they can be detected by the T cells. In this section we discuss how these fragments are generated and delivered to the class I MHC molecules, and we consider how cytotoxic T cells kill infected target cells.

Cytotoxic T Cells Recognize Fragments of Viral Proteins on the Surface of Virus-infected Cells [32]

The first clear evidence that MHC molecules present foreign antigens to T cells came from an experiment performed in 1974 that showed that cytotoxic T cells from a virus-infected mouse could kill cultured cells infected with the same virus only if these cells expressed some of the same class I MHC molecules as the infected mouse (Figure 23–49). This experiment demonstrated that the T cells of any individual that recognize a specific antigen will do so only when that antigen is associated with the allelic forms of MHC molecules expressed by that individual, a phenomenon known as *MHC restriction*. It was not until 10 years later,

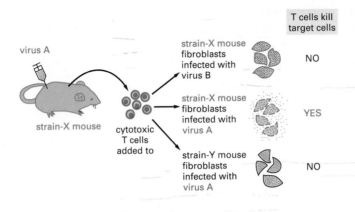

T cells kill
target cells

strain-X mouse fibroblasts infected with virus B	NO
strain-X mouse fibroblasts infected with virus A	YES
strain-Y mouse fibroblasts infected with virus A	NO

virus A

strain-X mouse

cytotoxic T cells added to

Figure 23–49 The classic experiment showing that a cytotoxic T cell recognizes some aspect of the surface of the host target cell in addition to a viral antigen. Mice of strain X are infected with virus A. Seven days later, the spleens of these mice contain active cytotoxic T cells able to kill virus-infected, strain-X fibroblasts in cell culture. As expected, they kill only fibroblasts infected with virus A and not those infected with virus B; thus the cytotoxic T cells are virus-specific. The same T cells, however, are also unable to kill fibroblasts from strain-Y mice infected with the same virus A, indicating that the cytotoxic T cells recognize a genetic difference between the two kinds of fibroblasts and not just the virus.

Pinning down the difference required the use of special strains of mice (known as *congenic strains*) that either were genetically identical except for the alleles at their class I MHC loci or were genetically different except for these alleles. In this way it was found that the killing of infected target cells required that they express at least one of the same class I MHC alleles expressed by the original infected mouse. This indicated that class I MHC proteins are necessary to present cell-surface-bound viral antigens to cytotoxic T cells.

however, that the chemical nature of the viral antigens recognized by cytotoxic T cells was discovered. In experiments on cells infected with influenza virus, it was unexpectedly found that some of the cytotoxic T cells activated by the virus specifically recognize internal proteins of the virus that would not be accessible in the intact virus particle. Subsequent evidence suggested that the T cells were recognizing degraded fragments of the internal viral proteins. Because a T cell can recognize tiny amounts of antigen (only a few hundred molecules), only a small fraction of the fragments generated from viral proteins have to get to the cell surface to attract an attack by a cytotoxic T cell. A mystery remained, however.

Since almost all proteins in a cell are known to be continually degraded (as discussed in Chapter 5), it is not difficult to understand how peptide fragments of internal viral proteins are produced in an infected cell. But it is less obvious how these fragments get exposed on the cell surface. Like other cellular proteins, the viral proteins are synthesized on cytosolic ribosomes, and, as explained in Chapter 12, special mechanisms are required if they are to leave the cytosol by crossing a membrane into some other compartment. Proteins destined for the cell surface usually begin their journey by crossing from the cytosol into the lumen of the endoplasmic reticulum (ER) (Figure 23–50). The puzzle of how viral protein fragments reach the cell surface was solved by the discovery of a remarkable transport system that vertebrate cells have for transporting such peptides from the cytosol into the ER lumen. Once inside the ER, the peptides can bind to MHC molecules that they encounter there and so can be carried to the cell surface with them.

MHC-encoded ABC Transporters Transfer Peptide Fragments from the Cytosol to the ER Lumen [33]

As noted in Chapter 5, proteolytic degradation in the cytosol is mainly mediated by an ATP- and ubiquitin-dependent mechanism that operates in *proteasomes*, which are large proteolytic enzyme complexes constructed from many different protein subunits. Although all proteasomes are probably able to generate peptide fragments that will bind to class I MHC molecules, some are thought to be specialized for this purpose because they contain two subunits that are encoded by genes located within the MHC chromosomal region.

The mechanism whereby peptides are delivered to the ER lumen from the cytosol was discovered through observations on mutant cells in which class I MHC molecules are not expressed at the cell surface but are instead degraded within the ER. The mutant genes in these cells proved to encode subunits of a protein belonging to the family of *ABC transporters*, which we encountered in Chapter 11. This transporter protein is located in the ER membrane and uses the energy of ATP hydrolysis to pump peptides from the cytosol into the ER lumen. The genes encoding its two subunits are in the MHC chromosomal region, and if either gene is inactivated by mutation, cells are unable to supply peptides to

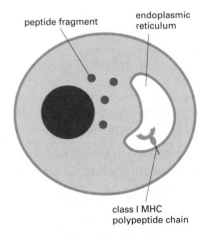

peptide fragment

endoplasmic reticulum

class I MHC polypeptide chain

Figure 23–50 The peptide-transport problem. How do peptide fragments get from the cytosol, where they are produced, into the ER lumen, where class I MHC molecules are made? A special transport process is required.

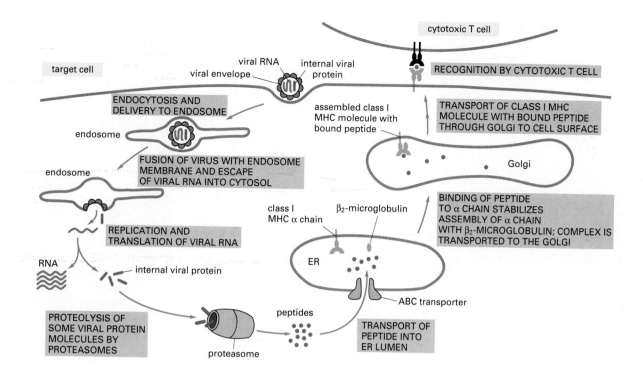

Figure 23–51 **The processing of viral protein for presentation to cytotoxic T cells.** A cytotoxic T cell will kill a virus-infected cell when it recognizes fragments of foreign protein bound to class I MHC molecules on the surface of the infected cell. Although not all viruses enter the cell in the way that this enveloped RNA virus does, fragments of internal viral proteins always follow the pathway shown. Only a very small proportion of the viral proteins synthesized in the cytosol are degraded, but it is a sufficient amount to attract an attack by a cytotoxic T cell.

class I MHC molecules. The fact that the class I MHC molecules in such mutant cells are degraded in the ER suggests that peptide binding is normally required for the proper folding and/or stable assembly of MHC molecules. In cells that are not infected by viruses, peptide fragments come from normal cytosolic and nuclear proteins that are degraded in the process of normal protein turnover; these peptides are carried to the cell surface by MHC molecules but are not antigenic because the cytotoxic T cells that could recognize them have been eliminated or inactivated during T cell development, as we discuss later. A current view of how viral proteins are processed for presentation to cytotoxic T cells is shown in Figure 23–51.

When T cells are activated by antigen, they secrete various signaling molecules, including **γ-interferon,** which greatly enhances anti-viral responses. The γ-interferon induces the expression of many genes within the MHC chromosomal region, including those that encode class I (and class II) MHC proteins, the two specialized proteasome subunits, and the two subunits of the peptide pump located in the ER. Thus all of the machinery required for presenting viral antigens to cytotoxic T cells is coordinately called into action by γ-interferon, creating a positive feedback that amplifies the immune response and culminates in the death of the infected cell.

Cytotoxic T Cells Induce Infected Target Cells to Kill Themselves [34]

Once a cytotoxic T cell has recognized a viral peptide bound to a class I MHC molecule on the surface of a target cell, its job is to destroy the cell before the virus that gave rise to the peptide can produce new viral particles that can escape from the infected cell. When a cytotoxic T cell, caught in the act of killing its target, is labeled with anti-tubulin antibodies, its centrosome is seen to be oriented toward the point of contact with the target cell (Figure 23–52). Moreover, antibody labeling shows that talin, a protein that helps link cell-surface receptors to cortical actin filaments, is concentrated in the cortex of the T cell at the contact site. It is thought that the aggregation of T cell receptors at the contact site leads to a local, talin-dependent alteration in the actin filaments in the cell cortex; a microtubule-dependent mechanism then moves the centrosome and its associ-

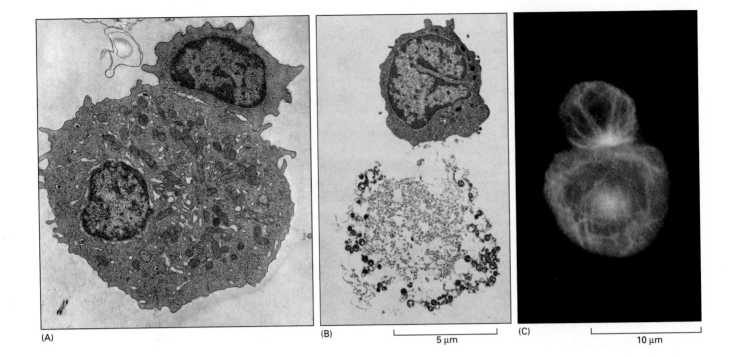

Figure 23–52 Cytotoxic T cells in the process of killing target cells in culture. The cells are visualized by electron microscopy in (A) and (B) and by immunofluorescence microscopy after staining with anti-tubulin antibodies in (C). The cytotoxic T cells were obtained from mice immunized with the target cells, which are foreign tumor cells. The T cells (the small cells) are shown binding to the target cell in (A) and (C) and having killed the target cell in (B). In an animal, as opposed to in a tissue culture dish, the killed target cell would be phagocytosed by neighboring cells before it disintegrated in the way that it has here. Note that the centrosome in the T cell and the microtubules radiating from it are oriented toward the point of cell-cell contact with the target cell (C). (A and B, from D. Zagury, J. Bernard, N. Thierness, M. Feldman, and G. Berke, *Eur. J. Immunol.* 5:818–822, 1975; C, reproduced from B. Geiger, D. Rosen, and G. Berke, *J. Cell Biol.* 95:137–143, 1982, by copyright permission of the Rockefeller University Press.)

ated Golgi apparatus toward the contact site, focusing the killing machinery on the target cell. A similar cytoskeletal polarization is seen when a helper T cell functionally interacts with a target cell.

The mechanism by which cytotoxic T cells kill their targets is not known for certain. They seem to employ at least two strategies, both of which are thought to operate by inducing the target cell to undergo *programmed cell death* (also called *apoptosis,* see p. 1169). In one strategy, binding to a target cell stimulates these cytotoxic T cells to release pore-forming protein called **perforin,** which is homologous to the complement component C9 and polymerizes in the target cell plasma membrane to form transmembrane channels. Perforin is stored in secretory vesicles and is released by local exocytosis at the point of contact with the target cell. The secretory vesicles also contain serine proteases and other proteins, which are also thought to play a part in killing the target cell, perhaps, by entering the target cell through the perforin channels and inducing programmed cell death. The second strategy, by contrast, involves the cytotoxic T cell activating a receptor on the surface of the target cell, thereby signaling the target cell to undergo programmed cell death.

Summary

Cytotoxic T cells directly kill infected target cells that display fragments of microbial protein on their surface. Some of the microbial proteins that are synthesized in the cytosol of the target cell are degraded by proteasomes, and some of the resulting peptide fragments are pumped by an ABC transporter into the lumen of the ER, where they bind to class I MHC molecules. The peptide-MHC complexes are then transported to the target cell surface, where they are recognized by cytotoxic T cells. These T cells are thought to kill infected target cells by inducing the target cell to undergo programmed cell death.

Helper T Cells and T Cell Activation [35]

Unlike cytotoxic T cells, **helper T cells** do not act directly to kill infected cells or to eliminate microorganisms. Instead, they stimulate macrophages to be more effective in destroying pathogens, and they help other types of lymphocytes to respond to antigen. The crucial importance of helper T cells in immunity is dramatically demonstrated by the devastating epidemic of *acquired immunodeficiency syndrome (AIDS)*. The disease is caused by a retrovirus (human immunodeficiency virus, HIV) that depletes the number of helper T cells by an unknown mechanism, thereby crippling the immune system and rendering the patient susceptible to infection by microorganisms that are normally not dangerous. As a result, most AIDS patients die of infection within several years of the onset of symptoms.

In this section we consider how class II MHC molecules on antigen-presenting cells acquire peptide fragments from endocytosed proteins and present them to helper T cells. We then discuss the multiple signals required to activate helper T cells and how these cells, once activated, help activate B cells and macrophages.

Helper T Cells Recognize Fragments of Endocytosed Foreign Protein Antigens in Association with Class II MHC Proteins [36]

Before they can help other lymphocytes respond to antigen, helper T cells must first be activated themselves. This activation occurs when helper T cells recognize foreign antigen bound to class II MHC proteins on the surface of specialized *antigen-presenting cells,* which are found in most tissues. We discuss these cells in more detail later.

Like the viral antigens presented to cytotoxic T cells, the antigens presented to helper T cells on antigen-presenting cells are degraded fragments of foreign protein that are bound to class II MHC molecules in much the same way that virus-derived peptides are bound to class I MHC molecules. But both the source of the peptide fragments presented and the route they take to find the MHC molecules are different from those of peptide fragments presented by class I MHC molecules to cytotoxic T cells. Rather than being derived from foreign protein synthesized inside the target cell, the peptides presented to helper T cells are derived from extracellular microbes or their products, which have been ingested by antigen-presenting cells and degraded in the acidic environment of endosomes. These peptides do not have to be pumped across a membrane because they do not originate in the cytosol; they are generated in a compartment that is topologically equivalent to the extracellular space. They never enter the lumen of the ER, where the class II MHC molecules are synthesized and assembled, but instead bind to preassembled class II heterodimers in a late endosomal compartment. Once the peptide binds, the class II MHC molecule alters its conformation, trapping the peptide in the binding groove for presentation at the cell surface to helper T cells.

If it is to function in the presentation of peptides derived from extracellular foreign proteins, a newly synthesized class II MHC molecule must avoid clogging its binding site prematurely with peptides derived from endogenously synthesized proteins in the ER lumen. A nonpolymorphic polypeptide, called the **invariant chain,** acts in at least two ways to help ensure this. First, it associates with newly synthesized class II MHC heterodimers in the ER in such a way as to prevent them from binding peptides in the lumen of the ER. Second, it directs class II MHC molecules from the *trans* Golgi network to the endosome compartment, where the invariant chain is released by proteolysis, freeing the class II molecules to bind peptide fragments derived from endocytosed proteins (Figure 23–53). In this way the functional differences between class I and class II MHC molecules are maintained—the former presenting molecules that come from the cytosol, the latter presenting molecules that come from the extracellular space.

Most of the class I and class II MHC molecules on the surface of a target cell have peptides derived from self proteins in their binding groove—for class I

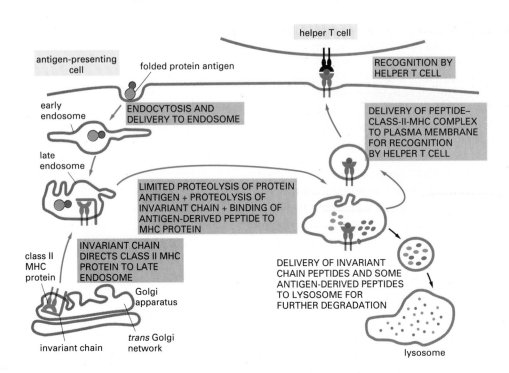

molecules, fragments of degraded cytosolic and nuclear proteins; for class II molecules, fragments of degraded membrane and serum proteins that pass through the endosome-lysosome system. Only a small fraction of the class II MHC molecules on the surface of an antigen-presenting cell will have foreign peptides bound to them. This, however, suffices to initiate an immune response, because only a few hundred such molecules are required to activate a helper T cell, just as only a few hundred peptide–class-I-MHC complexes on a target cell are required to activate a cytotoxic T cell.

Helper T Cells Are Activated by Antigen-presenting Cells [37]

Just as a B cell must be activated to proliferate and differentiate into an antibody-secreting cell before it can function, so a T cell must be activated to proliferate and differentiate before it can kill an infected target cell or help a macrophage or B cell. The initial activation of a T cell usually occurs when it recognizes a foreign peptide bound to an MHC molecule on the surface of an appropriate target cell. For a helper T cell the appropriate target is an **antigen-presenting cell.**

Antigen-presenting cells are derived from the bone marrow and comprise a heterogeneous set of cells, including *interdigitating dendritic cells* in lymphoid organs and *Langerhans cells* in skin, as well as the B cells and macrophages that will subsequently be the target of T cell help. Together with thymus epithelial cells, which have a special role in T cell development (discussed later), and activated T cells in some mammals, these specialized antigen-presenting cells are the only cell types that normally express class II MHC molecules (see Table 23–2). In addition to class II MHC molecules, antigen-presenting cells also express a second cell-surface molecule, called B7, that plays a crucial part in activating T cells, as we discuss below.

The T Cell Receptor Forms Part of a Large Signaling Complex in the Plasma Membrane [38]

The activation of a cytotoxic or helper T cell is a complicated process that is still incompletely understood. The T cell receptor heterodimer recognizes foreign peptides bound to MHC molecules on the surface of the target cell. As in the case of B cells, the receptor is associated with a set of invariant transmembrane

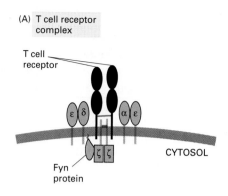

(A) T cell receptor complex

Figure 23–54 **Comparison of the antigen receptors on T and B cells.** In both cases the receptors (*black*) are associated with invariant transmembrane polypeptide chains (*green*) that act as signal transducers. The invariant chains associated with the T cell receptor form the CD3 complex, while those associated with the B cell receptor are called Ig-α, Ig-β, and Ig-γ. At least one of the invariant chains in both cases binds a Src-like tyrosine kinase (*red*). All of the chains shown, except for the CD3-ζ chains and the Src-like kinases, have extracellular Ig-like domains and are therefore members of the Ig superfamily.

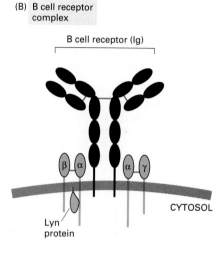

(B) B cell receptor complex

polypeptide chains (called the **CD3 complex**) that transduce the extracellular binding event into intracellular activating signals. The CD3 complex is thought to activate one or more members of the Src family of tyrosine kinases, including the *Fyn* protein, to phosphorylate various cellular proteins, including components of the CD3 complex itself and the enzyme phospholipase C-γ, which in turn activates the inositol phospholipid signaling pathway described in Chapter 15. The T cell and B cell receptors and their associated polypeptide chains are compared in Figure 23–54.

The T cell receptor and CD3 complex do not act on their own to activate a T cell. A variety of co-receptors also play important roles. We have already discussed the CD4 and CD8 co-receptors and their associated Lck kinase, and a number of others have also been defined. In the process of T cell activation, these receptors and co-receptors, as well as their associated Src-like tyrosine kinases, are thought to assemble into a large signaling complex in the plasma membrane of the T cell. Even this large signaling complex, however, is not enough on its own to activate a helper T cell: another, independent signaling pathway is also required.

Two Simultaneous Signals Are Required for Helper T Cell Activation [39]

To activate a helper T cell, an antigen-presenting cell must provide at least two signals. S*ignal 1* we have already discussed: it is provided by a foreign peptide bound to a class II MHC molecule on the surface of the presenting cell, which activates the T cell receptor complex illustrated in Figure 23–54A. Depending on the type of helper T cell (discussed below), *signal 2* is provided either by a secreted chemical signal, such as *interleukin-1 (IL-1)*, or by the plasma-membrane-bound signaling molecule **B7** on the surface of the antigen-presenting cell. B7 is recognized by a co-receptor protein called **CD28,** which is present on the surface of the helper T cell and is a member of the Ig superfamily. If helper T cells receive both signals, they are activated to proliferate and to secrete a variety of

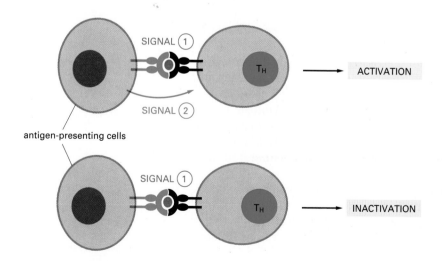

Figure 23–55 **Two signals are required for helper T cell activation.** Signal 2 can be provided by either a secreted signal such as interleukin-1 (IL-1) or the membrane-bound B7 protein on the antigen-presenting cell surface. Signal 1 without signal 2 can inactivate the T cell. The accessory proteins associated with the T cell receptor (illustrated in Figure 23–54A) that are required for generating signal 1 are not shown.

Table 23–3 Some Accessory Proteins on the Surface of T Cells

Protein*	Approximate Molecular Weight	Super-family	Expressed on	Ligand on Target Cell	Functions
CD3	γ chain = 25,000 δ chain = 20,000 ε chain = 20,000 ζ chain = 16,000	Ig Ig Ig —	all T cells	—	helps transduce signal when antigen-MHC complex binds to T cell receptors
CD4	55,000	Ig	helper T cells	class II MHC	promotes adhesion to antigen-presenting cells and to B cells; signals T cell
CD8	70,000 (homodimer or heterodimer)	Ig	cytotoxic T cells	class I MHC	promotes adhesion to infected target cells; signals T cell
CD28	80,000 (homodimer)	Ig	many helper and cytotoxic T cells	B7	provides signal 2 to some T cells
LFA-1	a_1 chain = 190,000 b_2 chain = 95,000	integrin	most white blood cells, including T cells	ICAM-1	promotes cell-cell adhesion

*CD stands for *cluster of differentiation*, as each of the CD proteins was originally defined as a T cell "differentiation antigen" recognized by multiple monoclonal antibodies. Their identification depended on large-scale collaborative studies in which hundreds of such antibodies, generated in many laboratories, were compared and found to consist of relatively few groups (or "clusters"), each recognizing a single cell-surface protein. Since these initial studies, however, about 100 CD proteins have been identified.

interleukins. In contrast, if they receive signal 1 without signal 2, they are altered so that they can no longer be activated even if they receive both signals (Figure 23–55). This has been suggested to be one mechanism whereby T cells become *tolerant,* as we discuss later.

Once a helper or cytotoxic T cell has been stimulated by antigen, other accessory proteins on its surface are called into play to increase the strength of T cell binding to the target cell. In Chapter 19, for example, we discussed how stimulation of a T cell activates the *lymphocyte-function-associated* protein *LFA-1,* a member of the integrin family of cell-adhesion proteins, so that LFA-1 can bind to its ligand on the target cell; the ligand is called *intercellular adhesion molecule 1 (ICAM-1),* and it is a member of the Ig superfamily.

Some of the co-receptors and other accessory proteins on the surface of T cells are summarized in Table 23–3.

Helper T Cells, Once Activated, Stimulate Themselves and Other T Cells to Proliferate by Secreting Interleukin-2 [40]

The combined action of signal 1 and signal 2 provokes helper T cell proliferation by a curiously indirect mechanism. It causes the T cells to stimulate their own proliferation by simultaneously secreting a growth factor called **interleukin-2 (IL-2)** and synthesizing cell-surface receptors that bind it. The binding of IL-2 to these IL-2 receptors then directly stimulates the T cells to proliferate. By this *autocrine mechanism,* helper T cells can continue to proliferate after they have left the surface of the antigen-presenting cell (Figure 23–56). The helper T cells can also stimulate the proliferation of any other nearby T cells, including cytotoxic T cells, that have first been induced by antigen to express IL-2 receptors. Because the expression of IL-2 receptors is strictly dependent on antigen stimulation, however, IL-2 causes the proliferation of only T cells that have encountered their specific antigen.

Once the requirements for T cell proliferation were discovered, it was possible to produce indefinitely proliferating, antigen-specific *T cell lines* in culture by administering IL-2 and periodically stimulating the cells with antigen to maintain the expression of IL-2 receptors. Single cells from such lines can be isolated

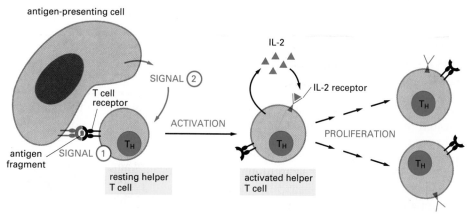

Figure 23–56 **The stimulation of T cell proliferation by IL-2.** Signals 1 and 2 activate the helper T cell to make IL-2 receptors and to secrete IL-2. The binding of IL-2 to its receptors stimulates the cell to grow and divide. When the antigen is eliminated, the T cells eventually stop producing IL-2 and IL-2 receptors, so cell proliferation stops. We shall see later that some helper T cells do not make IL-2; their proliferation, like that of cytotoxic T cells, is stimulated by IL-2 made by neighboring helper T cells.

to generate **T cell clones.** Such clones have been crucially important in T cell research. Together with *T cell hybridomas* (which are analogous to the mono-clonal-antibody-secreting B cell hybridomas mentioned earlier and are produced by fusing antigen-specific T cells with a T cell tumor line), T cell clones made it possible to isolate T cell receptors and their genes. They have also been widely used to study the mechanisms of T cell activation and the role of helper T cells in stimulating the responses of other cells, such as B cells and macrophages.

Helper T Cells Are Required for Most B Cells to Respond to Antigen [41]

The role of helper T cells in B cell antibody responses was first discovered in the mid-1960s through experiments in which either thymus cells or bone marrow cells were injected together with antigen into irradiated mice. Mice that had received only bone marrow or only thymus cells were unable to make antibody, but if a mixture of thymus and bone marrow cells was injected, large amounts of antibody were produced. It was later shown that the thymus provides T cells while the bone marrow provides B cells. The use of specific markers to distinguish between the injected T and B cells showed that the antibody-secreting cells are B cells, leading to the conclusion that T cells must help B cells respond to antigen.

There are some antigens, however, including many microbial polysaccharides, that can stimulate B cells to proliferate and differentiate into antibody-secreting cells without T cell help. Such *T-cell-independent antigens* are often large polymers with repeating, identical antigenic determinants (see Figure 23–25B); their multipoint binding to the membrane-bound antibody molecules that serve as antigen receptors on B cells may generate a signal strong enough to activate the B cells directly, but only some B cells can be activated in this way. Because these antigens do not activate helper T cells, they fail to induce memory B cells or antibody class switching (both of which require T cell help) and, therefore, mainly cause the production of low-affinity IgM antibodies.

The Activation of B Cells by Helper T Cells Is Mediated by Both Membrane-bound and Secreted Signals [42]

Whereas antigen-presenting cells such as dendritic cells and macrophages are omnivorous and ingest and present antigens nonspecifically, a B cell generally presents only an antigen that it specifically recognizes. In a primary antibody response helper T cells are normally activated by binding to a foreign antigen bound to a class II MHC protein on the surface of an antigen-presenting cell of the omnivorous type—such as an interdigitating dendritic cell in a secondary lymphoid organ. Once activated, the helper T cell can then help activate a B cell that specifically displays the same complex of foreign antigen and class II MHC protein on its surface. The display of antigen on the B cell surface reflects the selectivity with which it takes up foreign molecules from the extracellular fluid. These are selected by their binding to the specific membrane-bound antibodies (antigen receptors) on the surface of the B cell and are ingested by receptor-mediated endocytosis; they are then degraded and recycled to the cell surface in the form of peptides bound to class II MHC proteins. Thus the helper T cell activates those B cells that make membrane-bound antibodies that specifically recognize the antigen that initially activated the T cell. In secondary antibody responses memory B cells themselves may act as antigen-presenting cells and activate helper T cells, as well as being the subsequent targets of the helper T cells. The mutually reinforcing actions of T cells and B cells lead to an immune response that is both intense and highly selective.

Once a helper T cell has been activated and contacts a B cell, the contact initiates an internal rearrangement of the helper cell cytoplasm that orients the centrosome and Golgi apparatus toward the B cell, as described previously for a cytotoxic T cell contacting its target cell (see Figure 23–52). In this case, however, the orientation is thought to enable the helper T cell to direct membrane-bound and secreted signaling molecules onto the B cell surface. The membrane-bound signaling molecule is a transmembrane protein called **CD40 ligand**, which is expressed on the surface of activated, but not resting, helper T cells. It is recognized by the **CD40** transmembrane protein on the B cell surface. The interaction between CD40 ligand and CD40 is required for helper T cells to activate B cells to proliferate and mature into memory and antibody-secreting cells. This interaction is critical for T cell help: individuals whose T cells lack the CD40 ligand because of a mutation in the gene encoding the protein can make only IgM antibodies and are severely immunodeficient, being susceptible to the same infections that affect AIDS patients, whose helper T cells have been destroyed.

Secreted signals from the T cells provide additional help by activating B cells to proliferate and mature and, in some cases, to switch the class of antibody they produce. **Interleukin-4 (IL-4)** is one such signal. It stimulates B cell proliferation and maturation and promotes switching to IgE and IgG1 antibody production: if the *IL-4* gene is inactivated in a mouse by targeted genetic recombination, the mouse is unable to make IgE and makes very little IgG1.

Thus most B cells, like T cells, require multiple signals for activation, one provided by antigen binding to membrane-bound Ig molecules and the others provided by helper T cells; as in the case of T cells, if a B cell receives the first signal only, it may be functionally inactivated. The signals required for helper T and B cell activation are compared in Figure 23–57.

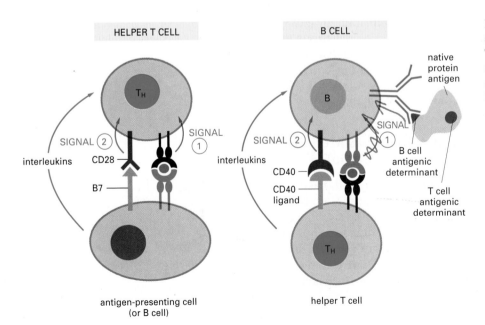

Figure 23–57 **Comparison of the signals required to activate a helper T cell and a B cell.** Note that signal 2 can be provided either by a secreted signaling molecule (an interleukin) or by a cell-cell contact interaction.

Some Helper T Cells Help Activate Cytotoxic T Cells and Macrophages by Secreting Interleukins [43]

There are at least two functionally distinct subclasses of helper T cells that can be distinguished by the interleukins that they secrete. **T$_H$1 cells** secrete IL-2 and γ-interferon and are concerned mainly with helping cytotoxic T cells and macrophages. **T$_H$2 cells** secrete IL-4 and IL-5 and are concerned mainly with helping B cells and eosinophils. We saw earlier that IL-2 secreted by activated helper T cells can stimulate activated cytotoxic T cells (as well as other activated helper T cells) to proliferate. Stimulated T cells also secrete other interleukins, such as *γ-interferon,* that help activate cytotoxic T cells to kill infected target cells and that enhance the ability of macrophages to phagocytose and destroy invading microorganisms. The activation of macrophages by T cells is especially important in defense against infections by microorganisms that can survive phagocytosis by nonactivated macrophages. Tuberculosis is one such infection.

The antigen-triggered secretion of interleukins underlies the familiar tuberculin skin test. If tuberculin (an extract of the bacterium responsible for tuberculosis) is injected into the skin of individuals who have had or have been immunized against tuberculosis (and therefore contain appropriate memory T cells), a characteristic immune response occurs in the skin. It is initiated at the site of injection by the secretion of interleukins by memory helper T cells that react to tuberculin. The interleukins attract macrophages (and lymphocytes) into the site, thereby causing the characteristic swelling of a positive reaction to tuberculin.

Another important effect of γ-interferon is to induce the expression of class II MHC proteins on the surface of some cells (such as endothelial cells) that do not normally express them, thereby enabling these cells to present antigen to helper T cells. As discussed earlier, γ-interferon also increases the efficiency with

Table 23–4 Properties of Some Interleukins*

Inter-leukin (IL)	Approximate Molecular Weight	Source	Target	Action
IL-1	15,000	antigen-presenting cells	helper T cells	helps activate
IL-2	15,000	some helper T cells	all activated T cells and B cells	stimulates proliferation
IL-3	25,000	some helper T cells	various hemo-poietic cells	stimulates proliferation
IL-4	20,000	some helper T cells	B cells	stimulates proliferation, maturation, and class switching to IgE and IgG1
IL-5	20,000	same helper T cells that make IL-4	B cells, eosinophils	promotes proliferation and maturation
IL-6	25,000	some helper T cells and macrophages	activated B cells, T cells	promotes B cell maturation to Ig-secreting cells; helps activate T cells
γ-Inter-feron	25,000 (dimer)	same helper T cells that make IL-2	B cells, macrophages, endothelial cells	activates various MHC genes and macrophages

Interleukins are secreted peptides and proteins that mainly mediate local interactions between white blood cells (leucocytes) but do not bind antigen; those secreted by lymphocytes are also called *lymphokines.* The amino acid sequence is known for all the proteins listed. The sources, target cells, and actions listed are those most relevant to the immune system, but most of the interleukins have many more sources, targets, and actions than are shown. This is especially the case for IL-1 and IL-6, which are also made by non-blood cells and act on many types of target cells other than blood cells; they are therefore more accurately called *cytokines.*

which target cells present viral peptide in association with class I MHC molecules for recognition by cytotoxic T cells.

Some of the interleukins secreted by helper T cells, along with some secreted by antigen-presenting cells, are listed in Table 23–4.

Summary

Helper T cells help activate B cells and macrophages. They are themselves initially activated when they recognize peptide fragments derived from foreign extracellular proteins that are endocytosed by specialized antigen-presenting cells. The ingested proteins are degraded in endosomes, and some of the resulting peptide fragments bind to class II MHC molecules, forming complexes that are carried to the cell surface, where the helper T cells recognize them. The activation of a helper T cell requires at least two signals: signal 1 is provided by the MHC-peptide complex, while signal 2 is provided by either the B7 protein on the surface of an antigen-presenting cell or a signal secreted by this cell. Once activated, helper T cells stimulate their own proliferation by secreting interleukin-2 and activate their target cells by a combination of membrane-bound and secreted signaling molecules.

Selection of the T Cell Repertoire

T cells develop in the thymus. Remarkably, more than 95% of the cells produced there die before they are able to mature and migrate to peripheral lymphoid organs. This waste is largely due to the stringent selection processes that operate on developing T cells to ensure that only cells with potentially useful receptors survive. In this section we examine both the positive and negative selection processes that help shape the T cell repertoire. This will lead us to consider why individuals differ genetically in their immune responsiveness and why the MHC molecules that present antigen to T cells show such extreme genetic variability.

Developing T Cells That Recognize Peptides in Association with Self MHC Molecules Are Positively Selected in the Thymus [44]

We have seen that T cells recognize antigen in association with self MHC molecules but not in association with foreign MHC molecules: that is, T cells show *MHC restriction*. This restriction reflects a process of **positive selection** during T cell development in the thymus, whereby those immature T cells that will be capable of recognizing foreign peptides presented by self MHC molecules are selected to survive, while the remainder, which would be of no use to the animal, die. Thus MHC restriction is an acquired property of the immune system that emerges as T cells develop in the thymus.

If, for example, a Y-strain thymus is transplanted into an X-strain mouse that has been irradiated to eliminate all of its mature T cells and then supplied with fresh bone marrow that can generate precursors of new ones, new X-strain T cells will develop in the Y-strain thymus. In most experiments of this type the mature X-strain T cells produced recognize foreign antigen in association with Y-strain but not X-strain MHC proteins, suggesting that the thymus dictates the specificity of MHC restriction. As T cells develop in the thymus, those with receptors that can recognize antigen in association with the types of MHC molecules expressed in the thymus are somehow selected to survive and proliferate. Similar experiments show that it is the epithelial cells of the thymus that are responsible for this positive selection process.

The most direct way to study the selection process is to follow the fate of a set of developing T cells of known specificity. This can be done using transgenic mice that express a specific pair of α and β T cell receptor genes derived from a

T cell clone of known antigen and MHC specificity. As is the case for immuno-globulin genes in B cells, the expression of the rearranged T cell receptor transgenes inhibits the rearrangement of endogenous T cell receptor genes to ensure allelic exclusion, so that a large proportion of the T cells in these transgenic mice express only the transgenic receptor. Thus the fate of T cells with a known specificity can be readily followed. Such experiments show that the transgenic T cells mature and populate the peripheral lymphoid tissues only if the transgenic mouse also expresses the same allelic form of MHC molecule as is recognized by the transgenic T cell receptor; if the mouse does not express the appropriate MHC molecule, the transgenic T cells die in the thymus. Thus, as suggested by the thymus transplantation experiments, the survival and matura-tion of a T cell depend on a match between its receptor and the MHC molecules expressed in the thymus. As part of this positive selection process, cytotoxic T cells are selected for recognition of class I MHC molecules while helper T cells are selected for recognition of class II MHC molecules. Thus genetically engi-neered mice that lack cell-surface class I MHC molecules specifically lack cyto-toxic T cells, whereas mice that lack class II MHC molecules specifically lack helper T cells.

Positive selection still leaves a large problem to be solved. If developing T cells with receptors that recognize self peptides associated with self MHC mol-ecules were to mature in the thymus and migrate to peripheral lymphoid tissues, they would wreak havoc. A second, *negative selection* process in the thymus is required to avoid this disaster.

Developing T Cells That React Strongly with Self Peptides Bound to Self MHC Molecules Are Eliminated in the Thymus [44, 45]

As discussed previously, a fundamental feature of the immune system is that it can distinguish self from nonself and normally does not react against self mol-ecules. This state of *immunological self tolerance* is acquired mainly during T cell development. Although some self-reactive B cells are eliminated or inactivated during B cell development, the main mechanism for self tolerance is thought to be the deletion in the thymus of developing self-reactive T cells—that is, T cells whose receptors bind strongly to the complex of a self peptide bound to a self MHC molecule. Because most B cells require helper T cells to respond to anti-gen, the elimination of self-reactive helper T cells also ensures that self-reactive B cells are harmless.

It is not enough, therefore, for the thymus to select *for* T cells that recognize self MHC molecules; it must also select *against* T cells that recognize self MHC molecules complexed with self peptides—in other words, it must pick out for survival just those T cells that will be capable of recognizing self MHC molecules complexed with foreign peptides, even though these peptides are not present in the developing thymus. It is thought that such T cells bind weakly in the thymus to self MHC molecules that are carrying self peptides mismatched to the T cell receptors. Thus the required goal can be achieved by (1) ensuring the death of T cells that bind *strongly* to the peptide-MHC complexes in the thymus while (2) promoting the survival of those that bind weakly and (3) permitting the death of those that do not bind at all. Process 2 is the positive selection we have just dis-cussed. Process 1 is called **negative selection** (Figure 23–58).

The most convincing evidence for negative selection derives once again from experiments with transgenic mice. After the introduction of T cell receptor transgenes encoding a receptor that recognizes a male-specific peptide antigen, for example, large numbers of mature T cells expressing the transgenic receptor are found in the thymus and peripheral lymphoid organs of female mice, but very few are found in males, where the cells die in the thymus before they have a chance to mature. Like positive selection, negative selection requires the inter-

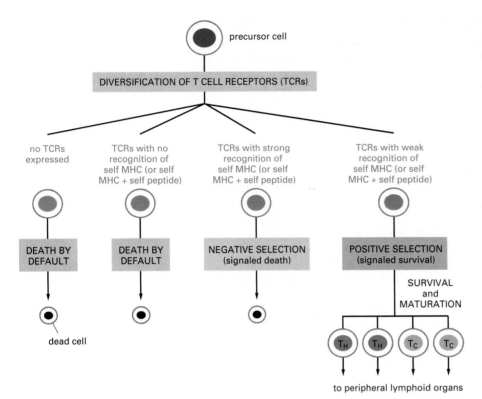

Figure 23–58 Positive and negative selection in the thymus. Only cells with receptors that recognize foreign peptides in association with self MHC molecules are selected to survive, mature, and migrate to peripheral lymphoid organs; all of the other cells undergo programmed cell death. During the process of positive selection, helper T cells (T_H) and cytotoxic T cells (T_C) diverge by an unknown mechanism such that helper cells express the CD4 co-receptor but not CD8 and recognize foreign peptides in association with class II MHC molecules, while cytotoxic cells express CD8 but not CD4 and recognize foreign peptides in association with class I MHC molecules (not shown).

action of a T cell receptor and a CD4 or CD8 co-receptor with an appropriate MHC molecule. Unlike positive selection, however, which occurs mainly on the surface of the thymus epithelial cells, negative selection occurs mainly on the surface of other cells, such as dendritic cells or macrophages, that originate in the bone marrow and migrate into the thymus. Like the epithelial cells, they have both class I and class II MHC molecules on their surface.

The mechanisms responsible for positive and negative T cell selection in the thymus are unknown, but in both cases programmed cell death (discussed in Chapters 21 and 22) is thought to be involved. During positive selection thymus epithelial cells seem to provide survival signals to weakly bound T cells, preventing the T cells from killing themselves. In negative selection, by contrast, the bone-marrow-derived cells signal the tightly bound T cells to kill themselves (Figure 23–58).

The deletion of self-reactive T cells in the thymus cannot eliminate all potentially self-reactive T cells, as some self molecules are not present in the thymus. Whereas at least some of the T cells with receptors that recognize such self molecules are presumably eliminated after they leave the thymus, others may be functionally inactivated by a process called *clonal anergy* (to distinguish it from *clonal deletion*, in which the self-reactive cells die). Although the molecular mechanism of clonal anergy is uncertain, it is postulated that the T cells recognize self peptides bound to MHC molecules on the surface of tissue cells that are unable to provide a signal 2; as discussed earlier, signal 1 without signal 2 can inactivate a T cell without killing it.

Some Allelic Forms of MHC Molecules Are Ineffective at Presenting Specific Antigens to T Cells: Immune Response (*Ir*) Genes [46]

Unlike class I MHC genes, which were first recognized by their effects on graft rejection, class II MHC genes were first recognized by their effects on T-cell-dependent immune responses to specific soluble antigens. When animals were immunized with a simple antigen, some made vigorous T-cell-dependent responses while others did not respond at all. Genetic studies indicated that the ability to respond to the antigen was controlled by a single gene, called an **immune response (*Ir*) gene,** and responses to different antigens were often controlled by different *Ir* genes. These, it transpired, are the MHC genes under a different name: *Ir* genes that control the response of helper T cells to an antigen map to one or other of the class II MHC loci; those that control the response of cytotoxic T cells to an antigen map to one or other of the class I MHC loci.

These observations initially were extremely puzzling, but once it was recognized that the MHC proteins play a crucial role in binding and presenting antigen to T cells, they could be readily explained: a genetic nonresponder to a simple antigen (usually one with only a single antigenic determinant) presumably lacks an MHC molecule that can bind and effectively present the antigenic determinant to an appropriate T cell. This explanation is supported by *in vitro* studies showing that purified class II MHC molecules from a *responder* animal can bind the relevant antigenic peptide, while those from a genetic *nonresponder* cannot.

Another mechanism, however, seems to be responsible for some cases of genetic nonresponsiveness to specific antigens. Certain combinations of self MHC molecules and foreign peptides are likely to resemble some combinations of self MHC molecules and self peptides. Because the helper T cells that react to such combinations are eliminated by negative selection during T cell development in the thymus, an animal may be genetically unable to respond to these foreign peptides.

The Role of MHC Proteins in Antigen Presentation to T Cells Provides an Explanation for Transplantation Reactions and MHC Polymorphism [47]

The role of MHC proteins in binding foreign antigens and presenting them to T cells provides an explanation for why so many T cells respond to foreign MHC molecules and thereby reject foreign organ grafts. Presumably, foreign MHC proteins with an endogenous peptide in their peptide-binding groove create complexes that can resemble self MHC molecules complexed with foreign peptides. Some cytotoxic T cell clones that react to a viral antigen in association with a self class I MHC molecule, for example, have been shown to react to a foreign class I MHC molecule in the absence of the viral antigen (Figure 23–59).

The antigen-presenting function of MHC proteins can also explain the extensive polymorphism of these molecules. In the evolutionary war between pathogenic microorganisms and the vertebrate immune system, microorganisms will tend to change their antigens to avoid associating with MHC molecules. When one succeeds, it will be able to sweep through a population as an epidemic. In such circumstances the few individuals that produce a new MHC molecule that can associate with an antigen of the altered microorganism will have a large selective advantage. In addition, individuals with two different alleles for each MHC molecule (heterozygotes) will have a better chance of resisting infection

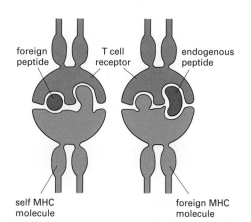

foreign peptide   T cell receptor   endogenous peptide

self MHC molecule   foreign MHC molecule

Figure 23–59 Foreign MHC mimicry. A foreign MHC molecule with an endogenous peptide bound in its binding groove can resemble a self MHC molecule with a foreign peptide (such as a viral peptide) bound to it. This is thought to explain why so many of our T cells can be activated by foreign MHC molecules in transplantation reactions.

than those with identical alleles at any given MHC locus, as they will have a greater capacity to present antigens from a wide range of pathogens. Thus selection will tend to promote and maintain a large diversity of MHC molecules in the population. This hypothesis that infectious diseases have provided the driving force for MHC polymorphism recently has received strong support from the finding that individuals in West Africa with a specific MHC allele have a reduced susceptibility to a severe form of malaria; while the allele is rare elsewhere, it is found in 25% of the West African population where this form of malaria is common.

If greater MHC diversity means greater resistance to infection, why do we each have so few MHC loci encoding these molecules, and why have we not evolved strategies for increasing their diversity—by alternative RNA splicing, for example, or by the genetic recombination mechanisms used to diversify antibodies? Presumably, this is because each time a new MHC molecule is added to the repertoire, the T cells that recognize self peptides in association with the new MHC molecule must be eliminated to maintain self tolerance. The elimination of these T cells would counteract the advantage of adding the new MHC molecule. Thus the number of MHC molecules we express may represent a balance between the advantages of presenting a wide diversity of foreign peptides to T cells against the disadvantages of restricting the T cell repertoire.

Immune Recognition Molecules Belong to an Ancient Superfamily [48]

Most of the proteins that mediate cell-cell recognition or antigen recognition in the immune system contain related structural elements, suggesting that the genes that encode them have a common evolutionary history. Included in this Ig superfamily are *antibodies, T cell receptors, MHC proteins,* the *CD4, CD8,* and *CD28 co-receptors,* most of the invariant polypeptide chains associated with B and T cell receptors, and the various *Fc receptors* on lymphocytes and other white blood

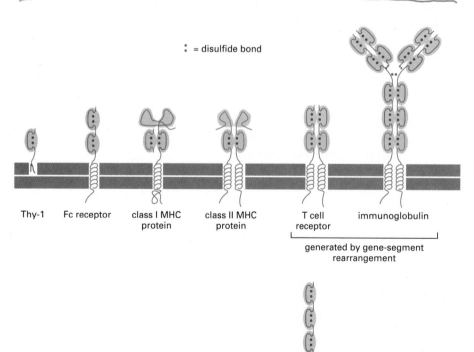

Figure 23–60 **Some of the membrane proteins belonging to the Ig superfamily.** The Ig domains are shaded in *gray*, and the antigen-binding domains, in *blue*. The immunoglobulin superfamily also includes many cell-surface proteins involved in cell-cell interactions outside the immune system, such as the neural cell-adhesion molecule (N-CAM) discussed in Chapter 19 and the receptors for various protein growth factors discussed in Chapters 15 and 17 (not shown).

cells—all of which contain one or more Ig or Ig-like domains (*Ig homology units*). In fact, 40% of the 150 or so polypeptides that have been characterized on the surface of white blood cells belong to this superfamily. Each of the Ig-like domains is 70–110 amino acids long and is thought to be folded into the characteristic sandwichlike structure made of two antiparallel β sheets, usually stabilized by a conserved disulfide bond. Many of these molecules are dimers or higher oligomers in which Ig-like domains of one chain interact with those in another (Figure 23–60).

Most of the amino acids in each Ig-like domain are usually encoded by a separate exon, and it seems likely that the entire supergene family evolved from a gene coding for a single Ig-like domain—similar to that encoding β_2-microglobulin (see Figure 23–45A) or the Thy-1 protein (see Figure 23–60)—that may have been involved in mediating cell-cell interactions. There is evidence that such a primordial gene arose before vertebrates diverged from their invertebrate ancestors some 400 million years ago. New family members presumably arose by exon and gene duplications, and similar duplication events probably gave rise to the multiple gene segments that encode antibodies and T cell receptors.

Summary

The T cell repertoire is shaped mainly by a combination of positive and negative selection processes that operate during T cell development in the thymus. These processes ensure that only T cells with potentially useful receptors survive and mature while the others undergo programmed cell death: T cells that will be able to recognize foreign peptides complexed with self MHC molecules are positively selected, while T cells that react strongly with self peptides complexed with self MHC molecules are eliminated.

Most of the proteins involved in cell-cell recognition and antigen recognition in the immune system, including antibodies, T cell receptors, and MHC molecules, as well as the various co-receptors discussed in this chapter, belong to the ancient Ig superfamily, which is thought to have evolved from a primordial gene encoding a single Ig-like domain.

References

General

Abbas, A.K.; Lichtman, A.H.; Pober, J.S. Cellular and Molecular Immunology. Philadelphia: Saunders, 1991.

Golub, E.S.; Green, D.R. Immunology: A Synthesis, 2nd ed. Sunderland, MA: Sinauer, 1991.

Janeway, C.; Travers, P. Immunobiology. London and New York: Current Science and Garland, 1994.

Kuby, J. Immunology. New York: W.H. Freeman, 1992.

Paul, W.E., ed. Fundamental Immunology. New York: Raven Press, 1984.

Cited

1. Gowans, J.L.; McGregor, D.D. The immunological activities of lymphocytes. *Prog. Allergy* 9:1–78, 1965.
2. Greaves, M.F.; Owen, J.J.T.; Raff, M.C. T and B Lymphocytes: Origins, Properties and Roles in Immune Responses. Amsterdam: Excerpta Medica, 1973.
3. Cooper, M.; Lawton, A. The development of the immune system. *Sci. Am.* 231(5):59–72, 1974.

   Ikuta, K.; Uchida, N.; Friedman, J.; Weissman, I.L. Lymphocyte development from stem cells. *Annu. Rev. Immunol.* 10:759–784, 1992.

   Owen, J.J.T. Ontogenesis of lymphocytes. In B and T Cells in Immune Recognition (F. Loor, G.E. Roelants, eds.), pp. 21–34. New York: Wiley, 1977.

4. Barclay, A.N., et al. The Leucocyte Antigens Facts Book. San Diego, CA: Academic Press, 1993.

   Möller, G., ed. Functional T Cell Subsets Defined by Monoclonal Antibodies. *Immunol. Rev.* 74, 1983.

   Raff, M.C. Cell-surface immunology. *Sci. Am.* 234(5):30–39, 1976.

   Reinherz, E.L.; Schlossman, S.F. The differentiation and function of human T lymphocytes. *Cell* 19:821–827, 1980.

5. Ada, G.L. Antigen binding cells in tolerance and immunity. *Transplant. Rev.* 5:105–129, 1970.

   Ada, G.L.; Nossal, G. The clonal selection theory. *Sci. Am.* 257(2):62–69, 1987.

   Burnet, F.M. The Clonal Selection Theory of Acquired Immunity. Nashville, TN: Vanderbilt University Press, 1959.

   Jerne, N.K. The immune system. *Sci. Am.* 229(1):52–60, 1973.

   Wigzell, H. Specific fractionation of immunocompetent cells. *Transplant. Rev.* 5:76–104, 1970.

6. Laver, W.G.; Air, G.M.; Webster, R.G.; Smith-Gill, S.J. Epitopes on protein antigens: misconceptions and realities. *Cell* 61:553–556, 1990.

   Pink, J.R.L.; Askonas, B.A. Diversity of antibodies to cross-reacting nitrophenyl haptens in inbred mice. *Eur. J. Immunol.* 4:426–429, 1974.

7. Bevilacqua, M.P. Endothelial-leukocyte adhesion molecules. *Annu. Rev. Immunol.* 11:767–804, 1993.

   Gowans, J.L.; Knight, E.J. The route of re-circulation of lymphocytes in the rat. *Proc. R. Soc. Lond. (Biol.)* 159:257–282, 1964.

   Mackay, C.R. Homing of naïve, memory and effector lymphocytes. *Curr. Opin. Immunol.* 5:423–427, 1993.

   Picker, L.J.; Butcher, E.C. Physiological and molecular mechanisms of lymphocyte homing. *Annu. Rev. Immunol.* 10:561–591, 1992.

   Springer, T.A. Adhesion receptors of the immune system. *Nature* 346:425–434, 1990.

8. Gray, D. Immunological memory. *Annu. Rev. Immunol.* 11:49–77, 1993.

   Mackay, C.R. Immunological memory. *Adv. Immunol.* 53:217–265, 1993.

   Sprent, J. Lifespans of naïve, memory and effector lymphocytes. *Curr. Opin. Immunol.* 5:433–438, 1993.

   Vitetta, E.S., et al. Memory B and T cells. *Annu. Rev. Immunol.* 9:193–217, 1991.

9. Billingham, R.E.; Brent, L.; Medawar, P.B. Quantitative studies on tissue transplantation immunity. III. Actively acquired tolerance. *Philos. Trans. R. Soc. Lond. (Biol.)* 239:357–414, 1956.

   Harris, D.E.; Cairns, L.; Rosen, F.S.; Borel, Y. A natural model of immunologic tolerance. Tolerance to murine C5 is mediated by T cells and antigen is required to maintain unresponsiveness. *J. Exp. Med.* 156:567–584, 1982.

   Lindstrom, J. Immunobiology of myasthenia gravis, experimental autoimmune myasthenia gravis and Lambert-Eaton syndrome. *Annu. Rev. Immunol.* 3:109–132, 1985.

   Nossal, G.J.V. Cellular and molecular mechanisms of B lymphocyte tolerance. *Adv. Immunol.* 52:283–331, 1992.

   Owen, R.D. Immunogenetic consequence of vascular anastomoses between bovine twins. *Science* 102:400–401, 1945.

   Schwartz, R.H. Acquisition of immunologic self-tolerance. *Cell* 57:1073–1081, 1989.

10. Davies, D.R.; Metzger, H. Structural basis of antibody function. *Annu. Rev. Immunol.* 1:87–118, 1983.

    Kabat, E.A. Structural Concepts in Immunology and Immunochemistry, 2nd ed. New York: Holt, Rinehart and Winston, 1976.

    Nisonoff, A.; Hopper, J.E.; Spring, S.B. The Antibody Molecule. New York: Academic Press, 1975.

11. DeFranco, A.L. Structure and function of the B cell antigen receptor. *Annu. Rev. Cell Biol.* 9:377–410, 1993.

    Möller, G., ed. Lymphocyte Immunoglobulin: Synthesis and Surface Representation. *Transplant. Rev.* 14, 1973.

    Reth, M. Antigen receptors on B lymphocytes. *Annu. Rev. Immunol.* 10:97–122, 1992.

12. Dutton, R.W.; Mishell, R.I. Cellular events in the immune response. The *in vitro* response of normal spleen cells to erythrocyte antigens. *Cold Spring Harb. Symp. Quant. Biol.* 32:407–414, 1967.

    Jerne, N.K., et al. Plaque forming cells: methodology and theory. *Transplant. Rev.* 18:130–191, 1974.

13. Edelman, G.M. The structure and function of antibodies. *Sci. Am.* 223(2):34–42, 1970.

    Porter, R.R. Structural studies of immunoglobulins. *Science* 180:713–716, 1973.

14. Burton, D.R.; Woof, J.M. Human antibody effector function. *Adv. Immunol.* 51:1–84, 1992.

    Ishizaka, T.; Ishizaka, K. Biology of immunoglobulin E. *Prog. Allergy* 19:60–121, 1975.

    Koshland, M.E. The coming of age of the immunoglobulin J chain. *Annu. Rev. Immunol.* 3:425–454, 1985.

    Kraehenbuhl, J.-P.; Neutra, M.R. Transepithelial transport and mucosal defense II: secretion of IgA. *Trends Cell Biol.* 2:170–174, 1992.

    Metzger, H. The receptor with high affinity for IgE. *Immunol. Rev.* 125:37–48, 1992.

    Morgan, E.L.; Weigle, W.O. Biological activities residing in the Fc region of immunoglobulin. *Adv. Immunol.* 40:61–134, 1987.

    Ravetch, J.V.; Kinet, J.-P. Fc receptors. *Annu. Rev. Immunol.* 9:457–492, 1991.

15. Berzofsky, J.A.; Berkover, I.J. Antigen-antibody interactions. In Fundamental Immunology (W.E. Paul, ed.), pp. 595–644. New York: Raven Press, 1984.

    Davies, D.R.; Sheriff, S.; Padlan, E.A. Antibody-antigen complexes. *J. Biol. Chem.* 263:10541–10544, 1988.

16. Müller-Eberhard, H.J. Molecular organization and function of the complement system. *Annu. Rev. Biochem.* 57:321–348, 1988.

    Reid, K.B. Activation and control of the complement system. *Essays Biochem.* 22:27–68, 1986.

    Reid, K.B.M.; Day, A.J. Structure-function relationships of the complement components. *Immunol. Today* 10:177–180, 1989.

    Tomlinson, S. Complement defense mechanisms. *Curr. Opin. Immunol.* 5:83–89, 1993.

17. Capra, J.D.; Edmundson, A.B. The antibody combining site. *Sci. Am.* 236(1):50–59, 1977.

    Wu, T.T.; Kabat, E.A. An analysis of the sequences of the variable regions of Bence Jones proteins and myeloma light chains and their implications for antibody complementarity. *J. Exp. Med.* 132:211–250, 1970.

18. Sakano, H., et al. Domains and the hinge region of an immunoglobulin heavy chain are encoded in separate DNA segments. *Nature* 277:627–633, 1979.

19. Alzari, P.M.; Lascombe, M.-B.; Poljak, R.J. Three-dimensional structure of antibodies. *Annu. Rev. Immunol.* 6:555–580, 1988.

    Davies, D.R.; Padlan, E.A.; Sheriff, S. Antibody-antigen complexes. *Annu. Rev. Biochem.* 59:439–473, 1991.

    Stanfield, R.L.; Fieser, T.M.; Lerner, R.A.; Wilson, I.A. Crystal structures of antibody to a peptide and its complex with peptide antigen at 2.8 Å. *Science* 248:712–719, 1990.

20. Hozumi, N.; Tonegawa, S. Evidence for somatic rearrangement of immunoglobulin genes coding for variable and constant regions. *Proc. Natl. Acad. Sci. USA* 73:3628–3632, 1976.

21. Alt, F.W.; Blackwell, T.K.; Vancopoulos, G.D. Development of the primary antibody repertoire. *Science* 238:1079–1087, 1987.

    Leder, P. The genetics of antibody diversity. *Sci. Am.* 246(5):72–83, 1982.

    Lieber, M.R. The mechanism of V(D)J recombination: a balance of diversity, specificity and stability. *Cell* 70:873–876, 1992.

    Schatz, D.G.; Oettinger, M.A.; Schlissel, M.S. V(D)J recombination: molecular biology and regulation. *Annu. Rev. Immunol.* 10:359–383, 1992.

Tonegawa, S. Somatic generation of antibody diversity. *Nature* 302:575–581, 1983.

22. Oettinger, M.A.; Shatz, D.G.; Gorka, C.; Baltimore, D. RAG-1 and RAG-2, adjacent genes that synergistically activate V(D)J recombination. *Science* 248:1517–1523, 1990.

   Thompson, C.B. RAG knockouts deliver a one/two punch. *Curr. Biol.* 2:180–182, 1992.

23. Berek, C. Somatic mutation and memory. *Curr. Opin. Immunol.* 5:218–222, 1993.

   French, D.L.; Laskov, R.; Scharff, M.D. The role of somatic hypermutation in the generation of antibody diversity. *Science* 244:1152–1157, 1989.

   Kocks, C.; Rajewsky, K. Stable expression and somatic hypermutation of antibody V regions in B-cell developmental pathways. *Annu. Rev. Immunol.* 7:537–559, 1989.

   Liu, Y.-J.; Johnson, G.D.; Gordon, J.; MacLennan, I.C.M. Germinal centres in T-cell-dependent antibody responses. *Immunol. Today* 13:17–21, 1992.

   Nossal, G.J.V. The molecular and cellular basis of affinity maturation in the antibody response. *Cell* 68:1–2, 1992.

24. Chen, J.; Alt, F.W. Gene rearrangement and B-cell development. *Curr. Opin. Immunol.* 5:194–200, 1993.

   Rolink, A.; Melchers, F. Molecular and cellular origins of B lymphocyte diversity. *Cell* 66:1081–1094, 1991.

25. Early, P., et al. Two mRNAs can be produced from a single immunoglobulin μ gene by alternative RNA processing pathways. *Cell* 20:313–319, 1980.

26. Esser, C.; Radbruch, A. Immunoglobulin class switching: molecular and cellular analysis. *Annu. Rev. Immunol.* 8:717–735, 1990.

   Finkelman, F.D., et al. Lymphokine control of *in vivo* immunoglobulin isotype selection. *Annu. Rev. Immunol.* 8:303–333, 1990.

   Harriman, W.H.; Völk, H.; Defranoux, N.; Wabl, M. Immunoglobulin class switch recombinations. *Annu. Rev. Immunol.* 11:361–384, 1993.

   Snapper, C.M.; Mond, J.J. Towards a comprehensive view of immunoglobulin class switching. *Immunol. Today* 14:15–17, 1993.

27. Allison, J.P.; Lanier, L.L. The structure, function and serology of the T cell antigen receptor complex. *Annu. Rev. Immunol.* 5:503–540, 1987.

   Chan, A.C.; Irving, B.A.; Weiss, A. New insights into T-cell antigen receptor structure and signal transduction. *Curr. Opin. Immunol.* 4:246–251, 1992.

   Davis, M.M. T cell receptor gene diversity and selection. *Annu. Rev. Biochem.* 59:475–496, 1990.

   Haas, W.; Pereira, P.; Tonegawa, S. Gamma/delta cells. *Annu. Rev. Immunol.* 11:637–685, 1993.

   Marrack, P.; Kappler, J. The T cell receptor. *Science* 238:1073–1079, 1987.

28. Kupfer, A.; Singer, S.J. Cell biology of cytotoxic and helper T-cell function. *Annu. Rev. Immunol.* 7:309–337, 1989.

   Rammensee, H.-G.; Falk, K.; Rotzschke, O. MHC molecules as peptide receptors. *Curr. Opin. Immunol.* 5:35–44, 1993.

   Swain, S.L., et al. Helper T-cell subsets: phenotype, function and the role of lymphokines in regulating their development. *Immunol. Rev.* 123:115–144, 1991.

29. Auffray, C.; Strominger, J.L. Molecular genetics of the human histocompatibility complex. *Adv. Human Genet.* 15:197–247, 1987.

   Bach, F.H.; Sacks, D.H. Transplantation immunology. *N. Engl. J. Med.* 317:489–492, 1987.

   Hood, L.; Steinmetz, M.; Malissen, B. Genes of the major histocompatibility complex of the mouse. *Annu. Rev. Immunol.* 1:529–568, 1983.

   Möller, G., ed. Molecular Genetics of Class I and II MHC Antigens. *Immunol. Rev.* 84 and 85, 1985.

30. Bjorkman, P.J.; Parham, P. Structure, function and diversity of class I major histocompatibility complex molecules. *Annu. Rev. Biochem.* 59:253–288, 1990.

   Brown, J.H., et al. Three-dimensional structure of the human class II histocompatibility antigen HLA-DR1. *Nature* 364:33–39, 1993.

   Matsumura, M.; Fremont, D.H.; Peterson, P.A.; Wilson, I.A. Emerging principles for the recognition of peptide antigens by MHC class I molecules. *Science* 257:927–934, 1992.

   Silver, M.L.; Guo, H.-C.; Strominger, J.L.; Wiley, D.C. Atomic structure of a human MHC molecule presenting an influenza virus peptide. *Nature* 360:367–369, 1992.

   Zhang, W., et al. Crystal structure of the major histocompatibility complex class I H-2K^b molecule containing a single viral peptide: implications for peptide binding and T-cell receptor recognition. *Proc. Natl. Acad. Sci. USA* 89:8403–8407, 1992.

31. Janeway, C.A., Jr. The T cell receptor as a multicomponent signalling machine: CD4/CD8 coreceptors and CD45 in T cell activation. *Annu. Rev. Immunol.* 10:645–674, 1992.

   Ledbetter, J.A., et al. CD4, CD8 and the role of CD45 in T cell activation. *Curr. Opin. Immunol.* 5:334–340, 1993.

   Miceli, M.C.; Parnes, J.R. Role of CD4 and CD8 in T cell activation and differentiation. *Adv. Immunol.* 53:59–122, 1993.

32. Townsend, A.R. Recognition of influenza virus proteins by cytotoxic T lymphocytes. *Immunol. Res.* 6:80–100, 1987.

   Yewdell, J.W.; Bennink, J.R. Cell biology of antigen processing and presentation to major histocompatibility complex class I molecule-restricted T lymphocytes. *Adv. Immunol.* 52:1–123, 1992.

   Zinkernagel, R.M.; Doherty, P.C. MHC-restricted cytotoxic T cells: studies on the biological role of polymorphic major transplantation antigens determining T-cell restriction-specificity, function and responsiveness. *Adv. Immunol.* 27:51–177, 1979.

33. Bijlmakers, M.-J.; Ploegh, H.L. Putting together an MHC class I molecule. *Curr. Opin. Immunol.* 5:21–26, 1993.

   De Maeyer, E.; De Maeyer-Guignard, J. Interferon-γ. *Curr. Opin. Immunol.* 4:321–326, 1992.

   Driscoll, J.; Finley, D. A controlled breakdown: antigen processing and the turnover of viral proteins. *Cell* 68:823–825, 1993.

   Goldberg, A.L.; Rock, K.L. Proteolysis, proteasomes and antigen presentation. *Nature* 357:375–379, 1992.

   Rammensee, H.-G.; Falk, K.; Rötzschke, O. Peptides naturally presented by MHC I molecules. *Annu. Rev. Immunol.* 11:213–244, 1992.

34. Schaerer, E.; Tschopp, J. Cytotoxic T cells keep their secrets. *Curr. Biol.* 3:167–169, 1993.

Taylor, M.K.; Cohen, J.J. Cell-mediated cytotoxicity. *Curr. Opin. Immunol.* 4:338–343, 1992.

Yagita, H., et al. Role of perforin in lymphocyte-mediated cytolysis. *Adv. Immunol.* 51:215–242, 1992.

35. Noelle, R.; Snow, E.C. T helper cells. *Curr. Opin. Immunol.* 4:333–337, 1992.

Pantaleo, G.; Graziosi, C.; Fauci, A.S. The immunopathogenesis of human immunodeficiency virus infection. *N. Engl. J. Med.* 328:327–335, 1993.

36. Cresswell, P. Chemistry and functional role of the invariant chain. *Curr. Opin. Immunol.* 4:87–92, 1992.

Germain, R.N.; Margulies, D.H. The biochemistry and cell biology of antigen processing and presentation. *Annu. Rev. Immunol.* 11:403–450, 1993.

Unanue, E.R. Cellular studies on antigen presentation by class II MHC molecules. *Curr. Opin. Immunol.* 4:63–69, 1992.

37. Knight, S.C.; Stagg, A.J. Antigen-presenting cell types. *Curr. Opin. Immunol.* 5:374–382, 1993.

Steinman, R.M. The dendritic cell system and its role in immunogenicity. *Annu. Rev. Immunol.* 9:271–296, 1991.

38. Cambier, J.C. Signal transduction by T- and B-cell antigen receptors: converging structures and concepts. *Curr. Opin. Immunol.* 4:257–264, 1992.

Izquierdo, M.; Cantrell, D.A. T-cell activation. *Trends Cell Biol.* 2:268–271, 1992.

Klausner, R.D.; Samelson, L.E. T cell antigen receptor activation pathways: the tyrosine kinase connection. *Cell* 64:875–878, 1991.

Malissen, B.; Schmitt-Verhulst, A.-M. Transmembrane signalling through the T-cell-receptor–CD3 complex. *Curr. Opin. Immunol.* 5:324–333, 1993.

Weiss, A. T cell antigen receptor signal transduction: a tale of tails and cytoplasmic protein-tyrosine kinases. *Cell* 73:209–212, 1993.

39. Dinarello, C.A. Interleukin-1 and its biologically related cytokines. *Adv. Immunol.* 44:153–205, 1989.

Hogg, N.; Landis, R.C. Adhesion molecules in cell interactions. *Curr. Opin. Immunol.* 5:383–390, 1993.

Jenkins, M.K.; Johnson, J.G. Molecules involved in T-cell costimulation. *Curr. Opin. Immunol.* 5:361–367, 1993.

Linsley, P.S.; Ledbetter, J.A. The role of the CD28 receptor during T cell responses to antigen. *Annu. Rev. Immunol.* 11:191–212, 1993.

Schwartz, R.H. Costimulation of T lymphocytes: the role of CD28, CTLA-4, and B7/BB1 in interleukin-2 production and immunotherapy. *Cell* 71:1065–1068, 1992.

Schwartz, R.H. T cell energy. *Sci. Am.* 269(2):62–71, 1993.

40. Fathman, C.G.; Frelinger, J.G. T-lymphocyte clones. *Annu. Rev. Immunol.* 1:633–656, 1983.

Minami, Y.; Kono, T.; Miyazaki, T.; Taniguchi, T. The IL-2 receptor complex: its structure, function and target genes. *Annu. Rev. Immunol.* 11:245–267, 1993.

Smith, K.A. Interleukin-2. *Curr. Opin. Immunol.* 4:271–276, 1992.

Taniguchi, T.; Minami, Y. The IL-2/IL-2 receptor system: a current overview. *Cell* 73:5–8, 1993.

41. Claman, H.N.; Chaperon, E.A. Immunologic complementation between thymus and marrow cells—a model for the two-cell theory of immunocompetence. *Transplant. Rev.* 1:92–113, 1969.

Davies, A.J.S. The thymus and the cellular basis of immunity. *Transplant. Rev.* 1:43–91, 1969.

42. Chesnut, R.W.; Grey, H.M. Antigen presentation by B cells and its significance in T-B interactions. *Adv. Immunol.* 39:51–94, 1986.

Lederman, S., et al. Non-antigen signals for B-cell growth and differentiation to antibody secretion. *Curr. Opin. Immunol.* 5:439–444, 1993.

Möller, G., ed. IL-4 and IL-5: Biology and Genetics. *Immunol. Rev.* 102, 1988.

Noelle, R.J.; Ledbetter, J.A.; Aruffo, A. CD40 and its ligand, an essential ligand-receptor pair for thymus-dependent B-cell activation. *Immunol. Today* 13:431–433, 1992.

Parker, D.C. T cell-dependent B cell activation. *Annu. Rev. Immunol.* 11:331–360, 1993.

43. Arai, K., et al. Cytokines: coordinators of immune and inflammatory responses. *Annu. Rev. Biochem.* 59:783–836, 1990.

Dawson, M.M. Lymphokines and Interleukins. Chichester, UK: Open University Press, 1991.

Mosmann, T.R.; Coffman, R.L. TH1 and TH2 cells: different patterns of lymphokine secretion lead to different functional properties. *Annu. Rev. Immunol.* 7:145–174, 1989.

Mosmann, T.R., et al. Diversity of cytokine synthesis and function of mouse CD4+ T cells. *Immunol. Rev.* 123:209–229, 1991.

Stout, R.D. Macrophage activation by T cells: cognate and non-cognate signals. *Curr. Opin. Immunol.* 5:398–403, 1993.

44. Benoist, C.; Mathis, D. Generation of the αβ T-cell repertoire. *Curr. Opin. Immunol.* 4:156–161, 1992.

Kruisbeek, A.M. Development of αβ T cells. *Curr. Opin. Immunol.* 5:227–332, 1993.

Rothenberg, E.V. The development of functionally responsive T cells. *Adv. Immunol.* 51:85–214, 1992.

von Boehmer, H. Thymic selection: a matter of life and death. *Immunol. Today* 13:454–458, 1992.

von Boehmer, H.; Kisielow, P. How the immune system learns about self. *Sci. Am.* 265(4):74–81, 1991.

45. Miller, J.F.A.P.; Morahan, G. Peripheral T cell tolerance. *Annu. Rev. Immunol.* 10:51–69, 1992.

46. Mengle-Gaw, L.; McDevitt, H.O. Genetics and expression of murine I_A antigens. *Annu. Rev. Immunol.* 3:367–396, 1985.

Schwartz, R.H. Immune response (Ir) genes in the murine major histocompatibility complex. *Adv. Immunol.* 39:31–201, 1986.

47. Benoist, C.; Mathis, D. Demystification of the allo-response. *Curr. Biol.* 1:143–144, 1991.

Hill, A.V.S., et al. Common West African HLA antigens are associated with protection from severe malaria. *Nature* 352:595–600, 1991.

48. Hunkapiller, T.; Hood, L. Diversity of the immunoglobulin gene superfamily. *Adv. Immunol.* 44:1–63, 1989.

Williams, A.F.; Barclay, A.N. The immunoglobulin superfamily—domains for cell surface recognition. *Annu. Rev. Immunol.* 6:381–406, 1988.

Cancer

Roughly one person in five, in the prosperous countries of the world, will die of cancer; but that is not the reason for devoting a chapter of this book to the subject. Heart disease causes more deaths, and in the world as a whole, other health problems, such as malnutrition and parasitic infections, are more serious. In the context of cell biology, however, cancer has a unique importance, for the family of diseases grouped under this heading reflect disturbances of the most fundamental rules of behavior of the cells in a multicellular organism. To understand cancer and to devise rational ways to treat it, we have to understand both the inner workings of cells and their social interactions in the tissues of the body. Thus the cancer research effort has profoundly benefited a much wider area of medical knowledge than that of cancer alone.

We have already discussed many offshoots of cancer research in the preceding chapters. In this concluding chapter we examine the disease itself. In the first section we consider the nature of cancer and the natural history of the disease from a cellular standpoint; in the second section we focus on its molecular basis.

Cancer as a Microevolutionary Process [1]

The body of an animal can be viewed as a society or ecosystem whose individual members are cells, reproducing by cell division and organized into collaborative assemblies or tissues. In our earlier discussion of the maintenance of tissues (in Chapter 22), our concerns were similar to those of the ecologist: cell births, deaths, habitats, territorial limitations, the maintenance of population sizes, and the like. The one ecological topic conspicuously absent was that of natural selection: we said nothing of competition or mutation among somatic cells. The reason is that a healthy body is in this respect a very peculiar society, where self-sacrifice, rather than competition, is the rule: all somatic cell lineages are committed to die, leaving no progeny but dedicating their existence to support of the germ cells, which alone have a chance of survival. There is no mystery in this, for the body is a clone, and the genome of the somatic cells is the same as the genome of the germ cells; by their self-sacrifice for the sake of the germ cells, the somatic cells help to propagate copies of their own genes.

Thus, unlike free-living cells such as bacteria, which compete to survive, the cells of a multicellular organism are committed to collaboration. Any mutation that gives rise to selfish behavior by individual members of the cooperative will jeopardize the future of the whole enterprise. Mutation, competition, and natural selection operating within the population of somatic cells are the basic ingredients of cancer: it is a disease in which individual mutant cells begin by prospering at the expense of their neighbors but in the end destroy the whole cellular society and die.

In this section we discuss the development of cancer as a microevolutionary process. This process occurs on a time scale of months or years in a population of cells in the body, and it is dependent on the same principles of mutation and natural selection that govern the long-term evolution of all living organisms.

Cancers Differ According to the Cell Type from Which They Derive [2]

Cancer cells are defined by two heritable properties: they and their progeny (1) reproduce in defiance of the normal restraints and (2) invade and colonize territories normally reserved for other cells. It is the combination of these features that makes cancers peculiarly dangerous. An isolated abnormal cell that does not proliferate more than its normal neighbors does no significant damage, no matter what other disagreeable properties it may have; but if its proliferation is out of control, it will give rise to a tumor, or *neoplasm*—a relentlessly growing mass of abnormal cells. As long as the neoplastic cells remain clustered together in a single mass, however, the tumor is said to be **benign,** and a complete cure can usually be achieved by removing the mass surgically. A tumor is counted as a cancer only if it is **malignant,** that is, only if its cells have the ability to invade surrounding tissue. Invasiveness usually implies an ability to break loose, enter the bloodstream or lymphatic vessels, and form secondary tumors, or **metastases,** at other sites in the body (Figure 24–1). The more widely a cancer metastasizes, the harder it becomes to eradicate.

Cancers are classified according to the tissue and cell type from which they arise. Cancers arising from epithelial cells are termed **carcinomas;** those arising from connective tissue or muscle cells are termed **sarcomas.** Cancers that do not fit in either of these two broad categories include the various **leukemias,** derived from hemopoietic cells, and cancers derived from cells of the nervous system. Table 24–1 lists the types of cancers that are common in the United States, together with their incidence and the death rate from them. Each of the broad categories has many subdivisions according to the specific cell type, the location in the body, and the structure of the tumor; many of the names used are fixed by tradition and have no modern rational basis. In parallel with the set of names for malignant tumors, there is a related set of names for benign tumors: an *adenoma,* for example, is a benign epithelial tumor with a glandular organization,

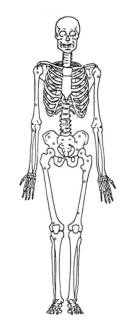

Figure 24–1 Metastasis. Malignant tumors typically give rise to metastases, making the cancer hard to eradicate. The drawing shows common sites in the bone marrow for metastases from carcinoma of the prostate gland. (From Union Internationale Contre le Cancer, TNM Atlas: Illustrated Guide to the Classification of Malignant Tumors, 2nd ed. Berlin: Springer-Verlag, 1986.)

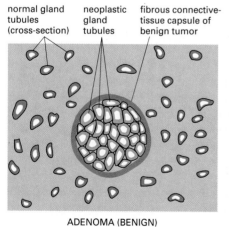

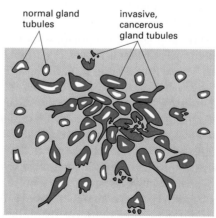

normal gland tubules (cross-section) neoplastic gland tubules fibrous connective-tissue capsule of benign tumor normal gland tubules invasive, cancerous gland tubules

ADENOMA (BENIGN) ADENOCARCINOMA (MALIGNANT)

Figure 24–2 The contrast between a benign glandular tumor (an adenoma) and a malignant glandular tumor (an adenocarcinoma). There are many forms that such tumors may take; the diagram illustrates types that might be found in the breast.

Table 24–1 **Cancer Incidence and Cancer Mortality in the United States, 1993**

Type of Cancer	New Cases per Year		Deaths per Year	
Total cancers	1,170,000		528,300	
Cancers of epithelia: carcinomas	992,700	(85%)	417,175	(79%)
Oral cavity and pharynx	29,800	(3%)	7,700	(1%)
Pancreas	27,700	(2%)	25,000	(5%)
Stomach	24,000	(2%)	13,600	(3%)
Liver and biliary system	15,800	(1%)	12,600	(2%)
Respiratory system (total)	187,100	(16%)	154,200	(29%)
Lung	170,000	(15%)	149,000	(28%)
Breast	183,000	(16%)	46,300	(9%)
Skin (total)	(>700,000)*		9,100	(2%)
Malignant melanoma	32,000	(3%)	6,800	(1%)
Reproductive tract (total)	244,400	(21%)	59,950	(11%)
Prostate gland	165,000	(14%)	35,000	(7%)
Ovary	22,000	(2%)	13,300	(3%)
Uterine cervix	13,500	(1%)	4,400	(1%)
Uterus (endometrium)	31,000	(3%)	5,700	(1%)
Urinary organs (total)	79,500	(7%)	20,800	(4%)
Bladder	52,300	(4%)	9,900	(2%)
Cancers of the hemopoietic and immune system: leukemias and lymphomas	93,000	(8%)	50,000	(9%)
Cancers of central nervous system and eye: gliomas, retinoblastoma, etc.	18,250	(2%)	12,350	(2%)
Cancers of connective tissues, muscles, and vasculature: sarcomas	8,000	(1%)	4,150	(1%)
All other cancers + unspecified sites	57,050	(5%)	43,425	(8%)

*Nonmelanoma skin cancers are not included in total of all cancers, since almost all are cured easily and many go unrecorded.

In the world as a whole, the five most common cancers are those of the lung, stomach, breast, colon/rectum, and uterine cervix, and the total number of new cancer cases per year is just over 6 million. Note that only about half the number of people who develop cancer die of it. (Data for USA from American Cancer Society, Cancer Facts and Figures, 1993.)

the corresponding type of malignant tumor being an *adenocarcinoma* (Figure 24–2); a *chondroma* and a *chondrosarcoma* are, respectively, benign and malignant tumors of cartilage. About 90% of human cancers are carcinomas, perhaps because most of the cell proliferation in the body occurs in epithelia or perhaps because epithelial tissues are most frequently exposed to the various forms of physical and chemical damage that favor the development of cancer.

Each cancer has characteristics that reflect its origin. Thus, for example, the cells of an epidermal *basal-cell carcinoma*, derived from a keratinocyte stem cell in the skin, will generally continue to synthesize cytokeratin intermediate filaments, whereas the cells of a *melanoma*, derived from a pigment cell in the skin, will often (but not always) continue to make pigment granules. Cancers originating from different cell types are, in general, very different diseases. The basal-cell carcinoma, for example, is only locally invasive and rarely forms metastases, whereas the melanoma is much more malignant and rapidly gives rise to many metastases (behavior that recalls the migratory tendencies of the normal pigment-cell precursors during development, discussed in Chapter 21). The basal-cell carcinoma is usually easy to remove by surgery, leading to complete cure; but the malignant melanoma, once it has metastasized, is often impossible to extirpate and consequently fatal.

Most Cancers Derive from a Single Abnormal Cell [3]

Even when a cancer has metastasized, its origins can usually be traced to a single **primary tumor,** arising in an identified organ and presumed to be derived by cell division from a single cell that has undergone some heritable change that enables it to outgrow its neighbors. By the time it is first detected, however, a typical tumor already contains about a billion cells or more (Figure 24–3), often including many normal cells—fibroblasts, for example, in the supporting connective tissue that is associated with a carcinoma. What evidence do we have that the cancer cells are indeed a clone descended from a single abnormal cell?

One type of demonstration comes from analysis of the cells' DNA. In almost all patients with *chronic myelogenous leukemia,* for example, the leukemic white blood cells are distinguished from the normal cells by a specific chromosomal abnormality (the so-called Philadelphia chromosome, created by a translocation between the long arms of chromosomes 9 and 22, as shown in Figure 24–4). When the DNA at the site of translocation is cloned and sequenced, it is found that the site of breakage and rejoining of the translocated fragments is identical in all the leukemic cells in any given patient but differs slightly (by a few hundred or thousand base pairs) from one patient to another, as expected if each case of the leukemia arises from a unique accident occurring in a single cell. We see later how Philadelphia translocations lead to leukemia by inappropriately activating a specific gene.

Another way to show that a cancer has a monoclonal origin is by exploiting the phenomenon of X-chromosome inactivation. A normal woman is a random mixture, or mosaic, of two classes of cells—those in which the paternal X chromosome is inactivated and those in which the maternal X chromosome is inactivated (as discussed in Chapter 9). The inactivation of one or the other of the X chromosomes occurs at random in each cell early in embryonic development, but once the choice has been made it is irreversible, so that when a cell divides it passes on its own state of X-inactivation to its daughters. Consequently, the state of X-chromosome inactivation—maternal or paternal—can be used as a heritable marker to trace the lineage of cells in the body. In the great majority of tumors that have been analyzed—both benign and malignant—all the tumor cells have been found to have the same X chromosome inactivated, strongly suggesting that they are derived from a single deranged cell (Figure 24–5).

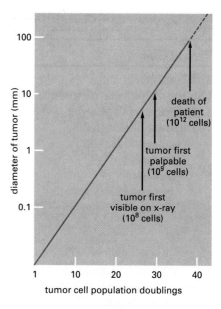

Figure 24–3 The growth of a typical human tumor such as a tumor of the breast. The diameter of the tumor is plotted on a logarithmic scale. Years may elapse before the tumor becomes noticeable.

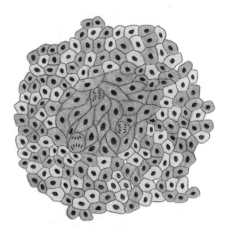

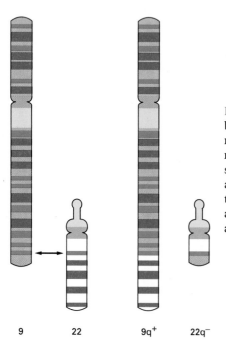

Figure 24–4 The translocation between chromosomes 9 and 22 responsible for chronic myelogenous leukemia. The smaller of the two resulting abnormal chromosomes is called the Philadelphia chromosome, after the city where the abnormality was first recorded.

Figure 24–5 Evidence from X-inactivation mosaics demonstrates the monoclonal origin of cancers. As a result of a random process that occurs in the early embryo, practically every normal tissue in a woman's body is a mixture of cells with different X chromosomes heritably inactivated (indicated here by the mixture of *red* cells and *gray* cells in the normal tissue). When the cells of a cancer are tested for their expression of an X-linked marker gene, however, they are usually all found to have the same X chromosome inactivated. This implies that they are all derived from a single cancerous founder cell.

Most Cancers Are Probably Initiated by a Change in the Cell's DNA Sequence [4]

If a single abnormal cell is to give rise to a tumor, it must pass on its abnormal-

memory, are a familiar feature of normal development, as manifest in the stability of the differentiated state and in such phenomena as X-chromosome inactivation; and there is no obvious a priori reason why they should not be involved in cancer. For one rare and extraordinary type of cancer—the teratocarcinoma (see Chapter 21)—the evidence does favor an epigenetic origin. There are, however, good reasons to think that most cancers are initiated by genetic change. Thus cells of a given cancer can often be shown to have a shared abnormality in their DNA sequence, as we have just seen for chronic myelogenous leukemia; many other examples are discussed in the second half of this chapter. Further evidence that genetic change can be a *cause* of cancer comes from a study of agents known to give rise to the disease. A correlation between **carcinogenesis** (the generation of cancer) and *mutagenesis* (the production of a change in the DNA sequence) is clear for three classes of agents: **chemical carcinogens** (which typically cause simple local changes in the nucleotide sequence), ionizing radiation such as x-rays (which typically cause chromosome breaks and translocations), and viruses (which introduce foreign DNA into the cell). The role of viruses in cancer is discussed later; we pause here to discuss chemical carcinogens.

In general, a given cancer cannot be blamed entirely on a single event or a single cause: as we shall see, cancers as a rule result from the chance occurrence in one cell of several independent accidents, with cumulative effects. There are, however, some unusually carcinogenic agents that increase the likelihood of the critical events to the point where it becomes virtually certain, given a high enough dosage, that at least one cell in the body will turn cancerous. The compound 2-naphthylamine, used in the chemical industry in the early part of this century, is one notorious example: in one British factory, all of the men who had been employed in distilling it (and were thereby subjected to prolonged exposure) eventually developed bladder cancer.

Many quite disparate chemicals have been shown to be likewise carcinogenic when they are fed to experimental animals or painted repeatedly on their skin. Some of these carcinogens act directly on the target cells; many others take effect only after they have been changed to a more reactive form by metabolic processes—notably by a set of intracellular enzymes known as the cytochrome P-450 oxidases. These enzymes normally help to convert ingested toxins and foreign lipid-soluble materials into harmless and easily excreted compounds, but they fail in this task with certain substances, converting them instead into direct carcinogens (Figure 24–6). Although the known chemical carcinogens are very diverse, most of them have at least one property in common—they cause muta-

Figure 24–6 Metabolic activation of a carcinogen. Many chemical carcinogens have to be activated by a metabolic transformation before they will cause mutations by reacting with DNA. The compound illustrated here is *aflatoxin B1*, a toxin from a mold (*Aspergillus flavus oryzae*) that grows on grain and peanuts when they are stored under humid tropical conditions. It is thought to be a contributory cause of liver cancer in the tropics and is associated with specific mutations of the *p53* gene (discussed later in this chapter).

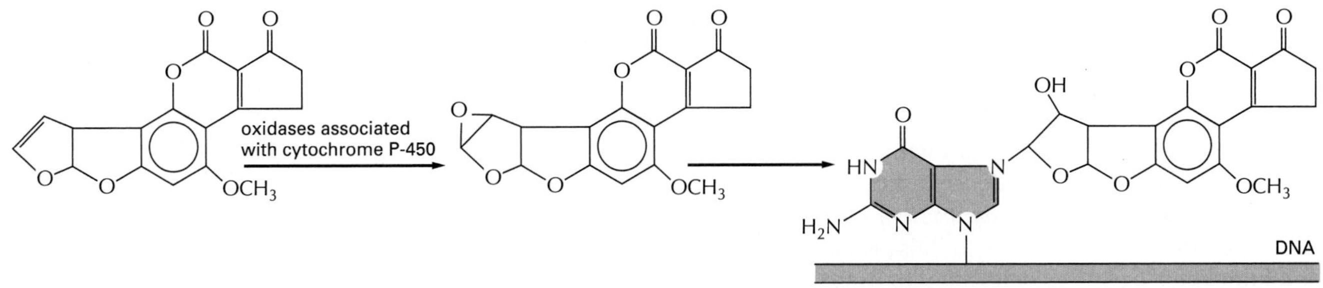

AFLATOXIN   AFLATOXIN-2,3-EPOXIDE   CARCINOGEN BOUND TO GUANINE IN DNA

Cancer as a Microevolutionary Process

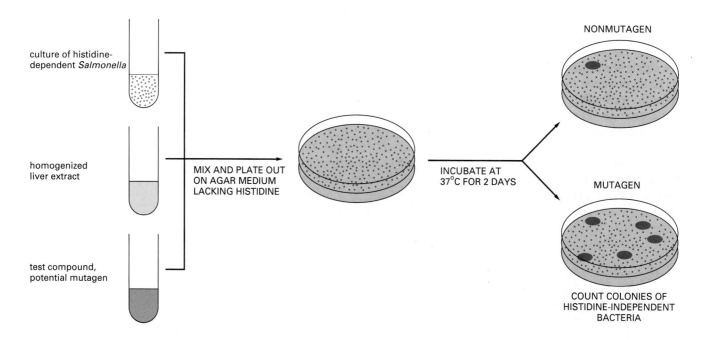

tions. In one popular test for mutagenicity, the carcinogen is mixed with an activating extract prepared from rat liver cells (to mimic the biochemical processing that occurs in an intact animal) and is added to a culture of specially designed test bacteria; the resulting mutation rate of the bacteria is then measured (Figure 24–7). Most of the compounds scored as mutagenic by this rapid and convenient bacterial assay also cause mutations and/or chromosome aberrations when tested on mammalian cells. When mutagenicity data from various sources are analyzed, one finds that the majority of carcinogens are mutagens.

There is, nevertheless, a significant minority of carcinogens that do not appear to be mutagenic. We shall discuss below how nonmutagenic substances may promote the development of cancer by affecting the behavior of preexisting mutant cells. But first we must consider how frequently such mutant cells are likely to arise in the normal course of events.

A Single Mutation Is Not Enough to Cause Cancer [1,5]

Something on the order of 10^{16} cell divisions take place in a human body in the course of a lifetime; in a mouse, with its smaller number of cells and its shorter life-span, the number is about 10^{12}. Even in an environment that is free of mutagens, mutations will occur spontaneously at an estimated rate of about 10^{-6} mutations per gene per cell division—a value set by fundamental limitations on the accuracy of DNA replication and repair. Thus, in a lifetime, every single gene is likely to have undergone mutation on about 10^{10} separate occasions in any individual human being, or about 10^6 occasions in a mouse. Among the resulting mutant cells one might expect that there would be many that have disturbances in genes involved in the regulation of cell division and that consequently disobey the normal restrictions on cell proliferation. From this point of view, the problem of cancer seems to be not why it occurs but why it occurs so infrequently.

Evidently, a single mutation is not enough to convert a typical healthy cell into a cancer cell that proliferates without restraint, or we would not be viable organisms. Many lines of evidence indicate that the genesis of a cancer as a rule requires that several independent rare accidents occur together in one cell. One such indication comes from epidemiological studies of the incidence of cancer as a function of age. If a single mutation were responsible, occurring with a fixed probability per year, the chance of developing cancer in any given year should be independent of age. In fact, for most types of cancer the chance goes up very

Figure 24–7 The Ames test for mutagenicity. The test uses a strain of *Salmonella* bacteria that require histidine in the medium because of a defect in a gene necessary for histidine synthesis. Mutagens can cause a further change in this gene that reverses the defect, creating revertant bacteria that do not require histidine. To increase the sensitivity of the test, the bacteria also have a defect in their DNA repair machinery that makes them especially susceptible to agents that damage DNA. A majority of compounds that are mutagenic by tests such as this are also carcinogenic and vice versa.

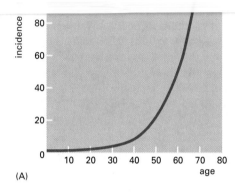

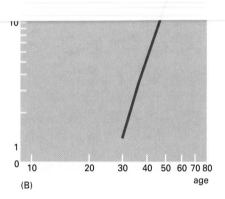

(A) age

(B) age

Figure 24–8 **Cancer incidence as a function of age.** The number of newly [...] scale (B). The incidence of cancer rises steeply as a function of age—roughly as the fifth power, in this example (that is, the slope of the log-log plot in (B) is about 5). If only a single mutation were required to trigger the cancer and this mutation had an equal chance of occurring at any time, the incidence would be independent of age (that is, the slope of the graph would be 0). The data suggest that a cell must accumulate the disruptive effects of about six independent rare accidents before it will give rise to a cancer of this type; the frequency of each of these events is a rate-limiting factor in the incidence of the cancer. (Data from C. Muir et al., Cancer Incidence in Five Continents, Vol. V. Lyon: International Agency for Research on Cancer, 1987.)

steeply with age—typically as the third, fourth, or fifth power (Figure 24–8). From such statistics it has been estimated that somewhere between three and seven independent random events, each of low probability, are typically required to turn a normal cell into a cancer cell; the smaller numbers apply to leukemias, the larger to carcinomas.

Now that specific mutations responsible for the development of cancer have been identified, it has become possible to test the effects of the mutant genes in transgenic mice; as we see later, the results give additional and more direct evidence for the hypothesis that a single mutation is insufficient to cause cancer. The hypothesis is also supported by many older studies of the phenomenon of **tumor progression,** whereby an initial mild disorder of cell behavior evolves gradually into a full-blown cancer. These observations of how tumors develop, moreover, provide insight into the nature of the multiple changes that must occur for a normal cell to become a cancer cell and into the factors that control their occurrence.

Cancers Develop in Slow Stages from Mildly Aberrant Cells [1, 5, 6]

For those cancers that have a discernible external cause, there is almost always a long delay between the causal event(s) and the onset of the disease: the incidence of lung cancer does not begin to rise steeply until after 10 or 20 years of heavy smoking; the incidence of leukemias in Hiroshima and Nagasaki did not show a marked rise until about 5 years after the explosion of the atomic bombs, and it did not reach its peak until 8 years had elapsed; industrial workers exposed for a limited period to chemical carcinogens do not usually develop the cancers characteristic of their occupation until 10, 20, or even more years after the exposure (Figure 24–9); and so on. During this long incubation period, the prospective cancer cells undergo a succession of changes. Chronic myelogenous leukemia, mentioned earlier, provides a clear and simple example. This disease begins as a disorder characterized by a nonlethal overproduction of white blood cells and continues as such for several years before changing into a much more rapidly progressing illness that usually ends in death within a few months. In the chronic early phase the leukemic cells in the body are distinguished simply by their possession of the chromosomal translocation mentioned previously. In the subsequent acute phase of the illness, the hemopoietic system is overrun by cells

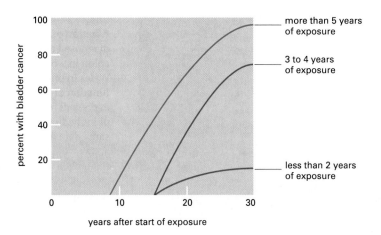

Figure 24–9 Delayed onset of cancer following exposure to a carcinogen. The graph shows the length of the delay before onset of bladder cancer in a set of 78 men who had been exposed to the carcinogen 2-naph-thylamine, grouped according to the duration of their exposure. (Modified from J. Cairns, Cancer: Science and Society. San Francisco: W.H. Freeman, 1978. After M.H.C. Williams, in Cancer, Vol. III (R.W. Raven, ed.). London: Butterfield, 1958.)

that show not only this chromosomal abnormality but also several others. It appears as though members of the initial mutant clone have undergone further mutations that make them proliferate more rapidly (or divide more times before they die or terminally differentiate), so that they come to outnumber both the normal hemopoietic cells and their cousins that have only the primary disorder.

Carcinomas and other solid tumors are thought to evolve in a similar way. Although most such cancers in humans are not diagnosed until a relatively late stage, in a few cases it is possible to observe the early steps in the development of the disease. We shall discuss one example—colorectal cancer—at the end of

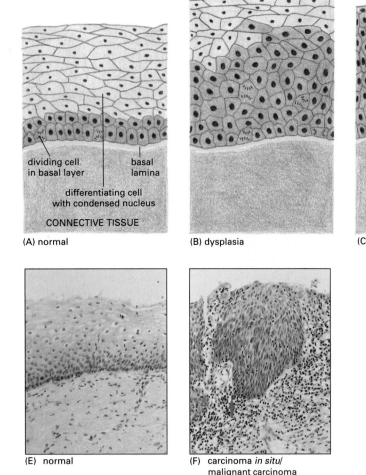

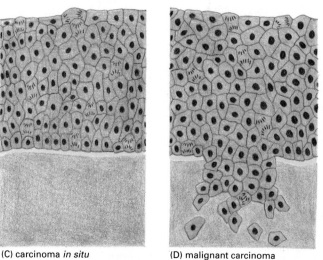

Figure 24–10 The stages of progression in the development of cancer of the epithelium of the uterine cervix. (A–D) Schematic diagrams. In dysplasia the most superficial cells still show some signs of differentiation, but this is incomplete, and proliferating cells are seen abnormally far above the basal layer. In carcinoma *in situ* the cells in all the layers are proliferating and apparently undifferentiated. True malignancy begins when the cells cross the basal lamina and begin to invade the underlying connective tissue. Several years may elapse from the first signs of dysplasia to the onset of full-blown malignant cancer. (E) Photograph of a section of normal cervical epithelium. (F) Photograph of a section of a cervical carcinoma that is just beginning to be invasive; the *arrow* points to cells that are in the process of escaping from the epithelium into the connective tissue below. (E and F, courtesy of Andrew Hanby.)

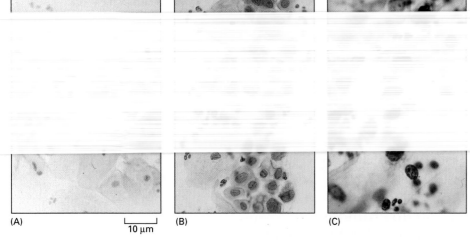

Figure 24-11 **Photographs of cells**

(A)   10 µm   (B)   (C)

undifferentiated, with scanty cytoplasm and a relatively large nucleus. Debris in the background includes some white blood cells. (Courtesy of Winifred Gray.)

this chapter. Another example is provided by cancers of the *uterine cervix* (the neck of the womb). These cancers derive from the multilayered cervical epithelium, which has an organization similar to that of the epidermis of the skin (discussed in Chapter 22). Normally, proliferation occurs only in the basal layer, generating cells that then move outward toward the surface, differentiating into flattened, keratin-rich, nondividing cells as they go, and finally being sloughed off from the surface (Figure 24-10A and E). When many specimens of this epithelium from different women are examined, however, it is not unusual to find patches of **dysplasia,** where dividing cells are no longer confined to the basal layer and there is some disorder in the process of differentiation (Figure 24-10B). Cells are sloughed from the surface in abnormally early stages of differentiation, and the presence of the dysplasia can be detected by scraping a sample of cells from the surface and viewing it under the microscope (the "Pap smear" technique—Figure 24-11). Left alone, the dysplastic patches will often remain harmless or even regress spontaneously; more rarely, however, they may progress, over a period of several years, to give rise to patches of so-called *carcinoma in situ* (Figure 24-10C). In these more serious lesions (somewhat misleadingly named, since they are not yet fully malignant), the usual pattern of cell division and differentiation is much more severely disrupted, and all the layers of the epithelium consist of undifferentiated proliferating cells, which are often highly variable in size and karyotype; the abnormal cells are still confined, however, to the epithelial side of the basal lamina. At this stage it is still easy to achieve a complete cure by destroying or removing the abnormal tissue surgically. Without such treatment the abnormal patch may still remain harmless or regress; but in an estimated 20–30% of cases it will develop, again over a period of several years, to give rise to a truly malignant cervical carcinoma (Figure 24-10D and F), whose cells break out of the epithelium by crossing the basal lamina and begin to invade the underlying connective tissue. Surgical cure becomes progressively more difficult as the invasive growth spreads.

Tumor Progression Involves Successive Rounds of Mutation and Natural Selection [6, 7]

As illustrated by the examples just discussed, cancers in general seem to arise by a process in which an initial population of slightly abnormal cells, descendants of a single mutant ancestor, evolves from bad to worse through successive cycles of mutation and natural selection. This evolution involves a large element of chance and usually takes many years; most of us die of other ailments before cancer has had time to develop. To understand the causation of cancer, it is essential to understand the factors that may speed up the process.

In general, the rate of evolution, whether in a population of cells exploiting the opportunities for cancerous behavior in the body or in a population of organisms adapting to a new environment on the surface of the earth, would be expected to depend on four main parameters: (1) the *mutation rate*, that is, the probability per gene per unit time that any given member of the population will undergo genetic change; (2) the *number of individuals in the population;* (3) the *rate of reproduction*, that is, the average number of generations of progeny produced per unit time; and (4) the *selective advantage* enjoyed by successful mutant individuals, that is, the ratio of the number of surviving fertile progeny they produce per unit time to the number of surviving fertile progeny produced by nonmutant individuals. Experimental studies on the induction of cancer in animals illustrate these evolutionary principles.

The Development of a Cancer Can Be Promoted by Factors That Do Not Alter the Cell's DNA Sequence [6, 8]

The stages by which an initial mild lesion progresses to become a cancer can be most easily observed in the skin. Skin cancers can be elicited in mice, for example, by repeatedly painting the skin with a mutagenic chemical carcinogen such as benzo[a]pyrene (a constituent of coal tar and tobacco smoke) or the related compound dimethylbenz[a]anthracene (DMBA). A single application of the carcinogen, however, usually does not by itself give rise to a tumor or any other obvious lasting abnormality. Yet it does cause latent genetic damage, and this can be detected through a greatly increased incidence of cancer when the cells are exposed either to further treatments with the same substance or to certain other, quite different, insults. A carcinogen that sows the seeds of cancer in this way is said to act as a **tumor initiator.** Simply wounding skin that has been exposed once to such an initiator can cause cancers to develop from some of the cells at the edge of the wound. Alternatively, repeated exposure over a period of months to certain substances known as **tumor promoters,** which are not themselves mutagenic, can cause cancer selectively in skin previously exposed to a tumor initiator. The most widely studied tumor promoters are *phorbol esters,* such as tetradecanoylphorbol acetate (TPA), which we have already encountered in another context as artificial activators of protein kinase C (and hence as agents that activate part of the phosphatidylinositol intracellular signaling pathway, discussed in Chapter 15). These substances cause cancers at high frequency only if they are applied *after* a treatment with a mutagenic initiator (Figure 24–12).

As one might expect for genetic damage, the hidden changes caused by a tumor initiator are irreversible: thus they can be uncovered by treatment with a tumor promoter even after a long delay. The immediate effect of the promoter

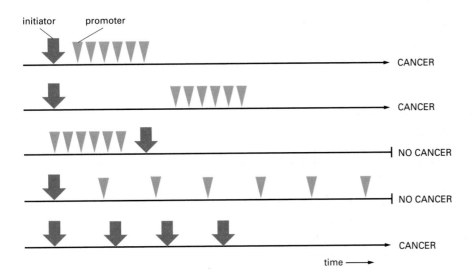

Figure 24–12 **Some possible schedules of exposure to a tumor initiator (mutagenic) and a tumor promoter (nonmutagenic) and their outcomes.** Cancer ensues only if the exposure to the promoter follows exposure to the initiator and only if the intensity of exposure to the promoter exceeds a certain threshold. Cancer can also occur as a result of repeated exposure to the initiator alone.

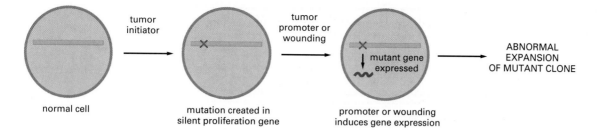

tumor
initiator

tumor
promoter or
wounding

ABNORMAL
EXPANSION
OF MUTANT CLONE

normal cell

mutation created in
silent proliferation gene

mutant gene
expressed

promoter or wounding
induces gene expression

is apparently to stimulate cell division (or to cause cells that would normally undergo terminal differentiation to continue dividing instead), and in the region that had previously been exposed to the initiator, this results in the growth of many small, benign, wartlike tumors called *papillomas*. The greater the prior dose of initiator, the larger the number of papillomas induced; it is thought that each papilloma (at least for low doses of the initiator) consists of a single clone of cells descended from a mutant cell that the initiator has engendered. Both wounding and the application of the promoter probably act by inducing the expression of some of the genes that directly or indirectly affect cell proliferation. Such genes may remain quiescent in the resting epithelium, so that any mutations they have undergone in response to the initiator go undetected; by inducing expression of the mutated genes, the promoter or the stimulus of wounding may enable them to begin influencing cell proliferation (Figure 24–13).

A typical papilloma might contain about 10^5 cells. If exposure to the tumor promoter is stopped, almost all the papillomas regress, and the skin regains a largely normal appearance—as expected from the hypothesis illustrated in Figure 24–13. In a few of the papillomas, however, further changes occur that enable growth to continue in an uncontrolled way, even after the promoter has been withdrawn. These changes seem to originate in occasional single papilloma cells, at about the frequency expected for spontaneous mutations. In this way a small proportion of the papillomas progress to become cancers. Thus the tumor promoter apparently favors the development of cancer, in this system at least, by expanding the population of cells that carry an initial mutation: the more such cells there are and the more times they divide, the greater the chance that at least one of them will undergo another mutation carrying it one more step along the road to malignancy. Although naturally occurring cancers do not necessarily arise through the specific sequence of distinct initiation and promotion steps just described, their evolution must be governed by similar principles. They too will evolve at a rate that depends both on the frequency of mutations and on influences affecting the survival, proliferation, and spread of certain types of mutant cells once they have been created.

Most Cancers Result from Avoidable Combinations of Environmental Causes [9]

The development of a cancer generally involves many steps, each governed by multiple factors, some dependent on the genetic constitution of the individual, others dependent on his or her environment and way of life. By changing our surroundings or our habits, therefore, we should, in principle, be able to reduce drastically our chance of developing almost any given type of cancer. This is demonstrated most clearly by a comparison of cancer incidence in different countries: for almost every cancer that is common in one country, there is another country where the incidence is several times lower (Table 24–2); and migrant populations tend to take on the pattern of cancer incidence typical of the host country, implying that the differences are due to environmental, not genetic, factors. From such data it is estimated that 80–90% of cancers should be avoidable. Unfortunately, different cancers have different environmental risk factors, and a country that happens to escape one such danger is no more likely than other countries to escape the rest; thus the incidence of all cancers combined

Figure 24–13 One hypothesis proposed to explain the observed effect of tumor promoters on the development of tumors. An initiator causes a mutation in a gene that is not normally expressed in resting cells. The promoter activates the mutant gene and, by causing cell division, increases the number of cells containing the mutation. Alternatively, the mutant gene might be constitutively expressed but have no effect until the promoter activates other genes required for cell proliferation.

Cancer as a Microevolutionary Process

Table 24–2 Variation Between Countries in the Incidence of Some Common Cancers

Site of Origin of Cancer	High-Incidence Population		Low-Incidence Population	
	Location	Incidence*	Location	Incidence*
Lung	USA (New Orleans, blacks)	110	India (Madras)	5.8
Breast	Hawaii (Hawaiians)	94	Israel (non-Jews)	14.0
Prostate	USA (Atlanta, blacks)	91	China (Tianjin)	1.3
Uterine cervix	Brazil (Recife)	83	Israel (non-Jews)	3.0
Stomach	Japan (Nagasaki)	82	Kuwait (Kuwaitis)	3.7
Liver	China (Shanghai)	34	Canada (Nova Scotia)	0.7
Colon	USA (Connecticut, whites)	34	India (Madras)	1.8
Melanoma	Australia (Queensland)	31	Japan (Osaka)	0.2
Nasopharynx	Hong Kong	30	UK (southwestern)	0.3
Esophagus	France (Calvados)	30	Romania (urban Cluj)	1.1
Bladder	Switzerland (Basel)	28	India (Nagpur)	1.7
Uterus	USA (San Francisco Bay Area, whites)	26	India (Nagpur)	1.2
Ovary	New Zealand (Polynesian Islanders)	26	Kuwait (Kuwaitis)	3.3
Rectum	Israel (European and USA born)	23	Kuwait (Kuwaitis)	3.0
Larynx	Brazil (São Paulo)	18	Japan (rural Miyagi)	2.1
Pancreas	USA (Los Angeles, Koreans)	16	India (Poona)	1.5
Lip	Canada (Newfoundland)	15	Japan (Osaka)	0.1
Kidney	Canada (NWT and Yukon)	15	India (Poona)	0.7
Oral cavity	France (Bas-Rhin)	14	India (Poona)	0.4
Leukemia	Canada (Ontario)	12	India (Nagpur)	2.2
Testis	Switzerland (urban Vaud)	10	China (Tianjin)	0.6

* Incidence = number of new cases per year per 100,000 population, adjusted for a standardized population age distribution (so as to eliminate effects due merely to differences of population age distribution). Figures for cancers of breast, uterine cervix, uterus, and ovary are for women; other figures are for men.

Adapted from V.T. DeVita, S. Hellman, and S.A. Rosenberg (eds.), Cancer: Principles and Practice of Oncology, 4th ed. Philadelphia: Lippincott, 1993; based on data from C. Muir et al., Cancer Incidence in Five Continents, Vol. 5. Lyon: International Agency for Research on Cancer, 1987.

(among individuals of a given age) is similar from country to country. There are, however, some subgroups whose abstinent way of life does seem to reduce the total cancer death rate: the incidence of cancer among strict Mormons in Utah, for example, is only about half that among Americans in general.

While such epidemiological observations indicate that cancer can be avoided, it remains difficult to identify the specific environmental risk factors or to establish how they act. Some certainly operate as mutagenic tumor initiators, directly provoking genetic change; others presumably serve as tumor promoters that help to enlarge the population of cells liable to progress, through further mutation, to full-blown cancer. The carcinogens in tobacco smoke, like the aflatoxin on tropical peanuts (see Figure 24–6), probably belong mostly in the first category, while the reproductive hormones that circulate in a woman's body at different stages of her life may belong in the second category. The importance of these hormones is indicated by the striking correlations that exist between a woman's reproductive history and her risk of developing breast cancer; the hormones presumably affect cancer incidence through their influence on cell proliferation in the breast (Figure 24–14). It is possible that some factors act in still other ways—for example, by causing heritable epigenetic changes. Of course, it is not necessary to under-

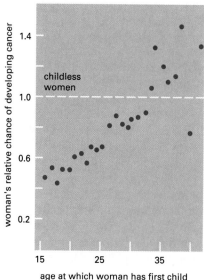

Figure 24–14 Effects of childbearing on the risk of breast cancer. The relative probability of breast cancer developing at some time in a woman's life is plotted as a function of the age at which she gives birth to her first child. The graph shows the value of the probability relative to that for a childless woman. The longer the delay before bearing the first child, the higher the probability of breast cancer, suggesting that exposure to certain combinations of reproductive hormones, especially estrogen, may promote development of the cancer. There is some evidence from laboratory studies that the first full-term pregnancy may result in a permanent epigenetic change in the cells of the breast, altering their subsequent responses to hormones. Other environmental factors may also be correlated with breast cancer, and some women carry genes that increase the risk of the disease. (From J. Cairns, Cancer: Science and Society. San Francisco: W.H. Freeman, 1978. After B. MacMahon, P. Cole, and J. Brown, *J. Natl. Cancer Inst.* 50:21–42, 1973.)

stand how cancer-causing agents act in order to identify them and show how to avoid them. In this task cancer epidemiology has had some notable successes and promises more to come. Simply by revealing the role of smoking, it has shown a way to reduce the total cancer death rate in North America and Europe by as much as 30%. The prevention of cancer is not only better than cure but seems also, given our present state of knowledge, to be much more readily attainable.

The Search for Cancer Cures Is Hard but Not Hopeless [10]

The difficulty of curing a cancer is like the difficulty of getting rid of weeds. Cancer cells can be removed surgically or destroyed with toxic chemicals or radiation; but it is hard to eradicate every single one of them. Surgery can rarely ferret out every metastasis, and treatments that kill cancer cells are generally toxic to normal cells as well. If even a few cancerous cells remain, they can proliferate to produce a resurgence of the disease; and unlike the normal cells they may evolve resistance to the poisons used against them. Yet the outlook is not hopeless. In spite of the difficulties, effective cures using anticancer drugs (alone or in combination with other treatments) have been devised for some formerly highly lethal cancers (notably Hodgkin's lymphoma, testicular cancer, choriocarcinoma, and some leukemias and other cancers of childhood). For several of the more common cancers, moreover, appropriate surgery or local radiotherapy enables a large proportion of patients to recover if the illness is diagnosed at a reasonably early stage. Effective treatments can sometimes be based on an understanding of the causes of a specific type of cancer. Estrogens, for example, appear to act as natural tumor promoters in cancer of the breast (see Figure 24–14), and treatment with an estrogen antagonist, such as the drug *tamoxifen,* is effective in many breast cancer patients in preventing or delaying recurrence of the disease. Even for types of cancer where a cure at present seems beyond our reach, there are treatments that will prolong life or at least relieve distress.

A great deal of clinical cancer research centers on the problem of how to kill cancer cells selectively. For the most part, current methods exploit relatively subtle differences between normal and neoplastic cells with respect to proliferation rate, metabolism, and radiosensitivity, and they have unpleasant toxic side effects. A few types of cancer cells are especially vulnerable to selective attack because they depend on specific hormones or because their surfaces have unusual chemical features that can be recognized by antibodies. In general, however, progress with the vexing problem of anticancer selectivity has been slow—a matter of trial and error and guesswork as much as rational calculation.

In the search for better ways of curbing the survival, proliferation, and spread of cancer cells, it is important to examine more closely the strategies by which they thrive and multiply.

Cancerous Growth Often Depends on Deranged Control of Cell Differentiation or Cell Death [11]

We have so far emphasized that cancer cells defy the normal controls on cell division: this is their central property. But there are other requirements too, if a tumor is to grow without limit. The tumor cells must, for example, stimulate the development of blood vessels to bring the nutrients and oxygen they require for growth, as discussed in Chapter 22. Moreover, many tissues are organized in such a way that even an uncontrolled increase in the frequency of cell division will not by itself produce a steadily growing tumor. The example of the uterine cervix, discussed above, illustrates this point. Like the epidermis of the skin and many other epithelia, the epithelium of the uterine cervix normally renews itself continually by shedding terminally differentiated cells from its outer surface and generating replacements from stem cells in the basal layer. On average, each normal stem cell division generates one daughter stem cell and one cell that is condemned to terminal differentiation and a cessation of cell division. If the stem cell simply divides more rapidly, terminally differentiated cells will be produced and shed more rapidly, and a balance of genesis and destruction will still be maintained. Thus if a transformed stem cell is to generate a steadily growing clone of progeny, the basic rules must be upset: either more than 50% of the daughter cells must remain as stem cells or the process of differentiation must be deranged so that daughter cells embarked on this route retain an ability to carry on dividing indefinitely and avoid dying or being discarded at the end of the production line (Figure 24–15).

Presumably, the development of such properties underlies the progression from a mild dysplasia of the uterine cervix to carcinoma *in situ* and malignant cancer (see Figure 24–10). Similar considerations apply to the development of cancer in other tissues that rely on stem cells, such as the skin, the lining of the gut, and the hemopoietic system. Several forms of leukemia, for example, seem to arise from a disruption of the normal program of differentiation, such that a committed progenitor of a particular type of blood cell continues to divide indefinitely, instead of differentiating terminally in the normal way and dying after a strictly limited number of division cycles (as discussed in Chapter 22). In general, changes that block the normal maturation of cells toward a nondividing, terminally differentiated state or prevent normal programmed cell death must play an essential part in many cancers. In the treatment of cancer, there-

Figure 24–15 Normal and deranged control of cell production from stem cells. (A) The normal strategy for producing new differentiated cells. (B and C) Two types of derangement that can give rise to the unbridled proliferation characteristic of cancer. Note that an excessive cell-division rate for the stem cells will not by itself have this effect, as long as each cell division produces only one daughter that is a stem cell.

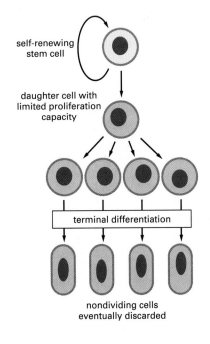

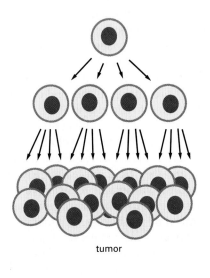

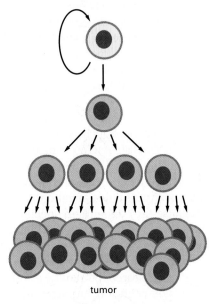

(A) NORMAL PATHWAY

(B) STEM CELL FAILS TO PRODUCE ONE NON-STEM-CELL DAUGHTER IN EACH DIVISION AND THEREBY PROLIFERATES TO FORM A TUMOR

(C) DAUGHTER CELLS FAIL TO DIFFERENTIATE NORMALLY AND THEREBY PROLIFERATE TO FORM A TUMOR fore, there is some prospect that drugs that promote cell differentiation may turn out to be a useful alternative to drugs that simply kill dividing cells.

To Metastasize, Cancer Cells Must Be Able to Cross Basal Laminae [12]

It is the ability to metastasize that makes cancers hard to eradicate surgically or by localized irradiation. To disseminate widely in the body, the cells of a typical solid tumor must be able to loosen their adhesion to their original neighbors, escape from the tissue of origin, burrow through other tissues until they reach a blood vessel or a lymphatic vessel, cross the basal lamina and endothelial lining of the vessel so as to enter the circulation, make an exit from the circulation elsewhere in the body, and survive and proliferate in the new environment in which they find themselves (Figure 24–16). Each of these steps requires different properties. For example, in a variety of carcinomas that have been studied, loss of adhesion to neighboring cells in an epithelium depends on loss of expression of the epithelial cell-cell adhesion molecule *E-cadherin*, but the ability to burrow through tissues seems to depend on the production of proteolytic enzymes that can break down extracellular matrix. The final steps in metastasis are probably the most difficult: many tumors release large numbers of cells into the circulation, but only a tiny proportion of these cells succeed in founding metastatic colonies.

A few types of normal cells—notably white blood cells—already have some or all of the properties needed to disseminate through the body, but for most cancers the ability to metastasize probably requires additional mutations or

Figure 24–16 Steps in the process of metastasis. This example illustrates the spread of a tumor from an organ such as the lung or bladder to the liver. Tumor cells may enter the bloodstream directly by crossing the wall of a blood vessel, as diagrammed here, or, more commonly perhaps, by crossing the wall of a lymphatic vessel that ultimately discharges its contents (lymph) into the bloodstream. Tumor cells that have entered a lymphatic vessel often become trapped in lymph nodes along the way, giving rise to lymph-node metastases. Studies in animals show that typically less than one in every thousand malignant tumor cells that enter the bloodstream will survive to produce a tumor at a new site.

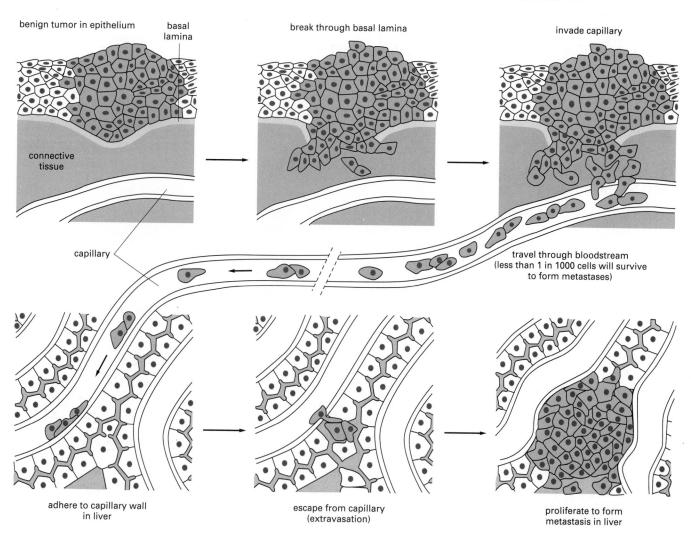

benign tumor in epithelium — basal lamina

break through basal lamina

invade capillary

connective tissue

capillary

travel through bloodstream (less than 1 in 1000 cells will survive to form metastases)

adhere to capillary wall in liver

escape from capillary (extravasation)

proliferate to form metastasis in liver

Cancer as a Microevolutionary Process

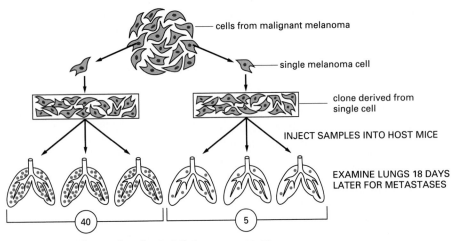

cells from malignant melanoma

single melanoma cell

clone derived from
single cell

INJECT SAMPLES INTO HOST MICE

EXAMINE LUNGS 18 DAYS
LATER FOR METASTASES

40

5

median number of metastatic tumors per mouse

Figure 24–17 Experiment showing that there are clonally heritable differences between the cells of a single tumor with respect to the ability to metastasize. Cells derived from a single cancer cell line are subcloned, and standard aliquots of each subclone are tested by injection into the bloodstream of host mice. The subclones differ markedly in the number of resulting metastases per mouse.

epigenetic changes. Such transformations, like the others involved in the development of cancer, are thought to occur at random in the initial tumor population: only those few cells that acquire the properties needed for metastasis and that happen to land in a suitable environment will be able to produce secondary tumors. In accordance with this concept of evolution through random variation and natural selection, one finds that the cells of a single tumor are heterogeneous in metastatic capacity (Figure 24–17).

An understanding of the molecular mechanisms of metastasis should eventually allow the design of treatments to block it. Some progress is being made along these lines. It has been shown, for example, that for tumor cells to cross a basal lamina they must have appropriate integrins to act as laminin receptors, which enable the cells to adhere to the lamina, and they must carry on their surface type-IV collagenase, which helps them digest the lamina (Figure 24–18). Antibodies or other reagents that block either laminin attachment or the activity of type-IV collagenase have been found to block metastasis in experimental animals. It remains to be seen whether human cancer patients can be helped by similar treatments.

Mutations That Increase the Mutation Rate Accelerate the Development of Cancer [1, 13]

As we have emphasized, the incidence of tumors and their rate of progression toward malignancy depends on the frequency of mutations. The mutation rate may be high because of mutagens in the environment or because of intracellular defects in the machinery governing replication, recombination, and repair of DNA. People with the rare genetic disorder *xeroderma pigmentosum*, for example, have a defect in the system of enzymes required to repair the type of damage done to DNA by ultraviolet irradiation; as a result, the slightest exposure of the skin to sunlight is liable to provoke skin cancers. A more general predisposition to cancer due to faults in DNA repair and replication occurs in the relatively common *HNPCC syndrome*, as we shall see on page 1290, and in *Bloom's syndrome, Fanconi's anemia,* and *ataxia-telangiectasia.* In all these rare genetic disorders the abnormality is inherited through the germ line and is therefore present in all the cells of the body. Similar genetic defects in DNA metabolism can also arise, however, through mutations originating in somatic cells. In fact, mutations that increase the mutation rate appear to be an important factor in the development of many cancers. Some of these mutations facilitate small local changes of DNA sequence. Others, especially common, facilitate gross disturbances of the genome.

Cancer cells often display an abnormal variability in the size and shape of their nuclei and in the number and structure of their chromosomes; indeed,

BINDING TO LAMININ

laminin receptor

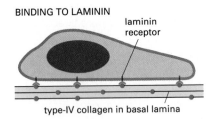

type-IV collagen in basal lamina

DIGESTION OF BASAL LAMINA BY TYPE-IV COLLAGENASE

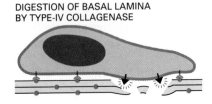

MOTILITY

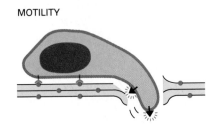

Figure 24–18 Three steps in crossing a basal lamina—a task that invasive tumor cells must be able to perform. Some of the experimental evidence for this mechanism is presented in Figure 19–59.

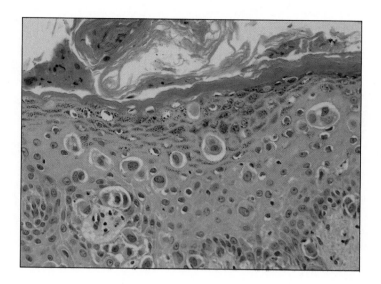

Figure 24–19 **Abnormal size and variability of cancer cell nuclei.** The photograph shows a section through the epidermis of the breast of a patient with *Paget's carcinoma,* in which the epidermis is invaded by cancer cells originating within the breast. The cancer cells can be distinguished from the ordinary epidermal cells by their abnormally large and variable nuclei and by the rim of clear space that appears to surround each of them. (Courtesy of Andrew Hanby.)

abnormal nuclear morphology is one of the key features used by pathologists to diagnose cancer (Figure 24–19). When cancer cells are grown in culture, they are often found to have an extraordinarily unstable karyotype: genes become amplified or deleted and chromosomes become lost, duplicated, or translocated with a far higher frequency than in normal cells in culture. Such chromosomal variability suggests that the cells have some heritable fault in the machinery or control of chromosome replication, repair, recombination, or segregation. Such a fault, arising by somatic mutation, would be liable to increase the likelihood of subsequent mutations in other classes of genes and so to provide a short cut to the accumulation of the multiple mutations required for cancerous behavior. Molecular genetic studies have revealed one mechanism for this destabilization of the karyotype in cancer cells, as we see in the last part of the chapter when we discuss the role of a protein known as the *p53 protein.*

The Enhanced Mutability of Cancer Cells Helps Them Evade Destruction by Anticancer Drugs [10, 14]

Because of the abnormally high mutability of many cancer cells, most malignant tumor cell populations are heterogeneous in many respects and capable of evolving at an alarming rate when subjected to new selection pressures. This aggravates the difficulties of cancer therapy. Repeated treatments with drugs that are selectively toxic to dividing cells can be used to kill the majority of neoplastic cells in a cancer patient, but it is rarely possible to kill them all. Usually some small proportion are drug-resistant, and the effect of the treatment is to favor the spread and evolution of cells with this trait.

To make matters worse, cells that are exposed to one drug often develop a resistance not only to that drug, but also to other drugs to which they have never been exposed. This phenomenon of **multidrug resistance** is frequently correlated with a curious change in the karyotype: the cell is seen to contain additional pairs of miniature chromosomes—so-called *double minute chromosomes*—or to have a *homogeneously staining region* interpolated in the normal banding pattern of one of its regular chromosomes. Both these aberrations consist of massively amplified numbers of copies of a small segment of the genome. The amplified DNA often contains a specific gene, known as the *multidrug resistance (mdr1)* gene, which codes for a plasma-membrane-bound transport ATPase (belonging to the ABC transporter superfamily discussed in Chapter 11) that is thought to prevent the intracellular accumulation of certain classes of lipophilic drugs by pumping them out of the cell. The amplification of other types of genes can also give the cancer cell a selective advantage: thus the gene for the enzyme dihydrofolate reductase (DHFR) often becomes amplified in response to cancer

Cancer as a Microevolutionary Process

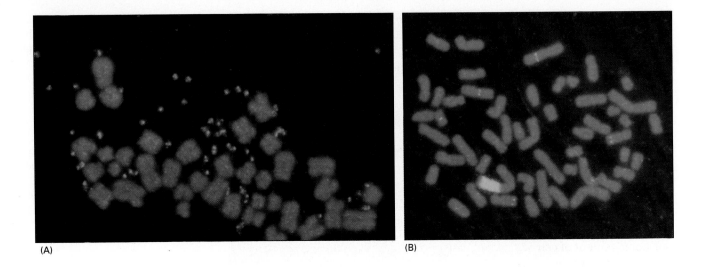

(A)   (B)

chemotherapy with the folic-acid antagonist methotrexate, and *myc* proto-oncogenes (to be discussed later), whose products stimulate cell proliferation, are similarly amplified in some cancers (Figure 24–20).

While defects in DNA replication, recombination, or repair may help cancer cells to evolve by increasing their mutability, they may also make the cells more vulnerable to certain types of attack. This may explain the observation—exploited in therapy—that the cells of many tumors are killed more easily than normal cells by irradiation or by exposure to specific drugs that interfere with DNA metabolism. As we learn more about the molecular mechanisms regulating DNA replication, recombination, and repair, it is beginning to be possible to pinpoint defects in these functions in individual cases of cancer, as the final section of this chapter will show. By using such information, we may be better able to kill the delinquent cells by designing drugs that exploit their particular weaknesses.

Summary

Cancer cells, by definition, proliferate in defiance of normal controls (that is, they are neoplastic) and are able to invade and colonize surrounding tissues (that is, they are malignant). By giving rise to secondary tumors, or metastases, they become hard to eradicate surgically. Most cancers are thought to originate from a single cell that has undergone a somatic mutation, but the progeny of this cell must undergo further changes, probably requiring several additional mutations, before they become cancerous. This phenomenon of tumor progression, which usually takes many years, reflects the operation of evolution by mutation and natural selection among somatic cells; the rate of the process is accelerated both by mutagenic agents (tumor initiators) and by certain nonmutagenic agents (tumor promoters) that affect gene expression, stimulate cell proliferation, and alter the ecological balance of mutant and nonmutant cells. Thus many factors contribute to the development of a given cancer, and since some of these factors are avoidable features of the environment, a large proportion of cancers are in principle preventable.

Much effort in cancer research has been devoted to the search for ways to cure the disease by exterminating cancer cells while sparing their normal neighbors. A rational approach to this problem requires an understanding of the special properties of cancer cells that enable them to evolve, multiply, and spread. Neoplastic cell proliferation often seems to be associated, for example, with a block in differentiation whereby the progeny of a stem cell are enabled to continue dividing instead of entering a terminal nondividing state or dying; in principle, the proliferation could be curbed by promoting cell differentiation. To become malignant, tumor cells must be able to cross basal laminae; antibodies can be designed that interfere with this ability, thereby hindering metastasis. Cancer cells are often found to be abnormally mutable.

Figure 24–20 Chromosomal changes in cancer cells reflecting gene amplification. In these examples the numbers of copies of a *myc* proto-oncogene have been amplified. The chromosomes are stained with a *red* fluorescent dye, while the multiple copies of the *myc* gene are detected by *in situ* hybridization with a *yellow* fluorescent probe. (A) Karyotype of a cell in which the *myc* gene copies are present as double minute chromosomes (paired *yellow specks*). (B) Karyotype of a cell in which the multiple *myc* gene copies appear as a homogeneously staining region (*yellow*) interpolated in one of the regular chromosomes. (Ordinary single-copy *myc* genes can be just seen as tiny *yellow dots* elsewhere in the genome.) Similar structures are seen in cancer cells that have other genes amplified. (Courtesy of Denise Sheer.)

This hastens evolution of the complex set of properties required for neoplasia and malignancy and helps the cancer cells develop resistance to anticancer drugs. At the same time, however, defects of DNA metabolism underlying such mutability may make the cancer cells uniquely vulnerable to a suitably designed therapeutic attack.

The Molecular Genetics of Cancer [15]

Because cancer is the outcome of a series of random genetic accidents subject to natural selection, no two cases even of the same variety of the disease are likely to be genetically identical. Nevertheless, all cancers can be expected to involve a disruption of the normal restraints on cell proliferation, and for each cell type there is a finite number of ways in which such disruption can occur. In fact, changes in a relatively small set of genes appear to be responsible for much of the derangement of cell behavior in cancer. The identification and characterization of many of these genes has been one of the great triumphs of molecular biology.

Cell proliferation can be regulated directly or indirectly—directly through the mechanism that determines whether a cell passes the restriction point, or "Start," of the cell-division cycle, as discussed in Chapter 17; or indirectly, for example, through regulation of the commitment to terminal differentiation or programmed cell death. In either case the normal regulatory genes can be loosely classified into those whose products help stimulate an increase in cell numbers and those whose products help inhibit it. Correspondingly, there are two mutational routes toward the uncontrolled cell proliferation and invasiveness that are characteristic of cancer. The first is to make a stimulatory gene hyperactive: this type of mutation has a dominant effect—only one of the cell's two gene copies need undergo the change—and the altered gene is called an **oncogene** (the normal allele being a **proto-oncogene;** from Greek *onkos,* a tumor). The second is to make an inhibitory gene inactive: this type of mutation usually has a recessive effect—both the cell's gene copies must be inactivated or deleted to free the cell of the inhibition—and the lost gene is called, for want of a better term, a **tumor suppressor gene.**

The mutant genes with a dominant effect—that is, the oncogenes—can be identified directly by taking DNA from the tumor cells and searching for fragments of it that, when introduced into normal cells, will cause these cells to behave like tumor cells. Techniques for achieving this feat were first devised in the late 1970s; their development followed earlier studies of a very similar process that occurs naturally, when viruses move their genetic material from cell to cell. This work paved the way for an explosion of discoveries of oncogenes and proto-oncogenes. More recently, progress has been made in the more difficult task of identifying and cloning tumor suppressor genes.

In this section we discuss oncogenes and tumor suppressor genes in turn. We conclude by presenting a case study of one common variety of cancer, where the steps of tumor progression can be related to a series of identified mutations.

Retroviruses Can Act as Vectors for Oncogenes That Transform Cell Behavior [16, 17, 18]

Viruses have played a remarkable part in the search for the genetic causes of human cancer. Although viruses have no role in the majority of common human cancers, they are more prominent as causes of cancer in some animal species, and analysis of animal tumor viruses has provided a key to the mechanisms of cancer in general.

The first animal tumor virus was discovered more than 80 years ago in chickens, which are subject to infections that cause connective-tissue tumors, or *sarcomas*. The infectious agent was characterized as a virus—the *Rous sarcoma virus,*

Table 24–3 Some Changes Commonly Observed When a Normal Tissue-Culture Cell Is Transformed by a Tumor Virus

1. **Plasma-membrane-related abnormalities**
   A. Enhanced transport of metabolites
   B. Excessive blebbing of plasma membrane
   C. Increased mobility of plasma membrane proteins

2. **Adherence abnormalities**
   A. Diminished adhesion to surfaces; therefore able to maintain a rounded morphology
   B. Failure of actin filaments to organize into stress fibers
   C. Reduced external coat of fibronectin
   D. High production of plasminogen activator, causing increased extracellular proteolysis

3. **Growth and division abnormalities**
   A. Growth to an unusually high cell density
   B. Lowered requirement for growth factors
   C. Less anchorage dependence (can grow even without attachment to rigid surface)
   D. "Immortal" (can continue proliferating indefinitely)
   E. Can cause tumors when injected into susceptible animals

which we now know to be an RNA virus. Like all the other RNA tumor viruses discovered since, it is a **retrovirus.** When it infects a cell, its RNA is copied into DNA by reverse transcription and the DNA is inserted into the host genome, where it can persist and be inherited by subsequent generations of cells. Figure 6–82 outlines the life cycle of a retrovirus and shows how its genome undergoes reverse transcription, integration into host DNA, and exit from and entry into host cells.

But how does the viral infection cause tumors? The solution to this problem, as to so many others in cell biology, depended on the development of a convenient assay by which different strains of virus could be rapidly tested for their tumor-causing capacity. The assay system, still widely used, consists simply of fibroblast cells proliferating in a culture dish. If active tumor virus is added to the culture medium, small colonies of abnormally proliferating **transformed** cells appear within a few days. Each such colony is a clone derived from a single cell that has been infected with the virus and has stably incorporated the viral genetic material. Released from the social controls on cell division, the transformed cells outgrow normal ones in the culture dish just as in the body and are therefore usually easy to select. The transformed cells commonly show a complex syndrome of abnormalities (summarized in Table 24–3). They tend not to be constrained by density-dependent inhibition of cell division (see Figure 17–39), for example, but pile up in layer upon layer as they proliferate (Figure 24–21). In addition, they often do not depend on anchorage for growth and are capable of

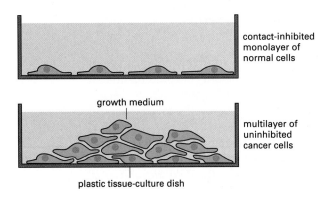

contact-inhibited monolayer of normal cells

growth medium

multilayer of uninhibited cancer cells

plastic tissue-culture dish

Figure 24–21 **Loss of contact inhibition.** Cancer cells, unlike most normal cells, usually continue to grow and pile up on top of one another after they have formed a confluent monolayer.

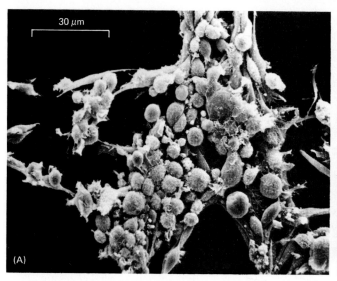
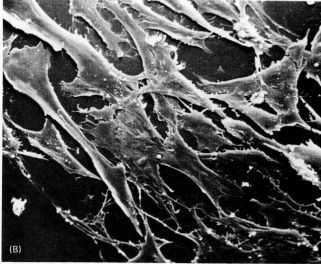

dividing even when held in suspension; they have an altered shape and adhere poorly to the substratum and to other cells, maintaining a rounded appearance reminiscent of a normal cell in mitosis; they may be able to proliferate even in the absence of growth factors; they are immortal and do not undergo senescence in culture; and when they are injected back into a suitable host animal, they can give rise to tumors.

The misbehavior of the transformed cells can be traced to an oncogene that is carried by the virus but is not necessary for the virus's own survival or reproduction. This was first demonstrated by the discovery of mutant Rous sarcoma viruses that multiply normally but no longer transform their host cells. The loss of transforming ability could be shown to correspond to loss or inactivation of a particular gene, which was given the name *src* (Figure 24–22). This specific gene in the Rous sarcoma virus is responsible for cell transformation *in vitro* and for tumor formation *in vivo*, but it is unnecessary baggage from the point of view of the virus's own propagation.

Retroviruses Pick Up Oncogenes by Accident [16, 17, 19]

If the viral *src* gene is bad for the animal and unnecessary to the virus, why is it present and where does it come from? When a radioactive DNA copy of the viral *src* gene sequence was used as a probe to search for related sequences by DNA-DNA hybridization, it was found that the genomes of normal vertebrate cells contain a sequence that is closely similar, but not identical, to the *src* gene of the Rous sarcoma virus. This normal cellular counterpart of the viral *src* gene (v-*src*) is called **c-*src*** (or just *src*). It is the proto-oncogene corresponding to the oncogene v-*src*. Evidently, the gene has been picked up accidentally by the retrovirus from the genome of a previous host cell but has undergone mutation in the process (Figure 24–23). The result is a perturbed gene function that leads to cancer and so brings the gene, and the virus that carries it, to the scientist's attention. The retrovirus has, in effect, cloned the gene for us. A large number of other oncogenes have been identified in other retroviruses and analyzed in similar ways (Table 24–4). Each has led to the discovery of a corresponding proto-oncogene that is present in every normal cell.

A Retrovirus Can Transform a Host Cell by Inserting Its DNA Next to a Proto-oncogene of the Host [20]

There are two ways in which a proto-oncogene can be converted into an oncogene upon incorporation into a retrovirus: the gene sequence may be altered

Figure 24–22 **Cell transformation by the Rous sarcoma virus.** The scanning electron micrographs show cells in culture infected with a form of the Rous sarcoma virus that carries a temperature-sensitive mutation in the gene responsible for transformation (the v-*src* oncogene). (A) The cells are transformed and have an abnormal rounded shape at low temperature (34°C), where the oncogene product is functional. (B) The same cells adhere strongly to the culture dish and thereby regain their normal flattened appearance when the oncogene product is inactivated by a shift to higher temperature (39°C). (Courtesy of G. Steven Martin.)

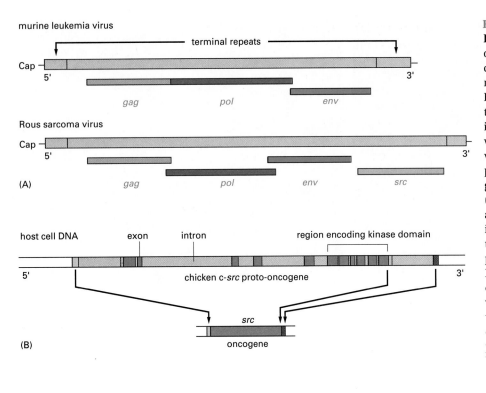

murine leukemia virus

Rous sarcoma virus

host cell DNA

chicken c-*src* proto-oncogene

Figure 24–23 The structure of the Rous sarcoma virus. (A) The organization of the viral genome as compared with that of a more typical retrovirus (murine leukemia virus). Rous sarcoma virus is unusual among the retroviruses that carry oncogenes in that it has retained all the three viral genes required for the ordinary viral life cycle: *gag* (which produces a polyprotein that is cleaved to generate the capsid proteins), *pol* (which produces reverse transcriptase and an enzyme involved in integrating the viral chromosome into the host genome), and *env* (which produces the envelope glycoprotein). In other oncogenic retroviruses one or more of these viral genes are wholly or partly lost in exchange for the acquisition of the transforming oncogene, and therefore infectious particles of the transforming virus can be generated only in a cell that is simultaneously infected with a nondefective, nontransforming *helper virus*, which supplies the missing functions. (Often the transforming oncogene is fused to a residual fragment of *gag*, leading to production of a hybrid oncogenic protein that includes part of the Gag sequence.) (B) The relationship between the v-*src* oncogene and the cellular *src* proto-oncogene from which it has been derived. The introns present in cellular *src* have been spliced out of v-*src*; in addition, v-*src* contains mutations that alter the amino acid sequence of the protein, making it hyperactive and unregulated as a tyrosine-specific protein kinase. Rous sarcoma virus has been highly selected (by cancer research workers) for its ability to transform cells to neoplasia, and it does this with unusual speed and efficiency.

Table 24–4 Some Oncogenes Originally Identified Through Their Presence in Transforming Retroviruses

Oncogene	Proto-oncogene Function	Source of Virus	Virus-induced Tumor
abl	protein kinase (tyrosine)	mouse cat	pre-B-cell leukemia sarcoma
erb-B	protein kinase (tyrosine): epidermal growth factor (EGF) receptor	chicken	erythroleukemia, fibrosarcoma
fes	protein kinase (tyrosine)	cat/chicken	sarcoma
fms	protein kinase (tyrosine): macrophage colony-stimulating factor (M-CSF) receptor	cat	sarcoma
fos *jun*	products associate to form AP-1 gene regulatory protein	mouse chicken	osteosarcoma fibrosarcoma
kit	protein kinase (tyrosine): Steel factor receptor	cat	sarcoma
raf	protein kinase (serine/threonine) activated by Ras	chicken/ mouse	sarcoma
myc	gene regulatory protein of the HLH family	chicken	sarcoma; myelocytoma, carcinoma
H-*ras*	GTP-binding protein	rat	sarcoma; erythroleukemia
K-*ras*	GTP-binding protein	rat	sarcoma; erythroleukemia
rel	gene regulatory protein related to NFκB	turkey	reticuloendotheliosis
sis	platelet-derived growth factor, B chain	monkey	sarcoma
src	protein kinase (tyrosine)	chicken	sarcoma

or truncated so that it codes for a protein with abnormal activity, or the gene may be brought under the control of powerful promoters and enhancers in the viral genome that cause its product to be made in excess or in inappropriate circumstances. Retroviruses can also exert similar oncogenic effects in a different way: DNA copies of the viral RNA may simply be inserted into the host cell genome at sites close to, or even within, proto-oncogenes. The resulting genetic disruption is called an **insertional mutation,** and the altered genome is inherited by all the progeny of the original host cell. More-or-less random insertion of DNA copies of the viral RNA into the host DNA occurs as part of the normal retroviral life cycle, and in at least one well-documented case, insertion anywhere within about 10,000 nucleotide pairs from a proto-oncogene can cause abnormal activation of that gene.

Insertional mutagenesis provides an important means of identifying proto-oncogenes, which can be tracked down by their proximity to the inserted viral DNA. Proto-oncogenes identified in this way often turn out to be the same as those discovered in the other way, as counterparts to oncogenes that retroviruses carry from cell to cell, but some new ones have been discovered as well (Table 24–5). An example is the *Wnt-1* gene, activated by insertional mutagenesis in breast cancers in mice infected with the mouse mammary tumor virus (Figure 24–24). This gene turns out to be closely homologous to the *Drosophila* gene *wingless,* which is involved in cell-cell communications that regulate details of the body pattern of the fly, as discussed in Chapter 21.

Different Searches for the Genetic Basis of Cancer Converge on Disturbances in the Same Proto-oncogenes [17, 21]

While some researchers pursued the line of investigation leading from retroviruses to oncogenes, others took a more direct approach and searched for DNA sequences in human cancer cells that would provoke uncontrolled proliferation when introduced into noncancerous cells. The assay was again done in cell culture, using an established line of mouse-derived fibroblast cells—NIH 3T3 cells—as the noncancerous hosts and transfecting them with DNA taken from human tumor cells. The findings were dramatic. Oncogenes were detected in many lines of human cancer cells, and in several cases these oncogenes turned out to be mutant alleles of some of the same proto-oncogenes that had been identified by the retroviral approach or of genes very closely related to them. About one in four human tumors, for example, was found to contain a mutated member of the *ras* gene family, first discovered as oncogenes carried by retroviruses that cause sarcomas in rats. Thus two independent lines of inquiry converged on the same genes.

Yet another approach that led to some of the same proto-oncogenes was based on the karyotyping of tumor cells. As mentioned earlier, in almost all patients with chronic myelogenous leukemia, the leukemic cells show the same chromosomal translocation, between chromosomes 9 and 22; likewise, in Burkitt's lymphoma there is regularly a translocation between chromosome 8 and one of the three chromosomes containing the genes that encode antibody molecules. In both these types of cancer the translocation breakpoint, where part of

Table 24–5 **Some Oncogenes Originally Identified by Means Other Than Their Presence in Transforming Retroviruses**

Means of Detection	Oncogenes
Insertional mutation	*Wnt*-1 (*int*-1), *fgf*-3 (*int*-2), *Notch*-1 (*int*-3), *lck*
Amplification	L-*myc*, N-*myc*
Transfection	*neu*, N-*ras*, *trk*, *ret*
Translocation	*bcl*-2, RARa

Figure 24–24 **Insertional mutagenesis.** In this example the process activates a gene called *Wnt*-1 (formerly called *int*-1) and produces breast cancer in mice infected with the mouse mammary tumor virus (MMTV). The sites of MMTV integration observed in 19 different tumor isolates are indicated by *arrows.* Note that the insertions can activate transcription of the *Wnt*-1 gene from distances of more than 10,000 nucleotide pairs away and from either side of the gene. This effect is attributed to a powerful enhancer DNA sequence present in the terminal repeats of the MMTV genome.

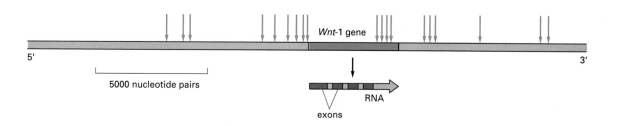

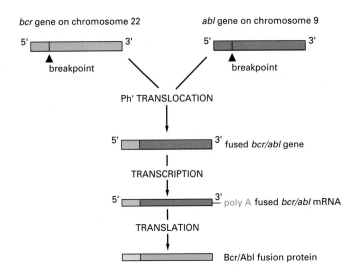

bcr gene on chromosome 22   abl gene on chromosome 9

breakpoint   breakpoint

Ph' TRANSLOCATION

5'   3' fused bcr/abl gene

TRANSCRIPTION

5'   3' poly A fused bcr/abl mRNA

TRANSLATION

Bcr/Abl fusion protein

Figure 24–25 The conversion of the *abl* proto-oncogene into an oncogene in patients with chronic myelogenous leukemia. The chromosome translocation responsible joins the *bcr* gene on chromosome 22 to the *abl* gene from chromosome 9, thereby generating a Philadelphia chromosome (see Figure 24–4). The resulting fusion protein has the amino terminus of the Bcr protein joined to the carboxyl terminus of the Abl tyrosine protein kinase. In consequence, the Abl kinase domain presumably becomes inappropriately active, driving excessive proliferation of a clone of hemopoietic cells in the bone marrow.

one chromosome is joined to another, was found to coincide exactly with the location of a proto-oncogene already known from retroviral studies—*abl* in chronic myelogenous leukemia, *myc* in Burkitt's lymphoma. Analogous chromosome translocations are similarly associated with some other types of cancer. From DNA sequencing studies it seems that in some cases the translocation turns a proto-oncogene into an oncogene by fusing the proto-oncogene to another gene in such a way that an altered protein is produced (Figure 24–25); in other cases the translocation moves a proto-oncogene into an inappropriate chromosomal environment that activates its transcription so that the normal protein is produced in excess.

A Proto-oncogene Can Be Made Oncogenic in Many Ways [17, 22]

So far, about 60 proto-oncogenes have been discovered (Tables 24–4 and 24–5 show a small selection); each of these can be converted into an oncogene that plays a dominant part in cancers of one sort or another. Most such genes have been encountered repeatedly, in a variety of mutant forms and in several kinds of cancer, suggesting that the majority of mammalian proto-oncogenes may already have been identified.

Figure 24–26 The activities and cellular locations of the products of the main classes of known proto-oncogenes. Some representative proto-oncogenes in each class are indicated in brackets.

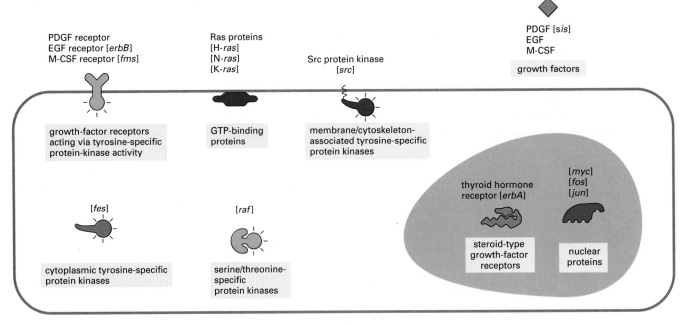

PDGF receptor
EGF receptor [erbB]
M-CSF receptor [fms]

Ras proteins
[H-ras]
[N-ras]
[K-ras]

Src protein kinase
[src]

PDGF [sis]
EGF
M-CSF

growth factors

growth-factor receptors acting via tyrosine-specific protein-kinase activity

GTP-binding proteins

membrane/cytoskeleton-associated tyrosine-specific protein kinases

[fes]

[raf]

thyroid hormone receptor [erbA]

[myc]
[fos]
[jun]

cytoplasmic tyrosine-specific protein kinases

serine/threonine-specific protein kinases

steroid-type growth-factor receptors

nuclear proteins

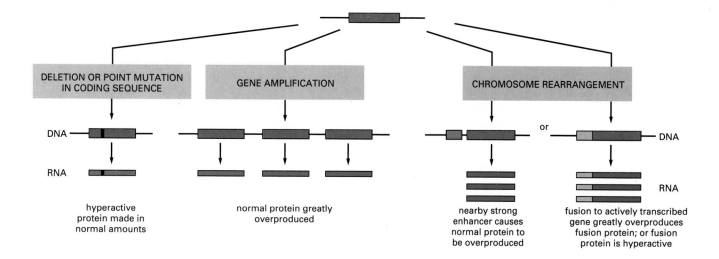

DELETION OR POINT MUTATION IN CODING SEQUENCE

DNA

RNA

hyperactive protein made in normal amounts

GENE AMPLIFICATION

normal protein greatly overproduced

CHROMOSOME REARRANGEMENT

or

DNA

RNA

nearby strong enhancer causes normal protein to be overproduced

fusion to actively transcribed gene greatly overproduces fusion protein; or fusion protein is hyperactive

But what functions do these genes have in a normal healthy cell, that mutations in them should be so dangerous? Most proto-oncogenes code for components of the mechanisms that regulate the social behavior of cells in the body—in particular, the mechanisms by which signals from a cell's neighbors can impel it to divide, differentiate, or die. In fact, many of the components of cell-signaling pathways were first identified through searches for oncogenes, and a full list of proto-oncogene products includes examples of practically every type of molecule involved in cell signaling—secreted proteins, transmembrane receptors, GTP-binding proteins, protein kinases, gene regulatory proteins, and so on, as summarized in Figure 24–26 and discussed in detail in Chapter 15. All these molecules normally serve in complex relay chains to deliver signals for the production of more cells when more cells are needed. But mutations can alter them so that they deliver the signals even when more cells are not needed. The proto-oncogene *erbB*, for example, codes for the receptor for epidermal growth factor (EGF); when EGF binds to the receptor's extracellular domain, the intracellular domain generates a stimulatory signal inside the cell. A mutation in c-*erbB* can turn it into an oncogene by deleting the extracellular EGF-binding domain in such a way that the intracellular stimulatory signal is produced constantly, even if no EGF is present. In a similar way a point mutation at an appropriate site in a *ras* gene can create a Ras protein that fails to hydrolyze its bound GTP and so persists abnormally in its active state, transmitting an intracellular signal for cell proliferation even when it should not. Innumerable other examples can be given.

The basic types of genetic accident that can convert a proto-oncogene into an oncogene are summarized in Figure 24–27. The gene may be altered by a point mutation, by a deletion, through a chromosomal translocation, or by insertion of a mobile genetic element such as retroviral DNA. The change can occur in the protein-coding region so as to yield a hyperactive product, or it can occur in adjacent control regions so that the gene is simply overexpressed. Alternatively, the gene may be overexpressed because it has been amplified to a high copy number through errors in the process of chromosome replication. (The mechanism is discussed later—see Figure 24–34.) Specific types of abnormality are characteristic of particular genes and of the responses to particular carcinogens. For example, 90% of the skin tumors evoked in mice by the tumor initiator dimethylbenz[a]anthracene (DMBA) have an A-to-T alteration at exactly the same site in a mutant *ras* gene; presumably, of the mutations caused by DMBA, it is only the ones at this site that efficiently activate skin cells to form a tumor. Members of the *myc* gene family, on the other hand, are frequently overexpressed or amplified. The Myc protein normally acts in the nucleus as a signal for cell proliferation, as discussed in Chapter 17; excessive quantities of Myc cause the cell to embark on the cell-division cycle in circumstances where a normal cell would halt.

Figure 24–27 **Three ways in which a proto-oncogene can be converted into an oncogene.** A fourth mechanism (not shown) involves recombination between retroviral DNA and a proto-oncogene (see Figure 24–24). This has effects similar to those of chromosome rearrangement, bringing the proto-oncogene under the control of a viral enhancer and/or fusing it to a viral gene that is actively transcribed.

The Molecular Genetics of Cancer

The Actions of Oncogenes Can Be Assayed Singly and in Combination in Transgenic Mice [23]

The concept of an oncogene is paradoxical. As we have argued at length in the first part of this chapter, a single mutation is not enough to cause a cancer. Yet an oncogene is defined as a dominantly acting gene and is typically assayed by its ability to cause neoplastic transformation of cultured cells on its own. This apparent contradiction reflects the gulf between the simplified models of cancer most widely studied by molecular biologists and the complexity of the actual human disease. The standard assay for identification of oncogenes does not test their effects on normal human somatic cells but on a mouse-derived fibroblast cell line, and the cells of this line, through establishment in culture, have already undergone mutations that make them abnormally easy to transform by a single further genetic change. Moreover, as we noted earlier, mice, with their shorter life-spans and smaller cell numbers, run an intrinsically smaller risk of cancer than do human beings, so that their cells may be less securely protected against the consequences of carcinogenic mutations than are human cells.

Even in a mouse, however, a single oncogene is not usually sufficient to turn a normal cell into a cancer cell. This can be strikingly demonstrated by studies of transgenic mice. An oncogene in the form of a DNA fragment, derived from either a virus or a tumor cell, can be linked to a suitable promoter DNA sequence and then injected into a mouse egg nucleus. Often this recombinant DNA molecule will become integrated into a mouse chromosome, leading to the generation of a strain of transgenic mice that carry the oncogene in all their cells. The oncogene introduced in this way may be expressed in many tissues or in only a select few, according to the tissue specificity of the associated promoter. Typically, in mice that are thus endowed with a *myc* or *ras* oncogene, some of the tissues that express the oncogene grow to an exaggerated size, and occasional cells with the passage of time undergo further changes and give rise to cancers. The vast majority of the cells in the transgenic mouse that express the *myc* or *ras* oncogene, however, do not give rise to cancers, showing that the single oncogene is not enough to cause neoplastic transformation. From the point of view of the whole animal, the inherited oncogene, nevertheless, is a serious menace because it increases the risk of developing cancer. Inherited oncogenes can act similarly in humans, although the phenomenon is rare. In the one well-documented example a mutant form of the *ret* proto-oncogene confers a hereditary predisposition to certain tumors of the thyroid and adrenal glands (a syndrome known as multiple endocrine neoplasia type 2A [MEN2A]).

The experimental analysis of oncogene action has been taken a step further by mating a pair of transgenic mice—one carrying a *myc* oncogene, the other carrying a *ras* oncogene—so as to obtain progeny that carry both oncogenes together. These offspring develop cancers at a much higher rate than either parental strain (Figure 24–28), but again the cancers originate as scattered isolated tumors among noncancerous cells. Thus, even with these two expressed oncogenes, the cells must undergo further, randomly generated changes to become cancerous.

The synergistic action of two or more specific oncogenes to make cells cancerous is known as *oncogene collaboration*, and it can be demonstrated *in vitro* as well as *in vivo*. If normal rat embryo fibroblasts are transfected with a *ras* oncogene alone or with a *myc* oncogene alone, they continue to behave normally. If they are transfected with both oncogenes together, they are transformed. In different cell types, different combinations of oncogenes are required for transformation. The lymphocyte-derived cancer known as B-cell lymphoma, for example, appears to depend on a collaboration between *myc* and the *bcl-2* gene (which we encountered in Chapter 21, in connection with programmed cell death). If *myc* alone is overexpressed, cells are driven round the division cycle inappropriately, but no cancer results because the progeny of such abnormally forced divisions are programmed to die. If *bcl-2* is overexpressed at the same

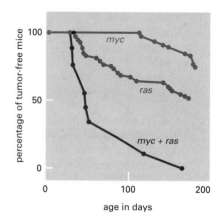

Oncogene collaboration in transgenic mice. The graphs show the incidence of tumors in three types of transgenic mice, one carrying a *myc* oncogene, one carrying a *ras* oncogene, and one carrying both oncogenes. For these experiments two lines of transgenic mice were first constructed. One carries an inserted copy of an oncogene created by fusing the proto-oncogene *myc* with the mouse mammary tumor virus promoter/ enhancer (which then drives *myc* overexpression in specific tissues such as the mammary gland). The other line carries an inserted copy of the oncogene v-H-*ras* under control of the same promoter/enhancer. Both strains of mice develop tumors much more frequently than normal, most often in the mammary or salivary glands. Mice that carry both oncogenes together were obtained by crossing the two strains. These hybrids develop tumors at a far higher rate still, much greater than the sum of the rates for the two oncogenes separately. Nevertheless, the tumors arise only after a delay and only from a small proportion of the cells in the tissues where the two genes are expressed. Some further accidental change, in addition to the two oncogenes, is apparently required for the development of cancer. (After E. Sinn et al., *Cell* 49:465–475, 1987. © Cell Press.)

time, however, the excess progeny cells survive and proliferate: *bcl*-2 acts as an oncogene because the Bcl-2 protein inhibits programmed cell death.

These phenomena reflect the fail-safe character of the complex network of controls that regulate cell proliferation and survival. It seems that multiple control mechanisms operate in parallel, so that a single faulty component is not enough to cause a disaster. In this control system, however, proto-oncogenes are only half the story. No less important are the tumor suppressor genes that act as counterbalance to them. As we see next, deletions and disturbances of tumor suppressor genes play as large a part in human cancer as do mutations of proto-oncogenes into oncogenes.

Loss of One Copy of a Tumor Suppressor Gene Can Create a Hereditary Predisposition to Cancer [24, 25]

Given a cancer cell, it is harder to pinpoint a normal gene that it lacks than it is to discover an abnormal gene that it possesses: one cannot take the DNA and use a cell transformation assay to identify something that simply is not there. Knowledge of tumor suppressor genes has thus been harder to obtain than knowledge of proto-oncogenes.

The difficulty is compounded because loss of a single tumor suppressor gene—even loss of both copies of the gene—usually is not sufficient by itself to cause a cancer; this makes it hard to incriminate specific mutations. We have seen that the same problem arises in tracking down oncogenes and can be resolved by assays using cultured cell lines that are unusually easy to transform with a single mutant gene. In the study of tumor suppressor genes the key insight came instead from a rare type of human cancer, **retinoblastoma,** that arises from cells in the body that are transformed by an unusually small number of mutations.

Retinoblastoma occurs in childhood; tumors develop from neural precursor cells in the immature retina. About one child in 20,000 is afflicted. There are two forms of the disease, one hereditary, the other not. In the hereditary form multiple tumors usually arise independently, affecting both eyes; in the non-hereditary form only one eye is affected and by only one tumor. Some hereditary sufferers from retinoblastoma are found to have a visibly abnormal karyotype, with a deletion of a specific band on chromosome 13, and deletions of this same locus are also encountered in tumor cells from patients with the nonhereditary disease. This suggests that the cancer may be caused by loss of a tumor suppressor gene rather than acquisition of an oncogene. Specifically, if all of the cells of patients with the hereditary disease lack one of the two normal copies of a tumor suppressor gene, then those cells will be predisposed to become cancerous: a single somatic mutation that eliminates the remaining good copy of the gene in one of the million or more cells in the growing retina will suffice to initiate a cancer. The gene whose loss thus appears to be critical for development of the cancer is called the retinoblastoma, or *Rb,* gene. In children without the hereditary predisposition retinoblastomas are very rare because they require the coincidence of two somatic mutations in a single retinal cell to destroy both copies of the *Rb* gene (Figure 24–29).

Using the known location of the chromosomal deletion associated with retinoblastoma, it was possible to clone and sequence the *Rb* gene and hence to analyze the genetic defects in detail in individual patients. Just as predicted, in those who suffer from the hereditary form of the disease, a deletion or loss-of-function mutation of the *Rb* gene is found in every cell of the body. The cells that are not cancerous are defective in only one of their two copies of the gene, while the cancerous cells are defective in both. In patients with the nonhereditary form of the disease, by contrast, the noncancerous cells show no defect in either copy of *Rb,* while the cancerous cells are again defective in both copies.

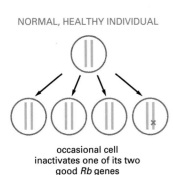

NORMAL, HEALTHY INDIVIDUAL

occasional cell inactivates one of its two good *Rb* genes

RESULT: NO TUMOR

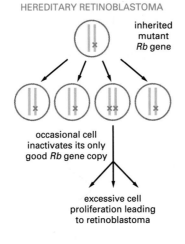

HEREDITARY RETINOBLASTOMA

inherited mutant *Rb* gene

occasional cell inactivates its only good *Rb* gene copy

excessive cell proliferation leading to retinoblastoma

RESULT: MOST PEOPLE WITH INHERITED MUTATION DEVELOP TUMOR

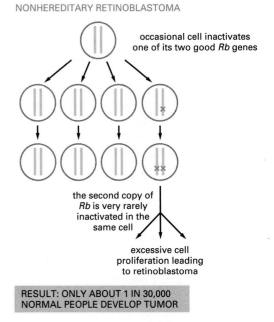

NONHEREDITARY RETINOBLASTOMA

occasional cell inactivates one of its two good *Rb* genes

the second copy of *Rb* is very rarely inactivated in the same cell

excessive cell proliferation leading to retinoblastoma

RESULT: ONLY ABOUT 1 IN 30,000 NORMAL PEOPLE DEVELOP TUMOR

Many different combinations of genetic mishaps can eliminate or cripple both copies of a gene. The first copy may, for example, be lost by a large-scale chromosomal deletion or inactivated by a point mutation; the second copy may be lost in a similar way or by mitotic recombination (explained in Figure 21–73) or by gene conversion (see Figure 6–66). The range of possibilities is summarized in Figure 24–30. It can be seen that most of these mechanisms are such that the second copy of the gene, and usually flanking regions of its chromosome as well, are either totally deleted or replaced by a copy of the corresponding region of the first defective chromosome; in other words, the usual heterozygous combination of maternal and paternal alleles of genes in that chromosomal region is lost. In fact, when tumor cells from many different retinoblastoma patients are analyzed, a **loss of heterozygosity** in the region of the *Rb* gene is observed in about 70% of cases. As we see below, loss of heterozygosity in the cells of a tumor provides the basis for a general method by which tumor suppressor genes can be identified and cloned.

Figure 24–29 The genetic mechanisms underlying retinoblastoma. In the hereditary form all cells in the body lack one of the normal two functional copies of a tumor suppressor gene, and tumors occur where the remaining copy is lost or inactivated by a somatic mutation. In the nonhereditary form all cells initially contain two functional copies of the gene, and the tumor arises because both copies are lost or inactivated through the coincidence of two somatic mutations in one cell.

Loss of the Retinoblastoma Tumor Suppressor Gene Plays a Part in Many Different Cancers [25, 26]

Given the *Rb* gene sequence, it is possible to test for its presence in cells of tumors other than retinoblastoma. The gene turns out to be frequently missing in several common types of cancer, including carcinomas of lung, breast, and bladder. These more common cancers arise by a more complex series of genetic changes than does retinoblastoma, and they make their appearance later in life and in other tissues of the body. But in all of them, it seems, loss of *Rb* is a major step in the progression toward malignancy. Thus although retinoblastoma is rare, cancers involving the *Rb* gene are not.

As discussed in Chapter 17, *Rb* is normally expressed in almost all the cells of the body, and its product appears to act as one of the main brakes on progress around the cell-division cycle. The braking action is normally regulated by phosphorylation. The Rb protein alternates between a phosphorylated and an unphosphorylated state in every cycle, and in cells that have withdrawn from cycling it is kept unphosphorylated. As long as it is unphosphorylated, Rb binds strongly to certain gene regulatory proteins and in this way prevents them from acting in the nucleus to promote DNA replication. Loss of the gene sets the cell free from this restraint.

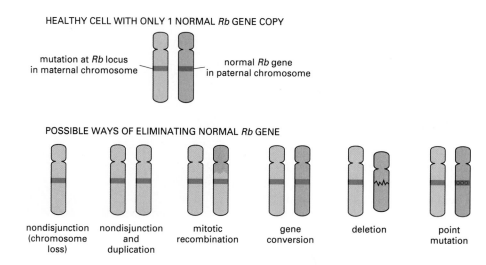

HEALTHY CELL WITH ONLY 1 NORMAL *Rb* GENE COPY

mutation at *Rb* locus in maternal chromosome — normal *Rb* gene in paternal chromosome

POSSIBLE WAYS OF ELIMINATING NORMAL *Rb* GENE

nondisjunction (chromosome loss)   nondisjunction and duplication   mitotic recombination   gene conversion   deletion   point mutation

Figure 24–30 Six ways of losing the remaining good copy of a tumor suppressor gene. A cell that is defective in only one of its two copies of a tumor suppressor gene usually behaves as a normal, healthy cell; the diagrams show how it may come to lose the function of the other gene copy as well and thereby progress toward cancer. Cloned DNA probes can be used in conjunction with restriction-fragment length polymorphisms (see Chapter 7) to analyze the tumor DNA and so to discover which type of event has occurred in a given patient. Note that most of the mechanisms result in a cell that totally lacks either the maternal or the paternal copy of the tumor suppressor gene, along with adjacent chromosomal regions. This is reflected in a *loss of heterozygosity* in the neighborhood of the genetic defect. Loss of heterozygosity at a specific site in the genome is a hallmark of a cancer dependent on loss of the function of a tumor suppressor gene. (After W.K. Cavenee et al., *Nature* 305:779–784, 1983. © 1983 Macmillan Magazines Ltd.)

The molecular interpretation of how Rb acts takes us a step further still in the understanding of cancer, for it throws light on yet another way in which cancer can be triggered—by DNA viruses. Before explaining how, we must pause to examine the role that viruses in general play in the causation of human cancer.

DNA Tumor Viruses Activate the Cell's DNA Replication Machinery as Part of Their Strategy for Survival [16]

About 15% of human cancers, in the world as a whole, are thought to arise by mechanisms that involve viruses. As shown in Table 24–6, the main culprits in humans are not retroviruses but DNA viruses. Evidence for their involvement comes partly from the detection of viruses in cancer patients and partly from

Table 24–6 Viruses Associated with Human Cancers

Virus	Associated Tumors	Areas of High Incidence
DNA viruses		
Papovavirus family		
Papillomavirus (many distinct strains)	warts (benign) carcinoma of uterine cervix	worldwide worldwide
Hepadnavirus family		
Hepatitis-B virus	liver cancer (hepatocellular carcinoma)	Southeast Asia, tropical Africa
Herpesvirus family		
Epstein-Barr virus	Burkitt's lymphoma (cancer of B lymphocytes) nasopharyngeal carcinoma	West Africa, Papua New Guinea Southern China, Greenland (Inuit)
RNA viruses		
Retrovirus family		
Human T-cell leukemia virus type I (HTLV-I)	adult T-cell leukemia/ lymphoma	Japan (Kyushu), West Indies
Human immunodeficiency virus (HIV-1, the AIDS virus)	Kaposi's sarcoma [cancer of endothelial cells of blood vessels or lymphatics (?)]	Central Africa

For all the above viruses, the number of people infected is much larger than the number who develop cancer: the viruses must act in conjunction with other factors. Moreover, some of the viruses probably contribute to cancer only indirectly; for example, HIV-1, by obliterating cell-mediated immune defenses, may allow endothelial cells transformed by some other agent to thrive as a tumor instead of being destroyed by the immune system.

epidemiology. Liver cancer, for example, is common in parts of the world (Africa and Southeast Asia) where hepatitis-B viral infections are common, and in those regions the cancer occurs almost exclusively in people who show signs of chronic hepatitis-B infection.

The precise role of a cancer-associated virus is often hard to decipher because there is a delay of many years from the initial viral infection to the development of the cancer. Moreover, the virus is responsible for only one of a series of steps in the progression to cancer, and other environmental factors and genetic accidents are also involved. In some cancers viruses seem to have indirect promoting actions; the hepatitis-B virus may, for example, favor the development of liver cancer by doing damage that provokes cell division in the liver. In some other human cancers, however, viruses help directly to cause neoplastic transformation of the cells they infect. Studies with cultured cells *in vitro* show how this can occur.

If a cell is to be stably transformed by a virus, a stable parasitic association must be established: the virus must not kill the cell, and the cell must retain the viral genes from one cell generation to the next—usually by integrating those genes into one or more of its own chromosomes, occasionally by retaining them as an extrachromosomal plasmid that replicates in step with the chromosomes. This applies equally to DNA viruses and retroviruses, but the two classes of virus differ fundamentally in the nature of the viral genes that cause neoplastic transformation.

As explained in Chapter 6, a **DNA tumor virus** normally propagates in the wild by a process that does not depend on the production of cancer. An SV40 virus (Figure 24–31), for example, produces a viral protein that rapidly activates the host cell's machinery for DNA replication. The virus then uses host proteins to replicate and transcribe its own genome; the infection continues until the host is killed, releasing a horde of new infectious virus particles. Much more rarely, however, the viral DNA fails to replicate and instead becomes stably incorporated into a host cell chromosome. If the viral gene that activates the host's machinery for DNA replication is transcribed, this gene can act as an oncogene, causing a cancerous transformation. Unlike the retroviral oncogenes discussed earlier, however, this oncogene is an essential part of the viral genome, and it has no counterpart in the normal host cell.

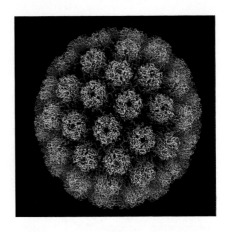

Figure 24–31 The SV40 virus. The structure of the capsid of this widely studied DNA virus that infects monkeys has been determined by x-ray diffraction. (Courtesy of Robert Grant, Stephen Crainic, and James M. Hogle.)

DNA Tumor Viruses Activate the Cell's Replication Machinery by Blocking the Action of Key Tumor Suppressor Genes [16, 27]

DNA viruses are a diverse group, but the general principles just described apply to most of those that are involved in cancer. The *papillomaviruses*, for example, are the cause of human warts and are also implicated in carcinomas of the uterine cervix. They are distantly related to the *polyomavirus* family that includes SV40, and the cancer that they cause in humans requires the integration of specific viral replication genes into a host chromosome, as illustrated in Figure 24–32.

Like SV40, papillomaviruses have to be able to commandeer the host cell's DNA synthesis machinery, and the viral genes that have this function can act as oncogenes. They are called the *E6* and *E7* genes of the papillomavirus, and their protein products are functionally equivalent to a single large dual-purpose protein called *large T antigen*, encoded by a corresponding viral oncogene in the SV40 genome. The mechanism of action is apparently simple: these viral proteins bind to the protein products of two key tumor suppressor genes of the host cell, putting them out of action and so permitting the cell to replicate its DNA and divide (Figure 24–33). One of these host proteins is Rb: by binding to it, the viral protein (E7 or large T) prevents it from binding to its normal associates in the cell.

The other tumor suppressor gene product that the viral proteins serve to inactivate is called **p53** (from its molecular mass of 53 kilodaltons). Like Rb, it

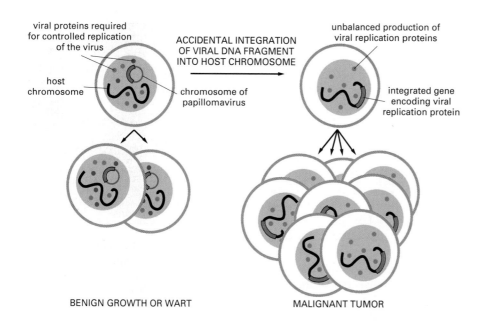

viral proteins required for controlled replication of the virus

host chromosome

chromosome of papillomavirus

ACCIDENTAL INTEGRATION OF VIRAL DNA FRAGMENT INTO HOST CHROMOSOME

unbalanced production of viral replication proteins

integrated gene encoding viral replication protein

BENIGN GROWTH OR WART

MALIGNANT TUMOR

Figure 24–32 How certain papillomaviruses are thought to give rise to cancer of the uterine cervix. Papillomaviruses have double-stranded circular DNA chromosomes of about 8000 nucleotide pairs. In a wart or other benign infection these chromosomes are stably maintained in the basal cells of the epithelium as plasmids whose replication is regulated so as to keep step with the chromosomes of the host (*left*). Rare accidents can cause the integration of a fragment of such a plasmid into a chromosome of the host, altering the environment of the viral genes and disrupting the control of their expression. The consequent unregulated production of viral replication proteins—in particular, the products of the viral E6 and E7 genes—tends to drive the host cell into S phase, thereby helping to generate a cancer (*right*).

plays a part in the development of many types of cancer, not only those dependent on viruses.

Mutations of the *p53* Gene Disable an Emergency Brake on Cell Proliferation and Lead to Genetic Instability [13, 14, 28]

People who inherit from their parents only one functional copy of the *p53* gene, like those who inherit only one functional copy of *Rb*, are predisposed to cancer. They are liable to develop several independent tumors in any of a wide variety of tissues at a relatively early age—typically as young adults. This propensity, running in families, is called *Li-Fraumeni* syndrome; like hereditary retinoblastoma, it is very rare. Tumor cells in the Li-Fraumeni patient have defects in both copies of *p53*, while the nontumor cells have a defect in only one.

These findings are typical of a tumor suppressor gene, and, in confirmation of this role for p53, it is found that artificially raised levels of normal p53 protein in an ordinary cultured cell will stop it from proliferating. The p53 protein binds to DNA and exerts its effect, in part at least, by inducing the transcription of another regulatory gene, whose product, a 21-kilodalton protein, binds to the complexes of G_1 cyclin with Cdk2 protein that normally serve to drive the cell past the G_1 checkpoint in the cell cycle, as discussed in Chapter 17. By blocking the kinase activity of these complexes, the 21-kilodalton protein prevents the cell from progressing into S phase and replicating its DNA.

In contrast with Rb, very little p53 protein is found in most of the cells of the body under normal conditions. In fact, p53 is not required for normal development: transgenic mice in which both copies of the gene have been knocked out

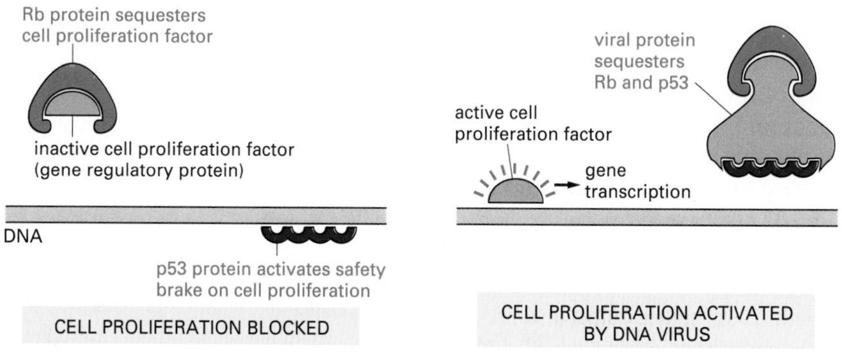

Rb protein sequesters cell proliferation factor

inactive cell proliferation factor (gene regulatory protein)

DNA

p53 protein activates safety brake on cell proliferation

CELL PROLIFERATION BLOCKED

viral protein sequesters Rb and p53

active cell proliferation factor

gene transcription

CELL PROLIFERATION ACTIVATED BY DNA VIRUS

Figure 24–33 Activation of cell proliferation by the SV40 DNA tumor virus. SV40 uses a single dual-purpose viral protein, called large T antigen, to sequester both Rb and p53; other related DNA tumor viruses use two separate viral proteins (E6 and E7 in the case of papillomavirus) for the same purpose.

The Molecular Genetics of Cancer

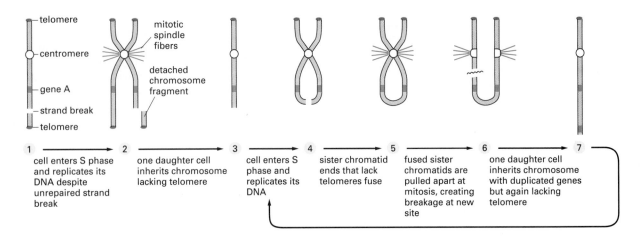

telomere

centromere

gene A

strand break

telomere

mitotic spindle fibers

detached chromosome fragment

1	2	3	4	5	6	7
cell enters S phase and replicates its DNA despite unrepaired strand break	one daughter cell inherits chromosome lacking telomere	cell enters S phase and replicates its DNA	sister chromatid ends that lack telomeres fuse	fused sister chromatids are pulled apart at mitosis, creating breakage at new site	one daughter cell inherits chromosome with duplicated genes but again lacking telomere	

appear normal in all respects except one—they usually develop cancer by the age of 3 months. These observations suggest that *p53* may serve a function that is required only occasionally or in special circumstances. Further evidence supports this idea. When normal cells are exposed to ultraviolet light or to gamma rays, they react by raising their concentration of p53 protein (by reducing the normally rapid rate of degradation of the molecule). The resulting high level of p53 protein blocks cell proliferation: the cells are barred from progressing into the S phase of the cell cycle, and either delay in the G_1 phase or die by apoptosis. Cells lacking the *p53* gene fail to show this response: they carry on dividing, plunging into DNA replication without pausing to repair the breaks and other DNA lesions caused by the damaging radiation. As a result, they either die or, worse, survive and proliferate but with a corrupted genome. A common consequence is that chromosomes become fragmented and rejoined, creating a highly unstable karyotype. This can lead to gene duplications and, through further rounds of cell division, to gene amplifications such as we described earlier for *mdr*1 and for *myc* in tumor cells (Figure 24–34).

It seems, therefore, that the normal function—or at least a normal function—of p53 is to enable cells to cope safely with DNA damage. It is possible that p53 may also act as a check on cell proliferation in other stressful circumstances. In fact, many cancer cells contain large quantities of p53 (of the mutant, ineffectual type), suggesting that the other genetic accidents that occur on the way to cancer are sufficient to call p53 into play. Loss or inactivation of p53 may thus be doubly dangerous in relation to cancer—first, by removing a block to the proliferation of cells that have suffered carcinogenic mutations and, second, by allowing further carcinogenic mutations to be generated when these cells divide. Mutations of the *p53* gene are the most common genetic lesion in human cancer, present in more than 50% of all cases of the disease.

Colorectal Cancers Develop Slowly Via a Succession of Visible Structural Changes [29, 30]

In the first part of this chapter we saw that most cancers develop gradually from a single aberrant cell, progressing from benign to malignant tumors by the accumulation of a half-dozen independent genetic accidents. In the second part of the chapter we have discussed what some of these accidents are in molecular terms. We must now consider how the general principles and the fragments of molecular genetic explanation fit together to give a coherent account of the development of one of the common human cancers. We take **colorectal cancer** as our example, because in this case the steps of tumor progression have been followed *in vivo* as a series of visible structural changes that can be related to specific molecular events.

Colorectal cancers arise from the epithelium lining the colon and rectum (the lower end of the gut). They are common, currently causing over 50,000 deaths

Figure 24–34 How replication of damaged DNA can lead to chromosome abnormalities and gene amplification. The diagram shows one of several possible mechanisms. The process begins with accidental DNA damage in a cell that lacks functional p53 protein. Instead of halting at the p53-dependent checkpoint in the G_1 phase of the division cycle, where a normal cell with damaged DNA would halt until the damage was repaired, the p53-defective cell enters S phase, with the consequences shown. Once a chromosome carrying a duplication and lacking a telomere has been generated, repeated rounds of replication, chromatid fusion, and unequal breakage can increase the number of copies of the duplicated region still further. Selection in favor of cells with increased numbers of copies of a gene in the affected chromosomal region will thus lead to mutants in which the gene is amplified to a high copy number. The multiple copies may eventually become visible as a homogeneously staining region in the chromosome, or they may—either through a recombination event or through unrepaired DNA strand breakage—become excised from their original locus and so appear as independent double minute chromosomes (see Figure 24–20).

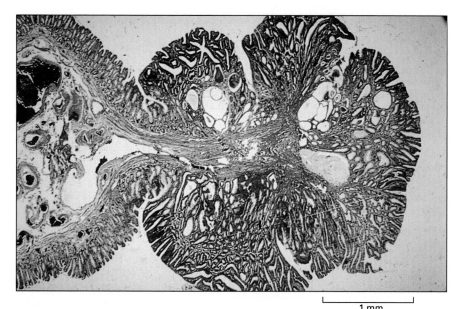

Figure 24–35 **Cross-section of an adenomatous polyp from the colon.** The polyp protrudes into the lumen of the colon. The rest of the wall of the colon is covered with normal colonic epithelium, forming typical short glands; the epithelium on the polyp appears mildly abnormal, forming longer glands. (Courtesy of Anne Campbell.)

1 mm

a year in the United States, or about 11% of total deaths from cancer. Like most cancers, they are not usually diagnosed until late in life (90% after the age of 55). However, routine examination of normal adults with a colonoscope (a fiber-optic device for viewing the interior of the colon and rectum) often reveals a small benign tumor, or *adenoma*, of the gut epithelium forming a protruding mass of tissue called a *polyp* (Figure 24–35). These adenomatous polyps are believed to be the precursors of a large proportion of colorectal cancers. Because progression of the disease is usually very slow, there is typically a period of 10–35 years in which the slowly growing tumor is detectable but has not yet turned malignant. Thus, when people are screened by colonoscopy in their fifties and the polyps are removed—a quick and easy surgical procedure—the subsequent incidence of colorectal cancer is very low—according to some studies, less than a quarter of what it would be otherwise.

Colon cancer provides a clear example of the phenomenon of tumor progression discussed previously. In polyps smaller than 1 cm in diameter, the cells individually and the local details of their arrangement in the epithelium usually appear almost normal. The larger the polyp, the more likely it is to contain cells that look abnormally undifferentiated and form abnormally organized structures. Sometimes, two or more sectors can be distinguished within a single polyp, the cells in one sector appearing relatively normal, those in the other appearing frankly cancerous, as though they have arisen as a mutant subclone within the original clone of adenomatous cells. At later stages in the disease the tumor cells become invasive, first breaking through the epithelial basal lamina, then spreading through the layer of muscle that surrounds the gut, and finally metastasizing to lymph nodes, liver, lung, and other tissues.

Mutations Leading to Colorectal Cancer Can Be Identified by Scanning the Cancer Cells and by Studying Families Prone to the Cancer [29, 31]

What are the mutations that accumulate with time to produce this type of cancer? One approach to answering this question is to use DNA probes for known proto-oncogenes and tumor suppressor genes to test whether any of these have undergone mutation in the colorectal cancer cells. In this way it has been shown that about 75% of colorectal cancers have inactivating mutations in the p53 tumor suppressor gene; about 50% have a point mutation in a *ras* proto-oncogene (usually a specific activating mutation, in codon 12 of the *K-ras* gene); a few percent have an amplified number of copies of a *myc* proto-oncogene; and another

Table 24–7 Some Genetic Abnormalities Detected in Colorectal Cancer Cells

Gene	Chromosome	Tumors with Mutations (%)	Class
K-ras	12	~50	oncogene
neu	17	2	oncogene
myc	8	2	oncogene
APC (adenomatous polyposis coli)	5	>70	tumor suppressor
DCC (deleted in colon carcinoma)	18	>70	tumor suppressor
p53	17	>70	tumor suppressor
HNPCC (hereditary non-polyposis colorectal cancer)	2	~15	tumor suppressor

Source: Modified from J.L. Marx, *Science* 260:751–752, 1992.

few percent have mutations in various other known proto-oncogenes (see Table 24–7).

Another approach is to track down the genetic defect in those rare families that show a hereditary predisposition to colorectal cancer. Chief among these hereditary cancer syndromes is a condition known as *familial adenomatous polyposis coli (APC)*, in which hundreds or thousands of polyps develop along the length of the colon. These make their appearance in early adult life, and if they are not removed, it is almost inevitable that one or more of them will progress to become malignant; on average, 12 years elapse from the first detection of polyps to the diagnosis of cancer. The disease can be traced to deletion or inactivating mutation of a specific gene—the **APC gene** on the long arm of chromosome 5. Individuals with the APC syndrome have inactivating mutations or deletions of the *APC* gene in all the cells of the body. Of the patients with colorectal cancer who do not have the APC syndrome, more than 65% have similar mutations in the cells of the cancer but not in their other tissues. Thus, by a route similar to that which we have discussed for retinoblastoma, mutation of *APC* has been identified as one of the central ingredients of colorectal cancer. The normal function of the APC protein is not known, but it has been found to bind to β-catenin (see Chapter 19), suggesting that it may be involved in some control mechanism based on the sites of anchorage of the cytoskeleton at cell-cell junctions.

Genetic Deletions in Colorectal Cancer Cells Reveal Sites of Loss of Tumor Suppressor Genes [29, 32]

Yet another approach has led to the discovery of further tumor suppressor genes. As explained earlier, the loss of both copies of a tumor suppressor gene often involves a loss of heterozygosity in the chromosomal region where the gene lies: in place of distinct maternal and paternal variants there is only one version of the chromosome sequence or sometimes none at all. Because there is a lot of genetic variability in human populations, it is possible in most individuals to find many differences of sequence between the maternal and the paternal copies of any given chromosome and to design genetic probes that will detect these differences. Using such probes, one can systematically scan the genome of the cells of a tumor looking for loss of a heterozygosity that is detectable in the normal cells of the individual. If a loss of heterozygosity is detected repeatedly at the same chromosomal site in many independently derived tumors, there is a strong presumption that that site normally harbors a tumor suppressor gene whose loss is instrumental in causing the tumors.

A systematic scan of a large number of colorectal cancers in this way reveals several distinct sites where loss of heterozygosity is so common as to suggest the presence of a tumor suppressor gene. The site of the *APC* gene is among them, as is the site of the *p53* gene; another site, showing loss of heterozygosity in more than 70% of the tumors, is occupied by a gene called *DCC* (*deleted in colon carcinoma*). This gene, which was discovered through this method of analysis, codes for what appears to be a transmembrane protein whose extracellular domain has similarities to that of the cell-adhesion molecule N-CAM, suggesting that it may be involved in cell-cell or cell-matrix adhesion or in the reception of signals from the cell's environment. Table 24–7 gives a summary of the main proto-oncogenes and tumor suppressor genes that have been implicated in colorectal cancer.

The Steps of Tumor Progression Can Be Correlated with Specific Mutations [29, 33]

In what sequence do these mutations occur, and what contribution does each of them make to the eventual unruly behavior of the cancer cell? Mutations inactivating the *APC* gene appear to be the first, or at least a very early, step. They can be detected already in small benign polyps at the same high frequency as in large malignant tumors. Their effect seems to be simply to increase the rate of cell proliferation relative to the rate of cell loss, without affecting the way the cells differentiate or the details of the histological pattern they form. Mutations activating the *ras* oncogene come a little later; they are rare in small polyps but common in larger ones that show disturbances of cell differentiation and histological pattern. When malignant colorectal carcinoma cells containing such *ras* mutations are grown in culture, they show typical features of transformed cells, such as the ability to proliferate without anchorage to a substratum; and the cells revert to a nontransformed character and reduce their rate of proliferation if their activated *ras* gene is eliminated.

Mutations in *DCC* and in *p53* come later still. They are rare in polyps but common in the malignant tumors that develop from them. Loss of *p53* seems to relieve the mutant cells of their last inhibitions, and if normal *p53* is transfected back into colorectal carcinoma cells, their proliferation is suppressed. According to one theory, oncogenic mutations of *ras*, like radiation-induced genetic damage, activate an emergency brake on proliferation that depends on p53 protein; only when this brake is disabled by mutation of *p53* can the cells finally give rise to a full-blown cancer. Loss of p53 function is thought to allow the cells not only to divide but also to accumulate further mutations at a rapid rate by progressing through the cell cycle when they are in no fit state to do so. Because these further mutations occur readily, they are probably not rate-limiting events in the incidence of the cancer (see Figure 24–8), but they may have important functional consequences. In this way one arrives at the final malignant tumor, with its many and varied genetic aberrations and its distressing ability to survive by evolving in the face of therapeutic attack (Figure 24–36).

Although mutations in *APC, ras, DCC,* and *p53* may be crucial rate-limiting steps in a large proportion of colorectal cancers, these mutations do not always occur in the same sequence, nor are they the only route to the disease. In about

Figure 24–36 Typical sequence of genetic changes underlying the development of a colorectal carcinoma. Note that the loss of *APC, DCC,* or *p53* generally requires two mutations, to eliminate both copies of the gene. Thus the changes shown here correspond to a total of seven mutations. (After E.R. Fearon and B. Vogelstein, *Cell* 61:759–767, 1990.)

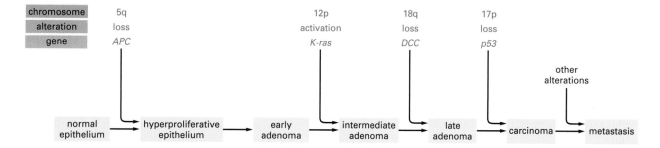

15% of cases, for example, it seems that the development of colorectal cancer is speeded from the outset by another type of mutation, which increases cell mutability by decreasing the fidelity of replication of specific classes of repetitive DNA sequences (such as dinucleotide repeats of the form CACACA . . .) that are scattered throughout the genome. Individuals who inherit this mutation that promotes mutations tend to develop colorectal (and some other) cancers at an early age; their condition is known as *hereditary non-polyposis colorectal cancer,* or *HNPCC.* The *HNPCC* gene is one with an ancient evolutionary history: it is a human homologue of a gene called *mutS* that is found in bacteria and yeast and is required for repair of regions in the DNA where there is a mismatch in the nucleotide sequences of the two strands of the double helix (see Figure 6–50). Such repair mechanisms are clearly of fundamental importance in all living organisms. Tumor cells with the *HNPCC* mutations usually do not show the gross chromosomal instability associated with *p53* mutations: they have found another and more subtle route toward rapid accumulation of the mutations required for cancer.

Other types of cancers that have been analyzed genetically often show an even larger variety of genetic lesions and even greater variation from one case of the disease to another. In the form of lung cancer known as small-cell lung cancer, for example, one finds mutations not only in *ras, p53,* and *APC* but also in *Rb,* in members of the *myc* gene family (in the form of amplification of the number of *myc* gene copies), and in at least five other known proto-oncogenes and tumor suppressor genes. Different combinations of mutations are encountered in different patients and correspond to cancers that react differently to treatment.

Each Case of Cancer Is Characterized by Its Own Array of Genetic Lesions

All medical progress depends on accurate diagnosis. If one cannot identify a disease correctly, one cannot discover its causes, predict its outcome, select the appropriate treatment for a given patient, or make trials on a population of patients to judge whether a proposed treatment is effective. As we have just seen, the traditional classification of cancers is simplistic: a single one of the conventional categories turns out on close scrutiny to be a heterogeneous collection of disorders, with some features in common but each characterized by its own array of genetic lesions (Figure 24–37). Molecular biology is beginning to provide tools to find out precisely which genes are amplified, which are deleted, and which are mutated in the tumor cells of any given patient. Such information may prove to be as important for the management and prevention of cancer as is the identification of microorganisms in patients with infectious diseases.

The discovery of oncogenes and, more recently, of tumor suppressor genes has marked the end of an era of groping in the dark for clues to the biochemical basis of cancer. It has been encouraging to find that there are, after all, some general principles and that some key genetic abnormalities are shared by many forms of the disease. But we are still far from fully understanding the common human cancers. We know the DNA sequences of many oncogenes and proto-oncogenes and of several crucial tumor suppressor genes but the precise physiological functions of only a few. In order to devise effective rational treatments, we need better understanding of how these and other molecules interact to govern the behavior of the individual cell, better understanding of the sociology of cells in tissues, and better understanding of the cell population genetics that govern the genesis of cancer cells through mutation and natural selection.

Looking back on the history of cell biology, we can be hopeful. The desire to understand that drives basic research will surely reveal new ways toward our humanitarian goals, not only in relation to cancer but also in wider areas that we can as yet scarcely foresee.

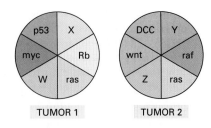

Figure 24–37 Each tumor will generally contain a different set of genetic lesions. In this schematic diagram, W, X, Y, and Z denote alterations in as yet undiscovered tumor suppressor genes or oncogenes. Tumors that arise from different tissues are generally more different in their genetic abnormalities than tumors of similar origin.

Summary

Two classes of genes are critical in the causation of cancer—tumor suppressor genes and proto-oncogenes. Loss-of-function mutations of tumor suppressor genes relieve cells of inhibitions that normally hold their numbers in check; gain-of-function mutations of proto-oncogenes stimulate cells to increase their numbers when they should not. These latter mutations have a dominant effect, and the mutant genes, known as oncogenes, can be identified by their ability to transform the behavior of cells into which they are introduced. Many oncogenes have been tracked down by their presence in transforming retroviruses, which can pick up such dangerously corrupted versions of host cell genes and carry them into other host cells. A large proportion of proto-oncogenes code for components of the pathways by which external signals stimulate cells to divide.

Mutations in tumor suppressor genes are usually recessive in their effects on the individual cell: there is no loss of control until both gene copies are put out of action. People who inherit one defective and one functional gene copy, however, are often strongly predisposed to cancer, since a single somatic mutation is enough, with the inherited mutation, to create a cell that totally lacks the tumor suppressor gene function. Cancers that run in families in this way are rare, but they provide one means to identify tumor suppressor genes whose loss turns out to be a feature of many common cancers. DNA viruses such as papillomaviruses and SV40 can promote the development of cancer by sequestering the products of tumor suppressor genes—in particular, the Retinoblastoma protein, which regulates progress through the cell division cycle in normal circumstances, and the p53 protein, which is thought to act as an emergency brake on cell division in cells that have suffered genetic damage.

The steps of tumor progression can be correlated with mutations that activate specific oncogenes and inactivate specific tumor suppressor genes. A loss of p53 function, for example, is a common late event and may be responsible for the genetic instability of many full-blown metastasizing cancers. Different combinations of mutations are found in different forms of cancer and even in patients that nominally have the same form of the disease, reflecting the random way in which mutations occur. Nevertheless, many of the same types of genetic lesions are encountered repeatedly, suggesting that there is only a limited number of ways in which our defenses against cancer can be breached.

References

General

Brugge, J.; Curran, T.; Harlow, E.; McCormick, F., eds. Origins of Human Cancer. Cold Spring Harbor, NY: Cold Spring Harbor Laboratory Press, 1991.

Cairns, J. Cancer: Science and Society. San Francisco: W.H. Freeman, 1978.

Cancer Biology: Readings from Scientific American. New York: W.H. Freeman, 1986.

De Vita, V.T.; Hellman, S.; Rosenberg, S.A., eds. Cancer: Principles and Practice of Oncology, 4th ed. Philadelphia: Lippincott, 1993.

Franks, L.M.; Teich, N.M., eds. Introduction to the Cellular and Molecular Biology of Cancer, 2nd ed. Oxford, UK: Oxford University Press, 1991.

Varmus, H.; Weinberg, R.A. Genes and the Biology of Cancer. New York: Scientific American Library, 1993.

Cited

1. Cairns, J. Mutation selection and the natural history of cancer. *Nature* 255:197–200, 1975.

   Nowell, P.C. The clonal evolution of tumor cell populations. *Science* 194:23–28, 1976.

2. Doll, R.; Peto, R. Epidemiology of cancer. In Oxford Textbook of Medicine (D.J. Weatherall, J.G.G. Ledingham, D.A. Warrell, eds.), 2nd ed., pp. 4.95–4.123. Oxford, UK: Oxford University Press, 1987.

   McGee, J.O'D.; Isaacson, P.G.; Wright, N.A., eds. Oxford Textbook of Pathology, Vol. 1, pp. 569–717. Oxford, UK: Oxford University Press, 1992.

   Parkin, D.M.; Laara, E.; Muir, C.S. Estimates of the worldwide frequency of sixteen major cancers in 1980. *Int. J. Cancer* 41:184–197, 1988.

   Robbins, S.L.; Cotran, R.S.; Kumar, V. Pathologic Basis of Disease, 4th ed., pp. 239–305. Philadelphia: Saunders, 1989.

3. Fearon, E.R.; Hamilton, S.R.; Vogelstein, B. Clonal analysis of human colorectal tumors. *Science* 238:193–197, 1987.

   Fialkow, P.J. Clonal origin of human tumors. *Biochim. Biophys. Acta* 458:283–321, 1976.

   Groffen, J.J., et al. Philadelphia chromosomal breakpoints are clustered within a limited region, *bcr*, on chromosome 22. *Cell* 36:93–99, 1984.

   Nowell, P.C.; Hungerford, D.A. A minute chromosome in human granulocytic leukemia. *Science* 132:1497, 1960.

4. Ames, B.; Durston, W.E.; Yamasaki, E.; Lee, F.D. Carcinogens are mutagens: a simple test system combining liver homogenates for activation and bacteria for detection. *Proc. Natl. Acad. Sci. USA* 70:2281–2285, 1973.

Ashby, J.; Tennant, R.W. Chemical structure, *Salmonella* mutagenicity and extent of carcinogenicity as indicators of genotoxic carcinogenesis among 222 chemicals tested in rodents by the U.S. NCI/NTP. *Mutation Res.* 204:17–115, 1988.

Case, R.M.P. Tumours of the urinary tract as an occupational disease in several industries. *Ann. R. Coll. Surg. Engl.* 39:213–235, 1966.

Doll, R. Strategy for detection of cancer hazards to man. *Nature* 265:589–597, 1977.

Harnden, D.G. Carcinogenesis. In Oxford Textbook of Pathology (J.O'D. McGee, P.G. Isaacson, N.A. Wright, eds.), Vol. 1, pp. 633–671. Oxford, UK: Oxford University Press, 1992.

5. Armitage, P.; Doll, R. The age distribution of cancer and a multi-stage theory of carcinogenesis. *Br. J. Cancer* 8:1–12, 1954.

Peto, R. Cancer epidemiology, multistage models, and short-term mutagenicity tests. In Origins of Human Cancer (H.H. Hiatt, J.D. Watson, J.A. Winsten, eds.), pp. 1403–1428. Cold Spring Harbor, NY: Cold Spring Harbor Laboratory, 1977.

Vogelstein, B.; Kinzler, K.W. The multistep nature of cancer. *Trends Genet.* 9:138–141, 1993.

6. Cairns, J. Cancer: Science and Society, pp. 144–156. San Francisco: W.H. Freeman, 1978.

Campion, M.J.; McCance, D.J.; Cuzick, J.; Singer, A. Progressive potential of mild cervical atypia: prospective, cytological, colposcopic, and virological study. *Lancet* 2:237–240, 1986.

Farber, E.; Cameron, C. The sequential analysis of cancer development. *Adv. Cancer Res.* 31:125–225, 1980.

Foulds, L. The natural history of cancer. *J. Chronic Dis.* 8:2–37, 1958.

McIndoe, W.A.; McLean, M.R.; Jones, R.W.; Mullins, P.R. The invasive potential of carcinoma *in situ* of the cervix. *Obstet. Gynecol.* 64:451–464, 1984.

7. Ames, B.N.; Gold, L.S. Too many rodent carcinogens: mitogenesis increases mutagenesis. *Science* 249:970–971, 1990.

Schnipper, L.E. Clinical implications of tumor-cell heterogeneity. *N. Engl. J. Med.* 314:1423–1431, 1986.

8. Berenblum, I. A speculative review: the probable nature of promoting action and its significance in the understanding of the mechanism of carcinogenesis. *Cancer Res.* 14:471–477, 1954.

Berenblum, I.; Shubik, P. The role of croton oil applications associated with a single painting of a carcinogen in tumour induction in the mouse's skin. *Br. J. Cancer* 1:379–383, 1947.

Leder, A.; Kuo, A.; Cardiff, R.D.; Sinn, E.; Leder, P. v-Ha-*ras* transgene abrogates the initiation step in mouse skin tumorigenesis: effects of phorbol esters and retinoic acid. *Proc. Natl. Acad. Sci. USA* 87:9178–9182, 1990.

Reddy, A.L.; Fialkow, P.J. Influence of dose of initiator on two-stage skin carcinogenesis in BALB/c mice with cellular mosaicism. *Carcinogenesis* 9:751–754, 1988.

9. Cairns, J. The treatment of diseases and the war against cancer. *Sci. Am.* 253(5):51–59, 1985.

Cohen, L.A. Diet and cancer. *Sci. Am.* 257(5):42–48, 1987.

Doll, R.; Peto, R. The causes of cancer: quantitative estimates of avoidable risks of cancer in the United States today. *J. Natl. Cancer Inst.* 66:1191–1308, 1981.

Enstrom, J.E. Cancer and mortality among active Mormons. *Cancer* 42:1943–1951, 1978.

Henderson, B.E.; Ross, R.K.; Pike, M.C. Hormonal chemoprevention of cancer in women. *Science* 259:633–639, 1993.

10. Boon, T. Teaching the immune system to fight cancer. *Sci. Am.* 266(3):82–89, 1993.

Fentiman, I.S. The local treatment of cancer. In Introduction to the Cellular and Molecular Biology of Cancer (L.M. Franks, N.M. Teich, eds.), 2nd ed., pp. 434–450. Oxford, UK: Oxford University Press, 1991.

Malpas, J. Chemotherapy. In Introduction to the Cellular and Molecular Biology of Cancer (L.M. Franks, N.M. Teich, eds.), 2nd ed., pp. 451–467. Oxford, UK: Oxford University Press, 1991.

Pastan, I.; FitzGerald, D. Recombinant toxins for cancer treatment. *Science* 254:1173–1177, 1991.

Wawrzynczak, E.J.; Thorpe, P.E. Monoclonal antibodies and therapy. In Introduction to the Cellular and Molecular Biology of Cancer (L.M. Franks, N.M. Teich, eds.), 2nd ed., pp. 468–509. Oxford, UK: Oxford University Press, 1991.

11. Sachs, L. Growth, differentiation, and the reversal of malignancy. *Sci. Am.* 254(1):40–47, 1986.

Sawyers, C.L.; Denny, C.T.; Witte, O.N. Leukemia and the disruption of normal hematopoiesis. *Cell* 64:337–350, 1991.

12. Fidler, I.; Hart, I. Biological diversity in metastatic neoplasms: origins and implications. *Science* 217:998–1003, 1982.

Frixen, U.H., et al. E-cadherin-mediated cell-cell adhesion prevents invasiveness of human carcinoma cells. *J. Cell Biol.* 113:173–185, 1991.

Stetler-Stevenson, W.G.; Aznavoorian, S.; Liotta, L.A. Tumor cell interactions with the extracellular matrix during invasion and metastasis. *Annu. Rev. Cell Biol.* 9:541–574, 1993.

Takeichi, M. Cadherins in cancer: implications for invasion and metastasis. *Curr. Opin. Cell Biol.* 5:806–811, 1993.

13. Barnes, D.E.; Lindahl, T.; Sedgwick, B. DNA repair. *Curr. Opin. Cell Biol.* 5:424–433, 1993.

Hanawalt, P.; Sarasin, A. Cancer-prone hereditary diseases with DNA processing abnormalities. *Trends Genet.* 2:124–129, 1986.

Kastan, M.B., et al. A mammalian cell cycle checkpoint pathway utilizing p53 and *GADD45* is defective in ataxia-telangiectasia. *Cell* 71:587–597, 1992.

Solomon, E.; Borrow, J.; Goddard, A.D. Chromosome aberrations and cancers. *Science* 254:1153–1160, 1991.

14. Gerlach, J.H., et al. Homology between P-glycoprotein and a bacterial haemolysin transport protein suggests a model for multidrug resistance. *Nature* 324:485–489, 1986.

Ma, C.; Martin, S.; Trask, B.; Hamlin, J.L. Sister chromatid fusion initiates amplification of the dihydrofolate reductase gene in Chinese hamster cells. *Genes Dev.* 7:605–620, 1993.

Roninson, I.B.; Abelson, H.T.; Housman, D.E.; Howell, N.; Varshavsky, A. Amplification of specific DNA sequences correlates with multi-drug resistance in Chinese hamster cells. *Nature* 309:626–628, 1984.

Smith, K.A.; Stark, M.B.; Gorman, P.A.; Stark, G.R. Fusions near telomeres occur early in the amplification of *CAD* genes in Syrian hamster cells. *Proc. Natl. Acad. Sci. USA* 89:5427–5431, 1992.

Wong, A.J., et al. Gene amplification of c-*myc* and N-*myc* in small cell carcinoma of the lung. *Science* 233:461–464, 1986.

15. Bishop, J.M. Molecular themes in oncogenesis. *Cell* 64:235–248, 1991.

Brugge, J.; Curran, T.; Harlow, E.; McCormick, F., eds. Origins of Human Cancer. Cold Spring Harbor, NY: Cold Spring Harbor Laboratory, 1991.

Harris, H. The genetic analysis of malignancy. *J. Cell Sci.* 4 (Suppl.):431–444, 1986.

Varmus, H.; Weinberg, R.A. Genes and the Biology of Cancer. New York: Scientific American Library, 1993.

16. Wyke, J.A. Viruses and Cancer. In Introduction to the Cellular and Molecular Biology of Cancer (L.M. Franks, N.M. Teich, eds.), 2nd ed., pp. 203–229. Oxford, UK: Oxford University Press, 1991.

zur Hausen, H. Viruses in human cancers. *Science* 254:1167–1173, 1991.

17. Cooper, G.M. Oncogenes. Boston: Jones and Bartlett, 1990.

Kahn, P.; Graf, T., eds. Oncogenes and Growth Control. Berlin: Springer, 1986.

Watson, J.D.; Hopkins, N.H.; Roberts, J.W.; Weiner, A.M. Molecular Biology of the Gene, 4th ed., pp. 961–1096. Menlo Park, CA: Benjamin-Cummings, 1987.

18. Martin, G.S. Rous sarcoma virus: a function required for the maintenance of the transformed state. *Nature* 227:1021–1023, 1970.

Temin, H.M.; Rubin, H. Characteristics of an assay for Rous sarcoma virus and Rous sarcoma cells in tissue culture. *Virology* 6:669–688, 1958.

19. Bishop, J.M. Viral oncogenes. *Cell* 42:23–38; 1985.

Bishop, J.M. Oncogenes. *Sci. Am.* 246(3):80–92, 1982.

Jove, R.; Hanafusa, H. Cell transformation by the viral *src* oncogene. *Annu. Rev. Cell Biol.* 3:31–56, 1987.

Swanstrom, R.; Parker, R.C.; Varmus, H.E.; Bishop, J.M. Transduction of a cellular oncogene—the genesis of Rous sarcoma virus. *Proc. Natl. Acad. Sci. USA* 80:2519–2523, 1983.

20. Rijsewijk, F., et al. The *Drosophila* homolog of the mouse mammary oncogene int-1 is identical to the segment polarity gene wingless. *Cell* 50:649–657, 1987.

Varmus, H.E. The molecular genetics of cellular oncogenes. *Annu. Rev. Genet.* 18:553–612, 1984.

21. Croce, C.M.; Klein, G. Chromosome translocations and human cancer. *Sci. Am.* 252(3):44–50, 1985.

Weinberg, R.A. A molecular basis of cancer. *Sci. Am.* 249(5):126–142, 1983.

22. Cantley, L.C., et al. Oncogenes and signal transduction. *Cell* 64:281–302, 1991.

Downward, J., et al. Close similarity of epidermal growth factor receptor and v-*erb-B* oncogene protein sequences. *Nature* 307:521–527, 1984.

Hunter, T. The proteins of oncogenes. *Sci. Am.* 251(2):70–79, 1984.

Quintanilla, M.; Brown, K.; Ramsden, M.; Balmain, A. Carcinogen-specific mutation and amplification of Ha-*ras* during mouse skin carcinogenesis. *Nature* 322:78–80, 1986.

23. Adams, J.M.; Cory, S. Transgenic models of tumor development. *Science* 254:1161–1167, 1991.

Bissonnette, R.P.; Echeverri, F.; Mahboubi, A.; Green, D.R. Apoptotic cell death induced by c-*myc* is inhibited by *bcl*-2. *Nature* 359:552–554, 1992.

Fanidi, A.; Harrington, E.A.; Evan, G.I. Cooperative interaction between c-*myc* and *bcl*-2 proto-oncogenes. *Nature* 359:554–556, 1992.

Hunter, T. Cooperation between oncogenes. *Cell* 64:249–270, 1991.

Mulligan, L.M., et al. Germ-line mutations of the *RET* proto-oncogene in multiple endocrine neoplasia type 2A. *Nature* 363:458–460, 1993.

24. Haber, D.A.; Housman, D.E. Rate-limiting steps: the genetics of pediatric cancers. *Cell* 64:5–8, 1991.

Knudson, A.G. Hereditary cancer, oncogenes, and antioncogenes. *Cancer Res.* 45:1437–1443, 1985.

Weinberg, R.A. Tumor suppressor genes. *Science* 254:1138–1146, 1991.

25. Hansen, M.F.; Cavenee, W.K. Retinoblastoma and the progression of tumor genetics. *Trends Genet.* 4:125–129, 1988.

Weinberg, R.A. The retinoblastoma gene and cell growth control. *Trends Biochem. Sci.* 15:199–202, 1990.

Weinberg, R.A. Finding the anti-oncogene. *Sci. Am.* 259(3):34–41, 1988.

26. Harbour, J.W., et al. Abnormalities in structure and expression of the human retinoblastoma gene in SCLC. *Science* 241:353–357, 1988. (SCLC = small-cell lung cancer.)

Hollingsworth, R.E.J.; Hensey, C.E.; Lee, W.-H. Retinoblastoma protein and the cell cycle. *Curr. Opin. Genet. Devel.* 3:55–62, 1993.

Lee, E.Y.-H.P., et al. Inactivation of the retinoblastoma susceptibility gene in human breast cancers. *Science* 241:218–221, 1988.

27. Gissman, L. Papillomaviruses and human oncogenesis. *Curr. Opin. Genet. Devel.* 2:97–102, 1992.

Lazo, P.A. Human papillomaviruses in oncogenesis. *Bioessays* 9:158–162, 1988.

Steinberg, B.M.; Brandsma, J.L.; Taichman, L.B. Cancer Cells 5 / Papillomaviruses. Cold Spring Harbor, NY: Cold Spring Harbor Laboratory, 1987.

28. El-Deiry, W.S.; et al. *WAF1*, a potential mediator of p53 tumor suppression. *Cell* 75:817–825, 1993.

Harper, J.W.; et al. The p21 CDK-interacting protein Cip1 is a potent inhibitor of G_1 cyclin-dependent kinases. *Cell* 75:805–816, 1993.

Hartwell, L. Defects in a cell cycle checkpoint may be responsible for the genomic instability of cancer cells. *Cell* 71:543–546, 1992.

Lane, D.P. Worrying about p53. *Curr. Biol.* 2:581–583, 1992.

Livingstone, L.R., et al. Altered cell cycle arrest and gene amplification potential accompany loss of wild-type p53. *Cell* 70:923–935, 1992.

Perry, M.E.; Levine, A.J. Tumor suppressor p53 and the cell cycle. *Curr. Opin. Genet. Devel.* 3:50–54, 1993.

Yin, Y.; Tainsky, M.A.; Bischoff, F.Z.; Strong, L.C.; Wahl, G.M. Wild-type p53 restores cell cycle control and inhibits gene amplification in cells with mutant p53 alleles. *Cell* 70:937–948, 1992.

29. Fearon, E.R.; Vogelstein, B. A genetic model for colorectal tumorigenesis. *Cell* 61:759–767, 1990.

30. Atkin, W.S.; Cuzick, J.; Northover, J.M.A.; Whynes, D.K. Prevention of colorectal cancer by once-only sigmoidoscopy. *Lancet* 341:736–740, 1993.

31. Nishisho, I., et al. Mutations of chromosome 5q21 genes in FAP and colorectal cancer patients. *Science* 253:665–669, 1991.

    Rubinfeld, B.; et al. Association of the APC gene product with β-catenin. *Science* 262:1731–1734, 1993.

    Su, L.-K.; Vogelstein, B.; Kinzler, K. Association of the APC tumor suppressor protein with catenins. *Science* 262:1734–1737, 1993.

32. Fearon, E.R., et al. Identification of a chromosome 18q gene that is altered in colorectal cancers. *Science* 247:49–56, 1990. (*DCC* gene.)

    Vogelstein, B., et al. Allelotype of colorectal carcinomas. *Science* 244:207–211, 1989.

33. Aaltonen, L.A., et al. Clues to the pathogenesis of familial colorectal cancer. *Science* 260:812–816, 1993.

    Baker, S.J.; Markowitz, S.; Fearon, E.R.; Willson, J.K.V.; Vogelstein, B. Suppression of human colorectal carcinoma cell growth by wild-type p53. *Science* 249:912–915, 1990.

    Fishel, R.; et al. The human mutator gene homolog *MSH2* and its association with hereditary nonpolyposis colon cancer. *Cell* 75:1027–1038, 1993.

    Leach, F.S.; et al. Mutations of *mutS* homolog in hereditary nonpolyposis colon cancer. *Cell* 75:1215–1235, 1993.

    Minna, J.D., et al. Molecular genetic analysis reveals chromosomal deletion, gene amplification, and autocrine growth factor production in the pathogenesis of human lung cancer. *Cold Spring Harbor Symp. Quant. Biol.* 51:843–853, 1986.

    Powell, S.M., et al. *APC* mutations occur early during colorectal tumorigenesis. *Nature* 359:235–237, 1992.

    Shirasawa, S.; Furuse, M.; Yokoyama, M.; Sasazuki, T. Altered growth of human colon cancer cell lines disrupted at activated Ki-*ras*. *Science* 260:85–88, 1993.

Glossary

A-kinase (cyclic AMP-dependent protein kinase)
Enzyme that phosphorylates target proteins in response to a rise in intracellular cyclic AMP. First identified in skeletal muscle as part of the pathway of regulation of glycogen breakdown in response to adrenaline.

ABC transporters
Members of a large superfamily of membrane transport proteins that hydrolyze ATP and transfer a diverse array of small molecules across a membrane.

acetyl
Chemical group derived from acetic acid. Acetyl groups are important in metabolism and often added as a covalent modification of proteins.

acetyl CoA
Small water-soluble molecule that carries acetyl groups in cells. Comprises an acetyl group linked to coenzyme A (CoA) by an easily hydrolyzable thioester bond. (*See* Figure 2–20.)

acetylcholine
Neurotransmitter that functions at cholinergic chemical synapses, found both in the brain and in the peripheral nervous system. It is the neurotransmitter at vertebrate neuromuscular junctions. (*See* Figure 15–9.)

acetylcholine receptor
Ion channel that opens in response to acetylcholine binding, thereby converting a chemical signal into an electrical one. Best understood example of a ligand-gated channel. Sometimes called the *nicotinic* acetylcholine receptor to distinguish it from a *muscarinic* receptor, which is a G-protein-linked cell-surface receptor.

acrosomal process
Long, thin, actin-containing spike produced from the head of certain sperm when they make contact with the egg. Seen in sea urchins and other marine invertebrates whose eggs are surrounded by a thick gelatinous coat.

acrosome
Region at the head end of a sperm cell that contains a sac of hydrolytic enzymes used to digest the protective coating of the egg.

actin
Abundant protein that forms actin filaments in all eucaryotic cells. The monomeric form is sometimes called globular or G-actin; the polymeric form is filamentous or F-actin.

actin-binding protein
Protein that associates with either actin monomers or actin filaments in cells and modifies their properties. Examples include myosin, α-actinin, and profilin.

actin filament (microfilament)
Helical protein filament formed by the polymerization of globular actin molecules. A major constituent of the cytoskeleton of all eucaryotic cells and part of the contractile apparatus of skeletal muscle.

action potential
Rapid, transient, self-propagating electrical excitation in the plasma membrane of a cell such as a neuron or muscle cell. Action potentials, or nerve impulses, allow long-distance signaling in the nervous system.

activation energy
Extra energy that must be possessed by atoms or molecules in addition to their ground-state energy in order to undergo a particular chemical reaction. (*See* Figure 2–15.)

active site
Region of an enzyme surface to which a substrate molecule binds in order to undergo a catalyzed reaction.

active transport
Movement of a molecule across a membrane or other barrier driven by energy other than that stored in the concentration gradient or electrochemical gradient of the transported molecule.

acyl group
Functional group derived from a carboxylic acid. (R represents an alkyl group, such as methyl.)

adaptation
Adjustment of sensitivity following repeated stimulation. This is the mechanism that allows a neuron, a photodetector, or a bacterium to react to small changes even against a high background level of stimulation.

adenylyl cyclase (adenylate cyclase)
Membrane-bound enzyme that catalyzes the formation of cyclic AMP from ATP. An important component of some intracellular signaling pathways.

adherens junction
Cell junction in which the cytoplasmic face of the plasma membrane is attached to actin filaments. Examples include the adhesion belts linking adjacent epithelial cells and the focal contacts on the lower surface of cultured fibroblasts.

adhesion belt (zonula adherens)
Beltlike adherens junction that encircles the apical end of an epithelial cell and attaches it to the adjoining cell. A contractile bundle of actin filaments runs along the cytoplasmic surface of the adhesion belt.

adhesion plaque—*see* **focal contact**

adipocyte
A fat cell.

ADP (adenosine 5′-diphosphate)
Nucleotide that is produced by hydrolysis of the terminal phosphate of ATP. It regenerates ATP when phosphorylated by an energy-generating process such as oxidative phosphorylation. (*See* Figure 2–18.)

adrenaline (epinephrine)
Hormone released by chromaffin cells (in the adrenal gland) and by some neurons in response to stress. Produces "fight or flight" responses, including increased heart rate and blood sugar levels.

aerobic
Describes a process that requires, or occurs in the presence of, gaseous oxygen (O_2).

alcohol
Polar organic molecule that contains a functional hydroxyl group (—OH) bound to a carbon atom that is not in an aromatic ring.
An example is ethyl alcohol. CH_3—CH_2—OH

aldehyde
Organic compound that contains the —CH=O group. An example is glyceraldehyde. Can be oxidized to an acid or reduced to an alcohol.

alga (plural algae)
Informal term used to describe a wide range of photosynthetic organisms, either procaryotic or eucaryotic. Eucaryotic examples include *Nitella, Volvox,* and *Fucus.*

alkaloid
Small but complex nitrogen-containing metabolite produced by plants as a defense against herbivores. Examples include caffeine, morphine, and colchicine.

alkane (adjective aliphatic)
Compound of carbon and hydrogen that has only single covalent bonds.
An example is ethane (CH_3CH_3).

alkene
Hydrocarbon with one or more carbon-carbon double bonds.
An example is ethylene.

alkyl group
General term for a group of covalently linked carbon and hydrogen atoms such as methyl (—CH_3) or ethyl (—CH_2CH_3) groups; these groups can be formed by removing a hydrogen atom from an alkane.

allele
One of a set of alternative forms of a gene. In a diploid cell each gene will have two alleles, each occupying the same position (locus) on homologous chromosomes.

allosteric protein
Protein that changes from one conformation to another when it binds another molecule or when it is covalently modified. The change in conformation alters the activity of the protein and can form the basis of directed movement

alpha helix (α helix)
Common structural motif of proteins in which a linear sequence of amino acids folds into a right-handed helix stabilized by internal hydrogen bonding between backbone atoms.

amide
Molecule containing a carbonyl group linked to an amine. Adjacent amino acids in a protein molecule are linked by amide groups.

amino acid
Organic molecule containing both an amino group and a carboxyl group. Those that serve as the building blocks of proteins are alpha amino acids, having both the amino and carboxyl groups linked to the same carbon atom. (*See* Panel 2–5, pp. 56–57.)

amino acyl tRNA
Activated form of amino acid used in protein synthesis. Consists of an amino acid linked through a labile ester bond from its carboxyl group to a hydroxyl group on tRNA. (*See* Figure 6–12.)

amino group
Weakly basic functional group derived from ammonia (NH_3) in which one or more hydrogen atoms are replaced by another atom. In aqueous solution it can accept a proton and carry a positive charge.

amino terminus (N terminus)
The end of a polypeptide chain that carries a free α-amino group.

amoeba (plural amoebae)
(1) Free-living single-celled eucaryote that crawls by changing its shape. (2) More narrowly, a particular genus of protozoa that move in this way.

Amoeba proteus
Species of giant freshwater amoeba widely used in studies of cell locomotion.

amoeboid locomotion
Distinctive form of cell crawling typified by *Amoeba proteus.* Associated with the extension of pseudopodia and with cytoplasmic streaming.

AMP (adenosine 5′-monophosphate)
One of the four nucleotides in an RNA molecule. Two phosphates are added to AMP to form ATP. (*See* Figure 2–30.)

amphipathic
Having both hydrophobic and hydrophilic regions, as in a phospholipid or a detergent molecule.

anabolism
System of biosynthetic reactions in a cell by which large molecules are made from smaller ones.

anaerobic
Describes a cell, organism, or metabolic process that functions in the absence of air or, more precisely, in the absence of molecular oxygen.

anaphase
Stage of mitosis during which the two sets of chromosomes separate and move away from each other. Composed of anaphase A (chromosomes move toward the two spindle poles) and anaphase B (spindle poles move apart).

Ångstrom (Å)
Unit of length used to measure atoms and molecules. Equal to 10^{-10} meter or 0.1 nanometer (nm).

animal pole
In yolky eggs, that end free of yolk and which cleaves more rapidly than the vegetal pole.

anterior
Situated toward the head end of the body.

antibiotic
Substance such as penicillin or streptomycin that is toxic to microorganisms. Usually a product of a specific microorganism or plant.

antibody (immunoglobulin)
Protein produced by B lymphocytes in response to a foreign molecule or invading organism. Often binds to the foreign molecule or cell extremely tightly, thereby inactivating it or marking it for destruction by phagocytosis or complement-induced lysis.

anticodon
Sequence of three nucleotides in a transfer RNA molecule that is complementary to the three-nucleotide codon on a messenger RNA molecule; the anticodon is matched to

a specific amino acid that is covalently attached to the transfer RNA molecule.

antigen
Molecule that provokes an immune response.

antigenic determinant (epitope)
Specific region of an antigenic molecule that binds to an antibody or a T cell receptor.

antiport
Membrane carrier protein that transports two different ions or small molecules across a membrane in opposite directions, either simultaneously or in sequence.

antisense RNA
RNA complementary to a specific RNA transcript of a gene that can hybridize to the specific RNA and block its function.

apical
Describes the tip of a cell, a structure, or an organ. The apical surface of an epithelial cell is the exposed free surface, opposite to the basal surface. The basal surface rests on the basal lamina that separates the epithelium from other tissue.

***Aplysia* (sea hare)**
Marine mollusk that is much used in studies of learning mechanisms.

apolar molecule—*see* **nonpolar molecule**

apoptosis
Programmed cell death.

aqueous
Pertaining to water, as in an aqueous solution.

aromatic
Refers to a molecule that contains carbon atoms in a ring, linked through alternating single and double bonds. Often a molecule related to benzene.

association constant (K_a)
Measure of the association of a complex. For the binding equilibrium $A + B \rightleftharpoons AB$, the association constant is given by $[AB]/[A][B]$, and it is larger the tighter the binding between A and B. (*See also* **dissociation constant.**)

aster
Star-shaped system of microtubules emanating from a centrosome or from a pole of a mitotic spindle.

astrocyte
Category of glial cell in the central nervous system that typically possesses long radial processes. Provides structural support to nerve cells and helps control their chemical and ionic extracellular environment.

ATP (adenosine 5′-triphosphate)
Nucleoside triphosphate composed of adenine, ribose, and three phosphate groups that is the principal carrier of chemical energy in cells. The terminal phosphate groups are highly reactive in the sense that their hydrolysis, or transfer to another molecule, takes place with release of a large amount of free energy. (*See* Figure 2–18.)

ATP synthase
Enzyme complex in the inner membrane of the mitochondrion and the thylakoid membrane of a chloroplast that catalyzes the formation of ATP from ADP and inorganic phosphate during oxidative phosphorylation and photosynthesis, respectively. Also present in the plasma membrane of bacteria.

ATPase
One of a large class of enzymes that catalyze a process that involves the hydrolysis of ATP.

autoantibody
Antibody that reacts with a molecule in the animal that produced the antibody; can cause autoimmune disease.

autocatalysis
Reaction that is catalyzed by one of its products, creating a positive feedback (self-amplifying) effect on the reaction rate.

autoradiography (radioautography)
Technique in which a radioactive object produces an image of itself on a photographic film. The image is called an autoradiograph or autoradiogram.

autosome
Any chromosome other than a sex chromosome.

auxin
Class of plant hormones.

axon
Long nerve cell process that is capable of rapidly conducting nerve impulses over long distances so as to deliver signals to other cells.

axonal transport
Directed transport of organelles and molecules along a nerve cell axon; can be anterograde (outward from the cell body) or retrograde (back toward the cell body).

axoneme
Bundle of microtubules and associated proteins that forms the core of cilia and flagella in a eucaryotic cell and is responsible for their movements.

axoplasm
Cytoplasm of the axon of a nerve cell, especially after it has been extruded from the axon.

B lymphocyte (B cell)
Type of lymphocyte that makes antibodies.

bacterium (plural bacteria)
Common name for any member of the diverse group of procaryotic organisms. Most are single cells, but multicellular forms also exist.

bacteriophage (phage)
Any virus that infects bacteria. They were the first organisms used for the study of molecular genetics and are now widely used as cloning vectors. (From Greek *phagein*, to eat.)

bacteriorhodopsin
Pigmented protein found in the plasma membrane of a salt-loving bacterium, *Halobacterium halobium;* it pumps protons out of the cell in response to light.

basal
Situated near the base. The basal surface of a cell is opposite the apical surface.

basal body
Short cylindrical array of microtubules plus their associated proteins found at the base of a eucaryotic cell cilium or flagellum. Serves as a nucleation site for the growth of the axoneme. Closely similar in structure to a centriole.

basal lamina (plural basal laminae)
Thin mat of extracellular matrix that separates epithelial sheets, and many types of cells such as muscle cells or fat

cells, from connective tissue. Sometimes called a basement membrane.

base
Molecule (usually containing nitrogen) that accepts a proton in solution. Often used to refer to the purines and pyrimidines in DNA and RNA.

base pair
Two nucleotides in an RNA or a DNA molecule that are paired by hydrogen bonds—for example, G with C and A with T or U.

basic
Having the properties of a base.

basophil
White blood cell that releases histamine in an inflammatory response. Closely related to a mast cell.

benzene
Compound composed of a six-membered ring of carbon atoms and containing three double bonds. Occurs as part of many biological molecules.

biosphere
The world of living organisms.

biotin
Low-molecular-weight compound used as a coenzyme. Useful technically as a covalent label for proteins, allowing them to be detected by the egg protein avidin, which binds extremely tightly to biotin. (*See* Figure 2–32.)

blastomere
One of the cells formed by the cleavage of a fertilized egg.

blastula
Early stage of an animal embryo, usually consisting of a hollow ball of cells, before gastrulation begins.

blotting
Biochemical technique in which macromolecules separated on an agarose or polyacrylamide gel are transferred to a sheet of paper, thereby immobilizing them for further analysis.

bond energy
Strength of the chemical linkage between two atoms, measured by the energy in kilocalories or kilojoules needed to break it.

brush border
Dense covering of microvilli on the apical surface of epithelial cells in the intestine and kidney; the microvilli aid absorption by increasing the surface area of the cell.

C-kinase
Ca^{2+}-dependent protein kinase that, when activated by diacylglycerol and an increase in the concentration of Ca^{2+}, phosphorylates target proteins on specific serine and threonine residues.

C terminus—*see* **carboxyl terminus**

Ca^{2+}-release channel
Ion channel in the membrane of the endoplasmic reticulum and sarcoplasmic reticulum (in muscle cells) that releases Ca^{2+} into the cytosol when activated.

cadherin
Member of a family of proteins that mediate Ca^{2+}-dependent cell-cell adhesion in animal tissues.

caged compound
Organic molecule designed to change into an active form when irradiated with light of a specific wavelength. An example is caged ATP.

calmodulin
Ubiquitous Ca^{2+}-binding protein whose binding to other proteins is governed by changes in intracellular Ca^{2+} concentration. Its binding modifies the activity of many target enzymes and membrane transport proteins.

calorie
Unit of heat. One calorie (small "c") is the amount of heat needed to raise the temperature of 1 gram of water by 1°C. A kilocalorie is the unit used to describe the energy content of foods.

Calvin cycle (Calvin-Benson cycle)
Major metabolic pathway by which CO_2 is fixed during photosynthesis.

CAM—*see* **cell-adhesion molecule**

cAMP—*see* **cyclic AMP**

CAP (catabolite gene activator protein)
Gene regulatory protein in procaryotes that, when glucose is absent, activates genes responsible for the breakdown of alternative carbon sources.

capsid
Protein coat of a virus, formed by the self-assembly of one or more protein subunits into a geometrically regular structure.

carbohydrate
General term for sugars and related compounds containing carbon, hydrogen, and oxygen, usually with the empirical formula $(CH_2O)_n$.

carbon fixation
Process by which green plants incorporate carbon atoms from atmospheric carbon dioxide into sugars. The second stage of photosynthesis.

carbonyl group
Pair of atoms consisting of a carbon atom linked to an oxygen atom by a double bond (C=O).

carboxyl group
Carbon atom linked both to an oxygen atom by a double bond and to a hydroxyl group. Molecules containing a carboxyl group are weak (carboxylic) acids.

carboxyl terminus (C terminus)
The end of a polypeptide chain that carries a free α-carbonyl group.

carcinogen
Agent, such as a chemical or a form of radiation, that causes cancer.

carcinoma
Cancer of epithelial cells; the most common form of human cancer.

cardiac muscle
Specialized form of striated muscle found in the heart, consisting of individual heart muscle cells linked together by cell junctions.

carrier protein
Membrane transport protein that binds to a solute and transports it across the membrane by undergoing a series of conformational changes.

cartilage
Form of connective tissue composed of cells (chondrocytes) embedded in a matrix rich in type II collagen and chondroitin sulfate.

catabolism
General term for the enzyme-catalyzed reactions in a cell by which complex molecules are degraded to simpler ones with release of energy. Intermediates in these reactions are sometimes called catabolites.

catalyst
Substance that accelerates a chemical reaction without itself undergoing a change. Enzymes are protein catalysts.

***cdc* gene (cell-division-cycle gene)**
Gene that controls a specific step or set of steps in the cell cycle. Originally identified in yeasts.

Cdk protein—*see* **cyclin-dependent protein kinase**

cDNA—*see* **complementary DNA**

cell-adhesion molecule (CAM)
Protein on the surface of an animal cell that mediates cell-cell binding.

cell body
Main part of a nerve cell that contains the nucleus. The other parts are axons and dendrites.

cell cortex
Specialized layer of cytoplasm on the inner face of the plasma membrane. In animal cells it is an actin-rich layer responsible for cell-surface movements.

cell cycle
Reproductive cycle of the cell: the orderly sequence of events by which the cell duplicates its contents and divides into two.

cell division
Separation of a cell into two daughter cells. In eucaryotic cells it entails division of the nucleus (mitosis) closely followed by division of the cytoplasm (cytokinesis).

cell fusion
Process in which the plasma membranes of two cells break down at the point of contact between them, allowing the two cytoplasms to mingle.

cell junction
Specialized region of connection between two cells or between a cell and the extracellular matrix.

cell line
Population of cells of plant or animal origin capable of dividing indefinitely in culture.

cell locomotion (cell migration)
Active movement of a cell from one location to another, particularly the migration of a cell over a surface.

cell-mediated immunity
Immune responses mediated by T lymphocytes.

cell plate
Flattened membrane-bounded structure that forms from fusing vesicles in the cytoplasm of a dividing plant cell and is the precursor of the new cell wall.

cell wall
Mechanically strong extracellular matrix deposited by a cell outside its plasma membrane. It is prominent in most plants, bacteria, algae, and fungi. Not present in most animal cells.

cellulose
Structural polysaccharide consisting of long chains of covalently linked glucose units. It provides tensile strength in plant cell walls.

central nervous system (CNS)
Main information-processing organ of the nervous system. In vertebrates it consists of the brain and spinal cord.

centriole
Short cylindrical array of microtubules, closely similar in structure to a basal body. A pair of centrioles is usually found at the center of a centrosome in animal cells.

centromere
Constricted region of a mitotic chromosome that holds sister chromatids together; also the site on the DNA where the kinetochore forms and then captures microtubules from the mitotic spindle.

centrosome (cell center)
Centrally located organelle of animal cells that is the primary microtubule organizing center and acts as the spindle pole during mitosis. In most animal cells it contains a pair of centrioles.

chaperone (molecular chaperone)
Protein that helps other proteins avoid misfolding pathways that produce inactive or aggregated states.

checkpoint
Point in the eucaryotic cell-division cycle where progress through the cycle can be halted until conditions are suitable for the cell to proceed to the next stage.

chelate
Combine reversibly, usually with high affinity, with a metal ion such as iron, calcium, or magnesium.

chemiosmotic coupling
Mechanism in which a gradient of hydrogen ions (a pH gradient) across a membrane is used to drive an energy-requiring process, such as ATP production or the rotation of bacterial flagella.

chemotaxis
Motile response of a cell or an organism that carries it toward or away from a diffusible chemical.

Chlamydomonas
Unicellular green alga with two flagella.

chlorophyll
Light-absorbing pigment that plays a central part in photosynthesis.

chloroplast
Specialized organelle in green algae and plants that contains chlorophyll and performs photosynthesis. It is a specialized form of plastid.

cholesterol
Lipid molecule with a characteristic four-ringed steroid structure that is an important component of the plasma membranes of animal cells. (*See* Figure 10–8.)

chromaffin cell
Cell that stores adrenaline in secretory vesicles and secretes it in times of stress when stimulated by the nervous system.

chromatid
One copy of a chromosome formed by DNA replication that is still joined at the centromere to the other copy.

chromatin
Complex of DNA, histones, and nonhistone proteins found in the nucleus of a eucaryotic cell. The material of which chromosomes are made.

chromatography
Biochemical technique in which a mixture of substances is separated by charge, size, or some other property by allowing it to partition between a moving phase and a stationary phase.

chromosome
Structure composed of a very long DNA molecule and associated proteins that carries part (or all) of the hereditary information of an organism. Especially evident in plant and animal cells undergoing mitosis or meiosis, where each chromosome becomes condensed into a compact, readily visible thread.

cilium (plural cilia)
Hairlike extension of a cell containing a core bundle of microtubules and capable of performing repeated beating movements. Cilia are found in large numbers on the surface of many eucaryotic cells, and they are responsible for the swimming of many single-celled organisms.

cisterna (plural cisternae)
Flattened membrane-bounded compartment, as found in the endoplasmic reticulum or Golgi apparatus.

citric acid cycle (TCA, or tricarboxylic acid cycle; Krebs cycle)
Central metabolic pathway found in all aerobic organisms. Oxidizes acetyl groups derived from food molecules to CO_2 and H_2O. Occurs in mitochondria in eucaryotic cells.

clathrin
Protein that assembles into a polyhedral cage on the cytoplasmic side of a membrane so as to form a clathrin-coated pit, which buds off to form a clathrin-coated vesicle.

cleavage
(1) Physical splitting of a cell into two. (2) Specialized type of cell division seen in many early embryos whereby a large cell becomes subdivided into many smaller cells without growth.

clone
Population of cells or organisms formed by repeated (asexual) division from a common cell or organism. Also used as a verb: "to clone a gene" means to produce many copies of a gene by repeated cycles of replication.

cloning vector
Genetic element, usually a bacteriophage or plasmid, that is used to carry a fragment of DNA into a recipient cell for the purpose of gene cloning.

coated pit
Invagination of the plasma membrane associated with a bristlelike layer of protein on its cytoplasmic surface. Pinches off to form a coated vesicle in the process of endocytosis.

coated vesicle
Small membrane-bounded organelle formed by the pinching off of a coated region of membrane. Some coats are made of clathrin, whereas others are made from other proteins.

codon
Sequence of three nucleotides in a DNA or messenger RNA molecule that represents the instruction for incorporation of a specific amino acid into a growing polypeptide chain.

coenzyme
Small molecule tightly associated with an enzyme that participates in the reaction that the enzyme catalyzes, often by forming a transient covalent bond to the substrate. Examples include biotin, NAD^+, and coenzyme A.

coenzyme A
Small molecule used in the enzymatic transfer of acyl groups in the cell. (*See also* **acetyl CoA** and Figure 2–20.)

cofactor
Inorganic ion or coenzyme that is required for an enzyme's activity.

coiled-coil
Especially stable rodlike protein structure formed by two α helices coiled around each other.

collagen
Fibrous protein rich in glycine and proline that is a major component of the extracellular matrix and connective tissues. Exists in many forms: type I, the most common, is found in skin, tendon, and bone; type II is found in cartilage; type IV is present in basal laminae.

combinatorial
Describes any process that is governed by a specific combination of factors (rather than by any single factor), with different combinations having different effects.

complement
System of serum proteins activated by antibody-antigen complexes or by microorganisms. Helps eliminate pathogenic microorganisms by directly causing their lysis or by promoting their phagocytosis.

complementary DNA (cDNA)
DNA molecule made as a copy of mRNA and therefore lacking the introns that are present in genomic DNA. Used to determine the amino acid sequence of a protein by DNA sequencing or to make the protein in large quantities by cloning followed by expression.

complementary nucleotide sequence
Two nucleic acid sequences are said to be complementary if they can form a perfect base-paired double helix with each other.

complex
Assembly of molecules that are held together by noncovalent bonds. Protein complexes perform most cell functions.

conformation
Spatial location of the atoms of a molecule—for example, the precise shape of a protein or other macromolecule in three dimensions.

connective tissue
Any supporting tissue that lies between other tissues and consists of cells embedded in a relatively large amount of extracellular matrix. Includes bone, cartilage, and loose connective tissue.

connexon
Water-filled pore in the plasma membrane formed by a ring of six protein subunits. Part of a gap junction: connexons from two adjoining cells join to form a continuous channel between the two cells.

consensus sequence
Average or most typical form of a sequence that is reproduced with minor variations in a group of related DNA, RNA, or protein sequences. The consensus sequence shows the nucleotide or amino acid most often found at each position. The preservation of a consensus implies that the sequence is functionally important.

constitutive
Produced in constant amount; opposite of regulated. Constitutive secretion, for example, occurs continuously without requiring an external stimulus.

cooperativity
Phenomenon in which the binding of one ligand molecule to a target molecule promotes the binding of successive ligand molecules. Seen in the assembly of large complexes, as well as in enzymes and receptors composed of multiple allosteric subunits, where it sharpens the response to a ligand. (*See* Figure 15–45.)

cosmid
Cloning vector used to carry large segments of DNA into and out of cells; derived from bacteriophage lambda.

co-transport (coupled transport)
Membrane transport process in which the transfer of one molecule depends on the simultaneous or sequential transfer of a second molecule.

coupled reaction
Linked pair of chemical reactions in which the free energy released by one of the reactions serves to drive the other.

covalent bond
Stable chemical link between two atoms produced by sharing one or more pairs of electrons.

crista (plural **cristae**)
(1) One of the folds of the inner mitochondrial membrane. (2) A sensory structure in the inner ear.

critical concentration
Concentration of an unassembled protein, such as actin or tubulin, that is in equilibrium with the assembled form of the protein. (*See* Panel 16–1, pp. 824–825.)

crossing over
Process whereby two homologous chromosomes break at corresponding sites and rejoin to produce two recombined chromosomes. (*See* Figure 6–56.)

cyclic AMP (cAMP)
Nucleotide that is generated from ATP in response to hormonal stimulation of cell-surface receptors. cAMP acts as a signaling molecule by activating A-kinase; it is hydrolyzed to AMP by a phosphodiesterase.

cyclin
Protein that periodically rises and falls in concentration in step with the eucaryotic cell cycle. Cyclins activate crucial protein kinases (called cyclin-dependent protein kinases) and thereby help control progression from one stage of the cell cycle to the next.

cyclin-dependent protein kinase (Cdk protein)
Protein kinase that has to be complexed with a cyclin protein in order to act; different Cdk-cyclin complexes are thought to trigger different steps in the cell-division cycle by phosphorylating specific target proteins.

cytochrome
Colored, heme-containing protein that transfers electrons during cellular respiration and photosynthesis.

cytokeratin
Member of a family of intermediate filament proteins characteristic of epithelial cells.

cytokine
Extracellular signaling protein or peptide that acts as a local mediator in cell-cell communication.

cytokinesis
Division of the cytoplasm of a plant or animal cell into two, as distinct from the division of its nucleus (which is mitosis).

cytokinin
One of a family of small molecules that regulate the growth and development of plant cells.

cytoplasm
Contents of a cell that are contained within its plasma membrane but, in the case of eucaryotic cells, outside the nucleus.

cytoskeleton
System of protein filaments in the cytoplasm of a eucaryotic cell that gives the cell shape and the capacity for directed movement. Its most abundant components are actin filaments, microtubules, and intermediate filaments.

cytosol
Contents of the main compartment of the cytoplasm, excluding membrane-bounded organelles such as endoplasmic reticulum and mitochondria. Originally defined operationally as the cell fraction remaining after membranes, cytoskeletal components, and other organelles have been removed by low-speed centrifugation.

cytotoxic T cell
Type of T lymphocyte responsible for killing infected cells.

dalton
Unit of molecular mass. Approximately equal to the mass of a hydrogen atom (1.66×10^{-24} g).

degenerate
Not a moral judgment but an adjective that describes multiple states that amount to the same thing: different triplet combinations of nucleotide bases (codons) that code for the same amino acid, for example.

denaturation
Dramatic change in conformation of a protein or nucleic acid caused by heating or by exposure to chemicals and usually resulting in the loss of biological function.

dendrite
Extension of a nerve cell, typically branched and relatively short, that receives stimuli from other nerve cells.

deoxyribonucleic acid—*see* **DNA**

desmosome
Specialized cell-cell junction, usually formed between two epithelial cells, characterized by dense plaques of protein into which intermediate filaments in the two adjoining cells insert.

detergent
Type of small amphipathic molecule that tends to coalesce in water, with its hydrophobic tails buried and its hydrophilic heads exposed; widely used to solubilize membrane proteins.

determination
Commitment by an embryonic cell to a particular specialized path of development; it reflects a change in the internal character of the cell.

development
Succession of changes that take place in an organism as a fertilized egg gives rise to an adult plant or animal.

diacylglycerol
Lipid produced by the cleavage of inositol phospholipids in response to extracellular signals. Composed of two fatty acid chains linked to glycerol, it serves as a signaling molecule to help activate protein kinase C.

Dictyostelium discoideum
Cellular slime mold widely used in the study of cell locomotion, chemotaxis, and differentiation.

differentiation
Process by which a cell undergoes a change to an overtly specialized cell type.

diffraction pattern
Pattern set up by wave interference between radiation transmitted or scattered by different parts of an object.

diffusion
Net drift of molecules in the direction of lower concentration due to random thermal movement.

diploid
Containing two sets of homologous chromosomes and hence two copies of each gene or genetic locus.

disaccharide
Carbohydrate molecule consisting of two covalently joined monosaccharide units.

dissociation constant (K_d)
Measure of the tendency of a complex to dissociate. For the binding equilibrium $A + B \rightleftharpoons AB$, the dissociation constant is given by $[A][B]/[AB]$, and it is smaller the tighter the binding between A and B. (*See also* **association constant.**)

disulfide bond (—S—S—)
Covalent linkage formed between two sulfhydryl groups on cysteines. Common way to join two proteins or to link together different parts of the same protein in the extracellular space.

DNA (deoxyribonucleic acid)
Polynucleotide formed from covalently linked deoxyribonucleotide units; serves as the carrier of genetic information.

DNA library
Collection of cloned DNA molecules, representing either an entire genome (genomic library) or DNA copies of the mRNA produced by a cell (cDNA library).

DNA sequencing
Determination of the order of nucleotides in a DNA molecule. (*See* Figures 7–7 and 7–8.)

DNA transcription—*see* transcription

domain
Portion of a protein that has a tertiary structure of its own. In larger proteins each domain is connected to other domains by short flexible regions of polypeptide.

dominant
Refers to the member of a pair of alleles that is expressed in the phenotype of the organism while the other allele is not, even though both alleles are present. Also refers to the phenotype expressed by a dominant allele. Opposite of recessive.

dominant negative mutation
Mutation that dominantly affects the phenotype by means of a defective protein or RNA molecule that interferes with the function of the normal gene product in the same cell.

dorsal
Relating to the back of an animal; also the upper surface of a leaf, wing, etc.

Drosophila melanogaster
Species of small fly, commonly called a fruit fly, much used in genetic studies of development.

dynein
Member of a family of large motor proteins that undergo ATP-dependent movement along microtubules. In the ciliary axoneme, dynein forms the side arms that cause adjacent microtubule doublets to slide past one another.

electrochemical gradient
Driving force that causes an ion to move across a membrane due to the combined influence of a difference in its concentration on the two sides of the membrane and the electrical charge difference across the membrane.

electron acceptor
Atom or molecule that takes up electrons readily, thereby gaining an electron and becoming reduced.

electron carrier
Molecule such as cytochrome c that transfers an electron from a donor molecule to an acceptor molecule.

electron donor
Molecule that easily gives up an electron, becoming oxidized in the process.

electron transport
Movement of electrons from a higher to a lower energy level along a series of electron carrier molecules, as in oxidative phosphorylation and photosynthesis.

elongation factor
Protein required for the addition of amino acids to growing polypeptide chains on ribosomes.

embryogenesis
Development of an embryo from a fertilized egg, or zygote.

endocrine cell
Specialized animal cell that secretes a hormone into the blood; usually part of a gland, such as the thyroid or pituitary gland.

endocytosis
Uptake of material into a cell by an invagination of the plasma membrane and its internalization in a membrane-bounded vesicle. (*See also* **pinocytosis** and **phagocytosis.**)

endoplasmic reticulum (ER)
Labyrinthine, membrane-bounded compartment in the cytoplasm of eucaryotic cells, where lipids are synthesized and membrane-bound proteins are made.

endosome
Membrane-bounded organelle in animal cells that carries materials newly ingested by endocytosis and passes many of them on to lysosomes for degradation.

endothelium
Single sheet of highly flattened cells (endothelial cells) that forms the lining of all blood vessels. Regulates exchanges between the bloodstream and surrounding tissues and is usually surrounded by a basal lamina.

enhancer
Regulatory DNA sequence to which gene regulatory proteins bind, influencing the rate of transcription of a structural gene that can be many thousands of base pairs away.

entropy
Thermodynamic quantity that measures the degree of disorder in a system; the higher the entropy, the more the disorder.

enzyme
Protein that catalyzes a specific chemical reaction.

epimerization
Reaction that alters the steric arrangement around one atom, as in a sugar molecule.

epinephrine—*see* **adrenaline**

epithelium
Coherent cell sheet formed from one or more layers of cells covering an external surface or lining a cavity.

epitope—*see* **antigenic determinant**

equilibrium constant (K)
Ratio of forward and reverse rate constants for a reaction and equal to the association constant. (*See* Figure 3–9.)

erythrocyte (red blood cell)
Small, hemoglobin-containing blood cell of vertebrates that transports oxygen and carbon dioxide to and from tissues. (From Greek *eruthros*, red.)

ER—*see* **endoplasmic reticulum**

Escherichia coli (E. coli)
Rodlike bacterium normally found in the colon of humans and other mammals and widely used in biomedical research.

ester
Molecule formed by the condensation reaction of an alcohol group with an acidic group. Most phosphate groups are esters. (*See* Panel 2–2, pp. 50–51.)

ethyl ($-CH_2CH_3$)
Hydrophobic chemical group derived from ethane (CH_3CH_3).

eucaryote (eukaryote)
Living organism composed of one or more cells with a distinct nucleus and cytoplasm. Includes all forms of life except viruses and bacteria (procaryotes).

euchromatin
Region of an interphase chromosome that stains diffusely; "normal" chromatin, as opposed to the more condensed heterochromatin.

exocytosis
Process by which most molecules are secreted from a eucaryotic cell. These molecules are packaged in membrane-bounded vesicles that fuse with the plasma membrane, releasing their contents to the outside.

exon
Segment of a eucaryotic gene that consists of DNA coding for a sequence of nucleotides in mRNA; an exon can encode amino acids in a protein. Usually adjacent to a noncoding DNA segment called an intron.

expression
Production of an observable phenotype by a gene—usually by the synthesis of a protein.

expression vector
A virus or plasmid that carries a DNA sequence into a suitable host cell and there directs the synthesis of a specific protein.

extracellular matrix (ECM)
Complex network of polysaccharides (such as glycosaminoglycans or cellulose) and proteins (such as collagen) secreted by cells. Serves as a structural element in tissues and also influences their development and physiology.

fatty acid
Compound such as palmitic acid that has a carboxylic acid attached to a long hydrocarbon chain. Used as a major source of energy during metabolism and as a starting point for the synthesis of phospholipids. (*See* Panel 2–4, pp. 54–55.)

fertilization
Fusion of a male and a female gamete (both haploid) to form a diploid zygote, which develops into a new individual.

fibroblast
Common cell type found in connective tissue. Secretes an extracellular matrix rich in collagen and other extracellular matrix macromolecules. Migrates and proliferates readily in wounded tissue and in tissue culture.

fixative
Chemical reagent such as formaldehyde or osmium tetroxide used to preserve cells for microscropy. Samples treated with these reagents are said to be "fixed," and the process is called fixation.

flagellum (plural flagella)
Long, whiplike protrusion whose undulations drive a cell through a fluid medium. Eucaryotic flagella are longer versions of cilia; bacterial flagella are completely different, being smaller and simpler in construction.

fluorescein
Fluorescent dye that fluoresces green when illuminated with blue light or ultraviolet light.

fluorescent dye
Molecule that absorbs light at one wavelength and responds by emitting light at another wavelength; the emitted light is of longer wavelength (and hence of lower energy) than the light absorbed.

focal contact (adhesion plaque)
Small region on the surface of a fibroblast or other cell that is anchored to the extracellular matrix. The attachment is mediated by transmembrane proteins such as integrins, which are linked, through other proteins, to actin filaments in the cytoplasm.

follicle cell
One of the cell types that surround a developing oocyte or egg.

free energy (G)
Energy that can be extracted from a system to drive reactions. Takes into account changes in both energy and entropy.

free-energy change (ΔG)

Change in the free energy during a reaction: the free energy of the product molecules minus the free energy of the starting molecules. A large negative value of ΔG indicates that the reaction has a strong tendency to occur. (*See* Panel 14–1, pp. 668–669.)

G₀ phase

(G-"zero" phase) State of withdrawal from the eucaroytic cell-division cycle by entry into a quiescent G_1 phase; often seen in differentiated cells.

G₁ phase

Gap 1 phase of the eucaryotic cell-division cycle, between the end of cytokinesis and the start of DNA synthesis.

G₂ phase

Gap 2 phase of the eucaryotic cell-division cycle, between the end of DNA synthesis and the beginning of mitosis.

G protein

One of a large family of heterotrimeric GTP-binding proteins that are important intermediaries in cell-signaling pathways. Usually activated by the binding of a hormone or other signaling ligand to a seven-pass transmembrane receptor protein.

GAG (glycosaminoglycan)

Long, linear, highly charged polysaccharide composed of a repeating pair of sugars, one of which is always an amino sugar. Mainly found covalently linked to a protein core in extracellular matrix proteoglycans. Examples include chondroitin sulfate, hyaluronic acid, and heparin.

gamete

Specialized haploid cell, either a sperm or an egg, serving for sexual reproduction.

ganglion (plural ganglia)

Cluster of nerve cells and associated glial cells located outside the central nervous system.

ganglioside

Any glycolipid having one or more sialic acid residues in its structure. Found in the plasma membrane of eucaryotic cells and especially abundant in nerve cells.

gap junction

Communicating cell-cell junction that allows ions and small molecules to pass from the cytoplasm of one cell to the cytoplasm of the next.

gastrula

Animal embryo at an early stage of development where cells are invaginating to form the rudiment of a gut cavity. (From Greek *gaster*, belly.)

gene

Region of DNA that controls a discrete hereditary characteristic, usually corresponding to a single protein or RNA. This definition includes the entire functional unit, encompassing coding DNA sequences, noncoding regulatory DNA sequences, and introns.

gene regulatory protein

General name for any protein that binds to a specific DNA sequence to alter the expression of a gene.

general transcription factor

Any of the proteins whose assembly around the TATA box is required for the initiation of transcription of most eucaryotic genes.

genetic code

Set of rules specifying the correspondence between nucleotide triplets (codons) in DNA or RNA and amino acids in proteins.

genome

The totality of genetic information belonging to a cell or an organism; the DNA that carries this information.

genomic DNA

DNA constituting the genome of a cell or an organism. Often used in contrast with cDNA (DNA prepared by reverse transcription from messenger RNA).

genotype

Genetic constitution of an individual cell or organism.

germ cells

Precursor cells that give rise to gametes.

germ line

The lineage of germ cells (which contribute to the formation of a new generation of organisms), as distinct from somatic cells (which form the body and leave no descendants).

giga-

Prefix denoting 10^9. (From Greek *gigas*, giant.)

glial cells

Supporting cells of the nervous system, including oligodendrocytes and astrocytes in the vertebrate central nervous system and Schwann cells in the peripheral nervous system.

globular protein

Any protein with an approximately rounded shape; contrasts with highly elongated, fibrous proteins such as collagen.

glucose

Six-carbon sugar that plays a major role in the metabolism of living cells. Stored in polymeric form as glycogen in animal cells and as starch in plant cells. (*See* Panel 2–3, pp. 52–53.)

glutaraldehyde

Small reactive molecule with two aldehyde groups that is often used as a cross-linking fixative.

glycerol

Small organic molecule that is the parent compound of many small molecules in the cell, including phospholipids.

CH₂OH
|
CHOH
|
CH₂OH

glycocalyx (cell coat)

Carbohydrate-rich layer that forms the outer coat of a eucaryotic cell. Composed of the oligosaccharides linked to intrinsic plasma membrane glycoproteins and glycolipids, as well as glycoproteins and proteoglycans that have been secreted and reabsorbed onto the cell surface.

glycogen

Polysaccharide composed exclusively of glucose units used to store energy in animal cells. Large granules of glycogen are especially abundant in liver and muscle cells.

glycolipid

Membrane lipid molecule with a short carbohydrate chain attached to a hydrophobic tail.

glycolysis

Ubiquitous metabolic pathway in the cytosol in which sugars are incompletely degraded with production of ATP. (Literally, "sugar splitting.")

glycoprotein

Any protein with one or more covalently linked oligosaccharide chains. Includes most secreted proteins and most proteins exposed on the outer surface of the plasma membrane.

glycosaminoglycan—*see* **GAG**

Golgi apparatus

Membrane-bounded organelle in eucaryotic cells where the proteins and lipids made in the endoplasmic reticulum are modified and sorted.

grana (singular **granum**)

Stacked membrane discs (thylakoids) in chloroplasts that contain chlorophyll and are the site of the light-trapping reactions of photosynthesis.

granulocyte

Category of white blood cell distinguished by conspicuous cytoplasmic granules. Includes neutrophils, basophils, and eosinophils.

gray crescent

Band of pale pigmentation that appears in the egg of some species of amphibian opposite the site of sperm entry following fertilization. Caused by rotation of the egg cortex and associated pigment granules. Marks the future dorsal site.

group (functional group)

Set of covalently linked atoms, such as a hydroxyl group (—OH) or an amino group (—NH$_2$), the chemical behavior of which is well characterized.

growth cone

Migrating motile tip of a growing nerve cell axon or dendrite.

growth factor

Extracellular polypeptide signaling molecule that stimulates a cell to grow or proliferate. Examples are epidermal growth factor (EGF) and platelet-derived growth factor (PDGF). Most growth factors have other actions besides the induction of cell growth or proliferation.

GTP (guanosine 5′-triphosphate)

Major nucleoside triphosphate used in the synthesis of RNA and in some energy-transfer reactions. Has a special role in microtubule assembly, protein synthesis, and cell signaling.

GTPase-activating protein (GAP)

Protein that binds to a Ras or Ras-related GTP-binding protein and inactivates it by stimulating its GTPase activity so that it hydrolyzes its bound GTP to GDP.

guanine nucleotide releasing protein (GNRP)

Protein that binds to a Ras or Ras-related GTP-binding protein and activates it by stimulating it to release its bound GDP and bind GTP in its place.

hair cell

Specialized sensory epithelial cell in the ear with bundles of giant microvilli (stereocilia) protruding from its apical surface. Sound vibrations tilt the stereocilia, evoking an electrical change in the hair cell, which thus acts as a sound detector.

haploid

Having only one set of chromosomes, as in a sperm cell or a bacterium, as distinct from diploid (having two sets of chromosomes).

heat shock protein (stress-response protein)

Protein synthesized in response to an elevated temperature or other stressful treatment; usually helps the cell to survive the stress.

HeLa cell

Line of human epithelial cells that grows vigorously in culture. Derived from a human cervical carcinoma.

α helix—*see* **alpha helix**

heme

Cyclic organic molecule containing an iron atom that carries oxygen in hemoglobin and carries an electron in cytochromes. (*See* Figure 14–28.)

hemidesmosome

Specialized cell junction between an epithelial cell and the underlying basal lamina.

hemoglobin

The major protein in red blood cells that associates with O$_2$ in the lungs by means of a bound heme group.

hemopoiesis (hematopoiesis)

Generation of blood cells, mainly in the bone marrow.

heterocaryon

Cell with two or more nuclei produced by the fusion of two or more different cells.

heterochromatin

Region of a chromosome that remains unusually condensed and transcriptionally inactive during interphase.

heterodimer

Protein complex composed of two different polypeptide chains.

heterozygote

Diploid cell or individual having two different alleles of a specified gene.

high-energy bond

Covalent bond whose hydrolysis releases an unusually large amount of free energy under the conditions existing in a cell. A group linked to a molecule by such a bond is readily transferred from one molecule to another. Examples include the phosphodiester bonds in ATP and the thioester linkage in acetyl CoA.

histamine

Small molecule derived from the amino acid histidine, released from mast cells and basophils in allergic reactions. Causes irritation, dilation of blood vessels, and contraction of smooth muscle.

histone

One of a group of small abundant proteins, rich in arginine and lysine, that are associated with DNA in eucaryotic chromosomes.

homeobox

Short (180 base pairs long) conserved DNA sequence that encodes a DNA-binding motif famous for its presence in genes that are involved in orchestrating the development of a wide range of organisms.

homeodomain

DNA-binding motif of 60 amino acids encoded by a homeobox.

homeotic mutation

Mutation that causes cells in one region of the body to behave as though they were located in another, causing bizarre disturbance of the body plan.

homologous chromosome (homologue)
One of two copies of a particular chromosome in a diploid cell, each copy being derived from a different parent.

homology
Similarity in structure of an organ or a molecule, reflecting a common evolutionary origin. Specifically, such a similarity in protein or nucleic acid sequence. Contrasted with analogy—a similarity that does not reflect a common evolutionary origin.

homozygote
Diploid cell or organism having two identical alleles of a specified gene.

housekeeping gene
Gene serving a function required in all the cell types of an organism, regardless of their specialized role.

hybridization
Process whereby two complementary nucleic acid strands form a double helix during an annealing period; a powerful technique for detecting specific nucleotide sequences.

hybridoma
Cell line used in the production of monoclonal antibodies; obtained by fusing antibody-secreting B lymphocytes with cells of a lymphocyte tumor.

hydrocarbon
Compound that has only carbon and hydrogen atoms.

hydrolysis (adjective **hydrolytic**)
Cleavage of a covalent bond with accompanying addition of water, —H being added to one product of the cleavage and —OH to the other.

hydrophilic
Polar molecule or part of a molecule that forms enough hydrogen bonds to water to dissolve readily in water. (Literally, "water loving.")

hydrophobic (lipophilic)
Nonpolar molecule or part of a molecule that cannot form favorable bonding interactions with water molecules and therefore does not dissolve in water. (Literally, "water hating.")

hydroxyl (—OH)
Chemical group consisting of a hydrogen atom linked to an oxygen, as in an alcohol.

hypertonic
Describes any medium with a sufficiently high concentration of solutes to cause water to move out of a cell due to osmosis. (From Greek *huper,* over.)

hypotonic
Describes any medium with a sufficiently low concentration of solutes to cause water to move into a cell due to osmosis. (From Greek *hupo,* under.)

immortalization
Production of a cell line capable of an unlimited number of cell divisions. Can be the result of a chemical or viral transformation or of fusion with cells of a tumor line.

immune response
Response made by the immune system of a vertebrate when a foreign substance or microorganism enters its body.

immune system
Population of lymphocytes and other white blood cells in the vertebrate body that defends it against infection.

immunoglobulin (Ig)
An antibody molecule. Higher vertebrates have five classes of immunoglobulin—IgA, IgD, IgE, IgG, and IgM—each with a different role in the immune response.

immunoglobulin like (Ig-like) domain
Characteristic protein domain of about 100 amino acids that is found in antibody molecules and in many other proteins that form the Ig superfamily.

***in situ* hybridization**
Technique in which a single-stranded RNA or DNA probe is used to locate a gene or an mRNA molecule in a cell or tissue. (*See also* **hybridization**.)

in vitro
Term used by biochemists to describe a process taking place in an isolated cell-free extract. Also used by cell biologists to refer to cells growing in culture (*in vitro*), as opposed to in an organism (*in vivo*). (Latin for "in glass.")

in vivo
In an intact cell or organism. (Latin for "in life.")

induction (embryonic)
Change in the developmental fate of one tissue caused by an interaction with another tissue.

inflammatory response
Local response of a tissue to injury or infection. Caused by invasion of white blood cells, which release various local mediators such as histamine.

initiation factor
Protein that promotes the proper association of ribosomes with mRNA and is required for the initiation of protein synthesis.

inositol
Cyclic molecule with six hydroxyl groups that forms the hydrophilic head group of inositol phospholipids.

inositol phospholipids (phosphoinositides)
One of a family of lipids containing phosphorylated inositol derivatives. Although minor components of the plasma membrane, they are important in signal transduction in eucaryotic cells. (*See* Figure 15–29.)

insulin
Polypeptide hormone that is secreted by β cells in the pancreas and helps regulate glucose metabolism in animals.

integrin
Member of the large family of transmembrane proteins involved in the adhesion of cells to the extracellular matrix.

interleukin
Secreted peptide or protein that mainly mediates local interactions between white blood cells (leucocytes).

intermediate filament
Fibrous protein filament (about 10 nm in diameter) that forms ropelike networks in animal cells. One of the three most prominent types of cytoskeletal filaments.

internal membrane
Eucaryotic cell membrane other than the plasma membrane. The membranes of the endoplasmic reticulum and the Golgi apparatus are examples.

interphase
Long period of the cell cycle between one mitosis and the next. Includes G_1 phase, S phase, and G_2 phase.

intron
Noncoding region of a eucaryotic gene that is transcribed into an RNA molecule but is then excised by RNA splicing when mRNA is produced.

ion channel
Transmembrane protein complex that forms a water-filled channel across the lipid bilayer through which specific inorganic ions can diffuse down their electrochemical gradients.

ionic bond
Cohesion between two atoms, one with a positive charge, the other with a negative charge. One type of noncovalent bond.

ionophore
Small hydrophobic molecule that dissolves in lipid bilayers and increases their permeability to specific inorganic ions.

IP$_3$ (inositol trisphosphate)
Small water-soluble molecule produced by the cleavage of the inositol phospholipid PIP_2 in response to extracellular signals; causes release of Ca^{2+} from the endoplasmic reticulum. (*See* Figure 15–30.)

isoelectric point
The pH at which a charged molecule in solution has no net electric charge and therefore does not move in an electric field.

isoforms
Multiple forms of the same protein that differ somewhat in their amino acid sequence. They can be produced by different genes or by alternative splicing of RNA transcripts from the same gene.

isomers
Molecules that are formed from the same atoms in the same chemical linkages but have different three-dimensional conformations.

isoprene
Small unsaturated hydrocarbon containing five carbon atoms. The parent compound of isoprenoids.

$$H_2C = C - CH = CH_2$$
$$\quad\ \ |$$
$$\quad\ \ CH_3$$

isoprenoid (polyisoprenoid)
Member of a large family of lipid molecules with a carbon skeleton based on multiple 5-carbon isoprene units. Examples include retinoic acid and cholesterol.

isotope
One of a number of forms of an atom that have the same chemistry but differ in atomic weight. May be either stable or radioactive.

joule
Standard unit of energy in the meter-kilogram-screened system. One joule is the energy delivered in one second by a one-watt power source. Approximately equal to 0.24 calories.

karyotype
Full set of chromosomes of a cell arranged with respect to size, shape, and number.

keratin (cytokeratin)
Member of the family of proteins that form keratin intermediate filaments, mainly in epithelial cells. Some specialized keratins are found in hair, nails, and feathers.

ketone
Organic molecule containing a carbonyl group linked to two alkyl groups.

kilo-
Prefix denoting 10^3.

kilocalorie (kcal)
Unit of heat energy equal to 1000 calories. Often used to express the energy content of food or molecules: bond strengths, for example, are measured in kcal/mole. An alternative unit in wide use is the kilojoule, equal to 0.24 kcal.

kilojoule
Standard unit of energy equal to 1000 joules, or 0.24 kilocalories.

kinesin
One type of motor protein that uses the energy of ATP hydrolysis to move along a microtubule.

kinetochore
Complex structure formed from proteins on a mitotic chromosome to which microtubules attach and which plays an active part in the movement of chromosomes to the pole. The kinetochore forms on the part of the chromosome known as the centromere.

Krebs cycle—*see* **citric acid cycle**

label
Chemical group or radioactive atom added to a molecule in order to follow its progress through a biochemical reaction or to locate it spatially. Also, as a verb, to add such a group or atom to a cell or molecule.

lagging strand
One of the two newly made strands of DNA found at a replication fork. The lagging strand is made in discontinuous lengths that are later joined covalently.

lambda bacteriophage (λ bacteriophage)
Virus that infects *E. coli;* widely used as a DNA cloning vector.

lamin
Extracellular matrix protein found in basal laminae.

lamins
Intermediate filament proteins that form the fibrous matrix (nuclear lamina) on the inner surface of the nuclear envelope.

leading strand
One of the two newly made strands of DNA found at a replication fork. The leading strand is made by continuous synthesis in the 5′-to-3′ direction.

lectin
Protein that binds tightly to a specific sugar. Abundant lectins derived from plant seeds are often used as affinity reagents to purify glycoproteins or to detect them on the surface of cells.

leucine zipper
Structural motif seen in many DNA-binding proteins in which two α helices from separate proteins are joined together in a coiled-coil, forming a protein dimer.

leucocyte—*see* **white blood cell**

ligand
Any molecule that binds to a specific site on a protein or other molecule. (From Latin *ligare*, to bind.)

ligase
Enzyme that joins together (ligates) two molecules in an energy-dependent process. DNA ligase, for example, links two DNA molecules together through a phosphodiester bond.

light chain
One of the smaller polypeptides of a multisubunit protein such as myosin or immunoglobulin.

lineage analysis
Tracing the ancestry of individual cells in a developing embryo.

linkage
(1) Mutual effect of the binding of one ligand on the binding of another that is a central feature of the behavior of all allosteric proteins. (2) Co-inheritance of two genetic loci that lie near each other on the same chromosome; the greater the linkage, the lower the frequency of recombination between the two loci.

lipase
Enzyme that catalyzes the cleavage of fatty acids from the glycerol moiety of a triglyceride.

lipid
Organic molecule that is insoluble in water but dissolves readily in nonpolar organic solvents. One class, the phospholipids, forms the structural basis of biological membranes.

lipid bilayer
Thin bimolecular sheet of mainly phospholipid molecules that forms the structural basis for all cell membranes. The two layers of lipid molecules are packed with their hydrophobic tails pointing inward and their hydrophilic heads outward, exposed to water.

lipophilic—*see* **hydrophobic**

liposome
Artificial phospholipid bilayer vesicle formed from an aqueous suspension of phospholipid molecules.

locus
In genetics, the position of a gene on a chromosome. Different alleles of the same gene all occupy the same locus. (From Latin *locus*, place.)

lumen
Cavity enclosed by an epithelial sheet (in a tissue) or by a membrane (in a cell).

lymph
Colorless fluid derived from blood by filtration through capillary walls. Carries lymphocytes in a special system of ducts and vessels—the lymphatic vessels.

lymphocyte
White blood cell that makes an immune response when activated by a foreign molecule (an antigen). T lymphocytes develop in the thymus and are responsible for cell-mediated immunity. B lymphocytes develop in the bone marrow in mammals and are responsible for the production of circulating antibodies.

lysis
Rupture of a cell's plasma membrane, leading to the release of cytoplasm and the death of the cell.

lysogeny
State of a bacterium in which it carries the DNA of an inactive virus integrated into its genome. The virus can subsequently be activated to replicate and lyse the cell.

lysosome
Membrane-bounded organelle in eucaryotic cells containing digestive enzymes, which are typically most active at the acid pH found in the lumen of lysosomes.

M phase
Period of the eucaryotic cell cycle during which the nucleus and cytoplasm divide.

M-phase-promoting factor—*see* **MPF**

macromolecule
Molecule such as a protein, nucleic acid, or polysaccharide with a molecular mass greater than a few thousand daltons. (Macro from Greek *makros*, large.)

macrophage
White blood cell that is specialized for the uptake of particulate material by phagocytosis.

malignant
Describes tumors and tumor cells that are invasive and/or able to undergo metasis; a malignant tumor is a cancer.

major histocompatibility complex—*see* **MHC**

MAP (microtubule-associated protein)
Any protein that binds to microtubules and modifies their properties. Many different kinds have been found, including structural proteins, such as MAP-2, and motor proteins, such as dynein.

MAP kinase (mitogen-activated protein kinase)
A protein kinase that performs a crucial step in relaying signals from the plasma membrane to the nucleus. Turned on by a wide range of proliferation- or differentiation-inducing signals.

mast cell
Widely distributed tissue cell that releases histamine as part of an inflammatory response. Closely related to blood basophils.

mega-
Prefix denoting 10^6. (From Greek *megas*, huge, powerful.)

meiosis
Special type of cell division by which eggs and sperm cells are produced, involving a diminution in the amount of genetic material. Comprises two successive nuclear divisions with only one round of DNA replication, which produces four haploid daughter cells from an initial diploid cell. (From Greek *meiosis*, diminution.)

melanocyte
Cell that produces the dark pigment melanin; responsible for the pigmentation of skin and hair.

membrane
Double layer of lipid molecules and associated proteins that encloses all cells and, in eucaryotic cells, many organelles; composed of a lipid bilayer and associated proteins.

membrane channel
Transmembrane protein complex that allows inorganic ions or other small molecules to diffuse passively across the lipid bilayer.

membrane potential
Voltage difference across a membrane due to a slight excess of positive ions on one side and of negative ions on the other. A typical membrane potential for an animal cell plasma membrane is –60 mV (inside negative relative to the surrounding fluid).

membrane protein
Protein that is normally closely associated with a cell membrane. (*See* Figure 10–13.)

membrane transport
Movement of molecules across a membrane mediated by a membrane transport protein.

meristem
An organized group of dividing cells whose derivatives give rise to the tissues and organs of a flowering plant. Key examples are the root apical meristem and shoot apical meristem.

mesenchyme
Immature, unspecialized form of connective tissue in animals, consisting of cells embedded in a tenuous extracellular matrix.

messenger RNA—*see* **mRNA**

metabolism
The sum total of the chemical processes that take place in living cells.

metaphase
Stage of mitosis at which chromosomes are firmly attached to the mitotic spindle at its equator but have not yet segregated toward opposite poles.

metaphase plate
Imaginary plane at right angles to the mitotic spindle and midway between the spindle poles; the plane in which chromosomes are positioned at metaphase.

metastasis
Spread of cancer cells from their site of origin to other sites in the body.

methyl (—CH$_3$)
Hydrophobic chemical group derived from methane (CH$_4$).

MHC (major histocompatibility complex)
Complex of vertebrate genes coding for a large family of cell-surface proteins that bind peptide fragments of foreign proteins and present them to T lymphocytes to induce an immune response. (*See* Figure 23–45.)

micro-
Prefix denoting 10^{-6}.

microelectrode (or micropipette)
Piece of fine glass tubing pulled to an even finer tip; used to penetrate a cell to study its physiology or to inject current or molecules.

microfilament—*see* **actin filament**

micrograph
Photograph of an image seen through a microscope. May be either a light micrograph or an electron micrograph depending on the type of microscope employed.

microinjection
Injection of molecules into a cell using a microelectrode.

micron (μm or micrometer)
Unit of measurement often applied to cells and organelles. Equal to 10^{-6} meter or 10^{-4} centimeter.

microtubule
Long, cylindrical structure composed of the protein tubulin. It is one of the three major classes of filaments of the cytoskeleton.

microtubule organizing center (MTOC)
Region in a cell, such as a centrosome or a basal body, from which microtubules grow.

microvillus (plural microvilli)
Thin cylindrical membrane-covered projection on the surface of an animal cell containing a core bundle of actin filaments. Present in especially large numbers on the absorptive surface of intestinal epithelial cells.

milli-
Prefix denoting 10^{-3}.

minus end
The end of a microtubule or actin filament at which the addition of monomers occurs least readily; the "slow-growing" end of the microtubule or actin filament. The minus end of an actin filament is also known as the pointed end. (*See* Panel 16–1, pp. 824–825.)

mitochondrion (plural mitochondria)
Membrane-bounded organelle, about the size of a bacterium, that carries out oxidative phosphorylation and produces most of the ATP in eucaryotic cells.

mitogen
Extracellular substances, such as a growth factor, that stimulates cell proliferation.

mitosis
Division of the nucleus of a eucaryotic cell, involving condensation of the DNA into visible chromosomes. (From Greek *mitos*, a thread, referring to the threadlike appearance of the condensed chromosomes.)

mitotic index
Percentage of cells in a population that are undergoing mitosis at any instant.

mitotic spindle
Array of microtubules and associated molecules that forms between the opposite poles of a eucaryotic cell during mitosis and serves to move the duplicated chromosomes apart.

module
In a protein or nucleic acid, a unit of structure or function that is used in a variety of different contexts.

mole
M grams of a substance, where M is its relative molecular mass (molecular weight); this will be 6×10^{23} molecules of the substance.

molecular weight
Numerically, the same as the relative molecular mass of a molecule expressed in daltons.

molecule
Group of atoms joined together by covalent bonds.

monoclonal antibody
Antibody secreted by a hybridoma clone. Because each such clone is derived from a single B cell, all of the antibody molecules it makes are identical.

monomer
Small molecular building block that can be linked to others of the same type to form a larger molecule (a polymer).

monosaccharide
Simple sugar with the general formula $(CH_2O)_n$, where $n = 3$ to 7.

mosaic
In genetics, an organism made of a mixture of cells with different genotypes.

motif
Element of structure or pattern that recurs in many contexts; specifically, a small structural domain that can be recognized in a variety of proteins.

motor protein
Protein that uses energy derived from nucleoside triphosphate hydrolysis to propel itself along a filament or polymeric molecule.

MPF (M-phase-promoting factor)
Protein complex containing cyclin and a protein kinase that triggers a cell to enter M phase. (Originally called maturation-promoting factor.)

mRNA (messenger RNA)
RNA molecule that specifies the amino acid sequence of a protein. Produced by RNA splicing (in eucaryotes) from a larger RNA molecule made by RNA polymerase as a complementary copy of DNA. It is translated into protein in a process catalyzed by ribosomes.

MTOC—*see* **microtubule-organizing center**

mutation
Heritable change in the nucleotide sequence of a chromosome.

myelin sheath
Insulating layer of specialized cell membrane wrapped around vertebrate axons. Produced by oligodendrocytes in the central nervous system and by Schwann cells in the peripheral nervous system.

myoblast
Mononucleated, undifferentiated muscle precursor cell. A skeletal muscle cell is formed by the fusion of multiple myoblasts.

myofibril
Long, highly organized bundle of actin, myosin, and other proteins in the cytoplasm of muscle cells that contracts by a sliding filament mechanism.

myosin
Type of motor protein that uses ATP to drive movements along actin filaments. Myosin II is a very large protein that forms the thick filaments of skeletal muscle that slide over actin filaments during contraction. Myosin I is smaller, more widely distributed, and not assembled into filaments; it is often membrane-bound.

N terminus—*see* **amino terminus**

Na⁺-K⁺ ATPase—*see* **sodium pump**

NAD⁺ (nicotine adenine dinucleotide)
Coenzyme that participates in an oxidation reaction by accepting a hydride ion (H^-) from a donor molecule. The NADH formed is an important carrier of electrons for oxidative phosphorylation.

NADP⁺ (nicotine adenine dinucleotide phosphate)
Coenzyme closely related to NAD^+ that is used extensively in biosynthetic, rather than catabolic, pathways.

nano-
Prefix denoting 10^{-9}.

nanometer (nm)
Unit of length commonly used to measure molecules and cell organelles. $1 \text{ nm} = 10^{-3} \text{ μm} = 10^{-9} \text{ m}$.

Nernst equation
Quantitative expression that relates the equilibrium ratio of concentrations of an ion on either side of a permeable membrane to the voltage difference across the membrane. (*See* Panel 11–2, p. 526.)

nerve cell—*see* **neuron**

neural crest
Group of embryonic cells derived from the roof of the neural tube that migrate to different locations and give rise to various types of adult cells, including nerve cells in peripheral ganglia, chromaffin cells, melanocytes, and Schwann cells.

neurite
Long process growing from a nerve cell in culture. A generic term that does not specify whether the process is an axon or a dendrite.

neurofilament
Type of intermediate filament found in nerve cells.

neuron (nerve cell)
Cell with long processes specialized to receive, conduct, and transmit signals in the nervous system.

neuropeptide
Peptide secreted by neurons as a signaling molecule either at synapsis or elsewhere.

neurotransmitter
Small signaling molecule secreted by the presynaptic nerve cell at a chemical synapse to relay the signal to the postsynaptic cell. Examples include acetylcholine, glutamate, GABA, glycine, and many neuropeptides.

nicotine adenine dinucleotide—*see* **NAD⁺**

nicotine adenine dinucleotide phosphate—*see* **NADP⁺**

Nitella
Green algae with giant multinucleated cells. Used in studies of plant physiology and actin-based cytoplasmic streaming.

nitrogen cycle
The natural circulation of nitrogen between organic molecules in living organisms and inorganic molecules in the soil.

nitrogen fixation
Biochemical process performed by specific bacteria that reduces atmospheric nitrogen (N_2) to ammonia and hence into various nitrogen-containing metabolites.

nitrogenase complex
Complex of enzymes in nitrogen-fixing bacteria that catalyzes the reduction of atmospheric N_2 to ammonia.

nm—*see* **nanometer**

NMR (nuclear magnetic resonance)
Resonant absorption of electromagnetic radiation at a specific frequency by atomic nuclei in a magnetic field, due to flipping of the orientation of their magnetic dipole moments. The NMR spectrum provides information about the chemical environment of the nuclei. Two-dimensional NMR is used widely to determine the three-dimensional structure of small proteins.

noncovalent bond
Chemical bond in which, in contrast to a covalent bond, no electrons are shared. Noncovalent bonds are relatively weak, but they can sum together to produce strong, highly specific interactions between molecules.

nonpolar molecule (apolar molecule)
Molecule lacking any asymmetric accumulation of positive or negative charge; such molecules are generally insoluble in water.

Northern blotting
Technique in which RNA fragments separated by electrophoresis are immobilized on a paper sheet; a specific molecule is then detected with a labeled nucleic acid probe.

nuclear envelope
Double membrane surrounding the nucleus. Consists of outer and inner membranes perforated by nuclear pores.

nuclear lamina
Fibrous layer on the inner surface of the inner nuclear membrane made up of a network of intermediate filaments formed from nuclear lamins.

nuclear magnetic resonance—*see* **NMR**

nuclear pore
Channel through the nuclear envelope that allows selected molecules to move between the nucleus and cytoplasm.

nucleation
Critical stage in the process of the assembly of a polymer at which a small cluster of monomers aggregates in the correct arrangement to initiate rapid polymerization; more generally, the rate-limiting step in an assembly process.

nucleic acid
RNA or DNA; consists of a chain of nucleotides joined together by phosphodiester bonds.

nucleolar organizer
Region of a chromosome containing a cluster of ribosomal RNA genes that gives rise to a nucleolus.

nucleolus
Structure in the nucleus where ribosomal RNA is transcribed and ribosomal subunits are assembled.

nucleoside
Compound composed of a purine or pyrimidine base linked to either a ribose or a deoxyribose sugar. (*See* Panel 2–6, pp. 58–59.)

nucleosome
Structural, beadlike unit of a eucaryotic chromosome composed of a short length of DNA wrapped around a core of histone proteins; the fundamental subunit of chromatin.

nucleotide
Nucleoside with one or more phosphate groups joined in ester linkages to the sugar moiety. DNA and RNA are polymers of nucleotides.

nucleus
Prominent membrane-bounded organelle in a eucaryotic cell, containing DNA organized into chromosomes.

nurse cell
Cell that is connected by cytoplasmic bridges to an oocyte and thereby provides macromolecules to the growing oocyte.

Okazaki fragments
Short lengths of DNA produced on the lagging strand during DNA replication, discovered by R. Okazaki. They are rapidly joined by DNA ligase to form a continuous DNA strand.

oligodendrocyte
Type of glial cell in the vertebrate central nervous system that forms a myelin sheath around axons.

oligomer
Short polymer, usually consisting (in a cell) of amino acids (oligopeptides), sugars (oligosaccharides), or nucleotides (oligonucleotides). (From Greek *oligos,* few, little.)

oncogene
One of a large number of genes that can help make a cell cancerous. Typically, a mutant form of a normal gene (proto-oncogene) involved in the control of cell growth or division.

oocyte
Developing egg; usually a large and immobile cell.

oogenesis
Formation and maturation of oocytes in the ovary.

operator
Short region of DNA in a bacterial chromosome that controls the transcription of an adjacent gene.

operon
In a bacterial chromosome, a group of contiguous genes that are transcribed into a single mRNA molecule.

Organizer—*see* **Spemann's Organizer**

osmium tetroxide
Inorganic compound (OsO_4) used as a fixative for electron microscopy.

osmosis
Net movement of water molecules across a semipermeable membrane driven by a difference in concentration of solute on either side. The membrane must be permeable to water but not to the solute molecules.

osmotic pressure
Pressure that must be exerted on the high solute concentration side of a semipermeable membrane to prevent the flow of water across the membrane due to osmosis.

oxidation (verb **oxidize**)
Loss of electron density from an atom, as occurs during the addition of oxygen to a molecule or when a hydrogen is removed. Opposite of reduction. (*See* Figure 2–14.)

oxidative phosphorylation
Process in bacteria and mitochondria in which ATP formation is driven by the transfer of electrons from food molecules to molecular oxygen. Involves the intermediate generation of a pH gradient across a membrane and chemiosmotic coupling.

palindromic sequence
Nucleotide sequence that is identical to its complementary strand when each is read in the same chemical direction —for example, GATC.

5′ XXXXGATCXXXX 3′
3′ XXXXCTAGXXXX 5′

patch-clamp recording
Electrophysiological technique in which a tiny electrode tip is sealed onto a patch of cell membrane, thereby

making it possible to record the flow of current through individual ion channels in the patch.

pathogen (adjective **pathogenic**)
An organism or other agent that causes diseases.

PCR (polymerase chain reaction)
Technique for amplifying specific regions of DNA by multiple cycles of DNA polymerization, each followed by a brief heat treatment to separate complementary strands.

peptide bond
Chemical bond between the carbonyl group of one amino acid and the amino group of a second amino acid—a special form of amide linkage. (*See* Figure 2–7.)

peptide map
Characteristic two-dimensional pattern (on paper or gel) formed by the separation of the mixture of peptides produced by the partial digestion of a protein.

peroxisome
Small membrane-bounded organelle that uses molecular oxygen to oxidize organic molecules. Contains some enzymes that produce and others that degrade hydrogen peroxide (H_2O_2).

pH
Common measure of the acidity of a solution: "p" refers to power of 10, "H" to hydrogen. Defined as the negative logarithm of the hydrogen ion concentration in moles per liter (M). Thus pH 3 (10^{-3} M H^+) is acidic and pH 9 (10^{-9} M H^+) is alkaline.

phage—*see* **bacteriophage**

phagocyte
General term for a professional phagocytic cell—that is, a cell such as a macrophage or neutrophil that is specialized to take up particles and microorganisms by phagocytosis.

phagocytosis
Process by which particulate material is endocytosed ("eaten") by a cell. Prominent in carnivorous cells, such as *Amoeba proteus*, and in vertebrate macrophages and neutrophils. (From Greek *phagein*, to eat.)

phenotype
The observable character of a cell or an organism.

phosphatase—*see* **phosphoprotein phosphatase**

phosphatidylinositol
An inositol phospholipid. (*See* Figure 15–29.)

phosphodiester bond
A covalent chemical bond formed when two hydroxyl groups are linked in ester linkage to the same phosphate group, such as adjacent nucleotides in RNA or DNA. (*See* Figure 2–10.)

phosphoinositide—*see* **inositol phospholipid**

phospholipid
The major category of lipid molecules used to construct biological membranes. Generally composed of two fatty acids linked through glycerol phosphate to one of a variety of polar groups.

phosphoprotein phosphatase
Enzyme that removes a phosphate group from a protein by hydrolysis.

phosphorylation
Reaction in which a phosphate group becomes covalently coupled to another molecule.

photon
Elementary particle of light and other electromagnetic radiation.

photosynthesis
Process by which plants and some bacteria use the energy of sunlight to drive the synthesis of organic molecules from carbon dioxide and water.

phylogeny
Evolutionary history of an organism or group of organisms, often presented in chart form as a phylogenetic tree.

pinocytosis
Type of endocytosis in which soluble materials are taken up from the environment and incorporated into vesicles for digestion. Literally, "cell drinking."

plasma membrane
Membrane that surrounds a living cell.

plasmid
Small circular DNA molecule that replicates independently of the genome. Used extensively as a vector for DNA cloning.

plasmodesma (plural **plasmodesmata**)
Communicating cell-cell junction in plants in which a channel of cytoplasm lined by plasma membrane connects two adjacent cells through a small pore in their cell walls.

plastid
Cytoplasmic organelle in plants, bounded by a double membrane, that carries its own DNA and is often pigmented. Chloroplasts are plastids.

platelet
Cell fragment, lacking a nucleus, that breaks off from a megakaryocyte in the bone marrow and is found in large numbers in the bloodstream. It helps initiate blood clotting when blood vessels are injured.

plus end
The end of a microtubule or actin filament at which addition of monomers occurs most readily; the "fast-growing" end of a microtubule or actin filament. The plus end of an actin filament is also known as the barbed end. (*See* Panel 16–1, pp. 824–825.)

point mutation
Change of a single nucleotide in DNA, especially in a region of DNA coding for protein.

polar bond
Covalent bond in which the electrons are attracted more strongly to one of the two atoms, creating a polarized distribution of electric charge.

polar molecule
Molecule in which there is a polarized distribution of positive and negative charges due to an uneven distribution of electrons. Polar molecules are likely to be soluble in water.

polymer
Large molecule made by forming a series of covalent bonds that link multiple identical or similar units (monomers).

polymerase chain reaction—*see* **PCR**

polypeptide
Linear polymer composed of multiple amino acids. Proteins are large polypeptides, and the two terms can be used interchangeably.

polyploid
Describes a cell or an organism that contains more than two sets of homologous chromosomes.

polyribosome (polysome)
mRNA molecule to which are attached a number of ribosomes engaged in protein synthesis.

polysaccharide
Linear or branched polymers of monosaccharides. These include glycogen, hyaluronic acid, and cellulose.

positional information
Information supplied to or possessed by cells according to their position in a multicellular organism. A cell's internal record of its positional information is called its positional value.

posterior
Situated toward the tail end of the body.

posttranslational modification
Enzyme-catalyzed change to a protein made after it is synthesized. Examples are cleavage, glycosylation, phosphorylation, methylation, and prenylation.

prenylation
Covalent attachment of an isoprenoid lipid group to a protein.

primary structure
Sequence of units in a linear polymer, such as the amino acid sequence of a protein.

probe
Defined fragment of RNA or DNA, radioactively or chemically labeled, used to locate specific nucleic acid sequences by hybridization.

procaryote (prokaryote)
Organism made of simple cells that lack a well-defined, membrane-enclosed nucleus: a bacterium or a cyanobacterium.

promoter
Nucleotide sequence in DNA to which RNA polymerase binds to begin transcription.

prophase
First stage of mitosis during which the chromosomes are condensed but not yet attached to a mitotic spindle.

protease (proteinase, proteolytic enzyme)
Enzyme such as trypsin that degrades proteins by hydrolyzing some of their peptide bonds.

proteasome
Type of large protein complex in the cytosol that is responsible for degrading proteins that have been marked for destruction by ubiquitination or by some other means.

protein
The major macromolecular constituent of cells. A linear polymer of amino acids linked together by peptide bonds in a specific sequence.

protein kinase
Enzyme that transfers the terminal phosphate group of ATP to a specific amino acid of a target protein.

protein phosphatase—*see* **phosphoprotein phosphatase**

proteinase—*see* **protease**

proteoglycan
Molecule consisting of one or more glycosaminoglycan (GAG) chains attached to a core protein.

proteolysis
Degradation of a protein, usually by hydrolysis at one or more of its peptide bonds.

proteolytic enzyme—*see* **protease**

proto-oncogene
Normal gene, usually concerned with the regulation of cell proliferation, that can be converted into a cancer-promoting oncogene by mutation.

protozoa
Free-living, nonphotosynthetic, single-celled, motile eucaryotic organisms, especially those, such as *Paramecium* or *Amoeba*, that live by feeding on other organisms.

pseudopodium (plural pseudopodia)
Large cell-surface protrusion formed by amoeboid cells as they crawl. More generally, any dynamic actin-rich extension of the surface of an animal cell.

pump
Transmembrane protein that drives the active transport of ions and small molecules across the lipid bilayer.

purine
One of the two categories of nitrogen-containing ring compounds found in DNA and RNA. Examples are adenine and guanine. (*See* Panel 2–6, pp. 58–59.)

pyrimidine
One of the two categories of nitrogen-containing ring compounds found in DNA and RNA. An example is cytosine. (*See* Panel 2–6, pp. 58–59.)

quaternary structure
Three-dimensional relationship of the different polypeptide chains in a protein complex.

radioactive isotope
Form of an atom with an unstable nucleus that emits radiation as it decays.

radioautography—*see* **autoradiography**

Ras protein
One of a large family of GTP-binding proteins that help relay signals from cell-surface receptors to the nucleus. Named for the *ras* gene, first identified in viruses that cause rat sarcomas.

reaction
In chemistry, any process in which the arrangement of atoms into molecules is changed.

reading frame
The phase in which nucleotides are read in sets of three to encode a protein; an mRNA molecule can be read in any one of three reading frames. (*See* Figure 3–17.)

receptor
Protein that binds a specific extracellular signaling molecule (ligand) and initiates a response in the cell. Cell-surface receptors, such as the acetylcholine receptor and the insulin receptor, are located in the plasma membrane, with their ligand-binding site exposed to the ex-

ternal medium. Intracellular receptors, such as steroid hormone receptors, bind ligands that diffuse into the cell across the plasma membrane.

recessive
Refers to the member of a pair of alleles that fails to be expressed in the phenotype of the organism when the dominant member is present. Also refers to the phenotype of an individual that has only the recessive allele.

recombinant DNA
Any DNA molecule formed by joining DNA segments from different sources. Recombinant DNAs are widely used in the cloning of genes, in the genetic modification of organisms, and in molecular biology generally.

recombination
Process in which chromosomes or DNA molecules are broken and the fragments are rejoined in new combinations. Can occur in the living cell—for example, through crossing-over during meiosis—or in the test tube using purified DNA and enzymes that break and ligate DNA strands.

red blood cell—*see* **erythrocyte**

reduction (verb **reduce**)
Addition of electron density to an atom, as occurs during the addition of hydrogen to a molecule or the removal of oxygen from it. Opposite of oxidation. (*See* Figure 2–14.)

relative molecular mass
Mass of a molecule expressed as a multiple of the mass of a hydrogen atom.

replication fork
Y-shaped region of a replicating DNA molecule at which the two daughter strands are formed and separate.

repressor
Protein that binds to a specific region of DNA to prevent transcription of an adjacent gene.

RER—*see* **rough endoplasmic reticulum**

residue
General term for the unit of a polymer. That portion of a sugar, amino acid, or nucleotide that is retained as part of the polymer chain during the process of polymerization.

respiration
General term for any process in a cell in which the uptake of O_2 molecules is coupled to the production of CO_2.

restriction enzyme (restriction nuclease)
One of a large number of nucleases that can cleave a DNA molecule at any site where a specific short sequence of nucleotides occurs. Extensively used in recombinant DNA technology.

restriction map
Diagrammatic representation of a DNA molecule indicating the sites of cleavage by various restriction enzymes.

retrovirus
RNA-containing virus that replicates in a cell by first making a double-stranded DNA intermediate.

reverse transcriptase
Enzyme, present in retroviruses, that makes a double-stranded DNA copy from a single-stranded RNA template molecule.

RGD
The amino acid sequence arginine-glycine-aspartate (RGD in the single-letter amino acid code), which is present in fibronectin and some other extracellular matrix proteins and is recognized by some integrins that bind these proteins.

ribonuclease
Enzyme that cuts an RNA molecule by hydrolyzing one or more of its phosphodiester bonds.

ribosomal RNA (rRNA)
Any one of a number of specific RNA molecules that form part of the structure of a ribosome and participate in the synthesis of proteins. Often distinguished by their sedimentation coefficient, such as 28S rRNA or 5S rRNA.

ribosome
Particle composed of ribosomal RNAs and ribosomal proteins that associates with messenger RNA and catalyzes the synthesis of protein.

RNA (ribonucleic acid)
Polymer formed from covalently linked ribonucleotide monomers.

RNA polymerase
Enzyme that catalyzes the synthesis of an RNA molecule on a DNA template from nucleoside triphosphate precursors. (*See* Figure 6–5.)

RNA splicing
Process in which intron sequences are excised from RNA molecules in the nucleus during formation of messenger RNA.

rough endoplasmic reticulum (RER)
Region of the endoplasmic reticulum associated with ribosomes; involved in the synthesis of secreted and membrane-bound proteins.

rRNA—*see* **ribosomal RNA**

S phase
Period of a eucaryotic cell cycle in which DNA is synthesized.

Saccharomyces
Genus of yeasts that reproduce asexually by budding or sexually by conjugation. Economically important in brewing and baking, they are also widely used in genetic engineering and as simple model organisms in the study of eucaryotic cell biology.

Salmonella
Rod-shaped, motile, aerobic genus of bacteria. Includes species that cause food poisoning.

sarcoma
Cancer of connective tissue.

sarcomere
Repeating unit of a myofibril in a muscle cell, composed of an array of overlapping thick (myosin) and thin (actin) filaments between two adjacent Z discs.

sarcoplasmic reticulum
Network of internal membranes in the cytoplasm of a muscle cell that contains high concentrations of sequestered Ca^{2+} that is released into the cytosol during muscle excitation.

satellite DNA

Regions of highly repetitive DNA from a eucaryotic chromosome, usually identifiable by its unusual nucleotide composition. Satellite DNA is not transcribed and has no known function.

saturated molecule

Molecule containing carbon-carbon bonds that has only single covalent bonds.

Schwann cell

Glial cell responsible for forming myelin sheaths in the peripheral nervous system.

second messenger

Small molecule that is formed in or released into the cytosol in response to an extracellular signal and helps to relay the signal to the interior of the cell. Examples include cAMP, IP_3, and Ca^{2+}.

secondary structure

Regular local folding pattern of a polymeric molecule; in proteins, α helices and β-pleated sheets.

secretory vesicle

Membrane-bounded organelle in which molecules destined for secretion are stored prior to release. Sometimes called secretory granule because darkly staining contents make the organelle visible as a small solid object.

SER—*see* **smooth endoplasmic reticulum**

serine/threonine kinase

Protein kinase that phosphorylates serines or threonines on its target protein.

Sertoli cell

Supporting cell of the mammalian testis that surrounds and nourishes developing sperm cells.

sex chromosome

Chromosome that may be present or absent, or present in a variable number of copies, according to the sex of the individual; in mammals, the X and Y chromosomes.

signal peptide

Short sequence of amino acids that determines the eventual location of a protein in the cell. An example is the N-terminal sequence of 20 or so amino acids that directs nascent secretory and transmembrane proteins to the endoplasmic reticulum.

signal transduction

Relaying of a signal by conversion from one physical or chemical form to another. In cell biology, the process by which a cell converts an extracellular signal into a response.

signaling molecule

Extracellular or intracellular molecule that cues the response of a cell to the behavior of other cells or objects in the environment.

sister chromatid—*see* **chromatid**

smooth endoplasmic reticulum (SER)

Region of the endoplasmic reticulum not associated with ribosomes; involved in the synthesis of lipids.

smooth muscle

Type of muscle found in the walls of arteries and of the intestine and other viscera, and in some other locations of the vertebrate body. Composed of long, spindle-shaped mononucleate cells. Called "smooth" because it lacks the striations caused by the sarcomeres in skeletal and cardiac muscle cells.

sodium pump (Na+-K+ ATPase)

Transmembrane carrier protein found in the plasma membrane of most animal cells that pumps Na+ out of and K+ into the cell, using the energy derived from ATP hydrolysis.

solute

Any molecule that is dissolved in a liquid. The liquid is called a solvent.

somatic cell

Any cell of a plant or animal other than a germ cell or germ-cell precursor. (From Greek *soma*, body.)

somite

One of a series of paired blocks of mesoderm that form during early development and lie on either side of the notochord in a vertebrate embryo. They give rise to the vertebral column; each somite produces the musculature of one vertebral segment, plus associated connective tissue including that forming the vertebrae to which that musculature is attached.

Southern blotting

Technique in which DNA fragments, separated by electrophoresis, are immobilized on a paper sheet; specific molecules are then detected with a labeled nucleic acid probe. (After E.M. Southern, inventor of the technique.)

Spemann's Organizer

Specialized tissue at the dorsal tip of the blastopore in an amphibian embryo; a source of signals that help to orchestrate formation of the embryonic body axis. (After H. Spemann and H. Mangold, co-discoverers.)

***src* gene**

Name of the first retroviral oncogene discovered (*v-src*) and its precursor proto-oncogene (*c-src*). The product of these genes is a membrane-associated protein kinase that phosphorylates many target proteins on tyrosine residues. (From *sarcoma*, the type of cancer that the *src* virus causes; pronounced "sark.")

starch

Polysaccharide composed exclusively of glucose units, used as an energy store in plant cells.

Start

Important checkpoint in the eucaryotic cell cycle. Passage through Start commits the cell to enter S phase.

stem cell

Relatively undifferentiated cell that can continue dividing indefinitely, throwing off daughter cells that can undergo terminal differentiation into particular cell types.

stereocilium

A large, rigid microvillus found in "organ pipe" arrays on the apical surface of hair cells in the ear. A stereocilium contains a bundle of actin filaments, rather than microtubules, and is thus not a true cilium.

steroid

Hydrophobic molecule related to cholesterol. Many important hormones such as estrogen and testosterone are steroids.

striated muscle

Muscle composed of transversely striped (striated) myofibrils. Skeletal and cardiac muscles of vertebrates are the best-known examples.

stroma

(1) The connective tissue in which a glandular or other epithelium is embedded. (2) The large interior space of a chloroplast, containing enzymes that incorporate CO_2 into sugars.

structural gene

Region of DNA that codes for a protein or for an RNA molecule that forms part of a structure or has an enzymatic function; as distinct from regions of DNA that regulate gene expression.

substrate

Molecule on which an enzyme acts.

substratum

Solid surface to which a cell adheres.

subunit

Component of a multicomponent complex—for example, one protein component of a protein complex.

sulfhydryl (thiol, —SH)

Chemical group containing sulfur and hydrogen found in the amino acid cysteine and other molecules. Two sulfhydryls can join to produce a disulfide bond.

supercoiled DNA

Region of DNA in which the double helix is further twisted on itself. (*See* Figure 9–55.)

symbiosis

Intimate association between two organisms of different species from which both derive a long-term selective advantage.

symport

Form of co-transport in which a membrane carrier protein transports two solute species across the membrane in the same direction. (*See also* **co-transport.**)

synapse

Communicating cell-cell junction that allows signals to pass from a nerve cell to another cell. In a chemical synapse the signal is carried by a diffusible neurotransmitter; in an electrical synapse a direct connection is made between the cytoplasms of the two cells via gap junctions.

synaptonemal complex

Structure that holds paired chromosomes together during prophase I of meiosis and that promotes genetic recombination.

syncytium

Mass of cytoplasm containing many nuclei enclosed by a single plasma membrane. Typically the result either of cell fusion or of a series of incomplete division cycles in which the nuclei divide but the cell does not.

T lymphocyte (T cell)

Type of lymphocyte responsible for cell-mediated immunity; includes both cytotoxic T cells and helper T cells.

TATA box

Consensus sequence in the promoter region of many eucaryotic genes that binds a general transcription factor and hence specifies the position where transcription is initiated.

telomere

End of a chromosome, associated with a characteristic DNA sequence that is replicated in a special way. Counteracts the tendency of the chromosome otherwise to

shorten with each round of replication. (From Greek *telos*, end.)

telophase

Final stage of mitosis in which the two sets of separated chromosomes decondense and become enclosed by nuclear envelopes.

temperature sensitive (ts) mutant

Organism or cell carrying a genetically altered protein (or RNA molecule) that performs normally at one temperature but is abnormal at another (usually higher) temperature.

tertiary structure

Complex three-dimensional form of a macromolecule, especially a protein.

Tetrahymena

Genus of ciliated protozoa used in studies of ciliary axonemes, self-splicing RNA, and telomere reproduction.

thioester bond

High-energy bond formed by a condensation reaction between an acid (acyl) group and a thiol group (—SH); seen, for example, in acetyl CoA and in many enzyme-substrate complexes.

thiol—*see* **sulfhydryl**

thylakoid

Flattened sac of membrane in a chloroplast that contains pigment and carries out the light-gathering reactions of photosynthesis. Stacks of thylakoids form the grana of chloroplasts.

tight junction

Cell-cell junction that seals adjacent epithelial cells together, preventing the passage of most dissolved molecules from one side of the epithelial sheet to the other.

topoisomerase (DNA topoisomerase)

Enzyme that makes reversible cuts in a double helical DNA molecule for the purpose of removing knots or unwinding excessive twists.

tracer

Molecule or atom that has been labeled either chemically or radioactively so that it can be followed in a biochemical process or readily located in a cell or tissue.

transcript

RNA product of DNA transcription.

transcription (DNA transcription)

Copying of one strand of DNA into a complementary RNA sequence by the enzyme RNA polymerase.

transcription factor

Term loosely applied to any protein required to initiate or regulate transcription in eucaryotes. Includes both gene regulatory proteins as well as the general transcription factors. (*See* Figure 9–34.)

transfection

Introduction of a foreign DNA molecule into a eucaryotic cell; usually followed by expression of one or more genes in the newly introduced DNA.

transfer ribonucleic acid—*see* **tRNA**

transformation

Heritable alteration in the properties of a eucaryotic cell. In the case of cultured animal cells, usually refers to the acquisition of cancerlike properties following treatment with a virus or a carcinogen.

transgenic
Describes a plant or animal that has stably incorporated one or more genes from another cell or organism and can pass them on to successive generations.

transition state
Structure that forms transiently in the course of a chemical reaction and has the highest free energy of any reaction intermediate; a rate-limiting step in the reaction.

translation (RNA translation)
Process by which the sequence of nucleotides in a messenger RNA molecule directs the incorporation of amino acids into protein; occurs on a ribosome.

transposable element
Segment of DNA that can move from one position in a genome to another.

triglyceride
Glycerol ester of fatty acids. The main constituent of fat droplets in animal tissues (where the fatty acids are saturated) and of vegetable oil (where the fatty acids are mainly unsaturated).

tRNA (transfer ribonucleic acid)
Set of small RNA molecules used in protein synthesis as an interface (adaptor) between mRNA and amino acids. Each type of tRNA molecule is covalently linked to a particular amino acid.

tyrosine kinase
Enzyme that transfers the terminal phosphate of ATP to a specific tyrosine residue on its target protein.

ubiquitin
Small, highly conserved protein present in all eucaryotic cells that becomes covalently attached to lysines of other proteins. Attachment of a chain of ubiquitins tags a protein for intracellular proteolytic destruction in a proteasome.

unsaturated
Describes a molecule that contains one or more double or triple carbon-carbon bonds, such as isoprene or benzene.

vector
In cell biology, an agent (virus or plasmid) used to transmit genetic material to a cell or organism. (*See also* **cloning vector.**)

vegetal pole
The end at which most of the yolk is located in an animal egg. The end opposite the animal pole.

ventral
Situated toward the belly surface of an animal.

vesicle
Small, membrane-bounded, spherical organelle in the cytoplasm of a eucaryotic cell.

virus
Particle consisting of nucleic acid (RNA or DNA) enclosed in a protein coat and capable of replicating within a host cell and spreading from cell to cell. Often the cause of disease.

V_{max}
Maximum rate of an enzymatic reaction when a substrate is present at saturation levels.

Western blotting
Technique by which proteins are separated and immobilized on a paper sheet and then analyzed, usually by means of a labeled antibody.

white blood cell (leucocyte)
Nucleated blood cell lacking hemoglobin; includes lymphocytes, neutrophils, eosinophils, basophils, and monocytes.

wild-type
Normal, nonmutant form of an organism; the form found in nature (in the wild).

x-ray crystallography
Technique for determining the three-dimensional arrangement of atoms in a molecule based on the diffraction pattern of x-rays passing through a crystal of the molecule.

Xenopus laevis (South African clawed toad)
Species of frog (not a toad) frequently used in studies of early vertebrate development.

yeast
Common term for several families of unicellular fungi. Includes species used for brewing beer and making bread, as well as pathogenic species (that is, species that cause disease).

Z disc (Z line)
Platelike region of a muscle sarcomere to which the plus ends of actin filaments are attached. Seen as a dark transverse line in micrographs.

zinc finger
Structural motif seen in many DNA-binding proteins, composed of a loop of polypeptide chain held in a hairpin bend bound to a zinc atom.

zonula adherens—*see* **adhesion belt**

zygote
Diploid cell produced by fusion of a male and female gamete. A fertilized egg.

Page numbers in **boldface** refer to a major text discussion of the entry; page numbers with an F refer to a figure, with an FF to figures that follow consecutively; page numbers with a T refer to a table; cf. means compare.

cures and treatments, **1267**, 1269, 1270
    multidrug resistance and gene amplifica-
      tion, 1271–1272, 1272F
defects in DNA replication, repair and
    recombination and, 1270–1271
dysplasia, 1262FF, 1263
epidemiology, 1265–1267, 1266T, 1267F
genetic lesions, heterogeneity of, 1290F
growth
    cell differentiation derangements in,
      1268–1269, 1268F
    rate of typical, 1258, 1258F
incidence
    with age, 1260–1261, 1261F
    in different countries, 1265–1266, 1266T
    and mortality, 1257T
incubation times, 1261, 1262F
initiation, 1260–1261, 1261F
metastasis, 1256, 1256F
    steps and mechanism, 1269–1270, 1269FF
as microevolutionary process, **1255–1272,**
    1256FF, 1257T, 1266T
molecular genetics, **1273–1290,** 1274FF,
    1274T, 1276T, 1277T, 1283T, 1288T
monoclonal origin, 1258, 1258F
multidrug resistance, 1271–1272, 1272F
mutation rate and, 244
mutation rate increases on, 1270–1271
mutations
    dominant with oncogenes, 1273
    recessive with tumor suppressor genes,
      1273, 1281, 1282F
p53 and, 889
prevention, 1265, 1267
progression
    cell changes, **1261–1264,** 1262FF
    colorectal cancer, 1287, 1287F, 1289, 1289F
    rate, factors affecting, 1263–1264
ras genes and, 763
Rb gene loss in, 1282
types, 1256, 1257T
    genetic lesions to classify, 1290
of uterine cervix, progression, 1262FF, 1263,
    1268
viruses, role in humans, 1283–1284, 1283T
Cancer cells
    see also Cancer; Transformation, neoplastic
defined, 1256
destabilization of karyotype, 1270–1272,
    1271FF, 1286, 1286F
gene amplification, 1271–1272, 1272F
high mutability of, 1270–1272
integrin phosphorylation, 998, 999F
metastasis and U-PA protease, 994F, 995
metastatic capacity, **1270,** 1270F
multidrug (MDR) resistance protein, 520,
    522T, 1271
during progression, 1261–1264, 1262FF
proliferation and antiproliferation genes,
    900–901, 900F
receptor tyrosine kinase signaling protein
    mutations and, 768–769
CAP, *see* Catabolite gene activator protein
5′ Cap
    in RNA degradation, 369
    RNA export and, 458
    synthesis, 368, 369F
Capacitation, of sperm, 1030
Capillary
    in bone remodeling, 1184, 1184F
    formation of new, 1152–1154, 1152FF
    structure, 1150, 1151F
Capping, 498–499, 499F
Capsid, 275–276, 275F
    genome size limits and, 286
    protein, synthesis and assembly, 279–280,
      279F
    of Semliki forest virus, 278–279, 278F
    structure, 126, 126F
Carbohydrates, *see* Polysaccharides; Sugars
Carbon
    double bonds, 50F
    important properties, 42, 50FF
    skeletons, 50F
Carbon compounds of cells, 42–47, 42FF, 43T,
    50FF, 60, 60F
Carbon cycle, 61, 62, 62F
Carbon dioxide
    band 3 protein and, 494
    in carbon cycle of biosphere, 61–62, 62F
    in carbon-fixation cycle, 688–689, 688FF
    as carbon source, 699–700, 700F
    in citric acid cycle, 71–72, 72F
    in origin of organic molecules, 4, 4F
Carbon dioxide pump, 690, 690F
Carbon fixation
    cycle, 688–689, 689FF, 690, 690F
    evolution, 15–16

reactions, 61, 62F, **687–690,** 688FF, 690, 690F
ribulose 1,5-bisphosphate, 688, 688F
ribulose bisphosphate carboxylase, 688, 688F
Carbon monoxide, as signaling molecule, 728–
    729
Carbon tetrachloride, liver poisoning by, 1150
Carboxyl group, 43, 51F
    in amino acids, 46, 46F, 56F
    transfer by biotin, 77, 78T, 79F, 130
Carboxylic acid group, in fatty acids, 45, 45F, 54F
Carcinogenesis
    see also Cancer; Cancer cells
    correlation with mutagenesis, 1259–1260,
      1259F
    initiation and promotion, **1264–1265,**
      1264FF, 1266
    multiple events for, 1260–1261, 1261F
    by oncogenes, **1280–1281,** 1280F
    p53 protein involvement, 1286, 1286F
Carcinogens
    chemical, 1259–1260, 1259FF
    ionizing radiation, 1259
    metabolic activation of, 1259, 1259F
    as tumor initiators, 1264–1265, 1264FF
    viruses, 1259
Carcinoma
    see also Cancer
    defined, 1256
    diagnosis on keratin type, 798–799
    growth rate of tumor, 1258, 1258F
    incidence and types, 1257, 1257T
    in situ, 1262F, 1263
    metastasis steps, 1269, 1269FF
    of prostate gland, metastases, 1256F
    Rb gene loss in, 1282
Cardiac muscle, 37F, 847, **856,** 856F
    acetylcholine effects, 752
    gap junctions, 960
Cardiac muscle cells, 1175–1176, 1176F
    acetylcholine effect, 727, 727F
    fat droplets in, 659, 659F
    mitochondria of, 655F, 656
    permanence, 1142–1143
    renewal of components, 1145
Cardiolipin, of mitochondria, 656, 714
Carotenoids, 691
Carpel, 1109F, 1116F
Carrier-mediated transport
    kinetics, 512, 512F
    types, 512–513, 513F
Carrier proteins, **512–522,** 512FF, 522T
    ABC transporter superfamily, **519–522,** 522F,
      522T
    asymmetrical distribution of some, 519, 520F
    Ca^{2+} pump, 516, 522T
    defined, 507, **509,** 510F
    as enzymes, 512–513, 512F
    ionophores, 511
    Na^+-driven, **518–519,** 520F, 522T
    Na^+-K^+ pump, **513–516,** 514FF, 522T
    schematic model of action, 513, 513F
    types, 512, 512F
Carrot cell, cellulose microfibrils and growth of,
    1003, 1004F
Cartilage, 975
    aggrecan aggregates in, 977, 977F
    embryonic origin, 1044
    extracellular matrix, 978F
    growth, 1182, 1182F
      erosion and bone deposition in
        development, 1182, 1185, 1185F
Cartilage cells, 1179, 1179F
    see also Chondrocytes
    fibroblasts, conversion to, 1141
Cassette mechanism, of gene regulation, 441–
    442, 442F
Catabolic, defined, 64
Catabolism
    anaerobic, 698–699, 699F
    compartmentalization of pathways, 87, 87F
    of food molecules, **66–74,** 67FF
    glycolysis, 68–70, 69F
    oxidative
      evolution and efficiency, 70
      outline, 70, 70F
    oxidative phosphorylation, 72–73, 73F
    regulation, **82–87,** 83FF
    reversal, 82, 84–86
    stages, outline, 66–67, 67F
Catabolite gene activator protein (CAP)
    cAMP binding, 129, 129F
    DNA bending by, 407F
    DNA sequence recognized, 407T
    domain homologies with other proteins,
      123F
    mechanism, 419–420, 419FF
    models, levels of organization, 118, 118F
    structure, 409, 409F

Catalase, 340T, 574
Catalysis
    allosteric proteins as switches, 198, 198FF
    by RNA, 7–8, 8F, 108–110, 109FF
Catalytic antibody, 132, 132F
Catecholamines, CaM-kinase II and, 750–751,
    751F
β-Catenin, 1288
Catenins, 954, 958T, 966F, 967, 1000T
Cation exchangers, 518–519, 522T
Cation-transporting ATPases, *see* Calcium ion
    ATPase; Sodium-potassium ion
    pump
Caveolae, 635, 636F
CD3 complex, 1240–1241, 1241F, 1242T, 1250F
CD4 protein, 1199, 1234–1235, 1235F, 1242T,
    1250–1251, 1250F
    HIV infection and, 284
CD8 protein, 1199, 1234–1235, 1235F, 1242T,
    1250–1251, 1250F
CD28 protein, 1241–1242, 1241F, 1242T
CD40 protein and ligand, 1244, 1244F
CD45 protein, 768
cdc genes and mutants
    see also individual genes
    defined, 882
    isolation, 883–884, 886
    list of some, with functions, 890T
    selection, 882, 883F
    terminology, 890T
*cdc*2 gene
    see also Cdc2 protein
    *CDC*28 gene interchangeability, 886, 886F
    fission yeast, **883–884,** 883F
Cdc2 protein, 874, 874F
    see also Cdk protein
    activation mechanism, 884, 884F
    cycle of activity in yeast, 886, 887F
    evolutionary relationships, 203F
    in fission vs. budding yeast, 886, 886F
    in MPF, activation and role, 883–884, 883F
    in Start kinase, activation and role, 886–887,
      886FF
    in vertebrate cell cycle, 892
Cdc7 protein, evolutionary relationships, 203F
*cdc*13 gene, 883
Cdc13 protein, homology to mitotic cyclin, 884
*cdc*25 gene, 883, 883F
Cdc25 protein, in MPF activation, 884–885, 884F
*CDC*28 gene, homology to *cdc*2 gene, 886, 886F
Cdk protein (cyclin-dependent protein kinase)
    activation, 205, 205F
    in control of cell cycle, 204, 869–870, 869F
    G_0 phase, lack in, 897
    induction by growth factors, 902
    as integrating switch, 204–205, 204FF, 211,
      211F
    in MPF, 874, 874F
    Rb protein and, 902
    structure, 204, 205F
Cdk2 protein, 203F, 892, 892F
cDNA, *see under* DNA
ced-3, ced-4, ced-9 genes and proteins, 1076, 1076F
Cell
    see also Eucaryotic cell; Procaryotic cell
    age in evolution, 11
    carbon compounds of, 42–47, 42FF, 43T,
      50FF, 60, 60F
    center, microtubules to define, 790, 791F
    chemical components, **41–60,** 42FF, 43T, 90T
    colony formation, 26–27, 27F
    components
      sizes, 140F
      turnover, 179, 1144–1146, 1146F
    defined, 3
    diffusion of molecules in, 95, 95FF
    earliest, 11
    energy from oxidations, 62–63, 63F
    evolution, **3–39,** 4FF
      allosteric changes and, 211
      membranes and, **9–10,** 10F
    fermentation in first, 697–698, 697F
    genetic information, amounts, 102
    glucose as food, 45
    hybrids *in vitro*, 160–162, 160F
    intracellular osmolarity, sources and control,
      515–516, 517F
    levels of organization, 788
    mechanical stability, intermediate filaments
      for, 801–802, 801FF
    metabolic cooperation, 87, 87F
    metabolism rates, 134
    multicellular organisms from, **26–38,** 27FF
    order and energy for, *see* Energy, for
      biological order
    organic compounds, need and sources, 684
    permanent
      in higher vertebrates, **1142–1146,** 1142FF

Page numbers in **boldface** refer to a major text discussion of the entry; F refers to a figure, FF to figures that follow consecutively; T refers to a table; cf. means compare.

I-9

Page numbers in **boldface** refer to a major text discussion of the entry; F refers to a figure, FF to figures that follow consecutively; T refers to a table; cf. means compare.

I-13

angiogenic factors, 1154
in blood vessels, 1150, 1151F
in bone, 1184, 1184F
in liver, 1147, 1149F, 1150
in lymphoid organs, 1202–1203, 1202FF
ICAMs on, 968
in inflammatory response, 1164, 1165F
interleukin effects on, 1245, 1245T
in platelet production, 1165, 1166F
proliferation in adult, 1151–1152
in tissues, 1139–1140, 1140F
Endothermic reactions, 668F
Energetically favorable/unfavorable reactions, 75
Energy
see also Free energy; ΔG
ATP as carrier, 65–66, 65F
for biological order, **60–66,** 61FF
chemical, 61–66, 62F
coupled metabolic reaction, 64–66, 65F
heat energy and, 60–61, 61F
photosynthesis, 61, 62F
thermodynamic analysis of cell and, 60–61, 61F
for biosynthesis, 74–75
ATP hydrolysis, 75–77, 76FF
group transfer reactions, 77, 79F
in polymerization, 80–81, 80FF
reducing power, 79–80, 79F
from catabolism of food molecules, **66–74,** 67FF
see also Catabolism; and individual processes
in coupled reactions, 75–77, 76FF
distribution, 64F
first and second law of thermodynamics and, 668FF
ΔG, **75**
plants to animals, 61–62, 62F
from respiratory chain, 665–671, 665F
Energy conversion, electron-transport chain in, 653–654, 654F
Engineered genes
see also Genetic engineering
to create dominant mutations, 326–327, 326FF
mutant genes, 325–326, 325F
techniques to alter, 323–324, 324F
transgenic animals with, 327–330, 328FF
engrailed gene, 1092, 1092F, 1100, 1100F
DNA sequence, control region, 432F
Engrailed protein, 121, 121F, 1092, 1092F
Enhancers, **422–423,** 422F, 424F
transposable elements as, 394
Enkephalins, 629, 773F
Entactin, 991, 991F
Enthalpy (H), 668FF
Entropy (S), 667, 668FF
env gene, 284F, 1276F
Envelope
proteins
Semliki forest virus, 278–279, 278F
sorting signals, 280, 280F
synthesis and assembly, 279–280, 279F
viral, 276, 275FF
Enzyme catalysis
see also Enzymes
activation energy, 63, 64F
ATP hydrolysis, coupled to, 133–134
coenzymes and, 130, 130FF
diffusion-limited, 134–135
equilibrium point, 133
feedback regulation, **82–86,** 84FF
K_m, 131, 131F
kinetics, 130–131, 131F
rate of, 97
acid and base catalysis, 132, 133F
compartmentalization and, 135, 135F
covalent intermediates with substrates, 133, 134F
multienzyme complexes, 135
transition states, 131–132, 132F
reaction pathways and, 63–64, 64F
regulation, by covalent modification, 86
by RNA, **108–110,** 109FF
substrate binding, 130–131, 131F
turnover number, 131
V_{max}, 131, 131F
Enzyme-linked immunosorbent assays, *see* ELISA
Enzyme-linked receptors, 537
classes, 759
growth factors, *see* Receptor tyrosine kinases
summary, **732–733,** 732FF
Enzymes
see also Active site; Enzyme catalysis; Regulatory site
as catalysts, 63–64, 64F
coenzyme binding to, 130, 130FF

complex, in sequential reactions, 86–87, 86F
concentrations in cell, 135
covalent bonds, making and breaking, 42
evolution of, 9–10, 10F, 13–14, 14F
in maintenance of life, 134
in prebiotic soup, 6
reaction pathways, specificity, 63–64, 64F
substrate, covalent bond with, 133, 134F
Eosinophils, 1162, 1163F, 1164T, 1168F, 1211
CSF effects on, 1171
immature, 1166F
interleukin effects on, 1245, 1245T
Epidermal growth factor, *see* EGF
Epidermis, 1140F
see also Skin
control of growth, 1159–1160
defined, 1154
embryonic origin, 1041, 1044–1045
glands of, 1160, 1161F
homeostatic control, 1159–1160, 1186
keratinocyte maturation, path, timing, molecular changes, 1158, 1158F
of leaf, 686F
maintenance and renewal, 1158–1160, 1159F
mutant keratins in, 801–802, 801F
plant, 28FF
proliferative unit, 1158F
reassembly in culture, 964–965
renewal, 1154–1155, **1158–1160,** 1157FF
stem cells, 1156–1159, 1158F
structure, 1156–1157, 1157FF
Epidermolysis bullosa simplex, 801–802, 801F
Epididymis, 1028
Epimerizations, of proteoglycans, 975
Epinephrine, *see* Adrenaline
Epithelia, 36FF, 950
see also Epithelial cell
actin filament bundles and tension in, 840
adherens junctions, 954–956, 955FF, 961F
basal lamina, 989–991, 989F
as cell filter, 992
blastula, formation of, 1039, 1040F
cancers and cell differentiation, 1268–1269, 1268F
catalogue of human cell types, 1188
cell junctions, summary, 961F
cell proliferation in, control of, 897
connective tissue underlying, 972F
desmosomes in, 954
development, cadherin regulation, 967–968
gap-junction functions in, 960
in gastrulation, sea urchin, 1040–1041, 1041F
hemidesmosomes, in, 956–957, 957F, 961F
mechanism of assembly, 963–964, 964F
in multicellular organisms, **30–32,** 30F
as permeability barrier, 951–953, 951FF
small intestine
cell types in, 1147, 1148F
renewal, 1154, 1155F
T cells in, 1228
tight junctions in, 952–953, 952F, 961F
tube formation in morphogenesis, 954, 955F
Epithelial cell
adhesion belt, 840, 840F
carcinomas, 1256–1257, 1257T
cell division and migration in gut, 895, 896F
endosomal compartments, 625, 626F
growth in culture, anchorage-dependence, 899–900, 899F
half-life, 896F
keratin filaments of, 798–799, 798T
keratins in desmosomes, 799–800, 800F
microvilli, 841–843, 842FF
nutrient transport carriers, 519, 520F
plasma membrane domain
asymmetric distribution, 500–501, 500F
creation, 632–633, 633F
polar structure, 631–632, 633F
semi-intact, preparations, 639F
transcytosis, 623F, 625
of IgA, 1210, 1211F
Epithelial tissue
general organization, 949–950
in skin, 1140F
Epitope, 1201
Epitope tagging, in genetic engineering, 325, 325F
Epstein-Barr virus, 1283T
Equilibrium, free-energy changes and, 667, 669F, 670, 671F
Equilibrium constant (K), 96, 96F
for tubulin or actin polymerization, 824F
Equilibrium point, of chemical reactions, 96, 96F, 97T, 133
Equilibrium potential, for ions, defined and calculated, 526F
Equilibrium sedimentation, 163–164, 164F, 165T
ER, *see* Endoplasmic reticulum

ER mannosidase, 605F
ER resident proteins, 589
ER retention signal, 589, 603, 603F
erbB gene, as proto-oncogene and oncogene, 768–769, 1276T, 1278F, 1279
ERKs, *see* MAP kinases
Erythroblast, 1169, 1169F
Erythrocyte, 37F, 1162, 1162FF, 1164T, 1168F
cell cortex of, 834
freeze-fracture electron microscopy, 494–495, 494F
function, 494
ghosts, 490–491, 490FF
inside-out vesicles, 491, 491F
lysis by complement, 1215, 1215F
membrane proteins, 489–491, 490FF
ankyrin, 492, 493F
band 3, 491F, 492, 493F, **494–495,** 495F
glycophorin, 488F, 491F, 492, 493–494, 495F
spectrin, 491–492, 491FF
osmotic behavior, 516F
precursors, 1166F
production, *see* Erythropoiesis
shape, 490F
spectrin-based cytoskeleton and, 492, 493F
Erythrocyte burst-forming cell (BFC-E), 1170, 1170F
Erythrocyte colony-forming cell (CFC-E), 1170, 1170F, 1171T
Erythromycin, 240T, 704, 704F
Erythropoiesis, **1169–1170,** 1170FF, 1171T
Erythropoietin, **1169–1170,** 1170F, 1171T
actions, specificity, 893, 894T
ES (embryonic stem) cells and cell lines, 1059, 1059F
esc gene, 1096F
Escherichia coli, see E. coli
Establishment methylase, 449
Esters, 51F
Estradiol, response rates to, 756, 756F
Estrogens, as tumor promoters, 1267, 1267F
Ethanol
in fermentation, 70
oxidation in peroxisomes, 575
Ethanolamine, in phospholipids, 45
Ethidium bromide, as DNA stain, 295, 295F
Ethylene, 1005, 1112, 1112F, 1114, 1114F
Ethylenediaminetetraacetic acid, *see* EDTA
Etioplasts, 685
Eubacteria, 13, 13F
Eucaryotes
evolutionary relationships, 38F
haploid and diploid stages, 1012, 1012F
procaryotes, comparison, 21, 22, 22T, 25
protein synthesis, rate, 238–239
protists as examples, 24–25, 24FF
mRNA, compared to procaryote, 237, 237F
transcription and translation, separation in cell, 238
translation, initiation, 461–462, 462FF
Eucaryotic cells
see also Cell; Eucaryotes
age in evolution, 12
chemotaxis, 831–832, 831F
chromosomes, packaging, 25, 26F
compartmentalization
and enzyme catalysis, 135, 135F
of genetic material, 25
cytoskeleton, 23–24, 23F
and evolution of, 787
diagram, 18F
membrane systems, 18FF
membranes, reasons for, 22
mitosis, 25, 26F
organelles, diagrams, micrographs, 17, 17FF
polarization, mechanism, 833–834, 833F
regulation
by GTP-binding proteins, 206–208, 206FF
by protein phosphorylations, 202, 202F, **204–205,** 204FF, 207F, 211, 211F
ribosome components, 233F
mRNA, overview, 105–106, 105F
symbiotic origin, **20–21,** 20FF
types, vertebrate, 34, 35F
Euchromatin, defined, 353
even-skipped (eve) gene and mutant, 426–429, 427FF, 1087F, 1088, 1088F, 1091, 1091FF, 1094
Evolution
see also Genome evolution; Origin of life
of aerobic life and cyanobacteria, 684
of antibody genes, 1218F
carbon and nitrogen fixation, 15–16
of cell compartmentalization, **554–556,** 554FF
of cells, **3–39,** 4FF
CG sequences, loss of, 452, 453F

Ig homology units, 1250F, 1251
IGF-1
    actions, targets, relatives, 894T
    effect on fat cell differentiation, 1181
IGF-1 receptor, 760–761, 760F, 770F
IGF-2, 894T
Igf-2 gene, genomic imprinting and, 451
Iκ-B, 748F
IL-1, 614, 1241, 1245T
IL-2, 767
    actions, targets, 894T
    T cell proliferation mechanism, 1242–1243, 1243F, 1245, 1245T
IL-2 receptors, 767, 768F, 1242–1243, 1243F, 1245T
IL-3
    actions, targets, related factors, 894T
    effect on hemopoietic cells, 1170–1171, 1171F, 1171T
    in erythropoiesis, 1170, 1170F, 1171F
IL-3 receptor, 1171, 1171F, 1171T
IL-4, 1244–1245, 1244F, 1245T
IL-5, 1245, 1245T
Image processing, light microscopy, 146, 147F
Image reconstruction, resolution, 155–156
Imaginal discs, 1097–1098, 1097F
    bristle formation in, 1101–1103, 1101FF
    morphogenetic signals in, 1100
Imago, 1078, 1078F
Imino acid, proline as an, 57F
Immune response
    ABC peptide pump in ER, 521, 522T
    affinity maturation, 1224
    amplification by γ-interferon, 1237, 1245, 1245T
    antibody responses, 1195–1196
    cell-mediated, *see* Cell-mediated immune responses
    clonal selection, **1199–1202**, 1200FF, 1204–1205, 1204F
    complement system, **1213–1215**, 1213FF
    genetic basis of antibody diversity, **1221–1227**, 1222FF
    genetic nonresponders, 1249
    immunological memory, cellular basis, 1203–1204, 1204F
    lymphocyte recirculation through lymphoid tissues, 1202–1203, 1202FF
    polyclonal, oligoclonal or monoclonal, 1201
    primary, 1203–1204, 1204F
        IgM in early stages, 1210, 1215F
    secondary, 1203–1204, 1204F
        IgG secretion in, 1210
    self-tolerance, development of, 1247–1248, 1248F
Immune response (*Ir*) genes, 1248–1249
Immune system, **1195–1251**, 1196FF
    catalogue of cells, 1189
    clonal selection, **1199–1202**, 1200FF, 1204–1205, 1204F
    fundamental features, 1195–1196
    self vs. nonself recognition, 1205–1206, 1205F
Immunity
    cellular basis, **1196–1206**, 1196FF
    discovery, 1195
Immunoblotting, *see* Western blotting
Immunofluorescence, 186, 186FF
Immunoglobulin, 1195
    see also Antibody; Antibody structure; Antibody synthesis; Antigen-antibody interaction; and individual immunoglobulins
    amounts in blood, 1206
    complement activation by, 1214
    domains, MHC molecule homology, 1230F, 1231
    fold, structure, 1219, 1219F
    module, 215, 216F
    ribbon model, light chain, 119F
    scheme, 389F
    structure, β sheets, 114FF
Immunoglobulin A (IgA)
    structure, properties, function, 1208, **1210–1211**, 1211F, 1212T
    synthesis, 1225–1227, 1227F
Immunoglobulin D (IgD)
    structure, properties, function, **1209, 1210**, 1212T
    synthesis, 1209, 1225–1227, 1226F
Immunoglobulin E (IgE)
    structure, properties, function, 1208, **1211**, 1212F, 1212T
    synthesis, 1225–1226, 1227F
    IL-4 effects, 1244, 1245T
Immunoglobulin E (IgE) receptors, in allergic response, 1211, 1212F
Immunoglobulin G (IgG)
    avidity, 1213

structure, properties, function, 1208, **1210**, 1210F, 1212T, 1218–1220, 1219FF
    synthesis, 1225–1227
Immunoglobulin G1 (IgG1), synthesis, IL-4 effects, 1244, 1245T
Immunoglobulin genes
    see also Antibody genes; and individual genes
    evolution by gene duplication, 389
Immunoglobulin M (IgM)
    avidity, 1213
    early production in pre-B cells, monospecificity control, 1225
    structure, properties, function, **1208–1210**, 1210F, 1212T
    synthesis, 1225, **1226–1227**, 1226FF
    synthesis in pre-B and virgin B cells, 1208–1209
Immunoglobulin superfamily, 1219, 1242, 1242T, **1250–1251**, 1250F
    gene evolution, 1250–1251
    in growth cone guidance, 1124
    N-CAM, 968–969, 968FF, 1000T
Immunological memory, cellular basis, **1203–1204**, 1204F
Immunological tolerance, **1205–1206**, 1205F
    acquired, 1205, 1205F
    of self, development of, 1247–1248, 1248F
Immunology, defined, 1195
In situ hybridization, 307, 307F
    identification of genes via, 312–313, 313FF
In vitro, defined, 158
In vivo, defined, 158
Inactivation center, of X chromosome, 447
Indirect immunocytochemistry, 186, 187F
Indole-3-acetic acid, 1114, 1114F
Inducible enzymes, SOS repair, 249
Induction
    in embryogenesis, **1052–1055**, 1052FF
    signaling molecules, 1053, 1054F, 1055
Infection
    bacterial vs. parasitic, 1165
    complement defense, 1213–1215, 1213FF
    defenses in vertebrates, 1195
    viral
        fusion proteins and, 645, 646F
        PCR to detect, 317
Infectious diseases, MHC polymorphism from, 1250
Inflammatory response, 1164–1165, 1165F
    angiogenesis in, 1153
    cell-cell adhesion processes in, 503–504, 503F
Influenza virus, 276F, 277
    budding, 280, 280F
    cellular defense against, 1236, 1236F
    uptake mechanism, 645, 646F
Inheritance
    cytoplasmic, 711–712, 711F
    maternal, 712
    maternal-effect mutation, 1082, 1082F
    Mendelian, 711F
Inhibitor-of-DNA-binding gene, 1062F
Inhibitory G protein, *see* G_i protein
Initiation factors (IFs), in protein synthesis, 234–235, 235F
Initiator tRNA, in protein synthesis, 235–237, 235FF
Inner cell mass, 1057–1058, 1057F
Inner membrane, of mitochondrion, 656, 657F, 658
Inner sheath, of axoneme, 816F
Inositol phospholipid signaling pathway, **744–749**, 744FF, 745T, 1241
    activation by integrins, 999
    amplification, 755–756
    cAMP interaction, 751
    cell responses mediated by, 745T
    diacylglycerol, 747–749, 748FF
    in egg cortical reaction, 1031–1032, 1032F
    in olfaction, 753
    in PDGF responses, 761, 762F
    summary, 748–749, 749F
Inositol phospholipids
    breakdown, 744–745, 745FF
    in membranes, 482
        asymmetry, 483
    profilin binding to, 832–833
Inositol trisphosphate, 736F
    Ca^{2+} propagation and oscillation, 746–747, 747F
    in Ca^{2+} release from ER, 745–746, 749F
    inactivation, 746
    as intracellular mediator, 744, **745–749**, 744FF, 745T
    production, 745, 746F, 749F
Input layer of neural network, 779, 779F
Insect, body plan, 1078–1079, 1078F
Insertional mutation, 1277, 1277T, 1277F
Instars, 1078, 1078F

Insulin, 894T
    amino acid sequence, 104, 104F
    conformation, role of disulfide bonds, 128, 128F
    gene, 340T
    genetic engineering, to produce, 320–321, 321F
    packaging as proinsulin, 628F
Insulin receptor, 760–761, 760F, 770F
Insulinlike growth factor-1, *see* IGF-1
Insulinlike growth factor-2, *see* IGF-2
Integral membrane proteins, defined, 486, 486F
Integral membrane proteoglycans, 502, 502F
Integrase, 467, 467F
    lambda, 271, 271F
    lambda-like, 272, 272F
    for retrovirus integration, 282, 283F
    transposases as, 284
    for transpositional recombination, 272–273, 273F
Integrins, 769, **841**, 842F, **995–999**, 996FF, 1000T, 1242, 1242T
    in adherens junctions, 955, 969F, 1000T
    affinity for ligand, 995–996
    in cell-cell adhesion, 967, 969F, 1000T
    cytoskeleton interaction with, 997
    as cytoskeleton-matrix mediator, 997, 998F
    in embryogenesis, 1047
    in focal contacts, 899F
    in growth cone guidance, 1124
    in hemidesmosomes, 957, 969F, 1000T
    inactivation, 998, 999F
    in inflammatory response, 504, 1164
    in lymphocyte adhesion in lymph nodes, 1203, 1203F
    in metastasis, 1270
    PH-30 binding to, 1033
    purification, 996
    structure, 996, 996F
    in transmembrane signaling, 998–999, 999F
    types, 996, 1000T
$β_2$ Integrins, 996
$β_3$ Integrins, 996, 998
Intercalary regeneration, 1065F, 1066
Intercalated disc, in cardiac muscle, 856, 856F
Intercellular adhesion molecules, 968, 1242, 1242T
Interferon, genetic engineering, to produce, 320–321, 321F
γ-Interferon, 1245–1246, 1245T
    action, 1237, 1245, 1245T
Interleukin-1, *see* IL-1
Interleukin-2, *see* IL-2
Interleukin-3, *see* IL-3
Interleukin-4, *see* IL-4
Interleukin-5, *see* IL-5
Interleukins, 1199, 1229, 1245T
    see also individual interleukins; γ-Interferon
Intermediary metabolism
    see also Biosynthesis; Metabolic pathways
    in plastids, 685
Intermediate filaments, 19F, 787, **796–802**, 796FF, 798T
    assembly, regulation, 797–798
    cytoplasmic, types, 798
    desmosomes and, 956, 957F, 958T, 961F, 969F, 1000T
    distribution, 800–801
    domains
        in filaments, 797
        of monomers, 796, 796F
    function, 787
        mechanical stability of cells, **801–802**, 801FF
        of variable regions, 802
    integrin binding, 997
    location, microtubule influences on, 794
    major types, 798T
    mechanical properties, 802, 802F
    in myofibrils, 855
    in nonepithelial cells, 799–800, 798T, 799F
    nucleus, surrounding, 335, 335F
    solubility, 796
    stress fibers, relationship, 840
    structure, 796–797, 796FF
    structure and function, overview, 788, 789F
    in terminal web, 843, 843F
Intermembrane space, 568, 569F
Internode, plant, 28F, 1108F
Interphase, defined, 864, 864F
Interspersed repeated DNAs, 391
Intestine
    adherens junctions, 954, 955F
    cell types in, 1148F
    epithelial sheet structure and permeability, 951–953, 951FF
    tissues in, 949F, 950
Intracellular digestion
    in lysosomes, *see* Lysosomes
    three pathways to lysosomes, 613, 614F

M

multienzyme complexes and compartmentalization, 135, 135F
sample scheme, 83F
spatial segregation in cells, 86–87, 86FF
Metabolic regulation, key enzymes, 82, 84–86, 84FF
Metabolism and storage, catalogue of specialized cells, 1188
Metabolites, concentrations, 135
Metalloproteases, 994
Metaphase, 864, 864F, **926–928,** 926FF
  arrest in, 929
  described, 915FF
  location of actin, tubulin, DNA, 145F
  mitotic spindle dynamics, 927–928, 927FF
  plate formation, 926–927, 926F
Metaphase arrest, of egg, 871, 872F
Metaphase checkpoint, defined, 868F, 869
Metaphase plate
  formation, 926–927, 926F
  in M phase, summary, 916F, 919
Metastasis
  defined, 994, 1256, 1256F
  genetic changes in colorectal cancer for, 1289, 1289F
  steps and mechanism, 1269–1270, 1269FF
Methane, oxidation to CO_2, 62, 63F
Methionine, 57F
  on initiator tRNA, 235
  on N-terminal of proteins, 220–221
Methotrexate, 1272
Methyl group, 43, 50F
  transfer by *S*-adenosylmethionine, 78T
5-Methyl cytosine, 449, 449F
  deamination, 250, 250F
Methyl transferase, 777, 777F
Methylation of DNA, role in bacteria, 293
Methylesterase, 777
7-Methylguanosine, on mRNA cap, 236F, 237
MHC molecules, **1229–1237,** 1230F, 1234T
  see also MHC molecules, class I; MHC molecules, class II
  discovery, in transplantation reactions, 1229–1230
  functional difference, class I vs. class II, 1239
  genetic loci, 1231, 1231F, 1234T, 1236
  polymorphism, 1230, 1230F, 1231, 1232
    evolution of, 1249–1250
  as products of immune response (*Ir*) genes, 1248–1249
  properties summarized, 1234T
  puzzling properties, 1230
  structure, 3-D, 1233
  in transplantation reactions, role, 1249, 1249F
MHC molecules, class I, 1250, 1250F
  assembly with antigen for presentation to cytotoxic T cells, 1236–1237, 1237F
  CD8 as co-receptor for, 1235, 1235F, 1242T
  discovery, in antigen presentation, 1235–1236, 1236F
  distribution, 1233, 1234T
  function, 1233–1234, 1233F, 1234T, 1235–1236, 1236F
  peptide binding, 1232–1233, 1232F
  structure, 1230–1231, 1230F, 1234T
    antigen-binding site, 1231–1233, 1232FF
MHC molecules, class II, 1250, 1250F
  assembly with antigen for presentation to helper T cells, 1239–1240, 1240F
  CD4 as co-receptor for, 1235, 1235F, 1242T
  distribution, 1233–1234, 1234T
  function, 1233–1234, 1233F, 1234T
  ineffective forms, 1249
  γ-interferon on, 1245
  invariant chain and, 1239, 1240F
  structure, 1230–1231, 1230F, 1234T
MHC restriction, 1235, 1236F
  development of, 1246–1247
Micelle, 55F, 479, 479F
  detergent, 488FF, 489
Microbody, *see* Peroxisomes
Microcinematography, 146
Micrococcal nuclease, 343, 346F
Microelectrodes, **181–182,** 181FF
  action potential studies, 531F
  patch-clamp recording with, 533, 533F
Microfibrils, in elastic fibers, 985–986
Microfilaments, defined, 789F
β₂-Microglobulin, 1230F, 1231, 1250F, 1251
Microinjection, into cells, 183–184, 184FF
Microsatellite
  defined, 303F
  in PCR, 318F
Microscopy, **139–156,** 140FF, 140T, 149T
  see also Electron microscopy; Light microscopy; Scanning microscopy
  limitations, 141

resolving power, 140F
specimen preparation, importance, 140
units of length, 140F
Microsomes, **580–581,** 581F
  isolation by ultracentrifugation, 162–164, 163FF, 165T
Microspikes, 793, 793F
  in bacterial infection, 830, 830F
  of growth cones, 1123, 1124F
  parallel actin bundles in, 835–836, 835FF
  protrusion in locomotion, 845
  size, components, 827, 827FF
Microsporidia, 20
Microtome, 142, 142F
Microtubule array, interphase, transformation at mitosis, 920
Microtubule-associated proteins, *see* MAPs
Microtubule organizing center
  see also Centrosome
  in formation of new cell center, 790, 791F
  microtubule orientation, 805, 806F
  variants, 808, 808F
Microtubules, 19F, 23–24, 787, **803–814,** 803FF
  see also Centrioles; Centrosomes; Cilia; Tubulin
  acetylation and detyrosination, 811–812
  actin filaments, comparison, 821
  in activated T cells, 1237–1238, 1238F
  antimitotic drug effects on, 804, 804F
  astral, *see* Microtubules, astral
  axonemal
    sliding, 817–818, 817FF
    structure, 816–817, 816F
  in basal bodies and centrioles, 818–819, 818F
  capture of kinetochores, 924, 924F
  in cell polarization, 794, 795F
  ciliary dynein binding, 817–818, 818F
  cytoplasmic dyneins, 813–814, 813F
  dynamic instability
    changes in, 810
    described, 808, 809F, 825F
    GTP hydrolysis, 809–810, 810F, 825F
    in cell morphogenesis, 811, 811F
    importance, 823
    in vitro, 809, 809F, 825F
    in vivo, **808–811,** 809FF, 825F
    M phase and interphase, 920, 920F
  function, 787
  growth from centrosome, 788–789, 790F
  half-life, 808–809
  kinesins, 813–814, 813F
  kinetochore, *see* Microtubules, kinetochore
  lability, 804
  mechanical properties, 802, 802F
  mitochondria, relationship to, 655, 655F
  in mitosis, *see* Microtubules, astral; Microtubules, kinetochore; Microtubules, polar; Mitotic spindle
  in mitotic spindle, *see* Mitotic spindle
  motor proteins and, 790, 792, 792F
  MPF effects on, transformation in mitosis, 877
  in neurulation, 1044, 1045F
  nucleation
    from centrosome, **807–808,** 807FF
    in vitro, **804–805,** 805F, 824FF
  in organelle movement
    mechanism, 813–814, 814F
    secretory vesicles, 630
  organelle positioning by, 792–793, 792F
  orientation in cells, 805, 806F
  pigment granule movement along, 789–790, 791F
  in plant
    cell division, 939, 939F
    direction of cell expansion and, 1112, 1112F
    orientation with cellulose microfibrils, 1003–1004, 1005F
  polar, *see* Microtubules, polar
  polarity, 788, 790F, 804, 824F
    defined, 805
    growth at plus end, 805, 806F, 824F
    hook-decoration method, 805, 806F
    MAPs in establishing, 813
  polymerization
    antimitotic drug effects on, 804, 804F
    GTP cap, 810, 810F, 825F
    in vitro, 804–805, 805F, 824FF
    MAP-2 and tau effects, 812
    on plus ends, 805, 806F, 824F
  protofilaments, 803–804, 803F
  stabilization
    by capping, 811, 811F, 825F
    by MAP-2 and tau, 812
  structure, **803–804,** 803F
    function, overview, 788, 789F
  in transport of endosomal vesicles, 625F

turnover, measurements, 811–812
visualization, 807, 807F
  by computer-assisted microscopy, 146, 147F
  in cell cycle, 915F
Microtubules, astral
  in anaphase B, 932, 932F
  creation, 920
  defined, 919F, 920
  in M phase, summary, 916FF
  mitotic spindle rotation, 936, 936F
  polarity, 928F
Microtubules, kinetochore
  in anaphase, 928F, 929
  in anaphase A
    depolymerization, 930–931, 931F
    shortening, 928F, **929–931,** 930FF
  chromosome binding, 921, 922F, 923
  dynamic flux, 924–925, 925F, 928F
  force on chromosomes, 925
  kinetochore capture, 924, 924F
  in M phase, summary, 916FF
  in mitotic spindle, dynamics, 927–928, 927FF
  polarity, 924, 928F
  stabilization by tension, 926
Microtubules, polar
  defined, 919–920, 919F
  dynamic properties, 920, 921F
  elongation in anaphase B, 929, 930F, **931–933,** 931FF
  in M phase, summary, 916FF
  in midbody, 937, 937F
  in mitotic spindle
    dynamics, 927–928, 927FF
    formation, 920–921, 921F
  overlap, 931F, 932
  polarity, 920, 921F, 928F
Microvilli, **841–843,** 842FF
  cytoskeleton, 842–843, 842F
  of epithelial cells, 519, 520F
  extracellular coat, 842
  formation and villin, 843
  size, function, 841–842
Mid-blastula transition, 1039
Midbody, 937, 937F
  in cytokinesis, summary, 915F, 917F
Midpiece, of sperm, 1026F, 1027
Milk, secretion, 1160, 1161F
Mismatch proofreading system, *see* Proofreading, mismatch proofreading system
Mitochondria, 19F, 552, 552F, 553T, 554F, **653–683,** 653FF
  see also Chemiosmotic coupling; Electrochemical proton gradients; Respiratory chain
  acetyl CoA production, 658–660, 658FF
  ATP synthase, **672–674,** 673FF
    see also ATP synthase
  biogenesis, *see* Mitochondrial biogenesis
  Ca^{2+} pump and Ca^{2+} concentration control, 743, 743F
  Ca^{2+} storage, 666–667
  in cell division, 918
  chloroplasts, structural comparison, 687, 687F
  citric acid cycle, 660–662, 661F
  compartments, internal (subcompartments), **656–658,** 657F
  contact sites for protein translocation, 570, 570FF
  division, 705, 705F
  electrochemical proton gradient, cf. chloroplasts, 695–696, 695F
  electron micrograph, 17F
  electron-transport chain, 654, 654F, 672, **675–683,** 675FF
    outline, 703F
  in energy conversion, overview, 653–654, 653FF
  energy metabolism, summary, 668F, **669**
  essential function, 655
  evolution, 555–556, 556F, 654, 701–702, 702F
    endosymbiont hypothesis, 17, 20–21, 21F, **714–715,** 714F
    from purple photosynthetic bacteria, 702F, 715
  fractionation, 656F, 672, 672F
  genome, *see* Mitochondrial genome
  half-life, 613
  isolation by ultracentrifugation, 162–164, 163FF, 165T
  locations in cell, 655–656, 654FF
  membranes
    lipid composition, 482T
    permeability, 657F, 663, 663T
  microtubules, relationship to, 655, 655F
  NADH, generation, 661–662, 661FF
  oxidative phosphorylation, defined, 662, 662F

Page numbers in **boldface** refer to a major text discussion of the entry; F refers to a figure, FF to figures that follow consecutively; T refers to a table; cf. means compare.

I-27

Page numbers in **boldface** refer to a major text discussion of the entry; F refers to a figure, FF to figures that follow consecutively; T refers to a table; cf. means compare.

I-29

cytokinesis mechanism, 938–941, 939FF
diagram, 18F
growth
    shape determination in, 1003–1005, 1004FF
    turgor pressure and, 1002
M phase, visualized, 918F
plastids, 684–685, 685F, 706T
size control by vacuoles, 612, 612F
special organelles, 19F
starch formation, 689
storage in vacuoles, 613
sucrose formation, 689
turgor pressure control, 612, 613
vacuoles, functions, 612–613, 612F
volume regulation of, 516, 517F
Plant development, **1108–1118,** 1108FF
and animal compared, 1108
apical meristem, 1110–1111, 1109FF
branching, regulation, 1113–1114
embryogenesis, 1108–1110, 1109F
flower development, 1115–1118, 1116F
germination, 1109F, 1110–1111
growth of root tip, 1111, 1111F
growth regulators, 1112, 1112F, **1114,** 1114F
homeotic selector genes, **1116–1117,** 1117FF
hormonal signals to coordinate, 1113–1114, 1114F
key genes, 1114–1115, **1116–1117,** 1117FF
molecular genetics, **1114–1118,** 1115FF
morphogenesis, 1111–1112, 1111F
pattern formation in shoot apex, 1113, 1113F
Plants
    see also Plant cells; Plant development
C$_3$ and C$_4$, 690, 690F
embryo, 1109F
evolution, early, 701–702, 702F
evolutionary age, 1108
flower, 1109F
germination, 1109F
glyoxysomes, 575, 576F
growth regulators, 1112, 1112F, **1114,** 1114F
membrane potential, 524
mitochondrial DNA, "junk," 710
modular construction, 1108, 1108F, 1112, 1112F
microtubule organizing centers in, 808
noncyclic photophosphorylation, 693–695, 694F
organs and tissue systems, 28FF
peroxisomes, 575, 576F
plasmodesmata, 961–962, 962F
repetitive patterning, 1112, 1112F
mRNA editing, 461
seed, 1109F
sexual reproduction, summary, 1108, 1109F
transpositional bursts in, 394
Plaques, in anchoring junctions, 954, 954F
Plasma cells, 1198–1199, 1199F
antibody selection, 1207, 1207F
Plasma membrane, 10, 18F
    see also Lipid bilayer; Membrane
action potentials, 528–530, 528FF
budding
    viral, 279FF, 280
capping, 498–499, 499F
in cell adhesion, *see* Focal contacts
cell coat, oligosaccharide functions, 502
cell cortex
    of animal cells, 834–835
    of erythrocytes, 834
cortical cytoskeleton and, 492–493
domains, formation of, 631–633, 633F
of eucaryotes, 489–491, 490FF
evolution, 20–21
exocytosis localization, 630, 631F
function, 477
glycolipids, **483–484,** 483FF
    amounts, 484
glycoprotein and glycolipid orientation in, 608, 608F
growth, constitutive secretory pathway and, 627, 627F
lateral diffusion, restrictions on, **500–501,** 500FF
lipid composition, 481–482, 482F, 482T
membrane potential
    basis, **524–527,** 526FF
    typical values, 527
MHC molecules, processing and insertion, 1236–1237, 1237F
in microvilli, 842–843, 842F
Na$^+$-driven carrier proteins, 518–519, 520F, 522T
Na$^+$-K$^+$ pump, **513–516,** 514FF, 522T
patch-clamp experiments, 181–182, 182F
patching, 498, 499F
permeabilization methods, 639F

pinocytosis of, rates, 619–620, 620F
of plant cells, 28FF
polarity, 280, 280F
protein, *see* Membrane proteins
Rab proteins, 644T
relative amount, 553F
shaping by cortical actin filaments, 793, 793F
structure, *see* Membrane structure
voltage gradient across, 534
Plasmids, 274, **286**
in nonpermissive cells, 282
replication, antisense RNA for control, 468, 468F
as tools in cell biology, 286
as vectors in DNA cloning, 308–309, 309F
Plasmin, 994
Plasminogen, 994
Plasmodesmata, 30, **961–962,** 962F
permeability, 962
Plasmodium falciparum, 520–521
Plastids, 653, **684–685,** 685F, 706T
biogenesis, lipid import, 593–594, 594F
evolution, 555–556, 556F
Plastocyanin, 694F
Plastoquinone, 676F, 694F
Platelet-derived growth factor, *see* PDGF
Platelets, 893, 893F, 1162, 1163F, 1164T, 1165, 1166F
activation via integrins, 998, 999F
Ploidy, cell size and, 885
Point mutations
in carcinogenesis, 1279, 1279F, 1282, 1283F
in colorectal cancer cells, 1287
pol gene, 284F, 467, 467F, 1276F
Pol protein, 467, 467F
Pol II, *see* RNA polymerase II
Polar body, 827F
production and fate, 1023, 1024F
Polar granules, 1085
Polar group, 48F, 478, 479F
Polar microtubules, *see* Microtubules, polar
Pole cells, *Drosophila*, 1079, 1080F
Poliovirus, 275FF, 278
Pollen grains, 1109F
in confocal scanning microscope, 148F
Poly-A addition
change in site, 456, 458F
control of gene expression by, 466, 466F
Poly-A polymerase, 369
Poly-A tail, 369, 370F, 371
effect on translation and mRNA stability, 466, 466F
function, 369
RNA export and, 458
Polyacrylamide-gel electrophoresis, 169–172, 169FF
    see also SDS polyacrylamide-gel electrophoresis
separating DNA molecules, 294, 295F
two-dimensional, 170–172, 171FF, 172T, 402
Polycistronic mRNA, defined, 237
Polycomb group genes, 1096–1097, 1096F
Polyisoprenoids, 55F
Polymer, biological, synthesis, 80–81, 80FF
Polymerase chain reaction, *see* PCR
Polymerization
of actin filaments, 821–823, 824FF
of biological polymers, 80–81, 80FF
of microtubules, *see* Microtubules, polymerization
Polymers
defined, 4
formation of earliest, 5, 5F
Polynucleotide kinase, labeling DNA molecules with, 296, 296F
Polynucleotides
biosynthesis, 77, 78F
formation, on template, 5–6, 6F
formation, spontaneous, 5, 5F
genetic code, 106, 106F
as templates, 5–6, 6F
Polyomaviruses, 276F
neoplastic transformation by, 282
Polypeptide
    see also Protein
bond types, 94F
as catalysts, in evolution, 8F, 9
disulfide bonds in, 112, 113F
formation, spontaneous, 5, 5F
steric limitations on bond angles, 91, 94F
Polyploid cell, defined, 349
Polyps
colorectal cancers from, 1287, 1287F, 1289, 1289F
mutation in, 1289, 1289F
Polyribosomes, 238, 238FF
in cytosol, 552F
defined, 108F

on ER, 578, 579F
Polysaccharides, 44–45, 53F
catabolism, outline, 66–67, 67F
cell weight, percent of, 90T
in extracellular matrix, 45
polymerization, 80–81, 80FF
Polysomes, *see* Polyribosomes
Polyspermy, 1031
blocks to, 1031–1032, 1032F
Polytene chromosomes, **348–351,** 348FF
bands, 348FF, 349–351
formation, 362F
gene location, 350–351
nonoverlap in nucleus, 383F, 384
puffs, gene activation in, 350, 350F
re-replication block, lack of, 879
Porin, 487, 496–497, 497F, 523
of mitochondria, 570, 656, 657F
Porphyrin ring, 675F, 676, 691F
Position effect
on gene expression, 435, 435F, 436–437, 437T
LCR override of, 436–437, 438F
variegation in *Drosophila*, 448, 448F
on X-chromosome inactivation, 447
Positional cloning, 315–316, 316F
Positional information, 1060, 1063–1066
in *Drosophila* embryo, 427, 427F
homeotic selector genes for memory of, **1098–1099,** 1099F
in imaginal discs of insects, 1097–1099, 1097FF
Positional labels, *Drosophila* segmentation genes and, 1091–1092, 1091FF
Positional values, **1063–1066,** 1065FF
Hox genes in vertebrates for, 1105–1106
limb patterning in vertebrates, 1106–1107, 1106F
neuronal development and, 1122
of neurons, for neural maps, 1126–1127, 1126F
rule of intercalation, 1064–1066, 1065F
Positive feedback
allosteric transition, 198–199, 199F
Ca^{2+}-release channels in ER, 746
in cell memory, 1062
in cell signaling, 757–758, 758F
in MPF activation, 884–885, 884F
in Start kinase activation, 887
Postcapillary high endothelial venules, 1202, 1202FF
Posterior group, egg-polarity genes, 1081, 1081F
Posterior system, 1085, 1085F
Postsynaptic cell, 536, 536F
grand PSP, input and output, 541–543, 542FF
Postsynaptic potential
excitatory, 541
grand, 542
inhibitory, 541
spatial and temporal summation, 542, 542F
Posttranscriptional controls
overview, 453, 453F
regulation of gene expression by, **453–468,** 453FF
Potassium ion
concentration, inside and outside cell, 508T
membrane potential and concentration gradient, 525–527, 526FF
transport by Na$^+$-K$^+$ pump, 514–516, 513FF
V_k, 526F
Potassium ion channels
Ca^{2+}-activated, at axon hillock, 544
delayed, 529, 531F
G$_i$ and, 752
membrane potential and leak channel, **524–527,** 526FF
voltage-gated, *see* Voltage-gated channels
Potato virus X, 276F
Poxvirus, 276F, 277
Pre-B cells
antibody gene rearrangements in, 1225, 1225F
antibody synthesis in, 1208–1209
Prebiotic conditions, on earth and simulated, 4–6, 4F
Prenyl groups, 485, 486FF
Preparative ultracentrifuge, 162FF, 163
Preprophase band, 940F, 941
Presynaptic cell, 536, 536F
    see also Neuromuscular junction; Synapses
Prickle cells, in epidermis, 1156, 1157F
Primary cilium, **820**
Primary cultures, 158
Primary follicle, 1025F
Primary lymphoid organs, 1196F, 1198, 1198F
Primary RNA transcript, 341, 341F, 368–369, 368FF, 379, 380F
    see also hnRNA
transcriptional control of synthesis, 403, 403F
Primary structure, of proteins, defined, 116, 118F

Page numbers in **boldface** refer to a major text discussion of the entry; F refers to a figure, FF to figures that follow consecutively; T refers to a table; cf. means compare.

I-33

Page numbers in **boldface** refer to a major text discussion of the entry; F refers to a figure, FF to figures that follow consecutively; T refers to a table; cf. means compare.

I-35

in protein synthesis, 106, 106F
RecA protein
  in general recombination, 266–267, 267F
  in SOS response, 249
  structure, 266F
RecBCD, in general recombination, 264F, 265
Receptor down-regulation, 624
Receptor guanylyl cyclases, **759**
Receptor proteins
  see also Receptors, cell surface; and
    individual receptors and types
  cargo receptors, 640, 640F
  in clathrin-coated vesicles, trapping, 640,
    640F
  fate in endosomal compartment, 623–624
  zinc finger family, 411, 412F
Receptor serine/threonine kinases, 203F, 759,
  769, 770F
  see also Protein kinases; and individual
    kinases
Receptor tyrosine kinases, 203F, 746F, 759, **760–**
  766, 760FF, **768–769**, 770F
  see also Protein kinases; and individual
    kinases
  activation
    dimerization and autophosphorylation,
      760–761, 761F
    of intracellular signaling proteins, 761–
      762, 761F
    long-lived via MAP kinases, 765–766, 766F
    short-lived, 765
  mutations in pathway and cancer, 768–769
  as proto-oncogene product, 1276T, 1278F,
    1279
  Ras proteins and, **763–765**, 763FF
  SH2 domain binding, 761–762, 762F
  subfamilies, 760, 760F
Receptor tyrosine phosphatases, 759, **767–768**
Receptors
  see also Receptor proteins; Receptors, cell
    surface; Receptors, intracellular
  antigen-specific, 1199–1202, 1200FF
  bacterial chemotaxis, 775–776, 775FF
  down-regulation, 771
  general principles, 722, 722F
  glycolipids as, 484
  for phagocytosis, 619
  phosphorylation in desensitization, 772, 772F
Receptors, cell surface
  see also Enzyme-linked receptors; G-protein-
    linked receptors; Transmitter-gated
    channels
  antibodies on B cells, 1206, 1207F
  antigen specific, on T cells, 1228, 1228F
  in chemotaxis
    cell-cortex actin polymerization, 831–832,
      832F
    G protein, 832–833
  to CSFs, 1170–1171, 1171F, 1171T
  for extracellular matrix (integrins), **995–999**,
    996FF, 1000T
  G-protein-linked, 734–735
  homing, 1202–1203
  modules, 216, 216F
  phosphorylation cascades on activation,
    733–734, 733F
  protease, 994–995, 994F
  proteoglycans as co-receptors, 977–978, 978F
  as signal transducers, 731
  three classes, **731–733**, 732F
Receptors, intracellular
  for NO, 728–729
  steroid-hormone receptor superfamily, **729–**
    731, 729FF
Recessive phenotype, defined, 1072F
Recognition helix, defined, 409, 409F
Recombinant DNA molecules
  defined, 293, 308
  formation, 308–309, 309F
  as vectors, 319, 319F
Recombinant DNA technology, **291–331**, 292FF,
  292T
  DNA cloning, **308–318**, 309FF
  DNA engineering, **319–331**, 319FF
  DNA footprinting, 297, 299F
  DNA fragmentation, **292–293**, 292FF
  DNA labeling, 295–296, 296F
  DNA probes, construction, 295–296, 296F
  DNA separation, **293–295**, 295FF
  DNA sequencing, **296–297**, 297FF
  history, major steps, 292T
  impact on cell biology, 291
  Northern blotting, 302, 302F
  nucleic acid hybridization, **300–307**, 300FF
  propagation of selected mutation in germ
    cells, chimeric mice, 1059
  reduced stringency hybridization, 306, 306F
  Southern blotting, 302–303, 302F

for study of membrane transport, 510–511
techniques, most important, 291–292, 292T
voltage-gated cation channel study by,
  534FF, 535
Recoverin, 754
Red blood cell, *see* Erythrocytes
Redox potentials, of respiratory chain, 680–681,
  680F
Reduced-stringency hybridization, *see* Nucleic
  acid hybridization, stringency
Reducing agents
  disulfide bonds and, 112, 113F
  evolution and development of new, 699–700,
    700F
Reducing power, in biosynthesis, 79–80, 79FF
Reduction, defined, 62, 63F
Regeneration
  of *Hydra*, 31–32
  intercalary regeneration, 1064–1066, 1065F
  of liver, 1150
Regulatory proteins, need for rapid turnover,
  727, 728F
Regulatory sequences of genes, eucaryote, **423**,
  424F
Regulatory site, allosteric regulation of
  metabolism and, 198
rel gene, 1276T
Release factors, in protein synthesis, 234, 235F
Reovirus, 276F
Replica methods, for EM, 153, 153F
Replicas, for *in situ* hybridization, 312, 313FF
Replicases, in RNA replication, 277–278
Replication bubble, 259, 259F
  mammalian, 358, 358FF, 360
Replication forks, mammalian, **358**, 358F
Replication origins, 259, 259FF
  on chromosome, 337–338, 338F
  replication units and, 359–360, 359F
  spacing, mammalian, 360
  in yeast
    core consensus sequence, 358
    isolation, identification, 357, 357F
Replication units, defined, 359–360, 360F
Reporter gene
  defined, 424F
  lacZ as, 424F
Reporter protein, to locate regulatory DNA, 321,
  322F
Reporter sequence, defined, 321
Repressor, in SOS response, 249
Reproduction, *see* Asexual reproduction;
  Parthenogenetic reproduction;
  Sexual reproduction
Residual bodies, 619
  in spermatogenesis, 1029F
Resolution
  electron microscopy, 148–149, 149F
  of light microscope, 141, 142F
Resolving power, of microscopes and eye, 140F
Resonance, 50F
Resonance energy transfer, in photosynthesis,
  691–692, 691FF
Respiration
  defined, 16–17
  efficiency of cellular, 670–671
Respiratory chain, 672, **675–683**, 675FF
  see also Electron-transport chain
  amounts of components, 679
  in bacteria, 683, 683F
  control mechanism for, **682**
  electrochemical proton gradient, *see*
    Electrochemical proton gradient
  electron carriers, **675–676**, 675FF
    mobility, 679
    order, methods to determine, 676, 677F
  electron flow, general, **663–666**, 664FF
  electron transfer
    path, 663, 664F, 677–678, 677F
    stoichiometries, 678
  energy conversion to electrochemical proton
    gradient, 665–666, 665F
  enzyme complexes, *see* Respiratory enzyme
    complexes
  Na⁺ based in some bacteria, 683
  NADH to O₂, 665
  proton and electron paths, 663, 664F
  proton pump, 675
    mechanism, **681–682**, 681F
  redox potentials of components and, 680,
    680F
  uncoupling, 682
Respiratory control, 682
Respiratory enzyme complexes, 665, **677–678**,
  677FF
  see also individual enzyme complexes
  location, 656, 657F
  proton pump, mechanism, 681–682, 681F
  redox potentials, **680–681**, 680F

Respiratory tract, cilia function, 815
Resting potential, *see* Membrane potential
Restriction fragment length polymorphisms, *see*
  RFLPs
Restriction fragments, defined, 293
Restriction maps
  construction, uses, 293, 294F
  of hemoglobin genes, 294F
Restriction nucleases, 292, **293**, 292FF
  cohesive ends, 293, 293F
  DNA libraries, use for, 308–309, 309F
  recognition sequences of some, 292F
Reticulocyte, 1169F
Retina
  development of connections, 1126–1130,
    1126FF
  embryonic origin, 1045
  neural, 1141F
  photoreceptor cells, 1144FF, 1145–1146
  pigment cells
    phagocytosis by, 1146, 1146F
    transdifferentiation, 1141
  pigment epithelium, 1141, 1141F
  reassembly after dissociation, 965, 965F
  structure, 1145F
Retinal
  in bacteriorhodopsin, 495, 496F
  in rhodopsin, activation, 753–754
Retinal ganglion cells, 1144F, 1145
Retinoblastoma, 1281–1282, 1282FF
  gene and protein, *see Rb* gene; Rb protein
  loss of heterozygosity, 1282, 1282F
Retinoic acid, effects on Hox gene expression,
  1107
Retinoids, **729–731**, 729FF
Retinotectal system, *see* Visual system
Retrotransposons, **284–285**, 283F
Retroviral vector, for hemopoietic cell clones,
  1167
Retroviruses, **282–284**, 283FF
  common genes, 284F
  endogenous, 286
  HIV, 283F, 284, 284F
  human cancer, 1283, 1283T
  insertional mutation, 1277, 1277F, 1277T
  life cycle, 282–283, 283F
  oncogene acquisition, 1275
  protein synthesis, 467, 467F
  transformation mechanism, 1275, **1277**,
    1277FF
  transpositional recombination, 273F
  as tumor viruses, **1273–1277**, 1274FF, 1274T,
    1276T
Reverse electron flow, evolution, 699, 700F
Reverse genetics, **323–331**, 324FF
  outline, 323
Reverse transcriptase, **282–283**, 283F, 467, 467F
  in cDNA synthesis, 310, 310F
  in loss of introns in evolution, 390–391
  of retrotransposons, 285
  of transposable elements, 392T
RFLPs
  in genetic mapping, 304–305, 304F
  as genetic markers, 303, 303FF
  high-resolution RFLP map, human genome,
    305
  in positional cloning, 316F
RGD sequence
  in fibronectin, 986F, 987
  integrins for, 996
Rho protein, 844
  stress fiber formation, 833
Rhodamine, 123, 144, 144FF, 499F, 655F
  as actin label, 829, 829F
Rhodobacter capsulatus, 497F
Rhodopsin, 37F, 498, 735
  synthesis in ER, 587–588, 588F
  in vertebrate vision, mechanism, 753–755,
    754FF
Rhombomeres, 1104, 1105F
Ribbon model, of proteins, 117F, 118
Ribonucleic acid, *see* RNA
Ribose
  in NADH and NAD⁺, 71F
  in nucleotides, 58F, 60
  in RNA, 60, 100F, 104
Ribosomal proteins
  see also RNPs; and individual proteins
  functions, 232
  structures, 232, 233F
Ribosomal RNA, *see* rRNA
Ribosomes, 223
  amounts in cells, 378
  assembly
    in nucleolus, **379–381**, 380F, 381F, 382
    on mRNA, 234–236, 235F
    self-assembly, 127
  of chloroplasts and mitochondria, 707

for B cell activation, 1244, 1244F, 1245T
breakdown pathway, 771
combinatorial action, 726, 726F
   integration, 733–734, 733F
concentration, 724, 727, 728F
different effects on different targets, 726–727, 727F, 731
for dorsoventral axis establishment in flies, 1083–1085, 1084FF
for fat cell differentiation, 1181
in inflammatory response, 1164–1165, 1165F
integrin activation pathways, 998, 999F
intracellular, for receptor tyrosine kinases, 761–762, 762F
intracellular mediators, **735–752**, 736FF
in mesoderm induction, 1053, 1054F, 1055
in nematode vulval induction, 1074, 1074F
NO (nitric oxide) gas, 728–729
receptors
   see also Receptors; Receptors, cell surface
   location and interaction, general
      principles, 722, 722F
regulation by proteoglycans, 976, 978, 978T
release, 722
response
   graded, 756, 756F
   nonlinear or all-or-none, 757–758, 758F
SH2 domains in, 761–762, 762F
steroid-hormone superfamily, **729–731,** 729F
in T cell activation
   cytotoxic, 1237, 1245
   helper, 1241–1242, 1241F, 1242T, 1245T
TGF-β superfamily, 769, 770F
turnover, rapid, need for, 727, 728F
types, 722–723, 722F
water-soluble vs. water-insoluble, 730
Signaling pathway
   see also Intracellular mediators; Signal
      transduction; Signaling molecules
Drosophila development, dorsoventral axis, **1083–1085,** 1084FF
for integrin activation, 998, 999F
nematode vulval induction, genetic analysis, 1071, 1071FF, 1074, 1074F
for stimulation by growth factor, 901–902, 901F
Silent mutation, 103
Silica gel, in chromatography, 166, 168
Simian virus 40, *see* SV40 virus
Single-strand DNA-binding proteins, *see* SSB
   proteins
Sinusoids, 1147, 1149F
sis gene, 768, 1276T, 1278F
Site-specific recombination, **271–273,** 271FF
of bacteriophage lambda, 271–272, 271FF, 281–282, 281F
conservative
   bacteriophage lambda, 271–272, 271FF
   with two DNA helices, 272
defined, 263
in generation of antibodies, 1223–1224
Ig gene assembly via, 1221–1222, 1222F
phase variation in bacteria and, 440–441, 440F
in transgenic animal studies, 272, 272F
transpositional, 272–273, 273F, 393, 393F
Skeletal muscle, 37F, **847–855,** 848FF, **1175–1178,** 1176FF
   see also Muscle; Muscle fibers
acetylcholine effect, 726–727, 727F
activity-dependent synapse elimination, 1128–1129, 1129F
contraction and glucose metabolism, 751–752, 752F
growth in adult, 1178
regeneration, from satellite cells, 1178, 1178F
Skeletal muscle cells
   see also Muscle cells; Muscle fibers;
      Myoblasts
development and differentiation, **1176–1178,** 1177F
modulations, 1177–1178, 1178F
myoblasts
   determination in embryo, 1061–1062, 1062F, 1176–1177
   fusion, 1177, 1177F
structure, summary, 1175–1176, 1176F
Skin, 983, 983F
   see also Dermis; Epidermis
cancers and cell differentiation, 1268–1269, 1268F
cellular architecture, 1140F
collagens in, 979, 979F, 980T
keratins of, 798
   mutant and blistering, 801–802, 801F
   outer layer, 802
Sliding clamp protein, 256–257, 257FF

Sliding filament model of muscle contraction, 849–851, 849FF, **853,** 856
Slime mold, *see Dictyostelium discoideum*
Small-cell lung cancer, 1290
Small intestine, *see* Gut; Intestine
Small nuclear ribonucleoproteins, *see* snRNPs
Smell, mechanism, 752F, 753
Smooth muscle, 37F, 847, **856–857,** 857F
of blood vessels, 1150, 1151F
contractile apparatus in, 856, 857F
control of contraction, 857, 857F
distribution, 856
gap junctions, 960
organization in gut, 949F
Smooth muscle cells, 1176, 1176F
SNAPs, 645, 645F
SNAREs
guiding vesicular transport, 642–643, 643FF
in vesicle fusion, 645, 645F
Sodium-bicarbonate ion exchanger, 519
Sodium-calcium ion exchanger, in Ca^{2+}
   concentration control, 743, 743F
Sodium dodecyl sulfate (SDS), 169, 169F, 489, 489F
Sodium-hydrogen ion exchanger, 518–519, 522T
Sodium ion
in action potential, 528–531, 529FF
concentration
   gradient and membrane potential, 526F, 529, 531F
   intracellular and extracellular, 508T
   transport by Na^+-K^+ pump, 514–516, 513FF
V_{Na}, 526F
Sodium ion channels
   see also Voltage-gated channels, Na^+
in action potential, 528–531, 528FF
cGMP-gated, 753–755, 754FF
Sodium ion-driven carrier proteins, 518–519, 520F, 522T
Sodium ion pump, in bacteria, 683
Sodium-potassium ion ATPase
   see Sodium-potassium ion pump
Sodium-potassium ion pump, **513–516,** 514FF, 522T
as ATPase, 209–210
in control of cell volume, **515–516,** 517F
distribution, 518
electrogenic, 525, 527
link to actin cortex, 835
membrane potential, role in, **524–525**
model of pumping cycle, 514, 515F
reversibility, 674
solubilization, purification, reconstitution, 489, 490F
structure, 514–515
Solation, of cell extracts, 837
Somatic cells
altruistic social behavior, function, 1012, 1255
defined, 1012
DNA stability, 244
vs. germ cells, 1012
size, 1022
telomerase deficiency and senescence, 904
Somatic hypermutation, in immune response, 1224
Somatic mutations, in nonhereditary
   retinoblastoma, 1281, 1282F
Somatostatin, 741
Somites
formation, 1045–1046, 1046FF
   cadherins in, 1047, 1047F
   cell migrations, 1048, 1048F
Sorting signals
   see also Signal peptides
for chloroplasts, 573, 574F
mannose 6-phosphate on lysosomal
   hydrolases, 615
for mitochondrial matrix, 568–569, 569F
for mitochondrial membranes and space, 572, 573F
for nucleus, **563–567**
for peroxisomes, 576–577
for protein sorting and transport, **556–560,** 557FF
for receptor-mediated endocytosis, 640, 640F
for secretory proteins, 628, 629, 629F
signal patch for M6P addition, 616–617, 616FF
sos (*son of sevenless*) gene, 765, 765F
Sos protein, 765, 765F
SOS response, in *E. coli*, 249
Southern blotting, 302–303, 302F
Sp1, DNA sequence recognized, 407T
Space-filling model
of cytochrome c, 131F
of lysozyme, 131F
of protein, 115F, 117F

Spacer DNA
eucaryote genes and control regions and, 423, 424F
in evolution, 387
Spatial summation, neuronal, 542
Species, morphological differences, reasons, 244
Spectrin, 491–492, 491FF, 834
modular construction, 836, 837F
structure, 492, 492F
summary, 845F
in terminal web, 843, 843F
Spemann's Organizer, *see* Organizer
Sperm, 37F, **1026–1029,** 1026FF
capacitation, 1030
centriole of, 1033F, 1034
defined, 1012, 1013F
development, *see* Spermatogenesis
differentiation step, 1027FF, 1028–1029
fertility assay, 1030, 1030F
flagella
   function, 815
   location of mitochondria in, 655F, 656
numbers in human ejaculate and fertilization, 1030
of plant, 1109F
plasma membrane domain, 501, 501F
size, 1026
special features, 1026–1027, 1026FF
structure of human, 1026, 1026FF
ZP3 as receptor, 1030, 1031F
in zygote formation, 1033–1034, 1033F
Spermatid, 1027FF, 1028–1029
Spermatocytes, primary and secondary, 1027FF, 1028–1029
Spermatogenesis, **1027–1029,** 1027FF
differences with oogenesis, 1027–1028, 1027FF
stages, 1027FF, 1028
syncytium, 1028–1029, 1029F
time scale, 1028F
Spermatogonia, 1027FF, 1028, 1156
Spermatozoon, *see* Sperm
Sphingomyelin, 593
in membranes, 481–482, 482T
   asymmetric distribution, 482, 483F
structure, 483F
Sphingosine, 593
Spindle pole body, 808, 808F, 833
in yeast, 882F
Spindle poles
astral exclusion force, 925, 925F
in M phase, summary, 915FF, 919
microtubule disassembly at, 928, 928F
microtubule orientation, 805, 806F
separation
   at anaphase, 931–932, 932F
   polar microtubules, 921
sister chromatid attachment to opposite, 925–926
Spleen, 1196F, 1198
hemopoiesis, site of, 1167, 1167F
Spliceosomes, 372, **373–375,** 373FF
evolution, 375–376, 377F
Splicing islands, 377–378, 378F
Split genes, *see* Introns
SPO1, control of transcription, 431, 431F
Spores, 35
of yeasts, 880F
Sporulation, in yeasts, 880F
Squames, of epidermis, 1157, 1157FF
Squid axon, experiments on action potentials
   using, 530, 531F
SRBCs, 1207
src gene, as proto-oncogene and oncogene, 769, 1275, 1276F, 1276T, 1278F
Src-like nonreceptor protein tyrosine kinases, 761, 762F
Src protein kinases, 203F, **767,** 1278F
CD4 and CD8 with, 1235
in focal contacts, 841
mutant, 769
synthesis of isoforms, 455, 455F
T cell receptor and, 1241, 1241F
SRF (serum response factor), 766, 766F
SRP, *see* Signal-recognition particle
SRP receptor, 583, 583F
SSB (single-strand DNA-binding) protein, **255–256,** 256FF
in general recombination in *E. coli*, 265–266, 266F
mammalian, 358F
in replication machine, 256–258, 257FF
in synapsis, 267
Staggered joint, *see* Heteroduplex joint
Staining of specimens
for light microscopy, 143, 143F, 145F
for TEM, 150–151, 151F
Stamen, 1109F, 1116F

Page numbers in **boldface** refer to a major text discussion of the entry; F refers to a figure, FF to figures that follow consecutively; T refers to a table; cf. means compare.

I-41

Transforming growth factor-βs, see TGF-β
    superfamily
Transgenes, defined, 327
Transgenic animals, **327–330**, 328FF
    study using site-specific recombination, 272,
        272F
Transgenic mice
    gene knockout, 329–330, 329FF
    genomic imprinting, 451
    globin gene expression, 436–437, 437T
    oncogene-carrying, 1280, 1280F
    in study of T cell development, 1246–1247
Transgenic organisms, 323
    defined, 327
Transgenic plants, **330**, 331F
Transition states, enzyme catalysis rate and,
        131–132, 132F
Transitional elements, cis Golgi network and,
        602, 602F
Translation
    see also Protein synthesis
    compartmentalization in eucaryotes, 25
    direction, 234
    efficiency in procaryotes and eucaryotes, 462
    in vitro systems, uses, 165
    initiation
        eIF-2 role, 462–463, 462F
        eucaryotes, **461–463,** 462FF
        internal ribosome entry sites, 462
        "leaky scanning," 461
        poly-A tails, 466, 466F
        procaryotes, 461
    mRNA
        direction on, 230
        selective degradation, 465–466
        start site, in eucaryotes, 461–462
        start site, in procaryotes, 461
    schematic, 107–108, 108F
Translation repressor proteins, 463
Translational control of protein synthesis, **461–
        464,** 462FF
    defined, 403, 403F
    eIF-2 phosphorylation, 462–463, 463F
    negative, 463–464, 463F
    poly-A addition, 466, 466F
    start site on mRNA, 461–462
    translational recording, 467, 467F
Translational frameshifting, 467, 467F
Translational recoding, **467**, 467F
Translocation apparatus
    in rough ER, 583, 583F
    for soluble proteins in ER, 584–585, 584FF
    for transmembrane proteins in ER, 586,
        586FF
Transmembrane proteins, **485–491**, 486FF, **493–
        501,** 494FF
    see also Membrane proteins
    acetylcholine receptor, 537–538, 538F
    α helix, 114, 487, 487FF, 496, 496F
    amphipathic properties, 485, 486F
    asymmetrical orientation, 486–488, 487FF
    bacteriorhodopsin, 495–496, 495FF
    band 3 protein, **494–495,** 491F, 495F
    β barrel, 487, 496–497, 497F
    chemotaxis receptors, 775–776, 775FF
    connexons, 959, 959F
    enzyme-linked receptors, **759–770,** 760FF
    freeze-fracture electron microscopy, 494–
        495, 494F
    glycophorin, 488F, 491F, 493–494, 495F
    glycosylation, 488, 488F
    hydropathy plots, 487, 488F
    integrins, **841,** 842F, 996, 996F
    ion channels, 523, 524F
    linker
        in anchoring junctions, 954, 954F
            cadherins, 966F, 967
            integrins, **995–999**, 996FF, 1000T
    multipass, 486F, **487, 494–497,** 495FF
        G-protein-linked receptor family, 735,
            735F
    synthesis in ER, **587–589,** 588FF
    PH-30, 1032–1033, 1033F
    photosynthetic reaction center, 497–498, 498F
    porin, 496–497, 497F
    receptors, in Drosophila development, 1083–
        1085, 1085F
    single-pass, 486F, **487**, 487FF, **493–494**, 495F
        cadherins, 966, 966F
        N-CAMs, 968, 968F
        synthesis in ER, **586,** 586FF
    solubilization, purification, reconstitution,
        488–489, 489FF
    synthesis in ER, 577, **586–591,** 586FF
    in tight junctions, 953, 953F
    as transport proteins, 509, 513
    tyrosine kinases, CSF receptors, 1171, 1171T
    vectorial labeling, 491

voltage-gated channels, 534–535, 534FF
    x-ray crystallography, 487
Transmembrane proteoglycan, 977–978, 978T
Transmembrane transport, of proteins, 556–557,
        557F
    see also Protein transport; Translocation
        apparatus
Transmission electron microscope (TEM), **149–
        152,** 149T, 149FF, **153–156**
Transmitter-gated channels, 523
    acetylcholine receptors
        neuromuscular junction, **537–538,** 537FF,
            546T
        vertebrate neurons, 539
    in cell signaling, 731, 732F
    at chemical synapses, **536–537,** 536F
    in LTP, 545–546, 545F
    pore size and subunits, 539, 539F
    psychoactive drugs, 539
    similarities and subtypes, 538–539
    subfamilies, 546T
Transplantation reactions
    MHC molecule discovery from, 1229–1230
    MHC role in, 1249, 1249F
Transport
    see also Active transport; Membrane
        transport
    paracellular, 953
    of proteins, see Protein transport
    transcellular, epithelial cell sheets, 951, 951F
Transport proteins
    anion antiporter, band 3, 494–495, 495F
    bacteriorhodopsin, 495–496, 495FF
    carriers, see Carrier proteins
    channels, see Channel proteins
    in chloroplast inner membrane, 696
    in epithelial cells, intestine, 951, 951F
    recombinant DNA technology to study, 511–
        512
    transmembrane protein structure, 509, 513
Transport vesicles, 18F, 555, 557, 557F, 599FF,
        600
    see also Secretory vesicles
    clathrin-coated, lysosomal enzyme
        packaging, 615–616, 615F
    coatomer-coated, 635, 635F, 640–642, 641FF
    docking mechanism, 643, 644F
    ER to Golgi, 602, 602F
    fusion with target, 644–645, 645F
    Golgi cisternae, between, 606, 607F
    Golgi to ER, 603–604, 603F
    H+ ATPase of, 622
    lysosomal enzymes, trans Golgi to lysosomes,
        615, 615F
    selectivity
        Rab protein and, 643, 644F, 644T
        SNAREs for, 642–643, 643FF
Transposable elements, 274, **284–286,** 285FF,
        392–395, 392T, 393FF
    see also Transposition
    DNA sequences, effects on, 393–394, 393F
    evolution and, 285–286
    gene regulation, effects on, 393–394
    L1 and Alu sequence in humans, 395, 392T,
        395F
    major families, characteristics, 392, 392T
    as mutagens, unusual features, 393–395
    rearrangement of host DNA, 285
    retrotransposons, 284–285, 283F
    transposition bursts and biological diversity,
        394–395
Transposase, 284–285, 285FF, 393, 393F
    alternative RNA splicing by, 455
    in transgenic animals, 328
    of transposable elements, 392T
Transposition
    defined, 284
    movement
        within chromosome, 285, 285F
        with DNA replication, 285, 286F
    of retrotransposons, **284–285,** 283F
    transgenic fruit flies and, 328
Transverse tubules, see T tubules
Treadmilling
    in actin filaments, 823, 825F
    in lamellipodium, 829, 829F
    in mitotic spindle microtubules, 928, 928F
    RecA protein in branch migration, 267, 267F
Triacylglycerols, 659, 659F
Tricarboxylic acid cycle, see Citric acid cycle
Trichomes, of plant, 29F
Triglycerides, 45, 54F, 659, 659F
Trimeric GTP-binding regulatory protein, see G
        protein
Triosephosphate isomerase, evolution, 389, 389F
Triple helix, of collagen, 116F, 979, 979F
Triskelion, clathrin, 638, 639F
Triton X-100, 489, 489F

Trophectoderm, 1057–1058, 1057F
Tropomyosin, **844,** 844F
    in regulation of muscle contraction, 854–855,
        854F
    summary, 845F
Troponin, **854–855,** 854F
trp operon, 417–418, 417FF
Trypanosomes
    RNA editing, 460, 460F
    trans RNA splicing, 460
Trypsin
    cleavage site, 173, 173T
    in tissue disruption, 157
Tryptophan, 57F
Tryptophan repressor, 417–418, 417FF
    structure, 409, 409F
Tuberculin skin test, 1245
Tubulin, 787, 789F, **803,** 803F
    see also Microtubules
    caged fluorescein label, 185, 186F
    in fly blastoderm, 1080F
    half-life, 809
    location in metaphase cells, 145F
    polymerization
        at centrosome, 805, 807–808, 807FF
        in vitro, **804–805,** 805F, 824FF
        at plus end, 805, 806F, 824F
    treadmilling in mitotic spindle, 928, 928F
γ-Tubulin, in centrioles, 808, 808F
Tumor
    see also Cancer; Cancer cells; Carcinogenesis;
        Transformed cells
    angiogenesis, 1153
    benign vs. malignant, 1256, 1256F
    primary, 1258
    progression, 1261–1264, 1262FF
Tumor initiators, 1264–1265, 1264FF
    carcinogens as, 1266
Tumor promoters, 1264–1265, 1264FF
    reproductive hormones as, 1266, 1267F
Tumor suppressor genes, **1281–1286,** 1282FF
    and cell-cycle control system, 900–901, 900F
    in colorectal cancer, 1287–1288, 1288T
    defined, 1273
    loss of heterozygosity, 1282, 1283F
    mutation, recessive effect, 1273, 1281
    p53 gene and p53 protein, 1284, **1285–1286,**
        1285F
    possible mechanisms for loss, 1282, 1283F
    Rb gene and product, **1282–1283,** 1282FF,
        1284, 1285F
    in retinoblastoma, 902–903, 902FF, **1281–
        1282,** 1282FF
    search for, using loss of heterozygosity, 1288
Tumor viruses
    DNA viruses, 282, **1283–1285,** 1283T, 1284FF
        human cancers, 1283–1284, 1283T, 1284FF
        life cycle, 1284
    retroviruses, **282–283,** 283F, **1273–1277,**
        1274FF, 1274T, 1276T
    Rous sarcoma virus, 1273–1275, 1275FF,
        1276T
    testing by cell transformation, 1274–1275,
        1274F, 1274T, 1277
Tumors, virus-induced, catalogue, 1276T
Turgor pressure
    of plants, 1002
    vacuole function in, 612–613, 612F
Turnover, of cell components, 179, 1144–1146
Turnover number, in enzyme catalysis, 131
Twins, graft tolerance, 1205
twist gene, 1084, 1084F
Two-dimensional gel electrophoresis, 170–172,
        171FF, 172T
Ty element, 392T
    life cycle, 285, 283F
Type III fibronectin repeat, 986–987, 986F
    in tenascin, 988
Type IV collagen, see Collagen, type IV
Tyrosine, 57F
Tyrosine hydroxylase, 750
Tyrosine-kinase-associated receptors, 759, **767,**
        767FF, 770
    see also Receptor tyrosine kinases
Tyrosine kinases, 734
    see also Protein kinases; Receptor tyrosine
        kinases

U-PA, 994–995, 994F
Ubiquinone
    as electron carrier, 677, 677F, 679
    in proton pumping, 681
    structure, 676, 676F
Ubiquitin
    functions, 219